SKY CATALOGUE 2000.0

Volume 1: Stars to Magnitude 8.0

SKY CATALOGUE 2000.0

Volume 1: Stars to Magnitude 8.0

Edited by

Alan Hirshfeld
Southeastern Massachusetts University

and

Roger W. Sinnott
Sky Publishing Corporation

1982

Cambridge University Press

Cambridge   London   New York   New Rochelle

Melbourne   Sydney

&

Sky Publishing Corporation
Cambridge, Massachusetts

Published by the Press Syndicate of the University of Cambridge
The Pitt Building, Trumpington Street, Cambridge CB2 1RP
32 East 57th Street, New York, New York 10022
296 Beaconsfield Parade, Middle Park, Melbourne 3206, Australia
and by Sky Publishing Corporation
49 Bay State Road, Cambridge, Massachusetts 02238-1290

First published 1982

Printed in the United States of America
by Essex Publishing Co., Inc.
Burlington, Vermont

Library of Congress Cataloging in Publication Data

Main entry under title:

Sky catalogue 2000.0. Volume 1: Stars to Magnitude 8.0

   1. Stars — Catalogs. I. Hirshfeld, Alan.

II. Sinnott, Roger W.

QB6.S54   523.8′908   81-17975
ISBN 0-521-24710-1 (v. 1)   AACR2
ISBN 0-521-28913-0 (pbk.: v. 1)

Preface

On a clear night, far away from city lights, the sky is filled with thousands of stars of all colors and brightnesses. Some form constellations, with mythological associations that hark back to ancient times, and others speckle the glowing band of the Milky Way. There are bright stars that guide mariners across the high seas and spacecraft across the solar system, while the fainter ones lead backyard astronomers to globular clusters and nebulae.

In 1981, Wil Tirion's *Sky Atlas 2000.0* was introduced by Sky Publishing Corp. as the first major new set of charts having its precessional epoch set for the turn of the century. This cartographic masterpiece conveniently records on 26 charts all stars down to 8th magnitude and meets the needs of observers with small or medium-size telescopes who seek galaxies, nebulae, clusters, comets, and asteroids against the starry vault. As with any atlas, however, an important element could not be included — detailed information about the stars themselves.

So intimately woven are the stars into many human endeavors that all luminaries visible to the naked eye, and many thousands of fainter ones, have been painstakingly measured, catalogued, and described by generations of astronomers. Their first goal was the careful determination of positions on the celestial sphere. With the advent of the spectroscope, photographic plate, and photometer, it has been possible to measure stellar motions accurately, as well as brightnesses, colors, and spectral types. For any star brighter than 9th magnitude, a great deal of information is likely to have been collected. Unfortunately, the data are commonly scattered throughout the astronomical literature.

Precise positions can be found in the massive four-volume *Smithsonian Astrophysical Observatory Star Catalog,* with its quarter-million entries. For the distances of nearby stars, L. F. Jenkins' *General Catalogue of Trigonometric Stellar Parallaxes* (1952) can be consulted. And for comprehensive data on the spectral types of stars, one might turn to the *Henry Draper Catalogue* (1918-24) or the *Bergedorf Spektraldurchmusterung* (1936-53). And so on. Even these sources have gaps in their coverage, or are outdated in some respects, and to learn the full story on a selected star may take hours of library research. Few amateurs have the necessary astronomical references.

For 30 years, the Skalnate Pleso *Atlas Catalogue* by Antonin Becvar has served as a fine general collection of star data. Including nearly 6,400 stars down to magnitude 6.25, it has been an indispensable reference for both amateur and professional astronomers. However, the *Atlas Catalogue* antedates some of the most modern references. And all too frequently a star of interest is just slightly too faint to be found in it; moreover, its positions are for the former epoch, 1950.0. A new comprehensive star catalogue is clearly needed.

The present volume was conceived by *Sky and Telescope* editor Leif J. Robinson when the Computer Sciences Corp. SKYMAP project documentation was published in 1978. Under project management of William E. Shawcross we have assembled *Sky Catalogue 2000.0,* Volume 1, mostly from computer-readable magnetic tapes that contain accurate modern stellar data as compiled for SKYMAP by David M. Gottlieb and Steven F. McLaughlin.

Specifically, the data herein have been derived from the character-coded computer tape of SKYMAP, Version 3.0, prepared by Wayne H. Warren, Jr., at the Astronomical Data Center, NASA Goddard Space Flight Center, from the original source catalogue compiled and supplied by D. M. Gottlieb and S. F. McLaughlin of Computer Sciences Corp. under contract to NASA GSFC. The SKYMAP catalogue and its sources of information are fully documented by David M. Gottlieb in "SKYMAP: A New Catalog of Stellar Data," *Astrophysical Journal Supplement Series,* Vol. 38, pages 287-308 (November, 1978). That paper should be consulted for specifics not covered here. SKYMAP includes some 248,000 stars to blue magnitude 9.0 — a compilation so huge that to publish it in printed form would require well over a dozen volumes the size of this one! To make this book practical and convenient, we have sought primarily the data of greatest value to the observer.

On these pages are summarized the essential characteristics of the 45,269 stars of visual magnitude 8.05 and brighter. Here, in short, is a written record of data for nearly every star plotted in *Sky Atlas 2000.0,* to which this volume serves as a companion. These are among its primary features: positional data are given for the epoch 2000.0; the modern **UBV** photometric system has been adopted to express stellar magnitudes and colors; complete spectral-type information is listed, including luminosity class and peculiarity code; distances are given along with the methods by which they were derived; multiple and variable stars have been flagged. In addition, each star has reference numbers from two comprehensive compendiums, the *SAO* and the *Henry Draper* catalogues.

Using computer facilities at Sky Publishing Corp., we have sorted the stars into strict order of right ascension and edited the data as described in the Introduction. To ensure the greatest freedom from error, the pages of the catalogue proper have been composed, formatted, and prepared in camera-ready sheets entirely by computer. Nevertheless, it is unthinkable that computers, their programs, magnetic media, and the humans who compiled the data always performed flawlessly. We are very interested in learning of errors that users may find in this catalogue. Write to Sky Publishing Corp., 49 Bay State Rd., Cambridge, Mass. 02238-1290.

Many of the 45,269 stars listed here have far more to tell us than can be summed up in a few neat columns of numbers. Those that are multiple or variable are more fully described in *Sky Catalogue 2000.0,* Volume 2, along with the many thousands of star clusters, nebulae, and galaxies deep in our night skies.

January, 1982

Alan Hirshfeld
Roger W. Sinnott

Introduction

ALL THE STARS that can be seen with the unaided eye, and some 37,000 other mostly nameless ones, are documented on the pages of *Sky Catalogue 2000.0*. Their sheer number poses a problem for the reader who turns to this book for information about a particular star. What is the best way to ferret out the page and line where such data are found?

If the star has a common name from antiquity, such as Vega or Rigel, its page number is listed in the index immediately following this Introduction, on page xviii. Nearly every prominent naked-eye star also has a Bayer designation (Greek letter), a Flamsteed number, or both, such as α Scorpii or 61 Cygni. When this is known, the index on page xix will help locate the star by its constellation. Finally, a variable star having its own special letter name, like T Centauri, will be found in the special index on page xxiv, where the variables within each constellation are arranged in order of discovery.

Superscripts are traditionally attached to some of the Bayer letters for stars appearing as naked-eye pairs, such as v^1 and v^2 Eridani, or strings of stars such as those in Orion's shield (which share the letter π with superscripts). Superscripts also distinguish the components of some wide, bright telescopic doubles (γ^1 and γ^2 Andromedae). But the components of other wide pairs, such as β Cygni, are not normally listed with superscripts; for them the same name is repeated in this catalogue, usually on the same page.

The catalogue lists star positions in the standard astronomical system of right ascension and declination. Specifically, its coordinate grid is defined by the mean equator and equinox at the beginning of the year 2000. Finding a star whose coordinates are known for this epoch is simply a matter of turning to the page that includes the star's right ascension, and then scanning down the declination column for a match. This is the procedure to use, for example, if the star's coordinates have been estimated from a chart of the Tirion *Sky Atlas 2000.0*, which is a useful companion to this catalogue.

Frequently, however, the star's coordinates will be known only for some other epoch, say 1950. Before searching for the star in this book, it is important to apply a rough mental correction for the effect of precession. This is the slow gyration of the Earth's rotational axis, which traces out a broad cone during 26,000 years. As a rule of thumb, a star's right ascension generally increases about 5 minutes in right ascension per century.

With a few keystrokes on a pocket calculator, one can find a much better correction for precession from these formulas, which give the increments to be added to a star's position to find its coordinates one year later:

$$\Delta\alpha = 3^s.07 + 1^s.34 \sin\alpha \tan\delta,$$
$$\Delta\delta = 20''.0 \cos\alpha.$$

For example, the 1950 coordinates of the bright star Deneb are $20^h 39^m 43^s.5$ and $+45° 06' 03''$. The rough values of these coordinates in degrees are 309°.93 and +45°.10, and the formulas tell us that the yearly changes are $\Delta\alpha = +2^s.04$ and $\Delta\delta = +12''.8$. Multiplying by 50 years gives a net change of $+1^m 42^s.0$ in right ascension and $+10' 40''$ in declination, implying that the coordinates of Deneb in the year 2000 are $20^h 41^m 25^s.5$ and $+45° 16' 43''$. Page 523 of the catalogue shows this to be not far from the truth.

These simple formulas for precession are not applicable to stars within a few degrees of the poles, nor across great time spans. More powerful techniques for dealing with such cases are discussed later.

Occasionally a user will want to look up a star knowing only the number from the *Henry Draper Catalogue* (HD) or the *Smithsonian Astrophysical Observatory Star Catalog* (SAO). A little ingenuity will go a long way toward finding the star within this volume. For example, the HD numbers were assigned in order of right ascension at epoch 1900. Precession has advanced the right ascensions of all stars by an average of 5 minutes of time since 1900, and has also given a slight twist to the coordinate grid. The HD star that was closest to the 0^h meridian in 1900 appears here at the foot of the second page. The numbers still run *almost* consecutively, especially for equatorial stars. But if the star being sought has a high northern or southern declination, be prepared to find it many pages ahead of or behind other stars with similar HD numbers. Some very large HD numbers, from the *Henry Draper Extension,* are interspersed throughout.

Finding a star by its SAO number is somewhat trickier because the SAO arrangement is in bands of declination 10° wide. Within each band, the numbers run in order of 1950 right ascension, so in our catalogue there are 18 overlapping sequences of SAO numbers. The key to finding a particular SAO star is to spot others in the same 10° band of declination with closely similar numbers.

POSITIONS AND PROPER MOTIONS

The transit instrument, meridian circle, and more recently the photographic zenith tube and prismatic astrolabe, have been used by astronomers to record star positions with great precision for over a century and a half. Modern knowledge of star positions and their individual proper motions comes from more than a hundred star catalogues produced within that

TABLE I. THE BRIGHTEST STARS IN THE SKY

Rank	Name	Bayer Designation	V Magnitude
1	Sirius	α Canis Majoris	− 1.46
2	Canopus	α Carinae	− 0.72
3	Rigil Kentaurus	α Centauri	− 0.27*
4	Arcturus	α Bootis	− 0.04
5	Vega	α Lyrae	+ 0.03
6	Capella	α Aurigae	+ 0.08
7	Rigel	β Orionis	+ 0.12
8	Procyon	α Canis Minoris	+ 0.38
9	Achernar	α Eridani	+ 0.46
10	Betelgeuse	α Orionis	+ 0.50 (var)
11	Hadar	β Centauri	+ 0.61
12	Altair	α Aquilae	+ 0.77
13	Aldebaran	α Tauri	+ 0.85 (var)
14	Acrux	α Crucis	+ 0.87*
15	Antares	α Scorpii	+ 0.96 (var)
16	Spica	α Virginis	+ 0.98
17	Pollux	β Geminorum	+ 1.14
18	Fomalhaut	α Piscis Austrini	+ 1.16
19	Deneb	α Cygni	+ 1.25
20	Mimosa	β Crucis	+ 1.25
21	Regulus	α Leonis	+ 1.35

*Combined magnitude for double stars whose components are listed separately in the *Catalogue*.

time period. Some measurements, still of value today, were made as early as the 1750's by the English astronomer J. Bradley.

The relative positions of selected, meticulously measured stars, such as the 1,535 listed in the *Fourth Fundamental Catalogue* (FK4), are considered known to within a small fraction of a second of arc. Widely distributed around the sky, such "fundamental" stars make up the reference frame to which the positions of the vast majority of others are referred. When the SAO catalogue was published in 1966, its compilers believed that the positions of its 259,000 stars had an average standard deviation of only 0".5 with respect to the FK4 system.

SKYMAP uses the SAO positions of stars whenever possible. In his 1978 paper, D. M. Gottlieb states that on the average these positions should still be reliable at epoch 2000 to better than 1", which is the tabular precision of *Sky Catalogue 2000.0*. For stars lacking an SAO number, the most notable being ς Sculptoris, positions were brought into SKYMAP from other sources. Whenever a proper motion is cited for a non-SAO star in this catalogue, the reader can assume its position is derived from the *Astronomische Gesellschaft Katalog 3* (AGK3) and compares in accuracy with an SAO position. But if a star has no listed proper motion or SAO number, the coordinates must be considered very approximate. Gottlieb estimates that such positions, derived from the HD, have an average uncertainty of 25".

MAGNITUDES AND COLORS

The science of measuring the brightnesses and colors of stars is called *photometry*. Over 2,000 years ago, the Greek astronomer Hipparchus had already recognized the importance of defining stellar brightnesses quantitatively. The brightest stars in the sky he called 1st magnitude and the faintest naked-eye stars 6th magnitude. Those intermediate in brilliance were assigned magnitudes 2 through 5.

In the 19th century, John Herschel realized that this arithmetic progression of magnitudes corresponds to equal *ratios* of brightness. Norman Pogson noted further in 1856 that a 1st magnitude star is about 100 times brighter than one of 6th magnitude. As a result, if a difference of five magnitudes corresponds to a ratio of 100, then each magnitude step represents a brightness ratio equal to the fifth root of 100, or about 2.512. Two steps denote a factor of 2.512^2, or about 6.3, and so on. In this catalogue, the visual magnitude of Capella is listed as 0.08, and that of Thuban (in Draco) as 3.65. The difference in their magnitudes is 3.57, showing that Capella is $2.512^{3.57}$ or almost 27 times brighter than Thuban.

Magnitudes on the Pogson scale were originally fixed by assigning Polaris, the North Star, a magnitude of exactly 2.00. When it was discovered that Polaris is a variable star of small amplitude, the system was standardized at Harvard and Mount Wilson observatories early in this century to a group of well-observed stars called the North Polar Sequence. Numerous other standard sequences have been added over the years. Incidentally, the magnitudes listed for Polaris and other variable stars in the present catalogue are the values when they are at their brightest.

Two factors complicate the measurement of stellar magnitudes:

(1) The light from a star is not monochromatic; that is, the light is composed of various colors, or wavelengths. The intensity peaks at a certain wavelength and falls off smoothly toward shorter and longer values. A star's surface temperature determines the wavelength of its maximum radiance. A red star appears that color because the major contribution to its overall light is in the red part of the spectrum. It does, however, emit some energy even in the blue. On the other hand, very hot bluish stars may give off the bulk of their radiation in the ultraviolet. The magnitude of a star clearly depends on where in the star's spectrum it is measured, unless it takes into account all wavelengths of emission (when it is called the *bolometric magnitude*).

(2) No light detector is uniformly sensitive across all parts of the spectrum. The sensitivity of the human eye peaks in the green region of the spectrum at a wavelength of about 5500 angstroms. (One angstrom is 10^{-8} centimeter.) Its sensitivity falls to zero in the ultraviolet (wavelengths shorter than 4000 A) and in the infrared (longer than 7000 A). The wavelength response also differs among the various photographic plates and photoelectric cells in use. A red star will appear bright on a red-sensitive plate, but relatively faint on a blue-sensitive one. Once again, it is clear that the wavelength at which the magnitude is measured must be specified.

Numerous photometric techniques have been introduced over the years. They fall into three broad categories: visual, photographic, and photoelectric. Most of the early visual photometers were constructed so the observer could directly compare the brightness of a star to that of an artificial light source of controllable intensity, or to another star. Instruments of this type were superseded in the late 1800's by the advent of photographic photometry. The magnitudes of large numbers of stars can be determined by measuring the sizes or densities of their images on photographic plates. The chief disadvantage of this method is the nonlinear response of the light-sensitive emulsion to brightness, color, and exposure time. Excellent discussions of visual and photographic photometry are those by H. F. Weaver in 1946 and by G. R. Miczaika and W. M. Sinton in 1961.

Since their astronomical adoption a half century ago, photoelectric photometers have grown increasingly sophisticated and easy to use. The photoelectric system most widely used by modern astronomers is the **UBV** system described in 1953 by H. Johnson and W. Morgan. Standard filters are placed in turn between the telescope and a photomultiplier tube to measure a star's energy within three distinct wavelength bands: **U**, centered at about 3500 A; **B**, 4350 A; and **V**, 5550 A. The **V** (also called the visual or yellow) band was selected to coincide with the peak sensitivity of the human eye. The filter used appears yellow, but the effective wavelength is actually in the green. The fact that the early photographic plates were most responsive in the blue region of the spectrum influenced the selection of the **B** band. The ultraviolet, or **U**, band provides further information about stars that was not obtained by the earlier visual or photographic methods; its wavelength characteristics were largely determined by instrumental limitations of the 1950's.

In 1963, Johnson published a list of **UBV** measurements for a set of stars intended to serve as standards. By always including these stars (or some from later lists) in an observing program, astronomers can determine how to correct their own measured magnitudes of other stars to the standard **UBV** system. The 10 primary standard stars are HD 12929 (α Arietis), 18331 (HR 875), 69267 (β Cancri), 74280 (η Hydrae), 135742 (β Librae), 140573 (α Serpentis), 143107 (ε Coronae Borealis), 147394 (τ Herculis), 214680 (10 Lacertae), and 219134 (HR 8832).

Table I lists the brightest stars in the sky and their **V** magnitudes as given in this catalogue. Similar lists have appeared in numerous astronomy books through the years, often showing great disagreement in the order of brightness. Yet the relative reliability and consistency of photoelectric photometry is illustrated by the fact that a list compiled by Johnson in 1957 agrees with our ranking, except that Hadar and Betelgeuse (an irregular variable) are switched. Apart from variables, the largest discrepancy is 0.07 in the case of Achernar; the **V** magnitudes of most other stars on the list agree to 0.03 or better with the values he assigned a quarter century ago.

Important information about a star's nature is revealed by computing a *color index,* such as **B − V** or **U − B**, from the separate magnitude values. Such a number, often referred to simply as the *color,* compares the energy in one region of the star's spectrum to that in another. It is a simple indicator of the spectral distribution of the light, which is related to the star's surface temperature. In other words, the color index is a measure of how hot the star is.

The relationship between surface temperature and **B − V** is plotted for main-sequence stars in Fig. 1. Main-sequence stars, by far the most common type in the galaxy, are stable stars that burn hydrogen in their cores; the Sun is an example. Note that the giants and supergiants, such as Aldebaran and Betelgeuse, tend to have slightly larger color indexes than the main-sequence stars of the same temperature. The values of **B − V** for most stars fall within the range −0.40 (hot, blue stars) to +2.00 (cool, red stars). For the Sun, the color index is +0.62, while for a white star such as Vega it is 0.00.

During the 1920's, several stars whose spectra clearly indicated that they were hot and blue were found to have "red" color indexes. In a series of notable papers, R. J. Trumpler demonstrated in 1930 that the anomaly did not stem from any peculiarity of the stars themselves. Rather, it was due to the effects of interstellar dust particles on their light. This material in the disk of the galaxy dims the light from distant stars, and does so more effectively at shorter wavelengths than at longer ones. As a result, starlight is increasingly reddened as it passes through the interstellar medium.

The reddening effect, small for nearby stars, is most noticeable for bright, blue supergiants. These stars lie near the plane of the galaxy and can be seen to great distances through the interstellar medium. For example HD 80077, which at 7,800 parsecs is one of the most distant stars in this catalogue (page 239), has an intrinsic **B − V** of −0.17 and an observed **B − V** of 1.30! The difference in the two values, called the *color excess,* is a measure of reddening, from which the amount of interstellar material can be inferred in various directions in the galaxy.

A value of **V** and **B − V** is listed in the catalogue for almost

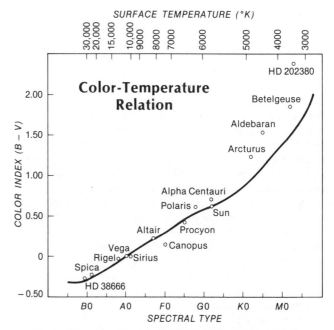

Fig. 1. *The curve shows the relationship between spectral type, color index, and surface temperature for main-sequence stars. Several well-known stars are plotted. HD 202380 (page 539) is the "reddest" type M star in the catalogue, with a color index of +2.39; HD 38666 (μ Columbae) is the "bluest" O type, with a color index of −0.28. The red stars falling well above the main-sequence relation are all giants or supergiants.*

every star. Their sum, of course, gives the value of **B**. The methods by which these magnitudes were measured or derived are explained in the Gottlieb paper, and for particular stars can be inferred from the number of decimal places printed. If both are listed to two places, then **V** and **B** were measured photoelectrically. When **V** was measured photoelectrically but **B** was derived from photographic photometry, **V** is given to two decimal places and **B − V** to one. When both are given to one place, then **B** and **V** were both derived from photographic photometry.

For photoelectric magnitudes, the expected uncertainties in **V** and **B − V** are about 0.02 and 0.01 magnitude, respectively. Otherwise, the uncertainties are typically about five times greater.

This catalogue includes all stars whose **V** magnitude (or **B** in the few cases when SKYMAP had no **V** magnitude) is brighter than 8.05. Their distribution by magnitude is given in Table II and illustrated in Fig. 2. Note that almost half of the stars in the catalogue are fainter than **V** = 7.49, even though the last magnitude class is terminated at 8.05.

Many wide double stars have their components listed individually in the catalogue, yet appear single to the unaided eye or in a low-power telescope. In such cases, there is frequently the need to know the *combined magnitude* of the pair. Logarithms to base 10 simplify such calculations because the log of 2.512 (the Pogson ratio) is exactly 0.4. Let us say the magnitude of the brighter component is m_1 and that of the fainter star m_2. To find their combined magnitude m, first compute the brightness ratio R, either by the method mentioned at the start of this section or by taking the antilog of $0.4\,(m_2 − m_1)$. Then, using this ratio and the magnitude of the *fainter* star, compute the combined magnitude from

$$m = m_2 − 2.5 \log (R + 1).$$

For example, the components of β Cygni (Albireo) are listed separately on page 485, where we see that $m_1 = 3.08$ and

---

TABLE II. STARS TO MAGNITUDE 8.05

Magnitude Class	Range Included	Number by Magnitude	Cumulative Total
−1	−1.50 to −0.51	2	2
0	−0.50 to +0.49	7	9
+1	+0.50 to +1.49	13	22
+2	+1.50 to +2.49	71	93
+3	+2.50 to +3.49	192	285
+4	+3.50 to +4.49	625	910
+5	+4.50 to +5.49	1,963	2,873
+6	+5.50 to +6.49	5,606	8,479
+7	+6.50 to +7.49	15,565	24,044
+8	+7.50 to +8.05	21,225	45,269

$m_2 = 5.11$. The brightness ratio of the stars is 6.487, and the formula shows the combined magnitude m to be equal to $5.11 - 2.5 \log (7.487)$, or 2.92. This is the naked-eye magnitude of Albireo.

Before anything was known about the true distances of the stars, it was natural to assume that the brightest stars were also the nearest. Actually, apparent brightness is an extremely poor clue to relative distance, because of the very wide range of luminosities, or intrinsic brightnesses, that stars can have. A high-luminosity star at a great distance can appear just as bright in the sky as one of low luminosity that is much nearer. In order to compare accurately the true properties of stars, the absolute visual magnitude M_v is used to describe their intrinsic brightnesses. By convention, M_v is the apparent magnitude a star would have if placed at the standard distance of 10 parsecs (32.6 light-years) from the Sun. The most luminous stars in the galaxy have absolute magnitudes of about -9. They are about 400,000 times brighter than the Sun, whose absolute magnitude is only about 4.8. Canopus, typical of the most luminous stars listed in the catalogue, is about 40 million times brighter than the least luminous one, HD 153336, which is listed on page 409.

SPECTRAL TYPES

The light from a star when spread into a spectrum reveals a wealth of information about the star's intrinsic properties, such as its composition, surface temperature, size, density, motion, and rate of rotation. Stellar spectra are generally characterized by the presence of dark absorption lines and bright emission lines cutting across the familiar continuum of colors. In practice, spectra are recorded on black-and-white photographic plates, since the color adds no extra information. The lines are not randomly placed; rather, each can be identified with a particular chemical element. In some sense, therefore, obtaining a spectrum of a distant star is like bringing a sample of its material back to Earth for analysis.

In 1863 the Italian astronomer Angelo Secchi demonstrated that stellar spectra can be classified according to the appearance of their absorption lines. He divided stars into four spectral categories. In 1886 E. C. Pickering set out to photograph spectra of all the brighter stars visible from the Harvard College Observatory. The first volume of his *Draper Catalogue of Stellar Spectra,* published in 1890, contained over 10,000 stars north of declination $-25°$ and down to magnitude 8. The massive project continued for decades, largely due to the extraordinary effort of Annie Jump Cannon and her co-workers. By 1949, the *Henry Draper Catalogue* and its two extensions contained spectral types for nearly 360,000 stars. The Harvard, or revised Henry Draper (HD), system categorizes stellar spectra primarily by the widths and darknesses ("strengths") of the absorption lines of hydrogen, helium, and numerous "metals" (the common astronomical term for any element other than hydrogen and helium). The sequence of spectral types *O, B, A, F, G, K,* and *M* corresponds to one of decreasing surface temperature. (The sequence was initially alphabetical, with *A* representing spectra with the strongest hydrogen lines and the now-obsolete *Q* the weakest. When the list was reordered by temperature, some of the original types were combined or deleted.) The properties of the principal spectral types are given in Table III. The HD types *R, N,* and *S* are still sometimes used. They refer to cool, giant stars with molecular absorption other than the titanium oxide lines typical of *K* and *M* stars.

Each spectral type is divided into 10 subclasses, indicated by an appended number (for example, *A6*). Occasionally, the subclasses have decimal extensions, such as *B3.5*. The change in surface temperature corresponding to adjacent subclasses depends on the spectral type. For instance, the temperature difference between types *B8* and *B9* is about 1000° K, whereas less than 100° separates *G8* and *G9*. The estimated uncertainty in an HD spectral type is three subclasses, according to D. M. Gottlieb.

Lower case letters are sometimes added to the *O* types to denote various emission-line patterns, and to the *M* types to indicate distinctive molecular absorption bands (spectral lines so closely spaced that they form wide bands of absorption). When astronomers recognized that stars of the same temperature can have vastly different luminosities, the prefix *d, g,* or *c* was added in some cases to represent dwarf (main sequence), giant, or supergiant, respectively.

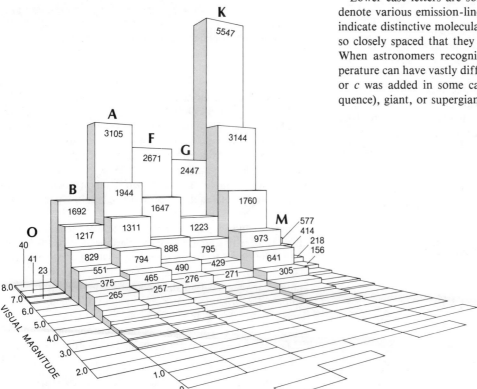

Fig. 2. The distribution of catalogue stars by spectral type, in half-magnitude steps. The height of each box is proportional to the number of stars in that particular category, with the exact number shown on the faces of the larger boxes. The two rectangles in the foreground represent Sirius and Canopus. Note the apparent deficiency of M stars. Actually, the intrinsic faintness of the numerous main-sequence stars of that type places most of them beyond the limiting magnitude of the present catalogue.

TABLE III. PROPERTIES OF THE PRINCIPAL SPECTRAL TYPES

Spectral Type	Apparent Color	Intrinsic $B - V$	Surface Temperature (°K)	Primary Absorption Lines in Spectrum	Examples
O	blue	less than −0.2	25,000 to 40,000	Strong lines of ionized helium and highly ionized metals; hydrogen lines weak	ζ Orionis (O9.5)
B	blue	−0.2 to 0.0	11,000 to 25,000	Lines of neutral helium prominent; hydrogen lines stronger than in type O	Spica (B1) Rigel (B8)
A	blue to white	0.0 to 0.3	7,500 to 11,000	Strong lines of hydrogen, ionized calcium, and other ionized metals; weak helium lines	Vega (A0) Sirius (A1) Deneb (A2)
F	white	0.3 to 0.6	6,000 to 7,500	Hydrogen lines weaker than in type A; ionized calcium strong; lines of neutral metals becoming prominent	Canopus (F0) Procyon (F5) Polaris (F8)
G	white to yellow	0.6 to 1.1	5,000 to 6,000	Numerous strong lines of ionized calcium and other ionized and neutral metals; hydrogen lines weaker than in type F	Sun (G2) Capella (G8)
K	orange to red	1.1 to 1.5	3,500 to 5,000	Numerous strong lines of neutral metals	Arcturus (K2) Aldebaran (K5)
M	red	greater than 1.5	3,000 to 3,500	Numerous strong lines of neutral metals; strong molecular bands (primarily titanium oxide)	Antares (M1) Betelgeuse (M2)

The HD spectral classification system has been superseded by the Morgan, Keenan, and Kellman (MK) system, developed in 1943. The MK classifications are based on a wider variety of line strengths and line-strength ratios, described fully by Keenan in 1963. It retains the major spectral-type designations of the HD system, but includes a more finely divided luminosity scale. The star's luminosity class is indicated by a Roman numeral I to VI, representing supergiant, bright giant, giant, subgiant, main sequence, and subdwarf, respectively. The supergiants are further divided into subclasses Ia-0, Ia, Iab, and Ib, in order of decreasing luminosity. When an MK spectral type has been assigned, the estimated uncertainty is one subclass and one third of a luminosity class.

There are a number of additional spectral types that occasionally appear in the catalogue, for special classes of stars. The superhot Wolf-Rayet stars, which even precede the blue O stars, have spectra characterized by broad emission lines from ejected matter, and are indicated by WR, WC, or WN. WC stars show strong emission lines of carbon and oxygen, while WN display prominent nitrogen emission. Those that fall in neither category are termed WR. The cool, red carbon stars, which follow the M class, were formerly called R or N in the HD system. They are now labeled type C. The spectra of some stars are so peculiar that they defy classification. These are simply designated by the letter P. Anomalies in chemical composition, rotation speed, atmospheric structure, or other properties are indicated by the codes in Table IV.

The overall character of the star population in any catalogue is wholly dependent on the selection criteria used. Table V gives the distribution of spectral types and luminosity classes for the stars in this catalogue. These numbers illustrate an important observational selection effect. Our sample of stars down to magnitude 8.05 is by no means a random one. That is, it does not give an accurate assessment of the true distribution of spectral types or luminosity classes among stars. The sample favors the hot, bright O and B main-sequence stars and the supergiants and giants, all of which can be seen to much greater distances (encompassing a far greater volume of space)

than less luminous stars. The results in Table V therefore overestimate the true percentages of O and B stars and supergiants and giants. Actually, supergiants and giants together are only about 1 percent as populous as main-sequence stars. Also, K and M stars are approximately 500 times more numerous in the solar neighborhood than types O and B, according to C. W. Allen. While the O and B stars contribute most of the light of the galaxy, it is the tremendous number of relatively inconspicuous stars that dominate its motion and stabilize it gravitationally.

TABLE IV. SPECTRAL PECULIARITY CODES

Code	Meaning
e	Emission lines
wl	Weak lines, possibly indicating metal-poor composition
m	Abnormally strong metallic lines, usually in type A stars; they imply cooler spectral type than do hydrogen lines
si	Abnormally strong silicon lines
cr	,, ,, chromium lines
sr-cr	,, ,, strontium and chromium lines
k	,, ,, calcium K line
he	,, ,, helium lines
s	Very narrow ("sharp") lines
n	Broad ("nebulous") lines indicating rapid rotation
nn	Very broad lines, indicating very rapid rotation
hb	Horizontal-branch star (old star that burns helium in its core)
sh	Shell star (type B-F main-sequence star with emission lines originating in an expelled gaseous shell)
neb	Star embedded in a nebula
comp	Composite spectrum; implies unresolved binary
var	Variable spectral type
p	General code for unspecified peculiarity, unless appended to type A, in which case it denotes abnormally strong metallic lines (related to Am stars)

DISTANCES

The distances to stars were computed in several ways. The most reliable ones are based on actual measurements of the trigonometric parallax. Because of the Earth's annual motion around the Sun, every six months finds the observer at an opposite extreme of the Earth's orbit. This changing viewpoint causes measurable shifts in the apparent positions of nearby stars. The trigonometric parallax is normally expressed in seconds of arc, and equals the maximum angle by which a star strays in the course of a year from its central or average apparent position. Specifically, the parallax is defined as the angle subtended at the star by the average radius of the Earth's orbit, 149,600,000 kilometers.

A second star with half the parallax of the first is twice as far away, and a star with a third the parallax is three times as far away. The reciprocal of the parallax in seconds of arc gives the distance of the star, expressed in a unit called the *parsec* (coined from *par*allax and *sec*ond). One parsec is about 3.26 light-years or over 30 trillion kilometers.

Measured stellar parallaxes are exceedingly small. Even the closest star system, α Centauri, has a parallax of only 0″.75, which is the apparent size of a postage stamp at a distance of nearly 7 kilometers. The reciprocal of 0″.75 gives the stars' distance, 1.3 parsecs, as listed on page 356. Their faint companion Proxima Centauri actually has the largest known parallax, 0″.77, but at 11th magnitude is much too faint to be included in this catalogue.

Stellar parallaxes become increasingly small the farther out we go. Beyond about 20 parsecs, this angle is so minute that it can no longer be reliably measured with today's techniques. The trigonometric parallax method has therefore been successfully applied to fewer than 1,000 stars. D. Mihalas and J. Binney recently pointed out that the 20-parsec maximum could be increased fivefold with the advent of astrometric satellites and space telescopes. Trigonometric distances in this catalogue are indicated by the letter "t."

The distances to the vast majority of stars cannot be measured in this fashion. Yet they can be deduced through application of the inverse-square law of brightness, which states that the observed intensity of light from a source decreases as the square of its distance. Thus, if a star were moved twice as far away, it would appear four times dimmer and its apparent magnitude would change by 1.51. It is straightforward to show that if m is the apparent magnitude and M is the absolute magnitude of a star, then

$$m - M = 5 \log d - 5,$$

where d is the distance in parsecs. (Notice that when d is 10 parsecs, the right side of this equation becomes zero and $m = M$, as it should.) The value of $m - M$ is called the distance modulus.

TABLE V. SPECTRAL-TYPE AND LUMINOSITY DISTRIBUTION

Spectral Type	Stars in Catalogue	Luminosity Class	Stars in Catalogue
O	0.4%	I	1.3%
B	13	II	1.1
A	20	III	35
F	16	IV	9
G	14	V	42
K	32	Undetermined	12
M	4		
Other	0.9		

TABLE VI. EFFECT OF ABSORPTION

Absorption (mag.):	0.005	0.05	0.25	0.5	1.5	2.5
Distance (pc)						
Actual:	10	100	500	1,000	2,500	5,000
Apparent:	10	102	560	1,300	4,500	16,000

Applying this equation to the most luminous stars in the galaxy, with absolute magnitudes on the order of -9, reveals that they could be as distant as 25,000 parsecs and still be included in this catalogue (neglecting the effects of interstellar absorption, which are discussed below). The Sun, on the other hand, would fall below the catalogue's limiting magnitude if it were moved beyond 45 parsecs.

If a star has a known spectral type and luminosity class, its approximate absolute magnitude can be inferred (see Fig. 3). Combining this information with its apparent magnitude in the equation above yields an estimate of the star's distance. This *spectroscopic parallax* method is extremely valuable, since it can be applied to the tremendous number of stars with reliable MK spectral types. Spectroscopic distances are marked in the catalogue with the letter "s."

Distances determined in this way are only estimates and must therefore be used with caution. They are highly sensitive to the assigned luminosity class. For example, if the spectrum of a main-sequence star (class V) is misinterpreted as that of a subgiant (class IV), the error in absolute magnitude could be as great as 1.0. The derived distance would then be overestimated by a factor of more than 1.5 (that is, the square root of 2.512). Also, the absolute magnitudes themselves are subject to some uncertainty, especially those of rare stars, such as supergiants. Gottlieb estimates the uncertainty to be either 0.2 magnitude or the difference in the values for adjacent spectral subclasses, whichever is greater.

A further complication with the spectroscopically derived distances arises from the dimming of light by interstellar dust particles along the line of sight. The dimming is small for stars closer than a few hundred parsecs. However, the apparent brightness of a remote supergiant can be reduced by several magnitudes, which could lead to a sizable overestimate of its distance. Interstellar absorption is most severe close to the plane of the Milky Way, especially toward the galactic center in Sagittarius. (In fact, were the dust not present, the combined radiance of the stars in the nucleus of the galaxy would rival that of the full moon!) There is essentially no absorption for stars with galactic latitudes greater than 20° from the plane of the Milky Way.

Near the galactic plane, the amount of absorption is typically 0.5 magnitude for each 1,000 parsecs of distance. That it cannot be ignored is shown clearly in Table VI. Here are listed the "actual" distances to a number of hypothetical stars, along with the amount of absorption in magnitudes and the apparent distances that would be computed if this absorption were neglected.

The magnitude-distance relation must therefore be revised to incorporate the effects of interstellar absorption:

$$m - M = 5 \log d - 5 + a.$$

But a varies with wavelength, so if the V magnitude is observed, then the absorption in the V band is the value that must be applied. Numerous studies have shown it generally to be equal to about three times the color excess of a star. As discussed in Gottlieb's 1978 paper, all stellar distances in SKYMAP that are based on a spectroscopic parallax have been adjusted for the effects of interstellar absorption, as a function of both the galactic longitude and latitude for each individual star.

Distances to *particular* stars can sometimes be refined by special techniques. The convergent point of a moving cluster, the period of a Cepheid variable star, the mass-luminosity relation as applied to a binary star orbit — these have led to some of the powerful methods in the astronomer's repertoire. Scattered throughout the literature, such special distance determinations could not be individually included in SKYMAP, although collectively they have had an important role in laying the foundations of Fig. 3. Thus, the distances listed for members of the Hyades cluster cannot be assumed to distinguish their individual locations within the cluster, even though their *average* is consistent with the recent determination by B. Hauck that the center of the Hyades lies about 42 to 48 parsecs from the Sun.

Reliable or even rough distances could not be determined for all stars in SKYMAP. Distances flagged with the "mn" code are minimum values assigned to stars with the smallest, hence most uncertain, trigonometric parallaxes. In some cases, only a maximum limit could be established (code "mx"). Greater distances would imply the existence of stars with space motions exceeding those considered likely from studies of galactic dynamics.

Distances are listed for 91 percent of the stars in this catalogue. Of these, 36 percent lie within 100 parsecs of the Sun, while fully 97 percent are closer than 500 parsecs. By comparison, the distance to the galactic center is about 8,500 parsecs.

RIGOROUS PRECESSION

The simple formulas for precession given on page vii are handy for telescopic observers who use setting circles, and are typical of the time-saving artifices that were worked out for rough pencil-and-paper computation in the days of logarithmic tables. But to take full advantage of the accuracy to which star positions are listed in this catalogue, the standard rigorous method should be used to reduce the mean place of a star from one epoch to another. The precessional constant now in use is that introduced in 1897 by the American astronomer Simon Newcomb, and the formulas are discussed in his 1906 book, *A Compendium of Spherical Astronomy*. They are similar to expressions published in Germany in 1830 by F. W. Bessel.

Three auxiliary constants are determined by the selection of initial and final epochs. Here, the initial epoch is 2000.0 and the final epoch is some number of tropical centuries T before or after that. (A tropical century contains 36,524.22 days.) Newcomb's expressions for these constants are

$$\zeta_0 = 2305''.646\,T + 0''.302\,T^2 + 0''.018\,T^3,$$
$$z = \zeta_0 + 0''.791\,T^2,$$
$$\theta = 2003''.829\,T - 0''.426\,T^2 - 0''.042\,T^3.$$

If the final epoch is before 2000, T is negative; otherwise, T is positive.

With these values, any star in the catalogue can be precessed to a chosen epoch. The first step is to correct the star's position for its proper motion during the interval. Simply multiply

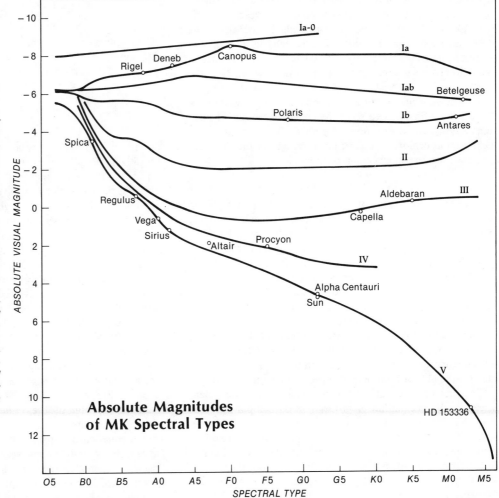

Fig. 3. *This is the relationship between absolute visual magnitude and MK spectral type, as adopted by the compilers of SKYMAP. The vertical scale shows the absolute magnitudes used for all stars in the catalogue except those closer than 10 parsecs (32.6 light-years), for which values from W. Gliese's 1969 catalogue are listed instead. The luminosity classes Ia-0, Ia, Iab, and Ib denote supergiants, in order of decreasing brightness, while II, III, IV, and V refer to bright giants, giants, subgiants, and main-sequence stars, respectively. Stars falling between adjacent luminosity classes (for example, Altair, with a spectral type A7 IV-V) are assigned an intermediate brightness. D. M. Gottlieb estimates the uncertainty in the absolute magnitudes to be at least 0.2 magnitude, and even greater where the curves are steep, as in the case of B stars on the main sequence.*

the tabulated values of $\mu(\alpha)$ and $\mu(\delta)$ by the number of years (negative if the final epoch is before 2000), and add the result to the star's right ascension and declination, respectively. Thus, to convert a position in this catalogue to epoch 1950, the proper motions must be multiplied by the factor -50. The star's coordinates, corrected for proper motion but not yet precessed, are denoted α_0 and δ_0, and the rigorous expressions for α and δ at the final epoch are

$$\cos \delta \sin (\alpha - z) = \cos \delta_0 \sin (\alpha_0 + \zeta_0),$$
$$\cos \delta \cos (\alpha - z) = \cos \theta \cos \delta_0 \cos (\alpha_0 + \zeta_0) - \sin \theta \sin \delta_0,$$
$$\sin \delta = \cos \theta \sin \delta_0 + \sin \theta \cos \delta_0 \cos (\alpha_0 + \zeta_0).$$

The first two equations are needed to resolve the ambiguity as to the quadrant in which α lies, but the third gives δ immediately. It is generally possible to write a single program for a pocket calculator to perform the entire computation.

The mean place given by these formulas differs from the star's apparent place on a given night by the small periodic effects of nutation, aberration, and stellar parallax. In many applications it is also important to correct for refraction by the Earth's atmosphere.

As an example, Polaris' position and proper motion can be taken from this catalogue and converted to its position at epoch 1755.0 by the use of these rigorous formulas. The result is $\alpha = 0^{\rm h}\, 43^{\rm m}\, 42^{\rm s}.2$, $\delta = +87° 59' 41''$, showing how much farther this star was from the pole a couple of centuries ago. This result compares very well with the position actually observed then by J. Bradley, and reduced to mean epoch 1755.0 by Bessel: $\alpha = 0^{\rm h}\, 43^{\rm m}\, 43^{\rm s}.0$, $\delta = +87° 59' 41''$.

Iterations of the same calculator program show that Polaris will be closest to the north celestial pole in the year 2102, at a distance of $27' 37''$. The same procedure, applied to the 2.08-magnitude star β Ursae Minoris, tells us that it was closest to the pole about the year 1200 B.C. Indeed, its ancient name Kochab is a contraction of the Arabic words for "the north star."

The precessional motion is complex, and Newcomb intended these formulas to serve the demanding needs of positional and geodetic astronomy for only a few centuries. The next section explores the long-term changes in the sky's appearance that are independent of precession.

MOTION IN SPACE

Over many thousands of years, the proper motions of the stars will noticeably alter the shapes of the constellations. Some stars are approaching the Sun and will become brighter; others are receding and must have been brighter in the remote past. To investigate these effects, it does not suffice to consider the proper motion of a star as carrying it uniformly across the sky. Rather, proper motion should be combined with radial velocity and distance to give a picture of the star's intrinsic motion through space, relative to that of the Sun.

This catalogue contains the information needed to study the individual space motions of the stars. In this context it is best to disregard precession, which is a shift of the Earth's coordinate system, not a motion of the stars themselves.

The first step is to take out of the catalogue the star's right ascension α, declination δ, proper motion components $\mu(\alpha)$ and $\mu(\delta)$, distance, and radial velocity. These six quantities describe the star's present position in space and direction of travel relative to the Sun. But to simplify the calculation it is best to transform them to a set of rectangular coordinates x, y, z and yearly changes Δx, Δy, Δz, the unit of length being the parsec.

TABLE VII. STARS OF LARGE PROPER MOTION

Name	Constellation	V Mag.	Annual Motion
Groombridge 1830	Ursa Major	6.45	7″.05
Lacaille 9352	Piscis Aus.*	7.34	6.90
61 Cygni	Cygnus	4.84	5.22
Lalande 21185	Ursa Major	7.49	4.77
ϵ Indi	Indus	4.69	4.70
o^2 Eridani	Eridanus	4.43	4.08
μ Cassiopeiae	Cassiopeia	5.17	3.75
α Centauri	Centaurus	-0.27	3.68
Lacaille 8760	Microscopium	6.68	3.46
82 G. Eridani	Eridanus	4.27	3.14
268 G. Ceti	Cetus	5.82	2.32
Arcturus	Bootes	-0.04	2.28

*In the year 2111, this star will reach the boundary of Piscis Austrinus and enter Sculptor.

The transformation is accomplished with the following equations, in which r is the star's distance in parsecs, and Δr its radial velocity in parsecs per year, found by dividing the radial velocity listed in the catalogue by the conversion factor 977,820. Also, $\Delta \alpha$ and $\Delta \delta$ are the proper-motion components in radians per year, found by dividing the listed $\mu(\alpha)$ by 13,751 and $\mu(\delta)$ by 206,265.

$$x = r \cos \delta \cos \alpha$$
$$y = r \cos \delta \sin \alpha$$
$$z = r \sin \delta,$$

and

$$\Delta x = (x/r)\Delta r - z\Delta\delta \cos \alpha - y\Delta\alpha$$
$$\Delta y = (y/r)\Delta r - z\Delta\delta \sin \alpha + x\Delta\alpha$$
$$\Delta z = (z/r)\Delta r + r\Delta\delta \cos \delta.$$

The six numerical constants may now be inserted in the following simple expressions for the star's rectangular coordinates x', y', z' at any other time:

$$x' = x + t\Delta x$$
$$y' = y + t\Delta y$$
$$z' = z + t\Delta z,$$

where t is the number of years in the past (negative) or future (positive). Finally, r' is the square root of the sum of the squares of these three numbers, and they can all be put back in the first three equations to solve for α and δ at time t.

Table VII lists the stars brighter than magnitude 8.05 that have the largest proper motions. At the head of the list is the 1,830th star in the catalogue of the English amateur Stephen Groombridge (1755-1832), whose ambitious project to measure positions with a Troughton transit circle was undertaken at age 51 and continued 20 years. Not until 1842 was the importance of this star recognized, when F. W. Argelander at Bonn Observatory noticed it had moved considerably. Even today, only two faint stars (Barnard's and Kapteyn's) are known to exhibit larger proper motions.

From the data for Groombridge 1830 on page 299, we see by its negative radial velocity that it is approaching. The star's rectangular coordinates in the year 2000 are $x = -6.958$, $y = +0.213$, and $z = +5.384$ parsecs. Note that the square root of the sum of the squares equals the star's current distance, 8.8 parsecs.

The equations for space motion show that it will continue to approach for another 8,800 years, when its distance will be 8.3 parsecs. Thereafter the star will draw away. In 100,000 years it will reach 30 parsecs and appear as a 9.1-magnitude star in Lupus. And as Joseph Ashbrook wrote in 1974, "One million years into the future, Groombridge's star will have receded into insignificance, being a 14th-magnitude object lost in the rich Milky Way of the southern constellation Norma.''

Other stars with negative radial velocities will also approach and pass the Sun. For example, α Centauri now lies at 1.3 parsecs, but 27,000 years hence it will appear twice as bright in Hydra and only 0.9 parsec (3 light-years) away.

When our Stone Age predecessors were making flint tools and campfires 330,000 years ago, Sirius stood near the Beehive cluster in Cancer, with only a ninth the radiance it has today. At that remote time, perhaps Aldebaran was the brightest star of the night sky, high in Cepheus at magnitude −1.4. Or might there be some other star, listed on a page of this catalogue, that outshone even Aldebaran?

We have assumed these stars remain constant in intrinsic brightness, and that they move in straight lines through space. After a few million years, however, their paths will increasingly show the effects of orbital motion around the Milky Way.

EDITING OF THE CATALOGUE

Apart from exceptions described here, the numerical values and star designations printed on these pages are directly traceable through a series of computer interfaces back to the SKYMAP magnetic tapes. As editors of this work, we have personally done all the programming needed to select the stars brighter than magnitude 8.05, sort them by right ascension, convert units and round off when appropriate, and finally to format the pages. In the course of these mechanical operations, we have inserted in the Notes column the 200 classical and other names of important nonvariable stars.

About 1,000 revisions have also been made to specific numerical values in the catalogue. These were suggested by computer scans of our disk file of the catalogue, during which we made specialized printouts to look for various kinds of "extremes."

For example, we have replaced all distances and absolute magnitudes of stars closer than 10 parsecs with the corresponding values from W. Gliese's 1969 *Catalogue of Nearby Stars.* For these particular stars, it appeared to us that a strictly trigonometric parallax was more reliable than SKYMAP's combination with a spectroscopic parallax.

We also made a listing of stars with the greatest and least absolute magnitudes, but found no unreasonable values. A list of all stars beyond 3,000 parsecs was checked for the presence of low-luminosity stars, but none were found.

A printout of **B − V** values greater than + 2.0 or less than − 0.2 turned up several discordant values, which we corrected. Most of the "unusual" values occurred for faint stars whose **B** and **V** had been derived from magnitudes determined by photographic photometry.

In another approach to data checking, we selected a realistic **B − V** range for each spectral class and had the computer print out exceptions to this scheme. For several of these stars, a more reasonable spectral type was inserted from the recent catalogues of N. Houk and A. P. Cowley (1975-78) and W. Buscombe (1980), and the absolute magnitude and distance revised accordingly. Most of the discordant entries, however, could be attributed to interstellar reddening.

SKYMAP Error Report No. 20, kindly provided by S. F. McLaughlin shortly before publication, indicated a problem with the hundreds digit in some of the radial velocities. Our printout of those exceeding ± 100 km/sec turned up a few dozen others with the same problem; all have been corrected using the 1953 Wilson catalogue.

The proper-motion components in SKYMAP are for the 1950 equator. For most stars these components are numerically the same in 2000, to the catalogue's tabular precision. But we have recomputed the positions and proper-motion components of stars within 2° of either pole, of selected stars within 10°, and of all stars whose proper motions exceed 2″ per year, and revised the catalogue where necessary.

Whenever the same ADS or variable star designation was attached in SKYMAP to two or more adjacent stars, we consulted various sources to resolve the confusion, if any. When two components of a clearly physical binary were listed as having somewhat different distances, we have used the simple mean for each.

The components of certain bright double stars, either fixed or showing very slow orbital motion, are traditionally catalogued separately, but were merged in SKYMAP. Such pairs usually had the correct combined magnitude, but often the HD number of one component and the SAO number of the other. We have "unmerged" the components of β Tucanae, γ Arietis, γ Andromedae, θ Eridani, θ¹ Orionis, γ Volantis, γ Leonis, α Crucis, μ Crucis, α Canum Venaticorum, ζ Ursae Majoris, α Centauri, π Bootis, ζ Coronae Borealis, ξ Lupi, α Herculis, γ Delphini, 61 Cygni, μ Cygni, and ζ Aquarii. A. Wallenquist's 1954 catalogue was of great help in assigning magnitudes to the components. The proper motions we give for α Centauri and 61 Cygni refer to the center of gravity.

The Bayer and Flamsteed designations of stars have a muddled history, and we found the SKYMAP tapes remarkably accurate in reflecting the consensus of earlier usage rather than the idiosyncrasies of any particular source. For discrepancies in nomenclature between SKYMAP and the Becvar *Atlas Catalogue* we followed D. Hoffleit's helpful 1979 paper.

We have not attempted to resolve the knotty problem of "duplicates," which afflicts any star catalogue that has been merged from two or more others. The earlier catalogues almost never agree *exactly* as to a star's position, and some algorithm must be used to decide whether two close matches are the same or two distinct stars. Complicating this process is the fact that the earlier catalogues seldom purport to be complete, or to reach the same magnitude limit.

For example, midway down on page 376 are the stars HD 138506 and SAO 225913. They are 1′ apart on the sky, but quite similar in magnitude and spectral type. It is highly suspicious that one should lack an SAO number and the other an HD number. No doubt somewhere down the road one of them was recorded in some earlier catalogue with an error in the position. A missing HD or SAO number for any star is a warning to treat the data cautiously.

In such cases it is tempting to adopt the SAO position and reject the other, but to have done so in wholesale fashion would only improve the present work cosmetically. Each instance of this kind calls for careful research or reobservation.

REFERENCES

SKYMAP sources. These works, in machine-readable form, were used to compile those portions of SKYMAP that have been excerpted for *Sky Catalogue 2000.0.*

Aitken, R. G., *New General Catalogue of Double Stars within 120° of the North Pole,* **1** and **2**, Washington, D. C., 1932: Carnegie Institution of Washington.

Blanco, V. M., S. Demers, G. G. Douglass, and M. P. Fitz-Gerald, *Publications* of the U. S. Naval Observatory, 2nd Series, **21**, Washington, D. C., 1968.

Cannon, A. J., and E. C. Pickering, *The Henry Draper Catalogue* and *The Henry Draper Extension,* Harvard College Observatory

Annals, **91-99**, **100**, and **112**, Cambridge, Mass., 1918-24, 1936, 1949. [HD]

Hoffleit, D., *Catalogue of Bright Stars*, New Haven, Conn., 1964: Yale University Observatory.

Jaschek, C., H. Conde, and C. C. de Sierra, *Publications* of La Plata Observatory, Ser. Astr., **28**, 1964.

Jenkins, L. F., *General Catalogue of Trigonometric Stellar Parallaxes*, New Haven, Conn., 1952: Yale University Observatory.

Kukarkin, B. V., *et al.*, *General Catalogue of Variable Stars*, 3rd edition, Moscow, 1969-71: Soviet Academy of Sciences.

Mermilliod, J., *Bulletin d'Information du Centre de Données Stellaires de Strasbourg*, **13**, 90, 1973.

Smith, C., *Astronomische Gesellschaft Katalog 3*, magnetic tape, 1977. [AGK3]

Smithsonian Institution, *Smithsonian Astrophysical Observatory Star Catalog*, Washington, D. C., 1966, 1971: Smithsonian Institution. [SAO]

U. S. Naval Observatory, *Index Catalogue of Visual Double Stars*, magnetic tape, 1974-75. [USNO]

Wilson, R. E., *General Catalogue of Stellar Radial Velocities*, Washington, D. C., 1953: Carnegie Institution of Washington.

The order of precedence adopted by the SKYMAP compilers is as follows, the chief source in each category being listed first:

Star numbers — HD, SAO
Star names (including variable star names) — Hoffleit, Kukarkin
Positions — SAO, AGK3, HD
Proper motions — SAO, AGK3
Photometric data — Mermilliod, Blanco, SAO, HD
Spectral types — Jaschek, Blanco, Mermilliod, HD, SAO
Radial velocities — Wilson
Trigonometric parallaxes — Jenkins
Multiple star data — USNO, Aitken

Other references. These books and articles cited in the Introduction can aid in interpreting and using the catalogue data.

Allen, C. W., *Astrophysical Quantities*, 3rd edition, London, 1973: Athlone Press.

Ashbrook, J., "The Story of Groombridge 1830," *Sky and Telescope*, **47**, 296, May, 1974.

Becvar, A., *Atlas of the Heavens — II: Catalogue 1950.0*, 4th edition, Prague and Cambridge, Mass., 1964: Czechoslovak Academy of Sciences.

Buscombe, W., *Fourth General Catalogue of MK Spectral Classifications*, Evanston, Ill., 1980: Dearborn Observatory.

Davis, G. A., Jr., "The Pronunciations, Derivations, and Meanings of a Selected List of Star Names," *Popular Astronomy*, **52**, 8, January, 1944.

Delporte, E., *Délimitation Scientifique des Constellations*, Cambridge, 1930: Cambridge University Press.

Fricke, W., and A. Kopff, *Fourth Fundamental Catalogue*, Veröffentlichungen des Astronomischen Rechen-Instituts, **10**, Karlsruhe, 1963: G. Braun. [FK4]

Gliese, W., *Catalogue of Nearby Stars*, Veröffentlichungen des Astronomischen Rechen-Instituts, **22**, Karlsruhe, 1969: G. Braun.

Gottlieb, D. M., "SKYMAP: A New Catalog of Stellar Data," *Astrophysical Journal Supplement Series*, **38**, 287, November, 1978.

Hack, M., and O. Struve, *Stellar Spectroscopy: Normal Stars*, and *Stellar Spectroscopy: Peculiar Stars*, Trieste, 1969-70: Osservatorio Astronomico de Trieste.

Harris, D. L., "The Stellar Temperature Scale and Bolometric Corrections," Chapter 14 in *Basic Astronomical Data*, K. Aa. Strand, editor, Chicago, 1963: University of Chicago Press.

Hauck, B., "The Distance Modulus of the Hyades, Coma Berenices, and Praesepe Clusters," *Astronomy and Astrophysics*, **99**, 207, June, 1981.

Hoffleit, D., "Discordances in Star Designations," *Bulletin d'Information du Centre de Données Stellaires de Strasbourg*, **17**, 38, 1979.

Houk, N., and A. P. Cowley, *University of Michigan Catalogue of Two-Dimensional Spectral Types for HD Stars*, **1** and **2**, Ann Arbor, Mich., 1975-78.

Jeffers, H. M., W. H. van den Bos, and F. M. Greeby, *Index Catalogue of Visual Double Stars, 1961.0*, Mount Hamilton, Calif., 1963: Lick Observatory.

Johnson, H. L., "The Fifty Brightest Stars," *Sky and Telescope*, **16**, 470, August, 1957.

Johnson, H. L., "Photometric Systems," Chapter 11 in *Basic Astronomical Data*, K. Aa. Strand, editor, Chicago, 1963: University of Chicago Press.

Johnson, H. L. "Infrared Photometry of M-Dwarf Stars," *Astrophysical Journal*, **141**, 170, January 1, 1965.

Johnson, H. L., and W. W. Morgan, "Fundamental Stellar Photometry for Standards of Spectral Type on the Revised System of the Yerkes Spectral *Atlas*," *Astrophysical Journal*, **117**, 313, May, 1953.

Keenan, P. C., "Classification of Stellar Spectra," Chapter 8 in *Basic Astronomical Data*, K. Aa. Strand, editor, Chicago, 1963: University of Chicago Press.

Miczaika, G. R., and W. M. Sinton, *Tools of the Astronomer*, Cambridge, Mass., 1961: Harvard University Press.

Mihalas, D., and J. Binney, *Galactic Astronomy*, 2nd edition, San Francisco, 1981: W. H. Freeman.

Morgan, W. W., P. C. Keenan, and E. Kellman, *An Atlas of Stellar Spectra*, Chicago, 1943: University of Chicago Press.

Newcomb, S., *A Compendium of Spherical Astronomy*, New York, 1906: Macmillan. (Also, Dover, 1960.)

Sharpless, S., "Interstellar Reddening," Chapter 12 in *Basic Astronomical Data*, K. Aa. Strand, editor, Chicago, 1963: University of Chicago Press.

Tirion, W., *Sky Atlas 2000.0*, Cambridge, Mass., 1981: Sky Publishing Corp.

Trumpler, R. J., *Lick Observatory Bulletin*, **14**, 154, 1930; also, *Publications* of the Astronomical Society of the Pacific, **42**, 214 and 267, 1930.

Wallenquist, A., *A General Catalogue of Differences in Magnitudes of Double Stars*, Uppsala Astronomiska Observatoriums Annaler, **4**, No. 2, Uppsala, 1954: Norblads Bokhandel.

Weaver, H. F., "The Development of Astronomical Photometry," *Popular Astronomy*, **54**, 211, 287, 339, 389, 451, 504, May-December, 1946.

LETTERS OF THE GREEK ALPHABET

A, α	Alpha	H, η	Eta	N, ν	Nu	T, τ	Tau
B, β	Beta	Θ, θ	Theta	Ξ, ξ	Xi	Y, υ	Upsilon
Γ, γ	Gamma	I, ι	Iota	O, o	Omicron	Φ, ø	Phi
Δ, δ	Delta	K, κ	Kappa	Π, π	Pi	X, χ	Chi
E, ε	Epsilon	Λ, λ	Lambda	P, ρ	Rho	Ψ, ψ	Psi
Z, ζ	Zeta	M, μ	Mu	Σ, σ	Sigma	Ω, ω	Omega

Indexes to Named Stars

KEY TO CONSTELLATION ABBREVIATIONS

And	Andromeda, Andromedae	Del	Delphinus, Delphini	Per	Perseus, Persei
Ant	Antlia, Antliae	Dor	Dorado, Doradus	Phe	Phoenix, Phoenicis
Aps	Apus, Apodis	Dra	Draco, Draconis	Pic	Pictor, Pictoris
Aql	Aquila, Aquilae	Equ	Equuleus, Equulei	PsA	Piscis Austrinus, Piscis Austrini
Aqr	Aquarius, Aquarii	Eri	Eridanus, Eridani	Psc	Pisces, Piscium
Ara	Ara, Arae	For	Fornax, Fornacis	Pup	Puppis, Puppis
Ari	Aries, Arietis	Gem	Gemini, Geminorum	Pyx	Pyxis, Pyxidis
Aur	Auriga, Aurigae	Gru	Grus, Gruis	Ret	Reticulum, Reticuli
Boo	Bootes, Bootis	Her	Hercules, Herculis	Scl	Sculptor, Sculptoris
Cae	Caelum, Caeli	Hor	Horologium, Horologii	Sco	Scorpius, Scorpii
Cam	Camelopardalis, Camelopardalis	Hya	Hydra, Hydrae	Sct	Scutum, Scuti
Cap	Capricornus, Capricorni	Hyi	Hydrus, Hydri	Ser	Serpens, Serpentis
Car	Carina, Carinae	Ind	Indus, Indi	Sex	Sextans, Sextantis
Cas	Cassiopeia, Cassiopeiae	Lac	Lacerta, Lacertae	Sge	Sagitta, Sagittae
Cen	Centaurus, Centauri	Leo	Leo, Leonis	Sgr	Sagittarius, Sagittarii
Cep	Cepheus, Cephei	Lep	Lepus, Leporis	Tau	Taurus, Tauri
Cet	Cetus, Ceti	Lib	Libra, Librae	Tel	Telescopium, Telescopii
Cha	Chamaeleon, Chamaeleontis	LMi	Leo Minor, Leonis Minoris	TrA	Triangulum Australe, Trianguli Australis
Cir	Circinus, Circini	Lup	Lupus, Lupi	Tri	Triangulum, Trianguli
CMa	Canis Major, Canis Majoris	Lyn	Lynx, Lyncis	Tuc	Tucana, Tucanae
CMi	Canis Minor, Canis Minoris	Lyr	Lyra, Lyrae	UMa	Ursa Major, Ursae Majoris
Cnc	Cancer, Cancri	Men	Mensa, Mensae	UMi	Ursa Minor, Ursae Minoris
Col	Columba, Columbae	Mic	Microscopium, Microscopii	Vel	Vela, Velorum
Com	Coma Berenices, Comae Berenices	Mon	Monoceros, Monocerotis	Vir	Virgo, Virginis
CrA	Corona Australis, Coronae Australis	Mus	Musca, Muscae	Vol	Volans, Volantis
CrB	Corona Borealis, Coronae Borealis	Nor	Norma, Normae	Vul	Vulpecula, Vulpeculae
Crt	Crater, Crateris	Oct	Octans, Octantis		
Cru	Crux, Crucis	Oph	Ophiuchus, Ophiuchi		
Crv	Corvus, Corvi	Ori	Orion, Orionis		
CVn	Canes Venatici, Canum Venaticorum	Pav	Pavo, Pavonis		
Cyg	Cygnus, Cygni	Peg	Pegasus, Pegasi		

(After each abbreviation is the full constellation name, followed by its genitive or possessive form, which is used with star designations. For example, the brightest star in Gemini is β Geminorum.)

Andromeda

1	o	584
2		584
3		584
4		586
5		586
6		587
7		587
8		589
9		589
10		590
11		590
12		590
13		592
14		594
15		595
16	λ	596
17	ι	596
18		597
19	κ	597
20	ψ	599
21	α	4
22		5
23		6
24	θ	7
25	σ	8
26		8
27	ρ	9
28		11
29	π	14
30	ε	14
31	δ	14
32		15
34	ζ	17
35	ν	18
36		19
37	μ	20
38	η	20
39		22
41		24
42	φ	24
43	β	24
44		24
45		25
46	ξ	28
47		29
48	ω	30
49		31
50	υ	33
51		33
52	χ	34
53	τ	34
55		37
56		38
57	γ¹	41
57	γ²	41
58		42
59		43
60		44
62		46
63		47
64		48
65		48
66		49

Antlia

	α	268
	δ	269
	ε	245
	ζ¹	246
	ζ²	246
	η	257
	θ	251
	ι	280

Apus

	α	359
	β	402
	γ	398
	δ¹	393
	δ²	394
	ε	350
	ζ	418
	η	348
	θ	344
	ι	418
	κ¹	375
	κ²	378

Aquarius

1		522
2	ε	526
3		526
4		528
5		529
6	μ	529
7		531
11		533
12		535
13	ν	537
15		541
16		543
17		544
18		544
19		545
20		545
21		545
22	β	548
23	ξ	550
25		551
26		552
28		560
29		560
31	o	561
32		561
33	ι	562
34	α	562
35		563
38		564
39		565
41		566
42		567
43	θ	567
44		567
45		567
46	ρ	568
47		568
48	γ	568
49		569
50		569
51		569
52	π	570
53		570
55	ζ¹	571
55	ζ²	571
56		572
57	σ	572
58		572
59	υ	573
60		573
62	η	573
63	κ	574
66		576
67		576
68		578
69		578
70		578
71	τ	579
73	λ	580
74		580
76	δ	581
77		581
78		581
81		583
82		584
83		585
86		585
88		586
89		586
90	φ	588
91	ψ¹	588
92	χ	589
93	ψ²	589
94		590
95	ψ³	590
96		590
97		591
98		591
99		592
100		594
101		595
102	ω¹	597
103		598
104		598
105	ω²	598
106		598
107		599
108		601

Aquila

4		460
5		461
8		463
10		467
11		468
12		469
13	ε	468
14		470
15		471
16	λ	472
17	ζ	471
18		472
19		473
20		475
21		476
22		478
23		479
24		479
25	ω	478
26		480
27		480
28		479
29		479
30	δ	483
31		482
32	ν	483
35		484
36		485
37		487
38	μ	487
39	κ	489
41	ι	488
42		489
44	σ	490
45		491
46		491
47	χ	492
48	ψ	493
49	υ	493
50	γ	493
51		496
52	π	494
53	α	496
54	o	496
55	η	497
56		497
57		498
58		498
59	ζ	498
60	β	498
61	φ	499
62		503
63	τ	503
64		505
65	θ	507
66		508
67	ρ	509
68		517
69		517
70		521
71		522

Ara

	α	422
	β	420
	γ	420
	δ	422
	ε¹	409
	ε²	410
	ζ	408
	η	404
	θ	439
	ι	419
	κ	420
	λ	426
	μ	428
	π	425
	σ	424

Aries

1		37
4		36
5	γ¹	38
5	γ²	38
6	β	38
7		38
8	ι	39
9	λ	39
10		41
11		42
12	κ	42
13	α	42
14		43
15		43
16		43
17	η	44
19		44
20		45
21		45
22	θ	46
24	ξ	48
26		50
27		50
29		51
30		52
31		52
32	ν	53
33		54
34	μ	55
35		55
36		55
37	o	55
38		56
39		56
40		57
41		57
42	π	57
43	σ	58
45		59
46	ρ	59
47		60
48	ε	60
49		61
52		62
53		63
54		63
55		63
56		64
57	δ	64
58	ζ	65
59		67
60		67
61	τ	68
62		68
63		68
64		69
65		69
66		70

Auriga

	ψ⁹	165
2		102
3	ι	104
4		104
5		105
6		105
7	ε	106
8	ζ	106
9		107
10	η	107
11	μ	110
13	α	112
14		111
15	λ	113
16		112
17		112
18		113
19		113
20	ρ	114
21	σ	115
22		115
24	φ	117
25	χ	119
26		123
27	o	127
29	τ	128
30	ξ	131
31	υ	129
32	ν	129
33	δ	133
34	β	133
35	π	133
36		134
37	θ	133
38		135
39		136
40		136
41		139
42		142
43		143
44	κ	141
45		145
46	ψ¹	147
47		150
48		149
49		153
50	ψ²	155
51		155
52	ψ³	155
53		155
54		156
55	ψ⁴	158
56	ψ⁵	159
57	ψ⁶	160
58	ψ⁷	162
59		163
60		163
61	ψ⁸	164
62		167
63		173
64		177
65		179
66		180

Bootes

1		336
2		336
3		338
4	τ	338
5	υ	339
6		339
7		340
8	η	341
9		342
10		342
11		343
12		346
13		345
14		347
15		347
16	α	348
17	κ	347
18		349
19	λ	348
20		349
21	ι	348
22		351
23	θ	351
24		352
25	ρ	353
26		354
27	γ	353
28	σ	354
29	π¹	356
29	π²	356
30	ζ	356
31		357
32		357
33		356
34		357
35	o	358
36	ε	358
37	ξ	360
38		359
39		360
40		363
41	ω	364
42	β	364
43	ψ	365
44		365
45		366
46		366
47		365
48	χ	369
49	δ	369
50		371
51	μ¹	372
51	μ²	372
52	ν¹	375
53	ν²	375
54	φ	377

Caelum

	α	97
	β	98
	γ	106
	δ	94
	ζ	100

Camelopard.

	γ	79
1		94
2		97
3		97
4		100
5		103
7		104
8		105
9	α	102
10	β	106
11		107
12		107
14		110
15		113
16		115
17		118
18		119
19		122
23		126
24		125
26		127
29		129
30		130
31		131
36		140
37		138
40		141
42		162
43		163
47		179
49		193
51		193
53		202

Cancer

1		199
2	ω	201
3		201
4		202
5		202
8		204
9		204
10	μ	205
12		205
14	ψ	206
15	ψ	208
16	ζ	207
17	β	209
18	χ	211
19	λ	211
20		213
21		213
22	φ¹	214
23	φ²	215
24		214
25		214
27		215
28		216
29		216
30	υ¹	217
31	θ	217
32	υ²	218
33	η	218
34		218
35		219
36		220
37		221
39		222
41	ε	222
43	γ	224
45		224
46		225
47	δ	224
48	ι	225
49		224
50		226
51	σ¹	228
53		228
54		228
55	ρ¹	228
57		229
58	ρ²	230
59	σ²	231
60		230
61		231
62	o	231
63		231
64	σ³	232
65	α	231
66	σ⁴	233
67		233
69	ν	233
70		234
72	τ	235
75		236
76	κ	235
77	ξ	236
79		236
81		237
82	π	239

Canes Venat.

2		307
3		308
4		310
5		310
6		311
7		312
8	β	314
9		315
10		317
11		318
12	α¹	321
12	α²	321
14		324
15		325
17		325
19		327
20		328
21		328
23		329
24		334
25		335

Canis Major

	λ	149
1	ζ	144
2	β	145
4	ξ¹	151
5	ξ²	153
6	ν¹	154
7	ν²	154
8	ν³	155
9	α	159
10		158
11		160
12		160
13	κ	161
14	θ	164
15		163
16	o¹	164
17		164
18	μ	165
19	π	165
20	ι	165
21	ε	166
22	σ	168
23	γ	169
24	o²	169
25	δ	172
26		174
27		175
28	ω	175
29		177
30	τ	177
31	η	180

Canis Minor

1		181
2	ε	181
3	β	182
4	γ	182
5	η	182
6		183
7	δ¹	185
8	δ²	185
9	δ³	185
10	α	188
11		192
13	ζ	196
14		199

Capricornus

1		508
2	ξ	508
3		510
4		511
5	α¹	511
6	α²	511
7	σ	512
8	ν	513
9	β	513
10	π	516
11	ρ	517
12	o	518
14	τ	522
15	υ	523
16	ψ	525
17		525
18	ω	528
19		530
20		532
22	η	535
23	θ	535
24		536
25	χ	537
27		537
28	φ	540
29		540
30		541

Capricornus

Fl.	Bayer	No.
31		541
32	ι	543
33		544
34	ζ	546
35		546
36		547
37		549
39	ε	550
40	γ	551
41		552
42		552
43	κ	552
44		553
45		553
46		553
47		554
48	λ	554
49	δ	554
51	μ	557

Carina

Bayer	No.
α	146
β	238
ε	212
η	275
θ	274
ι	239
υ	253
χ	198
ω	263

Cassiopeia

Fl.	Bayer	No.
	ι	49
1		585
2		586
4		592
5	τ	599
6		600
7	ρ	602
8	σ	604
9		2
10		3
11	β	4
12		10
13		12
14	λ	12
15	κ	12
16		13
17	ζ	14
18	α	15
19	ξ	15
20	π	16
21		16
22	ο	16
23		17
24	η	17
25	ν	17
26	υ¹	19
27	γ	20
28	υ²	20
30	μ	24
31		24
32		25
33	θ	25
34	φ	27
35		28
36	ψ	29
37	δ	29
38		31
39	χ	32
40		33
42		35
43		35
44		35
45	ε	38
46	ω	38
47		41
48		40
49		41
50		41
52		41
53		41
55		44

Centaurus

Bayer	No.
α¹	356
α²	356
β	344
γ	316
δ	304
ε	335
ζ	341
η	354
ι	329
κ	363
λ	293
μ	339
ν	339
ξ¹	323
ξ²	324
ο¹	292
ο²	292
π	288
ρ	305
σ	311
τ	315
υ¹	342
υ²	343
φ	342
χ	344
ψ	349
1	337
2	339
3	340
4	340
5 θ	345

Cepheus

Fl.	Bayer	No.
	μ	553
1	κ	506
2	θ	517
3	η	525
4		524
5	α	541
6		542
7		546
8	β	547
9		550
10	ν	554
11		552
12		554
13		557
14		560
16		559
17	ξ	561
18		561
19		561
20		561
21	ζ	564
22	λ	564
23	ε	566
24		563
25		567
26		570
27	δ	571
28		570
29	ρ	571
30		575
31		573
32	ι	579
33	π	586
34	ο	590
35	γ	597

Cetus

Fl.	Bayer	No.
1		603
2		2
3		2
6		5
7		6
8	ι	8
9		9
10		10
12		11
13		13
14		13
16	β	16
17	φ¹	16
18		16
19	φ²	18
20		19
21		19
22	φ³	20
23	φ⁴	20
25		22
26		22
27		23
28		23
30		24
31	η	24
32		24
33		24
34		25
37		26
38		26
39		26
42		27
43		28
44		29
45	θ	29
46		29
47		30
48		30
49		32
50		33
52	τ	35
53	χ	36
55	ζ	37
56		38
57		40
59	υ	40
60		41
61		41
63		44
64		43
65	ξ¹	44
66		44
67		45
68	ο	46
69		47
70		47
71		48
72	ρ	49
73	ξ²	49
75		50
76	σ	50
77		52
78	ν	52
80		52
81		53
82	δ	53
83	ε	53
84		54
86	γ	55
87	μ	56
89	π	55
91	λ	61
92	α	61
93		61
94		65
95		67
96	κ	67
97		68

Chamaeleon

Bayer	No.
θ	211
ι	242
κ	303
μ	258
ν	252
π	294
α	210
β	308
γ	271
δ¹	275
δ²	275
ε	301
ζ	247
η	223

Circinus

Bayer	No.
α	357
β	370
γ	372
δ	370
ε	370
ζ	361
η	365
θ	362

Columba

Bayer	No.
α	123
β	129
γ	132
δ	145
ε	118
η	133
θ	137
κ	142
λ	130
μ	127
ν¹	122
ν²	122
ξ	131
ο	112
π¹	137
π²	137
σ	132

Coma Beren.

Fl.	Bayer	No.
2		303
3		305
4		306
5		306
6		307
7		307
8		308
9		308
11		309
12		310
13		310
14		311
15	γ	311
16		311
17		312
18		312
20		312
21		312
22		314
23		314
24		314
25		315
26		315
27		318
28		318
29		319
30		319
31		319
32		320
35		320
36		322
37		322
38		323
39		324
40		324
41		324
42	α	325
43	β	326

Corona Aus.

Bayer	No.
α	474
β	474
γ	472
δ	473
ε	467
ζ	470
η¹	462
η²	462
θ	453
κ¹	453
κ²	453
λ	459
μ	461

Corona Bor.

Fl.	Bayer	No.
1	ο	371
2	η	372
3	β	374
4	θ	375
5	α	376
6	μ	376
7	ζ¹	378
7	ζ²	378
8	γ	379
9	π	380
10	δ	382
11	κ	382
12	λ	384
13	ε	385
14	ι	386
15	ρ	386
16	τ	389
17	σ	391
18	υ	392
19	ξ	394
20	ν¹	394
21	ν²	394

Corvus

Fl.	Bayer	No.
1	α	304
2	ε	305
3		305
4	γ	307
5	ζ	309
6		310
7	δ	312
8	η	313
9	β	314

Crater

Fl.	Bayer	No.
7	α	281
11	β	285
12	δ	288
13	λ	289
14	ε	289
15	γ	290
16	κ	290
21	θ	294
24	ι	294
27	ζ	297
30	η	300

Crux

Bayer	No.
α¹	311
α²	311
β	318
γ	313
δ	307
ε	309
ζ	308
η	304
θ¹	302
θ²	303
ι	317
κ	320
λ	321
μ¹	320
μ²	320

Cygnus

Fl.	Bayer	No.
	χ	495
1	κ	478
2		482
4		483
6	β	485
7		483
8		486
9		487
10	ι	485
11		488
12	φ	490
13	θ	488
14		490
15		492
16		491
17		493
18	δ	493
19		495
20		495
21	η	499
22		498
23		497
24	ψ	498
25		500
26		501
27		504
28		506
29		509
30		508
31	ο¹	509
32	ο²	510
33		508
34		511
35		512
36		512
37	γ	514
39		514
40		516
41		517
42		517
43		516
44		518
45	ω¹	518
46	ω²	518
47		520
48		521
49		523
50	α	523
51		524
52		525
53	ε	525
54	λ	526
55		527
56		528
57		529
58	ν	531
59		532
60		533
61		536
62	ξ	535
63		536
64	ζ	539
65	τ	540
66	υ	541
67	σ	541
68		541
69		545
70		546
71		547
72		549
73	ρ	549
74		550
75		551
76		552
77		552
78	μ¹	553
78	μ²	553
79		553
80	π¹	552
81	π²	554

Delphinus

Fl.	Bayer	No.
1		518
2	ε	519
3	η	520
4	ζ	520
5	ι	521
6	β	521
7	κ	522
8	θ	522
9	α	522
10		523
11	δ	524
12	γ¹	526
12	γ²	526
13		526
14		528
15		527
16		530
17		530
18		532

Dorado

Bayer	No.
α	95
β	120
γ	88
δ	126
ε	129
ζ	107
η¹	136
η²	139
θ	110
κ	99
λ	116
ν	138
π¹	145
π²	147

Draco

Fl.	Bayer	No.
1	λ	292
2		293
3		296
4		312
5	κ	314
6		314
7		318
8		321
9		322
10		339
11	α	344
12	ι	373
13	θ	386
14	η	395
15		396
16		399
17		399
18		401
19		407
20		408
21	μ	411
22	ζ	413
23	β	422
24	ν¹	423
25	ν²	423
26		424
27		422
28	ω	425
30		430
31	ψ	427
32	ξ	433
33	γ	434
34		433
35		430
36		443
37		444
39		448
40		436
41		436
42		449
43	φ	447
44	χ	447
45		453
46		458
47	ο	463
48		466
49		468
50		460
51		471
52	υ	465
53		475
54		476
55		474
57	δ	475
58	π	480
59		474
60	τ	477
61	σ	486
63	ε	494
64		502
65		502
66		504
67	ρ	502
68		507
69		500
71		512
73		518
74		517
75		517
76		524

Equuleus

Fl.	Bayer	No.
1		532
3		535
4		535
5	γ	538
6		538
7	δ	540
8	α	540
9		543
10	β	544

Eridanus

Fl.	Bayer	No.
	α	33
	θ¹	60
	θ²	60
	ι	54
	κ	49
	φ	45
	χ	38
1	τ¹	56
2	τ²	58
3	η	59
4		60
5		60
6		60
7		61
8	ρ¹	61
9	ρ²	61
10	ρ³	61
11	τ³	61
13	ζ	66
14		66
15		67
16	τ⁴	67
17		71
18	ε	72
19	τ⁵	72
20		73
21		74
22		75
23	δ	76
24		76
25		76
26	π	77
27	τ⁶	77
28	τ⁷	78
30		80
32		80
33	τ⁸	80
34	γ	81
35		82
36	τ⁹	82
37		85
38	ο¹	86

Mensa

γ	119
δ	88
ε	181
ζ	156
η	103
θ	165
ι	121
κ	129
μ	98
ν	90
ξ	104
π	122

Microscopium

α	528
β	529
γ	533
δ	535
ε	541
ζ	534
η	535
θ¹	543
θ²	544
ι	527
ν	520

Monoceros

1		133
2		133
3		134
5	γ	141
6		143
7		144
8		146
10		149
11	β	149
12		151
13		152
14		153
15		156
16		159
17		160
18		160
19		169
20		173
21		173
22	δ	173
24		175
25		187
26	α	190
27		200
28		201
29	ζ	205

Musca

α	315
β	318
γ	313
δ	323
ε	307
ζ¹	309
ζ²	309
η	327
θ	325
ι¹	330
ι²	331
λ	297
μ	298

Norma

γ¹	392
γ²	393
δ	388
ε	396
η	387
θ	392
ι¹	387

Norma (continued)

ι²	389
κ	391
λ	393
μ	399

Octans

α	535
β	577
γ¹	601
γ²	603
γ³	4
δ	351
ε	568
ζ	230
η	280
θ	1
ι	321
κ	336
λ	556
μ¹	524
μ²	523
ν	552
ξ	579
π¹	364
π²	365
ρ	379
σ	537
τ	593
υ	572
φ	448
χ	465
ψ	567
ω	367

Ophiuchus

1	δ	391
2	ε	393
3	υ	396
4	ψ	395
5	ρ	395
7	χ	396
8	φ	397
9	ω	398
10	λ	397
12		399
13	ζ	400
14		401
16		403
19		403
20		405
21		405
23		407
24		408
25	ι	406
26		409
27	κ	408
29		410
30		409
35	η	413
36		415
37		414
39	o	416
40	ξ	418
41		416
42	θ	418
43		419
44		420
45		421
49	σ	420
51		422
52		424
53		424
55	α	424
57	μ	425
58		428
60	β	428
61		428
62	γ	430
64	ν	435
66		436
67		436

Ophiuchus (continued)

68		437
69	τ	437
70		438
71		440
72		440
73		441
74		447

Orion

1	π³	101
2	π²	101
3	π⁴	101
4	o¹	102
5		102
6		103
7	π¹	103
8	π⁵	102
9	o²	103
10	π⁶	104
11		106
13		108
14		108
15		109
16		108
17	ρ	110
18		111
19	β	111
20	τ	112
21		113
22		114
23		114
24	γ	115
25		115
27		115
28	η	115
29		115
30	ψ	116
31		118
32		118
33		118
34	δ	119
35		120
36	υ	119
37	φ¹	120
38		120
39	λ	121
40	φ²	122
41	θ¹	121
42		121
43	θ²	121
44	ι	121
45		121
46	ε	121
47	ω	123
48	σ	123
49		123
50	ζ	124
51		125
52		128
53	κ	128
54	χ¹	131
55		129
56		130
57		131
58	α	131
59		133
60		133
61	μ	135
62	χ²	135
63		136
64		135
66		136
67	ν	137
68		140
69		140
70	ξ	140
71		141
72		141
73		142
74		142
75		142

Pavo

α		515
β		525
γ		545
δ		506
ε		501
ζ		459
η		429
θ		462
ι		441
κ		466
λ		463
μ¹		501
μ²		502
ν		452
ξ		448
o		539
π		440
ρ		521
σ		527
τ		478
υ		524
φ¹		520
φ²		523
ω		467

Pegasus

1		543
2		547
3		550
4		551
5		550
7		552
8	ε	553
9		553
10	κ	553
11		554
12		554
13		555
14		555
15		556
16		557
17		558
18		559
19		560
20		560
21		561
22	ν	562
23		562
24	ι	562
25		563
26	θ	564
27		563
28		564
29	π	564
30		568
31		568
32		568
33		569
34		570
35		570
36		571
37		571
38		571
39		572
40		575
41		575
42	ζ	576
43	o	576
44	η	576
45		577
46	ξ	578
47	λ	578
48	μ	579
49	σ	580
50	ρ	581
51		582
52		582
53	β	584
54	α	585
55		585
56		585
57		586
58		587

Pegasus (continued)

59		587
60		587
61		588
62	τ	590
63		590
64		591
65		591
66		591
67		592
68	υ	592
69		593
70		593
71		595
72		595
73		595
74		596
75		596
77		598
78		598
79		600
80		601
81	φ	601
82		601
84	ψ	603
85		1
86		3
87		4
88	γ	6
89	χ	6

Perseus

	ι	63
	φ	35
1		37
2		37
3		39
4		40
5		43
7		46
8		46
9		47
10		48
11		55
12		54
13	θ	55
14		55
15	η	57
16		57
17		58
18	τ	59
20		58
21		60
22	π	60
23	γ	62
24	ι	60
25	ρ	62
26	β	63
27	κ	63
28	ω	64
29		67
30		66
31		67
32		68
33	α	69
34		71
35	σ	71
36		72
37	ψ	73
38	o	76
39	δ	76
40		75
41	ν	76
42		78
43		81
44	ζ	80
45	ε	81
46	ξ	82
47	λ	84
48		85
49		84
50		85
51	μ	87
52		87

Perseus (continued)

53	90
54	89
55	91
56	91
57	95
58	96
59	98

Phoenix

α	10
β	23
γ	30
δ	31
ε	4
ζ	24
η	16
ι	595
κ	10
λ¹	12
λ²	13
μ	15
ν	26
ξ	15
π	604
ρ	18
σ	599
τ	1
υ	24
φ	38
χ	40
ψ	38

Pictor

α	160
β	127
γ	129
δ	139
ζ	113
η¹	106
η²	107
θ	115
ι	101
λ	98
μ	151
ν	146

Pisces

1		581
2		583
3		583
4	β	584
5		586
6	γ	589
7		590
8	κ	592
9		593
10	θ	593
13		594
14		595
16		596
17	ι	597
18	λ	598
19		599
20		600
21		600
22	π	601
23	γ	62
24		60
25	ρ	62
26	β	63
27	κ	63
28	ω	64
29		67
30		66
31		67
32		68
33	α	69
34		71
35	σ	71
36		72
37	ψ	73
38	o	76
39	δ	76
40		75
41	ν	76
42		78
43		81
44	ζ	80
45	ε	81
46	ξ	82
47	λ	84
48		85
49		84
50		85
51	μ	87
52		87

Pisces (continued)

41		9
42		9
44		10
47		11
48		11
51		12
52		12
53		14
54		14
55		14
57		17
58		17
59		17
60		17
61		17
62		17
63	δ	17
64		17
65		18
66		19
67		20
68		20
69	σ	22
71	ε	22
72		23
73		23
74	ψ¹	23
75		23
77		23
78		24
79	ψ²	24
80		24
81	ψ³	24
82		25
83	τ	25
84	χ	25
85	φ	25
86	ζ	25
87		26
88		26
89		27
90	υ	27
91		28
93	ρ	29
94		29
97		31
98	μ	31
99	η	31
101		32
102	π	33
105		34
106	ν	34
107		35
109		35
110	o	36
111	ξ	38
112		40
113	α	40

Piscis Aus.

	ζ	572
	π	584
	υ	563
5		547
6		548
7		550
8		550
9	ι	553
10	θ	554
12	η	560
13		561
14	μ	563
15	τ	564
16	λ	566
17	β	572
18	ε	575
19		576
21		579
22	γ	580
23	δ	581
24	α	582

Puppis

	ζ	203
	ν	155
	o	194
	π	176
	σ	183
	τ	161
1		191
2		192
3		191
4		192
5		193
6		195
7	ξ	194
8		196
9		196
10		196
11		198
12		200
14		203
15	ρ	205
16		206
18		207
19		207
20		208
21		210
22		213

Pyxis

α	224
β	222
γ	227
δ	230
ε	236
ζ	221
η	220
θ	241
κ	235
λ	242

Reticulum

α	87
β	76
γ	82
δ	82
ε	88
ζ¹	66
ζ²	67
η	90
ι	82
κ	71

Sagitta

1		477
2		482
4	ε	489
5	α	490
6	β	491
7	δ	494
8	ζ	495
9		497
10		498
11		499
12	γ	500
13		500
15		503
16	η	504
18		510

Sagittarius

α	482
β¹	481
β²	481
η	445
θ¹	500
θ²	500
ι	498
κ¹	514

Sagittarius

Fl.	Bayer	Page
	κ²	514
1		442
3		429
4		436
6		436
7		437
9		438
10	γ	439
13	μ	443
14		443
15		444
16		444
18		449
19	δ	447
20	ε	448
21		449
22	λ	450
24		454
25		454
26		458
27	φ	460
28		460
29		462
30		463
32	ν¹	464
33		464
34	σ	465
35	ν²	465
36	ξ¹	467
37	ξ²	467
38	ζ	470
39	ο	471
40	τ	472
41	π	474
42	ψ	477
43		478
44	ρ¹	480
45	ρ²	481
46	υ	480
47	χ¹	482
49	χ³	483
50		483
51		488
52		488
53		490
54		491
55		491
56		493
57		497
58	ω	498
59		499
60		500
61		499
62		502
63		502
65		504

Scorpius

Fl.	Bayer	Page
	ζ	407
	η	414
	θ	425
	ι¹	429
	ι²	431
	κ	427
	μ¹	405
	μ²	405
	ξ	387
1		382
2		383
3		384
4		384
5	ρ	384
6	π	385
7	δ	386
8	β¹	388
8	β²	388
9	ω¹	388
10	ω²	389
11		389
12		390
13		391
14	ν	390
15	ψ	390
16		390
17	χ	391
18		392
19	ο	394
20	σ	394
21	α	397
22		397
23	τ	399
25		403
26	ε	405
27		408
34	υ	422
35	λ	423

Sculptor

Bayer	Page
α	20
β	595
γ	590
δ	600
ε	36
ζ	1
η	11
θ	5
ι	9
κ¹	4
κ²	5
λ¹	15
λ²	16
μ	597
ξ	21
π	35
σ	22
τ	33

Scutum

Bayer	Page
α	454
β	461
γ	451
δ	458
ε	459
ζ	448
η	465

Serpens

Fl.	Bayer	Page
	φ	384
3		369
4		369
5		370
6		371
7		372
8		372
9	τ¹	373
10		374
11		375
12	τ²	375
13	δ	376
14		377
15	τ³	376
16		377
17	τ⁴	377
18	τ⁵	377
19	τ⁶	379
20	χ	379
21	ι	379
22	τ⁷	379
23	ψ	380
24	α	380
25		381
26	τ⁸	380
27	λ	381
28	β	381
31	υ	381
32	μ	382
34	ω	382
35	κ	382
36		382
37	ε	382
38	ρ	382
39		383
40		384
41	γ	384
43		387
44	π	386
45		389
47		389
50	σ	394
53	ν	418
55	ξ	425
56	ο	427
57	ζ	436
58	η	447
59		450
60		451
61		453
63	θ¹	466
63	θ²	466
64		467

Sextans

Fl.	Bayer	Page
4		254
6		254
7		255
8	γ	255
12		257
13		259
14		260
15	α	261
17		261
18		262
19		262
22	ε	264
23		266
25		267
29	δ	269
30	β	269
33		274
35		274
36		275
40		277
41		277

Taurus

Fl.	Bayer	Page
1	ο	69
2	ξ	70
4		71
5		71
6		72
7		73
10		73
11		75
12		75
13		75
14		76
16		76
17		76
18		76
19		76
20		77
21		77
22		77
23		77
25	η	77
27		78
28		78
29		77
30		78
31		79
32		81
33		81
35	λ	82
36		83
37		83
38	ν	83
40		83
41		84
42	ψ	84
43		85
44		86
45		86
46		86
47		87
48		87
49	μ	87
50	ω	88
51		88
52	φ	89
53		89
54	γ	89
56		89
57		89
58		89
59	χ	90
60		90
61	δ¹	90
62		91
63		91
64	δ²	91
65	κ	91
66		91
67		91
68	δ³	91
69	υ	92
70		92
71		92
72		92
73	π	92
74	ε	93
75		93
76		93
77	θ¹	93
78	θ²	93
79		93
80		93
81		94
83		94
85		94
86	ρ	95
87	α	96
88		95
89		96
90		96
91	σ¹	97
92	σ²	97
93		97
94	τ	98
95		98
96		101
97		101
98		104
99		104
102	ι	106
103		108
104		108
105		108
106		108
108		111
109		113
110		115
111		115
112	β	116
113		116
114		117
115		116
116		117
117		117
118		117
119		119
120		120
121		121
122		122
123	ζ	122
125		123
126		124
129		127
130		128
131		127
132		128
133		128
134		129
135		129
136		130
137		130
139		133

Telescopium

Bayer	Page
α	450
δ¹	452
δ²	453
ε	442
ζ	451
η	481
ι	488
κ	464
λ	467
ν	494
ξ	505
ρ	472

Triangulum

Fl.	Bayer	Page
2	α	37
3	ε	41
4	β	43
5		43
6		44
7		45
8	δ	45
9	γ	45
10		46
11		49
12		49
13		49
14		50
15		52

Triang. Aus.

Bayer	Page
α	404
β	384
γ	370
δ	392
ε	377
ζ	396
η	401
θ	399
ι	396
κ	384

Tucana

Bayer	Page
α	567
β¹	12
β²	12
γ	589
δ	570
ε	604
ζ	8
η	603
θ	12
ι	23
κ	26
λ¹	18
λ²	19
ν	572
π	9
ρ	15

Ursa Major

Fl.	Bayer	Page
1	ο	216
2		219
3	π¹	221
4	π²	222
5		229
6		230
8	ρ	233
9	ι	232
11	σ¹	236
12	κ	234
13	σ²	236
14	τ	237
15		236
16		238
17		239
18		239
22		248
23		246
24		247
25	θ	247
26		248
27		251
28		252
29	υ	254
30	φ	255
31		256
32		264
33	λ	264
34	μ	266
35		269
36		270
37		271
38		274
39		274
41		276
42		278
43		277
44		279
45	ω	279
46		279
47		281
48	β	282
49		281
50	α	282
51		282
52	ψ	284
53	ξ	287
54	ν	288
55		288
56		289
57		291
58		292
59		294
60		294
61		295
62		296
63	χ	297
64	γ	299
65		300
66		300
67		302
68		306
69	δ	307
70		309
71		310
73		311
74		312
75		312
76		316
77	ε	320
78		323
79	ζ	330
80		331
81		333
82		335
83		336
84		338
85	η	338
86		340

Ursa Minor

Fl.	Bayer	Page
	λ	416
1	α	50
3		345
4		345
5		352
7	β	360
11		371
13	γ	371
15	θ	375
16	ζ	380
19		390
20		391
21	η	392
22	ε	403
23	δ	423
24		422

Vela

Bayer	Page
γ	206
δ	224
κ	242
λ	235
μ	276
ο	222
φ	256
ψ	246

Virgo

Fl.	Bayer	Page
1	ω	294
2	ξ	297
3	ν	297
4		298
5	β	298
6		300
7		302
8	π	302
9	ο	303
10		305
11		305
12		306
13		308
15	η	308
16		309
17		310
20		313
21		314
25		315
26	χ	315
27		316
29	γ	316
30	ρ	316
31		316
32		317
33		318
34		318
35		318
37		319
38		320
40	ψ	320
41		320
43	δ	321
44		322
46		323
47	ε	323
48		323
49		325
50		325
53		326
54		326
55		327
57		327
59		328
60	σ	328
61		328
63		330
64		330
65		330
66		330
67	α	331
68		331
69		331
70		331
71		332
72		332
73		333
74		333
75		333
76		333
78		333
79	ζ	334
80		334
82		336
83		337
84		337
85		337
86		337
87		338
89		339
90		341
92		341
93	τ	343
95		345
96		345
98	κ	347
99	ι	348
100	λ	349
102	υ	349
104		352
105	φ	352
106		352
107	μ	357
108		358
109		358
110		364

Volans

Bayer	Page
α	233
β	214
γ¹	172
γ²	172
δ	176
ε	205
ζ	190
η	212
θ	221
ι	162
κ¹	211
κ²	211

Vulpecula

Fl.	Bayer	Page
1		477
2		478
3		481
4		483
5		483
6	α	484
7		484
8		484
9		487
10		492
12		496
13		497
14		500
15		501
16		502
17		505
18		507
19		508
20		508
21		509
22		510
23		510
24		510
25		514
26		521
27		521
28		522
29		522
30		525
31		529
32		530
33		532
35		546

Stars to Magnitude 8.0
for the Epoch 2000.0

Column Headings

HD — The star's number in the *Henry Draper Catalogue* (*Harvard Annals,* Vols. 91-99, 1918-24) and *Henry Draper Extension* (Vols. 100 and 102).

SAO — The star's number in the *Smithsonian Astrophysical Observatory Star Catalog* (1966).

Star Name — The Flamsteed number, Bayer letter, and constellation abbreviation, where assigned.

α 2000 — Right ascension for equator, equinox, and epoch 2000.0, in hours, minutes, and seconds of time.

δ 2000 — Declination for equator, equinox, and epoch 2000.0, in degrees, minutes, and seconds of arc.

$\mu(\alpha)$ — Proper motion in right ascension, expressed in seconds of time per year (positive means eastward motion).

$\mu(\delta)$ — Proper motion in declination, expressed in seconds of arc per year (positive means northward motion).

V — The star's visual **V** (yellow) magnitude in the **UBV** photometric system. Two decimal places indicate that the magnitude has been measured photoelectrically. One decimal place indicates that **V** has been derived from photographic photometry. However, if no visual magnitude is available, the star's **B** (blue) magnitude is listed instead, and followed by "B."

B − V — The color index in the **UBV** system. The greater the value of **B − V**, the redder the star's color. The bluest stars have negative color indexes.

M_v — The absolute visual magnitude. This would be the star's brightness if it were located at the standard distance of 10 parsecs (32.6 light-years) from our Sun.

Spec — The star's spectral type, luminosity class, and peculiarity code if any, on the MK system of Morgan, Keenan, and Kellman (1943). If the MK type is not available, the Henry Draper (HD) type is listed.

RV — Radial velocity in kilometers per second. Positive means receding (traveling away from our solar system).

d (pc) — Distance of the star in parsecs. Although a value is listed for almost every star, in most cases it must be regarded as a very rough estimate, even though shown to two significant figures. Appended letter codes tell how the distance was derived. The most reliable values are marked "t" to show that they are based on a trigonometric parallax. A letter "s" denotes spectroscopic parallax. If both of these methods were used, a "ts" appears. An "mn" code means the distance is a minimum value compatible with the star's negligible trigonometric parallax. The code "mx" means the distance is the maximum likely value, in view of the star's large space motion. To convert parsecs to light-years, multiply by 3.26.

ADS — The star's number in R. G. Aitken's *New General Catalogue of Double Stars* (1932). If the components of a binary or multiple system are bright and widely enough separated to be listed individually, the same ADS number appears at each entry.

Notes — Code letters qualify the data on certain stars. An "m" means the star is part of a binary or multiple system, or perhaps that it has a close optical neighbor (in the same line of sight but not physically associated). A "v" indicates the star is known to be variable, while "q" means its variability is questionable or suspected. The notation "d?" flags a few stars whose distance appears to be listed incorrectly, and "cs?" marks those whose color index seems incompatible with the spectral type.

Also appearing in this column are any special names by which a star is known. These include the designations of variable stars that had been assigned through the year 1968 (for example, RR Lyrae, V695 Cygni), and some of the old alphabetic names assigned by Lacaille to bright southern stars when Greek letters had been exhausted in a given constellation. A few stars of unusual interest, such as Groombridge 1830, are identified here. Finally, the classical names of over 180 stars (Sirius, Albireo, and so on) are transcribed here from the authoritative listing by George A. Davis, Jr., a noted star-name expert and Arabic scholar, as published in *Popular Astronomy* for January, 1944. Other common names, in use by navigators, are enclosed in brackets.

HD	SAO	Star Name	α 2000	δ 2000	μ(α)	μ(δ)	V	B−V	M_v	Spec	RV	d(pc)	ADS	Notes
			h m s	° ′ ″	s	″								
224699	73664		0 00 01.1	+38 51 34	0.000	+0.01	6.60	0.00	0.4	B9.5 V		170 s	17157	m
224720	53560		0 00 08.9	+46 56 27	+0.002	+0.03	7.3	0.1	1.4	A2 V	−24	150 s		
224726	147023		0 00 11.6	−0 21 35	+0.005	+0.02	7.3	1.1	0.2	K0 III		230 mx		
224721	73668		0 00 12.7	+38 18 14	0.000	−0.02	6.54	0.96	3.2	G5 IV		33 s		
224724	128523		0 00 15.8	+8 00 26	+0.004	0.00	7.8	1.1	−0.1	K2 III		310 mx		
224742	108955		0 00 17.8	+13 18 45	+0.004	+0.03	7.5	0.4	3.0	F2 V		80 s		
224750	231888		0 00 19.1	−44 17 26	+0.007	−0.11	6.29	0.76	3.0	G3 IV		41 s		m
224757	53562		0 00 23.0	+42 08 28	−0.001	−0.03	7.9	0.1	0.6	A0 V		280 s		
224743	165993		0 00 23.7	−10 27 44	−0.001	+0.04	8.0	0.4	3.0	F2 V		98 s		
224758	91648		0 00 23.8	+26 55 06	+0.003	−0.05	6.46	0.50	4.0	F8 V	0	31 s		
224759	108956		0 00 24.7	+12 16 02	+0.003	+0.02	7.6	1.1	0.2	K0 III		300 s		
224753	20991		0 00 24.9	+67 51 07	+0.002	−0.01	8.0	0.1	1.7	A3 V		170 s		
224763	165995		0 00 26.9	−16 41 49	+0.012	−0.03	7.5	0.5	3.4	F5 V		66 s		
224784	35983		0 00 30.8	+59 33 35	−0.010	−0.02	6.19	1.01		K0	−33			
224780	231892		0 00 32.0	−40 41 27	−0.002	−0.03	7.4	0.9	0.2	K0 III		280 s		
224782	248097		0 00 34.1	−53 05 51	+0.005	−0.01	6.80	0.6	4.4	G0 V		30 s		m
224783	255620		0 00 38.1	−66 40 57	+0.029	−0.01	7.60	0.5	3.4	F5 V		69 s		m
224792	20994		0 00 41.4	+62 10 33	−0.008	−0.05	7.05	0.49	3.7	F6 V		46 s		
224801	53568		0 00 43.6	+45 15 13	+0.002	+0.01	6.38	−0.07		A0 p	−3			CG And, v
224798	192314		0 00 44.2	−27 54 27	0.000	0.00	8.0	1.4	0.2	K0 III		220 s		
—	35986		0 00 50.1	+50 48 51	−0.002	+0.01	8.0	1.7	0.0	B8.5 V		35 s		
224826	20995		0 00 52.8	+66 50 53	−0.003	+0.02	7.2	1.1	−0.1	K2 III	−12	260 s		
224821	248099		0 00 53.0	−50 26 44	−0.002	+0.03	7.6	1.0	−0.1	K2 III		340 s		
224825	20996		0 00 57.4	+68 27 26	0.000	−0.01	8.0	0.1	0.6	A0 V		270 s		
224836	20998		0 01 02.4	+69 36 12	+0.002	0.00	8.0	0.0	0.4	B9.5 V		300 s		
—	21000		0 01 04.2	+69 34 32	−0.002	−0.01	7.80	0.00	0.4	B9.5 V		270 s		m
224834	231895	τ Phe	0 01 04.4	−48 48 36	−0.002	0.00	5.71	0.91	0.3	G8 III		120 s		
224842	231896		0 01 07.0	−41 53 16	+0.001	−0.05	8.0	0.4	3.4	F5 V		82 s		
224850	231898		0 01 17.0	−41 29 15	+0.003	−0.02	7.03	1.55		K5				
224854	10931		0 01 18.3	+70 55 45	−0.005	0.00	7.7	0.4	2.6	F0 V		100 s		
224870	53573		0 01 19.2	+49 58 54	+0.001	−0.01	6.22	0.97	0.2	K0 III	−20	160 s		
224865	248103		0 01 19.9	−50 20 14	0.000	+0.02	5.2	2.3	−0.5	M2 III	+2	53 s		
224862	192323		0 01 21.2	−28 43 31	−0.002	−0.02	7.2	0.8	0.2	K0 III		260 s		
224868	21005		0 01 21.6	+60 50 22	0.000	+0.01	7.26	0.13	−5.8	B0 Ib	−15	2600 s		
224882	73691		0 01 29.0	+30 44 09	−0.001	−0.03	8.00	0.6	2.8	G0 IV	−11	110 s	17167	m
224878	255624		0 01 29.6	−62 55 25	+0.002	0.00	7.6	0.1	1.4	A2 V		170 s		
224894	53579		0 01 33.5	+44 40 30	−0.001	−0.03	6.9	1.1	0.2	K0 III		220 s		
224895	91659		0 01 34.9	+28 25 26	−0.002	−0.03	7.0	1.1	0.0	K1 III	−13	250 s		
224889	258207	θ Oct	0 01 35.8	−77 03 57	−0.016	−0.17	4.78	1.27	−0.1	K2 III	+24	78 s		
224891	10933		0 01 35.9	+72 14 14	−0.010	+0.02	7.5	1.1	0.2	K0 III		260 s		
224893	21009		0 01 36.8	+61 13 23	−0.001	+0.01	5.55	0.41	−2.1	A5 II	−23	260 s		
224896	108967		0 01 38.2	+14 37 56	+0.005	+0.01	7.9	0.9	3.2	G5 IV		88 s		
224890	10934		0 01 39.2	+73 36 45	+0.016	+0.01	6.5	0.2		A m	−8			
224914	231901		0 01 41.5	−40 08 56	+0.003	−0.02	6.8	0.6	3.4	F5 V		40 s		
224907	91660		0 01 42.9	+24 15 12	+0.002	−0.02	6.8	1.1	0.2	K0 III		210 s		
224906	53580		0 01 43.7	+42 22 02	0.000	0.00	6.1	0.0		B9	−11			
224910	166002		0 01 44.9	−16 31 53	+0.006	+0.04	7.9	1.1	−0.1	K2 III		180 mx		
224919	21014		0 01 48.8	+65 27 21	−0.008	+0.07	7.9	0.5	3.4	F5 V		77 s		
224918	21013		0 01 49.1	+66 18 19	+0.004	−0.01	7.4	1.1	0.2	K0 III	−18	250 s		
224926	147041	29 Psc	0 01 49.3	−3 01 39	+0.001	0.00	5.10	−0.12	−1.2	B8 III	+23	180 s		
224936	214931		0 01 55.0	−37 13 48	−0.003	−0.04	7.0	0.3	0.2	K0 III		230 s		
224937	231903		0 01 56.2	−41 36 51	+0.008	−0.03	7.75	0.42	1.9	F3 IV		110 mx		m
224935	147042	30 Psc	0 01 57.5	−6 00 51	+0.003	−0.04	4.41	1.63		M3 IV	−12	30 mn		
224938	21016		0 01 58.3	+66 26 18	−0.002	+0.01	7.27	0.04	0.4	B9.5 V		210 s		
224949	214933		0 02 00.0	−37 54 08	+0.009	−0.04	7.0	1.2	0.2	K0 III		140 mx		
224945	147045		0 02 02.4	−2 45 59	+0.002	−0.01	6.9	0.1	1.7	A3 V		110 s		
224955	53586		0 02 07.1	+44 31 54	−0.001	−0.03	7.8	1.1	−0.1	K2 III		380 s		
224960	166005		0 02 07.2	−14 40 34	−0.001	−0.02	7.40	1.9		S7.3 e	+13			W Cet, v
224930	91669	85 Peg	0 02 10.1	+27 04 56	+0.063	−0.98	5.75	0.67	5.4	G2 V	−36	12 t	17175	m
224969	53589		0 02 11.8	+47 34 39	0.000	+0.01	7.9	0.1	1.4	A2 V		190 s		
224968	53590		0 02 12.5	+49 46 16	0.000	−0.01	8.0	0.8	3.2	G5 IV		88 s		
224974	166007		0 02 16.7	−13 24 27	+0.019	+0.04	7.1	0.5	4.0	F8 V		41 s		
224980	21020		0 02 17.5	+60 42 12	−0.002	−0.01	6.73	1.82	−0.4	K8 III	−25	180 s		
224990	—	ζ Scl	0 02 19.8	−29 42 36			5.02	−0.15	−1.1	B5 V	0	170 s		m
224993	36003		0 02 21.8	+50 30 00	−0.002	+0.01	7.4	0.1	0.6	A0 V		220 s		
224995	128544	31 Psc	0 02 24.1	+8 57 25	0.000	0.00	6.32	0.18		A4 n	+11			
224992	21022		0 02 24.4	+67 09 48	−0.005	−0.01	7.8	0.7	4.4	G0 V		48 s		
225004	128546		0 02 28.6	+8 23 38	−0.002	−0.03	8.0	0.9	3.2	G5 IV		93 s		
225003	128547	32 Psc	0 02 29.6	+8 29 08	−0.006	−0.04	5.63	0.29	3.0	F2 V	+10	34 s		
224999	53592		0 02 30.1	+47 25 24	+0.002	0.00	7.6	0.5	4.0	F8 V		53 s		
225001	108979		0 02 32.1	+16 15 25	−0.003	+0.02	7.2	0.1	0.6	A0 V	−11	210 s		
225009	10937		0 02 36.0	+66 05 56	+0.002	+0.01	5.86	1.09	0.3	G8 III	−18	100 s	1	m
225010	10938		0 02 38.3	+66 06 00	+0.001	−0.01	7.34	0.08	1.4	A2 V	−7	150 s		
225021	21026		0 02 40.0	+51 28 21	+0.004	0.00	8.0	0.9	3.2	G5 IV		91 s		
225019	1		0 02 41.6	+82 58 24	−0.008	0.00	7.2	0.1	0.6	A0 V		210 s		

HD	SAO	Star Name	α 2000	δ 2000	μ(α)	μ(δ)	V	B-V	M$_v$	Spec	RV	d(pc)	ADS	Notes
			h m s	° ′ ″	s	″								
225022	36007		0 02 42.2	+42 55 15	−0.004	−0.01	7.8	0.1	0.6	A0 V		270 s		
225023	53596		0 02 45.8	+35 48 56	0.000	0.00	7.30	0.1	0.6	A0 V	−2	220 s	8	m
225020	2		0 02 46.8	+80 16 57	+0.021	+0.03	7.90	0.4	3.0	F2 V		93 s	5	m
225028	108983		0 02 47.1	+2 07 48	+0.005	−0.10	7.80	0.6	4.4	G0 V		48 s	9	m
225059	128552		0 02 57.0	−0 53 37	+0.001	−0.01	8.00	1.1	0.2	K0 III		360 s		m
225045	166014		0 02 57.5	−20 02 46	+0.008	+0.07	6.25	0.53	4.0	F8 V		27 s		
225061	192337		0 03 00.2	−34 44 41	−0.001	+0.01	7.7	0.5	3.2	G5 IV		81 s		
225062	248107		0 03 00.8	−60 41 31	−0.001	0.00	7.9	0.5	0.2	K0 III		340 s		
225069	166016		0 03 07.6	−24 08 43	+0.003	0.00	6.5	0.5	3.2	G5 IV		46 s		
225068	36018		0 03 08.3	+42 44 53	+0.004	0.00	6.7	0.1	1.4	A2 V		120 s		
225076	166017		0 03 10.6	−24 35 19	+0.004	0.00	7.96	0.42	3.1	F3 V		89 s		
225072	36020		0 03 10.9	+43 38 35	+0.001	−0.01	8.0	1.1	0.2	K0 III		370 s		
225073	91678		0 03 13.3	+17 33 09	+0.005	−0.05	6.6	1.1	0.2	K0 III		190 s		
225086	128557		0 03 20.4	−4 08 45	−0.002	−0.01	8.00	0.4	3.0	F2 V		100 s		m
—	21039		0 03 21.9	+59 04 44	0.000	+0.02	7.8	2.0	−0.3	K5 III		220 s		
225093	4020		0 03 23.6	+73 10 29	−0.008	+0.02	7.6	0.1	1.4	A2 V	−14	160 s		
225094	10942		0 03 25.6	+63 38 32	−0.001	+0.06	6.24	0.33	−6.8	B3 Ia	−43	2100 s		
225097	73716		0 03 26.0	+20 39 57	+0.005	−0.15	7.6	0.7	4.4	G0 V		44 s		
225101	192342		0 03 26.4	−36 15 03	+0.008	+0.02	6.9	0.4	2.6	F0 V		62 s		
225095	21040		0 03 27.3	+55 33 03	+0.003	−0.01	7.91	−0.02		B1 e				
225106	91679		0 03 30.3	+19 22 53	+0.001	−0.03	7.9	1.1	0.2	K0 III		340 s		
225107	91680		0 03 30.5	+12 21 24	−0.001	0.00	7.2	0.2	2.1	A5 V		110 s		
225105	36023		0 03 33.2	+42 21 07	+0.007	−0.05	7.6	1.1	0.2	K0 III		210 mx		
225120	—		0 03 37.4	−31 05 36			7.7	0.1	3.4	F5 V		73 s		
225127	108992		0 03 40.1	+3 54 24	−0.001	−0.02	8.0	0.4	2.6	F0 V		120 s		
225124	36025		0 03 41.9	+48 26 14	−0.001	0.00	6.8	0.1	1.4	A2 V		120 s		
225126	91681		0 03 42.5	+13 21 55	+0.002	−0.01	8.0	1.1	0.2	K0 III		360 s		
225132	147059	2 Cet	0 03 44.3	−17 20 10	+0.002	0.00	4.55	−0.05	−0.3	B9 IV	−5	91 s		
225125	36027		0 03 44.6	+42 01 25	+0.010	+0.02	7.8	0.1	1.7	A3 V		67 mx		
225136	10948		0 03 51.8	+66 42 44	+0.003	0.00	6.29	1.67	−0.5	M2 III	+15	200 s		
225137	21046		0 03 52.5	+57 22 42	+0.003	+0.01	8.00	0.2		A m				m
225155	166025		0 03 53.4	−28 23 38	+0.010	−0.02	7.6	0.4	3.2	G5 IV		77 s		
225151	91682		0 03 56.2	+14 22 45	+0.007	−0.01	7.3	0.4	2.6	F0 V		87 s		
225167	214947		0 03 57.8	−48 07 30	+0.002	+0.02	7.80	0.4	2.6	F0 V		100 s		m
225161	91683		0 04 00.2	+12 08 44	+0.001	−0.02	7.20	0.4	2.6	F0 V		83 s	26	m
225169	258208		0 04 00.7	−80 23 40	+0.020	+0.02	7.9	0.1	2.1	A5 V		140 s		
225171	36029		0 04 09.2	+49 59 18	0.000	0.00	8.00	1.1	−0.1	K2 III		380 s		m
225181	91684		0 04 10.1	+17 26 11	−0.003	−0.02	7.9	0.5	3.4	F5 V		78 s		
225172	36030		0 04 10.1	+49 52 14	−0.001	+0.01	7.4	1.1	0.2	K0 III		260 s		
225187	192345		0 04 11.7	−30 08 05	0.000	+0.01	7.15	−0.09		B8				
225180	10954	9 Cas	0 04 13.5	+62 17 16	−0.001	+0.01	5.88	0.30	−0.6	A0 III	−18	120 s		m
225179	10955		0 04 15.8	+65 09 14	0.000	+0.05	7.59	1.68	−0.3	K5 III		300 s		
225197	147064		0 04 19.6	−16 31 44	+0.002	−0.06	5.78	1.10	0.2	K0 III	−27	110 s		
225200	166031		0 04 20.2	−29 16 07	+0.002	+0.02	6.39	0.01	0.6	A0 V n		140 s		
225190	21050		0 04 21.2	+54 16 29	0.000	0.00	7.81	−0.08	−1.6	B3.5 V		660 s		
225206	166032		0 04 22.9	−29 22 53	+0.002	+0.01	7.76	−0.05	0.2	B9 V n		330 s		
225212	147066	3 Cet	0 04 30.0	−10 30 35	−0.001	−0.01	5.1	1.1	−4.4	K3 Ib	−42	780 s		
225233	255629		0 04 30.7	−72 53 52	+0.004	−0.05	7.31	0.44		F5				
225218	36037		0 04 36.6	+42 05 32	0.000	−0.02	6.10	0.1	1.4	A2 V	−8	87 s	30	m
225229	—		0 04 37.1	−31 10 36			7.8	1.9	0.2	K0 III		97 s		
225251	231916		0 04 38.2	−51 34 59	+0.001	+0.04	7.4	1.0	0.2	K0 III		270 s		
225252	255630		0 04 38.9	−70 24 15	+0.011	−0.01	7.7	0.8	3.2	G5 IV		78 s		
225220	53617		0 04 40.0	+34 15 53	−0.001	−0.03	8.00	1.1	0.2	K0 III		360 s	32	m
225241	91690		0 04 41.3	+14 57 55	−0.002	−0.01	7.3	0.5	3.4	F5 V		59 s		
225253	255631		0 04 41.4	−71 26 13	+0.008	−0.01	5.59	−0.12	−1.2	B8 III	−3	230 s		
225216	10956		0 04 41.9	+67 10 00	+0.016	+0.04	5.67	1.07	0.0	gK1	−27	140 s		
225215	4028		0 04 43.3	+70 29 08	+0.004	+0.01	8.0	1.1	0.2	K0 III		330 s		
225268	231917		0 04 43.9	−56 50 28	+0.001	+0.04	7.8	1.0	0.2	K0 III		330 s		
225239	53622		0 04 53.5	+34 39 36	+0.061	+0.10	6.12	0.62	4.7	G2 V	+4	21 ts		
225257	21062		0 04 54.9	+58 31 55	+0.001	−0.01	6.62	0.04	−1.6	B3.5 V	+11	340 s	36	m
225283	—		0 04 55.0	−32 00 36			7.9	1.1	0.2	K0 III		320 s		
225276	73731		0 04 55.9	+26 38 56	+0.008	−0.01	6.25	1.40		K2	−5		42	m
225261	73730		0 04 56.3	+23 16 11	+0.028	0.00	7.9	0.9	3.2	G5 IV		42 mx		
225272	10960		0 05 00.0	+65 25 33	+0.015	+0.07	7.5	1.1	0.2	K0 III	+22	170 mx		
225292	73733		0 05 01.0	+27 40 30	+0.005	0.00	6.5	0.9	0.3	G8 III	+12	180 s		
225291	—		0 05 02.0	+45 40 24			7.7	0.5	4.0	F8 V	−17	54 s	41	m
225297	192349		0 05 02.5	−36 00 54	+0.017	+0.07	7.74	0.54	3.4	F5 V		65 s		
6	128569		0 05 03.7	−0 30 12	+0.003	−0.05	6.29	1.10	0.2	gG9	+14	130 s		
225289	10962		0 05 06.1	+61 18 51	+0.002	+0.01	5.80	−0.09	−1.9	B6 III	+14	300 s		
23	231920		0 05 07.4	−52 09 07	−0.013	−0.13	7.1	0.5	4.4	G0 V		35 s		
1	10963		0 05 08.7	+67 50 22	0.000	−0.04	7.7	1.1	0.2	K0 III		280 s		
4	53627		0 05 09.0	+30 19 45	−0.004	−0.02	7.78	0.43		F0			47	m
3	36042		0 05 09.8	+45 13 45	−0.001	−0.01	6.50	0.1	0.6	A0 V	−18	150 s	46	m
24	248112		0 05 09.9	−62 50 43	+0.011	+0.02	8.0	0.7	4.4	G0 V		53 s		
21	192350		0 05 10.2	−32 28 08	−0.005	−0.06	7.1	1.2	0.2	K0 III		190 s		

HD	SAO	Star Name	α 2000	δ 2000	μ(α)	μ(δ)	V	B-V	M_v	Spec	RV	d(pc)	ADS	Notes
17	53630		0ʰ05ᵐ13.0ˢ	+35°34'17"	+0.001ˢ	-0.02"	6.87	0.16	0.0	A3 III		220 s		
15	36043		0 05 17.4	+48 28 41	-0.006	+0.03	8.0	0.9	3.2	G5 IV		91 s		
32	192351		0 05 19.5	-38 51 26	+0.003	+0.04	7.0	0.6	3.2	G5 IV		58 s		
14	10965		0 05 20.0	+68 53 05	+0.008	+0.02	7.0	0.0	0.4	B9.5 V		120 mx		
28	128572	33 Psc	0 05 20.1	-5 42 27	-0.001	+0.09	4.61	1.04	0.0	K1 III	-6	84 s		
53	231922		0 05 21.5	-56 57 21	+0.003	-0.02	7.4	0.1	0.6	A0 V		210 s		
43	109004		0 05 22.2	+3 36 23	-0.001	+0.01	6.9	0.9	3.2	G5 IV		55 s		
25	36044		0 05 22.3	+49 46 12	-0.003	-0.07	7.6	0.4	2.6	F0 V		97 s		
46	128574		0 05 25.4	-3 51 07	-0.001	-0.04	7.7	1.1	0.2	K0 III		320 s		
69	—		0 05 26.1	-79 15 36			7.8	1.1	0.2	K0 III		270 s		
48	147073		0 05 26.4	-14 25 20	+0.002	-0.03	7.6	1.1	0.2	K0 III		300 s		
68	248114		0 05 26.7	-67 57 37	+0.003	-0.01	6.9	1.2	0.2	K0 III		170 s		
65	192352		0 05 27.8	-37 18 10	+0.005	0.00	7.6	1.3	-0.1	K2 III		290 s		
62	128575		0 05 28.0	-7 39 59	0.000	-0.07	7.1	1.4	-0.3	K5 III		220 mx		
66	214959		0 05 28.1	-42 53 16	0.000	-0.01	7.5	1.6	-0.5	M4 III		400 s		
63	128577		0 05 31.1	-9 37 02	+0.006	-0.06	7.1	0.5	3.4	F5 V		56 s		
58	21074		0 05 34.8	+53 10 18	+0.001	+0.01	7.3	1.1	0.2	K0 III	+6	250 s		
80	248116		0 05 35.8	-61 18 37	+0.002	0.00	7.6	0.9	3.2	G5 IV		75 s		
79	231923		0 05 36.2	-51 15 36	+0.001	+0.01	7.4	0.8	-0.1	K2 III		310 s		
71	21076		0 05 39.5	+55 42 37	-0.001	+0.02	7.1	1.1	0.2	K0 III	-10	230 s		
87	91701	86 Peg	0 05 41.9	+13 23 47	+0.003	0.00	5.51	0.90	0.2	K0 III	+2	120 s		
85	91704		0 05 44.6	+17 50 19	+0.004	-0.01	8.0	0.4	3.0	F2 V		98 s	56	m
105	214961		0 05 52.5	-41 45 13	+0.010	-0.09	7.53	0.60	4.4	G0 V		41 s		
104	—		0 05 54.6	-30 21 36			7.9	1.0	4.4	G0 V		28 s		
101	91706		0 05 54.8	+18 14 06	-0.010	-0.15	7.8	0.5	4.0	F8 V		57 s		
100	73737		0 05 56.6	+24 34 08	+0.002	-0.01	7.1	1.4	-0.3	K5 III		300 s		
98	36054		0 05 57.0	+49 37 21	0.000	+0.01	7.60	0.1	1.7	A3 V		150 s	59	m
99	36056		0 05 58.0	+44 44 20	-0.003	-0.03	7.9	0.1	1.7	A3 V		180 s		
111	73738		0 06 00.6	+28 16 38	+0.001	+0.01	6.9	1.4	-0.3	K5 III		270 s		
114	91711		0 06 01.0	+16 27 08	0.000	+0.01	7.9	0.9	3.2	G5 IV		88 s		
112	73739		0 06 03.2	+24 54 59	+0.007	-0.02	7.4	1.1	0.2	K0 III		210 mx		
108	10973		0 06 03.2	+63 40 49	-0.002	+0.02	7.40	0.17		08.8 p	-63			
110	36057		0 06 04.3	+40 25 07	+0.001	0.00	6.7	0.8	3.2	G5 IV		50 s		
126	109006		0 06 07.9	+9 42 53	+0.001	-0.02	7.8	0.5	3.4	F5 V		76 s		
141	166043		0 06 09.1	-29 09 06	+0.003	-0.01	7.89	-0.02	0.6	A0 V n		290 s		m
125	36059		0 06 11.3	+44 04 03	-0.002	-0.03	7.9	1.4	-0.3	K5 III		430 s		
160	248117		0 06 14.8	-64 14 29	-0.012	-0.01	7.84	0.46	3.0	F2 V		80 s		m
123	21085		0 06 15.9	+58 26 12	+0.033	+0.04	5.96	0.67		G5	-12	24 t	61	m
—	—		0 06 17.2	+61 05 24			7.66	1.91		K5				
—	10975		0 06 18.0	+61 04 34	+0.003	0.00	7.71	1.90						
146	128584		0 06 18.6	-1 14 19	-0.002	-0.05	8.0	1.1	-0.1	K2 III		410 s		
142	214963		0 06 19.0	-49 04 31	+0.058	-0.03	5.70	0.52	2.9	G1 IV	+1	28 ts		m
135	10977		0 06 19.9	+60 27 24	+0.003	-0.02	7.95	0.23		F0				
145	36061		0 06 20.1	+40 53 55	+0.007	+0.01	7.0	1.1	0.2	K0 III		230 s		
—	10976		0 06 20.5	+68 38 10	0.000	-0.01	8.0	1.3						
156	147084		0 06 24.9	-18 02 17	+0.001	+0.01	7.5	0.4	2.6	F0 V		97 s		
153	36062		0 06 25.7	+42 45 12	+0.004	+0.03	8.00	0.6	4.5	G1 V	-32	50 s		
144	10978	10 Cas	0 06 26.5	+64 11 47	+0.001	+0.01	5.59	-0.03	-1.2	B8 III	0	200 s		
167	73744		0 06 35.3	+28 -33 13	-0.003	0.00	6.7	0.9	0.2	G9 III	+3	200 s		
166	73743		0 06 36.7	+29 01 17	+0.029	-0.18	6.13	0.75	5.9	K0 V	-8	14 ts	69	m
190	192358		0 06 38.7	-36 05 05	+0.001	-0.04	7.6	1.1	3.2	G5 IV		46 s		
203	166053		0 06 50.0	-23 06 27	+0.007	-0.04	6.18	0.38	2.4	A7 V	-2	47 mx		
213	192359		0 06 53.3	-38 52 41	0.000	0.00	6.9	1.3	0.2	K0 III		140 s		
—	214967		0 06 56.0	-43 59 07	+0.002	+0.03	7.6	0.9	3.2	G5 IV		67 s		
222	73750		0 07 01.9	+22 50 41	+0.001	0.00	8.0	0.4	2.6	F0 V		120 s		
231	109014		0 07 04.0	+6 52 31	-0.003	-0.03	7.8	0.2	2.1	A5 V		140 s		
204	4037		0 07 08.7	+72 46 28	0.000	-0.05	7.9	0.1	1.7	A3 V		160 s		
230	36074		0 07 10.8	+46 45 26	-0.001	-0.01	7.7	1.1	0.2	K0 III		290 s		
276	255632		0 07 13.8	-76 43 50	+0.026	+0.02	7.54	0.41	2.8	F5 IV-V		91 s		
236	73752		0 07 14.4	+20 33 21	+0.001	+0.01	8.0	1.1	0.2	K0 III		360 s		
256	147090		0 07 18.1	-17 23 11	-0.001	+0.04	6.19	0.14	1.7	A3 V		74 s		
273	231930		0 07 22.1	-56 50 13	+0.002	-0.01	7.6	1.1	-0.1	K2 III		350 s		
249	73755		0 07 22.4	+26 27 03	+0.009	-0.10	7.3	1.1		K1 IV	+13	100 mx		
268	166063		0 07 22.6	-25 21 22	+0.017	-0.11	7.06	0.45	3.4	F5 V		44 mx		
267	147092		0 07 23.1	-11 36 44	-0.001	+0.03	7.8	0.1	1.4	A2 V		190 s		
285	231932		0 07 28.0	-56 00 41	+0.008	+0.04	7.4	0.2	3.4	F5 V		64 s		
261	10989		0 07 28.6	+69 04 48	+0.002	+0.02	7.9	1.1	0.2	K0 III		310 s		
—	53656		0 07 32.4	+32 00 32	-0.007	-0.01	8.0	0.9						
281	53659		0 07 35.1	+31 40 08	+0.001	-0.03	7.00	0.9	3.2	G5 IV		58 s	83	m
310	248120		0 07 36.6	-62 18 55	+0.003	-0.01	7.9	1.6	-0.5	M2 III		470 s		
278	21108		0 07 37.4	+52 46 23	-0.001	-0.01	7.60	0.8	3.2	G5 IV		75 s	81	m
279	53660		0 07 37.8	+40 04 42	+0.005	-0.01	7.5	1.1	0.2	K0 III		280 s		
290	53661		0 07 37.9	+40 08 54	-0.002	-0.01	7.00	0.9	3.2	G5 IV		58 s	82	m
385	258214		0 07 38.3	-86 02 19	+0.035	+0.02	7.4	-0.2	0.4	B9.5 V		230 mx		
291	53662		0 07 40.2	+39 02 06	-0.009	-0.04	8.01	0.42	3.4	F5 V		84 s		

3

HD	SAO	Star Name	α 2000	δ 2000	μ(α)	μ(δ)	V	B-V	M_v	Spec	RV	d(pc)	ADS	Notes
			h m s	° ′ ″	s	″								
323	248123		0 07 40.7	−65 55 58	+0.002	−0.07	6.9	1.1	3.2	G5 IV		34 s		
292	53665		0 07 42.9	+37 11 09	0.000	0.00	7.60	0.5	3.4	F5 V		69 s	86	m
315	128595		0 07 44.0	−2 32 56	+0.002	0.00	6.43	−0.14		B8 p	+13			
319	166066		0 07 46.7	−22 30 32	+0.004	−0.04	5.94	0.14	1.7	A3 V		67 mx	89	m
298	21112		0 07 49.6	+57 10 55	+0.002	−0.01	8.0	1.1	0.2	K0 III		340 s		
313	53666		0 07 50.8	+38 43 17	−0.002	−0.04	7.66	0.16	2.1	A5 V		130 s		
299	21114		0 07 52.0	+55 34 38	+0.023	+0.02	7.9	0.6	4.4	G0 V		49 s		
228	214975		0 07 58.8	−43 27 20	−0.003	−0.01	7.6	0.9	3.2	G5 IV		59 s		
369	255636		0 08 00.5	−78 35 57	+0.001	−0.01	6.8	0.1	1.4	A2 V		120 s		
333	73760		0 08 00.7	+29 50 15	0.000	0.00	8.0	0.4	3.0	F2 V		98 s		
332	53668		0 08 01.6	+31 23 30	−0.002	−0.01	7.10	0.2	2.1	A5 V		100 s		m
344	192367		0 08 03.4	−33 31 46	−0.003	+0.01	5.68	1.12		K0				
341	53670		0 08 09.4	+31 22 43	+0.002	−0.02	7.3	0.1	1.7	A3 V		130 s		
352	128602		0 08 12.0	−2 26 52	0.000	0.00	6.07	1.38	−0.1	gK2	+1	140 s		
365	231937		0 08 13.3	−49 43 20	−0.001	0.00	7.8	2.8	3.4	F5 V		76 s		
348	36087		0 08 14.8	+46 42 31	−0.002	+0.01	7.7	1.1	−0.1	K2 III		330 s		
361	147100		0 08 16.1	−14 49 27	−0.002	0.00	7.0	0.6	4.4	G0 V		34 s		
360	128604		0 08 17.3	−8 49 26	−0.004	−0.03	5.99	1.04	0.3	gG8	+20	120 s		
358	73765	21 α And	0 08 23.2	+29 05 26	+0.010	−0.16	2.06	−0.11		A0 p	−12	22 mn	94	Alpheratz, m
377	109027		0 08 25.7	+6 37 00	+0.006	0.00	8.00	0.6	4.7	G2 V	−4	46 s		
392	166073		0 08 27.6	−24 05 36	+0.004	+0.03	7.8	−0.4	3.4	F5 V		76 s		
375	53674		0 08 28.3	+34 56 04	+0.009	+0.01	7.5	0.5	4.0	F8 V		49 s		
373	21125		0 08 30.1	+52 20 11	+0.009	+0.02	7.9	0.5	3.4	F5 V		77 s		
391	128605		0 08 30.9	−6 47 32	+0.004	0.00	8.0	1.1	−0.1	K2 III		320 mx		
371	10999		0 08 32.3	+63 12 16	−0.007	+0.02	6.42	1.03	−2.1	G3 II		130 mx		
402	147103		0 08 33.3	−17 34 39	0.000	−0.02	6.06	1.67		Ma III				
442	255638		0 08 39.0	−74 14 20	+0.003	−0.02	7.7	1.1	−0.1	K2 III		370 s		
400	53677		0 08 41.0	+36 37 36	−0.008	−0.14	6.19	0.48	3.4	dF5	−14	25 ts		
406	128607		0 08 43.0	−2 13 22	+0.002	−0.01	7.3	0.8	3.2	G5 IV		67 s		
370	4043		0 08 43.9	+74 12 54	+0.011	+0.03	7.5	0.1	0.6	A0 V	+6	160 mx		
405	91731		0 08 45.8	+14 08 32	+0.001	−0.02	7.8	1.1	0.2	K0 III		330 s		
427	192369		0 08 46.2	−34 47 39	+0.002	+0.01	7.9	0.4	3.4	F5 V		81 s		
439	214978		0 08 48.4	−47 03 39	+0.010	−0.05	8.0	0.1	4.0	F8 V		62 s		
418	109033		0 08 48.7	+8 01 18	0.000	+0.05	7.30	0.4	2.6	F0 V		87 s	101	m
414	36097		0 08 49.8	+40 29 39	0.000	−0.01	6.9	1.4	−0.3	K5 III		280 s		
413	36098		0 08 50.1	+40 50 27	+0.002	−0.02	8.00	0.9	3.2	G5 IV		77 s	105	m
417	73769		0 08 52.1	+25 27 46	+0.008	+0.04	6.23	0.97	0.2	K0 III	+15	140 mx		
422	109035		0 08 52.2	+0 41 33	+0.001	+0.01	7.6	1.1	−0.1	K2 III		350 s		
437	166077		0 08 53.4	−22 10 46	−0.003	0.00	7.2	1.2	−0.5	M2 III		350 s		
---	---		0 08 55.2	+62 43 35			8.00	1.6		R7.7				
433	36103		0 08 59.5	+49 51 29	−0.007	0.00	7.8	0.1	1.7	A3 V		160 s		
457	214980		0 08 59.7	−39 44 15	+0.011	−0.03	7.8	0.7	4.4	G0 V		40 s		
452	---		0 09 00.2	−21 11 37			7.20	1.1	0.2	K0 III		250 s		m
434	73772		0 09 00.2	+28 14 53	+0.002	+0.01	6.5	0.1	1.4	A2 V	+2	110 s		
469	231943		0 09 02.3	−54 00 07	+0.006	+0.02	6.33	0.74	3.1	G4 IV		43 s		m
448	91734	87 Peg	0 09 02.3	+18 12 43	+0.010	−0.02	5.53	1.04	0.2	K0 III	−23	110 mx		
466	192372		0 09 02.6	−35 05 31	+0.003	−0.01	7.8	0.5	2.6	F0 V		82 s		
445	36104		0 09 03.7	+40 50 36	+0.013	−0.15	7.20	0.9	3.2	G5 IV		77 s	105	m
447	91736		0 09 04.2	+19 55 30	0.000	+0.03	7.3	0.5	3.4	F5 V		60 s		
444	36105		0 09 07.7	+45 31 29	+0.002	+0.03	8.0	1.1	0.2	K0 III		340 s		
480	231947		0 09 08.5	−50 10 01	+0.002	−0.02	7.20	0.00	0.0	B8.5 V		280 s		m
432	21133	11 β Cas	0 09 10.6	+59 08 59	+0.068	−0.18	2.27	0.34	1.9	F2 IV	+12	13 ts	107	Caph, m
471	73776		0 09 15.8	+25 16 56	+0.014	−0.14	7.77	0.68		G0				m
443	11005		0 09 18.6	+65 04 16	+0.043	+0.05	7.01	0.92	5.7	dG9	+7	15 mn		
483	91740		0 09 19.3	+17 32 02	0.000	−0.13	7.2	0.7	4.4	G0 V		36 s		
431	4048		0 09 20.1	+79 42 54	+0.038	−0.02	6.01	0.19		A3	+1	26 mn	102	m
493	166083	κ¹ Scl	0 09 20.9	−27 59 16	+0.005	0.00	5.42	0.42	3.0	dF2	+9	28 s	111	m
496	214983	ε Phe	0 09 24.6	−45 44 51	+0.012	−0.18	3.88	1.03	0.2	K0 III	−9	23 ts		
487	36109		0 09 27.4	+46 23 17	0.000	−0.01	7.0	0.0	0.4	B9.5 V	−6	200 s		
489	91742		0 09 28.0	+19 06 57	+0.019	+0.03	7.96	0.67	4.9	G3 V	−24	38 s		m
491	109039		0 09 28.0	+1 14 38	+0.003	0.00	7.29	1.23	3.2	G5 IV		29 s		
502	91744		0 09 30.9	+19 33 33	+0.002	−0.05	7.80	0.8	3.1	G4 IV	+2	87 s		
509	21145		0 09 44.2	+52 01 40	+0.002	0.00	8.00	1.1	0.2	K0 III		340 s	116	m
565	248130		0 09 52.2	−62 17 47	+0.004	+0.04	6.6	−0.1	1.4	A2 V		110 s		
516	21150		0 09 52.3	+51 25 23	+0.001	0.00	8.0	1.1	0.2	K0 III		340 s		
529	91749		0 09 52.4	+12 42 40	0.000	+0.02	8.0	1.1	0.2	K0 III		360 s		
545	128618		0 09 54.9	−2 33 40	0.000	−0.01	7.2	1.1	−0.1	K2 III	+18	280 s		
543	---		0 09 55.0	+4 49 57	+0.002	+0.02	8.0	0.9	3.2	G5 IV		93 s		
563	214987		0 09 56.3	−44 39 59	0.000	+0.03	7.6	0.8	3.2	G5 IV		74 s		
569	248131		0 09 58.1	−65 21 28	+0.006	−0.02	8.0	0.8	3.2	G5 IV		90 s		
562	166089		0 09 59.1	−25 52 32	+0.005	−0.01	7.65	0.13	2.1	A5 V n		110 mx		
527	11011		0 09 59.3	+65 44 37	+0.002	+0.01	8.0	0.1	0.6	A0 V		270 s		
560	91750	34 Psc	0 10 02.2	+11 08 45	+0.003	0.00	5.51	−0.07	−0.2	B8 V	+14	140 s	122	m
636	258215	γ³ Oct	0 10 02.3	−82 13 27	−0.010	−0.01	5.28	1.05	0.3	gG8	+15	81 s		
561	128619		0 10 03.2	−8 58 27	+0.005	+0.01	7.1	0.4	2.6	F0 V		81 s		

4

HD	SAO	Star Name	α 2000	δ 2000	μ(α)	μ(δ)	V	B-V	M_V	Spec	RV	d(pc)	ADS	Notes
610	255639		0^h 10^m 04^s.3	−79° 00′ 07″	+0^s.004	0″.00	7.9	1.7	3.4	F5 V		13 s		
626	255640		0 10 07.8	−78 18 03	+0.001	−0.01	7.6	1.1	0.2	K0 III		300 s		
553	11013		0 10 10.0	+64 38 48	−0.009	0.00	8.0	1.1	0.2	K0 III		320 s		
567	21156		0 10 16.4	+52 15 18	+0.001	−0.01	7.1	0.0	0.4	B9.5 V		210 s		
587	128621		0 10 18.8	−5 14 55	+0.002	−0.03	5.84	0.98	0.2	gG9	+24	130 s		
571	36123	22 And	0 10 19.2	+46 04 20	+0.001	+0.01	5.03	0.40	−2.0	F2 II	−5	240 s		
598	73785		0 10 26.7	+28 39 12	0.000	0.00	8.0	1.6	−0.5	M4 III	−11	510 s		
584	21162		0 10 29.7	+57 09 57	+0.003	+0.01	6.5	−0.1		B8				
613	53694		0 10 32.2	+33 07 48	−0.003	−0.03	6.83	1.43	−0.3	K4 III	−14	270 s		
595	21165		0 10 34.2	+52 19 00	0.000	−0.01	8.0	0.1	0.6	A0 V		280 s		
593	21164		0 10 34.6	+59 40 24	+0.001	0.00	6.70	0.02	−3.5	B1 V	−3	810 s		
646	192380		0 10 37.3	−31 16 41	−0.002	0.00	7.1	0.8	0.2	K0 III		240 s		
661	255642		0 10 38.7	−73 13 29	+0.031	+0.02	6.64	0.37	2.1	A5 V		60 mx		
632	109047		0 10 39.1	+2 18 03	−0.001	+0.01	8.0	1.1	−0.1	K2 III		410 s		
633	109048		0 10 40.7	+2 03 20	−0.001	+0.02	7.9	0.5	4.0	F8 V		60 s		
645	147127		0 10 42.8	−12 34 48	+0.010	−0.03	5.85	1.01		K1 IV	+4	120 mx		
628	21170		0 10 44.9	+58 01 06	−0.001	0.00	8.0	0.0	0.4	B9.5 V		300 s		
655	166095		0 10 46.4	−21 55 59	−0.001	+0.02	7.6	1.1	0.2	K0 III		270 s		
656	192382		0 10 46.6	−34 51 37	+0.005	+0.02	6.90	1.1	0.2	K0 III		220 s		m
639	53697		0 10 46.6	+37 28 26	+0.003	+0.02	7.8	0.4	3.0	F2 V		91 s		
630	36132		0 10 48.3	+49 00 39	+0.002	0.00	7.9	0.0	0.4	B9.5 V		290 s		
641	53699		0 10 48.4	+30 41 09	0.000	0.00	8.0	0.1	0.6	A0 V		300 s		
638	21175		0 10 52.8	+56 58 22	+0.002	−0.02	7.7	1.1	0.2	K0 III		300 s		
666	147129		0 10 53.4	−13 54 09	+0.001	+0.03	7.7	1.1	0.2	K0 III		320 s		
672	147130		0 10 57.9	−18 34 22	−0.001	−0.02	7.7	1.6	−0.5	M4 III		440 s		
663	21178		0 11 07.7	+57 35 15	−0.006	0.00	6.7	1.1	0.2	K0 III		180 s		
688	128628		0 11 09.5	−3 19 16	+0.001	0.00	6.8	0.9	3.2	G5 IV		53 s		
--	21180		0 11 12.7	+53 24 56	+0.003	0.00	8.0	1.6	−0.1	K2 III		230 s		
685	109054		0 11 15.6	+7 56 54	0.000	0.00	7.9	1.1	−0.1	K2 III		390 s		
693	147133	6 Cet	0 11 15.8	−15 28 05	−0.006	−0.26	4.89	0.49	3.7	F6 V	+15	18 ts		
706	214994		0 11 17.7	−40 22 23	+0.001	−0.01	7.60	0.8	3.2	G5 IV		76 s		m
678	4054		0 11 22.1	+74 29 05	+0.017	−0.01	7.1	1.1	0.2	K0 III		220 s		
733	192387		0 11 31.4	−34 27 36	+0.002	0.00	7.8	1.2	3.2	G5 IV		48 s		
697	21188		0 11 33.1	+59 45 44	−0.002	+0.01	8.00	0.9	3.2	G5 IV		88 s	140	m
720	166103	κ² Scl	0 11 34.4	−27 47 59	+0.001	+0.02	7.20	1.4	−0.3	K5 III	−6	320 s		m
717	128631		0 11 35.1	−3 04 41	+0.010	0.00	7.80	0.5	4.0	F8 V		58 s	144	m
731	147136		0 11 35.5	−18 36 02	+0.005	+0.02	6.9	1.3	3.2	G5 IV		27 s		
698	21190		0 11 37.0	+58 12 43	−0.001	0.00	7.08	0.18	−3.7	B5 II	−24	1000 s		
783	258216		0 11 37.5	−80 10 27	+0.004	+0.01	7.2	0.9	3.2	G5 IV		54 s		
709	21191		0 11 38.8	+55 57 43	0.000	+0.01	7.90	0.00	0.0	B8.5 V		360 s	143	m
710	21192		0 11 41.8	+54 20 08	+0.002	0.00	8.0	0.0	0.4	B9.5 V		320 s		
739	192388	θ Scl	0 11 43.9	−35 07 59	+0.014	+0.13	5.25	0.44	3.3	F4 V	−2	24 s		
725	21193		0 11 44.8	+57 16 16	+0.001	−0.01	7.09	0.63	−3.3	F5 Ib-II		1000 s		
745	109059		0 11 47.5	+9 08 24	−0.001	−0.01	7.7	0.9	3.2	G5 IV		80 s		
724	11030		0 11 50.4	+66 07 35	+0.002	+0.01	7.30	0.1	0.6	A0 V		200 s	145	
770	214999		0 11 55.9	−42 10 20	−0.008	−0.04	6.5	1.1	0.2	K0 III		170 s		
793	255644		0 11 56.5	−75 28 35	+0.011	+0.03	7.40	1.1	0.2	K0 III		280 s		m
743	36148		0 11 59.0	+48 09 09	+0.006	+0.02	6.16	1.45	0.2	K0 III	+16	84 s		
756	73802		0 12 01.3	+23 28 20	+0.007	+0.02	6.6	0.9	3.2	G5 IV		49 s		
790	231966		0 12 01.4	−54 45 01	+0.001	+0.02	7.9	1.4	−0.3	K5 III		430 s		
755	73803		0 12 01.5	+28 36 22	0.000	−0.10	7.2	0.4	2.6	F0 V		83 s		
754	36149		0 12 01.9	+48 40 48	+0.002	+0.03	7.6	1.1	0.2	K0 III		280 s		
780	147142		0 12 02.7	−13 48 46	−0.003	0.00	6.6	0.8	3.2	G5 IV		48 s		
763	36150		0 12 06.5	+47 29 28	−0.005	−0.02	7.4	1.1	0.2	K0 III		260 s		
761	21202		0 12 08.0	+53 37 25	+0.004	−0.01	6.81	0.25		F0			148	m
762	36151		0 12 08.7	+48 10 56	+0.002	0.00	7.4	1.1	0.2	K0 III		260 s		
787	147144		0 12 09.9	−17 56 18	+0.004	−0.03	5.25	1.48	−0.3	K5 III	−8	130 s		
776	53715		0 12 14.0	+39 54 08	+0.013	+0.03	8.0	0.4	2.6	F0 V		110 mx		
784	73809		0 12 15.7	+22 33 24	+0.001	+0.01	7.8	1.6	−0.5	M2 III		450 s		
775	36153		0 12 15.7	+43 49 04	0.000	−0.02	7.9	1.6	−0.5	M4 III		490 s		
774	11036		0 12 21.6	+62 53 36	+0.001	+0.02	7.49	1.72	−2.2	K1 II		440 s		
804	91771		0 12 28.2	+20 14 03	+0.015	−0.02	8.0	0.9	3.2	G5 IV		79 mx		
801	36156		0 12 33.4	+44 35 26	−0.001	−0.01	7.5	1.1	0.2	K0 III		300 s		
800	36157		0 12 34.1	+44 42 26	+0.009	−0.02	6.6	1.1	0.2	K0 III		190 s		
1348	258218		0 12 34.4	−88 21 47	+0.044	0.00	7.3	−0.1	0.6	A0 V		210 s		
802	36158		0 12 35.9	+44 18 58	+0.005	+0.01	7.2	1.1	0.2	K0 III		250 s		
834	166114		0 12 39.0	−26 51 12	+0.022	+0.12	7.97	0.95	5.9	K0 V		21 s		
820	128639		0 12 40.1	−1 13 39	+0.004	−0.01	7.2	1.1	0.2	K0 III		260 s		
845	215002		0 12 40.3	−43 09 53	+0.003	−0.01	7.4	1.1	3.2	G5 IV		45 s		
809	21208		0 12 45.0	+57 13 38	−0.001	−0.01	7.8	1.1	−0.1	K2 III		350 s		
818	36160		0 12 48.0	+46 05 45	+0.009	0.00	6.7	0.4	3.0	F2 V		53 s		
870	231973		0 12 50.2	−57 54 47	−0.015	+0.02	7.4	0.0	3.2	G5 IV		70 s		
829	53725		0 12 50.4	+37 41 37	+0.002	−0.01	6.73	−0.13	−2.5	B2 V	−9	650 s		
828	21213		0 12 52.2	+51 29 04	+0.001	+0.01	8.0	1.4	−0.3	K5 III		420 s		
853	128642		0 12 55.1	−3 22 53	0.000	0.00	7.5	1.6	−0.5	M2 III		400 s		

HD	SAO	Star Name	α 2000	δ 2000	μ(α)	μ(δ)	V	B–V	M$_v$	Spec	RV	d(pc)	ADS	Notes
			h m s	° ′ ″	s	″								
852	91776		0 13 00.0	+12 19 51	+0.004	0.00	7.9	0.4	2.6	F0 V		110 s		
842	21214		0 13 00.2	+55 51 32	0.000	0.00	7.95	0.51	-6.7	A9 I		5500 s		
841	11040		0 13 01.1	+62 43 41	+0.002	0.00	7.73	1.13	3.0	F2 V		30 s		
849	36164		0 13 01.8	+49 55 36	+0.001	+0.01	7.1	0.1	1.4	A2 V		130 s		
877	166120		0 13 03.9	-22 28 13	0.000	+0.02	6.6	1.1	0.2	K0 III	-16	170 s		
863	53729		0 13 04.2	+35 17 29	-0.002	-0.06	7.71	0.95	3.2	G5 IV		58 s		
848	11041		0 13 07.9	+68 10 23	-0.002	+0.03	7.9	0.5	3.4	F5 V		76 s		
876	128643		0 13 08.2	-7 40 58	+0.006	+0.03	8.0	1.1	0.2	K0 III		190 mx		
874	91779		0 13 09.4	+16 55 17	-0.002	-0.08	6.6	0.8	3.2	G5 IV	+10	48 s		
903	215005		0 13 09.6	-43 03 54	+0.004	+0.01	7.3	0.7	2.6	F0 V		49 s		
861	11044		0 13 12.4	+62 02 28	+0.007	-0.02	6.64	0.19		A m				
873	73821		0 13 13.0	+20 45 09	+0.001	-0.03	6.90	1.1		K0				
902	192406		0 13 13.7	-37 49 23	-0.001	0.00	7.19	1.53		K2				
890	147153		0 13 13.7	-17 11 05	+0.001	+0.01	7.53	1.24		K0				m
886	91781	88 γ Peg	0 13 14.1	+15 11 01	0.000	-0.01	2.83	-0.23	-3.0	B2 IV	+4	150 s		Algenib, m,v
901	166124		0 13 15.4	-26 19 31	+0.001	0.00	7.5	1.0	0.2	K0 III		280 s		
871	21217		0 13 15.8	+51 49 59	+0.008	+0.02	8.00	1.4	-0.3	K5 III		260 mx	158	m
899	166126		0 13 17.9	-20 37 07	+0.001	+0.06	7.30	0.4	2.6	F0 V		87 s		m
896	109069		0 13 19.3	+8 55 47	+0.002	-0.01	8.0	0.4	2.6	F0 V		120 s		
1032	258217		0 13 19.8	-84 59 38	+0.008	+0.03	5.77	1.72	-0.3	gK5		120 s		
883	36168		0 13 20.5	+49 43 24	+0.005	-0.01	7.1	0.1	0.6	A0 V		130 mx		
895	73823		0 13 23.9	+26 59 14	-0.001	-0.04	6.30	0.65		F5	-13		161	m
960	255645		0 13 26.5	-76 44 29	-0.001	-0.02	8.0	1.1	0.2	K0 III		360 s		
905	36173	23 And	0 13 30.8	+41 02 08	-0.010	-0.14	5.72	0.31	2.1	A5 V	-29	47 s		
915	36178		0 13 38.5	+42 56 57	+0.001	0.00	7.4	1.1	0.2	K0 III		280 s		
942	166130		0 13 42.1	-26 01 19	+0.002	-0.06	5.94	1.55		K2				
943	166131		0 13 44.2	-26 17 06	-0.001	+0.02	6.3	1.5	-0.3	K5 III		210 s		
957	215009		0 13 44.9	-48 41 04	-0.003	-0.03	7.1	0.1	1.4	A2 V		130 s		
941	166132		0 13 45.5	-23 12 52	+0.010	-0.06	6.9	0.8	0.2	K0 III		100 mx		
931	109077		0 13 45.8	+8 11 36	+0.010	-0.01	7.9	0.4	3.0	F2 V		94 s		
912	11050		0 13 46.7	+64 51 54	+0.016	+0.02	8.0	0.6	4.4	G0 V		51 s		
940	109078		0 13 47.5	+1 23 01	-0.002	-0.01	6.8	1.1	0.2	K0 III		210 s		
980	231976		0 13 53.1	-56 59 58	+0.013	+0.06	6.8	0.8	4.0	F8 V		27 s		
955	—		0 13 53.5	-17 32 38			7.34	-0.15		B5	-34			
936	21224		0 13 56.4	+59 59 58	+0.005	0.00	6.88	1.12	-2.1	G8 II	0	170 mx		
966	128648		0 14 01.2	-3 54 32	+0.003	-0.02	7.5	0.8	0.3	G6 III	-23	280 s		
952	53744		0 14 02.2	+33 12 21	-0.001	-0.02	6.10	0.02	1.2	A1 V	+1	96 s		
919	4062		0 14 02.7	+76 01 38	+0.008	+0.01	7.60	1.6	-0.5	M2 III	-1	380 s		m
1025	255647		0 14 04.5	-74 54 42	+0.017	+0.04	7.0	1.1	0.2	K0 III		230 s		
962	11054		0 14 06.4	+60 43 13	-0.002	-0.01	7.71	0.70	3.0	F2 V		51 s		
976	73830		0 14 09.0	+26 15 31	+0.005	-0.03	7.1	0.5	4.0	F8 V	-25	41 s		
991	192412		0 14 13.1	-3 48 29	+0.004	-0.03	7.8	1.4	3.2	G5 IV		34 s		
973	21229		0 14 14.4	+55 17 00	+0.002	+0.01	7.7	1.1	-0.1	K2 III		340 s		
1004	231979		0 14 14.6	-55 04 11	0.000	-0.03	6.7	1.0	3.4	F5 V		22 s		
989	128652		0 14 16.5	-8 49 59	0.000	-0.04	8.0	1.1	0.2	K0 III		360 s		
1023	248140		0 14 18.5	-68 55 13	+0.003	-0.04	7.6	0.1	1.7	A3 V		150 s		
974	21232		0 14 19.3	+52 51 40	+0.001	+0.01	8.0	0.1	0.6	A0 V		280 s		
1002	166137		0 14 20.2	-27 31 54	0.000	-0.16	7.2	0.8	4.4	G0 V		27 s		
1022	248139		0 14 20.7	-62 46 18	+0.001	+0.02	6.5	2.3	-0.5	M2 III		91 s		
1000	166138		0 14 20.8	-21 11 49	-0.005	-0.23	6.91	0.47	4.0	F8 V		38 s		
972	11056		0 14 21.0	+66 12 18	+0.007	+0.01	7.6	1.1	0.2	K0 III		270 s		
947	4064		0 14 22.4	+76 01 21	0.000	+0.02	7.9	0.9	3.2	G5 IV	-16	85 s		
999	128654		0 14 24.3	-2 11 51	-0.002	-0.01	7.3	1.1	0.2	K0 III		270 s		
1014	128655		0 14 27.5	-7 46 50	+0.004	+0.01	5.12	1.62	-0.5	M3 III	-2	130 s	180	m
1015	147167		0 14 28.0	-14 25 55	-0.005	-0.12	7.0	0.5	3.4	F5 V	-1	52 s		
995	36190		0 14 33.6	+41 01 58	+0.004	+0.03	7.0	1.1	0.2	K0 III		230 s		
1013	91792	89 χ Peg	0 14 36.1	+20 12 24	+0.007	+0.01	4.80	1.57	-0.5	M2 III	-46	120 s		
1031	109084		0 14 36.4	+1 17 48	0.000	-0.01	7.1	0.5	3.4	F5 V		54 s		
1038	147169	7 Cet	0 14 38.4	-18 55 58	-0.002	-0.06	4.44	1.66	-0.5	gM1	-23	86 s		
1037	147170		0 14 40.2	-14 48 18	+0.007	-0.03	6.9	0.9	3.2	G5 IV	-30	55 s		
1036	147171		0 14 40.9	-12 44 53	+0.001	+0.01	8.00	1.1	0.2	K0 III		360 s	185	m
1066	215018		0 14 44.0	-46 02 01	+0.004	+0.02	7.3	1.2	0.2	K0 III		210 s		
1051	147175		0 14 48.2	-14 10 42	-0.005	-0.06	6.9	0.1	1.4	A2 V	-9	130 s		
1026	11063		0 14 49.4	+62 50 10	+0.003	+0.03	8.04	0.18	0.6	A0 V		240 s	183	m
1064	128660		0 14 54.4	-9 34 11	+0.002	-0.01	5.75	-0.08	0.2	B9 V		130 s		
1063	128661		0 14 55.5	-3 01 39	-0.005	-0.03	6.9	1.6	-0.5	M2 III		270 mx		
1048	73838		0 14 56.0	+22 17 03	+0.005	-0.01	6.00	-0.01	0.6	A0 V	-15	45 mx		
1059	73839		0 14 56.2	+21 32 32	0.000	-0.03	7.40	1.1	-0.1	K2 III		320 s	190	m
1089	192418		0 14 58.1	-34 54 17	+0.006	-0.02	6.17	1.34		K0				
1061	109087	35 Psc	0 14 58.7	+8 49 15	+0.006	-0.02	5.79	0.31	2.6	F0 V		43 s	191	UU Psc, m,v
1090	192420		0 14 59.9	-36 09 37	+0.001	+0.03	7.8	2.2	-0.3	K5 III		150 s		
1044	21247		0 15 01.9	+54 04 42	+0.001	-0.01	7.9	0.9	3.2	G5 IV		85 s		
1045	21248		0 15 04.0	+52 38 23	+0.001	0.00	7.6	0.0	0.4	B9.5 V		260 s		
1074	53754		0 15 06.7	+33 01 49	-0.001	-0.02	7.35	1.02		K0				
1075	53755		0 15 06.9	+31 32 09	+0.003	0.00	6.5	1.4	-0.3	K5 III	+2	230 s		

HD	SAO	Star Name	α 2000	δ 2000	μ(α)	μ(δ)	V	B−V	M_v	Spec	RV	d(pc)	ADS	Notes
1069	11069		0ʰ15ᵐ09.5ˢ	+60°46'20"	+0.003	0.00	7.67	1.43	3.2	G8 IV		37 s		
1083	73842		0 15 10.5	+27 16 59	+0.002	−0.02	6.10	−0.02	1.2	A1 V	−7	96 s		m
1071	21250		0 15 11.8	+53 34 05	+0.001	−0.01	7.9	1.1	0.2	K0 III		320 s		
1070	21249		0 15 12.7	+59 46 47	−0.001	+0.02	7.99	0.59	−2.1	A5 II		620 s	192	m
1116	215021		0 15 14.9	−40 05 42	+0.002	+0.02	7.39	0.98		G5				m
1102	166150		0 15 15.5	−22 53 28	+0.001	−0.02	8.0	1.1	0.2	K0 III		330 s		
1101	147179		0 15 16.3	−16 45 22	+0.002	−0.01	7.8	0.4	3.0	F2 V		89 s		
1082	36202		0 15 17.6	+44 12 12	−0.001	0.00	7.30	0.00	0.4	B9.5 V		240 s	i97	m
1115	192424		0 15 22.6	−32 02 41	+0.007	+0.02	6.2	1.6		M5 III	+35	220 mx		S Sc1, v
—	192426		0 15 29.3	−39 27 21	−0.005	−0.11	7.7	1.3	0.2	K0 III		160 mx		
1138	215023		0 15 29.4	−40 27 15	−0.002	−0.07	7.7	1.3	0.2	K0 III		220 s		
1109	36206		0 15 33.6	+50 03 00	0.000	−0.02	7.6	1.1	0.2	K0 III		290 s		
1133	109091		0 15 38.6	+5 50 36	+0.003	−0.02	6.96	1.1	0.2	K0 III		250 s		
1157	215024		0 15 41.2	−46 14 23	−0.001	−0.01	7.7	1.5	0.2	K0 III		140 s		
1184	248143		0 15 46.6	−66 55 34	−0.012	−0.07	8.0	0.5	4.0	F8 V		63 s		
1128	21256		0 15 48.4	+59 26 26	0.000	+0.01	7.88	0.03		B8				
1153	128668		0 15 51.6	−5 36 06	+0.001	−0.01	7.70	1.1	0.2	K0 III		320 s	205	m
1142	11074		0 15 55.1	+61 00 01	0.000	−0.01	6.45	0.83	0.3	G8 III		170 s		
1221	255650		0 15 55.2	−75 54 41	−0.001	+0.01	6.49	1.00	0.3	gG5		140 s		
1160	109094		0 15 57.2	+4 15 04	+0.001	−0.01	7.1	0.1	0.6	A0 V		200 s		
1169	109096		0 16 05.3	+8 06 56	+0.003	−0.03	8.00	0.2	2.1	A5 V		150 s	211	m
1187	192430		0 16 08.9	−31 26 47	+0.011	−0.03	5.67	1.35	−0.3	K5 III		130 mx		m
1237	258219		0 16 13.7	−79 51 07	+0.176	−0.07	6.7	0.6	4.4	G0 V		28 s		
1141	4071		0 16 13.9	+76 57 03	+0.005	0.00	6.70	−0.07		B9	−8		207	m
1185	36221		0 16 21.5	+43 35 42	+0.004	−0.03	6.15	0.05		A0	+3		215	m
1205	166163		0 16 23.4	−22 35 17	+0.002	−0.04	7.7	0.6	3.2	G5 IV		80 s		
1235	215028		0 16 23.9	−43 46 50	+0.001	−0.01	7.6	1.1	0.2	K0 III		86 s		
1261	255651		0 16 24.7	−70 23 32	+0.005	+0.02	7.3	−0.1	1.4	A2 V		150 s		
1192	21261		0 16 25.7	+53 56 46	+0.001	+0.01	7.4	0.8	3.2	G5 IV		67 s		
1260	248149		0 16 31.0	−66 41 23	+0.006	+0.04	7.4	1.1	0.2	K0 III		270 s		
1250	231986		0 16 32.1	−55 28 09	+0.001	+0.01	7.6	1.2	0.2	K0 III		240 s		
1233	192437		0 16 32.4	−32 45 38	+0.003	−0.01	7.4	0.5	0.2	K0 III		280 s		
1231	147195		0 16 34.0	−13 24 13	−0.002	−0.01	7.2	0.1	1.7	A3 V		120 s		
1227	109100	36 Psc	0 16 34.0	+8 14 25	−0.002	−0.01	6.11	0.92	0.3	gG6	+1	140 s		
1228	109101		0 16 39.3	+1 51 03	0.000	+0.02	7.30	1.6		M5 III	−6			
1226	91811		0 16 39.7	+11 56 18	0.000	−0.04	8.0	1.1	−0.1	K2 III		410 s		
1210	21263		0 16 41.3	+54 39 38	−0.002	+0.01	7.80	0.1	1.7	A3 V	+11	160 s	218	m
1259	192442		0 16 42.0	−32 41 16	−0.002	0.00	6.7	0.7	0.2	K0 III		200 s		
1256	166167		0 16 42.4	−20 12 37	+0.001	−0.01	6.47	−0.09	−0.2	B8 V		220 s		
1223	53772		0 16 43.0	+36 37 47	−0.002	−0.03	6.99	0.05	1.4	A2 V	−1	130 s	220	m
1224	53773		0 16 44.5	+36 29 30	+0.008	+0.06	7.72	0.46	3.8	dF7	+7	61 s	221	m
1243	91812		0 16 45.6	+13 55 01	+0.002	+0.01	7.5	0.1	1.7	A3 V	−1	150 s		
1255	109102		0 16 45.6	+10 14 42	+0.001	+0.02	6.7	1.6	−0.5	M2 III	+11	280 s		
1271	215032		0 16 47.2	−46 34 25	+0.012	−0.05	8.0	0.2	3.4	F5 V		82 s		
1324	255652		0 16 49.2	−78 46 51	+0.029	−0.04	6.77	0.46	1.9	F2 IV		82 s		
1241	36233		0 16 51.6	+46 36 51	−0.002	−0.02	7.4	0.9	3.2	G5 IV		67 s		
1273	231987		0 16 53.6	−52 39 03	+0.032	+0.21	6.85	0.64	4.7	G2 V		22 ts		
1240	36234		0 16 53.9	+49 27 42	−0.002	−0.01	6.7	1.6	−0.5	M2 III		260 s		
1254	73868		0 16 54.9	+23 15 23	+0.003	−0.01	7.1	1.1	0.2	K0 III		240 s		
1239	11084		0 16 57.0	+61 32 00	0.000	0.00	5.74	0.88	0.3	G8 III	−4	120 s	222	m
1238	11083		0 16 57.4	+63 33 11	+0.003	0.00	7.87	0.07	−0.6	B7 V		400 s		
1265	21268		0 17 00.2	+53 14 36	0.000	0.00	7.8	0.0	0.4	B9.5 V		290 s		
1283	91815		0 17 01.7	+10 23 20	+0.001	−0.02	7.6	1.1	−0.1	K2 III		350 s		
1119	43		0 17 02.8	+84 57 19	+0.010	−0.01	8.0	0.1	1.4	A2 V		210 s		
1264	21269		0 17 04.7	+56 54 35	−0.003	−0.01	7.4	0.1	0.6	A0 V		220 s		
1280	53777	24 θ And	0 17 05.4	+38 40 54	−0.004	−0.01	4.61	0.06	1.4	A2 V	+1	44 s		
1279	36236		0 17 09.0	+47 56 53	+0.001	+0.02	5.89	−0.09		B9				
1321	215035		0 17 12.1	−44 21 52	+0.003	−0.01	8.0	1.0	3.0	F2 V		18 s		
1278	21270		0 17 15.8	+54 58 51	+0.001	0.00	7.8	0.0	0.4	B9.5 V		280 s		
1320	215036		0 17 16.5	−43 51 12	+0.037	−0.04	7.98	0.65	4.2	G5 IV-V		41 mx		
1317	109111		0 17 24.5	+8 52 36	+0.006	+0.08	6.67	0.59		F5	+36		238	m
1329	109113		0 17 28.7	+0 19 18	+0.001	+0.12	7.9	0.5	4.0	F8 V		60 s	241	m
1306	—		0 17 32.0	+50 17 18	0.000	+0.04	7.50	2.0		N7.7	−43			
1343	147205		0 17 32.5	−19 03 04	+0.001	+0.01	6.45	0.37	2.6	F0 V		54 s		
1315	53785		0 17 32.9	+36 33 03	+0.001	0.00	7.58	0.95	3.2	G5 IV		55 s		
1391	255654		0 17 33.4	−69 52 49	−0.012	−0.04	7.6	0.7	4.4	G0 V		44 s		
1382	248154		0 17 41.2	−60 47 57	+0.012	−0.01	7.3	0.4	2.1	A5 V		67 mx		
1337	21273		0 17 43.0	+51 25 59	−0.001	0.00	5.90	0.03	−6.0	O9 III	−35	1700 s		AO Cas, m,v
1397	248155		0 17 47.3	−66 21 32	+0.011	0.00	6.9	1.1	−4.4	K0 Ib		91 mx		
1367	109119		0 17 47.6	+1 41 19	+0.006	+0.01	6.17	0.94	0.3	gG6	−9	140 s		
1369	128685		0 17 48.2	−1 51 48	−0.001	−0.01	7.2	0.9	3.2	G5 IV		62 s		
1455	255655		0 17 49.3	−79 13 57	−0.004	−0.02	8.0	0.3	4.4	G0 V		53 s		
1352	91825		0 17 49.9	+16 19 52	+0.015	−0.02	7.5	0.5	3.4	F5 V	+8	66 s		
1334	21277		0 17 50.1	+59 03 27	−0.002	0.00	7.69	0.01		B5	−4			
1364	91826		0 17 51.2	+20 13 37	+0.002	−0.01	7.30	1.6	−1.4	M4 II-III	+9	560 s		

HD	SAO	Star Name	α 2000	δ 2000	μ(α)	μ(δ)	V	B-V	M_V	Spec	RV	d(pc)	ADS	Notes
			$0^h 17^m 52.8^s$	$+7° 51' 58''$	$+0.001^s$	$0.00''$								
1365	109121		0 17 52.8	+7 51 58	+0.001	0.00	7.52	0.31		A5				
1350	36242		0 17 52.9	+44 34 40	+0.004	0.00	7.5	1.1	0.2	K0 III		290 s		
1375	91828		0 17 56.2	+12 46 18	-0.001	-0.01	6.6	0.8	3.2	G5 IV	+3	48 s		
1388	147208		0 17 58.8	-13 27 19	+0.028	+0.02	6.50	0.59	4.4	G0 V	+28	25 s		
1385	109126		0 18 03.4	+0 37 02	-0.001	+0.03	7.4	0.9	3.2	G5 IV		69 s		
1393	109127		0 18 07.8	+3 47 48	-0.002	+0.01	7.2	0.4	2.6	F0 V		81 s		
1374	36247		0 18 09.0	+46 12 51	+0.002	-0.01	7.0	1.1	0.2	K0 III		210 s		
1431	166174		0 18 16.8	-21 08 18	0.000	+0.01	6.68	0.00	0.6	A0 V n		160 s	251	m
1419	91832		0 18 17.1	+11 12 21	-0.002	-0.03	6.05	1.03	0.2	K0 III	+9	140 s		
1383	11092		0 18 17.4	+61 43 37	-0.003	0.00	7.63	0.26	-5.1	B1 II	-40	1900 s		
1434	192453		0 18 18.2	-36 30 36	+0.001	+0.01	6.7	1.5	0.2	K0 III		100 s		
1421	128688		0 18 18.5	-2 00 54	+0.002	-0.01	7.4	1.1	-0.1	K2 III		320 s		
1392	36250		0 18 19.6	+44 33 31	-0.002	-0.01	7.5	1.1	-0.1	K2 III		340 s		
1404	53798	25 σ And	0 18 19.6	+36 47 07	-0.005	-0.03	4.52	0.05	1.4	A2 V	-8	42 s		
1444	231993		0 18 20.4	-49 54 23	-0.004	-0.06	7.8	0.8	0.2	K0 III		330 s		
1402	--		0 18 23.1	+49 30 20			7.80	0.5	4.0	F8 V		57 s	248	m
1466	248159		0 18 26.0	-63 28 39	+0.014	-0.05	7.4	0.5	4.0	F8 V		47 s		
1359	4080		0 18 27.7	+76 16 35	+0.008	-0.02	7.1	0.0	0.4	B9.5 V	-2	210 s		
1400	11095		0 18 28.7	+62 12 13	-0.001	+0.02	6.96	1.54	8.0	dK5	-30			
1452	215044		0 18 29.4	-46 31 56	-0.007	+0.02	7.6	0.2	3.4	F5 V		69 s		
1429	73883		0 18 30.4	+26 08 24	-0.002	-0.04	7.60	0.1	1.4	A2 V	-11	170 s	252	m
1427	21288		0 18 37.1	+52 55 36	+0.001	0.00	7.9	1.1	0.2	K0 III		320 s		
1439	53803		0 18 38.2	+31 31 02	+0.005	0.00	5.80	-0.01	1.2	A1 V	-5	83 s		
1449	73884		0 18 39.6	+22 52 49	0.000	+0.05	7.1	0.6	4.4	G0 V	+12	35 s		
1448	36255		0 18 40.9	+44 37 47	+0.001	-0.02	7.2	0.1	1.4	A2 V		150 s		
1461	128690		0 18 41.7	-8 03 10	+0.028	-0.14	6.46	0.68	3.6	G0 IV-V		27 ts		
1438	36256	26 And	0 18 42.0	+43 47 28	+0.002	0.00	6.00	-0.08		B9	+7		254	m
1483	215047		0 18 42.5	-43 14 07	+0.005	+0.02	6.33	1.21	-0.1	gK2		180 s		
1458	21292		0 18 56.4	+52 10 20	+0.001	-0.02	7.8	1.1	0.2	K0 III		310 s		
1447	11101		0 18 56.7	+63 17 17	+0.006	-0.02	7.6	0.4	2.6	F0 V		97 s		
---	11100		0 18 56.9	+68 25 21	+0.001	+0.03	7.8	1.7						
1468	21293		0 18 58.5	+50 41 39	-0.002	0.00	7.9	0.2	2.1	A5 V		140 s		
1457	11104		0 19 00.2	+60 20 07	-0.002	0.00	7.82	0.62	-6.6	F0 I		4600 s		
1558	255657		0 19 05.2	-77 59 19	+0.011	+0.03	7.9	0.4	2.6	F0 V		120 s		
1479	--		0 19 14.4	+59 42 17	+0.010	-0.02	7.82	0.34		F0			263	m
1513	109135		0 19 16.5	+0 31 14	-0.001	0.00	7.9	0.7	4.4	G0 V		51 s		
1467	4086		0 19 17.7	+73 07 23	-0.007	0.00	7.3	1.1	0.2	K0 III		240 s		
1486	21296		0 19 18.5	+59 08 21	+0.003	+0.01	7.30	0.02		B9	+2			TV Cas, m,v
1531	215052		0 19 20.4	-46 29 56	+0.001	+0.02	7.3	1.7	-0.5	M2 III		340 s		
1501	73889		0 19 21.9	+26 27 15	-0.001	-0.01	7.70	0.9	0.3	G8 III	-11	300 s		
1532	231998		0 19 22.5	-53 50 39	+0.002	+0.01	7.9	1.6	-0.3	K5 III		370 s		
1478	11110		0 19 23.1	+67 19 25	+0.002	0.00	8.0	0.1	0.6	A0 V		270 s		
1510	91846		0 19 23.4	+19 59 55	+0.002	-0.05	7.7	1.1	0.2	K0 III		310 s		
1542	231999		0 19 24.1	-59 41 28	+0.008	-0.08	7.6	1.1	3.2	G5 IV		48 s		
1522	128694	8 ι Cet	0 19 25.6	-8 49 26	-0.001	-0.03	3.56	1.22	-0.1	K2 III	+19	50 s		m
1572	255658		0 19 26.3	-70 39 46	+0.007	-0.03	7.8	1.1	0.2	K0 III		270 mx		
1529	147216		0 19 28.8	-17 40 43	+0.002	+0.05	8.0	0.9	3.2	G5 IV		93 s		
1540	166190		0 19 34.2	-21 48 24	0.000	+0.03	7.8	1.4	0.2	K0 III		190 s		
1550	232001		0 19 35.0	-50 25 46	+0.004	+0.01	7.7	1.2	-0.3	K5 III		390 s		
1539	128696		0 19 36.3	-7 13 26	+0.001	-0.12	7.2	0.5	3.4	F5 V		57 s		
1527	36269		0 19 41.5	+40 43 47	-0.002	0.00	6.33	1.18	0.0	K1 III	-38	170 s		
1537	53812		0 19 46.6	+36 57 46	-0.004	-0.01	7.4	1.1	0.2	K0 III		270 s		
1567	128698		0 19 50.9	-2 28 44	0.000	+0.04	7.1	0.9	3.2	G5 IV		59 s		
1536	21305		0 19 53.0	+55 42 53	+0.001	+0.01	7.9	1.1	-0.1	K2 III		370 s		
1563	91850		0 19 56.2	+16 15 04	-0.001	0.00	6.60	0.98	0.3	gG8	+20	180 s		
1565	109142		0 19 57.0	+1 34 56	-0.002	0.00	7.8	0.8	3.2	G5 IV		84 s		
1589	166193		0 19 58.0	-25 07 20	+0.002	-0.12	7.5	1.1	0.2	K0 III		95 mx		
1562	53817		0 20 00.3	+38 13 39	-0.012	-0.27	7.20	0.7	4.4	G0 V	+9	36 s	271	m
1588	147221		0 20 02.9	17 42 02	+0.003	0.00	6.8	1.1	0.2	K0 III		210 s		
1581	248163	ζ Tuc	0 20 04.4	-64 52 30	+0.271	+1.17	4.23	0.58	5.0	G0 V	+9	7.1 t		
1585	109145		0 20 05.0	+6 17 34	0.000	+0.02	7.20	1.1	0.2	K0 III		250 s		m
1561	36272		0 20 05.1	+48 51 56	0.000	-0.01	6.3	0.1	0.6	A0 V	-2	130 s		
1560	36271		0 20 05.2	+50 06 41	-0.002	-0.02	7.9	1.1	0.2	K0 III		320 s		
1575	36273		0 20 07.3	+43 55 32	-0.001	-0.03	7.8	1.1	-0.1	K2 III		380 s		
1586	109146		0 20 09.5	+3 02 01	+0.001	0.00	7.9	1.6	-0.5	M2 III		490 s		
1594	147223		0 20 10.5	-13 03 41	+0.003	+0.02	6.9	0.8	3.2	G5 IV		54 s		
1583	73898		0 20 14.4	+24 39 57	-0.002	0.00	7.6	0.9	3.2	G5 IV		77 s		
1608	109149		0 20 21.5	+4 46 43	-0.002	-0.02	8.0	0.9	3.2	G5 IV		93 s		
1559	11118		0 20 22.2	+68 50 49	+0.017	0.00	7.7	0.4	3.0	F2 V		84 s		
1606	53820		0 20 24.4	+30 56 09	+0.001	+0.01	5.90	-0.10	-1.3	B6 IV	+4	280 s		
1627	91854		0 20 30.3	+10 55 04	0.000	-0.02	7.1	1.1	0.2	K0 III	+6	240 s		
1601	36277		0 20 30.7	+48 58 07	0.000	0.00	6.6	0.6	4.4	G0 V		28 s		
1605	53821		0 20 31.2	+30 58 29	+0.004	+0.06	7.60	1.1		K1 IV	+10	250 mx		
1614	36280		0 20 33.0	+45 30 37	+0.007	+0.04	7.20	0.5	3.4	F5 V		58 s	277	m
1598	21313		0 20 33.8	+56 33 23	+0.003	-0.03	6.8	0.8	3.2	G5 IV		52 s		

HD	SAO	Star Name	α 2000	δ 2000	μ(α)	μ(δ)	V	B-V	M_V	Spec	RV	d(pc)	ADS	Notes
			h m s	° ′ ″	s	″								
1635	109152	41 Psc	0 20 35.8	+8 11 25	0.000	+0.01	5.37	1.34	-0.2	gK3	+16	120 s		
1685	248167	π Tuc	0 20 38.9	-69 37 30	-0.002	0.00	5.51	-0.05	0.4	B9.5 V	+12	110 s		
1613	11123		0 20 43.6	+61 52 46	0.000	0.00	7.10	1.6	-2.4	M2 II	-29	620 s		
1632	53825		0 20 45.5	+32 54 41	-0.002	-0.01	5.79	1.59	-0.3	K5 III	-36	150 s		
1667	166201		0 20 48.4	-23 37 49	-0.001	+0.01	6.8	0.1	2.6	F0 V		68 s		
1642	91857		0 20 50.1	+18 28 57	+0.005	+0.04	7.7	0.9	3.2	G5 IV		81 s		
1683	192474		0 20 53.5	-39 14 24	-0.005	-0.04	7.2	0.4	3.4	F5 V		58 s		
1641	53827		0 20 54.0	+32 58 42	+0.001	-0.06	7.50	0.5	3.4	F5 V	-4	66 s	285	m
1674	166204		0 20 54.1	-29 41 26	-0.007	-0.10	7.7	0.7	4.4	G0 V		38 s		
1663	91858		0 20 54.5	+10 58 37	-0.003	-0.02	7.00	0.1	0.6	A0 V	-18	190 s	287	m
1624	11124		0 20 57.3	+67 40 04	+0.005	+0.09	7.91	0.82		G5			283	
1662	91859		0 20 59.4	+12 46 18	+0.001	+0.01	7.5	0.1	1.7	A3 V	-19	140 s		
1679	128707		0 21 04.1	-2 54 33	+0.005	+0.02	7.0	1.1	0.2	K0 III		220 mx		
1671	53828	27 ρ And	0 21 07.2	+37 58 07	+0.005	-0.03	5.18	0.42	2.1	F5 IV	+9	40 s		
1677	53831		0 21 09.0	+35 52 43	-0.003	-0.02	7.37	0.24	0.6	A0 V		160 s		
1672	73911		0 21 10.0	+25 20 18	+0.002	0.00	7.8	1.6	-0.5	M2 III		460 s		
1670	36291		0 21 10.5	+44 56 14	-0.001	-0.01	8.0	1.1	-0.1	K2 III		410 s		
1721	192477		0 21 14.5	-35 47 56	-0.001	-0.04	7.2	1.3	-0.5	M2 III		350 s		
1658	11128		0 21 22.1	+67 00 19	+0.004	+0.01	7.23	0.01		A0			293	m
---	53834		0 21 22.2	+33 40 47	0.000	-0.01	8.0	1.5						
1801	255663		0 21 28.6	-77 25 37	+0.003	0.00	5.97	1.40	0.2	K0 III		75 s		
1767	248169		0 21 30.3	-65 45 10	+0.006	-0.02	7.4	0.9	3.2	G5 IV		68 s		
1737	166207	ι Scl	0 21 31.2	-28 58 54	+0.003	-0.07	5.18	1.00	0.3	gG5	+21	76 s		
1777	248170		0 21 33.4	-66 18 17	+0.005	-0.02	7.1	0.0	0.4	B9.5 V		110 mx		
1712	53835		0 21 35.0	+38 11 14	+0.003	-0.02	6.9	1.1	0.2	K0 III		220 s		
1697	11134		0 21 41.7	+61 41 30	+0.021	-0.04	7.26	0.51	4.0	F8 V		45 s		
1695	11130		0 21 42.3	+67 49 23	-0.004	0.00	6.7	1.1	0.2	K0 III		180 s		
1760	166210		0 21 46.2	-20 03 28	+0.005	0.00	5.12	1.82		M5 II e	+29			T Cet, v
1727	21325		0 21 51.3	+57 58 06	+0.002	-0.01	7.9	0.1	0.6	A0 V		260 s		
1766	166213		0 21 52.3	-23 00 28	+0.009	-0.10	7.60	0.6	4.4	G0 V		44 s	302	m
1764	91863		0 21 57.2	+16 52 38	+0.004	-0.05	7.9	0.5	3.4	F5 V		78 s		
1732	11141		0 22 08.3	+66 27 40	+0.003	+0.01	7.7	0.2	2.1	A5 V		130 s		
1788	128718		0 22 14.1	-5 11 29	-0.002	0.00	7.1	0.1	1.4	A2 V		140 s		
1731	4101		0 22 14.4	+71 30 27	+0.005	-0.03	7.8	1.1	-0.1	K2 III		330 s		
1786	73929		0 22 16.9	+24 40 46	-0.001	-0.01	7.7	0.1	0.6	A0 V		270 s		
1770	53842		0 22 17.2	+37 45 41	+0.006	-0.04	7.7	0.9	3.2	G5 IV		78 s		
1803	166222		0 22 20.0	-23 50 24	+0.002	-0.03	7.7	1.0	0.2	K0 III		310 s		
1841	248172		0 22 20.6	-67 37 32	+0.009	+0.04	7.8	1.1	0.2	K0 III		310 mx		
1817	232008		0 22 21.3	-50 59 33	+0.008	+0.05	6.6	0.9	0.2	K0 III		190 s		
1795	73930		0 22 23.4	+26 59 51	+0.004	+0.05	8.0	1.6		M5 III	-90			T And, v
1794	53844		0 22 23.5	+35 32 12	0.000	0.00	7.6	1.4	-0.3	K5 III		380 s		
1796	91866	42 Psc	0 22 25.4	+13 28 58	+0.004	+0.03	6.23	1.22	0.2	K0 III	+3	120 s	303	m
1860	248174		0 22 34.1	-61 02 06	+0.010	+0.01	7.3	-0.1	0.6	A0 V		110 mx		
1854	192490		0 22 46.4	-35 23 11	+0.011	+0.04	7.6	1.2	4.4	G0 V		20 s		
1811	21334		0 22 47.8	+54 38 43	-0.001	0.00	7.3	0.9	3.2	G5 IV		66 s		
1856	192491		0 22 48.1	-39 15 45	+0.010	-0.03	6.9	0.2	2.6	F0 V		71 s		
1826	73938		0 22 49.6	+29 27 17	+0.005	+0.01	6.9	0.1	1.4	A3 V	+5	110 s		
1835	147237	9 Cet	0 22 51.7	-12 12 34	+0.027	+0.06	6.39	0.66	4.7	dG2	-7	23 ts		m
1855	192493		0 22 51.7	-37 24 15	-0.001	-0.02	7.1	1.5	-0.3	K5 III		300 s		
1831	53849		0 22 57.3	+38 45 15	+0.001	0.00	6.9	1.6	-0.5	M2 III	-23	300 s		
1868	128728		0 22 59.9	-6 27 47	-0.003	-0.01	7.7	1.1	0.2	K0 III		310 s		
1832	73940		0 23 00.0	+22 22 31	+0.013	-0.21	7.62	0.64	4.0	F8 V		45 s		
1848	36309		0 23 03.0	+44 31 32	-0.002	+0.01	7.2	0.1	1.4	A2 V		150 s		
1879	147241		0 23 04.3	-15 56 33	+0.005	+0.03	6.45	1.60	-0.5	M2 III	-22	230 mx		
1926	248177		0 23 04.9	-65 07 17	-0.014	-0.06	7.4	0.6	4.4	G0 V		40 s		
1876	91871		0 23 12.0	+20 05 09	+0.002	-0.02	7.9	0.4	2.6	F0 V		110 s		
1909	192495		0 23 12.6	-31 02 10	+0.002	0.00	6.55	0.00	0.4	B9.5 V		170 s		
1845	21343		0 23 14.5	+55 47 33	+0.006	0.00	6.6	1.6		M5 III	-12	14 mn		T Cas, v
1844	21345		0 23 17.6	+60 14 08	-0.002	-0.01	7.62	0.30	0.6	A0 V		160 s		
1842	11152		0 23 18.4	+65 21 09	+0.004	+0.01	7.4	0.4	2.6	F0 V		89 s		
1843	11153		0 23 19.3	+62 18 50	+0.001	-0.01	8.0	1.4	-0.3	K5 III		390 s		
1923	192498		0 23 22.2	-29 50 50	+0.003	-0.03	6.9	1.5	-0.5	M2 III		310 s		
---	21348		0 23 22.4	+59 35 44	0.000	+0.01	8.0	2.2	-0.3	K7 III		180 s		
1873	21351		0 23 24.1	+55 48 08	+0.002	+0.01	8.0	0.0	0.4	B9.5 V		310 s		
1921	147245		0 23 25.1	-18 53 30	+0.004	-0.03	7.2	1.1	3.2	G5 IV		37 s		
1948	232016		0 23 27.4	-53 50 22	0.000	+0.02	8.0	0.4	1.4	A2 V		120 s		
1907	91874		0 23 27.4	+11 50 12	+0.001	+0.03	8.0	1.1	0.2	K0 III		360 s		
1963	232017		0 23 32.8	-54 03 15	-0.006	+0.02	7.7	0.8	0.2	K0 III		310 s		
1919	36320		0 23 38.1	+42 03 21	-0.004	-0.04	8.0	0.5	3.4	F5 V		82 s		
1935	109173		0 23 38.4	+2 44 34	+0.002	-0.01	7.8	0.4	2.6	F0 V		110 s		
1985	248178		0 23 42.7	-62 25 30	+0.031	+0.09	7.1	0.8	3.2	G5 IV		59 s		
1918	36322		0 23 43.1	+45 05 21	+0.002	+0.03	7.61	0.96	0.2	G9 III	+36	300 s		
1973	215079		0 23 43.7	-49 21 29	+0.001	-0.01	7.9	1.4	-0.5	M2 III		470 s		
1917	11157		0 23 52.8	+60 57 45	-0.020	-0.06	6.58	0.36	3.0	F2 V		51 s		
1978	128739		0 23 59.8	-3 28 31	+0.002	-0.01	7.80	0.1	0.6	A0 V		280 s	326	m

HD	SAO	Star Name	α 2000	δ 2000	μ(α)	μ(δ)	V	B−V	M$_v$	Spec	RV	d(pc)	ADS	Notes
1967	53860		0^{h}24^{m}01.9^s	+38°34'38"	−0.001^s	−0.02"	7.39	1.97		S6.6 e	−11			R And, m,v
1987	128740		0 24 03.5	−9 19 37	+0.001	+0.04	7.2	1.6		M5 III	+33			S Cet, v
1952	36330		0 24 05.2	+44 15 53	−0.001	0.00	6.6	0.4	3.0	F2 V	−4	53 s		
1951	21361		0 24 07.8	+57 05 42	−0.002	−0.05	7.7	1.1	−0.1	K2 III		340 s		
2029	232024		0 24 12.1	−54 34 02	+0.005	−0.07	7.8	−0.4	2.6	F0 V		110 mx		
2008	166239		0 24 13.2	−23 24 12	+0.001	−0.02	7.7	0.6	3.2	G5 IV		80 s		
1966	21363		0 24 14.7	+53 49 20	+0.007	−0.04	7.5	0.5	3.4	F5 V		65 s		
1976	21366		0 24 15.6	+52 01 12	+0.001	0.00	5.57	−0.12	−1.6	B5 IV	−12	270 s	328	m
1950	11167		0 24 17.1	+60 27 11	−0.002	0.00	7.58	0.15	0.4	B9.5 V		210 s		
2025	166242		0 24 25.9	−27 01 36	+0.051	+0.09	7.93	0.95	6.9	K3 V	+6	15 s		
2023	128743		0 24 29.5	−2 13 08	−0.002	−0.03	6.07	1.22	0.0	gK1	+15	130 s		
2024	128744		0 24 32.8	−4 38 53	−0.001	−0.01	7.5	1.6	−0.5	M2 III		390 s		
2035	91881		0 24 38.1	+14 18 56	+0.001	0.00	6.7	1.1	0.2	K0 III	−16	200 s		
2019	53870		0 24 39.6	+31 22 24	0.000	0.00	6.8	0.0	0.4	B9.5 V	+6	190 s		
2071	232028		0 24 42.5	−53 59 03	+0.025	−0.03	7.40	0.7	4.4	G0 V		40 s		m
2034	91882		0 24 44.7	+20 04 14	−0.001	−0.03	8.0	1.1	0.2	K0 III		360 s		
2070	232027		0 24 45.0	−51 02 39	+0.061	−0.26	6.82	0.60	4.1	G4 IV−V		36 s		
2011	11172	12 Cas	0 24 47.4	+61 49 52	+0.002	0.00	5.40	0.00	0.4	B9.5 V	−6	98 s		
2066	147258		0 24 49.6	−18 28 18	0.000	−0.05	7.0	1.2	3.2	G5 IV		34 s		
2088	215085		0 24 56.4	−41 53 09	+0.009	+0.02	7.5	0.9	3.2	G5 IV		61 s		
2121	248179		0 24 58.0	−62 12 23	+0.017	0.00	8.0	0.5	4.0	F8 V		63 s		
2098	192509		0 25 01.3	−30 41 54	−0.013	−0.31	7.58	0.62	4.4	G0 V		39 s		
−−	53874		0 25 04.9	+39 08 46	+0.002	+0.01	8.0	1.5						
2054	21381		0 25 06.3	+53 02 48	+0.002	0.00	5.7	−0.1	0.4	B9.5 V		110 s		
2057	36347		0 25 08.5	+48 02 53	+0.027	+0.03	7.7	0.5	4.0	F8 V		54 s		
2148	−−		0 25 09.3	−65 24 45			7.8	0.5	3.4	F5 V		75 s		
2093	53876		0 25 17.6	+35 02 14	−0.001	−0.01	7.8	1.4	−0.3	K5 III		410 s		
2132	192512		0 25 21.9	−39 15 36	+0.001	−0.08	6.9	1.0	0.2	K0 III		160 mx		
2129	147263		0 25 22.8	−17 25 39	+0.001	−0.05	7.3	1.1	−0.1	K2 III		300 s		
2114	109192	44 Psc	0 25 24.1	+1 56 23	−0.001	−0.01	5.77	0.86	0.3	gG5	−4	120 s		
2092	53879		0 25 24.6	+39 49 48	+0.005	−0.01	7.9	1.1	0.2	K0 III		340 mx		
2228	−−		0 25 35.2	−67 44 45			7.9	1.1	0.2	K0 III		350 s		
2161	166259		0 25 36.6	−19 56 03	−0.001	−0.05	7.9	0.4	3.4	F5 V		78 s		
2203	232035		0 25 38.1	−51 24 11	0.000	+0.02	7.8	0.8	3.2	G5 IV		81 s		
2128	73973		0 25 40.2	+22 56 42	+0.002	+0.01	8.0	0.1	1.4	A2 V		210 s		
2112	21387		0 25 41.7	+54 59 40	0.000	+0.01	6.9	1.1	−0.1	K2 III		240 s		
2140	109195		0 25 41.9	+7 41 28	+0.003	−0.04	7.1	1.1	0.2	K0 III	−18	240 s		
2178	166260		0 25 42.3	−21 37 54	+0.001	0.00	7.63	0.06	1.2	A1 V n		190 s		
2102	21388		0 25 44.5	+57 57 50	+0.004	0.00	7.51	1.68	−0.5	M2 III		350 s		
2199	166262		0 25 45.5	−27 42 39	+0.002	−0.05	7.0	1.3	0.2	K0 III		140 s		
2151	255670	β Hyi	0 25 46.0	−77 15 15	+0.688	+0.33	2.80	0.62	3.8	G1 IV	+23	6.3 t		
2137	36358		0 25 49.6	+49 39 25	−0.001	0.00	8.0	0.0	0.0	B8.5 V		370 s		
2083	4114		0 25 51.1	+71 48 26	0.000	+0.01	6.89	0.03	−3.5	B1 V	−5	900 s		
2110	11180		0 25 52.0	+65 23 02	−0.001	+0.02	7.6	0.0	0.4	B9.5 V		250 s		
2195	128758		0 25 52.5	−8 21 06	0.000	−0.01	7.8	1.1	0.2	K0 III		330 s		
2154	53888		0 25 53.4	+35 18 21	−0.003	0.00	8.03	0.13	1.4	A2 V		200 s		
2224	192520		0 25 56.2	−30 44 53	+0.001	+0.01	7.6	0.7	3.2	G5 IV		75 s		
2189	53890		0 26 02.6	+35 49 10	+0.002	+0.01	7.51	0.14	1.7	A3 V		140 s		
2191	91892		0 26 03.7	+20 08 49	0.000	+0.01	6.5	0.0	0.0	B8.5 V		200 s		
2152	21392		0 26 06.5	+55 29 29	+0.002	+0.01	7.3	1.1	0.2	K0 III	−10	250 s		
2170	21395		0 26 11.8	+56 46 46	−0.007	−0.04	6.68	0.87	0.3	G5 III		190 s	350	m
2262	215092	κ Phe	0 26 12.1	−43 40 48	+0.010	+0.04	3.94	0.17	1.7	A3 V	+9	19 ts		
2188	36363		0 26 12.9	+49 23 20	−0.001	−0.02	8.0	0.1	0.6	A0 V		280 s		
2259	192523		0 26 13.4	−39 38 56	−0.002	−0.05	7.2	0.7	0.2	K0 III		250 s		
2257	166267		0 26 16.0	−21 40 23	+0.002	−0.01	7.5	1.2	3.2	G5 IV		41 s		
2235	109206		0 26 16.4	+3 49 33	+0.001	0.00	6.90	0.00	0.0	B8.5 V		240 s		m
2261	215093	α Phe	0 26 17.0	−42 18 22	+0.019	−0.39	2.39	1.09	0.2	K0 III	+75	24 s		[Ankaa], m
2207	21398		0 26 19.3	+51 16 49	0.000	0.00	6.90	0.5	−2.0	F4 II var	−22	540 s		TU Cas, v
2268	147275		0 26 21.5	−18 41 37	0.000	0.00	6.4	1.5	−0.5	M2 III		240 s		
2219	36365		0 26 21.6	+49 02 55	−0.001	−0.01	7.1	1.1	0.2	K0 III		220 s		
2234	73985		0 26 21.9	+24 44 06	−0.001	−0.02	7.5	0.8	3.2	G5 IV		72 s		
2218	21401		0 26 30.9	+56 06 48	+0.002	0.00	8.0	0.0	0.4	B9.5 V		310 s		
2273	128760	10 Cet	0 26 37.3	−0 02 59	+0.005	0.00	6.19	0.90	0.3	gG4	−23	140 s		
2244	21405		0 26 39.6	+56 38 28	+0.004	0.00	6.6	0.1	0.6	A0 V		150 s		
2254	36370		0 26 40.2	+47 05 00	−0.001	−0.02	7.8	1.1	−0.1	K2 III		350 s		
2265	53899		0 26 40.6	+34 11 03	+0.001	+0.01	7.7	1.6	−0.5	M2 III		440 s		
2354	248186		0 26 42.0	−69 18 26	+0.011	−0.02	7.1	2.1	0.2	K0 III		52 s		
2304	109212		0 26 59.5	+7 07 22	+0.001	+0.02	8.0	1.1	0.2	K0 III		360 s		
2302	73997		0 27 04.7	+25 02 32	+0.009	−0.01	6.7	0.5	3.4	F5 V		47 s		
2324	128766		0 27 05.4	−5 00 13	+0.001	−0.02	7.1	1.1	0.2	K0 III		240 s		
2335	147282		0 27 06.9	−9 52 23	−0.001	−0.03	7.4	0.9	3.2	G5 IV		70 s		
2169	4118		0 27 07.2	+80 03 08	+0.008	+0.01	6.5	0.0	0.4	B9.5 V	+6	160 s		
2313	53905		0 27 07.8	+31 10 39	+0.001	+0.02	7.5	1.6	−0.5	M2 III	+30	410 s		
2312	53904		0 27 07.8	+38 46 42	0.000	0.00	7.8	1.1	0.2	K0 III		330 s		
2291	11206		0 27 10.7	+60 40 01	0.000	0.00	7.90	1.1	0.2	K0 III		310 s		m

HD	SAO	Star Name	α 2000	δ 2000	μ(α)	μ(δ)	V	B–V	M_V	Spec	RV	d(pc)	ADS	Notes
			h m s	° ′ ″	s	″								
2316	74000		0 27 11.0	+21 15 29	+0.002	0.00	8.00	0.1	1.7	A3 V		180 s	362	m
2315	74001		0 27 12.4	+25 34 42	−0.003	−0.03	7.90	1.1	−0.2	K3 III	−36	420 s		
2301	36374		0 27 12.5	+49 59 08	+0.002	0.00	6.90	0.00	0.4	B9.5 V		190 s	361	m
2363	166282		0 27 14.6	−25 32 50	+0.002	−0.01	5.98	1.03	3.2	G5 IV		23 s		
2366	192533		0 27 17.7	−33 52 55	0.000	0.00	7.9	1.9	−0.3	K5 III		240 s		
2344	109216		0 27 20.2	+2 48 51	−0.001	0.00	7.7	0.9	0.3	G4 III	0	310 s		
2361	128769		0 27 20.5	−8 39 31	+0.002	−0.04	7.9	0.4	3.0	F2 V		96 s		
2376	166284		0 27 21.9	−28 14 36	−0.003	0.00	6.9	1.1	3.2	G5 IV		37 s		
2300	21412		0 27 22.1	+59 45 27	−0.002	0.00	7.4	1.1	0.2	K0 III	−32	250 s		
2330	74006		0 27 26.8	+26 16 43	+0.007	−0.07	7.8	0.5	4.0	F8 V		57 s		
2381	192537		0 27 30.1	−31 49 43	+0.008	+0.02	7.5	0.4	3.0	F2 V		79 s		
2358	91903		0 27 31.0	+16 01 32	+0.006	+0.01	6.45	0.24		A5	−4			m
2357	53910		0 27 35.6	+34 02 05	+0.002	−0.01	8.00	0.9	3.2	G8 IV	+1	91 s		
2405	215101		0 27 37.4	−41 22 57	−0.001	0.00	8.0	0.8	2.6	F0 V		63 s		
2395	166285		0 27 37.5	−20 08 05	+0.004	0.00	6.80	0.22	1.5	A7 IV		110 s		
2394	147286		0 27 40.3	−16 24 34	+0.013	+0.04	7.30	0.5	4.0	F8 V		46 s	366	m
2329	21415		0 27 40.5	+58 33 14	+0.001	0.00	7.43	0.07	−1.7	B3 V	−14	500 s		
2429	192545	η Sc1	0 27 55.7	−33 00 26	−0.001	−0.04	4.81	1.64		gM5	+11	47 mn		
2410	91909		0 27 58.4	+19 30 51	−0.001	−0.01	6.6	1.1	0.2	K0 III	+8	190 s		
2438	147289		0 28 00.5	−11 39 32	+0.003	−0.01	7.5	1.6	−0.5	M4 III		390 s		
2458	232052		0 28 02.2	−52 57 09	+0.007	+0.01	7.4	1.0	0.2	K0 III		270 s		
2411	91910	47 Psc	0 28 02.9	+17 53 36	+0.008	+0.02	5.2	1.6	−0.5	M3 III	+6	140 s		TV Psc, v
2423	74020		0 28 10.2	+20 47 46	0.000	0.00	7.6	1.1	0.2	K0 III		300 s		
2436	91912	48 Psc	0 28 12.6	+16 26 42	+0.001	−0.01	6.06	1.58	−0.1	gK2	−7	82 s		
2421	36390		0 28 13.7	+44 23 40	+0.009	−0.01	5.17	0.03		A2	+2	28 mn		
2455	166295		0 28 13.9	−29 02 25	−0.004	−0.02	6.8	1.4	3.2	G5 IV		21 s		
2454	109224		0 28 19.9	+10 11 23	+0.002	−0.20	6.04	0.43	3.0	F2 V	−10	38 s		
2516	248194		0 28 20.8	−64 43 39	+0.022	+0.05	7.8	0.4	2.6	F0 V		110 s		
2475	166296		0 28 21.0	−20 20 06	−0.008	−0.10	6.43	0.59	3.1	G4 IV		46 s		m
2433	36396		0 28 21.9	+45 13 38	+0.003	−0.01	7.9	1.1	0.2	K0 III		340 s		
2476	166297		0 28 22.8	−22 51 18	+0.001	−0.01	7.7	1.7	0.2	K0 III		120 s		
2490	215103		0 28 26.4	−39 54 54	+0.011	−0.03	5.43	1.56	−0.3	gK5	+32	130 s		
2477	166299		0 28 26.4	−29 17 05	+0.004	−0.03	7.2	0.0	3.4	F5 V		59 s		
2452	53920		0 28 27.7	+33 47 57	0.000	−0.01	7.3	1.1	0.2	K0 III		260 s		
2453	53921		0 28 28.5	+32 26 15	−0.002	−0.02	6.92	0.07		A2 p	−18			
2443	36400		0 28 30.5	+47 43 31	0.000	−0.02	8.0	1.1	−0.1	K2 III		380 s		
2488	147292		0 28 33.3	−11 14 14	+0.008	0.00	6.90	0.5	3.4	F5 V		50 s		m
2472	53924		0 28 36.8	+33 38 02	+0.002	+0.01	7.5	1.1	0.2	K0 III		290 s		
2530	232057		0 28 41.3	−54 25 52	+0.003	0.00	8.0	0.8	0.2	K0 III		360 s		
2529	232055		0 28 43.0	−50 31 58	+0.016	0.00	6.3	0.7	0.2	K0 III		140 mx		
2527	166303		0 28 50.7	−24 38 12	+0.001	+0.01	7.13	0.31	0.6	F0 III n		200 s		m
2469	21428		0 28 54.0	+57 12 36	−0.002	0.00	7.3	0.6	4.4	G0 V		38 s		
2507	53928		0 28 56.5	+36 54 00	−0.001	+0.01	6.26	0.92	0.3	G5 III	+10	150 s		
2486	36408		0 28 57.0	+48 24 50	+0.001	+0.01	7.4	1.6	−0.5	M2 III		360 s		
2523	91918		0 29 08.5	+11 19 13	+0.001	+0.01	8.00	0.4	2.6	F0 V		120 s	392	m
2553	147299		0 29 10.3	−11 35 54	+0.001	+0.03	7.6	1.1	−0.1	K2 III		350 s		
2506	21431		0 29 14.0	+59 28 11	−0.007	+0.01	7.80	0.8	0.3	G4 III	−55	280 s		
2552	74035		0 29 20.8	+28 49 42	+0.004	+0.03	7.31	1.47	−0.2	K3 III	+32	240 s		
2608	215112		0 29 22.9	−40 39 56	0.000	−0.01	7.4	0.2	0.2	K0 III		270 s		
2593	128784		0 29 29.3	−3 28 09	+0.002	+0.02	7.0	1.4	−0.3	K5 III		290 s		
2564	74038		0 29 31.0	+24 52 51	−0.002	−0.08	7.9	0.5	4.0	F8 V		59 s		
2592	91923		0 29 32.0	+10 23 15	+0.003	+0.03	8.0	1.1	0.2	K0 III		360 s		
2615	192565		0 29 34.7	−32 34 08	+0.006	+0.19	7.62	0.43	3.4	F5 V		70 s		
2535	11225		0 29 36.1	+62 03 51	+0.007	0.00	7.00	1.06	0.2	K0 III		200 s		
2612	128787		0 29 38.9	−2 50 26	+0.005	−0.04	7.0	1.1	−0.1	K2 III		170 mx		
2591	53935		0 29 39.1	+32 59 13	−0.001	−0.01	7.8	0.8	3.2	G5 IV		83 s		
2559	21441		0 29 47.7	+57 01 50	+0.001	−0.01	7.2	0.0	0.4	B9.5 V		220 s		
2632	192567		0 29 48.8	−32 07 01	0.000	−0.05	6.57	1.34		K0				
2622	109234		0 29 50.7	+9 41 51	+0.001	−0.03	8.0	1.1	0.2	K0 III		360 s		
2624	128788		0 29 50.9	−0 19 15	+0.003	+0.03	7.70	0.9	0.3	G6 III	+14	300 s		
2611	53939		0 29 51.6	+38 19 27	−0.003	−0.08	7.5	0.5	4.0	F8 V		51 s		
2630	147307		0 29 51.8	−14 51 51	+0.010	−0.03	6.14	0.37	2.6	F4 IV–V		50 s		
2629	128789		0 29 55.3	−1 07 04	+0.010	−0.07	7.6	0.4	2.6	F0 V	0	67 mx		
2637	128791	12 Cet	0 30 02.3	−3 57 26	+0.001	−0.01	5.72	1.56	−0.4	M0 III	+5	170 s	410	m
2628	74041	28 And	0 30 07.2	+29 45 06	+0.003	−0.05	5.23	0.24		A m	−10	35 mn	409	m
2648	109238		0 30 08.6	+4 51 35	−0.002	+0.01	6.6	0.5	3.4	F5 V		44 s		
2610	11228		0 30 16.2	+64 44 57	+0.003	+0.01	7.27	−0.02		A0				
2656	91927		0 30 18.2	+15 47 43	+0.002	+0.02	7.1	1.1	0.2	K0 III	+10	240 s		
2626	21457		0 30 19.8	+59 58 39	+0.002	−0.01	5.95	−0.01		B9	−20		412	m
2696	166318		0 30 22.6	−23 47 16	−0.002	+0.02	5.19	0.12	1.7	A3 V	+1	48 s		
2726	215119		0 30 26.0	−48 12 54	+0.013	−0.08	5.69	0.35	2.6	dF0	+2	39 s		
2724	215120		0 30 27.7	−40 56 22	0.000	+0.03	6.19	0.33	2.6	dF0		50 s		
2723	192574		0 30 29.3	−39 31 07	+0.005	−0.07	6.6	0.9	0.2	K0 III		120 mx		
2666	53944		0 30 29.3	+32 10 36	+0.001	−0.02	7.6	0.5	3.4	F5 V	−8	70 s		
2751	232070		0 30 39.5	−54 58 12	+0.004	+0.04	7.8	0.1	1.7	A3 V		170 s		

11

HD	SAO	Star Name	α 2000	δ 2000	μ(α)	μ(δ)	V	B-V	M_v	Spec	RV	d(pc)	ADS	Notes
			h m s	° ' "	s	"								
2688	53947		0 30 39.8	+32 08 11	0.000	−0.02	7.60	0.4	3.0	F2 V		83 s		m
2654	11235		0 30 41.4	+62 21 13	−0.001	−0.01	7.38	−0.01		B3	−1			
2665	21465		0 30 45.2	+57 03 54	+0.004	−0.06	7.7	0.9	3.2	G5 IV		77 s		
2714	91934		0 30 47.1	+16 02 15	−0.002	0.00	6.90	0.1	0.6	A0 V		180 s	420	m
2702	36429		0 30 52.3	+41 31 16	−0.001	−0.05	7.6	0.1	0.6	A0 V		250 s		
2589	4130		0 30 54.8	+77 01 11	+0.098	−0.02	6.21	0.84	3.2	K0 IV	+19	40 s		
2760	147317		0 30 59.8	−10 05 03	−0.002	0.00	6.86	0.20		A3			426	m
--	21468		0 31 00.1	+55 47 29	+0.002	+0.01	7.9	1.9	−0.1	K2 III		150 s		
2712	36432		0 31 02.4	+50 15 15	+0.001	+0.02	7.4	1.1	0.2	K0 III		260 s		
2710	21470		0 31 03.7	+57 21 54	−0.005	0.00	7.7	0.1	0.6	A0 V		250 s		
2748	109247		0 31 03.8	+9 18 19	−0.003	−0.04	7.9	1.1	−0.1	K2 III		390 s		
2701	21471		0 31 04.0	+55 42 08	+0.012	−0.10	6.8	0.9	3.2	G5 IV		51 s		
2739	36436		0 31 12.8	+43 56 48	+0.001	0.00	6.6	0.0	0.0	B8.5 V	−6	210 s		
2663	11239		0 31 12.9	+69 47 08	+0.058	−0.07	7.5	0.5	4.0	F8 V	+11	49 s		
2811	215129		0 31 18.3	−43 36 25	−0.001	−0.02	7.6	−0.1	1.4	A2 V		180 s		
2822	215130		0 31 23.0	−45 54 06	0.000	0.00	7.7	0.9	3.2	G5 IV		67 s		
2834	215131	λ¹ Phe	0 31 24.9	−48 48 13	+0.014	+0.03	4.77	0.02	0.6	A0 V	−5	52 mx		m
2766	53955		0 31 25.1	+38 58 51	+0.002	0.00	7.32	0.11	1.7	A3 V		130 s		
2729	11243	13 Cas	0 31 25.2	+66 31 11	+0.004	0.00	6.18	−0.10	−0.9	B6 V	−10	180 mx		
2767	53956		0 31 25.5	+33 34 54	+0.004	−0.01	5.87	1.14	0.0	K1 III	+9	140 s		m
2792	91941		0 31 30.2	+11 31 56	−0.002	−0.04	7.3	0.2	2.1	A5 V		110 s		
2779	74061		0 31 30.8	+20 49 44	+0.003	−0.02	7.40	1.4	−0.3	K4 III	+4	350 s		
2778	74062		0 31 31.2	+21 06 37	+0.001	+0.01	7.9	1.4	−0.3	K5 III		430 s		
2808	128805		0 31 31.5	−4 16 56	−0.005	−0.15	7.7	0.5	4.0	F8 V		55 s		
2884	248201	β¹ Tuc	0 31 32.7	−62 57 30	+0.014	−0.05	4.37	−0.07	−0.2	B8 V	+10	33 mx		m
2885	248202	β² Tuc	0 31 33.6	−62 57 57	+0.015	−0.05	4.53	0.15		A2 V	+10			m
2871	232078		0 31 35.8	−53 07 05	+0.004	0.00	7.2	0.1	1.4	A2 V		130 s		
2806	91942		0 31 36.0	+16 01 15	0.000	−0.04	7.0	1.1	0.2	K0 III	−6	230 s		
2807	109252		0 31 37.1	+9 09 40	−0.002	−0.01	7.2	1.1	−0.1	K2 III		300 s		
2830	128806		0 31 40.6	−1 47 37	+0.001	−0.01	7.03	0.08		A0				
2774	21486		0 31 41.1	+52 50 22	−0.006	−0.02	5.60	1.15	−0.1	K2 III	−52	140 s		
2816	91945		0 31 41.4	+12 54 49	−0.005	−0.15	7.4	1.1	0.2	K0 III		130 mx		
2815	--		0 31 43.3	+19 28 55	−0.003	−0.02	8.0	0.9	3.2	G5 IV		93 s		
2772	21489	14 λ Cas	0 31 46.3	+54 31 20	+0.005	−0.01	4.73	−0.10		B8	−12	13 mn	434	m
2805	36443		0 31 47.4	+45 55 35	+0.004	−0.03	7.6	0.9	3.2	G5 IV		74 s		
2804	36444		0 31 52.8	+48 39 29	+0.002	0.00	7.6	0.1	1.4	A2 V		170 s		
2841	91949		0 31 54.7	+19 38 33	−0.001	−0.01	7.30	1.4	−0.3	K5 III	+4	330 s		
2824	21492		0 32 04.1	+55 59 39	−0.001	−0.01	7.4	1.1	0.2	K0 III	−18	260 s		
2894	147325		0 32 04.7	−18 13 20	+0.003	−0.03	8.00	1.1	0.2	K0 III		360 s		m
2826	21494		0 32 06.9	+53 00 11	−0.002	+0.03	7.8	1.1	−0.1	K2 III		350 s		
2825	21496		0 32 11.1	+53 48 55	−0.003	0.00	7.0	1.1	−0.1	K2 III	−28	250 s		
2866	53969		0 32 13.9	+34 59 39	0.000	−0.02	6.73	0.17	2.1	A5 V	+3	84 s		
2851	21498		0 32 21.4	+58 20 19	−0.002	−0.02	7.00	0.4	3.0	F2 V		62 s	443	m
2916	166339		0 32 22.0	−24 38 49	+0.009	0.00	7.3	0.7	3.0	F2 V		44 s		
2913	109262	51 Psc	0 32 23.7	+6 57 20	+0.002	+0.01	5.67	0.00	0.2	B9 V	+19	120 s	449	m
2888	36453		0 32 26.7	+43 29 42	+0.002	−0.01	6.4	0.1		A0 p	−21			
2873	21502		0 32 29.9	+51 43 43	−0.002	+0.01	7.1	1.1	0.2	K0 III		230 s		
2911	91953		0 32 31.8	+18 47 40	−0.002	−0.09	7.0	1.4	−0.3	K5 III		200 mx		
2924	74083		0 32 34.4	+27 34 50	−0.001	−0.01	6.6	0.1	0.6	A0 V	+2	160 s		
2910	74084	52 Psc	0 32 35.4	+20 17 40	+0.009	−0.04	5.38	1.08	0.2	gK0	−13	95 s	452	m
2925	74086		0 32 39.4	+23 11 35	+0.008	+0.02	6.84	0.92	0.2	K0 III	−112	210 s	451	m
2947	166344		0 32 39.4	−25 21 34	0.000	−0.04	6.90	1.1	0.2	K0 III		220 s	456	m
2923	53974		0 32 43.2	+38 50 38	0.000	−0.02	7.1	0.4	2.6	F0 V		79 s		
3003	248208		0 32 43.8	−63 01 53	+0.013	−0.03	5.09	0.04		A2	+5	13 mn		m
2908	36455		0 32 44.3	+43 13 15	+0.005	−0.01	8.0	1.1	−0.1	K2 III		360 mx		
2900	21505		0 32 44.6	+55 15 37	+0.001	0.00	8.0	1.1	0.2	K0 III		340 s		
2901	21506		0 32 47.4	+54 07 10	+0.009	−0.05	6.89	1.25	−0.1	K2 III	−107	220 s		
2942	74090		0 32 49.0	+28 16 49	−0.001	+0.01	6.30	1.00	0.2	K0 III	−12	170 s	455	m
2933	53976		0 32 51.7	+35 51 29	−0.001	−0.01	7.94	−0.02	0.2	B9 V		350 s	454	m
2954	91956		0 32 55.6	+16 10 00	+0.002	+0.01	6.90	0.5	3.4	F5 V		50 s	458	m
2953	91958		0 32 59.2	+18 23 42	−0.001	−0.03	7.6	0.4	3.0	F2 V		83 s		
2905	11256	15 κ Cas	0 32 59.9	+62 55 55	0.000	0.00	4.16	0.14	−6.6	B1 Ia	−2	930 s		
3013	215138		0 33 07.1	−39 45 35	−0.002	−0.03	7.9	1.0	0.2	K0 III		340 s		
2952	21512		0 33 10.3	+54 53 42	+0.008	−0.04	5.93	1.04	0.2	K0 III	−35	130 mx		
2995	128820		0 33 11.3	−3 50 53	−0.001	0.00	7.4	1.1	0.2	K0 III		280 s		
3025	215139		0 33 11.7	−46 57 07	+0.002	−0.01	7.6	−0.2	2.1	A5 V		130 s		
2994	91965		0 33 18.4	+13 50 26	+0.007	+0.03	7.8	0.5	3.4	F5 V		75 s		
2904	4142		0 33 19.4	+70 58 55	+0.008	+0.01	6.4	0.1	0.6	A0 V	−10	100 mx		
3112	255679	θ Tuc	0 33 23.4	−71 16 00	+0.017	−0.02	6.13	0.23		A5	+2			
--	192605		0 33 27.7	−36 47 21	+0.005	+0.01	7.9	0.9	4.4	G0 V		33 s		
3024	128823		0 33 29.1	−0 36 31	0.000	−0.02	7.1	0.1	1.7	A3 V		120 s		
3075	232087		0 33 29.4	−55 19 43	−0.001	0.00	7.90	0.1	1.4	A2 V		200 s		m
2993	53983		0 33 30.1	+40 06 23	+0.002	+0.04	8.00	0.5	4.0	F8 V		63 s	463	m
3047	192606		0 33 32.1	−38 06 50	+0.003	−0.03	7.3	0.7	4.4	G0 V		32 s		
3093	248212		0 33 32.6	−61 08 50	+0.005	+0.01	8.00	0.1	0.6	A0 V		210 mx		m

HD	SAO	Star Name	α 2000	δ 2000	μ(α)	μ(δ)	V	B-V	M_V	Spec	RV	d(pc)	ADS	Notes
			h m s	° ′ ″	s	″								
2974	11260		0 33 34.3	+60 32 56	+0.001	+0.01	7.90	-0.01		B9	+4			
3045	166362		0 33 36.4	-26 05 34	0.000	-0.03	7.50	1.1	0.2	K0 III		290 s	466	m
3029	74098		0 33 39.5	+20 26 02	0.000	+0.01	7.84	0.27		A3				
3059	166367		0 33 41.0	-29 33 31	-0.002	-0.03	5.55	1.27	-0.1	K2 III		120 s		
3074	192609		0 33 43.8	-35 00 07	-0.004	-0.51	6.41	0.61	3.0	G3 IV	+29	22 mx		m
3085	215143		0 33 48.0	-39 55 28	+0.004	+0.02	7.5	-0.1	0.0	B8.5 V		320 s		
--	53989		0 33 48.6	+32 51 58	-0.001	0.00	7.9	0.6						
3070	109278		0 33 54.1	+3 19 10	-0.002	-0.03	7.80	0.8	0.3	G4 III	-10	320 s		
3128	232088		0 33 59.7	-57 05 08	-0.001	0.00	7.3	1.5	3.2	G5 IV		24 s		
3027	21522		0 34 00.6	+54 55 48	+0.007	0.00	8.0	1.1	0.2	K0 III		320 mx		
3041	36474		0 34 02.7	+43 01 37	+0.005	+0.08	7.9	1.1	0.2	K0 III		140 mx		
3137	232090		0 34 03.6	-58 12 06	+0.010	+0.02	7.4	2.4	0.2	K0 III		39 s		
3089	109282		0 34 08.3	+5 57 27	+0.002	+0.02	7.8	0.4	3.0	F2 V		92 s		
3039	21525		0 34 08.8	+59 53 57	-0.013	-0.02	7.96	0.53		G0				
3081	53993		0 34 10.5	+31 38 43	+0.003	-0.01	7.7	0.4	3.0	F2 V		88 s		
3088	91971		0 34 10.8	+10 17 58	+0.005	-0.15	6.8	0.8	3.2	G5 IV		52 s		
3051	21528		0 34 12.6	+54 03 59	-0.001	-0.01	7.6	0.1	0.6	A0 V		240 s		
3136	215146		0 34 19.4	-42 25 53	+0.008	+0.02	6.9	0.1	1.4	A2 V		88 mx		
3078	36480		0 34 23.3	+48 05 11	-0.002	-0.01	7.9	0.1	0.6	A0 V		270 s		
3038	11265	16 Cas	0 34 24.8	+66 45 02	+0.004	+0.01	6.4	-0.1		B9	-21			
3145	215148		0 34 25.3	-45 25 34	+0.003	+0.01	6.7	1.3	0.2	K0 III		140 s		
3126	128830		0 34 27.0	-6 30 15	-0.006	-0.09	6.9	0.4	3.0	F2 V		59 s		
3158	232091		0 34 27.7	-52 22 24	+0.025	+0.04	5.57	0.47	3.4	dF5	+35	26 s		
3125	128831		0 34 29.7	-4 32 48	+0.006	+0.01	7.01	0.73	4.4	dG0	+10	25 s	475	m
3079	36481		0 34 29.8	+47 54 56	+0.041	+0.06	7.37	0.54	4.0	F8 V		40 mx		
3067	11269		0 34 33.7	+62 54 14	-0.003	-0.03	7.67	0.48		F5			470	m
3066	11268		0 34 37.4	+67 31 01	+0.007	+0.02	7.3	0.1	1.7	A3 V		130 s		
3173	192621		0 34 40.4	-37 59 45	+0.006	+0.01	7.4	1.0	0.2	K0 III		240 mx		
3166	91980		0 34 55.2	+13 22 16	-0.006	-0.05	6.4	1.1	0.2	K0 III		180 s		
--	36488		0 35 00.3	+41 22 20	+0.003	-0.01	8.0	1.4						
3141	36489		0 35 00.6	+42 41 40	-0.017	-0.08	7.70	1.1	3.2	K0 IV	+1	79 s		
3187	166381		0 35 02.0	-26 07 33	+0.001	-0.01	7.64	0.98	5.3	G6 V		18 s		
3165	53999		0 35 09.6	+36 49 57	-0.001	-0.01	6.60	1.45	-0.3	K4 III	-9	240 s	486	m
3123	11273		0 35 11.3	+62 59 23	0.000	+0.01	7.27	0.59	2.6	F0 V		53 s		
3149	21537		0 35 13.7	+51 50 17	+0.005	-0.13	7.9	0.5	4.0	F8 V		59 s		
3196	128839	13 Cet	0 35 14.8	-3 35 34	+0.027	-0.02	5.20	0.56	4.0	F8 V	+9	17 ts	490	m
3232	147349		0 35 19.4	-18 33 52	+0.001	-0.07	8.0	0.9	0.2	K0 III		180 mx		
3234	166386		0 35 25.1	-24 51 50	+0.003	+0.01	7.8	1.0	0.2	K0 III		330 s		
3228	109297		0 35 28.5	+4 17 39	-0.001	-0.01	7.9	0.4	2.6	F0 V		110 s		
3184	21541		0 35 29.0	+54 12 13	+0.007	+0.01	7.1	0.1	1.4	A2 V		100 mx		
3203	54005		0 35 30.3	+35 39 33	+0.001	+0.01	7.71	0.07	1.4	A2 V		180 s		
3229	128843	14 Cet	0 35 32.7	-0 30 20	+0.009	-0.06	5.93	0.44	3.0	dF2	+6	35 s		
3303	232099		0 35 33.4	-54 49 19	+0.003	-0.05	6.06	1.01	0.2	gK0		150 s		
3277	215155		0 35 34.2	-39 44 47	+0.010	-0.17	7.8	0.6	3.2	G5 IV		45 mx		
3192	21543		0 35 35.3	+53 29 20	0.000	0.00	7.2	0.1	0.6	A0 V		200 s		
3147	11279		0 35 37.3	+67 55 37	+0.002	+0.04	6.91	2.05	-3.3	K2 Ib-II	-18	350 s		
3302	215157	λ² Phe	0 35 41.0	-48 00 04	+0.004	-0.10	5.51	0.44	4.0	F8 V	+8	20 s		
3395	--		0 35 44.6	-73 13 56			7.85	0.91		G5				
3200	21547		0 35 47.5	+55 42 26	0.000	0.00	7.8	0.0	0.4	B9.5 V		280 s		
3253	74127		0 35 48.1	+21 46 25	-0.001	-0.01	8.0	1.1	-0.1	K2 III		410 s		
3211	36497		0 35 48.7	+49 01 16	0.000	-0.01	6.9	1.1	0.0	K1 III		230 s	493	m
3255	91988		0 35 49.9	+16 34 09	-0.004	-0.13	7.6	0.4	3.0	F2 V		85 s		
3268	91990		0 35 54.7	+13 12 24	-0.010	-0.18	6.41	0.52	3.4	F5 V	-25	37 s		
3375	232105		0 36 01.0	-59 43 02	+0.012	-0.01	6.9	0.5	2.1	A5 V		56 mx		
3296	128852		0 36 01.8	-5 34 15	+0.015	-0.10	6.6	0.5	3.4	F5 V		44 s		
3266	74132		0 36 02.3	+29 59 36	+0.014	-0.40	7.94	0.70	4.7	G2 V	-49	40 s	497	m
3325	147360		0 36 02.9	-14 58 25	-0.006	-0.08	6.45	1.06	0.2	K0 III		160 s		
3294	109306		0 36 03.6	+7 09 22	+0.004	0.00	7.9	0.5	3.4	F5 V		78 s		
3239	21549		0 36 05.8	+54 50 55	+0.003	+0.01	7.8	0.1	1.4	A2 V		180 s		
3265	54015		0 36 06.6	+38 15 08	+0.002	-0.01	7.40	1.6	-0.4	M0 III	-16	360 s		
3326	166400		0 36 06.7	-22 50 32	-0.006	-0.04	6.06	0.29		A7 p				
3337	166399		0 36 07.5	-20 45 31	+0.007	+0.02	7.90	0.4	2.6	F0 V		120 s		m
3240	21551		0 36 08.2	+54 10 07	+0.002	+0.01	5.08	-0.11	-0.2	B8 V	+1	110 s		
3293	74134		0 36 09.6	+24 01 31	+0.001	-0.01	7.30	0.8	3.2	G5 IV		66 s	499	m
3250	21552		0 36 14.9	+57 18 27	+0.003	-0.02	7.6	1.1	-0.1	K2 III		320 s		
3264	36504		0 36 15.2	+48 33 23	-0.002	+0.02	7.52	-0.04	-2.5	B2 V	-5	830 s		
3322	74136		0 36 19.9	+27 15 17	+0.001	-0.01	6.2	-0.1	-6.5	B8 I pe	+1	3500 s		
3291	36506		0 36 20.3	+44 38 21	0.000	0.00	7.3	0.0	0.0	B8.5 V	-9	290 s		
3335	109308		0 36 24.2	+8 52 27	0.000	-0.02	7.9	0.9	3.2	G5 IV		88 s		
3283	11291		0 36 27.2	+60 19 35	0.000	0.00	5.79	0.29	-4.8	A3 Ib	-9	890 s		
3444	248225		0 36 37.4	-65 07 29	+0.005	-0.02	6.42	1.26		K0				
3405	215165		0 36 37.8	-49 07 57	+0.038	-0.13	6.78	0.64		G0		12 mn		m
3389	166410		0 36 42.9	-27 25 08	-0.001	+0.03	7.90	0.4	3.0	F2 V		96 s	514	m
3345	21562		0 36 46.4	+54 57 27	-0.005	-0.01	7.2	1.1	-0.1	K2 III	-30	270 s		
3346	36509		0 36 46.5	+44 29 19	-0.002	+0.04	5.13	1.60	-0.3	K5 III	-33	110 s		

HD	SAO	Star Name	α 2000	δ 2000	μ(α)	μ(δ)	V	B-V	M_V	Spec	RV	d(pc)	ADS	Notes
			$0^h36^m47^s.3$	$+15°13'54''$	$0^s.000$	$-0''.01$								
3379	91995	53 Psc	0 36 47.3	+15 13 54	0.000	−0.01	5.89	−0.15	−1.7	B3 V	−12	330 s		
3528	255687		0 36 50.0	−76 18 35	+0.015	0.00	6.8	0.2	2.6	F0 V		71 s		
3400	147370		0 36 51.6	−11 07 40	+0.002	−0.02	7.9	0.9	3.2	G5 IV		86 s		
3369	54033	29 π And	0 36 52.8	+33 43 09	+0.001	0.00	4.36	−0.14	−1.1	B5 V	+9	120 s	513	m
3360	21566	17 ζ Cas	0 36 58.2	+53 53 49	+0.002	−0.01	3.67	−0.19	−2.5	B2 V	+2	170 s		
3368	21569		0 37 01.5	+55 13 03	+0.001	0.00	7.8	1.1	−0.1	K2 III		350 s		
3411	74148		0 37 07.2	+24 00 51	−0.001	−0.04	6.4	1.1	0.2	K0 III	−1	170 s		
3913	258234		0 37 10.6	−85 15 10	+0.060	−0.04	8.0	1.9	−0.3	K5 III		170 mx		
3434	92001		0 37 17.9	+11 26 11	+0.003	0.00	7.3	0.1	1.4	A2 V		150 s		
3488	232114		0 37 18.1	−54 23 39	+0.007	−0.01	6.41	1.00	0.2	gK0		180 s		
3459	192642		0 37 18.7	−35 25 15	+0.003	−0.02	7.9	1.1	4.4	G0 V		24 s		
3443	166418		0 37 20.6	−24 46 02	+0.102	−0.01	5.57	0.72	5.2	G5 V	+17	13 ts	520	m
3421	54038		0 37 21.1	+35 23 58	−0.001	−0.01	5.48	0.88	0.3	G5 III	0	110 s		
3460	192643		0 37 22.4	−37 17 17	+0.046	−0.03	7.01	0.75	5.2	G5 V		22 s		m
3601	255688		0 37 23.4	−75 36 22	+0.003	+0.05	7.98	0.21		A0				
3431	36517		0 37 26.4	+40 19 57	−0.001	+0.01	6.86	0.02	0.6	A0 V		170 s		
3457	109315		0 37 30.4	+3 08 07	+0.006	−0.06	6.39	1.33	−0.3	K4 III	+4	140 mx		
3410	21573		0 37 31.2	+55 13 31	+0.003	0.00	7.7	0.6	4.4	G0 V		46 s		
3509	232117		0 37 33.1	−53 43 16	−0.003	0.00	7.6	1.5	0.2	K0 III		160 s		
3409	21575		0 37 36.9	+59 10 13	+0.002	−0.02	8.0	1.1	0.2	K0 III		320 s		
3468	74154		0 37 40.6	+21 05 56	+0.004	−0.05	7.0	0.9	3.2	G5 IV		57 s		
3447	36523		0 37 44.3	+42 31 28	+0.005	−0.02	7.92	0.45		F5				
3366	4165		0 37 45.9	+72 53 43	+0.005	+0.02	6.97	0.01	−1.6	B3.5 V	−15	86 mx	516	m
−−	36526		0 37 50.8	+40 52 18	+0.002	+0.01	7.8	1.5						
3408	11301		0 37 51.3	+67 38 56	+0.001	−0.01	7.40	1.68	−0.3	K5 III		270 s		
3454	36529		0 37 56.9	+47 24 28	+0.011	−0.08	7.8	0.5	3.4	F5 V		75 s		
3507	166424		0 37 58.0	−22 20 16	+0.004	+0.03	7.8	1.5	3.2	G5 IV		30 s		
3503	109323		0 38 00.4	+2 45 49	0.000	0.00	7.62	1.08		K0				
3512	128868		0 38 04.4	−0 30 13	−0.003	−0.01	6.68	1.26	−0.1	K2 III	−56	190 s	534	m
3552	215173		0 38 06.4	−41 35 34	+0.003	−0.02	7.5	0.4	4.4	G0 V		42 s		
3522	166432		0 38 06.7	−27 37 15	−0.001	+0.04	7.9	1.1	0.2	K0 III		300 s		
3611	248228		0 38 08.1	−66 02 36	+0.019	+0.04	7.5	0.6	4.4	G0 V		42 s		
3537	147375		0 38 11.7	−16 14 42	+0.001	0.00	8.0	0.9	3.2	G5 IV		93 s		
3161	102		0 38 14.0	+84 40 30	−0.021	−0.02	7.2	0.5	4.0	F8 V		43 s		
3689	255689		0 38 17.7	−73 57 45	−0.032	−0.10	7.42	0.48	2.8	F5 IV−V		71 mx		
3532	92004		0 38 21.2	+19 42 18	+0.003	−0.02	7.9	0.5	3.4	F5 V		78 s		
3533	109333		0 38 24.3	+8 36 18	+0.001	+0.01	8.0	0.9	3.2	G5 IV		93 s		
3497	21588		0 38 24.8	+51 16 07	+0.003	−0.01	7.9	0.1	1.4	A2 V		190 s		
3598	215174		0 38 27.1	−45 54 06	−0.004	+0.05	7.5	0.0	4.0	F8 V		49 s		
3580	166438		0 38 31.6	−20 17 50	0.000	0.00	6.74	−0.11		B8				
3568	147377		0 38 33.2	−10 41 24	+0.002	+0.03	8.0	1.1	0.2	K0 III		360 s		
3546	74164	30 ε And	0 38 33.3	+29 18 42	−0.017	−0.25	4.37	0.87	0.3	G8 III	−84	41 ts		
3489	11306		0 38 33.5	+60 19 29	0.000	0.00	6.75	1.70	−3.4	K3 Ib−II	−28	630 s		
3496	21589		0 38 35.3	+57 31 00	−0.002	0.00	7.6	0.0	0.4	B9.5 V		260 s		
3581	166439		0 38 36.0	−23 35 59	−0.005	−0.06	7.2	0.2	3.4	F5 V		56 s		
3719	255690		0 38 40.9	−73 08 14	−0.002	+0.02	6.85	0.11		A0				
3531	36542		0 38 41.6	+46 57 25	−0.003	+0.01	6.90	0.9	3.2	G5 IV		54 s	538	m
4152	258237		0 38 47.2	−85 24 45	−0.057	−0.02	7.70	0.9	3.2	G5 IV		79 s		m
3605	166443		0 38 48.8	−25 06 28	+0.005	−0.02	6.60	1.1	0.2	K0 III		190 s		m
3519	21592		0 38 49.4	+59 49 34	+0.004	0.00	6.74	0.01		A0				
3622	166446		0 38 52.9	−25 35 44	−0.003	−0.01	7.77	0.22	2.4	A7 V		120 s		m
3646	215178		0 38 53.3	−43 17 45	0.000	−0.01	7.67	0.95	0.3	gG8		280 s		
3669	232130		0 38 55.8	−52 20 13	−0.001	+0.01	7.9	1.3	0.2	K0 III		240 s		
3590	74168		0 38 57.5	+26 19 24	0.000	−0.03	7.2	1.1	−0.2	K3 III	0	310 s		
3565	36547		0 38 58.7	+44 39 46	+0.005	−0.02	8.0	0.1	0.6	A0 V		100 mx		
−−	215180		0 39 03.9	−41 53 59	0.000	+0.01	7.8	1.4	0.2	K0 III		190 s		
3666	215182		0 39 04.2	−41 53 56	+0.003	+0.03	7.8	1.1	0.2	K0 III		300 s		
3574	36550		0 39 09.8	+49 21 16	+0.001	−0.01	5.43	1.64	−0.1	K2 III	−10	67 s	546	m
3698	232134		0 39 11.2	−57 58 06	+0.008	−0.03	6.9	0.2	4.0	F8 V		38 s		
3628	109348		0 39 13.2	+3 08 01	+0.052	+0.29	7.32	0.63	4.7	G2 V	−28	32 s		
3627	54058	31 δ And	0 39 19.6	+30 51 40	+0.011	−0.08	3.27	1.28	−0.2	K3 III	−7	49 s	548	m
3651	74175	54 Psc	0 39 21.7	+21 15 02	−0.033	−0.37	5.87	0.85	5.8	K0 V	−34	11 t		m
3695	192659		0 39 33.0	−29 49 14	+0.002	−0.02	7.7	1.7	0.2	K0 III		120 s		
3638	36555		0 39 34.6	+49 34 22	+0.001	−0.02	8.0	0.9	3.2	G5 IV		90 s		
3626	54061		0 39 35.2	+36 47 26	−0.003	−0.02	7.90	0.4	2.6	F0 V		120 s	553	m
3674	92011		0 39 40.6	+11 31 57	+0.010	0.00	7.6	0.5	4.0	F8 V		51 s		
4229	258238		0 39 41.3	−85 42 03	+0.176	0.00	6.80	1.26	−0.3	K5 III		100 mx		
3708	147387		0 39 43.5	−11 09 02	−0.004	−0.11	6.8	0.5	3.4	F5 V		48 s		
3703	109355		0 39 46.8	+3 39 08	+0.002	+0.02	7.8	0.8	3.2	G5 IV		84 s		
3440	106		0 39 47.3	+82 29 38	−0.055	+0.09	6.40	0.55	4.0	dF8	−33	30 s		
3948	258236		0 39 48.5	−80 39 24	+0.034	+0.01	7.90	0.74		G0				
3750	215185		0 39 51.9	−44 47 48	+0.004	+0.01	6.01	1.14	0.2	gK0		110 s		
3637	11320		0 39 51.9	+63 13 38	−0.005	−0.03	7.74	0.50	2.2	F6 IV	−28	120 s		
3690	74182	55 Psc	0 39 55.5	+21 26 18	+0.002	−0.03	5.36	1.16	−0.9	K0 II−III	−17	160 s	558	m
3735	192663		0 39 57.8	−33 57 42	+0.027	−0.11	6.69	0.51	4.0	F8 V		35 s		Z Scl, q

HD	SAO	Star Name	α 2000	δ 2000	μ(α)	μ(δ)	V	B-V	M_v	Spec	RV	d(pc)	ADS	Notes
3553	4176		0 40 00.0	+76 52 17	+0.006	0.00	7.00	1.1	0.2	K0 III		220 s		m
3682	21602		0 40 05.3	+51 52 45	+0.005	+0.01	7.7	1.1	0.2	K0 III		290 s		
3700	36564		0 40 12.8	+47 15 20	+0.001	-0.01	8.00	0.6	4.4	G0 V		52 s	559	m
3727	92016		0 40 16.7	+15 49 28	+0.001	-0.01	7.8	0.8	3.2	G5 IV		84 s		
3743	74184		0 40 19.2	+24 03 00	+0.003	+0.02	7.21	0.19	2.1	dA5	-3	100 s	562	m
3681	21604		0 40 19.7	+59 27 53	+0.001	-0.02	7.04	1.11	0.2	K0 III p	-12	220 s		
3757	74186		0 40 23.0	+23 46 12	+0.006	+0.02	8.0	1.1	0.2	K0 III		230 mx		
3770	128888		0 40 24.3	-7 39 23	+0.008	-0.09	7.9	0.5	4.0	F8 V		60 s		
3699	21608		0 40 24.6	+59 30 51	-0.001	-0.01	7.26	1.92	-0.1	K2 III		110 s		
3660	4181		0 40 25.8	+70 42 31	+0.002	0.00	7.2	0.1	0.6	A0 V		190 s		
3823	232143		0 40 26.4	-59 27 16	+0.118	+0.45	5.89	0.55	4.5	G1 V		20 ts		
3794	147395		0 40 28.5	-16 31 00	+0.002	-0.04	6.49	0.92		G5				m
3742	54068		0 40 29.0	+36 27 03	-0.001	+0.01	8.0	0.1	0.6	A0 V		300 s		
3853	232145		0 40 30.1	-55 30 40	+0.005	+0.05	6.8	2.0	-0.1	K2 III		79 s		
3712	21609	18 α Cas	0 40 30.4	+56 32 15	+0.006	-0.03	2.23	1.17	-0.9	K0 II-III	-4	37 s	561	Schedar, m,q
3795	166475		0 40 32.8	-23 48 16	+0.047	-0.33	6.14	0.70	4.9	dG3	-53	21 ts		
3809	166477		0 40 33.2	-29 25 14	+0.001	+0.02	6.8	-0.2	3.2	G5 IV		54 s		
3807	128891		0 40 42.3	-4 21 07	-0.001	-0.02	5.91	1.10	0.3	gG7	+35	100 s		m
3845	192674		0 40 47.0	-30 51 35	+0.001	-0.02	7.7	1.9	3.2	G5 IV		16 s		
3821	128892		0 40 47.4	-7 13 58	+0.001	-0.10	6.97	0.65	4.9	dG3	+5	29 ts	566	m
3765	54074		0 40 49.1	+40 11 14	+0.030	-0.67	7.36	0.94	6.6	K2 V	-63	14 t		
3819	109365		0 40 49.7	+4 28 42	-0.003	-0.02	8.0	0.9	3.2	G5 IV		93 s		
3888	232148		0 40 51.8	-53 12 36	+0.015	-0.04	7.4	0.5	4.0	F8 V		48 s		
3828	74195		0 40 59.0	+21 01 13	-0.001	-0.01	7.3	1.1	0.2	K0 III		260 s		
3817	54079	32 And	0 41 07.1	+39 27 31	-0.001	0.00	5.33	0.89	0.3	G8 III	-5	100 s		
—	36572		0 41 09.4	+49 58 02	+0.006	-0.02	7.9	0.9						
3861	109369		0 41 11.7	+9 21 19	-0.009	-0.09	6.5	0.5	3.4	F5 V	-19	43 s		
3827	54080		0 41 12.1	+39 36 14	0.000	0.00	8.01	-0.24	-1.6	B3.5 V		820 s		
3802	21621		0 41 12.6	+52 11 52	-0.004	-0.02	6.9	0.1	0.6	A0 V		170 s		
3919	215194	μ Phe	0 41 19.5	-46 05 06	-0.002	+0.01	4.59	0.97	0.3	G8 III	+17	70 s		
3838	36576		0 41 26.7	+44 48 52	+0.010	+0.06	8.0	0.9	3.2	G5 IV		93 s		
3909	166489		0 41 29.7	-20 17 59	+0.001	0.00	7.9	1.4	-0.1	K2 III		280 s		
3884	92030		0 41 31.7	+12 57 34	-0.003	-0.08	7.9	0.7	4.4	G0 V		51 s		
3917	166490		0 41 34.2	-25 11 47	-0.002	-0.03	6.7	1.1	0.2	K0 III		170 s		
3801	11331		0 41 34.7	+65 52 06	+0.013	-0.06	7.2	0.5	4.0	F8 V		42 s		
3883	74200		0 41 35.9	+24 37 44	+0.008	-0.02	6.00	0.2		A m	-15			
3894	92031		0 41 41.9	+20 08 13	+0.006	-0.06	7.8	0.5	4.0	F8 V		57 s		
3872	36580		0 41 42.3	+49 50 00	+0.008	+0.01	7.6	0.5	3.4	F5 V		70 s		
3980	232152	ξ Phe	0 41 46.3	-56 30 06	+0.011	+0.05	5.70	0.20		A5 p	+10			m
3816	11334		0 41 49.5	+67 46 49	+0.001	+0.02	7.9	0.4	2.6	F0 V		110 s		
3892	54085		0 41 51.1	+38 53 47	+0.001	0.00	7.67	-0.06	0.0	B8.5 V		340 s		
—	—		0 41 52.7	+44 30 34	+0.007	+0.07	7.7	0.6	4.4	G0 V		47 s		
—	54087		0 41 54.7	+39 02 56	-0.003	-0.01	8.0	1.6						
3979	215199		0 41 57.5	-43 07 30	+0.004	0.00	7.07	0.36		F5				
3881	21632		0 41 57.6	+59 54 54	-0.001	+0.02	7.4	0.1	1.7	A3 V	+33	130 s		
4001	232155		0 41 59.9	-55 47 00	-0.005	0.00	7.3	1.0	4.4	G0 V		21 s		m
3914	36585		0 42 01.3	+40 41 18	-0.002	-0.06	7.1	0.5	3.4	F5 V		55 s		
3945	109383		0 42 02.0	+1 01 26	+0.002	0.00	8.0	1.1	-0.1	K2 III		410 s		
3856	11336		0 42 03.5	+66 08 52	-0.001	0.00	5.83	1.04	1.7	G9 III-IV	-3	56 s		
3901	21637	19 ξ Cas	0 42 03.8	+50 30 45	+0.001	0.00	4.80	-0.11	-2.5	B2 V	-8	260 s		
3900	21638		0 42 06.0	+51 58 12	+0.003	+0.01	8.0	1.1	0.2	K0 III		330 s		
3975	147408		0 42 14.4	-11 48 18	+0.005	-0.02	6.9	1.1	0.2	K0 III		190 mx		
3934	36588		0 42 21.9	+43 56 17	+0.001	0.00	7.6	1.1	0.2	K0 III		300 s		
3972	109392		0 42 23.1	+4 09 58	+0.001	-0.05	7.59	0.51	4.0	F8 V	+8	52 s	588	m
3933	21641		0 42 26.6	+58 34 12	-0.001	-0.01	7.9	1.1	0.2	K0 III		310 s		
4089	248237	ρ Tuc	0 42 28.4	-65 28 05	+0.009	+0.05	5.39	0.50	3.3	dF4	+14	24 s		
3970	74215		0 42 28.9	+27 38 35	+0.003	0.00	7.5	0.4	2.6	F0 V		93 s		
3924	21642		0 42 31.0	+58 45 13	+0.005	0.00	6.1	0.0		B9	-2			
3992	109393		0 42 32.7	+5 40 30	0.000	-0.01	7.8	0.1	1.7	A3 V		160 s		
4066	215203		0 42 36.4	-47 11 38	+0.001	0.00	7.8	1.4	3.2	G5 IV		27 s		
4053	192689		0 42 37.4	-36 01 20	0.000	+0.02	7.1	1.6	-0.5	M2 III		310 s		
3950	21646		0 42 37.9	+52 20 13	-0.001	-0.01	6.91	0.12	-4.4	B1 III	-92	1200 s		
3941	21645		0 42 38.8	+57 35 46	-0.001	0.00	7.80	0.00	0.4	B9.5 V		280 s	587	m
3891	4191		0 42 40.2	+71 21 58	+0.006	0.00	7.3	0.1	0.6	A0 V	-15	72 mx	582	m
4036	166503		0 42 40.5	-20 11 38	0.000	-0.01	7.5	0.9	3.2	G5 IV		61 s		
3923	11341		0 42 40.8	+68 09 52	-0.001	+0.03	8.0	1.1	0.2	K0 III		320 s		
4125	248239		0 42 41.2	-65 36 32	+0.005	-0.01	7.90	0.1	1.7	A3 V		120 mx		m
4088	248238		0 42 41.8	-60 15 45	+0.035	-0.04	5.98	1.32	-0.3	K5 III		64 mx		
4065	192690	λ¹ Scl	0 42 42.8	-38 27 48	0.000	0.00	6.06	-0.03	0.6	A0 V		120 s		m
4006	74217		0 42 47.6	+30 06 08	0.000	-0.05	7.90	0.9	1.7	G9 III-IV	-22	170 s		
3949	11342		0 42 48.0	+62 45 56	-0.002	-0.01	7.79	0.15		A0				
3989	36596		0 42 48.8	+45 55 49	-0.004	+0.02	7.04	1.52	-0.3	K5 III	-21	290 s		
3940	11343		0 42 50.0	+64 17 28	0.000	-0.01	7.26	0.73	-7.3	A1 Ia		3400 s		
4048	147415		0 42 50.9	-9 55 18	0.000	-0.02	6.60	0.5	4.0	F8 V		33 s		m
4014	92047		0 42 52.6	+16 39 43	-0.003	-0.10	6.6	1.1	0.2	K0 III		180 mx		

HD	SAO	Star Name	α 2000	δ 2000	μ(α)	μ(δ)	V	B-V	M_V	Spec	RV	d(pc)	ADS	Notes
			$^h\ ^m\ ^s$	$^\circ\ '\ ''$	^{s}	$''$								
3890	4192		0 42 52.8	+74 03 55	−0.002	0.00	8.0	0.0	0.4	B9.5 V		300 s		
3826	4189		0 42 57.3	+79 11 49	+0.001	+0.02	8.0	1.1	0.2	K0 III		340 s		
4061	128916		0 43 01.7	−3 51 23	+0.004	0.00	7.4	0.4	3.0	F2 V		76 s		
4113	192693		0 43 12.8	−37 58 57	+0.007	−0.10	7.7	0.7	3.2	G5 IV		73 mx		
4130	215208		0 43 17.4	−45 10 52	+0.004	+0.02	7.10	0.8	3.2	G5 IV		60 s		m
4086	109406		0 43 17.4	+2 03 12	+0.001	−0.01	7.9	0.4	2.6	F0 V		110 s		
4097	128917		0 43 18.8	−6 20 15	+0.002	−0.02	7.5	1.1	0.2	K0 III		290 s		
4150	232162	η Phe	0 43 21.2	−57 27 48	0.000	0.00	4.36	0.00	0.0	A0 IV	+10	71 s		m
4094	109408		0 43 23.6	+2 35 03	0.000	+0.03	7.9	0.4	2.6	F0 V		110 s		
4029	21657		0 43 26.2	+56 32 03	−0.002	+0.02	7.8	1.1	0.2	K0 III		300 s		
4058	36602	20 π Cas	0 43 28.0	+47 01 29	−0.002	−0.03	4.94	0.18		A5	+13	22 mn		
4128	147420	16 β Cet	0 43 35.3	−17 59 12	+0.016	+0.04	2.04	1.02	0.2	K0 III	+13	21 ts		Deneb Kaitos
4169	192698		0 43 48.6	−37 28 21	0.000	0.00	7.6	0.4	3.4	F5 V		69 s		
4145	147422		0 43 50.1	−12 00 42	0.000	−0.20	6.02	1.10	3.2	G5 IV		20 s		
4119	54111		0 43 52.2	+33 18 39	0.000	+0.02	7.0	0.2	2.1	A5 V		97 s		
4102	36606		0 43 59.4	+49 04 11	0.000	−0.03	7.3	1.1	0.2	K0 III		240 s		
4042	4202		0 44 03.4	+70 49 25	+0.002	−0.01	6.9	1.1	0.2	K0 III	−3	200 s		
4309	255698		0 44 07.8	−74 15 57	+0.026	−0.04	7.57	0.49	4.0	F8 V		52 s		
4134	36611		0 44 11.2	+46 14 07	+0.007	−0.03	7.90	0.5	3.4	F5 V	+9	78 s	616	m
4188	147423	17 ϕ¹ Cet	0 44 11.3	−10 36 34	−0.001	−0.11	4.75	1.01	0.2	K0 III	+1	81 s		
4211	192703	λ² Scl	0 44 12.0	−38 25 18	+0.020	+0.12	5.90	1.16	0.2	gK0		100 s		
4186	128926		0 44 13.4	−7 53 18	0.000	−0.01	7.8	0.1	1.7	A3 V		160 s		
4228	215215		0 44 14.0	−49 23 10	+0.001	−0.01	7.0	0.0	1.4	A2 V		130 s		
4251	215219		0 44 21.7	−49 42 25	+0.002	0.00	7.70	0.03	0.6	A0 V		240 s		
4133	21668		0 44 23.8	+54 09 09	+0.001	+0.01	7.4	0.1	0.6	A0 V		210 s		
4142	36617		0 44 26.3	+47 51 51	−0.003	+0.02	5.67	−0.12	−1.1	B5 V	−60	230 s		
4208	166526		0 44 26.7	−26 30 58	+0.005	+0.14	7.79	0.67	4.4	G0 V		40 mx		
4219	147425		0 44 27.6	−13 24 57	+0.005	−0.02	7.7	0.5	4.0	F8 V		55 s		
4263	232165		0 44 28.4	−52 50 18	+0.001	0.00	7.8	1.9	−0.1	K2 III		140 s		
4294	248243		0 44 32.3	−62 29 53	+0.021	0.00	6.07	0.44	3.4	F5 V		33 s		m
4438	255701		0 44 35.4	−78 04 56	+0.006	0.00	6.6	1.3	3.2	G5 IV		25 s		
4174	36618		0 44 37.1	+40 40 47	+0.001	0.00	7.50	1.6		M2 e	−101			
4308	248244		0 44 39.1	−65 38 59	+0.025	−0.74	6.54	0.65	5.2	G5 V		19 mx		
4126	11358		0 44 40.7	+67 33 18	+0.007	−0.01	7.6	1.1	0.2	K0 III		270 s		
4180	36620	22 o Cas	0 44 43.4	+48 17 04	+0.002	0.00	4.54	−0.07	−2.5	B2 V	−8	210 s	622	m
4247	166528		0 44 44.3	−22 00 22	−0.005	+0.09	5.24	0.33	3.0	F2 V	+10	28 s		
4261	192712		0 44 45.7	−31 23 22	+0.003	0.00	7.6	0.9	3.2	G5 IV		65 s		
4194	36623		0 44 47.7	+42 29 47	−0.003	−0.03	8.0	0.9	3.2	G5 IV		93 s		
4132	11360		0 44 48.2	+67 09 30	+0.039	−0.05	6.9	0.5	4.0	F8 V		37 s		
4332	248245		0 44 54.6	−63 11 46	+0.002	−0.02	7.7	1.1	−0.1	K2 III		360 s		
4246	109427		0 44 55.1	+3 12 04	+0.001	+0.02	7.6	1.6	−0.5	M2 III		420 s		
4293	215221		0 44 57.0	−42 40 36	−0.008	−0.10	5.94	0.28	0.5	A7 III		120 s		
4272	147430		0 44 59.5	−12 08 39	−0.001	0.00	8.0	0.9	3.2	G5 IV		93 s		
4304	232168		0 44 59.8	−53 42 55	+0.025	0.00	6.15	0.53	3.2	F8 IV–V		38 s		
4179	11366		0 45 02.3	+62 31 43	+0.001	−0.01	6.8	1.1	0.2	K0 III		190 s		
4075	4208		0 45 04.6	+75 56 17	+0.106	−0.09	7.19	0.75	5.0	dG4	−8	23 s		
4256	109432		0 45 04.9	+1 47 09	−0.003	−0.56	8.03	1.00	6.3	K2 V	+6	18 s		
4231	54124		0 45 10.1	+37 22 58	−0.001	0.00	8.0	1.1	0.2	K0 III		360 s		
4271	128932		0 45 10.8	+0·15 11	+0.017	−0.05	7.1	0.9	5.6	dG		20 s		m
4269	74240		0 45 12.8	+23 35 27	+0.001	0.00	7.3	0.0	0.0	B8.5 V	+18	280 s		
4222	21677		0 45 17.1	+55 13 18	−0.003	+0.01	5.50	0.1	0.6	A0 V	−9	93 s	625	m
4343	215225		0 45 20.1	−48 50 11	+0.006	+0.01	7.5	0.0	1.7	A3 V		150 s		
4244	36626		0 45 21.6	+46 18 16	+0.003	−0.02	7.8	1.1	−0.1	K2 III		340 s		
4301	128935		0 45 24.0	−4 37 45	+0.002	+0.04	6.15	1.62	−0.4	gM0	+7	180 s		
4307	147432	18 Cet	0 45 28.6	−12 52 51	−0.003	−0.20	6.15	0.61	4.4	G0 V	−13	21 s		m
4161	4216	21 Cas	0 45 39.0	+74 59 17	−0.004	−0.02	5.66	0.05		A2	+11		624	YZ Cas, m,v
4313	109437		0 45 40.3	+7 50 42	0.000	0.00	8.0	0.9	3.2	G5 IV		93 s		
4338	147436		0 45 41.6	−16 25 27	+0.001	0.00	6.47	0.31	1.7	F0 IV		89 s	636	m
4266	21681		0 45 45.2	+56 46 30	−0.006	0.00	7.6	0.4	−6.6	F2 I	−25	190 mx		
4391	215232		0 45 45.5	−47 33 06	+0.018	+0.09	5.80	0.64	3.2	K0 IV		33 s		m
4312	74251		0 45 46.3	+26 10 23	−0.001	0.00	7.90	1.4	−2.3	K5 II	−19	1100 s		
4243	11374		0 45 47.6	+66 29 32	0.000	−0.07	7.6	0.7	4.4	G0 V		44 s		
4378	215231		0 45 48.0	−41 54 33	+0.025	−0.07	7.90	1.4	8.0	K5 V		11 ts		m
4230	11375		0 45 50.5	+67 51 56	0.000	+0.02	8.0	1.1	−0.1	K2 III		360 s		
4285	21683		0 45 52.7	+54 39 10	+0.001	+0.03	8.0	1.1	−0.1	K2 III		380 s		
4352	109441		0 45 54.6	+0 34 44	+0.007	+0.01	7.9	0.8	3.2	G5 IV		87 s		
4390	192721		0 45 57.3	−30 42 35	+0.001	−0.01	7.7	0.4	4.4	G0 V		47 s		
4375	147437		0 45 59.1	−16 39 08	+0.002	+0.04	8.00	0.4	2.6	F0 V		120 s		m
4322	36638		0 46 03.2	+40 48 32	+0.008	−0.06	7.7	0.5	3.4	F5 V		73 s		
4335	36640		0 46 10.8	+44 51 41	+0.003	0.00	6.05	−0.07		A p	0			
4398	166555		0 46 11.7	−22 31 19	+0.014	0.00	5.50	0.98	0.3	gG6	−15	100 s		
4395	147438		0 46 13.6	−11 27 10	−0.008	−0.06	7.6	0.9	3.2	G5 IV		77 s		
4321	21689		0 46 15.0	+55 18 19	+0.002	−0.01	6.5	0.1	1.4	A2 V	−8	100 s		
4510	248248		0 46 16.8	−62 40 31	+0.010	+0.02	7.4	0.4	2.6	F0 V		91 s		
4350	—		0 46 21.3	+48 14 40	0.000	+0.01	8.00	1.9		S5.5 e	−45			U Cas, v

HD	SAO	Star Name	α 2000	δ 2000	μ(α)	μ(δ)	V	B-V	M_V	Spec	RV	d(pc)	ADS	Notes
4372	54138		0 46 24.3	+30 56 34	-0.001	-0.05	7.40	1.1		K1 IV	+13		639	m
4496	232177		0 46 25.2	-57 55 40	+0.009	0.00	7.1	1.1	0.2	K0 III		210 s		
4388	54139		0 46 27.0	+30 57 06	-0.003	-0.02	7.60	1.1	-0.2	K3 III	-26	360 s		
4364	36646		0 46 27.0	+45 25 34	-0.003	-0.02	7.9	0.1	1.7	A3 V	+10	170 s		
4371	36647		0 46 27.6	+44 17 09	+0.001	-0.03	7.9	1.1	0.2	K0 III		350 s		
4409	92071		0 46 29.6	+13 09 08	+0.001	0.00	7.7	1.1	0.2	K0 III		310 s		
4408	92072	57 Psc	0 46 32.9	+15 28 32	-0.002	-0.04	5.38	1.65	-0.5	M2 III	-27	140 s		
4521	232180		0 46 35.3	-58 02 56	+0.003	-0.01	7.2	2.0	-0.5	M2 III		210 s		
4471	192731		0 46 37.2	-30 11 29	-0.002	-0.01	6.8	0.7	3.2	G5 IV		53 s		
4385	36650		0 46 37.9	+46 53 37	+0.006	+0.01	7.3	1.1	0.2	K0 III		250 s		
4520	232179		0 46 38.0	-54 06 05	-0.001	+0.08	7.90	0.5	3.4	F5 V		79 s		m
4295	11380		0 46 39.0	+69 19 31	+0.037	+0.01	6.4	0.4	3.0	F2 V	-14	48 s		
4362	21693		0 46 42.3	+59 34 29	-0.001	0.00	6.39	1.09	-4.5	G0 Ib	-15	1100 s		
4432	74261		0 46 44.9	+21 37 26	+0.005	-0.03	7.1	0.5	4.0	F8 V		42 s		
4406	36652		0 46 51.6	+46 21 49	-0.004	-0.02	7.6	0.9	3.2	G5 IV	+1	75 s		
4507	166568		0 46 59.9	-24 13 01	+0.005	-0.03	7.7	0.2	2.1	A5 V		64 mx		
4482	92080	58 Psc	0 47 01.4	+11 58 26	+0.004	-0.03	5.50	0.97	3.2	G5 IV	-1	22 s		
4310	4224		0 47 02.5	+73 05 18	+0.005	0.00	7.7	1.1	0.2	K0 III		280 s		
4404	21697		0 47 09.5	+60 16 03	+0.002	+0.01	7.6	1.1	-0.1	K2 III		310 s		
4490	92082	59 Psc	0 47 13.5	+19 34 43	+0.007	+0.01	6.13	0.27	2.1	A5 V	0	55 s		
4479	54147		0 47 20.1	+39 01 47	+0.003	+0.01	7.52	1.30	0.2	K0 III		170 s		
4502	74267	34 ζ And	0 47 20.3	+24 16 02	-0.007	-0.08	4.06	1.12	-2.2	K1 II	-24	48 ts		m,v
4526	109461	60 Psc	0 47 23.5	+6 44 28	+0.001	-0.01	5.99	0.94	0.3	G8 III	+14	140 s		
4477	36661		0 47 27.7	+49 39 06	+0.004	+0.02	7.8	0.0	0.4	B9.5 V		200 mx		
4525	92084		0 47 29.5	+11 38 33	+0.001	+0.01	7.7	0.9	3.2	G5 IV		80 s		
---	---		0 47 30.3	+51 06 01	+0.002	+0.02	8.0	1.1	-0.1	K2 III		390 s		
4597	192744		0 47 30.7	-36 56 24	-0.015	-0.35	7.85	0.54		G0		14 mn		
4538	92085		0 47 33.9	+18 54 04	+0.001	0.00	7.6	0.1	1.4	A2 V		170 s		
4565	128957		0 47 36.9	-2 19 20	+0.001	0.00	7.30	1.6	-0.5	M1 III	+20	360 s		
4595	192746		0 47 41.0	-31 21 11	-0.001	-0.01	7.1	1.8	0.2	K0 III		83 s		
4585	147451		0 47 43.2	-18 03 41	+0.003	+0.04	5.70	1.30	0.2	K0 III	+2	72 s		
4550	74278		0 47 45.0	+26 17 30	+0.008	+0.01	7.0	1.1	0.2	K0 III	-6	190 mx		
4549	74279		0 47 45.1	+27 05 50	0.000	+0.01	7.80	0.8	0.3	G4 III	-29	320 s		
4382	4226	23 Cas	0 47 46.0	+74 50 51	+0.004	0.00	5.3	-0.1	0.0	B8.5 V	-3	110 s		
4568	74280	61 Psc	0 47 54.7	+20 55 31	+0.011	+0.01	6.54	0.50	4.0	F8 V	+1	32 s		
4623	166584		0 47 56.7	-29 20 38	+0.005	+0.02	7.58	0.32	0.6	F0 III		220 mx	664	m
4607	128961		0 47 57.3	-9 20 46	+0.013	+0.22	7.74	0.51	3.4	F5 V		53 mx		
4536	21716		0 47 58.9	+51 26 42	-0.001	0.00	7.20	0.1	0.6	A0 V	-5	200 s	659	m
4563	54151		0 48 00.5	+36 29 13	+0.002	-0.05	7.8	0.5	4.0	F8 V		57 s		
4622	166585		0 48 01.0	-21 43 21	+0.002	-0.01	5.57	-0.06	0.2	B9 V	+21	120 s		
4440	4229		0 48 09.0	+72 40 30	+0.029	+0.03	5.87	1.01	0.2	gK0	+1	140 s		
4523	11400		0 48 11.4	+65 07 11	+0.001	0.00	7.10	0.8	3.2	G5 IV		59 s		m
4627	109470	62 Psc	0 48 17.3	+7 18 00	+0.007	+0.01	5.93	1.10	0.3	G8 III	-1	100 s		
4639	128965		0 48 20.2	-0 29 23	+0.001	-0.04	8.0	1.1	0.2	K0 III		360 s		
4628	109471		0 48 22.9	+5 16 50	+0.051	-1.14	5.75	0.88	6.6	K2 V	-13	6.9 t		96 G. Piscium, m
4591	36676		0 48 25.0	+49 42 59	+0.003	+0.02	7.9	0.1	1.4	A2 V		190 s		
4815	255710	λ Hyi	0 48 35.5	-74 55 24	+0.036	-0.03	5.07	1.37	-0.5	M1 III	-9	130 s		
4603	21725		0 48 39.5	+55 47 56	-0.002	-0.01	7.9	0.1	1.4	A2 V		190 s		
4656	109474	63 δ Psc	0 48 40.9	+7 35 06	+0.006	-0.05	4.43	1.50	-0.3	K5 III	+32	88 s		m
4691	166593		0 48 41.3	-28 29 42	+0.008	-0.03	6.7	0.6	2.6	F0 V		45 s		
4616	21727		0 48 41.3	+52 06 04	+0.004	0.00	6.9	0.1	1.7	A3 V		110 s		
4678	147459		0 48 43.9	-17 32 48	+0.004	0.00	7.8	1.1	0.2	K0 III		340 s		
4429	4231		0 48 47.1	+78 28 02	+0.014	-0.03	7.2	1.1	0.2	K0 III		240 s		
4670	92097		0 48 48.0	+18 18 51	-0.001	+0.01	7.6	0.0	0.4	B9.5 V		280 s		
4654	92096		0 48 48.3	+20 12 51	+0.001	-0.03	8.0	1.1	0.2	K0 III		360 s		
4636	21729	25 ν Cas	0 48 50.0	+50 58 06	+0.004	0.00	4.89	-0.11		B9	+1			
4774	248257		0 48 52.6	-60 22 12	+0.001	+0.02	7.6	0.6	1.7	A3 V		72 s		
4653	54164		0 48 54.4	+32 55 27	-0.003	-0.02	7.7	0.2	2.1	A5 V		130 s		
4737	215245		0 48 56.6	-46 41 52	-0.001	+0.02	6.27	0.90		K0				
4676	92099	64 Psc	0 48 58.6	+16 56 26	0.000	-0.20	5.07	0.51	4.0	F8 V	+2	20 ts		m
4723	192759		0 49 00.5	-36 15 48	-0.004	+0.02	7.6	0.5	3.2	G5 IV		77 s		
4669	36685		0 49 02.1	+41 07 55	+0.003	-0.02	7.69	1.16	0.2	K0 III		250 s		
4751	215246		0 49 03.7	-42 33 44	+0.001	-0.01	7.7	-0.5	0.0	B8.5 V		340 s		
4698	92100		0 49 04.5	+14 48 33	-0.001	0.00	7.0	1.4	-0.3	K5 III		290 s		
4614	21732	24 η Cas	0 49 05.9	+57 48 58	+0.137	-0.52	3.44	0.57	4.6	G0 V	+9	5.9 t	671	m
4714	166601		0 49 05.9	-21 09 04	-0.004	-0.02	7.1	0.9	3.2	G5 IV		53 s		
4668	36686		0 49 06.0	+44 07 44	-0.004	-0.08	7.7	0.9	3.2	G5 IV		80 s		
4686	74288		0 49 09.5	+28 43 09	-0.001	-0.01	7.3	0.9	0.3	G8 III	-2	250 s		
4785	232195		0 49 11.8	-53 46 28	-0.006	-0.04	7.9	0.0	4.0	F8 V		60 s		
4647	21738		0 49 13.7	+57 04 30	-0.001	0.00	7.07	1.77	-0.5	M2 III	-36	260 s		
4732	166602		0 49 13.8	-24 08 11	+0.006	-0.05	5.90	0.95	3.2	G5 IV	+23	25 s	679	m
4685	36690		0 49 16.0	+41 04 54	-0.002	-0.02	7.12	1.05	3.2	G5 IV		41 s		
4612	11413		0 49 16.2	+68 28 03	0.000	+0.01	8.0	0.0	0.0	B8.5 V		350 s		
4730	147464		0 49 25.5	-13 33 40	+0.007	-0.09	5.80	1.1	-0.1	K2 III	+4	88 mx	680	m
4747	166607		0 49 26.7	-23 12 47	+0.038	+0.11	7.17	0.77	5.5	dG7	+5	19 ts		

HD	SAO	Star Name	α 2000	δ 2000	μ(α)	μ(δ)	V	B−V	M_v	Spec	RV	d(pc)	ADS	Notes
			h m s	° ′ ″	s	″								
4695	36693		0 49 30.9	+46 30 42	+0.004	+0.05	7.3	1.1	0.2	K0 III		180 mx		
4772	166608		0 49 33.3	−23 21 42	−0.001	0.00	6.28	0.13	1.7	A3 V		77 s		
4702	54169		0 49 33.6	+38 12 17	+0.007	0.00	8.0	0.5	4.0	F8 V		62 s		
4701	36696		0 49 37.1	+47 45 50	−0.005	−0.02	7.0	0.1	1.4	A2 V	−14	130 s		
4674	11417		0 49 39.4	+60 45 33	+0.004	0.00	8.02	0.50		F5			677	m
4760	109485		0 49 39.9	+6 24 26	−0.001	+0.01	8.0	1.1	−0.1	K2 III		410 s		
4781	128971		0 49 44.4	−7 50 58	0.000	0.00	8.0	1.1	0.2	K0 III		360 s		
4635	4239		0 49 46.4	+70 26 59	+0.073	+0.22	7.76	0.90	6.3	dK2	−29	20 mn		m
4727	36699	35 ν And	0 49 48.8	+41 04 44	+0.002	−0.02	4.53	−0.15	−1.1	B5 V	−24	130 s		
4744	54175		0 49 52.5	+30 27 01	+0.017	−0.02	7.62	1.07	3.2	G8 IV		57 s		m
4757	74296	65 Psc	0 49 53.1	+27 42 37	+0.007	0.00	6.3	0.4	2.6	F0 V	+5	55 s	683	m
4790	128974		0 49 54.9	−0 13 25	+0.002	+0.01	6.7	1.1	0.2	K0 III		200 s		
4652	4241		0 49 55.3	+71 10 32	−0.006	−0.01	7.4	0.0	0.4	B9.5 V		230 s		
4860	232201		0 49 56.1	−57 23 11	−0.003	0.00	7.2	1.9	−0.5	M4 III		230 s		
4808	166610		0 49 56.4	−22 06 07	+0.004	+0.02	8.0	0.9	−0.3	K5 III		280 mx		
4756	54179		0 49 58.3	+34 58 27	−0.007	−0.01	8.0	0.5	3.4	F5 V		82 s		
4667	4244		0 50 02.6	+70 46 28	+0.008	0.00	7.0	0.0	0.4	B9.5 V		140 mx		
4849	215254		0 50 03.6	−43 23 41	+0.001	+0.02	6.48	0.29		F0				
--	--		0 50 04.8	+81 58 03			7.50							RX Cep, v
4813	147470	19 φ² Cet	0 50 07.5	−10 38 39	−0.016	−0.22	5.19	0.50	4.0	F8 V	+8	17 ts		
5278	258243		0 50 09.9	−83 44 38	+0.086	+0.04	8.0	0.3	3.4	F5 V		84 s		
4798	74301		0 50 14.7	+28 22 08	−0.002	−0.02	7.80	1.1	0.0	K1 III	−10	360 s		
4778	36702		0 50 18.2	+45 00 08	+0.007	+0.01	6.15	0.01		A0 p	+2			
4777	21759		0 50 25.0	+50 37 50	−0.004	0.00	7.82	0.47	3.4	F5 V		75 s	684	m
4806	36704		0 50 26.7	+40 48 27	+0.003	−0.03	8.0	0.0	0.0	B8.5 V		400 s		
4768	21757		0 50 27.7	+59 40 18	−0.001	0.00	7.57	0.38	−5.7	B5 Ib	−39	2300 s		
4833	74303		0 50 32.2	+21 09 49	−0.001	0.00	7.8	0.8	3.2	G5 IV		84 s		
4831	74305		0 50 38.5	+25 35 04	−0.002	−0.03	7.4	0.9	0.3	G8 III	−12	260 s		
4919	232203	ρ Phe	0 50 41.0	−50 59 13	+0.006	+0.05	5.22	0.36	2.6	dF0 n	+22	33 s		
4775	11424		0 50 43.5	+64 14 51	+0.004	−0.01	5.39	0.49	0.6	G0 III	+3	91 s		
4797	21763		0 50 47.4	+58 45 15	0.000	0.00	8.0	0.1	0.6	A0 V		270 s		
5028	255713		0 50 52.3	−71 09 04	−0.003	+0.05	7.1	−0.3	3.0	F2 V		66 s		
4666	4247		0 50 56.0	+77 57 14	+0.023	0.00	6.74	0.27		A5	−15			
4818	21767		0 50 57.3	+51 30 29	+0.014	0.00	6.39	0.28	1.9	F2 IV	+2	79 s		
4726	4248		0 50 57.6	+73 34 02	0.000	−0.01	7.9	1.4	−0.3	K5 III		380 s		
4843	36711		0 50 59.4	+46 43 17	+0.001	+0.01	7.8	1.1	0.2	K0 III		310 s		
4899	109506		0 51 04.3	+2 44 36	−0.001	−0.07	7.60	0.61		G0				
4811	11427		0 51 07.7	+63 06 08	−0.004	+0.02	8.0	1.1	−0.1	K2 III		360 s		
4855	36712		0 51 09.9	+48 12 26	+0.005	+0.03	7.9	1.1	−0.1	K2 III		330 mx		
4915	128986		0 51 10.7	−5 02 23	+0.016	−0.13	6.8	0.7	4.4	G0 V		35 ts		
4869	54198		0 51 12.1	+38 02 31	+0.002	−0.05	7.3	1.1	0.2	K0 III		250 mx		
4817	11430		0 51 16.3	+61 48 21	0.000	+0.01	6.07	1.88	−8.0	K3 Ia	−21	3100 s		
4928	109507		0 51 18.2	+3 23 06	+0.001	−0.06	6.37	1.07	0.2	K0 III	+6	160 s		
4939	128989		0 51 19.8	−9 24 21	−0.002	0.00	7.1	0.4	3.0	F2 V		66 s		
4975	192779		0 51 22.0	−38 31 45	0.000	−0.04	7.2	0.5	4.0	F8 V		42 s		
4903	54203		0 51 24.8	+32 39 26	−0.008	−0.08	7.4	0.5	4.0	F8 V		47 s		
4829	11431		0 51 25.4	+65 58 47	−0.003	−0.04	8.0	1.1	0.2	K0 III		320 s		
4841	11433		0 51 25.5	+63 46 52	−0.004	0.00	6.86	0.57	−7.0	B5 Ia	−26	2500 s		
4902	36718		0 51 30.1	+41 13 55	+0.002	+0.01	7.29	−0.04	0.6	A0 V	+4	220 s		
4973	166630		0 51 31.3	−23 34 47	+0.001	0.00	7.93	1.02		G5				
4965	128991		0 51 32.7	−3 08 37	+0.001	−0.02	7.28	0.13		A0				
4881	21775		0 51 33.6	+51 34 16	−0.001	−0.01	6.2	0.1	0.6	A0 V	−14	130 s		
4935	92123		0 51 34.9	+12 47 05	+0.006	−0.01	6.72	0.33	2.6	F0 V	−6	65 s	702	m
4901	36719		0 51 35.4	+44 12 28	+0.005	0.00	7.9	0.4	2.6	F0 V		110 s		
4971	128993		0 51 37.2	−8 49 06	+0.001	+0.01	7.4	1.1	0.2	K0 III		280 s		
4912	36721		0 51 46.8	+48 03 26	−0.003	+0.03	7.60	1.1	0.2	K0 III		280 s	699	m
4925	36722		0 51 50.1	+48 59 00	+0.002	−0.01	7.6	1.1	−0.1	K2 III		320 s		
5080	232211		0 51 51.2	−56 34 40	+0.004	+0.08	7.5	0.8	4.0	F8 V		35 s		
5042	215270		0 51 52.0	−43 42 33	+0.002	−0.01	6.90	0.35		F2				m
4924	21780		0 51 54.5	+50 46 29	+0.003	0.00	7.7	0.9	3.2	G5 IV		77 s		
4961	54211		0 51 57.0	+33 53 24	0.000	−0.01	7.00	1.1	−0.1	K2 III		260 s		m
4932	36723		0 51 58.2	+50 01 07	+0.006	−0.03	7.9	0.1	1.4	A2 V		92 mx		
5058	166640		0 52 12.7	−22 37 02	−0.008	−0.20	7.14	0.53	4.4	G0 V	−22	35 s	716	m
5057	166638		0 52 13.0	−21 06 31	−0.004	−0.07	7.6	0.6	3.4	F5 V		56 s		
5039	--		0 52 14.3	−9 50 21			7.8	1.1	−0.1	K2 III		380 s		
4880	4254		0 52 18.7	+71 38 20	0.000	+0.02	8.0	0.9	3.2	G5 IV		89 s		
5007	74332		0 52 23.2	+25 46 50	+0.007	0.00	7.70	1.1	0.0	K1 III	+13	210 mx		
5190	248269	λ¹ Tuc	0 52 24.4	−69 30 16	0.000	−0.08	6.22	0.56		F8				m
5036	74333		0 52 28.1	+21 24 36	+0.004	−0.05	7.3	0.5	3.4	F5 V		60 s		
5208	248271		0 52 28.4	−69 30 09	+0.002	−0.05	7.6	0.5	4.0	F8 V		52 s		
5038	92128		0 52 28.9	+13 38 56	+0.002	0.00	7.4	0.1	0.6	A0 V		230 s		
4796	4253		0 52 29.0	+79 50 26	−0.002	+0.01	7.8	0.1	0.6	A0 V		260 s		
5077	166643		0 52 31.4	−19 53 44	+0.002	−0.03	8.0	1.1	−0.1	K2 III		410 s		
5168	248270		0 52 35.8	−63 44 29	+0.009	+0.04	7.1	0.4	3.4	F5 V		54 s		
5097	--		0 52 38.0	−21 25 21			8.0	0.8	0.2	K0 III		360 s		

HD	SAO	Star Name	α 2000	δ 2000	μ(α)	μ(δ)	V	B-V	M_v	Spec	RV	d(pc)	ADS	Notes
5096	147492		$0^h 52^m 38^s.9$	$-12° 56' 14''$	$-0^s.002$	$0''.00$	7.90	0.38		F0				
5135	232214		0 52 40.4	−51 33 01	+0.003	+0.03	7.88	0.72	4.9	G3 V		35 s		
5098	166647		0 52 40.5	−24 00 21	+0.002	+0.04	5.46	1.24	−0.1	gK2	+34	120 s	726	m
––	166646		0 52 40.5	−21 26 20	+0.002	−0.07	7.5	1.1	0.2	K0 III		160 mx		
4996	21794		0 52 43.6	+52 16 42	+0.001	0.00	8.0	0.1	0.6	A0 V		280 s		
4947	11440		0 52 45.4	+68 51 58	+0.003	0.00	8.00	0.1	1.4	A2 V		190 s	710	m
5005	––		0 52 49.1	+56 37 31	+0.002	+0.01	7.76	0.09		O6	−24		719	m
5065	54224		0 52 51.6	+40 14 44	−0.008	+0.05	7.0	0.6	4.4	G0 V		34 s		
5066	54225		0 52 53.3	+38 32 55	+0.002	−0.02	6.50	0.02	1.2	A1 V	+16	110 s		
5049	36740		0 52 56.1	+47 05 54	−0.001	−0.01	8.0	0.1	0.6	A0 V		280 s		
4994	11442		0 52 58.2	+63 55 29	+0.001	−0.02	7.99	0.36		F0			718	m
5072	54227		0 52 59.3	+39 02 37	+0.005	+0.16	7.76	0.55	2.3	F7 IV	−11	24 mx		
5112	129009	20 Cet	0 53 00.4	−1 08 39	0.000	−0.01	4.77	1.57	−0.4	M0 III	+16	110 s		
5133	192793		0 53 01.1	−30 21 25	+0.048	+0.04	7.16	0.94	6.9	K3 V	−2	13 ts		
5015	11444		0 53 04.1	+61 07 27	−0.010	+0.18	4.82	0.53	2.4	F8 IV	+21	21 ts	721	m
5032	21799		0 53 05.5	+57 13 08	+0.002	−0.02	7.10	0.00	0.4	B9.5 V	−9	210 s	723	m
5132	147496		0 53 07.7	−17 39 06	−0.001	+0.01	7.64	0.32	1.7	F0 IV		150 s		
5092	54231		0 53 07.7	+30 20 57	+0.002	−0.01	7.70	1.1	−0.2	K3 III	+22	380 s		
5303	255716		0 53 07.9	−74 39 06	+0.063	+0.03	7.8	0.1		G0				
5156	166651		0 53 12.3	−24 46 37	+0.007	+0.04	6.46	0.44	2.4	F2 IV-V		57 s	733	m
5119	92136		0 53 18.0	+19 45 58	+0.009	−0.04	7.4	0.5	3.4	F5 V		62 s		
4959	4256		0 53 19.0	+74 18 46	+0.010	+0.01	7.6	0.2	2.1	A5 V		120 s		
5083	36745		0 53 19.2	+49 39 19	0.000	−0.02	7.2	0.0	0.0	B8.5 V	+6	250 s		
5143	109522		0 53 19.4	+4 05 10	−0.004	−0.03	7.40	0.4	2.6	F0 V		91 s	732	m
5091	36746		0 53 21.1	+46 35 44	+0.003	+0.01	7.9	0.1	0.6	A0 V		260 s		
4993	11446		0 53 22.1	+69 57 24	+0.001	+0.04	7.3	1.6	−0.5	M4 III		320 s		
5071	11449		0 53 27.9	+60 39 57	+0.004	0.00	7.80	0.11	0.6	A0 V		200 mx		
5118	54237		0 53 28.2	+37 25 05	+0.001	−0.04	6.06	1.14	0.2	K0 III	−6	120 s		
5111	36747		0 53 33.2	+48 31 08	−0.003	−0.02	7.6	1.4	−0.3	K5 III		350 s		
5138	74340		0 53 35.7	+29 04 49	−0.002	0.00	7.9	0.1	1.4	A2 V		200 s		
5137	74342		0 53 36.9	+29 29 21	0.000	−0.05	6.7	1.1	0.2	K0 III	−12	200 s		
5276	248276		0 53 37.8	−62 52 18	+0.010	0.00	5.5	2.6	−0.5	M4 III	−12	41 s		
5129	36751		0 53 40.4	+43 21 50	+0.024	−0.11	7.3	0.6	4.4	G0 V	−10	38 s		
5181	129015		0 53 41.2	−4 31 32	+0.006	0.00	6.6	0.9	3.2	G5 IV		49 s		
5202	147503		0 53 42.1	−13 55 13	+0.003	+0.01	7.6	0.6	4.4	G0 V		43 s		
5128	21814		0 53 47.5	+52 41 21	+0.008	−0.02	6.30	0.1	0.6	A0 V	−1	82 mx	735	m
5164	74346		0 53 49.6	+28 33 30	+0.003	−0.01	7.90	1.1	0.0	K1 III		380 s		
5283	232221		0 53 56.9	−54 35 30	+0.004	+0.05	7.80	0.8	3.2	G5 IV		83 s		m
5275	215283		0 53 59.1	−48 47 29	+0.010	+0.01	7.0	0.4	3.2	G5 IV		57 s		
5197	74349		0 54 00.1	+25 29 17	+0.005	0.00	7.4	1.1	0.2	K0 III		270 s		
5271	192804		0 54 03.9	−32 20 06	+0.002	0.00	7.9	0.3	3.0	F2 V		97 s		
5136	21820		0 54 04.9	+57 00 51	+0.004	0.00	7.3	0.4	2.6	F0 V		87 s		
5110	11453		0 54 06.1	+66 26 09	+0.011	+0.01	7.06	0.41	2.6	F0 V		70 s		
5196	54248		0 54 09.9	+32 53 18	0.000	−0.01	7.3	1.1	−0.1	K2 III		300 s		
5322	232222		0 54 14.9	−57 24 12	+0.001	+0.01	7.3	1.5	0.2	K0 III		130 s		
5268	129019	21 Cet	0 54 17.5	−8 44 27	+0.001	−0.05	6.16	0.92	0.4	gG3	+45	120 s		
5311	215285		0 54 23.4	−48 56 57	+0.002	−0.06	7.9	0.2	3.0	F2 V		94 s		
5177	21827		0 54 30.0	+56 19 10	+0.009	−0.02	7.2	0.2	2.1	A5 V		72 mx		
5267	92145	66 Psc	0 54 35.2	+19 11 18	+0.001	−0.01	5.80	−0.01	1.2	A1 V	+12	83 s	746	m
4853	143		0 54 52.8	+83 42 27	+0.035	−0.01	5.5	0.1	1.4	A2 V	+28	67 s		
5235	––		0 54 54.4	+58 33 36			7.80	1.8		C7.7 e	−39			W Cas, v
5286	74359	36 And	0 54 58.0	+23 37 42	+0.010	−0.03	5.47	1.00	0.0	gK1	+2	120 s	755	m
5366	215287		0 54 58.5	−45 13 12	0.000	−0.02	7.4	1.4	0.2	K0 III		180 s		
5294	74361		0 54 59.1	+24 06 01	−0.014	−0.18	7.38	0.65	3.2	G5 IV		69 s		m
5499	255721		0 54 59.4	−74 18 12	+0.029	+0.09	6.69	0.99	3.2	K0 IV		48 s		
5234	21832	26 υ¹ Cas	0 55 00.0	+58 58 22	−0.004	−0.04	4.83	1.21	−0.1	K2 III	−23	91 s	748	m
5457	248281	λ² Tuc	0 55 00.3	−69 31 38	+0.002	−0.04	5.45	1.09	0.3	G7 III	+5	80 s		
5273	36763		0 55 05.1	+48 40 43	−0.002	0.00	6.5	1.6	−0.5	M2 III	−52	240 s		
5258	21836		0 55 09.6	+57 16 34	−0.001	0.00	7.9	1.4	−0.3	K5 III		390 s		
5279	36764		0 55 09.7	+46 52 59	+0.005	0.00	7.7	1.1	0.2	K0 III		300 s		
5285	36765		0 55 10.2	+42 15 00	+0.002	−0.02	7.9	0.1	1.7	A3 V		170 s		
5388	215291		0 55 11.5	−47 24 22	−0.011	−0.18	6.9	1.0	4.0	F8 V		19 s		
––	21837		0 55 12.4	+56 29 00	+0.002	+0.02	7.8	1.4	3.2	G5 IV		34 s		
5316	74365		0 55 14.6	+24 33 26	+0.002	−0.01	6.20	1.62	−0.5	M4 III	−10	220 s		
5244	11466		0 55 14.7	+64 32 45	−0.002	−0.01	7.59	0.23		A3				
5336	109533		0 55 19.6	+6 51 15	+0.003	+0.01	7.7	1.1	0.2	K0 III		310 s		
5488	248282		0 55 22.3	−68 37 43	+0.005	−0.02	8.0	0.5	4.0	F8 V		63 s		
5377	166682		0 55 22.7	−24 39 37	+0.004	−0.01	7.4	0.5	3.2	G5 IV		68 s		
5328	74367		0 55 25.2	+28 33 34	+0.004	−0.04	7.3	0.2	2.1	A5 V		72 mx		
5306	36770		0 55 25.7	+45 05 28	−0.006	−0.05	8.0	1.1	−0.1	K2 III		270 mx		
5314	36771		0 55 26.4	+40 22 56	−0.001	−0.02	6.9	0.1	1.4	A2 V		130 s	766	m
5376	147512		0 55 30.6	−12 33 09	+0.009	+0.02	7.9	0.5	3.4	F5 V		78 s		
5362	129029		0 55 33.1	+0 01 11	0.000	−0.04	7.60	1.4	−0.3	K4 III	−29	380 s		
5384	129032		0 55 42.3	−7 20 50	0.000	−0.04	5.85	1.52	−0.3	gK5	+2	160 s		
5344	36775		0 55 50.4	+46 52 33	+0.001	−0.01	7.99	−0.02	0.6	A0 V		300 s		

HD	SAO	Star Name	α 2000	δ 2000	μ(α)	μ(δ)	V	B-V	M_V	Spec	RV	d(pc)	ADS	Notes
5474	232227		0 55 52.8	-53 11 33	+0.002	-0.04	6.5	2.1	0.2	K0 III		42 s		
5445	166686		0 55 55.4	-27 46 32	0.000	+0.01	6.10	1.67		M				
5382	74373	67 Psc	0 55 58.4	+27 12 33	-0.001	+0.01	5.90	0.11	1.7	A3 V	-8	66 s		
5437	147519	22 φ³ Cet	0 56 01.4	-11 16 00	-0.002	0.00	5.31	1.52	-0.3	K4 III	-26	120 s		
5435	129035		0 56 02.0	-3 44 18	+0.001	-0.02	7.5	0.5	3.4	F5 V		65 s		
--	36780		0 56 05.8	+40 28 15	0.000	-0.02	7.0	2.9						
5442	129038		0 56 08.2	-8 01 24	0.000	-0.01	7.5	0.5	4.0	F8 V		51 s		
5418	92159		0 56 09.1	+13 57 07	0.000	0.00	6.8	0.9	3.2	G5 IV		53 s		
5342	11475		0 56 12.7	+61 15 15	-0.003	+0.02	8.03	0.12	-3.4	B8 II		1700 s		
5343	21846		0 56 12.8	+57 59 48	+0.005	-0.01	6.21	1.37	-0.2	K3 III	-30	170 s		
5373	36781		0 56 13.0	+46 39 14	-0.001	0.00	7.6	0.4	3.0	F2 V		83 s		
5397	54277		0 56 14.7	+35 13 43	+0.001	0.00	6.8	0.1	0.6	A0 V	+7	180 s		
5473	166692		0 56 18.6	-25 33 15	0.000	-0.03	7.7	2.5	-0.5	M2 III		130 s		
5562	248283		0 56 21.2	-63 57 32	+0.048	+0.14	7.4	0.6	4.4	G0 V		40 s		
5380	21853		0 56 29.9	+57 17 11	0.000	0.00	7.5	0.1	0.6	A0 V		230 s		
5431	36785		0 56 30.6	+42 58 44	-0.001	0.00	6.5	1.1	0.2	K0 III		190 s		
5465	109540		0 56 32.9	+9 13 44	+0.001	-0.03	8.0	1.1	0.2	K0 III		360 s		
5395	21855	28 υ² Cas	0 56 39.7	+59 10 52	-0.012	-0.04	4.63	0.96	1.8	G8 III-IV	-47	29 ts		
5394	11482	27 γ Cas	0 56 42.4	+60 43 00	+0.003	0.00	2.47	-0.15	-4.6	B0 IV e	-7	240 s	782	m,v
5448	54281	37 μ And	0 56 45.1	+38 29 58	+0.013	+0.04	3.87	0.13	2.1	A5 V	+8	25 ts	788	m
5409	21857		0 56 45.6	+60 02 00	+0.003	0.00	7.86	0.05	0.2	B9 V		300 s		
5506	129044		0 56 46.9	-2 43 47	-0.001	-0.03	7.0	0.4	2.6	F0 V		76 s		
5408	11484		0 56 47.0	+60 21 46	+0.004	0.00	5.55	-0.07		B9	-2		784	m
5524	166699		0 56 49.1	-25 21 48	+0.003	+0.01	7.19	0.14	2.1	A5 V		100 s		
5569	215296		0 56 52.9	-46 51 52	+0.002	-0.01	7.9	1.5	-0.1	K2 III		170 s		
5392	11486		0 56 55.0	+64 32 00	-0.004	+0.05	7.01	0.99	-6.5	F4 I		71 mx		
5357	11481		0 56 55.5	+68 46 34	+0.025	-0.01	6.37	0.38	2.6	F0 V	-8	53 s		
5479	54284		0 56 59.2	+31 53 46	+0.001	-0.01	8.0	1.6	-0.5	M2 III		510 s		
5516	74388	38 η And	0 57 12.4	+23 25 04	-0.002	-0.04	4.42	0.94	1.8	G8 III-IV	-10	35 mn		m
5544	109552		0 57 12.9	+0 20 32	-0.001	-0.01	7.70	1.1	0.2	K0 III p	-13	320 s		
--	--		0 57 14.0	-74 59 59			8.00							U Tuc, v
5468	21866		0 57 14.6	+54 57 08	+0.007	-0.02	7.2	0.1	0.6	A0 V		93 mx		
5459	11491		0 57 19.5	+61 25 20	+0.008	0.00	6.43	0.91	3.2	G8 IV	-9	42 s		
5492	21871		0 57 20.2	+52 14 25	0.000	+0.02	7.50	1.1	-0.1	K2 III	-38	310 s	792	m
5477	21870		0 57 20.5	+53 18 09	+0.001	0.00	7.9	1.1	0.2	K0 III		320 s		
5556	74393		0 57 29.7	+21 29 10	+0.007	-0.03	7.1	0.8	3.2	G5 IV		59 s		
5601	147533		0 57 32.5	-10 28 32	+0.002	+0.02	7.65	-0.06	0.6	A0 V		260 s		
5617	147537		0 57 37.6	-18 59 56	0.000	0.00	6.90	0.09	1.4	A2 V n		130 s	799	m
5671	232237		0 57 38.0	-52 15 58	+0.003	+0.01	7.8	1.4	3.2	G5 IV		35 s		
5588	109556		0 57 39.6	+1 47 08	+0.001	+0.01	6.97	1.19	-0.1	gK2		260 s		
5526	36795		0 57 39.6	+45 50 23	+0.001	+0.01	6.12	1.02	0.2	gK0	+5	150 s		
5600	109557		0 57 42.1	+2 05 37	+0.004	0.00	8.00	0.46		F8				
5575	74395	68 Psc	0 57 50.1	+28 59 32	+0.001	-0.01	5.42	1.08	0.2	K0 III	-1	99 s		
5624	129050		0 57 50.2	-1 45 59	+0.002	-0.01	6.51	1.04		G5				
5756	248291		0 57 51.9	-66 33 42	-0.004	-0.03	7.60	0.6	4.4	G0 V		44 s		m
5612	92183		0 57 54.4	+13 41 45	-0.001	-0.01	6.32	0.89	0.3	G5 III	+15	160 s		
5554	21883		0 57 55.8	+52 08 11	+0.003	-0.01	7.2	0.1	1.4	A2 V		140 s		
5643	166710		0 57 57.8	-20 40 35	-0.001	-0.04	8.0	0.5	2.6	F0 V		87 s		
--	36800		0 57 58.6	+47 30 58	-0.002	+0.01	7.8	2.6						
5623	109561		0 57 59.0	+2 29 11	+0.002	-0.01	7.8	1.1	0.2	K0 III		330 s		
5660	166711		0 58 00.2	-22 35 46	-0.003	-0.03	8.0	1.8	-0.3	K5 III		290 s		
5582	54297		0 58 02.5	+37 29 59	0.000	-0.01	7.3	0.4	2.6	F0 V		89 s		
5739	232243		0 58 05.9	-57 42 02	+0.012	+0.04	7.80	0.5	4.0	F8 V		58 s		m
5659	147542		0 58 09.4	-15 41 05	+0.002	-0.01	7.08	0.46	3.8	dF7	+1	45 s	806	m
5597	54299		0 58 09.7	+39 28 32	-0.001	-0.02	6.52	1.16	0.2	K0 III		140 s		
5676	166713		0 58 12.3	-25 52 34	+0.001	+0.02	7.9	1.0	0.2	K0 III		350 s		
5490	4278		0 58 12.9	+70 29 48	+0.004	+0.01	6.8	0.0	0.4	B9.5 V		180 s		
5608	54306		0 58 14.3	+33 57 05	+0.004	-0.05	5.98	1.00	0.2	K0 III	-17	140 s		
5581	36805		0 58 16.4	+48 13 09	0.000	-0.03	7.6	1.1	0.2	K0 III		280 s		
5596	36806		0 58 16.5	+45 35 48	-0.001	-0.04	7.2	0.1	1.7	A3 V	-19	130 s		
5641	74402		0 58 18.9	+21 24 16	0.000	0.00	6.40	0.09	1.4	A2 V	-6	94 s	805	m
5654	109563		0 58 19.4	+6 50 40	+0.002	0.00	6.9	0.8	3.2	G5 IV		54 s		
5771	248293		0 58 22.4	-60 41 47	+0.007	+0.02	6.23	0.10		A3 m				
5551	11503		0 58 26.5	+63 42 50	-0.004	+0.01	7.71	0.62	-5.7	B1.5 Ib	-51	1600 s		
5650	74405		0 58 28.1	+26 47 08	-0.004	-0.01	7.50	1.4	-0.3	K5 III	-22	360 s		
5550	11502		0 58 31.0	+66 21 08	+0.007	0.00	6.0	0.0		B9	-10			
5675	92188		0 58 31.7	+16 18 49	+0.001	+0.01	8.0	1.1	0.2	K0 III		360 s		
5737	166716	α Scl	0 58 36.3	-29 21 27	+0.001	0.00	4.31	-0.16	-1.2	B8 III	+10	130 s		
5637	36815		0 58 42.8	+49 10 37	+0.002	-0.06	7.9	0.1	0.6	A0 V		120 mx		
5638	36817		0 58 43.2	+47 02 12	+0.002	0.00	6.85	-0.11	0.6	A0 V	-13	180 s		
5735	147549		0 58 43.7	-19 38 00	+0.003	-0.03	7.2	1.7	-0.5	M2 III	+29	300 s		
5722	147546	23 φ⁴ Cet	0 58 43.8	-11 22 48	-0.003	-0.02	5.61	0.97	3.2	G5 IV	-19	22 s		
5783	232245		0 58 44.1	-51 15 56	+0.002	0.00	7.7	1.1	-0.1	K2 III		360 s		
--	36818		0 58 45.0	+48 56 18	+0.001	-0.01	8.0	1.6						
5720	129056		0 58 45.7	-5 52 58	0.000	-0.08	6.6	1.1	0.2	K0 III		190 mx		

HD	SAO	Star Name	α 2000	δ 2000	μ(α)	μ(δ)	V	B-V	M_V	Spec	RV	d(pc)	ADS	Notes
			h m s	° ′ ″	s	″								
5663	21894		0 58 57.4	+52 31 10	−0.001	0.00	7.9	0.1	1.4	A2 V		190 s		
5761	147556		0 59 02.5	−19 19 35	+0.002	0.00	7.9	1.0	3.2	G5 IV		65 s		
5705	74415		0 59 02.8	+27 39 26	+0.001	+0.01	7.1	1.1	−0.2	K3 III	−9	290 s		
5754	129059		0 59 02.9	−2 27 13	−0.001	−0.02	8.0	1.1	−0.1	K2 III		410 s		
5704	54316		0 59 06.5	+32 27 20	0.000	−0.01	7.3	0.1	1.4	A2 V		150 s		
5877	232248		0 59 19.4	−58 24 18	+0.018	−0.04	7.5	2.1	0.2	K0 III		63 s		
5781	129063		0 59 20.5	−0 40 29	+0.003	−0.04	7.62	0.51	3.4	F5 V		64 s	819	m
5780	109577		0 59 23.2	+0 46 44	−0.002	−0.10	7.61	1.42	−0.3	K4 III	−104	380 s	818	m
5825	192850		0 59 25.4	−34 38 11	+0.001	+0.02	7.7	1.1	0.2	K0 III		280 s		
5741	54321		0 59 26.2	+40 09 19	+0.003	−0.01	7.4	0.1	0.6	A0 V		230 s		
--	54323		0 59 30.4	+32 52 40	+0.002	+0.03	8.0	0.9						
5580	4282		0 59 30.5	+73 59 16	−0.004	−0.01	8.0	1.1	0.2	K0 III		320 s		
5728	36825		0 59 32.3	+48 57 58	+0.003	−0.03	7.1	0.5	3.4	F5 V		55 s		
5750	54328		0 59 35.8	+32 29 31	+0.028	−0.05	7.1	0.5	3.4	F5 V	+23	45 mx		
5822	129069		0 59 42.1	−4 19 17	0.000	0.00	7.8	1.1	0.2	K0 III		340 s		
5869	192853		0 59 44.4	−31 20 15	+0.003	−0.07	7.5	1.6	0.2	K0 III		130 s		
5820	109581		0 59 49.6	+6 29 00	+0.001	0.00	6.11	1.67		Ma	−15	20 mn		m
5835	129070		0 59 50.7	−8 19 25	+0.004	+0.02	7.5	0.8	3.2	G5 IV		73 s		
5904	232251		0 59 51.7	−49 45 02	+0.002	+0.01	7.5	1.2	−0.5	M2 III		400 s		
5801	54330		0 59 57.9	+31 51 14	−0.002	−0.02	7.9	0.9	3.2	G5 IV		88 s		
5764	36829		0 59 59.7	+48 01 12	+0.001	−0.01	7.15	−0.12	0.0	B8.5 V	−8	270 s		
4449	128952		0 66 40.3	−4 25 36	0.000	−0.25	7.7	0.9	3.2	G5 IV		49 mx		
5747	11515		1 00 02.7	+60 30 37	+0.003	−0.01	7.07	0.96	−2.1	G8 II	+16	680 s		
5861	129072		1 00 03.5	−2 01 10	+0.004	−0.06	7.30	0.5	3.4	F5 V		60 s	825	m
5788	36833		1 00 03.5	+44 42 48	+0.001	−0.02	6.00	−0.01		A1 n	+17		824	m
5859	109587		1 00 05.5	+2 37 59	−0.002	−0.01	7.9	1.1	−0.1	K2 III		400 s		
6045	255728		1 00 09.8	−70 10 45	−0.004	−0.07	8.00	0.43	3.3	F4 V				
5936	232254		1 00 10.8	−50 31 28	+0.005	+0.03	7.8	1.0	0.2	K0 III		310 mx		
5799	36835		1 00 13.2	+47 19 08	+0.004	0.00	7.5	1.1	−0.1	K2 III		310 s		
5800	36836		1 00 13.6	+46 52 36	+0.007	+0.02	8.0	0.5	3.4	F5 V		80 s		
5857	92194		1 00 15.7	+18 11 56	+0.002	−0.07	7.40	0.9	3.2	G5 IV		69 s	827	m
5831	54337		1 00 19.6	+35 41 29	+0.001	+0.01	7.7	0.1	1.4	A2 V		180 s		
5798	21912		1 00 21.0	+52 08 21	+0.002	−0.01	7.8	1.1	0.2	K0 III		300 s		
5873	92195		1 00 21.1	+17 13 34	+0.002	−0.03	8.00	1.1	−0.1	K2 III		420 s		
5871	74426		1 00 27.9	+21 14 57	0.000	−0.01	7.2	1.1	−0.1	K2 III		290 s		
5911	147563		1 00 28.0	−19 23 20	+0.010	−0.04	7.7	1.5	4.4	G0 V		13 s		
5883	92196		1 00 29.7	+18 41 29	+0.006	−0.01	7.3	0.9	3.2	G5 IV		65 s		
5534	162		1 00 29.8	+80 32 45	+0.035	+0.03	6.6	0.4	3.0	F2 V		53 s		
5715	4288		1 00 31.1	+70 59 00	+0.018	0.00	6.5	0.1	0.6	A0 V	+6	92 mx		
5854	54339		1 00 31.5	+37 47 29	+0.003	0.00	7.19	0.13		A0	+7			
5813	21914		1 00 32.6	+58 21 50	0.000	−0.01	7.2	0.1	1.7	A3 V	−4	120 s		
5934	166740		1 00 33.4	−25 44 57	+0.004	−0.05	7.34	1.03		G5				
5963	192863		1 00 35.2	−36 14 18	+0.003	0.00	6.9	1.2	0.2	K0 III		180 s		
5932	147566		1 00 38.3	−18 07 07	+0.003	+0.02	7.9	0.4	2.6	F0 V		110 s		
6013	232258		1 00 40.0	−52 34 56	+0.001	+0.03	7.80	0.4	2.6	F0 V		110 s		m
5933	166742		1 00 40.0	−20 05 26	+0.001	0.00	7.2	1.6	0.2	K0 III		120 s		
5919	129076		1 00 40.3	−1 39 32	+0.001	−0.03	7.0	1.6	−0.5	M4 III		310 s		
5852	36841		1 00 41.6	+43 06 46	+0.001	−0.03	7.8	1.1	0.2	K0 III		340 s		
5959	129078		1 00 50.1	−9 22 26	+0.001	0.00	7.6	0.1	1.7	A3 V		160 s		
5897	54346		1 00 53.5	+36 39 22	−0.005	−0.12	7.5	1.1	0.2	K0 III		110 mx		
6172	255729		1 00 53.9	−72 41 51	+0.006	0.00	7.68	1.33		K0				
5882	21918		1 00 56.4	+50 52 45	+0.001	0.00	7.70	−0.05	−1.0	B5.5 V		520 s		
5975	129081		1 01 01.6	−4 38 55	−0.001	−0.06	7.4	0.5	4.0	F8 V		49 s		
5896	21924		1 01 14.3	+52 03 19	+0.002	−0.01	7.6	0.1	0.6	A0 V		240 s		
6055	192870	ξ Scl	1 01 18.3	−38 54 59	+0.007	+0.06	5.59	1.18	0.2	gK0	−31	93 s		
5916	36851		1 01 19.0	+45 27 08	+0.010	−0.02	6.85	0.90	1.8	G8 III-IV	−71	110 s		
6107	248301		1 01 19.2	−60 51 36	+0.013	+0.11	6.9	0.6	4.4	G0 V		31 s		
5988	92207		1 01 21.7	+11 54 46	+0.001	+0.02	7.9	0.5	3.4	F5 V		79 s	828	m
6041	192871		1 01 21.9	−33 21 37	0.000	+0.01	7.7	1.4	0.2	K0 III		180 s		
5986	92208		1 01 25.8	+18 32 14	+0.001	0.00	8.0	0.9	3.2	G5 IV		93 s		
5927	36853		1 01 26.9	+49 32 41	0.000	+0.02	6.90	0.9	3.2	G5 IV		54 s	842	m
5839	11526		1 01 30.2	+69 21 31	0.000	0.00	7.50	0.00	0.4	B9.5 V	+5	240 s	836	m
6222	255730		1 01 33.1	−71 33 01	+0.003	−0.06	7.74	1.10		K0				
6080	192874		1 01 34.1	−38 12 53	−0.005	−0.08	6.6	1.1	0.2	K0 III		160 s		
6031	129088		1 01 34.3	−6 48 00	+0.002	+0.02	7.2	1.1	−0.1	K2 III		290 s		
5953	36856		1 01 37.0	+48 20 50	0.000	−0.04	8.0	1.1	−0.1	K2 III		370 s		
5838	4294		1 01 37.8	+70 31 18	+0.003	+0.02	7.9	0.2	2.1	A5 V		140 s		
6037	147572		1 01 38.6	−16 15 56	+0.005	−0.08	6.6	0.8	3.2	G5 IV		47 s		
6030	109610		1 01 42.8	+9 44 45	0.000	+0.03	8.0	0.9	3.2	G5 IV		93 s		
6009	74439		1 01 43.5	+25 17 32	+0.010	−0.01	6.8	0.8	3.2	G5 IV		51 s		
5944	21931		1 01 45.5	+57 49 00	−0.003	0.00	6.8	0.1	1.4	A2 V	−8	120 s		
5982	36857		1 01 46.3	+47 42 20	+0.015	−0.04	7.9	0.5	3.4	F5 V		76 s		
6192	232268		1 02 01.7	−57 00 09	+0.001	+0.02	6.11	0.94	0.3	G8 III		150 s		
5966	21935		1 02 01.7	+59 32 46	+0.002	+0.01	8.0	0.4	0.6	F0 III		270 s		
5906	11534		1 02 05.4	+68 40 44	+0.002	−0.02	8.0	0.9	3.2	G5 IV		89 s		

HD	SAO	Star Name	α 2000	δ 2000	μ(α)	μ(δ)	V	B-V	M_V	Spec	RV	d(pc)	ADS	Notes
6077	109614		1ʰ02ᵐ06.8ˢ	+7°56'26"	+0.001	+0.04	7.9	0.9	3.2	G5 IV	+24	86 s		
5981	21937		1 02 07.4	+56 19 41	+0.004	-0.03	6.9	1.1	0.2	K0 III		200 s		
6124	166759		1 02 10.9	-21 36 32	0.000	+0.01	8.0	1.2	0.2	K0 III		290 s		
6179	215337		1 02 15.5	-45 10 20	-0.001	-0.02	7.6	0.4	2.6	F0 V		72 s		
5679	168		1 02 18.3	+81 52 33	+0.011	0.00	6.60	0.6	0.4	G2 III	+7	170 s	830	U Cep, m,v
6028	21938		1 02 18.4	+51 02 06	0.000	-0.01	6.6	0.1	1.7	A3 V	+6	95 s		
6178	192884	σ Scl	1 02 26.3	-31 33 08	+0.006	+0.02	5.50	0.08	1.4	A2 V	-21	66 s		
6156	166761		1 02 27.6	-21 36 36	+0.017	+0.01	7.93	0.80	3.2	G5 IV		73 mx	860	m
6120	109618		1 02 29.5	+8 49 17	-0.002	-0.01	6.9	0.4	2.6	F0 V		71 s		
6093	74443		1 02 35.5	+27 44 54	+0.005	+0.02	6.70	0.5	3.4	F5 V		46 s	857	m
6446	255731		1 02 36.9	-77 33 01	+0.012	-0.01	7.18	1.38	-2.2	K2 II p				
6027	21942		1 02 37.1	+59 17 12	0.000	-0.02	7.70	1.1	6.3	K2 V		19 s		
6059	21943		1 02 39.6	+52 51 03	+0.001	-0.01	7.3	0.1	0.6	A0 V		210 s		
6152	109623		1 02 42.6	+9 07 58	+0.004	-0.03	6.7	0.5	3.4	F5 V		46 s		
6311	248308		1 02 42.9	-65 27 22	+0.001	0.00	6.21	1.64		Mb				
---	---		1 02 47.0	+61 52 04			8.00	1.6		R7.7				
6118	54374	69 σ Psc	1 02 49.0	+31 48 16	+0.001	-0.02	5.50	-0.05	0.2	B9 V	+10	120 s		
6245	215343		1 02 49.1	-46 23 51	0.000	+0.01	5.36	0.90	0.3	gG6	-1	100 s		
6092	---		1 02 49.9	+47 41 19			8.00	0.1	0.6	A0 V		280 s	856	m
6026	11540		1 02 50.1	+62 36 32	-0.005	-0.01	7.20	0.95		K0				
6133	74445		1 02 50.3	+26 17 56	+0.002	+0.01	6.9	0.4	3.0	F2 V	+9	60 s		
6282	232274		1 02 52.3	-56 03 15	0.000	+0.02	7.9	1.7	-0.1	K2 III		200 s		
6132	74446		1 02 53.2	+29 59 09	-0.003	0.00	8.00	1.1	-0.1	K2 III		420 s		
6084	21949		1 02 53.7	+51 47 52	+0.001	-0.01	6.83	0.02	-1.0	B5.5 V	-17	310 s	859	m
6116	36874	39 And	1 02 54.2	+41 20 43	-0.002	-0.01	5.98	0.16	2.4	A7 V	+4	52 s	863	m
6186	109627	71 ε Psc	1 02 56.5	+7 53 24	-0.005	+0.03	4.28	0.96	0.2	K0 III	+7	66 s		
6114	36875		1 03 01.4	+47 22 34	+0.009	-0.01	6.45	0.25		A3	+3	18 mn	862	m
6203	129094	25 Cet	1 03 02.5	-4 50 12	-0.008	-0.10	5.43	1.11	1.7	K0 III-IV	+15	43 s		m
6015	4307		1 03 07.5	+70 19 20	+0.003	+0.02	8.0	0.1	0.6	A0 V		270 s		
6148	36878		1 03 09.9	+47 51 14	+0.001	-0.04	7.7	0.0	-1.0	B5.5 V		490 s		
6112	21951		1 03 09.9	+53 08 26	+0.003	0.00	7.7	1.6	-0.5	M4 III		390 s		
6405	---		1 03 12.4	-70 50 46			6.8 B							
6254	166771		1 03 15.7	-26 10 48	+0.006	-0.06	8.01	0.99	3.2	G8 IV		79 s		
6269	166774		1 03 17.5	-29 31 33	-0.009	-0.03	6.29	0.93	3.2	G5 IV		32 s		
6334	248309		1 03 17.9	-60 05 52	-0.015	-0.10	6.83	0.47	3.4	F5 V		46 s		m
6290	215346		1 03 18.4	-41 01 16	+0.003	0.00	7.1	1.2	-0.3	K5 III		300 s		
6098	21953		1 03 20.3	+59 05 32	-0.002	0.00	8.00	1.1	0.2	K0 III		320 s		
6289	192893		1 03 24.2	-32 04 40	+0.002	+0.02	7.5	1.9	0.2	K0 III		82 s		
6242	109637		1 03 25.9	+3 15 18	+0.002	-0.03	8.0	1.1	-0.1	K2 III		410 s		
6229	74458		1 03 36.4	+23 46 06	+0.001	-0.03	7.73	1.10		G0				
6130	11551		1 03 36.9	+61 04 30	-0.001	+0.01	5.92	0.49		F0	-1		868	m
6147	21956		1 03 37.0	+58 54 34	+0.005	-0.01	7.0	1.4	-0.3	K5 III	-12	260 s		
5621	171		1 03 40.1	+84 36 26	+0.048	0.00	6.7	0.5	3.4	F5 V		47 s		
6266	109641		1 03 40.7	+5 14 03	-0.001	-0.05	7.8	0.2	2.1	A5 V		140 s		
6277	109642		1 03 47.1	+6 45 54	+0.001	-0.01	7.00	0.1	0.6	A0 V		190 s	874	m
6288	109643	26 Cet	1 03 48.9	+1 22 00	+0.008	-0.04	6.04	0.27	2.6	dF0	+6	49 s	875	m
6226	36891		1 03 53.3	+47 38 33	0.000	0.00	6.82	-0.03	0.0	B8.5 V	-55	230 s		
6354	215348		1 03 55.4	-40 38 52	-0.001	+0.02	7.60	0.1	1.4	A2 V		170 s		m
6224	36892		1 03 56.3	+49 43 47	-0.002	-0.02	7.6	1.1	0.2	K0 III		280 s		
6200	21964		1 03 56.6	+53 05 16	+0.003	0.00	8.0	0.4	2.6	F0 V		110 s		
6225	36893		1 03 57.5	+48 51 39	+0.013	+0.06	7.2	0.5	4.0	F8 V		43 s		
6264	54390		1 03 58.6	+35 28 09	0.000	-0.02	8.00	0.1	1.7	A3 V		180 s	873	m
---	54391		1 04 00.1	+36 49 59	+0.002	-0.02	7.9	0.7						
6367	215350		1 04 02.3	-40 31 53	+0.008	-0.06	7.50	0.5	3.4	F5 V		66 s		
6211	21967		1 04 02.4	+52 30 08	+0.001	-0.05	5.99	1.47	-0.1	K2 III	-7	110 s		
6262	54393		1 04 04.5	+38 41 18	-0.001	-0.01	7.30	1.6	-0.5	M3 III	-30	360 s		
6273	54394		1 04 06.1	+31 30 07	+0.002	-0.03	8.0	1.1	0.2	K0 III		360 s		
6238	21971		1 04 15.1	+57 07 02	0.000	0.00	7.1	1.1	0.2	K0 III	-2	220 s		
6250	21974		1 04 16.4	+54 12 15	-0.016	-0.06	6.8	0.6	4.4	G0 V		30 s		
6210	11557		1 04 19.5	+61 34 49	-0.010	-0.02	5.84	0.54	4.0	dF8	-16	23 s		
6413	215353		1 04 24.0	-47 56 27	+0.008	+0.01	6.83	1.59	-0.5	M2 III		270 mx		
6249	21976		1 04 24.4	+57 58 52	-0.001	-0.01	7.56	0.12	0.4	B9.5 V		220 s		
6272	36900		1 04 24.6	+49 12 16	0.000	-0.05	7.9	0.4	2.6	F0 V		110 s		
6301	74471		1 04 27.6	+29 39 32	+0.006	-0.11	6.19	0.43	3.4	F5 V	0	36 s		
6403	192907		1 04 32.6	-33 31 58	-0.002	-0.01	6.43	1.09		G5				m
6495	248313		1 04 34.0	-64 42 05	+0.004	0.00	7.3	1.1	0.2	K0 III		260 s		
6314	54398		1 04 36.3	+39 59 28	+0.007	-0.02	6.72	0.31	2.6	F0 V	+11	66 s		
6434	192911		1 04 40.0	-39 29 19	-0.016	-0.54	7.72	0.60	3.0	G3 IV		28 mx		
6328	54400		1 04 40.3	+33 10 57	0.000	-0.03	7.9	1.1	-0.1	K2 III		390 s		
6375	109653		1 04 43.5	+2 18 49	0.000	-0.04	7.3	1.1	0.2	K0 III		260 s		
6444	215354		1 04 46.7	-43 03 51	+0.004	+0.02	7.9	0.9	3.0	F2 V		34 s		
6300	21988		1 04 46.8	+51 00 36	+0.002	0.00	6.54	-0.08	-1.7	B3 V	-5	400 s		
6299	21987		1 04 49.2	+55 46 43	-0.001	+0.01	7.8	0.1	0.6	A0 V		250 s		
6412	166785		1 04 49.3	-20 43 58	-0.003	0.00	7.8	0.6	3.4	F5 V		59 s		
6360	92228		1 04 49.9	+18 53 42	+0.005	-0.02	7.7	0.9	3.2	G5 IV		79 s		

HD	SAO	Star Name	α 2000	δ 2000	μ(α)	μ(δ)	V	B-V	M$_v$	Spec	RV	d(pc)	ADS	Notes
6163	4316		1^{h}04^{m}50^s.0	+74°22'20"	+0^s.006	0".00	6.8	0.1	0.6	A0 V		160 s		
6374	109654		1 04 50.4	+7 02 45	+0.010	-0.10	7.9	0.5	4.0	F8 V		60 s		
6386	109656	73 Psc	1 04 52.5	+5 39 23	+0.002	-0.01	6.00	1.51	-0.3	gK5	-15	180 s		
6384	92229		1 04 55.1	+16 14 33	0.000	+0.02	8.00	1.6	-0.5	M2 III		500 s		
6426	147597		1 05 03.8	-13 45 34	0.000	-0.03	7.5	1.1	0.2	K0 III		290 s		
6397	92230	72 Psc	1 05 05.3	+14 56 46	0.000	+0.06	5.68	0.41	3.0	F2 V	+4	33 s	889	m
6623	255735		1 05 08.4	-71 44 02	+0.027	+0.06	7.36	1.10	-0.2	K3 III		140 mx		
6466	192913		1 05 08.5	-37 33 51	-0.001	0.00	7.7	1.2	4.4	G0 V		18 s		
7070	258255		1 05 08.7	-83 35 27	-0.004	+0.01	7.4	1.6	0.2	K0 III		120 s		
6409	92233		1 05 12.1	+19 11 53	0.000	0.00	7.4	1.6	-0.5	M4 III		390 s		
6521	232290		1 05 14.3	-57 47 42	+0.002	0.00	7.6	0.3	2.1	A5 V		110 s		
6493	215355		1 05 22.0	-40 16 12	0.000	-0.01	7.4	-0.1	2.6	F0 V		90 s		
6461	147600		1 05 25.4	-12 54 11	+0.005	+0.07	7.7	0.6	4.4	G0 V		45 s		
6662	255738		1 05 27.2	-70 56 00	0.000	+0.01	8.0	1.1	0.2	K0 III		360 s		
6006	179		1 05 28.1	+81 57 42	+0.003	0.00	7.90	0.02	0.6	A0 V		280 s		
6783	255739		1 05 31.2	-77 35 01	+0.002	+0.03	7.4	0.0	0.0	B8.5 V		300 s		
6491	166790		1 05 34.6	-26 57 19	+0.001	-0.05	7.8	0.4	2.6	F0 V		91 s		
6482	147601	27 Cet	1 05 36.7	-9 58 45	-0.003	-0.03	6.12	1.01	0.2	K0 III	+12	150 s		
6456	74482	74 ψ¹ Psc	1 05 40.9	+21 28 24	+0.004	-0.01	5.34	-0.03	-0.1	B9.5 IV	-3	120 s	899	m
6457	74483	74 ψ¹ Psc	1 05 41.6	+21 27 55	+0.003	-0.01	5.56	-0.07	0.2	B9 V	-4	120 s	899	m
6479	109666	77 Psc	1 05 49.1	+4 54 30	+0.001	-0.12	6.35	0.38	3.4	dF5	-7	48 s	903	m
6480	109667	77 Psc	1 05 51.3	+4 54 35	+0.002	-0.11	7.25	0.49	3.3	dF4	-10	48 s	903	m
6343	11564		1 05 52.9	+65 58 16	+0.001	0.00	7.26	0.16		B7 e	-8			
6478	92243		1 05 56.5	+15 22 57	+0.002	-0.07	7.3	0.4	3.0	F2 V		74 s		
6571	215363		1 06 03.1	-41 50 10	+0.002	+0.02	7.5	1.5	-0.1	K2 III		140 s		
6417	22000		1 06 03.7	+57 45 22	-0.001	-0.01	7.13	0.04	-1.0	B5.5 V	-24	350 s		
6595	215365	β Phe	1 06 05.0	-46 43 07	-0.003	+0.01	3.31	0.89	0.3	G8 III	-1	40 s		m
6530	147606	28 Cet	1 06 05.0	-9 50 22	+0.002	+0.01	5.58	0.01	0.0	A0 IV		130 s		
6546	147608		1 06 06.3	-10 59 05	+0.001	-0.01	6.9	0.8	3.2	G5 IV		55 s		
6559	166799		1 06 07.7	-23 59 34	-0.002	-0.04	6.14	1.09	3.2	G5 IV		22 s		
6469	54420		1 06 08.2	+35 30 58	-0.002	+0.01	7.6	0.4	2.6	F0 V		98 s		
6575	192920		1 06 09.6	-38 13 58	+0.004	+0.02	7.9	2.0	0.2	K0 III		88 s		
6476	54421		1 06 11.1	+32 10 53	+0.001	-0.02	6.6	1.1	0.2	K0 III	+28	190 s		m
6416	11571		1 06 22.7	+62 45 42	+0.015	-0.02	6.54	0.19		A3	+11			
6619	192925		1 06 26.4	-35 39 39	+0.005	-0.02	6.61	0.14		A m				
6573	147610		1 06 26.9	-9 45 59	+0.002	+0.01	7.5	0.1	1.4	A2 V		160 s		
6629	215367		1 06 27.5	-39 51 24	0.000	+0.01	6.7	1.4	0.2	K0 III		110 s		
6544	92247		1 06 30.0	+19 24 24	+0.002	-0.06	8.0	0.5	3.4	F5 V		82 s		
6557	92250	75 Psc	1 06 33.6	+12 57 23	+0.001	+0.04	6.12	0.96	3.2	G5 IV	+8	30 s		
6566	92251		1 06 36.5	+13 53 04	+0.004	-0.05	7.4	0.4	3.0	F2 V		75 s		
6568	109677		1 06 37.4	+8 21 38	-0.005	-0.02	6.9	0.4	2.6	F0 V		73 s		
6447	11576		1 06 37.5	+64 23 36	-0.009	-0.05	7.52	0.57		G0				m
6567	109678		1 06 41.9	+8 52 08	+0.001	+0.05	7.8	0.1	1.7	A3 V		160 s		
6616	147612		1 06 44.4	-15 30 37	+0.001	-0.02	7.3	0.6	4.4	G0 V		39 s		
6554	54431		1 06 47.4	+35 08 43	-0.003	-0.02	7.3	0.1	1.4	A2 V		150 s		
6628	166806		1 06 48.9	-22 51 22	-0.003	-0.09	7.9	0.9	3.2	G5 IV		75 s		
--	22013		1 06 51.0	+57 05 57	-0.003	+0.02	7.9	1.2	0.2	K0 III		270 s		
7101	258257		1 06 52.0	-81 38 55	+0.044	-0.01	7.80	1.1	0.2	K0 III		170 mx		m
6475	22012		1 06 53.8	+59 51 32	-0.001	+0.01	6.87	0.00	1.4	A2 V	0	120 s		
6591	109681		1 06 53.9	+8 39 07	-0.001	-0.03	7.8	0.1	1.7	A3 V		160 s		
6414	4327		1 06 55.2	+70 55 54	+0.018	0.00	6.7	0.1	1.4	A2 V	-6	56 mx		
6523	22016		1 06 58.3	+50 35 10	0.000	0.00	8.0	1.1	0.2	K0 III		340 s		
6626	129125		1 06 58.5	-4 54 23	+0.001	-0.03	8.03	0.09	0.6	A0 V		260 s		
6541	36933		1 06 58.7	+48 14 59	+0.001	-0.03	7.9	1.4	-0.3	K5 III		380 s		
6474	11581		1 06 59.4	+63 46 25	-0.003	+0.02	7.61	1.62	-8.0	G0 Ia		4000 s		
6497	22015		1 07 00.2	+56 56 04	+0.013	-0.13	6.43	1.18	-0.1	K2 III	-96	200 s		
6614	92255		1 07 03.9	+11 33 07	+0.004	0.00	7.20	0.4	2.6	F0 V		83 s	920	m
6613	92256		1 07 07.2	+20 09 02	+0.001	-0.01	7.3	0.9	3.2	G5 IV		65 s		
6586	54434		1 07 09.1	+38 39 01	+0.003	-0.02	7.57	0.50	4.0	dF8	+14	52 s	918	m
6540	22021		1 07 09.4	+53 29 54	+0.002	0.00	6.50	1.1	0.2	K0 III	+7	170 s	915	m
6539	22019		1 07 10.5	+56 22 29	-0.002	+0.01	7.4	0.1	0.6	A0 V		220 s		
6651	129128		1 07 10.7	-1 43 53	+0.004	-0.01	7.40	0.5	3.4	F5 V	-6	63 s	923	m
6669	166813		1 07 11.2	-24 51 20	+0.001	-0.01	7.5	1.2	0.2	K0 III		230 s		
6668	166814		1 07 13.0	-23 59 47	+0.007	-0.03	6.37	0.23	2.4	A7 V		47 mx		
6564	36936		1 07 14.0	+49 33 21	+0.004	0.00	6.82	-0.04	0.6	A0 V		180 s		
6584	36938		1 07 15.0	+44 36 59	+0.002	-0.02	7.7	1.1	0.2	K0 III		310 s		
6793	248324	ι Tuc	1 07 18.6	-61 46 31	+0.011	0.00	5.37	0.88	0.3	G5 III	-8	100 s		
6585	36940		1 07 21.2	+43 31 21	+0.010	-0.06	7.3	0.5	3.4	F5 V		60 s		
6612	54438		1 07 22.5	+38 01 44	0.000	-0.01	7.0	0.0	0.6	B9.5 V	-5	210 s		
6583	36939		1 07 23.4	+48 03 14	+0.002	0.00	7.6	0.5	3.4	F5 V		69 s		
6611	36943		1 07 28.7	+41 15 32	+0.010	-0.05	7.3	0.5	3.4	F5 V		61 s		
6735	215372		1 07 31.9	-41 44 50	+0.017	-0.07	7.3	0.3	4.4	G0 V		38 s		
6538	11584		1 07 36.9	+63 41 39	0.000	+0.02	7.67	0.99		G5				
6722	166819		1 07 38.1	-27 43 17	+0.004	+0.05	7.99	1.09	3.2	K0 IV		73 s		
6742	192938		1 07 41.4	-32 48 52	+0.008	-0.05	7.9	1.8	3.2	G5 IV		21 s		

HD	SAO	Star Name	α 2000	δ 2000	μ(α)	μ(δ)	V	B-V	M$_v$	Spec	RV	d(pc)	ADS	Notes
6706	147622	30 Cet	1^{h}07^{m}46^s.1	-9°47'08"	+0^s.010	+0".02	5.82	0.43	3.0	F2 V	+22	34 s		
6646	36948		1 07 46.7	+41 30 48	0.000	0.00	7.3	0.1	1.4	A2 V		150 s		
6767	215374	υ Phe	1 07 47.8	-41 29 14	+0.003	+0.01	5.21	0.16	1.7	A3 V	+9	47 s		m
6819	232301		1 07 48.6	-55 58 12	+0.003	-0.05	7.6	1.4	4.4	G0 V		14 s		
6645	—		1 07 51.8	+46 51 08			7.60	1.1	-0.9	K0 II-III	-26	450 s	926	m
6580	11586		1 07 52.0	+62 39 13	-0.004	+0.01	7.83	1.95	-0.5	M2 III		280 s		
6696	92263		1 07 56.3	+17 24 44	+0.003	-0.03	7.7	0.4	2.6	F0 V		100 s		
6695	74506	79 ψ² Psc	1 07 57.1	+20 44 21	+0.006	-0.09	5.60	0.13	1.7	A3 V	-2	35 mx		
6664	54443		1 07 58.8	+39 15 08	-0.005	-0.02	7.82	0.60	-8.0	G0 Ia	+6	240 mx		
6734	109694		1 07 59.5	+1 59 35	+0.009	-0.42	6.46	0.85	3.2	K0 IV	-95	45 s		
6658	36950	41 And	1 08 00.8	+43 56 32	+0.015	-0.05	5.03	0.11		A2 m	+9	17 mn		
6680	54445	78 Psc	1 08 01.2	+32 00 43	+0.016	-0.03	6.25	0.40	2.1	F5 IV	+14	68 s		
6870	248325		1 08 04.0	-61 52 18	+0.019	-0.01	7.45	0.26	0.3	A5 III		66 mx		
6678	36951		1 08 05.9	+42 50 52	+0.001	0.00	7.7	0.1	0.6	A0 V		260 s		
6688	36953		1 08 10.4	+44 47 45	0.000	-0.01	7.47	-0.09	0.0	B8.5 V		310 s		
6656	22029		1 08 11.7	+55 09 21	-0.001	-0.02	7.8	0.4	3.0	F2 V		87 s		
6715	74509		1 08 12.3	+21 58 35	+0.028	-0.07	7.67	0.67	3.2	G5 IV		38 mx		
6714	74510		1 08 15.4	+28 52 04	-0.002	-0.09	6.8	0.4	2.6	F0 V		69 s		
6582	22024	30 μ Cas	1 08 16.2	+54 55 15	+0.395	-1.58	5.17	0.69	5.8	G5 V p	-97	7.7 t		m
6634	22028		1 08 17.2	+60 08 11	+0.004	+0.01	8.00	1.1	0.2	K0 III		320 s		m
6761	109696		1 08 21.7	+9 54 30	0.000	+0.01	6.8	1.6	-0.5	M2 III		290 s		
6763	109697	80 Psc	1 08 22.1	+5 38 59	-0.018	-0.18	5.52	0.34	2.6	F0 V	+7	37 s		m
6677	22031		1 08 22.5	+54 13 55	+0.001	+0.01	7.0	0.1	0.6	A0 V		180 s		
6882	232306	ζ Phe	1 08 23.0	-55 14 45	+0.002	+0.03	3.92	-0.08	-0.2	B8 V	+18	67 s		m,v
6788	129140		1 08 26.7	-6 10 31	-0.002	-0.01	7.9	0.4	2.6	F0 V		120 s		
6676	22032		1 08 33.3	+58 15 49	+0.001	0.00	5.79	-0.01		B8	-4			
6869	215379		1 08 33.6	-46 40 04	+0.001	+0.04	6.89	0.29		A5				m
6713	36960		1 08 34.8	+48 25 17	-0.001	0.00	7.9	1.4	-0.3	K5 III		390 s		
6805	147632	31 η Cet	1 08 35.3	-10 10 56	+0.015	-0.13	3.45	1.16	-0.1	K2 III	+12	36 ts		m
6610	4334		1 08 40.8	+70 40 57	+0.001	+0.03	7.6	0.1	0.6	A0 V		230 s		
6854	192949		1 08 42.3	-32 09 41	-0.004	-0.04	7.9	0.7	0.2	K0 III		340 mx		
5848	181		1 08 44.6	+86 15 26	+0.077	-0.01	4.25	1.21	-0.1	K2 III	+9	68 s		
6785	54460		1 08 48.5	+30 24 28	0.000	-0.01	7.5	0.2	2.1	A5 V		120 s		
6815	109700		1 08 55.8	+9 43 48	+0.001	-0.03	7.2	0.0	0.4	B9.5 V		230 s		
6910	215382		1 09 02.4	-42 24 07	+0.003	0.00	7.4	0.7	4.4	G0 V		31 s		
6473	4331		1 09 12.3	+80 00 42	-0.010	-0.03	6.3	1.1	0.2	K0 III	-27	160 s		
6675	11595		1 09 13.0	+69 41 10	+0.001	-0.01	6.90	0.31		B0.5 III				
6813	54462		1 09 13.1	+33 56 29	-0.001	-0.02	7.3	0.1	1.4	A2 V		150 s		
6812	54463		1 09 17.2	+34 48 22	+0.003	0.00	7.3	0.1	1.7	A3 V		130 s		
6996	232315		1 09 22.1	-56 35 43	+0.015	-0.02	7.13	0.46	2.1	F5 IV		78 mx		m
6847	74521		1 09 23.1	+22 54 35	+0.007	+0.06	7.5	0.4	3.0	F2 V		80 s		
6811	36972	42 φ And	1 09 30.1	+47 14 31	+0.001	0.00	4.25	-0.07	-1.2	B8 III	0	120 s	940	m
6795	22040		1 09 34.5	+57 21 00	-0.012	-0.03	6.7	0.4	3.0	F2 V		54 s		
6927	147641		1 09 35.9	-11 40 45	0.000	-0.03	7.4	0.1	1.7	A3 V		140 s		
6886	74523		1 09 39.0	+23 47 38	0.000	-0.03	7.30	0.4	2.6	F0 V	-21	87 s	955	m
6755	11602		1 09 42.8	+61 32 50	+0.086	+0.07	7.68	0.72	4.2	F9 V	-320	41 s		
6860	54471	43 β And	1 09 43.8	+35 37 14	+0.015	-0.11	2.06	1.58	-0.4	M0 III	0	27 ts	949	Mirach, m
6859	54473		1 09 45.2	+38 07 29	+0.003	0.00	7.60	0.4	2.6	F0 V		100 s	948	m
6810	22043		1 09 45.4	+52 51 17	+0.001	+0.02	7.8	1.1	0.2	K0 III		300 s		
6903	92283	81 ψ³ Psc	1 09 49.1	+19 39 31	0.000	+0.01	5.55	0.69	3.2	G5 IV	-8	30 s		
6844	36976		1 09 51.8	+49 03 16	0.000	0.00	7.1	0.1	0.6	A0 V		190 s		
6833	22044		1 09 52.3	+54 44 21	+0.003	+0.05	6.74	1.17	0.0	K1 III	-245	220 s		
6893	54476		1 09 57.0	+33 52 53	+0.007	0.00	7.1	0.5	3.4	F5 V		54 s		
6843	22048		1 10 00.6	+52 02 00	-0.001	0.00	7.94	0.77		F5			950	m
7187	255748		1 10 01.5	-72 57 27	+0.007	0.00	7.14	0.08		A0				m
7006	166846		1 10 03.2	-26 11 38	-0.002	-0.01	7.4	0.9	0.2	K0 III		270 s		
7082	232324		1 10 07.2	-57 41 40	-0.002	-0.11	6.5	0.3	3.2	G5 IV		45 s		
6966	92288		1 10 11.4	+15 40 27	+0.002	-0.02	6.8	1.49	-0.4	M0 III	-3	200 s		
6976	129154	32 Cet	1 10 11.9	-8 54 22	-0.001	-0.03	6.40	1.04	0.3	gG6	-20	130 s		
6901	36982		1 10 12.9	+40 56 05	-0.001	-0.03	7.9	0.1	1.4	A2 V		200 s		
7097	232325		1 10 15.5	-56 51 47	-0.005	-0.06	7.3	-0.1	2.6	F0 V		86 s		
6920	36984	44 And	1 10 18.6	+42 04 53	-0.012	-0.04	5.65	0.60	4.0	F8 V	-11	19 s		
6953	74530		1 10 19.3	+25 27 28	0.000	-0.11	5.80	1.47	-0.3	K5 III	+5	130 mx		
7026	147650		1 10 23.2	-16 18 26	-0.002	-0.02	7.6	0.8	3.2	G5 IV		76 s		
7025	129157		1 10 30.8	-6 05 08	+0.004	-0.01	7.5	0.1	1.4	A2 V		85 mx		
6841	11613		1 10 32.1	+64 10 29	+0.007	-0.02	7.29	0.22		A3				
6917	22057		1 10 32.5	+52 07 45	0.000	-0.02	7.5	0.1	1.4	A2 V		160 s		
7014	109715	33 Cet	1 10 33.5	+2 26 45	0.000	-0.01	5.95	1.51	-0.3	gK4	-3	150 s		
6918	22060		1 10 34.3	+51 00 48	+0.001	+0.03	6.86	0.45		F5			963	m
7174	248332		1 10 35.2	-65 55 21	+0.012	+0.01	7.8	0.5	3.4	F5 V		75 s		
6829	11612	31 Cas	1 10 39.3	+68 46 44	+0.007	-0.02	5.29	-0.02		A0	+1	22 mn		m
6963	36995		1 10 41.7	+42 55 55	-0.015	-0.19	7.80	1.1	0.2	K0 III		64 mx		m
7048	129159		1 10 42.9	-4 50 33	-0.002	0.00	7.2	0.4	2.6	F0 V		83 s		
7199	248334		1 10 47.0	-66 11 18	+0.017	-0.13	8.0	0.8	3.2	G5 IV		62 mx		
6840	11614		1 10 50.6	+67 46 50	+0.042	+0.04	6.7	0.6	4.4	G0 V	-10	28 s		

HD	SAO	Star Name	α 2000	δ 2000	μ(α)	μ(δ)	V	B–V	M$_V$	Spec	RV	d(pc)	ADS	Notes
6962	36996		1^{h}10^{m}52^s.2	+49°09'50"	0^s.000	–0".02	7.4	1.1	–0.1	K2 III		290 s		
7047	109718		1 10 54.3	+9 33 51	+0.007	+0.25	7.23	0.57	4.4	G0 V		37 s	972	m
6948	22068		1 10 58.6	+56 46 29	0.000	0.00	7.26	–0.03	0.4	B9.5 V		240 s		
7093	147654		1 11 01.8	–18 48 18	+0.007	0.00	7.5	1.7	–0.1	K2 III		160 s		
6961	22070	33 θ Cas	1 11 06.1	+55 09 00	+0.027	–0.02	4.33	0.17	2.4	A7 V	+9	37 mn		m
7034	54493	82 Psc	1 11 06.7	+31 25 29	–0.001	–0.01	5.16	0.23		A7	+2			
7019	54494	45 And	1 11 10.2	+37 43 27	–0.001	0.00	5.80	–0.10	–1.6	B7 III		290 s		m
7011	36999		1 11 10.6	+46 11 02	+0.001	–0.02	6.8	1.4	–0.3	K5 III		240 s		
7089	129164		1 11 12.6	–1 16 28	–0.002	–0.03	8.0	0.4	3.0	F2 V		98 s		
7018	37001		1 11 14.4	+41 13 19	+0.003	0.00	7.9	0.9	3.2	G5 IV		87 s		
7017	37000		1 11 14.8	+46 37 48	+0.001	0.00	7.2	0.0	0.4	B9.5 V		210 s		
7076	92302		1 11 16.2	+14 41 30	–0.001	–0.02	8.0	1.1	0.2	K0 III		370 s		
7122	147656		1 11 17.2	–13 30 13	+0.001	0.00	7.6	1.6	–0.5	M4 III		420 s		
6960	11615		1 11 25.5	+64 12 11	+0.006	–0.01	5.55	–0.07		B9	–10			
7087	74544	84 χ Psc	1 11 27.1	+21 02 05	+0.003	–0.01	4.66	1.03	0.2	K0 III	+16	75 s		
7107	92304		1 11 28.9	+10 17 32	+0.001	+0.01	6.6	0.9	3.2	G5 IV		49 s		
7134	147658		1 11 32.0	–12 50 36	–0.002	–0.07	7.4	0.6	4.4	G0 V		41 s		
7106	74546	83 τ Psc	1 11 39.5	+30 05 23	+0.006	–0.03	4.51	1.09	1.7	K0 III–IV	+30	29 s		
7055	37004		1 11 39.8	+49 37 25	+0.002	–0.02	7.6	1.1	0.2	K0 III		280 s		
7119	74548		1 11 40.0	+22 43 20	–0.003	–0.02	7.10	0.1	1.7	A3 V		120 s		m
6972	11617	32 Cas	1 11 41.4	+65 01 08	+0.004	–0.01	5.57	–0.09		B8	–2			RU Cas, q
7147	129169	34 Cet	1 11 43.4	–2 15 04	–0.004	–0.02	5.94	1.40	–0.3	K4 III	–9	180 s		
7085	54500		1 11 44.3	+36 24 35	–0.002	+0.02	8.0	0.1	1.4	A2 V		210 s		
7084	37008		1 11 50.5	+43 45 08	+0.003	–0.02	8.0	1.1	–0.1	K2 III		410 s		
7221	192974		1 12 00.1	–39 09 43	+0.008	–0.05	7.9	1.2	3.2	G5 IV		48 s		
7442	255751		1 12 10.7	–73 54 26	+0.025	+0.04	7.0	0.5	4.0	F8 V		39 s		
7202	147669		1 12 15.9	–10 56 25	+0.001	–0.01	7.9	1.1	0.2	K0 III		340 s		
6798	4341		1 12 16.6	+79 40 27	+0.034	0.00	5.7	0.1	0.6	A0 V	+18	71 mx		
7138	37020		1 12 16.8	+40 54 39	+0.002	–0.04	7.6	0.4	2.6	F0 V		100 s		
7193	92310		1 12 19.0	+12 16 53	–0.006	–0.07	7.1	0.5	3.4	F5 V		55 s		
7259	192977		1 12 23.4	–30 48 09	+0.005	–0.07	6.52	0.48	2.3	F7 IV		70 s		
7145	37022		1 12 24.5	+41 18 31	–0.001	–0.01	8.0	0.1	0.6	A0 V		300 s		
7182	74554		1 12 28.6	+23 15 21	+0.003	–0.01	7.6	0.4	2.6	F0 V		98 s		
7234	129175		1 12 30.0	–9 13 25	+0.027	0.00	7.80	0.6	4.4	G0 V		43 mx		m
7218	109727		1 12 30.6	+2 28 17	–0.012	–0.10	6.9	0.5	4.0	F8 V	+3	37 s		m
––	––		1 12 31.3	+1 57 54			6.90							m
7083	11629		1 12 31.6	+64 37 20	–0.002	–0.02	8.03	0.05		A0				
7158	37023		1 12 33.9	+45 20 16	+0.002	+0.03	6.5	1.4	–0.3	K5 III	+22	230 s		
7257	166876		1 12 34.5	–20 27 30	0.000	–0.02	7.6	0.6	3.4	F5 V		52 s		
7102	11631		1 12 41.2	+65 00 32	–0.001	–0.02	7.90	0.4	2.6	F0 V		110 s	983	m
7312	192980		1 12 45.3	–37 51 24	+0.007	–0.03	5.92	0.28	0.5	A7 III		83 mx		
7311	192981		1 12 46.7	–35 12 20	+0.001	–0.01	6.9	0.6	0.2	K0 III		220 s		
7268	129177		1 12 47.6	–6 47 00	–0.004	–0.01	6.9	0.8	3.2	G5 IV	–11	56 s		
7215	54514		1 12 52.8	+32 04 32	+0.001	0.00	6.98	0.05		A2	–2		988	m
7169	22094		1 12 53.9	+51 36 09	+0.005	–0.06	7.3	0.2	2.1	A5 V		84 mx		
7323	192983		1 12 54.9	–35 44 44	–0.002	+0.01	7.83	0.07	1.4	A2 V n		190 s		
7189	37024		1 12 57.6	+47 11 41	–0.001	–0.03	7.70	0.9	0.3	G6 III	–21	280 s		
7229	74561		1 12 59.4	+30 03 51	+0.001	–0.03	6.19	1.00	0.3	G8 III	+36	140 s	990	m
7205	37026		1 13 05.8	+41 39 14	+0.026	–0.06	7.29	0.78	3.2	G5 IV		50 mx		
7157	11637		1 13 09.9	+61 42 21	+0.005	–0.01	6.41	0.01		B9	–2		987	m
7177	22099		1 13 14.4	+57 35 20	–0.006	–0.01	8.0	1.4	–0.3	K5 III		400 s		
7254	54523		1 13 16.1	+34 05 55	+0.001	0.00	6.75	–0.15	0.0	B8.5 V	+1	220 s		
7156	11640		1 13 17.5	+62 49 48	+0.001	+0.01	7.7	1.1	0.2	K0 III		280 s		
7243	37031		1 13 20.5	+41 14 12	+0.003	+0.02	8.0	1.4	–0.3	K5 III		460 s		
7214	37029		1 13 22.1	+49 16 49	+0.002	–0.01	7.8	0.1	1.4	A2 V		180 s		
7321	129182		1 13 23.0	–2 33 03	+0.004	–0.02	7.2	1.1	0.2	K0 III		250 mx		
7071	4347		1 13 24.5	+73 55 15	–0.007	0.00	7.8	0.6	4.4	G0 V		46 s		
7382	192991		1 13 26.5	–39 32 26	+0.014	+0.02	7.4	0.2	2.6	F0 V		93 s		
7266	37034		1 13 32.4	+44 10 19	–0.001	–0.07	7.8	1.1	–0.1	K2 III		180 mx		
7299	74566		1 13 32.7	+29 44 09	–0.001	–0.05	6.8	0.9	1.8	G8 III–IV	–12	100 s		
7300	74567		1 13 33.8	+26 27 07	+0.002	–0.02	7.90	1.1	–0.1	K2 III		400 s		
7267	37035		1 13 34.2	+41 09 03	+0.004	0.00	8.00	0.9	3.2	G5 IV		91 s	994	m
––	––		1 13 37.6	+62 57 38			7.50	1.6		R7.7				
7344	109739	86 ζ Psc	1 13 43.8	+7 34 31	+0.010	–0.05	4.86	0.32	2.1	F0 IV–V	+9	33 s	996	m
7402	192994		1 13 44.5	–31 07 02	+0.003	–0.04	7.73	1.07	1.7	K0 III–IV		140 s		m
7318	74571	85 φ Psc	1 13 44.8	+24 35 01	+0.001	–0.03	4.65	1.04	0.2	K0 III	+6	74 s	995	m
7308	74570		1 13 45.0	+26 14 30	–0.002	–0.02	7.90	1.4	–0.3	K5 III	–46	440 s		
7345	109740	86 ζ Psc	1 13 45.2	+7 34 42	+0.010	–0.05	6.30	0.49	3.7	F6 V	+11	33 s	996	m
7430	215418		1 13 49.2	–42 22 20	0.000	–0.01	7.8	1.6	0.2	K0 III		94 s		
7376	129187		1 13 51.9	–4 50 50	+0.003	–0.02	7.2	0.4	3.0	F2 V		68 s		
7398	147682		1 14 01.0	–13 28 33	+0.002	+0.02	6.9	0.2	2.1	A5 V		89 s		
7334	54534		1 14 01.3	+34 29 32	+0.004	–0.02	7.1	0.5	3.4	F5 V		54 s		
7528	248338		1 14 03.0	–61 00 35	–0.006	–0.02	8.0	0.4	3.4	F5 V		82 s		
7252	11644		1 14 03.1	+60 53 02	–0.005	+0.03	7.12	0.09	–3.5	B1 V	–3	90 mx		
7351	74576		1 14 04.9	+28 31 47	+0.006	–0.04	6.5	1.6		M2 s	+2			

25

HD	SAO	Star Name	α 2000	δ 2000	μ(α)	μ(δ)	V	B-V	M_v	Spec	RV	d(pc)	ADS	Notes
			$h \quad m \quad s$	$\circ \quad ' \quad ''$	s	$''$								
--	22114		1 14 07.1	+56 00 48	+0.005	+0.02	7.8	1.8	-0.3	K5 III		270 s		
7374	92326	87 Psc	1 14 07.6	+16 08 01	-0.002	-0.02	5.98	-0.08	-1.2	B8 III	-16	270 s		
7284	22113		1 14 10.3	+58 48 45	+0.002	-0.04	7.2	0.5	3.4	F5 V		57 s		
7516	232350		1 14 20.7	-50 24 59	0.000	0.00	7.34	1.13	0.0	K1 III		280 s		
7384	54541		1 14 20.8	+30 32 31	+0.003	-0.02	7.50	0.1	0.6	A0 V		240 s	1000	m
7527	232352		1 14 21.7	-53 07 27	+0.005	+0.01	7.4	1.7	0.2	K0 III		110 s		
7438	129192		1 14 22.4	-7 54 39	+0.009	+0.28	7.87	0.78	5.5	dG7	+16	28 s		
7439	129193	37 Cet	1 14 23.9	-7 55 23	+0.008	+0.28	5.13	0.46	3.0	dF2	+22	23 ts	1003	m
7449	129195		1 14 29.2	-5 02 50	-0.011	-0.13	7.5	0.5	4.0	F8 V		51 s		
7477	166898		1 14 29.9	-22 26 51	0.000	-0.03	7.9	2.1	-0.1	K2 III		110 s		
7349	37044		1 14 35.9	+49 25 45	+0.005	-0.02	7.2	0.0	0.4	B9.5 V		99 mx		
7239	4352		1 14 39.6	+71 06 03	-0.001	-0.02	7.9	1.1	0.2	K0 III		300 s		
7363	37045		1 14 41.3	+50 00 52	0.000	0.00	8.0	1.1	-0.1	K2 III		370 s		
7582	232353		1 14 41.6	-57 23 02	+0.005	0.00	7.8	1.8	0.2	K0 III		110 s		
7446	109753	88 Psc	1 14 42.3	+6 59 43	-0.001	-0.02	6.03	1.08	0.3	gG6	-9	100 s		
7497	166902		1 14 42.9	-22 31 37	0.000	-0.02	7.6	2.1	-0.1	K2 III		96 s		
7392	54543		1 14 44.0	+39 26 58	+0.001	-0.02	7.71	0.96	3.2	G5 IV		59 s		
7331	11652		1 14 45.4	+60 56 23	+0.005	-0.02	7.24	0.45	2.3	F7 IV		97 s	999	m
7476	129196	38 Cet	1 14 49.1	-0 58 26	-0.001	+0.21	5.70	0.42	3.1	dF3	+26	32 s		m
7463	109756		1 14 51.1	+9 44 28	-0.001	-0.02	7.9	0.5	3.4	F5 V		78 s		
7495	147692		1 14 54.1	-15 49 09	+0.007	-0.03	7.4	0.5	4.0	F8 V		47 s		
7597	232355		1 14 55.6	-55 22 25	+0.001	-0.02	7.6	2.3	0.2	K0 III		300 s		
7418	37054		1 15 00.1	+46 37 47	-0.003	0.00	8.0	1.1	-0.1	K2 III		370 s		
7608	232356		1 15 00.6	-55 37 58	+0.003	0.00	7.60	1.4	-0.3	K5 III		380 s		m
7693	248342		1 15 00.9	-68 49 10	+0.073	+0.10	7.21	0.97		G5		18 mn		m
7434	54547		1 15 01.6	+34 53 04	+0.002	0.00	7.9	1.1	0.2	K0 III		340 s		
7570	215428	ν Phe	1 15 11.1	-45 31 53	+0.064	+0.19	4.96	0.58	4.0	F8 V	+12	15 ts		
7645	--		1 15 12.0	-58 01 15			8.0	2.8	-0.5	M2 III		99 s		
7544	166905		1 15 14.4	-22 17 49	+0.001	+0.02	7.5	0.8	3.2	G5 IV		73 s		
7475	54552		1 15 25.6	+38 29 11	+0.003	-0.01	7.90	0.9	3.2	G5 IV		87 s	1009	m
7444	22130		1 15 27.6	+51 44 28	+0.001	0.00	7.5	1.1	0.2	K0 III		270 s		
7511	74591		1 15 28.1	+20 24 54	-0.002	0.00	7.1	0.1	1.4	A2 V		140 s		
7416	11659		1 15 31.6	+60 31 11	+0.005	+0.02	8.00	0.9	0.3	G8 III		310 s		
7551	109764		1 15 35.6	+0 54 44	+0.003	0.00	6.7	0.1	1.4	A2 V		120 s		
--	22133		1 15 40.4	+57 42 24	0.000	-0.01	8.0	1.7	-0.3	K5 III		330 s		
7788	248346	κ Tuc	1 15 46.2	-68 52 34	+0.076	+0.11	4.86	0.47	3.7	F6 V	+9	18 ts		
7564	109766		1 15 46.8	+9 47 05	-0.004	+0.04	7.20	0.4	3.0	F2 V		69 s	1016	m
7550	74592		1 15 53.3	+20 30 09	+0.007	-0.02	7.5	0.8	3.2	G5 IV		73 s		
7629	166915		1 15 53.7	-23 58 23	-0.002	+0.01	7.13	0.30	0.6	F0 III		200 s		
7563	92337		1 15 55.9	+16 11 24	+0.004	-0.02	7.7	0.1	1.4	A2 V		180 s		
7507	37062		1 16 03.8	+49 53 46	+0.001	-0.01	7.6	1.4	-0.3	K5 III		340 s		
7458	11666		1 16 04.7	+61 54 27	-0.004	-0.05	7.30	0.38	2.6	F0 V		78 s		
7561	74593		1 16 04.9	+25 46 09	-0.001	-0.01	6.84	2.62		N7.7	+18			Z Psc, v
7676	193014		1 16 06.6	-34 08 58	0.000	-0.02	8.0	0.3		A5 p				
7706	215433		1 16 11.8	-42 00 33	-0.007	0.00	6.58	1.20		M5				
7389	4358		1 16 12.0	+71 44 38	+0.002	+0.01	7.83	0.78	0.2	K0 III	-17	340 s		
6319	193		1 16 13.4	+87 08 44	+0.069	-0.02	6.25	1.12	0.2	gK0	-5	130 s		
--	22138		1 16 16.7	+55 16 45	0.000	+0.01	7.9	1.8	-0.5	M2 III		350 s		
7578	54567		1 16 18.7	+33 06 53	+0.001	-0.03	6.02	1.15	0.0	K1 III	+6	150 s		
7661	147702		1 16 24.1	-12 05 51	+0.009	-0.02	7.5	0.9	3.2	G5 IV		72 s		
7546	37067		1 16 24.5	+48 04 56	+0.002	0.00	6.61	-0.05		A p				
7615	74596		1 16 27.4	+23 35 22	-0.004	-0.01	6.69	0.03		A0				
7858	248347		1 16 29.1	-67 25 52	+0.023	+0.02	6.9	0.6	3.0	F2 V		42 s		
7590	37069		1 16 29.2	+42 56 21	-0.010	-0.03	6.7	0.6	4.4	G0 V		29 s		
7238	4354		1 16 30.7	+79 54 36	-0.020	+0.06	6.26	0.42	3.0	F2 V	-43	43 s		
7660	109775		1 16 35.9	+2 44 05	+0.004	-0.03	7.54	0.36		F0				
7672	129204	39 Cet	1 16 36.2	-2 30 01	-0.007	-0.06	5.41	0.90	0.3	gG5	-20	110 s		m
7406	4360		1 16 36.8	+74 01 36	+0.005	-0.02	7.10	0.00	0.4	B9.5 V		200 s		
7730	166921		1 16 39.4	-25 40 29	+0.004	-0.03	8.0	1.6	3.2	G5 IV		29 s		
7753	193020		1 16 39.9	-38 07 12	+0.001	0.00	7.8	1.8	0.2	K0 III		110 s		
7767	215435		1 16 42.2	-41 10 50	+0.003	+0.01	7.5	1.3	0.2	K0 III		200 s		
7659	74601		1 16 50.3	+21 03 08	+0.006	-0.06	7.2	0.5	3.4	F5 V		57 s		
7777	193022		1 16 50.5	-39 32 05	-0.001	+0.03	7.7	1.3	4.4	G0 V		16 s		
7727	129210		1 16 58.9	-2 16 44	+0.017	-0.13	6.52	0.56	4.0	F8 V	+9	30 s		
7795	215438		1 17 00.6	-42 31 57	+0.003	-0.01	7.86	-0.09		B9				
7916	248350		1 17 03.6	-66 23 53	+0.008	+0.01	6.24	0.05		A0				m
7684	74604		1 17 05.0	+29 59 32	+0.001	-0.04	8.0	0.4	3.0	F2 V		98 s		
7647	37077		1 17 05.0	+44 54 07	+0.001	-0.04	6.4	1.4	-0.3	K5 III	-52	220 s		
7736	109783		1 17 09.7	+2 00 54	0.000	+0.01	8.00	0.9	3.2	G5 IV		91 s		m
7735	92347		1 17 13.6	+10 35 48	+0.001	0.00	8.00	0.5	3.4	F5 V		83 s	1035	m
7669	54579		1 17 16.5	+39 28 51	-0.001	+0.01	6.68	0.03	0.4	B9.5 V		160 s		
7763	109787		1 17 19.4	+1 50 28	0.000	-0.01	7.90	1.1	0.2	K0 III		350 s	1042	m
--	22151		1 17 20.1	+56 24 52	+0.004	+0.01	7.8	2.2	-0.3	K5 III		160 s		
7819	193030		1 17 21.7	-37 16 12	-0.002	0.00	7.70	1.1	0.2	K0 III		320 s		m
7724	54580		1 17 24.0	+31 44 41	-0.004	+0.01	6.8	1.1	0.2	K0 III	-34	210 s		

HD	SAO	Star Name	α 2000	δ 2000	μ(α)	μ(δ)	V	B-V	M_V	Spec	RV	d(pc)	ADS	Notes
7636	22154		1^{h}17^{m}26.2^s	+57°37'55"	+0.001^s	-0.01"	6.61	0.14		B1.5 e	-14			
7772	92350		1 17 32.9	+14 14 33	+0.001	-0.02	7.4	0.8	3.2	G5 IV		68 s		
7974	248354		1 17 34.9	-65 12 44	+0.004	0.00	7.3	1.6	-0.5	M4 III		360 s		
7899	232365		1 17 41.0	-51 02 07	-0.003	-0.03	8.0	0.5	1.4	A2 V		100 s		
7722	37086		1 17 41.7	+44 37 54	+0.001	0.00	7.40	1.4	-0.3	K5 III		350 s		m
7812	129217		1 17 42.2	-7 39 35	+0.001	+0.02	7.0	0.1	1.4	A2 V		130 s		
7886	215443		1 17 43.6	-46 10 20	0.000	0.00	8.01	1.02		G5				
7666	22159		1 17 44.9	+56 37 54	+0.005	-0.01	6.7	1.1	0.2	K0 III		190 s		
7804	109793	89 Psc	1 17 47.9	+3 36 52	-0.003	-0.02	5.16	0.07	1.7	A3 V	+5	49 s		
7710	37087		1 17 47.9	+49 00 30	+0.003	-0.02	7.13	0.07		A0			1040	m
7694	22160		1 17 50.3	+55 26 43	0.000	0.00	7.4	0.1	-3.5	B1 V	-9	1100 s		
8025	248356		1 17 50.9	-67 06 37	+0.016	+0.03	7.2	1.1	0.2	K0 III		200 mx		
7734	37091		1 17 57.6	+47 40 50	+0.001	-0.02	7.6	1.1	-0.1	K2 III		320 s		
7934	232371		1 17 59.1	-52 06 21	+0.006	+0.01	7.4	0.7	2.1	A5 V		57 s		
7898	193037		1 18 03.1	-34 08 14	0.000	0.00	7.74	0.26		A7 p				
7681	11682		1 18 07.1	+61 43 04	+0.002	-0.02	7.76	1.76		M				
7802	74617		1 18 08.7	+26 27 03	-0.001	-0.02	7.9	0.1	1.7	A3 V		170 s		
7758	37096		1 18 10.1	+47 25 12	+0.002	0.00	6.25	1.35	0.2	K0 III	-1	100 s		
7340	211		1 18 10.8	+81 33 38	-0.009	0.00	7.7	0.9	3.2	G5 IV		80 s		
7733	22165		1 18 14.0	+57 48 12	+0.001	+0.01	6.7	1.6	-0.5	M4 III		240 s		
7909	193040		1 18 14.9	-33 08 15	-0.001	+0.02	7.67	1.01	3.2	K0 IV		73 s		
7665	11681		1 18 18.7	+67 48 59	-0.001	0.00	6.7	0.1	0.6	A0 V		160 s		
7847	92359		1 18 23.1	+18 34 39	+0.001	+0.02	7.9	0.9	3.2	G5 IV		87 s		
7908	166940		1 18 25.3	-23 00 49	0.000	-0.05	7.27	0.28	0.5	A7 III		210 s		
7907	166941		1 18 28.7	-20 11 27	0.000	0.00	7.8	1.2	0.2	K0 III		240 s		
7931	166944		1 18 32.1	-28 43 58	+0.005	+0.03	7.89	1.03	4.6	K0 IV-V		39 s		
7824	37103		1 18 32.9	+43 16 55	0.000	+0.02	7.4	1.4	-0.3	K5 III		350 s		
7895	129219		1 18 41.0	-0 52 05	+0.029	-0.26	8.00	0.81	6.1	K1 V	+13	25 ts	1057	m
---	---		1 18 41.7	+32 25 16			8.00							m
7646	4371		1 18 42.1	+72 52 41	+0.006	-0.02	7.2	0.8	3.2	G5 IV		61 s		
8001	215450		1 18 44.1	-43 20 00	0.000	0.00	6.76	1.48		K2				
7741	11685		1 18 45.3	+65 36 22	+0.004	-0.03	8.0	1.1	0.2	K0 III		320 s		
7655	4372		1 18 45.4	+72 24 09	+0.002	+0.04	7.1	0.9	3.2	G5 IV		58 s		
7853	54592		1 18 46.9	+37 23 10	-0.001	-0.01	6.40	0.1	1.7	A3 V	+5	87 s	1055	m
7830	37107		1 18 50.3	+45 40 05	-0.001	+0.02	7.7	1.1	0.2	K0 III		300 s		
7864	54593		1 18 53.2	+39 57 48	+0.002	-0.03	7.50	0.9	3.2	G5 IV		72 s	1053	m
7871	54594		1 18 54.0	+33 30 56	-0.001	-0.02	7.9	0.2	2.1	A5 V		150 s		
7918	109802		1 18 57.4	+7 25 49	0.000	0.00	7.9	0.9	3.2	G5 IV		88 s		
8306	255762		1 19 01.7	-77 01 49	+0.039	+0.05	8.0	0.5	4.0	F8 V		63 s		
7862	37108		1 19 03.0	+45 42 23	-0.001	+0.04	7.4	1.1	-0.1	K2 III		290 s		
7471	218		1 19 06.6	+80 51 44	+0.012	+0.02	7.17	0.38		F0			1030	m
7623	4373		1 19 11.9	+76 47 47	-0.007	0.00	7.5	1.1	-0.1	K2 III	-78	320 s		
8094	232377		1 19 20.3	-53 38 10	+0.007	+0.04	7.3	1.8	0.2	K0 III		88 s		
8316	255764		1 19 20.9	-76 08 56	+0.011	+0.02	7.3	0.5	4.0	F8 V		45 s		
7980	92365		1 19 24.1	+18 08 15	-0.005	-0.06	7.8	1.1	0.2	K0 III		230 mx		
7964	74637	90 υ Psc	1 19 27.9	+27 15 51	+0.002	-0.01	4.76	0.03	1.4	A2 V	+8	47 s		
7990	109807		1 19 28.2	+8 23 47	-0.002	+0.05	8.0	1.1	0.2	K0 III		360 s		
8038	166958		1 19 29.0	-24 57 04	-0.001	-0.10	7.7	1.4	3.2	G5 IV		33 s		
7505	221		1 19 29.4	+80 53 38	-0.007	0.00	6.65	0.03		A0				
8076	193052		1 19 33.1	-39 21 46	-0.002	+0.04	7.8	0.6	4.0	F8 V		52 s		
8106	215460		1 19 33.7	-47 17 33	-0.002	-0.01	7.2	2.1	-0.5	M4 III		150 s		
7943	54601		1 19 35.3	+34 45 09	+0.001	-0.01	7.7	1.1	-0.1	K2 III		360 s		
8241	248360		1 19 45.2	-68 53 55	+0.001	+0.11	7.1	0.5	3.4	F5 V		54 s		
8036	129235	42 Cet	1 19 48.2	-0 30 32	+0.001	-0.01	5.87	0.64	0.3	G8 III	+14	130 s	1081	m
7902	22187		1 19 51.6	+58 12 28	0.000	0.00	6.96	0.41	-5.7	B6 Ib	-31	1800 s		
7979	37120		1 19 53.0	+43 12 37	+0.002	+0.01	8.0	1.1	0.2	K0 III		360 s		
8071	147741		1 19 57.7	-15 48 51	+0.004	-0.10	7.50	0.6	4.4	G0 V		42 s	1087	
8178	232381		1 19 58.1	-57 20 59	+0.004	+0.03	7.43	0.44		F5				
8070	147742		1 20 02.9	-9 53 01	+0.002	-0.07	6.7	0.4	3.0	F2 V		54 s		
7959	22192		1 20 03.1	+52 39 58	+0.001	-0.02	8.00	0.1	0.6	A0 V		280 s	1076	m
7927	22191	34 φ Cas	1 20 04.8	+58 13 54	0.000	0.00	4.98	0.68	-8.5	F0 Ia	-24	2900 s	1073	m
8007	37123		1 20 04.9	+41 58 15	+0.002	-0.05	7.0	0.4	2.6	F0 V		74 s		
8103	166967		1 20 05.3	-20 28 16	+0.001	0.00	7.6	1.5	-0.1	K2 III		230 s		
7978	22194		1 20 11.5	+52 07 54	-0.002	+0.01	7.3	0.0	0.4	B9.5 V		220 s		
8214	248365		1 20 12.6	-60 04 28	+0.008	0.00	7.4	1.9	0.2	K0 III		80 s		
8030	54610		1 20 14.1	+35 07 18	+0.004	-0.03	7.9	0.9	3.2	G5 IV		88 s		
8353	255765		1 20 14.4	-72 44 57	+0.001	+0.03	7.48	0.33	2.6	F0 V		92 s		
8130	193059		1 20 15.6	-36 14 36	-0.002	-0.03	7.48	0.04	1.2	A1 V		180 s		
8016	37130		1 20 17.9	+43 30 36	+0.001	+0.01	7.2	1.1	0.2	K0 III		250 s		
7732	4376		1 20 19.4	+77 34 14	-0.005	+0.09	6.31	0.92	0.3	G5 III	-74	150 s		
8188	215464		1 20 19.9	-48 40 35	+0.009	+0.06	7.57	0.54	3.4	F5 V		58 s		m
8144	166971		1 20 22.5	-28 59 14	+0.002	-0.01	7.37	1.08	1.7	K0 III-IV		120 s		
8005	22199		1 20 24.6	+51 49 59	+0.002	-0.01	7.5	0.0	0.0	B8.5 V		300 s		
8121	147746		1 20 27.8	-11 14 20	-0.003	-0.07	6.15	1.11	0.2	K0 III		130 s		
8224	232385		1 20 27.9	-57 20 20	+0.020	+0.08	7.00	0.54	3.8	F7 V		41 s		m

HD	SAO	Star Name	α 2000	δ 2000	μ(α)	μ(δ)	V	B-V	M_V	Spec	RV	d(pc)	ADS	Notes
			^{h} ^{m} ^{s}	$^\circ$ $'$ $''$	^{s}	$''$								
8129	166973		1 20 30.0	-19 56 58	+0.011	-0.24	7.59	0.70	4.4	G0 V		33 mx		
8120	129239		1 20 34.4	-3 14 49	-0.001	+0.01	6.23	1.02		G5				
8004	22200		1 20 36.8	+54 57 46	-0.011	-0.06	7.3	0.6	4.4	G0 V		37 s		
8027	22201		1 20 40.6	+51 35 41	+0.002	-0.01	7.20	0.00	0.4	B9.5 V		210 s	1083	m
8142	147750		1 20 40.7	-13 53 23	+0.004	+0.02	7.0	0.8	3.2	G5 IV	+8	58 s		
8110	92380		1 20 41.0	+15 41 45	+0.003	-0.01	7.5	0.9	3.2	G5 IV	+8	74 s		
8176	166976		1 20 45.8	-25 06 14	-0.003	-0.03	7.5	1.2	0.2	K0 III		210 s		
7976	11705		1 20 48.3	+65 05 04	-0.004	+0.02	8.0	1.1	0.2	K0 III		320 s		
8013	11708		1 20 52.0	+60 56 59	+0.003	+0.02	7.53	-0.02		B8				
8153	129241		1 20 53.0	-1 18 38	+0.003	+0.01	7.3	0.1	0.6	A0 V		220 s		
8054	22205		1 21 03.5	+51 59 02	+0.032	-0.09	7.5	0.5	3.4	F5 V		44 mx		
8117	74646		1 21 03.6	+25 09 30	+0.008	-0.10	8.0	0.5	3.4	F5 V		76 mx		
8100	54620		1 21 04.0	+38 02 03	+0.025	+0.03	7.89	0.63	4.4	G0 V		46 s		
8286	232387		1 21 04.1	-54 57 31	+0.003	-0.02	7.8	2.1	0.2	K0 III		73 s		
8053	22204		1 21 04.3	+54 37 43	+0.002	-0.02	7.2	0.0	0.0	B8.5 V		250 s		
8003	11712	35 Cas	1 21 05.1	+64 39 31	+0.009	-0.01	6.34	0.09		A0	-15		1088	m
8099	37139		1 21 06.4	+43 14 01	+0.001	-0.02	7.4	1.1	0.2	K0 III		270 s		
8126	74647	91 Psc	1 21 07.3	+28 44 18	+0.002	-0.07	5.23	1.39	0.2	K0 III	-36	59 s		
8151	74651		1 21 14.9	+22 22 29	+0.001	+0.01	7.2	1.4	-0.3	K5 III		310 s		
8202	129246		1 21 17.4	-6 09 34	+0.002	-0.03	7.0	0.6	4.4	G0 V		33 s		
8296	232389		1 21 18.4	-49 44 30	+0.005	+0.04	6.9	0.5	3.0	F2 V		46 s		
8187	92388		1 21 19.2	+11 32 14	+0.004	+0.02	7.40	0.4	2.6	F0 V		91 s	1097	m
8002	11713		1 21 23.1	+68 04 14	-0.003	+0.02	8.0	1.1	0.2	K0 III		320 s		
8315	232390		1 21 24.5	-50 56 13	+0.003	0.00	6.6	0.5	0.2	K0 III		190 s		
7926	4384		1 21 31.4	+74 34 50	-0.001	-0.01	7.3	0.1	1.4	A2 V		140 s		
8305	215473		1 21 31.9	-45 08 13	+0.013	+0.08	7.45	0.37		F0				
8294	193070		1 21 39.2	-33 13 56	+0.006	+0.03	7.2	1.4	3.2	G5 IV		28 s		
8437	248369		1 21 48.8	-66 23 04	-0.006	-0.01	7.1	0.0	0.6	A0 V		200 s		
8160	37148		1 21 51.9	+50 07 19	+0.001	0.00	6.7	1.1	0.2	K0 III		190 s		
8161	37149		1 21 55.7	+47 16 57	+0.003	+0.01	7.70	0.1	1.4	A2 V		170 s	1099	m
8250	92395		1 21 58.4	+12 36 15	+0.004	0.00	7.0	1.1	0.2	K0 III		230 s		
8519	248370		1 21 58.4	-69 43 08	-0.002	+0.01	7.24	0.42		F2				m
7925	4387		1 21 59.2	+76 14 21	+0.020	-0.02	6.5	0.1	1.7	A3 V	-16	31 mx		
7924	4386		1 21 59.5	+76 42 37	-0.004	-0.03	7.11	0.82		K0	-23	12 mn		
8248	92396		1 22 02.5	+15 47 21	+0.005	-0.03	7.5	0.5	3.4	F5 V	+4	67 s		
7851	4383		1 22 03.1	+79 01 31	+0.030	-0.05	7.3	1.1	0.2	K0 III		220 mx		
8351	193072		1 22 09.2	-37 03 06	+0.002	-0.01	6.9	0.2	2.6	F0 V		71 s		
8436	232398		1 22 09.5	-59 07 35	-0.001	-0.03	7.53	0.83		G5				
8230	54634		1 22 11.4	+31 42 20	+0.004	-0.01	7.7	1.1	0.2	K0 III		310 s		
8209	37154		1 22 12.5	+43 35 03	0.000	0.00	6.6	0.0	-1.0	B5.5 V	+17	330 s		
8262	92398		1 22 17.7	+18 40 57	+0.038	-0.01	6.96	0.62	4.9	G3 V	+2	27 ts		
8207	37155	46 ξ And	1 22 20.3	+45 31 44	+0.003	+0.01	4.88	1.08	1.7	K0 III-IV	-12	35 s		
8463	248372		1 22 21.2	-61 19 47	+0.005	+0.01	7.5	2.1	-0.5	M2 III		210 s		
8275	92400		1 22 22.2	+16 49 30	-0.006	-0.07	7.8	0.8	3.2	G5 IV		84 s		
8391	215478		1 22 24.3	-43 36 08	+0.013	+0.05	7.02	0.33	1.7	F0 IV		110 mx		
8274	74668		1 22 24.6	+22 05 12	-0.002	-0.01	7.2	0.5	4.0	F8 V	-47	44 s		
8381	193076		1 22 28.1	-34 39 52	+0.003	-0.01	7.5	1.0	3.2	G5 IV		55 s		
8206	22223		1 22 29.5	+51 46 38	+0.003	0.00	7.7	1.4	-0.3	K5 III		350 s		
8350	147767		1 22 30.4	-19 04 53	-0.004	-0.07	6.50	0.5	3.4	F5 V		42 s	1106	m
8337	129263		1 22 34.0	-3 48 03	+0.005	-0.01	7.90	1.1	0.2	K0 III		210 mx		m
8335	129262	43 Cet	1 22 34.7	-0 26 59	+0.001	-0.01	6.49	1.08	0.2	gK0	+14	160 s		
8334	109834		1 22 36.8	+1 43 34	+0.004	-0.04	6.20	1.52	-0.4	gM0	-15	210 s		
8410	193079		1 22 42.9	-36 35 46	+0.004	-0.01	6.8	0.9	3.2	G5 IV		40 s		
8333	109835		1 22 43.4	+4 44 18	-0.002	-0.02	7.0	0.9	3.2	G5 IV		57 s		
8331	92406		1 22 43.4	+10 22 09	-0.006	-0.10	7.5	0.8	3.2	G5 IV		72 s		
8593	248375		1 22 48.1	-69 04 50	+0.021	+0.05	7.7	0.9	3.2	G5 IV		62 s		
8356	92407		1 22 53.7	+11 21 59	+0.005	-0.05	6.9	1.1	0.2	K0 III		160 mx		
8320	54641		1 22 56.3	+31 50 29	+0.001	+0.01	7.6	1.6	-0.5	M4 III		420 s		
8357	109841		1 22 56.7	+7 25 11	+0.006	+0.25	7.3	0.9	3.2	G5 IV		60 mx		
8406	147769		1 23 00.8	-16 29 12	-0.009	-0.14	7.93	0.65	4.4	G0 V		45 s		
8449	193083		1 23 02.3	-37 42 09	0.000	-0.01	7.4	1.4	-0.3	K5 III		350 s		
8407	147771		1 23 02.5	-16 33 36	+0.007	+0.08	7.8	0.7	4.4	G0 V		48 s		
8389	147768		1 23 02.6	-12 57 59	+0.033	-0.04	7.85	0.90	5.9	dK0	+31	21 s		m
8347	54647		1 23 12.0	+33 17 24	-0.001	+0.01	7.8	1.4	-0.3	K5 III		410 s		
8290	22231		1 23 15.7	+52 44 16	+0.003	+0.01	8.0	0.1	0.6	A0 V		270 s		
8474	193087		1 23 17.0	-33 32 51	+0.003	+0.02	6.9	1.3	0.2	K0 III		140 s		
8473	193086		1 23 17.5	-31 15 33	+0.001	-0.04	8.0	0.8	3.2	G5 IV		89 s		
8346	54648		1 23 18.0	+36 30 53	+0.002	0.00	6.7	0.0	0.0	B8.5 V		220 s		
8317	37166		1 23 18.2	+45 19 58	+0.001	-0.01	7.3	0.0	0.4	B9.5 V	+4	240 s		
8402	92410		1 23 19.1	+10 50 33	-0.002	-0.06	7.9	0.4	2.6	F0 V		110 s		
8272	22230		1 23 21.5	+58 08 35	+0.017	-0.09	7.10	0.5	3.4	F5 V	+7	54 s	1105	m
8447	147773		1 23 21.6	-17 56 06	+0.002	-0.03	7.20	1.6	-0.5	M3 III	0	350 s		
8388	74682		1 23 24.9	+20 28 08	-0.001	0.00	5.97	1.71	-0.3	K5 III	-11	130 s		
8307	22232		1 23 25.9	+52 48 27	+0.010	+0.02	7.8	0.7	4.4	G0 V		48 s		
8328	37167		1 23 28.3	+46 53 28	+0.008	-0.02	8.0	0.9	3.2	G5 IV		90 s		

HD	SAO	Star Name	α 2000	δ 2000	μ(α)	μ(δ)	V	B-V	M_v	Spec	RV	d(pc)	ADS	Notes
8498	193090		1^{h}23^{m}30^s.9	-30°56'44"	-0^s.001	-0".04	5.84	1.61	-0.3	K5 III		150 s		
8487	167011		1 23 34.8	-24 21 11	-0.002	0.00	6.65	0.24	2.4	A7 V n		71 s	1113	m
8535	215484		1 23 37.2	-41 16 12	+0.005	-0.07	7.7	0.9	4.0	F8 V		34 s		
8375	54654		1 23 37.4	+34 14 45	+0.018	+0.12	6.29	0.83	3.2	G8 IV	+3	42 s		
8374	54655	47 And	1 23 40.5	+37 42 54	+0.007	-0.01	5.50	0.26		A m	+13			
8534	215486		1 23 40.6	-40 57 20	-0.001	-0.06	7.40	1.1	0.2	K0 III		270 mx		m
8373	37171		1 23 46.4	+48 22 35	+0.002	0.00	7.7	1.1	-0.1	K2 III		330 s		
8065	4391		1 23 46.7	+78 43 34	-0.002	0.00	6.07	0.39	1.4	A2 V	-75	62 s		
--	--		1 23 47.2	+30 33 41	-0.001	-0.01	7.9	1.1	-0.1	K2 III		400 s		
8483	129272		1 23 50.7	-5 49 09	-0.004	-0.19	7.5	0.5	4.0	F8 V		51 s		
8442	92413		1 23 50.9	+17 49 06	+0.006	-0.03	6.8	0.4	2.6	F0 V	-15	70 s		
8652	248376		1 23 56.8	-60 37 43	-0.002	+0.02	7.7	2.3	-0.5	M2 III		170 s		
8420	54659		1 23 58.3	+39 55 26	+0.003	-0.02	8.0	0.1	1.4	A2 V		210 s		
8783	255773		1 24 00.5	-72 19 29	+0.005	0.00	7.80	0.16		A2				
8512	129274	45 θ Cet	1 24 01.3	-8 11 01	-0.006	-0.22	3.60	1.06	0.2	K0 III	+17	35 ts	1118	m
8511	129275	44 Cet	1 24 02.4	-8 00 27	+0.011	-0.07	6.21	0.23		A5				m
8365	22237		1 24 03.4	+54 21 33	-0.010	-0.02	8.0	0.9	3.2	G5 IV		90 s		
8226	4393		1 24 07.9	+72 50 48	-0.002	0.00	7.30	0.1	0.6	A0 V		200 s	1107	m
8398	22242		1 24 11.7	+51 10 24	+0.005	-0.01	7.7	0.1	1.4	A2 V		110 mx		
8581	193096		1 24 12.5	-31 48 44	+0.016	-0.07	6.86	0.58	4.4	G0 V		30 s		
8441	37177		1 24 18.7	+43 08 32	+0.001	-0.01	6.72	0.01		A2 p				
8372	22240		1 24 20.0	+58 56 47	0.000	0.00	8.0	1.1	0.2	K0 III		320 s		
8556	129277		1 24 20.5	-6 54 53	+0.002	+0.01	5.91	0.41	2.6	dF0	+29	42 s	1123	m
8452	54665		1 24 23.6	+36 53 00	+0.002	-0.02	7.1	0.0	0.4	B9.5 V	+14	220 s		
8592	167022		1 24 24.6	-28 29 52	+0.006	+0.05	7.6	1.6	0.2	K0 III		140 s		
8397	22244		1 24 25.1	+57 11 51	-0.002	-0.03	7.7	1.4	-0.3	K5 III		350 s		
8663	215491		1 24 33.7	-45 43 14	0.000	+0.04	7.8	1.9	0.2	K0 III		97 s		
--	54669		1 24 37.5	+32 00 17	0.000	0.00	8.0	1.0		K0				
8589	147793		1 24 39.7	-15 39 37	+0.001	0.00	6.14	0.91		G5				
8629	193101		1 24 40.4	-34 08 23	+0.009	+0.05	6.7	0.9	3.2	G5 IV		41 s		
8651	215492		1 24 40.7	-41 29 33	+0.001	-0.03	5.42	1.04	0.3	G8 III	+73	90 s		
8578	129278		1 24 41.1	-0 58 13	0.000	-0.01	7.9	0.2	2.1	A5 V		150 s		
8681	215494		1 24 41.8	-44 31 43	-0.001	-0.01	6.26	1.14		K0				
8599	129280		1 24 48.6	-2 50 55	0.000	-0.03	6.15	0.96		G5				
8587	109873		1 24 54.5	+3 20 56	+0.001	-0.04	8.0	0.1	0.6	A0 V		300 s		
8507	37181		1 24 58.1	+47 10 09	-0.002	+0.01	7.80	0.8	-2.1	G5 II	-28	760 s		
8531	54675		1 24 59.1	+37 27 18	+0.001	0.00	8.0	0.1	1.4	A2 V		210 s		
8627	129283		1 25 00.5	-5 56 46	+0.001	+0.01	6.80	0.1	0.6	A0 V	+16	170 s	1131	m
8810	248381		1 25 05.3	-64 22 11	+0.004	-0.02	5.93	1.56	-0.4	M0 III		180 s		
8696	193107		1 25 07.4	-39 22 43	+0.007	-0.07	7.80	0.8	3.2	G5 IV		83 s		m
8561	54679		1 25 08.2	+35 44 00	-0.001	0.00	7.9	0.9	3.2	G5 IV		87 s		
8574	74702		1 25 12.6	+28 34 00	+0.020	-0.16	7.9	0.5	4.0	F8 V		40 mx		
8612	74705		1 25 18.9	+20 21 00	+0.001	+0.01	7.8	0.1	0.6	A0 V		280 s		
8626	92428		1 25 19.1	+16 15 26	0.000	+0.01	7.40	1.4	-0.3	K5 III	-4	350 s		
8787	232418		1 25 20.0	-59 29 56	+0.011	+0.01	7.05	0.37		F0				m
8344	4400		1 25 28.6	+74 06 57	0.000	+0.03	7.9	0.1	0.6	A0 V		260 s		
8715	167039		1 25 30.0	-24 43 09	0.000	-0.09	7.6	0.9	3.4	F5 V		37 s		
8634	74707		1 25 35.5	+23 30 40	+0.001	-0.03	6.18	0.43	3.4	F5 V	-16	36 s		
8705	147803	46 Cet	1 25 37.1	-14 35 56	+0.002	-0.01	4.90	1.23	-0.2	K3 III	-23	110 s		
8686	109883		1 25 40.7	+2 58 19	+0.003	-0.01	7.00	0.4	2.6	F0 V		76 s	1138	m
8583	37195		1 25 40.8	+47 07 07	+0.002	+0.01	7.8	0.9	0.3	G4 III	-10	290 s		
8594	37196		1 25 41.4	+44 47 10	+0.013	-0.03	7.7	1.1	0.2	K0 III		130 mx		
8622	54691		1 25 43.4	+36 21 27	+0.004	-0.04	7.80	0.5	4.0	F8 V		58 s	1136	m
8424	4401		1 25 46.4	+70 58 48	+0.004	-0.01	6.5	0.1	0.6	A0 V	+11	150 s		
8538	22268	37 δ Cas	1 25 48.9	+60 14 07	+0.040	-0.04	2.68	0.13	2.1	A5 V	+7	19 ts		Ruchbah, m,q
8713	129292		1 25 52.4	-3 55 38	+0.005	-0.02	6.7	1.1	0.2	K0 III		200 mx		
8728	129293		1 25 55.0	-8 30 15	0.000	-0.01	6.9	0.9	3.2	G5 IV		55 s		
8491	11751	36 ψ Cas	1 25 55.9	+68 07 48	+0.014	+0.03	4.74	1.05	0.2	K0 III	-12	74 s	1129	m
8821	215503		1 25 56.1	-47 53 55	-0.002	-0.11	7.86	0.75	3.2	G5 IV		57 s		m
8703	92434		1 25 58.2	+10 24 25	+0.003	+0.01	7.20	0.1	0.6	A0 V		210 s	1139	m
8767	167043		1 26 04.9	-22 47 54	+0.002	+0.01	7.1	1.2	0.2	K0 III		170 s		
8734	129294		1 26 06.2	-2 12 05	0.000	-0.03	7.9	0.7	4.4	G0 V		51 s		
8673	54695		1 26 08.6	+34 34 47	+0.018	-0.08	6.31	0.47	3.4	F5 V	+17	37 s		
8723	92436	93 ρ Psc	1 26 15.2	+19 10 20	-0.002	+0.01	5.38	0.39	3.0	F2 V	-9	30 ts		
8672	37203		1 26 16.8	+42 18 40	+0.007	-0.02	7.2	0.4	2.6	F0 V		82 s		
8671	37204		1 26 18.5	+43 27 28	+0.009	-0.06	5.96	0.49	4.0	F8 V	+31	25 s		
8733	92439		1 26 23.5	+20 04 15	+0.004	-0.04	6.5	1.1	0.2	K0 III		190 s		
8779	129300		1 26 27.2	-0 23 55	+0.003	0.00	6.41	1.24	0.2	gK0	-6	110 s		
8364	4402		1 26 27.3	+77 40 45	-0.013	+0.01	7.73	0.49		F5	-11			
8832	193116		1 26 29.1	-30 16 30	+0.003	+0.03	7.4	1.1	0.2	K0 III		240 s		
8654	22277		1 26 32.3	+53 07 49	+0.001	0.00	7.7	0.8	3.2	G5 IV		78 s		
--	--		1 26 34.6	+60 17 25			7.90	0.1	-6.8	B3 Ia		4000 s		
8763	92444	94 Psc	1 26 41.6	+19 14 25	+0.003	-0.06	5.50	1.11	0.0	gK1	-42	120 s		
8710	37214		1 26 43.1	+43 41 31	+0.013	-0.04	7.0	0.4	3.0	F2 V	+7	62 s		
8869	215506		1 26 43.7	-40 29 01	0.000	-0.01	7.7	1.4	-0.3	K5 III		390 s		

HD	SAO	Star Name	α 2000	δ 2000	μ(α)	μ(δ)	V	B-V	M$_v$	Spec	RV	d(pc)	ADS	Notes
8747	74721		1^h 26^m 43.7^s	+27° 14' 48"	0$.^s$000	-0$.^{''}$05	6.8	1.1	0.2	K0 III	-5	210 s		
8731	37218		1 26 50.1	+45 40 39	-0.003	0.00	7.8	0.0	0.4	B9.5 V		280 s		
8829	147812	47 Cet	1 26 51.5	-13 03 24	+0.001	+0.01	5.66	0.35	2.8	F1 V	+10	37 s		
8803	109895		1 26 53.4	+3 32 07	0.000	-0.03	6.58	-0.04		B8 n	+15		1148	m
8746	54702		1 26 53.9	+37 17 48	+0.005	+0.03	7.8	0.8	3.2	G5 IV		85 s		
8901	215507		1 26 58.0	-41 44 52	+0.005	-0.02	6.6	1.2	-0.1	K2 III		210 s		
8879	193122		1 26 58.0	-32 32 36	-0.001	-0.03	5.79	3.86		N7.7 p	-8			R Scl, m,v
8828	129305		1 27 01.5	-0 09 27	+0.020	-0.35	7.93	0.74	3.2	G5 IV		22 mx		m
8953	232426		1 27 02.0	-55 04 55	-0.012	-0.04	7.5	1.2	3.4	F5 V		23 s		
8761	54703		1 27 02.3	+34 38 19	-0.001	-0.03	7.9	0.9	3.2	G5 IV		88 s		
8866	167049		1 27 03.3	-20 29 31	0.000	-0.01	7.1	1.7	0.2	K0 III		94 s		
8887	193123		1 27 04.9	-30 14 09	-0.001	-0.03	6.90	1.1	0.2	K0 III		220 s		m
8791	74730		1 27 05.4	+25 26 23	-0.001	-0.01	7.11	1.54	-2.3	K3 II	-17	660 s		
8774	54705		1 27 06.1	+34 22 39	+0.012	-0.01	6.27	0.46	3.4	F5 V	+15	36 s		
8667	11762		1 27 08.9	+64 42 43	+0.003	+0.03	7.8	0.1	1.7	A3 V		150 s		
8786	54706		1 27 14.3	+39 39 15	-0.003	-0.07	7.9	0.4	2.6	F0 V		110 s		
8815	74732		1 27 20.3	+29 46 15	+0.006	-0.06	7.2	0.4	3.0	F2 V	-6	70 s		
--	167055		1 27 23.2	-22 20 18	0.000	-0.08	6.5	1.1	0.2	K0 III		180 s		
8895	167056		1 27 25.2	-20 21 06	+0.002	+0.03	6.9	0.9	3.0	F2 V		29 s		
236740	11765		1 27 26.2	+60 17 04	-0.001	0.00	7.89	0.46	-6.8	B3 Ia		3700 s	1145	m
8801	37227		1 27 26.5	+41 06 01	0.000	0.00	6.40	0.1	1.7	A3 V	+1	87 s	1151	m
8877	129311		1 27 28.0	-9 17 50	-0.001	-0.03	7.3	1.1	0.2	K0 III		260 s		
8963	215510		1 27 29.4	-45 50 46	+0.002	+0.03	6.97	1.48		K0				
8977	215513		1 27 35.0	-46 09 05	+0.002	+0.02	7.71	0.08	1.4	A2 V		170 s		
8784	22292		1 27 36.8	+51 48 32	-0.001	-0.01	7.8	1.1	0.2	K0 III		300 s		
8701	11764		1 27 37.4	+66 04 41	-0.003	+0.05	6.98	1.92	-2.2	K2 II p	0	550 s		
9163	255779		1 27 38.9	-71 42 16	0.000	+0.05	7.5	0.8	3.2	G5 IV		71 s		
8799	37228	48 ω And	1 27 39.2	+45 24 24	+0.033	-0.10	4.83	0.42	2.0	F4 IV	+11	36 s	1152	m
8875	109905		1 27 39.7	+5 21 10	-0.001	-0.13	7.00	0.62		G0	-15		1158	m
--	37229		1 27 42.5	+48 08 23	-0.008	-0.05	8.00						1153	m
8847	54711		1 27 42.6	+31 58 00	-0.002	-0.08	6.8	1.1	0.2	K0 III		180 mx		
9248	255781		1 27 45.3	-75 02 36	+0.005	+0.03	7.4	1.6	-0.5	M2 III		390 s		
8921	147819		1 27 46.5	-10 54 06	+0.012	+0.02	6.13	1.32	0.2	K0 III		89 s	1162	m
8837	37234		1 27 46.9	+40 20 08	+0.001	-0.02	6.40	-0.04	-0.8	B9 III		260 s		
8959	167062		1 27 59.4	-22 02 12	0.000	+0.02	6.7	1.0	3.2	G5 IV		40 s		
8943	129315		1 28 01.0	-2 02 05	0.000	0.00	6.9	0.4	2.6	F0 V		72 s		
8973	167063		1 28 02.6	-20 26 01	-0.002	-0.04	7.4	1.0	3.2	G5 IV		47 s		
8957	147822		1 28 05.8	-17 15 39	+0.003	+0.04	7.30	0.4	2.6	F0 V		87 s	1171	m
8846	37236		1 28 06.6	+47 45 09	+0.003	-0.02	7.5	0.1	0.6	A0 V		220 s		
--	37240		1 28 06.9	+42 28 01	0.000	-0.02	7.9	2.2						
8985	167068		1 28 07.4	-28 03 03	+0.010	+0.09	7.6	0.3	3.4	F5 V		71 s		
8862	37241		1 28 12.4	+44 02 57	+0.001	-0.01	6.5	0.0	0.4	B9.5 V	-2	170 s		
8972	147824		1 28 16.2	-12 14 45	+0.003	0.00	6.9	0.2	2.1	A5 V		90 s		
9067	215517		1 28 18.6	-49 00 22	0.000	0.00	7.1	0.5	4.4	G0 V		35 s		
8909	54719		1 28 19.1	+30 33 27	+0.004	-0.06	6.9	0.5	3.4	F5 V	-16	50 s		
8884	37243		1 28 19.4	+42 46 55	+0.001	-0.03	7.70	0.9	0.3	G7 III	-17	300 s	1161	m
9053	215516	γ Phe	1 28 21.9	-43 19 06	-0.001	-0.20	3.41	1.57	-4.4	K5 Ib	+26	280 s		
8949	109907		1 28 22.8	+7 57 41	+0.007	0.00	6.20	1.12	0.0	K1 III	+2	170 s		m
8941	92453		1 28 24.3	+17 04 45	+0.008	-0.03	6.8	0.5	3.4	F5 V	+9	47 s		
8908	54720		1 28 25.8	+37 04 14	+0.001	-0.01	7.3	0.0	0.4	B9.5 V	-22	240 s		
8956	109908		1 28 27.3	+7 57 28	+0.006	-0.01	8.02	0.53	3.2	F8 IV-V		91 s		
8881	37246		1 28 28.7	+48 19 14	+0.002	+0.01	8.0	0.0	0.4	B9.5 V		300 s		
9066	193133		1 28 34.0	-37 57 37	+0.002	-0.04	7.9	1.1	0.2	K0 III		320 s		
8907	37248		1 28 34.2	+42 16 02	+0.003	-0.12	6.7	0.5	4.0	F8 V		34 s		
9014	147827		1 28 36.4	-11 31 58	-0.001	-0.03	8.0	0.4	2.6	F0 V		120 s		
9231	248387		1 28 41.9	-68 10 33	-0.004	-0.05	7.3	1.1	0.2	K0 III		260 s		
9065	193136		1 28 43.2	-33 45 49	-0.005	-0.02	6.58	0.30	1.7	F0 IV		95 s		
9063	167078		1 28 47.7	-24 47 51	-0.002	+0.02	7.02	0.25	2.4	A7 V		78 s		
9085	193137		1 28 49.6	-36 49 59	+0.001	+0.01	7.4	1.4	-0.5	M2 III		390 s		
9061	147831		1 28 57.3	-17 56 16	+0.004	-0.01	6.7	0.5	3.4	F5 V		46 s		
8929	22317		1 29 01.0	+51 41 10	+0.003	+0.02	7.2	0.8	3.2	G5 IV		63 s		
8730	4407		1 29 02.1	+74 12 18	+0.047	-0.12	7.30	0.8	3.2	G5 IV	+24	63 mx		m
9024	109916		1 29 03.0	+7 17 39	-0.002	-0.05	6.7	0.4	3.0	F2 V	-8	55 s		
8997	74742		1 29 04.8	+21 43 22	+0.033	-0.19	7.74	0.96	6.3	K2 V	+44	18 s	1176	m
9265	--		1 29 05.9	-67 28 58			7.7	1.6	-0.5	M2 III		430 s		
8996	54733		1 29 12.8	+36 37 40	+0.002	-0.01	7.2	1.1	0.2	K0 III		260 s		
8906	22318		1 29 15.4	+60 02 10	+0.001	+0.01	7.11	0.74	-4.6	F3 Ib		1400 s		
9033	74747		1 29 23.4	+25 16 32	+0.004	+0.01	6.7	1.1	0.2	K0 III		200 s		
8995	37264		1 29 26.3	+42 16 55	+0.011	-0.05	7.3	0.5	3.4	F5 V		59 s		
9195	232433		1 29 28.4	-51 30 07	+0.007	-0.27	7.9	0.5	3.2	G5 IV		39 mx		
9184	215525		1 29 30.3	-46 45 23	-0.002	+0.02	6.31	1.66	-0.5	gM4		220 s		
9101	129325		1 29 32.1	-9 29 21	0.000	+0.02	7.4	0.8	3.2	G5 IV		68 s		
9207	232437		1 29 32.6	-52 22 01	+0.002	-0.02	7.1	1.7	-0.3	K5 III		230 s		
9132	167086	48 Cet	1 29 36.0	-21 37 45	+0.004	+0.01	5.12	0.02	1.2	A1 V	-8	61 s	1184	m
9118	147841		1 29 40.9	-13 13 26	0.000	-0.08	7.3	0.4	3.0	F2 V		72 s		

HD	SAO	Star Name	α 2000	δ 2000	μ(α)	μ(δ)	V	B-V	M$_v$	Spec	RV	d(pc)	ADS	Notes
			h m s	° ′ ″	s	″								
9071	74751		1 29 41.2	+22 49 34	+0.002	−0.02	7.50	0.1	0.6	A0 V		240 s	1183	m
9151	167087		1 29 42.7	−25 37 04	+0.003	+0.02	6.5	1.2	0.2	K0 III		130 s		
8928	11776		1 29 46.5	+65 10 20	−0.002	+0.04	6.9	1.1	−0.1	K2 III		230 s		
9091	74753		1 29 48.5	+23 41 25	−0.004	−0.05	7.6	0.5	4.0	F8 V		54 s		
8965	22327		1 29 51.2	+60 15 05	0.000	0.00	7.27	0.03		B0.5 V	+1			
9100	92463	97 Psc	1 29 52.8	+18 21 21	+0.004	0.00	6.00	0.14	0.1	A4 III	+4	150 s		
9045	37269		1 29 52.8	+42 37 40	+0.001	0.00	7.9	0.4	3.0	F2 V		94 s		
9139	129331		1 29 53.5	−5 35 47	+0.003	−0.02	6.8	0.5	3.4	F5 V		47 s		
8992	22328		1 29 53.8	+58 45 45	−0.001	0.00	7.79	0.89	−4.6	F6 Ib		2100 s		
9158	147844		1 29 58.3	−15 43 44	0.000	−0.02	7.8	1.1	−0.1	K2 III		390 s		
9175	167088		1 29 59.9	−22 31 28	+0.009	−0.06	7.80	0.6	4.4	G0 V		48 s	1186	m
9057	37275	49 And	1 30 06.1	+47 00 26	+0.001	−0.04	5.27	1.00	0.2	K0 III	−11	100 s		
9138	109926	98 μ Psc	1 30 11.1	+6 08 38	+0.020	−0.04	4.84	1.37	−0.3	K4 III	+35	67 mx		m
9218	167092		1 30 13.7	−28 51 57	+0.002	−0.03	8.0	0.6	0.2	K0 III		360 s		
9022	22338		1 30 18.2	+59 46 55	+0.002	+0.01	6.90	1.44	−0.2	K3 III	+40	210 s		
9293	232443		1 30 19.4	−52 29 37	+0.004	−0.01	6.9	1.5	−0.3	K5 III		280 s		
8991	11784		1 30 20.7	+63 51 32	+0.003	+0.01	8.03	0.07	0.6	A0 V		270 s		
9228	167093		1 30 22.8	−26 12 28	+0.003	−0.06	5.93	1.34	−0.3	gK4	−1	180 s	1193	m
—	22341		1 30 23.2	+54 08 51	+0.001	0.00	8.0	1.7	4.0	F8 V		12 s		
9107	37282		1 30 33.8	+44 55 42	+0.001	0.00	8.0	1.1	0.2	K0 III		360 s		
9405	—		1 30 35.4	−65 07 02			7.9	0.4	2.6	F0 V		120 s		
9203	109934		1 30 38.3	+2 52 55	+0.001	0.00	6.9	1.6		M5 III	−45			R Psc, v
9172	74763		1 30 39.0	+25 54 54	−0.001	+0.06	7.24	0.38	2.6	F0 V		81 s		
9379	232444		1 30 46.3	−59 39 35	+0.017	−0.02	7.94	0.60	4.4	G0 V		49 s		
9202	92471		1 30 47.0	+15 01 17	0.000	−0.04	8.0	1.6	−0.5	M4 III		510 s		
9030	11788		1 30 52.2	+66 05 54	+0.014	0.00	6.2	0.1	0.6	A0 V	+9	91 mx		
9137	37285		1 30 58.9	+49 20 37	−0.004	−0.04	7.1	1.1	0.2	K0 III		220 s		
9261	129342		1 30 59.7	−4 57 36	+0.002	+0.01	7.0	1.1	0.2	K0 III		230 s		
9351	232445		1 31 01.7	−51 12 59	+0.017	−0.11	7.7	0.2	4.4	G0 V		47 s		
9177	54750		1 31 02.3	+39 29 39	−0.001	−0.03	7.58	−0.07	0.4	B9.5 V		270 s		
9478	—		1 31 03.3	−66 29 04			7.9	0.1	0.6	A0 V		290 s		
9021	4422	38 Cas	1 31 13.7	+70 15 53	+0.027	−0.07	5.81	0.47	3.4	dF5	+5	30 s		
9236	92478		1 31 13.9	+17 09 08	+0.002	−0.02	8.0	0.4	2.6	F0 V		120 s		
9362	215536	δ Phe	1 31 15.0	−49 04 22	+0.014	+0.16	3.95	0.99	1.7	K0 III-IV	−7	28 s		
9477	248394		1 31 15.5	−65 07 07	+0.045	+0.10	7.3	0.8	0.2	K0 III		63 mx		
9378	232446		1 31 16.3	−49 54 15	−0.004	−0.08	7.6	0.4	2.6	F0 V		93 s		
9339	193159		1 31 16.8	−38 25 46	0.000	0.00	7.8	1.9	−0.3	K5 III		250 s		
9105	11793		1 31 18.2	+63 20 52	+0.003	+0.01	7.47	0.55	−6.3	B5 Iab	−37	2400 s		
9224	74767		1 31 19.3	+29 24 47	+0.012	−0.09	7.22	0.60	4.4	G0 V	+14	35 s		
9439	232448		1 31 25.8	−55 58 04	+0.002	0.00	7.5	1.3	0.2	K0 III		200 s		
9223	54754		1 31 26.1	+38 46 00	0.000	−0.02	7.1	0.0	0.0	B8.5 V	+11	270 s		
9270	92484	99 η Psc	1 31 28.9	+15 20 45	+0.002	0.00	3.62	0.97	0.3	G8 III	+15	44 s	1199	m
9404	215537		1 31 29.7	−44 39 20	−0.005	−0.01	7.86	0.40		F2				
9349	193161		1 31 29.8	−29 59 05	+0.006	+0.04	6.62	0.89	3.2	G8 IV		48 s		
9136	11794		1 31 32.9	+61 33 08	+0.001	−0.02	7.65	0.14	1.2	A1 V		170 s		
9468	232451		1 31 33.6	−59 35 35	+0.021	0.00	7.99	0.47	3.4	F5 V		79 s		
9336	147861		1 31 34.0	−19 01 25	−0.001	−0.05	6.82	0.26	0.5	A7 III		180 s		m
9438	232450		1 31 35.3	−53 22 09	+0.008	+0.04	7.78	0.45	3.4	F5 V		74 s		m
9315	147860		1 31 36.7	−12 15 29	+0.001	+0.01	7.2	0.4	2.6	F0 V		83 s		
9499	248395		1 31 37.4	−61 14 58	−0.003	+0.01	7.2	0.2	2.6	F0 V		85 s		
9414	215539		1 31 39.0	−45 34 32	0.000	+0.01	6.17	0.06		A0				
9337	147862		1 31 39.1	−19 01 57	+0.002	−0.02	7.4	1.0	0.2	K0 III		260 s		
9145	11798		1 31 40.9	+61 02 34	−0.003	0.00	7.80	0.00	−1.6	B7 III		590 s	1196	m
9307	92485		1 31 42.7	+10 53 22	+0.004	+0.02	7.6	0.9	3.2	G5 IV		75 s		
9377	193163		1 31 43.2	−30 17 00	0.000	−0.06	5.82	1.08	0.2	gK0		120 s		
9335	147863		1 31 47.5	−10 15 26	−0.004	−0.06	7.4	0.1	1.4	A2 V		160 s		
9306	92487		1 31 48.9	+15 37 31	−0.001	+0.01	7.8	1.6	−0.5	M4 III		450 s		
9154	11800		1 31 49.7	+61 22 39	0.000	+0.01	7.70	0.9	0.3	G8 III		270 s		
9167	11801		1 31 52.4	+61 31 26	−0.001	+0.02	8.0	0.4	−2.0	F1 II		730 s		
9388	167115		1 31 53.4	−27 42 11	+0.001	+0.01	7.3	1.0	0.2	K0 III		250 s		
9235	22362		1 31 58.6	+52 28 12	−0.003	0.00	7.9	0.4	2.6	F0 V		110 s		
9359	129352		1 32 00.8	−6 43 08	0.000	0.00	7.7	1.1	−0.1	K2 III		360 s		
9312	92492		1 32 03.0	+16 56 49	+0.009	−0.20	6.90	0.9	3.2	G5 IV	−5	44 mx	1202	m
9234	22364		1 32 05.5	+54 01 08	+0.001	+0.01	7.8	0.0	0.0	B8.5 V		330 s		IZ Per, v
9411	167116		1 32 07.4	−23 38 54	−0.001	−0.08	7.24	0.28	2.6	F0 V		85 s		
9298	54762		1 32 07.5	+34 48 00	−0.001	0.00	6.39	−0.12	−0.9	B6 V		280 s		
9371	109952		1 32 14.9	+3 41 06	+0.001	−0.01	7.8	1.1	0.2	K0 III		330 s		
9200	11805		1 32 14.9	+63 37 15	0.000	0.00	7.85	0.13	1.2	A1 V		180 s		
9304	54765		1 32 15.3	+38 16 17	−0.003	−0.02	7.3	0.9	3.2	G5 IV		65 s		
9421	147869		1 32 16.2	−18 33 58	+0.004	−0.05	8.00	0.9	3.2	G5 IV		91 s		m
9451	167119		1 32 19.4	−26 33 10	−0.001	−0.09	7.66	0.30	1.5	A7 IV		95 mx	1212	m
9166	11803		1 32 24.2	+68 24 33	+0.018	−0.05	6.76	1.23	−0.2	K3 III	−14	150 mx		
9165	11804		1 32 32.6	+70 01 10	+0.002	−0.01	6.9	0.0	0.4	B9.5 V		190 s		
9528	215545		1 32 33.0	−49 31 40	+0.017	−0.04	8.0	1.4	4.4	G0 V		16 s		
9463	147875		1 32 36.1	−18 58 38	−0.002	−0.02	8.0	1.0	−0.1	K2 III		420 s		

HD	SAO	Star Name	α 2000	δ 2000	μ(α)	μ(δ)	V	B-V	M$_v$	Spec	RV	d(pc)	ADS	Notes
9544	215547		1^h 32^m 36.2^s	-49° 43' 40"	-0$.^s$003	-0$.^{\prime\prime}$07	6.28	0.46	3.7	dF6		33 s		
--	248397		1 32 40.2	-68 22 03	-0.004	-0.01	8.0							
9250	11813		1 32 43.3	+63 35 38	+0.001	+0.01	7.21	1.42	-4.5	G0 Ib	-19	930 s		
9370	54771		1 32 47.6	+35 50 33	-0.001	-0.05	6.80	1.1	0.2	K0 III		210 s	1211	m
9369	54773		1 32 50.3	+38 07 16	-0.003	-0.04	8.0	0.5	3.4	F5 V		82 s		
9525	193173		1 32 56.0	-36 51 55	0.000	-0.02	5.51	1.02	0.3	gG8	+13	100 s		
9386	54775		1 32 57.7	+33 51 09	+0.005	+0.02	7.6	0.5	3.4	F5 V		70 s		
9484	129363		1 33 03.4	-9 00 53	+0.002	+0.01	6.59	-0.04		A0				
9514	147879		1 33 04.5	-19 09 18	+0.001	-0.02	7.8	1.0	-0.1	K2 III		380 s		
9303	11820		1 33 08.8	+60 26 59	0.000	+0.01	7.64	0.00		A0				
10062	258268		1 33 13.3	-79 55 39	+0.015	-0.01	7.3	-0.2	0.6	A0 V		98 mx		
9311	11822		1 33 13.8	+60 41 09	-0.001	-0.03	7.15	0.30	-5.7	B5 Ib	-39	2100 s	1209	m
9540	167134		1 33 15.9	-24 10 40	+0.021	-0.15	6.97	0.76	5.6	G8 V	0	17 ts		m
9554	167135		1 33 15.9	-25 40 04	+0.006	-0.06	7.9	1.3	0.2	K0 III		120 mx		
9496	109964		1 33 18.2	+8 12 31	+0.001	-0.03	6.60	1.1	0.2	K0 III		190 s	1214	m
9472	74789		1 33 18.9	+23 58 31	0.000	+0.02	7.6	0.6	4.4	G0 V		43 s		
9354	22387		1 33 21.2	+55 49 05	0.000	+0.01	7.3	1.1	0.2	K0 III	-15	240 s		
9352	22389		1 33 25.7	+58 19 39	+0.001	+0.01	5.70	1.52	0.2	K0 III	-1	63 s		
9366	22391		1 33 26.2	+54 56 43	-0.001	+0.02	7.1	1.1	-4.4	K3 Ib	-32	1400 s		
9619	215556		1 33 26.4	-43 54 06	+0.018	+0.05	7.83	0.83		K0				
9493	74792		1 33 28.6	+23 35 25	-0.002	-0.03	7.6	0.9	3.2	G5 IV		74 s		
9850	255792		1 33 30.4	-70 52 04	-0.009	-0.02	7.8	0.5	3.4	F5 V		75 s		
9329	11825		1 33 32.6	+62 31 37	-0.003	0.00	7.2	1.1	0.2	K0 III	-23	230 s		
9443	37310		1 33 33.8	+40 54 06	+0.003	-0.02	7.6	0.2	2.1	A5 V		120 s		
--	37311		1 33 36.3	+40 59 42	0.000	-0.02	7.8	1.5						
10042	255794		1 33 39.4	-78 30 17	-0.008	-0.13	6.11	0.97	3.0	G3 IV		26 s		m
9562	129371		1 33 42.8	-7 01 31	+0.012	-0.08	5.76	0.64	4.7	dG2	-15	15 s		
5914	209		1 33 48.9	+89 00 57	+0.214	0.11	6.46	0.11		A2	-10			
9500	54785		1 33 54.2	+35 36 31	+0.004	+0.01	7.30	1.6	-0.5	M4 III	+1	360 s		
9383	11831		1 33 55.5	+62 09 00	-0.004	+0.01	7.90	0.14	1.4	A2 V		180 s		
9408	22397	39 χ Cas	1 33 55.8	+59 13 56	-0.005	-0.02	4.71	1.00	0.2	K0 III	+6	80 s		
9798	248401		1 33 59.8	-62 28 34	-0.004	0.00	7.2	-0.2	2.1	A5 V		100 s		
9650	167144		1 34 05.3	-28 47 01	0.000	+0.01	7.0	1.0	-0.1	K2 III		270 s		
9771	232465		1 34 13.0	-57 00 05	+0.002	+0.01	6.90	0.00	0.4	B9.5 V		200 s		m
9531	54788		1 34 16.5	+37 14 14	+0.001	-0.01	5.80	-0.07	-0.2	B8 V	-4	160 s		
9519	37320		1 34 17.0	+42 22 50	-0.002	-0.02	7.8	1.1	-0.1	K2 III		380 s		
9673	167147		1 34 18.3	-27 21 46	+0.001	+0.04	7.92	0.23	2.1	A5 V		140 s		
9733	215567		1 34 20.7	-45 41 27	+0.011	+0.10	6.91	0.90		G5				m
9692	167148		1 34 23.3	-28 14 13	-0.002	-0.02	6.9	1.4	-0.5	M2 III		300 s		
9508	37321		1 34 27.9	+49 33 51	0.000	0.00	7.8	0.1	1.4	A2 V		180 s		
9407	11834		1 34 33.4	+68 56 54	-0.068	+0.12	6.53	0.68	5.3	G6 V	-31	18 s		
9585	54793		1 34 37.0	+35 06 47	0.000	+0.04	7.3	1.1	0.2	K0 III		260 s		
9672	147886	49 Cet	1 34 37.7	-15 40 35	+0.006	+0.01	5.63	0.07	1.4	A2 V		69 s		
9743	193187		1 34 43.5	-35 20 45	-0.002	-0.04	8.0	0.4	0.2	K0 III		360 s		
9671	129384		1 34 46.4	-2 02 02	-0.005	-0.05	7.6	0.9	3.2	G5 IV		77 s		
9670	109989		1 34 48.7	+0 56 41	+0.013	-0.27	6.90	0.53	4.0	F8 V	-18	26 mx		
9640	92520		1 34 49.0	+18 27 38	+0.002	-0.07	5.89	1.52	-0.5	M2 III	-26	170 mx		
9742	193188		1 34 50.6	-31 53 32	-0.006	-0.03	6.12	1.11		G5				
9690	129387		1 34 51.0	-3 31 26	-0.001	-0.03	6.7	1.1	0.2	K0 III		200 s		
9731	167153		1 34 51.5	-23 41 57	+0.002	+0.04	6.50	1.1	0.2	K0 III		180 s		m
9656	92521		1 34 51.5	+12 33 32	0.000	-0.01	7.30	0.1	1.7	A3 V		130 s		m
9657	109990		1 34 52.4	+8 16 29	-0.001	-0.01	7.1	1.1	0.2	K0 III		250 s		
9655	92523		1 34 54.8	+14 23 08	-0.001	+0.01	7.9	0.9	3.2	G5 IV		88 s		
9616	54801		1 34 58.1	+33 07 05	-0.002	-0.13	6.6	0.7	4.4	G0 V	-25	27 s		
9770	193189		1 35 00.8	-29 54 36	+0.010	+0.12	7.10	0.94	6.9	K3 V	-20	15 ts		m
9638	74806		1 35 01.9	+29 06 02	0.000	-0.01	7.76	1.14	-2.2	K2 II	-20	980 s		
9716	129389		1 35 06.2	-2 20 09	+0.003	-0.03	7.5	0.1	0.0	A0 V		240 s		
9896	232470		1 35 15.2	-58 08 22	+0.036	-0.03	6.01	0.38	3.0	F2 V		38 s		
9604	22414		1 35 24.0	+53 20 44	0.000	0.00	6.8	0.0	0.0	B8.5 V		210 s		
9546	11846		1 35 24.3	+63 04 54	-0.001	-0.01	6.80	1.1	6.1	K1 V		14 s	1233	m
9590	22413		1 35 25.7	+56 02 39	-0.003	+0.01	7.1	0.0	0.4	B9.5 V	+6	200 s		
9782	147893		1 35 26.3	-13 22 53	+0.016	+0.04	7.17	0.60	4.4	G0 V		34 s		
9701	74810		1 35 26.4	+26 41 02	0.000	-0.08	7.8	0.5	4.0	F8 V		57 s		
9635	22417		1 35 37.0	+52 09 50	+0.004	-0.01	7.7	1.1	-0.1	K2 III		320 s		
9714	74812		1 35 38.2	+28 16 09	+0.001	0.00	6.81	1.03	0.2	K1 III		230 s		
9934	232472		1 35 41.9	-55 14 17	+0.017	+0.19	8.0	0.6	4.4	G0 V		51 s		
9925	232471		1 35 43.1	-53 12 01	+0.012	+0.03	7.80	1.1	0.2	K0 III		170 mx		m
9962	248407		1 35 44.8	-60 57 59	+0.009	+0.01	7.5	0.3	1.4	A2 V		90 mx		
9766	92530	101 Psc	1 35 46.4	+14 39 41	0.000	-0.01	6.22	-0.04	-0.8	B9 III	-16	240 s		
9907	215574		1 35 50.4	-43 00 08	+0.003	-0.13	7.9	1.4	3.2	G5 IV		25 s		
9895	215573		1 35 50.5	-39 56 52	-0.004	-0.07	6.6	0.8	3.0	F2 V		28 s		
9712	37344		1 35 52.4	+41 04 35	+0.012	0.00	6.38	1.11	0.0	K1 III	+65	190 s		
9847	147900		1 35 52.9	-17 31 58	+0.019	-0.21	7.17	0.70	3.0	dF2	+8	42 s		m
9780	92531		1 35 54.7	+17 26 01	+0.011	+0.01	5.9	0.2	2.1	A5 V	+3	49 mx		
9727	54818		1 35 56.6	+39 52 53	+0.005	-0.02	7.9	1.1	0.2	K0 III		240 mx		

HD	SAO	Star Name	α 2000	δ 2000	μ(α)	μ(δ)	V	B–V	M_v	Spec	RV	d(pc)	ADS	Notes
			h m s	° ′ ″	s	″								
9856	147901	50 Cet	1 35 58.8	−15 24 00	+0.001	+0.02	5.4	1.1	−0.1	K2 III	+24	130 s		
9894	193200		1 35 58.8	−36 27 13	−0.002	0.00	7.7	1.3	−0.5	M2 III		440 s		
9817	110001		1 36 02.8	+7 38 43	−0.002	+0.02	7.70	0.5	4.0	F8 V		55 s	1254	m
9709	37345		1 36 03.3	+47 06 51	+0.004	−0.01	7.0	0.0	0.4	B9.5 V	−10	73 mx		
9634	22421		1 36 05.1	+60 12 47	+0.004	+0.01	7.90	1.72	−0.3	K5 III		320 s		
9906	193201	τ Scl	1 36 08.3	−29 54 27	+0.007	+0.04	5.69	0.33	3.3	dF4	+5	30 s		m
9764	54820		1 36 08.4	+36 32 16	−0.001	+0.01	7.9	0.2	2.1	A5 V		150 s		
9666	22426		1 36 16.2	+59 27 34	+0.002	−0.01	7.71	0.45	0.7	F5 III		250 s		
9855	110004		1 36 16.8	+4 18 51	+0.001	+0.01	7.90	1.1	0.2	K0 III		350 s	1257	m
9747	37350		1 36 22.7	+47 19 33	−0.001	−0.01	7.5	0.9	3.2	G5 IV		69 s		
9614	11853		1 36 23.4	+67 36 55	+0.004	+0.01	7.1	0.1	0.6	A0 V		180 s		
9815	54827		1 36 25.8	+35 08 02	+0.004	−0.02	8.0	0.9	3.2	G5 IV		93 s		
9746	37351		1 36 27.1	+48 43 22	−0.001	−0.01	5.92	1.21	0.2	K0 III	−43	100 s		
9723	22428		1 36 28.1	+54 41 47	+0.001	+0.01	7.2	0.1	0.6	A0 V		200 s		
9759	22430		1 36 38.3	+51 44 55	+0.001	0.00	7.9	0.2	2.1	A5 V		140 s		
9882	110009		1 36 41.1	+6 43 04	−0.001	−0.03	7.1	1.1	0.2	K0 III		240 s		
9889	110011		1 36 43.4	+7 49 53	0.000	+0.01	6.6	1.6	−0.5	M2 III		260 s		
10052	232477		1 36 44.7	−58 16 15	+0.003	+0.01	6.0	1.8	−0.5	M2 III		160 s		
—	255796		1 36 46.4	−69 57 10	−0.017	−0.06	8.0							
9695	11866		1 36 46.7	+63 24 52	−0.001	−0.01	7.62	0.09	−1.2	B8 III		450 s		
9800	37355		1 36 47.0	+48 18 54	+0.004	−0.01	7.3	0.4	2.6	F0 V	−6	85 s		
9826	37362	50 υ And	1 36 47.8	+41 24 20	−0.015	−0.38	4.09	0.54	4.0	F8 V	−28	13 ts		m
10019	215581		1 36 48.1	−48 48 14	−0.002	+0.04	6.9	1.0	3.2	G5 IV		37 s		
9737	22429		1 36 48.3	+59 37 35	+0.005	−0.01	7.2	0.4	0.6	F0 III		190 s		
9854	54834		1 36 48.4	+32 51 12	0.000	−0.01	8.0	1.1	0.2	K0 III		360 s		
9961	193207		1 36 48.9	−32 17 48	+0.002	+0.02	7.2	1.1	3.2	G5 IV		37 s		
9842	54833		1 36 50.6	+38 27 10	+0.001	−0.02	7.9	1.1	0.2	K0 III		340 s		
9919	92536	102 π Psc	1 37 05.9	+12 08 30	−0.005	+0.05	5.57	0.35	2.6	F0 V	−1	38 s		
9822	37366		1 37 06.3	+49 04 32	+0.002	−0.04	7.5	0.4	3.0	F2 V		79 s		
10039	215583		1 37 07.3	−48 17 29	−0.005	−0.09	7.8	0.4	3.4	F5 V		40 s		
9959	129406		1 37 13.1	−4 08 34	−0.003	+0.01	7.7	1.1	0.2	K0 III		320 s		
9958	129407		1 37 16.1	−0 21 02	−0.001	−0.07	7.1	0.7	4.4	G0 V		34 s		
9861	37369		1 37 16.6	+46 27 35	−0.001	+0.01	7.6	0.5	3.4	F5 V		69 s		
9860	37368		1 37 17.3	+46 56 55	+0.005	−0.01	7.8	0.1	0.6	A0 V		100 mx		
9612	4448		1 37 22.4	+74 18 04	+0.007	0.00	6.4	0.0		B8				
10693	258270		1 37 22.6	−82 16 54	+0.014	−0.02	7.60	0.8	3.2	G5 IV		76 s		m
9812	22441		1 37 22.7	+58 38 14	0.000	−0.01	7.40	0.00	0.0	B8.5 V		270 s	1262	m
9939	74830		1 37 24.9	+25 10 02	−0.017	−0.22	6.99	0.91	0.2	K0 III		54 mx		
9839	22444		1 37 25.0	+55 04 16	+0.002	+0.02	7.6	1.1	0.2	K0 III		270 s		
11025	258273		1 37 28.2	−84 46 11	+0.028	+0.02	5.69	0.94	0.2	K0 III	+18	130 s		
10049	193214		1 37 32.4	−38 08 06	+0.006	+0.03	7.2	1.0	−0.1	K2 III		220 mx		
10009	129412		1 37 37.6	−9 24 15	+0.017	+0.09	6.24	0.53		F5		18 mn		m
9986	92543		1 37 40.8	+12 04 42	+0.009	0.00	6.9	0.6	4.4	G0 V		32 s		
10144	232481	α Eri	1 37 42.9	−57 14 12	+0.013	−0.03	0.46	−0.16	−1.6	B5 IV	+19	26 s		Achernar
9777	11873		1 37 44.1	+67 35 23	+0.001	−0.07	8.0	1.4	−0.3	K5 III		190 mx		
9901	37371		1 37 44.5	+48 24 45	+0.016	+0.01	6.8	0.5	3.4	F5 V		47 s		
10101	215588		1 37 45.6	−47 10 41	−0.004	+0.02	7.58	1.02		K0				
9811	11875		1 37 46.8	+64 44 23	−0.004	+0.02	6.6	0.2	−6.8	A6 Iab		210 mx		
9938	54847		1 37 50.5	+38 01 19	−0.001	−0.11	7.9	0.7	4.4	G0 V		51 s		
9852	11878		1 37 51.1	+61 51 39	0.000	−0.04	7.93	1.46	0.3	G8 III		150 s		
10024	129415		1 37 52.0	−3 26 27	+0.001	0.00	6.7	1.4	−0.3	K5 III		250 s		
10180	248411		1 37 53.6	−60 30 41	−0.001	+0.02	7.4	0.8	3.2	G5 IV		67 s		
10800	258271		1 37 55.8	−82 53 30	+0.068	+0.13	5.87	0.61	4.7	G2 V		17 s		
9869	22453		1 37 57.0	+59 17 57	+0.003	0.00	7.9	0.1	1.4	A2 V		190 s		
9927	37375	51 And	1 37 59.5	+48 37 42	+0.007	−0.11	3.57	1.28	−0.2	K3 III	+16	57 s		
10007	74841		1 38 02.3	+21 53 09	+0.006	−0.08	7.9	0.5	3.4	F5 V		78 s		
10121	215589		1 38 03.5	−46 05 02	+0.004	+0.01	6.97	1.16		K0				
9838	11880		1 38 06.1	+65 32 07	+0.004	0.00	7.8	1.1	0.2	K0 III		290 s		
9900	22456		1 38 07.3	+57 58 40	−0.001	0.00	5.56	1.38	−2.1	G5 II	−8	170 s		
9983	54855		1 38 10.9	+35 20 20	+0.002	+0.01	7.7	1.1	−0.1	K2 III		360 s		
9878	11883		1 38 12.6	+62 21 08	−0.001	+0.02	6.72	−0.04	−0.6	B7 V		270 s		
—	215591		1 38 16.0	−47 56 05	+0.001	+0.03	7.4	0.4	1.7	A3 V		88 s		
10162	—		1 38 18.8	−47 55 26			7.26	0.34	1.7	F0 IV		120 s		
10057	110024		1 38 19.8	+2 35 10	0.000	0.00	7.0	1.1	0.2	K0 III		230 s		
10100	147921		1 38 25.7	−15 52 20	+0.006	−0.03	7.50	1.1	0.0	K1 III	+28	170 mx		
10870	258272		1 38 26.1	−83 14 52	+0.040	−0.03	7.3	1.6	0.2	K0 III		110 s		
10142	193224		1 38 27.3	−36 31 42	−0.002	−0.12	5.94	1.05	3.2	G5 IV		21 s		
10167	215592		1 38 30.6	−42 55 42	−0.004	−0.03	6.67	0.33		F0				
9774	4453	40 Cas	1 38 30.9	+73 02 24	−0.002	−0.01	5.28	0.96	−0.9	G8 II-III	−4	170 s	1268	m
9996	37393		1 38 31.6	+45 24 01	−0.003	+0.02	6.36	0.04		A0 p	+3			
10291	248413		1 38 35.6	−65 36 24	+0.011	−0.03	7.1	0.8	3.2	G5 IV		59 s		
10045	74845		1 38 37.3	+21 23 53	+0.004	0.00	6.8	1.1	−0.1	K2 III		240 s		
10149	193226		1 38 38.1	−32 10 09	−0.001	0.00	7.4	0.6	4.4	G0 V		38 s		
10014	37396		1 38 38.9	+44 08 39	−0.001	−0.02	7.8	0.4	2.6	F0 V		110 s		
10006	37397		1 38 42.6	+46 20 49	−0.003	+0.02	7.5	1.4	−0.3	K5 III		330 s		

HD	SAO	Star Name	α 2000	δ 2000	μ(α)	μ(δ)	V	B-V	M$_v$	Spec	RV	d(pc)	ADS	Notes
10116	129423		1^{h}38^{m}44^s.6	−5°44′07″	−0^s.009	−0″.04	7.4	0.5	3.4	F5 V		63 s		
10033	54863		1 38 46.7	+38 39 08	+0.001	+0.03	7.80	0.8	3.2	G5 IV		83 s	1284	m
10241	232484		1 38 48.9	−53 26 23	−0.007	−0.05	6.84	0.44	2.6	dF0		59 s		m
10161	167199		1 38 49.9	−25 01 20	0.000	+0.02	6.70	−0.08	0.2	B9 V n		200 s	1298	m
10148	167198		1 38 51.6	−21 16 32	+0.008	+0.04	5.7	0.4	2.6	F0 V	+18	41 s		
10056	54866		1 38 51.7	+31 38 58	−0.001	+0.01	7.9	1.1	0.2	K0 III		340 s		
10088	74848		1 38 56.6	+21 55 07	+0.003	0.00	7.7	0.1	0.6	A0 V		270 s		
10055	37403		1 39 05.9	+41 09 41	+0.001	+0.01	7.2	0.1	0.6	A0 V		210 s		
9973	11893		1 39 06.7	+61 04 44	−0.004	0.00	6.88	0.87	−8.4	F2 Ia		440 mx		
10269	232488		1 39 07.3	−56 25 47	+0.008	−0.03	7.3	0.4	4.0	F8 V		45 s		
10074	54870		1 39 07.5	+36 32 36	−0.002	−0.01	7.1	0.1	0.6	A0 V		200 s		
10240	215601		1 39 08.1	−48 46 30	+0.003	+0.04	7.30	0.08	1.4	A2 V		150 s		
10095	74853		1 39 14.3	+27 45 23	+0.003	−0.05	7.07	1.20	−0.2	K3 III	−36	280 s		
10113	92555		1 39 15.1	+16 37 32	0.000	−0.02	7.10	0.8	3.2	G5 IV	+4	60 s	1300	m
10135	92556		1 39 15.4	+14 17 09	+0.006	−0.02	6.9	1.1	0.2	K0 III	−3	170 mx		
10209	167202		1 39 16.5	−29 01 22	+0.003	+0.01	7.41	0.33	0.6	F0 III		220 s		
10031	22475		1 39 17.3	+54 36 02	−0.002	+0.01	7.80	0.1	0.6	A0 V	−3	250 s	1286	m
10072	37406	52 χ And	1 39 20.9	+44 23 10	−0.002	+0.02	4.98	0.89	0.3	G8 III	+7	86 s		
10186	147935		1 39 21.0	−17 47 37	+0.002	0.00	7.62	0.28	0.6	F0 III		250 s	1303	m
10216	193229		1 39 21.6	−33 33 50	+0.002	+0.03	6.6	0.8	3.2	G5 IV		44 s		
10165	110033		1 39 25.0	+0 36 44	−0.002	+0.02	7.7	0.1	1.7	A3 V		160 s		
10054	22479		1 39 30.5	+52 16 02	−0.002	0.00	7.6	0.0	0.4	B9.5 V		250 s		
10126	74857		1 39 35.9	+28 06 41	+0.038	+0.16	7.72	0.74	5.6	G8 V	+55	27 s		
10086	37410		1 39 36.0	+45 52 40	+0.022	−0.22	6.6	0.9	3.2	G5 IV	+6	48 s		
--	54879		1 39 37.2	+36 12 26	+0.002	0.00	8.0	1.3						
10164	92558	105 Psc	1 39 40.7	+16 24 21	+0.005	−0.01	5.97	1.12	0.2	K0 III	+18	120 s		
10268	193237		1 39 41.4	−37 28 24	−0.001	−0.01	7.20	1.1	0.2	K0 III		250 s		m
10063	22481		1 39 42.2	+55 47 04	−0.001	+0.01	7.39	0.25	−6.5	B8 Iab	−32	4300 s		
10361	232490		1 39 47.7	−56 11 41	+0.034	+0.03	5.07	0.88	6.7	K0 V	+23	6.5 t		p Eri, m
10372	232491		1 39 49.2	−55 08 59	+0.009	+0.02	7.5	1.6	0.2	K0 III		130 s		
10608	255801		1 39 53.4	−74 59 01	+0.010	+0.05	8.0	0.5	3.4	F5 V		82 s		
10254	167208		1 39 53.7	−22 54 49	−0.001	−0.03	7.26	1.64		M5 III				
10226	147941		1 39 54.5	−9 58 21	+0.016	+0.08	7.86	0.61	4.0	F8 V		51 s		
10132	37415		1 39 57.7	+41 40 00	+0.001	0.00	6.9	0.9	3.2	G5 IV		55 s		
10155	37416		1 40 04.4	+40 41 01	−0.004	−0.03	6.8	0.9	3.2	G5 IV		52 s		
10358	215607		1 40 06.2	−48 45 39	−0.001	+0.01	7.2	1.6	−0.5	M2 III		350 s		
10232	129435		1 40 06.9	−0 14 33	+0.001	−0.02	7.2	0.5	3.4	F5 V		57 s		
10109	22492		1 40 12.4	+54 26 44	−0.002	0.00	7.8	0.0	0.0	B8.5 V		330 s		
10110	22493		1 40 13.1	+53 52 06	−0.001	0.00	6.6	1.1	−0.1	K2 III	−62	200 s		
10329	193242		1 40 16.8	−34 28 36	0.000	+0.01	7.8	1.0	3.2	G5 IV		57 s		
10169	54887		1 40 17.2	+39 35 16	+0.003	−0.02	7.2	1.1	0.2	K0 III		260 s		
10472	248416		1 40 24.0	−60 59 55	+0.009	+0.01	8.0	0.3	2.6	F0 V		120 s		
10107	22496		1 40 30.7	+59 03 13	−0.001	+0.02	6.98	−0.04		B9				
9653	263		1 40 31.7	+81 25 48	−0.003	0.00	7.1	0.1	0.6	A0 V		200 s		
10205	37418	53 τ And	1 40 34.7	+40 34 37	+0.001	−0.02	4.94	−0.09	−0.6	B8 IV	−14	130 s	1311	m
10262	110046		1 40 34.9	+8 45 39	+0.005	+0.01	6.7	0.4	3.0	F2 V		55 s		
10146	22501		1 40 35.0	+54 56 32	0.000	+0.01	8.00	0.1	0.6	A0 V		280 s	1309	m
10318	147952		1 40 35.5	−16 17 36	+0.005	−0.13	8.0	0.8	3.2	G5 IV		66 mx		
10370	193248		1 40 36.4	−36 29 00	+0.009	−0.14	7.8	0.7	3.2	G5 IV		53 mx		
10204	37419		1 40 39.7	+43 17 52	+0.011	−0.03	5.61	0.21	2.6	F0 V	+17	40 s		
10315	129442		1 40 46.2	−2 37 20	0.000	−0.05	7.0	0.9	3.2	G5 IV		57 s		
10181	22505		1 40 52.6	+53 34 16	+0.001	+0.01	7.5	1.1	−0.1	K2 III		310 s		
9899	4461		1 40 53.0	+77 58 08	+0.005	−0.01	6.60	0.00	0.4	B9.5 V		170 s	1288	m
10211	37425		1 41 00.6	+49 06 14	+0.003	0.00	7.8	0.5	4.0	F8 V		56 s		
10309	74866		1 41 07.2	+23 01 34	+0.002	+0.02	7.6	0.9	3.2	G5 IV		75 s		
10447	193253		1 41 12.2	−33 23 52	+0.004	+0.02	7.3	1.0	3.2	G5 IV		45 s		
10308	74870		1 41 18.3	+25 44 44	+0.009	−0.04	6.17	0.44	3.4	F5 V	+5	36 s	1326	m
10859	255806	τ¹ Hyi	1 41 21.4	−79 08 54	+0.021	+0.02	6.33	0.94		K0				m
10392	129449		1 41 22.1	−0 30 24	−0.002	−0.01	7.9	0.9	3.2	G5 IV		88 s		
10380	110065	106 ν Psc	1 41 25.8	+5 29 15	−0.002	+0.01	4.44	1.36	−0.2	K3 III	0	42 ts		
10481	193255		1 41 27.2	−38 07 59	+0.004	+0.06	6.17	0.42	3.0	F2 V		39 s		
10196	11914		1 41 33.6	+62 40 37	0.000	−0.01	7.80	0.9	5.6	G8 V		27 s	1318	m
10145	11913		1 41 37.6	+66 54 38	+0.116	−0.24	7.68	0.68	5.2	G5 V	+16	31 s		
10925	258274		1 41 38.3	−80 03 08	−0.001	−0.01	7.8	−0.2	1.7	A3 V		160 s		
10348	74872		1 41 39.1	+30 02 50	−0.001	0.00	5.99	1.01	0.2	K0 III	+5	140 s		
10553	232497		1 41 41.0	−50 02 20	−0.003	−0.01	6.64	0.12		A2				
10453	147962		1 41 44.7	−11 19 29	+0.003	−0.41	5.75	0.44	3.4	F5 V	−10	25 mx	1339	m
10307	37434		1 41 47.1	+42 36 49	+0.073	−0.15	4.95	0.62	4.7	G2 V	+4	11 ts		
10615	248421		1 41 47.9	−60 47 22	+0.002	−0.04	5.71	1.27	0.2	gK0	+2	87 s		
10510	147966		1 42 01.9	−19 11 34	−0.002	−0.03	7.69	0.40		F5				
10552	193262		1 42 02.4	−39 08 51	+0.003	0.00	7.9	0.9	3.2	G5 IV		74 s		
10538	193261		1 42 02.9	−36 49 57	−0.003	−0.02	5.72	−0.01	0.6	A0 V	+20	100 s		
10260	11920		1 42 02.9	+61 02 18	+0.002	0.00	6.71	−0.03	−0.6	B7 V		270 s		
10390	54912		1 42 03.4	+35 14 44	+0.004	−0.03	5.40	−0.09	0.2	B9 V	−2	85 mx		
10363	37443		1 42 03.9	+43 38 25	+0.001	0.00	7.1	0.1	0.6	A0 V	0	200 s		

HD	SAO	Star Name	α 2000	δ 2000	μ(α)	μ(δ)	V	B-V	M_v	Spec	RV	d(pc)	ADS	Notes
			h m s	° ′ ″	s	″								
10407	74880		1 42 05.8	+29 30 21	-0.001	0.00	7.4	0.1	1.4	A2 V	+7	160 s		
10614	232500		1 42 06.4	-55 51 52	0.000	+0.05	7.4	2.0	0.2	K0 III		69 s		
10537	193263	π Scl	1 42 08.5	-32 19 37	-0.006	-0.02	5.26	1.04	0.3	gG8	+10	87 s		
10222	11918		1 42 09.5	+66 47 30	+0.004	-0.03	7.0	1.1	0.2	K0 III	+22	210 s		
10970	258275		1 42 11.8	-79 54 46	+0.016	+0.03	7.9	0.2	2.6	F0 V		120 s		
10293	22520		1 42 17.7	+58 37 40	+0.002	-0.01	6.30	-0.02		B9			1334	m
10221	11919	43 Cas	1 42 20.4	+68 02 35	+0.010	0.00	5.59	-0.07		A0 p	+5			
10439	54918		1 42 24.0	+32 57 41	0.000	-0.03	7.7	1.1	0.2	K2 III		360 s		
10506	129461		1 42 25.1	-2 09 36	+0.006	-0.01	8.0	0.4	3.0	F2 V		98 s		
9993	4464		1 42 26.0	+79 19 56	+0.025	-0.08	8.0	0.9	3.2	G5 IV		91 s		
10477	92588		1 42 26.9	+15 46 43	+0.002	0.00	7.6	1.1	0.2	K0 III		300 s		
10647	232501		1 42 29.3	-53 44 26	+0.019	-0.09	5.52	0.53	4.4	G0 V	+13	17 s		
10476	74883	107 Psc	1 42 29.7	+20 16 07	-0.021	-0.67	5.24	0.84	5.9	K1 V	-34	7.5 t		m
10530	129464		1 42 32.0	-8 39 35	+0.004	+0.02	7.1	0.9	3.2	G5 IV		59 s		
10572	167242		1 42 35.6	-20 10 26	-0.001	-0.05	7.30	1.4	-0.3	K4 III	+25	330 s		
10502	110082		1 42 35.9	+10 14 40	0.000	-0.02	7.5	0.1	1.4	A2 V		170 s		
10404	37451		1 42 36.6	+48 24 59	+0.001	-0.01	8.0	0.0	0.0	B8.5 V		360 s		
10332	11927		1 42 36.7	+60 32 54	0.000	-0.01	7.40	1.1	0.0	K1 III	+4	270 s	1337	m
10438	37452		1 42 37.1	+40 22 39	-0.001	-0.01	6.8	1.1	0.2	K0 III		210 s		
10625	215624		1 42 42.8	-41 38 51	-0.001	-0.03	7.9	1.1	0.2	K0 III		200 s		
10550	129465		1 42 43.4	-3 41 25	-0.001	-0.03	4.99	1.38	-1.3	K3 II-III	-34	170 s		
10304	11928		1 42 44.6	+63 53 28	-0.006	-0.04	7.7	1.1	0.2	K0 III	-9	250 mx		
10671	232503		1 42 49.3	-54 41 48	+0.005	+0.01	7.9	0.6	1.4	A2 V		86 s		
10500	54926		1 42 55.7	+31 01 38	0.000	-0.01	8.0	1.1	-0.1	K2 III		420 s		
10250	4470	42 Cas	1 42 55.8	+70 37 22	+0.016	-0.01	5.18	-0.04		A0	+6	21 mn		
11359	258276		1 42 56.9	-83 22 00	+0.003	-0.01	7.4	0.9	3.2	G5 IV		68 s		
10387	22535		1 42 57.6	+58 06 55	0.000	-0.03	7.1	0.4	3.0	F2 V		65 s		
10362	11931		1 42 58.3	+61 25 18	+0.001	+0.02	6.34	0.01	-5.7	B5 Ib		2100 s		
10465	37456		1 43 11.0	+48 30 59	-0.001	-0.01	6.9	1.6	-0.5	M2 III	-70	280 s		
10611	167249		1 43 14.3	-21 37 11	+0.004	-0.01	8.00	0.9	3.2	G5 IV		91 s	1358	m
10486	37460		1 43 16.5	+45 19 20	+0.014	-0.01	6.34	1.01		K2 IV	+12	140 mx		
10425	11941	44 Cas	1 43 19.6	+60 33 05	+0.001	-0.01	5.78	-0.02		B9	-37		1344	m
10437	22541		1 43 21.5	+59 37 58	+0.013	-0.04	6.7	1.1	0.2	K0 III		170 mx		
10731	232508		1 43 28.5	-56 14 04	+0.007	-0.02	7.6	1.6	0.2	K0 III		140 s		
10229	4471		1 43 29.7	+74 44 32	+0.003	0.00	8.0	1.4	-0.3	K5 III		420 s		
10498	22551		1 43 36.1	+51 30 57	0.000	0.00	6.7	1.4	-0.3	K5 III		230 s		
10609	110088		1 43 38.4	+5 44 44	+0.001	+0.01	8.0	1.1	0.2	K0 III		370 s		
10497	22552		1 43 38.6	+52 53 09	-0.001	-0.01	6.80	0.2	-2.0	A7 II		490 s	1354	m
10516	22554	φ Per	1 43 39.6	+50 41 20	+0.003	-0.01	4.07	-0.04	-3.9	B1 IV pe	+1	350 s		v
10495	22555		1 43 46.9	+55 52 39	-0.003	-0.01	7.20	0.1	1.4	A2 V	-19	140 s	1353	m
10259	4473		1 43 48.4	+74 36 15	+0.015	-0.03	6.8	0.8	3.2	G5 IV		52 s		
10588	54939		1 43 49.9	+32 11 30	-0.002	0.00	6.4	0.9	1.8	G8 III-IV	-5	84 s		
10653	129476		1 43 52.3	-7 28 47	-0.003	-0.06	7.74	0.05	0.6	A0 V		180 mx		
10546	37466		1 43 52.8	+49 39 21	-0.003	-0.02	7.44	0.00	0.4	B9.5 V		260 s		
10658	129477		1 43 54.7	-4 45 56	-0.002	-0.03	6.19	1.54		K0				
10474	11946		1 43 54.8	+60 26 12	-0.001	0.00	7.80	0.20	1.7	A3 V		140 s		
10691	193273		1 43 57.0	-30 43 40	+0.005	-0.03	8.0	0.4	1.7	A3 V		53 mx		
10700	147986	52 τ Cet	1 44 04.0	-15 56 15	-0.119	+0.86	3.50	0.72	5.7	G8 V p	-16	3.6 t		m
10583	37468		1 44 05.4	+43 14 23	-0.003	-0.04	7.7	1.1	0.2	K0 III		310 s		
10600	54942		1 44 09.3	+38 31 19	0.000	0.00	7.9	0.9	3.2	G5 IV		88 s		
10494	11950		1 44 11.3	+61 51 01	-0.003	+0.03	7.31	1.22	-8.2	F5 Ia		4800 s		
10577	37469		1 44 13.8	+48 12 41	+0.003	-0.01	7.02	0.02	0.4	B9.5 V		210 s		
10556	22567		1 44 14.0	+54 53 11	-0.006	-0.01	7.3	0.5	4.0	F8 V		44 s		
10543	22566		1 44 17.8	+57 32 11	+0.005	-0.03	6.40	0.1	1.4	A2 V	+5	96 s	1359	m
10638	54945		1 44 22.7	+32 30 58	+0.004	-0.06	6.8	0.1	1.7	A3 V	0	58 mx		
10617	54944		1 44 22.9	+37 40 07	-0.001	0.00	6.9	0.1	0.6	A0 V		190 s		
10707	147988		1 44 25.3	-11 17 53	+0.004	+0.06	7.5	0.5	4.0	F8 V		52 s		
10597	37475		1 44 26.4	+46 08 24	-0.001	-0.02	6.4	1.4	-0.3	K5 III	-19	200 s		
10840	248427		1 44 32.9	-61 01 07	+0.003	0.00	7.0	-0.4	0.4	B9.5 V		210 s		
10698	110104		1 44 35.9	+3 13 25	-0.002	-0.01	7.0	1.1	0.2	K0 III		230 s		
10762	193280		1 44 36.6	-34 24 24	+0.001	-0.04	6.8	1.6	0.2	K0 III		95 s		
10649	54948		1 44 38.4	+36 56 18	-0.002	-0.03	7.4	0.1	1.4	A2 V		160 s		
10725	129482		1 44 43.5	-6 45 58	0.000	+0.02	6.60	0.6	4.4	G0 V		28 s	1376	m
10344	4477		1 44 45.1	+75 52 21	-0.002	-0.01	7.2	0.4	2.6	F0 V		80 s		
10587	22578		1 44 46.0	+57 05 21	+0.002	0.00	6.2	0.1	0.6	A0 V	+5	130 s		
10681	74899		1 44 51.1	+28 27 28	+0.002	-0.01	7.3	0.1	0.6	A0 V	-1	220 s		
10697	92611	109 Psc	1 44 55.8	+20 04 59	-0.003	-0.11	6.27	0.75	5.0	dG4	-44	16 s		
10778	167265		1 44 57.4	-28 43 37	+0.001	-0.08	6.9	1.4	-0.5	M2 III		290 mx		
10524	11952		1 45 00.8	+69 30 10	+0.002	+0.02	7.3	0.8	3.2	G5 IV		64 s		
10786	167268		1 45 04.2	-28 32 31	0.000	-0.01	7.9	0.8	3.2	G5 IV		85 s		
10743	110109		1 45 04.8	+2 30 12	+0.001	0.00	7.8	1.1	0.2	K0 III		330 s		
10696	74904		1 45 06.5	+28 44 05	+0.002	-0.04	7.8	1.6	-0.5	M4 III		470 s		
10777	147996		1 45 08.0	-14 33 45	-0.001	-0.03	7.5	0.8	3.2	G5 IV		73 s		
10713	54950		1 45 20.5	+35 42 47	0.000	+0.03	7.5	0.2	2.1	A5 V		120 s		
11068	255813		1 45 20.7	-73 33 12	+0.012	-0.01	7.8	0.4	2.6	F0 V		110 s		

HD	SAO	Star Name	α 2000	δ 2000	μ(α)	μ(δ)	V	B-V	M_V	Spec	RV	d(pc)	ADS	Notes
			$\mathrm{^h \ ^m \ ^s}$	$\mathrm{^{\circ} \ ' \ ''}$	$\mathrm{^s}$	$''$								
10776	129485		1 45 21.3	-3 10 08	+0.001	+0.02	7.7	1.6	-0.5	M2 III		430 s		
10761	110110	110 o Psc	1 45 23.5	+9 09 28	+0.005	+0.05	4.26	0.96	0.2	K0 III	+14	65 s		
11185	255814		1 45 23.9	-77 21 26	+0.006	0.00	7.30	0.5	3.4	F5 V		60 s		m
10596	11956		1 45 25.0	+64 33 34	+0.002	-0.02	7.8	1.1	-0.1	K2 III		330 s		
10831	167273		1 45 32.9	-26 23 49	-0.003	+0.01	7.6	2.2	-0.5	M2 III		170 s		
10830	167275	ε Scl	1 45 38.7	-25 03 10	+0.012	-0.06	5.31	0.39	2.8	F1 V	+15	30 s	1394	m
10783	110111		1 45 42.6	+8 33 34	+0.002	-0.01	6.58	-0.05		A2 p	+19			
10795	110112		1 45 43.7	+3 40 01	-0.002	-0.02	6.9	0.1	1.7	A3 V		110 s		
10849	167277		1 45 44.5	-26 18 30	0.000	-0.05	7.9	1.7	-0.1	K2 III		200 s		
10809	110114		1 45 54.9	+3 25 01	-0.001	-0.02	6.9	0.2	2.1	A5 V		93 s		
10824	129490		1 45 59.1	-5 44 00	-0.001	-0.03	5.34	1.52	-0.3	gK4	+11	110 s		
10680	22600		1 45 59.3	+59 43 59	+0.004	+0.01	7.80	1.1	0.2	K0 III		290 s		
10863	167279		1 46 01.0	-27 20 57	+0.007	-0.05	6.39	0.36	3.0	F2 V n		48 s		
10934	232519		1 46 05.8	-50 48 59	+0.003	-0.02	5.49	1.60	-0.5	gM4	-2	160 s		
10939	232520		1 46 06.3	-53 31 18	+0.015	+0.07	5.04	0.04		A0	+10	18 mn		
10292	277		1 46 09.2	+80 15 14	+0.009	+0.01	8.0	1.4	-0.3	K5 III		420 s		
10910	215640		1 46 13.6	-42 22 54	-0.001	-0.01	7.8	2.2	-0.3	K5 III		150 s		
10773	--		1 46 19.2	+43 42 21	+0.002	0.00	7.4	0.1	0.6	A0 V	-14	230 s		
10878	148007		1 46 30.9	-10 35 53	-0.002	-0.01	7.5	0.8	3.2	G5 IV		74 s		
10845	92622		1 46 35.3	+17 24 46	+0.004	+0.01	6.55	0.25	2.6	F0 V	-1	62 s		
10808	37497		1 46 35.8	+41 33 07	+0.003	-0.01	7.9	0.9	3.2	G5 IV		88 s		
10757	22612		1 46 46.7	+59 09 23	-0.013	-0.04	7.5	0.6	4.4	G0 V		41 s		
10969	215647		1 46 46.9	-42 07 57	+0.005	-0.04	7.5	1.4	-0.1	K2 III		160 mx		
10821	54974		1 46 50.6	+38 44 56	0.000	-0.02	7.6	0.4	2.6	F0 V		99 s		
10756	11975		1 46 56.2	+60 40 11	-0.002	0.00	7.54	0.44	-7.1	B8 Ia		4800 s		
10920	148008		1 46 58.3	-13 53 20	+0.002	+0.01	6.7	0.0	0.4	B9.5 V		190 s		
10866	74926		1 47 03.3	+26 10 06	+0.001	-0.01	7.80	1.1	-0.2	K3 III	+22	400 s		
10894	92628		1 47 09.1	+10 50 40	+0.001	0.00	7.0	0.0	0.4	B9.5 V	+10	210 s		
10755	11979		1 47 09.5	+63 39 13	+0.007	-0.03	8.00	0.9	0.3	G5 III		280 mx		
10958	148011		1 47 15.7	-17 29 03	-0.001	-0.03	7.4	0.1	1.4	A2 V		160 s		
11022	215650		1 47 16.7	-41 45 37	+0.002	+0.05	6.18	1.54	-0.1	K2 III		93 s		
10806	22620		1 47 17.8	+58 01 17	0.000	0.00	7.00	1.1	0.2	K0 III		210 s		m
11023	215651		1 47 18.0	-42 38 20	-0.001	0.00	7.7	0.2	2.1	A5 V		130 s		
10930	110128		1 47 21.5	+0 48 44	+0.001	+0.02	8.0	0.8	3.2	G5 IV		89 s		
10987	167295		1 47 26.5	-28 41 05	-0.001	-0.03	7.9	1.0	3.2	G5 IV		64 s		
10860	37507		1 47 27.1	+43 04 47	0.000	-0.04	7.7	0.1	0.6	A0 V		260 s		
11058	215654		1 47 37.6	-46 15 27	+0.002	-0.01	6.7	1.5	-0.5	M2 III		280 s		
10998	167298		1 47 38.8	-20 50 43	+0.002	-0.04	6.55	1.39		K0				
10852	22627		1 47 42.2	+54 00 16	0.000	0.00	7.2	0.0	0.4	B9.5 V		210 s		
10780	11983		1 47 44.7	+63 51 09	+0.088	-0.25	5.63	0.81	5.9	K0 V	+2	8.8 t		m
11604	258280	τ² Hyi	1 47 46.6	-80 10 36	-0.028	-0.06	6.07	0.33	2.6	dF0		47 s		m
11050	193303		1 47 47.7	-37 09 35	0.000	+0.02	6.32	1.02		K0				
10874	37513		1 47 47.9	+46 13 47	+0.001	-0.05	6.32	0.43	3.4	F5 V	-3	38 s		
10872	22630		1 48 02.5	+53 53 24	0.000	+0.01	8.00	0.1	0.6	A0 V		270 s	1416	m
10982	92637	4 Ari	1 48 10.8	+16 57 20	+0.004	-0.03	5.70	-0.03	0.4	B9.5 V	+10	110 s		
--	22629		1 48 11.9	+59 39 06	-0.001	0.00	8.0	1.9	-0.1	K2 III		150 s		
--	--		1 48 18.5	+59 40 09			8.00	1.1	-0.1	K2 III		360 s		
11112	215660		1 48 20.3	-41 29 43	+0.035	+0.15	7.13	0.65	5.0	G4 V		23 ts		
10995	92638		1 48 20.7	+17 01 22	-0.002	+0.05	7.4	0.6	4.4	G0 V	+12	39 s		
11037	110138		1 48 26.0	+3 41 07	0.000	+0.02	5.91	0.97	0.3	gG6	+3	120 s		
10898	22639		1 48 35.0	+58 27 29	0.000	+0.02	7.40	0.35	-5.7	B2 Ib		2000 s		
11100	167312		1 48 35.8	-26 15 13	-0.003	+0.01	7.09	0.28	2.6	F0 V		79 s		
11080	148020		1 48 36.7	-14 45 56	0.000	-0.02	7.5	1.1	0.2	K0 III		280 s		
10975	54991		1 48 38.8	+37 57 10	+0.009	-0.03	5.94	0.97	0.2	K0 III	+37	140 s		
11049	110142		1 48 40.6	+7 41 01	+0.003	-0.03	7.30	1.1	0.2	K0 III		260 s	1435	m
11137	193312		1 48 40.8	-38 39 28	-0.003	-0.10	7.80	0.4	3.0	F2 V		91 s		
11007	54994		1 48 41.5	+32 41 25	-0.014	+0.30	5.79	0.55	3.4	F5 V	-27	29 ts		
11211	232530		1 48 49.9	-51 37 58	+0.005	-0.03	7.5	0.4	0.2	K0 III		220 mx		
11013	54996		1 48 50.6	+34 08 33	0.000	-0.01	8.0	1.6	-0.5	M4 III		510 s		
11210	232531		1 48 52.0	-50 29 00	+0.004	+0.02	7.8	-0.7	0.4	B9.5 V		180 mx		
11054	92645		1 48 54.5	+16 22 38	-0.001	-0.01	7.7	0.4	2.6	F0 V		100 s		
11046	55000		1 49 12.8	+34 54 13	+0.002	-0.03	7.9	0.9	3.2	G5 IV		88 s		
16477	258291		1 49 13.2	-88 21 27	+0.051	+0.02	7.95	0.08		B8				m
11031	37536		1 49 15.6	+47 53 50	-0.001	0.00	5.82	0.29		A2	-2		1438	m
11219	215666		1 49 16.8	-44 14 30	+0.001	0.00	7.9	1.1	-0.1	K2 III		390 s		
11183	193320		1 49 19.4	-31 04 22	+0.008	-0.02	6.4	1.2	0.2	K0 III		140 s		
11462	255818		1 49 22.8	-72 24 43	+0.003	+0.01	7.7	0.0	0.4	B9.5 V		290 s		
11131	148033		1 49 23.3	-10 42 13	-0.010	-0.09	6.76	0.61	4.4	dG0	-3	28 s		
11079	74953		1 49 27.1	+26 28 22	0.000	0.00	6.7	0.0	0.0	B8.5 V	+13	220 s		
11171	148036	53 χ Cet	1 49 35.0	-10 41 11	-0.010	-0.09	4.67	0.33	1.9	F2 IV	-1	30 ts		m
11864	258282		1 49 36.9	-81 21 34	+0.033	-0.03	7.5	1.5	0.2	K0 III		150 s		
11394	--		1 49 47.3	-64 22 12			7.40	1.26	-0.1	K2 III		280 s		
11262	193326		1 49 48.7	-38 24 14	-0.001	+0.24	6.37	0.52	4.0	F8 V	+15	30 s		m
11071	37546		1 49 51.1	+45 16 18	+0.001	-0.02	8.0	0.0	0.4	B9.5 V		300 s		
10648	291		1 49 53.1	+80 53 11	-0.005	0.00	7.90	0.1	0.6	A0 V		280 s	1411	m

HD	SAO	Star Name	α 2000	δ 2000	μ(α)	μ(δ)	V	B-V	M$_V$	Spec	RV	d(pc)	ADS	Notes
			$1^h49^m53^s.2$	$+85°12'56''$	$+0^s.017$	$-0''.01$								
10124	282		1 49 53.2	+85 12 56	+0.017	−0.01	8.00	0.2	2.1	A5 V		150 s	1366	m
11170	110154		1 49 55.9	+7 13 25	−0.001	−0.06	7.9	0.7	2.9	G1 IV	−14	100 s		
10790	4496		1 49 56.4	+76 59 34	+0.029	+0.07	8.0	0.9	3.2	G5 IV		91 s		
11249	167334		1 49 57.6	−29 02 27	+0.002	0.00	7.4	1.3	0.2	K0 III		170 s		
11343	232538		1 50 06.1	−54 27 52	+0.012	−0.03	8.0	1.2	0.2	K0 III		130 mx		
11155	74966	1 Ari	1 50 08.5	+22 16 31	−0.001	0.00	5.86	0.74	0.2	K0 III	+1	54 s	1457	m
11399	232540		1 50 14.3	−58 56 22	0.000	+0.02	7.4	1.1	0.2	K0 III		250 s		
11226	129526		1 50 15.3	−6 42 15	+0.014	+0.04	7.3	0.5	4.0	F8 V		45 s		
11247	148049		1 50 15.7	−12 53 14	0.000	−0.01	7.0	1.6	−0.5	M2 III		310 s		
11332	215673		1 50 20.2	−47 48 59	+0.012	+0.06	6.14	1.01	3.2	K0 IV		36 s		
11166	55017		1 50 23.7	+30 48 13	0.000	−0.03	7.9	0.5	3.4	F5 V		78 s		
11440	248447		1 50 25.6	−62 57 29	+0.004	−0.01	7.5	1.1	0.2	K0 III		290 s		
11323	215674		1 50 28.7	−42 30 29	+0.003	−0.01	7.5	1.2	0.2	K0 III		230 s		
11094	22663		1 50 29.3	+53 44 35	+0.002	+0.01	8.00	1.6		M5 II−III				TT Per, v
11152	55016		1 50 29.8	+38 18 11	+0.006	−0.08	7.7	0.4	2.6	F0 V		110 s		
11190	55023		1 50 44.7	+36 03 05	0.000	−0.02	8.0	0.1	0.6	A0 V		310 s		
11103	22668		1 50 46.2	+56 45 55	+0.006	0.00	7.9	1.1	−0.1	K2 III		350 mx		
11093	22669		1 50 51.4	+58 28 04	+0.001	+0.01	7.9	1.4	−0.3	K5 III		370 s		
11257	92659		1 50 51.9	+11 02 36	−0.005	−0.02	5.94	0.30	2.6	F0 V	+11	47 s		
11202	55026		1 50 53.1	+35 50 45	+0.003	+0.02	7.3	0.5	3.4	F5 V		59 s		
11413	232542		1 50 54.3	−50 12 23	−0.006	0.00	5.94	0.14		A0				
11151	22678		1 50 57.0	+51 56 00	+0.004	−0.12	5.90	0.43	3.4	F5 V	−17	32 s		
11439	232544		1 50 59.2	−54 16 49	+0.002	−0.03	7.5	1.2	−0.1	K2 III		320 s		
11216	55028		1 51 01.3	+37 33 15	+0.001	0.00	7.9	1.1	0.2	K0 III		340 s		
11188	37556		1 51 02.5	+47 25 11	+0.001	−0.01	7.28	−0.02		B8	−9			
11201	37557		1 51 07.3	+43 31 30	−0.001	−0.02	7.8	1.1	0.2	K0 III		330 s		
11224	55030		1 51 08.9	+38 01 28	−0.004	+0.01	7.4	1.6	−0.5	M4 III		390 s		
11285	74983		1 51 09.4	+20 30 53	+0.001	0.00	6.8	0.4	2.6	F0 V		69 s		
11284	74984		1 51 13.2	+24 39 06	−0.001	−0.04	7.60	0.2	2.1	A5 V		130 s	1473	m
10733	294		1 51 13.7	+81 22 29	0.000	0.00	7.5	0.1	0.6	A0 V		240 s		
11379	193335		1 51 14.6	−30 54 10	+0.016	+0.07	7.21	0.47	4.0	F8 V		44 s		
—	22680		1 51 16.7	+55 24 23	+0.002	0.00	8.0	1.1	0.2	K0 III		300 s		
11092	12006		1 51 17.0	+64 51 16	+0.001	−0.02	6.57	2.09	−5.1	K5 Iab−Ib		780 s	1459	m
11236	55032		1 51 17.8	+38 50 08	+0.003	0.00	7.8	0.1	0.6	A0 V		270 s		
11126	12009		1 51 19.2	+60 21 13	0.000	−0.01	7.98	0.13	−1.2	B8 III		500 s	1461	m
11421	215679		1 51 26.2	−39 50 09	+0.001	−0.02	6.5	1.1	0.2	K0 III		150 s		
11437	215681		1 51 26.4	−45 28 17	0.000	−0.02	7.2	0.7	0.2	K0 III		250 s		
11187	22682		1 51 26.6	+54 55 28	+0.002	−0.02	7.94	0.26		A0 p	+5			
11353	148059	55 ζ Cet	1 51 27.6	−10 20 06	+0.002	−0.04	3.73	1.14	−0.1	K2 III	+9	58 s		Baten Kaitos, m
11339	129535		1 51 28.8	−4 13 13	+0.009	−0.06	7.3	0.2	2.1	A5 V		30 mx		
11223	37561		1 51 33.0	+49 52 10	0.000	+0.01	7.7	0.0	0.4	B9.5 V		260 s		
11351	129537		1 51 36.2	−2 38 18	+0.004	−0.04	7.1	0.4	3.0	F2 V		67 s		
11252	37562		1 51 37.1	+47 05 11	−0.009	−0.01	7.50	0.9	3.2	G5 IV		71 s	1469	m
11365	129539		1 51 37.8	−6 52 29	+0.001	−0.02	6.5	1.1	0.2	K0 III		180 s		
11389	148066		1 51 39.9	−13 13 58	−0.002	+0.04	7.0	1.1	0.2	K0 III		230 s		
11253	37568		1 51 43.5	+46 20 18	+0.001	−0.02	8.0	0.0	0.4	B9.5 V		300 s		
11396	148069		1 51 45.9	−15 38 52	−0.001	0.00	6.5	1.6	−0.5	M2 III		250 s		
11326	92666		1 51 47.1	+18 17 23	−0.002	−0.12	6.7	0.8	3.2	G5 IV		51 s		
10971	4503		1 51 48.0	+75 35 23	+0.019	−0.03	6.9	0.5	3.4	F5 V		51 s		
11459	193342		1 51 52.0	−37 02 32	0.000	+0.03	7.2	0.3	4.0	F8 V		43 s		
11241	22690	1 Per	1 51 59.4	+55 08 51	+0.002	0.00	5.52	−0.18	−2.5	B2 V	−3	400 s		
11574	232553		1 52 06.3	−55 45 36	+0.003	+0.02	7.3	1.1	3.4	F5 V		24 s		
11291	22696	2 Per	1 52 09.3	+50 47 34	+0.002	−0.02	5.79	−0.07		B9				
11363	74989		1 52 10.4	+24 07 02	0.000	+0.02	6.8	1.1	0.2	K0 III		210 s		
11114	4506		1 52 10.4	+70 45 08	+0.003	+0.05	7.8	0.5	4.0	F8 V		57 s		
11481	193344		1 52 11.3	−32 32 25	−0.003	−0.02	7.96	0.10	2.1	A5 V n		150 s		
11526	193349		1 52 21.5	−39 25 00	0.000	−0.05	8.0	0.8	3.2	G5 IV		86 s		
11348	55047		1 52 22.2	+37 25 48	−0.002	−0.12	6.8	0.5	4.0	F8 V		37 s		
11361	55049		1 52 32.7	+37 59 03	+0.013	+0.05	8.0	0.5	4.0	F8 V		62 s		
11336	37578		1 52 33.4	+44 48 44	+0.002	−0.03	7.5	0.0	0.0	B8.5 V	−20	290 s		
11733	248455	η¹ Hyi	1 52 34.8	−67 56 40	+0.005	0.00	6.7	0.1	0.6	A0 V	+15	100 mx		q
11565	193358		1 52 49.0	−34 51 12	+0.008	−0.03	7.9	1.1	3.2	G5 IV		56 s		
11506	148079		1 52 50.5	−19 30 27	+0.003	−0.11	7.8	1.2	4.4	G0 V		20 s		
11335	22705		1 52 50.7	+51 28 29	−0.001	0.00	6.2	0.1	0.6	A0 V	+6	130 s		
11522	148078		1 52 52.0	−16 55 45	+0.003	−0.05	5.80	0.27		A6 m				
11573	193361		1 52 59.7	−32 46 38	+0.008	−0.02	8.03	0.24	0.5	A7 III		76 mx		
11443	74996	2 α Tri	1 53 04.8	+29 34 44	+0.001	−0.23	3.41	0.49	2.2	F6 IV	−13	18 ts		m
11505	129551		1 53 06.2	−1 19 38	−0.012	−0.36	7.42	0.63	4.4	G0 V		37 s		
11453	74997		1 53 10.6	+28 49 11	+0.003	0.00	7.0	1.4	−0.3	K5 III	+4	290 s		
11430	55058		1 53 10.7	+37 19 17	0.000	−0.05	7.40	0.5	3.4	F5 V	−13	63 s	1500	m
11428	37587	55 And	1 53 17.3	+40 43 47	0.000	0.00	5.40	1.32	0.0	gK1	−7	81 s		m
11643	193368		1 53 23.1	−38 35 41	+0.011	+0.03	6.10	1.12	−2.1	K0 II		50 mx		
11629	193369		1 53 26.7	−33 33 54	−0.004	−0.02	7.1	1.3	0.2	K0 III		160 s		
11418	37588		1 53 28.9	+49 26 42	−0.002	+0.01	7.0	0.8	3.2	G5 IV		57 s		
11520	92682		1 53 29.9	+12 40 57	−0.001	−0.03	7.1	0.5	3.4	F5 V		56 s		

HD	SAO	Star Name	α 2000	δ 2000	μ(α)	μ(δ)	V	B-V	M$_v$	Spec	RV	d(pc)	ADS	Notes
11502	92680	5 γ¹ Ari	1ʰ53ᵐ31.7ˢ	+19°17′45″	+0.006ˢ	−0.11″	4.68	0.0		A si	+4	36 mx	1507	m
11503	92681	5 γ² Ari	1 53 31.8	+19 17 37	+0.006	−0.10	4.59	0.0	0.2	B9 V	−1	36 mx	1507	Mesartim, m
11559	110206	111 ξ Psc	1 53 33.3	+3 11 15	+0.001	+0.03	4.62	0.94	0.2	K0 III	+30	77 s		
11594	148086		1 53 38.2	−10 24 57	−0.004	−0.03	7.8	1.1	0.2	K0 III		330 s		
11558	110207		1 53 38.3	+7 38 02	+0.001	−0.01	7.9	1.1	−0.1	K2 III		390 s		
11695	215696	ψ Phe	1 53 38.7	−46 18 09	−0.008	−0.08	4.41	1.59	−0.5	M4 III	+1	96 s		
8395	258		1 53 47.6	+88 33 07	+0.106	+0.01	8.0	1.1	0.2	K0 III		370 s		
11408	22716		1 53 48.4	+55 35 53	+0.006	+0.01	6.5	0.1	1.4	A2 V	+8	100 s		
11451	37593		1 53 54.1	+49 56 35	−0.001	0.00	6.9	1.1	0.2	K0 III		210 s		
11547	75004		1 53 55.6	+25 46 33	+0.007	−0.06	7.49	0.39		F0				
11592	92688		1 53 57.7	+10 36 50	−0.012	−0.29	6.78	0.46	3.4	F5 V		47 s		
11401	22719		1 54 00.3	+60 09 12	−0.001	+0.01	7.89	1.99	−0.5	M3 III		280 s		
12121	255831		1 54 05.8	−77 29 28	+0.001	+0.05	7.90	0.4	2.6	F0 V		120 s		m
11809	232565		1 54 08.8	−57 55 09	+0.004	+0.01	7.4	1.6	−0.1	K2 III		180 s		
11347	12028		1 54 20.5	+69 10 56	+0.002	−0.03	7.9	1.6	−0.5	M2 III		410 s		
11753	215697	φ Phe	1 54 21.9	−42 29 50	−0.003	−0.03	5.11	−0.06	1.7	A3 V	+12	48 s		
11640	110214		1 54 22.1	+8 46 52	+0.001	+0.01	7.0	1.6	−0.5	M2 III		310 s		
11415	12031	45 ε Cas	1 54 23.6	+63 40 13	+0.005	−0.02	3.38	−0.15	−2.9	B3 III	−8	160 s		
11637	92689		1 54 24.5	+11 24 10	+0.002	+0.02	7.4	1.1	0.2	K0 III		270 s		
11395	12030		1 54 34.2	+68 09 06	+0.002	0.00	7.6	0.1	0.6	A0 V		230 s		
11691	129560		1 54 37.2	−6 16 17	+0.002	+0.04	8.0	0.8	3.2	G5 IV		89 s		
11579	37604		1 54 37.8	+40 39 23	0.000	+0.01	7.2	1.1	−0.1	K2 III		290 s		
11636	75012	6 β Ari	1 54 38.3	+20 48 29	+0.007	−0.11	2.64	0.13	2.1	A5 V	−2	14 ts		Sheratan
11160	4509		1 54 41.2	+78 11 48	+0.017	−0.03	6.9	0.8	3.2	G5 IV		54 s		
11607	55077		1 54 42.3	+39 56 55	−0.001	−0.02	7.5	0.9	3.2	G5 IV		71 s		
11571	22736		1 54 44.6	+50 17 46	−0.002	−0.01	7.1	1.1	−0.1	K2 III		250 s		
11673	92693		1 54 47.1	+13 45 27	0.000	+0.02	8.0	0.9	3.2	G5 IV		93 s		
11613	37607		1 54 53.7	+40 42 08	+0.004	−0.06	6.24	1.25	−0.1	K2 III	+32	170 s		
11650	75017		1 54 55.6	+27 50 31	+0.003	−0.01	7.5	0.9	0.2	G9 III		290 s		
11977	248460	η² Hyi	1 54 56.1	−67 38 50	+0.014	+0.08	4.69	0.95	0.3	G5 III	−16	69 s		
11544	22740		1 54 56.3	+56 34 57	+0.001	0.00	6.82	1.15	−4.5	G2 Ib		1300 s		
11624	55082		1 54 57.4	+37 07 42	+0.002	−0.01	6.26	1.17	0.2	K0 III	−2	130 s		
11555	22743		1 54 58.9	+54 11 20	0.000	0.00	7.5	0.8	3.2	G5 IV		71 s		
11785	167392		1 55 04.4	−24 53 22	+0.001	+0.03	7.1	1.1	0.2	K0 III		190 s		
11672	75019		1 55 04.8	+27 33 15	−0.002	0.00	8.0	0.1	0.6	A0 V		300 s		
11671	75020		1 55 07.3	+28 47 52	+0.002	−0.04	7.80	0.4	3.0	F2 V		91 s	1522	m
11578	22748		1 55 15.8	+57 01 36	+0.006	−0.05	7.6	0.7	4.4	G0 V		44 s		
11577	22749		1 55 17.2	+57 05 36	−0.003	+0.04	7.7	0.1	−2.8	A0 II		860 s		
11944	248462		1 55 19.9	−60 18 44	−0.005	−0.04	7.10	0.4	2.6	F0 V		79 s		m
11442	4515		1 55 26.1	+70 29 11	−0.002	+0.02	8.0	0.9	3.2	G5 IV		89 s		
11316	4512		1 55 26.2	+76 13 27	−0.013	−0.01	7.40	0.2	2.1	A5 V		110 s	1504	m
11764	92700		1 55 39.4	+12 34 24	+0.005	0.00	7.7	0.8	3.2	G5 IV		79 s		
13121	258285		1 55 41.8	−84 45 05	+0.008	0.00	7.1	1.5	−0.1	K2 III		180 s		
11053	310		1 55 42.0	+81 57 33	+0.015	0.00	7.1	1.4	−0.3	K5 III		290 s		
11606	22753		1 55 42.7	+59 16 24	0.000	0.00	7.02	0.06		B3 e	+13			
11689	37616		1 55 45.8	+42 50 42	+0.004	−0.01	7.54	0.41		F2				
11995	248464		1 55 46.3	−60 51 41	+0.001	+0.06	6.06	0.37		F0				
11658	22760		1 55 46.9	+51 41 18	+0.010	−0.03	7.2	1.1	0.2	K0 III	+9	190 mx		
11956	232572		1 55 48.2	−55 04 19	+0.006	+0.01	6.9	0.6	1.4	A2 V		60 s		
11736	55100		1 55 48.3	+30 21 19	+0.007	−0.03	6.8	0.2	2.1	A5 V		58 mx		
11835	167401		1 55 49.0	−20 33 47	0.000	−0.10	8.0	0.3	3.0	F2 V		100 s		
11854	167404		1 55 49.8	−29 06 49	+0.003	+0.01	7.4	0.8	0.2	K0 III		280 s		
12363	255835	σ Hyi	1 55 50.5	−78 20 55	+0.033	+0.05	6.16	0.44		F2				
11763	75030	7 Ari	1 55 51.0	+23 34 39	+0.001	0.00	5.74	1.19	0.2	K0 III	+14	99 s		RR Ari, q
12087	248466		1 55 52.7	−68 23 29	−0.003	+0.02	7.10	0.91		K0				
11803	110235		1 55 53.7	+1 50 59	+0.010	+0.19	6.01	0.56		G0	+30	23 mn	1538	m
11605	12042		1 55 54.1	+62 21 49	+0.002	0.00	7.58	0.12	−1.9	B6 III		560 s		
11727	55102		1 55 54.3	+37 16 40	0.000	0.00	5.89	1.63	−0.1	K2 III	+7	75 s	1534	m
11878	193387		1 55 56.1	−36 14 43	−0.001	0.00	7.94	0.23	3.0	F2 V		97 s		
11937	232573	χ Eri	1 55 57.5	−51 36 32	+0.073	+0.30	3.70	0.85	3.2	G5 IV	−6	15 ts		m
11529	12038	46 ω Cas	1 55 59.9	+68 41 07	+0.002	−0.01	4.99	−0.10		B8	−24			
11719	37618		1 56 00.6	+43 03 37	+0.002	−0.03	7.50	1.4	−0.3	K4 III	+2	360 s		
11150	312		1 56 02.9	+80 54 30	+0.048	−0.08	7.6	1.1	0.2	K0 III		160 mx		
11965	232574		1 56 09.2	−49 50 11	0.000	+0.02	6.5	0.4	0.0	B8.5 V		100 s		
11749	55107	56 And	1 56 09.3	+37 15 06	+0.015	+0.01	5.67	1.06	0.2	K0 III	+59	75 s	1534	m
11888	167413		1 56 18.5	−24 14 03	+0.001	−0.02	7.9	1.5	−0.1	K2 III		250 s		
11669	12046		1 56 23.8	+61 16 50	+0.002	−0.02	7.44	0.09	−1.9	B6 III		550 s	1531	m
11570	12043		1 56 31.6	+70 12 14	0.000	+0.01	8.0	1.4	−0.3	K5 III		380 s		
11780	37625		1 56 32.2	+41 53 32	+0.001	0.00	6.7	0.0	0.4	B9.5 V		190 s		
11943	193396		1 56 34.6	−32 08 18	+0.002	+0.01	6.5	1.0	3.2	G5 IV		34 s		
12057	232579		1 56 37.2	−58 28 18	+0.003	0.00	7.4	0.7	3.4	F5 V		45 s		
11735	22772		1 56 37.5	+53 18 17	0.000	0.00	7.0	0.1	0.6	A0 V		180 s		
12681	258284		1 56 37.8	−81 29 58	+0.028	+0.02	7.1	0.3	2.6	F0 V		78 s		
11930	167416	56 Cet	1 56 40.1	−22 31 36	+0.004	−0.02	5.1	1.6	−0.3	K5 III	+27	98 s		
12003	215711		1 56 50.3	−43 10 43	+0.005	−0.02	7.87	0.23	2.4	A7 V n		62 mx		

HD	SAO	Star Name	α 2000	δ 2000	μ(α)	μ(δ)	V	B-V	M$_v$	Spec	RV	d(pc)	ADS	Notes
11773	22777		1^h 56^m 52.2^s	+52°29'23"	+0$.^s$002	+0$.^{\prime\prime}$01	7.7	1.4	-0.3	K5 III		340 s		
12362	255836		1 56 54.0	-75 21 57	+0.007	+0.03	7.9	1.1	0.2	K0 III		350 s		
11975	167422		1 56 59.9	-25 37 15	+0.002	-0.01	7.4	1.4	0.2	K0 III		150 s		
12042	232581		1 57 00.0	-51 45 58	+0.039	+0.26	6.10	0.48	4.0	F8 V		24 ts		
12002	193401		1 57 03.1	-33 56 29	-0.003	-0.07	8.0	0.0	2.6	F0 V		120 s		
11993	193404		1 57 08.3	-31 05 03	+0.004	-0.03	7.3	0.8	0.2	K0 III		260 s		
11964	148123		1 57 09.5	-10 14 32	-0.026	-0.23	6.41	0.82		G5		13 mn		m
12055	215715		1 57 10.0	-47 23 06	+0.010	+0.02	4.83	0.88		G5	+12	17 mn		
11609	4522		1 57 11.2	+71 43 28	+0.003	-0.04	7.2	0.1	1.4	A2 V		140 s		
--	55124		1 57 11.9	+31 03 48	+0.001	+0.01	8.0	1.2						
11952	129578		1 57 12.8	-7 48 49	+0.002	+0.02	7.7	0.9	3.2	G5 IV		78 s		
11992	167425		1 57 13.1	-26 38 41	-0.001	+0.02	7.9	1.0	0.2	K0 III		340 s		
12033	193407		1 57 16.1	-37 29 47	+0.001	+0.03	7.91	1.23		K0				
11985	148125		1 57 16.2	-17 40 37	+0.004	0.00	7.7	1.4	-0.3	K5 III		350 mx		
11909	92721	8 ι Ari	1 57 21.0	+17 49 03	+0.002	-0.02	5.10	0.92		K1 p	-5			
12040	193408		1 57 24.1	-36 03 25	+0.002	+0.04	7.1	0.1	2.1	A5 V		98 s		
12108	232585		1 57 29.3	-52 12 01	+0.002	-0.05	7.90	0.5	3.4	F5 V		79 s		m
11982	129580		1 57 31.1	-6 49 39	+0.004	+0.03	8.0	0.1	1.4	A2 V		210 s		
11744	12051		1 57 36.3	+65 24 37	-0.004	+0.01	7.81	0.37	-1.6	B3.5 V		370 s		
11800	22785		1 57 40.0	+60 13 08	-0.001	0.00	7.79	2.04	-4.4	K5 Ib		1400 s		
11928	75048		1 57 43.7	+27 48 16	+0.002	-0.05	5.82	1.60	-0.5	M4 III	-3	180 s		
12603	255838		1 57 48.6	-78 45 50	+0.012	+0.01	7.7	0.5	3.4	F5 V		71 s		
11927	55144		1 57 51.1	+33 48 22	0.000	-0.01	8.0	0.0	0.4	B9.5 V		330 s		
--	4518		1 57 53.1	+79 30 43	-0.005	+0.01	8.0	2.2						
11588	4524		1 57 53.2	+75 20 10	-0.003	-0.03	6.7	0.9	3.2	G5 IV		50 s		
12270	248472		1 57 53.6	-65 25 29	-0.001	-0.01	6.37	0.90		G5				
11907	37652		1 57 55.4	+40 45 36	+0.011	-0.10	6.9	0.5	3.4	F5 V		49 s		
11973	75051	9 λ Ari	1 57 55.7	+23 35 46	-0.007	-0.01	4.79	0.28	1.7	F0 IV	-1	42 s	1563	m
11905	37653		1 57 56.4	+41 41 40	+0.001	-0.01	6.6	-0.1		B9				
11821	12057		1 57 56.7	+61 11 34	0.000	-0.01	7.95	1.44	-0.1	K2 III		160 s		
--	75054		1 57 57.6	+23 36 13	-0.008	0.00	7.32	0.56	4.4	G0 V		38 s		
11884	37648		1 57 59.1	+47 05 45	0.000	0.00	6.5	1.1	0.2	K0 III	-7	170 s		
12020	129588		1 57 59.3	-2 03 33	+0.002	+0.02	6.60	0.1	0.6	A0 V		160 s	1567	m
11962	75052		1 58 01.0	+28 31 11	+0.001	-0.02	7.9	0.1	0.6	A0 V		280 s		
11883	37650		1 58 03.1	+49 38 35	0.000	-0.02	7.0	1.1	-0.1	K2 III		240 s		
11961	55147		1 58 03.8	+31 08 05	-0.002	-0.03	7.1	1.6	-0.5	M4 III	-45	340 s		
11926	37658		1 58 04.5	+41 23 10	-0.004	+0.03	7.60	0.8	3.2	G5 IV		76 s	1560	m
--	--		1 58 05.2	+23 36 15			7.32	0.57						
11575	4526		1 58 05.4	+76 22 12	+0.022	-0.05	7.9	0.9	3.2	G5 IV		88 s		
11860	22793		1 58 18.2	+59 37 34	+0.003	-0.02	6.65	0.08	0.6	A0 V		140 s		
11866	22799		1 58 19.7	+57 52 03	+0.001	0.00	7.94	0.07	1.2	A1 V		210 s		
12135	193422		1 58 26.7	-33 04 01	+0.002	-0.02	6.35	1.02		G5				m
12076	129596		1 58 27.6	-7 04 43	+0.001	+0.02	7.0	1.6	-0.5	M2 III		310 s		
12157	215718		1 58 27.9	-41 10 13	+0.006	-0.03	7.30	0.4	2.6	F0 V		87 s		m
11922	22804		1 58 29.0	+53 04 40	-0.001	+0.01	7.8	0.8	3.2	G5 IV		82 s		
11991	55156		1 58 30.8	+36 15 06	-0.002	0.00	7.9	0.9	3.2	G5 IV		88 s		
11969	37668		1 58 32.4	+45 07 21	0.000	-0.01	8.0	1.4	-0.3	K5 III		390 s		
11857	12065		1 58 33.2	+61 41 53	+0.002	0.00	6.02	-0.04		B8				
11949	37665	3 Per	1 58 33.5	+49 12 15	+0.001	+0.04	5.69	1.01	3.2	G5 IV	0	23 s		
--	37669		1 58 35.7	+47 02 21	+0.002	-0.03	7.8	0.7						
11865	12066		1 58 38.0	+61 32 50	-0.003	+0.02	7.44	1.28	0.3	G8 III	-11	170 s		
11979	37673		1 58 44.2	+45 26 05	+0.003	-0.02	7.7	1.6	-0.5	M4 III		390 s		
12116	148139		1 58 45.9	-11 18 04	+0.004	-0.09	6.6	0.8	3.2	G5 IV		48 s		
12311	248474	α Hyi	1 58 46.2	-61 34 12	+0.038	+0.03	2.86	0.28	2.6	F0 V	+1	11 s		
12007	55161		1 58 46.3	+36 32 10	+0.001	-0.01	7.2	0.1	1.4	A2 V		140 s		
11755	4536		1 58 50.2	+73 09 08	-0.012	-0.06	7.1	0.9	3.2	G5 IV		59 s		
12155	167448		1 58 54.4	-21 07 57	+0.003	+0.08	7.2	0.7	3.4	F5 V		38 s		
12082	92735		1 58 54.5	+18 21 29	+0.005	-0.02	7.8	0.8	3.2	G5 IV		84 s		
12215	215720		1 58 56.5	-42 56 57	-0.001	0.00	7.4	1.0	-0.1	K2 III		310 s		
12050	55166		1 59 00.0	+34 20 21	0.000	-0.04	7.7	0.9	3.2	G5 IV	-41	78 s		
12180	167451		1 59 00.7	-22 55 09	+0.007	+0.06	6.8	-0.2	3.0	F2 V		59 s	1581	m
12278	232592		1 59 02.7	-52 14 49	+0.003	+0.01	7.6	0.7	0.2	K0 III		300 s		
12051	55167		1 59 06.5	+33 12 34	+0.019	-0.36	7.14	0.78	3.2	G5 IV	-37	15 mx		m
12178	148142		1 59 13.4	-15 07 19	-0.001	-0.01	7.5	0.9	3.2	G5 IV		71 s		
11874	12071		1 59 14.9	+67 02 37	+0.007	-0.01	7.4	0.1	1.7	A3 V		130 mx		
12206	167461		1 59 19.5	-26 25 55	+0.005	+0.03	6.79	0.02	1.2	A1 V		110 mx		
12140	92739		1 59 25.8	+12 17 42	0.000	-0.03	6.10	0.19	2.2	A6 V	-12	60 s		
12268	215725		1 59 26.0	-40 43 38	-0.006	-0.03	7.3	0.9	4.4	G0 V		24 s		
12204	148144		1 59 30.9	-13 52 22	+0.001	+0.01	7.0	1.6	-0.5	M2 III	+29	310 s		
--	22816		1 59 33.7	+55 42 52	-0.002	-0.02	7.9	0.8	3.2	G5 IV		86 s		
12025	22817		1 59 35.1	+54 49 21	+0.004	+0.01	6.9	1.6		M5 III	+17			U Per, v
12452	--		1 59 35.3	-65 27 54			7.2	0.4	3.0	F2 V		69 s		
12139	75077		1 59 35.6	+21 03 30	+0.010	-0.02	5.87	1.03	0.2	K0 III	-2	110 mx		
11946	12076		1 59 37.9	+64 37 18	+0.005	-0.01	5.26	0.01		A0	+5		1571	m
11996	22813		1 59 38.4	+60 00 28	-0.008	0.00	7.29	1.79	-0.2	K3 III		160 s		

HD	SAO	Star Name	α 2000	δ 2000	μ(α)	μ(δ)	V	B-V	M_V	Spec	RV	d(pc)	ADS	Notes
12296	215726		1 59 38.7	-42 01 50	-0.005	-0.10	5.57	1.06	-0.2	gK3	+27	140 s		
12477	248476		1 59 41.0	-66 03 59	+0.004	+0.02	6.10	1.17	-0.1	K2 III		170 s		
12333	215729		1 59 41.2	-48 36 50	+0.026	-0.01	8.0	0.5	3.2	G5 IV		51 mx		
12284	193436		1 59 41.9	-35 25 10	-0.001	+0.04	7.5	0.6	2.6	F0 V		61 s		
12100	37685		1 59 43.7	+41 20 59	+0.004	-0.02	7.5	0.8	3.2	G5 IV		72 s		
12255	167466	57 Cet	1 59 46.0	-20 49 28	0.000	+0.02	5.41	1.65		Ma	-15			
12014	22820		1 59 50.8	+59 09 40	0.000	+0.01	7.58	1.97	-4.4	K0 Ib		1000 s		
12203	110264		1 59 58.3	+6 02 08	+0.004	-0.01	7.1	0.9	3.2	G5 IV		61 s		
12274	167471	59 υ Cet	2 00 00.2	-21 04 40	+0.009	-0.02	4.00	1.57	-0.5	M1 III	+18	79 s		
12254	129620		2 00 08.3	-4 17 01	+0.001	+0.02	8.0	1.1	0.2	K0 III		370 s		
12235	110266	112 Psc	2 00 09.1	+3 05 50	+0.015	-0.25	5.88	0.62		G0	-17	31 t		
12233	110268		2 00 12.7	+6 55 04	0.000	-0.02	7.8	1.4	-0.3	K5 III		410 s		
11978	12078		2 00 14.4	+67 22 56	+0.001	0.00	8.0	1.1	0.2	K0 III		320 s		
12262	129622		2 00 14.6	-3 22 07	-0.004	-0.02	6.9	0.1	0.6	A0 V		180 s		
12131	37690		2 00 14.9	+44 25 12	+0.002	-0.03	7.1	0.8	3.2	G5 IV		58 s		
12431	232600		2 00 16.9	-55 00 45	+0.008	+0.02	6.6	0.9	3.2	G5 IV		40 s		
12175	55176		2 00 17.4	+38 35 51	-0.001	-0.01	7.3	0.1	0.6	A0 V		210 s		
--	55178		2 00 18.8	+36 16 24	+0.001	+0.01	7.8	1.6						
12137	37692		2 00 21.5	+44 06 10	+0.004	0.00	7.6	0.9	3.2	G5 IV		77 s		
12252	110272		2 00 21.7	+4 23 23	-0.001	+0.02	7.1	0.9	3.2	G5 IV		61 s		
12249	92746		2 00 22.7	+10 37 49	0.000	+0.01	8.0	0.9	3.2	G5 IV		91 s		
12292	129624		2 00 26.8	-8 31 25	+0.006	-0.01	5.51	1.52		Mb	+6			m
12388	215733		2 00 28.1	-41 46 35	+0.001	-0.02	7.3	1.3	-0.1	K2 III		260 s		
12508	248478		2 00 30.2	-62 45 45	+0.004	+0.04	7.80	0.4	3.0	F2 V		91 s		m
12186	55182		2 00 30.3	+39 58 00	-0.001	-0.02	8.0	0.9	3.2	G5 IV		93 s		
12387	215734		2 00 32.0	-40 43 52	+0.039	-0.42	7.36	0.66	4.9	G3 V		20 mx		
11844	4541		2 00 38.2	+75 57 11	-0.004	0.00	7.4	1.1	0.0	K1 III		290 s		
13255	258287		2 00 38.6	-82 30 30	+0.013	0.00	7.9	0.1	2.1	A5 V		150 s		
12369	167481		2 00 42.4	-24 25 18	+0.002	-0.08	7.04	1.42		K2				
12201	37697		2 00 45.8	+41 21 39	-0.001	-0.02	7.7	0.9	3.2	G5 IV		80 s		
12210	55186		2 00 47.4	+39 37 31	-0.005	-0.05	8.0	0.5	4.0	F8 V		62 s		
12112	22835		2 00 48.5	+59 57 33	+0.007	-0.03	6.73	0.14	-7.5	A2 Ia		120 mx		
12328	129630		2 00 49.3	-8 28 19	+0.005	-0.06	7.1	0.9	3.2	G5 IV		62 s		
12231	55188		2 00 56.3	+38 41 03	-0.001	-0.02	8.0	1.1	0.2	K0 III		360 s		
12951	255839		2 00 56.5	-78 52 13	+0.035	-0.05	6.9	1.5	4.0	F8 V		38 s		
12247	55189		2 00 57.9	+34 16 54	+0.009	-0.01	8.0	0.9	3.2	G5 IV		93 s		
12273	75087		2 00 59.6	+29 14 17	+0.007	-0.02	7.9	0.5	3.4	F5 V		78 s		
12601	--		2 01 00.1	-65 07 00			7.9	1.1	-0.1	K2 III		400 s		
12356	129633		2 01 04.1	-2 29 45	+0.003	+0.02	7.9	0.7	4.4	G0 V		51 s		
12379	148153		2 01 05.4	-11 02 57	+0.005	+0.08	7.8	0.5	4.0	F8 V		56 s		
12634	--		2 01 09.1	-66 21 01			7.4	1.6	-0.5	M2 III		390 s		
12315	92750		2 01 09.7	+16 33 55	-0.003	-0.01	7.70	0.4	3.0	F2 V		87 s	1605	m
12306	75090		2 01 10.2	+24 26 38	-0.001	-0.03	7.4	0.9	3.2	G5 IV		70 s		
12245	37701		2 01 13.4	+41 12 44	0.000	0.00	6.9	1.1	0.2	K0 III		220 s		
12438	193455	π For	2 01 14.7	-30 00 06	-0.008	-0.11	5.35	0.88	0.3	G5 III	+24	100 s		
12161	12086		2 01 20.2	+60 30 43	-0.004	-0.01	7.88	0.27	0.5	A8 III		300 s	1599	m
12367	92755		2 01 30.4	+17 57 15	+0.003	-0.02	8.0	1.6	-0.5	M2 III		510 s		
12354	75093		2 01 33.2	+23 23 43	-0.007	-0.01	6.6	0.4	2.6	F0 V	-18	62 s		
12193	22848		2 01 33.4	+59 58 20	-0.001	-0.02	7.60	0.19	1.4	A2 V		150 s		
12524	215739	χ Phe	2 01 42.3	-44 42 49	-0.003	-0.04	5.14	1.49	-0.3	gK5	-31	120 s		
12680	248482		2 01 50.2	-63 37 22	+0.007	+0.04	6.9	0.6	4.0	F8 V		33 s		
12390	92756		2 01 50.3	+15 04 02	0.000	+0.01	6.7	1.1	0.2	K0 III		200 s		
12414	110286		2 01 52.5	+7 51 51	+0.009	-0.05	7.2	0.4	3.0	F2 V		69 s		
12853	255841		2 01 52.7	-74 26 45	+0.004	-0.01	7.0	2.1	-0.5	M4 III		160 s		
12480	148161		2 01 56.6	-16 34 09	+0.002	-0.01	7.8	1.1	-0.1	K2 III		380 s		
12111	4554	48 Cas	2 01 57.3	+70 54 26	-0.013	+0.01	4.48	0.17	1.9	A4 V	-5	33 ts	1598	m
12208	12090		2 01 58.9	+61 54 19	0.000	0.00	7.44	1.69	7.4	K5 V	-12	10 s		
12447	110291	113 α Psc	2 02 02.7	+2 45 49	+0.002	0.00	3.79	0.03	1.4	A2 V	+9	30 s	1615	Alrescha, m
12243	12094		2 02 06.5	+60 17 41	-0.001	-0.02	7.97	0.10	-0.6	B7 V		380 s		
12413	75101		2 02 07.1	+22 06 14	0.000	-0.06	7.3	0.5	3.4	F5 V		61 s		
12402	75100		2 02 07.7	+28 24 00	+0.004	0.00	6.7	1.1	0.0	K1 III	+16	220 s		
12013	4550		2 02 09.1	+75 30 08	+0.004	-0.01	6.70	0.1	0.6	A0 V	-43	160 s	1588	m
12586	215741		2 02 10.8	-45 50 07	+0.007	+0.01	7.8	0.1	3.4	F5 V		74 s		
12695	248484		2 02 12.2	-62 35 32	+0.010	+0.06	7.9	0.5	4.0	F8 V		60 s		
12619	215742		2 02 15.1	-49 24 33	-0.002	-0.01	7.4	1.5	-0.1	K2 III		210 s		
12303	22859	4 Per	2 02 18.0	+54 29 15	+0.004	0.00	5.04	-0.08	-0.2	B8 V	-2	110 s		
12352	37713		2 02 20.0	+46 50 33	+0.003	-0.01	7.1	0.0	0.0	B8.5 V		240 s		
12314	22860		2 02 21.9	+53 37 44	-0.002	0.00	7.9	0.1	0.6	A0 V		260 s		
12456	92761		2 02 24.9	+12 41 07	+0.006	-0.02	7.1	1.1	0.2	K0 III		160 mx		
12375	37720		2 02 25.1	+42 50 57	+0.002	0.00	7.7	1.1	0.2	K0 III		310 s		
12563	167511		2 02 28.0	-29 39 55	-0.003	+0.03	6.42	0.14		A3 p	+12			
12600	215744		2 02 30.2	-40 31 15	-0.003	-0.02	8.0	0.3	3.4	F5 V		82 s		
12671	232610		2 02 34.8	-52 31 58	-0.001	+0.04	8.0	0.9	1.7	A3 V		55 s		
12561	167512		2 02 35.1	-21 57 58	-0.001	0.00	6.8	-0.4	0.4	B9.5 V		190 s		
12479	92763		2 02 35.1	+13 28 36	+0.001	-0.01	5.94	1.59	-0.5	M4 III	-7	190 s		

HD	SAO	Star Name	α 2000	δ 2000	μ(α)	μ(δ)	V	B-V	M$_V$	Spec	RV	d(pc)	ADS	Notes
			h m s	° ′ ″	s	″								
12562	167513		2 02 35.3	−26 44 27	0.000	+0.04	7.6	1.6	0.2	K0 III		140 s		
12483	110295		2 02 35.4	+9 04 53	0.000	−0.02	7.8	0.8	3.2	G5 IV		84 s		
12851	255842		2 02 37.9	−70 25 11	−0.002	+0.02	6.9	1.2	3.2	G5 IV		29 s		
12160	4557		2 02 39.6	+71 17 54	+0.008	+0.03	7.7	1.6	−0.5	M2 III		220 mx		
12778	248487		2 02 41.6	−65 53 05	−0.005	0.00	7.8	1.1	−0.1	K2 III		380 s		
12596	167516		2 02 51.7	−23 53 12	−0.002	−0.03	6.4	1.4	0.2	K0 III		110 s		
12538	129649		2 02 52.3	−2 22 40	+0.005	−0.03	7.0	0.1	1.7	A3 V		61 mx		
12279	12095	52 Cas	2 02 52.6	+64 54 06	0.000	−0.01	6.00	0.03	1.4	A2 V	−25	83 s		
12005	4555		2 02 57.1	+77 54 59	0.000	0.00	6.04	1.14	0.2	K0 III	−3	120 s		
12512	92768		2 02 57.6	+11 01 08	+0.001	+0.01	6.8	1.1	0.2	K0 III		210 s		
12471	55218	3 ε Tri	2 02 57.9	+33 17 02	−0.001	−0.01	5.50	0.03	−0.2	A2 III	+3	140 s	1621	m
12712	232614		2 02 58.3	−53 02 10	+0.001	+0.01	7.5	0.0	1.4	A2 V		170 s		
12583	148170		2 02 58.4	−15 18 22	+0.001	0.00	5.86	0.98	3.2	G5 IV	+6	24 s		
12301	12097	53 Cas	2 03 00.2	+64 23 24	+0.001	0.00	5.58	0.38	−5.6	B8 Ib	−20	1000 s		
12536	110302		2 03 01.6	+3 21 30	+0.014	+0.15	7.1	0.6	4.4	G0 V		35 s		
12434	37730		2 03 03.9	+46 21 05	+0.001	−0.05	7.8	1.1	0.2	K0 III		300 s		
12424	37731		2 03 09.4	+48 37 15	+0.017	−0.02	7.90	0.7	4.4	G0 V		49 s	1620	m
12173	4559		2 03 10.4	+73 51 02	−0.006	0.00	6.30	0.1	1.7	A3 V	−5	80 s	1606	m
12573	129655	60 Cet	2 03 11.6	+0 07 42	+0.005	+0.03	5.43	0.15		A3 p				
12365	12100		2 03 11.7	+60 42 14	−0.001	−0.03	7.49	0.08	−1.6	B7 III		460 s		
12433	22877		2 03 25.2	+54 41 37	−0.005	−0.02	7.7	0.5	4.0	F8 V	−14	55 s		
12628	148173		2 03 25.5	−17 01 58	+0.006	+0.03	7.9	0.2	2.1	A5 V		110 mx		
12608	129661		2 03 25.8	−4 19 45	+0.001	+0.03	7.7	0.9	3.2	G5 IV		80 s		
12216	4560	50 Cas	2 03 26.0	+72 25 17	−0.009	+0.03	3.98	−0.01	1.2	A1 V	−14	36 s		
12453	22880		2 03 28.4	+51 39 08	0.000	+0.02	7.6	0.1	0.6	A0 V		240 s		
12704	193482		2 03 32.6	−37 12 01	+0.001	−0.04	7.9	0.5	2.6	F0 V		86 s		
12535	75112		2 03 32.8	+27 28 56	+0.006	0.00	7.40	1.1	−0.1	K2 III		210 mx		
12627	129663		2 03 35.1	−9 27 38	−0.002	−0.01	7.4	0.9	3.2	G5 IV		70 s		
12558	75114	10 Ari	2 03 39.3	+25 56 09	+0.010	+0.03	5.63	0.54	3.4	F5 V	+16	25 s	1631	m
12642	129665		2 03 40.5	−4 06 13	+0.001	−0.06	5.62	1.59	−5.9	cK5	+25	1400 s		
12594	92774		2 03 42.5	+18 15 12	−0.001	−0.02	6.21	1.42	−0.3	K5 III	+10	200 s		
12545	55233		2 03 47.0	+35 35 28	−0.005	−0.02	7.6	0.9	3.2	G5 IV		77 s		
12641	129667	61 Cet	2 03 48.1	−0 20 25	+0.005	−0.04	5.93	0.88	−0.9	G5 II-III	+24	61 mx	1634	m
12399	12102		2 03 49.2	+64 14 14	+0.003	−0.02	7.49	1.79	−8.0	G5 Ia		4000 s		
12546	55234		2 03 50.6	+33 41 53	+0.006	+0.01	7.9	0.4	3.0	F2 V		94 s		
12685	148178		2 03 53.9	−17 30 50	0.000	−0.02	7.2	1.4	−0.3	K5 III		320 s		
12533	37734	57 γ¹ And	2 03 53.9	+42 19 47	+0.004	−0.05	2.18	1.20	−0.1	K2 III	−12	37 mn	1630	Almach, m
12534	37735	57 γ² And	2 03 54.7	+42 19 51	+0.003	−0.05	5.03			A0 p	−14	37 mn	1630	m
12759	215754		2 03 55.4	−45 24 46	+0.033	+0.07	7.31	0.68		G0		10 mn		m
12640	110316		2 04 00.1	+8 05 41	+0.003	−0.05	7.4	0.5	3.4	F5 V		63 s		
12702	148180		2 04 12.4	−11 51 35	+0.003	+0.01	6.6	1.1	0.2	K0 III		190 s		
12518	22897		2 04 13.2	+51 58 07	+0.004	+0.02	6.66	−0.03	0.0	B8.5 V		170 mx		
12691	129675		2 04 14.6	−5 42 39	+0.001	+0.03	7.8	1.1	0.2	K0 III		330 s		
12482	12103		2 04 16.4	+60 14 39	−0.005	−0.08	7.39	0.50	2.2	F6 IV		84 mx		
12350	4563		2 04 19.0	+71 12 43	−0.006	0.00	7.7	0.4	2.6	F0 V	−9	100 s		
12638	75122		2 04 21.0	+25 55 13	+0.002	0.00	7.13	1.04	0.3	G8 III	−19	200 s		
12808	215758		2 04 24.5	−43 30 29	+0.005	−0.01	7.8	0.5	4.4	G0 V		49 s		
12767	167532	ν For	2 04 29.4	−29 17 49	+0.001	−0.01	4.69	−0.17	−0.6	A0 III	+19	110 s		
12745	148183		2 04 30.3	−16 13 00	−0.002	−0.05	7.5	0.5	3.4	F5 V		67 s		
12905	232622		2 04 30.6	−58 20 40	+0.001	+0.02	7.4	0.3	1.4	A2 V		110 s		
12661	75125		2 04 34.2	+25 24 52	−0.008	−0.17	7.44	0.72	0.2	K0 III		73 mx		
12894	232620		2 04 35.3	−54 52 54	+0.012	−0.02	6.6	0.7	3.0	F2 V		32 s		
12992	248489		2 04 35.6	−65 08 14	+0.019	+0.09	7.50	0.4	2.6	F0 V		96 s		m
12557	22901		2 04 39.8	+54 45 23	0.000	0.00	8.0	0.1	0.6	A0 V		280 s		
12468	12105		2 04 40.1	+65 06 13	+0.008	−0.03	6.5	0.1	0.6	A0 V	−4	85 mx		
12807	193493		2 04 41.1	−35 16 26	+0.002	−0.03	7.6	1.5	0.2	K0 III		160 s		
13713	258290		2 04 44.6	−82 18 00	+0.013	0.00	7.9	0.1	0.0	B8.5 V		170 mx		
12776	148188		2 04 46.7	−17 17 16	+0.003	−0.03	7.5	1.6	−0.5	M2 III		400 s		
12730	110325		2 04 51.0	+7 44 08	+0.003	0.00	6.5	1.1	0.2	K0 III		180 s		
12806	167538		2 04 55.1	−28 21 31	0.000	+0.01	7.7	0.9	0.2	K0 III		310 s		
12736	110326		2 05 00.4	+10 06 58	−0.001	−0.01	7.1	1.1	0.2	K0 III		240 s		
12509	12110		2 05 01.3	+64 23 08	−0.001	0.00	7.09	0.34	−4.4	B1 III	−17	940 s	1635	m
12591	22907		2 05 03.7	+55 37 20	+0.005	−0.04	6.6	0.4	2.6	F0 V		63 s		
12569	12116		2 05 06.7	+60 20 43	−0.002	−0.16	7.51	0.58	2.3	F7 IV		44 mx		
12915	215763		2 05 07.0	−49 40 56	+0.002	+0.02	7.80	0.2	2.1	A5 V		93 s		
12230	4562	47 Cas	2 05 07.2	+77 16 53	+0.037	−0.05	5.40	0.4	2.6	F0 V	−26	36 s		m
12835	167540		2 05 16.2	−28 33 31	+0.001	−0.04	7.34	1.49		K5				
12834	167541		2 05 16.6	−28 22 40	+0.004	+0.03	7.2	0.9	3.2	G5 IV		57 s		
12728	75133		2 05 16.8	+29 06 15	+0.002	0.00	7.90	1.1	0.0	K1 III		380 s		
12783	110329		2 05 19.0	+0 19 21	−0.001	−0.04	8.00	0.9	5.2	G5 V	+22	36 s		
12773	110328		2 05 23.2	+10 04 36	0.000	0.00	7.2	1.4	−0.3	K5 III		310 s		
12675	22914		2 05 26.9	+50 38 37	+0.001	−0.01	7.2	0.5	3.4	F5 V		55 s		
12568	12117		2 05 27.6	+62 57 20	+0.002	−0.01	8.00	0.6	−2.1	G1 II	+12	760 s		
12962	215765		2 05 29.1	−48 29 27	−0.002	−0.01	7.9	1.3	−0.1	K2 III		220 s		
12339	4565	49 Cas	2 05 31.2	+76 06 55	−0.005	−0.01	5.30	0.9	0.3	G8 III	0	99 s	1625	m

HD	SAO	Star Name	α 2000	δ 2000	μ(α)	μ(δ)	V	B-V	M$_v$	Spec	RV	d(pc)	ADS	Notes
12979	232629		2^{h}05^{m}35^s.2	-51°59'27"	+0^s.003	+0".04	7.3	0.4	3.0	F2 V		67 s		
13220	255847		2 05 41.3	-71 36 25	+0.021	+0.06	7.2	0.8	3.2	G5 IV		62 s		
12743	55257		2 05 46.2	+39 45 35	-0.004	-0.02	8.0	0.4	2.6	F0 V		120 s		
12772	75136		2 05 46.4	+25 08 42	-0.001	-0.01	8.0	0.9	3.2	G5 IV		93 s		
12805	92789		2 05 47.0	+13 20 33	+0.006	+0.01	7.80	0.5	3.4	F5 V		76 s		
12814	110332		2 05 48.4	+7 01 47	+0.005	+0.06	6.9	0.4	2.6	F0 V		73 s		
12825	110334		2 05 54.9	+6 40 39	-0.001	0.00	7.9	0.4	3.0	F2 V		94 s		
12623	12122		2 06 01.0	+63 10 22	+0.015	-0.01	7.52	1.17	5.9	K0 V		12 s		
12751	37763		2 06 01.7	+41 31 56	+0.001	-0.04	6.8	0.2	2.1	A5 V		85 s		
12441	4570		2 06 02.6	+74 35 00	-0.001	-0.01	7.5	0.1	0.6	A0 V	-11	240 s		
12901	148199		2 06 10.6	-10 16 33	+0.002	-0.01	6.7	0.4	2.6	F0 V		66 s		
12741	37765		2 06 10.9	+46 51 36	+0.008	-0.04	7.5	0.5	4.0	F8 V		50 s		
12740	37764		2 06 11.5	+49 09 23	+0.003	0.00	7.95	-0.05	-2.6	B2.5 IV		1100 s		
12872	110337		2 06 12.2	+8 14 51	-0.001	-0.03	6.31	1.65	-0.5	gM2	-26	220 s		
12986	193506		2 06 14.2	-39 32 21	-0.001	-0.02	7.2	1.4	-0.5	M2 III		350 s		
12909	129693		2 06 14.5	-4 21 54	-0.004	-0.01	7.3	0.4	3.0	F2 V		74 s		
12831	75141		2 06 16.0	+20 35 33	-0.004	0.00	7.7	0.5	4.0	F8 V		56 s		
12709	22919		2 06 16.3	+57 18 26	0.000	0.00	7.96	0.09	-2.6	B4 III	-20	910 s		
12944	148201		2 06 20.2	-15 26 48	+0.001	-0.03	7.1	1.1	0.2	K0 III		240 s		
13109	232634		2 06 28.3	-55 04 58	+0.004	+0.02	6.9	1.0	3.2	G5 IV		40 s		
12923	129695		2 06 29.3	+0 02 06	+0.004	-0.01	6.28	0.90		K0				
12886	92795		2 06 29.6	+13 27 43	+0.003	-0.01	7.2	1.1	0.2	K0 III		250 s		
12846	75144		2 06 29.9	+24 20 02	-0.002	-0.15	7.0	0.6	4.4	G0 V		34 s		
12975	167556		2 06 30.5	-23 34 01	+0.001	-0.01	7.7	1.6	0.2	K0 III		140 s		
12869	75146	12 κ Ari	2 06 33.8	+22 38 54	+0.001	-0.03	5.03	0.11		A m	+12			
13032	193509		2 06 36.7	-37 07 08	-0.003	-0.02	7.50	1.1	0.2	K0 III		290 s		m
12907	92798		2 06 39.5	+10 28 40	-0.004	0.00	7.70	0.5	4.0	F8 V		55 s		m
12899	92797		2 06 41.5	+14 35 02	-0.001	0.00	8.0	0.4	2.6	F0 V		120 s		
12868	75147		2 06 42.5	+25 49 51	0.000	-0.01	7.4	0.1	1.7	A3 V		140 s		
12942	110345		2 06 48.2	+1 26 32	+0.001	+0.02	8.0	1.1	0.2	K0 III		360 s		
12884	75148		2 06 49.1	+29 16 10	0.000	0.00	6.5	0.1	1.4	A2 V	-6	110 s		
12885	75149	11 Ari	2 06 49.2	+25 42 17	+0.001	-0.01	6.00	-0.03	-0.2	B8 V	-9	160 s	1658	m
13004	148210		2 06 51.9	-19 08 19	+0.009	-0.02	6.5	1.4	0.2	K0 III		110 s		
12897	75150		2 06 55.8	+26 19 25	+0.007	+0.03	7.3	1.1	0.0	K1 III		190 mx		
13263	248497		2 06 58.9	-65 56 45	+0.001	-0.02	6.7	0.6	0.2	K0 III		200 s		
13065	193513		2 06 59.8	-34 45 15	+0.001	-0.04	7.8	1.4	0.2	K0 III		200 s		
12984	129700		2 07 01.7	-4 13 04	-0.004	-0.02	7.6	0.4	2.6	F0 V		100 s		
12998	129701		2 07 02.3	-8 36 23	+0.002	-0.01	6.8	0.8	3.2	G5 IV		52 s		
12527	4575		2 07 09.1	+74 52 09	+0.008	-0.05	7.7	0.1	1.4	A2 V		150 mx		
12929	75151	13 α Ari	2 07 10.3	+23 27 45	+0.014	-0.14	2.00	1.15	-0.1	K2 III	-14	26 ts		Hamal
13242	248495		2 07 11.5	-60 59 00	+0.018	+0.11	7.0	1.2	3.4	F5 V		18 s		
12954	75153		2 07 20.9	+27 14 17	+0.003	+0.01	6.8	0.1	1.4	A2 V		120 s		
13083	167564		2 07 25.5	-28 53 47	+0.001	0.00	7.68	1.07	0.2	K0 III		280 s		
13246	232642		2 07 26.1	-59 40 48	+0.013	-0.03	7.60	0.5	3.4	F5 V		69 s		m
12312	342		2 07 32.0	+80 39 54	+0.006	-0.01	7.5	0.1	1.4	A2 V		170 s		
13043	129706		2 07 34.3	-0 37 04	-0.016	-0.35	6.91	0.61	4.7	G2 V	-40	28 ts		
12972	75154		2 07 37.4	+29 44 30	+0.004	0.00	8.0	0.1	1.7	A3 V		140 mx		
13018	92801		2 07 46.0	+18 01 45	0.000	+0.01	6.5	0.1	1.7	A3 V		92 s		
13042	110355		2 07 47.0	+5 59 06	0.000	-0.02	7.5	1.4	-0.3	K5 III		360 s		
12941	37787		2 07 49.9	+44 21 41	+0.001	-0.03	8.0	0.9	3.2	G5 IV		89 s		
12857	22948		2 07 50.8	+57 01 39	+0.005	+0.01	8.0	1.1	0.2	K0 III		320 s		
12983	55281		2 07 52.6	+35 33 13	-0.004	-0.03	7.5	0.4	2.6	F0 V		96 s		
12965	37789		2 07 56.6	+43 09 32	-0.002	-0.01	7.8	0.0	0.4	B9.5 V		300 s		
13057	92803		2 08 03.8	+15 48 17	+0.002	-0.02	7.5	1.1	0.2	K0 III		280 s		
12715	4583		2 08 04.8	+71 50 26	+0.007	-0.02	8.0	1.1	0.2	K0 III		320 s		
13218	215777		2 08 05.6	-41 52 50	-0.001	-0.01	6.8	-0.1	1.4	A2 V		120 s		
13168	167576		2 08 08.7	-27 34 18	+0.005	-0.01	7.2	0.7	2.1	A5 V		54 s		
13307	232645		2 08 14.2	-55 31 34	+0.002	-0.02	7.70	1.1	0.2	K0 III		320 s		m
13055	75159		2 08 16.9	+24 49 09	+0.001	+0.02	7.6	1.6	-0.5	M2 III		420 s		
13091	110359		2 08 19.3	+8 50 59	+0.001	-0.02	7.9	0.1	0.6	A0 V		280 s		
13141	148231		2 08 19.7	-10 02 29	+0.003	+0.01	6.9	0.1	1.7	A3 V		110 s		
13014	37791		2 08 26.0	+43 11 27	+0.008	-0.06	7.6	0.5	3.4	F5 V		71 s		
13139	129714		2 08 27.2	-0 25 11	+0.001	+0.01	7.7	1.1	-0.1	K2 III		360 s		
13041	55289	58 And	2 08 29.2	+37 51 33	+0.013	-0.04	4.82	0.12	1.9	A4 V	+8	38 s		
12928	22956		2 08 30.8	+58 51 55	0.000	0.00	7.84	0.20	-3.6	B7 II		1500 s		
13155	129717		2 08 31.8	-6 40 42	+0.001	0.00	6.6	0.8	3.2	G5 IV		47 s		
13013	37794		2 08 33.5	+44 27 34	+0.001	-0.05	6.4	0.9	0.3	G8 III	+24	160 s		
12953	22959		2 08 40.4	+58 25 26	-0.001	+0.01	5.67	0.61	-7.3	A1 Ia	-36	1900 s		
13179	129719		2 08 42.3	-0 26 13	-0.004	-0.01	7.9	0.4	2.6	F0 V		110 s	1673	m
12882	12145		2 08 45.2	+65 02 15	-0.002	0.00	7.51	0.38	-7.1	B6 Ia		4900 s		
13215	148237		2 08 45.6	-17 46 46	-0.001	-0.03	6.1	1.6	-0.5	M2 III		210 s		
13279	193533		2 08 49.0	-35 49 24	+0.004	+0.03	7.8	1.0	0.2	K0 III		320 s		
13071	55295		2 08 52.8	+39 20 55	-0.001	+0.01	7.8	0.1	1.4	A2 V		190 s		
13239	148238		2 08 55.4	-16 00 36	+0.003	+0.01	7.9	0.4	2.6	F0 V		110 s		
12555	4579		2 08 56.2	+79 20 46	+0.001	+0.01	7.1	0.4	2.6	F0 V		80 s		

HD	SAO	Star Name	α 2000	δ 2000	μ(α)	μ(δ)	V	B–V	M_v	Spec	RV	d(pc)	ADS	Notes
13278	193534		2^{h}08^{m}56^s.8	−31°44′44″	0^s.000	−0″.01	7.0	1.3	0.2	K0 III		160 s		
12971	22963		2 08 58.8	+60 11 38	−0.003	−0.04	7.97	0.13	1.4	A2 V		190 s		
13077	37804		2 09 03.4	+41 57 53	−0.002	−0.03	6.9	0.0	0.4	B9.5 V		200 s		
13260	167589		2 09 04.8	−22 59 25	+0.002	0.00	7.6	1.5	−0.1	K2 III		230 s		
12800	4588		2 09 08.2	+71 33 08	+0.064	−0.23	6.8	0.5	4.0	F8 V	−1	36 s		
13228	129722		2 09 09.0	−2 19 54	−0.007	−0.04	7.1	0.5	4.0	F8 V	−11	42 s		
13336	215785		2 09 09.2	−43 31 00	−0.005	−0.05	5.7	0.8	0.2	K0 III		130 s		
13076	37806		2 09 11.3	+45 50 25	−0.001	−0.01	7.7	0.5	3.4	F5 V		71 s		
13300	193537		2 09 12.3	−31 48 39	0.000	−0.07	7.7	1.7	−0.3	K5 III		210 mx		
13276	148240		2 09 17.0	−19 34 13	+0.002	0.00	7.3	1.2	0.2	K0 III		210 s		
13227	110371		2 09 22.2	+5 59 10	−0.001	+0.04	7.1	0.4	2.6	F0 V		79 s		
13201	92810		2 09 23.1	+17 13 27	+0.010	−0.18	6.4	0.5	3.4	F5 V	+11	41 s		
12467	344		2 09 25.2	+81 17 45	−0.015	+0.01	6.0	0.1	0.6	A0 V	−13	120 s		
13174	75171	14 Ari	2 09 25.3	+25 56 24	+0.006	−0.03	4.98	0.33	0.6	F2 III	+1	75 s		m
13070	22975		2 09 25.3	+53 14 35	−0.002	+0.02	7.9	1.1	0.2	K0 III		310 s		
13161	55306	4 β Tri	2 09 32.5	+34 59 14	+0.012	−0.04	3.00	0.14	0.3	A5 III	+10	35 s		
13388	215789		2 09 33.7	−45 28 02	+0.004	−0.04	7.3	−0.2	2.6	F0 V		85 s		
13305	167599		2 09 34.7	−24 20 44	−0.003	−0.02	6.5	0.8	2.6	F0 V		30 s		
13426	232655		2 09 36.6	−53 16 34	0.000	+0.03	7.6	1.2	3.2	G5 IV		42 s		
13304	148244		2 09 37.1	−16 07 35	+0.004	−0.03	7.7	0.9	3.2	G5 IV		81 s		
13160	55308		2 09 37.4	+37 06 58	+0.001	−0.01	7.9	0.1	0.6	A0 V		280 s		
13397	215792		2 09 38.1	−48 35 58	+0.005	0.00	7.7	0.8	3.2	G5 IV		80 s		
13151	37817		2 09 38.4	+42 51 30	+0.014	−0.09	7.20	0.5	3.4	F5 V		58 s	1675	m
13189	55309		2 09 40.1	+32 18 59	0.000	+0.01	7.8	1.1	−0.1	K2 III		380 s		
13249	110374		2 09 40.2	+4 13 56	+0.007	−0.01	7.5	0.9	3.2	G5 IV		72 s		
13138	37819		2 09 40.2	+42 00 28	+0.003	−0.03	7.75	−0.06	0.4	B9.5 V		300 s		
13335	167600		2 09 42.8	−29 00 27	0.000	−0.02	7.44	1.59	−0.1	K2 III		150 s		
13424	232656		2 09 45.2	−51 43 49	+0.018	+0.07	7.2	0.3	0.2	K0 III		100 mx		
13248	92812		2 09 46.4	+13 10 33	−0.001	0.00	7.6	0.0	0.4	B9.5 V		280 s		
13315	148247		2 09 46.5	−14 53 06	0.000	−0.02	7.7	1.1	0.2	K0 III		320 s		
13387	215793		2 09 48.4	−40 51 59	+0.004	−0.02	6.7	0.0	3.0	F2 V		56 s		
13285	110378		2 09 51.2	+3 46 08	+0.001	−0.01	6.9	0.1	1.4	A2 V		130 s		
13538	248502		2 09 53.8	−62 15 59	+0.008	+0.01	7.9	0.6	4.4	G0 V		50 s		
13211	55314		2 09 54.5	+32 32 57	−0.001	−0.01	7.50	0.9	3.2	G5 IV		72 s		m
13423	215794		2 10 04.8	−43 48 56	+0.005	0.00	6.32	0.90	5.9	dK0		11 s		
13137	22993		2 10 07.7	+53 50 35	+0.003	−0.04	6.31	0.95	3.2	G5 IV	+10	33 s		
13578	248504		2 10 08.4	−64 21 19	+0.057	+0.01	7.49	0.62		G0		18 mn		m
13570	248505		2 10 15.5	−61 05 50	+0.002	0.00	7.6	1.9	−0.1	K2 III		130 s		
13136	22994		2 10 15.7	+56 33 35	0.000	+0.02	7.75	2.25	−4.8	M2 Ib		1100 s		KK Per, v
13247	55321		2 10 16.0	+33 21 55	0.000	0.00	7.76	0.01	0.6	A0 V		270 s	1681	m
13149	22997		2 10 23.9	+56 17 50	+0.005	−0.03	7.60	1.1	0.2	K0 III	+10	270 s		m
13928	255857		2 10 24.2	−76 37 26	+0.033	+0.01	6.7	0.6	2.6	F0 V		45 s		
13122	22995		2 10 25.4	+59 58 48	+0.006	−0.05	6.65	0.34	−2.0	F5 II		86 mx		
13445	232658		2 10 25.6	−50 49 28	+0.224	+0.65	6.12	0.82	6.5	K0 V	+50	11 t		
12543	345		2 10 30.1	+81 28 55	+0.017	−0.01	6.90	0.1	1.7	A3 V		110 s	1659	m
13490	215798		2 10 30.3	−45 26 50	+0.006	+0.01	7.2	0.9	3.2	G5 IV		53 s		
13422	193551		2 10 30.7	−30 47 43	+0.001	−0.06	8.0	1.5	0.2	K0 III		190 s		
13345	110382		2 10 32.4	+6 05 55	+0.008	+0.01	7.4	0.6	4.4	G0 V		41 s		
13515	215800		2 10 33.2	−48 05 41	+0.004	−0.01	7.8	1.0	3.2	G5 IV		55 s		
13461	193552		2 10 35.2	−35 30 37	−0.002	0.00	7.6	1.4	0.2	K0 III		170 s		
13325	92822	15 Ari	2 10 37.5	+19 30 01	+0.006	−0.02	5.70	1.65	−0.5	M2 III	+61	160 s		
13462	193553		2 10 40.0	−38 22 04	0.000	−0.06	7.7	1.8	−0.1	K2 III		150 s		
13223	23006		2 10 40.2	+51 03 16	+0.002	−0.02	7.5	0.1	0.6	A0 V		220 s		
13435	167613		2 10 41.6	−28 13 09	+0.007	0.00	7.06	1.02	6.3	K2 V		12 s		
13406	148253		2 10 45.5	−12 55 37	0.000	−0.01	7.43	0.32	0.6	A0 V		150 s		
13294	55330	59 And	2 10 52.7	+39 02 22	−0.001	−0.01	5.63	−0.02	0.4	B9.5 V	+1	110 s	1683	m
13951	255858		2 10 53.4	−75 57 33	+0.003	0.00	6.9	1.2	3.2	G5 IV		28 s		
13344	75183		2 10 54.3	+23 47 57	+0.002	+0.01	7.3	0.4	2.6	F0 V		86 s		
13442	148257		2 11 03.9	−15 04 14	−0.002	−0.02	6.73	0.07	1.7	A3 V		100 s		
13459	148259		2 11 08.4	−19 18 27	−0.002	+0.01	7.4	1.4	0.2	K0 III		160 s		
13668	232665		2 11 09.4	−59 09 08	+0.003	0.00	7.4	1.4	0.2	K0 III		160 s		
13457	148258		2 11 10.1	−14 42 08	−0.001	−0.10	7.8	0.5	4.0	F8 V		57 s		
14024	255861		2 11 11.6	−76 17 57	+0.006	−0.02	8.0	0.5	3.4	F5 V		82 s		
13363	75188	16 Ari	2 11 12.0	+25 56 13	−0.001	0.00	6.02	1.36	−0.1	K2 III	−19	130 s		
13588	215805		2 11 16.7	−46 35 07	+0.007	−0.01	7.93	0.18		A m				
13950	255859		2 11 17.9	−74 30 05	+0.007	+0.01	7.2	1.1	0.2	K0 III		260 s		
13421	110390	64 Cet	2 11 21.0	+8 34 12	−0.010	−0.11	5.63	0.56	4.0	dF8	−18	19 s		
13456	148262		2 11 22.2	−10 03 08	−0.002	−0.17	6.01	0.39	3.0	dF2	+11	38 s		
13382	75191		2 11 23.2	+21 22 38	+0.021	−0.01	7.2	0.7	4.4	G0 V		37 s		
13372	55338	5 Tri	2 11 25.0	+31 31 35	+0.003	−0.01	6.20	0.11		A m	+11			
13355	55337		2 11 26.3	+37 21 08	0.000	0.00	7.9	0.0	0.0	B8.5 V		380 s		
13236	12163		2 11 27.4	+61 12 23	+0.002	+0.02	8.0	1.1	0.2	K0 III		320 s		
13267	23011	5 Per	2 11 28.9	+57 38 45	−0.002	+0.01	6.36	0.33	−7.0	B5 Ia	−34	2700 s	1685	m
−−	23013		2 11 29.3	+56 07 45	0.000	+0.01	7.4	1.5	0.0	K1 III		180 s		
13323	37844		2 11 31.9	+47 12 33	+0.002	+0.01	8.0	0.0	0.0	B8.5 V		350 s		

HD	SAO	Star Name	α 2000	δ 2000	μ(α)	μ(δ)	V	B-V	M$_V$	Spec	RV	d(pc)	ADS	Notes
13523	148265		2^{h}11^{m}32.4^s	-18°58'34"	0$.^s$000	+0$.''$02	6.8	1.1	3.2	G5 IV		32 s		
13511	148264		2 11 33.5	-17 44 32	+0.004	+0.04	6.7	1.4	-0.3	K5 III		250 s		
13468	129739	63 Cet	2 11 35.8	-1 49 32	-0.001	-0.03	5.93	0.97	0.2	gG9	+32	140 s		
13467	110395		2 11 43.3	+3 27 11	+0.001	+0.01	6.7	0.2	2.1	A5 V		84 s		
13696	232670		2 11 48.2	-54 02 56	+0.001	0.00	7.7	0.7	2.1	A5 V		60 s		
13666	215812		2 11 53.9	-47 10 16	-0.005	-0.06	7.4	0.3	0.2	K0 III		270 s		
13907	255860		2 11 55.3	-70 57 04	+0.002	0.00	7.10	0.8	3.2	G5 IV		60 s		m
13354	37848		2 11 55.4	+49 35 49	-0.001	-0.02	7.9	0.5	3.4	F5 V		75 s		
13616	193568		2 12 01.2	-32 19 46	+0.003	0.00	8.00	0.4	2.6	F0 V		120 s		m
13615	--		2 12 03.8	-32 16 44			8.00	0.6	4.4	G0 V		53 s		m
13568	148270		2 12 04.2	-17 43 45	+0.002	-0.03	7.4	1.1	0.2	K0 III		280 s		
13415	37853		2 12 09.6	+42 08 55	0.000	+0.01	8.0	1.1	0.2	K0 III		360 s		
13614	167629		2 12 10.7	-27 30 42	+0.001	+0.03	7.6	1.2	0.2	K0 III		240 s		
13482	75199		2 12 15.3	+23 57 31	+0.009	-0.15	7.80	0.8	3.2	G5 IV		58 mx	1696	m
13546	110403		2 12 15.6	+2 44 43	+0.001	+0.02	6.7	0.9	3.2	G5 IV		50 s		
13480	55347	6 Tri	2 12 22.2	+30 18 11	-0.005	-0.06	4.94	0.78	0.3	G5 III	-18	85 s	1697	m
12881	4592		2 12 29.8	+79 41 38	-0.002	+0.04	7.17	0.29		A m				m
13652	167634		2 12 34.3	-26 19 22	-0.003	-0.04	7.92	1.08	0.2	K0 III		310 s		
13522	75203		2 12 37.5	+24 10 04	+0.003	0.00	5.96	1.37	0.2	K0 III	-1	85 s		
13926	248512		2 12 42.0	-66 09 16	-0.004	-0.04	7.30	1.1	0.2	K0 III		260 s		m
13567	92837		2 12 42.5	+13 55 06	0.000	-0.01	7.3	1.1	0.2	K0 III		270 s		
13755	215818		2 12 46.9	-44 29 22	0.000	0.00	7.8	0.7	2.1	A5 V		74 s		
13612	129752	66 Cet	2 12 47.4	-2 23 37	+0.024	-0.06	5.54	0.57	4.2	dF9	-3	22 ts	1703	m
13555	75204	17 η Ari	2 12 48.0	+21 12 39	+0.011	+0.01	5.27	0.43	3.4	F5 V	+6	24 s		
12927	4594		2 12 50.1	+79 41 30	-0.005	+0.03	6.47	0.23	0.3	A5 III		160 s		m
13723	193574		2 12 52.0	-34 32 20	+0.005	-0.01	7.1	1.4	0.2	K0 III		140 s		
13566	92838		2 12 53.5	+19 49 16	+0.001	-0.01	8.0	1.1	0.2	K0 III		360 s		
13453	37862		2 12 53.5	+48 34 18	0.000	0.00	7.5	1.4	-0.3	K5 III		330 s		
13709	193573	μ For	2 12 54.4	-30 43 26	+0.001	-0.01	5.28	-0.02	1.2	A1 V	+17	27 ts		
13507	37865		2 12 54.7	+40 40 04	+0.003	-0.12	7.4	0.6	4.4	G0 V		39 s		
13403	23033		2 12 56.0	+57 12 17	+0.033	-0.21	7.02	0.62	4.7	G2 V	-34	29 s		m
13611	110408	65 ξ¹ Cet	2 12 59.9	+8 50 48	-0.002	0.00	4.37	0.89	-2.1	G8 II	-4	200 s		
13692	167637		2 13 00.8	-21 00 01	+0.003	+0.04	5.86	1.01	4.4	gG6	+38	20 s		d?
13596	92841	19 Ari	2 13 03.2	+15 16 47	+0.007	-0.02	5.71	1.55	-0.4	M0 III	+23	170 s		
13477	37864		2 13 04.8	+49 23 02	-0.001	+0.02	7.8	1.1	0.2	K0 III		300 s		
13412	23034		2 13 06.6	+58 47 58	-0.002	+0.01	7.95	0.26	0.6	A9 III m		260 s		
13565	55358		2 13 09.6	+30 33 33	0.000	+0.02	7.8	0.9	0.3	G4 III	+17	310 s		
13520	37867	60 And	2 13 13.2	+44 13 54	-0.002	-0.01	4.83	1.48	-0.3	K4 III	-46	97 s		
13531	37868		2 13 13.3	+40 30 28	+0.006	-0.08	7.5	0.7	4.4	G0 V		42 s		
13730	167641		2 13 13.7	-22 52 18	-0.001	0.00	7.21	0.97	3.2	G8 IV		56 s		
13778	193580		2 13 17.7	-36 03 40	+0.001	-0.01	7.5	1.3	0.2	K0 III		200 s		
13722	148274		2 13 19.1	-14 54 27	+0.001	0.00	8.0	0.4	2.6	F0 V		120 s		
13320	12171		2 13 20.8	+69 22 02	+0.004	-0.01	8.0	0.1	0.6	A0 V		230 mx		
13222	4599		2 13 21.2	+74 01 40	+0.013	-0.03	6.29	0.91	3.2	G5 IV	-37	34 s		
--	110414		2 13 26.9	+4 12 18	+0.002	-0.01	5.6	1.1	0.2	K0 III		120 s		
13437	23041		2 13 28.6	+59 11 45	+0.001	-0.03	7.60	1.18	-2.1	G5 II		650 s		
13564	37871		2 13 29.3	+40 47 24	+0.003	-0.03	7.6	0.1	1.7	A3 V		150 s		
13683	110415		2 13 29.6	+5 00 47	+0.004	-0.04	6.6	0.4	2.6	F0 V	+1	62 s		
13794	193583		2 13 30.5	-33 20 08	+0.003	0.00	6.9	1.3	0.2	K0 III		150 s		
13519	23046		2 13 33.2	+53 03 31	-0.002	+0.01	6.8	1.4	-0.3	K5 III		240 s		
13765	167645		2 13 36.0	-24 47 24	0.000	+0.06	7.8	0.6	3.4	F5 V		59 s		
13530	23047		2 13 36.3	+51 03 57	+0.037	-0.17	5.31	0.93	0.2	K0 III	+27	51 mx		m
13649	75214		2 13 41.3	+25 35 46	0.000	-0.07	7.1	0.5	3.4	F5 V		55 s		
13476	23044		2 13 41.4	+58 33 41	-0.001	+0.03	6.44	0.60	-6.8	A3 Iab	-41	2200 s		
13729	129762		2 13 42.1	-3 01 56	+0.001	-0.02	7.30							m
13427	12175		2 13 47.8	+64 51 17	0.000	-0.02	7.4	0.1	1.4	A2 V		150 s		
13682	92846		2 13 51.7	+19 36 53	+0.002	+0.02	7.1	1.1	0.2	K0 III		240 s		
13763	129765		2 13 53.4	-9 03 52	-0.001	+0.03	6.6	1.1	0.2	K0 III		190 s		
13691	75215		2 14 02.1	+26 37 36	+0.001	+0.01	7.3	1.1	0.0	K1 III	-9	280 s		
13594	37878		2 14 02.5	+47 29 03	-0.006	-0.05	6.06	0.40	3.4	F5 V	-8	34 s	1709	m
13504	12179		2 14 06.3	+61 41 05	-0.006	-0.02	7.6	0.5	0.7	F6 III		220 s		
13739	92847		2 14 12.5	+12 17 02	-0.002	-0.01	7.6	0.1	0.6	A0 V		260 s		
14090	248515		2 14 13.6	-63 47 46	0.000	+0.01	7.5	0.8	3.2	G5 IV		71 s		
13607	37882		2 14 14.6	+49 05 29	+0.003	-0.03	8.0	1.1	0.2	K0 III		320 s		
14141	248518	π¹ Hyi	2 14 14.8	-67 50 30	+0.007	+0.04	5.55	1.55		gMa	+26	20 mn		
13449	12177		2 14 17.3	+66 43 42	+0.020	-0.03	7.7	1.1	0.2	K0 III	-28	85 mx		
14401	255864		2 14 19.1	-76 21 28	+0.006	+0.02	6.5	1.1	3.2	G5 IV		31 s		
13847	167651		2 14 19.5	-27 54 26	+0.000	-0.06	7.7	0.2	3.0	F2 V		86 s		
13704	55373		2 14 20.0	+30 43 29	-0.004	-0.04	7.3	0.4	2.6	F0 V		89 s		
13804	129770		2 14 21.9	-5 16 12	-0.002	-0.03	8.0	0.4	3.0	F2 V		98 s		
13474	12180	55 Cas	2 14 29.0	+66 31 29	-0.001	0.00	6.2	0.1	1.4	A2 V	-12	86 s		
13648	37886		2 14 30.4	+49 05 05	+0.001	-0.01	8.0	1.1	-0.1	K2 III		360 s		
13940	215831		2 14 31.8	-41 10 00	-0.002	-0.02	5.91	0.97	0.2	G9 III		140 s		
13803	110422		2 14 32.4	+6 39 04	+0.003	-0.01	8.0	1.1	0.2	K0 III		360 s		
13819	110424		2 14 36.8	+1 40 38	-0.002	-0.01	7.5	1.1	0.2	K0 III		280 s		

HD	SAO	Star Name	α 2000	δ 2000	μ(α)	μ(δ)	V	B-V	M_v	Spec	RV	d(pc)	ADS	Notes
13747	75223		2 14 37.9	+28 41 28	+0.012	-0.10	6.5	1.1	0.2	K0 III	+15	80 mx		
13746	55383		2 14 42.0	+30 23 42	0.000	0.00	8.00	0.00	0.0	B8.5 V		400 s	1723	m
13679	37889		2 14 46.3	+46 41 15	+0.002	-0.02	6.7	0.0	0.0	B8.5 V		200 s		
13946	193597		2 14 47.8	-37 32 00	-0.002	-0.04	7.5	1.7	0.2	K0 III		120 s		
13690	37891		2 14 52.4	+48 46 57	0.000	-0.01	6.9	0.8	3.2	G5 IV		54 s		
13661	23073		2 14 53.0	+54 31 54	-0.001	0.00	7.79	-0.01	0.0	B8.5 V	-50	360 s		
13540	12185		2 14 55.1	+66 19 16	-0.006	0.00	7.4	0.5	3.4	F5 V		61 s		
13633	23071		2 14 57.0	+58 29 25	-0.001	-0.01	7.85	0.17	-1.9	B6 III		560 s	1717	m
13801	75228		2 14 57.1	+22 37 44	-0.002	-0.05	7.9	0.4	2.6	F0 V		120 s		
13920	167654		2 15 02.0	-23 32 25	+0.003	-0.04	7.2	0.6	4.0	F8 V		40 s		
13669	23079		2 15 02.5	+55 47 36	0.000	-0.01	7.90	0.35		B2 e				
13678	23085		2 15 07.7	+55 05 19	+0.001	+0.03	7.1	1.1	0.2	K0 III	-8	220 s		
13885	110428		2 15 09.5	+0 43 12	+0.001	-0.01	6.8	0.1	1.4	A2 V		120 s		
13992	193600		2 15 11.3	-35 07 05	+0.016	0.00	7.9	0.9	3.4	F5 V		41 s		
13917	148290		2 15 11.9	-12 03 09	0.000	-0.03	7.81	0.12	0.6	A0 V		230 s		
13590	12188		2 15 12.9	+64 01 30	-0.001	+0.03	7.90	0.35	-3.6	B2 III		940 s		
14004	193604		2 15 13.8	-36 19 14	+0.007	+0.02	7.9	1.3	0.2	K0 III		230 s		
13825	75231		2 15 24.3	+24 16 18	+0.033	-0.18	6.81	0.69	5.5	dG7	-2	18 s		
14112	232686		2 15 24.7	-50 53 12	-0.002	-0.03	7.6	1.3	3.2	G5 IV		36 s		
14057	215838		2 15 25.0	-41 04 08	0.000	+0.03	6.97	1.62		K5				
13717	23091		2 15 26.8	+55 35 44	-0.002	+0.02	7.86	0.11	-2.8	A0 II	-44	1400 s		
13936	129781		2 15 28.2	-9 27 56	-0.001	+0.03	6.55	-0.02		A0				
14287	248521	π² Hyi	2 15 28.6	-67 44 47	+0.007	-0.01	5.69	1.30	-0.3	gK5	+17	160 s		
13738	23093		2 15 28.7	+52 30 41	-0.001	0.00	7.2	1.1	-0.1	K2 III	-66	260 s		
13884	92859		2 15 35.1	+16 11 43	+0.002	0.00	8.0	1.4	-0.3	K5 III		450 s		
14021	167665		2 15 37.8	-26 31 25	-0.002	-0.01	7.6	0.7	3.2	G5 IV		77 s		
13872	75238	21 Ari	2 15 42.7	+25 02 35	-0.007	-0.08	5.58	0.49	3.4	F5 V	-44	26 s		m
13579	12190		2 15 42.8	+67 40 20	+0.093	-0.30	7.18	0.90	6.3	K2 V	-14	22 ts		m
14151	232689		2 15 43.6	-52 12 24	+0.009	+0.06	7.2	0.1	4.0	F8 V		43 s		
14733	--		2 15 44.3	-79 11 23			7.9	1.4	0.2	K0 III		190 s		
13745	23099		2 15 45.8	+55 59 47	0.000	0.00	7.87	0.17	-5.0	B0 III	-30	2200 s		
13871	75239	20 Ari	2 15 46.0	+25 46 59	+0.013	-0.06	5.79	0.44	3.0	dF2	+26	32 s		
14001	148298		2 15 46.0	-18 14 24	-0.005	-0.18	7.91	1.02	0.2	K0 III	+5	120 mx	1733	m
13785	37902		2 15 47.6	+49 42 05	-0.003	+0.02	8.0	0.9	3.2	G5 IV		89 s		
14181	232690		2 15 50.8	-55 16 46	0.000	-0.02	7.6	0.5	2.6	F0 V		75 s		
14355	248522		2 15 51.8	-69 25 56	+0.031	+0.06	7.6	0.9	3.2	G5 IV		67 s		
13857	55396		2 15 54.4	+36 46 13	+0.001	-0.03	7.9	0.2	2.1	A5 V		150 s		
13686	12195		2 15 55.1	+63 14 13	-0.003	+0.02	7.01	1.87	-4.4	K3 Ib	+3	860 s		
13869	55397	7 Tri	2 15 56.2	+33 21 32	-0.001	-0.03	5.28	-0.01	0.4	B9.5 V	-1	89 s		
13283	4601		2 15 56.8	+79 10 46	-0.001	+0.01	7.1	0.1	1.4	A2 V		130 s		
13818	37905		2 15 57.8	+47 48 42	+0.006	-0.07	6.4	1.1	0.2	K0 III	+16	160 s		
13744	23101		2 15 58.5	+58 17 38	-0.002	+0.01	7.59	0.74	-6.6	A0 Iab	-52	2600 s		
14119	193611		2 16 01.7	-36 28 20	+0.008	-0.13	7.5	0.8	3.4	F5 V		38 s		
13843	37908		2 16 01.8	+42 41 49	0.000	0.00	7.6	0.0	0.4	B9.5 V		280 s		
14196	232692		2 16 02.8	-52 53 22	-0.002	-0.01	7.3	0.3	2.6	F0 V		87 s		
--	37906		2 16 04.4	+48 28 57	-0.004	-0.01	8.0	1.6		M5 III	+114			R Ari, v
13913	75245		2 16 07.1	+25 03 24	+0.002	-0.01	7.2	1.6						
14099	167672		2 16 07.6	-28 35 02	0.000	+0.03	7.9	1.2	0.2	K0 III		270 s		
14044	148304		2 16 15.7	-9 49 17	+0.017	-0.04	7.30	0.6	4.4	G0 V	+32	38 s		m
13933	55406		2 16 22.1	+31 02 37	+0.005	0.00	7.9	0.4	2.6	F0 V		120 s		
13997	92865		2 16 27.6	+12 22 47	+0.015	-0.23	8.01	0.80	3.2	G5 IV		33 mx		
14042	110441		2 16 28.2	+2 14 16	+0.002	-0.01	8.0	0.9	3.2	G5 IV		93 s		
14180	215844		2 16 29.4	-43 15 42	+0.004	-0.02	7.5	1.2	3.2	G5 IV		40 s		
14228	232696	φ Eri	2 16 30.6	-51 30 44	+0.010	-0.02	3.56	-0.12	-0.2	B8 V	+10	37 mx		m
13868	37913		2 16 32.1	+48 52 43	-0.001	0.00	7.6	0.9	3.2	G5 IV		74 s		
13867	37915		2 16 36.0	+49 49 11	+0.002	0.00	7.57	0.02	-1.1	B5 V e	0	450 s		
13855	23112		2 16 37.5	+52 43 19	-0.001	+0.01	8.0	0.4	3.0	F2 V		94 s		
14207	215846		2 16 38.5	-42 55 35	+0.002	+0.06	7.8	1.6	0.2	K0 III		150 s		
13725	12199		2 16 41.8	+67 17 01	+0.002	-0.02	7.01	1.86	-2.3	K4 II		410 s		
14379	248526		2 16 44.3	-63 23 33	+0.004	+0.06	7.6	0.5	3.4	F5 V		68 s		
12918	359		2 16 45.6	+83 33 41	+0.036	-0.04	6.8	1.1	0.2	K0 III		210 s		
13841	23113		2 16 46.3	+57 01 46	0.000	0.00	7.39	0.23	-5.7	B2 Ib	-39	2400 s		
13931	37918		2 16 47.4	+43 46 22	+0.010	-0.19	7.7	0.6	4.4	G0 V		46 s		
13882	23118		2 16 49.2	+51 31 44	0.000	0.00	7.4	0.9	3.2	G5 IV		66 s		
13854	23115		2 16 51.7	+57 03 19	0.000	0.00	6.48	0.28	-6.2	B1 Iab	-40	1800 s		m
13866	23117		2 16 57.0	+56 43 09	-0.005	+0.02	7.50	0.19	-5.7	B2 Ib	-47	2700 s		
14115	129797		2 16 57.2	-6 34 42	0.000	0.00	7.1	0.1	0.6	A0 V		200 s		
14129	129798	67 Cet	2 16 59.0	-6 25 20	+0.006	-0.10	5.51	0.96	0.3	G8 III	+7	70 mx		
13974	55420	8 δ Tri	2 17 03.2	+34 13 27	+0.093	-0.24	4.87	0.61	4.4	G0 V	-6	10 mx	1739	m
13955	37920		2 17 07.5	+45 50 19	0.000	+0.01	7.6	0.1	0.6	A0 V		230 s		
14067	75262		2 17 10.3	+23 46 05	-0.003	-0.03	6.4	0.9	0.2	G9 III	-13	180 s		
13865	23123		2 17 11.8	+58 38 11	+0.008	-0.01	8.04	0.44		F0				
14128	110452		2 17 13.5	+1 51 37	+0.001	0.00	7.8	0.4	2.6	F0 V		110 s		
14055	55427	9 γ Tri	2 17 18.8	+33 50 50	+0.004	-0.05	4.01	0.02	0.6	A0 V	+14	46 s		
14247	193622		2 17 19.8	-35 58 58	+0.014	+0.05	6.7	1.2	3.2	G5 IV		29 s		

HD	SAO	Star Name	α 2000	δ 2000	μ(α)	μ(δ)	V	B-V	M_V	Spec	RV	d(pc)	ADS	Notes
14082	75265		$2^h 17^m 25.1^s$	$+28°44'43''$	$+0\overset{s}{.}006$	$-0\overset{''}{.}06$	7.02	0.50		F5	+6		1752	m
13829	12206		2 17 25.6	+65 15 44	−0.006	−0.01	7.7	0.5	4.0	F8 V		54 s		
14028	−−		2 17 33.3	+44 18 26	+0.004	+0.09	7.50	1.6		M7 s	−29			W And, v
13929	23133		2 17 33.8	+58 01 13	−0.001	−0.02	7.45	0.25		A m	−8			
14398	232701		2 17 40.4	−56 15 06	+0.004	−0.13	7.9	0.6	4.4	G0 V		48 s		
14064	37927		2 17 41.7	+42 35 23	−0.002	−0.02	8.0	0.4	2.6	F0 V	−44	120 s		
14147	92873		2 17 45.4	+18 27 18	+0.003	+0.02	7.4	0.4	2.6	F0 V		93 s		
14105	37930		2 17 52.1	+40 33 01	−0.003	0.00	7.7	1.1	0.2	K0 III		320 s		
14146	75269		2 17 57.0	+29 00 27	+0.001	0.00	6.8	1.6	−0.4	M0 III	+27	270 s		
14294	167688		2 17 59.1	−28 22 22	−0.001	+0.02	7.8	0.6	3.4	F5 V		59 s		
13982	23143	8 Per	2 17 59.8	+57 54 00	+0.008	+0.01	5.75	1.17	0.3	G8 III	+3	91 s		
14095	37931		2 18 00.6	+42 54 04	−0.001	0.00	7.1	0.9	3.2	G5 IV		60 s		
14214	110456		2 18 01.3	+1 45 28	+0.024	+0.38	5.58	0.60	4.2	F9 V	+27	20 ts		
14192	92878		2 18 03.6	+14 28 00	−0.001	−0.02	8.00	0.4	2.6	F0 V		120 s		
14285	167689		2 18 03.8	−23 37 43	+0.001	−0.02	7.4	0.7	2.6	F0 V		53 s		
13994	23149	7 Per	2 18 04.4	+57 31 00	−0.002	+0.01	5.98	1.05	−2.1	G5 II	−11	360 s		
14203	92879		2 18 06.0	+13 25 47	0.000	−0.02	7.8	0.2	2.1	A5 V		140 s		
14155	75270		2 18 06.6	+29 50 49	−0.002	−0.02	7.4	0.1	1.7	A3 V		140 s		
14191	92877	22 θ Ari	2 18 07.5	+19 54 04	−0.001	0.00	5.62	0.01	1.2	A1 V	+6	77 s		
14284	148319		2 18 08.5	−14 08 00	0.000	+0.04	8.0	1.6	−0.5	M4 III	+28	510 s		
14094	37932		2 18 11.2	+48 08 54	0.000	−0.01	7.2	0.1	1.7	A3 V		120 s		
14577	−−		2 18 12.4	−67 47 22			7.8	1.6	−0.5	M2 III		450 s		
13908	12210		2 18 14.3	+65 35 40	+0.002	−0.03	7.7	0.5	4.0	F8 V		53 s		
14340	193631		2 18 14.7	−30 47 21	−0.002	+0.01	7.7	1.6	0.2	K0 III		140 s		
13589	4610		2 18 18.1	+77 44 30	+0.013	−0.05	7.9	1.1	0.2	K0 III		330 s		
14062	23163		2 18 24.5	+54 16 46	0.000	0.00	7.6	1.1	0.2	K0 III		270 s		
14293	129815		2 18 27.8	−8 21 32	−0.001	+0.03	7.6	0.8	3.2	G5 IV		77 s		
14222	75273		2 18 34.4	+21 53 52	+0.002	−0.01	7.7	0.4	2.6	F0 V		110 s		
14190	55444		2 18 35.9	+37 04 01	−0.002	+0.01	8.0	1.1	−0.1	K2 III		420 s		
14189	37940		2 18 36.2	+40 16 43	−0.002	−0.01	7.24	0.43		F2			1763	m
14280	110465		2 18 40.8	+4 11 46	−0.001	−0.04	8.00	0.9	3.2	G5 IV		91 s	1772	m
14279	110464		2 18 41.6	+8 10 52	+0.002	+0.02	7.0	1.1	0.2	K0 III		230 s		
14010	12213		2 18 44.7	+64 25 30	+0.001	0.00	7.11	0.60	−7.1	B9 Ia	−48	3300 s		
14389	167696		2 18 48.8	−26 26 25	+0.006	+0.02	8.0	1.2	0.2	K0 III		260 mx		
14338	129819		2 18 51.6	−7 27 31	−0.003	−0.03	7.6	0.5	3.4	F5 V		69 s		
14278	92882		2 18 52.2	+12 59 32	+0.014	+0.03	7.1	0.9	3.2	G5 IV		61 s		
14376	148325		2 18 53.7	−19 32 04	−0.002	−0.01	7.0	1.6	−0.4	M0 III	−7	290 s		
14252	75276	10 Tri	2 18 57.0	+28 38 33	+0.001	0.00	5.03	0.04	1.4	A2 V	+3	53 s	1770	m
14037	12215		2 18 57.2	+63 56 50	+0.007	−0.04	7.6	0.1	1.7	A3 V		91 mx		
14262	75277		2 18 58.0	+23 10 04	0.000	0.00	6.46	0.34	4.4	G0 V	−13	26 s		
14412	167697		2 18 58.4	−25 56 44	−0.016	+0.45	6.34	0.73	5.2	G5 V	+5	12 ts		
14518	232706		2 18 58.5	−53 56 15	+0.001	+0.01	7.8	1.3	−0.1	K2 III		330 s		
12648	356		2 18 59.6	+85 44 11	−0.008	+0.10	7.2	0.9	3.2	G5 IV		65 s		
14420	167698		2 19 03.8	−28 55 26	+0.006	0.00	8.0	1.1	0.2	K0 III		210 mx		
14134	23178		2 19 04.4	+57 08 07	0.000	0.00	6.56	0.48	−6.8	B3 Ia	−44	2000 s		m
15532	258297		2 19 07.1	−82 57 15	+0.014	−0.01	7.8	1.1	0.2	K0 III		290 s		
14305	92884		2 19 08.7	+19 41 16	−0.001	−0.11	6.9	0.5	4.0	F8 V	+2	38 s		
14350	110470		2 19 10.3	+3 32 12	0.000	−0.02	8.0	0.9	3.2	G5 IV		93 s		
14213	37947		2 19 10.8	+46. 28 21	−0.001	−0.01	6.1	0.1	1.7	A3 V	−15	75 s		
14143	23182		2 19 13.9	+57 10 10	0.000	+0.01	6.65	0.44	−6.8	B2 Ia	−42	2100 s	1766	m
14188	37946		2 19 14.0	+50 08 38	+0.004	−0.02	7.30	1.4	−0.3	K5 III	−3	290 s		m
14212	37948	62 And	2 19 16.7	+47 22 48	−0.006	0.00	5.30	−0.01		A0	−30	26 mn		
14386	129825	68 o Cet	2 19 20.6	−2 58 39	−0.001	−0.23	3.04	1.42		Md	+64	29 mn	1778	Mira, m,v
14124	12220		2 19 21.4	+60 29 10	+0.001	−0.02	8.00	0.2	1.6	A9 IV		180 s	1765	m
14221	37949		2 19 22.6	+48 57 19	−0.009	+0.07	6.4	0.4	2.6	F0 V	−19	57 s		
15946	258299		2 19 23.3	−84 36 04	−0.004	−0.01	8.0	1.6	−0.5	M4 III		500 s		
14509	215860		2 19 24.6	−41 50 54	0.000	+0.03	6.37	1.16	0.3	gG5		110 s		
14642	248531		2 19 25.3	−61 56 17	+0.023	+0.16	7.7	0.5	3.4	F5 V		67 mx		
14498	193640		2 19 25.7	−38 58 39	+0.007	−0.01	7.8	1.2	0.2	K0 III		180 mx		
14385	110474		2 19 28.7	+2 49 13	+0.002	+0.01	7.80	0.8	0.3	G5 III	+11	320 s		
14333	75284		2 19 34.9	+25 30 46	+0.001	+0.01	7.9	0.1	0.6	A0 V		290 s		
14272	55453		2 19 37.3	+39 50 06	+0.003	−0.01	6.50	−0.09	−0.6	B8 IV		260 s		
14220	23199		2 19 37.6	+52 33 40	0.000	0.00	7.00	−0.03	−1.0	B5.5 V	−46	370 s		
14508	193642		2 19 40.0	−33 37 09	−0.002	+0.02	8.0	2.0	−0.5	M2 III		310 s		
14462	148333		2 19 40.6	−19 09 18	+0.002	−0.05	7.7	1.1	0.2	K0 III		260 s		
14417	129830		2 19 40.7	−4 20 44	+0.002	+0.01	6.50	0.08		A2				
14173	23192		2 19 41.0	+60 00 45	−0.005	−0.03	7.20	0.95	−2.1	G5 II		160 mx		m
14172	23194		2 19 44.2	+60 01 46	−0.004	−0.01	6.92	0.23	1.4	A2 V		99 s		m
14793	248535		2 19 46.7	−68 05 05	−0.006	+0.02	7.0	0.1	1.7	A3 V		110 s		
14348	55460		2 19 53.1	+31 20 15	+0.012	−0.09	7.3	0.5	3.4	F5 V		61 mx		
14641	232717		2 19 54.2	−55 56 41	+0.003	+0.03	5.81	1.55	−0.3	K5 III	+49	150 s		m
14447	129833		2 19 55.9	−6 19 20	+0.004	−0.03	7.9	0.9	3.2	G5 IV		86 s		
14373	75287		2 20 04.3	+30 11 19	+0.001	−0.02	6.5	1.1	0.2	K0 III	−1	190 s		
14304	23205		2 20 09.0	+50 57 30	+0.004	−0.02	6.7	0.5	3.4	F5 V		46 s		
14171	12226		2 20 12.9	+64 20 15	−0.003	+0.02	6.60	−0.03	0.6	A0 V	−26	160 s		

HD	SAO	Star Name	α 2000	δ 2000	μ(α)	μ(δ)	V	B−V	M_v	Spec	RV	d(pc)	ADS	Notes
			h m s	° ′ ″	s	″								
14438	92890		2 20 13.5	+14 17 52	+0.001	−0.02	8.0	1.1	0.2	K0 III		360 s		
14682	232720		2 20 15.4	−56 05 50	+0.003	+0.02	7.1	1.7	−0.1	K2 III		130 s		
14640	232718		2 20 15.9	−50 18 09	+0.004	0.00	7.8	0.4	0.2	K0 III		330 s		
14614	215866		2 20 20.7	−44 03 28	+0.001	+0.04	8.0	0.8	0.2	K0 III		360 s		
14548	167715		2 20 21.9	−27 16 23	−0.001	+0.01	8.0	1.3	0.2	K0 III		250 s		
14703	232722		2 20 22.8	−56 47 03	−0.003	−0.01	6.5	1.1	3.2	G5 IV		28 s		
14270	23207		2 20 29.0	+56 59 36	+0.001	+0.01	7.84	2.27	−5.6	M3 Iab	−44	1600 s		AD Per, v
14217	12230		2 20 36.4	+63 52 26	−0.003	0.00	7.8	1.1	0.2	K0 III		290 s		
14409	55471		2 20 39.8	+38 22 11	+0.001	0.00	7.9	1.1	0.2	K0 III		350 s		
14372	37955		2 20 41.4	+47 18 40	+0.001	0.00	6.11	−0.08	−1.1	B5 V	+2	260 s		
14322	23214		2 20 42.7	+55 54 33	−0.002	0.00	6.79	0.32	−5.6	B8 Ib	−35	2000 s		
14629	193652		2 20 42.9	−39 02 02	+0.018	−0.03	7.6	2.5	0.2	K0 III		40 s		
14805	248540		2 20 43.6	−62 32 44	+0.025	+0.04	7.7	1.1	0.2	K0 III		110 mx		
14547	148340		2 20 43.7	−11 36 25	+0.001	+0.01	7.9	0.9	3.2	G5 IV		85 s		
14456	75293		2 20 47.0	+28 31 25	+0.002	−0.01	7.90	0.9	0.3	G8 III		330 s		
14455	75294		2 20 49.5	+28 44 11	−0.002	+0.01	7.3	0.4	2.6	F0 V		89 s		
14745	232725		2 20 54.0	−52 57 49	−0.008	+0.01	7.5	0.8	3.4	F5 V		41 s		
14392	37960	63 And	2 20 58.1	+50 09 05	+0.003	−0.02	5.5	−0.1		A0	−2			
14330	23217		2 20 59.6	+57 09 31	+0.001	+0.01	7.95	2.26	−5.7	M1 Iab	−44	1600 s		FZ Per, v
14359	23220		2 20 59.7	+55 24 30	+0.001	+0.01	7.6	1.1	0.2	K0 III		270 s		
14445	55475		2 21 02.5	+37 04 13	+0.010	−0.04	7.6	0.4	2.6	F0 V		92 mx		
14437	37963		2 21 02.7	+42 56 37	+0.001	−0.02	7.1	0.0	0.4	B9.5 V		200 s		
13714	372		2 21 06.2	+80 43 37	−0.001	+0.01	7.60	0.1	0.6	A0 V		240 s	1754	m
14384	23229		2 21 12.9	+54 30 36	+0.005	0.00	6.9	0.5	3.4	F5 V		48 s		
14628	167722		2 21 13.7	−19 54 55	0.000	+0.01	6.6	1.2	3.2	G5 IV		27 s		
14512	75300		2 21 14.4	+23 25 36	0.000	−0.01	7.7	1.6	−0.5	M2 III		430 s		
14346	23225		2 21 15.1	+59 14 53	−0.002	−0.01	7.6	1.1	0.2	K0 III	+11	270 s		
14471	——		2 21 31.3	+46 25 38			7.90	1.4	−0.3	K5 III		380 s	1793	m
14587	110493		2 21 31.9	+7 45 08	+0.002	+0.01	7.5	0.2	2.1	A5 V		120 s		
14792	215872		2 21 33.6	−48 18 44	0.000	−0.01	7.50	0.9	3.2	G5 IV		72 s		m
——	——		2 21 34.7	+27 36 28			8.00	0.9	0.2	G9 III		360 s		
14477	37971		2 21 35.7	+44 36 00	−0.001	0.00	7.20	0.1	1.4	A2 V		140 s	1795	m
15066	——		2 21 40.3	−71 39 42			7.9	1.1	0.2	K0 III		350 s		
14625	129848		2 21 40.4	−0 08 57	−0.001	−0.03	7.6	0.9	0.3	G8 III	+5	280 s		
14636	129850		2 21 40.8	−7 32 37	+0.001	−0.01	7.5	0.8	3.2	G5 IV		74 s		
16701	258302		2 21 40.9	−85 43 05	+0.003	−0.02	7.8	0.2	2.6	F0 V		110 s		
14404	23234		2 21 42.3	+57 51 47	0.000	+0.01	7.84	2.30	−4.8	M2 Ib	−39	970 s		
14626	129849		2 21 43.4	−0 21 06	+0.001	−0.04	7.3	1.1	0.2	K0 III		260 s		
15008	248545	δ Hyi	2 21 45.0	−68 39 34	−0.008	+0.01	4.09	0.03	1.4	A2 V	+11	35 s		
14502	37973		2 21 46.3	+43 49 03	0.000	−0.05	7.8	0.8	3.2	G5 IV		81 s		
14597	92900		2 21 47.1	+15 31 09	+0.001	−0.01	8.0	0.1	0.6	A0 V	+36	300 s		
13965	4619		2 21 47.2	+78 15 16	+0.006	+0.03	7.9	1.1	−0.1	K2 III		290 mx		
14692	148353		2 21 49.7	−14 17 05	0.000	−0.04	7.5	0.4	2.6	F0 V	+6	93 s		
14596	92901		2 21 50.2	+16 09 45	+0.003	0.00	8.0	1.1	0.2	K0 III		360 s		
14433	23243		2 21 55.3	+57 14 34	−0.001	−0.01	6.39	0.57	−7.3	A1 Ia	−47	2800 s		
14652	110495	69 Cet	2 21 56.5	+0 23 45	−0.001	0.00	5.28	1.65		Ma	+23			
14834	215875		2 21 57.0	−49 31 15	−0.001	+0.03	7.5	0.3	3.2	G5 IV		74 s		
14758	167730		2 21 59.0	−29 21 08	+0.001	−0.18	7.90	0.7	4.4	G0 V		50 s	1816	m
14443	23244		2 22 00.5	+57 08 44	+0.001	+0.03	8.05	0.34	−5.7	B2 Ib	−40	2800 s		
14691	148354		2 22 01.3	−10 46 40	+0.009	−0.08	5.48	0.34	2.6	F0 V	+12	36 s		
14757	167731		2 22 02.5	−27 51 40	0.000	+0.04	7.5	0.3	3.2	G5 IV		72 s		
14728	148356		2 22 04.9	−17 39 44	+0.001	−0.06	5.87	1.23		K0				
14595	75308		2 22 06.6	+22 52 25	−0.001	+0.01	6.6	0.9	3.2	G5 IV	+23	48 s		
14469	23250		2 22 06.8	+56 36 16	0.000	+0.01	7.63	2.17	−5.6	M3 Iab	−40	1700 s		SU Per, v
——	75307		2 22 06.9	+27 36 15	−0.006	−0.02	8.0	0.7	3.2	G5 IV		93 s		
14791	193667		2 22 07.0	−36 06 24	+0.006	−0.06	7.8	1.3	0.2	K0 III		120 mx		
14928	232732		2 22 11.7	−59 45 35	−0.004	−0.01	7.6	0.5	1.4	A2 V		91 s		
14832	215878		2 22 11.9	−43 11 59	+0.007	+0.05	6.31	1.00		G5				
14690	129858	70 Cet	2 22 12.3	−0 53 06	−0.002	−0.05	5.42	0.31	2.6	F0 V n	+20	37 s		
14468	23253		2 22 18.1	+59 09 45	+0.004	+0.03	7.8	1.4	−0.3	K5 III		260 mx		
14743	148358		2 22 18.5	−16 30 26	+0.003	−0.01	7.3	1.1	0.2	K0 III		270 s		
14489	23256	9 Per	2 22 21.3	+55 50 45	0.000	0.00	5.17	0.37	−7.5	A2 Ia	−15	2400 s	1802	m
14585	55494		2 22 23.4	+37 41 55	+0.002	−0.01	8.0	0.9	3.2	G5 IV		93 s		
14608	55496		2 22 24.1	+30 19 30	0.000	+0.03	7.80	1.1	−0.1	K2 III		380 s		
14414	12247		2 22 26.4	+65 14 35	0.000	+0.02	7.3	0.5	4.0	F8 V		45 s		
14802	167736	κ For	2 22 32.5	−23 48 59	+0.014	−0.06	5.20	0.60	4.5	G1 V	+18	14 ts		
14688	92905		2 22 33.2	+16 52 14	+0.004	0.00	6.8	0.1	0.6	A0 V	+15	140 mx		
14846	193674		2 22 35.2	−39 24 50	0.000	0.00	7.8	1.1	0.2	K0 III		290 s		
14649	55500		2 22 45.0	+33 42 13	−0.001	0.00	7.9	0.0	0.4	B9.5 V		320 s		
14788	148362		2 22 45.6	−14 54 05	+0.004	0.00	7.68	0.03	1.2	A1 V		110 mx		
14622	37986		2 22 50.1	+41 23 46	−0.008	−0.10	5.82	0.27	1.8	F1 IV	−35	64 s		m
14606	37984		2 22 50.4	+43 31 05	−0.002	+0.01	7.5	0.1	1.4	A2 V		150 s		
15248	255880	κ Hyi	2 22 52.3	−73 38 45	−0.019	+0.01	5.01	1.09	0.2	K0 III		77 s		
14402	12250		2 22 53.0	+68 45 43	+0.016	0.00	7.4	0.9	3.2	G5 IV	+12	68 s		
14535	23263		2 22 53.4	+57 14 43	+0.001	+0.01	7.45	0.71	−7.5	A2 Ia p	−53	4200 s		

HD	SAO	Star Name	α 2000	δ 2000	μ(α)	μ(δ)	V	B−V	M_v	Spec	RV	d(pc)	ADS	Notes
14633	37987		$2^h22^m54\overset{s}{.}2$	$+41°28'49''$	$0\overset{s}{.}000$	$0\overset{''}{.}00$	7.46	−0.21		O8	−36			
14943	232736		2 22 54.6	−51 05 32	+0.002	+0.07	5.92	0.22		A3				
14851	193678		2 22 55.1	−31 24 06	+0.001	−0.04	7.9	1.3	0.2	K0 III		230 s		
14830	148366		2 22 57.7	−18 21 16	+0.010	−0.11	6.22	0.94		G5				
14739	92911		2 22 59.6	+17 35 52	+0.002	0.00	7.3	0.1	1.4	A2 V	−9	150 s		
14542	23266		2 23 00.4	+57 23 14	0.000	0.00	6.98	0.61	−7.1	B8 Ia	−47	2900 s		
14890	193679		2 23 06.4	−37 34 35	−0.002	−0.03	6.53	1.61		K2				
14519	12259		2 23 11.9	+63 36 09	−0.004	+0.03	7.79	1.18	3.2	G5 IV		40 s		
14954	215888		2 23 12.7	−48 16 21	+0.003	0.00	7.9	0.3	−0.1	K2 III		390 s		
14787	92914		2 23 13.9	+10 50 12	−0.001	+0.02	7.9	0.9	3.2	G5 IV		89 s		
14882	193680		2 23 14.4	−29 52 11	−0.006	−0.11	6.95	0.56	4.0	F8 V		37 s		m
14571	23272		2 23 15.8	+58 05 19	+0.003	+0.01	8.0	1.1	0.2	K0 III		310 s		
14663	37991		2 23 16.2	+46 08 24	−0.003	+0.01	7.9	0.1	1.4	A2 V		180 s		
14685	55503		2 23 17.4	+38 15 09	+0.002	−0.01	7.1	0.1	0.6	A0 V		200 s		
14619	23278		2 23 20.1	+52 04 59	+0.005	−0.01	6.6	0.1	1.7	A3 V		92 s		
14840	129868		2 23 22.3	−6 11 36	−0.001	−0.05	7.00	0.30	2.6	F0 V		75 s		
14880	167744		2 23 23.3	−26 59 45	+0.001	−0.06	7.0	1.0	3.2	G5 IV		42 s		
15051	232745		2 23 23.8	−56 48 48	+0.003	+0.04	6.7	1.5	0.6	A0 V		20 s		
14648	37994		2 23 25.4	+49 10 53	+0.002	−0.03	8.0	0.9	3.2	G5 IV		89 s		
14735	55505		2 23 28.5	+35 26 27	−0.002	−0.02	6.8	0.8	3.2	G5 IV		54 s		
14552	12262		2 23 30.9	+61 46 27	+0.007	−0.01	7.9	0.5	3.4	F5 V		75 s		
15060	232746		2 23 35.5	−55 37 00	+0.005	+0.08	6.9	0.7	3.4	F5 V		35 s		
14381	4637		2 23 40.9	+73 48 35	−0.005	−0.03	8.0	0.1	1.7	A3 V		170 s		
14662	23283		2 23 51.7	+55 21 52	+0.001	−0.02	6.28	0.86	−4.6	F7 Ib	−26	1100 s	1820	m
14783	55511		2 23 52.9	+33 51 30	−0.002	+0.01	7.6	1.1	0.2	K0 III	−2	300 s		
14800	55512		2 23 54.6	+33 30 26	−0.002	−0.03	7.38	0.12		A0			1824	m
15194	--		2 23 57.5	−65 19 50			7.9	1.6	−0.5	M2 III		470 s		
14771	38002		2 24 02.5	+42 06 08	−0.005	−0.01	6.7	1.1	0.2	K0 III		200 s		
15088	232753		2 24 03.2	−54 21 42	0.000	−0.06	7.9	0.8	0.2	K0 III		300 mx		
14940	148373		2 24 09.5	−16 15 17	−0.001	+0.01	6.6	0.4	2.6	F0 V		63 s		
14866	92922		2 24 10.3	+10 16 20	+0.003	−0.02	7.2	1.1	0.2	K0 III		260 s		
14799	55514		2 24 11.0	+39 20 45	+0.004	−0.01	7.82	−0.06	0.4	B9.5 V		310 s		
15429	255881		2 24 11.4	−74 26 09	+0.014	+0.07	7.9	0.5	0.2	K0 III		210 mx		
14152	4630		2 24 15.5	+79 46 15	+0.009	−0.01	7.80	0.4	2.6	F0 V		110 s		
14807	55516		2 24 16.6	+39 49 43	0.000	−0.03	7.3	0.5	3.4	F5 V		60 s		
14632	12268		2 24 17.5	+63 02 35	0.000	+0.02	7.65	0.16	0.6	A0 V		200 s		
14617	12267		2 24 18.5	+63 32 13	−0.003	+0.01	7.54	1.58	−1.1	K2 II-III		330 s		
14988	167757		2 24 20.0	−25 50 50	+0.002	−0.02	6.5	1.5	0.2	K0 III		92 s		
14938	129879		2 24 23.4	−3 06 18	+0.005	−0.13	7.20	0.51	3.4	F5 V		53 s		
14770	38005	64 And	2 24 24.8	+50 00 24	+0.002	−0.03	5.19	0.98	0.3	G8 III	−13	89 s		
15048	193690		2 24 26.1	−37 21 40	−0.003	−0.03	7.0	0.2	3.4	F5 V		53 s		
14887	92924		2 24 27.2	+15 31 30	−0.001	−0.02	7.8	0.4	2.6	F0 V	−39	110 s		
14971	148375		2 24 28.1	−13 17 35	−0.004	−0.02	7.7	0.5	3.4	F5 V		72 s		
14797	38006		2 24 30.1	+47 22 24	+0.003	+0.02	7.60	1.6	−0.4	M0 III	+26	350 s		
14903	110516		2 24 30.6	+9 42 49	+0.004	−0.05	7.6	1.1	0.2	K0 III		150 mx		
15099	215893		2 24 31.6	−45 08 30	−0.001	−0.03	8.0	1.2	0.2	K0 III		290 s		
15064	215892		2 24 33.7	−40 50 26	+0.019	+0.12	6.18	0.66		G0				
14875	75342		2 24 41.4	+29 14 40	+0.002	−0.01	7.1	1.1	−0.2	K3 III		280 s		
15047	167766		2 24 45.7	−28 39 06	+0.011	−0.04	8.0	0.7	3.4	F5 V		57 s		
14951	92932	24 ξ Ari	2 24 49.0	+10 36 39	+0.001	−0.01	5.47	−0.10	−1.0	B7 IV	+4	200 s		
14413	4640		2 24 50.3	+76 07 57	−0.003	0.00	7.6	0.1	0.6	A0 V		250 s		
15122	215894		2 24 53.2	−42 09 35	+0.004	+0.02	7.8	−0.5	1.4	A2 V		150 mx		
15233	248555	λ Hor	2 24 53.9	−60 18 43	−0.009	−0.13	5.35	0.39	0.6	F2 III p	+27	89 s		
15005	129887		2 24 56.0	−3 53 34	−0.002	−0.06	6.90	1.02	0.2	K0 III		210 s	1840	m
15004	129888	71 Cet	2 24 58.3	−2 46 48	0.000	0.00	6.33	0.00		A0				
15098	193694		2 24 58.5	−35 53 03	−0.001	−0.02	7.8	1.5	0.2	K0 III		170 s		
14918	75348		2 25 01.8	+25 29 12	−0.001	0.00	7.75	0.83	0.3	G5 III		310 s		m
14855	38012		2 25 03.4	+45 38 51	−0.005	−0.07	7.6	0.9	3.2	G5 IV		73 s		
14901	55527		2 25 09.9	+33 52 00	0.000	−0.01	7.4	1.6	−0.5	M4 III		380 s		
14827	23305		2 25 13.6	+55 15 15	+0.006	−0.02	7.63	0.06	0.4	B9.5 V		97 mx	1829	m
14893	55530		2 25 15.5	+37 07 07	+0.001	−0.04	7.36	−0.02	0.4	B9.5 V		240 s		
14818	23304	10 Per	2 25 15.9	+56 36 37	0.000	+0.01	6.25	0.31	−6.8	B2 Ia	−46	2200 s		
15097	167774		2 25 24.1	−21 48 36	−0.002	−0.06	7.8	0.2	4.0	F8 V		58 s		
14825	23311		2 25 25.8	+58 12 09	+0.003	−0.01	7.90	0.1	1.4	A2 V	+2	180 s	1832	m
15379	248556		2 25 26.3	−66 29 41	−0.004	−0.02	6.41	1.54		Mb				
14935	55537		2 25 26.6	+35 02 32	0.000	+0.01	6.9	0.8	3.2	G5 IV		55 s		
14795	23307		2 25 26.6	+60 00 22	+0.003	+0.02	7.68	0.00	−0.9	B6 V		450 s		
15205	215897		2 25 27.6	−48 29 29	+0.002	−0.01	7.8	1.4	0.2	K0 III		110 s		
14948	55539		2 25 27.9	+32 24 27	+0.002	−0.02	7.5	1.1	−0.1	K2 III		330 s		
15439	255883		2 25 30.5	−70 06 07	+0.006	+0.02	8.0	0.2	2.1	A5 V		150 s		
14969	75355		2 25 31.2	+29 52 49	−0.001	−0.04	7.90	1.1	−0.2	K3 III		420 s		
15029	92941		2 25 33.9	+11 58 17	−0.008	−0.28	7.6	0.5	3.4	F5 V		52 mx		
14872	23319	65 And	2 25 37.3	+50 16 43	+0.002	−0.01	4.71	1.53	−0.3	K4 III	−5	82 s		m
15014	92940		2 25 39.7	+19 33 17	+0.003	+0.01	8.0	1.1	0.2	K0 III		360 s		
14817	12277		2 25 40.1	+61 32 59	+0.003	−0.01	7.01	0.22	0.2	B9 V		180 s	1833	m

HD	SAO	Star Name	α 2000	δ 2000	μ(α)	μ(δ)	V	B-V	M$_V$	Spec	RV	d(pc)	ADS	Notes
			^{h} ^{m} ^{s}	^{o} ' "	^{s}	"								
15042	92942		2 25 41.4	+10 30 21	+0.001	0.00	7.6	0.0	0.4	B9.5 V		280 s		
15028	75359		2 25 47.1	+20 16 49	+0.008	-0.05	8.0	0.9	3.2	G5 IV		90 mx		
14968	55544		2 25 49.7	+37 00 49	-0.002	+0.02	7.4	1.6	-0.5	M2 III		380 s		
15027	75360		2 25 51.9	+25 09 35	-0.001	0.00	7.0	0.1	1.7	A3 V		110 s		
15154	167785		2 25 52.8	-23 50 33	+0.006	+0.01	7.8	0.2	1.7	A3 V		67 mx		
14343	4641		2 25 55.0	+79 34 36	-0.001	+0.01	8.0	1.1	-0.1	K2 III		390 s		
15130	148385	72 ρ Cet	2 25 56.9	-12 17 26	-0.001	-0.01	4.89	-0.03	0.2	B9 V	+10	85 s		
14871	23321		2 25 57.9	+56 06 10	0.000	0.00	7.95	0.10	0.4	B9.5 V	-32	310 s		DM Per, m,v
15144	148386		2 26 00.2	-15 20 28	-0.004	-0.05	5.83	0.15		A4 p	-8	21 mn	1849	m
15096	110531		2 26 01.7	+5 46 48	+0.025	+0.13	7.95	0.81	3.2	G5 IV		43 mx		
15105	129898		2 26 02.2	-0 10 42	+0.001	+0.01	6.9	1.6		M5 III	+42			R Cet, v
14342	4642		2 26 02.7	+79 38 21	+0.020	-0.09	7.9	1.1	-0.1	K2 III		200 mx		
15026	55550		2 26 05.0	+30 54 23	0.000	-0.07	7.8	1.1	0.2	K0 III		210 mx		
15013	55552		2 26 09.5	+34 27 59	+0.010	-0.21	7.9	0.9	3.2	G5 IV		51 mx		
15095	92948		2 26 10.9	+10 38 16	+0.005	-0.25	7.6	0.7	4.4	G0 V		43 s		
15414	248563		2 26 12.6	-62 55 06	+0.003	-0.05	7.7	1.1	0.2	K0 III		320 s		
15115	110532		2 26 16.2	+6 17 33	+0.006	-0.05	6.7	0.4	3.0	F2 V		55 s		
14899	23326		2 26 18.2	+57 13 42	-0.002	0.00	7.39	0.44	-5.6	B8 Ib	-42	2100 s		
15508	248564		2 26 23.5	-68 38 08	+0.007	+0.01	7.9	1.1	0.2	K0 III		350 s		
15114	92950		2 26 26.9	+12 53 55	+0.001	-0.02	7.8	0.0	0.4	B9.5 V		300 s		
14914	23327		2 26 33.7	+59 39 32	+0.012	-0.03	7.1	1.1	0.2	K0 III	+14	190 mx		
15220	167795		2 26 35.1	-20 02 34	+0.005	+0.10	5.88	1.26		K0	+42			
15362	232769		2 26 37.5	-51 42 18	-0.005	-0.06	7.6	0.5	3.4	F5 V		65 s		
15339	215905		2 26 45.0	-45 59 59	+0.001	+0.01	7.15	1.10	0.2	K0 III		210 s		
14956	23335		2 26 45.2	+57 40 44	-0.004	-0.01	7.20	0.72	-6.8	B2 Ia	-24	320 mx		
15165	92952		2 26 45.6	+10 33 55	+0.002	-0.01	6.80	0.1	1.7	A3 V		110 s		m
14947	23334		2 26 46.7	+58 52 34	-0.002	+0.01	8.01	0.45		06.8	-54			
15093	55560		2 26 48.4	+35 36 51	+0.002	-0.02	7.0	0.4	2.6	F0 V		76 s		
14694	4646		2 26 55.9	+74 35 02	-0.002	+0.01	7.8	1.4	-0.3	K5 III		390 s		
15141	75367		2 26 57.7	+26 02 27	-0.004	-0.03	6.97	0.58	3.4	F5 V		43 s		
15371	215906	κ Eri	2 26 59.1	-47 42 14	+0.002	-0.01	4.25	-0.14	-2.2	B5 III	+29	190 s		
15127	55565		2 27 02.1	+33 55 43	-0.002	0.00	7.8	0.7	4.4	G0 V		49 s		
15152	75370		2 27 07.0	+27 00 48	-0.004	-0.06	6.18	1.43		K5	-48			
15126	55568		2 27 14.8	+36 58 00	+0.001	-0.03	7.7	1.6	-0.5	M4 III		440 s		
15228	110537		2 27 23.4	+10 11 54	-0.020	-0.20	6.49	0.46	3.4	F5 V	-41	42 s		
15000	23340		2 27 25.1	+59 06 21	+0.015	-0.09	7.3	0.4	3.0	F2 V		70 s		
15176	55570	11 Tri	2 27 27.7	+31 48 05	-0.002	-0.02	5.54	1.12	0.2	K0 III	-39	99 s		
15151	38039		2 27 28.6	+41 02 53	+0.001	-0.03	8.0	0.1	0.6	A0 V		300 s		
15841	255890		2 27 28.6	-75 44 26	+0.002	+0.03	7.6	1.1	0.2	K0 III		300 s		
15275	129913		2 27 29.3	-0 45 07	+0.001	-0.01	7.2	1.4	-0.3	K5 III		310 s		
15227	92957		2 27 32.0	+16 38 37	+0.001	-0.04	7.3	0.4	2.6	F0 V	+15	87 s		
15113	38036		2 27 34.4	+48 13 56	0.000	+0.01	7.7	0.0	0.4	B9.5 V		260 s		
15299	129918		2 27 38.6	-6 55 43	+0.002	-0.03	7.3	0.9	3.2	G5 IV		66 s		
15394	193720		2 27 42.4	-33 53 47	+0.002	+0.02	7.8	-0.2	4.0	F8 V		57 s		
15433	215909		2 27 43.6	-45 02 52	-0.002	-0.03	7.8	1.4	0.2	K0 III		170 s		
15701	248570		2 27 47.1	-69 31 27	0.000	+0.04	7.2	1.6	-0.5	M4 III		350 s		
14863	4652		2 27 50.8	+72 07 59	-0.001	-0.01	7.7	0.0	0.0	B8.5 V		300 s		
15138	23353	66 And	2 27 51.7	+50 34 12	+0.004	-0.09	6.12	0.41	2.6	F0 V	-4	44 s		
14836	4649		2 27 55.0	+73 42 45	+0.021	-0.04	7.3	0.4	2.6	F0 V		84 s		
15412	193722		2 27 59.3	-30 26 30	+0.002	-0.02	7.8	1.5	-0.5	M2 III		450 s		
15137	23354		2 27 59.8	+52 32 58	0.000	0.00	7.88	0.04	-4.4	09.5 V		2000 s		
15328	110542		2 27 59.9	+1 57 39	0.000	-0.01	6.45	0.97	0.2	K0 III	+18	180 s		m
15427	193723	φ For	2 28 01.6	-33 48 40	+0.001	+0.01	5.14	0.10	0.6	A2 IV	+16	76 s		
15198	38046		2 28 02.3	+41 15 10	+0.003	+0.02	7.9	1.1	0.2	K0 III		340 s		
15646	248567		2 28 04.1	-64 18 00	+0.003	+0.01	6.37	-0.04		B9				
15653	248571		2 28 07.5	-65 33 50	-0.009	-0.05	7.3							
15318	110543	73 ξ² Cet	2 28 09.5	+8 27 36	+0.002	0.00	4.28	-0.06	-0.8	B9 III	+11	98 s		
15218	55578		2 28 09.9	+38 50 28	+0.005	-0.01	6.90	0.4	2.6	F0 V		72 s	1864	m
15257	75382	12 Tri	2 28 09.9	+29 40 10	-0.001	-0.08	5.4	0.4	2.6	F0 V	-25	36 s		
15520	232779		2 28 10.8	-50 18 11	+0.008	+0.02	7.1	1.1	0.2	K0 III		200 s		
15256	75383		2 28 11.6	+29 52 23	-0.001	-0.01	7.80	0.8	0.3	G5 III		320 s	1868	m
15243	55579		2 28 14.9	+34 11 44	+0.001	0.00	7.8	1.1	0.2	K0 III		320 s		
15426	167808		2 28 15.6	-26 25 51	+0.001	+0.04	7.8	0.8	0.2	K0 III		330 s		
15447	193724		2 28 16.2	-36 20 28	+0.002	+0.02	7.2	1.1	-0.1	K2 III		300 s		
15069	12295		2 28 20.4	+62 13 08	+0.022	+0.05	7.9	0.7	4.5	G1 V		48 s		
15178	23363		2 28 28.3	+53 04 07	+0.004	-0.01	7.5	1.1	-0.1	K2 III	+26	290 s		
15209	38051		2 28 33.6	+48 17 38	+0.003	-0.03	7.9	0.5	4.0	F8 V		58 s		
15471	193727		2 28 35.4	-31 06 09	-0.003	-0.02	6.11	1.11	3.2	G5 IV		21 s		
15102	12296		2 28 38.0	+63 11 15	0.000	-0.03	7.5	0.1	1.4	A2 V		150 s		
15628	232787		2 28 41.2	-56 26 10	+0.006	+0.05	7.9	1.2	2.6	F0 V		31 s		
15432	148408		2 28 46.4	-11 20 52	+0.005	-0.08	7.0	0.4	3.0	F2 V		64 s		
15335	75391	13 Tri	2 28 48.4	+29 55 55	-0.005	+0.08	5.89	0.58	4.4	G0 V	+40	20 s		
15544	215914		2 28 52.8	-40 26 08	+0.001	-0.01	7.6	1.5	3.2	G5 IV		27 s		
15496	167821		2 29 02.6	-23 12 29	+0.002	-0.04	8.0	1.7	-0.3	K5 III		370 s		
15089	12298	ι Cas	2 29 03.9	+67 24 09	-0.002	+0.02	4.52	0.12		A5 p	+1	20 mn	1860	m,v

HD	SAO	Star Name	α 2000	δ 2000	μ(α)	μ(δ)	V	B-V	M_v	Spec	RV	d(pc)	ADS	Notes
15333	55594		2h 29m 07.3s	+37°19'46"	+0.002	-0.03	6.60	0.2	2.1	A5 V		79 s	1881	m
15590	215917		2 29 11.9	-42 04 33	+0.009	-0.06	7.99	0.65	3.2	G5 IV		80 mx		
15385	75398		2 29 13.6	+23 28 08	+0.006	-0.02	6.1	0.2	2.1	A5 V	+21	63 s		
15348	55597		2 29 17.3	+38 13 30	+0.001	0.00	7.4	0.1	1.7	A3 V		140 s		
15398	75400		2 29 17.4	+21 35 38	+0.007	-0.03	8.0	0.5	3.4	F5 V		82 s		
15515	167823		2 29 18.7	-21 02 18	+0.011	-0.04	7.20	0.5	3.4	F5 V		58 s		m
15377	55600		2 29 21.5	+30 51 54	+0.001	-0.02	7.7	1.1	0.2	K0 III		320 s		
15576	193738		2 29 21.6	-35 56 30	-0.004	-0.02	7.6	1.1	0.2	K0 III		280 s		
15505	148418		2 29 23.4	-12 54 39	-0.001	+0.02	7.3	0.9	3.2	G5 IV		65 s		
15253	23369		2 29 24.9	+55 32 11	+0.004	-0.01	6.51	0.09	1.4	A2 V	+2	100 s	1878	m
15555	167829		2 29 34.0	-24 06 16	+0.026	+0.03	7.34	1.05	5.9	dK0	+29	18 mn		
15453	110565		2 29 35.2	+9 33 57	-0.001	+0.02	6.07	1.02	0.2	K0 III	-11	150 s	1896	m
15365	38067		2 29 46.0	+46 01 53	+0.003	-0.09	6.8	0.9	3.2	G5 IV	+34	51 s		
15251	12307		2 29 46.7	+60 42 20	-0.002	+0.02	8.00	0.00	0.0	B8.5 V		340 s		
15376	38068		2 29 49.4	+44 09 22	+0.001	+0.02	8.0	1.1	-0.1	K2 III		360 s		
15588	167832		2 29 55.3	-22 40 58	+0.005	+0.01	6.77	0.19		A2			1906	m
15420	55605		2 29 55.8	+39 08 14	-0.001	-0.01	6.75	1.10	0.2	K0 III		170 s		
15316	23374		2 29 58.8	+57 49 16	+0.003	+0.01	7.23	0.77	-6.8	A3 Iab	-44	2400 s		
15346	23381		2 30 00.8	+53 51 09	+0.001	+0.02	8.0	0.9	3.2	G5 IV		89 s		
16148	255895		2 30 05.2	-75 47 03	+0.014	+0.05	7.7	0.5	3.4	F5 V		71 s		
15345	23383		2 30 11.8	+55 20 39	+0.003	+0.01	7.9	1.1	0.2	K0 III		310 s		
15442	55606		2 30 12.4	+38 57 07	+0.001	-0.06	7.7	0.1	1.4	A2 V		130 mx		
15634	167837		2 30 13.7	-25 11 11	+0.006	+0.04	6.51	0.29		F0	+25			
15397	38070		2 30 13.8	+48 30 08	+0.001	+0.02	7.9	0.6	4.4	G0 V		49 s		
15793	232795		2 30 15.5	-57 48 34	+0.005	+0.03	6.7	1.9		Mb				
15464	55611		2 30 16.6	+33 50 02	+0.006	-0.05	6.25	1.07	0.0	K1 III	+7	160 mx		
15773	232793		2 30 22.9	-51 22 36	-0.004	-0.02	7.8	0.1	3.0	F2 V		89 s		
15436	38075		2 30 24.8	+45 25 53	-0.001	-0.02	7.30	0.8	3.2	G5 IV		64 s	1899	m
15418	38074		2 30 25.0	+47 51 12	-0.002	-0.02	7.6	1.1	0.2	K0 III		270 s		
15598	129945		2 30 26.3	-5 01 52	-0.002	-0.01	7.66	0.06	0.6	A0 V		230 s		
15524	75407		2 30 32.3	+25 14 06	+0.005	-0.08	5.92	0.41	3.4	F5 V	-11	32 s	1904	m
15652	167839		2 30 32.7	-22 32 44	0.000	-0.03	6.10	1.61		K5	-19			
15435	38077		2 30 35.3	+47 48 36	-0.002	-0.04	7.6	0.9	3.2	G5 IV		72 s		
15699	193745		2 30 37.4	-31 16 45	+0.003	-0.02	8.0	1.7	-0.5	M2 III		470 s		
15550	92979	26 Ari	2 30 38.4	+19 51 19	+0.006	-0.03	6.15	0.25	2.6	F0 V	+19	51 s		
15451	38078		2 30 39.1	+43 40 44	0.000	+0.05	7.8	0.7	4.4	G0 V		48 s		
15473	55614		2 30 40.8	+40 10 03	+0.006	-0.03	6.9	1.1	-0.1	K2 III		190 mx		
15463	38079		2 30 42.0	+45 03 24	+0.004	-0.01	8.0	0.9	3.2	G5 IV		89 s		
15633	110583		2 30 45.1	+0 15 19	-0.004	-0.07	6.00	0.17		A3 m				
15740	193747		2 30 46.3	-33 00 26	-0.001	0.00	7.4	0.1	2.6	F0 V		92 s		
15512	55615		2 30 47.2	+36 07 02	+0.001	-0.01	7.6	0.1	1.7	A3 V		150 s		
15407	23389		2 30 50.7	+55 32 53	+0.010	-0.10	7.40	0.5	3.4	F5 V		61 s	1901	m
15936	248583		2 30 52.8	-64 02 47	+0.001	-0.01	7.7	0.1	1.4	A2 V		180 s		
15596	92983	27 Ari	2 30 54.3	+17 42 14	+0.002	-0.08	6.23	0.90	1.8	G5 III-IV	-116	71 s		
15657	129952		2 30 55.4	-7 59 53	-0.001	-0.02	7.9	0.1	0.6	A0 V		280 s		
15792	215928		2 30 59.4	-45 40 07	+0.004	-0.01	7.9	1.5	0.2	K0 III		180 s		
15697	148434		2 31 03.7	-15 08 00	-0.001	+0.05	7.6	0.6	4.4	G0 V		43 s		
16034	248587		2 31 11.3	-67·58 06	+0.004	-0.02	7.5	0.6	4.4	G0 V		42 s		
15510	38084		2 31 15.4	+46 23 41	-0.004	-0.02	7.0	0.2	2.1	A5 V	-11	92 s		
15375	12315		2 31 22.0	+64 31 53	-0.004	+0.01	8.0	1.1	-0.1	K2 III		360 s		
15449	23394		2 31 26.4	+58 01 28	-0.009	-0.02	6.8	1.4	-0.3	K5 III		240 s		
15582	55625		2 31 27.9	+37 27 15	+0.002	-0.02	7.70	0.4	2.6	F0 V		110 s	1914	m
15695	110591		2 31 29.7	+1 05 39	-0.003	-0.02	7.30	0.1	1.4	A2 V		150 s	1924	m
15694	110589		2 31 30.0	+2 16 02	+0.001	0.00	5.25	1.27	-0.2	K3 III	+26	120 s		
15631	75419		2 31 31.9	+27 34 10	-0.001	+0.01	7.2	0.4	3.0	F2 V		70 s		
15783	167853		2 31 33.4	-27 30 15	+0.001	+0.04	7.60	0.1	1.4	A2 V		170 s	1931	m
15934	232802		2 31 34.0	-56 36 05	+0.001	0.00	7.5	2.2	0.2	K0 III		55 s		
15960	232804		2 31 37.9	-58 52 14	+0.007	+0.04	7.7	0.8	4.0	F8 V		39 s		
16522	255898	μ Hyi	2 31 40.7	-79 06 33	+0.046	-0.04	5.28	0.98	0.3	G4 III	-15	78 s		
15624	55629		2 31 45.5	+39 57 48	0.000	0.00	7.8	1.1	0.2	K0 III		330 s		
15673	75422		2 31 47.7	+21 22 31	0.000	-0.01	7.5	1.1	0.2	K0 III		290 s		
15625	55630		2 31 50.2	+38 07 18	+0.004	-0.03	7.20	1.1	0.2	K0 III		250 s	1919	m
8890	308	1 α UMi	2 31 50.4	+89 15 51	+0.232	-0.01	2.02	0.60	-4.6	F8 Ib	-17		1477	Polaris, m,v
15498	23402		2 31 50.6	+57 31 47	+0.001	-0.03	7.3	1.1	0.2	K0 III	+11	240 s		
15497	23403		2 31 52.8	+57 41 51	-0.005	0.00	7.02	0.78	-7.1	B6 Ia	-39	2100 s	1911	m
15579	38089		2 31 54.1	+46 35 09	+0.008	-0.02	7.1	0.4	3.0	F2 V	+23	64 s		
15623	38096		2 31 59.4	+42 23 28	+0.001	-0.01	8.0	1.1	0.2	K0 III		320 s		
15807	167856		2 32 00.2	-23 05 05	+0.001	-0.01	8.0	-0.1	1.7	A3 V		190 s		
15798	148445	76 σ Cet	2 32 05.1	-15 14 41	-0.005	-0.12	4.75	0.45	2.1	F5 IV	-29	33 s		
15656	55635	14 Tri	2 32 06.1	+36 08 50	+0.004	+0.02	5.15	1.47	-0.3	K5 III	-36	120 s		
15594	38092		2 32 06.3	+50 10 50	+0.001	0.00	7.3	1.6	-0.5	M2 III		330 s		
15672	55637		2 32 06.5	+34 40 42	+0.004	-0.03	8.0	0.5	3.4	F5 V		82 s		
15780	129958		2 32 08.7	-1 11 42	+0.003	-0.03	7.9	1.1	-0.1	K2 III		410 s		
15593	38093		2 32 08.8	+50 13 00	0.000	-0.01	7.72	0.06	0.6	A0 V		260 s		
15779	129959	75 Cet	2 32 09.3	-1 02 06	-0.002	-0.03	5.35	1.02	0.4	gG3	-5	66 s		

HD	SAO	Star Name	α 2000	δ 2000	μ(α)	μ(δ)	V	B-V	M_v	Spec	RV	d(pc)	ADS	Notes
15875	193757		2 32 11.3	-32 36 22	+0.006	+0.01	6.6	1.3	0.2	K0 III		130 s		
15747	92991		2 32 11.8	+12 41 14	+0.004	+0.01	7.6	0.1	1.4	A2 V		120 mx		
15559	23407		2 32 11.8	+55 27 31	+0.013	-0.02	6.7	0.4	3.0	F2 V		54 s		
15717	75427		2 32 12.1	+22 00 25	+0.005	-0.01	7.2	0.4	2.6	F0 V		83 s		
15771	110596		2 32 12.1	+6 43 02	-0.001	-0.01	7.8	1.1	0.2	K0 III		330 s		
15847	167860		2 32 13.6	-27 47 20	+0.006	+0.07	7.6	1.0	-0.1	K2 III		170 mx		
15889	193759		2 32 14.7	-36 25 39	+0.005	+0.01	6.30	1.02	0.3	G8 III		140 s		
15734	75428		2 32 14.7	+21 07 06	-0.002	+0.01	7.6	0.9	3.2	G5 IV		75 s		
15929	215936		2 32 14.9	-45 52 16	-0.004	-0.02	7.3	0.6	3.2	G5 IV		66 s		
15704	55641		2 32 15.3	+32 05 55	-0.002	0.00	7.7	0.6	4.4	G0 V		46 s		
15683	38099		2 32 22.4	+40 16 21	+0.001	-0.01	6.7	0.1	0.6	A0 V		170 s		
15733	75430		2 32 22.9	+22 20 03	+0.002	+0.01	7.8	0.4	2.6	F0 V		110 s		
15684	55640		2 32 23.3	+40 05 07	+0.002	-0.01	7.6	0.0	0.4	B9.5 V		270 s		
15887	167866		2 32 30.0	-27 00 13	+0.001	-0.01	7.4	0.3	1.7	A3 V		100 s		
15557	12325		2 32 40.3	+61 44 05	0.000	+0.01	7.4	0.4	3.1	F3 V	-15	72 s		
15821	110603		2 32 42.0	+1 02 00	-0.001	-0.01	7.5	0.8	3.2	G5 IV		72 s		
15558	12326		2 32 42.1	+61 27 21	-0.005	0.00	7.83	0.52		O6	-50		1920	m
15714	55644		2 32 44.3	+39 32 57	+0.001	-0.04	7.7	0.0	0.0	B8.5 V		340 s		
15873	148450		2 32 49.3	-11 24 48	0.000	0.00	7.5	1.1	0.2	K0 III		290 s		
15755	55650		2 32 52.4	+34 32 33	-0.005	-0.01	5.83	1.07	3.2	G5 IV	-2	22 s		
15814	92998	29 Ari	2 32 54.1	+15 02 05	-0.001	+0.04	6.04	0.54	3.4	F5 V	+6	30 s		
15745	55649		2 32 55.8	+37 19 59	+0.006	-0.08	7.5	0.4	2.6	F0 V		95 s		
15649	23418		2 32 55.9	+54 32 40	0.000	-0.01	7.8	0.1	1.7	A3 V		160 s		
16667	255907		2 33 02.2	-79 02 09	+0.004	+0.02	7.7	1.1	0.2	K0 III		320 s		
15975	193763	λ¹ For	2 33 07.0	-34 39 00	-0.002	-0.02	5.90	1.06	0.2	gK0		130 s		
15788	75436		2 33 07.7	+29 57 52	-0.001	-0.03	7.8	1.1	0.2	K0 III	+11	330 s		
15754	38113		2 33 09.2	+41 49 03	+0.004	-0.01	8.0	0.5	4.0	F8 V		62 s		
15753	38112		2 33 09.3	+42 53 45	-0.014	-0.08	7.1	0.6	4.4	G0 V		35 s		
15640	23420		2 33 13.3	+59 59 58	-0.002	-0.02	7.59	0.05	-1.1	B5 V		380 s	1932	m
16493	255904		2 33 16.6	-75 53 36	+0.025	+0.04	7.00	0.4	3.0	F2 V		63 s		m
15958	167870		2 33 17.0	-26 17 28	0.000	-0.01	8.0	1.0	-0.1	K2 III		420 s		
15703	23425		2 33 18.9	+52 18 32	-0.001	-0.01	7.00	0.1	1.4	A2 V	-11	130 s	1938	m
16408	--		2 33 19.2	-72 39 45			7.8	0.5	3.4	F5 V		75 s		
15472	4673		2 33 26.6	+70 57 24	0.000	+0.01	7.92	0.06		B3 e	-45			
15904	110612		2 33 27.6	+5 42 00	0.000	0.00	7.9	1.1	0.2	K0 III		350 s		
15690	23426		2 33 32.6	+57 32 14	0.000	-0.01	8.01	0.66	-5.7	B1.5 Ib	-35	1700 s	1937	m
15832	55658		2 33 33.5	+31 24 30	-0.003	-0.03	7.40	1.1	-0.1	K2 III		320 s	1947	m
16226	248595		2 33 33.6	-62 35 13	+0.004	+0.01	6.77	-0.06		B9	+9			
16048	193770		2 33 35.5	-37 24 24	+0.003	+0.04	7.7	1.7	-0.3	K5 III		320 s		
15869	93002		2 33 36.4	+18 52 48	+0.006	+0.01	6.8	0.2	2.1	A5 V	+17	86 s		
16032	193768		2 33 36.7	-32 49 37	-0.001	-0.03	7.8	0.9	0.2	K0 III		330 s		
15996	167878		2 33 40.1	-20 00 07	-0.003	-0.08	6.21	1.10		K0				
15292	4672		2 33 40.3	+77 40 01	+0.007	+0.12	7.8	0.8	3.2	G5 IV		77 mx		
16105	215942		2 33 40.8	-48 23 21	-0.006	+0.02	8.00	0.5	4.0	F8 V		63 s		m
15831	55659		2 33 42.6	+34 20 29	+0.001	-0.01	7.6	0.1	0.6	A0 V		250 s		
15971	148467		2 33 43.7	-13 08 56	+0.002	-0.02	6.5	1.6		M5 III	-27			U Cet, v
16427	255901		2 33 43.9	-72 20 57	+0.029	-0.01	7.0	0.4	4.0	F8 V		40 s		
16047	193772		2 33 50.5	-29 55 58	+0.013	+0.09	7.3	0.2	3.4	F5 V		61 s		
16046	167882	ω For	2 33 50.6	-28 13 57	-0.002	-0.01	4.90	-0.05		B9	+10		1954	m
16241	248598		2 33 51.3	-60 58 59	+0.018	+0.01	6.9	0.2	3.0	F2 V		61 s		
15866	55660		2 33 54.2	+31 34 49	-0.004	-0.07	8.01	0.66		G5				
16170	232818		2 33 54.6	-51 05 37	-0.001	-0.03	6.24	0.52	3.4	F5 V		33 s		
16419	255902		2 33 55.0	-70 46 06	0.000	+0.01	7.8	1.6	-0.5	M2 III		450 s		
15995	148469		2 33 55.3	-12 20 13	0.000	-0.03	7.0	0.5	4.0	F8 V		40 s		
16065	193773		2 33 55.8	-32 11 52	+0.009	+0.05	8.0	0.7	4.4	G0 V		45 s		
15728	23430		2 33 58.1	+59 12 45	-0.004	-0.01	7.1	1.1	0.2	K0 III	-23	220 s		
15533	4677		2 33 58.5	+71 17 40	+0.014	-0.06	6.7	1.1	0.2	K0 III		180 s		
15883	55661		2 33 59.4	+31 24 23	-0.001	+0.01	7.7	1.1	0.2	K0 III		310 s		
15942	93004		2 34 03.4	+12 10 51	+0.011	-0.03	7.7	0.7	4.4	G0 V		45 s		
15896	55664		2 34 05.2	+32 27 08	-0.001	-0.01	7.5	1.1	0.2	K0 III		290 s		
16530	255906		2 34 07.2	-74 01 25	+0.007	+0.01	7.3	0.1	1.4	A2 V		150 s		
15994	129981		2 34 07.3	-5 38 05	+0.001	+0.04	7.15	1.08	0.0	K1 III		270 s	1953	m
15865	55662		2 34 08.3	+38 44 06	+0.002	-0.01	6.7	1.1	0.2	K0 III		200 s		
16087	193774		2 34 10.4	-31 31 28	-0.003	-0.05	7.50	0.4	2.6	F0 V		96 s		m
15830	38127		2 34 10.6	+42 47 06	+0.037	-0.21	7.63	0.67	4.4	G0 V	+16	31 mx		
16095	193775		2 34 11.7	-33 07 05	+0.022	-0.11	7.6	0.3	4.0	F8 V		34 mx		
16199	110616		2 34 13.0	+8 59 20	+0.002	-0.02	7.8	0.4	3.0	F2 V		89 s		
15895	55666		2 34 13.9	+33 47 30	-0.003	-0.04	6.9	1.1	0.2	K0 III		210 s		
15894	55667		2 34 17.6	+35 08 59	+0.003	-0.06	7.3	0.1	1.4	A2 V		74 mx		
15983	93006		2 34 25.8	+11 36 16	+0.002	-0.04	7.0	0.1	0.6	A0 V		190 s		
15417	4675		2 34 30.8	+76 43 05	+0.022	-0.05	6.8	1.6	-0.5	M2 III		230 mx		
11696	362		2 34 32.7	+88 28 17	+0.140	-0.03	7.9	0.1	1.7	A3 V		82 mx		
15682	12337		2 34 35.1	+67 44 33	-0.004	-0.05	7.8	0.8	3.2	G5 IV		81 s		
16093	167885		2 34 35.7	-21 55 49	-0.003	-0.07	7.1	1.1	0.2	K0 III		190 s		
15863	38133		2 34 40.2	+50 03 56	-0.001	-0.01	6.8	0.0	0.4	B9.5 V		180 s		

HD	SAO	Star Name	α 2000	δ 2000	μ(α)	μ(δ)	V	B-V	M$_V$	Spec	RV	d(pc)	ADS	Notes
16074	129984	77 Cet	2^{h}34^{m}42^s.6	-7°51'34"	+0^s.004	-0".06	5.75	1.40	-0.3	gK4	+25	160 s		m
15939	55672		2 34 43.0	+34 05 03	+0.004	+0.02	7.8	0.4	2.6	F0 V		110 s		
16062	110624		2 34 52.9	+1 14 41	0.000	-0.01	8.0	1.1	0.2	K0 III		360 s		
16240	215950		2 34 59.6	-42 06 47	0.000	-0.02	7.3	0.9	-0.5	M4 III		370 s		
17216	258306		2 35 01.7	-82 14 08	+0.012	-0.10	7.5	0.4	4.4	G0 V		42 s		
16145	148481		2 35 04.0	-17 17 23	0.000	+0.01	7.9	0.1	0.6	A0 V		290 s		
16060	110625		2 35 04.1	+7 28 17	-0.001	-0.10	6.18	1.06	0.3	gG6	-25	120 s		
15769	12342		2 35 07.4	+66 02 56	+0.005	-0.03	7.1	1.1	0.2	K0 III	-14	210 s		
16195	167893		2 35 10.8	-28 41 51	-0.002	+0.01	8.0	-0.1	0.6	A0 V		300 s		
15922	38140		2 35 11.7	+49 51 39	+0.007	+0.01	7.9	0.1	0.6	A0 V		100 mx		
15980	38143		2 35 13.9	+40 35 49	0.000	0.00	7.8	0.0	0.4	B9.5 V		310 s		
16015	55679		2 35 14.8	+32 51 50	0.000	0.00	7.9	0.7	4.4	G0 V		51 s		
16141	129992		2 35 19.9	-3 33 37	-0.010	-0.43	6.83	0.66	3.2	G5 IV	-53	53 s		
16084	93012		2 35 20.8	+13 08 33	-0.001	-0.03	7.8	0.8	3.2	G5 IV		84 s		
16152	129994		2 35 24.4	-9 21 03	0.000	-0.01	7.2	0.1	0.6	A0 V		210 s		
16004	55680		2 35 27.9	+39 39 52	+0.002	-0.02	6.30	-0.10	-0.6	B8 IV		240 s	1961	m
15829	12346		2 35 30.1	+63 14 35	-0.007	0.00	7.4	0.9	5.2	G5 V		27 s		
16346	232825		2 35 31.3	-50 51 12	+0.002	-0.01	7.1	1.0	3.2	G5 IV		43 s		
15992	38147		2 35 33.4	+44 38 24	-0.005	-0.01	7.4	0.0	0.4	B9.5 V	0	220 mx		
16202	148487		2 35 38.5	-14 56 42	-0.001	-0.07	6.8	1.1	0.2	K0 III		210 s		
16028	55684		2 35 38.7	+37 18 45	0.000	-0.01	5.71	1.39	0.2	K0 III	-6	74 s	1964	m
15912	23446		2 35 39.4	+58 30 31	-0.001	-0.03	8.0	0.5	4.0	F8 V		61 s		
15726	4685		2 35 42.2	+70 55 03	+0.001	-0.01	8.0	1.1	-0.1	K2 III		360 s		
15850	12347		2 35 42.8	+63 38 03	-0.005	-0.05	7.8	0.5	3.4	F5 V		73 s		
17946	258311		2 35 43.2	-85 01 03	+0.001	-0.02	7.2	1.5	0.2	K0 III		130 s		
16027	38152		2 35 43.4	+43 35 57	-0.001	+0.03	7.9	1.1	-0.1	K2 III		340 s		
15784	12345		2 35 43.8	+68 22 06	+0.008	+0.01	6.8	0.5	-2.0	F4 II		110 mx		
16058	55687	15 Tri	2 35 46.7	+34 41 15	+0.002	-0.05	5.35	1.66	-0.5	M4 III	-10	140 s		m
16070	55688		2 35 50.0	+34 43 32	0.000	-0.01	6.8	0.2	2.1	A5 V	+7	86 s		
16161	110635	78 v Cet	2 35 52.4	+5 35 36	-0.002	-0.02	4.86	0.87	0.3	G8 III	+5	82 s	1971	m
16090	55691		2 35 54.5	+31 08 24	+0.001	-0.07	7.9	0.7	2.9	G1 IV	-4	100 s		
16057	55690		2 35 54.8	+37 03 28	-0.004	-0.02	7.6	0.9	3.2	G5 IV		75 s		
16252	148492		2 35 56.5	-17 37 58	-0.004	-0.02	7.1	0.6	4.4	G0 V		35 s		
16263	167903		2 35 59.2	-21 24 13	-0.001	+0.02	7.40	0.7	4.4	G0 V		40 s	1978	m
16212	130004	80 Cet	2 35 59.9	-7 49 54	-0.003	-0.06	5.53	1.59	-0.4	M0 III	+14	150 s		m
16150	93016		2 36 01.1	+13 45 18	0.000	-0.01	7.4	0.5	3.4	F5 V		64 s		
16189	110637		2 36 02.5	+6 25 57	+0.008	-0.03	7.0	0.2	2.1	A5 V		42 mx		
16099	75459		2 36 02.5	+29 52 27	-0.002	-0.02	8.00	1.1	-0.2	K3 III		440 s		
16111	75460		2 36 04.0	+29 24 13	-0.001	-0.02	7.1	0.0	0.0	B8.5 V	+4	260 s		
16160	110636		2 36 04.8	+6 53 13	+0.121	+1.46	5.82	0.98	6.5	K3 V	+23	7.2 t		268 G. Ceti, m
15416	4679		2 36 07.8	+79 43 43	+0.008	-0.04	7.6	0.8	3.2	G5 IV		77 s		
16307	193795	z¹ For	2 36 09.2	-30 02 41	-0.001	+0.01	5.75	1.02		G5				
16128	75461		2 36 09.7	+27 10 01	0.000	+0.01	7.9	0.9	3.2	G5 IV		88 s		
--	23457		2 36 09.7	+52 29 15	+0.001	+0.01	7.9	2.5	-0.5	M2 III		150 s		
15963	23451		2 36 12.2	+58 04 32	-0.001	0.00	8.03	0.06	-5.0	A2 Ib		3900 s		
16269	148495		2 36 16.7	-11 09 58	+0.003	+0.01	7.4	1.1	0.2	K0 III		270 s		
15849	12351		2 36 16.9	+67 30 48	+0.003	+0.05	7.4	0.4	3.0	F2 V		74 s		
16097	55695		2 36 18.9	+40 12 18	+0.002	0.00	7.4	0.0	0.4	B9.5 V		250 s		
16277	148496		2 36 20.2	-12 53 48	0.000	+0.02	7.7	0.9	3.2	G5 IV		78 s		
16261	130008		2 36 26.5	-0 38 25	+0.001	-0.04	7.8	0.1	1.7	A3 V		160 s		
16317	167914		2 36 29.3	-23 55 10	+0.001	-0.02	7.5	1.1	0.2	K0 III		250 s		
15937	12356		2 36 33.3	+63 29 48	+0.001	-0.03	7.6	1.1	0.2	K0 III		270 s		
16295	148499		2 36 33.9	-14 39 24	-0.003	-0.03	6.9	1.1	0.2	K0 III		220 s		
16247	110640		2 36 35.0	+7 43 47	-0.004	-0.03	5.81	1.04	0.2	gK0	-25	130 s		
16108	38161		2 36 36.0	+42 24 04	+0.002	0.00	6.7	0.0	0.0	B8.5 V	-15	200 s		
16234	93022	31 Ari	2 36 37.8	+12 26 52	+0.019	-0.08	5.68	0.49	3.4	F5 V	+7	32 ts		
16039	23464		2 36 40.4	+56 11 49	-0.002	0.00	7.36	1.59	-2.1	K0 II		400 s		
16187	55703		2 36 42.8	+31 36 28	-0.003	0.00	6.10	1.05		K0	+3			
16388	193805		2 36 46.1	-35 49 45	-0.001	-0.05	7.5	1.5	-0.5	M2 III		400 s		
16358	167920		2 36 47.7	-25 01 26	-0.005	-0.07	7.9	0.4	4.4	G0 V		51 s		
16082	23470		2 36 48.2	+51 57 40	0.000	-0.01	7.4	0.8	3.2	G5 IV	-16	66 s		
16356	167918		2 36 50.0	-20 07 20	-0.003	-0.04	7.0	0.9	0.2	K0 III		230 s		
17137	258308		2 36 51.4	-79 48 51	-0.007	0.00	7.8	1.0	3.2	G5 IV		63 s		
16068	23469		2 36 52.8	+55 54 56	+0.001	0.00	7.7	1.1	-0.1	K2 III		320 s		
16404	193808		2 36 53.6	-35 19 13	+0.005	-0.01	7.5	1.2	0.2	K0 III		210 s		
16038	23467		2 36 55.1	+59 53 01	0.000	-0.01	7.60	0.00	-0.2	B8 V		320 s	1970	m
16176	55705		2 36 57.1	+38 43 58	+0.012	-0.19	5.90	0.48	3.4	F5 V	+1	30 s		m
16232	75470	30 Ari	2 36 57.7	+24 38 55	+0.011	+0.01	6.50	0.41	2.1	F5 IV	+17	61 s	1982	m
16417	193811	λ² For	2 36 58.5	-34 34 42	-0.002	-0.27	5.79	0.66	3.2	G5 IV	+4	23 ts		
16246	75471	30 Ari	2 37 00.5	+24 38 51	+0.010	0.00	7.09	0.50	3.7	F6 V	+15	61 s	1982	m
16175	38170		2 37 02.1	+42 03 44	0.000	-0.07	7.2	0.7	4.4	G0 V		36 s		
16403	167922		2 37 05.2	-28 59 38	+0.005	-0.03	7.2	0.6	4.4	G0 V		33 s		
16220	55711		2 37 06.4	+32 53 32	+0.005	+0.07	6.25	0.48	3.4	F5 V	0	35 s		
16329	130020		2 37 07.9	-2 20 03	+0.002	+0.01	7.5	0.4	3.0	F2 V		80 s		
16245	55716		2 37 10.5	+30 24 26	+0.001	-0.03	7.80	0.1	0.6	A0 V	+7	280 s		m

HD	SAO	Star Name	α 2000	δ 2000	μ(α)	μ(δ)	V	B-V	M$_v$	Spec	RV	d(pc)	ADS	Notes
			h m s	° ′ ″	s	″								
16219	55715		2 37 20.7	+39 53 46	+0.001	-0.02	6.40	-0.12	-1.1	B5 V	+8	310 s		
16555	232835	η Hor	2 37 24.2	-52 32 36	+0.009	-0.01	5.31	0.27		A5	-3	15 mn		
16088	23474		2 37 26.0	+60 05 19	0.000	0.00	7.5	0.4	0.6	F0 III		220 s		
16520	215969		2 37 27.3	-46 38 47	+0.005	+0.01	7.8	0.7	1.4	A2 V		63 s		
15036	404		2 37 27.6	+83 50 06	-0.022	+0.03	6.8	1.1	0.2	K0 III		210 s		
16037	12360		2 37 27.8	+64 06 49	+0.001	0.00	7.6	0.5	4.0	F8 V		51 s		
16302	75476		2 37 34.3	+20 34 30	+0.002	0.00	6.9	0.8	3.2	G5 IV		56 s		
15948	12357		2 37 34.4	+69 03 24	+0.006	-0.02	7.4	1.1	0.2	K0 III	-44	190 mx		
16024	12361		2 37 36.0	+65 44 43	+0.008	-0.01	5.78	1.56	-5.9	cK5	+41	1700 s		
16184	--		2 37 36.1	+50 14 14	+0.002	-0.01	7.30	0.00	0.0	B8.5 V		260 s		
16400	130026	81 Cet	2 37 41.7	-3 23 46	+0.002	-0.04	5.65	1.02	0.3	gG5	+8	91 s		
16446	167931		2 37 44.3	-22 59 31	-0.001	+0.05	6.9	1.0	3.2	G5 IV	+24	39 s		
16353	93029		2 37 45.2	+12 16 12	-0.003	+0.02	7.58	1.01		G5				
16285	55723		2 37 50.5	+35 09 39	+0.001	-0.06	7.7	0.9	3.2	G5 IV		79 s		
16434	148511		2 37 50.6	-13 07 47	+0.003	+0.07	7.9	0.5	4.0	F8 V		59 s		
16067	12363		2 37 52.7	+65 19 07	-0.003	-0.07	7.6	0.5	4.0	F8 V		52 s		
16503	193826		2 37 53.0	-33 40 39	-0.004	-0.01	7.4	1.1	0.2	K0 III		230 s		
16284	55722		2 37 54.4	+38 55 06	-0.002	-0.04	8.0	0.1	1.4	A2 V		210 s		
16777	248611		2 37 58.6	-67 18 03	+0.001	0.00	6.0	1.1	0.2	K0 III		140 s		
16399	110653		2 38 00.7	+7 41 43	+0.006	-0.04	6.39	0.44		F5	+13			
15920	4694		2 38 01.9	+72 49 06	-0.006	+0.02	5.16	0.88	0.3	G8 III	-2	94 s		
15842	4691		2 38 09.7	+75 31 39	-0.003	-0.03	7.8	0.1	1.7	A3 V		160 s		
16591	215977		2 38 12.0	-41 54 14	+0.011	+0.06	7.26	0.94	5.9	K0 V		17 s		
16258	23486		2 38 12.6	+50 29 29	-0.001	-0.01	7.1	1.1	-0.1	K2 III	-65	250 s		
16590	215979		2 38 15.4	-41 33 10	+0.005	+0.02	7.2	1.1	0.2	K0 III		230 s		
16327	55729		2 38 17.7	+37 43 37	-0.003	-0.04	6.18	0.47	3.7	dF6	+9	30 s	1996	m
16538	193829	ι² For	2 38 18.6	-30 11 40	+0.008	-0.08	5.83	0.48	3.4	F5 V		29 s		
16066	12364		2 38 23.0	+68 03 44	-0.005	+0.04	7.8	0.4	-2.0	F2 II		130 mx		
16589	193834		2 38 24.6	-37 59 27	+0.008	-0.06	6.49	0.52	3.4	dF5		37 s		
236979	23484		2 38 25.3	+57 02 47	0.000	+0.01	7.91	2.32	-5.6	M3 Iab		1400 s		YZ Per, v
16350	55735		2 38 27.7	+38 05 21	-0.001	0.00	6.3	0.0	-0.6	A0 III	+2	230 s		
16397	55739		2 38 27.8	+30 49 00	-0.038	-0.38	7.34	0.60	4.5	G1 V	-100	37 s		
16635	215982		2 38 30.0	-44 43 58	-0.001	+0.03	7.4	0.1	2.1	A5 V		110 s		
16387	75490		2 38 30.1	+27 30 57	+0.001	+0.02	8.0	1.6	-0.5	M4 III		510 s		
16634	215983		2 38 30.7	-44 42 30	-0.004	-0.02	8.0	0.7	3.2	G5 IV		90 s		
16526	167939		2 38 33.1	-20 27 23	-0.004	-0.05	7.0	1.3	0.2	K0 III		160 s		
16467	110655		2 38 36.8	+3 26 35	+0.005	+0.01	6.21	1.00	0.2	G9 III	+2	150 s		m
16423	93033		2 38 37.3	+19 43 39	+0.001	-0.01	7.50	1.1	0.2	K0 III		290 s		m
16376	55742		2 38 38.0	+36 20 18	0.000	-0.04	8.0	0.9	3.2	G5 IV		92 s		
16699	232842		2 38 44.4	-52 57 10	+0.004	+0.04	7.40	0.5	4.0	F8 V		48 s		
16396	55748		2 38 45.7	+33 25 07	0.000	-0.02	7.50	1.1	-0.1	K2 III	-3	320 s	2004	m
16602	193836		2 38 46.3	-30 35 58	+0.003	0.00	7.5	1.4	0.2	K0 III		160 s		
16432	75495	32 ν Ari	2 38 48.9	+21 57 41	-0.001	-0.01	5.30	0.16	2.4	A7 V	+8	38 s		
16386	55746		2 38 51.5	+39 09 00	-0.001	-0.01	7.86	-0.06	0.4	B9.5 V		310 s		
16940	255910		2 38 53.5	-70 40 48	+0.005	+0.01	6.6	1.6	0.2	K0 III		86 s		
16600	167945		2 38 54.5	-25 44 44	0.000	-0.09	7.14	1.20	-0.1	K2 III		190 mx		
16744	232845		2 38 57.0	-54 50 16	+0.006	-0.03	7.50	0.5	3.4	F5 V		66 s		m
16218	12369		2 38 58.1	+62 35 30	+0.003	-0.02	6.66	-0.02	0.4	B9.5 V		180 s		m
16500	110658		2 38 59.5	+8 55 11	0.000	-0.02	8.0	0.1	0.6	A0 V		300 s		m
16480	--		2 39 00.2	+14 51 40	+0.001	-0.01	7.40	1.1	-1.1	K2 II-III	+1	510 s	2008	m
16548	130036		2 39 00.7	-8 55 39	+0.017	-0.03	7.1	0.6	4.4	G0 V		35 s		
16499	93034		2 39 03.3	+10 38 14	+0.003	-0.06	6.8	1.1	0.2	K0 III		170 mx		
16230	12371		2 39 03.3	+62 34 30	+0.003	-0.05	7.62	0.11	0.6	A0 V		130 mx		
16624	167951		2 39 05.5	-27 49 20	+0.003	-0.06	7.2	1.1	0.2	K0 III		140 mx		
16743	232847		2 39 06.6	-52 56 05	-0.006	+0.06	6.9	0.3	2.1	A5 V		76 s		
16569	148523		2 39 08.8	-9 49 53	-0.002	-0.03	6.7	0.8	3.2	G5 IV		51 s		
16385	38195		2 39 08.8	+46 12 13	+0.001	-0.01	7.9	0.1	0.6	A0 V		260 s		
16293	23496		2 39 12.9	+58 50 35	+0.001	-0.03	7.8	1.1	0.2	K0 III		290 s		
16816	248613		2 39 16.9	-60 32 50	0.000	0.00	7.8	1.4	0.2	K0 III		200 s		
16755	215994		2 39 24.9	-49 08 55	-0.001	-0.01	7.8	1.4	-0.3	K5 III		410 s		
16650	167955		2 39 26.1	-25 34 09	+0.002	+0.01	7.8	0.9	0.2	K0 III		330 s		
16582	110665	82 δ Cet	2 39 28.9	+0 19 43	+0.001	0.00	4.07	-0.22	-3.0	B2 IV	+13	260 s		v
16891	248616		2 39 31.7	-64 16 55	+0.004	+0.02	6.55	-0.08		B9				
16620	148528	83 ε Cet	2 39 33.7	-11 52 20	+0.010	-0.23	4.84	0.45	2.8	F5 IV-V	+15	22 ts		m
16939	248618		2 39 34.6	-66 57 26	+0.006	-0.02	7.2	1.5	0.2	K0 III		130 s		
16690	193841		2 39 35.0	-34 30 39	-0.002	-0.03	8.0	1.6	0.2	K0 III		150 s		
16978	248621	ε Hyi	2 39 35.5	-68 16 00	+0.017	+0.01	4.11	-0.06	-0.8	B9 III	+6	91 s		
16864	248615		2 39 36.7	-61 32 00	+0.002	-0.02	7.9	0.0	0.4	B9.5 V		320 s		
16649	167957		2 39 39.4	-20 25 43	-0.008	-0.20	7.4	0.7	4.4	G0 V		32 s		
16852	232851		2 39 40.0	-59 34 04	+0.005	-0.01	7.20	1.1	0.2	K0 III		250 s		m
16899	248617		2 39 44.6	-62 41 47	+0.001	-0.01	7.6	1.1	0.2	K0 III		300 s		
16853	232852		2 39 44.8	-59 34 06	+0.005	+0.02	8.00	0.2	2.1	A5 V		150 s		m
16715	193844		2 39 46.8	-35 01 31	-0.001	-0.02	7.3	1.6	0.2	K0 III		110 s		
16754	215996		2 39 47.9	-42 53 30	+0.009	-0.02	4.75	0.06	1.4	A2 V	+20	25 mx		m
16512	55760		2 39 48.3	+33 29 42	0.000	-0.04	7.9	1.1	0.2	K0 III		330 s		

HD	SAO	Star Name	α 2000	δ 2000	μ(α)	μ(δ)	V	B-V	M_V	Spec	RV	d(pc)	ADS	Notes
			$2^h39^m50^s.5$	$+33°56'56''$	$-0^s.001$	$-0''.03$								
16511	55761		2 39 50.5	+33 56 56	-0.001	-0.03	7.80	1.1	0.2	K0 III		320 s	2020	m
16963	248622		2 39 50.6	-66 07 07	+0.010	-0.06	7.3	0.5	4.0	F8 V		45 s		
16619	130042		2 39 51.8	+0 08 48	+0.005	-0.16	7.84	0.65	5.0	dG4	+40	37 s	2028	m
16173	4701		2 39 52.8	+71 13 27	-0.001	-0.02	8.0	1.1	0.2	K0 III		320 s		
16675	148534		2 40 00.6	-16 18 31	+0.001	0.00	7.3	0.1	1.4	A2 V		150 s		
16349	12377		2 40 00.9	+62 36 08	+0.003	-0.02	7.3	0.1	0.6	A0 V		200 s		
16733	193846		2 40 02.4	-30 38 02	-0.001	-0.06	6.52	1.04		G5				
16607	93039		2 40 09.0	+16 43 37	+0.002	-0.04	7.9	0.1	1.4	A2 V		200 s		
16673	130047		2 40 12.2	-9 27 11	-0.011	-0.08	5.78	0.52	3.7	dF6	-4	25 s		
16647	110673		2 40 15.6	+6 06 43	+0.003	0.00	6.25	0.40	3.0	dF2	+18	43 s		
16580	75506		2 40 18.4	+29 47 48	+0.001	0.00	7.5	0.0	0.4	B9.5 V	+14	260 s		
16485	38207		2 40 20.9	+49 33 37	+0.001	-0.01	6.5	0.1	0.6	A0 V		140 s		
16461	23513		2 40 21.8	+52 48 15	+0.002	-0.01	7.49	0.14		A0				
16594	55773		2 40 26.1	+32 19 13	+0.002	0.00	7.5	0.1	1.4	A2 V	+2	170 s		
16723	148538		2 40 26.4	-14 26 55	+0.003	+0.02	6.6	0.2		A m				
16639	93041		2 40 27.1	+13 31 38	+0.001	-0.01	7.7	1.1	0.2	K0 III		310 s		
16785	193848		2 40 30.9	-31 28 02	+0.006	+0.01	7.6	0.1	3.4	F5 V		68 s		
16753	167965		2 40 31.4	-24 08 07	+0.001	+0.02	8.00	0.5	3.4	F5 V		83 s	2044	m
16629	75509		2 40 31.6	+21 11 16	+0.002	+0.01	7.7	1.6	-0.5	M2 III		440 s		
16789	193850		2 40 31.8	-36 29 39	+0.002	0.00	7.8	1.2	0.2	K0 III		260 s		
16410	12380		2 40 31.8	+61 29 02	0.000	-0.01	7.80	1.1	-0.9	K0 II-III		460 s	2014	m
16878	232856		2 40 32.0	-51 47 19	+0.003	-0.09	8.0	0.3	3.4	F5 V		82 s		
16646	93042		2 40 33.0	+18 35 59	-0.002	+0.03	7.9	0.5	3.4	F5 V		78 s		
16460	23515		2 40 33.4	+56 16 17	-0.001	0.00	7.6	0.4	-6.6	F0 I		3200 s		
16545	38212		2 40 36.9	+44 05 29	+0.002	0.00	7.3	0.1	0.6	A0 V	+5	200 s		
16784	193849		2 40 38.9	-30 08 05	+0.046	+0.10	8.02	0.57	4.4	G0 V	+34	23 mx		
16920	232857	ζ Hor	2 40 39.6	-54 33 00	+0.005	+0.01	5.21	0.40	3.0	dF2	-1	26 s		
16644	75511		2 40 39.8	+21 13 21	+0.001	-0.01	7.9	0.9	3.2	G5 IV		88 s		
16815	215999	ι Eri	2 40 40.0	-39 51 19	+0.012	-0.02	4.11	1.02	0.2	K0 III	-9	61 s		
17072	248626		2 40 40.0	-69 14 02	+0.012	-0.03	6.7	0.5	4.4	G0 V		28 s		
16448	23518		2 40 40.4	+57 18 31	+0.005	-0.06	7.06	1.15	-0.1	K2 III	-15	190 mx		
16628	75510	33 Ari	2 40 41.0	+27 03 39	+0.005	-0.03	5.30	0.09	1.7	A3 V	+17	51 s	2033	m
16638	—		2 40 42.3	+26 37 19	+0.006	-0.01	7.55	0.54	3.8	F7 V	+1	53 s	2034	m
16950	232858		2 40 43.6	-58 08 14	-0.003	-0.08	8.0	0.6	4.0	F8 V		60 s		
16429	12383		2 40 45.0	+61 16 58	+0.002	+0.03	7.67	0.62	-5.5	09.5 III		1300 s	2018	m
16708	110677		2 40 47.3	+2 53 54	+0.003	+0.02	7.80	0.8	3.2	G5 IV	+65	83 s		
16851	216002		2 40 51.8	-41 17 59	+0.001	0.00	7.66	1.62		K5				
16721	110680		2 40 56.5	+4 52 15	0.000	0.00	7.90	0.4	2.6	F0 V		120 s	2043	m
16694	93052		2 41 06.6	+18 47 57	0.000	-0.03	7.04	0.06		B9	+20		2042	m
16524	23524		2 41 07.6	+56 16 12	+0.001	-0.01	7.5	0.0	0.4	B9.5 V		240 s		
16740	110683		2 41 07.7	+6 04 20	-0.001	+0.01	7.3	1.1	-0.1	K2 III		300 s		
16604	38223		2 41 13.0	+47 34 18	+0.001	+0.01	7.8	0.1	0.6	A0 V		240 s		
16765	130055	84 Cet	2 41 13.9	-0 41 45	+0.014	-0.13	5.71	0.52	3.7	dF6	+8	29 ts	2046	m
16763	110688		2 41 21.6	+4 25 40	+0.001	-0.06	6.90	0.1	0.6	A0 V		73 mx	2050	m
16762	110686		2 41 23.6	+8 31 38	-0.001	-0.03	7.7	1.1	0.2	K0 III		310 s		
—	—		2 41 23.6	+44 16 22	-0.001	-0.03	8.0	0.5	4.0	F8 V		63 s		
16990	232860		2 41 26.2	-53 19 12	+0.002	+0.02	7.9	1.2	-0.1	K2 III		390 s		
17215	255913		2 41 28.0	-71 27 46	+0.037	-0.07	7.76	0.72	5.2	G5 V		31 s		m
16826	148549		2 41 29.9	-15 06 05	+0.004	-0.09	7.3	0.5	3.4	F5 V		61 s		
16786	110689		2 41 30.3	+0 32 41	+0.001	-0.06	6.8	0.9	3.2	G5 IV		53 s		
16825	148550		2 41 33.9	-14 32 59	-0.003	+0.04	5.98	0.43	3.4	F5 V	+2	33 s		
17014	232863		2 41 36.6	-56 17 48	+0.007	0.00	7.4	1.1	0.2	K0 III		250 s		
16393	12386		2 41 36.8	+69 42 37	-0.001	+0.03	8.0	0.0	0.0	B8.5 V		350 s		
16897	193863		2 41 39.5	-30 03 41	+0.002	-0.03	7.3	1.5	-0.3	K5 III		330 s		
16556	23534		2 41 40.8	+58 58 38	+0.005	-0.01	7.2	0.5	3.4	F5 V		57 s		
16977	216009		2 41 41.4	-47 07 43	+0.001	+0.02	8.0	0.8	0.2	K0 III		360 s		
16576	23537		2 41 42.8	+55 28 54	+0.001	0.00	7.90	0.1	1.4	A2 V		190 s	2040	m
16824	130061		2 41 48.2	-3 12 49	-0.002	+0.01	6.05	1.17	0.2	gG9	+4	100 s		
16575	23538		2 41 49.6	+57 27 15	+0.003	-0.01	8.0	0.9	3.2	G5 IV		89 s		
16440	12387		2 41 50.9	+68 28 12	+0.002	+0.02	8.0	0.0	-3.6	B7 II		1200 s		
16916	193867		2 41 52.2	-30 24 15	+0.005	+0.02	7.7	0.4	3.2	G5 IV		78 s		
16669	38238		2 41 52.3	+45 00 06	-0.002	-0.01	7.8	1.4	-0.3	K5 III		360 s		
16383	4709		2 41 56.5	+71 37 35	+0.001	0.00	7.4	1.1	0.2	K0 III		250 s		
16896	167977		2 41 57.5	-22 36 22	+0.003	-0.02	7.9	1.6		M6 III				
16835	110692		2 41 58.9	+0 33 00	+0.001	0.00	7.8	0.3	2.6	F0 V		110 s		
16802	93059		2 42 00.0	+10 32 41	-0.001	-0.03	6.6	1.1	0.2	K0 III		190 s		
16706	38247		2 42 00.4	+41 23 39	-0.001	0.00	7.8	1.1	0.2	K0 III		300 s		
16761	55792		2 42 00.7	+30 56 09	+0.008	-0.04	7.2	1.1	0.2	K0 III		120 mx		
16791	93057		2 42 02.4	+17 45 25	+0.001	-0.09	7.8	0.8	3.2	G5 IV		84 s		
16975	193873		2 42 06.5	-38 23 02	+0.001	+0.01	6.01	0.92		G5				
17006	216013		2 42 08.4	-46 31 28	+0.002	-0.09	6.10	0.88	3.2	G8 IV		38 s		
16989	216011		2 42 09.0	-42 31 54	+0.003	+0.03	7.4	1.2	0.2	K0 III		220 s		
16728	38254		2 42 13.2	+42 41 54	+0.002	-0.05	7.94	0.00		B9			2052	m
16739	55793	12 Per	2 42 14.9	+40 11 38	-0.001	-0.18	4.91	0.59	4.2	F9 V	-23	15 ts		
16773	55800		2 42 18.2	+34 17 41	-0.001	+0.02	7.1	0.9	3.2	G5 IV		59 s		

HD	SAO	Star Name	α 2000	δ 2000	μ(α)	μ(δ)	V	B–V	M$_V$	Spec	RV	d(pc)	ADS	Notes
			h m s	° ′ ″	s	″								
16912	148558		2 42 18.6	-16 01 18	+0.004	0.00	7.4	0.2	2.1	A5 V		95 mx		
16811	93062	34 μ Ari	2 42 21.9	+20 00 41	+0.002	-0.04	5.70	-0.02	0.3	A0 IV-V	-7	120 s	2062	m
16505	12393		2 42 22.7	+68 03 51	0.000	-0.01	7.04	1.24	-0.2	K3 III	-50	280 s		
16616	12399		2 42 23.7	+62 21 30	-0.010	0.00	7.3	0.1	0.6	A0 V		160 mx		
16847	93066		2 42 28.7	+15 13 00	-0.001	-0.03	8.0	0.0	0.4	B9.5 V		330 s		
16861	93067		2 42 28.8	+10 44 30	-0.002	-0.02	6.30	0.06	1.4	A2 V	+6	94 s		
16737	38261		2 42 31.3	+47 01 48	0.000	0.00	7.6	0.4	2.6	F0 V		98 s		
17051	232864	ι Hor	2 42 33.4	-50 48 01	+0.035	+0.23	5.41	0.56	3.0	G3 IV	+17	17 ts		
16626	12402		2 42 35.6	+61 35 45	+0.000	+0.03	7.0	0.4	2.6	F0 V		74 s		
16771	38268		2 42 35.7	+42 35 28	+0.004	-0.02	7.33	0.94		G5				
16810	55804		2 42 41.3	+32 54 05	+0.005	-0.05	7.6	0.4	2.6	F0 V		100 s		
17070	216017		2 42 41.4	-49 43 46	+0.001	-0.02	7.63	0.43	3.4	F5 V		70 s		
17005	167996		2 42 50.3	-28 09 05	-0.001	-0.01	6.8	1.1	4.4	G0 V		15 s		
17367	255915		2 42 50.7	-71 12 19	+0.007	0.00	7.6	1.6	-0.5	M4 III		410 s		
16809	55807		2 42 52.9	+37 10 25	+0.001	-0.02	8.0	0.1	1.7	A3 V		180 s		
16994	167995		2 42 57.6	-20 17 21	+0.008	+0.03	7.50	0.6	4.4	G0 V		42 s	2079	m
16735	23556		2 42 59.6	+53 31 34	+0.008	-0.03	5.84	1.12	0.2	K0 III	-12	110 s	2059	m
16883	75531		2 43 01.0	+20 13 50	+0.001	-0.01	8.0	0.9	3.2	G5 IV		91 s	2072	m
16780	38274		2 43 01.8	+48 15 56	+0.001	+0.01	6.60	0.8	3.2	G5 IV	-5	47 s	2064	m
16727	23555	11 Per	2 43 02.8	+55 06 22	+0.004	-0.02	5.6	-0.1	-0.2	B8 V		120 mx		
17025	193876		2 43 03.5	-31 04 12	+0.001	-0.04	7.5	0.9	4.4	G0 V		27 s		
16869	75530		2 43 04.7	+27 00 14	-0.001	-0.02	7.8	0.1	0.6	A0 V		270 s		
16972	130072		2 43 07.3	-6 38 34	+0.004	-0.04	7.8	1.1	0.2	K0 III		160 mx		
16718	23554		2 43 08.9	+58 53 22	-0.003	0.00	7.58	0.15	0.6	A0 V		200 s	2058	m
16970	110707	86 γ Cet	2 43 18.0	+3 14 09	-0.010	-0.15	3.47	0.09	1.4	A2 V	-5	23 ts	2080	m
16971	110708		2 43 18.4	+1 43 47	0.000	0.00	8.0	1.6	-0.5	M2 III		510 s		
17098	216019		2 43 20.2	-40 31 39	+0.001	+0.03	6.36	-0.02		A0				m
17125	216020		2 43 22.6	-44 39 13	-0.004	-0.02	7.5	1.0	0.2	K0 III		280 s		
17060	193877		2 43 24.2	-34 05 32	+0.002	-0.03	7.2	1.4	0.2	K0 III		150 s		
17374	248634		2 43 24.3	-69 09 35	-0.003	+0.07	6.6	1.4	0.2	K0 III		120 s		
17326	248632		2 43 26.6	-66 42 52	+0.019	-0.07	6.26	0.53		F6	-20			m
16770	23560		2 43 26.9	+56 30 13	0.000	-0.01	7.9	0.1	1.7	A3 V		160 s		
16908	75532	35 Ari	2 43 27.0	+27 42 26	+0.001	-0.01	4.66	-0.13	-1.7	B3 V	+19	180 s		
17001	130079		2 43 28.8	-2 31 54	+0.001	+0.01	6.6	1.1	0.2	K0 III		190 s		
16833	38280		2 43 32.6	+46 58 53	-0.001	-0.01	8.0	0.1	0.6	A0 V		270 s		
16855	38285		2 43 35.4	+43 32 16	+0.008	-0.03	6.7	0.1	1.4	A2 V	+20	72 mx		
17289	248631		2 43 35.5	-62 55 11	-0.002	-0.05	7.4	0.6	4.4	G0 V		40 s		
16944	75535		2 43 37.9	+24 04 22	+0.001	-0.02	7.7	1.1	0.2	K0 III		320 s		
16969	93073		2 43 38.9	+17 32 54	+0.002	-0.01	7.8	1.6	-0.5	M2 III		470 s		
17083	193880		2 43 39.5	-33 02 45	-0.002	+0.01	6.8	0.9	3.2	G5 IV		47 s		
16956	75537		2 43 41.0	+21 08 50	0.000	-0.01	7.7	0.1		A2 mp				
--	--		2 43 43.5	-27 53 50			7.90							m
16882	38287		2 43 43.9	+41 29 51	+0.001	-0.03	7.5	0.1	0.6	A0 V		220 s		
16981	93075		2 43 44.8	+13 17 25	0.000	+0.01	8.0	1.1	0.2	K0 III		370 s		
17082	168005		2 43 45.5	-27 53 59	+0.006	+0.01	7.90	0.9	3.2	G5 IV		87 s	2092	m
16984	93077		2 43 48.1	+12 01 47	+0.003	0.00	8.0	1.1	0.2	K0 III		360 s		
17097	168006		2 43 48.6	-28 48 23	+0.006	+0.03	8.0	1.4	0.2	K0 III		220 s		
16955	75539		2 43 51.2	+25 38 17	-0.001	0.00	6.35	0.08	1.7	A3 V	-11	85 s	2082	m
16778	23564		2 43 53.4	+59 49 22	-0.002	0.00	7.71	0.90	-7.5	A2 Ia	-36	3600 s		
17039	130083		2 43 55.6	-7 54 42	+0.002	-0.04	6.6	1.1	0.2	K0 III		190 s		
17135	193884		2 44 00.4	-33 10 18	+0.002	+0.02	7.8	1.0	3.2	G5 IV		60 s		
17037	130084		2 44 00.8	-6 00 32	+0.009	+0.03	7.1	0.5	3.4	F5 V		54 s		
16954	55822		2 44 01.3	+32 22 14	0.000	-0.02	7.7	1.1	0.2	K0 III		300 s		
16901	38289	14 Per	2 44 05.1	+44 17 50	0.000	0.00	5.43	0.90	-4.5	G0 Ib	-3	880 s		
17081	148575	89 π Cet	2 44 07.3	-13 51 32	-0.001	-0.01	4.25	-0.14	-0.6	B7 V	+15	93 s		
17080	148576		2 44 10.2	-12 29 32	+0.003	-0.07	7.6	1.1	0.2	K0 III		140 mx		
17254	232871		2 44 10.7	-52 34 14	-0.002	-0.01	6.15	0.09		A2				
16968	55824		2 44 11.4	+35 06 54	0.000	-0.01	7.30	0.00	-1.0	B5.5 V		430 s	2084	m
16895	38288	13 θ Per	2 44 11.9	+49 13 43	+0.034	-0.08	4.12	0.49	3.8	F7 V	+25	13 ts	2081	m
17169	193887		2 44 13.0	-36 18 27	+0.004	-0.01	7.2	0.7	4.4	G0 V		31 s		
17134	168011		2 44 14.4	-25 29 54	+0.012	+0.08	7.00	0.6	4.4	G0 V		33 s	2098	m
17017	93081	36 Ari	2 44 19.1	+17 45 50	+0.003	-0.03	6.46	1.07	-0.1	gK2	-32	210 s		
17168	193888		2 44 20.4	-32 31 30	+0.001	-0.03	6.22	0.04		A0	+21			
16843	23575		2 44 23.6	+57 48 11	+0.010	-0.03	8.0	1.1	0.2	K0 III		190 mx		
—	—		2 44 28.5	+48 42 25	0.000	-0.01	7.4	0.0	0.0	B8.5 V		270 s		
17152	168014		2 44 28.8	-24 24 56	-0.003	-0.25	8.0	0.9	3.2	G5 IV		59 mx		
17008	75543		2 44 30.0	+28 00 54	+0.003	-0.03	8.0	0.1	1.4	A2 V		210 s		
17065	110717		2 44 32.7	+5 39 28	+0.002	-0.02	8.0	0.8	3.2	G5 IV		91 s		
17036	93082	37 o Ari	2 44 32.9	+15 18 42	0.000	-0.02	5.80	-0.01	0.2	B9 V	-7	130 s		
16933	38294		2 44 33.1	+46 50 29	+0.008	-0.10	7.0	0.5	3.4	F5 V	+25	52 s		
17007	75544		2 44 36.6	+29 27 37	+0.002	-0.06	7.80	0.4	2.6	F0 V	-8	110 s	2091	m
17122	130091		2 44 42.7	-0 50 45	0.000	+0.01	8.0	1.1	0.2	K0 III		360 s		
17035	75546		2 44 45.0	+24 11 04	+0.004	-0.04	7.8	0.4	2.6	F0 V		110 s		
17046	75547		2 44 48.0	+21 36 51	0.000	-0.01	7.8	0.8	3.2	G5 IV		84 s		
16769	12421		2 44 49.6	+67 49 29	+0.003	-0.03	5.8	0.1	1.4	A2 V	+5	75 s		

HD	SAO	Star Name	α 2000	δ 2000	μ(α)	μ(δ)	V	B−V	M_v	Spec	RV	d(pc)	ADS	Notes
			h m s	$\circ$ $'$ $''$	s	$''$								
17094	110723	87 μ Cet	2 44 56.4	+10 06 51	+0.019	−0.03	4.27	0.31	1.7	F0 IV	+29	30 ts		
17302	216030		2 44 56.4	−48 40 11	−0.001	+0.01	7.5	1.3	−0.3	K5 III		370 s		
17093	93083	38 Ari	2 44 57.5	+12 26 45	+0.008	−0.08	5.18	0.24	1.5	A7 IV	−2	46 mx		
17207	168021		2 45 01.2	−22 10 00	−0.003	−0.10	7.3	0.0	4.0	F8 V		45 s		
17104	93085		2 45 02.3	+14 14 15	−0.002	−0.04	7.80	0.5	3.4	F5 V		76 s	2101	m
17277	216029		2 45 04.2	−41 17 43	−0.001	−0.03	7.95	1.42		K5				
17206	148584	1 τ¹ Eri	2 45 06.1	−18 34 21	+0.023	+0.04	4.47	0.48	3.7	F6 V	+26	15 ts		
17105	93086		2 45 07.6	+13 09 55	0.000	0.00	8.0	1.1	−0.1	K2 III		410 s		
16894	12428		2 45 11.3	+61 23 09	−0.006	−0.01	8.0	0.9	3.2	G5 IV		89 s		
17205	148586		2 45 12.5	−15 25 38	+0.003	+0.02	7.6	0.9	3.2	G5 IV		74 s		
17325	216036		2 45 16.4	−46 17 14	−0.001	−0.01	6.85	1.36		K0				
17224	168025		2 45 18.2	−20 24 05	−0.002	+0.02	7.1	0.1	0.6	A0 V		180 s		
17163	110730		2 45 20.8	+4 42 42	+0.004	−0.04	6.03	0.31	0.6	gF0 n	+20	120 s		
17242	168027		2 45 22.7	−24 39 07	+0.004	+0.04	7.8	0.9	3.0	F2 V		40 s		
16964	23588		2 45 24.6	+56 33 48	+0.001	−0.01	7.8	0.1	0.6	A0 V		240 s	2094	m
17504	248642		2 45 27.5	−63 42 16	+0.004	0.00	5.74	0.93	0.2	K0 III	−11	130 s		m
17566	248644	ζ Hyi	2 45 32.6	−67 37 00	+0.013	+0.04	4.84	0.06		A2	+4	23 mn		
17324	216038		2 45 32.9	−42 50 19	0.000	−0.06	6.9	1.1	3.2	G5 IV		32 s		
17653	255919		2 45 37.0	−71 14 11	+0.014	0.00	6.9	0.4	3.0	F2 V		57 s		
17322	193895		2 45 40.9	−38 09 33	+0.016	+0.24	7.9	−0.3	4.0	F8 V		46 mx		
17176	93089		2 45 43.0	+17 01 16	+0.001	+0.01	7.8	1.1	0.2	K0 III		330 s		
−−	193896		2 45 44.6	−39 43 03	+0.006	+0.04	8.0	0.1	3.2	G5 IV		93 s		
17354	193898		2 45 44.8	−39 43 02	+0.007	+0.05	8.00	0.9	3.2	G5 IV		91 s		m
17118	55849		2 45 47.4	+34 25 47	+0.002	0.00	7.9	1.1	−0.1	K2 III		370 s		
17160	75558		2 45 47.7	+20 41 34	−0.001	−0.01	7.6	0.1	0.6	A0 V		250 s		
17538	248643		2 45 49.4	−62 43 28	−0.001	−0.03	7.4	1.1	0.2	K0 III		270 s		
17076	38306		2 45 51.0	+40 54 40	+0.002	−0.01	7.7	0.1	0.6	A0 V		240 s		
17321	193897		2 45 55.2	−30 28 53	+0.004	−0.05	7.8	0.4	4.4	G0 V		47 s		
17102	38311		2 45 59.3	+40 36 42	0.000	−0.04	7.5	0.0	0.4	B9.5 V		260 s		
17251	130100		2 45 59.4	−4 57 24	+0.002	0.00	8.00	0.4	3.0	F2 V		100 s	2111	m
17148	55855		2 46 03.0	+35 27 03	+0.003	−0.01	7.7	1.1	0.2	K0 III		310 s		
17033	23594		2 46 06.0	+53 14 09	−0.002	0.00	8.0	1.1	−0.1	K2 III		360 s		
17320	168034		2 46 08.0	−25 20 05	+0.001	−0.04	7.4	1.2	0.2	K0 III		200 s		
17229	93094		2 46 16.4	+15 00 27	+0.001	−0.03	7.9	0.1	0.6	A0 V		280 s		
17063	23597		2 46 21.2	+53 09 48	+0.002	+0.01	8.00	1.4	−0.3	K5 III		390 s	2102	m
17092	38313		2 46 22.0	+49 39 12	+0.003	0.00	7.8	1.1	0.2	K0 III		290 s		
17350	168038		2 46 26.1	−21 39 41	+0.007	+0.04	7.1	0.3	2.6	F0 V		79 s		
17146	38320		2 46 31.9	+43 45 55	+0.008	−0.01	7.8	0.5	3.4	F5 V		73 s		
17915	255921		2 46 32.3	−76 11 37	+0.020	+0.05	7.4	1.1	0.2	K0 III		230 mx		
17090	23599		2 46 35.8	+54 04 41	+0.002	+0.01	7.3	0.4	3.0	F2 V		71 s		
16758	4723		2 46 40.8	+75 24 26	+0.001	−0.06	7.10	0.1	1.7	A3 V		120 s	2087	m
17390	168045		2 46 45.1	−21 38 23	+0.006	+0.01	6.49	0.38	1.9	F3 IV		82 s		
17461	216044		2 46 45.5	−40 57 35	+0.006	0.00	7.9	0.3	4.0	F8 V		61 s		
17088	23601		2 46 51.3	+57 44 02	−0.001	+0.01	7.50	0.82	−7.1	B9 Ia	−42	2900 s		
17477	193911		2 46 56.6	−37 20 44	−0.003	0.00	7.9	1.4	−0.3	K5 III		440 s		
17228	55868		2 46 58.2	+35 59 01	+0.004	0.00	6.25	0.93	0.3	G8 III	+21	160 s		
17722	248650		2 47 02.0	−67 16 39	+0.003	0.00	7.7	0.5	4.0	F8 V		55 s		
17240	55872		2 47 03.5	+35 33 18	−0.004	−0.04	6.40	0.4	3.0	F2 V	−4	48 s	2117	m
17259	55875		2 47 03.7	+31 23 42	+0.002	−0.03	8.0	0.0	0.0	B8.5 V		400 s		
17219	38329		2 47 08.4	+42 10 30	−0.001	−0.02	7.7	0.1	0.6	A0 V		240 s		
17086	12438		2 47 09.7	+60 34 11	+0.001	−0.03	6.7	0.2	−4.8	A7 Ib	−14	1200 s		
17438	168051		2 47 11.1	−22 29 09	−0.002	−0.03	6.47	0.39		F2				
17309	75571		2 47 15.4	+22 57 38	−0.002	−0.02	7.3	0.9	3.2	G5 IV		65 s		
17238	38333		2 47 22.3	+43 24 08	−0.003	−0.01	7.60	0.9	0.2	G9 III	−27	270 s		
17364	110745		2 47 23.0	+9 18 28	+0.001	−0.04	7.6	1.1	0.2	K0 III		310 s		
17211	38330		2 47 23.4	+49 04 28	−0.003	+0.01	8.0	0.1	1.4	A2 V		200 s		
17332	93105		2 47 27.4	+19 22 20	+0.009	−0.15	7.40	0.7	4.4	G0 V	+5	31 ts	2122	m
17269	55879		2 47 27.5	+37 47 28	−0.001	−0.02	7.0	1.1	0.2	K0 III		230 s		
17331	75575		2 47 30.2	+23 38 14	−0.002	+0.03	7.4	0.5	3.4	F5 V		62 s		
17245	38335		2 47 31.1	+44 16 21	0.000	0.00	6.8	0.5	3.4	F5 V	−14	46 s		
17528	193916	η¹ For	2 47 33.6	−35 33 03	−0.004	−0.03	6.51	0.96		K0				
17369	93108		2 47 37.9	+15 13 52	0.000	−0.03	7.7	0.8	3.2	G5 IV		78 s		
17330	75576		2 47 38.4	+29 40 42	−0.001	−0.02	7.20	0.00	0.4	B9.5 V	−3	230 s	2126	m
17198	23611		2 47 38.7	+53 56 38	+0.001	+0.02	7.80	0.1	0.6	A0 V		250 s	2115	m
16458	433		2 47 47.6	+81 26 55	+0.006	−0.07	5.78	1.30	0.2	K0 III	+18	86 s		
17361	75578	39 Ari	2 47 54.5	+29 14 50	+0.011	−0.12	4.51	1.11	0.0	K1 III	−15	58 mx		
17280	38341		2 47 54.8	+44 12 21	+0.001	−0.03	7.9	0.0	0.4	B9.5 V		280 s		
17217	23614		2 47 55.5	+55 36 49	+0.005	−0.02	6.9	0.1	0.6	A0 V		100 mx		
17491	148612		2 47 55.9	−12 27 38	0.000	−0.04	6.40	1.6	−0.5	M4 III	−14	240 s		Z Eri, v
17527	168059		2 47 57.6	−24 48 00	+0.005	−0.05	7.9	1.1	0.2	K0 III		140 mx		
17414	93113		2 47 59.3	+15 30 29	0.000	−0.02	6.8	1.1	0.2	K0 III		210 s		
17755	248652		2 48 01.2	−62 57 51	−0.009	−0.09	7.8	0.5	3.4	F5 V		75 s		
17257	23618		2 48 03.0	+51 33 06	−0.002	0.00	7.6	0.1	0.6	A0 V		230 s		
17272	38344		2 48 05.0	+48 11 53	0.000	+0.01	7.8	1.1	0.2	K0 III		290 s		
17576	193923		2 48 07.8	−36 58 54	0.000	+0.02	8.00	0.6	4.4	G0 V		53 s		m

HD	SAO	Star Name	α 2000	δ 2000	μ(α)	μ(δ)	V	B-V	M_v	Spec	RV	d(pc)	ADS	Notes
			h m s	° ′ ″	s	″								
17382	75580		2 48 09.0	+27 04 08	+0.020	-0.11	7.62	0.84	6.1	K1 V	+6	17 ts		m
17563	193921		2 48 09.1	-31 25 24	+0.001	-0.09	7.1	1.0	3.2	G5 IV		44 s		
17490	130117		2 48 13.1	-3 37 08	+0.001	-0.01	6.7	0.1	0.6	A0 V		170 s		
17316	38346		2 48 14.1	+43 37 21	0.000	-0.01	7.4	0.0	0.4	B9.5 V	-29	220 s		
17244	23620		2 48 14.5	+55 46 24	-0.002	-0.02	7.7	1.1	0.2	K0 III		280 s		
17446	93115		2 48 19.5	+17 30 35	-0.003	0.00	7.3	1.6		M6 III	+7			T Ari, v
17235	23622		2 48 23.8	+58 12 00	0.000	+0.01	7.8	1.1	-0.1	K2 III		330 s		
17234	23621		2 48 24.5	+58 49 56	0.000	-0.01	7.5	0.8	3.2	G5 IV		70 s		
17524	130123		2 48 29.2	-5 44 20	-0.001	-0.03	7.1	1.1	0.2	K0 III		240 s		
17315	38348		2 48 29.9	+48 27 14	+0.001	-0.02	7.7	0.0	0.4	B9.5 V		260 s		
17307	23626		2 48 30.6	+50 53 51	0.000	+0.01	8.0	0.0	0.4	B9.5 V		290 s		
17610	193924		2 48 30.8	-33 46 59	-0.001	-0.03	6.8	0.8	3.2	G5 IV		48 s		
17459	93118	40 Ari	2 48 32.0	+18 17 01	+0.003	-0.03	5.82	1.20	0.2	K0 III	+47	100 s		
17627	193926		2 48 37.9	-37 24 03	+0.006	+0.07	6.7	0.1	3.0	F2 V		56 s		
17561	168066		2 48 38.3	-20 49 16	+0.002	-0.01	7.9	0.2	3.2	G5 IV		88 s		
17433	55899		2 48 43.7	+31 06 55	+0.017	-0.17	6.76	0.96	0.2	K0 III		48 mx		
18050	255926		2 48 44.2	-73 50 30	-0.002	+0.03	7.5	1.4	-0.3	K5 III		360 s		
17306	23627		2 48 44.6	+54 10 10	0.000	+0.01	7.93	1.30	-6.3	G0 I		3700 s		
17471	75588		2 48 45.8	+25 11 17	+0.004	0.00	5.90	-0.07	0.2	B9 V	+14	120 mx		
17359	38352		2 48 50.2	+49 11 03	+0.002	0.00	7.56	0.02	0.4	B9.5 V		250 s	2139	m
17702	216058		2 48 52.2	-46 20 50	+0.001	+0.01	7.4	0.6	3.4	F5 V		51 s		
17138	12445		2 48 55.4	+69 38 03	0.000	+0.04	6.40	0.1	1.4	A2 V	-39	96 s		RZ Cas, m,v
17556	130126		2 48 55.6	-5 10 10	+0.001	-0.04	8.0	0.4	2.6	F0 V		120 s		
17715	216059		2 48 59.1	-45 05 22	-0.001	+0.01	6.78	1.01	0.2	K0 III		200 s		
17848	248656	ν Hor	2 49 01.5	-62 48 24	+0.014	+0.03	5.26	0.10		A0	+31	23 mn		
17290	23629		2 49 01.9	+58 01 44	+0.015	-0.11	7.8	0.7	4.4	G0 V		48 s		
17559	130127		2 49 02.4	-4 13 30	-0.001	-0.02	7.1	1.4	-0.3	K5 III		300 s		
17394	38355		2 49 02.7	+47 37 31	+0.001	-0.03	8.0	1.1	0.2	K0 III		320 s		
17470	55904		2 49 05.2	+32 50 41	+0.001	-0.01	8.0	0.1	0.6	A0 V		280 s		
17652	193931	β For	2 49 05.4	-32 24 22	+0.007	+0.16	4.46	0.99	0.3	G6 III	+17	61 s		m
17469	55905		2 49 05.8	+33 10 32	0.000	-0.01	7.9	1.1	0.2	K0 III		320 s		
17412	38357		2 49 10.9	+47 13 01	+0.006	-0.06	6.7	1.1	0.2	K0 III		180 mx		
17346	23632		2 49 14.3	+57 02 01	+0.002	0.00	6.82	1.26	0.2	K0 III		150 s		
17458	38363		2 49 15.2	+41 56 07	+0.003	-0.01	8.0	1.1	0.2	K0 III		320 s		
17467	38364		2 49 16.3	+40 28 45	-0.002	-0.03	8.0	0.1	0.6	A0 V		300 s		
17543	93127	42 π Ari	2 49 17.4	+17 27 51	0.000	-0.01	5.22	-0.06	-1.3	B6 IV	+9	190 s	2151	m
17693	193933		2 49 25.7	-30 48 48	+0.003	+0.02	7.6	0.6	1.4	A2 V		85 s		
17497	55911		2 49 26.6	+33 56 11	+0.001	+0.01	7.30	0.1	1.4	A2 V		150 s	2150	m
17484	55910		2 49 27.0	+37 19 34	+0.001	-0.01	6.45	0.43	3.4	F5 V	+12	39 s		
17378	23637		2 49 30.7	+57 05 03	+0.001	-0.01	6.25	0.89	-7.7	A5 Ia	-38	2100 s		
17445	38366		2 49 34.5	+47 40 43	+0.001	-0.01	7.9	1.1	0.2	K0 III		300 s		
17456	38369		2 49 35.0	+47 08 32	+0.002	-0.03	7.9	0.1	1.4	A2 V		180 s		
17616	110775		2 49 37.9	+0 55 15	+0.002	-0.03	7.1	0.9	3.2	G5 IV		60 s		
17530	55914		2 49 43.9	+35 10 43	+0.001	-0.01	8.0	0.4	2.6	F0 V		120 s		
17156	4737		2 49 44.1	+71 45 13	+0.015	-0.02	7.7	0.9	3.2	G5 IV		78 s		
17700	168080		2 49 47.3	-20 14 45	0.000	+0.03	7.80	1.4	-0.3	K5 III		420 s		m
17713	168081	γ¹ For	2 49 50.9	-24 33 37	-0.003	-0.12	6.14	1.07	0.3	G5 III		110 s	2167	m
17608	93132		2 49 53.4	+12 39 21	-0.001	-0.02	7.6	0.4	3.0	F2 V		85 s		
17729	168082	γ² For	2 49 54.1	-27 56 30	+0.003	+0.03	5.39	0.02		A0	+24	15 mn		
17480	38374		2 49 57.8	+48 48 08	+0.003	0.00	7.80						2154	m
17691	148630		2 49 57.9	-13 50 46	+0.002	+0.02	8.0	1.1	0.2	K0 III		360 s		
17327	12455		2 49 58.2	+64 37 51	+0.003	0.00	7.43	0.34	-3.4	B8 II		950 s	2142	m
17573	75596	41 Ari	2 49 58.9	+27 15 38	+0.005	-0.11	3.63	-0.10	-0.2	B8 V	+4	36 mx	2159	m
17429	23643		2 50 00.2	+55 29 47	0.000	-0.01	8.0	0.1	0.6	A0 V		270 s		
17572	55920		2 50 00.4	+30 31 35	+0.004	-0.04	6.80	0.4	2.6	F0 V	-7	69 s	2158	
17496	38377		2 50 07.2	+48 08 42	+0.003	-0.06	7.6	0.4	3.0	F2 V		82 s		
17179	4738		2 50 12.2	+72 54 49	+0.002	-0.04	7.90	0.00	0.0	B8.5 V		360 s	2135	m
17793	193940	η² For	2 50 14.7	-35 50 37	+0.004	+0.02	5.92	0.90		K0				m
17711	130140		2 50 20.2	-6 48 23	0.000	0.00	7.4	1.1	0.2	K0 III		280 s		
17699	130139		2 50 21.3	-4 59 11	-0.002	0.00	7.10	0.1	0.6	A0 V		200 s	2170	m
17606	55927		2 50 25.9	+31 58 23	+0.002	-0.04	6.61	0.03		A0				
18293	255929	ν Hyi	2 50 28.7	-75 04 00	-0.007	-0.02	4.75	1.33	-0.3	gK6	+5	100 s		
17521	23654		2 50 33.2	+52 16 58	+0.002	-0.01	7.4	0.0	0.4	B9.5 V		230 s		
17584	55928	16 Per	2 50 34.9	+38 19 07	+0.016	-0.10	4.23	0.34	0.6	F2 III	+14	50 mx		m
17605	55930		2 50 35.4	+36 56 51	+0.001	-0.16	6.6	0.6	4.4	G0 V		28 s		
17553	38382		2 50 39.7	+48 17 25	0.000	-0.02	7.8	0.1	0.6	A0 V		250 s		
17829	193944	η³ For	2 50 40.3	-35 40 34	0.000	-0.06	5.47	1.25	0.2	gK0	+12	80 s		
17506	23655	15 η Per	2 50 41.8	+55 53 44	+0.002	-0.01	3.76	1.68	-4.4	K3 Ib	-1	250 s	2157	m
17659	93140		2 50 42.6	+19 09 39	+0.014	-0.06	6.6	0.5	4.0	F8 V		34 s		
17864	216065		2 50 47.8	-39 55 54	+0.004	+0.01	6.36	0.05		B9				
17799	168092		2 50 50.1	-23 01 39	+0.005	+0.01	7.5	0.2	1.4	A2 V		93 mx		
17541	23656		2 50 50.7	+54 39 47	-0.001	+0.01	7.9	0.9	3.2	G5 IV		85 s		
18121	248663		2 50 52.5	-67 31 21	+0.004	0.00	6.6	1.0	0.2	K0 III		190 s		
17624	55936		2 50 52.7	+35 23 39	+0.004	+0.01	6.6	1.1	-0.1	K2 III		210 s		
17454	12467		2 50 53.1	+62 22 02	-0.004	-0.02	7.8	0.0	0.0	B8.5 V		220 mx		

HD	SAO	Star Name	α 2000	δ 2000	μ(α)	μ(δ)	V	B−V	M_v	Spec	RV	d(pc)	ADS	Notes
17675	75602		2ʰ50ᵐ53.6ˢ	+23°44′29″	0.000	−0.″02	7.9	1.1	0.2	K0 III		340 s		
17824	168094	2 τ² Eri	2 51 02.2	−21 00 15	−0.004	−0.02	4.75	0.91	0.2	K0 III	−9	81 s	2179	m
17674	55940		2 51 04.1	+30 17 12	+0.009	−0.03	7.52	0.59	4.4	G0 V	+5	41 s		
17910	216069		2 51 06.2	−40 57 48	0.000	+0.01	7.0	1.3	−0.3	K5 III		290 s		
17673	55942		2 51 06.7	+30 30 36	+0.002	0.00	7.63	1.20	0.0	K1 III	−22	290 s		
18004	232916		2 51 07.8	−57 11 34	+0.003	0.00	7.3	1.5	−0.3	K5 III		340 s		
17505	12470		2 51 08.1	+60 25 02	+0.002	−0.03	7.06	0.40		O7	−17		2161	m
—	12464		2 51 14.2	+68 49 52	+0.012	0.00	7.95	1.46						
17780	110789		2 51 17.4	+1 42 08	0.000	−0.01	7.93	0.25	1.7	A3 V		150 s		m
17791	110791		2 51 20.3	+2 10 22	0.000	−0.02	7.1	1.4	−0.3	K5 III		300 s		
17808	130145		2 51 22.3	−2 59 52	0.000	−0.01	8.0	0.1	0.6	A0 V		300 s		
17884	193948		2 51 22.5	−30 26 15	+0.006	−0.04	7.6	0.3	3.4	F5 V		68 s		
17797	130148		2 51 26.8	−0 41 15	+0.005	+0.04	8.00	0.4	3.0	F2 V		100 s		m
17790	110793		2 51 29.3	+5 50 30	0.000	−0.04	8.0	0.2	2.1	A5 V		150 s		
17806	130149		2 51 29.3	+0 05 11	0.000	+0.03	8.0	1.1	0.2	K0 III		360 s		
17769	93144	43 σ Ari	2 51 29.5	+15 04 55	+0.002	−0.02	5.49	−0.09	−0.6	B7 V	+17	170 s		
17709	55946	17 Per	2 51 30.8	+35 03 35	+0.001	−0.06	4.53	1.56	−0.3	K5 III	+14	85 s		
18120	248667		2 51 36.0	−63 32 28	−0.002	−0.04	6.6	1.2	0.2	K0 III		150 s		
17656	38397		2 51 41.6	+46 50 31	−0.003	−0.02	5.88	0.89	0.3	gG5	−12	130 s		
17671	38398		2 51 43.2	+45 03 33	+0.002	−0.02	7.8	1.4	−0.3	K5 III		350 s		
17581	23662		2 51 45.4	+58 18 53	−0.008	+0.02	6.45	0.10		A m	−5			m
17622	23665		2 51 47.3	+53 01 55	+0.004	0.00	7.0	1.1	−0.1	K2 III	+50	240 s		
17880	148642		2 51 49.7	−17 15 01	+0.001	+0.04	8.0	1.1	0.2	K0 III		370 s		
17689	38400		2 51 52.7	+44 53 38	−0.001	−0.01	6.7	1.1	0.2	K0 III		180 s		
17926	193951		2 51 55.2	−30 48 52	−0.009	+0.11	6.40	0.48	3.2	F8 IV–V		44 s		
18023	232921		2 51 58.0	−49 52 14	+0.003	−0.05	7.06	1.04	3.2	G5 IV		38 s		
17463	12472		2 51 58.7	+68 53 18	+0.001	−0.01	5.80	0.64	−5.5	F6 I–II var	−7	1100 s		SU Cas, v
18035	232922		2 52 01.1	−50 12 45	+0.002	−0.01	7.80	1.29		K0				
17540	12476		2 52 06.5	+65 38 07	−0.002	−0.01	7.1	0.1	1.4	A2 V		130 s		
17758	55957		2 52 11.1	+37 05 38	−0.001	−0.01	7.0	1.1	0.2	K0 III		230 s		
17873	130155		2 52 11.7	−0 39 01	+0.001	−0.03	7.81	0.14	1.4	A2 V		170 s		
17895	130157		2 52 14.1	−8 16 02	0.000	−0.02	7.20	1.6		M5 III				RR Eri, v
18185	248673		2 52 19.2	−62 54 35	+0.011	+0.01	6.03	1.25	0.2	K0 III		100 s		
17591	12478		2 52 22.5	+63 24 35	+0.024	−0.12	7.0	0.6	4.4	G0 V	−11	32 s		
17756	38406		2 52 22.5	+42 01 15	+0.001	−0.01	7.2	1.1	0.2	K0 III		230 s		
17980	168114		2 52 31.7	−27 57 27	−0.001	+0.03	7.3	0.7	3.2	G5 IV		66 s		
17925	148647		2 52 32.0	−12 46 10	+0.027	−0.17	6.04	0.87	6.6	K0 V	+19	7.9 t		
—	23670		2 52 32.3	+53 22 39	0.000	+0.01	7.9	1.4	−0.3	K5 III		430 s		
17907	110807		2 52 39.4	+6 28 28	+0.001	−0.02	7.9	0.0	0.4	B9.5 V		320 s	2193	m
17744	38409		2 52 39.8	+48 48 59	+0.002	−0.03	7.3	0.1	0.6	A0 V		210 s		
17377	4747		2 52 42.4	+74 44 17	+0.009	−0.04	7.6	0.1	1.4	A2 V		110 mx		
18055	216081		2 52 43.5	−41 23 32	+0.004	−0.01	6.70	0.1	1.4	A2 V		79 mx		m
18048	193958		2 52 45.0	−36 50 47	−0.001	+0.02	7.9	1.3	−0.1	K2 III		330 s		
17943	130160		2 52 50.4	−9 26 28	+0.005	+0.05	6.32	0.19		A2				
17743	23674		2 52 51.9	+52 59 52	0.000	−0.01	6.50	0.07		B9	+1	28 mn	2185	m
17441	4750		2 52 54.7	+73 16 40	+0.019	+0.02	7.3	0.9	3.2	G5 IV		65 s		
18292	248676		2 52 54.9	−65 27 20	+0.001	−0.02	6.6	1.2	0.2	K0 III		140 s		
17688	12485		2 52 57.4	+60 28 17	−0.001	−0.01	7.40	0.1	1.7	A3 V	+2	130 s	2184	m
17631	12481		2 52 58.4	+64 40 12	−0.001	0.00	7.7	0.4	3.0	F2 V		85 s		
18696	255934		2 53 04.9	−77 51 46	−0.002	+0.03	7.5	1.7	−0.3	K5 III		260 s		
17734	23675		2 53 09.7	+58 45 57	−0.002	+0.01	8.0	0.0	0.4	B9.5 V		300 s		
17918	93164		2 53 11.6	+16 29 00	+0.004	−0.06	6.4	0.4	3.0	F2 V	+9	48 s		
17976	130164		2 53 12.5	−5 14 57	−0.003	0.00	7.3	1.1	−0.1	K2 III		300 s		
18083	193963		2 53 16.5	−34 29 18	+0.011	+0.10	8.0	0.4	4.4	G0 V		53 s		
17225	4745		2 53 17.3	+78 40 13	+0.049	+0.01	8.0	0.9	3.2	G5 IV		91 s		
18046	168120		2 53 18.1	−22 05 27	−0.001	−0.01	6.7	1.0	3.2	G5 IV		39 s		
17818	38418		2 53 21.1	+48 34 10	+0.001	−0.02	6.50	1.1	0.2	K0 III	−1	170 s	2192	m
17905	55972		2 53 24.3	+31 38 41	−0.001	−0.01	6.5	0.5	3.4	F5 V		42 s		
18015	130170		2 53 27.1	−8 50 54	+0.004	−0.01	7.80	0.6	4.4	G0 V		48 s		m
17707	12492		2 53 29.9	+63 38 48	−0.002	−0.01	8.0	0.9	3.2	G5 IV		89 s		
18149	193965	ψ For	2 53 34.2	−38 26 13	+0.004	+0.04	5.92	0.44	3.0	dF2		34 s		
18031	130173		2 53 34.7	−7 45 02	−0.002	−0.07	7.2	0.4	3.0	F2 V		71 s		
18071	168121		2 53 35.3	−22 22 35	+0.006	−0.08	5.95	1.04		G5				
18012	110816		2 53 38.1	+1 58 08	−0.003	−0.16	6.7	0.9	3.2	G5 IV		50 s		
17973	93166		2 53 38.3	+14 40 12	−0.001	−0.01	7.5	1.6	−0.5	M2 III		400 s		
18229	232930		2 53 40.0	−51 30 29	−0.008	−0.02	7.6	0.7	3.0	F2 V		51 s		
17817	23680		2 53 40.7	+53 48 10	+0.004	−0.03	7.9	0.1	1.4	A2 V		180 s		
17904	55975	20 Per	2 53 42.5	+38 20 15	+0.004	−0.07	5.33	0.41	3.3	F4 V	+6	25 s	2200	m
18030	130176		2 53 43.2	−4 14 52	0.000	−0.03	7.8	1.1	0.2	K0 III		320 s		
18099	168125		2 53 44.5	−23 08 45	−0.001	−0.02	8.0	1.9	−0.5	M2 III		350 s		
17986	110817		2 53 46.2	+9 20 09	+0.004	−0.02	6.7	1.6	−0.5	M4 III		240 mx		
18010	93169		2 53 56.9	+14 13 08	−0.001	−0.01	8.0	0.9	3.2	G5 IV		93 s		
17964	75625		2 53 59.1	+26 44 52	0.000	−0.01	7.9	0.4	2.6	F0 V		120 s		
17891	38424		2 54 01.0	+47 09 40	+0.002	−0.01	6.70	0.00	0.4	B9.5 V	+6	170 s	2199	m
18390	248679		2 54 01.1	−64 54 00	+0.017	+0.07	7.0	0.5	3.4	F5 V		51 s		

HD	SAO	Star Name	α 2000	δ 2000	μ(α)	μ(δ)	V	B−V	M$_v$	Spec	RV	d(pc)	ADS	Notes
			h m s	° ′ ″	s	″								
18265	232933		2 54 06.3	−50 52 21	−0.001	−0.03	6.21	1.58	0.2	K0 III		58 s		
18095	148660		2 54 07.1	−10 26 33	−0.001	−0.01	6.9	0.5	3.4	F5 V		51 s		
—	232934		2 54 09.5	−50 50 46	+0.001	+0.01	7.5	1.2						
18060	110822		2 54 12.5	+5 02 52	−0.001	−0.03	7.9	0.1	1.7	A3 V		170 s		
18107	148661		2 54 13.8	−11 05 23	+0.001	−0.01	7.9	0.1	0.6	A0 V		280 s		
17922	38428		2 54 14.0	+42 35 19	+0.018	−0.09	7.0	0.5	3.4	F5 V	+26	52 s		
17878	23685	18 τ Per	2 54 15.4	+52 45 45	0.000	0.00	3.95	0.74	0.3	G4 III	+2	54 s	2202	m
18019	75627		2 54 17.6	+20 33 56	0.000	−0.01	7.0	1.1	0.2	K0 III		230 s		
18239	216090		2 54 19.4	−42 34 06	−0.002	+0.02	7.4	1.3	3.2	G5 IV		40 s		
18325	232937		2 54 20.0	−55 52 46	+0.006	+0.06	6.7	0.6	2.6	F0 V		44 s		
18059	110823		2 54 20.9	+7 40 49	+0.001	−0.01	7.4	0.1	0.6	A0 V		230 s		
18423	248681		2 54 20.9	−64 26 08	+0.002	0.00	6.56	1.39		K0				
17902	23686		2 54 22.7	+51 10 00	−0.006	−0.02	6.5	1.1	0.2	K0 III		170 s		
17921	38430		2 54 32.3	+48 08 43	+0.001	0.00	7.6	0.1	0.6	A0 V		230 s		
18227	193975		2 54 36.6	−30 53 28	−0.002	0.00	6.9	1.0	4.4	G0 V		17 s		
18220	193974		2 54 36.9	−29 50 36	0.000	−0.02	7.7	1.6	0.2	K0 III		140 s		
17985	55989		2 54 36.9	+37 44 32	0.000	0.00	7.90	0.9	3.2	G5 IV		86 s		
18067	93176		2 54 37.5	+12 56 19	−0.001	−0.05	7.9	0.1	1.7	A3 V		170 s		
18131	130180		2 54 38.6	−5 19 51	+0.001	−0.02	7.3	1.1	0.2	K0 III		260 s		
18145	130181		2 54 47.0	−0 02 54	0.000	+0.03	6.52	1.05	0.2	K0 III	+6	170 s		
18448	248682		2 54 47.7	−64 00 08	+0.020	+0.10	6.6	1.5	3.2	G5 IV		18 s		
18252	193981		2 54 49.6	−33 31 30	+0.002	+0.01	6.9	0.6	1.7	A3 V		52 s		
18066	75633		2 54 49.8	+20 22 21	+0.003	−0.02	7.3	1.1	0.2	K0 III		260 s		
18183	148663		2 54 54.1	−14 01 29	−0.001	0.00	7.2	0.4	2.6	F0 V		83 s		
18091	93178		2 54 55.1	+17 44 05	+0.003	−0.02	7.0	0.1	1.7	A3 V		110 s		
18162	110833		2 55 08.1	+3 29 12	+0.001	−0.01	6.8	0.4	2.6	F0 V		68 s		
17911	23692		2 55 12.5	+59 49 46	−0.002	0.00	8.0	0.1	1.7	A3 V		170 s		
18175	110838		2 55 14.0	+0 26 11	0.000	−0.06	7.02	1.14	0.3	G8 III	−34	150 s		
18144	93185		2 55 17.4	+16 18 32	+0.015	−0.05	7.8	0.8	3.2	G5 IV		56 mx		
17345	4753		2 55 18.7	+80 06 23	+0.005	0.00	7.3	0.1	0.6	A0 V		210 s		
17857	12509		2 55 20.9	+64 09 29	0.000	+0.02	7.68	0.75	−5.6	B8 Ib		1700 s		
18104	75638		2 55 21.8	+27 43 21	−0.002	−0.01	6.70	0.00	0.4	B9.5 V		180 s	2215	m
18173	93186		2 55 23.2	+10 27 27	−0.001	−0.01	7.7	0.4	2.6	F0 V		110 s		
18276	168148		2 55 24.7	−25 56 02	+0.003	+0.01	7.4	1.7	−0.5	M2 III		350 s		
17696	4763		2 55 28.2	+73 22 17	+0.006	0.00	6.8	0.9	3.2	G5 IV		52 s		
18389	216099		2 55 28.3	−48 56 00	+0.002	+0.01	7.9	0.4	0.2	K0 III		340 s		
18008	23702		2 55 28.4	+51 04 44	0.000	+0.02	8.0	0.9	3.2	G5 IV		89 s		
18290	168150		2 55 29.5	−25 18 16	+0.003	−0.06	6.7	1.2	0.2	K0 III		150 mx		
18216	110843		2 55 31.5	+2 01 07	0.000	0.00	6.60	−0.04	0.6	A0 V		160 s		
17580	4759		2 55 32.9	+76 31 45	−0.003	0.00	7.70	0.9	3.2	G5 IV		79 s	2187	m
18041	38442		2 55 33.9	+47 18 17	0.000	+0.01	7.4	1.1	0.2	K0 III	−11	250 s		
18040	38443		2 55 38.0	+48 20 24	0.000	0.00	7.2	0.1	1.4	A2 V	−8	140 s		
18143	75644		2 55 38.9	+26 52 24	+0.020	−0.18	7.58	0.92	6.3	dK2	+33	22 ts	2218	m
18192	93188		2 55 40.1	+14 42 24	−0.002	−0.02	7.8	0.1	0.6	A0 V		270 s		
18142	56009		2 55 41.1	+31 02 35	−0.001	−0.01	7.20	1.6	−1.4	M3 II-III	−24	480 s		
18299	148671		2 55 42.7	−19 45 23	+0.002	+0.01	7.8	1.2	0.2	K0 III		250 s		
18191	93189	45 Ari	2 55 48.4	+18 19 54	−0.001	−0.01	5.91	1.47		M6 III	+46	20 mn		RZ Ari, q
18377	193987		2 55 50.3	−38 39 04	−0.001	+0.01	7.10	1.1	0.2	K0 III		240 s		m
18157	56013		2 55 53.8	+31 34 46	+0.002	−0.04	8.00	0.5	3.4	F5 V		82 s	2219	m
17971	12518		2 55 54.2	+60 23 36	−0.001	0.00	7.75	1.07	−8.2	F5 Ia		7300 s		
18112	38446		2 55 55.4	+41 48 25	+0.002	−0.01	8.0	0.4	2.6	F0 V		120 s		
17948	12517		2 55 56.9	+61 31 17	+0.020	+0.04	5.61	0.44	3.4	F5 V	+29	28 s		m
17916	12514		2 56 02.6	+65 48 25	+0.004	−0.03	7.6	0.1	0.6	A0 V		140 mx		
18258	110849		2 56 04.1	+6 10 30	+0.001	+0.01	7.5	1.1	0.2	K0 III		290 s		
19151	255939		2 56 06.0	−79 35 34	−0.010	+0.01	7.9	1.1	0.2	K0 III		310 s		
18350	168156		2 56 07.4	−26 12 00	+0.001	+0.03	7.4	0.1	1.7	A3 V		140 s		
17785	4776		2 56 11.5	+72 53 10	+0.012	−0.05	8.00	0.6	4.4	G0 V	−3	52 s	2204	m
18262	110851		2 56 13.7	+8 22 54	+0.004	−0.08	5.97	0.48	3.8	dF7	+29	27 s		
17705	4768		2 56 15.3	+75 12 05	+0.016	−0.08	7.9	0.5	3.4	F5 V		80 s		
18202	75651		2 56 15.8	+29 09 52	+0.002	−0.01	6.5	0.9	0.3	G8 III	+29	170 s		
17695	4767		2 56 16.7	+75 39 08	+0.019	−0.03	7.8	0.4	3.0	F2 V		91 s		
18039	23712		2 56 21.2	+57 55 55	−0.001	+0.01	8.0	1.1	0.2	K0 III		320 s		
17958	12519		2 56 24.8	+64 19 57	+0.001	0.00	6.24	2.03	−4.4	K3 Ib	−22	460 s		m
18322	130197	3 η Eri	2 56 25.6	−8 53 54	+0.005	−0.22	3.89	1.11		K1 III-IV	−20	23 mn		Azha
18256	93195	46 ρ Ari	2 56 26.1	+18 01 23	+0.019	−0.21	5.63	0.43	3.4	F5 V	+15	30 mx		
18346	148677		2 56 28.7	−15 01 13	0.000	−0.10	7.7	0.1	1.4	A2 V		46 mx		
17993	12521		2 56 29.3	+62 36 37	−0.004	+0.03	7.50	1.6	−0.5	M1 III		340 s		
18125	38451		2 56 31.0	+49 47 34	+0.002	0.00	7.8	0.1	1.4	A2 V		170 s		
18155	38455		2 56 33.3	+47 09 50	0.000	0.00	6.02	1.34	0.2	K0 III	−13	91 s		m
18331	130199		2 56 37.4	−3 42 44	−0.002	−0.04	5.17	0.08	1.2	A1 V	−15	58 s		
—	130201		2 56 41.2	−5 21 06	−0.001	+0.01	8.0	1.1	−0.1	K2 III		420 s		
18447	193997		2 56 44.2	−36 41 59	+0.005	+0.01	7.3	1.7	−0.1	K2 III		150 s		
18446	193998		2 56 48.3	−35 22 41	−0.002	0.00	6.6	1.3	3.2	G5 IV		22 s		
17929	12520		2 56 48.9	+68 50 10	+0.003	0.00	7.89	0.28	−1.1	B5 V		350 s		
18153	23721		2 56 50.5	+51 15 40	−0.001	−0.04	6.4	1.4	−0.3	K5 III	+5	210 s		

HD	SAO	Star Name	α 2000	δ 2000	μ(α)	μ(δ)	V	B-V	M_V	Spec	RV	d(pc)	ADS	Notes
			h m s	° ′ ″	s	″								
18456	193999		2 56 52.3	−36 17 49	−0.001	0.00	7.5	0.5	3.4	F5 V		60 s		
18180	38460		2 56 54.9	+48 42 16	+0.003	+0.02	7.8	0.5	3.4	F5 V		74 s		
17947	12522		2 56 55.8	+68 10 41	+0.002	0.00	7.7	0.6	4.4	G0 V		46 s		
18139	23719		2 56 58.1	+55 29 25	+0.004	−0.04	8.0	0.5	3.4	F5 V		81 s		
18330	110862		2 57 00.5	+10 09 42	−0.005	−0.08	8.0	0.5	3.4	F5 V		84 s		
18675	248692		2 57 01.2	−63 02 06	0.000	+0.01	7.50	0.5	4.0	F8 V		50 s		m
18345	110865		2 57 04.4	+4 30 04	0.000	+0.02	6.11	1.69		M2	+52			
18359	110866		2 57 09.5	+3 40 42	0.000	−0.03	7.9	0.5	3.4	F5 V		78 s		
18428	168176		2 57 09.6	−21 14 51	+0.002	+0.03	7.8	0.9	0.2	K0 III		330 s		
18369	110867		2 57 10.3	+0 26 50	+0.001	−0.02	6.62	0.32	2.1	A5 V	−4	69 s		
18384	130205		2 57 10.6	−0 34 28	−0.001	+0.02	7.30	0.9	0.3	G8 III	+10	250 s	2237	m
18487	194005		2 57 12.5	−36 26 00	+0.001	0.00	7.5	0.3	3.0	F2 V		78 s		
18445	168180		2 57 12.9	−24 58 30	−0.001	−0.02	7.84	0.96		G5			2242	m
18466	194002		2 57 13.0	−29 51 19	+0.001	0.00	6.3	0.0	2.1	A5 V		70 s		
18455	168181		2 57 14.5	−24 58 08	+0.001	−0.02	7.36	0.86		G5			2242	m
18368	110868		2 57 14.9	+1 53 16	+0.007	−0.08	7.55	0.57		G0			2236	m
18231	38466		2 57 15.3	+46 01 12	−0.001	0.00	8.0	0.0	0.0	B8.5 V		350 s		
18296	56031	21 Per	2 57 17.2	+31 56 03	0.000	−0.03	5.11	−0.01		A0 p	+8			
18200	23729		2 57 21.5	+52 30 05	+0.002	0.00	8.0	1.1	0.2	K0 III	−41	320 s		
18454	168183	4 Eri	2 57 23.6	−23 51 43	+0.007	−0.03	5.45	0.23		A5	+29	15 mn		
18357	93205		2 57 28.5	+16 17 37	+0.004	−0.04	6.8	1.1	0.2	K0 III		180 mx		
18546	194007		2 57 32.6	−38 11 28	0.000	+0.01	6.41	−0.03		A0				
18637	232950		2 57 36.7	−55 00 47	+0.005	+0.06	6.8	0.5	2.6	F0 V		53 s		
18545	194008		2 57 37.7	−37 31 28	−0.002	+0.04	7.7	1.2	0.2	K0 III		250 s		
18152	12532		2 57 37.8	+61 07 44	+0.002	0.00	7.67	0.14		B9				
18295	38472		2 57 54.1	+46 39 39	−0.002	−0.04	7.8	0.2	2.1	A5 V		130 s		
18405	93209		2 57 54.8	+17 48 48	+0.002	−0.02	7.1	1.1	0.2	K0 III		240 s		
18151	12538		2 57 58.2	+63 16 51	+0.001	+0.01	7.8	1.1	0.2	K0 III		290 s		
18137	12536		2 57 58.7	+64 25 09	0.000	+0.02	6.9	0.8	3.2	G5 IV		54 s		
18367	56039		2 58 02.0	+31 31 01	+0.004	0.00	7.9	1.1	0.2	K0 III		320 s		
18339	56036		2 58 02.3	+38 36 54	−0.001	−0.01	6.04	1.41	0.2	K0 III	−41	84 s		
18498	148689		2 58 02.4	−15 50 19	+0.011	+0.13	7.80	0.6	4.4	G0 V		48 s	2247	m
18404	75662	47 Ari	2 58 05.1	+20 40 08	+0.017	−0.03	5.80	0.41	2.6	dF0	+29	39 s		
18535	168191	6 Eri	2 58 05.6	−23 36 22	+0.004	+0.05	5.84	1.33	−0.1	gK2	+7	120 s		
18056	12530		2 58 07.2	+69 11 34	+0.020	−0.11	7.71	0.40	0.8	A1 IV-V		62 mx	2226	m
17982	4783		2 58 07.9	+72 09 14	−0.004	0.00	7.9	0.0	0.4	B9.5 V		280 s		
18511	148690		2 58 12.5	−12 00 22	+0.001	−0.02	6.7	0.9	3.2	G5 IV		50 s		
18268	23738		2 58 13.3	+53 47 37	+0.003	0.00	7.7	0.0	0.0	B8.5 V		310 s		
18622	216113	θ¹ Eri	2 58 15.6	−40 18 17	−0.004	+0.02	3.24	0.12	1.7	A3 V	+12	17 s		Acamar, m
18623	216114	θ² Eri	2 58 16.2	−40 18 16	−0.006	+0.02	4.35			A2	+19	17 s		m
18393	56043		2 58 18.6	+32 27 40	+0.001	+0.07	6.7	1.1	0.2	K0 III		170 mx		
17992	4784		2 58 23.1	+72 40 23	+0.007	−0.02	7.9	0.9	3.2	G5 IV		88 s		
18462	93212		2 58 24.0	+10 12 18	+0.001	+0.01	7.9	1.1	−0.1	K2 III		410 s		
18636	194015		2 58 33.3	−37 59 31	+0.001	+0.02	7.6	1.0	3.2	G5 IV		52 s		
18342	38480		2 58 34.8	+47 41 01	−0.001	−0.03	8.0	1.1	−0.1	K2 III		360 s		
18509	110878		2 58 37.3	+2 07 15	+0.001	−0.01	7.5	1.1	0.2	K0 III		300 s		
18543	130215		2 58 42.0	−2 46 57	−0.002	−0.05	5.23	0.00		A2	−7	30 mn		m
18495	93216		2 58 45.5	+13 36 17	+0.002	−0.07	7.4	0.8	3.2	G5 IV		68 s		
18411	56047	22 π Per	2 58 45.6	+39 39 46	+0.003	−0.04	4.70	0.06	1.4	A2 V	+14	44 s		
18557	130216		2 58 47.3	−9 46 35	0.000	−0.01	6.14	0.22		A2				
18866	248701	β Hor	2 58 47.8	−64 04 16	+0.003	+0.02	4.99	0.13	0.3	A5 III	+24	87 s		
18484	75671		2 58 53.0	+21 37 04	+0.003	−0.02	7.50	0.1	1.7	A3 V		150 s	2253	m
18418	56051		2 58 54.5	+38 24 13	−0.001	0.00	7.4	1.1	0.2	K0 III		260 s		
17889	4782		2 58 56.9	+77 04 51	−0.010	0.00	7.0	0.7	4.4	G0 V		34 s		
18763	232956		2 58 57.2	−53 41 20	+0.002	+0.03	7.8	0.0	0.2	K0 III		330 s		
18709	216120		2 58 58.9	−43 44 54	+0.004	−0.16	7.3	0.6	4.4	G0 V		38 s		
18400	38487		2 58 59.3	+43 57 18	−0.001	−0.06	8.0	1.1	0.2	K0 III		220 mx		
18660	194018		2 59 01.3	−34 11 21	0.000	−0.01	6.6	1.5	−0.1	K2 III		140 s		
18449	56052	24 Per	2 59 03.6	+35 11 00	−0.004	+0.01	4.93	1.23	−0.1	K2 III	−36	92 s		
18650	168202		2 59 06.5	−28 54 25	+0.001	−0.04	6.14	1.04		G5				
18572	130218		2 59 08.3	−4 47 00	−0.001	0.00	8.0	0.1	0.6	A0 V		300 s		
18795	232958		2 59 08.5	−55 33 57	+0.004	−0.04	6.9	0.9	0.2	K0 III		220 s		
18519	75673	48 ε Ari	2 59 12.6	+21 20 25	−0.001	0.00	4.63	0.04	1.4	A2 V	−6	48 mn	2257	m
18570	110883		2 59 22.3	+6 39 06	+0.001	−0.05	7.90	0.4	2.6	F0 V		120 s	2261	m
18326	12552		2 59 23.0	+60 33 59	−0.001	−0.01	7.84	0.37		O8	−40			
18582	110884		2 59 25.8	+1 39 03	−0.002	−0.04	7.9	1.1	0.2	K0 III		350 s		
18337	23748		2 59 26.6	+59 35 32	+0.005	−0.01	7.69	0.26		B9				
18618	130223		2 59 34.4	−4 06 58	+0.014	0.00	7.8	0.5	4.0	F8 V		58 s		
18829	232961		2 59 34.6	−53 54 30	0.000	+0.02	6.7	0.5	0.2	K0 III		200 s		
18692	168209	ζ For	2 59 36.1	−25 16 27	+0.013	+0.09	5.71	0.40	0.6	gA9 n	+27	110 s		
18634	130227		2 59 36.4	−7 10 48	+0.001	−0.04	6.6	1.1	0.2	K0 III		190 s		
18735	194023		2 59 38.2	−32 30 27	0.000	0.00	6.31	0.00		A0				
18482	38492		2 59 39.8	+41 02 00	+0.002	−0.04	5.89	1.44	−0.1	K2 III	+32	110 s		
18633	130228	5 Eri	2 59 41.0	−2 27 54	−0.001	−0.02	5.56	−0.08		B9 n	+18			
18959	248705		2 59 42.4	−63 37 43	+0.006	+0.01	7.6	1.1	0.2	K0 III		300 s		

HD	SAO	Star Name	α 2000	δ 2000	μ(α)	μ(δ)	V	B−V	M_v	Spec	RV	d(pc)	ADS	Notes
18604	110889	91 λ Cet	2^{h}59^{m}42^s.8	+8°54′27″	0^s.000	−0″.01	4.70	−0.12	−2.2	B5 III	+10	220 s		
18352	12554		2 59 47.1	+61 17 25	−0.002	+0.02	6.81	0.22	−3.5	B1 V	−2	660 s		
18743	194024		2 59 47.7	−30 10 20	0.000	−0.03	7.8	0.7	3.4	F5 V		52 s		
18391	23749		2 59 48.6	+57 39 49	−0.001	+0.01	6.89	1.94	−8.0	G0 Ia	−33	1900 s		
18474	38493		2 59 49.8	+47 13 15	+0.002	+0.03	5.47	0.89		G4 p	+7	26 mn		
18690	148706		2 59 55.1	−13 41 00	−0.001	+0.02	7.4	0.9	3.2	G5 IV		70 s		
18439	23753		3 00 04.0	+55 11 17	+0.005	−0.03	7.1	0.1	0.6	A0 V	−4	80 mx		
18627	93228		3 00 05.1	+10 39 22	+0.009	−0.07	7.4	0.5	3.4	F5 V		63 s		
18851	216131		3 00 07.7	−47 25 00	+0.003	+0.05	8.0	−0.3	1.7	A3 V		180 s		
18810	216128		3 00 07.8	−41 52 28	+0.001	−0.01	6.9	1.4	0.2	K0 III		130 s		
18958	248706		3 00 10.3	−59 49 10	+0.004	+0.02	7.9	1.3	3.2	G5 IV		42 s		
18742	168212		3 00 10.4	−20 48 09	−0.004	−0.01	7.9	0.9	4.4	G0 V		34 s		
18528	38497		3 00 11.2	+43 07 36	+0.003	−0.01	7.6	0.1	1.4	A2 V		170 s		
18552	56067		3 00 11.7	+38 07 54	0.000	−0.03	5.90	−0.06	−0.2	B8 V e	−16	160 s		
19733	258323		3 00 13.3	−81 05 21	+0.026	+0.04	7.7	0.5	2.6	F0 V		83 s		
18267	12553		3 00 16.3	+69 30 07	+0.005	−0.03	8.0	0.1	1.4	A2 V		150 mx		
18579	56076		3 00 17.0	+31 07 30	+0.003	−0.04	7.4	0.5	3.4	F5 V		63 s		
18741	148712		3 00 19.0	−11 27 32	+0.002	+0.01	7.4	0.9	3.2	G5 IV		70 s		
18654	93229		3 00 31.4	+18 00 18	0.000	−0.02	7.00	0.1	0.6	A0 V		190 s	2279	m
18643	75681		3 00 36.4	+22 49 45	+0.001	+0.03	7.4	0.9	3.2	G5 IV		69 s		
18539	38503		3 00 36.9	+48 05 33	+0.001	−0.05	7.7	1.1	0.2	K0 III		280 s		
18549	38504		3 00 37.8	+47 53 05	−0.003	−0.02	8.00	0.1	1.4	A2 V		190 s	2271	m
18819	168221		3 00 43.5	−27 38 29	−0.002	−0.11	7.61	0.59	4.4	G0 V		43 s		
18700	93232		3 00 44.1	+10 52 13	+0.005	−0.03	5.95	1.59	−0.3	K5 III	+18	160 s		
18791	168220		3 00 48.5	−20 41 37	+0.002	+0.01	7.4	1.3	−0.5	M2 III		390 s		
18718	93234		3 00 49.0	+10 14 38	0.000	−0.02	7.8	1.1	−0.1	K2 III		390 s		
18760	130242	7 Eri	3 00 50.9	−2 52 43	+0.001	+0.01	6.11	1.77	−0.5	gM1	+81	160 s		
––	––		3 00 51.8	−50 38 32			7.20							T Hor, v
18717	93235		3 00 53.3	+15 01 52	−0.006	−0.07	7.3	0.5	4.0	F8 V		46 s		
18537	23765		3 00 53.3	+52 21 08	+0.002	−0.02	5.28	−0.05	−1.0	B7 IV	−4	170 s	2270	m
18473	23761		3 00 53.8	+59 39 57	0.000	−0.02	7.40	0.00		B9 p	−1			m
18614	38508		3 00 55.5	+40 25 06	−0.001	+0.02	7.9	0.1	1.7	A3 V		160 s		
18837	194033		3 00 57.5	−30 21 45	+0.002	−0.05	7.4	0.7	0.2	K0 III		280 s		
––	216136		3 01 03.7	−41 21 06	0.000	0.00	7.6	1.6	−0.5	M2 III		430 s		
18864	194035		3 01 04.3	−33 06 34	+0.009	+0.03	7.3	0.0	3.4	F5 V		61 s		
18784	130243	8 ρ¹ Eri	3 01 09.9	−7 39 46	+0.006	−0.07	5.75	1.05	5.3	dG6	+14	12 s		
18641	38512		3 01 11.0	+40 47 47	−0.006	−0.03	7.2	1.4	−0.3	K5 III		190 mx		
18665	56092		3 01 15.3	+36 07 04	+0.001	−0.01	7.5	1.1	−0.1	K2 III		310 s		
18789	110908		3 01 25.4	+1 09 22	0.000	0.00	7.37	0.21	1.7	A3 V		120 s		
18896	168235		3 01 26.7	−28 26 08	+0.002	+0.01	7.8	0.3	2.6	F0 V		110 s		
18715	56095		3 01 28.9	+32 24 45	+0.001	−0.02	6.7	0.8	3.2	G5 IV		50 s	2286	m
18907	168238	ε For	3 01 37.6	−28 05 29	+0.021	−0.42	5.89	0.79	3.2	G5 IV	+31	16 mx		
19299	248711		3 01 52.1	−69 41 03	+0.015	0.00	7.8	1.5	0.2	K0 III		160 s		
18832	110915		3 01 52.2	+5 20 10	+0.002	+0.02	6.25	1.05		G8	−59			
19029	216143		3 01 52.6	−47 33 33	−0.008	0.00	7.8	0.6	3.4	F5 V		59 s		
18730	56099		3 01 53.2	+38 52 43	−0.001	+0.02	6.6	0.1	1.7	A3 V		96 s		
18769	75693	49 Ari	3 01 54.1	+26 27 45	−0.001	+0.01	5.90	0.14		A m	−4			
18885	148721		3 01 56.0	−9 57 42	+0.003	−0.01	5.83	1.11	0.3	gG6	+12	98 s		
19007	216142		3 02 01.6	−41 50 14	−0.002	−0.02	8.0	1.7	−0.1	K2 III		130 s		
18921	148724		3 02 02.4	−18 12 28	−0.003	−0.04	7.4	0.4	2.6	F0 V		93 s		
18894	130251		3 02 09.2	−6 29 41	+0.006	−0.15	6.19	0.60		F8				
19400	255945	θ Hyi	3 02 15.5	−71 54 09	+0.007	+0.02	5.53	−0.14		B8	+12			m
18884	110920	92 α Cet	3 02 16.7	+4 05 23	−0.001	−0.07	2.53	1.64	−0.5	M2 III	−26	40 mn		Menkar
––	––		3 02 18.1	−71 59 18			6.30							m
18883	110921	93 Cet	3 02 22.4	+4 21 10	+0.001	+0.01	5.61	−0.10	−1.6	B7 III	+12	270 s		
18978	168249	11 τ³ Eri	3 02 23.4	−23 37 28	−0.011	−0.05	4.09	0.16	2.1	A5 V	−10	23 ts		
18678	23775		3 02 24.3	+54 21 48	+0.002	−0.02	7.4	0.2	2.1	A5 V		110 s		
18803	75696		3 02 26.0	+26 36 34	+0.018	−0.16	6.7	0.6	4.4	G0 V	+11	29 s		
19242	248712		3 02 29.9	−63 52 09	+0.001	+0.03	8.0	0.1	1.4	A2 V		210 s		
18953	130254	9 ρ² Eri	3 02 42.2	−7 41 07	+0.003	0.00	5.32	0.94	0.3	gG5	+25	90 s	2312	m
18909	110927		3 02 48.6	+8 28 19	0.000	−0.02	7.7	1.4	−0.3	K5 III		400 s		
18768	38527		3 02 52.1	+47 06 39	+0.006	−0.05	6.8	0.5	4.0	F8 V		37 s		
19012	168256		3 02 55.4	−21 50 40	+0.005	+0.06	8.0	0.1	1.7	A3 V		120 mx		
19141	216150		3 02 55.8	−46 58 30	+0.002	+0.01	5.82	1.30	0.3	gG8	+17	78 s		
19241	248714		3 02 55.9	−60 47 53	+0.004	0.00	7.3	1.9	0.2	K0 III		76 s		
18975	130256		3 03 01.7	−2 05 12	+0.007	0.00	7.51	0.52	3.8	dF7	+35	53 s	2316	m
18974	110933		3 03 10.5	+1 22 05	+0.002	−0.02	8.0	1.1	0.2	K0 III		360 s		
18859	38535		3 03 25.4	+41 33 35	0.000	−0.03	7.6	1.1	−0.1	K2 III		300 s		
18995	110934		3 03 28.0	+6 13 35	−0.002	−0.06	6.6	0.4	2.6	F0 V		64 s		
18940	75711		3 03 28.5	+23 03 43	+0.008	+0.02	7.1	0.6	4.4	G0 V		35 s		
19285	232977		3 03 28.9	−58 55 59	+0.008	−0.01	5.4	1.6	−0.5	M4 III		150 s		V Hor, v
18928	75709		3 03 30.2	+28 16 11	+0.006	−0.02	6.3	0.2	2.1	A5 V	+11	71 s		
18881	56114		3 03 31.8	+38 24 35	0.000	−0.03	7.0	0.1	0.6	A0 V	+14	190 s		
19063	148739		3 03 32.0	−10 58 30	−0.001	−0.10	7.20	0.5	4.0	F8 V		44 s	2323	m
18972	93256		3 03 33.2	+14 28 14	−0.001	−0.04	8.00	1.1	3.2	K0 IV		91 s		

HD	SAO	Star Name	α 2000	δ 2000	μ(α)	μ(δ)	V	B-V	M$_v$	Spec	RV	d(pc)	ADS	Notes
18900	56117		3^{h}03^{m}35.5^s	+36°26'32"	+0.013	-0.03	7.60	0.5	4.0	F8 V		52 s		m
19319	232981	μ Hor	3 03 36.8	-59 44 16	-0.009	-0.06	5.11	0.34	2.6	dF0	+17	30 s		
18749	23781		3 03 39.4	+59 26 58	-0.005	+0.05	7.8	0.8	3.2	G5 IV		81 s		
18880	38540		3 03 41.8	+42 25 00	0.000	-0.01	7.9	0.5	3.4	F5 V		75 s		
19096	168268		3 03 46.9	-21 21 41	+0.013	-0.02	7.70	0.7	4.4	G0 V		46 s	2326	m
18766	12585		3 03 48.4	+60 18 43	-0.006	-0.02	7.2	0.5	3.4	F5 V	-50	56 s		
19838	56121		3 03 51.8	+34 27 17	-0.002	-0.01	7.8	0.2	2.1	A5 V		140 s		
19215	216156		3 03 55.1	-43 53 53	0.000	+0.03	6.6	1.2	0.2	K0 III		150 s		
18878	38543		3 03 56.4	+47 50 56	-0.003	+0.01	6.6	0.4	2.6	F0 V		63 s		
18908	38545		3 04 03.1	+44 06 46	0.000	0.00	7.9	0.1	0.6	A0 V		250 s		
19370	248721		3 04 03.4	-59 58 07	+0.002	-0.05	8.0	0.9	3.0	F2 V		44 s		
19082	130265		3 04 05.9	-5 14 36	-0.003	0.00	7.4	1.4	-0.3	K5 III		340 s		
19263	216158		3 04 08.3	-47 51 21	0.000	+0.06	7.6	0.6	0.2	K0 III		300 s		
18757	12587		3 04 09.5	+61 42 22	+0.101	-0.68	6.62	0.63	5.0	G4 V	-7	16 mx		m
18950	56124		3 04 09.6	+38 04 45	+0.003	-0.02	6.9	0.0	0.4	B9.5 V	-5	190 s		
19018	75715		3 04 15.1	+21 28 24	0.000	+0.01	7.8	1.1	0.2	K0 III		330 s		
19107	130269	10 ρ³ Eri	3 04 16.3	-7 36 03	+0.003	+0.02	5.26	0.20		A3	+15	16 mn		
18503	4804		3 04 16.7	+75 08 57	-0.002	-0.02	7.0	0.0	0.4	B9.5 V		200 s		
18971	38551		3 04 27.8	+40 17 37	+0.001	+0.02	7.3	1.1	0.2	K0 III		240 s		
19330	232983		3 04 33.0	-51 19 17	+0.009	+0.09	7.50	0.6	4.4	G0 V		42 s		m
19121	110945		3 04 38.0	+1 51 49	+0.002	+0.01	6.05	1.04	0.2	K0 III	+1	140 s		
19080	93260		3 04 40.7	+15 51 22	0.000	-0.09	6.5	1.1	-0.2	K3 III	-32	160 mx		
19437	248725		3 04 43.5	-60 11 48	+0.005	-0.01	7.6	1.9	0.2	K0 III		88 s		
19225	168285		3 04 43.6	-25 46 46	+0.004	-0.03	7.6	1.1	3.2	G5 IV		46 s		
18925	23789	23 γ Per	3 04 47.7	+53 30 23	0.000	0.00	2.93	0.70	0.3	G8 III	+3	34 s	2324	m
19112	93261		3 04 49.5	+13 47 52	+0.001	0.00	7.4	0.9	0.3	G8 III		270 s		
19524	248726		3 04 50.0	-65 11 07	-0.005	+0.03	8.0	0.5	3.4	F5 V		82 s		
19423	232989		3 04 55.2	-57 10 01	-0.001	-0.04	7.8	0.2	4.4	G0 V		49 s		
19755	255951		3 04 59.7	-74 19 12	+0.019	+0.06	6.9	1.5	0.2	K0 III		120 s		
18964	23790		3 05 03.6	+52 37 42	+0.001	-0.01	7.9	1.4	-0.3	K5 III		370 s		
19058	56138	25 ρ Per	3 05 10.5	+38 50 25	+0.011	-0.10	3.39	1.65	-0.5	M4 III	+28	60 s		v
19210	130277		3 05 11.1	-8 16 27	+0.003	+0.03	6.5	1.1	0.2	K0 III		180 s		
19251	148756		3 05 12.8	-19 04 52	-0.001	-0.06	7.6	1.1	0.2	K0 III		280 s		
18876	12595		3 05 16.1	+63 01 47	-0.001	0.00	7.47	0.04		B8	-3			
19337	216162		3 05 18.3	-40 59 04	+0.003	0.00	7.4	0.9	3.0	F2 V		35 s		
19057	38558		3 05 19.1	+41 07 13	+0.002	-0.01	7.28	0.07	0.6	A0 V		190 s		
19066	38559		3 05 20.7	+40 34 57	-0.005	0.00	6.05	1.01	0.2	K0 III	-34	150 s		
19783	255953		3 05 22.2	-73 28 19	+0.006	+0.02	7.4	0.9	3.2	G5 IV		68 s		
18639	4811		3 05 23.3	+73 56 41	+0.002	-0.02	7.5	0.1	0.6	A0 V		240 s		
19718	255952		3 05 25.7	-72 00 14	+0.009	+0.02	8.0	1.1	0.2	K0 III		360 s		
19135	75723	52 Ari	3 05 26.6	+25 15 19	0.000	0.00	6.1	0.0	0.0	B8.5 V	+9	160 s	2336	m
19153	75727		3 05 30.2	+20 54 09	+0.002	-0.02	7.9	0.1	1.4	A2 V		200 s		
19272	148759		3 05 31.5	-16 36 30	-0.001	-0.02	7.4	0.4	3.0	F2 V		75 s		
18970	23791		3 05 32.4	+56 42 21	-0.002	+0.08	4.76	1.02	-0.9	K0 II-III	-45	140 s		
19070	38562		3 05 35.0	+44 24 36	-0.001	+0.01	7.7	0.1	0.6	A0 V		240 s		
19102	56144		3 05 35.4	+36 47 54	+0.001	-0.03	7.1	0.1	1.4	A2 V		130 s		
19219	110960		3 05 38.3	+6 07 03	0.000	0.00	7.7	1.1	-0.1	K2 III		360 s		
18991	23793		3 05 39.9	+56 04 07	0.000	-0.04	6.11	1.02	0.2	K0 III	-11	150 s		
18892	12598		3 05 40.8	+65 09 25	+0.002	-0.02	7.2	1.1	0.2	K0 III	-10	230 s		
19166	75729		3 05 41.0	+23 05 38	0.000	0.00	7.7	1.4	-0.3	K5 III		390 s		
19091	38564		3 05 45.9	+43 42 08	-0.004	-0.01	7.3	0.5	3.4	F5 V		60 s	2334	m
19395	194080		3 05 52.2	-37 20 21	-0.002	+0.01	7.5	-0.3	1.4	A2 V		170 s		
19152	56151		3 05 57.9	+34 00 34	-0.001	-0.01	7.8	0.1	0.6	A0 V		260 s		
18438	4810		3 06 07.8	+79 25 07	0.013	+0.01	5.49	1.57	-0.5	M1 III	-38	160 s	2294	m
19283	110965		3 06 11.0	+4 49 25	-0.002	-0.01	7.9	1.1	0.2	K0 III		340 s		
19111	38573		3 06 11.7	+45 46 11	-0.002	+0.01	8.0	1.1	0.2	K0 III		320 s		
19258	93275		3 06 17.1	+11 39 56	+0.001	0.00	7.30	1.6	-0.5	M1 III	-74	360 s		
19101	38574		3 06 18.3	+49 15 11	-0.006	+0.03	7.9	0.4	3.0	F2 V		90 s		
19039	23801		3 06 18.5	+57 30 32	0.000	0.00	7.93	0.65	-6.6	F0 I		4800 s		
19294	110968		3 06 22.5	+3 48 35	+0.001	-0.01	7.9	0.1	1.4	A2 V		200 s		
19270	93276		3 06 23.6	+13 11 14	0.000	-0.06	5.62	1.08	3.2	G5 IV	-15	20 s		
19304	110969		3 06 25.5	+3 15 50	0.000	-0.07	8.0	0.4	2.6	F0 V		120 s		
19207	56155		3 06 26.8	+34 06 51	+0.002	-0.02	7.9	0.9	3.2	G5 IV		84 s		
19377	148770		3 06 27.3	-16 12 08	-0.001	-0.01	8.0	1.1	0.2	K0 III		360 s		
19216	56158		3 06 32.1	+33 37 31	+0.001	-0.01	7.8	0.0	0.4	B9.5 V		280 s		
19366	148771		3 06 33.2	-10 15 09	0.000	-0.02	6.8	0.8	3.2	G5 IV		52 s		
19480	194091		3 06 33.3	-39 46 48	0.000	+0.02	7.4	1.4	-0.3	K5 III		340 s		
19312	110972		3 06 33.3	+5 37 29	+0.001	-0.02	8.0	0.9	3.2	G5 IV		93 s		
19349	130284		3 06 33.4	-6 05 19	0.000	-0.01	5.27	1.60	-0.5	gM3	+17	140 s		m
19436	194088		3 06 39.5	-29 59 15	-0.002	-0.04	7.2	0.7	0.2	K0 III		250 s		
19077	--		3 06 41.7	+57 28 16	+0.001	-0.02	8.0	1.1	0.2	K0 III		320 s		
19890	255957		3 06 43.7	-72 01 31	0.000	+0.02	7.1	1.1	0.2	K0 III		240 s		
19174	--		3 06 44.1	+45 45 27			7.80	0.00	0.4	B9.5 V		270 s	2341	m
19346	110978		3 06 46.9	+2 42 53	-0.002	0.00	7.97	0.11	0.6	A0 V		250 s		
18962	12607		3 06 49.2	+67 24 40	-0.006	+0.02	7.88	0.60		G0				

HD	SAO	Star Name	α 2000	δ 2000	μ(α)	μ(δ)	V	B-V	M_V	Spec	RV	d(pc)	ADS	Notes
19257	56166		3 06 51.8	+30 31 38	+0.002	-0.04	7.1	0.2	2.1	A5 V		98 s		
19411	130294		3 07 12.7	-1 48 12	-0.002	-0.03	7.1	0.9	3.2	G5 IV		60 s		
19515	194097		3 07 13.0	-32 21 05	+0.003	0.00	7.8	1.2	0.2	K0 III		260 s		
19361	93283		3 07 16.7	+15 12 11	0.000	-0.02	7.90	1.1	-0.2	K3 III		420 s		
19310	75740		3 07 16.8	+29 07 24	-0.001	-0.01	7.6	0.0	0.4	B9.5 V		280 s		
19467	148780		3 07 18.5	-13 45 42	0.000	-0.26	7.00	0.65	5.2	G5 V	+12	23 s		
19065	12608		3 07 18.9	+64 03 28	-0.002	+0.01	5.89	-0.03	0.2	B9 V	-2	140 s		
19421	130295		3 07 19.2	-5 23 06	0.000	+0.01	8.0	1.4	-0.3	K5 III		450 s		
--	38586		3 07 19.5	+40 12 09	+0.001	0.00	8.0	1.5						
18875	4825		3 07 21.8	+72 59 10	+0.017	-0.09	7.7	0.9	3.2	G5 IV		79 s		
19374	93284	53 Ari	3 07 25.6	+17 52 48	-0.002	+0.01	6.11	-0.12	-2.5	B2 V	+28	500 s		
19195	23817		3 07 29.1	+54 06 41	+0.001	-0.01	8.0	0.0	0.4	B9.5 V		300 s		
19301	56176		3 07 32.1	+39 05 13	-0.001	-0.02	7.82	0.32	3.1	F3 V	-2	88 s		
20313	255962		3 07 32.2	-78 59 22	+0.026	+0.07	5.57	0.30	-2.0	F0 II	+3	63 mx		
18787	4824		3 07 35.1	+75 48 17	+0.001	+0.03	7.60	1.6	-0.5	M2 III		390 s	2339	m
19308	56178		3 07 39.3	+36 37 04	+0.020	-0.21	7.90	0.7	4.4	G0 V		35 mx		m
19450	110988		3 07 39.7	+1 24 21	0.000	-0.01	7.3	1.1	0.2	K0 III		260 s		
19743	248736		3 07 40.0	-61 42 57	+0.016	+0.11	7.07	0.93	3.2	G5 IV		46 s		
19829	248739		3 07 44.6	-65 55 13	+0.002	0.00	7.7	0.4	2.6	F0 V		110 s		
19279	38587		3 07 47.3	+47 18 31	0.000	0.00	6.41	0.12	0.6	A0 V	-10	120 s		
19940	248742		3 07 49.2	-69 15 57	-0.004	-0.01	6.15	1.02		G5				
19431	110990		3 07 49.9	+9 55 24	+0.001	+0.01	7.92	0.29	2.1	A5 V		130 s		
19545	168321		3 07 50.8	-27 49 52	+0.005	-0.02	6.19	0.16	1.4	A2 V		50 mx		
19781	248737		3 07 50.8	-61 51 17	+0.001	0.00	7.4	0.6	4.4	G0 V		40 s		
19142	12613		3 07 51.0	+62 38 24	-0.001	+0.01	8.0	1.1	0.2	K0 III		320 s		
19389	75744		3 07 52.2	+23 41 16	+0.002	-0.05	7.9	0.2	2.1	A5 V		150 s		
19462	110994		3 07 56.2	+7 15 22	+0.001	0.00	8.0	0.1	0.6	A0 V		300 s		
19563	194103		3 07 57.1	-30 18 50	+0.004	-0.01	7.3	0.3	1.4	A2 V		68 mx		
19522	148785		3 08 01.2	-12 02 27	+0.009	+0.08	8.00	0.6	3.0	G3 IV	+50	65 mx		
19268	23825		3 08 03.9	+52 12 48	+0.003	-0.02	6.31	-0.01	-1.1	B5 V	+6	260 s		
19447	93290		3 08 06.3	+17 01 38	-0.003	-0.01	8.0	0.4	2.6	F0 V		120 s		
19801	248740		3 08 08.1	-61 09 02	+0.001	+0.01	7.4	1.8	0.2	K0 III		92 s		
19554	148787		3 08 09.2	-19 18 10	-0.001	-0.01	7.8	1.7	-0.5	M2 III		370 s		
19356	38592	26 β Per	3 08 10.1	+40 57 21	0.000	0.00	2.12	-0.05	-0.2	B8 V	+4	29 ts	2362	Algol, m,v
19344	38591		3 08 10.9	+41 35 04	-0.001	-0.05	7.9	0.0	-1.0	B5.5 V		490 s		
19477	93292		3 08 12.7	+13 10 13	+0.002	+0.02	7.9	1.1	0.2	K0 III		340 s		
19343	38594		3 08 15.7	+44 12 18	+0.002	-0.11	7.9	0.5	3.4	F5 V		75 s		
19460	93293	54 Ari	3 08 21.1	+18 47 42	+0.003	-0.01	6.27	1.58	-0.4	gM0	+43	210 s		
19511	110999		3 08 24.8	+6 43 24	+0.002	+0.02	7.80	0.8	0.3	G4 III	+20	320 s		
19504	93296		3 08 25.1	+10 47 45	-0.004	-0.03	7.5	0.5	3.4	F5 V		65 s		
19256	23829		3 08 25.5	+57 00 49	+0.013	-0.08	6.65	0.59	4.4	G0 V		27 s		
19322	23831		3 08 26.9	+51 10 02	+0.001	0.00	7.0	0.1	1.4	A2 V		130 s		
19521	111000		3 08 27.7	+7 38 34	-0.002	-0.03	7.6	0.4	2.6	F0 V		100 s		
19525	111002		3 08 38.6	+8 28 15	-0.001	+0.07	6.28	1.06	0.2	G9 III	+38	140 s		
19861	248743		3 08 42.2	-60 28 37	+0.003	+0.01	7.6	0.9	0.6	A0 V		67 s		
19659	194109		3 08 42.4	-35 25 58	-0.003	-0.11	7.1	0.6	4.4	G0 V		33 s		
19444	56196		3 08 45.9	+35 27 32	+0.003	-0.02	7.70	0.1	0.6	A0 V		250 s	2364	m
19427	56195		3 08 47.6	+37 02 46	+0.002	+0.01	8.0	1.6	-0.5	M2 III		460 s		
19939	248744		3 08 50.3	-64 33 55	+0.013	+0.08	6.9	0.1	2.1	A5 V		75 mx		
19288	23832		3 08 50.6	+58 20 52	+0.006	-0.03	7.46	0.30	2.6	F0 V		94 s		
19632	168331		3 08 52.3	-24 53 16	+0.016	+0.13	7.4	0.7	3.2	G5 IV		60 mx		
19243	12619		3 08 53.9	+62 23 02	-0.003	-0.02	6.62	0.26	-3.5	B1 V e	-25	760 s		
19622	148791		3 08 54.1	-18 58 09	+0.001	0.00	7.1	0.7	0.6	A0 V		70 s		
19552	111003		3 09 00.1	+9 42 43	+0.001	0.00	8.0	1.4	-0.3	K5 III		450 s		
19551	93301		3 09 01.0	+12 51 36	+0.001	+0.02	7.1	0.8	3.2	G5 IV		59 s		
19332	23836		3 09 01.9	+56 02 42	0.000	0.00	8.0	1.1	0.2	K0 III		320 s		
19373	38597	ι Per	3 09 03.9	+49 36 49	+0.130	-0.08	4.05	0.61	3.7	G0 V	+50	12 t		m
19193	12618		3 09 07.2	+66 55 04	+0.018	-0.02	8.0	0.9	3.2	G5 IV		89 s		
19658	168335		3 09 13.2	-24 08 13	+0.001	+0.04	7.7	1.8	-0.3	K5 III		260 s		
19678	168336		3 09 13.5	-26 42 59	+0.001	-0.01	7.5	0.5	0.6	A0 V		120 s		
19549	75755		3 09 20.0	+20 45 40	+0.003	-0.01	6.6	1.1	-0.1	K2 III		220 s		
19267	12621		3 09 22.0	+64 18 02	-0.001	0.00	6.9	0.8	3.2	G5 IV		54 s		
19509	56202		3 09 28.1	+37 18 06	-0.004	-0.03	7.1	1.1	0.2	K0 III		220 s		
19476	38609	27 κ Per	3 09 29.7	+44 51 27	+0.017	-0.15	3.80	0.98	0.2	K0 III	+29	53 s	2368	m
19729	194121		3 09 33.6	-31 15 13	0.000	+0.02	8.0	1.0	3.2	G5 IV		68 s		
19568	93304		3 09 34.6	+19 22 51	+0.002	0.00	7.0	0.1	0.6	A0 V		190 s		
19620	111013		3 09 36.3	+5 12 10	+0.006	-0.02	7.99	0.58	4.4	dG0	+36	52 s	2373	m
19548	75757	55 Ari	3 09 36.7	+29 04 38	+0.002	-0.01	5.72	0.12	-0.6	B7 V	-2	130 s		
19676	168339		3 09 36.9	-20 00 45	+0.002	-0.03	8.0	1.7	-0.5	M2 III		470 s		
19583	93307		3 09 40.6	+17 15 38	0.000	-0.05	7.7	1.1	-0.1	K2 III		360 s		
19593	93309		3 09 45.6	+20 00 03	+0.002	-0.04	7.9	1.1	0.2	K0 III		340 s		
19701	148801		3 09 48.6	-19 27 34	0.000	-0.07	7.8	1.2	0.2	K0 III		250 mx		
19649	111016		3 09 51.3	+1 29 02	+0.001	-0.04	7.65	0.36	2.6	F0 V		96 s		
21190	258330		3 09 54.0	-83 31 53	-0.004	+0.03	7.60	0.4	2.6	F0 V		100 s		m
19648	111018		3 09 56.7	+3 03 39	+0.002	-0.06	7.6	0.4	2.6	F0 V		100 s		

HD	SAO	Star Name	α 2000	δ 2000	μ(α)	μ(δ)	V	B-V	M$_v$	Spec	RV	d(pc)	ADS	Notes
20060	248748		3^h 10^m 03.4^s	-63° 54' 49"	+0.007^s	+0.01"	6.65	0.13	1.4	A2 V		80 mx		m
19916	233015		3 10 03.8	-50 50 00	+0.010	0.00	7.7	0.4	4.4	G0 V		47 s		
19015	4834		3 10 04.0	+76 05 54	-0.009	+0.02	7.7	0.1	0.6	A0 V		250 s		
19600	75762		3 10 08.7	+27 49 12	0.000	-0.04	6.42	0.01	0.6	A0 V	-5	140 s		
19628	75768		3 10 14.5	+21 16 19	+0.002	0.00	7.9	1.1	0.2	K0 III		340 s		
19647	93315		3 10 17.8	+16 48 48	+0.001	-0.03	8.0	0.1	0.6	A0 V		300 s		
19712	130319		3 10 18.0	-1 41 41	0.000	-0.01	7.35	-0.03	0.4	B9.5 V		240 s		
19441	23846		3 10 18.4	+59 31 28	-0.001	0.00	7.92	0.36		B5				
20335	255963		3 10 19.2	-74 17 51	+0.011	+0.26	7.1	0.3	3.4	F5 V		54 s		
19540	38620		3 10 20.3	+45 56 23	-0.003	-0.04	7.5	0.8	3.2	G5 IV		71 s		
20094	248750		3 10 22.5	-63 15 14	-0.002	-0.01	8.0	0.4	2.6	F0 V		120 s		
20171	248752		3 10 25.3	-67 30 22	0.000	0.00	8.0	0.0	0.4	B9.5 V		340 s		
19637	75771		3 10 27.0	+26 53 47	0.000	+0.07	6.02	1.28	-0.2	K3 III	-16	140 mx		
19948	216197		3 10 27.4	-48 44 03	+0.003	-0.02	6.12	1.12	0.2	gK0		130 s		
19666	93318		3 10 27.7	+15 59 26	+0.001	+0.05	7.9	0.2	2.1	A5 V		150 s		
19853	194131		3 10 27.9	-33 03 01	+0.005	-0.01	7.9	1.3	-0.5	M2 III		220 mx		
19826	168354		3 10 35.3	-23 44 18	+0.004	+0.04	6.38	0.93	3.2	K0 IV		43 s		
19739	111027		3 10 35.9	+0 12 45	-0.001	-0.02	7.29	0.24	1.7	A3 V		110 s		
19754	130323		3 10 38.4	-5 23 38	+0.001	-0.02	8.0	1.1	0.2	K0 III		360 s		
19698	93320		3 10 38.8	+11 52 21	+0.003	-0.02	5.90	-0.06	-0.2	B8 V		160 s		
20037	233023		3 10 39.2	-57 48 35	+0.001	0.00	6.7	0.6	3.2	G5 IV		50 s		
19665	75773		3 10 39.8	+21 53 33	-0.003	-0.08	7.8	1.1	0.2	K0 III		160 mx		
20146	--		3 10 40.2	-64 57 23			7.9	1.1	0.2	K0 III		350 s		
19814	168355		3 10 41.4	-20 06 41	+0.003	+0.01	7.6	1.4	3.2	G5 IV		30 s		
19904	194134		3 10 42.5	-39 03 06	-0.004	+0.01	7.1	0.3	1.4	A2 V		93 s		
19673	93319		3 10 42.9	+18 44 36	+0.002	-0.03	7.9	1.1	0.2	K0 III		340 s		
19723	111028		3 10 49.2	+10 00 23	+0.004	+0.03	7.1	1.1	-0.1	K2 III		280 s		
19440	12635		3 10 50.3	+63 47 14	+0.008	-0.10	7.4	0.5	4.0	F8 V		48 s	2371	m
19810	148810		3 10 51.5	-11 07 31	+0.007	-0.11	7.20	0.9	3.2	G5 IV		62 mx		m
19645	56223		3 10 52.2	+32 13 35	+0.003	-0.02	7.3	1.1	0.2	K0 III		250 s		
19760	111030		3 10 55.4	+2 18 57	0.000	-0.03	6.79	1.05	0.2	K0 III		190 s		
19558	23857		3 11 00.1	+53 08 37	-0.001	0.00	7.5	0.2	2.1	A5 V		120 s		
19439	12636		3 11 00.6	+64 53 47	-0.009	-0.01	6.5	0.2	2.1	A5 V		74 s		
19008	4836		3 11 04.1	+78 22 35	+0.002	-0.02	8.0	0.9	3.2	G5 IV		91 s		
19989	216205		3 11 04.3	-46 21 01	+0.001	+0.01	8.0	0.8	0.2	K0 III		360 s		
19850	148816		3 11 05.1	-13 15 49	+0.001	+0.03	6.5	0.6	4.4	G0 V		27 s		
19759	111031		3 11 06.4	+8 43 20	+0.005	-0.03	7.9	0.2	2.1	A5 V		53 mx		
19899	168364		3 11 06.9	-24 50 23	-0.001	-0.03	8.0	1.2	-0.3	K5 III		450 s		
20506	--		3 11 13.8	-75 48 42			7.9	0.4	2.6	F0 V		120 s		
19627	38624		3 11 14.2	+45 01 08	+0.001	-0.02	7.5	0.1	1.4	A2 V		160 s		
19887	148821		3 11 16.7	-16 01 31	-0.001	-0.02	6.26	1.20		K0				
19656	56224	28 ω Per	3 11 17.3	+39 36 42	-0.002	+0.01	4.63	1.11	0.0	K1 III	+7	83 s		m
19836	130328		3 11 18.7	-3 48 44	-0.001	-0.03	6.05	1.66	-0.5	gM1	+24	180 s	2389	m
19789	93327		3 11 21.9	+13 02 52	-0.001	+0.02	6.4	0.8	3.2	G5 IV	+11	44 s		
19664	56225		3 11 23.5	+38 04 13	-0.002	+0.04	7.4	1.6	-0.5	M2 III		350 s		
19557	23858		3 11 25.3	+57 54 12	+0.002	+0.01	7.59	2.07		R5	-7			
20465	255966		3 11 28.1	-74 39 55	+0.006	+0.03	7.6	1.2	0.2	K0 III		240 s		
19897	148822		3 11 30.8	-13 31 26	0.000	-0.09	7.9	0.5	3.4	F5 V		78 s		
19536	12644		3 11 31.2	+60 38 06	-0.005	-0.03	7.30	0.05		A0	+13			
19787	93328	57 δ Ari	3 11 37.7	+19 43 36	+0.011	-0.01	4.35	1.03	-0.1	K2 III	+25	78 s		
19866	130333		3 11 40.2	-0 29 40	-0.003	-0.04	7.7	1.1	0.2	K0 III		310 s		
18778	500		3 11 42.3	+81 28 15	-0.022	0.00	5.95	0.15		A m	-3	17 mn	2348	m
19624	23867		3 11 42.8	+52 09 49	+0.002	-0.01	6.88	0.02		B5				
19708	56236		3 11 44.5	+37 35 02	+0.004	0.00	7.3	1.1	0.2	K0 III		250 s		
19934	148824		3 11 46.7	-14 55 34	+0.003	-0.01	7.9	1.1	0.2	K0 III	+35	340 s		
19987	168368		3 11 48.9	-29 09 45	+0.001	-0.05	6.9	1.1	3.2	G5 IV		36 s		
20029	194148		3 11 52.4	-39 01 23	+0.006	-0.02	7.0	0.9	3.4	F5 V		28 s		
18591	498		3 11 53.3	+82 54 14	+0.013	-0.02	7.2	0.9	3.2	G5 IV		64 s		
20486	255967		3 11 55.5	-74 00 57	+0.014	+0.07	7.7	0.2	2.1	A5 V		99 mx		
19833	93331		3 11 56.3	+16 09 58	0.000	0.00	7.5	1.4	-0.3	K5 III		370 s		
19275	4840		3 11 56.3	+74 23 37	+0.005	-0.09	4.87	0.02	0.6	A0 V	+10	67 s		
19500	12643		3 11 57.5	+66 44 15	+0.006	+0.01	8.0	1.1	0.2	K0 III		310 s		
20035	194149		3 11 57.6	-39 21 56	+0.007	0.00	6.9	1.0	0.2	K0 III		190 mx		
19684	38631		3 11 59.0	+46 07 40	+0.008	-0.05	6.9	0.4	2.6	F0 V		69 s		
19976	168369		3 12 01.2	-20 47 28	+0.002	+0.02	8.0	1.5	-0.3	K5 III		450 s		
20177	233033		3 12 01.6	-56 24 06	+0.003	+0.04	7.19	1.47		K5				
19903	130338		3 12 02.1	+0 06 27	0.000	0.00	7.2	1.1	0.2	K0 III		250 s		
20010	168373	α For	3 12 04.2	-28 59 13	+0.025	+0.64	3.87	0.52	3.3	F8 IV	-21	14 t	2402	
19736	38635		3 12 09.6	+42 22 34	+0.003	-0.01	6.0	-0.1		B8				
19928	130344		3 12 09.7	-0 14 45	+0.002	-0.03	7.68	1.08	0.2	K0 III		270 s		
19706	38634		3 12 11.2	+46 23 11	-0.005	-0.02	7.1	0.5	3.4	F5 V		55 s		
20081	216207		3 12 13.9	-40 24 33	-0.003	+0.03	7.4	0.2	1.7	A3 V		120 s		
19832	75788	56 Ari	3 12 14.2	+27 15 26	+0.001	-0.01	5.79	-0.12		A0 p	+11			SX Ari, v
20001	168376		3 12 15.3	-20 37 08	0.000	+0.03	6.9	0.1	0.4	B9.5 V		150 s		
19986	148831		3 12 16.1	-16 03 30	+0.003	-0.02	8.0	1.1	0.2	K0 III		360 s		

HD	SAO	Star Name	α 2000	δ 2000	μ(α)	μ(δ)	V	B-V	M_v	Spec	RV	d(pc)	ADS	Notes
			h m s	° ' "	s	"								
19896	93335		3 12 24.8	+16 31 02	−0.001	0.00	7.4	0.1	1.4	A2 V	−6	160 s		
20121	216209		3 12 25.7	−44 25 10	+0.008	0.00	5.93	0.44	0.7	F6 III	+34	66 mx		m
19926	111044		3 12 26.3	+6 39 39	−0.001	0.00	5.56	1.08	3.2	G5 IV	+5	19 s		
19735	38638		3 12 26.3	+47 43 34	+0.007	−0.07	6.35	1.43	−0.3	K5 III	−36	83 mx		m
19769	38642		3 12 30.4	+43 51 37	+0.016	−0.03	7.7	0.6	4.4	G0 V		46 s		
20071	194153		3 12 30.8	−32 23 48	+0.007	+0.02	6.9	1.1	0.2	K0 III		190 s		
19683	23875		3 12 31.9	+56 08 57	+0.002	−0.01	7.9	1.6	−0.5	M2 III		410 s		
20234	233037		3 12 33.1	−57 19 18	+0.002	+0.01	5.74	2.28		N7.7	+14			
19944	111046		3 12 35.2	+8 35 39	0.000	+0.02	7.9	0.9	3.2	G5 IV		88 s		
20070	194154		3 12 36.7	−30 09 20	+0.006	0.00	7.7	0.8	0.2	K0 III		190 mx		
20080	194157		3 12 37.5	−31 07 44	+0.001	+0.01	7.1	1.3	0.2	K0 III		160 s		
19340	4844		3 12 42.4	+74 56 10	−0.002	−0.03	8.0	0.9	3.2	G5 IV		91 s		
19994	130355	94 Cet	3 12 46.4	−1 11 46	+0.013	−0.06	5.06	0.57	4.0	F8 V	+18	18 ts	2406	m
19984	111050		3 12 48.5	+2 41 42	0.000	0.00	8.03	0.15	0.6	A0 V		250 s		
19705	23878		3 12 50.1	+55 20 25	−0.001	+0.01	8.0	1.1	−0.1	K2 III		360 s		
19963	93339		3 12 54.3	+11 07 57	−0.001	−0.03	7.98	0.33		A p				
20201	216214		3 12 54.4	−47 09 20	−0.008	−0.03	7.5	0.0	4.4	G0 V		42 s		
20102	168386		3 12 55.5	−26 15 35	−0.002	0.00	7.6	1.1	3.2	G5 IV		48 s		
20170	216213		3 12 56.2	−42 22 28	0.000	+0.03	7.5	0.7	2.6	F0 V		34 s		
19507	4851		3 12 59.3	+71 12 25	−0.006	−0.02	7.9	1.1	0.2	K0 III		300 s		
20069	168384		3 13 00.0	−20 01 38	+0.009	+0.05	6.8	0.2	0.2	K0 III		150 mx		
20144	194163		3 13 01.4	−35 56 38	0.000	0.00	6.27	−0.08		B9				
19902	56256		3 13 02.7	+32 53 46	+0.014	−0.07	8.0	0.9	3.2	G5 IV		60 mx		
19983	93341		3 13 03.1	+11 16 08	0.000	0.00	8.0	0.5	3.4	F5 V		82 s		
19805	38647		3 13 05.2	+49 00 33	+0.003	−0.04	7.94	0.12		A0				
20217	216216		3 13 09.9	−47 47 03	+0.008	+0.04	8.0	1.1	0.2	K0 III		220 mx		
19925	56259		3 13 10.4	+31 04 07	0.000	−0.04	7.3	1.1	−0.1	K2 III		270 s		
20233	216217		3 13 13.0	−49 19 46	+0.002	+0.01	7.7	1.3	0.2	K0 III		220 s		
20200	216215		3 13 15.7	−41 22 47	0.000	+0.01	7.3	1.2	−0.1	K2 III		300 s		
20142	168390		3 13 21.8	−27 34 24	+0.002	+0.06	7.1	0.1	1.4	A2 V		110 mx		
19845	38652		3 13 23.9	+48 10 37	+0.003	−0.02	5.90	0.97	0.3	G8 III	−7	120 s		
19438	4849		3 13 27.5	+74 17 48	+0.004	−0.01	7.5	0.5	4.0	F8 V		50 s		
19266	4843		3 13 31.1	+78 11 28	−0.005	+0.02	7.3	1.1	0.2	K0 III		250 s		
20118	148845		3 13 37.1	−13 38 21	+0.003	+0.03	7.6	0.1	1.4	A2 V		170 s		
20176	168397		3 13 37.9	−29 48 15	+0.002	0.00	6.16	1.05		G5				
19962	56270		3 13 45.6	+36 08 25	+0.009	−0.02	8.0	0.9	3.2	G5 IV		90 s		
20927	255969		3 13 47.5	−78 49 36	−0.008	+0.01	7.2	0.4	2.6	F0 V		83 s		
19634	12661		3 13 49.3	+68 31 31	−0.002	−0.04	7.7	0.9	3.2	G5 IV		78 s		
19901	38657		3 13 49.4	+47 55 06	−0.001	−0.02	7.8	1.4	−0.3	K5 III		350 s		
19864	23887		3 13 50.1	+51 47 06	+0.011	−0.02	7.9	0.5	3.4	F5 V		75 s		
20232	194172		3 13 50.1	−38 48 34	−0.003	−0.04	6.9	0.1	0.6	A0 V		160 s		
19893	38655		3 13 50.3	+49 34 08	+0.003	−0.03	7.15	0.04		B9				
19942	38659		3 13 51.2	+43 51 46	+0.005	−0.08	7.3	0.9	3.2	G5 IV		65 s		
20216	194170		3 13 53.5	−32 49 47	+0.003	0.00	7.8	0.9	0.2	K0 III		330 s		
20043	93345		3 13 54.8	+18 58 24	+0.002	+0.01	6.6	1.4	−0.3	K5 III		240 s		
20127	130367		3 13 54.9	−6 39 53	+0.006	−0.13	7.9	0.4	3.0	F2 V		56 mx		
20115	111062		3 14 02.8	+0 44 22	+0.005	0.00	7.35	0.51	4.0	F8 V	+24	43 s	2416	m
19820	23886		3 14 05.3	+59 33 49	0.000	0.00	7.11	0.51	−5.4	O9 IV	−4	1600 s		CC Cas, m,v
20086	93350		3 14 08.3	+15 35 25	+0.001	−0.02	7.4	0.1	0.6	A0 V	+17	230 s		
19981	56274		3 14 08.4	+40 07 11	+0.002	+0.01	7.1	0.0	0.4	B9.5 V		200 s		
20270	216223		3 14 10.9	−40 15 21	−0.001	+0.01	7.4	0.9	0.2	K0 III		270 s		
20078	75806		3 14 17.1	+22 57 14	+0.005	−0.01	6.8	1.1	0.2	K0 III		210 s	2414	m
19534	4855		3 14 17.6	+74 14 37	+0.005	0.00	7.20	1.6	−0.5	M2 III	+13	330 s		
20310	216224		3 14 25.2	−42 40 23	0.000	+0.02	7.8	1.4	−0.1	K2 III		180 s		
20205	130375		3 14 32.9	−9 41 15	+0.002	−0.03	7.7	0.5	3.4	F5 V		72 s		
19844	12667		3 14 33.5	+61 42 56	−0.003	0.00	7.87	0.06		B9				
20196	130374		3 14 36.2	−2 19 58	+0.001	+0.01	7.3	1.6	−0.5	M2 III		370 s		
20301	194176		3 14 40.0	−35 33 28	+0.002	+0.01	6.87	0.74		G0				
20215	130376		3 14 42.1	−6 42 10	+0.005	−0.01	7.2	1.1	0.2	K0 III		180 mx		
20165	111070		3 14 47.2	+8 58 51	+0.027	−0.40	7.83	0.86	6.1	K1 V	−22	20 mx		
20099	56279		3 14 48.2	+32 39 31	+0.001	−0.19	7.97	0.96		G5				
20239	148855		3 14 49.8	−14 26 36	+0.002	−0.01	8.00	1.1	0.2	K0 III		360 s		m
20408	233049		3 14 52.2	−50 33 05	+0.004	+0.03	7.5	0.8	3.4	F5 V		41 s		
20485	233050		3 14 52.9	−59 30 49	+0.004	+0.01	7.00	1.1	0.2	K0 III		230 s		m
20150	75810	58 ζ Ari	3 14 54.0	+21 02 40	−0.002	−0.07	4.89	−0.01	0.0	A0 IV	+7	91 s		
20017	38672		3 14 55.2	+48 41 45	+0.001	0.00	7.91	0.29		B7 e	−28			
20063	38674		3 14 56.6	+42 30 14	+0.006	+0.02	6.06	1.07	3.2	G5 IV	+22	25 s		
20380	216227		3 14 57.0	−44 07 13	+0.002	+0.02	7.4	1.3	3.2	G5 IV		33 s		
20659	248765		3 14 57.8	−69 46 57	+0.001	+0.01	7.3	0.8	0.2	K0 III		270 s		
20293	168406		3 15 00.1	−26 06 01	−0.002	−0.01	6.25	0.04	−2.8	A0 II		650 s		
20022	38673		3 15 00.3	+48 16 49	+0.001	+0.01	8.0	0.1	0.6	A0 V		270 s		
20268	148858		3 15 04.8	−13 49 41	+0.004	−0.02	7.5	0.9	3.2	G5 IV	+26	72 s		
20407	216230		3 15 06.2	−45 39 54	−0.014	+0.14	6.77	0.58	4.9	G3 V		24 s		
19615	4858		3 15 07.6	+73 52 06	0.000	−0.02	6.80	1.1	0.2	K0 III		200 s		m
20586	248764		3 15 11.0	−64 26 38	−0.004	−0.06	6.90	0.5	3.4	F5 V		50 s		m

HD	SAO	Star Name	α 2000	δ 2000	μ(α)	μ(δ)	V	B−V	M_V	Spec	RV	d(pc)	ADS	Notes
			h m s	° ' "	s	"								
20322	168411		3 15 13.2	−28 20 42	0.000	+0.01	7.8	1.2	0.2	K0 III		240 s		
20105	56282		3 15 13.6	+37 16 09	+0.002	0.00	7.8	0.1	0.6	A0 V		260 s		
20291	168410		3 15 15.0	−20 01 07	+0.001	+0.01	6.9	0.0	0.6	A0 V		180 s		
20113	56284		3 15 17.1	+36 28 35	+0.003	+0.01	7.6	0.0	0.0	B8.5 V		310 s		
20149	56285		3 15 20.4	+30 33 24	−0.001	+0.01	5.50	0.01	−0.4	A1 III	−3	150 s		
20388	194188		3 15 23.8	−38 59 59	−0.002	+0.03	7.4	1.0	3.0	F2 V		30 s		
20398	194190		3 15 26.8	−39 45 46	+0.001	−0.01	6.9	1.3	0.2	K0 III		150 s		
20378	194187		3 15 28.0	−34 50 41	+0.001	−0.04	8.0	1.2	−0.3	K5 III		450 s		
20321	148861		3 15 29.5	−18 10 27	+0.001	−0.01	7.3	0.4	2.6	F0 V		89 s		
20290	130386		3 15 31.7	−8 46 12	0.000	−0.01	6.7	1.1	0.2	K0 III	+15	200 s		
20076	38681		3 15 33.0	+49 09 17	−0.002	−0.03	8.0	1.1	0.2	K0 III		320 s		
20433	216234		3 15 36.7	−43 17 07	−0.002	+0.03	7.6	1.8	−0.5	M2 III		340 s		
19968	12673		3 15 38.6	+61 07 40	+0.001	−0.01	7.56	0.08		B9	−7			
20163	56289		3 15 41.4	+38 31 03	+0.001	+0.01	7.9	0.0	0.4	B9.5 V		280 s		
20413	194196		3 15 43.8	−39 03 48	−0.001	−0.03	7.6	1.8	−0.3	K5 III		250 s		
20340	148864		3 15 45.7	−16 49 46	−0.001	−0.04	7.99	−0.12	−1.6	B3.5 V	−25	770 s		
19321	4852		3 15 45.8	+80 07 50	−0.004	−0.05	6.9	1.1	0.2	K0 III		220 s		
20193	56293		3 15 46.9	+32 51 23	−0.003	+0.01	6.31	0.37	2.6	F0 V	+14	52 s	2431	m
20041	23903		3 15 47.9	+57 08 27	0.000	+0.01	5.79	0.73	−7.1	A0 Ia	−12	1500 s	2424	m
−−	56292		3 15 48.4	+36 41 03	+0.004	−0.02	8.0	1.8						
20096	23907		3 15 48.5	+50 57 22	+0.002	−0.02	6.80	0.1	0.6	A0 V		160 s	2425	m
20320	130387	13 ζ Eri	3 15 49.9	−8 49 11	−0.001	+0.05	4.80	0.23		A m	−7	16 mn		
21024	255973	ι Hyi	3 15 57.8	−77 23 19	+0.035	+0.07	5.52	0.44		F2	+19			
20319	130388		3 16 00.8	−5 55 07	0.000	0.00	6.17	−0.02		B9 n	+7	13 mn	2440	m
20209	56297		3 16 01.7	+35 02 43	−0.013	−0.04	7.6	0.5	3.4	F5 V		67 s		
20210	56296		3 16 01.8	+34 41 20	+0.003	−0.03	6.25	0.28		A m	+25		2433	m
20278	93365		3 16 03.0	+11 37 43	+0.012	−0.01	8.03	0.61		G5	+44			
20375	148870		3 16 04.4	−19 21 17	0.000	+0.02	8.0	0.5	0.2	K0 III		370 s		
20162	38685		3 16 04.6	+45 20 45	+0.005	−0.04	6.16	1.67	−0.5	M2 III	−30	160 mx		
20337	130390		3 16 06.1	−7 55 37	−0.004	−0.05	7.7	1.1	−0.1	K2 III		200 mx		
20112	23911		3 16 07.7	+51 34 01	+0.001	−0.07	7.9	0.5	4.0	F8 V		58 s		
20880	255970		3 16 08.0	−73 32 56	+0.001	+0.03	7.9	−0.1	1.7	A3 V		170 s		
20040	23906		3 16 10.6	+60 06 56	−0.003	−0.01	7.7	1.1	0.2	K0 III	−35	290 s		
20423	194197		3 16 11.3	−30 49 40	+0.002	+0.01	6.65	−0.07		B9				
20123	23914		3 16 12.1	+50 56 16	−0.001	−0.01	5.03	1.15	−2.1	G5 II	+22	210 s		
−−	38690		3 16 13.9	+43 39 27	+0.003	−0.02	8.0	2.4						
20358	148873		3 16 15.3	−11 59 09	+0.001	0.00	7.50	1.4	−0.3	K4 III	−24	360 s		
20356	130391		3 16 21.6	−5 43 49	+0.001	−0.01	6.5	1.4	−0.3	K5 III		230 s		
19967	12677		3 16 25.1	+66 22 50	+0.008	−0.01	6.7	0.1	0.6	A0 V		100 mx		
20355	130392		3 16 27.4	−4 17 13	0.000	−0.01	6.9	0.1	1.4	A2 V		120 s		
20484	194202		3 16 32.6	−35 41 27	+0.002	+0.01	7.0	0.4	1.4	A2 V		86 s		
20236	56306		3 16 32.8	+38 38 03	+0.004	−0.06	7.9	1.1	−0.1	K2 III		190 mx		
20277	56308		3 16 35.0	+32 11 01	−0.001	−0.10	6.06	0.99	0.3	gG8	+19	140 s		
20395	130395	14 Eri	3 16 35.6	−9 09 17	−0.001	+0.05	6.14	0.40	3.3	dF4	−5	37 s		
20192	38691		3 16 36.3	+48 16 56	0.000	+0.03	7.80	0.9	−2.1	G9 II		710 s		
19906	12676		3 16 39.5	+69 58 45	+0.001	−0.02	7.8	0.1	0.6	A0 V		240 s		
20385	130394		3 16 40.5	−3 31 49	+0.004	−0.05	7.5	0.5	3.4	F5 V		67 s		
20542	216239		3 16 44.5	−41 16 13	0.000	+0.01	7.9	0.6	3.4	F5 V		62 s		
20235	38694		3 16 48.6	+43 44 54	0.000	−0.02	8.0	1.1	−0.1	K2 III		360 s		
20191	23924		3 16 49.0	+51 13 05	+0.002	−0.03	7.19	0.03		B9				
19580	4860		3 16 50.1	+78 30 02	+0.013	−0.02	8.00	1.1	0.2	K0 III		350 s	2403	m
20504	194206		3 16 55.0	−31 21 08	+0.001	−0.02	7.14	0.20		A0				
20328	75832		3 16 56.9	+23 07 36	+0.001	−0.06	7.4	1.1	0.2	K0 III		240 mx		
20134	23922		3 16 59.6	+60 04 02	−0.002	−0.01	7.43	0.13	−1.7	B3 V	−15	460 s		
20402	111096		3 17 08.9	+6 48 00	+0.003	−0.02	7.0	0.1	1.7	A3 V		120 s		
20431	130400		3 17 10.0	−4 08 22	+0.002	0.00	7.3	0.2	2.1	A5 V		110 s		
20283	38700		3 17 11.4	+40 29 00	+0.001	−0.02	6.70	0.1	0.6	A0 V	−8	160 s	2443	m
20421	111098		3 17 11.7	+1 33 28	+0.002	0.00	7.78	0.24	1.4	A2 V		150 s		
20368	93370		3 17 12.3	+17 34 22	+0.001	−0.01	7.7	1.4	−0.3	K5 III		390 s		
20520	168441		3 17 17.4	−23 31 13	0.000	+0.01	6.7	1.1	0.2	K0 III		170 s		
20317	56317		3 17 21.2	+36 21 17	+0.001	−0.03	7.9	0.1	0.6	A0 V		270 s		
19597	4863		3 17 21.8	+78 52 31	0.000	+0.02	6.8	0.4	2.6	F0 V		67 s		
20430	111104		3 17 26.2	+7 39 23	+0.010	+0.02	7.40	0.56	4.0	dF8	+32	44 s	2451	m
20640	216246		3 17 26.4	−47 45 06	−0.002	+0.03	5.85	1.24	0.2	gK0		97 s		
20104	12686		3 17 31.5	+65 39 31	−0.001	+0.01	6.36	0.09	1.4	A2 V	−65	94 s	2436	m
20439	111105		3 17 32.7	+7 41 25	+0.011	0.00	7.78	0.62	4.4	dG0	+32	43 s		
20316	38703		3 17 35.0	+42 40 30	+0.004	−0.02	7.70	0.4	2.6	F0 V		100 s	2445	m
20296	38702		3 17 39.4	+46 56 53	+0.001	−0.05	7.5	0.1	0.6	A0 V		220 s		
20367	56323		3 17 40.0	+31 07 36	−0.008	−0.06	6.5	0.6	4.4	G0 V	+5	27 s		
20282	−−		3 17 43.1	+50 21 40	+0.002	−0.03	7.7	0.1	0.6	A0 V		240 s		
20347	56320		3 17 43.9	+38 38 21	+0.010	−0.04	7.28	0.61	4.4	G0 V		35 s	2446	m
20889	248775		3 17 44.2	−68 09 37	−0.001	+0.01	8.0	0.5	3.4	F5 V		82 s		
20346	56322		3 17 45.7	+39 17 00	+0.002	−0.01	5.96	0.07	1.7	A3 V	+27	71 s		
20766	248770	ζ¹ Ret	3 17 46.2	−62 34 31	+0.195	+0.66	5.54	0.64	4.7	G2 V	+12	13 ts		
20315	38704	30 Per	3 17 47.3	+44 01 30	+0.003	−0.03	5.47	−0.06	−0.2	B8 V	0	140 s		

HD	SAO	Star Name	α 2000	δ 2000	μ(α)	μ(δ)	V	B-V	M_V	Spec	RV	d(pc)	ADS	Notes
			$\overset{h}{3}\ \overset{m}{17}\ \overset{s}{49.7}$	$+22°49'55''$	$+0\overset{s}{.}002$	$-0\overset{''}{.}05$								
20420	75845		3 17 49.7	+22 49 55	+0.002	-0.05	7.6	0.1	1.7	A3 V		150 s		
20584	168446		3 17 50.8	-26 21 09	+0.004	-0.01	7.4	0.6	4.0	F8 V		40 s		
20458	93373		3 17 52.3	+13 50 50	-0.001	-0.01	7.5	0.1	0.6	A0 V	+6	240 s		
20550	148889		3 17 59.1	-10 26 35	+0.006	-0.03	7.1	0.9	3.2	G5 IV		62 s		
20888	248776		3 17 59.3	-66 55 37	+0.012	+0.01	6.05	0.13	1.7	A3 V		54 mx		
20606	168449		3 18 02.8	-28 47 49	+0.014	-0.01	5.91	0.33	2.6	dF0		44 s		
20429	75847		3 18 04.1	+24 52 38	+0.001	-0.07	7.7	0.4	2.6	F0 V		100 s		
20807	248774	ζ² Ret	3 18 12.9	-62 30 23	+0.194	+0.66	5.24	0.60	4.7	G2 V	+12	12 ts		m
20500	93375		3 18 15.0	+12 49 27	0.000	0.00	7.6	0.0	0.0	B8.5 V	+16	340 s		
20457	75850		3 18 19.9	+24 04 54	-0.002	-0.03	7.3	1.1	0.2	K0 III		270 s		
20477	93376		3 18 19.9	+18 10 18	-0.006	-0.10	7.6	0.7	4.4	G0 V		43 s		
20610	168452	15 Eri	3 18 22.0	-22 30 41	0.000	0.00	4.88	0.90	0.3	gG6	+24	82 s	2463	
20559	130408	95 Cet	3 18 22.3	-0 55 49	+0.017	-0.05	5.38	1.04		K1 IV	+28	36 mn	2459	m
20512	93377		3 18 27.0	+15 10 38	0.000	-0.29	7.42	0.78	5.2	dG5	+11	23 s		
20428	56332		3 18 30.3	+37 56 09	+0.005	-0.04	7.4	1.1	0.2	K0 III		190 mx		
20427	56333		3 18 31.8	+38 27 32	+0.010	-0.07	7.5	0.5	3.4	F5 V		63 s		
20393	38709		3 18 32.3	+44 43 08	0.000	-0.05	8.0	0.1	0.6	A0 V		270 s		
20765	233069		3 18 34.4	-51 17 49	-0.004	-0.01	7.8	0.6	2.6	F0 V		69 s		
20392	38710		3 18 37.4	+46 53 15	+0.006	-0.05	7.8	0.7	4.4	G0 V		48 s		
20365	23944	29 Per	3 18 37.8	+50 13 20	+0.003	-0.02	5.15	-0.06	-1.7	B3 V	-5	200 s		
20631	148897		3 18 41.1	-18 33 35	+0.009	-0.05	5.71	0.37	3.0	dF2	+18	34 s	2465	m
20622	148896		3 18 41.1	-14 15 15	+0.001	+0.01	7.9	1.1	0.2	K0 III	+84	350 s		
20468	56340		3 18 43.7	+34 13 21	0.000	0.00	4.82	1.49	-2.2	K2 II	+2	200 s		
20391	38711		3 18 44.7	+49 46 12	+0.002	-0.02	7.94	0.12	1.2	A1 V		200 s		
20637	148898		3 18 50.0	-15 33 06	+0.009	-0.01	7.8	1.1	-0.1	K2 III		120 mx		
20855	233071		3 18 54.8	-57 58 46	+0.009	+0.11	7.45	1.05	3.2	K0 IV		59 s		
21166	255979		3 18 58.2	-73 59 11	0.000	-0.10	7.24	0.41		F2				
20958	248778		3 19 00.4	-65 53 18	0.000	-0.07	7.8	0.5	3.4	F5 V		75 s		
20511	56342		3 19 01.3	+32 41 17	-0.001	0.00	8.0	0.1	0.6	A0 V		270 s		
20619	130415		3 19 01.8	-2 50 37	+0.017	-0.11	7.04	0.65	4.7	dG2	+21	28 s		
20418	38714	31 Per	3 19 07.6	+50 05 42	+0.002	-0.02	5.04	-0.06	-1.1	B5 V	+3	160 s		
20966	248779		3 19 09.7	-65 43 40	-0.006	+0.01	7.5	0.1	1.4	A2 V		160 s		
20749	194229		3 19 17.6	-35 00 15	-0.001	-0.03	7.0	0.1	2.6	F0 V		76 s		
20499	56345		3 19 19.2	+39 03 21	+0.005	-0.06	7.3	0.4	3.0	F2 V		71 s		
20630	111120	96 κ Cet	3 19 21.6	+3 22 13	+0.018	+0.10	4.83	0.68	5.0	G5 V	+19	9.3 t		m
20759	194231		3 19 23.6	-36 33 53	+0.011	+0.07	7.4	0.7	3.0	F2 V		49 s		
20720	168460	16 τ⁴ Eri	3 19 30.9	-21 45 28	+0.003	+0.04	3.69	1.62	-0.5	gM3	+42	69 s	2472	m
20805	216258		3 19 31.1	-42 34 13	+0.002	+0.04	7.5	1.2	3.2	G5 IV		40 s		
20498	38720		3 19 33.1	+44 00 15	+0.001	-0.01	8.0	0.9	3.2	G5 IV		89 s		
21139	--		3 19 33.9	-71 27 29			8.0	1.1	0.2	K0 III		360 s		
20729	168462		3 19 34.8	-24 07 23	0.000	-0.02	5.61	1.66	-0.5	gM2	+15	150 s		
20693	148903		3 19 35.4	-11 24 33	+0.004	-0.02	8.0	0.9	3.2	G5 IV		93 s		
20579	75860		3 19 37.4	+27 32 37	-0.002	+0.03	7.9	0.1	0.6	A0 V		260 s		
20672	111127		3 19 44.8	+1 11 56	+0.001	+0.01	7.4	1.1	0.2	K0 III		270 s		
20728	148904		3 19 46.4	-18 50 39	+0.001	+0.02	7.0	0.6	0.6	A0 V		84 s		
20487	38723		3 19 47.2	+48 37 40	+0.002	-0.03	7.64	0.07		A0				
20629	93386		3 19 47.7	+19 04 35	0.000	0.00	7.4	0.1	0.6	A0 V		230 s		
20656	93387		3 19 51.9	+13 22 37	-0.001	-0.01	7.3	0.5	4.0	F8 V		45 s		
20794	216263		3 19 55.7	-43 04 10	+0.278	+0.75	4.27	0.71	5.3	G5 V	+87	6.2 t		82 G. Eridani
20618	75863	59 Ari	3 19 55.7	+27 04 16	-0.002	-0.07	5.90	0.86	0.3	gG5	0	130 s		
20665	111128		3 19 55.9	+8 41 59	-0.001	0.00	7.6	0.0	0.4	B9.5 V		280 s		
20085	4878		3 19 56.9	+75 14 15	+0.003	-0.02	7.5	0.0	0.4	B9.5 V		250 s		
20804	194236		3 19 57.4	-34 00 42	0.000	+0.01	7.5	0.7	3.2	G5 IV		72 s		
20774	168467		3 19 58.3	-28 59 36	+0.002	-0.06	7.7	1.3	-0.1	K2 III		200 mx		
20645	93389		3 19 59.2	+17 30 04	+0.002	+0.06	7.9	0.4	2.6	F0 V		110 s		
20336	12704		3 19 59.3	+65 39 09	+0.003	-0.01	4.84	-0.15	-2.5	B2 V e	+20	260 s		m
20699	111131		3 20 01.8	+1 29 01	0.000	+0.01	6.8	1.1	0.2	K0 III		210 s		
20703	130423		3 20 02.1	-1 35 32	-0.002	0.00	7.6	0.5	3.4	F5 V		71 s		
20782	168469		3 20 03.5	-28 51 13	+0.027	-0.04	7.38	0.65		K0		18 mn		m
--	56354		3 20 06.0	+32 01 18	+0.001	0.00	7.80	1.6		M5 II-III				UZ Per, v
20510	23958		3 20 06.2	+50 58 09	+0.003	-0.01	7.0	0.0	0.4	B9.5 V		190 s		
20273	12703		3 20 07.0	+69 43 53	+0.004	0.00	6.70	0.1	0.6	A0 V	-9	160 s	2455	m
20548	38731		3 20 07.2	+47 16 55	+0.010	-0.07	8.0	0.9	3.2	G5 IV		89 s		
20558	38733		3 20 13.9	+46 46 50	0.000	-0.02	7.8	1.1	0.2	K0 III		290 s		
19978	4875		3 20 19.8	+77 44 06	+0.020	-0.05	5.45	0.19		A4 n	+4	30 mn	2450	m
20682	93391		3 20 20.1	+19 43 47	+0.002	-0.01	7.7	0.1	1.4	A2 V		180 s	2475	m
20644	75871		3 20 20.3	+29 02 55	0.000	-0.01	4.47	1.55	-1.1	K2 II-III	-2	81 s		
20568	38735		3 20 20.9	+45 52 24	+0.001	-0.03	7.8	1.4	-0.3	K5 III		360 s		
20655	75873		3 20 21.1	+23 41 20	-0.001	-0.07	7.50	0.2	2.1	A5 V		71 mx	2473	m
20509	23960		3 20 22.4	+55 32 14	-0.001	+0.01	7.3	1.1	0.2	K0 III	+5	240 s		
20681	93394		3 20 23.2	+19 52 22	0.000	+0.01	7.1	1.1	0.2	K0 III		240 s		
20537	23962		3 20 23.7	+51 37 06	+0.004	-0.03	7.6	0.0	0.4	B9.5 V		100 mx		
20663	75875	60 Ari	3 20 25.5	+25 39 46	+0.001	-0.08	6.12	1.21	0.2	K0 III		110 s		
20664	75876		3 20 26.8	+24 32 36	-0.001	-0.01	8.0	1.1	0.2	K0 III		360 s		
20818	168470		3 20 28.0	-24 06 53	+0.006	-0.03	7.5	1.0	3.4	F5 V		31 s		

HD	SAO	Star Name	α 2000	δ 2000	μ(α)	μ(δ)	V	B−V	M$_v$	Spec	RV	d(pc)	ADS	Notes
			h m s	° ′ ″	s	″								
20577	38739		3 20 28.7	+46 09 58	+0.001	−0.02	7.7	1.1	−0.1	K2 III		320 s		
20717	93398		3 20 33.7	+12 20 48	+0.010	+0.02	7.4	0.5	3.4	F5 V	+45	63 s		
21115	248782		3 20 34.4	−65 47 09	+0.002	+0.09	8.0	0.1	1.7	A3 V		140 mx		
20680	75879		3 20 38.2	+26 55 39	−0.001	+0.01	8.00	1.1	−0.1	K2 III		420 s		
20852	168474		3 20 44.5	−26 17 16	+0.002	+0.14	6.8	0.5	2.6	F0 V		54 s		
20853	168475		3 20 45.1	−26 36 23	+0.003	+0.03	6.39	0.54	3.7	dF6		31 s		
20524	23967		3 20 51.3	+57 52 29	−0.002	0.00	8.0	1.1	0.2	K0 III		320 s		
20716	75883		3 20 51.5	+20 30 34	+0.001	−0.02	6.8	1.1	−0.1	K2 III		240 s		
20771	111138		3 20 53.4	+1 09 54	+0.009	−0.04	8.00	0.6	4.4	G0 V		53 s	2484	m
20653	56365		3 20 53.5	+38 53 07	+0.001	−0.01	8.0	0.1	0.6	A0 V		270 s		
20363	12710		3 20 53.7	+68 27 24	+0.002	−0.04	7.5	1.1	0.2	K0 III		260 s		
20914	194245		3 20 58.8	−37 27 00	+0.002	−0.01	7.5	1.9	−0.3	K5 III		210 s		
20678	56366		3 20 59.3	+33 13 05	+0.013	−0.13	7.90	1.1	0.2	K0 III		64 mx		
21223	248785		3 21 02.9	−68 32 02	+0.018	+0.07	7.1	0.0	1.7	A3 V		45 mx		
20255	4884		3 21 03.5	+73 41 58	+0.005	0.00	7.1	1.1	0.2	K0 III		230 s		
20791	111142	97 Cet	3 21 06.7	+3 40 32	+0.003	−0.02	5.69	0.97	0.3	gG8	+11	110 s		
21114	248783		3 21 11.6	−62 04 29	+0.004	0.00	8.0	1.1	0.2	K0 III		360 s		
20756	75886	61 τ Ari	3 21 13.6	+21 08 49	+0.002	−0.02	5.28	−0.07	−1.1	B5 V p	+14	190 s		m
20789	111143		3 21 17.3	+9 39 37	0.000	0.00	7.8	0.1	1.4	A2 V		190 s		
20536	12716		3 21 18.9	+61 59 54	−0.001	−0.05	6.6	0.0	0.6	B8.5 V	−7	200 s		
21475	255982		3 21 23.4	−74 35 39	+0.002	+0.02	8.04	0.25		A3				
20894	168482		3 21 24.0	−23 38 07	−0.002	−0.03	5.52	0.88	0.3	G5 III	+8	110 s		
20677	38750	32 Per	3 21 26.5	+43 19 47	−0.005	0.00	4.95	0.04	1.4	A2 V	−7	51 s		
20676	38749		3 21 29.7	+45 23 15	0.000	+0.02	7.50	0.00	0.0	B8.5 V		280 s	2483	m
21011	216278		3 21 33.2	−47 46 37	−0.001	−0.02	6.39	1.00		K0				
20911	168485		3 21 41.8	−20 19 27	+0.002	0.00	6.60	0.1	0.6	A0 V		160 s		m
20953	194251		3 21 44.6	−31 06 49	+0.001	−0.01	7.8	0.6	2.6	F0 V		69 s		
20662	23977		3 21 46.4	+52 43 53	+0.002	0.00	7.9	0.1	0.6	A0 V		260 s		
20768	56380		3 21 51.5	+31 39 25	0.000	+0.01	8.0	0.5	3.4	F5 V		79 s		
20675	38753		3 21 52.6	+49 04 15	+0.019	−0.07	6.20	0.5	3.4	F5 V	+25	36 s		m
20670	23979		3 21 56.7	+52 37 17	+0.018	−0.09	7.9	0.9	3.2	G5 IV		65 mx		
20924	148921		3 21 58.7	−15 27 30	+0.002	+0.03	7.3	1.1	0.2	K0 III		260 s		
20588	12721		3 22 04.6	+62 44 27	+0.001	−0.06	7.70	0.5	4.0	F8 V		54 s		m
20884	111155		3 22 05.0	+4 12 38	+0.002	+0.01	7.9	1.1	0.2	K0 III		340 s		
20425	4887		3 22 08.4	+71 06 02	+0.004	+0.01	7.9	1.1	0.2	K0 III		300 s		
21068	216282		3 22 09.4	−46 45 48	−0.001	−0.07	7.3	0.3	0.2	K0 III		260 s		
20906	130446		3 22 11.2	−5 08 15	+0.001	+0.01	7.8	1.1	0.2	K0 III		330 s		
21177	233086		3 22 11.6	−58 41 55	+0.003	−0.02	7.5	2.1	−0.1	K2 III		92 s		
20825	75892	62 Ari	3 22 11.9	+27 36 27	+0.001	−0.01	5.52	1.10	0.2	gK0	+6	100 s		
20980	168493		3 22 16.3	−25 35 16	+0.001	+0.01	6.35	0.01		A0				
20965	168492		3 22 16.4	−21 06 01	−0.001	−0.03	6.5	1.7	3.2	G5 IV		14 s		
20812	56387		3 22 18.2	+31 24 51	0.000	+0.01	7.5	1.1	0.2	K0 III		260 s		
20614	12724		3 22 20.4	+62 43 47	0.000	−0.03	7.9	0.1	0.6	A0 V		250 s		
21210	233088		3 22 25.3	−59 42 13	−0.001	−0.02	7.8	1.4	0.2	K0 III		200 s		
20844	56389		3 22 31.0	+32 14 23	0.000	−0.02	7.6	1.6	−0.5	M2 III		370 s		
20893	75899	63 Ari	3 22 45.2	+20 44 31	−0.003	−0.02	5.09	1.24	−0.2	K3 III	+2	110 s		
20873	—		3 22 52.6	+29 49 04	+0.002	−0.02	7.9	0.1	1.4	A2 V	+15	180 s	2499	m
21176	233089		3 22 52.7	−51 18 44	+0.001	+0.04	7.8	0.3	4.0	F8 V		57 s		
21057	168499		3 22 54.5	−27 18 09	+0.004	−0.03	8.0	1.2	−0.1	K2 III		180 mx		
20849	56392		3 22 54.8	+37 39 37	+0.005	−0.02	7.8	1.1	0.2	K0 III		220 mx		
21113	216287		3 22 55.9	−41 15 20	+0.001	0.00	7.7	1.4	−0.3	K5 III		390 s		
20523	4891		3 22 57.5	+71 16 48	+0.004	−0.06	7.6	0.4	2.6	F0 V		95 s		
20900	75903		3 23 04.8	+26 54 04	−0.001	−0.02	7.3	0.1	1.4	A2 V		140 s		
20750	23991		3 23 05.3	+55 13 48	+0.003	−0.05	7.6	0.2	2.1	A5 V		120 s		
20891	56399		3 23 12.1	+33 04 42	+0.009	−0.06	7.7	0.6	4.4	G0 V		46 s		
20809	38768		3 23 13.1	+49 12 48	+0.002	−0.02	5.29	−0.07	−1.1	B5 V	+5	180 s		
21020	148930		3 23 16.0	−11 20 50	+0.006	0.00	7.70	0.5	3.4	F5 V		72 s		m
21066	148933		3 23 17.1	−17 26 20	−0.003	+0.06	6.6	1.1	0.2	K0 III		190 s		
21019	130457		3 23 17.7	−7 47 39	0.000	−0.22	6.20	0.70		G0		15 mn	2507	m
20947	75906		3 23 19.3	+20 58 14	0.000	−0.01	7.60	0.8	3.2	G5 IV		76 s	2504	m
20905	56402		3 23 23.8	+30 59 01	+0.001	−0.04	7.4	1.1	0.2	K0 III		250 s		
7868	216291		3 23 24.2	−41 21 50	−0.002	+0.02	8.0	2.0						
21064	148934		3 23 24.9	−14 52 34	0.000	+0.08	7.5	0.6	4.4	G0 V		42 s		
20871	38773		3 23 26.5	+41 42 30	0.000	−0.01	8.0	0.0	0.0	B8.5 V		350 s		
21063	148936		3 23 27.6	−14 13 16	−0.001	+0.03	8.00	0.9	3.2	G5 IV		91 s		m
20904	56404		3 23 28.3	+33 19 55	+0.002	−0.08	7.7	0.5	4.0	F8 V		53 s		
20872	38776		3 23 30.7	+41 12 23	+0.004	−0.05	7.8	0.1	1.4	A2 V		64 mx		
19855	536		3 23 31.3	+82 09 00	+0.013	−0.01	7.30	1.6	−0.5	M2 III	−7	360 s		
20762	23996		3 23 32.7	+58 43 23	0.000	+0.02	6.7	1.1	0.2	K0 III		190 s		
20566	4896		3 23 33.0	+71 25 08	−0.002	+0.01	7.7	0.0	0.0	B8.5 V		330 s		
21032	111162		3 23 36.2	+0 54 35	−0.001	−0.11	6.6	1.1	0.2	K0 III		150 mx		
21327	233104		3 23 36.5	−58 59 39	+0.003	−0.03	7.5	1.7	−0.3	K5 III		280 s		
20933	56407		3 23 36.8	+32 59 41	−0.001	−0.01	8.0	0.4	3.0	F2 V		95 s		
21018	111161		3 23 39.0	+4 52 55	0.000	0.00	6.38	0.86		F8	+3		2509	m
21131	168507		3 23 39.7	−27 55 48	+0.001	+0.01	7.2	1.2	0.2	K0 III		190 s		

HD	SAO	Star Name	α 2000	δ 2000	μ(α)	μ(δ)	V	B-V	M$_V$	Spec	RV	d(pc)	ADS	Notes
			h m s	° ′ ″	s	″								
20842	24004		3 23 43.1	+51 46 13	+0.003	−0.03	7.85	0.10		A0				
21149	194268		3 23 44.5	−32 42 26	+0.001	0.00	6.50	1.37		K0				
20863	38779		3 23 47.0	+48 36 14	0.000	−0.04	6.99	0.02		B9				
21052	111164		3 23 54.8	+3 02 36	+0.001	−0.01	7.9	0.0	0.4	B9.5 V		320 s		
21208	194272		3 23 59.5	−37 10 15	0.000	−0.06	7.7	1.0	0.2	K0 III		260 mx		
21006	93412		3 24 00.9	+19 54 24	+0.001	−0.01	7.2	1.4	−0.3	K5 III		310 s		
21563	248797		3 24 02.5	−69 37 29	+0.006	+0.02	6.15	0.48		A3				
20507	4897		3 24 04.8	+74 10 17	+0.049	−0.13	7.0	0.5	3.4	F5 V	−28	28 mx		
21246	216297		3 24 06.4	−41 13 52	−0.002	−0.03	7.2	0.2	0.2	K0 III		260 s		
21144	168514		3 24 07.8	−20 55 51	−0.001	0.00	7.5	0.3	3.2	G5 IV		71 s		
20972	56414		3 24 08.5	+34 00 01	+0.001	0.00	8.0	1.1	0.2	K0 III		320 s		
21051	93416		3 24 10.0	+12 37 47	+0.001	−0.02	6.04	1.23		G5	+21			
20711	12738		3 24 14.9	+67 27 19	+0.010	−0.01	7.80	0.5	4.0	F8 V		56 s	2494	m
21174	168518		3 24 16.3	−24 18 03	+0.005	+0.01	7.1	1.1	0.2	K0 III		210 s		
21273	216299		3 24 17.1	−45 05 48	+0.008	+0.04	7.8	0.4	3.4	F5 V		62 s		
21017	75912	64 Ari	3 24 18.4	+24 43 27	+0.001	−0.04	5.50	1.19	0.2	K0 III	+13	88 s		
20902	38787	33 α Per	3 24 19.3	+49 51 40	+0.003	−0.02	1.80	0.48	−4.6	F5 Ib	−2	190 s		Mirfak, m
21220	194275		3 24 23.4	−31 07 38	+0.002	+0.10	7.3	0.6	3.2	G5 IV		67 s		
21413	233109		3 24 23.5	−57 33 05	+0.001	+0.02	7.3	0.5	3.2	G5 IV		66 s		
21160	148943		3 24 24.5	−13 59 34	0.000	+0.02	6.90	0.1	0.6	A0 V		180 s	2523	m
21161	148944		3 24 24.8	−15 39 13	+0.017	−0.10	7.51	0.61		G0			2524	m
21050	75915	65 Ari	3 24 26.0	+20 48 13	0.000	0.00	5.90	−0.04	0.6	A0 V	−9	120 s		
20971	56417		3 24 26.4	+40 02 37	0.000	−0.03	7.6	1.4	−0.3	K5 III		330 s		
20994	56418		3 24 29.6	+34 17 10	−0.001	−0.01	8.0	0.1	1.4	A2 V		190 s		
20995	56419		3 24 29.7	+33 32 10	+0.003	−0.03	5.60	−0.03	0.4	B9.5 V	+2	110 s	2514	m
20931	38791		3 24 30.0	+49 08 22	+0.004	−0.04	7.87	0.09		A2				
21319	216302		3 24 34.5	−45 39 49	0.000	+0.03	7.40	0.1	1.4	A2 V		160 s		m
21049	75917		3 24 34.5	+22 02 25	+0.003	−0.02	7.0	0.1	0.6	A0 V		190 s		
21360	233106		3 24 36.2	−51 03 45	−0.001	+0.02	6.70	0.1	0.6	A0 V		170 s		m
21940	255988		3 24 37.3	−76 44 47	−0.009	−0.07	6.81	0.20		A0				
20797	12743		3 24 40.5	+64 35 10	−0.001	+0.01	5.23	2.08	−2.4	M0 II	−21	190 s		
20987	38798		3 24 44.7	+40 45 39	−0.002	−0.01	7.80	0.00	0.4	B9.5 V		270 s	2513	m
21111	111171		3 24 46.3	+9 48 29	+0.003	−0.02	7.9	0.9	3.2	G5 IV		88 s		
21120	111172	1 o Tau	3 24 48.7	+9 01 44	−0.005	−0.07	3.60	0.89	0.3	G8 III	−21	46 s		
21062	75918		3 24 48.9	+28 39 09	+0.002	−0.05	7.0	0.1	0.6	A0 V	+6	180 s		
20961	38796		3 24 52.1	+47 54 54	+0.004	−0.03	7.64	0.12	0.6	A0 V		100 mx		
21197	130471		3 24 59.5	−5 21 50	−0.017	−0.77	7.86	1.16	8.0	K5 V	−12	11 ts		
21218	148951		3 25 00.7	−14 06 59	+0.006	+0.01	7.9	0.5	3.4	F5 V		78 s		
21134	93421		3 25 04.6	+10 58 35	+0.002	−0.02	8.0	0.1	0.6	A0 V		300 s		
21038	38809		3 25 09.4	+41 15 26	0.000	0.00	6.4	0.1	0.6	A0 V	−19	140 s		
21196	130473		3 25 09.6	−4 28 42	+0.003	+0.06	7.4	0.9	3.2	G5 IV		70 s		
21142	93422		3 25 10.5	+12 29 08	+0.002	−0.02	7.6	0.1	0.6	A0 V		260 s		
21298	194282		3 25 11.7	−29 50 37	+0.004	−0.01	7.2	1.2	3.2	G5 IV		37 s		
21341	194283		3 25 12.8	−37 09 12	+0.003	−0.02	7.0	−0.1	1.7	A3 V		120 s		
--	75923		3 25 13.9	+21 12 52	−0.001	0.00	7.9	1.9		M3				
21252	148954		3 25 14.7	−15 02 21	−0.014	−0.27	7.97	0.67	4.4	dG0	+45	45 s		
20898	12750		3 25 16.0	+60 29 00	−0.002	−0.02	7.94	0.44	−3.6	B2 III		870 s	2510	m
21061	56428		3 25 17.1	+37 55 07	0.000	−0.01	7.9	0.9	3.2	G5 IV		85 s		
20930	24017		3 25 18.1	+58 41 39	−0.005	0.00	7.1	1.1	0.2	K0 III	−32	220 s		
21047	38812		3 25 19.3	+40 57 11	−0.001	+0.02	7.3	0.1	0.6	A0 V		200 s		
21215	130476		3 25 20.2	−4 39 17	0.000	+0.02	7.1	0.9	3.2	G5 IV		60 s		
21157	93425		3 25 23.3	+15 49 35	+0.003	−0.03	7.9	1.1	−0.1	K2 III		390 s		
21110	56436		3 25 23.9	+31 43 52	+0.002	0.00	7.29	1.51		K4 III-IV	+19			
21169	93427		3 25 24.0	+14 05 13	0.000	−0.04	8.0	0.1	1.4	A2 V		210 s		
21462	233112		3 25 24.8	−52 22 54	+0.001	+0.02	7.8	1.1	0.2	K0 III		300 s		
20710	4902		3 25 26.5	+71 41 06	−0.001	+0.01	7.6	0.0	0.0	B8.5 V		310 s		
21385	194285		3 25 34.8	−38 18 46	+0.011	+0.02	7.6	0.6	3.4	F5 V		59 s		
21722	248804		3 25 36.2	−69 20 12	−0.001	+0.06	5.96	0.42	2.1	F5 IV		58 s		
21340	168536		3 25 36.4	−27 16 20	+0.005	+0.01	7.7	0.7	3.2	G5 IV		80 s		
21060	38815		3 25 37.6	+45 30 55	−0.001	−0.02	7.70	0.00	0.0	B8.5 V		310 s	2525	m
21425	216312		3 25 42.4	−41 19 32	+0.003	+0.02	8.00	0.4	2.6	F0 V		120 s		m
21243	111181		3 25 44.2	+5 56 07	+0.004	−0.02	8.0	0.1	0.6	A0 V		68 mx		
20959	24021		3 25 47.7	+59 25 57	−0.002	+0.01	8.01	0.27	−2.9	B3 III		830 s		
21004	24024		3 25 48.3	+53 55 18	+0.010	−0.03	6.51	0.29	1.1	A9 III-IV	−4	82 mx		
20709	4903		3 25 49.9	+73 12 28	+0.013	−0.06	7.02	1.32	−0.2	K3 III	−20	170 mx		
21028	24026		3 25 55.4	+53 19 20	+0.001	−0.03	7.3	0.1	0.6	A0 V		210 s		
21423	194289	χ¹ For	3 25 55.7	−35 55 15	+0.002	0.00	6.39	0.08		A2				
21071	38817		3 25 57.4	+49 07 16	+0.003	−0.02	6.07	−0.07	−0.9	B6 V	+10	240 s		
21045	24028		3 26 01.8	+51 25 06	−0.007	+0.07	7.7	0.5	4.0	F8 V		53 s		
21085	38818		3 26 04.6	+49 44 43	+0.004	−0.01	7.2	0.1	−2.3	A3 II		130 mx		
21091	38824		3 26 10.9	+48 23 02	+0.004	−0.03	7.49	0.04	0.6	A0 V		110 mx		
21411	194293		3 26 11.1	−30 37 06	+0.017	+0.20	7.89	0.71	5.6	G8 V		29 s		m
21473	216316		3 26 11.6	−41 38 13	+0.001	+0.03	6.32	0.06	1.2	A1 V		100 s		
21037	24029		3 26 16.0	+56 08 30	−0.002	+0.02	7.8	1.4	−0.3	K5 III		360 s		
21504	216319		3 26 21.9	−42 50 12	+0.002	+0.04	7.5	1.0	0.2	K0 III		260 s		

HD	SAO	Star Name	α 2000	δ 2000	μ(α)	μ(δ)	V	B-V	M_v	Spec	RV	d(pc)	ADS	Notes
21430	168544		3^{h}26^{m}22^s.4	-27°19'03"	0^s.000	+0".05	5.93	0.94		G5				
21383	148959		3 26 24.0	-15 21 50	0.000	-0.03	7.6	0.1	0.6	A0 V		250 s		
21229	56450		3 26 30.6	+31 49 18	-0.001	+0.01	7.9	0.1	1.7	A3 V		160 s		
21014	12763		3 26 34.0	+60 16 53	-0.001	-0.03	8.0	0.9	3.2	G5 IV		89 s		
20984	12758		3 26 34.5	+63 11 43	-0.001	-0.01	7.3	0.9	3.2	G5 IV		65 s		
21242	75927		3 26 35.2	+28 42 54	+0.002	-0.11	6.5	0.9	3.2	G5 IV	+16	45 s		
21117	24036		3 26 39.4	+50 50 47	+0.002	-0.02	7.7	0.0	0.0	B8.5 V		310 s		
20831	4905		3 26 40.4	+72 12 21	+0.007	-0.07	7.2	0.4	2.6	F0 V		83 s		
21460	168549		3 26 44.3	-28 33 36	+0.004	+0.05	7.70	1.1	0.2	K0 III		250 mx	2543	m
21044	12764		3 26 48.5	+60 24 18	0.000	-0.01	7.7	0.1	1.4	A2 V		170 s		
21367	130495		3 26 49.1	+0 01 34	+0.001	0.00	7.1	1.1	0.2	K0 III		240 s		
21152	38834		3 26 50.3	+47 54 58	+0.003	-0.01	7.72	0.11		B9				
20967	12759		3 26 52.8	+66 12 40	+0.002	-0.04	7.2	1.1	0.2	K0 III	+20	230 s		
21192	38837		3 26 53.7	+43 39 21	-0.002	-0.03	7.6	1.1	0.2	K0 III		270 s		
21419	130497		3 26 55.3	-7 58 57	+0.001	-0.02	8.0	1.1	0.2	K0 III		360 s		
--	--		3 26 56.6	+49 01 03	0.000	+0.02	6.9	1.4	-0.3	K5 III		250 s		
21269	56455		3 27 02.2	+34 25 47	-0.001	+0.01	6.8	0.7	4.4	G0 V		29 s		
21335	93436		3 27 03.2	+18 45 24	+0.003	0.00	6.57	0.13	0.0	A3 III	+31	200 s		
21181	38838		3 27 05.2	+48 12 20	-0.002	-0.02	6.84	-0.01	0.2	B9 V		210 s		
21534	194307		3 27 06.1	-37 53 27	+0.002	+0.07	8.00	0.5	3.4	F5 V		83 s		m
21364	111195	2 ξ Tau	3 27 10.1	+9 43 58	+0.004	-0.03	3.74	-0.09		B8 p	-2	27 mn		
21607	216327		3 27 12.2	-48 51 46	+0.005	-0.06	7.4	1.3	3.2	G5 IV		30 s		
21405	111197		3 27 13.8	+2 16 48	+0.002	-0.02	7.40	0.12	0.6	A0 V		190 s		
21765	248810		3 27 16.9	-61 15 31	+0.007	+0.08	8.0	0.3	3.0	F2 V		98 s		
21379	93439		3 27 18.6	+12 44 06	0.000	-0.02	6.20	-0.02	0.4	B9.5 V	+15	140 s		
21292	56462		3 27 26.1	+38 02 32	+0.002	-0.10	7.8	0.5	4.0	F8 V		56 s		
21116	24040		3 27 26.3	+59 11 44	-0.002	-0.01	7.8	0.0	0.4	B9.5 V		270 s		
21440	111200		3 27 31.9	+5 52 13	+0.001	-0.04	7.3	1.1	0.2	K0 III		260 s		
21574	194312	χ² For	3 27 33.3	-35 40 54	+0.006	-0.01	5.71	1.29		K0	+30			
21586	216330		3 27 34.6	-39 52 51	-0.001	-0.01	7.1	1.7	0.2	K0 III		93 s		
21439	111202		3 27 37.5	+7 45 03	+0.002	0.00	8.0	0.4	2.6	F0 V		120 s		
21239	38846		3 27 37.6	+48 16 25	+0.003	0.00	7.95	0.10		A0				
21238	38845		3 27 38.8	+49 35 59	+0.001	-0.03	7.8	0.1	0.6	A0 V		240 s		
21594	216331		3 27 39.6	-39 50 02	-0.002	-0.03	7.6	0.9	3.2	G5 IV		62 s		
21438	111204		3 27 39.7	+8 45 55	+0.002	-0.03	7.9	0.2	2.1	A5 V		150 s		
21513	148969		3 27 41.3	-14 21 59	-0.002	0.00	7.20	0.1	0.6	A0 V		210 s	2551	m
21626	216332		3 27 43.7	-43 51 21	+0.003	+0.03	6.72	0.50	2.8	G0 IV		61 s		
20030	540		3 27 45.3	+83 31 42	+0.005	0.00	7.4	1.1	0.2	K0 III		280 s		
21416	75938		3 27 51.1	+20 55 31	0.000	-0.02	7.6	1.1	0.2	K0 III		310 s		
21279	38847		3 27 55.7	+47 44 09	+0.002	-0.03	7.26	0.05		B9				
21659	216335		3 27 58.5	-45 33 40	0.000	+0.02	7.7	1.1	0.2	K0 III		280 s		
22387	255993		3 27 59.3	-77 36 48	+0.015	+0.02	7.70	1.1	0.2	K0 III		320 s		m
21530	148972		3 28 00.9	-11 17 12	0.000	-0.05	5.73	1.10	-0.1	gK2	-2	150 s		
21437	75941		3 28 01.5	+20 27 48	+0.001	+0.04	7.40	0.1	1.7	A3 V		140 s		m
21278	38849		3 28 03.0	+49 03 46	+0.003	-0.03	4.98	-0.10	-1.7	B3 V	+7	200 s		
20674	4904		3 28 03.0	+78 28 44	+0.010	-0.02	7.9	0.5	3.4	F5 V		77 s		
21545	148973		3 28 09.4	-12 45 13	0.000	+0.03	7.3	1.1	0.2	K0 III		270 s		
21403	56471		3 28 10.3	+32 48 41	+0.002	-0.01	7.6	1.1	0.2	K0 III		280 s		
21635	194318	χ³ For	3 28 11.4	-35 51 12	+0.002	0.00	6.50	0.13		A0				m
21826	233133		3 28 17.4	-57 52 56	+0.001	-0.01	7.2	0.7	-0.3	K5 III		320 s		
21346	38855		3 28 18.4	+44 22 38	+0.001	0.00	6.9	1.1	0.2	K0 III		200 s		
21402	56475		3 28 20.7	+33 48 27	+0.003	-0.05	5.60	0.02	1.4	A2 V	+6	69 s		
21203	12770		3 28 23.5	+60 15 21	+0.003	-0.02	6.80	0.02	-0.2	B8 V	+5	220 s	2538	m
21467	75945	66 Ari	3 28 26.5	+22 48 15	0.000	-0.10	6.03	0.95	3.2	G5 IV	+49	29 s	2552	m
21571	148974		3 28 27.4	-13 12 35	0.000	-0.05	7.2	1.1	0.2	K0 III		250 s		
21224	--		3 28 28.7	+59 54 24	-0.002	+0.01	7.54	1.06	4.0	F8 V		23 s	2540	m
21400	56476		3 28 30.1	+40 10 18	+0.001	0.00	7.3	1.1	0.2	K0 III		240 s		
21377	38860		3 28 33.2	+43 45 11	+0.001	-0.03	7.3	0.1	0.6	A0 V		200 s		
22773	258340		3 28 33.8	-80 42 33	-0.016	+0.01	7.7	1.1	0.2	K0 III		280 s		
21524	93448		3 28 36.3	+11 23 27	-0.001	-0.02	6.80	0.8	3.2	G5 IV		53 s	2561	m
21483	56488		3 28 46.5	+30 22 32	-0.001	0.00	7.06	0.36	-2.9	B3 III	-5	460 s		
21511	75946		3 28 48.1	+20 37 27	0.000	0.00	7.2	0.1	0.6	A0 V		210 s		
21450	56486		3 28 49.0	+35 48 29	+0.004	-0.14	7.3	0.9	3.2	G5 IV		65 s		
21376	38866		3 28 51.5	+47 49 49	-0.010	-0.01	7.7	0.6	4.4	G0 V		46 s		
21362	38862		3 28 52.3	+49 50 54	+0.003	-0.02	5.59	-0.04	-0.9	B6 V	0	180 s		
21375	38865		3 28 53.7	+49 04 13	+0.005	-0.02	7.47	0.11	1.2	A1 V		100 mx		
21106	12769		3 28 54.4	+69 17 01	+0.001	+0.06	7.6	0.5	3.4	F5 V		66 s		
21704	194326		3 28 54.7	-33 39 20	+0.005	-0.01	7.1	0.9		K0 III		220 mx		
21449	38870		3 28 57.3	+40 11 11	+0.005	+0.01	7.60	0.8	3.2	G5 IV		74 s	2553	m
21541	93449		3 28 58.6	+14 59 46	+0.003	-0.04	7.4	0.1	0.6	A0 V	+8	230 s		
21585	111215		3 29 03.7	+3 14 55	0.000	+0.01	6.5	0.9	3.2	G5 IV		46 s		
21291	24054		3 29 04.1	+59 56 25	0.000	0.00	4.21	0.41	-7.1	B9 Ia	-7	1100 s	2544	m
21398	38868		3 29 07.6	+48 18 11	+0.002	-0.02	7.36	0.01	0.2	B9 V		260 s		
21448	38873		3 29 13.9	+45 02 57	+0.001	-0.01	7.17	0.27	-3.5	B1 V	-12	730 s	2559	m
21679	168579		3 29 17.2	-21 22 15	0.000	+0.02	8.0	1.3	-0.1	K2 III		360 s		

HD	SAO	Star Name	α 2000	δ 2000	μ(α)	μ(δ)	V	B–V	M$_V$	Spec	RV	d(pc)	ADS	Notes
21568	93452		3^h 29^m 19^s.6	+16° 11' 55"	+0^s.002	−0".03	8.0	0.4	2.6	F0 V		120 s		
21428	38872	34 Per	3 29 22.0	+49 30 32	+0.003	−0.02	4.67	−0.09	−2.3	B3 IV	−1	210 s	2558	m
22001	248819	κ Ret	3 29 22.7	−62 56 15	+0.056	+0.38	4.72	0.40	3.4	F5 V	+12	20 ts		m
21654	—		3 29 23.4	−10 33 11			7.8	0.5	3.4	F5 V		74 s		
21140	12774		3 29 24.9	+69 27 35	−0.001	0.00	7.7	0.1	0.6	A0 V		240 s		
21689	148984		3 29 25.8	−19 33 18	+0.004	+0.02	7.4	0.9	3.2	G5 IV		60 s		
21455	38874		3 29 26.2	+46 56 16	+0.002	−0.03	6.24	0.13	−1.1	B5 V	−1	200 s	2560	m
21825	216346		3 29 26.6	−42 28 09	+0.000	−0.03	7.4	0.8	2.1	A5 V		44 s		
21737	168582		3 29 35.1	−23 28 33	+0.006	+0.03	6.9	0.4	3.4	F5 V		49 s		
21688	148985		3 29 36.0	−12 40 30	0.000	0.00	5.59	0.17		A1 n	+15			
21665	130520		3 29 39.0	−6 48 18	−0.007	−0.10	5.99	1.02		G5				
21590	93454		3 29 42.5	+16 45 44	+0.002	−0.02	7.1	0.1		A0 p	+2			
21482	38880		3 29 46.0	+47 29 46	−0.003	0.00	8.0	0.1	0.6	A0 V		270 s		
21760	168587		3 29 48.9	−22 30 19	0.000	+0.03	6.8	0.9	0.2	K0 III		210 s		
21481	38882		3 29 49.8	+47 58 36	+0.001	−0.03	7.67	0.10	0.6	A0 V		220 s		
21479	38881		3 29 51.9	+49 12 47	+0.003	−0.04	7.28	0.10	1.4	A2 V		150 s		
21389	24061		3 29 54.8	+58 52 44	+0.001	0.00	4.54	0.56	−7.1	A0 Ia	−6	1100 s		
21882	216347		3 29 54.9	−42 38 03	−0.007	0.00	5.78	0.22		A3	+12			
21852	194339		3 29 58.0	−35 44 55	0.000	−0.02	7.1	0.5	1.4	A2 V		73 s		
22676	255998		3 29 58.9	−78 21 07	−0.004	−0.02	5.70	0.93		K0	+10			
21447	24064		3 30 00.2	+55 27 07	−0.005	−0.01	5.09	0.05	1.2	A1 V	+1	58 s	2565	m
21567	56496		3 30 03.2	+35 40 17	0.000	−0.01	7.8	1.6		M5 III	−79			R Per, v
21026	4912		3 30 06.3	+75 15 30	+0.001	−0.04	8.0	0.1	0.6	A0 V		290 s		
22449	255996		3 30 08.0	−74 35 40	+0.003	+0.04	7.52	0.16		A0				
21579	56498		3 30 10.8	+36 38 18	+0.002	−0.01	7.8	0.5	3.4	F5 V		75 s		
21427	24062		3 30 11.2	+59 21 58	+0.005	−0.04	6.13	0.08	0.6	A0 V	+11	110 s	2563	m
21267	12780		3 30 13.0	+68 25 59	+0.007	−0.05	7.98	0.02	0.0	B8.5 V		80 mx		
21465	24065		3 30 13.1	+55 22 28	+0.002	−0.02	7.10	2.00	−0.3	K5 III	−13	150 s		
21779	148994		3 30 13.2	−11 38 47	+0.006	−0.09	6.9	0.9	3.2	G5 IV		55 s		
21899	216350		3 30 13.5	−41 22 12	−0.001	−0.17	6.12	0.48	4.0	F8 V		27 s		
21611	75953		3 30 16.2	+30 02 28	0.000	0.00	7.6	0.1	0.6	A0 V	+10	230 s		
21488	24068		3 30 17.8	+52 53 42	+0.005	−0.03	7.29	1.39		K2	+22		2566	m
21179	4917		3 30 19.5	+71 51 51	+0.003	+0.01	6.7	1.6	−0.5	M2 III	−23	270 s		
22014	233139		3 30 21.7	−54 48 28	+0.003	+0.05	7.9	0.7	3.2	G5 IV		89 s		
21686	93463	4 Tau	3 30 24.4	+11 20 11	−0.001	−0.02	5.14	−0.03	0.4	B9.5 V	0	87 s		
21565	38892		3 30 25.2	+42 58 46	−0.002	−0.02	7.5	0.0	0.4	B9.5 V		240 s		
21687	93464		3 30 26.3	+10 27 31	0.000	−0.02	7.2	1.1	0.2	K0 III		250 s		
21540	38887		3 30 26.9	+47 03 45	+0.001	−0.04	7.0	0.0	0.0	B8.5 V	+6	230 s		
21552	38890	35 σ Per	3 30 34.4	+47 59 43	0.000	−0.02	4.35	1.37	−0.2	K3 III	+16	71 s		
21981	216357		3 30 37.0	−47 22 30	+0.008	+0.02	5.99	0.11	1.4	A2 V		56 mx		
21790	130528	17 Eri	3 30 37.0	−5 04 30	+0.001	+0.01	4.73	−0.09	−0.2	B8 V	+15	97 s		
21551	38893		3 30 37.0	+48 06 13	+0.003	−0.03	5.82	−0.04	−0.6	B8 IV	+9	180 s		
21841	149001		3 30 39.9	−19 26 08	−0.002	−0.01	7.5	0.7	4.4	G0 V		33 s		
21871	168598		3 30 40.2	−27 55 20	+0.003	+0.10	7.80	0.5	3.4	F5 V		76 s		m
21811	130532		3 30 40.5	−9 43 51	+0.002	+0.01	7.0	1.1	0.2	K0 III		230 s		
21777	111231		3 30 44.6	+5 30 46	−0.001	−0.01	7.8	0.1	1.4	A2 V		190 s		
21589	38898		3 30 44.7	+42 12 07	+0.005	−0.04	6.94	0.39	3.0	F2 V		59 s		
21755	111230		3 30 45.2	+6 11 20	+0.002	−0.01	5.94	0.96	0.3	gG5	+11	120 s		
21778	111232		3 30 46.9	+5 22 00	+0.005	−0.01	7.2	0.8	3.2	G5 IV		62 s		
21861	149002		3 30 49.1	−18 27 56	−0.001	−0.02	7.5	1.1	0.2	K0 III		300 s		
21938	194347		3 30 50.2	−37 22 20	−0.018	−0.02	7.5	0.0	4.4	G0 V		42 s		
22252	248825		3 30 51.6	−66 29 24	+0.002	0.00	5.83	−0.06	−0.6	B7 V		180 s		
21754	93469	5 Tau	3 30 52.3	+12 56 12	+0.001	0.00	4.11	1.12	−0.9	K0 II-III	+15	97 s		
22166	248821		3 30 52.6	−62 00 29	+0.005	+0.02	6.7	0.9	4.4	G0 V		18 s		
21685	75962		3 31 00.6	+27 43 29	+0.001	−0.01	7.87	0.14	1.7	A3 V		160 s		
21797	111235		3 31 01.4	+7 11 54	−0.001	−0.01	7.9	1.1	0.2	K0 III		340 s		
21920	168603		3 31 02.3	−26 37 18	+0.001	−0.07	7.7	1.5	0.2	K0 III		160 s		
21700	75964		3 31 03.4	+27 43 54	+0.002	−0.02	7.44	0.03	1.2	A1 V	+8	180 s		m
21822	130537		3 31 03.9	−0 28 51	−0.001	−0.01	6.6	0.1	0.6	A0 V		160 s		
22462	255997		3 31 04.8	−72 22 26	−0.008	−0.08	7.9	0.5	3.4	F5 V		79 s		
21584	24074		3 31 06.7	+50 28 55	−0.001	0.00	7.4	0.2		A m				
21651	56505		3 31 08.9	+39 08 52	+0.002	−0.03	7.29	0.35	3.4	F5 V		60 s		
21875	149008		3 31 10.9	−13 29 56	+0.001	−0.01	7.6	0.1	0.6	A0 V		250 s		
21650	38907		3 31 15.6	+41 43 35	+0.001	0.00	7.33	0.03		B7 e	−15			
22120	233144		3 31 17.3	−55 14 56	+0.006	+0.03	7.2	1.1	−0.3	K5 III		320 s		
—	38902		3 31 20.2	+49 27 23	+0.001	−0.02	7.7	0.9						
21743	75970		3 31 20.7	+27 34 19	+0.003	−0.03	6.40	0.1	0.6	A0 V	+6	140 s	2582	m
21787	93473		3 31 20.9	+17 56 15	−0.002	−0.02	7.2	1.1	−0.1	K2 III		290 s		
—	—		3 31 21.5	+49 26 35			7.90	0.9	−6.1	G8 I				
21988	194353		3 31 24.4	−33 32 46	0.000	+0.02	7.2	0.7	3.2	G5 IV		63 s		
22189	248827		3 31 24.6	−59 51 27	0.000	−0.02	7.7	−0.1	1.4	A2 V		180 s		
21919	149013		3 31 25.8	−16 46 42	+0.009	+0.04	7.6	0.5	3.4	F5 V		69 s		
21620	38906		3 31 29.3	+49 12 36	0.000	0.00	6.3	0.1	0.6	A0 V	−23	130 s		
21730	56510		3 31 30.0	+34 22 18	+0.002	−0.01	7.8	1.1	0.2	K0 III		290 s		
21684	38910		3 31 30.2	+40 45 37	+0.002	−0.03	6.6	0.2	2.1	A5 V		76 s		

HD	SAO	Star Name	α 2000	δ 2000	μ(α)	μ(δ)	V	B-V	M_v	Spec	RV	d(pc)	ADS	Notes
			$3^h 31^m 33^s.2$	$+47°51'44''$	$+0^s.003$	$-0''.03$								
21641	38908		3 31 33.2	+47 51 44	+0.003	-0.03	6.76	-0.02	0.2	B9 V	+15	210 s	2579	m
21999	194355		3 31 34.7	-31 40 28	-0.003	-0.06	7.9	0.1	0.6	A0 V		190 mx		
21887	130544		3 31 40.0	-3 29 53	0.000	0.00	7.1	0.9	3.2	G5 IV		61 s		
21025	4916		3 31 41.2	+78 00 52	-0.008	+0.02	7.6	1.1	0.2	K0 III		290 s		
21649	38909		3 31 44.8	+48 57 08	+0.002	-0.03	7.8	0.0	0.4	B9.5 V		270 s		
21753	56515		3 31 47.0	+36 29 02	0.000	0.00	7.82	0.20		A3				
21661	38912		3 31 49.1	+49 24 04	+0.001	0.00	6.3	0.1	0.2	B9 V		150 s		
21997	168612		3 31 53.8	-25 36 51	+0.006	-0.02	6.38	0.12		A0				
21672	38913		3 31 53.9	+48 44 04	+0.003	-0.05	6.63	-0.02	-0.2	B8 V		220 s		
21477	12790		3 31 57.2	+67 25 31	+0.004	-0.05	7.9	1.1	0.2	K0 III		300 s		
21835	75979		3 32 00.6	+23 38 51	+0.004	0.00	7.9	1.1	0.2	K0 III		300 mx		
21476	12792		3 32 01.2	+67 35 08	+0.002	-0.02	7.50	0.4	2.6	F0 V		92 s	2576	m
22177	233148		3 32 01.7	-52 28 25	+0.017	-0.05	7.4	0.7	3.2	G5 IV		64 mx		
21948	130548		3 32 02.5	-7 50 24	-0.002	-0.01	7.4	0.1	0.6	A0 V		230 s		
22165	216370		3 32 05.4	-49 09 33	-0.002	+0.03	7.60	1.1	0.2	K0 III		160 s		m
21509	12794		3 32 06.5	+66 18 07	-0.003	-0.05	8.0	1.1	0.2	K0 III		320 s		
21699	38917		3 32 08.5	+48 01 25	+0.002	-0.02	5.47	-0.10	-1.2	B8 III	+1	220 s		
21577	12799		3 32 08.8	+62 16 05	+0.002	0.00	7.2	1.1	0.2	K0 III	+6	230 s		
21751	38922		3 32 09.3	+42 44 59	-0.001	0.00	7.8	0.0	0.4	B9.5 V		270 s		
21834	75980		3 32 09.6	+30 00 22	0.000	+0.01	8.0	0.1	0.6	A0 V	+3	270 s		
21915	93479		3 32 12.0	+11 32 33	+0.003	-0.04	6.90	0.1	0.6	A0 V		180 s	2591	m
22054	194360		3 32 13.1	-31 34 15	+0.002	0.00	8.0	0.4	3.0	F2 V		98 s		
21978	130551		3 32 15.4	-7 05 24	-0.002	-0.04	7.80	0.5	4.0	F8 V		58 s	2596	m
21771	38923		3 32 16.3	+44 50 26	0.000	-0.01	7.3	1.1	0.2	K0 III	-10	240 s		
21795	38925		3 32 17.5	+40 55 16	+0.002	0.00	6.7	0.0	0.4	B9.5 V		170 s		
21809	56525		3 32 19.8	+38 03 44	-0.005	-0.05	7.4	0.9	3.2	G5 IV		67 s		
20084	550		3 32 20.4	+84 54 40	+0.054	-0.13	5.61	0.92	-0.9	G8 II-III	+33	200 s		
21770	38924	36 Per	3 32 26.2	+46 03 26	-0.005	-0.07	5.31	0.40	0.7	F4 III	-45	84 s		
22052	168617		3 32 26.3	-24 36 59	0.000	+0.02	7.2	-0.2	1.4	A2 V		150 s		
21864	56529		3 32 31.9	+31 38 52	-0.001	-0.03	7.9	1.1	0.2	K0 III		310 s		
22231	233152		3 32 34.7	-50 22 44	+0.008	+0.08	5.68	1.10	-0.2	K3 III	+40	150 mx		
21933	111246	6 Tau	3 32 35.9	+9 22 25	+0.002	-0.04	5.77	-0.08		B8				
21803	38927		3 32 39.0	+44 51 20	+0.001	-0.01	6.41	0.03	-3.0	B2 IV	+4	570 s		KP Per, v
21847	56530		3 32 39.9	+35 39 35	+0.012	-0.04	7.4	0.5	4.0	F8 V		47 s		
21856	56531		3 32 39.9	+35 27 42	-0.001	0.00	5.90	-0.06	-3.5	B1 V	+25	640 s		
22366	248833		3 32 43.3	-61 23 38	+0.001	-0.06	7.5	1.0	3.2	G5 IV		53 s		
22359	248832		3 32 44.1	-60 34 29	+0.019	+0.05	7.55	0.50	3.2	F8 IV-V		74 s		m
22007	130560		3 32 48.3	-0 30 11	-0.001	+0.02	8.00	0.9	3.2	G5 IV	+18	91 s		
21993	111249		3 32 51.0	+5 15 17	+0.001	-0.02	7.4	0.9	3.2	G5 IV		70 s		
22382	248834		3 32 51.6	-61 01 00	+0.011	+0.05	6.41	1.04	3.2	G5 IV		27 s		
22049	130564	18 ε Eri	3 32 55.8	-9 27 30	-0.066	+0.02	3.73	0.88	6.1	K2 V	+15	3.3 t		
21962	93484		3 32 57.6	+16 35 54	-0.001	-0.09	7.1	0.5	3.4	F5 V		54 s		
22527	248836		3 32 59.9	-67 25 09	-0.009	-0.06	7.3	0.5	3.4	F5 V		59 s		
21904	56533		3 33 07.7	+35 23 57	+0.001	0.00	7.5	0.8	3.2	G5 IV		70 s		
22047	130565		3 33 12.1	-1 51 38	0.000	+0.01	7.6	0.9	3.2	G5 IV		77 s		
21913	56534		3 33 12.2	+33 21 33	+0.002	+0.01	7.6	0.4	3.0	F2 V		82 s		
21844	38937		3 33 21.1	+47 57 00	+0.013	-0.02	6.7	0.4	2.6	F0 V	+38	65 s		
22017	93487		3 33 21.1	+13 46 56	-0.002	-0.02	7.3	0.8	3.2	G5 IV		67 s		
21695	12807		3 33 21.1	+62 13 49	0.000	0.00	8.0	0.1	0.6	A0 V		270 s		
21742	24091		3 33 26.6	+59 25 00	+0.019	-0.31	8.00	1.1		K1 IV		50 mx		
21639	12805		3 33 26.9	+65 33 05	+0.002	-0.02	7.1	0.1	1.4	A2 V		130 s		
237150	24092		3 33 29.7	+58 45 46	0.000	-0.03	7.90	0.38	3.0	F2 V		92 s		
21802	24096		3 33 30.6	+55 33 33	+0.001	-0.01	7.89	0.16	0.6	A0 V		220 s		
21769	24093		3 33 32.1	+58 45 54	+0.001	-0.05	6.40	0.13	1.9	A4 V	+7	78 s	2592	m
22497	248837		3 33 34.7	-62 38 26	+0.012	+0.08	7.2	0.4	2.6	F0 V		83 s		
21912	56538		3 33 34.9	+39 53 59	+0.001	-0.04	5.80	0.11		A m	+4			IW Per, v
21784	24094		3 33 36.8	+58 15 48	+0.001	-0.02	8.00	0.1	0.6	A0 V		270 s	2593	m
22128	130572		3 33 37.8	-7 24 54	0.000	+0.01	7.8	0.2	2.1	A5 V		140 s		
21819	24099		3 33 38.9	+54 58 29	-0.005	0.00	5.98	0.11	1.4	A2 V	+14	76 s		
21943	56539		3 33 39.0	+38 00 46	+0.001	-0.03	7.24	-0.03		B8				
21794	24098		3 33 40.9	+57 52 04	-0.002	0.00	6.4	0.5	3.4	F5 V	-72	40 s		
21944	56540		3 33 43.5	+37 47 46	0.000	-0.01	7.9	1.1	-0.1	K2 III		350 s		
22074	111259		3 33 44.8	+6 55 34	0.000	+0.01	7.9	0.9	3.2	G5 IV		88 s		
22173	149040		3 33 45.3	-15 07 40	-0.003	-0.12	7.1	0.5	3.4	F5 V		55 s		
21708	12811		3 33 46.8	+63 26 28	0.000	-0.01	8.0	0.0	0.4	B9.5 V		300 s		
22203	168634	19 τ⁵ Eri	3 33 47.2	-21 37 58	+0.003	-0.02	4.27	-0.11	-0.2	B8 V	+14	78 s		
22045	93491		3 33 48.4	+18 13 40	+0.003	+0.01	7.3	1.1	0.2	K0 III		260 s		
22024	75990		3 33 51.3	+23 49 52	-0.001	-0.01	7.9	1.1	0.2	K0 III		340 s		
22023	75991		3 33 53.2	+24 14 35	+0.001	-0.01	7.9	0.0	0.4	B9.5 V		320 s		
22248	194373		3 33 54.5	-30 37 40	+0.004	+0.01	7.15	1.08		G5				
22262	194375		3 33 56.8	-31 04 49	-0.002	+0.06	6.20	0.48	3.4	F5 V		34 s		m
21974	56544		3 34 02.2	+39 18 22	+0.001	-0.02	8.0	1.4	-0.3	K5 III		390 s		
22323	216386		3 34 04.3	-41 20 19	+0.001	+0.04	7.9	-0.4	2.6	F0 V		110 s		
22072	93494		3 34 08.3	+17 49 57	+0.006	-0.31	6.17	0.89	5.5	dG7	+11	18 mn		
22448	233163		3 34 10.0	-54 30 43	+0.002	0.00	7.5	1.1	0.2	K0 III		260 s		

72

HD	SAO	Star Name	α 2000	δ 2000	μ(α)	μ(δ)	V	B-V	M_v	Spec	RV	d(pc)	ADS	Notes
			h m s	° ' "	s	"								
21931	38946		3 34 12.7	+48 37 04	+0.001	-0.01	7.34	0.03	0.2	B9 V		250 s	2609	m
22149	111264		3 34 13.6	+3 25 55	-0.001	-0.05	8.0	0.9	3.2	G5 IV		93 s		
22201	130581		3 34 15.8	-6 31 15	-0.001	+0.01	8.0	1.1	0.2	K0 III		360 s		
21843	24105		3 34 19.0	+59 44 06	-0.001	0.00	7.77	0.73	-2.9	B3 III		320 s		
22634	248842		3 34 24.8	-65 45 52	+0.007	+0.01	6.75	0.17		A2				
22091	75999	7 Tau	3 34 26.6	+24 27 53	+0.001	-0.02	5.92	0.13	1.7	A3 V	+29	65 s	2616	m
22347	194380		3 34 30.1	-34 50 00	+0.002	-0.08	7.1	0.4	2.6	F0 V		67 s		
22624	248844		3 34 33.2	-64 42 37	-0.004	-0.01	8.0	0.3	2.6	F0 V		120 s		
22322	194379		3 34 33.5	-31 52 29	+0.001	0.00	6.40	1.40		K0				m
22243	149047		3 34 37.4	-9 52 07	+0.001	0.00	6.25	0.02		A1				
22597	248841		3 34 38.6	-62 33 25	+0.008	+0.04	7.1	-0.2	1.4	A2 V		88 mx		
21806	12820		3 34 38.6	+63 53 20	+0.001	0.00	7.67	0.30	-1.6	B3.5 V		410 s		
21894	24109		3 34 41.7	+58 35 08	-0.001	0.00	7.94	0.21	0.0	B8.5 V		330 s		
22022	38959		3 34 42.6	+44 48 11	+0.002	-0.01	7.7	1.1	0.2	K0 III		290 s		
22515	233169		3 34 42.7	-54 19 23	+0.005	+0.03	6.9	0.1	3.4	F5 V		51 s		
22044	56554		3 34 44.9	+38 48 20	+0.006	-0.05	7.9	0.4	2.6	F0 V		110 s		
22182	111271		3 34 45.5	+9 09 07	+0.002	-0.10	7.7	0.1	1.4	A2 V		41 mx		
22211	111273		3 34 49.1	+6 25 04	-0.001	-0.01	6.49	0.63		F5 n	-11			
22333	168648		3 34 52.3	-25 34 57	+0.003	+0.05	6.6	1.4	3.2	G5 IV		19 s		
22240	130587		3 34 53.8	-3 24 44	+0.002	-0.02	7.58	0.14		A2				
21903	24111		3 35 00.7	+60 02 28	-0.004	+0.02	6.46	0.39	3.3	dF4	+21	43 s	2612	m
22124	56559		3 35 01.1	+32 01 00	+0.004	-0.04	6.60	0.4	2.4	F2 IV-V	-4	66 s	2622	IX Per, m,v
22039	38962		3 35 03.5	+45 26 57	-0.001	-0.02	7.8	1.1	-0.1	K2 III		330 s		
22356	168651		3 35 03.8	-24 18 28	+0.002	+0.01	7.5	0.7	0.2	K0 III		290 s		
22425	194389		3 35 06.0	-35 56 15	-0.002	+0.01	7.8	1.2	-0.1	K2 III		380 s		
22090	56558		3 35 06.0	+38 35 09	+0.001	-0.02	7.0	0.4	2.6	F0 V		74 s		
22587	233172		3 35 08.2	-57 48 37	+0.004	+0.06	7.6	0.4	0.2	K0 III		270 mx		
22255	111275		3 35 10.1	+5 42 51	0.000	-0.01	8.0	0.5	3.4	F5 V		82 s		
22114	56561		3 35 11.7	+38 00 54	-0.001	-0.02	7.58	0.07		B8				
21610	4936		3 35 12.3	+73 20 50	+0.003	-0.02	6.57	0.03		A0	-9			
22532	216395		3 35 15.3	-49 25 04	+0.005	+0.02	8.0	0.7	3.2	G5 IV		91 s		
22225	93501		3 35 18.4	+18 54 09	+0.001	-0.06	7.8	1.4	-0.3	K5 III	-9	230 mx		
22088	38966		3 35 20.7	+43 50 53	-0.002	0.00	8.00	1.1	-0.1	K2 III		360 s	2621	m
21960	24115		3 35 28.3	+59 31 43	-0.001	0.00	8.0	0.0	0.0	B8.5 V		350 s		
22122	38968		3 35 35.7	+42 53 16	+0.013	-0.14	7.40	0.5	4.0	F8 V		47 s		m
22195	56569		3 35 37.7	+31 40 49	+0.004	0.00	7.60	0.4	2.6	F0 V	+22	96 s	2628	m
22586	233173		3 35 37.9	-52 33 22	0.000	+0.03	8.04	-0.18	-3.6	B2 III		2100 s		
22596	233174		3 35 42.6	-53 12 22	+0.011	-0.04	7.9	0.7	3.0	F2 V		56 s		
22541	216398		3 35 49.8	-43 42 59	+0.001	+0.02	7.2	0.4	0.6	A0 V		130 s		
22379	130598		3 35 57.2	-5 07 30	-0.001	+0.01	6.6	1.1	0.2	K0 III		190 s		
22409	149057		3 35 57.6	-11 11 37	+0.002	+0.08	5.57	0.91	0.3	gG7	+37	110 s		
22136	38971		3 35 58.5	+47 05 28	+0.003	-0.01	6.89	-0.02	-0.2	B8 V	+12	250 s		
22224	56572		3 35 59.6	+34 55 26	0.000	+0.01	7.7	0.2	2.1	A5 V		120 s		
22748	248850		3 36 00.5	-61 32 33	+0.006	+0.06	7.1	0.6	2.6	F0 V		51 s		
22309	93504		3 36 03.1	+16 28 03	-0.020	-0.29	7.8	0.7	4.4	G0 V		37 mx		
21971	12830		3 36 03.7	+63 17 19	+0.001	+0.02	7.60	1.4	-0.3	K4 III	-22	330 s	2617	m
22156	38975		3 36 04.6	+46 35 04	-0.001	0.00	7.70	0.9	0.3	G6 III	-26	270 s		
22280	76017		3 36 11.3	+29 58 59	0.000	-0.02	7.80	0.1	0.6	A0 V		250 s	2633	m
22354	111283		3 36 11.6	+8 53 11	0.000	-0.01	7.9	1.1	0.2	K0 III		340 s		
23128	256002		3 36 12.1	-73 58 29	+0.005	+0.04	7.62	1.14		K0				
---	216401		3 36 14.7	-47 12 02	+0.003	+0.05	7.9	1.4	0.2	K0 III				
22609	---		3 36 16.4	-47 12 07			7.9	1.0	0.2	K0 III				
22470	149063	20 Eri	3 36 17.3	-17 28 02	+0.001	-0.01	5.23	-0.13	1.4	A2 V p	+14	58 s		
22329	93506		3 36 21.4	+19 33 34	0.000	-0.02	8.0	0.1	1.4	A2 V		210 s		
22105	24119		3 36 22.1	+55 53 14	+0.004	0.00	6.82	0.11	0.0	B8.5 V		180 s		
22328	93507		3 36 24.2	+20 04 04	+0.006	-0.04	7.6	0.5	3.4	F5 V	+35	69 s		
23549	256007		3 36 24.6	-79 05 55	-0.010	-0.02	6.7	1.2	0.2	K0 III		160 s		
22494	149064		3 36 25.1	-19 22 37	+0.002	0.00	7.2	0.3	0.6	A0 V		140 s		
21922	12829		3 36 25.8	+67 57 09	+0.009	+0.04	7.9	0.5	4.0	F8 V		58 s		
22135	24120		3 36 26.4	+52 55 53	+0.003	-0.01	7.27	1.86	-0.3	K5 III		200 s		
22688	233178		3 36 29.3	-52 46 39	+0.004	+0.03	7.66	0.95		K0				
22192	38980	37 ψ Per	3 36 29.3	+48 11 34	+0.002	-0.02	4.23	-0.06		B5 e	0			
23474	256005		3 36 30.4	-78 19 23	-0.004	+0.01	6.29	1.15		K0				
22317	76024		3 36 39.8	+29 13 26	+0.003	-0.01	6.6	0.2	2.1	A5 V	+20	78 s		
22306	56580		3 36 46.8	+34 20 51	+0.002	-0.03	6.9	1.1	0.2	K0 III		210 s		
22468	111291		3 36 47.2	+0 35 16	-0.002	-0.16	5.71	0.92	5.7	dG9	-23	14 ts	2644	m
22594	194399		3 36 51.3	-33 46 49	+0.003	+0.02	7.0	0.5	3.0	F2 V		52 s		
22484	111292	10 Tau	3 36 52.3	+0 24 06	-0.016	-0.48	4.28	0.58	4.0	F8 V	+28	13 ts		
22705	233181		3 36 53.1	-49 57 29	0.000	0.00	7.50	0.6	4.4	G0 V		42 s		m
22327	56581		3 36 54.9	+35 03 08	0.000	0.00	7.44	0.14		B8				
23004	248854		3 36 56.5	-68 26 18	+0.001	+0.01	7.7	0.8	3.2	G5 IV		78 s		
22374	76029		3 36 58.0	+23 12 40	-0.001	-0.02	6.72	0.13		B9 p				
22291	38986		3 37 00.9	+43 15 32	+0.001	-0.01	8.0	1.1	-0.1	K2 III		360 s		
22431	93517		3 37 02.3	+18 21 51	+0.001	0.00	8.0	0.1	1.4	A2 V		210 s		
22621	194404		3 37 04.5	-36 17 32	-0.001	+0.01	7.3	0.1	0.6	A0 V		180 s		

HD	SAO	Star Name	α 2000	δ 2000	μ(α)	μ(δ)	V	B−V	M$_v$	Spec	RV	d(pc)	ADS	Notes
			h m s	° ′ ″	s	″								
22267	38985		3 37 04.6	+47 15 41	−0.001	0.00	7.68	1.77		K5				
22663	216405		3 37 05.6	−40 16 29	−0.001	−0.03	4.58	1.04	0.2	K0 III	+12	74 s		
22360	56584		3 37 07.0	+31 07 30	−0.002	−0.02	8.0	0.9	3.2	G5 IV		89 s		
22593	168685		3 37 09.1	−29 49 50	−0.001	−0.03	7.3	−0.4	0.6	A0 V		220 s		
--	--		3 37 13.3	+49 04 53	0.000	−0.02	4.1 B							
22113	12835		3 37 20.4	+62 56 04	0.000	0.00	7.9	0.1	0.6	A0 V		260 s		
22352	38988		3 37 22.6	+40 32 16	+0.001	−0.04	7.8	1.1	0.2	K0 III		300 s		
22353	56585		3 37 23.4	+40 05 23	+0.002	−0.02	7.2	0.0	0.4	B9.5 V		210 s		
22418	56588		3 37 27.7	+31 07 12	−0.012	+0.02	7.0	0.5	3.4	F5 V	−38	52 s		
23358	256006		3 37 41.7	−74 49 22	−0.006	−0.01	6.89	0.32		A5				
22253	24128		3 37 43.9	+56 44 22	−0.002	+0.01	6.53	0.34		B0.5 III	+5			
22619	149071		3 37 45.2	−15 29 06	−0.001	−0.02	7.2	0.8	3.2	G5 IV		64 s		
22810	216414		3 37 45.4	−49 00 19	−0.002	+0.01	7.3	1.8	−0.1	K2 III		110 s		
22522	93524		3 37 47.7	+15 25 51	+0.001	−0.02	6.5	0.1	1.7	A3 V	+33	91 s		
22557	111304		3 37 49.8	+4 08 43	0.000	+0.02	7.0	0.2	2.1	A5 V		97 s		
22572	130614		3 37 50.3	+0 08 27	+0.002	−0.02	7.9	0.9	3.2	G5 IV		88 s		
22556	111305		3 37 52.5	+5 38 29	−0.001	+0.03	8.0	0.5	4.0	F8 V		63 s		
22545	111303		3 37 54.9	+9 55 58	+0.001	−0.01	7.7	1.1	−0.1	K2 III		360 s		
22442	56593		3 37 56.2	+37 25 52	+0.002	0.00	7.4	0.8	3.2	G5 IV		67 s		
22086	12836		3 37 59.1	+67 59 39	+0.001	−0.02	8.0	0.9	3.2	G5 IV		89 s		
22402	38999		3 38 00.2	+42 34 59	+0.002	−0.02	6.3	−0.1		B8	−1			
22298	24131		3 38 01.0	+55 10 15	0.000	0.00	7.60	0.53		B2 e				
22936	233194		3 38 05.3	−57 17 09	+0.002	−0.01	7.3	1.6	−0.3	K5 III		290 s		
22728	194416		3 38 09.9	−31 17 59	0.000	+0.02	7.25	1.57	−0.3	K5 III		290 s		
22989	233197		3 38 10.2	−59 46 35	+0.003	+0.05	6.94	0.40		F2				m
22630	130620		3 38 10.4	−7 29 00	0.000	+0.03	7.8	0.8	3.2	G5 IV		84 s		
22538	93526		3 38 10.8	+19 21 39	+0.001	−0.01	7.7	0.2		A m				
22389	39000		3 38 12.9	+49 05 00	−0.002	−0.02	7.0	0.0	0.4	B9.5 V		200 s		
22401	39001		3 38 15.5	+47 34 36	+0.002	−0.05	7.45	0.01	0.6	A0 V	−2	240 s		
22866	216420		3 38 16.4	−49 22 05	−0.001	+0.02	7.6	0.1	2.1	A5 V		130 s		
22428	39005		3 38 18.7	+44 48 06	+0.002	−0.03	8.00	0.1	0.6	A0 V	+1	270 s		m
22316	24133		3 38 19.6	+56 55 58	+0.003	−0.03	6.30	−0.13		B9 p				
22657	149077		3 38 20.4	−11 12 04	+0.003	+0.05	8.0	0.9	3.2	G5 IV		93 s		
22865	216421		3 38 26.1	−46 14 38	+0.001	−0.08	8.0	0.6	3.4	F5 V		65 s		
22579	93528		3 38 26.3	+15 08 59	0.000	−0.06	7.9	0.9	3.2	G5 IV		88 s		
22417	39004		3 38 26.8	+49 12 22	+0.001	0.00	6.7	0.0	0.4	B9.5 V		170 s		
22675	130628		3 38 29.1	−7 23 30	−0.001	−0.05	5.85	0.98		G5	−30			
22973	233199		3 38 31.6	−56 08 36	0.000	+0.02	7.96	0.96	5.3	G6 V		24 s		
21910	4951		3 38 36.2	+74 44 17	+0.001	+0.05	7.46	1.03	0.3	G8 III	−104	240 s		
23521	256008		3 38 39.7	−75 45 44	+0.031	+0.12	7.68	1.19		K0				
22578	76043		3 38 40.7	+22 39 35	+0.002	−0.04	6.78	0.1		A0	0			
22826	194426		3 38 42.0	−35 12 17	0.000	0.00	7.0	0.9	0.2	K0 III		230 s		
22653	111314		3 38 42.8	+2 43 42	+0.001	0.00	8.0	0.1	1.4	A2 V		210 s		
22788	168700		3 38 47.6	−25 36 31	+0.001	0.00	7.7	1.8	−0.1	K2 III		150 s		
22789	168701	τ For	3 38 47.6	−27 56 35	+0.001	+0.03	6.01	−0.02	0.6	A0 V		120 s		
22755	168699		3 38 49.2	−20 12 56	+0.001	+0.02	7.9	0.9	0.2	K0 III		350 s		
22714	130632		3 38 50.0	−8 30 25	−0.002	−0.02	7.50	0.9	3.2	G5 IV		72 s	2660	m
22697	130631		3 38 54.5	−1 31 32	+0.003	−0.02	7.4	0.4	2.6	F0 V		93 s		
--	--		3 38 54.5	−1 31 31	+0.003	−0.01	7.3	0.4	2.6	F0 V		87 s		
22686	111318		3 38 55.0	+2 45 49	+0.001	−0.01	7.1	0.1	0.6	A0 V		200 s		
22521	39015		3 38 57.2	+42 32 19	−0.017	−0.12	7.0	0.6	4.4	G0 V	−39	32 s		
22615	76045		3 39 00.0	+20 54 58	0.000	−0.02	6.51	0.15	0.1	A4 III	−5	180 s		
22713	130634	21 Eri	3 39 01.0	−5 37 34	−0.001	−0.20	5.96	0.92	5.2	dG5	+40	21 mn		
22451	24139		3 39 06.4	+52 49 09	0.000	+0.02	7.78	0.52	3.8	F7 V		60 s		
22614	76046		3 39 06.7	+24 42 11	+0.002	−0.04	7.09	0.04		A0	−1			
22265	12841		3 39 07.5	+65 39 37	+0.003	+0.02	7.4	0.1	0.6	A0 V		210 s		
22684	111323		3 39 08.1	+9 08 37	0.000	−0.03	7.8	0.1	1.7	A3 V		160 s		
22637	76050		3 39 13.1	+21 50 36	+0.001	−0.04	7.26	0.08		A0	+28			
22682	93532		3 39 17.0	+13 53 30	+0.001	−0.09	6.9	0.9	3.2	G5 IV		55 s		
21970	4953		3 39 24.8	+75 44 23	−0.003	+0.01	6.27	0.97		G5	+28			
22799	149082		3 39 25.3	−10 26 15	−0.002	−0.10	6.3	0.9	3.2	G5 IV		42 s		
22695	93536		3 39 25.7	+16 32 12	+0.003	−0.03	6.16	1.00		G5	+14		2661	m
22427	24140		3 39 32.4	+59 26 44	−0.002	+0.01	6.89	1.40	0.2	K0 III	−32	110 s		
22681	76056		3 39 32.8	+22 50 24	−0.001	0.00	8.0	0.1	0.6	A0 V		300 s		
22836	149085		3 39 36.7	−17 21 56	0.000	−0.03	7.0	1.1	0.2	K0 III		230 s		
22439	24141		3 39 37.0	+59 22 12	−0.002	−0.02	7.8	0.7	4.4	G0 V		48 s		
22798	130639		3 39 38.2	−3 23 35	−0.002	−0.07	6.23	1.04		G5				
22987	216429		3 39 38.2	−43 30 36	−0.003	+0.06	7.8	1.1	0.2	K0 III		300 s		
23100	233209		3 39 39.4	−56 14 05	+0.003	+0.01	6.7	1.5	−0.3	K5 III		250 s		
23079	233208		3 39 43.0	−52 54 57	−0.021	−0.09	7.3	0.7	4.4	G0 V		30 s		
22797	--		3 39 44.3	−1 35 33	−0.001	−0.02	8.0	0.9	3.2	G5 IV		93 s		
22489	24145		3 39 44.9	+56 23 00	0.000	+0.01	6.6	0.1	1.4	A2 V		100 s		
22986	216431		3 39 45.4	−40 21 08	+0.001	+0.02	7.0	−0.2	1.4	A2 V		130 s		
22820	130642		3 39 47.2	−6 46 37	+0.003	+0.03	7.0	0.1	1.4	A2 V		130 s		
22702	76060		3 39 51.1	+25 11 43	+0.002	−0.03	7.80	0.35		A2				

HD	SAO	Star Name	α 2000	δ 2000	μ(α)	μ(δ)	V	B-V	M$_v$	Spec	RV	d(pc)	ADS	Notes
			h m s	° ′ ″	s	″								
22796	111334	12 Tau	3 39 51.1	+3 03 24	−0.003	+0.01	5.57	0.94	0.3	gG6	+21	110 s		
22854	149087		3 39 53.4	−10 56 45	−0.008	+0.01	7.1	0.9	3.2	G5 IV		60 s		
22924	168720		3 39 53.4	−28 30 42	−0.001	−0.12	6.9	0.4	4.0	F8 V		39 s		
22399	12854		3 39 58.7	+63 52 12	+0.020	−0.16	6.80	0.46		F5			2650	m
22369	12853		3 39 58.7	+64 17 06	−0.003	−0.05	7.4	0.5	4.0	F8 V		46 s		
22819	130645		3 39 59.4	−1 07 14	+0.001	+0.02	6.12	1.00		G5				
23137	233215		3 40 01.7	−55 32 54	+0.005	−0.03	8.0	0.6	0.6	A0 V		60 mx		
22985	194440		3 40 02.9	−34 32 59	0.000	−0.11	8.0	0.1	3.4	F5 V		82 s		
22692	56613		3 40 07.8	+34 06 58	−0.006	−0.01	7.60	0.5	3.4	F5 V		68 s	2668	m
22290	12847		3 40 08.0	+69 42 47	−0.004	−0.03	5.55	−0.15		A p				
22892	130648		3 40 10.6	−9 02 22	+0.001	0.00	7.0	0.5	3.4	F5 V		52 s		
22905	149094		3 40 11.4	−15 13 34	0.000	0.00	6.33	0.88		G5				
22921	149095		3 40 12.2	−19 35 36	−0.002	0.00	7.0	1.4	3.2	G5 IV		24 s		
22818	111338		3 40 15.5	+7 35 03	−0.005	−0.08	7.6	0.1	1.4	A2 V		67 mx		
23071	216434		3 40 19.3	−43 14 27	−0.001	−0.02	7.2	1.7	−0.5	M2 III		290 s		
22767	76070		3 40 21.5	+21 24 21	−0.002	+0.01	7.20	0.2	2.1	A5 V		110 s	2673	m
22879	130649		3 40 22.0	−3 13 02	+0.047	−0.21	6.68	0.55	4.2	F9 V	+114	26 ts		
22878	111340		3 40 27.8	+5 07 33	0.000	0.00	6.7	0.8	3.2	G5 IV		51 s	2681	m
22833	93548		3 40 33.6	+13 33 11	−0.001	+0.02	7.9	0.9	3.2	G5 IV		88 s		
22920	130652	22 Eri	3 40 38.3	−5 12 39	0.000	−0.01	5.53	−0.15		B8	+16			
22766	76071		3 40 38.8	+28 46 24	+0.003	−0.04	7.30	0.1	0.6	A0 V	+11	200 s	2679	m
23202	233221		3 40 40.3	−53 44 18	+0.011	−0.01	7.6	0.6	3.2	G5 IV		77 s		
22983	149100		3 40 40.3	−19 28 21	+0.008	+0.05	6.9	1.2	3.2	G5 IV		30 s		
22691	39033		3 40 43.5	+44 41 12	+0.003	−0.05	7.8	1.1	0.2	K0 III		290 s		
22805	76073	11 Tau	3 40 46.2	+25 19 47	+0.001	−0.01	6.20	0.06	1.4	A2 V	+6	87 s		
22918	130653		3 40 47.4	−2 19 57	+0.025	−0.21	6.95	0.96		K0		24 mn		
23024	168732		3 40 49.1	−27 58 12	−0.004	−0.03	7.1	0.7	0.2	K0 III		240 s		
22721	39034		3 40 49.8	+41 50 29	−0.003	−0.04	7.9	0.4	3.0	F2 V		90 s		
22733	56623		3 40 50.3	+39 07 02	+0.005	−0.01	7.84	0.20	1.7	A3 V		95 mx	2677	m
22968	149104		3 40 51.5	−12 36 59	+0.006	+0.03	7.50	0.4	3.0	F2 V		79 s	2690	m
22720	39035		3 40 55.5	+43 42 08	0.000	+0.02	8.0	0.4	2.6	F0 V		110 s		
22732	39038		3 40 56.0	+41 11 19	−0.003	+0.08	7.1	0.4	3.0	F2 V		65 s		
22929	111350		3 41 04.8	+2 57 01	−0.001	−0.01	7.9	1.1	0.2	K0 III		340 s		
22780	56628		3 41 07.8	+37 34 48	+0.002	−0.03	5.57	−0.07	−0.6	B7 V n	−1	160 s		
23136	216439		3 41 09.4	−40 46 06	−0.001	0.00	7.4	0.7	1.7	A3 V		57 s		
22601	24155		3 41 11.8	+58 50 40	+0.004	−0.08	7.1	0.4	2.6	F0 V		78 s		
23010	149107		3 41 13.7	−11 48 11	+0.005	+0.01	6.49	0.36	1.9	F2 IV		82 s		m
22860	76080		3 41 18.3	+28 42 12	0.000	+0.01	6.9	0.0	0.4	B9.5 V	+8	180 s		
23055	149114		3 41 22.3	−19 35 05	−0.001	0.00	6.59	0.09		A0				
22814	56630		3 41 22.6	+36 22 46	0.000	+0.01	7.8	0.9	3.2	G5 IV		80 s		
22916	93552		3 41 26.3	+19 23 19	0.000	0.00	7.8	0.1	0.6	A0 V		270 s		
22952	93553		3 41 28.2	+12 36 28	+0.003	−0.08	8.0	0.5	4.0	F8 V		62 s		
22965	111356		3 41 32.0	+10 04 52	−0.002	−0.05	7.3	0.9	3.2	G5 IV		67 s		
23067	149116		3 41 34.3	−14 17 57	+0.003	−0.01	7.4	1.1	0.2	K0 III		270 s		
23009	130668		3 41 38.1	−0 09 49	0.000	−0.01	7.8	0.1	0.6	A0 V		270 s		
22915	76084		3 41 41.5	+25 22 12	+0.001	−0.01	8.0	1.1	−0.1	K2 III		360 s		
22859	56637		3 41 43.4	+37 35 13	+0.004	−0.04	7.77	0.08		A0				
23274	216445		3 41 45.5	−47 46 30	−0.003	+0.06	7.1	0.1	1.4	A2 V		130 s		
22611	12870		3 41 48.3	+62 38 56	+0.003	+0.02	7.55	4.29		N7.7	−3			U Cam, m, v
23285	216446		3 41 49.4	−48 14 22	−0.001	−0.08	8.00	0.4	2.6	F0 V		120 s		m
23095	149120		3 41 54.3	−14 21 48	0.000	−0.01	7.0	0.1	0.6	A0 V		190 s		
22846	39047		3 41 54.7	+43 30 52	−0.002	−0.04	7.90	0.9	3.2	G5 IV		84 s	2688	m
23148	168751		3 41 59.2	−24 39 08	+0.012	−0.03	6.8	0.7	3.4	F5 V		32 s		
23121	—		3 42 03.2	−17 08 41			7.90	0.5	3.4	F5 V		79 s	2706	m
25254	258353		3 42 03.4	−84 04 58	−0.009	+0.06	7.9	0.2	3.0	F2 V		94 s		
22649	12874		3 42 09.3	+63 13 00	−0.003	+0.02	5.10	1.63		S5.3	−22	34 mn		
23308	216449		3 42 09.8	−45 57 28	−0.007	+0.01	6.49	0.52	4.0	F8 V		32 s		
23344	216452		3 42 13.4	−49 09 06	+0.001	+0.04	7.5	1.0	0.2	K0 III		280 s		
23145	149123		3 42 13.7	−15 57 49	+0.003	−0.03	8.00	1.1	−0.1	K2 III		420 s	2707	m
23227	194467	δ For	3 42 14.9	−31 56 18	0.000	+0.02	5.00	−0.16		B5 n	+26			
22992	76088		3 42 15.8	+22 47 11	+0.001	−0.08	7.4	0.4	3.0	F2 V		75 s		
24428	—		3 42 17.5	−80 01 38			7.9	2.2	4.0	F8 V		60 s		
23016	93557	13 Tau	3 42 18.9	+19 42 01	0.000	−0.01	5.50	−0.01	−0.2	B8 V e	−10	140 s		
22951	56646	40 Per	3 42 22.5	+33 57 54	0.000	−0.01	4.97	−0.01		B0.5 V	+19		2699	m
23368	233231		3 42 22.6	−50 47 43	−0.003	+0.02	7.3	0.9	0.2	K0 III		260 s		
24427	258350		3 42 26.0	−80 01 13	−0.009	0.00	7.90	0.5	4.0	F8 V		60 s		m
22963	56650		3 42 27.6	+32 56 21	+0.002	−0.10	6.70	0.5	4.0	F8 V	−33	34 s	2701	m
23007	76089		3 42 29.7	+26 34 22	+0.005	−0.03	7.8	1.1	0.2	K0 III	+33	200 mx		
23028	93560		3 42 29.8	+20 09 01	+0.001	0.00	8.0	0.1	1.4	A2 V		210 s		
25887	258356		3 42 32.4	−85 15 45	+0.010	+0.01	6.41	−0.01		B9				m
23593	248870		3 42 34.3	−64 11 43	+0.005	+0.03	7.5	1.9	0.2	K0 III		84 s		
22220	4964		3 42 35.6	+77 10 11	−0.008	−0.03	8.0	0.1	0.6	A0 V		190 mx		
22845	24172		3 42 36.6	+52 17 56	+0.002	−0.03	7.8	1.1	0.2	K0 III		300 s		
22976	56652		3 42 36.7	+33 33 21	+0.002	−0.07	7.9	1.1	−0.1	K2 III		180 mx		
23052	93561		3 42 36.9	+17 17 37	−0.008	+0.04	7.2	0.7	4.4	G0 V		36 s		

HD	SAO	Star Name	α 2000	δ 2000	μ(α)	μ(δ)	V	B-V	M$_v$	Spec	RV	d(pc)	ADS	Notes
			h m s	° ' "	s	"								
23083	111363		3 42 37.2	+8 38 04	-0.002	-0.03	7.1	0.1	1.4	A2 V		140 s		
22872	24174		3 42 41.6	+51 10 24	+0.021	-0.17	8.0	0.5	4.2	F9 V	+56	57 s		
22764	24169		3 42 42.6	+59 58 10	0.000	+0.01	5.76	1.71	-4.4	K0 Ib	-10	730 s	2691	m
23144	130678		3 42 43.5	-4 36 21	0.000	+0.01	7.7	0.0	0.4	B9.5 V		290 s		
23207	149128		3 42 44.5	-18 42 49	0.000	-0.05	7.4	0.3	1.4	A2 V		120 s		
22553	12873		3 42 45.2	+69 50 46	-0.003	-0.01	7.90	0.1	1.7	A3 V		160 s	2678	m
23250	168766		3 42 49.1	-22 54 37	0.000	0.00	7.9	1.5	0.2	K0 III		180 s		
23319	194475		3 42 50.0	-37 18 49	-0.008	-0.07	4.59	1.20	-0.3	gK5	+10	95 s		m
22928	39053	39 δ Per	3 42 55.4	+47 47 15	+0.003	-0.03	3.01	-0.13	-2.2	B5 III	-9	100 s		m
23456	233238		3 43 04.6	-50 38 37	+0.014	+0.49	6.97	0.52	4.5	G1 V		31 s		
23075	76094		3 43 06.3	+25 40 54	-0.005	0.00	7.30	0.1	3.4	F5 V		58 s	2708	m
23000	39056		3 43 07.4	+40 59 38	+0.007	-0.04	8.0	0.4	2.6	F0 V		110 s		
23618	248873		3 43 09.0	-61 30 31	-0.001	+0.02	7.0	0.6	1.4	A2 V		59 s		
23318	194477		3 43 10.1	-31 01 10	+0.002	0.00	7.1	1.7	-0.5	M4 III		300 s		
23495	233240		3 43 12.3	-52 14 52	0.000	+0.02	7.8	0.9	3.2	G5 IV		70 s		
23249	130686	23 δ Eri	3 43 14.8	-9 45 48	-0.007	+0.75	3.54	0.92	3.8	K0 IV	-6	9.0 t		
23142	93565		3 43 21.3	+16 25 13	0.000	-0.02	7.8	0.0	0.4	B9.5 V		300 s		
23223	130685		3 43 22.3	-2 04 11	0.000	-0.07	7.2	1.1	0.2	K0 III		210 mx		
23060	56662		3 43 24.0	+34 06 59	+0.001	0.00	7.50	0.11	-2.5	B2 V p		780 s		
24063	256019		3 43 24.6	-74 00 33	-0.001	-0.01	7.54	0.24		A0				
23026	39057		3 43 29.4	+43 10 03	-0.003	-0.03	7.8	1.1	-0.1	K2 III		330 s		
23672	248875		3 43 30.6	-61 37 36	-0.003	-0.03	7.6	0.9	3.2	G5 IV		75 s		
23281	149132		3 43 33.7	-10 29 09	-0.001	-0.02	5.60	0.22		A m	+16			
23262	130687		3 43 36.3	-2 04 41	+0.003	0.00	8.0	1.1	-0.1	K2 III		410 s		
23157	76103		3 43 41.5	+23 38 58	+0.002	-0.04	7.90	0.34	2.5	A9 V	-1	110 s		
23155	76102		3 43 43.1	+25 04 52	+0.002	-0.04	7.51	0.15		A2	+1			
23304	130691		3 43 45.6	-9 36 12	+0.004	-0.02	7.2	1.1	0.2	K0 III		190 mx		
23141	76105		3 43 45.9	+26 22 52	-0.002	0.00	7.28	1.16	0.2	K0 III	-28	210 s		
23183	93568	14 Tau	3 43 47.1	+19 39 54	+0.008	-0.05	6.14	1.01	0.2	K0 III	+78	150 s		
23050	39061		3 43 47.6	+42 36 13	+0.033	-0.24	7.47	0.59	4.7	G2 V	+32	36 s		m
23317	130693		3 43 54.7	-7 52 56	-0.002	0.00	7.8	0.8	3.2	G5 IV	-16	84 s		
23356	149134		3 43 55.4	-19 06 40	+0.023	+0.15	7.2	1.1	0.2	K0 III		50 mx		
23405	168781		3 43 58.0	-28 37 44	+0.002	-0.03	7.2	0.4	1.4	A2 V		88 s		
23107	56667		3 43 58.9	+38 22 26	+0.001	0.00	7.40	1.4	-0.3	K5 III	+14	310 s	2717	m
23082	39066		3 44 05.5	+44 53 04	-0.002	-0.01	7.8	1.4	-0.3	K5 III	+8	360 s		
23508	216459		3 44 06.1	-40 39 37	0.000	-0.08	6.45	1.06	0.0	K1 III		200 s		m
23049	39063		3 44 06.3	+48 31 25	0.000	-0.01	6.06	1.55	-0.3	K4 III	-12	160 s	2712	m
23484	194491		3 44 09.1	-38 16 54	+0.017	+0.29	7.00	0.88	5.9	K0 V		15 s		
23817	248877	β Ret	3 44 12.0	-64 48 26	+0.049	+0.08	3.85	1.13	3.2	K0 IV	+51	17 ts		m
23534	216461		3 44 13.1	-43 14 11	-0.002	-0.01	7.8	1.7	-0.5	M2 III		270 s		
23660	233250		3 44 16.5	-54 28 40	+0.006	+0.12	7.2	1.1	0.2	K0 III		130 mx		
23548	216463		3 44 18.4	-41 53 52	0.000	-0.02	7.2	1.7	-0.3	K5 III		240 s		
23180	56673	38 o Per	3 44 19.1	+32 17 18	+0.001	-0.01	3.83	0.05	-4.4	B1 III	+19	310 s	2726	m,v
23435	168785		3 44 21.6	-21 06 23	+0.002	0.00	8.0	1.8	-0.3	K5 III		290 s		
23258	76121		3 44 28.1	+20 55 43	+0.001	-0.02	6.00	0.01	0.6	A0 V	+12	110 s		
23363	130698	24 Eri	3 44 30.4	-1 09 47	0.000	0.00	5.25	-0.10	-1.0	B7 IV	+39	180 s		
23193	56675		3 44 31.3	+36 27 37	+0.004	-0.03	5.60	0.06	0.0	A3 III	+22	130 s		
23697	233252		3 44 33.8	-54 16 26	+0.002	+0.07	6.30	1.04		K0				m
23245	76122		3 44 37.2	+27 53 51	-0.002	-0.07	6.70	0.4	2.6	F0 V		42 s	2735	m
23362	111388		3 44 40.1	+1 56 18	+0.002	-0.01	7.9	1.1	-0.1	K2 III		390 s		
23139	39071		3 44 40.8	+46 05 58	-0.001	-0.03	6.11	0.28	1.5	A7 IV	+9	75 s		
23104	24187		3 44 41.1	+50 32 34	+0.002	-0.03	7.6	0.1	0.6	A0 V		230 s		
23257	76124		3 44 44.9	+27 55 19	+0.012	-0.12	6.77	0.64	5.2	G5 V	+49	42 s	2735	m
23576	194501		3 44 45.2	-38 49 06	-0.006	+0.05	7.3	0.6	4.4	G0 V		38 s		
23647	216467		3 44 47.1	-45 55 34	+0.006	+0.06	8.0	0.9	4.0	F8 V		35 s		
24062	256022		3 44 47.1	-70 01 37	+0.001	-0.11	7.2	0.6	4.4	G0 V		36 s		
23288	76126	16 Tau	3 44 48.1	+24 17 22	+0.001	-0.04	5.45	-0.05	-1.0	B7 IV	+3	180 s		Celaeno
23140	39072		3 44 48.7	+46 02 09	+0.030	-0.11	7.71	0.86	-0.1	K2 III		43 mx		
23670	216469		3 44 50.5	-48 03 41	+0.002	-0.08	6.49	1.02		K0				
23759	233257		3 44 50.9	-56 20 04	+0.004	-0.02	7.9	1.5	-0.3	K5 III		310 mx		
23507	168790		3 44 52.4	-23 22 51	0.000	-0.04	7.6	0.6	0.2	K0 III		300 s		
23302	76131	17 Tau	3 44 52.5	+24 06 48	+0.001	-0.04	3.70	-0.11	-1.9	B6 III	+12	120 s		Electra
23413	130704	25 Eri	3 44 56.4	-0 17 48	+0.004	0.00	5.55	1.42	-0.3	gK5	+70	150 s		
23721	233255		3 44 56.6	-52 04 53	0.000	-0.03	7.9	1.2	-0.1	K2 III		390 s		
23013	12888		3 44 59.2	+61 27 50	+0.001	-0.01	7.6	0.1	1.4	A2 V		160 s		
24085	256023		3 45 01.8	-70 01 28	+0.001	-0.10	7.5	0.6	4.4	G0 V		42 s		
23412	111393		3 45 02.7	+2 37 17	0.000	-0.05	6.7	0.5	3.4	F5 V		46 s		
23389	111391		3 45 02.8	+6 38 11	+0.001	+0.01	8.0	0.1	1.4	A2 V		210 s		
23231	56681		3 45 04.4	+40 00 45	+0.002	-0.01	7.60	0.8	3.2	G5 IV		74 s		m
22648	4972		3 45 07.4	+74 32 36	+0.006	+0.04	6.8	0.9	3.2	G5 IV		52 s		
23931	248881		3 45 08.1	-63 43 46	-0.001	-0.04	7.4	1.6	3.2	G5 IV		21 s		
23324	76137	18 Tau	3 45 09.7	+24 50 21	+0.002	-0.04	5.65	-0.08	-0.2	B8 V	+4	150 s		
23230	39078	41 ν Per	3 45 11.6	+42 34 43	-0.001	0.00	3.77	0.42	-2.0	F5 II	-13	140 s	2738	m
23338	76140	19 Tau	3 45 12.4	+24 28 02	+0.001	-0.04	4.30	-0.11	-0.9	B6 V	+6	110 s		Taygeta, m
23749	233259		3 45 12.8	-51 15 36	+0.002	+0.02	7.7	0.8	3.2	G5 IV		77 s		

HD	SAO	Star Name	α 2000	δ 2000	μ(α)	μ(δ)	V	B-V	M_v	Spec	RV	d(pc)	ADS	Notes
			^{h} ^{m} ^{s}	^{o} $'$ $''$	^{s}	$''$								
23719	216470		3 45 15.7	-47 21 35	-0.002	-0.02	5.73	0.96	0.3	gG8	-2	120 s		
23542	168802		3 45 20.0	-20 31 25	-0.003	-0.01	7.5	-0.2	2.1	A5 V		120 s		
24188	256025		3 45 24.2	-71 39 30	+0.008	+0.04	6.6	-0.5	0.6	A0 V		120 mx		
23361	76145		3 45 26.1	+24 02 07	+0.002	-0.04	8.04	0.21	1.7	A3 V	+10	160 s		
23219	39079		3 45 27.5	+47 39 37	+0.002	-0.02	7.18	-0.01	0.4	B9.5 V		220 s		
23267	39082		3 45 30.9	+42 00 26	+0.002	-0.02	6.8	0.1	0.6	A0 V		170 s		
23541	149150		3 45 31.3	-15 22 24	0.000	-0.02	7.5	1.1	0.2	K0 III		290 s		
23388	76150		3 45 31.9	+21 14 48	+0.001	-0.05	7.73	0.18		A3	0			
23616	168809		3 45 33.1	-25 54 55	-0.002	+0.01	6.9	0.3	1.7	A3 V		82 s		
23256	39081		3 45 33.7	+45 21 15	+0.007	-0.01	7.7	0.4	3.0	F2 V	+14	85 s		
23467	111398		3 45 34.5	+3 14 23	0.000	0.00	7.9	1.1	-0.1	K2 III		390 s		
23411	93585		3 45 34.8	+16 15 47	0.000	0.00	8.0	0.1	0.6	A0 V		300 s		
23301	56685		3 45 36.9	+38 40 35	-0.004	+0.02	6.5	1.1	0.2	K0 III		170 s		
23387	76152		3 45 37.7	+24 20 08	+0.001	-0.04	7.18	0.16	1.2	A1 V	+4	130 s		m
23402	76154		3 45 39.8	+22 41 40	+0.002	-0.04	7.81	0.19	0.6	A0 V	+9	240 s		m
23466	111400	29 Tau	3 45 40.4	+6 03 00	+0.001	-0.01	5.35	-0.12	-1.7	B3 V	+13	240 s	2750	m
23481	111401		3 45 43.9	+7 11 19	-0.002	0.00	7.2	0.1	1.4	A2 V		150 s		
23410	76156		3 45 48.9	+23 08 48	+0.003	-0.06	6.85	0.04		A0	+3		2748	m
23408	76155	20 Tau	3 45 49.5	+24 22 04	+0.002	-0.04	3.88	-0.07	-1.6	B7 III	+8	120 s		Maia
23190	24195		3 45 50.5	+55 03 50	+0.001	-0.03	6.8	0.2	2.1	A5 V		85 s		
23409	76158		3 45 51.6	+24 02 20	+0.002	-0.04	7.85	0.20	1.4	A2 V	+6	160 s		
23287	39084		3 45 53.3	+45 35 59	+0.004	-0.04	7.5	0.1	0.6	A0 V	+2	220 s		
23527	111403		3 45 53.4	+0 13 22	0.000	+0.01	7.9	1.1	-0.1	K2 III		400 s		
23432	76159	21 Tau	3 45 54.3	+24 33 17	+0.001	-0.04	5.76	-0.04	-0.2	B8 V	0	150 s		Sterope
23430	76161		3 45 59.0	+25 23 56	+0.001	-0.03	8.02	0.22		A0	+7			
23300	39085		3 45 59.2	+45 40 55	+0.002	-0.02	5.60	-0.07	0.2	B9 V	+2	120 s	2746	m
23005	12890		3 46 01.1	+67 12 06	+0.017	-0.11	5.80	0.35	3.0	F2 V	+6	36 s		
23089	12891		3 46 02.3	+63 20 42	0.000	-0.01	4.80	0.80	0.6	G0 III	-2	59 s		
23441	76164	22 Tau	3 46 02.8	+24 31 40	+0.001	-0.04	6.42	-0.02	0.2	B9 V	0	180 s		
23502	93588		3 46 06.2	+10 32 55	0.000	0.00	7.70	0.9	0.3	G8 III	+3	300 s		
23614	149158	26 π Eri	3 46 08.4	-12 06 06	+0.003	+0.06	4.42	1.63		gMa	+46	50 mn		
23526	111407		3 46 09.4	+6 48 11	+0.001	-0.07	5.91	0.99	0.2	G9 III	-26	140 s		
23635	149160		3 46 13.1	-9 53 01	0.000	-0.01	6.9	1.1	0.2	K0 III		220 s		
22912	4984		3 46 14.1	+71 37 07	+0.002	+0.02	7.10	0.8	3.2	G5 IV		60 s	2718	m
24136	248886		3 46 15.7	-66 30 13	-0.002	-0.01	7.9	0.1	1.7	A3 V		170 s		
23479	76169		3 46 15.9	+24 11 24	+0.001	-0.04	7.96	0.32	2.4	A7 V	-2	120 s	2755	m
23480	76172	23 Tau	3 46 19.5	+23 56 54	+0.002	-0.04	4.18	-0.06	-1.3	B6 IV	+6	120 s		Merope
23501	93589		3 46 20.6	+16 22 27	+0.004	0.00	8.0	0.4	2.6	F0 V		120 s		
23570	111410		3 46 25.2	+7 14 39	+0.001	-0.03	7.9	1.1	0.2	K0 III		340 s		
23538	93590		3 46 26.2	+13 30 31	+0.002	-0.04	6.8	0.1	0.6	A0 V		180 s		
23489	76173		3 46 27.2	+24 15 18	+0.001	-0.04	7.35	0.10	1.4	A2 V	+4	140 s		
23738	168820	σ For	3 46 27.3	-29 20 17	0.000	+0.01	5.90	0.12		A2				
24579	256028		3 46 33.7	-76 42 56	+0.004	+0.03	7.8	0.0	0.0	B8.5 V		260 mx		
24450	256026		3 46 37.0	-74 30 26	+0.012	+0.11	8.02	0.39	2.6	F0 V		110 s		
23478	56695		3 46 40.7	+32 17 23	-0.001	-0.02	6.67	0.08	-2.3	B3 IV	+15	450 s		
23757	168823		3 46 41.6	-25 21 27	+0.001	+0.02	7.3	1.2	0.2	K0 III		190 s		
23756	168825		3 46 44.6	-24 51 22	-0.001	-0.01	7.7	1.2	-0.3	K5 III		400 s		
23385	39095		3 46 49.5	+47 36 14	0.000	0.00	7.9	1.1	0.2	K0 III		300 s		
23754	168827	27 τ⁶ Eri	3 46 50.8	-23 14 59	-0.012	-0.53	4.23	0.42	3.1	F3 V	+7	17 ts		
23634	111414		3 46 56.0	+9 31 51	-0.001	0.00	7.6	0.1	1.4	A2 V		180 s		
23737	149166		3 46 56.1	-14 28 43	0.000	-0.02	7.7	1.1	0.2	K0 III		310 s		
23906	216487		3 46 57.1	-43 48 25	0.000	+0.01	7.8	1.5	-0.3	K5 III		220 s		
23568	76183		3 46 59.3	+24 31 13	+0.001	-0.04	6.81	0.02	0.4	B9.5 V	-4	180 s		
23426	39097		3 46 59.4	+46 37 04	-0.003	0.00	7.8	1.1	-0.1	K2 III		330 s		
23829	194525		3 47 00.9	-30 03 19	+0.002	-0.04	7.2	1.5	-0.3	K5 III		310 s		
23439	39100		3 47 01.9	+41 25 39	+0.052	-1.23	7.65	0.81	6.1	K1 V	+50	14 mx	2757	m
23644	111416		3 47 03.5	+8 57 18	-0.002	+0.04	7.4	0.9	3.2	G5 IV		69 s		
23384	24213		3 47 10.6	+51 42 23	+0.006	-0.07	6.85	0.37		F0				
24512	256029	γ Hyi	3 47 14.5	-74 14 20	+0.013	+0.12	3.24	1.62	-0.4	M0 III	+16	49 s		
23477	39101		3 47 15.9	+44 04 27	+0.001	-0.02	7.1	0.0	0.4	B9.5 V	+8	200 s		
23254	12900		3 47 17.4	+62 28 30	+0.001	0.00	7.8	0.0	-1.0	B5.5 V		480 s		
24023	233270		3 47 19.5	-50 45 15	-0.004	-0.07	6.5	1.2	0.2	K0 III		150 s		
23856	194531		3 47 20.0	-29 54 07	+0.014	-0.07	6.55	0.49	3.4	F5 V		40 s		
23632	76193		3 47 21.0	+23 48 13	+0.003	-0.03	6.99	0.03	1.2	A1 V	+2	140 s		
23629	76192		3 47 21.0	+24 06 58	+0.002	-0.04	6.30	0.02	0.6	A0 V	+5	130 s		m
23869	168835		3 47 22.2	-29 28 09	+0.002	-0.07	6.9	0.8	0.2	K0 III		170 mx		
23628	76194		3 47 24.0	+24 35 20	+0.002	-0.03	7.65	0.21	1.9	A4 V	0	130 s		
23631	76197		3 47 24.2	+23 54 54	0.000	-0.03	7.26	0.05	1.4	A2 V	+3	150 s	2767	m
23643	76198		3 47 26.8	+23 40 42	+0.001	-0.04	7.77	0.15	1.7	A3 V	+9	150 s		
24022	233271		3 47 28.3	-50 03 53	+0.008	+0.02	7.8	1.1	0.2	K0 III		240 mx		
23630	76199	25 η Tau	3 47 29.0	+24 06 18	+0.001	-0.04	2.87	-0.09	-1.6	B7 III	+10	73 s		Alcyone, m
23642	76200		3 47 29.4	+24 17 19	+0.001	-0.04	6.81	0.06	0.6	A0 V	+15	160 s		
23813	149173		3 47 30.4	-17 31 26	-0.002	-0.02	7.8	1.1	-0.1	K2 III		380 s		
23383	24215		3 47 32.1	+55 55 21	+0.004	-0.01	6.10	-0.03		B9				
23458	24223		3 47 38.2	+50 21 51	+0.001	+0.01	8.0	1.1	-0.1	K2 III		370 s		

HD	SAO	Star Name	α 2000	δ 2000	μ(α)	μ(δ)	V	B−V	M_v	Spec	RV	d(pc)	ADS	Notes
23878	168836	28 τ⁷ Eri	3ʰ47ᵐ39.5ˢ	−23°52′29″	+0.003	+0.05	5.24	0.07		A2	+29	17 mn		
23452	24219		3 47 40.1	+51 31 41	+0.003	−0.05	7.29	0.17	0.6	A0 V		170 s		
24394	256027		3 47 40.9	−69 53 20	−0.004	0.00	7.6	0.2	1.7	A3 V		140 s		
23795	130725		3 47 47.4	−4 50 14	−0.001	−0.01	7.5	1.1	0.2	K0 III		290 s		
23626	56707		3 47 48.9	+32 11 42	−0.002	−0.04	6.25	0.47	4.4	G0 V	−4	23 s		
23958	194537		3 47 49.5	−36 06 22	+0.001	+0.01	6.21	−0.10		B8 n	+5			
23625	56709		3 47 52.5	+33 36 00	−0.001	0.00	6.57	0.08	−2.5	B2 V	+34	470 s	2772	m
23940	194535	ρ For	3 47 56.0	−30 10 06	+0.002	−0.24	5.54	0.98	0.3	gG5	+53	97 s		
23827	130730		3 47 58.7	−2 51 42	−0.002	0.00	6.9	0.1	1.4	A2 V		130 s		
23920	168841		3 47 59.0	−26 19 43	−0.001	+0.01	6.9	0.7	1.4	A2 V		51 s		
23596	39110		3 48 00.3	+40 31 49	+0.005	+0.01	7.2	0.5	4.0	F8 V		42 s		
24293	248888		3 48 01.1	−64 50 14	+0.054	+0.07	7.85	0.65	5.2	G5 V		34 s		
23605	56710		3 48 01.1	+38 20 39	−0.006	+0.02	7.9	0.4	3.0	F2 V		91 s		m
23779	111424		3 48 02.1	+7 07 45	+0.003	−0.01	7.8	1.1	0.2	K0 III		330 s		
23566	39107		3 48 04.3	+45 21 16	0.000	−0.04	7.7	0.2	2.1	A5 V	−6	130 s		
23712	76206		3 48 06.5	+24 59 19	−0.002	−0.02	6.46	1.70		K5				
24636	256034		3 48 11.9	−74 41 39	+0.019	+0.03	7.13	0.40		F2				
23793	93611	30 Tau	3 48 16.2	+11 08 36	+0.002	−0.02	5.07	−0.13	−1.7	B3 V	+19	220 s	2778	m
23552	24231		3 48 18.2	+50 44 12	+0.002	0.00	6.14	0.06	−0.6	B7 V		180 s	2769	m
23753	76215		3 48 20.8	+23 25 16	+0.002	−0.05	5.44	−0.07	−0.2	B8 V	−2	130 s		
23825	93612		3 48 21.9	+10 47 35	−0.002	−0.13	7.90	0.7	3.0	G3 IV	−13	96 s		
23805	111428		3 48 22.4	+8 34 49	+0.007	−0.01	7.42	1.11		G5				
23624	39114		3 48 22.4	+42 52 51	−0.001	−0.02	7.8	0.0	0.0	B8.5 V		320 s		
24185	233280		3 48 25.0	−54 17 39	+0.004	+0.01	7.0	0.3	3.4	F5 V		53 s		
23917	149178		3 48 25.7	−13 18 24	−0.002	−0.01	7.6	1.1	0.2	K0 III		300 s		
23763	76216		3 48 30.0	+24 20 43	+0.002	−0.05	6.94	0.12	1.2	A1 V	+8	120 s		
23841	111430		3 48 30.7	+9 38 46	+0.005	+0.01	6.69	1.22	−0.1	K2 III	−80	220 s		
24033	194544		3 48 32.4	−31 46 42	+0.002	+0.01	7.6	0.3	1.7	A3 V		110 s		m
24071	194550		3 48 35.3	−37 37 20	+0.004	−0.01	4.27	−0.01	0.6	A0 V	+16	50 mx		m
23978	168851		3 48 35.7	−20 54 12	−0.001	−0.02	5.81	1.64	−0.3	gK5	+3	140 s		
23565	24234		3 48 37.2	+51 49 26	−0.010	−0.01	7.60	0.66	5.2	G5 V		30 s		
24502	256032		3 48 37.5	−70 21 32	+0.002	+0.02	7.33	1.08	0.2	K0 III		240 s		
23901	130740		3 48 37.8	−4 25 16	+0.001	−0.01	7.7	0.6	4.4	G0 V		47 s		
23887	111433		3 48 38.8	+0 13 40	+0.004	0.00	5.91	1.23		K0	+66			
23581	24235		3 48 39.5	+51 22 31	+0.002	−0.04	7.21	1.22	0.2	K0 III		190 s		
22828	––		3 48 40.8	+78 07 12	+0.010	+0.02	7.1	0.0	0.0	B8.5 V		160 mx		
23730	56719		3 48 43.6	+34 18 36	0.000	−0.02	7.2	0.5	4.0	F8 V		43 s		
23915	130742		3 48 44.9	−2 25 44	+0.001	0.00	8.0	0.9	3.2	G5 IV		93 s		
24112	216499		3 48 46.7	−40 23 58	+0.007	+0.05	7.3	0.1	4.0	F8 V		45 s		
23937	130743		3 48 47.6	−7 00 54	+0.001	−0.02	7.7	1.6	−0.5	M4 III		440 s		BR Eri, v
24250	233283		3 48 51.2	−55 00 00	+0.004	+0.04	7.3	1.3	0.2	K0 III		180 s		
23623	24241		3 48 54.1	+50 50 25	+0.003	−0.08	7.33	0.41	3.0	F2 V		68 s		
23854	93615		3 48 56.1	+16 27 34	−0.003	−0.10	7.5	0.5	3.4	F5 V		68 s		
23406	12909		3 48 56.3	+64 44 53	+0.001	−0.08	7.67	0.42		F5			2765	m
23822	76225		3 48 56.9	+23 51 26	+0.004	−0.05	6.6	0.4	2.6	F0 V	+19	61 s		
23729	39126		3 48 57.2	+40 40 15	0.000	−0.04	7.8	0.1	1.7	A3 V		150 s		
24070	168858		3 49 00.1	−26 01 57	−0.004	−0.02	6.9	0.8	3.0	F2 V		33 s		
23802	56722		3 49 07.3	+32 15 51	+0.001	−0.01	7.44	0.17		B7			2784	m
23728	39128		3 49 08.1	+43 57 47	0.000	+0.02	5.9	0.4	2.6	F0 V	−15	44 s		
23850	76228	27 Tau	3 49 09.7	+24 03 12	+0.001	−0.04	3.63	−0.08	−1.2	B8 III	+9	90 s	2786	Atlas, m
23862	76229	28 Tau	3 49 11.1	+24 08 12	+0.001	−0.05	5.06	−0.08		B8 p	+4	29 mn		BU Tau/Pleione, v
23852	76232		3 49 11.2	+22 36 33	+0.001	−0.05	7.72	0.18		A0	+2			
23277	5000		3 49 13.8	+70 52 16	+0.006	−0.06	5.44	0.09		A p	+17	26 mn		
23872	76234		3 49 16.7	+24 23 46	+0.002	−0.04	7.52	0.10	1.4	A2 V	+3	160 s		
23992	130746		3 49 19.0	−1 27 13	+0.005	−0.05	6.90	0.9	3.2	G5 IV		55 s		m
23594	24244		3 49 19.6	+57 07 06	+0.002	−0.02	6.50	0.06	0.2	B9 V		160 s		m
24150	194557		3 49 21.6	−36 05 32	+0.003	+0.01	6.7	1.7	0.2	K0 III		77 s		
23873	76236		3 49 21.6	+24 22 50	+0.001	−0.05	6.62	−0.03	0.4	B9.5 V	+1	78 mx		m
24307	233290		3 49 23.8	−53 14 35	−0.001	−0.03	7.6	0.6	3.0	F2 V		58 s		
23689	24249		3 49 24.6	+50 56 12	0.000	0.00	8.0	1.1	0.2	K0 III		320 s		
23886	76237		3 49 25.9	+24 14 52	+0.001	−0.04	7.97	0.18	1.7	A3 V	+9	160 s		
23603	24245		3 49 26.7	+57 07 20	+0.004	−0.03	7.3	0.1	0.6	A0 V		120 mx		
24160	194559		3 49 27.2	−36 12 01	−0.004	−0.05	4.17	0.95	0.3	G5 III	+2	53 s		
23675	24248		3 49 27.4	+52 39 20	−0.001	+0.01	6.70	0.44		B0.5 III	+2		2783	m
22689	594		3 49 30.0	+80 19 21	−0.002	+0.02	6.70	1.6		M5 III				SS Cep, v
23475	12916		3 49 31.2	+65 31 34	−0.001	−0.01	4.47	1.88	−4.8	M2 Ib	−3	390 s		
23848	56727	42 Per	3 49 32.6	+33 05 29	−0.002	0.00	5.11	0.07	1.4	A2 V	−14	53 s		
23741	39129		3 49 33.4	+47 58 57	+0.003	−0.02	7.2	0.9	3.2	G5 IV		61 s		
24306	233291		3 49 36.1	−52 04 49	+0.005	+0.01	6.9	2.2	−0.5	M4 III		150 s		
23523	12917		3 49 36.6	+63 17 50	−0.002	−0.01	5.85	0.18		A3	−14			
23913	76242		3 49 38.1	+22 32 01	+0.001	−0.03	7.00	0.03		B9	−7			
23450	12915		3 49 38.2	+67 09 55	+0.003	−0.03	7.70	0.9	0.3	G8 III	+6	270 s		
24031	130752		3 49 38.6	−2 19 45	−0.004	−0.06	7.1	0.6	4.4	G0 V		35 s		
23988	93625		3 49 42.7	+10 33 33	−0.002	0.00	7.7	0.9	3.2	G5 IV		81 s		
23923	76244		3 49 43.5	+23 42 42	+0.001	−0.05	6.16	−0.05	0.2	B9 V	+2	160 s		

HD	SAO	Star Name	α 2000	δ 2000	μ(α)	μ(δ)	V	B-V	M$_v$	Spec	RV	d(pc)	ADS	Notes
23990	111441		3^h 49^m 46.4^s	+9°24'29"	+0.002	0.00	6.80	0.00	0.4	B9.5 V		190 s	2796	m
23989	111442		3 49 47.2	+9 52 40	+0.001	-0.05	7.7	1.1	0.2	K0 III		310 s		
24249	216509		3 49 49.1	-42 43 40	-0.003	0.00	7.36	0.17		A2				
23950	76250		3 49 55.0	+22 14 40	+0.001	-0.04	6.07	-0.01	-1.2	B8 III	+14	250 s		
23948	76249		3 49 56.4	+24 20 56	+0.001	-0.05	7.54	0.08		A0	+5			
23964	76251		3 49 58.1	+23 50 55	+0.002	-0.05	6.74	0.06		A0	+6		2795	m
24291	216513		3 50 02.7	-45 23 07	+0.001	+0.01	6.94	0.94	0.3	G8 III		210 s		
23621	12922		3 50 03.1	+61 47 52	+0.008	-0.14	7.4	1.1	0.2	K0 III	+38	120 mx		
23965	76254		3 50 03.4	+22 35 30	+0.012	-0.06	8.0	0.5	3.4	F5 V	+12	64 mx		
23885	56730		3 50 03.5	+37 52 26	+0.007	-0.05	6.6	0.4	2.6	F0 V		61 s		
23838	39134		3 50 04.5	+44 58 04	-0.003	-0.03	5.66	0.76	4.4	G0 V	+14	14 s		
24053	111446		3 50 08.8	+6 37 15	+0.003	-0.03	7.8	0.6	4.4	G0 V		47 s		
23946	76253		3 50 09.8	+29 38 40	-0.001	-0.01	7.90	0.1	0.6	A0 V		260 s		m
23922	56735		3 50 13.3	+34 49 14	+0.004	-0.07	6.80	0.4	2.6	F0 V		68 s	2794	m
24157	149199		3 50 13.9	-18 44 43	0.000	+0.02	7.3	-0.1	0.6	A0 V		220 s		
24098	130762		3 50 16.2	-1 31 22	+0.007	+0.02	6.50	0.4	3.0	F2 V		50 s	2803	m
24248	194564		3 50 16.7	-32 17 09	+0.004	0.00	6.92	1.54	-0.3	K5 III		260 s		
24107	130763		3 50 18.3	-3 53 19	0.000	-0.01	7.50	1.1	-2.2	K1 II	+39	870 s		
23985	76256		3 50 18.8	+25 34 46	+0.003	-0.11	5.26	0.21		A3	+4		2799	m
24940	256038		3 50 19.6	-75 53 41	+0.014	+0.04	7.37	1.61		K2				
24120	130764		3 50 20.1	-5 04 34	-0.001	-0.01	8.0	1.1	0.2	K0 III		360 s		
23401	5006	γ Cam	3 50 21.5	+71 19 57	+0.004	-0.04	4.63	0.03	0.9	A3 IV	-1	56 s		m
24040	93630		3 50 22.9	+17 28 35	+0.008	-0.25	7.52	0.66	4.4	G0 V		35 s		
23800	24255		3 50 25.1	+52 28 54	0.000	0.00	6.89	0.35	-3.9	B1 IV	-18	660 s		
23898	39138		3 50 27.7	+42 35 18	+0.001	-0.01	7.9	1.6	-0.5	M2 III		410 s		
24013	76259		3 50 28.1	+24 29 43	+0.002	-0.03	7.3	0.1	1.4	A2 V		140 s		
24501	248891		3 50 29.8	-60 25 42	0.000	0.00	7.7	2.4	-0.5	M2 III		150 s		
23962	56737		3 50 30.1	+34 03 06	+0.002	-0.05	7.40	1.4	-0.3	K5 III	+16	320 s		
24267	194569		3 50 34.8	-30 41 24	0.000	-0.02	7.2	1.1	0.2	K0 III		210 s		
24305	194570		3 50 37.5	-36 25 31	0.000	+0.02	6.86	-0.04		A0 n	+6			
23649	12925		3 50 38.7	+63 42 27	-0.001	+0.01	7.9	1.6	-0.5	M4 III		410 s		
24133	111453		3 50 41.9	+1 33 52	0.000	+0.02	6.7	0.4	2.6	F0 V		66 s		
23036	4999		3 50 42.8	+78 19 43	+0.001	-0.01	8.0	0.0	0.0	B8.5 V		380 s		
24330	194573		3 50 44.3	-38 58 56	+0.002	+0.03	7.0	0.8	-0.5	M4 III		320 s		
24012	56741		3 50 51.2	+35 05 58	0.000	-0.02	7.6	0.0	0.4	B9.5 V		250 s		
--	--		3 50 51.9	+33 01 32	-0.001	-0.02	4.7	0.5	4.0	F8 V		14 s		
24076	76264		3 50 52.4	+23 57 43	+0.002	-0.03	6.93	0.09	1.4	A2 V	+5	120 s		
24304	168886		3 50 59.5	-29 24 05	+0.001	-0.03	7.8	1.7	-0.1	K2 III		180 s		
24426	216522		3 50 59.7	-48 31 40	-0.001	+0.02	7.6	0.7	3.4	F5 V		33 s		
23551	12924		3 51 00.3	+69 06 11	+0.002	-0.01	7.2	1.1	0.2	K0 III	-4	230 s		
24406	216520		3 51 03.1	-44 21 55	+0.003	+0.01	8.01	0.39		F0				
24181	111460		3 51 03.7	+1 43 31	+0.001	+0.02	7.9	0.9	3.2	G5 IV		86 s		
24349	194578		3 51 12.6	-31 39 16	-0.001	-0.05	7.8	1.1	0.2	K0 III		300 s		
24155	93637		3 51 15.8	+13 02 45	+0.001	-0.03	6.20	-0.07	-1.9	B9 II-III	+16	430 s		
24361	194579		3 51 22.1	-30 52 19	+0.012	+0.03	7.5	1.6	0.2	K0 III		120 mx		
24118	76272		3 51 25.2	+25 09 47	-0.001	0.00	6.9	0.1	1.4	A2 V		120 s		
24075	56750		3 51 26.3	+34 36 17	+0.008	-0.06	7.9	1.1	-0.1	K2 III		120 mx		
24416	216523		3 51 26.3	-40 05 08	-0.002	0.00	7.7	1.3	-0.5	M4 III		440 s		
24328	168888		3 51 31.1	-21 16 51	0.000	-0.05	6.6	1.0	0.2	K0 III		180 s		
24316	149212		3 51 32.5	-17 10 00	-0.005	-0.06	7.6	1.1	0.2	K0 III		170 mx		
24566	233301		3 51 32.6	-56 17 37	+0.001	+0.03	7.46	0.92	3.2	G8 IV		67 s		
24010	39153		3 51 36.5	+47 32 33	-0.003	+0.04	8.0	1.1	0.2	K0 III		320 s		
24154	76275		3 51 36.6	+22 01 54	+0.001	-0.02	6.8	1.1	0.2	K0 III	+63	190 s		
24500	216527		3 51 41.4	-47 19 00	0.000	+0.04	6.71	0.28		A3				
23662	12929		3 51 41.8	+68 30 27	+0.004	-0.01	6.3	-0.1		B8				
24359	168894		3 51 42.0	-22 55 51	+0.003	-0.06	7.70	0.9	3.2	G5 IV		79 s	2825	m
24650	248898		3 51 43.0	-60 21 36	+0.001	-0.03	8.0	0.6	4.0	F8 V		60 s		
23602	5015		3 51 43.8	+70 30 28	+0.002	-0.03	7.40	0.1	1.7	A3 V		140 s		m
24499	216529		3 51 51.1	-46 09 45	0.000	+0.02	7.86	0.40	1.9	F2 IV		150 s		
24131	56761		3 51 53.7	+34 21 33	0.000	0.00	5.77	0.00	-3.5	B1 V	+18	550 s		m
24338	149218		3 51 55.3	-11 58 03	+0.001	+0.05	7.5	1.6	-0.5	M2 III		400 s		
24178	76279		3 51 57.3	+25 59 56	+0.001	-0.05	7.64	0.16		A0	0			
23894	12939		3 51 59.8	+60 20 12	-0.001	+0.02	7.0	1.1	0.2	K0 III	+38	210 s		
24263	111469	31 Tau	3 52 00.2	+6 32 05	+0.001	0.00	5.67	0.06		B9	+16			m
24193	76281		3 52 00.8	+25 20 30	0.000	+0.02	8.0	1.1	0.2	K0 III		310 s		
24167	56762		3 52 04.3	+31 10 06	-0.002	-0.04	6.25	0.20		A3	-38			
24117	39161		3 52 04.5	+40 47 50	-0.003	-0.01	8.00	0.1	1.4	A2 V		190 s	2815	m
24206	76283		3 52 05.4	+22 40 18	+0.014	-0.33	7.58	0.69	5.2	dG5	+8	29 s		
24325	130780		3 52 08.6	-1 08 56	-0.001	-0.01	6.80	0.00	0.4	B9.5 V		190 s		m
24213	76286		3 52 11.4	+25 09 47	-0.009	-0.16	6.8	0.6	4.4	G0 V		30 s		
24336	130784		3 52 17.0	-0 39 22	+0.003	-0.01	7.0	0.1	1.4	A2 V		130 s		
24262	93643		3 52 18.5	+18 55 04	+0.001	-0.01	7.9	0.1	0.6	A0 V		280 s		
24190	56768		3 52 18.9	+34 13 19	0.000	-0.01	7.43	0.05	-2.5	B2 V	+18	720 s		
23909	12942		3 52 20.6	+62 20 41	-0.001	0.00	6.8	0.1	1.4	A2 V		120 s		
24278	93644		3 52 23.0	+18 35 53	-0.002	-0.02	7.9	0.1	0.6	A0 V		280 s		

HD	SAO	Star Name	α 2000	δ 2000	μ(α)	μ(δ)	V	B−V	M_v	Spec	RV	d(pc)	ADS	Notes
			h m s	° ′ ″	s	″								
24371	130785		3 52 30.8	−5 03 15	+0.001	+0.03	7.1	0.4	2.6	F0 V		79 s		
24165	39167		3 52 33.0	+40 21 56	−0.001	0.00	7.3	0.9	3.2	G5 IV		65 s		
24425	149223		3 52 34.6	−15 44 29	−0.001	+0.01	7.6	1.1	0.2	K0 III		310 s		
24228	56775		3 52 39.5	+32 24 31	+0.002	−0.01	6.6	1.1	0.2	K0 III		180 s		
24576	216534		3 52 39.8	−44 56 47	−0.001	0.00	7.61	1.30		K0				
24388	130789	30 Eri	3 52 41.5	−5 21 41	−0.001	−0.01	5.48	−0.10		B7 n	+15		2832	m
23883	12944		3 52 41.8	+65 31 54	0.000	0.00	7.4	0.6	4.4	G0 V		39 s		
24380	111474		3 52 47.6	+1 44 52	+0.001	−0.03	7.90	1.1	0.2	K0 III		350 s	2831	m
24472	168913		3 52 49.6	−22 16 34	−0.002	+0.04	7.3	0.2	2.6	F0 V		85 s		
24345	93648		3 52 55.4	+14 22 58	−0.001	+0.01	8.00	0.9	3.2	G5 IV		91 s	2829	m
24471	149227		3 52 57.1	−18 27 35	+0.001	0.00	7.8	1.1	−0.1	K2 III		380 s		
24129	24274		3 52 58.1	+51 03 02	−0.002	−0.02	7.83	0.13	−3.1	B9 II		1300 s		
24301	76294		3 52 59.3	+26 40 43	+0.009	−0.12	8.00	0.6	2.8	G0 IV	+25	67 mx		
24424	130793		3 52 59.8	−4 22 15	+0.002	−0.04	7.6	0.5	3.4	F5 V		69 s		
24177	39169		3 53 01.8	+45 57 27	−0.007	+0.03	8.0	0.0	0.4	B9.5 V		120 mx		
24446	130796		3 53 08.3	−6 38 02	0.000	0.00	6.5	0.0	0.4	B9.5 V		170 s		
24357	93650		3 53 10.0	+17 19 38	+0.010	−0.03	5.97	0.34	3.3	F4 V	+35	34 s		
24497	149229		3 53 12.8	−18 26 03	0.000	+0.01	6.22	0.88		F2				
24400	111476		3 53 13.1	+7 46 12	+0.004	−0.11	6.6	0.5	3.4	F5 V	+13	44 s		
24757	233314		3 53 22.7	−54 51 02	+0.001	0.00	7.76	−0.12	−0.9	B6 V		540 s		
24486	149230		3 53 24.0	−10 32 43	+0.001	0.00	8.0	1.1	0.2	K0 III		360 s		
24649	216539		3 53 26.9	−41 13 21	0.000	+0.07	7.5	0.1	4.0	F8 V		49 s		
24241	39173		3 53 28.1	+46 54 15	−0.005	+0.02	8.0	0.9	3.2	G5 IV		89 s		
24456	111480		3 53 30.0	+2 07 08	0.000	−0.02	6.7	0.0	0.4	B9.5 V		180 s		
24522	149232		3 53 30.6	−14 58 11	+0.001	+0.03	7.2	1.1	−0.1	K2 III		280 s		
24706	216540		3 53 33.3	−46 53 37	+0.003	−0.03	5.93	1.24	−0.2	K3 III		170 s		
24368	76305		3 53 34.4	+25 40 59	+0.001	−0.03	6.36	0.12		A0	+8			
24365	76304		3 53 37.6	+28 08 54	+0.006	−0.01	7.84	0.84	5.6	G8 V	+20	25 s		
24240	39175		3 53 38.7	+48 39 02	+0.004	−0.03	5.76	1.05	0.2	K0 III	+8	120 s		
24626	194608		3 53 38.9	−34 43 57	+0.003	−0.01	5.11	−0.13	−1.3	B6 IV	+18	190 s		
24587	168925	33 τ⁸ Eri	3 53 42.6	−24 36 45	+0.001	−0.01	4.65	−0.13	−1.1	B5 V	+23	140 s		
24141	24276		3 53 43.2	+57 58 31	+0.011	−0.09	5.8	0.2		A m	−5			
24443	93655		3 53 45.3	+13 30 18	+0.005	0.00	8.0	0.5	3.4	F5 V		82 s		
24065	12951		3 53 45.4	+63 23 23	−0.001	−0.02	8.0	1.4	−0.3	K5 III		390 s		
24539	149234		3 53 46.6	−12 44 01	+0.001	−0.02	8.0	1.1	0.2	K0 III		360 s		
24696	216543		3 53 47.9	−43 09 45	0.000	+0.03	7.93	1.10		K0				
24399	76307		3 53 50.2	+26 53 46	0.000	+0.01	7.5	0.9	−2.1	G8 II		610 s		
23688	5021		3 53 55.9	+74 39 58	−0.003	−0.02	7.9	0.9	3.2	G5 IV		88 s		
24616	168929		3 53 59.3	−23 08 09	+0.024	−0.26	6.70	0.82	4.4	dG0	+100	20 s		
24561	149236		3 53 59.6	−12 26 57	−0.004	0.00	7.5	0.9	3.2	G5 IV		71 s		
24310	39181		3 54 00.8	+45 30 43	0.000	−0.02	7.9	1.1	0.2	K0 III		310 s		
24434	76311		3 54 01.9	+21 56 43	0.000	0.00	7.1	0.0	0.0	B8.5 V	+14	230 s		
24398	56799	44 ζ Per	3 54 07.8	+31 53 01	+0.001	−0.01	2.85	0.12	−5.7	B1 Ib	+21	340 s	2843	Atik, m
24661	168933		3 54 17.3	−27 40 08	+0.003	+0.04	6.9	0.4	1.4	A2 V		82 s		
24555	130806	32 Eri	3 54 17.4	−2 57 17	+0.002	+0.01	4.46	0.68	0.3	G8 III	+27	68 s	2850	m
24552	111490		3 54 22.5	+1 15 41	0.000	−0.13	8.0	0.6	4.4	G0 V		53 s		
24745	216546		3 54 23.0	−40 21 26	−0.002	+0.01	5.71	0.60	3.4	F5 V		48 mn		m
24550	111492		3 54 27.2	+5 10 29	−0.002	−0.02	7.44	0.39	−0.7	F3 II-III	+15	420 s	2849	m
24496	93662		3 54 27.9	+16.36 55	+0.015	−0.19	6.9	0.7	4.4	G0 V		32 s		
24527	93664		3 54 33.4	+13 46 15	0.000	−0.06	7.8	0.8	3.2	G5 IV		85 s		
24863	233321		3 54 33.9	−52 41 26	+0.003	−0.04	6.46	0.16		A2				m
24623	130812		3 54 34.8	−9 31 15	−0.002	−0.02	7.1	0.4	3.0	F2 V		66 s		
24805	216548		3 54 37.4	−46 25 01	−0.002	−0.01	6.91	0.16		A0				
24585	111497		3 54 42.5	+2 05 25	+0.001	+0.03	6.9	0.9	3.2	G5 IV		56 s		
23999	5031		3 54 43.5	+70 11 56	−0.002	+0.06	7.9	1.4	−0.3	K5 III		250 mx		
24573	111494		3 54 45.4	+9 10 40	+0.001	−0.02	7.4	0.1	0.6	A0 V		230 s		
22677	599		3 54 46.9	+83 32 42	+0.008	−0.01	7.5	0.0	0.4	B9.5 V		270 s		
24755	194620		3 54 49.0	−35 42 13	+0.001	+0.20	7.69	0.45		F8		15 mn		
24341	24291		3 54 51.1	+52 25 11	+0.012	−0.15	7.83	0.67	4.5	G1 V		42 s		
--	56807		3 54 56.0	+39 51 35	+0.002	−0.01	7.9	1.5						
24693	149250		3 54 58.7	−14 54 21	+0.002	0.00	7.3	1.6	−0.5	M2 III		370 s		
24505	76315		3 54 59.8	+28 11 17	+0.001	−0.06	7.97	0.69	0.3	G5 III		190 mx		
24238	12955		3 55 03.9	+61 10 00	+0.062	−0.25	7.85	0.83	5.9	K0 V	+48	25 ts		
24571	93670		3 55 05.7	+18 35 26	+0.002	0.00	8.0	0.5	3.4	F5 V		82 s		
24787	194624		3 55 07.6	−35 33 53	+0.004	−0.01	7.9	1.4	−0.1	K2 III		290 s		
24754	168950		3 55 13.9	−24 01 58	+0.003	−0.01	7.3	1.6		M5 III	+42			T Eri, v
24825	194625		3 55 15.5	−38 45 34	−0.001	−0.01	6.8	0.0	0.4	B9.5 V		170 s		
24712	149251		3 55 16.2	−12 05 57	−0.004	−0.04	6.00	0.32		A7 p	+22			
24534	56815		3 55 22.9	+31 02 45	−0.001	−0.01	6.10	0.29		O pe	+17	21 mn	2859	X Per, m,v
25054	248907		3 55 23.7	−60 54 26	+0.010	+0.02	7.5	0.5	3.4	F5 V		62 s		
24432	39191		3 55 25.0	+49 02 28	+0.002	+0.02	6.82	0.55	−3.9	B3 II	−11	530 s		
24620	93676		3 55 26.0	+17 37 57	+0.001	−0.01	8.00	0.9	3.2	G5 IV		91 s	2864	m
24835	194628		3 55 26.8	−37 58 43	−0.001	−0.02	7.7	0.7	3.2	G5 IV		80 s		
24824	194630		3 55 29.8	−36 18 36	+0.004	+0.04	7.5	0.4	1.7	A3 V		99 s		
23998	5033		3 55 32.3	+72 37 15	+0.009	−0.06	8.0	1.1	0.2	K0 III		240 mx		

HD	SAO	Star Name	α 2000	δ 2000	μ(α)	μ(δ)	V	B-V	M_V	Spec	RV	d(pc)	ADS	Notes
24823	194629		3^{h}55^{m}34.1^s	−33°12′52″	0.000	+0″.01	7.9	1.0	−0.1	K2 III		410 s		
24797	168960		3 55 35.7	−25 55 39	+0.002	+0.03	7.4	1.1	−0.1	K2 III		320 s		
24798	168961		3 55 36.3	−26 12 53	−0.001	−0.01	8.0	0.2	0.2	K0 III		360 s		
24421	24301		3 55 37.1	+52 13 37	−0.012	+0.11	6.79	0.51	3.4	F5 V		43 s		
24431	24300		3 55 38.5	+52 38 28	+0.001	−0.01	6.73	0.38	−5.1	O9 IV-V	−10	1400 s		
24977	233330		3 55 40.1	−51 47 58	0.000	+0.02	7.8	0.8	2.1	A5 V		60 s		
24395	24298		3 55 46.2	+56 55 08	+0.002	−0.03	6.92	0.28	−2.0	A7 II		590 s		
24784	149260		3 55 46.3	−19 26 01	+0.001	+0.02	7.5	1.2	0.2	K0 III		210 s		
24727	130824		3 55 49.2	−4 31 37	−0.002	−0.05	7.1	0.5	3.4	F5 V		55 s		
25171	248911		3 55 49.4	−65 11 15	+0.022	+0.06	7.7	0.5	4.0	F8 V		55 s		
24504	39195		3 55 58.1	+47 52 16	+0.002	−0.02	5.37	−0.08	−0.9	B6 V	+10	170 s		
25170	248912		3 56 04.0	−63 27 49	+0.011	+0.05	6.14	1.10	0.2	K0 III		130 s		
24990	216558		3 56 05.2	−47 46 25	−0.013	−0.12	7.60	0.5	4.0	F8 V		53 s		m
24782	130827		3 56 05.9	−7 38 32	+0.001	−0.02	7.7	1.1	0.2	K0 III		310 s		
24875	168973		3 56 11.1	−29 39 36	+0.001	−0.01	7.5	0.9	0.2	K0 III		280 s		
24409	24307		3 56 11.3	+59 38 32	−0.036	+0.18	6.70	0.7	4.4	G0 V		29 s		m
24967	216559		3 56 15.5	−43 06 57	+0.001	−0.02	7.54	1.24		K0				
24763	111507		3 56 18.6	+2 12 27	−0.002	−0.01	7.9	1.1	0.2	K0 III		340 s		
24834	149267		3 56 27.6	−13 35 51	0.000	+0.01	6.41	1.69	−0.5	gM3	+38	200 s		
24772	111510		3 56 28.2	+5 51 05	+0.003	0.00	7.9	0.9	3.2	G5 IV		88 s		
24702	76334		3 56 28.6	+22 40 28	+0.012	−0.23	7.86	0.69	4.4	G0 V		35 mx		
24640	56824		3 56 28.6	+35 04 52	0.000	0.00	5.49	−0.03	−2.5	B2 V	+17	340 s		
24892	168977		3 56 28.7	−25 10 52	+0.009	−0.27	6.9	1.0	4.4	G0 V		18 s		
24966	194639		3 56 29.3	−38 57 44	+0.002	0.00	6.8	−0.2	0.4	B9.5 V		190 s		
24164	5040		3 56 30.2	+71 49 20	−0.010	+0.01	6.4	0.4	2.6	F0 V	−2	57 s		
24833	149268		3 56 31.7	−10 51 08	+0.006	+0.06	6.6	0.5	3.4	F5 V		44 s		
24779	111512		3 56 36.0	+6 42 39	+0.005	−0.04	7.8	0.4	2.6	F0 V		110 mx		
24546	24314	43 Per	3 56 36.4	+50 41 43	+0.010	−0.13	5.28	0.41	3.4	F5 V	+27	24 s		m
24064	5038		3 56 36.6	+74 04 49	+0.001	−0.01	6.9	1.1	0.2	K0 III		210 s		
24792	111514		3 56 37.6	+3 03 28	+0.002	0.00	7.4	1.1	0.2	K0 III		270 s		
24832	130833		3 56 37.8	−9 45 02	+0.003	+0.02	6.19	0.28		F0				
24819	130832		3 56 42.8	−1 34 20	−0.002	−0.03	7.90	0.1	1.4	A2 V		200 s	2886	m
25074	216570		3 56 51.4	−48 28 42	−0.001	−0.03	7.5	0.5	3.2	G5 IV		73 s		
24740	76339	32 Tau	3 56 52.0	+22 28 41	+0.005	−0.11	5.63	0.30	2.6	F0 V	+32	39 ts		
24952	168981		3 56 57.0	−27 30 58	−0.001	+0.01	7.8	1.4	4.0	F8 V		18 s		
24817	111516		3 57 01.7	+6 02 24	+0.002	−0.06	6.09	0.06		A0	+8			
25038	216568		3 57 01.7	−40 02 14	+0.001	0.00	7.2	1.0	0.2	K0 III		250 s		
24769	76343	33 Tau	3 57 03.8	+23 10 31	+0.001	−0.02	6.00	0.02		B0.5 IV				
24480	12968		3 57 08.2	+61 06 32	0.000	−0.02	5.00	1.45	−0.3	gK4	−2	120 s	2867	m
24777	76347		3 57 11.2	+21 19 24	−0.001	−0.02	6.9	1.1	0.2	K0 III		200 s		
24768	76345		3 57 11.7	+25 16 58	0.000	+0.01	7.50	0.9	0.3	G8 III		240 s		
24975	168983		3 57 12.1	−24 37 53	+0.001	+0.02	7.3	0.2	1.4	A2 V		130 s		
25149	233337		3 57 12.4	−52 44 52	+0.001	+0.01	7.6	1.6	0.2	K0 III		140 s		
25797	256050		3 57 12.7	−76 30 40	+0.006	+0.04	7.7	0.4	3.0	F2 V		87 s		
24689	39212		3 57 16.3	+41 52 49	0.000	−0.02	6.80	0.1	1.4	A2 V		120 s	2884	m
24479	12969		3 57 25.3	+63 04 20	0.000	+0.01	5.03	−0.09	−0.8	B9 III	+5	150 s		
24701	39213		3 57 25.8	+43 18 59	+0.004	−0.04	7.2	0.1	0.6	A0 V	0	200 s		
24802	76350		3 57 26.3	+24 27 43	+0.009	−0.01	6.16	1.37	0.2	K0 III	−13	93 s		
24916	130840		3 57 28.6	−1 09 35	−0.013	−0.16	8.00	1.14	8.0	dK5	+6	10 ts	2894	m
25210	233340		3 57 33.9	−54 21 27	−0.004	−0.02	8.0	0.6	3.0	F2 V		67 s		
24747	56838		3 57 35.1	+36 29 30	+0.006	−0.09	6.9	0.4	2.6	F0 V		71 s		
24964	130847		3 57 38.2	−9 42 00	0.000	0.00	8.0	1.1	0.2	K0 III		370 s		
24844	76355		3 57 47.2	+22 55 29	+0.010	+0.03	8.00	1.1	0.2	K0 III	+27	100 mx		m
−−	56841		3 57 50.7	+36 36 33	+0.003	−0.02	7.0	2.4						
24760	56840	45 ε Per	3 57 51.1	+40 00 37	+0.002	−0.02	2.89	−0.18		B0.5 V	−1	40 mn	2888	m
25169	216579		3 57 51.6	−46 22 57	+0.001	−0.07	8.02	0.47		F5				
24688	24319		3 57 57.9	+52 07 42	−0.002	0.00	7.96	1.02	0.2	K0 III		350 s		
25025	149283	34 γ Eri	3 58 01.7	−13 30 31	+0.004	−0.11	2.95	1.59	−0.4	M0 III	+62	44 s	2904	Zaurak, m
24809	56847		3 58 03.0	+34 48 52	+0.001	+0.02	6.4	0.2	2.1	A5 V	−2	71 s		
24296	5046		3 58 03.1	+72 43 29	+0.005	+0.01	6.7	1.1	0.2	K0 III		190 s		
25245	216584		3 58 15.1	−49 36 36	+0.001	+0.04	7.1	1.1	0.2	K0 III		220 s		
24962	111529		3 58 15.3	+7 36 18	0.000	−0.01	7.8	0.5	3.4	F5 V		74 s		
25222	216582		3 58 15.4	−48 10 14	−0.001	+0.01	8.0	1.1	0.2	K0 III		210 s		
24899	76358		3 58 20.8	+24 04 53	+0.001	−0.03	7.19	0.04		B9	+8			
25001	111532		3 58 21.1	+1 26 45	−0.001	0.00	7.8	1.4	−0.3	K5 III	+35	430 s		
24973	111531		3 58 23.2	+7 28 25	−0.001	−0.02	8.0	1.1	0.2	K0 III		360 s		
24843	56849		3 58 29.1	+38 50 25	+0.003	−0.04	6.30	1.08	0.2	K0 III	+22	150 s		
25549	248920		3 58 35.5	−68 20 37	+0.002	+0.01	8.0	1.3	0.2	K0 III		230 s		
25000	93699		3 58 37.3	+11 01 39	0.000	−0.03	7.5	1.6	−0.5	M2 III		400 s		
25040	130858		3 58 37.9	−2 39 06	−0.002	−0.02	7.10	0.8	3.2	G5 IV		60 s	2909	m
24733	24324		3 58 38.1	+53 59 20	0.000	−0.01	7.02	0.25	2.4	A7 V		78 s		
25041	130859		3 58 40.0	−3 39 55	−0.004	−0.01	7.3	0.9	3.2	G5 IV		66 s		
25346	233347		3 58 42.9	−57 06 09	+0.004	+0.01	6.05	0.44	0.6	F2 III		110 s		
24569	12976		3 58 43.8	+65 34 33	+0.001	−0.01	8.0	0.1	0.6	A0 V		270 s		
24513	12975		3 58 43.9	+68 01 25	−0.001	−0.02	7.4	0.9	3.2	G5 IV		67 s		

HD	SAO	Star Name	α 2000	δ 2000	μ(α)	μ(δ)	V	B-V	M_V	Spec	RV	d(pc)	ADS	Notes
25422	248918	δ Ret	3 58 44.7	-61 24 01	+0.001	-0.02	4.56	1.62	-0.5	M2 III	-1	97 s		
24717	24325		3 58 49.9	+57 27 03	+0.005	-0.04	6.9	0.1	0.6	A0 V		90 mx		
24775	24326		3 58 51.2	+51 29 57	0.000	0.00	7.58	1.64	-4.4	K2 Ib		1900 s	2896	m
25069	130860		3 58 52.2	-5 28 12	-0.004	-0.18	5.83	1.00	5.7	dG9	+36	11 s		d?
25527	248921		3 58 52.2	-65 52 53	-0.002	0.00	7.4	1.6	-0.5	M2 III		390 s		
25316	233346		3 58 53.7	-51 05 25	0.000	+0.01	8.0	1.0	3.2	G5 IV		67 s		
24912	56856	46 ξ Per	3 58 57.8	+35 47 28	0.000	0.00	4.04	0.01		O7	+70	46 mn		Menkib
24970	76365		3 59 05.2	+27 11 51	-0.001	-0.01	7.4	0.1	0.6	A0 V		210 s		
25315	216589		3 59 08.4	-48 54 53	+0.001	+0.01	7.8	0.7	0.6	A0 V		170 s		
25049	93705		3 59 19.7	+13 58 30	0.000	-0.06	7.3	0.8	3.2	G5 IV		67 s		
25301	216590		3 59 20.2	-43 55 01	+0.001	0.00	6.84	1.10		G5				
24854	39225		3 59 21.2	+48 45 44	-0.001	+0.01	7.6	1.1	-0.1	K2 III		310 s		
24898	39228		3 59 24.5	+44 28 53	+0.002	-0.05	7.8	0.1	1.7	A3 V		150 s		
25189	169008		3 59 25.2	-20 19 59	-0.001	-0.03	7.6	0.9	-2.2	K2 II	+14	930 s		
25165	149299		3 59 30.0	-12 34 27	-0.001	-0.03	5.60	1.48	-0.3	gK5	-5	150 s		
25449	233350		3 59 37.6	-56 28 22	+0.005	+0.05	7.0	0.2	0.6	A0 V		120 mx		
24982	56866		3 59 39.9	+38 49 14	0.000	0.00	6.40	0.10	1.2	A1 V	-2	97 s	2910	m
25102	93710		3 59 40.6	+10 19 51	+0.011	+0.01	6.37	0.42	3.4	F5 V	+40	39 s		
25231	149303		3 59 46.5	-17 54 44	+0.001	+0.02	7.7	1.4	-0.3	K5 III		390 s		
25448	233351		3 59 54.0	-54 09 39	0.000	+0.05	7.5	0.7	3.0	F2 V		48 s		
25267	169017	36 τ⁹ Eri	3 59 55.4	-24 00 59	0.000	+0.01	4.66	-0.13	-0.6	A0 III	+24	110 s		
25581	248922		3 59 59.4	-61 53 42	+0.003	-0.01	6.6	1.6	-0.3	K5 III		190 s		
25250	149305		4 00 00.0	-19 14 51	0.000	-0.04	7.1	0.7	4.0	F8 V		35 s		
25358	216594		4 00 01.8	-42 23 25	+0.001	+0.01	7.5	0.7	0.2	K0 III		280 s		
25022	56876		4 00 10.4	+40 00 50	0.000	+0.01	6.9	0.1	0.6	A0 V		170 s		
25470	233354		4 00 15.7	-51 33 53	+0.002	+0.02	6.51	1.64		Ma				
25153	93714		4 00 16.6	+14 18 21	+0.003	-0.02	7.8	0.5	3.4	F5 V	+40	74 s		
25334	194682		4 00 19.0	-31 26 44	+0.003	-0.04	8.0	2.3	-0.5	M2 III		200 s		
25057	39244		4 00 34.8	+42 13 35	+0.002	+0.01	8.0	0.1	0.6	A0 V		270 s		
25175	93716		4 00 36.8	+17 17 48	0.000	-0.03	6.30	0.06	0.4	B9.5 V		130 s		
25186	93718		4 00 37.0	+13 53 33	+0.002	-0.01	7.5	1.1	0.2	K0 III		290 s		
25371	194689		4 00 40.6	-30 29 26	+0.004	+0.01	5.93	0.04	0.6	A0 V		93 mx		
25204	93719	35 λ Tau	4 00 40.7	+12 29 25	0.000	-0.01	3.47	-0.12	-1.7	B3 V	+15	100 s		v
25938	256053		4 00 43.8	-71 10 00	+0.006	+0.04	6.58	0.08		A0				
25203	93720		4 00 44.4	+15 28 29	-0.003	-0.06	7.8	0.9	3.2	G5 IV		82 s		
25614	233362		4 00 44.8	-58 39 43	+0.001	+0.01	7.2	0.0	0.6	A0 V		210 s		
25202	93721		4 00 48.7	+18 11 38	+0.009	-0.03	5.89	0.32	3.3	F4 V	+25	33 s		m
25705	248925	γ Ret	4 00 53.8	-62 09 34	0.000	+0.03	4.51	1.65		Mb	-7			
25344	169029		4 00 54.7	-20 42 44	0.000	+0.01	8.0	0.9	0.2	K0 III		370 s		
25201	76388		4 00 56.8	+23 12 06	+0.001	-0.03	6.5	0.0	0.4	B9.5 V		150 s	2926	m
25590	233363		4 00 59.0	-54 23 31	-0.002	+0.03	7.70	0.1	1.4	A2 V		180 s		m
25136	56890		4 01 04.6	+38 22 38	+0.002	-0.04	7.6	1.4	-0.3	K5 III		330 s		
25230	76393		4 01 06.9	+20 11 57	+0.002	-0.07	6.8	0.9	3.2	G5 IV	+21	50 s		
25385	169036		4 01 12.1	-22 16 32	0.000	+0.02	7.2	0.0	2.6	F0 V		84 s		
--	56897		4 01 12.3	+37 07 37	-0.001	+0.01	7.8	0.6	4.4	G0 V		48 s		
25152	56899		4 01 14.7	+36 59 24	-0.002	+0.02	6.30	-0.01	0.4	B9.5 V		140 s		
25142	56896		4 01 16.6	+40 05 42	+0.005	-0.04	7.3	0.4	3.0	F2 V		69 s		
25537	216602		4 01 17.5	-46 08 14	-0.001	+0.02	7.23	1.02		G5				
25728	248927	z Ret	4 01 18.2	-61 04 44	+0.009	+0.09	4.97	1.42	-0.4	gM0	+61	120 s		
25401	169040		4 01 20.9	-26 30 59	+0.006	+0.02	7.5	0.1	3.4	F5 V		66 s		
25021	24348		4 01 20.9	+56 02 23	-0.002	+0.01	7.1	0.1	0.6	A0 V		190 s		
25228	76395		4 01 25.9	+28 30 09	+0.003	-0.01	7.06	0.17	0.6	A0 V		160 s		
25099	39249		4 01 25.9	+47 27 02	-0.002	0.00	7.3	1.1	0.2	K0 III	-27	240 s		
25340	130878	35 Eri	4 01 31.9	-1 32 59	+0.001	-0.01	5.28	-0.15	-1.1	B5 V	+16	190 s		
25184	56902		4 01 34.7	+38 39 55	-0.001	-0.01	7.90						2932	m
25056	24350		4 01 37.1	+53 51 58	-0.001	0.00	7.03	1.20	-4.5	G0 Ib	-5	1200 s		
25330	111566		4 01 46.1	+9 59 52	0.000	0.00	5.67	0.02		B8	+3		2938	m
25311	93728		4 01 46.1	+13 50 14	-0.002	-0.09	7.8	0.5	4.0	F8 V		57 s		
25200	56907		4 01 46.5	+39 59 21	-0.001	0.00	7.3	0.1	1.4	A2 V		140 s		
25109	24353		4 01 49.9	+50 38 17	+0.002	-0.03	8.03	0.14	0.6	A0 V		250 s		
25436	149324		4 01 50.0	-18 22 03	0.000	+0.02	7.8	0.4	3.0	F2 V		93 s		
25467	169046		4 01 55.3	-21 01 15	-0.001	+0.01	7.5	1.0	-0.1	K2 III		330 s		
25535	194709		4 02 03.5	-34 28 53	+0.031	+0.01	6.74	0.62		G1		14 mn		m
25296	76403		4 02 14.9	+28 07 37	+0.001	+0.01	7.31	0.95	0.3	G8 III		250 s	2944	m
25414	130884		4 02 16.7	-1 37 58	-0.004	-0.03	8.0	0.5	3.4	F5 V		82 s		
25874	248932		4 02 27.1	-61 21 26	+0.049	+0.10	6.73	0.68	4.2	G5 IV-V		32 s		m
26595	256059		4 02 27.8	-78 37 59	+0.002	-0.01	6.6	1.3	0.2	K0 III		130 s		
25412	111573		4 02 29.1	+3 50 50	-0.001	0.00	6.9	0.0	0.0	B8.5 V		240 s		
25271	56916		4 02 29.4	+39 05 50	+0.003	-0.04	7.70	0.26	1.7	A3 V		130 s		
25522	149329		4 02 30.6	-19 29 55	+0.001	-0.03	8.0	0.4	4.0	F8 V		62 s		
25457	130893		4 02 36.6	-0 16 08	+0.010	-0.25	5.38	0.50	3.7	F6 V	+18	20 ts		
26088	248935		4 02 36.9	-68 56 51	+0.008	+0.01	7.5	1.9	-0.1	K2 III		130 s		
25466	130894		4 02 38.2	-0 48 03	-0.001	0.00	6.9	0.1	0.6	A0 V		180 s		
25391	93735		4 02 38.5	+15 03 43	0.000	-0.06	8.00	0.6	4.4	G0 V	-39	53 s		
25587	169060		4 02 42.8	-27 29 00	+0.008	+0.08	7.39	0.54	4.0	F8 V		46 s		

HD	SAO	Star Name	α 2000	δ 2000	μ(α)	μ(δ)	V	B−V	M_v	Spec	RV	d(pc)	ADS	Notes
			h m s	° ′ ″	s	″								
25132	24360		4 02 43.4	+58 56 57	−0.003	0.00	7.55	0.05	−1.6	B3.5 V	−16	500 s		
29034	258368		4 02 44.3	−86 14 43	+0.021	+0.02	8.0	0.1	1.7	A3 V		170 mx		
25575	169058		4 02 47.1	−22 35 41	+0.001	−0.08	6.8	1.0	0.2	K0 III		160 mx		
25574	169061		4 02 53.9	−20 30 22	+0.002	−0.01	7.8	1.2	−0.1	K2 III		380 s		
25090	12998		4 02 54.7	+62 25 17	−0.005	0.00	7.34	0.30	−1.0	B5.5 V	−3	240 mx		
25703	194717		4 02 55.8	−39 48 20	−0.001	−0.01	7.6	0.1	1.4	A2 V		170 s		
25174	24367		4 02 55.9	+56 15 11	+0.003	−0.06	7.8	0.4	0.6	F0 III		160 mx		
25464	111576		4 02 57.9	+8 08 24	0.000	−0.01	7.8	0.4	2.6	F0 V		110 s		
25463	111577		4 02 59.6	+9 12 30	+0.007	−0.17	6.9	0.5	4.0	F8 V		38 s		
25771	216618		4 02 59.7	−48 22 09	0.000	+0.01	7.0	1.5	0.2	K0 III		110 s		
25816	233375		4 03 03.0	−51 38 48	−0.002	+0.03	7.7	1.1	0.2	K0 III		290 s		
25456	93740		4 03 06.6	+14 29 00	+0.002	−0.01	8.0	0.5	4.0	F8 V		62 s		
25320	39271		4 03 06.6	+41 55 56	0.000	+0.01	7.7	0.1	0.6	A0 V		240 s		
25713	194719		4 03 08.7	−39 21 54	−0.002	0.00	7.5	1.3	0.2	K0 III		200 s		
25490	111579	38 v Tau	4 03 09.3	+5 59 21	0.000	0.00	3.91	0.03	1.2	A1 V	−6	34 s		
25338	56926		4 03 09.5	+37 07 18	+0.001	+0.04	7.5	0.8	3.2	G5 IV		69 s		
25477	111578		4 03 10.4	+8 52 57	−0.001	0.00	6.9	1.1	0.2	K0 III		220 s		
24545	5062		4 03 11.5	+78 12 09	+0.001	−0.03	7.0	1.1	0.2	K0 III		220 s		
25373	56929		4 03 12.9	+34 50 00	+0.001	−0.02	7.6	1.1	0.2	K0 III		280 s		
24841	5066		4 03 14.3	+72 49 44	+0.014	−0.03	8.0	0.9	3.2	G5 IV		91 s		
25622	169068		4 03 15.6	−21 25 04	−0.001	+0.01	7.7	0.9	0.2	K0 III		310 s		
25521	111580		4 03 18.9	+3 04 39	−0.002	+0.02	7.1	0.4	2.6	F0 V		79 s		
25520	111581		4 03 21.0	+3 11 15	−0.001	0.00	7.3	1.1	−0.1	K2 III		300 s		
25293	39272		4 03 22.2	+48 50 28	+0.001	0.00	7.0	0.5	2.3	F7 IV n		83 s		
25631	169071		4 03 24.7	−20 08 39	0.000	−0.01	6.46	−0.18		B3 n	+20			
25795	216619		4 03 29.0	−44 08 17	−0.001	+0.01	8.03	1.1		K0				
25675	169075		4 03 29.2	−24 27 38	+0.004	−0.04	7.27	1.53		Mb				
24163	637		4 03 30.8	+81 34 35	−0.013	−0.06	7.7	0.5	3.4	F5 V		72 s		
25367	39278		4 03 31.3	+42 10 56	0.000	+0.01	8.00	0.1	1.4	A2 V		190 s	2950	m
25676	169080		4 03 33.9	−25 50 53	0.000	−0.01	6.8	2.0	−0.3	K5 III		130 s		
24226	638		4 03 34.0	+81 13 35	−0.006	−0.02	7.8	1.1	0.2	K0 III		340 s		
25661	169078		4 03 36.7	−20 09 30	+0.001	+0.01	7.01	1.22	−2.2	K2 II	+24	700 s		
25455	76414		4 03 38.7	+28 39 43	−0.002	0.00	8.0	0.1	1.7	A3 V		160 s		
25860	216625		4 03 39.2	−47 51 44	−0.002	0.00	6.63	0.20		A5				
25913	233384		4 03 40.3	−52 55 46	+0.005	−0.01	7.5	−0.1	2.6	F0 V		98 s		
24126	636		4 03 43.8	+81 52 30	+0.021	−0.07	7.2	1.1	0.2	K0 III		230 mx		
25786	194726		4 03 44.3	−37 04 14	−0.001	−0.05	7.1	0.6	1.7	A3 V		63 s		
25558	111585	40 Tau	4 03 44.5	+5 26 08	0.000	−0.01	5.33	−0.08	−1.7	B3 V	+12	230 s		
25843	216624		4 03 44.7	−44 43 58	−0.003	+0.03	7.65	0.18		A0				
25411	56934		4 03 48.5	+37 58 18	+0.001	−0.03	7.80	0.1	0.6	A0 V		250 s	2956	m
25292	24381		4 03 51.1	+53 16 39	+0.002	−0.08	7.79	0.58	4.0	F8 V		53 s		
25968	233386		4 03 53.2	−55 23 59	+0.002	+0.01	7.4	0.6	1.4	A2 V		71 s		
25487	76418		4 03 54.4	+28 07 33	+0.002	−0.02	8.00	0.00	−0.2	B8 V e	−20	360 s		RW Tau, v
25350	39281		4 03 55.8	+48 39 43	−0.002	−0.01	7.9	1.1	0.2	K0 III		310 s		
25570	111586		4 03 56.6	+8 11 50	+0.011	+0.03	5.45	0.38	3.0	F2 V	+36	30 s		
25389	39286		4 03 58.9	+43 59 19	−0.004	−0.01	7.8	0.1	1.7	A3 V		150 s		
25647	130909		4 04 02.3	−4 36 02	+0.001	0.00	7.5	1.4	−0.3	K5 III		370 s		
25444	56936		4 04 07.1	+39 30 38	+0.014	−0.05	7.13	0.68	3.2	G5 IV	+23	61 s	2959	m
25700	149345		4 04 08.7	−16 35 21	+0.007	−0.07	6.39	1.26		K2				
25621	111590		4 04 09.8	+2 49 37	+0.010	−0.12	5.36	0.50	2.2	F6 IV	−18	40 s		
25410	39289		4 04 09.9	+42 31 41	+0.014	+0.08	7.9	0.5	4.0	F8 V		58 s		
25584	93746		4 04 12.9	+15 01 36	−0.001	−0.04	7.9	0.9	3.2	G5 IV		88 s		
25754	149353		4 04 17.9	−19 28 09	+0.001	0.00	7.4	0.8	0.6	A0 V		78 s		
25605	93747		4 04 18.8	+12 30 27	+0.004	0.00	7.5	1.6	−0.5	M2 III		240 mx		
25555	76425	36 Tau	4 04 21.6	+24 06 21	0.000	−0.01	5.47	0.86	0.6	G0 III	+18	71 s	2965	m
25723	149351		4 04 22.6	−12 47 33	0.000	+0.01	5.61	1.08	0.2	gK0	+32	110 s		
25476	56942		4 04 23.4	+38 47 48	+0.002	−0.06	7.6	0.1	1.7	A3 V		76 mx		
24800	5069		4 04 26.8	+76 08 44	0.000	0.00	8.0	1.1	0.2	K0 III		340 s		
25291	24384		4 04 27.1	+59 09 20	0.000	0.00	5.06	0.50	−2.0	F0 II	−20	190 s		
25508	56946		4 04 33.9	+35 55 14	+0.002	−0.01	8.0	0.1	0.6	A0 V		280 s		
25956	216629		4 04 35.3	−45 54 07	−0.001	+0.02	7.8	1.4	3.2	G5 IV		24 s		
25362	24391		4 04 36.3	+55 03 59	−0.003	−0.04	6.60	0.4	1.9	F2 IV		85 s	2957	m
25347	24390		4 04 39.1	+56 44 31	+0.002	−0.05	8.00	0.9	0.3	G5 III		310 s		
25627	93749		4 04 39.5	+17 31 04	−0.002	+0.01	6.8	0.9	3.2	G5 IV		53 s		
25803	169095		4 04 40.9	−20 22 54	+0.002	−0.01	6.13	1.16		K0				
25604	76430	37 Tau	4 04 41.7	+22 04 55	+0.007	−0.06	4.36	1.07	0.2	K0 III	+9	60 s		m
25539	56951		4 04 43.1	+32 34 16	+0.001	−0.01	6.87	0.06	−1.7	B3 V	+30	400 s		
25966	216632		4 04 47.5	−44 40 06	+0.001	+0.01	8.01	1.1		K2				
25883	194739		4 04 52.5	−30 13 17	+0.001	+0.03	7.8	0.4	2.6	F0 V		90 s		
25290	13004		4 04 53.0	+62 19 20	−0.005	−0.01	7.7	0.5	3.4	F5 V		71 s		
25752	130925		4 04 53.1	−2 25 39	−0.004	−0.02	7.2	0.1	0.6	A0 V		210 s		
25538	56953		4 04 57.8	+37 05 14	+0.003	−0.06	8.0	0.1	0.6	A0 V		68 mx		m
25707	93750		4 05 11.1	+13 17 47	0.000	−0.02	7.6	0.1	1.4	A2 V		180 s		
25626	76435		4 05 12.2	+27 36 35	+0.001	−0.03	7.91	0.09		A3				
25790	130932		4 05 14.2	−1 00 42	0.000	−0.03	7.6	0.9	3.2	G5 IV		77 s		

HD	SAO	Star Name	α 2000	δ 2000	μ(α)	μ(δ)	V	B−V	M$_v$	Spec	RV	d(pc)	ADS	Notes
			h m s	° ′ ″	s	″								
25669	76436		4 05 15.2	+23 47 51	−0.001	−0.01	8.0	0.5	4.0	F8 V		60 s		
25789	130934		4 05 15.9	−0 34 10	0.000	−0.07	7.1	0.5	3.4	F5 V		55 s		
25680	76438	39 Tau	4 05 20.2	+22 00 32	+0.012	−0.13	5.90	0.62	5.2	dG5	+26	15 ts		m
26097	233395		4 05 20.8	−51 22 20	+0.010	+0.09	7.8	0.9	3.2	G5 IV		70 s		
25595	39304		4 05 28.6	+40 18 06	−0.001	−0.01	7.8	1.1	0.2	K0 III		300 s		
25679	76440		4 05 28.7	+27 52 47	0.000	−0.04	8.0	0.1	1.4	A2 V		190 s		
25749	93754		4 05 31.1	+14 17 11	+0.002	−0.03	7.60	0.9	−0.9	G9 II−III	−41	510 s		
25658	56961		4 05 33.5	+31 29 51	+0.001	0.00	7.4	1.4	−0.3	K5 III		320 s		
26074	216638		4 05 34.1	−45 40 01	+0.001	−0.01	7.57	0.23		A5				
25657	56960		4 05 35.9	+34 14 21	+0.003	−0.06	6.6	0.4	3.0	F2 V	+29	52 s		
25983	194754		4 05 36.7	−31 03 03	+0.004	−0.02	6.8	0.6	4.0	F8 V		31 s		
25945	169110		4 05 37.4	−27 39 06	+0.015	+0.10	5.59	0.31	2.6	F0 V	+61	39 s		
25643	56958		4 05 39.5	+37 45 18	+0.002	−0.04	7.8	0.4	2.6	F0 V		110 s		
25800	111597		4 05 41.5	+10 02 15	−0.001	+0.02	6.8	0.9	3.2	G5 IV		52 s		
25944	169111		4 05 46.7	−20 30 44	−0.002	+0.01	6.34	0.92		G5				
25408	13009		4 05 53.6	+61 47 39	−0.002	−0.01	7.60	2.26		R8	−9			UV Cam, q
25921	149368		4 05 54.0	−10 17 45	+0.002	0.00	7.30	1.6	−0.5	M4 III	+49	360 s		
26025	194759		4 05 54.6	−35 04 12	+0.002	−0.01	8.0	1.2	0.2	K0 III		290 s		
25910	130948		4 05 56.5	−8 51 21	+0.002	+0.01	6.26	0.06	1.4	A2 V		92 s		
25427	13011		4 05 58.1	+61 37 58	−0.005	0.00	7.8	0.1	0.6	A0 V		250 s		
25616	39308		4 06 02.1	+46 55 32	+0.005	−0.05	6.6	0.1	1.4	A2 V	+44	110 s		
26492	256060		4 06 02.9	−69 52 16	+0.002	+0.04	7.7	0.5	2.1	A5 V		83 s		
25274	13006		4 06 03.2	+68 40 48	+0.002	+0.01	5.87	1.54	−0.1	K2 III	−47	94 s		
25768	76449		4 06 04.4	+26 12 43	−0.004	−0.03	7.6	0.5	4.0	F8 V		50 s		
25443	13012		4 06 07.9	+62 06 07	−0.001	0.00	6.74	0.33		B0.5 III	−2			
24978	5080		4 06 11.7	+76 24 56	+0.021	−0.07	7.7	0.4	3.0	F2 V		87 s		
25869	111604		4 06 12.9	+8 30 10	+0.001	+0.10	7.9	0.5	4.0	F8 V		60 s		
25825	93760		4 06 16.1	+15 41 53	+0.009	−0.02	7.85	0.59		G0	+36			
26004	149373		4 06 16.2	−19 30 48	+0.001	−0.02	7.60	1.1	−2.1	K0 II	−12	870 s		
25849	93761		4 06 22.8	+14 24 08	+0.004	+0.06	7.9	0.4	3.0	F2 V		94 s		
25453	13018		4 06 30.5	+63 44 43	−0.009	−0.02	7.5	1.1	0.2	K0 III		230 mx		
25642	24412	47 λ Per	4 06 35.0	+50 21 05	−0.001	−0.03	4.29	−0.01	0.2	B9 V	+6	62 s		
25823	76455	41 Tau	4 06 36.3	+27 36 00	+0.002	−0.05	5.20	−0.13		A0 p	−2			
25602	24411		4 06 36.6	+54 00 31	+0.006	−0.10	6.31	0.99	1.7	K0 III−IV	−8	82 s		
26024	149377		4 06 36.9	−18 03 04	+0.001	−0.02	6.7	1.4	−0.3	K5 III		250 s		
25868	93763		4 06 37.4	+19 09 11	+0.002	+0.08	7.7	0.4	3.0	F2 V		82 s		
25799	56968		4 06 38.9	+32 23 07	−0.001	0.00	7.02	0.09	−1.7	B3 V	+20	410 s		
25425	13015		4 06 38.9	+65 31 16	+0.005	−0.02	6.1	0.1	1.4	A2 V	−3	83 s		
25835	76460		4 06 41.4	+25 43 21	0.000	+0.01	7.96	0.18	0.6	A0 V		220 s		
25498	13022		4 06 43.1	+62 20 38	+0.002	−0.03	8.0	1.1	0.2	K1 III		350 s		
26246	216651		4 06 51.2	−49 37 44	+0.005	+0.02	7.1	0.7	2.6	F0 V		46 s		
25834	56974		4 06 52.1	+30 16 25	+0.001	−0.02	7.51	1.54	−2.2	K1 II		570 s		
25833	56973		4 06 55.8	+33 26 47	+0.001	0.00	6.69	0.04	−1.1	B5 V p	+16	330 s	2990	AG Per, m,v
26087	169135		4 06 58.0	−21 59 39	0.000	−0.01	7.10	0.1	1.7	A3 V		120 s	3000	m
26054	149382		4 06 58.8	−15 59 23	+0.001	−0.03	8.0	0.4	3.0	F2 V		98 s		
25867	76461	42 ψ Tau	4 07 00.4	+29 00 05	−0.007	+0.01	5.23	0.34	2.6	F0 V	+9	33 s		
26040	149379		4 07 00.6	−10 00 01	−0.007	−0.14	7.1	0.7	4.4	G0 V		34 s		
26263	216654		4 07 06.1	−46 17 53	+0.001	+0.05	8.0	0.4	3.0	F2 V		42 s		
25225	5088		4 07 10.7	+74 00 01	+0.006	−0.07	6.7	1.1	0.2	K0 III		190 s		
25978	93771		4 07 11.3	+12 16 05	+0.004	−0.02	7.4	0.0	0.4	B9.5 V	+22	73 mx		
25641	24416		4 07 12.0	+56 44 15	+0.014	−0.20	7.60	1.1	3.2	K0 IV		63 mx		
25847	56977		4 07 13.2	+37 17 10	+0.004	−0.04	7.7	1.1	0.2	K0 III		200 mx		
26068	130961		4 07 15.5	−9 45 27	−0.002	+0.01	7.0	1.1	0.2	K0 III		230 s		
25977	93772		4 07 18.7	+13 32 18	−0.002	0.00	7.9	0.1	1.7	A3 V		170 s		
26203	194776		4 07 19.4	−36 08 36	−0.001	0.00	7.8	0.0	1.7	A3 V		160 s		
26491	248945		4 07 21.6	−64 13 21	+0.032	+0.33	6.38	0.64	4.9	G3 V		20 s		
25866	56979		4 07 23.0	+38 01 25	−0.001	−0.01	8.00	1.1	0.2	K0 III		320 s	2992	m
25173	5087		4 07 24.2	+75 10 32	+0.046	−0.29	7.16	0.54	4.0	F8 V	+36	36 mx		
25213	5089		4 07 24.9	+74 38 13	−0.001	−0.03	6.8	1.1	0.2	K0 III		200 s		
26262	216655		4 07 25.1	−42 55 01	−0.001	+0.01	6.59	0.93		G5				
26066	130963		4 07 29.3	−6 00 26	+0.001	+0.02	6.9	0.1	1.4	A2 V		130 s		
25893	56982		4 07 34.2	+38 04 29	+0.015	−0.22	7.13	0.86	6.3	dK2	+27	17 mn	2995	m
26242	194782		4 07 37.3	−37 03 47	−0.001	−0.04	6.6	0.6	3.2	G5 IV		48 s		
26015	93775		4 07 42.0	+15 09 46	+0.009	−0.02	6.01	0.40	3.0	dF2	+36	38 s	2999	m
25787	24424		4 07 45.1	+51 27 10	0.000	0.00	7.67	0.03	−1.6	B3.5 V	+4	590 s		
25638	13031		4 07 51.0	+62 19 48	−0.003	−0.01	7.0	0.1	−5.3	B0 II−III	−9	1600 s	2984	SZ Cam, m,v
26039	93776		4 07 54.6	+16 31 50	0.000	−0.01	7.5	0.0		B9 m	+16			
26382	233406		4 07 57.9	−50 10 33	−0.001	+0.01	7.5	1.6	−0.3	K5 III		320 s		
26038	93777		4 07 59.5	+17 20 23	+0.001	−0.01	5.89	1.50	0.2	K0 III	−31	69 s	3006	m
25907	39327		4 08 01.5	+43 10 48	+0.003	−0.07	6.70	1.1	0.2	K0 III	+46	190 s		m
26616	248950		4 08 10.7	−65 34 38	+0.004	+0.01	7.8	1.1	0.2	K0 III		330 s		
26392	216664		4 08 11.4	−48 20 45	−0.001	−0.01	7.8	0.7	3.2	G5 IV		84 s		
25932	39331		4 08 13.0	+43 11 28	0.000	−0.02	6.6	0.0	0.0	B8.5 V		200 s		
25918	39329		4 08 14.1	+44 39 44	−0.029	−0.16	7.7	0.8	3.2	G5 IV		33 mx		
25975	57000	49 Per	4 08 15.2	+37 43 40	−0.009	−0.19	6.09	0.95	0.0	K1 III	−40	64 mx		

HD	SAO	Star Name	α 2000	δ 2000	μ(α)	μ(δ)	V	B−V	M$_v$	Spec	RV	d(pc)	ADS	Notes
26301	194790		4 08 17.3	−32 51 26	+0.002	+0.02	7.43	0.24	1.7	A3 V		110 s		m
25999	57004		4 08 18.1	+32 27 36	0.000	0.00	7.4	0.1	1.4	A2 V		160 s		
25949	39334		4 08 19.4	+41 29 25	+0.004	−0.02	7.60	0.1	0.6	A0 V		100 mx	3001	m
26063	93778		4 08 20.5	+19 44 15	−0.002	−0.01	7.90	0.1	1.4	A2 V		180 s	3010	m
25734	13036		4 08 24.2	+60 52 36	0.000	0.00	7.5	0.0	0.4	B9.5 V		240 s		
25889	24429		4 08 27.9	+50 56 33	+0.002	−0.01	7.7	0.0	0.0	B8.5 V		320 s		
26413	216667		4 08 33.9	−45 51 54	+0.007	+0.02	6.59	0.38	2.6	dF0		58 s		m
25998	57006	50 Per	4 08 36.5	+38 02 23	+0.014	−0.20	5.51	0.46	3.8	dF7	+25	23 ts		
26032	57009		4 08 36.5	+33 43 44	−0.003	0.00	7.8	0.8	3.2	G5 IV		82 s		
26340	194795		4 08 37.5	−34 07 08	−0.001	+0.02	8.0	1.4	0.2	K0 III		210 s		
25878	24431		4 08 38.6	+53 21 39	−0.001	0.00	7.0	1.1	0.2	K0 III	+16	220 s		XX Cam, v
26081	76472		4 08 38.9	+25 52 40	−0.001	0.00	7.4	0.9	−2.1	G8 II	−13	570 s		
25940	39336	48 Per	4 08 39.6	+47 42 45	+0.002	−0.03	4.04	−0.03	−1.7	B3 V p	+3	140 s		
25997	57007		4 08 39.9	+39 42 44	+0.002	−0.01	8.0	1.1	0.2	K0 III		320 s		
25973	39338		4 08 41.2	+43 41 03	−0.002	+0.04	7.9	1.1	0.2	K0 III		310 s		
26163	111624		4 08 42.0	+10 05 58	−0.001	−0.04	7.0	1.6	−0.5	M2 III		320 s		
26429	216668		4 08 44.0	−47 23 39	−0.002	+0.01	7.5	1.3	−0.1	K2 III		160 s		
26237	130981		4 08 46.5	−8 40 12	0.000	+0.01	7.6	0.0	0.4	B9.5 V		280 s		
25929	24433		4 08 47.2	+50 12 02	−0.002	0.00	7.0	0.1	0.6	A0 V		180 s		
26141	93781		4 08 49.4	+17 17 31	+0.001	−0.03	7.6	0.1	0.6	A0 V		220 s		
26116	76474		4 08 51.6	+23 52 16	+0.003	−0.02	6.95	1.20	−0.1	K2 III		260 s		
26128	76475		4 08 53.3	+23 05 51	0.000	0.00	6.8	0.1	1.7	A3 V		100 s	3019	m
25987	39339		4 08 54.6	+46 13 40	+0.003	−0.03	7.70	0.1	0.6	A0 V		250 s	3007	m
26031	39341		4 08 56.5	+41 30 14	−0.003	0.00	6.9	0.1	1.7	A3 V		110 s		
26349	169171		4 08 57.1	−28 54 34	−0.002	−0.04	7.8	0.3	3.0	F2 V		89 s		
26339	169170		4 09 00.4	−25 01 04	+0.003	+0.01	7.6	2.3	−0.1	K2 III		73 s		
26299	149406		4 09 01.1	−17 36 49	−0.004	−0.05	7.8	0.4	3.0	F2 V		89 s		
26171	93784		4 09 01.5	+13 23 54	+0.001	−0.01	5.95	0.05	0.4	B9.5 V	−25	120 s		
25528	5105		4 09 03.1	+71 20 39	−0.001	+0.02	7.5	0.0	0.4	B9.5 V		250 s		
26297	149407		4 09 03.3	−15 53 29	+0.002	−0.02	7.7	0.6	4.5	G1 V	+15	45 s		
26257	111629		4 09 08.9	+0 10 44	+0.003	+0.02	7.9	0.5	4.0	F8 V		60 s		
26162	93785	43 Tau	4 09 09.9	+19 36 33	+0.008	−0.03	5.50	1.07	3.2	G5 IV	+24	19 s		
26051	39342		4 09 10.2	+40 09 42	−0.002	−0.11	7.60	0.5	4.0	F8 V		51 s	3015	m
25098	5095		4 09 11.1	+78 57 22	−0.003	0.00	8.0	0.0	0.4	B9.5 V		320 s		
26080	57018		4 09 11.8	+36 25 39	+0.001	0.00	7.6	1.6	−0.5	M2 III		370 s		
26326	149412		4 09 17.8	−16 23 09	0.000	+0.01	5.37	−0.15	−1.7	B3 V	+14	250 s		
25948	24440		4 09 22.3	+54 49 44	+0.010	−0.09	6.3	0.5	3.4	F5 V	−5	37 s		
25473	5104		4 09 22.5	+73 34 14	−0.010	+0.04	6.9	0.5	3.4	F5 V	−29	50 s		
25914	24438		4 09 23.1	+57 05 28	−0.001	−0.01	7.99	0.60	−7.1	B6 Ia	−26	4400 s		
26103	57019		4 09 23.3	+36 02 05	+0.003	−0.01	7.8	0.5	4.0	F8 V		56 s		
25906	24437		4 09 24.3	+58 33 36	−0.004	−0.02	7.1	0.4	2.6	F0 V		76 s		
26256	111631		4 09 27.3	+6 43 30	+0.003	−0.06	6.7	0.0	0.4	B9.5 V	+15	46 mx		
25877	24436		4 09 27.6	+59 54 29	0.000	0.00	6.28	1.14	−2.1	G8 II	−14	430 s		
26443	194801		4 09 27.8	−36 39 18	−0.001	−0.05	7.7	1.4	3.2	G5 IV		33 s		
26522	216673		4 09 31.8	−47 25 49	0.000	+0.01	8.0	1.4	0.2	K0 III		210 s		
26388	169179		4 09 31.9	−26 01 18	+0.004	−0.04	7.6	1.7	−0.1	K2 III		170 s		
26442	194803		4 09 35.2	−32 58 31	+0.005	−0.05	7.4	1.4	0.2	K0 III		140 mx		
−−	−−		4 09 35.3	−81 51 24			8.00B							U Men, v
26161	57026		4 09 38.7	+31 39 07	+0.005	+0.06	6.90	0.5	3.4	F5 V		50 s	3029	m
26337	130994		4 09 40.7	−7 53 32	+0.001	+0.13	7.1	0.6	4.4	G0 V		35 s		
26292	111634		4 09 42.9	+3 19 21	−0.004	−0.06	6.5	0.4	3.0	F2 V		51 s		
26140	57025		4 09 43.2	+35 58 38	+0.004	−0.07	7.6	1.1	0.2	K0 III		150 mx		
−−	57027		4 09 43.4	+33 59 50	−0.001	−0.01	8.0	2.2						
26212	76480		4 09 43.5	+24 04 24	−0.001	−0.01	7.3	0.1	1.7	A3 V		120 s		
26374	149419		4 09 44.9	−12 33 38	−0.001	+0.01	7.5	0.1	0.6	A0 V		250 s		
26336	130995		4 09 49.1	−3 34 26	0.000	0.00	6.8	0.1	0.6	A0 V		180 s		
26412	149421		4 09 52.3	−19 17 15	−0.002	−0.01	7.0	1.1	0.2	K0 III		210 s		
26411	149422		4 09 58.6	−19 00 13	−0.002	−0.04	7.2	0.9	0.2	K0 III		250 s		
26428	169185		4 09 59.2	−25 02 32	+0.006	+0.01	7.50	1.1	0.2	K0 III		200 mx		m
25765	13043		4 10 01.9	+68 22 25	+0.011	−0.03	7.4	0.4	2.6	F0 V		88 s		
22701	623		4 10 02.0	+86 37 35	+0.167	−0.08	5.8	0.5	3.4	F5 V	−4	31 s		
26312	93795		4 10 02.3	+10 50 47	+0.001	0.00	8.0	0.9	3.2	G5 IV		93 s		
25008	650		4 10 02.8	+80 41 55	−0.005	0.00	5.2	0.1	1.4	A2 V	+4	58 s	2963	m
26487	194806		4 10 03.4	−34 29 51	+0.002	+0.03	7.2	1.7	−0.5	M2 III		290 s		
26754	248957		4 10 07.4	−61 35 56	+0.001	−0.07	6.9	0.6	4.4	G0 V		30 s		
26560	216679		4 10 12.6	−42 37 21	−0.001	+0.01	7.7	1.4	3.2	G5 IV		46 s		
26277	76483		4 10 16.7	+23 58 51	−0.001	−0.01	7.3	0.9	3.2	G5 IV		64 s		
26200	57033		4 10 21.2	+39 13 32	+0.001	−0.03	6.99	0.39	2.6	F0 V		70 s		
26409	131001	37 Eri	4 10 22.4	−6 55 26	−0.001	−0.01	5.44	0.94	0.3	gG6	−10	100 s		
26199	39349		4 10 25.0	+42 10 25	0.000	−0.02	7.75	1.96	−0.5	M2 III		270 s		
26450	149429		4 10 28.6	−15 01 45	−0.004	+0.01	7.5	0.4	2.6	F0 V		97 s		
26586	216680		4 10 29.5	−40 32 09	−0.003	+0.01	7.8	0.4	1.7	A3 V		100 s		
26235	57040		4 10 36.7	+38 29 12	−0.001	−0.02	7.9	0.1	1.4	A2 V		180 s		
26519	169196		4 10 41.4	−25 25 34	−0.002	−0.15	7.8	0.9	3.4	F5 V		39 s		
26345	93801		4 10 42.2	+18 25 22	+0.008	−0.04	6.62	0.42	3.7	F6 V	+33	38 s		

HD	SAO	Star Name	α 2000	δ 2000	μ(α)	μ(δ)	V	B-V	M_v	Spec	RV	d(pc)	ADS	Notes
26465	149430		4^{h}10^{m}44^s.1	−11°19′52″	−0^s.001	+0″.03	7.9	1.1	−0.1	K2 III		390 s		
26441	131003		4 10 44.4	−4 52 28	+0.004	−0.13	7.36	0.65		G0			3041	m
26518	169198		4 10 44.5	−25 05 45	−0.004	−0.05	7.3	0.7	4.0	F8 V		37 s		
26575	194810		4 10 45.8	−35 16 26	−0.002	−0.03	6.44	1.07		G5				
26464	131005		4 10 47.6	−8 49 11	+0.002	+0.01	5.70	1.06	0.2	gG9	+30	110 s		
26322	76485	44 Tau	4 10 49.8	+26 28 51	−0.002	−0.04	5.41	0.34	2.6	F0 V	+19	35 s		
26612	216682	δ Hor	4 10 50.5	−41 59 37	+0.017	+0.07	4.93	0.33	2.6	dF0	+37	28 s		
26380	93803		4 10 55.1	+15 56 49	0.000	0.00	7.2	0.1	0.6	A0 V	+11	210 s		
26504	149436		4 10 57.5	−14 21 48	+0.001	−0.03	7.9	1.4	−0.3	K5 III		430 s		
26311	57047		4 10 59.0	+33 35 12	0.000	−0.01	5.72	1.44	−5.9	cK5	+20	2100 s		
26665	216683		4 11 03.1	−46 56 31	−0.002	−0.13	7.0	0.8	3.4	F5 V		32 s		
26398	93805		4 11 03.3	+16 38 49	0.000	0.00	7.0	0.0	0.0	B8.5 V	+32	220 s		
26473	131008		4 11 04.8	−5 16 14	−0.003	−0.05	7.9	1.1	0.2	K0 III		340 s		
26253	39359		4 11 07.8	+45 24 08	−0.002	−0.03	7.8	0.8	3.2	G5 IV		81 s		
26448	111646		4 11 09.3	+2 19 06	−0.001	−0.01	7.9	1.6	−0.5	M2 III		490 s		
26399	93806		4 11 10.4	+15 37 34	−0.001	−0.02	7.9	0.1	0.6	A0 V		280 s		
26285	39362		4 11 13.3	+41 44 50	+0.004	−0.08	7.0	1.1	0.2	K0 III		140 mx		
26462	111648	45 Tau	4 11 20.2	+5 31 23	+0.010	+0.01	5.72	0.35	3.3	F4 V	+37	30 s		m
26112	24456		4 11 21.0	+60 04 55	−0.002	−0.02	8.0	0.1	0.6	A0 V		280 s		
26549	149440		4 11 25.5	−15 58 07	−0.001	+0.01	7.1	1.1	−0.1	K2 III		270 s		
26321	39366		4 11 26.7	+40 54 45	+0.001	−0.05	7.2	0.1	1.4	A2 V		140 s		
26516	131012		4 11 31.6	−0 25 22	0.000	−0.01	6.7	1.1	0.2	K0 III		200 s		
26590	169205		4 11 34.2	−20 20 27	+0.003	+0.04	7.8	0.5	3.0	F2 V		72 s		
26591	169206		4 11 36.0	−20 21 23	+0.002	+0.04	5.79	0.18		A1 pe				
26385	57056		4 11 41.2	+31 33 02	−0.001	−0.02	7.90	0.1	0.6	A0 V		270 s		m
26753	216685		4 11 42.7	−47 05 08	−0.002	0.00	7.1	0.7	3.2	G5 IV		60 s		
26890	233432		4 11 45.1	−59 10 12	+0.008	+0.01	8.00	0.5	3.4	F5 V		83 s		m
26574	131019	38 o¹ Eri	4 11 51.8	−6 50 15	0.000	+0.08	4.04	0.33	−0.7	F2 II−III	+11	85 mx		Beid
26111	13058		4 11 54.0	+63 02 29	+0.011	−0.03	7.4	0.9	3.2	G5 IV		68 s		
26584	131020		4 11 56.1	−8 50 14	+0.004	+0.01	6.80	0.8	3.2	G5 IV		53 s		m
26833	233431		4 11 58.9	−53 24 41	−0.001	+0.01	6.8	1.2	0.2	K0 III		170 s		
26625	149446		4 12 00.1	−17 16 28	0.000	−0.04	6.5	1.1	0.2	K0 III		180 s		
26407	57058		4 12 03.7	+35 21 34	0.000	0.00	7.7	0.1	0.6	A0 V		240 s		
26624	149448		4 12 05.2	−13 13 24	−0.002	−0.01	8.0	0.1	0.6	A0 V		300 s		
26573	111659		4 12 09.2	+0 44 06	−0.001	0.00	6.8	0.8	3.2	G5 IV		51 s	3054	m
26697	194822		4 12 10.3	−30 26 11	+0.001	+0.02	7.9	1.2	0.2	K0 III		260 s		
25406	5109		4 12 14.2	+79 01 48	+0.020	+0.01	7.1	1.1	0.2	K0 III		230 mx		
26608	131025		4 12 20.3	−4 24 38	−0.001	−0.01	6.9	1.4	−0.3	K5 III		270 s		
26729	194827		4 12 23.8	−31 34 54	+0.014	−0.13	7.5	0.7	3.2	G5 IV		42 mx		
26514	76499		4 12 24.9	+23 34 29	−0.001	−0.02	7.19	1.06	0.3	G6 III		200 s		m
26758	194831		4 12 30.5	−36 09 09	+0.002	+0.02	7.10	0.5	3.4	F5 V		55 s		m
26546	93810		4 12 31.3	+17 16 39	+0.004	−0.02	6.09	1.08	0.3	gG5	+28	100 s		
26820	216694		4 12 31.6	−44 22 06	+0.003	+0.01	6.71	1.48		K0				
26633	131028		4 12 33.0	−2 30 42	+0.002	+0.12	7.6	1.1	0.2	K0 III		140 mx		
26420	39384		4 12 36.5	+42 07 07	+0.001	0.00	7.84	0.32		B3 e	+3			
26855	216697		4 12 38.9	−45 22 04	+0.002	+0.01	8.0	1.1	−0.1	K2 III		310 s		
26446	39385		4 12 44.5	+40 39 13	−0.002	−0.01	8.0	1.1	−0.1	K2 III		360 s		
26545	76502		4 12 46.6	+23 31 16	−0.001	0.00	8.0	1.1	−0.1	K2 III		340 s		
26742	169229		4 12 47.9	−23 49 17	−0.001	−0.01	6.7	1.7	−0.1	K2 III		120 s		
26342	24468		4 12 48.8	+54 24 46	−0.002	0.00	7.8	0.4	2.6	F0 V		110 s		
26572	93814		4 12 49.0	+19 32 20	−0.001	−0.01	7.9	0.1	1.4	A2 V		180 s		
26771	194834		4 12 50.5	−30 18 51	+0.001	+0.01	8.0	1.8	−0.1	K2 III		180 s		
26571	76505		4 12 51.2	+22 24 49	0.000	−0.01	6.20	0.19	−2.3	B8 II−III	+8	380 s		
26101	13061		4 12 51.6	+68 30 06	−0.008	+0.03	6.32	1.18	0.2	K0 III	−24	130 s		
26663	131033		4 12 56.0	−6 23 20	−0.001	−0.13	7.1	1.1	0.2	K0 III		120 mx		
26640	111666		4 12 56.1	+1 55 25	+0.001	0.00	7.6	1.1	0.2	K0 III		300 s		
26526	57072		4 12 58.3	+32 31 57	−0.001	−0.02	6.8	1.1	0.2	K0 III		200 s		
26770	169233		4 13 00.6	−28 32 24	+0.007	+0.06	7.45	0.52	4.4	G0 V		41 s	3069	m
--	57073		4 13 00.8	+33 44 32	+0.002	−0.02	7.8	1.4						
26769	169232		4 13 02.2	−26 26 06	−0.002	+0.10	7.8	1.2	4.0	F8 V		23 s		
26799	194836		4 13 02.4	−32 47 40	+0.001	+0.04	7.1	0.9	1.7	A3 V		40 s		
26662	131036		4 13 02.4	−4 18 43	0.000	+0.01	8.0	0.9	3.2	G5 IV		90 s		
26757	169231		4 13 03.2	−23 07 42	−0.002	−0.02	6.80	0.4	2.6	F0 V		69 s	3067	m
27415	256074		4 13 06.7	−72 49 14	+0.010	+0.03	7.0	0.5	4.0	F8 V		39 s		
26832	194838		4 13 07.3	−36 32 41	+0.003	+0.02	7.6	2.5	−0.5	M2 III		130 s		
26692	131037		4 13 12.5	−3 53 05	+0.001	+0.01	7.7	1.6	−0.5	M2 III		430 s		
27346	256073		4 13 15.3	−70 25 13	+0.011	+0.06	7.1	0.9	2.6	F0 V		34 s		
26544	57075		4 13 23.4	+35 28 37	+0.001	0.00	7.6	0.1	0.6	A0 V		230 s		
26750	149466		4 13 29.6	−10 23 14	+0.001	−0.03	7.6	1.6	−0.5	M4 III		420 s		BM Eri, v
26677	111671		4 13 31.3	+8 53 25	+0.001	−0.02	6.51	0.16		A3	+8		3063	m
26690	111672	46 Tau	4 13 33.0	+7 42 58	0.000	+0.01	5.29	0.36	3.1	F3 V	+4	27 s	3064	m
26877	194846		4 13 33.6	−33 30 20	−0.001	−0.03	7.8	1.8	4.0	F8 V		58 s		
26676	93821		4 13 34.7	+10 12 45	+0.003	−0.02	6.23	0.05		B8				
26927	216705		4 13 35.6	−40 21 29	0.000	+0.02	6.37	1.46		G5				
26817	169239		4 13 36.9	−25 31 44	+0.003	+0.03	8.00	1.1	0.2	K0 III		360 s	3075	m

HD	SAO	Star Name	α 2000	δ 2000	μ(α)	μ(δ)	V	B-V	M$_v$	Spec	RV	d(pc)	ADS	Notes
			h m s	° ′ ″	s	″								
26739	131041		4 13 38.1	-1 08 59	0.000	0.00	6.44	-0.13	-1.1	B5 V	+15	320 s		
27100	233441		4 13 39.7	-58 01 19	+0.002	+0.04	7.0	0.1	0.6	A0 V		160 s		
26482	39392		4 13 41.0	+49 05 39	+0.001	0.00	7.3	0.8	3.2	G5 IV		64 s		
26570	57079		4 13 43.7	+39 40 49	0.000	-0.01	7.6	1.4	-0.3	K5 III		340 s		
26076	5125		4 13 44.9	+72 07 35	+0.004	-0.02	6.03	1.01	0.0	K1 III	-4	160 s		
29138	258373		4 13 48.0	-84 29 07	+0.006	+0.03	7.21	-0.06		B1 k				
27043	233440		4 13 48.2	-51 35 36	+0.003	+0.03	7.4	1.4	0.2	K0 III		160 s		
26703	93823		4 13 49.8	+12 45 12	+0.001	-0.03	6.25	1.16		K0	+48			
26495	24489		4 13 54.5	+50 28 42	0.000	0.00	7.79	1.28	0.2	K0 III		220 s		
26722	111674	47 Tau	4 13 56.3	+9 15 49	-0.001	-0.04	4.84	0.80	0.3	gG5	-7	81 s	3072	m
26934	194852		4 13 57.1	-37 01 50	-0.001	-0.02	6.7	1.2	0.2	K0 III		160 s		
26921	194848		4 13 58.6	-32 02 32	-0.002	-0.02	7.1	0.8	3.2	G5 IV		62 s		
26605	57081		4 13 59.5	+37 58 01	-0.005	0.00	6.5	0.9	0.2	G9 III	+29	170 s		
26967	216710	α Hor	4 14 00.0	-42 17 40	+0.004	-0.21	3.86	1.10	0.0	K1 III	+22	59 s		
26721	93824		4 14 00.9	+11 01 39	+0.001	0.00	7.9	0.1	1.4	A2 V		200 s		
26920	194849		4 14 03.4	-30 06 46	+0.003	+0.01	7.1	1.5	0.2	K0 III		120 s		
26829	149472		4 14 03.4	-15 59 35	-0.001	+0.01	7.6	1.6	-0.5	M2 III		410 s		
26604	57083		4 14 07.0	+38 28 12	+0.005	0.00	7.5	0.4	2.6	F0 V		91 s		
27019	216711		4 14 09.6	-46 07 57	+0.015	-0.10	6.78	0.58		G0				m
27142	233444		4 14 11.2	-56 10 35	+0.002	+0.07	7.3	0.5	3.4	F5 V		56 s		
26903	169249		4 14 11.7	-24 12 18	0.000	+0.02	7.6	1.1	2.1	A5 V		36 s		
26587	39398		4 14 12.8	+44 44 57	+0.001	-0.05	7.7	0.2	2.1	A5 V		130 s		
26711	93825		4 14 14.9	+18 53 39	+0.001	+0.01	7.9	0.4	2.6	F0 V		110 s		
26749	93827		4 14 18.2	+12 20 54	+0.023	-0.05	6.9	0.7	4.4	G0 V		32 s		
27305	--		4 14 19.3	-64 58 03			7.8	0.5	3.4	F5 V		75 s		
26748	93828		4 14 19.6	+14 32 54	+0.001	-0.03	7.9	1.1	0.2	K0 III		350 s		
26979	194858		4 14 19.7	-38 15 35	+0.002	+0.07	6.8	0.3	3.2	G5 IV		52 s		
26232	13067		4 14 20.5	+69 30 24	+0.002	0.00	7.2	1.1	0.2	K0 III		240 s		
27212	248968		4 14 20.7	-60 12 31	+0.003	+0.02	7.8	2.0	2.6	F0 V		11 s		
26512	24493		4 14 22.2	+54 31 28	-0.003	+0.01	7.1	0.1	0.6	A0 V		190 s		
26846	149478	39 Eri	4 14 23.6	-10 15 23	-0.001	-0.16	4.87	1.17	-0.2	K3 III	+7	100 s	3079	m
26796	111678		4 14 24.9	+1 39 33	-0.002	-0.03	7.8	0.8	3.2	G5 IV		84 s		
27256	248969	α Ret	4 14 25.5	-62 28 26	+0.007	+0.05	3.35	0.91	-2.1	G6 II	+36	120 s		m
26719	76512		4 14 26.3	+23 41 47	0.000	0.00	7.0	1.1	0.2	K0 III		200 s		
27018	216716		4 14 30.3	-40 53 40	+0.002	-0.06	7.5	0.7	3.2	G5 IV		73 s		
26737	76515		4 14 30.4	+22 27 07	+0.007	-0.03	7.06	0.42	3.4	F5 V	+38	54 s		
27728	256076		4 14 31.1	-75 48 27	+0.007	+0.04	7.23	1.64	-0.3	K4 III		220 s		
26710	76514		4 14 32.2	+26 15 20	-0.005	-0.11	7.16	0.64	4.7	G2 V	-9	30 s		
26736	76516		4 14 32.3	+23 34 31	+0.009	-0.03	8.0	0.9	3.2	G5 IV	+42	88 s		
26784	93831		4 14 34.3	+10 42 06	+0.008	-0.02	7.12	0.51		F5	+37			
26793	111680		4 14 36.2	+10 00 40	0.000	-0.02	5.22	-0.10		B8	+7			
26845	131050		4 14 36.9	-2 08 23	0.000	+0.01	7.9	1.4	-0.3	K5 III		430 s		
26220	5131		4 14 43.6	+71 12 55	+0.002	-0.02	8.0	0.1	0.6	A0 V		280 s		
27304	248973		4 14 48.7	-62 11 32	+0.001	+0.09	5.45	1.10	0.0	gK1	+36	120 s		m
26702	57094		4 14 51.9	+37 32 34	-0.001	+0.01	6.31	0.97		G5				
26673	39409	52 Per	4 14 53.3	+40 29 01	+0.001	-0.02	4.71	1.01	-4.5	G5 Ib var	-2	560 s		
26630	39404	51 μ Per	4 14 53.8	+48 24 33	+0.001	-0.02	4.14	0.95	-4.5	G0 Ib	+8	460 s	3071	m
26708	57096		4 14 58.3	+36 30 08	+0.003	-0.01	7.6	0.1	1.4	A2 V		160 s		
26940	149487		4 15 00.3	-16 10 54	-0.002	-0.06	7.4	0.9	3.2	G5 IV		68 s		
26843	93832		4 15 00.9	+10 44 54	-0.001	0.00	8.0	0.1	1.4	A2 V		210 s		
26620	24498		4 15 01.7	+50 41 13	+0.001	-0.01	7.45	1.59	0.2	K0 III		130 s		
26553	24495		4 15 01.8	+57 27 36	0.000	-0.01	6.1	0.1	1.4	A2 V	-23	84 s		
27016	194866		4 15 06.3	-30 04 18	+0.004	+0.01	7.70	0.1	1.4	A2 V		84 mx		m
26766	76520		4 15 10.1	+29 54 07	+0.009	-0.17	7.10	1.07		K1 IV		26 mn		
27078	194870		4 15 11.2	-37 11 48	0.000	+0.02	6.7	-0.3	0.6	A0 V		160 s		
27097	194872		4 15 11.3	-38 17 41	-0.004	+0.08	6.9	1.1	4.4	G0 V		14 s		
26596	24497		4 15 13.7	+54 57 50	-0.001	-0.04	8.00	0.5	2.6	F9 IV	+2	120 s		
26686	39413		4 15 13.8	+44 46 03	-0.002	-0.03	7.3	0.1	0.6	A0 V		200 s		
26965	131063	40 o² Eri	4 15 16.2	-7 39 10	-0.151	-3.41	4.43	0.82	6.0	K1 V	-42	4.9 t	3093	Keid, m
27096	194873		4 15 24.8	-36 12 14	0.000	+0.02	7.5	1.4	3.2	G5 IV		32 s		
27242	233454		4 15 25.4	-53 36 08	+0.010	+0.04	7.7	1.3	3.2	G5 IV		38 s		
26913	111695		4 15 25.7	+6 11 59	-0.007	-0.10	6.93	0.70	4.9	G3 V	-8	23 s		
26994	149491		4 15 26.9	-16 26 38	+0.001	+0.05	6.73	-0.07		B8				
26923	111698		4 15 29.1	+6 11 12	-0.004	-0.11	6.31	0.59	2.8	G0 IV	-8	50 s	3085	m
26912	111696	49 μ Tau	4 15 32.0	+8 53 32	+0.001	-0.02	4.29	-0.06	-1.7	B3 V	+18	140 s		
26947	131066		4 15 37.8	+0 00 13	0.000	+0.01	7.9	0.2	2.1	A5 V		150 s		
26874	76525		4 15 42.3	+20 49 09	+0.007	-0.05	7.84	0.70	5.0	G4 V	+27	34 s		
26993	131071		4 15 43.1	-8 38 12	+0.002	0.00	7.9	1.1	0.2	K0 III		340 s		
26815	57106		4 15 45.6	+31 23 50	+0.002	-0.02	7.9	0.1	0.6	A0 V		270 s		
26745	39423		4 15 45.7	+44 41 03	+0.002	-0.01	8.0	0.0	0.4	B9.5 V		310 s		
26911	93836	48 Tau	4 15 46.2	+15 24 02	+0.008	-0.03	6.32	0.40	3.4	F5 V	+37	38 s		m
27066	169273		4 15 46.3	-23 14 18	+0.002	+0.02	6.6	-0.1	1.4	A2 V		110 s		
27040	149494		4 15 46.8	-18 38 27	-0.001	+0.01	7.1	1.1	0.2	K0 III		240 s		
27076	169274		4 15 51.5	-22 09 05	-0.002	-0.03	6.8	0.4	1.4	A2 V		71 s		
26842	57110		4 15 55.6	+31 41 36	-0.001	-0.04	7.33	0.56	4.4	dG0	-26	38 mn	3082	m

HD	SAO	Star Name	α 2000	δ 2000	μ(α)	μ(δ)	V	B-V	M_V	Spec	RV	d(pc)	ADS	Notes
27274	233456		4 15 56.6	-53 18 33	+0.085	+0.42	7.65	1.11	8.0	K5 V		12 ts		
29245	258377		4 15 59.9	-84 10 53	+0.051	+0.04	7.1	1.2	0.2	K0 III		170 s		
26991	111705		4 16 01.4	+0 27 13	+0.002	-0.02	7.40	0.1	0.6	A0 V		230 s	3095	m
27290	233457	γ Dor	4 16 01.6	-51 29 12	+0.011	+0.19	4.25	0.30	2.6	F0 V	+27	21 ts		
26873	--		4 16 02.1	+30 02 07	+0.001	-0.06	7.70	0.7	4.4	G0 V		45 s	3089	m
27231	216727		4 16 08.9	-42 37 33	-0.002	+0.06	7.8	0.4	2.6	F0 V		77 s		
26860	57112		4 16 10.5	+34 52 08	+0.002	-0.02	7.70	0.1	1.7	A3 V		150 s	3088	m
26990	111707		4 16 16.2	+7 09 33	-0.008	-0.06	7.6	0.7	4.4	G0 V		44 s		
27115	149500		4 16 18.8	-17 52 30	+0.002	+0.01	7.7	1.1	0.2	K0 III		310 s		
27463	248977		4 16 21.1	-60 56 55	+0.006	+0.02	6.37	0.07		A0				m
27442	233463	ε Ret	4 16 28.9	-59 18 07	-0.007	-0.16	4.44	1.08	-0.3	gK5	+29	21 ts		
27241	194886		4 16 29.1	-38 53 03	-0.001	-0.04	7.1	0.1	0.6	A0 V		160 s		
27063	131080		4 16 30.1	-0 34 24	-0.004	-0.19	7.8	0.7	4.4	G0 V		49 s		
27093	149503		4 16 36.0	-10 05 07	+0.002	+0.06	7.50	0.1	1.4	A2 V		110 mx	3103	m
26883	57119		4 16 39.0	+40 01 35	0.000	-0.01	7.9	1.1	0.2	K0 III		320 s		
27545	248978		4 16 39.6	-64 18 57	+0.007	+0.02	7.6	0.4	2.6	F0 V		100 s		
26764	24512		4 16 43.1	+53 36 42	-0.001	0.00	5.19	0.05		A2	-3	29 mn		
26824	39430		4 16 44.7	+46 46 41	-0.003	-0.01	7.9	0.0	0.4	B9.5 V		290 s		
27621	248981		4 16 46.0	-66 46 50	+0.010	+0.02	7.9	1.1	0.2	K0 III		280 mx		
27693	248983		4 16 46.4	-69 43 37	+0.009	+0.04	7.4	1.2	3.2	G5 IV		39 s		
27029	93838		4 16 48.5	+16 12 51	-0.003	-0.02	6.8	0.9	3.2	G5 IV		52 s		
26670	13075		4 16 53.6	+61 51 01	+0.002	-0.01	5.6	-0.1		B8 nn	-2			
27230	169299		4 16 56.3	-28 53 41	+0.003	-0.06	7.8	0.7	1.4	A2 V		36 mx		
27028	93840		4 17 01.2	+19 40 31	0.000	-0.11	7.10	0.40	3.4	dF5	-3	55 s	3102	m
26494	5142		4 17 03.6	+70 27 07	0.000	-0.01	7.8	1.1	-0.1	K2 III		350 s		
26755	24514		4 17 08.1	+57 51 38	+0.004	-0.03	5.71	1.09	0.2	K0 III	-38	110 s		
26857	24518		4 17 12.3	+50 51 57	0.000	0.00	6.8	1.1	-0.1	K2 III		230 s		
27045	76532	50 ω Tau	4 17 15.6	+20 34 43	-0.003	-0.06	4.94	0.26		A m	+16			
26897	39435		4 17 16.7	+46 41 47	-0.001	-0.02	7.9	1.1	-0.1	K2 III		360 s		
27146	111722		4 17 17.3	+2 31 48	-0.001	0.00	7.5	0.8	3.2	G5 IV		73 s		
27179	131092		4 17 19.1	-6 28 19	-0.002	0.00	5.94	1.08	0.2	gK0	-2	120 s		
26907	39438		4 17 19.6	+46 13 04	-0.001	+0.01	7.20	0.1	1.4	A2 V		140 s		m
26718	13080		4 17 19.9	+61 47 49	-0.001	0.00	7.9	1.1	0.2	K0 III		310 s		
26792	24516		4 17 20.8	+57 10 39	-0.005	-0.04	6.5	0.0	0.0	B8.5 V		130 mx		
27109	93844		4 17 38.1	+19 33 38	+0.001	-0.03	7.8	1.1	0.2	K0 III		280 s		
27657	248985		4 17 39.7	-63 15 19	-0.004	-0.01	6.20	0.00	0.4	B9.5 V		150 s		m
27145	93847		4 17 43.1	+13 50 12	+0.006	+0.05	7.0	0.9	3.2	G5 IV		58 s		
26763	13084		4 17 43.5	+62 20 48	-0.004	+0.01	7.3	1.6	-0.5	M2 III		320 s		ZZ Cam, v
27556	233469		4 17 45.2	-57 18 30	+0.002	0.00	8.0	1.5	-0.3	K5 III		450 s		
27377	194901		4 17 50.7	-34 11 12	0.000	0.00	7.3	0.4	2.6	F0 V		72 s		
26669	13081		4 17 51.0	+67 05 21	+0.006	-0.01	6.90	0.00	0.0	B8.5 V		140 mx	3086	m
25904	670		4 17 51.7	+80 32 08	-0.005	0.00	6.6	1.1	0.2	K0 III		190 s		
27376	194902	41 Eri	4 17 53.6	-33 47 54	+0.005	-0.01	3.56	-0.12	0.0	B8.5 V	+18	40 mx		m
26839	24521		4 17 55.5	+58 47 25	-0.003	-0.02	7.40	0.1	0.6	A0 V		210 s	3098	m
28525	258372	δ Men	4 17 59.2	-80 12 51	+0.006	+0.06	5.69	0.84		K0	-20			
27471	216737		4 18 00.2	-45 39 03	+0.006	+0.07	7.54	0.64		G0				
27004	39445		4 18 01.1	+43 41 03	-0.001	0.00	7.78	0.16	0.4	B9.5 V		250 s		
27149	93849		4 18 01.7	+18 15 24	+0.008	-0.03	7.52	0.67		G5	+44			
27026	39447		4 18 08.1	+42 08 29	+0.003	-0.03	6.1	-0.1	-0.2	B8 V		170 s		
26801	13088		4 18 11.1	+61 54 46	+0.001	+0.01	7.74	0.00		B9			3097	m
27325	149524		4 18 14.4	-14 38 19	+0.001	+0.01	6.8	1.1	0.2	K0 III	+15	210 s		
27128	57138		4 18 14.5	+32 23 25	-0.001	-0.07	7.6	0.2	2.1	A5 V		100 mx		
26961	24531		4 18 14.6	+50 17 44	+0.005	-0.05	4.62	0.05		A2	+20			b Per, v
27553	233472		4 18 15.4	-52 54 07	+0.007	+0.08	7.8	0.2	3.4	F5 V		76 s		
27362	169317		4 18 15.9	-20 42 56	-0.001	+0.01	6.3	1.9		Mb				
27454	194912		4 18 18.5	-38 06 35	-0.001	-0.01	7.9	1.7	3.2	G5 IV		24 s		
27127	57140		4 18 19.7	+33 50 52	+0.002	-0.04	7.9	0.5	4.0	F8 V		58 s		
27176	76541	51 Tau	4 18 23.1	+21 34 45	+0.007	-0.03	5.65	0.28	2.4	A8 V	+35	43 s		m
27236	111729		4 18 24.5	+9 29 12	-0.003	-0.03	6.54	0.16		A2	+28			
27125	57141		4 18 29.7	+36 51 41	-0.001	-0.02	8.0	0.1	1.4	A2 V		190 s		
26580	5154		4 18 30.7	+72 24 33	+0.002	-0.02	7.8	1.1	-0.1	K2 III		350 s		
27044	39452		4 18 31.0	+43 43 39	+0.004	-0.01	7.5	0.1	1.4	A2 V		120 mx		
26972	24533		4 18 32.3	+52 01 32	+0.001	0.00	7.7	0.9	3.2	G5 IV		78 s		
27271	111735		4 18 33.8	+2 28 13	0.000	-0.07	8.0	0.9	3.2	G5 IV		93 s		
26882	13090		4 18 36.0	+60 29 37	+0.004	-0.02	8.00	0.00	0.4	B9.5 V		120 mx	3105	m
27411	169325		4 18 37.4	-22 58 13	+0.003	+0.03	6.07	0.30		A m				
26985	24534		4 18 39.8	+53 30 55	-0.001	-0.01	8.0	0.9	3.2	G5 IV		90 s		
27604	233476		4 18 40.1	-52 51 36	+0.006	+0.07	6.09	0.49	2.1	F5 IV		56 s		m
27399	149534		4 18 47.0	-17 42 38	+0.008	-0.07	7.7	1.4	-0.3	K5 III		85 mx		
27647	233480		4 18 48.2	-54 08 05	0.000	+0.05	7.2	0.7	0.6	A0 V		74 s		
28203	--		4 18 52.0	-75 44 01			8.0	0.7	4.4	G0 V		53 s		
27452	169329		4 18 54.9	-24 09 17	+0.002	-0.03	8.0	1.4	0.2	K0 III		220 s		
27424	149535		4 19 00.8	-17 51 03	0.000	+0.01	7.5	0.1	1.4	A2 V		170 s		
27144	39458		4 19 00.8	+41 32 26	-0.001	-0.03	8.0	1.1	-0.1	K2 III		360 s		
27490	194923		4 19 02.9	-33 54 18	+0.001	0.00	6.37	0.13		A2				

HD	SAO	Star Name	α 2000	δ 2000	μ(α)	μ(δ)	V	B-V	M_V	Spec	RV	d(pc)	ADS	Notes
			^{h} ^{m} ^{s}	$^{\circ}$ $^{\prime}$ $^{\prime\prime}$	^{s}	$^{\prime\prime}$								
27248	76545		4 19 04.1	+24 01 41	+0.001	-0.01	7.8	0.8	3.2	G5 IV		80 s		
27489	194921		4 19 04.5	-31 19 39	+0.001	+0.05	7.7	1.0	3.2	G5 IV		58 s		
28566	256083		4 19 05.4	-79 31 13	-0.016	+0.04	7.1	-0.3	1.4	A2 V		120 mx		
27603	216748		4 19 07.2	-47 34 05	-0.001	+0.01	6.9	1.0	0.2	K0 III		200 s		
27774	248991		4 19 08.6	-62 26 38	+0.001	+0.01	7.8	1.1	0.2	K0 III		330 s		
27296	93861		4 19 09.1	+13 43 17	+0.001	+0.01	7.8	0.4	3.0	F2 V		90 s		
27157	39461		4 19 13.0	+42 22 52	-0.002	-0.06	6.9	0.5	3.4	F5 V		49 s		
27084	39457		4 19 13.2	+50 02 56	+0.007	-0.05	5.5	0.2	2.1	A5 V	-17	31 mx		
26936	13094		4 19 13.8	+61 35 01	+0.001	-0.02	8.00	0.1	1.7	A3 V		170 s	3109	m
27175	39462		4 19 14.3	+41 00 20	+0.002	+0.02	8.0	1.4	-0.3	K5 III		390 s		
27588	216749		4 19 16.6	-44 16 06	+0.005	-0.04	5.34	1.08		K2	+24	27 mn		m
27488	169337		4 19 16.9	-27 16 22	+0.004	-0.04	8.0	2.0	-0.3	K5 III		170 mx		
27295	76548	53 Tau	4 19 26.1	+21 08 32	+0.002	-0.04	5.35	-0.09	0.2	B9 V p	+10	100 s		
27361	93864		4 19 28.5	+11 57 59	0.000	+0.01	7.7	0.5	3.4	F5 V		72 s		
27214	57151		4 19 30.1	+37 59 53	-0.003	-0.03	7.4	0.4	2.6	F0 V		87 s		
27259	57155		4 19 30.6	+30 52 59	+0.001	-0.04	7.8	1.4	-0.3	K5 III		380 s		
27436	131115		4 19 30.9	-8 07 09	-0.001	+0.01	7.0	0.1	0.6	A0 V		190 s		
27350	93865		4 19 33.9	+14 41 31	0.000	-0.01	7.9	0.1	1.4	A2 V		180 s		
27509	169340		4 19 34.7	-25 57 26	+0.008	+0.02	7.7	1.0	2.6	F0 V		42 s		
27309	76551	56 Tau	4 19 36.6	+21 46 25	+0.002	-0.04	5.38	-0.14		A0 p	+12			
27386	93866		4 19 37.5	+10 07 17	0.000	-0.04	6.31	1.43		K0	-27			
27408	111748		4 19 38.4	+2 59 31	+0.002	-0.02	7.9	0.1	1.7	A3 V		170 s		
27646	216752		4 19 41.6	-43 28 46	0.000	+0.02	7.8	1.3	0.2	K0 III		300 s		
27518	169344		4 19 42.4	-25 01 28	+0.002	+0.02	6.87	1.44	-0.3	K5 III	-29	270 s		
27225	57154		4 19 43.3	+39 54 55	+0.003	-0.02	8.00	0.4	2.6	F0 V		120 s	3122	m
27372	93867		4 19 44.5	+14 16 27	+0.005	-0.19	7.55	0.98	0.3	G7 III	-17	48 mx		
27631	216753		4 19 45.5	-41 57 37	-0.003	-0.09	7.7	1.2	4.4	G0 V		19 s		
27335	76552		4 19 45.7	+23 44 19	0.000	+0.01	7.70	0.1	0.6	A0 V		230 s	3131	m
27385	93869		4 19 47.2	+13 05 09	-0.002	-0.01	7.8	1.1	0.2	K0 III		330 s		
27371	93868	54 γ Tau	4 19 47.5	+15 37 39	+0.008	-0.02	3.63	0.99	0.2	K0 III	+39	49 s		
27500	149544		4 19 47.9	-18 31 11	+0.005	+0.07	7.8	0.4	3.0	F2 V		89 s		
27234	57158		4 19 48.5	+39 17 51	0.000	-0.02	8.0	1.4	-0.3	K5 III		390 s		
27499	149546		4 19 53.3	-14 35 30	+0.002	0.00	7.8	1.1	0.2	K0 III		340 s		
27508	149547		4 19 53.3	-17 27 53	+0.001	-0.03	7.9	1.1	-0.1	K2 III		400 s		
27487	149543		4 19 53.8	-12 23 56	+0.003	+0.01	6.9	0.8	3.2	G5 IV		55 s		
27383	93870		4 19 54.8	+16 31 21	+0.008	-0.03	6.88	0.56	4.2	F9 V	+37	33 s	3135	m
27466	131119		4 19 56.9	-4 26 19	-0.006	-0.03	7.9	0.9	3.2	G5 IV		86 s		
27397	93872	57 Tau	4 19 57.6	+14 02 07	+0.008	-0.02	5.59	0.28	3.1	F3 V	+42	32 s		m
28032	249000		4 20 00.0	-69 23 57	+0.005	-0.02	7.8	1.7	-0.1	K2 III		180 s		
26415	5153		4 20 03.3	+78 04 47	+0.007	0.00	7.1	0.9	3.2	G5 IV		59 s		
27370	76555		4 20 04.0	+23 35 55	+0.011	-0.04	7.5	0.9	3.2	G5 IV	+8	69 s		
27528	149554		4 20 09.0	-16 26 14	+0.001	-0.01	6.80	-0.03		B9			3140	m
27349	57166		4 20 09.9	+31 57 12	0.000	-0.01	6.16	1.71	-0.3	K5 III	-18	150 s		
27192	24544		4 20 11.5	+50 55 16	+0.001	0.00	5.55	-0.01	-3.0	B2 IV	-18	410 s		
27308	57164		4 20 12.0	+37 12 03	+0.002	-0.01	7.6	1.6	-0.5	M2 III		370 s		
27406	93873		4 20 12.9	+19 14 00	+0.008	-0.03	7.48	0.56	4.0	dF8	+39	47 s		
27279	57163		4 20 14.1	+39 49 52	+0.001	-0.01	7.8	1.1	-0.1	K2 III		330 s		
27278	39468		4 20 14.3	+41 48 29	+0.001	-0.02	5.92	0.94	3.2	G5 IV	+24	28 s		
27485	131123		4 20 14.4	-3 45 04	-0.003	-0.29	7.86	0.65		G0	-36	22 mn		
27498	131122		4 20 15.2	-2 37 44	+0.001	-0.01	7.08	1.55	-0.5	M4 III	+86	330 s		
27460	111755		4 20 16.6	+7 51 14	0.000	+0.01	7.6	0.0	0.0	B8.5 V		330 s		
27917	248997		4 20 19.1	-63 10 26	+0.004	-0.01	7.91	0.43	2.6	F0 V		99 s		
27382	76558	52 φ Tau	4 20 21.2	+27 21 02	-0.002	-0.08	4.95	1.15		K1 III	+3	90 s	3137	m
27348	57171	54 Per	4 20 24.6	+34 34 00	-0.002	0.00	4.93	0.94	0.3	G8 III	-27	84 s		m
27429	93874		4 20 25.0	+18 44 33	+0.008	-0.04	6.11	0.37	3.1	F3 V	+42	40 s		
27584	169350		4 20 25.9	-21 20 08	+0.004	-0.02	7.98	1.02		G5				
27516	131126		4 20 28.8	-1 18 57	-0.002	-0.01	7.50	0.1	1.7	A3 V		150 s		m
27405	76559		4 20 30.3	+25 49 39	0.000	0.00	7.79	0.18	0.4	B9.5 V		250 s		
27293	39472		4 20 31.7	+43 14 20	0.000	+0.04	7.31	1.80		K2	+1			
27459	93876	58 Tau	4 20 36.3	+15 05 43	+0.007	-0.02	5.26	0.22	2.6	F0 V	+36	34 s		
27536	131129		4 20 38.6	-6 14 45	+0.007	-0.04	6.27	0.91		G5				
27616	169354		4 20 38.9	-20 38 23	+0.001	-0.01	5.38	-0.02	1.4	A2 V	+32	63 s		
27022	13098		4 20 40.2	+65 08 26	-0.004	0.00	5.27	0.81	0.3	G5 III	-19	99 s		
27598	149559		4 20 41.1	-16 49 52	0.000	-0.04	7.30	1.6		M5 II	+99			
27497	111756		4 20 41.2	+6 07 51	-0.001	-0.05	5.77	0.92	0.3	gG6	+7	120 s		
27563	131132		4 20 42.8	-7 35 33	0.000	-0.01	5.85	-0.13	-2.2	B5 III		380 s		
27369	57173		4 20 43.9	+35 48 00	0.000	0.00	8.0	1.1	-0.1	K2 III		360 s		
27233	24546		4 20 44.4	+53 20 53	+0.001	+0.01	8.0	0.0	0.4	B9.5 V		300 s		
---	24545		4 20 46.7	+56 06 09	+0.001	0.00	8.0	2.1	-0.3	K5 III		200 s		
27505	111759		4 20 49.0	+9 13 33	+0.004	+0.04	6.53	0.15		A3	+39			
27550	131131		4 20 49.6	+0 04 34	+0.001	+0.02	7.9	0.5	3.4	F5 V		78 s		
27629	149563		4 20 50.6	-19 20 07	+0.001	+0.02	7.08	0.86		G0			3145	
27483	93878		4 20 52.7	+13 51 51	+0.008	-0.02	6.16	0.45	3.7	F6 V	+37	31 s		
27332	39475		4 20 53.1	+45 27 54	+0.002	-0.01	7.7	0.1	0.6	A0 V		240 s		
27276	24554		4 20 53.7	+50 22 39	+0.001	+0.01	7.90	1.1	-0.1	K2 III		360 s	3136	m

HD	SAO	Star Name	α 2000	δ 2000	μ(α)	μ(δ)	V	B-V	M_v	Spec	RV	d(pc)	ADS	Notes
			h m s	° ' "	s	"								
26012	673		4 20 54.1	+81 58 39	-0.008	+0.02	7.4	0.1	0.6	A0 V		230 s		
27703	194943		4 20 54.2	-34 44 00	0.000	-0.08	8.0	0.6	2.1	A5 V		48 mx		
27381	57175		4 20 55.1	+38 34 37	+0.001	-0.01	7.7	0.4	3.0	F2 V		84 s		
27826	233489		4 20 55.7	-51 02 39	+0.001	+0.02	7.90	0.9	3.2	G5 IV		87 s		m
27292	24556		4 20 57.5	+50 15 18	+0.001	+0.01	7.30	1.1	-0.1	K2 III	-18	280 s		m
26267	679		4 20 57.6	+80 24 45	-0.005	0.00	8.0	0.9	3.2	G5 IV		93 s		
29116	258378	ν Men	4 20 58.1	-81 34 47	0.000	+0.13	5.79	0.35		A5 p				
27403	57177		4 20 58.9	+36 34 26	0.000	-0.03	7.6	0.1	1.4	A2 V		170 s		
27574	131137		4 21 01.5	-3 43 08	0.000	+0.01	7.6	0.9	3.2	G5 IV		77 s		
27428	57179		4 21 02.5	+33 22 19	+0.001	-0.08	7.8	1.1	0.2	K0 III		180 mx		
27232	24552		4 21 03.0	+55 31 54	-0.003	+0.02	8.00	0.1	1.7	A3 V		170 s		m
27815	216758		4 21 03.5	-48 48 41	-0.004	+0.01	7.4	-0.1	1.4	A2 V		160 s		
27224	24550		4 21 03.6	+56 42 44	+0.001	0.00	7.06	1.45	0.2	K0 III		110 s		
27549	111764		4 21 04.6	+1 11 39	0.000	0.00	7.8	1.1	-0.1	K2 III		390 s		
27666	169359		4 21 05.5	-21 14 54	0.000	0.00	7.7	0.9	3.2	G5 IV		62 s		
27747	194948		4 21 10.9	-37 32 05	-0.003	+0.02	7.8	1.4	3.2	G5 IV		36 s		
34652	258386		4 21 14.5	-87 48 45	+0.067	+0.09	6.5	0.5	3.4	F5 V		43 s		
27805	216760		4 21 15.0	-45 02 38	+0.001	+0.03	7.2	1.1	3.2	G5 IV		27 s		
27482	76563		4 21 15.1	+27 21 00	-0.001	-0.01	7.28	2.05	-0.1	K2 III		73 s		
27427	57180		4 21 16.6	+38 14 38	-0.001	-0.04	7.6	0.4	2.6	F0 V		110 s		
26684	5164		4 21 19.8	+76 06 21	+0.006	-0.02	6.6	0.0	-1.0	B5.5 V	+6	150 mx		
27174	13105		4 21 22.8	+62 14 43	+0.004	-0.14	8.0	0.9	3.2	G5 IV		89 s		
27525	93880		4 21 23.8	+18 15 59	0.000	0.00	7.9	0.1	1.4	A2 V		180 s		
27458	57184		4 21 26.7	+34 34 15	0.000	-0.01	7.6	1.4	-0.3	K5 III		340 s		
27611	131140		4 21 27.0	-0 05 53	-0.001	-0.12	5.86	1.32		K2		21 mn	3152	m
27651	131141		4 21 30.2	-6 04 25	-0.003	-0.05	7.2	1.1	0.2	K0 III		260 s		
27676	149575		4 21 31.0	-14 16 25	+0.003	-0.01	8.0	1.4	-0.3	K5 III		450 s		
27710	169368		4 21 31.2	-25 43 43	+0.003	-0.05	6.09	0.35	3.0	dF2	+24	28 ts	3159	m
27524	76564		4 21 31.6	+21 02 24	+0.008	-0.03	6.80	0.44		F8	+37			
27534	93883		4 21 32.1	+18 25 04	+0.007	-0.03	6.80	0.44		F5	+37			
27396	39483	53 Per	4 21 33.1	+46 29 56	+0.002	-0.04	4.85	-0.03	-1.7	B3 V	+1	170 s		
27561	93885		4 21 34.8	+14 24 34	+0.008	-0.03	6.61	0.41	3.4	F5 V	+38	44 s		
27665	131142		4 21 35.1	-8 06 31	-0.001	-0.01	8.0	0.1	0.6	A0 V		300 s		
27533	76565		4 21 35.1	+21 11 20	0.000	+0.03	7.1	0.5	4.0	F8 V		41 s		
27481	57185		4 21 36.1	+31 36 53	-0.002	-0.02	8.0	1.4	-0.3	K5 III		410 s		
27660	131143		4 21 37.4	-6 17 05	-0.001	-0.02	6.6	0.1	0.6	A0 V		160 s		
27448	39486		4 21 41.8	+41 24 26	0.000	0.00	8.0	0.0	0.4	B9.5 V		300 s		
27579	93889		4 21 45.3	+13 35 06	+0.001	0.00	7.4	0.2	2.1	A5 V	+11	110 s		
27395	39484		4 21 45.4	+50 02 05	0.000	-0.02	7.50	0.6	4.4	G0 V		41 s	3141	m
27804	194956		4 21 45.5	-37 46 54	-0.002	-0.04	6.8	0.8	3.2	G5 IV		47 s		
27642	111773		4 21 46.2	+2 23 23	+0.001	-0.15	6.9	0.5	3.4	F5 V		51 s		
27843	216764		4 21 46.6	-42 47 16	+0.002	+0.08	7.70	0.9	3.2	G5 IV		79 s		m
27723	169372		4 21 47.3	-21 46 14	+0.005	0.00	7.50	0.56		G0				
27245	13113		4 21 47.6	+60 44 09	+0.008	-0.11	5.39	1.49	0.2	K0 III	+29	56 s		
27700	149579		4 21 50.1	-13 52 56	-0.002	-0.01	8.0	0.1	0.6	A0 V		300 s		
27322	24563		4 21 51.8	+56 30 24	-0.002	+0.02	5.9	0.1	1.4	A2 V	-18	77 s		
28093	249009	η Ret	4 21 53.4	-63 23 11	+0.013	+0.17	5.24	0.96	0.3	G7 III	+45	76 mx		
27842	216765		4 21 54.5	-41 13 12	+0.004	-0.07	7.90	0.5	3.4	F5 V		79 s		m
27532	76566		4 21 55.5	+29 39 01	0.000	-0.02	7.6	0.1	1.4	A2 V		170 s		
27495	57186		4 22 00.3	+39 56 01	-0.001	-0.13	7.30	0.5	4.0	F8 V		45 s	3155	m
27628	93892	60 Tau	4 22 03.4	+14 04 38	+0.008	-0.02	5.72	0.31		A m	+41			m
27136	13107		4 22 03.4	+67 43 38	+0.003	+0.02	7.8	0.8	3.2	G5 IV		83 s		
26209	680		4 22 06.0	+81 37 55	+0.007	-0.03	7.5	0.1	0.6	A0 V		170 mx		
27699	131150		4 22 10.1	-4 40 39	-0.001	-0.01	8.00	0.2	2.1	A5 V		150 s	3163	m
27639	76571		4 22 22.7	+20 49 17	0.000	0.00	5.91	1.66	-0.3	K5 III	-9	140 s	3158	m
27719	131152		4 22 28.0	-0 33 08	+0.001	-0.03	7.8	1.1	0.2	K0 III		330 s		
27596	57194		4 22 33.8	+32 28 04	0.000	-0.01	7.4	1.4	-0.3	K5 III		330 s		
27638	76573	59 χ Tau	4 22 34.8	+25 37 45	+0.001	-0.02	5.39	-0.03	0.4	B9.5 V	+20	97 s	3161	m
27743	131156		4 22 35.3	-4 54 21	-0.002	-0.03	8.0	0.4	2.6	F0 V		120 s		
27523	39491		4 22 37.2	+43 26 56	-0.003	-0.04	7.8	0.5	3.4	F5 V		73 s		
27456	24579		4 22 37.5	+51 56 31	+0.016	-0.08	7.8	0.8	3.2	G5 IV		68 mx		
27504	39490		4 22 39.8	+47 39 35	+0.002	-0.02	7.85	1.83	-0.3	K5 III		270 s		
27691	93896		4 22 44.1	+15 03 22	+0.007	-0.03	6.99	0.56		G0	+38	18 mn	3169	m
27685	93895		4 22 44.7	+16 47 26	+0.010	-0.05	7.85	0.67	5.0	G4 V	+33	36 s		
27823	169383		4 22 46.3	-20 14 34	+0.001	+0.01	8.0	0.4	1.4	A2 V		120 s		
29598	258379		4 22 51.0	-82 53 57	-0.010	0.00	6.76	0.20		A2				m
27595	57195		4 22 53.5	+37 57 04	+0.001	-0.02	7.8	1.1	-0.1	K2 III		330 s		
27697	93897	61 δ¹ Tau	4 22 56.0	+17 32 33	+0.008	-0.03	3.76	0.98	0.2	K0 III	+38	51 s		
27750	111782		4 22 57.7	+2 53 58	+0.001	0.00	7.6	0.1	1.4	A2 V		180 s		
27402	24577		4 22 57.9	+59 36 59	+0.004	-0.03	6.2	0.1	0.6	A0 V	+12	120 s	3146	m
--	13119		4 23 00.0	+63 20 46	0.000	-0.03	8.0	1.6						
27942	194968		4 23 02.7	-36 29 29	+0.006	-0.03	7.7	1.4	3.2	G5 IV		33 s		
27881	169391		4 23 05.6	-24 53 32	+0.001	-0.02	5.83	1.51		K5				
27941	194971		4 23 07.6	-35 32 42	-0.001	0.00	6.39	1.24		G5				m
27762	93899		4 23 16.8	+11 22 39	+0.002	-0.05	7.10	0.1	1.4	A2 V		140 s	3174	m

HD	SAO	Star Name	α 2000	δ 2000	μ(α)	μ(δ)	V	B-V	M_V	Spec	RV	d(pc)	ADS	Notes
27950	194972		4^{h}23^{m}17.9^s	-32°09'10"	0.000^s	-0.03"	7.1	0.3	1.4	A2 V		97 s		
27841	131170		4 23 19.6	-7 42 18	-0.001	-0.01	7.50	1.1	0.2	K0 III		290 s	3178	m
27905	169396		4 23 20.1	-25 23 45	+0.003	-0.30	7.81	0.63		G0		12 mn		
27749	93900	63 Tau	4 23 25.0	+16 46 38	+0.007	-0.03	5.64	0.30		A m	+35			
27761	93901		4 23 26.9	+16 37 47	0.000	0.00	7.9	0.1	0.6	A0 V		240 s		
27731	76582		4 23 30.3	+24 24 19	+0.007	-0.04	7.19	0.45		F5	+34			
27742	76585		4 23 32.3	+20 58 56	+0.001	-0.03	5.90	0.03	0.2	B9 V		140 s		
27464	24580		4 23 34.5	+59 43 57	-0.001	-0.01	7.8	0.2	2.1	A5 V		130 s		
27683	57200		4 23 35.0	+35 54 11	0.000	0.00	7.3	0.8	3.2	G5 IV		64 s		
27650	39501		4 23 35.8	+42 25 41	+0.002	-0.03	6.50	0.00	-3.1	B9 II		830 s	3172	m
27927	169398		4 23 36.0	-20 16 28	-0.004	-0.06	8.0	1.0	3.4	F5 V		37 s		
27957	169405		4 23 40.3	-27 49 47	0.000	+0.04	8.0	1.8	-0.5	M4 III		390 s		
27861	131176	42 ξ Eri	4 23 40.8	-3 44 44	-0.004	-0.05	5.17	0.08	1.2	A1 V	-11	57 s		
27760	76588		4 23 44.1	+22 57 53	0.000	-0.01	7.9	1.6	-0.5	M2 III		380 s		
28057	216775		4 23 44.6	-43 27 21	0.000	-0.01	7.9	0.5	2.6	F0 V		83 s		
27820	111791	66 Tau	4 23 51.8	+9 27 40	-0.001	-0.01	5.12	0.07		A2	-4	23 mn	3182	m
28398	249015		4 23 55.8	-66 44 12	+0.004	+0.07	7.10	0.1	1.7	A3 V		120 s		m
27778	76591	62 Tau	4 23 59.7	+24 18 04	+0.001	-0.01	6.17	0.20	-1.7	B3 V	+13	220 s	3179	m
27716	57206		4 23 59.7	+35 12 57	-0.002	-0.01	7.8	0.4	2.6	F0 V		110 s		
28028	194984	43 Eri	4 24 02.1	-34 01 01	+0.005	+0.05	3.96	1.49	-0.5	M1 III	+24	78 s		
28131	216781		4 24 03.4	-47 09 24	+0.001	0.00	7.7	1.9	0.2	K0 III		160 s		
28778	--		4 24 03.7	-75 36 43			8.0	0.8	3.2	G5 IV		90 s		
27837	93908		4 24 04.4	+12 58 21	+0.004	-0.03	7.7	1.1	0.2	K0 III		210 mx		
27609	24585		4 24 05.4	+53 29 42	+0.005	-0.07	7.0	0.4	2.6	F0 V		75 s		
27819	93907	64 δ² Tau	4 24 05.7	+17 26 38	+0.008	-0.04	4.80	0.16	2.4	A7 V	+38	30 s		m
28248	233504		4 24 06.9	-57 15 11	+0.002	+0.01	7.2	0.3	0.6	A0 V		130 s		
27493	13130		4 24 10.0	+62 01 58	+0.004	-0.01	7.0	0.4	2.6	F0 V		73 s		
28255	233506		4 24 12.1	-57 04 17	-0.012	-0.08	6.29	0.66	5.0	G4 V comp		18 s		m
27836	93910		4 24 12.5	+14 45 29	+0.004	-0.03	7.62	0.60		G0	+38			
28107	216780		4 24 12.9	-40 03 10	0.000	+0.02	7.30	0.00	0.4	B9.5 V		240 s		m
27636	24588		4 24 13.9	+51 56 14	-0.002	-0.01	6.9	0.1	1.4	A2 V		120 s		
27808	76593		4 24 14.4	+21 44 09	+0.008	-0.06	7.14	0.51		F5	+34			
27860	93912		4 24 14.4	+12 09 30	+0.001	-0.01	6.6	1.1	0.2	K0 III		190 s		
28154	216783		4 24 15.9	-46 38 42	-0.003	-0.02	7.1	0.4	3.4	F5 V		56 s		
27878	111793		4 24 16.3	+7 19 12	-0.001	-0.02	7.8	1.1	0.2	K0 III		330 s		
27190	5186		4 24 17.4	+73 04 27	+0.004	-0.01	7.9	1.1	-0.1	K2 III		370 s		
28247	233507		4 24 20.0	-55 50 02	+0.003	+0.02	7.7	1.6	0.2	K0 III		140 s		
27759	57210		4 24 20.0	+35 14 20	+0.001	-0.03	7.7	1.6	-0.5	M2 III		380 s		
27848	93913		4 24 22.2	+17 04 44	+0.007	-0.03	6.97	0.44		F8	+43			
27770	57211		4 24 24.7	+34 18 53	+0.001	-0.03	7.20	0.00	0.4	B9.5 V	+11	210 s	3185	m
27796	76594		4 24 25.5	+29 01 11	+0.001	-0.01	7.9	1.6	-0.5	M2 III		450 s		
27911	111797		4 24 26.6	+2 29 27	+0.004	-0.06	7.9	0.4	2.6	F0 V		100 mx		
27729	39509		4 24 27.5	+41 43 47	+0.001	-0.04	6.9	0.1	1.4	A2 V		120 s		
27859	93914		4 24 28.2	+16 53 10	+0.008	-0.03	7.80	0.60	4.7	G2 V	+44	42 s		
27777	57212	55 Per	4 24 29.1	+34 07 50	+0.002	-0.04	5.60	-0.06	-0.6	B7 V		160 s		
--	149612		4 24 29.6	-14 36 40	-0.001	+0.01	8.0	0.9	3.2	G5 IV		93 s		
27673	24589		4 24 32.8	+50 50 42	+0.002	-0.06	6.9	0.4	3.0	F2 V		59 s		
27786	57216	56 Per	4 24 37.4	+33 57 35	+0.003	-0.07	5.76	0.40	3.4	F5 V	-32	30 s	3188	m
27902	93916		4 24 40.3	+12 54 15	-0.001	+0.01	7.8	0.4	2.6	F0 V		100 s		
27924	111798		4 24 42.0	+10 02 48	+0.002	-0.05	7.22	0.21	0.6	A0 V		160 s		
27877	93915		4 24 42.9	+18 54 46	+0.001	-0.02	7.4	0.0	0.4	B9.5 V		220 s		
27984	131192		4 24 45.7	-9 13 25	-0.002	0.00	7.8	0.1	0.6	A0 V		270 s		
28254	233508		4 24 50.6	-50 37 20	-0.008	-0.14	7.90	0.9	3.2	G5 IV		87 s		m
28090	169421		4 24 51.8	-28 12 29	-0.001	+0.02	7.8	1.5	0.2	K0 III		170 s		
29058	256092		4 24 53.0	-77 41 06	-0.014	-0.01	7.1	0.5	4.0	F8 V		41 s		
28143	194996		4 24 56.4	-34 45 28	-0.002	-0.11	6.55	0.44	2.1	F5 IV n		78 s		m
27901	93918		4 24 57.1	+19 02 30	+0.007	-0.04	5.97	0.37	3.3	F4 V	+37	34 s		
28075	169422		4 25 01.4	-21 12 40	+0.002	+0.01	7.7	1.5	3.2	G5 IV		29 s		
28413	249016		4 25 05.3	-61 14 19	-0.003	+0.02	5.94	1.54		K5	-19			
27784	39517		4 25 09.5	+44 21 54	+0.001	0.00	7.8	1.1	-0.1	K2 III		340 s		
28471	249019		4 25 09.6	-64 04 48	-0.005	+0.33	7.91	0.65	5.2	G5 V		35 s		m
27887	76599		4 25 10.7	+25 44 57	+0.004	-0.02	7.84	0.44	3.4	F5 V		77 s		
28037	131197		4 25 17.4	-5 39 08	-0.003	+0.02	7.5	0.9	3.2	G5 IV		73 s		
28246	216790		4 25 19.1	-44 09 39	+0.002	+0.07	6.39	0.44	3.7	F6 V		35 s		
27831	57219		4 25 19.8	+39 03 08	-0.001	-0.08	7.8	0.7	4.4	G0 V		48 s		
27934	76601	65 κ Tau	4 25 22.1	+22 17 38	+0.007	-0.05	4.22	0.13	2.4	A7 V	+40	23 s	3201	m
--	--		4 25 22.6	+22 17 48			5.40							m
28187	195001		4 25 23.7	-35 40 33	+0.005	-0.08	7.7	0.8	4.4	G0 V		33 s		
28329	233511		4 25 24.5	-51 56 07	+0.001	+0.01	8.0	1.3	0.2	K0 III		250 s		
27946	76602	67 Tau	4 25 24.9	+22 11 59	+0.008	-0.05	5.28	0.25	2.4	A7 V	+32	36 s		m
27795	39520		4 25 29.1	+46 08 38	0.000	+0.01	7.41	0.27	-1.6	B3.5 V	-21	360 s		
27962	93923	68 δ³ Tau	4 25 29.3	+17 55 41	+0.008	-0.03	4.30	0.05	1.4	A2 V	+35	38 s	3206	m
28349	233512		4 25 29.6	-53 06 42	+0.004	+0.04	7.10	0.4	2.6	F0 V		79 s		
27875	57220		4 25 31.1	+37 16 24	-0.001	-0.01	7.8	0.2	2.1	A5 V		130 s		
28115	149621		4 25 34.6	-16 25 07	+0.002	+0.01	7.7	1.1	0.2	K0 III		310 s		

HD	SAO	Star Name	α 2000	δ 2000	μ(α)	μ(δ)	V	B-V	M_V	Spec	RV	d(pc)	ADS	Notes
			^{h} ^{m} ^{s}	$^\circ$ $'$ $''$	^{s}	$''$								
28054	131201		4 25 36.2	−5 05 29	+0.005	−0.03	8.0	1.1	−0.1	K2 III		150 mx		
28026	111807		4 25 36.8	+5 38 44	0.000	+0.01	7.3	1.1	0.2	K0 III		260 s		
27991	93925	70 Tau	4 25 37.3	+15 56 27	+0.007	−0.03	6.46	0.49	3.4	dF5	+36	38 s		m
27933	76603		4 25 38.3	+29 31 36	+0.002	0.00	8.0	1.1	0.2	K0 III		340 s		
28053	131200		4 25 39.6	−1 24 38	−0.001	0.00	7.00	0.9	3.2	G5 IV		58 s	3217	m
28072	--		4 25 41.9	−2 13 57	+0.005	+0.01	8.00	0.5	4.0	F8 V		63 s	3214	m
28102	149622		4 25 42.3	−9 55 36	0.000	−0.01	6.9	1.4	−0.3	K5 III		270 s		
28088	131204		4 25 43.7	−5 09 12	0.000	−0.01	7.5	1.1	0.2	K0 III		290 s		
28701	256088		4 25 45.7	−69 59 38	+0.027	+0.01	7.85	0.65	4.7	G2 V		40 s		
28299	216798		4 25 46.5	−44 51 17	+0.002	0.00	7.5	−0.1	0.0	B8.5 V		320 s		
28141	149624		4 25 49.9	−15 39 37	+0.003	−0.01	7.9	1.1	−0.1	K2 III		390 s		
27989	93926		4 25 51.8	+18 51 51	+0.009	−0.03	7.53	0.68		G0	+41		3210	m
27858	39525		4 25 52.9	+47 03 30	−0.001	−0.01	8.0	1.4	−0.3	K5 III		400 s		
28007	93927		4 25 53.4	+17 26 51	+0.003	−0.01	7.9	0.4	2.6	F0 V	+30	110 s		
28087	111813		4 25 56.1	+1 01 56	0.000	+0.01	8.0	0.0	0.0	B8.5 V		400 s		
28069	111811		4 25 57.4	+5 09 00	+0.007	−0.01	7.36	0.51	3.4	dF5	+32	57 s		
28172	149629		4 26 00.5	−17 52 01	+0.003	−0.04	7.7	1.1	0.2	K0 III		310 s		
28086	111814		4 26 01.2	+4 22 25	0.000	−0.01	6.5	1.1	0.2	K0 III		180 s		
27899	39529		4 26 04.7	+42 13 11	0.000	−0.01	7.6	0.1	1.4	A2 V		170 s		
27898	39528		4 26 05.2	+42 38 40	−0.001	−0.04	8.00	0.1	0.6	A0 V		280 s	3205	m
28034	93928		4 26 05.8	+15 31 26	+0.008	−0.04	7.49	0.54		G0	+41			
27971	57229		4 26 06.2	+31 26 20	+0.006	−0.12	5.28	0.97	0.0	K1 III	+28	95 mx		
28024	76608	69 υ Tau	4 26 18.4	+22 48 49	+0.008	−0.05	4.29	0.26	1.1	F0 III–IV	+35	42 s		m
28033	76609		4 26 18.4	+21 28 14	+0.007	−0.03	7.38	0.54		F8	+42			
28052	93932	71 Tau	4 26 20.7	+15 37 06	+0.008	−0.02	4.49	0.25	2.6	F0 V	+41	24 s		m
28114	111817		4 26 21.0	+8 35 24	0.000	−0.01	6.06	0.02	−1.3	B6 IV	+14	260 s		
28185	149631		4 26 26.1	−10 33 05	+0.003	−0.08	7.9	0.9	3.2	G5 IV		88 s		
27857	24610		4 26 26.2	+52 30 17	+0.010	−0.08	7.9	0.9	3.2	G5 IV		86 s		
28159	131207		4 26 27.0	−0 30 41	−0.001	−0.03	7.50	1.6	−2.4	M1 II	−7	960 s		
28313	195019		4 26 30.5	−32 48 59	+0.003	0.00	6.7	0.7	2.6	F0 V		38 s		
28085	93934		4 26 31.2	+17 09 17	0.000	−0.01	7.7	0.9	3.2	G5 IV		77 s		
28100	93935	73 π Tau	4 26 36.4	+14 42 49	0.000	−0.03	4.69	0.98	0.3	G8 III	+32	72 s		
28140	111818		4 26 37.4	+9 23 38	−0.001	−0.03	8.0	0.0	0.0	B8.5 V		400 s		
28358	216803		4 26 37.7	−40 31 45	+0.001	+0.03	7.10	0.1	0.6	A0 V		200 s		m
27970	39535		4 26 42.4	+41 07 22	−0.002	−0.03	7.9	0.0	0.4	B9.5 V		290 s		
28208	131213		4 26 46.0	−6 52 56	0.000	0.00	7.3	0.0	0.4	B9.5 V		240 s		
27856	24614		4 26 46.1	+55 38 38	−0.002	+0.01	7.60	0.4	2.6	F0 V		97 s	3207	m
27969	39537		4 26 49.5	+42 54 34	+0.013	−0.18	8.0	0.9	3.2	G5 IV		47 mx		
28171	111824		4 26 50.6	+7 05 41	−0.003	+0.07	7.2	0.8	3.2	G5 IV		62 s		
28124	93938		4 26 50.9	+17 01 42	0.000	−0.06	7.1	1.1	0.2	K0 III		210 s		
28264	149639		4 26 53.0	−18 39 21	+0.005	+0.03	7.2	0.5	4.0	F8 V		44 s		
28312	169455		4 26 56.9	−24 04 53	−0.001	−0.01	6.11	0.14		A2			3230	m
28939	256096		4 26 57.7	−72 38 00	−0.004	+0.01	7.90	0.9	3.2	G5 IV		87 s		m
27921	24618		4 26 59.6	+51 12 23	−0.002	0.00	7.9	0.0	0.4	B9.5 V		290 s		
28139	93940		4 27 00.6	+19 07 03	+0.006	−0.13	7.90	0.7	4.4	G0 V	+10	49 s		m
28191	111827		4 27 00.6	+2 04 47	+0.004	−0.04	6.23	1.09	0.0	K1 III	+21	150 mx		
27920	24617		4 27 00.7	+52 00 09	+0.002	0.00	7.7	1.4	−0.3	K5 III		350 s		
27855	24615		4 27 00.8	+57 35 07	+0.002	−0.02	6.20	0.1	0.6	A0 V	−1	130 s	3203	m
28138	93941		4 27 01.5	+19 50 37	0.000	0.00	7.8	0.1	0.6	A0 V		230 s		
26836	691		4 27 02.7	+80 49 27	+0.004	−0.02	5.43	1.17	0.2	K0 III	−9	88 s		
27546	5198		4 27 03.6	+71 46 52	+0.004	−0.03	7.5	0.0	0.0	B8.5 V		150 mx		
28150	93942		4 27 04.7	+18 12 27	−0.001	−0.01	6.90	0.1	0.6	A0 V	+20	160 s	3226	m
28454	216809		4 27 05.9	−46 56 51	+0.006	−0.27	6.10	0.46	4.0	F8 V		26 s		
28667	249027		4 27 06.3	−62 47 59	−0.003	+0.01	6.87	−0.01	0.6	A0 V		180 s		
28149	76613	72 Tau	4 27 17.4	+22 59 47	+0.001	−0.02	5.53	−0.10	−0.9	B6 V	+5	190 s		
28067	57239		4 27 18.4	+36 31 13	+0.002	−0.03	6.8	0.8	3.2	G5 IV		53 s		
28218	111831		4 27 21.8	+7 03 45	−0.002	+0.01	7.6	0.5	3.4	F5 V		70 s		
27981	39539		4 27 22.7	+48 47 21	−0.001	−0.01	8.0	0.1	0.6	A0 V		280 s		
28574	233528		4 27 24.5	−53 24 17	+0.005	0.00	7.8	0.5	0.4	B9.5 V		56 mx		
28005	39540		4 27 24.6	+46 51 11	+0.008	−0.31	6.7	0.7	4.4	G0 V	+39	29 s		
28388	169470		4 27 26.7	−29 11 47	+0.005	+0.13	7.7	0.7	3.2	G5 IV		80 s		
28217	93943		4 27 28.7	+11 12 45	0.000	0.00	5.88	0.05	−1.6	B7 III		250 s	3228	m
28453	216811		4 27 32.5	−40 28 22	−0.002	+0.02	7.3	0.4	4.0	F8 V		45 s		
28004	39541		4 27 35.6	+48 17 51	0.000	0.00	8.0	0.9	3.2	G5 IV		90 s		
28205	93944		4 27 35.8	+15 35 20	+0.007	−0.03	7.42	0.53	4.4	dG0	+41	40 s		
28428	195034		4 27 40.7	−32 24 32	+0.003	+0.01	6.9	0.4	3.4	F5 V		51 s		
28065	39545		4 27 43.1	+43 11 27	−0.001	−0.01	7.50	0.9	3.2	G5 IV		71 s	3225	m
28732	249029		4 27 46.0	−62 31 16	−0.003	+0.01	5.75	1.01		K0				m
28237	93945		4 27 46.1	+11 44 12	+0.008	0.00	7.51	0.56		F8	+43			
28357	149650		4 27 47.5	−11 07 27	0.000	+0.02	7.3	0.9	3.2	G5 IV		65 s		
28083	39547		4 27 54.7	+43 19 52	0.000	−0.02	7.74	0.03	0.6	A0 V		270 s		
28396	169475		4 27 55.3	−21 30 14	−0.001	−0.04	7.30	0.5	3.4	F5 V		60 s	3247	m
28387	149659		4 27 58.0	−17 50 29	0.000	−0.03	7.50	1.4	−2.3	K5 II	+26	910 s		
28226	76618		4 28 00.7	+21 37 12	+0.007	−0.04	5.72	0.27		A m	+36			m
28538	216818		4 28 02.0	−42 45 15	−0.001	−0.01	8.0	0.7	1.4	A2 V		110 s		

HD	SAO	Star Name	α 2000	δ 2000	μ(α)	μ(δ)	V	B−V	M_v	Spec	RV	d(pc)	ADS	Notes
			h m s	° ′ ″	s	″								
28283	111837		4 28 02.9	+10 04 20	−0.001	−0.02	7.46	0.22	0.6	A0 V		170 s	3238	m
28322	111840		4 28 03.6	+1 51 31	+0.002	+0.01	6.15	1.02	0.2	G9 III	+30	140 s		
28169	57243		4 28 04.8	+35 01 13	+0.001	−0.02	8.0	1.1	0.2	K0 III		320 s		
28284	111838		4 28 05.3	+8 09 19	0.000	+0.01	6.9	0.0	0.0	B8.5 V		250 s		
28386	149661		4 28 07.2	−15 10 49	0.000	+0.04	7.1	0.9	3.2	G5 IV		60 s		
28552	216821		4 28 09.4	−41 57 36	−0.002	0.00	6.44	1.64		Ma				
26356	685		4 28 13.1	+83 48 28	−0.005	+0.01	5.57	−0.13	−1.1	B5 V	−7	220 s		
28438	169483		4 28 14.8	−22 46 07	+0.002	−0.01	6.9	1.4	3.2	G5 IV		24 s		
28535	195046		4 28 14.9	−38 49 05	−0.003	−0.01	8.0	1.2	3.2	G5 IV		49 s		
28225	76619		4 28 17.8	+27 46 51	−0.001	0.00	7.76	0.51	0.6	A0 V		130 s		
28309	93949		4 28 18.1	+10 09 45	+0.001	0.00	7.3	1.6		M5 III	+32			R Tau, v
27134	5194		4 28 20.2	+79 38 43	+0.010	+0.03	8.0	0.5	4.0	F8 V		62 s		
28098	39549		4 28 23.0	+49 13 54	+0.004	−0.02	8.0	1.1	−0.1	K2 III		330 mx		
28294	93948	76 Tau	4 28 23.4	+14 44 27	+0.007	−0.02	5.90	0.32	2.6	dF0	+44	45 s		
28321	93951		4 28 23.7	+11 39 58	0.000	0.00	7.5	1.1	0.2	K0 III		250 s		
28292	93950	75 Tau	4 28 26.3	+16 21 35	0.000	+0.03	4.97	1.13	−0.1	K2 III	+18	100 s		
28550	195052		4 28 27.7	−37 36 49	+0.003	−0.13	8.0	0.3	3.2	G5 IV		90 mx		
28215	57247		4 28 31.8	+34 11 56	0.000	−0.01	7.7	0.9	3.2	G5 IV		78 s		
28375	111845		4 28 32.0	+1 22 51	+0.001	−0.02	5.55	−0.10		B8	+18			
28307	93955	77 θ¹ Tau	4 28 34.4	+15 57 44	+0.007	−0.03	3.85	0.96	0.2	K0 III	+40	54 s		
28178	39554		4 28 34.6	+41 48 24	+0.001	−0.01	8.0	0.9	3.2	G5 IV		90 s		
28305	93954	74 ε Tau	4 28 36.9	+19 10 49	+0.008	−0.04	3.54	1.02	0.2	K0 III	+39	45 s		m
28479	149668		4 28 38.9	−19 27 30	+0.002	−0.09	5.96	1.22	0.0	gK1	+26	130 s		
28319	93957	78 θ² Tau	4 28 39.6	+15 52 15	+0.007	−0.02	3.42	0.18	0.5	A7 III	+40	38 s		m
28437	—		4 28 41.1	−11 57 34			7.90	0.9	3.2	G5 IV		87 s	3252	m
28304	76625		4 28 42.1	+20 40 39	+0.001	+0.01	7.80	0.00	0.0	B8.5 V		300 s		m
28344	93958		4 28 48.2	+17 17 06	+0.007	−0.04	7.85	0.60	4.7	G2 V	+41	43 s		
28355	93960	79 Tau	4 28 50.2	+13 02 52	+0.007	−0.01	5.02	0.22	2.4	A7 V	+33	32 s		
28521	—		4 28 51.3	−25 11 39			7.80	0.8	3.2	G5 IV		83 s	3257	m
28271	57250		4 28 52.5	+30 21 53	+0.001	−0.02	6.40	0.52	3.3	dF4	−37	36 s	3243	m
28134	24641		4 28 52.6	+52 23 22	−0.001	0.00	7.6	0.0	−1.0	B5.5 V		460 s		
28097	24640		4 28 54.2	+55 00 02	−0.004	−0.02	7.8	0.5	4.0	F8 V		56 s		
28363	93961		4 28 59.7	+16 09 31	+0.007	−0.04	6.59	0.53		F8	+45	18 mn	3248	m
27758	5210		4 29 01.3	+71 39 48	+0.001	−0.03	8.00	0.4	2.6	F0 V		120 s	3215	m
28213	39557		4 29 01.9	+42 42 56	−0.003	−0.02	7.3	0.4	3.0	F2 V		71 s		
28497	149674		4 29 06.9	−13 02 55	0.000	−0.01	5.60	−0.23	−3.5	B1 V n	+12	660 s		
28281	57254		4 29 12.3	+35 15 57	+0.001	−0.01	7.5	0.1	1.4	A2 V		160 s		
28666	216828		4 29 16.2	−42 11 55	0.000	−0.01	7.8	0.6	1.7	A3 V		76 s		
28354	76627		4 29 19.8	+27 24 16	+0.001	−0.02	6.6	0.1	0.6	A0 V	+20	160 s		
28700	216832		4 29 19.9	−46 30 55	+0.004	+0.03	6.16	1.06	0.3	gG8		130 s		
28394	93962		4 29 20.5	+17 32 42	+0.007	−0.03	7.05	0.50		G0	+35			
28488	131243		4 29 26.2	−2 24 51	+0.001	−0.08	8.0	1.1	0.2	K0 III		160 mx		
28406	93963		4 29 30.4	+17 51 47	+0.008	−0.03	6.91	0.45		F8	+35			m
—	24647		4 29 31.2	+50 36 36	+0.001	0.00	8.0	2.5	−0.3	K5 III		110 s		
28424	93965		4 29 32.6	+13 53 41	+0.012	−0.01	7.72	1.22	0.0	K1 III	+96	300 s		
28599	169503		4 29 32.7	−26 16 34	+0.003	+0.02	6.9	2.3	0.2	K0 III		36 s		
27133	701		4 29 36.5	+80 55 54	−0.008	+0.01	7.2	0.1	1.4	A2 V		150 s		
28687	195067		4 29 37.7	−39 05 28	−0.001	−0.03	7.3	0.6	1.7	A3 V		62 s		
28487	111851		4 29 38.9	+5 09 51	+0.001	−0.03	7.20	1.6	−2.4	M3 II	−6	830 s		
—	—		4 29 41.8	+42 16 53	−0.001	−0.01	7.5	0.0	0.0	B8.5 V		290 s		
28475	93968		4 29 42.8	+10 31 18	0.000	0.00	6.6	0.1		B9				
28639	169505		4 29 43.4	−29 01 48	−0.002	−0.03	7.7	1.1	0.2	K0 III		290 s		
28710	195069		4 29 48.5	−39 25 01	+0.003	−0.01	7.7	1.5	3.2	G5 IV		29 s		
—	—		4 29 49.6	+42 16 04			7.4	0.1	1.4	A2 V		150 s		
28826	233540		4 29 50.2	−52 59 09	+0.001	−0.08	7.8	0.6	3.0	F2 V		61 s		
27932	13157		4 29 51.9	+69 22 42	+0.003	−0.03	7.0	1.1	0.2	K0 III		220 s		
28854	233541		4 29 54.3	−53 55 51	−0.001	−0.01	7.8	0.9	3.2	G5 IV		70 s		
28280	39564		4 29 55.7	+46 26 22	0.000	0.00	7.9	0.1	0.6	A0 V		260 s		
28665	169509		4 29 57.7	−28 52 47	+0.002	+0.01	7.5	0.7	2.6	F0 V		54 s		
28096	13159		4 29 58.2	+62 42 24	+0.001	−0.03	8.0	0.1	1.7	A3 V		170 s		
26659	693		4 30 00.1	+83 20 26	−0.030	+0.11	5.46	0.87	0.3	G8 III	−38	110 s		
28505	93971		4 30 02.3	+10 15 45	−0.001	−0.06	6.5	0.9	0.3	G8 III	−63	170 s		
28405	57258		4 30 03.5	+32 30 42	−0.002	0.00	7.8	0.4	2.6	F0 V		100 s		
28484	93969		4 30 05.2	+16 10 18	+0.002	+0.02	7.9	1.6	−0.5	M2 III		380 s		
28485	93970	80 Tau	4 30 08.5	+15 38 17	+0.007	−0.02	5.58	0.32	2.6	F0 V	+30	38 s	3264	m
28625	149682		4 30 09.6	−13 35 32	−0.002	+0.01	6.24	1.00		G5				
29137	249044		4 30 11.3	−67 52 40	+0.038	+0.42	7.68	0.71		G0		20 mn		
28825	216841		4 30 13.2	−47 58 42	−0.001	+0.04	7.8	0.6	0.6	A0 V		230 s		
28756	195081		4 30 13.8	−38 59 20	+0.004	+0.06	7.2	1.5	0.2	K0 III		130 s		
28766	216839		4 30 15.2	−41 39 34	−0.003	+0.01	7.6	1.4	−0.3	K5 III		380 s		
28483	93973		4 30 17.9	+19 50 25	+0.006	−0.04	7.10	0.47	3.4	dF5	+38	53 s		
28755	195083		4 30 18.1	−38 16 29	−0.001	0.00	7.2	1.1	3.2	G5 IV		38 s		
28447	76634		4 30 20.0	+28 07 56	+0.011	+0.06	6.6	0.9	3.2	G5 IV	+24	48 s		
28720	195076		4 30 22.4	−30 26 40	+0.001	+0.01	7.2	0.5	0.6	A0 V		110 s		
28482	76635		4 30 22.4	+23 35 19	+0.001	−0.01	7.1	0.1	0.6	A0 V		200 s		

HD	SAO	Star Name	α 2000	δ 2000	μ(α)	μ(δ)	V	B–V	M_v	Spec	RV	d(pc)	ADS	Notes
28393	57261		$4^h 30^m 24\overset{s}{.}8$	$+39°52'24''$	$+0\overset{s}{.}010$	$-0\overset{''}{.}20$	8.0	1.1	0.2	K0 III		55 mx		
27807	5216		4 30 26.7	+73 57 57	+0.021	−0.08	8.0	1.1	−0.1	K2 III		130 mx		
28812	216842		4 30 26.8	−43 12 45	+0.001	+0.03	6.9	1.0	0.4	B9.5 V		50 s		
28353	39567		4 30 29.2	+46 09 01	+0.005	−0.06	7.8	0.2	2.1	A5 V		55 mx		
28813	216844		4 30 29.3	−43 24 36	−0.002	+0.04	7.2	0.7	0.4	B9.5 V		88 s		
28527	93975		4 30 33.6	+16 11 38	+0.007	−0.03	4.78	0.17	2.4	A7 V	+38	24 ts		m
28708	169521		4 30 33.7	−23 01 29	−0.001	0.00	7.1	0.2	1.4	A2 V		110 s		
28798	216845		4 30 36.5	−40 23 43	+0.005	+0.04	7.7	0.9	4.4	G0 V		41 s		
28556	93979	83 Tau	4 30 37.3	+13 43 28	+0.007	−0.02	5.43	0.26	2.6	dF0	+39	37 s		m
28459	57269		4 30 38.3	+32 27 29	0.000	−0.01	6.21	−0.04	0.4	B9.5 V	+20	150 s		
28546	93978	81 Tau	4 30 38.8	+15 41 31	+0.007	−0.02	5.48	0.26		A m	+39			m
28776	195085		4 30 40.3	−35 39 13	0.000	+0.04	5.96	1.00		K0				m
28336	24659		4 30 41.2	+52 18 22	+0.001	−0.02	8.0	0.0	0.4	B9.5 V		310 s		
28257	24656		4 30 41.7	+57 24 42	0.000	0.00	7.90	1.6	−1.4	M4 II–III	−21	600 s		RV Cam, v
28435	39576		4 30 46.2	+40 08 48	+0.001	0.00	8.0	0.9	3.2	G5 IV		88 s		
28568	93981		4 30 46.7	+16 08 53	+0.007	−0.04	6.51	0.43		F2	+44			
28707	169523		4 30 47.5	−21 22 28	−0.001	−0.01	7.6	0.2	1.7	A3 V		130 s		
28874	216849		4 30 48.5	−45 35 47	−0.001	+0.02	8.0	2.0	−0.3	K5 III		450 s		
28168	13165		4 30 49.9	+64 26 29	+0.001	−0.03	7.60	1.6	−0.5	M3 III	−23	360 s		RY Cam, v
28873	216850	δ Cae	4 30 50.1	−44 57 13	0.000	0.00	5.07	−0.19	−1.7	B3 V	+15	230 s		
28719	169525		4 30 51.2	−20 49 48	−0.002	+0.01	8.0	0.4	0.6	A0 V		160 s		
28416	39575		4 30 53.1	+44 36 10	+0.002	−0.05	7.1	1.1	0.2	K0 III	−48	220 s		
28837	216847		4 30 53.3	−41 10 27	+0.002	−0.01	7.1	0.4	2.6	F0 V		66 s		
28810	195089		4 30 53.7	−36 52 36	+0.002	−0.02	7.2	1.4	0.2	K0 III		140 s		
28608	93982		4 30 57.1	+10 45 07	+0.007	0.00	7.04	0.47	3.4	dF5	+37	52 s		
--	--		4 31 00.8	+40 32 55	+0.002	−0.01	7.9	0.5	4.0	F8 V		60 s		
28630	111863		4 31 03.9	+6 47 29	−0.006	−0.09	7.30	0.8	3.2	G5 IV		66 s	3279	m
28595	93983		4 31 07.2	+15 06 19	+0.003	−0.04	6.34	1.70	−0.5	M3 III	+38	200 s		
28020	5226		4 31 07.6	+70 20 21	0.000	0.00	7.9	1.1	0.2	K0 III		320 s		
28629	111864		4 31 07.9	+7 19 56	0.000	0.00	7.9	0.1	1.7	A3 V		170 s		
--	--		4 31 10.8	+45 14 54			8.00	1.6		R7.7				
28581	76636		4 31 13.1	+23 20 47	+0.001	−0.02	7.02	1.73	−0.1	K2 III		110 s		
28649	111868		4 31 16.2	+7 27 09	0.000	−0.02	7.8	0.8	3.2	G5 IV		84 s		
28823	169533		4 31 20.3	−28 52 53	−0.001	+0.02	7.5	0.7	2.6	F0 V		56 s		
28622	93987		4 31 20.7	+15 48 56	−0.002	−0.02	7.84	0.45	3.0	F2 V		85 s		
29115	233550		4 31 23.6	−59 46 01	+0.002	+0.01	7.0	0.2	2.1	A5 V		86 s		
28503	57278		4 31 24.0	+40 00 36	−0.001	−0.01	7.00	0.04	0.0	B8.5 V		230 s	3273	m
28763	149702		4 31 25.8	−13 38 42	−0.001	−0.06	6.21	0.12		A2			3284	m
28635	93988		4 31 29.4	+13 54 13	+0.007	−0.03	7.78	0.54		F8	+43			
28621	76640		4 31 35.7	+22 50 42	+0.002	−0.03	7.9	0.1	0.6	A0 V		240 s		
28869	195103		4 31 35.9	−31 31 11	−0.004	−0.13	7.6	0.7	3.0	F2 V		55 s		
28737	131268		4 31 44.5	−1 21 09	0.000	−0.02	7.8	0.8	3.2	G5 IV		84 s		
30054	--		4 31 45.8	−80 17 11			7.9	1.1	0.2	K0 III		350 s		
28715	111876		4 31 50.4	+5 45 53	+0.001	−0.01	6.6	0.0	0.4	B9.5 V		180 s		
28677	93993	85 Tau	4 31 51.7	+15 51 06	+0.007	−0.03	6.02	0.33	3.3	F4 V	+36	35 s		
28749	131270	45 Eri	4 31 52.6	−0 02 38	0.000	−0.01	4.91	1.32	−1.3	K3 II–III	+17	170 s		
28807	149706		4 31 56.8	−12 37 55	+0.006	+0.03	7.1	0.5	3.4	F5 V		54 s		
28121	5230		4 31 56.8	+70 34 34	+0.003	−0.02	7.7	1.1	0.2	K0 III		300 s		
28591	57285		4 31 56.9	+36 44 34	+0.001	−0.06	6.7	0.9	3.2	G5 IV		50 s		
28775	131272		4 31 59.4	−4 22 53	+0.001	+0.01	7.9	1.1	0.2	K0 III		350 s		
28913	195108		4 31 59.6	−33 53 55	−0.001	0.00	8.0	−0.2	0.4	B9.5 V		330 s		
28859	169542		4 31 59.7	−20 41 57	−0.001	−0.02	7.0	2.1	−0.3	K5 III		130 s		
28446	24672	1 Cam	4 32 01.8	+53 54 39	0.000	0.00	5.4	0.1	−5.0	B0 III	−7	890 s	3274	m
28403	24667		4 32 03.7	+57 50 27	+0.003	−0.02	8.00	1.38	−0.1	K2 III		300 s		
29078	233551		4 32 04.6	−49 56 54	0.000	+0.01	8.0	1.0	2.6	F0 V		43 s		
28736	111879		4 32 04.8	+5 24 36	+0.007	+0.01	6.38	0.42	3.1	dF3	+40	43 s		
28901	169545		4 32 06.9	−28 48 23	+0.013	−0.11	7.42	1.09		K5				
28735	93994		4 32 10.6	+10 45 39	+0.003	−0.05	7.9	1.1	−0.1	K2 III		320 s		
27757	5221		4 32 16.3	+77 37 28	+0.066	−0.12	7.8	0.8	3.2	G5 IV	+38	61 mx		
28620	57288		4 32 17.1	+37 02 35	+0.002	−0.01	6.8	0.5	3.4	F5 V		48 s		
28697	76646		4 32 21.7	+25 11 08	0.000	−0.03	7.42	0.16	1.4	A2 V		140 s		
28796	111884		4 32 24.3	+0 58 49	0.000	−0.01	7.8	0.0	0.0	B8.5 V		370 s		
28596	39592		4 32 32.5	+44 54 07	−0.002	−0.02	7.84	0.12	0.6	A0 V		240 s		
29046	216863		4 32 35.3	−40 12 28	0.000	+0.02	8.0	1.9	−0.5	M2 III		170 s		
29076	216866		4 32 36.9	−43 26 03	+0.002	+0.01	7.9	2.1	−0.3	K5 III		370 s		
28843	131279		4 32 37.3	−3 12 33	−0.001	0.00	5.81	−0.14		B9				m
29187	233559		4 32 43.2	−55 35 51	−0.001	+0.09	7.1	1.6	2.6	F0 V		13 s		
29028	195114		4 32 56.9	−33 05 49	0.000	+0.07	8.0	−0.2	0.4	B9.5 V		93 mx		
28795	93998		4 33 00.8	+19 20 54	+0.001	−0.03	7.5	1.1	0.2	K0 III		240 s		
29087	195118		4 33 01.3	−38 17 01	+0.001	+0.02	8.00	0.1	1.4	A2 V		210 s		m
28646	39595		4 33 03.5	+44 55 39	0.000	−0.01	7.9	1.4	−0.3	K5 III		390 s		
28947	149722		4 33 06.2	−12 32 36	−0.003	−0.03	6.6	1.1	0.2	K0 III		190 s		
28342	13176		4 33 07.1	+67 37 54	+0.003	−0.01	6.9	0.1	1.4	A2 V		120 s		
29013	169560		4 33 09.7	−25 33 37	−0.002	−0.02	7.4	0.5	3.0	F2 V		65 s		
29112	195122		4 33 09.9	−39 29 12	+0.006	0.00	7.8	0.3	3.0	F2 V		89 s		

HD	SAO	Star Name	α 2000	δ 2000	μ(α)	μ(δ)	V	B-V	M_v	Spec	RV	d(pc)	ADS	Notes
28401	13180		4 33 15.2	+64 59 18	+0.002	-0.06	7.4	0.5	3.4	F5 V		63 s		
28579	24680		4 33 15.9	+55 27 44	0.000	-0.10	7.7	0.5	4.0	F8 V		54 s		
28604	24682		4 33 16.8	+52 48 19	+0.001	0.00	7.22	1.93	-0.3	K5 III		180 s		m
28433	13181		4 33 19.0	+64 10 08	+0.001	0.00	7.0	0.8	3.2	G5 IV		57 s		
28693	39602		4 33 21.5	+43 01 54	+0.002	-0.06	6.80	0.4	2.6	F0 V	+2	68 s		m
28970	149725		4 33 22.0	-10 47 09	-0.001	+0.02	6.06	1.38	0.2	K0 III		78 s		
28704	39604	57 Per	4 33 24.8	+43 03 50	+0.000	+0.01	6.09	0.38	2.6	F0 V	-23	46 s		m
29085	169570	50 υ¹ Eri	4 33 30.6	-29 46 00	-0.008	-0.27	4.51	0.98	0.3	gG6	+20	63 s		
28204	5238		4 33 30.7	+72 31 44	+0.007	-0.08	6.0	0.2	2.1	A5 V	+10	58 s	3267	m
28961	131298		4 33 31.7	-6 59 16	0.000	+0.04	6.7	1.1	-0.1	K2 III		230 s		
--	249053		4 33 32.4	-63 01 43	-0.002	+0.04	6.7	1.6		M4				R Ret, v
28867	94002		4 33 32.9	+18 01 00	+0.001	-0.02	6.20	0.07	0.2	B9 V n		150 s	3297	m
29399	249054		4 33 34.0	-62 49 26	-0.017	-0.02	5.79	1.04	0.2	K0 III		120 s		m
28879	94004		4 33 38.0	+16 19 23	+0.001	-0.02	6.59	0.18	2.6	F0 V	+24	63 s		
28980	131301		4 33 38.6	-8 15 19	0.000	+0.01	6.7	0.1	0.6	A0 V		170 s		
28958	131299		4 33 39.2	-1 49 53	+0.001	+0.02	7.7	1.1	0.2	K0 III		310 s		
28979	131303		4 33 44.1	-5 18 38	-0.002	-0.10	6.8	1.1	0.2	K0 III		160 mx		
28922	94008		4 33 46.4	+10 31 04	0.000	0.00	7.86	0.17	0.6	A0 V		220 s		
28911	94006		4 33 46.7	+13 15 07	+0.008	-0.01	6.62	0.43		F2	+35			
28930	111890		4 33 48.0	+9 24 48	-0.001	-0.04	6.01	1.06	0.3	G8 III	-26	120 s		
28946	111893		4 33 50.3	+5 23 04	-0.006	-0.30	7.93	0.79	0.2	K0 III		51 mx		
28910	94007	86 ρ Tau	4 33 50.8	+14 50 40	+0.007	-0.02	4.65	0.25	2.6	F0 V	+38	26 s		
29009	131309	46 Eri	4 33 54.7	-6 44 20	0.000	-0.01	5.72	-0.13		B9	+2		3305	m
29073	149732		4 33 54.8	-18 10 47	+0.003	+0.10	7.7	0.6	4.4	G0 V		47 s		
29305	233564	α Dor	4 33 59.8	-55 02 42	+0.006	0.00	3.27	-0.10	-0.6	A0 III	+26	59 s		m
28954	111894		4 34 00.2	+7 16 21	-0.003	0.00	7.9	0.1	1.7	A3 V		170 s		
--	216876		4 34 01.1	-43 31 30	+0.004	-0.08	8.0	2.3	-0.1	K2 III		86 s		
29134	169574		4 34 02.7	-28 26 59	-0.001	-0.04	6.9	0.6	3.4	F5 V		39 s		
28747	39609		4 34 04.3	+45 38 06	-0.002	-0.04	7.7	0.0	0.4	B9.5 V		270 s		
28782	39612		4 34 04.5	+43 05 15	+0.001	-0.02	7.7	0.2	2.1	A5 V		130 s		
29143	195129		4 34 07.0	-30 34 29	-0.001	-0.06	7.8	0.8	2.6	F0 V		52 s		
28978	111896		4 34 08.2	+5 34 07	-0.001	-0.01	5.68	0.05		A2	-7			
29064	131315	47 Eri	4 34 11.5	-8 13 53	-0.002	+0.01	5.11	1.70	-0.5	gM3	-12	110 s		
29065	131316		4 34 11.7	-8 58 13	-0.003	-0.11	5.26	1.47	-1.3	K4 II-III	-27	200 s		
29063	131317		4 34 14.2	-6 50 16	-0.002	-0.04	6.09	1.38		K2				
29005	111898		4 34 14.8	+3 44 59	-0.001	0.00	7.8	1.1	0.2	K0 III		330 s		
29751	256106		4 34 16.1	-73 12 32	+0.005	+0.05	6.82	0.97		K0				
28772	39615		4 34 16.6	+45 54 49	+0.004	-0.04	7.3	0.5	3.4	F5 V		58 s		
29052	111902		4 34 26.8	+0 24 26	0.000	-0.09	7.5	0.8	3.2	G5 IV		72 s		
29769	256107		4 34 34.8	-72 44 01	+0.005	+0.06	7.01	0.22		A2				
28992	94014		4 34 35.2	+15 30 16	+0.007	-0.04	7.94	0.63		F8	+42			
26774	702		4 34 36.6	+84 28 19	+0.017	-0.03	7.4	1.1	-0.1	K2 III		310 s		
29211	195133		4 34 36.7	-33 38 47	0.000	+0.02	7.3	1.1	3.2	G5 IV		43 s		
29163	169581		4 34 37.9	-24 02 24	0.000	+0.04	6.6	2.3	3.2	G5 IV		48 s		
28929	76654		4 34 37.9	+28 57 40	0.000	-0.02	5.70	-0.06		B6 pe	+13		3304	m
29231	195135		4 34 38.6	-35 39 29	-0.006	-0.23	7.6	0.6	3.2	G5 V		78 s		
28771	24694		4 34 38.9	+51 00 11	0.000	-0.05	7.80	0.4	2.6	F0 V		110 s	3298	m
28991	94015		4 34 42.1	+17 44 53	+0.004	-0.04	7.9	0.4	3.0	F2 V		88 s		
29230	195138		4 34 43.7	-34 49 09	0.000	-0.03	7.1	1.5	0.2	K0 III		130 s		
29447	233577		4 34 44.3	-59 09 06	+0.006	+0.03	8.0	1.4	0.2	K0 III		220 s		
29277	216881		4 34 44.9	-42 01 46	-0.001	-0.01	7.6	2.1	-0.5	M2 III		190 s		
28976	76658		4 34 45.3	+22 41 33	+0.002	-0.02	6.80	0.4	3.0	F2 V		58 s	3311	m
28794	--		4 34 45.9	+50 07 28	+0.002	-0.02	7.1	0.1	0.6	A0 V		190 s		m
29198	169585		4 34 54.6	-25 02 23	0.000	-0.05	7.40	0.5	3.4	F5 V		63 s		
29251	195141		4 34 55.0	-35 34 36	-0.001	-0.01	7.1	0.7	3.2	G5 IV		60 s		
29240	195140		4 34 56.9	-33 20 42	+0.002	+0.05	7.7	1.3	3.2	G5 IV		38 s		
29038	94018		4 34 58.6	+16 59 41	0.000	-0.08	7.22	1.18	-0.2	K3 III	+43	190 mx		
29184	169584		4 35 00.5	-19 55 15	+0.005	+0.08	6.13	1.17	0.2	K0 III		120 s		
29051	94019		4 35 02.3	+17 12 06	-0.001	+0.01	7.1	1.4	-0.3	K5 III		250 s		
29197	169586		4 35 04.4	-20 44 34	+0.002	+0.03	7.8	0.4	3.2	G5 IV		83 s		
28734	24695		4 35 05.1	+57 25 13	+0.005	-0.11	6.7	0.4	2.6	F0 V		65 s		
29172	131334		4 35 13.1	-9 44 13	+0.001	-0.01	7.7	0.1	0.6	A0 V		260 s		
29173	131335		4 35 14.0	-9 44 12	+0.001	-0.02	6.37	0.11	0.6	A0 V		120 s	3318	m
29239	169589		4 35 17.7	-24 32 00	+0.006	-0.01	6.6	0.9	4.0	F8 V		21 s		
26367	695		4 35 24.3	+85 31 38	+0.017	+0.03	6.7	0.5	4.0	F8 V	-47	35 s		
29022	57327		4 35 30.4	+33 16 21	+0.002	-0.03	7.6	0.1	1.4	A2 V		170 s		
29291	195148	52 υ² Eri	4 35 33.3	-30 33 45	-0.004	-0.01	3.82	0.98	0.3	K0 III	-4	53 s		
29117	94020		4 35 33.7	+16 08 41	+0.001	-0.04	7.5	0.8	3.2	G5 IV		70 s		
--	94024		4 35 35.0	+10 10 13	+0.002	-0.05	7.82	0.54	4.4	G0 V		48 s		
29037	57328		4 35 35.4	+32 34 48	+0.002	-0.02	7.7	0.5	4.0	F8 V		53 s		
29472	233583		4 35 35.7	-53 32 32	0.000	-0.01	7.8	1.7	0.2	K0 III		130 s		
29140	94026	88 Tau	4 35 39.2	+10 09 39	+0.004	-0.04	4.25	0.18	2.4	A V m	+29	25 ts	3317	m
29103	94021		4 35 40.0	+19 58 07	-0.001	-0.05	8.00	0.5	4.0	F8 V	+12	61 s		m
29532	233586		4 35 41.3	-57 25 33	+0.005	+0.05	7.6	1.0	4.0	F8 V		28 s		
29104	94022		4 35 42.6	+19 52 54	0.000	-0.02	7.10	0.5	4.0	F8 V	-2	41 s	3316	m

HD	SAO	Star Name	α 2000	δ 2000	μ(α)	μ(δ)	V	B-V	M_v	Spec	RV	d(pc)	ADS	Notes
			$4^h 35^m 43^s.1$	$+59°24'03''$	$-0^s.004$	$-0''.02$								
28793	24702		4 35 43.1	+59 24 03	-0.004	-0.02	6.7	0.1	0.6	A0 V		160 s		
28943	39630		4 35 44.7	+45 12 53	+0.002	0.00	7.7	1.4	-0.3	K5 III		360 s		
29139	94027	87 α Tau	4 35 55.2	+16 30 33	+0.005	-0.19	0.85	1.54	-0.3	K5 III	+54	21 ts	3321	Aldebaran, m
28877	24705		4 36 00.4	+55 30 46	-0.001	-0.03	7.3	0.5	4.0	F8 V		45 s		
28988	39632		4 36 00.6	+42 50 14	-0.001	-0.04	8.0	0.9	3.2	G5 IV		90 s		
29227	131344		4 36 01.6	-3 36 43	-0.001	-0.01	6.33	-0.10		B9	+20		3328	m
29226	131343		4 36 01.7	-2 13 50	-0.001	-0.01	7.69	-0.10	0.4	B9.5 V		290 s		
29446	216890		4 36 04.5	-45 08 10	+0.006	+0.03	7.29	0.43	3.2	F8 IV-V		66 s		
28987	39633		4 36 12.9	+47 21 23	0.000	-0.01	7.6	0.4		F0				
29150	76668		4 36 13.9	+21 32 11	+0.002	-0.04	7.4	0.9	3.2	G5 IV		67 s		
28986	39634		4 36 14.9	+47 22 15	0.000	-0.04	7.50	0.5	3.4	F5 V		65 s		m
28926	24711		4 36 15.8	+52 52 45	+0.001	+0.01	7.9	0.9	3.2	G5 IV		86 s		
29158	76669		4 36 17.5	+21 41 21	+0.004	-0.06	7.9	0.5	3.4	F5 V		78 s		
29248	131346	48 ν Eri	4 36 19.1	-3 21 09	0.000	0.00	3.93	-0.21	-3.6	B2 III	+15	320 s		m,v
29302	149757		4 36 19.1	-17 26 38	-0.001	-0.01	7.9	0.4	2.6	F0 V		110 s		
29207	94030		4 36 19.7	+11 24 44	+0.001	-0.02	6.71	0.31	0.6	A0 V		110 s		
28828	13199		4 36 20.6	+61 21 47	+0.002	+0.03	7.2	1.1	-0.1	K2 III		260 s		
28801	13198		4 36 22.8	+62 54 34	-0.001	-0.03	7.9	0.1	1.4	A2 V		180 s		
29471	216892		4 36 23.8	-43 27 03	-0.003	-0.03	7.9	2.0	0.2	K0 III		160 s		
28780	13196		4 36 24.1	+64 15 41	-0.004	-0.01	5.9	0.1	0.6	A0 V	-16	110 s		
29427	195155		4 36 25.6	-37 49 33	+0.001	-0.07	7.2	0.5	2.6	F0 V		62 s		
29169	76670		4 36 29.1	+23 20 27	+0.008	-0.05	6.02	0.38		F2	+43			
--	39636		4 36 29.1	+44 58 40	0.000	0.00	7.9	2.4						YY Per, q
29377	169609		4 36 35.6	-27 02 54	+0.001	-0.03	7.20	0.5	3.4	F5 V		58 s	3342	m
29122	57337		4 36 37.9	+36 57 10	+0.001	-0.06	6.62	1.23	0.2	K0 III		120 s		
29096	39638		4 36 39.1	+41 04 37	+0.001	0.00	8.00	0.1	1.4	A2 V		200 s	3324	m
29225	94033		4 36 40.7	+15 52 09	+0.007	-0.03	6.65	0.43		F8	+11			
29094	39639	58 Per	4 36 41.3	+41 15 53	-0.001	-0.02	4.25	1.22	-2.1	G8 II	+5	160 s		
29712	249066		4 36 45.6	-62 04 39	-0.009	-0.08	5.40	1.58		M6 III	+26	230 mx		R Dor, m,v
31396	258384		4 36 46.7	-83 29 48	-0.012	+0.07	7.3	1.7	3.2	G5 IV		19 s		
29323	131351		4 36 49.9	-8 27 37	-0.002	0.00	6.9	0.1	0.6	A0 V		190 s		
29132	57338		4 36 50.6	+37 26 22	+0.003	-0.01	7.62	0.07		A0				
29435	195163		4 36 50.9	-30 43 00	-0.001	+0.01	6.30	-0.10	-0.1	B9 IV-V		190 s		
27909	721		4 36 52.7	+80 52 55	-0.010	+0.03	7.3	0.4	2.6	F0 V		87 s		
29247	94034		4 36 54.9	+15 15 40	+0.003	-0.04	7.3	0.6	4.4	G0 V		37 s		
29483	195167		4 36 59.9	-35 37 09	-0.001	+0.03	7.9	1.3	0.2	K0 III		230 s		
29453	195164		4 37 05.6	-30 25 46	0.000	+0.02	6.8	1.6	0.2	K0 III		92 s		
29286	94037		4 37 08.2	+10 50 32	0.000	+0.01	7.51	0.24	0.6	A0 V		170 s		
29093	39641		4 37 08.2	+48 23 59	+0.002	0.00	7.90	0.00	-1.0	B5.5 V		520 s	3325	m
29224	76675		4 37 09.0	+27 55 33	-0.002	-0.01	7.4	0.1	0.6	A0 V	-2	230 s		
29271	94035		4 37 10.5	+15 56 15	+0.007	-0.13	7.8	1.1	0.2	K0 III		69 mx		
29433	169617		4 37 13.5	-22 36 58	+0.002	-0.04	7.6	0.1	0.4	B9.5 V		230 s		
29335	111928		4 37 13.6	+0 59 54	-0.001	0.00	5.31	-0.12	-0.6	B7 V	+24	150 s		
29214	57341		4 37 14.5	+30 24 00	+0.003	-0.11	8.0	0.5	3.4	F5 V		80 s		
29260	94036		4 37 14.6	+18 32 35	-0.001	-0.01	6.80	0.5	-4.6	F7 Ib var	-3	1000 s		SZ Tau, v
--	39645		4 37 15.7	+44 48 21	-0.003	0.00	7.7	2.0						
30554	256114		4 37 15.7	-79 27 31	+0.006	+0.06	7.98	1.08	-1.1	K1 II-III		110 mx		
29443	169619		4 37 17.1	-23 02 56	+0.001	+0.01	7.9	-0.5	0.0	B8.5 V		380 s		
29559	216899		4 37 19.2	-41 52 23	+0.002	+0.05	6.6	0.5	2.1	A5 V		51 s		
28779	13201		4 37 23.5	+68 16 58	-0.001	+0.03	7.4	0.0	0.0	B8.5 V		280 s		
29246	76676		4 37 24.6	+25 43 38	+0.003	-0.04	7.46	1.36	-0.1	K2 III	+38	250 s		
29156	39649		4 37 27.0	+43 01 16	+0.003	-0.02	7.9	0.9	3.2	G5 IV		86 s		
29310	94040		4 37 31.9	+15 08 45	+0.007	-0.04	7.55	0.60	4.4	dG0	+40	40 s		
28973	24717		4 37 32.4	+60 05 25	+0.001	0.00	7.5	0.1	0.6	A0 V		220 s		
29391	131358	51 Eri	4 37 36.0	-2 28 25	+0.002	-0.06	5.23	0.28		A5	+21	14 mn	3350	m
29514	195174		4 37 38.1	-29 54 15	-0.001	-0.10	8.00	0.5	3.4	F5 V		83 s		m
29405	131361		4 37 42.7	-2 38 47	0.000	-0.02	7.0	1.1	0.2	K0 III		230 s		
28256	5252		4 37 43.6	+78 59 09	-0.003	-0.08	7.4	0.4	2.6	F0 V		90 s		
29465	149773		4 37 44.2	-18 19 52	-0.001	-0.02	7.7	1.1	0.2	K0 III		320 s		
29543	195176		4 37 44.4	-31 24 38	-0.001	-0.02	7.7	1.3	3.2	G5 IV		38 s		
29021	13210		4 37 52.1	+60 40 34	+0.007	+0.02	7.9	0.9	3.2	G5 IV		84 s		
29376	111940		4 37 54.3	+7 19 03	+0.001	0.00	7.02	-0.05	-1.0	B5.5 V	+25	380 s		
29506	169624		4 37 59.0	-20 41 14	-0.001	+0.03	7.1	-0.6	0.0	B8.5 V		260 s		
29482	149776		4 37 59.5	-13 01 46	0.000	-0.04	6.80	0.1	0.6	A0 V		170 s	3355	m
--	149777		4 37 59.9	-13 01 54	+0.002	+0.01	7.8	0.1	0.6	A0 V		280 s		
29542	169626		4 38 01.4	-27 15 41	+0.005	-0.03	7.8	2.3	-0.1	K2 III		86 s		
29203	39658		4 38 05.6	+46 14 01	0.000	-0.03	7.1	0.9	5.6	G8 V	-6	20 s		
29235	39660		4 38 07.1	+42 07 01	+0.005	-0.08	7.50	1.1	0.2	K0 III	+16	130 mx	3338	m
29781	233601		4 38 07.7	-58 12 14	-0.002	+0.03	6.6	1.4	3.2	G5 IV		21 s		
29388	94044	90 Tau	4 38 09.4	+12 30 39	+0.007	-0.01	4.27	0.13	2.1	A5 V	+45	27 s		m
29375	94043	89 Tau	4 38 09.4	+16 02 00	+0.006	-0.02	5.79	0.31	2.4	dA8	+38	44 s		m
29504	149782		4 38 09.6	-16 43 39	-0.003	-0.07	7.7	1.1	-0.1	K2 III		200 mx		
29585	195183		4 38 09.6	-34 06 35	-0.001	0.00	7.7	1.0	3.2	G5 IV		58 s		
29667	216908		4 38 10.1	-44 37 37	0.000	+0.01	7.80	0.2	2.1	A5 V		140 s		m
29503	149781	53 Eri	4 38 10.7	-14 18 15	-0.005	-0.16	3.87	1.09	-0.1	K2 III	+42	44 ts		m

96

HD	SAO	Star Name	α 2000	δ 2000	μ(α)	μ(δ)	V	B-V	M_v	Spec	RV	d(pc)	ADS	Notes
			h m s	° ′ ″	s	″								
29502	149780		4 38 11.4	-13 15 29	+0.001	+0.01	8.0	0.5	4.0	F8 V		62 s		
29994	249073		4 38 13.0	-68 48 43	0.000	0.00	7.7	0.1	0.6	A0 V		260 s		
29060	13215		4 38 14.9	+60 53 03	0.000	-0.02	6.7	1.1	0.2	K0 III		190 s		
29309	57353		4 38 15.2	+31 59 55	0.000	-0.02	7.12	0.33	-1.6	B3.5 V	+18	280 s		
29410	111943		4 38 15.4	+8 41 39	-0.001	-0.03	7.8	1.1	0.2	K0 III		280 s		
29365	76680		4 38 15.7	+20 41 05	-0.001	-0.01	5.70	-0.05	-0.2	B8 V	-14	150 s		
30479	256116		4 38 21.1	-77 39 21	-0.010	+0.01	6.05	1.10		K0				
29556	149787		4 38 28.8	-19 38 28	+0.001	-0.01	7.1	0.2	1.7	A3 V		100 s		
29364	76682		4 38 29.5	+26 56 24	+0.004	-0.06	7.30	0.4	2.6	F0 V	+4	86 s	3353	m
--	13217		4 38 33.8	+61 12 56	+0.005	-0.08	7.6	0.6						
29521	131375		4 38 34.1	-9 20 35	0.000	+0.01	7.8	0.5	4.0	F8 V		57 s		
29441	111944		4 38 36.2	+8 10 32	0.000	+0.01	7.64	0.1		B3	-19			
29308	57354		4 38 38.7	+38 26 55	+0.001	-0.01	7.7	0.0	0.0	B8.5 V		310 s		
29420	94047		4 38 39.2	+17 24 28	+0.001	-0.02	7.8	0.4	2.6	F0 V		100 s		
29920	249072		4 38 49.4	-62 22 43	+0.001	+0.03	6.5	0.1	1.4	A2 V		110 s		
29419	76683		4 38 51.2	+23 09 00	+0.008	-0.05	7.53	0.57		F5	+40			
29554	131380		4 38 52.4	-8 13 30	+0.001	-0.02	7.7	0.0	0.4	B9.5 V		290 s		
29297	39666		4 38 53.3	+42 14 24	0.000	-0.03	7.0	1.1	0.2	K0 III	+3	210 s		
29573	149789		4 38 53.5	-12 07 23	-0.004	-0.01	5.01	0.07		A m	+7	12 mn		
29877	233609		4 38 54.1	-58 58 51	+0.003	0.00	7.8	0.0	1.4	A2 V		190 s		
29626	169635		4 38 56.0	-25 42 24	+0.001	-0.06	7.3	1.9	0.2	K0 III		76 s		
29637	169637		4 38 56.4	-27 54 42	+0.001	0.00	7.7	1.2	1.7	A3 V		34 s		
29461	94049		4 38 57.2	+14 06 21	+0.006	-0.01	7.96	0.65		G5	+40			
29191	24737		4 38 57.8	+56 38 18	-0.001	0.00	8.0	0.9	3.2	G5 IV		90 s		
29363	57358		4 38 58.9	+36 32 48	+0.002	-0.01	7.7	0.1	1.7	A3 V		150 s		
28567	5259		4 39 03.7	+76 58 29	-0.003	+0.02	8.0	1.1	0.2	K0 III		340 s		
29805	233605		4 39 04.2	-51 40 22	+0.001	+0.02	6.3	2.0	0.2	K0 III		41 s		
29499	111954		4 39 06.0	+7 52 16	+0.006	0.00	5.39	0.26	2.5	dA9	+36	37 s		m
29497	94053		4 39 08.8	+13 00 11	-0.001	-0.07	7.19	0.43	3.4	F5 V		57 s		
29479	94051	91 σ¹ Tau	4 39 09.2	+15 47 59	+0.003	-0.01	5.07	0.15		A m	+19			
28723	5266		4 39 09.7	+74 28 07	+0.005	-0.10	7.8	1.1	0.2	K0 III		190 mx		
29517	111955		4 39 14.3	+7 59 14	0.000	-0.01	7.8	0.8	3.2	G5 IV		81 s		
29418	57361		4 39 14.4	+31 44 30	+0.003	-0.01	7.9	0.4	2.6	F0 V		110 s		
29488	94054	92 σ² Tau	4 39 16.4	+15 55 05	+0.006	-0.02	4.68	0.16	2.1	A5 V	+37	33 s		m
29613	149797		4 39 19.6	-14 21 33	+0.008	-0.13	5.45	1.06	0.0	gK1	+56	120 s		
29711	195205		4 39 19.6	-33 25 48	-0.001	+0.01	7.8	0.5	3.4	F5 V		68 s		
29459	76689		4 39 23.1	+25 13 07	+0.001	0.00	6.3	0.1	1.7	A3 V	+21	82 s		
29409	57362		4 39 25.0	+35 57 17	+0.001	-0.01	7.8	0.1	0.6	A0 V		250 s		
29899	233613		4 39 25.4	-56 53 20	0.000	+0.01	8.0	1.5	-0.1	K2 III		270 s		
29718	195206		4 39 27.8	-33 59 52	0.000	+0.17	7.7	1.0	3.2	G5 IV		60 s		
29583	131386		4 39 28.2	-0 05 05	-0.001	-0.05	7.6	1.1	0.2	K0 III		300 s		
29373	39673		4 39 33.1	+43 39 33	+0.002	0.00	8.0	0.0	-0.9	B6 V		540 s		
29663	149804		4 39 37.0	-14 35 44	+0.002	+0.07	7.9	0.5	4.0	F8 V		60 s		
29674	169640		4 39 37.0	-21 14 52	+0.001	+0.01	7.20	0.9	3.2	G5 IV		63 s	3375	m
29670	149808		4 39 43.8	-16 57 48	0.000	+0.01	7.8	1.4	-0.3	K5 III		410 s		
--	--		4 39 45.9	+41 37 44			7.30	2.0		N7.7				
29610	131392		4 39 47.2	-1 03 10	+0.001	-0.02	6.11	0.94		K0				
29767	195209		4 39 51.4	-35 38 38	0.000	+0.04	7.00	0.20	3.4	F5 V		53 s		
29550	94059		4 39 52.1	+17 29 20	-0.001	-0.02	8.0	1.1	0.2	K0 III		300 s		
29317	24743	3 Cam	4 39 54.6	+53 04 47	0.000	-0.01	5.05	1.07	0.2	K0 III	-41	85 s	3359	m
29316	24744	2 Cam	4 39 58.1	+53 28 23	+0.005	-0.09	5.35	0.32	2.6	F0 V	+20	34 s	3358	m
30306	--		4 40 00.6	-71 29 42			7.3	0.8	3.2	G5 IV		65 s		
29589	94063	93 Tau	4 40 03.3	+12 11 52	0.000	-0.01	5.40	-0.12	-1.0	B7 IV	+23	190 s		
29458	57364		4 40 05.1	+39 47 17	+0.003	-0.04	7.5	1.1	-0.1	K2 III		310 s		
29737	169650		4 40 06.7	-24 28 57	-0.005	+0.02	5.58	0.92	0.3	G6 III	-18	110 s		
29793	195215		4 40 07.9	-35 01 03	0.000	+0.02	8.0	1.3	0.2	K0 III		250 s		
29147	--		4 40 08.4	+66 08 38	-0.007	-0.13	7.70	1.9		S4.7 e	-2			T Cam, v
29457	39679		4 40 11.0	+40 47 20	-0.006	-0.02	6.53	0.41	3.0	F2 V		48 s		
29824	195216		4 40 14.2	-38 48 26	-0.001	-0.01	7.4	0.9	0.2	K0 III		270 s		
29372	24747		4 40 14.8	+51 06 26	+0.001	0.00	8.0	1.1	0.2	K0 III		330 s		
30003	233622		4 40 18.3	-58 56 37	+0.009	+0.19	6.53	0.68	5.2	G5 V		18 s		m
29202	13224		4 40 21.1	+64 33 27	+0.002	+0.01	8.0	0.9	3.2	G5 IV		89 s		
29632	111969		4 40 22.4	+7 47 49	+0.007	-0.07	7.8	1.1	0.2	K0 III		91 mx		
29537	76697		4 40 22.5	+29 58 19	+0.002	-0.04	6.9	0.4	2.6	F0 V	+25	72 s		
29755	149818	54 Eri	4 40 26.4	-19 40 18	+0.001	-0.10	4.32	1.61	-0.5	gM4	-34	92 s	3380	m
28745	5270		4 40 27.3	+76 34 02	-0.009	+0.05	7.7	0.5	4.0	F8 V		54 s		
29875	216926	α Cae	4 40 33.6	-41 51 50	-0.014	-0.08	4.45	0.34	3.0	F2 V	-1	20 s		m
29930	216928		4 40 34.1	-48 32 16	-0.001	+0.04	6.91	1.09	3.2	G5 IV		31 s		
29487	39683		4 40 36.1	+44 05 50	-0.001	+0.01	7.43	0.21	-0.2	B8 V	+8	230 s		
29813	169663		4 40 42.4	-28 12 06	-0.001	-0.28	7.74	0.62		G0				
29889	195222		4 40 48.5	-39 50 39	+0.003	+0.03	7.9	1.1	0.6	A0 V		61 s		
29890	195223		4 40 48.5	-39 51 09	+0.002	+0.06	7.9	0.8	2.1	A5 V		58 s		
29929	216929		4 40 48.7	-45 47 34	-0.001	+0.01	7.3	1.1	1.4	A2 V		39 s		
30030	233624		4 40 48.9	-56 01 51	+0.005	+0.04	7.80	0.4	2.6	F0 V		110 s		m
29371	24751		4 40 49.5	+57 52 47	+0.002	-0.01	7.1	0.0	0.4	B9.5 V		200 s		

HD	SAO	Star Name	α 2000	δ 2000	μ(α)	μ(δ)	V	B−V	M$_v$	Spec	RV	d(pc)	ADS	Notes
29728	111975		4^h 40^m 56^s.1	+0° 57' 46"	−0^s.001	0".00	8.00	0.1	0.6	A0 V		300 s	3383	m
29717	111973		4 40 57.5	+2 30 21	−0.002	0.00	7.9	0.4	2.6	F0 V		110 s		
29765	131405		4 41 04.9	−6 44 51	+0.001	0.00	7.1	1.1	0.2	K0 III		240 s		
29697	76708		4 41 18.8	+20 54 05	−0.017	−0.26	7.98	1.11	6.9	K3 V		14 ts		
29646	76707		4 41 19.7	+28 36 54	+0.003	−0.03	5.70	−0.02	1.4	A2 V	+25	72 s	3379	m
30053	233627		4 41 19.8	−52 37 32	−0.004	−0.21	7.5	1.1	3.4	F5 V		27 s		
29620	57374		4 41 20.4	+34 39 56	+0.001	−0.04	7.9	1.1	0.2	K0 III		320 s		
29526	39688		4 41 24.0	+48 18 04	+0.004	−0.04	5.67	−0.02	0.6	A0 V	+23	100 s		
29872	169670		4 41 31.0	−23 58 36	+0.001	0.00	7.4	2.4	−0.5	M2 III		130 s		
29580	39689		4 41 33.7	+44 29 48	−0.004	−0.01	8.0	0.0		B9 p				
29841	149829		4 41 33.9	−14 41 16	+0.001	−0.02	7.4	0.9	3.2	G5 IV		70 s		
29789	131410		4 41 34.8	−0 44 39	+0.005	+0.07	7.98	0.37		F2	+36			
29788	111984		4 41 35.8	+0 33 37	−0.001	+0.03	8.0	0.0	0.4	B9.5 V		330 s		
29587	39690		4 41 36.3	+42 07 06	+0.049	−0.41	7.29	0.64	4.7	G2 V	+113	31 mx		
29851	149831		4 41 41.8	−12 28 30	−0.002	+0.04	6.6	0.1	1.4	A2 V		110 s		
30042	216939		4 41 46.1	−47 15 37	+0.005	+0.08	6.76	0.76		G0				m
29645	57377		4 41 50.2	+38 16 49	+0.020	−0.10	5.99	0.57	3.4	F5 V	+47	28 s		
30065	216942		4 41 51.5	−47 49 33	−0.005	−0.03	7.40	0.4	2.6	F0 V		91 s		m
30143	233632		4 41 59.9	−54 48 02	+0.001	+0.05	8.00	1.1	0.2	K0 III		360 s		m
29992	195239	β Cae	4 42 03.4	−37 08 40	+0.003	+0.19	5.05	0.37	4.0	F8 V	+31	17 ts		
29268	13229		4 42 03.8	+69 05 37	+0.006	−0.07	7.8	0.8	3.2	G5 IV		81 s		
29745	76717		4 42 04.0	+24 00 39	−0.001	0.00	7.1	0.5	3.4	F5 V		54 s		
30011	195241		4 42 08.1	−37 36 48	−0.001	+0.01	6.9	−0.3	0.6	A0 V		180 s		
30187	233634		4 42 10.4	−56 07 08	+0.005	+0.06	7.9	1.0	3.4	F5 V		35 s		
29600	39695		4 42 10.4	+49 55 11	+0.001	−0.03	7.7	0.1	0.6	A0 V		240 s		
29330	13230		4 42 11.6	+67 17 37	+0.004	+0.01	7.9	0.1	0.6	A0 V		180 mx		
−−	76720		4 42 12.0	+22 56 31	−0.001	−0.02	7.16	0.05	0.6	A0 V		190 s		
29763	76721	94 τ Tau	4 42 14.6	+22 57 25	0.000	−0.02	4.28	−0.13	−1.7	B3 V	+15	150 s		m
29819	111989		4 42 14.9	+9 38 03	+0.002	−0.05	6.74	0.35	3.0	F2 V		55 s		
29630	39696		4 42 17.6	+47 35 11	+0.004	−0.06	7.8	1.1	0.2	K0 III		180 mx		
30324	249086		4 42 21.1	−63 12 55	+0.003	+0.04	7.7	0.4	2.6	F0 V		110 s		
29963	169680		4 42 22.0	−23 10 43	−0.003	−0.03	7.0	1.6	−0.3	K5 III		260 s		
29870	131419		4 42 22.4	−0 55 50	−0.002	−0.04	6.8	0.4	2.6	F0 V		71 s		
29961	169681		4 42 31.8	−20 58 23	+0.002	+0.01	7.7	0.8	3.4	F5 V		45 s		
29189	5278		4 42 35.7	+73 05 57	−0.001	−0.01	8.0	1.4	−0.3	K5 III		410 s		
29296	13232		4 42 37.5	+69 55 44	+0.004	+0.01	8.0	0.0	0.4	B9.5 V		310 s		
29743	57384		4 42 37.7	+36 18 42	+0.002	0.00	7.8	0.0	0.4	B9.5 V		270 s		
29926	131422		4 42 40.2	−5 45 25	−0.002	0.00	7.1	0.1	0.6	A0 V		200 s		
30185	233638	λ Pic	4 42 46.3	−50 28 52	−0.005	+0.04	5.31	0.98	0.3	gG7	+5	93 s		
29742	57385		4 42 47.6	+38 25 35	−0.003	−0.01	6.7	0.2	2.1	A5 V		82 s		
29862	94079		4 42 48.7	+12 12 35	+0.005	−0.23	7.6	0.9	3.2	G5 IV		42 mx		
30062	195247		4 42 51.4	−34 12 57	0.000	+0.04	7.6	1.0	0.2	K0 III		310 s		
29836	94078		4 42 51.7	+18 43 14	+0.008	−0.09	7.50	0.9	3.2	G5 IV	+15	69 s		m
29722	39699	59 Per	4 42 54.3	+43 21 55	+0.004	−0.05	5.29	0.00	0.6	A0 V	+9	84 s		
30060	195248		4 42 55.7	−31 42 33	+0.003	−0.01	7.4	0.6	2.6	F0 V		62 s		
29936	131424		4 42 56.6	−0 35 27	+0.005	−0.07	7.9	0.5	4.2	F9 V	+3	54 s		
30023	169689		4 42 58.8	−24 18 05	+0.004	+0.05	7.6	0.8	2.6	F0 V		49 s		
29869	94082		4 42 59.5	+14 48 52	+0.001	+0.01	6.8	0.9	3.2	G5 IV		50 s		
29925	112007		4 43 00.3	+1 06 28	0.000	0.00	8.0	0.0	0.4	B9.5 V		330 s		
30612	256122	μ Men	4 43 04.1	−70 55 52	+0.003	+0.03	5.54	−0.12	−0.3	B9 IV	−26	150 s		
29785	57387		4 43 04.9	+33 55 47	+0.001	−0.01	7.60	1.1	0.2	K0 III	−5	280 s	3397	m
29971	131432		4 43 07.6	−6 28 15	+0.002	−0.01	8.0	0.1	0.6	A0 V		300 s		
30080	195250		4 43 09.2	−30 45 56	−0.003	−0.07	5.68	1.41	0.2	gK0	−4	71 s		
30001	149843		4 43 10.3	−15 37 15	+0.003	−0.03	8.0	1.4	−0.3	K5 III		450 s		
30278	233644		4 43 13.5	−54 35 36	+0.004	−0.19	7.8	0.8	3.2	G5 IV		82 s		
29859	76727	95 Tau	4 43 13.7	+24 05 20	+0.001	−0.02	6.13	0.54	4.4	G0 V	+8	22 s		
30051	169694		4 43 17.1	−23 37 42	+0.003	−0.01	7.1	0.4	3.0	F2 V		61 s		
29606	24777		4 43 18.1	+59 31 16	+0.001	−0.05	7.30	0.1	1.7	A3 V	+10	99 mx	3391	m
30218	216956		4 43 20.2	−47 13 17	+0.001	−0.01	7.4	1.3	0.2	K0 III		270 s		
29721	39702		4 43 21.5	+49 58 25	0.000	−0.02	5.80	0.02		B8				m
30294	233646		4 43 29.7	−52 32 47	+0.004	0.00	7.5	1.2	0.2	K0 III		230 s		
30194	216955		4 43 31.5	−41 45 13	+0.001	−0.01	7.9	0.4	0.2	K0 III		340 s		
−−	39703		4 43 34.2	+49 47 59	+0.003	−0.09	8.0	1.2						
30020	131442	55 Eri	4 43 34.6	−8 47 38	+0.001	−0.01	6.79	0.63	0.3	G6 III	+40	100 s	3409	m
30021	131443	55 Eri	4 43 35.1	−8 47 46	+0.002	−0.02	6.72	0.89	1.8	G6 III−IV	+48	100 s	3409	m
29883	76728		4 43 35.4	+27 41 17	+0.005	−0.24	8.01	0.91		K4		23 t		
30259	216963		4 43 36.5	−48 18 15	+0.001	+0.04	7.9	1.0	3.2	G5 IV		64 s		
30202	216961		4 43 44.1	−41 03 53	0.000	+0.01	6.25	1.46		K5				
30317	233647		4 43 44.3	−52 15 48	+0.005	+0.05	6.62	0.37	2.6	F0 V		57 s		
30050	149847		4 43 45.8	−10 40 57	+0.001	−0.01	7.79	0.65		A m	+32	16 mn		RZ Eri, v
30105	169702		4 43 46.9	−20 21 08	−0.003	−0.02	7.6	0.8	3.2	G5 IV		78 s		
29867	57393		4 43 48.2	+32 51 55	−0.003	−0.04	6.5	0.1	1.7	A3 V	0	87 s		
30292	216967		4 43 49.4	−48 19 38	−0.001	+0.28	6.91	0.50		F5		30 mn		
29935	76729		4 43 53.9	+22 56 39	0.000	−0.02	8.0	0.0	0.4	B9.5 V		330 s		
30149	169707		4 44 02.0	−25 55 09	+0.003	−0.05	8.0	1.3	2.6	F0 V		28 s		

HD	SAO	Star Name	α 2000	δ 2000	μ(α)	μ(δ)	V	B-V	M_V	Spec	RV	d(pc)	ADS	Notes
			h m s	° ′ ″	s	″								
31230	256126		4 44 02.1	−78 01 21	+0.004	−0.06	8.0	0.1	0.6	A0 V		190 mx		
29833	39712		4 44 03.1	+42 25 08	+0.001	−0.01	7.51	0.31	0.4	B9.5 V		190 s	3400	m
29977	94090		4 44 03.6	+12 59 29	−0.001	+0.04	7.9	0.5	3.4	F5 V		74 s		
30193	195257		4 44 04.9	−31 29 11	+0.001	−0.05	7.6	0.9	3.2	G5 IV		59 s		
30076	131451	56 Eri	4 44 05.2	−8 30 13	0.000	0.00	5.90	−0.11	−2.5	B2 V	+15	460 s		
31081	256124		4 44 05.3	−76 18 19	+0.033	+0.11	7.74	0.50	3.2	F8 IV-V		75 mx		
30127	149856		4 44 07.9	−18 39 59	+0.004	−0.01	5.53	0.02	0.6	A0 V		60 mx		
29549	13243		4 44 08.2	+66 45 55	0.000	+0.02	8.0	1.1	0.2	K0 III		330 s		
30305	216969		4 44 10.1	−46 16 04	+0.002	0.00	7.9	1.1	0.2	K0 III		320 s		
29866	39715		4 44 12.9	+40 47 13	0.000	−0.01	6.08	0.06	−0.6	B8 IV n	+41	210 s		
29901	57395		4 44 13.7	+36 10 51	0.000	−0.01	8.0	1.4	−0.3	K5 III		390 s		
30182	169712		4 44 15.3	−27 34 40	+0.002	−0.02	7.0	2.2	−0.1	K2 III		68 s		
30478	233664	κ Dor	4 44 21.2	−59 43 58	+0.005	+0.04	5.27	0.20		A3	+2	16 mn		
29827	39713		4 44 21.5	+48 10 31	+0.004	−0.09	6.9	1.1	0.2	K0 III		140 mx		
30034	94095		4 44 25.8	+11 08 46	+0.007	−0.01	5.40	0.25	2.2	dA6	+39	41 s		m
30216	169714		4 44 28.8	−27 56 55	−0.002	+0.01	7.4	1.8	3.2	G5 IV		17 s		
30157	149859		4 44 33.9	−16 03 54	+0.004	+0.04	7.6	0.7	4.4	G0 V		44 s		
29882	39716		4 44 34.9	+44 46 01	0.000	−0.04	7.8	0.2	2.4	A7 V	+24	120 s		
29923	57401		4 44 35.4	+38 15 17	+0.011	+0.01	6.8	0.6	4.4	G0 V		30 s		
30112	112033		4 44 42.1	+0 34 06	−0.003	−0.02	7.22	−0.17	−1.6	B3.5 V	+9	570 s		
30047	94097		4 44 44.7	+14 37 35	+0.001	−0.03	7.8	0.7	4.4	G0 V		47 s		
30240	169725		4 44 46.9	−26 46 07	−0.003	0.00	8.0	0.7	0.2	K0 III		360 s		
29846	24798		4 44 54.9	+52 20 22	−0.002	0.00	7.9	0.0	−1.0	B5.5 V		510 s		
30610	249095		4 44 58.0	−63 13 47	−0.001	0.00	6.46	1.08		K0				
30490	233666		4 44 58.5	−56 17 40	−0.003	−0.03	7.7	1.5	0.2	K0 III		160 s		
30238	169727		4 45 04.1	−21 17 01	+0.001	−0.02	5.72	1.47	−0.1	gK2	+22	96 s		
30316	195267		4 45 07.5	−33 01 52	−0.001	+0.04	7.8	0.8	4.0	F8 V		41 s		
28791	739		4 45 08.2	+80 51 10	+0.005	−0.01	7.9	0.9	3.2	G5 IV		87 s		
29998	57407		4 45 12.2	+35 09 29	0.000	−0.03	7.9	0.4	3.0	F2 V		94 s		
28985	5281		4 45 14.3	+79 39 25	−0.005	+0.02	6.6	0.1	0.6	A0 V		160 s		
29957	39725		4 45 21.1	+43 47 09	+0.001	−0.01	7.90	1.4	−0.3	K5 III		400 s	3414	m
29951	39724		4 45 27.7	+47 28 35	+0.002	0.00	7.3	0.9	3.2	G5 IV		66 s		
30211	131468	57 μ Eri	4 45 30.1	−3 15 17	+0.001	−0.01	4.02	−0.15	−1.6	B5 IV	+7	130 s		
30424	216980		4 45 35.0	−42 33 17	+0.001	+0.01	7.2	0.5	0.6	A0 V		98 s		
30501	233668		4 45 38.4	−50 04 30	−0.047	−0.35	7.59	0.89	5.9	K0 V		23 ts		
30122	76737		4 45 42.4	+23 37 41	+0.001	−0.01	6.20	0.06	−2.2	B5 III		360 s		
--	--		4 45 42.6	+75 06 08			7.40							X Cam, v
29796	13255		4 45 45.0	+63 15 36	+0.004	−0.01	7.7	0.9	3.2	G5 IV		78 s		
30397	195275		4 45 49.6	−34 00 18	+0.003	+0.02	6.86	0.00		B9				
30111	76738		4 45 50.1	+28 39 39	+0.001	−0.03	7.10	0.9	0.3	G8 III	+22	210 s		m
--	149867		4 45 50.9	−11 56 57	−0.002	0.00	7.70	0.4	2.6	F0 V		110 s	3428	m
30432	195278		4 45 55.3	−39 21 24	−0.006	−0.02	6.05	1.07		K0				
30234	112051		4 45 56.2	+4 21 36	−0.001	+0.01	8.0	0.0	0.4	B9.5 V		330 s		
30805	249104		4 45 59.1	−66 04 44	+0.004	+0.03	7.6	0.4	2.6	F0 V		100 s		
29329	5289		4 46 00.4	+76 36 40	+0.020	−0.13	6.49	0.51	3.4	F5 V	−6	38 s		
28941	743		4 46 00.8	+80 32 41	−0.001	−0.01	7.8	1.1	0.2	K0 III		330 s		
30210	94111		4 46 01.7	+11 42 21	+0.005	0.00	5.37	0.20		A m	+41			m
30369	169744		4 46 04.6	−25 09 33	+0.004	+0.05	7.5	0.4	3.0	F2 V		73 s		
30179	94110		4 46 05.1	+18 48 03	0.000	−0.01	7.8	1.1	0.2	K0 III		330 s		
--	149869		4 46 06.4	−12 58 30	−0.002	+0.01	7.9	0.1	1.7	A3 V		180 s		
28167	736		4 46 08.4	+84 01 46	+0.007	−0.04	7.3	1.1	0.2	K0 III		270 s		
30332	149871		4 46 09.3	−11 33 41	−0.002	0.00	7.6	0.0	0.4	B9.5 V		270 s		
30551	216987		4 46 09.6	−49 14 45	+0.004	+0.04	6.4	1.6		M5 III	+207	18 mn		R Pic, v
30168	76740		4 46 12.1	+26 02 07	+0.002	−0.04	7.5	0.1	0.6	A0 V		240 s		
30286	112058		4 46 16.4	+3 16 07	+0.003	+0.03	7.9	0.7	4.4	G0 V	+20	51 s		
30197	94112		4 46 16.8	+18 44 05	+0.005	−0.06	6.01	1.21	−0.3	gK4	+38	180 s		
30475	195282		4 46 19.7	−38 06 50	−0.001	−0.01	7.4	0.5	0.2	K0 III		280 s		
29950	24808		4 46 20.1	+55 39 18	−0.003	−0.02	7.5	1.1	0.2	K0 III		270 s		
30321	131481		4 46 24.0	−2 57 16	+0.004	−0.05	6.33	0.04		A2				
29826	13260		4 46 24.4	+64 48 36	+0.004	−0.06	7.0	0.9	3.2	G5 IV		57 s		
30422	169752		4 46 25.7	−28 05 15	0.000	+0.01	6.19	0.19	0.4	A3 III-IV		120 s		
30462	195283		4 46 26.3	−35 48 23	0.000	+0.03	7.8	1.3	0.2	K0 III		230 s		
30015	24815		4 46 26.4	+51 05 46	+0.001	−0.01	7.7	1.1	−0.1	K2 III		330 s		
30385	149874		4 46 26.8	−17 49 22	+0.004	+0.06	7.7	1.1	0.2	K0 III		200 mx		
30090	39741		4 46 26.8	+42 20 56	−0.001	+0.08	6.7	0.6	4.4	G0 V	+29	28 s		
30500	195286		4 46 31.8	−37 27 21	−0.001	0.00	7.7	1.7	−0.3	K5 III		320 s		
30233	94115		4 46 35.9	+19 29 38	0.000	−0.01	8.0	1.1	−0.1	K2 III		410 s		
30285	112066		4 46 36.2	+8 19 04	+0.001	−0.05	8.0	1.1	−0.1	K2 III		370 s		
30969	249109		4 46 37.3	−68 32 45	−0.004	+0.05	7.4	0.5	4.0	F8 V		47 s		
30540	216990		4 46 38.9	−42 05 32	−0.001	+0.03	7.7	0.2	2.1	A5 V		130 s		
30138	39746		4 46 44.3	+40 18 46	0.000	−0.03	5.97	0.93	0.2	G9 III	+34	140 s		
30311	112069		4 46 45.5	+9 01 04	+0.006	+0.01	7.26	0.55	3.4	dF5	+40	51 s		
30447	169759		4 46 49.5	−26 18 10	+0.002	−0.02	7.8	0.6	2.6	F0 V		77 s		
30420	149879		4 46 50.2	−18 23 57	0.000	−0.01	7.8	0.5	3.4	F5 V		74 s		
30791	249105		4 46 52.5	−60 36 14	+0.009	+0.12	7.3	1.0	4.4	G0 V		21 s		

HD	SAO	Star Name	α 2000	δ 2000	μ(α)	μ(δ)	V	B-V	M$_v$	Spec	RV	d(pc)	ADS	Notes
			h m s	° ′ ″	s	″								
30137	39747		4 46 53.9	+42 59 43	0.000	−0.02	8.0	0.0	0.4	B9.5 V		310 s		
30439	169760		4 46 57.2	−21 12 42	+0.002	+0.01	7.20	0.9	3.2	G5 IV		63 s		m
30734	233680		4 46 57.8	−55 46 47	+0.004	+0.05	8.0	1.5	0.2	K0 III		190 s		
30497	169764		4 46 58.9	−29 24 28	−0.001	0.00	7.5	0.8	0.2	K0 III		280 s		
30566	195291		4 47 01.0	−38 18 31	−0.001	+0.01	7.3	0.8	3.2	G5 IV		67 s		
30152	39748		4 47 02.1	+41 18 18	+0.001	−0.01	7.18	0.07	0.6	A0 V		210 s		
30342	112074		4 47 06.6	+10 03 07	0.000	+0.02	7.19	0.16	1.4	A2 V		130 s	3448	m
30790	233682		4 47 06.9	−59 08 12	+0.002	+0.03	6.7	1.3	0.2	K0 III		130 s		
30310	94119		4 47 07.0	+18 29 15	−0.001	−0.04	7.8	1.1	0.2	K0 III		330 s		
30880	249110		4 47 15.4	−62 49 45	−0.001	−0.03	7.8	0.2	2.1	A5 V		140 s		
30365	112077		4 47 17.9	+5 47 18	+0.001	−0.01	6.7	0.1	0.6	A0 V		160 s		
30593	195295		4 47 19.1	−36 12 34	+0.004	+0.01	7.73	2.32		N7.7	−7			T Cae, v
30865	249108		4 47 22.0	−61 28 35	+0.009	−0.04	7.50	0.4	2.6	F0 V		96 s		m
30178	39752		4 47 35.7	+45 59 58	+0.003	+0.02	7.78	1.88	−4.8	M2 Ib		2300 s		
30495	149888	58 Eri	4 47 36.2	−16 56 04	+0.009	+0.17	5.51	0.63	4.5	dG1	+17	15 ts		
30436	131489		4 47 41.2	−0 05 20	−0.001	−0.03	7.98	0.42	0.7	F4 III	−12	290 s		
30282	57431		4 47 46.3	+36 43 22	+0.001	−0.01	8.0	0.4	2.6	F0 V	+14	120 s		AW Per, v
30136	24830		4 47 49.4	+53 18 05	+0.002	−0.07	7.00	0.4	3.0	F2 V		62 s	3434	m
30608	195300	ζ Cae	4 47 49.6	−30 01 13	+0.003	+0.10	6.37	1.07		K1 IV		180 mx		
30221	39758		4 47 53.4	+45 29 22	+0.002	−0.03	7.8	0.1	−0.6	A0 III	+8	430 s		
30121	24829	4 Cam	4 48 00.2	+56 45 26	+0.007	−0.15	5.26	0.25		A m	+19	24 mn	3432	m
30167	24835		4 48 01.7	+53 07 18	0.000	−0.01	7.40	0.1	1.7	A3 V		130 s	3442	m
30110	24826		4 48 02.9	+59 15 41	0.000	+0.01	8.00	0.1	1.4	A2 V		200 s	3436	m
32195	258391		4 48 04.2	−80 46 54	+0.008	−0.04	7.8	0.7	4.4	G0 V		48 s		
30486	112088		4 48 05.7	+0 40 32	−0.004	−0.01	7.48	0.44	3.0	F2 V		74 s		
31010	249111		4 48 06.7	−63 28 49	0.000	−0.01	7.1	1.1	0.2	K0 III		240 s		
30144	24834		4 48 06.9	+55 36 09	+0.009	−0.10	6.4	0.4	2.6	F0 V	+22	55 s		
30895	233690		4 48 11.7	−56 49 03	+0.004	−0.01	7.5	1.8	−0.1	K2 III		140 s		
30535	131500		4 48 13.7	−9 30 24	−0.001	0.00	7.2	0.1	0.6	A0 V		210 s		
30379	76761		4 48 18.9	+29 14 09	+0.002	−0.05	7.0	0.5	3.4	F5 V		53 s		
29949	13267		4 48 19.6	+67 45 49	+0.011	−0.07	7.0	0.4	2.6	F0 V		75 s		
30402	76764		4 48 20.4	+26 57 59	0.000	−0.01	7.8	1.1	0.2	K0 III		300 s		
30378	76762		4 48 22.7	+29 46 22	+0.001	−0.03	7.41	0.03	0.4	B9.5 V		240 s		
30622	149899		4 48 23.3	−19 51 07	0.000	+0.02	7.8	1.6	−0.5	M2 III		460 s		
30308	39767		4 48 25.7	+43 20 18	+0.003	−0.03	7.90	1.1	0.2	K0 III		320 s		
30508	112091		4 48 27.4	+2 42 54	−0.002	+0.07	6.6	0.9	3.2	G5 IV		49 s		
30850	217006		4 48 31.6	−49 47 34	−0.001	−0.02	7.36	0.97	3.2	G5 IV		49 s		
30606	149901	59 Eri	4 48 32.4	−16 19 47	0.000	+0.04	5.77	0.55	3.7	dF6	+35	23 s		
30469	94134		4 48 33.2	+12 55 31	−0.001	−0.02	7.8	0.8	3.2	G5 IV		80 s		
30788	217004		4 48 33.7	−43 58 47	+0.001	+0.03	6.72	0.95		G5				
30456	94132		4 48 35.3	+17 48 23	0.000	−0.04	8.02	0.42		F2				m
30562	131504		4 48 36.3	−5 40 26	+0.020	−0.24	5.78	0.62	4.4	dG0	+78	22 ts		
30544	112096		4 48 39.3	+3 38 57	0.000	0.00	7.32	−0.05		B9	+33			
30507	112095		4 48 40.9	+7 50 24	−0.001	+0.03	7.4	1.4	−0.3	K5 III		330 s		
30483	94138		4 48 41.2	+16 38 02	+0.002	−0.09	7.6	0.5	3.4	F5 V		65 s		
30455	94136		4 48 42.1	+18 42 33	+0.014	−0.39	6.97	0.62	4.7	G2 V	+55	28 mx		
30468	76767		4 48 43.8	+21 18 56	−0.001	−0.02	6.9	0.1	0.6	A0 V		190 s		
30545	112098		4 48 44.5	+3 35 18	0.000	−0.01	6.03	1.19	0.0	K1 III	−19	140 s		
31028	249114		4 48 47.0	−60 15 32	0.000	0.00	6.9	1.4	0.2	K0 III		120 s		
30692	169788		4 48 49.9	−23 16 45	+0.003	−0.18	7.8	0.4	4.0	F8 V		57 s		
29678	5309		4 48 50.4	+75 56 29	+0.011	−0.13	6.06	0.28	2.6	F0 V	−6	49 s		m
30353	39773		4 48 53.7	+43 16 32	+0.006	0.00	7.76	0.46		A pe	−10			KS Per, v
30573	112100		4 48 56.1	+3 41 22	−0.002	−0.02	7.1	0.1	0.6	A0 V		200 s		
30861	217011		4 48 57.2	−47 08 04	+0.002	+0.04	7.4	0.4	0.6	A0 V		120 s		
31249	249120		4 48 57.8	−67 42 48	−0.004	+0.01	7.64	0.19	1.7	A3 V		140 s		
41301	258410		4 49 00.8	−88 16 17	−0.020	+0.01	7.15	1.68	−0.5	M4 III		320 s		
31009	233694		4 49 04.6	−56 40 00	−0.002	+0.02	6.9	1.7		Ma				
30165	13277		4 49 05.1	+61 30 27	−0.002	−0.01	7.6	1.6	−0.5	M4 III	+53	370 s		
30467	76770		4 49 05.5	+27 00 43	−0.001	−0.03	7.7	0.5	2.4	F8 IV		110 s		
30787	195322		4 49 09.3	−35 05 00	+0.004	+0.14	7.9	0.2	3.4	F5 V		78 s		
30719	169793		4 49 10.0	−21 53 29	−0.003	−0.02	8.0	1.1	0.2	K0 III		330 s		
30848	217012		4 49 10.7	−42 23 06	−0.001	0.00	7.27	0.17		A2				
30732	169794		4 49 11.2	−23 03 42	−0.001	0.00	7.9	1.2	−0.1	K2 III		390 s		
30786	195323		4 49 11.6	−34 18 43	0.000	−0.03	7.0	1.0	3.2	G5 IV		40 s		
30454	57441		4 49 12.8	+31 26 15	+0.001	−0.10	5.58	1.12	0.2	K0 III	+23	97 s		
30637	131515		4 49 13.4	−4 59 12	−0.001	−0.01	7.0	0.0	0.0	B8.5 V		260 s		
30847	217013		4 49 14.0	−41 35 13	−0.003	−0.02	7.5	1.4	0.2	K0 III		280 s		
30691	149910		4 49 14.2	−15 43 06	0.000	−0.02	7.9	1.1	−0.1	K2 III		400 s		
30482	76772		4 49 15.7	+28 21 04	0.000	0.00	7.5	1.1	0.2	K0 III	−9	260 s		
30466	76771		4 49 15.9	+29 34 16	0.000	−0.03	7.29	0.05		A0 p	+17			
30679	131521		4 49 18.8	−9 00 34	−0.002	−0.01	7.8	0.1	1.4	A2 V		190 s		
30453	57444		4 49 19.0	+32 35 18	+0.002	−0.03	5.9	0.1	1.7	A3 V	+21	69 s		
30877	217016		4 49 20.3	−42 24 50	−0.002	0.00	7.32	0.20		A0				
30410	39779		4 49 26.1	+43 32 31	+0.002	0.00	8.00	0.9	5.6	G8 V		30 s		
30452	57446		4 49 30.0	+37 20 05	+0.002	−0.02	8.0	1.1	0.2	K0 III		340 s		

HD	SAO	Star Name	α 2000	δ 2000	μ(α)	μ(δ)	V	B-V	M_v	Spec	RV	d(pc)	ADS	Notes
			$4^h49^m31.2^s$	$-41°57'39''$	$+0.006^s$	$+0.04''$								
30893	217021		4 49 31.2	-41 57 39	+0.006	+0.04	7.8	0.5	3.4	F5 V		68 s		
30589	94148		4 49 32.0	+15 53 19	+0.005	-0.02	7.74	0.57	4.0	dF8	+40	51 s		
30339	24853		4 49 33.3	+53 21 57	+0.007	-0.06	7.9	0.5	4.0	F8 V		59 s		
30389	39780		4 49 40.9	+49 35 19	0.000	+0.02	7.8	0.1	1.7	A3 V		160 s		
30919	217025		4 49 42.0	-41 51 14	+0.002	+0.01	7.5	1.7	-0.1	K2 III		160 s		
30743	149916		4 49 42.2	-13 46 10	-0.008	-0.16	6.26	0.44	3.4	F5 V		37 s		
30700	131526		4 49 42.4	-6 25 01	-0.001	-0.03	7.9	0.9	3.2	G5 IV		88 s		
30605	94151	96 Tau	4 49 44.0	+15 54 15	0.000	-0.01	6.08	1.60	-0.2	gK3	+13	110 s	3464	m
30999	217027		4 49 47.3	-48 18 41	-0.001	+0.02	7.9	1.0	3.2	G5 IV		78 s		
31065	233699		4 49 48.0	-54 27 17	-0.003	-0.01	7.80	0.00	0.0	B8.5 V		360 s		m
30741	149917		4 49 48.2	-11 44 51	+0.001	+0.01	8.0	1.4	-0.3	K5 III		450 s		
30652	112106	1 π³ Ori	4 49 50.3	+6 57 41	+0.031	+0.02	3.19	0.45	3.8	F6 V	+24	7.7 t		m
30876	195332		4 49 52.5	-35 06 29	0.000	-0.13	7.5	0.8	0.2	K0 III		160 mx		
30504	57447		4 49 54.6	+37 29 18	-0.003	+0.04	4.88	1.44	-2.3	K4 II	-23	270 s		
30634	94153		4 49 54.8	+13 41 20	-0.001	-0.02	7.5	1.1	-0.1	K2 III		270 s		
30859	195331		4 49 56.1	-30 58 14	-0.002	+0.02	7.7	1.2	0.2	K0 III		250 s		
30633	94152		4 49 57.0	+15 31 52	0.000	-0.02	7.9	0.1	1.7	A3 V		160 s		
30677	112111		4 50 03.6	+8 24 28	+0.001	-0.01	6.86	-0.04	-3.5	B1 V	+5	960 s		
30164	13278		4 50 05.5	+66 32 09	+0.003	-0.02	7.80	1.1	0.2	K0 III		310 s	3452	m
30801	149923		4 50 06.7	-15 48 15	-0.001	-0.03	6.8	1.1	0.2	K0 III		210 s		RX Eri, v
30891	195335		4 50 08.3	-30 15 38	-0.001	0.00	8.0	1.3	-0.1	K2 III		360 s		
30687	94156		4 50 09.5	+10 36 18	+0.002	-0.05	7.90	0.40	3.0	F2 V		91 s		
30244	13283		4 50 10.0	+63 37 33	+0.003	-0.01	6.7	0.1	0.6	A0 V		160 s		
30814	149924	60 Eri	4 50 11.5	-16 13 02	+0.002	+0.05	5.1	1.1	0.2	K0 III	+37	94 s		
30985	217032		4 50 16.2	-41 19 15	0.000	+0.07	6.07	0.37	1.7	F0 IV		68 s		m
30714	112117		4 50 17.2	+6 56 59	-0.001	+0.01	7.20	0.1	1.4	A2 V		140 s		m
30856	169805		4 50 17.7	-24 22 08	0.000	-0.03	8.0	1.1	3.2	G5 IV		58 s		
30676	94158		4 50 23.8	+17 12 09	+0.007	-0.04	7.12	0.56		F8	+42			
30781	131533		4 50 27.7	-3 51 55	-0.001	-0.01	7.6	1.1	0.2	K0 III		310 s		
30997	195342		4 50 28.4	-38 33 36	+0.002	+0.12	7.30	0.5	3.4	F5 V		60 s		m
30918	169810		4 50 31.8	-27 06 02	+0.001	+0.02	7.6	1.3	0.2	K0 III		210 s		
30712	94159		4 50 33.8	+15 05 00	+0.007	-0.02	7.71	0.72		G5	+43			
30739	112124	2 π² Ori	4 50 36.6	+8 54 01	0.000	-0.03	4.36	0.01	0.6	A0 V	+24	55 s		
30085	5326		4 50 36.6	+70 56 30	+0.003	-0.01	6.4	-0.1		B9				
30465	24863		4 50 39.3	+50 24 33	+0.001	0.00	8.0	1.6	-0.5	M2 III		450 s		
31518	249123		4 50 43.9	-69 49 50	-0.004	+0.03	7.7	1.7	-0.5	M2 III		380 s		
30902	149930		4 50 44.4	-18 54 01	+0.002	-0.04	7.6	1.4	-0.1	K2 III		250 s		
30796	112132		4 50 45.3	+1 08 53	0.000	-0.01	6.69	0.05	1.4	A2 V		110 s		
30427	24861		4 50 48.0	+56 31 49	+0.002	-0.01	7.8	0.5	3.4	F5 V		74 s		
30651	57454		4 50 48.1	+31 33 56	+0.001	-0.03	7.8	1.1	0.2	K0 III		300 s		
30738	94162		4 50 48.5	+16 12 38	+0.006	-0.02	7.29	0.50		F8	+42			
30812	131538		4 50 49.2	-0 05 41	+0.001	-0.04	7.26	1.06	0.0	K1 III	-9	280 s		
30769	112129		4 50 50.4	+7 59 41	+0.001	-0.06	7.3	0.5	3.4	F5 V		61 s		
30675	76784		4 50 52.0	+28 18 51	+0.001	-0.01	7.55	0.35	-1.6	B3.5 V	+20	330 s		
30951	169813		4 50 53.7	-26 07 46	-0.001	-0.05	7.7	1.5	0.2	K0 III		160 s		
31203	233710	ι Pic	4 50 56.3	-53 27 35	-0.011	+0.08	5.23	0.33	1.7	F0 IV	+5	49 s		m
30983	169815		4 51 07.6	-25 18 37	+0.001	+0.01	7.0	0.4	0.6	A0 V		110 s		
30557	39799		4 51 09.3	+48 44 26	-0.004	-0.04	5.66	0.99	0.3	G5 III	+29	97 s		
31172	217042		4 51 10.5	-48 03 45	-0.001	+0.01	8.0	1.6	0.2	K0 III		190 s		
30836	112142	3 π⁴ Ori	4 51 12.3	+5 36 18	0.000	0.00	3.69	-0.17	-3.6	B2 III	+23	280 s		
30810	94163		4 51 12.6	+11 04 07	+0.007	+0.01	6.77	0.52	3.7	F6 V	+39	36 ts	3475	m
--	217041		4 51 12.9	-43 03 06	+0.003	+0.07	7.6	1.1	4.0	F8 V		22 s		
30243	13285		4 51 13.3	+68 10 09	0.000	+0.01	7.00	2.0		N7.7	-12			ST Cam, v
31131	--		4 51 14.0	-43 02 48			7.80							m
30686	57456		4 51 16.3	+34 05 36	-0.001	+0.01	7.8	0.1	1.7	A3 V		150 s		
31532	249125		4 51 20.8	-68 04 11	-0.011	+0.06	7.60	0.6	4.4	G0 V		44 s		m
30780	94164	97 Tau	4 51 22.4	+18 50 23	+0.006	-0.03	5.13	0.21	2.6	dF0 n	+39	32 s		m
30809	94165		4 51 23.2	+15 26 00	+0.005	-0.03	7.90	0.53		F8				
31244	233715		4 51 26.9	-51 43 33	-0.001	0.00	6.6	0.7	-1.3	K3 II-III		370 s		
31229	217044		4 51 27.8	-49 23 43	-0.003	+0.08	7.4	0.9	4.4	G0 V		40 s		
31093	195357		4 51 28.1	-34 54 23	+0.002	-0.03	5.86	0.08	1.2	A1 V		79 s		m
30898	131543		4 51 28.2	-0 32 49	0.000	-0.01	7.86	0.27	1.4	A2 V		150 s		
30708	57460		4 51 29.5	+35 48 55	+0.004	-0.11	6.8	0.9	3.2	G5 IV		51 s		
30914	131545		4 51 30.5	-3 39 33	-0.001	+0.01	7.4	1.1	-0.1	K2 III		320 s		
30755	76788		4 51 31.3	+28 31 38	+0.001	0.00	8.02	2.93		N7.7	+16			TT Tau, v
30650	39807		4 51 32.8	+43 34 41	+0.003	-0.01	7.51	-0.03	-0.9	B6 V	+34	430 s		
--	149938		4 51 35.3	-14 35 21	-0.003	-0.03	7.80	0.5	3.4	F5 V		76 s	3485	m
30963	149937		4 51 39.0	-10 17 26	-0.003	-0.01	7.1	0.0	0.4	B9.5 V		220 s		
30962	131549		4 51 40.4	-8 24 05	-0.001	+0.01	7.1	0.1	0.6	A0 V		200 s		
31163	195363		4 51 42.8	-38 36 38	+0.001	+0.02	7.6	0.2	1.7	A3 V		140 s		
30870	112150		4 51 43.3	+9 58 30	0.000	-0.01	6.11	0.08	-1.1	B5 V	+11	210 s		
30649	39808		4 51 43.5	+45 50 03	+0.037	-0.55	6.97	0.58		G1 V-VI	+27	29 t		m
31143	195362		4 51 45.7	-35 50 26	-0.007	-0.05	8.0	1.0	0.2	K0 III		150 mx		
31129	195360		4 51 48.3	-33 16 07	-0.002	+0.03	8.0	0.9	0.2	K0 III		360 s		
30869	94171		4 51 50.1	+13 39 18	+0.008	-0.03	6.27	0.50	3.7	dF6	+39	31 s	3483	m

HD	SAO	Star Name	α 2000	δ 2000	μ(α)	μ(δ)	V	B-V	M_v	Spec	RV	d(pc)	ADS	Notes
			$4^h 51^m 50^s.8$	$+42°02'49''$										
30707	39812		4 51 50.8	+42 02 49	$+0^s.003$	$-0''.02$	7.60	0.98	0.3	G8 III		280 s		
31142	195364		4 51 54.1	-34 14 20	+0.006	0.00	6.6	0.8	3.4	F5 V		27 s		
30376	13288		4 51 57.1	+66 29 46	+0.011	-0.10	8.0	0.9	3.2	G5 IV		90 s		
30913	112155		4 51 57.2	+9 52 24	+0.002	-0.01	6.79	0.42	3.0	F2 V		54 s		
31841	--		4 51 59.4	-73 00 26			7.90	0.64	2.9	G1 IV		97 s		
31274	217050		4 52 01.8	-46 51 07	-0.002	0.00	7.12	0.96	0.3	G4 III		190 s		
30582	24878		4 52 04.1	+56 25 24	0.000	-0.06	7.6	0.5	3.4	F5 V		69 s		
30442	13291		4 52 05.2	+63 30 19	+0.006	-0.10	5.44	1.57	-0.5	M2 III	-36	110 mx		m
30825	57468		4 52 10.5	+31 09 52	-0.002	-0.09	6.8	0.9	3.2	G5 IV		52 s		
30977	112163		4 52 12.2	+1 30 33	-0.001	-0.04	8.0	1.1	-0.1	K2 III		410 s		
30795	57467		4 52 13.9	+34 26 37	-0.001	0.00	7.8	0.1	1.4	A2 V		180 s		
30794	57469		4 52 20.3	+36 38 31	+0.003	-0.04	6.8	1.1	0.2	K0 III	-37	200 s		
31056	131560		4 52 20.8	-7 42 35	-0.001	-0.03	7.5	0.8	3.2	G5 IV		74 s		
30960	112162		4 52 21.0	+9 20 18	+0.001	+0.02	7.9	1.1	0.2	K0 III		310 s		
30736	39819		4 52 21.5	+45 56 23	+0.005	-0.05	6.69	0.55	3.8	F7 V	+24	34 s		
31091	149951		4 52 24.4	-13 16 16	-0.002	-0.01	7.6	0.1	1.7	A3 V		160 s		
30842	57474		4 52 27.7	+31 59 10	-0.001	-0.01	7.5	0.1	1.7	A3 V		140 s		
30793	57470		4 52 27.8	+39 17 04	+0.005	-0.04	7.91	1.32	-0.3	K5 III		190 mx		
31407	233719		4 52 28.1	-55 41 51	0.000	0.00	7.5	0.7	0.0	B8.5 V		120 s		
30990	112166		4 52 29.7	+9 51 50	-0.002	-0.02	7.7	1.1	0.2	K0 III		280 s		
30959	94176	4 o¹ Ori	4 52 31.9	+14 15 02	0.000	-0.06	4.74	1.84		M3 s	-7	32 mn		
31089	149955		4 52 34.7	-10 29 20	-0.001	-0.01	7.9	0.0	0.4	B9.5 V		320 s		
31311	217051		4 52 35.3	-43 03 37	+0.002	+0.06	7.83	1.63		Ma				
30834	57475	2 Aur	4 52 37.9	+36 42 11	-0.002	-0.01	4.78	1.41	-0.2	K3 III	-17	82 s		
31215	195372		4 52 38.3	-30 39 39	+0.002	-0.02	7.80	0.1	1.4	A2 V		190 s		m
30989	94180		4 52 41.0	+12 23 05	-0.001	-0.02	7.2	0.0	0.0	B8.5 V		240 s		
31193	169837		4 52 41.4	-25 46 31	+0.001	-0.08	7.7	0.9	3.4	F5 V		38 s		
30974	94181		4 52 45.1	+14 37 21	+0.014	-0.05	7.8	0.7	4.4	G0 V		47 s		
30912	76802		4 52 47.1	+27 53 51	+0.004	-0.03	5.97	0.37	0.6	F2 III n	+38	110 s		
30823	39826		4 52 47.7	+42 35 12	-0.001	0.00	5.71	0.11	-0.6	A0 III	-2	140 s		
31109	131568	61 ω Eri	4 52 53.6	-5 27 10	-0.002	+0.02	4.39	0.25	1.6	A9 IV	-9	36 s		
30097	5335		4 52 56.1	+75 42 36	+0.024	-0.17	7.2	0.8	3.2	G5 IV	-43	28 mx		
32495	256143		4 53 02.3	-78 24 06	0.000	+0.02	7.5	0.1	0.6	A0 V		240 s		
31975	256139		4 53 05.6	-72 24 27	-0.012	+0.27	6.28	0.52	4.0	F8 V		29 s		
30752	24894		4 53 09.8	+52 50 27	-0.001	-0.02	6.4	0.1	1.4	A2 V	-13	95 s		
31225	169847		4 53 12.0	-20 46 22	-0.002	-0.03	6.90	0.1	1.7	A3 V		110 s	3511	m
31087	112173		4 53 14.4	+8 33 21	+0.002	0.00	7.80	0.1	1.4	A2 V		180 s	3502	m
31349	195381		4 53 16.9	-38 08 40	+0.003	0.00	6.9	1.6	0.2	K0 III		99 s		
31018	76810		4 53 22.0	+24 10 38	-0.002	-0.02	7.8	0.5	4.0	F8 V		57 s		
31139	112179	5 Ori	4 53 22.7	+2 30 30	+0.002	-0.02	5.33	1.64	-0.5	gM1	+13	130 s		
30556	13294		4 53 25.5	+66 00 22	-0.002	+0.01	7.6	0.9	3.2	G5 IV		74 s		
31155	112182		4 53 28.2	+1 07 48	0.000	0.00	7.36	1.54	0.2	K0 III		110 s		
32035	256140		4 53 28.4	-72 45 53	+0.004	-0.01	7.2	0.5	0.6	A0 V		110 s		
31754	249138		4 53 30.7	-66 40 32	+0.002	0.00	6.41	1.63		K5				
31039	76812		4 53 33.9	+23 18 56	-0.001	0.00	6.6	1.1	0.2	K0 III		190 s		
30943	57483		4 53 34.8	+36 48 22	+0.002	-0.02	6.8	0.9	3.2	G5 IV		52 s		
31072	94183		4 53 35.5	+19 04 22	+0.001	+0.01	7.5	1.4	-0.3	K5 III		370 s		
31033	76811		4 53 36.3	+25 21 59	+0.001	-0.01	7.30	0.1	0.6	A0 V		210 s	3501	m
31574	233727		4 53 36.5	-57 19 56	+0.001	-0.02	7.9	0.7	3.0	F2 V		56 s		
31507	233726		4 53 36.6	-52 19 44	-0.001	+0.03	7.5	0.8	0.4	B9.5 V		88 s		
31310	169859		4 53 37.4	-26 04 53	-0.001	+0.05	7.9	1.9	-0.1	K2 III		150 s		
30685	13300		4 53 39.6	+61 28 55	+0.003	-0.04	6.6	0.1	1.4	A2 V		110 s		
31168	112185		4 53 41.7	+6 01 01	0.000	0.00	7.6	0.4	2.6	F0 V		100 s		
31071	76814		4 53 43.9	+23 33 16	+0.003	0.00	7.50	1.1	0.2	K0 III		290 s	3504	m
30530	13295		4 53 50.5	+68 13 09	+0.007	-0.06	7.4	1.1	0.2	K0 III		200 mx		
30555	13296		4 53 51.8	+67 29 17	-0.009	-0.04	7.50	0.4	3.0	F2 V		78 s	3479	m
30055	5336		4 53 52.7	+77 33 58	+0.001	-0.01	7.9	0.1	0.6	A0 V		280 s		
31138	94185		4 53 55.6	+15 55 55	0.000	-0.09	7.9	0.5	4.0	F8 V		57 s		
31209	112191		4 53 55.8	+1 34 10	0.000	0.00	6.61	0.04		A2 n	+21			
--	57486		4 53 56.0	+40 00 23	0.000	+0.01	8.0	1.7						
31000	57487		4 53 56.1	+36 45 27	0.000	+0.01	7.5	0.8	3.2	G5 IV		73 s		
31301	149968		4 54 00.6	-12 27 45	+0.001	+0.01	8.0	1.6	-0.5	M4 III		510 s		
31153	94187		4 54 03.0	+17 01 36	+0.005	0.00	7.24	0.53		F8	+54			
30614	13298	9 α Cam	4 54 03.0	+66 20 34	+0.001	+0.01	4.29	0.03	-6.2	O9.5 Ia	+6	860 s		
31392	195386		4 54 03.9	-35 24 16	+0.008	+0.16	7.7	0.9	3.2	G5 IV		66 s		
31430	195387		4 54 04.3	-38 09 47	+0.001	0.00	6.7	1.5	0.2	K0 III		98 s		
30988	39841		4 54 04.7	+42 11 51	+0.001	+0.01	7.94	1.55	0.0	K1 III		210 s		
31137	76820		4 54 12.8	+23 59 56	-0.004	-0.04	8.0	0.4	3.0	F2 V		98 s		
30924	39837		4 54 13.5	+49 34 53	-0.001	+0.01	7.9	0.7	4.4	G0 V		50 s	3498	m
30792	24897		4 54 14.8	+58 38 14	0.000	0.00	8.0	0.0	0.4	B9.5 V		300 s		
31237	112197	8 π⁵ Ori	4 54 15.0	+2 26 26	0.000	0.00	3.72	-0.18	-3.6	B2 III	+23	290 s		v
31308	131588		4 54 27.2	-5 25 15	+0.001	0.00	7.9	1.1	0.2	K0 III		340 s		
31297	131587		4 54 27.8	-3 13 35	+0.001	-0.02	7.40	0.5	3.4	F5 V		63 s		m
31222	94190		4 54 28.9	+12 27 04	+0.007	-0.04	7.8	0.8	3.2	G5 IV		80 s		
31406	169876		4 54 32.7	-26 43 45	0.000	+0.02	8.0	1.8	0.2	K0 III		120 s		

HD	SAO	Star Name	α 2000	δ 2000	μ(α)	μ(δ)	V	B-V	M_v	Spec	RV	d(pc)	ADS	Notes
31640	233729		4 54 33.3	-52 31 25	-0.003	-0.02	7.5	0.3	1.7	A3 V		110 s		
31495	195395		4 54 35.4	-39 05 21	-0.002	+0.02	7.7	0.9	3.2	G5 IV		70 s		
31255	94191		4 54 39.7	+11 56 09	0.000	-0.02	8.0	0.1	0.6	A0 V		260 s		
31445	169879		4 54 40.8	-28 52 48	+0.004	+0.11	7.7	1.5	0.2	K0 III		160 mx		
31254	94192		4 54 41.0	+12 01 33	-0.001	-0.01	7.45	0.14	0.6	A0 V		220 s		
31460	195394		4 54 42.5	-31 26 03	0.000	-0.29	8.04	0.58	4.4	G0 V		53 s		
31253	94193		4 54 43.8	+12 21 06	+0.008	-0.06	7.2	0.5	4.0	F8 V		42 s		
31283	94197	6 Ori	4 54 46.8	+11 25 34	-0.001	+0.02	5.19	0.12		A3	+9			
31296	112203		4 54 47.7	+7 46 44	-0.001	-0.03	5.33	1.22	0.0	gK1	-5	95 s		
31429	169882		4 54 48.9	-25 39 10	-0.005	0.00	8.0	1.3	0.2	K0 III		210 mx		
31331	112206		4 54 50.6	+0 28 03	0.000	0.00	5.99	-0.12	-1.1	B5 V	+17	260 s		
31069	39851		4 54 51.2	+44 03 39	+0.003	-0.05	6.08	-0.02	0.2	B9 V	+1	150 s		
31746	233733		4 54 53.0	-58 32 50	+0.013	+0.08	6.12	0.44		F5				
31295	94201	7 π¹ Ori	4 54 53.7	+10 09 03	+0.003	-0.13	4.65	0.09		A0 p	+13			m
31529	195400		4 54 54.7	-39 37 43	-0.001	+0.02	6.10	1.42		K0				
31268	94200		4 54 54.8	+16 23 11	+0.001	-0.01	7.1	0.1	0.6	A0 V		200 s		
31306	112205		4 54 56.1	+8 36 01	-0.003	0.00	6.94	0.15	0.6	A0 V		150 s	3517	m
31236	94199		4 54 58.3	+19 29 06	+0.004	-0.04	6.37	0.29	2.6	dF0	+35	57 s		
31282	94202		4 55 00.2	+14 35 37	+0.001	-0.01	7.5	0.4	2.6	F0 V		92 s		
31098	39855		4 55 01.8	+42 55 54	+0.002	-0.02	7.54	1.42	-0.1	K2 III		240 s		
30958	24904	5 Cam	4 55 03.1	+55 15 33	-0.001	-0.01	5.60	0.1	0.6	A0 V	+2	97 s	3508	m
31414	149985		4 55 06.7	-16 44 26	0.000	0.00	5.70	0.96	0.2	gG9	+10	130 s		
31875	249144		4 55 06.9	-62 47 42	+0.002	+0.09	8.04	0.11	1.7	A3 V		110 mx		
31506	195399		4 55 08.1	-30 17 11	-0.001	+0.01	7.4	0.1	0.6	A0 V		190 s		
31721	233734		4 55 08.7	-54 25 51	-0.001	-0.01	7.4	1.8	0.2	K0 III		92 s		
31032	24910		4 55 10.2	+51 36 17	+0.003	-0.02	6.9	0.1	0.6	A0 V		170 s		
31930	249146		4 55 10.8	-63 57 33	-0.001	+0.01	7.9	0.5	4.0	F8 V		60 s		
32440	256145	η Men	4 55 11.3	-74 56 13	+0.007	+0.06	5.47	1.52	-0.3	K6 III	+26	140 s		
31118	39859		4 55 14.8	+43 24 59	0.000	-0.04	7.01	1.81	-4.4	K5 Ib	+19	1400 s		
31444	149988		4 55 18.6	-16 25 04	0.000	+0.04	5.72	0.88	0.3	gG4	+32	110 s		R Eri, q
31315	94206		4 55 19.5	+16 37 11	0.000	-0.03	7.2	0.1	0.6	A0 V		210 s		
31603	195405		4 55 20.0	-39 45 33	0.000	-0.01	7.5	2.3	3.2	G5 IV		72 s		
30934	24907		4 55 22.2	+58 50 10	+0.003	-0.03	7.9	1.1	0.2	K0 III		310 s		
31604	195407		4 55 25.2	-39 45 14	0.000	+0.10	7.4	1.7	0.2	K0 III		100 s		
31234	57503		4 55 28.1	+32 09 18	+0.001	-0.01	7.4	0.1	0.6	A0 V		210 s		
30751	13303		4 55 28.2	+67 32 53	0.000	-0.01	7.9	1.4	-0.3	K5 III		400 s		
31517	169889		4 55 30.1	-25 43 40	+0.001	+0.03	6.72	0.27	2.6	dF0		67 s		
31233	57504		4 55 30.4	+32 47 03	-0.002	0.00	7.3	1.1	0.2	K0 III	-50	240 s		
31294	76841		4 55 34.6	+27 12 10	-0.002	-0.01	7.4	1.1	0.2	K0 III	-2	260 s		
31527	169891		4 55 38.3	-23 14 31	-0.005	+0.13	7.6	0.5	4.4	G0 V		44 s		
31753	233737		4 55 38.6	-51 58 51	-0.002	+0.01	7.5	2.4	-0.1	K2 III		60 s		
31338	94210		4 55 41.6	+20 00 06	+0.012	-0.33	7.98	0.80	5.9	dK0	+27	26 s		
31293	57506		4 55 45.8	+30 33 04	+0.001	-0.03	7.05	0.11		A0 pe	+8	21 mn		AB Aur, v
31305	57507		4 55 48.1	+30 20 16	0.000	-0.03	7.56	0.11	0.6	A0 V		210 s		
31373	94212		4 55 50.2	+15 02 24	+0.001	-0.02	5.70	-0.08	-1.2	B8 III	+9	240 s		
31374	94213		4 55 52.1	+13 37 51	0.000	0.00	7.81	0.12	0.6	B9.5 V		270 s		
31195	39872		4 55 54.8	+45 02 04	+0.002	-0.02	7.86	0.03	-0.6	B7 V		410 s		
31412	112219		4 55 55.8	+4 40 13	+0.009	-0.19	7.02	0.55	4.0	F8 V		38 s		
31411	112220		4 55 58.3	+5 23 57	-0.001	-0.01	6.50	0.02		A0 n	+22			
31516	150000		4 55 58.5	-14 52 26	+0.001	+0.06	7.8	0.1	1.7	A3 V		160 s		
31437	112223		4 56 01.0	+1 37 31	0.000	+0.04	6.9	0.2	2.1	A5 V		93 s		
31220	39875		4 56 03.2	+43 29 37	-0.002	-0.01	7.38	1.80	-4.7	M0 Ib		2000 s		
31363	76846		4 56 06.5	+22 34 35	+0.002	-0.06	7.4	1.1	0.2	K0 III		180 mx		
31134	24919		4 56 07.1	+52 52 11	0.000	+0.01	5.7	0.1	1.4	A2 V	-22	72 s		
31423	112225		4 56 08.8	+7 54 17	-0.007	-0.05	6.5	0.5	3.4	F5 V		42 s		
31688	195413		4 56 09.4	-38 56 27	-0.001	-0.02	7.8	1.4	-0.1	K2 III		280 s		
31453	112231		4 56 10.7	+2 02 36	-0.001	-0.01	7.9	0.1	1.7	A3 V		170 s		
31571	150006		4 56 12.6	-17 44 11	0.000	0.00	7.50	0.9	3.2	G5 IV		72 s		m
31382	76850		4 56 14.2	+21 34 21	+0.002	-0.02	7.3	0.4	2.6	F0 V		87 s		
31362	76848		4 56 15.5	+24 35 32	-0.002	-0.02	6.37	0.33	2.6	F0 V	-9	57 s		
31166	24922		4 56 16.6	+52 05 50	0.000	-0.05	7.8	0.5	4.0	F8 V		56 s	3520	
31906	233742		4 56 17.1	-56 25 36	+0.004	+0.04	7.3	1.6	0.2	K0 III		120 s		
31327	57511		4 56 19.9	+36 10 08	-0.001	+0.01	6.07	0.41	-5.7	B2 Ib	-5	1000 s		
31421	94218	9 o² Ori	4 56 22.2	+13 30 52	-0.005	-0.05	4.07	1.15	-0.1	K2 III	+1	68 s	3540	m
31512	131614	62 Eri	4 56 24.1	-5 10 17	-0.001	-0.01	5.51	-0.13		B9	+24			m
31585	150008		4 56 26.1	-15 22 50	-0.001	+0.03	7.6	0.2	2.1	A5 V		130 s		
31790	217069		4 56 34.8	-43 25 57	0.000	0.00	7.5	2.3	-0.3	K5 III		98 s		
31489	112240		4 56 43.7	+8 49 20	+0.001	0.00	7.49	0.22	1.4	A2 V		130 s		
31569	131619		4 56 44.6	-4 46 53	-0.002	-0.01	7.8	0.0	0.4	B9.5 V		300 s		
31476	94220		4 56 47.5	+16 44 03	+0.001	-0.03	7.4	1.1	0.2	K0 III		280 s		
31636	150011		4 56 47.7	-16 08 07	-0.002	-0.02	7.80	0.1	0.6	A0 V		280 s	3556	m
31189	24925		4 56 49.5	+55 49 26	-0.001	-0.03	6.9	1.4	-0.3	K5 III	-1	250 s		
31826	217072		4 56 50.9	-44 11 35	+0.005	+0.01	7.81	1.15		K1 III-IV		280 mx		
31381	57518		4 56 51.7	+33 58 38	+0.001	-0.01	8.0	0.9	3.2	G5 IV		90 s		
31998	233753		4 56 52.2	-56 23 26	-0.001	-0.03	7.5	1.6	3.2	G5 IV		23 s		

HD	SAO	Star Name	α 2000	δ 2000	μ(α)	μ(δ)	V	B-V	M_V	Spec	RV	d(pc)	ADS	Notes
31152	24924		4 56 52.6	+58 05 57	+0.001	0.00	8.0	1.1	0.2	K0 III		320 s		
31946	233750		4 56 52.9	-52 51 01	-0.001	-0.02	7.5	1.1	1.4	A2 V		42 s		
31555	112248		4 56 57.1	+2 46 47	+0.001	-0.05	7.7	1.1	0.2	K0 III		320 s		
31625	131622		4 56 58.5	-8 27 32	-0.001	0.00	6.8	0.2	2.1	A5 V		89 s		
31825	217074		4 56 58.7	-41 59 02	+0.002	+0.01	7.6	0.5	1.4	A2 V		90 s		
31398	57522	3 ι Aur	4 56 59.5	+33 09 58	0.000	-0.02	2.69	1.53	-2.3	K3 II	+18	82 s		
31292	39884		4 57 02.6	+49 55 18	+0.007	-0.13	6.9	0.4	2.6	F0 V		71 mx		
31326	39886		4 57 04.7	+47 52 45	+0.002	+0.02	7.89	0.00	0.6	A0 V		290 s		
31594	112255		4 57 04.8	+0 13 17	0.000	-0.03	7.81	1.12	0.2	K0 III		290 s		
32012	233756		4 57 05.8	-55 36 44	-0.001	-0.01	7.8	1.2	1.4	A2 V		41 s		
32255	249154		4 57 06.7	-65 00 57	+0.004	+0.04	7.3	-0.4	0.6	A0 V		220 s		
32065	233757		4 57 07.4	-58 04 11	-0.001	+0.09	6.9	0.3	3.0	F2 V		60 s		
31177	24927		4 57 09.5	+59 07 14	-0.001	-0.02	6.8	1.6	-0.5	M2 III		270 s		
31554	94225		4 57 13.1	+10 48 37	-0.001	+0.02	7.89	0.29	0.6	A0 V		190 s		
31151	13317		4 57 15.2	+61 45 11	-0.001	0.00	7.10	0.6	4.4	G0 V	-21	34 s	3526	m
31623	131625		4 57 17.1	-1 04 03	-0.003	-0.04	6.23	0.42		F2	+13			
31278	24929	7 Cam	4 57 17.1	+53 45 08	-0.003	+0.01	4.47	-0.02	1.2	A1 V	-8	45 s	3536	m
31539	94227		4 57 22.3	+17 09 13	-0.001	-0.01	5.48	1.31	0.2	K0 III	+25	74 s		
31067	13316		4 57 31.6	+65 16 43	+0.003	-0.03	6.8	0.9	3.2	G5 IV		51 s		
31943	217082		4 57 40.7	-43 01 56	+0.009	-0.05	7.4	1.3	0.6	A0 V		21 mx		
31743	150028		4 57 44.5	-13 42 19	-0.001	-0.02	7.7	0.0	0.4	B9.5 V		290 s		
31851	195435		4 57 44.6	-33 04 07	0.000	-0.01	7.8	1.3	-0.1	K2 III		310 s		
31726	150029		4 57 44.7	-14 13 53	0.000	+0.01	6.15	-0.21	-3.5	B1 V	+11	850 s		
31860	195436		4 57 46.2	-34 53 31	+0.003	+0.04	7.7	0.9	0.2	K0 III		310 s		
31014	13315		4 57 47.4	+67 46 34	0.000	-0.03	7.1	0.0	0.0	B8.5 V		250 s		
31693	131634		4 57 48.3	-4 39 17	+0.001	-0.01	7.6	0.9	3.2	G5 IV		75 s		
31553	76858	99 Tau	4 57 48.6	+23 56 55	+0.001	-0.01	5.79	1.11	0.2	K0 III	+4	110 s	3557	m
--	--		4 57 49.3	+48 00 48	-0.001	-0.03	7.8	0.1	0.6	A0 V		260 s		
31813	169923		4 57 49.4	-25 03 37	+0.001	+0.01	8.0	1.5	-0.3	K5 III		450 s		
31451	57529		4 57 50.6	+39 37 56	+0.001	-0.02	8.0	0.9	3.2	G5 IV		91 s		
31712	131635		4 57 54.4	-6 06 09	0.000	-0.04	7.2	1.1	0.2	K0 III		260 s		
32064	233762		4 57 58.1	-50 34 32	-0.001	+0.01	7.60	0.8	3.2	G5 IV		76 s		m
31552	57535		4 58 01.3	+30 07 33	+0.002	-0.02	7.9	0.1	0.6	A0 V		270 s		
32227	233767		4 58 02.3	-58 12 29	+0.002	+0.01	6.86	1.50	-0.3	K4 III		240 s		
31836	169926		4 58 04.3	-22 01 09	0.000	+0.02	7.4	0.4	0.6	A0 V		120 s		
31592	76862	98 Tau	4 58 09.4	+25 03 01	+0.002	-0.05	5.60	0.00	0.4	B9.5 V	+26	100 s	3547	m
31810	150038		4 58 10.3	-16 46 49	0.000	+0.01	6.6	0.1	1.4	A2 V		110 s		
31739	131640		4 58 10.7	-2 12 45	-0.001	+0.03	6.35	0.10		A0			3570	m
31500	57533		4 58 13.1	+40 03 24	+0.007	-0.11	7.4	0.5	3.4	F5 V		61 s		
31904	195443		4 58 14.7	-30 57 33	+0.002	+0.02	8.0	1.5	0.2	K0 III		190 s		
31738	112278		4 58 16.8	+0 27 13	-0.011	-0.05	7.10	0.71	3.2	G5 IV		60 mx		
32363	249156		4 58 21.9	-63 12 48	-0.006	-0.04	7.9	0.5	4.0	F8 V		60 s		
31324	24934		4 58 22.3	+58 36 56	-0.001	-0.03	7.80	0.8	0.3	G7 III	+35	290 s		
31768	131645		4 58 26.1	-3 44 02	+0.004	0.00	7.0	0.5	3.4	F5 V		53 s		
31550	57539		4 58 28.3	+37 19 45	+0.002	-0.05	6.7	0.2	2.1	A5 V		83 s		
33245	256148		4 58 30.7	-78 11 10	+0.007	+0.04	7.6	0.5	4.0	F8 V		52 s		
32208	233768		4 58 32.5	-54 10 18	-0.001	-0.07	7.5	1.0	2.6	F0 V		35 s		
32299	233771		4 58 32.6	-59 07 55	+0.003	-0.05	8.0	1.0	-0.1	K2 III		410 s		
31767	112281	10 π⁶ Ori	4 58 32.8	+1 42 51	0.000	0.00	4.47	1.40	-2.2	K2 II	+14	190 s		
31939	169937		4 58 33.3	-28 53 18	+0.002	-0.02	7.4	1.9	-0.3	K5 III		200 s		
31848	150044		4 58 36.3	-14 47 02	+0.002	+0.02	7.5	0.1	1.4	A2 V		160 s		
31887	150045		4 58 44.6	-17 48 26	0.000	+0.02	7.1	1.4	-0.3	K5 III		310 s		
31648	76866		4 58 45.9	+29 50 36	-0.003	-0.04	7.66	0.17	1.4	A2 V		160 s		
32138	217093		4 58 48.6	-45 17 25	-0.001	+0.03	7.6	0.9	0.6	A0 V		140 s		
34172	258395	ξ Men	4 58 50.9	-82 28 14	-0.008	+0.01	5.85	0.93	0.3	G8 III		130 s		
31679	76868		4 58 52.7	+24 29 45	+0.001	0.00	8.0	0.0	-1.0	B5.5 V		550 s		
32154	217097		4 58 55.9	-43 41 36	+0.003	+0.07	7.8	1.4	4.0	F8 V		10 s		
31747	94239		4 58 57.1	+14 32 59	0.000	+0.01	7.60	0.05	-0.9	B6 V		390 s		
31149	13325		4 58 58.1	+69 06 42	+0.001	+0.01	8.0	1.1	0.2	K0 III		330 s		
31764	94240		4 58 59.3	+14 32 34	0.000	-0.02	5.85	0.04	-0.9	B6 V		170 s	3579	m
32169	217099		4 59 00.2	-43 23 32	-0.002	-0.03	7.4	1.4	3.4	F5 V		20 s		
31925	150052		4 59 01.2	-16 22 33	-0.010	+0.15	5.66	0.43	3.0	dF2	+31	32 s	3588	m
31487	24941		4 59 03.7	+51 56 32	-0.001	0.00	8.0	1.1	0.2	K0 III		330 s		
31858	131657		4 59 05.6	-2 08 12	-0.002	+0.02	8.0	0.1	1.4	A2 V		210 s		
32030	169947		4 59 08.6	-28 26 27	-0.001	+0.04	7.5	1.0	3.2	G5 IV		56 s		
31647	57548	4 Aur	4 59 15.3	+37 53 24	+0.004	-0.10	4.94	0.04	0.6	A0 V	+5	69 s	3572	m
31510	24942		4 59 17.6	+52 29 34	+0.002	-0.06	8.0	1.1	0.2	K0 III		210 mx		
32079	195455		4 59 19.2	-32 14 21	+0.003	+0.03	6.7	1.5	0.2	K0 III		94 s		
32152	195458		4 59 20.3	-36 52 15	-0.006	+0.14	7.1	0.1	3.4	F5 V		55 s		
31617	39918		4 59 20.8	+43 19 25	-0.004	0.00	7.40	0.00	-3.0	B2 IV	+4	190 mx		
31936	131664		4 59 26.5	-9 09 57	0.000	+0.01	8.0	0.1	1.4	A2 V		210 s		
32278	217101		4 59 29.5	-49 27 29	+0.001	0.00	7.1	0.2	2.6	F0 V		80 s		
32078	169950		4 59 35.8	-25 22 09	0.000	0.00	8.0	0.7	1.7	A3 V		72 s		
31996	150058		4 59 36.4	-14 48 21	+0.001	+0.02	5.9	1.8		C6 II	+32			R Lep, v
31782	76876		4 59 36.7	+25 56 10	-0.004	-0.07	7.29	0.80	3.2	K0 IV		66 s		

HD	SAO	Star Name	α 2000	δ 2000	μ(α)	μ(δ)	V	B-V	M$_v$	Spec	RV	d(pc)	ADS	Notes
			h m s	° ′ ″	s	″								
32062	169949		4 59 38.0	-22 23 59	+0.007	-0.08	7.9	0.7	3.2	G5 IV		76 mx		
32313	217103		4 59 41.4	-49 07 25	-0.002	+0.01	7.9	1.3	3.2	G5 IV		60 s		
31664	39921		4 59 42.3	+41 52 24	-0.007	+0.01	6.72	0.96	0.2	K0 III		200 s		
31845	94248		4 59 44.3	+15 55 01	+0.006	-0.02	6.75	0.45	3.4	F5 V	+44	46 s		m
31579	24943	8 Cam	4 59 46.3	+53 09 20	-0.001	-0.01	6.08	1.46	-0.1	K2 III	-2	120 s		
32008	150060	63 Eri	4 59 50.4	-10 15 48	+0.001	-0.13	5.38	0.80	0.2	K0 III	-12	110 s		
32224	195467		4 59 51.3	-36 37 27	0.000	-0.01	8.0	1.2	0.2	K0 III		290 s		
32762	249164		4 59 52.2	-68 34 55	-0.001	+0.02	8.03	0.24	1.7	A3 V		150 s		
31806	76880		4 59 53.6	+27 19 30	+0.001	-0.05	6.80	0.00	0.4	B9.5 V		180 s	3587	m
32045	150064	64 Eri	4 59 55.7	-12 32 15	+0.003	-0.09	4.79	0.26	1.7	F0 IV	-15	42 s		S Eri, q
32182	169959		4 59 56.3	-29 34 38	-0.002	-0.04	8.0	1.5	0.2	K0 III		190 s		
32135	169957		4 59 57.8	-25 03 15	+0.001	+0.05	7.4	0.6	2.6	F0 V		64 s		
32292	217104		5 00 00.6	-43 17 35	0.000	+0.06	7.8	1.7	3.4	F5 V		48 s		
32206	195469		5 00 01.1	-34 10 01	-0.001	+0.01	7.8	0.0	0.6	A0 V		250 s		
31918	94249		5 00 01.5	+10 23 28	0.000	+0.01	8.00	0.30	1.7	A3 V		140 s		m
31691	39925		5 00 01.8	+43 59 06	+0.001	-0.04	8.00	1.17	-2.0	G0 II		580 s		
32290	217106		5 00 07.9	-41 02 53	+0.002	+0.01	7.8	0.4	1.4	A2 V		110 s		
31264	13331		5 00 08.9	+69 57 32	+0.001	+0.01	8.00	0.1	0.6	A0 V		280 s		m
33519	256153		5 00 13.3	-78 18 01	-0.007	-0.01	6.29	1.51		K0				m
32507	233781		5 00 13.7	-57 36 28	+0.001	+0.02	7.9	0.2	0.6	A0 V		200 s		
32413	233779		5 00 14.5	-51 51 28	-0.001	0.00	7.7	1.9	0.2	K0 III		92 s		
31867	76883		5 00 17.5	+25 08 12	+0.005	+0.01	7.96	0.68	4.7	G2 V	-27	41 s		
31761	57559	5 Aur	5 00 18.2	+39 23 41	-0.001	0.00	5.95	0.41	3.4	F5 V	+6	32 s	3589	m
30338	783		5 00 20.5	+81 11 39	-0.001	+0.03	5.2	1.1	-0.2	K3 III	-8	120 s		
31780	57560	6 Aur	5 00 23.2	+39 39 18	0.000	+0.01	6.7	1.4	-0.3	K5 III	-24	230 s		
32252	195478		5 00 24.1	-31 48 50	-0.003	-0.01	8.0	0.3	0.6	A0 V		180 s		
32022	112303		5 00 29.5	+5 05 56	+0.002	-0.03	8.02	0.57	3.2	G5 IV		92 s	3596	m
32039	112304		5 00 32.5	+3 36 55	+0.001	0.00	6.07	-0.07	0.6	A0 V	+31	120 s		
32714	249165		5 00 32.9	-64 23 40	+0.009	-0.06	8.0	1.1	0.2	K0 III		190 mx		
31817	57564		5 00 33.3	+36 37 38	0.000	+0.01	7.2	0.1	0.6	A0 V		200 s		
31966	94253		5 00 33.8	+14 23 02	+0.003	+0.05	6.7	0.9	5.2	G5 V		20 s		
32040	112305		5 00 33.8	+3 36 58	0.000	-0.01	6.60	0.1	0.6	A0 V	+42	160 s	3597	m
32115	131684		5 00 39.7	-2 03 57	-0.002	+0.01	6.32	0.28		A5				
32179	150076		5 00 40.3	-13 30 15	-0.002	-0.02	7.50	0.1	0.6	A0 V		240 s	3606	m
31816	39934		5 00 41.5	+40 13 57	+0.002	+0.01	7.4	1.4	-0.3	K5 III	-3	330 s		
32021	94256		5 00 42.6	+10 54 50	+0.001	+0.01	6.81	0.04	0.4	B9.5 V		170 s		
32426	217115		5 00 45.8	-47 36 14	+0.001	-0.02	7.7	1.6	-0.1	K2 III		240 s		
32073	112308		5 00 47.1	+4 34 03	+0.005	-0.04	7.1	1.1	0.2	K0 III		140 mx		
32147	131688		5 00 49.0	-5 45 12	+0.037	-1.09	6.22	1.06	6.4	K3 V	+27	9.2 t		
30841	5373		5 00 53.6	+77 46 31	-0.009	0.00	7.7	0.5	4.0	F8 V		55 s		
32400	217114		5 00 56.2	-42 00 11	-0.002	+0.01	7.5	2.0	-0.5	M2 III		400 s		
32145	112316		5 01 05.9	+3 43 02	-0.001	-0.01	6.9	0.0	0.0	B8.5 V		240 s		
31646	24949		5 01 06.9	+58 39 57	+0.001	-0.04	7.4	0.8	3.2	G5 IV		68 s		
31690	24953		5 01 07.7	+56 16 11	+0.001	-0.04	7.8	0.8	3.2	G5 IV		82 s		
33244	256152		5 01 09.8	-74 20 25	+0.004	+0.04	7.50	0.1	0.6	A0 V		160 mx		m
32310	169979		5 01 18.3	-23 42 27	+0.002	+0.08	7.60	0.4	3.0	F2 V		83 s	3618	m
31844	39943		5 01 18.5	+45 26 46	+0.002	-0.01	7.93	0.04	0.4	B9.5 V		310 s		
32287	150082		5 01 18.9	-17 33 29	0.000	-0.05	8.0	0.5	3.4	F5 V		82 s		
32018	76895		5 01 18.9	+26 32 04	0.000	0.00	7.8	0.0	-1.0	B5.5 V		500 s	3602	m
32096	94267		5 01 19.6	+14 07 02	+0.002	-0.04	7.9	0.9	3.2	G5 IV		88 s		
32112	94265		5 01 20.7	+15 44 41	0.000	+0.01	8.0	1.1	-0.1	K2 III		410 s		
32370	195491		5 01 21.2	-31 23 38	+0.003	0.00	8.0	1.5	0.2	K0 III		190 s		
31855	39945		5 01 21.4	+45 53 25	-0.001	-0.02	7.5	0.4	3.0	F2 V		77 s		
31895	39947		5 01 22.2	+42 00 23	-0.002	+0.02	7.84	1.34	-2.2	K1 II		880 s		
31913	57573		5 01 23.1	+39 57 38	+0.001	+0.01	8.00	0.5	-6.3	F8 I p	-21	3500 s		RX Aur, v
32309	169981		5 01 25.5	-20 03 07	+0.002	-0.01	4.91	-0.05	0.2	B9 V	+24	88 s		
32219	131697		5 01 25.6	-0 41 50	0.000	-0.03	6.6	0.1	1.7	A3 V		96 s		
32249	131700	65 ψ Eri	5 01 26.3	-7 10 26	0.000	+0.01	4.81	-0.19	-2.5	B2 V	+25	290 s		
32369	195494		5 01 28.0	-30 14 06	-0.001	-0.04	7.7	1.2	0.2	K0 III		250 s		
31866	39948		5 01 28.2	+44 23 54	-0.001	+0.02	7.31	0.22	1.7	A3 V		110 s		
32453	195501		5 01 34.5	-39 43 05	-0.001	+0.03	6.03	0.88		G5				
31662	13341		5 01 35.9	+61 04 41	+0.002	-0.17	6.03	0.41	3.4	F5 V	+11	34 s	3590	m
32517	217124		5 01 36.2	-44 49 50	+0.001	-0.08	8.0	0.8	3.2	G5 IV		86 s		
33701	256157		5 01 41.6	-78 10 59	+0.016	+0.06	7.6	1.1	0.2	K0 III		260 mx		
29475	772		5 01 42.5	+84 52 20	0.000	-0.01	7.6	0.1	1.4	A2 V		170 s		
31977	57580		5 01 42.6	+38 47 21	+0.001	+0.01	7.8	1.1	0.2	K0 III		310 s		
31779	24958		5 01 43.4	+55 34 30	+0.002	-0.02	7.5	0.1	1.7	A3 V		140 s		
32092	76903		5 01 44.2	+26 40 15	-0.001	+0.02	6.81	0.48	4.7	G2 V		26 s	3608	m
32036	57582		5 01 45.9	+31 46 42	-0.001	0.00	7.70	0.15	0.4	B9.5 V		230 s		
32093	76904		5 01 46.3	+26 39 03	+0.001	+0.03	6.74	0.48	4.7	G2 V		26 s		
31757	24957		5 01 46.5	+57 08 01	+0.002	0.00	8.0	0.9	3.2	G5 IV		90 s		
32202	94274		5 01 47.5	+11 22 29	-0.001	-0.03	7.19	0.03	0.0	B8.5 V		240 s		m
32308	150086		5 01 47.8	-10 56 03	+0.002	+0.01	6.8	0.8	3.2	G5 IV		54 s		
32263	112334		5 01 50.4	+0 43 19	+0.001	-0.03	5.92	1.27		K0	+21			
32491	195503		5 01 54.4	-36 58 43	+0.002	-0.06	7.6	0.6	4.4	G0 V	.	41 s		

HD	SAO	Star Name	α 2000	δ 2000	μ(α)	μ(δ)	V	B-V	M$_v$	Spec	RV	d(pc)	ADS	Notes
33563	256155		5 01 56.3	-76 37 41	-0.008	-0.12	7.52	0.48	3.4	F5 V		63 s		m
32368	--		5 01 57.0	-19 40 10			7.7	0.3	0.6	A0 V		180 s		
32111	76906		5 01 57.3	+28 17 29	-0.004	+0.01	8.00	0.1	0.6	A0 V		190 mx	3613	m
31964	39955	7 ε Aur	5 01 58.1	+43 49 24	0.000	0.00	2.99	0.54	-8.5	F0 Ia	-3	1400 s	3605	m,v
32273	112340		5 01 59.9	+1 36 32	-0.001	+0.01	6.24	-0.04		B8			3623	m
--	112341		5 02 00.6	+1 36 42	-0.001	+0.02	6.50	0.1	0.6	A0 V		150 s	3623	m
32306	131711		5 02 01.0	-5 30 03	+0.002	-0.03	6.7	0.4	2.6	F0 V		66 s		
32588	217130		5 02 03.5	-45 35 57	-0.002	+0.02	7.8	0.8	0.6	A0 V		150 s		
31499	13338		5 02 05.7	+69 09 38	+0.017	-0.08	7.10	0.5	3.4	F5 V	+18	54 s		m
32489	195505		5 02 07.5	-31 20 55	+0.002	-0.02	7.2	1.4	0.2	K0 III		150 s		
32436	169997		5 02 09.8	-26 16 30	+0.006	-0.08	5.02	1.07	0.2	gK0	+27	81 s		
32778	233796		5 02 16.8	-56 05 00	-0.008	+0.62	7.03	0.64	5.2	G5 V		24 ts		m
31312	5385		5 02 20.1	+74 16 09	+0.006	+0.04	6.06	1.57	-0.3	K5 III	-52	180 s		
32259	94280		5 02 20.2	+13 54 37	-0.002	-0.10	7.8	0.7	4.4	G0 V		49 s		
32515	195509		5 02 22.7	-31 46 17	-0.001	+0.08	5.94	1.17	0.3	G8 III		99 s		
32068	39966	8 ζ Aur	5 02 28.6	+41 04 33	+0.001	-0.02	3.75	1.22	-2.3	K4 II	+13	160 s		v
32304	112350		5 02 31.5	+7 25 28	-0.001	0.00	7.0	0.8	3.2	G5 IV		59 s		
32711	233795		5 02 31.8	-50 26 43	+0.003	+0.02	8.0	0.3	1.7	A3 V		130 s		
31990	39962		5 02 36.7	+49 31 50	+0.002	0.00	8.0	1.1	0.2	K0 III		330 s		
32395	131714		5 02 37.7	-8 53 40	-0.001	-0.01	7.3	1.1	0.2	K0 III		260 s		
33285	256154	β Men	5 02 43.2	-71 18 50	+0.001	+0.02	5.31	1.00		G8	-11	17 mn		
32359	112356		5 02 44.5	+3 27 27	0.000	-0.01	8.0	0.0	0.4	B9.5 V		330 s		
32503	170010	1 Lep	5 02 44.7	-22 47 43	0.000	0.00	5.75	1.20	0.0	gK1	+33	120 s		
--	170010		5 02 44.9	-22 47 42	+0.004	+0.02	5.8	1.1	0.2	K0 III		130 s		
32393	131715		5 02 45.4	-4 12 35	+0.003	+0.02	5.85	1.21	-6.0	cK3	+38	2300 s		
32050	39967		5 02 46.5	+46 39 36	0.000	-0.03	7.6	0.0	0.4	B9.5 V		260 s		
--	24969		5 02 47.0	+53 45 33	-0.001	+0.01	8.0	2.4	-0.3	K5 III		130 s		
32743	217140	η¹ Pic	5 02 48.6	-49 09 05	-0.005	+0.03	5.38	0.42	3.3	dF4	+21	26 s		m
31675	13344		5 02 50.4	+66 49 22	+0.012	-0.34	6.19	0.48	4.0	F8 V	+17	32 ts		
32760	217144		5 02 55.8	-49 42 12	-0.005	+0.07	7.3	1.1	0.2	K0 III		220 s		
32678	217136		5 02 57.3	-41 36 28	+0.002	+0.02	7.9	1.6	-0.1	K2 III		230 s		
32143	39974		5 02 57.3	+42 34 08	-0.001	-0.01	7.45	-0.06		B8				
32348	94285		5 03 01.7	+12 03 34	0.000	+0.05	7.25	0.35	3.0	F2 V		71 s		
32468	131720		5 03 01.8	-8 39 48	-0.001	-0.01	7.10	0.1	0.6	A0 V		200 s	3640	m
32693	217138		5 03 01.9	-42 29 54	-0.001	-0.03	6.9	1.4	-0.5	M2 III		300 s		
32366	112360		5 03 02.1	+9 22 56	+0.001	-0.01	7.59	-0.02	0.6	A0 V		250 s		
32301	76920	102 ι Tau	5 03 05.7	+21 35 24	+0.005	-0.04	4.64	0.15	2.4	A7 V	+42	28 mn		
32189	39975		5 03 07.2	+40 33 12	+0.001	0.00	7.7	0.4	2.6	F0 V		100 s		
32821	233803		5 03 11.6	-50 19 19	-0.002	-0.01	7.4	1.3	2.6	F0 V		22 s		
32392	112362		5 03 12.7	+7 12 38	0.000	-0.04	7.9	0.5	4.0	F8 V		60 s		
32806	217150		5 03 15.7	-49 29 34	+0.002	+0.01	7.1	1.1	0.2	K0 III		220 s		
32188	39979		5 03 18.6	+41 26 30	0.000	0.00	6.14	0.16	-0.6	A0 III	-1	170 s		
31911	13349		5 03 19.5	+60 25 19	-0.002	-0.04	7.43	0.29		A5			3615	m
31910	13351	10 β Cam	5 03 25.1	+60 26 32	0.000	-0.01	4.03	0.92	-4.5	G0 Ib	-2	460 s	3615	m
32270	57596		5 03 38.2	+37 16 05	+0.001	-0.04	7.5	0.0	0.0	B8.5 V	+9	290 s		
32526	131731		5 03 42.7	-2 32 24	+0.001	+0.01	6.80	1.1	0.2	K0 III		210 s	3650	m
32483	112370		5 03 45.5	+6 38 25	0.000	-0.03	7.6	1.1	0.2	K0 III		300 s		
33000	233810		5 03 47.9	-56 18 09	+0.002	+0.06	7.9	0.7	2.6	F0 V		63 s		
32387	76929		5 03 51.9	+24 58 22	+0.009	-0.11	8.00	0.9	5.6	G8 V	+57	30 s		
32612	150109		5 03 51.9	-14 22 10	-0.001	+0.03	6.41	-0.18		B3	+16			
32667	170029		5 03 53.2	-24 23 16	+0.001	-0.03	5.61	0.10		A3	+7			
32820	217153		5 03 53.9	-41 44 42	+0.002	+0.15	6.31	0.53	4.5	G1 V		23 s		
32282	39986		5 03 54.0	+40 41 01	+0.001	0.00	8.04	0.01	0.0	B8.5 V		380 s		
32677	170030		5 03 54.5	-25 53 30	0.000	+0.02	8.0	1.2	0.2	K0 III		290 s		
32613	150113		5 03 56.6	-14 33 23	+0.003	0.00	6.8	1.1	-0.1	K2 III		250 s		
32610	150111		5 03 59.3	-10 38 04	-0.002	+0.04	7.4	1.1	0.2	K0 III		270 s		
32212	39982		5 04 02.2	+49 30 34	-0.002	-0.03	7.9	0.0	0.4	B9.5 V		300 s		
32233	39984		5 04 03.1	+47 39 52	+0.001	-0.03	7.0	1.1	0.2	K0 III		220 s		
31563	5403		5 04 13.0	+73 45 50	+0.006	-0.01	6.7	1.1	0.2	K0 III	+22	190 s		
32406	57610		5 04 14.4	+30 29 41	0.000	0.00	6.14	1.21	0.2	K0 III	+18	120 s		
32296	39988		5 04 16.3	+45 46 46	+0.004	-0.06	6.5	0.1	1.4	A2 V		62 mx		
32622	131738		5 04 17.9	-6 01 53	0.000	0.00	7.10	0.00	0.4	B9.5 V		220 s	3662	m
32461	76938		5 04 20.9	+23 31 37	0.000	0.00	8.0	1.4	-0.3	K5 III		400 s		
32482	76939		5 04 21.6	+21 16 42	+0.001	-0.01	6.19	1.32	0.2	K0 III	+48	100 s		
32481	76940		5 04 22.0	+21 38 37	0.000	0.00	8.00	0.00	-1.0	B5.5 V		550 s		m
32971	217159		5 04 23.2	-49 36 42	0.000	+0.05	7.8	1.0	2.1	A5 V		41 s		
32831	195532	γ Cae	5 04 24.3	-35 29 00	+0.010	-0.05	4.55	1.20	0.2	gK0	+10	52 s		m
32846	195534		5 04 26.0	-35 42 19	+0.002	+0.04	6.34	0.30		F0				
32577	112385		5 04 28.4	+7 49 49	+0.001	0.00	7.1	1.1	0.2	K0 III		240 s		
32328	39995		5 04 32.8	+43 43 40	+0.002	-0.02	7.58	-0.06	-0.2	B8 V		360 s		
32549	94290	11 Ori	5 04 34.1	+15 24 14	+0.001	-0.03	4.68	-0.06		A0 p	+17	26 mn		
--	39991		5 04 36.1	+49 43 57	+0.001	-0.01	8.0	1.1						
32707	150121		5 04 36.4	-14 56 52	+0.001	-0.01	7.4	1.4	-0.3	K5 III		350 s		
32428	57614		5 04 36.8	+32 19 13	-0.001	-0.07	6.4	0.2		A m	-8			
32739	150122		5 04 37.1	-17 53 06	-0.001	+0.01	7.8	0.1	1.7	A3 V		170 s		

HD	SAO	Star Name	α 2000	δ 2000	μ(α)	μ(δ)	V	B−V	M_v	Spec	RV	d(pc)	ADS	Notes
			h m s	° ′ ″	s	″								
32706	150120		5 04 37.7	−12 55 41	−0.003	+0.03	7.3	0.1	1.7	A3 V		130 s		
32480	76941		5 04 37.8	+27 41 46	+0.001	−0.03	6.50	0.24	2.4	A7 V	+22	63 s		
31590	5407		5 04 39.8	+74 04 02	+0.004	−0.01	6.40	0.1	1.4	A2 V	−9	98 s		m
32776	170039		5 04 41.5	−21 13 16	−0.002	−0.03	7.6	0.9	0.2	K0 III		300 s		
32576	94293		5 04 41.8	+14 51 31	0.000	−0.04	6.7	0.1		A3 p				
33293	249182		5 04 45.2	−64 32 47	+0.002	−0.01	8.0	1.2	3.2	G5 IV		52 s		
32724	150123		5 04 47.4	−11 41 59	+0.004	−0.24	7.8	0.7	4.4	G0 V		43 mx		
32595	94296		5 04 49.1	+13 18 31	+0.001	−0.01	7.5	0.0	0.0	B8.5 V		320 s		
32789	170040		5 04 49.8	−20 14 47	+0.003	+0.06	7.3	0.9	3.2	G5 IV		55 s		
32509	76945		5 04 50.0	+26 43 17	0.000	−0.01	7.5	0.1	1.4	A2 V		160 s		
32561	94294		5 04 50.0	+18 15 14	−0.001	−0.02	7.8	0.1	1.4	A2 V		190 s		
32803	170041		5 04 50.7	−21 54 17	−0.001	−0.04	7.4	1.6		M5 III	−4			T Lep, v
33075	233811		5 04 51.0	−52 10 52	−0.001	−0.06	7.9	0.6	4.0	F8 V		57 s		
32929	195543		5 04 53.0	−39 33 42	0.000	−0.04	7.6	1.3	0.2	K0 III		210 s		
32686	131742		5 04 54.5	−3 02 23	0.000	0.00	6.05	−0.11	−1.6	B5 IV	+27	330 s		
33652	256158		5 04 56.0	−72 25 15	−0.013	−0.03	7.8	1.1	0.2	K0 III		270 mx		
32773	150129		5 04 56.7	−14 13 21	−0.003	−0.07	8.0	1.1	0.2	K0 III		170 mx		
33042	217164	η² Pic	5 04 57.9	−49 34 40	+0.006	0.00	5.03	1.49	−0.5	M2 III	+36	130 s		
32186	13360		5 04 57.9	+60 10 45	−0.001	−0.04	7.9	1.1	0.2	K0 III		320 s		
32520	76948		5 04 58.1	+29 58 32	−0.002	+0.01	8.00	1.1	0.2	K0 III		330 s	3660	m
32418	40001		5 04 58.7	+41 52 49	+0.002	−0.02	7.23	0.16	1.9	A4 V		110 s		
33116	233814		5 05 00.6	−54 24 28	+0.001	−0.01	6.27	1.54	−0.3	K5 III		190 s		
33146	233817		5 05 04.0	−55 26 59	0.000	+0.03	7.56	1.01	0.2	K0 III		290 s		
32662	112399		5 05 06.5	+6 07 48	+0.004	−0.02	7.6	0.7	4.4	G0 V		44 s		
33486	249185		5 05 06.7	−68 05 10	+0.002	+0.01	7.86	−0.03	0.6	A0 V		280 s		
32575	76952		5 05 12.0	+23 47 55	−0.001	−0.01	8.00	1.1	0.2	K0 III		330 s		m
32660	112400		5 05 12.8	+8 56 34	−0.002	+0.02	7.46	−0.03	0.4	B9.5 V		260 s		
32890	170050		5 05 16.1	−26 09 09	0.000	−0.07	5.73	1.17	0.2	gK0		100 s		
32643	94302		5 05 16.6	+15 14 22	+0.001	−0.03	7.8	0.1		A0 p				
32889	170049		5 05 16.9	−25 30 27	+0.003	−0.02	7.9	1.3	3.2	G5 IV		42 s		
32855	170047		5 05 17.2	−21 15 32	−0.002	−0.03	7.6	0.3	0.6	A0 V		160 s		
33229	233820		5 05 18.9	−56 46 52	0.000	+0.01	7.9	0.7	3.0	F2 V		56 s		
32685	112403		5 05 20.0	+7 33 44	−0.001	−0.02	8.0	0.0	0.4	B9.5 V		330 s		
33230	233821		5 05 23.7	−57 11 41	+0.004	0.00	7.7	1.8	−0.1	K2 III		150 s		
32736	112406		5 05 23.7	+1 10 39	+0.001	−0.01	6.17	3.45		C6 II	+17			W Ori, v
32416	40008		5 05 27.3	+46 54 59	0.000	0.00	6.80	0.5	4.0	F8 V		36 s	3659	m
32887	170051	2 ε Lep	5 05 27.6	−22 22 16	+0.001	−0.07	3.19	1.46	−0.3	K5 III	+1	50 s		
33262	233822	ζ Dor	5 05 30.6	−57 28 22	−0.004	+0.12	4.72	0.52	4.0	F8 V	−2	13 ts		
32642	94306		5 05 32.0	+19 48 23	−0.001	−0.02	6.44	0.21	0.5	A7 III	−17	150 s	3672	m
32641	76954		5 05 37.7	+23 03 39	0.000	−0.01	6.70	0.00	−1.0	B5.5 V	+1	320 s		m
32574	57626		5 05 49.1	+35 50 46	+0.001	−0.01	7.80	0.5	3.4	F5 V		75 s	3670	m
32734	94311		5 05 49.2	+13 17 10	+0.003	−0.10	7.7	1.1	0.2	K0 III		110 mx		
32656	76955		5 05 53.4	+26 25 48	0.000	−0.01	6.5	0.0	−1.0	B5.5 V	+17	300 s		
33115	217175		5 05 53.9	−44 46 44	+0.001	−0.01	7.8	1.3	0.2	K0 III		200 s		
32457	40013		5 05 54.8	+50 02 45	0.000	−0.02	7.2	0.0	0.4	B9.5 V		210 s		
33180	217177		5 05 55.5	−48 40 40	0.000	+0.01	7.7	2.3	−0.3	K5 III		160 s		
32608	57629		5 06 00.8	+35 56 12	−0.001	0.00	6.4	0.1	1.7	A3 V	+14	85 s		
32839	112418		5 06 05.1	+3 47 18	0.000	−0.01	7.6	1.1	0.2	K0 III		300 s		
32499	40015		5 06 05.3	+48 39 35	−0.001	0.00	8.0	0.1	0.6	A0 V		280 s		
32049	13362		5 06 06.2	+68 58 22	+0.001	−0.01	6.8	0.4	3.0	F2 V		57 s		
33354	233824		5 06 07.9	−58 11 56	0.000	+0.02	7.4	1.7	−0.1	K2 III		150 s		
32633	57631		5 06 08.2	+33 55 08	0.000	0.00	7.09	−0.06		B9 p				
32343	25001	11 Cam	5 06 08.4	+58 58 21	0.000	−0.01	5.08	−0.08	−1.7	B3 V e	−11	210 s		m
33012	170065		5 06 09.5	−26 29 17	−0.001	+0.03	7.9	1.1	0.2	K0 III		320 s		
33875	256160		5 06 09.6	−73 02 15	+0.007	+0.07	6.27	−0.01		A0				
32868	131763		5 06 10.0	−1 14 38	−0.002	0.00	7.7	0.9	3.2	G5 IV		80 s		
32913	150139		5 06 11.0	−10 25 22	−0.002	−0.07	8.0	0.5	3.4	F5 V		82 s		
33476	249188		5 06 11.4	−62 47 22	−0.004	+0.31	7.7	0.5	3.4	F5 V		49 mx		
32357	25003	12 Cam	5 06 12.1	+59 01 16	0.000	−0.03	6.40	1.1	0.2	K0 III	−8	160 s		m
32884	131766		5 06 12.9	−3 29 05	−0.001	−0.01	8.0	0.0	0.4	B9.5 V		330 s		
32508	40017		5 06 15.5	+49 58 29	+0.003	−0.08	7.6	0.5	3.4	F5 V		67 s		
32445	25007		5 06 21.9	+54 24 21	−0.001	+0.03	7.30	0.8	3.2	G5 IV		65 s	3667	m
32966	150144		5 06 22.1	−14 41 48	0.000	0.00	7.0	0.0	0.4	B9.5 V		210 s		
32683	57637		5 06 27.5	+34 51 16	+0.002	−0.07	8.0	1.6	−0.5	M2 III		190 mx		
32356	13369		5 06 29.6	+61 10 12	+0.005	−0.07	6.04	1.37		K0	−40			
32673	57639		5 06 29.8	+35 33 44	+0.001	−0.01	7.7	0.6	4.4	G0 V		46 s		
32630	40026	10 η Aur	5 06 30.8	+41 14 04	+0.003	−0.07	3.17	−0.18	−1.7	B3 V	+7	61 mx		
32456	25009		5 06 31.7	+55 21 13	+0.002	0.00	7.9	0.8	3.2	G5 IV		86 s		
32672	57638		5 06 34.1	+38 31 20	−0.001	−0.01	7.76	0.17	−1.6	B3.5 V	+5	490 s		
34297	256162		5 06 35.5	−77 34 02	+0.043	−0.40	7.31	0.66		G0		18 mn		
32996	150149		5 06 36.6	−13 07 20	+0.001	+0.03	6.05	−0.06		A0				
32619	40025		5 06 36.7	+44 43 56	−0.001	−0.01	7.48	0.16	1.9	A4 V		130 s		
33539	249193		5 06 38.5	−63 08 56	+0.002	−0.03	8.0	0.4	2.6	F0 V		120 s		
32881	94324		5 06 39.5	+10 41 48	0.000	+0.01	7.8	1.1	0.2	K0 III		310 s		
32537	25019	9 Aur	5 06 40.6	+51 35 52	−0.003	−0.17	5.00	0.33	2.6	F0 V	−1	30 s	3675	m

HD	SAO	Star Name	α 2000	δ 2000	μ(α)	μ(δ)	V	B–V	M_v	Spec	RV	d(pc)	ADS	Notes
			h m s	° ′ ″	s	″								
32811	76962		5 06 40.9	+22 30 40	+0.001	0.00	7.0	0.0	0.4	B9.5 V		200 s		
32850	94322		5 06 42.3	+14 26 45	+0.021	−0.26	7.76	0.81	5.9	K0 V		24 s		
32851	94323		5 06 43.0	+14 05 04	−0.001	−0.14	7.5	0.5	4.0	F8 V		50 s		
33475	233832		5 06 44.9	−59 52 40	+0.003	+0.02	7.9	1.0	0.2	K0 III		330 s		
32964	131777	66 Eri	5 06 45.6	−4 39 18	+0.001	+0.01	5.12	−0.06		B9	+31		3698	m
32733	57642		5 06 47.3	+34 32 07	+0.003	−0.02	7.9	0.5	3.4	F5 V		77 s		
32655	40029		5 06 49.6	+43 10 29	+0.001	0.00	6.20	0.43	−2.0	F2 II p	−12	400 s		
32994	131780		5 06 55.3	−5 09 56	+0.004	−0.02	7.2	0.5	3.4	F5 V		58 s		
32906	94330		5 07 04.4	+15 56 15	+0.001	+0.02	7.5	1.1	−0.1	K2 III		330 s		
33081	150154		5 07 08.7	−17 18 02	+0.001	−0.18	7.2	0.5	3.4	F5 V		57 s		
33095	150156		5 07 09.7	−19 23 32	−0.003	+0.26	6.44	0.62	4.4	G0 V		24 s		m
33197	195566		5 07 14.8	−31 18 37	+0.001	0.00	8.0	2.6	0.2	K0 III		40 s		
32087	5425		5 07 17.0	+72 05 07	−0.001	−0.02	8.0	1.1	−0.1	K2 III		380 s		
35877	258401		5 07 18.5	−83 51 35	+0.024	+0.15	6.8	0.6	3.4	F5 V		36 s		
32653	25031		5 07 22.2	+50 18 21	+0.001	+0.01	7.40	0.00	0.4	B9.5 V		240 s	3684	m
33093	150159		5 07 24.9	−12 29 26	+0.009	−0.08	5.97	0.60		F8	+50			
33331	217185		5 07 26.0	−44 49 18	+0.001	+0.02	6.90	−0.09		A0				
33069	131791		5 07 26.3	−8 39 13	−0.001	−0.02	7.02	−0.11	0.0	B8.5 V		250 s		
32923	94332	104 Tau	5 07 27.0	+18 38 42	+0.038	+0.02	4.92	0.65	5.0	G4 V	+20	13 ts	3701	m
32606	25029		5 07 30.6	+55 32 24	−0.001	−0.01	7.60	0.5	4.0	F8 V		52 s	3683	m
33684	249198		5 07 34.1	−63 23 59	+0.003	−0.04	5.1	2.2	−0.5	M4 III	+19	59 s		
33514	233837		5 07 34.1	−54 59 22	−0.006	+0.05	7.5	0.1	3.0	F2 V		80 s		
33038	112439		5 07 34.8	+2 28 08	0.000	−0.01	7.9	0.0	0.4	B9.5 V		320 s		
32863	57655		5 07 37.1	+32 45 50	+0.003	−0.05	7.5	1.1	0.2	K0 III	+56	270 s		
33021	112436	13 Ori	5 07 38.3	+9 28 19	0.000	−0.38	6.17	0.62	3.7	G1 IV–V	−24	33 ts		m
33378	217186		5 07 38.9	−43 58 11	0.000	+0.01	8.0	2.2	−0.3	K5 III		140 s		
33163	150165		5 07 42.2	−17 17 20	0.000	−0.08	7.3	1.1	0.2	K0 III		200 mx		
33020	94338		5 07 46.2	+10 53 43	+0.001	+0.02	7.1	0.1	0.6	A0 V		190 s		
32629	25032		5 07 46.3	+55 45 32	+0.005	−0.03	7.1	1.1	0.2	K0 III	+10	220 s		
32977	76971	106 Tau	5 07 48.3	+20 25 06	−0.003	−0.03	5.3	0.1	1.7	A3 V	−1	51 s		
33162	150166		5 07 49.7	−12 35 31	−0.002	−0.07	6.6	1.6	−0.5	M2 III		220 mx		
33111	131794	67 β Eri	5 07 50.9	−5 05 11	−0.007	−0.08	2.79	0.13	0.0	A3 III	−8	28 ts		Cursa, m
32681	25034		5 07 51.6	+53 12 42	+0.004	−0.02	8.0	1.1	−0.1	K2 III		340 mx		
32849	57659		5 07 52.7	+39 51 05	0.000	−0.02	8.0	0.0	0.4	B9.5 V		310 s		
33054	112440	14 Ori	5 07 52.8	+8 29 54	+0.002	−0.06	5.34	0.33		A m	+6	28 mn	3711	m
––	––		5 07 54.0	+28 57 59	+0.001	−0.01	6.3	1.6	−0.5	M2 III		220 s		
32991	76972	105 Tau	5 07 55.4	+21 42 17	0.000	−0.01	5.89	0.19	−2.5	B2 V e	+25	430 s		m
32963	76970		5 07 55.7	+26 19 42	−0.006	−0.05	7.58	0.66	3.2	G5 IV	−63	75 s		
33239	170097		5 08 00.2	−20 07 16	−0.001	−0.06	7.30	0.9	−2.1	G9 II	−11	130 mx		
33283	170100		5 08 01.0	−26 47 52	+0.004	−0.06	7.9	0.5	3.2	G5 IV		88 s		
33158	131799		5 08 03.5	−6 13 07	+0.002	−0.02	8.0	1.1	0.2	K0 III		360 s		
33066	112450		5 08 04.9	+7 53 52	−0.002	−0.17	6.7	1.1	0.2	K0 III		96 mx		
33045	94340		5 08 05.8	+15 51 18	+0.002	−0.02	7.5	1.1	−0.1	K2 III		330 s		
33053	94341		5 08 06.3	+14 31 57	0.000	−0.02	7.7	0.9	0.3	G5 III	−18	310 s		
33110	112452		5 08 06.3	+3 44 54	−0.001	−0.02	7.8	0.1	0.6	A0 V		270 s		
32990	76974	103 Tau	5 08 06.6	+24 15 55	0.000	0.00	5.50	0.06	−2.5	B2 V	+16	290 s	3709	m
33498	217196		5 08 06.9	−49 17 50	+0.003	+0.02	7.6	−0.1	0.6	A0 V		140 s		
33238	150170		5 08 13.1	−15 05 44	0.000	0.00	6.6	0.8	3.2	G5 IV		48 s		
33377	195581		5 08 14.7	−35 43 06	0.000	0.00	6.52	1.08		G5				
32904	57667		5 08 20.0	+39 43 41	+0.002	−0.01	8.0	1.1	−0.1	K2 III		380 s		
33224	131806		5 08 20.1	−8 39 54	0.000	−0.01	5.78	−0.06		B8			3722	m
––	––		5 08 22.7	+47 20 42	−0.023	−0.01	8.0	1.1	0.2	K0 III		64 mx		
33473	217198		5 08 29.8	−41 12 50	−0.007	+0.28	6.72	0.66	4.7	dG2	+40	24 s		m
33292	150172		5 08 31.2	−18 07 12	0.000	0.00	7.60	1.4	−0.3	K5 III		380 s		m
33207	131809		5 08 31.8	−2 49 28	−0.004	−0.01	6.9	0.4	3.0	F2 V		60 s		
33208	131810		5 08 33.1	−3 00 35	0.000	0.00	7.3	0.5	4.0	F8 V		45 s		
33034	76978		5 08 35.2	+28 16 25	−0.001	0.00	7.0	0.1	1.4	A2 V		130 s		
33017	57678		5 08 42.9	+32 41 52	+0.002	−0.08	8.0	0.9	3.2	G5 IV		91 s		
33256	131813	68 Eri	5 08 43.6	−4 27 22	+0.003	+0.02	5.12	0.44	3.4	F5 V	+9	22 s		
32747	25042		5 08 46.0	+57 24 05	+0.003	−0.02	7.81	0.11		B9				
33155	94347		5 08 46.1	+11 45 09	0.000	−0.01	7.4	1.1	0.2	K0 III		270 s		
33121	94345		5 08 50.3	+19 51 35	0.000	−0.01	6.5	0.8	3.2	G5 IV	+7	46 s		
32989	57677		5 08 53.9	+39 29 16	0.000	−0.02	8.04	0.06	−1.0	B5.5 V		540 s	3718	m
32948	40053		5 08 55.3	+44 07 14	+0.004	+0.02	8.0	0.9	3.2	G5 IV		91 s		
33236	112465		5 08 56.1	+3 13 07	−0.001	−0.03	7.00	0.1	0.6	A0 V		190 s	3728	m
32903	40052		5 09 04.2	+49 07 18	−0.003	0.00	6.60	0.1	1.7	A3 V		93 s	3715	m
33562	217201		5 09 07.4	−42 01 08	−0.001	−0.04	7.6	0.2	2.6	F0 V		100 s		
33328	131824	69 λ Eri	5 09 08.7	−8 45 15	0.000	0.00	4.27	−0.19	−3.0	B2 IV	+3	280 s		
33172	94351		5 09 11.8	+18 57 32	+0.001	−0.02	7.9	0.9	3.2	G5 IV		88 s		
33221	94353		5 09 12.6	+11 29 44	+0.001	−0.01	7.8	0.4	2.6	F0 V		110 s		
33650	217204		5 09 12.7	−48 58 48	+0.002	−0.02	7.3	1.0	0.2	K0 III		260 s		
33234	94355		5 09 14.6	+11 29 36	+0.001	−0.01	7.8	0.4	2.6	F0 V		110 s		
33254	112467	16 Ori	5 09 19.6	+9 49 47	+0.004	−0.01	5.43	0.24		A m	+37	20 mn		m
33450	170126		5 09 25.0	−23 07 08	−0.002	−0.01	7.3	1.2	0.2	K0 III		200 s		
33345	131829		5 09 29.6	−2 08 06	−0.001	+0.02	6.52	1.02		G5				

HD	SAO	Star Name	α 2000	δ 2000	μ(α)	μ(δ)	V	B-V	M_v	Spec	RV	d(pc)	ADS	Notes
			$5^h 09^m 29^s.7$	$+14°21'21''$	$0^s.000$	$-0''.01$								
33253	94357		5 09 29.7	+14 21 21	0.000	-0.01	7.8	1.6	-0.5	M2 III		470 s		
33089	57687		5 09 36.4	+38 41 09	+0.003	+0.01	7.0	0.9	3.2	G5 IV		57 s		
32518	13382		5 09 36.8	+69 38 22	+0.013	-0.06	6.5	1.1	0.0	K1 III	-8	140 mx		
34002	249205		5 09 37.3	-63 46 39	+0.002	+0.08	7.8	0.1	2.6	F0 V		110 s		
33649	217205		5 09 39.6	-44 20 07	+0.001	+0.11	7.3	1.2	3.0	F2 V		22 s		
32784	13390		5 09 40.1	+62 28 47	-0.003	-0.05	6.8	0.2	2.1	A5 V	-3	83 s		
33276	94359	15 Ori	5 09 41.9	+15 35 50	0.000	-0.01	4.82	0.32	1.9	F2 IV	+31	41 mn		m
32651	13385		5 09 42.8	+66 00 16	+0.003	0.00	8.0	1.1	0.2	K0 III		330 s		
33495	170131		5 09 43.2	-22 29 38	-0.002	0.00	7.10	0.4	2.6	F0 V		79 s	3742	m
33185	76989		5 09 43.7	+29 47 57	+0.006	+0.02	6.80	0.5	4.0	F8 V		36 s		m
33561	195610		5 09 44.7	-33 31 16	-0.001	-0.02	7.8	1.7	-0.3	K5 III		300 s		
33204	76990		5 09 45.0	+28 01 49	+0.004	-0.06	6.01	0.27	2.6	dF0	+41	48 s	3730	m
32715	13388		5 09 45.0	+64 55 11	+0.004	-0.17	6.4	0.4	3.0	F2 V	0	48 s		
33313	112475		5 09 45.9	+7 10 53	+0.003	+0.02	7.5	0.5	4.0	F8 V		49 s		
32231	5439		5 09 49.7	+75 28 52	+0.005	-0.05	7.9	0.9	3.2	G5 IV		89 s		
33339	112478		5 09 57.6	+9 13 56	-0.002	-0.03	7.9	1.1	0.2	K0 III		340 s		
33340	112481		5 09 58.2	+8 10 23	-0.003	-0.11	7.20	0.5	4.0	F8 V	-64	44 s	3737	m
32230	5442		5 10 01.4	+75 40 56	+0.002	-0.04	7.30	0.1	0.6	A0 V		220 s	3681	m
34074	--		5 10 02.0	-65 02 44			7.4	2.5		K0 III		36 s		
33680	217208		5 10 02.6	-40 33 33	+0.003	+0.03	7.4	0.9	-0.1	K2 III		310 s		
33419	131834		5 10 03.2	-0 33 55	0.000	-0.05	6.10	1.10	-0.1	K2 III		170 s		
33252	76998		5 10 03.8	+27 33 22	+0.014	-0.10	7.30	0.5	3.4	F5 V		47 mx		m
33217	57703		5 10 05.1	+31 55 49	+0.001	-0.05	7.8	0.2	2.1	A5 V		140 s		
33368	112486		5 10 07.9	+9 57 45	0.000	0.00	7.70	-0.01	0.6	A0 V		260 s		
33613	195614		5 10 08.8	-31 20 56	-0.003	0.00	7.8	0.2	0.6	A0 V		200 s		
32959	25054		5 10 08.9	+56 42 53	+0.002	0.00	8.0	1.1	-0.1	K2 III		370 s		
33431	131838		5 10 09.5	-3 51 08	-0.002	0.00	7.7	0.0	0.4	B9.5 V		290 s		
33336	94363		5 10 10.4	+13 32 54	-0.004	-0.04	6.8	0.4	3.0	F2 V	0	56 s		
--	--		5 10 11.3	+46 09 46	0.000	-0.01	7.7	1.4	-0.3	K5 III		370 s		
33269	57706		5 10 14.4	+30 34 29	+0.003	-0.13	7.7	0.5	4.0	F8 V		55 s		
32973	25057		5 10 17.9	+57 04 20	+0.001	-0.01	8.00	0.00	0.4	B9.5 V		300 s	3725	m
33203	57704		5 10 18.9	+37 18 07	0.000	-0.01	6.02	0.72		K3	+9		3734	m
35267	258400		5 10 25.2	-80 24 29	+0.018	+0.08	7.8	0.6	4.0	F8 V		52 s		
34802	256167		5 10 26.6	-77 13 01	-0.008	+0.01	7.8	0.9	3.2	G5 IV		82 s		
33668	195616		5 10 27.6	-32 03 32	-0.002	+0.03	8.0	1.2	-0.1	K2 III		410 s		
33594	150198		5 10 32.8	-18 39 46	-0.002	0.00	7.9	0.1	0.6	A0 V		280 s		
33447	112490		5 10 32.9	+5 18 54	0.000	0.00	7.3	0.1	0.6	A0 V		220 s		
33550	150194		5 10 34.0	-11 38 55	-0.001	0.00	7.0	0.0	0.0	B8.5 V		250 s		
33202	40078		5 10 34.0	+40 50 39	-0.004	-0.01	7.67	0.53	3.4	F5 V		65 s		
33299	57713		5 10 34.9	+30 47 52	0.000	0.00	6.69	1.58	-4.4	K1 Ib		1100 s		
33402	94366		5 10 39.9	+17 26 24	0.000	-0.03	7.9	0.0	0.4	B8.5 V		380 s		
33740	195621		5 10 41.7	-36 16 29	-0.001	0.00	7.8	0.4	0.4	B9.5 V		160 s		
32893	13394		5 10 42.7	+63 35 48	+0.006	-0.09	6.70	0.4	2.6	F0 V		65 s	3723	m
33167	40077		5 10 42.8	+46 57 44	+0.006	-0.15	5.68	0.42	3.4	F5 V	+33	29 s		
33858	217221		5 10 43.3	-48 27 18	+0.003	+0.03	7.70	0.1	1.4	A2 V		180 s		S Pic, m,v
33628	170149		5 10 43.4	-20 44 54	-0.001	+0.01	7.30	0.00	0.4	B9.5 V		240 s		m
33712	195619		5 10 44.1	-32 50 27	0.000	0.00	8.0	0.9	0.2	K0 III		360 s		
33667	170151		5 10 44.4	-25 54 34	-0.002	+0.06	6.41	1.25		K0				
33400	77012		5 10 47.5	+20 34 10	+0.003	-0.01	7.8	0.4	2.6	F0 V	+45	110 s		
33555	131847		5 10 57.9	-2 15 14	+0.005	-0.14	6.25	0.98	3.2	K0 IV		38 s		
33467	94373		5 10 58.9	+14 55 18	-0.001	-0.01	7.8	0.0	0.4	B9.5 V		300 s		
33770	195625		5 10 59.1	-34 54 14	+0.003	+0.01	7.2	1.0	-0.1	K2 III		290 s		
33610	131850		5 11 08.5	-6 01 24	-0.001	+0.02	7.91	0.12	0.6	A0 V		240 s		
33709	170161		5 11 10.4	-24 20 57	+0.001	0.00	7.9	1.0	3.2	G5 IV		64 s		
33736	170164		5 11 10.6	-26 42 14	0.000	-0.07	7.8	1.2	0.2	K0 III		260 s		
33333	57723		5 11 10.8	+35 21 27	+0.001	-0.04	7.8	0.4	2.6	F0 V		110 s		
33759	170168		5 11 12.3	-29 35 45	+0.003	-0.03	8.0	0.7	2.6	F0 V		66 s		
33545	112500		5 11 17.9	+7 00 18	-0.001	-0.02	7.9	1.1	0.2	K0 III		340 s		
33413	77017		5 11 18.0	+29 43 49	0.000	-0.01	7.0	0.1	0.6	A0 V		180 s		
33608	131852		5 11 19.1	-2 29 26	+0.004	+0.01	5.90	0.46	2.1	F5 IV	+31	54 s		
33835	195633		5 11 21.4	-36 28 46	0.000	-0.02	8.0	2.1	-0.1	K2 III		120 s		
33664	150206		5 11 22.8	-11 50 57	+0.001	+0.06	5.68	1.46	-0.5	M4 III	+46	170 s		RX Lep, v
33849	195637		5 11 25.9	-38 56 01	-0.003	+0.02	8.0	1.8	-0.1	K2 III		180 s		
33427	57729		5 11 27.5	+30 48 22	-0.001	+0.02	6.7	1.1	0.2	K0 III		190 s		
33921	217226		5 11 28.8	-43 56 47	+0.002	+0.01	7.7	2.1	-0.1	K2 III		63 s		
33324	40090		5 11 28.8	+42 44 55	+0.001	0.00	7.9	0.0	0.4	B9.5 V		300 s		
33201	25064		5 11 30.0	+53 14 05	+0.002	-0.03	8.0	0.4	2.6	F0 V		110 s		
33807	195631		5 11 30.1	-30 13 31	+0.002	+0.02	7.1	1.1	3.4	F5 V		23 s		
33364	57726		5 11 33.2	+37 18 00	-0.001	-0.04	7.50	1.43		K0				
33398	57727		5 11 34.0	+35 57 26	-0.001	0.00	7.3	1.1	-0.1	K2 III	-37	280 s		
35391	258402		5 11 35.3	-80 11 33	+0.005	-0.10	7.9	2.2	0.2	K0 III		65 s		
33872	195639		5 11 35.9	-37 23 43	0.000	0.00	6.57	1.62		K5				
33463	77025		5 11 38.2	+29 54 12	0.000	-0.03	6.38	1.78	-0.5	M2 III	+10	180 s		
33647	112505		5 11 41.3	+0 30 53	+0.001	0.00	6.67	-0.07	-0.2	B8 V		240 s	3767	m
33554	94377		5 11 41.5	+16 02 44	+0.001	+0.01	5.18	1.49	-0.3	K5 III	-6	130 s		

HD	SAO	Star Name	α 2000	δ 2000	μ(α)	μ(δ)	V	B-V	M_V	Spec	RV	d(pc)	ADS	Notes
			h m s	° ′ ″	s	″								
33297	40092	.	5 11 42.2	+46 56 33	+0.001	+0.01	7.82	0.06	0.6	A0 V		260 s		
33646	112509		5 11 45.2	+1 02 13	−0.001	−0.01	5.89	0.66		F5	−19		3764	m
33636	112506		5 11 46.3	+4 24 13	+0.012	−0.13	7.06	0.58	4.4	G0 V		33 s		
33725	131861		5 11 54.2	−9 06 48	−0.004	−0.56	8.04	0.81	6.1	K1 V	+6	27 ts		m
33267	25067		5 11 55.6	+53 27 07	+0.001	0.00	7.1	1.1	0.2	K0 III	−5	220 s		
33755	150215		5 11 57.1	−10 57 41	−0.002	−0.02	7.9	0.1	0.6	A0 V		280 s		
33645	112511		5 12 01.3	+6 50 17	−0.001	+0.01	7.70	1.1	0.2	K0 III		320 s		m
34851	256168		5 12 01.8	−75 21 43	+0.006	−0.01	7.8	1.1	0.2	K0 III		330 s		
34060	217234		5 12 03.0	−49 03 41	−0.003	−0.03	7.83	−0.08	0.6	A0 V		280 s		
34349	249218		5 12 03.3	−65 10 33	+0.009	+0.06	7.04	0.43		F5				
33982	217232		5 12 03.5	−41 40 05	+0.005	+0.10	7.4	0.4	0.2	K0 III		170 mx		
33411	40101		5 12 03.5	+42 23 59	0.000	+0.02	7.9	0.0	0.0	B8.5 V		360 s		
33845	170182		5 12 05.0	−26 35 29	+0.002	−0.04	8.0	1.3	0.2	K0 III		250 s		
33662	112512		5 12 07.8	+6 51 36	0.000	+0.01	7.9	1.1	0.2	K0 III	+22	340 s		
33503	57737		5 12 09.6	+32 53 42	+0.001	0.00	7.6	0.4	−2.0	F2 II		740 s		
33461	40108		5 12 14.4	+41 12 54	0.000	−0.02	7.78	0.27	−2.5	B2 V nne		940 s		
33462	40110		5 12 14.8	+40 06 07	−0.001	0.00	6.98	0.06	0.0	B8.5 V		220 s		
33981	195651		5 12 16.7	−39 03 34	+0.001	+0.01	8.0	1.8	−0.1	K2 III		180 s		
33802	150223	3 ι Lep	5 12 17.8	−11 52 09	+0.001	−0.02	4.45	−0.10	−0.2	B8 V	+25	85 s	3778	m
33585	77028		5 12 21.4	+26 27 16	−0.006	−0.16	6.8	0.9	0.3	G5 III	−12	74 mx		
34142	233871		5 12 22.0	−51 33 00	+0.003	+0.06	7.5	1.0	0.2	K0 III		260 mx		
32650	5455		5 12 22.5	+73 56 48	+0.002	−0.03	5.4	0.1		A0 p	+9	15 mn		
33001	13400		5 12 23.3	+67 40 47	+0.006	−0.04	7.3	0.4	3.0	F2 V		72 s		
33751	131865		5 12 23.6	+0 01 26	0.000	−0.01	7.8	1.1	0.2	K0 III		330 s		
35798	258405		5 12 26.1	−81 32 31	+0.008	+0.05	6.51	1.11		G5				
32765	5457		5 12 26.9	+72 44 37	+0.002	−0.03	7.9	0.0	0.4	B9.5 V		300 s		
33844	150228		5 12 35.9	−14 57 04	0.000	+0.03	7.4	0.9	3.2	G5 IV		70 s		
33779	112522		5 12 37.1	+0 04 49	0.000	+0.01	7.5	0.5	3.4	F5 V		65 s		
33459	--		5 12 37.6	+44 26 08	+0.001	+0.02	7.5	0.0	0.0	B8.5 V		300 s		
33833	131873		5 12 48.1	−6 03 26	+0.001	−0.03	5.91	0.96	0.2	G9 III	+23	140 s		
33980	170194		5 12 51.7	−27 10 38	−0.001	+0.02	7.0	1.3	−0.1	K2 III		220 s		
33568	57745		5 12 53.4	+37 44 37	−0.001	−0.01	8.0	0.9	3.2	G5 IV		91 s		
33904	150237	5 μ Lep	5 12 55.8	−16 12 20	+0.002	−0.03	3.31	−0.11	−0.8	B9 III p	+28	66 s		
33979	170196		5 12 55.8	−27 09 31	0.000	+0.01	7.5	1.0	0.2	K0 III		280 s		
33266	13409		5 13 03.1	+61 51 00	+0.002	+0.01	6.17	0.02		A0	−4			
33542	40128		5 13 07.9	+44 34 01	−0.001	0.00	7.28	0.08	0.0	B8.5 V		260 s		
33501	40122		5 13 12.9	+49 51 41	+0.003	−0.03	7.7	0.4	2.6	F0 V		100 s		
33604	40133		5 13 13.3	+40 11 37	0.000	−0.01	7.34	0.04	−2.5	B2 V pe	+7	790 s		
33949	150239	4 κ Lep	5 13 13.8	−12 56 30	−0.001	−0.01	4.36	−0.10	−1.2	B8 III	+18	130 s	3800	m
33356	25071		5 13 14.7	+58 39 47	+0.005	−0.06	8.0	0.9	3.2	G5 IV		90 s		
33632	57754		5 13 17.3	+37 20 13	−0.013	−0.14	6.50	0.5	4.0	F8 V		31 s		m
33856	112528	17 ρ Ori	5 13 17.4	+2 51 40	0.000	−0.01	4.46	1.19	−0.2	K3 III	+41	86 s	3797	m
34395	233883		5 13 18.5	−59 35 09	+0.001	+0.02	7.1	1.3	−0.1	K2 III		220 s		
34376	233881		5 13 21.9	−58 33 27	0.000	0.00	8.0	0.2	0.0	B8.5 V		260 s		
--	150249		5 13 22.9	−16 54 33	−0.001	−0.01	7.7	1.4	−0.3	K5 III		390 s		
33641	57755	11 μ Aur	5 13 25.6	+38 29 04	−0.002	−0.08	4.86	0.18		A m	+23	22 mn		
34327	233880		5 13 28.0	−55 34 06	+0.007	−0.08	7.10	0.6	4.4	G0 V		35 s		m
33918	131884		5 13 28.6	−4 39 10	−0.001	−0.01	7.70	0.17	0.6	A0 V		210 s		m
33296	13413	14 Cam	5 13 31.2	+62 41 29	−0.005	+0.01	6.50	0.21	2.4	A7 V n	−4	65 s		
33883	112535		5 13 31.5	+1 58 05	0.000	+0.01	6.09	0.42	1.4	A2 V	+7	53 s	3799	m
33671	57760		5 13 31.9	+35 39 11	+0.001	0.00	7.84	0.10	0.6	A0 V		250 s		
34012	150253		5 13 32.6	−17 06 25	−0.002	−0.03	7.6	0.5	4.0	F8 V		51 s		
34088	195662		5 13 32.7	−31 54 17	0.000	−0.02	7.7	0.0	3.0	F2 V		88 s		m
33948	131887		5 13 33.2	−8 08 52	0.000	0.00	6.37	−0.13		A0				
34104	195665		5 13 37.1	−32 04 00	−0.002	+0.03	7.2	0.1	1.7	A3 V		130 s		
33928	131889		5 13 38.9	−3 37 19	−0.001	0.00	7.6	0.0	0.0	B8.5 V		330 s		
34348	233884		5 13 43.6	−52 58 32	+0.001	0.00	7.5	0.9	3.2	G5 IV		61 s		
33601	--		5 13 44.5	+46 59 29			7.70	0.00	0.0	B8.5 V		320 s	3781	m
34649	249225	θ Dor	5 13 45.4	−67 11 07	+0.002	+0.03	4.83	1.28	−0.1	K2 III	+11	80 s		
34167	195669		5 13 46.4	−35 49 33	+0.002	−0.05	6.9	1.6	0.2	K0 III		95 s		
33265	13415		5 13 46.9	+64 56 57	+0.005	−0.02	8.0	1.4	−0.3	K5 III		370 mx		
33946	112543		5 13 47.2	+0 33 37	0.000	−0.02	6.32	1.46	−1.3	K4 II-III	−11	330 s		
33994	131893		5 13 47.7	−6 44 54	+0.001	0.00	7.34	−0.06	0.0	B8.5 V		290 s		
33704	57767		5 13 49.2	+37 01 56	0.000	−0.05	6.85	0.01	0.6	A0 V		170 s		
34347	233886		5 13 53.2	−52 01 52	−0.001	−0.02	6.05	1.39		K5				
34087	170216		5 13 55.8	−22 59 19	+0.001	+0.02	7.40	0.9	3.2	G5 IV		69 s	3819	m
33500	25080		5 13 57.6	+56 30 28	−0.001	−0.04	7.9	0.1	1.4	A2 V		180 s		
34045	150255		5 13 59.8	−14 36 24	−0.001	+0.01	6.21	0.37		F2				
33749	57770		5 14 02.5	+36 54 58	+0.001	0.00	7.65	0.08	0.0	B8.5 V		290 s		
34071	150258		5 14 04.0	−16 39 57	+0.002	−0.02	8.0	1.1	0.2	K0 III		360 s		
34122	170218		5 14 04.4	−21 38 59	+0.001	+0.03	7.7	0.2	1.4	A2 V		150 s		
33275	13417		5 14 04.8	+66 03 25	−0.004	+0.05	7.9	0.4	3.0	F2 V		95 s		
33864	77043		5 14 05.7	+21 13 26	+0.001	0.00	7.8	0.1	1.4	A2 V		180 s		
33231	13416		5 14 07.5	+67 28 54	+0.001	−0.07	7.2	0.1	0.6	A0 V		89 mx		
33927	94409		5 14 11.9	+12 04 13	−0.001	−0.02	7.8	1.1	−0.1	K2 III		380 s		

HD	SAO	Star Name	α 2000	δ 2000	μ(α)	μ(δ)	V	B−V	M_v	Spec	RV	d(pc)	ADS	Notes
			h m s	° ′ ″	s	″								
33733	40144		5 14 12.8	+40 14 32	+0.001	0.00	7.4	1.1	0.2	K0 III	−25	260 s		
34068	150260		5 14 13.3	−13 03 16	−0.001	−0.02	8.0	1.1	0.2	K0 III		360 s		
33829	57780		5 14 14.3	+30 24 00	−0.002	0.00	7.2	0.4	2.6	F0 V		84 s		
33164	13412		5 14 19.6	+69 49 26	+0.004	−0.10	7.60	1.1		K1 IV		23 mn	3759	m
34101	150263		5 14 20.0	−15 49 36	+0.015	−0.23	7.45	0.73	5.3	dG6	+33	26 s		
34183	170227		5 14 20.9	−26 47 41	+0.001	−0.02	7.3	1.4	0.2	K0 III		160 s		
33772	57776		5 14 24.1	+39 21 48	+0.002	−0.05	8.0	0.9	3.2	G5 IV		91 s		
34086	150264		5 14 27.4	−10 25 00	−0.003	+0.01	7.6	0.4	2.6	F0 V		100 s		
34153	150267		5 14 28.7	−18 23 05	−0.004	+0.03	7.70	1.1	0.2	K0 III		290 mx		m
34266	195683		5 14 28.8	−35 58 38	+0.001	0.00	5.76	1.01	0.3	gG8	+13	110 s		
34198	170230		5 14 30.5	−26 12 30	+0.004	−0.05	7.0	1.3	0.2	K0 III		150 s		
33897	77046		5 14 30.9	+25 55 03	−0.001	−0.02	7.9	0.1	1.4	A2 V		190 s		
34085	131907	19 β Ori	5 14 32.2	−8 12 06	0.000	0.00	0.12	−0.03	−7.1	B8 Ia	+21	280 s	3823	Rigel, m
32781	5464		5 14 35.6	+76 28 22	−0.001	−0.01	6.30	−0.02		B9			3738	m
34055	131905		5 14 36.8	−0 33 46	0.000	0.00	6.8	1.6	−0.5	M4 III		290 s		
34214	170232		5 14 37.6	−26 57 42	+0.003	+0.04	7.9	0.4	3.2	G5 IV		88 s		
33863	57790		5 14 38.4	+30 40 19	+0.001	−0.01	8.0	0.1	0.6	A0 V		280 s		
33441	13422		5 14 38.9	+63 07 38	+0.002	+0.01	6.8	0.4	2.6	F0 V	−16	67 s		
33862	57789		5 14 39.3	+31 24 05	−0.001	−0.06	6.7	1.1	0.2	K0 III		200 s		
34121	131910		5 14 41.2	−7 04 17	−0.001	−0.01	7.10	1.1	−0.1	K2 III		280 s	3825	m
34043	112556		5 14 44.0	+5 09 22	−0.001	+0.01	5.50	1.37	−0.2	K3 III	−8	120 s		
33654	25087		5 14 44.4	+53 12 50	+0.001	0.00	6.2	0.1	0.6	A0 V	−5	130 s		
34021	---		5 14 45.0	+12 31 31	0.000	−0.01	7.90	0.1	0.6	A0 V	+17	290 s	3822	m
33323	13421		5 14 46.3	+67 21 37	+0.005	−0.02	7.6	0.4	2.6	F0 V		99 s		
33827	40157		5 15 02.0	+42 28 33	+0.001	−0.03	8.00	0.9	3.2	G5 IV		90 s	3813	m
---	---		5 15 06.1	−46 55 06			7.90							T Pic, v
34358	195689		5 15 08.8	−36 39 10	+0.001	0.00	6.80	0.8	3.2	G5 IV		53 s		m
34054	94417		5 15 09.0	+15 03 45	0.000	−0.01	7.3	0.0	0.4	B9.5 V	+14	240 s		
34182	131916		5 15 09.0	−6 48 21	−0.001	0.00	7.6	0.4	3.0	F2 V		83 s		
33861	40163		5 15 09.8	+40 07 58	−0.001	+0.02	7.70	1.6	−0.5	M3 III	+21	400 s		UZ Aur, v
34081	---		5 15 10.5	+8 25 54	0.000	−0.04	7.70	0.4	2.6	F0 V		110 s	3827	m
34295	170235		5 15 10.5	−28 20 07	−0.003	+0.02	7.1	0.1	1.4	A2 V		130 s		
33618	25088		5 15 11.3	+59 24 21	+0.002	−0.01	6.15	1.18	−0.1	K2 III	+3	170 s		
34031	77054		5 15 11.7	+20 03 21	+0.007	−0.04	7.7	0.6	4.4	G0 V	+22	46 s		
34137	112568		5 15 11.8	+1 33 21	−0.001	−0.01	7.9	1.1	0.2	K0 III		340 s		
34136	112567		5 15 12.1	+4 16 13	0.000	+0.02	7.8	0.1	0.6	A0 V		270 s		
33798	40158		5 15 15.5	+47 10 13	+0.010	−0.12	7.10	0.8	3.2	G5 IV		59 s	3812	m
34541	233893		5 15 16.3	−52 39 54	+0.002	+0.02	7.2	0.4	0.2	K0 III		250 s		
34180	131917		5 15 18.4	−1 24 33	−0.003	+0.04	6.15	0.39	2.8	F1 V		45 s		
34325	170239		5 15 20.6	−29 45 32	0.000	−0.01	7.3	1.3	0.2	K0 III		180 s		
34179	112572		5 15 21.7	+0 03 37	+0.001	0.00	8.02	−0.03	−0.2	B8 V		420 s		
34310	170238		5 15 24.3	−26 56 36	0.000	−0.02	5.07	−0.10		B9	+29			
34274	150277		5 15 24.3	−19 03 37	+0.001	−0.03	7.9	0.5	2.6	F0 V		83 s		
33959	57799	14 Aur	5 15 24.3	+32 41 16	−0.002	+0.01	5.02	0.23	2.5	A9 V	−10	32 s	3824	m
35184	256172		5 15 25.4	−73 35 19	−0.001	0.00	6.6	0.2	1.4	A2 V		91 s		
34135	112571		5 15 26.5	+7 03 18	−0.001	−0.01	8.0	0.5	3.4	F5 V		82 s		
34053	77057	108 Tau	5 15 27.6	+22 17 05	−0.001	−0.01	6.20	0.08	0.6	A2 IV	−7	120 s		m
35107	256170		5 15 29.0	−72 23 58	+0.001	−0.03	7.9	1.1	0.2	K0 III		350 s		
33853	40167		5 15 29.5	+46 24 22	−0.003	+0.01	7.94	0.01	0.4	B9.5 V		320 s		
33896	40172		5 15 31.9	+42 31 44	−0.002	−0.01	8.0	1.1	0.2	K0 III		340 s		
34149	112574		5 15 35.3	+8 26 28	−0.001	−0.02	7.8	0.1	0.6	A0 V		270 s		
34587	233895		5 15 38.8	−52 10 55	+0.002	+0.03	6.49	1.21		K0				
34540	217257		5 15 42.9	−49 26 44	+0.010	−0.07	7.7	1.2	3.2	G5 IV		44 s		
34324	170243		5 15 43.7	−22 53 39	−0.002	+0.01	6.9	0.2	0.6	A0 V		140 s		
33841	25097		5 15 44.0	+50 33 44	+0.002	−0.02	7.5	0.2	2.1	A5 V		110 s		
34239	131922		5 15 44.8	−4 47 48	+0.002	+0.04	7.2	0.7	4.4	G0 V		37 s		
34435	195694		5 15 47.0	−34 55 36	−0.002	+0.03	6.66	0.15		A				m
34030	57808		5 15 53.2	+34 25 22	+0.002	−0.14	7.90	0.5	4.0	F8 V		60 s	3834	m
34148	77064		5 15 57.8	+20 07 32	+0.001	−0.03	8.0	1.1	−0.1	K2 III		380 s		
34096	77063		5 16 00.3	+25 57 28	0.000	0.00	8.00	0.1	0.6	A0 V		280 s	3839	m
34309	150282		5 16 00.6	−14 30 21	−0.003	−0.01	7.7	0.0	0.4	B9.5 V	+46	290 s		
34203	94426	18 Ori	5 16 04.1	+11 20 29	0.000	−0.01	5.50	−0.02	−0.6	A0 III	−8	150 s		
33852	25100		5 16 04.4	+51 57 24	+0.001	−0.01	8.00	0.00	0.0	B8.5 V		360 s		m
34517	217258		5 16 10.0	−41 14 57	+0.001	+0.01	7.2	1.4	−0.1	K2 III		210 s		
34318	150286		5 16 10.5	−11 21 10	0.000	0.00	6.5	0.6	4.4	G0 V		27 s		
34263	112584		5 16 11.3	+1 45 24	−0.001	−0.01	7.9	1.1	0.2	K0 III		340 s		
34530	217259		5 16 11.5	−41 04 22	−0.006	+0.06	8.00	0.9	3.2	G5 IV		91 s		m
34007	40180		5 16 12.4	+42 45 23	−0.003	−0.01	7.8	0.1	0.6	A0 V		260 s		
34280	131928		5 16 14.4	−3 29 00	−0.001	−0.01	7.78	0.10	0.4	B9.5 V		240 s	3856	m
34078	57816		5 16 18.2	+34 18 43	+0.001	+0.03	5.96	0.22	−4.4	O9.5 V	+59	630 s	3843	AE Aur, m,v
34496	195705		5 16 23.8	−33 32 16	−0.001	−0.02	6.97	1.00		G5				
34484	195704		5 16 24.1	−32 30 49	+0.002	−0.02	7.1	0.3	4.0	F8 V		42 s		
34433	170258		5 16 25.4	−23 03 35	+0.004	0.00	7.6	0.5	3.2	G5 IV		77 s		
33988	40183		5 16 27.1	+46 24 58	0.000	+0.01	6.88	0.25		B2 e	−26			
34474	170262		5 16 28.6	−28 08 21	0.000	−0.03	7.5	1.8	−0.3	K5 III		240 s		

HD	SAO	Star Name	α 2000	δ 2000	μ(α)	μ(δ)	V	B-V	M_v	Spec	RV	d(pc)	ADS	Notes
			^{h} ^{m} ^{s}	^{o} ' "	^{s}	"								
34095	57819		5 16 29.1	+36 04 17	+0.002	-0.02	7.2	0.1	1.4	A2 V		140 s		
--	40184		5 16 32.0	+46 08 28	+0.004	-0.05	8.0	0.3						
34371	150291		5 16 32.0	-11 15 44	-0.001	+0.01	8.0	1.1	0.2	K0 III		360 s		
34040	40185		5 16 32.9	+42 40 02	-0.002	-0.01	8.00	0.4	2.6	F0 V		120 s	3842	m
34473	170265		5 16 35.6	-25 19 56	+0.002	+0.01	7.7	0.7	0.6	A0 V		100 s		
34631	217265		5 16 37.9	-45 54 49	+0.001	+0.02	7.0	0.6	0.4	B9.5 V		83 s		
34317	112588		5 16 41.0	+1 56 50	-0.001	-0.01	6.42	-0.02	0.6	A0 V		150 s		
34029	40186	13 α Aur	5 16 41.3	+45 59 53	+0.008	-0.42	0.08	0.80	0.3	G8 III	+30	13 t	3841	Capella, m
34251	94431		5 16 43.8	+18 26 21	0.000	0.00	7.18	0.10	-1.6	B3.5 V	+25	390 s	3854	m
34694	217270		5 16 47.0	-49 36 02	0.000	-0.04	7.8	1.3	0.2	K0 III		230 s		
34447	150295		5 16 48.0	-17 08 30	-0.001	+0.02	6.56	-0.16		B2	+12			
34260	77076		5 16 49.3	+20 07 05	+0.001	-0.03	7.6	0.1	1.7	A3 V		150 s		
34482	170270		5 16 51.4	-22 16 29	-0.002	-0.10	8.0	0.3	0.2	K0 III		210 mx		
34291	94436		5 16 52.0	+12 24 34	0.000	-0.03	7.9	1.1	0.2	K0 III		340 s		
34092	40188		5 16 52.9	+42 40 39	-0.002	0.00	8.0	1.1	0.2	K0 III		340 s		
34708	217271		5 16 54.2	-48 11 31	0.000	-0.03	7.5	1.5	-0.3	K5 III		370 s		
34768	233907		5 16 54.4	-52 27 09	0.000	+0.01	7.5	1.2	2.6	F0 V		27 s		
33973	25108		5 16 58.9	+54 10 26	+0.002	0.00	7.8	1.1	0.2	K0 III		300 s		
34554	195714		5 17 00.2	-31 16 36	0.000	+0.16	7.47	0.46	3.7	F6 V		57 s		
34338	112591		5 17 05.3	+9 55 40	0.000	+0.01	7.5	0.0	0.0	B8.5 V		320 s		
34316	94439		5 17 05.3	+12 33 57	+0.001	0.00	7.2	1.1	0.2	K0 III		250 s		
34304	94437		5 17 05.6	+16 21 05	+0.001	-0.03	7.5	0.8	3.2	G5 IV		73 s		
34175	57832		5 17 12.2	+39 27 50	-0.001	0.00	7.50	1.1	0.2	K0 III	-28	270 s	3851	m
34645	195718		5 17 14.9	-39 17 17	-0.002	+0.02	7.0	0.1	1.4	A2 V		120 s		
34250	77079		5 17 15.5	+28 54 17	+0.001	-0.01	6.9	0.4	2.6	F0 V	+2	71 s		
34566	170279		5 17 15.6	-27 14 45	+0.001	0.00	7.8	1.5	-0.1	K2 III		250 s		
34019	25112		5 17 17.8	+53 35 10	+0.002	-0.02	6.4	1.6		M5 III	+8		3845	R Aur, m,v
35390	256174		5 17 19.1	-72 42 06	+0.009	0.00	8.0	1.4	-0.3	K5 III		330 mx		
34513	150300		5 17 22.2	-17 46 29	-0.001	+0.01	6.8	0.2	2.1	A5 V		89 s		
34387	94443		5 17 26.2	+11 43 25	0.000	-0.03	7.9	0.1	0.6	A0 V		280 s		
35324	256173		5 17 26.2	-71 31 57	+0.003	-0.06	7.80	1.1	0.2	K0 III		330 s		m
34400	112597		5 17 27.1	+8 49 43	0.000	-0.04	7.8	1.1	0.2	K0 III		330 s		
33924	13444		5 17 27.5	+60 10 55	0.000	-0.13	6.98	0.45		F5	+17			
34642	195721	o Col	5 17 29.0	-34 53 43	+0.007	-0.34	4.83	1.00	3.2	K0 IV	+21	21 mn		
34110	40195		5 17 30.0	+48 56 48	+0.001	-0.01	8.0	1.1	0.2	K0 III		330 s		
34258	57843		5 17 30.9	+31 49 25	0.000	-0.02	8.0	1.4	-0.3	K5 III		420 s		
34335	77084		5 17 31.2	+20 07 53	-0.001	-0.10	6.77	0.49		F5	-27		3866	CD Tau, m,v
33555	5478		5 17 31.5	+74 32 06	+0.001	-0.07	7.3	0.1	1.4	A2 V		86 mx		
34303	77082		5 17 31.7	+24 00 35	-0.001	-0.03	6.8	1.1	0.2	K0 III		200 s		
34428	112598		5 17 32.2	+5 28 55	+0.002	-0.02	7.3	0.9	3.2	G5 IV		65 s		
34527	150303		5 17 35.3	-15 13 12	-0.001	-0.01	6.70	0.00	0.0	B8.5 V	+67	220 s	3883	m
35230	249241		5 17 35.3	-68 35 37	-0.006	-0.01	7.57	0.88		K0				
34028	25115		5 17 36.0	+55 31 01	+0.001	+0.01	7.5	0.1	0.6	A0 V		220 s		
34503	131952	20 τ Ori	5 17 36.3	-6 50 40	-0.001	-0.01	3.60	-0.11	-2.2	B5 III	+20	130 s	3877	m
34538	150304		5 17 40.1	-13 31 11	-0.001	-0.05	5.50	0.93	0.2	gG9	+75	120 s		
34445	112601		5 17 40.9	+7 21 12	0.000	-0.14	7.7	0.6	4.4	G0 V		47 s		
34914	233916		5 17 43.4	-54 28 22	-0.003	+0.01	6.9	1.4	0.2	K0 III		120 s		
34781	217274		5 17 45.0	-43 36 12	-0.002	+0.12	7.4	0.8	3.4	F5 V		39 s		
34677	195723		5 17 45.2	-33 25 57	0.000	+0.03	7.87	-0.07	0.0	B8.5 V		370 s		
34415	94447		5 17 46.1	+13 38 24	+0.002	0.00	7.9	0.4	2.6	F0 V		110 s		
34190	40203		5 17 48.7	+46 07 40	+0.001	-0.02	7.25	1.48	-0.2	K3 III		230 s		
34471	112604		5 17 56.9	+8 06 17	-0.002	+0.06	7.5	1.1	0.2	K0 III		270 mx		
34618	150316		5 18 00.7	-18 36 48	-0.001	-0.04	7.9	1.6	-0.5	M2 III		490 t		
34552	131962		5 18 00.8	-8 13 44	0.000	-0.03	8.0	1.4	-0.3	K5 III		450 s		
34511	131959		5 18 00.9	-0 02 14	0.000	+0.02	7.39	-0.10	-1.1	B5 V		490 s		
34454	--		5 18 04.0	+13 25 03	-0.001	-0.02	7.84	1.75	-1.4	M2 II-III	+10	720 s		
34616	150319		5 18 08.7	-16 11 14	-0.008	+0.02	7.60	0.9	0.2	G9 III	+24	92 mx		
34302	57850		5 18 10.1	+37 39 00	0.000	-0.01	7.88	0.09	0.0	B8.5 V		300 s	3871	m
34334	57853	16 Aur	5 18 10.6	+33 22 18	+0.004	-0.16	4.54	1.27	-0.2	K3 III	-28	68 mx	3872	m
34500	112606		5 18 11.1	+9 12 59	0.000	-0.04	7.5	0.1	1.4	A2 V		170 s		
33541	5483		5 18 13.3	+73 16 05	0.000	-0.03	5.8	0.1	0.6	A0 V	0	110 s		
34690	170291		5 18 15.3	-25 15 07	-0.004	-0.02	7.5	1.3	4.4	G0 V		16 s		
34384	77089		5 18 15.3	+28 46 48	-0.004	-0.03	7.3	0.2		A m	-23			
35666	256177		5 18 15.7	-74 41 53	0.000	+0.02	7.8	1.1	0.2	K0 III		280 s		
34269	40214		5 18 15.8	+42 47 32	+0.004	-0.02	5.48	1.65	-0.5	M4 III	-38	160 s		
34364	57858	17 Aur	5 18 18.8	+33 46 02	+0.001	-0.03	6.14	-0.06	0.2	B9 V	+25	150 s		AR Aur, v
34676	170290		5 18 18.9	-21 49 29	0.000	0.00	7.6	0.3	1.4	A2 V		120 s		
34333	57857		5 18 20.9	+36 37 54	-0.001	-0.02	7.71	0.08	-1.0	B5.5 V	-1	460 s		EO Aur, v
34248	40213		5 18 24.4	+47 02 00	+0.002	0.00	8.0	0.9	3.2	G5 IV		91 s		
34596	131967		5 18 25.4	-5 39 40	0.000	-0.09	7.0	1.1	0.2	K0 III		170 mx		
34536	112609		5 18 27.8	+6 44 47	0.000	-0.02	7.7	1.1	0.2	K0 III		310 s		
34469	77092		5 18 31.6	+21 47 33	0.000	-0.01	7.8	0.0	0.0	B8.5 V		340 s		
34247	40215		5 18 32.3	+48 55 23	0.000	-0.04	7.1	1.1	0.2	K0 III	+26	230 s		
34615	131971		5 18 34.1	-6 14 35	+0.002	-0.09	7.6	0.6	4.4	G0 V		44 s		
34522	94453		5 18 37.8	+12 05 18	+0.004	-0.08	8.0	0.5	3.4	F5 V		82 s		

HD	SAO	Star Name	α 2000	δ 2000	μ(α)	μ(δ)	V	B−V	M$_V$	Spec	RV	d(pc)	ADS	Notes
34332	40224		5^h 18^m 40.4^s	+40° 27′ 54″	+0.002^s	−0.01″	6.18	1.37	0.2	K0 III	−17	94 s		
35199	249244		5 18 43.9	−62 59 06	−0.001	−0.01	8.0	1.1	0.2	K0 III		360 s		
34721	150326		5 18 50.4	−18 07 48	+0.026	+0.06	5.96	0.58	4.4	dG0	+40	20 s	3899	m
34547	94455		5 18 52.9	+13 34 02	0.000	−0.01	7.5	0.0	0.4	B9.5 V	+9	260 s		
35024	233925		5 18 55.0	−51 34 46	+0.001	+0.02	7.8	0.3	0.2	K0 III		330 s		
34381	40229		5 18 55.4	+41 05 43	−0.001	−0.03	6.7	1.4	−0.3	K5 III		240 s		
34452	57884		5 19 00.0	+33 44 54	+0.001	−0.03	5.41	−0.19		B9 p	+29			
34411	40233	15 λ Aur	5 19 08.4	+40 05 57	+0.046	−0.66	4.71	0.63	4.4	G0 V	+66	13 ts	3886	m
34658	112624	21 Ori	5 19 11.2	+2 35 45	−0.001	−0.05	5.34	0.41	−2.0	F5 II	+11	290 s		
34673	131977		5 19 12.8	−3 04 26	+0.048	+0.13	7.74	1.04	6.9	K3 V	+85	14 ts	3900	m
34579	77098		5 19 14.6	+20 08 05	−0.003	−0.03	6.08	1.01	−0.9	G8 II-III	−47	240 s	3894	m
34477	57889		5 19 15.4	+34 53 26	+0.001	−0.01	7.01	0.18	0.0	B8.5 V		190 s	3888	m
34559	77097	109 Tau	5 19 16.5	+22 05 47	+0.001	−0.08	4.94	0.93	0.3	G8 III	+19	85 s		
34798	150335		5 19 17.4	−18 31 12	0.000	+0.01	6.36	−0.16	−0.2	B8 V		210 s	3910	m
34897	195748		5 19 17.5	−33 42 27	+0.004	+0.03	6.7	1.6		M5 III	+67			T Col, v
34797	150336		5 19 18.2	−18 30 36	0.000	0.00	6.54	−0.11		B8 pe				
34736	131980		5 19 21.3	−7 20 49	+0.001	+0.02	7.86	−0.08	0.4	B9.5 V		310 s		
35072	233926	ζ Pic	5 19 22.1	−50 36 22	+0.002	+0.23	5.45	0.51	0.6	F8 III	+45	61 mx		
34499	57893	18 Aur	5 19 23.5	+33 59 08	+0.001	−0.01	6.49	0.24	2.1	A5 V	+7	69 s	3893	m
34868	170311		5 19 23.7	−27 22 08	0.000	−0.01	5.99	−0.04	0.0	A0 IV		160 s		
34233	25125	15 Cam	5 19 27.8	+58 07 02	+0.001	−0.02	6.13	−0.03	−2.3	B3 IV	−3	380 s		
34867	170313		5 19 31.2	−25 07 26	0.000	0.00	7.1	1.8	−0.5	M4 III		260 s		
34776	131988		5 19 34.2	−9 42 02	−0.001	+0.02	8.0	0.1	0.6	A0 V		310 s		
34816	150340	6 λ Lep	5 19 34.4	−13 10 36	−0.001	0.00	4.29	−0.26		B0.5 IV	+20	25 mn		
34748	131983		5 19 35.2	−1 24 44	−0.001	−0.01	6.34	−0.11	−3.0	B1.5 V	+19	630 s		
34636	94461		5 19 38.9	+15 47 09	−0.001	−0.01	8.0	0.1	0.6	A0 V	−7	300 s		
34774	131989		5 19 42.1	−4 52 38	−0.002	−0.01	7.36	0.16	0.6	A0 V		170 s		
34018	13454		5 19 42.8	+67 59 45	−0.001	−0.03	7.40	0.1	1.4	A2 V		150 s		m
34912	170319		5 19 46.1	−25 55 39	+0.001	+0.08	7.2	−0.1	3.0	F2 V		68 s		
34745	112633		5 19 48.9	+2 30 49	+0.001	−0.10	7.8	0.5	4.0	F8 V		57 s		
34864	150344		5 19 54.1	−15 36 49	+0.001	−0.02	8.0	1.1	0.2	K0 III		360 s		
34863	150345	7 ν Lep	5 19 59.0	−12 18 56	−0.001	+0.01	5.30	−0.12	−0.6	B7 V nn	+16	150 s		
35116	217288		5 20 00.4	−47 02 57	0.000	−0.02	7.4	1.4	−0.5	M2 III		390 s		
35474	249255		5 20 00.4	−66 03 28	−0.003	+0.05	7.2	0.4	3.0	F2 V		69 s		
34590	57907		5 20 00.8	+32 47 08	+0.003	−0.04	7.0	0.7	4.4	G0 V		33 s		
34578	57906	19 Aur	5 20 00.8	+33 57 28	0.000	−0.01	5.03	0.27	−2.1	A5 II	−4	230 s		
34498	40242		5 20 02.3	+44 25 32	0.000	−0.01	6.7	1.1	0.2	K0 III	+13	190 s		
34827	131994		5 20 03.1	−5 12 30	−0.001	0.00	7.21	−0.02	0.4	B9.5 V		230 s		
34720	94466		5 20 03.5	+13 32 41	−0.001	−0.10	7.8	0.5	4.0	F8 V		57 s		
34535	57901		5 20 04.6	+39 47 21	+0.001	−0.07	7.2	0.1	0.6	A0 V		67 mx		
34545	57903		5 20 06.4	+39 20 34	+0.002	−0.01	7.80	1.4	−0.3	K5 III		390 s	3898	m
34577	57908		5 20 08.2	+35 47 10	0.000	−0.01	7.35	1.83	−0.5	M2 III		270 s		
34576	57911		5 20 11.3	+36 40 28	−0.001	−0.01	7.50	−0.02	−1.6	B3.5 V	0	540 s		
34557	40248		5 20 14.8	+41 05 10	0.000	−0.06	5.52	0.11	1.7	A3 V	+13	58 s		
34719	94467		5 20 18.2	+19 34 41	0.000	−0.02	6.8	0.1		A0 p	+17			
35046	195763		5 20 20.5	−34 41 56	−0.001	0.00	6.34	0.33	2.8	dF1		49 s		
35159	217290		5 20 21.1	−46 33 50	−0.002	−0.01	8.0	1.3	−0.1	K2 III		270 s		
34255	13460		5 20 22.6	+62 39 14	0.000	0.00	5.61	1.75	−5.9	cK4	−6	1100 s		
34793	94472		5 20 24.4	+10 53 25	−0.001	−0.03	7.5	0.1	0.6	A0 V		240 s		
34880	132004		5 20 26.3	−5 22 00	−0.001	+0.03	6.39	−0.03		B9			3926	m
34968	170327		5 20 26.8	−21 14 23	0.000	0.00	4.71	−0.05	0.6	A0 V	+30	66 s	3930	m
35402	249254		5 20 28.2	−60 46 47	−0.001	+0.03	6.9	0.2	1.4	A2 V		99 s		
34950	150355		5 20 32.8	−13 23 11	−0.001	−0.05	8.0	0.9	3.2	G5 IV		93 s		
35158	217292		5 20 36.7	−43 32 03	0.000	+0.03	6.80	1.66	−0.5	M2 III		260 s		
35114	195766		5 20 38.0	−39 45 17	+0.004	+0.02	7.3	0.6	3.4	F5 V		46 s		
34004	5501		5 20 38.4	+71 42 52	−0.004	−0.02	6.8	0.8	3.2	G5 IV		51 s		
34533	40251		5 20 39.2	+46 57 50	0.000	0.00	6.54	0.60		F2	+17		3903	m
34200	13458		5 20 41.0	+66 44 49	+0.002	+0.03	6.6	0.9	3.2	G5 IV		47 s		
34656	57919		5 20 43.0	+37 26 18	−0.001	−0.01	6.79	0.02		O7	0			m
34811	94477		5 20 43.2	+15 38 12	0.000	−0.01	7.7	0.2	2.1	A5 V	+29	130 s		
34791	94476		5 20 43.5	+17 14 34	−0.001	−0.01	7.7	0.0	0.4	B9.5 V		270 s		
34878	112653		5 20 43.6	+2 32 41	+0.001	+0.02	6.6	0.9	3.2	G5 IV		49 s		
35071	170336		5 20 44.5	−29 12 21	0.000	+0.02	7.2	0.3	0.6	A0 V		130 s		
34698	57921		5 20 49.9	+34 16 03	+0.001	0.00	7.9	1.1	−0.1	K2 III		370 s		
34888	112655		5 20 52.2	+2 55 45	0.000	+0.04	6.8	0.2	2.1	A5 V		85 s		
34855	112654		5 20 53.1	+9 43 31	0.000	+0.05	6.5	0.1	1.4	A2 V		110 s		
34624	40260		5 20 54.3	+42 21 55	−0.002	+0.02	7.80	0.8	1.8	G6 III-IV	+9	160 s		
34625	40261		5 20 54.8	+40 53 07	+0.001	+0.01	7.30	−0.01	0.4	B9.5 V		230 s		
35021	170335		5 20 55.0	−21 02 24	−0.001	+0.04	8.0	−0.4	0.4	B9.5 V		330 s		
35009	150364		5 20 55.3	−19 41 14	−0.001	+0.04	7.4	1.2	0.2	K0 III		220 s		
34854	94481		5 20 56.4	+11 41 46	−0.002	−0.01	7.8	1.1	−0.1	K2 III		380 s		
34810	94478		5 20 56.5	+19 48 51	0.000	−0.02	6.18	1.23	0.2	K0 III	0	110 s		
34762	77121		5 20 59.3	+27 57 26	−0.001	−0.02	6.30	0.04	−0.2	B8 V	+7	170 s		
34635	40262		5 21 04.1	+42 30 17	−0.001	+0.02	7.73	0.01	0.4	B9.5 V		280 s		
34655	40263		5 21 07.3	+42 21 53	+0.002	−0.04	7.7	1.1	0.2	K0 III		300 s		

HD	SAO	Star Name	α 2000	δ 2000	μ(α)	μ(δ)	V	B-V	M_v	Spec	RV	d(pc)	ADS	Notes
35129	195769		5ʰ21ᵐ11ˢ.0	-30°04'21"	0ˢ.000	-0".02	7.4	1.5	0.2	K0 III		150 s		
34907	94484		5 21 12.0	+11 06 34	0.000	-0.05	7.5	0.4	2.6	F0 V		96 s		
34790	77124		5 21 12.7	+29 34 12	0.000	0.00	5.70	0.06	1.4	A2 V	-19	70 s		
36192	256179		5 21 13.2	-75 55 00	-0.004	+0.01	7.9	0.2	2.1	A5 V		150 s		
35165	195770		5 21 16.8	-34 20 43	0.000	0.00	6.09	-0.19		B5 pne	+20			m
34544	25137		5 21 17.3	+54 15 10	-0.001	+0.03	6.7	0.9	3.2	G5 IV		50 s		
34991	132021		5 21 17.9	-5 48 41	0.000	-0.01	7.4	0.5	4.0	F8 V		48 s		
35043	150368		5 21 18.0	-15 09 05	-0.001	0.00	7.7	0.2	2.1	A5 V		130 s		
34959	112660		5 21 19.3	+4 00 43	0.000	-0.01	6.57	-0.09		B5 p	+5			
35042	150372		5 21 21.4	-14 33 20	-0.002	-0.01	7.2	0.0	0.0	B8.5 V	+21	280 s		
35128	170348		5 21 22.6	-27 56 59	0.000	-0.01	8.0	1.6	0.2	K0 III		160 s		
35356	233940		5 21 23.1	-51 34 31	+0.003	+0.03	7.4	0.4	0.2	K0 III		270 s		
35229	195778		5 21 28.1	-39 29 52	0.000	+0.05	7.4	1.5	0.2	K0 III		140 s		
35008	132023		5 21 28.4	-1 32 44	0.000	+0.03	7.12	-0.10		B8				
35288	217304		5 21 29.8	-44 22 25	-0.003	+0.01	7.4	1.2	0.6	A0 V		46 s		
35007	132024		5 21 31.7	-0 24 59	-0.001	0.00	5.68	-0.12	-1.7	B3 V	+7	290 s	3941	m
35417	233945		5 21 36.2	-53 18 51	0.000	+0.02	7.4	1.1	0.2	K0 III		250 s		
34937	94489		5 21 39.1	+15 50 49	-0.002	-0.05	7.4	0.4	3.0	F2 V		76 s		
34926	94488		5 21 39.8	+18 54 24	-0.001	-0.02	7.50	0.6	4.4	G0 V		41 s		m
34989	112667		5 21 43.5	+8 25 43	0.000	0.00	5.80	-0.13	-3.5	B1 V	+26	670 s		
35039	132028	22 Ori	5 21 45.7	-0 22 57	0.000	0.00	4.73	-0.17	-3.0	B2 IV	+29	340 s		
35162	170351		5 21 46.2	-24 46 23	-0.002	-0.01	5.06	0.67	1.3	A3 IV-V	+5	25 s	3954	m
34759	40269	20 ρ Aur	5 21 48.4	+41 48 17	+0.002	-0.03	5.23	-0.15	-1.1	B5 V	+5	190 s		
34634	25143		5 21 49.3	+51 29 11	+0.026	-0.22	7.9	0.4	3.0	F2 V		30 mx		
35104	150375		5 21 51.0	-13 45 22	0.000	-0.01	6.56	-0.09		B8				
34330	13463		5 21 52.5	+67 01 29	+0.008	-0.07	7.8	1.1	0.2	K0 III		160 mx		
35473	233949		5 21 55.2	-54 16 22	+0.001	+0.04	7.1	0.1	1.4	A2 V		130 s		
35137	150379		5 21 55.3	-17 36 14	+0.001	-0.01	6.9	1.1	0.2	K0 III		220 s		
34807	57938		5 21 55.4	+39 34 22	+0.001	-0.02	7.4	0.1	1.4	A2 V		160 s		
35227	195779		5 21 56.4	-31 35 24	+0.001	+0.01	8.0	2.1	-0.1	K2 III		120 s		
34769	40275		5 21 56.8	+42 17 10	-0.003	0.00	7.1	1.1	0.2	K0 III	-4	230 s		
35079	132032		5 21 58.2	-2 57 50	0.000	+0.01	7.06	-0.03	-1.7	B3 V		470 s		
35301	195782		5 21 58.4	-38 29 20	0.000	+0.02	7.9	0.9	0.2	K0 III		350 s		
35017	112673		5 22 01.1	+7 00 51	-0.001	-0.04	7.8	1.1	0.2	K0 III		330 s		
33940	5503		5 22 03.8	+75 11 43	+0.007	-0.05	8.0	0.5	4.0	F8 V		65 s		
35067	112679		5 22 05.5	+3 34 18	0.000	0.00	7.7	1.4	-0.3	K5 III	+51	390 s		
35038	112676		5 22 05.9	+7 05 23	-0.003	0.00	7.6	0.4	3.0	F2 V		85 s		
35037	112677		5 22 08.1	+8 00 19	+0.001	0.00	7.8	0.1	1.4	A2 V		190 s		
37134	258412		5 22 08.1	-81 02 21	-0.023	+0.01	7.90	0.1	1.4	A2 V		130 mx		m
35274	195784		5 22 10.7	-32 32 22	+0.001	+0.01	7.9	0.8	0.2	K0 III		340 s		
34788	40280		5 22 11.1	+45 04 54	+0.001	0.00	7.70	0.1	0.6	A0 V		250 s	3932	m
35066	112681		5 22 11.2	+5 23 43	+0.003	-0.05	7.20	0.5	3.4	F5 V		58 s		m
34875	57946		5 22 11.9	+36 11 57	+0.001	-0.03	7.7	1.1	0.2	K0 III		300 s		
35064	112683		5 22 17.0	+7 11 57	0.000	-0.01	7.9	1.1	0.2	K0 III		340 s		
35155	132035		5 22 18.5	-8 39 58	+0.002	-0.01	6.9	1.9		S4.1				
34575	25144		5 22 20.0	+59 16 41	+0.032	-0.28	7.07	0.75	5.3	dG6	-23	27 ts		
35580	233952		5 22 22.2	-56 08 03	0.000	+0.03	6.11	-0.10		B9				
35036	94498		5 22 23.4	+16 07 21	+0.001	-0.05	7.4	0.1	0.6	A0 V	+48	220 s		
36062	256180		5 22 25.8	-71 30 23	+0.003	+0.08	7.44	0.19		A3				
35245	--		5 22 31.7	-19 52 06			7.9	-0.3	0.6	A0 V		290 s		
36208	256182		5 22 32.5	-73 48 05	-0.003	+0.02	7.2	1.1	0.2	K0 III		250 s		
33564	5496		5 22 33.5	+79 13 52	-0.028	+0.16	5.05	0.47	3.7	F6 V	-10	19 ts	3864	m
34921	57950		5 22 35.2	+37 40 35	0.000	+0.02	7.50	0.15	-4.6	B0 IV pe	-9	1900 s		
35757	249259		5 22 38.3	-62 53 44	0.000	0.00	7.7	1.1	0.2	K0 III		320 s		
35284	170364		5 22 41.1	-23 15 33	0.000	0.00	7.6	0.8	0.2	K0 III		300 s		
35062	94501		5 22 45.5	+19 00 42	+0.001	0.00	7.6	0.7	0.6	G0 III	+39	240 s		
35134	112693		5 22 45.5	+2 47 45	-0.002	-0.06	6.74	0.07	0.6	A0 V		88 mx		
35149	112697	23 Ori	5 22 49.9	+3 32 40	0.000	0.00	5.00	-0.15	-3.5	B1 V	+18	470 s	3962	m
34904	40290		5 22 50.3	+41 01 45	-0.001	+0.01	5.54	0.12	1.7	A3 V	-14	56 s		
35148	112699		5 22 50.8	+3 33 06	-0.001	-0.01	7.18	-0.12		A				
34623	13476		5 22 54.8	+61 45 05	-0.005	-0.01	7.8	0.8	3.2	G5 IV		82 s		
35110	94505		5 22 56.8	+14 21 41	0.000	-0.01	7.40	1.1	0.2	K0 III		280 s	3961	m
35034	77136		5 23 00.1	+29 43 45	0.000	-0.03	8.02	0.03	0.4	B9.5 V		330 s		
35014	77137		5 23 00.2	+29 05 29	0.000	+0.01	8.0	1.1	0.2	K0 III		340 s		
35035	77138		5 23 01.1	+28 28 08	-0.002	-0.03	7.4	0.2		A m	+46			
35261	132046		5 23 04.2	-8 06 21	0.000	0.00	7.57	0.00	0.6	A0 V		250 s	3974	m
35192	112704		5 23 05.4	+1 03 24	0.000	-0.01	7.70	0.1	0.6	A0 V		260 s	3968	m
35191	112705		5 23 06.0	+1 17 23	0.000	0.00	7.70	1.1	-0.1	K2 III		360 s		m
35203	112707		5 23 10.0	+1 08 22	-0.001	-0.01	7.97	-0.09	-0.9	B6 V		580 s		
35386	170375		5 23 11.9	-26 42 21	+0.002	+0.01	6.49	0.50	3.7	dF6		35 s		
35813	249261		5 23 12.1	-62 13 51	0.000	0.00	8.0	1.1	-0.1	K2 III		420 s		
35416	195796		5 23 12.3	-31 44 55	-0.007	-0.06	7.60	0.4	2.6	F0 V		100 s		m
34903	40294		5 23 12.5	+47 01 17	+0.001	-0.02	7.10	1.4	-0.3	K5 III	-9	280 s	3948	m
35281	132053		5 23 18.5	-8 24 57	0.000	-0.02	5.99	-0.03		A0			3978	m
35307	150392		5 23 18.7	-14 49 58	-0.001	+0.01	7.9	1.1	0.2	K0 III	+50	350 s		

HD	SAO	Star Name	α 2000	δ 2000	μ(α)	μ(δ)	V	B–V	M$_V$	Spec	RV	d(pc)	ADS	Notes
35076	77139	22 Aur	5 23 22.8	+28 56 12	+0.001	-0.03	6.40	-0.04	0.2	B9 V	+9	170 s		
35515	195807		5 23 23.9	-39 40 43	+0.001	0.00	5.71	1.62		Ma				
35147	94509		5 23 26.4	+16 58 54	0.000	-0.15	8.0	0.5	4.0	F8 V		62 s		
35604	217317		5 23 26.7	-49 22 31	+0.001	-0.01	7.8	1.0	0.2	K0 III		330 s		
34787	25161	16 Cam	5 23 27.6	+57 32 39	-0.004	-0.06	5.20	0.1	0.6	A0 V	+10	81 s		m
35337	150396	8 Lep	5 23 30.1	-13 55 38	0.000	0.00	5.25	-0.21	-3.0	B2 IV	+18	450 s		
35146	94510		5 23 30.6	+18 36 03	0.000	-0.01	7.50	1.4	-0.3	K5 III	+9	330 s		
35242	112708		5 23 31.0	+5 19 21	-0.002	0.00	6.35	0.12		A0	+9			
35173	94512		5 23 31.6	+16 02 26	-0.001	0.00	6.90	0.00	0.0	B8.5 V	+21	230 s	3969	m
35189	94514	110 Tau	5 23 37.6	+16 41 57	-0.002	-0.02	6.10	0.14	1.4	A2 V	+21	75 s		
--	170383		5 23 37.9	-22 18 28	-0.002	-0.01	7.6	0.1	1.7	A3 V		150 s		
35171	94513		5 23 38.4	+17 19 27	+0.019	0.00	7.95	1.10	-0.1	K2 III	+37	21 ts		
35528	195809		5 23 39.0	-37 20 12	+0.002	+0.01	6.82	1.04		G5				
35190	94515		5 23 41.7	+16 29 50	-0.003	-0.02	7.7	0.6	4.4	G0 V		46 s		
34786	25163		5 23 42.0	+58 56 28	+0.001	-0.01	7.80	0.9	0.3	G8 III	-17	290 s		
35299	132057		5 23 42.2	-0 09 35	-0.001	0.00	5.70	-0.21	-2.5	B2 V	+22	440 s		
35430	--		5 23 43.2	-22 17 17			7.40	0.1	1.7	A3 V		140 s	3993	m
34853	25166		5 23 43.4	+56 05 20	+0.003	+0.01	6.9	0.9	3.2	G5 IV		55 s		
35353	132061		5 23 43.5	-8 17 21	-0.001	+0.01	7.69	0.22	0.6	A0 V		200 s		
35298	112714		5 23 50.3	+2 04 56	0.000	+0.01	7.89	-0.14	0.2	B9 V		350 s		
35089	57972		5 23 51.3	+36 23 54	+0.001	-0.01	6.9	0.8	3.2	G5 IV		54 s		
35317	132060		5 23 51.3	-0 52 02	-0.001	-0.01	6.11	0.50		F5			3991	m
35514	195810		5 23 51.5	-32 09 04	0.000	+0.01	8.0	1.5	0.2	K0 III		190 s		
35441	170386		5 23 52.5	-20 43 39	+0.002	+0.03	7.80	1.1	-0.1	K2 III	+15	380 s		
35270	94519		5 23 53.7	+11 05 01	-0.002	0.00	7.5	0.5	3.4	F5 V		66 s		
35369	132067	29 Ori	5 23 56.7	-7 48 29	-0.001	-0.04	4.14	0.96	0.3	G8 III	-18	58 s		
35567	195816		5 24 08.1	-34 24 55	+0.002	+0.06	7.5	0.2	1.7	A3 V		120 s		
35102	57974		5 24 09.3	+39 21 56	+0.001	-0.01	8.0	1.1	0.2	K0 III		340 s		
35782	233962		5 24 09.4	-53 25 43	-0.002	-0.01	7.3	0.2	1.4	A2 V		120 s		
35603	195820		5 24 10.9	-38 03 27	0.000	-0.01	8.0	1.8	0.2	K0 III		120 s		
35578	195818		5 24 12.6	-35 09 07	+0.001	-0.04	8.0	0.9	-0.1	K2 III		410 s		
30881	818		5 24 14.3	+86 17 23	+0.007	-0.02	7.96	-0.01	0.6	A0 V		220 mx		
35678	217321		5 24 15.8	-43 36 32	-0.001	0.00	7.3	1.5	-0.3	K5 III		340 s		
35101	40311		5 24 22.7	+42 36 37	-0.001	0.00	8.00	0.1	1.4	A2 V		200 s	3975	m
35296	94526	111 Tau	5 24 25.4	+17 23 00	+0.017	-0.01	4.98	0.52	4.0	F8 V	+37	16 ts		m
35012	25177		5 24 27.6	+52 13 49	+0.001	-0.02	7.8	1.1	-0.1	K2 III		340 s		
35505	150416		5 24 28.3	-16 58 34	+0.001	-0.03	5.65	0.00		A0				
35411	132071	28 η Ori	5 24 28.6	-2 23 50	0.000	-0.01	3.36	-0.17	-3.5	B1 V	+20	230 s	4002	m,v
35410	132070	27 Ori	5 24 28.8	-0 53 29	-0.001	+0.13	5.08	0.96	0.2	K0 III	+21	95 s		
35120	40317		5 24 30.5	+41 49 37	+0.001	-0.01	7.80	-0.05	-1.0	B5.5 V		490 s		
35396	112726		5 24 32.3	+3 31 05	0.000	0.00	7.7	0.4	3.0	F2 V		87 s		
--	57979		5 24 34.8	+37 39 42	+0.002	-0.04	8.0	1.4						
35407	112729		5 24 36.1	+2 21 10	+0.001	-0.01	6.32	-0.15	-1.1	B5 V	-8	310 s		
35238	57987		5 24 38.3	+31 13 26	-0.003	-0.01	6.28	1.24	0.0	K1 III	+40	150 s		
35239	57988		5 24 38.4	+31 08 35	-0.001	-0.01	5.94	0.04	-0.8	B9 III	+8	190 s		
35186	57981	21 σ Aur	5 24 39.1	+37 23 07	0.000	-0.01	4.99	1.42	-0.3	K4 III	-19	110 s	3984	m
35456	132075		5 24 40.2	-2 29 52	-0.002	+0.01	6.94	-0.05		B8			4007	m
35364	94530		5 24 41.5	+11 31 47	-0.002	-0.10	7.3	0.9	3.2	G5 IV		66 s		
35859	233964		5 24 42.2	-52 18 46	-0.001	-0.02	6.79	0.07	0.6	A0 V		150 s		m
34839	13482		5 24 44.5	+63 23 10	-0.016	-0.06	7.90	0.7	4.4	G0 V		49 s	3956	m
35439	112734	25 Ori	5 24 44.8	+1 50 47	0.000	0.00	4.95	-0.20	-3.5	B1 V	+19	490 s		
35202	57985		5 24 44.8	+36 11 59	0.000	-0.02	6.7	1.1	-0.1	K2 III		220 s		
35860	233965	θ Pic	5 24 46.1	-52 18 59	-0.002	-0.03	6.27	0.07	0.6	A0 V		120 s		m
34804	13481		5 24 46.8	+64 43 53	-0.002	+0.02	8.00	0.5	4.0	F8 V		62 s	3955	m
35739	217323		5 24 48.5	-42 26 23	+0.001	+0.02	8.0	1.5	0.2	K0 III		360 s		
35349	94531		5 24 50.5	+17 11 48	0.000	-0.01	7.60	0.00	-1.0	B5.5 V	+14	450 s		
35765	217325		5 24 55.5	-44 13 33	-0.001	+0.01	6.08	1.20		K0				
35234	57994		5 24 58.0	+37 30 12	0.000	0.00	8.0	0.0	0.4	B9.5 V		310 s		
34450	5517		5 24 59.1	+73 42 24	-0.005	-0.02	6.9	1.6	-0.5	M2 III		310 s		
35405	94534		5 24 59.8	+11 07 38	-0.001	-0.03	7.6	1.1	0.2	K0 III		300 s		
35593	150422		5 25 00.0	-19 22 29	0.000	-0.03	6.90	0.1	0.6	A0 V		180 s		m
34666	13479		5 25 00.4	+69 13 25	+0.002	-0.02	8.0	1.1	0.2	K0 III		330 s		
34917	13484		5 25 01.0	+61 49 33	-0.001	-0.02	8.0	0.0	0.4	B9.5 V		310 s		
35502	132081		5 25 01.1	-2 48 57	-0.001	-0.02	7.36	-0.04	-1.1	B5 V		450 s		m
35536	150420		5 25 01.5	-10 19 45	-0.002	-0.02	5.61	1.56	-0.3	K5 III	+57	150 s		
36061	249267		5 25 01.9	-61 06 20	+0.003	0.00	8.0	0.9	0.2	K0 III		360 s		
35468	112740	24 γ Ori	5 25 07.8	+6 20 59	-0.001	-0.01	1.64	-0.22	-3.6	B2 III	+18	110 s		Bellatrix, m
35395	77157		5 25 10.3	+20 35 02	0.000	0.00	6.77	0.23		B0.5 III	+12			
35949	233973		5 25 11.2	-54 19 05	+0.002	+0.04	7.2	0.7	3.0	F2 V		42 s		
35501	112744		5 25 11.3	+1 55 24	-0.001	0.00	7.42	-0.06	-0.2	B8 V		330 s	4012	m
35295	57999		5 25 12.9	+34 51 19	0.000	-0.04	6.55	1.11		K1 III-IV p	-15		4000	m
35934	233972		5 25 13.6	-52 40 36	0.000	+0.02	7.2	1.3	-0.1	K2 III		250 s		
35756	195836		5 25 16.9	-36 29 44	-0.001	-0.01	7.6	1.2	3.2	G5 IV		42 s		
36767	256188		5 25 18.1	-75 41 30	-0.036	-0.29	7.18	0.53		G0		11 mn		
35737	195835		5 25 19.0	-34 50 49	-0.001	+0.03	7.4	0.5	3.2	G5 IV		68 s		

HD	SAO	Star Name	α 2000	δ 2000	μ(α)	μ(δ)	V	B-V	M_V	Spec	RV	d(pc)	ADS	Notes
35563	132084		5 25 19.3	−7 41 46	−0.001	+0.02	8.0	1.4	−0.3	K5 III		450 s		
35591	150427		5 25 20.0	−12 33 14	−0.001	−0.02	6.7	0.9	3.2	G5 IV		50 s		
35326	58003		5 25 28.0	+34 26 40	0.000	−0.01	8.0	1.1	−0.1	K2 III		390 s		
35826	217328		5 25 28.5	−41 20 20	0.000	+0.02	7.6	1.7	0.2	K0 III		300 s		
35327	58004		5 25 30.0	+34 11 11	0.000	0.00	6.9	0.4	3.0	F2 V		58 s		
35548	132086		5 25 31.1	−0 32 39	+0.001	−0.01	6.57	−0.05		B9			4020	m
35575	132088		5 25 36.4	−1 29 29	−0.001	0.00	6.43	−0.17	−1.7	B3 V	+2	420 s		
35313	58006		5 25 38.4	+38 02 40	−0.001	−0.02	7.9	0.1	0.6	A0 V		270 s	4004	m
35615	132092		5 25 40.7	−9 33 09	0.000	−0.01	8.0	1.1	0.2	K0 III		360 s		
36008	233974		5 25 42.6	−53 28 54	+0.008	+0.04	7.8	0.6	3.4	F5 V		59 s		
35588	112752		5 25 46.9	+0 31 15	−0.001	+0.02	6.16	−0.18	−1.7	B3 V	−24	370 s		
35559	112751		5 25 50.4	+5 57 22	−0.001	−0.02	7.9	1.1	0.2	K0 III		340 s		
35304	40327		5 25 51.5	+43 13 56	−0.001	0.00	7.89	1.73	−0.3	K5 III		320 s		
35824	195846		5 25 53.1	−35 21 11	+0.002	−0.04	7.70	0.5	3.4	F5 V		72 s		
35627	132095		5 25 53.1	−6 21 46	−0.010	−0.03	7.6	0.5	4.0	F8 V		51 s		
35522	94539		5 25 54.4	+15 27 13	+0.001	−0.01	7.0	0.0	0.4	B9.5 V	+18	200 s		
35641	132097		5 25 56.3	−5 59 30	−0.001	−0.02	8.0	1.4	−0.3	K5 III		450 s		
35736	150442		5 25 59.8	−19 41 44	0.000	−0.02	5.65	0.44	3.3	dF4	+6	29 s	4034	m
35811	195848		5 26 01.0	−32 12 34	+0.002	0.00	7.00	1.1	0.2	K0 III		230 s		m
35640	132100		5 26 02.3	−5 31 06	+0.001	+0.01	6.23	−0.06	0.2	B9 V		160 s		
35533	94542		5 26 02.6	+15 40 19	0.000	0.00	7.5	0.1	0.6	A0 V	+25	230 s		
35685	150437		5 26 04.2	−10 45 28	−0.001	0.00	8.0	0.0	0.4	B9.5 V		330 s		
36155	233977		5 26 04.5	−58 33 42	+0.004	0.00	7.5	−0.4	0.6	A0 V		85 mx		
35698	150440		5 26 04.6	−12 54 18	0.000	0.00	7.8	1.1	0.2	K0 III		330 s		
35532	94543	113 Tau	5 26 05.6	+16 42 01	−0.001	0.00	6.25	−0.08	−2.5	B2 V n	+31	500 s		
--	--		5 26 05.6	−86 23 21			6.40							R Oct, v
34188	5516		5 26 10.8	+78 24 54	+0.004	+0.02	6.8	0.9	3.2	G5 IV		52 s		
35312	40329		5 26 14.5	+46 49 07	+0.001	−0.04	8.0	1.1	0.2	K0 III		330 s		
35854	195854		5 26 14.9	−32 30 17	+0.021	−0.09	7.5	0.7	0.2	K0 III		49 mx		
35497	77168	112 β Tau	5 26 17.5	+28 36 27	+0.002	−0.17	1.65	−0.13	−1.6	B7 III	+8	40 mx		Elnath, m
36189	233981	λ Dor	5 26 19.2	−58 54 46	−0.002	+0.03	5.14	1.00		G6	+10	20 mn		
35622	112759		5 26 22.5	+8 59 59	−0.001	−0.02	7.8	0.1	0.6	A0 V		270 s		
35638	112761		5 26 22.9	+3 51 22	+0.002	+0.01	7.8	0.5	3.4	F5 V		74 s		
35796	170424		5 26 23.9	−20 42 56	−0.003	+0.01	7.40	0.1	1.7	A3 V		140 s	4042	m
35673	112765		5 26 31.2	+2 56 10	0.000	0.00	6.50	0.00	0.2	B9 V		170 s	4033	m
35623	112764		5 26 32.1	+8 15 19	0.000	−0.01	7.9	0.1	0.6	A0 V		280 s		
35611	94549		5 26 37.8	+16 13 10	+0.001	−0.02	7.4	0.4	3.0	F2 V		73 s		
36791	256192		5 26 38.5	−73 39 55	+0.009	+0.07	7.6	0.9	3.2	G5 IV		75 s		
35656	112767		5 26 38.7	+6 52 09	0.000	−0.01	6.42	−0.02		B9				
35794	150452		5 26 39.2	−13 51 35	0.000	0.00	7.1	0.1	0.6	A0 V		200 s		
36369	249276		5 26 41.3	−64 19 28	0.000	+0.02	7.6	−0.3	0.6	A0 V		250 s		
36752	256191		5 26 41.6	−72 57 00	+0.004	−0.01	7.9	1.4	−0.3	K5 III		440 s		
35975	195861		5 26 44.8	−36 35 07	−0.001	0.00	7.3	−0.2	0.6	A0 V		220 s		
35696	112770		5 26 44.8	+4 04 54	0.000	−0.04	7.5	0.7	4.4	G0 V		42 s		
34886	5533		5 26 47.2	+70 13 42	+0.004	−0.02	7.0	0.0	0.4	B9.5 V		140 mx		
35635	94551		5 26 48.2	+13 34 57	0.000	0.00	7.7	0.0	0.4	B9.5 V		270 s		
35520	58028		5 26 48.8	+34 23 30	+0.001	−0.01	5.94	0.14		A1 p	+7			
35715	112775	30 ψ Ori	5 26 50.2	+3 05 44	0.000	0.00	4.59	−0.21	−3.0	B2 IV	+12	330 s	4039	m
35521	58030		5 26 51.3	+33 15 45	+0.001	0.00	6.15	1.15	0.2	K0 III	−9	130 s		
--	--		5 26 53.0	+34 10 10	0.000	−0.02	5.8	0.1	0.6	A0 V		110 s		
35519	58029		5 26 54.2	+35 27 26	−0.002	−0.01	6.2	1.1	−0.1	K2 III	−21	180 s		
35730	112777		5 26 54.3	+3 36 52	+0.001	−0.01	7.20	−0.15		B5 p	−3			
35777	132116		5 26 59.1	−2 21 39	0.000	0.00	6.62	−0.18	−2.5	B2 V	+19	670 s		
35694	112776		5 26 59.1	+8 04 40	+0.001	−0.01	7.5	1.1	0.2	K0 III		280 s		
35714	112779		5 27 00.0	+7 10 13	+0.001	−0.03	7.6	0.1	0.6	A0 V		260 s		
36584	249281		5 27 00.1	−68 37 22	+0.001	−0.02	6.03	0.34		F0	+1			m
35916	170436		5 27 01.2	−24 54 13	+0.002	+0.03	8.0	1.3	0.2	K0 III		250 s		
35586	77177		5 27 01.2	+27 36 36	0.000	−0.03	7.88	0.49		F8			4032	m
35850	150461		5 27 04.7	−11 54 04	0.000	−0.05	6.35	0.51	3.4	F5 V	+19	36 s		
36060	217340		5 27 05.2	−40 56 37	0.000	+0.09	5.87	0.23		A m				m
36095	217341		5 27 06.3	−44 51 41	0.000	+0.05	6.8	1.6	−0.1	K2 III		140 s		
35974	195865		5 27 06.5	−31 33 59	−0.005	+0.04	7.3	0.5	4.4	G0 V		38 s		
35776	132119		5 27 07.2	−1 04 59	−0.002	+0.05	8.0	0.4	3.0	F2 V		98 s		
35762	112781		5 27 08.2	+3 51 19	−0.001	−0.01	6.74	−0.18	−2.5	B2 V	+17	710 s		
35600	58040		5 27 08.3	+30 12 31	+0.001	−0.01	5.74	0.16	−5.5	B9 Ib	+17	1400 s		
35792	132121		5 27 09.4	−1 22 02	0.000	+0.01	7.22	−0.15	−1.7	B3 V		600 s		
35887	150463		5 27 10.0	−16 31 27	0.000	−0.14	7.5	0.5	3.4	F5 V		65 s		
35671	94554	115 Tau	5 27 10.0	+17 57 44	0.000	−0.02	5.42	−0.10	−1.1	B5 V	+19	200 s	4038	m
35601	77179		5 27 10.2	+29 55 15	+0.001	−0.02	7.35	2.20	−4.7	M1 Ib	−1	1100 s		
35712	94557		5 27 11.0	+11 31 33	+0.002	0.00	7.9	0.5	3.4	F5 V		78 s		
35693	94556		5 27 13.8	+15 15 28	−0.001	0.00	6.16	0.08	1.4	A2 V p	+25	88 s		
35476	40340		5 27 13.9	+44 00 20	+0.002	−0.02	7.4	1.1	0.2	K0 III	−47	270 s		
36038	195868		5 27 14.4	−34 36 51	−0.004	0.00	6.9	1.0	3.2	G5 IV		42 s		
35775	112784		5 27 15.2	+2 20 28	−0.001	0.00	6.5	1.1	0.2	K0 III		180 s		
36137	217342		5 27 15.6	−46 05 40	+0.001	−0.04	7.71	0.40	3.1	F3 V		81 s		

HD	SAO	Star Name	α 2000	δ 2000	μ(α)	μ(δ)	V	B−V	M$_v$	Spec	RV	d(pc)	ADS	Notes
35788	112787		5^{h}27^{m}23.7^s	+4°15'41"	−0^{s}001	+0."01	7.7	1.1	0.2	K0 III		310 s		
35994	170440		5 27 24.7	−26 35 06	0.000	−0.04	7.1	−0.1	1.4	A2 V		140 s		
35496	40341		5 27 27.1	+47 19 11	+0.001	−0.04	8.0	1.1	−0.1	K2 III		380 s		
35746	94564		5 27 30.0	+16 02 24	−0.001	−0.01	7.7	0.0	0.4	B9.5 V		270 s		
35930	150469		5 27 32.3	−13 34 44	+0.002	−0.23	7.3	0.5	3.4	F5 V		45 mx		
35475	40343		5 27 34.3	+49 15 33	−0.003	+0.01	7.8	0.1	0.6	A0 V		250 s		
35555	40346		5 27 35.8	+40 39 01	+0.001	+0.01	7.8	0.1	0.6	A0 V		260 s		
35991	170445		5 27 36.4	−21 22 32	0.000	+0.04	6.07	1.04	0.3	gG7	+34	120 s		
35834	112792		5 27 36.9	+1 06 26	+0.001	−0.01	7.67	−0.05	−0.2	B8 V		370 s	4056	m
35708	77184	114 Tau	5 27 38.0	+21 56 13	0.000	−0.01	4.88	−0.15	−1.7	B3 V	+14	210 s	4048	m
35620	58051	24 φ Aur	5 27 38.8	+34 28 32	0.000	−0.05	5.07	1.40		K3 p	+31	27 mn		m
35544	40345		5 27 39.2	+43 22 06	+0.001	+0.01	6.79	0.00	0.6	A0 V	−1	170 s		
36435	249280		5 27 39.3	−60 24 59	−0.021	−0.10	6.97	0.76	5.2	G5 V		20 s		
34740	5531		5 27 39.7	+74 33 22	+0.003	−0.01	7.20	0.1		A0 p			3982	m
35786	94568		5 27 42.1	+11 28 15	+0.001	−0.05	7.9	1.1	0.2	K0 III		340 s		
35633	58053		5 27 43.3	+34 31 56	+0.001	−0.01	8.04	0.32		B0.5 IV				
35882	132134		5 27 44.6	−1 48 47	−0.001	+0.01	7.80	−0.07		B5				
36037	170450		5 27 44.7	−25 38 12	+0.001	+0.01	7.5	−0.1	1.4	A2 V		160 s		
36136	195878		5 27 44.8	−39 13 18	0.000	+0.01	6.8	0.9	3.2	G5 IV		42 s		
35770	94566	116 Tau	5 27 45.6	+15 52 27	0.000	−0.02	5.50	0.01	0.2	B9 V	+15	110 s		
35899	132135		5 27 45.7	−2 08 42	0.000	+0.01	7.52	−0.14	−1.1	B5 V	+25	530 s		
35653	58056		5 27 45.8	+33 56 46	−0.001	0.00	7.44	0.12		B0.5 V	+3			
35573	40349		5 27 47.2	+44 09 56	−0.001	−0.02	7.8	1.1	−0.1	K2 III		350 s		
36094	195875		5 27 50.4	−32 25 04	−0.001	−0.03	6.90	0.9	3.2	G5 IV		55 s		m
35421	25203		5 27 52.9	+56 16 43	+0.004	−0.08	8.0	0.5	3.4	F5 V		82 s		
35881	112793		5 27 54.1	+1 06 18	−0.001	0.00	7.78	−0.09	−0.2	B8 V		400 s		
35543	40347		5 27 54.5	+48 13 36	−0.001	−0.04	6.6	1.4	−0.3	K5 III		230 s		
36018	150476		5 27 56.8	−18 00 15	−0.001	0.00	7.6	0.7	4.4	G0 V		44 s		
35585	40352		5 27 57.9	+44 54 39	+0.001	−0.04	7.7	1.1	0.2	K0 III		300 s		
35542	40350		5 27 58.3	+48 22 54	−0.001	0.00	7.2	1.1	0.2	K0 III	+8	230 s		
36154	195883		5 27 58.9	−38 54 25	−0.002	+0.01	7.70	0.9	3.2	G5 IV		79 s		m
35818	94575		5 27 59.1	+12 54 42	+0.001	−0.01	7.8	0.1	0.6	A0 V		250 s		
35681	58065		5 28 00.8	+33 45 50	+0.001	−0.20	6.80	0.5	3.4	F5 V		48 s	4050	m
35912	112794		5 28 01.5	+1 17 54	+0.001	0.00	6.41	−0.18	−2.5	B2 V	+34	610 s		
35802	94573	117 Tau	5 28 01.6	+17 14 20	+0.001	−0.05	5.77	1.63	−0.3	gK5	−23	130 s		
35947	132138		5 28 06.2	−1 40 10	−0.004	−0.05	7.8	0.5	3.4	F5 V		74 s		
35910	112795		5 28 06.8	+3 32 09	−0.001	0.00	7.58	−0.10	−0.9	B6 V	+18	490 s		
35833	94577		5 28 09.2	+16 26 20	−0.003	−0.05	7.1	0.6	4.4	G0 V		35 s		
36142	195884		5 28 13.1	−32 49 33	−0.001	−0.14	7.9	0.3	0.6	A0 V		28 mx		
36079	170457	9 β Lep	5 28 14.7	−20 45 35	−0.001	−0.09	2.84	0.82	−2.1	G2 II	−14	97 s	4066	Nihal, m
36187	195887		5 28 15.2	−37 13 51	0.000	+0.07	5.57	0.02		A1	+50			
36141	195885		5 28 17.9	−31 50 27	+0.002	−0.04	8.0	0.7	3.2	G5 IV		93 s		
36108	170461		5 28 20.9	−22 26 04	−0.011	+0.13	6.8	0.7	4.4	G0 V		26 s		
35865	94579		5 28 24.9	+16 36 17	+0.001	−0.01	7.9	0.0	0.4	B9.5 V		300 s		
35971	132144		5 28 25.6	+0 01 12	−0.001	−0.02	6.67	−0.06		B9				
35832	77191		5 28 25.7	+22 44 33	0.000	−0.04	7.7	1.1	0.2	K0 III		290 s		
36003	132145		5 28 25.9	−3 29 59	−0.022	−0.80	7.64	1.12	8.0	K5 V	−58	11 ts		m
35742	58076		5 28 26.3	+34 41 48	0.000	−0.03	6.7	1.1	0.2	K0 III		190 s		
34885	5539		5 28 30.0	+74 18 27	+0.005	−0.02	6.9	0.0	0.4	B9.5 V		150 mx		
36017	132146		5 28 33.1	−4 41 49	−0.002	0.00	7.53	0.26		A3				
35909	94580		5 28 34.8	+13 40 44	+0.001	−0.02	6.30	0.15	1.9	A4 V	+27	73 s		
36035	132149		5 28 36.8	−5 21 43	+0.002	−0.07	8.0	0.5	3.4	F5 V		85 s		
36002	112808		5 28 41.9	+1 13 38	−0.002	0.00	7.6	0.1	1.4	A2 V		180 s		
36059	132154		5 28 42.4	−8 22 37	−0.001	0.00	7.00	0.9	3.2	G5 IV		58 s	4071	m
36705	249286		5 28 44.9	−65 26 55	+0.008	+0.13	6.80	0.6	4.4	G0 V		30 s		m
36013	112813		5 28 45.2	+1 38 38	0.000	0.00	6.88	−0.13	−3.0	B1.5 V	+11	860 s		m
35169	13503		5 28 45.6	+69 20 14	+0.003	+0.01	7.8	1.1	−0.1	K2 III		350 s		
36153	170469		5 28 47.5	−24 15 40	0.000	−0.03	7.9	0.5	1.4	A2 V		99 s		
34531	5530		5 28 47.9	+78 18 02	+0.006	−0.08	6.8	0.4	2.6	F0 V	+16	70 s		
36012	112814		5 28 48.3	+2 09 54	−0.001	+0.01	7.24	−0.10		B3	+21			
35945	94582		5 28 50.0	+16 25 36	0.000	−0.01	7.65	0.02	0.4	B9.5 V		260 s		
35956	94583		5 28 51.7	+12 33 03	+0.007	−0.21	6.9	0.5	4.0	F8 V	+9	30 ts		
36417	233991		5 28 51.7	−50 42 56	−0.001	−0.02	7.4	1.8	−0.1	K2 III		130 s		
36058	132157		5 28 56.6	−3 18 27	−0.003	−0.01	6.39	−0.01		B9			4078	m
35944	77196		5 28 58.1	+20 26 31	+0.002	−0.03	7.3	1.1	0.2	K0 III	+40	250 s		
36000	94585		5 28 58.4	+10 08 09	−0.001	−0.01	7.8	0.1	0.6	A0 V		280 s		
35968	94584		5 28 59.6	+13 08 55	−0.002	−0.04	7.9	0.9	3.2	G5 IV		87 s		
36090	132163		5 29 01.1	−4 41 31	+0.003	+0.01	7.4	1.6		M5 III	+22			S Ori, m,v
35761	40360		5 29 01.2	+42 16 15	+0.005	−0.02	6.8	0.8	3.2	G5 IV	−11	51 s		
36255	195898		5 29 06.7	−30 07 00	−0.001	+0.01	6.75	1.06	0.2	gK0		190 s		
36106	132164		5 29 06.7	−5 26 42	−0.001	0.00	8.0	1.1	−0.1	K2 III		410 s		
36120	132167		5 29 08.9	−5 47 27	0.000	+0.01	7.96	−0.03		B9				
36660	249287		5 29 13.7	−61 55 18	+0.001	−0.01	8.0	1.1	0.2	K0 III		360 s		
35943	77200	118 Tau	5 29 16.2	+25 08 57	0.000	−0.03	5.80	−0.04			+16		4068	m
35898	58091		5 29 16.6	+32 11 58	0.000	−0.05	7.2	0.5	4.0	F8 V		43 s		

HD	SAO	Star Name	α 2000	δ 2000	μ(α)	μ(δ)	V	B−V	M$_V$	Spec	RV	d(pc)	ADS	Notes
			h m s	° ′ ″	s	″								
36689	249288		5 29 17.4	−62 18 52	+0.001	−0.02	6.59	1.53		K2				
35706	40361		5 29 18.6	+49 10 32	−0.001	−0.01	7.8	1.1	0.2	K0 III		300 s		
35985	94586		5 29 19.1	+18 21 55	−0.001	−0.02	6.74	0.12	1.4	A2 V		110 s	4073	m
36134	132170		5 29 23.5	−3 26 47	−0.003	−0.01	5.79	1.15	0.0	K1 III	+23	140 s		
36296	195902		5 29 24.4	−30 57 22	0.000	0.00	8.0	1.3	−0.5	M2 III		510 s		
36151	132172		5 29 25.2	−7 15 42	−0.001	−0.04	6.71	−0.13	−1.1	B5 V	+19	370 s		m
34653	5535		5 29 25.6	+77 58 39	+0.003	−0.01	6.5	0.2	2.1	A5 V	−16	78 s		
35760	40362		5 29 26.3	+48 00 03	+0.004	−0.05	7.8	0.8	3.2	G5 IV		82 s		
35667	25216		5 29 27.0	+54 09 11	−0.004	−0.01	8.0	1.1	0.2	K0 III		330 s		
36045	94588		5 29 27.2	+13 25 39	−0.001	−0.02	8.0	0.5	3.4	F5 V		80 s		
36117	132169		5 29 27.3	−0 02 32	+0.001	−0.01	7.97	0.10		A0				
36237	170480		5 29 31.0	−21 02 22	+0.002	+0.03	7.6	0.0	1.4	A2 V		180 s		
36133	112824		5 29 33.5	+3 08 51	0.000	−0.01	6.94	−0.09	−2.5	B2 V	+23	690 s	4088	m
36447	217361		5 29 34.6	−44 46 34	−0.001	+0.01	7.3	1.9	−0.5	M2 III		370 s		
36139	132174		5 29 37.6	−0 01 17	+0.001	−0.02	6.39	−0.01		A0				
36274	170482		5 29 39.7	−22 43 09	0.000	−0.02	7.7	0.9	3.2	G5 IV		70 s		
35984	77205		5 29 40.6	+29 11 10	+0.002	−0.05	6.24	0.45	3.4	F5 V	+13	36 s		
36150	132175		5 29 41.5	−0 48 09	+0.003	−0.05	7.90	0.1	1.4	A2 V		200 s	4096	m
35921	58105		5 29 42.6	+35 22 30	0.000	0.00	6.81	0.20	−5.5	O9.5 III	−29	1600 s	4072	m
35940	58108		5 29 43.7	+35 06 32	+0.001	−0.01	7.0	1.1	0.2	K0 III	+18	220 s		
36167	132176	31 Ori	5 29 43.9	−1 05 32	0.000	−0.02	4.71	1.57	−0.3	K5 III	+14	93 s		CI Ori, m,q
35953	58111		5 29 44.9	+33 20 41	0.000	0.00	7.8	0.4	2.6	F0 V		110 s		
36104	94592		5 29 47.5	+12 16 11	0.000	−0.03	7.0	0.0	0.0	B8.5 V	+14	240 s		
36073	94589		5 29 49.8	+18 24 59	+0.001	−0.04	7.60	0.1	0.6	A0 V		240 s	4087	m
36520	217365		5 29 50.3	−47 04 36	+0.004	−0.02	5.50	0.4	2.6	F0 V		38 s		m
36072	94590		5 29 51.2	+18 25 50	+0.001	−0.03	7.60	0.4		F0			4087	m
35607	13517		5 29 51.5	+60 16 05	+0.002	−0.02	6.9	0.1	0.6	A0 V	+7	170 s		
35741	25217		5 29 52.9	+53 25 58	−0.005	+0.03	7.1	0.4	2.6	F0 V		77 s		
36166	112830		5 29 54.7	+1 47 21	0.000	0.00	5.78	−0.20	−3.0	B1.5 V	+12	560 s		
36103	94593		5 29 57.9	+16 09 02	+0.001	−0.01	7.3	1.6	−0.5	M2 III		340 s		
36054	77211		5 30 03.2	+23 45 31	0.000	0.00	8.0	0.9	3.2	G5 IV		90 s		
36219	132184		5 30 04.2	−1 44 57	−0.001	+0.01	7.63	−0.07		B9				
36044	77210		5 30 06.0	+29 32 56	+0.001	−0.01	7.11	0.85	0.3	gG6	+47	230 s	4086	m
36519	217366		5 30 07.1	−43 34 45	−0.002	+0.04	7.73	1.46	−0.2	K3 III		300 s		
36553	217368		5 30 09.4	−47 04 40	+0.002	−0.13	5.46	0.62	3.0	G3 IV	+16	31 s		m
35583	13518	17 Cam	5 30 10.2	+63 04 02	0.000	0.00	5.42	1.71	−0.3	K5 III	−19	100 s		
39780	258418		5 30 14.2	−84 47 06	−0.007	+0.05	6.26	−0.02	1.2	A1 V		100 s		
36113	77215		5 30 15.0	+20 33 06	−0.001	−0.01	7.08	−0.05	−1.0	B5.5 V	+19	360 s		
36876	249297		5 30 15.9	−63 55 40	+0.001	+0.05	6.19	0.22		F0				m
36271	132190		5 30 16.0	−5 59 16	−0.002	0.00	8.0	0.9	3.2	G5 IV		93 s		
35518	13516		5 30 19.0	+66 14 06	+0.002	−0.05	8.0	0.9	3.2	G5 IV		90 s		
36552	217369		5 30 19.5	−43 41 37	−0.001	+0.03	8.0	0.7	3.2	G5 IV		35 s		
36217	112837		5 30 19.8	+4 12 15	0.000	−0.05	6.21	1.27	−0.1	K2 III		150 s		CK Ori, q
36285	132192		5 30 20.6	−7 26 06	−0.001	0.00	6.33	−0.19	−3.0	B1.5 V	+11	710 s		
36043	58125		5 30 21.1	+31 48 10	0.000	−0.01	7.8	0.1	0.6	A0 V		260 s		
35964	40381		5 30 21.3	+40 30 31	0.000	−0.01	7.3	1.1	0.2	K0 III		250 s		
36042	58124		5 30 23.4	+34 12 01	0.000	−0.02	7.80	0.8	0.3	G7 III	+7	300 s		
36162	94596		5 30 26.0	+15 21 37	−0.002	−0.05	5.80	0.14	1.7	A3 V	−12	61 s		
36071	58128		5 30 28.6	+31 16 56	+0.001	0.00	7.89	0.09	0.6	A0 V		270 s		
36635	234002		5 30 30.2	−50 06 01	+0.001	0.00	7.9	1.4	−0.3	K5 III		440 s		
36269	112844		5 30 34.6	+0 21 56	0.000	0.00	7.8	0.2	2.1	A5 V		140 s		
36085	58130		5 30 35.9	+31 30 40	+0.001	−0.01	7.54	1.77	−0.3	K5 III		260 s		m
36027	58126		5 30 38.6	+39 49 42	+0.003	−0.02	7.7	1.1	0.2	K0 III		300 s		
36160	77220		5 30 43.3	+22 27 44	+0.003	−0.02	6.29	1.18		K0	+2			
36789	234009		5 30 44.8	−57 09 36	−0.004	0.00	7.0	0.6	2.1	A5 V		49 s		
36041	58129		5 30 45.0	+39 49 33	+0.002	−0.04	6.50	0.1	0.2	G9 III	+12	180 s		m
35864	25224		5 30 45.0	+53 09 43	−0.001	−0.02	7.8	0.4	2.6	F0 V		100 s		
36267	112849	32 Ori	5 30 47.0	+5 56 53	+0.001	−0.03	4.20	−0.14	−1.6	B5 IV	+21	150 s	4115	m
36248	112846		5 30 47.4	+8 23 56	−0.001	+0.05	7.70	0.5	4.0	F8 V		55 s	4114	m
36040	40387		5 30 48.4	+41 27 43	−0.002	−0.04	6.00	1.11	0.2	K0 III p	+14	140 s		
36432	170497		5 30 48.7	−21 24 28	+0.001	0.00	8.0	1.9	−0.1	K2 III		160 s		
36263	94602		5 30 51.7	+10 15 16	+0.001	0.00	7.47	−0.08	0.4	B9.5 V		260 s	4113	m
36342	132200		5 30 52.8	−4 15 17	−0.001	−0.01	7.52	0.14		A2				m
36264	94603		5 30 53.7	+10 09 46	−0.001	−0.01	7.09	−0.01	0.6	A0 V		200 s		
36310	112857		5 30 57.6	+4 40 07	−0.001	0.00	7.95	−0.02	−0.9	B6 V		530 s	4117	m
35362	5553		5 30 57.9	+71 55 27	0.000	−0.01	7.7	1.1	0.2	K0 III		290 s		
36379	150522		5 31 00.0	−10 04 53	+0.011	−0.32	6.91	0.56		G0				
36516	170508		5 31 00.8	−28 31 44	−0.006	+0.13	7.9	0.5	3.2	G5 IV		88 s		
36262	94605		5 31 01.4	+12 05 49	−0.001	0.00	7.63	−0.11	−1.6	B3.5 V	+20	650 s		
36340	112859		5 31 04.8	+3 21 12	−0.001	0.00	7.99	−0.14	−1.6	B3.5 V	+17	810 s		
36473	170506	10 Lep	5 31 07.5	−20 51 49	0.000	−0.04	5.55	0.00		A0	−11	21 mn		
35962	25228		5 31 08.7	+51 16 58	−0.002	−0.03	7.7	0.8	3.2	G5 IV		77 s		
36648	217374		5 31 10.2	−42 18 00	0.000	+0.06	7.0	0.5	2.1	A5 V		60 s		m
36597	195924	ε Col	5 31 12.7	−35 28 15	+0.002	−0.04	3.87	1.14	0.2	gK0	−5	42 s		
36351	112861	33 Ori	5 31 14.4	+3 17 32	0.000	0.00	5.46	−0.18	−3.0	B1.5 V	+20	470 s	4123	m

118

HD	SAO	Star Name	α 2000	δ 2000	μ(α)	μ(δ)	V	B-V	M_v	Spec	RV	d(pc)	ADS	Notes
			h m s	° ′ ″	s	″								
36430	132210		5 31 20.9	−6 42 30	0.000	0.00	6.22	−0.17	−2.5	B2 V	+23	550 s		
36215	77224		5 31 22.6	+27 46 09	+0.002	−0.04	7.6	0.5	4.0	F8 V		51 s		
36149	58141		5 31 24.7	+36 46 45	0.000	0.00	7.80	1.1	−4.4	K3 Ib		1900 s		
36148	58139		5 31 26.7	+38 19 09	+0.001	−0.06	7.5	1.4	−0.3	K5 III		230 mx		
36395	132211		5 31 27.3	−3 40 39	+0.051	−2.10	7.97	1.47	9.1	M1 V	+11	5.9 t		m
35961	25232		5 31 29.0	+54 39 14	−0.014	−0.39	7.53	0.64	4.5	G1 V	+3	37 s		m
36392	112867		5 31 29.8	+1 41 23	0.000	−0.02	7.56	−0.14	−1.7	B3 V	+17	700 s		
36337	94618		5 31 35.8	+14 55 44	+0.001	0.00	6.6	0.0	−1.0	B5.5 V	+17	310 s		
36734	217382		5 31 36.0	−45 55 31	+0.002	+0.04	5.86	1.35		K2				
37066	249302		5 31 36.3	−61 32 28	+0.002	+0.01	7.3	−0.5	0.4	B9.5 V		240 s		
36320	94617		5 31 39.1	+18 14 35	0.000	−0.08	6.9	1.4	−0.3	K5 III		160 mx		
36376	112868		5 31 40.9	+9 13 33	−0.001	−0.01	7.79	0.09	0.0	B8.5 V		340 s		
36429	112873		5 31 41.1	+2 49 58	−0.001	+0.01	7.56	−0.13	−1.1	B5 V		540 s		m
36487	132218		5 31 41.4	−7 02 52	0.000	+0.04	7.81	−0.11	−1.1	B5 V		600 s		
36634	195930		5 31 42.5	−33 23 05	−0.001	+0.06	8.0	0.7	3.2	G5 IV		93 s		
36457	132217		5 31 43.6	−3 12 58	0.000	+0.01	6.8	0.9	3.2	G5 IV		52 s		
36804	217384		5 31 43.8	−49 01 06	0.000	+0.01	7.5	−0.1	0.0	B8.5 V		320 s		
36212	58145		5 31 45.3	+34 52 55	−0.002	0.00	7.77	0.24	−3.9	B3 II		1300 s		
36410	112874		5 31 45.7	+5 18 30	+0.001	−0.02	6.64	0.9	3.2	G5 IV		80 s		
35815	13525		5 31 46.0	+62 59 07	0.000	−0.01	7.5	0.1	0.6	A0 V	+13	230 s		
36083	40393		5 31 47.5	+48 56 07	0.000	+0.02	7.3	0.0	0.0	B8.5 V		270 s		
32196	843		5 31 48.6	+85 56 18	+0.023	−0.08	6.5	0.2	2.1	A5 V	−6	59 mx		
36067	25237		5 31 52.9	+51 01 41	+0.001	−0.01	7.60	1.1	−2.2	K1 II	−15	750 s		
37763	256201	γ Men	5 31 53.1	−76 20 28	+0.032	+0.29	5.19	1.13	−0.3	K4 III	+57	130 s		m
36512	132222	36 υ Ori	5 31 55.8	−7 18 05	−0.001	−0.01	4.62	−0.26	−4.1	B0 V	+17	560 s		
36561	150535		5 31 56.1	−16 40 35	+0.004	−0.23	7.2	0.8	3.2	G5 IV		44 mx		
36307	77229		5 31 56.2	+26 41 07	−0.001	0.00	7.7	1.1	−0.1	K2 III		340 s		
36082	25238		5 31 56.7	+51 14 15	−0.001	−0.01	7.9	0.8	3.2	G5 IV		85 s		
36173	40399		5 31 59.5	+42 29 51	+0.002	−0.01	8.0	0.0	0.4	B9.5 V		310 s		
36486	132220	34 δ Ori	5 32 00.3	−0 17 57	0.000	0.00	2.23	−0.22		09.5 II	+16		4134	Mintaka, m,v
36485	132221	34 δ Ori	5 32 00.4	−0 17 05	0.000	0.00	6.85	−0.16	−2.5	B2 V	+21	720 s	4134	m
36375	94625		5 32 02.9	+17 31 50	−0.001	−0.03	8.0	0.1	0.6	A0 V		280 s		
36541	132225		5 32 06.9	−6 42 30	−0.001	0.00	7.69	−0.08	−0.9	B6 V		500 s		
36157	40398		5 32 09.9	+47 11 33	−0.001	−0.03	7.79	0.04	0.4	B9.5 V		270 s		
36260	58151		5 32 10.5	+38 04 49	+0.002	−0.04	8.0	0.9	3.2	G5 IV		91 s		
36389	94628	119 Tau	5 32 12.7	+18 35 39	0.000	0.00	4.38	2.07	−4.8	M2 Ib	+23	290 s		CE Tau, v
36335	77233		5 32 14.0	+29 11 43	−0.003	−0.04	7.9	0.5	3.4	F5 V	−31	77 s		
36408	94631		5 32 14.5	+17 03 23	−0.001	+0.01	6.00	−0.04	−1.0	B7 IV	+15	220 s	4131	m
36407	94633		5 32 18.0	+17 44 37	+0.001	−0.04	7.99	0.46		F0				
36889	217387		5 32 20.6	−47 28 31	0.000	+0.04	7.6	0.5	3.2	G5 IV		77 s		
36146	40401		5 32 21.3	+49 23 40	+0.001	−0.01	7.50	0.5	3.4	F5 V		65 s	4119	m
36406	94634		5 32 22.0	+19 07 20	−0.001	−0.07	7.8	0.5	2.3	F7 IV		120 s		
36763	195940		5 32 22.5	−36 05 22	+0.002	−0.02	7.9	1.5	−0.3	K5 III		430 s		
36374	77237		5 32 27.2	+26 58 54	−0.003	−0.02	7.1	0.0	−1.0	B5.5 V	+20	390 s		
36066	25241	18 Cam	5 32 33.8	+57 13 16	+0.015	−0.22	6.5	0.7	4.4	G0 V	+37	26 s		
36245	40408		5 32 35.3	+44 21 37	−0.002	0.00	7.95	0.07	0.0	B8.5 V		340 s		
36558	132233		5 32 37.7	+0 00 42	−0.001	−0.02	6.5	1.4	−0.3	K5 III		230 s		
36243	40409		5 32 38.6	+45 32 13	−0.002	−0.03	7.9	0.5	4.0	F8 V		59 s		
36591	132234		5 32 41.3	−1 35 32	0.000	−0.01	5.34	−0.19	−3.5	B1 V	+34	590 s	4141	m
36656	150546		5 32 41.9	−16 19 12	+0.001	+0.01	6.8	1.1	−0.1	K2 III		240 s		
37167	249306		5 32 42.6	−60 07 29	−0.002	−0.01	7.2	1.0	0.2	K0 III		250 s		
36371	58164	25 χ Aur	5 32 43.6	+32 11 31	0.000	0.00	4.76	0.34	−6.3	B5 Iab	0	930 s		
36673	150547	11 α Lep	5 32 43.7	−17 49 20	0.000	0.00	2.58	0.21	−4.7	F0 Ib	+25	290 s	4146	Arneb, m
36244	40411		5 32 44.2	+45 29 28	+0.002	−0.02	7.9	0.9	3.2	G5 IV		87 s		
36372	58166		5 32 46.0	+32 01 59	+0.001	−0.01	7.6	1.1	−0.1	K2 III		320 s		
36605	132238		5 32 49.5	−0 42 47	−0.001	0.00	7.96	0.08		B9				
36848	195948		5 32 51.3	−38 30 47	+0.003	0.00	5.48	1.22		K5	−1	34 mn		
37279	249308		5 32 53.0	−63 27 23	+0.003	0.00	7.3	0.1	0.6	A0 V		220 s		
36360	58165		5 32 53.3	+36 19 11	+0.003	−0.05	7.1	0.2		A m				
36557	112889		5 32 55.2	+7 19 30	0.000	0.00	7.7	0.9	3.2	G5 IV		80 s		
37004	217394		5 32 56.3	−47 41 20	−0.001	−0.05	7.6	−0.4	0.6	A0 V		180 s		
36629	132244		5 32 57.1	−4 33 58	0.000	+0.02	7.64	0.02	−2.5	B2 V	+21	840 s		
36469	77244		5 32 57.4	+25 53 38	−0.001	−0.02	8.0	1.1	0.2	K0 III		340 s		
36425	58170		5 32 59.5	+31 52 18	−0.002	−0.03	7.36	0.07		A2	+6			
37297	249309		5 32 59.6	−64 13 40	+0.006	−0.01	5.34	1.04	0.3	gG8	+10	87 s		
36715	150550		5 33 00.7	−17 15 31	−0.001	−0.01	8.0	1.4	−0.3	K5 III		450 s		
36762	170542		5 33 01.4	−25 21 59	−0.003	+0.04	7.3	1.7	−0.5	M2 III		310 s		
35863	13527		5 33 03.1	+68 00 48	+0.003	−0.18	6.9	0.4	2.6	F0 V	+32	73 s		
36628	132245		5 33 03.5	−1 14 27	+0.000	+0.01	7.98	−0.04		B8				
36672	150549		5 33 04.0	−10 28 32	+0.004	+0.01	8.0	1.1	−0.1	K2 III		230 mx		
36501	77246		5 33 05.3	+23 08 32	0.000	−0.02	7.8	1.1	0.2	K0 III		300 s		
36646	132247		5 33 07.1	−1 43 07	−0.002	−0.04	6.47	−0.10	−1.7	B3 V	+37	400 s	4150	m
36874	195952		5 33 07.3	−35 08 23	+0.007	−0.04	5.7	1.2	0.2	K0 III	+15	91 s		
36304	40417		5 33 08.3	+46 52 51	+0.004	−0.07	7.6	0.1	0.6	A0 V		56 mx		
36627	112894		5 33 08.8	+3 07 52	−0.001	0.00	7.56	−0.11	−0.9	B6 V		490 s		

HD	SAO	Star Name	α 2000	δ 2000	μ(α)	μ(δ)	V	B-V	M_v	Spec	RV	d(pc)	ADS	Notes
36242	25247		5 33 13.6	+51 28 14	-0.001	0.00	7.93	0.05	0.4	B9.5 V		300 s		
36602	—		5 33 13.7	+7 09 12			7.98	2.85		N7.7	+5			RT Ori, v
36453	58177		5 33 15.5	+32 17 23	+0.001	-0.01	6.61	-0.04		B9				
36888	195955		5 33 16.1	-34 18 14	-0.001	+0.01	6.8	0.5	2.1	A5 V		59 s		
36387	58174		5 33 16.9	+39 30 21	-0.004	-0.14	7.8	0.7	4.4	G0 V		48 s		
37119	234021		5 33 17.3	-51 38 24	0.000	+0.06	8.0	-0.3	2.6	F0 V		120 s		
36601	112896		5 33 19.6	+8 32 19	+0.001	+0.01	7.9	1.1	0.2	K0 III		340 s		
37106	217400		5 33 26.9	-49 22 40	+0.003	-0.11	7.2	0.6	3.4	F5 V		47 s		
36484	58179		5 33 27.4	+32 48 04	+0.001	-0.06	6.48	0.09		A m	+34			
36404	40426		5 33 28.6	+42 06 32	0.000		6.55	-0.02		B8	+1			
36278	25250		5 33 30.2	+52 39 11	-0.001	-0.02	8.0	0.1	0.6	A0 V		280 s		
36846	170549		5 33 30.3	-24 20 16	+0.001	-0.17	7.70	0.5	4.0	F8 V		55 s	4157	m
36546	77252		5 33 30.6	+24 37 45	0.000	-0.03	6.9	0.0	0.0	B8.5 V		230 s		
35303	5557		5 33 30.8	+77 12 17	-0.002	-0.01	8.0	1.1	-0.1	K2 III		410 s		
36695	132255		5 33 31.3	-1 09 22	-0.001	0.00	5.32	-0.18	-3.5	B1 V	+22	570 s		VV Ori, m,v
36576	94649	120 Tau	5 33 31.6	+18 32 25	0.000	0.00	5.69	0.01		B p	+44			
37227	234025		5 33 34.4	-57 05 01	+0.002	+0.03	6.8	0.8	2.6	F0 V		36 s		
37350	249311	β Dor	5 33 37.5	-62 29 24	0.000	+0.01	3.8	0.5	-8.0	F9 Ia var	+7	2300 s		v
36710	132259		5 33 37.6	-3 27 48	-0.001	-0.01	6.9	1.1	-0.1	K2 III		260 s		
36499	58182		5 33 38.0	+34 43 32	-0.003	-0.01	6.27	0.15	1.9	A4 V	-14	71 s		
36589	77255		5 33 38.8	+20 28 26	0.000	-0.01	6.10	-0.07	-0.9	B6 V		230 s		
36358	40425		5 33 40.2	+48 16 40	+0.004	+0.02	7.7	0.9	3.2	G5 IV		79 s		
36506	58184		5 33 43.1	+32 44 57	+0.003	0.00	6.49	1.58		M0				
37226	234026		5 33 44.3	-54 54 08	+0.006	+0.01	6.43	0.55	3.4	F5 V		35 s		m
36943	195964		5 33 47.6	-31 27 05	+0.003	-0.02	8.0	1.8	0.2	K0 III		120 s		
36942	170560		5 33 50.4	-29 56 25	+0.001	+0.03	7.9	0.4	1.4	A2 V		120 s		
36422	40429		5 33 50.8	+44 47 07	0.000	-0.02	8.00	0.5	3.4	F5 V		82 s	4137	m
36965	170561		5 33 52.0	-29 50 56	+0.001	+0.02	6.4	0.3	0.6	A0 V		91 s		
36653	94652	35 Ori	5 33 54.2	+14 18 21	-0.001	0.00	5.64	-0.14	-1.7	B3 V	+19	290 s		
36575	77258		5 33 54.3	+27 09 51	0.000	0.00	8.0	1.1	0.2	K0 III		340 s		
36941	170562		5 33 56.1	-28 08 59	0.000	-0.03	7.9	1.8	-0.3	K5 III		280 s		
36741	112901		5 33 57.6	+1 24 28	0.000	0.00	6.59	-0.17	-2.5	B2 V	+14	660 s		
36667	94653		5 33 59.9	+15 35 03	-0.004	-0.05	7.3	0.5	4.0	F8 V		46 s		
35029	5554		5 34 02.1	+79 51 09	-0.006	-0.02	7.3	0.1	0.6	A0 V		220 s		
36760	132267		5 34 02.2	-0 28 36	0.000	-0.01	7.63	-0.10		B8				
36814	132271		5 34 02.3	-7 01 27	-0.002	-0.05	6.5	1.1	0.2	K0 III		190 s		
36779	132269		5 34 03.8	-1 02 08	0.000	0.00	6.23	-0.18	-2.5	B2 V	+4	560 s	4159	m
36780	132270		5 34 04.0	-1 28 14	-0.001	-0.03	5.93	1.55	-0.3	K5 III		170 s		
35919	13531		5 34 04.0	+69 39 39	-0.001	+0.08	8.0	0.5	3.4	F5 V		82 s	4118	m
35249	5558		5 34 09.3	+78 19 21	-0.004	-0.01	7.8	0.1	0.6	A0 V		270 s		
36468	40434		5 34 09.6	+43 56 14	+0.003	-0.03	7.23	0.02	0.4	B9.5 V	+40	210 s		
36626	77262		5 34 12.4	+26 52 55	-0.002	-0.01	7.8	0.1	1.7	A3 V		160 s		
36922	150577		5 34 13.6	-18 20 19	-0.001	0.00	8.0	0.1	0.6	A0 V		300 s		
36827	132274		5 34 14.7	-2 52 53	-0.001	+0.02	6.69	-0.17		B5	+5			
36923	150578		5 34 15.6	-18 48 45	-0.002	-0.02	7.1	1.4	-0.3	K5 III		300 s		
36811	—		5 34 15.7	-1 52 52	+0.001	0.00	7.08	0.18		B9				
36332	25255		5 34 16.3	+56 16 17	+0.001	+0.01	8.0	0.0	0.4	B9.5 V		310 s		
36777	112904	38 Ori	5 34 16.7	+3 46 00	-0.002	-0.02	5.36	0.05		A2	-9	19 mn		m
36535	40439		5 34 16.9	+41 07 22	-0.001	-0.03	7.0	1.4	-0.3	K5 III	0	270 s		
36643	77266		5 34 21.3	+27 19 50	-0.004	-0.02	7.9	0.4	2.6	F0 V		110 s		
36843	132278		5 34 24.1	-4 48 17	+0.001	0.00	6.80	0.18		A3				
37278	234031		5 34 24.5	-52 37 43	+0.001	+0.04	7.3	0.8	0.2	K0 III		260 s		
37105	195976		5 34 27.3	-35 51 17	-0.001	-0.02	7.6	0.8	2.1	A5 V		55 s		
36886	150580		5 34 28.7	-10 30 03	+0.001	+0.01	7.5	1.4	-0.3	K5 III		360 s		
36840	132277		5 34 29.2	-0 00 44	-0.001	0.00	6.5	0.9	3.2	G5 IV		47 s		
36865	132282		5 34 32.4	-4 29 18	0.000	-0.03	7.40	-0.07		B9			4172	m
—	—		5 34 34.8	+48 29 27	+0.001	0.00	7.9	0.6	4.4	G0 V		49 s		
36823	112910		5 34 38.4	+6 07 36	0.000	-0.04	7.9	0.1	1.4	A2 V		200 s		
37260	217408		5 34 40.3	-48 56 52	-0.001	-0.03	7.4	0.4	1.4	A2 V		94 s		
36824	112912		5 34 43.2	+5 39 39	-0.001		6.69	-0.15	-1.7	B3 V	+14	480 s		
36883	132285		5 34 43.3	-4 23 32	+0.002	0.00	7.22	-0.08		B8			4176	m
37993	256208		5 34 44.7	-73 44 28	-0.003	+0.04	5.79	1.71		Mb				
36917	132288		5 34 46.8	-5 34 17	-0.001	-0.04	7.96	0.16		B9				V372 Ori, v
36920	132290		5 34 47.9	-7 12 00	0.000	0.00	6.9	1.1	0.2	K0 III		220 s		
36822	112914	37 φ¹ Ori	5 34 49.2	+9 29 22	0.000	0.00	4.41	-0.16	-4.6	B0 IV	+33	570 s		
36916	132292		5 34 53.7	-4 06 37	-0.003	+0.01	6.73	-0.05		B9				
37132	195979		5 34 55.3	-31 41 40	-0.002	+0.01	7.3	-0.5	0.4	B9.5 V		240 s		
36522	40445		5 34 55.5	+49 03 49	-0.002	-0.06	7.8	0.5	3.4	F5 V		74 s		
36194	13540		5 34 55.6	+65 02 55	+0.002	-0.04	7.5	0.1	1.4	A2 V		160 s		
36898	132291		5 34 56.3	-0 07 22	-0.001		7.06	-0.07		B5	+14			m
37080	170576		5 34 57.6	-21 04 04	-0.001	-0.01	7.4	1.5	0.2	K0 III		140 s		
37501	249314		5 34 57.6	-61 10 33	+0.001	-0.02	6.32	0.85	3.2	K0 IV		42 s		
36936	132296		5 34 58.8	-4 21 16	-0.002	-0.01	7.52	-0.11		B8				
36915	132293		5 35 00.1	-0 48 57	-0.002	0.00	8.02	-0.01		B9				
36959	132298		5 35 00.8	-6 00 33	-0.001	0.00	5.67	-0.22	-3.5	B1 V	+30	680 s		

HD	SAO	Star Name	α 2000	δ 2000	μ(α)	μ(δ)	V	B-V	M$_v$	Spec	RV	d(pc)	ADS	Notes
			h m s	o ′ ″	s	″								
37462	234037		5 35 02.4	-58 52 16	-0.005	+0.02	6.4	1.4	-0.1	K2 III		150 s		
36960	132301		5 35 02.6	-6 00 07	0.000	0.00	4.78	-0.25	-4.1	B0 V	+28	600 s	4182	m
36758	77276		5 35 03.5	+24 17 26	0.000	-0.02	7.8	1.1	0.2	K0 III		310 s		
36820	94665		5 35 04.6	+15 38 20	0.000	-0.01	7.3	0.0	0.4	B9.5 V		230 s		
36958	132302		5 35 04.7	-4 43 54	0.000	0.00	7.36	-0.07	-1.7	B3 V	+23	580 s		KX Ori, v
36981	132303		5 35 05.9	-5 12 15	-0.003	+0.01	7.82	-0.11		B8				
36861	112921	39 λ Ori	5 35 08.2	+9 56 02	0.000	-0.01	3.39	-0.18		O8	+34		4179	Meissa, m
36935	132300		5 35 09.1	-0 16 10	0.000	+0.01	7.51	-0.12		B8				
37000	132311		5 35 10.9	-5 55 37	-0.001	-0.01	7.49	-0.13	-1.1	B5 V	+19	520 s		
36954	132305		5 35 12.6	-0 44 07	-0.001	0.00	6.94	-0.10	-1.7	B3 V	+1	500 s		
36895	112922		5 35 12.8	+9 36 48	+0.001	0.00	6.74	-0.20	-2.5	B2 V	+8	710 s		
36881	94671		5 35 13.3	+10 14 24	+0.001	-0.01	5.60	0.12	-1.2	B8 III		160 s	4181	m
245203	112923		5 35 13.7	+9 41 50	0.000	0.00	7.9	-0.2	0.0	B8.5 V		380 s		
36896	112925		5 35 14.1	+9 11 01	-0.001	-0.01	7.9	1.4	-0.3	K5 III		430 s		
36722	58208		5 35 14.4	+32 44 57	+0.002	+0.01	7.97	0.11		A0				
37192	195986		5 35 15.3	-33 04 48	+0.001	+0.11	5.78	1.12	-2.1	K0 II		45 mx		
37899	256207		5 35 15.5	-71 08 18	+0.003	-0.01	7.80	1.1	-0.1	K2 III		380 s		m
37025	132315		5 35 15.6	-6 01 59	-0.002	-0.02	7.17	-0.12		B5	+22			
37020	---	41 θ¹ Ori	5 35 15.7	-5 23 14			6.73	0.03		Pec	+33		4186	A component, m,v
37021	---	41 θ¹ Ori	5 35 16.1	-5 23 07			7.95			O	+24		4186	BM Ori, m,v
37022	132314	41 θ¹ Ori	5 35 16.4	-5 23 23	0.000	0.00	5.14	0.03		O6 p	+28		4186	C component, m
36914	112927		5 35 16.6	+8 42 03	-0.001	-0.01	7.5	1.6	-0.5	M4 III		400 s		
36894	112926		5 35 16.7	+9 46 39	+0.001	-0.01	7.7	-0.2		B9				
36793	77281		5 35 17.1	+25 49 44	-0.001	-0.01	7.8	0.1	0.6	A0 V		260 s		
37023	---	41 θ¹ Ori	5 35 17.2	-5 23 16			6.70	0.10		B0.5 V p	+31		4186	D component, m
36641	40453		5 35 19.9	+43 14 20	+0.003	-0.03	7.4	0.8	3.2	G5 IV		67 s		
37017	132317		5 35 21.7	-4 29 37	-0.001	+0.03	6.55	-0.14	-3.0	B1.5 V	+29	720 s		
37016	132319		5 35 22.4	-4 25 31	+0.001	-0.03	6.23	-0.16	-1.7	B3 V	+31	390 s		m
37041	132321	43 θ² Ori	5 35 22.8	-5 24 58	0.000	0.00	5.08	-0.08	-4.4	O9.5 V p	+36	800 s	4188	m
37018	132320	42 Ori	5 35 23.1	-4 50 18	0.000	0.00	4.59	-0.19	-3.6	B2 III	+30	440 s	4187	m
36588	40450		5 35 24.3	+49 35 52	+0.004	-0.05	7.5	1.1	0.2	K0 III		210 mx		
37043	132323	44 ι Ori	5 35 25.9	-5 54 36	0.000	0.00	2.76	-0.24	-6.0	O9 III	+22	570 s	4193	m
37042	132322		5 35 26.2	-5 24 58	-0.001	+0.03	6.39	-0.08	-3.5	B1 V	+29	840 s		
37104	150588		5 35 26.3	-15 44 15	-0.002	+0.02	6.6	0.0	0.0	B8.5 V		210 s		
37224	195992		5 35 26.8	-33 16 10	-0.002	+0.09	7.30	0.5	3.4	F5 V		60 s		m
36819	77285	121 Tau	5 35 27.0	+24 02 23	+0.001	-0.01	5.38	-0.09	-1.7	B3 V	+23	240 s		
36706	58210		5 35 27.7	+37 54 06	-0.001	-0.02	7.50	0.2	2.1	A5 V		120 s	4168	m
37040	132325		5 35 31.0	-4 21 53	0.000	-0.02	6.31	-0.13		B5	+30		4192	m
37061	132328		5 35 31.3	-5 16 03	0.000	-0.01	6.85	0.28	-3.5	B1 V	+40	630 s		NU Ori, v
37058	132331		5 35 33.3	-4 50 16	+0.001	0.00	7.34	-0.16	-2.5	B2 V p	+23	930 s		V359 Ori, v
36721	58211		5 35 33.9	+38 01 12	+0.001	+0.01	7.80	0.1	1.4	A2 V		180 s		
37055	132332		5 35 35.8	-3 15 10	-0.001	0.00	6.40	-0.12	-1.7	B3 V	+24	400 s		m
38602	256214	ι Men	5 35 36.5	-78 49 15	+0.003	+0.02	6.05	-0.02		B9				
37630	249319		5 35 37.5	-62 03 07	+0.003	-0.06	7.9	0.8	3.2	G5 IV		86 s		
36625	40457		5 35 38.2	+48 04 41	+0.003	-0.03	7.5	1.1	-0.1	K2 III		320 s		
39197	258420		5 35 38.8	-81 36 42	-0.024	+0.04	7.3	0.5	3.4	F5 V		58 s		
37077	132336	45 Ori	5 35 39.5	-4 51 22	0.000	+0.01	5.26	0.24	0.6	gF0	-9	86 s	4196	m
36879	77293		5 35 40.4	+21 24 12	-0.001	0.00	7.57	0.19		O6				
37213	170581		5 35 41.0	-27 34 18	+0.001	-0.26	7.9	0.7	3.2	G5 IV		55 mx		
37212	170582		5 35 47.5	-25 44 18	-0.002	+0.02	7.66	2.12		N7.7	+40			
37162	150595		5 35 47.6	-19 56 48	-0.001	0.00	8.0	1.5	-0.3	K5 III		460 s		
37053	112939		5 35 51.5	+5 01 43	0.000	-0.05	7.8	0.5	3.4	F5 V		76 s		
37152	150594		5 35 53.5	-13 14 43	0.000	0.00	8.0	1.1	0.2	K0 III		360 s		
37115	---		5 35 54.2	-5 36 58			7.07	-0.05	-0.9	B6 V	+21	380 s	4202	m
36859	77295		5 35 55.4	+27 39 44	-0.002	-0.04	6.27	1.54		K0	-9			
37276	195999		5 35 56.2	-30 32 10	0.000	0.00	7.6	0.9	3.2	G5 IV		66 s		
36975	94684		5 36 02.5	+19 32 56	0.000	-0.02	7.2	0.0	0.4	B9.5 V		210 s		
37434	217422		5 36 02.7	-47 18 50	-0.002	-0.02	6.11	1.16		K0				
37112	132342		5 36 03.4	-0 46 49	0.000	-0.01	8.02	-0.09		B9				
37129	132345		5 36 06.1	-4 25 33	-0.001	0.00	7.13	-0.14	-2.5	B2 V p	+28	840 s		
37151	132347		5 36 06.2	-7 23 47	+0.001	+0.01	7.40	-0.08	-0.2	B8 V		330 s		
37033	94689		5 36 08.5	+12 26 12	0.000	-0.01	7.7	0.9	3.2	G5 IV		79 s		
37286	170587		5 36 10.2	-28 42 28	+0.001	0.00	6.26	0.16		A0				
37101	112947		5 36 10.3	+4 37 46	-0.001	-0.03	7.8	0.4	2.6	F0 V		110 s		
36756	40468		5 36 11.2	+44 19 34	-0.008	-0.01	7.2	0.5	3.4	F5 V	+22	57 s		
37211	150603		5 36 12.0	-15 49 18	-0.002	-0.03	7.9	1.1	0.2	K0 III		340 s		
37128	132346	46 ε Ori	5 36 12.7	-1 12 07	0.000	0.00	1.70	-0.19	-6.2	B0 Ia	+26	370 s		Alnilam, m
37150	132351		5 36 14.9	-5 38 53	0.000	0.00	6.55	-0.19	-1.7	B3 V	+11	450 s		
36719	40466		5 36 15.9	+47 42 56	-0.001	-0.01	6.11	0.26	2.6	F0 V	+14	50 s		
36755	40469		5 36 16.7	+44 46 11	-0.002	-0.02	8.0	0.0	0.4	B9.5 V		310 s		
37149	132350		5 36 17.7	-1 38 08	-0.001	0.00	8.03	-0.10		B8				
37629	234047		5 36 21.7	-57 27 29	0.000	+0.05	7.4	1.5	-0.5	M2 III		390 s		
37433	217425		5 36 24.4	-44 02 17	0.000	-0.01	7.6	1.6	0.2	K0 III		140 s		
36909	58238		5 36 25.0	+32 00 13	+0.001	-0.04	7.98	0.51	4.7	G2 V		45 s		
36931	58240		5 36 26.1	+30 35 27	0.000	+0.01	7.92	1.10	-2.0	F8 II		500 s		

HD	SAO	Star Name	α 2000	δ 2000	μ(α)	μ(δ)	V	B-V	M_v	Spec	RV	d(pc)	ADS	Notes
			h m s	° ′ ″	s	″								
37013	77313		5 36 26.4	+21 59 36	-0.003	-0.07	7.20	0.5	4.0	F8 V	+26	43 s	4200	m
36994	77310		5 36 30.1	+25 56 22	+0.001	0.00	6.49	0.43	3.4	F5 V	+3	42 s		
37173	132353		5 36 30.5	-1 59 02	0.000	0.00	7.86	-0.06		B8				
37109	94696		5 36 33.3	+12 06 03	-0.001	-0.01	7.9	0.9	3.2	G5 IV		87 s		
36678	25276		5 36 35.2	+54 25 45	0.000	0.00	5.73	1.63	-0.3	K5 III	+1	130 s		
245459	112954		5 36 35.4	+9 51 56	0.000	+0.01	8.0	1.1	0.2	K0 III		340 s		
37209	132359		5 36 35.6	-6 03 54	-0.001	-0.01	5.70	-0.23	-3.5	B1 V	+29	690 s	4212	m
37377	196012		5 36 36.2	-35 03 43	0.000	+0.04	6.8	1.9	0.2	K0 III		61 s		
37511	217427		5 36 36.3	-48 17 05	+0.003	+0.01	7.6	0.0	3.4	F5 V		71 s		
36555	13547		5 36 39.1	+61 24 25	-0.001	+0.02	8.0	0.9	3.2	G5 IV		90 s		
37327	170594		5 36 44.1	-24 42 10	+0.001	+0.05	8.0	0.7	0.4	B9.5 V		130 s		
37125	94698		5 36 49.2	+15 37 06	0.000	+0.01	7.6	1.1	0.2	K0 III		280 s		
36891	40481		5 36 52.4	+40 10 56	0.000	-0.01	6.09	1.03	-4.5	G3 Ib	-18	1200 s		
36949	58249		5 36 53.3	+38 20 32	0.000	-0.02	8.0	1.1	0.2	K0 III		340 s		
37160	112958	40 φ² Ori	5 36 54.3	+9 17 27	+0.006	-0.30	4.09	0.95	0.2	K0 III	+99	60 s		
37935	249322		5 36 54.7	-66 33 37	-0.003	+0.01	6.31	-0.07		A0				
36065	5578		5 36 55.9	+74 19 08	+0.012	-0.11	7.0	0.4	2.6	F0 V	-18	74 mx		
38283	256213		5 37 02.2	-73 41 58	+0.034	-0.11	6.7	0.6		WN7				
37124	77323		5 37 02.4	+20 43 51	-0.005	-0.42	7.68	0.67	4.1	G4 IV-V		30 mx		
37147	94700	122 Tau	5 37 03.7	+17 02 25	+0.003	-0.03	5.55	0.22		A5	+41	18 mn		
37171	94702		5 37 04.3	+11 02 06	+0.003	-0.01	5.94	1.58	-0.3	K5 III	-112	160 s		
37345	150623		5 37 06.5	-18 54 28	-0.002	-0.02	8.0	1.1	-0.1	K2 III		410 s		
36929	40485		5 37 06.9	+41 49 42	-0.002	-0.03	7.40	0.5	4.0	F8 V		47 s	4204	m
37628	234050		5 37 08.2	-50 24 33	-0.002	0.00	7.8	1.3	0.2	K0 III		230 s		
37306	150617		5 37 08.7	-11 46 33	+0.001	-0.03	6.11	0.05		A1				
37098	77322		5 37 08.8	+26 55 27	+0.001	-0.03	5.70	-0.07	-1.2	B8 III	+10	230 s	4208	m
39091	258421	π Men	5 37 10.0	-80 28 09	+0.123	+1.06	5.65	0.60	3.0	G3 IV	+12	15 mx		
37234	112965		5 37 13.2	+4 46 07	0.000	+0.01	7.75	0.04	0.4	B9.5 V		280 s	4217	m
37272	132369		5 37 14.4	-1 40 03	-0.001	+0.01	7.91	-0.11		B9				
36570	13550	19 Cam	5 37 15.1	+64 09 17	+0.001	-0.07	6.00	0.01		B9			4177	m
34360	866		5 37 15.8	+83 51 40	+0.013	-0.04	7.3	0.1	0.6	A0 V		99 mx		
36496	13548		5 37 16.2	+66 41 46	-0.001	-0.03	6.50	0.2	2.1	A5 V	-24	74 s		m
37430	170601	ν¹ Col	5 37 16.5	-27 52 17	+0.001	-0.06	5.9	1.1	2.1	A5 V		15 s		
36948	40487		5 37 16.9	+44 04 14	+0.001	-0.02	7.5	0.1	0.6	A0 V	-18	230 s		
37232	112966		5 37 19.3	+8 57 06	0.000	0.00	6.12	-0.17	-3.0	B1.5 V	+42	630 s		
36878	40484		5 37 20.3	+47 56 24	0.000	+0.01	7.4	0.8	3.2	G5 IV		67 s		
37496	196029		5 37 21.8	-38 01 23	+0.001	+0.01	6.9	1.2	0.2	K0 III		170 s		
36927	40488		5 37 26.3	+46 49 14	+0.003	-0.07	8.0	1.1	0.2	K0 III		180 mx		
36770	25277		5 37 27.0	+56 29 16	0.000	-0.01	7.5	1.1	0.2	K0 III		270 s		
37303	132375		5 37 27.3	-5 56 18	+0.001	0.00	6.02	-0.22	-3.5	B1 V	+29	800 s		
36973	40491		5 37 29.3	+44 35 23	-0.002	-0.04	7.4	0.8	3.2	G5 IV		67 s		
36569	13551		5 37 29.4	+65 08 29	0.000	-0.02	7.9	0.1	1.4	A2 V		190 s		
37473	196028		5 37 31.5	-31 03 42	+0.002	0.00	7.3	1.4	3.2	G5 IV		26 s		
37071	58258		5 37 32.4	+37 05 12	-0.003	+0.03	7.7	0.6	4.4	G0 V		46 s		
36130	5581		5 37 33.4	+74 41 20	-0.005	-0.18	7.5	0.7	4.4	G0 V	-61	42 s		
37321	132376		5 37 34.7	-1 25 18	-0.001	+0.03	7.09	-0.08	-1.7	B3 V		510 s	4222	m
37359	150627		5 37 36.3	-10 23 49	-0.002	+0.02	8.0	1.1	0.2	K0 III		360 s		
37334	132378		5 37 36.6	-4 56 03	0.000	0.00	7.19	-0.17		B3	+27			
37271	112970		5 37 37.1	+6 06 43	+0.006	-0.09	7.4	0.5	3.4	F5 V		64 s		
37231	94713		5 37 37.7	+15 14 40	0.000	-0.02	7.7	0.8	3.2	G5 IV		78 s		
37202	77336	123 ζ Tau	5 37 38.6	+21 08 33	0.000	-0.02	3.00	-0.19	-3.0	B2 IV p	+24	150 s		
37484	170610		5 37 39.4	-28 37 34	0.000	0.00	7.2	0.7	3.4	F5 V		39 s		
37720	234054		5 37 39.6	-53 32 07	+0.001	+0.02	7.3	0.2	1.4	A2 V		120 s		
37472	170609		5 37 41.4	-26 43 07	-0.001	0.00	7.5	0.5	2.6	F0 V		72 s		
37495	170613	ν² Col	5 37 44.6	-28 41 22	-0.003	+0.05	5.31	0.46	3.0	F2 V	+36	25 s		
37138	58263		5 37 45.7	+33 33 33	0.000	0.00	6.33	1.31	0.2	K0 III	+29	110 s		
37332	132382		5 37 45.8	-0 46 42	0.000	+0.01	7.60	-0.13		B8				
37610	217437		5 37 47.9	-41 09 49	+0.002	+0.03	7.6	0.5	4.4	G0 V		44 s		
37549	196033		5 37 48.0	-34 41 23	0.000	-0.01	6.5	1.4	0.2	K0 III		110 s		
37761	234056		5 37 48.3	-54 33 50	+0.001	+0.11	7.70	0.9	3.2	G5 IV		79 s		m
37510	170614		5 37 50.7	-28 56 18	+0.002	-0.04	7.3	1.5	0.2	K0 III		140 s		
36357	5587		5 37 52.2	+71 39 13	+0.002	-0.07	6.8	0.9	3.2	G5 IV		53 s		
37330	112980		5 37 53.3	+0 58 07	-0.001	0.00	7.38	-0.07	-0.9	B6 V		450 s		m
37356	132387		5 37 53.4	-4 48 49	0.000	+0.02	6.19	-0.04	-3.0	B1.5 V	+29	540 s		m
37342	112981		5 37 56.2	+0 59 15	0.000	+0.01	8.00	-0.13	-1.1	B5 V		660 s		m
37146	58265		5 37 57.3	+35 36 15	-0.003	-0.01	7.3	1.1	0.2	K0 III	+9	250 s		
37531	196034		5 37 59.4	-30 30 58	-0.001	+0.02	8.0	1.4	0.2	K0 III		220 s		
36806	25281		5 38 00.6	+59 21 52	-0.001	-0.01	7.6	0.0	0.4	B9.5 V		260 s		
35783	5577		5 38 01.0	+78 21 30	+0.022	-0.27	7.69	0.46	3.7	F6 V	+19	59 mx		
37320	112979		5 38 01.0	+7 32 29	0.000	-0.02	5.88	-0.07		B8	+19			
37655	217439		5 38 01.8	-42 57 50	+0.009	+0.25	7.44	0.60	4.4	G0 V		39 s		
36850	25284		5 38 05.5	+58 04 17	-0.005	-0.03	7.8	1.1	0.2	K0 III		280 mx		
37370	—		5 38 06.4	-0 11 05	-0.001	-0.02	7.46	-0.04		B9			4234	m
37530	170618		5 38 06.9	-27 12 39	0.000	-0.01	6.7	1.9	-0.1	K2 III		79 s		
37719	217446		5 38 08.6	-47 42 52	-0.003	+0.01	7.9	0.7	0.2	K0 III		340 s		

HD	SAO	Star Name	α 2000	δ 2000	μ(α)	μ(δ)	V	B-V	M_v	Spec	RV	d(pc)	ADS	Notes
37706	217444		5 38 09.3	-46 06 22	-0.013	-0.47	7.34	0.78	5.2	G5 V		23 s		m
37371	---		5 38 09.8	-0 11 03	0.000	-0.03	7.95	0.10		B9				
37341	112984		5 38 11.9	+6 00 11	0.000	-0.09	7.5	1.4	-0.3	K5 III		150 mx		
37008	25290		5 38 12.0	+51 26 45	-0.057	+0.11	7.73	0.83	6.3	K2 V	-44	24 ts		
37397	132390		5 38 13.7	-1 10 10	0.000	-0.01	6.84	-0.15	-1.7	B3 V	+23	510 s		
37410	132392		5 38 15.2	-4 06 31	0.000	-0.04	6.85	0.12		A2				
37169	58269		5 38 15.8	+37 44 39	0.000	-0.02	7.56	0.04	0.4	B9.5 V		260 s		
37369	112987		5 38 16.0	+1 59 15	-0.001	0.00	7.7	0.9	3.2	G5 IV		80 s		
37368	112985		5 38 16.2	+4 56 01	0.000	-0.02	7.9	0.1	1.4	A2 V		200 s		
37933	234065		5 38 16.5	-59 07 00	+0.001	+0.01	7.1	-0.5	1.4	A2 V		140 s		
37781	234059		5 38 17.2	-50 38 28	-0.001	+0.01	6.6	0.5	0.6	A0 V		76 s		
37355	112986		5 38 21.5	+8 29 15	-0.002	0.00	7.0	0.8	3.2	G5 IV		59 s		
37877	234063		5 38 25.4	-54 57 14	-0.006	0.00	7.0	1.4	-0.1	K2 III		200 s		
38004	249327		5 38 26.3	-61 10 03	-0.002	+0.01	7.9	0.0	0.6	A0 V		270 s		
37444	132396		5 38 26.9	-5 02 41	-0.001	-0.01	7.64	0.33		A2				
37718	217448		5 38 36.3	-41 17 49	-0.002	-0.03	7.90	0.5	3.4	F5 V		79 s		m
37548	150643		5 38 37.6	-16 55 09	+0.004	-0.10	7.5	0.7	4.4	G0 V		41 s		
37481	132405		5 38 37.8	-6 34 26	-0.001	0.00	5.96	-0.23	-3.5	B1 V	+15	780 s		
37269	58280	26 Aur	5 38 38.0	+30 29 33	-0.001	-0.01	5.40	0.41	0.3	G5 III	+2	110 s	4229	m
37251	58279		5 38 38.5	+32 40 31	0.000	-0.03	7.87	0.94	0.3	G8 III		330 s		
37492	132407		5 38 39.6	-8 28 17	0.000	-0.02	6.7	0.0	0.4	B9.5 V		190 s		
37252	58281		5 38 41.6	+32 28 30	+0.003	-0.05	7.83	0.47	3.7	F6 V		67 s		m
38268	249329		5 38 43.1	-69 06 03	+0.006	0.00	6.0	0.2		O				m
294271	---		5 38 43.4	-2 31 33			7.91	-0.11		B8				
37717	217450		5 38 43.5	-40 42 27	-0.001	+0.01	5.8	0.6	0.0	B8.5 V		60 s		
37184	40507		5 38 43.7	+41 50 21	+0.004	-0.04	6.8	1.4	-0.3	K5 III		220 mx		
36554	5594		5 38 44.4	+70 21 54	+0.002	0.00	6.9	0.1	0.6	A0 V		170 s		
37468	132406	48 σ Ori	5 38 44.7	-2 36 00	0.000	0.00	3.73	-0.24	-4.4	O9.5 V	+29	550 s	4241	m
37569	150646		5 38 46.1	-17 04 14	-0.001	-0.02	8.0	0.0	0.6	B9.5 V		330 s		
37479	132408		5 38 47.0	-2 35 39	0.000	+0.01	6.65	-0.19	-2.5	B2 V p	+29	550 s	4241	m
37467	112994		5 38 52.6	+2 51 41	0.000	-0.02	7.89	-0.10	0.4	B9.5 V		320 s		
37507	132411	49 Ori	5 38 53.0	-7 12 47	-0.001	-0.05	4.80	0.13	1.0	A4 IV	+4	57 s		
37491	132410		5 38 55.1	-1 10 13	+0.001	-0.02	7.9	1.1	0.2	K0 III		350 s		
37145	40505		5 38 56.9	+49 24 58	+0.004	-0.01	7.1	0.9	3.2	G5 IV		59 s		
37329	77350		5 38 57.4	+26 37 04	+0.002	-0.02	6.4	0.9	0.2	G9 III	+15	170 s		
37526	132414		5 39 02.3	-5 11 39	0.000	0.00	7.61	-0.12	-1.7	B3 V		680 s		
37266	58287		5 39 04.2	+37 58 36	-0.001	0.00	7.53	0.04	0.4	B9.5 V		260 s		
37070	25298		5 39 05.4	+56 21 36	+0.002	-0.13	6.9	0.5	3.4	F5 V	+20	50 s		
38042	234073		5 39 07.1	-58 57 28	0.000	+0.04	6.7	1.1	0.2	K0 III		200 s		
37441	94726		5 39 07.6	+14 33 24	0.000	0.00	7.3	0.1	0.6	A0 V		210 s		
37406	94723		5 39 08.1	+19 46 07	+0.001	0.00	7.9	1.1	0.2	K0 III		320 s		
37250	40512		5 39 08.5	+41 21 30	-0.004	-0.13	6.5	1.1	0.2	K0 III		110 mx		
37490	113001	47 ω Ori	5 39 11.0	+4 07 17	0.000	0.00	4.57	-0.11	-2.9	B3 III e	+22	310 s		
37665	170639		5 39 11.1	-24 52 51	0.000	-0.01	7.8	1.1	-0.1	K2 III		390 s		
37387	77355		5 39 14.8	+23 19 24	0.000	-0.01	7.48	1.94	-4.4	K1 Ib	+6	900 s	4239	m
37352	58292		5 39 15.0	+30 09 02	-0.002	0.00	7.68	0.10	-0.2	B8 V		300 s		
37643	150652		5 39 16.2	-17 50 58	0.000	0.00	6.38	-0.12		B9			4254	m
---	150654		5 39 17.0	-17 49 44	+0.002	-0.02	8.0							
37367	77354		5 39 18.4	+29 12 55	+0.001	0.00	5.96	0.16	-2.5	B2 V	+30	310 s		
37688	170649		5 39 24.2	-25 14 34	0.000	+0.04	7.5	1.1	3.4	F5 V		27 s		
37366	58297		5 39 24.8	+30 53 28	0.000	+0.01	7.62	0.09		B3	+21			
37622	150653		5 39 25.1	-11 12 39	0.000	-0.02	8.00	-0.08	-1.7	B3 V n		870 s		
37283	40515		5 39 25.8	+42 40 49	-0.003	-0.03	7.4	1.6	-0.5	M2 III	+17	350 s		
37439	77358		5 39 27.0	+21 45 46	0.000	-0.04	6.30	0.06	1.4	A2 V	+37	93 s		
37488	94731		5 39 27.4	+11 30 11	+0.001	0.00	7.6	1.6	-0.5	M2 III		390 s		
37635	132425		5 39 30.8	-9 42 24	0.000	-0.01	6.50	-0.10	-0.6	B7 V	+21	260 s		
37594	132424		5 39 31.1	-3 33 53	0.000	0.00	6.00	0.27		A9 m				
37487	94732		5 39 32.4	+12 00 14	+0.001	-0.04	8.0	0.4	2.6	F0 V		120 s		
37664	150657		5 39 34.4	-15 52 05	-0.001	+0.01	7.3	1.1	0.2	K0 III		260 s		
37522	113005		5 39 34.5	+9 50 55	-0.001	-0.01	7.3	1.1	0.2	K0 III		250 s		
37339	58296		5 39 36.6	+37 59 19	-0.001	0.00	6.96	0.00	0.4	B9.5 V		210 s		
37593	113011		5 39 37.9	+0 50 02	0.000	-0.01	7.5	1.1	0.2	K0 III		280 s		
37795	196059	α Col	5 39 38.9	-34 04 27	0.000	-0.03	2.64	-0.12	-0.2	B8 V e	+35	37 s		Phact
36384	5593		5 39 43.7	+75 02 38	-0.001	+0.03	6.17	1.57	-0.3	K5 III	-3	180 s		
37438	77360	125 Tau	5 39 44.2	+25 53 50	+0.001	-0.02	5.18	-0.15	-2.5	B2 V	+15	330 s		
37702	170652		5 39 44.5	-20 26 07	-0.002	0.00	6.90	0.00	0.0	B8.5 V		240 s	4260	m
37591	113012		5 39 45.2	+4 26 05	+0.001	+0.01	7.9	0.0	0.4	B9.5 V		320 s		
37385	58300		5 39 48.7	+35 39 19	+0.002	-0.04	7.80	1.1	0.2	K0 III		310 s	4242	m
37606	113014		5 39 49.0	+1 29 31	0.000	+0.01	6.90	-0.07	-0.2	B8 V		260 s		
37811	196061		5 39 49.7	-32 37 45	-0.002	-0.03	5.45	0.92	0.2	gK0	-8	110 s		
37466	77363		5 39 51.5	+24 13 31	0.000	+0.04	7.1	0.4	3.0	F2 V		65 s		
37384	58301		5 39 54.3	+37 57 27	0.000	-0.02	7.32	0.10	0.4	B9.5 V		210 s	4243	m
37560	94737		5 39 54.7	+13 01 22	+0.001	-0.01	7.20	0.1	0.6	A0 V		200 s	4253	m
37919	217465		5 39 56.1	-41 30 50	+0.001	-0.01	7.5	1.2	0.2	K0 III		230 s		
37641	132428		5 39 56.4	-1 55 36	-0.001	0.00	7.56	-0.06		B7				

HD	SAO	Star Name	α 2000	δ 2000	μ(α)	μ(δ)	V	B-V	M_v	Spec	RV	d(pc)	ADS	Notes
37832	196062		5 39 57.3	−30 38 10	+0.002	0.00	8.0	1.4	−0.1	K2 III		310 s		
42556	258427		5 39 59.9	−85 54 52	−0.019	+0.01	6.7	1.3	0.2	K0 III		130 s		
37338	40517		5 40 02.9	+43 19 28	−0.001	−0.04	7.1	1.4	−0.3	K5 III	−22	280 s		
37422	58307		5 40 03.2	+34 17 20	+0.001	0.00	8.0	1.1	0.2	K0 III		340 s		
37778	170663		5 40 05.9	−23 43 24	−0.002	−0.03	7.2	1.0	0.2	K0 III		260 s		
36289	5591		5 40 06.0	+76 46 00	−0.005	−0.05	7.9	0.4	3.0	F2 V		94 s		
37850	196064		5 40 06.3	−31 14 43	−0.002	+0.02	7.6	0.1	0.6	A0 V		200 s		
37559	94739		5 40 09.5	+19 40 48	0.000	−0.01	7.6	0.0	0.4	B9.5 V		250 s		
37632	113017		5 40 10.5	+5 39 52	+0.001	−0.03	7.9	1.1	−0.1	K2 III		390 s		
37972	217470		5 40 11.4	−43 35 04	−0.001	+0.01	7.8	1.9	−0.1	K2 III		100 s		
37674	132436		5 40 13.4	−1 27 45	−0.001	0.00	7.68	−0.08		B8	+15			
36768	13559		5 40 13.4	+69 58 56	+0.005	+0.02	7.0	1.1	0.2	K0 III		220 s		
37588	94743		5 40 17.6	+16 23 33	+0.005	−0.17	7.5	0.5	3.4	F5 V		47 mx		
37539	77375		5 40 18.1	+24 32 16	0.000	−0.02	7.1	1.1	0.2	K0 III	+34	230 s		
37603	94746		5 40 18.8	+15 21 00	0.000	−0.03	7.00	0.4	2.6	F0 V	+21	75 s	4256	m
37687	132439		5 40 19.8	−3 25 38	+0.001	−0.01	7.05	0.03		B8				
37699	132438		5 40 20.0	−2 26 09	−0.001	0.00	7.62	−0.13		B5	+15			
37700	132440		5 40 25.2	−4 25 17	0.000	0.00	7.96	−0.09	−1.1	B5 V		610 s		
37136	13564		5 40 29.1	+61 56 57	0.000	+0.02	6.7	0.2	2.1	A5 V	−16	80 s		
37436	58314		5 40 32.2	+39 50 05	+0.001	−0.04	7.9	0.4	2.6	F0 V		110 s		
37747	132442		5 40 32.4	−7 39 36	0.000	−0.01	7.5	1.1	0.2	K0 III		290 s		
37282	25313		5 40 34.9	+55 05 56	0.000	−0.11	7.9	0.4	3.0	F2 V		95 s		
37557	77381		5 40 35.5	+28 58 38	−0.002	+0.01	7.02	1.16		G5	−21			
37519	58319		5 40 36.0	+31 21 30	+0.001	−0.01	6.04	0.04	−0.6	B7 V	−7	170 s		
37744	132441		5 40 37.2	−2 49 30	0.000	+0.01	6.22	−0.20	−3.5	B1 V	+29	870 s		
37847	170678		5 40 39.6	−20 17 56	+0.001	−0.01	7.00	0.9	0.3	G4 III	+36	220 s		
37404	40525		5 40 41.2	+45 42 28	+0.001	−0.02	8.0	0.1	1.7	A3 V		170 s		
37536	58322		5 40 42.1	+31 55 15	+0.001	+0.01	6.11	2.12	−5.6	M2 Iab	+5	1100 s		
37328	25315		5 40 42.3	+52 34 13	−0.001	0.00	7.2	0.8	3.2	G5 IV		63 s		
38149	234085		5 40 42.6	−53 13 02	+0.001	+0.02	7.4	0.4	1.4	A2 V		94 s		
37809	150681		5 40 44.2	−12 13 45	−0.002	+0.01	8.0	0.9	3.2	G5 IV		93 s		
37698	113029		5 40 45.2	+5 04 39	0.000	−0.01	7.2	0.9	3.2	G5 IV		62 s		
37742	132444	50 ζ Ori	5 40 45.5	−1 56 34	0.000	0.00	1.77	−0.21	−5.9	09.5 Ib	+18	340 s	4263	Alnitak, m
37808	150680		5 40 45.9	−10 24 35	−0.003	−0.03	6.52	−0.16		A p				
37685	113028		5 40 46.2	+9 15 55	−0.004	−0.07	7.7	0.6	4.4	G0 V		46 s		
37505	58320		5 40 47.3	+37 19 55	0.000	−0.01	7.9	0.1	0.6	A0 V		270 s		
37756	132445		5 40 50.6	−1 07 44	−0.001	−0.01	4.95	−0.21	−3.0	B2 IV	+26	390 s		
37962	196074		5 40 52.0	−31 21 02	−0.004	−0.33	7.85	0.65	5.0	G4 V		37 s		
37961	170689		5 40 52.4	−29 43 15	0.000	0.00	7.0	1.4	3.2	G5 IV		24 s		
37828	150685		5 40 54.8	−11 12 01	+0.005	−0.06	7.0	1.1	0.2	K0 III		130 mx		
37776	132446		5 40 56.2	−1 30 26	−0.001	0.00	6.98	−0.14	−2.5	B2 V	+27	790 s		
37777	132449		5 40 57.5	−3 50 34	−0.001	+0.02	7.5	1.1	−0.1	K2 III		330 s		
37574	58327		5 40 57.8	+32 53 45	0.000	−0.01	6.8	0.5	4.0	F8 V	−10	37 s		
--	170690		5 40 58.7	−27 57 06	0.000	+0.02	7.9	1.6	−0.5	M2 III		490 s		
37905	170686		5 40 59.4	−21 32 37	−0.002	+0.02	7.9	0.5	2.6	F0 V		83 s		
37518	58325		5 40 59.7	+38 30 57	0.000	−0.01	7.7	1.1	0.2	K0 III		300 s		
38214	234088		5 41 01.5	−54 44 23	0.000	−0.03	8.0	1.7	−0.3	K5 III		340 s		
37806	132452		5 41 02.1	−2 43 01	−0.001	0.00	7.90	0.03		A0				
37949	170691		5 41 02.3	−25 30 27	−0.001	+0.01	8.0	0.0	0.4	B9.5 V		300 s		
37788	113033		5 41 05.5	+0 20 16	+0.001	+0.03	5.93	0.30		A5	−12			
37516	58326		5 41 05.6	+39 38 07	−0.005	−0.02	7.9	0.4	2.6	F0 V		110 s		
37515	40534		5 41 06.0	+40 52 57	−0.001	−0.01	7.7	0.1	1.7	A3 V		160 s		
37807	132455		5 41 08.0	−3 37 55	−0.001	+0.02	7.90	−0.10		B8				
38244	234090		5 41 10.8	−54 27 50	−0.001	−0.01	8.0	1.2	−0.3	K5 III		450 s		
37617	58329		5 41 15.0	+33 55 08	+0.001	−0.02	7.9	1.1	0.2	K0 III	+39	320 s		
38617	249334		5 41 15.2	−68 46 11	+0.003	+0.01	7.9	1.4	−0.3	K5 III		440 s		
37987	170697		5 41 16.7	−26 20 46	−0.011	+0.03	7.80	0.5	3.4	F5 V		76 s	4281	m
37711	94759	126 Tau	5 41 17.7	+16 32 02	+0.001	−0.02	4.86	−0.13	−2.3	B3 IV	+21	250 s	4265	m
37639	58335		5 41 19.9	+31 11 11	−0.001	−0.01	7.45	−0.01	0.6	A0 V		230 s		
37394	25319		5 41 20.3	+53 28 52	+0.001	−0.52	6.23	0.84	6.1	K1 V	+1	11 ts		m
37646	77393		5 41 20.9	+29 29 15	+0.001	−0.03	6.43	−0.11	0.6	A0 V	+18	150 s	4262	m
37647	77394		5 41 21.3	+29 29 38	0.000	−0.05	7.17	−0.05	2.4	A V	+19	90 s		
37889	132458		5 41 22.3	−6 56 06	0.000	+0.02	7.67	−0.12	−2.5	B2 V		1000 s		
37710	94760		5 41 22.5	+17 31 20	0.000	−0.01	6.89	1.85	−0.5	M2 III		210 s		
37786	113036		5 41 25.7	+9 11 56	−0.001	+0.02	7.48	−0.14	0.0	B8.5 V		310 s		
37824	113040		5 41 26.7	+3 46 40	+0.001	−0.01	7.2	0.8	3.2	G5 IV		63 s		
38056	196088		5 41 26.9	−33 24 02	+0.001	0.00	6.4	−0.4	0.6	A0 V		140 s		
38212	217479		5 41 32.3	−48 15 02	−0.002	+0.02	7.20	0.1	0.6	A0 V		210 s		m
37773	94765		5 41 34.3	+15 12 49	+0.005	−0.21	7.7	0.6	4.4	G0 V		43 mx		
37946	150700		5 41 35.0	−15 20 15	−0.003	−0.01	7.2	1.1	0.2	K0 III		250 s		
37887	132463		5 41 35.9	−3 43 53	0.000	−0.02	7.70	−0.01	0.6	A0 V		260 s		
37315	13568		5 41 36.5	+60 37 22	0.000	−0.04	7.0	0.1	1.7	A3 V		110 s		
37696	77398		5 41 36.6	+26 36 56	−0.002	0.00	8.0	0.0	0.4	B9.5 V		320 s		
37903	132464		5 41 38.3	−2 15 33	0.000	−0.01	7.83	0.10	−3.0	B1.5 V	+7	950 s		
37904	132465		5 41 40.2	−2 53 46	+0.002	+0.03	6.42	0.30	0.6	gF0		150 s	4279	m

HD	SAO	Star Name	α 2000	δ 2000	μ(α)	μ(δ)	V	B−V	M_V	Spec	RV	d(pc)	ADS	Notes
37971	150703		5^h 41^m 41^s.4	−16° 43′ 32″	−0^s.001	+0″.01	6.21	−0.13		B5	+16			
38616	249336		5 41 42.5	−67 24 11	−0.002	0.00	7.05	−0.03		A0				
37771	94767		5 41 44.1	+19 21 24	−0.001	−0.02	7.8	0.1	0.6	A0 V		250 s		
37514	40537		5 41 46.0	+48 27 55	−0.001	0.00	7.8	0.1	1.4	A2 V		180 s		
37281	13567		5 41 47.4	+63 17 46	−0.001	−0.22	7.2	0.8	3.2	G5 IV		62 s		
37970	150705		5 41 50.3	−12 18 36	−0.002	+0.05	7.3	0.1	1.4	A2 V		150 s		
37986	150708		5 41 53.5	−15 37 49	+0.015	−0.11	7.37	0.79		G5				
37884	113046		5 41 53.6	+4 59 17	0.000	−0.02	7.8	0.2	2.1	A5 V		140 s		
37752	77413		5 41 54.5	+23 19 34	−0.001	−0.01	6.50	−0.06	−1.6	B7 III		400 s		
37670	58343		5 41 56.3	+35 37 53	−0.004	+0.01	6.8	0.1	0.6	A0 V		170 s		
37768	77414		5 41 57.1	+24 19 07	0.000	−0.01	7.9	0.1	0.6	A0 V		270 s		
38511	249335		5 41 57.6	−62 54 52	0.000	−0.01	6.8	0.1	0.6	A0 V		180 s		
37784	77420		5 42 03.9	+22 39 37	−0.001	−0.02	6.4	1.1	−0.1	K2 III	−21	190 s		
38139	196095		5 42 07.6	−31 54 28	0.000	+0.03	7.6	2.4	−0.1	K2 III		63 s		
38243	217482		5 42 08.5	−43 30 30	−0.002	+0.03	7.4	1.4	0.2	K0 III		160 s		
38025	150713		5 42 08.7	−12 59 27	−0.001	−0.03	7.5	0.9	3.2	G5 IV		71 s		
38138	196096		5 42 11.5	−30 32 07	−0.001	+0.01	6.19	0.01		A0				
38090	170715	12 Lep	5 42 13.9	−22 22 25	−0.001	+0.02	5.87	0.11		A2				
38054	150716		5 42 14.3	−17 31 49	−0.003	+0.01	6.3	1.1	0.2	K0 III		160 s		
37926	113050		5 42 14.3	+8 22 24	−0.002	−0.02	7.8	1.1	0.2	K0 III		310 s		
38170	196098		5 42 15.0	−34 40 04	−0.001	+0.05	5.2	0.0	0.4	B9.5 V	+34	93 s		
37958	113053		5 42 17.2	+2 22 02	−0.004	0.00	6.6	0.0	0.0	B8.5 V		120 mx		
37708	58347		5 42 18.7	+38 11 50	−0.001	−0.01	7.8	0.1	0.6	A0 V		260 s		
37657	40544		5 42 19.8	+43 03 36	0.000	0.00	7.23	0.04		B2 e	+48			
38137	170718		5 42 20.0	−27 42 28	0.000	−0.02	7.9	0.5	1.7	A3 V		110 s		
38211	196100		5 42 20.4	−37 27 29	−0.001	+0.01	7.81	1.13	0.0	K1 III		350 s		
37944	113052		5 42 22.3	+6 39 07	0.000	0.00	7.9	0.5	3.4	F5 V		78 s		
38024	132473		5 42 24.7	−9 35 27	0.000	−0.08	7.7	1.1	0.2	K0 III		180 mx		
37289	13570		5 42 26.4	+65 41 51	0.000	−0.01	5.60	1.24	0.2	K0 III	−19	86 s		
37984	113056	51 Ori	5 42 28.6	+1 28 29	−0.004	−0.01	4.91	1.17	0.0	K1 III	+88	87 s		
37881	94787		5 42 28.7	+18 59 14	0.000	−0.03	7.4	1.1	0.2	K0 III	+12	260 s		
37800	77430		5 42 28.9	+29 51 00	−0.003	−0.04	7.27	0.55	2.4	F8 IV	+3	90 s	4280	m
37943	94789		5 42 30.0	+10 32 19	0.000	−0.02	7.1	0.2	2.1	A5 V		98 s		
37737	58352		5 42 31.0	+36 12 01	−0.001	0.00	8.00	0.31	−5.6	B0 II		2500 s		
38072	150721		5 42 32.3	−11 38 42	−0.003	−0.03	7.1	1.1	0.2	K0 III		250 s		
38002	132472		5 42 32.8	−0 00 58	0.000	−0.02	7.0	1.1	0.2	K0 III		230 s		
38053	150722		5 42 33.6	−10 37 45	−0.001	−0.02	7.1	0.5	4.0	F8 V		43 s		
37420	13574		5 42 38.4	+61 36 11	+0.001	−0.02	8.0	0.5	3.4	F5 V		80 s		
38746	249339		5 42 41.9	−67 08 20	+0.007	−0.03	7.80	0.5	4.0	F8 V		58 s		m
38000	113058		5 42 45.7	+5 50 38	+0.001	−0.10	7.8	1.1	0.2	K0 III		120 mx		
37766	58355		5 42 47.3	+38 11 53	−0.002	0.00	7.2	1.1	0.2	K0 III	+75	240 s		
38242	196106		5 42 49.6	−34 04 52	−0.002	+0.02	8.0	0.9	0.2	K0 III		360 s		
38343	217485		5 42 50.3	−45 14 07	−0.003	+0.01	7.5	0.2	0.4	B9.5 V		190 s		
38121	150729		5 42 52.3	−12 21 06	0.000	−0.04	8.0	1.4	−0.3	K5 III		450 s		
38253	196107		5 42 53.3	−33 57 30	−0.001	−0.02	7.30	0.8	3.2	G5 IV		66 s		m
37940	94793		5 42 53.6	+18 58 50	0.000	−0.01	6.80	0.00	0.4	B9.5 V		180 s		m
38089	132477		5 42 53.8	−6 47 46	−0.001	+0.07	5.97	0.44	3.3	F4 V	−11	33 s	4299	m
38022	113063		5 42 56.2	+5 14 08	−0.001	−0.02	7.9	0.0	0.4	B9.5 V		310 s		
37981	94795		5 42 58.1	+14 10 43	−0.003	−0.01	6.73	1.08		K1 IV	+63			
37859	58359		5 42 58.2	+33 18 51	0.000	−0.04	6.80	0.5	3.4	F5 V		48 s	4287	m
37736	40552		5 42 59.9	+44 51 06	0.000	+0.01	7.7	0.1	1.4	A2 V	+6	180 s		
36971	5620		5 43 01.1	+73 59 09	+0.003	+0.01	6.8	0.4	2.6	F0 V		70 s	4246	m
--	--	24 Cam	5 43 01.3	+73 58 58			6.80							
37601	25333	24 Cam	5 43 01.6	+56 34 54	+0.002	+0.03	6.05	0.95	0.2	gG9	−29	150 s		
38099	132480		5 43 09.2	−1 36 47	0.000	−0.03	6.31	1.47	−0.3	K4 III		190 s		
--	217489		5 43 13.7	−46 27 14	−0.001	+0.01	7.6							
37799	40559		5 43 13.8	+41 41 47	+0.006	−0.10	6.8	0.7	4.4	G0 V		29 s		
38297	196114		5 43 14.5	−33 25 30	0.000	+0.01	6.9	2.0	−0.5	M2 III		180 s		
37841	40560		5 43 16.9	+41 07 21	−0.003	−0.01	7.47	0.01		B8.5 V		300 s	4288	m
38046	94800		5 43 19.1	+10 30 04	−0.004	−0.02	7.8	1.1	−0.1	K2 III		210 mx		
37967	77450		5 43 19.4	+23 12 15	0.000	−0.02	6.21	−0.06	−1.7	B3 V e	+19	360 s		
38206	150739		5 43 21.5	−18 33 27	+0.001	−0.02	5.73	−0.01		A0				
37956	77449		5 43 22.6	+29 12 00	−0.001	−0.02	6.59	1.11	0.0	K1 III		210 s		
37723	25342		5 43 23.1	+50 42 07	−0.001	−0.05	7.3	1.1	0.2	K0 III		240 s		
38098	113074		5 43 23.9	+5 21 30	0.000	−0.02	6.75	−0.03	0.0	B8.5 V		210 s		
37979	77452		5 43 25.8	+24 05 09	0.000	−0.01	7.9	1.1	0.2	K0 III		330 s		
38471	217496		5 43 26.4	−49 50 34	−0.002	+0.01	7.60	0.00	0.4	B9.5 V		280 s		m
37693	25339		5 43 26.8	+52 29 19	+0.027	+0.02	7.0	0.6	4.4	G0 V		33 s		
38266	170735		5 43 27.5	−26 08 41	+0.001	0.00	7.8	1.2	0.2	K0 III		260 s		
38385	196116		5 43 30.0	−39 24 25	+0.003	0.00	6.3	0.9	2.6	F0 V		23 s		
38469	217495		5 43 30.9	−47 32 47	−0.001	−0.01	7.5	1.7	−0.5	M2 III		370 s		
--	58369		5 43 32.3	+38 21 13	+0.002	−0.03	8.0	1.9						
37585	13581		5 43 33.8	+61 02 17	+0.002	−0.06	7.9	0.5	3.4	F5 V		76 s		
38108	113080		5 43 36.4	+6 53 18	−0.001	−0.02	7.24	−0.06	0.0	B8.5 V		280 s		
37681	25341		5 43 38.0	+55 19 43	+0.002	−0.04	7.5	0.9	3.2	G5 IV		71 s		

HD	SAO	Star Name	α 2000	δ 2000	μ(α)	μ(δ)	V	B-V	M$_v$	Spec	RV	d(pc)	ADS	Notes
			h m s	° ′ ″	s	″								
38942	249341		5 43 38.0	-68 04 53	+0.003	+0.04	7.8	1.1	0.2	K0 III		330 s		
38205	150740		5 43 38.4	-11 17 51	0.000	-0.02	8.0	1.1	0.2	K0 III		360 s		
38010	77459		5 43 38.9	+25 26 23	-0.001	+0.01	6.84	0.03	-3.5	B1 V pe	+19	1000 s		
38185	132490		5 43 39.2	-8 56 13	-0.001	+0.04	7.61	-0.11	0.4	B9.5 V		280 s		
38458	217497		5 43 41.1	-45 49 58	+0.002	+0.09	6.39	0.29	0.6	F0 III n		140 s		
38279	170739		5 43 41.6	-22 46 49	0.000	+0.01	8.0	1.3	0.2	K0 III		250 s		
38295	170740		5 43 46.4	-21 25 46	-0.001	-0.03	7.0	1.2	3.2	G5 IV		33 s		
38264	150745		5 43 48.1	-19 39 05	-0.003	+0.04	7.5	0.6	3.2	G5 IV		71 s		
38009	77466		5 43 48.8	+27 17 46	-0.002	+0.01	7.8	0.0	0.4	B9.5 V		290 s		
34109	873		5 43 49.0	+85 40 05	-0.009	0.00	6.6	0.1	0.6	A0 V	-14	160 s		
37840	40563		5 43 50.2	+47 42 49	+0.002	-0.02	7.3	1.1	0.2	K0 III		250 s		
38117	94811		5 43 51.0	+11 24 22	+0.001	+0.03	7.7	1.1	0.2	K0 III		290 s		
38294	170742		5 43 52.7	-21 16 49	+0.013	-0.01	7.2	0.7	4.4	G0 V		31 s		
37465	13578		5 43 56.6	+66 37 34	+0.009	-0.08	7.8	1.1	0.2	K0 III		140 mx		
37735	25344		5 43 58.6	+54 51 54	0.000	-0.01	6.8	0.1	0.6	A0 V		170 s		
38225	132493		5 44 02.6	-4 33 32	-0.002	+0.01	8.0	1.1	0.2	K0 III		360 s		
38164	113089		5 44 03.2	+6 30 55	-0.001	-0.03	7.7	0.9	0.3	G5 III	+36	290 s		
38313	150749		5 44 03.9	-18 17 19	-0.004	-0.09	7.3	1.1	0.2	K0 III		140 mx		
38384	170751		5 44 07.8	-28 13 35	+0.001	+0.04	7.8	1.9	-0.3	K5 III		230 s		
37638	13590	23 Cam	5 44 08.5	+61 28 36	0.000	0.00	6.15	0.90	3.2	G5 IV	-4	33 s		
38488	217502		5 44 09.1	-41 06 24	0.000	-0.01	8.0	1.6	-0.1	K2 III		270 s		
37839	25346		5 44 09.2	+50 48 12	+0.001	-0.01	8.0	1.1	0.2	K0 III		330 s		
38032	58379		5 44 11.7	+33 06 01	-0.001	-0.01	7.8	0.1	1.4	A2 V		180 s		
38383	170754		5 44 12.1	-27 32 41	-0.004	+0.03	7.5	0.8	1.7	A3 V		52 s		
38033	58380		5 44 12.6	+32 00 15	+0.002	-0.02	7.82	1.05	4.7	G2 V		23 s		
38683	234112		5 44 13.5	-55 41 44	-0.002	+0.05	7.9	0.4	4.0	F8 V		60 s		
38580	217505		5 44 15.5	-49 10 26	0.000	0.00	7.4	1.7	0.2	K0 III		190 s		
38133	94814		5 44 16.4	+18 50 03	-0.001	-0.01	7.5	0.1	0.6	A0 V		230 s		
37937	40570		5 44 17.0	+43 33 23	0.000	-0.02	7.7	0.9	3.2	G5 IV	+6	79 s		
38277	150750		5 44 17.7	-10 01 00	+0.006	-0.14	7.2	0.6	4.4	G0 V		36 s		
37838	25347		5 44 19.2	+52 41 43	0.000	0.00	8.0	0.0	0.0	B8.5 V		370 s		
38719	234115		5 44 19.9	-56 54 59	-0.003	+0.02	7.3	0.2	0.6	A0 V		160 s		
38084	77478		5 44 21.0	+27 43 53	-0.001	0.00	8.00	0.1	0.6	A0 V		290 s	4308	m
38273	132502		5 44 25.3	-5 27 53	-0.001	-0.01	7.9	0.4	3.0	F2 V		94 s		
38392	170757		5 44 26.4	-22 25 18	-0.023	-0.36	6.15	0.94	6.6	G5 V	-10	8.1 t	4334	m
38393	170759	13 γ Lep	5 44 27.7	-22 26 55	-0.021	-0.37	3.60	0.47	4.1	F6 V	-10	8.1 t	4334	m
38382	170756		5 44 28.3	-20 07 35	-0.002	+0.05	6.34	0.58		G0				
38237	113092		5 44 30.5	+4 20 21	-0.002	-0.01	7.8	0.1	1.7	A3 V		160 s		
38094	58385		5 44 30.8	+30 29 17	+0.006	0.00	7.35	0.87	4.7	G2 V		24 s		
38182	94816		5 44 31.3	+15 03 49	0.000	-0.02	7.50	0.6	4.4	G0 V	+48	41 s	4323	m
38292	132504		5 44 31.5	-4 41 45	-0.002	0.00	7.2	0.1	0.6	A0 V		210 s		
38312	132505		5 44 33.0	-6 51 52	0.000	0.00	6.78	0.17	1.4	A2 V		100 s		
38116	77483		5 44 37.6	+29 01 00	-0.001	0.00	7.87	0.20		B5			4314	m
38142	77485		5 44 37.8	+24 54 58	0.000	-0.03	7.77	1.02	0.3	G8 III	+22	280 s		
38368	150756		5 44 39.1	-12 46 35	-0.002	+0.02	7.8	0.1	0.6	A0 V		280 s		
38221	94819		5 44 41.3	+12 24 55	0.000	-0.02	7.8	0.8	3.2	G5 IV		83 s		
38270	113093		5 44 43.4	+3 49 52	0.000	-0.02	7.5	0.0	0.4	B9.5 V		260 s	4329	m
39014	249346	δ Dor	5 44 46.5	-65 44 08	-0.004	+0.01	4.35	0.21	2.4	A7 V	-3	24 s		
38426	170770		5 44 50.6	-21 39 35	-0.001	0.00	6.70	0.00	-1.6	B3.5 V	+19	450 s	4339	m
37837	25349		5 44 50.6	+57 14 36	+0.001	-0.01	7.8	1.1	-0.1	K2 III		350 s		
38113	58391		5 44 52.6	+33 43 22	0.000	-0.03	7.4	0.8	3.2	G5 IV		68 s		
38219	94823		5 44 53.2	+16 05 10	0.000	-0.01	6.8	0.1	0.6	A0 V		170 s		
38161	77494		5 44 55.4	+26 20 34	-0.001	0.00	7.70	0.00	0.4	B9.5 V		280 s	4324	m
38092	58388		5 44 56.9	+38 29 55	+0.001	-0.04	7.5	0.9	0.2	G9 III	+30	270 s		
38153	77495		5 44 59.0	+28 11 41	+0.001	+0.01	8.00	0.1	0.6	A0 V		290 s		
38291	113097		5 45 00.4	+6 21 01	-0.002	-0.01	7.18	-0.07	0.0	B8.5 V		270 s		
38263	94827		5 45 01.2	+12 53 19	-0.002	0.00	6.6	0.1	1.7	A3 V		94 s		
38309	113099		5 45 01.8	+4 00 29	+0.001	-0.02	6.09	0.31	1.7	F0 IV	+8	75 s	4333	m
38631	217510		5 45 05.1	-41 46 54	+0.003	+0.02	7.2	0.2	1.7	A3 V		110 s		
38612	217509		5 45 06.5	-40 11 57	+0.001	-0.01	7.5	0.9	0.2	K0 III		290 s		
38141	58397		5 45 06.6	+33 37 36	-0.001	-0.06	6.8	0.5	3.4	F5 V		48 s		
--	58396		5 45 07.2	+35 07 31	0.000	-0.01	7.2	2.2						
38247	94830		5 45 11.5	+18 42 16	0.000	0.00	6.62	1.62	-6.1	G8 Iab		1700 s		
38536	196136		5 45 13.1	-31 40 10	-0.001	0.00	6.8	0.6	3.0	F2 V		43 s		
38509	170789		5 45 20.2	-23 50 08	-0.006	+0.01	8.0	0.5	0.2	K0 III		150 mx		
38308	94835		5 45 22.5	+12 47 23	0.000	0.00	7.44	0.77		G0				
38643	196142		5 45 24.0	-39 18 44	+0.001	+0.04	7.0	0.7	2.1	A5 V		47 s		
38350	113107		5 45 25.4	+6 17 52	+0.001	0.00	7.21	0.06	1.4	A2 V		140 s		
38455	150771		5 45 25.6	-12 26 04	0.000	-0.11	7.6	0.9	3.2	G7 IV	-33	77 s		
38200	58404		5 45 29.6	+31 19 28	+0.001	-0.03	6.46	1.14		K0				
38873	234123		5 45 30.4	-55 34 40	-0.002	0.00	7.1	0.4	2.6	F0 V		68 s		
38567	170796		5 45 30.6	-27 29 07	+0.001	-0.02	7.5	2.1	-0.5	M2 III		210 s		
38232	77506		5 45 31.7	+29 17 54	0.000	-0.01	7.41	0.69	-2.0	F2 II		490 s		
38261	77513		5 45 32.9	+25 06 59	0.000	-0.03	7.46	1.16	-0.1	K2 III		330 s		
38059	40579		5 45 35.2	+48 59 07	+0.005	+0.01	8.0	1.1	0.2	K0 III		210 mx		

HD	SAO	Star Name	α 2000	δ 2000	μ(α)	μ(δ)	V	B-V	M_v	Spec	RV	d(pc)	ADS	Notes
37995	25357		5 45 36.0	+54 05 18	0.000	0.00	8.0	0.9	3.2	G5 IV		92 s		
38130	40582		5 45 36.3	+42 51 49	0.000	-0.01	8.0	1.4	-0.3	K5 III	+12	410 s		
38307	77516		5 45 39.3	+20 41 42	0.000	0.00	6.95	3.03		N2	+17			Y Tau, v
38610	196146		5 45 40.7	-31 30 55	-0.003	+0.01	7.6	1.1	2.6	F0 V		32 s		
38199	58406		5 45 47.7	+37 39 32	0.000	-0.01	7.47	1.74	-0.5	M2 III		320 s		
38336	94839		5 45 49.2	+18 47 32	0.000	-0.01	7.9	0.0	0.4	B9.5 V		300 s		
38189	40584		5 45 49.4	+40 30 26	-0.003	-0.01	6.5	0.1	1.7	A3 V	-4	89 s		
38792	217517		5 45 49.6	-47 00 24	-0.003	-0.02	7.6	-0.4	2.6	F0 V		100 s		
38104	40583	27 o Aur	5 45 54.0	+49 49 35	-0.001	0.00	5.47	0.03		A0 p	-6	20 mn		
38579	170799		5 45 54.5	-23 08 25	-0.001	-0.04	7.3	1.3	0.2	K0 III		180 s		
37857	13601		5 45 54.9	+63 17 45	0.000	-0.05	7.6	0.4	2.6	F0 V		98 s		
38306	77523		5 45 55.2	+25 54 48	-0.001	-0.02	7.7	0.0	0.4	B9.5 V		280 s	4343	m
38319	77526		5 45 58.6	+23 44 50	0.000	-0.02	7.9	0.1	0.6	A0 V		270 s		
38666	196149	μ Col	5 45 59.9	-32 18 23	0.000	-0.02	5.17	-0.28	-4.4	O9.5 V	+110	840 s		
39062	249350		5 46 01.2	-61 13 47	0.000	+0.02	7.40	1.30		K0				
38532	150784		5 46 01.4	-12 25 24	-0.002	-0.02	7.8	1.4	-0.3	K5 III		410 s		
38230	58412		5 46 01.7	+37 17 05	+0.040	-0.51	7.36	0.83	6.1	K1 V	-31	16 ts		m
36737	5626		5 46 02.4	+79 37 13	+0.017	-0.05	7.9	0.9	3.2	G5 IV		87 s		
38495	132515		5 46 02.7	-4 16 07	0.000	-0.02	6.25	1.03	0.0	K1 III		180 s	4361	m
38364	77530		5 46 03.8	+21 17 07	-0.001	-0.05	7.9	0.1	1.4	A2 V		200 s		
38188	40588		5 46 11.5	+44 46 59	0.000	-0.02	7.93	0.14	0.6	A0 V	+7	270 s		
37856	13602		5 46 13.1	+64 46 12	+0.005	+0.04	6.8	1.1	0.2	K0 III		200 s		
38246	58414		5 46 13.1	+39 04 30	0.000	-0.03	7.6	0.9	3.2	G5 IV		75 s		
38628	170809		5 46 13.9	-23 38 41	0.000	+0.01	6.9	1.3	-0.5	M4 III		310 s		
38452	94846		5 46 24.2	+12 31 15	+0.001	-0.01	7.9	0.0	0.4	B9.5 V		300 s		
38179	40591		5 46 25.0	+47 54 10	+0.004	-0.01	7.98	0.10	0.4	B9.5 V		120 mx		
38333	58419		5 46 26.8	+32 38 59	+0.002	-0.02	7.8	1.1	0.2	K0 III		310 s		
38871	217521		5 46 27.3	-46 35 49	0.000	+0.02	5.31	1.04	0.2	gG9	+11	96 s		m
38973	234128		5 46 28.0	-53 13 09	+0.005	-0.13	6.6	0.7	4.4	G0 V		23 s		
38178	40592		5 46 29.5	+49 07 35	0.000	-0.01	7.8	0.4	2.6	F0 V		110 s		
37393	5641		5 46 29.7	+74 36 37	+0.041	-0.18	7.3	0.6	4.4	G0 V	+25	38 s		
38091	25362	26 Cam	5 46 30.4	+56 06 57	+0.003	-0.06	5.94	0.17	1.9	A4 V n	+26	64 s		
38593	150793		5 46 30.6	-11 01 33	+0.003	-0.02	8.0	1.4	-0.3	K5 III		450 s		
38609	150794		5 46 30.9	-12 43 44	+0.001	0.00	8.0	0.0	0.4	B9.5 V		330 s		
38575	132524		5 46 32.9	-6 56 41	+0.002	-0.02	7.9	0.5	3.4	F5 V		78 s		
38529	113119		5 46 34.9	+1 10 05	-0.005	-0.15	5.95	0.78	5.0	dG4	+29	18 mn		
38494	94849		5 46 40.3	+10 55 55	-0.001	-0.03	7.3	1.1	-0.1	K2 III		280 s		
38805	196159		5 46 41.9	-36 13 54	-0.002	-0.03	6.9	0.5	0.6	A0 V		87 s		
38528	113123		5 46 43.1	+2 42 26	0.000	-0.01	8.0	0.1	0.6	A0 V		300 s		
38478	94848	129 Tau	5 46 45.4	+15 49 21	0.000	-0.01	5.90	-0.06	-1.6	B7 III p		300 s		
38465	77544		5 46 45.7	+20 10 25	-0.002	-0.01	8.0	0.4	3.0	F2 V		97 s		
38216	40597		5 46 45.9	+48 40 10	+0.001	-0.01	7.7	0.9	3.2	G5 IV		80 s		
38526	113120		5 46 49.7	+9 39 47	-0.001	-0.02	7.9	1.1	0.2	K0 III		320 s		
41217	—		5 46 49.8	-82 26 21			7.6	0.1	1.4	A2 V		170 s		
38258	40600		5 46 51.4	+47 28 00	+0.001	-0.05	7.46	0.02	0.0	B8.5 V	-3	270 s		
38527	113124		5 46 52.1	+9 31 20	-0.002	-0.06	5.79	0.88	0.3	G8 III	-26	130 s	4369	m
38772	170829		5 46 52.6	-28 53 25	-0.005	-0.12	7.6	0.6	3.4	F5 V		56 s		
38129	25364		5 46 54.7	+56 55 26	+0.003	-0.02	6.8	0.1	0.6	A0 V	+20	170 s		
38477	77547		5 46 56.6	+20 16 55	-0.001	0.00	7.2	1.1	0.2	K0 III	+11	240 s		
38678	150801	14 ζ Lep	5 46 57.2	-14 49 20	-0.002	0.00	3.55	0.10	1.7	A3 V	+20	24 ts		
38651	132529		5 46 57.5	-9 50 20	0.000	-0.02	7.9	1.1	-0.1	K2 III		400 s		
38940	217530		5 46 57.9	-45 38 49	+0.007	+0.14	7.40	0.50	2.9	F6 IV-V		74 s		
38699	150803		5 46 58.7	-16 38 47	0.000	-0.01	7.60	1.4	-0.3	K4 III	+30	380 s		
38229	25371		5 47 00.0	+51 31 16	-0.001	-0.03	6.5	0.9	3.2	G5 IV		45 s		
38939	217531		5 47 00.3	-44 48 00	-0.002	-0.01	8.00	0.1	1.4	A2 V		210 s		m
38402	58426		5 47 00.5	+34 18 42	0.000	0.00	7.7	0.0	0.0	B8.5 V		330 s		
38804	170836		5 47 04.6	-28 38 21	0.000	+0.01	6.0	0.1	0.0	B8.5 V		130 s		
38713	150806		5 47 07.6	-16 14 16	-0.003	0.00	6.3	0.6	4.4	G0 V		24 s		
38450	77548		5 47 09.9	+29 25 51	0.000	-0.01	8.0	0.0	0.4	B9.5 V		320 s		
38128	25366		5 47 10.2	+58 47 06	+0.002	-0.01	6.5	1.1	0.2	K0 III		180 s		
38559	94856		5 47 11.4	+12 20 23	-0.002	-0.02	7.1	1.1	0.2	K0 III	+7	230 s		
38758	170834		5 47 12.5	-21 32 36	-0.001	+0.01	7.8	1.7	-0.5	M2 III		430 s		
39110	234137		5 47 13.0	-54 21 39	-0.002	-0.01	6.18	1.40		K5				
38545	94855	131 Tau	5 47 13.1	+14 29 18	+0.001	-0.04	5.70	0.04	1.7	A3 V	+21	63 s		
38712	150808		5 47 13.9	-12 48 17	0.000	0.00	7.7	1.6	-0.5	M2 III		430 s		
38463	77551		5 47 14.1	+28 37 26	-0.002	-0.01	8.00	0.00	0.4	B9.5 V		310 s		m
38441	58431		5 47 14.3	+31 39 34	+0.001	-0.06	7.64	0.03	0.6	A0 V		62 mx		
38358	40609		5 47 14.7	+42 31 36	+0.001	-0.08	6.29	1.35		K0	-16			
39060	234134	β Pic	5 47 17.1	-51 03 59	+0.001	+0.09	3.85	0.17	0.3	A5 III	+28	24 ts		
38127	25367		5 47 17.3	+58 58 31	+0.002	-0.06	7.5	0.5	3.4	F5 V		66 s		
38197	25373		5 47 17.5	+55 38 52	-0.001	-0.06	8.0	1.1	0.2	K0 III		300 mx		
38885	196171		5 47 18.6	-35 40 29	-0.001	+0.04	6.32	1.18		K0				
38286	25376		5 47 19.9	+50 05 19	+0.004	-0.01	7.8	1.1	0.2	K0 III		310 s		
38676	132533		5 47 20.9	-5 35 43	-0.001	-0.03	8.0	1.1	0.2	K0 III		360 s		
38827	170841		5 47 21.8	-27 08 03	+0.002	-0.01	7.3	0.7	4.0	F8 V		33 s		

HD	SAO	Star Name	α 2000	δ 2000	μ(α)	μ(δ)	V	B-V	M$_v$	Spec	RV	d(pc)	ADS	Notes
38921	196174		5^h 47^m 22.1^s	-38°13'53"	-0.001^s	-0.02"	7.6	-0.3	0.6	A0 V		260 s		
38491	77555		5 47 24.8	+29 39 28	0.000	-0.11	8.00	0.5	4.0	F8 V		63 s	4371	m
38558	94858	130 Tau	5 47 26.1	+17 43 44	0.000	-0.01	5.49	0.30	2.6	F0 V	+9	35 s		
38543	94857		5 47 26.6	+18 43 46	+0.001	-0.03	7.77	1.09	0.2	K0 III		290 s		
38920	196175		5 47 26.7	-36 29 17	-0.002	-0.05	7.10	1.1	-0.1	K2 III		280 s		m
38735	150814		5 47 26.7	-10 32 00	-0.002	-0.03	6.03	0.16		A3				
38757	150817		5 47 27.4	-15 15 11	0.000	-0.04	7.0	0.0	0.4	B9.5 V		210 s		
38525	77559		5 47 29.3	+22 55 21	0.000	0.00	8.0	1.1	0.2	K0 III		340 s		
38989	217537		5 47 29.4	-41 35 26	-0.001	-0.02	6.80	1.56		M5 III				
38650	113142		5 47 29.9	+4 06 01	0.000	+0.01	7.67	-0.06	0.4	B9.5 V		280 s		
38058	13608		5 47 32.7	+63 35 55	+0.003	0.00	6.6	0.8	3.2	G5 IV		47 s		
38524	77560		5 47 34.6	+25 34 07	0.000	-0.01	6.42	1.19	0.0	K1 III	-18	170 s		
38257	25377		5 47 37.0	+54 29 40	-0.003	-0.02	6.5	1.1	0.2	K0 III		180 s		
38674	113145		5 47 39.8	+4 51 08	+0.001	-0.03	8.0	1.1	-0.1	K2 III		390 s		
38621	94863		5 47 40.1	+15 01 52	+0.002	-0.01	8.0	0.2	2.1	A5 V		150 s		
38622	94864	133 Tau	5 47 42.9	+13 53 58	+0.001	-0.01	5.29	-0.17	-2.5	B2 V	+28	350 s	4381	m
38971	196180		5 47 43.8	-36 40 20	+0.006	-0.08	8.0	1.5	0.2	K0 III		130 mx		
38770	132541		5 47 45.3	-9 31 15	0.000	+0.02	7.8	1.1	0.2	K0 III		330 s		
38771	132542	53 κ Ori	5 47 45.3	-9 40 11	0.000	-0.01	2.06	-0.17		B0.5 Ia	+21	21 mn		Saiph
38755	132539		5 47 46.7	-6 26 06	-0.001	0.00	7.68	-0.11	-0.9	B6 V		520 s		
38734	132537		5 47 47.5	-2 17 51	+0.001	-0.02	8.00	1.1		K0 III		360 s	4391	m
39810	256239		5 47 48.9	-72 42 09	+0.005	+0.02	6.53	1.08		K0				
38695	113149		5 47 54.8	+7 57 38	-0.002	0.00	7.90	0.5	4.0	F8 V		60 s	4388	m
38584	77569		5 47 56.1	+24 41 12	0.000	-0.01	7.20	1.1	-0.1	K2 III	+15	280 s		m
38503	58442		5 47 56.1	+35 09 35	-0.002	+0.01	6.60	0.5	-3.3	F8 Ib-II		830 s	4377	m
38605	77571		5 47 57.7	+24 39 41	0.000	-0.01	8.0	1.1	0.0	K1 III		370 s		
39040	217543		5 47 58.0	-40 39 09	0.000	+0.03	6.4	1.6	0.2	K0 III		76 s		
38710	113150	52 Ori	5 48 00.2	+6 27 15	0.000	-0.02	5.27	0.23	2.1	dA5 n	+42	43 s	4390	m
38986	196184		5 48 01.0	-34 28 12	+0.003	+0.01	7.6	1.5	0.2	K0 III		160 s		
38672	94868		5 48 02.0	+12 25 07	-0.001	0.00	6.68	-0.10	-1.0	B5.5 V	+27	320 s	4386	m
38824	132549		5 48 10.4	-8 23 01	0.000	0.00	7.29	-0.11	0.4	B9.5 V		240 s	4402	m
38583	58451		5 48 11.5	+30 32 06	+0.002	-0.03	7.00	1.50		K4	+24		4383	m
39177	217547		5 48 13.1	-48 55 05	-0.002	+0.02	7.50	0.1	0.6	A0 V		240 s		
38449	40612		5 48 14.6	+46 43 19	+0.002	-0.01	7.1	1.1	0.2	K0 III		220 s		
38767	113155		5 48 14.9	+3 53 56	-0.004	-0.06	7.80	0.5	4.0	F8 V		58 s	4395	m
39027	196189		5 48 19.7	-33 25 53	-0.001	-0.02	6.85	1.38	-0.3	K4 III		270 s		
38935	170857		5 48 20.8	-22 03 12	-0.005	+0.01	8.0	1.0	-0.1	K2 III		180 mx		
39039	196190		5 48 21.6	-34 56 12	-0.001	-0.02	7.28	0.94	0.3	G6 III		240 s		
38670	77578		5 48 22.3	+20 52 10	+0.001	-0.02	6.07	-0.08	-0.6	B7 V	+7	210 s	4392	m
--	217550		5 48 22.6	-48 58 55	-0.001	+0.02	8.0	1.3						
38823	132550		5 48 25.4	-0 45 35	-0.002	0.00	7.7	0.2	2.1	A5 V		130 s		
38904	150823		5 48 27.5	-13 21 35	0.000	-0.05	7.40	1.1	0.2	K0 III		280 s	4410	m
38798	113161		5 48 28.1	+4 42 11	-0.001	-0.04	7.29	0.03	0.6	A0 V		210 s	4401	m
38709	94872		5 48 28.8	+17 26 00	+0.001	-0.01	7.29	0.06	0.4	B9.5 V		200 s		
38779	113160		5 48 30.7	+7 33 05	0.000	-0.01	7.2	0.4	3.0	F2 V		67 s		
39281	234145		5 48 34.1	-53 40 35	+0.006	-0.12	7.7	1.4	0.2	K0 III		190 s		
38858	132554		5 48 34.8	-4 05 40	+0.004	-0.22	5.97	0.64	4.1	G4 IV-V	+29	24 s		
39567	249362		5 48 35.2	-65 10 05	+0.002	-0.02	7.6	1.8	0.2	K0 III		100 s		
38931	150830		5 48 39.4	-14 18 51	-0.003	+0.01	6.9	0.8	3.2	G5 IV		54 s		
38867	132556		5 48 40.9	-1 47 09	0.000	-0.01	7.1	1.1	-0.1	K2 III		280 s		
38753	94874		5 48 44.6	+18 14 34	+0.001	-0.05	7.3	0.4	3.0	F2 V		70 s		
38809	113165		5 48 45.4	+7 52 35	0.000	-0.04	7.9	1.1	0.2	K0 III		320 s		
38866	132557		5 48 45.4	-1 30 07	+0.001	+0.02	7.5	0.2	2.1	A5 V		120 s		
38856	113167		5 48 46.0	+0 43 32	0.000	+0.01	7.25	-0.14	0.0	B8.5 V		280 s		
38797	94875		5 48 47.2	+11 59 49	-0.001	-0.01	6.93	-0.10	0.0	B8.5 V		240 s		
39086	196200		5 48 51.2	-32 25 38	-0.002	-0.02	7.5	0.9	3.2	G5 IV		58 s		
38604	58460		5 48 51.7	+39 32 01	0.000	-0.02	6.9	0.7	4.4	G0 V	+20	32 s		
38751	77592	132 Tau	5 49 00.0	+24 34 03	0.000	-0.03	4.86	1.01	0.3	G8 III	+16	73 s		
39123	196204		5 49 03.9	-31 41 34	+0.001	-0.05	7.7	0.8	3.0	F2 V		48 s		
38284	13618		5 49 04.8	+62 48 30	-0.001	-0.01	6.50	0.1	1.4	A2 V	-6	100 s	4376	m
38750	77593		5 49 06.9	+25 39 02	+0.001	0.00	7.20	1.44	-2.2	K2 II	-9	670 s		
39148	196206		5 49 07.3	-32 30 14	0.000	+0.01	6.80	1.1	0.2	K0 III		210 s		m
38900	113171		5 49 07.6	+4 12 46	-0.002	+0.01	7.82	-0.06	0.4	B9.5 V		310 s		
38656	58465	29 τ Aur	5 49 10.4	+39 10 52	-0.002	-0.02	4.52	0.94	0.3	G8 III	-20	70 s	4398	m
39055	150841		5 49 12.5	-19 23 59	+0.001	+0.01	7.3	0.7	0.6	A0 V		83 s		
38796	77596		5 49 12.7	+21 08 10	0.000	0.00	7.32	-0.07	0.4	B9.5 V		240 s		
38749	77594		5 49 17.4	+29 43 37	+0.002	-0.01	7.8	0.2	2.1	A5 V	-11	140 s		
38808	77597		5 49 20.4	+24 13 22	0.000	0.00	8.00	0.6	-3.3	G3 Ib-II		1500 s		
38686	58472		5 49 21.7	+38 44 17	0.000	-0.04	7.3	1.1	-0.1	K2 III	+31	290 s		
39280	217557		5 49 22.5	-44 40 57	+0.001	+0.02	7.75	0.94	3.2	G8 IV		75 s		
39176	170883		5 49 23.4	-29 33 40	-0.001	-0.02	8.0	1.5	0.2	K0 III		190 s		
38835	94883		5 49 25.1	+18 37 05	+0.002	-0.04	7.6	0.4	2.6	F0 V		100 s		
39235	196213		5 49 26.1	-37 11 06	+0.002	+0.01	7.6	-0.4	0.4	B9.5 V		280 s		
39054	150844		5 49 32.0	-12 13 29	+0.001	+0.01	8.00	1.4	-0.3	K5 III		460 s	4430	m
39000	132570		5 49 32.5	-0 40 55	-0.001	-0.01	7.62	-0.07	0.6	A0 V		250 s		

HD	SAO	Star Name	α 2000	δ 2000	μ(α)	μ(δ)	V	B-V	M_V	Spec	RV	d(pc)	ADS	Notes
38899	94888	134 Tau	5 49 32.9	+12 39 04	-0.001	-0.02	4.91	-0.07	-0.3	B9 IV	+19	110 s		m
39312	217559		5 49 34.0	-44 52 31	-0.001	+0.01	6.3	1.5	-0.1	K2 III		130 s		
38926	94891		5 49 35.3	+10 43 27	+0.001	+0.02	8.0	0.4	3.0	F2 V		96 s		
39174	170887		5 49 36.0	-26 20 57	-0.001	-0.07	8.0	0.4	0.6	A0 V		66 mx		
39608	249364		5 49 36.4	-60 40 35	+0.001	+0.01	7.2	2.4	-0.1	K2 III		52 s		
39070	150845		5 49 36.4	-14 29 01	-0.002	-0.05	5.49	0.88	0.3	gG6	-2	110 s	4432	m
38807	77604		5 49 38.1	+27 33 03	+0.001	-0.03	7.2	1.1	0.2	K0 III	+48	250 s		
38731	58478		5 49 41.5	+38 41 51	+0.001	-0.01	7.9	0.2	2.1	A5 V		140 s		
39008	132574		5 49 42.8	-0 21 05	0.000	-0.02	7.40	1.1	-0.2	K3 III	-12	330 s		
39523	234154	γ Pic	5 49 49.6	-56 10 00	+0.009	-0.07	4.51	1.10	0.0	K1 III	+16	80 s		
39353	217563		5 49 50.8	-43 21 52	-0.003	+0.03	7.3	0.0	0.0	B8.5 V		250 s		
39190	170892		5 49 53.4	-22 58 18	-0.001	+0.02	5.87	0.06		A2	+44			
39844	249368	ε Dor	5 49 53.6	-66 54 05	-0.003	+0.01	5.11	-0.14		B5	+16	18 mn		
39241	196219		5 49 54.1	-30 37 16	+0.001	0.00	6.7	1.0	3.2	G5 IV		40 s		
39463	234152		5 49 55.5	-51 02 20	+0.002	+0.01	7.7	0.2	0.2	K0 III		310 s		
38819	58484		5 49 55.5	+31 47 08	+0.002	-0.02	7.00	0.1	1.7	A3 V		110 s	4421	m
39172	170891		5 49 56.0	-20 20 36	+0.001	0.00	8.0	1.3	0.2	K0 III		250 s		
39033	113182		5 49 58.5	+0 09 12	+0.002	-0.02	7.7	0.0	0.4	B9.5 V		290 s		
39007	113179		5 50 02.5	+9 52 16	0.000	0.00	5.80	0.87	0.3	G7 III	+44	130 s		
39295	196221		5 50 03.1	-34 33 13	0.000	+0.01	7.30	1.4	-0.3	K5 III		330 s		m
39104	132577		5 50 03.2	-5 53 54	-0.001	-0.04	8.00	0.9	3.2	G5 IV		91 s		m
38960	94897		5 50 08.3	+19 09 58	-0.001	-0.01	7.8	0.2	2.1	A5 V		140 s		
39294	196225		5 50 09.1	-32 48 32	-0.003	-0.01	6.8	0.4	0.6	A0 V		100 s		
39565	234158		5 50 11.7	-55 04 52	-0.001	+0.08	7.5	0.5	2.6	F0 V		74 s		
39051	113186		5 50 13.1	+4 25 24	+0.001	-0.04	5.97	1.36	-0.2	K3 III	+27	150 s		
39577	234160		5 50 16.3	-55 03 13	-0.001	+0.06	7.4	0.7	-0.1	K2 III		310 mx		
40953	256248	κ Men	5 50 16.6	-79 21 41	-0.007	+0.07	5.47	-0.08		B9 n	+5			
39755	249366		5 50 18.8	-62 04 49	-0.004	-0.04	7.82	0.40	3.0	F2 V		86 s		
39488	217566		5 50 23.3	-48 59 10	0.000	+0.02	7.2	-0.3	3.2	G5 IV		63 s		
39082	113191		5 50 23.8	+4 57 23	0.000	-0.02	7.42	0.02	0.4	B9.5 V		230 s		
39102	113195		5 50 25.4	+3 12 06	-0.001	+0.02	7.9	1.4	-0.3	K5 III		430 s		
37879	5665		5 50 26.9	+75 42 53	-0.002	-0.02	7.9	0.5	4.0	F8 V		60 s		
39368	196231		5 50 28.4	-37 37 17	-0.001	+0.02	7.00	1.4	-0.3	K5 III		290 s		m
39547	234162		5 50 28.5	-52 46 04	-0.003	+0.02	6.35	0.76	3.4	F5 V		25 s		
39019	94904	135 Tau	5 50 28.8	+14 18 21	0.000	-0.04	5.52	1.01	0.2	K0 III	+46	110 s		
39118	113198		5 50 29.9	+2 01 28	0.000	-0.02	5.98	0.91	4.4	G0 V	+7	13 s		
39018	94903		5 50 32.3	+18 01 34	+0.001	-0.01	7.72	-0.08	0.4	B9.5 V		290 s		
38618	25403	29 Cam	5 50 34.0	+56 55 08	-0.001	-0.01	6.50	0.1	1.4	A2 V	+4	100 s	4412	m
38817	40642		5 50 37.1	+44 00 44	-0.005	+0.04	7.5	0.1	1.4	A2 V	+33	160 s		
39169	132583		5 50 37.7	-1 25 48	0.000	-0.04	7.85	1.06	0.2	gK0	+3	310 s	4442	m
39902	249373		5 50 39.0	-65 15 38	0.000	+0.01	7.91	0.05	0.6	A0 V		260 s		
38980	77619		5 50 39.7	+27 30 16	-0.001	-0.06	7.2	0.5	3.4	F5 V		56 s		
38850	40644		5 50 44.9	+43 41 21	+0.002	0.00	8.0	0.0	0.4	B9.5 V		310 s		
39005	77622		5 50 47.7	+26 03 16	+0.001	-0.03	7.9	0.1	1.7	A3 V		170 s		
38998	77621		5 50 48.2	+27 41 10	+0.001	0.00	7.49	1.68	-0.5	M2 III	+28	350 s		
39098	94914		5 50 48.6	+14 26 35	0.000	-0.02	6.75	-0.07	0.4	B9.5 V		190 s	4441	m
39099	94916		5 50 51.3	+14 02 53	+0.001	+0.01	6.60	1.06	0.0	K1 III	-48	210 s		
39640	234169		5 50 53.3	-52 06 32	+0.001	-0.08	5.17	0.99	0.3	G8 III	+1	89 s		
38765	25411		5 50 56.3	+51 30 52	+0.018	-0.04	6.29	1.05	0.2	K1 III	+26	66 mx		
39425	196240	β Col	5 50 57.5	-35 46 06	+0.004	+0.40	3.12	1.16	-0.1	K2 III	+89	44 s		Wazn
39004	77625		5 50 58.0	+27 58 04	-0.001	+0.01	5.56	0.97	0.2	K0 III	+8	120 s	4474	m
40156	249378		5 51 00.2	-69 54 09	-0.005	-0.06	7.80	0.4	3.0	F2 V		91 s		m
38475	13627		5 51 01.0	+65 45 11	-0.001	-0.03	6.90	0.5	4.0	F8 V		38 s	4405	m
38944	58496	31 υ Aur	5 51 02.4	+37 18 20	+0.003	-0.04	4.74	1.62	-0.5	gM1	+38	110 s		
39156	94924		5 51 03.1	+11 29 56	-0.001	+0.01	6.9	0.9	3.2	G5 IV		55 s		
39097	94920		5 51 04.7	+18 33 06	+0.001	0.00	6.9	1.1	0.2	K0 III		210 s		
39443	196244		5 51 16.8	-32 17 56	-0.002	-0.01	8.0	1.0	0.2	K0 III		360 s		
39364	170926	15 δ Lep	5 51 19.2	-20 52 45	+0.016	-0.65	3.81	0.99	0.3	G8 III	+99	48 s		
39319	150869		5 51 21.7	-11 38 47	-0.001	-0.01	7.30	0.00	0.4	B9.5 V		240 s	4458	m
39291	132591	55 Ori	5 51 21.9	-7 31 05	0.000	0.00	5.35	-0.20	-3.6	B2 III	+20	620 s		
39963	249376		5 51 23.1	-64 02 01	-0.002	+0.02	6.36	0.86	0.2	K0 III		170 s		
39045	58504		5 51 25.6	+32 07 30	0.000	+0.01	6.25	1.75	-0.5	M3 III	+103	180 s	4443	m
39168	94927		5 51 27.8	+18 26 38	0.000	-0.03	7.1	0.9	3.2	G5 IV		59 s		
39385	170931		5 51 28.6	-22 55 34	0.000	-0.02	6.17	1.02		K0				
39003	58502	32 ν Aur	5 51 29.3	+39 08 55	0.000	+0.01	3.97	1.13	0.2	K0 III	+10	45 s	4440	m
39602	--		5 51 37.3	-41 40 20			8.01	0.83	5.2	G5 V		30 s		m
39167	77642		5 51 40.1	+22 54 53	+0.002	-0.04	7.8	1.1	0.2	K0 III		310 s		
39115	58515		5 51 43.5	+30 57 25	+0.001	-0.02	7.4	0.1	0.6	A0 V		220 s		
39361	150877		5 51 45.1	-10 26 12	-0.001	-0.03	8.0	1.4	-0.3	K5 III		450 s		
39601	217577		5 51 45.6	-40 08 49	-0.001	0.00	7.9	1.0	3.2	G5 IV		88 s		
39442	150882		5 51 48.1	-19 23 13	-0.001	+0.02	7.5	1.7	-0.3	K5 III		280 s		
39184	77647		5 51 50.6	+23 22 58	-0.001	-0.02	7.0	1.4	-0.3	K5 III		270 s		
39227	94934		5 51 51.9	+19 31 17	-0.001	-0.01	7.31	-0.03	0.4	B9.5 V		240 s		
39484	170941		5 51 54.1	-25 13 09	-0.002	-0.02	7.1	0.5	1.4	A2 V		79 s		
39654	217583		5 51 55.4	-41 06 19	+0.001	+0.01	6.9	1.5	-0.3	K5 III		280 s		

HD	SAO	Star Name	α 2000	δ 2000	μ(α)	μ(δ)	V	B-V	M$_v$	Spec	RV	d(pc)	ADS	Notes
39065	58512		5^{h}51^{m}56^s.4	+39°33'46"	+0^s.002	-0".05	7.9	0.1	1.4	A2 V		72 mx		
39376	132599		5 51 57.2	-7 18 36	+0.001	0.00	7.88	-0.08	0.4	B9.5 V		310 s	4462	m
39543	170945		5 51 59.5	-29 26 55	-0.001	+0.02	6.45	1.48		K0				
39441	150884		5 52 01.1	-15 30 21	-0.004	+0.02	7.2	0.1	0.6	A0 V		140 mx		
39403	132602		5 52 01.7	-9 48 46	-0.003	-0.01	7.4	0.4	2.6	F0 V		91 s		
39621	196258		5 52 04.2	-36 17 19	-0.002	-0.07	8.0	1.7	0.2	K0 III		140 s		
39421	132603		5 52 07.5	-9 02 29	-0.002	+0.05	5.97	0.10		A0				
39533	170946		5 52 07.7	-25 56 35	+0.015	+0.05	6.86	0.92		G5				
38655	13632		5 52 07.8	+65 09 30	0.000	+0.01	7.7	1.1	0.2	K0 III		290 s		
39076	40663		5 52 10.0	+40 57 47	+0.004	-0.03	7.9	0.5	4.0	F8 V		59 s		
37463	906		5 52 10.9	+80 36 31	-0.006	0.00	8.0	0.1	1.4	A2 V		210 s		
39482	150887		5 52 11.4	-17 33 10	-0.001	+0.10	7.8	0.8	3.2	G5 IV		84 s		
39500	150890		5 52 12.5	-19 03 14	-0.001	+0.01	7.2	1.4	0.2	K0 III		150 s		
39114	58520		5 52 13.8	+38 33 36	+0.001	-0.03	7.10	0.00	0.0	B8.5 V		250 s	4452	m
38374	5670		5 52 15.7	+72 28 40	-0.015	+0.02	7.8	0.5	4.0	F8 V		57 s		
39344	113217		5 52 16.9	+7 19 46	+0.003	-0.02	7.8	1.1	0.2	K0 III		310 s		
38831	25419	30 Cam	5 52 17.3	+58 57 51	+0.001	-0.02	6.0	0.0		B9				
38438	5673		5 52 17.6	+71 17 22	+0.004	+0.01	7.2	0.1	1.7	A3 V		130 s		
39919	234180		5 52 17.6	-56 27 46	0.000	0.00	7.4	1.0	0.2	K0 III		270 s		
39937	234181		5 52 20.3	-57 09 22	+0.003	-0.08	5.94	0.66		F5				
39440	132607		5 52 21.7	-8 57 41	+0.001	+0.02	8.0	1.1	0.2	K0 III		360 s		
39317	94945	137 Tau	5 52 22.2	+14 10 18	-0.001	-0.01	5.60	-0.05		B9 p	-4			
248587	94943		5 52 22.8	+19 08 59	0.000	0.00	7.94	0.65	-6.6	A0 Iab		3500 s		
39044	--		5 52 23.3	+46 47 56			7.90	1.1	0.2	K0 III		330 s	4449	m
39286	94942		5 52 23.4	+19 52 05	0.000	0.00	6.0	0.5		B9 comp				
39575	170953		5 52 23.8	-26 17 29	0.000	-0.01	8.0	0.5	0.6	A0 V		140 s		
39875	234179		5 52 24.3	-52 03 19	-0.001	-0.02	7.0	1.2	-0.5	M2 III		320 s		
39400	113220	56 Ori	5 52 26.4	+1 51 19	-0.001	-0.01	4.78	1.38	-2.2	K2 II	+10	240 s	4467	m
39374	113219		5 52 26.8	+6 12 33	-0.001	0.00	7.0	0.9	3.2	G5 IV		57 s		
39438	132608		5 52 29.3	-2 17 07	-0.008	-0.01	7.9	0.4	3.4	F5 V		78 s		
39823	217591		5 52 30.0	-49 05 05	+0.003	-0.06	7.9	0.7	3.2	G5 IV		68 s		
39305	94948		5 52 30.9	+19 45 17	-0.002	-0.02	7.6	1.1	0.2	K0 III		290 s		
39514	150896		5 52 32.3	-14 33 24	-0.002	+0.02	8.0	0.9	3.2	G5 IV		93 s		
39720	196266		5 52 33.1	-37 37 52	+0.002	-0.03	5.63	1.04	0.2	K0 III	+32	120 s		m
39182	58524		5 52 39.4	+39 34 28	-0.002	-0.02	6.50	0.09	0.0	A3 III	-19	200 s		
39225	58528		5 52 39.9	+33 55 02	+0.001	+0.01	5.98	1.60	-2.4	M1 II	+100	470 s		
39752	196274		5 52 47.6	-38 31 33	0.000	0.00	6.80	1.1	0.2	K0 III		210 s		m
39152	40670		5 52 47.8	+44 22 41	+0.003	-0.01	7.9	1.1	0.2	K0 III		320 s		
39718	196270		5 52 49.0	-34 33 39	-0.003	-0.06	6.6	0.8	4.4	G0 V		20 s		
39706	196268		5 52 50.1	-32 20 21	-0.002	+0.02	7.8	0.8	3.2	G5 IV		81 s		
39251	58531		5 52 52.6	+34 26 42	-0.002	-0.08	7.90	0.5	4.0	F8 V		60 s	4463	m
38645	13635		5 52 55.6	+68 28 18	+0.003	-0.04	6.20	0.95	0.2	G9 III	-1	160 s		
39358	77667		5 52 58.2	+20 46 40	+0.001	-0.02	7.7	0.0	0.4	B9.5 V		280 s		
39735	196273		5 53 01.3	-30 34 19	+0.002	0.00	7.9	1.8	-0.3	K5 III		300 s		
39764	196276	λ Col	5 53 06.8	-33 48 05	-0.001	+0.03	4.87	-0.15	-1.1	B5 V	+30	160 s		
39918	217598		5 53 09.4	-47 57 20	-0.001	+0.01	6.60	1.1	-0.1	K2 III		220 s		m
39456	94960		5 53 12.6	+11 48 52	+0.002	-0.10	7.2	0.5	3.4	F5 V		58 s		
40201	249381		5 53 14.9	-61 50 23	+0.001	-0.01	7.50	0.1	0.6	A0 V		240 s		m
39479	113234		5 53 17.9	+9 33 32	-0.002	-0.02	7.7	1.1	-0.1	K2 III		340 s		
39417	77680		5 53 19.0	+20 17 57	0.000	-0.01	6.5	-0.1		B9	-6			
39357	77675	136 Tau	5 53 19.6	+27 36 44	0.000	-0.01	4.58	-0.02	0.4	B9.5 V	-16	64 s		
38924	13642		5 53 20.2	+61 40 35	+0.001	+0.01	8.0	0.1	1.4	A2 V		190 s		
39901	217599		5 53 22.8	-42 55 17	0.000	+0.01	6.3	1.4	0.2	K0 III		89 s		
39029	25430		5 53 25.8	+57 57 50	+0.004	-0.04	7.1	1.1	-0.1	K2 III		220 mx		
39529	113239		5 53 26.2	+6 15 18	-0.003	-0.01	7.3	0.7	4.4	G0 V		38 s		
39455	94961		5 53 26.8	+18 10 12	-0.001	-0.01	7.4	0.4	-2.0	F2 II		700 s		
39917	217600		5 53 27.4	-43 33 33	0.000	-0.08	7.91	0.81		G0				
39315	58535		5 53 28.7	+37 20 20	-0.003	+0.07	6.90	0.5	3.4	F5 V		50 s	4472	m
--	--		5 53 29.7	+42 49 01	0.000	0.00	8.0	1.1	0.2	K0 III		340 s		
40308	249385		5 53 30.1	-63 19 00	0.000	-0.01	8.0	0.1	0.6	A0 V		300 s		
39416	77683		5 53 30.3	+25 04 24	0.000	-0.01	7.7	0.7	-3.3	G3 Ib-II		1300 s		
39688	150919		5 53 30.8	-16 15 56	-0.003	-0.11	8.0	0.5	3.8	F7 V	+3	68 s		
39436	77684		5 53 33.6	+24 17 12	0.000	-0.01	7.8	0.0	0.0	B8.5 V		350 s		
39634	132619		5 53 34.6	-6 16 09	0.000	+0.02	7.9	1.6	-0.5	M2 III		490 s		
39330	58539		5 53 36.6	+36 07 49	0.000	-0.01	7.3	1.1	0.2	K0 III	+2	250 s		
248712	58544		5 53 39.7	+33 44 16	0.000	-0.06	8.0	0.7	4.4	G0 V		52 s		
39647	132622		5 53 41.6	-5 42 18	-0.001	-0.02	7.09	-0.01	0.4	B9.5 V		210 s		
39962	217604		5 53 51.5	-42 13 49	0.000	+0.10	7.97	0.41	3.0	F2 V		91 s		
39915	196294		5 53 52.9	-37 32 00	-0.002	-0.03	7.4	0.3	1.4	A2 V		110 s		
39733	150922		5 53 54.4	-12 23 38	-0.001	+0.04	8.0	1.1	0.2	K0 III		360 s		
39789	150925		5 53 57.4	-19 38 18	-0.002	0.00	6.5	0.9	0.6	A0 V		42 s		
39761	150924		5 53 58.7	-14 07 36	+0.001	+0.05	7.4	1.1	0.2	K0 III		280 s		
39570	94968		5 53 58.7	+12 25 09	-0.001	-0.27	7.76	0.58		F8				m
39166	25439		5 54 03.5	+57 02 44	+0.002	-0.04	7.9	0.4	2.6	F0 V		110 s		
40307	249388		5 54 04.4	-60 01 24	-0.005	-0.05	7.1	1.2	3.2	G5 IV		35 s		

HD	SAO	Star Name	α 2000	δ 2000	μ(α)	μ(δ)	V	B-V	M_v	Spec	RV	d(pc)	ADS	Notes
248996	94972		5^{h}54^{m}04.7^s	+12°03'50"	-0.001	-0.01	7.8	1.1	0.2	K0 III		310 s		
40409	249390		5 54 06.1	-63 05 23	+0.021	+0.54	4.65	1.05		K0	+25	13 mn		
39932	196296		5 54 07.0	-35 54 12	-0.002	0.00	6.8	0.5	3.0	F2 V		44 s		
39612	94975		5 54 09.6	+10 14 41	+0.002	0.00	7.20	0.9	3.2	G5 IV		62 s	4491	m
40105	234191		5 54 10.6	-50 21 43	+0.008	+0.57	6.52	0.90	6.1	K1 V		17 mn		
40455	249391		5 54 11.9	-64 28 56	-0.001	+0.06	6.63	0.38		F2				
40079	217610		5 54 12.2	-47 11 09	-0.001	-0.02	7.6	0.7	0.2	K0 III		300 s		
39632	94977		5 54 13.3	+10 35 12	0.000	+0.01	6.12	1.47	-2.1	G9 II	+13	250 s		
39891	170993		5 54 14.0	-29 08 52	-0.002	-0.05	6.36	0.37		F2				m
39685	113253		5 54 15.6	+3 13 31	+0.003	-0.06	6.31	1.29		K0	-4			
39477	58556		5 54 15.8	+30 29 38	-0.001	0.00	7.69	0.17	-1.0	B5.5 V	+4	350 s	4483	m
39732	132630		5 54 22.3	-1 04 37	-0.002	-0.01	7.8	1.6	-0.5	M2 III		470 s		
39587	77705	54 χ¹ Ori	5 54 22.9	+20 16 34	-0.013	-0.09	4.41	0.59	4.4	G0 V	-14	9.9 t		
39930	196298		5 54 23.4	-31 08 10	-0.002	+0.04	8.0	1.6	0.2	K0 III		160 s		
39817	150927		5 54 23.6	-15 43 56	0.000	-0.01	7.70	0.00	0.4	B9.5 V		290 s	4500	m
39683	113256		5 54 28.6	+8 03 15	0.000	-0.01	7.14	-0.03	0.4	B9.5 V		220 s		
39787	132634		5 54 29.6	-8 24 46	+0.004	-0.04	6.7	1.1	0.2	K0 III		200 s		
39855	150931		5 54 30.0	-19 42 16	+0.005	-0.03	7.40	0.9	3.2	G5 IV		69 s	4503	m
39662	94979		5 54 32.1	+11 45 45	+0.001	-0.07	6.50	0.03	1.4	A2 V		52 mx		m
39777	132635		5 54 34.6	-4 03 50	0.000	+0.02	6.57	-0.18	-2.5	B2 V	+25	650 s		
39731	113262		5 54 35.7	+5 21 14	-0.001	0.00	8.00	0.9	3.2	G5 IV		90 s	4496	m
39681	94982		5 54 36.6	+11 29 07	+0.003	-0.03	7.6	0.0	0.4	B9.5 V		260 s		
39661	94980		5 54 37.4	+15 30 27	0.000	-0.02	8.02	-0.01	0.4	B9.5 V		310 s	4495	m
39945	171001		5 54 39.2	-26 39 35	-0.001	+0.06	6.8	0.9	1.7	A3 V		33 s		
40200	217612		5 54 41.2	-49 37 37	+0.001	+0.01	6.10	-0.13		B5	+12			
39853	150932		5 54 43.5	-11 46 26	+0.004	+0.04	5.66	1.53	-0.3	K5 III	+87	150 s		
39775	113265		5 54 44.0	+0 58 06	0.000	0.00	6.00	1.34		K0	+22			
39527	58565		5 54 44.3	+35 04 17	+0.002	0.00	7.9	1.1	0.2	K0 III		330 s		
39680	94983		5 54 44.7	+13 51 17	0.000	0.00	7.85	0.06		O6 pe	-18			m
40292	234199		5 54 50.1	-52 38 07	-0.001	+0.25	5.29	0.31	2.6	F0 V	+24	34 s		
39413	40690		5 54 50.2	+47 07 51	+0.004	-0.02	7.9	0.1	1.4	A2 V		190 s		
39283	25450	30 ξ Aur	5 54 50.7	+55 42 25	-0.001	+0.02	4.99	0.05		A2 p	-12	26 mn		
39526	58566		5 54 52.1	+36 14 54	0.000	-0.01	7.3	1.1	0.2	K0 III	-20	260 s		
40091	196309		5 54 52.4	-39 57 29	-0.002	+0.01	5.57	1.51	-0.3	K5 III		150 s		
39645	77710		5 54 53.9	+22 31 12	0.000	+0.01	7.6	0.9	0.3	G7 III	+19	280 s		
39729	94990		5 54 55.0	+14 13 10	-0.002	-0.04	6.80	1.1	0.2	K0 III		200 s	4497	m
39773	113267		5 54 55.5	+5 51 36	+0.001	-0.04	6.80	0.00	0.4	B9.5 V		190 s	4499	m
40434	234205		5 54 56.3	-59 36 29	+0.001	+0.03	7.2	1.5	0.2	K0 III		130 s		
39774	113268		5 54 56.5	+5 21 13	0.000	-0.01	7.7	0.9	3.2	G5 IV		79 s		
39698	94986	57 Ori	5 54 56.6	+19 44 58	0.000	-0.01	5.92	-0.17	-2.5	B2 V	+7	480 s		
39899	150935		5 54 57.3	-12 56 23	-0.004	-0.08	7.9	0.5	4.0	F8 V		59 s		
39220	25447	31 Cam	5 54 57.7	+59 53 18	0.000	-0.02	5.2	0.1	0.6	A0 V	-3	83 s		TU Cam, m,v
39699	94989		5 54 58.4	+17 24 06	0.000	-0.02	7.40	1.4	-0.3	K5 III	+31	330 s		
39586	58569		5 54 59.0	+31 42 06	-0.003	-0.18	5.8	0.1	1.7	A3 V	-21	28 mx		
40028	196305		5 54 59.5	-30 18 21	0.000	+0.02	7.5	0.6	0.4	B9.5 V		120 s		
249092	94988		5 54 59.6	+19 36 16	+0.001	-0.01	8.0	1.1	0.2	K0 III		350 s		
40273	234200		5 55 00.3	-50 59 18	+0.004	+0.02	8.0	-0.2	3.0	F2 V		98 s		
39678	77714		5 55 00.5	+21 24 22	+0.001	0.00	7.8	0.0	0.4	B9.5 V		290 s		
39833	132641		5 55 01.9	-0 30 30	-0.007	-0.06	7.8	0.7	4.4	G0 V		49 s		
39888	132646		5 55 07.2	-9 48 37	+0.001	+0.01	7.5	0.4	3.0	F2 V		80 s		
39849	132645		5 55 07.3	-1 55 38	0.000	-0.04	7.4	0.1	1.4	A2 V		160 s		
40992	--		5 55 07.4	-72 48 31			8.0	0.8	3.2	G5 IV		90 s		
40484	234207		5 55 09.0	-59 54 36	+0.001	+0.01	8.0	1.7	0.2	K0 III		130 s		
39801	113271	58 α Ori	5 55 10.2	+7 24 26	+0.002	+0.01	0.50	1.85	-5.6	M2 Iab	+21	95 s	4506	Betelgeuse, m,v
39771	94993		5 55 15.7	+12 53 44	0.000	0.00	7.9	0.7	4.4	G0 V		51 s		
39677	77715		5 55 22.2	+29 57 50	-0.001	-0.01	7.2	0.4	2.6	F0 V		82 s		
39911	132648		5 55 24.8	-4 49 29	-0.001	0.00	7.1	0.1	0.6	A0 V		200 s		
39713	77721		5 55 29.0	+29 10 25	-0.001	-0.03	7.64	1.01	0.3	G5 III	+69	250 s		
40176	196316	ξ Col	5 55 29.8	-37 07 15	+0.002	-0.03	4.97	1.11	0.0	gK1	+60	97 s		
39910	132652		5 55 30.2	-4 36 59	+0.002	0.00	5.87	1.18	-0.1	K2 III	+26	150 s		
40101	171018		5 55 34.4	-28 57 13	-0.001	+0.01	7.8	1.8	-0.5	M4 III		360 s		
40046	150940		5 55 35.1	-18 57 37	0.000	-0.02	7.60	1.1	0.2	K0 III		300 s		m
39927	132653		5 55 35.3	-4 47 18	-0.001	-0.02	6.28	0.06		A0				
39958	132655		5 55 35.4	-7 40 15	0.000	0.00	7.8	1.1	0.2	K0 III		330 s		
40126	171020		5 55 36.6	-29 54 58	+0.001	+0.09	8.0	0.3	4.0	F8 V		62 s		
39430	25454		5 55 37.6	+53 27 35	+0.003	0.00	8.00	1.6	-0.5	M2 III		500 s		m
39828	94997		5 55 38.9	+13 00 08	+0.002	0.00	7.7	0.9	3.2	G5 IV		79 s		
40125	171021		5 55 39.2	-29 08 15	0.000	-0.03	8.03	0.42	3.4	F5 V		84 s		
39746	77724		5 55 40.2	+27 42 57	0.000	+0.01	7.04	0.22	-5.1	B1 II		1500 s		
39448	25457		5 55 41.2	+52 58 39	+0.002	-0.05	8.0	0.9	3.2	G5 IV		93 s		
40216	196320		5 55 43.1	-38 06 18	+0.002	-0.01	7.5	0.3	3.4	F5 V		65 s		
40152	196317		5 55 43.4	-31 52 30	-0.002	+0.02	7.9	1.1	0.2	K0 III		320 s		
--	77730		5 55 49.2	+20 10 30	0.000	-0.01	5.4							U Ori, v
39697	58575		5 55 50.1	+36 56 13	-0.002	-0.02	7.70	0.2	2.1	A5 V		130 s	4505	m
40100	171023		5 55 52.6	-21 41 23	-0.001	-0.12	6.80	0.6	4.4	G0 V		30 s	4527	m

HD	SAO	Star Name	α 2000	δ 2000	μ(α)	μ(δ)	V	B-V	M_V	Spec	RV	d(pc)	ADS	Notes
40508	234212		5h 55m 54s.5	-56° 33' 18"	-0.004	+0".04	7.4	1.3	0.2	K0 III		190 s		
40566	234215		5 55 55.1	-59 11 55	+0.001	+0.03	7.8	1.1	2.6	F0 V		34 s		
39660	40707		5 55 56.6	+41 19 24	-0.002	-0.02	6.6	0.4	2.6	F0 V		61 s		
39953	113276		5 55 57.2	+0 50 09	0.000	-0.01	7.6	0.1	0.6	A0 V		250 s		
40195	196322		5 55 58.8	-31 31 55	-0.001	+0.01	6.7	-0.1	0.4	B9.5 V		190 s		
39907	95006		5 56 01.8	+11 31 19	-0.001	0.00	7.90	0.00	0.4	B9.5 V		300 s	4520	m
39881	95004		5 56 03.5	+13 55 31	+0.027	-0.47	6.60	0.65	4.4	dG0	-2	23 ts	4519	m
40071	150950		5 56 06.0	-13 08 07	-0.001	0.00	7.9	0.0	0.4	B9.5 V		310 s		
39301	13654		5 56 07.9	+63 16 53	+0.002	-0.06	7.9	0.2	2.1	A5 V		81 mx		
39745	58582		5 56 08.9	+35 14 12	0.000	-0.01	7.8	1.1	0.2	K0 III		310 s		
40138	171031		5 56 10.6	-21 07 19	+0.001	+0.05	7.0	-0.2	0.6	A0 V		190 s		
40151	171034		5 56 14.2	-22 50 25	+0.008	+0.02	5.96	1.11	5.9	dK0	+34	10 s		d?
39551	25467		5 56 14.4	+51 48 14	-0.001	-0.01	6.5	0.1	1.7	A3 V	-12	91 s		
40173	171037		5 56 18.1	-24 05 21	-0.001	-0.01	7.6	1.6	-0.3	K5 III		350 s		
40340	196331		5 56 20.7	-39 09 21	+0.001	0.00	7.2	1.3	0.2	K0 III		170 s		
40248	196330	σ Col	5 56 20.9	-31 22 57	0.000	+0.01	5.5	0.4	2.6	F0 V	+19	34 s		
39643	40712		5 56 23.2	+47 42 58	0.000	-0.02	7.6	0.0	0.4	B9.5 V		260 s		
40136	150957	16 η Lep	5 56 24.2	-14 10 04	-0.003	+0.14	3.71	0.33	1.7	F0 IV	-2	20 ts		
39784	58585		5 56 25.0	+36 53 11	+0.001	-0.02	7.9	0.1	1.4	A2 V		190 s		
39985	113284		5 56 28.0	+9 30 35	0.000	0.00	5.99	-0.07		B9				
40120	150958		5 56 28.2	-11 16 42	-0.001	-0.06	8.0	1.1	-0.1	K2 III		270 mx		
40319	196332		5 56 31.9	-33 26 05	0.000	+0.01	7.2	0.2	0.6	A0 V		150 s		
39866	77744		5 56 33.7	+28 56 32	0.000	0.00	6.40	0.30	-5.0	A2 Ib	+19	1300 s		
40235	171045		5 56 34.4	-23 12 56	+0.001	+0.03	6.36	1.07		K0				
40192	150962		5 56 44.1	-13 09 30	+0.001	-0.01	7.9	0.5	4.0	F8 V		60 s		
40234	150965		5 56 45.5	-18 17 36	-0.001	-0.02	7.9	0.1	1.4	A2 V		200 s		
39814	58591		5 56 46.6	+38 34 40	0.000	-0.01	7.9	0.1	0.6	A0 V		270 s		
40006	95024		5 56 48.6	+15 04 50	-0.001	-0.01	7.6	1.1	0.2	K0 III		290 s		
40147	132679		5 56 48.8	-9 18 53	-0.001	-0.01	8.0	0.9	3.2	G5 IV		93 s		
40359	196335		5 56 48.9	-31 58 34	0.000	+0.01	6.44	1.07		K0				
40020	95027		5 56 49.3	+11 31 16	+0.007	-0.05	5.87	1.10	3.2	G5 IV	+21	22 s		
40005	95025		5 56 50.6	+16 21 19	+0.001	0.00	7.23	-0.13	-1.6	B3.5 V	+32	570 s		
40168	132683		5 56 54.6	-8 23 02	-0.001	+0.01	6.9	0.8	3.2	G5 IV		54 s		
40483	217634		5 56 54.8	-44 00 32	+0.005	+0.08	6.63	0.52	3.6	G0 IV-V		40 s		
39864	58596		5 56 54.8	+35 34 44	-0.001	-0.02	7.4	1.6	-0.5	M4 III		360 s		
39845	58593		5 56 55.4	+38 17 13	+0.003	-0.05	7.2	1.4	-0.3	K5 III	+31	300 s		
39970	77750		5 56 56.1	+24 14 59	0.000	0.00	6.02	0.39	-7.1	A0 Ia	+1	2700 s		
40358	171057		5 56 58.9	-28 45 33	0.000	-0.03	8.0	1.0	0.2	K0 III		360 s		
39925	58601		5 57 00.2	+30 36 33	+0.001	-0.03	7.8	1.4	-0.3	K5 III		380 s		
40040	95030		5 57 01.7	+15 44 28	+0.005	-0.25	7.9	0.7	4.4	G0 V	-24	42 mx		
40233	150966		5 57 03.3	-10 51 17	+0.004	-0.04	7.8	1.1	0.2	K0 III		180 mx		
39743	40720		5 57 04.8	+49 01 46	+0.002	-0.02	6.47	0.99	0.3	G8 III	-2	160 s		
39949	77753		5 57 05.5	+27 19 00	0.000	-0.01	7.21	1.08	-2.0	G0 II	+14	450 s		
39783	40721		5 57 06.7	+45 30 32	0.000	-0.03	8.00	1.6	-0.5	M4 III		460 s		TW Aur, v
39983	77756		5 57 07.3	+22 50 21	-0.001	0.00	6.90	1.6		M5 III	+30			BQ Ori, v
39628	25477		5 57 07.8	+55 56 52	+0.005	-0.09	6.97	1.20		K2 IV		120 mx		
40039	95031		5 57 08.3	+19 13 00	0.000	-0.01	7.86	-0.07	0.4	B9.5 V		310 s		
40643	234222		5 57 09.1	-53 25 04	0.000	-0.04	7.60	1.1	0.2	K0 III		300 s		m
40665	234224		5 57 14.4	-53 25 34	+0.001	-0.01	6.45	1.48		K2				m
39939	58604		5 57 16.3	+32 41 36	0.000	-0.01	8.0	0.0	0.4	B9.5 V		320 s		
40002	77759		5 57 19.3	+26 51 13	+0.001	+0.01	8.0	1.1	-0.1	K2 III	-32	390 s		
40210	113296		5 57 25.1	+0 01 41	-0.002	+0.01	6.90	0.00	0.6	A0 V		180 s	4542	m
40188	113292		5 57 27.7	+5 00 04	0.000	+0.01	7.8	0.2	2.1	A5 V		130 s		
39923	58606		5 57 28.7	+38 38 14	+0.003	-0.07	7.8	1.4	-0.3	K5 III		150 mx		
39938	58609		5 57 29.4	+35 46 37	0.000	-0.03	7.9	0.9	3.2	G5 IV		87 s		
40113	95045		5 57 29.7	+14 03 36	-0.002	-0.01	7.0	1.1	-0.1	K2 III	+7	250 s		
40166	113294		5 57 30.0	+6 23 28	-0.001	0.00	8.0	0.1	0.6	A0 V		280 s		
40494	196352	γ Col	5 57 32.2	-35 17 00	0.000	+0.01	4.36	-0.18	-2.3	B3 IV	+24	210 s		m
40379	150972		5 57 33.3	-18 03 19	-0.001	-0.01	6.9	0.5	3.4	F5 V		51 s		
39429	13662		5 57 34.9	+66 05 45	+0.007	-0.02	6.5	1.1	0.2	K0 III	-22	180 s		
40557	217641		5 57 35.6	-41 06 19	-0.003	0.00	8.0	-0.3	0.4	B9.5 V		330 s		
40065	77769		5 57 36.8	+24 36 46	0.000	-0.01	7.3	0.0	0.0	B8.5 V		280 s		
39863	40726		5 57 37.2	+45 54 06	-0.001	0.00	6.5	0.9	3.2	G5 IV	+3	46 s		
40355	150973		5 57 39.6	-15 24 23	0.000	-0.01	6.9	0.0	0.0	B8.5 V		240 s		
40001	58611		5 57 39.7	+33 15 37	-0.002	-0.04	7.5	0.5	4.0	F8 V		50 s		
39724	25485		5 57 40.6	+55 39 24	+0.001	+0.01	7.1	0.1	0.6	A0 V		200 s		
40301	132692		5 57 41.3	-6 05 41	0.000	+0.01	7.5	1.6	-0.5	M2 III	+40	400 s		
40418	150977		5 57 44.1	-19 23 13	-0.003	-0.03	7.8	0.5	-0.1	K2 III		380 s		
40376	150974		5 57 46.8	-14 12 34	-0.001	+0.01	8.00	1.1	-0.1	K2 III		420 s	4549	m
40186	95054		5 57 48.9	+14 41 04	0.000	+0.03	7.8	0.7	4.4	G0 V		49 s		
39982	58612		5 57 50.1	+37 31 06	-0.002	0.00	7.5	1.1	-0.1	K2 III		320 s		
40540	196357		5 57 52.6	-34 28 34	+0.001	-0.01	7.6	0.3	1.4	A2 V		120 s		
40282	113306		5 57 54.1	+1 13 28	-0.005	+0.01	6.22	1.48		K5 III-IV	+37	170 mx		
39905	40729		5 57 55.2	+46 42 23	+0.002	-0.01	7.9	0.5	3.4	F5 V		77 s		
40522	196356		5 57 56.5	-31 55 17	+0.002	-0.02	7.5	1.4	-0.1	K2 III		240 s		

HD	SAO	Star Name	α 2000	δ 2000	μ(α)	μ(δ)	V	B-V	M$_v$	Spec	RV	d(pc)	ADS	Notes
40111	77775	139 Tau	5h57m59.6s	+25°57'14"	0.000	0.00	4.82	-0.06	-5.7	B1 Ib	+8	1100 s		
39354	13663		5 58 07.1	+69 36 16	+0.010	+0.02	7.1	0.4	2.6	F0 V		78 s		
39723	25488		5 58 07.5	+59 23 36	+0.003	-0.05	7.7	1.1	0.2	K0 III		290 s		
38126	920		5 58 11.4	+81 21 50	+0.005	-0.02	8.0	0.9	3.2	G5 IV		93 s		
40347	132700		5 58 11.6	-0 59 39	0.000	-0.02	6.22	1.14		K0				
40335	113311		5 58 13.3	+1 51 22	-0.001	-0.02	7.2	0.1	0.6	A0 V		210 s		
40555	171088		5 58 14.4	-29 06 35	-0.002	0.00	7.8	0.5	0.2	K0 III		330 s		
40554	171089		5 58 18.5	-27 58 58	-0.003	+0.01	8.0	1.0	3.2	G5 IV		67 s		
40037	40735		5 58 19.0	+40 47 12	+0.001	-0.02	7.89	0.08	1.4	A2 V		190 s		
40397	132703		5 58 21.5	-4 39 03	+0.005	-0.21	6.80	0.6	4.4	G0 V		30 s	4557	m
40163	77781		5 58 23.9	+29 37 39	-0.001	+0.01	8.00	0.00		B9.5 V		310 s	4544	m
40372	113315	59 Ori	5 58 24.4	+1 50 13	0.000	-0.01	5.90	0.21		A5	+14			m
40207	77785		5 58 24.5	+23 09 20	0.000	-0.03	7.5	1.1	0.2	K0 III		280 s		
40333	113313		5 58 27.6	+7 51 37	+0.001	-0.01	7.3	1.1	0.2	K0 III	+17	250 s		
40464	150988		5 58 30.7	-10 50 26	-0.004	-0.04	8.0	0.0	0.4	B9.5 V		94 mx		
40085	58619		5 58 33.1	+38 53 37	-0.002	+0.03	7.19	1.17	-0.1	K2 III	+10	290 s		
40537	171090		5 58 35.4	-21 32 38	+0.007	+0.04	7.8	0.2	3.2	G5 IV		84 s		
40241	77790		5 58 36.7	+24 47 52	+0.002	-0.04	7.6	0.8	3.2	G5 IV		76 s		
40733	217648		5 58 37.6	-44 02 04	-0.001	+0.01	5.81	1.07	0.2	K0 III		120 s		
40242	77792		5 58 38.1	+23 39 57			7.81	1.68	-0.5	M2 III		410 s		
40316	95068		5 58 39.9	+16 35 52	-0.001	0.00	7.39	0.00	0.4	B9.5 V		230 s		
40868	234234		5 58 46.1	-52 39 17	-0.003	-0.01	8.0	0.5	0.2	K0 III		360 s		
39781	13668		5 58 46.4	+60 22 47	-0.002	-0.11	7.0	0.4	2.6	F0 V		76 s		
40446	113321	60 Ori	5 58 49.5	+0 33 10	-0.001	0.00	5.22	0.01		A1	+34	40 mn		m
40280	77800		5 58 52.8	+25 46 38	-0.004	-0.02	6.60	0.99	0.2	G9 III	0	190 s		
40369	95075		5 58 53.1	+12 48 31	-0.001	+0.02	5.70	0.89	-0.1	K2 III	+12	150 s	4562	m
40331	95070		5 58 54.8	+18 49 12	+0.001	-0.02	7.1	1.1	0.2	K0 III	-34	230 s		
40255	77795		5 58 55.8	+29 59 03	+0.001	-0.01	7.1	0.1	1.7	A3 V		120 s		
40720	196377		5 58 56.0	-36 59 41	0.000	-0.04	7.9	0.9	3.2	G5 IV		74 s		
40064	40742		5 58 59.3	+45 51 50	+0.002	+0.01	7.8	0.8	3.2	G5 IV		83 s		
40535	132714	1 Mon	5 59 01.0	-9 22 56	+0.001	+0.01	6.12	0.29	-2.0	F2 II	+15	420 s		
40909	234236		5 59 01.7	-51 13 24	0.000	+0.01	7.1	0.0	0.0	B8.5 V		230 s		
40536	132715	2 Mon	5 59 04.3	-9 33 31	+0.001	-0.05	5.03	0.19		A m	+22	32 mn		
40297	77804		5 59 05.4	+27 33 43	-0.001	0.00	7.27	0.29	-5.2	A0 Ib		2100 s		
40808	217650	η Col	5 59 08.8	-42 48 55	+0.001	-0.02	3.96	1.14	0.2	K0 III	+17	44 s		
40718	196380		5 59 16.2	-32 12 34	0.000	+0.01	7.4	1.0	3.2	G5 IV		52 s		
40184	40747		5 59 17.2	+40 02 02	+0.005	+0.01	7.4	0.5	3.4	F5 V		63 s		
40131	40745		5 59 20.3	+46 55 06	-0.001	-0.01	8.00	0.03	0.6	A0 V		300 s		
250019	95089		5 59 20.3	+11 45 46	-0.001	0.00	7.7	1.1	0.2	K0 III		300 s		
40084	40743		5 59 21.7	+49 55 27	0.000	-0.01	5.89	1.23	3.2	G5 IV	-4	18 s		
40605	150994		5 59 21.9	-12 50 07	-0.001	+0.01	7.3	1.1	-0.1	K2 III		300 s		
40206	40749		5 59 23.8	+41 55 02	-0.001	-0.02	7.9	0.1	0.6	A0 V		270 s		
40412	95084		5 59 24.5	+17 48 53	+0.002	-0.09	7.90	0.5	3.4	F5 V		79 s	4571	m
40143	40746		5 59 25.1	+45 37 09	+0.003	-0.02	6.60	0.1	0.6	A0 V	-12	150 s	4551	m
40491	113327		5 59 27.4	+8 24 57	-0.002	0.00	6.8	1.1	0.2	K0 III		200 s		
40463	95092		5 59 27.7	+12 28 09	+0.001	-0.06	7.8	1.1	0.2	K0 III		190 mx		
--	58629		5 59 28.6	+38 37 27	+0.001	-0.03	7.7	1.2						
40035	25502	33 δ Aur	5 59 31.6	+54 17 05	+0.009	-0.13	3.72	1.00	0.2	K0 III	+8	50 s		m
40426	95090		5 59 31.7	+18 51 08	0.000	-0.01	7.5	0.1	0.6	A0 V		230 s		
40183	40750	34 β Aur	5 59 31.7	+44 56 51	-0.005	0.00	1.90	0.03	0.6	A2 IV	-18	22 ts	4556	Menkalinan, m,v
40160	40748		5 59 33.0	+46 32 05	+0.002	-0.01	7.48	-0.07	-1.0	B5.5 V	+5	490 s		
40205	40751		5 59 36.2	+44 06 41	-0.003	-0.02	7.7	1.1	0.2	K0 III		290 s		
40574	132723		5 59 37.6	-1 26 40	-0.001	-0.02	6.63	-0.07		B9				
40313	58634		5 59 38.4	+36 48 51	-0.001	-0.04	7.9	1.1	0.2	K0 III		320 s		
40805	196387		5 59 38.7	-35 21 10	-0.003	+0.03	7.0	1.5	0.2	K0 III		120 s		
40443	77813		5 59 39.5	+21 36 15	0.000	-0.02	6.7	0.1	0.6	A0 V		160 s		
40423	--		5 59 42.4	+22 28 15	-0.001	0.00	7.54	0.23		A0			4577	m
40312	58636	37 θ Aur	5 59 43.2	+37 12 45	+0.004	-0.08	2.62	-0.08		A0 p	+29	25 mn	4566	m
40062	25504		5 59 45.6	+55 19 15	-0.001	-0.09	6.5	0.1	1.4	A2 V	+45	100 s		
40240	40754		5 59 47.9	+43 59 09	+0.002	0.00	7.90	1.1	0.2	K0 III		320 s	4563	m
40083	25505		5 59 48.0	+54 32 50	+0.001	-0.03	6.14	1.20	0.2	K0 III	-6	100 s		
40637	132728		5 59 48.3	-6 35 51	0.000	-0.01	7.5	1.1	0.2	K0 III		280 s		
40590	113339		5 59 51.6	+0 03 21	+0.002	-0.04	8.0	0.5	3.4	F5 V		84 s		
40573	113337		5 59 53.5	+3 59 27	-0.001	-0.02	7.8	0.1	0.6	A0 V		260 s		
40239	40756	35 π Aur	5 59 56.1	+45 56 12	0.000	-0.01	4.26	1.72	-2.4	M3 II	+1	200 s		
40672	132734		6 00 01.1	-8 06 29	0.000	-0.02	6.9	1.1	0.2	K0 III		220 s		
40825	171114		6 00 02.7	-29 12 21	-0.001	+0.06	7.6	1.3	3.2	G5 IV		36 s		
40657	132732		6 00 03.3	-3 04 27	0.000	-0.07	4.53	1.22	-0.1	K2 III	+26	80 s		
40441	77818		6 00 04.9	+28 07 33	0.000	-0.02	7.0	1.4	-0.3	K5 III	+4	270 s		
40460	77819		6 00 06.0	+27 16 20	-0.001	-0.08	6.60	1.02	0.2	K1 III	+96	210 s		
40635	132733		6 00 07.8	-0 30 09	0.000	-0.02	7.80	0.00	0.4	B9.5 V	+31	290 s		m
40531	95098		6 00 09.7	+19 59 27	0.000	-0.01	7.90	-0.13	0.0	B8.5 V		380 s		
40204	25513		6 00 09.8	+51 04 51	+0.002	-0.03	6.6	0.1	0.6	A0 V		160 s		
40616	113346		6 00 13.6	+4 50 09	-0.005	+0.15	7.6	0.6	4.4	G0 V		43 s		
40326	40767		6 00 14.1	+43 10 52	+0.002	-0.01	7.90	0.00	0.4	B9.5 V		300 s	4574	m

HD	SAO	Star Name	α 2000	δ 2000	μ(α)	μ(δ)	V	B−V	M$_v$	Spec	RV	d(pc)	ADS	Notes
			h m s	° ' "	s	"								
40439	58645		6 00 14.4	+33 08 13	0.000	0.00	6.8	0.1	1.4	A2 V		120 s		
40570	95103		6 00 14.9	+15 07 52	−0.001	0.00	7.70	1.1	−4.4	K3 Ib		2000 s		
40602	113344		6 00 15.4	+8 57 23	−0.001	−0.01	7.90	0.37		A m				
40571	95105		6 00 17.1	+12 36 42	0.000	+0.01	6.5	0.0	0.4	B9.5 V		160 s		
40745	151011		6 00 17.6	−12 53 59	−0.001	−0.04	6.22	0.36		F0				
41002	217664		6 00 17.7	−44 00 23	0.000	−0.04	7.6	−0.3	1.7	A3 V		160 s		
40325	40769		6 00 18.9	+44 35 31	−0.002	−0.04	6.40	1.1	−0.1	K2 III	+2	190 s	4576	m
40459	58646		6 00 19.6	+31 56 27	0.000	−0.01	7.2	1.1	0.2	K0 III	−20	240 s		
40382	58643		6 00 20.7	+39 52 38	−0.002	−0.01	7.6	0.9	3.2	G5 IV		76 s		
40887	—		6 00 23.5	−31 02 28			7.85	1.14		K4 p		25 t		m
40569	95106		6 00 24.1	+16 17 52	0.000	−0.05	6.7	1.1	−0.1	K2 III		220 s		
40458	58647		6 00 24.2	+34 30 29	−0.003	0.00	6.8	1.1	0.2	K0 III		200 s		
40545	77831		6 00 28.3	+22 53 59	+0.001	−0.02	6.9	0.1	1.4	A2 V		130 s		
40457	58649		6 00 28.4	+35 18 43	−0.001	−0.01	7.9	0.5	−4.6	F5 Ib		2200 s		CO Aur, q
40886	171122		6 00 28.5	−27 53 23	0.000	−0.06	8.0	0.1	0.6	A0 V		55 mx		
40744	132740		6 00 31.3	−7 28 10	+0.001	−0.01	7.3	1.1	0.2	K0 III		260 s		
41296	234252		6 00 35.6	−58 06 08	0.000	+0.02	6.8	0.5	0.6	A0 V		81 s		
40082	25512		6 00 36.7	+59 53 07	0.000	−0.01	8.0	1.1	−0.1	K2 III		380 s		
40530	77830		6 00 36.8	+29 00 19	0.000	−0.02	7.4	0.4	2.6	F0 V		88 s		
40141	25516		6 00 36.9	+56 55 25	+0.003	0.00	7.6	0.9	3.2	G5 IV		75 s		
40311	40772		6 00 38.0	+49 58 44	−0.001	−0.02	7.9	0.1	1.4	A2 V		200 s		
39894	13674		6 00 48.5	+67 00 51	+0.001	−0.03	6.9	0.1	0.6	A0 V	−16	180 s		
41214	234251		6 00 49.2	−51 12 59	−0.003	+0.10	5.67	0.20		A0	+5			
40726	113362		6 00 49.3	+2 54 57	−0.001	−0.01	6.97	0.9	3.2	G5 IV		66 s		
41089	217668		6 00 50.9	−42 52 14	−0.001	+0.01	6.6	−0.1	0.0	B8.5 V		210 s		
40529	58654		6 00 52.8	+31 57 43	0.000	−0.02	7.8	0.1	1.7	A3 V		160 s		
40394	40778	36 Aur	6 00 58.5	+47 54 07	+0.001	−0.02	5.73	−0.01		A p	+16			
40589	77837		6 01 00.4	+27 34 20	0.000	0.00	6.05	0.25	−6.5	B9 Iab	+17	2400 s	4589	m
41070	196410		6 01 02.6	−37 04 18	−0.004	+0.01	7.20	0.9	3.2	G5 IV		63 s		m
41155	217674		6 01 05.6	−43 54 10	+0.004	+0.03	7.4	0.1	3.4	F5 V		64 s		
40681	95117		6 01 05.6	+16 59 29	0.000	−0.02	7.69	−0.03	0.4	B9.5 V		280 s		
40488	40782		6 01 06.1	+41 56 20	−0.002	−0.05	8.0	0.5	3.4	F5 V		80 s		
250290	77842		6 01 07.6	+23 18 17	0.000	−0.01	7.38	0.61	−5.7	B3 Ib		1400 s		
40108	13681		6 01 08.2	+62 19 00	0.000	0.00	8.0	1.1	0.2	K0 III		330 s		
41069	196411		6 01 08.9	−36 28 09	−0.001	−0.01	7.6	0.1	0.6	A0 V		230 s		
40588	58657		6 01 10.1	+31 02 05	0.000	+0.01	6.00	0.08	0.6	A0 V	−11	100 s		
40679	95119		6 01 10.5	+19 25 13	0.000	−0.01	7.5	0.1	0.6	A0 V		240 s		
40972	171138		6 01 13.1	−25 25 03	−0.001	−0.02	6.05	0.02	1.2	A1 V		93 s		
40487	40787		6 01 13.2	+43 16 42	0.000	−0.01	7.8	1.4	−0.3	K5 III		380 s		
41451	249421		6 01 14.7	−60 29 08	+0.003	+0.01	7.2	0.9	3.2	G5 IV		53 s		
40034	13675		6 01 15.1	+65 31 33	0.000	−0.04	6.70	0.1	1.7	A3 V		97 s		m
41047	196413		6 01 16.3	−33 54 42	+0.001	−0.02	5.55	1.58		K5	+19			
40774	113367		6 01 17.2	+9 04 22	−0.002	+0.01	7.7	0.9	3.2	G5 IV		79 s		
40470	40788		6 01 17.6	+45 09 37	+0.001	−0.01	7.6	0.1	0.6	A0 V		240 s		
36905	914		6 01 20.4	+85 10 56	+0.013	0.00	6.11	1.57	0.2	K0 III	−46	69 s		
40697	95127		6 01 20.9	+18 57 40	0.000	−0.02	7.3	0.1	0.6	A0 V		220 s		
40600	58662		6 01 25.2	+32 38 25	+0.002	−0.03	7.4	0.8	3.2	G5 IV		67 s		
40756	95129		6 01 28.9	+13 07 53	−0.001	−0.02	7.5	0.0	0.4	B9.5 V		260 s		
40678	77851		6 01 29.1	+23 42 14	−0.001	−0.01	7.35	0.12		F2				
40990	171140		6 01 29.9	−20 38 33	0.000	+0.04	8.0	1.0	0.2	K0 III		360 s		
40837	113377		6 01 31.1	+2 54 34	−0.003	0.00	7.5	0.1	0.6	A0 V		240 s		
40836	113376		6 01 31.3	+3 11 10	0.000	−0.02	7.3	0.0	0.0	B8.5 V		280 s		
40970	151032		6 01 34.1	−16 23 56	−0.001	0.00	7.1	1.1	0.2	K0 III		240 s		
40969	151033		6 01 38.9	−15 35 21	−0.003	−0.01	6.7	1.1	0.2	K0 III		200 s		
40773	95133		6 01 39.6	+16 56 02	−0.001	−0.01	7.4	1.1	0.2	K0 III	−17	260 s		
41183	196424		6 01 41.1	−38 10 02	−0.004	−0.01	8.0	0.7	3.2	G5 IV		93 s		
40724	77858		6 01 41.6	+22 24 03	−0.001	−0.02	6.30	−0.07	−0.2	B8 V		200 s		
40486	40789		6 01 42.9	+48 57 35	−0.001	0.00	5.96	1.45	0.2	K0 III	+11	76 s		
41017	151038		6 01 44.1	−18 03 39	−0.004	+0.02	8.0	0.9	3.2	G5 IV		93 s		
40967	151037	3 Mon	6 01 50.3	−10 35 53	−0.001	0.00	4.95	−0.12	−1.6	B5 IV	+39	200 s	4615	m
40676	58670		6 01 50.5	+31 19 35	+0.001	−0.03	8.0	0.9	3.2	G5 IV		91 s		
41687	249425		6 01 53.6	−64 51 38	−0.008	−0.05	7.1	0.8	4.4	G0 V		26 s		
40669	58671		6 02 05.3	+38 43 13	0.000	0.00	7.20	0.9	3.2	G5 IV		62 s	4597	m
40935	113387		6 02 05.5	+1 45 39	−0.001	−0.01	7.9	0.1	0.6	A0 V		270 s		
40438	25527		6 02 05.6	+55 18 55	+0.001	−0.01	8.0	1.4	−0.3	K5 III		430 s		
41245	196432		6 02 06.2	−38 59 23	+0.003	+0.05	6.9	0.4	3.2	G5 IV		55 s		
41172	171158		6 02 06.2	−27 25 39	−0.002	−0.10	7.12	0.46	2.8	F5 IV−V		71 s		m
40182	13684		6 02 06.4	+65 03 35	+0.002	−0.02	7.4	1.1	0.2	K0 III		260 s		
40985	132761		6 02 08.7	−5 08 14	0.000	−0.01	6.9	1.1	−0.1	K2 III		250 s		
41586	249424		6 02 09.2	−60 05 49	+0.001	0.00	6.4	2.0	−0.5	M2 III		140 s		
40998	132764		6 02 10.8	−7 17 22	0.000	0.00	6.9	0.1	0.6	A0 V		180 s		
41085	151045		6 02 11.6	−19 41 29	−0.001	−0.02	7.7	−0.3	0.6	A0 V		260 s		
41369	217687		6 02 14.0	−47 24 59	−0.001	+0.02	8.0	0.5	1.7	A3 V		160 s		
41555	234265		6 02 14.9	−58 13 17	+0.003	0.00	8.0	1.3	−0.1	K2 III		360 s		
40140	13683		6 02 16.3	+66 53 59	0.000	−0.01	7.7	1.4	−0.3	K5 III		390 s		

HD	SAO	Star Name	α 2000	δ 2000	μ(α)	μ(δ)	V	B-V	M$_v$	Spec	RV	d(pc)	ADS	Notes
			h m s	° ′ ″	s	″								
40964	113392		6 02 17.1	+1 41 39	0.000	+0.01	6.59	-0.05		A0	+3			
21175	217684		6 02 18.6	-40 15 01	-0.001	+0.03	6.80	1.1	0.2	K0 III		210 s		
40898	95149		6 02 22.2	+12 57 11	0.000	-0.01	7.9	1.1	0.2	K0 III		330 s		
40932	113389	61 μ Ori	6 02 22.9	+9 38 51	+0.001	-0.03	4.12	0.15		A m	+45	37 t	4617	m
41146	151048		6 02 23.3	-18 34 03	-0.001	-0.05	8.0	0.9	3.2	G5 IV		93 s		
40912	95150		6 02 30.6	+13 17 22	+0.002	-0.07	8.0	0.5	3.4	F5 V		81 s		
40931	95153		6 02 33.1	+13 01 31	-0.002	-0.03	7.40	1.6	-0.4	M0 III	+16	350 s		
41125	151052		6 02 33.8	-14 29 50	0.000	+0.01	6.20	0.96		G5				
--	171166		6 02 33.9	-22 42 48	-0.004	-0.02	8.00	0.1	0.6	A0 V		100 mx		m
40787	58683		6 02 37.7	+32 35 29	+0.003	+0.02	6.77	0.40	3.4	F5 V		47 s		
40947	95155		6 02 38.8	+12 40 17	+0.001	0.00	7.9	0.0	0.4	B9.5 V		310 s		
40650	40802		6 02 45.0	+47 48 34	+0.002	+0.09	6.5	0.5	3.4	F5 V		42 s		
41274	171175		6 02 47.5	-29 59 48	-0.001	+0.03	7.9	1.4	0.2	K0 III		210 s		
41553	234271		6 02 47.9	-53 34 58	-0.003	+0.01	6.9	1.4	0.2	K0 III		130 s		
40626	40801		6 02 48.6	+49 54 20	+0.003	-0.05	6.05	-0.06	0.6	A0 V	+22	120 s		
40722	40808		6 02 53.5	+43 22 43	+0.002	-0.02	6.5	1.1	0.2	K0 III	-19	170 s		
40832	58688		6 02 55.1	+32 38 09	+0.006	-0.21	6.24	0.42	3.4	F5 V	+34	37 s		
38387	929		6 02 57.1	+82 45 04	+0.027	+0.02	8.00	0.5	4.0	F8 V		63 s	4492	m
40960	95158		6 02 58.9	+18 00 26	+0.001	-0.04	7.80	1.1	0.0	K1 III	+42	350 s		
40894	77883		6 02 59.3	+28 40 37	0.000	0.00	7.56	-0.08	-2.5	B2 V		920 s		
41256	171176		6 03 00.6	-23 13 07	-0.001	+0.06	7.4	0.9	0.2	K0 III		280 s		
41393	196443		6 03 03.0	-37 32 51	0.000	+0.01	6.9	0.5	1.4	A2 V		73 s		
40568	25530		6 03 04.3	+57 01 07	+0.001	-0.04	6.6	0.8	3.2	G5 IV		47 s		
40981	77889		6 03 06.9	+20 35 11	-0.001	0.00	7.2	0.8	3.2	G5 IV		62 s		
40710	40810		6 03 08.0	+47 59 17	0.000	-0.08	7.3	1.1	0.2	K0 III		180 mx		
42029	249435		6 03 08.0	-67 18 34	-0.003	+0.02	7.7	0.1	1.4	A2 V		180 s		
41392	196445		6 03 08.3	-36 52 24	-0.002	-0.01	6.9	-0.1	0.4	B9.5 V		200 s		
40649	25532		6 03 09.3	+53 32 29	+0.001	0.00	6.9	0.4	2.6	F0 V		72 s		
41079	113402		6 03 10.0	+7 37 38	0.000	0.00	6.9	1.1	-0.1	K2 III		240 s		
41143	132778		6 03 14.6	-0 36 08	0.000	-0.02	7.7	1.1	0.2	K0 III		300 s		
41011	95163		6 03 14.6	+16 55 19	0.000	0.00	7.9	0.1	0.6	A0 V		270 s		
41042	95165		6 03 15.1	+12 37 04	-0.001	0.00	7.7	0.0	0.4	B9.5 V		280 s		
41312	171180		6 03 15.5	-26 17 04	+0.004	+0.09	5.04	1.34	-0.2	gK3	+183	110 s	4645	m
40801	40818	38 Aur	6 03 17.9	+42 54 41	+0.011	-0.14	6.10	0.97	0.2	K0 III	+38	71 mx		
41465	217693		6 03 18.2	-41 41 27	-0.001	0.00	8.0	1.0	3.2	G5 IV		84 s		
41176	132782		6 03 18.9	-2 28 44	-0.002	-0.02	7.1	0.1	0.6	A0 V		190 s		
41924	249432		6 03 20.0	-63 21 20	+0.002	0.00	8.0	1.1	0.2	K0 III		360 s		
40930	58695		6 03 21.7	+31 23 25	-0.003	+0.03	7.2	0.5	4.0	F8 V		43 s		
41652	234275		6 03 24.0	-53 53 26	+0.003	0.00	7.1	1.6	-0.1	K2 III		150 s		
41076	95170		6 03 24.7	+11 40 51	-0.001	-0.01	6.00	-0.04	0.4	B9.5 V	-11	130 s		
41311	171183		6 03 24.9	-21 47 25	-0.001	-0.09	7.5	0.2	2.6	F0 V		98 s		
41165	113406		6 03 26.1	+3 51 44	-0.001	-0.07	7.3	1.1	0.2	K0 III	+25	180 mx		
41040	95166	64 Ori	6 03 27.3	+19 41 26	0.000	-0.02	5.14	-0.11	-1.2	B8 III	+12	190 s		
41041	95168		6 03 28.4	+18 10 59	0.000	-0.01	8.0	0.1	0.6	A0 V		290 s		
40000	5718		6 03 29.5	+73 04 18	-0.001	+0.01	4.03	0.86	5.9	K0 V				
41142	113407		6 03 32.2	+8 35 23	-0.001	-0.03	7.70	0.32	2.6	F0 V		100 s		
40996	77900		6 03 33.8	+26 31 47	-0.001	+0.01	7.0	0.0	0.0	B8.5 V		240 s		
40945	58696		6 03 35.5	+33 52 57	+0.001	-0.02	8.0	1.1	0.2	K0 III		340 s		
41101	95174		6 03 48.1	+18 15 52	0.000	-0.02	7.7	0.0	0.0	B8.5 V		330 s		
41118	95176		6 03 49.3	+17 07 34	+0.002	-0.04	7.5	1.1	-0.1	K2 III		320 s		
41119	95177		6 03 49.3	+16 43 55	-0.001	-0.01	7.8	0.1	1.4	A2 V		180 s		
40615	13691		6 03 52.3	+60 27 10	+0.003	-0.01	8.0	1.1	0.2	K0 III		330 s		
41117	77911	62 χ² Ori	6 03 55.1	+20 08 18	0.000	-0.01	4.63	0.28	-6.8	B2 Ia	+17	1100 s		
41253	113416		6 03 58.1	+2 51 51	0.000	-0.02	7.31	-0.02	-1.0	B5.5 V	+34	420 s		
41702	234282		6 04 00.0	-51 05 56	-0.001	-0.02	8.0	-0.4	1.4	A2 V		210 s		
41056	77909		6 04 01.0	+27 34 14	+0.001	-0.03	7.9	0.1	0.6	A0 V		270 s		m
41574	196460		6 04 06.9	-39 50 11	-0.001	0.00	8.0	1.3	0.2	K0 III		250 s		
41473	171201		6 04 07.1	-29 20 19	-0.002	+0.01	7.8	0.0	0.6	A0 V		260 s		
41116	77915	1 Gem	6 04 07.2	+23 15 48	0.000	-0.10	4.16	0.82	0.3	gG5	+20	59 s		m
41383	151074		6 04 11.0	-13 41 23	-0.002	0.00	8.0	0.0	0.4	B9.5 V		330 s		
41202	95188		6 04 11.1	+14 04 25	+0.001	-0.03	7.9	0.4	3.0	F2 V		92 s		
41139	77918		6 04 12.0	+25 26 38	+0.001	-0.04	7.0	1.1	0.2	K0 III	+19	220 s		
41366	151075		6 04 12.8	-14 05 03	-0.001	+0.04	6.9	0.5	3.4	F5 V		51 s		
41335	132793		6 04 13.4	-6 42 33	-0.001	0.00	5.21	-0.06	-2.8	B2 IV-V nne	+51	390 s		
41534	196459		6 04 20.2	-32 10 20	-0.001	+0.13	5.65	-0.19	-1.7	B3 V	+94	300 s		
41864	234291		6 04 23.9	-54 23 03	-0.006	+0.01	7.4	0.9	3.4	F5 V		34 s		
41700	217702		6 04 28.4	-45 02 12	-0.009	+0.24	6.35	0.52	3.6	G0 IV-V		36 s		
40979	40830		6 04 30.0	+44 15 38	+0.010	-0.15	6.8	0.5	4.0	F8 V		35 s		
40873	25549		6 04 30.0	+51 35 02	0.000	0.00	6.45	0.18			+20		4633	m
41701	217703		6 04 30.5	-45 11 48	+0.002	0.00	6.7	1.0	0.2	K0 III		200 s		
40957	40831		6 04 33.6	+45 35 10	0.000	-0.05	7.40	0.1	1.4	A2 V		150 s	4639	m
41631	196466		6 04 36.0	-36 42 36	+0.001	-0.03	8.0	2.9	0.0	K1 III		33 s		
41630	196467		6 04 36.8	-36 39 12	-0.001	-0.01	7.8	1.0	0.4	B9.5 V		77 s		
40978	40834		6 04 38.3	+46 35 06	-0.001	-0.01	7.21	-0.05		B2.5 e	-3			
41572	171210		6 04 39.2	-28 37 03	+0.002	+0.01	7.8	0.9	3.2	G5 IV		67 s		

HD	SAO	Star Name	α 2000	δ 2000	μ(α)	μ(δ)	V	B−V	M_V	Spec	RV	d(pc)	ADS	Notes
41075	58709		6^{h}04^{m}39.7^s	+38°34′31″	0.000^s	0.00″	6.8	0.0	0.0	B8.5 V		220 s		
41306	95197		6 04 40.1	+10 30 36	−0.001	0.00	8.0	0.1	0.6	A0 V		280 s		
41742	217706		6 04 40.1	−45 04 44	−0.007	+0.26	5.93	0.49	3.4	F5 V		30 s		m
41381	132803		6 04 42.7	−1 34 49	−0.002	−0.01	6.8	0.1	0.6	A0 V		170 s		
41571	171211		6 04 43.4	−28 03 40	+0.001	−0.01	7.2	0.3	0.6	A0 V		130 s		
41488	151086		6 04 46.3	−17 57 26	0.000	−0.02	7.8	0.0	0.0	B8.5 V		370 s		
41824	217708		6 04 46.9	−48 27 31	−0.010	−0.04	6.58	0.72		G5		15 mn		m
41512	151089		6 04 49.9	−18 04 42	0.000	0.00	7.9	1.1	0.2	K0 III		350 s		
41285	95199		6 04 50.9	+16 39 34	−0.001	−0.02	7.80	−0.11	−1.0	B5.5 V	−5	580 s		
41434	132807		6 04 56.2	−5 20 37	−0.001	−0.04	7.90	0.00	0.4	B9.5 V		320 s		m
41361	113429	63 Ori	6 04 58.1	+5 25 12	0.000	0.00	5.67	1.04	0.3	G8 III	+20	100 s		
41304	95203		6 04 58.3	+14 23 20	+0.007	−0.17	6.8	0.5	3.4	F5 V	+37	35 mx		
41380	113430	66 Ori	6 04 58.3	+4 09 31	0.000	−0.01	5.63	1.04	−2.1	G7 II	+33	350 s		
41511	151093	17 Lep	6 04 59.0	−16 29 04	−0.001	0.00	4.93	0.24		A2	+20	15 mn		
41162	58716		6 05 02.6	+37 57 51	0.000	−0.01	6.40	0.5	4.0	F8 V	+5	30 s	4649	m
41074	40840	39 Aur	6 05 03.4	+42 58 54	−0.003	−0.14	5.87	0.30	2.6	F0 V	+34	45 s		
41054	40839		6 05 04.7	+45 29 16	+0.001	−0.01	8.0	1.1	−0.3	K5 III		410 s		
42026	234298		6 05 06.4	−56 13 57	−0.003	+0.05	6.5	1.3	−0.1	K2 III		180 s		
40827	25551		6 05 08.2	+59 23 36	+0.001	−0.05	6.34	1.10	−0.1	K2 III	+31	190 s		
41433	113435		6 05 08.5	+0 51 58	−0.001	−0.02	7.04	0.98		G7	−14		4662	m
40055	5730		6 05 09.5	+75 35 09	+0.005	−0.02	6.4	1.4	−0.3	K5 III	+4	220 s		
41344	95208		6 05 10.9	+15 26 57	0.000	−0.01	7.6	0.1	1.4	A2 V		170 s		
42025	234299		6 05 15.3	−54 24 31	−0.001	0.00	6.9	1.4	0.2	K0 III		130 s		
40225	5732		6 05 16.0	+73 59 57	−0.006	−0.06	7.60	0.5	4.0	F8 V		53 s	4603	m
41903	217713		6 05 19.0	−46 56 57	−0.004	−0.06	7.6	1.0	0.2	K0 III		250 mx		
40737	13700		6 05 19.5	+63 43 11	+0.005	−0.06	7.6	1.1	0.2	K0 III		200 mx		
41376	95212		6 05 22.2	+15 11 22	+0.001	−0.05	7.7	1.1	0.2	K0 III		300 s		
41460	113439		6 05 22.5	+0 36 46	−0.002	−0.02	7.02	1.60	−0.3	K5 III	0	250 s		
40872	25555		6 05 25.2	+58 44 10	+0.001	−0.03	8.0	1.1	0.2	K0 III		340 s		
41547	151098		6 05 26.9	−10 14 34	+0.001	+0.02	5.87	0.37		F0	+33			
41759	196483		6 05 27.0	−35 30 49	0.000	+0.02	5.80	0.02		A0				
41741	196482		6 05 28.2	−33 30 09	−0.002	−0.02	7.3	0.0	0.4	B9.5 V		230 s		
41821	217711		6 05 29.8	−40 03 40	−0.001	−0.04	7.8	1.2	−0.1	K2 III		380 s		
41269	58727		6 05 33.9	+33 35 56	+0.001	−0.01	6.10	−0.08	−0.8	B9 III p	+25	230 s		
41446	95222		6 05 41.1	+11 13 08	0.000	−0.02	8.0	0.1	1.4	A2 V		200 s		
41698	171222		6 05 45.6	−24 11 44	+0.001	−0.02	6.95	1.63		M6 III	+12			S Lep, v
41201	40845		6 05 48.2	+44 41 22	−0.002	−0.03	8.0	1.1	−0.1	K2 III		380 s		
41649	151104		6 05 48.3	−16 57 33	0.000	+0.01	7.8	0.1	0.6	A0 V		270 s		
41937	217715		6 05 50.1	−43 56 09	−0.001	−0.02	7.3	0.1	2.1	A5 V		110 s		
41161	40844		6 05 52.5	+48 14 58	+0.001	+0.01	6.79	−0.10		O9 n	+5		4655	m
41501	113442		6 05 52.5	+5 44 16	−0.001	−0.01	7.91	−0.12	0.4	B9.5 V		320 s		
41666	151106		6 05 55.6	−13 14 54	0.000	−0.02	7.70	0.4	2.6	F0 V		110 s	4685	m
41609	132827		6 05 57.7	−6 37 50	−0.001	0.00	8.0	0.1	1.4	A2 V		210 s		
41331	58734		6 05 58.2	+31 57 03	0.000	−0.03	7.9	0.9	3.2	G5 IV		87 s		
41481	95225		6 06 00.7	+12 28 46	−0.001	−0.01	8.00	0.00	0.4	B9.5 V		320 s	4675	m
42211	234308		6 06 01.7	−55 58 08	−0.006	+0.03	8.00	1.1	0.2	K0 III		260 mx		m
41608	132829		6 06 02.2	−5 52 50	+0.001	−0.02	7.20	1.6	−0.5	M1 III	+7	350 s		
41644	132833		6 06 04.2	−7 18 55	−0.002	+0.02	7.9	1.1	0.2	K0 III		340 s		
41843	171231		6 06 05.5	−29 45 31	0.000	−0.04	5.81	0.05	1.5	A IV		73 s		
41863	196490		6 06 05.6	−33 13 09	+0.003	+0.03	6.8	0.3	3.4	F5 V		48 s		
41645	132834		6 06 05.7	−7 37 19	0.000	+0.01	7.2	1.1	0.2	K0 III		260 s		
41398	77949		6 06 06.4	+28 56 04	−0.002	−0.02	7.46	0.32	−5.7	B2 Ib	+18	2300 s		
41568	113445		6 06 07.7	+6 01 25	−0.001	−0.02	8.0	0.1	0.6	A0 V		280 s		
41330	58739		6 06 08.4	+35 23 15	−0.011	−0.30	6.12	0.60	4.4	G0 V	−12	23 ts		m
41695	151110	18 θ Lep	6 06 09.3	−14 56 07	−0.001	+0.02	4.67	0.05		A0 n	+32	18 mn		
42525	249448	η¹ Dor	6 06 09.5	−66 02 23	+0.003	+0.02	5.71	−0.03		B9				
42024	217718		6 06 11.4	−45 48 58	+0.004	−0.06	7.6	0.6	4.0	F8 V		49 s		
40858	13713		6 06 12.1	+64 04 58	+0.007	+0.05	8.0	0.5	3.4	F5 V		80 s		
41594	113449		6 06 14.6	+3 20 07	+0.001	−0.02	8.00	1.46		K0				
41375	58743		6 06 15.7	+31 59 49	0.000	−0.01	7.1	0.0	0.0	B8.5 V		250 s		
41430	77956		6 06 17.0	+29 05 50	+0.002	−0.04	7.51	1.24	−0.2	K3 III		350 s		
41415	58748		6 06 21.8	+31 24 48	−0.001	0.00	8.0	0.9	3.2	G5 IV		91 s		
41429	77958		6 06 22.4	+29 30 45	+0.001	−0.01	6.03	1.68	−2.4	M3 II	−36	490 s	4673	m
41456	77960		6 06 23.9	+26 31 33	−0.001	0.00	7.6	0.9	0.3	G5 III	−20	280 s		
41661	132838		6 06 24.4	−0 57 51	0.000	−0.05	7.70	0.4	3.0	F2 V		86 s	4690	m
41580	95234		6 06 28.1	+10 45 02	+0.001	−0.02	7.19	−0.06	0.6	A0 V		210 s	4687	m
41841	171236		6 06 31.9	−23 06 39	−0.002	−0.02	5.47	0.08		A m	−15	21 mn	4704	m
41357	58749	40 Aur	6 06 35.0	+38 28 57	+0.001	−0.05	5.35	0.23		A3	+18	38 mn		
41818	151122		6 06 38.2	−16 31 05	−0.003	−0.03	6.7	0.9	3.2	G5 IV		49 s		
41692	132841		6 06 38.6	−4 11 38	−0.001	0.00	5.38	−0.13	−1.6	B5 IV	+20	250 s	4698	m
41879	171241		6 06 38.8	−23 32 27	0.000	−0.01	8.0	0.7	0.2	K0 III		360 s		
40956	13715		6 06 39.2	+63 27 14	−0.004	+0.01	6.4	1.1	0.2	K0 III	−15	170 s		
42058	217719		6 06 39.4	−41 13 12	0.000	0.00	7.0	−0.2	0.6	A0 V		190 s		
41641	113456		6 06 40.3	+6 43 52	−0.002	+0.02	7.9	0.2	2.1	A5 V		140 s		
41593	95235		6 06 40.4	+15 32 31	−0.008	−0.11	6.76	0.81	5.9	dK0	−12	17 ts		

HD	SAO	Star Name	α 2000	δ 2000	μ(α)	μ(δ)	V	B-V	M_v	Spec	RV	d(pc)	ADS	Notes
			h m s	° ′ ″	s	″								
37599	925		6 06 41.0	+85 17 04	0.000	-0.01	7.6	0.8	3.2	G5 IV		77 s		
42078	217720	π¹ Col	6 06 41.0	-42 17 55	-0.004	0.00	6.12	0.25		A m				
42169	217721		6 06 43.1	-49 13 22	0.000	+0.08	7.2	0.1	2.1	A5 V		100 s		
41898	171247		6 06 44.6	-24 55 35	-0.001	+0.05	7.1	0.7	2.6	F0 V		48 s		
41977	196499		6 06 45.5	-31 25 02	-0.001	-0.01	7.4	1.2	3.2	G5 IV		37 s		
41543	77971		6 06 48.6	+23 38 19	0.000	0.00	6.8	1.1	0.2	K0 III	-15	210 s		
41616	95240		6 06 50.0	+14 12 41	0.000	-0.03	7.8	0.4	2.6	F0 V		110 s		
42701	249453		6 06 51.0	-67 17 00	+0.002	0.00	7.6	1.1	0.2	K0 III		300 s		
41897	171248		6 06 51.2	-23 05 36	-0.003	-0.01	7.50	0.1	0.6	A0 V		240 s		m
41814	151126		6 06 51.8	-11 10 25	0.000	0.00	6.66	-0.15		B5	+13			
41715	113458		6 06 54.7	+2 10 23	0.000	-0.02	7.8	0.1	0.6	A0 V		260 s		
41976	171253		6 06 54.9	-28 31 06	-0.001	+0.05	8.0	1.3	0.2	K0 III		250 s		
41839	151127		6 06 55.8	-12 03 54	-0.001	0.00	7.9	1.1	0.2	K0 III		340 s		
41602	95244		6 06 55.8	+16 22 00	+0.001	-0.01	6.4	1.1	0.2	K0 III		170 s		
41615	95245		6 06 56.0	+16 10 37	-0.001	-0.03	7.8	0.1	1.7	A3 V		160 s		
41756	132845		6 06 56.2	-3 20 29	-0.001	0.00	6.93	-0.11		B5	+19			
41933	171251		6 06 57.5	-21 48 46	0.000	-0.02	6.0	1.7	-0.5	M4 III		180 s		
41794	132848		6 06 58.5	-6 12 08	-0.001	-0.01	6.6	0.2	2.1	A5 V		79 s		
41733	113465		6 06 59.3	+0 04 36	0.000	-0.02	7.4	1.1	-0.1	K2 III	+35	300 s		
42190	217725		6 06 59.7	-46 12 21	-0.009	-0.03	6.9	0.0	3.0	F2 V		59 s		
42168	217724		6 07 01.8	-45 05 29	+0.007	+0.05	6.51	1.16		K2				
40955	13717		6 07 02.5	+65 24 06	-0.003	-0.04	7.6	0.1	1.7	A3 V		160 s		
41563	77974		6 07 02.7	+26 40 16	0.000	+0.01	7.5	0.8	0.3	G6 III	-6	270 s		
42540	249451		6 07 03.4	-62 09 17	+0.003	-0.08	5.05	1.25	0.2	K0 III	+22	59 s		
42054	196503		6 07 03.6	-34 18 43	0.000	0.00	5.9	0.0	-1.0	B5.5 V	+18	240 s		
44327	258434		6 07 05.3	-81 03 47	-0.002	+0.04	7.3	0.2	2.6	F0 V		89 s		
41479	58756		6 07 06.7	+35 12 49	+0.001	-0.01	7.4	1.1	0.2	K0 III	-4	260 s		
42045	196505		6 07 07.7	-32 30 56	-0.001	+0.03	7.6	1.7	0.2	K0 III		120 s		
42044	196504		6 07 07.8	-31 53 51	+0.001	+0.05	7.9	0.3	3.4	F5 V		78 s		
41676	95248		6 07 08.9	+14 26 58	+0.001	0.00	7.7	1.1	-0.1	K2 III		340 s		
41640	95247		6 07 09.0	+16 33 24	+0.004	-0.06	7.2	0.5	3.4	F5 V		58 s		
41712	113463		6 07 09.4	+9 09 59	+0.005	-0.05	7.8	0.4	3.0	F2 V		88 s		
41770	113471		6 07 14.3	+1 30 31	-0.001	-0.03	7.9	0.4	3.0	F2 V		92 s		
42137	196509		6 07 17.1	-37 02 13	0.000	-0.02	7.5	1.6	-0.3	K5 III		320 s		
41675	95255		6 07 17.7	+16 43 09	0.000	-0.01	8.0	0.1	0.6	A0 V		290 s		
41657	77983		6 07 17.9	+20 49 12	-0.001	-0.01	7.9	1.1	-0.1	K2 III		370 s		
42138	196510		6 07 18.7	-37 12 01	-0.004	-0.01	7.5	0.7	3.2	G5 IV		73 s		
41600	77980		6 07 19.9	+26 40 56	-0.001	0.00	7.0	0.0	0.4	B9.5 V		200 s		
41808	113475		6 07 23.2	+3 05 29	-0.001	-0.01	7.8	0.1	0.6	A0 V		260 s		
41754	113470		6 07 23.3	+9 11 36	0.000	0.00	7.90	-0.05	0.0	B8.5 V		380 s		
42524	234328		6 07 24.3	-58 46 26	-0.001	0.00	7.6	1.7	0.2	K0 III		120 s		
42099	196508		6 07 25.8	-30 37 40	+0.001	-0.03	7.9	2.1	-0.3	K5 III		180 s		
41523	58762		6 07 25.8	+36 16 29	+0.001	0.00	7.00	0.00	0.4	B9.5 V		200 s	4691	m
41467	40868		6 07 26.8	+41 51 15	+0.001	-0.02	6.12	1.22	0.2	K0 III	+6	110 s		
41791	113473		6 07 27.1	+8 16 15	-0.002	0.00	8.01	-0.04	0.4	B9.5 V		330 s		
42136	196511		6 07 29.1	-35 13 30	0.000	-0.01	7.9	-0.1	0.6	A0 V		290 s		
42166	196512		6 07 30.7	-36 18 36	0.000	-0.03	7.7	0.1	0.6	A0 V		220 s		
41542	58763		6 07 31.4	+38 04 56	0.000	-0.02	7.0	1.1	0.2	K0 III		220 s		
42167	196514	θ Col	6 07 31.6	-37 15 10	0.000	0.00	5.02	-0.11		B9	+45			
41753	95259	67 v Ori	6 07 34.2	+14 46 06	0.000	-0.02	4.42	-0.17	-1.7	B3 V	+22	170 s		
42490	234327		6 07 34.7	-55 42 20	-0.012	-0.02	7.9	0.8	3.2	G5 IV		85 s		
41790	95260		6 07 35.9	+10 27 25	+0.001	+0.02	6.8	1.1	0.2	K0 III		200 s		
41789	95261		6 07 36.6	+11 00 30	-0.001	+0.02	7.5	1.1	0.2	K0 III		280 s		
41711	95258		6 07 37.8	+18 54 46	-0.008	-0.05	7.8	0.5	4.0	F8 V		56 s		
41690	77990		6 07 38.7	+21 52 23	-0.001	-0.01	7.71	0.21	-3.5	B1 V	+16	1000 s		m
42165	196516		6 07 41.4	-34 38 23	+0.001	+0.01	7.3	1.0	0.2	K0 III		260 s		
42042	151142	19 Lep	6 07 41.5	-19 09 57	0.000	+0.05	5.31	1.68	-0.5	gM2	+29	130 s		
41710	77996		6 07 46.9	+22 42 29	+0.001	+0.01	7.3	0.9	3.2	G5 IV		66 s		
42304	217729		6 07 47.0	-44 43 45	0.000	+0.03	6.8	0.6	2.1	A5 V		51 s		
41541	40872		6 07 50.7	+42 39 56	+0.001	-0.01	7.20	0.00	-1.0	B5.5 V	+5	410 s	4696	m
42303	217730	π² Col	6 07 52.9	-42 09 14	-0.001	-0.01	6.20	0.1	0.6	A0 V	+31	130 s		m
41832	95264		6 07 55.7	+14 20 31	0.000	-0.02	7.9	1.1	0.2	K0 III		330 s		
41599	58769		6 07 55.9	+36 03 54	+0.002	-0.05	7.1	1.1	-0.1	K2 III	+44	260 s		
42188	196517		6 07 56.7	-32 56 32	-0.002	+0.04	8.0	1.5	3.2	G5 IV		33 s		
42504	234330		6 07 56.8	-54 26 21	-0.001	+0.01	7.1	1.6	0.2	K0 III		110 s		
41709	77999		6 07 57.1	+26 24 01	-0.004	0.00	7.9	0.1	0.6	A0 V		270 s		
41638	58772		6 07 57.3	+34 09 49	-0.001	0.00	7.7	0.0	0.4	B9.5 V		280 s		
45575	--		6 07 58.7	-83 46 49			7.5	1.1	0.2	K0 III		290 s		
41727	78002		6 07 59.0	+25 27 21	-0.004	-0.03	8.0	0.0	0.4	B9.5 V		110 mx		
44828	258437		6 08 01.3	-82 02 39	-0.007	+0.02	7.5	0.0	2.6	F0 V		95 s		
41656	58774		6 08 02.2	+34 50 15	+0.001	-0.03	7.8	1.1	-0.1	K2 III		360 s		
41786	78006		6 08 02.3	+21 17 42	-0.003	-0.01	7.90	0.4	2.6	F0 V		110 s	4716	m
42074	151148		6 08 06.4	-15 54 03	-0.003	-0.02	7.8	0.5	3.4	F5 V		74 s		
42041	151147		6 08 08.0	-10 34 15	0.000	-0.01	7.8	0.1	0.6	A0 V		270 s		
41637	58776		6 08 11.6	+37 58 56	-0.001	-0.02	7.10	0.1	1.4	A2 V		140 s	4709	m

HD	SAO	Star Name	α 2000	δ 2000	μ(α)	μ(δ)	V	B−V	M_v	Spec	RV	d(pc)	ADS	Notes
			$h\ m\ s$	$°\ '\ ''$	s	$''$								
41237	13724		6 08 12.9	+62 58 49	−0.004	−0.01	7.9	0.2	2.1	A5 V		140 s		
41986	113493		6 08 17.3	+3 23 18	−0.001	0.00	8.0	0.1	0.6	A0 V		280 s		
42449	217733		6 08 17.7	−46 50 58	0.000	−0.02	8.0	1.4	0.2	K0 III		250 s		
41578	40877		6 08 18.9	+45 33 17	0.000	0.00	7.4	0.1	0.6	A0 V		230 s		
41636	40881		6 08 22.9	+41 03 20	−0.001	−0.05	6.36	1.05	0.2	K0 III	−87	160 s		
42116	151154		6 08 25.4	−11 08 46	+0.001	−0.01	6.80	0.1	1.4	A2 V		120 s	4741	m
---	217732		6 08 27.9	−41 16 25	−0.001	−0.03	8.0	0.4	3.2	G5 IV				
42391	--		6 08 28.1	−41 16 53			7.6	1.2	3.2	G5 IV				
41725	58782		6 08 29.3	+35 08 23	−0.001	−0.06	7.7	0.6	4.4	G0 V		46 s		
41943	95282		6 08 30.4	+13 58 13	+0.001	−0.02	7.4	0.1	−2.6	B2.5 IV	+15	910 s	4728	m
41785	58784		6 08 30.4	+30 34 21	0.000	−0.05	8.0	0.9	3.2	G5 IV		91 s		
42448	217735		6 08 34.5	−44 21 22	−0.002	+0.01	6.30	0.00	0.4	B9.5 V		150 s		m
41073	13721		6 08 37.1	+68 42 44	0.000	−0.04	7.9	0.1	0.6	A0 V		280 s		
41998	113499		6 08 37.7	+9 38 04	0.000	−0.01	7.78	0.00	0.0	B8.5 V		340 s		
42114	132880		6 08 43.0	−5 20 03	−0.001	−0.12	7.9	1.1	0.2	K0 III		130 mx		
43107	249461	v Dor	6 08 44.3	−68 50 36	−0.009	+0.02	5.06	−0.08	−0.2	B8 V	+18	110 s		
42133	132883		6 08 45.6	−7 56 08	−0.001	+0.01	6.7	0.1	1.4	A2 V		110 s		
42463	217737		6 08 46.4	−43 48 45	−0.001	+0.01	7.1	−0.1	0.6	A0 V		200 s		
42035	113503		6 08 47.2	+8 40 12	+0.002	+0.01	6.55	−0.07		B9				
42132	132884		6 08 47.5	−6 49 13	−0.003	+0.05	6.6	0.9	3.2	G5 IV		49 s		
41890	78029		6 08 47.6	+25 38 44	−0.002	−0.05	7.5	1.6	−0.5	M2 III		380 s		
42764	234348		6 08 48.1	−59 27 57	+0.001	+0.02	7.9	0.7	0.2	K0 III		340 s		
41807	58787		6 08 53.9	+36 40 42	+0.002	−0.03	7.7	1.4	−0.3	K5 III		360 s		
42131	132885		6 08 56.7	−2 02 14	+0.002	−0.03	7.2	0.8	3.2	G5 IV		63 s		
42301	171293		6 08 57.8	−22 25 39	0.000	−0.04	5.50	−0.01	0.0	A0 IV	+44	120 s		
42426	196532		6 08 57.8	−38 03 53	0.000	0.00	7.6	0.3	0.6	A0 V		160 s		
42111	113507		6 08 57.8	+2 29 58	0.000	−0.02	5.43	0.06	−0.3	A0 III−IV	+34	120 s	4749	m
42184	132886		6 08 58.0	−9 38 52	−0.003	+0.01	8.0	0.1	1.4	A2 V		210 s		
42092	113509		6 08 59.5	+2 29 46	−0.002	−0.01	6.92	0.03	0.6	A0 V		180 s		
45040	258438		6 09 00.0	−82 09 25	−0.013	+0.04	7.6	−0.7	0.0	B8.5 V		180 mx		
42503	217738		6 09 02.6	−41 07 07	0.000	−0.03	7.4	0.2	0.6	A0 V		160 s		
41849	58789		6 09 03.6	+34 53 37	−0.001	−0.02	7.8	1.6	−0.5	M4 III		420 s		
41765	40889		6 09 05.4	+41 32 58	−0.002	−0.04	7.7	0.0	0.0	B8.5 V		330 s		
41889	58791		6 09 06.6	+31 15 55	0.000	0.00	7.50	0.1	1.4	A2 V		160 s	4734	m
42205	132888		6 09 06.8	−8 18 08	+0.001	−0.02	6.8	1.1	0.2	K0 III		210 s		
42263	151166		6 09 07.1	−15 42 07	0.000	−0.01	8.0	0.1	0.6	A0 V		300 s		
42089	113512		6 09 07.2	+7 30 50	0.000	+0.02	6.7	0.6	4.4	G0 V		29 s		
42282	151168		6 09 15.2	−13 13 16	+0.001	−0.01	8.0	0.4	2.6	F0 V		120 s		
42326	151171		6 09 16.9	−17 17 30	+0.001	+0.01	7.7	0.1	0.6	A0 V		260 s		
42327	151173		6 09 20.2	−18 07 34	0.000	−0.02	6.35	−0.03		A0				m
41589	25592		6 09 23.2	+55 57 48	+0.005	−0.03	7.8	1.1	0.2	K0 III		270 mx		
42683	217747		6 09 23.2	−49 33 45	−0.003	+0.08	6.49	0.52	4.0	F8 V		32 s		
41994	78041		6 09 26.5	+27 11 38	0.000	0.00	8.00	0.9	−2.1	G5 II		920 s		
42719	234350		6 09 27.4	−51 23 23	−0.004	+0.03	7.5	0.5	4.4	G0 V		42 s		
42461	196536		6 09 29.9	−31 01 27	0.000	0.00	7.5	1.5	0.2	K0 III		150 s		
42279	132894		6 09 30.6	−6 32 11	−0.001	−0.02	7.6	0.0	0.4	B9.5 V		270 s		
41867	40894		6 09 30.8	+41 14 41	+0.002	+0.03	7.9	1.1	−0.1	K2 III		370 s		
42049	78045		6 09 32.3	+22 11 24	−0.001	−0.01	5.93	1.63	0.2	K2 III	+8	85 s		
42416	171311		6 09 32.5	−25 16 33	0.000	0.00	8.0	0.4	4.0	F8 V		62 s		
41905	58801		6 09 34.2	+38 17 45	−0.001	−0.01	7.6	0.1	0.6	A0 V		240 s		
42341	151178		6 09 34.2	−14 35 03	−0.004	+0.04	5.56	1.16	−0.1	gK2	+31	140 s		
42278	132900		6 09 36.1	−5 42 40	+0.003	+0.01	6.17	0.34		F0				
42203	113523		6 09 37.5	+4 43 01	−0.001	−0.03	7.8	0.0	0.0	B8.5 V		340 s		
42918	234356		6 09 37.8	−59 31 20	−0.001	+0.02	7.6	0.7	0.0	B8.5 V		120 s		
41764	40892		6 09 39.0	+49 03 47	+0.001	−0.02	7.9	1.4	−0.3	K5 III		430 s		
42625	217743		6 09 39.3	−41 58 49	−0.007	+0.03	7.6	−0.4	4.0	F8 V		52 s		
---	217746		6 09 39.4	−41 58 49	−0.005	+0.02	7.6	0.0	4.0	F8 V		51 s		
42088	78049		6 09 39.5	+20 29 16	0.000	0.00	7.55	0.06		O6	+23			
41847	40898		6 09 39.7	+43 08 27	+0.001	−0.01	7.20	0.1	0.6	A0 V		200 s		m
42423	171314		6 09 41.7	−22 07 09	−0.001	0.00	8.0	0.7	0.2	K0 III		360 s		
42087	78050	3 Gem	6 09 44.0	+23 06 48	+0.001	0.00	5.75	0.21	−5.7	B2.5 Ib	+16	1200 s	4751	m
42500	171319		6 09 45.2	−29 27 36	0.000	+0.01	7.9	1.7	0.2	K0 III		140 s		
42414	171316		6 09 46.2	−20 30 18	−0.001	−0.04	7.9	1.2	−0.1	K2 III		390 s		
42257	113528		6 09 46.6	+0 58 49	−0.001	−0.02	7.5	0.0	0.4	B9.5 V		260 s		
42486	171318		6 09 47.1	−26 42 03	−0.001	+0.02	6.27	1.01	0.2	K0 III		160 s		
42443	171317		6 09 47.8	−22 46 27	+0.005	+0.07	5.71	0.44	3.7	dF6	+22	25 s		m
42256	113530		6 09 49.2	+2 52 09	+0.001	−0.08	6.7	1.1	0.2	K0 III		150 mx		
42180	95308		6 09 53.5	+14 51 10	0.000	0.00	7.30	−0.05	0.4	B9.5 V	+16	240 s		
42499	171321		6 09 56.1	−25 37 32	−0.002	+0.05	7.66	0.57	4.0	F8 V		50 s	4775	m
42235	113529		6 09 56.8	+9 28 03	+0.001	−0.03	7.29	−0.11	0.6	A0 V		220 s		
41597	25597	37 Cam	6 09 59.0	+58 56 09	+0.003	+0.02	5.36	1.09	0.3	G8 III	+31	81 s		
42816	234354		6 09 59.1	−51 24 22	0.000	+0.01	7.9	1.5	−0.3	K5 III		430 s		
43224	249462		6 10 06.8	−65 18 43	+0.001	+0.03	7.6	1.0	0.2	K0 III		300 s		
42569	171325		6 10 07.5	−29 49 06	0.000	+0.02	7.4	0.2	0.4	B9.5 V		190 s		
42317	113538		6 10 09.2	+3 18 45	−0.001	−0.01	7.9	0.9	3.2	G5 IV		87 s		

HD	SAO	Star Name	α 2000	δ 2000	μ(α)	μ(δ)	V	B-V	M$_v$	Spec	RV	d(pc)	ADS	Notes
			h m s	° ' "	s	"								
42299	113536		6 10 09.9	+6 19 14	-0.001	+0.01	7.9	0.1	1.7	A3 V		170 s		
42682	217753		6 10 10.3	-40 21 13	-0.003	+0.08	5.4	2.0	-0.5	M2 III	-19	85 s		
41929	40907		6 10 11.8	+46 45 33	+0.002	0.00	7.1	1.1	0.2	K0 III		240 s		
42106	58818		6 10 14.5	+30 32 59	+0.004	-0.04	7.80	0.8	0.3	G7 III	+24	150 mx		
43834	256274	α Men	6 10 14.6	-74 45 11	+0.031	-0.21	5.09	0.72	5.4	G5 V	+35	8.7 t		
42233	95321		6 10 15.4	+15 54 23	0.000	-0.03	7.40	0.9	3.2	G5 IV		69 s		m
42459	151187		6 10 16.4	-15 03 35	-0.003	0.00	6.7	0.1	0.6	A0 V		170 s		
42129	58820		6 10 16.9	+30 25 31	+0.002	-0.04	8.0	0.9	3.2	G5 IV		91 s		
42933	234359	δ Pic	6 10 17.9	-54 58 07	-0.001	+0.01	4.81	-0.23		B1 n	-2			
42275	95323		6 10 19.8	+13 48 40	+0.002	-0.12	7.5	0.1	3.0	F2 V		74 mx		
42553	171328		6 10 19.9	-25 29 05	+0.001	-0.03	7.7	1.4	0.2	K0 III		190 s		
41928	40909		6 10 21.4	+47 44 21	+0.003	-0.01	7.9	1.6	-0.5	M4 III		490 s		
42409	132912		6 10 21.7	-2 00 27	+0.001	-0.02	8.0	0.9	3.2	G5 IV		90 s		
42568	171330		6 10 21.8	-25 37 56	0.000	0.00	8.0	1.2	3.2	G5 IV		51 s		
42353	113542		6 10 23.8	+2 54 08	0.000	+0.01	6.8	0.0	0.4	B9.5 V		190 s		
44148	256277		6 10 26.6	-77 06 19	-0.003	+0.03	6.9	0.3	0.6	A0 V		110 s		
42853	217757		6 10 27.9	-48 57 25	-0.004	-0.04	7.5	0.2	3.0	F2 V		81 s		
42216	78063		6 10 29.9	+22 59 52	0.000	-0.01	7.30	0.00	0.4	B9.5 V	-5	230 s	4768	m
42273	95328		6 10 31.6	+17 41 39	-0.001	-0.02	7.90	1.1	6.3	K2 V		21 s		
42698	196559		6 10 31.9	-34 49 00	0.000	-0.01	6.7	0.7	4.4	G0 V		23 s		
42621	171339		6 10 34.5	-27 09 15	-0.002	-0.05	5.72	1.07	0.2	K0 III	+1	110 s		
43092	234361		6 10 36.0	-58 51 17	0.000	-0.01	7.4	1.5	-0.1	K2 III		200 s		
41413	13734		6 10 37.8	+68 24 46	-0.003	-0.14	7.6	0.4	2.6	F0 V		100 mx		
42065	40912		6 10 37.9	+40 34 39	+0.001	-0.01	7.8	0.5	4.0	F8 V		56 s		
42157	58828		6 10 39.4	+31 57 33	-0.001	-0.04	7.8	1.1	0.0	K1 III		340 s		
42156	58827		6 10 39.5	+32 02 53	+0.002	-0.03	7.6	0.1	0.6	A0 V		240 s		
42834	217758		6 10 39.8	-45 16 55	-0.001	0.00	6.31	-0.03		A0				
43770	256276		6 10 42.5	-73 00 47	-0.008	+0.03	7.8	0.1	0.6	A0 V		180 mx		
42294	95332		6 10 42.6	+18 42 10	-0.001	0.00	8.0	0.1	0.6	A2 IV		290 s		
42154	58829		6 10 44.5	+34 15 38	-0.001	-0.01	7.9	0.1	0.6	A0 V		270 s		
42352	95335		6 10 47.5	+13 39 35	+0.002	0.00	6.95	0.00	-1.0	B5.5 V	+14	380 s		
43144	234362		6 10 49.8	-59 48 01	0.000	+0.12	7.8	0.4	4.0	F8 V		57 s		
43199	249464		6 10 52.7	-61 29 59	0.000	+0.03	7.4	0.5	2.6	F0 V		69 s		
42272	78066		6 10 53.1	+26 00 54	+0.001	-0.01	7.29	2.77		N7.7	+48			TU Gem, v
41783	25603		6 10 53.2	+59 10 09	-0.001	-0.01	7.2	1.1	0.2	K0 III		230 s		
42032	40913		6 10 55.8	+47 30 15	-0.002	-0.07	7.8	0.1	1.7	A3 V		81 mx		
42401	95343		6 10 59.1	+11 59 41	0.000	-0.01	7.35	-0.06		B5	+8			
42536	132924		6 11 01.2	-6 45 15	0.000	0.00	6.15	0.01		A0				
42731	196563		6 11 01.5	-31 46 29	-0.001	-0.04	7.5	1.8	0.2	K0 III		97 s		
42351	95337		6 11 01.7	+18 07 45	-0.002	-0.05	6.33	1.35	-2.2	K1 II	-3	430 s		
42551	132926		6 11 02.0	-7 17 04	-0.002	-0.01	7.6	0.0	0.0	B8.5 V		330 s		
42481	113557		6 11 02.9	+2 19 50	-0.002	0.00	7.7	1.1	-0.1	K2 III		340 s		
42141	40920		6 11 03.6	+41 13 57	0.000	-0.01	7.6	0.0	0.4	B9.5 V		260 s		
41965	25608		6 11 05.1	+53 37 11	0.000	0.00	8.0	1.1	0.2	K0 III		360 s		
42152	40922		6 11 05.2	+40 32 06	0.000	-0.04	8.0	0.1	0.6	A0 V		280 s		
42533	132925		6 11 06.9	-1 53 02	0.000	-0.01	8.0	0.1	0.6	A0 V		290 s		
42577	132929		6 11 09.5	-7 41 49	-0.001	+0.04	7.8	0.4	3.0	F2 V		89 s		
42128	40921		6 11 13.2	+44 57 09	+0.001	+0.01	7.5	0.8	3.2	G5 IV		74 s		
42151	40923		6 11 13.2	+41 55 42	-0.001	-0.01	7.8	1.1	0.2	K0 III		310 s		
42729	171356		6 11 13.5	-26 28 56	-0.002	0.00	6.09	-0.04		A0				
41782	13746		6 11 13.6	+60 27 12	-0.002	-0.02	6.7	1.6	-0.5	M2 III		260 s		
42314	78072		6 11 14.8	+28 54 28	-0.001	+0.01	7.6	0.0	0.0	B8.5 V		320 s		
43455	249469	η² Dor	6 11 15.0	-65 35 23	-0.004	+0.11	5.01	1.62		Mb	+35	21 mn		
42917	217764		6 11 15.3	-44 54 52	-0.001	0.00	7.1	1.4	-0.1	K2 III		200 s		
42801	196568		6 11 15.7	-33 52 15	+0.002	0.00	7.4	0.6	-0.1	K2 III		310 s		
42084	40919		6 11 15.8	+47 54 24	-0.001	-0.03	6.9	0.1	1.4	A2 V		120 s		
42379	78074		6 11 18.0	+21 33 50	0.000	+0.01	7.39	0.35	-5.1	B1 II		1500 s		
42604	132934		6 11 18.6	-9 38 09	-0.001	0.00	8.0	0.1	0.6	A0 V		300 s		
42313	58838		6 11 19.4	+30 40 23	0.000	-0.02	8.00	0.1	0.6	A0 V		290 s	4779	m
42659	151199		6 11 21.6	-15 47 34	-0.001	0.00	6.7	0.1	1.7	A3 V		100 s		
42400	78077		6 11 23.1	+20 54 19	0.000	0.00	6.83	0.18	-3.7	B5 II	+10	920 s		
42619	132935		6 11 23.4	-7 15 02	0.000	0.00	7.8	1.6	-0.5	M2 III		470 s		
42779	171359		6 11 27.7	-27 03 29	-0.002	-0.02	7.4	0.6	1.4	A2 V		71 s		
42477	95352		6 11 27.9	+13 38 19	0.000	-0.02	5.90	0.00	0.4	B9.5 V	+13	110 s		
42196	40928		6 11 30.0	+43 09 54	+0.001	+0.02	7.10	0.8	3.2	G5 IV		60 s	4774	m
42173	40927		6 11 30.6	+44 08 04	+0.001	-0.02	7.6	0.1	1.4	A2 V		170 s		
42195	40929		6 11 32.0	+43 47 59	+0.002	-0.03	7.2	1.1	-0.1	K2 III		280 s		
42398	78079	5 Gem	6 11 32.2	+24 25 12	0.000	-0.06	5.80	1.11	0.2	K0 III	+22	110 s		
42640	132938		6 11 32.6	-6 58 18	0.000	-0.01	8.0	0.4	3.0	F2 V		98 s		
42213	40930		6 11 33.2	+42 09 05	-0.003	-0.02	7.9	0.0	0.4	B9.5 V		300 s		
42397	78081		6 11 34.7	+25 00 35	+0.002	-0.02	7.8	0.6	2.8	G0 IV		99 s		
42127	40924	41 Aur	6 11 36.4	+48 42 46	+0.001	-0.07	5.78	0.10	0.6	A0 V	+33	92 s	4773	m
42815	171364		6 11 37.0	-27 55 40	+0.003	+0.07	7.60	0.4	3.0	F2 V		83 s	4806	m
42476	95354		6 11 39.0	+17 22 38	0.000	-0.03	7.20	0.03	0.0	A0 IV	+29	250 s	4789	m
42679	151201		6 11 40.7	-10 15 03	+0.003	0.00	7.2	1.1	0.2	K0 III		250 s		

HD	SAO	Star Name	α 2000	δ 2000	μ(α)	μ(δ)	V	B-V	M$_V$	Spec	RV	d(pc)	ADS	Notes
			h m s	° ′ ″	s	″								
42711	151204		6 11 40.8	-12 35 55	0.000	+0.04	7.95	0.35		F2				
42548	113569		6 11 41.4	+7 46 43	0.000	+0.07	7.6	0.4	2.6	F0 V		100 s		
42814	171367		6 11 41.7	-27 13 52	-0.001	-0.06	7.90	1.1	0.2	K0 III		350 s		m
42367	58842		6 11 42.2	+31 10 24	-0.001	+0.01	8.0	0.1	1.4	A2 V		200 s		
42747	151208		6 11 43.1	-18 11 10	0.000	+0.02	6.7	1.1	-0.1	K2 III		230 s		
42657	132941		6 11 43.6	-4 39 56	-0.001	-0.01	6.18	-0.08		B9			4799	m
42083	25613		6 11 46.0	+52 38 50	+0.001	-0.07	6.3	0.1	1.4	A2 V	+13	72 mx		
42832	171369		6 11 46.4	-26 43 59	-0.002	-0.02	7.9	1.1	3.2	G5 IV		56 s		
42475	78092		6 11 51.4	+21 52 07	+0.001	+0.01	6.60	1.6	-5.7	M1 Iab	+17	2200 s		TV Gem, v
42690	132944		6 11 51.7	-6 33 01	-0.001	-0.01	5.05	-0.20	-2.5	B2 V	+29	320 s		
43140	234367		6 11 52.1	-50 47 34	0.000	-0.02	7.5	0.1	0.6	A0 V		200 s		
41476	5774		6 11 54.7	+71 07 51	-0.001	+0.02	7.5	0.5	3.4	F5 V		67 s		
42677	132943		6 11 56.1	-1 19 25	0.000	0.00	7.8	0.0	0.0	B8.5 V		350 s		
42474	78094		6 11 56.2	+23 12 25	0.000	0.00	7.30	1.63	-5.6	M2 Iab pe	+16	2600 s		WY Gem, v
42560	95362	70 ξ Ori	6 11 56.3	+14 12 31	0.000	-0.02	4.48	-0.18	-1.7	B3 V	+24	170 s		m
44165	256278		6 11 58.4	-75 05 00	0.000	+0.04	7.4	0.8	3.2	G5 IV		71 s		
42597	113578		6 11 58.9	+7 23 30	+0.002	0.00	7.05	-0.09	-2.6	B2.5 IV	+23	760 s		
42618	113580		6 12 00.4	+6 46 58	+0.013	-0.27	6.87	0.63	5.0	G4 V	-54	26 ts		
42509	95359	68 Ori	6 12 01.3	+19 47 26	0.000	-0.01	5.70	-0.07	0.4	B9.5 V	+31	120 s		m
42596	95368		6 12 01.8	+11 30 45	-0.001	-0.01	7.6	1.1	-0.1	K2 III		330 s		
42544	95361		6 12 02.9	+19 31 00	0.000	-0.02	7.90	1.1	-0.1	K2 III		380 s		
42545	95365	69 Ori	6 12 03.2	+16 07 49	+0.001	-0.02	4.95	-0.14	-1.1	B5 V	+22	160 s		
42454	78095		6 12 05.5	+29 29 33	+0.001	+0.01	7.32	1.32	-4.5	G2 Ib		1300 s	4790	m
42889	171378		6 12 07.2	-26 03 01	-0.005	0.00	7.2	1.7	0.2	K0 III		94 s		
43618	249477		6 12 11.0	-65 31 52	+0.002	+0.15	6.83	0.46	3.4	F5 V		46 s		m
42675	113584		6 12 12.2	+4 40 26	0.000	0.00	7.9	0.1	0.6	A0 V		270 s		
42594	95372		6 12 13.7	+13 41 12	-0.001	-0.01	7.8	0.0	0.4	B9.5 V		290 s		
42655	95375		6 12 15.8	+10 20 00	-0.002	-0.01	7.50	-0.05	-1.0	B5.5 V	+7	470 s		
42543	78098	6 Gem	6 12 19.1	+22 54 30	0.000	-0.01	6.2	1.6	-7.2	M1 Ia	+22	3100 s		BU Gem, q
42471	58852		6 12 20.2	+32 41 36	0.000	0.00	5.78	1.66	-0.1	K2 III	-51	76 s		
42968	196580		6 12 22.2	-30 28 54	-0.004	0.00	7.9	1.1	0.4	B9.5 V		70 s		
42811	151218		6 12 27.0	-10 29 21	-0.001	-0.04	7.4	1.1	0.2	K0 III		280 s		
43071	196585		6 12 29.0	-36 33 51	-0.001	-0.01	6.89	-0.14	-1.7	B3 V		520 s		
42849	151220		6 12 29.4	-13 37 24	+0.001	+0.03	7.4	0.0	0.4	B9.5 V		250 s		
42988	171389		6 12 30.6	-29 01 05	-0.008	-0.06	7.44	0.51		F5				
43519	249475		6 12 32.1	-61 28 26	-0.001	+0.01	7.30	0.00	0.4	B9.5 V		240 s		m
42350	40940		6 12 35.6	+47 25 15	+0.002	+0.02	7.54	0.03		B9				
42530	58856		6 12 36.0	+30 46 43	+0.001	-0.04	7.8	0.1	1.7	A3 V		160 s		
42529	58857		6 12 39.7	+32 58 21	-0.001	-0.01	8.0	0.0	0.4	B9.5 V		320 s		
42987	171394		6 12 40.7	-27 43 35	-0.001	-0.01	7.6	0.8	0.2	K0 III		310 s		
43010	171396		6 12 41.8	-27 44 48	-0.004	-0.04	7.7	1.4	0.0	K1 III		190 mx		
42824	132965		6 12 44.4	-2 30 16	-0.003	+0.01	6.62	0.05		A0				
42927	151227		6 12 46.2	-17 45 47	-0.001	+0.02	6.53	-0.16		B4 n	+8			
43518	234379		6 12 47.8	-59 56 07	+0.005	+0.03	7.7	0.9	0.2	K0 III		280 mx		
43122	196591		6 12 49.7	-37 04 54	-0.002	-0.01	8.0	-0.4	0.0	B8.5 V		400 s		
42883	151225		6 12 50.6	-10 17 24	-0.003	-0.04	7.6	1.1	-0.1	K2 III		350 s		
42863	132970		6 12 50.8	-7 17 56	0.000	+0.04	8.0	1.6	-0.5	M2 III		480 s		
41927	13756	36 Cam	6 12 51.0	+65 43 06	+0.002	-0.03	5.32	1.34	-1.1	K2 II-III	+7	170 s		
43056	171401		6 12 54.8	-27 44 07	-0.001	-0.01	8.0	0.7	3.2	G5 IV		93 s		
42771	113590		6 12 54.9	+9 02 11	-0.001	+0.02	7.97	-0.08	0.4	B9.5 V		330 s		
42881	132972		6 12 55.5	-8 43 47	+0.001	-0.03	6.7	0.0	0.4	B9.5 V		180 s		
42773	113591		6 12 58.2	+7 16 41	-0.002	-0.01	7.00	1.56		K2				
41096	5770		6 12 58.5	+77 17 12	+0.002	0.00	7.5	0.9	3.2	G5 IV		71 s		
43070	171406		6 12 59.3	-28 27 49	+0.001	+0.03	7.3	0.5	0.6	A0 V		110 s		
42770	95389		6 12 59.3	+10 17 18	-0.001	+0.02	6.59	-0.06	0.4	B9.5 V		170 s		
42787	113592		6 12 59.5	+6 00 58	0.000	-0.01	6.46	1.63	-0.5	M2 III		230 s		
43120	196594		6 13 01.3	-34 34 35	+0.001	-0.07	8.0	0.5	-0.1	K2 III		280 mx		
42786	113593		6 13 01.9	+7 32 03	+0.001	-0.02	8.0	0.1	0.6	A0 V		280 s		
42879	132973		6 13 02.8	-4 56 07	+0.001	-0.05	6.9	1.1	0.2	K0 III		210 s		
42845	113596		6 13 03.7	+3 30 31	-0.001	-0.02	7.51	-0.13	0.0	B8.5 V		320 s		
42758	95390		6 13 10.8	+19 00 32	0.000	-0.03	7.44	-0.10	-1.2	B8 III		520 s		
42911	132977		6 13 12.5	-5 00 35	-0.001	0.00	7.34	1.02	0.3	G7 III	+42	230 s		
42807	95394		6 13 12.6	+10 37 39	+0.006	-0.28	6.45	0.67	5.3	G6 V	+3	18 ts		
42821	113597		6 13 15.7	+9 35 54	0.000	+0.02	8.0	0.1	0.6	A0 V		290 s		
42963	151234		6 13 18.4	-10 38 39	-0.001	-0.02	7.8	0.4	3.0	F2 V		89 s		
43028	151241		6 13 24.6	-15 23 01	0.000	+0.01	6.8	1.1	0.2	K0 III	-3	210 s		
42349	25623		6 13 27.5	+55 35 06	0.000	0.00	7.8	0.8	3.2	G5 IV		84 s		
42860	113603		6 13 30.4	+9 37 43	0.000	0.00	7.58	-0.13	0.0	B8.5 V		330 s		
42617	58866		6 13 30.5	+37 09 31	0.000	-0.02	6.6	0.1	0.6	A0 V		150 s		
42784	95397		6 13 33.3	+18 40 49	+0.002	-0.02	6.20	-0.08	-0.2	B8 V		190 s		
43179	171425		6 13 33.3	-29 23 46	-0.002	-0.01	6.4	0.1	0.0	B8.5 V		140 s		
42959	132986		6 13 35.6	-2 15 11	-0.002	0.00	7.79	0.00	0.0	B8.5 V		350 s		
43598	234387		6 13 40.3	-56 55 05	-0.001	-0.01	6.7	1.1	0.2	K0 III		190 s		
43921	249486		6 13 40.8	-66 40 47	-0.003	+0.07	7.6	0.5	3.4	F5 V		68 s		
43504	234384		6 13 41.8	-51 40 04	-0.001	-0.04	6.9	0.5	0.6	A0 V		90 s		

HD	SAO	Star Name	α 2000	δ 2000	μ(α)	μ(δ)	V	B−V	M$_v$	Spec	RV	d(pc)	ADS	Notes
42616	40959		6 13 43.1	+41 41 51	0.000	0.00	7.15	0.09		A2 p				
42841	95402		6 13 44.1	+19 19 59	0.000	−0.02	7.5	0.8	−2.1	G5 II		760 s		
43162	171428		6 13 45.0	−23 51 43	−0.005	+0.12	6.39	0.72		G5				
42466	25630		6 13 45.2	+51 10 21	−0.001	−0.07	6.04	1.06	0.0	K1 III	+11	160 s		
42722	58871		6 13 48.6	+33 39 06	−0.001	−0.01	7.9	1.1	0.2	K0 III		330 s		
42589	40958		6 13 49.4	+44 45 09	+0.001	−0.02	7.7	1.1	0.2	K0 III		320 s		
42983	113610		6 13 52.6	+2 48 30	+0.004	−0.28	7.9	1.1	0.2	K0 III	−26	58 mx		
43023	132994		6 13 54.2	−3 44 29	−0.001	+0.02	5.83	0.94	5.5	dG7	+49	12 s		d?
43065	132995		6 13 56.2	−6 49 24	−0.001	−0.01	7.9	0.0	0.0	B8.5 V		370 s		
43066	132997		6 13 56.9	−7 14 51	−0.002	+0.03	6.7	0.0	0.4	B9.5 V		180 s		
42508	25631		6 13 59.0	+52 00 41	+0.002	−0.09	7.8	0.4	3.0	F2 V		89 s		
42767	58874		6 14 00.5	+33 14 38	0.000	−0.03	6.7	1.4	−0.3	K5 III		240 s		
42737	58873		6 14 03.0	+37 19 13	−0.002	−0.01	7.9	0.0	0.4	B9.5 V		300 s		
42062	13759		6 14 08.4	+67 50 18	0.000	+0.02	7.66	1.43		K2				
42871	78121		6 14 08.8	+23 59 12	−0.001	−0.01	8.00	0.4	2.6	F0 V		120 s		m
−−	78122		6 14 09.1	+23 57 17	0.000	0.00	8.0							
43021	113614		6 14 10.3	+2 34 30	0.000	+0.02	8.0	0.1	0.6	A0 V		280 s		
43452	217794		6 14 11.8	−41 39 37	−0.004	+0.05	7.6	0.0	3.0	F2 V		85 s		
43047	113615		6 14 12.5	+3 54 07	−0.001	−0.02	7.31	−0.05	0.4	B9.5 V		240 s		
42955	95417		6 14 18.4	+14 29 37	0.000	0.00	7.70	0.1	1.2	A1 V		200 s	4835	m
43236	151255		6 14 18.9	−19 31 31	0.000	+0.01	7.90	1.64	−0.5	M3 III	+72	460 s		
42736	58877		6 14 20.5	+39 40 21	+0.003	−0.06	7.8	0.1	0.6	A0 V		60 mx		
42999	95423		6 14 26.5	+11 48 49	0.000	0.00	7.66	0.06	0.6	A0 V		230 s		
42939	78129		6 14 28.4	+21 46 54	−0.001	0.00	7.8	1.1	0.2	K0 III		310 s		
42954	95419		6 14 28.5	+17 54 22	−0.001	−0.01	6.50	0.2		A m	+24			m
43083	113618		6 14 30.1	+4 32 34	+0.001	−0.02	6.9	1.1	0.2	K0 III		220 s		
43369	171448		6 14 30.3	−29 36 13	0.000	+0.02	7.10	0.1	0.6	A0 V		200 s	4858	m
43020	95425		6 14 31.1	+11 47 50	−0.001	−0.01	7.30	1.1	0.2	K0 III	−48	260 s		m
43157	133003		6 14 36.6	−4 34 06	−0.001	0.00	5.83	−0.17		A0			4846	m
43113	113626		6 14 37.2	+2 17 23	−0.001	−0.01	7.50	0.00	0.4	B9.5 V		250 s		
44533	256286		6 14 41.9	−73 37 36	0.000	+0.02	6.8	−0.5	0.4	B9.5 V		190 s		
43044	95431		6 14 42.6	+14 35 11	0.000	−0.02	7.03	0.01	−0.2	B8 V	+16	240 s	4840	m
43043	95429		6 14 43.3	+16 02 14	+0.001	+0.03	6.7	0.9	0.3	G8 III	+32	190 s		
43195	133007		6 14 43.9	−4 26 57	−0.001	−0.01	7.30	1.1	0.2	K0 III		260 s	4851	m
42527	25633		6 14 46.3	+57 46 27	+0.006	−0.02	7.7	1.1	0.2	K0 III		260 mx		
43042	95432	71 Ori	6 14 50.8	+19 09 23	−0.007	−0.19	5.20	0.44	3.7	F6 V	+36	20 s	4842	m
43232	133012	5 γ Mon	6 14 51.3	−6 16 29	−0.001	−0.02	3.98	1.32	−0.2	K3 III	−5	66 s	4853	m
42995	78135	7 η Gem	6 14 52.6	+22 30 24	−0.004	−0.01	3.28	1.60	−0.5	M3 III	+19	57 s	4841	Propus, m,v
43251	133015		6 14 54.5	−8 48 14	−0.003	+0.02	7.24	−0.10	0.0	B8.5 V		280 s		
43450	196623		6 14 54.8	−30 51 11	−0.002	−0.01	7.5	0.6	1.4	A2 V		84 s		
43080	95437		6 14 57.5	+16 44 46	−0.002	−0.03	7.8	0.1	0.6	A0 V		260 s		
42783	40967		6 14 58.6	+46 23 43	0.000	−0.02	7.3	0.0	0.0	B8.5 V		290 s		
42061	5785		6 15 00.3	+70 40 03	−0.001	−0.04	7.80	0.8	0.3	G7 III	+13	320 s		
41497	5779		6 15 01.5	+76 30 18	−0.002	−0.02	8.00	0.5	4.0	F8 V	−17	63 s	4771	m
43471	171461		6 15 01.6	−29 21 57	0.000	−0.03	7.3	0.1	1.4	A2 V		150 s		
43288	133019		6 15 03.8	−8 37 36	−0.007	+0.03	7.8	0.5	3.4	F5 V		74 s		
44447	256285		6 15 05.9	−71 42 10	−0.005	+0.06	6.64	0.56		F8				
43396	171458		6 15 08.3	−20 16 20	0.000	+0.03	5.91	1.32		K0				
43112	95444		6 15 08.5	+13 51 04	+0.002	+0.01	5.91	−0.23	−3.5	B1 V	+36	760 s		m
42782	40969		6 15 10.2	+48 50 03	0.000	−0.02	7.19	−0.15	−1.0	B5.5 V		440 s		
43188	113633		6 15 11.9	+8 26 47	0.000	+0.01	7.7	1.1	0.2	K0 III		300 s		
42980	58898		6 15 15.6	+31 28 19	+0.002	−0.04	8.0	0.9	3.2	G5 IV		91 s		
42903	58894		6 15 17.3	+39 56 46	+0.001	+0.01	7.9	1.6	−0.5	M4 III		440 s		
43429	151274		6 15 17.5	−18 28 36	+0.001	−0.05	6.00	1.05	0.2	K0 III		130 s		
43152	95445		6 15 21.9	+16 25 51	+0.001	−0.02	7.50	1.4	−4.4	K5 Ib	−10	1900 s		
43039	78143	44 κ Aur	6 15 22.6	+29 29 53	−0.005	−0.26	4.35	1.02	0.3	G8 III	+20	46 mx		
43393	151272		6 15 23.6	−11 54 16	0.000	+0.02	6.6	0.1	1.4	A2 V		110 s		
43095	78146		6 15 24.7	+25 20 02	0.000	+0.01	7.7	1.4	−0.3	K5 III		370 s		
43153	95447	72 Ori	6 15 25.1	+16 08 35	0.000	−0.01	5.30	−0.14	−0.6	B7 V	+23	150 s		
43362	133029		6 15 25.9	−9 02 07	−0.003	+0.01	6.10	−0.08		B9			4866	m
43302	133025		6 15 26.8	−1 24 00	0.000	−0.02	7.5	0.8	3.2	G5 IV		72 s		
43574	196631		6 15 28.9	−32 32 05	−0.003	+0.03	8.0	0.0	0.4	B9.5 V		300 s		
43319	133027		6 15 29.6	−4 54 55	+0.001	−0.05	5.99	0.09		A2			4865	m
43286	113645		6 15 30.2	+3 57 29	0.000	−0.01	7.00	−0.13	−1.0	B5.5 V	+39	400 s	4863	m
43301	113649		6 15 31.7	+0 50 06	−0.002	−0.01	7.2	0.0	−1.0	B5.5 V	+9	410 s		
43318	133028		6 15 34.2	−0 30 44	−0.011	−0.22	5.65	0.50	3.7	F6 V	−36	26 ts		
43489	151280		6 15 34.9	−19 29 41	0.000	−0.02	7.7	0.2	0.4	B9.5 V		200 s		
43264	113648		6 15 35.6	+7 39 09	0.000	0.00	7.54	−0.03	0.4	B9.5 V		270 s		
43017	58905		6 15 38.9	+36 08 55	−0.005	0.00	6.92	0.45	3.3	dF4	+7	51 s	4849	m
44247	249497		6 15 39.4	−66 17 33	0.000	+0.01	7.34	−0.10	0.4	B9.5 V		240 s		
43285	113650		6 15 40.1	+6 03 58	−0.001	−0.02	6.07	−0.13	−0.9	B6 V	+26	250 s		
42633	13772	40 Cam	6 15 40.6	+59 59 57	+0.005	−0.02	5.35	1.34	0.2	K0 III	+12	67 s		m
43185	95456		6 15 41.1	+18 17 50	−0.004	−0.02	6.8	1.1	−0.1	K2 III		230 s		
43502	171468		6 15 41.8	−20 12 05	+0.001	+0.02	7.4	1.1	0.2	K0 III	+49	250 s		
43445	151283		6 15 44.8	−13 43 06	0.000	−0.01	5.01	−0.08	−0.2	B8 V	+38	110 s		

HD	SAO	Star Name	α 2000	δ 2000	μ(α)	μ(δ)	V	B-V	M_v	Spec	RV	d(pc)	ADS	Notes
			h m s	° ′ ″	s	″								
43247	95457	73 Ori	6 15 44.9	+12 33 04	0.000	0.00	5.33	-0.02	-1.9	B9 II-III	+13	280 s		
43411	133039		6 15 46.3	-6 12 31	-0.001	-0.01	7.9	0.0	0.0	B8.5 V		370 s		
43317	113653		6 15 46.8	+4 17 01	-0.001	0.00	6.64	-0.17	-2.3	B3 IV	+13	600 s		
43358	113656		6 15 53.9	+1 10 09	-0.001	+0.03	6.37	0.46		F5	+3			m
43148	78149		6 15 54.6	+27 51 42	0.000	-0.03	8.0	1.1	-0.1	K2 III		390 s		
43516	151290		6 15 55.6	-17 01 45	-0.001	+0.01	7.7	0.9	3.2	G5 IV		80 s		
43299	95460		6 15 56.3	+10 16 54	-0.002	-0.04	6.84	1.22	0.2	K0 III		140 s		
43636	171482		6 15 57.1	-29 47 18	0.000	+0.01	6.67	1.55		K2				
43129	58921		6 16 01.5	+33 11 55	-0.003	-0.02	7.8	0.1	1.4	A2 V		180 s		
43173	78154		6 16 03.1	+28 12 05	0.000	-0.02	7.9	0.1	0.6	A0 V		270 s		
42250	5793		6 16 03.2	+70 46 54	+0.004	-0.44	7.43	0.77	5.5	dG7	+25	23 s		
42721	25636		6 16 04.1	+59 12 54	+0.002	-0.02	6.8	0.9	3.2	G5 IV		51 s		
42869	25640		6 16 04.9	+52 11 09	0.000	+0.01	7.8	1.1	-0.1	K2 III		380 s		
43408	113661		6 16 05.0	+0 52 32	-0.001	-0.03	7.8	0.0	0.4	B9.5 V		280 s		
43094	58915		6 16 05.8	+37 38 10	-0.001	-0.04	7.1	1.1	0.2	K0 III		230 s		
43544	151294		6 16 07.6	-16 37 04	-0.001	+0.01	5.92	-0.17		B5 n	+14			
43282	95461		6 16 09.2	+19 03 19	0.000	0.00	7.90	0.9	-2.1	G5 II		890 s		
42819	25638		6 16 12.1	+56 56 04	-0.026	-0.19	7.4	0.6	4.4	G0 V		40 s		
43718	196638		6 16 12.8	-34 03 13	-0.002	+0.05	7.5	1.4	-0.3	K5 III		370 s		
43406	113664		6 16 13.7	+5 06 51	-0.001	0.00	7.16	-0.07	0.4	B9.5 V		230 s		
43443	113666		6 16 14.8	+0 00 53	+0.001	-0.03	7.3	0.0	0.0	B8.5 V		280 s		
44105	234417		6 16 15.1	-59 13 10	+0.002	-0.22	6.40	0.5	4.0	F8 V		30 s		
43405	113663		6 16 17.9	+9 01 40	0.000	-0.01	7.8	0.0	0.4	B9.5 V		290 s		
44120	234418		6 16 18.5	-59 12 49	-0.007	-0.33	6.43	0.59	4.9	G3 V		20 s		m
43261	78168	8 Gem	6 16 19.0	+23 58 12	-0.001	-0.01	6.08	0.90	0.3	G5 III	-21	140 s		
43298	95469		6 16 19.6	+18 09 21	+0.001	-0.02	8.0	0.1	1.4	A2 V		200 s		
43228	58927		6 16 20.6	+30 06 09	+0.001	-0.03	6.9	0.0	0.4	B9.5 V		190 s		
43461	113668		6 16 21.1	+1 04 49	+0.001	-0.01	6.63	-0.05		B8				
42839	25641		6 16 22.0	+56 44 05	+0.001	-0.01	7.7	1.6	-0.5	M4 III		410 s		
43246	78165		6 16 22.1	+28 51 06	0.000	-0.03	7.3	0.1	1.4	A2 V	-1	150 s		
43531	133051		6 16 22.8	-9 56 05	-0.002	+0.01	7.8	1.1	0.2	K0 III		330 s		
43335	95473		6 16 23.7	+17 10 53	0.000	-0.02	6.39	1.58	-2.3	K5 II	+38	510 s		
43386	95476	74 Ori	6 16 26.6	+12 16 20	+0.006	+0.19	5.04	0.42	2.8	F5 IV-V	+9	28 ts		m
43765	196641		6 16 27.1	-35 25 52	-0.002	+0.02	7.2	0.4	0.2	K0 III		250 s		
43355	95475		6 16 31.6	+19 00 46	+0.001	+0.05	7.30	0.5	2.3	F7 IV		99 s	4881	m
43785	196643	κ Col	6 16 33.0	-35 08 26	-0.001	+0.08	4.37	1.00	0.3	G8 III	+24	58 s		
43245	58928		6 16 34.8	+31 32 53	+0.002	-0.08	7.8	1.1	0.2	K0 III		150 mx		
43847	196646		6 16 35.5	-39 15 52	-0.002	-0.02	6.00	0.16		A0				
43628	151303		6 16 41.5	-12 02 50	-0.001	-0.03	7.00	0.4	3.0	F2 V		63 s	4891	m
44578	249501		6 16 41.9	-69 04 35	+0.001	-0.01	7.4	1.1	-0.1	K2 III		320 s		
43941	217826		6 16 44.7	-43 38 37	-0.003	+0.01	7.2	0.1	0.4	B9.5 V		180 s		
44104	234419		6 16 48.1	-54 37 01	+0.001	+0.02	7.1	1.7	-0.1	K2 III		130 s		
43670	151307		6 16 50.1	-15 07 33	-0.001	-0.01	7.89	1.21	-0.2	gK3	+39	420 s		
43717	151311		6 16 52.7	-18 36 26	-0.010	0.00	7.6	0.5	4.0	F8 V		52 s		
43806	171504		6 16 56.4	-29 52 13	+0.001	+0.06	7.9	1.1	3.4	F5 V		31 s		
43526	113673		6 16 58.3	+7 03 11	-0.001	+0.01	6.57	-0.12		B8				
43384	78176	9 Gem	6 16 58.7	+23 44 27	0.000	0.00	6.25	0.45	-6.3	B3 Iab	+13	1400 s		
43899	196653		6 17 01.1	-37 44 15	-0.001	-0.04	5.53	1.14	0.2	gK0	+70	90 s		
43745	171502		6 17 03.5	-22 42 54	+0.008	-0.25	6.07	0.56	4.4	dG0	-3	22 s		m
43588	113676		6 17 03.7	+1 41 13	-0.001	0.00	7.8	1.1	-0.1	K2 III		360 s		
42507	13775		6 17 04.3	+69 28 42	+0.002	0.00	7.6	0.1	0.6	A0 V		250 s		
44164	234420		6 17 04.5	-55 24 53	-0.003	0.00	7.2	0.4	0.0	B8.5 V		130 s		
43496	95486		6 17 06.2	+15 51 03	0.000	0.00	7.33	-0.08	-3.4	B8 II	+12	1400 s		m
43525	113675	75 Ori	6 17 06.5	+9 56 32	0.000	-0.06	5.39	0.10		A2 n	+13	29 mn	4890	m
43762	171506		6 17 07.8	-22 22 26	-0.001	0.00	6.9	0.7	0.6	A0 V		71 s		
43458	95485		6 17 09.2	+18 54 53	-0.002	-0.02	7.1	0.9	0.3	G5 III	+35	220 s		
43940	196659		6 17 09.5	-37 15 11	-0.002	-0.01	5.87	0.14		A2				
43804	171508		6 17 12.7	-24 04 29	+0.001	-0.03	6.8	0.9	3.2	G5 IV		47 s		
43828	171513		6 17 13.8	-26 55 59	-0.002	+0.01	6.8	1.0	3.2	G5 IV		39 s		
43480	95487		6 17 14.0	+18 40 59	0.000	0.00	7.80	0.8	-2.1	G5 II		860 s		
43074	25650		6 17 14.8	+52 32 32	-0.002	+0.01	7.9	1.1	0.2	K0 III		340 s		
43184	40994		6 17 14.9	+46 01 41	+0.003	-0.06	7.5	0.2	2.1	A5 V		57 mx		
43665	133062		6 17 15.4	-5 20 37	+0.001	-0.01	8.0	0.1	1.4	A2 V		200 s		
43587	113683		6 17 16.0	+5 06 01	-0.015	+0.16	5.71	0.61	4.4	dG0	+13	21 ts		m
42756	13779		6 17 18.9	+64 49 56	+0.005	-0.02	7.5	1.1	0.2	K0 III		290 s		
43281	40998		6 17 19.2	+40 53 13	+0.002	-0.05	7.6	0.1	0.6	A0 V		250 s		
43479	78182		6 17 20.3	+22 08 38	-0.001	-0.01	8.0	0.9	3.2	G5 IV		91 s		
43171	40993		6 17 21.0	+49 29 09	+0.001	0.00	7.4	0.1	1.4	A2 V		160 s		
43204	40997		6 17 21.8	+45 51 39	+0.001	-0.02	7.6	1.1	-0.1	K2 III		350 s		
42994	25647		6 17 23.5	+57 24 53	-0.004	-0.06	7.1	0.5	4.0	F8 V		41 s		
42974	25646		6 17 26.4	+58 49 36	-0.001	0.00	7.6	0.1	1.4	A2 V		170 s		
43583	95494		6 17 33.2	+14 03 30	0.000	-0.01	6.59	-0.03	0.4	B9.5 V	+10	170 s		
43382	58944		6 17 33.6	+35 08 56	+0.005	+0.03	6.6	0.5	3.7	F6 V		39 s		
41496	5782		6 17 33.9	+79 19 48	-0.006	0.00	7.9	0.1	0.6	A0 V		280 s		
43244	40999	42 Aur	6 17 34.6	+46 25 27	-0.004	+0.02	6.52	0.27	2.6	F0 V	-8	61 s		

HD	SAO	Star Name	α 2000	δ 2000	μ(α)	μ(δ)	V	B-V	M$_v$	Spec	RV	d(pc)	ADS	Notes
			h m s	° ′ ″	s	″								
43760	151318	6 Mon	6 17 35.1	−10 43 31	−0.001	−0.01	6.74	0.38	0.6	gF2	+27	160 s		
43333	58942		6 17 36.1	+39 51 31	+0.001	0.00	7.5	1.1	0.2	K0 III		270 s		
43879	171521		6 17 37.5	−24 26 41	+0.005	−0.12	7.30	0.5	3.4	F5 V		60 s	4908	m
43331	41003		6 17 38.9	+41 28 37	+0.001	+0.01	8.0	1.4	−0.3	K5 III		420 s		
43332	41004		6 17 39.2	+40 18 39	+0.003	+0.03	7.8	0.6	4.4	G0 V		47 s		
44246	234427		6 17 39.5	−53 58 44	+0.001	+0.02	6.9	1.1	0.2	K0 III		200 s		
44058	217830		6 17 41.0	−40 39 32	+0.001	−0.01	7.5	0.4	0.6	A0 V		130 s		
43827	151322		6 17 41.7	−16 48 57	−0.001	+0.01	5.14	1.30	−0.2	K3 III	−8	110 s		
43861	151323		6 17 44.3	−19 11 23	0.000	−0.03	6.5	0.1	0.4	B9.5 V		150 s		
43381	58946		6 17 45.0	+39 28 26	+0.001	−0.01	7.1	1.1	−0.1	K2 III		260 s		
43495	78186		6 17 45.6	+29 13 11	0.000	−0.01	7.8	0.1	0.6	A0 V		260 s		
43353	41007		6 17 46.7	+40 11 17	0.000	−0.04	7.5	0.0	0.4	B9.5 V		250 s		
43280	41002		6 17 47.7	+45 58 04	−0.001	−0.01	7.8	0.1	0.6	A0 V		270 s		
44577	249502		6 17 49.7	−64 01 51	−0.001	−0.01	7.54	−0.08	0.4	B9.5 V		270 s		
43607	95495		6 17 50.8	+19 27 25	0.000	−0.02	7.53	−0.03	0.6	A0 V		240 s		
44267	234430		6 17 51.6	−52 44 00	0.000	−0.01	6.41	1.46		K0				
43974	196665		6 17 51.7	−31 00 34	+0.001	−0.01	7.5	1.5	0.2	K0 III		150 s		
43757	133071		6 17 53.2	−4 22 58	−0.001	−0.04	6.8	1.1	0.2	K0 III		200 s		
42973	13787	1 Lyn	6 17 54.9	+61 30 55	0.000	0.00	5.2	1.6	−0.5	M2 III	+11	130 s		
43743	113701		6 17 58.6	+0 13 27	−0.001	−0.03	7.7	0.0	0.4	B9.5 V		280 s		
43421	58953		6 17 59.8	+39 05 29	+0.004	−0.06	8.0	1.1	−0.1	K2 III		170 mx		
43537	78191		6 17 59.9	+28 00 24	+0.001	−0.01	7.5	0.1	0.6	A0 V	+12	230 s		
43683	95502		6 18 05.4	+14 22 58	−0.001	0.00	6.15	0.05	1.4	A2 V	+11	88 s	4901	m
44295	234434		6 18 05.7	−53 33 31	+0.001	+0.05	6.8	0.4	0.6	A0 V		100 s		
43581	78195		6 18 06.3	+26 26 04	−0.001	−0.02	8.00	1.1	−2.1	K0 II		920 s		
43227	25654		6 18 07.6	+53 40 17	−0.002	0.00	7.6	0.1	0.6	A0 V		250 s		
43777	133075		6 18 07.6	−0 22 16	0.000	−0.02	7.77	0.16		B3 n	+20			
43843	133076		6 18 08.3	−6 43 33	+0.002	+0.02	6.8	1.1	0.2	K0 III		210 s		
44407	234439		6 18 10.4	−57 20 11	0.000	+0.03	7.9	0.9	0.2	K0 III		340 s		
43955	151334		6 18 13.6	−19 58 01	−0.001	+0.01	5.2	0.1	−2.5	B2 V	+23	350 s		
43380	41010	43 Aur	6 18 16.8	+46 21 38	+0.001	−0.13	6.38	1.11	−0.1	K2 III	0	98 mx		
43776	113704		6 18 16.8	+3 59 03	+0.001	−0.02	7.5	1.1	0.2	K0 III		270 s		
43857	133078		6 18 17.4	−5 39 17	0.000	−0.01	6.70	0.1	0.6	A0 V		160 s	4910	m
44864	249506		6 18 20.7	−69 09 44	0.000	−0.05	7.9	0.4	3.0	F2 V		96 s		
43624	78196		6 18 20.7	+27 12 36	+0.001	−0.07	6.66	1.09	0.0	K1 III	−49	180 mx		
42506	5807		6 18 22.2	+72 57 00	+0.010	−0.12	7.8	0.4	3.0	F2 V		84 mx		
43379	41011		6 18 26.4	+49 11 09	0.000	−0.07	8.0	0.9	3.2	G5 IV		93 s		
−−	217838		6 18 26.8	−48 02 20	−0.002	−0.01	8.0	0.3	−0.3	K5 III				
44277	−−		6 18 27.3	−48 02 24			7.2	1.2	−0.3	K5 III		320 s		
44446	234444		6 18 28.6	−57 00 00	−0.001	0.00	8.0	−0.2	0.6	A0 V		300 s		
43823	113709		6 18 31.8	+4 30 34	−0.002	−0.02	7.8	0.4	3.0	F2 V		88 s		
44382	234441		6 18 32.5	−53 37 49	0.000	+0.03	7.3	0.9	0.2	K0 III		260 s		
44163	196675		6 18 35.7	−38 53 38	+0.002	−0.02	7.3	1.0	0.2	K0 III		260 s		
43646	78201		6 18 37.6	+29 46 53	0.000	−0.05	6.9	0.1	0.6	A0 V	+12	180 s		
43558	58968		6 18 39.2	+38 14 09	+0.001	−0.01	7.8	1.4	−0.3	K5 III		380 s		
43821	113710		6 18 40.5	+9 02 50	+0.002	−0.04	6.24	0.87		K0	−14			
43702	78207		6 18 43.8	+25 20 55	+0.001	0.00	8.0	0.0	0.0	B8.5 V		380 s		
44184	196676		6 18 44.8	−35 54 37	−0.002	+0.01	7.9	0.6	−0.1	K2 III		400 s		
44362	234442		6 18 46.8	−50 21 33	0.000	0.00	7.04	0.83		G5				
43693	78206		6 18 47.2	+28 03 19	0.000	−0.01	8.00	1.1	−0.1	K2 III		390 s		
44021	151346		6 18 48.7	−15 01 29	−0.001	+0.02	6.06	1.66	−0.5	gM1	+51	190 s		
43841	95516		6 18 49.4	+11 09 54	0.000	−0.02	7.09	0.02	0.6	A0 V		190 s	4918	m
43993	133091		6 18 50.5	−9 23 25	0.000	−0.03	5.36	1.24	0.0	gK1	+7	90 s		
42818	13788		6 18 50.8	+69 19 11	+0.001	−0.11	4.80	0.03	0.6	A0 V	−17	66 s		
43950	133089		6 18 51.0	−5 03 51	−0.001	−0.03	7.9	0.2	2.1	A5 V		140 s		
44135	171553		6 18 52.5	−28 07 20	0.000	+0.12	7.9	0.8	4.4	G0 V		37 s		
43644	58975		6 18 53.8	+35 12 28	0.000	−0.05	6.5	1.1	0.2	K0 III		180 s		
43740	78210		6 18 54.4	+23 36 12	0.000	−0.01	6.6	0.9	3.2	G5 IV	+42	47 s		
44102	171552		6 18 55.4	−25 00 46	−0.002	0.00	7.1	0.4	0.6	B9.5 V		120 s		
43913	113718		6 18 58.4	+4 11 28	0.000	−0.01	7.5	0.1	0.6	A0 V		240 s		
44081	171549		6 18 58.9	−20 55 34	−0.001	−0.01	5.81	−0.15		B5	+31			
43840	95520		6 18 59.6	+12 42 05	0.000	−0.01	7.44	0.02	0.6	A0 V		220 s	4921	m
43753	78211		6 18 59.6	+23 00 03	0.000	−0.01	7.90	0.30		B0.5 III				
43819	95519		6 19 01.8	+17 19 30	−0.001	−0.01	6.32	−0.07		B9.5 pe	+4			
43855	95521		6 19 02.9	+14 05 40	+0.001	−0.01	7.8	0.5	3.4	F5 V		74 s		
43259	25660		6 19 03.1	+59 33 34	+0.003	−0.05	7.5	0.9	3.2	G5 IV		72 s		
44116	171556		6 19 03.1	−23 32 02	0.000	−0.02	8.0	0.5	0.2	K0 III		360 s		
43523	41016		6 19 04.8	+47 23 38	−0.002	−0.03	7.0	0.5	3.4	F5 V		53 s		
44293	217843		6 19 05.5	−41 04 29	0.000	0.00	7.4	−0.2	0.6	A0 V		230 s		
43873	95524		6 19 06.2	+12 44 48	0.000	−0.01	6.74	−0.12	0.0	B8.5 V		220 s		
43990	133094		6 19 06.2	−3 44 49	−0.001	−0.01	8.0	0.1	0.6	A0 V		290 s		
44037	133098		6 19 07.8	−8 35 11	−0.002	−0.01	6.22	−0.04		B9				
43988	133093		6 19 07.9	−1 22 03	0.000	−0.03	7.6	1.1	0.2	K0 III		290 s		
43692	58979		6 19 07.9	+33 47 43	−0.002	−0.01	7.9	0.4	2.6	F0 V		110 s		
43352	25662		6 19 10.0	+56 31 36	−0.004	−0.03	7.0	1.1	0.2	K0 III		230 s		

HD	SAO	Star Name	α 2000	δ 2000	μ(α)	μ(δ)	V	B-V	M_V	Spec	RV	d(pc)	ADS	Notes
44183	171565		6 19 10.2	-29 00 55	+0.002	-0.01	7.9	1.9	-0.3	K5 III		240 s		
45080	249514		6 19 15.1	-69 47 03	-0.004	+0.08	7.0	0.4	4.0	F8 V		40 s		
44144	171562		6 19 15.2	-24 58 29	-0.001	+0.02	7.20	0.1	0.6	A0 V		210 s		m
--	--		6 19 17.7	-24 58 20			7.10							
43818	78222		6 19 19.2	+23 28 09	0.000	-0.01	6.92	0.29	-5.6	B0 II	+18	1600 s		
44080	--		6 19 19.8	-12 31 18			8.00	0.00	0.4	B9.5 V		330 s	4941	m
44019	133100		6 19 21.0	-0 54 54	-0.001	-0.02	7.60	1.1		K2 IV	+39			
43910	95530		6 19 22.0	+12 59 24	+0.001	0.00	7.8	0.1	-7.5	A2 Ia		5700 s		
45538	256294		6 19 22.2	-75 42 00	+0.016	-0.02	7.8	1.1	0.2	K0 III		190 mx		
43836	78225		6 19 22.5	+23 16 28	0.000	-0.01	6.95	0.48	-3.1	B9 II		550 s		m
43931	95534		6 19 25.5	+13 26 03	+0.003	0.00	6.92	0.47	3.8	F7 V	+43	42 s	4924	m
44405	217847		6 19 28.2	-42 53 13	-0.003	-0.01	8.0	-0.4	0.4	B9.5 V		330 s		
43930	95536		6 19 28.7	+13 26 54	0.000	-0.01	7.67	1.13	6.1	K1 V		14 s		
43437	25666		6 19 30.8	+56 15 26	+0.001	-0.01	8.0	1.1	0.2	K0 III		350 s		
43929	95535		6 19 31.6	+16 10 41	0.000	-0.02	8.03	0.41	3.4	F5 V		83 s		
43751	58984		6 19 33.1	+34 38 23	0.000	-0.01	7.7	0.0	0.4	B9.5 V		270 s		
42330	5809		6 19 35.8	+76 49 40	0.000	-0.07	7.9	1.4	-0.3	K5 III		250 mx		
43700	41026		6 19 35.8	+40 39 03	+0.002	-0.02	8.0	1.1	0.2	K0 III		340 s		
42937	13796		6 19 35.8	+69 34 09	-0.005	-0.03	6.8	0.2	2.1	A5 V		89 s		
43378	25665	2 Lyn	6 19 37.3	+59 00 39	-0.001	+0.02	4.48	0.01	1.4	A2 V	-4	35 ts		
44225	171571		6 19 37.9	-22 06 10	-0.005	+0.01	6.5	1.8	-0.1	K2 III		92 s		
44880	249509		6 19 38.5	-63 49 25	+0.002	+0.04	6.9	1.1	0.2	K0 III		220 s		
42992	13799		6 19 38.8	+68 40 38	-0.001	+0.02	6.9	0.6	4.4	G0 V		32 s		
44264	171576		6 19 39.6	-26 59 02	-0.003	-0.07	7.3	1.4	-0.1	K2 III		180 mx		
43947	95538		6 19 40.1	+16 00 48	-0.001	-0.02	6.6	0.5	4.0	F8 V	+43	33 s		
44323	196686		6 19 40.9	-34 23 48	+0.001	0.00	5.78	-0.08		B9 n	+26			
44182	151357		6 19 40.9	-17 31 33	-0.003	+0.01	7.27	-0.08	0.0	B8.5 V		280 s		
43983	95540		6 19 41.2	+12 17 33	+0.001	-0.02	7.80	0.00		B8.5 V		350 s	4939	m
44112	133114	7 Mon	6 19 42.7	-7 49 23	-0.001	0.00	5.27	-0.19	-2.5	B2 V	+29	360 s		
43739	58985		6 19 43.6	+38 26 08	+0.003	-0.02	7.10	0.4	2.6	F0 V		78 s	4922	m
44404	196693		6 19 47.9	-39 29 11	0.000	-0.01	6.70	1.1	0.2	K0 III		200 s		
44263	171579		6 19 48.0	-25 24 22	-0.001	-0.01	7.9	1.3	0.2	K0 III		240 s		
43906	78231		6 19 51.7	+25 00 48	+0.001	-0.06	7.60	1.1	-0.1	K2 III	-60	230 mx	4930	m
44200	151365		6 19 57.2	-13 59 35	-0.002	0.00	6.6	0.1	0.6	A0 V		160 s		
43885	78233		6 19 58.9	+28 25 36	0.000	-0.04	7.27	0.24		A3	+5	37 mn	4929	m
44242	151366		6 19 59.2	-17 41 00	-0.001	+0.02	7.90	0.1	1.4	A2 V		200 s	4960	m
44131	133118		6 19 59.5	-2 56 40	-0.001	0.00	4.90	1.60	-0.5	gM1	+47	120 s		
44846	249513		6 19 59.7	-61 19 49	-0.003	+0.05	6.8	1.5	3.2	G5 IV		19 s		
44033	95543		6 20 04.2	+14 39 04	0.000	-0.02	5.69	1.61	-4.4	K3 Ib	+33	670 s		
44291	171586		6 20 05.3	-23 06 19	-0.001	0.00	7.8	0.2	2.1	A5 V		140 s		
44594	217861		6 20 06.1	-48 44 28	+0.023	-0.27	6.60	0.66		G0		12 mn		
44178	133124		6 20 06.6	-7 52 18	-0.002	-0.01	6.9	0.2	2.1	A5 V		91 s		
44018	95544		6 20 12.1	+19 52 26	+0.001	-0.02	7.7	0.9	3.2	G5 IV		79 s		
44380	171596		6 20 13.6	-30 01 05	0.000	+0.01	7.8	1.2	0.2	K0 III		260 s		
44219	151367		6 20 14.1	-10 43 30	+0.003	-0.01	7.9	0.9	3.2	G5 IV		87 s		
43926	58993		6 20 14.5	+30 22 09	-0.001	-0.06	7.9	0.4	2.6	F0 V		110 s		
43795	41030		6 20 16.1	+42 48 00	+0.003	-0.01	7.80	0.8	0.3	G6 III	-50	320 s		
44176	133125		6 20 16.2	-2 03 18	0.000	-0.02	7.7	0.1	0.6	A0 V		250 s		
44241	151368		6 20 16.9	-10 52 54	-0.001	0.00	7.3	0.1	1.4	A2 V		150 s		
44109	113733		6 20 17.0	+7 43 09	-0.001	0.00	7.40	0.00	0.4	B9.5 V		240 s	4951	m
44465	196699		6 20 17.4	-36 21 11	-0.001	+0.02	7.6	0.3	0.6	A0 V		160 s		
44402	196698	1 ζ CMa	6 20 18.7	-30 03 48	0.000	0.00	3.02	-0.19	-1.7	B3 V	+32	88 s		Furud
43621	25672		6 20 18.9	+54 14 59	-0.001	-0.09	7.3	0.4	2.6	F0 V		87 s		
41804	985		6 20 24.3	+80 21 23	-0.005	-0.10	7.53	0.45		F4				
44666	217867		6 20 27.1	-49 06 15	+0.002	0.00	6.5	1.2	-0.1	K2 III		200 s		
44064	78252		6 20 32.0	+21 05 27	+0.001	-0.01	8.0	1.1	0.2	K0 III		340 s		
44073	95554		6 20 34.4	+18 02 56	0.000	-0.02	7.60	0.8	0.3	G7 III	+16	280 s		m
44379	171602		6 20 34.9	-23 38 16	-0.003	-0.06	7.9	1.5	0.2	K0 III		94 s		
44030	78250		6 20 35.2	+25 36 30	0.000	-0.04	7.58	1.48	-0.3	K4 III		340 s		
44765	234456		6 20 35.4	-52 04 12	+0.002	-0.01	6.6	1.7	-0.1	K2 III		110 s		
44506	196707		6 20 36.2	-34 08 38	0.000	+0.03	5.53	-0.19	-2.5	B2 V n	+73	400 s		
44050	78251		6 20 36.3	+25 11 20	-0.001	0.00	7.20	0.00	0.4	B9.5 V		220 s		m
44288	133135		6 20 39.9	-9 53 26	-0.002	+0.02	7.0	0.1	1.4	A2 V		130 s		
44688	217870		6 20 40.5	-46 40 40	-0.005	-0.02	7.9	0.9	-0.3	K5 III		230 mx		
42951	5817		6 20 42.0	+72 08 45	+0.001	-0.02	6.80	0.8	3.2	G5 IV		53 s		m
44427	171607		6 20 46.7	-24 17 58	-0.002	+0.04	7.3	0.5	0.0	B8.5 V		130 s		
44213	113741		6 20 48.3	+5 44 23	0.000	-0.03	7.98	1.95	-5.6	M2 Iab	+17	3300 s		m
43750	25674		6 20 49.9	+52 58 13	-0.002	-0.02	7.8	0.1	0.6	A0 V		270 s		
44525	196713		6 20 50.2	-30 38 12	-0.001	-0.01	7.7	1.6	-0.5	M2 III		440 s		
44286	133136		6 20 50.3	-4 35 45	+0.001	-0.05	6.6	0.2	2.1	A5 V		78 s		
44016	58997		6 20 51.3	+31 38 22	-0.001	-0.05	7.8	0.5	3.4	F5 V		73 s		
44173	95563		6 20 52.2	+11 45 23	+0.002	0.00	6.54	-0.10	-2.2	B5 III	+19	520 s		
46262	256300		6 20 53.9	-79 03 59	-0.012	+0.04	7.1	1.1	0.2	K0 III		240 s		
44396	151389		6 20 54.5	-16 59 04	-0.005	-0.16	7.9	0.9	3.2	G5 IV		75 mx		
44610	196717		6 20 54.7	-37 29 37	+0.004	-0.01	7.8	0.8	3.2	G5 IV		82 s		

HD	SAO	Star Name	α 2000	δ 2000	μ(α)	μ(δ)	V	B-V	M_v	Spec	RV	d(pc)	ADS	Notes
			h m s	° ′ ″	s	″								
45038	249519		6 20 55.3	−62 21 15	+0.001	−0.02	7.7	0.1	1.4	A2 V		180 s		
44524	171609		6 20 56.4	−29 40 15	+0.003	−0.07	7.4	0.4	2.6	F0 V		79 s		
44937	234460		6 20 58.0	−58 21 07	0.000	+0.01	7.7	0.8	0.4	B9.5 V		97 s		
44321	133142		6 20 58.2	−6 41 31	−0.001	0.00	7.64	−0.05	0.4	B9.5 V		280 s		
46175	256299		6 20 59.3	−78 24 44	−0.018	−0.02	8.0	0.8	3.2	G5 IV		90 s		
44303	133140		6 21 00.7	−3 30 11	0.000	−0.02	8.0	0.4	3.0	F2 V		97 s		
44172	95567		6 21 00.7	+14 42 18	0.000	−0.01	7.34	−0.10	−0.9	B6 V	−30	440 s		
44442	151394		6 21 02.6	−17 57 27	−0.002	0.00	8.0	0.1	0.6	A0 V		300 s		
44274	113750		6 21 02.8	+2 34 07	−0.001	−0.04	7.50	1.6	−0.5	M4 III		380 s	4966	m
44005	59000		6 21 03.1	+36 18 47	+0.001	−0.02	7.90	1.1	0.2	K0 III		330 s	4955	m
44235	113746		6 21 03.9	+9 45 44	−0.001	+0.01	7.99	−0.16	0.0	B8.5 V		400 s		
44071	78257		6 21 04.0	+29 22 18	−0.003	+0.01	6.9	0.4	2.6	F0 V	−10	73 s		
44394	151391		6 21 05.3	−11 49 03	−0.001	0.00	6.6	1.4	−0.3	K5 III		240 s		
44236	113748		6 21 05.9	+7 32 09	−0.002	−0.02	8.0	0.1	1.4	A2 V		200 s		
44015	59001		6 21 10.4	+37 56 19	0.000	−0.01	7.8	1.1	0.2	K0 III		310 s		
44335	133146		6 21 10.7	−4 21 01	−0.002	−0.02	8.0	0.1	0.6	A0 V		290 s		
44737	217873		6 21 11.1	−44 45 30	−0.002	+0.01	7.6	0.2	0.0	B8.5 V		240 s		
44092	78259		6 21 12.0	+29 32 28	+0.003	−0.04	6.30	0.06	1.2	A1 V	+25	100 s		
44171	78264		6 21 14.3	+21 07 57	−0.001	0.00	7.30	−0.08	0.0	B8.5 V		290 s	4962	m
44004	41039		6 21 17.8	+42 20 24	0.000	−0.04	6.8	0.1	0.6	A0 V		170 s		
44485	151402		6 21 19.5	−16 48 14	0.000	0.00	7.9	0.4	0.4	B9.5 V		320 s		
44462	151400		6 21 20.0	−14 02 47	−0.001	0.00	8.0	1.6	−0.5	M2 III		510 s		
44354	133148		6 21 20.2	−1 13 17	−0.001	−0.03	7.7	0.0	0.4	B9.5 V		280 s		
44193	78267		6 21 22.1	+21 11 59	0.000	−0.01	7.59	0.06	0.6	A0 V		230 s		
44458	151401		6 21 24.6	−11 46 24	−0.001	−0.01	5.64	−0.02	−3.5	B1 V pe	+21	630 s	4978	m
45058	234464		6 21 24.8	−60 01 20	−0.004	−0.01	7.2	0.7	2.1	A5 V		51 s		
44918	234461		6 21 25.0	−52 39 44	−0.002	0.00	7.4	1.2	3.2	G5 IV		38 s		
44643	196721		6 21 25.2	−33 04 17	−0.007	+0.02	7.6	1.0	0.2	K0 III		160 mx		
44333	113758		6 21 25.7	+2 16 07	−0.001	−0.03	6.31	0.24		A5	−26		4971	m
44234	95572		6 21 25.9	+17 45 49	+0.001	−0.04	6.4	1.1	0.2	K0 III	+10	170 s		
44642	171623		6 21 31.3	−29 09 03	+0.001	+0.04	7.7	0.6	3.4	F5 V		56 s		
44253	95576		6 21 35.6	+19 53 31	0.000	−0.01	7.3	0.9	3.2	G5 IV	+36	66 s		
44623	171626		6 21 40.1	−25 21 06	0.000	−0.05	7.4	1.1	2.6	F0 V		29 s		
45002	234465		6 21 40.7	−56 21 45	−0.001	−0.02	6.9	1.4	−0.5	M2 III		300 s		
44480	133156		6 21 41.6	−8 14 10	−0.002	−0.02	7.3	1.6	−0.5	M2 III		350 s		
44482	133158		6 21 42.8	−9 40 58	−0.001	+0.01	7.38	−0.08	0.6	A0 V		230 s		
44420	133153		6 21 43.4	−0 32 22	−0.001	−0.17	7.7	0.9	5.0	G4 V	+5	35 s		
44318	95580		6 21 44.9	+12 51 42	0.000	−0.01	6.9	0.8	3.2	G5 IV		54 s		
44029	41041		6 21 45.4	+45 36 42	+0.002	0.00	7.3	1.1	0.2	K0 III		270 s		
43905	25681	45 Aur	6 21 46.2	+53 27 08	+0.003	−0.09	5.36	0.43	0.7	F5 III	−1	83 mx		
44418	113766		6 21 46.9	+2 20 32	−0.001	−0.01	7.3	1.1	0.2	K0 III	−9	260 s		
44251	78272		6 21 49.2	+23 45 39	−0.001	−0.03	7.7	1.1	0.2	K0 III		300 s		
44958	234463		6 21 50.1	−51 14 13	+0.002	+0.03	6.8	0.5	1.4	A2 V		67 s		
44749	196732		6 21 50.9	−34 45 46	−0.001	0.00	7.6	1.1	0.2	K0 III		270 s		
44547	133164		6 21 58.4	−8 13 49	−0.002	−0.03	7.77	−0.02	0.6	A0 V		270 s		
44091	41044		6 22 01.3	+44 03 30	+0.001	−0.03	7.0	0.8	3.2	G5 IV		59 s		
44192	59016		6 22 03.0	+34 35 49	0.000	−0.08	7.5	0.5	4.0	F8 V		51 s		
44746	196733		6 22 03.1	−30 32 02	+0.001	+0.01	7.7	1.4	−0.3	K5 III		400 s		
43812	25678	4 Lyn	6 22 03.6	+59 22 20	0.000	0.00	6.20	0.1	1.4	A2 V	−24	90 s	4950	m
44250	78275		6 22 03.8	+29 57 50	0.000	−0.01	7.1	0.1	0.6	A0 V	+8	190 s		
44776	196736		6 22 04.4	−35 02 37	+0.001	0.00	7.4	1.2	0.2	K0 III		220 s		
43771	13819		6 22 04.6	+60 46 41	−0.002	−0.06	7.4	0.1	1.4	A2 V		100 mx		
44621	151419		6 22 05.9	−15 39 46	−0.006	−0.03	6.7	0.8	3.2	G5 IV		50 s		
44762	196735	δ Col	6 22 06.7	−33 26 11	−0.002	−0.06	3.85	0.88	0.5	gG1	−3	36 s		
43749	13818		6 22 08.1	+61 45 37	0.000	−0.11	7.2	0.4	2.6	F0 V	+8	83 s		
44436	113773		6 22 08.8	+9 09 30	0.000	−0.02	7.7	1.1	0.2	K0 III		300 s		
44546	133167		6 22 10.4	−4 45 13	0.000	0.00	8.0	0.1	0.6	A0 V		290 s		
44844	196742		6 22 13.0	−39 13 09	+0.004	−0.03	7.8	1.1	0.2	K0 III		300 s		
44372	95586		6 22 14.4	+19 42 09	+0.001	−0.02	7.8	0.1	1.4	A2 V		180 s		
44736	171642		6 22 15.7	−26 16 39	−0.001	−0.03	7.9	1.2	−0.1	K2 III		390 s		
44125	41047		6 22 18.3	+44 24 44	+0.001	0.00	7.7	0.2	2.1	A5 V		130 s		
45057	234468		6 22 18.9	−53 20 06	−0.001	0.00	6.86	−0.14	−2.2	B5 III		650 s		
43793	13820		6 22 22.5	+62 06 14	−0.004	−0.01	7.6	1.1	0.2	K0 III		310 s		
44972	217889		6 22 23.4	−45 57 40	+0.001	−0.02	6.9	0.3	0.4	B9.5 V		130 s		
44124	41048		6 22 25.3	+46 11 40	0.000	−0.01	7.40	0.1	1.4	A2 V		160 s	4970	m
43979	25687		6 22 26.0	+56 00 46	−0.002	−0.03	7.9	0.4	2.6	F0 V		110 s		
44316	78282		6 22 28.0	+28 54 44	0.000	−0.01	7.70	1.1	0.0	K1 III		330 s		
44300	59021		6 22 28.3	+30 56 42	0.000	−0.04	7.9	0.4	2.6	F0 V		110 s		
44605	133175		6 22 29.3	−2 42 35	−0.001	−0.03	7.7	1.1	0.2	K0 III		300 s		
44843	196744		6 22 30.5	−33 52 11	−0.001	+0.01	7.90	0.7	4.4	G0 V		50 s		m
45270	249526		6 22 30.5	−60 13 09	−0.006	+0.04	6.60	0.6	4.4	G0 V		28 s		m
44299	59022		6 22 34.9	+32 45 38	+0.001	−0.01	8.00	0.4	3.4	F5 V		82 s	4982	m
44497	95599		6 22 36.5	+12 34 12	−0.002	−0.01	6.0	0.4	2.6	F0 V	+21	47 s		
45669	249532	π¹ Dor	6 22 38.3	−69 59 03	+0.004	+0.03	5.56	1.51		K5	+16			
44743	151428	2 β CMa	6 22 41.9	−17 57 22	−0.001	0.00	1.98	−0.23	−4.8	B1 II-III	+34	220 s		Mirzam, m,v

HD	SAO	Star Name	α 2000	δ 2000	μ(α)	μ(δ)	V	B−V	M$_v$	Spec	RV	d(pc)	ADS	Notes
45077	234471		6^{h}22^{m}43.1^s	−50°06'06"	−0.001^s	−0.01"	8.02	0.88		G5				
44639	133182		6 22 43.7	−2 11 47	+0.002	−0.07	5.9	1.6		M5 III	+30	160 mx		V Mon, v
44602	113790		6 22 45.0	+4 12 36	−0.002	−0.04	6.85	0.10	1.7	A3 V		110 s		
44391	78291		6 22 47.7	+27 59 12	−0.002	0.00	7.74	1.40	−2.1	G9 II		630 s		
44654	133183		6 22 49.5	−0 50 01	0.000	−0.03	7.7	0.0	0.4	B9.5 V		280 s		
44496	95602		6 22 49.6	+17 34 27	0.000	−0.02	6.91	0.02	0.6	A0 V	+21	180 s	4991	m
45034	217892		6 22 51.8	−45 55 14	−0.002	+0.01	8.0	1.3	−0.1	K2 III		130 s		
44896	196752		6 22 53.7	−33 36 54	0.000	−0.02	7.1	1.3	−0.5	M2 III		330 s		
45056	217893		6 22 55.1	−45 37 47	−0.002	+0.01	6.70	1.1	0.2	K0 III		200 s		m
44249	41054		6 22 55.2	+42 54 13	−0.002	0.00	7.7	0.0	0.4	B9.5 V		290 s		
44638	113795		6 22 55.4	+3 45 38	−0.001	−0.01	7.5	1.1	0.2	K0 III	−1	280 s		
45229	234473	ν Pic	6 22 55.8	−56 22 12	−0.004	−0.02	5.61	0.24		A m	+7			
44478	78297	13 μ Gem	6 22 57.6	+22 30 49	+0.004	−0.11	2.88	1.64	−0.5	M3 III	+55	46 s	4990	m
44821	171659		6 22 57.6	−24 33 23	0.000	−0.01	7.5	0.8	3.2	G5 IV		71 s		
44434	78294		6 23 00.3	+28 37 16	+0.001	−0.03	7.8	0.1	0.6	A0 V		260 s		
45461	249531		6 23 01.3	−63 41 00	−0.004	+0.01	6.1	2.9	−0.5	M2 III		36 s		
44701	133189		6 23 01.4	−3 16 37	0.000	+0.01	6.57	−0.15	−1.0	B5.5 V	+8	330 s		IM Mon, m,v
44451	78299		6 23 05.6	+27 43 53	−0.002	+0.01	7.9	0.1	0.6	A0 V		270 s		
44720	133192		6 23 06.0	−3 30 57	−0.001	0.00	7.22	−0.12	0.0	B8.5 V		280 s		
43167	5826		6 23 06.4	+74 50 47	+0.020	−0.01	7.9	0.5	4.0	F8 V		62 s		
44061	25693		6 23 06.4	+56 58 35	+0.001	−0.06	6.6	1.1	0.2	K0 III		190 s		
42702	5818		6 23 07.5	+78 23 45	−0.005	−0.10	8.0	0.9	3.2	G5 IV		93 s		
45797	256298		6 23 07.8	−70 56 42	0.000	+0.01	8.0	0.1	1.4	A2 V		210 s		
44584	95604		6 23 09.4	+16 30 56	+0.001	+0.01	8.00	0.01	0.4	B9.5 V		330 s	4997	m
44281	41056		6 23 11.3	+43 14 10	+0.001	−0.01	7.9	0.1	0.6	A0 V		280 s		
44390	59028		6 23 13.0	+35 31 14	−0.002	0.00	7.3	0.1	0.6	A0 V		210 s		
44979	196761		6 23 13.0	−34 59 48	0.000	0.00	6.49	−0.05	0.4	B9.5 V		170 s		
44956	196759		6 23 14.3	−31 47 24	−0.005	+0.01	6.4	1.0	3.2	G5 IV		34 s		
44934	196758		6 23 15.7	−30 56 44	−0.001	0.00	7.6	0.8	2.6	F0 V		52 s		
44800	151438		6 23 18.3	−14 50 11	0.000	−0.05	6.9	0.1	1.7	A3 V		110 s		
44700	113801		6 23 18.4	+3 45 52	0.000	−0.01	6.40	−0.15	−2.3	B3 IV	+29	520 s		
44210	25697		6 23 20.8	+50 44 29	−0.002	−0.01	7.9	1.1	0.2	K0 III		340 s		
44389	59030		6 23 21.8	+38 02 08	0.000	−0.03	7.8	1.1	0.2	K0 III		310 s		
44756	133199		6 23 22.6	−4 41 15	−0.001	−0.01	6.67	0.06		B9				
44637	95612		6 23 24.6	+15 06 06	0.000	−0.01	8.03	0.24	−2.5	B2 V pe	+20	1100 s		
44933	171672		6 23 25.7	−26 22 43	+0.001	+0.03	7.3	1.9	−0.3	K5 III		190 s		
44189	25699		6 23 29.4	+53 04 38	0.000	−0.02	7.7	0.1	0.6	A0 V		260 s		
44978	171677		6 23 29.4	−29 51 39	−0.001	0.00	7.00	1.1	0.2	K0 III		230 s		m
44541	78309		6 23 30.2	+27 07 04	0.000	−0.01	7.90	0.1	0.6	A0 V		280 s		m
44893	151443		6 23 30.5	−18 59 55	−0.001	−0.04	7.4	1.1	0.2	K0 III		270 s		
44977	171678		6 23 32.8	−29 16 11	−0.003	+0.04	7.9	1.1	3.2	G5 IV		56 s		
44676	95617		6 23 32.9	+14 02 30	+0.001	0.00	7.80	1.1	0.0	K1 III		350 s		
44123	25696		6 23 34.3	+57 21 18	+0.001	0.00	7.76	1.26	0.0	K1 III		270 s		
44913	171673		6 23 35.9	−20 10 37	0.000	+0.03	7.7	0.2	0.4	B9.5 V		210 s		
44816	133203		6 23 35.9	−9 52 28	0.000	+0.02	6.19	1.84		K5				
44712	95620		6 23 36.5	+10 56 01	−0.001	−0.01	7.76	0.00	0.6	A0 V		270 s		
45291	234475		6 23 37.7	−52 10 52	−0.003	−0.02	5.98	1.04		G5				
44413	41062		6 23 38.8	+40 31 17	+0.003	−0.01	6.9	0.9	3.2	G5 IV		54 s		
44892	151449		6 23 42.8	−16 27 57	0.000	+0.05	6.6	0.2	2.1	A5 V		78 s		
45098	196769		6 23 45.1	−37 00 53	−0.002	0.00	6.90	−0.10	0.0	B8.5 V		240 s		
44891	151450		6 23 46.0	−15 04 18	−0.002	−0.01	6.3	1.1	−0.1	K2 III		190 s		
44769	113810	8 Mon	6 23 46.0	+4 35 34	−0.001	+0.01	4.33	0.20	0.3	A5 III	+16	54 s	5012	m
44770	113811	8 Mon	6 23 46.4	+4 35 44	−0.002	−0.01	6.72	0.45	3.3	dF4	+16	54 s	5012	m
45450	234482		6 23 46.9	−58 32 38	+0.001	+0.02	6.48	0.12		A2				
44953	151453		6 23 47.5	−19 47 07	−0.002	+0.01	6.60	−0.15		B8			5023	
44771	113812		6 23 47.6	+2 39 54	0.000	−0.03	7.10	1.1	0.2	K0 III	+19	230 s	5013	m
44841	133206		6 23 53.5	−4 43 45	−0.001	−0.02	7.4	1.1	0.2	K0 III		260 s		
44474	41066		6 23 53.8	+40 48 30	−0.001	−0.01	7.7	1.4	−0.3	K5 III		390 s		
44738	95625		6 23 54.0	+14 06 48	0.000	−0.02	7.3	0.1		A0 p	+22			
45018	171688		6 23 55.8	−25 34 39	−0.001	−0.03	5.63	1.56	−0.3	gK5	+34	140 s		
45348	234480	α Car	6 23 57.1	−52 41 44	+0.003	+0.02	−0.72	0.15	−8.5	F0 Ia	+21	360 s		Canopus
45228	217904		6 23 57.9	−45 56 51	+0.030	+0.15	7.88	0.70		G8		16 mn		m
45145	196774		6 24 00.9	−36 42 28	−0.002	+0.05	5.62	1.04		G5				m
44782	113815		6 24 01.2	+9 10 19	−0.001	+0.01	7.38	−0.07	0.4	B9.5 V		250 s		
45895	256301		6 24 01.8	−70 11 15	−0.002	0.00	7.9	0.3	2.6	F0 V		110 s		
44783	113817		6 24 02.3	+8 53 05	−0.001	−0.02	6.26	−0.08		A0	+9			
45171	196778		6 24 03.5	−39 42 08	−0.001	+0.02	7.8	1.0	−0.1	K2 III		390 s		
45158	196776		6 24 05.9	−36 42 10	−0.003	+0.02	5.70	0.9	3.2	G5 IV		32 s		m
45347	234481		6 24 07.3	−50 31 45	0.000	+0.04	7.90	0.4	2.6	F0 V		120 s		m
44951	151458		6 24 10.2	−11 31 49	−0.004	−0.04	5.22	1.24	−0.2	K3 III	−26	120 s		
44150	13826		6 24 10.7	+60 05 02	−0.003	−0.01	7.2	1.1	0.2	K0 III		250 s		
44768	95627		6 24 11.7	+15 51 03	+0.001	−0.02	7.87	−0.07	−0.6	A0 III		490 s		
45074	171693		6 24 12.9	−26 26 40	−0.001	+0.01	6.9	1.3	3.2	G5 IV		28 s		
45557	249535		6 24 13.7	−60 16 52	−0.006	+0.04	5.80	0.00		A0				
44813	113823		6 24 14.3	+7 53 42	−0.001	0.00	7.90	0.1	1.7	A3 V		170 s	5018	m

HD	SAO	Star Name	α 2000	δ 2000	μ(α)	μ(δ)	V	B-V	M_v	Spec	RV	d(pc)	ADS	Notes
44495	41070		$6^h24^m17.4^s$	$+43°13'52''$	$+0.002$	-0.01	7.1	0.9	3.2	G5 IV		61 s		
44996	151461		6 24 20.5	-12 57 45	-0.001	-0.01	6.12	-0.08		B8			5030	m
45016	151462		6 24 20.9	-16 13 27	-0.003	0.00	6.70	0.1	1.7	A3 V		100 s	5034	m
44386	25710		6 24 23.5	+51 04 38	-0.001	0.00	7.2	1.4	-0.3	K5 III		320 s		
45289	217907		6 24 24.2	-42 50 52	-0.009	+0.77	6.67	0.68	5.2	G5 V	+48	20 s		
45112	171703		6 24 24.5	-27 18 52	+0.002	+0.15	8.0	0.8	3.4	F5 V		49 s		
45701	249539		6 24 26.4	-63 25 44	-0.001	-0.10	6.46	0.66		G0				
44812	95631		6 24 28.0	+13 00 55	0.000	0.00	7.4	0.9	-4.5	G5 Ib		1900 s		
44633	59047		6 24 29.3	+37 19 00	+0.002	0.00	7.2	1.1	0.2	K0 III		240 s		
44869	113829		6 24 29.5	+8 23 50	-0.002	0.00	7.3	0.2	2.1	A5 V		110 s		
44614	59046		6 24 30.8	+38 34 08	+0.008	-0.10	7.7	0.6	4.4	G0 V		46 s		
44494	41071		6 24 31.4	+47 07 12	+0.004	-0.06	7.9	0.5	3.4	F5 V		78 s		
44907	113830		6 24 32.3	+4 11 10	0.000	-0.02	7.30	-0.08	0.4	B9.5 V		240 s		
44538	41073		6 24 32.9	+45 10 01	-0.001	-0.02	8.0	1.4	-0.3	K5 III		450 s		
44229	25707		6 24 33.2	+59 41 56	+0.001	-0.06	7.5	0.1	1.7	A3 V		76 mx		
44853	95637		6 24 33.8	+11 15 24	+0.001	0.00	6.9	1.1	0.2	K0 III	+52	220 s		
44948	133215		6 24 33.9	-1 25 07	0.000	-0.02	6.73	-0.10	0.0	B8.5 V		220 s	5029	m
44851	95638		6 24 36.5	+12 51 38	0.000	-0.03	7.90	1.1	5.9	K0 V		25 s		
44694	59050		6 24 38.8	+35 15 19	0.000	-0.02	6.8	1.1	0.2	K0 III		200 s		
44780	78331		6 24 43.7	+25 02 55	0.000	-0.01	6.5	1.1	0.0	K1 III	+7	190 s		
45184	171711		6 24 43.8	-28 46 48	-0.012	-0.12	6.39	0.62		G0				
44271	25709		6 24 43.9	+59 40 09	0.000	-0.06	8.00	1.1	0.2	K0 III		240 mx		m
45306	217910		6 24 44.4	-40 17 02	-0.001	-0.01	6.31	-0.05		B9				
44612	41080		6 24 46.5	+43 32 54	0.000	-0.03	7.3	0.1	0.6	A0 V		210 s		
45268	196795		6 24 46.9	-36 35 53	+0.004	-0.01	8.0	1.3	-0.3	K5 III		270 mx		
45013	133220		6 24 47.4	-5 30 15	-0.003	-0.06	8.0	0.5	3.4	F5 V		81 s		
45509	234488		6 24 48.1	-52 48 23	-0.001	0.00	6.51	1.70		K0				
44966	113840		6 24 50.1	+4 52 20	-0.003	-0.03	7.1	0.5	3.4	F5 V		54 s		
44766	78333		6 24 52.5	+29 42 26	-0.002	-0.01	6.5	-0.1		B9	+29			
44867	95641		6 24 52.8	+16 03 25	+0.003	-0.05	6.33	1.08	3.2	G8 IV	+73	30 s		
44537	41076	46 ψ¹ Aur	6 24 53.8	+49 17 16	0.000	0.00	4.91	1.97	-5.7	M0 Iab	+5	600 s		q
44711	59055		6 24 54.6	+36 02 23	+0.001	-0.02	8.0	1.1	0.2	K0 III		340 s		
45796	249542		6 24 55.7	-63 49 40	-0.002	+0.02	6.27	-0.13		B7	-1			
44796	78334		6 24 59.4	+28 01 56	0.000	-0.01	7.8	0.1	0.6	A0 V		260 s		
44944	95649		6 25 00.8	+10 31 02	-0.002	-0.02	7.90	0.00	0.4	B9.5 V		310 s	5035	m
44904	95648		6 25 05.7	+16 59 42	0.000	+0.01	7.00	-0.05	0.4	B9.5 V	+3	210 s		
45614	234493		6 25 05.9	-55 43 31	-0.001	+0.06	7.2	0.6	3.0	F2 V		48 s		
45143	151477		6 25 06.7	-15 08 24	-0.002	0.00	7.9	0.1	1.4	A2 V		200 s		
44556	25716		6 25 12.0	+51 37 59	+0.001	-0.07	7.1	1.1	0.2	K0 III		200 mx		
44794	59063		6 25 12.8	+32 16 18	0.000	0.00	6.9	1.1	0.2	K0 III		210 s		
44990	113845		6 25 13.0	+7 05 09	+0.001	0.00	5.98	1.22	-6.2	G5 Iab var	+32	2200 s		T Mon, v
44965	95653		6 25 13.1	+11 41 02	0.000	0.00	7.82	0.31	-3.9	B3 II	+27	1200 s		
45407	—		6 25 13.7	-40 34 19			8.0	2.2	-0.3	K5 III		170 s		
49836	258446		6 25 14.1	-84 40 49	+0.013	0.00	8.0	1.6	-0.5	M2 III		490 s		
45140	151478		6 25 14.8	-10 55 56	+0.002	-0.03	7.80	1.1	0.2	K0 III		320 s	5048	m
44987	113846		6 25 15.8	+9 09 56	-0.001	0.00	7.5	1.1	0.2	K0 III	+35	280 s		
45067	133229		6 25 16.4	-0 56 46	+0.015	-0.22	5.87	0.56	2.6	dF0	+45	33 s		
45050	113855		6 25 18.3	+1 30 04	-0.001	-0.01	6.66	-0.04		A0	+7			m
44883	78344		6 25 18.6	+23 42 33	0.000	-0.03	7.1	0.4	2.6	F0 V		79 s		
45328	196799		6 25 20.0	-30 52 51	-0.004	0.00	8.0	1.3	3.4	F5 V		24 s		
44985	95656		6 25 20.2	+11 14 00	-0.008	-0.05	7.8	0.5	4.0	F8 V		56 s		
45265	171725		6 25 21.3	-25 59 24	-0.001	+0.01	8.0	1.2	0.2	K0 III		290 s		
44986	95658		6 25 21.5	+10 48 33	0.000	+0.01	7.5	0.0	0.0	B8.5 V		310 s		
45124	133233		6 25 21.8	-9 05 29	0.000	-0.01	7.3	1.1	0.2	K0 III		260 s		
45817	249544		6 25 25.0	-62 25 21	+0.001	+0.07	7.9	0.4	2.6	F0 V		120 s		
44984	95659		6 25 28.0	+14 43 19	-0.001	0.00	6.24	2.34		N7.7	+13			BL Ori, v
46116	249550	π² Dor	6 25 28.7	-69 41 25	-0.003	+0.20	5.38	0.97	0.3	gG4	+9	84 s		
45383	196805		6 25 29.9	-35 03 50	-0.001	-0.04	6.25	1.36		K0				m
44926	78348		6 25 32.6	+23 26 31	+0.001	-0.02	6.8	0.9	3.2	G5 IV		51 s		
44927	78349		6 25 32.9	+23 19 37	-0.001	-0.02	6.00	-0.03	0.6	A0 V	-32	120 s		
44941	78352		6 25 34.1	+22 27 29	0.000	-0.01	7.80	0.1	1.7	A3 V		160 s	5042	m
45139	133234		6 25 34.7	-2 59 27	-0.001	+0.02	6.7	0.1	0.6	A0 V		160 s		
45424	196808		6 25 37.9	-35 02 18	+0.001	+0.02	7.30	0.1	1.7	A3 V		130 s		m
44473	25717		6 25 38.2	+58 25 31	+0.001	-0.10	7.7	0.5	4.0	F8 V		54 s		
45364	196806		6 25 38.5	-31 28 52	+0.005	-0.02	7.9	0.7	4.4	G0 V		42 s		
45153	133236		6 25 38.8	-4 49 57	-0.002	-0.01	7.9	0.0	0.0	B8.5 V		370 s		
45382	171731		6 25 40.3	-29 42 10	-0.001	+0.01	6.89	-0.02	0.6	A0 V		180 s		
45047	95669		6 25 41.0	+11 13 13	-0.002	-0.02	7.6	1.1	0.2	K0 III		290 s		
45045	95666		6 25 41.9	+14 05 34	-0.002	0.00	7.4	1.1	1.4	A2 V		160 s		
45439	196810		6 25 42.3	-35 41 49	-0.002	+0.03	7.88	-0.06	0.4	B9.5 V		310 s		
44974	78355		6 25 42.7	+21 38 42	-0.001	-0.01	6.6	0.9	3.2	G5 IV	-24	48 s		
44834	59068		6 25 43.0	+38 03 01	0.000	+0.02	7.2	0.8	3.2	G5 IV		62 s		
45572	217922		6 25 43.6	-48 10 38	-0.001	-0.03	5.76	-0.06		B9				m
45304	171729		6 25 45.7	-20 58 23	+0.001	0.00	7.2	1.0	0.2	K0 III		240 s		
45501	217921		6 25 46.3	-40 58 31	-0.001	-0.01	6.90	0.1	1.7	A3 V		110 s		m

HD	SAO	Star Name	α 2000	δ 2000	μ(α)	μ(δ)	V	B-V	M_v	Spec	RV	d(pc)	ADS	Notes
45137	113868		6 25 46.5	+2 16 20	0.000	0.00	6.51	-0.03		B9				
45168	133238		6 25 47.0	-3 53 21	-0.002	-0.01	6.35	1.02		G5				
43810	5835		6 25 48.4	+73 26 08	-0.001	+0.01	7.7	1.1	-0.1	K2 III		370 s		
45481	196812		6 25 49.8	-38 56 09	-0.002	0.00	7.3	2.1	-0.3	K5 III		140 s		
45152	113872		6 25 51.5	+0 48 45	+0.001	-0.01	6.8	0.9	3.2	G5 IV		53 s		
45090	95672		6 25 52.1	+11 07 33	-0.002	-0.02	7.2	0.1	0.6	A0 V		210 s		
45136	113871		6 25 54.3	+4 31 24	-0.001	0.00	7.97	0.02	0.6	A0 V		280 s	5057	m
45459	196813		6 25 57.4	-35 20 15	-0.001	-0.01	7.0	1.6	-0.3	K5 III		240 s		
45151	113873		6 25 58.1	+4 25 17	-0.001	-0.01	7.89	-0.10	0.0	B8.5 V		380 s		
45239	133246		6 25 58.7	-7 53 39	-0.001	+0.02	6.40	0.14		A2				
45089	95674		6 26 01.0	+15 09 00	0.000	-0.01	7.0	1.1	0.2	K0 III	-34	220 s		
44412	13834		6 26 05.6	+62 41 26	+0.004	+0.01	7.6	1.4	-0.3	K5 III		340 mx		
44866	59073		6 26 08.3	+39 55 44	0.000	-0.02	7.9	0.0	0.4	B9.5 V		320 s		
44648	25727		6 26 08.7	+55 52 21	-0.001	-0.03	7.9	1.1	0.2	K0 III		340 s		
45215	133248		6 26 09.5	-3 30 56	-0.007	+0.07	6.5	0.9	3.2	G5 IV		46 s		
44649	25728		6 26 10.1	+55 42 13	+0.001	+0.01	7.8	0.8	3.2	G5 IV		84 s		
44832	41091		6 26 10.2	+44 43 59	+0.002	+0.01	7.1	0.0	0.4	B9.5 V		220 s		
45088	95677		6 26 10.2	+18 45 24	-0.008	-0.18	6.76	0.94	6.9	dK3		12 ts	5054	m
45284	133252		6 26 13.1	-7 21 42	-0.001	-0.01	7.37	-0.10	0.0	B8.5 V		300 s		
45298	151501		6 26 18.0	-10 02 30	0.000	-0.01	8.0	0.1	1.4	A2 V		210 s		
45212	113876		6 26 20.2	+2 19 03	0.000	-0.01	7.7	1.1	-0.1	K2 III		350 s		
44901	41099		6 26 21.6	+41 57 34	0.000	-0.01	7.1	0.9	3.2	G5 IV		62 s		
44902	41101		6 26 22.9	+40 21 29	-0.002	-0.02	7.8	0.1	0.6	A0 V		280 s		
45404	171742		6 26 23.5	-20 10 27	+0.001	+0.01	7.5	1.5	-0.5	M2 III		400 s		
45211	113877		6 26 24.0	+4 06 12	-0.001	-0.01	7.9	1.1	0.2	K0 III		330 s		
45257	133254		6 26 25.0	-0 46 37	0.000	-0.01	7.6	0.0	0.4	B9.5 V		270 s		
44691	25731		6 26 25.8	+56 17 06	-0.003	+0.02	5.64	0.24		A m	-13	20 mn		RR Lyn, m,v
45437	151509		6 26 27.7	-19 19 29	0.000	-0.01	7.1	0.2	1.4	A2 V		110 s		
45005	59083		6 26 29.4	+34 00 37	0.000	-0.01	7.6	1.4	-0.3	K5 III		350 s		
45422	151508		6 26 30.6	-16 36 05	-0.005	+0.01	7.7	1.1	0.2	K0 III		180 mx		
45479	171751		6 26 32.6	-26 06 13	-0.002	+0.01	7.7	1.8	-0.5	M2 III		330 s		
45321	133257		6 26 34.4	-4 35 51	0.000	0.00	6.15	-0.15		B3	+10			
44647	25732		6 26 36.6	+58 25 05	-0.001	-0.03	7.90	1.06	0.3	G8 III		280 s		
44962	59082		6 26 36.9	+39 40 27	+0.001	-0.01	7.7	0.9	3.2	G5 IV		79 s		
45180	95684		6 26 37.6	+15 31 24	-0.001	-0.01	6.88	-0.03	0.2	B9 V	+10	220 s	5062	m
45553	196825		6 26 38.7	-33 24 58	0.000	+0.01	7.3	0.4	3.2	G5 IV		67 s		
45194	95686		6 26 39.2	+13 06 04	+0.002	-0.10	6.6	0.5	3.8	F7 V	-6	36 s		
45320	133260		6 26 39.5	-1 30 27	0.000	-0.04	5.87	0.08	0.5	A7 III p		120 s		
45420	151512		6 26 42.5	-14 36 11	-0.001	0.00	6.5	1.1	0.2	K0 III		180 s		
45317	113892		6 26 43.4	+0 27 15	0.000	-0.02	7.70	0.9	3.2	G5 IV		79 s	5069	m
45681	217931		6 26 43.6	-44 04 02	+0.001	+0.02	7.1	0.2	0.6	A0 V		200 s		
45380	133263		6 26 44.8	-7 30 41	0.000	0.00	6.27	-0.04		A0			5070	m
44708	25733	5 Lyn	6 26 48.8	+58 25 02	0.000	-0.01	5.21	1.53	-0.3	K4 III	-3	100 s	5036	m
65322	258465		6 26 49.4	-88 44 37	-0.003	+0.02	7.8	-1.3	0.6	A0 V		280 s		
45552	196828		6 26 49.6	-31 52 18	0.000	+0.01	8.0	0.4	2.6	F0 V		100 s		
44690	25734		6 26 50.5	+58 47 01	+0.001	-0.02	8.0	1.1	-0.1	K2 III		390 s		
45210	95690		6 26 54.9	+17 27 05	-0.004	-0.01	7.9	0.7	4.4	G0 V		51 s		
44671	13840		6 26 55.6	+60 37 58	-0.001	-0.05	7.9	0.9	3.2	G5 IV		88 s		
45398	133265		6 26 58.1	-4 27 33	-0.002	0.00	6.8	1.1	0.2	K0 III		200 s		
45357	113896		6 26 58.6	+0 50 28	-0.002	-0.01	6.71	0.04		A0	+10			
45738	217933		6 26 59.2	-44 07 24	-0.001	+0.01	7.6	0.7	3.0	F2 V		75 s		
45085	59088		6 27 00.6	+35 25 46	+0.001	0.00	7.9	0.0	0.4	B9.5 V		300 s		
45418	133267		6 27 00.8	-4 21 21	0.000	-0.01	6.50	-0.15		B5 n	+15			
45984	234519		6 27 04.0	-58 00 08	-0.002	-0.01	5.82	1.28		K0	+13			
45610	196834		6 27 06.5	-32 10 00	+0.004	-0.08	7.2	0.9	4.4	G0 V		23 s		
45680	196842		6 27 07.4	-37 53 44	-0.002	+0.04	6.50	0.4	2.6	F0 V		60 s		m
45536	171764		6 27 08.7	-21 33 08	+0.002	+0.01	8.0	0.0	0.6	A0 V		280 s		
45397	113900		6 27 09.1	+0 08 05	-0.001	-0.02	7.9	0.0	0.0	B8.5 V		360 s		m
45551	171769		6 27 09.6	-24 14 19	0.000	0.00	7.8	1.7	-0.3	K5 III		310 s		
45588	171774		6 27 11.2	-25 51 24	-0.014	-0.22	6.07	0.53	4.2	dF9		24 s		m
45535	171767		6 27 12.1	-20 25 14	0.000	-0.02	7.4	0.9	0.6	A0 V		60 s		
46069	249552		6 27 12.1	-60 43 43	-0.005	+0.01	7.2	1.1	0.2	K0 III		230 s		
45495	151519		6 27 12.7	-12 50 17	-0.002	+0.01	7.9	0.0	0.4	B9.5 V		310 s		
45416	113905		6 27 13.7	+0 17 57	0.000	-0.01	5.20	1.18	-2.2	K1 II	+33	300 s		
45433	133269		6 27 15.5	-0 16 34	0.000	-0.01	5.55	1.38	-0.3	gK5	+39	150 s		
45314	95697		6 27 15.8	+14 53 23	+0.001	+0.01	6.64	0.15		O9 pe	+10			
257331	78387		6 27 15.9	+21 59 27	0.000	0.00	8.0	1.1	0.2	K0 III		350 s		
45251	78390		6 27 19.0	+21 17 37	+0.003	-0.04	7.7	0.9	3.2	G5 IV		79 s		
45923	234518		6 27 19.8	-53 33 59	-0.001	+0.03	8.0	0.5	2.6	F0 V		87 s		
45415	113906		6 27 20.3	+2 54 29	-0.003	+0.01	5.55	1.04	0.2	gG9	+53	100 s		
45377	113901		6 27 20.4	+9 13 01	-0.001	0.00	8.0	0.1	1.4	A2 V		200 s		
45522	151523		6 27 21.3	-15 18 58	-0.003	-0.01	8.0	0.0	0.4	B9.5 V		330 s		
45431	113911		6 27 27.9	+5 10 29	0.000	-0.02	6.7	0.4	2.6	F0 V		66 s		
45607	171780		6 27 28.3	-22 07 26	0.000	+0.03	7.6	0.5	2.1	A5 V		78 s		
43884	5843		6 27 32.4	+76 01 13	-0.004	+0.03	7.9	0.9	3.2	G5 IV		87 s		

HD	SAO	Star Name	α 2000	δ 2000	μ(α)	μ(δ)	V	B−V	M$_V$	Spec	RV	d(pc)	ADS	Notes
45192	59096		6^h 27^m 35^s.4	+32° 33′ 47″	0^s.000	−0″.06	6.43	1.26	0.2	K0 III	+57	120 s		
45604	151527		6 27 36.2	−19 27 17	+0.002	−0.03	7.7	0.9	−0.1	K2 III		360 s		
45566	151525		6 27 36.6	−15 18 22	−0.002	+0.01	7.40	0.00	0.0	B8.5 V		300 s	5087	m
45004	41106		6 27 38.3	+48 46 37	+0.002	+0.05	7.3	0.5	3.4	F5 V		61 s		
45237	59100		6 27 40.2	+30 38 43	+0.001	−0.01	6.96	1.16	0.2	K0 III		160 s		
45353	95703		6 27 40.5	+18 13 25	−0.003	+0.01	7.5	0.5	3.4	F5 V		67 s		
45445	113916		6 27 42.9	+9 37 35	−0.002	−0.01	7.8	0.1	1.7	A3 V		160 s		
45515	133280		6 27 45.3	−2 39 35	−0.001	0.00	7.88	−0.03	0.0	B8.5 V		340 s		
45352	78395		6 27 46.5	+20 47 22	−0.003	−0.05	6.60	1.1	0.2	K0 III	−30	190 s	5080	m
45517	133286		6 27 47.6	−4 55 54	0.000	0.00	7.7	0.1	0.6	A0 V		260 s		
45834	217940		6 27 47.9	−41 07 54	+0.001	+0.10	7.4	0.9	0.2	K0 III		190 mx		
46190	249554		6 27 48.6	−62 08 59	0.000	+0.03	6.8	0.4	0.6	A0 V		96 s		
45532	133287		6 27 50.8	−5 09 18	0.000	0.00	8.0	0.1	0.6	A0 V		290 s		
45105	41109		6 27 51.0	+47 24 19	0.000	+0.02	6.3	0.0		B9				
45394	78402	16 Gem	6 27 56.6	+20 29 46	−0.002	+0.01	6.10	0.01	1.4	A2 V	+39	87 s		
45546	133290	10 Mon	6 27 57.5	−4 45 44	−0.001	0.00	5.05	−0.18	−2.5	B2 V	+25	320 s		m
45413	95707		6 27 58.5	+18 48 49	−0.001	−0.03	7.8	0.4	2.6	F0 V		110 s		
45629	151530		6 27 58.8	−13 09 36	−0.003	+0.01	7.10	−0.04	0.0	B8.5 V		260 s		
45235	59104		6 27 59.5	+36 12 08	0.000	−0.01	7.70	0.7	4.4	G0 V		45 s		m
45507	113922		6 28 02.8	+5 55 21	−0.001	0.00	7.5	0.8	3.2	G5 IV		72 s		
45748	171803		6 28 05.7	−27 08 38	+0.001	−0.01	7.9	0.5	0.4	B9.5 V		140 s		
45514	113925		6 28 09.4	+5 57 31	−0.002	−0.01	7.6	1.1	−0.1	K2 III		330 s		
45813	196857	λ CMa	6 28 10.1	−32 34 48	−0.002	+0.02	4.48	−0.17		B5	+41	17 mn		
45735	171802		6 28 12.7	−22 00 22	0.000	−0.04	7.9	1.1	−0.3	K5 III		430 s		
45530	113926		6 28 13.9	+5 16 18	0.000	−0.01	7.41	−0.04		A1 p			5097	m
44472	5861		6 28 14.5	+70 32 08	+0.001	+0.02	6.00	0.1	1.4	A2 V	−32	83 s	5039	m
47332	256308		6 28 14.8	−77 17 24	0.000	+0.06	7.0	0.3	0.6	A0 V		120 mx		
45715	151537		6 28 15.9	−19 47 49	+0.001	−0.01	7.5	1.1	0.2	K0 III		240 s		
45563	113929		6 28 16.7	+1 54 44	0.000	−0.03	6.48	−0.07		B9				
45677	151534		6 28 17.5	−13 03 09	+0.001	+0.02	7.55	0.03		B3 pe	+25			
45452	95715		6 28 17.8	+18 28 07	−0.001	−0.01	7.5	0.8	3.2	G5 IV		72 s		
46214	234527		6 28 17.9	−59 37 05	0.000	−0.01	7.3	0.7	3.4	F5 V		42 s		
45843	196859		6 28 18.0	−34 22 50	−0.002	+0.01	8.0	1.6	−0.1	K2 III		240 s		
45512	95718		6 28 18.8	+10 18 14	+0.002	−0.04	6.15	1.15	0.2	K0 III	−20	120 s		
44849	13848		6 28 21.7	+62 42 44	−0.004	−0.06	7.9	0.4	2.6	F0 V		110 s		
45659	133303		6 28 26.1	−6 32 52	0.000	0.00	8.04	−0.04	0.0	B8.5 V		380 s		
45506	95719		6 28 27.9	+16 14 18	−0.007	−0.05	6.23	0.90	3.2	G5 IV	+41	32 s		
45351	59121		6 28 28.8	+34 29 32	−0.001	−0.01	7.7	0.1	1.7	A3 V		150 s		
45311	59117		6 28 31.7	+39 06 37	+0.001	−0.03	7.6	0.4	2.6	F0 V		100 s		
43748	5846		6 28 33.6	+77 57 32	+0.002	−0.06	7.80	0.8	1.8	G5 III-IV	+8	160 s		
45412	59128	48 Aur	6 28 34.0	+30 29 35	0.000	−0.01	5.55	0.68		G0	+22	37 mn		RT Aur, v
45733	151542		6 28 34.5	−13 02 58	−0.002	−0.01	7.52	0.32	1.4	A2 V		120 s		
45692	133311		6 28 35.6	−8 27 45	−0.004	+0.09	7.8	0.4	3.0	F2 V		88 s		
45655	133305		6 28 35.8	−1 35 23	−0.002	0.00	7.9	0.1	0.6	A0 V		270 s		
45765	151546		6 28 37.2	−17 27 58	−0.001	0.00	5.77	1.12		G5				
47674	256310		6 28 37.4	−78 54 17	−0.009	+0.07	7.40	0.4	2.6	F0 V		91 s		m
45654	133307		6 28 38.1	−0 43 18	+0.002	−0.02	6.9	0.5	4.0	F8 V		38 s		
45871	196861		6 28 39.1	−32 22 16	−0.001	+0.02	5.74	−0.17	−1.6	B3.5 V	+23	290 s		m
45983	217951		6 28 42.2	−41 04 28	+0.002	+0.03	6.32	0.40		F2				
45393	59129		6 28 42.6	+34 51 51	−0.001	−0.02	7.3	0.1	0.6	A0 V		210 s		
46085	217955		6 28 43.1	−49 17 07	0.000	0.00	7.9	0.3	0.4	B9.5 V		220 s		
45861	171819		6 28 45.3	−28 00 01	−0.002	−0.02	7.7	1.5	−0.1	K2 III		240 s		
46212	234529		6 28 45.7	−55 55 03	0.000	+0.01	8.0	0.2	0.4	B9.5 V		230 s		
45350	59126		6 28 45.7	+38 57 44	−0.003	−0.08	7.86	0.74	3.2	G5 IV		86 s		
45391	59131		6 28 46.0	+36 28 48	−0.024	−0.22	7.1	0.6	4.4	G0 V	−4	35 s		
45674	133312		6 28 47.0	−0 34 21	0.000	−0.01	6.7	0.4	2.6	F0 V		65 s		
45725	133316	11 β Mon	6 28 48.9	−7 01 58	−0.002	0.00	3.76	−0.15	−1.7	B3 V e	+22	220 s	5107	m
45726	133317	11 β Mon	6 28 49.4	−7 02 03	−0.002	0.00	5.40	−0.07	−2.6	B2.5 IV	+18	220 s	5107	m
45175	25752		6 28 50.7	+52 26 55	+0.003	+0.03	7.20	0.9	3.2	G5 IV		63 s	5078	m
46068	217956		6 28 52.1	−45 46 36	−0.001	+0.05	7.70	0.1	0.6	A0 V		260 s		m
45689	113938		6 28 52.4	+0 07 40	+0.001	−0.01	6.7	0.1	0.6	A0 V		160 s		
45372	41123		6 28 53.6	+40 07 10	−0.001	−0.03	8.00	0.1	0.6	A0 V		300 s	5088	m
257937	78420		6 28 53.8	+20 14 21	0.000	−0.01	8.00	0.10	0.6	A0 V		260 s		
43853	5848		6 28 54.9	+77 54 24	+0.005	−0.07	7.8	0.4	3.0	F2 V		89 s		
45764	151550		6 28 56.0	−11 15 48	−0.004	−0.01	7.79	−0.03	0.0	B8.5 V		110 mx		
45504	78417		6 28 56.6	+26 58 04	+0.009	−0.07	6.5	0.5	3.4	F5 V	−7	42 s		
45542	78423	18 ν Gem	6 28 57.7	+20 12 43	0.000	−0.02	4.15	−0.13	−1.0	B7 IV	+39	110 s	5103	m
45638	95732		6 29 00.0	+11 01 10	−0.002	−0.01	6.4	0.4	2.6	F0 V	+41	58 s		
45580	95730		6 29 03.5	+17 44 42	−0.010	−0.17	7.6	0.6	4.4	G0 V		43 s		
46039	217958		6 29 07.1	−40 22 20	+0.004	+0.01	7.40	1.1	0.2	K0 III		240 mx		
––	217962		6 29 08.4	−42 40 26	−0.003	+0.03	8.0	0.4	0.2	K0 III		360 s		
45904	171825		6 29 08.6	−25 53 06	−0.001	+0.01	8.0	0.8	0.2	K0 III		360 s		
45173	25758		6 29 09.3	+54 36 43	0.000	+0.01	7.7	0.9	3.2	G5 IV		81 s		
45597	95733		6 29 09.5	+17 29 46	−0.001	−0.03	7.80	1.57	0.2	K0 III		130 s		
46083	––		6 29 11.2	−42 40 54			7.9	0.3	0.2	K0 III				

HD	SAO	Star Name	α 2000	δ 2000	μ(α)	μ(δ)	V	B-V	M$_v$	Spec	RV	d(pc)	ADS	Notes
			h m s	° ′ ″	s	″								
45919	171826		6 29 11.9	-26 07 08	0.000	0.00	8.0	1.6	-0.3	K5 III		390 s		
46040	217961		6 29 12.0	-40 22 50	-0.002	-0.01	8.0	1.2	0.2	K0 III		290 s		
45724	113940		6 29 14.8	+2 38 46	0.000	-0.04	6.16	1.54		Ma	+9			m
46437	249561		6 29 18.9	-61 00 21	-0.009	+0.06	8.0	0.2	4.0	F8 V		62 s		
45889	151559		6 29 21.2	-17 48 45	-0.001	+0.02	8.00	0.1	0.6	A0 V		300 s		m
45941	171831		6 29 25.5	-22 35 28	+0.001	+0.01	6.90	0.1	1.4	A2 V		130 s	5128	m
45760	113943		6 29 26.4	+4 15 10	-0.001	-0.03	7.9	0.0	0.4	B9.5 V		310 s		
46025	196874		6 29 27.6	-34 05 38	-0.002	+0.04	7.8	0.2	1.7	A3 V		140 s		
45857	151558		6 29 28.1	-11 27 52	-0.003	-0.02	8.0	0.1	1.4	A2 V		200 s		
46355	234541		6 29 28.4	-56 51 11	-0.005	+0.02	5.22	1.09	0.3	gG8	+13	79 s		
45831	133332		6 29 36.7	-3 33 05	0.000	-0.01	7.8	0.8	3.2	G5 IV		83 s		
45759	113947		6 29 36.7	+8 29 34	-0.003	+0.01	7.6	0.5	4.0	F8 V		51 s		
45688	95741		6 29 37.4	+16 58 37	+0.001	-0.05	7.88	1.10	3.2	G5 IV		49 s	5112	m
45916	151561		6 29 37.7	-17 00 42	-0.002	-0.02	7.0	0.0	0.4	B9.5 V		210 s		
45758	95746		6 29 41.7	+10 39 57	-0.001	+0.02	7.8	0.5	3.4	F5 V		74 s		
46273	234539		6 29 49.1	-50 14 21	-0.006	-0.06	5.30	0.4	3.0	F2 V	+2	29 s		m
45722	95745		6 29 49.8	+17 37 27	0.000	0.00	7.8	0.1	1.4	A2 V		180 s		
46479	234549		6 29 55.1	-59 35 26	-0.001	+0.01	6.8	0.2	2.6	F0 V		68 s		
45789	113953		6 29 56.0	+7 06 43	+0.001	0.00	7.10	-0.13		B3	+5		5124	m
46208	217974		6 29 57.2	-42 16 40	-0.002	+0.04	7.5	1.2	-0.3	K5 III		370 s		
258394	95749		6 29 58.3	+13 36 55	0.000	0.00	8.00	1.4	-0.3	K5 III		430 s		BK Gem, v
--	--		6 30 00.0	+29 46 05	0.000	-0.02	7.0	1.4	-0.3	K7 III		280 s		
46095	196879		6 30 00.6	-31 10 27	0.000	+0.01	7.4	0.1	0.6	A0 V		200 s		
45757	95748		6 30 01.3	+17 54 23	-0.001	-0.02	7.55	0.06	0.6	A0 V	+36	230 s	5121	m
45829	113956		6 30 02.3	+7 55 17	0.000	+0.01	6.63	1.58	-6.1	K0 Iab		2100 s		
45466	41130	47 Aur	6 30 02.9	+46 41 08	0.000	+0.01	5.90	1.44	0.2	K0 III	-47	75 s		
46730	249568		6 30 03.1	-65 34 06	+0.001	+0.05	6.29	0.33		F2				
45787	95754		6 30 05.1	+14 30 03	+0.001	-0.02	8.0	0.5	3.4	F5 V		81 s		
46094	171852		6 30 05.2	-28 39 27	0.000	-0.02	7.8	1.0	3.4	F5 V		34 s		
45827	113957		6 30 05.4	+9 01 44	-0.001	-0.03	6.57	0.14		A0 p	+14			
45976	151573		6 30 11.1	-10 04 54	-0.002	-0.01	5.93	1.38		K0				
46037	151577		6 30 11.9	-19 12 54	+0.002	-0.03	6.6	1.6	-0.5	M2 III		240 s		
45953	133344		6 30 17.9	-6 25 50	-0.001	-0.03	7.97	-0.04	0.4	B9.5 V		330 s		
45769	78445		6 30 21.8	+22 32 42	-0.001	-0.01	6.80	0.1	0.6	A0 V		170 s	5129	m
45721	78440		6 30 21.9	+28 12 43	0.000	-0.01	6.9	0.1	1.4	A2 V	-4	120 s		
45271	13855		6 30 22.6	+60 08 44	0.000	-0.03	6.7	0.0	0.0	B8.5 V		220 s		
45622	41134		6 30 25.1	+41 24 05	-0.001	-0.02	6.6	0.0	0.0	B8.5 V		210 s		
45826	95759		6 30 25.2	+17 56 59	0.000	-0.01	7.60	0.8	3.2	G5 IV		75 s	5132	m
45411	25770		6 30 26.7	+55 41 47	0.000	0.00	7.7	1.1	-0.1	K2 III		360 s		
45974	133347		6 30 26.8	-3 01 18	0.000	+0.01	7.10	0.8	3.2	G5 IV		60 s		m
46976	249573		6 30 27.8	-68 56 54	0.000	0.00	7.9	0.0	0.4	B9.5 V		320 s		
46819	--		6 30 28.4	-66 01 25			7.5	1.1	0.2	K0 III		290 s		
45911	113973		6 30 29.1	+4 19 56	-0.001	-0.01	7.31	-0.14	-2.5	B2 V		880 s		m
46035	151583		6 30 29.7	-14 57 14	-0.001	+0.01	6.90	0.00	0.0	B8.5 V		230 s	5144	m
45910	113974		6 30 32.9	+5 51 59	+0.001	-0.03	6.77	0.33	-3.6	B2 III sh	-1	1100 s		AX Mon, q
46034	151582		6 30 32.9	-12 04 53	-0.002	-0.01	6.6	1.1	0.2	K0 III		190 s		
46383	217984		6 30 33.6	-48 14 48	-0.001	-0.01	7.9	1.5	-0.3	K5 III		230 s		
45465	25774		6 30 34.3	+52 48 33	+0.001	-0.02	7.8	0.4	2.6	F0 V		110 s		
46064	151585		6 30 34.7	-13 08 54	-0.002	0.00	6.16	-0.15		B2	+2		5148	m
46131	171861		6 30 38.4	-22 19 18	-0.002	0.00	7.2	0.3	0.0	B8.5 V		160 s		
46114	151587		6 30 39.3	-17 55 20	-0.012	-0.22	7.72	0.83	5.0	dG4	+1	28 s		
46288	217981		6 30 40.7	-40 26 54	0.000	0.00	7.30	0.00	0.0	B8.5 V		290 s		m
45851	78453		6 30 40.7	+20 25 17	-0.001	-0.01	8.00	0.1	0.6	A0 V		290 s		m
45637	41138		6 30 45.2	+43 52 50	+0.002	-0.03	7.6	1.1	0.2	K0 III		300 s		
46189	171866		6 30 46.3	-27 46 12	0.000	+0.01	5.93	-0.15	-2.0	B4 IV		390 s		
45410	25771	6 Lyn	6 30 47.1	+58 09 46	-0.003	-0.34	5.88	0.94	3.2	K0 IV	+36	34 s		m
46021	133354		6 30 47.3	-3 42 47	-0.001	-0.01	7.9	0.1	1.7	A3 V		170 s		
45824	78455		6 30 51.1	+26 38 38	+0.002	0.00	8.00	0.9	0.3	G8 III		330 s		
46272	196890		6 30 51.6	-35 00 02	-0.001	+0.02	7.1	0.5	2.1	A5 V		66 s		
46353	217985		6 30 53.7	-41 20 24	-0.001	0.00	7.8	1.3	-0.3	K5 III		410 s		
45783	59166		6 30 59.7	+32 48 18	-0.001	-0.03	8.0	1.6	-0.5	M2 III	+48	480 s		
46365	217986		6 30 59.7	-40 54 59	-0.002	0.00	6.20	1.40		K2				
46975	249575		6 31 01.3	-66 52 15	+0.003	+0.01	8.0	1.6	-0.5	M4 III		500 s		
45720	59164		6 31 02.9	+39 45 16	-0.002	-0.03	6.6	0.1	0.6	A0 V		160 s		
46415	217987		6 31 04.4	-43 43 03	-0.004	+0.01	6.68	1.00	3.2	G8 IV		40 s		
46697	234554		6 31 05.8	-59 00 16	+0.005	+0.05	7.4	1.2	-0.1	K2 III		240 mx		
46364	196899		6 31 07.1	-39 02 27	-0.004	+0.01	8.0	1.7	-0.3	K5 III		300 mx		
46005	113984		6 31 09.2	+9 56 24	-0.001	0.00	7.70	0.00	0.0	B8.5 V		330 s	5154	m
45995	95766		6 31 09.4	+11 15 04	0.000	-0.02	6.14	-0.08	-2.5	B2 V nne	-20	500 s	5153	m
45951	95765		6 31 10.0	+16 56 19	0.000	-0.05	6.20	1.16	-0.1	K2 III	+27	180 s	5146	m
46792	249572		6 31 10.6	-61 52 47	-0.002	+0.01	6.15	-0.15		B5	+34			
45823	59169		6 31 12.0	+33 45 22	-0.001	-0.01	7.8	1.4	-0.3	K5 III		380 s		
47213	256311		6 31 12.9	-70 30 59	-0.001	+0.02	7.4	0.5	4.0	F8 V		47 s		
46349	196897		6 31 13.0	-35 15 33	-0.001	0.00	5.8	1.7	1.7	A3 V		66 s		
46169	151599		6 31 15.4	-13 36 08	-0.002	+0.01	7.5	0.4	2.6	F0 V		97 s		

HD	SAO	Star Name	α 2000	δ 2000	μ(α)	μ(δ)	V	B-V	M_v	Spec	RV	d(pc)	ADS	Notes
			h m s	° ' "	s	"								
46505	217992		6 31 15.7	−49 37 00	+0.001	+0.01	7.5	0.8	0.2	K0 III		280 s		
45994	−−		6 31 17.7	+15 44 14	0.000	−0.02	7.90	0.9	3.2	G5 IV		86 s	5152	m
45971	95768		6 31 18.0	+16 17 26	−0.004	−0.04	7.9	0.9	3.2	G5 IV		87 s		
46569	234551		6 31 18.2	−51 49 34	+0.011	+0.10	5.60	0.54	4.2	dF9	+16	19 s		
45576	25782		6 31 19.6	+53 46 32	+0.002	−0.02	7.9	0.9	3.2	G5 IV		88 s		
46090	113993		6 31 21.4	+2 54 41	−0.008	+0.01	7.70	0.7	4.4	G0 V		46 s	5161	m
46696	234556		6 31 22.3	−57 36 48	+0.003	−0.02	8.00	1.1	0.2	K0 III		360 s		m
46184	151602		6 31 22.9	−12 23 30	+0.002	−0.02	5.30	1.1	−0.2	K3 III	+17	120 s		m
45993	95771		6 31 23.5	+17 25 07	0.000	−0.01	7.3	1.1	0.2	K0 III	+49	250 s		
46347	196905		6 31 24.1	−32 52 03	0.000	+0.04	6.5	1.6	−0.5	M2 III		250 s		
46185	151604		6 31 24.4	−12 33 33	+0.002	+0.01	6.8	0.0	−1.0	B5.5 V	+7	340 s		
46270	171884		6 31 29.5	−21 24 18	0.000	+0.04	6.7	0.5	2.1	A5 V		52 s		
45443	13866		6 31 31.1	+61 21 01	+0.002	+0.01	8.0	0.9	3.2	G5 IV		93 s		
46018	95772		6 31 31.3	+15 51 12	+0.001	−0.02	7.6	0.1	1.4	A2 V		170 s		
45848	59173		6 31 31.4	+37 10 43	−0.002	+0.01	6.6	1.1	0.2	K0 III		190 s		
46019	95774		6 31 33.3	+15 25 19	−0.001	−0.02	7.6	0.2	2.1	A5 V		130 s		
46627	234555		6 31 33.7	−51 55 28	−0.001	+0.01	7.8	1.5	−0.1	K2 III		250 s		
46431	196906		6 31 34.8	−36 56 26	−0.001	+0.05	6.2	1.2	−0.5	M2 III		220 s		
46017	95773		6 31 35.1	+16 49 43	+0.001	+0.01	7.0	1.1	−0.1	K2 III	+4	260 s		
46165	133365		6 31 36.0	−5 22 07	−0.001	0.00	7.46	−0.07	0.4	B9.5 V		260 s		
45899	59177		6 31 36.7	+32 09 59	+0.001	−0.02	7.00	0.00	0.4	B9.5 V	−3	200 s	5151	m
46031	95778	19 Gem	6 31 37.3	+15 54 12	−0.001	−0.02	6.4	0.2	2.1	A5 V	+21	71 s		
46269	171886		6 31 38.4	−21 03 59	+0.002	−0.05	7.9	1.0	0.2	K0 III		340 s		
46106	114001		6 31 38.4	+5 01 35	0.000	−0.02	7.92	0.14	−3.5	B1 V	+13	1200 s		
46075	95783		6 31 39.2	+11 47 32	−0.002	−0.01	6.50	−0.12	−0.9	B6 V		300 s		
46105	114002		6 31 39.8	+5 46 12	+0.001	−0.01	7.13	−0.03		A1 p			5162	m
45799	41151		6 31 41.5	+43 42 51	−0.003	0.00	8.0	0.9	3.2	G5 IV		93 s		
46138	114007		6 31 44.8	+2 00 58	0.000	−0.01	7.5	0.0	0.4	B9.5 V		260 s		
46462	196909		6 31 45.0	−37 10 21	−0.002	+0.03	7.56	−0.10	0.4	B9.5 V		270 s		
46122	114005		6 31 46.0	+6 47 08	−0.004	−0.03	7.7	0.6	3.0	G3 IV	−8	88 s		
45636	25786		6 31 47.5	+55 37 39	+0.002	−0.04	8.0	1.1	0.2	K0 III		360 s		
46089	95788		6 31 48.2	+11 32 40	+0.001	+0.03	5.23	0.15	1.9	A4 V	−3	45 s		
46229	133369		6 31 50.0	−8 09 28	−0.001	−0.01	5.43	1.38	−0.1	gK2	+3	90 s		
46328	171895	4 ξ¹ CMa	6 31 51.3	−23 25 06	−0.001	+0.01	4.34	−0.25	−3.9	B1 IV	+27	440 s	5176	m,v
46149	114009		6 31 52.4	+5 01 57	−0.001	−0.03	7.59	0.17		O8	+33			
46150	114010		6 31 55.5	+4 56 35	0.000	0.00	6.76	0.13		O6	+36		5165	m
46308	151614		6 31 55.6	−18 03 46	−0.001	−0.01	6.7	1.1	0.2	K0 III		200 s		
46413	171899		6 31 56.2	−29 18 35	−0.001	−0.03	7.5	1.1	−0.1	K2 III		330 s		
46860	234564	μ Pic	6 31 58.4	−58 45 15	0.000	−0.01	5.70	−0.06	0.4	B9.5 V		120 s		m
46030	78467		6 31 59.4	+22 11 09	0.000	0.00	7.2	0.1	0.6	A0 V		200 s		
46412	171901		6 32 00.2	−28 01 54	−0.001	−0.03	7.8	1.4	−0.3	K5 III		410 s		
45909	59186		6 32 01.2	+36 35 52	0.000	−0.02	7.8	0.2	2.1	A5 V		140 s		
46087	95790		6 32 03.2	+19 43 10	−0.001	0.00	7.9	0.9	3.2	G5 IV		87 s		
46179	114013		6 32 07.1	+6 02 07	0.000	−0.02	6.69	−0.04	0.2	B9 V		200 s		
46223	114017		6 32 09.1	+4 49 23	−0.002	−0.02	7.28	0.22		O5	+43			
47121	249579		6 32 10.0	−65 19 02	−0.007	+0.05	7.8	0.6	2.6	F0 V		74 s		
46283	133381		6 32 15.1	−7 24 27	−0.001	+0.04	7.27	0.25		A m				
45528	13871		6 32 16.9	+62 45 54	+0.004	−0.06	7.60	0.9	0.2	G9 III	−47	93 mx		
46199	114019		6 32 17.4	+8 49 07	−0.001	+0.01	6.9	0.1	1.7	A3 V		110 s		
46136	95795		6 32 18.5	+17 47 04	+0.002	+0.03	6.28	0.53	3.7	F6 V	0	30 s	5166	m
46241	114023	12 Mon	6 32 19.2	+4 51 21	−0.003	−0.01	5.86	0.99	0.2	K0 III	+21	140 s		
46652	218002		6 32 19.6	−45 18 37	+0.003	−0.03	7.17	1.09	3.2	K0 IV		50 s		
45743	25794		6 32 19.9	+54 51 13	−0.004	−0.11	7.7	0.5	4.0	F8 V		54 s		
46148	95800		6 32 20.7	+15 42 23	+0.001	0.00	7.20	0.5	3.4	F5 V	−13	57 s	5168	m
46568	196917		6 32 21.3	−37 41 48	+0.005	−0.08	5.3	1.4	3.2	G5 IV	+39	19 mn		
46304	133382		6 32 23.0	−5 52 08	0.000	−0.04	5.60	0.25		A5				m
46178	95803		6 32 23.3	+11 40 25	+0.001	−0.03	6.03	1.07	0.2	K0 III	−21	130 s	5170	m
46052	59194		6 32 27.1	+32 27 17	−0.002	−0.02	6.0	0.2	2.4	A7 V	−9	51 s		WW Aur, m,v
46446	171913		6 32 32.4	−21 01 19	0.000	0.00	7.10	0.00	−1.0	B5.5 V	−2	420 s		m
46894	234566		6 32 34.3	−55 38 56	+0.013	−0.10	7.9	0.8	3.2	G5 IV		85 s		
45836	25801		6 32 35.3	+50 57 27	−0.001	−0.01	7.7	1.1	0.2	K0 III		320 s		
46682	218005		6 32 36.5	−44 13 16	−0.001	−0.02	7.3	0.0	1.7	A3 V		130 s		
46547	196918		6 32 36.8	−32 01 57	−0.002	−0.01	5.69	−0.17		B3	+20			m
46754	218007		6 32 38.5	−48 08 25	−0.001	−0.03	7.60	1.1	−0.1	K2 III		350 s		
46222	95811		6 32 39.6	+15 01 15	0.000	−0.02	7.2	0.9	3.2	G5 IV		64 s		
46430	151624		6 32 40.1	−15 50 19	0.000	+0.01	8.0	0.1	1.4	A2 V		200 s		
46050	59199		6 32 40.2	+35 17 37	+0.001	−0.02	6.9	1.1	0.2	K0 III		210 s		
46198	95810		6 32 41.4	+17 01 43	0.000	0.00	7.8	0.1	0.6	A0 V		260 s		
46301	114031		6 32 42.6	+3 48 51	−0.001	−0.01	7.6	0.5	3.4	F5 V		70 s		
46380	133389		6 32 43.2	−7 30 33	−0.001	−0.01	8.00	0.43		B1 e	+15			
46072	59203		6 32 44.9	+35 03 46	0.000	−0.03	7.2	1.1	0.2	K0 III		240 s		
46523	171925		6 32 46.0	−27 24 15	+0.007	+0.10	7.00	0.5	3.4	F5 V		53 s		m
46460	151630		6 32 46.1	−20 01 16	+0.001	−0.02	8.0	0.8	−0.1	K2 III		420 s		
47095	249583		6 32 46.3	−61 49 40	−0.002	+0.03	7.2	1.3	0.2	K0 III		170 s		
46407	151625		6 32 46.8	−11 09 59	−0.001	0.00	6.24	1.11	−0.9	K0 II pe		270 s		

151

HD	SAO	Star Name	α 2000	δ 2000	μ(α)	μ(δ)	V	B-V	M_v	Spec	RV	d(pc)	ADS	Notes
			h m s	° ′ ″	s	″								
46459	151629		6 32 47.4	−17 42 36	−0.003	−0.02	7.7	0.0	0.4	B9.5 V		280 s		
45950	41162		6 32 49.6	+46 55 23	+0.003	−0.01	7.1	1.4	−0.3	K5 III		300 s		
46837	234565		6 32 49.8	−50 28 22	+0.001	+0.02	7.5	0.7	4.4	G0 V		35 s		
46161	78477		6 32 51.8	+25 50 51	+0.001	0.00	8.0	1.1	−0.1	K2 III		390 s		
46300	114034	13 Mon	6 32 54.2	+7 19 58	0.000	−0.01	4.50	0.00	−5.2	A0 Ib	+12	860 s		
46742	218009		6 32 56.1	−43 34 33	+0.002	+0.03	8.00	1.6	−0.5	M2 III		500 s		
46836	218012		6 32 57.6	−48 32 39	+0.001	−0.12	6.9	0.8	3.4	F5 V		30 s		
46389	133395		6 32 58.4	−3 49 14	0.000	−0.03	8.0	1.1	0.2	K0 III		350 s		
46426	133399		6 32 59.2	−8 57 51	−0.002	−0.01	7.39	0.07	0.6	A0 V		200 s		
46049	59207		6 32 59.8	+39 47 05	−0.001	−0.01	7.20	0.1	1.4	A2 V		150 s	5167	m
46160	78480		6 33 02.0	+27 49 31	0.000	−0.02	7.42	1.51	−0.3	K5 III		350 s		
44385	5874		6 33 04.1	+78 58 47	0.000	−0.02	6.6	1.1	0.2	K0 III		190 s		
46177	78483		6 33 04.7	+28 04 55	−0.004	−0.06	7.6	0.1	1.4	A2 V		100 mx		
46264	95819		6 33 05.5	+16 56 56	0.000	0.00	7.73	−0.02		B5	−4			
46377	114039		6 33 06.0	+1 16 22	−0.002	−0.02	7.40	1.4	−0.3	K4 III	+15	340 s		
46016	41168		6 33 09.6	+46 28 35	0.000	+0.01	6.8	0.0	0.0	B8.5 V	−5	230 s		
46727	196930		6 33 10.2	−38 37 31	0.000	−0.02	6.3	0.9	0.2	K0 III		170 s		
46726	196931		6 33 11.8	−37 49 20	−0.001	0.00	7.2	0.2	2.6	F0 V		84 s		
47673	−−		6 33 12.5	−71 38 59			8.0	1.1	0.2	K0 III		360 s		
46651	171939		6 33 14.1	−29 45 35	0.000	−0.02	7.8	1.6	−0.3	K5 III		360 s		
46582	171934		6 33 16.0	−23 04 00	−0.001	+0.01	7.9	1.1	−0.3	K5 III		430 s		
46280	95822		6 33 16.4	+19 50 37	0.000	−0.01	7.7	0.0	0.0	B8.5 V		330 s		
46014	41173		6 33 19.8	+47 13 07	+0.004	+0.01	7.9	1.6	−0.5	M2 III		200 mx		
44579	5877		6 33 24.6	+78 10 06	+0.004	−0.01	7.1	0.0	0.4	B9.5 V		220 mx		
47001	234574		6 33 26.1	−52 19 44	−0.005	+0.04	6.19	1.09		G5				
46602	171940		6 33 26.5	−20 55 26	−0.002	−0.03	6.4	1.5	3.2	G5 IV		17 s		
46694	196933		6 33 27.1	−32 13 36	−0.001	−0.01	7.1	1.7	−0.3	K5 III		230 s		
46013	41174		6 33 27.3	+48 22 30	+0.007	+0.05	7.8	0.8	3.2	G5 IV		84 s		
46680	171945		6 33 27.9	−29 12 35	+0.001	+0.02	7.9	0.2	0.6	A0 V		200 s		
46373	95829		6 33 31.9	+14 35 26	0.000	−0.01	7.9	1.1	−0.1	K2 III		380 s		
46915	218023		6 33 32.9	−47 29 53	+0.001	0.00	8.0	0.9	3.2	G5 IV		82 s		
46601	171941		6 33 33.6	−20 08 20	−0.001	0.00	7.5	0.5	3.0	F2 V		65 s		
46374	95830		6 33 36.1	+14 09 19	−0.001	−0.09	5.53	1.11	0.2	K0 III	−12	100 s		
46487	133404		6 33 37.8	−1 13 13	−0.001	−0.02	5.10	−0.14	−0.9	B6 V	+25	160 s		
46816	196944		6 33 38.4	−39 38 32	+0.009	−0.04	8.00	1.1	−0.1	K2 III		140 mx		
47064	234577		6 33 39.2	−54 07 01	−0.002	+0.01	8.0	0.5	0.6	A0 V		140 s		
46679	171950		6 33 39.9	−26 19 45	+0.001	+0.02	7.4	0.3	0.6	A0 V		140 s		
46541	133408		6 33 41.1	−7 12 12	0.000	+0.01	7.9	0.0	0.4	B9.5 V		310 s		
46277	78496		6 33 41.3	+27 58 49	−0.001	0.00	7.44	1.23	0.2	G9 III		200 s		
46157	59221		6 33 41.9	+39 26 45	+0.002	−0.05	7.2	1.1	−0.1	K2 III		290 s		
46251	59222		6 33 42.6	+33 01 27	+0.001	+0.01	6.40	0.03	1.4	A2 V	−9	100 s		
46914	218024		6 33 45.7	−43 51 33	−0.001	+0.07	8.0	1.0	0.2	K0 III		260 mx		
259654	114049		6 33 45.8	+9 10 08	0.000	−0.02	7.7	1.1	−0.1	K2 III		350 s		
46753	196941		6 33 46.0	−30 34 59	−0.001	+0.01	7.8	2.1	−0.5	M2 III		240 s		
46815	196945		6 33 49.4	−36 13 56	−0.002	+0.09	5.42	1.44		M1	+32			
46484	114058		6 33 54.3	+4 39 44	−0.001	−0.01	7.74	0.36	−3.5	B1 V		860 s		
46048	25811		6 33 57.6	+52 27 41	+0.001	−0.08	7.20	0.1	1.7	A3 V	+9	62 mx	5178	m
46336	78501		6 33 57.9	+27 02 41	+0.001	−0.02	7.76	1.05	0.2	G9 III		290 s		
48584	256317		6 34 02.7	−77 41 14	−0.004	−0.04	6.7	1.0	3.4	F5 V		21 s		
46814	196946		6 34 03.7	−30 20 13	+0.001	0.00	7.8	0.9	3.0	F2 V		40 s		
46721	171959		6 34 06.8	−21 32 10	−0.001	+0.01	7.9	−0.1	0.6	A0 V		280 s		
46401	78508		6 34 07.1	+22 07 26	0.000	−0.04	8.00	0.1	1.7	A3 V		180 s		m
46813	171964		6 34 08.3	−29 37 49	+0.001	−0.05	6.8	0.6	1.4	A2 V		57 s	5214	m
46558	114065		6 34 08.4	+2 53 54	0.000	+0.02	6.8	0.4	2.6	F0 V		69 s		m
46974	218028		6 34 10.1	−42 05 56	−0.001	−0.05	7.30	0.8	3.2	G5 IV		66 s		m
45767	13879		6 34 13.6	+65 17 15	−0.005	+0.08	7.6	0.4	3.0	F2 V		83 s		
46616	133421		6 34 18.1	−4 53 59	0.000	+0.02	7.26	−0.14	0.0	B8.5 V		280 s		
46296	59230		6 34 19.3	+38 04 34	+0.001	−0.01	6.60	1.1	0.2	K0 III		190 s	5188	m
46719	151652		6 34 19.8	−16 00 40	−0.002	+0.04	7.5	0.5	3.4	F5 V		67 s		
46597	114073		6 34 21.5	+3 18 22	−0.001	−0.03	7.50	0.9	0.3	G5 III	+13	270 s	5202	m
46644	133424		6 34 21.5	−3 04 27	−0.001	0.00	7.30	1.1	−0.1	K2 III	+19	290 s		m
46557	114071		6 34 22.6	+6 41 43	−0.001	−0.01	7.55	0.12	1.4	A2 V		160 s		
46573	114075		6 34 23.4	+2 32 02	−0.002	−0.01	7.93	0.34		O7	+14			
46555	114070		6 34 24.0	+9 05 07	−0.001	−0.01	7.7	1.4	−0.3	K5 III		370 s		
46135	25816		6 34 24.8	+51 05 41	−0.002	−0.02	8.0	1.1	0.2	K0 III		360 s		
46716	151653		6 34 32.0	−11 13 48	−0.001	−0.02	7.44	0.40	2.6	F0 V		81 s	5212	m
46101	25814	7 Lyn	6 34 32.8	+55 21 11	0.000	0.00	6.5	1.1	0.2	K0 III	−20	180 s		
−−	218031		6 34 33.1	−48 16 58	−0.001	−0.02	8.0	1.5						
47447	249588		6 34 33.8	−61 29 46	+0.004	+0.05	7.30	1.1	0.2	K0 III		260 s		m
46936	196955		6 34 35.2	−32 42 58	0.000	+0.01	5.62	−0.08	−0.2	B8 V	+42	140 s		
46612	114079		6 34 35.6	+4 58 04	+0.002	−0.04	7.10	1.6	−0.4	M0 III		310 s		
259919	114077		6 34 35.6	+7 46 38	−0.001	+0.01	7.7	1.1	0.2	K0 III		310 s		
46276	41188		6 34 38.4	+43 03 57	+0.001	−0.03	7.6	0.4	2.6	F0 V		100 s		
46516	95852		6 34 39.2	+19 25 44	0.000	−0.01	6.9	0.1	0.6	A0 V	+30	180 s		
46359	59239		6 34 41.7	+38 32 25	+0.001	−0.01	7.60	0.5	3.4	F5 V		69 s	5191	m

HD	SAO	Star Name	α 2000	δ 2000	μ(α)	μ(δ)	V	B-V	M_v	Spec	RV	d(pc)	ADS	Notes
			h m s	° ′ ″	s	″								
46973	196959		6 34 43.3	-34 00 33	0.000	+0.04	7.40	0.4	2.6	F0 V		91 s		m
47520	249592		6 34 45.6	-62 33 42	-0.003	+0.03	8.0	1.1	0.2	K0 III		360 s		
46854	171973		6 34 45.7	-20 33 27	-0.002	0.00	6.9	0.7	0.6	A0 V		67 s		
46833	151661		6 34 46.2	-19 57 07	+0.001	+0.06	7.8	0.2	0.0	B8.5 V		120 mx		
46554	95855		6 34 46.2	+16 22 54	0.000	0.00	8.0	0.1	1.4	A2 V		200 s		
46642	114085	14 Mon	6 34 46.3	+7 34 21	-0.001	-0.01	6.45	-0.02		A0	+38		5211	m
47545	249593		6 34 47.0	-63 15 43	-0.003	-0.01	7.6	0.1	1.4	A2 V		170 s		
47169	218037		6 34 51.0	-47 22 35	+0.002	-0.02	7.3	1.1	0.2	K0 III		240 s		
46853	151664		6 34 51.1	-19 38 50	0.000	-0.02	7.4	1.5	-0.3	K5 III		350 s		
47286	234588		6 34 51.8	-53 16 55	-0.003	0.00	6.9	1.9	-0.3	K5 III		160 s		
46737	133434		6 34 52.4	-4 40 52	+0.001	-0.01	7.9	1.1	0.2	K0 III		330 s		
46594	95860		6 34 53.6	+16 12 21	0.000	0.00	6.7	0.9	3.2	G5 IV		49 s		
46971	196962		6 34 54.2	-30 05 28	-0.001	+0.04	6.9	1.9	-0.3	K5 III		170 s		
46997	196963		6 34 55.9	-32 18 24	-0.004	+0.01	7.90	-0.12	0.0	B8.5 V		160 mx		m
46593	95861		6 34 56.7	+16 45 49	0.000	-0.01	7.22	-0.13	0.4	B9.5 V		230 s		
46595	95862		6 34 57.4	+15 19 51	+0.001	0.00	6.7	0.8	3.2	G5 IV	0	50 s		W Gem, v
46972	196964		6 34 57.8	-31 17 30	0.000	0.00	7.4	1.9	-0.3	K5 III		200 s		
47306	234589		6 34 58.5	-52 58 32	-0.001	+0.01	4.39	-0.02	-0.8	B9 III	+23	100 s		
47025	196970		6 35 00.1	-35 02 59	-0.003	-0.01	7.4	0.7	3.2	G5 IV		68 s		
46660	95865		6 35 00.3	+11 07 27	0.000	-0.01	8.04	0.31	-3.5	B1 V		1000 s		m
46464	59257		6 35 01.9	+31 26 05	-0.001	-0.01	7.1	1.1	0.2	K0 III		230 s		
46047	13885		6 35 02.7	+60 42 16	-0.002	-0.03	8.0	1.1	0.2	K0 III		360 s		
46828	151665		6 35 02.8	-12 36 28	+0.006	-0.16	7.8	0.5	4.0	F8 V		55 mx		
46933	171982	5 ξ² CMa	6 35 03.3	-22 57 53	0.000	+0.01	4.54	-0.05	0.6	A0 V	+32	61 s		
47013	196967		6 35 04.2	-30 56 12	-0.003	+0.04	8.0	1.1	1.7	A3 V		47 s		
46451	59255		6 35 06.2	+37 03 51	+0.002	-0.01	7.40	0.9	3.2	G5 IV		69 s	5201	m
46659	95867		6 35 07.1	+12 40 08	-0.001	-0.03	7.6	0.1	1.7	A3 V		150 s		
47567	249594		6 35 09.5	-61 10 24	+0.001	+0.02	7.4	0.3	1.7	A3 V		100 s		
46250	25822		6 35 10.2	+52 51 50	-0.001	-0.01	7.3	0.1	1.4	A2 V		150 s		
46553	78524	49 Aur	6 35 12.0	+28 01 20	0.000	-0.02	5.27	-0.03	0.4	B9.5 V	+17	92 s		
46785	133437		6 35 12.3	-1 30 37	-0.001	-0.03	8.0	1.6	-0.5	M4 III		490 s		
46994	171989		6 35 13.1	-25 50 24	0.000	+0.01	7.8	-0.3	0.0	B8.5 V		370 s		
46320	41191		6 35 14.0	+47 56 57	+0.001	-0.01	7.9	0.4	2.6	F0 V		110 s		
46932	171988		6 35 15.6	-20 23 31	0.000	+0.01	7.9	0.0	0.4	B9.5 V		290 s		
46769	114097		6 35 15.8	+0 53 24	0.000	-0.01	5.80	0.00	-5.6	B8 Ib	+10	1900 s		
46802	133439		6 35 17.1	-4 47 25	-0.001	-0.02	7.4	1.1	0.2	K0 III		270 s		
46709	95870		6 35 17.4	+9 59 19	-0.001	0.00	5.88	1.51	-0.3	K5 III	+39	170 s		
47145	196977		6 35 21.0	-37 26 57	+0.004	-0.05	8.0	1.0	-0.3	K5 III		450 s		
46275	25825		6 35 21.7	+51 58 29	-0.002	-0.08	7.8	0.4	2.6	F0 V		110 s		
46482	59259		6 35 22.1	+37 42 54	+0.001	-0.05	7.20	0.9	3.2	G5 IV		63 s	5208	m
47062	196975		6 35 22.4	-31 47 36	0.000	+0.03	8.0	1.3	-0.1	K2 III		360 s		
47144	196978		6 35 24.1	-36 46 48	-0.002	+0.02	5.59	-0.14	0.4	B9.5 V	+20	110 s		m
44490	1026		6 35 26.8	+80 34 24	-0.002	-0.02	7.3	1.1	0.2	K0 III		260 s		
47061	196976		6 35 27.1	-31 35 30	0.000	-0.01	7.89	-0.13	0.0	B8.5 V		380 s		
46784	114100		6 35 28.7	+5 30 45	-0.001	0.00	7.96	1.63	-0.4	M0 III		420 s		
47012	171992		6 35 30.8	-22 06 31	0.000	-0.01	6.5	1.2	0.2	K0 III		130 s		
46707	95875		6 35 31.3	+13 13 26	0.000	0.00	7.8	0.1	0.6	A0 V		260 s		
47187	196983		6 35 31.6	-37 51 29	0.000	-0.04	7.6	0.7	-0.1	K2 III		340 s		
47168	196982		6 35 33.8	-36 15 32	+0.002	-0.04	7.1	0.0	0.6	A0 V		190 s		
47371	234592		6 35 34.5	-50 17 46	-0.001	+0.01	6.7	1.4	0.2	K0 III		120 s		
46193	25824		6 35 35.2	+57 59 11	+0.002	-0.02	7.9	0.2	2.1	A5 V		150 s		
46871	133445		6 35 35.8	-2 14 34	+0.007	-0.11	8.0	0.5	4.0	F8 V		62 s		
46706	95873		6 35 36.1	+19 00 09	+0.001	-0.02	7.5	1.1	0.2	K0 III	-23	270 s		
46421	41193		6 35 36.9	+45 37 24	-0.001	0.00	7.70	1.6	-0.5	M4 III	+8	440 s		TU Aur, v
46552	59265		6 35 37.6	+32 34 38	0.000	+0.01	7.8	0.0	0.0	B8.5 V		350 s		
47352	218048		6 35 39.3	-47 26 00	-0.003	-0.01	7.2	1.1	0.2	K0 III		230 s		
46931	133452		6 35 42.4	-8 59 59	-0.001	-0.01	7.9	1.1	0.2	K0 III		330 s		
47265	218046		6 35 48.2	-41 03 17	-0.001	+0.03	7.87	1.03		K0				
47011	151681		6 35 50.6	-16 06 10	-0.002	0.00	7.39	-0.04	0.4	B9.5 V		250 s	5242	m
46764	95881		6 35 51.5	+17 11 59	0.000	-0.03	7.44	0.01	0.6	A0 V		230 s		
46765	95882		6 35 51.6	+16 26 47	-0.001	-0.01	7.4	1.1	0.2	K0 III	+15	260 s		
46927	133454		6 35 51.8	-3 11 19	0.000	-0.01	7.5	1.1	-0.1	K2 III		320 s		
47566	234600		6 35 51.9	-56 13 39	-0.002	+0.04	7.9	0.5	0.2	K0 III		340 s		
47408	218052		6 35 52.7	-48 16 48	+0.001	-0.02	7.50	1.1	0.2	K0 III		290 s		m
46782	95887		6 35 52.7	+13 10 39	0.000	+0.01	7.7	1.1	0.2	K0 III		300 s		
46747	95883		6 35 53.7	+17 41 32	-0.001	-0.02	7.4	1.1	0.2	K0 III	+23	260 s		
47230	196987		6 35 53.9	-36 05 20	-0.005	-0.10	6.35	0.48	4.4	G0 V		25 s		m
46990	151680		6 35 54.0	-11 22 29	-0.001	-0.02	7.1	1.1	0.2	K0 III		240 s		
46493	41198		6 35 54.7	+43 42 19	0.000	-0.03	7.1	0.9	3.2	G5 IV		60 s		
46949	133457		6 35 57.6	-3 58 39	-0.001	0.00	8.0	0.1	1.4	A2 V		200 s		
46641	59271		6 35 59.2	+32 17 27	+0.001	-0.03	7.3	0.1	1.4	A2 V	+16	150 s		
46657	59272		6 35 59.3	+31 28 46	0.000	-0.01	7.2	1.1	-0.1	K2 III		270 s		
47229	196988		6 35 59.4	-35 10 16	+0.002	+0.08	7.60	0.5	3.4	F5 V		69 s		m
46885	114115		6 35 59.9	+4 29 51	-0.002	-0.01	6.55	-0.06	0.2	B9 V		190 s		
46640	59269		6 36 01.3	+34 11 54	+0.001	+0.02	8.0	0.9	3.2	G5 IV		91 s		

HD	SAO	Star Name	α 2000	δ 2000	μ(α)	μ(δ)	V	B-V	M_v	Spec	RV	d(pc)	ADS	Notes
46781	95889		6 36 02.1	+16 47 49	+0.001	-0.03	6.7	0.5	3.4	F5 V	+31	46 s		
46318	25829		6 36 02.2	+56 23 11	-0.004	-0.05	6.5	0.4	3.0	F2 V	+2	51 s		
46825	95892		6 36 04.4	+13 41 38	-0.001	-0.01	7.00	0.2		A m			5231	m
47116	172002		6 36 04.9	-24 07 08	-0.002	-0.01	7.90	0.00	0.4	B9.5 V		320 s		
47141	172003		6 36 05.3	-24 51 59	-0.002	-0.03	7.6	1.2	-0.1	K2 III		320 s		
47186	172008		6 36 08.7	-27 37 20	+0.001	-0.26	7.9	0.8	3.2	G5 IV		59 mx		
47209	172010		6 36 09.4	-28 50 08	-0.001	0.00	7.1	1.6	-0.1	K2 III		150 s		
46883	95893		6 36 09.6	+10 17 05	-0.001	0.00	7.79	0.40	-2.5	B2 V n	+15	1000 s		
47283	196990		6 36 12.6	-36 08 21	-0.004	-0.11	7.90	0.7	4.4	G0 V		50 s		m
46316	25830		6 36 14.8	+58 49 38	+0.002	0.00	8.0	1.1	0.2	K0 III		360 s		
46317	25832		6 36 15.2	+58 06 25	+0.017	-0.19	6.7	0.5	3.4	F5 V		41 mx		
47207	172016		6 36 19.3	-26 32 08	0.000	-0.06	7.6	1.6	-0.3	K5 III		320 s		
47139	151692		6 36 20.4	-19 29 29	-0.001	+0.03	7.3	0.1	0.6	A0 V		200 s		
--	151693		6 36 21.4	-18 39 36	-0.002	+0.03	7.9	0.7	4.4	G0 V		51 s		
47138	151694	6 v¹ CMa	6 36 22.7	-18 39 36	-0.001	+0.01	5.52	0.76	0.3	G8 III	+25	110 s	5253	m
46294	25833		6 36 24.2	+59 39 50	+0.003	-0.03	6.8	0.1	0.6	A0 V		180 s		
46966	114120		6 36 25.8	+6 05 00	0.000	+0.01	6.87	-0.04		O8	+43			
46780	78540		6 36 26.3	+27 16 43	+0.001	-0.09	6.89	0.65	4.7	dG2	+29	31 ts	5234	m
47699	234607		6 36 27.1	-57 32 21	+0.004	+0.06	7.0	0.6	3.0	F2 V		44 s		
47540	234604		6 36 27.4	-50 19 22	+0.003	+0.14	7.2	0.6	3.2	G5 IV		65 s		
47282	196995		6 36 27.9	-31 53 12	-0.002	+0.02	6.8	0.4	3.0	F2 V		53 s		
46900	95895		6 36 28.8	+15 45 00	-0.001	-0.06	7.2	0.1	0.6	A0 V		75 mx		
45821	5898		6 36 30.5	+72 00 34	-0.010	-0.23	7.79	0.65	4.4	G0 V	+41	41 s		
46687	59280		6 36 32.8	+38 26 43	0.000	-0.03	5.29	2.61		C5 II	+12	22 mn		UU Aur, m,v
46763	59287		6 36 33.6	+31 21 32	0.000	0.00	7.9	0.1	1.4	A2 V		190 s		
47054	133469		6 36 35.3	-5 12 40	0.000	-0.02	5.52	-0.08		B8 pne	+27			
46762	59285		6 36 36.0	+33 23 25	-0.001	-0.02	7.9	0.9	3.2	G5 IV		87 s		
45190	5891		6 36 36.2	+77 55 31	+0.007	-0.03	7.6	0.1	0.6	A0 V		120 mx		
47247	172021		6 36 40.8	-22 36 54	-0.003	-0.01	6.35	-0.11		B8			5260	m
47205	151702	7 v² CMa	6 36 41.0	-19 15 22	+0.004	-0.08	3.95	1.06	0.0	K1 III	+3	25 ts		
46866	78545		6 36 42.3	+23 05 52	0.000	-0.02	6.8	0.1	0.6	A0 V		170 s		
47182	151700		6 36 46.6	-13 19 15	0.000	-0.03	5.97	1.56		K2				
46704	41209		6 36 47.1	+41 08 06	-0.002	-0.01	7.9	0.1	1.4	A2 V		200 s	5233	m
47475	218060		6 36 51.2	-41 33 25	0.000	+0.01	6.34	1.15		K0				
46760	59290		6 36 52.9	+37 16 32	0.000	-0.03	8.0	1.1	0.2	K0 III		360 s		
45967	5903		6 36 53.7	+70 30 10	0.000	-0.10	7.8	0.1	1.4	A2 V		50 mx		
47279	172031		6 36 55.4	-22 15 59	-0.001	-0.01	7.90	1.6	-0.5	M4 III		480 s		m
47246	151707		6 37 01.0	-16 21 24	+0.002	0.00	7.81	-0.02	0.4	B9.5 V		300 s		
47463	197011		6 37 01.9	-38 08 48	0.000	+0.04	6.04	1.03	3.2	G5 IV		23 s		m
46984	95902		6 37 02.0	+19 09 36	+0.001	-0.04	7.5	1.1	0.2	K0 III	-37	270 s		
47517	218062		6 37 02.5	-41 15 00	-0.002	+0.01	7.5	1.3	-0.3	K5 III		360 s		
47348	--		6 37 03.0	-26 00 56			7.8	1.9	0.2	K0 III		97 s		
47390	172037		6 37 03.3	-28 42 23	-0.001	-0.03	7.3	1.4	0.2	K0 III		150 s		
47072	114136		6 37 03.4	+5 31 10	-0.001	+0.01	7.15	0.2		A m			5256	m
47391	197008		6 37 03.7	-32 13 31	+0.025	-0.05	7.66	0.69		G5		12 mn		
47658	234608		6 37 04.9	-50 28 57	-0.002	0.00	8.00	1.6	-0.5	M2 III		500 s		m
47160	133484		6 37 05.6	-2 10 45	-0.002	-0.01	7.0	0.0	0.4	B9.5 V		210 s		
47721	234614		6 37 07.4	-52 20 50	0.000	-0.02	8.0	-0.1	0.6	A0 V		300 s		
48467	256320		6 37 09.0	-70 55 06	-0.007	-0.01	7.9	0.1	0.6	A0 V		180 mx		
47088	114140		6 37 10.5	+6 03 33	+0.001	0.00	7.61	-0.03	-4.4	B1 III		2000 s		
47537	218063		6 37 13.2	-40 46 13	-0.004	+0.02	7.4	1.3	0.2	K0 III		270 s		
47500	197014		6 37 13.7	-36 59 26	-0.002	+0.02	5.71	-0.12	0.4	B9.5 V	+29	120 s		m
47107	114142		6 37 14.3	+5 48 21	-0.001	-0.01	8.00	-0.03	-6.7	B1.5 Ia		6700 s		
47601	218064		6 37 19.8	-43 27 06	-0.001	-0.01	6.87	-0.16	-2.2	B5 III		650 s		
46924	59300		6 37 21.1	+31 42 38	+0.001	-0.02	7.8	1.1	0.2	K0 III		310 s		
47087	95904		6 37 23.0	+11 07 26	0.000	-0.01	8.0	0.1	1.4	A2 V		200 s		
47129	114146		6 37 24.0	+6 08 07	0.000	0.00	6.06	0.05		O8	+25			
47369	151721		6 37 26.9	-19 47 41	0.000	+0.02	7.5	0.1	0.0	B8.5 V		260 s		
47020	78557		6 37 27.1	+24 35 27	-0.001	+0.01	6.40	0.10	1.7	A3 V	-2	87 s		
47031	78558		6 37 29.2	+22 08 55	0.000	-0.02	7.9	0.1	0.6	A0 V		270 s		
47157	114147		6 37 31.7	+9 08 39	-0.005	-0.09	7.50	0.9	3.2	G5 IV		72 s	5262	m
47278	133491		6 37 32.0	-8 14 07	0.000	+0.01	7.23	0.00	0.4	B9.5 V		230 s		
47127	95907		6 37 33.7	+12 10 51	-0.007	-0.27	7.60	0.8	3.2	G5 IV	+53	75 s		m
47499	197016		6 37 34.3	-30 29 33	-0.002	+0.04	7.9	1.1	0.4	B9.5 V		70 s		
47910	234623		6 37 36.0	-55 20 37	-0.001	+0.03	6.90	1.1	0.2	K0 III		220 s		m
47179	114150		6 37 36.4	+5 39 32	-0.001	0.00	8.0	0.5	2.3	F7 IV		130 s		
47156	95914		6 37 36.8	+10 51 11	-0.003	-0.04	6.5	1.1	0.2	K0 III	+2	180 s		
47198	114151		6 37 37.4	+4 55 20	-0.002	-0.06	8.00	0.9	3.2	G8 IV		90 s		
47768	218075		6 37 37.9	-49 31 39	-0.007	+0.03	7.1	0.5	2.1	A5 V		66 s		
46590	25842	11 Lyn	6 37 38.4	+56 51 27	0.000	+0.01	5.7	0.1	0.6	A0 V	0	110 s		
48527	256321		6 37 38.5	-70 36 15	-0.003	-0.06	7.3	0.5	4.0	F8 V		45 s		
47051	78561		6 37 38.7	+22 42 59	0.000	-0.01	7.7	1.1	-0.1	K2 III		340 s		
47221	114155		6 37 39.8	+1 48 41	-0.001	-0.01	7.9	0.0	0.4	B9.5 V		310 s		
47600	197021		6 37 40.0	-38 37 42	-0.001	+0.01	7.91	-0.12	0.0	B8.5 V		380 s		
47220	114154		6 37 40.3	+2 42 15	-0.002	-0.05	6.17	1.08	0.0	K1 III	-8	170 s		

HD	SAO	Star Name	α 2000	δ 2000	μ(α)	μ(δ)	V	B-V	M$_v$	Spec	RV	d(pc)	ADS	Notes
46608	25843		6^{h}37^{m}40.5	+55°21'33"	-0.002	+0.01	7.4	0.1	1.4	A2 V		160 s		
47366	151725		6 37 40.8	-12 59 06	+0.002	-0.13	6.12	1.00		K0				
46480	13897	8 Lyn	6 37 41.3	+61 28 52	-0.028	-0.28	5.94	0.89	0.3	gG7	-46	130 s		m
47241	114156		6 37 41.8	+2 06 02	0.000	-0.02	8.0	0.1	0.6	A0 V		290 s		
47105	95912	24 γ Gem	6 37 42.7	+16 23 57	+0.003	-0.04	1.93	0.00	0.0	A0 IV	-13	26 ts		Alhena, m
47671	218070		6 37 42.7	-44 29 45	+0.001	0.00	7.69	-0.02		A0				
47670	218071	ν Pup	6 37 45.6	-43 11 45	-0.001	-0.01	3.17	-0.11	-1.2	B8 III	+28	75 s		
47104	95913		6 37 45.8	+18 24 05	0.000	-0.01	7.4	0.0	0.0	B8.5 V		300 s		
47536	197019		6 37 47.5	-32 20 23	+0.008	+0.07	5.27	1.18	-0.1	K2 III	+79	120 s		
47619	197025		6 37 50.1	-36 06 50	-0.003	-0.07	7.7	0.3	4.0	F8 V		54 s		
47274	114163		6 37 50.7	+0 31 38	0.000	0.00	7.4	0.1	0.6	A0 V		230 s		
47364	133500		6 37 51.1	-8 46 57	0.000	-0.06	7.40	1.1	0.2	K0 III	-20	270 s	5277	m
47086	78568		6 37 52.5	+23 36 15	+0.001	0.00	6.80	1.1	0.2	K0 III		200 s		m
47240	114162		6 37 52.6	+4 57 26	0.000	+0.02	6.15	0.14	-5.7	B1 Ib	+36	1500 s		
47257	114164		6 37 53.0	+2 16 09	0.000	-0.01	7.2	0.0	0.4	B9.5 V		220 s		
47564	197020		6 37 53.1	-30 38 02	-0.004	-0.03	8.0	0.7	1.7	A3 V		72 s		
47442	151730	8 ν³ CMa	6 37 53.3	-18 14 15	-0.001	-0.01	4.43	1.15	0.0	K1 III	-2	72 s		
45947	5909		6 37 54.9	+73 41 44	-0.035	-0.02	6.24	0.38	3.0	F2 V	+6	43 s		
47563	172063		6 37 59.3	-28 29 45	0.000	+0.03	7.7	1.1	2.1	A5 V		38 s		
48189	249604		6 38 00.6	-61 31 59	-0.002	+0.08	6.18	0.62		G0				m
47050	59309		6 38 04.8	+30 53 20	0.000	-0.03	7.3	0.1	1.4	A2 V	-12	150 s		
46776	25850		6 38 06.4	+51 12 08	+0.002	+0.01	7.9	1.6	-0.5	M4 III		490 s		
46607	25844		6 38 06.9	+59 26 57	-0.003	+0.01	7.60	0.8	3.2	G5 IV		76 s		m
47292	114174		6 38 11.8	+7 08 54	0.000	-0.01	7.49	0.11	0.0	B8.5 V		270 s		
46463	13901		6 38 18.8	+64 44 05	-0.001	-0.01	7.40	0.5	3.4	F5 V		63 s	5236	m
46839	25852		6 38 18.8	+50 32 45	0.000	0.00	7.9	1.1	0.2	K0 III		340 s		
47176	78572		6 38 18.8	+24 27 01	0.000	-0.04	6.70	0.2	2.1	A5 V		82 s	5270	m
46702	25849		6 38 20.2	+57 11 29	+0.001	-0.03	6.8	0.9	3.2	G5 IV		53 s		
47420	133511		6 38 20.4	-2 32 37	-0.002	+0.01	6.14	1.48		K2				
47339	114179		6 38 22.3	+6 19 03	-0.001	-0.01	7.90	0.8	0.3	G8 III		320 s		
47152	78571	53 Aur	6 38 22.9	+28 59 03	-0.002	-0.02	5.79	0.00		A0 p	+14			
47579	172068		6 38 23.0	-23 34 47	+0.001	+0.01	6.6	0.9	4.4	G0 V		17 s		
48024	234627		6 38 23.0	-52 55 57	-0.001	+0.01	7.0	0.9	-0.3	K5 III		290 s		
48447	249609		6 38 25.4	-66 05 07	-0.003	+0.05	7.2	1.1	0.2	K0 III		250 s		
46981	41227		6 38 26.3	+44 18 46	+0.001	-0.01	7.9	0.4	3.0	F2 V	-19	94 s		
47382	114183		6 38 28.5	+4 36 27	-0.001	+0.01	7.14	0.17	-5.0	B0 III	+15	1600 s		
47471	133518		6 38 31.9	-9 19 18	+0.002	-0.02	7.48	1.64	-0.5	M2 III		370 s		
47313	95934		6 38 33.0	+12 44 18	+0.001	0.00	8.0	0.4	2.6	F0 V		120 s		
47048	59313		6 38 34.1	+39 07 10	0.000	-0.01	8.0	1.4	-0.3	K5 III		450 s		
47561	151737		6 38 35.3	-16 52 25	-0.001	-0.02	6.03	0.03	0.6	A0 V		110 s		
47047	41230		6 38 36.3	+40 20 11	-0.002	-0.24	7.70	1.1	0.2	K0 III		55 mx		m
46635	13910		6 38 36.5	+61 28 41	+0.002	-0.02	7.1	0.1	1.7	A3 V		120 s		
47973	218093		6 38 37.6	-48 13 13	0.000	+0.01	4.93	0.87		G5	+28	17 mn		m
47432	114191		6 38 38.1	+1 36 49	-0.001	-0.01	6.21	0.15		09.5 II	+58			
48246	249607		6 38 38.5	-60 10 59	0.000	+0.01	7.2	-0.3	0.4	B9.5 V		220 s		
47070	59316	51 Aur	6 38 39.5	+39 23 27	-0.002	-0.11	5.69	1.35	0.2	K0 III	+33	77 s		
47805	197035		6 38 39.5	-37 59 46	+0.001	-0.03	7.16	-0.10	1.4	A2 V		140 s		
47046	41232		6 38 40.8	+41 34 54	0.000	0.00	6.8	0.0	0.4	B9.5 V		190 s	5269	m
48002	218096		6 38 43.7	-48 43 09	+0.001	0.00	8.0	-0.2	1.4	A2 V		190 s		
47924	218088		6 38 43.8	-45 03 51	-0.001	+0.02	7.70	0.00		B8.5 V		350 s		m
47596	151742		6 38 45.5	-16 15 39	-0.001	0.00	6.9	0.1	0.6	A0 V		180 s		
47417	114192		6 38 47.9	+6 54 07	0.000	0.00	6.92	0.05	-4.6	B0 IV	+32	1400 s		
47923	218091		6 38 47.9	-43 49 29	-0.002	+0.02	7.57	-0.04		A0				
47100	59319	52 ψ³ Aur	6 38 49.1	+39 54 09	-0.001	-0.01	5.20	-0.07	-1.2	B8 III	+9	180 s		
47431	114194		6 38 49.4	+4 42 02	+0.001	-0.01	6.57	-0.07	-0.9	B6 V		280 s		
47416	114193		6 38 50.9	+7 58 20	-0.002	+0.02	7.79	-0.11	0.6	A0 V		270 s		
47216	59326		6 38 52.6	+33 01 05	-0.005	-0.01	7.2	0.4	2.6	F0 V		82 s		
47256	78580		6 38 53.6	+27 48 14	0.000	-0.04	7.30	0.1	1.4	A2 V	+42	150 s		m
47908	218094		6 38 55.1	-40 35 10	0.000	-0.01	7.90	0.5	4.0	F8 V		60 s		
47381	95940		6 38 55.8	+15 12 23	-0.002	-0.07	7.9	0.5	3.4	F5 V		79 s		
46606	13911		6 38 57.2	+64 09 19	+0.001	-0.08	8.0	1.1	-0.1	K2 III	+28	210 mx		
47743	172086		6 38 57.5	-27 07 12	-0.003	0.00	8.0	0.8	1.7	A3 V		63 s		
47099	41236		6 38 58.9	+41 29 39	+0.001	-0.04	7.7	1.4	-0.3	K5 III		390 s		
47379	95943		6 39 03.2	+19 35 45	0.000	+0.01	8.0	0.4	2.6	F0 V		120 s		
47358	78586		6 39 05.2	+22 01 51	0.000	-0.03	6.04	1.03	0.2	G9 III	-9	140 s		
47255	59330		6 39 05.7	+31 45 59	-0.002	+0.01	7.3	0.1	0.6	A0 V	+6	220 s		
47907	197046		6 39 05.7	-37 32 00	-0.002	-0.03	7.9	1.1	0.2	K0 III		320 s		
47530	114213		6 39 10.4	+0 24 11	0.000	+0.01	7.7	0.9	3.2	G5 IV		80 s		
47667	151751		6 39 16.6	-14 08 45	0.000	0.00	4.82	1.50	-2.2	K2 II	+29	190 s		
---	---		6 39 16.8	-31 49 47			7.80							m
47451	95951		6 39 16.9	+13 41 10	-0.001	-0.03	7.6	1.1	0.2	K0 III		290 s		
48405	249614		6 39 17.2	-60 58 03	+0.001	0.00	7.9	2.2	-0.3	K5 III		160 s		
47174	41239	50 ψ² Aur	6 39 19.8	+42 29 20	+0.001	-0.06	4.79	1.23	-0.2	K3 III	+17	100 s		m
48000	---		6 39 24.8	-39 31 21			8.0	1.3	0.2	K0 III		250 s		
47850	172101		6 39 26.6	-28 26 28	+0.002	-0.02	7.6	1.9	-0.5	M2 III		290 s		

HD	SAO	Star Name	α 2000	δ 2000	μ(α)	μ(δ)	V	B-V	M_V	Spec	RV	d(pc)	ADS	Notes
			^{h} ^{m} ^{s}	$^\circ$ ′ ″	^{s}	″								
48069	218102		6 39 26.6	−43 25 36	−0.001	+0.01	7.17	1.55		K5				
47690	151754		6 39 27.2	−12 02 08	−0.001	−0.01	7.9	1.1	0.2	K0 III		330 s		
46633	13912		6 39 27.9	+66 09 44	0.000	0.00	7.60	0.00	0.0	B8.5 V		330 s	5255	m
47415	78596		6 39 31.3	+24 36 00	+0.001	+0.08	6.38	0.53		F5	+18			
47905	172106		6 39 31.4	−30 02 19	−0.003	+0.02	7.8	3.4	−0.3	K5 III		28 s		
47395	78593	54 Aur	6 39 33.1	+28 15 47	0.000	−0.02	6.03	−0.08	−1.9	B6 III	+19	350 s	5289	m
47553	95957		6 39 34.9	+10 14 26	−0.001	−0.01	8.0	0.1	0.6	A0 V		290 s		
47554	114220		6 39 35.4	+9 45 08	−0.001	−0.02	7.9	0.4	2.6	F0 V		110 s		
48446	249615		6 39 35.7	−60 24 03	+0.001	+0.06	7.8	0.6	3.2	G5 IV		83 s		
47827	172102		6 39 36.2	−23 41 44	+0.001	0.00	6.05	−0.03		A0				m
48267	234643		6 39 37.3	−52 49 57	0.000	0.00	7.3	1.0	−0.3	K5 III		340 s		
47946	197057		6 39 42.6	−30 28 13	−0.001	−0.18	5.71	1.14		K0				
48292	234644		6 39 43.2	−53 27 05	+0.001	−0.01	7.2	0.4	1.4	A2 V		86 s		
47663	133536		6 39 46.3	−1 22 50	−0.001	−0.02	7.4	1.1	0.2	K0 III		260 s		
48150	218109		6 39 46.6	−43 24 09	−0.003	+0.02	7.41	−0.13		B8				
47575	95963		6 39 47.5	+12 58 59	−0.002	0.00	5.90	0.06	1.7	A3 V	−16	69 s	5302	m
47549	95965		6 39 56.5	+19 18 44	0.000	−0.01	7.7	1.1	0.2	K0 III		300 s		
48087	197064		6 39 56.8	−38 09 32	+0.004	−0.03	6.58	1.18		K0				
47608	95972		6 39 57.0	+11 40 52	0.000	0.00	7.8	0.0	0.4	B9.5 V		290 s		
47270	41245		6 39 57.9	+44 00 50	+0.003	0.00	6.4	1.1	0.0	K1 III	−30	200 s		
47548	78609		6 39 59.1	+20 29 12	−0.001	−0.02	7.9	1.6	−0.5	M2 III		460 s		
47574	95968		6 40 02.1	+19 39 28	0.000	−0.07	7.3	1.1	0.2	K0 III		170 mx		
50506	258451	ζ Men	6 40 02.9	−80 48 49	−0.007	+0.06	5.64	0.20	1.0	A4 IV	+9	76 s		
47868	151771		6 40 04.2	−17 17 37	0.000	−0.01	7.8	0.4	2.6	F0 V		110 s		
47525	78610		6 40 05.3	+23 40 25	0.000	−0.02	7.9	0.5	3.4	F5 V		77 s		
47760	133543		6 40 05.5	−2 21 42	−0.003	−0.01	7.8	0.1	0.6	A0 V		260 s		
261683	114234		6 40 07.3	+9 18 01	0.000	−0.01	7.5	1.3	−0.4	K8 III		360 s		
46701	13914		6 40 10.3	+67 19 24	+0.004	−0.04	8.00	0.00	0.4	B9.5 V		100 mx	5267	m
47335	41251		6 40 12.8	+44 19 57	+0.004	−0.03	6.8	0.9	3.2	G5 IV	−13	53 s		
47942	151775		6 40 15.1	−19 05 46	−0.001	+0.01	8.0	0.1	1.4	A2 V		200 s		
47799	133548		6 40 15.8	−4 27 39	0.000	+0.03	7.50	1.4	−0.3	K5 III	+35	350 s		m
45560	5911		6 40 16.8	+79 35 58	+0.002	+0.02	6.5	0.1	0.6	A0 V	−7	150 s		
47758	114239		6 40 18.1	+3 32 26	−0.001	−0.03	7.80	1.1	−2.2	K2 II		910 s		m
47821	133552		6 40 18.8	−6 20 47	+0.001	+0.01	6.85	1.65	−0.5	M3 III	−14	280 s		
47412	41253		6 40 21.3	+40 58 09	−0.002	−0.02	7.30	0.4	3.0	F2 V		72 s	5296	m
48559	234651		6 40 21.9	−59 07 22	−0.001	+0.02	7.6	0.0	0.0	B8.5 V		290 s		
47759	114243		6 40 25.0	+2 33 28	−0.001	−0.03	7.7	0.0	0.4	B9.5 V		280 s		
47732	114241		6 40 28.5	+9 49 03	0.000	−0.02	7.90	0.00	0.0	B8.5 V		370 s	5316	m
45866	5919		6 40 28.9	+77 59 45	+0.005	0.00	5.73	1.47	0.2	K0 III	−14	67 s		
47005	13917		6 40 31.4	+61 55 22	−0.003	+0.02	6.6	0.1	1.4	A2 V		110 s		
47756	114244		6 40 31.7	+6 22 17	−0.002	+0.01	6.51	−0.14	0.2	B9 V		180 s		
46509	5925		6 40 32.6	+71 44 55	+0.004	+0.01	5.92	1.19	0.2	K0 III	−23	110 s		
48166	197076		6 40 33.4	−34 12 44	−0.003	+0.05	7.6	0.4	1.4	A2 V		100 s		
48112	172145		6 40 34.4	−28 15 54	0.000	−0.01	7.2	0.9	3.2	G5 IV		52 s		
261783	114245		6 40 37.2	+9 35 43	−0.001	0.00	7.9	1.4	−0.3	K5 III		420 s		
47573	59350		6 40 40.5	+31 06 59	0.000	−0.01	7.7	1.1	0.2	K0 III		300 s		
47777	114250		6 40 42.3	+9 39 22	0.000	+0.01	7.94	−0.16		B3	+6			
47467	41261		6 40 42.6	+40 10 05	−0.002	−0.02	7.2	0.5	3.4	F5 V		58 s		
47590	59351		6 40 46.8	+33 01 30	−0.002	−0.03	7.7	0.1	1.4	A2 V		180 s		
48402	218123		6 40 47.3	−47 37 16	−0.001	+0.02	6.65	−0.14		B8				
47309	25875		6 40 49.0	+51 58 42	+0.001	−0.29	7.6	0.7	4.4	G0 V		38 mx		
48403	218124		6 40 49.3	−47 40 29	−0.001	−0.01	6.4	1.4	−0.5	M2 III		240 s		
47629	59354		6 40 52.5	+30 12 41	0.000	−0.01	8.0	0.0	0.4	B9.5 V		320 s		
48038	151785		6 40 57.3	−12 10 57	0.000	0.00	6.93	0.04	−1.6	B3.5 V	−10	400 s		
49268	256326		6 40 57.9	−71 46 32	+0.008	−0.05	6.51	1.11		K0				m
47839	114258	15 Mon	6 40 58.6	+9 53 45	0.000	−0.01	4.66	−0.25		O7	+33	22 mn	5322	S Mon, m,q
47254	25874		6 40 59.5	+56 05 57	+0.002	−0.02	7.0	1.1	0.2	K0 III		230 s		
46962	13919		6 40 59.6	+64 59 13	0.000	0.00	7.6	0.9	3.2	G5 IV		75 s		
47705	78629		6 40 59.9	+25 19 25	+0.001	−0.01	7.9	1.6	−0.5	M2 III		460 s		
48016	133566		6 41 01.7	−7 34 21	−0.002	+0.01	7.1	1.1	−0.1	K2 III	+46	260 s		
46419	5924		6 41 01.9	+74 03 00	−0.002	−0.01	7.9	0.4	3.0	F2 V		94 s		
49194	256325		6 41 02.8	−70 39 05	−0.002	+0.07	7.0	0.2	2.1	A5 V		95 s		
47753	78632		6 41 03.2	+20 38 44	0.000	+0.01	7.60	0.8	3.2	G5 IV		75 s	5320	m
47650	59359		6 41 03.7	+31 21 05	−0.001	−0.02	8.0	0.9	3.2	G5 IV		91 s		
47355	25878		6 41 04.2	+53 23 55	+0.001	+0.01	7.6	0.5	3.4	F5 V		71 s		
47964	114269		6 41 05.3	+0 29 43	−0.001	0.00	5.79	−0.10	−0.8	B9 III		210 s		
48261	197084		6 41 07.8	−30 38 43	−0.001	+0.01	7.23	−0.06	0.4	B9.5 V		230 s		
48013	133567		6 41 09.0	−5 42 17	+0.001	+0.03	7.5	1.1	0.2	K0 III	−7	280 s		
47888	114265		6 41 09.0	+8 59 08	−0.001	−0.06	6.7	0.5	3.4	F5 V		45 s	5328	m
47572	59356		6 41 09.5	+39 14 05	+0.002	+0.01	7.3	1.1	0.2	K0 III		270 s		
47887	114264		6 41 09.6	+9 27 54	0.000	−0.03	7.14	−0.18	−3.6	B2 III	+19	1400 s	5327	m
48060	133568		6 41 10.0	−7 59 36	−0.001	+0.01	7.90	−0.09	0.4	B9.5 V		320 s	5336	m
48305	197086		6 41 10.9	−34 30 37	+0.001	−0.05	7.3	1.4	−0.3	K5 III		330 s		
48383	218126		6 41 14.0	−40 20 59	−0.002	+0.01	6.12	−0.14	−1.7	B3 V nn		370 s		m
47886	95997		6 41 17.1	+11 00 12	0.000	+0.01	6.11	1.68	−0.5	M2 III	+16	190 s		

HD	SAO	Star Name	α 2000	δ 2000	μ(α)	μ(δ)	V	B-V	M_v	Spec	RV	d(pc)	ADS	Notes
			h m s	° ' "	s	"								
48382	218127		6 41 18.1	−40 17 35	0.000	−0.01	8.0	0.0	0.6	A0 V		300 s		
47730	78633		6 41 18.5	+29 42 48	0.000	−0.03	7.3	1.1	0.0	K1 III		280 s		
48609	234655		6 41 19.9	−52 41 48	0.000	−0.01	7.8	0.0	2.6	F0 V		110 s		
47731	78636	25 Gem	6 41 20.9	+28 11 47	+0.001	−0.01	6.42	1.09	−4.5	G5 Ib	−6	1500 s		m
47863	95996		6 41 21.8	+16 23 51	−0.001	−0.01	6.2	0.0	0.6	A0 V	+17	130 s		
48106	133574		6 41 22.2	−7 10 08	−0.002	−0.01	6.80	0.00	0.0	B8.5 V		220 s	5340	m
47885	95998		6 41 22.8	+15 25 38	−0.001	0.00	7.50	0.2	2.1	A5 V		120 s	5330	m
48056	133573		6 41 23.3	−2 21 00	+0.005	−0.14	8.0	0.5	4.0	F8 V		62 mx		
48035	114278		6 41 25.8	+0 57 25	0.000	−0.01	7.50	0.1	0.6	A0 V		230 s		m
48289	172170		6 41 25.8	−27 35 12	0.000	+0.04	8.0	−0.1	0.6	A0 V		300 s		
47961	114274		6 41 27.3	+9 51 15	0.000	0.00	7.3	0.0	−1.0	B5.5 V	+23	430 s		
48240	172169		6 41 27.5	−24 22 59	0.000	−0.01	7.9	0.2	0.4	B9.5 V		220 s		
48220	172166		6 41 28.0	−22 06 14	−0.001	0.00	7.7	0.3	0.2	K0 III		310 s		
46896	13920		6 41 28.5	+68 38 51	−0.006	0.00	8.0	1.1	0.2	K0 III		360 s		
48164	151796		6 41 29.1	−13 56 43	−0.001	+0.03	8.0	0.1	0.6	A0 V		290 s		
47984	114277		6 41 29.9	+6 06 39	+0.001	−0.01	6.8	0.0	0.2	B9 V		210 s		
49006	249624		6 41 30.2	−65 28 08	0.000	−0.07	7.8	0.4	3.0	F2 V		91 s		
48203	172167		6 41 30.7	−20 19 06	+0.004	−0.06	6.8	1.2	3.2	G5 IV		23 s		
48990	−−		6 41 37.3	−64 36 59			7.8	1.1	0.2	K0 III		330 s		
47703	59365		6 41 37.5	+35 55 55	−0.003	−0.02	6.46	0.49	3.4	F5 V	+86	38 s		
47836	78643		6 41 42.7	+27 04 53	−0.001	0.00	7.52	0.89	0.3	G8 III		280 s		
48338	172177		6 41 45.2	−27 38 01	−0.002	+0.01	7.7	1.1	0.2	K0 III		290 s		
48010	96004		6 41 47.3	+14 24 49	−0.001	−0.02	7.7	0.4	2.6	F0 V		100 s		
47773	59371		6 41 47.8	+32 38 13	0.000	−0.01	7.7	0.5	3.4	F5 V		73 s		
49398	−−		6 41 49.0	−71 13 12			8.0	1.1	0.2	K0 III		360 s		
48011	96007		6 41 50.7	+14 13 01	+0.002	−0.02	6.8	0.1	1.4	A2 V		120 s		
48503	197101		6 41 53.8	−39 47 17	−0.002	−0.02	7.75	−0.06	0.6	A0 V		270 s		
47607	41272		6 41 54.1	+47 47 37	−0.001	−0.02	7.6	0.1	1.4	A2 V		170 s		
48101	114292		6 41 54.6	+2 34 53	0.000	−0.02	7.4	1.1	0.2	K0 III	+19	260 s		
48217	133585		6 41 56.4	−9 10 03	+0.003	−0.04	5.19	1.53	−0.4	gM0	+1	130 s		
262096	78650		6 41 56.6	+20 07 23	−0.001	+0.02	8.0	0.5	4.0	F8 V		62 s		
48315	172178		6 41 57.0	−20 15 00	−0.002	−0.06	6.9	1.3	0.2	K0 III		150 s		
48099	114293		6 41 59.2	+6 20 42	+0.001	−0.01	6.37	−0.05		O6	+31			
48287	151807		6 41 59.6	−16 00 25	−0.001	+0.02	6.95	−0.03	0.4	B9.5 V		200 s	5358	m
48157	114297		6 42 04.7	+3 14 55	0.000	−0.01	7.80	0.1	1.4	A2 V		190 s		m
49359	249630		6 42 04.9	−69 42 25	−0.003	0.00	7.6	0.1	0.6	A0 V		250 s		
48286	151809		6 42 05.0	−15 12 55	−0.013	−0.13	7.2	0.7	4.4	G0 V		37 s		
48215	133587		6 42 07.4	−6 06 52	+0.001	−0.01	7.03	−0.14	−1.0	B5.5 V	+23	400 s		
−−	−−		6 42 08.0	+25 28 03			7.80							m
49532	256328		6 42 08.4	−71 53 29	+0.008	−0.02	8.0	1.1	−0.1	K2 III		350 mx		
48543	197108		6 42 16.3	−38 23 55	−0.002	0.00	6.29	0.34		A3				
48144	114299		6 42 17.1	+7 23 51	0.000	−0.02	6.86	1.63	−0.3	K5 III	−10	230 s		
48485	197104		6 42 19.0	−31 39 59	−0.001	−0.01	7.80	−0.10	0.4	B9.5 V		300 s		
48631	218137		6 42 20.3	−41 43 43	0.000	0.00	7.9	0.8	3.2	G5 IV		85 s		
48212	114302		6 42 21.4	+2 52 08	−0.001	0.00	7.70	0.00	0.4	B9.5 V		280 s	5359	m
47862	59384		6 42 23.0	+37 05 47	0.000	0.00	7.8	0.8	3.2	G5 IV		84 s		
48097	96015	26 Gem	6 42 24.3	+17 38 43	0.000	−0.09	5.21	0.06	1.4	A2 V	+14	37 mx		
49306	249631		6 42 25.6	−67 50 34	−0.004	+0.05	6.9	−0.2	0.6	A0 V		140 mx		
48460	172199		6 42 26.1	−25 41 12	0.000	−0.03	8.0	0.6	0.6	A0 V		120 s		
48676	218138		6 42 27.0	−42 34 11	−0.005	+0.10	7.95	0.59	3.6	G0 IV-V		73 s		m
48425	172196		6 42 27.3	−23 13 56	−0.002	+0.01	7.40	0.00	−1.0	B5.5 V		480 s	5371	m
49076	249627		6 42 28.6	−61 45 22	−0.002	+0.01	7.80	0.5	3.4	F5 V		76 s		m
47215	13929		6 42 29.3	+66 11 51	−0.001	−0.04	7.10	0.5	4.0	F8 V		42 s		m
48007	78658		6 42 32.9	+29 22 31	+0.001	0.00	7.5	0.0	0.4	B9.5 V		250 s		
48606	197113		6 42 34.1	−36 13 22	+0.002	−0.03	8.0	1.4	0.2	K0 III		220 s		
47959	59391		6 42 34.1	+34 05 47	−0.002	−0.02	7.8	0.0	0.4	B9.5 V		300 s		
47882	59387		6 42 34.2	+38 53 31	−0.002	−0.09	7.00	0.35	3.0	F2 V		63 s		
48674	218139		6 42 34.5	−41 48 54	+0.001	−0.01	7.4	1.1	0.2	K0 III		250 s		
48352	151818		6 42 37.0	−10 45 04	+0.002	−0.06	7.8	0.1	1.4	A2 V		37 mx		
48301	133598		6 42 37.5	−3 50 11	−0.001	−0.02	7.5	0.1	1.4	A2 V		170 s		
48630	197119		6 42 40.3	−36 37 51	−0.001	−0.01	7.86	−0.01	0.6	A0 V		280 s		
47748	41282		6 42 40.4	+47 44 09	+0.004	−0.08	7.3	0.5	3.4	F5 V		61 s		
48279	114307		6 42 40.4	+1 42 57	−0.001	−0.02	7.93	0.12		O8	+10		5364	m
48605	197117		6 42 41.4	−34 46 43	+0.001	+0.04	8.0	1.2	−0.3	K5 III		450 s		
48581	197115		6 42 43.2	−32 42 46	0.000	0.00	6.8	0.6	2.6	F0 V		48 s		
48199	96027		6 42 43.7	+11 46 18	+0.001	0.00	7.9	0.1	0.6	A0 V		270 s		
262337	96023		6 42 44.4	+19 35 17	0.000	0.00	7.9	1.1	0.2	K0 III		330 s		
48006	59397		6 42 45.0	+32 33 31	0.000	−0.02	6.9	0.8	3.2	G5 IV		54 s		
48501	172204		6 42 45.7	−22 26 57	−0.007	+0.08	6.13	0.34		F0			5377	m
47373	13935		6 42 46.2	+64 05 44	0.000	−0.05	7.9	1.1	0.2	K0 III		340 s		
48332	133603		6 42 47.4	−4 08 28	−0.004	−0.01	6.7	0.5	3.4	F5 V		45 s		
48580	197116		6 42 47.4	−30 54 08	−0.002	+0.05	8.0	0.7	1.4	A2 V		79 s		
48857	234665		6 42 48.5	−50 27 05	−0.003	+0.02	7.00	−0.15	−1.1	B5 V		420 s		m
48971	234669		6 42 50.5	−56 00 59	−0.002	−0.01	7.8	0.3	0.6	A0 V		170 s		
48439	151825		6 42 56.8	−11 45 57	−0.001	+0.01	8.0	0.1	0.6	A0 V		290 s		

HD	SAO	Star Name	α 2000	δ 2000	μ(α)	μ(δ)	V	B-V	M_V	Spec	RV	d(pc)	ADS	Notes
			^{h} ^{m} ^{s}	$^{\circ}$ $'$ $''$	^{s}	$''$								
48458	151828		6 42 57.0	-15 44 05	+0.003	-0.03	7.5	0.4	3.0	F2 V		80 s		
48054	59401		6 42 58.7	+34 07 40	+0.001	-0.01	7.9	0.0	0.4	B9.5 V		320 s		
48481	151829		6 43 02.2	-13 41 54	0.000	-0.04	7.0	0.1	0.6	A0 V		190 s		
48725	197131		6 43 02.2	-37 22 12	-0.001	0.00	8.00	-0.09	0.4	B9.5 V		330 s		
48706	197127		6 43 02.9	-34 43 51	-0.004	+0.01	8.0	1.4	-0.1	K2 III		270 mx		
47914	41288	55 ψ⁴ Aur	6 43 04.9	+44 31 28	-0.004	-0.03	5.02	1.48	-0.3	K5 III	-73	120 s		
262496	96036		6 43 05.9	+13 00 51	+0.001	-0.02	8.0	1.1	0.2	K0 III		350 s		
48629	172224		6 43 05.9	-29 14 14	-0.001	+0.01	7.30	-0.01	0.6	A0 V		220 s		
48348	114312		6 43 06.5	+3 02 00	0.000	-0.02	6.19	1.37		K2	+31			
48231	96034		6 43 08.9	+17 57 48	0.000	-0.02	7.8	0.1	1.7	A3 V		160 s		
48628	172226		6 43 09.9	-28 21 08	0.000	+0.01	7.1	0.9	0.6	A0 V		54 s		
48602	172225		6 43 11.4	-26 07 36	+0.001	+0.02	7.8	1.6	-0.5	M2 III		470 s		
48518	151838		6 43 12.2	-15 50 02	-0.005	-0.02	7.9	1.1	-0.1	K2 III		160 mx		
48394	114317		6 43 13.1	-0 01 17	0.000	-0.01	7.70	0.9	3.2	G5 IV		79 s		
48073	59406		6 43 13.8	+37 08 48	+0.003	-0.04	6.19	1.03	0.2	K0 III	-41	150 s		
48907	218155		6 43 13.8	-48 13 24	-0.002	0.00	7.5	1.4	0.2	K0 III		280 s		
47726	25899		6 43 14.0	+55 32 04	-0.005	0.00	8.0	0.9	3.2	G5 IV		93 s		
48574	172221		6 43 14.3	-21 51 44	0.000	+0.01	7.3	-0.4	0.0	B8.5 V		280 s		
47747	25901		6 43 15.0	+54 06 47	0.000	-0.01	7.9	0.4	3.0	F2 V		97 s		
48276	96038		6 43 18.3	+18 08 43	0.000	-0.01	7.6	0.0	0.4	B9.5 V		270 s		
47879	41290		6 43 19.4	+49 24 15	+0.006	+0.01	7.7	1.1	0.2	K0 III		180 mx		
48600	172227		6 43 19.6	-21 43 57	+0.001	0.00	8.0	0.3	0.2	K0 III		360 s		
48815	218151		6 43 21.2	-40 54 33	-0.001	-0.01	7.4	1.7	-0.3	K5 III		350 s		
48797	197134		6 43 23.3	-39 11 35	-0.001	-0.01	6.30	0.26		A3				
48393	114321		6 43 26.7	+5 50 57	-0.003	0.00	7.2	0.8	0.3	G5 III	+9	240 s		
48142	59410		6 43 28.7	+37 00 41	0.000	-0.01	8.0	1.1	-0.1	K2 III		410 s		
48195	59414		6 43 32.4	+31 22 25	0.000	-0.01	7.9	0.1	0.6	A0 V		280 s		
48497	133616		6 43 36.0	-1 20 48	-0.001	0.00	7.5	0.0	0.0	B8.5 V		310 s		
48836	197139		6 43 37.1	-37 46 26	-0.002	+0.02	7.2	1.1	0.2	K0 III		220 s		
48330	96044		6 43 37.3	+16 46 58	+0.001	0.00	8.0	0.1	0.6	A0 V		290 s		
49947	256330		6 43 37.4	-73 07 07	+0.008	-0.11	6.37	0.96		K0				
48434	114324		6 43 38.6	+3 55 56	0.000	-0.01	5.90	-0.02	-5.0	B0 III	+35	1200 s		
48625	151850		6 43 41.6	-16 03 00	-0.002	-0.02	7.7	1.1	-0.1	K2 III		350 s		
48875	—		6 43 42.0	-39 50 59			7.6	0.3	2.6	F0 V		100 s		
48345	96047		6 43 47.2	+18 39 22	-0.004	0.00	7.6	0.9	3.2	G5 IV		76 s		
48758	172238		6 43 53.5	-25 32 07	-0.003	0.00	6.8	1.1	3.2	G5 IV		32 s		
48777	172240		6 43 53.7	-27 35 27	+0.001	+0.03	7.4	0.2	0.6	A0 V		180 s		
48329	78682	27 ε Gem	6 43 55.9	+25 07 52	0.000	-0.01	2.98	1.40	-4.5	G8 Ib	+10	210 s	5381	Mebsuta, m
48835	197142		6 43 58.3	-32 25 54	-0.001	-0.03	7.8	1.3	-0.3	K5 III		410 s		
48433	96051	30 Gem	6 43 59.2	+13 13 40	0.000	-0.06	4.49	1.16	0.0	K1 III	+14	70 s	5387	m
48004	25914		6 44 00.5	+51 27 10	0.000	-0.03	7.6	0.1	0.6	A0 V		250 s		
47659	13945		6 44 05.2	+62 57 46	+0.001	-0.03	6.8	1.1	-0.1	K2 III		240 s		
49192	234682		6 44 09.7	-54 43 52	-0.002	0.00	6.8	0.3	0.6	A0 V		110 s		
47979	25916		6 44 11.6	+53 17 47	+0.006	-0.18	6.27	1.08	0.2	K0 III	+19	71 mx		
47522	13943		6 44 11.7	+66 17 48	+0.004	+0.01	7.8	0.5	4.0	F8 V		57 s		
48271	59422		6 44 11.7	+36 59 38	+0.001	-0.02	7.9	0.4	2.6	F0 V		110 s		
48272	59425		6 44 12.5	+36 06 34	+0.001	+0.01	6.30	0.06	1.4	A2 V	-10	93 s		
49219	234683		6 44 12.6	-54 41 43	-0.003	+0.01	6.60	0.1	0.6	A0 V		160 s		m
48228	41305		6 44 17.4	+40 37 22	-0.001	-0.16	6.84	1.59	-0.5	M4 III	+41	98 mx	5379	m
47930	25918		6 44 24.5	+56 55 56	+0.001	-0.04	7.20	1.6	-0.5	M2 III	+22	350 s		
48568	114344		6 44 25.1	+6 16 55	0.000	-0.04	8.0	0.4	3.0	F2 V		97 s		
48616	114347		6 44 25.4	+3 08 29	0.000	-0.01	6.86	0.79	-2.0	F2 II		320 s		
48594	114346		6 44 26.7	+5 12 20	-0.001	-0.02	7.4	1.1	0.2	K0 III	+7	270 s		
48917	197149	10 CMa	6 44 28.3	-31 04 13	-0.002	+0.01	5.20	0.00	-1.6	B3.5 V	+34	220 s		m
46758	5943		6 44 28.4	+76 35 09	-0.006	-0.04	7.2	0.9	3.2	G5 IV		65 s		
45618	1043		6 44 29.9	+82 06 55	+0.003	-0.05	6.4	0.1	1.4	A2 V	+6	100 s		
48494	78691		6 44 30.0	+20 41 35	0.000	-0.01	7.1	0.1	0.6	A0 V		190 s		
48813	151863		6 44 31.1	-19 39 31	+0.001	+0.02	7.60	0.1	1.7	A3 V		150 s		m
49133	—		6 44 36.1	-46 46 09			7.67	1.51		K2				
48691	114352		6 44 36.5	+0 35 20	0.000	-0.02	7.84	0.07		B0.5 IV	-4			
49518	249636		6 44 44.3	-61 13 28	-0.001	0.00	6.8	1.3	0.2	K0 III		130 s		
48450	78692	28 Gem	6 44 45.4	+28 58 15	-0.001	-0.03	5.44	1.45	0.2	gK0	+16	51 s		
48872	—		6 44 46.2	-19 03 01			7.2	0.1	0.6	A0 V		200 s		
48270	41308		6 44 46.6	+44 30 21	+0.001	0.00	6.8	0.9	3.2	G5 IV	+20	52 s		
48940	172262		6 44 47.5	-28 53 43	-0.002	+0.07	8.0	0.5	2.6	F0 V		87 s		
48717	114355		6 44 49.5	+3 40 46	-0.001	-0.02	7.7	0.0	-1.0	B5.5 V	+40	530 s		m
48890	151873		6 44 50.9	-18 34 14	-0.001	-0.03	7.0	0.1	1.4	A2 V		130 s		
48938	172264		6 44 51.8	-27 20 29	-0.001	+0.31	6.45	0.54	4.7	G2 V		22 s		
48566	96058		6 44 52.5	+19 31 41	0.000	+0.02	8.0	0.1	2.6	F0 V		120 s		
48565	78698		6 44 55.0	+20 51 38	+0.007	+0.01	7.1	0.5	4.0	F8 V		42 s		
50002	256331		6 44 56.2	-70 26 02	+0.002	0.00	6.11	1.33		K2				
48548	78696		6 44 56.3	+22 50 15	-0.001	0.00	6.8	0.1	1.4	A2 V		120 s		
48809	133636		6 44 59.6	-5 56 33	-0.002	+0.01	7.9	0.0	0.0	B8.5 V		370 s		
49028	197163		6 45 02.3	-30 35 10	-0.002	+0.02	6.52	-0.13	-0.6	B8 IV		270 s		m
262990	78702		6 45 03.7	+22 47 23	0.000	-0.01	7.6	1.1	0.2	K0 III		290 s		

HD	SAO	Star Name	α 2000	δ 2000	μ(α)	μ(δ)	V	B−V	M$_V$	Spec	RV	d(pc)	ADS	Notes
			h m s	° ′ ″	s	″								
49337	234689		6 45 05.4	−52 21 44	−0.001	+0.05	7.6	1.5	3.2	G5 IV		27 s		
49234	218168		6 45 05.7	−46 56 10	0.000	0.00	7.76	−0.14	0.0	B8.5 V		360 s		
48808	133638		6 45 06.6	−1 01 16	0.000	0.00	7.7	0.0	0.4	B9.5 V		280 s		
49029	197165		6 45 07.2	−31 39 17	−0.001	0.00	7.61	−0.10	0.6	A0 V		250 s		
48915	151881	9 α CMa	6 45 08.9	−16 42 58	−0.038	−1.21	−1.46	0.01	1.4	A1 V	−8	2.7 t	5423	Sirius, m
52792	258452		6 45 09.3	−82 53 45	−0.022	0.00	7.7	1.1	2.6	F0 V		34 s		
48412	59441		6 45 09.6	+39 22 04	−0.001	−0.15	7.0	0.4	2.6	F0 V		70 mx		
48510	59444		6 45 10.0	+30 49 31	−0.001	−0.04	7.30	0.1	0.6	A0 V		220 s	5403	m
48923	151882		6 45 10.0	−18 16 08	0.000	−0.05	7.0	0.9	3.2	G5 IV		57 s		
49260	218170		6 45 10.4	−47 13 22	−0.001	0.00	7.22	−0.16	−1.7	B3 V k		610 s		
48807	114363		6 45 11.3	−0 02 35	+0.002	0.00	7.02	0.21	−5.6	B7 Ib		2400 s		
49259	218172		6 45 13.1	−46 51 13	−0.003	−0.01	7.48	−0.14	0.0	B8.5 V		310 s		
48771	114362		6 45 13.8	+5 12 45	0.000	−0.04	7.5	0.1	0.6	A0 V		240 s		
48207	25932		6 45 14.7	+53 05 09	+0.002	−0.03	7.2	1.4	−0.3	K5 III		320 s		
48737	96074	31 ξ Gem	6 45 17.3	+12 53 44	−0.008	−0.19	3.36	0.43	0.7	F5 III	+25	23 ts		
49095	197171		6 45 22.8	−31 47 37	−0.017	−0.32	5.92	0.48	3.7	dF6	+32	25 ts		
48641	78707		6 45 23.2	+23 38 46	+0.001	−0.01	7.6	1.1	−0.1	K2 III		330 s		
49001	172273		6 45 23.2	−23 27 43	−0.003	+0.04	6.05	1.20	0.2	K0 III		100 s		
48640	78706		6 45 23.4	+24 40 20	−0.001	−0.01	7.30	1.1	−4.4	K1 Ib		1700 s	5409	m
49800	249640		6 45 25.4	−65 02 39	0.000	−0.01	7.3	1.7	−0.5	M2 III		300 s		
49396	234693		6 45 26.0	−52 12 04	0.000	−0.01	6.3	1.2	3.2	G5 IV		23 s		
48784	114366		6 45 26.4	+9 45 55	+0.002	−0.04	6.7	0.4	2.6	F0 V		66 s		
48493	59447		6 45 27.6	+38 22 57	−0.002	−0.03	7.9	0.2	2.1	A5 V		150 s		
49169	197181		6 45 28.6	−36 50 57	−0.002	0.00	7.6	0.4	0.6	A0 V		140 s		
48410	41315		6 45 28.7	+44 14 00	+0.012	−0.19	7.8	0.7	4.4	G0 V	+5	43 mx		
49131	197177		6 45 31.2	−30 56 56	0.000	0.00	5.56	−0.11	−1.7	B3 V	+17	280 s		
48845	114371		6 45 33.1	+4 35 38	0.000	−0.02	7.70	0.1	0.6	A0 V		260 s		m
48638	78710		6 45 35.0	+27 40 25	0.000	+0.01	6.7	1.1	−0.2	K3 III	−34	230 s		
48999	151892		6 45 40.6	−17 10 37	−0.003	+0.02	6.94	−0.07	0.4	B9.5 V		200 s		
49068	172286		6 45 42.9	−20 51 11	−0.001	−0.02	7.43	1.26	−4.4	K0 Ib	+22	2300 s		
50160	256335		6 45 49.7	−70 31 21	+0.003	−0.01	7.5	1.1	0.2	K0 III		290 s		
48914	114378		6 45 49.9	+2 29 56	0.000	−0.02	7.50	−0.04	−5.7	B5 Ib	+9	3900 s		
48922	133645		6 45 49.9	−0 43 01	−0.002	−0.02	6.7	0.1	0.6	A0 V		160 s		
48050	13949		6 45 50.9	+62 38 58	0.000	−0.01	7.8	0.0	0.0	B8.5 V		370 s		
49517	234699		6 45 53.7	−52 24 36	−0.001	0.00	5.80	1.53	0.2	K0 III	+36	52 s		
48843	96084	32 Gem	6 45 54.1	+12 41 37	0.000	0.00	6.4	0.4	2.6	F0 V	+9	58 s		
49091	172290		6 45 57.2	−20 46 31	−0.003	−0.01	7.4	1.0	−0.1	K2 III		310 s		
48979	133651		6 45 57.5	−7 20 20	−0.003	−0.02	6.9	0.1	0.6	A0 V		180 s		
49188	172297		6 45 57.8	−28 48 42	−0.001	+0.01	7.64	−0.11	−1.0	B5.5 V		540 s		
48589	59457		6 45 59.1	+38 58 14	0.000	+0.01	7.27	−0.04	0.6	A0 V		220 s		
49048	151895		6 45 59.2	−14 47 46	−0.002	−0.02	5.32	0.07	1.4	A2 V	−19	58 s		
48684	59463		6 46 02.8	+33 01 45	+0.013	−0.03	7.9	0.5	4.0	F8 V		60 s		
49319	197192		6 46 03.3	−39 32 24	0.000	+0.01	6.62	−0.13		B8				
48563	41324		6 46 04.0	+41 46 29	+0.003	−0.02	7.6	1.1	0.2	K0 III		300 s		
49126	172296		6 46 06.9	−20 45 16	−0.003	−0.01	8.00	0.1	0.6	A0 V		290 s	5437	m
48805	78717		6 46 10.3	+23 22 17	−0.001	+0.05	6.5	0.9	3.2	G5 IV		45 s		
48587	41327		6 46 11.3	+42 15 54	+0.001	−0.03	7.2	1.1	0.2	K0 III		250 s		
49336	197195		6 46 12.0	−37 46 32	−0.001	0.00	6.21	−0.13	−1.7	B3 V		370 s		
50295	−−		6 46 12.3	−71 02 49			7.9	0.1	0.6	A0 V		290 s		
53842	258455		6 46 13.1	−83 59 28	+0.007	+0.08	7.8	0.4	3.4	F5 V		75 s		
49302	197193		6 46 13.6	−34 14 34	−0.001	+0.02	6.9	0.3	0.6	A0 V		110 s		
48250	25939	12 Lyn	6 46 14.1	+59 26 30	−0.003	0.00	4.87	0.08		A2 n	−4		5400	m
49019	133655		6 46 14.2	−5 15 02	0.000	−0.01	7.8	1.4	−0.3	K5 III		390 s		
46588	5946		6 46 14.4	+79 33 53	−0.030	−0.61	5.45	0.50	4.0	F8 V	+13	22 ts		
49258	172304		6 46 16.2	−29 33 41	−0.001	−0.01	6.9	0.7	3.2	G5 IV		55 s		
49257	172303		6 46 17.8	−28 11 39	−0.001	+0.01	7.8	1.6	−0.5	M2 III		470 s		
48489	25946		6 46 20.6	+50 22 48	−0.001	−0.04	7.2	0.1	0.6	A0 V		210 s		
48864	96089		6 46 21.1	+18 50 14	+0.001	−0.02	6.8	0.0	0.4	B9.5 V	+24	190 s		
49301	197194		6 46 21.2	−30 44 39	+0.004	−0.03	7.6	0.8	4.0	F8 V		37 s		
48883	78726		6 46 26.0	+20 42 15	0.000	−0.01	8.0	0.0	0.4	B9.5 V		320 s		
49183	151907		6 46 26.1	−19 52 40	+0.004	0.00	6.8	1.1	0.2	K0 III		190 s		
49123	133660		6 46 31.8	−8 55 07	+0.001	−0.01	7.22	−0.01	0.4	B9.5 V		220 s		
48977	114388	16 Mon	6 46 32.4	+8 35 14	0.000	−0.01	5.93	−0.17	−2.1	B2.5 V	+10	390 s		
48976	114387		6 46 32.8	+9 12 14	0.000	−0.02	7.40	1.4	−0.3	K5 III	+21	330 s		
49148	151909		6 46 33.7	−14 31 24	+0.008	+0.03	6.6	0.1	1.7	A3 V		42 mx		
49147	151911		6 46 39.0	−10 06 27	−0.001	0.00	5.66	−0.05	0.0	A0 IV	+21	140 s		
49015	114392		6 46 40.7	+8 21 47	0.000	+0.01	7.5	0.4	2.6	F0 V		97 s		
49705	234704		6 46 41.5	−54 41 41	−0.006	+0.02	6.46	0.86		G5				m
49420	197199		6 46 41.6	−36 35 45	0.000	+0.02	7.6	1.2		K0 III		240 s		
48682	41330	56 ψ⁵ Aur	6 46 44.3	+43 34 39	0.000	+0.16	5.22	0.57	4.4	G0 V	−24	15 ts	5425	m
49614	218186		6 46 44.5	−47 48 17	−0.001	0.00	6.94	1.40	−0.2	K3 III		230 s		m
49145	133664		6 46 46.4	−6 08 41	−0.001	+0.01	7.90	0.1	0.6	A0 V		280 s	5445	m
49559	218185		6 46 46.4	−44 58 26	−0.001	0.00	7.96	1.49	−0.3	K4 III		450 s		
49014	96099		6 46 47.2	+10 44 05	−0.001	−0.01	7.4	0.0	0.4	B9.5 V		250 s		
48954	96097		6 46 49.2	+16 46 22	0.000	0.00	6.7	0.2	2.1	A5 V		82 s		

HD	SAO	Star Name	α 2000	δ 2000	μ(α)	μ(δ)	V	B-V	M_v	Spec	RV	d(pc)	ADS	Notes
			h m s	° ' "	s	"								
48432	25947	13 Lyn	6 46 49.5	+57 10 09	+0.003	-0.04	5.35	0.96	0.2	K0 III	+19	110 s		
49229	151919	11 CMa	6 46 51.0	-14 25 34	-0.001	+0.01	5.29	-0.04		B8 n	+17			
49689	234705		6 46 52.8	-51 15 57	-0.001	-0.10	5.40	1.34	0.0	gK1	-5	86 s		
58805	258460		6 46 59.6	-87 01 30	+0.007	0.00	6.47	0.42		F2				
46550	1054		6 47 01.2	+80 14 33	+0.011	+0.02	7.5	1.1	0.2	K0 III		280 s		
49333	172318	12 CMa	6 47 01.4	-21 00 55	-0.001	0.00	6.04	-0.15		B8				
49557	218188		6 47 01.5	-40 45 13	-0.001	-0.02	7.4	1.7	-0.1	K2 III		310 s		
49334	172320		6 47 05.6	-21 06 07	-0.001	-0.01	8.0	1.0	0.2	K0 III		360 s		
49274	151924		6 47 07.0	-12 11 56	-0.001	+0.01	7.7	1.1	0.2	K0 III		300 s		
49013	96102		6 47 07.6	+17 07 12	+0.001	0.00	7.4	0.0	0.4	B9.5 V		250 s		
49468	197208		6 47 09.3	-30 57 26	-0.005	+0.04	7.90	0.5	3.4	F5 V		79 s		m
48994	78742		6 47 13.2	+22 11 40	0.000	-0.09	7.0	1.1	0.2	K0 III		130 mx		
49042	96108		6 47 13.6	+16 55 35	-0.003	-0.02	7.8	0.5	3.4	F5 V		74 s		
49315	151931		6 47 14.2	-16 04 01	-0.003	-0.02	7.84	-0.07		B5	+9			
49877	234710		6 47 18.7	-55 32 24	0.000	+0.02	5.61	1.52		K2				
49058	96110		6 47 19.2	+18 19 22	-0.001	-0.01	7.5	0.1	1.4	A2 V		170 s		
49161	114410	17 Mon	6 47 19.8	+8 02 14	-0.002	-0.01	4.77	1.40	-0.3	K4 III	+47	100 s		
49592	197216		6 47 21.3	-38 55 17	+0.001	0.00	7.6	0.1	0.6	A0 V		210 s		
49591	197215		6 47 21.3	-37 55 47	-0.001	-0.02	5.26	-0.08	-0.3	B9 IV	+47	130 s		m
49758	218197		6 47 22.6	-48 33 48	-0.004	+0.04	6.80	0.6	4.4	G0 V		30 s		m
49296	133676		6 47 22.7	-9 57 45	-0.001	-0.02	7.94	1.69	-2.4	M2 II		1100 s		
49086	96112		6 47 23.1	+16 58 20	+0.001	-0.02	7.7	1.1	-0.1	K2 III		340 s		
49512	197212		6 47 23.1	-32 01 31	0.000	-0.02	7.9	1.7	0.2	K0 III		140 s		
49059	96111		6 47 23.4	+18 11 36	0.000	-0.05	6.20	0.07	1.4	A2 V	+16	91 s	5447	m
48881	59483		6 47 24.0	+38 32 03	+0.002	-0.09	7.3	1.1	0.2	K0 III		130 mx		
48388	13955		6 47 28.1	+63 03 18	+0.002	-0.03	8.0	0.9	3.2	G5 IV		93 s		
49488	172334		6 47 28.4	-28 34 24	+0.002	+0.03	7.7	-0.3	-1.6	B3.5 V		720 s		
49555	197214		6 47 28.9	-33 30 48	-0.002	-0.02	8.0	1.0	0.2	K0 III		360 s		
51320	256339		6 47 29.0	-76 51 05	-0.004	0.00	6.90	0.9	3.2	G5 IV		55 s		m
49463	172333		6 47 31.7	-25 41 04	0.000	+0.01	8.0	0.3	0.6	A0 V		180 s		
49415	151935		6 47 36.2	-17 30 29	-0.001	+0.02	7.07	-0.03	0.4	B9.5 V		210 s		
49331	133679		6 47 37.1	-8 59 54	-0.001	0.00	5.07	1.80	-2.4	M1 II	+24	260 s		
52494	258454		6 47 39.3	-81 03 57	-0.006	+0.07	7.40	1.4	-0.3	K5 III		270 mx		m
48781	41346	57 ψ⁶ Aur	6 47 39.5	+48 47 22	0.000	+0.01	5.22	1.12	0.0	K1 III	-8	110 s		
49778	218203		6 47 41.6	-47 18 00	+0.002	0.00	7.11	1.04	0.3	G5 III		190 s		
49728	218202		6 47 43.7	-44 43 09	+0.001	+0.01	7.77	1.14		K2				
49041	59490		6 47 44.7	+30 06 57	-0.002	+0.01	8.0	0.9	3.2	G5 IV		93 s		
49273	114423		6 47 45.6	+4 26 44	-0.002	0.00	7.9	1.1	-0.1	K2 III		380 s		
49294	114426		6 47 47.3	+0 20 23	-0.002	-0.01	7.00	0.16	1.4	A2 V		120 s	5455	m
49120	78750		6 47 47.5	+21 41 29	+0.001	-0.02	7.5	1.6	-0.5	M2 III		380 s		
49292	114425		6 47 49.8	+2 55 18	-0.001	-0.02	7.78	1.65	-0.5	M2 III		420 s		
49245	114424		6 47 51.1	+7 52 59	+0.001	+0.01	7.9	0.1	0.6	A0 V		270 s		
49293	114428	18 Mon	6 47 51.6	+2 24 44	-0.001	-0.01	4.47	1.11	0.2	K0 III	+11	59 s		
49852	218206		6 47 51.9	-49 03 36	-0.002	-0.01	7.5	-0.2	2.6	F0 V		98 s		
49370	--		6 47 55.7	-5 09 24			7.5	0.0	0.0	B8.5 V		300 s		
292296	--		6 47 56.2	+0 19 26			7.93	0.50		A0				
49057	59491		6 47 56.2	+31 02 48	-0.001	0.00	7.9	0.1	1.4	A2 V		200 s		
48863	41349		6 47 57.3	+48 00 18	-0.002	0.00	8.0	1.1	-0.1	K2 III		410 s		
49459	151938		6 47 57.7	-13 42 12	-0.001	0.00	7.8	0.1	0.6	A0 V		260 s		
49291	114430		6 47 58.5	+6 45 39	-0.002	+0.03	7.8	1.1	0.2	K0 III		320 s		
49369	133682		6 47 59.3	-4 15 06	-0.001	+0.03	8.0	0.1	1.7	A3 V		180 s		
49438	--		6 47 59.8	-11 06 26			7.80	0.00	0.4	B9.5 V		290 s	5463	m
49243	96122		6 48 00.0	+11 29 00	+0.001	+0.01	8.0	0.1	0.6	A0 V		290 s		
49529	151943		6 48 09.7	-17 13 49	-0.003	-0.04	7.8	0.1	0.6	A0 V		260 s		
49385	114435		6 48 11.4	+0 18 17	-0.004	+0.03	7.89	0.47	4.4	G0 V		49 s		m
50241	249647	α Pic	6 48 11.4	-61 56 29	-0.010	+0.27	3.27	0.21	2.1	A5 V	+21	16 s		
48766	25963		6 48 12.8	+55 42 16	+0.007	-0.10	5.52	0.47	3.4	F5 V	+9	26 s	5436	m
49084	59495		6 48 15.7	+35 16 07	0.000	-0.01	8.0	0.0	0.4	B9.5 V		330 s		
50127	234726		6 48 15.8	-57 39 21	-0.001	+0.05	7.1	1.3	0.2	K0 III		160 s		
49775	197232		6 48 16.2	-38 39 22	-0.002	+0.03	8.0	0.7	0.2	K0 III		360 s		
49290	96128		6 48 16.3	+14 35 33	-0.001	0.00	7.5	0.1	0.6	A0 V		240 s		
49201	78758		6 48 17.6	+23 13 27	-0.001	+0.02	8.00	0.9	3.2	G5 IV		90 s		m
49850	218208		6 48 18.1	-43 48 02	-0.002	+0.05	7.41	0.16		A3				
49434	133687		6 48 19.0	-1 19 09	-0.003	-0.04	5.75	0.28		A5				
49548	--		6 48 19.7	-15 19 30			7.96	0.15	0.0	B8.5 V		310 s	5473	m
49009	41354		6 48 21.5	+41 18 09	-0.003	0.00	8.0	1.1	-0.1	K2 III	+21	410 s		
49481	133696		6 48 21.5	-7 45 17	-0.001	0.00	6.7	0.0	0.0	B8.5 V		220 s		
49435	133690		6 48 21.9	-1 49 05	-0.001	0.00	7.60	1.48	-0.3	K5 III	+71	380 s		
49368	114437		6 48 22.2	+5 32 30	-0.001	0.00	7.70	1.78		S5.1				
49367	114438		6 48 23.9	+6 12 46	-0.001	0.00	7.70	1.1	-2.2	K1 II	+2	880 s		
49573	151949		6 48 27.2	-12 50 27	-0.002	-0.01	6.9	0.0	0.0	B8.5 V		230 s		
49478	133698		6 48 29.0	-1 39 43	-0.001	0.00	7.64	0.93		K0				
49139	59500		6 48 30.2	+34 06 53	0.000	-0.01	8.0	0.1	1.4	A2 V		210 s		
49702	172356		6 48 34.7	-27 22 52	-0.001	0.00	8.0	-0.3	-1.0	B5.5 V		640 s		
50198	234729		6 48 36.0	-58 28 02	0.000	+0.01	8.0	0.3	1.4	A2 V		140 s		

HD	SAO	Star Name	α 2000	δ 2000	μ(α)	μ(δ)	V	B−V	M_v	Spec	RV	d(pc)	ADS	Notes
			h m s	° ′ ″	s	″								
49723	172357		6 48 37.2	−28 12 55	−0.001	−0.01	7.3	0.4	0.6	A0 V		120 s		
49750	197235		6 48 38.0	−31 21 26	−0.003	+0.01	8.0	0.4	2.6	F0 V		100 s		
49724	172358		6 48 38.2	−28 22 07	+0.001	+0.04	8.0	0.6	0.6	A0 V		120 s		
50099	234727		6 48 38.6	−54 00 51	+0.002	+0.02	7.6	0.0	2.6	F0 V		100 s		
49409	114448		6 48 39.4	+7 37 22	−0.003	−0.37	7.93	0.60	4.4	G0 V	−87	48 s	5469	m
49942	218218		6 48 42.9	−43 48 06	−0.003	0.00	7.34	−0.10		B8				
49119	41358		6 48 43.2	+40 43 37	+0.002	−0.03	7.8	1.1	0.2	K0 III		330 s		
49834	197238		6 48 43.8	−35 46 07	−0.002	0.00	8.0	0.8	0.2	K0 III		360 s		
49429	114449		6 48 43.9	+8 51 59	−0.001	−0.01	6.9	1.1	0.2	K0 III	+21	220 s		
49749	172365		6 48 44.0	−28 44 57	0.000	+0.02	7.9	1.2	0.2	K0 III		270 s		
49700	172361		6 48 47.2	−23 12 10	−0.011	−0.03	7.5	0.8	3.4	F5 V		39 s		
49649	151957		6 48 48.7	−16 12 41	0.000	−0.05	7.70	1.1	0.2	K0 III		310 s	5484	m
49833	197239		6 48 49.1	−34 56 11	−0.002	+0.02	6.92	−0.04	0.4	B9.5 V		200 s		
49793	172368		6 48 50.4	−29 44 36	−0.002	+0.02	7.0	1.6	−0.1	K2 III		140 s		
49647	151958		6 48 51.5	−14 23 21	−0.002	+0.01	7.9	1.1	−0.1	K2 III		380 s		
49199	59507		6 48 53.8	+36 57 52	−0.001	−0.04	7.9	0.1	1.7	A3 V		180 s		
49178	59506		6 48 53.8	+37 30 18	−0.009	−0.07	7.8	0.6	4.4	G0 V		47 s		
49366	78769		6 48 57.2	+20 33 51	0.000	+0.01	7.5	0.0	0.0	B8.5 V		300 s		
49662	151962		6 48 57.7	−15 08 41	0.000	−0.01	5.39	−0.10	−0.9	B6 V	+23	180 s	5487	m
50050	218222		6 48 59.0	−48 05 56	−0.003	+0.03	7.7	0.2	3.2	G5 IV		80 s		
49625	133715		6 48 59.3	−9 48 09	0.000	−0.01	7.9	1.1	0.2	K0 III		330 s		
49567	114465		6 49 03.6	+1 00 07	0.000	−0.01	6.15	−0.13	−3.4	B3 II−III	+23	770 s		
49790	172373		6 49 06.5	−26 03 48	0.000	−0.02	7.40	1.1	−0.1	K2 III		320 s	5492	m
49381	78771		6 49 08.5	+23 12 06	−0.001	−0.02	7.6	0.1	0.6	A0 V		250 s		
49345	78770		6 49 10.2	+27 11 25	0.000	−0.04	6.6	0.1	1.4	A2 V		110 s		
50080	218225		6 49 13.2	−46 42 30	−0.003	+0.05	7.36	0.01	0.4	B9.5 V		230 s		
49406	78774		6 49 15.3	+22 18 39	0.000	−0.01	7.1	0.0	0.4	B9.5 V	+46	210 s		
49699	151964		6 49 15.8	−12 40 04	0.000	0.00	7.5	0.0	0.4	B9.5 V		260 s		
49872	172375		6 49 16.1	−28 58 25	−0.004	0.00	7.0	0.6	2.6	F0 V		51 s		
49643	133718		6 49 16.3	−2 16 19	−0.001	0.00	5.74	−0.10	−0.2	B8 V		120 s		m
264311	78775		6 49 18.9	+22 09 40	0.000	0.00	8.0	1.1	0.2	K0 III		350 s		
49197	41363		6 49 21.2	+43 45 33	−0.004	−0.05	7.6	0.5	3.4	F5 V		71 s		
49894	172380		6 49 26.7	−28 39 08	+0.001	+0.02	7.40	1.1	0.2	K0 III		280 s		m
49565	96152		6 49 28.1	+12 05 18	−0.002	+0.09	7.9	0.5	4.0	F8 V		59 s		
−−	41360		6 49 28.8	+49 41 34	0.000	+0.02	8.0	2.0	−0.3	K5 III		220 s		
49641	114472		6 49 29.3	+3 41 30	−0.001	−0.01	7.8	1.1	0.2	K0 III		320 s		
50126	218228		6 49 31.4	−45 34 01	−0.001	0.00	7.11	−0.03		B8				m
49961	197253		6 49 31.7	−32 32 33	−0.001	−0.01	6.8	0.6	1.4	A2 V		60 s		
50078	218227		6 49 33.8	−41 25 46	−0.001	+0.09	6.9	0.7	3.4	F5 V		35 s		
49741	133729		6 49 35.6	−7 24 45	−0.002	0.00	7.1	1.1	−0.1	K2 III	+19	270 s		
49539	96153		6 49 37.3	+19 10 04	−0.001	−0.02	7.6	0.1	0.6	A0 V		250 s		
49868	172383		6 49 37.9	−24 08 36	−0.001	−0.01	7.8	0.1	1.4	A2 V		180 s		
49892	172384		6 49 38.2	−24 08 09	+0.001	0.00	7.20	0.1	0.6	A0 V		210 s		m
49082	25973		6 49 38.7	+53 01 52	−0.003	−0.07	8.00	0.4	2.6	F0 V		120 s	5462	m
49812	151975		6 49 40.0	−15 02 41	−0.001	−0.01	7.56	0.01	0.4	B9.5 V		250 s		
49380	59521		6 49 41.2	+32 36 25	−0.003	−0.04	5.71	1.29	0.2	K0 III	−16	85 s		
49866	151978		6 49 43.1	−19 23 01	+0.001	+0.07	7.7	0.9	3.2	G5 IV		67 s		
49500	78778		6 49 43.2	+25 29 05	+0.002	−0.03	6.92	1.60	0.2	K0 III		81 s		
49891	172389		6 49 43.9	−24 04 33	0.000	−0.01	6.33	0.0	0.6	A0 V		130 s	5498	m
49713	133730		6 49 44.2	−1 20 23	−0.001	+0.01	7.7	0.0		B9 p				
49364	59518		6 49 44.3	+37 30 46	−0.001	+0.01	6.5	1.4	−0.3	K5 III		230 s		
49845	151977		6 49 44.8	−16 18 36	0.000	0.00	6.8	1.4	−0.3	K5 III		260 s		
49582	96160		6 49 49.7	+17 59 21	−0.002	−0.05	7.6	0.4	2.6	F0 V		100 s		
49606	96161	33 Gem	6 49 49.8	+16 12 10	−0.001	−0.01	5.70	−0.13	−1.2	B8 III	+13	240 s		m
49344	59519		6 49 49.8	+39 28 45	−0.002	+0.01	7.60	0.00	0.4	B9.5 V		280 s		
50013	197258	13 κ CMa	6 49 50.4	−32 30 31	−0.001	0.00	3.96	−0.23	−2.5	B2 V e	+14	200 s		
50337	234737		6 49 51.3	−53 37 20	0.000	+0.03	4.40	0.92	0.4	gG3	+26	52 s		
49677	114480		6 49 53.5	+8 38 45	+0.001	+0.01	7.7	0.0	0.4	B9.5 V		280 s		
50223	218235		6 49 54.5	−46 36 52	−0.001	+0.37	5.14	0.45	0.7	F5 III	+19	37 ts		
49787	133737		6 49 55.4	−5 30 48	−0.002	0.00	7.54	−0.05	−3.5	B1 V pe	+13	1400 s		
49639	96166		6 49 55.7	+11 54 48	0.000	−0.01	7.7	1.1	0.2	K0 III		300 s		
50310	234735	τ Pup	6 49 56.1	−50 36 53	+0.003	−0.07	2.93	1.20	0.2	K0 III	+36	25 s		
50196	218236		6 49 57.7	−45 27 00	+0.001	−0.04	6.55	1.51	−0.3	K5 III		230 s		m
49270	41370		6 49 58.2	+47 02 34	−0.001	−0.04	8.0	1.4	−0.3	K5 III		450 s		
48680	13964		6 49 58.7	+67 30 48	−0.002	+0.01	7.67	1.56		K5				
49288	41371		6 50 00.9	+46 11 05	−0.001	−0.02	7.30	1.1	0.2	K0 III		260 s	5480	m
50571	249654		6 50 01.1	−60 14 57	+0.001	+0.11	6.11	0.46		F5				
49676	96170		6 50 02.1	+12 03 30	−0.001	0.00	7.3	1.6	−0.5	M2 III		350 s		
50450	234739		6 50 02.2	−55 54 12	−0.002	0.00	7.4	0.2	2.1	A5 V		120 s		
50862	249658		6 50 02.7	−66 17 34	+0.001	+0.04	7.0	1.1	3.2	G5 IV		35 s		
49711	114484		6 50 05.7	+6 15 23	−0.001	0.00	7.5	0.0	0.0	B8.5 V		310 s		
50012	172403		6 50 05.9	−27 20 03	0.000	0.00	7.04	−0.18	−2.3	B3 IV		740 s		
49888	151986		6 50 07.9	−12 35 05	−0.003	0.00	7.23	−0.11		B5 e	+3			
50096	197262		6 50 10.6	−33 53 06	−0.002	+0.04	7.41	−0.05	0.6	A0 V		230 s		
50336	218239		6 50 10.8	−48 52 53	+0.001	+0.02	7.5	0.6	1.4	A2 V		170 s		

HD	SAO	Star Name	α 2000	δ 2000	μ(α)	μ(δ)	V	B–V	M$_V$	Spec	RV	d(pc)	ADS	Notes
49659	96171		6^h 50^m 11^s.0	+16°27'13"	+0^s.002	−0".03	7.7	0.9	3.2	G5 IV		80 s		
50157	197264		6 50 11.1	−39 03 18	0.000	−0.03	7.2	0.8	0.2	K0 III		260 s		
49499	59530		6 50 12.1	+35 04 07	−0.004	−0.04	7.1	0.9	3.2	G5 IV		60 s		
49739	114487		6 50 12.2	+5 40 25	−0.004	0.00	6.7	0.2	2.1	A5 V		81 s		
49808	114491		6 50 13.1	+0 08 12	0.000	−0.01	7.98	0.38	2.6	F0 V		110 s		
49636	96172		6 50 13.9	+18 47 52	−0.001	−0.01	7.8	0.1	1.7	A3 V	+28	160 s		
50415	234740		6 50 14.0	−52 47 58	0.000	+0.03	7.9	−0.7	0.4	B9.5 V		320 s		
49955	151990		6 50 16.4	−17 17 32	−0.002	−0.03	6.63	0.08	1.4	A2 V		110 s		
49980	151992		6 50 21.8	−17 05 02	+0.001	+0.01	5.79	1.44		K0				
50123	197263		6 50 23.3	−31 42 22	−0.001	+0.01	5.56	0.14	0.0	B8.5 V		110 s		m
49738	96179	35 Gem	6 50 25.4	+13 24 48	0.000	0.00	5.65	1.34	0.2	K0 III	+26	77 s		
51211	249662		6 50 25.5	−69 59 11	0.000	−0.01	6.7	1.2	4.4	G0 V		13 s		
49977	151996		6 50 26.4	−14 06 48	−0.001	+0.01	7.97	0.20	−3.0	B1.5 V pne	+16	1400 s		
264717	96178		6 50 30.6	+18 30 22	0.000	−0.01	7.9	1.1	0.2	K0 III		330 s		
49806	114494		6 50 31.8	+7 11 20	0.000	+0.01	7.5	1.1	0.2	K0 III	+8	280 s		
49710	96180		6 50 33.1	+19 17 52	0.000	−0.02	8.0	0.1	0.6	A0 V		290 s		
50093	172420		6 50 36.9	−25 46 41	−0.001	+0.01	6.33	−0.18	−1.7	B3 V		400 s		
50178	197270		6 50 38.6	−31 16 34	−0.002	+0.03	7.2	0.8	3.2	G5 IV		61 s		
49976	133761		6 50 42.3	−8 02 28	−0.001	0.00	6.29	0.00		A0 p				
49935	133758		6 50 43.0	−4 16 00	−0.002	−0.02	6.6	0.0	0.0	B8.5 V		210 s		
50072	172422		6 50 45.1	−20 50 09	−0.001	0.00	7.60	−0.11	0.0	B8.5 V		330 s		
50177	172427		6 50 45.4	−29 26 51	0.000	−0.01	7.9	0.4	3.2	G5 IV		88 s		
49520	41380	58 ψ⁷ Aur	6 50 45.9	+41 46 53	−0.002	−0.13	5.02	1.27	−0.2	K3 III	+61	110 s		m
50195	197272		6 50 46.6	−30 34 38	0.000	+0.02	8.0	2.3	−0.3	K5 III		150 s		
49696	78792		6 50 47.9	+26 17 58	0.000	−0.01	8.0	1.1	0.2	K0 III		340 s		
49933	133760		6 50 49.8	−0 32 27	+0.002	−0.19	5.77	0.39	3.0	dF2	−15	35 s	5505	m
50235	197277		6 50 52.3	−34 22 02	0.000	0.00	4.99	1.38		K5	+30	34 mn		
50118	172425		6 50 52.5	−20 54 38	−0.001	−0.01	7.1	−0.3	−1.6	B3.5 V	+18	540 s		
49840	96190		6 50 54.2	+11 39 50	−0.003	−0.01	7.4	0.9	3.2	G5 IV		69 s		
48879	13973	42 Cam	6 50 57.1	+67 34 19	+0.001	0.00	5.14	−0.17	−2.3	B3 IV	+5	300 s		
50117	152007		6 50 58.4	−17 51 17	+0.003	+0.08	7.9	0.9	3.2	G5 IV		87 s		
49932	114502		6 50 59.4	+5 50 32	−0.002	−0.04	7.0	0.5	3.4	F5 V		52 s		
49736	78795		6 51 00.2	+25 45 36	−0.001	−0.21	7.0	0.5	4.0	F8 V		39 s		
50291	197282		6 51 05.3	−34 16 34	−0.004	+0.03	8.0	1.0	−0.1	K2 III		310 mx		
49603	41385		6 51 05.4	+40 23 08	+0.002	−0.02	8.0	0.1	0.6	A0 V		310 s		
50414	218250		6 51 05.9	−42 12 14	−0.001	+0.02	7.3	1.1	−0.1	K2 III		300 s		
50503	218251		6 51 07.1	−47 19 33	−0.001	+0.05	7.30	1.15	−0.1	K2 III		300 s		
52365	256347		6 51 07.8	−77 40 26	−0.007	+0.09	7.7	0.1	1.4	A2 V		85 mx		
50153	152014		6 51 08.2	−19 00 49	−0.002	−0.01	6.9	0.1	1.7	A3 V		110 s		
49361	25993		6 51 08.3	+54 25 08	−0.002	−0.03	8.0	0.5	4.0	F8 V		63 s		
50233	172441		6 51 10.7	−27 15 38	−0.003	+0.02	7.9	1.6	−0.1	K2 III		230 s		
50067	152011		6 51 12.0	−10 05 06	0.000	−0.01	7.40	1.4	−0.3	K4 III	+7	340 s	5516	m
49909	96193		6 51 14.9	+15 04 35	+0.002	−0.04	7.2	1.4	−0.3	K5 III	+40	310 s		
50041	133773		6 51 18.3	−3 30 50	−0.002	−0.01	7.9	0.0	0.0	B8.5 V		370 s		
50115	152015		6 51 19.5	−11 07 10	−0.003	−0.04	7.9	1.1	0.2	K0 III		330 s		
49733	59548		6 51 21.0	+32 09 34	−0.001	0.00	7.9	0.1	0.6	A0 V		280 s		
50446	218253		6 51 25.8	−40 33 00	−0.002	0.00	7.45	−0.05	0.4	B9.5 V		260 s		
51557	256344	ı Vol	6 51 27.0	−70 57 49	0.000	+0.02	5.40	−0.11	−1.3	B6 IV	+18	210 s		
50622	218258		6 51 27.7	−49 17 36	0.000	+0.01	7.5	0.2	1.4	A2 V		170 s		
49674	41390		6 51 30.4	+40 52 02	+0.003	−0.15	7.7	0.6	4.4	G0 V		47 s		
50358	172453		6 51 32.6	−29 22 31	−0.001	0.00	7.9	0.8	0.2	K0 III		340 s		
50621	218261		6 51 32.7	−48 17 33	+0.003	−0.01	6.42	1.21	0.2	K0 III		120 s		
49908	78805	36 Gem	6 51 32.9	+21 45 40	−0.001	−0.03	5.27	−0.02	1.4	A2 V	+34	59 s	5511	m
50138	133781		6 51 33.3	−6 58 00	−0.001	−0.01	6.67	0.01	0.4	B9.5 V	+34	180 s		
50903	234754		6 51 33.6	−58 19 42	+0.001	+0.01	8.0	−0.2	0.4	B9.5 V		330 s		
50005	114515		6 51 35.7	+9 26 24	0.000	−0.03	6.9	1.4	−0.3	K5 III	+13	270 s		
50039	114523		6 51 36.3	+2 39 03	+0.002	−0.05	7.2	0.5	3.4	F5 V		56 s		
50330	172454		6 51 37.1	−26 32 06	−0.003	0.00	7.4	0.4	1.4	A2 V		94 s		
50472	197290		6 51 37.9	−38 12 54	0.000	+0.01	7.88	−0.11	0.0	B8.5 V		380 s		
50062	114525		6 51 39.2	+3 02 30	−0.001	−0.04	6.38	0.04		A0	+45			
50134	133785		6 51 41.0	−2 24 09	−0.002	0.00	6.8	1.1	−0.1	K2 III		240 s		
50568	218259		6 51 41.1	−44 18 22	+0.011	+0.15	7.98	0.90	0.2	K0 III		68 mx		m
50304	172455		6 51 41.3	−23 48 09	0.000	+0.02	8.0	−0.7	0.6	A0 V		300 s		
50445	197291		6 51 42.3	−36 13 49	−0.004	−0.07	5.96	0.18		A2				
50083	114528		6 51 45.7	+5 05 05	0.000	+0.01	6.92	0.06		B2 e	+10			
49633	41392		6 51 47.7	+46 29 59	0.000	0.00	7.5	0.9	−2.1	G8 II	+6	850 s		
42855	1037		6 51 48.1	+86 41 03	+0.016	−0.11	6.60	1.20	−0.2	K3 III	+26	180 mx		
50598	218266		6 51 49.0	−43 06 05	0.000	0.00	7.9	0.7	3.2	G5 IV		88 s		
50379	172461		6 51 49.2	−26 34 54	−0.001	+0.01	8.00	0.00	−1.0	B5.5 V		600 s	5539	m
49732	41394		6 51 51.0	+41 34 36	−0.004	+0.10	7.9	0.5	4.0	F8 V		60 s		
50170	133789		6 51 51.8	−2 10 33	0.000	0.00	6.9	0.4	3.0	F2 V		60 s		
50082	114534		6 51 53.9	+6 36 24	0.000	0.00	7.7	0.6	4.4	G0 V		46 s		
50352	172460		6 51 55.0	−22 18 13	−0.002	−0.02	8.0	0.3	3.2	G5 IV		93 s		
50060	96210		6 51 55.7	+10 48 03	−0.006	+0.04	7.9	0.5	4.2	F9 V	+71	54 s		
49951	78813		6 51 58.1	+25 39 48	−0.002	+0.02	6.75	0.16	1.4	A2 V		100 s		

HD	SAO	Star Name	α 2000	δ 2000	μ(α)	μ(δ)	V	B-V	M_v	Spec	RV	d(pc)	ADS	Notes
265200	96208		6ʰ51ᵐ58ˢ.4	+19°40'28"	0ˢ.000	−0".02	7.9	1.1	0.2	K0 III		330 s		
49600	26003		6 51 59.6	+51 59 26	+0.001	−0.01	8.0	1.1	−0.1	K2 III		410 s		
49968	78816		6 52 00.0	+23 36 06	−0.003	−0.01	5.65	1.45	−0.3	K5 III	+40	160 s		
50922	234756		6 52 01.4	−55 57 15	0.000	−0.01	8.0	1.7	−0.3	K5 III		340 s		
50499	197294		6 52 02.1	−33 54 54	−0.005	+0.09	7.2	0.8	4.4	G0 V		26 s		
50255	152024		6 52 02.1	−11 12 17	−0.007	+0.01	7.9	0.9	3.2	G5 IV		88 s		
49534	26000		6 52 04.1	+55 46 34	+0.002	−0.02	8.0	0.9	3.2	G5 IV		93 s		
50167	114536		6 52 04.1	+1 15 04	0.000	−0.02	7.86	1.54		K5			5533	m
50498	197295		6 52 09.7	−30 25 35	+0.001	+0.08	7.9	0.2	2.6	F0 V		110 s		
50251	133799		6 52 12.0	−5 38 45	−0.001	−0.02	7.1	0.0	0.0	B8.5 V		260 s		
50252	133802		6 52 13.4	−6 27 57	0.000	−0.01	8.0	0.0	0.4	B9.5 V		330 s		
48268	5987		6 52 13.8	+77 13 30	−0.001	−0.03	7.9	0.1	0.6	A0 V		280 s		
50281	133805		6 52 18.0	−5 10 25	−0.036	0.00	6.59	1.06	6.7	dK2	−10	9.6 t		
50147	96219		6 52 18.9	+11 19 08	−0.001	0.00	7.8	0.1	0.6	A0 V		260 s		
50518	197298		6 52 19.7	−32 00 53	−0.001	+0.01	7.5	0.2	2.1	A5 V		120 s		
50282	133807		6 52 22.8	−5 18 58	−0.002	−0.01	6.30	0.99		K0				
50206	114544		6 52 24.8	+7 53 46	0.000	−0.06	8.0	0.5	3.4	F5 V		81 s		
50543	172485		6 52 28.1	−29 05 54	+0.001	0.00	8.0	1.7	−0.5	M2 III		470 s		
49884	41400		6 52 35.1	+42 37 12	−0.001	+0.02	7.9	0.1	0.6	A0 V		280 s		
49904	59566		6 52 36.2	+39 55 43	0.000	−0.02	7.3	1.1	0.2	K0 III		270 s		
51801	256346		6 52 37.0	−70 30 37	+0.002	+0.03	7.4	1.1	0.2	K0 III		270 s		
50187	96222		6 52 37.4	+11 35 51	−0.001	−0.01	7.7	1.1	0.2	K0 III		300 s		
50228	114551		6 52 39.0	+7 49 20	+0.001	0.00	7.8	0.0	−1.0	B5.5 V	−1	550 s		
50785	218272		6 52 39.6	−42 30 16	−0.001	−0.01	6.52	0.42		F2				
50883	218277		6 52 40.1	−46 48 33	−0.001	+0.01	8.04	0.96		G5				
50491	172486		6 52 40.6	−21 11 04	−0.001	−0.01	7.47	−0.12	0.0	B8.5 V		310 s		
50538	172494		6 52 41.0	−26 07 04	−0.003	−0.06	7.50	1.4	−0.3	K5 III		210 mx		m
50058	78820		6 52 41.4	+29 26 37	+0.002	−0.09	7.7	0.5	3.4	F5 V		73 s		
51210	234764		6 52 45.4	−59 20 28	−0.002	0.00	6.41	0.18		A2				
50463	152041		6 52 46.2	−16 12 46	+0.001	−0.02	7.14	−0.10		B3	+17			
51043	234762		6 52 46.8	−54 05 24	−0.003	+0.02	6.4	1.3	0.2	K0 III		120 s		
--	--		6 52 46.9	+28 58 18	−0.002	+0.01	7.1	1.1	0.0	K1 III		260 s		
50860	218278		6 52 47.0	−43 58 32	−0.001	0.00	6.46	−0.10		B9				m
50019	59570	34 θ Gem	6 52 47.3	+33 57 40	0.000	−0.05	3.60	0.11	0.0	A3 III	+20	51 s	5532	m
50535	172497		6 52 47.5	−23 02 50	−0.001	0.00	7.86	−0.10	0.4	B9.5 V		310 s		
49858	41402		6 52 48.4	+47 12 20	+0.001	−0.01	8.00	0.4	3.0	F2 V		100 s	5525	m
50808	218276		6 52 49.0	−40 41 36	−0.003	0.00	7.4	0.7	0.2	K0 III		270 s		
50277	114556		6 52 49.3	+8 22 49	−0.004	−0.03	5.77	0.26		A5 n	+27			
50132	78824		6 52 54.2	+25 45 52	−0.002	−0.01	7.7	0.4	2.6	F0 V		100 s		
50462	152042		6 52 54.5	−12 09 15	−0.002	+0.02	7.14	0.26		A m				
50300	114558		6 52 57.0	+8 39 52	−0.001	−0.02	7.5	1.1	0.2	K0 III	−23	280 s		
50562	172499		6 52 58.5	−21 50 08	0.000	−0.01	8.03	0.18	−1.6	B3.5 V	+25	430 s		
50372	114561		6 52 59.2	+2 44 37	−0.001	0.00	7.70	0.9	−2.1	G6 II	+15	840 s		
50882	218280		6 52 59.6	−42 37 39	−0.002	0.00	7.6	−0.4	3.0	F2 V		85 s		
50648	172504		6 53 00.0	−26 57 27	−0.003	−0.06	6.50	1.53		M			5548	m
50018	59571	59 Aur	6 53 01.4	+38 52 09	0.000	+0.01	6.12	0.37	3.0	F2 V	+1	40 s	5534	m
48840	5996		6 53 02.5	+74 08 13	−0.002	+0.01	7.6	0.9	3.2	G5 IV		77 s		
50056	59574		6 53 03.3	+35 47 19	−0.002	+0.01	6.01	1.45	3.2	G5 IV	+6	14 s		
49618	26012	14 Lyn	6 53 05.0	+59 26 55	−0.001	−0.04	5.33	0.65	0.3	G4 III	+13	100 s	5514	m
49881	41403		6 53 05.8	+49 32 03	0.000	−0.01	7.90	1.1	−0.1	K2 III		400 s		
50186	78827		6 53 06.0	+25 18 41	−0.001	−0.04	7.4	0.2		A m				
50857	197319		6 53 06.5	−39 41 26	−0.002	−0.01	8.0	1.6	−0.1	K2 III		240 s		
49949	41406		6 53 07.6	+44 50 21	+0.001	−0.08	6.26	0.21		A5	+3			
50461	133826		6 53 07.9	−7 45 55	−0.001	+0.01	7.8	0.1		A0 p				
50105	59579		6 53 11.9	+34 57 38	+0.001	+0.01	7.8	1.1	0.2	K0 III		340 s		
50037	59576	60 Aur	6 53 13.3	+38 26 16	+0.003	−0.18	6.30	0.49	3.4	F5 V	+32	43 mn		
50646	172511		6 53 13.3	−24 10 03	+0.005	−0.02	7.4	0.1	−1.6	B3.5 V		36 mx		
50343	96238		6 53 14.0	+10 35 26	−0.002	−0.01	7.4	0.0	0.4	B9.5 V		250 s		
52449	256349		6 53 16.3	−74 43 30	−0.018	+0.19	7.65	0.52		F5		11 mn		
50513	133831		6 53 18.4	−9 41 26	−0.002	−0.02	7.9	0.0	0.0	B8.5 V		370 s		
50644	152055		6 53 18.7	−19 01 58	−0.004	0.00	5.64	0.28	2.6	F0 V		40 s		
50643	152056		6 53 21.6	−18 56 00	−0.002	+0.02	6.11	0.16		A2				
50711	172516		6 53 22.1	−24 41 03	−0.001	−0.02	7.5	−0.2	1.4	A2 V		170 s		
50371	96241		6 53 22.4	+10 59 47	+0.001	−0.14	6.24	0.97	0.2	G9 III	−34	100 mx		
49902	26023		6 53 26.1	+51 31 04	−0.001	−0.03	7.00	0.5	3.4	F5 V		53 s	5531	m
50319	96239		6 53 26.9	+19 15 57	−0.001	−0.04	7.5	0.2	2.1	A5 V		120 s		
50707	172520	15 CMa	6 53 32.8	−20 13 27	−0.001	0.00	4.82	−0.22	−3.9	B1 IV	+31	560 s		
51268	234770		6 53 33.7	−54 52 59	+0.007	+0.08	7.6	1.4	0.2	K0 III		140 mx		
50806	172524		6 53 33.8	−28 32 23	+0.021	−0.44	6.04	0.71	3.2	G5 IV	+72	37 s		
49671	13997		6 53 37.6	+61 00 55	−0.006	+0.01	7.9	1.6	−0.5	M4 III		270 mx		
50245	78835		6 53 38.0	+29 31 04	+0.002	−0.02	8.0	1.1	0.2	K0 III		360 s		
49340	13986	43 Cam	6 53 42.2	+68 53 18	+0.001	+0.01	5.12	−0.13	−1.0	B7 IV	−21	170 s		
50163	59585		6 53 42.6	+38 18 45	+0.001	0.00	7.7	1.4	−0.3	K5 III		390 s		
50639	133840		6 53 45.1	−9 30 57	0.000	−0.05	7.0	0.5	3.4	F5 V		52 s		
50674	152065		6 53 47.6	−11 15 49	−0.002	+0.03	7.7	1.1	0.2	K0 III		300 s		

HD	SAO	Star Name	α 2000	δ 2000	μ(α)	μ(δ)	V	B−V	M_v	Spec	RV	d(pc)	ADS	Notes
			$^h\ ^m\ ^s$	$^{\circ}\ '\ ''$	^{s}	$''$								
50705	152066		6 53 48.3	−13 11 28	−0.001	+0.02	7.7	0.0	0.0	B8.5 V		340 s		
51138	218291		6 53 49.1	−46 25 36	0.000	0.00	8.0	0.4	1.4	A2 V		120 s		
50804	172529		6 53 49.8	−24 13 48	0.000	+0.01	7.3	0.6	1.7	A3 V		64 s		
50802	172528		6 53 50.0	−21 52 52	0.000	+0.01	8.0	0.4	1.4	A2 V		120 s		
50434	96244		6 53 51.3	+14 45 06	−0.001	−0.02	7.80	0.8	0.3	G5 III	−5	310 s		
50185	59588		6 53 53.0	+39 10 11	0.000	−0.01	7.6	1.4	−0.3	K5 III		390 s		
50483	96247		6 53 54.0	+11 01 24	−0.002	−0.03	7.7	0.6	4.4	G0 V		46 s		
50920	197329		6 53 54.3	−31 36 58	0.000	+0.02	7.9	0.2	0.6	A0 V		200 s		
50853	172531		6 53 55.2	−24 32 21	−0.002	−0.01	6.2	0.6	0.6	A0 V		52 s		
50826	172530		6 53 56.0	−22 19 43	−0.002	+0.01	7.7	0.8	−0.1	K2 III		360 s		
50204	59589	61 ψ⁸ Aur	6 53 56.9	+38 30 18	−0.001	−0.03	6.2	0.0		B9 pe	+27			
51015	197331		6 53 56.9	−36 33 23	0.000	+0.02	7.6	1.0	3.2	G5 IV		54 s		
50299	59591		6 53 57.1	+31 07 18	+0.001	−0.02	7.8	0.8	3.2	G5 IV		84 s		
51266	234774		6 54 02.2	−50 36 42	−0.006	+0.22	6.26	0.99	0.2	gK0		160 s		
53144	256352		6 54 03.6	−77 46 48	−0.001	+0.06	7.70	0.1	1.4	A2 V		110 mx		m
50557	114580		6 54 06.0	+6 41 16	0.000	−0.01	7.20	0.9	3.2	G5 IV		63 s		m
50525	96252		6 54 06.1	+10 31 30	−0.001	0.00	7.9	1.4	−0.3	K5 III		410 s		
50316	59594		6 54 07.4	+32 30 42	−0.002	+0.01	6.9	0.0	0.4	B9.5 V		200 s		
50877	172542	16 o¹ CMa	6 54 07.8	−24 11 02	−0.001	+0.01	3.86	1.73	−6.0	K3 Iab	+36	520 s		
50700	133855		6 54 08.4	−5 51 09	−0.002	0.00	6.41	0.17		A3 n	+17		5557	m
50990	197333		6 54 10.9	−30 36 48	−0.004	0.00	7.9	−0.1	0.6	A0 V		130 mx		
48506	5994		6 54 11.0	+78 00 28	+0.003	−0.08	7.8	0.5	4.0	F8 V		58 s		
50226	41417		6 54 11.2	+40 53 24	+0.007	−0.08	8.0	0.6	4.4	G0 V		52 s		
50778	152071	14 θ CMa	6 54 11.3	−12 02 19	−0.010	−0.02	4.07	1.43	−0.3	K4 III	+97	73 s		
50896	172546		6 54 12.9	−23 55 42	−0.001	0.00	6.86	−0.30		WN5	0			
50581	114586		6 54 13.8	+6 41 17	0.000	0.00	7.4	0.1	0.6	A0 V		220 s		
50730	133864		6 54 13.8	−6 01 07	0.000	−0.01	7.9	1.1	0.2	K0 III		330 s		
51042	197334		6 54 14.8	−34 13 21	0.000	+0.02	7.35	−0.03	0.6	A0 V		220 s		m
50636	114587		6 54 18.2	+5 06 36	−0.001	−0.02	7.8	0.0	0.0	B8.5 V		350 s		
50751	133871		6 54 18.6	−6 23 22	−0.002	−0.02	7.9	0.0	0.0	B8.5 V		370 s		
50777	133876		6 54 19.8	−9 36 45	+0.001	−0.02	7.8	1.4	−0.3	K5 III		400 s		
51014	197335		6 54 20.0	−31 48 04	−0.002	+0.02	6.9	−0.2	0.6	A0 V		180 s		
50482	78852		6 54 21.3	+21 09 41	+0.002	−0.05	6.70	0.9	3.2	G5 IV		50 s	5553	m
50747	133870		6 54 24.6	−1 07 38	0.000	−0.01	5.45	0.18		A2	−9	19 mn		
50876	152078		6 54 25.6	−14 27 03	−0.002	+0.01	8.0	0.1	1.4	A2 V		200 s		
51208	218296		6 54 26.7	−42 21 55	0.000	+0.02	6.33	2.25		N7.7	+32			
50580	96256		6 54 27.1	+14 37 32	−0.001	0.00	7.8	0.0	0.4	B9.5 V		290 s		
50608	96259		6 54 28.7	+11 03 20	0.000	+0.01	7.9	0.1	0.6	A0 V		270 s		
50456	78853		6 54 29.1	+26 31 51	0.000	−0.01	7.6	0.1	0.6	A0 V		250 s		
50161	26033		6 54 32.5	+50 02 42	−0.002	−0.02	7.2	0.1	0.6	A0 V		210 s		
50938	152085		6 54 35.7	−17 55 02	−0.002	+0.01	7.66	−0.12	0.0	B8.5 V		340 s		
50635	96265	38 Gem	6 54 38.6	+13 10 40	+0.005	−0.08	4.65	0.30	2.6	F0 V p	+22	26 ts	5559	m
50607	96262		6 54 39.9	+15 50 19	+0.001	+0.01	6.5	0.9	3.2	G5 IV		47 s		
50725	114596		6 54 40.2	+6 43 26	+0.001	0.00	7.1	0.1	1.7	A3 V		120 s		
51315	218303		6 54 40.9	−43 59 43	−0.001	0.00	7.5	1.1	0.2	K0 III		260 s		
50402	59602		6 54 41.5	+36 56 20	+0.002	−0.04	7.5	1.4	−0.3	K5 III		370 s		
50820	133881		6 54 42.0	−1 45 23	0.000	+0.01	6.21	0.56	−1.7	B3 V e	+13	370 s		
52302	−−		6 54 42.2	−70 01 59			7.9	1.4	−0.3	K5 III		440 s		
51038	172560		6 54 42.4	−24 55 27	−0.002	+0.03	7.9	0.9	−1.0	B5 V p		160 s		
50554	78855		6 54 42.6	+24 14 42	−0.005	−0.11	6.9	0.5	4.0	F8 V		38 s		
51136	197341		6 54 43.1	−32 47 24	+0.004	+0.03	7.8	1.3	−0.3	K5 III		180 mx		
50202	41422		6 54 45.9	+48 54 17	−0.002	−0.01	6.9	0.1	1.7	A3 V		110 s		
50315	41423		6 54 46.5	+42 56 51	−0.002	−0.02	7.3	0.1	1.7	A3 V	+3	130 s		
50481	78854		6 54 46.9	+29 49 38	0.000	−0.01	7.9	0.0	0.0	B8.5 V		390 s		
50958	152087		6 54 47.1	−14 09 14	−0.002	+0.04	7.9	1.6	−0.5	M2 III		460 s		
50959	152090		6 54 53.2	−14 51 45	+0.001	−0.05	7.4	0.5	4.0	F8 V		48 s		
50479	59605		6 54 55.7	+32 29 10	+0.001	0.00	8.0	1.1	0.2	K0 III		360 s		
50634	78858		6 54 56.6	+21 34 16	−0.001	−0.01	6.80	0.00	0.4	B9.5 V	−4	190 s	5564	m
51258	197355		6 54 57.9	−36 19 07	+0.001	+0.04	7.5	1.8	−0.5	M2 III		320 s		
50890	133890		6 54 58.8	−2 48 13	−0.001	0.00	6.04	1.10	0.3	gG6	+20	110 s		
50793	114608		6 55 01.8	+8 31 14	0.000	−0.01	7.9	1.1	0.2	K0 III		330 s		
51055	172569	17 CMa	6 55 02.6	−20 24 17	−0.001	−0.01	5.80	0.1	1.4	A2 V		75 s	5585	m
50633	78861		6 55 04.3	+22 33 37	0.000	−0.01	8.0	0.1	1.4	A2 V		200 s		
50981	133896		6 55 07.1	−8 29 58	−0.002	−0.02	7.76	0.04	0.4	B9.5 V		280 s		m
51312	197363		6 55 07.1	−38 33 22	−0.002	0.00	7.40	1.1	0.2	K0 III		280 s		m
50818	114612		6 55 07.8	+7 16 08	−0.002	+0.01	7.7	1.1	0.2	K0 III		300 s		
50933	133894		6 55 08.5	−2 28 52	−0.002	0.00	7.7	1.1	0.2	K0 III		300 s		
50982	133898		6 55 11.8	−9 19 17	−0.001	0.00	7.73	−0.02	0.0	B8.5 V		350 s		
51054	152100		6 55 11.9	−15 42 14	−0.002	−0.03	7.5	1.1	−0.1	K2 III		320 s		
50868	114617		6 55 13.6	+5 26 02	0.000	−0.01	7.92	−0.20	−2.5	B2 V ne	−16	1100 s		
51202	172574		6 55 14.6	−29 02 17	0.000	−0.07	7.10	0.5	3.4	F5 V		55 s		m
50420	41429		6 55 14.7	+43 54 36	+0.001	−0.01	6.13	0.32	2.6	F0 V	−7	47 s		
50384	41426		6 55 15.1	+45 49 35	−0.003	−0.07	6.34	0.94		K0	+31			
52493	256351		6 55 16.0	−70 29 20	+0.001	+0.02	7.3	−0.6	1.4	A2 V		150 s		
51085	152102		6 55 16.0	−17 12 55	−0.002	0.00	7.89	0.00	0.4	B9.5 V		320 s		

HD	SAO	Star Name	α 2000	δ 2000	μ(α)	μ(δ)	V	B-V	M_V	Spec	RV	d(pc)	ADS	Notes
50692	78866	37 Gem	6 55 18.6	+25 22 32	-0.003	+0.02	5.73	0.57	4.4	dG0	-11	19 s		
51028	152101		6 55 20.6	-10 11 21	-0.002	-0.02	7.8	0.1	1.4	A2 V		190 s		
51079	133902		6 55 31.5	-8 25 03	0.000	+0.01	7.93	-0.03	0.0	B8.5 V		370 s		
50576	59609		6 55 31.7	+37 54 55	-0.004	-0.04	6.80	1.1	0.2	K0 III		210 s	5574	m
51200	172578		6 55 32.7	-22 02 16	-0.002	-0.01	6.81	-0.20	-2.3	B3 IV	+10	660 s	5601	m
50931	114626		6 55 34.6	+8 19 29	-0.001	+0.02	6.29	0.04		A0	+33			
51367	197366		6 55 36.3	-33 48 24	-0.001	+0.01	7.1	1.3	-0.1	K2 III		220 s		
51434	197368		6 55 36.9	-38 40 00	-0.007	+0.05	7.6	0.3	3.4	F5 V		70 s		
50767	78870		6 55 37.1	+24 00 28	0.000	-0.02	7.7	0.0	-1.6	B3.5 V		650 s		
51199	172579	19 π CMa	6 55 37.3	-20 08 11	+0.003	+0.04	4.68	0.37	0.6	gF2	+8	65 s	5602	m
52418	249686		6 55 41.1	-68 30 19	+0.001	+0.05	7.1	1.1	0.5	K0 III		240 s		
54481	258458		6 55 41.5	-80 50 16	+0.004	+0.04	7.9	1.6	-0.3	K5 III		380 s		
51283	172588		6 55 46.7	-22 56 29	-0.002	0.00	5.28	-0.19	-3.4	B3 II-III	+38	550 s		
50975	114633		6 55 48.2	+8 52 39	0.000	0.00	7.5	0.5	4.0	F8 V		49 s		
50605	59612		6 55 50.7	+39 33 11	-0.001	0.00	8.0	0.1	0.6	A0 V		310 s		
51553	218317		6 55 50.8	-41 19 13	-0.001	0.00	8.0	1.6	0.2	K0 III		360 s		
51282	152122		6 55 52.5	-19 45 41	0.000	-0.01	7.9	1.3	-0.3	K5 III		430 s		
51343	172591		6 55 52.9	-25 31 07	+0.002	+0.01	7.40	0.4	2.6	F0 V		90 s	5606	m
51929	234792		6 55 53.1	-56 56 34	-0.002	+0.60	7.43	0.59	3.0	G2 IV		24 mx		
50887	96277		6 55 53.4	+18 29 30	-0.001	-0.04	7.6	0.2	2.1	A5 V		130 s		
51706	234787		6 55 53.6	-50 05 10	-0.001	+0.01	7.8	1.9	-0.5	M2 III		430 s		
51106	133907		6 55 54.1	-1 35 07	-0.001	+0.01	7.36	0.23		A3 m				
51411	197371		6 55 54.7	-31 47 24	-0.001	+0.01	6.40	0.00	0.0	B8.5 V		190 s		m
50816	78876		6 55 56.2	+27 17 10	0.000	+0.01	7.0	0.0	0.4	B9.5 V		210 s		
49632	6006		6 55 58.5	+71 52 57	-0.001	-0.02	8.0	1.1	0.2	K0 III		360 s		
51251	--		6 56 05.1	-14 02 36			5.2	0.1	1.4	A2 V		56 s		
51250	152123	18 μ CMa	6 56 06.6	-14 02 37	-0.001	0.00	5.00	1.18		G5	+20	27 mn	5605	m
51309	152126	20 ι CMa	6 56 08.1	-17 03 14	-0.001	+0.01	4.38	-0.06	-3.9	B3 II	+41	410 s		
50951	96283		6 56 11.8	+18 48 00	0.000	0.00	7.4	0.0	0.4	B9.5 V		240 s		
51001	96287		6 56 11.9	+15 15 56	0.000	-0.02	7.7	1.4	-0.3	K5 III		370 s		
51024	96289		6 56 12.1	+12 34 50	0.000	0.00	7.7	0.9	3.2	G5 IV		80 s		
51074	114642		6 56 13.6	+8 44 43	0.000	0.00	7.0	1.1	-0.1	K2 III	-18	260 s		
50720	41445		6 56 14.9	+40 32 10	+0.002	-0.06	7.6	0.5	3.4	F5 V		69 s		
--	59620		6 56 15.0	+36 56 52	+0.002	-0.04	8.0	1.1						
51799	218324		6 56 16.0	-48 43 16	0.000	+0.01	4.95	1.69	-0.5	gM3	+22	110 s		
50974	96288		6 56 16.6	+17 44 13	0.000	-0.05	6.8	1.1	0.2	K0 III		200 s		
51607	197382		6 56 17.8	-38 36 35	-0.002	0.00	7.68	-0.01	0.6	A0 V		260 s		
50928	78885		6 56 18.7	+25 00 33	-0.002	+0.07	7.7	0.1	1.7	A3 V		100 mx		
52063	234799		6 56 23.4	-57 34 35	+0.005	-0.16	7.4	0.9	4.4	G0 V		25 s		
51104	96294		6 56 25.7	+9 57 23	-0.002	-0.02	5.90	-0.08	-0.6	B7 V	+33	190 s		
51684	218321		6 56 29.7	-40 59 25	-0.002	+0.01	7.6	0.5	2.6	F0 V		83 s		
50658	41446	ψ⁹ Aur	6 56 32.2	+46 16 27	+0.002	+0.01	5.80	-0.06	-0.6	B8 IV	-41	190 s		
51220	114660		6 56 33.1	+0 06 21	-0.003	-0.01	7.7	1.1	0.2	K0 III	+47	300 s		
51518	172609		6 56 33.3	-25 54 42	0.000	-0.06	7.8	0.9	0.2	K0 III		310 mx		
51457	172606		6 56 34.1	-22 38 42	+0.001	0.00	8.00	1.1	0.2	K0 III		350 s		m
51219	114659		6 56 34.1	+1 09 45	-0.001	-0.57	7.42	0.69	5.6	G8 V	-12	23 s		
54239	256355	θ Men	6 56 34.5	-79 25 13	-0.003	0.00	5.45	0.05		A0	+5			
51335	133924		6 56 34.8	-8 20 03	-0.003	-0.02	6.93	0.24	1.7	A3 V		92 s		
51604	197384		6 56 36.5	-31 22 53	-0.001	-0.03	7.9	1.7	-0.1	K2 III		200 s		
51360	133927		6 56 39.3	-8 10 57	-0.001	+0.02	7.60	-0.07	0.6	A0 V		250 s		
51682	197388		6 56 45.6	-35 20 30	-0.001	-0.01	6.29	1.29		K0				
51968	234800		6 56 46.5	-51 00 57	0.000	0.00	7.6	0.6	0.2	K0 III		300 s		
51277	114669		6 56 47.6	+0 48 07	-0.001	-0.02	7.45	1.65	-0.3	K5 III		290 s		
51102	78896		6 56 48.5	+20 58 01	0.000	-0.01	7.8	0.0	0.0	B8.5 V		350 s		
51188	96302		6 56 50.0	+11 23 31	-0.001	0.00	7.7	0.0	0.4	B9.5 V		280 s		
51332	133933		6 56 51.5	-3 41 33	-0.003	-0.03	7.4	0.4	2.6	F0 V		92 s		
51549	172617		6 56 55.2	-21 06 04	0.000	+0.02	7.81	-0.15	-1.0	B5.5 V	+17	580 s		
50763	41450		6 56 55.9	+46 42 18	-0.010	-0.09	5.86	1.08	0.2	K0 III	+39	120 s		
51101	78897		6 56 59.6	+24 38 34	-0.002	-0.03	6.9	1.1	0.2	K0 III		220 s		
51424	133937		6 57 00.0	-8 10 44	-0.001	0.00	6.34	0.64	3.4	F5 V		28 s		
51000	59631		6 57 00.5	+33 40 51	-0.002	0.00	5.89	0.88	4.4	G0 V	-10	13 s		
51127	78898		6 57 01.2	+24 57 28	-0.002	-0.01	8.00	0.1	0.6	A0 V		300 s	5608	m
51573	152151		6 57 01.5	-19 26 20	-0.002	-0.08	6.80	0.5	3.4	F5 V		48 s	5623	m
51546	152150		6 57 03.1	-17 05 54	+0.001	-0.01	7.0	1.1	0.2	K0 III		230 s		
266822	96305		6 57 03.7	+15 35 22	-0.002	-0.01	8.0	0.7	4.4	G0 V		52 s		
52198	234807		6 57 05.8	-55 02 52	-0.003	0.00	7.7	1.7	-0.3	K5 III		310 s		
51572	152155		6 57 09.0	-17 33 36	+0.001	0.00	7.88	0.00	0.4	B9.5 V		300 s		
51480	152149		6 57 09.4	-10 49 26	+0.001	+0.02	6.93	0.31	-1.0	B5.5 V	+58	240 s		
51826	197400		6 57 12.1	-36 53 09	-0.001	+0.03	7.60	-0.05	-1.6	B5 IV		640 s		
50551	26050		6 57 13.1	+57 33 48	+0.002	+0.02	6.05	1.49	-0.1	gK2	-54	98 s		
51680	172632		6 57 14.7	-25 49 05	0.000	+0.01	8.0	0.8	0.2	K0 III		360 s		
51630	172631		6 57 14.7	-22 12 12	-0.001	-0.01	6.59	-0.19	-2.0	B2 IV-V				
50522	26051	15 Lyn	6 57 16.5	+58 25 21	0.000	-0.14	4.35	0.85	1.8	G5 III-IV	+9	32 s	5586	m
51825	197402		6 57 17.5	-35 30 27	-0.004	+0.01	6.23	0.46	3.2	F8 IV-V		25 ts		
50865	41455		6 57 18.3	+47 16 40	-0.001	0.00	8.0	0.1	1.4	A2 V		210 s		

HD	SAO	Star Name	α 2000	δ 2000	μ(α)	μ(δ)	V	B-V	M_v	Spec	RV	d(pc)	ADS	Notes
			h m s	° ′ ″	s	″								
51021	59634		6 57 18.6	+37 23 39	+0.002	-0.02	6.6	1.6	-0.5	M4 III		260 s		
50521	26052		6 57 21.8	+59 11 07	0.000	-0.03	7.9	0.1	1.4	A2 V		200 s		
52301	234812		6 57 24.0	-56 06 05	-0.003	+0.03	7.7	0.6	3.2	G5 IV		80 s		
51330	96314		6 57 25.6	+11 54 27	+0.001	0.00	6.2	0.4	2.6	F0 V	+9	51 s		
52447	249689		6 57 26.5	-60 51 05	+0.002	+0.02	8.0	0.8	3.2	G5 IV		89 s		
50947	41461		6 57 27.8	+43 13 45	-0.001	-0.02	8.0	1.1	0.2	K0 III		360 s		
51356	96316		6 57 27.9	+10 58 26	-0.001	-0.02	7.8	0.5	4.0	F8 V		56 s		
52300	234813		6 57 28.1	-55 40 03	-0.003	+0.01	7.7	-0.2	0.4	B9.5 V		290 s		
52024	218336		6 57 31.5	-43 47 19	+0.001	0.00	7.29	-0.05	0.4	B9.5 V		240 s		
51296	96312		6 57 33.2	+19 43 43	+0.001	-0.12	7.3	0.2	2.1	A5 V		28 mx		
51676	172639		6 57 33.4	-21 02 52	+0.002	+0.02	8.0	0.4	4.0	F8 V		62 s		
51733	172644		6 57 33.8	-24 37 50	-0.005	+0.10	5.46	0.36	2.6	dF0	+20	35 s	5629	m
51732	172643		6 57 34.7	-23 09 26	0.000	+0.02	7.9	0.1	4.4	G0 V		51 s		
51901	197407		6 57 35.7	-35 28 00	+0.001	-0.06	7.0	1.3	0.2	K0 III		150 s		
50973	41463	16 Lyn	6 57 37.0	+45 05 39	-0.002	0.00	4.90	0.03	1.4	A2 V	-9	50 s		
51507	114686		6 57 40.9	+1 29 28	-0.001	-0.01	8.02	-0.10	-1.6	B3.5 V	-19	760 s		
51823	172650		6 57 42.4	-27 32 15	-0.003	0.00	6.1	0.5	-1.6	B3.5 V		140 s		
51354	96318		6 57 42.8	+17 54 06	-0.001	-0.01	7.12	-0.18		B3 e	+8			
51047	41464		6 57 43.5	+40 49 44	+0.004	-0.03	7.7	1.1	0.2	K0 III		250 mx		
52298	234815		6 57 45.4	-52 38 54	-0.001	-0.37	6.94	0.46		F5		15 mn		
51353	78915		6 57 45.9	+20 28 00	-0.001	+0.01	7.8	0.0	0.4	B9.5 V		290 s		
51295	78912		6 57 46.3	+22 53 32	-0.010	-0.15	7.5	0.6	4.4	G0 V		42 s		
52060	218340		6 57 46.5	-42 01 39	-0.002	+0.07	8.00	0.2	2.1	A5 V		130 s		m
52096	218342		6 57 49.3	-42 36 23	-0.001	0.00	7.6	-0.2	1.7	A3 V		160 s		
51098	41469		6 57 54.0	+41 42 19	+0.004	-0.06	6.8	1.1	0.2	K0 III		160 mx		
51506	114689		6 57 55.9	+6 32 56	-0.001	-0.01	7.7	0.0	-1.0	B5.5 V	-10	530 s		
52279	218354		6 57 56.9	-49 08 47	-0.001	+0.03	7.5	-0.2	0.6	A0 V		250 s		
51566	114692		6 57 57.7	+2 17 33	-0.001	0.00	7.7	0.6	4.4	G0 V	-20	47 s		
51700	152171		6 57 59.0	-10 47 11	+0.001	-0.02	6.7	1.1	0.2	K0 III		200 s		
52196	218350		6 58 00.6	-46 05 45	-0.001	+0.02	6.83	-0.06		A0				
50520	14021		6 58 00.7	+63 02 27	+0.001	0.00	7.5	0.2	2.1	A5 V		120 s		
50972	41465		6 58 00.7	+48 29 47	0.000	0.00	7.9	0.1	0.6	A0 V		290 s		
51596	114694		6 58 03.7	+0 22 08	0.000	-0.04	8.0	0.1	1.4	A2 V		200 s		
50297	14016		6 58 06.1	+68 21 56	+0.001	+0.01	7.9	1.1	0.2	K0 III		340 s		
51502	96325		6 58 06.5	+14 13 42	0.000	-0.02	7.30	0.00	0.4	B9.5 V		230 s		m
51725	133965		6 58 07.0	-9 01 38	-0.001	+0.02	7.4	1.6	-0.5	M4 III		370 s		
51925	172656		6 58 07.6	-27 09 53	0.000	-0.01	6.37	-0.19	-1.7	B3		410 s		m
51964	197415		6 58 08.4	-30 44 48	0.000	-0.06	7.1	0.9	3.2	G5 IV		53 s		
51726	152175		6 58 08.4	-10 18 56	-0.001	+0.01	6.8	1.1	-0.1	K2 III		230 s		
50688	14024		6 58 10.6	+60 47 22	-0.004	-0.14	7.7	0.5	4.0	F8 V		55 s		
51419	78921		6 58 11.7	+22 28 32	+0.004	+0.09	6.9	0.8	3.2	G5 IV		54 s		
51533	96331		6 58 13.3	+14 14 57	-0.002	-0.01	7.30	0.2	2.1	A5 V		110 s		m
51850	152177		6 58 13.8	-17 46 45	+0.001	-0.01	7.9	1.1	0.2	K0 III		330 s		
52093	197422		6 58 16.0	-35 25 09	+0.001	0.00	7.7	0.0	0.4	B9.5 V		270 s		
51645	114701		6 58 17.4	+3 53 45	0.000	0.00	7.4	1.1	0.2	K0 III	+40	260 s		
51620	114704		6 58 21.3	+6 10 03	-0.002	+0.03	7.06	2.65		C5 II	+16			RV Mon, v
51272	59649		6 58 22.8	+37 59 16	-0.003	-0.03	8.0	1.1	0.2	K0 III		360 s		
51722	133970		6 58 24.1	-4 31 30	0.000	+0.01	7.9	0.2	2.1	A5 V		150 s		
51876	152182		6 58 24.1	-16 03 17	-0.001	+0.01	7.15	-0.08	0.4	B9.5 V		220 s		
52092	197427		6 58 25.0	-34 06 41	-0.002	+0.01	5.06	-0.16		B4	+19			
49801	6017		6 58 27.0	+75 11 54	-0.006	+0.02	7.9	0.5	3.4	F5 V		80 s		
51756	133972		6 58 28.0	-3 01 26	-0.002	-0.01	7.23	-0.07		B0.5 IV	+25		5640	m
51531	96335		6 58 31.9	+19 13 20	0.000	0.00	7.4	1.1	-0.1	K2 III		300 s		
52760	249694		6 58 32.2	-61 05 55	-0.001	+0.01	7.9	1.3	-0.1	K2 III		350 s		
51986	172667		6 58 35.6	-23 52 51	-0.001	+0.01	6.73	0.03	0.6	A0 V		160 s		
52054	172671		6 58 35.7	-26 42 50	-0.003	+0.01	7.9	1.6	-0.3	K5 III		370 s		
52018	172669		6 58 35.8	-25 24 50	-0.001	+0.01	5.58	-0.17		B3	+28		5651	m
52622	234825		6 58 36.2	-56 23 41	-0.004	-0.03	6.45	0.39		F2				
51845	133978		6 58 36.6	-7 11 08	+0.002	-0.06	7.12	0.81	4.4	G0 V		24 s		
52089	172676	21 ε CMa	6 58 37.5	-28 58 20	0.000	0.00	1.50	-0.21	-4.4	B2 II	+27	150 s	5654	Adhara, m
51046	26074		6 58 37.7	+51 52 01	-0.017	-0.22	7.7	0.6	4.4	G0 V		47 s		
51693	114713		6 58 38.9	+7 37 19	-0.001	-0.03	6.27	0.11		A2	-27			
52141	197431		6 58 39.1	-33 08 43	-0.004	+0.04	7.5	2.0	-0.3	K5 III		180 s		
52603	234826		6 58 39.6	-55 43 45	-0.009	-0.09	6.27	1.16		K0				
52362	218360		6 58 41.7	-45 46 04	-0.001	-0.02	6.22	0.00		A0				
52140	197432		6 58 43.7	-30 59 52	0.000	+0.02	6.42	-0.14		B6	+14			m
51530	78929	39 Gem	6 58 47.3	+26 04 52	-0.012	+0.09	6.10	0.46	3.4	F5 V	+6	34 s		m
51351	59655		6 58 50.6	+39 32 10	-0.001	+0.01	7.6	0.1	0.6	A0 V		250 s		
51594	78935		6 58 51.8	+21 36 51	-0.002	-0.13	7.9	1.1	0.2	K0 III		100 mx		
51814	114722		6 58 56.9	+3 36 08	-0.001	-0.01	5.97	1.06	0.3	G8 III	+17	110 s	5648	m
51871	133986		6 58 57.0	-4 01 06	-0.002	+0.01	7.90	0.1	0.6	A0 V		280 s		m
50273	6024		6 58 58.7	+71 55 55	-0.001	0.00	7.7	1.1	0.2	K0 III		310 s		
52470	218362		6 59 00.0	-47 01 24	0.000	-0.01	7.27	-0.01	0.4	B9.5 V		230 s		
52220	197436		6 59 00.3	-32 43 21	-0.001	0.00	6.9	1.1	3.2	G5 IV		35 s		
52138	172685		6 59 00.6	-26 28 26	0.000	+0.02	7.16	-0.18	0.0	B8.5 V		270 s		m

HD	SAO	Star Name	α 2000	δ 2000	μ(α)	μ(δ)	V	B-V	M_v	Spec	RV	d(pc)	ADS	Notes
51440	59658	62 Aur	6 59 02.8	+38 03 08	-0.003	-0.12	6.00	1.23	-0.1	K2 III	+25	120 mx		
52219	172690		6 59 04.7	-28 57 44	-0.001	0.00	8.0	-0.2	2.6	F0 V		120 s		
51185	26079		6 59 04.8	+50 48 39	-0.001	-0.05	7.9	1.1	-0.1	K2 III		390 s		
51717	96341		6 59 05.3	+16 02 26	0.000	-0.02	7.6	1.1	-0.1	K2 III		330 s		
51716	96342		6 59 06.9	+16 07 25	0.000	0.00	7.7	0.9	3.2	G5 IV		80 s		
52904	249695		6 59 07.3	-60 50 33	-0.001	-0.02	7.1	1.8	-0.1	K2 III		120 s		
50476	14025		6 59 07.8	+68 45 14	+0.004	+0.05	7.7	0.5	4.0	F8 V		55 s		
51839	114727		6 59 09.1	+5 17 06	-0.001	-0.01	7.9	0.9	3.2	G5 IV		87 s		
52332	197440		6 59 11.2	-36 48 55	-0.001	+0.02	7.6	1.0	3.0	F2 V		33 s		
52492	218369		6 59 14.7	-45 13 41	0.000	0.00	7.9	0.4	-0.3	K5 III		440 s		
52854	234843		6 59 19.3	-58 00 08	+0.001	+0.02	8.0	1.6	-0.1	K2 III		240 s		
51418	41475		6 59 20.0	+42 18 53	-0.001	0.00	6.6	0.1	0.6	A0 V	-23	160 s		
51892	114731		6 59 20.1	+7 19 01	-0.001	-0.01	6.35	-0.11		B8				
51689	78945		6 59 22.6	+25 14 08	-0.001	-0.02	7.63	0.52	4.0	F8 V		52 s		
51560	59665		6 59 27.0	+37 05 53	-0.001	-0.01	7.90	0.1		B1	+8		5642	m
51688	78947	40 Gem	6 59 27.9	+25 54 51	-0.001	-0.01	6.30	-0.11	-1.2	B8 III		320 s		
51582	59666		6 59 28.9	+37 06 08	-0.001	0.00	7.8	0.1	0.6	A0 V		270 s		
51956	114736		6 59 31.6	+0 54 59	-0.001	-0.02	7.7	1.1	0.2	K0 III		300 s		
52162	152211		6 59 36.1	-13 00 09	0.000	-0.01	7.88	-0.06		B3 n	+21			
52273	172700		6 59 39.2	-21 36 12	-0.001	-0.01	6.2	-0.6		B8				
51397	41476		6 59 40.4	+48 37 21	-0.001	-0.04	7.4	1.1	0.2	K0 III		280 s		
52414	197446		6 59 41.5	-33 28 30	0.000	+0.03	6.6	1.0	1.7	A3 V		28 s		
52395	172707		6 59 43.6	-29 42 27	+0.002	-0.12	7.79	0.53	4.4	G0 V		48 s		
51324	26083		6 59 44.2	+52 34 18	-0.002	-0.04	6.7	1.1	0.2	K0 III		200 s		
52356	172706		6 59 44.4	-28 23 57	-0.001	+0.02	6.99	-0.17	-1.0	B5.5 V		400 s		
56938	258461		6 59 47.1	-83 45 10	+0.003	+0.09	7.6	1.5	0.2	K0 III		150 s		
51911	96359		6 59 50.4	+15 56 27	0.000	-0.01	7.20	0.7	4.4	G0 V		36 s	5660	m
53501	249704		6 59 50.5	-67 54 59	-0.006	+0.24	5.17	1.40	-0.2	K3 III	+39	78 mx		
52574	218374		6 59 50.8	-41 00 05	+0.002	+0.06	6.9	1.3	0.2	K2 III		150 s		
51955	96360		6 59 52.1	+11 48 08	0.000	0.00	7.7	1.1	-0.1	K2 III		350 s		
52242	152218		6 59 52.4	-15 01 16	+0.001	+0.04	7.41	0.40	3.0	F2 V		73 s		
52078	114749		6 59 57.1	+4 04 59	+0.002	-0.04	7.8	1.1	0.2	K0 III		320 s		
51592	41481		6 59 57.4	+41 52 01	-0.001	-0.07	7.6	0.5	3.4	F5 V		71 s		
53143	249700		6 59 59.5	-61 20 11	-0.024	+0.25	6.81	0.80	4.6	K0 IV-V		28 s		
52465	197448		7 00 02.9	-30 04 41	+0.003	-0.03	7.1	1.2	-0.1	K2 III		250 s		
49878	6022		7 00 04.0	+76 58 39	+0.022	-0.01	4.55	1.36	-0.3	K4 III	-26	93 s		
51612	41482		7 00 05.1	+42 11 04	0.000	+0.01	7.8	0.4	2.6	F0 V		110 s		
52348	172718		7 00 08.1	-20 09 32	-0.002	-0.01	6.1	-0.2		B8				
52516	197456		7 00 09.7	-31 08 28	+0.002	+0.05	6.7	0.5	3.4	F5 V		44 s		
51833	59678		7 00 10.2	+30 17 10	+0.001	-0.03	7.80	0.9	0.3	G8 III	+55	320 s		
53048	234852		7 00 10.5	-56 54 23	0.000	+0.02	7.6	0.3	0.0	B8.5 V		190 s		
51323	26085		7 00 11.6	+56 07 08	+0.002	0.00	8.0	1.1	-0.1	K2 III		410 s		
52548	197457		7 00 14.0	-31 04 07	-0.001	+0.03	7.4	2.2	-0.3	K5 III		130 s		
52388	152228		7 00 15.0	-19 48 04	0.000	-0.01	6.8	1.3	0.2	K0 III		150 s		
52005	96363	41 Gem	7 00 15.7	+16 04 44	0.000	-0.01	5.68	1.66	-5.9	cK4	+22	1400 s		
52184	114756		7 00 15.8	+2 30 44	0.000	0.00	7.8	1.1	-0.1	K2 III		360 s		
52547	197459		7 00 16.1	-30 40 42	0.000	+0.02	7.5	1.5	0.2	K0 III		150 s		
53191	249701		7 00 16.7	-60 51 47	0.000	-0.02	7.90	0.00	0.0	B8.5 V		380 s		m
52265	134031		7 00 17.9	-5 22 01	-0.008	+0.08	6.30	0.57		F8				
52037	96365		7 00 18.1	+15 32 04	+0.001	-0.04	6.7	0.0	0.0	B8.5 V		210 s		
52437	172725		7 00 19.3	-22 07 11	-0.001	-0.02	6.53	-0.17	-1.4	B4 V ne	+9	370 s	5687	m
50904	14037		7 00 21.0	+67 19 51	+0.004	-0.01	6.5	0.9	3.2	G5 IV		46 s		
52266	134034		7 00 21.0	-5 49 38	0.000	-0.02	7.23	-0.01	-4.8	O9 V	+5	2100 s		
52347	152229		7 00 23.3	-15 15 35	-0.001	-0.03	7.91	-0.07	0.6	A0 V		290 s		
52902	218381		7 00 23.4	-49 08 07	-0.002	-0.03	6.5	0.8	3.2	G5 IV		46 s		
52312	134036		7 00 23.6	-8 24 25	-0.001	0.00	5.96	-0.08	-0.8	B9 III		230 s		m
52074	96368		7 00 23.9	+15 17 16	0.000	-0.03	6.7	1.4	-0.3	K5 III		240 s		
52206	114762		7 00 25.0	+1 55 54	0.000	+0.01	6.6	0.0	0.0	B8.5 V		200 s		
52462	172732		7 00 28.1	-22 38 43	0.000	-0.01	7.27	1.55		K5				
51710	41490		7 00 29.1	+42 05 27	-0.005	-0.02	7.80	0.8	3.2	G7 IV	+12	83 s		
52756	218379		7 00 29.3	-41 22 47	+0.001	+0.21	8.0	1.2	0.2	K0 III		86 mx		
52485	172736		7 00 35.1	-20 38 23	+0.001	+0.02	6.60	0.1	1.7	A3 V		95 s	5692	m
52155	96372		7 00 36.9	+12 43 25	-0.002	+0.01	7.80	0.1	1.7	A3 V		160 s	5676	m
52704	197471		7 00 37.8	-35 50 18	-0.003	-0.01	8.0	0.0	0.4	B9.5 V		300 s		
52436	152235		7 00 38.5	-15 23 57	-0.001	-0.02	7.1	1.4	-0.3	K5 III		300 s		
52705	197474		7 00 38.9	-36 02 09	-0.002	-0.02	8.0	0.9	2.6	F0 V		50 s		
52124	96371		7 00 39.1	+16 57 32	+0.001	-0.03	7.5	0.0	0.4	B9.5 V		260 s		
52382	134041		7 00 39.2	-9 12 12	-0.001	-0.01	6.49	0.21	-5.7	B1 Ib	+51	1600 s		
52035	78963		7 00 40.3	+23 26 24	-0.001	-0.02	7.0	0.1	0.6	A0 V		190 s		
51805	41495		7 00 41.1	+40 20 59	+0.001	-0.01	7.7	0.1	0.6	A0 V		270 s		
52619	172749		7 00 42.6	-28 29 21	-0.001	-0.04	6.4	1.0	4.0	F8 V		16 s		
53327	249706		7 00 43.1	-61 08 32	-0.001	+0.01	7.6	0.2	0.4	B9.5 V		190 s		
53989	256359		7 00 47.0	-71 23 49	+0.007	+0.02	7.7	0.5	4.0	F8 V		55 s		
52596	172750		7 00 47.4	-25 38 37	-0.001	+0.01	7.36	-0.13	-1.6	B3.5 V		600 s	5699	m
52617	172755		7 00 47.4	-27 46 26	-0.002	+0.01	8.0	1.0	-0.1	K2 III		410 s		

HD	SAO	Star Name	α 2000	δ 2000	μ(α)	μ(δ)	V	B-V	M_v	Spec	RV	d(pc)	ADS	Notes
			h m s	o ' "	s	"								
52597	172753		7 00 47.6	-26 05 36	+0.001	+0.01	7.87	-0.14	0.0	B8.5 V		380 s		
52703	197475		7 00 49.7	-33 27 55	+0.001	+0.09	6.5	1.0	0.2	K0 III		170 s		
52852	218384		7 00 51.1	-40 56 55	+0.002	+0.04	7.4	1.4	0.2	K0 III		160 s		
53047	234854		7 00 51.5	-51 24 09	-0.002	+0.02	4.9	2.3	-0.5	M2 III	+5	47 s		
52540	152239		7 00 51.8	-19 26 37	-0.003	+0.02	7.0	0.2	0.6	A0 V		140 s		
52595	172757		7 00 54.1	-24 29 24	-0.002	-0.01	7.5	0.3	1.7	A3 V		100 s		
52237	96379		7 00 56.4	+12 24 01	-0.001	0.00	7.70	1.1	0.2	K0 III		310 s	5684	m
54154	256360		7 00 57.7	-72 33 37	-0.001	+0.03	7.5	0.4	2.6	F0 V		95 s		
52071	78968		7 00 58.0	+27 09 25	0.000	-0.01	7.12	1.25		K2 IV				
52287	96381		7 00 58.6	+10 37 32	0.000	0.00	6.9	1.4	-0.3	K5 III		270 s		
52432	134049		7 01 01.9	-3 15 07	0.000	+0.02	7.15	1.71		R5	+22			
267805	96378		7 01 02.1	+18 52 44	+0.001	-0.01	8.0	1.4	-0.3	K5 III		440 s		
52818	197482		7 01 03.2	-36 37 57	-0.003	+0.03	8.0	1.3	0.2	K0 III		250 s		
53046	218386		7 01 04.3	-49 42 30	-0.002	0.00	7.8	1.4	0.2	K0 III		53 s		
53349	234863		7 01 05.1	-58 56 24	-0.007	+0.14	6.02	0.30	2.6	F0 V		48 s		
52899	218385		7 01 05.5	-40 47 42	-0.003	0.00	7.35	-0.02	0.6	A0 V		220 s		
53658	249715		7 01 05.7	-65 42 23	-0.015	+0.07	7.33	0.46	3.7	F6 V		53 s		
52670	172763		7 01 05.9	-25 12 55	-0.001	+0.02	5.63	-0.17		B3	+6			
53279	234862		7 01 08.6	-57 11 31	+0.001	-0.01	8.0	1.1	3.4	F5 V		82 s		
52230	96382		7 01 09.1	+19 12 53	+0.001	-0.03	7.7	1.1	0.2	K0 III		300 s		
52309	96385		7 01 09.8	+11 46 28	0.000	-0.02	7.10	0.1	1.4	A2 V		140 s	5689	m
52782	197481		7 01 10.7	-30 39 51	+0.001	+0.05	7.40	0.4	2.6	F0 V		91 s		m
52283	96384		7 01 11.7	+16 06 19	-0.001	-0.02	7.8	1.1	0.2	K0 III		320 s		
52147	78972		7 01 12.6	+29 13 01	0.000	0.00	7.7	0.9	0.3	G5 III		310 s		
51638	26094		7 01 13.6	+54 51 24	0.000	-0.03	6.7	0.0	0.0	B8.5 V		220 s		
52698	172768		7 01 13.7	-25 56 55	+0.015	+0.04	6.71	0.90	5.9	K0 V		13 s		
52431	114789		7 01 14.8	+0 00 37	-0.006	-0.01	7.58	0.41	3.4	F5 V		69 s		
52100	59697		7 01 17.1	+32 24 52	+0.001	-0.02	6.59	0.27	2.6	F0 V	-28	63 s	5680	m
52536	134056		7 01 17.4	-8 12 11	0.000	+0.01	8.02	0.04	0.0	B8.5 V		360 s		
51610	26097		7 01 17.9	+55 19 51	-0.001	+0.01	7.80	1.9		S7.7 e	+28			R Lyn, v
53142	234861		7 01 18.3	-50 28 00	0.000	0.00	7.3	0.1	0.6	A0 V		200 s		
50885	6041		7 01 21.5	+70 48 30	+0.004	-0.02	5.68	1.33	0.2	K0 III	-17	79 s		m
52815	172781		7 01 22.6	-29 43 44	-0.002	-0.02	8.0	0.7	0.2	K0 III		360 s		
267827	78979		7 01 25.3	+26 14 08	-0.001	+0.01	7.8	0.9	3.2	G5 IV		82 s		
52533	134061		7 01 26.9	-3 07 04	0.000	-0.01	7.70	-0.09	-4.8	O9 V	+6	2500 s	5705	m
52642	152248		7 01 31.2	-13 07 56	+0.001	-0.02	7.88	1.69	-0.5	M2 III		410 s		
52813	172786		7 01 32.0	-27 51 34	0.000	0.00	8.0	-0.1	1.4	A2 V		210 s		
52946	197490		7 01 33.0	-35 20 54	-0.001	0.00	7.7	1.2	0.2	K0 III		240 s		
52812	172788		7 01 33.5	-27 13 23	-0.001	0.00	6.6	0.8	-1.7	B3 V		130 s		
51866	41500		7 01 38.6	+48 22 42	+0.055	-0.44	8.00	0.99	6.9	K3 V	-23	19 mx		
52778	172789		7 01 39.6	-23 30 05	0.000	+0.01	7.0	1.5	-0.3	K5 III		280 s		
52479	114798		7 01 41.3	+4 49 05	-0.001	0.00	6.63	0.06		A0	-11			
52201	59703		7 01 42.7	+33 41 07	-0.001	-0.01	7.4	0.1	1.7	A3 V	+7	140 s		
52877	172797	22 σ CMa	7 01 43.1	-27 56 06	-0.001	0.00	3.46	1.73	-5.7	M0 Iab	+22	460 s	5719	m
51068	14044		7 01 45.2	+69 38 46	0.000	-0.01	6.7	0.8	3.2	G5 IV		51 s		
52372	96393		7 01 45.5	+17 58 31	0.000	-0.01	7.4	0.1	1.4	A2 V		150 s		
53071	218388		7 01 46.7	-40 53 48	-0.001	+0.01	7.91	-0.13	-2.5	B2 V		1100 s		
52993	197497		7 01 47.1	-35 32 52	-0.001	+0.01	6.59	-0.18	0.4	B9.5 V		170 s		
52721	152255		7 01 49.3	-11 18 03	-0.002	+0.01	6.59	0.06		B2 e	+16		5713	m
52694	152256		7 01 49.9	-10 52 56	0.000	+0.01	6.59	0.73	4.0	F8 V		23 s	5712	m
52611	134073		7 01 52.8	-1 20 44	0.000	-0.03	6.17	1.28		K0				m
52370	78987		7 01 54.1	+22 02 22	-0.001	-0.01	7.8	0.5	3.4	F5 V		74 s		
52559	114801		7 01 55.0	+5 33 27	-0.003	+0.01	6.59	-0.02	-2.8	B2 IV-V	+34	580 s		
52666	134076		7 01 56.3	-5 43 20	-0.001	0.00	5.20	1.68	-0.5	gM2	+3	120 s		
52930	172802		7 01 56.8	-26 41 32	-0.001	+0.04	7.4	0.6	0.6	A0 V		93 s		
52806	152260		7 01 57.0	-17 38 40	-0.001	-0.03	7.5	0.1	1.4	A2 V		170 s		
51495	14049		7 01 57.1	+63 40 37	+0.002	-0.02	6.6	1.4	-0.3	K5 III		240 s		
52099	41509		7 02 01.2	+44 26 55	+0.001	-0.02	7.1	1.1	0.2	K0 III		240 s		
53632	234876		7 02 01.8	-59 27 38	0.000	-0.01	7.1	1.8	-0.3	K5 III		200 s		
53526	234873		7 02 02.4	-57 18 53	-0.001	+0.03	7.1	0.5	3.0	F2 V		53 s		
52745	134084		7 02 04.7	-9 53 13	-0.002	0.00	7.79	0.26	2.1	A5 V		130 s		
52422	78992		7 02 05.4	+21 47 53	-0.001	0.00	8.0	0.0	0.4	B9.5 V		320 s		
52690	134082		7 02 06.6	-3 45 18	-0.001	-0.01	6.7	1.6	-4.7	M1 Ib		1700 s		
53348	234868		7 02 06.9	-50 25 49	-0.001	-0.02	7.80	0.1	0.6	A0 V		280 s		m
52403	78990		7 02 07.3	+25 25 33	0.000	0.00	7.04	0.11	1.4	A2 V		120 s		
52719	134083		7 02 07.5	-6 01 20	-0.001	-0.01	7.0	1.4	-0.3	K5 III	-31	280 s		
52925	172808		7 02 08.5	-23 29 51	-0.001	0.00	8.00	0.9	3.2	G5 IV		90 s	5728	m
51416	14050		7 02 10.9	+65 45 06	-0.003	0.00	7.7	0.1	0.6	A0 V		260 s		
52474	78993		7 02 13.4	+20 49 42	0.000	0.00	8.0	0.1	1.4	A2 V		200 s		
53253	218396		7 02 15.5	-43 24 15	+0.001	-0.01	6.43	-0.04		A0				
52473	78996		7 02 16.5	+22 05 45	0.000	-0.04	7.3	0.8	3.2	G5 IV		66 s		
51970	26107		7 02 17.0	+51 34 04	-0.003	-0.07	7.5	1.1	0.2	K0 III		230 mx		
52556	96403		7 02 17.4	+15 20 10	0.000	-0.02	5.8	1.1	0.2	K0 III	-14	130 s		
53041	172821		7 02 20.1	-28 16 54	0.000	-0.02	7.5	0.7	0.0	B8.5 V		120 s		
52472	78995		7 02 20.3	+25 21 51	-0.003	-0.05	7.5	0.1	1.7	A3 V		140 s		

HD	SAO	Star Name	α 2000	δ 2000	μ(α)	μ(δ)	V	B-V	M_v	Spec	RV	d(pc)	ADS	Notes
			h m s	° ′ ″	s	″								
53252	218397		7 02 20.6	−42 37 57	−0.001	0.00	7.00	−0.14	0.0	B8.5 V		250 s		
52452	78998		7 02 23.2	+25 50 46	−0.002	0.00	8.0	0.9	3.2	G5 IV		93 s		
52497	78999	42 ω Gem	7 02 24.7	+24 12 55	−0.001	0.00	5.18	0.94	−2.1	G5 II	−9	290 s		
52554	96407		7 02 25.5	+17 45 21	+0.001	+0.04	5.94	1.63	−0.5	M2 III	+23	180 s		
51999	26110		7 02 28.2	+53 23 01	−0.001	−0.01	7.4	0.0	0.0	B8.5 V		310 s		
52685	114818		7 02 30.2	+8 40 08	−0.001	−0.01	7.9	0.7	4.4	G0 V		51 s		
52609	96409		7 02 33.4	+16 40 27	+0.001	−0.02	5.82	1.68	−0.3	K5 III	+35	130 s		
53015	152280		7 02 35.8	−19 46 08	−0.002	+0.01	8.0	0.1	1.4	A2 V		180 s		
52451	59713		7 02 39.0	+31 20 24	−0.001	−0.01	7.8	1.4	−0.3	K5 III		410 s		
52968	152276		7 02 40.2	−13 11 22	−0.002	0.00	7.86	−0.03	0.4	B9.5 V		310 s		
51349	14051		7 02 40.5	+69 12 28	−0.004	−0.03	7.4	1.4	−0.3	K5 III	+13	350 s		
52986	152281		7 02 41.3	−17 34 54	0.000	−0.01	7.45	−0.03	0.4	B9.5 V		250 s		
52254	41512		7 02 42.9	+45 57 27	−0.002	0.00	7.8	0.1	0.6	A0 V		270 s		
53494	218407		7 02 45.8	−49 55 40	+0.001	+0.01	7.1	−0.3	1.4	A2 V		140 s		
52938	134103		7 02 46.0	−8 27 09	−0.001	−0.01	7.8	1.1	−0.1	K2 III		360 s		
52471	59715		7 02 46.4	+31 32 45	+0.001	−0.02	8.0	1.1	0.2	K0 III		360 s		
52984	152282		7 02 48.1	−13 05 45	−0.002	−0.01	7.79	0.00	0.6	A0 V		270 s		
53013	152285		7 02 49.4	−16 41 59	0.000	−0.01	6.90	1.1	0.2	K0 III		210 s	5736	m
51832	14056		7 02 51.3	+61 18 27	−0.001	−0.02	7.8	0.4	3.0	F2 V		90 s		
52838	114828		7 02 52.6	+3 47 19	−0.001	+0.01	7.6	1.1	0.2	K0 III		290 s		
52918	134106	19 Mon	7 02 54.6	−4 14 21	−0.001	0.00	4.99	−0.20	−3.5	B1 V	+25	500 s		
53123	172837		7 02 54.7	−24 35 01	+0.001	0.00	7.14	−0.07	0.6	A0 V		200 s		
53012	152286		7 02 55.1	−13 13 04	0.000	+0.03	8.0	0.5	3.4	F5 V		81 s		
53138	172839	24 o² CMa	7 03 01.4	−23 50 00	−0.001	0.00	3.03	−0.09	−6.8	B3 Ia	+48	860 s		
53139	172840		7 03 01.8	−24 13 59	0.000	−0.02	7.4	1.0	0.2	K0 III		270 s		
53405	218406		7 03 04.8	−40 21 26	−0.004	+0.06	7.5	0.3	1.7	A3 V		110 s		
52145	26119		7 03 05.3	+54 10 24	−0.003	−0.03	7.70	0.7	4.4	G0 V		46 s	5706	m
53088	152293		7 03 05.7	−16 00 22	+0.001	0.00	7.9	1.4	−0.3	K5 III		410 s		
53035	152288		7 03 07.7	−11 11 58	+0.001	−0.02	7.87	−0.06	−1.0	B5.5 V	+19	560 s		
−−	59717		7 03 10.8	+39 25 40	+0.001	+0.01	8.0	2.1						
52886	114831		7 03 12.8	+9 01 25	−0.003	−0.08	7.4	0.5	3.4	F5 V		64 s		
52836	96421		7 03 14.2	+14 20 10	+0.001	−0.01	7.1	0.1	1.7	A3 V		120 s		
53060	152295		7 03 14.2	−11 08 17	0.000	−0.03	7.99	0.01	0.6	A0 V		300 s		
52068	26116		7 03 14.3	+57 55 47	+0.002	0.00	7.7	1.1	−0.1	K2 III		360 s		
53921	234890		7 03 15.5	−59 10 41	+0.003	+0.01	5.50	−0.11		B9	−5			m
53869	234888		7 03 17.0	−57 30 10	+0.001	−0.02	7.70	1.1	−0.1	K2 III		360 s		m
52913	114835		7 03 17.8	+9 08 18	−0.002	+0.01	5.97	0.12		A2 n	−11			
53004	114842		7 03 23.5	+0 20 37	−0.002	−0.03	7.9	0.0	0.4	B9.5 V		310 s		
53762	234883		7 03 24.5	−52 25 31	−0.001	+0.01	6.6	0.2	0.0	B8.5 V		140 s		
53056	134119		7 03 25.2	−3 01 32	−0.002	−0.01	7.48	0.14		A3		12 mn		
52937	114839		7 03 27.3	+8 43 31	−0.002	+0.02	8.0	0.1	0.6	A0 V		290 s		
268381	79012		7 03 28.3	+25 20 16	+0.001	−0.02	7.7	1.1	0.2	K0 III		320 s		
53868	234891		7 03 29.6	−55 56 55	0.000	−0.01	7.4	1.3	0.2	K0 III		190 s		
52828	96426		7 03 29.7	+19 13 18	−0.001	−0.04	7.8	0.1	0.6	A0 V		260 s		
53402	197521		7 03 30.1	−33 35 21	−0.006	+0.01	7.9	1.1	−0.3	K5 III		180 mx		
52711	79009		7 03 30.4	+29 20 14	+0.012	−0.83	5.93	0.60	3.2	G8 IV	+22	11 mx		
51370	6054		7 03 33.5	+71 45 28	+0.006	+0.05	7.8	0.5	4.0	F8 V		57 s		
53003	114844		7 03 34.5	+2 26 14	0.000	−0.01	7.1	0.7	4.4	G0 V		34 s		
53522	197528		7 03 35.3	−39 57 53	−0.002	+0.05	8.0	0.7	3.2	G5 IV		92 s		
53440	197523		7 03 36.8	−32 58 51	−0.005	+0.05	7.6	0.7	3.4	F5 V		49 s		
52960	96429		7 03 38.0	+10 57 06	−0.001	−0.02	5.13	1.39	−0.2	K3 III	+21	96 s		
53083	134125		7 03 41.6	−0 33 25	−0.001	+0.01	7.1	0.0	0.0	B8.5 V		250 s		
53344	172872		7 03 42.8	−25 05 03	−0.001	−0.01	6.87	−0.16	−2.3	B3 IV		680 s		
52709	59730		7 03 43.0	+33 53 34	−0.003	−0.05	7.4	1.1	0.2	K0 III		280 s		
53244	152303	23 γ CMa	7 03 45.4	−15 38 00	0.000	−0.01	4.11	−0.12	−3.4	B8 II	+30	320 s		
52496	41529		7 03 45.9	+47 15 33	−0.001	+0.03	7.7	0.5	4.0	F8 V		55 s		
53373	172876		7 03 47.3	−25 39 11	−0.001	+0.02	7.97	−0.17	−1.0	B5.5 V		620 s		
53631	218417		7 03 50.3	−41 12 26	−0.003	−0.01	7.3	1.3	0.2	K0 III		180 s		
52976	96432		7 03 51.6	+12 35 40	0.000	0.00	5.98	1.58	−0.3	K5 III	−16	160 s		
53267	152307		7 03 53.6	−13 05 46	+0.002	+0.03	7.02	1.41	−0.1	K2 III		190 s		
53811	218427		7 03 53.7	−49 35 02	−0.005	+0.15	4.93	0.13		A2	+25	17 mn		
53840	234895		7 03 54.4	−50 39 41	0.000	0.00	7.4	1.2	0.2	K0 III		220 s		
53548	197529		7 03 56.1	−34 16 01	+0.001	0.00	7.8	1.3	0.2	K0 III		220 s		
53240	152308		7 03 57.2	−10 07 27	−0.002	−0.01	6.45	−0.08		B8				
53705	218421		7 03 57.2	−43 36 29	−0.011	+0.39	5.28	0.66	4.9	G3 V		13 ts		m
52654	41536		7 03 58.3	+40 34 57	−0.002	+0.02	7.0	1.1	0.2	K0 III		230 s		
53706	218423		7 03 58.8	−43 36 42	−0.011	+0.37	6.79	0.80	5.9	K0 V		15 s		
52552	41531		7 03 59.0	+47 03 00	−0.008	+0.02	7.1	0.5	3.4	F5 V		54 s		
53053	114857		7 04 01.4	+8 15 00	−0.002	−0.01	7.7	0.9	3.2	G5 IV		80 s		
53113	114858		7 04 02.2	+5 07 54	0.000	−0.02	7.4	1.1	0.2	K0 III		270 s		
52450	26131		7 04 02.6	+51 17 09	0.000	−0.01	7.7	1.1	0.2	K0 III		310 s		
53303	152311		7 04 02.7	−12 17 18	+0.001	0.00	7.42	−0.11	0.4	B9.5 V		250 s		
53704	218424		7 04 02.7	−42 20 15	0.000	0.00	5.20	0.20		A m	+28	18 mn		
52737	59738		7 04 03.5	+36 18 48	+0.003	−0.02	6.7	0.9	3.2	G5 IV		50 s		
53208	134133		7 04 05.2	−5 19 25	−0.001	+0.01	5.62	1.29	−0.2	gK3	+40	150 s		

HD	SAO	Star Name	α 2000	δ 2000	μ(α)	μ(δ)	V	B-V	M_v	Spec	RV	d(pc)	ADS	Notes
51067	6052		7 04 05.8	+75 13 49	-0.020	-0.26	7.12	0.56	4.0	dF8	+23	39 s	5669	m
52973	79031	43 ζ Gem	7 04 06.5	+20 34 13	-0.001	0.00	3.79	0.79	-4.5	G0 Ib var	+7	430 s	5742	Mekbuda, m,v
52607	41535		7 04 06.7	+45 03 48	-0.002	-0.03	7.7	0.4	2.6	F0 V		100 s		
268518	--		7 04 06.9	+20 33 59			7.58	0.62		G			5742	
52996	96439		7 04 09.5	+18 40 07	+0.001	-0.01	7.1	0.1	1.4	A2 V		130 s		
53433	172890		7 04 13.5	-21 15 59	0.000	-0.02	7.65	-0.01	0.6	A0 V		260 s		
51066	6053		7 04 15.5	+75 24 41	+0.002	0.00	7.3	0.9	3.2	G5 IV		65 s		
53204	114863		7 04 16.2	+1 30 26	0.000	-0.01	7.4	0.0	0.4	B9.5 V		250 s		
52823	59741		7 04 16.9	+37 33 51	-0.001	0.00	6.6	0.1	0.6	A0 V		160 s		
54118	234902		7 04 18.3	-56 44 59	0.000	0.00	5.17	-0.04		A0 p	+30	41 mn		
53203	114865		7 04 19.4	+4 01 25	-0.002	+0.02	7.7	0.9	3.2	G5 IV		80 s		
53205	114867		7 04 20.2	+1 29 18	0.000	-0.01	6.57	0.01		B9				m
53602	197537		7 04 20.4	-30 46 26	-0.002	+0.06	7.92	-0.08	0.4	B9.5 V		100 mx		
53110	96443		7 04 20.5	+15 14 19	-0.001	-0.02	7.2	0.1	1.4	A2 V		140 s		
53111	96446		7 04 20.5	+14 52 33	-0.001	-0.01	7.40	1.1	-0.1	K2 III		310 s		
53202	114866		7 04 22.5	+4 34 19	-0.003	-0.01	7.40	0.00	0.4	B9.5 V		250 s	5747	m
53367	152320		7 04 25.3	-10 27 19	-0.002	-0.04	6.96	0.44	-4.6	B0 IV e	+19	1800 s		m
53955	234899		7 04 27.6	-50 22 20	+0.004	+0.01	7.5	1.2	0.2	K0 III		210 s		
53173	114869		7 04 30.6	+8 14 55	-0.001	-0.01	8.0	0.1	1.4	A2 V		200 s		
51396	6059		7 04 32.6	+73 25 40	-0.007	-0.02	7.5	0.8	3.2	G5 IV		73 s		
53395	134147		7 04 33.2	-8 26 29	-0.001	-0.02	7.33	1.14	0.2	K0 III		220 s		
53676	197540		7 04 35.6	-32 00 31	-0.001	-0.02	7.9	-0.3	0.0	B8.5 V		380 s		
53456	152324		7 04 38.3	-11 31 27	0.000	-0.01	7.21	0.01		B3	+18		5761	m
53834	218432		7 04 41.4	-40 29 02	+0.001	+0.01	7.0	1.4	0.2	K0 III		130 s		
53261	114874		7 04 41.7	+5 55 33	-0.002	-0.02	7.90						5755	m
53698	197542		7 04 44.3	-31 56 44	-0.003	0.00	7.1	0.2	3.0	F2 V		66 s		
52935	59745		7 04 45.2	+36 35 58	0.000	-0.06	7.8	0.8	3.2	G5 IV		84 s		
53629	172906		7 04 47.0	-22 01 56	+0.001	-0.06	6.09	1.22	0.2	K0 III		98 s		
53545	152337		7 04 49.5	-16 17 06	+0.003	-0.09	8.0	0.5	3.4	F5 V		81 s		
53598	172911		7 04 53.6	-20 49 48	0.000	+0.02	7.1	1.6	-2.4	M0 II	+54	800 s		
53078	59755		7 04 59.1	+30 21 44	-0.001	-0.02	7.2	0.9	3.2	G5 IV		65 s		
53781	197545		7 05 00.0	-31 31 36	-0.002	-0.03	7.4	1.6	-0.3	K5 III		310 s		
53732	172925		7 05 00.2	-27 06 50	-0.001	+0.02	8.0	0.1	0.6	A0 V		240 s		
53596	152342		7 05 01.5	-13 58 01	+0.001	+0.02	7.9	1.4	-0.3	K5 III		410 s		
53670	172920		7 05 05.3	-20 12 08	-0.002	+0.02	6.9	1.4	0.2	K0 III		130 s		
53672	172922		7 05 06.2	-21 27 03	+0.002	+0.01	6.8	1.6	0.2	K0 III		95 s		
52860	41538		7 05 08.9	+47 46 30	0.000	-0.01	6.3	0.0		B9				
53451	114891		7 05 08.9	+0 19 46	-0.002	-0.03	7.90	1.1	-1.1	K1 II-III		600 s	5768	m
52628	26137		7 05 15.6	+55 42 29	0.000	-0.03	8.0	0.1	1.4	A2 V		210 s		
54179	234907		7 05 16.4	-50 21 37	-0.002	+0.03	6.4	2.0	0.2	K0 III		44 s		
53623	152349		7 05 16.7	-12 19 33	-0.001	+0.03	7.97	-0.06	-2.5	B2 V		1100 s		
53891	197550		7 05 16.7	-34 08 34	-0.006	+0.07	7.90	0.5	3.4	F5 V		79 s		m
53257	79042	44 Gem	7 05 18.3	+22 38 14	0.000	-0.01	5.90	-0.03	0.4	B9.5 V	-8	120 s		
55302	256370		7 05 18.8	-72 59 28	+0.002	+0.01	8.0	1.1	-0.1	K2 III		420 s		
53330	96468		7 05 21.5	+16 22 17	0.000	-0.02	7.7	1.1	0.2	K0 III		300 s		
54290	234909		7 05 21.7	-52 36 34	0.000	+0.01	7.70	1.4	-0.3	K5 III		400 s		m
53621	134169		7 05 21.8	-8 08 00	0.000	-0.01	7.1	1.1	0.2	K0 III	+3	240 s		
53983	197554		7 05 24.2	-38 14 04	+0.001	-0.04	7.79	0.03	0.6	A0 V		260 s		
54009	--		7 05 26.4	-39 52 03			7.9	1.2	0.2	K0 III		270 s		
54612	249729		7 05 27.9	-61 24 09	-0.001	+0.04	7.9	1.3	0.2	K0 III		240 s		
54343	234913		7 05 29.5	-54 07 10	-0.001	+0.02	7.3	2.1	0.4	B9.5 V		13 s		
--	--		7 05 31.2	-34 47 09			6.50							m
53952	197558		7 05 32.3	-34 46 40	-0.002	+0.07	6.14	0.37		F0				
53590	134171		7 05 32.6	-0 47 16	+0.001	-0.03	7.6	0.9	3.2	G8 IV	+15	75 s		
54038	218440		7 05 32.7	-40 38 23	0.000	+0.08	6.9	1.2	4.4	G0 V		13 s		
53416	96473		7 05 34.6	+14 28 22	-0.001	-0.02	6.8	0.0	0.0	B8.5 V		220 s		
53418	96475		7 05 35.2	+13 42 35	+0.001	-0.02	7.3	1.6	-0.5	M2 III		350 s		
53667	134175		7 05 35.2	-8 43 46	0.000	-0.03	7.74	0.24		B0.5 III	+38			
53566	114900		7 05 38.1	+3 56 56	+0.001	-0.02	7.8	0.5	4.0	F8 V		57 s		
53510	114899		7 05 39.0	+9 11 09	+0.003	-0.03	5.78	1.52	0.2	gK0	+46	51 s		
52859	26145		7 05 40.2	+52 45 27	-0.001	-0.07	6.90	0.1	1.4	A2 V	+27	81 mx	5746	m
53756	152360		7 05 42.0	-12 48 42	-0.001	0.00	7.33	-0.08	-3.0	B2 IV	+19	1000 s		
53540	114902		7 05 44.6	+7 50 38	-0.002	-0.01	7.8	1.1	0.2	K0 III		320 s		
53287	59766		7 05 47.3	+31 23 35	+0.001	-0.01	6.7	1.6	-0.5	M2 III		280 s		
53538	96481		7 05 48.1	+11 13 27	-0.001	0.00	7.8	0.1	0.6	A0 V		260 s		
53777	152367		7 05 48.8	-13 49 22	0.000	+0.02	8.0	0.5	3.4	F5 V		81 s		
53755	152363		7 05 49.6	-10 39 40	0.000	-0.03	6.49	-0.05	-4.1	B0 V	+16	1000 s	5782	m
52030	6074		7 05 51.8	+70 43 55	+0.008	-0.01	6.5	1.1	0.2	K0 III	+20	190 s		
53665	134179		7 05 52.6	-1 01 14	+0.001	-0.19	8.0	0.5	4.0	F8 V		54 mx		
53536	96483		7 05 55.8	+15 10 52	0.000	-0.03	7.0	1.1	0.2	K0 III		230 s		
53228	59767		7 05 57.6	+36 34 23	+0.001	-0.04	7.0	0.9	3.2	G5 IV		57 s		
54208	218447		7 05 57.7	-42 19 39	-0.002	+0.03	7.20	-0.02	0.4	B9.5 V		220 s		m
53314	59769		7 05 58.7	+33 44 32	0.000	-0.01	7.9	0.9	3.2	G5 IV		88 s		
54031	197566		7 06 00.5	-30 39 20	0.000	+0.01	6.34	-0.16	-2.3	B3 IV	+14	510 s		
52708	26141		7 06 01.2	+59 48 07	+0.004	0.00	6.5	1.1	0.2	K0 III	+22	180 s		

HD	SAO	Star Name	α 2000	δ 2000	μ(α)	μ(δ)	V	B-V	M$_v$	Spec	RV	d(pc)	ADS	Notes
			h m s	° ′ ″	s	″								
54234	218448		7 06 01.7	−42 18 47	−0.004	0.00	7.10	0.5	3.4	F5 V		55 s		m
54153	197572		7 06 02.2	−38 22 58	+0.001	+0.02	6.11	0.70	0.6	gG0	+22	120 s		
53561	96485		7 06 05.0	+13 59 08	0.000	−0.01	7.40	1.4	−0.3	K5 III	+42	330 s		
54315	218451		7 06 05.4	−45 58 07	−0.001	−0.01	7.60	1.4	−0.3	K5 III		380 s		m
54207	218449		7 06 05.5	−40 57 02	−0.001	0.00	8.0	1.4	0.2	K0 III		160 s		
53414	79053		7 06 10.7	+28 11 32	−0.001	+0.01	7.9	0.1	1.4	A2 V		200 s		
53329	59773		7 06 11.5	+34 28 26	−0.004	−0.05	5.55	0.91	0.4	gG3	+5	91 s		
53824	152379		7 06 12.3	−10 09 59	−0.001	0.00	8.0	0.0	0.4	B9.5 V		320 s		
55151	249740		7 06 14.2	−68 50 15	0.000	+0.01	6.47	1.04	0.3	gG5		130 s		
53588	96487		7 06 17.3	+17 44 38	−0.001	−0.02	7.1	0.0	0.4	B9.5 V		220 s		
54232	197579		7 06 20.4	−37 42 35	−0.003	0.00	7.50	−0.02	0.6	A0 V		240 s		
54341	218453		7 06 21.0	−43 36 39	+0.001	+0.01	6.7	0.4	0.6	A0 V		93 s		
54203	197578		7 06 21.0	−35 44 59	−0.003	−0.01	7.7	1.5	−0.1	K2 III		240 s		
53851	134190		7 06 27.6	−3 06 58	−0.001	−0.01	7.4	0.0	0.4	B9.5 V		250 s		
53907	152386		7 06 28.2	−10 13 08	+0.001	0.00	7.9	1.1	−0.1	K2 III		390 s		
54713	234930		7 06 30.7	−57 46 43	0.000	−0.01	7.0	1.1	1.7	A3 V		31 s		
51558	6072		7 06 35.5	+76 10 18	−0.006	−0.05	8.0	1.1	0.2	K0 III		360 s		
53975	152393		7 06 35.8	−12 23 38	−0.002	0.00	6.48	−0.10		O8	+33			
53640	96493		7 06 36.4	+19 06 32	−0.002	−0.04	7.7	1.1	0.2	K0 III		300 s		
53974	152394		7 06 40.7	−11 17 39	−0.001	−0.01	5.39	0.05		B0.5 IV	+31		5795	m
55150	249742		7 06 47.0	−67 00 58	−0.004	−0.02	7.9	1.1	0.2	K0 III		350 s		
53850	114929		7 06 47.2	+4 42 58	−0.001	−0.03	7.8	1.1	0.2	K0 III		330 s		
53946	134201		7 06 50.3	−2 44 52	−0.001	0.00	7.9	0.0	0.0	B8.5 V		370 s		
54173	172986		7 06 52.3	−24 57 38	−0.003	+0.01	6.08	1.34	−0.1	K2 III		130 s		
53145	26155		7 06 54.4	+53 04 23	0.000	−0.03	7.50	1.1	0.0	K1 III		320 s		
---	---		7 06 55.3	−7 33 17			7.70	2.0		N7.7				RY Mon, v
54025	152399		7 06 56.3	−11 19 39	−0.002	0.00	7.62	−0.03	−1.0	B5.5 V	+17	530 s	5799	m
53794	96499		7 06 57.6	+13 24 10	+0.001	−0.01	8.0	0.1	0.6	A0 V		290 s		
53972	134207		7 06 59.2	−5 28 20	−0.002	−0.03	8.0	0.4	3.0	F2 V		97 s		
54224	172989		7 07 00.1	−26 39 28	+0.001	−0.03	6.62	−0.17	−3.5	B1 V		970 s		
53283	41557		7 07 01.0	+49 28 17	+0.001	+0.01	7.8	0.4	2.6	F0 V		110 s		
54084	152402		7 07 01.1	−12 14 08	0.000	−0.01	8.00	0.17	1.4	A2 V		180 s		
54257	172991		7 07 02.7	−25 46 51	−0.002	0.00	7.3	1.5	−0.3	K5 III		340 s		
53878	114934		7 07 05.6	+9 52 45	+0.002	−0.03	8.0	0.4	2.6	F0 V		120 s		
53929	114935		7 07 06.3	+4 54 37	−0.002	−0.02	6.11	−0.13		B8				
54475	218465		7 07 07.0	−40 53 36	−0.001	0.00	5.79	−0.17		B9				
54339	172999		7 07 09.6	−29 32 22	+0.002	+0.02	7.3	1.0	−0.5	M2 III		370 s		
54364	197593		7 07 11.4	−30 47 42	+0.001	+0.01	8.0	1.2	−0.1	K2 III		410 s		
54503	218468		7 07 11.5	−40 32 15	0.000	−0.01	7.9	1.5	−0.1	K2 III		260 s		
54732	234940		7 07 13.3	−51 58 04	−0.001	+0.06	5.96	1.00	3.2	G5 IV		24 s		
53874	96506		7 07 14.9	+15 48 10	0.000	+0.02	7.9	0.9	3.2	G5 IV		87 s		
54967	234944		7 07 16.3	−59 42 59	0.000	0.00	6.47	−0.12	−1.7	B3 V		400 s		
53766	79065		7 07 16.8	+24 10 05	0.000	−0.04	6.90	1.6	−0.5	M2 III	−12	300 s		m
53792	79067		7 07 17.9	+22 30 31	0.000	−0.01	7.70	1.6	−0.4	M0 III		420 s		
53791	79070		7 07 21.3	+22 42 13	0.000	0.00	7.10	1.9		S7.7 e	−41			R Gem, m,v
54141	134220		7 07 21.8	−9 59 07	+0.001	−0.01	6.9	0.1	1.4	A2 V		130 s		
53686	59794		7 07 22.2	+34 00 34	−0.002	−0.03	5.91	1.51	0.2	K0 III	+14	69 s		
54309	173002		7 07 22.5	−23 50 25	−0.001	+0.02	5.71	−0.10	−3.5	B1 V e		650 s		
54197	152414		7 07 23.4	−14 03 08	−0.003	−0.02	8.01	−0.07	0.0	B8.5 V		400 s		
53744	79066		7 07 24.9	+28 10 38	0.000	+0.02	6.2	−0.1	0.2	B9 V		160 s		
54792	234943		7 07 27.6	−52 12 15	−0.004	+0.06	6.5	1.4	0.2	K0 III		110 s		
54446	197600		7 07 30.8	−31 48 26	−0.002	−0.02	8.0	1.1	−0.3	K5 III		450 s		
53966	96514		7 07 35.8	+14 20 33	+0.001	+0.02	7.9	0.9	3.2	G5 IV		87 s		
54678	218473		7 07 40.0	−43 58 42	−0.001	+0.03	7.6	0.9	1.4	A2 V		69 s		
53742	59796		7 07 41.7	+33 11 44	+0.002	−0.01	8.0	0.9	3.2	G5 IV		93 s		
56241	256375		7 07 42.7	−76 02 56	0.000	−0.01	8.0	1.1	−0.1	K2 III		420 s		
55478	249747		7 07 46.7	−67 56 12	−0.003	+0.01	8.00	0.00	0.0	B8.5 V		400 s		
54219	134228		7 07 46.9	−3 22 16	0.000	0.00	7.90	1.76	−0.5	M2 III		380 s		
54079	114947		7 07 49.4	+7 28 16	0.000	−0.04	5.75	1.18	0.2	gK0	+24	96 s		
54581	197610		7 07 49.6	−35 48 55	0.000	−0.06	7.5	1.1	0.2	K0 III		260 s		
54522	173028		7 07 50.8	−29 25 58	−0.002	−0.03	7.9	0.7	−0.1	K2 III		390 s		
54336	152425		7 07 52.4	−15 42 03	0.000	−0.03	7.60	0.6	4.4	G0 V		44 s	5814	m
54579	197613		7 07 52.8	−34 50 01	−0.003	+0.13	7.9	0.6	4.4	G0 V		48 s		
54049	96525		7 07 56.7	+12 34 17	0.000	−0.02	7.8	1.1	−0.1	K2 III		360 s		
53714	41578		7 07 57.1	+40 46 57	0.000	−0.04	7.9	0.1	1.4	A2 V		200 s		
54046	96526		7 08 00.2	+15 31 43	−0.003	−0.21	7.82	0.57	4.4	G0 V	−13	48 s		m
54361	152427		7 08 03.3	−11 55 22	−0.002	+0.02	6.56	2.42		N7.7	+21			W CMa, v
53685	41579		7 08 06.2	+44 02 23	0.000	−0.01	6.9	0.9	3.2	G5 IV		56 s		
55105	234952		7 08 06.6	−57 19 53	−0.002	+0.02	6.7	0.8	2.6	F0 V		32 s		
53409	26168		7 08 09.3	+55 51 06	−0.001	−0.07	7.9	0.4	2.6	F0 V		110 s		
54626	197616		7 08 11.0	−31 25 39	+0.003	−0.03	7.0	0.6	3.4	F5 V		41 s		
54100	96530		7 08 11.9	+15 31 16	−0.004	−0.20	7.72	0.53	3.8	dF7	−11	59 s		m
53899	59806		7 08 13.1	+33 49 56	−0.001	−0.04	6.4	1.1	0.2	K0 III	−3	170 s		
54643	197619		7 08 16.3	−33 22 24	−0.003	+0.02	8.0	−0.3	1.7	A3 V		180 s		
54918	218479		7 08 19.2	−47 50 23	−0.001	+0.04	8.0	0.4	2.6	F0 V		63 s		

HD	SAO	Star Name	α 2000	δ 2000	μ(α)	μ(δ)	V	B−V	M_v	Spec	RV	d(pc)	ADS	Notes
			$7^h 08^m 21.8^s$	$+16°08'05''$	$+0.002^s$	$-0.04''$								
54130	96534		7 08 21.8	+16 08 05	+0.002	−0.04	7.0	0.9	3.2	G5 IV		58 s		
54519	173041		7 08 22.0	−20 51 39	+0.001	−0.05	6.90	1.4	−2.3	K5 II	+15	650 s	5823	m
54131	96535	45 Gem	7 08 22.0	+15 55 51	0.000	−0.10	5.44	1.03	0.2	K0 III	−17	110 s	5812	m
54217	114963		7 08 22.4	+7 52 53	0.000	−0.01	7.3	0.1	0.6	A0 V		220 s		
54966	234950		7 08 22.6	−50 05 59	+0.001	−0.02	7.9	0.6	3.4	F5 V		67 s		
54439	152438		7 08 23.2	−11 51 10	0.000	−0.02	7.71	0.04	−3.6	B2 III n		1600 s		
54605	173047	25 δ CMa	7 08 23.4	−26 23 36	−0.001	0.00	1.86	0.65	−8.0	F8 Ia	+34	940 s		
55196	234957		7 08 27.5	−58 22 21	+0.002	0.00	8.0	0.4	0.6	A0 V		160 s		
54547	152448		7 08 27.6	−18 36 12	−0.001	+0.01	7.9	1.4	−0.3	K5 III		420 s		
54246	96540		7 08 29.9	+10 55 31	−0.002	0.00	7.1	0.1	1.4	A2 V		140 s		
53634	26172		7 08 30.2	+51 22 23	0.000	0.00	7.8	1.4	−0.3	K5 III		410 s		
54493	152446		7 08 32.2	−12 53 07	0.000	+0.03	7.21	0.00	−3.6	B2 III		1100 s		
54672	173054		7 08 33.6	−28 36 46	−0.003	+0.05	7.5	0.5	4.0	F8 V		48 s		
53925	59812		7 08 36.2	+37 26 42	−0.001	−0.02	6.16	1.21	0.0	K1 III	+10	150 s		
54355	114976		7 08 36.8	+4 10 33	0.000	−0.03	7.4	1.6	−0.5	M2 III		370 s		
54183	96538		7 08 37.2	+18 18 04	0.000	−0.01	7.4	1.1	0.2	K0 III		260 s		
54184	96541		7 08 38.0	+17 21 16	0.000	0.00	7.9	0.2	2.1	A5 V		150 s		
55864	256373	γ¹ Vol	7 08 42.3	−70 29 50	+0.002	+0.10	5.7			dF4	−3	23 mn		m
54300	96548		7 08 42.6	+10 01 25	−0.001	−0.01	7.00	3.23		C7.7 e	+48			R CMi, v
54211	96543		7 08 43.7	+17 39 26	0.000	0.00	7.6	0.5	3.4	F5 V		69 s		
55865	256374	γ² Vol	7 08 45.0	−70 29 57	+0.006	+0.10	3.8			G8 III	+3	23 mn		m
54670	173062		7 08 46.0	−25 37 04	−0.002	−0.01	8.0	0.2	0.6	A0 V		210 s		
54245	96546		7 08 46.1	+16 29 41	0.000	−0.02	7.8	0.1	0.6	A0 V		260 s		
54244	96547		7 08 46.7	+16 54 29	0.000	−0.02	7.60	1.1	−0.2	K3 III	+38	350 s	5816	m
54669	173064		7 08 49.3	−24 02 39	−0.001	−0.01	6.4	0.6	−1.7	B3 V		160 s		
54893	197632		7 08 51.0	−39 39 21	−0.001	0.00	4.83	−0.18	−1.7	B3 V	+20	200 s		
54044	59818		7 08 52.3	+33 56 03	−0.001	+0.01	6.6	1.1	0.2	K0 III		190 s		
54637	152463		7 08 53.8	−18 11 05	0.000	−0.02	7.7	0.9	3.2	G5 IV		80 s		
55000	218484		7 08 58.0	−45 19 52	−0.002	+0.01	7.00	−0.07	0.0	B8.5 V		250 s		
54771	173073		7 09 00.4	−28 09 31	−0.002	+0.02	7.75	−0.16	0.0	B8.5 V		360 s		m
54351	96559		7 09 04.8	+15 25 17	−0.012	−0.30	8.02	0.63	4.4	G0 V		44 mx		
52762	6092		7 09 05.3	+71 57 34	−0.002	−0.01	7.10	1.1	0.2	K0 III		240 s		m
53924	41586		7 09 05.7	+44 15 34	+0.002	−0.03	7.5	0.4	2.6	F0 V		95 s		
54840	197635		7 09 05.9	−32 54 14	−0.002	+0.02	7.6	1.0	−0.1	K2 III		340 s		
54489	114995		7 09 07.7	+2 15 12	0.000	−0.02	7.5	0.9	0.2	G9 III	+20	280 s		
54839	197637		7 09 08.5	−31 56 02	−0.002	−0.01	7.5	0.0	0.0	B8.5 V		270 s		
54321	96558		7 09 08.7	+19 32 47	0.000	−0.03	7.2	1.4	−0.3	K5 III		320 s		
54182	59823		7 09 10.3	+30 23 11	+0.001	−0.09	7.9	0.5	4.0	F8 V		60 s		
54073	59821		7 09 13.2	+38 36 42	−0.001	−0.02	7.93	0.31	2.1	A5 V		130 s		
53764	26176		7 09 13.8	+54 01 35	−0.002	−0.03	8.0	1.1	0.2	K0 III		360 s		
54242	59825		7 09 18.9	+30 08 51	0.000	−0.03	7.4	1.1	0.2	K0 III		270 s		
54662	152470		7 09 20.2	−10 20 51	0.000	−0.02	6.21	0.03		O6	+58			
54660	134264		7 09 20.3	−8 17 03	−0.003	0.00	7.4	0.8	3.2	G5 IV		68 s		
54405	96566		7 09 22.0	+16 32 56	−0.010	−0.16	7.9	0.5	4.0	F8 V		59 s		
55527	249754		7 09 25.0	−60 22 49	0.000	+0.05	7.40	1.1	0.2	K0 III		280 s		m
54159	59824		7 09 25.7	+36 33 48	0.000	−0.02	6.80	0.1	0.6	A0 V		170 s	5820	m
54596	115006		7 09 29.7	+1 55 11	−0.001	−0.01	7.8	0.0	−1.0	B5.5 V	−4	550 s		
57336	256381		7 09 30.3	−79 25 52	−0.006	+0.04	7.4	0.1	0.6	A0 V		180 mx		
54980	197643		7 09 30.4	−35 59 28	−0.001	+0.01	7.3	0.6	2.6	F0 V		58 s		
54658	134269		7 09 32.3	−6 02 36	0.000	+0.03	7.9	0.4	2.6	F0 V		110 s		
54764	152477		7 09 33.1	−16 14 05	−0.002	−0.01	6.03	0.06	−5.1	B1 II	+6	1200 s	5837	m
53105	14086		7 09 34.2	+69 21 34	−0.006	−0.04	8.0	1.4	−0.3	K5 III		390 mx		
54862	173091		7 09 34.2	−24 18 28	−0.001	+0.02	8.0	1.5	−0.3	K5 III		450 s		
54814	152480		7 09 34.9	−19 06 54	−0.003	+0.02	8.00	−0.09	0.0	B8.5 V		400 s		
54371	79121		7 09 35.3	+25 43 42	−0.010	−0.18	7.10	0.70	5.3	dG6	+19	23 s	5827	m
54763	152478		7 09 35.5	−15 14 36	−0.002	−0.01	7.1	0.1	1.4	A2 V		140 s		
54568	115007		7 09 37.3	+6 50 47	−0.002	0.00	7.8	0.1	1.7	A3 V		160 s		
53633	14095		7 09 37.7	+60 47 31	−0.001	−0.05	6.7	1.1	0.2	K0 III	+2	200 s		
54937	173107		7 09 42.9	−27 53 01	−0.001	−0.04	7.4	1.4	−0.5	M4 III		370 s		
54912	173101		7 09 42.9	−25 13 52	−0.001	0.00	5.69	−0.16	−1.6	B3.5 V	+28	280 s		m
54403	79124		7 09 44.1	+25 44 04	−0.001	−0.04	7.8	0.4	2.6	F0 V		110 s		
54427	79126		7 09 44.3	+22 21 53	+0.001	−0.01	7.8	1.1	−0.1	K2 III		390 s		
54619	115009		7 09 50.9	+9 18 32	−0.001	−0.02	6.8	1.1	−0.1	K2 III		240 s		
54834	152484		7 09 52.1	−16 14 05	−0.003	0.00	6.7	0.2	2.1	A5 V		83 s		
54565	96577		7 09 53.9	+13 08 51	0.000	0.00	7.2	1.1	0.2	K0 III		240 s		
54511	96575		7 09 55.5	+18 50 39	0.000	−0.02	7.8	0.1	0.6	A0 V		270 s		
54859	152487		7 09 57.1	−13 36 57	+0.001	0.00	7.2	1.1	0.2	K0 III		250 s		
54632	115012		7 09 57.9	+7 43 29	0.000	−0.01	6.7	1.1	−0.1	K2 III		230 s		
55019	173116		7 10 00.5	−28 44 54	0.000	0.00	7.31	−0.17	−2.6	B4 III		960 s		
54935	—		7 10 04.2	−19 34 36			7.5	0.6	−1.0	B5.5 V	+20	200 s		
54563	79131		7 10 06.7	+21 14 49	−0.012	−0.48	6.43	0.89	5.7	G9 V	−15	22 mn		
54879	152491		7 10 07.9	−11 48 08	−0.003	+0.04	7.65	0.00	−4.4	O9.5 V	+16	1800 s		
53711	14101		7 10 08.4	+62 08 07	−0.002	−0.04	7.5	1.4	−0.3	K5 III		370 s		
54911	152493		7 10 08.7	−15 41 05	−0.001	0.00	7.32	−0.08	−4.4	B2 II	+16	2000 s		
54958	152497		7 10 09.2	−18 41 07	−0.001	+0.01	6.23	0.40		F0				

HD	SAO	Star Name	α 2000	δ 2000	μ(α)	μ(δ)	V	B-V	M_v	Spec	RV	d(pc)	ADS	Notes
54957	152499		7^{h}10^{m}11.4^s	−18°33'13"	−0.004^s	−0.01"	7.9	1.1	0.2	K0 III		230 mx		
54959	152500		7 10 12.2	−19 11 53	−0.004	+0.05	7.3	0.5	2.1	A5 V		70 s		
54810	134282	20 Mon	7 10 13.6	−4 14 14	0.000	+0.21	4.92	1.03	0.2	K0 III	+79	87 s		m
54857	134286		7 10 15.7	−9 18 46	−0.002	−0.03	7.78	0.20	1.4	A2 V		160 s		
54878	152495		7 10 16.0	−11 07 23	+0.003	+0.01	7.9	1.4	−0.3	K5 III		410 s		
55070	173122		7 10 19.3	−27 29 29	−0.001	+0.01	5.46	1.00	0.3	gG7	+15	99 s		
55598	234990		7 10 24.0	−55 35 19	+0.002	0.00	7.50	1.1	0.2	K0 III		290 s		m
55247	197658		7 10 26.4	−37 12 59	0.000	−0.05	7.7	1.5	0.2	K0 III		170 s		
53957	26186		7 10 28.5	+57 21 14	+0.001	−0.01	7.8	1.1	0.2	K0 III		330 s		
54616	79133		7 10 29.5	+24 06 29	−0.006	−0.12	7.9	0.5	3.4	F5 V		78 s		
55039	173123		7 10 30.1	−20 16 30	0.000	−0.01	7.6	0.3	1.7	A3 V		110 s		
54930	152507		7 10 30.2	−10 09 53	+0.001	−0.02	7.7	1.1	0.2	K0 III		310 s		
54854	134290		7 10 30.3	−0 09 28	−0.001	+0.01	7.8	1.1	0.2	K0 III		320 s		
54929	134293		7 10 33.2	−7 52 12	0.000	+0.04	7.34	0.00	0.4	B9.5 V		230 s		
54691	96588		7 10 33.9	+19 48 10	0.000	−0.01	7.9	0.5	3.4	F5 V		80 s		
55173	—		7 10 34.0	−30 38 43			7.4	0.3	−1.6	B3.5 V		320 s		
54805	115032		7 10 37.2	+8 47 05	−0.001	−0.01	7.8	0.0	0.4	B9.5 V		300 s		
55400	218507		7 10 37.8	−45 34 50	−0.002	+0.04	7.9	−0.2	1.4	A2 V		160 s		
54908	134292		7 10 38.9	−1 50 35	−0.002	−0.01	8.0	0.1	0.6	A0 V		290 s		
55348	218505		7 10 39.6	−41 15 53	0.000	+0.01	7.70	0.1	1.4	A2 V		180 s		m
54562	59846		7 10 43.2	+32 37 09	+0.001	+0.01	7.2	1.1	0.2	K0 III		250 s		
55246	197660		7 10 43.6	−31 23 55	−0.005	−0.04	7.8	0.8	−0.1	K2 III		160 mx		
55014	152512		7 10 45.9	−11 34 26	−0.002	+0.01	7.77	−0.07	0.0	B8.5 V		360 s		
54995	134299		7 10 46.8	−9 20 10	−0.001	−0.01	7.41	−0.11	0.0	B8.5 V		300 s		
55526	218514		7 10 47.4	−48 55 54	−0.002	+0.21	5.14	1.24	−0.3	gK4	+64	120 s		
54592	59850		7 10 48.7	+32 13 22	−0.003	−0.08	7.7	0.9	3.2	G5 IV		78 s		
54851	115038		7 10 49.8	+9 22 39	0.000	−0.02	7.5	0.0	0.4	B9.5 V		260 s		
54402	41601		7 10 51.9	+45 14 50	0.000	−0.01	7.7	1.1	0.2	K0 III		320 s		
55095	152519		7 10 52.3	−18 03 04	+0.009	−0.14	7.9	0.9	3.2	G5 IV		60 mx		
54990	134306		7 11 03.0	−3 53 54	−0.001	−0.01	6.6	1.1	0.2	K0 III		180 s		
54401	41605		7 11 07.6	+47 54 38	+0.008	−0.19	8.0	1.1	−0.1	K2 III		56 mx		
54719	59858	46 τ Gem	7 11 08.3	+30 14 43	−0.002	−0.05	4.41	1.26	−0.1	K2 III	+22	67 s	5846	m
54591	41608		7 11 09.5	+40 02 34	−0.005	−0.10	7.3	0.5	3.4	F5 V		61 s		
55036	134309		7 11 10.4	−4 42 13	−0.001	0.00	7.9	0.1	−4.8	A3 Ib		2700 s		
54615	59853		7 11 10.7	+38 11 42	0.000	−0.02	7.9	1.1	0.2	K0 III		340 s		
55090	134311		7 11 10.9	−7 42 16	−0.002	−0.02	7.7	1.1	0.2	K0 III		300 s		
54989	115050		7 11 11.5	+3 11 30	−0.001	−0.01	6.8	0.8	3.2	G5 IV		52 s		
55091	134315		7 11 12.7	−9 08 03	0.000	−0.02	7.58	0.36	2.6	F0 V		88 s		
55449	197675		7 11 15.0	−39 16 10	−0.002	+0.02	7.8	0.4	0.4	B9.5 V		160 s		
54099	26197		7 11 15.3	+59 03 52	+0.001	−0.03	7.5	1.1	0.2	K0 III	+12	300 s		
54901	96603		7 11 16.7	+15 19 56	+0.001	−0.04	7.3	0.4	0.6	F2 III	+27	210 s		
55397	197670		7 11 17.0	−34 23 24	0.000	−0.02	7.82	−0.12	0.0	B8.5 V		370 s		
56478	256378		7 11 17.3	−70 07 09	−0.001	+0.09	7.0	1.4	0.2	K0 III		140 s		
55137	152527		7 11 17.5	−12 02 32	−0.001	−0.11	7.2	0.9	3.2	G5 IV		64 s		
54717	59864		7 11 19.4	+33 06 43	−0.009	−0.01	7.2	0.5	3.4	F5 V		57 s		
56708	256379		7 11 19.9	−72 45 19	0.000	+0.01	8.0	1.1	0.2	K0 III		360 s		
55271	173152		7 11 20.5	−21 48 13	−0.004	−0.01	7.00	0.00	0.0	B8.5 V		120 mx	5863	m
55135	152528		7 11 20.9	−10 25 45	+0.001	−0.01	7.31	−0.07		B6 e	+14			
55011	115053		7 11 21.7	+1 38 19	−0.001	0.00	7.9	1.1	−0.1	K2 III		380 s		
55474	197678		7 11 22.4	−39 06 06	−0.004	+0.05	7.60	0.1	1.4	A2 V		110 mx		m
54801	79141	47 Gem	7 11 23.0	+26 51 23	−0.002	−0.04	5.60	0.12	1.9	A4 V	+39	55 s		
54871	79143		7 11 23.4	+20 15 33	+0.001	−0.02	8.0	0.9	3.2	G5 IV		93 s		
55057	134316	21 Mon	7 11 23.5	−0 18 07	−0.002	−0.01	5.45	0.29	2.6	dF0 n	+30	37 s		
55213	152532		7 11 25.1	−17 19 51	−0.001	+0.01	6.67	−0.04	0.4	B9.5 V		180 s		
55345	173161		7 11 26.5	−27 43 56	−0.004	−0.02	7.1	1.0	0.2	K0 III		220 s		
55211	152531		7 11 28.1	−13 04 13	+0.002	+0.02	7.67	−0.01	0.6	A0 V		260 s		
54122	14109		7 11 28.6	+60 13 29	−0.001	−0.02	7.3	1.1	0.2	K0 III	+5	260 s		
54777	59869		7 11 28.8	+30 28 21	−0.002	+0.01	7.9	0.4	3.0	F2 V		94 s		
54240	26199		7 11 29.8	+58 00 55	0.000	−0.13	8.0	0.9	3.2	G5 IV		85 mx		
54825	79142		7 11 29.9	+26 24 24	−0.001	+0.01	6.7	1.1	−2.1	K0 II	+41	570 s		
54319	26201		7 11 31.6	+55 36 44	+0.004	−0.04	7.3	0.4	2.6	F0 V		86 s		
55720	218519		7 11 31.7	−49 25 30	0.000	+0.79	7.53	0.71	5.3	G6 V		28 s		
60102	258468		7 11 39.1	−84 28 09	−0.005	+0.02	7.60	0.00	0.4	B9.5 V		280 s		m
54716	59866	63 Aur	7 11 39.3	+39 19 14	+0.004	0.00	4.90	1.45	−1.3	K4 II-III	−27	170 s		
55344	173168		7 11 41.5	−20 52 58	−0.002	+0.03	5.84	−0.04	0.6	A0 V		110 s		
54986	96611		7 11 46.9	+16 58 36	+0.002	−0.02	6.7	1.4	−0.3	K5 III		250 s		
54985	96610		7 11 48.9	+19 53 08	0.000	−0.02	8.00	0.1	1.4	A2 V		210 s		m
55054	96616		7 11 51.2	+10 30 43	+0.003	−0.06	8.0	0.5	3.8	F7 V	−45	67 s		
55111	115062		7 11 51.3	+5 39 17	0.000	−0.01	6.09	−0.02		A0	+46			
55185	134330	22 δ Mon	7 11 51.8	−0 29 34	0.000	0.00	4.15	−0.01	0.0	A0 IV	+15	64 s	5864	m
55964	235005		7 11 55.5	−56 09 20	−0.002	+0.04	6.6	0.9	0.2	K0 III		190 s		
54590	41612		7 11 57.0	+49 47 24	−0.001	−0.02	6.8	1.6	−0.5	M2 III		290 s		
54715	41614		7 12 01.1	+43 50 34	−0.001	−0.03	6.8	0.1	0.6	A0 V		180 s		
56239	249774		7 12 02.0	−63 11 24	0.000	0.00	6.02	−0.02		A0				m
55568	197686		7 12 04.1	−30 49 18	−0.003	+0.01	6.10	0.27		A5				

HD	SAO	Star Name	α 2000	δ 2000	μ(α)	μ(δ)	V	B-V	M_V	Spec	RV	d(pc)	ADS	Notes
			$7^h 12^m 05^s.3$	$-27°20'05''$	$-0^s.001$	$+0''.01$								
55523	173189		7 12 05.3	-27 20 05	-0.001	+0.01	6.91	-0.14	-1.6	B3.5 V		480 s		m
55184	115065		7 12 07.3	+5 28 29	-0.003	0.00	6.16	1.14		K0	+20			
56146	235009		7 12 10.3	-59 50 22	+0.001	0.00	7.9	0.2	0.0	B8.5 V		250 s		
55522	173193	26 CMa	7 12 12.1	-25 56 33	-0.001	+0.01	5.92	-0.17		B3	+22			
55696	197692		7 12 15.0	-38 10 28	-0.002	-0.02	7.5	1.0	3.2	G5 IV		53 s		
55719	218525		7 12 15.7	-40 29 56	-0.002	-0.02	5.31	0.06		A3 p	-7	23 mn		
56560	249780		7 12 17.6	-67 46 59	+0.035	-0.21	7.8	0.7	4.4	G0 V		48 s		
54896	59882		7 12 18.5	+36 07 36	-0.001	-0.03	7.9	0.1	1.4	A2 V		200 s		
56117	235010		7 12 18.5	-58 24 13	-0.002	+0.04	7.6	1.0	2.6	F0 V		37 s		
55595	173200		7 12 24.0	-27 28 27	-0.002	+0.01	6.5	0.6	1.4	A2 V		48 s		
55663	197691		7 12 24.1	-31 58 59	+0.001	-0.05	7.13	1.67	-0.5	M4 III		320 s		m
55912	218535		7 12 25.1	-49 19 37	-0.002	+0.02	7.8	1.1	-0.1	K2 III		370 s		
55718	197694		7 12 25.6	-36 32 40	-0.002	0.00	5.96	-0.14		B5	+17			m
55940	218536		7 12 26.0	-49 49 31	-0.001	0.00	7.8	1.1	0.2	K0 III		330 s		
55052	79162	48 Gem	7 12 26.3	+24 07 42	-0.001	-0.05	5.85	0.36	0.7	F5 III	+13	110 s		
55204	96623		7 12 27.5	+12 31 57	0.000	-0.02	7.9	0.0	0.4	B9.5 V		310 s		
55417	134351		7 12 27.6	-8 12 43	-0.001	+0.01	8.03	-0.04		A0			5877	m
54318	14112		7 12 28.2	+62 15 44	+0.004	-0.05	7.8	0.0	0.4	B9.5 V		72 mx		
56266	249778		7 12 29.2	-61 22 54	+0.002	-0.03	7.3	2.0	-0.3	K5 III		170 s		
55941	235008		7 12 29.7	-50 19 15	0.000	-0.03	7.9	1.2	-0.5	M4 III		490 s		
55233	96624		7 12 29.9	+10 11 52	+0.001	-0.01	7.7	1.1	0.2	K0 III		300 s		
54822	41619		7 12 30.0	+43 59 19	+0.001	-0.02	7.8	1.4	-0.3	K5 III		410 s		
55892	218537		7 12 33.6	-46 45 34	-0.014	+0.10	4.49	0.32	2.6	F0 V	-1	24 s		
53609	6105		7 12 34.0	+72 40 29	0.000	-0.04	7.5	0.1	0.6	A0 V		250 s		
56023	235012		7 12 45.3	-51 12 42	-0.001	+0.01	7.20	-0.14	-1.7	B3 V k		600 s		
55538	152558		7 12 45.4	-15 30 08	-0.001	-0.03	7.82	-0.03	-3.6	B2 III	+17	1500 s		
55201	79174		7 12 46.2	+21 21 08	-0.003	-0.04	7.70	0.9	3.2	G5 IV		79 s	5873	m
56216	235017		7 12 46.3	-57 29 28	0.000	-0.03	6.80	0.1	0.6	A0 V		170 s		m
55283	96630		7 12 48.0	+15 10 42	0.000	+0.01	7.90	0.1	0.6	A0 V	-16	280 s	5875	m
54155	14111		7 12 48.1	+66 37 00	+0.002	0.00	7.7	1.4	-0.3	K5 III		390 s		
55130	79170		7 12 49.0	+27 13 29	+0.001	-0.11	6.43	0.49	3.7	dF6	-13	34 s	5871	m
55156	79172		7 12 49.3	+25 44 55	-0.001	-0.01	7.07	0.03		A0	-7			
55489	134360		7 12 49.4	-7 09 00	+0.001	-0.03	7.90	0.4	3.0	F2 V		95 s	5888	m
56145	235016		7 12 52.6	-53 53 40	+0.006	-0.02	7.83	0.58	4.0	F8 V		54 s		
55910	218541		7 12 52.8	-43 53 05	-0.002	+0.01	7.4	-0.4	0.6	A0 V		230 s		
55412	115082		7 12 55.9	+3 44 38	-0.002	-0.03	7.9	0.9	3.2	G5 IV		88 s		
55860	218539		7 12 56.9	-40 23 28	0.000	+0.03	8.0	1.1	0.2	K0 III		360 s		
55693	173219		7 13 02.5	-24 13 32	-0.009	+0.10	7.1	1.0	3.2	G5 IV		46 s		
55129	59892		7 13 04.1	+31 57 56	-0.002	-0.01	6.7	0.1	0.6	A0 V		160 s		
55589	152570		7 13 07.1	-11 15 05	-0.001	+0.01	5.78	1.51	0.2	K0 III		54 s		
54776	26219		7 13 09.5	+54 29 37	+0.002	+0.01	7.0	0.1	1.4	A2 V		130 s		
55838	197709		7 13 12.1	-34 49 00	+0.002	-0.01	7.9	2.3	-0.3	K5 III		140 s		
54944	41625		7 13 12.6	+47 36 25	-0.003	-0.08	6.6	0.2	2.1	A5 V		64 mx		
55817	197706		7 13 13.1	-30 58 02	0.000	-0.04	7.90	0.00	0.4	B9.5 V		310 s		m
56022	218546		7 13 13.3	-45 10 59	-0.002	-0.10	4.89	-0.02		A0 p	+4	23 mn		L¹ Pup
55333	96635		7 13 14.4	+18 55 17	0.000	-0.02	7.5	0.1	0.6	A0 V		250 s		
56644	249788		7 13 15.9	-64 52 25	-0.001	0.00	7.6	0.0	0.6	B8.5 V		330 s		
55533	134365		7 13 18.2	-0 15 35	-0.001	-0.01	6.8	0.4	2.6	F0 V		68 s		
55633	134371		7 13 20.7	-8 51 46	0.000	-0.01	7.5	0.2	2.1	A5 V		120 s		
55383	96638	51 Gem	7 13 22.2	+16 09 32	+0.001	-0.04	5.00	1.66	-0.5	M4 III	-9	120 s		BQ Gem, m,q
54895	26223		7 13 23.3	+51 25 43	+0.001	+0.01	5.47	1.67	-0.5	M2 III	-51	140 s		
55411	96639		7 13 23.6	+13 45 45	-0.002	-0.02	6.6	1.1	0.2	K0 III		190 s		
55762	173230		7 13 23.8	-22 40 24	0.000	-0.01	6.1	1.7	-0.1	K2 III		85 s		
55484	115092		7 13 24.0	+9 00 16	-0.001	-0.01	7.9	1.1	0.2	K0 III		330 s		
55836	173237		7 13 27.4	-29 10 58	-0.004	-0.01	7.8	-0.2	1.7	A3 V		120 mx		
55987	197716		7 13 29.3	-38 29 28	0.000	+0.02	7.60	0.1	1.4	A2 V		170 s		m
55225	59896		7 13 30.8	+32 08 43	-0.002	-0.03	7.30	0.2	2.1	A5 V		110 s	5884	m
56096	218549		7 13 32.3	-44 38 23	+0.010	+0.33	5.10	1.56		Md	+53	20 mn		L² Pup, m,v
55857	173244		7 13 36.2	-27 21 23	-0.002	-0.01	6.11	-0.26	-2.3	B3 IV		480 s		
55003	41628		7 13 37.1	+47 16 00	+0.004	0.00	7.1	0.4	3.0	F2 V		66 s		
56350	235022		7 13 39.7	-53 40 02	-0.002	+0.02	6.8	0.5	0.6	A0 V		84 s		
55685	134377		7 13 44.5	-5 09 21	0.000	-0.02	7.5	0.8	3.2	G5 IV	+3	73 s		
55958	197719		7 13 47.1	-31 05 02	-0.002	0.00	6.5	0.4		B3	+28			
55856	173247		7 13 48.3	-22 54 23	-0.001	-0.02	6.27	-0.13		B3	+17		5912	m
55557	96654		7 13 49.7	+10 41 28	0.000	0.00	7.1	1.1	0.2	K0 III		230 s		
55356	79185		7 13 50.1	+28 29 39	-0.001	0.00	7.90	0.1	0.6	A0 V		290 s	5893	m
55509	96651		7 13 50.9	+15 51 06	-0.002	-0.05	7.8	0.5	3.4	F5 V		74 s		
55583	115101		7 13 51.4	+9 28 05	-0.002	-0.04	7.7	1.1	0.2	K0 III		300 s		
56287	218556		7 13 52.0	-49 56 10	-0.002	+0.01	7.4	2.0	-0.3	K5 III		170 s		
55078	41630		7 13 52.7	+48 29 56	+0.001	0.00	7.70	0.1	0.6	A0 V		260 s	5879	m
55508	96652		7 13 53.1	+16 34 49	+0.001	-0.02	7.4	0.4	3.0	F2 V		77 s		
52881	6102		7 13 55.0	+78 45 06	+0.004	0.00	6.9	0.2	2.1	A5 V		92 s		
55709	134385		7 13 56.0	-3 43 27	-0.001	+0.01	7.90	1.1	0.2	K0 III		340 s	5906	m
55754	134390		7 13 56.1	-9 27 21	0.000	-0.01	7.84	-0.06	0.4	B9.5 V		310 s		
56017	197724		7 13 56.1	-33 20 03	-0.001	+0.01	7.6	1.0	-0.1	K2 III		350 s		

HD	SAO	Star Name	α 2000	δ 2000	μ(α)	μ(δ)	V	B-V	M$_v$	Spec	RV	d(pc)	ADS	Notes
55777	—		7^h 13^m 56.2^s	-10° 26′ 06″	^{s}	$''$	7.77	0.27	2.6	F0 V		110 s		
55985	197722		7 13 57.2	-30 20 24	-0.002	-0.01	6.3	0.3	-1.6	B3.5 V	+19	190 s		
54070	6115		7 13 57.9	+71 49 00	+0.004	+0.02	6.35	1.12	0.2	K0 III	-68	140 s		
56318	235026		7 13 58.1	-50 28 18	0.000	+0.02	6.9	-0.1	0.4	B9.5 V		200 s		
56265	218557		7 13 58.9	-48 18 06	+0.006	-0.05	7.9	-0.1	2.6	F0 V		110 s		
55555	96655		7 13 59.3	+13 58 27	-0.002	-0.03	7.6	0.5	4.0	F8 V		51 s		
55905	—		7 13 59.9	-23 40 10			7.50	1.1	0.2	K0 III		280 s	5914	m
56114	218554		7 14 00.4	-40 11 35	-0.001	+0.04	8.0	1.4	0.2	K0 III		220 s		
56046	197727		7 14 02.6	-31 38 37	0.000	-0.05	7.4	0.1	0.4	B9.5 V		200 s		
55680	115108		7 14 03.1	+4 34 39	0.000	-0.04	7.6	0.9	3.2	G5 IV		76 s		
55354	59902		7 14 08.8	+35 32 52	+0.003	+0.02	7.7	1.1	0.2	K0 III		320 s		
56264	218560		7 14 10.2	-46 19 31	-0.002	+0.03	7.5	0.7	0.2	K0 III		280 s		
56068	197729		7 14 10.9	-32 26 30	-0.001	-0.05	7.9	1.0	0.2	K0 III		340 s		
55775	134391		7 14 10.9	-3 54 05	-0.001	+0.01	5.75	1.58	-0.3	gK5	+22	150 s	5911	m
55854	152591		7 14 11.3	-12 48 36	-0.002	0.00	7.74	-0.03	0.0	B8.5 V		330 s		
55904	—		7 14 12.0	-18 58 10			8.0	0.4	3.0	F2 V		97 s		
56477	235030		7 14 12.2	-54 48 03	+0.001	+0.03	8.0	1.1	0.2	K0 III		330 s		
55679	115113		7 14 13.6	+7 13 45	-0.002	-0.02	7.8	0.1	0.6	A0 V		260 s		
56142	197732		7 14 14.1	-36 25 37	0.000	+0.01	7.2	1.2	3.4	F5 V		21 s		
56014	173264	27 CMa	7 14 15.1	-26 21 09	-0.001	0.00	4.66	-0.19	-2.9	B3 III pe	0	320 s		m
55832	134395		7 14 15.5	-9 56 51	-0.001	0.00	5.90	1.52	-0.2	gK3	+43	120 s		
56044	173268		7 14 16.7	-28 10 12	-0.001	-0.02	7.92	-0.16	0.0	B8.5 V		380 s		
55751	115119		7 14 20.0	+3 06 41	-0.001	0.00	5.35	1.19	0.2	gK0	+37	76 s		
56737	249792		7 14 22.5	-61 04 12	-0.016	+0.01	7.15	0.43	3.1	F3 V		60 s		
55579	79191		7 14 26.5	+24 42 39	0.000	-0.02	6.70	-0.03		B9	+3			m
55879	152598		7 14 28.2	-10 19 00	-0.001	-0.01	6.00	-0.18	-4.6	B0 IV	+33	1200 s		
56013	152608		7 14 32.1	-19 59 29	-0.001	0.00	7.36	-0.16	-1.6	B3.5 V	+20	610 s		m
55730	96672		7 14 32.5	+12 06 57	-0.004	-0.02	5.62	1.01	0.2	gK0	+30	120 s		
55750	115120		7 14 33.3	+8 02 00	-0.002	-0.01	8.0	0.1	0.6	A0 V		290 s		
55793	115124		7 14 33.7	+5 16 13	-0.002	-0.03	7.7	0.0	0.6	B8.5 V		340 s		
55649	79197		7 14 34.0	+20 31 00	-0.001	-0.02	8.0	1.1	0.2	K0 III		360 s		
56067	173276		7 14 34.1	-25 18 09	-0.001	-0.01	7.40	0.00	0.0	B8.5 V		290 s		m
56162	197736		7 14 34.8	-31 29 37	-0.002	-0.01	8.00	0.00	0.4	B9.5 V		320 s		m
56161	197738		7 14 36.6	-30 39 17	-0.008	+0.05	6.93	0.77	3.2	G5 IV		56 s		
55774	115123		7 14 36.7	+8 56 42	-0.002	-0.08	7.9	1.1	0.2	K0 III		180 mx		
56456	218567		7 14 38.1	-48 16 18	0.000	0.00	4.76	-0.10		B8	+44			m
56186	197740		7 14 40.8	-32 33 46	-0.003	+0.03	7.9	0.4	4.4	G0 V		51 s		
55621	79199	52 Gem	7 14 41.9	+24 53 06	+0.004	-0.09	5.82	1.56	-0.5	M1 III	+47	100 mx	5909	m
56094	173280		7 14 41.9	-23 29 21	-0.001	-0.01	7.7	-0.1	-1.6	B3.5 V		720 s		
56455	218570		7 14 45.8	-46 50 59	-0.002	-0.01	5.72	-0.11		A0 p				
56284	197745		7 14 46.9	-37 14 43	0.000	0.00	7.22	-0.18	0.0	B8.5 V		280 s		
55199	26240		7 14 48.0	+52 32 42	+0.001	-0.02	7.70	0.4	2.6	F0 V		110 s	5896	m
56012	152609		7 14 48.2	-15 28 34	+0.001	-0.04	7.27	0.43		F8			5925	m
56139	173282	28 ω CMa	7 14 48.6	-26 46 22	-0.001	0.00	3.85	-0.17	-2.3	B3 IV e	+26	170 s		
55727	96676		7 14 50.6	+18 33 43	-0.001	0.00	7.8	1.1	-0.1	K2 III		380 s		
56378	218566		7 14 50.7	-41 10 05	0.000	+0.03	7.9	-0.1	0.0	B8.5 V		380 s		
55805	115132		7 14 51.0	+7 13 31	0.000	-0.01	7.9	1.1	-0.1	K2 III		380 s		
56160	173283		7 14 51.0	-27 02 17	-0.004	-0.03	5.7	1.7	-0.3	K5 III	+15	120 s		
55672	79203		7 14 52.3	+23 06 39	-0.002	-0.01	7.8	1.1	0.0	K1 III		360 s		
54943	14126		7 14 53.9	+62 08 11	-0.003	-0.06	7.8	0.7	4.4	G0 V		49 s		
56211	173286		7 14 55.0	-28 57 54	-0.001	-0.03	7.40	-0.17	-2.9	B3 III		1100 s		
56410	218568		7 14 57.1	-41 25 33	0.000	+0.02	5.94	-0.16		B8				
56006	134412		7 14 59.1	-7 40 47	-0.001	+0.02	7.69	-0.09	0.0	B8.5 V		350 s		
55951	134409		7 15 00.6	-2 49 25	0.000	-0.01	7.90	0.1	1.4	A2 V		200 s	5924	m
56137	173285		7 15 00.6	-21 53 22	-0.001	0.00	7.7	1.1	0.2	K0 III		310 s		
55876	115140		7 15 00.8	+4 48 22	+0.001	-0.01	7.4	0.9	3.2	G5 IV		69 s		
56159	173287		7 15 04.2	-22 46 36	0.000	+0.01	7.8	0.8	-0.5	M2 III		450 s		
56110	152616		7 15 08.7	-14 36 35	-0.001	-0.02	7.1	1.1	0.2	K0 III		240 s		
56376	197755		7 15 11.8	-36 13 34	-0.001	-0.01	7.38	-0.08	0.4	B9.5 V		250 s		
55922	115144		7 15 14.9	+5 49 06	-0.001	0.00	7.4	0.1	0.6	A0 V		230 s		
55848	96690		7 15 17.9	+15 14 18	0.000	-0.02	7.9	1.1	0.2	K0 III		340 s		
56282	173305		7 15 19.0	-28 21 54	0.000	-0.02	7.8	0.7	0.2	K0 III		330 s		
56003	134414	24 Mon	7 15 19.2	-0 09 41	-0.002	0.00	6.41	0.90	0.3	gG5	-10	170 s	5933	m
55823	96687		7 15 19.8	+18 48 53	0.000	-0.01	7.8	0.1	1.7	A3 V		160 s		
56342	197756		7 15 21.0	-30 41 11	-0.002	+0.01	5.36	-0.17		B5	+33			
56705	235040		7 15 21.0	-52 29 59	-0.007	+0.10	5.97	1.10		G5				
55198	26244		7 15 21.6	+57 17 57	+0.001	+0.01	7.8	1.6	-0.5	M2 III		470 s		
55822	96691		7 15 22.5	+19 03 43	-0.002	-0.04	7.70	0.2	2.1	A5 V		130 s	5922	m
56531	218572		7 15 23.9	-42 05 19	-0.001	-0.02	6.99	-0.06	0.6	A0 V		190 s		
56453	197762		7 15 28.7	-36 33 38	+0.002	+0.01	7.4	1.8	-0.3	K5 III		230 s		
56910	235051		7 15 29.0	-58 32 33	-0.004	0.00	7.0	0.8	2.1	A5 V		43 s		
56662	218576		7 15 29.0	-48 13 47	-0.015	+0.17	7.65	0.61		G0		16 mn		
55801	79214		7 15 29.5	+23 51 23	-0.001	-0.02	7.7	0.9	3.2	G5 IV		80 s		
55973	115150		7 15 30.0	+9 29 52	-0.004	-0.02	7.1	0.4	3.0	F2 V		67 s		
55405	26249		7 15 31.6	+51 26 57	0.000	-0.02	7.2	1.1	0.2	K0 III		260 s		

HD	SAO	Star Name	α 2000	δ 2000	μ(α)	μ(δ)	V	B-V	M$_v$	Spec	RV	d(pc)	ADS	Notes
55178	26245		7^{h}15^{m}32^s.4	+59°46'30"	+0^s.001	-0".03	7.20	0.93		K0	+8			
55847	79216		7 15 33.6	+21 58 00	-0.001	0.00	7.40	1.4	-0.3	K5 III	+62	350 s		m
--	173315		7 15 34.5	-27 25 24	-0.003	-0.02	7.8	1.4	-0.3	K5 III		400 s		
56279	173310		7 15 34.7	-20 07 44	-0.001	-0.02	8.02	-0.08	0.4	B9.5 V		330 s		
56585	218575		7 15 37.5	-42 09 17	-0.003	-0.02	7.3	1.8	0.2	K0 III		88 s		
55576	41643		7 15 37.8	+45 24 35	0.000	-0.02	6.6	1.1	0.2	K0 III		190 s		
56703	235042		7 15 37.9	-50 29 56	-0.001	+0.01	7.5	1.3	0.2	K0 III		180 s		
56031	115159		7 15 39.4	+7 58 40	+0.002	+0.01	5.82	1.60	-0.5	gM4	-9	180 s		
56961	235056		7 15 39.4	-58 48 21	-0.002	0.00	7.2	2.3	0.2	K5 III		100 s		
55648	41645		7 15 41.0	+42 05 37	+0.002	-0.05	8.0	1.4	-0.3	K5 III		450 s		
55919	96697		7 15 42.9	+19 50 51	+0.001	0.00	7.5	0.0	0.4	B9.5 V		260 s		
56207	152621		7 15 43.0	-10 35 02	0.000	-0.01	5.95	1.18	0.2	K0 III		100 s		
56341	173319		7 15 47.5	-23 44 26	+0.001	0.00	6.3	0.1	0.6	A0 V		120 s		
54684	6125		7 15 48.1	+70 30 23	+0.005	-0.07	7.8	0.7	3.0	G2 IV	-38	80 mx		
55575	41644		7 15 50.1	+47 14 24	+0.003	-0.18	5.58	0.58	4.4	G0 V	+85	19 ts		
56274	152626		7 15 50.8	-13 02 58	-0.034	+0.17	7.74	0.62	4.7	G2 V	+60	36 ts		
55703	41654		7 15 54.8	+41 55 47	-0.003	-0.18	7.7	0.6	4.4	G0 V		47 s		
55280	26248	18 Lyn	7 15 54.8	+59 38 15	-0.012	-0.26	5.20	1.07	-0.1	K2 III	+24	67 mx		
56641	218578		7 15 54.9	-42 08 24	-0.002	+0.02	7.4	0.1	0.0	B8.5 V		230 s		
56451	173325		7 15 55.5	-27 26 55	+0.015	-0.06	8.0	0.3	3.2	G5 IV		56 mx		
56310	152630		7 15 56.1	-16 14 08	-0.003	+0.01	6.86	-0.17	-3.5	B1 V	+24	1100 s		
55870	79221	53 Gem	7 15 57.1	+27 53 51	-0.001	0.00	5.71	1.63		gMa	+24	19 mn		
56079	115165		7 15 57.1	+9 42 50	0.000	-0.02	7.9	1.1	-0.1	K2 III		390 s		
56935	235058		7 15 57.1	-55 44 00	-0.008	+0.05	7.7	0.9	0.2	K0 III		250 mx		
56273	152629		7 15 57.3	-12 32 36	0.000	0.00	7.92	-0.08	0.4	B9.5 V		320 s		
52029	1144		7 15 59.8	+82 26 25	-0.014	-0.01	7.4	1.4	-0.3	K5 III		350 s		
54451	6123		7 15 59.9	+73 19 22	-0.004	+0.01	7.8	0.5	4.0	F8 V		57 s		
56909	235057		7 15 59.9	-54 57 41	+0.001	+0.03	7.5	0.6	0.2	K0 III		290 s		
55869	79222		7 16 01.9	+29 15 17	+0.001	-0.03	8.0	0.0	0.0	B8.5 V		400 s		
55077	14129		7 16 02.1	+64 46 52	+0.002	0.00	7.3	0.5	4.0	F8 V		46 s		
56367	152636		7 16 05.8	-14 30 36	-0.002	0.00	6.9	0.1	1.4	A2 V		120 s		
56472	173334		7 16 06.0	-26 49 23	+0.001	-0.01	7.9	-0.2	0.4	B9.5 V		320 s		
56554	197770		7 16 06.8	-30 26 05	-0.001	+0.02	7.15	-0.17	-1.4	B4 V		510 s		
56960	235059		7 16 09.9	-54 02 30	-0.004	+0.01	7.20	1.1	0.2	K0 III		250 s		m
57125	249801		7 16 11.6	-60 31 29	0.000	+0.05	8.0	0.4	1.4	A2 V		120 s		
56430	152642		7 16 13.1	-17 17 49	+0.002	-0.03	7.83	-0.03	0.0	B8.5 V		360 s		
56305	134437		7 16 13.4	-7 52 51	-0.001	0.00	7.9	0.9	3.2	G5 IV		88 s		
56405	152641		7 16 14.3	-15 35 08	-0.005	-0.01	5.46	0.08	1.4	A2 V	+10	63 s		
56813	218589		7 16 15.5	-46 46 28	0.000	+0.04	5.66	1.44		K5	+20			
56429	152646		7 16 19.2	-16 43 03	+0.001	-0.05	8.0	0.1	0.6	A0 V		290 s		
56621	197777		7 16 19.3	-32 02 34	-0.001	-0.06	7.0	1.3	-0.3	K5 III		290 s		
56907	218595		7 16 19.4	-50 02 53	-0.003	+0.02	7.6	-0.7	0.6	A0 V		250 s		
56733	197785		7 16 31.7	-38 19 08	-0.002	0.00	5.80	-0.13	-1.6	B5 IV		300 s		
55076	14131		7 16 32.8	+67 05 14	+0.006	+0.02	7.8	0.8	3.2	G5 IV		84 s		
56363	134444		7 16 33.7	-5 02 07	-0.002	-0.01	7.9	1.1	0.2	K0 III		330 s		
56618	173360		7 16 34.9	-27 52 52	-0.001	+0.04	4.64	1.60	-0.5	gM3	+42	110 s		
55668	26253		7 16 35.6	+52 02 06	+0.002	-0.08	7.7	0.9	3.2	G5 IV		80 s		
56579	173350		7 16 36.3	-23 49 35	0.000	-0.02	7.2	0.1	-1.0	B5.5 V		310 s		
56303	115185		7 16 36.4	+1 52 44	+0.004	-0.11	7.6	0.7	4.4	G0 V		44 s		
56577	173349		7 16 36.7	-23 18 56	-0.001	0.00	4.51	1.15	-0.4	gM0	+28	96 s	5951	m
57375	249805		7 16 37.3	-64 07 32	+0.005	+0.17	8.0	0.5	3.4	F5 V		82 s		
56578	173353		7 16 38.1	-23 18 38	-0.004	+0.06	7.0	-0.1	2.6	F0 V	+37	77 s		
56501	152651		7 16 42.3	-13 27 17	-0.001	-0.02	7.74	-0.04	0.0	B8.5 V		350 s		
56200	96712		7 16 44.5	+16 08 49	-0.001	+0.03	6.9	0.5	3.4	F5 V	+22	49 s		
56525	152654		7 16 47.0	-13 40 54	-0.001	-0.02	7.20	0.05	0.6	A0 V		190 s		
56152	79238		7 16 48.4	+24 32 09	-0.001	-0.03	7.10	0.1	0.6	A0 V		200 s	5945	m
56779	197790		7 16 49.4	-36 35 34	-0.001	0.00	5.03	-0.17	-2.5	B2 V	+8	310 s		
57623	249809	δ Vol	7 16 49.8	-67 57 27	-0.001	0.00	3.98	0.79	-5.4	F8 I-II	+23	730 s		
56637	173371		7 16 50.3	-25 09 32	+0.001	0.00	7.2	-0.1	2.1	A5 V		110 s		
56575	152656		7 16 57.1	-14 06 47	+0.001	-0.01	7.9	1.1	0.2	K0 III		330 s		
--	197788		7 16 57.1	-30 54 26	+0.001	-0.02	8.0							
56731	197789		7 16 57.1	-30 53 48	0.000	-0.02	6.32	0.20		A m				m
56495	134454		7 16 57.6	-7 31 37	-0.002	-0.01	7.59	0.34		A3 p				
57920	256390		7 16 57.7	-71 44 59	-0.006	+0.08	7.8	0.5	3.4	F5 V		75 s		
56730	173376		7 16 58.9	-29 43 48	-0.004	0.00	7.8	0.2	1.7	A3 V		130 s		
56359	115194		7 17 00.1	+9 27 48	0.000	-0.02	7.1	0.1	1.4	A2 V		140 s		
56752	197791		7 17 00.5	-33 02 46	-0.003	+0.01	7.80	-0.03	0.4	B9.5 V		300 s		
56492	134455		7 17 01.1	-6 14 18	-0.001	0.00	7.9	0.1	1.4	A2 V		190 s		
56176	79241		7 17 03.3	+26 41 22	+0.003	-0.14	6.40	0.89	3.2	G7 IV	-6	42 s		
56634	173373		7 17 04.2	-21 11 23	-0.004	-0.11	7.1	1.4	0.2	K0 III		120 mx		
55549	14136		7 17 05.5	+59 55 00	+0.001	-0.01	7.84	0.66		F5 pe				
56593	152660		7 17 06.9	-12 02 04	+0.002	-0.08	6.72	0.43		F5			5956	m
56751	173381		7 17 07.2	-28 36 02	0.000	-0.01	7.0	0.2	0.6	A0 V		140 s		
56597	152665		7 17 07.4	-13 59 24	+0.002	-0.01	7.75	-0.02	0.0	B8.5 V		360 s	5957	m
56855	197795	π Pup	7 17 08.5	-37 05 51	-0.001	0.00	2.70	1.62	-0.3	K5 III	+16	40 s		m

HD	SAO	Star Name	α 2000	δ 2000	μ(α)	μ(δ)	V	B−V	M$_v$	Spec	RV	d(pc)	ADS	Notes
			$7^h17^m09^s.5$	$+33°05'30''$	$-0^s.005$	$-0''.18$								
56124	59945		7 17 09.5	+33 05 30	−0.005	−0.18	6.9	0.7	4.4	G0 V		32 s		
57001	218606		7 17 09.7	−46 41 42	−0.001	+0.02	7.60	1.1	0.2	K0 III		300 s		m
56329	96722		7 17 11.4	+13 51 55	0.000	−0.05	7.5	0.5	4.0	F8 V		49 s		
56391	96725		7 17 12.2	+11 07 38	−0.002	+0.01	7.8	0.1	0.6	A0 V		260 s		
56151	59947		7 17 13.2	+31 57 41	−0.002	−0.04	7.8	1.1	0.2	K0 III		330 s		
56224	79243		7 17 15.5	+26 21 50	−0.003	0.00	7.25	1.17	0.0	K1 III		240 s		
56777	173392		7 17 16.5	−29 02 39	−0.001	−0.01	7.8	0.5	−0.1	K2 III		380 s		
56957	218604		7 17 17.7	−40 39 23	+0.002	+0.13	7.3	1.4	4.4	G0 V		12 s		
56446	115204		7 17 17.8	+6 40 50	0.000	−0.02	6.65	−0.11		B9				
55894	41661		7 17 17.9	+48 27 53	+0.001	−0.01	7.9	0.1	1.4	A2 V		200 s		
56750	173388		7 17 18.6	−25 46 25	0.000	0.00	8.0	1.3	−0.1	K2 III		360 s		
56123	59948		7 17 19.8	+35 59 50	−0.001	−0.04	7.7	1.1	0.2	K0 III		310 s		
55700	26257		7 17 20.2	+58 02 58	−0.001	−0.04	7.4	0.4	2.6	F0 V		91 s		
56518	115206		7 17 21.1	+0 38 52	−0.001	−0.04	7.9	0.9	3.2	G5 IV		87 s		
56809	173396		7 17 22.7	−29 07 54	+0.004	−0.04	7.0	0.2	0.4	B9.5 V		160 s		
56978	197806		7 17 25.7	−39 01 29	−0.002	+0.01	7.38	−0.11	0.0	B8.5 V		300 s		
56977	197807		7 17 26.4	−38 54 49	−0.003	+0.05	6.8	1.1	−0.1	K2 III		240 s		
55969	41664		7 17 26.8	+46 19 27	−0.004	0.00	8.0	1.1	0.2	K0 III		330 mx		
56932	197804		7 17 27.2	−36 59 11	−0.001	−0.01	7.95	−0.06	0.0	B8.5 V		380 s		
57358	235076		7 17 28.1	−57 53 38	+0.006	−0.04	7.7	0.9	0.2	K0 III		310 s		
57095	218611		7 17 29.5	−46 58 45	−0.002	+0.59	6.68	0.99	6.3	K2 V		13 ts		m
56222	59954		7 17 30.5	+31 41 55	0.000	0.00	6.7	0.0	0.0	B8.5 V	+23	220 s		
56613	134473		7 17 31.6	−5 49 22	−0.001	−0.02	7.0	0.0	0.4	B9.5 V		210 s		
56614	134474		7 17 31.7	−6 40 48	−0.001	+0.01	6.29	1.62		K2				
56747	152672		7 17 32.9	−18 47 43	0.000	−0.04	7.84	1.51	−0.1	K2 III		240 s		
55866	26260		7 17 33.7	+52 07 51	0.000	−0.03	5.92	1.26	0.2	K0 III	−7	97 s		
56610	134475		7 17 35.2	−3 44 52	−0.001	0.00	7.8	0.0	0.4	B9.5 V		300 s		
56515	115211		7 17 38.4	+9 17 32	−0.002	−0.11	7.60	0.6	4.4	G0 V		44 s	5958	m
57446	235079		7 17 39.1	−59 15 43	−0.002	+0.03	7.4	0.4	1.7	A3 V		87 s		
56122	41670		7 17 40.4	+40 56 42	−0.002	−0.03	7.4	1.1	0.2	K0 III		280 s		
56171	59955		7 17 42.6	+38 52 36	−0.001	−0.02	6.5	1.1	0.2	K0 III		180 s		
56875	173409		7 17 46.8	−25 59 09	−0.003	+0.02	8.00	0.4	2.6	F0 V		120 s	5966	m
56876	173411		7 17 47.9	−26 47 51	−0.001	0.00	6.46	−0.15	−1.1	B5 V n		320 s		
56418	79253		7 17 54.4	+26 20 00	−0.002	0.00	7.6	0.9	0.3	G7 III	−4	290 s		
56803	152681		7 17 57.5	−13 53 16	−0.006	+0.06	7.53	0.43	3.0	F2 V		76 s		
56246	59958		7 18 00.6	+38 21 49	+0.001	−0.04	8.0	1.4	−0.3	K5 III		450 s		
56149	41676		7 18 01.2	+45 07 43	−0.001	−0.06	7.7	0.4	3.0	F2 V		86 s		
56221	41679	64 Aur	7 18 02.1	+40 53 00	−0.001	+0.01	5.7	0.1	1.7	A3 V	−15	64 s		
56998	197816		7 18 02.5	−30 56 14	−0.002	+0.04	7.9	0.2	0.6	A0 V		200 s		
56719	134481		7 18 03.1	−3 06 53	+0.001	−0.01	8.0	0.1	0.6	A0 V		290 s		
56386	59964		7 18 03.9	+30 57 22	−0.002	−0.02	6.24	−0.03	0.4	B9.5 V	+34	150 s		
57197	218618		7 18 04.1	−43 59 11	−0.002	0.00	5.85	−0.12		B9				
56537	96746	54 λ Gem	7 18 05.5	+16 32 25	−0.003	−0.04	3.58	0.11	1.7	A3 V	−9	25 ts	5961	m
57357	235081		7 18 10.1	−52 01 59	−0.001	−0.01	7.9	0.4	0.6	A0 V		150 s		
56718	134482		7 18 11.2	−0 10 43	0.000	−0.01	7.9	0.1	0.6	A0 V		290 s		
56955	173421		7 18 12.0	−22 20 26	−0.003	−0.05	6.7	1.3	0.2	K0 III		130 s		
57334	235080		7 18 12.3	−51 03 04	+0.003	−0.18	7.6	0.3	4.4	G0 V		44 s		
57150	197824		7 18 18.4	−36 44 03	−0.001	0.00	4.66	−0.10	−1.7	B3 V e	+19	180 s		m
56119	26267		7 18 19.1	+50 30 17	0.000	+0.01	7.9	0.1	0.6	A0 V		280 s		
57555	235087		7 18 26.9	−57 21 07	−0.001	0.00	7.90	0.7	4.4	G0 V		50 s		m
57241	218623		7 18 28.4	−41 57 30	−0.001	0.00	7.8	1.7	0.2	K0 III		130 s		
56764	115232		7 18 30.7	+4 03 02	−0.003	0.00	8.0	0.1	0.6	A0 V		290 s		
56169	41681		7 18 31.9	+49 27 54	0.000	+0.01	5.05	0.08	0.4	A3 III−IV	−12	83 s		
57122	197828		7 18 31.9	−32 20 48	−0.002	+0.01	6.98	−0.09	0.4	B9.5 V		210 s		
57120	197827		7 18 33.1	−30 47 56	−0.003	0.00	7.70	0.00	−1.6	B3.5 V		670 s		
57240	197836		7 18 33.6	−39 12 37	0.000	+0.01	5.25	0.01		A2	+32	21 mn		
56821	115237		7 18 36.0	+0 16 43	−0.001	−0.01	7.7	0.2	2.1	A5 V		130 s		
55944	26264		7 18 36.2	+58 55 07	−0.001	−0.01	7.40	0.4	2.6	F0 V		91 s		m
56971	152698		7 18 37.0	−16 38 44	−0.002	−0.01	7.9	1.1	−0.1	K2 III		380 s		
56245	41684		7 18 37.2	+47 08 11	+0.002	−0.03	7.8	0.8	3.2	G5 IV		84 s		
56716	96756		7 18 38.0	+11 04 13	−0.001	−0.05	7.5	1.1	0.2	K0 III		280 s		
57219	197837		7 18 38.2	−36 44 34	−0.001	0.00	5.11	−0.16	−1.7	B3 V	+23	230 s		m
56894	134493		7 18 38.5	−7 34 51	−0.001	+0.02	7.7	1.4	−0.3	K5 III		380 s		
57808	249817		7 18 38.9	−63 23 12	−0.002	+0.01	7.6	0.0	0.4	B9.5 V		280 s		
56244	41685		7 18 39.7	+48 07 24	+0.001	+0.01	8.0	1.1	0.2	K0 III		360 s		
56714	96755		7 18 40.3	+14 21 48	−0.001	0.00	7.7	0.0	0.4	B9.5 V	+31	290 s		
57060	173444	29 CMa	7 18 40.3	−24 33 32	−0.001	0.00	4.98	−0.15		O7.8	−11			UW CMa, m,v
56922	134495		7 18 40.7	−9 07 30	0.000	−0.03	7.5	1.1	−0.1	K2 III		330 s		
57061	173446	30 τ CMa	7 18 42.4	−24 57 15	−0.001	+0.01	4.39	−0.16	−6.0	O9 III	+40	1100 s	5977	m
57554	235088		7 18 45.8	−55 16 50	0.000	+0.04	7.8	0.3	0.4	B9.5 V		190 s		
56793	115238		7 18 46.1	+8 29 12	−0.001	−0.03	7.4	1.1	0.2	K0 III		260 s		
56715	96758		7 18 47.0	+13 31 43	0.000	−0.01	7.40	1.1	−0.1	K2 III		320 s		m
57333	218629		7 18 49.2	−42 26 41	−0.002	−0.02	7.8	1.7	0.2	K0 III		190 s		
57146	173453		7 18 51.1	−26 35 09	−0.002	+0.01	5.28	0.96	−2.0	G0 II	+32	210 s		
56482	59972		7 18 52.5	+38 40 26	0.000	+0.01	6.90	0.96	0.2	K0 III		220 s		

HD	SAO	Star Name	α 2000	δ 2000	μ(α)	μ(δ)	V	B-V	M_V	Spec	RV	d(pc)	ADS	Notes
56891	115247		7^h 18^m 55^s.5	+0° 04′ 45″	+0^s.001	−0″.07	6.7	1.1	0.2	K0 III		160 mx		
57415	218631		7 18 56.3	−45 17 49	−0.022	+0.12	7.15	0.51		F5				
56792	96762		7 18 57.6	+11 50 32	−0.001	0.00	8.0	0.4	2.6	F0 V		120 s		
57553	235090		7 18 58.5	−52 32 25	−0.004	−0.06	8.00	0.5	3.4	F5 V		83 s		m
56563	59975		7 19 00.6	+35 09 29	−0.001	+0.01	7.1	1.1	0.2	K0 III		240 s		
57118	152710		7 19 01.9	−19 16 49	0.000	0.00	6.09	0.62		F0				
57087	152709		7 19 03.7	−17 12 51	+0.002	−0.09	6.8	1.1	0.2	K0 III		160 mx		
56844	96764		7 19 05.8	+10 18 52	+0.001	−0.02	7.4	0.1	1.4	A2 V		160 s		
57301	197840		7 19 06.5	−35 11 05	−0.001	−0.01	7.82	−0.02	0.4	B9.5 V		310 s		
57355	197843		7 19 08.4	−38 02 17	−0.002	−0.01	6.9	1.3	0.2	K0 III		140 s		
55745	14145		7 19 08.7	+66 44 23	−0.010	−0.15	7.60	0.5	4.0	F8 V		53 s		m
57193	173462		7 19 09.8	−25 33 56	0.000	+0.01	7.47	−0.16	−1.0	B5.5 V		490 s		
57238	173468		7 19 11.1	−29 00 48	−0.001	−0.02	7.5	1.2	−0.1	K2 III		310 s		
57353	197846		7 19 12.0	−37 19 13	−0.003	−0.01	8.0	0.3	0.6	A0 V		180 s		
56099	26269		7 19 12.4	+59 07 32	+0.010	−0.03	7.7	0.5	4.0	F8 V	−54	55 s		
57192	173464		7 19 12.7	−24 57 19	−0.001	+0.02	7.30	0.00	−1.6	B3.5 V	+26	560 s	5985	m
57299	197842		7 19 13.6	−33 43 38	−0.002	−0.01	6.4	1.6	0.2	K0 III		80 s		
57582	218639		7 19 15.8	−49 58 40	−0.009	+0.13	7.4	0.6	0.2	K0 III		120 mx		
55893	14146		7 19 17.7	+64 06 56	+0.009	−0.05	8.0	1.1	0.2	K0 III		190 mx		
56647	59977		7 19 17.9	+34 32 47	−0.003	−0.03	7.2	1.4	−0.3	K5 III		320 s		
57767	235101		7 19 18.1	−58 03 09	+0.002	+0.05	6.9	0.1	0.6	A0 V		150 s		
57139	152723		7 19 18.2	−17 31 29	−0.002	+0.01	6.6	0.0	0.0	B8.5 V		200 s		
57487	218637		7 19 18.9	−46 14 07	+0.001	−0.03	7.2	0.8	0.2	K0 III		250 s		
56990	134507		7 19 18.9	−0 20 03	−0.002	−0.01	7.8	0.6	4.4	G0 V		48 s		
57190	173466		7 19 19.8	−22 02 59	−0.003	−0.03	7.50	0.1	1.4	A2 V		160 s	5986	m
56354	26272		7 19 20.7	+51 29 37	−0.003	−0.04	7.8	0.5	4.0	F8 V		57 s		
57719	235100		7 19 21.8	−55 58 13	−0.001	+0.05	7.6	1.0	−0.1	K2 III		340 s		
57078	134510		7 19 22.1	−8 47 01	+0.001	−0.02	7.15	1.71	−0.5	M2 III		290 s		
56989	115263		7 19 22.3	+2 44 26	−0.001	−0.02	5.89	1.07	0.2	G9 III	+24	120 s		
56628	59979		7 19 22.4	+36 40 32	0.000	−0.02	7.8	0.8	3.2	G5 IV		84 s		
56790	79285		7 19 23.8	+22 01 29	−0.001	0.00	8.0	0.9	3.2	G5 IV	+6	93 s		
57411	197850		7 19 27.1	−37 30 26	−0.001	−0.03	7.81	−0.04	0.0	B8.5 V		350 s		
56916	96772		7 19 27.2	+10 26 37	0.000	−0.01	7.4	0.1	1.4	A2 V		160 s		
57167	152724		7 19 28.1	−16 23 42	+0.011	−0.13	5.70	0.35		F0	−39	14 mn		R CMa, m,v
56761	79286		7 19 30.7	+26 49 22	0.000	−0.01	6.8	0.9	0.3	G8 III		200 s		
57919	249822		7 19 32.5	−61 08 41	+0.001	+0.02	6.9	1.3	0.2	K0 III		150 s		
57807	235104		7 19 32.8	−57 26 56	−0.002	+0.02	6.8	1.2	0.6	A0 V		35 s		
56965	96775		7 19 34.7	+10 23 57	0.000	−0.01	7.40	1.6	−0.4	M0 III	+27	350 s		
−−	6136		7 19 44.9	+77 46 22	0.000	−0.05	7.9	1.3						
57006	115269		7 19 47.7	+7 08 36	+0.005	−0.05	5.91	0.53	4.0	dF8	+22	23 s		
56606	41694		7 19 50.2	+45 09 09	0.000	−0.03	8.0	1.1	0.2	K0 III		360 s		
56385	26279		7 19 51.1	+54 55 18	−0.001	−0.03	7.6	0.1	0.6	A0 V		250 s	5968	m
57764	235105		7 19 51.6	−52 33 52	0.000	+0.03	7.9	0.6	3.2	G5 IV		88 s		
56243	26274		7 19 52.9	+59 15 10	−0.001	−0.03	6.9	1.1	−0.1	K2 III	+8	250 s		
56586	41695		7 19 54.8	+46 58 19	−0.003	−0.16	7.6	0.9	3.2	G5 IV		72 mx		
56789	59983		7 19 55.7	+33 36 51	−0.001	−0.02	7.1	0.1	1.4	A2 V		140 s		
57184	134522		7 19 57.0	−4 35 01	−0.002	−0.02	7.8	0.1	1.4	A2 V		180 s		
56987	96781		7 20 02.6	+19 31 15	+0.003	−0.03	7.3	1.1	0.2	K0 III		260 s		
57346	173487		7 20 04.5	−20 42 13	−0.003	+0.01	6.8	0.1	0.6	A0 V		140 s		
57213	134526		7 20 04.5	−4 59 30	−0.005	−0.02	7.10	0.2	2.1	A5 V		92 mx	5992	m
57049	96783		7 20 06.8	+15 08 34	+0.001	−0.02	6.50	−0.02	1.4	A2 V	+13	110 s		
57212	134527		7 20 06.8	−3 54 51	0.000	+0.01	6.8	1.1	0.2	K0 III		200 s		
56986	79294	55 δ Gem	7 20 07.3	+21 58 56	−0.002	−0.01	3.53	0.34	1.9	F2 IV	+3	18 ts	5983	Wasat, m
57048	96786		7 20 09.5	+16 08 24	0.000	−0.03	7.7	1.1	0.2	K0 III		310 s		
57252	134529		7 20 13.1	−6 08 17	−0.001	+0.01	7.7	1.1	0.2	K0 III		300 s		
57321	134531		7 20 18.7	−8 32 58	−0.001	+0.01	7.1	0.4	2.6	F0 V		79 s		
57852	235111		7 20 21.7	−52 18 35	−0.005	+0.12	5.52	0.49		F2		12 mn		m
57437	173498		7 20 22.8	−22 05 06	0.000	−0.01	7.1	0.9	3.2	G5 IV		52 s		
57969	235114		7 20 22.8	−56 17 42	−0.004	+0.05	6.7	0.7	0.6	A0 V		60 s		
56964	79295		7 20 22.9	+28 07 21	+0.001	+0.01	7.6	0.1	1.4	A2 V		180 s		
57319	134538		7 20 30.2	−6 36 13	+0.002	−0.06	7.9	0.9	3.2	G5 IV		88 s		
55943	14151		7 20 30.7	+68 32 25	+0.001	−0.02	7.90	0.2	2.1	A5 V		150 s	5962	m
57275	115294		7 20 31.0	+0 24 12	−0.001	−0.01	6.90	−0.01		B9			5996	m
57108	96791		7 20 31.2	+17 51 51	0.000	−0.02	7.9	1.1	0.2	K0 III		340 s		
57527	173505		7 20 32.4	−26 42 02	−0.003	+0.03	6.6	0.7	2.6	F0 V		34 s		
57917	235116		7 20 38.7	−52 05 10	−0.002	−0.01	5.4	0.0	0.4	B9.5 V	+21	100 s		
57918	235117		7 20 38.7	−52 11 55	−0.004	0.00	7.20	0.1	0.6	A0 V		190 mx		m
58092	235128		7 20 39.2	−58 40 24	0.000	0.00	7.3	0.5	0.6	A0 V		100 s		
57386	134543		7 20 42.8	−8 25 16	−0.001	−0.01	8.02	−0.01	−3.0	B1.5 V nnpe		1400 s		
57291	115301		7 20 43.6	+3 34 54	−0.001	−0.01	6.8	0.0	−1.0	B5.5 V	+7	360 s		
57343	134541		7 20 46.3	−0 50 50	0.000	−0.01	7.7	0.2	2.1	A5 V		130 s		
57270	115300		7 20 49.3	+9 39 13	−0.002	−0.06	7.40	0.5	3.4	F5 V		63 s	6000	m
57069	79300		7 20 49.6	+29 44 03	+0.001	0.00	7.1	0.1	1.4	A2 V	+5	140 s		
57949	235121		7 20 49.7	−51 24 15	−0.005	+0.05	7.0	0.1	1.4	A2 V		110 mx		
57385	134545		7 20 49.8	−6 08 09	0.000	−0.01	7.1	0.1	0.6	A0 V		190 s		

HD	SAO	Star Name	α 2000	δ 2000	μ(α)	μ(δ)	V	B-V	M_v	Spec	RV	d(pc)	ADS	Notes
57618	173520		7h 20m 49.9s	-29° 51' 51"	-0.001	-0.03	7.88	-0.15	-1.1	B5 V n		590 s		
57637	197876		7 20 50.2	-32 28 37	-0.002	-0.01	7.3	1.0	0.2	K0 III		260 s		
57573	173517		7 20 53.2	-22 51 06	0.000	-0.04	6.61	-0.15	-1.7	B3 V	+10	460 s		
57544	173515		7 20 54.0	-20 38 58	-0.002	-0.03	7.50	0.1	1.7	A3 V		140 s		m
58116	235131		7 20 54.9	-56 46 48	0.000	+0.01	8.00	0.00	0.4	B9.5 V		330 s		m
57593	173522		7 20 54.9	-26 57 49	0.000	+0.02	6.01	-0.17	-1.7	B3 V		350 s	6015	m
56711	26290		7 20 55.3	+52 16 04	+0.001	-0.02	7.93	1.27	-0.1	K2 III		330 s		
57313	115307		7 20 57.6	+6 14 16	-0.001	+0.01	7.6	0.9	3.2	G5 IV		76 s		
57478	152749		7 20 58.2	-14 21 37	-0.002	+0.01	5.45	0.98	0.2	gK0	+13	110 s		
57290	96803		7 21 00.6	+10 11 43	0.000	-0.03	7.80	0.1	0.6	A0 V		270 s	6002	m
57269	96802		7 21 01.6	+13 04 44	0.000	-0.04	7.8	0.1	1.7	A3 V		170 s		
57968	235126		7 21 01.8	-50 51 16	0.000	0.00	7.3	0.1	0.0	B8.5 V		220 s		
56941	41705		7 21 03.0	+42 39 20	-0.002	-0.05	6.5	1.1	0.2	K0 III	+46	180 s		
57615	173529		7 21 04.3	-25 53 29	-0.001	+0.02	5.87	1.64	-0.5	gM4	+23	180 s		
57829	218661		7 21 05.0	-41 11 51	-0.002	0.00	7.8	1.1	0.2	K0 III		330 s		
57761	197881		7 21 10.9	-34 15 22	0.000	-0.01	7.9	1.6	-0.1	K2 III		230 s		
57785	197883		7 21 15.0	-35 29 49	-0.001	-0.04	8.0	0.1	0.6	A0 V		240 s		
57517	134563		7 21 16.8	-8 52 42	+0.002	-0.16	6.55	0.54		F5				
56913	41708		7 21 17.5	+45 13 42	-0.004	+0.01	5.77	0.32	2.6	F0 V	+25	43 s		
57896	218663		7 21 20.6	-41 59 27	-0.001	0.00	7.3	0.5	0.6	A0 V		79 s		
57340	96809		7 21 21.0	+13 09 53	0.000	-0.01	7.5	0.1	1.7	A3 V		150 s		
58017	218666		7 21 22.0	-48 31 37	-0.002	+0.02	7.10	0.1	0.6	A0 V		200 s		m
57498	134562		7 21 22.5	-3 17 18	0.000	+0.01	7.9	1.4	-0.3	K5 III		410 s		
57383	96811		7 21 27.4	+11 00 42	-0.002	-0.01	7.4	1.1	0.2	K0 III		260 s		
57758	173544		7 21 28.3	-28 11 30	-0.003	-0.02	7.4	0.7	0.2	K0 III		270 s		
57104	59999		7 21 30.8	+37 39 44	-0.001	-0.13	6.8	1.1	0.2	K0 III		95 mx		
57267	79316		7 21 32.6	+26 09 33	-0.002	-0.01	7.90	0.7	4.7	G2 V	+15	44 s		
57539	134570		7 21 34.6	-5 53 50	0.000	0.00	6.47	-0.10		B8 n	+17			
58089	218670		7 21 35.5	-49 24 17	-0.002	+0.06	7.8	1.1	0.2	K0 III		280 mx		
57847	197890		7 21 35.6	-34 00 36	-0.003	+0.03	7.8	0.5	0.2	K0 III		330 s		
59973	256406		7 21 35.7	-79 09 52	-0.022	+0.11	7.70	1.1	0.2	K0 III		150 mx		m
57228	--		7 21 36.4	+32 03 31			7.9	1.6	-0.5	M2 III		470 s		
57892	197892		7 21 36.6	-36 29 39	0.000	+0.01	7.9	0.3	0.0	B8.5 V		220 s		
58448	249832		7 21 41.1	-61 57 05	-0.004	+0.02	7.10	0.00	0.0	B8.5 V		230 mx		m
57757	173548		7 21 41.2	-24 13 57	0.000	-0.01	7.3	1.2	-0.5	M2 III		370 s		
57891	197893		7 21 43.6	-35 26 16	-0.001	-0.02	8.0	1.3	0.2	K0 III		250 s		
55667	6156		7 21 50.8	+75 05 21	-0.008	-0.06	7.1	0.1	1.4	A2 V		96 mx		
58292	235143		7 21 51.5	-55 46 17	0.000	-0.01	7.4	0.8	0.6	A0 V		70 s		
57068	41713		7 21 55.5	+46 13 34	+0.002	-0.04	8.00	1.1	-0.1	K2 III		420 s		m
58015	197899		7 21 56.6	-38 56 12	-0.002	+0.02	8.0	0.1	4.0	F8 V		62 s		
57423	79328	56 Gem	7 21 56.8	+20 26 37	-0.005	-0.03	5.10	1.52	-0.4	gM0	+4	130 s	6016	m
57180	41715		7 22 00.1	+41 18 37	-0.004	-0.06	7.7	0.8	3.2	G5 IV		78 s		
57682	134580		7 22 01.9	-8 58 45	0.000	+0.02	6.43	-0.19	-4.8	O9 V	+23	1600 s		
57264	60010	65 Aur	7 22 02.6	+36 45 38	-0.007	-0.03	5.13	1.08	0.2	K0 III	+23	85 s	6009	m
57608	115335		7 22 03.4	+0 10 38	0.000	0.00	5.99	-0.07		B8				
57945	197900		7 22 08.3	-32 35 14	0.000	0.00	7.0	0.7	0.6	A0 V		71 s		
57962	197902		7 22 10.4	-32 01 54	-0.004	+0.04	7.9	1.4	-0.1	K2 III		300 s		
57631	115338		7 22 11.8	+2 08 35	0.000	0.00	7.8	0.1	0.6	A0 V		260 s		
57263	60012		7 22 13.3	+38 59 45	-0.001	-0.04	6.4	1.1	0.2	K0 III	+3	180 s		
57821	152776		7 22 13.5	-19 00 59	0.000	-0.01	4.96	-0.04		B8	+27			
57066	26306		7 22 15.1	+50 08 55	-0.002	-0.04	7.30	0.4	2.6	F0 V	+7	87 s	6004	m
57838	197906		7 22 16.3	-35 55 04	-0.002	+0.11	7.50							
56820	26298	47 Cam	7 22 17.2	+59 54 07	+0.001	+0.01	6.30	0.2	2.1	A5 V	+7	69 s	5995	m
57708	134585		7 22 18.4	-2 58 44	-0.001	0.00	6.23	0.68		F5				
57820	152778		7 22 19.6	-16 18 42	0.000	0.00	7.0	1.1	-0.1	K2 III		260 s		
58266	218676		7 22 20.0	-49 20 38	-0.001	-0.01	7.9	0.7	2.6	F0 V		59 s		
57890	173565		7 22 22.9	-20 30 23	-0.001	+0.02	7.40	1.6		M6 III	+36			
57911	173569		7 22 23.4	-24 54 14	-0.001	-0.01	7.4	1.6	-0.1	K2 III		180 s		
57749	134588		7 22 25.3	-5 58 58	-0.001	-0.01	5.82	0.35	0.6	gF2	+11	110 s		
57707	115351		7 22 28.2	+0 42 02	+0.002	-0.06	6.6	0.9	3.2	G5 IV		49 s		
57733	134587		7 22 30.9	-0 15 32	-0.001	+0.01	6.7	0.0	0.4	B9.5 V		180 s		
56788	14171		7 22 35.7	+62 51 56	-0.003	-0.03	7.9	1.1	-0.1	K2 III		390 s		
58063	197914		7 22 35.7	-32 02 46	0.000	0.00	6.6	0.2	-1.0	B5.5 V		210 s		
57990	173577		7 22 35.9	-27 08 15	+0.003	+0.01	7.9	0.8	0.2	K0 III		340 s		
57129	26309		7 22 38.0	+52 50 18	-0.002	-0.06	7.72	1.26		K2				
58037	173579		7 22 38.3	-28 35 07	-0.003	0.00	8.0	0.6	-0.3	K5 III		450 s		
58238	218678		7 22 38.7	-44 37 06	-0.001	-0.01	7.71	-0.10	0.0	B8.5 V		300 s		
58062	197915		7 22 38.9	-31 14 42	-0.001	+0.03	8.0	0.9	0.2	K0 III		360 s		
58011	173581		7 22 43.2	-26 00 39	-0.001	-0.01	7.11	-0.07		B1 e				
55403	6155		7 22 44.4	+77 53 47	-0.011	-0.01	7.6	0.4	3.0	F2 V		83 s		
57337	41728		7 22 44.4	+42 39 54	+0.002	+0.02	7.7	0.2	2.1	A5 V		130 s		
58518	235154		7 22 46.2	-56 40 54	-0.001	-0.06	8.0	1.1	3.2	G5 IV		58 s		
57863	134593		7 22 48.1	-9 37 56	-0.003	-0.09	7.8	0.5	4.0	F8 V		57 s		
57705	115355		7 22 48.5	+9 50 44	+0.001	-0.02	7.9	0.1	1.4	A2 V		190 s		
58424	235152		7 22 48.7	-52 31 11	-0.003	-0.03	7.9	0.6	3.2	G5 IV		88 s		

HD	SAO	Star Name	α 2000	δ 2000	μ(α)	μ(δ)	V	B−V	M_v	Spec	RV	d(pc)	ADS	Notes
			$7^h22^m48.8^s$	$-16°43'35''$	-0.001^s	$+0.01''$								
57939	152790		7 22 48.8	-16 43 35	-0.001	+0.01	7.7	1.1	0.2	K0 III		310 s		
57989	173582		7 22 49.2	-21 29 09	-0.002	0.00	7.9	1.0	0.2	K0 III		340 s		
57103	26312	19 Lyn	7 22 52.0	+55 16 53	0.000	-0.03	5.6	0.0	0.0	B8.5 V	+5	130 s	6012	m
57746	115361		7 22 53.7	+6 22 25	-0.001	+0.03	7.9	0.5	4.0	F8 V		59 s		
58540	235155		7 22 57.0	-55 34 36	-0.009	+0.12	6.90	1.1	0.2	K0 III		120 mx		m
58134	173596		7 22 57.6	-29 45 04	-0.004	-0.01	7.7	0.6	0.2	K0 III		210 mx		
58061	173591		7 22 58.0	-25 46 01	-0.003	+0.02	7.50	1.6		M5 Ib p			6033	VY CMa, m,v
57861	134595		7 22 58.0	-3 53 32	-0.003	-0.04	7.90	0.5	3.4	F5 V		79 s		m
58155	197925		7 23 00.6	-31 55 26	-0.001	0.00	5.42	-0.15		B5 n	+24			
57728	96838		7 23 01.3	+15 21 31	-0.001	-0.01	6.7	0.8	3.2	G5 IV	0	50 s		
57703	96837		7 23 04.5	+18 16 25	-0.003	-0.04	6.8	0.4	3.0	F2 V		59 s		
57704	96839		7 23 05.5	+17 24 35	-0.001	-0.01	7.4	0.0	0.4	B9.5 V		250 s		
58774	249839		7 23 06.2	-62 16 18	-0.002	+0.03	7.08	0.98		K0				
59239	256402		7 23 07.1	-70 38 13	-0.009	+0.01	7.3	1.1	0.2	K0 III		260 s		
--	197932		7 23 11.0	-39 05 28	-0.003	0.00	7.9	0.4	2.6	F0 V		120 s		
58034	152800		7 23 12.5	-17 46 19	0.000	-0.06	7.7	1.1	0.2	K0 III		300 mx		
57984	134601		7 23 15.5	-10 04 08	0.000	-0.09	8.0	0.5	4.0	F8 V		62 s		
57262	26317		7 23 16.1	+52 16 45	+0.006	+0.02	7.9	0.5	3.4	F5 V		78 s		
58260	197934		7 23 19.6	-36 20 26	-0.001	-0.01	6.73	-0.12	-2.9	B3 III k		790 s		
58591	235159		7 23 22.0	-53 30 30	-0.002	+0.01	6.9	1.2	0.2	K0 III		180 s		
57902	115374		7 23 22.8	+0 33 04	0.000	0.00	7.6	0.1	1.4	A2 V		180 s		
58056	152802		7 23 24.1	-13 21 01	0.000	-0.01	8.0	0.1	0.6	A0 V		290 s		
58131	173609		7 23 24.4	-20 13 48	-0.001	+0.02	7.37	0.34		B2 n	+32			
57744	79354	58 Gem	7 23 28.1	+22 56 43	-0.001	-0.04	6.00	-0.01	1.2	A1 V	+18	91 s		
57727	79352	57 Gem	7 23 28.4	+25 03 02	-0.005	-0.02	5.03	0.90	0.3	G8 III	+6	88 s		
58215	173622		7 23 29.1	-27 50 02	+0.003	0.00	5.38	1.53		K2	+48	20 mn		
57511	41734		7 23 29.1	+41 21 48	-0.001	-0.04	7.8	0.1	0.6	A0 V		270 s		
58492	218692		7 23 30.9	-47 57 17	+0.004	-0.01	7.9	1.1	3.2	G5 IV		29 s		
58286	197938		7 23 31.8	-32 12 08	-0.001	-0.02	5.38	-0.19		B3	+21			
58287	197940		7 23 32.1	-34 31 04	+0.001	-0.03	7.4	1.2	0.2	K0 III		220 s		
58285	197939		7 23 34.5	-31 18 01	-0.001	-0.02	7.98	-0.08	0.4	B9.5 V		330 s		
58127	152810		7 23 37.7	-14 06 00	-0.001	+0.01	7.58	0.03		B5 e				
58473	218691		7 23 39.0	-44 05 22	-0.002	+0.01	7.6	1.8	-0.1	K2 III		150 s		
58283	197941		7 23 40.1	-30 26 57	-0.004	+0.02	7.0	0.3	1.4	A2 V		100 s		
57309	26320		7 23 41.5	+54 05 30	-0.001	-0.03	8.0	0.0	0.4	B9.5 V		330 s		
58657	235163		7 23 42.0	-53 40 09	-0.003	+0.03	7.23	-0.07	0.2	B9 V		260 s		
54942	1159		7 23 47.3	+80 37 34	+0.001	-0.01	7.3	1.1	0.2	K0 III		260 s		
57881	96848		7 23 47.8	+16 40 30	-0.002	0.00	7.8	1.1	0.2	K0 III		330 s		
57930	96855		7 23 48.3	+10 42 02	-0.001	-0.04	8.0	0.1	1.7	A3 V		180 s	6035	m
58170	152818		7 23 49.8	-15 52 03	0.000	+0.01	7.8	1.4	-0.3	K5 III		400 s		
58444	197950		7 23 51.0	-39 32 16	-0.002	-0.01	7.0	0.7	0.2	K0 III		230 s		
58074	134616		7 23 51.8	-5 16 17	+0.004	0.00	8.0	0.9	3.2	G5 IV		92 s		
58324	173640		7 23 52.3	-29 17 23	-0.003	0.00	7.4	-0.1	0.6	A0 V		230 s		
58192	152819		7 23 53.7	-17 24 55	-0.005	-0.10	6.9	0.5	4.0	F8 V		38 s		
58325	197944		7 23 54.3	-30 13 00	-0.002	+0.01	6.59	-0.20	-1.6	B3.5 V	+7	430 s		
58941	249844		7 23 55.9	-60 50 18	-0.001	+0.01	7.8	0.5	0.6	A0 V		120 s		
57726	60046		7 23 56.0	+33 40 03	0.000	-0.02	8.0	1.1	-0.1	K2 III		410 s		
57791	79358		7 23 56.6	+27 47 41	+0.001	-0.03	7.67	1.32		K2				
58420	197951		7 23 58.2	-35 50 15	-0.002	+0.02	6.31	-0.15		B8				m
57979	96860		7 23 59.9	+10 54 22	-0.002	-0.02	7.4	0.1	1.4	A2 V		160 s		
58098	134618		7 24 01.2	-3 58 46	0.000	-0.02	7.10	1.1	0.2	K0 III	+8	240 s	6047	m
58052	115387		7 24 01.3	+1 59 12	0.000	-0.02	7.8	0.1	1.4	A2 V		180 s		
58394	197949		7 24 03.8	-31 53 35	-0.002	+0.01	7.9	0.8	0.0	K1 III		380 s		
58350	173651	31 η CMa	7 24 05.6	-29 18 11	-0.001	0.00	2.44	-0.07	-7.0	B5 Ia	+41	760 s		Aludra, m
58377	173654		7 24 07.9	-28 49 41	-0.001	0.00	6.81	-0.17	-1.0	B5.5 V		370 s		
57669	41738	66 Aur	7 24 08.4	+40 40 20	-0.001	-0.03	5.19	1.23	0.2	K0 III	+21	66 s		
58895	235171		7 24 09.4	-58 29 31	-0.010	+0.14	6.58	0.71	3.2	G5 IV		47 s		m
58349	173653		7 24 09.7	-27 14 14	0.000	+0.01	8.0	1.0	0.2	K0 III		360 s		
58306	173649		7 24 12.3	-22 21 45	0.000	0.00	7.3	1.7	-0.3	K5 III		250 s		
57101	14179		7 24 12.8	+64 08 51	-0.003	-0.04	7.4	0.1	1.4	A2 V		160 s		
57769	60049		7 24 16.9	+36 18 38	0.000	-0.05	7.0	0.4	3.0	F2 V		63 s		
58346	173656		7 24 17.1	-22 54 46	-0.001	0.00	6.20	-0.08		B9				
58305	152827		7 24 17.5	-18 20 56	0.000	+0.02	7.9	0.2	2.1	A5 V		150 s		
58122	115394		7 24 18.1	+3 10 39	-0.001	-0.03	7.7	1.1	0.2	K0 III		300 s		
57647	41739		7 24 19.8	+45 44 36	+0.001	-0.08	8.0	1.4	-0.3	K5 III		150 mx		
58072	96868		7 24 24.6	+10 14 44	0.000	-0.01	6.9	0.9	3.2	G5 IV		56 s		
57836	60050		7 24 25.2	+34 14 10	-0.001	-0.02	8.0	0.9	3.2	G5 IV		93 s		
58050	96866		7 24 27.7	+15 31 02	+0.001	0.00	6.41	-0.18	-2.9	B3 III	+38	730 s		
58487	197960		7 24 27.7	-34 21 31	-0.002	-0.05	7.7	1.2	0.2	K0 III		240 s		
58121	115397		7 24 28.3	+6 09 01	-0.001	0.00	7.70	0.9	0.3	G7 III	+14	290 s		
56818	6175		7 24 33.3	+70 41 11	-0.003	-0.01	8.0	0.1	0.6	A0 V		300 s		
57927	79366	59 Gem	7 24 33.4	+27 38 16	+0.001	+0.01	5.76	0.34	2.6	F0 V	-5	40 s		
58144	115398		7 24 34.4	+9 05 44	-0.002	-0.01	7.8	0.1	1.7	A3 V		160 s		
58441	173671		7 24 36.7	-26 40 45	+0.001	+0.03	8.04	-0.15	0.6	A0 V		310 s		
59640	256408		7 24 37.0	-71 28 13	0.000	+0.05	6.5	0.5	1.4	A2 V		55 s		

HD	SAO	Star Name	α 2000	δ 2000	μ(α)	μ(δ)	V	B-V	M_V	Spec	RV	d(pc)	ADS	Notes
			$7^h 24^m 37^s.5$	$-38°16'25''$	$-0^s.005$	$-0''.02$								
58618	197969		7 24 37.5	-38 16 25	-0.005	-0.02	7.4	0.2	2.6	F0 V		91 s		
58343	152834		7 24 40.0	-16 12 04	-0.001	-0.01	5.33	-0.05	-1.7	B3 V	-5	220 s		
57155	14180		7 24 40.8	+65 50 20	-0.001	-0.04	7.4	0.2	2.1	A5 V		110 s		
58534	197962		7 24 41.3	-31 46 58	-0.003	0.00	5.50	0.1	0.6	A0 V		67 s		m
58535	197964		7 24 43.8	-31 48 32	-0.002	+0.01	5.33	1.08	-0.1	gK2	+20	120 s		m
58635	197974		7 24 47.1	-37 17 24	+0.001	+0.01	6.16	0.25	1.7	A3 V		63 s		
57814	41749		7 24 48.6	+41 58 11	0.000	-0.02	7.8	1.4	-0.3	K5 III		410 s		
58071	79370		7 24 49.7	+20 00 03	0.000	-0.01	7.6	0.1	0.6	A0 V		260 s		
58439	152837		7 24 50.7	-19 00 44	-0.002	0.00	6.30	0.1	-5.0	A2 Ib		1600 s		m
57790	41748		7 24 51.9	+44 51 06	-0.003	-0.05	7.80	1.1	0.2	K0 III		330 s	6044	m
57646	26333		7 24 57.1	+51 53 14	+0.002	-0.04	5.80	1.61	-0.1	K2 III	+18	82 s		
--	218712		7 24 57.2	-47 06 41	-0.001	-0.01	8.0	1.1	0.2	K0 III		370 s		
58187	96871	1 CMi	7 24 58.1	+11 40 11	-0.001	-0.02	5.30	0.10	0.1	A4 III	0	110 s		
58705	197981		7 25 03.7	-37 58 02	0.000	+0.06	7.5	-0.5	2.1	A5 V		120 s		
58770	218711		7 25 04.6	-43 23 12	0.000	-0.01	8.00	0.1	0.6	A0 V		300 s		
58462	152839		7 25 06.6	-14 52 58	-0.003	+0.01	6.77	0.04	0.6	A0 V		160 s		
58981	235179		7 25 07.4	-53 06 49	-0.002	+0.06	7.6	0.6	2.6	F0 V		66 s		
58510	173684		7 25 07.8	-21 10 29	-0.001	-0.04	6.80	0.12		B2 n	+20		6065	m
58461	152840		7 25 08.2	-13 45 08	-0.015	-0.01	5.78	0.42	2.6	F0 V	+7	37 s		
58162	96872		7 25 09.6	+19 00 19	-0.001	-0.02	7.6	1.1	0.2	K0 III		300 s		
57177	14182		7 25 10.6	+67 08 50	0.000	-0.01	7.6	0.1	0.6	A0 V		260 s		
58937	218715		7 25 11.2	-49 54 45	-0.001	+0.01	7.8	1.5	0.2	K0 III		130 s		
58630	173692		7 25 16.2	-28 11 14	-0.002	-0.02	7.5	-0.4	0.0	B8.5 V		320 s		
57789	41751		7 25 17.9	+49 39 26	+0.002	+0.01	7.9	1.1	0.2	K0 III		350 s		
57625	26335		7 25 17.9	+56 34 09	-0.007	-0.06	7.70	0.5	4.0	F8 V		55 s		m
58800	218713		7 25 18.9	-41 09 27	+0.002	+0.06	7.8	-0.1	3.0	F2 V		89 s		
58585	173691		7 25 19.8	-21 58 58	-0.001	0.00	6.05	0.23		A5				m
55075	1164		7 25 22.0	+81 15 27	0.000	-0.03	6.2	0.0		B9	-8			
58612	173697		7 25 25.1	-25 13 03	-0.002	-0.01	5.77	-0.11		B9				
57667	26337		7 25 26.0	+56 32 39	-0.002	-0.01	7.60	1.1	-0.1	K2 III		350 s		m
55967	6169		7 25 26.4	+78 15 15	+0.004	-0.07	7.3	1.1	0.2	K0 III		190 mx		
57880	41753		7 25 27.6	+46 48 36	-0.008	-0.06	7.8	1.1	0.2	K0 III		150 mx		
58673	173703		7 25 33.4	-26 37 30	-0.001	+0.01	7.3	1.8	1.4	A2 V		13 s		
58609	152853		7 25 33.6	-19 26 14	-0.001	+0.02	7.0	1.2	0.2	K0 III		180 s		
58297	96886		7 25 34.3	+13 04 22	-0.001	-0.01	7.6	1.1	-0.1	K2 III		350 s		
58246	79376		7 25 35.6	+20 29 48	0.000	-0.02	7.20	0.2	2.1	A5 V	-27	110 s	6060	m
60816	256415	ε Men	7 25 38.2	-79 05 38	-0.011	+0.01	5.53	1.28		K5	+11	24 mn		
58367	115425	2 ε CMi	7 25 38.8	+9 16 34	-0.001	-0.01	4.99	1.01	0.3	G8 III	-8	78 s		
58766	197986		7 25 43.0	-31 44 19	0.000	0.00	6.30	-0.18		B4	+10			
58207	79374	60 ζ Gem	7 25 43.5	+27 47 53	-0.009	-0.09	3.79	1.03	0.2	K0 III	+8	50 s		
58432	115429		7 25 45.2	+4 26 45	-0.002	-0.11	7.7	1.1	0.2	K0 III		130 mx		
58526	134654		7 25 51.0	-5 46 30	-0.002	-0.01	5.97	0.92	-6.3	G0 I	+14	2600 s		
57950	41759		7 25 51.8	+48 32 54	-0.001	0.00	6.75	0.00	0.4	B9.5 V		290 s		
58431	115432		7 25 53.0	+6 42 48	-0.003	-0.02	7.8	0.4	2.6	F0 V		110 s		
58338	96888		7 25 54.1	+18 08 48	0.000	-0.04	6.90	0.9	3.2	G5 IV		55 s	6064	m
58206	60069		7 25 55.5	+32 51 41	-0.003	-0.01	7.4	0.0	0.4	B9.5 V		250 s		
56842	6180		7 25 56.0	+73 51 03	-0.006	-0.01	7.4	0.1	0.6	A0 V		230 s		
58647	152860		7 25 56.3	-14 10 42	+0.002	+0.01	6.81	0.04	0.4	B9.5 V		180 s		
58556	134655		7 25 57.2	-2 14 54	-0.001	-0.04	7.1	0.6	4.4	G0 V		35 s		
58962	218717		7 25 59.3	-42 06 30	+0.001	+0.07	7.30	1.4	-0.3	K5 III		230 mx		m
58093	41763		7 25 59.6	+42 10 01	-0.009	0.00	7.6	0.4	3.0	F2 V		85 s		
58698	--		7 26 02.3	-18 20 48			7.90	0.1	0.6	A0 V		280 s	6078	m
58383	96892		7 26 02.5	+14 06 09	+0.001	-0.02	7.41	-0.09		B9			6068	m
58580	134658		7 26 03.4	-4 32 15	0.000	-0.02	6.76	-0.01		B9				
58696	152865		7 26 03.4	-15 39 56	-0.007	-0.09	7.0	0.9	3.2	G5 IV		58 s		
58500	115437		7 26 04.4	+5 34 49	0.000	-0.03	7.3	1.1	0.2	K0 III		250 s		
58797	173724		7 26 05.2	-26 47 24	-0.001	+0.03	7.9	0.5	0.2	K0 III		340 s		
59103	218720		7 26 06.0	-48 10 03	-0.002	-0.01	7.9	1.2	0.2	K0 III		340 s		
58244	60073		7 26 06.5	+31 53 48	0.000	0.00	6.8	0.1	0.6	A0 V	-14	180 s		
59102	218719		7 26 06.8	-47 46 24	0.000	+0.01	8.00	1.1	0.2	K0 III		360 s		m
58722	152868		7 26 07.2	-18 07 18	-0.002	+0.01	7.32	-0.02	0.0	B8.5 V		290 s		
58888	197992		7 26 10.0	-31 38 37	-0.001	0.00	8.0	-0.6	1.4	A2 V		210 s		
60378	256413		7 26 12.2	-75 35 26	+0.008	0.00	8.0	0.9	0.2	K0 III		360 s		
56195	6173		7 26 12.6	+78 02 40	-0.004	-0.08	7.1	0.1	0.6	A0 V		59 mx		
58764	173725		7 26 13.5	-20 54 21	+0.002	+0.02	7.0	-0.1	0.4	B9.5 V		210 s		
59219	235192		7 26 21.8	-51 01 07	0.000	0.00	5.10	1.06	-0.3	gK6	+8	120 s		
58961	197996		7 26 22.2	-35 10 37	-0.002	+0.03	7.8	0.4	0.2	K0 III		330 s		
58693	134666		7 26 25.1	-10 05 19	-0.002	+0.03	6.9	0.1	1.4	A2 V		120 s		
58691	134664		7 26 25.9	-7 22 25	-0.001	0.00	8.0	1.1	0.2	K0 III		350 s		
58453	96897		7 26 27.3	+18 31 00	-0.001	-0.03	7.26	0.33		F0			6073	m
58552	96899		7 26 27.9	+10 36 30	-0.002	0.00	6.42	0.09	1.4	A2 V	-12	97 s		
58689	134667		7 26 33.2	-3 02 58	0.000	-0.04	8.0	1.1	-0.1	K2 III		400 s		
58666	134665		7 26 34.1	-0 28 10	-0.002	-0.01	6.9	0.1	1.4	A2 V		130 s		
57044	6187		7 26 35.1	+73 04 57	+0.006	+0.01	7.36	0.31	1.9	F2 IV	-34	150 s	6028	m
58791	152878		7 26 35.5	-13 11 21	-0.003	0.00	6.93	0.00	0.4	B9.5 V		200 s		

HD	SAO	Star Name	α 2000	δ 2000	μ(α)	μ(δ)	V	B-V	M_v	Spec	RV	d(pc)	ADS	Notes
			h m s	° ′ ″	s	″								
58886	173741		7 26 38.7	-23 04 54	0.000	+0.01	7.41	-0.01	0.6	A0 V		230 s		
58690	134671		7 26 38.9	-4 19 06	-0.002	+0.01	8.0	1.1	0.2	K0 III		350 s		
59946	249856		7 26 39.2	-69 34 33	-0.016	+0.04	7.8	0.5	3.4	F5 V		75 s		
58382	60080		7 26 39.6	+31 37 12	-0.001	-0.02	7.30	0.8	3.2	G5 IV		66 s	6072	m
58907	173743		7 26 40.6	-23 42 44	-0.001	+0.02	6.5	0.3	0.6	A0 V		93 s		
59027	197999		7 26 40.7	-34 56 18	-0.003	+0.03	8.03	0.96	3.2	G6 IV		76 s		
58599	96901		7 26 41.2	+11 00 33	0.000	0.00	6.30	-0.13	-1.3	B6 IV		330 s		
59026	198000		7 26 42.2	-34 08 27	-0.003	0.00	5.90	-0.17		B3 n	+7			
57128	6190		7 26 42.4	+73 05 01	+0.006	+0.02	7.82	0.29	1.7	F0 IV	-29	150 s	6028	m
58142	41764	21 Lyn	7 26 42.8	+49 12 42	-0.001	-0.05	4.64	-0.02	0.3	A1 IV	+26	74 s		
58755	134674		7 26 46.0	-5 52 08	-0.002	+0.01	7.7	1.1	-0.1	K2 III		350 s		
58205	41767		7 26 46.1	+44 28 42	0.000	-0.04	7.8	1.1	0.2	K0 III		340 s		
57666	14191		7 26 46.4	+64 56 50	0.000	+0.03	8.00	0.6	4.4	G0 V		53 s	6053	m
58551	79386		7 26 50.2	+21 32 09	-0.022	-0.02	6.54	0.46	3.7	F6 V	+47	31 ts		
58579	79391	61 Gem	7 26 56.2	+20 15 26	0.000	-0.02	5.93	0.30	2.6	F0 V	+9	45 s		
57925	26343		7 26 57.1	+59 09 07	-0.002	-0.01	7.8	1.1	-0.1	K2 III		380 s		
56863	6184		7 26 57.7	+75 36 45	+0.002	0.00	7.3	0.1	0.6	A0 V		210 s		
58295	41772		7 26 58.5	+43 15 30	0.000	0.00	6.7	1.1	-0.1	K2 III		230 s		
59099	198005		7 26 58.6	-34 18 41	-0.025	+0.12	7.03	0.48		F8				m
59197	218729		7 26 59.4	-41 37 45	-0.001	0.00	7.9	0.2	0.0	B8.5 V		330 s		
58978	173752		7 26 59.4	-23 05 10	-0.001	0.00	5.4	0.1	-4.6	B0 IV pe	+48	940 s		
59579	235207		7 26 59.9	-59 34 43	+0.001	+0.02	7.4	1.6	0.2	K0 III		120 s		
58714	115455		7 27 05.0	+9 02 36	-0.001	-0.01	7.7	1.1	0.2	K0 III		300 s		
59995	249857		7 27 05.9	-68 36 55	-0.004	+0.10	7.5	0.4	3.0	F2 V		79 s		
58271	41771		7 27 07.8	+47 38 46	-0.002	+0.01	7.9	0.1	0.6	A0 V		280 s		
58954	152894		7 27 07.9	-17 51 52	0.000	+0.01	5.63	0.32		F0	-29		6093	m
58686	96908		7 27 08.9	+11 57 20	-0.002	+0.01	6.6	1.1	0.2	K0 III		190 s		
58715	115456	3 β CMi	7 27 09.0	+8 17 21	-0.004	-0.04	2.90	-0.09	-1.0	B8 V	+22	42 s		Gomeisa
59140	198009		7 27 17.1	-33 53 24	+0.001	-0.02	7.3	1.0	3.2	G5 IV		50 s		
58810	115466		7 27 18.9	+1 27 08	0.000	-0.03	7.4	1.1	0.2	K0 III		260 s		
59368	218735		7 27 20.4	-47 46 56	-0.002	+0.03	7.9	0.4	2.1	A5 V		93 s		
58451	60086		7 27 24.0	+39 18 54	-0.007	-0.05	6.9	1.1	0.2	K0 III		160 mx		
59468	235206		7 27 25.1	-51 24 08	-0.034	0.00	6.74	0.70	4.2	G5 IV-V		32 s		
58729	96914		7 27 25.8	+15 18 58	-0.001	-0.02	7.40	0.00	0.0	B8.5 V	+25	300 s	6086	m
57742	14195		7 27 26.0	+66 19 54	+0.001	-0.02	6.3	-0.1		B9				
59076	173770		7 27 33.8	-21 08 46	-0.001	-0.01	7.70	0.7	0.6	G0 III	+20	260 s		
58497	60088		7 27 34.1	+38 39 59	0.000	-0.02	6.87	0.90	3.2	G5 IV		45 s		
59074	152905		7 27 34.3	-18 29 31	-0.002	-0.01	6.95	-0.09	0.6	A0 V		190 s		m
59075	152906		7 27 37.0	-18 29 46	+0.002	+0.03	7.65	0.34	-6.5	B8 I		4200 s		
58809	96917		7 27 40.0	+14 40 35	+0.001	-0.01	7.6	1.1	-0.1	K2 III		340 s		
58712	79401		7 27 40.5	+22 08 28	-0.001	-0.02	6.8	1.4	-0.3	K5 III		270 s	6087	m
59446	218743		7 27 42.7	-47 24 51	0.000	0.00	7.59	0.09	-0.6	B7 V		350 s		
58973	134698		7 27 42.7	-3 04 17	-0.001	0.00	7.84	-0.15		B5	-3			
59136	173778		7 27 42.8	-22 51 35	-0.001	-0.02	5.95	-0.09		B8				
59192	173787		7 27 44.1	-27 04 31	-0.003	0.00	7.9	0.9	0.2	K0 III		340 s		
58728	79403	63 Gem	7 27 44.3	+21 26 42	-0.004	-0.12	5.22	0.39	2.8	F5 IV-V	+25	31 s	6089	m
58683	79402		7 27 48.8	+27 17 33	-0.005	-0.10	7.37	0.92	0.3	G8 III	+55	59 s		
59215	173793		7 27 49.3	-28 22 17	-0.001	0.00	7.10	-0.12	-1.0	B5.5 V		420 s		m
59067	152908		7 27 50.6	-11 33 11	-0.001	0.00	5.79	0.58	-3.3	G8 Ib-II	+15	660 s	6104	m
58781	96918		7 27 50.7	+19 02 41	+0.008	-0.07	7.2	0.8	3.2	G5 IV		63 s		
59502	218746		7 27 58.8	-47 06 27	-0.006	+0.06	7.8	1.0	0.2	K0 III		220 mx		
59256	173799		7 27 59.1	-29 09 21	-0.001	+0.01	5.54	-0.08		B9 p	+4			
59018	134707		7 27 59.6	-2 52 55	-0.001	-0.01	6.8	1.1	0.2	K0 III		210 s		
58747	79405		7 28 00.8	+26 13 40	-0.002	+0.01	7.2	0.1	1.4	A2 V		140 s		
59235	173797		7 28 01.8	-26 50 20	0.000	+0.01	6.50	1.4	-0.3	K5 III		230 s		m
58923	115477	5 η CMi	7 28 02.0	+6 56 31	0.000	-0.05	5.25	0.22	0.6	gF0	+18	85 s	6101	m
57878	14201		7 28 04.0	+66 29 28	-0.003	-0.05	7.6	0.2	2.1	A5 V		120 s		
59189	152918		7 28 05.2	-19 07 05	-0.001	+0.01	7.2	0.0	0.0	B8.5 V		280 s		
58520	41783		7 28 08.6	+46 31 17	-0.001	-0.06	6.8	0.1	1.4	A2 V		76 mx		
58972	115478	4 γ CMi	7 28 09.7	+8 55 33	-0.004	+0.02	4.32	1.43	-0.2	K3 III	+47	65 s	6100	m
59467	218745		7 28 09.8	-41 30 56	-0.001	-0.02	7.9	0.7	0.2	K0 III		340 s		
58595	41789		7 28 10.5	+40 39 26	+0.008	-0.01	7.3	0.8	3.2	G5 IV		67 s		
59780	235221		7 28 11.4	-58 04 22	0.000	0.00	6.5	1.2	0.2	K0 III		140 s		
58521	41784		7 28 11.6	+45 59 27	0.000	0.00	6.30	1.6		M5 Ib-II				Y Lyn, v
58746	79408		7 28 12.1	+29 25 29	-0.001	-0.02	7.4	0.2	2.1	A5 V	+12	120 s		
58549	41787		7 28 15.9	+46 17 09	-0.005	-0.09	7.8	0.5	3.4	F5 V		77 s		
59129	152919		7 28 16.9	-11 55 57	0.000	-0.02	7.9	0.0	0.4	B9.5 V		310 s		
59530	218748		7 28 17.5	-43 57 47	-0.001	0.00	7.57	-0.05	0.4	B9.5 V		270 s		
59618	218750		7 28 18.6	-48 48 46	-0.001	+0.01	8.0	-0.4	0.0	B8.5 V		400 s		
58548	41788		7 28 21.2	+48 03 35	-0.001	-0.03	7.9	0.8	3.2	G5 IV		87 s		
59128	134717		7 28 22.1	-7 09 32	-0.003	-0.02	7.9	0.5	3.4	F5 V		77 s		
59466	198031		7 28 22.8	-37 48 37	-0.002	-0.05	6.58	0.06		A0				
59416	198029		7 28 22.9	-35 15 37	-0.003	-0.04	7.5	0.5	1.7	A3 V		81 s		
58184	14206		7 28 24.1	+61 52 38	+0.004	-0.05	7.4	1.1	0.2	K0 III		210 mx		
58899	79410		7 28 24.7	+21 32 59	-0.009	-0.02	7.16	0.92		G5				

HD	SAO	Star Name	α 2000	δ 2000	μ(α)	μ(δ)	V	B-V	M$_v$	Spec	RV	d(pc)	ADS	Notes
			h m s	° ′ ″	s	″								
59063	115495		7 28 25.6	+3 23 34	−0.002	−0.03	7.9	0.9	3.2	G5 IV		88 s		
59212	152924		7 28 26.2	−12 57 21	−0.003	+0.02	7.95	0.05	0.6	A0 V		280 s		m
58662	41794		7 28 31.2	+42 45 10	−0.003	−0.01	7.90	1.1	0.2	K0 III		350 s	6094	m
59211	134724		7 28 34.2	−10 02 39	−0.001	−0.02	6.62	−0.12		B5	+30			
59061	115496		7 28 37.7	+8 30 26	0.000	−0.01	7.9	1.1	0.2	K0 III		340 s		
58898	79414		7 28 39.6	+27 33 08	−0.001	+0.01	6.6	1.1	−0.1	K2 III		220 s		
58919	79416		7 28 40.5	+24 38 55	0.000	−0.07	7.40	0.5	4.0	F8 V		48 s		m
59411	173829		7 28 41.0	−29 07 07	+0.004	−0.02	6.8	1.3	0.2	K0 III		150 s		
59344	173823		7 28 41.0	−21 48 16	−0.002	+0.01	7.7	1.0	0.2	K0 III		310 s		
59110	115499		7 28 41.7	+2 16 59	−0.002	−0.04	7.5	0.1	0.6	A0 V		240 s		
58183	14208		7 28 43.1	+63 52 29	+0.010	+0.01	7.8	0.5	3.4	F5 V		74 s		
58640	41795		7 28 43.4	+47 55 13	−0.005	−0.08	6.8	1.6	−0.5	M4 III	+26	210 mx		
58829	60107		7 28 44.0	+34 26 13	−0.002	−0.01	7.4	0.9	3.2	G5 IV		69 s		
59090	115497		7 28 46.7	+7 55 39	−0.003	−0.05	8.0	0.5	4.0	F8 V		62 s		
59059	96930		7 28 47.3	+15 06 34	0.000	−0.03	6.22	−0.05	0.2	B9 V	+34	160 s		
59499	198039		7 28 51.4	−31 50 49	−0.003	0.00	5.95	−0.15	−1.6	B3.5 V	+2	320 s		
58661	41797		7 28 51.5	+48 11 01	+0.001	−0.05	5.60	−0.10		B9 pe			6095	m
60150	249864		7 28 51.5	−64 30 36	+0.001	−0.01	6.39	1.56		K5				
58990	79425		7 28 52.3	+21 49 16	0.000	−0.02	8.0	1.1	0.2	K0 III		360 s		
58681	41799		7 28 52.8	+46 18 52	−0.002	−0.08	6.6	1.1	0.2	K0 III		180 mx		
59230	134731		7 28 55.9	−3 50 00	−0.001	−0.01	7.7	1.4	−0.3	K5 III		380 s		
59527	198041		7 28 56.6	−34 55 07	−0.001	0.00	6.91	−0.12	0.0	B8.5 V		240 s		
57509	6200		7 28 58.6	+74 08 54	+0.002	0.00	7.8	0.2	2.1	A5 V		140 s		
−−	60116		7 29 01.7	+31 59 38	+0.012	+0.18	7.73	0.94		K0				
59228	115509		7 29 04.8	+3 48 17	−0.002	−0.02	7.9	0.9	3.2	G5 IV		88 s		
59550	198042		7 29 04.8	−31 27 23	−0.002	0.00	5.77	−0.19	−2.3	B3 IV	+8	410 s		
59635	198045		7 29 05.6	−38 48 43	−0.003	+0.02	5.43	−0.16	−2.9	B3 III	+26	460 s		
59408	152936		7 29 05.8	−18 52 33	0.000	−0.02	7.20	1.1	0.2	K0 III		250 s		m
58946	60118	62 ρ Gem	7 29 06.6	+31 47 03	+0.012	+0.17	4.18	0.32	2.6	F0 V	−6	19 ts	6109	m
59152	96937		7 29 11.2	+12 45 36	−0.001	−0.02	6.6	0.2	2.1	A5 V	+26	79 s		
59180	96939		7 29 12.7	+11 34 52	−0.002	0.00	6.9	1.1	0.2	K0 III	+82	220 s		
59150	96938		7 29 13.3	+14 21 30	−0.001	−0.01	7.00	0.1	1.7	A3 V	−6	120 s		m
59717	218755	σ Pup	7 29 13.8	−43 18 05	−0.005	+0.19	3.25	1.52	−0.3	K5 III	+88	51 s		
59573	198043		7 29 14.7	−30 38 38	+0.001	+0.01	7.9	1.2	−0.3	K5 III		430 s		
59313	134739		7 29 14.7	−4 32 30	−0.001	−0.02	7.7	1.1	0.2	K0 III		300 s		
59594	198048		7 29 17.1	−34 10 16	−0.002	+0.01	7.2	0.2	1.4	A2 V		120 s		
59271	115514		7 29 17.8	+3 22 48	−0.001	−0.02	7.70	0.2	2.1	A5 V		130 s	6121	m
59179	96940		7 29 18.5	+12 26 56	0.000	−0.03	7.8	0.4	3.0	F2 V		89 s		
59311	134740		7 29 18.6	−1 54 19	−0.001	−0.01	5.59	1.50	−0.3	gK5	−5	150 s		
59227	115513		7 29 18.8	+6 34 57	−0.003	+0.03	6.7	0.8	3.2	G5 IV		50 s		
59404	152942		7 29 19.3	−13 08 23	−0.002	−0.03	7.30	1.1	0.2	K0 III		260 s		m
59441	152944		7 29 20.3	−17 20 44	0.000	−0.04	7.4	0.6	4.4	G0 V		39 s		
59037	79427	64 Gem	7 29 20.3	+28 07 05	−0.003	−0.06	5.05	0.11	2.2	A6 V	+35	37 s		
59438	152943		7 29 21.8	−14 59 57	−0.013	−0.26	6.05	0.47	3.3	dF4	−6	34 s	6126	m
59381	152941		7 29 22.0	−10 19 39	−0.001	−0.03	5.75	1.62	−0.3	gK5	−7	140 s		
59895	235231		7 29 24.9	−52 22 46	−0.002	+0.05	7.9	1.0	0.2	K0 III		340 s		
59380	134742		7 29 25.6	−7 33 04	+0.004	+0.13	5.86	0.48	4.2	dF9	+9	22 s		
58987	60125		7 29 27.0	+34 48 12	0.000	−0.01	7.90	0.1	0.6	A0 V		290 s	6111	m
59203	96947		7 29 28.1	+16 09 50	0.000	−0.01	7.9	0.1	1.4	A2 V		200 s		
59675	198053		7 29 30.0	−36 41 07	−0.005	−0.03	7.4	1.3	−0.5	M2 III		170 mx		
59149	96945		7 29 30.7	+19 38 00	+0.002	0.00	6.7	1.1	0.2	K0 III		200 s		
59014	60126		7 29 30.8	+35 09 50	−0.001	−0.04	7.2	1.1	−0.1	K2 III		290 s		
59704	198055		7 29 31.4	−38 07 22	−0.002	+0.07	7.6	0.3	4.0	F8 V		53 s		
59295	115518		7 29 35.4	+7 50 27	0.000	+0.01	7.5	1.1	0.2	K0 III		290 s		
59435	134747		7 29 35.4	−9 15 34	−0.001	−0.01	7.9	0.2		A5 p				
59869	218762		7 29 45.0	−46 41 22	−0.003	+0.01	7.40	0.03	0.6	A0 V		230 s		
−−	−−		7 29 46.2	−73 22 46			7.70							S Vol, v
59294	96952	6 CMi	7 29 47.7	+12 00 24	0.000	−0.02	4.54	1.28	−0.1	K2 III	−15	70 s		
59868	218761		7 29 47.9	−44 54 43	−0.001	0.00	7.84	−0.04	−1.0	B5.5 V		490 s		
59148	79434	65 Gem	7 29 48.7	+27 54 58	−0.003	−0.03	5.01	1.11	−0.1	K2 III	+36	110 s	6119	m
59612	173864		7 29 51.4	−23 01 28	−0.001	−0.01	4.85	0.23	−4.8	A5 Ib	+36	690 s		m
60131	235243		7 29 52.8	−56 43 39	+0.001	+0.02	8.0	0.9	0.6	A0 V		78 s		
59544	152962		7 29 53.4	−14 21 30	−0.005	0.00	7.5	0.4	2.6	F0 V		97 s		
59543	152963		7 29 54.3	−13 59 17	0.000	+0.05	7.19	−0.07	−1.0	B5.5 V	+5	430 s		
58855	41808	22 Lyn	7 29 55.9	+49 40 21	+0.012	−0.08	5.36	0.45	3.7	F6 V	−27	23 ts		m
59991	235238		7 29 56.5	−51 12 07	−0.004	+0.08	7.9	0.4	2.6	F0 V		96 s		
60290	249866		7 29 57.0	−61 14 58	+0.006	−0.06	7.3	1.0	3.4	F5 V		28 s		
59058	60128		7 29 57.7	+38 26 17	−0.004	−0.07	7.7	0.9	5.2	G5 V	+14	32 s		
60060	235239		7 29 59.6	−52 39 03	−0.004	+0.06	5.87	1.01	3.2	G5 V		22 s		
59292	96953		7 30 01.1	+18 37 30	0.000	−0.02	7.8	0.1	1.4	A2 V		190 s		
59652	173873		7 30 02.1	−25 08 47	−0.001	−0.01	7.11	0.19	1.7	A3 V		110 s		
59610	152968		7 30 02.2	−19 31 37	−0.001	−0.01	6.9	1.4	0.2	K0 III		130 s		
59541	134759		7 30 05.0	−8 31 06	0.000	−0.04	8.0	0.5	3.4	F5 V		81 s		
59375	96958		7 30 06.1	+10 02 50	0.000	−0.04	7.6	1.6	−0.5	M2 III		420 s		
60173	235246		7 30 07.4	−56 19 06	−0.001	+0.03	8.0	1.3	−0.3	K5 III		450 s		

HD	SAO	Star Name	α 2000	δ 2000	μ(α)	μ(δ)	V	B-V	M_v	Spec	RV	d(pc)	ADS	Notes
			h m s	° ′ ″	s	″								
59106	60131		7 30 10.3	+38 10 09	+0.002	−0.07	6.9	0.6	4.4	G0 V		31 s		
59631	152971		7 30 11.3	−17 04 05	+0.001	+0.02	7.7	1.1	0.2	K0 III		310 s		
59084	41817		7 30 13.5	+40 50 44	−0.003	−0.05	7.0	0.1	0.6	A0 V		190 s		
59333	79443		7 30 14.7	+20 47 04	0.000	0.00	7.4	0.1	0.6	A0 V		230 s		
60250	235250		7 30 17.9	−57 12 37	−0.001	+0.02	7.0	1.6	−0.1	K2 III		150 s		
59894	218764		7 30 18.4	−40 59 21	−0.002	−0.02	6.6	0.9	3.2	G5 IV		38 s		
59083	41819		7 30 18.9	+41 35 55	0.000	−0.03	7.80	1.1	0.2	K0 III		330 s	6122	m
59268	79441		7 30 22.3	+29 11 28	+0.001	−0.03	8.0	0.9	3.2	G5 IV		93 s		
59360	96960		7 30 23.4	+19 48 51	+0.004	−0.11	7.1	0.6	4.4	G0 V		35 s		
59267	60139		7 30 25.0	+31 08 43	0.000	−0.02	8.0	0.9	3.2	G5 IV		93 s		
59539	115530		7 30 25.7	+3 18 25	−0.001	−0.02	7.7	1.1	−0.1	K2 III		350 s		
59432	96964		7 30 26.1	+13 51 53	−0.003	−0.03	7.5	0.5	3.4	F5 V		68 s	6135	m
59737	173885		7 30 26.8	−20 48 27	−0.001	+0.03	7.5	0.1	0.6	A0 V		200 s		
60130	235247		7 30 28.6	−50 36 32	0.000	+0.01	6.7	0.3	0.6	A0 V		110 s		
59511	115531		7 30 28.9	+2 15 42	−0.001	−0.03	7.0	1.1	0.2	K0 III		220 s		
60228	235252		7 30 30.9	−54 23 56	+0.001	+0.04	5.96	1.56	−0.3	K5 III		160 s		
59034	41818		7 30 33.4	+48 52 47	−0.001	−0.02	7.2	0.1	1.4	A2 V		140 s		
59538	115532		7 30 33.5	+5 15 15	−0.001	−0.02	7.70	0.00	0.4	B9.5 V		280 s	6140	m
59864	198063		7 30 34.2	−34 05 25	+0.001	+0.02	7.64	−0.09		B0.5 II				
59671	134769		7 30 36.0	−9 03 23	−0.001	−0.01	7.8	1.1	0.2	K0 III		320 s		
59563	115535		7 30 36.3	+4 33 14	+0.001	−0.02	7.9	1.1	0.2	K0 III		340 s		
57810	6208		7 30 37.2	+74 47 56	−0.007	0.00	7.8	0.8	3.2	G5 IV		85 s		
59430	79451		7 30 38.8	+20 29 31	−0.001	+0.01	7.1	0.1	0.6	A0 V	−3	200 s		
59173	41822		7 30 39.1	+41 53 27	−0.001	−0.01	7.3	1.1	0.2	K0 III		270 s		
59012	26380		7 30 39.3	+49 56 21	0.000	0.00	7.9	0.9	3.2	G5 IV		89 s		
59967	198069		7 30 42.3	−37 20 23	−0.009	+0.04	6.65	0.63	5.2	G5 V		20 s		
59890	198064		7 30 42.5	−30 57 44	−0.002	0.00	4.65	0.93	−4.5	G1 Ib	+14	640 s		
59692	134772		7 30 42.9	−9 18 36	0.000	+0.01	7.8	0.1	0.6	A0 V		260 s		
59670	134771		7 30 45.2	−6 43 05	0.000	−0.01	8.0	0.1	1.4	A2 V		200 s		
59693	134775		7 30 47.3	−9 46 38	−0.002	+0.01	6.10	0.8	−4.5	G4 Ib var	+35	1200 s		U Mon, v
60201	235257		7 30 49.7	−51 09 31	−0.005	−0.02	7.2	1.7	0.2	K0 III		96 s		
59669	134774		7 30 51.0	−5 13 36	−0.001	−0.01	6.24	1.18		K0				
59732	152984		7 30 51.4	−13 10 06	−0.002	+0.01	8.0	0.1	0.6	A0 V		290 s		
58425	14211		7 30 52.6	+68 27 57	−0.001	−0.04	5.64	1.11	0.2	K0 III	+56	110 s		
59605	115545		7 30 56.6	+8 33 08	0.000	0.00	7.2	0.6	4.4	G0 V	+24	36 s		
59057	26382		7 30 56.8	+51 59 56	0.000	−0.01	6.6	0.2	2.1	A5 V		81 s		
59645	115549		7 30 57.4	+0 09 18	0.000	−0.03	7.8	0.1	0.6	A0 V		260 s		
59730	152986		7 30 58.5	−10 06 28	−0.002	−0.02	6.57	1.62		K2				
59690	134776		7 30 59.2	−2 09 54	+0.001	−0.01	7.30	0.8	3.2	G5 IV		66 s	6147	m
59989	198072		7 31 02.0	−34 59 02	0.000	+0.01	7.7	0.9	−0.3	K5 III		390 s		
59105	26385		7 31 03.5	+50 02 56	−0.001	−0.02	7.7	1.6	−0.5	M2 III		450 s		
55966	1179		7 31 04.5	+82 24 41	0.000	−0.04	4.96	1.66		Mb	+14	44 mn		VZ Cam, v
59603	96973		7 31 08.7	+11 12 10	−0.001	−0.02	6.5	1.1	0.2	K0 III		180 s		
60434	235266		7 31 10.9	−56 37 27	−0.003	+0.03	7.9	1.0	0.2	K0 III		340 s		
57644	6207		7 31 11.7	+76 57 17	−0.005	0.00	7.2	0.8	3.2	G5 IV		63 s		
61314	256419		7 31 12.7	−74 41 25	+0.003	+0.03	7.7	0.1	1.7	A3 V		160 s		
59688	115559		7 31 13.5	+2 10 13	+0.001	−0.03	7.5	0.7	4.4	G0 V		42 s		
59726	134781		7 31 17.2	−1 07 32	−0.001	0.00	6.60	0.8	3.2	G5 IV		48 s		m
59856	153001		7 31 19.7	−14 57 57	−0.002	−0.03	7.2	1.1	0.2	K0 III		250 s		
59765	134784		7 31 21.3	−4 40 44	−0.001	+0.02	7.9	1.1	0.2	K0 III		330 s		
59806	134785		7 31 23.9	−8 44 25	0.000	0.00	7.7	1.1	−0.1	K2 III		350 s		
61313	256420		7 31 24.0	−74 17 25	−0.004	+0.03	7.7	0.4	2.6	F0 V		110 s		
58679	14217		7 31 24.3	+66 08 04	−0.001	−0.02	7.0	1.1	0.2	K0 III		230 s		
60499	235274		7 31 25.1	−57 56 24	−0.003	+0.03	6.9	1.0	0.2	K0 III		220 s		
59509	79464		7 31 25.2	+27 37 22	−0.001	0.00	7.95	0.17		F8				
60098	198086		7 31 25.7	−36 09 10	−0.001	+0.02	6.68	−0.12	−2.2	B5 III		570 s		
59936	153009		7 31 30.2	−19 30 42	0.000	−0.02	7.3	1.6	−0.3	K5 III		290 s		
58917	14221		7 31 36.2	+62 30 05	0.000	−0.05	6.90	0.1	0.6	A0 V	+2	180 s	6125	m
59371	41833		7 31 37.9	+42 38 27	0.000	−0.03	7.52	1.67		K5				
59934	153011		7 31 38.1	−17 11 50	0.000	−0.02	7.8	0.0	−1.0	B5.5 V	+16	560 s		
58381	6221		7 31 38.6	+71 49 18	−0.004	0.00	8.0	0.9	3.2	G5 IV		93 s		
59600	79470		7 31 39.1	+24 00 00	0.000	+0.01	8.0	1.1	0.2	K0 III		360 s		
60168	198093		7 31 42.8	−35 53 16	0.000	+0.01	6.61	−0.08		A0				m
60287	218780		7 31 43.9	−43 17 23	−0.003	+0.02	8.00	1.1	0.2	K0 III		350 s		m
59749	115569		7 31 47.5	+9 49 57	−0.001	−0.03	7.8	1.4	−0.3	K5 III		410 s		
59686	96985		7 31 48.3	+17 05 09	+0.003	−0.08	5.42	1.13	0.2	K0 III	−40	93 s		
59805	115572		7 31 49.1	+1 09 53	+0.002	−0.05	7.5	1.1	0.2	K0 III		280 s		
59830	115575		7 31 52.7	+0 27 26	−0.002	−0.03	7.9	1.1	0.2	K0 III		330 s		
59643	79474		7 31 54.5	+24 30 13	0.000	0.00	7.80	2.24		R6	+41			
59507	60148		7 31 55.6	+38 53 47	−0.003	+0.01	6.54	0.07	1.4	A2 V	+7	110 s		
59929	134797		7 31 58.1	−5 56 34	−0.001	−0.02	7.0	1.1	0.2	K0 III	+25	220 s		
59764	96990		7 31 59.1	+12 39 41	0.000	−0.04	6.6	0.9	3.2	G5 IV	+59	47 s		
59557	60152		7 32 01.1	+35 36 11	0.000	+0.01	7.1	1.4	−0.3	K5 III		300 s		
59033	14226		7 32 02.3	+61 45 29	0.000	−0.11	6.7	0.9	3.2	G5 IV	−2	51 s		
59880	115577		7 32 02.6	+2 43 32	−0.002	−0.01	6.8	0.1	0.6	A0 V		170 s		

HD	SAO	Star Name	α 2000	δ 2000	μ(α)	μ(δ)	V	B-V	M_v	Spec	RV	d(pc)	ADS	Notes
			h m s	° ′ ″	s	″								
59984	134806		7 32 05.7	−8 52 52	−0.006	−0.16	5.90	0.54		F5			6158	m
59881	115581	7 δ¹ CMi	7 32 05.8	+1 54 52	−0.001	0.00	5.25	0.22		A8	+29	30 mn		
60197	173956		7 32 07.8	−29 38 06	−0.001	+0.01	7.8	1.2	−0.3	K5 III		410 s		
58545	6226		7 32 11.6	+71 21 51	−0.004	+0.02	7.3	1.1	−0.1	K2 III		310 s		
60092	153028		7 32 16.2	−18 31 04	−0.002	−0.02	7.5	0.4	2.6	F0 V		97 s		
60312	198104		7 32 22.2	−35 57 40	−0.001	0.00	6.35	−0.10		B9				m
59784	79480		7 32 22.7	+21 06 15	0.000	−0.02	8.00	0.1	1.4	A2 V		210 s		m
59800	96995		7 32 26.9	+18 21 58	0.000	−0.03	7.80	0.5	3.4	F5 V		76 s	6156	m
59828	79483		7 32 34.4	+21 24 31	−0.001	−0.01	6.8	0.9	3.2	G5 IV		52 s		
59555	41841		7 32 39.8	+46 32 05	−0.005	−0.04	8.0	0.9	3.2	G5 IV		93 s		
60114	134821		7 32 41.7	−9 40 10	−0.001	−0.01	7.5	0.0	0.0	B8.5 V		310 s		
59950	115591		7 32 43.0	+8 19 06	−0.001	−0.01	6.9	1.6		M5 III	+68			S CMi, v
60139	134823		7 32 44.8	−9 24 40	0.000	−0.02	7.36	1.42		K2				
60649	235283		7 32 46.3	−53 33 19	0.000	0.00	7.1	0.7	0.0	B8.5 V		96 s		
60266	173973		7 32 47.1	−20 54 32	+0.001	+0.02	8.0	1.1	0.2	K0 III	+62	350 s		X Pup, v
60086	134820		7 32 49.3	−2 02 04	−0.002	−0.02	6.8	0.1	0.6	A0 V		170 s		
60138	134824		7 32 49.6	−7 17 05	−0.001	−0.01	7.7	0.0	0.4	B9.5 V		280 s		
59878	79489		7 32 50.5	+22 53 15	−0.002	−0.01	6.54	1.01	0.2	G9 III	+30	180 s	6160	m
60044	115601		7 32 52.5	+5 02 07	0.000	−0.02	7.7	0.0	0.0	B8.5 V		340 s		
75261	−−		7 32 54.8	−88 53 57			7.8	0.0	0.4	B9.5 V		310 s		
59011	14229		7 32 58.4	+66 56 16	−0.004	−0.01	8.0	1.4	−0.3	K5 III		450 s		
60156	134826		7 32 59.1	−6 51 57	0.000	−0.05	7.29	1.12		K0	−52			
59747	60168		7 33 00.6	+37 01 48	−0.003	+0.01	7.8	0.8	3.2	G5 IV		84 s		
60344	173980		7 33 02.2	−23 56 03	+0.001	+0.01	7.71	−0.17	−1.0	B5.5 V		550 s		
57724	6213		7 33 05.0	+78 41 12	−0.012	−0.11	7.4	0.6	4.4	G0 V		40 s		
59997	97008		7 33 06.5	+15 58 56	0.000	−0.01	7.8	0.1	0.6	A0 V		270 s		
59798	60170		7 33 09.5	+33 56 46	−0.001	−0.01	7.4	0.1	0.6	A0 V		230 s		
60345	173987		7 33 09.7	−24 42 39	−0.001	+0.01	5.85	0.17		A3				
60111	115610	8 δ² CMi	7 33 11.6	+3 17 25	−0.001	+0.04	5.59	0.31		F0 n	+1			
59925	79495		7 33 12.6	+26 06 50	−0.001	−0.01	7.9	0.1	1.4	A2 V		200 s		
60574	218792		7 33 13.3	−43 05 11	0.000	−0.04	6.52	0.92		K0				
60559	198120		7 33 13.4	−40 03 32	+0.001	−0.01	6.26	−0.12		B8				
59848	60171		7 33 14.7	+32 51 43	−0.003	+0.01	7.90	0.9	3.2	G5 IV		87 s	6162	m
60341	153062		7 33 19.5	−19 24 45	+0.001	−0.07	5.66	1.12	0.2	gK0	+16	110 s		
60108	115612		7 33 19.7	+9 19 34	−0.002	−0.01	7.8	0.6	4.4	G0 V		47 s		
59827	60172		7 33 21.0	+35 47 55	+0.005	−0.02	8.0	0.5	3.4	F5 V		82 s		
61312	249881		7 33 21.8	−68 46 04	−0.001	+0.01	6.7	0.3	0.6	A0 V		110 s		
60325	153061		7 33 22.0	−14 20 19	−0.001	−0.01	6.21	−0.04	−3.5	B1 V	+22	680 s		
60081	97012		7 33 24.6	+15 38 21	−0.001	−0.01	6.6	1.1	0.2	K0 III	+55	190 s		
59826	60173		7 33 24.7	+37 11 17	0.000	+0.02	6.4	1.1	0.2	K0 III		180 s		
60498	198121		7 33 25.2	−33 24 00	0.000	−0.01	7.36	−0.13	−1.6	B3.5 V		580 s		
60753	235291		7 33 27.2	−50 35 03	−0.001	0.00	6.70	−0.09	−1.3	B6 IV		400 s		
60136	97014		7 33 27.3	+11 00 42	−0.003	−0.02	7.7	1.6	−0.5	M2 III		440 s		
62093	256426		7 33 27.8	−77 04 57	+0.003	+0.03	7.3	−0.1	0.6	A0 V		220 s		
60931	235299		7 33 29.6	−59 11 55	+0.003	−0.01	7.1	2.0	−0.3	K5 III		150 s		
59947	60183		7 33 31.9	+30 37 43	0.000	+0.02	7.4	1.1	0.2	K0 III		280 s		
59924	60181		7 33 33.2	+32 37 34	0.000	−0.04	8.0	1.1	0.2	K0 III		360 s		
60276	134835		7 33 33.4	−6 13 38	−0.002	−0.02	7.3	1.1	0.2	K0 III	−9	260 s		
−−	60184		7 33 34.4	+30 51 32	0.000	0.00	7.8	1.9						
60107	97016	68 Gem	7 33 36.4	+15 49 36	−0.001	−0.02	5.25	0.05	1.2	A1 V	+13	62 s		
60207	115626		7 33 41.3	+6 41 10	0.000	+0.03	7.7	0.2	2.1	A5 V		130 s		
59847	41850		7 33 45.6	+41 10 26	+0.001	0.00	7.9	1.4	−0.3	K5 III		430 s		
60414	153072		7 33 47.8	−14 31 26	−0.001	0.00	4.9	1.6	−5.6	M2 Iab pe	+22	1200 s		
60687	218801		7 33 50.2	−43 06 50	0.000	−0.01	6.8	0.9	−0.3	K5 III		270 s		
60606	198130		7 33 50.9	−36 20 18	−0.001	0.00	5.54	−0.06	−1.7	B3 V k	−10	270 s		
59721	26407		7 33 51.3	+51 18 51	−0.002	−0.03	6.5	1.1	0.2	K0 III		180 s		
60555	174008		7 33 52.2	−28 34 04	−0.003	+0.02	6.6	0.5	2.6	F0 V		48 s		
60752	218802		7 33 52.8	−46 44 18	−0.002	−0.02	8.0	0.8	3.4	F5 V		49 s		
59506	14236		7 33 56.7	+60 00 58	−0.005	−0.03	8.0	1.1	0.2	K0 III		320 mx		
60686	198137		7 33 58.4	−39 54 21	−0.002	+0.04	6.6	1.3	0.2	K0 III		130 s		
60668	198136		7 34 02.3	−37 19 09	−0.002	+0.01	7.81	−0.04	0.0	B8.5 V		370 s		
59642	26406		7 34 03.0	+57 15 31	−0.004	−0.04	7.6	1.1	0.2	K0 III		300 s		
60411	134844		7 34 03.1	−7 37 49	−0.003	+0.03	7.6	1.1	0.2	K0 III		300 s		
60532	174009		7 34 03.1	−22 17 46	−0.003	+0.04	4.45	0.51	2.3	F7 IV	+61	26 s		
60275	97021		7 34 05.0	+10 34 06	0.000	−0.01	6.20	−0.01	0.4	B9.5 V	0	140 s		
60907	235303		7 34 08.5	−50 50 26	−0.001	+0.07	7.70	1.6	−0.5	M2 III		270 mx		m
59876	41853		7 34 09.2	+46 09 00	−0.003	−0.07	7.30	1.4	−0.3	K5 III		190 mx	6167	m
59923	41854		7 34 09.5	+42 05 15	−0.007	0.00	7.9	0.4	3.0	F2 V		94 s		
60553	174014		7 34 12.8	−20 08 16	−0.001	−0.01	6.91	−0.17	−3.9	B3 II	+28	1500 s		
60646	198138		7 34 12.8	−33 27 48	−0.005	+0.07	6.11	0.29		F0				
60513	153080		7 34 13.3	−16 11 16	−0.004	+0.13	6.9	0.9	3.2	G5 IV		55 s		
58985	6238		7 34 14.6	+71 35 24	−0.002	−0.05	7.6	0.1	1.7	A3 V		150 s		
60357	115644	9 δ³ CMi	7 34 15.7	+3 22 17	−0.001	−0.02	5.81	−0.02		A0 n	+34			m
60584	174019		7 34 18.5	−23 28 25	−0.007	0.00	5.09	0.44	3.0	F2 V	−5	24 s	6190	m
60442	134847		7 34 19.5	−1 19 07	−0.002	−0.02	7.9	0.9	3.2	G5 IV		88 s		

HD	SAO	Star Name	α 2000	δ 2000	μ(α)	μ(δ)	V	B-V	M_v	Spec	RV	d(pc)	ADS	Notes
			$7^h 34^m 19.6^s$	$+14°45'10''$	0.000^s	$-0.03''$								
60302	97024		7 34 19.6	+14 45 10	0.000	-0.03	8.0	0.4	2.6	F0 V		120 s		
60839	218814		7 34 19.6	-46 45 37	+0.001	-0.03	7.9	1.2	0.2	K0 III		270 s		
60791	218812		7 34 22.9	-42 04 25	-0.003	0.00	7.40	0.1	0.6	A0 V		230 s		m
60106	60193		7 34 24.6	+33 07 20	-0.001	-0.04	7.70	0.1	1.4	A2 V		180 s	6173	m
60303	97028		7 34 25.6	+14 06 04	+0.001	-0.01	7.6	1.6	-0.5	M2 III		420 s		
60491	134849		7 34 26.1	-6 53 47	-0.005	-0.03	8.0	1.1	0.2	K0 III		150 mx		
60813	218813		7 34 27.3	-42 23 56	+0.001	+0.02	8.0	-0.7	1.7	A3 V		180 s		
61056	235309		7 34 28.1	-54 03 55	0.000	+0.02	7.2	0.6	3.0	F2 V		48 s		
60180	79510		7 34 28.2	+28 24 39	0.000	0.00	7.7	0.0	0.4	B9.5 V		290 s		
60552	153083		7 34 28.7	-13 52 09	-0.004	+0.12	6.73	0.48	3.4	F5 V		45 s	6189	m
60629	174028		7 34 28.8	-26 07 00	-0.002	-0.03	6.65	-0.01		A0				
60301	97030		7 34 29.1	+16 57 14	0.000	-0.01	7.7	0.9	3.2	G5 IV		80 s		
59976	41858		7 34 30.6	+44 55 31	+0.005	-0.01	7.8	1.1	0.2	K0 III		200 mx		
60204	79512		7 34 31.5	+28 41 10	0.000	-0.02	6.7	0.8	3.2	G5 IV	-12	51 s		
60355	97033		7 34 32.0	+12 18 16	-0.001	-0.02	7.90	0.00	0.0	B8.5 V		380 s	6180	m
60601	174026		7 34 34.3	-21 18 23	-0.001	0.00	7.5	0.1	1.4	A2 V		150 s		
60666	174033		7 34 34.8	-27 00 43	-0.003	+0.08	5.77	1.06	0.2	gK0		120 s		
60178	60198	66 α Gem	7 34 35.9	+31 53 18	-0.013	-0.10	1.58	0.04	1.2	A1 V	-1	14 ts	6175	Castor, m
59975	41857		7 34 36.0	+48 11 28	-0.002	-0.01	7.2	0.1	1.7	A3 V		130 s		
60475	115650		7 34 37.8	+0 10 27	0.000	0.00	7.8	0.0	0.4	B9.5 V		300 s		
61031	235310		7 34 39.5	-51 28 29	-0.001	0.00	6.3	-0.1	0.6	A0 V		140 s		
57508	1194		7 34 39.7	+80 53 48	-0.003	0.00	6.4	1.1	0.2	K0 III	-2	180 s		
60235	79517		7 34 40.5	+28 30 07	+0.001	-0.04	7.38	1.37	-0.2	K3 III		290 s		
61267	249884		7 34 42.3	-61 04 57	-0.001	+0.01	8.0	0.2	0.6	A0 V		210 s		
60906	218817		7 34 43.7	-44 41 30	-0.002	+0.04	6.8	1.9	-0.1	K2 III		91 s		
60683	174037		7 34 43.7	-26 02 20	-0.004	-0.13	8.0	0.3	3.2	G5 IV		93 s		
60299	79519		7 34 44.4	+22 57 11	+0.002	-0.01	7.9	0.9	3.2	G5 IV		88 s		
60490	115652		7 34 45.3	+2 25 15	-0.001	-0.01	7.7	1.1	0.2	K0 III		300 s		
60489	115653		7 34 46.0	+2 43 30	-0.003	-0.03	6.55	0.22		A m	+46			
60725	174040		7 34 47.1	-29 24 37	-0.008	-0.03	7.4	1.2	0.2	K0 III		110 mx		
59596	14239		7 34 49.0	+63 10 18	+0.001	+0.01	8.0	1.1	0.2	K0 III		360 s		
61142	235319		7 34 49.3	-55 46 25	0.000	-0.02	7.6	1.3	3.2	G5 IV		36 s		
60105	41863		7 34 49.4	+40 52 37	-0.002	0.00	7.6	0.1	1.4	A2 V		180 s		
60298	79521		7 34 50.4	+24 57 15	+0.010	-0.36	7.35	0.67	4.7	G2 V	-130	31 s		
60929	218818		7 34 51.0	-44 17 47	-0.001	0.00	6.81	-0.03	0.6	A0 V		180 s		
60504	115655		7 34 53.1	+3 20 34	+0.001	+0.01	7.7	0.9	3.2	G5 IV		80 s		
60297	79522		7 34 55.4	+28 15 51	-0.001	-0.01	7.9	0.1	0.6	A0 V		280 s		
60526	115658		7 34 57.4	+3 16 34	-0.003	+0.02	7.7	0.9	3.2	G5 IV		80 s		
60503	115660		7 34 59.4	+4 26 23	-0.001	0.00	6.9	1.1	0.2	K0 III		220 s		
--	--		7 35 00.3	-2 20 32	-0.001	-0.02	7.9	1.1	0.2	K0 III		340 s		
60501	115661		7 35 02.8	+6 11 41	-0.002	0.00	7.6	1.6	-0.5	M2 III		420 s		
60502	115662		7 35 03.0	+5 17 48	-0.001	-0.02	7.3	0.0	0.4	B9.5 V		240 s		
60765	174050		7 35 06.5	-26 07 09	-0.001	+0.01	7.90	1.1	0.2	K0 III		340 s	6195	m
60063	41864		7 35 07.3	+48 40 51	+0.001	+0.04	6.9	0.4	2.6	F0 V		74 s		
60318	60204		7 35 08.7	+30 57 40	-0.002	+0.01	5.33	1.01	0.2	K0 III	-6	100 s	6185	m
60408	79527		7 35 11.2	+23 02 46	+0.003	0.00	7.2	0.8	3.2	G5 IV		64 s		
60700	153099		7 35 11.9	-17 32 54	-0.007	+0.15	7.5	0.8	3.2	G5 IV		73 s		
60234	41869		7 35 16.2	+40 30 22	+0.001	-0.01	7.6	0.6	4.4	G0 V		44 s		
61077	218821		7 35 16.8	-46 34 39	0.000	+0.02	7.61	0.06	0.6	A0 V		230 s		
60580	115665		7 35 17.1	+0 32 42	-0.001	-0.04	7.8	0.9	3.2	G5 IV		81 s		
60718	153101		7 35 20.5	-14 02 50	-0.007	-0.07	7.9	0.9	3.2	G5 IV		88 s		
62153	256428		7 35 21.9	-74 16 31	-0.001	+0.01	6.34	-0.02	0.4	B9.5 V		150 s		m
60863	174058		7 35 22.8	-28 22 10	-0.006	-0.02	4.64	-0.11		B8	+13		6205	m
60742	153104		7 35 23.4	-15 22 59	-0.002	-0.02	7.2	1.4	-0.3	K5 III		320 s		
60383	79531		7 35 23.9	+28 37 39	-0.001	0.00	7.1	0.1	1.7	A3 V	-8	120 s		
60487	97054		7 35 24.4	+17 22 59	0.000	-0.01	7.6	0.9	3.2	G5 IV		77 s		
60563	115666		7 35 25.0	+7 34 50	-0.003	+0.01	6.6	0.1	0.6	A0 V		160 s		
61168	218823		7 35 27.4	-49 49 53	-0.006	-0.03	7.8	1.5	0.2	K0 III		180 mx		
61029	198161		7 35 27.9	-39 24 12	+0.001	+0.02	8.0	0.5	0.6	A0 V		140 s		
60861	174063		7 35 32.5	-26 09 15	0.000	-0.01	8.0	-0.6	0.4	B9.5 V		330 s		
60620	115670		7 35 33.1	+3 59 15	-0.001	0.00	7.7	0.0	0.4	B9.5 V		280 s		
60862	174065		7 35 33.4	-26 37 40	-0.002	+0.05	6.5	0.2	1.4	A2 V		83 s		
61008	198163		7 35 37.2	-36 09 46	0.000	0.00	7.85	-0.13	0.0	B8.5 V		370 s		
62364	256429		7 35 38.1	-75 32 37	-0.031	+0.01	7.6	0.5	4.0	F8 V		54 s		
61248	235336		7 35 39.6	-52 32 02	+0.002	-0.02	4.94	1.40	-0.4	gM0	+62	120 s		
61904	256427		7 35 40.2	-70 31 25	0.000	-0.01	7.3	0.7	0.2	K0 III		270 s		
60759	134871		7 35 47.6	-8 20 49	-0.001	0.00	8.0	0.1	1.4	A2 V		200 s		
61006	198168		7 35 50.7	-33 23 33	+0.001	+0.01	7.22	-0.12	-2.6	B2.5 IV		830 s		
61194	218827		7 35 52.7	-47 09 21	0.000	0.00	7.2	1.1	-0.1	K2 III		280 s		
60922	174083		7 35 54.3	-23 32 52	0.000	+0.02	7.5	0.7	0.2	K0 III		280 s		
61246	218828		7 35 54.5	-49 56 15	+0.001	+0.03	7.8	1.7	-0.1	K2 III		140 s		
60522	79533	69 υ Gem	7 35 55.3	+26 53 45	-0.002	-0.11	4.06	1.54	-0.4	M0 III	-21	78 s		m
61087	198174		7 35 55.7	-37 26 44	+0.001	+0.01	7.26	-0.08		B9				
60335	41877		7 35 55.9	+43 01 52	-0.002	-0.05	6.50	0.4	2.6	F0 V	+21	60 s	6191	m
60856	153115		7 35 56.9	-14 42 40	0.000	-0.02	7.98	0.00		B9				

HD	SAO	Star Name	α 2000	δ 2000	μ(α)	μ(δ)	V	B-V	M$_v$	Spec	RV	d(pc)	ADS	Notes
60875	153117		7^{h}35^{m}57^s.0	-18°50'47"	-0^s.001	0".00	8.0	0.4	2.6	F0 V		120 s		
60807	134877		7 35 59.9	-9 18 47	0.000	-0.03	7.7	0.9	3.2	G5 IV		80 s		
60779	134875		7 36 01.4	-3 09 07	-0.019	+0.10	7.18	0.56		G0		11 mn		
61394	235343		7 36 01.7	-55 53 15	-0.001	+0.01	6.39	1.18		G5				
58777	6244		7 36 01.8	+76 50 01	-0.005	+0.04	8.0	1.1	-0.1	K2 III		410 mx		
60921	174087		7 36 02.5	-22 17 37	-0.001	0.00	7.76	-0.15	0.6	A0 V		270 s		m
61436	235349		7 36 03.6	-57 17 44	+0.004	+0.08	6.7	1.4	0.2	K0 III		120 s		
60855	153118		7 36 03.8	-14 29 34	-0.001	0.00	5.65	-0.10	-3.0	B2 IV	+21	470 s	6208	m
62689	256431		7 36 04.1	-77 38 01	-0.001	+0.04	6.18	1.73		K2				
59720	14247		7 36 04.3	+66 28 14	-0.003	+0.01	7.4	1.6	-0.5	M2 III	+19	390 s		
60951	174092		7 36 07.7	-22 09 38	+0.003	-0.02	6.34	0.99		G5				
60900	153124		7 36 07.8	-16 53 42	0.000	-0.04	7.8	1.1	0.2	K0 III		320 s		
60899	153120		7 36 08.4	-14 41 53	+0.002	-0.01	7.91	1.03		K0				
61026	174097		7 36 08.4	-27 30 15	+0.002	-0.02	7.00	1.49	-0.1	K2 III		160 s		
60898	153121		7 36 09.2	-14 24 31	-0.001	-0.03	7.80	1.21		K2				
60973	174094		7 36 12.3	-21 55 00	0.000	+0.02	7.9	1.0	3.2	G5 IV		64 s		
61245	218831		7 36 13.9	-44 57 27	+0.004	-0.01	6.9	0.9	0.2	K0 III		220 s		
61392	235347		7 36 14.6	-53 49 12	+0.001	+0.02	7.70	1.4	-0.3	K5 III		400 s		m
60777	115683		7 36 15.2	+2 46 57	0.000	-0.02	7.9	1.1	0.2	K0 III		330 s		
60853	134883		7 36 16.6	-8 18 41	+0.002	0.00	6.27	1.54		K2				
61834	249893		7 36 17.3	-67 22 40	+0.001	+0.01	7.9	0.1	1.4	A2 V		200 s		
61192	218830		7 36 17.3	-40 31 54	0.000	-0.05	7.5	0.6	1.4	A2 V		100 s		
61950	249895		7 36 19.1	-69 03 23	-0.003	+0.01	7.0	0.2	0.0	B8.5 V		180 s		
60944	153132		7 36 21.1	-16 16 52	0.000	0.00	7.72	2.17	-0.5	M2 III		200 s		
61290	218835		7 36 22.9	-47 14 31	-0.002	+0.02	8.0	-0.1	0.4	B9.5 V		330 s		
60676	97072		7 36 24.6	+18 10 54	-0.001	-0.04	7.8	0.1	1.4	A2 V		190 s		
61435	235351		7 36 26.2	-54 11 14	+0.002	-0.01	7.50	0.1	0.6	A0 V		240 s		m
60969	153133		7 36 27.4	-14 35 43	-0.001	-0.01	7.01	-0.08		B9				
60825	115691		7 36 28.3	+2 52 03	-0.002	-0.02	8.0	0.1	0.6	A0 V		290 s		
60437	41879		7 36 31.6	+46 10 49	-0.003	-0.03	5.65	1.54	-0.3	gK5	+29	150 s		
60776	115690		7 36 31.6	+7 21 02	-0.001	-0.01	7.8	0.1	1.4	A2 V		190 s		
61071	174112		7 36 31.8	-25 19 58	0.000	0.00	6.84	-0.07	-0.9	B6 V		350 s		m
61576	235358		7 36 33.3	-58 22 40	-0.001	+0.01	7.8	1.0	2.6	F0 V		39 s		
61243	198184		7 36 33.3	-39 51 43	0.000	+0.03	8.0	0.2	1.4	A2 V		160 s		
61138	198181		7 36 33.9	-31 19 14	+0.002	0.00	7.3	1.9	-0.5	M2 III		250 s		
59897	14250		7 36 34.2	+65 05 47	+0.001	+0.01	7.3	1.1	0.2	K0 III		260 s		
60803	115693		7 36 34.6	+5 51 42	-0.008	+0.01	5.91	0.60		F8	+4			
60997	153142		7 36 35.6	-14 29 00	-0.001	0.00	7.76	-0.05	0.4	B9.5 V		290 s	6216	m
60998	153143		7 36 36.0	-14 29 05	-0.001	0.00	7.9	0.0	0.4	B9.5 V		310 s		
61333	218837		7 36 37.7	-44 59 14	-0.001	+0.01	7.20	-0.13	-1.7	B3 V		590 s		
61068	153149		7 36 41.0	-19 42 08	-0.001	0.00	5.74	-0.19	-4.4	B2 II	+22	1100 s		
61017	153145		7 36 41.3	-14 26 36	0.000	0.00	6.68	-0.06		B9				m
61391	218841		7 36 43.9	-48 49 48	-0.001	+0.01	5.72	-0.06		B9				
60801	97078		7 36 45.8	+12 51 55	-0.002	-0.03	7.7	1.6	-0.5	M2 III		440 s		
60294	26423		7 36 46.9	+55 45 18	-0.002	-0.04	5.92	1.12	0.2	K0 III	+1	120 s		
61045	153150		7 36 47.0	-14 33 41	-0.002	+0.02	8.02	-0.12		A0				
60868	115701		7 36 47.6	+6 04 29	-0.003	0.00	7.9	0.0	0.4	B9.5 V		320 s		
60252	26422		7 36 48.6	+56 51 30	+0.002	-0.06	7.7	0.9	3.2	G5 IV		81 s		
61377	218840		7 36 50.8	-45 25 53	0.000	+0.02	7.1	1.5	0.2	K0 III		120 s		
59874	14252		7 36 51.4	+66 40 24	+0.001	-0.04	8.0	1.1	0.2	K0 III		360 s		
60821	97080		7 36 51.8	+12 04 44	-0.001	0.00	8.0	0.4	2.6	F0 V		120 s		
61390	218842		7 36 54.5	-45 01 41	-0.004	+0.01	7.9	0.9	3.2	G5 IV		76 s		
61160	174125		7 36 57.2	-24 17 48	0.000	+0.01	7.4	0.4	1.4	A2 V		94 s		
61332	218839		7 36 57.6	-40 11 33	-0.002	-0.03	7.0	1.4	-0.1	K2 III		180 s		
60272	26424		7 36 58.2	+57 32 59	+0.010	-0.18	8.0	0.9	3.2	G5 IV		52 mx		
61135	174123		7 37 00.0	-21 23 37	-0.005	+0.04	6.7	1.4	3.2	G5 IV		21 s		
61191	174134		7 37 03.5	-26 48 59	+0.001	-0.07	6.88	1.06	0.2	K0 III		190 s		
61209	174137		7 37 04.7	-27 25 18	0.000	+0.01	7.13	0.32	0.0	B8.5 V		170 s		m
60820	79553		7 37 05.4	+20 09 24	+0.001	-0.11	6.7	1.1	0.2	K0 III		98 mx		
60103	14257		7 37 05.5	+63 20 23	+0.004	-0.01	7.8	0.4	3.0	F2 V		89 s		
60848	97083		7 37 05.6	+16 54 15	0.000	-0.01	6.87	-0.20	-5.2	O8 V pe	+15	2600 s		BN Gem, q
61114	153158		7 37 08.4	-14 19 29	-0.001	0.00	7.79	1.68		K5				
60334	26428		7 37 14.0	+57 26 32	+0.002	0.00	8.0	1.1	0.2	K0 III		360 s		
60436	26430		7 37 15.0	+53 50 26	-0.003	-0.05	7.3	1.1	0.2	K0 III		270 s		
61064	134899	25 Mon	7 37 16.6	-4 06 40	-0.005	+0.01	5.13	0.44	0.7	F5 III	+46	53 ts		m
61227	174141		7 37 16.8	-23 46 30	-0.001	+0.02	6.37	0.54		F0				
60654	41887		7 37 17.8	+40 01 33	-0.001	-0.04	6.5	1.6	-0.5	M2 III	+31	250 s		
61308	198193		7 37 20.1	-30 47 32	+0.001	0.00	7.7	-0.4	0.0	B8.5 V		350 s		
60633	41888		7 37 20.8	+40 50 43	-0.003	+0.02	7.8	0.8	3.2	G5 IV		84 s		
60989	115709		7 37 20.8	+5 24 18	0.000	0.00	7.7	1.1	-0.1	K2 III		360 s		
61330	198195		7 37 22.0	-34 58 07	-0.002	+0.01	4.53	-0.09	-0.2	B8 V	+24	88 s		m
61373	198201		7 37 23.5	-36 25 03	+0.001	-0.01	7.1	1.3	0.2	K0 III		150 s		
60694	60229		7 37 25.8	+38 52 06	0.000	0.00	7.94	0.15	0.6	A0 V		240 s	6211	m
61095	134902		7 37 28.8	-2 02 14	-0.001	-0.01	7.70	1.4	-0.3	K5 III		390 s		m
60915	97087		7 37 30.4	+18 55 05	-0.001	-0.08	6.8	0.4	2.6	F0 V		70 s		

HD	SAO	Star Name	α 2000	δ 2000	μ(α)	μ(δ)	V	B-V	M_v	Spec	RV	d(pc)	ADS	Notes
60800	60231		7ʰ37ᵐ31ˢ.0	+31°36'49"	-0ˢ.001	-0".02	7.6	0.0	0.4	B9.5 V	-1	280 s		
61597	235368		7 37 31.9	-51 25 18	-0.002	-0.03	7.3	1.8	3.2	G5 IV		16 s		
60293	14261		7 37 32.6	+60 32 17	+0.002	-0.02	6.9	0.1	0.6	A0 V	-12	180 s		
61741	235369		7 37 34.3	-57 13 00	-0.001	+0.01	8.0	0.4	1.4	A2 V		120 s		
61153	134907		7 37 37.9	-9 49 30	-0.001	0.00	7.8	1.1	0.2	K0 III		320 s		
61224	153172		7 37 38.9	-14 26 28	+0.001	-0.01	6.52	-0.04		B9				m
60914	79558		7 37 40.0	+22 20 25	-0.001	-0.04	6.96	1.05		G5				
61093	115719		7 37 41.4	+1 55 40	-0.001	-0.02	6.8	0.0	0.4	B9.5 V		180 s		
61176	134911		7 37 43.1	-7 43 38	-0.001	0.00	7.1	0.1	1.7	A3 V		120 s		
61409	198205		7 37 44.7	-35 16 38	-0.003	+0.01	6.60	1.14		K0				
61223	153174		7 37 44.9	-13 25 54	0.000	-0.02	7.8	1.1	-0.1	K2 III		370 s		
61453	198210		7 37 45.1	-38 00 38	0.000	+0.06	6.38	1.48		K2				
60771	60232		7 37 48.1	+38 52 43	-0.002	-0.01	7.19	0.27	2.1	A5 V		94 s		
61037	97093		7 37 50.1	+14 20 26	-0.003	-0.01	6.6	0.0	0.4	B9.5 V		180 s		
61038	97095		7 37 50.8	+14 02 31	+0.001	-0.10	6.6	0.0	0.4	B9.5 V		32 mx		
61202	134918		7 37 51.5	-6 57 29	-0.001	0.00	6.7	0.0	0.4	B9.5 V		180 s		
60652	41893		7 37 53.9	+48 46 25	-0.003	-0.03	5.92	0.22		A3	+10			
61431	198212		7 37 57.5	-32 07 54	0.000	-0.05	6.90	0.5	3.4	F5 V		50 s		m
61236	134924		7 38 01.6	-8 07 33	0.000	0.00	6.9	0.8	3.2	G5 IV		55 s		
60406	14263		7 38 04.5	+61 32 01	+0.002	+0.03	7.2	0.5	3.4	F5 V	-41	58 s		
60847	60234		7 38 07.3	+38 15 32	-0.001	-0.02	7.5	1.6	-0.5	M4 III		400 s		
61323	153180		7 38 08.8	-16 02 38	+0.002	+0.03	7.3	1.1	0.2	K0 III		260 s		
61470	198216		7 38 11.1	-31 27 14	+0.002	-0.02	7.2	0.5	2.6	F0 V		68 s		
60936	60236		7 38 11.3	+32 26 18	+0.002	-0.01	7.6	0.9	3.2	G5 IV		77 s		
61740	235374		7 38 11.7	-51 38 41	-0.002	+0.01	7.3	0.3	0.6	A0 V		140 s		
61277	153179		7 38 11.7	-10 36 48	-0.002	-0.02	7.2	1.4	-0.3	K5 III		320 s		
61035	79562		7 38 14.4	+24 21 37	+0.002	+0.01	6.3	0.4	2.6	F0 V	+7	56 s		
60737	41897		7 38 16.3	+47 44 55	-0.002	-0.17	7.6	0.6	4.4	G0 V		44 s		
61429	174175		7 38 17.9	-25 21 53	-0.001	-0.01	4.67	-0.10		B8	+41		6246	m
61715	218852		7 38 18.2	-48 36 04	0.000	+0.03	5.68	0.65	-6.5	F4 Iab	+11	2200 s		
60614	26436		7 38 19.9	+55 12 20	-0.001	-0.06	7.9	0.5	3.4	F5 V		78 s		
60935	60238		7 38 20.6	+33 11 11	-0.001	-0.04	6.9	0.9	3.2	G5 IV		55 s		
61623	198226		7 38 24.2	-39 59 29	0.000	+0.02	6.59	-0.05		A0				
61856	235378		7 38 24.6	-54 06 58	-0.001	+0.03	7.7	1.6	-0.3	K5 III		340 s		
61558	198223		7 38 25.4	-32 45 38	0.000	-0.01	7.9	1.0	0.2	K0 III		340 s		
60693	26438		7 38 26.8	+51 52 32	-0.002	0.00	7.8	0.5	3.4	F5 V		74 s		
61557	198224		7 38 27.6	-31 51 39	-0.005	+0.03	7.5	1.1	0.2	K0 III		260 s		
60292	14262		7 38 28.9	+66 14 05	+0.002	-0.04	7.0	1.1	0.2	K0 III		230 s		
61275	115733		7 38 30.4	+0 30 18	0.000	-0.01	7.10	0.00	0.4	B9.5 V		220 s	6240	m
61642	198229		7 38 32.5	-38 46 52	+0.002	-0.03	6.19	1.02		G5				
60986	60243	70 Gem	7 38 32.8	+35 02 55	+0.003	+0.03	5.56	0.93	3.2	G5 IV	-36	24 s	6229	m
59762	6259		7 38 33.1	+73 35 45	-0.013	-0.05	7.6	0.5	3.4	F5 V		70 s		
62402	249913		7 38 36.1	-68 16 18	+0.002	-0.05	8.0	0.1	1.4	A2 V		210 s		
61949	235385		7 38 37.2	-56 13 47	-0.002	0.00	6.9	1.4	0.2	K0 III		130 s		
62091	249907		7 38 38.1	-60 37 46	0.000	+0.02	7.1	0.5	0.6	A0 V		93 s		
61512	174190		7 38 39.7	-26 19 00	+0.001	0.00	7.15	-0.11	0.4	B9.5 V		220 s		
61641	198237		7 38 43.8	-36 29 49	-0.001	0.00	5.80	-0.16		B5	+19			
61595	198230		7 38 44.3	-31 43 54	-0.001	-0.06	7.9	0.7	3.4	F5 V		54 s		
61640	198238		7 38 46.6	-35 48 14	-0.002	-0.04	8.00	1.1	0.2	K0 III		350 s		m
61763	218860		7 38 46.8	-44 49 49	-0.002	-0.01	7.9	-0.1	0.0	B8.5 V		380 s		
61593	198234		7 38 48.3	-30 27 44	-0.001	0.00	7.9	0.7	0.4	B9.5 V		120 s		
61320	115740		7 38 49.7	+3 31 49	0.000	-0.02	7.9	1.1	0.2	K0 III		340 s		
61555	174199		7 38 49.7	-26 48 13	-0.002	+0.02	3.82	-0.17		B8	+24		6255	m
61383	134936		7 38 51.3	-5 27 56	+0.015	-0.13	7.7	0.6	4.4	G0 V		47 s		
61273	115739		7 38 51.3	+7 57 58	+0.009	-0.07	7.2	0.0	0.4	B9.5 V		30 mx		
61881	235384		7 38 51.8	-50 34 36	-0.004	+0.06	7.7	0.6	3.2	G5 IV		78 s		
60960	41906		7 38 52.4	+42 28 31	-0.001	-0.03	7.9	0.5	3.4	F5 V		80 s		
60961	41907		7 38 52.5	+42 27 29	-0.002	0.00	7.50	0.5	3.4	F5 V		66 s		m
61367	134937		7 38 56.7	-0 15 39	-0.001	-0.02	7.30	1.6	-0.5	M2 III	+17	350 s		
61532	—		7 38 58.7	-20 16 33			7.90	0.1	1.7	A3 V		170 s	6254	m
61966	235391		7 39 00.4	-53 16 24	-0.001	+0.01	6.06	-0.11		A0				m
61366	115749		7 39 00.5	+0 57 40	-0.002	-0.02	7.90	0.9	3.2	G5 IV		86 s	6249	m
59641	6258		7 39 02.7	+75 47 10	-0.007	-0.13	7.9	0.5	3.4	F5 V		79 s		
61297	97115		7 39 04.3	+11 52 16	-0.002	-0.01	7.4	1.4	-0.3	K5 III		350 s		
61340	115748		7 39 05.9	+8 55 35	-0.002	+0.02	7.4	0.9	3.2	G5 IV		70 s		
61554	153195		7 39 07.0	-18 40 46	-0.002	-0.03	6.4	-0.1		B9				
61341	115753		7 39 09.7	+7 49 51	+0.001	-0.01	7.9	0.1	1.4	A2 V		200 s		
61110	60247	71 o Gem	7 39 09.9	+34 35 04	-0.003	-0.11	4.90	0.40	0.6	F3 III	+7	70 s		
61653	174213		7 39 10.9	-29 28 30	+0.001	+0.01	7.0	0.0	0.0	B8.5 V		300 s		
61219	79580		7 39 11.9	+24 13 21	-0.001	0.00	6.00	0.01	1.4	A2 V	-11	83 s		
60818	26443		7 39 12.3	+53 54 03	-0.002	-0.07	6.6	0.0	0.4	B9.5 V		60 mx		
60845	26446		7 39 14.8	+52 33 59	-0.002	-0.06	7.5	1.1	0.2	K0 III		210 mx		
61947	218872		7 39 16.9	-49 03 05	+0.001	+0.01	7.20	0.00	0.0	B8.5 V		280 s		m
61272	79583		7 39 17.8	+20 08 55	-0.001	+0.01	7.5	0.4	2.6	F0 V		95 s		
61421	115756	10 α CMi	7 39 18.1	+5 13 30	-0.047	-1.03	0.38	0.42	2.6	F5 IV	-3	3.5 t	6251	Procyon, m

HD	SAO	Star Name	α 2000	δ 2000	μ(α)	μ(δ)	V	B-V	M_v	Spec	RV	d(pc)	ADS	Notes
			^{h} ^{m} ^{s}	^{o} $'$ $''$	^{s}	$''$								
61553	153199		7 39 18.4	-16 19 03	-0.001	0.00	7.9	1.1	0.2	K0 III		340 s		
60912	26449		7 39 21.6	+50 31 43	-0.008	-0.06	6.9	0.5	3.4	F5 V		49 s		
--	198251		7 39 22.1	-38 26 42	-0.004	-0.02	8.0	1.1	-0.1	K2 III		220 mx		
61589	153204		7 39 24.2	-16 50 50	0.000	-0.01	6.5	1.1	-0.1	K2 III		200 s		
61107	41915		7 39 25.8	+40 40 16	0.000	-0.01	7.4	0.9	3.2	G5 IV		69 s		
61672	174219		7 39 26.9	-26 51 47	-0.001	0.00	6.2	0.3		B8				
61551	153200		7 39 27.1	-11 33 51	-0.001	-0.01	7.0	0.0	0.0	B8.5 V		250 s		
61831	198253		7 39 27.3	-38 18 30	-0.002	0.00	4.84	-0.20		B3 n	+26			
61338	97120	74 Gem	7 39 28.6	+17 40 29	0.000	0.00	5.05	1.56	-0.4	gM0	+28	120 s		
61832	198258		7 39 29.5	-38 51 49	-0.001	+0.04	7.7	1.1	0.2	K0 III		290 s		
61687	174223		7 39 29.7	-26 55 09	-0.002	-0.02	6.80	-0.12	-1.0	B7 IV		360 s	6264	m
56862	1195		7 39 34.7	+84 11 50	-0.009	-0.01	7.8	1.1	0.2	K0 III		330 s		
61829	198260		7 39 37.2	-36 30 13	0.000	-0.01	7.9	0.7	0.0	B8.5 V		140 s		
61650	153209		7 39 37.8	-17 52 17	-0.001	-0.01	7.4	0.1	0.6	A0 V		230 s		
61759	174230		7 39 39.3	-29 11 25	+0.003	+0.03	7.3	0.7	-0.1	K2 III		300 s		
61828	198263		7 39 40.7	-35 41 34	0.000	-0.06	7.9	0.6	2.6	F0 V		72 s		
61946	218877		7 39 41.9	-43 16 47	0.000	+0.01	7.16	0.01	0.6	A0 V		210 s		m
61799	198262		7 39 42.4	-34 21 44	+0.001	0.00	8.0	0.5	-0.1	K2 III		410 s		
61649	153211		7 39 43.2	-15 49 40	0.000	-0.01	7.2	1.4	-0.3	K5 III		310 s		
61878	198265		7 39 43.7	-38 08 22	-0.002	+0.01	5.73	-0.13		B5	+30			m
61459	115764		7 39 46.7	+7 50 27	+0.001	-0.01	7.9	1.1	0.2	K0 III		340 s		
61899	198268		7 39 47.8	-38 15 39	-0.002	0.00	5.75	-0.07		B3 n	+23			
61583	134950		7 39 48.7	-6 14 40	-0.006	0.00	7.90	0.9	3.2	G5 IV		87 s	6262	m
61827	198264		7 39 49.3	-32 34 44	-0.001	-0.03	7.65	0.62		O8				
61926	198272		7 39 52.4	-38 32 54	-0.004	0.00	7.97	-0.06		B8				
61295	60254		7 39 54.0	+32 00 35	-0.001	-0.05	6.17	0.35	2.6	F0 V	+25	47 s		
61732	174236		7 39 54.5	-23 17 00	-0.001	-0.03	7.9	0.4	3.2	G5 IV		88 s		
62036	218882		7 39 54.6	-45 58 10	0.000	-0.01	7.35	1.71	-1.4	M3 II-III		510 s		
61755	174239		7 39 55.7	-23 21 44	-0.002	0.00	7.9	0.8	0.2	K0 III		340 s		
61925	198273		7 39 57.9	-37 34 46	-0.001	-0.01	5.99	-0.04	-2.3	B3 IV	+23	360 s		
60844	26450		7 39 58.6	+59 33 48	0.000	0.00	7.1	0.1	1.4	A2 V		140 s		
61605	134953		7 39 59.2	-1 52 28	-0.003	+0.01	7.7	0.0	0.0	B8.5 V		340 s		
61606	134954		7 39 59.3	-3 35 52	+0.005	-0.29	7.20	0.95	6.3	K2 V	-21	13 ts		m
61777	174245		7 40 01.6	-24 43 22	0.000	-0.02	7.4	1.7	-0.3	K5 III		270 s		
61501	97129		7 40 02.1	+11 05 22	-0.003	+0.01	7.5	1.1	0.2	K0 III		280 s		
64723	258478		7 40 03.4	-82 34 52	-0.018	+0.05	7.5	1.5	-0.1	K2 III		220 s		
61850	198271		7 40 04.3	-30 10 08	-0.001	+0.01	7.07	1.09	0.2	K0 III		220 s		
61563	115773		7 40 06.9	+5 13 51	-0.001	-0.02	6.02	-0.04		A0	+16	19 mn	6263	m
62061	218886		7 40 07.9	-44 25 50	+0.005	+0.09	7.5	0.4	3.2	G5 IV		73 s		
61825	174247		7 40 08.3	-24 32 26	0.000	-0.03	7.5	1.3	0.2	K0 III		200 s		
60959	26453		7 40 09.6	+57 15 43	+0.001	-0.03	8.0	0.1	0.6	A0 V		300 s		
62088	218887		7 40 09.9	-45 20 57	-0.002	+0.02	7.4	1.1	-0.1	K2 III		310 s		
61774	153225		7 40 13.5	-19 39 39	0.000	+0.01	5.93	1.16	0.2	K0 III		110 s	6273	m
61252	41921		7 40 13.7	+41 09 50	+0.001	+0.03	6.8	0.1	0.6	A0 V		170 s		
61294	60257		7 40 14.5	+38 20 40	-0.004	-0.01	5.73	1.63	-0.3	K5 III	+46	130 s		
61874	174254		7 40 18.4	-26 35 25	-0.002	+0.02	8.0	0.2	1.7	A3 V		150 s		
63670	256440		7 40 19.6	-78 23 13	-0.016	+0.09	6.9	-0.7	1.4	A2 V		81 mx		
62034	198282		7 40 20.0	-39 25 19	0.000	-0.11	6.5	1.3	0.2	K0 III		130 s		
61750	153224		7 40 21.0	-11 45 08	0.000	0.00	6.5	1.1	-0.1	K2 III		210 s		
61772	153227		7 40 23.1	-15 15 50	-0.001	-0.03	5.0	1.1	-2.3	K3 II	0	290 s		
61500	97130		7 40 23.9	+19 21 27	-0.003	-0.05	7.8	0.5	3.4	F5 V		74 s		
61846	174256		7 40 24.9	-22 02 46	-0.002	+0.02	7.1	1.2	0.2	K0 III		170 s		
61944	174266		7 40 26.3	-30 04 43	-0.001	+0.05	6.92	0.00	-1.0	B5.5 V		290 s		
60982	26455		7 40 28.0	+59 05 52	0.000	-0.01	8.0	1.1	0.2	K0 III		360 s		
61943	174269		7 40 28.8	-30 02 55	-0.003	+0.03	7.4	0.3	0.4	B9.5 V		150 s		
62851	249919		7 40 29.4	-69 50 47	-0.001	+0.04	7.8	1.1	0.2	K0 III		330 s		
--	--		7 40 29.6	+34 41 31	0.000	-0.07	4.1	1.1	0.2	K0 III		60 s		
61919	174268		7 40 30.2	-28 13 10	0.000	+0.01	7.9	1.1	-0.1	K2 III		390 s		
61989	198283		7 40 30.4	-33 25 35	-0.005	+0.13	7.9	0.0	4.0	F8 V		60 s		
51802	1168		7 40 31.2	+87 01 12	-0.048	-0.03	5.07	1.63	-0.5	M2 III	-25	120 s		
61749	134969		7 40 35.4	-8 11 09	+0.001	-0.04	6.01	0.15		A2				
62377	235411		7 40 40.6	-55 39 05	-0.003	0.00	7.3	0.2	0.6	A0 V		160 s		
61987	174273		7 40 43.4	-27 56 44	0.000	-0.02	6.76	-0.16		B8				
61630	97136		7 40 47.2	+13 46 15	+0.001	-0.02	6.4	1.1	0.2	K0 III	+5	180 s		
63056	256435		7 40 47.4	-71 45 49	0.000	-0.01	7.1	1.1	0.2	K0 III		240 s		
61721	115795		7 40 48.8	+4 04 19	-0.002	-0.09	7.7	1.1	0.2	K0 III		150 mx		
61106	26459	23 Lyn	7 40 49.5	+57 04 58	-0.002	-0.01	6.06	1.46	0.2	gK0	-13	67 s		
61984	174275		7 40 50.4	-24 16 26	+0.001	+0.01	7.70	1.36	0.2	K0 III		180 s		
62485	235415		7 40 50.4	-58 52 15	-0.002	+0.01	7.3	0.6	0.0	B8.5 V		120 s		
62400	235414		7 40 52.4	-54 42 25	-0.001	+0.02	6.8	0.7	2.1	A5 V		43 s		
62058	198286		7 40 52.7	-31 39 39	+0.001	+0.01	6.65	1.20	-8.0	G0 Ia		5200 s		R Pup, q
62296	218898		7 40 52.9	-49 41 51	+0.002	-0.01	7.9	0.7	3.0	F2 V		40 s		
61337	41928		7 40 56.8	+46 36 56	+0.002	-0.03	7.9	1.1	0.2	K0 III		350 s		
62212	218896		7 40 57.5	-43 11 04	0.000	0.00	7.33	-0.04	0.4	B9.5 V		240 s		
61603	79607		7 40 58.4	+23 01 08	-0.001	0.00	5.89	1.58	-0.3	K5 III	+39	160 s		

HD	SAO	Star Name	α 2000	δ 2000	μ(α)	μ(δ)	V	B–V	M$_v$	Spec	RV	d(pc)	ADS	Notes
62151	198291		7^{h}41^{m}00^s.2	−36°52'08"	0^s.000	−0".02	7.5	1.4	−0.1	K2 III		250 s		
62295	218899		7 41 01.4	−47 49 50	+0.003	+0.02	7.6	−0.4	1.4	A2 V		180 s		
61957	153246		7 41 09.8	−17 08 42	−0.002	0.00	8.01	−0.06	0.0	B8.5 V		400 s		
62027	174289		7 41 10.3	−21 14 50	0.000	0.00	7.7	0.9	−0.3	K5 III		390 s		
61363	41934		7 41 12.4	+48 07 53	−0.005	−0.13	5.56	1.01	3.2	G5 IV	+40	25 mn		
61499	60267		7 41 13.1	+39 12 42	−0.001	−0.02	7.8	0.1	0.6	A0 V		270 s		
62150	198293		7 41 13.2	−32 38 40	−0.002	−0.02	7.67	0.53	−6.8	B3 Ia		3100 s		
62227	198297		7 41 14.2	−39 14 10	−0.001	+0.01	7.73	−0.05		B9				
61935	134986	26 α Mon	7 41 14.8	−9 33 04	−0.005	−0.02	3.93	1.02	0.2	K0 III	+11	54 s		
62226	198298		7 41 15.7	−38 32 01	−0.002	+0.01	5.41	−0.15		B3	+40			
62279	218903		7 41 21.0	−41 53 21	−0.005	+0.11	7.3	0.8	3.4	F5 V		44 s		
62318	218905		7 41 21.7	−44 37 56	0.000	0.00	6.41	0.00	0.4	B9.5 V		160 s		
62278	218902		7 41 21.8	−41 09 53	−0.001	0.00	7.40	−0.06	0.4	B9.5 V		200 s		
62277	218901		7 41 22.8	−40 16 20	−0.001	−0.02	7.9	1.2	−0.3	K5 III		450 s		
62082	174298		7 41 23.6	−22 20 14	+0.001	−0.01	6.3	2.0	−0.5	M2 III		120 s		
62317	218906		7 41 25.4	−43 38 09	−0.001	−0.01	7.04	0.01	0.6	A0 V		190 s		
61717	79615		7 41 34.6	+22 07 03	0.000	−0.01	7.4	0.1	1.4	A2 V		160 s		
62225	198302		7 41 35.1	−33 44 27	+0.002	−0.01	7.8	0.1	0.0	B8.5 V		270 s		
61887	115813		7 41 35.2	+3 37 29	+0.001	−0.03	5.95	−0.04		A0 n	−24			
61747	79616		7 41 37.0	+23 05 27	0.000	0.00	8.0	0.1	0.6	A0 V		310 s		
62191	174311		7 41 39.7	−27 13 58	0.000	+0.01	7.92	−0.11	0.0	B8.5 V		380 s		
61600	60275		7 41 41.5	+37 25 47	−0.001	−0.01	8.00	0.1	1.7	A3 V		180 s	6276	m
62018	134995		7 41 45.0	−7 09 02	−0.003	+0.01	8.0	0.4	2.6	F0 V		120 s		
62171	174310		7 41 45.6	−22 24 15	−0.003	+0.04	6.8	1.0	1.7	A3 V		32 s		
61837	97150		7 41 47.1	+15 34 57	−0.001	0.00	7.5	1.1	0.2	K0 III		280 s		
61933	115815		7 41 47.2	+4 23 25	−0.001	−0.02	7.9	0.0	0.4	B9.5 V		320 s		
62612	235421		7 41 47.8	−56 18 53	−0.002	+0.01	6.7	0.6	0.0	B8.5 V		91 s		
61361	26467		7 41 47.9	+54 45 57	−0.001	0.00	6.8	1.1	0.2	K0 III		210 s		
63295	256438	ζ Vol	7 41 49.3	−72 36 22	+0.007	+0.02	3.95	1.04	0.2	K0 III	+48	54 s		m
61885	97154		7 41 51.7	+13 28 50	−0.003	−0.02	5.74	1.67	−0.5	M2 III	+7	160 s		
65001	258480		7 41 57.2	−82 20 30	+0.017	−0.03	7.9	1.0	0.2	K0 III		340 s		
62376	198315		7 41 57.9	−38 31 43	−0.002	0.00	6.54	−0.10		B8				
62375	198317		7 41 59.9	−38 20 55	−0.003	+0.01	7.9	0.6	3.2	G5 IV		88 s		
62316	198311		7 42 01.2	−32 04 44	+0.001	0.00	7.0	0.2	2.6	F0 V		77 s		
62128	153264		7 42 01.9	−11 51 10	−0.002	+0.04	7.9	0.9	3.2	G5 IV		88 s		
61913	97157		7 42 03.1	+14 12 31	0.000	−0.01	5.56	1.64		M3 s	−16	18 mn		
61836	79621		7 42 05.0	+22 55 43	−0.001	−0.03	7.6	1.1	0.2	K0 III		300 s		
62315	198312		7 42 06.6	−30 11 13	−0.001	−0.01	6.96	−0.15	−1.1	B5 V		410 s		
62758	235430		7 42 10.2	−58 37 51	−0.001	+0.01	6.4	−0.1	−1.6	B5 IV	−4	390 s		
62663	235428		7 42 11.6	−55 10 43	−0.001	+0.02	6.9	0.5	0.0	B8.5 V		110 s		
60405	6273		7 42 14.7	+75 33 51	−0.011	−0.08	7.4	0.9	3.2	G5 IV		69 s		m
61912	79623		7 42 15.4	+20 58 35	−0.001	−0.03	7.90	1.15	3.2	G5 IV		43 s		
62164	153273		7 42 17.4	−10 52 47	0.000	−0.01	7.70	1.9		S3.6				SU Mon, v
61599	41945		7 42 18.5	+47 13 26	0.000	0.00	8.0	0.0	0.4	B9.5 V		330 s		
63088	249934		7 42 18.5	−66 54 14	0.000	−0.04	7.9	1.1	0.2	K0 III		350 s		
62312	174337		7 42 19.2	−27 11 00	0.000	+0.02	7.88	−0.12	0.0	B8.5 V		380 s		
61805	60283		7 42 19.3	+30 10 05	0.000	0.00	7.1	1.1	−0.1	K2 III		270 s		
62241	153280		7 42 20.8	−19 06 24	0.000	−0.04	7.8	0.5	3.4	F5 V		74 s		
62805	235434		7 42 20.9	−58 40 12	−0.001	0.00	7.60	0.00	0.2	B9 V		300 s		m
62714	235432		7 42 23.8	−56 08 47	−0.001	0.00	6.7	1.1	0.4	B9.5 V		39 s		
--	26472		7 42 24.5	+52 18 32	−0.003	−0.11	8.0	1.3	3.2	G5 IV		43 s		
62217	153278		7 42 25.0	−13 50 56	+0.003	−0.05	8.0	0.1	0.6	A0 V		290 s		
62358	174342		7 42 26.0	−27 43 48	−0.001	−0.01	7.40	−0.06	0.4	B9.5 V		250 s		
61768	60281		7 42 26.7	+37 06 20	0.000	−0.03	7.8	0.4	2.6	F0 V		110 s		
62503	198332		7 42 30.7	−39 12 05	−0.001	+0.01	7.26	−0.08		B9				
62123	115827		7 42 34.4	+4 49 22	−0.001	−0.04	7.9	0.8	3.2	G5 IV		89 s		
62447	198331		7 42 34.6	−36 22 30	−0.001	−0.01	7.61	−0.11	0.0	B8.5 V		330 s		
62850	235436		7 42 36.2	−59 17 48	−0.006	+0.18	7.19	0.64	3.0	G2 IV		69 s		
62202	153288		7 42 36.7	−16 04 09	−0.002	+0.02	7.2	1.1	0.2	K0 III		240 s		
62542	218917		7 42 37.1	−42 13 46	−0.001	+0.02	8.04	0.17	−1.1	B5 V		290 s		
62560	--		7 42 40.9	−42 33 09			7.80	0.00	0.0	B8.5 V		350 s		m
61997	79628		7 42 43.0	+22 13 24	−0.001	−0.07	7.13	0.41		F5				
61859	60291		7 42 43.5	+34 00 01	−0.006	−0.01	6.0	0.4	2.6	F0 V	−11	48 s		
62161	115831		7 42 45.9	+4 04 13	−0.002	−0.04	8.0	0.5	3.4	F5 V		82 s		
62412	174356		7 42 48.0	−26 21 04	−0.001	−0.03	5.64	0.99	0.2	K0 III	−18	120 s		
62181	115834		7 42 50.1	+1 36 35	0.000	+0.02	7.6	0.1	1.7	A3 V		160 s		
61396	14296		7 42 50.3	+61 09 26	−0.002	−0.02	8.0	1.1	0.2	K0 III		360 s		
62826	235437		7 42 51.3	−55 18 45	−0.001	+0.01	7.6	0.4	−1.0	B5.5 V		260 s		
63429	256445		7 42 52.2	−71 32 29	−0.001	+0.03	7.6	1.4	−0.3	K5 III		380 s		
62897	235440		7 42 53.3	−58 13 48	−0.004	+0.03	6.21	1.06	0.2	K0 III		150 s		
61745	41955		7 42 54.2	+45 49 03	+0.001	−0.07	7.8	1.1	0.2	K0 III		170 mx		
62559	--		7 42 54.3	−38 54 09			8.0	−0.3	2.6	F0 V		120 s		
62393	174360		7 42 56.2	−21 59 43	+0.005	−0.14	6.9	0.9	0.2	K0 III		110 mx		
62644	218923		7 42 57.1	−45 10 24	−0.006	−0.56	5.06	0.78	3.2	G5 IV	+23	23 ts		
62428	174365		7 42 57.9	−25 50 23	+0.001	0.00	8.0	−0.5	1.4	A2 V		210 s		

HD	SAO	Star Name	α 2000	δ 2000	μ(α)	μ(δ)	V	B-V	M_v	Spec	RV	d(pc)	ADS	Notes
62351	153301		7^{h}42^{m}58.8	−17°03′44″	−0.001	+0.01	6.54	0.81		G5			6315	m
61497	26474	24 Lyn	7 43 00.4	+58 42 37	−0.005	−0.05	4.99	0.08	0.0	A3 III	+9	100 s	6285	m
62539	198339		7 43 03.0	−33 18 46	−0.007	+0.04	7.9	0.1	2.6	F0 V		110 s		
62475	174367		7 43 03.2	−27 40 08	0.000	−0.01	6.9	0.1	1.7	A3 V		110 s		
62264	115839		7 43 05.4	+0 11 21	0.000	−0.01	6.19	1.02	0.2	K0 III	+8	160 s	6313	m
62756	218928		7 43 06.8	−49 59 34	−0.001	−0.01	6.6	0.4	0.6	A0 V		94 s		
62595	198342		7 43 06.9	−38 51 51	0.000	−0.02	6.7	1.2	3.2	G5 IV		29 s		
62578	198343		7 43 11.9	−36 03 01	−0.002	+0.01	5.60	−0.13		B8 n	−1			
62305	135023		7 43 14.3	−4 07 53	0.000	−0.06	7.1	0.4	2.6	F0 V		80 s		
61144	14289		7 43 16.5	+69 20 11	0.000	−0.05	7.2	1.1	0.2	K0 III		260 s		
62577	198345		7 43 16.7	−33 48 09	−0.002	+0.03	7.5	0.5	0.6	A0 V		120 s		
62044	79638	75 σ Gem	7 43 18.7	+28 53 01	+0.005	−0.23	4.28	1.12	0.0	K1 III	+46	40 mx		m
63428	249940		7 43 19.8	−70 03 50	+0.004	−0.10	7.7	0.6	4.4	G0 V		46 s		
62469	174379		7 43 20.3	−22 17 55	+0.007	−0.01	7.7	0.3	3.2	G5 IV		80 s		
62286	115846		7 43 21.1	+4 56 41	−0.002	−0.05	6.6	1.1	0.2	K0 III		190 s		
62848	235442		7 43 21.4	−52 09 49	−0.007	+0.17	6.6	0.4	4.4	G0 V		28 s		
62141	79641		7 43 22.1	+22 23 58	−0.002	+0.01	6.21	0.93	0.2	K0 III	−3	160 s		
62642	−−		7 43 28.4	−37 44 13			7.59	0.01		A0				
62943	235446		7 43 29.4	−54 14 30	−0.001	0.00	6.90	0.9	3.2	G5 IV		55 s		m
62069	60302		7 43 29.7	+32 26 48	0.000	−0.03	7.7	1.6	−0.5	M2 III		440 s		
62323	115851		7 43 31.0	+3 29 09	+0.001	−0.09	7.50	0.5	4.0	F8 V		50 s		m
62367	135030		7 43 32.1	−4 40 51	−0.001	−0.01	7.14	−0.08		B6 e				
62576	174391	1 Pup	7 43 32.3	−28 24 40	−0.001	+0.02	4.59	1.63	−0.3	gK5	+33	79 s	6324	m
62641	198346		7 43 32.6	−36 41 29	0.000	+0.01	7.4	2.0	−0.3	K5 III		170 s		
62783	218933		7 43 33.6	−45 51 51	−0.001	+0.01	7.23	1.72		K5				
62423	153309		7 43 35.2	−10 08 33	−0.001	−0.03	7.6	1.1	−0.1	K2 III		340 s		
62535	174386		7 43 35.6	−21 16 55	−0.002	−0.01	7.96	−0.04	0.6	A0 V		300 s		
62385	135031		7 43 36.5	−1 20 55	0.000	−0.03	7.9	1.1	−0.1	K2 III		380 s		
62386	135034		7 43 37.2	−4 42 30	−0.003	−0.02	6.8	0.9	3.2	G5 IV		53 s		
62739	218931		7 43 39.0	−41 28 04	0.000	−0.01	7.0	1.8	−0.3	K5 III		140 s		
62575	174397		7 43 39.1	−27 14 07	+0.001	−0.02	8.0	0.4	0.2	K0 III		360 s		
62555	174395		7 43 39.1	−25 30 14	0.000	−0.01	6.55	0.06		A3				m
62713	218932		7 43 41.9	−40 56 03	+0.011	−0.19	5.17	1.10	−0.2	gK3	+53	120 s		
62712	198352		7 43 42.8	−38 12 07	−0.003	0.00	6.41	−0.15		B9 VI p	−6			
61657	26483		7 43 46.5	+59 17 20	+0.001	0.00	7.7	0.4	2.6	F0 V		100 s		
61995	41963		7 43 47.2	+43 47 41	−0.008	−0.02	7.2	0.4	3.2	G5 IV		63 s		
62753	218935		7 43 48.0	−40 18 33	+0.001	+0.01	6.63	−0.12	−2.5	B2 V ne		640 s		
62686	198353		7 43 48.4	−36 13 06	−0.001	0.00	7.9	1.0	−0.1	K2 III		390 s		
62623	174400	3 Pup	7 43 48.4	−28 57 18	−0.001	0.00	3.96	0.18	−7.5	A2 Ia	+25	1700 s		
62347	115858		7 43 49.9	+9 36 13	−0.003	−0.02	7.9	0.1	1.4	A2 V		200 s		
62321	97192		7 43 53.4	+15 15 27	−0.001	−0.02	7.4	0.4	2.6	F0 V		93 s		
62530	153319		7 43 55.4	−16 10 41	−0.002	0.00	7.5	0.0	0.4	B9.5 V		260 s		
62230	79649		7 43 57.9	+24 14 38	−0.002	−0.02	6.9	0.2	2.1	A5 V	+28	89 s		
63406	249941		7 44 00.7	−67 12 08	0.000	+0.03	8.0	1.6	−0.5	M2 III		490 s		
62803	198364		7 44 02.6	−38 49 05	0.000	+0.01	7.38	−0.06		B9				
62846	218940		7 44 03.2	−42 58 40	−0.002	+0.01	7.4	1.1	−0.5	M2 III		390 s		
61931	26488		7 44 04.1	+50 26 01	−0.001	−0.03	5.27	0.1		A0	+6	24 mn		
62711	198358		7 44 05.1	−31 19 57	0.000	+0.01	7.5	1.1	3.2	G5 IV		46 s		
62710	198359		7 44 05.2	−31 12 59	−0.002	+0.03	7.9	0.7	3.2	G5 IV		88 s		
62285	79650	76 Gem	7 44 06.8	+25 47 03	−0.001	−0.02	5.31	1.54	−0.3	K5 III	+3	130 s		
62437	115864		7 44 07.3	+2 24 18	−0.003	−0.02	6.47	0.20		F0	+15			
62158	60304		7 44 09.5	+38 50 16	+0.001	−0.02	7.6	1.6	−0.5	M4 III		420 s		
62781	198366		7 44 09.5	−36 03 46	−0.007	+0.07	5.80	0.31	2.1	dA5		48 s		
62366	97196		7 44 09.6	+14 10 24	−0.002	−0.01	7.5	1.6	−0.5	M4 III		400 s		
62735	198360		7 44 09.6	−31 40 48	−0.001	+0.02	7.5	0.8	2.6	F0 V		48 s		
62513	135042		7 44 10.8	−3 15 06	−0.003	−0.03	8.0	0.5	4.0	F8 V		62 s		
63008	235458		7 44 12.4	−50 27 24	−0.013	+0.14	6.64	0.52		F5		14 mn		m
63584	249943		7 44 13.1	−69 49 17	−0.009	+0.01	6.18	−0.06		A0				
62346	79652		7 44 13.8	+20 12 06	+0.005	−0.11	7.3	0.9	3.2	G5 IV		67 s		
62407	97199		7 44 13.9	+12 51 33	−0.002	−0.05	6.4	1.1	0.2	K0 III	+26	180 s		
−−	235462		7 44 16.6	−50 27 59	−0.011	+0.14	7.9	0.7	4.4	G0 V		51 s		
62749	198365		7 44 18.1	−30 14 45	+0.002	0.00	7.8	1.3	0.6	A0 V		46 s		
62549	135048		7 44 19.8	−5 03 18	−0.001	+0.19	7.69	0.62	4.5	G1 V	+85	42 s		
62068	41969		7 44 20.4	+45 22 08	−0.001	0.00	7.60	1.1	0.2	K0 III		300 s	6316	m
62587	135052		7 44 23.3	−9 33 51	0.000	−0.01	8.0	0.1	1.4	A2 V		200 s		
62912	218946		7 44 26.7	−40 33 02	−0.001	0.00	7.9	0.7	0.2	K0 III		340 s		
62345	79653	77 κ Gem	7 44 26.8	+24 23 52	−0.002	−0.06	3.57	0.93	0.3	G8 III	+21	45 s	6321	m
62365	79657		7 44 32.0	+23 20 42	+0.001	−0.03	7.7	1.4	−0.3	K5 III		400 s		
62747	174433		7 44 33.9	−24 40 26	−0.002	+0.01	5.62	−0.19	−3.5	B1 V		660 s		
62893	198379		7 44 34.1	−37 56 36	−0.002	0.00	5.87	−0.12	−0.2	B8 V n	+37	160 s		
62874	198377		7 44 36.3	−35 07 59	−0.002	−0.06	8.0	0.5	1.4	A2 V		100 s		
62319	60312		7 44 38.0	+31 06 50	0.000	−0.01	7.80	0.2	2.1	A5 V		140 s	6323	m
63007	218949		7 44 40.1	−47 00 30	−0.001	+0.01	6.8	−0.1	0.4	B9.5 V		190 s		
63178	235476		7 44 40.7	−54 38 23	0.000	+0.02	7.3	1.7	−0.3	K5 III		250 s		
62420	79660		7 44 42.7	+22 27 20	0.000	0.00	8.0	1.1	0.2	K0 III		360 s		

HD	SAO	Star Name	α 2000	δ 2000	μ(α)	μ(δ)	V	B−V	M_v	Spec	RV	d(pc)	ADS	Notes
			h m s	° ′ ″	s	″								
62702	153337		7 44 43.1	−17 41 37	0.000	0.00	7.7	1.1	0.2	K0 III		310 s		
63513	249944		7 44 43.9	−66 04 19	−0.002	+0.04	6.38	0.95		G5				
63365	249942		7 44 44.7	−62 27 06	−0.001	+0.04	7.7	1.1	−0.1	K2 III		360 s		
62938	198386		7 44 44.8	−38 03 13	−0.002	+0.02	7.59	−0.02		A0				
60843	6288		7 44 45.9	+76 04 35	−0.003	+0.02	8.0	1.1	0.2	K0 III		360 s		
62728	153338		7 44 47.7	−15 05 06	−0.006	+0.03	8.0	0.4	3.0	F2 V		97 s		
62157	41971		7 44 53.4	+48 43 42	−0.003	−0.01	7.9	0.4	3.0	F2 V		94 s		
62301	60313		7 44 56.2	+39 33 23	+0.004	−0.69	6.77	0.54	4.0	F8 V	−3	36 s		
62523	97211		7 45 00.6	+15 53 41	0.000	−0.03	7.8	0.1	0.6	A0 V		270 s		
62992	198389		7 45 01.6	−38 09 30	0.000	0.00	7.88	0.12		A0				
62457	79663		7 45 01.9	+23 01 10	+0.001	−0.01	7.5	1.1	0.2	K0 III		290 s		
62284	41977		7 45 02.7	+41 11 41	−0.003	+0.02	8.0	1.1	0.2	K0 III		360 s		
62196	41974		7 45 04.2	+48 23 41	−0.007	+0.05	7.6	0.4	2.6	F0 V		100 s		
62991	198390		7 45 04.6	−37 53 16	−0.003	+0.01	6.51	−0.11	−2.2	B5 III n	+24	530 s		
63176	235480		7 45 07.3	−50 17 16	+0.002	+0.02	7.50	0.4	3.0	F2 V		79 s		m
63154	218957		7 45 08.1	−48 54 56	0.000	−0.02	7.6	0.6	0.6	A0 V		100 s		
63033	198396		7 45 08.4	−39 50 55	0.000	+0.01	7.9	0.7	−0.1	K2 III		390 s		
62510	79665	79 Gem	7 45 09.2	+20 18 58	−0.001	0.00	6.30	0.00	0.6	A0 V	−12	140 s		
62888	174454		7 45 09.8	−24 39 10	+0.001	0.00	7.6	0.8	0.0	B8.5 V		100 s		
62699	135062		7 45 12.2	−4 27 16	0.000	−0.06	6.9	1.4	−0.3	K5 III		240 mx		
65517	258483		7 45 15.2	−81 58 03	+0.004	0.00	7.9	−0.4	1.4	A2 V		200 s		
63032	198398		7 45 15.2	−37 58 07	−0.001	0.00	3.59	1.72	−5.9	cK	+17	530 s		
62937	174459		7 45 17.9	−27 38 29	+0.002	−0.03	8.0	0.4	0.2	K0 III		360 s		
63118	218955		7 45 18.0	−43 45 08	−0.001	0.00	6.03	−0.07		B5				
62697	135063		7 45 18.5	−0 26 19	−0.001	−0.02	7.80	0.1	0.6	A0 V		270 s		m
62509	79666	78 β Gem	7 45 18.9	+28 01 34	−0.047	−0.05	1.14	1.00	0.2	K0 III	+3	11 ts	6335	Pollux, m
62342	41979		7 45 21.1	+43 01 51	+0.002	−0.03	8.0	0.9	3.2	G5 IV		93 s		
63080	198400		7 45 21.4	−39 20 42	−0.003	+0.02	7.18	−0.03		A0				
62724	135064		7 45 21.7	−0 30 49	−0.003	−0.01	7.7	0.4	3.0	F2 V		88 s		
63153	218961		7 45 24.9	−44 52 58	−0.003	−0.01	7.57	0.09	0.6	A0 V		220 s		
63079	198401		7 45 27.2	−37 35 44	−0.003	+0.03	6.96	−0.07		A0				
63117	218960		7 45 28.1	−40 44 21	+0.001	−0.01	7.9	0.9	0.2	K0 III		310 s		
62863	153362	2 Pup	7 45 28.6	−14 41 10	−0.002	−0.02	6.8	0.1	0.6	A0 V		170 s	6348	m
62864	153363	2 Pup	7 45 29.0	−14 41 27	−0.001	−0.03	6.09	0.11		A0			6348	m
62816	153359		7 45 29.2	−10 09 39	+0.001	0.00	7.7	0.5	4.0	F8 V		54 s		
63427	235492		7 45 31.2	−58 57 17	−0.002	+0.01	7.84	0.90	0.3	G5 III		310 s		m
62567	79672		7 45 32.1	+25 59 20	+0.001	−0.01	7.35	1.53	−0.3	K5 III		330 s	6340	m
62671	115890		7 45 34.1	+7 43 53	−0.001	−0.01	7.8	1.1	0.2	K0 III		330 s		
63077	198404		7 45 34.8	−34 10 22	−0.024	+1.66	5.37	0.60	4.4	G0 V	+102	17 ts		
63382	235490		7 45 35.6	−56 43 21	0.000	+0.02	6.12	0.39		F0				m
63219	218965		7 45 36.1	−47 35 09	−0.004	−0.04	8.0	0.5	3.4	F5 V		51 s		
62488	60322		7 45 41.3	+34 33 19	−0.001	−0.01	8.00	0.1	1.7	A3 V		180 s	6337	m
62454	60320		7 45 42.2	+39 32 50	−0.004	−0.01	7.1	0.4	2.6	F0 V		80 s		
62834	135074		7 45 42.7	−5 40 43	−0.001	0.00	7.1	0.1	0.6	A0 V		190 s		
61883	14310		7 45 42.8	+64 43 00	−0.006	−0.03	7.9	0.5	3.4	F5 V		80 s		
63255	218966		7 45 42.9	−48 21 18	−0.001	+0.04	7.6	1.3	−0.1	K2 III		300 s		
63050	174477		7 45 46.6	−29 57 01	+0.005	−0.01	8.0	0.6	0.6	A0 V		57 mx		
62883	135076		7 45 50.8	−7 31 46	−0.002	−0.03	7.6	0.5	4.0	F8 V		52 s		
63028	174475		7 45 51.8	−24 15 21	−0.001	+0.01	6.74	−0.14	−2.3	B3 IV		600 s		
61907	14311		7 45 52.2	+65 09 13	−0.002	−0.06	7.0	0.4	3.0	F2 V		63 s	6319	m
62615	79677		7 45 52.9	+26 43 32	−0.001	−0.01	8.0	0.0	0.0	B8.5 V		400 s		
63252	218971		7 45 54.2	−44 45 42	0.000	−0.04	7.5	0.8	0.2	K0 III		280 s		
62952	153372	4 Pup	7 45 56.8	−14 33 50	−0.001	0.00	5.04	0.33		F0	−2	16 mn		
63046	174482		7 46 01.4	−23 15 35	+0.001	+0.02	7.4	−0.2	0.0	B8.5 V		300 s		
63345	218975		7 46 01.6	−49 31 54	−0.002	−0.02	8.0	1.7	3.2	G5 IV		53 s		
62257	26496		7 46 01.7	+55 38 31	−0.001	−0.04	7.4	0.1	0.6	A0 V		230 s		
62902	135079		7 46 02.2	−6 46 21	+0.003	−0.10	5.49	1.38	−0.3	K5 III	−33	140 s		
63493	235498		7 46 02.2	−57 14 18	−0.003	+0.03	7.9	1.4	−0.1	K2 III		300 s		
63045	174487		7 46 03.6	−23 08 31	−0.007	−0.04	6.9	0.7	4.0	F8 V		31 s		
62721	97221	81 Gem	7 46 07.4	+18 30 36	−0.005	−0.06	4.88	1.45	−0.3	K5 III	+81	110 s		
62720	79680		7 46 07.9	+21 07 20	+0.002	0.00	7.41	0.38		F2			6347	m
63403	235494		7 46 09.0	−51 33 02	−0.002	0.00	7.9	0.7	4.0	F8 V		50 s		
62767	97222		7 46 09.3	+16 26 18	0.000	−0.01	7.4	1.1	0.2	K0 III		270 s		
62901	135081		7 46 10.0	−3 57 37	−0.001	+0.03	8.0	0.1	0.6	A0 V		290 s		
62695	79679		7 46 10.3	+23 32 55	0.000	−0.01	7.5	0.0	0.4	B9.5 V		260 s		
63215	198416		7 46 10.3	−37 56 02	−0.004	0.00	5.85	−0.11	0.2	B9 V n	+28	130 s		
63344	218978		7 46 13.2	−47 41 25	−0.001	0.00	8.0	0.4	0.0	B8.5 V		230 s		
62378	26498		7 46 13.9	+51 01 32	+0.001	0.00	7.0	1.4	−0.3	K5 III		290 s		
62832	97224	11 CMi	7 46 16.2	+10 46 06	−0.002	−0.02	5.30	0.01	1.2	A1 V	+31	66 s		
62766	97223		7 46 16.4	+17 00 50	+0.002	−0.03	7.80	0.5	3.4	F5 V		76 s	6349	m
62997	153378		7 46 17.2	−12 24 54	−0.001	−0.04	7.9	1.1	0.2	K0 III		330 s		
64142	256447		7 46 18.4	−71 43 15	+0.002	+0.04	7.7	0.4	3.0	F2 V		87 s		
63343	218979		7 46 18.8	−46 09 36	−0.002	−0.01	7.60	−0.06	−0.9	B6 V		500 s		
63251	198419		7 46 19.0	−37 53 49	+0.001	+0.03	7.69	0.02		B9				
63148	174500		7 46 19.4	−28 54 37	−0.001	+0.02	7.85	0.25	2.6	F0 V		110 s		

HD	SAO	Star Name	α 2000	δ 2000	μ(α)	μ(δ)	V	B-V	M_V	Spec	RV	d(pc)	ADS	Notes
63214	198418		7 46 21.9	-33 25 33	-0.002	-0.02	7.9	0.0	0.4	B9.5 V		290 s		
63342	218980		7 46 21.9	-45 05 16	-0.002	-0.01	7.8	1.5	-0.1	K2 III		230 s		
63685	249956		7 46 23.1	-61 25 49	-0.011	+0.27	7.39	0.76	5.2	G5 V		25 s		
63402	218982		7 46 24.1	-49 14 28	-0.002	+0.04	7.2	-0.2	1.4	A2 V		140 s		
63127	174499		7 46 26.9	-21 32 46	0.000	-0.02	7.8	1.2	-0.5	M2 III		470 s		
63235	198420		7 46 26.9	-33 36 01	-0.003	+0.03	7.4	1.5	-0.3	K5 III		350 s		
63627	235505		7 46 27.1	-59 05 21	-0.001	+0.02	7.3	0.9	4.0	F8 V		28 s		
62140	14322	49 Cam	7 46 27.4	+62 49 50	-0.005	-0.06	6.49	0.26		F0 p	+2			
63308	198424		7 46 33.3	-40 03 35	-0.001	0.00	6.57	-0.13	-1.7	B3 V		440 s		
63291	198423		7 46 33.3	-37 46 22	-0.005	-0.04	6.35	1.34		K0				
63025	135086		7 46 34.1	-6 20 25	-0.003	-0.01	8.0	0.1	1.4	A2 V		200 s		
63068	153385		7 46 35.0	-11 41 20	+0.002	-0.04	7.8	0.1	0.6	A0 V		270 s		
62969	115912		7 46 37.1	+1 51 27	0.000	-0.02	7.40	0.1	0.6	A0 V		230 s	6361	m
62647	60328		7 46 39.2	+37 31 03	+0.002	+0.01	5.3	1.6	-0.5	M2 III	-35	150 s		
62858	97226		7 46 39.4	+16 38 55	-0.001	0.00	7.5	0.1	0.6	A0 V		250 s		
63490	235500		7 46 39.6	-50 59 51	-0.006	+0.04	7.1	0.4	1.4	A2 V		82 s		
62066	14321	51 Cam	7 46 40.1	+65 27 21	+0.006	+0.02	5.92	1.18	0.2	K0 III	-29	110 s		
62979	115913		7 46 43.3	+3 37 49	-0.002	-0.01	7.9	0.0	0.4	B9.5 V		320 s		
63451	218988		7 46 43.9	-48 05 44	-0.002	+0.02	7.2	0.1	1.4	A2 V		150 s		S Pup, q
63112	153389		7 46 44.8	-12 40 31	+0.001	0.00	6.39	-0.02		B9				
63165	174516		7 46 45.9	-20 49 51	-0.001	+0.02	7.20	0.00	0.4	B9.5 V		230 s	6368	m
63449	218987		7 46 46.1	-46 48 05	0.000	+0.01	7.40	0.00	0.0	B8.5 V		300 s		m
63095	135093		7 46 49.9	-8 49 13	0.000	-0.01	7.9	0.9	3.2	G5 IV		88 s		
63274	174522		7 46 49.9	-29 56 55	-0.002	-0.01	8.00	-0.12	0.0	B8.5 V		400 s		
64409	256448		7 46 50.7	-74 05 25	0.000	+0.04	8.0	0.1	0.6	A0 V		300 s		
60062	1222		7 46 52.0	+81 40 55	-0.012	-0.04	7.6	0.0	0.4	B9.5 V		160 mx		
63709	235513		7 46 54.7	-58 54 24	+0.001	+0.01	8.0	-0.2	0.4	B9.5 V		340 s		
63563	235508		7 46 56.7	-52 20 47	-0.002	+0.02	6.94	-0.13	0.4	B9.5 V		200 s		
63531	235507		7 46 57.7	-50 11 03	-0.001	0.00	7.09	0.05	-1.1	B5 V n		440 s		
63020	115917		7 47 00.0	+6 02 59	-0.002	-0.01	7.9	1.1	-0.1	K2 III		390 s		
63021	115919		7 47 00.3	+4 55 30	+0.001	-0.03	7.0	0.0	0.4	B9.5 V		210 s		
62452	26504		7 47 00.4	+55 32 29	0.000	-0.01	8.0	1.6	-0.5	M2 III		510 s		
62195	14326		7 47 00.8	+64 03 06	+0.002	-0.04	7.00	0.1	1.4	A2 V		130 s	6336	m
61951	14320		7 47 04.6	+69 09 15	-0.005	-0.08	7.2	0.4	2.6	F0 V		84 s		
62765	60331		7 47 04.9	+34 57 42	-0.001	-0.03	7.6	0.1	1.4	A2 V		170 s		
63401	198435		7 47 05.8	-39 19 53	-0.002	-0.01	6.31	-0.17		B9				
63425	218992		7 47 07.0	-41 30 12	-0.002	+0.01	6.94	-0.18		B1				m
--	218993		7 47 07.3	-41 31 03	-0.001	-0.01	7.8							
63466	218995		7 47 08.0	-43 20 28	-0.001	0.00	7.5	-0.3	0.4	B9.5 V		260 s		
62926	79688		7 47 09.0	+23 39 49	+0.001	-0.03	7.5	1.1	0.2	K0 III		290 s		
63093	115930		7 47 09.8	+0 55 06	0.000	-0.02	7.5	0.8	3.2	G5 IV		73 s		
63424	198439		7 47 10.1	-39 06 11	-0.003	0.00	7.1	0.4	3.0	F2 V		61 s		
63467	218996		7 47 10.9	-43 23 41	-0.001	+0.01	7.58	-0.05	0.0	B8.5 V		330 s		m
63064	115929		7 47 11.2	+2 27 53	-0.001	-0.02	7.9	0.2	2.1	A5 V		150 s		
63400	198438		7 47 12.0	-37 58 17	+0.003	-0.01	7.49	0.98		K0				
63271	174533		7 47 12.5	-22 31 11	-0.001	0.00	5.90	-0.19	-3.9	B1 IV	+7	840 s		
63399	198437		7 47 14.5	-36 04 23	-0.004	+0.04	6.5	1.0	0.2	K0 III		180 s		
62830	60335		7 47 15.0	+33 00 40	-0.001	-0.01	7.2	0.8	3.2	G5 IV		64 s		
62668	41995		7 47 20.7	+47 20 19	-0.001	+0.02	7.6	1.1	0.2	K0 III		310 s		
62856	60336		7 47 22.1	+33 33 52	0.000	-0.07	8.0	0.9	3.2	G5 IV		93 s		
63018	97243		7 47 22.4	+15 31 14	-0.002	-0.06	7.7	0.9	3.2	G5 IV		80 s		
63666	235515		7 47 23.5	-53 19 58	-0.001	-0.01	7.62	0.02	0.4	B9.5 V		270 s		m
63465	198442		7 47 24.9	-38 30 40	-0.001	0.00	5.08	-0.11		B3	+12			m
63373	174542		7 47 28.7	-29 41 37	-0.002	-0.01	7.7	0.9	4.4	G0 V		29 s		
62898	60340	80 π Gem	7 47 30.3	+33 24 56	-0.001	-0.03	5.14	1.60	-0.4	gM0	-12	120 s	6364	m
61994	6310		7 47 30.4	+70 12 24	-0.018	-0.14	7.2	0.7	4.4	G0 V	-24	27 ts		
65836	258485		7 47 31.0	-81 35 47	-0.012	+0.04	7.0	-0.4	1.4	A2 V		130 s		
63578	219000		7 47 31.5	-46 36 31	-0.001	0.00	5.23	-0.14	-3.5	B1 V	+36	510 s		
63107	115932		7 47 32.5	+9 37 56	-0.003	+0.13	7.3	0.5	4.0	F8 V		45 s		
63108	115933		7 47 33.1	+8 58 21	+0.003	+0.01	7.0	0.4	2.6	F0 V		75 s		
63062	97247		7 47 34.3	+15 42 19	-0.001	-0.02	7.9	0.0	0.4	B9.5 V		320 s		
63579	219001		7 47 35.0	-47 00 47	-0.003	0.00	6.98	-0.11	0.4	B8.5 V		250 s		
63423	198440		7 47 35.1	-30 32 53	-0.004	+0.04	7.86	0.29		B0.5 II		80 mx		
63605	219002		7 47 37.4	-47 04 16	+0.002	0.00	7.90	0.7	4.4	G0 V		50 s		m
63302	153404		7 47 38.4	-15 59 27	-0.001	0.00	6.35	1.78	-6.1	G8 Iab		1200 s		
61690	6306		7 47 39.6	+73 59 28	0.000	-0.04	7.7	0.6	4.4	G0 V		47 s		
63464	198445		7 47 43.1	-32 40 21	-0.004	-0.05	7.4	0.6	2.1	A5 V		66 s		
63323	153409		7 47 45.2	-16 00 52	0.000	-0.01	6.70	1.1	-0.1	K2 III		230 s		m
63358	153411		7 47 45.9	-19 59 06	-0.001	-0.01	8.01	-0.10	0.0	B8.5 V		400 s		
63161	115942		7 47 50.0	+7 20 28	0.000	-0.02	7.5	0.0	0.4	B9.5 V		260 s		
63105	79700		7 47 55.8	+21 02 05	0.000	-0.01	7.9	1.1	-0.1	K2 III		390 s		
63336	153414	5 Pup	7 47 56.6	-12 11 36	-0.008	+0.05	5.48	0.48	3.4	F5 V	+27	25 s	6381	m
63725	219010		7 47 56.8	-49 45 31	-0.005	+0.01	7.9	0.7	3.0	F2 V		53 s		
62522	14330		7 47 57.1	+60 17 45	-0.006	-0.09	7.02	0.57		F5			6354	m
--	--		7 48 00.7	-19 24 01			7.80							m

HD	SAO	Star Name	α 2000	δ 2000	μ(α)	μ(δ)	V	B-V	M_v	Spec	RV	d(pc)	ADS	Notes
59664	1221		7 48 01.8	+83 04 08	-0.003	+0.01	8.0	1.1	0.2	K0 III		360 s		
63241	115947		7 48 03.3	+5 24 35	-0.001	-0.04	7.00	1.1	0.2	K0 III		230 s		m
63661	219008		7 48 04.2	-44 00 06	0.000	+0.02	7.5	1.3	-0.3	K5 III		370 s		
63462	174558	o Pup	7 48 05.1	-25 56 14	-0.001	0.00	4.50	-0.05	-3.5	B1 V e	+16	390 s	6384	m
63015	60345		7 48 06.8	+32 51 26	+0.001	-0.05	7.20	0.9	3.2	G5 IV		63 s		m
63640	219006		7 48 08.4	-40 39 08	-0.001	-0.03	6.14	1.58		Ma				T Pup, q
63641	219009		7 48 09.8	-42 30 28	-0.001	0.00	7.77	-0.03	0.6	A0 V		270 s		
63485	174563		7 48 12.4	-24 34 37	-0.003	+0.01	7.8	-0.3	0.6	A0 V		270 s		
63394	153421		7 48 14.7	-14 05 48	-0.004	0.00	7.9	1.1	0.2	K0 III		340 s		
63870	235529		7 48 15.2	-54 42 50	+0.002	+0.01	6.7	0.8	0.2	K0 III		200 s		
63355	135118		7 48 15.8	-6 46 24	-0.001	+0.03	7.0	0.1	1.7	A3 V		110 s		
62854	42005		7 48 16.6	+49 22 14	-0.007	+0.02	7.8	0.5	3.4	F5 V		77 s		
63547	174570		7 48 19.2	-28 21 29	+0.004	+0.03	7.76	1.00	0.2	K0 III		170 mx		
63926	235532		7 48 19.2	-56 28 16	-0.003	0.00	6.33	1.01	0.2	K0 III		52 s		
63744	219018		7 48 20.2	-47 04 39	-0.009	-0.08	4.71	1.06	0.2	K0 III	-1	72 s		
62853	42007		7 48 20.4	+49 45 24	-0.002	-0.02	7.4	1.1	0.2	K0 III		270 s		
--	--		7 48 20.6	+28 33 13			7.3	1.1	0.0	K1 III		290 s		
63210	97260		7 48 22.6	+18 20 13	+0.001	-0.02	7.70	0.9	0.3	G8 III	+13	300 s	6376	m
61496	6305		7 48 23.0	+76 59 11	-0.006	-0.01	8.0	0.9	3.2	G5 IV		93 s		
62995	42010		7 48 27.5	+41 36 51	-0.002	0.00	8.0	1.1	0.2	K0 III		360 s		
63316	115959		7 48 27.6	+2 02 35	-0.001	+0.01	7.8	0.1	0.6	A0 V		270 s		
62667	26511		7 48 28.7	+59 05 05	+0.001	0.00	8.00	0.4	2.6	F0 V		120 s	6363	m
63138	79703		7 48 28.7	+28 45 52	0.000	0.00	7.10	1.1	0.2	K0 III		240 s		m
63809	219023		7 48 31.7	-48 02 07	-0.004	+0.02	7.9	0.8	3.2	G5 IV		59 s		
63208	79704	82 Gem	7 48 33.6	+23 08 28	-0.001	0.00	6.2	0.6	0.4	G2 III	-5	150 s	6378	m
63886	235534		7 48 33.7	-52 41 03	-0.001	+0.01	7.2	-0.1	1.4	A2 V		150 s		
63543	174578		7 48 35.6	-22 00 34	0.000	-0.01	7.28	1.18	0.2	K0 III		200 s		
61906	6315		7 48 36.0	+74 02 33	+0.005	-0.10	7.1	1.1	0.2	K0 III		140 mx		
63625	198463		7 48 36.5	-30 45 18	-0.001	+0.05	7.21	-0.15	-0.9	B6 V		420 s		
63638	198469		7 48 41.6	-31 37 04	-0.002	+0.01	6.94	-0.14	0.0	B8.5 V		240 s		
63637	198468		7 48 42.6	-30 15 15	-0.019	+0.21	7.54	0.65		G5		14 mn		
63541	153434		7 48 43.8	-19 10 29	-0.002	0.00	7.5	1.1	-0.1	K2 III		320 s		
63261	79710		7 48 45.5	+20 47 32	+0.001	-0.05	8.0	1.1	0.2	K0 III		360 s		
64141	249971		7 48 47.6	-61 09 46	-0.005	+0.06	7.6	0.2	2.6	F0 V		100 s		
63738	198475		7 48 49.3	-38 12 14	-0.010	-0.04	7.2	-0.1	3.4	F5 V		57 s		
63705	198473		7 48 49.4	-35 44 25	0.000	-0.03	7.8	0.7	-0.5	M2 III		470 s		
63806	219024		7 48 51.4	-43 18 41	-0.001	+0.01	7.12	-0.15	-2.9	B3 III		1000 s		
62808	26517		7 48 51.8	+56 46 52	+0.004	-0.01	8.0	0.9	3.2	G5 IV		93 s		
63598	174589		7 48 53.5	-24 58 26	+0.011	-0.25	7.93	0.54		G0		12 mn		
63299	97268		7 48 53.6	+18 11 24	-0.001	-0.05	7.7	1.1	0.2	K0 III		310 s		
63436	115967		7 48 58.1	+0 39 42	-0.001	-0.03	7.60	0.4	3.0	F2 V		83 s		m
63435	115966		7 48 58.8	+4 19 58	-0.003	-0.03	6.53	0.78		G0	-6			
63660	174592		7 49 01.6	-24 54 44	-0.002	+0.02	5.33	0.76	0.4	gG3	+2	97 s		
63352	97273		7 49 01.9	+13 22 15	+0.004	-0.05	6.04	1.38	0.2	K0 III	-57	87 s		
63540	153441		7 49 05.9	-11 10 52	-0.007	+0.02	6.8	0.4	2.6	F0 V		70 s		
64067	235539		7 49 06.7	-56 24 37	-0.001	+0.01	5.59	1.13	0.2	K0 III	+22	100 s		m
63520	135135		7 49 12.0	-4 39 32	0.000	-0.01	7.8	1.1	0.2	K0 III		320 s		
63949	219034		7 49 12.8	-46 51 27	-0.001	+0.01	5.84	-0.14		B2	+25			
64185	249975		7 49 12.8	-60 17 01	-0.008	+0.15	5.78	0.42		F2				m
63555	153443		7 49 13.2	-10 29 57	-0.001	0.00	7.5	0.1	0.6	A0 V		230 s		
63351	97275		7 49 13.4	+19 46 11	+0.001	-0.01	7.8	1.1	0.2	K0 III		330 s		
63922	219035		7 49 14.3	-46 22 24	-0.001	0.00	4.11	-0.18	-5.6	B0 II	+24	820 s		m
63868	219029		7 49 14.5	-40 42 04	+0.001	0.00	6.50	-0.15	-1.1	B5 V		330 s		
63786	198480		7 49 14.6	-35 14 36	0.000	+0.01	5.93	-0.05	0.6	A0 V		120 s		
63700	174601	7 ξ Pup	7 49 17.6	-24 51 35	-0.001	0.00	3.34	1.24	-4.5	G3 Ib	+3	230 s	6393	m
63333	79718		7 49 18.8	+24 29 20	-0.009	-0.01	7.1	0.5	3.4	F5 V		55 s		
63804	198482		7 49 19.5	-33 19 50	-0.001	-0.04	7.69	1.21	-7.1	B9.5 Ia				m
63593	153447		7 49 19.8	-11 25 46	-0.002	0.00	8.0	0.1	0.6	A0 V		290 s		
63679	174603		7 49 21.9	-22 33 54	-0.001	-0.04	7.92	1.59	-0.5	M2 III		480 s		
63883	219033		7 49 25.2	-40 15 23	0.000	-0.02	7.4	1.0	-0.5	M2 III		390 s		
64184	235545		7 49 26.8	-59 22 51	+0.041	-0.22	7.52	0.67		G5		10 mn		
64121	235540		7 49 27.6	-54 48 42	0.000	-0.07	7.4	0.6	3.2	G5 IV		70 s		
63948	219040		7 49 28.1	-44 45 06	-0.005	0.00	6.32	0.96	0.2	K0 III		170 s		
63655	153449		7 49 28.5	-13 21 12	-0.003	-0.02	6.1	-0.1		B9				
63656	153450		7 49 28.7	-15 20 44	-0.001	-0.01	8.0	1.1	-0.1	K2 III		410 s		
63759	174613		7 49 29.3	-27 50 03	+0.001	-0.03	7.9	0.8	1.4	A2 V		65 s		
63536	115981		7 49 29.8	+3 13 13	+0.001	-0.03	7.7	0.6	4.4	G0 V		47 s	6391	m
63882	198489		7 49 30.6	-39 00 04	0.000	+0.01	7.9	-0.1	0.2	K0 III		340 s		
63475	97280		7 49 32.3	+12 48 50	0.000	-0.03	6.5	1.1	0.2	K0 III		180 s		
63719	174609		7 49 32.4	-21 17 06	-0.001	0.00	7.8	-0.4	0.0	B8.5 V		370 s		
63852	198487		7 49 35.3	-33 17 20	-0.004	0.00	5.6	1.6	-0.3	K5 III		130 s		
63473	97281		7 49 36.5	+14 07 52	-0.003	-0.02	7.9	0.9	3.2	G5 IV		85 s		
64162	235543		7 49 38.3	-55 04 43	-0.004	+0.01	7.60	0.1	0.6	A0 V		250 s		m
63696	153453		7 49 39.5	-14 05 07	-0.005	0.00	6.6	1.4	-0.3	K5 III	+33	210 mx		
63497	97284		7 49 40.1	+12 08 56	-0.002	-0.04	7.3	1.1	0.2	K0 III		260 s		

HD	SAO	Star Name	α 2000	δ 2000	μ(α)	μ(δ)	V	B-V	M_v	Spec	RV	d(pc)	ADS	Notes
63474	97283		7^{h}49^{m}40.7^s	+13°51'30"	+0^s.002	-0".05	7.6	0.8	3.2	G5 IV		78 s		
63734	153456		7 49 40.8	-19 30 14	-0.004	+0.03	6.7	1.3	0.2	K0 III		140 s		
64484	249978		7 49 41.0	-66 11 46	0.000	-0.01	5.79	-0.04		B9				
63697	153454	6 Pup	7 49 41.1	-17 13 43	+0.003	-0.12	5.18	1.28	-0.2	K3 III	+44	51 mx		
63516	115983		7 49 41.6	+8 10 52	0.000	-0.06	7.9	0.7	4.4	G0 V		51 s		
63280	60367		7 49 44.3	+38 01 19	+0.001	-0.02	8.0	0.0	0.4	B9.5 V		330 s		
63754	174617		7 49 45.1	-20 12 25	-0.004	-0.11	6.56	0.58		G0			6398	m
63496	97286		7 49 46.3	+15 06 22	-0.001	-0.02	7.9	0.9	3.2	G5 IV		86 s		
63410	79726		7 49 47.6	+26 15 49	-0.003	-0.02	6.80	0.95	0.3	G8 III		200 s		
63898	198492		7 49 50.2	-36 05 38	-0.002	-0.03	7.00	1.1	0.2	K0 III		230 s		m
63694	135149		7 49 50.6	-8 34 47	-0.001	+0.01	7.9	1.6	-0.5	M2 III		470 s		
63988	219045		7 49 52.0	-40 54 53	-0.001	+0.01	7.07	-0.07	0.0	B8.5 V		260 s		
64320	235553		7 49 54.8	-60 03 04	0.000	+0.02	6.72	1.24		K0				
63433	79729		7 49 54.9	+27 21 49	-0.002	+0.01	7.7	0.9	3.2	G5 IV		80 s		
63652	135147		7 49 55.3	-2 41 25	-0.001	+0.01	8.0	0.1	0.6	A0 V		290 s		
63432	79730		7 49 56.9	+28 11 56	-0.001	-0.01	6.7	0.1	1.4	A2 V		110 s		
63847	174625		7 49 58.9	-23 47 53	-0.007	0.00	6.7	0.7	2.6	F0 V		38 s		
63944	198496		7 49 59.9	-35 06 28	0.000	-0.01	8.0	0.0	3.4	F5 V		82 s		
63822	153468		7 50 05.6	-19 31 25	-0.001	-0.01	6.12	1.26		K0				
63692	135154		7 50 06.8	-1 40 56	+0.001	-0.03	7.7	1.1	0.2	K0 III		310 s		
63515	79732		7 50 06.9	+22 43 54	+0.002	-0.04	8.0	1.1	0.2	K0 III		360 s		
62976	14338		7 50 07.9	+60 20 32	0.000	+0.02	6.78	0.09		A2				
63752	135158		7 50 10.5	-9 11 00	0.000	0.00	5.61	1.44	-0.2	gK3	-7	120 s		
63366	60369		7 50 13.5	+39 46 13	-0.001	-0.04	6.9	1.6	-0.5	M2 III		300 s		
63714	135157		7 50 13.9	-1 27 33	-0.001	-0.02	8.0	0.1	0.6	A0 V		300 s		
63915	174636		7 50 19.4	-27 58 21	-0.001	-0.01	7.9	1.0	-0.5	M4 III		490 s		
64008	198504		7 50 20.6	-35 47 26	-0.002	+0.01	7.3	0.6	-0.1	K2 III		300 s		
63691	116003		7 50 22.6	+5 45 54	-0.002	-0.02	7.7	0.0	0.4	B9.5 V		290 s		
63312	42027		7 50 22.7	+45 47 30	-0.001	-0.01	7.1	0.1	1.4	A2 V	+4	140 s		
64225	235552		7 50 23.8	-50 30 35	-0.007	-0.05	5.7	1.3	-0.3	K5 III		150 mx		
64028	198506		7 50 26.0	-36 14 31	-0.002	+0.01	7.26	-0.08	0.4	B9.5 V		240 s		
63675	116004		7 50 28.8	+9 45 55	+0.001	-0.05	8.0	0.4	3.0	F2 V		99 s		
63800	135160		7 50 30.4	-3 35 52	-0.003	-0.01	7.1	0.1	1.7	A3 V		120 s		
63648	97298		7 50 31.6	+14 50 30	-0.004	-0.03	7.3	0.4	3.0	F2 V		72 s		
63349	42029		7 50 33.0	+45 20 26	+0.001	-0.01	8.0	1.1	-0.1	K2 III		410 s		
63123	26527		7 50 33.8	+59 03 10	0.000	-0.05	7.9	0.4	2.6	F0 V		110 s		
63647	97301		7 50 35.7	+16 52 45	-0.001	-0.02	7.8	0.8	3.2	G5 IV		84 s		
64066	198509		7 50 35.8	-36 15 29	-0.002	+0.02	7.9	-0.4	0.4	B9.5 V		320 s		
--	116008		7 50 36.2	+1 54 54	-0.001	-0.01	8.0	1.1	0.2	K0 III		370 s		
63820	135164		7 50 36.9	-5 48 10	-0.001	0.00	7.5	0.1	0.6	A0 V		240 s		
63590	79739		7 50 39.6	+24 09 39	-0.001	-0.01	7.0	0.4	2.6	F0 V		77 s		
64249	219062		7 50 40.2	-48 32 38	-0.005	-0.03	7.92	-0.01	0.6	A0 V		120 mx		
64181	219059		7 50 42.3	-44 34 47	-0.003	+0.02	6.45	0.90		K0				
63961	174647		7 50 43.8	-22 03 20	-0.001	0.00	7.5	1.5	0.2	K0 III		150 s		
63799	116014		7 50 47.3	+3 16 38	+0.003	-0.03	6.18	1.12	0.0	K1 III	-48	170 s	6405	m
63774	116013		7 50 48.5	+5 51 33	-0.003	+0.03	7.4	0.1	1.4	A2 V		160 s		
63798	116016		7 50 50.9	+4 27 33	-0.002	+0.04	6.6	0.8	3.2	G5 IV		48 s		
63838	116019		7 50 53.3	+0 04 45	0.000	-0.01	6.5	1.1	0.2	K0 III		180 s		
63894	153479		7 50 55.2	-11 07 44	0.000	-0.04	6.3	1.1	0.2	K0 III		160 s		
64100	198513		7 50 55.3	-33 03 21	-0.003	0.00	7.4	1.1	3.2	G5 IV		44 s		
64507	235562		7 50 56.9	-59 38 48	0.000	-0.02	7.1	0.2	0.4	B9.5 V		150 s		
64042	174658		7 50 59.9	-24 31 43	-0.001	-0.03	6.45	-0.01		A0			6414	m
63610	60380		7 51 00.1	+31 36 48	-0.001	-0.04	6.86	0.17	0.9	A6 III-IV	+6	160 s		
63611	60382		7 51 01.3	+30 15 44	0.000	+0.01	7.6	0.9	3.2	G5 IV		77 s		
63589	60379		7 51 02.3	+33 14 01	-0.001	0.00	6.00	0.15		A m	-10			m
63797	116021		7 51 02.6	+7 26 16	0.000	0.00	7.3	0.1	0.6	A0 V		220 s		
64304	219068		7 51 02.6	-47 56 10	+0.001	+0.01	7.7	1.1	0.2	K0 III		290 s		
63859	116025		7 51 03.0	+0 40 29	-0.001	+0.05	7.9	1.1	0.2	K0 III		340 s		
63332	26535		7 51 05.6	+54 07 45	-0.005	+0.05	6.02	0.46	3.4	F5 V	-2	32 s		
63570	60377		7 51 06.1	+37 12 41	+0.002	-0.02	7.70	0.4	2.6	F0 V		110 s	6399	m
63631	60383		7 51 06.1	+31 36 57	-0.001	-0.05	7.75	0.17	1.3	A6 IV		200 s		
64000	153488		7 51 06.4	-18 14 51	0.000	+0.01	6.6	1.1	0.2	K0 III		190 s		
64158	198518		7 51 07.5	-33 37 23	+0.001	-0.06	7.9	0.5	0.2	K0 III		340 s		
63978	153489		7 51 13.7	-12 36 39	+0.001	-0.05	8.0	1.1	-0.1	K2 III		400 s		
64677	249984		7 51 13.9	-62 55 49	-0.004	+0.04	7.3	0.8	3.2	G5 IV		65 s		
64318	219071		7 51 14.1	-47 12 59	+0.001	0.00	6.53	-0.12	-2.9	B3 III		750 s		
63588	60384		7 51 15.4	+37 04 54	+0.001	-0.04	6.6	1.1	-0.1	K2 III		220 s		
63103	14340		7 51 16.7	+64 05 22	-0.011	-0.10	8.0	1.1	0.2	K0 III	+12	160 mx		
63977	153490		7 51 17.7	-11 10 20	-0.001	-0.01	7.9	1.1	0.2	K0 III		330 s		
63348	26536		7 51 19.2	+54 43 57	-0.001	-0.01	7.40	1.6	-0.4	M0 III	-5	360 s		
63976	135180		7 51 20.3	-9 24 17	-0.002	+0.01	7.30	0.1	0.6	A0 V		220 s	6412	m
64287	219069		7 51 20.4	-43 05 44	0.000	-0.01	6.32	-0.17		B3	+14			
63836	97313		7 51 24.1	+14 10 14	0.000	-0.01	7.9	1.1	0.2	K0 III		340 s		
63712	79747		7 51 26.5	+29 10 04	-0.001	-0.01	6.9	0.9	0.3	G8 III		210 s		
63711	60390		7 51 30.1	+30 31 49	0.000	0.00	8.0	0.0	0.4	B9.5 V		330 s		

195

HD	SAO	Star Name	α 2000	δ 2000	μ(α)	μ(δ)	V	B−V	M$_V$	Spec	RV	d(pc)	ADS	Notes
			h m s	° ′ ″	s	″								
63996	135181		7 51 31.6	−6 59 00	−0.002	−0.01	7.4	0.1	1.4	A2 V		160 s		
64222	198526		7 51 33.0	−32 42 57	−0.002	0.00	7.5	1.0	−0.1	K2 III		330 s		
64365	219076		7 51 40.2	−42 53 18	−0.002	0.00	6.04	−0.18		B3	+32			
64077	153499	8 Pup	7 51 40.8	−12 49 10	0.000	−0.02	6.5	0.4	3.0	F2 V	+21	50 s		
63975	116043	13 ζ CMi	7 51 41.9	+1 46 01	−0.001	−0.01	5.14	−0.12		B8	+32	25 mn		
64152	174679		7 51 42.8	−21 10 26	−0.005	+0.02	5.63	0.96	0.3	gG8	+32	120 s		
64244	198531		7 51 43.6	−30 19 17	0.000	−0.02	7.9	0.7	0.2	K0 III		350 s		
64133	153506		7 51 45.6	−18 21 26	0.000	+0.01	7.8	0.6	0.4	B9.5 V		300 s		
64096	153500	9 Pup	7 51 46.2	−13 53 53	−0.005	−0.34	5.17	0.60	4.5	G1 V	−18	15 ts	6420	m
64624	235572		7 51 48.5	−55 42 09	−0.002	+0.01	7.6	0.8	0.2	K0 III		300 s		
64172	174682		7 51 49.2	−20 43 17	+0.004	−0.06	6.9	0.4	3.4	F5 V		50 s		
64114	153509		7 51 55.9	−11 01 59	−0.004	−0.18	7.7	0.9	3.2	G5 IV		78 mx		
64113	135189		7 51 56.5	−9 39 22	−0.003	−0.02	7.5	0.1	1.7	A3 V		150 s		
63889	97318		7 51 56.7	+19 19 30	−0.004	−0.04	5.99	1.13	0.2	K0 III	+39	120 s		
64301	198533		7 51 57.2	−31 38 12	−0.001	−0.01	7.72	−0.10	0.4	B9.5 V		290 s		m
64021	116050		7 51 58.9	+5 20 59	+0.001	+0.01	7.9	0.5	3.4	F5 V		78 s		
64403	219079		7 52 00.8	−41 47 12	−0.002	+0.01	7.4	1.6	−0.1	K2 III		150 s		
63385	26539		7 52 03.9	+59 30 01	+0.002	−0.07	7.2	1.1	0.2	K0 III		170 mx		
63630	42034		7 52 03.9	+45 55 58	−0.002	−0.05	6.5	0.1	1.7	A3 V	+23	93 s		
64052	116054		7 52 07.1	+3 16 38	+0.003	−0.09	6.31	1.59		Ma	−62			
64110	135190		7 52 08.4	−3 03 16	−0.002	−0.01	7.00	0.2	2.1	A5 V		96 s	6421	m
63793	60396		7 52 08.6	+36 10 08	+0.002	−0.04	6.9	1.6	−0.5	M2 III		300 s		
64440	219082		7 52 13.0	−40 34 33	−0.001	0.00	3.73	1.04	0.3	G5 III	+24	38 s		
64784	235582		7 52 13.2	−59 36 49	−0.002	0.00	7.50	0.00	0.4	B9.5 V		260 s		m
64458	219083		7 52 14.4	−42 19 22	−0.002	−0.01	7.9	−0.6	0.0	B8.5 V		380 s		
63814	60398		7 52 15.5	+36 16 10	+0.004	−0.04	7.0	0.5	4.0	F8 V		40 s		
64379	198540		7 52 15.6	−34 42 19	−0.016	+0.24	5.01	0.44	3.4	F5 V	+28	17 ts		m
64238	153520	10 Pup	7 52 18.8	−14 50 47	−0.001	−0.08	5.69	0.37	−6.5	cF3	+17	2700 s		
64831	235585		7 52 22.4	−59 29 34	0.000	+0.02	7.2	0.6	0.6	A0 V		85 s		
64259	153522		7 52 28.2	−13 51 46	+0.002	−0.08	6.61	1.11	−0.1	K2 III	−29	220 s	6426	m
63534	26541		7 52 28.3	+57 01 04	+0.005	+0.01	7.5	1.1	−0.1	K2 III		280 mx		
64722	235579		7 52 29.7	−54 22 02	−0.001	0.00	5.70	−0.15	−3.6	B2 III		700 s		
63792	42042		7 52 31.2	+41 32 29	+0.002	0.00	8.0	1.1	0.2	K0 III		360 s		
64192	135198		7 52 34.7	−2 37 17	−0.001	0.00	7.6	1.6	−0.5	M2 III		420 s		
63586	26543		7 52 36.5	+55 12 34	0.000	−0.04	6.2	0.1	0.6	A0 V	+16	140 s		
64294	153527		7 52 37.8	−15 25 43	−0.001	−0.03	7.5	0.0	0.4	B9.5 V		260 s		
64439	198543		7 52 37.8	−32 07 04	−0.002	−0.02	7.9	0.8	0.2	K0 III		340 s		
62965	6336		7 52 38.0	+71 47 02	−0.001	−0.06	7.6	0.9	3.2	G5 IV		76 s		
64503	198545		7 52 38.6	−38 51 47	−0.001	−0.01	4.49	−0.19	−2.3	B3 IV	−21	230 s		
64336	153530		7 52 38.7	−19 47 15	−0.001	0.00	7.9	0.4	0.0	B8.5 V		190 s		
63874	60399		7 52 39.9	+39 32 18	−0.002	−0.02	7.1	0.4	2.6	F0 V		80 s		
64165	116064		7 52 42.1	+3 23 02	0.000	−0.03	7.50	0.12		A2			6425	m
64358	174714		7 52 44.9	−22 13 27	−0.001	0.00	7.5	1.5	−0.1	K2 III		210 s		
64719	—		7 52 45.3	−51 34 31			7.9	0.6	0.6	A0 V		110 s		
64718	—		7 52 45.4	−51 33 31			7.9	0.5	0.6	A0 V		130 s		
64235	135205		7 52 47.8	−5 25 42	−0.002	−0.03	5.76	0.41	0.7	gF5	−2	100 s		m
64501	198548		7 52 49.1	−36 54 48	+0.002	−0.04	7.9	1.1	0.2	K0 III		320 s		
64717	235584		7 52 50.0	−50 31 16	−0.002	+0.02	7.10	0.03	−1.7	B3 V		460 s		
64761	235588		7 52 50.9	−52 58 17	0.000	+0.01	7.33	0.06	0.4	B9.5 V		210 s		
63455	14352		7 52 52.1	+62 53 06	−0.002	−0.01	7.8	0.8	3.2	G5 IV		84 s		
63384	14349		7 52 55.4	+64 54 13	−0.001	−0.02	7.60	1.4	−0.3	K5 III		380 s	6406	m
64419	174721		7 52 57.3	−25 42 43	+0.003	−0.03	7.9	0.8	0.2	K0 III		340 s		
64092	79768		7 53 00.9	+22 20 04	0.000	−0.01	7.1	1.1	0.2	K0 III	−9	240 s		
64572	198553		7 53 03.5	−36 21 50	−0.001	−0.01	5.43	1.16	−0.3	gK5	+12	140 s		
64740	219106		7 53 03.7	−49 36 47	0.000	+0.01	4.63	−0.23	−3.6	B2 III	+8	440 s		
64455	174730		7 53 04.7	−26 24 50	0.000	+0.02	7.75	−0.10	−1.2	B8 III		620 s		
64332	153536		7 53 05.1	−11 37 32	−0.001	−0.05	8.0	1.9		S6.2				
64164	97330		7 53 05.3	+15 22 37	−0.001	−0.01	7.9	0.1	1.4	A2 V		200 s		
64546	198555		7 53 13.1	−31 48 28	+0.002	0.00	7.4	1.3	−0.1	K2 III		260 s		
64571	198559		7 53 13.3	−34 56 17	−0.003	+0.02	6.64	0.83	4.4	G0 V		18 s		
64570	198560		7 53 15.8	−34 02 21	−0.002	−0.01	7.3	1.4	−0.1	K2 III		220 s		
64188	97332		7 53 15.9	+15 35 56	−0.001	−0.03	7.7	1.1	0.2	K0 III		310 s		
65077	249993		7 53 16.2	−62 21 32	−0.002	−0.02	7.9	1.1	0.2	K0 III		350 s		
64760	219111		7 53 18.2	−48 06 11	0.000	0.00	4.24	−0.14	−5.7	B3 Ib	+41	970 s		
64211	97334		7 53 18.3	+15 44 07	+0.001	−0.03	7.50	1.1	0.2	K0 III		290 s	6430	m
64189	97335		7 53 20.0	+14 55 02	0.000	0.00	7.9	0.6	4.4	G0 V		50 s		
64107	79772		7 53 20.9	+28 03 09	−0.001	0.00	6.8	0.1	0.6	A0 V		170 s		
63903	42049		7 53 22.8	+45 44 32	+0.002	+0.04	7.6	0.9	3.2	G5 IV		75 s		
64019	60412		7 53 24.2	+39 28 56	−0.002	−0.03	7.6	1.1	−0.1	K2 III		350 s		
64210	97338		7 53 25.8	+16 50 45	−0.001	−0.01	7.5	1.1	0.2	K0 III		290 s		
64595	174741		7 53 28.3	−29 49 29	+0.001	−0.04	7.9	0.2	2.6	F0 V		110 s		
65094	249995		7 53 29.4	−61 03 17	+0.002	−0.01	8.0	−0.2	0.6	A0 V		300 s		
64145	79774	83 φ Gem	7 53 29.7	+26 45 57	−0.003	−0.03	4.97	0.09	1.7	A3 V	+8	45 s		
64674	198565		7 53 33.5	−36 56 10	−0.004	+0.03	7.4	0.4	2.1	A5 V		89 s		
64291	97344		7 53 36.9	+10 41 14	−0.004	−0.05	7.70	0.9	0.3	G6 III	+67	300 s		

HD	SAO	Star Name	α 2000	δ 2000	μ(α)	μ(δ)	V	B-V	M_v	Spec	RV	d(pc)	ADS	Notes
64566	174742		7^{h}53^{m}38$.^s$7	−23°18′51″	+0$.^s$001	−0$.^{\prime\prime}$01	6.7	0.1	0.6	A0 V		140 s		
64565	174744		7 53 40.0	−23 11 16	−0.001	−0.02	7.1	0.3	0.4	B9.5 V		140 s		
63769	26554		7 53 43.4	+58 12 49	−0.006	−0.02	7.4	0.4	2.6	F0 V		91 s		
64353	116080		7 53 44.9	+9 00 04	+0.002	−0.05	7.6	1.1	−0.1	K2 III		200 mx		
64207	79777		7 53 45.3	+26 33 51	+0.008	−0.18	8.00	0.5	4.2	F9 V	+25	42 mx		
65191	250001		7 53 48.3	−62 22 48	−0.001	+0.05	7.6	0.9	3.2	G5 IV		75 s		
64451	135225		7 53 48.8	−6 03 33	0.000	−0.02	7.8	0.1	0.6	A0 V		260 s		
64616	174755		7 53 49.3	−26 15 42	−0.002	+0.02	6.92	0.93		G5				
64163	60421		7 53 50.9	+32 20 04	−0.003	−0.05	8.0	0.9	3.2	G5 IV		93 s		
65152	250000		7 53 54.5	−60 35 01	−0.002	−0.01	7.2	1.9	−0.3	K5 III		180 s		
64515	135231		7 53 57.4	−9 19 26	−0.002	−0.04	7.7	1.1	0.2	K0 III		310 s		
64330	97348		7 53 57.7	+16 02 10	0.000	−0.02	7.10	0.1	1.4	A2 V		140 s	6440	m
64657	174762		7 53 58.7	−26 20 56	+0.001	0.00	6.85	1.55		Mb V				
64827	219122		7 54 02.2	−41 42 10	−0.001	0.00	6.85	−0.11	0.4	B9.5 V		200 s		
64516	153560		7 54 03.0	−10 33 40	0.000	−0.01	7.7	1.1	−0.1	K2 III		350 s		
64777	198570		7 54 04.6	−34 01 21	−0.004	−0.02	7.46	−0.09	0.6	A0 V		140 mx		
64611	153563		7 54 07.2	−18 19 50	+0.001	+0.01	7.10	1.4	−0.3	K5 III		300 s	6452	m
64697	174767		7 54 10.8	−26 09 08	−0.001	0.00	7.8	1.3	−0.3	K5 III		410 s		
64802	198575		7 54 10.9	−35 52 39	−0.001	−0.01	5.49	−0.19		B5	+28			
64187	60422		7 54 11.1	+36 44 29	−0.001	−0.01	7.1	0.9	3.2	G5 IV		61 s		
64391	97352		7 54 11.8	+13 25 15	0.000	−0.06	7.7	0.9	3.2	G5 IV		80 s		
64539	135233		7 54 12.5	−3 32 42	−0.002	+0.01	7.8	0.1	1.7	A3 V		160 s		
65170	235606		7 54 13.0	−58 34 30	−0.006	−0.11	7.80	0.6	4.4	G0 V		48 s		m
64351	79782		7 54 13.2	+21 06 19	0.000	−0.01	6.9	1.6	−0.5	M2 III	+9	310 s		
63932	26561		7 54 14.3	+54 42 53	+0.008	−0.10	7.2	0.9	3.2	G5 IV		64 s		
64903	219129		7 54 18.9	−43 04 38	0.000	−0.04	7.90	1.1	0.2	K0 III		350 s		m
63347	6349		7 54 20.9	+71 04 46	−0.003	−0.03	7.2	0.0	0.0	B8.5 V		270 s		
64610	135235		7 54 23.5	−9 47 09	−0.001	−0.01	7.9	0.9	3.2	G5 IV		88 s		
65038	235604		7 54 24.5	−50 10 15	0.000	0.00	7.49	−0.12	−1.1	B5 V		520 s		
64756	174779		7 54 25.6	−28 06 19	0.000	+0.01	6.61	0.07	0.0	A3 III		210 s		
64513	116094		7 54 27.1	+5 30 37	−0.001	+0.01	7.9	0.0	0.4	B9.5 V		320 s		
64902	219131		7 54 27.4	−41 39 29	−0.001	−0.03	7.3	1.2	0.2	K0 III		210 s		
64106	42055	25 Lyn	7 54 29.2	+47 23 10	−0.002	0.00	6.25	1.15	0.2	gK0	−63	130 s		
65321	250006		7 54 29.6	−61 52 19	+0.005	−0.06	7.2	2.0	−0.5	M2 III		210 s		
64607	135238		7 54 32.9	−2 47 44	−0.002	0.00	8.00	0.4	2.6	F0 V		120 s	6454	m
64927	198580		7 54 33.1	−40 07 51	+0.001	−0.02	8.0	0.7	−0.1	K2 III		410 s		
64606	135239		7 54 34.0	−1 24 45	−0.018	−0.06	7.44	0.73	5.6	G8 V	+93	23 s		
64954	219133		7 54 35.1	−42 28 21	−0.002	−0.01	7.5	1.5	−0.1	K2 III		320 s		
64206	42059		7 54 38.6	+41 49 07	0.000	0.00	7.6	1.1	−0.1	K2 III		360 s		
64876	198579		7 54 39.8	−34 50 49	−0.001	−0.02	6.15	1.53	−0.1	K2 III		94 s		
64144	42058	26 Lyn	7 54 42.6	+47 33 52	−0.005	0.00	5.45	1.46	0.2	K0 III	+17	59 s		
65013	219136		7 54 44.6	−44 21 20	−0.002	+0.01	7.0	0.0	0.4	B9.5 V		190 s		
64372	60437		7 54 45.8	+30 20 30	−0.002	+0.01	7.80	0.8	0.3	G7 III	+82	320 s		
64124	42057		7 54 45.9	+49 47 05	0.000	−0.01	7.8	1.1	0.2	K0 III		330 s		
65037	219137		7 54 47.5	−44 22 42	0.000	−0.02	7.8	−0.2	0.4	B9.5 V		300 s		
64820	174784		7 54 48.1	−24 18 18	−0.003	+0.01	7.7	1.9	0.2	K0 III		92 s		
64324	60434		7 54 48.4	+34 37 11	−0.011	−0.17	7.8	0.6	4.4	G0 V	+15	47 s		
64493	97357		7 54 49.2	+18 05 59	+0.001	−0.03	7.40	1.4	−0.3	K4 III	−12	350 s		
65074	219140		7 54 50.3	−45 39 52	−0.002	−0.01	7.34	−0.10	0.6	A0 V		220 s		
64898	198583		7 54 51.0	−31 32 03	−0.003	+0.02	7.19	−0.11	0.0	B8.5 V		270 s		
64975	198588		7 54 52.3	−38 25 12	−0.005	−0.02	7.2	0.6	0.6	A0 V		91 s		
65273	235615		7 54 53.2	−52 34 59	−0.010	−0.02	5.63	1.30	−0.1	gK2	+26	120 s		
64468	97359		7 54 54.1	+19 14 10	+0.007	−0.47	7.79	0.95	8.2	dK6	−19	16 ts		m
64253	42062		7 54 54.9	+43 59 39	0.000	+0.02	8.0	0.9	3.2	G5 IV		93 s		
64512	97360		7 54 55.5	+17 57 27	−0.001	0.00	7.5	0.8	0.3	G6 III	+5	280 s		
64751	153580		7 54 58.1	−13 17 06	+0.002	−0.06	8.00	0.5	3.4	F5 V		83 s	6461	m
64323	60436		7 54 59.0	+39 42 47	+0.003	−0.05	7.7	0.1	1.4	A2 V		180 s		
65189	235612		7 55 00.4	−52 34 59	−0.001	+0.02	6.38	−0.01		B9				
64857	174790		7 55 00.4	−23 27 54	−0.002	+0.02	6.9	0.4	0.6	A0 V		110 s		
64584	97364		7 55 01.9	+12 34 10	−0.001	0.00	7.1	1.1	0.2	K0 III		240 s		
65250	235616		7 55 03.4	−54 38 29	−0.005	+0.01	7.9	0.8	0.2	K0 III		340 mx		
64309	42066		7 55 07.6	+41 57 38	+0.003	0.00	7.9	0.1	1.4	A2 V		200 s		
64630	116110		7 55 07.8	+6 59 45	−0.001	−0.02	8.0	0.4	2.6	F0 V		120 s		
64467	79791		7 55 07.8	+28 23 02	−0.010	−0.07	7.3	0.5	3.4	F5 V		59 s		
64732	135253		7 55 10.2	−4 34 20	−0.002	−0.05	8.0	0.1	1.4	A2 V		200 s		
64388	60445		7 55 10.7	+37 11 40	−0.003	−0.04	7.9	0.4	2.6	F0 V		110 s		
64649	116112		7 55 10.9	+9 20 59	−0.002	−0.01	7.7	0.9	3.2	G5 IV		80 s		
64974	198591		7 55 13.7	−38 43 18	+0.001	+0.04	6.5	1.9	0.2	K0 III		280 s		
64252	42065		7 55 13.9	+48 08 52	+0.001	−0.02	7.80	0.1	0.6	A0 V		280 s	6447	m
65249	235617		7 55 15.5	−52 54 08	−0.004	+0.04	7.8	−0.8	3.0	F2 V		89 s		
64923	174797		7 55 16.0	−26 25 31	+0.001	0.00	8.0	0.0	0.4	B9.5 V		300 s		
64794	153583		7 55 17.1	−11 09 06	0.000	−0.01	7.7	0.0	0.0	B8.5 V		340 s		
65089	−−		7 55 17.3	−41 50 00			7.54	0.69		G0				m
−−	219146		7 55 17.9	−41 49 51	−0.010	+0.17	7.5	0.9	4.4	G0 V				
64922	174799		7 55 18.3	−25 55 56	−0.002	0.00	7.5	1.5	−0.1	K2 III		210 s		

HD	SAO	Star Name	α 2000	δ 2000	μ(α)	μ(δ)	V	B-V	M$_V$	Spec	RV	d(pc)	ADS	Notes
			h m s	° ' "	s	"								
64387	60446		7 55 18.7	+38 12 36	+0.002	−0.02	8.0	1.1	0.2	K0 III		360 s		
64972	174802		7 55 21.9	−28 17 01	+0.001	+0.02	7.19	−0.12	−3.7	B6 II		1500 s		
65188	219150		7 55 23.9	−47 18 48	−0.001	−0.01	7.2	1.4	−0.3	K5 III		320 s		
65297	235621		7 55 24.2	−53 37 21	0.000	+0.01	7.80	1.1	0.2	K0 III		330 s		m
65187	219151		7 55 25.6	−46 39 58	−0.001	0.00	8.0	0.8	0.2	K0 III		360 s		
65012	198594		7 55 27.3	−30 21 10	−0.001	−0.04	7.1	0.6	3.2	G5 IV		62 s		
64894	153591		7 55 29.3	−19 02 54	+0.001	−0.04	6.8	0.1	1.4	A2 V		120 s		
64447	60447		7 55 30.3	+36 58 50	0.000	−0.02	8.0	1.1	0.2	K0 III		360 s		
64018	14368		7 55 30.7	+62 02 38	0.000	0.00	7.3	0.9	3.2	G5 IV		66 s		
64685	116120		7 55 31.3	+8 51 46	−0.001	−0.09	5.86	0.35	3.3	dF4	+22	33 s		
64837	153589		7 55 32.1	−10 39 44	−0.001	−0.03	7.4	0.5	3.4	F5 V		63 s		
64793	135260		7 55 33.1	−2 11 23	−0.002	+0.02	8.0	0.1	0.6	A0 V		300 s		
64921	153593		7 55 35.6	−15 34 43	−0.009	−0.03	7.0	1.1	0.2	K0 III		98 mx		
64446	42075		7 55 38.4	+40 18 00	−0.001	−0.03	7.8	1.1	0.2	K0 III		330 s		
65186	219154		7 55 38.8	−43 46 56	0.000	−0.03	7.50	1.1	0.2	K0 III		280 s		m
64648	79799	85 Gem	7 55 39.8	+19 53 02	−0.001	−0.04	5.35	−0.04	0.4	B9.5 V	+13	96 s		
64491	60453		7 55 40.7	+35 24 47	−0.005	−0.01	6.1	0.3		A pe	+28			
64920	153594		7 55 41.1	−13 43 58	−0.004	0.00	7.8	0.4	3.0	F2 V		89 s		
64730	97374		7 55 45.4	+13 19 54	0.000	−0.06	7.9	0.9	3.2	G5 IV		88 s		
65211	219157		7 55 46.4	−43 50 42	−0.002	0.00	6.02	−0.11		B5 n	+14			m
64745	97378		7 55 47.6	+10 26 37	0.000	−0.01	7.5	0.0	0.4	B9.5 V		260 s		
65270	219158		7 55 49.8	−45 58 47	−0.003	+0.01	6.75	−0.12	−1.0	B5.5 V		360 s		
64488	60456		7 55 54.7	+39 17 13	+0.001	+0.01	7.2	0.1	1.4	A2 V		140 s		
65248	219161		7 55 55.7	−44 37 13	0.000	+0.01	7.7	−0.2	0.0	B8.5 V		350 s		
64942	135272		7 55 58.1	−9 47 50	+0.001	−0.03	8.0	0.9	3.2	G5 IV		92 s		
64967	153599		7 55 58.7	−12 36 30	−0.001	+0.07	6.7	1.1	0.2	K0 III		200 s		
64815	116135		7 56 00.5	+6 30 44	+0.012	−0.11	7.9	0.5	4.0	F8 V		60 s		
64682	79803		7 56 01.5	+24 39 56	0.000	0.00	7.80	1.1	0.2	K0 III		330 s		m
64602	60462		7 56 01.8	+34 22 10	−0.002	−0.02	7.3	1.1	0.2	K0 III		260 s		
64704	79804		7 56 02.5	+23 41 33	−0.011	−0.15	7.36	0.64	4.4	G0 V		35 s		m
65085	——		7 56 02.6	−26 43 48			8.0	0.1	1.4	A2 V		180 s		
65494	235631		7 56 05.0	−55 03 26	−0.003	+0.01	8.0	1.0	0.2	K0 III		360 s		
64705	79805		7 56 06.4	+23 37 25	−0.001	−0.03	6.7	0.0	0.0	B8.5 V		220 s		
63873	14367		7 56 06.7	+68 15 18	0.000	−0.02	8.0	1.4	−0.3	K5 III		450 s		
64988	135277		7 56 08.9	−9 51 48	0.000	−0.01	7.8	0.1	0.6	A0 V		260 s		
65425	235630		7 56 11.7	−51 20 18	+0.002	+0.01	6.5	1.1	0.2	K0 III		170 s		
65131	174837		7 56 13.7	−28 10 57	+0.001	−0.03	7.41	−0.07	−0.8	B9 III		430 s		
64681	79806		7 56 15.4	+29 51 08	−0.001	+0.01	6.8	1.1	0.2	K0 III		210 s		
65663	250018		7 56 15.9	−61 05 59	+0.001	0.00	7.0	−0.1	0.4	B9.5 V		210 s		
62613	1254		7 56 17.1	+80 15 55	−0.190	+0.08	6.56	0.72	5.6	dG8	−8	15 ts		
65662	250019		7 56 18.6	−60 31 35	−0.001	+0.01	5.73	1.55		K2				
66168	——		7 56 19.1	−71 59 19			7.9	0.5	3.4	F5 V		79 s		
64767	79810		7 56 19.6	+22 34 17	0.000	−0.16	7.7	1.1	0.2	K0 III		75 mx		
64852	97386		7 56 19.9	+10 51 54	−0.001	−0.01	7.60	0.1	0.6	A0 V		250 s	6473	m
65424	219177		7 56 20.7	−49 58 54	+0.001	0.00	7.86	3.05		C5 II				
65005	135280		7 56 21.7	−7 59 28	0.000	0.00	7.7	0.1	1.4	A2 V		180 s		
64250	26574		7 56 21.8	+59 40 34	−0.002	−0.05	7.9	0.5	3.4	F5 V		78 s		
65102	174836		7 56 21.8	−20 27 48	+0.001	−0.01	6.8	0.5	0.6	A0 V		86 s		
65183	198609		7 56 22.7	−30 17 07	0.000	−0.03	6.33	1.66		M6 III				
64917	116143		7 56 22.8	+4 15 15	−0.001	+0.01	7.90	0.1	1.7	A3 V		170 s		m
64938	116145		7 56 23.8	+4 29 09	−0.001	0.00	6.17	0.98	0.3	G8 III	+17	140 s		
65315	219169		7 56 24.2	−40 44 11	+0.001	0.00	6.78	−0.18		B3 n	+14			
64729	60464		7 56 24.6	+31 41 45	+0.002	0.00	7.6	0.4	2.6	F0 V		99 s		
64347	26579		7 56 26.8	+56 30 16	0.000	−0.03	6.5	0.1	0.6	A0 V	+28	150 s		
64868	97387		7 56 27.7	+11 44 56	−0.001	−0.01	7.7	0.0	0.4	B9.5 V		290 s		
65143	174842		7 56 30.3	−23 00 16	0.000	−0.01	7.9	−0.2	0.4	B9.5 V		320 s		
64937	116148		7 56 31.6	+6 24 24	−0.002	+0.01	7.9	1.4	−0.3	K5 III		430 s		
65098	153611		7 56 32.0	−15 31 27	0.000	+0.01	7.3	1.1	−0.1	K2 III		300 s		
64916	97390		7 56 33.1	+11 02 43	−0.002	−0.03	7.6	1.6	−0.5	M2 III		420 s		
63748	6359		7 56 34.3	+71 50 50	+0.010	+0.03	7.3	0.5	3.4	F5 V		60 s		
62807	6348		7 56 35.5	+79 31 09	+0.001	+0.01	7.8	1.1	0.2	K0 III		330 s		
65181	174847		7 56 37.7	−25 24 52	+0.001	0.00	8.0	1.0	−0.1	K2 III		410 s		
65127	153612		7 56 38.8	−16 37 40	+0.001	−0.01	7.7	1.1	−0.1	K2 III		350 s		
65141	153616		7 56 40.0	−16 31 52	−0.004	−0.03	7.2	0.5	3.4	F5 V		58 s		
64726	60466		7 56 43.0	+36 32 32	−0.004	−0.07	7.7	0.9	3.2	G5 IV		80 s		
65125	153618		7 56 45.0	−12 56 07	−0.004	−0.01	7.3	1.1	−0.1	K2 III		240 mx		
——	250023		7 56 45.3	−60 48 54	+0.002	0.00	7.58	0.04		B8				m
65575	235635	χ Car	7 56 46.7	−52 58 56	−0.004	−0.18	3.47	−0.18	−3.0	B2 IV	+19	180 s		
65750	235638		7 56 50.7	−59 07 35	−0.003	−0.02	6.25	1.93		K5				
65228	174852	11 Pup	7 56 51.5	−22 52 48	−0.002	+0.01	4.20	0.72	−2.0	F8 II	+14	150 s		
65082	135287		7 56 54.4	−3 27 24	−0.002	−0.07	7.80	1.1	0.2	K0 III		210 mx		m
64983	97400		7 56 56.8	+11 41 18	−0.001	−0.02	7.9	0.0	0.4	B9.5 V		320 s		
65710	235639		7 56 57.4	−57 25 48	+0.002	+0.02	7.3	1.1	0.2	K0 III		240 s		
65442	219186		7 56 57.8	−42 24 22	+0.002	+0.02	6.09	1.36	−0.1	K2 III		120 s		
65460	219189		7 56 57.8	−43 30 01	0.000	+0.01	5.35	−0.18	−1.6	B3.5 V	+28	240 s		

HD	SAO	Star Name	α 2000	δ 2000	μ(α)	μ(δ)	V	B-V	M_V	Spec	RV	d(pc)	ADS	Notes
65553	219194		7ʰ56ᵐ59ˢ.1	-49°15'55"	-0ˢ.001	-0".01	7.1	1.6	0.2	K0 III		110 s		
64960	97399	1 Cnc	7 56 59.4	+15 47 25	-0.002	-0.05	5.78	1.28	0.2	K0 III	+10	89 s		
65441	219188		7 57 00.4	-41 38 33	-0.001	-0.01	7.82	0.00	0.6	A0 V		180 s		
65354	198622		7 57 03.1	-34 22 43	0.000	0.00	6.8	1.2	-0.1	K2 III		230 s		
65079	116160		7 57 03.8	+2 57 03	-0.002	-0.01	7.83	-0.14		B3 pne	-11			
64959	79818		7 57 05.7	+20 58 02	-0.001	+0.01	7.20	0.9	3.2	G5 IV		63 s	6478	m
65379	198627		7 57 06.7	-35 12 54	-0.009	+0.08	8.0	0.3	0.2	K0 III		140 mx		
63887	6364		7 57 07.5	+71 40 49	-0.002	-0.01	7.6	0.1	0.6	A0 V	-10	250 s		
65491	219193		7 57 07.6	-43 29 03	-0.005	-0.02	7.6	0.2	3.2	G5 IV		77 s		
65378	198625		7 57 08.6	-33 14 58	-0.001	+0.03	7.30	-0.17	0.4	B9.5 V		240 s		
65401	198628		7 57 11.7	-33 08 20	-0.001	-0.01	7.3	1.4	-0.3	K5 III		340 s		
65908	250030		7 57 12.5	-63 17 49	-0.003	+0.01	6.14	-0.10		B9 n	+23			
65243	153625		7 57 12.5	-14 35 29	-0.018	+0.19	7.99	0.63		G0		12 mn		
62434	1255		7 57 15.4	+81 11 24	-0.018	-0.03	7.1	1.1	-0.1	K2 III		270 s		
65066	116162		7 57 15.9	+8 38 29	0.000	-0.02	6.05	1.00	0.2	K0 III	-36	150 s		
65123	116165		7 57 16.1	+1 07 37	-0.012	-0.01	6.35	0.50	3.7	dF6	0	33 s	6483	m
65044	97410		7 57 18.5	+13 16 08	-0.001	-0.03	7.5	1.1	0.2	K0 III		280 s		
65551	219197		7 57 18.5	-44 06 35	0.000	+0.01	5.09	-0.17		B3	+16			m
65199	135294		7 57 19.7	-7 47 59	0.000	+0.01	7.5	0.1	0.6	A0 V		240 s		
65598	219200		7 57 19.9	-47 53 25	-0.002	+0.01	6.22	-0.10		B5 n	+12			m
---	250028		7 57 20.0	-60 48 45	+0.002	+0.02	7.8	-0.2	0.4	B9.5 V		300 s		
65869	250029		7 57 21.1	-60 46 03	-0.001	+0.01	7.71	0.03	0.2	B9 V		290 s		
65065	97411		7 57 22.6	+13 12 25	-0.001	-0.02	7.90	0.9	3.2	G5 IV		87 s	6482	m
65488	198635		7 57 24.9	-37 20 14	-0.001	-0.01	7.5	1.8	-0.1	K2 III		140 s		
65121	116168		7 57 25.4	+6 51 05	-0.002	-0.01	7.3	0.9	3.2	G5 IV		66 s		
65158	135292		7 57 25.7	-0 38 10	-0.003	-0.01	7.17	0.01		A2	+32			
65198	135295		7 57 27.3	-2 56 52	0.000	-0.02	7.7	1.4	-0.3	K5 III		390 s		
65622	219203		7 57 28.1	-46 19 37	0.000	-0.01	7.07	-0.11	-1.1	B5 V nn		420 s		
65139	116170		7 57 28.1	+5 17 06	-0.003	-0.01	8.0	0.5	4.0	F8 V		62 s		
65175	135296		7 57 28.9	-1 18 42	-0.002	+0.01	7.8	1.6	-0.5	M2 III		470 s		
65063	97412		7 57 31.5	+18 06 03	-0.002	-0.05	7.8	0.4	3.0	F2 V		89 s		
64934	60474		7 57 32.6	+32 39 23	-0.003	-0.02	7.0	0.1	1.4	A2 V		130 s		
64914	60473		7 57 33.3	+34 03 27	+0.001	-0.02	8.0	0.9	3.2	G5 IV		93 s		
65658	219209		7 57 38.1	-46 35 34	0.000	+0.02	7.24	-0.13	-1.0	B5.5 V		450 s		m
65550	---		7 57 39.5	-38 55 04			7.9	1.0	3.2	G5 IV		64 s		
63383	6357		7 57 39.9	+77 34 36	+0.002	-0.01	6.8	0.1	1.7	A3 V		110 s		
65456	198636		7 57 40.1	-30 20 04	-0.001	+0.01	4.79	0.15	1.4	A2 V	+28	41 s		
66110	250041		7 57 41.6	-65 54 30	+0.001	+0.02	7.7	1.4	-0.3	K5 III		400 s		
65023	79822		7 57 41.7	+27 29 18	+0.001	-0.04	7.7	1.1	0.2	K0 III		310 s		
65852	235645		7 57 42.5	-57 28 45	-0.001	+0.01	7.41	0.05	0.6	A0 V		230 s		
65950	250036		7 57 45.4	-60 55 35	+0.001	+0.01	6.86	0.00	-0.8	B9 III		340 s		
65592	198641		7 57 45.5	-40 07 24	-0.003	+0.01	7.5	0.5	4.4	G0 V	+42	42 s		AP Pup, v
65907	250035		7 57 46.9	-60 18 11	+0.069	+0.12	5.59	0.57	4.7	G2 V	+13	17 ts		m
65616	219208		7 57 47.0	-41 32 09	-0.001	-0.02	7.3	1.4	0.2	K0 III		230 s		
65279	135300		7 57 48.9	-6 36 01	0.000	-0.03	7.9	1.4	-0.3	K5 III		410 s		
65685	219218		7 57 51.7	-45 34 40	-0.001	+0.02	5.17	1.27	-0.4	gM0	+51	130 s		
65412	174884		7 57 54.1	-20 25 42	+0.002	+0.02	7.8	1.5	-0.5	M2 III		450 s		
65656	219212		7 57 54.5	-41 38 28	-0.001	+0.01	7.53	-0.11	-1.0	B5.5 V		510 s		
65638	219210		7 57 54.9	-40 47 06	+0.003	-0.03	6.8	1.0	0.2	K0 III		210 s		
65021	60480		7 57 58.4	+32 56 27	0.000	-0.03	8.0	1.1	0.2	K0 III		360 s		
65820	235644		7 57 58.4	-52 17 38	-0.001	-0.02	7.6	-0.2	4.0	F8 V		52 s		
65511	174887		7 57 58.9	-27 50 12	-0.001	0.00	7.9	0.5	0.0	B8.5 V		160 s		
65987	250040		7 58 03.0	-60 36 54	0.000	+0.01	7.61	-0.06		A0 p				
---	---		7 58 04.0	-60 48 21			6.68	1.27		A9				
65375	153641		7 58 04.2	-10 53 01	-0.002	0.00	8.00	1.1	0.2	K0 III		350 s	6490	m
65615	198643		7 58 04.7	-37 34 48	-0.002	+0.01	7.47	-0.10	0.0	B8.5 V		310 s		
65241	116179		7 58 05.8	+7 12 49	-0.001	-0.01	6.41	-0.04		B9				
65258	116180		7 58 06.1	+5 00 41	-0.002	+0.01	7.9	0.9	3.2	G5 IV		88 s		
65411	153643		7 58 06.5	-14 38 34	0.000	0.00	7.7	1.4	-0.3	K5 III		380 s		
---	250042		7 58 08.3	-60 46 50	+0.002	+0.01	7.22	0.05		B9				
---	250043		7 58 09.4	-60 48 59	+0.001	+0.02	7.6	0.0	0.4	B9.5 V		280 s		
65220	97423		7 58 11.1	+13 19 46	+0.001	-0.05	7.7	0.9	3.2	G5 IV		80 s		
65683	198646		7 58 12.1	-39 36 49	0.000	-0.03	7.99	-0.03	0.6	A0 V		300 s		
---	250045		7 58 13.0	-60 48 04	0.000	+0.01	7.20	0.00	0.2	B9 V		250 s		
65818	219226		7 58 14.3	-49 14 42	-0.002	+0.01	4.41	-0.17	-4.0	B1.5 III	+20	470 s		V Pup, m,v
64958	42097		7 58 16.5	+43 58 39	+0.004	+0.01	6.4	1.1	0.2	K0 III	-49	170 s		
66109	250049		7 58 17.2	-63 07 54	-0.005	0.00	7.11	0.46	0.6	A0 V		110 s		
65345	116182	14 CMi	7 58 20.6	+2 13 29	-0.011	+0.10	5.29	0.92	0.2	K0 III	+46	82 mx		m
65867	235646		7 58 21.3	-51 26 55	-0.005	+0.03	6.4	1.0	2.6	F0 V		23 s		
66042	235650		7 58 22.0	-59 41 51	-0.001	+0.02	7.7	0.8	0.2	K0 III		310 s		
65652	198648		7 58 22.3	-33 36 32	-0.003	+0.03	7.2	1.1	-0.1	K2 III		290 s		
66066	250047		7 58 22.4	-60 51 23	+0.001	-0.02	7.40	0.00	0.0	B8.5 V		300 s		m
65410	135315		7 58 23.9	-6 24 18	-0.001	-0.02	7.8	0.1	0.6	A0 V		260 s		
65948	235648		7 58 30.1	-54 34 08	-0.002	-0.04	7.3	1.4	0.2	K0 III		160 s		
65409	135317		7 58 30.2	-3 38 16	-0.002	-0.01	8.0	0.1	0.6	A0 V		300 s		

HD	SAO	Star Name	α 2000	δ 2000	μ(α)	μ(δ)	V	B-V	M_v	Spec	RV	d(pc)	ADS	Notes
			h m s	° ′ ″	s	″								
65372	116186		7 58 30.3	+2 56 09	+0.001	-0.02	6.67	0.14		A3				
65257	97429		7 58 31.4	+16 31 07	0.000	-0.01	5.99	1.47	0.2	K0 III	-1	75 s		
65705	198653		7 58 31.6	-36 25 27	-0.004	0.00	8.0	0.2	1.4	A2 V		160 s		
--	135318		7 58 34.6	-3 52 16	-0.001	-0.01	8.0	1.1	-0.1	K2 III		420 s		
65041	42099		7 58 34.8	+43 30 15	0.000	+0.01	7.33	-0.18	-1.6	B3.5 V	-9	600 s		
65721	198656		7 58 36.8	-34 56 43	-0.030	+0.18	7.96	0.74	5.3	G6 V		31 ts		
65723	198657		7 58 37.8	-37 01 14	+0.007	-0.19	7.01	0.98		G5		14 mn		
65636	174914		7 58 38.4	-27 34 39	0.000	-0.02	7.1	1.5	0.2	K0 III		110 s		
65563	153655		7 58 43.0	-15 39 19	-0.002	-0.01	7.7	1.1	0.2	K0 III		310 s		
66137	250053		7 58 43.7	-60 42 05	-0.001	+0.01	7.82	0.05	0.6	A0 V		250 s		
65396	116192		7 58 45.1	+5 37 23	-0.001	-0.01	6.80	0.00	0.4	B9.5 V		190 s	6492	m
65343	97433		7 58 45.9	+12 07 14	0.000	-0.02	7.9	0.9	3.2	G5 IV		88 s		
65905	219237		7 58 50.4	-47 40 56	0.000	+0.02	7.27	-0.09	0.0	B8.5 V		280 s		m
66194	250055		7 58 50.5	-60 49 27	-0.001	+0.01	5.80	-0.10	-1.7	B3 V	-3	300 s		
65395	116193		7 58 52.9	+8 27 13	+0.003	+0.02	7.4	0.8	3.2	G5 IV		68 s		
65701	174926		7 58 53.4	-27 54 26	+0.001	0.00	7.5	1.2	3.2	G5 IV		40 s		
65587	153662		7 58 55.0	-17 03 27	+0.001	-0.01	7.7	1.4	-0.3	K5 III		390 s		
65019	42101		7 58 57.2	+48 37 33	-0.001	0.00	7.4	1.1	0.2	K0 III		270 s		
65930	219244		7 59 00.7	-48 22 34	0.000	+0.01	6.83	-0.12	-2.5	B2 V		690 s		m
65904	219240		7 59 01.7	-45 12 58	-0.002	0.00	5.99	-0.14		B6 n	-3			
65929	219243		7 59 01.8	-47 08 28	-0.002	-0.01	7.3	0.9	3.2	G5 IV		55 s		
66260	250058		7 59 02.5	-61 35 00	-0.003	+0.03	7.2	0.6	2.6	F0 V		54 s		
65610	153666		7 59 02.6	-13 53 47	-0.004	+0.01	8.0	0.4	2.6	F0 V		120 s		
65526	135329		7 59 03.9	-4 19 56	+0.002	-0.04	6.8	0.1	1.7	A3 V		100 s		
66064	235658		7 59 04.3	-54 32 47	-0.005	+0.02	7.6	1.2	0.2	K0 III		240 s		
65699	174932	12 Pup	7 59 05.6	-23 18 37	-0.001	0.00	5.11	1.12	-6.0	cK2	+11	1700 s		
65848	198666		7 59 07.4	-39 25 15	-0.002	-0.02	7.24	-0.09	0.6	A0 V		210 s		
65831	198665		7 59 07.7	-38 03 03	-0.002	+0.05	7.1	0.5	2.6	F0 V		58 s		
66027	235655		7 59 10.2	-51 27 57	-0.001	0.00	7.3	-0.1	0.4	B9.5 V		240 s		
65698	174935		7 59 10.8	-22 05 39	-0.001	-0.01	7.9	1.4	-0.1	K2 III		300 s		
66005	219249		7 59 12.2	-49 58 36	-0.001	+0.01	6.50	0.00	-1.6	B3.5 V	+13	400 s		m
65254	60497		7 59 12.5	+33 52 26	0.000	-0.01	7.8	0.1	0.6	A0 V		280 s		
65525	116204		7 59 12.6	+5 15 05	-0.002	-0.02	7.8	0.8	3.2	G5 IV		84 s		
66006	219250		7 59 13.4	-49 58 25	-0.001	+0.01	6.6	0.0	-1.6	B3.5 V	+23	420 s		
66920	256463		7 59 16.2	-73 14 41	+0.002	-0.04	6.34	0.14		A2				
65673	153675		7 59 17.6	-16 40 59	-0.002	-0.01	7.9	1.4	-0.3	K5 III		420 s		
65431	97439		7 59 17.7	+16 49 07	-0.004	-0.04	7.40	0.5	4.0	F8 V		48 s	6493	m
65078	26603		7 59 21.2	+50 20 38	-0.001	-0.03	7.5	1.1	-0.1	K2 III		330 s		
65718	153679		7 59 21.2	-19 27 58	0.000	0.00	8.0	-0.2	0.6	A0 V		300 s		
66080	235661		7 59 22.2	-52 32 35	-0.003	+0.01	7.7	0.8	0.2	K0 III		310 s		
65450	97442		7 59 22.7	+15 21 56	-0.001	-0.02	7.8	1.1	0.2	K0 III		330 s		
65473	97443		7 59 23.9	+13 49 38	-0.001	-0.05	7.9	0.9	3.2	G5 IV		88 s		
65275	60500		7 59 24.9	+34 40 47	-0.001	-0.01	7.6	1.6	-0.5	M2 III	-25	420 s		
65524	97444		7 59 25.3	+9 56 07	-0.008	-0.02	8.0	0.5	3.4	F5 V		82 s		
65926	219247		7 59 26.9	-41 56 25	-0.003	+0.03	7.9	1.1	0.2	K0 III		150 s		
65925	198668		7 59 28.3	-39 17 50	-0.008	-0.05	5.24	0.39	2.6	dF0	-8	30 s		
65813	174945		7 59 30.3	-28 26 00	+0.001	-0.04	7.71	0.32	2.6	F0 V		100 s		
65888	198667		7 59 30.6	-37 18 08	0.000	0.00	7.43	-0.13	0.0	B8.5 V		310 s		
65430	79846		7 59 33.9	+20 50 38	+0.013	-0.55	7.70	0.83	5.9	K0 V	-28	26 mx		
65812	174946		7 59 34.1	-27 12 10	-0.001	-0.01	8.0	0.9	3.2	G5 IV		78 s		
65522	97445		7 59 35.0	+13 14 32	-0.001	-0.01	6.02	1.32	-0.3	K5 III	+27	180 s		
63331	6363		7 59 36.7	+79 51 41	-0.004	-0.01	6.9	1.1	-0.1	K2 III		250 s		
66342	250063		7 59 37.6	-60 35 13	0.000	+0.01	5.16	1.72	-2.4	M0 II	+24	300 s		
66341	250064		7 59 40.2	-60 12 27	+0.001	0.00	6.32	-0.06		B8	+23			
65219	42109		7 59 40.7	+46 37 11	-0.002	-0.05	8.00	0.6	4.4	G0 V		53 s	6491	m
66061	219263		7 59 41.7	-48 16 48	-0.001	-0.02	7.9	0.2	4.4	G0 V		51 s		
65471	79847		7 59 42.7	+23 10 58	0.000	-0.03	6.8	1.1	0.2	K0 III		210 s		
--	--		7 59 43.0	-47 18 17			7.40							m
--	--		7 59 43.2	-61 05 46			6.70	1.1	0.0	K1 III		220 s		
65847	--		7 59 43.7	-27 17 16			8.0	1.4	-0.3	K5 III		450 s		
65695	135345	27 Mon	7 59 44.1	-3 40 47	-0.004	-0.01	4.93	1.21	-0.1	K2 III	-29	97 s		
65763	153682		7 59 45.0	-12 52 02	+0.002	0.00	7.4	0.0	0.4	B9.5 V		250 s		
66079	219265		7 59 45.7	-47 18 13	-0.002	0.00	6.71	-0.09	0.4	B9.5 V		180 s		m
65864	174951		7 59 46.0	-28 06 28	0.000	-0.03	7.7	1.6	0.2	K0 III		140 s		
65740	135347		7 59 46.1	-9 06 31	-0.002	-0.02	7.9	0.1	1.7	A3 V		170 s		
65887	198670		7 59 49.0	-30 24 46	-0.002	-0.02	7.69	1.26	0.2	K0 III		210 s		
65392	60505		7 59 51.4	+36 04 55	-0.001	-0.01	6.4	1.4	-0.3	K5 III		220 s		
65810	153687		7 59 52.0	-18 23 58	-0.001	-0.05	4.61	0.08	1.7	A3 V	-12	38 s		
65629	97452		7 59 53.3	+9 53 55	+0.007	-0.16	7.9	0.9	3.2	G5 IV		53 mx		
66078	219267		7 59 54.2	-46 10 05	+0.002	-0.08	7.99	0.78		G0				
64626	14393		7 59 57.3	+68 56 44	+0.001	+0.01	8.0	1.1	0.2	K0 III		360 s		
65923	198673		7 59 57.3	-30 22 35	+0.005	-0.03	7.73	0.55	4.4	G0 V		46 s		m
66039	219266		7 59 58.5	-41 42 15	-0.002	-0.07	7.9	0.6	4.0	F8 V		34 s		
65717	135349		7 59 58.8	-1 47 51	-0.003	0.00	7.7	1.1	0.2	K0 III		310 s		
65239	42111		7 59 58.9	+47 47 07	+0.004	-0.05	8.0	0.9	3.2	G5 IV		93 s		

HD	SAO	Star Name	α 2000	δ 2000	μ(α)	μ(δ)	V	B-V	M$_v$	Spec	RV	d(pc)	ADS	Notes
			h m s	° ′ ″	s	″								
65862	174954		8 00 02.5	-21 19 54	+0.002	+0.05	8.0	1.2	-0.5	M2 III		510 s		
66105	219268		8 00 02.9	-45 01 32	-0.001	+0.01	7.1	0.4	0.0	B8.5 V		130 s		
65863	174955		8 00 03.0	-23 21 03	-0.002	+0.01	7.8	-0.1	0.6	A0 V		270 s		
63855	6373		8 00 04.8	+77 38 45	+0.001	-0.06	7.7	1.1	0.2	K0 III		250 mx		
65369	42115		8 00 07.9	+41 25 21	-0.002	-0.04	7.7	1.1	-0.1	K2 III		360 s		
65846	153689		8 00 10.0	-16 57 10	-0.004	0.00	7.0	1.1	0.2	K0 III		230 s		
64307	6378		8 00 11.7	+73 55 04	-0.002	-0.04	5.41	1.42	-0.2	K3 III	+35	110 s		
66339	235669		8 00 12.5	-55 12 56	-0.003	+0.03	7.1	1.2	0.2	K0 III		170 s		
66019	198684		8 00 14.7	-35 59 31	-0.004	0.00	7.2	0.1	1.4	A2 V		130 s		
66210	219283		8 00 14.8	-48 58 52	-0.002	+0.05	6.02	0.04		A2				
66162	219276		8 00 15.0	-45 47 37	-0.006	+0.06	7.4	0.7	3.2	G5 IV		70 s		
65407	42118		8 00 16.5	+40 47 35	0.000	-0.02	8.0	0.9	3.2	G5 IV		93 s		
65944	174961		8 00 18.4	-26 35 31	-0.001	-0.01	7.92	0.91	0.2	K0 III		350 s		
65715	116219		8 00 18.7	+8 58 28	-0.002	+0.02	8.0	0.1	1.7	A3 V		180 s		
65882	153698		8 00 19.3	-18 00 17	-0.001	-0.01	7.9	1.4	-0.3	K5 III		420 s		
66190	219280		8 00 19.6	-45 27 25	0.000	0.00	6.5	1.3	0.2	K0 III		120 s		
66591	250069		8 00 20.0	-63 34 03	0.000	+0.02	4.82	-0.17	-2.3	B3 IV	+21	260 s		
66037	198686		8 00 20.9	-34 16 06	-0.003	+0.02	7.4	1.2	-0.1	K2 III		280 s		
65694	97462		8 00 21.5	+12 12 50	0.000	+0.01	7.8	0.1	0.6	A0 V		270 s		
65778	116223		8 00 22.5	+3 14 55	-0.002	-0.02	7.9	0.2	2.1	A5 V		150 s		
64911	14394		8 00 23.8	+64 44 50	-0.002	-0.03	7.4	0.4	2.6	F0 V		90 s		
65789	116225		8 00 24.9	-0 03 59	-0.001	-0.01	7.7	1.1	0.2	K0 III		310 s		
66059	198689		8 00 28.0	-34 07 26	-0.001	+0.02	7.48	0.05	0.6	A0 V		220 s		
66255	219292		8 00 28.8	-48 52 17	-0.002	+0.01	6.12	-0.11		A0 p				
65826	135360		8 00 28.9	-4 52 48	-0.001	+0.01	7.1	0.5	4.0	F8 V		42 s		
65919	153704		8 00 29.3	-17 55 49	0.000	0.00	7.9	1.1	-0.1	K2 III		380 s		
65903	153703		8 00 30.3	-17 02 03	+0.002	-0.06	7.4	1.1	0.2	K0 III		230 mx		
65583	79856		8 00 32.2	+29 12 44	-0.011	-1.17	7.00	0.71	5.6	G8 V	+13	11 mx		
65666	97463		8 00 32.3	+18 50 34	-0.001	-0.02	6.9	0.1	0.6	A0 V		180 s		
66235	219290		8 00 35.3	-45 20 11	-0.002	+0.02	7.4	-0.2	0.4	B9.5 V		200 s		
65737	97469		8 00 36.3	+12 38 30	-0.001	-0.02	6.6	1.1	0.2	K0 III		190 s		
65839	135365		8 00 38.4	-4 35 28	-0.001	-0.01	7.9	1.1	-0.1	K2 III		390 s		
65736	97468		8 00 41.2	+18 14 44	-0.001	-0.01	7.1	0.9	3.2	G5 IV		61 s		
66187	219287		8 00 41.7	-40 11 47	-0.001	+0.01	7.4	0.5	0.6	A0 V		120 s		
68039	256469		8 00 43.8	-79 09 24	-0.002	+0.06	7.4	1.1	0.2	K0 III		270 s		
65760	97470		8 00 43.8	+16 35 16	-0.001	-0.02	7.7	1.4	-0.3	K5 III		390 s		
65875	135368		8 00 44.0	-2 52 54	-0.001	+0.01	6.51	-0.07	-2.1	B2.5 V e	+33	530 s		
65876	135370		8 00 46.5	-4 31 13	0.000	+0.02	7.9	1.4	-0.3	K5 III		430 s		
66569	235681		8 00 46.8	-59 39 30	-0.002	+0.03	7.8	0.7	2.6	F0 V		60 s		
65877	135372		8 00 46.9	-4 45 30	-0.001	+0.01	7.8	0.1	0.6	A0 V		270 s		
65759	97472	3 Cnc	8 00 47.3	+17 18 31	-0.001	-0.01	5.55	1.32	0.2	K0 III	+41	76 s		
65735	97471		8 00 48.0	+19 48 57	0.000	-0.02	6.25	1.11	0.0	K1 III	+28	180 s		
65804	97474		8 00 49.2	+12 40 42	+0.001	-0.01	6.6	0.0	0.4	B9.5 V		180 s		
66441	235680		8 00 49.9	-54 09 05	-0.002	0.00	5.87	-0.14	0.0	B8.5 V	0	150 s		
66231	219294		8 00 50.4	-40 31 38	+0.001	-0.02	7.9	1.1	0.2	K0 III		340 s		
65996	153714		8 00 54.3	-15 44 21	+0.002	-0.06	7.4	0.2	2.1	A5 V		39 mx		
66207	198700		8 00 55.0	-38 43 30	+0.003	0.00	7.7	1.1	-0.1	K2 III		360 s		
65542	42123		8 00 55.6	+39 54 36	+0.002	-0.01	7.6	1.1	0.2	K0 III		310 s		
65714	79861	2 ω Cnc	8 00 55.8	+25 23 34	+0.001	0.00	5.83	1.02	0.2	K0 III	+2	130 s		
66155	198698		8 00 57.3	-32 21 45	-0.001	-0.04	7.9	1.5	-0.3	K5 III		430 s		
66523	235683		8 00 58.4	-57 26 58	-0.002	+0.03	7.6	1.4	-0.1	K2 III		260 s		
65757	79864		8 01 00.7	+23 34 59	-0.002	-0.03	6.40	1.1		K1 III-IV	+25		6513	m
65994	153718		8 01 02.2	-14 08 16	+0.001	-0.01	7.2	0.1	1.7	A3 V		130 s		
65938	135379		8 01 03.3	-6 25 09	-0.005	-0.05	6.5	0.8	3.2	G5 IV		47 s		
66311	219302		8 01 08.9	-41 45 32	-0.003	-0.01	7.88	-0.10	0.4	B9.5 V		310 s		
66152	174992		8 01 11.7	-29 09 38	-0.002	+0.02	7.9	-0.2	-1.0	B5.5 V		610 s		
65390	26620		8 01 12.1	+54 27 38	-0.005	-0.03	6.9	0.1	1.7	A3 V		110 s		
65953	135380	28 Mon	8 01 13.2	-1 23 33	+0.004	-0.08	4.68	1.49	-0.3	K4 III	+27	88 s		
65972	135381		8 01 13.4	-3 31 10	-0.001	+0.01	7.9	1.1	0.2	K0 III		340 s		
65874	116241		8 01 13.4	+8 22 34	-0.002	-0.05	7.4	0.6	4.4	G0 V		40 s		
65900	116244		8 01 13.8	+4 52 47	-0.003	+0.01	5.65	0.00		A0	+46			m
66097	---		8 01 14.1	-22 19 27			7.80	0.1	1.4	A2 V		120 s	6524	m
66149	174988		8 01 14.3	-25 24 48	+0.001	+0.01	7.3	0.9	1.7	A3 V		41 s		
66750	250074		8 01 18.6	-61 52 38	+0.001	+0.03	8.0	0.9	0.2	K0 III		360 s		
66279	198703		8 01 18.9	-35 54 15	-0.001	+0.01	7.39	1.62	-0.3	K5 III		340 s		
66464	219314		8 01 20.1	-49 37 14	0.000	0.00	7.24	-0.11	-2.9	B3 III		980 s		
65301	26618		8 01 20.7	+59 02 50	+0.002	+0.03	5.77	0.39	3.0	F2 V	-40	35 s		
---	174995		8 01 21.5	-22 20 00	0.000	+0.01	7.80	0.6	4.4	G0 V		120 s	6524	m
65641	42128		8 01 22.4	+40 51 24	-0.002	-0.02	7.6	0.9	3.2	G5 IV		75 s		
66181	174998		8 01 22.5	-24 57 13	-0.001	0.00	8.0	-0.6	-1.0	B5.5 V		640 s		
66546	235686		8 01 22.8	-54 30 54	-0.002	+0.02	5.95	-0.04	-1.4	B4 V	+19	270 s		m
65857	97483		8 01 23.0	+16 56 57	-0.003	-0.03	7.7	0.6	4.4	G0 V		47 s	6519	m
66507	235685		8 01 23.7	-51 26 23	+0.001	0.00	7.1	0.6	0.4	B9.5 V		96 s		
66655	235692		8 01 28.2	-58 07 09	-0.003	-0.02	7.8	0.3	0.6	A0 V		170 s		
66293	198709		8 01 29.4	-35 22 46	-0.001	-0.02	7.51	-0.08	0.4	B9.5 V		260 s		

HD	SAO	Star Name	α 2000	δ 2000	μ(α)	μ(δ)	V	B-V	M_v	Spec	RV	d(pc)	ADS	Notes
			h m s	° ' "	s	"								
65873	97485	5 Cnc	8 01 30.2	+16 27 19	0.000	−0.01	5.90	−0.02	0.2	B9 V	−16	130 s		
65823	79868		8 01 30.2	+23 54 45	−0.001	−0.03	7.40	1.1	0.2	K0 III		280 s	6518	m
66607	235690		8 01 31.5	−55 27 18	−0.002	−0.01	6.28	−0.15	0.0	B8.5 V		180 s		
67200	250085		8 01 31.9	−70 01 25	−0.004	+0.11	6.8	1.5	4.4	G0 V		30 s		
66438	219315		8 01 32.3	−45 13 52	−0.001	+0.02	7.8	0.1	0.4	B9.5 V		200 s		
65664	42130		8 01 32.4	+41 23 14	−0.003	−0.10	7.2	0.5	3.4	F5 V		59 s		
66146	153731		8 01 34.4	−18 38 27	−0.001	−0.01	7.7	1.1	−0.1	K2 III		350 s		
66126	153729		8 01 35.1	−15 15 19	+0.004	−0.01	7.5	1.1	0.2	K0 III		220 mx		
66522	235688		8 01 35.1	−50 36 22	−0.001	−0.02	7.21	0.05	−3.6	B2 III p		1300 s		
66277	198706		8 01 35.2	−30 35 33	+0.003	−0.01	7.63	1.52	−0.1	K2 III		220 s		
66278	198708		8 01 35.8	−30 37 14	0.000	0.00	8.0	0.7	0.2	K0 III		360 s		
66309	198711		8 01 36.2	−35 47 20	−0.001	+0.01	7.84	−0.04	0.6	A0 V		280 s		
66358	198714		8 01 37.3	−37 17 01	−0.002	+0.01	5.95	0.14		A2				
66308	198712		8 01 39.8	−34 41 45	+0.006	−0.14	8.0	−0.3	3.4	F5 V		82 s		
65339	14402	53 Cam	8 01 42.3	+60 19 27	−0.006	−0.02	6.01	0.14		A2 p	−5			
65856	79869	4 Cnc	8 01 43.7	+25 05 23	−0.001	+0.02	6.20	0.01	1.2	A1 V	−9	100 s		m
66094	135392		8 01 44.4	−8 35 37	−0.004	−0.01	7.2	0.5	3.4	F5 V		59 s	6526	m
66223	175008		8 01 44.5	−21 09 36	0.000	−0.02	7.8	1.5	−0.3	K5 III		420 s		
65755	42132		8 01 47.4	+40 08 40	−0.003	−0.09	8.0	0.5	4.0	F8 V		64 s		
66384	198716		8 01 47.9	−36 10 53	−0.003	−0.06	7.72	0.25	2.6	F0 V		110 s		
66011	116251		8 01 50.6	+8 54 51	0.000	+0.03	6.22	0.57		F5	+4			
66478	219321		8 01 51.1	−44 36 00	−0.002	0.00	6.55	1.33	0.2	K0 III		110 s		
67279	250086		8 01 51.9	−70 01 34	−0.001	+0.06	8.0	1.1	−0.1	K2 III		350 mx		
66382	198717		8 01 54.1	−33 10 42	−0.003	−0.01	7.9	1.5	−0.3	K5 III		430 s		
65801	60523		8 01 55.0	+35 24 47	−0.004	−0.01	6.34	1.56	0.2	K0 III	−16	78 s		
66050	116254		8 01 55.4	+4 32 27	−0.003	−0.03	7.3	1.1	0.2	K0 III		270 s		
66306	175018		8 01 56.9	−27 12 54	−0.003	−0.01	6.60	0.00	0.4	B9.5 V		170 s	6535	m
66706	235694		8 01 59.7	−54 39 44	−0.005	+0.06	7.0	0.6	2.1	A5 V		54 s		
66179	135400		8 02 00.2	−9 21 26	−0.005	0.00	6.9	0.1	1.7	A3 V		110 s		
66381	198719		8 02 01.8	−31 08 12	0.000	−0.02	7.90	0.10	1.4	A2 V		190 s		
65970	97491		8 02 04.7	+19 44 14	0.000	−0.03	7.5	0.0	0.4	B9.5 V		260 s		
66435	198725		8 02 06.2	−37 03 02	+0.001	0.00	6.34	1.62		Ma				
66749	235696		8 02 06.8	−55 36 15	−0.002	+0.01	7.37	−0.03	0.6	A0 V		230 s		
66090	116258		8 02 07.4	+2 57 37	−0.001	−0.01	7.5	1.1	−0.1	K2 III		330 s		
67315	250090		8 02 08.5	−69 46 31	−0.003	+0.02	7.7	1.1	−0.1	K2 III		360 s		
66768	235698		8 02 15.3	−55 32 27	−0.003	−0.01	6.69	−0.03	0.6	A0 V		170 s		
66141	116260		8 02 15.8	+2 20 04	−0.002	+0.10	4.39	1.25	−0.1	K2 III	+71	70 s		m
65788	42136		8 02 15.9	+44 01 19	−0.003	−0.01	7.9	0.4	2.6	F0 V		120 s		
65429	14406		8 02 16.6	+60 59 37	0.000	+0.03	6.7	0.4	3.0	F2 V	−20	56 s		
65899	60524		8 02 21.3	+35 53 18	−0.001	−0.05	8.0	1.1	0.2	K0 III		360 s		
--	235697		8 02 23.7	−51 26 25	+0.001	+0.02	7.27	1.09	3.2	G5 IV		37 s		
66500	198730		8 02 24.7	−36 50 15	0.000	−0.03	7.85	−0.04	0.4	B9.5 V		310 s		
66456	198728		8 02 25.6	−32 18 28	−0.002	+0.01	7.9	0.4	0.4	B9.5 V		160 s		
66242	135406		8 02 25.9	−6 20 14	0.000	−0.02	6.33	0.62		G0		16 mn		
66653	219334		8 02 26.9	−46 20 12	−0.012	+0.22	7.53	0.65		G0		15 mn		
66605	219333		8 02 27.3	−44 40 02	+0.002	0.00	6.70	0.1		A0 p				m
66376	175036		8 02 28.6	−20 19 06	−0.002	−0.02	7.50	0.1	0.6	A0 V		240 s	6539	m
66199	116264		8 02 30.3	+3 04 35	0.000	−0.03	7.80	1.1	0.2	K0 III		330 s	6533	m
65448	14407		8 02 30.7	+63 05 24	−0.002	−0.02	6.40	0.59	4.0	F8 V	+20	25 s		m
67199	250088		8 02 31.4	−66 01 17	−0.023	−0.15	7.9	0.8	3.2	G5 IV		54 mx		
66117	97498		8 02 32.6	+12 27 08	−0.002	0.00	7.50	1.1	0.2	K0 III		290 s	6531	m
66118	97499		8 02 32.9	+12 14 00	+0.001	−0.05	7.7	1.1	−0.1	K2 III		360 s		
66048	79884		8 02 33.5	+21 52 19	+0.001	0.00	8.0	1.1	0.2	K0 III		360 s		
66433	175048		8 02 35.7	−25 02 14	+0.004	−0.04	7.4	0.7	3.0	F2 V		48 s		
65626	26634		8 02 35.8	+57 16 25	−0.003	−0.07	6.5	0.6	4.4	G0 V	+26	27 s		
66115	97500		8 02 36.9	+15 54 54	0.000	−0.01	7.8	1.1	0.2	K0 III		340 s		
66454	175052		8 02 37.5	−27 32 42	+0.001	−0.03	7.10	0.8	3.2	G5 IV		60 s	6544	m
66540	198734		8 02 37.7	−36 24 04	0.000	+0.01	6.57	0.24	2.6	F0 V		62 s		
66268	135413		8 02 40.2	−2 23 49	+0.001	0.00	7.6	0.5	3.4	F5 V		70 s		
66747	219345		8 02 40.9	−49 57 03	−0.003	+0.02	7.4	−0.2	2.6	F0 V		93 s		
65800	42137		8 02 42.3	+48 35 33	−0.002	−0.01	7.5	1.4	−0.3	K5 III		370 s		
66116	97503		8 02 42.6	+14 56 58	0.000	0.00	8.0	0.1	2.1	A5 V		100 s		
66624	219339		8 02 44.7	−41 18 36	−0.001	+0.01	5.47	−0.11	0.4	B9.5 V		100 s		m
66351	135416		8 02 46.3	−6 19 21	−0.002	+0.01	6.7	1.1	0.2	K0 III		200 s		
65734	26637		8 02 46.5	+54 07 53	−0.001	0.00	7.5	0.2	2.1	A5 V	−27	120 s		
66539	198735		8 02 48.0	−31 12 57	−0.007	0.00	7.70	0.00	−1.6	B3.5 V		66 mx		m
66583	198738		8 02 50.3	−34 32 45	−0.005	0.00	7.4	1.2	−0.5	M2 III		210 mx		
66582	198739		8 02 53.5	−33 09 12	−0.001	+0.02	7.34	−0.14	−1.6	B3.5 V		590 s		
66650	219342		8 02 54.6	−40 13 37	−0.003	−0.03	7.8	0.0	0.6	A0 V		280 s		
66600	198741		8 02 54.7	−36 13 13	−0.002	−0.01	7.9	0.8	3.2	G5 IV		85 s		
66765	219347		8 02 55.5	−48 19 30	−0.003	0.00	6.62	−0.16	−2.2	B5 III		580 s		
66676	219343		8 02 55.9	−40 24 08	−0.004	+0.03	7.1	1.4	0.2	K0 III		160 s		
66598	198743		8 03 04.1	−32 27 50	−0.002	0.00	5.76	1.24	−0.1	K2 III		130 s		m
66623	198744		8 03 06.1	−35 05 00	−0.006	+0.13	7.9	0.7	4.4	G0 V		42 s		
66726	219346		8 03 07.0	−41 07 36	−0.003	0.00	7.3	0.4	0.2	K0 III		260 s		

HD	SAO	Star Name	α 2000	δ 2000	μ(α)	μ(δ)	V	B–V	M_v	Spec	RV	d(pc)	ADS	Notes
67170	250091		8^{h}03^{m}07.7^s	−62°07′58″	+0.001	0.00	7.4	0.3	0.4	B9.5 V		150 s		
65497	14413		8 03 09.4	+65 08 08	−0.001	−0.06	7.6	0.4	3.0	F2 V		83 s		
66555	175065		8 03 10.9	−24 36 23	+0.010	−0.08	7.9	0.7	3.2	G5 IV		88 s		
65914	42139		8 03 12.8	+46 13 04	+0.002	−0.01	7.6	0.5	3.4	F5 V		71 s		
66303	97510		8 03 14.4	+9 56 33	0.000	−0.02	7.70	1.1	−0.1	K2 III		360 s	6542	m
---	---		8 03 17.2	+26 16 02			7.50						6538	m
66176	79893		8 03 18.5	+26 16 03	0.000	−0.01	7.0	0.1	1.4	A2 V		130 s		
67559	250098		8 03 21.3	−69 58 16	−0.001	+0.02	6.8	−0.3	0.6	A0 V		170 s		
66742	198749		8 03 22.8	−38 38 19	−0.001	−0.01	7.8	1.5	−0.3	K5 III		420 s		
66301	97513		8 03 24.5	+13 40 08	0.000	−0.02	7.60	0.6	4.4	G0 V		44 s	6543	m
66085	60531		8 03 26.6	+37 08 13	0.000	−0.01	7.1	1.4	−0.3	K5 III		300 s		
66300	97512		8 03 27.6	+16 50 50	+0.002	−0.07	7.5	1.1	0.2	K0 III		150 mx		
66348	97514		8 03 29.0	+12 10 51	−0.006	−0.07	6.70	0.5	4.0	F8 V		35 s	6546	m
66812	219355		8 03 29.4	−42 56 55	−0.003	−0.02	6.29	1.01	0.2	K0 III		170 s		
65872	26641		8 03 29.6	+52 45 27	+0.001	−0.01	7.7	1.4	−0.3	K5 III		390 s		
66940	219360		8 03 29.7	−49 29 56	−0.005	+0.01	6.64	1.02	0.2	K0 III		190 s		
66216	79896	χ Gem	8 03 31.0	+27 47 39	−0.002	−0.04	4.94	1.12	−0.1	K2 III	−11	100 s		m
66488	135430		8 03 31.9	−8 10 24	−0.003	+0.04	7.40	0.5	4.0	F8 V		48 s		m
66321	97516		8 03 33.8	+14 05 24	0.000	−0.03	7.4	1.1	0.2	K0 III		270 s		
66287	79899		8 03 34.1	+20 20 19	+0.002	−0.02	7.1	0.8	3.2	G5 IV		60 s		
66811	198752	ζ Pup	8 03 35.0	−40 00 12	−0.003	+0.01	2.25	−0.26		O5.8	−24			
66175	60532		8 03 39.9	+36 20 40	0.000	−0.03	6.7	1.6	−0.5	M4 III	−1	280 s		SV Lyn, q
66320	97518		8 03 40.0	+17 43 32	−0.001	−0.04	7.7	1.1	0.2	K0 III		310 s		
66698	175087		8 03 42.1	−29 25 09	+0.002	−0.01	7.4	0.6	0.6	A0 V		93 s		
66347	79903		8 03 50.4	+22 04 14	−0.003	−0.01	6.8	1.1	0.2	K0 III	−4	210 s		
66084	42146		8 03 51.2	+44 12 57	0.000	−0.05	8.0	1.1	0.2	K0 III		360 s		
67706	256471		8 03 51.3	−70 20 14	+0.001	+0.03	7.3	−0.1	0.6	A0 V		210 s		
66446	116287		8 03 52.1	+6 19 26	0.000	0.00	7.9	0.0	0.0	B8.5 V		380 s		
67624	250099		8 03 53.3	−68 56 14	−0.002	−0.01	6.8	1.1	0.2	K0 III		210 s		
66513	116290		8 03 55.3	+3 23 56	−0.002	0.00	7.7	0.0	0.4	B9.5 V		290 s		
67041	219367		8 03 56.2	−47 48 56	+0.002	+0.02	7.4	−0.3	0.4	B9.5 V		250 s		
66113	42148		8 03 57.5	+43 50 24	−0.001	+0.03	7.3	0.4	2.6	F0 V		88 s		
66644	153775		8 03 57.5	−16 20 33	−0.001	−0.03	8.0	1.1	0.2	K0 III		350 s		
66740	175094		8 04 01.2	−26 58 23	−0.002	+0.05	7.5	0.2	3.0	F2 V		81 s		
66392	79909		8 04 01.4	+21 44 10	+0.001	−0.05	7.6	1.1	0.2	K0 III		310 s		
66789	175099		8 04 04.8	−29 10 25	0.000	−0.01	7.37	0.08	1.4	A2 V		150 s		
66722	175092		8 04 05.0	−21 22 35	+0.001	−0.02	7.7	0.4	0.6	A0 V		150 s		
66739	175093		8 04 05.7	−21 49 00	0.000	−0.03	7.4	1.0	3.0	F2 V		30 s		
66444	97526		8 04 06.0	+13 41 15	−0.007	+0.05	7.4	1.1	0.2	K0 III		200 mx		
66299	60538		8 04 08.3	+33 01 51	−0.001	+0.01	7.10	0.1	0.6	A0 V	−9	200 s	6549	m
66643	153777		8 04 08.6	−10 25 44	+0.001	−0.04	8.0	1.1	0.2	K0 III		360 s		
66594	135444		8 04 10.1	−4 49 37	−0.001	0.00	7.56	−0.13		B5	+10			
65580	14419		8 04 11.2	+67 40 20	−0.004	+0.01	7.8	1.1	0.2	K0 III		330 s		
67221	235727		8 04 14.5	−54 28 35	0.000	−0.01	7.7	1.0	0.2	K0 III		310 s		
66888	198764		8 04 16.1	−32 40 30	−0.001	0.00	5.31	1.91	−0.5	M2 III	+36	100 s		
66510	97527		8 04 16.3	+10 51 25	−0.003	0.00	7.7	0.2	2.1	A5 V		130 s		
66112	42150		8 04 16.7	+47 56 53	−0.003	−0.03	8.0	0.9	3.2	G5 IV		93 s		
66530	97531		8 04 21.1	+10 14 59	−0.001	+0.02	7.5	0.8	3.2	G5 IV		73 s		
67061	219375		8 04 22.2	−44 14 16	−0.002	−0.02	7.41	1.01	3.2	G5 IV		47 s		
66509	97530		8 04 23.1	+12 17 23	+0.007	−0.14	7.78	0.85	6.3	dK2	−12	25 ts	6554	m
67276	235728		8 04 23.1	−54 59 42	−0.003	+0.02	7.9	0.1	0.4	B9.5 V		250 s		
66613	116304		8 04 26.5	+0 57 41	−0.001	0.00	7.7	1.1	0.2	K0 III		310 s		
66640	135448		8 04 27.5	−1 35 44	−0.004	−0.02	7.8	0.1	0.6	A0 V		120 mx		
66443	79913		8 04 28.0	+21 00 29	−0.002	+0.01	7.4	0.1	1.4	A2 V		160 s		
66573	116302		8 04 29.7	+9 16 06	+0.017	+0.04	7.4	0.6	4.4	G0 V		40 s		
67059	219382		8 04 30.1	−42 45 30	0.000	+0.01	7.40	0.00	−1.6	B3.5 V		590 s		m
---	250100		8 04 30.8	−62 50 25	+0.004	−0.03	7.8	1.1	0.2	K0 III		320 s		
66264	42156		8 04 31.1	+42 03 39	−0.001	−0.02	8.0	1.1	0.2	K0 III		360 s		
66784	153791		8 04 31.2	−18 59 02	−0.003	−0.03	8.0	1.1	−0.1	K2 III		400 s		
67688	250104		8 04 31.4	−67 12 29	−0.003	+0.03	8.0	0.1	1.4	A2 V		210 s		
66783	153790		8 04 32.5	−17 39 55	−0.001	−0.01	6.5	1.4	−0.3	K5 III	−7	230 s		
66885	175114		8 04 32.6	−25 41 58	−0.001	−0.01	6.50	1.1	−0.1	K2 III		210 s	6566	m
67127	219383		8 04 33.8	−46 09 06	−0.001	+0.01	7.68	1.13	0.2	K0 III		310 s		
66069	26650		8 04 35.8	+53 40 42	+0.001	−0.01	7.1	0.1	1.4	A2 V		140 s		
66067	26648		8 04 36.9	+54 44 46	−0.002	−0.02	7.70	0.1	0.6	A0 V		260 s	6545	m
66884	175116		8 04 38.3	−24 21 50	−0.003	+0.01	7.5	0.9	2.1	A5 V		45 s		
66932	175122		8 04 38.5	−29 57 57	−0.003	−0.01	7.39	1.59	−0.5	M2 III		380 s		
67023	198771		8 04 39.6	−34 53 51	−0.004	+0.01	8.0	−0.1	2.6	F0 V		120 s		
66319	60543		8 04 40.2	+39 44 28	−0.002	−0.01	7.1	1.1	0.2	K0 III		240 s		
66834	153796	14 Pup	8 04 41.4	−19 43 41	0.000	0.00	6.13	−0.16	−1.7	B3 V	+14	370 s		
67249	235730		8 04 42.4	−50 35 27	0.000	−0.01	5.95	1.21	0.2	K0 III		100 s		
67536	250101		8 04 42.9	−62 50 10	−0.001	+0.02	6.30	−0.10	−1.4	B4 V n	0	350 s		m
66211	26657		8 04 44.6	+50 14 18	+0.001	−0.01	8.0	1.1	0.2	K0 III		360 s		
66552	97537		8 04 45.2	+18 50 31	−0.002	−0.02	6.10	−0.05	0.4	B9.5 V	+31	140 s		
64486	6392		8 04 47.1	+79 28 47	−0.010	−0.05	5.30	0.1	0.6	A0 V	+3	87 s		m

HD	SAO	Star Name	α 2000	δ 2000	μ(α)	μ(δ)	V	B-V	M$_V$	Spec	RV	d(pc)	ADS	Notes
			h m s	° ′ ″	s	″								
66083	26651		8 04 47.6	+55 29 05	−0.001	−0.01	7.8	0.8	3.2	G5 IV		84 s		
66665	116308		8 04 47.8	+6 11 08	−0.001	−0.03	7.80	−0.26		B0.5 III	+13			
66735	135461		8 04 49.0	−5 24 42	−0.002	+0.01	7.7	1.6	−0.5	M2 III		430 s		
67125	219385		8 04 50.7	−42 06 39	−0.003	−0.01	7.9	1.4	−0.1	K2 III		170 s		
67365	235733		8 04 51.0	−54 59 20	−0.005	+0.06	7.40	0.1	1.4	A2 V		99 mx		m
66713	116311		8 04 51.5	+4 08 29	−0.002	0.00	7.7	0.0	0.0	B8.5 V		350 s		
66833	153797		8 04 52.1	−13 56 38	−0.002	+0.03	7.9	1.6	−0.5	M2 III		470 s		
66550	79917		8 04 52.3	+23 36 49	−0.008	−0.09	7.9	0.9	3.2	G5 IV		88 s		
66714	116312		8 04 56.0	+4 01 55	−0.003	0.00	7.9	0.9	3.2	G5 IV		88 s		
66778	135465		8 04 58.5	−3 29 45	−0.002	+0.01	6.7	0.1	0.6	A0 V		170 s		
66686	97541		8 05 00.9	+10 30 00	0.000	−0.01	7.7	0.9	0.3	G5 III	+17	310 s		
66929	175130		8 05 01.2	−22 26 05	0.000	−0.02	7.80	0.6	4.4	G0 V		48 s		m
67515	250102		8 05 02.6	−60 23 16	+0.001	+0.01	7.70	0.1	0.6	A0 V		260 s		m
66831	135472		8 05 02.7	−8 53 27	−0.001	−0.03	8.0	0.1	1.4	A2 V		200 s		
67078	198776		8 05 03.2	−33 35 28	−0.003	+0.02	6.6	0.8	3.4	F5 V		25 s		
67364	235735		8 05 04.0	−53 06 28	+0.003	−0.01	5.53	1.34		M0	+18	17 mn		
66664	97542	8 Cnc	8 05 04.4	+13 07 05	−0.002	−0.07	5.12	0.01	0.0	A0 IV	+22	79 mx		
66905	153803		8 05 07.7	−17 01 26	0.000	0.00	7.8	1.1	0.2	K0 III		320 s		
66138	26658		8 05 10.3	+57 46 23	−0.010	−0.09	6.9	0.5	3.4	F5 V	+11	50 s		
67191	198780		8 05 15.1	−39 03 12	−0.001	+0.01	7.5	1.3	2.1	A5 V		26 s		
67297	219397		8 05 20.2	−44 13 30	−0.001	−0.01	8.0	−0.2	0.4	B9.5 V		330 s		
67341	219400		8 05 20.3	−46 58 44	0.000	−0.01	6.19	−0.15	−1.1	B5 V n		280 s		
67074	175145		8 05 21.6	−26 45 40	+0.001	0.00	7.9	1.3	−0.5	M2 III		490 s		
66711	97545		8 05 21.7	+17 37 19	−0.001	0.00	7.7	1.4	−0.3	K5 III		390 s		
67363	219402		8 05 22.0	−48 48 21	−0.002	+0.02	7.3	0.0	0.6	A0 V		210 s		
66469	42167		8 05 24.3	+41 41 10	0.000	−0.01	7.9	0.5	3.4	F5 V		78 s		
66801	116324		8 05 24.3	+5 49 38	−0.001	−0.02	8.00	0.00	0.4	B9.5 V		330 s	6571	m
66776	116323		8 05 24.7	+8 11 55	0.000	−0.01	8.00	0.9	3.2	G5 IV		91 s	6570	m
66470	42168		8 05 25.2	+41 11 24	+0.001	−0.01	7.7	0.1	1.7	A3 V		160 s		
66956	135483		8 05 29.0	−10 07 23	−0.001	0.00	6.9	0.0	0.4	B9.5 V		200 s		
67015	153815		8 05 31.5	−13 34 21	−0.003	0.00	7.40	0.1	0.6	A0 V		230 s	6579	m
66775	97548		8 05 33.2	+13 30 14	0.000	−0.02	7.5	1.1	0.2	K0 III		280 s		
66684	79928		8 05 36.9	+27 31 47	0.000	−0.01	6.20	0.01	0.6	A0 V		130 s	6569	m
67385	219406		8 05 43.0	−45 10 16	−0.001	+0.01	7.07	−0.13	−1.0	B5.5 V		410 s		m
65871	14426		8 05 44.0	+68 22 54	−0.038	−0.22	7.70	0.5	3.4	F5 V	−5	56 mx		m
67243	198791		8 05 44.8	−33 34 09	−0.001	+0.01	6.14	1.11	3.2	G5 IV		22 s		m
67242	198792		8 05 46.2	−33 19 58	−0.004	+0.02	7.8	0.3	0.6	A0 V		160 mx		
66925	116336		8 05 47.6	+2 09 54	−0.001	−0.01	6.8	0.1	0.6	A0 V	−6	170 s		
66950	135488		8 05 49.5	−0 34 25	−0.002	−0.02	6.41	1.05		K0				
67118	153827		8 05 54.1	−16 24 01	0.000	+0.01	7.8	1.4	−0.3	K5 III		400 s		
67140	153833		8 05 58.8	−19 53 30	−0.001	+0.04	6.7	0.5	0.4	B9.5 V		89 s		
65754	6415		8 05 59.5	+71 47 25	+0.002	−0.04	7.40	1.1	0.2	K0 III		280 s		m
66682	60556		8 06 02.2	+36 16 20	0.000	−0.05	7.0	0.4	3.0	F2 V		63 s		
66286	26662		8 06 02.5	+59 14 54	−0.001	0.00	6.60	0.1	0.6	A0 V		160 s		m
67893	250109		8 06 03.4	−64 22 37	−0.004	+0.02	7.9	0.5	3.4	F5 V		79 s		
68139	—		8 06 05.3	−69 33 27			7.7	1.1	−0.1	K2 III		360 s		
66979	116342		8 06 07.2	+5 33 35	+0.001	+0.01	7.8	1.1	0.2	K0 III		330 s		
67262	175173		8 06 08.2	−28 14 26	0.000	−0.03	7.6	0.8	2.6	F0 V		49 s		
66922	97556		8 06 09.0	+15 20 48	−0.003	−0.01	7.6	0.5	3.4	F5 V		71 s		
67441	219415		8 06 10.8	−40 21 00	−0.003	0.00	8.0	0.3	0.4	B9.5 V		200 s		
67029	116345		8 06 13.9	+1 10 38	0.000	−0.01	7.7	1.4	−0.3	K5 III		390 s		
67440	219416		8 06 15.1	−40 20 38	−0.001	−0.01	8.0	1.0	0.4	B9.5 V		84 s		
67439	198805		8 06 15.8	−40 06 00	0.000	−0.01	7.50	0.9	3.2	G5 IV		72 s		m
67138	135501		8 06 16.9	−9 39 49	+0.002	−0.09	8.0	1.1	0.2	K0 III		160 mx		
67213	153837		8 06 17.3	−18 22 24	−0.001	−0.03	7.9	0.1	0.6	A0 V		280 s		
66875	79940	9 Cnc	8 06 18.3	+22 38 08	−0.001	−0.01	5.99	1.66	−0.5	M4 III	+26	190 s		
67411	198806		8 06 19.1	−38 53 46	0.000	+0.07	7.0	1.0	0.2	K0 III		210 s		
67644	235745		8 06 20.3	−54 02 45	0.000	−0.02	7.99	1.18	−0.3	K5 III		460 s		
67158	135504		8 06 26.3	−9 14 15	0.000	0.00	7.93	−0.04	0.0	A0 IV		280 s	6588	m
67111	135502		8 06 26.3	−2 23 38	−0.001	−0.01	7.4	0.9	3.2	G5 IV		69 s		
67159	135505		8 06 27.4	−9 14 42	0.000	−0.02	6.01	−0.04	−0.3	B9 IV		280 s	6588	m
67261	153845		8 06 27.4	−19 49 36	−0.002	−0.03	8.0	0.3	0.2	K0 III		360 s		
67290	175178		8 06 30.9	−20 20 04	−0.001	−0.02	8.0	0.7	1.7	A3 V		77 s		
67612	219424		8 06 32.2	−49 12 06	−0.003	+0.01	7.9	0.2	0.4	B9.5 V		220 s		
66948	79948		8 06 34.4	+22 27 24	−0.008	−0.05	7.2	0.7	4.4	G0 V		37 s		
67990	250111		8 06 36.3	−63 30 57	−0.003	+0.05	7.0	1.1	0.2	K0 III		230 s		
67288	153848		8 06 39.1	−17 17 51	−0.001	0.00	8.0	0.9	3.2	G5 IV		92 s		
67582	219422		8 06 40.3	−45 15 58	−0.001	0.00	5.05	1.50		K5	+25	37 mn		
67621	219427		8 06 41.3	−48 29 51	−0.004	0.00	6.34	−0.19	−2.9	B3 III		220 mx		
67109	116353		8 06 48.3	+9 46 01	0.000	−0.03	7.9	0.1	0.6	A0 V		290 s		
67620	219428		8 06 48.9	−45 50 05	−0.001	0.00	7.24	1.07	0.2	K0 III		260 s		
67642	219432		8 06 50.0	−47 33 39	−0.002	−0.01	7.9	−0.3	0.0	B8.5 V		380 s		
67409	175196		8 06 51.4	−27 06 52	−0.002	−0.01	7.00	0.00	0.4	B9.5 V		210 s		m
67208	135514		8 06 51.7	−1 38 51	−0.002	+0.01	7.8	1.1	0.2	K0 III		330 s		
67047	79951		8 06 56.3	+22 46 30	0.000	−0.05	7.5	0.8	3.2	G5 IV		74 s		

HD	SAO	Star Name	α 2000	δ 2000	μ(α)	μ(δ)	V	B-V	M_V	Spec	RV	d(pc)	ADS	Notes
67763	235749		8h 07m 00.2s	-51°56'20"	-0.001s	+0.03"	7.13	1.26	0.2	K0 III		150 s		
67458	175200		8 07 00.5	-29 24 12	+0.027	-0.37	6.81	0.60	4.1	G4 IV-V	-19	26 ts		
67048	79953		8 07 01.3	+20 53 30	-0.003	-0.01	8.00	0.9	3.2	G5 IV		91 s	6590	m
67352	153857		8 07 04.2	-14 19 17	-0.004	-0.02	7.6	1.1	0.2	K0 III		240 mx		
67705	219439		8 07 05.5	-48 11 23	-0.002	+0.01	7.80	0.1	0.6	A0 V		280 s		m
67704	219438		8 07 05.6	-47 47 41	-0.001	-0.01	7.4	-0.4	0.4	B9.5 V		250 s		
67177	97566		8 07 06.3	+9 58 49	-0.002	0.00	6.8	0.9	3.2	G5 IV		53 s		
67556	198815		8 07 08.8	-36 22 53	-0.002	+0.09	7.3	0.4	3.4	F5 V		62 s		
66824	42174	28 Lyn	8 07 09.9	+43 15 38	-0.001	-0.03	6.2	0.1	0.6	A0 V	+9	140 s		
67408	153860		8 07 15.9	-17 22 07	-0.003	+0.02	6.5	1.1	-0.1	K2 III		200 s		
67655	219437		8 07 16.9	-41 44 09	-0.001	-0.02	7.4	1.4	-0.3	K5 III		220 s		
67230	116366		8 07 17.6	+7 16 59	-0.003	-0.01	7.6	0.4	3.0	F2 V		85 s		
67456	175206		8 07 17.9	-20 33 17	-0.002	-0.01	5.38	0.10		A m	+12	30 mn		
67009	60573		8 07 18.0	+33 14 50	+0.001	-0.02	8.0	0.5	3.4	F5 V		82 s		
67930	235755		8 07 18.4	-56 38 52	+0.001	0.00	7.1	1.9	-0.3	K5 III		170 s		
67762	219444		8 07 19.1	-48 23 11	-0.016	+0.19	6.73	0.96		K0		14 mn		
68057	250116		8 07 24.5	-60 20 53	-0.001	+0.03	8.0	-0.1	0.6	A0 V		310 s		
66660	26674		8 07 25.2	+55 18 06	+0.001	0.00	7.20	1.08		K1 III-IV				
68036	235760		8 07 25.7	-59 26 16	+0.001	+0.01	7.0	0.5	0.4	B9.5 V		110 s		
67150	97570		8 07 26.0	+19 49 03	-0.007	-0.04	7.8	0.5	4.0	F8 V		58 s		
67848	235753		8 07 30.8	-51 06 39	-0.004	+0.01	7.87	0.93	3.2	G5 IV		66 s		
67523	175217	15 ρ Pup	8 07 32.6	-24 18 15	-0.006	+0.05	2.81	0.43	-2.0	F6 II	+47	92 s		m,v
67821	219448		8 07 33.4	-49 16 11	-0.002	+0.02	7.7	1.5	-0.5	M4 III		440 s		
67929	235758		8 07 38.8	-54 03 53	-0.010	+0.01	7.60	1.01	0.2	K0 III		160 mx		
67087	60580		8 07 39.7	+31 33 04	-0.001	-0.04	7.9	0.5	4.0	F8 V		60 s		
67404	135530		8 07 41.3	-3 24 24	-0.001	0.00	6.83	1.60	-0.5	gM4	+24	290 s		
67847	219451		8 07 42.0	-49 00 39	+0.001	-0.01	6.6	0.6	0.4	B9.5 V		74 s		
67604	175222		8 07 42.7	-28 19 16	+0.001	+0.03	7.9	1.5	-0.3	K5 III		430 s		
67322	116371		8 07 43.6	+9 26 19	+0.001	0.00	7.9	0.0	0.4	B9.5 V		320 s		
67758	219447		8 07 43.8	-41 48 42	0.000	+0.02	7.15	-0.11	-1.6	B3.5 V		350 s		
67175	79956		8 07 44.1	+27 00 48	+0.001	0.00	7.9	0.1	0.6	A0 V		280 s		
67066	60579		8 07 44.5	+35 37 31	-0.001	-0.01	7.8	0.2	2.1	A5 V		140 s		
67702	198826		8 07 44.9	-37 25 31	-0.004	+0.01	7.9	1.1	0.2	K0 III		320 s		
68499	250125		8 07 45.1	-68 53 56	-0.005	+0.03	7.3	0.1	1.4	A2 V		150 s		
67228	79959	10 μ Cnc	8 07 45.8	+21 34 54	+0.001	-0.07	5.30	0.63	3.0	G2 IV	-36	31 ts		
67086	60582		8 07 48.4	+36 14 35	0.000	-0.03	7.9	0.1	1.4	A2 V		200 s		
67065	60578		8 07 48.4	+38 34 50	-0.001	-0.02	7.08	0.33	2.6	F0 V		74 s		
68561	250130		8 07 51.2	-69 53 55	-0.002	+0.04	7.6	0.0	0.4	B9.5 V		280 s		
68520	250128	ε Vol	8 07 55.9	-68 37 02	-0.004	+0.02	4.35	-0.11	-1.1	B5 V	+10	120 s		m
67452	—		8 07 56.3	-1 20 58			8.0	0.1	1.4	A2 V		210 s	6611	m
67735	198832		8 07 59.7	-34 01 09	-0.002	0.00	7.92	0.04	0.6	A0 V		290 s		
67487	135536		8 08 00.1	-3 43 47	-0.002	-0.02	8.0	1.1	0.2	K0 III		360 s		
67736	198833		8 08 00.2	-35 12 41	0.000	-0.02	7.20	0.2		A m				
67551	153869		8 08 01.9	-13 30 28	-0.002	-0.02	7.2	0.5	4.0	F8 V		44 s		
66947	26683		8 08 02.9	+50 08 13	0.000	0.00	7.3	1.1	0.2	K0 III		260 s		
68768	—		8 08 03.1	-72 07 46			8.0	1.1	0.2	K0 III		360 s		
67780	198838		8 08 08.0	-36 25 07	-0.005	+0.02	7.7	0.0	2.6	F0 V		110 s		
67673	175236		8 08 08.4	-25 48 54	-0.001	0.00	8.0	1.1	-0.3	K5 III		450 s		
67346	97579		8 08 08.7	+19 12 57	-0.002	-0.10	8.0	0.5	4.0	F8 V		62 s		
69547	256482		8 08 10.0	-78 41 39	+0.015	0.00	7.04	0.98	0.2	K0 III		230 s		
67778	198839		8 08 14.3	-33 20 23	-0.004	+0.02	7.47	0.00	0.0	B8.5 V		290 s		
68279	250121		8 08 14.4	-61 04 40	-0.002	-0.01	6.60							
67652	—		8 08 15.4	-22 08 19			7.90	0.9	3.2	G5 IV		87 s	6614	m
67174	60590		8 08 18.6	+38 31 30	-0.001	0.00	7.2	1.4	-0.3	K5 III		310 s		
67225	60592		8 08 19.0	+34 01 57	+0.001	-0.04	7.5	0.1	1.4	A2 V		170 s		
67698	175239		8 08 19.3	-23 37 03	-0.001	0.00	6.63	-0.09	-1.1	B5 V e		340 s	6618	m
67204	60591		8 08 21.4	+37 52 03	0.000	-0.01	7.8	1.1	-0.1	K2 III		380 s		
67425	97587		8 08 23.2	+13 09 12	+0.001	+0.01	7.8	1.1	0.2	K0 III		330 s		
67044	42187		8 08 24.6	+49 01 15	-0.001	-0.04	7.3	0.0	0.0	B8.5 V		280 s		
68423	250127		8 08 24.6	-63 48 03	-0.002	0.00	6.28	-0.06		B8 pne	+30			
67670	—		8 08 26.6	-19 51 21			7.50	0.1	0.6	A0 V		300 s	6622	m
67006	26687	27 Lyn	8 08 27.4	+51 30 24	-0.006	-0.01	4.84	0.05	1.4	A2 V	+5	60 mn	6600	m
67503	116388		8 08 27.4	+5 22 00	-0.002	-0.04	7.6	1.1	0.2	K0 III		300 s		
68117	235768		8 08 28.1	-52 37 03	-0.006	+0.03	7.40	0.4	3.0	F2 V		75 s		m
66285	6425		8 08 31.4	+71 16 09	-0.005	-0.04	7.8	0.8	3.2	G5 IV		84 s		
67594	135551	29 ζ Mon	8 08 35.6	-2 59 02	-0.001	-0.01	4.34	0.97	-4.5	G2 Ib	+30	560 s	6617	m
67905	198849		8 08 36.3	-39 29 59	+0.001	-0.01	6.8	1.0	0.6	A0 V		41 s		
67888	198848		8 08 37.5	-37 40 54	-0.001	-0.01	6.37	-0.04	-2.2	B5 III		450 s		
67728	153881		8 08 37.6	-19 50 24	0.000	-0.01	8.00	1.1	0.2	K0 III		300 s	6622	m
67547	116390		8 08 37.8	+1 53 43	-0.002	0.00	7.6	0.0	0.4	B9.5 V		280 s		
67483	97594	12 Cnc	8 08 42.4	+13 38 27	0.000	-0.02	6.27	0.43	3.4	F5 V	-10	38 s		
67751	175250		8 08 43.5	-20 21 48	-0.001	0.00	6.36	0.16		A3 p				m
68372	250129		8 08 44.3	-60 37 33	-0.002	0.00	7.4	0.3	0.6	A0 V		150 s		
67592	116396		8 08 46.2	+0 45 14	-0.001	-0.03	7.8	1.1	0.2	K0 III		330 s		
67614	116398		8 08 48.6	+0 27 33	-0.002	-0.02	7.6	1.4	-0.3	K5 III		380 s		

HD	SAO	Star Name	α 2000	δ 2000	μ(α)	μ(δ)	V	B-V	M_V	Spec	RV	d(pc)	ADS	Notes
			h m s	° ′ ″	s	″								
67402	79974		8 08 49.5	+27 28 49	-0.001	-0.03	6.90	0.9	0.2	G9 III	+14	220 s	6612	m
--	219484		8 08 49.9	-44 44 25	-0.006	+0.10	8.0	-0.5						
68092	219488		8 08 51.1	-47 10 29	-0.001	-0.01	7.4	-0.6	0.0	B8.5 V		300 s		
67924	198853		8 08 55.7	-33 14 46	-0.001	+0.02	7.76	-0.04	-1.0	B5.5 V		530 s		
68767	250138		8 08 56.5	-68 59 30	-0.003	+0.01	7.50	0.1	0.6	A0 V		240 s		m
67725	153887		8 08 56.8	-11 20 24	-0.001	-0.02	6.32	0.01		B9				
68950	256476		8 08 57.2	-71 43 16	+0.003	+0.03	7.5	1.1	-0.1	K2 III		330 s		
68456	250131		8 09 00.6	-61 18 07	-0.022	-0.28	4.76	0.43	3.4	dF5	+25	21 ts		
67589	116400		8 09 01.1	+8 02 22	-0.007	0.00	7.6	0.4	2.6	F0 V		100 s		
67797	153890	16 Pup	8 09 01.5	-19 14 42	-0.002	-0.01	4.40	-0.15	-1.1	B5 V	+19	130 s		
68005	198858		8 09 02.5	-39 30 51	-0.001	-0.04	7.1	1.2	0.2	K0 III		170 s		
67816	175262		8 09 02.6	-20 29 57	-0.002	-0.09	7.5	0.8	4.4	G0 V		29 s		
67921	198854		8 09 06.7	-30 19 21	+0.002	0.00	6.65	1.40		K5				m
67920	175272		8 09 07.0	-29 52 20	0.000	-0.02	7.8	1.1	3.2	G5 IV		53 s		
67884	175270		8 09 07.7	-26 13 57	+0.005	-0.06	8.00	1.1	-0.1	K2 III		250 mx		m
67796	153891		8 09 09.0	-14 34 58	0.000	-0.04	7.9	1.6	-0.5	M2 III		470 s		
67947	198855		8 09 09.4	-31 38 45	0.000	0.00	7.4	1.0	3.2	G5 IV		52 s		
68161	219493		8 09 09.5	-48 41 04	0.000	-0.01	5.8	-0.2	0.0	B8.5 V		150 s		
67977	198859		8 09 10.1	-35 27 18	0.000	+0.01	6.20	0.89		G5				
68635	250135		8 09 12.3	-64 46 35	+0.001	0.00	7.6	1.1	0.2	K0 III		300 s		
67882	175271		8 09 15.5	-21 34 36	-0.005	-0.09	7.5	0.9	3.2	G5 IV		63 s		
67722	135562		8 09 15.6	-4 33 10	-0.001	+0.03	7.4	1.6	-0.5	M2 III		380 s		
67835	153895		8 09 16.5	-15 34 12	-0.001	+0.01	7.2	0.0	0.0	B8.5 V		270 s		
68276	235778		8 09 18.0	-51 00 44	-0.006	+0.01	7.60	0.06	0.6	A0 V		150 mx		
67720	135561		8 09 18.2	-0 26 59	-0.001	-0.03	7.3	0.1	0.6	A0 V		220 s		
67318	42197		8 09 20.6	+45 46 01	-0.001	-0.06	7.6	0.9	3.2	G5 IV		77 s		
67717	116407		8 09 21.9	+1 18 50	-0.002	-0.02	7.9	0.1	0.6	A0 V		280 s		
67370	42199		8 09 23.0	+42 25 50	-0.001	-0.08	6.27	1.27	-0.1	K2 III	+38	160 mx		
67880	153898		8 09 28.4	-16 14 56	-0.002	-0.01	5.68	-0.17		B3 s	+33		6632	m
67147	26696		8 09 28.8	+55 15 21	-0.002	-0.03	6.9	0.4	2.6	F0 V		73 s		
68243	219501		8 09 29.2	-47 20 44	-0.001	-0.01	4.27	-0.23	-3.9	B1 IV	+20	430 s		
68346	235781		8 09 30.0	-51 50 34	-0.006	+0.03	6.53	0.60	4.4	G0 V		25 s		
67501	60604		8 09 30.4	+32 13 18	+0.004	+0.01	7.10	0.4	3.0	F2 V	+11	66 s	6623	m
68518	235786		8 09 30.5	-59 32 29	-0.002	+0.01	6.6	0.2	0.4	B9.5 V		130 s		
68371	235782		8 09 31.2	-52 34 27	-0.001	+0.01	7.25	-0.02	-1.1	B5 V		420 s		
68003	175282		8 09 31.6	-27 27 04	-0.001	0.00	7.8	0.9	0.2	K0 III		330 s		
68273	219504	γ Vel	8 09 31.9	-47 20 12	-0.001	0.00	1.78	-0.22		WC7	+35			m
67715	116414		8 09 32.2	+5 55 35	0.000	-0.11	7.5	0.5	3.4	F5 V		68 s		
66823	14442		8 09 33.5	+65 39 59	+0.002	-0.05	7.30	1.4	-0.3	K5 III	-10	210 mx		
68434	235784		8 09 33.6	-56 05 07	-0.001	+0.03	5.8	0.1	1.4	A2 V		76 s		
67542	79982		8 09 35.4	+29 05 33	+0.002	-0.03	6.5	0.9	-2.1	G5 II	+16	540 s		
68217	219502		8 09 35.9	-44 07 23	-0.001	-0.01	5.21	-0.19		B3	+8			
67482	60606		8 09 39.7	+35 42 09	+0.001	-0.05	7.30	1.1	0.2	K0 III		260 s		m
67541	60608		8 09 42.3	+33 22 42	-0.001	-0.04	7.6	1.1	0.2	K0 III		310 s		
--	219513		8 09 43.1	-47 56 16	0.000	-0.02	5.3	-0.6	-1.6	B3.5 V		240 s		
68324	219515		8 09 43.1	-47 56 14	0.000	0.00	5.23	-0.21	-1.7	B3 V	+5	240 s		
67369	42202		8 09 45.9	+48 22 28	-0.007	-0.07	7.9	0.5	3.4	F5 V		78 s		
68242	219507		8 09 47.6	-42 38 26	-0.001	+0.01	6.26	-0.04		A0				
67854	135574		8 09 49.8	-4 09 07	-0.002	-0.02	8.0	1.1	0.2	K0 III		360 s		
68241	219510		8 09 51.5	-41 16 35	-0.002	+0.01	7.9	1.5	-0.3	K5 III		430 s		
68270	219514		8 09 52.0	-43 31 44	-0.001	-0.01	7.17	1.46	0.2	K0 III		110 s		
68154	198868		8 09 53.5	-35 57 23	-0.001	-0.03	7.9	1.1	0.2	K0 III		320 s		
67712	97610		8 09 54.6	+17 15 06	0.000	-0.01	7.8	1.1	0.2	K0 III		330 s		
67628	79988		8 09 57.9	+29 07 30	-0.001	-0.04	7.50	1.4	-0.3	K5 III		360 s		
67852	116425		8 09 58.4	+1 01 13	0.000	-0.02	7.9	0.4	2.6	F0 V		110 s		
67743	97614		8 10 00.8	+17 00 59	-0.001	-0.02	7.4	1.6	-0.5	M2 III	+22	370 s		
67224	26701		8 10 03.7	+58 14 53	-0.003	-0.08	6.0	1.1	0.2	K0 III	+34	140 s		
--	198871		8 10 06.2	-32 55 06	-0.001	0.00	7.7	2.1	-0.3	K5 III		170 s		
67939	135580		8 10 06.3	-6 44 43	-0.001	-0.03	7.6	1.6	-0.5	M2 III		420 s		
67789	97618		8 10 06.6	+14 30 31	-0.004	-0.08	7.6	0.5	3.4	F5 V		70 s		
67587	60611		8 10 11.4	+35 27 17	+0.017	-0.25	6.60	0.5	4.0	F8 V	-55	33 s		m
67317	26702		8 10 12.5	+55 48 15	-0.002	-0.07	8.00	0.1	1.4	A2 V		65 mx	6620	m
67690	79991		8 10 13.1	+25 50 40	-0.001	-0.03	6.80	1.1	0.2	K0 III	+6	210 s		m
67709	79990		8 10 13.9	+27 05 22	0.000	+0.01	7.90	1.1	0.0	K1 III		380 s		
68109	175306		8 10 15.5	-22 32 01	-0.003	-0.03	6.7	1.1	2.6	F0 V		22 s		
68451	219536		8 10 16.0	-49 02 07	-0.001	-0.01	7.33	-0.14	-3.6	B2 III		1400 s		
67873	116428		8 10 17.8	+9 10 04	0.000	-0.01	6.9	0.8	3.2	G5 IV		54 s		
66751	14446		8 10 20.4	+69 43 29	+0.031	+0.11	6.6	0.5	4.0	F8 V	-6	33 s		
68478	219538		8 10 20.4	-49 14 16	-0.001	-0.01	6.47	-0.15		B3				
68298	198882		8 10 23.7	-36 13 57	0.000	0.00	7.9	1.3	-0.3	K5 III		430 s		
67767	79995	14 ψ Cnc	8 10 27.1	+25 30 26	-0.005	-0.35	5.73	0.81	5.3	dG6	-43	17 mn		m
68496	219541		8 10 27.2	-49 09 50	-0.001	+0.01	7.93	-0.10	-0.9	B6 V		570 s		
68083	135587		8 10 30.5	-9 36 19	-0.001	-0.03	7.7	1.6	-0.5	M4 III		440 s		
68391	198886		8 10 34.5	-39 39 53	-0.001	+0.01	8.0	-0.2	3.4	F5 V		82 s		
71676	--		8 10 38.3	-84 27 12			7.9	1.6	-0.5	M2 III		470 s		

HD	SAO	Star Name	α 2000	δ 2000	μ(α)	μ(δ)	V	B−V	M_v	Spec	RV	d(pc)	ADS	Notes
68146	153924	18 Pup	8 10 39.8	-13°47'57"	-0.017	+0.05	5.54	0.49	3.8	dF7	+38	24 ts		m
68018	116436		8 10 40.2	+2 56 59	-0.004	-0.02	7.1	0.9	3.2	G5 IV		60 s		
67481	26712		8 10 44.6	+52 08 32	+0.004	-0.06	7.6	0.5	4.0	F8 V		52 s		
68416	198892		8 10 47.7	-38 16 10	0.000	-0.03	7.6	0.2	0.6	A0 V		190 s		
68363	198891		8 10 51.2	-32 22 44	-0.003	+0.06	7.20	1.34		K2				
68633	235791		8 10 53.0	-51 11 27	-0.001	-0.01	8.00	0.29	-1.0	B5.5 V		370 s		
66170	6428		8 10 56.5	+77 38 30	-0.043	-0.11	8.0	0.4	3.0	F2 V		100 s		m
68474	198897		8 10 57.4	-39 04 41	0.000	0.00	7.37	-0.09	0.4	B9.5 V		250 s		m
68145	135592		8 10 58.3	-6 44 52	-0.004	+0.03	6.6	1.1	0.2	K0 III		190 s		
--	219555		8 10 58.4	-48 30 04	-0.008	-0.01	7.53	0.56	4.4	G0 V		42 s		
67959	97628		8 10 58.8	+14 37 46	-0.002	-0.02	6.23	0.02	1.4	A2 V	+24	93 s		
68608	219559		8 10 59.3	-49 17 05	-0.002	-0.01	7.89	-0.11	-2.2	B5 III		1000 s		
68450	198898		8 11 01.5	-37 17 32	-0.002	0.00	6.44	-0.02	-5.6	B0 II		1900 s		m
68261	153934		8 11 02.1	-18 21 11	-0.005	-0.02	8.0	0.9	3.2	G5 IV		92 s		
68062	116439		8 11 02.9	+9 34 36	0.000	0.00	8.0	0.4	2.6	F0 V		120 s		
68292	153936		8 11 03.4	-18 58 27	0.000	-0.01	7.5	0.0	0.4	B9.5 V		260 s		
68512	219552		8 11 07.2	-40 45 29	-0.002	+0.02	6.6	1.7	-0.1	K2 III		100 s		
63834	1278		8 11 08.6	+84 24 26	-0.018	-0.01	7.4	0.1	0.6	A0 V		230 s		
67562	26716		8 11 09.7	+53 14 43	+0.002	-0.07	6.8	0.5	3.4	F5 V		49 s		
68657	219565		8 11 11.1	-48 27 43	+0.001	0.00	5.82	-0.15	-1.7	B3 V	+15	320 s		
68173	135599		8 11 13.1	-1 33 58	0.000	-0.01	7.9	0.1	0.6	A0 V		280 s		
--	116449		8 11 15.0	+1 16 36	-0.005	-0.01	7.80							m
66344	6434		8 11 15.2	+76 38 22	-0.005	-0.03	8.0	1.1	0.2	K0 III		360 s		
68290	153942	19 Pup	8 11 16.2	-12 55 37	-0.002	+0.01	4.72	0.95	0.2	K0 III	+36	45 ts	6647	m
68099	116444		8 11 16.5	+9 49 16	-0.001	-0.02	6.10	-0.10	-1.6	B7 III		330 s		
--	--		8 11 17.7	-7 06 59			8.00						6645	m
67539	26717		8 11 18.0	+54 32 24	+0.005	-0.10	6.5	0.9	3.2	G5 IV		46 s		
68627	219566		8 11 18.7	-45 49 06	-0.001	0.00	6.94	0.97	3.2	G5 IV		56 s		
68313	153944		8 11 20.5	-14 22 23	-0.002	-0.02	8.0	1.1	0.2	K0 III		350 s		
68553	198908		8 11 21.5	-39 37 07	-0.001	-0.01	4.45	1.62	-5.9	cK	+16	850 s		
68510	198905		8 11 21.5	-36 48 04	-0.002	-0.02	7.2	0.9	0.2	K0 III		250 s		
67827	60625		8 11 21.6	+38 43 54	-0.009	-0.06	6.58	0.59	4.4	G0 V	+26	26 s		
68601	219569		8 11 25.9	-42 59 14	0.000	0.00	4.75	0.18		A3	+19	23 mn		m
68312	135611		8 11 32.9	-7 46 21	-0.003	-0.03	5.36	0.89	0.3	G8 III	-11	100 s		
68287	135610		8 11 33.5	-6 00 56	-0.008	+0.07	7.8	0.6	4.4	G0 V		48 s		
68122	97636		8 11 36.7	+16 13 03	0.000	-0.01	7.4	1.1	0.2	K0 III		280 s		
68968	235802		8 11 36.9	-59 14 57	0.000	+0.06	7.3	1.3	3.2	G5 IV		31 s		
68200	116460		8 11 38.0	+6 10 07	-0.001	-0.04	8.0	0.9	3.2	G5 IV		93 s		
68017	60631		8 11 38.6	+32 27 25	-0.037	-0.65	6.81	0.69	5.0	G4 V	+27	20 mx		m
68765	219579		8 11 39.3	-50 04 05	-0.002	0.00	7.34	-0.01	0.4	B9.5 V		220 s		
67824	42216		8 11 40.1	+45 25 33	+0.001	0.00	7.8	1.1	0.2	K0 III		330 s		
69009	235804		8 11 40.9	-60 04 33	-0.003	+0.04	7.2	1.8	0.2	K0 III		87 s		
67500	14455		8 11 43.1	+60 19 50	-0.003	-0.01	8.0	1.1	0.2	K0 III		360 s		
68198	97642		8 11 44.6	+11 11 01	-0.001	0.00	7.9	0.1	1.4	A2 V		200 s		
67825	42219		8 11 46.1	+45 19 25	-0.001	0.00	7.8	1.4	-0.3	K5 III		410 s		
67345	14454		8 11 47.8	+66 01 03	-0.008	-0.02	7.9	0.4	2.6	F0 V		110 s		
68168	97640		8 11 49.2	+16 31 23	-0.003	-0.29	7.3	0.6	4.4	G0 V	+9	35 mx		
68197	97643		8 11 49.3	+13 26 45	-0.009	-0.07	7.5	0.8	3.2	G5 IV		73 s		
68284	116465		8 11 49.4	+4 16 27	-0.003	+0.03	7.9	0.5	4.0	F8 V		60 s		
68535	175365		8 11 50.4	-27 31 27	+0.003	-0.03	7.6	0.8	-1.6	B3.5 V		180 s		
68623	198918		8 11 53.4	-35 46 25	+0.001	-0.02	6.6	1.5	-0.5	M4 III		270 s		
68716	219577		8 11 54.8	-41 47 20	+0.001	-0.01	6.5	1.9	-0.1	K2 III		78 s		
67872	42220		8 11 54.9	+45 12 34	+0.002	+0.01	7.8	1.1	0.2	K0 III		330 s		
68570	198916		8 11 54.9	-30 12 44	+0.001	0.00	7.8	0.6	0.0	B8.5 V		140 s		
68868	235801		8 11 57.1	-51 57 44	-0.003	+0.02	8.0	0.1	0.6	A0 V		240 s		
68808	219587		8 11 59.9	-46 38 39	-0.001	+0.01	5.76	0.59		F8 p				AH Vel, v
69455	--		8 12 01.2	-69 45 11			7.4	1.1	0.2	K0 III		270 s		
68569	175370		8 12 02.7	-29 00 08	+0.002	-0.01	8.0	-0.2	-1.0	B5.5 V		640 s		
68806	219589		8 12 02.9	-45 59 37	-0.001	+0.03	7.39	1.05	3.2	G5 IV		69 s		
67871	42222		8 12 05.0	+46 56 30	-0.001	-0.03	7.7	0.1	1.4	A2 V		190 s		
68763	219586		8 12 05.7	-43 24 03	-0.003	-0.01	6.46	1.71	-0.1	K2 III		85 s		
68426	135622		8 12 08.2	-6 05 38	-0.002	-0.02	8.0	1.1	0.2	K0 III		360 s		
67870	42223		8 12 11.6	+48 57 33	-0.002	0.00	7.0	1.1	0.2	K0 III		230 s		
68467	135624		8 12 12.4	-8 32 10	-0.002	-0.03	8.0	0.1	1.4	A2 V		210 s		
68257	97645	16 ζ Cnc	8 12 12.6	+17 38 52	+0.005	-0.14	4.67	0.53			-8	16 ts	6650	m
69600	256485		8 12 15.6	-71 16 06	+0.009	-0.14	7.6	0.4	3.0	F2 V		83 s		
69134	235813		8 12 15.7	-60 02 22	-0.001	0.00	7.9	2.1	-0.3	K5 III		180 s		
68332	97647		8 12 22.1	+14 00 14	-0.002	-0.01	6.4	0.2	2.1	A5 V	-9	73 s		
68761	198931		8 12 27.0	-36 59 20	-0.003	+0.01	6.55	-0.07		B0.5 III				
68895	219602		8 12 30.8	-46 15 51	0.000	+0.01	6.03	-0.11	-1.1	B5 V	+13	270 s		m
69051	235810		8 12 32.6	-54 08 47	-0.003	+0.03	6.9	0.2	2.6	F0 V		73 s		
68892	219603		8 12 33.0	-45 03 35	-0.006	+0.05	7.97	0.85	3.2	G5 IV		90 s		
67935	26724		8 12 33.6	+50 00 33	+0.009	-0.01	7.7	0.5	4.0	F8 V		55 s		
68784	198935		8 12 34.9	-36 40 06	-0.001	-0.04	7.5	1.5	-0.1	K2 III		210 s		
68254	80012		8 12 36.2	+28 49 11	+0.001	-0.02	8.0	0.1	0.6	A0 V		300 s		

HD	SAO	Star Name	α 2000	δ 2000	μ(α)	μ(δ)	V	B-V	M$_v$	Spec	RV	d(pc)	ADS	Notes
			h m s	° ′ ″	s	″								
68425	97650		8 12 38.4	+11 11 28	-0.001	-0.01	7.4	1.4	-0.3	K5 III		350 s		
68758	175390		8 12 45.9	-29 54 39	-0.002	+0.03	6.52	0.06		A2				
68439	97652		8 12 48.0	+11 11 05	-0.001	-0.01	7.2	0.0	0.0	B8.5 V		270 s		
67447	14456		8 12 48.7	+68 28 26	0.000	+0.01	5.32	1.04	-2.1	G8 II	-9	310 s		
68504	116481		8 12 50.6	+4 55 42	-0.001	-0.03	7.2	0.0	0.4	B9.5 V		230 s		
68843	198938		8 12 50.6	-36 13 44	-0.002	-0.03	7.63	-0.12	-0.9	B6 V nn		490 s		
68862	198942		8 12 51.4	-37 55 27	-0.002	0.00	6.42	0.10		A3				
68586	135634		8 12 51.9	-7 12 15	+0.009	-0.08	7.93	0.90	0.2	K0 III		120 mx		
68483	116479		8 12 52.5	+9 34 40	-0.003	-0.03	7.80	0.2	2.1	A5 V	+32	140 s	6659	m
68165	60644		8 12 54.8	+39 28 24	+0.007	-0.01	7.7	1.1	0.2	K0 III		140 mx		
68461	97653		8 12 59.7	+16 30 51	-0.001	-0.02	6.01	0.89	0.2	K0 III	-20	150 s		
68860	198944		8 13 04.0	-34 34 41	-0.002	+0.02	6.70	1.30	0.2	K0 III	+19	140 s		RS Pup, v
68687	153987		8 13 07.0	-12 43 11	-0.001	-0.03	7.6	0.1	1.4	A2 V		180 s		
68351	80016	15 ψ Cnc	8 13 08.8	+29 39 23	-0.001	-0.02	5.64	-0.07		A0 p	+20			
68798	175401		8 13 12.9	-23 30 42	-0.003	+0.03	7.6	-0.2	1.7	A3 V		160 s		
69171	235818		8 13 14.9	-52 14 10	-0.002	0.00	7.0	-0.2	0.0	B8.5 V		250 s		
68886	198947		8 13 15.0	-32 09 20	0.000	+0.02	7.62	-0.12	-1.6	B3.5 V		650 s		
68821	--		8 13 17.1	-24 16 57			7.40	0.4	2.6	F0 V		91 s	6670	m
68944	198949		8 13 17.9	-36 20 32	-0.003	-0.01	7.33	-0.14		B8				
68780	153994		8 13 19.5	-18 08 24	0.000	-0.05	7.9	0.1	1.4	A2 V		190 s		
68752	153993	20 Pup	8 13 19.9	-15 47 18	-0.001	-0.01	4.99	1.07	-2.1	G5 II	+17	230 s		
67991	26727		8 13 21.7	+56 31 39	+0.001	-0.07	6.8	0.4	3.0	F2 V		57 s		
68667	135643		8 13 21.7	-1 09 57	-0.002	0.00	6.4	1.1	0.2	K0 III		180 s		
68982	198954		8 13 22.0	-38 26 19	-0.004	-0.01	7.54	0.10	-1.7	B3 V		510 s		
68962	198953		8 13 22.3	-36 18 37	-0.004	+0.01	7.34	-0.15		B8				
68194	42232		8 13 27.8	+48 16 50	-0.002	-0.02	6.7	0.0	0.4	B9.5 V		190 s		
68980	198957		8 13 29.5	-35 53 59	0.000	0.00	4.78	-0.11		B3 p	+35			
68979	198956		8 13 32.6	-33 18 00	+0.004	-0.04	7.8	0.9	-0.1	K2 III		380 s		
68615	97662		8 13 34.1	+10 50 49	0.000	-0.08	7.40	0.5	4.0	F8 V		48 s	6663	m
69194	235819		8 13 34.1	-50 11 46	+0.001	-0.01	5.51	1.61	-0.3	K5 III	-7	120 s		
68730	135649		8 13 34.1	-6 35 40	-0.002	0.00	7.2	1.1	0.2	K0 III	+20	250 s		
68978	198958		8 13 34.4	-31 44 08	-0.027	+0.14	6.75	0.61	4.2	G5 IV-V		32 s		m
68643	97664		8 13 35.9	+10 22 51	0.000	-0.01	7.8	0.7	4.4	G0 V		49 s	6665	m
69144	219629		8 13 36.2	-46 59 31	0.000	0.00	5.13	-0.14	-2.6	B2.5 IV	+25	320 s		m
69002	198960		8 13 41.0	-33 34 09	+0.002	+0.02	6.40	1.1	-0.1	K2 III		200 s		m
68543	80024		8 13 41.6	+23 08 16	-0.003	-0.01	6.40	0.11	0.0	A3 III	-2	190 s		
69168	219631		8 13 45.6	-46 34 44	-0.001	-0.01	6.50	-0.15	-2.3	B3 IV		570 s		
68077	26732		8 13 50.1	+56 27 08	-0.002	-0.03	5.85	1.01	0.2	K0 III	+7	130 s		
69066	198967		8 13 50.4	-35 12 32	-0.003	-0.05	6.89	0.79	0.2	K0 III		220 s		
68705	116503		8 13 51.9	+6 18 10	+0.001	-0.03	7.6	0.1	0.6	A0 V		250 s		
68542	80025		8 13 53.5	+27 46 43	0.000	-0.02	7.60	0.4	3.0	F2 V		83 s	6662	m
68704	116504		8 13 53.7	+7 43 58	-0.001	-0.01	7.6	0.1	0.6	A0 V		260 s		
69283	235821		8 13 57.0	-50 18 44	-0.011	+0.02	7.80	0.96	0.2	K0 III		160 mx		
69191	219639		8 13 58.0	-44 45 04	-0.005	+0.04	7.42	1.11	0.2	K0 III		240 s		
68856	135663		8 13 58.3	-10 01 55	-0.001	0.00	8.0	0.0	0.4	B9.5 V		330 s		
69081	198969		8 13 58.3	-36 19 21	0.000	-0.01	5.08	-0.19	-1.7	dB3 n	+18	220 s		m
69082	198970		8 13 58.8	-36 20 28	0.000	-0.01	6.11	-0.18		B8				
69142	219635		8 14 02.8	-40 20 52	+0.003	-0.07	4.44	1.17	-0.1	gK2	+14	81 s		m
69106	198971		8 14 03.8	-36 57 09	0.000	-0.02	7.14	-0.09		B0.5 II				
69213	219640		8 14 05.2	-44 34 29	+0.004	+0.06	7.0	0.6		A7 var	+20			AI Vel, v
69257	219642		8 14 05.2	-46 55 06	-0.009	+0.06	7.31	0.47	3.4	F5 V		58 s		
69105	198973		8 14 05.5	-36 25 34	-0.002	-0.01	7.67	-0.09	0.0	B8.5 V		340 s		
68881	135666		8 14 06.8	-8 51 35	-0.001	-0.03	7.8	1.1	0.2	K0 III		330 s		
69576	250157		8 14 07.5	-61 26 32	-0.006	+0.04	7.1	1.4	1.4	A2 V		23 s		
69080	198972		8 14 10.9	-32 08 28	-0.001	-0.01	6.06	-0.16		B5 n	-54			m
68703	97669		8 14 11.1	+17 40 33	0.000	+0.01	6.40	0.4	2.6	F0 V	-3	58 s	6673	m
69040	175425		8 14 13.0	-24 55 57	-0.001	-0.03	7.7	0.5	2.1	A5 V		76 s		
69123	198979		8 14 13.1	-35 29 25	-0.005	+0.01	5.78	1.02	0.2	gK0		130 s		
68563	60662		8 14 17.1	+33 33 14	+0.004	-0.07	7.7	1.1	0.2	K0 III		150 mx		
68776	97671		8 14 20.9	+13 02 54	-0.001	-0.02	6.50	0.9	0.3	G8 III	+25	170 s		m
--	219652		8 14 21.0	-45 50 17	+0.001	-0.02	6.00	1.1	0.2	K0 III		140 s		m
67201	6457		8 14 23.7	+75 50 05	-0.001	-0.03	7.9	0.1	0.6	A0 V		280 s		
69302	219657		8 14 23.8	-45 50 04	-0.001	+0.01	5.80	-0.13		B3	+20			m
68937	135670		8 14 23.8	-3 49 48	-0.003	-0.02	7.5	1.1	-0.1	K2 III		330 s		
68725	80030		8 14 27.1	+20 42 28	+0.001	-0.01	6.9	0.4	3.0	F2 V		59 s		
69235	219647		8 14 27.2	-40 30 28	+0.003	-0.03	7.95	0.07	0.6	A0 V		300 s		
--	42238		8 14 28.4	+48 26 17	+0.001	-0.01	7.9	1.4						
68612	60667		8 14 28.8	+33 56 37	-0.002	-0.03	7.6	1.1	-0.1	K2 III		350 s		
69517	235831		8 14 28.9	-56 45 24	+0.002	-0.02	7.4	0.2	0.6	A0 V		160 s		
69253	219651		8 14 29.5	-40 49 33	-0.002	0.00	6.62	-0.15	-1.7	B3 V		460 s		
69210	198983		8 14 29.8	-38 30 32	-0.003	+0.02	7.5	0.5	1.4	A2 V		86 s		
68835	97672		8 14 31.9	+10 11 18	0.000	0.00	8.0	0.9	3.2	G5 IV		93 s		
68905	116524		8 14 38.9	+1 09 26	0.000	0.00	7.6	0.1	1.4	A2 V		180 s		
68724	80032		8 14 39.3	+26 43 49	+0.003	-0.03	7.70	1.1	0.2	K0 III		320 s		
70270	256489		8 14 42.6	-73 48 24	-0.009	+0.06	7.40	0.2	2.1	A5 V		120 s		m

HD	SAO	Star Name	α 2000	δ 2000	μ(α)	μ(δ)	V	B−V	M$_V$	Spec	RV	d(pc)	ADS	Notes
68562	42242		8^{h}14^{m}49.0^s	+43°02′06″	−0.001	−0.03	6.80	0.1	0.6	A0 V		170 s	6675	m
70049	250167		8 14 50.7	−69 37 25	−0.001	−0.03	7.9	0.1	0.6	A0 V		290 s		
69404	219674		8 14 51.0	−46 29 08	−0.002	+0.02	6.43	−0.15	−1.7	B3 V nne		410 s		
68640	42243		8 14 54.0	+40 34 44	−0.001	0.00	7.7	0.1	1.4	A2 V		180 s		
69340	219669		8 14 55.1	−41 00 13	0.000	−0.02	7.9	0.5	3.2	G5 IV		88 s		
69101	154023		8 14 55.7	−17 16 33	−0.001	+0.07	6.9	0.5	3.4	F5 V		50 s		
69280	198991		8 14 56.6	−35 41 20	−0.002	+0.02	6.70	1.1	−0.1	K2 III		230 s		m
68903	97677		8 14 59.5	+16 04 42	0.000	−0.01	7.2	0.0	0.0	B8.5 V	+1	270 s		
68659	42244		8 14 59.8	+41 08 27	−0.002	−0.06	7.6	0.4	3.0	F2 V		85 s		
69160	175445		8 15 00.4	−21 40 27	0.000	+0.02	8.00	1.1	0.2	K0 III		350 s		m
69249	198988		8 15 00.5	−30 25 11	−0.002	−0.01	8.00	−0.02	0.6	A0 V		300 s		
69250	198989		8 15 00.6	−30 51 24	0.000	+0.02	7.4	0.6	1.7	A3 V		66 s		
70953	256491		8 15 02.7	−79 19 06	+0.002	+0.04	7.3	0.1	0.6	A0 V		220 s		
69514	219681		8 15 15.2	−49 10 50	−0.005	−0.01	8.0	−0.5	0.6	A0 V		140 mx		
69863	250164		8 15 15.9	−62 54 57	−0.003	−0.03	5.16	0.09		A2	+4	15 mn		
69185	154031		8 15 17.2	−16 25 43	−0.001	+0.01	8.0	1.6	−0.5	M2 III		490 s		
69596	235838		8 15 23.2	−50 26 58	−0.002	−0.02	6.44	1.49	−0.1	K2 III		110 s		
69156	154032		8 15 24.7	−11 17 35	−0.002	0.00	8.0	1.1	0.2	K0 III		360 s		
69655	235839		8 15 25.5	−52 03 37	+0.002	−0.06	6.63	0.58	4.4	G0 V		28 s		
69402	199000		8 15 28.0	−37 22 19	0.000	0.00	7.50	0.00	0.0	B8.5 V		310 s		m
69113	116544		8 15 31.3	+2 49 56	0.000	−0.01	7.9	0.1	0.6	A0 V		280 s		
69056	97681		8 15 33.1	+11 25 51	−0.014	−0.25	7.8	0.8	3.2	G5 IV		44 mx		
69229	154035		8 15 33.4	−13 37 28	0.000	−0.01	7.20	1.6	−0.5	M2 III	−4	340 s		
69653	219694		8 15 37.3	−49 59 35	−0.002	−0.01	7.60	0.00	0.0	B8.5 V		310 s		
69153	135689		8 15 39.1	−1 14.51	−0.001	0.00	8.0	0.9	3.2	G5 IV		93 s		
69031	97680		8 15 39.4	+17 34 20	−0.001	−0.03	7.8	1.1	0.2	K0 III		330 s		
68192	14472		8 15 42.2	+66 10 32	−0.002	−0.09	7.1	0.4	3.0	F2 V		65 s		
69652	219695		8 15 44.1	−49 15 52	−0.001	0.00	7.8	0.0	1.4	A2 V		190 s		
69860	235844		8 15 44.4	−60 03 26	−0.005	+0.02	7.9	1.8	−0.3	K5 III		290 s		
69178	116551		8 15 48.4	+2 48 04	−0.001	−0.01	7.1	0.1	0.6	A0 V		200 s	6698	m
68933	60688		8 15 49.2	+32 10 38	−0.001	−0.01	7.5	0.5	3.4	F5 V		65 s		
68457	14479		8 15 50.5	+60 22 50	−0.002	+0.01	6.45	0.20	2.4	dA7	−16	65 s	6680	m
69445	199010		8 15 52.5	−30 55 33	−0.004	−0.01	6.21	0.78		G5				m
69650	219696		8 15 52.9	−46 18 44	−0.002	+0.04	6.8	0.7	1.4	A2 V		49 s		
69072	97684		8 15 58.3	+18 41 38	−0.002	−0.02	7.50	1.1	0.2	K0 III		290 s	6696	m
69511	199015		8 15 58.8	−35 54 10	0.000	+0.01	6.16	1.55	−0.1	K2 III		98 s		
69444	199014		8 15 59.0	−30 11 02	0.000	+0.01	7.0	1.0	3.2	G5 IV		44 s		
69354	154045		8 15 59.3	−19 24 31	0.000	−0.03	7.2	1.3	−0.3	K5 III		320 s		
69293	135699		8 16 00.2	−9 46 35	−0.003	−0.01	8.0	1.1	0.2	K0 III		360 s		
69891	235847		8 16 05.7	−57 16 11	−0.001	−0.01	7.83	−0.02	0.0	B8.5 V		360 s		
68638	26742		8 16 06.2	+57 05 37	−0.039	−0.24	7.50	0.75	5.3	dG6	+17	26 s		m
68831	42257		8 16 09.9	+46 48 07	−0.001	−0.01	6.9	1.4	−0.3	K5 III		280 s		
69371	154048		8 16 11.9	−15 41 10	−0.002	0.00	7.36	1.19	0.2	gK0	+28	210 s		
69565	199018		8 16 12.5	−35 52 55	+0.001	+0.03	7.00	0.9	3.2	G5 IV		58 s		m
68745	26747		8 16 14.5	+52 20 32	+0.001	−0.01	7.7	1.1	0.2	K0 III		310 s		
69222	116560		8 16 15.1	+9 24 21	−0.002	−0.03	7.7	1.1	−0.1	K2 III		360 s		
74566	258497		8 16 17.6	−86 33 27	−0.038	+0.04	7.0	−0.1	0.6	A0 V		150 mx		
69620	199020		8 16 25.0	−36 12 07	0.000	+0.02	7.17	−0.14	−0.9	B6 V		410 s		
69674	199024		8 16 25.0	−39 45 10	−0.001	+0.01	6.9	1.1	0.2	K0 III		200 s		
69028	60695		8 16 25.9	+35 33 01	0.000	−0.06	7.7	0.0	0.4	B9.5 V		68 mx		
69351	135709		8 16 26.5	−3 13 32	−0.007	+0.03	7.10	0.5	4.0	F8 V		42 s	6707	m
—	135706		8 16 26.6	−0 15 54	−0.006	−0.06	7.9	1.1	0.2	K0 III		120 mx		
69438	154055		8 16 28.2	−16 19 14	−0.003	−0.04	7.60	0.9	0.2	G9 III	+26	300 s	6711	m
69267	116569	17 β Cnc	8 16 30.9	+9 11 08	−0.003	−0.05	3.52	1.48	−0.3	K4 III	+22	52 s	6704	m
69887	235850		8 16 31.3	−52 45 37	−0.002	0.00	7.14	−0.08	0.0	B8.5 V		270 s		
69266	97693		8 16 32.8	+11 02 37	0.000	−0.04	7.4	1.1	−0.1	K2 III		320 s		
69692	199029		8 16 33.1	−38 47 55	−0.001	0.00	7.9	1.1	−0.1	K2 III		390 s		
68953	42261		8 16 33.6	+44 56 08	0.000	−0.04	7.7	0.2	2.1	A5 V		130 s		
69243	97694		8 16 33.9	+11 43 35	+0.001	−0.01	5.9	1.6		M5 III	+32	26 mn		R Cnc, v
69690	199030		8 16 41.0	−36 38 23	−0.001	−0.03	8.0	1.0	−0.1	K2 III		410 s		
69314	116573		8 16 42.4	+7 23 21	−0.001	−0.05	7.9	0.1	1.7	A3 V		170 s		
69437	135715		8 16 42.5	−9 03 14	−0.003	−0.02	7.6	0.9	3.2	G5 IV		77 s		
69562	175492		8 16 44.5	−21 53 00	0.000	−0.05	6.96	−0.15		B4	+12			
69221	80063		8 16 45.4	+21 13 52	−0.002	−0.01	7.63	0.95		G5				
69333	116574		8 16 45.7	+7 22 21	−0.003	−0.09	7.6	0.5	3.4	F5 V		71 s		
70201	250172		8 16 47.0	−64 24 10	−0.003	+0.03	7.9	0.1	0.6	A0 V		290 s		
69460	135717		8 16 47.5	−9 00 50	−0.001	−0.01	7.00	0.1	0.6	A0 V		190 s	6714	m
67707	6472		8 16 48.0	+75 44 34	−0.010	−0.11	7.8	0.5	3.4	F5 V		75 s		
69530	154061		8 16 52.0	−15 09 06	−0.001	+0.01	7.17	1.48	−0.3	gK5	−5	310 s		m
—	235856		8 16 53.0	−53 37 35	−0.003	+0.01	7.36	1.10	0.2	K0 III		220 s		
70200	250174		8 16 53.2	−64 11 13	−0.001	+0.02	6.8	−0.1	0.4	B9.5 V		190 s		
65299	1300		8 16 53.8	+84 03 27	−0.007	−0.02	6.4	0.1	0.6	A0 V	−3	150 s		
69589	175497		8 16 54.1	−21 19 13	−0.002	+0.02	6.60	0.02		A0				
69992	235857		8 16 57.2	−53 42 08	−0.001	+0.02	7.76	0.05	0.6	A0 V		250 s		
69459	135718		8 16 57.2	−5 25 08	−0.002	+0.01	7.8	1.1	0.2	K0 III		340 s		

HD	SAO	Star Name	α 2000	δ 2000	μ(α)	μ(δ)	V	B-V	M$_V$	Spec	RV	d(pc)	ADS	Notes
			h m s	° ' "	s	"								
68771	26750		8 16 58.0	+59 11 20	0.000	-0.03	6.70	1.1	0.2	K0 III	-29	200 s		m
70896	—		8 16 58.2	-76 26 00			8.0	0.8	3.2	G5 IV		90 s		
69856	219726		8 17 01.2	-45 45 06	-0.003	+0.02	8.00	1.07	0.2	K0 III		330 s		
69760	199036		8 17 06.0	-34 29 50	-0.004	+0.02	7.60	0.04	0.6	A0 V		170 mx		m
69365	97705		8 17 07.3	+11 21 04	0.000	-0.03	7.8	1.1	0.2	K0 III		320 s		
69502	135724		8 17 09.4	-3 21 50	-0.002	-0.02	6.6	0.0	0.4	B9.5 V		170 s		
69837	219728		8 17 14.5	-40 14 30	-0.002	0.00	7.80	0.03	0.6	A0 V		260 s		
69526	135730		8 17 15.6	-5 22 09	-0.003	+0.02	7.73	0.39		F2		19 mn	6719	m
69952	219736		8 17 16.5	-46 53 35	-0.002	0.00	7.9	-0.3	0.6	A0 V		280 s		
69882	219733		8 17 17.5	-42 31 17	-0.001	0.00	7.15	0.35	-4.4	B1 III k		1800 s		m
69479	116582		8 17 18.4	+4 13 09	-0.001	+0.01	6.7	0.7	4.4	G0 V	0	29 s		
69973	219740		8 17 19.8	-47 55 12	+0.002	+0.01	6.64	-0.09	-1.1	B5 V n		340 s		
69312	80068		8 17 22.9	+27 02 57	+0.003	-0.01	7.4	1.1	0.0	K1 III		300 s		
69665	154076	21 Pup	8 17 23.0	-16 17 06	-0.002	-0.01	6.16	0.01		A0				
69757	175521		8 17 26.7	-29 07 58	+0.002	-0.04	7.6	0.9	0.2	K0 III		300 s		
69818	199047		8 17 28.3	-32 52 10	-0.002	+0.04	6.95	0.03	1.4	A2 V		130 s		
69756	175520		8 17 28.7	-26 18 08	-0.003	-0.02	7.2	0.5	3.4	F5 V		52 s		
69611	135736		8 17 29.1	-3 59 23	-0.012	-0.45	7.69	0.58	4.0	F8 V		27 mx		
69582	135734		8 17 29.5	-3 06 26	-0.001	-0.01	7.5	0.9	3.2	G5 IV		74 s		
69478	116585		8 17 31.6	+8 51 58	-0.001	-0.02	6.29	0.98	0.3	G8 III	+29	150 s		
69349	80069		8 17 32.3	+27 24 08	-0.002	-0.01	7.80	1.1	0.0	K1 III		360 s		
70685	—		8 17 33.2	-72 17 55			8.0	1.1	0.2	K0 III		360 s		
68435	6482		8 17 34.0	+70 10 38	+0.003	-0.02	7.3	1.1	0.2	K0 III		270 s		
70175	235869		8 17 34.4	-58 41 51	0.000	+0.01	6.8	0.3	0.4	B9.5 V		110 s		
69755	175522		8 17 34.4	-25 00 10	-0.001	-0.03	7.5	1.4	-0.5	M2 III		410 s		
69554	116588		8 17 35.4	+1 08 45	-0.001	-0.03	7.6	1.1	0.2	K0 III		300 s		
69364	80070		8 17 36.2	+24 50 13	+0.001	-0.03	7.4	1.1	0.2	K0 III		270 s		
69523	116587		8 17 37.8	+6 14 42	0.000	-0.02	7.5	1.1	-0.1	K2 III		330 s		
69581	116591		8 17 38.8	+0 28 44	-0.001	-0.02	7.8	0.1	1.4	A2 V		190 s		
69817	175528		8 17 40.2	-29 41 35	-0.001	0.00	7.1	0.6	0.6	A0 V		85 s		
69608	116592		8 17 44.0	+2 47 37	-0.001	0.00	7.6	1.1	-0.1	K2 III		350 s		
70164	235870		8 17 44.0	-55 53 31	-0.007	+0.07	7.30	1.1	0.2	K0 III		190 mx		m
70163	235871		8 17 44.9	-55 46 32	-0.001	+0.01	7.10	1.1	0.2	K0 III		240 s		m
69904	199051		8 17 46.1	-36 04 00	0.000	+0.03	6.4	0.7	-0.3	K5 III		220 s		
69633	116593		8 17 49.5	+2 55 11	-0.001	-0.02	7.7	1.1	0.2	K0 III		320 s		
69330	60706		8 17 50.2	+35 43 43	-0.003	0.00	6.9	0.4	2.6	F0 V		74 s		
68930	26756	29 Lyn	8 17 50.4	+59 34 16	0.000	0.00	5.5	0.2	2.1	A5 V	-17	48 s		
69663	135745		8 17 52.0	-1 48 15	-0.005	-0.23	7.45	0.86		G5				
69242	42275		8 17 53.5	+43 24 35	-0.002	+0.01	7.5	0.4	2.6	F0 V		96 s		
69834	175537		8 17 54.5	-25 21 41	-0.001	-0.01	8.0	0.7	3.4	F5 V		62 s		
69311	60707		8 17 55.4	+38 48 03	+0.001	-0.01	7.84	0.91	3.2	G5 IV		69 s		
70267	235872		8 17 55.8	-59 10 01	-0.006	-0.03	6.42	0.39		F8				m
69879	175543		8 17 58.3	-30 00 12	+0.003	+0.01	6.45	1.04	0.3	gG6	-12	140 s		
69632	116597		8 17 59.1	+6 14 01	-0.002	0.00	7.0	1.1	0.2	K0 III	-2	230 s		
69791	154084		8 18 00.3	-16 51 39	-0.001	-0.01	8.0	1.1	0.2	K0 III		350 s		
69772	154083		8 18 00.9	-14 58 31	-0.003	-0.04	8.0	0.0	0.0	B8.5 V	+27	390 s		
70084	219751		8 18 00.9	-47 05 31	-0.002	0.00	7.07	-0.13	-1.6	B7 III		480 s		
70333	250181		8 18 05.1	-60 18 51	-0.001	-0.01	7.7	-0.4	0.4	B9.5 V		290 s		
69477	80076		8 18 06.5	+26 25 29	-0.003	0.00	7.1	0.1	1.7	A3 V		120 s		
69630	97716		8 18 08.4	+11 39 25	+0.001	-0.03	8.0	0.1	1.4	A2 V		210 s		
69391	60708		8 18 10.5	+37 10 16	+0.001	-0.02	7.9	1.4	-0.3	K5 III		430 s		
70003	199059		8 18 12.6	-37 22 27	-0.002	+0.02	6.70	0.23		A0				m
69629	97718		8 18 14.4	+15 40 42	-0.001	0.00	6.5	1.1	0.2	K0 III		190 s		
69149	26760		8 18 15.7	+54 08 36	-0.003	-0.04	6.27	1.54	-0.3	K5 III	+25	200 s		
70002	199061		8 18 17.3	-35 27 06	+0.001	0.00	5.58	1.24	-0.1	gK2		120 s		
70514	250186		8 18 18.9	-65 36 48	+0.005	+0.02	5.07	1.15		K0	0	22 mn		
69947	175554		8 18 20.1	-26 20 42	+0.001	-0.07	7.4	1.3	-0.3	K5 III		310 mx		
69811	135750		8 18 20.6	-10 07 34	-0.001	-0.01	7.9	1.1	0.2	K0 III		350 s		
69720	116603		8 18 21.3	+5 04 26	0.000	-0.02	7.8	1.1	0.2	K0 III		330 s		
69924	175555		8 18 22.8	-24 36 59	0.000	+0.01	8.0	-0.2	0.6	A0 V		300 s		
69686	116602		8 18 22.9	+9 09 05	-0.006	0.00	7.0	0.0	0.0	B8.5 V		100 mx		
69434	60713		8 18 23.2	+37 03 20	0.000	-0.05	7.4	1.1	0.2	K0 III		280 s		
69830	154093		8 18 23.8	-12 37 55	+0.018	-0.98	5.98	0.76	5.6	G8 V	+30	11 mx		
69789	135752		8 18 24.7	-7 34 14	-0.001	+0.04	7.6	0.4	2.6	F0 V		99 s		
69174	26763		8 18 29.0	+54 33 49	+0.003	-0.03	6.6	0.2	2.1	A5 V		78 s		
71243	256496	α Cha	8 18 31.7	-76 55 10	+0.033	+0.11	4.07	0.39	2.2	F6 IV	-14	24 s		
70265	235875		8 18 32.0	-52 42 22	-0.001	+0.04	8.0	-0.1	0.6	A0 V		300 s		
69752	116608		8 18 32.6	+5 11 05	-0.001	-0.06	7.9	0.5	3.4	F5 V		78 s		
70060	199070		8 18 33.2	-36 39 34	-0.009	+0.09	4.45	0.22	0.5	A7 III	+5	62 s		
69348	42281		8 18 34.8	+47 02 02	-0.002	-0.04	7.5	0.9	3.2	G5 IV		72 s		
69901	154098		8 18 37.3	-14 33 59	-0.001	-0.01	6.8	0.1	1.4	A2 V		120 s		
70119	219765		8 18 37.3	-40 19 49	-0.001	-0.05	7.94	1.03	0.2	K0 III		350 s		
70100	199072		8 18 39.8	-38 41 40	0.000	-0.01	7.9	-0.6	1.4	A2 V		200 s		
69551	60720		8 18 41.6	+33 20 09	-0.001	-0.01	7.8	0.1	0.6	A0 V		270 s		
69498	60716		8 18 42.7	+37 44 14	+0.001	0.00	8.0	1.1	0.2	K0 III		360 s		

HD	SAO	Star Name	α 2000	δ 2000	μ(α)	μ(δ)	V	B-V	M_V	Spec	RV	d(pc)	ADS	Notes
			h m s	° ′ ″	s	″								
69010	14495		8 18 44.9	+63 15 31	−0.009	−0.11	7.5	0.6	4.4	G0 V		43 s		
71086	256495		8 18 45.0	−75 00 53	+0.003	0.00	7.6	0.1	1.4	A2 V		170 s		
70195	219772		8 18 53.7	−42 31 38	−0.001	−0.02	7.07	0.96	0.2	K0 III		240 s		
69698	80087		8 18 54.5	+21 29 39	0.000	−0.01	7.88	1.23	0.2	K0 III		220 s		
69684	80086		8 18 54.6	+26 20 07	−0.002	−0.03	7.1	0.4	2.6	F0 V		78 s		
70157	199074		8 18 57.3	−39 04 44	0.000	0.00	7.10	−0.06	0.6	A0 V		200 s		
70251	219781		8 18 58.2	−47 12 40	−0.002	−0.01	7.5	0.0	0.0	B8.5 V		270 s		
69263	26766		8 18 58.9	+56 15 17	+0.001	0.00	7.1	1.4	−0.3	K5 III		300 s		
70309	219785		8 19 05.6	−48 11 52	−0.001	0.00	6.45	−0.14	−1.7	B3 V		420 s		m
––	14497		8 19 07.3	+63 33 44	−0.007	0.00	7.7	1.6						
70193	199077		8 19 08.0	−38 10 38	−0.003	+0.02	8.0	−0.5	0.6	A0 V		300 s		
69788	97725		8 19 08.7	+16 07 19	−0.001	−0.02	6.8	0.1	0.6	A0 V	+26	180 s		
70156	199076		8 19 08.8	−33 06 20	−0.004	+0.01	7.97	−0.07	−1.0	B5.5 V		570 s		
70264	219783		8 19 09.0	−46 12 19	−0.002	−0.01	7.16	1.36	0.2	K0 III		130 s		m
69734	80089		8 19 09.1	+24 10 28	−0.001	−0.02	7.1	0.9	3.2	G5 IV		61 s		
69966	135772		8 19 14.3	−7 33 30	−0.001	+0.01	7.2	1.4	−0.3	K5 III	+40	310 s		
69965	135771		8 19 15.0	−5 37 09	0.000	+0.05	7.6	0.6	4.4	G0 V		43 s		
69997	154105		8 19 15.0	−10 09 57	−0.003	+0.03	6.32	0.33	0.6	F2 III		140 s		
70040	154110		8 19 16.9	−18 28 07	−0.001	−0.06	7.9	1.6	−0.5	M2 III		290 mx		
69148	14500		8 19 17.2	+62 30 26	−0.002	0.00	5.71	0.89	4.4	G0 V	−2	12 s		
70097	175590		8 19 18.3	−23 47 48	+0.001	+0.01	7.8	1.0	−0.3	K5 III		410 s		
70075	175589		8 19 22.2	−20 09 51	−0.001	+0.01	6.7	−0.5	0.6	A0 V		170 s		
69787	80094		8 19 23.9	+21 15 27	+0.001	−0.02	8.03	1.18	3.2	G5 IV		42 s		
70215	199080		8 19 25.2	−33 44 15	−0.003	−0.02	7.9	1.5	−0.3	K5 III		430 s		
70235	199084		8 19 29.3	−34 35 25	−0.002	0.00	6.43	−0.08		B9				
69661	60726		8 19 29.8	+39 35 43	+0.001	−0.03	7.9	1.1	0.2	K0 III		360 s		
70115	175598		8 19 31.5	−21 15 21	+0.001	+0.01	8.0	1.1	−0.3	K5 III		450 s		
69715	60729		8 19 31.7	+35 02 43	−0.004	−0.02	7.10	0.2	2.1	A5 V		100 s		m
68375	6487		8 19 32.2	+75 45 25	+0.008	+0.02	5.54	0.90	3.2	G5 IV	+7	25 s		
69604	42288		8 19 34.9	+45 28 01	+0.001	−0.03	7.7	1.1	−0.1	K2 III		360 s		
70186	175604		8 19 40.9	−25 25 25	+0.001	+0.02	8.0	1.0	−0.3	K5 III		460 s		
70188	175606		8 19 42.8	−27 04 55	+0.001	−0.01	7.1	1.5	0.2	K0 III		110 s		
70622	235894		8 19 44.2	−60 06 12	0.000	−0.01	7.53	−0.05		B9				
70414	219797		8 19 44.6	−49 32 12	−0.005	0.00	7.14	1.19	0.2	K0 III		170 s		
70289	199090		8 19 45.3	−37 25 04	−0.001	0.00	6.9	−0.2	2.6	F0 V		73 s		
69767	60731		8 19 47.4	+35 12 04	0.000	−0.01	7.2	1.1	0.2	K0 III		250 s		
70367	219795		8 19 47.4	−42 43 21	−0.002	+0.01	7.85	1.21	0.2	K0 III		340 s		
71046	256497	κ¹ Vol	8 19 49.1	−71 30 54	−0.002	+0.03	5.37	−0.06		B9	+36			m
70013	116630		8 19 49.8	+3 56 52	+0.002	−0.03	6.05	0.97		G5	−47			m
70136	154119		8 19 53.9	−12 51 26	−0.003	−0.01	6.9	1.1	−0.1	K2 III		250 s		
69750	42293		8 19 55.8	+39 53 02	−0.001	−0.04	6.7	1.1	0.2	K0 III		200 s		
70248	175614		8 19 56.7	−29 20 50	+0.003	+0.01	7.9	0.8	−0.5	M2 III		490 s		
70510	235888		8 19 57.8	−51 56 30	−0.004	+0.05	6.8	0.6	2.1	A5 V		49 s		
70327	199094		8 19 58.3	−36 36 44	−0.002	−0.04	7.58	0.02	0.6	A0 V		240 s		
71066	256499	κ² Vol	8 20 00.8	−71 30 19	−0.001	+0.04	5.65	−0.10		A0 p				m
69978	97731		8 20 01.3	+16 13 23	+0.001	−0.02	8.0	1.4	−0.3	K5 III		450 s		
70605	235895		8 20 01.7	−56 51 27	−0.001	+0.03	8.02	−0.01	0.4	B9.5 V		330 s		
69897	80104	18 χ Cnc	8 20 03.8	+27 13 03	−0.001	−0.38	5.14	0.47	3.7	F6 V	+33	17 ts		
69713	42292		8 20 03.8	+44 39 47	−0.007	−0.09	8.0	0.5	3.4	F5 V		84 s		
70305	199093		8 20 04.6	−31 05 52	−0.003	−0.01	7.1	1.6	0.2	K0 III		110 s		
70245	175616		8 20 04.7	−23 27 43	−0.002	0.00	7.1	1.1	−0.3	K5 III		300 s		
69602	26782		8 20 07.4	+53 11 51	+0.002	−0.01	7.7	1.1	−0.1	K2 III		360 s		
70801	250193		8 20 11.7	−63 30 15	−0.005	−0.03	7.9	1.1	0.2	K0 III		280 mx		
70424	219800		8 20 12.2	−42 03 33	0.000	−0.02	7.2	1.4	0.2	K0 III		170 s		
70110	135783		8 20 13.1	−0 54 34	+0.003	−0.10	6.18	0.60		G0				
70148	135787		8 20 17.0	−5 19 45	+0.001	−0.03	6.13	1.33		K2				
70325	175633		8 20 18.7	−29 32 26	−0.001	+0.03	7.5	−0.2	0.6	A0 V		250 s		
77149	258501		8 20 19.2	−87 43 41	−0.005	+0.08	6.7	0.5	3.4	F5 V		47 s		
69549	26781		8 20 19.2	+57 18 17	0.000	0.00	7.6	0.9	3.2	G5 IV		77 s		
71007	250199		8 20 19.4	−69 03 42	+0.001	−0.03	6.9	1.0	0.2	K0 III		210 s		
69994	80112		8 20 20.9	+20 44 51	+0.005	−0.05	5.83	1.13	0.3	gG5	−17	90 s		
70093	116642		8 20 23.4	+8 00 28	0.000	+0.01	7.5	1.1	0.2	K0 III		290 s		
69548	26784	30 Lyn	8 20 26.0	+57 44 36	+0.008	+0.01	5.89	0.39	3.0	F2 V	−15	38 s		
70302	175634		8 20 27.3	−22 55 29	−0.001	0.00	6.13	1.04		K0				m
70092	116645		8 20 29.0	+9 24 01	−0.003	−0.02	6.9	1.1	−0.1	K2 III		250 s		
69682	26788		8 20 29.1	+53 34 28	−0.007	−0.10	6.4	0.4	2.6	F0 V	+11	57 s		
70183	135790		8 20 29.4	−6 37 38	−0.001	−0.01	7.3	1.1	−0.1	K2 III		300 s		
70011	80113	19 λ Cnc	8 20 32.1	+24 01 20	−0.002	−0.02	5.30	−0.03	0.2	B9 V	+24	100 s		
70210	135792		8 20 34.6	−6 30 10	−0.001	−0.02	8.0	1.1	0.2	K2 III		410 s		
70506	219811		8 20 34.7	−44 15 17	0.000	+0.01	7.4	0.0	0.6	A0 V		220 s		
71701	256503	θ Cha	8 20 38.7	−77 29 04	−0.039	+0.04	4.35	1.16	1.7	K0 III-IV	+22	24 s		m
68951	6504		8 20 40.3	+72 24 26	+0.001	−0.03	5.98	1.54	−0.3	K5 III	+12	170 s	6724	m
70030	80114		8 20 40.9	+25 20 10	−0.001	−0.02	7.40	1.1	−0.2	K3 III		330 s		
70181	116649		8 20 41.2	+2 39 53	−0.005	−0.02	7.9	0.4	3.0	F2 V		94 s		
70916	250197		8 20 42.6	−64 03 59	−0.003	+0.11	7.9	1.1	0.2	K0 III		170 mx		

HD	SAO	Star Name	α 2000	δ 2000	μ(α)	μ(δ)	V	B–V	M_v	Spec	RV	d(pc)	ADS	Notes
			$h \quad m \quad s$	$\circ \quad ' \quad ''$	s	$''$								
70300	154134		8 20 45.3	-14 21 48	-0.004	+0.01	7.5	0.1	1.7	A3 V		140 s		
69961	60744		8 20 47.8	+35 46 42	-0.004	-0.04	7.0	0.1	0.6	A0 V		100 mx		
69285	14507		8 20 48.0	+67 31 35	0.000	-0.03	7.20	1.6	-0.5	M3 III	+10	350 s		
70531	219813		8 20 50.3	-41 04 17	+0.001	-0.03	7.5	0.2	1.7	A3 V		110 s		
70255	135796		8 20 52.7	-4 52 00	0.000	-0.02	7.9	0.1	1.7	A3 V		170 s		
70875	250198		8 20 53.7	-62 23 35	0.000	+0.02	7.7	0.0	0.4	B9.5 V		290 s		
70359	154139		8 20 54.4	-17 21 29	-0.001	+0.05	8.0	1.4	-0.3	K5 III		440 s		
70391	175646		8 20 54.5	-22 27 03	-0.001	-0.03	8.00	0.1	1.7	A3 V		180 s		m
70703	235908		8 21 00.5	-52 13 41	-0.002	+0.01	6.6	0.4	0.6	A0 V		88 s		
69433	14509		8 21 03.6	+65 26 34	+0.001	+0.03	8.03	0.73		G0				
70341	135801		8 21 04.0	-8 53 01	-0.001	0.00	7.7	1.1	0.2	K0 III		320 s		
70530	199113		8 21 07.0	-34 53 38	-0.005	+0.01	7.90	-0.03	0.4	B9.5 V		160 mx		
70662	219819		8 21 07.3	-47 19 27	-0.002	-0.02	7.78	1.30	-0.1	K2 III		380 s		
70982	250202		8 21 07.7	-64 06 22	0.000	-0.03	6.12	0.93	0.3	gG5		140 s		
70731	235911		8 21 09.1	-52 35 15	-0.004	-0.02	7.0	0.3	2.1	A5 V		83 s		
70602	219817		8 21 11.2	-42 16 25	-0.002	-0.01	7.8	1.3	-0.5	M2 III		410 s		
70051	60748		8 21 11.3	+35 19 46	+0.001	-0.03	8.0	1.6	-0.5	M2 III		510 s		
70839	235917		8 21 12.0	-57 58 23	-0.001	+0.01	6.97	-0.09	-2.9	B3 III	+15	860 s		
71321	—		8 21 13.4	-71 58 20			7.6	0.4	2.6	F0 V		100 s		
70358	135806		8 21 15.7	-8 03 31	+0.001	-0.03	7.3	1.1	0.2	K0 III	+35	260 s		
70298	116665		8 21 15.9	+1 03 49	0.000	-0.16	7.6	0.4	3.0	F2 V		60 mx		
70409	154145		8 21 18.6	-14 58 34	-0.001	-0.02	7.0	1.1	0.2	K0 III		230 s		
70340	135804		8 21 20.1	-1 36 09	-0.001	-0.04	6.50	0.02		A0	+29		6762	m
70556	199119		8 21 21.0	-36 29 04	-0.001	-0.01	5.20	-0.19		B3	+16			m
70442	154150		8 21 21.1	-20 04 45	-0.001	-0.02	5.58	0.77	-2.1	G3 II	-8	140 s		
70555	199118		8 21 23.0	-33 03 16	-0.001	0.00	4.83	1.45	-0.4	M0 III	+33	110 s		
70612	199123		8 21 24.1	-39 37 15	0.000	-0.01	6.16	0.17		A5				
70682	219824		8 21 25.2	-44 59 25	-0.004	-0.01	7.90	0.1	0.6	A0 V		150 mx		m
70204	80125		8 21 28.1	+21 54 33	0.000	-0.04	7.4	1.1	0.2	K0 III		270 s		
70642	199126		8 21 28.1	-39 42 17	-0.018	+0.24	7.18	0.71	4.3	G6 IV-V		39 s		
70482	175660		8 21 30.1	-20 47 39	-0.004	+0.02	7.2	0.6	0.6	A0 V		87 s		
70660	219823		8 21 31.8	-41 08 01	-0.002	0.00	8.0	0.7	0.2	K0 III		360 s		
70407	135811		8 21 32.5	-8 11 30	-0.001	+0.01	8.0	1.1	0.2	K0 III		360 s		
70422	135814		8 21 34.8	-8 40 12	-0.002	+0.02	7.1	0.1	1.4	A2 V		140 s		
69389	14511		8 21 35.5	+69 01 48	-0.009	-0.06	7.9	0.4	3.0	F2 V		94 s		
70438	154152		8 21 36.3	-11 02 46	0.000	-0.02	6.9	0.4	2.6	F0 V		73 s		
70145	60759		8 21 36.9	+33 56 12	-0.002	-0.01	7.20	1.1	0.2	K0 III		250 s	6757	m
70680	219826		8 21 37.5	-40 27 17	0.000	-0.01	7.6	0.1	1.4	A2 V		160 s		
70717	219833		8 21 38.2	-45 00 11	-0.003	0.00	7.92	1.06	0.2	K0 III		320 s		
70715	219830		8 21 41.4	-42 50 10	-0.002	0.00	7.8	0.1	0.4	B9.5 V		240 s		
70276	97753		8 21 42.9	+17 17 08	0.000	+0.01	6.80	1.9		S7.7 e	-1		6763	V Cnc, m,v
70459	135821		8 21 46.9	-9 57 56	-0.001	-0.01	8.0	1.1	-0.1	K2 III		430 s		
--	--		8 21 48.0	-76 21 18			7.50							R Cha, v
70319	97756		8 21 48.4	+14 19 09	-0.004	+0.02	7.9	1.1	0.2	K0 III	+73	340 s		
71308	—		8 21 50.2	-69 28 21			8.0	1.6	-0.5	M2 III		490 s		
70338	97761		8 21 53.3	+13 37 26	-0.002	-0.03	7.1	0.1	1.4	A2 V	+25	140 s		
70374	97763		8 21 53.5	+10 03 53	-0.008	-0.08	8.0	0.9	3.2	G5 IV		93 s		
70523	154159		8 21 54.5	-17 35 11	-0.007	-0.02	5.75	1.05	0.0	K1 III	+68	140 s		
71242	250206		8 21 55.1	-67 20 04	-0.005	+0.05	7.52	0.96	0.0	K1 III		320 mx		
69893	26796		8 21 55.9	+56 37 27	-0.001	-0.05	7.9	0.9	3.2	G5 IV		88 s		
70499	154158		8 21 56.8	-10 44 54	-0.001	0.00	6.7	0.4	3.0	F2 V		56 s		
70337	97762		8 21 57.0	+14 46 04	-0.001	-0.03	7.5	1.1	0.2	K0 III		280 s		
70522	154160		8 22 00.1	-13 46 08	-0.004	-0.02	8.00	0.5	3.4	F5 V		83 s		m
70435	116677		8 22 03.6	-0 09 37	-0.002	-0.02	7.5	0.8	3.2	G5 IV		72 s		
70336	80131		8 22 03.9	+19 57 34	+0.002	-0.03	7.4	1.1	0.2	K0 III		270 s		
71576	256505	η Vol	8 22 04.7	-73 24 01	-0.006	+0.01	5.29	0.01		A2	+20	20 mn		m
71532	—		8 22 06.5	-72 57 28			8.0	0.1	0.6	A0 V		300 s		
70657	175688		8 22 06.6	-28 37 25	0.000	+0.01	7.6	0.4	0.2	K0 III		300 s		
70765	219840		8 22 07.3	-41 41 36	0.000	-0.02	7.9	1.7	-0.3	K5 III		180 s		
70764	219839		8 22 07.9	-40 59 32	-0.001	+0.01	7.70	0.5	3.4	F5 V		72 s		m
70780	219841		8 22 08.3	-43 39 46	-0.002	+0.02	8.03	1.57	-0.3	K5 III		420 s		
69054	6511		8 22 09.6	+74 49 13	+0.013	+0.04	6.37	1.01	0.2	K0 III	-32	170 s	6736	m
70458	116680		8 22 11.4	+2 09 05	0.000	-0.02	6.8	1.4	-0.3	K5 III		270 s		
70851	219843		8 22 12.5	-47 46 59	0.000	0.00	7.58	-0.06	0.6	A0 V		250 s		
70714	199136		8 22 15.5	-31 55 17	-0.003	-0.01	7.6	1.5	0.2	K0 III		160 s		
69960	26800		8 22 16.2	+56 50 11	-0.001	-0.03	7.9	0.9	3.2	G5 IV		88 s		
71065	235926		8 22 16.3	-59 46 55	+0.001	+0.02	8.0	1.6	0.2	K0 III		160 s		
70456	97770		8 22 23.5	+10 39 42	0.000	0.00	7.8	1.1	0.2	K0 III		330 s		
70131	42314		8 22 26.0	+49 28 44	0.000	0.00	7.9	1.1	0.2	K0 III		360 s		
70494	116685		8 22 27.0	+5 00 59	-0.001	0.00	7.4	1.4	-0.3	K5 III		350 s		
70574	135832		8 22 30.1	-6 10 45	-0.004	0.00	6.15	0.22		A3				
71130	235932	ε Car	8 22 30.8	-59 30 34	-0.003	+0.01	1.86	1.27	-2.1	K0 II	+12	62 s		[Avior]
70930	219848		8 22 31.5	-48 29 25	-0.002	+0.01	4.82	-0.15	-3.6	B2 III	+26	460 s		m
70433	97773		8 22 31.9	+16 09 45	-0.001	-0.01	6.7	1.1	0.2	K0 III		200 s		
70725	175704		8 22 33.0	-29 41 32	0.000	0.00	7.30	0.4	2.6	F0 V		87 s		m

HD	SAO	Star Name	α 2000	δ 2000	μ(α)	μ(δ)	V	B-V	M_v	Spec	RV	d(pc)	ADS	Notes
			h m s	° ' "	s	"								
70549	135831		8 22 33.7	−0 32 13	−0.002	−0.01	7.9	0.1	0.6	A0 V		290 s		
69733	14518		8 22 33.9	+66 32 41	−0.008	−0.03	7.4	1.1	0.2	K0 III		280 s		
70951	219851		8 22 36.4	−49 03 04	−0.005	+0.02	7.76	−0.06	0.6	A0 V		170 mx		
70892	219847		8 22 37.4	−44 51 55	−0.006	+0.02	7.3	0.5	2.6	F0 V		65 s		
70273	42318		8 22 38.9	+40 26 38	−0.002	−0.04	7.0	0.4	2.6	F0 V		77 s		
69976	14522		8 22 44.0	+60 37 52	+0.001	0.00	6.5	0.9	3.2	G5 IV	−6	45 s		
––	––		8 22 44.4	+33 30 41			8.00	2.0		N7.7 e				
70815	199149		8 22 44.9	−34 19 33	−0.002	+0.03	7.9	1.4	0.2	K0 III		210 s		
70796	199148		8 22 45.6	−31 36 29	−0.001	−0.02	7.23	−0.11	0.0	B8.5 V		280 s		
70673	154177	22 Pup	8 22 46.7	−13 03 17	−0.004	−0.05	6.11	1.00	0.2	gK0	−17	150 s		
70596	135835		8 22 47.3	−0 52 05	−0.003	−0.03	7.0	1.1	0.2	K0 III		230 s		
71972	256507		8 22 48.0	−76 25 49	−0.005	+0.04	7.10	0.1	1.4	A2 V		130 mx		
70631	135836		8 22 49.0	−6 41 39	−0.001	+0.01	7.9	0.1	0.6	A0 V		280 s		
70761	175709		8 22 49.8	−26 20 53	−0.001	0.00	5.90	0.37	−4.6	F2 Ib	+65	1300 s	6782	m
70272	42319	31 Lyn	8 22 50.1	+43 11 17	−0.002	−0.10	4.25	1.55	−0.3	K5 III	+24	76 s		
70871	199156		8 22 51.4	−39 53 45	−0.002	−0.06	8.0	0.3	3.4	F5 V		82 s		
70652	135840		8 22 53.9	−7 32 36	−0.002	0.00	5.96	1.67		Ma				
70629	135837		8 22 55.2	−2 57 31	−0.001	−0.04	8.0	1.1	0.2	K0 III		360 s		
71043	235933		8 22 55.2	−52 07 26	−0.001	+0.01	5.9	0.3	0.4	B9.5 V		76 s		
70948	219857		8 22 57.1	−43 13 40	−0.002	+0.01	7.0	0.2	0.0	B8.5 V		160 s		
70401	60768		8 22 57.7	+31 35 07	0.000	−0.01	7.7	0.9	3.2	G5 IV		80 s		
70519	97775		8 22 59.9	+17 40 06	0.000	−0.05	8.0	1.1	0.2	K0 III		360 s		
70947	219858		8 23 08.1	−40 22 54	−0.002	−0.01	7.4	0.3	0.0	B8.5 V		260 s		
70203	26811		8 23 15.8	+54 36 00	−0.005	−0.04	7.9	0.9	3.2	G5 IV		87 s		
70946	199166		8 23 17.1	−38 17 09	+0.002	0.00	6.32	1.63		Ma				
––	219863		8 23 19.9	−44 57 58	−0.001	−0.02	7.8	0.0	−1.0	B5.5 V		550 s		
70569	97781	20 Cnc	8 23 21.8	+18 19 56	−0.004	−0.03	5.9	0.4	2.6	F0 V	+36	45 s		
71613	256506		8 23 25.4	−70 14 54	0.000	+0.01	7.0	1.0	−0.1	K2 III		270 s		
70593	97783		8 23 25.7	+18 08 11	0.000	−0.01	7.9	1.1	−0.1	K2 III		400 s		
71060	219869		8 23 27.9	−45 36 42	−0.002	+0.02	7.5	0.4	0.6	A0 V		130 s		
65172	1312		8 23 30.1	+85 41 31	−0.025	−0.05	7.4	0.8	3.2	G5 IV		68 s		
70928	199168		8 23 30.5	−30 48 03	−0.003	+0.01	7.00	1.1	−0.1	K2 III		260 s		m
70889	175729		8 23 31.8	−27 49 21	−0.009	−0.04	7.1	0.9	4.4	G0 V		22 s		
71922	256510		8 23 38.0	−74 09 57	−0.009	+0.04	7.7	0.1	1.7	A3 V		160 s		
70945	199171		8 23 41.7	−30 23 32	−0.003	0.00	7.3	0.1	0.6	A0 V		190 s		
71257	235943		8 23 42.2	−55 19 13	−0.002	−0.04	7.6	−0.2	2.6	F0 V		98 s		
70757	116711		8 23 45.5	+0 03 33	−0.003	−0.03	7.4	0.4	2.6	F0 V		91 s		
71077	219871		8 23 47.7	−42 02 02	−0.001	−0.01	7.43	−0.06	0.2	B9 V		280 s		
70313	26819		8 23 48.5	+53 13 10	−0.003	−0.10	5.6	0.1	1.4	A2 V	+21	48 mx		
70963	175743		8 23 49.4	−28 58 12	0.000	+0.01	6.70	0.1	0.6	A0 V		160 s	6794	m
70756	116712		8 23 50.5	+5 09 29	0.000	0.00	8.0	1.4	−0.3	K5 III		450 s		
70028	14530		8 23 51.3	+65 30 01	−0.003	0.00	7.7	1.1	0.2	K0 III		320 s		
70566	60784		8 23 52.0	+32 17 38	−0.001	0.00	7.5	0.1	1.7	A3 V	−2	150 s		
70734	97788	21 Cnc	8 23 55.2	+10 37 55	0.000	−0.03	6.08	1.49	−0.3	K5 III	+3	190 s	6787	m
70804	135861		8 23 56.1	−1 29 52	−0.001	−0.02	6.8	0.9	3.2	G5 IV		52 s		
71015	199176		8 23 58.7	−32 53 50	−0.002	−0.01	7.14	−0.14	−2.9	B3 III		980 s		
69892	14528		8 24 00.2	+68 55 28	+0.004	+0.02	7.2	0.5	3.4	F5 V		57 s		
71057	199178		8 24 00.7	−36 43 00	−0.001	−0.02	7.46	−0.14	0.4	B9.5 V		260 s		
70627	60787		8 24 05.9	+31 18 02	−0.001	−0.04	7.4	1.1	0.2	K0 III		280 s		
71181	219876		8 24 07.4	−45 14 10	−0.001	0.00	7.61	1.17	−0.3	K5 III		380 s		
70942	154213		8 24 08.3	−19 08 06	−0.005	−0.05	6.9	1.1	0.2	K0 III		140 mx		
71104	199179		8 24 08.9	−35 30 35	−0.001	−0.02	7.40	0.1	1.4	A2 V		160 s		m
71076	—		8 24 09.8	−34 58 18			7.4	0.6	1.7	A3 V		66 s		
70859	135871		8 24 11.7	−5 40 45	0.000	−0.01	7.6	1.4	−0.3	K5 III		380 s		
70177	14531		8 24 12.0	+62 18 00	−0.001	0.00	7.8	0.0	0.4	B9.5 V		300 s		
70516	42337		8 24 15.6	+44 56 59	−0.006	−0.18	7.70	0.64	4.7	G2 V		37 s	6783	m
72922	258496		8 24 20.0	−80 54 51	−0.063	+0.22	5.69	1.02		K0				
71368	235947		8 24 22.4	−55 47 22	−0.001	−0.02	7.0	1.6	−0.3	K5 III		250 s		
70844	116721		8 24 27.7	+7 38 16	+0.003	−0.03	7.8	1.1	0.2	K0 III		330 s		
70938	135876		8 24 27.8	−8 31 13	−0.002	+0.02	7.7	1.6	−0.5	M4 III		440 s		
71367	235949		8 24 31.7	−53 44 11	−0.001	−0.02	7.6	−0.3	0.6	A0 V		260 s		
67934	1319		8 24 33.2	+82 25 51	−0.007	−0.03	6.2	0.1	0.6	A0 V	−16	80 mx		
70923	135874		8 24 33.2	−1 08 57	+0.006	−0.21	7.05	0.60	4.4	dG0	+9	31 ts		m
82390	—		8 24 33.9	−89 01 47			7.3	0.4	2.6	F0 V		87 s		
71216	219883		8 24 34.9	−40 44 48	−0.003	+0.02	7.10	−0.15	−0.2	B8 V		160 s		
70958	135877	1 Hya	8 24 34.9	−3 45 04	−0.014	−0.03	5.61	0.46	3.4	dF5	+72	21 ts		
69220	6520		8 24 36.0	+77 56 19	+0.004	0.00	8.0	0.1	0.6	A0 V		310 s		
70937	135882		8 24 36.3	−4 43 01	+0.001	−0.05	6.01	0.46	3.3	dF4	−35	33 s		
71010	154221		8 24 38.3	−15 27 36	+0.002	−0.02	7.9	1.1	0.2	K0 III		340 s		
70995	154220		8 24 39.6	−11 31 31	+0.001	−0.04	7.4	1.1	−0.1	K2 III		310 s		
71160	199188		8 24 39.6	−32 55 27	−0.004	−0.02	8.00	1.4	−0.3	K5 III		220 mx		m
71405	235952		8 24 40.8	−55 21 53	−0.004	0.00	6.8	−0.2	0.0	B8.5 V		230 s		
70647	42342		8 24 42.7	+42 00 18	+0.001	0.00	6.02	1.59	−0.3	K5 III	+27	160 s		m
70843	97797		8 24 45.4	+17 10 55	+0.001	−0.12	7.1	0.5	3.4	F5 V	−16	54 s		
70592	42340		8 24 46.0	+47 10 21	+0.003	0.00	7.9	1.1	0.2	K0 III		340 s		

HD	SAO	Star Name	α 2000	δ 2000	μ(α)	μ(δ)	V	B-V	M$_v$	Spec	RV	d(pc)	ADS	Notes
70826	80164		8 24 49.1	+20 09 11	0.000	0.00	7.20	0.2	2.1	A5 V		110 s		m
71271	219886		8 24 51.2	-41 24 48	-0.001	-0.01	7.9	1.7	-0.3	K5 III		180 s		
71142	175775		8 24 52.7	-23 41 23	-0.004	+0.03	6.8	0.8	3.0	F2 V		31 s		
69408	6523		8 24 53.2	+76 57 53	+0.009	+0.01	7.1	1.1	0.2	K0 III		240 s		
71141	175777		8 24 55.0	-23 09 13	-0.003	+0.04	5.68	0.06		A2				
70825	80165		8 24 55.1	+23 56 42	-0.004	+0.03	7.3	0.4	1.9	F2 IV	-1	120 s		
70899	97801		8 24 55.8	+12 14 10	+0.001	-0.01	8.0	1.1	0.2	K0 III		360 s		
70755	60792		8 24 56.4	+32 15 21	+0.001	-0.02	7.9	1.1	0.2	K0 III		340 s		
71302	219890		8 24 57.1	-42 46 10	-0.002	+0.02	5.98	-0.17		B5 n	+23			m
71099	154231		8 24 58.3	-15 17 39	-0.003	+0.02	6.7	0.1	1.7	A3 V		100 s		
71072	154230		8 25 00.2	-12 45 51	-0.002	+0.01	6.9	0.0	0.0	B8.5 V		240 s		
72234	256511		8 25 00.4	-74 54 46	+0.005	+0.02	7.5	0.8	3.2	G5 IV		71 s		
70935	116737		8 25 01.1	+9 25 34	+0.001	+0.02	6.90	1.1	0.2	K0 III		220 s	6798	m
71176	175783		8 25 03.7	-24 02 46	-0.002	+0.02	5.28	1.48	-0.3	K5 III	+26	130 s	6800	m
71386	219900		8 25 04.1	-49 09 36	+0.006	-0.11	7.66	0.78	3.2	G5 IV		78 s		
70771	60794		8 25 04.9	+35 00 41	0.000	-0.02	6.06	1.27	0.2	K0 III	+33	100 s		
70991	116740		8 25 06.3	+4 29 50	0.000	-0.02	7.8	1.6	-0.5	M4 III		470 s		
71336	219898		8 25 07.0	-43 21 54	0.000	-0.01	7.99	-0.10	-1.7	B3 V		870 s		
71491	235957		8 25 09.1	-55 28 24	-0.003	-0.01	7.10	0.00		B8.5 V		260 s		m
71097	135890		8 25 09.2	-9 47 10	0.000	-0.02	7.8	0.0	0.0	B8.5 V		360 s		
70547	26826		8 25 09.4	+53 55 55	-0.004	+0.01	7.3	0.4	3.0	F2 V		72 s		
71255	199198		8 25 10.7	-31 56 26	-0.001	-0.01	6.7	0.2	0.4	B9.5 V		120 s		
71071	135893		8 25 14.1	-7 10 14	-0.007	-0.03	8.0	0.9	3.2	G5 IV		93 s		
70312	14539		8 25 15.0	+64 42 49	-0.004	-0.05	7.5	0.5	3.4	F5 V		68 s		
71285	199201		8 25 17.5	-34 21 59	-0.002	-0.01	7.5	0.9	3.2	G5 IV		61 s		
71197	175793		8 25 18.4	-22 08 27	0.000	0.00	6.9	1.1	0.2	K0 III		200 s		
71196	175792		8 25 19.1	-21 02 45	-0.002	+0.06	6.01	0.40		F2				
71549	235963		8 25 22.2	-55 44 56	+0.002	0.00	7.00	0.1	1.4	A2 V		130 s		m
71634	235968		8 25 30.2	-58 07 59	-0.002	0.00	6.67	-0.02	-1.0	B7 IV		310 s		
71726	250221		8 25 30.3	-61 20 56	-0.001	-0.03	8.0	0.4	1.7	A3 V		110 s		
71510	235962		8 25 31.0	-51 43 41	-0.005	-0.01	5.20	0.00	-1.6	B3.5 V	+18	100 mx		m
71095	116747		8 25 35.5	+2 06 07	-0.001	-0.02	5.73	1.53	-0.3	gK5	+12	160 s		
71172	135897		8 25 37.2	-8 57 23	-0.003	-0.01	7.6	1.1	0.2	K0 III		300 s		
71231	154240		8 25 39.3	-17 26 22	-0.001	-0.01	6.4	1.1	0.2	K0 III		170 s		
71155	135896		8 25 39.5	-3 54 23	-0.005	-0.03	3.90	-0.02	0.6	A0 V	+10	46 s		
72072	—		8 25 40.7	-71 02 50			7.4	1.1	-0.1	K2 III		320 s		
70857	60801		8 25 41.8	+36 30 12	0.000	-0.03	8.0	1.1	0.2	K0 III		360 s		
71878	250228	β Vol	8 25 44.3	-66 08 13	-0.005	-0.16	3.77	1.13	-0.1	K2 III	+27	59 s		
71137	135898		8 25 45.9	-0 24 44	-0.001	-0.08	7.3	1.1	0.2	K0 III	-2	200 mx		m
71136	116751		8 25 48.3	+1 34 25	-0.001	-0.07	6.7	1.1	0.2	K0 III		190 mx		
71171	135901		8 25 48.5	-4 10 39	-0.001	+0.01	7.2	1.4	-0.3	K5 III		320 s		
71334	175807		8 25 49.5	-29 55 51	+0.011	-0.30	7.82	0.67	5.0	G4 V		36 s		
71030	97806	25 Cnc	8 25 49.8	+17 02 46	-0.013	-0.16	6.14	0.41	3.0	F2 V	+38	40 s		
71863	250226		8 25 51.7	-64 36 03	-0.001	0.00	5.97	0.96		G5				
71459	219910		8 25 51.9	-42 09 12	-0.001	0.00	5.47	-0.15	-1.1	B5 V	+28	210 s		
71115	116752		8 25 54.7	+7 33 53	-0.002	-0.01	5.13	0.94	-2.1	G8 II	+15	280 s	6805	m
71053	97807		8 25 54.8	+17 50 01	-0.006	-0.03	8.0	0.5	4.2	F9 V	+28	57 s		
71267	154244		8 25 55.5	-14 55 48	-0.001	+0.03	5.98	0.17		A2				
71230	—		8 25 57.1	-8 46 26			7.80	1.1	0.2	K0 III		330 s		m
71381	199215		8 26 00.6	-32 10 12	-0.004	+0.02	7.20	0.4	2.6	F0 V		83 s		m
71253	154247		8 26 01.1	-10 31 01	-0.001	-0.02	6.6	0.1	1.4	A2 V		110 s		
70920	60804		8 26 02.4	+36 58 56	+0.001	-0.05	7.9	1.1	0.2	K0 III		350 s		
71252	135908		8 26 06.0	-8 30 54	-0.005	-0.02	6.9	0.4	2.6	F0 V		73 s		
71587	219917		8 26 07.7	-48 25 56	-0.005	+0.04	7.5	-0.3	2.6	F0 V		98 s		
71489	219913		8 26 08.4	-40 47 11	0.000	0.00	7.5	1.1	0.2	K0 III		220 s		
71528	219915		8 26 15.7	-42 08 15	-0.001	0.00	7.89	0.03	-2.1	B2.5 V		970 s		
71399	175821		8 26 17.4	-27 08 42	+0.002	-0.07	6.9	1.1	3.4	F5 V		21 s		
71487	199222		8 26 17.5	-39 03 33	-0.002	-0.01	6.70							m
71458	199218		8 26 17.6	-32 56 48	-0.003	+0.03	8.0	1.0	-0.1	K2 III		410 s		
70311	6535		8 26 19.3	+70 07 40	-0.019	+0.01	7.5	0.8	3.2	G5 IV	-63	73 s		
71228	116761		8 26 20.1	+2 29 23	+0.003	-0.07	7.60	1.1	0.0	K1 III	+2	190 mx		
71630	219920		8 26 22.9	-48 06 05	+0.002	-0.03	7.5	0.0	0.6	A0 V		230 s		
71544	219918		8 26 24.5	-41 11 17	-0.002	-0.03	7.5	0.4	0.4	B9.5 V		130 s		
71722	235976		8 26 25.4	-52 48 28	-0.002	-0.01	6.1	0.8	0.6	A0 V		41 s		
71631	219922		8 26 26.9	-48 11 07	-0.001	0.00	7.9	1.2	-0.1	K2 III		140 s		
71297	135916	2 Hya	8 26 27.1	-3 59 15	-0.004	-0.06	5.59	0.22		A5 m	+27	22 mn		m
71093	80181	22 φ¹ Cnc	8 26 27.6	+27 53 36	-0.002	-0.12	5.57	1.40	-0.3	K5 III	+24	89 mx		m
71485	199225		8 26 31.2	-33 12 47	-0.003	+0.02	7.8	1.6	-0.3	K5 III		360 s		
71695	219929		8 26 35.8	-49 29 50	+0.001	0.00	7.25	-0.06	0.6	A0 V		210 s		
71609	219923		8 26 38.2	-43 24 31	+0.001	0.00	7.81	0.09	-2.6	B2.5 IV		860 s		
71152	80185	24 Cnc	8 26 40.0	+24 32 06	-0.003	-0.09	7.02	0.32	1.7	dA3	+15	91 s	6811	m
71607	219924		8 26 40.4	-42 31 27	-0.002	-0.01	7.21	0.14	1.4	A2 V		130 s		
71357	154255		8 26 40.5	-11 46 29	-0.001	+0.01	8.0	0.9	3.2	G5 IV		93 s		
71377	154257		8 26 41.9	-12 32 04	-0.007	-0.02	5.54	1.19	-0.1	K2 III	+65	130 s		
71262	116762		8 26 43.5	+7 39 04	-0.003	+0.01	6.8	0.1	0.6	A0 V	+16	170 s		

HD	SAO	Star Name	α 2000	δ 2000	μ(α)	μ(δ)	V	B−V	M_V	Spec	RV	d(pc)	ADS	Notes
71250	97819	27 Cnc	$8^h 26^m 43\overset{s}{.}8$	$+12°39'17''$	$-0\overset{s}{.}002$	$-0\overset{''}{.}10$	5.50	1.60	−0.5	M2 III	−7	150 mx		
71417	154262		8 26 43.8	−18 50 13	−0.001	−0.01	8.0	0.0	0.4	B9.5 V		330 s		
71721	219935		8 26 45.7	−48 57 22	−0.001	−0.01	7.6	−0.2	0.4	B9.5 V		280 s		
71151	80188	23 φ² Cnc	8 26 46.9	+26 56 07	−0.001	−0.01	6.30	0.2	0.1	A4 III	−31	170 s	6815	m
71027	42360		8 26 47.1	+39 53 32	+0.002	−0.02	7.2	1.1	0.2	K0 III		250 s		
71807	235986		8 26 47.2	−55 21 47	−0.002	0.00	7.4	0.0	0.4	B9.5 V		240 s		
71416	154266		8 26 48.0	−18 10 25	−0.001	−0.04	7.4	0.1	0.6	A0 V		230 s		
71113	60816		8 26 48.2	+34 20 16	−0.001	0.00	7.8	0.1	0.6	A0 V		270 s		
71572	199234		8 26 48.6	−34 35 51	−0.003	−0.01	8.0	0.1	1.4	A2 V		180 s		
71524	199233		8 26 49.5	−32 01 36	−0.002	+0.02	8.0	1.0	0.2	K0 III		360 s		
71542	199235		8 26 49.5	−34 10 27	−0.003	+0.02	7.5	1.1	0.2	K0 III		240 s		
71523	175851		8 26 50.7	−29 12 55	−0.001	+0.01	6.5	0.4	0.4	B9.5 V		95 s		
71343	135921		8 26 51.9	−1 12 57	−0.001	−0.01	7.8	1.1	0.2	K0 III		330 s		
71374	135925		8 26 54.7	−6 12 12	0.000	−0.01	7.8	1.1	−0.1	K2 III		380 s		
71584	199238		8 26 56.4	−36 40 52	−0.002	−0.01	7.4	1.6	−0.3	K5 III		310 s		
71805	235988		8 26 57.6	−52 42 19	−0.012	−0.06	6.50	0.42	3.7	F6 V		36 s		m
71841	235993		8 26 58.5	−55 19 57	−0.002	+0.02	7.7	−0.3	0.6	A0 V		260 s		
70968	42358		8 26 58.7	+49 33 55	0.000	−0.03	7.9	1.4	−0.3	K5 III		430 s		
71504	175852		8 26 59.1	−23 15 09	−0.004	−0.01	7.6	−0.3	3.4	F5 V		71 s		
70919	26837		8 26 59.3	+53 37 44	−0.002	−0.06	7.9	0.4	2.6	F0 V		110 s		
71310	116767		8 27 00.1	+7 13 15	0.000	−0.01	7.0	0.1	1.4	A2 V		130 s		
70545	14549		8 27 01.0	+67 17 10	+0.002	+0.01	7.6	1.1	0.2	K0 III		300 s		
71790	235987		8 27 01.4	−51 07 39	−0.005	+0.02	7.4	0.3	1.7	A3 V		100 s		
72353	256515		8 27 01.5	−71 32 33	+0.002	−0.03	7.5	1.1	0.2	K0 III		260 s		
71803	235990		8 27 02.6	−51 28 25	−0.008	+0.02	7.9	0.2	2.6	F0 V		110 s		
71625	199242		8 27 03.4	−37 56 33	−0.002	+0.04	7.8	1.0	−0.1	K2 III		380 s		
71583	199239		8 27 04.0	−32 11 01	−0.002	−0.01	6.8	0.8	3.2	G5 IV		51 s		
71465	154270		8 27 05.3	−15 38 33	0.000	0.00	7.8	1.4	−0.3	K5 III		410 s		
71624	199245		8 27 09.0	−36 55 27	−0.001	−0.02	7.6	1.3	−0.3	K5 III		370 s		
71520	175858		8 27 10.0	−23 26 16	+0.001	−0.01	7.2	1.0	−0.3	K5 III		310 s		
71622	199246		8 27 16.4	−31 40 22	−0.002	−0.01	6.33	0.90		K0				
72337	250235		8 27 17.0	−70 05 36	−0.001	+0.04	5.53	−0.03		B9	+20			
71433	135931		8 27 17.2	−6 24 35	−0.003	−0.03	6.59	0.51		F5				
71187	60822		8 27 18.6	+35 49 02	−0.006	−0.04	7.9	0.9	3.2	G5 IV		88 s		
71670	199248		8 27 20.5	−34 41 50	−0.002	0.00	7.3	1.7	−0.3	K5 III		250 s		
71411	116777		8 27 23.9	+3 41 32	0.000	−0.04	8.0	1.1	0.2	K0 III		360 s		
71518	154273		8 27 26.7	−14 56 09	−0.003	−0.02	6.67	−0.17	−1.0	B5.5 V	+12	340 s		
71919	236001		8 27 27.4	−55 00 42	−0.002	−0.01	6.53	−0.02		A0				m
71621	175874		8 27 28.4	−28 17 06	−0.003	+0.01	7.9	0.8	0.2	K0 III		340 s		
71431	116780		8 27 31.5	+0 49 40	+0.007	+0.02	7.8	0.7	4.4	G0 V		49 s		
71581	175870		8 27 33.2	−20 50 38	0.000	−0.01	6.56	0.04		A0				m
71619	175875		8 27 33.2	−25 04 52	−0.002	−0.04	7.6	0.7	0.6	A0 V		89 s		
72719	256516		8 27 34.2	−75 21 30	−0.005	+0.08	6.9	0.3	1.4	A2 V		66 mx		
71479	135935		8 27 35.1	−5 08 55	−0.009	−0.09	7.3	0.6	4.4	G0 V		39 s		
71691	199250		8 27 36.5	−32 41 29	−0.004	−0.01	8.0	1.3	0.2	K0 III		250 s		
71935	236002		8 27 36.5	−53 05 19	−0.008	+0.01	5.09	0.25	2.6	dF0	+25	32 s		
71148	42369		8 27 36.7	+45 39 10	−0.002	−0.35	6.32	0.62	4.4	G0 V	−34	23 ts		
71690	199251		8 27 37.7	−31 59 38	0.000	−0.01	7.8	1.2	−0.1	K2 III		380 s		
70985	26845		8 27 38.8	+55 11 20	+0.002	−0.01	7.7	1.1	0.2	K0 III		310 s		
71499	135938		8 27 40.8	−4 24 54	+0.002	−0.05	6.84	0.34		F0			6825	m
71740	199258		8 27 42.3	−33 22 59	−0.002	−0.02	7.8	1.2	0.2	K0 III		260 s		
71538	135941		8 27 46.1	−7 13 12	+0.001	+0.01	7.8	1.4	−0.3	K5 III		410 s		
71668	175880		8 27 46.3	−24 53 30	−0.003	−0.02	7.5	0.6	−0.1	K2 III		340 s		
71579	154278		8 27 46.7	−13 35 36	−0.001	0.00	7.5	1.4	−0.3	K5 III		370 s		
71934	219964		8 27 50.4	−50 06 01	−0.001	−0.01	7.54	0.08	0.0	B8.5 V		320 s		
71276	60829		8 27 52.3	+36 07 28	0.000	−0.05	7.7	0.1	0.6	A0 V		270 s		
71617	154280		8 27 53.0	−16 52 30	0.000	−0.01	8.0	1.1	0.2	K0 III		360 s		
71688	175883		8 27 53.4	−26 07 57	−0.004	−0.01	6.62	0.11		A2				
71407	97833		8 27 53.8	+18 04 08	+0.001	−0.02	7.9	0.5	3.4	F5 V		80 s		
71090	26849		8 27 55.2	+53 27 26	−0.001	−0.02	7.7	1.1	−0.1	K2 III		360 s		
71515	116787		8 27 55.5	+3 13 06	−0.001	−0.01	7.5	1.1	0.2	K0 III		280 s		
71497	116786		8 27 56.1	+4 49 17	0.000	−0.01	7.8	0.8	3.2	G5 IV		84 s		
71801	199260		8 27 59.2	−35 06 50	−0.002	0.00	5.75	−0.15	−1.7	B3 V	+23	310 s		m
71857	219962		8 28 01.3	−41 14 44	−0.001	0.00	7.80	0.1	0.6	A0 V		270 s		m
71475	97838		8 28 02.6	+11 42 24	0.000	0.00	7.6	1.1	−0.1	K2 III		340 s		
71323	60830		8 28 03.9	+35 53 32	−0.005	0.00	6.9	1.1	0.2	K0 III		220 s		
71969	236005		8 28 04.2	−51 14 08	−0.002	−0.01	8.0	−0.6	0.4	B9.5 V		330 s		
71245	42376		8 28 07.0	+46 14 56	−0.005	−0.03	7.0	1.1	0.2	K0 III		230 s		
71428	80201		8 28 07.1	+21 08 55	+0.001	−0.13	7.0	0.9	3.2	G5 IV		57 s		
71354	60832		8 28 08.4	+33 31 44	−0.002	−0.04	7.60	1.1	0.2	K0 III		300 s	6821	m
71771	175896		8 28 12.6	−27 33 51	−0.002	+0.02	7.52	−0.11	−1.0	B5.5 V		510 s		
71915	219968		8 28 14.3	−42 13 50	−0.002	+0.02	7.58	1.54	−0.1	K2 III		130 s		
71597	116796		8 28 14.4	+0 14 36	−0.002	−0.05	7.31	1.16	−0.1	K2 III	+110	300 s		
72019	236008		8 28 15.3	−51 46 13	−0.005	0.00	7.6	−0.2	0.4	B9.5 V		160 mx		
71856	199268		8 28 19.2	−35 10 02	−0.001	+0.01	7.58	1.18	−0.1	K2 III		340 s		

HD	SAO	Star Name	α 2000	δ 2000	μ(α)	μ(δ)	V	B-V	M_v	Spec	RV	d(pc)	ADS	Notes
71665	135956		8 28 19.6	-8 48 58	0.000	-0.04	6.43	1.20		K0				
71389	60840		8 28 20.4	+33 41 07	0.000	-0.01	7.2	1.1	-0.1	K2 III		290 s		
71226	42378		8 28 20.5	+49 41 14	-0.006	-0.05	7.3	0.4	2.6	F0 V		87 s		
71664	135957		8 28 21.8	-8 13 02	-0.001	-0.03	8.0	1.4	-0.3	K5 III		450 s		
71353	60838		8 28 22.4	+37 00 42	+0.004	-0.04	7.7	1.1	0.2	K0 III		240 mx		
71663	135958		8 28 29.1	-2 31 02	-0.001	-0.02	6.39	0.33	2.6	dF0	-14	55 s	6828	m
71595	116800		8 28 33.1	+6 57 54	+0.004	-0.08	7.9	0.5	3.4	F5 V		78 s		
71894	199273		8 28 33.6	-33 50 49	-0.003	+0.02	8.0	0.5	1.4	A2 V		100 s		
71640	116802		8 28 35.9	+1 14 50	-0.002	-0.02	7.5	0.5	3.4	F5 V		65 s		
71815	175902		8 28 35.9	-23 04 18	-0.003	-0.03	6.51	0.05		A0				
71496	80204	28 Cnc	8 28 36.7	+24 08 41	-0.002	-0.06	6.1	0.2	2.1	A5 V	+12	57 mx		
71967	219972		8 28 36.8	-41 10 28	-0.002	0.00	7.5	0.4	0.6	A0 V		130 s		
72473	250243		8 28 36.9	-68 37 02	-0.008	-0.02	7.9	0.1	1.7	A3 V		140 mx		
72201	236019		8 28 37.0	-58 22 12	-0.002	0.00	8.0	0.5	0.2	K0 III		360 s		
71555	97843	29 Cnc	8 28 37.3	+14 12 39	-0.001	-0.02	5.90	0.19	2.1	A5 V	+2	55 s		
71834	175907		8 28 41.0	-23 49 47	-0.006	-0.04	7.9	-0.4	1.4	A2 V		78 mx		
72142	236017		8 28 41.8	-54 29 56	+0.002	+0.01	8.0	-0.3	0.4	B9.5 V		330 s		
71913	199276		8 28 42.9	-34 43 56	-0.001	-0.03	7.68	-0.11	-1.0	B5.5 V		550 s		
71661	116806		8 28 43.5	+4 19 26	-0.001	-0.05	7.9	1.1	0.2	K0 III		340 s		
72275	236021		8 28 43.7	-60 07 22	0.000	-0.01	7.4	0.9	3.2	G5 IV		70 s		V Car, v
---	154299		8 28 49.8	-16 03 46	-0.001	+0.07	8.0	0.9	3.2	G5 IV		93 s		
71766	135965		8 28 50.9	-9 44 54	-0.001	0.00	6.00	0.42		F2				
73291	256521		8 28 51.4	-78 55 33	+0.012	-0.01	8.0	1.1	0.2	K0 III		360 s		
71244	26860		8 28 51.7	+53 51 43	+0.009	-0.07	7.0	0.6	4.4	G0 V		34 s		
72014	219977		8 28 51.9	-42 35 14	-0.002	+0.01	6.26	-0.06	-1.7	B3 V nne		380 s		m
71912	175919		8 28 52.3	-29 57 05	0.000	-0.01	7.9	0.5	0.6	A0 V		130 s		
71833	175912		8 28 53.3	-20 57 01	-0.001	-0.01	6.67	-0.06		B8				m
72740	256518		8 28 53.6	-72 04 53	-0.006	+0.06	7.1	0.1	1.4	A2 V		100 mx		
71224	26857		8 28 55.7	+56 40 01	+0.002	-0.02	7.2	1.1	0.2	K0 III		250 s		
71930	199279		8 28 57.4	-30 51 01	-0.001	-0.04	7.6	0.4	0.6	A0 V		140 s		
72108	219985		8 29 04.6	-47 55 44	-0.002	0.00	5.33	-0.14	-3.6	B2 III	+14	570 s		m
---	---		8 29 04.8	-47 56 04			5.70							m
72066	219981		8 29 04.8	-43 55 07	-0.008	+0.01	6.75	1.39	-0.3	K5 III		180 mx		
71782	135967		8 29 05.6	-5 52 34	-0.002	0.00	7.3	1.1	0.2	K0 III	-30	270 s		
72089	219983		8 29 06.8	-45 33 26	-0.001	+0.02	7.8	-0.2	0.4	B9.5 V		300 s		
72067	219982		8 29 07.6	-44 09 37	0.000	+0.01	5.79	-0.16	-1.7	B3 V	+3	320 s		m
71765	135966		8 29 07.8	-0 18 02	-0.002	-0.03	7.4	0.1	0.6	A0 V		230 s		
71681	97848		8 29 09.1	+12 33 06	-0.001	-0.02	8.0	1.1	0.2	K0 III		360 s		
71537	60846		8 29 12.9	+32 41 33	+0.002	0.00	6.92	-0.09		A0	-11			
71494	60843		8 29 13.1	+38 52 26	0.000	0.00	7.93	0.20	0.6	A0 V		220 s		
70351	6545		8 29 24.8	+77 03 32	-0.032	-0.07	7.8	0.4	2.6	F0 V		110 s		
72127	219996		8 29 27.5	-44 43 30	0.000	-0.01	4.99	-0.16	-1.0	B5.5 V	+20	160 s		
71997	175939		8 29 27.5	-27 19 57	0.000	-0.01	6.70	-0.13		B9				
72389	250244		8 29 29.4	-60 10 46	-0.001	+0.01	7.85	0.98		G5				
72303	236028		8 29 29.8	-54 12 43	-0.002	0.00	6.60	0.00	0.4	B9.5 V		170 s		m
71830	135971		8 29 30.4	-0 57 35	-0.001	0.00	6.9	1.1	0.2	K0 III		220 s		
72047	199286		8 29 30.9	-34 15 44	0.000	0.00	7.7	1.3	0.2	K0 III		220 s		
71945	154313		8 29 31.5	-18 45 26	-0.001	+0.01	7.70	0.00	0.0	B8.5 V		340 s		m
72323	236031		8 29 35.3	-55 25 01	-0.002	+0.02	7.7	-0.3	0.6	A0 V		260 s		
72322	236032		8 29 36.3	-55 11 27	-0.002	-0.02	6.36	0.80		G0				m
71795	116818		8 29 36.4	+8 25 02	0.000	0.00	6.7	1.1	0.2	K0 III		200 s		
72139	219998		8 29 37.0	-41 31 31	-0.002	-0.01	7.9	0.3	0.4	B9.5 V		190 s		
71638	60852		8 29 37.3	+32 03 10	+0.001	0.00	7.9	1.1	0.2	K0 III		350 s		
71730	80209		8 29 39.9	+24 20 41	-0.002	-0.03	7.20	1.00	0.2	K0 III		250 s		
72086	199293		8 29 42.7	-34 03 53	0.000	-0.01	7.9	0.0	0.6	A0 V		270 s		
72634	250250		8 29 43.1	-67 08 23	0.000	+0.02	7.50	0.1	0.6	A0 V		240 s		m
72232	220007		8 29 45.5	-46 19 55	-0.003	0.00	6.1	-0.1	0.0	B8.5 V	+9	160 s		
71088	14568		8 29 46.2	+67 17 51	-0.010	+0.01	5.88	0.97	3.2	G5 IV	-3	26 s		
71884	116822		8 29 47.5	+0 16 24	-0.002	-0.03	7.5	1.1	0.2	K0 III		290 s		
71729	80210		8 29 49.4	+26 11 34	-0.002	-0.01	7.00	0.1	0.6	A0 V		190 s		m
71956	154316		8 29 50.5	-10 28 01	-0.006	+0.04	7.4	1.1	0.2	K0 III		220 mx		
72085	199294		8 29 51.7	-30 19 31	-0.002	+0.01	8.0	1.4	-0.3	K5 III		460 s		U Pyx, v
71926	116824		8 29 55.1	+0 01 08	-0.001	-0.04	7.2	0.6	4.4	G0 V		36 s		
72158	199299		8 30 09.4	-33 04 27	-0.002	0.00	7.7	1.3	-0.1	K2 III		310 s		
72372	236039		8 30 15.6	-51 08 04	+0.007	+0.03	7.0	0.2	0.6	A0 V		70 mx		
71369	14573	1 o UMa	8 30 15.8	+60 43 05	-0.018	-0.11	3.36	0.84	-0.9	G4 II-III	+20	71 s	6830	Muscida, m
71991	135985		8 30 18.9	-2 22 49	-0.001	-0.01	7.9	0.0	0.4	B9.5 V		320 s		
---	---		8 30 22.8	-26 02 07			8.00							m
72025	135988		8 30 24.4	-7 02 25	0.000	-0.02	7.5	0.1	0.6	A0 V		250 s		
72060	154326		8 30 25.1	-13 28 07	-0.002	-0.01	7.2	0.1	1.4	A2 V		140 s		
72227	199308		8 30 28.5	-32 09 34	-0.002	-0.01	5.65	1.49		K2				
72268	199310		8 30 29.5	-36 43 16	-0.002	+0.01	6.69	1.95	-0.1	K2 III		70 s		
72056	154327		8 30 30.3	-10 17 03	-0.002	-0.02	8.0	1.1	0.2	K0 III		360 s		
71554	26876		8 30 34.2	+54 07 10	-0.004	0.00	6.7	0.0	0.4	B9.5 V		180 s		
72318	220020		8 30 34.2	-40 30 42	-0.002	-0.01	7.20	0.00	0.4	B9.5 V		230 s		m

HD	SAO	Star Name	α 2000	δ 2000	μ(α)	μ(δ)	V	B-V	M$_v$	Spec	RV	d(pc)	ADS	Notes
			h m s	° ' "	s	"								
72076	154332		8 30 35.6	-10 11 54	-0.001	-0.02	7.8	0.4	2.6	F0 V		110 s		
72195	175976		8 30 36.7	-27 50 04	-0.004	-0.02	7.83	0.38	3.0	F2 V		89 s		
72317	220024		8 30 38.4	-40 30 00	-0.002	0.00	7.8	1.4	-0.3	K5 III		420 s		
72350	220025		8 30 39.1	-44 44 14	-0.002	+0.01	6.30	-0.02	-1.6	B5 IV	+24	330 s		m
71705	42400		8 30 41.6	+46 44 58	-0.003	+0.01	7.4	0.5	3.4	F5 V		63 s		
72979	256520		8 30 43.2	-70 50 39	-0.005	-0.05	7.6	0.1	1.4	A2 V		140 mx		
71223	14572		8 30 44.5	+68 30 52	+0.002	-0.05	6.9	0.1	0.6	A0 V		180 s		
72118	154333		8 30 46.4	-11 48 00	-0.002	-0.01	7.6	1.1	0.2	K0 III		300 s		
72348	220026		8 30 46.5	-41 30 54	-0.003	+0.01	6.80	1.1	-0.1	K2 III		240 s		m
72267	199313		8 30 48.6	-31 08 05	-0.002	-0.01	7.2	-0.4	0.4	B9.5 V		230 s		
72300	199315		8 30 50.5	-33 38 41	-0.005	-0.01	7.1	0.4	2.6	F0 V		70 s		
70470	6549		8 30 52.1	+78 13 42	0.000	-0.03	7.1	0.9	3.2	G5 IV		62 s		
72173	154340		8 30 53.4	-15 50 03	-0.001	+0.01	6.8	1.1	-0.1	K2 III		240 s		
71988	97869		8 31 00.3	+18 58 06	-0.003	-0.01	7.4	0.1	1.7	A3 V		140 s		
72097	116850		8 31 02.4	+1 20 13	-0.003	-0.02	7.9	0.4	2.6	F0 V		110 s		
71844	42406		8 31 03.7	+40 03 19	-0.002	-0.02	7.0	0.1	1.7	A3 V		110 s		
72191	154342		8 31 03.8	-14 00 43	0.000	+0.01	7.9	0.1	1.7	A3 V		170 s		
71882	60863		8 31 04.3	+35 21 46	0.000	0.00	7.9	1.1	0.2	K0 III		340 s		
72245	175984		8 31 04.4	-20 18 15	-0.002	0.00	7.3	0.4	1.4	A2 V		100 s		
71867	60862		8 31 04.5	+38 17 28	+0.001	-0.03	7.5	1.4	-0.3	K5 III		360 s		
72882	250255		8 31 04.8	-66 51 56	-0.008	+0.03	8.0	0.4	2.6	F0 V		120 s		
72384	199324		8 31 05.8	-39 03 40	-0.002	0.00	7.1	0.0	0.4	B9.5 V		210 s		
71534	14576		8 31 07.2	+59 54 45	-0.003	-0.04	7.1	0.5	4.0	F8 V		42 s		
71658	26885		8 31 09.9	+54 57 35	+0.001	-0.01	7.6	0.1	0.6	A0 V		250 s		
71866	42408		8 31 10.7	+40 13 28	-0.001	-0.04	6.73	0.08		A0 p				
72485	220039		8 31 10.7	-47 52 00	0.000	0.00	6.39	-0.15		B4	+11			
71987	80225		8 31 15.9	+25 44 25	0.000	-0.03	7.5	0.1	1.4	A2 V		170 s		
72075	97875		8 31 16.2	+11 38 18	+0.001	-0.04	7.8	1.1	0.2	K0 III		330 s		
72484	--		8 31 16.2	-46 31 12			8.0	1.8	-0.3	K5 III		290 s		
--	--		8 31 16.7	-59 13 38			8.00B							X Car, v
72265	154353		8 31 16.9	-18 47 50	-0.001	-0.01	8.0	1.1	0.2	K0 III		360 s		
71906	60866		8 31 19.9	+37 15 56	-0.001	0.00	6.10	-0.03	-0.2	B8 V		170 s		
72881	250257		8 31 21.9	-65 31 22	+0.001	-0.01	7.5	-0.1	0.4	B9.5 V		260 s		
72539	220046		8 31 22.3	-48 44 57	-0.001	0.00	7.97	-0.07	0.6	A0 V		300 s		
72297	154358		8 31 23.9	-19 30 19	-0.001	-0.01	7.6	0.1	2.6	F0 V		100 s		
--	--		8 31 24.4	-36 41 42			7.70							m
72436	199329		8 31 24.5	-39 03 51	-0.002	0.00	6.28	-0.14	-1.1	B5 V n		290 s		m
72435	199328		8 31 25.8	-36 41 37	+0.005	-0.08	7.70	1.1	0.2	K0 III		240 mx		m
71222	6563		8 31 26.4	+71 01 27	-0.005	-0.04	7.8	0.5	4.0	F8 V		57 s		
71147	6562		8 31 27.6	+72 17 29	+0.003	0.00	7.8	1.1	0.2	K0 III		320 s		
72346	--		8 31 28.7	-25 41 08			7.70	0.1	0.6	A0 V		260 s	6863	m
72650	236055		8 31 29.5	-54 23 38	+0.002	-0.05	6.34	1.31		K2				
72041	80229	30 υ¹ Cnc	8 31 30.4	+24 04 52	-0.006	-0.05	5.75	0.28	2.6	F0 V	+19	43 s		
72310	154359		8 31 30.9	-19 34 39	-0.003	-0.01	5.42	-0.06		A0	+12	25 mn	6862	m
72095	97878		8 31 31.2	+17 20 08	+0.001	-0.03	7.8	1.1	-0.1	K2 III		380 s		
72332	175996		8 31 32.5	-21 04 49	+0.002	-0.09	7.8	0.5	3.2	G5 IV		85 s		
71974	60870		8 31 34.8	+34 57 58	-0.002	+0.01	7.50	0.9	3.2	G5 IV		72 s	6851	m
72094	97881	31 θ Cnc	8 31 35.7	+18 05 39	-0.004	-0.06	5.35	1.56	-0.3	K5 III	+45	120 s		m
72537	220048		8 31 36.5	-45 47 06	-0.003	0.00	7.1	0.3	0.0	B8.5 V		150 s		
72187	116857		8 31 37.3	+1 08 52	0.000	0.00	7.4	1.1	0.2	K0 III		270 s		
72843	250258		8 31 38.4	-62 20 20	-0.001	0.00	7.9	-0.4	0.0	B8.5 V		380 s		
72555	220051		8 31 39.8	-47 14 26	+0.001	+0.02	7.03	-0.14	-1.4	B4 V		490 s		
72696	236058		8 31 40.6	-55 48 03	-0.001	-0.02	8.0	1.0	0.2	K0 III		360 s		
72115	97884		8 31 41.3	+18 59 15	-0.001	-0.03	6.7	1.1	0.2	K0 III		200 s		
--	26888		8 31 42.8	+55 46 35	+0.002	0.00	7.9	1.6	0.2	K0 III		150 s		
71829	26891		8 31 43.0	+50 42 31	-0.003	-0.01	7.8	0.8	3.2	G5 IV		84 s		
73391	256522		8 31 44.6	-74 51 17	-0.002	0.00	6.7	1.7	0.2	K0 III		80 s		
72295	136008		8 31 45.6	-8 51 16	0.000	-0.02	7.8	0.1	0.6	A0 V		280 s		
71207	6564		8 31 53.5	+72 41 30	0.000	-0.02	7.9	1.4	-0.3	K5 III		430 s		
72208	116863		8 31 54.5	+9 48 52	0.000	-0.01	6.83	-0.07		A p	+8			
71881	26892		8 31 55.0	+50 37 01	-0.008	-0.33	7.43	0.63	4.5	G1 V	+17	36 s		
72434	176020		8 31 56.0	-26 42 10	-0.003	+0.05	7.9	0.3	0.4	B9.5 V		190 s		
72514	199340		8 31 56.9	-39 03 59	-0.001	-0.01	7.2	0.4	0.4	B9.5 V		130 s		
72379	154363		8 31 59.0	-18 23 27	-0.004	-0.15	7.1	0.8	3.2	G5 IV		59 s		
72774	236063		8 31 59.4	-56 22 17	0.000	-0.02	7.3	1.7	0.2	K0 III		100 s		
72467	176026		8 32 01.9	-29 22 17	+0.001	-0.16	7.8	0.4	0.2	K0 III		150 mx		
72631	220059		8 32 01.9	-47 19 43	+0.001	-0.03	7.98	1.72	-0.3	K5 III		330 s		
72480	199339		8 32 03.0	-31 42 38	-0.002	+0.03	7.71	-0.10	-1.0	B5.5 V		540 s		
72737	236062		8 32 04.8	-53 12 44	-0.003	+0.01	5.69	0.58	0.2	K0 III	+19	130 s		
72573	199341		8 32 09.0	-39 57 11	-0.002	+0.01	7.37	-0.04	0.6	A0 V		230 s		
72256	97890		8 32 11.3	+12 48 10	0.000	-0.01	7.26	1.15		K0				
72114	60881		8 32 14.2	+32 10 15	0.000	-0.01	7.6	0.4	3.0	F2 V		85 s		
73744	256525		8 32 16.3	-76 55 40	-0.050	+0.18	7.61	0.60	4.4	G0 V		42 s		
72611	220062		8 32 16.8	-41 49 57	-0.003	+0.02	7.1	0.2	0.6	A0 V		110 s		
72146	80232		8 32 17.3	+29 19 09	-0.001	-0.02	7.1	0.8	3.2	G5 IV	+2	59 s		

HD	SAO	Star Name	α 2000	δ 2000	μ(α)	μ(δ)	V	B−V	M$_v$	Spec	RV	d(pc)	ADS	Notes
			h m s	° ′ ″	s	″								
72363	136012		8 32 18.2	−5 06 29	−0.002	+0.01	7.1	1.1	0.2	K0 III		240 s		
72648	220063		8 32 18.9	−43 55 53	0.000	0.00	7.62	0.13	−1.6	B3.5 V		490 s		m
72754	220069		8 32 23.3	−49 36 07	−0.001	−0.02	6.89	0.19						
72361	116871		8 32 27.8	+3 40 23	−0.001	−0.04	7.9	0.1	1.4	A2 V		200 s		
72820	236069		8 32 28.7	−55 24 33	+0.002	0.00	7.9	−0.5	0.6	A0 V		290 s		
72570	199344		8 32 29.7	−32 14 02	−0.001	−0.01	7.5	0.6	0.2	K0 III		290 s		
72647	220066		8 32 30.9	−40 12 38	−0.022	+0.16	7.9	0.6	3.2	G5 IV		50 mx		
72462	154373		8 32 33.2	−15 01 46	−0.004	+0.05	6.38	0.27		A5				m
71952	26896		8 32 33.4	+53 06 53	+0.001	−0.08	6.24	1.01	0.2	gK0	+44	160 s		
72412	136017		8 32 35.3	−5 53 58	+0.001	0.00	6.8	0.9	3.2	G5 IV		52 s		
72858	236073		8 32 35.4	−55 07 24	−0.004	+0.02	7.35	−0.02	0.4	B9.5 V		240 s		
72360	116874		8 32 37.1	+5 43 03	−0.001	+0.02	7.40	1.1	0.2	K0 III		280 s		
72474	154374		8 32 37.5	−14 25 05	−0.002	0.00	6.7	1.1	0.2	K0 III		200 s		
72430	136018		8 32 38.1	−3 10 44	−0.004	0.00	7.3	0.1	1.4	A2 V		150 mx		
72306	97900		8 32 38.2	+15 44 21	0.000	−0.02	7.8	1.1	0.2	K0 III		330 s		
72359	97902	34 Cnc	8 32 39.8	+10 03 58	0.000	−0.01	6.30	−0.02	0.6	A0 V	+4	140 s		
72752	220080		8 32 40.5	−47 02 30	−0.001	+0.02	6.49	1.29	−0.1	K2 III		170 s		
72878	236075		8 32 41.1	−55 28 12	0.000	0.00	7.50	0.1	0.6	A0 V		240 s		m
73468	256524		8 32 42.3	−73 21 23	−0.008	+0.08	6.12	0.95		K0				
72292	80243	33 η Cnc	8 32 42.4	+20 26 28	−0.003	−0.05	5.33	1.25	−0.2	K3 III	+24	130 s		
72645	199350		8 32 42.9	−37 28 11	−0.003	0.00	7.9	0.7	3.2	G5 IV		88 s		
72325	97901		8 32 43.4	+15 16 28	0.000	−0.01	7.5	1.1	−0.1	K2 III		330 s		
72132	42419		8 32 43.5	+42 08 23	0.000	−0.02	7.7	0.9	0.3	G5 III	−28	310 s		
73373	256523		8 32 44.2	−71 47 42	−0.008	+0.03	7.9	0.0	0.4	B9.5 V		180 mx		
72628	199349		8 32 46.7	−33 33 33	−0.001	0.00	7.9	1.1	0.2	K0 III		320 s		
72673	199352		8 32 51.4	−31 30 03	−0.087	+0.76	6.38	0.79	5.9	K0 V	+18	13 ts		
71553	14582		8 32 53.4	+69 19 11	+0.006	−0.03	6.31	1.35	0.2	K0 III	−30	100 s		
72800	220085		8 32 53.9	−47 36 21	+0.001	−0.01	6.61	0.14	−6.5	B9 I		3300 s		
72237	60891		8 32 54.2	+34 23 03	−0.003	−0.10	7.7	1.1	−0.1	K2 III		120 mx		
72690	199355		8 32 54.3	−38 10 31	−0.001	+0.01	7.90	1.1	0.2	K0 III		340 s		m
72184	60890		8 32 54.9	+38 00 59	−0.009	−0.17	5.90	1.11	−0.1	K2 III	+15	70 mx		
72838	220087		8 32 56.2	−49 09 18	−0.002	0.00	7.30	1.81	−0.3	K5 III		230 s		
72688	199353		8 32 58.5	−34 38 02	−0.001	0.00	6.36	0.95		K0				
72324	80245	32 υ² Cnc	8 33 00.0	+24 05 05	−0.005	−0.05	6.36	1.02	0.2	G9 III	+75	150 s		
72798	220086		8 33 01.6	−45 45 10	−0.003	+0.01	6.45	−0.14	−2.2	B5 III		540 s		m
71704	14585		8 33 04.7	+67 17 35	−0.007	−0.05	7.8	1.1	0.2	K0 III	+6	310 mx		
72626	176061		8 33 04.8	−24 36 23	−0.001	−0.02	6.19	0.27		A5	−8	16 mn	6871	m
72429	97905		8 33 04.9	+11 16 05	−0.004	−0.04	7.8	0.7	4.4	G0 V		49 s		
72003	26899		8 33 05.2	+55 21 16	−0.003	−0.08	8.00						6858	m
72976	236081		8 33 05.6	−56 23 47	−0.001	0.00	7.51	−0.11	0.6	A0 V		240 s		
72568	154383		8 33 09.1	−15 45 53	−0.002	−0.01	7.5	1.1	0.2	K0 III		280 s		
72816	220090		8 33 09.6	−43 41 03	−0.001	+0.04	7.11	0.99	0.2	K0 III		240 s		
72625	176063		8 33 09.7	−22 19 41	+0.001	0.00	8.0	1.3	−0.1	K2 III		360 s		
72280	60893		8 33 10.4	+33 25 52	+0.001	0.00	7.70	0.1	0.6	A0 V		260 s	6866	m
72506	136026		8 33 11.0	−5 13 21	+0.001	−0.01	6.6	0.1	1.4	A2 V		110 s		
72787	199363		8 33 19.9	−38 22 16	0.000	−0.01	6.49	−0.18	−1.4	dB4	+5	380 s		
72600	199361		8 33 20.3	−36 39 41	−0.002	−0.02	7.65	−0.04	−1.7	B3 V		640 s		
72589	154388		8 33 20.5	−12 44 12	0.000	+0.01	7.9	0.1	1.7	A3 V		170 s		
72235	42425		8 33 21.6	+41 45 28	+0.004	−0.01	7.6	0.9	3.2	G5 IV		77 s		
72291	60896	32 Lyn	8 33 21.7	+36 26 11	−0.012	0.00	6.24	0.36	3.0	F2 V	0	38 ts		
72771	199362		8 33 24.1	−34 38 56	+0.001	−0.01	7.86	−0.09	−1.4	B4 V		710 s		
72526	136030		8 33 25.2	−0 18 31	−0.001	−0.03	7.9	0.0	0.0	B8.5 V		370 s		
73406	—		8 33 27.6	−70 06 42			7.6	1.4	−0.3	K5 III		380 s		
72834	220098		8 33 28.3	−41 36 59	0.000	−0.14	7.0	1.2	3.2	G5 IV		31 s		
72900	220103		8 33 30.3	−46 58 16	0.000	−0.01	6.24	1.56		K2				
72899	220102		8 33 31.2	−46 14 52	−0.003	0.00	7.62	1.03	−0.1	K2 III		350 s		
72565	136035		8 33 31.5	−2 58 39	−0.001	−0.03	7.9	1.4	−0.3	K5 III		430 s		
72814	199366		8 33 32.0	−38 12 16	0.000	−0.03	7.8	1.3	−0.3	K5 III		410 s		
72665	154395		8 33 33.4	−16 06 35	−0.001	0.00	6.6	1.1	0.2	K0 III		190 s		
72562	136036		8 33 35.3	−0 45 03	−0.002	−0.05	8.0	1.1	0.2	K0 III		360 s		
74343	256526		8 33 36.0	−80 09 30	−0.019	+0.01	7.0	1.9	−0.3	K5 III		180 s		
72341	60899		8 33 36.5	+35 47 53	+0.001	−0.04	7.9	1.1	0.2	K0 III		340 s		
72490	97910		8 33 36.6	+13 33 03	+0.002	−0.02	7.6	0.9	3.2	G5 IV		77 s		
72832	199370		8 33 38.2	−38 50 56	−0.003	0.00	5.96	−0.14		B8				
72750	176078		8 33 38.3	−25 44 41	0.000	−0.02	8.0	0.9	0.2	K0 III		360 s		
72561	116890		8 33 43.5	+4 45 24	0.000	−0.02	5.87	1.07	0.3	gG5	+1	95 s		
72505	97913		8 33 45.0	+13 15 26	−0.002	−0.05	6.28	1.17		K0	+28			
72769	176080		8 33 45.9	−23 21 18	−0.021	+0.18	7.21	0.74	5.2	dG5	+18	23 s		
72664	136040		8 33 46.7	−9 59 53	−0.001	+0.01	7.6	1.6	−0.5	M2 III		420 s		
72831	199371		8 33 53.1	−33 29 28	−0.002	−0.05	8.0	0.7	0.6	A0 V		100 s		
72605	116891		8 33 53.8	+1 35 05	0.000	−0.03	7.70	0.1	0.6	A0 V		260 s	6874	m
72113	26907		8 33 55.1	+58 36 25	−0.001	−0.04	6.8	0.1	0.6	A0 V		170 s		
72766	176085		8 33 55.6	−22 02 01	−0.003	+0.01	8.0	0.4	1.4	A2 V		120 s		
72583	97915		8 33 58.0	+10 35 11	−0.003	−0.02	7.60	0.4	2.6	F0 V		100 s	6873	m
73064	236087		8 33 58.2	−52 09 13	−0.001	+0.01	8.0	0.2	1.4	A2 V		160 s		

HD	SAO	Star Name	α 2000	δ 2000	μ(α)	μ(δ)	V	B-V	M_v	Spec	RV	d(pc)	ADS	Notes
72795	176089		8^{h}34^{m}01^s.5	-25°05'56"	-0^s.001	+0".01	7.0	0.2	1.4	A2 V		110 s		
72660	136044		8 34 01.5	-2 09 05	-0.003	+0.02	5.81	0.00		A0				
72958	220113		8 34 04.1	-41 40 45	-0.002	0.00	6.8	1.4	-0.1	K2 III		250 s		
72503	80252		8 34 06.5	+25 14 01	-0.003	-0.01	8.01	0.22	1.7	A3 V		160 s		
72997	220117		8 34 07.8	-44 32 41	-0.003	0.00	7.6	0.0	-1.0	B5.5 V		530 s		
72746	136051		8 34 09.3	-9 57 09	-0.001	+0.01	7.5	0.4	3.0	F2 V		78 s		
73105	236091		8 34 09.8	-53 04 15	+0.001	+0.02	6.80	-0.11	-1.0	B5.5 V		360 s		
73010	220121		8 34 10.8	-45 38 11	-0.004	+0.01	7.5	0.1	0.0	B8.5 V		220 mx		
72957	199383		8 34 11.1	-39 43 45	+0.002	+0.03	6.9	1.3	-0.1	K2 III		210 s		
72936	199381		8 34 11.2	-36 54 22	-0.008	+0.06	6.70	0.40	3.0	F2 V		53 s		
72183	26911		8 34 12.3	+56 55 24	-0.002	-0.03	8.00	0.1	0.6	A0 V		300 s	6867	m
72996	220118		8 34 12.6	-42 06 08	-0.003	-0.01	6.9	1.4	0.2	K0 III		180 s		
72617	116894		8 34 13.3	+8 27 07	0.000	-0.03	6.03	0.33		F0	+16			m
72356	42432		8 34 15.8	+45 11 45	-0.003	-0.05	7.9	0.4	2.6	F0 V		110 s		
72604	97918		8 34 23.3	+18 23 51	-0.001	-0.01	7.6	1.1	-0.1	K2 III		340 s		
73169	236094		8 34 24.3	-54 40 02	-0.002	+0.01	8.0	0.2	0.4	B9.5 V		230 s		
75001	258500		8 34 24.5	-82 34 00	-0.016	+0.06	6.9	0.6	2.1	A5 V		48 s		
72869	176098		8 34 25.7	-23 49 38	-0.004	-0.06	7.8	0.3	2.1	A5 V		71 mx		
73077	220127		8 34 26.4	-47 07 34	-0.003	+0.02	8.01	1.12		K0				
72955	199386		8 34 26.6	-34 11 34	-0.001	+0.02	7.7	1.6	-0.1	K2 III		210 s		
73127	236093		8 34 28.9	-51 05 36	0.000	-0.01	6.59	-0.15	-1.1	B5 V n		340 s		
72993	199389		8 34 29.2	-37 36 41	-0.001	0.00	6.40	1.4	-0.3	K5 III		220 s		m
72913	176103		8 34 29.4	-27 05 54	-0.005	+0.01	6.6	1.4	-0.1	K2 III		170 s		
72392	42433		8 34 29.7	+47 08 22	+0.001	0.00	6.70	0.01		A0	-19			
72760	136057		8 34 31.6	-0 43 34	-0.011	+0.01	7.5	0.8	3.2	G5 IV		72 s		
72954	199388		8 34 31.7	-32 35 55	-0.004	-0.14	6.43	0.75	4.2	G5 IV-V		27 s		m
72722	116903		8 34 33.3	+5 30 12	+0.001	-0.02	6.9	1.4	-0.3	K5 III		270 s		
73075	220129		8 34 35.7	-44 16 13	0.000	-0.03	7.3	0.5	0.2	K0 III		260 s		
72037	14590	2 UMa	8 34 36.2	+65 08 42	-0.008	-0.07	5.47	0.18		A m	-16	22 mn		
73155	220138		8 34 43.7	-49 56 39	0.000	+0.02	5.01	1.33		K0	+4	24 mn		
72524	60907	33 Lyn	8 34 43.8	+36 25 10	-0.003	-0.05	5.80	0.04	-0.2	A2 III	+25	160 s		
72759	116908		8 34 45.0	+1 29 14	-0.003	+0.01	7.8	1.1	0.2	K0 III		330 s		
73041	199392		8 34 47.2	-36 58 56	-0.003	+0.04	7.94	1.34	-0.3	K5 III		450 s		
72932	—		8 34 50.5	-21 27 30			7.9	0.5	3.2	G5 IV		89 s		
73308	236099		8 34 53.0	-58 47 47	+0.004	-0.08	7.8	0.3	2.6	F0 V		110 s		
72441	42434		8 34 53.5	+48 32 01	-0.003	-0.03	6.8	0.1	0.6	A0 V		170 s		
71703	6579		8 34 55.2	+73 28 47	+0.018	+0.02	7.4	1.1	0.2	K0 III		180 mx		
71828	6583		8 34 58.2	+72 00 14	-0.002	-0.06	7.9	0.1	1.7	A3 V		73 mx		
73054	199397		8 34 58.8	-37 20 33	-0.002	0.00	7.94	0.16	2.1	A5 V		150 s		
72543	42436		8 35 02.8	+41 01 26	-0.005	-0.02	7.6	0.8	3.2	G5 IV		76 s		
72780	97927		8 35 04.1	+11 17 01	-0.004	-0.06	7.5	0.5	4.0	F8 V		50 s		
72991	176121		8 35 05.7	-22 15 17	+0.004	-0.05	7.5	1.1	0.2	K0 III		250 s		
72848	116915		8 35 09.3	+0 21 43	+0.002	-0.06	7.0	0.9	3.2	G5 IV		58 s		
73287	236100		8 35 10.0	-54 12 21	-0.003	+0.02	7.07	-0.11		B9				
73089	199398		8 35 11.3	-34 01 27	-0.002	+0.04	6.8	1.3	-0.1	K2 III		190 s		
73121	199402		8 35 12.6	-39 58 12	+0.007	+0.05	6.47	0.58	4.4	dG0		26 s		
72458	26917		8 35 13.4	+52 12 00	-0.002	-0.02	7.7	0.9	3.2	G5 IV		81 s		
73119	199403		8 35 14.4	-39 40 54	+0.002	-0.01	7.63	1.64	-0.3	K5 III		310 s		
73073	199399		8 35 14.8	-31 17 07	-0.003	+0.02	7.6	1.5	0.2	K0 III		160 s		
73390	236105		8 35 15.4	-58 13 29	-0.004	0.00	5.26	-0.14	-1.7	B3 V n	+28	240 s		
73118	199405		8 35 15.7	-39 15 13	-0.003	0.00	7.7	1.6	-0.3	K5 III		340 s		
73007	154429		8 35 19.2	-19 02 29	+0.001	-0.01	7.8	1.4	-0.3	K5 III		410 s		
72779	97928	35 Cnc	8 35 19.4	+19 35 24	-0.003	-0.01	6.58	0.68	0.6	G0 III	+36	150 s		
73389	236106		8 35 19.6	-58 00 33	+0.005	+0.02	4.86	1.00	0.3	gG6	+24	70 s		
73273	236102		8 35 21.9	-51 49 13	-0.003	+0.01	7.57	1.13	0.2	K0 III		240 s		
72908	116920		8 35 24.9	+2 44 37	-0.001	0.00	6.33	1.02	0.2	G9 III	-4	150 s		
73244	220154		8 35 26.1	-47 34 18	-0.003	0.00	7.04	1.35		K5				
72968	136076	3 Hya	8 35 28.1	-7 58 56	-0.002	+0.02	5.72	-0.03		A p	+24			
73072	176131		8 35 28.7	-26 50 37	-0.002	+0.01	5.96	0.37		A2				
73088	176133		8 35 29.7	-29 31 27	0.000	-0.01	7.4	0.1	0.6	A0 V		200 s		
73369	236107		8 35 29.7	-55 06 12	-0.003	-0.02	7.9	0.0	0.4	B9.5 V		290 s		
72558	42438		8 35 33.0	+49 15 29	+0.001	-0.02	7.8	0.4	2.6	F0 V		110 s		
73286	220158		8 35 36.4	-48 32 53	-0.002	+0.01	7.9	0.1	0.4	B9.5 V		250 s		
73218	220155		8 35 37.0	-42 33 39	-0.003	0.00	7.46	0.43	3.4	F5 V		65 s		
73341	236108		8 35 41.0	-52 19 21	0.000	0.00	7.4	1.5	-0.5	M4 III		390 s		
73151	199408		8 35 42.9	-32 12 51	+0.003	-0.02	7.9	2.1	-0.3	K5 III		180 s		
73217	220157		8 35 43.3	-40 52 07	0.000	-0.04	7.38	0.51	3.4	F5 V		45 s		
73116	176139		8 35 46.3	-28 28 42	0.000	-0.02	7.1	1.4	0.2	K0 III		130 s		
74305	256529		8 35 47.7	-77 00 05	-0.007	+0.02	7.7	1.4	-0.3	K5 III		400 s		
72945	116929		8 35 50.9	+6 37 12	-0.009	-0.15	5.63	0.56	3.7	dF6	+25	22 ts	6886	m
73048	154434		8 35 51.8	-11 17 31	-0.001	-0.05	7.5	1.6	-0.5	M2 III		400 s		
73340	236110		8 35 52.1	-50 58 12	-0.001	0.00	5.80	-0.13		B9 p				
73465	236112		8 35 52.4	-56 22 24	-0.015	+0.06	7.1	0.1	3.0	F2 V		67 s		
73326	220167		8 36 02.1	-46 30 07	-0.002	-0.01	7.29	-0.02	-1.0	B5.5 V		450 s		
72884	80275		8 36 02.4	+24 03 03	0.000	-0.01	6.9	0.1	0.6	A0 V		180 s		

HD	SAO	Star Name	α 2000	δ 2000	μ(α)	μ(δ)	V	B−V	M$_V$	Spec	RV	d(pc)	ADS	Notes
			h m s	° ′ ″	s	″								
73019	––		8 36 05.4	+2 02 56	−0.001	0.00	7.8	0.0	0.4	B9.5 V		300 s		
72943	97950		8 36 07.7	+15 18 49	+0.001	−0.03	6.3	0.2	2.1	A5 V	+4	69 s		
73268	199416		8 36 09.9	−36 15 49	−0.004	+0.03	7.5	0.8	0.6	A0 V		74 s		
73504	236117		8 36 10.3	−56 42 45	−0.001	−0.05	7.2	0.0	3.4	F5 V		59 s		
72965	97952		8 36 10.9	+13 46 38	+0.001	−0.01	7.80	0.1	0.6	A0 V		280 s		m
73215	199414		8 36 14.0	−30 35 52	−0.002	−0.01	7.9	0.2	−1.0	B5.5 V		400 s		
72778	42443		8 36 15.6	+42 34 46	+0.002	−0.02	7.0	0.1	1.4	A2 V	−30	130 s		
73137	154442		8 36 15.6	−15 07 27	−0.001	+0.01	8.0	1.1	−0.1	K2 III		410 s		
73241	199417		8 36 16.7	−31 20 14	−0.001	+0.04	7.59	−0.10	0.0	B8.5 V		330 s		
73781	250280		8 36 19.5	−64 39 33	+0.008	−0.02	8.0	0.1	0.6	A0 V		150 mx		
73990	256527		8 36 20.9	−70 40 36	−0.001	0.00	7.0	0.1	0.4	B9.5 V		170 s		
74046	256528		8 36 21.7	−71 39 44	−0.003	−0.05	8.0	1.1	0.2	K0 III		360 s		
73257	199419		8 36 21.7	−32 15 41	−0.005	+0.05	6.7	1.7	0.2	K0 III		79 s		
72984	97956		8 36 21.8	+16 19 03	−0.001	−0.01	8.0	0.5	4.0	F8 V		62 s		
73301	199421		8 36 23.0	−38 27 23	+0.001	+0.02	7.1	0.6	3.4	F5 V		47 s		
73145	154443		8 36 23.0	−13 09 04	−0.001	+0.03	7.9	0.4	3.0	F2 V		94 s		
73238	176158		8 36 23.5	−28 04 55	−0.007	+0.02	7.9	1.1	−0.1	K2 III		210 mx		
73741	250281		8 36 31.3	−61 57 47	−0.002	+0.01	8.0	0.4	1.4	A2 V		120 s		
73178	136086		8 36 39.5	−9 42 53	−0.001	−0.02	7.5	1.1	0.2	K0 III		290 s		
73000	80284		8 36 39.9	+22 10 23	−0.001	−0.02	8.00	0.6	4.4	G0 V		53 s		m
72861	60918		8 36 40.4	+39 46 14	−0.001	−0.02	7.5	1.1	0.2	K0 III		290 s		
73701	250282		8 36 40.5	−60 19 39	0.000	0.00	7.86	−0.04		B9				
72777	42444		8 36 40.9	+49 22 35	+0.002	0.00	7.3	1.6	−0.5	M2 III		360 s		
73461	220181		8 36 40.9	−47 29 51	0.000	0.00	7.38	0.32		A5				m
72982	80283		8 36 42.8	+25 56 14	+0.001	−0.01	8.0	0.1	0.6	A0 V		300 s		
73478	220182		8 36 42.8	−47 59 55	+0.001	0.00	7.37	−0.10	0.0	B8.5 V		300 s		
73569	236123		8 36 45.7	−52 30 57	−0.001	0.00	7.3	0.7	2.6	F0 V		49 s		
71973	6592		8 36 48.7	+74 43 24	−0.004	−0.03	6.30	0.2	2.1	A5 V	−6	69 s	6872	m
––	––		8 36 48.9	+48 20 37	−0.001	0.00	7.4	1.1	0.2	K0 III		280 s		
73383	199424		8 36 51.1	−35 05 28	−0.003	−0.04	7.9	−0.3	0.6	A0 V		280 s		
73364	199426		8 36 57.6	−31 43 21	−0.003	0.00	8.0	0.7	0.2	K0 III		360 s		
73266	154459		8 36 57.7	−18 40 50	0.000	−0.03	7.1	0.8	3.2	G5 IV		59 s		
73502	220186		8 37 01.1	−44 06 52	−0.002	0.00	6.99	1.24	0.2	K0 III		160 s		RZ Vel, v
72720	26924		8 37 03.5	+55 50 21	−0.003	0.00	8.0	1.1	−0.1	K2 III		410 s		
73335	176170		8 37 03.6	−26 24 59	−0.002	−0.03	6.7	1.2	0.2	K0 III		160 s		
73501	220189		8 37 04.2	−43 48 42	−0.002	+0.01	7.57	0.89	3.2	G5 IV		61 s		m
73143	116953	36 Cnc	8 37 05.7	+9 39 19	−0.002	−0.01	6.00	0.09	1.7	A3 V	+17	72 s		
73363	176173		8 37 08.2	−26 04 23	−0.004	0.00	7.7	1.2	−0.1	K2 III		340 mx		
73381	199428		8 37 09.3	−30 39 44	0.000	+0.03	7.9	0.5	0.4	B9.5 V		140 s		
73609	220195		8 37 09.5	−49 25 31	−0.002	0.00	7.40	0.1	0.6	A0 V		230 s		m
73282	154461		8 37 09.6	−16 01 11	0.000	0.00	7.7	1.1	0.2	K0 III		320 s		
72338	14600		8 37 10.0	+69 41 54	−0.003	−0.05	7.1	1.1	0.2	K0 III		240 s		
73681	236125		8 37 10.5	−53 15 32	−0.004	+0.03	7.87	0.08	1.2	A1 V		150 mx		
71827	6590		8 37 13.9	+77 02 48	+0.004	+0.06	7.5	0.5	4.0	F8 V		49 s		
73525	220192		8 37 18.4	−40 41 30	−0.003	0.00	7.29	0.60	0.6	A0 V		89 s		
73887	250288		8 37 19.0	−62 51 12	0.000	−0.02	5.47	1.02	0.3	gG8	+21	95 s		m
73524	199438		8 37 19.8	−40 08 51	−0.027	+0.03	6.55	0.60	4.1	G4 IV–V	0	32 s		
73080	80292		8 37 22.1	+28 17 40	+0.001	−0.05	6.7	0.9	3.2	G5 IV	−27	50 s		
73499	199434		8 37 22.7	−36 08 13	−0.005	−0.01	7.6	1.5	−0.3	K5 III		200 mx		
72792	26928		8 37 25.8	+55 59 09	−0.003	−0.03	7.7	0.4	2.6	F0 V		100 s		
73281	136103		8 37 27.1	−4 56 01	−0.001	+0.02	6.19	1.05		K0				
73522	––		8 37 29.5	−37 41 52			7.9	1.5	−0.3	K5 III		430 s		
73476	199436		8 37 29.6	−33 44 45	−0.002	−0.05	6.48	0.33		A5				m
73438	176183		8 37 31.0	−28 10 43	−0.005	−0.06	7.9	0.3	0.2	K0 III		140 mx		
73457	199435		8 37 32.0	−30 29 55	−0.001	+0.01	8.0	−0.3	0.4	B9.5 V		330 s		
73332	154464		8 37 33.8	−12 35 10	−0.004	+0.01	6.6	0.4	2.6	F0 V		63 s		
73634	220204		8 37 38.6	−42 59 21	−0.001	+0.01	4.14	0.11	−5.8	A9 I–II	+19	950 s		
73262	116965	4 δ Hya	8 37 39.3	+5 42 13	−0.005	−0.01	4.16	0.00	0.6	A0 V	+11	43 ts		m
73588	220203		8 37 39.7	−40 26 10	0.000	+0.01	7.0	1.6	−0.5	M2 III		320 s		
73658	220206		8 37 40.0	−46 16 58	0.000	−0.01	6.88	0.05	−1.0	B5.5 V		340 s		
73814	236134		8 37 42.8	−54 59 39	−0.001	+0.01	7.2	2.0	−0.3	K5 III		160 s		
72923	26933		8 37 43.4	+52 01 51	0.000	−0.11	7.7	1.1	0.2	K0 III		110 mx		
74623	256534		8 37 46.6	−76 44 18	+0.001	+0.02	7.6	1.1	0.2	K0 III		300 s		
73210	97976		8 37 46.7	+19 16 02	−0.002	−0.01	6.76	0.19	2.1	A5 V	+28	82 s		
73657	220207		8 37 47.9	−42 19 12	−0.002	+0.01	7.84	1.21	0.2	K0 III		270 s		
73378	154469		8 37 48.0	−12 32 12	−0.001	0.00	7.9	0.4	3.0	F2 V		94 s		
73133	60932		8 37 49.2	+32 17 18	−0.001	−0.01	7.9	0.1	1.4	A2 V		200 s		
73227	97978		8 37 49.7	+14 28 21	+0.001	−0.03	7.8	1.1	0.2	K0 III		330 s		
73350	136111		8 37 50.3	−6 48 26	−0.019	+0.03	6.77	0.65		G0			6900	m
73495	176189	η Pyx	8 37 52.1	−26 15 18	−0.002	−0.02	5.27	−0.04		A0	+31	15 mn		m
73414	154470		8 37 53.0	−12 42 22	−0.001	−0.01	7.2	1.1	0.2	K0 III		250 s		
73456	176186		8 37 54.7	−20 18 44	−0.004	+0.01	8.0	1.6	−0.1	K2 III		240 s		
73632	220209		8 37 55.1	−40 22 24	−0.001	+0.01	7.9	1.0	−0.1	K2 III		390 s		
74461	256532		8 37 56.0	−74 34 30	−0.005	−0.05	6.7	0.6	2.1	A5 V		48 s		
73094	60930		8 37 56.1	+39 38 40	0.000	−0.03	7.5	0.1	1.4	A2 V		160 s		

HD	SAO	Star Name	α 2000	δ 2000	μ(α)	μ(δ)	V	B-V	M$_V$	Spec	RV	d(pc)	ADS	Notes
			h m s	o ' "	s	"								
73776	220217		8 37 56.6	-50 04 49	-0.002	+0.04	7.68	1.09	0.2	K0 III		270 s		
73400	136112		8 37 57.5	-8 52 54	-0.011	-0.03	7.2	0.5	3.4	F5 V		58 s		
73399	136113		8 37 59.0	-8 44 06	+0.003	+0.05	8.0	0.9	3.2	G5 IV		93 s		
73280	97981		8 37 59.3	+11 26 59	0.000	-0.02	8.0	0.1	1.4	A2 V		210 s		
73261	97980		8 38 00.6	+15 08 05	-0.004	-0.06	7.8	0.7	4.4	G0 V		49 s		
73739	220216		8 38 00.8	-46 54 14	+0.001	+0.01	7.42	1.58		Ma				
73431	154475		8 38 01.7	-11 44 38	0.000	-0.02	6.5	0.1	0.6	A0 V		160 s		
73316	116975	37 Cnc	8 38 05.1	+9 34 29	-0.003	-0.01	6.53	-0.02		A0	+28			
73653	199447		8 38 05.7	-38 00 33	0.000	-0.02	7.21	-0.17	0.0	B8.5 V		280 s		
73226	80303		8 38 08.4	+26 02 55	-0.010	-0.19	7.56	0.62	4.7	dG2	+24	36 s		
73699	199448		8 38 09.0	-40 04 53	0.000	-0.01	7.59	0.05	-1.0	B5.5 V		460 s		
73696	199451		8 38 11.3	-39 32 10	+0.001	-0.01	7.9	1.9	0.0	K1 III		130 s		
73413	136116		8 38 12.4	-1 52 07	-0.003	+0.03	7.8	0.8	3.2	G5 IV		84 s		
73757	220222		8 38 13.4	-45 35 36	-0.002	+0.02	7.78	1.03	-0.1	K2 III		380 s		
73586	176208		8 38 15.4	-26 38 29	-0.005	0.00	8.0	1.2	0.2	K0 III		230 mx		
73172	60937		8 38 17.6	+37 01 13	-0.001	-0.04	7.5	0.4	2.6	F0 V		97 s		
73192	60939		8 38 18.9	+32 48 06	-0.002	-0.01	5.94	1.12	0.2	K0 III	+4	120 s		
73604	176209		8 38 19.1	-26 26 04	-0.001	-0.06	8.0	1.6	-0.3	K5 III		290 mx		
73451	136117		8 38 20.3	-6 39 45	0.000	-0.01	6.51	0.44		A2				
73412	116983		8 38 20.4	+0 41 27	-0.003	-0.01	7.2	1.1	0.2	K0 III	-10	250 s		
73516	154479		8 38 20.6	-15 16 31	-0.003	0.00	8.0	1.1	0.2	K0 III		360 s		
73017	26935		8 38 22.1	+53 24 05	-0.009	-0.02	5.66	0.96	3.2	G5 IV	-43	24 s		
73294	80308		8 38 23.1	+20 12 26	+0.001	-0.01	7.84	0.48		F5	-9			
73904	236140		8 38 23.9	-53 43 17	-0.002	+0.04	7.67	0.08		A0				
73813	220224		8 38 24.8	-46 46 38	-0.004	0.00	7.59	-0.02		B9				m
74088	250290		8 38 27.8	-62 50 36	+0.001	-0.01	6.5	3.2	-0.3	K5 III		23 s		
73346	97990		8 38 28.9	+17 03 35	-0.001	0.00	8.0	0.5	3.4	F5 V		82 s		
73347	97992		8 38 29.0	+14 51 28	+0.001	-0.04	7.7	0.4	2.6	F0 V		110 s		
74543	256536		8 38 37.2	-74 04 27	+0.007	+0.15	6.82	1.04	3.2	K0 IV		47 s		
73279	60945		8 38 38.1	+30 42 45	0.000	0.00	7.6	1.1	-0.1	K2 III		350 s		
73603	154492		8 38 40.2	-19 44 14	0.000	-0.02	6.33	1.59		K5			6903	m
73884	220233		8 38 43.1	-47 47 16	-0.002	0.00	7.85	1.86		K2				
73921	236141		8 38 44.1	-51 16 27	0.000	+0.01	7.77	0.89	3.2	G5 IV		67 s		
73952	236142		8 38 44.9	-53 05 26	-0.001	+0.01	6.46	-0.10	0.2	B9 V		180 s		m
73471	116988	5 σ Hya	8 38 45.4	+3 20 29	-0.001	-0.02	4.44	1.21	-0.1	K2 III	+25	75 s		
73344	80310		8 38 45.5	+23 41 09	-0.003	-0.16	6.90	0.54		F8				
73968	236143		8 38 47.9	-53 31 21	0.000	+0.02	7.1	2.1	-0.5	M2 III		170 s		
73601	136129		8 38 53.2	-10 01 55	0.000	-0.07	8.0	1.1	0.2	K0 III		220 mx		
73580	136130		8 38 56.3	-6 50 54	-0.001	-0.01	7.10	0.1	0.6	A0 V		200 s		m
73131	26940		8 38 59.9	+52 55 30	-0.002	-0.04	6.5	1.1	0.2	K0 III	+39	180 s		
74045	236148		8 39 04.2	-55 56 32	-0.001	+0.01	7.90	0.5	3.4	F5 V		79 s		m
74405	256535	θ Vol	8 39 05.3	-70 23 12	+0.004	-0.05	5.20	0.01		A0	+13	22 mn		m
73449	97999		8 39 06.1	+19 40 36	-0.002	-0.02	7.45	0.26	2.5	A9 V n	+30	98 s		m
73752	176226		8 39 07.9	-22 39 42	-0.017	+0.43	5.05	0.73	4.0	F8 V comp	+43	18 ts	6914	m
---	220236		8 39 09.4	-40 25 09	-0.001	+0.01	7.1	0.4						
73882	---		8 39 09.9	-40 25 04			7.19	0.40	-5.2	O8 V		2500 s		m
73029	14612		8 39 10.2	+59 56 21	-0.003	-0.04	6.4	0.1	0.6	A0 V	-14	150 s		
72790	14608		8 39 11.6	+68 57 11	+0.003	-0.01	8.0	1.1	0.2	K0 III		360 s		
72905	14609	3 π¹ UMa	8 39 11.7	+65 01 15	-0.004	+0.09	5.64	0.62	4.4	G0 V	-12	17 s		
73827	199469		8 39 12.4	-33 36 39	-0.001	+0.03	7.4	1.1	0.2	K0 III		250 s		
74148	250291		8 39 13.2	-60 19 02	-0.004	+0.02	6.36	0.1		A0				m
74388	250296		8 39 14.2	-69 16 27	-0.003	0.00	7.1	0.0	0.0	B8.5 V		260 s		
73881	199470		8 39 14.2	-37 28 13	-0.001	+0.02	7.0	1.4	-0.3	K5 III		290 s		
73687	154504		8 39 14.8	-12 44 14	-0.001	-0.01	6.6	0.1	0.6	A0 V		160 s		
73171	26946		8 39 17.5	+52 42 42	-0.003	-0.03	5.91	1.17	0.2	K0 III	+27	110 s		
73901	199474		8 39 18.8	-39 47 59	-0.001	-0.04	7.28	0.15	1.4	A2 V		130 s		
73900	199473		8 39 22.0	-36 36 25	-0.015	+0.04	6.13	0.42	1.9	F3 IV		60 mx		m
73806	176238		8 39 23.1	-28 22 38	-0.003	+0.02	8.00	0.4	2.6	F0 V		120 s	6918	m
74071	236151		8 39 23.8	-53 26 23	-0.004	+0.02	5.49	-0.16	-0.9	B6 V	+9	190 s		
73669	136137		8 39 24.1	-4 51 36	0.000	+0.01	8.0	1.1	-0.1	K2 III		410 s		
73599	116997		8 39 24.5	+8 01 02	-0.002	-0.04	6.45	1.08	0.0	K1 III	+17	200 s		
73395	60960		8 39 27.3	+35 31 11	+0.001	-0.01	7.3	0.1	1.4	A2 V		150 s		
73791	154508		8 39 28.8	-18 15 58	-0.006	-0.02	7.1	0.1	1.4	A2 V		91 mx		
73427	60963		8 39 31.8	+32 30 56	-0.001	-0.02	6.8	1.4	-0.3	K5 III		260 s		
73578	98007		8 39 39.5	+16 08 29	-0.002	-0.04	7.8	1.1	-0.1	K2 III		380 s		
73747	136143		8 39 40.6	-9 11 56	-0.002	-0.01	8.0	1.1	-0.1	K2 III		410 s		
73879	176249		8 39 40.8	-29 04 57	+0.001	-0.02	7.1	0.8	0.2	K0 III		240 s		
74043	220243		8 39 42.3	-46 23 00	-0.004	+0.02	7.81	1.65	-0.3	K5 III		400 s		
73898	176253	ζ Pyx	8 39 42.5	-29 33 40	-0.001	-0.10	4.89	0.90	0.3	gG4	-32	75 s	6923	m
72582	6605		8 39 42.6	+73 37 47	-0.003	-0.10	6.15	1.02	0.2	K0 III	+1	140 mx		
73575	98006		8 39 42.6	+19 46 42	-0.003	-0.02	6.67	0.25	0.6	F0 III	+31	160 s		
73574	80327		8 39 42.8	+20 05 10	-0.002	-0.02	7.73	0.22	2.1	A5 V	+37	130 s		m
73766	136144		8 39 43.6	-9 35 13	0.000	-0.02	7.60	1.6		M5 II				RV Hya, v
73668	117000		8 39 43.7	+5 45 50	+0.012	-0.32	7.24	0.60	4.5	G1 V	-18	35 s	6913	m
73576	98009		8 39 44.5	+19 16 30	-0.003	-0.02	7.67	0.20	2.4	A7 V n	+33	110 s		

HD	SAO	Star Name	α 2000	δ 2000	μ(α)	μ(δ)	V	B-V	M$_v$	Spec	RV	d(pc)	ADS	Notes
73947	199480		8 39 50.0	-34 00 36	-0.001	-0.03	7.2	0.9	-0.1	K2 III		290 s		
73598	98010		8 39 50.6	+19 32 27	-0.003	-0.01	6.58	0.96	0.2	K0 III	+34	190 s	6915	m
73407	42475		8 39 50.7	+41 21 58	-0.003	0.00	8.0	1.1	-0.1	K2 III		410 s		
73667	98015		8 39 50.8	+11 31 21	-0.007	-0.51	7.64	0.83	6.1	K1 V	-13	21 ts		m
73897	176257		8 39 52.5	-24 44 04	-0.001	-0.01	7.9	0.1	0.6	A0 V		230 s		
73844	154514		8 39 53.5	-17 18 11	0.000	-0.16	6.61	1.53	-0.5	M4 III	+31	84 mx		AK Hya, v
73508	60968		8 39 56.3	+32 43 38	-0.002	-0.04	6.8	1.1	0.2	K0 III	+14	210 s		
73618	98013		8 39 56.4	+19 33 11	-0.003	-0.01	7.32	0.19		A m	+40		6915	m
73619	98014		8 39 57.7	+19 32 31	-0.003	+0.01	7.54	0.25		A m	+33			
74146	236158		8 39 58.6	-53 03 03	-0.002	+0.01	5.20	-0.15	-1.1	B5 V	+36	180 s		
74236	236166		8 39 59.3	-59 32 50	0.000	-0.01	6.7	-0.2	0.4	B9.5 V		180 s		
74169	236160		8 39 59.3	-53 15 40	-0.003	+0.01	7.25	-0.03		A0 m				
73840	154515	6 Hya	8 40 01.4	-12 28 31	-0.006	-0.01	4.98	1.42	-0.3	K4 III	-11	110 s		
73984	199486		8 40 02.5	-34 01 11	-0.004	+0.01	7.4	0.6	1.7	A3 V		71 s		
74182	236162		8 40 03.1	-53 51 07	0.000	-0.02	7.36	0.29		A2				m
74086	220250		8 40 03.3	-44 42 55	-0.002	-0.01	7.99	0.90	0.2	K0 III		360 s		
73859	154518		8 40 03.7	-15 02 11	-0.001	0.00	7.8	1.6	-0.5	M2 III		470 s		
74006	199490	β Pyx	8 40 06.1	-35 18 30	+0.001	-0.02	3.97	0.94	0.3	G4 III	-15	46 s		m
73665	80333	39 Cnc	8 40 06.4	+20 00 27	-0.003	0.00	6.39	0.98	0.2	K0 III	+35	170 s		
74168	236163		8 40 08.0	-51 56 30	-0.004	0.00	7.4	-0.1	0.0	B8.5 V		220 mx		
73666	80336		8 40 11.4	+19 58 16	-0.003	-0.02	6.61	0.01	1.2	A1 V	+34	120 s		m
74130	220258		8 40 12.2	-48 26 45	-0.004	+0.01	7.71	1.37	0.2	K0 III		180 s		
73108	14616	4 π² UMa	8 40 12.9	+64 19 40	-0.008	+0.02	4.60	1.17	-0.1	K2 III	+15	87 s		
73764	117006		8 40 13.3	+6 33 55	-0.001	+0.03	6.6	1.1	0.2	K0 III		190 s		
74196	236165		8 40 17.3	-53 00 55	-0.003	+0.03	5.58	-0.14	-0.6	B8 IV	+14	170 s		
74195	236164	o Vel	8 40 17.6	-52 55 19	-0.002	+0.02	3.62	-0.18	-1.7	B3 V	+17	120 s		
74698	256538		8 40 17.8	-71 52 25	+0.025	+0.18	7.5	0.8	3.2	G5 IV		47 mx		
73801	117010		8 40 18.0	+1 55 12	+0.001	-0.05	6.8	1.1	-0.1	K2 III		240 s		
73711	98018		8 40 18.0	+19 31 55	-0.003	-0.01	7.54	0.16		A m	+39			
73596	60970		8 40 18.3	+31 56 31	-0.003	-0.03	6.1	0.4	3.0	F2 V	+13	43 s		
74067	220252		8 40 19.2	-40 15 51	-0.004	-0.01	5.20	-0.01	0.2	B9 V	+21	99 s		m
73712	98019		8 40 20.0	+19 20 55	-0.003	-0.03	6.78	0.26	2.5	A9 V	+31	71 s		
73709	98020		8 40 20.7	+19 41 11	-0.002	-0.03	7.70	0.20		A m	+17		6921	m
74104	220256		8 40 20.7	-42 21 10	-0.003	0.00	7.73	0.75	4.4	G0 V		33 s		
74105	220257		8 40 21.0	-42 23 21	-0.003	+0.03	6.89	0.16	1.4	A2 V		110 s		m
73246	26951		8 40 22.0	+59 35 06	-0.003	-0.05	7.7	1.1	0.2	K0 III		310 s		
73710	98021		8 40 22.0	+19 40 11	-0.002	-0.02	6.44	1.02	0.2	K0 III	+35	170 s	6921	m
73394	26954		8 40 22.5	+51 45 06	+0.005	-0.10	7.80	1.1	0.2	K0 III		110 mx	6906	m
73730	80340		8 40 23.4	+19 50 05	-0.002	-0.02	8.02	0.19		A m	+28			
74541	---		8 40 23.5	-67 48 26			7.89	1.15	-0.3	K5 III		440 s		
74024	176271		8 40 26.7	-27 15 28	-0.001	-0.01	8.0	0.3	1.4	A2 V		140 s		
73731	98024	41 ε Cnc	8 40 26.9	+19 32 42	-0.003	-0.01	6.30	0.17	0.4	A6 III	+30	150 s		m
73962	176268		8 40 28.7	-21 00 46	0.000	+0.01	7.8	1.4	-0.1	K2 III		260 s		
74115	220260		8 40 32.4	-41 04 13	-0.003	+0.01	8.00	0.00	0.4	B9.5 V		300 s		
73799	98028		8 40 32.6	+10 55 27	0.000	-0.01	6.9	1.1	0.2	K0 III		220 s		
73638	60972		8 40 33.6	+33 16 39	-0.002	-0.01	6.8	1.1	0.2	K0 III		210 s		
74167	220261		8 40 35.3	-45 11 29	+0.001	-0.01	5.71	1.65		K5				
74224	236172		8 40 35.3	-51 00 49	-0.003	+0.02	7.54	1.54	-0.3	K5 III		350 s		
74375	236181		8 40 37.0	-59 45 40	-0.001	0.00	4.33	-0.11	-3.6	B2 III	+13	350 s		m
74180	220265		8 40 37.6	-46 38 55	0.000	0.00	3.84	0.71	-8.4	F2 Ia	+25	1900 s		m
73763	98027		8 40 39.2	+19 13 42	-0.002	-0.01	7.80	0.22	2.5	A9 V	+38	120 s		
73293	14621		8 40 40.4	+60 37 23	-0.005	-0.01	7.8	0.4	2.6	F0 V		110 s		
73393	26956		8 40 42.2	+55 40 03	-0.030	-0.37	8.04	0.68	4.9	G3 V	+37	38 s		
74166	220262		8 40 42.5	-42 20 24	-0.001	+0.02	7.67	1.30	-0.3	K5 III		390 s		
73447	26959		8 40 42.9	+51 25 51	-0.003	+0.05	7.0	0.9	3.2	G5 IV		57 s		
73785	98030		8 40 43.2	+19 43 09	-0.002	-0.02	6.85	0.20	0.6	A9 III	+31	180 s		
74341	236179		8 40 43.6	-57 32 42	-0.002	+0.02	6.34	0.20		A2				m
73615	60974		8 40 43.7	+38 09 32	-0.002	-0.03	7.8	1.1	0.2	K0 III		330 s		
73940	154533		8 40 43.8	-11 55 39	-0.001	-0.01	7.90	0.9	3.2	G5 IV		87 s		m
74022	176277		8 40 45.4	-20 49 56	+0.003	+0.02	8.0	1.2	0.2	K0 III		260 s		
74194	220270		8 40 47.6	-45 03 31	-0.003	-0.01	7.54	0.23		O9 k				
74322	236180		8 40 47.7	-56 10 04	-0.003	0.00	7.8	-0.2	1.7	A3 V		160 s		
74275	236177		8 40 48.5	-52 48 06	-0.002	+0.02	7.30	-0.01	0.6	A0 V		220 s		
73637	42488		8 40 50.2	+40 01 04	-0.001	0.00	7.6	0.4	2.6	F0 V		100 s		
73976	154535		8 40 51.0	-10 26 48	-0.004	-0.02	7.8	1.1	0.2	K0 III		220 mx		
73857	98035		8 40 52.1	+9 49 27	-0.002	-0.03	7.6	0.2	0.6	A9 III var	+24	250 s		VZ Cnc, v
74210	220274		8 40 53.4	-45 45 48	-0.005	-0.03	7.96	1.24	-0.1	K2 III		220 mx		
74234	220276		8 40 53.5	-48 13 30	+0.001	+0.02	6.95	-0.16	-2.5	B2 V k		730 s		
73821	98033		8 40 54.5	+16 29 58	0.000	-0.01	7.6	0.4	2.6	F0 V		100 s		
73819	98032		8 40 56.2	+19 34 49	-0.003	-0.01	6.78	0.17	2.2	A6 V n	+28	82 s		
---	42489		8 40 58.8	+44 53 36	+0.002	-0.01	7.9	1.8						
73593	42490	34 Lyn	8 41 01.0	+45 50 02	+0.002	+0.09	5.37	0.99	2.8	G0 IV	-37	25 mn		
74251	220279		8 41 01.6	-48 04 06	-0.001	-0.03	7.76	-0.11		B5				
73997	136176		8 41 01.6	-9 03 08	-0.002	0.00	6.63	-0.02		A0				
74562	250303		8 41 03.6	-64 20 05	-0.004	+0.02	7.8	0.1	0.6	A0 V		250 mx		

HD	SAO	Star Name	α 2000	δ 2000	μ(α)	μ(δ)	V	B-V	M_V	Spec	RV	d(pc)	ADS	Notes
74638	250304		8 41 04.5	-66 49 02	-0.005	-0.03	6.8	-0.2	0.0	B8.5 V		170 mx		
74273	220282		8 41 05.3	-48 55 21	-0.001	+0.01	5.90	-0.21	-2.5	B2 V n		460 s		
74063	154545		8 41 08.1	-19 54 57	-0.003	+0.01	7.7	0.4	1.7	A3 V		98 s		
74192	199505		8 41 12.6	-36 18 38	-0.001	-0.06	8.0	0.4	3.0	F2 V		90 s		
74272	220284		8 41 13.0	-47 19 01	-0.002	+0.01	4.77	0.12	-2.3	A3 II	+17	260 s		
74126	176302		8 41 13.5	-27 26 28	-0.001	0.00	8.0	1.3	-0.1	K2 III		360 s		
73636	42492		8 41 14.5	+47 28 49	-0.002	-0.07	7.6	0.7	4.4	G0 V		44 s		
73871	80354		8 41 15.3	+20 28 38	0.000	+0.01	6.73	0.10		A2	0		6930	m
73975	117025		8 41 16.4	+0 57 28	0.000	-0.02	7.8	1.1	-0.1	K2 III		380 s		
73938	98042		8 41 16.8	+11 02 55	-0.002	-0.01	7.56	0.11	0.9	A3 IV		210 s		
74175	199506		8 41 18.0	-33 14 21	-0.002	-0.05	7.9	0.9	0.2	K0 III		340 s		
73890	--		8 41 18.2	+19 15 38	-0.004	-0.03	7.92	0.24	2.4	A7 V n	+40	27 mx	6931	m
74014	136179		8 41 18.5	-4 48 34	+0.004	-0.07	7.8	1.1	0.2	K0 III		200 mx		
75416	256543	η Cha	8 41 19.9	-78 57 48	-0.008	+0.02	5.47	-0.10	-0.3	B9 IV	+18	140 s		
74923	256539		8 41 23.6	-72 35 20	-0.014	+0.07	6.9	0.4	2.6	F0 V		72 s		
73469	26961		8 41 25.0	+59 04 14	-0.011	-0.10	7.1	0.8	3.2	G5 IV		59 s		
73995	117027		8 41 29.2	+2 36 31	-0.001	0.00	7.8	1.1	0.2	K0 III		330 s		
74985	256540		8 41 32.6	-73 36 51	-0.001	+0.02	7.1	-0.2	0.6	A0 V		200 s		
74319	220293		8 41 34.9	-44 59 30	-0.001	+0.02	6.69	-0.10	0.4	B9.5 V		180 s		m
74162	176314		8 41 35.1	-22 11 39	0.000	0.00	7.9	0.2	0.6	A0 V		200 s		
73329	14624		8 41 36.4	+64 42 21	0.000	-0.08	8.0	1.1	0.2	K0 III		160 mx		
74137	154552	9 Hya	8 41 43.2	-15 56 36	-0.001	-0.10	4.88	1.06	0.0	K1 III	-2	95 s	6937	m
75505	256544		8 41 45.1	-79 02 50	-0.008	+0.06	7.5	0.1	0.6	A0 V		84 mx		
73759	42496		8 41 45.6	+44 51 49	-0.002	-0.01	7.9	0.4	3.0	F2 V	+22	97 s		
74438	236190		8 41 46.1	-53 03 46	-0.007	+0.24	7.60	0.24	1.4	A2 V		140 mx		
74123	136188		8 41 49.2	-8 33 36	-0.001	-0.01	7.7	0.1	1.4	A2 V		190 s		
73974	80361		8 41 50.0	+19 52 27	-0.003	-0.02	6.90	0.96	0.2	K0 III	+31	220 s		
74093	136187		8 41 50.3	-5 36 39	+0.002	-0.03	7.9	1.6	-0.5	M2 III		480 s		
73483	14627		8 41 54.2	+62 12 35	+0.005	-0.02	7.8	1.1	0.2	K0 III		320 mx		
74203	176323		8 41 54.7	-22 01 41	0.000	0.00	7.5	1.8	-0.3	K5 III		240 s		
74371	220300		8 41 56.8	-45 24 39	-0.001	0.00	5.23	0.21	-6.3	B5 Iab	+25	1400 s		
90105	258546		8 41 59.5	-89 27 39	+0.040	-0.03	7.5	1.4	-0.1	K2 III		230 s		
73446	14628		8 42 01.4	+63 27 18	+0.001	-0.04	7.8	1.4	-0.3	K5 III		410 s		
75134	256542		8 42 04.7	-74 36 31	+0.002	+0.03	7.5	0.0	0.4	B9.5 V		270 s		
73614	26966		8 42 06.0	+59 01 45	-0.005	-0.03	7.9	0.4	2.6	F0 V		110 s		
74028	98053		8 42 06.5	+19 24 42	-0.002	0.00	7.96	0.21	2.4	A7 V	+31	130 s		
74497	236196		8 42 06.7	-52 45 23	-0.003	+0.02	7.85	0.65		G5				m
74157	136193		8 42 09.2	-3 03 00	-0.002	-0.01	7.4	1.6	-0.5	M2 III		390 s		
74190	154558		8 42 09.7	-11 57 59	-0.002	-0.03	6.45	0.16		A2				
74516	236198		8 42 09.7	-52 58 03	-0.005	+0.02	7.40	0.1	1.2	A1 V		170 mx		
74536	236201		8 42 12.3	-53 22 11	+0.002	0.00	7.97	0.02	-0.6	B7 V		470 s		
74300	199516		8 42 12.5	-32 13 54	-0.001	-0.01	7.31	-0.02	0.4	B9.5 V		240 s		
74189	136195		8 42 14.0	-8 34 54	-0.002	+0.02	7.8	1.1	0.2	K0 III		330 s		
74455	220313		8 42 15.9	-48 05 57	-0.003	-0.01	5.51	-0.18		B3 n	+58			m
74535	236202		8 42 18.9	-53 06 00	-0.003	+0.02	5.53	-0.16		A0 p	+20			m
74622	236206		8 42 20.9	-55 46 27	-0.009	+0.05	6.29	1.18		K0				
74454	220315		8 42 22.8	-46 34 47	+0.002	-0.01	7.87	-0.02	0.4	B9.5 V		310 s		
74156	117040		8 42 25.0	+4 34 43	+0.001	-0.18	7.7	0.6	4.4	G0 V		46 s		
74560	236205		8 42 25.4	-53 06 50	-0.002	+0.02	4.86	-0.18	-1.6	B5 IV	+22	200 s		m
--	--		8 42 26.1	-53 06 50			5.80							m
74496	220317		8 42 28.2	-47 31 17	-0.002	-0.05	6.71	0.24	1.4	A2 V		120 s		
74681	236210		8 42 29.3	-59 31 38	0.000	+0.01	7.84	0.06	0.6	A0 V		260 s		
74217	136201		8 42 30.6	-8 29 55	-0.002	-0.01	7.00	0.00	0.4	B9.5 V		210 s		m
74011	60997		8 42 30.7	+34 11 17	+0.002	-0.26	7.4	0.5	4.0	F8 V	+45	32 mx		
74298	--		8 42 31.7	-22 22 23			7.6	0.3	2.1	A5 V		110 s		
74314	176338		8 42 33.1	-24 12 43	0.000	0.00	7.8	1.5	0.2	K0 III		170 s		
74531	220319		8 42 34.6	-48 09 49	-0.002	0.00	7.26	-0.17		B9				m
74530	220321		8 42 36.4	-48 04 32	-0.001	-0.01	8.00	0.00	0.0	B8.5 V		400 s		
74216	136202		8 42 36.8	-4 24 38	-0.001	-0.04	6.9	1.1	-0.1	K2 III		250 s		
74417	199528		8 42 37.9	-37 22 58	+0.002	-0.04	7.9	0.1	-0.1	K2 III		390 s		
73992	60998		8 42 41.8	+39 03 42	+0.001	+0.02	7.5	1.6	-0.5	M2 III		410 s		
74368	176343		8 42 46.1	-25 52 40	-0.001	-0.01	7.5	0.7	1.4	A2 V		65 s		
74057	61003		8 42 46.3	+31 51 46	-0.001	-0.03	7.1	0.5	4.0	F8 V	0	42 s		
74558	220325		8 42 47.4	-46 48 27	-0.009	+0.07	6.91	0.28	0.5	A7 III		100 mx		
74416	199532		8 42 48.1	-34 44 02	+0.003	0.00	7.8	1.8	-0.5	M2 III		370 s		
73190	6618		8 42 51.1	+73 10 08	-0.003	-0.02	7.0	0.1	0.6	A0 V	+4	190 s		
74294	136210		8 42 53.8	-9 48 16	-0.002	+0.02	6.8	0.1	1.4	A2 V		120 s		
74155	98062		8 42 54.1	+16 26 59	0.000	-0.01	7.9	0.1	0.6	A0 V		280 s		
73933	42504		8 42 56.2	+48 11 33	-0.004	-0.11	7.6	0.7	4.4	G0 V		44 s		
74475	199535		8 42 57.0	-35 56 36	-0.001	+0.01	6.42	0.02		A1				
75485	256546		8 42 57.9	-77 07 01	+0.009	+0.04	8.0	0.1	0.6	A0 V		98 mx		
73971	42508		8 43 00.1	+46 54 04	-0.004	-0.05	6.22	0.96	3.2	G5 IV	-7	31 s		
73392	6620		8 43 00.7	+70 09 53	-0.004	-0.03	7.1	1.1	0.2	K0 III		250 s		
73745	14637		8 43 01.8	+60 55 48	-0.003	-0.03	7.4	1.1	0.2	K0 III		280 mx		
74451	199536		8 43 03.0	-31 35 45	-0.004	-0.03	7.9	1.2	0.2	K0 III		200 mx		

HD	SAO	Star Name	α 2000	δ 2000	μ(α)	μ(δ)	V	B−V	M$_V$	Spec	RV	d(pc)	ADS	Notes
			^{h} ^{m} ^{s}	$^\circ$ ′ ″	^{s}	″								
74678	236217		8 43 03.4	−53 04 43	−0.004	−0.01	7.69	0.07	1.2	A1 V		190 s		
74429	176352		8 43 03.7	−29 31 28	−0.002	0.00	7.5	0.6	0.6	A0 V		98 s		
74601	220337		8 43 04.5	−46 13 35	+0.001	−0.07	7.98	1.06	0.2	K0 III		330 s		
74599	220339		8 43 06.7	−45 33 44	−0.002	−0.01	6.84	1.51	0.2	K0 III		88 s		
74200	98066		8 43 08.3	+18 09 03	−0.001	0.00	7.9	1.1	0.2	K0 III		340 s		
74228	98069	45 Cnc	8 43 12.3	+12 40 51	0.000	0.00	5.6	0.7	4.4	G0 V	−13	18 s		
75747	256549		8 43 12.6	−79 04 11	−0.009	+0.04	6.05	0.23		A5				
74280	117050	7 η Hya	8 43 13.4	+3 23 55	−0.001	−0.01	4.30	−0.20	−1.7	B3 V	+21	160 s		
74474	176355		8 43 14.2	−28 40 51	−0.001	+0.01	8.00	1.1	0.2	K0 III		350 s	6953	m
74650	220346		8 43 16.2	−47 48 24	−0.005	0.00	7.35	−0.05	0.4	B9.5 V		180 mx		
74556	199542		8 43 16.4	−39 08 13	+0.002	−0.06	8.0	0.0	3.4	F5 V		82 s		
74198	80378	43 γ Cnc	8 43 17.1	+21 28 06	−0.007	−0.04	4.66	0.02	1.2	A1 V	+29	48 mx		Asellus Bor., m
74576	199544		8 43 18.0	−38 52 57	−0.026	+0.33	6.56	0.93	6.1	K1 V	+18	11 ts		
74713	236220		8 43 22.4	−52 12 32	+0.001	+0.01	7.80	1.08	0.2	K0 III		290 s		
74010	42511		8 43 24.8	+48 51 51	−0.008	−0.05	6.9	0.4	2.6	F0 V		73 s	6945	m
74888	250308		8 43 30.2	−61 41 19	+0.015	−0.07	7.0	−0.1	0.4	B9.5 V		96 mx		
74153	61011		8 43 32.2	+36 20 52	+0.002	−0.02	7.7	1.1	0.2	K0 III		320 s		
74617	199549		8 43 32.3	−39 35 59	+0.001	+0.01	7.76	1.55		K5				
74409	154576		8 43 32.4	−14 02 31	−0.002	0.00	7.60	0.1	0.6	A0 V		250 s		m
74152	61012		8 43 33.8	+36 42 52	−0.002	−0.05	7.0	1.1	0.2	K0 III		230 s		
−−	199545		8 43 34.9	−33 09 27	−0.001	−0.04	8.0	0.5						
74762	236223		8 43 35.4	−53 31 30	−0.005	−0.03	7.79	0.20		A2				
74575	199546	α Pyx	8 43 35.5	−33 11 11	−0.001	+0.01	3.68	−0.18	−4.4	B2 II	+15	410 s		
75591	256548		8 43 36.7	−77 02 57	−0.004	+0.01	7.9	0.1	0.6	A0 V		290 s		
74712	220356		8 43 38.4	−47 24 12	−0.004	−0.01	7.96	1.24	−0.1	K2 III		350 mx		SW Vel, v
74119	42513		8 43 39.5	+42 40 54	−0.004	−0.03	7.7	0.4	2.6	F0 V		100 s		
74395	136221		8 43 40.4	−7 14 01	−0.001	0.00	4.62	0.84	−4.5	G2 Ib	+31	670 s		m
74753	220361		8 43 40.5	−49 49 22	+0.002	0.00	5.16	−0.20	−4.1	B0 V n	+28	670 s		
74330	117058		8 43 41.4	+9 20 23	−0.001	−0.03	7.3	1.1	0.2	K0 III		260 s		
74711	220362		8 43 47.3	−46 47 55	−0.002	+0.01	7.11	0.08		B3				
74774	236224		8 43 51.1	−50 39 19	+0.003	−0.10	7.25	0.37	3.4	F5 V		59 s		
74491	154582		8 43 51.6	−15 57 15	0.000	0.00	8.0	1.1	0.2	K0 III		360 s		
75116	250315		8 43 54.4	−68 12 41	+0.001	+0.02	6.32	1.50		K2				
74446	136225		8 43 57.5	−6 30 17	0.000	−0.03	8.0	1.1	0.2	K0 III		360 s		
74393	117069		8 43 59.6	+4 20 05	−0.002	−0.01	6.37	−0.06		B8				
74674	199562		8 44 00.6	−38 46 47	−0.002	+0.04	7.00	1.1	0.2	K0 III		230 s		m
75428	256547		8 44 01.5	−74 18 09	+0.004	+0.01	7.7	0.1	0.6	A0 V		260 s		
74773	220368		8 44 09.7	−47 06 55	−0.001	+0.02	7.24	−0.11	−1.4	B4 V		540 s		
74243	61018		8 44 10.0	+36 55 05	−0.003	−0.10	6.4	0.4	3.0	F2 V	+4	47 s		
74292	61019		8 44 14.9	+32 03 45	−0.003	+0.01	6.9	0.1	1.4	A2 V	−8	130 s		
74214	42518		8 44 15.5	+42 06 02	0.000	+0.02	7.0	1.1	−0.1	K2 III		260 s		
74612	176381		8 44 18.1	−20 16 02	−0.004	−0.02	7.50	1.1	0.2	K0 III		230 mx	6969	m
74731	199564		8 44 18.6	−37 31 37	−0.003	−0.02	6.7	0.5	−0.1	K2 III		230 s		
74630	176384		8 44 19.1	−21 38 34	0.000	−0.04	8.0	1.3	0.2	K0 III		250 s		
74772	220371		8 44 23.9	−42 38 57	−0.002	+0.02	4.06	0.87	0.3	gG5	−2	56 s		m
74710	199567		8 44 27.9	−32 01 16	−0.002	+0.01	7.5	1.5	−0.1	K2 III		200 s		
75171	250317		8 44 30.0	−65 49 32	−0.010	+0.10	6.05	0.20	1.4	A2 V		68 mx		
74486	117081		8 44 31.1	+5 17 08	−0.004	0.00	7.9	0.1	0.6	A0 V		280 s		
74853	220382		8 44 31.5	−47 59 11	−0.001	−0.02	7.73	1.72	−0.1	K2 III		170 s		
74790	220376		8 44 33.1	−41 43 18	−0.001	−0.01	8.0	0.6	0.6	A0 V		160 s		
74771	199569		8 44 34.1	−37 57 51	+0.001	+0.03	8.00	1.1	0.2	K0 III		350 s		m
74804	220381		8 44 40.3	−41 16 38	0.000	0.00	7.20	0.00	−1.0	B5.5 V		220 s		m
74442	98087	47 δ Cnc	8 44 41.0	+18 09 15	−0.001	−0.23	3.94	1.08	0.2	K0 III	+17	47 mx	6967	Asellus Aus., m
74690	176395		8 44 41.4	−23 47 21	+0.003	−0.07	6.80	1.1	0.2	K0 III		210 s	6974	m
74956	236232	δ Vel	8 44 42.2	−54 42 30	+0.003	−0.08	1.96	0.04	0.6	A0 V	+2	21 ts		m
75590	256550		8 44 42.8	−74 47 16	−0.002	+0.01	6.5	1.9	−0.1	K2 III		73 s		
74842	220383		8 44 44.9	−42 38 03	0.000	−0.28	7.22	0.73	5.2	G5 V		23 s		
74521	98089	49 Cnc	8 44 45.0	+10 04 54	−0.001	−0.02	5.66	−0.11		A0 p	+24			
74708	176397		8 44 46.7	−22 17 51	−0.001	+0.01	7.8	1.6	−0.3	K5 III		360 s		
74955	236233		8 44 48.7	−53 04 44	−0.006	0.00	7.58	0.08		A0				
74568	117086		8 44 49.5	+1 42 30	−0.001	−0.02	8.0	1.1	0.2	K0 III		360 s		
74868	220387		8 44 50.9	−44 32 33	−0.017	+0.13	6.56	0.55	3.0	G3 IV		52 s		
74824	199573		8 44 51.8	−37 08 50	−0.001	0.00	5.76	−0.15		B8				
74643	136237		8 44 52.4	−8 15 23	+0.011	−0.10	7.5	0.5	3.4	F5 V		68 s		
74706	176404		8 44 55.1	−21 10 04	−0.001	0.00	6.11	0.22		A2				
74787	199575		8 45 00.5	−30 58 40	−0.002	+0.01	7.1	0.4	0.6	A0 V		120 s		
74591	117088	10 Hya	8 45 01.2	+5 40 50	0.000	−0.01	6.13	0.20		A3 n	−6			
74704	154603		8 45 04.4	−15 25 30	−0.007	−0.07	7.9	0.5	3.4	F5 V		80 s		
75086	236241		8 45 05.3	−58 43 30	−0.004	−0.02	6.21	−0.09		B8				m
74822	199578		8 45 08.8	−32 14 58	+0.004	−0.09	6.90	0.4	3.0	F2 V		60 s		m
74920	220397		8 45 10.5	−46 02 18	+0.001	+0.01	7.54	0.03		B3				
74361	42525		8 45 10.8	+43 38 59	−0.004	−0.02	7.3	0.5	3.4	F5 V		59 s		
74567	98092		8 45 12.0	+15 20 32	−0.001	+0.01	7.8	0.8	3.2	G5 IV		84 s		
74900	220395		8 45 12.9	−41 52 44	−0.004	0.00	7.4	0.9	−0.1	K2 III		230 s		
74702	136244		8 45 17.4	−6 58 42	0.000	−0.01	6.9	1.1	0.2	K0 III		220 s		

HD	SAO	Star Name	α 2000	δ 2000	μ(α)	μ(δ)	V	B-V	M$_v$	Spec	RV	d(pc)	ADS	Notes
74590	98096		8^h 45^m 19^s.7	+15°02'48"	-0^s.001	-0".05	7.7	0.5	4.0	F8 V		54 s		
74688	136242		8 45 20.4	-2 36 06	0.000	-0.02	6.41	0.52	3.4	dF5	-18	36 s	6977	m
74952	220402		8 45 21.3	-45 54 57	-0.001	-0.02	7.91	0.21		A3				
74485	61029	46 Cnc	8 45 21.4	+30 41 52	0.000	-0.01	6.13	0.94	0.3	G5 III	-12	140 s		
74390	42530		8 45 21.5	+42 24 13	+0.004	-0.07	8.0	1.1	0.2	K0 III		150 mx		
74686	117093		8 45 23.2	+0 11 52	-0.003	0.00	6.8	0.8	3.2	G5 IV		54 s		
74607	98098		8 45 29.5	+18 49 03	-0.008	-0.01	6.9	0.4	3.0	F2 V		59 s		
74484	61032		8 45 30.3	+33 14 23	+0.001	0.00	7.9	1.1	-0.1	K2 III	-28	410 s		
--	176430		8 45 30.5	-28 12 03	-0.002	0.00	7.8							R Pyx, v
74546	80398		8 45 31.2	+28 26 43	-0.008	+0.03	7.3	0.4	3.0	F2 V	+3	72 s		
74935	199585		8 45 32.5	-39 45 06	-0.009	+0.04	7.9	0.1	3.2	G5 IV		88 s		
74685	117095		8 45 34.8	+4 39 50	+0.003	-0.10	6.4	1.1	0.2	K0 III		130 mx		
74483	61033		8 45 37.8	+37 17 27	-0.004	-0.08	6.9	0.4	2.6	F0 V		74 s		
75893	256551		8 45 38.9	-77 04 31	+0.005	-0.02	6.9	0.3	0.6	A0 V		120 s		
74799	154609		8 45 43.3	-13 16 53	-0.002	+0.04	8.0	1.1	0.2	K0 III		360 s		
75105	236249		8 45 43.4	-52 58 18	-0.003	-0.01	7.7	0.0	0.4	B9.5 V		260 s		
75152	236251		8 45 43.8	-55 51 17	-0.009	+0.10	7.2	0.1	2.1	A5 V		83 mx		
74389	42532		8 45 47.0	+48 52 42	+0.001	0.00	7.4	0.1	0.6	A0 V		230 s		
74258	26983		8 45 47.2	+58 12 14	+0.001	-0.01	8.0	0.5	3.4	F5 V		82 s		
74979	220409		8 45 47.5	-40 36 57	+0.001	-0.01	7.24	-0.04	0.0	B8.5 V		280 s		
75009	220412		8 45 47.5	-44 14 52	0.000	+0.01	6.70	-0.09	0.4	B9.5 V		180 s		
74463	42536		8 45 48.2	+43 38 11	0.000	0.00	8.0	1.1	0.2	K0 III		360 s		
74966	199588		8 45 48.4	-36 44 35	0.000	0.00	7.44	-0.14	0.0	B8.5 V		310 s		
74879	176439		8 45 49.2	-25 23 15	-0.002	+0.03	6.10	0.08		A0				
74895	176440		8 45 49.7	-27 51 22	-0.001	+0.03	7.6	0.9	1.4	A2 V		51 s		
74978	199590		8 45 51.2	-37 52 38	+0.001	+0.01	6.7	0.4	1.4	A2 V		72 s		
76236	256552		8 45 55.4	-79 30 15	-0.010	+0.08	5.79	1.60		K0				
74948	199589		8 45 56.3	-32 09 50	-0.004	+0.01	7.1	1.1	0.2	K0 III		220 s		
73797	6629		8 45 57.0	+73 17 32	+0.002	-0.01	7.4	0.1	1.4	A2 V	-14	160 s		
75168	236254		8 45 58.2	-54 57 38	-0.002	0.00	7.2	0.5	3.2	G5 IV		63 s		
74834	154612		8 45 58.6	-10 14 35	-0.003	-0.01	7.8	1.1	0.2	K0 III		340 s		
75104	220424		8 45 58.8	-48 41 15	-0.002	-0.04	7.45	1.16		K5				
74326	26986		8 46 00.9	+57 26 06	-0.003	0.00	8.01	0.14		A0				
75063	220422		8 46 01.7	-46 02 30	-0.001	0.00	3.91	0.00	-0.6	A0 III	+24	75 s		
74794	136254		8 46 02.4	-2 02 56	+0.003	+0.04	5.70	1.10	0.2	gK0	+10	100 s		
74745	98106		8 46 02.8	+9 57 06	0.000	-0.01	7.8	1.4	-0.3	K5 III		410 s		
74670	80407		8 46 03.9	+25 21 13	0.000	0.00	7.90	0.1	1.4	A2 V		200 s	6981	m
75062	220421		8 46 04.0	-43 45 05	-0.001	-0.01	7.92	-0.06	0.4	B9.5 V		320 s		
74605	61041		8 46 04.3	+33 35 12	0.000	-0.04	7.1	1.1	-0.1	K2 III		280 s		
74833	136257		8 46 06.5	-7 35 03	-0.002	+0.01	8.0	1.1	-0.1	K2 III		410 s		
74860	154615		8 46 06.9	-11 00 23	-0.002	+0.02	6.25	1.61		K5			7026	m
74859	154616		8 46 09.5	-10 49 52	+0.001	-0.06	7.3	1.6	-0.5	M4 III		290 mx		
74946	176444		8 46 10.0	-26 36 47	-0.001	-0.01	7.3	1.2	0.2	K0 III		190 s		
73991	6632		8 46 12.1	+70 56 39	-0.001	-0.01	7.8	1.1	-0.1	K2 III		380 s		
74669	80409		8 46 14.0	+27 35 43	-0.003	-0.07	7.23	0.94	0.2	K0 III		180 mx		
--	220426		8 46 14.9	-42 02 07	-0.003	0.00	7.9	0.2	0.0	B8.5 V		250 s		
74977	199595		8 46 16.0	-31 28 19	+0.001	-0.02	8.0	1.2	0.2	K0 III		290 s		
74930	176445		8 46 17.3	-20 25 40	-0.014	+0.02	7.8	0.7	4.4	G0 V		40 s		
75129	220436		8 46 19.3	-47 33 00	-0.001	-0.01	6.87	0.26		B8.5 V		160 s		
75060	220428		8 46 19.7	-40 23 48	-0.001	0.00	7.98	-0.10	0.4	B9.5 V		330 s		
74831	117102		8 46 20.0	+0 38 48	-0.001	-0.02	7.00	0.2	2.1	A5 V		96 s		m
74741	80412		8 46 20.3	+20 36 59	-0.001	-0.03	6.9	0.1	1.4	A2 V		130 s		
75025	199598		8 46 21.5	-33 54 08	-0.004	+0.01	7.3	1.1	0.2	K0 III		210 s		
74858	136265		8 46 21.7	-6 24 41	-0.002	-0.03	8.0	1.1	-0.1	K2 III		410 s		
75201	236260		8 46 21.7	-52 31 26	-0.003	-0.02	8.0	1.6	-0.1	K2 III		240 s		
74918	154622	12 Hya	8 46 22.4	-13 32 52	+0.001	-0.02	4.32	0.90	0.3	G8 III	-8	64 s		m
72520	1378		8 46 22.6	+82 14 21	-0.005	-0.02	6.7	0.1	0.6	A0 V		170 s		
75059	--		8 46 22.8	-38 50 52			7.7	0.0	0.2	K0 III		310 s		
75081	220432		8 46 23.7	-41 07 31	-0.002	0.00	6.21	-0.05		A0				
74793	98113		8 46 24.2	+10 35 30	-0.004	-0.04	7.6	0.5	4.0	F8 V		52 s		
74993	176451		8 46 26.7	-26 37 38	-0.001	+0.03	8.0	0.7	0.2	K0 III		360 s		
74740	80414		8 46 28.9	+22 21 08	-0.003	-0.01	8.0	0.1	1.4	A2 V		210 s		
74781	98112		8 46 29.4	+17 29 45	-0.002	-0.03	7.9	0.5	3.4	F5 V		78 s		
75149	220442		8 46 30.6	-45 54 46	0.000	-0.01	5.46	0.27	-4.4	B2 II	+25	510 s		
74815	117106		8 46 32.5	+8 28 42	-0.001	+0.01	7.0	0.0	0.4	B9.5 V	+25	210 s		
75126	220439		8 46 32.9	-42 33 58	-0.003	0.00	7.08	-0.13		B9				m
75022	176457		8 46 34.9	-29 45 30	0.000	-0.01	7.58	1.45	0.2	K0 III		140 s		
74915	136268		8 46 35.3	-8 38 28	-0.002	-0.02	7.6	1.1	0.2	K0 III		300 s		
75021	176458		8 46 36.1	-29 43 42	-0.002	-0.01	7.09	1.95		R8	+11			
75057	199605		8 46 36.7	-33 14 43	-0.008	+0.03	7.88	0.41	3.0	F2 V		87 s		
75041	176459		8 46 37.7	-28 38 22	-0.002	0.00	8.0	1.8	0.2	K0 III		120 s		
74738	80415	48 ι Cnc	8 46 40.0	+28 45 54	-0.001	-0.04	6.57	0.04	1.7	A3 V		130 s	6988	m
74739	80416	48 ι Cnc	8 46 41.8	+28 45 36	-0.002	-0.04	4.02	1.01	-2.1	G8 II	+16	130 s	6988	m
75311	236268		8 46 42.6	-56 46 11	0.000	+0.01	4.49	-0.17	-2.5	B2 V n	+27	250 s		
75451	250325		8 46 42.7	-63 11 30	-0.008	-0.04	7.6	1.1	0.2	K0 III		180 mx		

HD	SAO	Star Name	α 2000	δ 2000	μ(α)	μ(δ)	V	B-V	M_v	Spec	RV	d(pc)	ADS	Notes
			h m s	° ′ ″	s	″								
75125	199609		8 46 43.7	−39 54 28	−0.001	0.00	7.06	−0.08	0.4	B9.5 V		220 s		
74991	154628		8 46 45.5	−18 45 28	0.000	−0.01	6.5	1.1	0.2	K0 III		180 s		
74912	136270		8 46 45.8	−0 21 13	−0.001	−0.04	8.0	1.1	0.2	K0 III		360 s		
74874	117112	11 ε Hya	8 46 46.5	+6 25 07	−0.013	−0.05	3.38	0.68	0.6	G0 III	+36	34 s	6993	m
74700	61046		8 46 49.1	+38 02 31	−0.002	−0.04	7.4	1.4	−0.3	K5 III		350 s		
75112	199607		8 46 49.1	−34 37 22	−0.002	0.00	6.37	−0.13	−1.4	B4 V		350 s		
75147	199613		8 46 49.5	−39 29 47	−0.001	0.00	7.60	0.1	0.6	A0 V		250 s		m
75111	199608		8 46 51.4	−34 04 01	−0.004	0.00	7.5	0.4	0.6	A0 V		130 s		
74873	98117	50 Cnc	8 46 55.9	+12 06 36	−0.005	−0.06	5.70	0.11	1.2	A1 V	+23	32 mx		
75211	220448		8 47 01.4	−44 04 29	−0.002	0.00	7.51	0.41		B5				
75241	220452		8 47 05.4	−45 04 31	0.000	−0.02	6.59	−0.12	−2.2	B5 III		540 s		
75399	236274		8 47 05.7	−58 04 39	+0.002	−0.02	7.9	0.4	0.2	K0 III		340 s		
74811	80423		8 47 07.2	+28 09 42	−0.003	−0.06	6.58	0.63	3.0	G2 IV		52 s		
74910	98118		8 47 07.8	+11 25 10	+0.002	−0.02	7.5	1.1	−0.1	K2 III		330 s		
74072	6638		8 47 08.3	+72 23 09	−0.021	−0.01	7.6	0.4	3.0	F2 V		83 s		
74944	117119		8 47 10.3	+4 21 28	−0.001	−0.02	7.2	0.0	0.4	B9.5 V		230 s		
74926	98120		8 47 10.9	+11 09 33	−0.002	−0.02	7.6	0.1	0.6	A0 V		260 s	6995	m
75351	236272		8 47 14.8	−51 55 19	−0.002	−0.06	7.82	0.62	4.4	G0 V		43 s		m
74988	136276		8 47 14.9	−1 53 50	−0.002	0.00	5.29	0.04		A0 n	+2	26 mn		
74943	117121		8 47 15.3	+7 36 33	+0.001	−0.04	8.0	1.1	−0.1	K2 III		410 s		
74909	98121		8 47 18.1	+15 19 45	−0.001	−0.01	7.7	0.1	0.6	A0 V		270 s		
75276	220464		8 47 18.8	−46 09 20	0.000	−0.01	5.75	0.56	−6.6	F2 Iab	+32	2300 s		
75098	154636		8 47 21.6	−17 03 13	0.000	−0.04	6.6	0.7	4.4	G0 V		28 s	6999	m
75223	199622		8 47 21.9	−39 47 58	−0.002	+0.02	7.5	−0.3	0.6	A0 V		250 s		
75199	199619		8 47 24.0	−34 36 02	0.000	+0.01	6.70	1.4	−0.3	K5 III		250 s		m
75222	199621		8 47 25.1	−36 45 03	0.000	0.00	7.41	0.38	−6.2	B0 I k		4000 s		
75258	220466		8 47 25.9	−42 27 09	−0.003	+0.01	7.19	−0.12		B9				
75274	220471		8 47 27.3	−42 46 04	−0.005	+0.03	8.02	1.02		K0				
75309	220477		8 47 27.8	−46 27 05	−0.002	−0.01	7.85	0.02		B2				
75256	220473		8 47 33.7	−40 13 48	−0.003	+0.02	7.80	1.4	−0.3	K5 III		410 s		m
75012	117127		8 47 34.7	+0 04 38	−0.001	−0.03	7.83	0.09		B9				m
75178	176478		8 47 38.2	−24 23 35	0.000	+0.02	8.0	0.7	0.6	A0 V		100 s		
75272	199626		8 47 40.4	−38 59 31	0.000	0.00	6.98	−0.15	0.4	B9.5 V		210 s		
75289	220481		8 47 40.4	−41 44 14	0.000	−0.24	6.36	0.58		G0		13 mn		
75271	199628		8 47 44.0	−38 56 30	0.000	−0.01	7.20	0.1	0.6	A0 V		210 s		m
75324	220484		8 47 47.7	−42 16 22	−0.002	0.00	7.12	−0.13		B9				
74828	42549		8 47 48.4	+41 43 08	+0.001	−0.05	8.0	0.9	3.2	G5 IV		93 s		
75161	154643		8 47 48.7	−19 47 01	−0.004	−0.03	8.0	1.5	−0.3	K5 III		220 mx		
74425	14663		8 47 51.0	+66 12 38	−0.005	−0.09	7.8	0.5	4.0	F8 V		57 s		
75254	199629		8 47 51.5	−33 23 53	+0.002	0.00	6.8	1.0	0.2	K0 III		210 s		
75364	220489		8 47 54.1	−43 21 14	−0.001	+0.04	7.23	0.85	0.2	K0 III		260 s		m
75466	236284		8 48 00.2	−52 51 01	−0.003	+0.01	6.28	−0.11		B9				
75426	220498		8 48 00.7	−49 34 31	−0.003	−0.09	6.8	0.8	3.0	F2 V		31 s		
73796	6635		8 48 01.7	+77 41 18	−0.010	−0.07	7.7	0.4	2.6	F0 V		100 s		
75932	256554		8 48 03.8	−71 18 59	+0.003	+0.02	7.7	0.1	0.6	A0 V		260 s		
75140	136287		8 48 04.8	−6 33 31	0.000	−0.02	6.09	1.28		K0				
75035	98133		8 48 06.3	+17 23 41	0.000	−0.04	7.50	1.1	−0.1	K2 III		330 s		m
75387	220495		8 48 08.8	−42 27 50	−0.001	−0.01	6.42	−0.20		B8				
75121	117140		8 48 12.2	+2 34 48	−0.002	−0.05	7.50	0.5	3.4	F5 V		66 s	7003	m
75588	236291		8 48 14.0	−57 39 30	−0.002	+0.01	7.3	1.4	−0.1	K2 III		220 s		
75306	176496		8 48 14.4	−28 38 19	−0.003	−0.01	6.7	1.5	−0.5	M2 III		280 s		
75197	136288		8 48 15.2	−7 58 21	−0.001	−0.01	7.9	0.1	1.4	A2 V		200 s		
75138	117144		8 48 18.2	+0 33 17	−0.001	−0.02	7.26	1.48		K2			7004	m
75863	250334		8 48 19.0	−68 39 54	−0.003	+0.03	8.0	1.1	−0.1	K2 III		420 s		
75234	154645		8 48 20.7	−19 01 38	+0.001	0.00	8.0	1.1	−0.1	K2 III		410 s		
75795	250332		8 48 23.1	−65 25 50	+0.003	0.00	7.50	1.1	0.2	K0 III		290 s		m
75137	117146	13 ρ Hya	8 48 25.9	+5 50 16	−0.001	−0.04	4.36	−0.04	0.6	A0 V	+33	57 s	7006	m
75193	136292		8 48 26.1	−2 31 16	−0.002	−0.02	7.5	0.8	3.2	G5 IV		72 s		
75611	236295		8 48 28.7	−55 26 57	+0.001	0.00	7.68	−0.05	0.6	A0 V		260 s		
75446	220504		8 48 30.6	−42 24 01	−0.003	+0.01	7.38	−0.13		B9				
75175	117149		8 48 32.7	+2 33 49	−0.001	−0.03	7.1	1.1	0.2	K0 III		240 s		
75535	220510		8 48 36.8	−49 13 43	−0.003	0.00	7.68	1.09	0.2	K0 III		270 s		
75157	98142		8 48 37.0	+10 26 00	+0.001	+0.03	7.1	1.6	−0.5	M2 III	−13	320 s		
75217	136294		8 48 37.3	−1 02 42	−0.004	−0.01	6.42	1.14	0.2	K0 III		140 s		
75344	176508		8 48 37.7	−25 39 15	0.000	−0.01	7.5	1.1	1.4	A2 V		42 s		
75712	236301		8 48 39.1	−59 57 09	−0.002	+0.01	7.9	1.4	0.2	K0 III		210 s		
75156	98143		8 48 39.4	+12 32 49	−0.001	0.00	6.61	1.67		Ma	+67			
75445	199654		8 48 42.6	−39 14 03	−0.005	+0.03	7.0	−0.2	1.7	A3 V		120 s		
75443	—		8 48 44.7	−37 44 07			7.9	1.9	−0.3	K5 III		240 s		
75534	220513		8 48 44.7	−47 45 49	−0.001	−0.01	7.85	0.37		B5				
75174	98144		8 48 44.9	+12 35 46	−0.002	0.00	8.0	0.9	3.2	G5 IV		93 s		
74604	14667		8 48 49.3	+66 42 29	−0.002	−0.04	6.1	−0.1		B8				
75476	220509		8 48 51.7	−40 17 34	−0.003	−0.07	7.96	−0.12	0.0	B8.5 V		100 mx		
74358	6649		8 48 54.2	+72 01 14	+0.003	+0.01	8.00	1.1	0.2	K0 III		360 s		m
75475	199659		8 48 54.3	−38 58 13	−0.001	0.00	6.9	0.3	0.2	K0 III		220 s		

HD	SAO	Star Name	α 2000	δ 2000	μ(α)	μ(δ)	V	B-V	M_v	Spec	RV	d(pc)	ADS	Notes
75232	117157		8ʰ48ᵐ54.5ˢ	+3°45′21″	-0.001	-0.01	7.9	0.5	3.4	F5 V		78 s		
74717	14668		8 48 59.3	+63 19 26	+0.001	+0.01	8.0	0.9	3.2	G5 IV		93 s		
75052	61060		8 49 00.3	+38 20 43	-0.001	0.00	7.40	0.1	1.4	A2 V		160 s	7005	m
75032	42559		8 49 00.5	+41 06 28	-0.001	-0.01	8.0	1.4	-0.3	K5 III		450 s		
75500	199662		8 49 02.1	-39 24 54	-0.001	-0.01	7.5	-0.3	0.4	B9.5 V		260 s		
75549	220517		8 49 03.1	-43 45 38	-0.003	+0.02	7.31	-0.13	-1.7	B3 V		600 s		
75282	136301		8 49 03.2	-5 14 22	-0.002	-0.01	7.90	1.1	0.2	K0 III		350 s		m
75519	199666		8 49 05.8	-39 57 14	-0.009	+0.14	7.7	-0.1	3.2	G5 IV		78 s		
75302	117163		8 49 12.4	+3 29 06	-0.011	+0.07	7.8	0.7	4.4	G0 V		49 s		
75393	154658		8 49 15.2	-15 33 55	+0.002	-0.07	7.5	0.5	4.0	F8 V		50 s		
75608	220523		8 49 21.2	-43 22 13	-0.001	+0.01	7.45	-0.09	0.0	B8.5 V		310 s		
75333	136308	14 Hya	8 49 21.6	-3 26 35	-0.002	-0.03	5.31	-0.09		A p	+33			
75391	154659		8 49 22.3	-10 53 05	0.000	+0.03	7.3	1.1	0.2	K0 III		270 s		
75410	154661		8 49 27.0	-11 01 08	-0.002	+0.02	7.5	1.1	-0.1	K2 III		340 s		
75514	199668		8 49 27.2	-30 56 01	-0.008	-0.03	7.7	0.8	3.4	F5 V		46 s		
75300	117168		8 49 29.1	+8 51 53	-0.001	0.00	7.90	0.2	2.1	A5 V		150 s	7021	m
75584	199671		8 49 34.3	-36 57 45	-0.008	+0.01	8.0	0.4	2.6	F0 V		100 s		
75537	117170		8 49 36.9	+1 28 56	0.000	-0.01	6.9	0.0	0.4	B9.5 V		200 s		
75908	250339		8 49 37.7	-62 30 30	+0.006	-0.05	7.9	0.4	3.0	F2 V		95 s		
75630	220531		8 49 39.1	-40 19 13	-0.001	-0.02	5.48	0.06		A2	+17	14 mn		
75929	250341		8 49 39.7	-63 10 56	+0.001	-0.01	7.2	1.1	0.2	K0 III		250 s		
75657	220537		8 49 40.2	-42 49 25	-0.002	0.00	7.48	-0.13	0.0	B8.5 V		310 s		
75422	136316		8 49 40.5	-8 38 18	0.000	-0.06	6.8	0.2	2.1	A5 V		52 mx		
74810	14672		8 49 40.9	+64 15 51	-0.004	-0.02	7.2	1.1	0.2	K0 III		250 s		
75547	176536		8 49 42.1	-28 15 24	-0.004	-0.04	7.4	-0.2	0.6	A0 V		100 mx		
75408	136315		8 49 42.9	-5 41 40	-0.001	-0.01	7.3	0.9	3.2	G5 IV		65 s		
75117	42563		8 49 44.5	+44 58 40	-0.007	-0.02	7.6	0.7	4.4	G0 V	+20	44 s		
75495	176535		8 49 44.9	-21 02 55	0.000	-0.07	6.47	0.23		A3				
75135	42564		8 49 47.2	+44 42 17	-0.002	+0.01	7.8	0.1	0.6	A0 V	+3	270 s		
75655	220538		8 49 47.4	-41 44 36	-0.001	-0.01	7.91	-0.02	0.0	B8.5 V		380 s		
75710	220540		8 49 47.7	-45 18 28	0.000	+0.01	4.93	0.05		A2	+5	22 mn		
75374	117172		8 49 48.7	+7 00 52	+0.001	-0.02	6.7	1.1	0.2	K0 III		200 s		
76270	256556		8 49 50.4	-72 33 03	+0.001	+0.02	6.11	0.20		A2				
75605	199678		8 49 51.5	-32 46 50	0.000	-0.05	5.21	0.87	0.4	gG3	-8	82 s		
75215	61068		8 49 51.9	+33 43 13	-0.003	0.00	7.4	1.1	0.2	K0 III		270 s		
75654	199682		8 49 52.2	-39 08 30	-0.007	+0.03	6.39	0.23	-0.2	A2 III		150 mx		
75407	117176		8 49 54.2	+2 21 46	-0.001	-0.09	7.3	0.8	3.2	G5 IV		67 s		
75653	199681		8 49 55.4	-36 37 00	-0.011	+0.02	7.1	0.9	3.4	F5 V		29 s		
74905	14674		8 49 57.2	+61 14 59	+0.005	-0.03	7.9	0.4	3.0	F2 V		94 s		
75390	117175		8 49 57.4	+6 32 50	-0.001	-0.03	6.7	0.0	0.4	B9.5 V	-10	190 s		
75564	176539		8 49 58.8	-23 57 08	0.000	+0.01	8.0	1.4	-0.1	K2 III		310 s		
75852	236323		8 50 01.7	-55 04 45	0.000	-0.01	8.0	0.3	1.4	A2 V		140 s		
75629	176546		8 50 02.2	-29 27 46	-0.001	0.00	5.87	0.95	0.3	gG7	-10	130 s		
75725	220544		8 50 05.2	-42 33 38	-0.002	-0.02	7.81	0.92	0.2	K0 III		330 s		
75724	220543		8 50 05.7	-42 06 25	+0.002	-0.02	7.70	-0.05		B0.5 V				
74940	14676		8 50 07.4	+61 49 59	-0.001	+0.01	7.75	1.07		K0				
75671	199683		8 50 08.9	-34 13 46	-0.002	+0.01	7.9	0.3	0.4	B9.5 V		190 s		
75824	236322		8 50 09.9	-50 35 29	-0.004	0.00	8.01	0.55	4.4	G0 V		53 s		
75810	220556		8 50 10.2	-49 55 27	-0.002	+0.02	7.60	0.4	2.6	F0 V		81 s		m
75708	199687		8 50 12.7	-36 40 32	-0.002	-0.01	7.8	1.0	0.2	K0 III		330 s		
75650	199685		8 50 15.2	-30 25 00	0.000	0.00	7.6	0.6	0.6	A0 V		100 s		
75790	220555		8 50 19.0	-44 43 15	-0.001	+0.02	7.50	0.98	3.2	G5 IV		50 s		m
75669	176551		8 50 19.7	-28 51 39	-0.001	+0.01	7.6	0.7	1.7	A3 V		64 s		
75759	220552		8 50 20.9	-42 05 24	-0.002	-0.01	6.00	-0.10	-4.8	O9 V		1300 s		
75649	176554		8 50 21.5	-28 37 05	-0.002	-0.03	6.17	-0.08		B9				
75722	199690		8 50 23.2	-35 55 49	-0.002	0.00	7.00	0.00	0.4	B9.5 V		210 s		m
75529	136325		8 50 24.7	-4 11 53	+0.005	-0.15	7.00	1.1	0.2	K0 III		89 mx	7040	m
75628	176549		8 50 24.7	-21 08 39	-0.001	+0.02	8.0	0.3	0.4	B9.5 V		200 s		
75874	236325		8 50 26.0	-51 53 52	-0.004	-0.03	7.60	1.28	-0.1	K2 III		250 mx		
75600	154678		8 50 28.2	-14 36 49	-0.001	-0.01	7.9	0.1	1.4	A2 V		200 s		
75692	176558		8 50 28.2	-28 23 24	0.000	-0.01	8.0	1.9	-0.3	K5 III		260 s		
75789	220558		8 50 30.4	-43 18 44	-0.004	+0.01	7.53	1.55	-0.3	K5 III		350 s		
75691	176559	γ Pyx	8 50 31.9	-27 42 36	-0.010	+0.08	4.01	1.27	-0.3	K4 III	+25	73 s		
75332	61074		8 50 32.1	+33 17 06	-0.005	-0.08	6.25	0.49	4.0	F8 V	+5	28 s		
75431	98159		8 50 33.3	+17 01 55	-0.001	-0.03	7.9	0.1	1.7	A3 V		170 s		
75821	220561		8 50 33.4	-46 31 45	-0.001	0.00	5.10	-0.21		O9.5 II	+8			m
76143	250347		8 50 34.8	-66 47 34	+0.015	+0.09	5.35	0.42		F0 n	+42	14 mn		m
75561	136327		8 50 36.3	-4 41 57	-0.003	+0.04	7.2	0.1	1.4	A2 V		140 s		
75331	61075		8 50 38.2	+36 52 35	-0.004	-0.07	7.9	1.1	-0.1	K2 III		180 mx		
75470	98161		8 50 40.3	+18 00 11	-0.001	0.00	6.72	0.86		G5			7039	m
75353	61077		8 50 44.0	+35 04 16	-0.015	+0.11	7.50	0.5	4.0	F8 V		50 s	7034	m
75469	98162		8 50 45.1	+18 49 56	-0.001	-0.03	6.10	-0.01	0.6	A0 V	+19	130 s		
75419	80454		8 50 46.1	+26 49 36	-0.005	-0.03	7.4	0.4	2.6	F0 V		93 s		
74225	6656		8 50 46.8	+78 09 54	+0.002	-0.02	7.30	1.6		M5 III	-43			
75721	176571		8 50 47.2	-26 07 07	-0.001	-0.02	7.5	1.1	3.2	G5 IV		47 s		

HD	SAO	Star Name	α 2000	δ 2000	μ(α)	μ(δ)	V	B-V	M_v	Spec	RV	d(pc)	ADS	Notes
			h m s	° ′ ″	s	″								
75807	199696		8 50 47.3	-37 17 20	-0.005	-0.02	7.9	1.1	-0.1	K2 III		220 mx		
75860	220566		8 50 53.1	-43 45 06	-0.002	-0.01	7.60	0.73	-6.3	B1.5 Iab		1700 s		m
71986	1381		8 50 54.5	+85 03 03	-0.087	-0.09	7.5	0.4	3.0	F2 V	+1	78 s		
74653	6661		8 51 00.3	+73 45 05	-0.002	-0.05	7.7	1.1	0.2	K0 III		320 s		
75372	42568		8 51 00.4	+40 56 27	0.000	-0.01	7.8	0.5	3.4	F5 V		75 s		
75528	98168	54 Cnc	8 51 01.4	+15 21 03	-0.008	+0.07	6.38	0.64	4.4	dG0	+45	23 s		
76032	236329		8 51 03.3	-57 46 07	+0.004	-0.05	7.9	1.4	-0.3	K5 III		420 s		
75620	117197		8 51 04.1	+0 28 09	-0.003	-0.02	8.01	0.08	0.6	A0 V		270 s		
75739	176576		8 51 04.5	-22 21 28	+0.002	+0.01	7.6	1.7	0.2	K0 III		120 s		
75818	199702		8 51 05.2	-34 25 12	+0.002	0.00	7.9	-0.1	0.4	B9.5 V		320 s		
75689	136335		8 51 05.6	-9 25 31	0.000	-0.01	7.6	0.4	2.6	F0 V		100 s		
75704	154689		8 51 08.3	-13 43 55	-0.002	-0.03	7.2	0.1	0.6	A0 V		210 s		
75665	136336		8 51 10.1	-6 14 18	-0.001	0.00	8.0	1.1	0.2	K0 III		360 s		
75558	98169		8 51 10.8	+15 59 58	-0.002	+0.01	7.2	0.9	3.2	G5 IV	+50	64 s		
75871	220575		8 51 11.0	-41 39 42	-0.001	-0.01	7.80	0.01	-1.0	B5.5 V		300 s		
75869	199708		8 51 13.8	-38 38 01	-0.001	-0.02	7.01	-0.20	-2.5	B2 V		800 s		
76092	236333		8 51 14.9	-59 29 01	-0.001	+0.01	8.0	0.0	0.4	B9.5 V		300 s		
75775	176586		8 51 18.8	-20 31 05	-0.001	+0.01	8.00	0.5	2.3	F7 IV	+19	140 s		
75870	199710		8 51 19.4	-39 44 41	0.000	+0.03	7.72	-0.15	0.0	B8.5 V		350 s		
75720	136341		8 51 19.9	-9 07 24	0.000	-0.01	7.8	1.1	-0.1	K2 III		380 s		
75388	42571		8 51 23.3	+45 55 59	0.000	-0.04	7.8	1.4	-0.3	K5 III		410 s		
75955	220582		8 51 26.0	-45 37 22	-0.001	+0.02	7.73	0.01	0.6	A0 V		260 s		
75638	98174		8 51 27.1	+12 07 40	-0.003	-0.04	7.90	0.4	2.6	F0 V		120 s	7049	m
75926	220581		8 51 27.8	-42 30 15	-0.003	+0.03	6.55	0.04	1.4	A2 V n		110 s		
76188	250349		8 51 28.1	-61 21 46	-0.024	+0.15	7.0	1.2	4.4	G0 V		13 s		
74904	6667		8 51 29.0	+70 17 57	0.000	0.00	7.13	1.64	-0.5	M2 III		310 s		
75618	98173		8 51 29.7	+17 45 14	-0.002	-0.03	8.0	0.9	3.2	G5 IV		93 s		
75229	27016		8 51 30.4	+58 22 06	+0.001	+0.01	8.0	1.1	0.2	K0 III		360 s		
76006	220589		8 51 32.3	-49 33 28	0.000	0.00	7.33	0.66	4.0	F8 V		35 s		
75905	199712		8 51 32.7	-37 28 47	-0.001	+0.02	8.0	0.7	3.0	F2 V		59 s		
75737	136345	15 Hya	8 51 34.4	-7 10 38	-0.003	0.00	5.54	0.15		A p	+37		7050	m
75954	220583		8 51 35.8	-43 23 46	-0.001	+0.02	7.8	1.4	-0.5	M2 III		470 s		
76113	236339		8 51 36.6	-57 38 01	-0.002	+0.01	5.59	-0.11		B8	+8			
75845	176588		8 51 37.5	-26 12 23	-0.003	-0.01	7.60	1.4	-0.3	K5 III		370 s		m
76234	250352		8 51 44.2	-61 25 11	-0.001	+0.01	7.3	0.3	0.4	B9.5 V		140 s		
76131	236340		8 51 45.5	-55 48 35	-0.002	-0.03	6.69	-0.05	0.0	B8.5 V		220 s		
75700	98178		8 51 49.3	+11 53 39	-0.001	+0.01	7.8	1.1	0.2	K0 III	-5	330 s		
76004	220591		8 51 49.9	-44 09 05	-0.002	-0.02	6.38	-0.14	-1.7	B3 V		400 s		
75107	14683		8 51 51.7	+66 26 55	-0.001	+0.01	7.80	0.9	0.3	G8 III	-57	320 s		
76162	236347		8 51 52.1	-57 17 38	-0.001	-0.01	8.0	0.1	1.7	A3 V		170 s		
75298	27017		8 51 54.1	+58 07 36	-0.002	0.00	7.9	0.4	3.0	F2 V		94 s		
75989	220592		8 51 55.6	-40 59 12	0.000	-0.01	6.49	-0.11	0.4	B9.5 V		170 s		
75718	98181		8 51 56.3	+11 08 10	+0.001	-0.01	7.5	1.1	0.2	K0 III		290 s		
75506	42576	35 Lyn	8 51 56.8	+43 43 35	-0.001	+0.04	5.15	0.98	0.2	K0 III	+15	98 s		
75921	176599		8 51 58.7	-27 38 31	-0.001	-0.02	7.5	0.4	2.6	F0 V		83 s		
75646	80467		8 52 00.4	+25 43 07	0.000	-0.04	7.5	0.9	0.2	G9 III	-8	290 s		m
76060	220597		8 52 02.4	-46 17 19	-0.001	+0.01	7.87	-0.04	0.0	B8.5 V		380 s		
75768	117208		8 52 03.3	+4 27 57	0.000	-0.02	7.50	0.1	1.4	A2 V		170 s	7055	m
75488	42578		8 52 05.5	+47 33 50	+0.005	-0.21	8.00	0.9	3.2	G5 IV		46 mx		m
75637	61097		8 52 09.6	+29 51 14	-0.002	0.00	6.9	1.1	0.2	K0 III		220 s		
75556	42581		8 52 09.9	+42 00 09	-0.003	-0.08	5.99	1.25	0.2	K0 III	+57	100 s		
75523	42580		8 52 11.6	+45 18 45	-0.003	-0.04	5.99	1.26	0.2	gK0	+12	90 s		
75918	154702		8 52 16.2	-18 36 58	-0.004	-0.02	7.62	0.46		F5				
75767	117212		8 52 16.5	+8 03 46	+0.012	-0.25	6.6	0.7	4.4	G0 V	+4	28 s		
75881	154701		8 52 20.5	-12 48 29	0.000	-0.09	6.9	0.6	4.4	G0 V		32 s		
76233	236357		8 52 21.7	-56 14 32	-0.001	0.00	7.21	-0.08	0.4	B9.5 V		230 s		
76187	236353		8 52 23.5	-53 19 33	-0.002	0.00	7.97	0.06	0.6	A0 V		270 s		
75811	117214		8 52 24.1	+5 20 24	0.000	-0.02	6.33	0.12		A3	-6		7061	m
76214	236358		8 52 24.7	-55 20 31	-0.001	0.00	6.78	0.01	0.6	A0 V		170 s		m
76001	199728		8 52 26.1	-32 30 33	-0.001	-0.01	6.50	1.46		K2				
75716	80476	53 Cnc	8 52 28.5	+28 15 33	-0.001	-0.01	6.23	1.62	-0.5	M2 III	+11	210 s		m
75916	154704		8 52 30.6	-13 14 01	+0.001	-0.02	6.13	1.14	0.2	K0 III		120 s	7062	m
76017	199729		8 52 32.4	-31 10 04	0.000	+0.03	8.0	1.4	-0.1	K2 III		310 s		
76015	176616		8 52 32.4	-28 59 26	-0.002	-0.02	8.0	1.9	-0.1	K2 III		160 s		
76382	250362		8 52 32.7	-62 51 39	0.000	+0.01	7.9	0.1	0.6	A0 V		280 s		
75698	61102	51 σ¹ Cnc	8 52 34.6	+32 28 27	0.000	+0.01	5.7	0.1	1.7	A3 V	-23	64 s	7057	m
--	--		8 52 34.7	-12 43 51			6.20							
75800	98190		8 52 34.9	+18 11 29	-0.001	+0.02	7.2	0.4	2.6	F0 V		84 s		
75941	154706		8 52 35.3	-15 45 08	-0.002	-0.03	7.6	0.1	0.6	A0 V		250 s		
75732	80478	55 ρ¹ Cnc	8 52 35.8	+28 19 51	-0.037	-0.24	5.95	0.87	5.6	G8 V	+27	13 ts		
76016	176622		8 52 38.5	-29 59 24	-0.002	+0.01	7.5	1.9	0.2	K0 III		84 s		
76161	220609		8 52 38.6	-48 21 33	-0.001	-0.01	5.91	-0.15	-0.9	B6 V n	+3	230 s		
76072	199731		8 52 38.6	-36 32 45	-0.002	-0.02	6.42	0.56	0.3	G8 III		170 s		m
75554	27025		8 52 39.9	+49 56 42	+0.001	+0.01	7.8	0.5	4.0	F8 V		59 s		
76230	236362		8 52 40.7	-52 07 44	-0.002	0.00	6.39	0.00		A0				m

HD	SAO	Star Name	α 2000	δ 2000	μ(α)	μ(δ)	V	B-V	M$_v$	Spec	RV	d(pc)	ADS	Notes
76111	220603		8^{h}52^{m}41.2^s	−40°58'58"	−0.003^s	+0.02"	7.5	1.4	0.2	K0 III		170 s		
76186	220612		8 52 47.6	−47 23 38	−0.001	−0.01	6.81	−0.03	0.4	B9.5 V		190 s		
76110	199737		8 52 47.9	−38 43 27	−0.001	+0.01	5.82	1.50		gMa				
76614	250368		8 52 49.1	−69 49 49	−0.001	+0.01	7.45	−0.01	0.6	A0 V		230 s		
76056	199735		8 52 49.3	−32 34 02	−0.003	0.00	7.9	1.3	−0.3	K5 III		430 s		
75963	154710		8 52 52.3	−11 43 06	0.000	−0.02	8.0	0.9	3.2	G5 IV		93 s		
76160	220613		8 52 54.0	−43 45 12	−0.001	−0.01	7.59	1.08	0.2	K0 III		260 s		
75731	61107		8 52 54.4	+34 44 36	−0.003	−0.11	7.8	1.1	0.2	K0 III		110 mx		
76381	236369		8 52 56.4	−59 50 15	0.000	+0.01	7.8	0.3	0.6	A0 V		170 s		
75783	61109		8 53 00.0	+29 57 40	−0.002	−0.06	7.5	1.1	−0.1	K2 III		190 mx		
76346	236368		8 53 03.8	−56 38 58	−0.002	+0.03	6.03	−0.02		B9				
76027	154714		8 53 05.3	−16 57 10	−0.004	−0.07	6.6	1.1	0.2	K0 III		170 mx		
75487	27026		8 53 05.9	+59 03 22	+0.002	0.00	6.1	0.4	2.6	F0 V	+9	50 s		
75616	27028		8 53 06.0	+52 23 24	−0.004	+0.03	7.0	0.5	3.4	F5 V		53 s		
76043	176631		8 53 06.2	−22 07 00	−0.002	0.00	7.6	0.7	3.4	F5 V		49 s		
76086	176635		8 53 07.3	−28 21 29	−0.001	−0.01	7.6	0.4	0.2	K0 III		300 s		
75782	61111		8 53 11.4	+37 04 12	−0.009	−0.16	7.2	0.6	4.4	G0 V		36 s		
75864	98196		8 53 11.9	+17 22 07	−0.004	−0.04	6.87	0.40		F2				
76105	176637		8 53 19.3	−25 14 47	−0.002	+0.03	8.0	1.1	0.2	K0 III		330 s		
75899	98198		8 53 20.6	+19 19 20	0.000	−0.02	7.9	1.4	−0.3	K5 III		430 s		
75486	14691	5 UMa	8 53 22.5	+61 57 44	−0.001	+0.02	5.7	0.4	2.6	F0 V	−31	42 s		
76268	220618		8 53 22.6	−44 20 02	+0.003	−0.05	7.42	0.11	0.6	A0 V		190 s		
76104	176638		8 53 22.8	−22 59 35	−0.002	+0.01	8.0	0.3	0.4	B9.5 V		200 s		
76013	136370		8 53 24.7	−6 12 45	−0.001	−0.03	7.8	1.1	0.2	K0 III		330 s		
76181	199750		8 53 25.4	−34 49 10	−0.003	−0.02	7.9	2.3	−0.1	K2 III		84 s		V Pyx, v
76413	236377		8 53 26.9	−56 31 09	−0.003	+0.02	8.0	0.0	1.4	A2 V		210 s		
76053	154719		8 53 29.3	−11 23 31	0.000	+0.01	7.1	1.4	−0.3	K5 III		300 s		
76011	117236		8 53 33.8	+3 04 07	−0.001	+0.01	7.2	1.6		M5 III	+74			S Hya, v
76102	154721		8 53 36.3	−17 21 34	+0.001	0.00	7.4	0.5	3.4	F5 V		63 s		
76282	220624		8 53 36.6	−43 31 58	−0.002	+0.01	7.57	0.04	0.6	A0 V		230 s		
75976	98202		8 53 46.0	+13 49 50	+0.001	0.00	7.1	0.1	1.7	A3 V		120 s		
76538	250369		8 53 48.7	−60 21 15	−0.001	0.00	5.78	−0.08		B5	+2			
76360	220636		8 53 50.6	−47 31 15	0.000	−0.03	5.33	0.26		A m	−1	19 mn		m
75995	98204		8 53 50.7	+17 32 42	−0.001	−0.02	7.2	0.1	0.6	A0 V		210 s		
76305	220628		8 53 50.7	−40 26 51	−0.001	0.00	6.5	0.7	2.1	A5 V		34 s		
76050	117242		8 53 51.9	+1 49 27	+0.001	−0.05	7.30	0.4	2.6	F0 V		87 s	7074	m
76358	220637		8 53 54.9	−45 22 37	−0.001	+0.01	7.45	0.86	0.2	K0 III		280 s		
75974	80491		8 53 55.2	+19 58 02	−0.002	−0.01	6.68	0.70		F8				m
75896	61117		8 53 55.6	+35 32 17	−0.002	−0.03	6.00	0.04	0.0	A3 III	+22	160 s		
75994	98207		8 53 56.1	+18 14 00	+0.003	−0.05	7.80	0.8	3.2	G5 IV	−34	83 s		
76441	236383		8 53 56.4	−52 28 21	−0.002	+0.01	7.57	−0.09	0.0	B8.5 V		330 s		
76323	220634		8 53 56.5	−41 49 38	−0.004	0.00	7.40	0.00	0.4	B9.5 V		170 s		m
76037	117244		8 53 57.6	+8 25 18	−0.005	+0.02	8.00	0.5	4.0	F8 V		63 s	7073	m
76155	154727		8 53 58.7	−15 35 10	−0.003	−0.02	7.9	1.1	−0.1	K2 III		400 s		
75576	14693		8 53 59.0	+62 35 01	−0.020	−0.05	7.8	0.7	4.4	G0 V		49 s		
77049	256565		8 54 00.8	−74 53 05	−0.003	+0.03	7.8	1.1	−0.1	K2 III		380 s		
76341	220638		8 54 00.8	−42 29 10	+0.002	−0.01	7.17	0.30	−6.2	O9 I		3700 s		
76300	199763		8 54 01.0	−35 39 04	−0.007	+0.11	7.4	0.3	3.4	F5 V		63 s		
76263	199761		8 54 06.3	−31 14 39	0.000	0.00	7.5	1.6	−0.1	K2 III		190 s		
76357	220640		8 54 06.4	−43 25 17	−0.001	−0.02	7.96	1.44	−0.3	K5 III		450 s		
76458	236386		8 54 07.3	−51 42 00	−0.003	−0.01	7.70	−0.03	0.4	B9.5 V		290 s		
75895	61121		8 54 09.3	+39 15 17	−0.002	+0.01	7.5	0.4	3.0	F2 V		78 s		
76278	199765		8 54 09.7	−32 09 20	−0.004	+0.02	7.2	1.5	0.2	K0 III		120 s		
76380	220642		8 54 12.7	−43 29 54	−0.001	+0.02	6.97	1.64	−0.3	K5 III		230 s		
75959	61125	57 Cnc	8 54 14.6	+30 34 45	+0.003	−0.02	5.39	1.05	0.2	K0 III	−60	100 s	7071	m
76175	136388		8 54 14.8	−8 45 40	−0.001	0.00	7.5	0.1	1.4	A2 V		160 s	7080	m
76151	136389		8 54 17.9	−5 26 04	−0.028	+0.03	6.00	0.67	4.9	dG3	+26	12 ts		
76025	80496		8 54 18.6	+22 12 40	−0.009	−0.22	7.6	0.9	3.2	G5 IV		45 mx		
76246	154730		8 54 19.3	−18 51 49	−0.003	−0.02	8.0	1.1	−0.1	K2 III		410 s		
76439	220646		8 54 24.4	−45 27 15	−0.003	+0.01	7.94	−0.14	0.4	B9.5 V		320 s		
76081	98219		8 54 27.4	+14 14 41	−0.001	+0.02	8.0	0.9	3.2	G5 IV		93 s		
76149	117253		8 54 29.0	+2 05 09	−0.005	−0.09	7.9	0.5	4.0	F8 V		60 s		
76296	176679		8 54 29.5	−23 31 17	+0.001	+0.05	7.4	0.7	4.4	G0 V		33 s		
75933	42592		8 54 30.0	+40 07 59	+0.003	−0.16	7.5	0.8	3.2	G5 IV		57 mx		
76148	117252		8 54 30.7	+5 27 51	−0.001	−0.01	7.7	1.4	−0.3	K5 III		390 s		
76243	154732		8 54 38.5	−11 22 38	+0.002	+0.01	6.8	1.4	−0.3	K5 III		260 s		
76438	199779		8 54 49.4	−38 24 17	−0.001	−0.01	7.9	1.8	−0.3	K5 III		280 s		
77375	256567		8 54 52.1	−77 36 22	−0.003	0.00	8.0	0.1	0.6	A0 V		300 s		
76423	199778		8 54 52.3	−36 03 27	0.000	−0.08	7.40	1.1	0.2	K0 III		240 mx		m
76640	236402		8 54 53.9	−58 14 24	−0.003	0.00	6.38	−0.11	−2.2	B5 III		480 s		
76274	136400		8 54 54.7	−7 49 07	0.000	−0.02	7.90	0.2	2.1	A5 V		150 s	7086	m
76222	117259		8 54 55.4	+2 01 00	−0.002	−0.01	8.0	1.1	0.2	K0 III		360 s		
76095	80503		8 54 56.0	+26 11 52	+0.005	−0.46	6.75	0.67	4.5	dG1	+37	25 s	7082	m
——	27042		8 54 56.2	+52 57 22	−0.001	0.00	7.6	1.3	0.2	K0 III		200 s		
75894	27044		8 54 57.4	+52 47 39	−0.001	−0.04	7.4	0.0	0.4	B9.5 V		250 s		

HD	SAO	Star Name	α 2000	δ 2000	μ(α)	μ(δ)	V	B-V	M_V	Spec	RV	d(pc)	ADS	Notes
			h m s	° ′ ″	s	″								
76321	154737		8 54 57.6	-15 46 46	-0.008	+0.05	7.2	1.1	0.2	K0 III		160 mx		
76273	136401		8 54 58.3	-3 33 20	-0.001	+0.02	7.02	0.20		A2				
76354	154740		8 54 58.7	-18 48 44	-0.013	+0.01	7.8	0.8	3.2	G5 IV		84 s		
76069	61131		8 54 58.8	+32 44 37	-0.001	-0.04	7.9	0.1	0.6	A0 V		280 s		
76312	154735		8 54 58.8	-10 59 09	0.000	-0.02	7.6	1.1	0.2	K0 III		310 s		
76469	199784		8 55 00.0	-39 47 12	-0.002	-0.02	7.5	1.2	-0.1	K2 III		330 s		
76612	236400		8 55 00.4	-54 41 59	-0.002	+0.01	7.2	0.4	4.0	F8 V		43 s		
76437	199780		8 55 01.4	-34 08 36	-0.001	-0.07	7.2	1.2	0.2	K0 III		200 s		
76516	220656		8 55 02.2	-43 49 52	-0.004	+0.01	7.98	0.03	0.6	A0 V		230 mx		
76403	176686		8 55 02.3	-28 04 59	-0.002	-0.02	7.6	0.9	0.2	K0 III		300 s		
76728	250374		8 55 02.8	-60 38 41	-0.003	+0.04	3.84	-0.10	-1.0	B8.5 III	+25	93 s		m
75764	14696		8 55 08.5	+63 28 06	-0.009	-0.05	8.0	0.9	3.2	G5 IV		93 s		
76534	220660		8 55 08.5	-43 28 01	-0.001	-0.01	8.04	0.13	-1.7	B3 V ne		890 s		m
76567	220661		8 55 08.8	-46 21 39	-0.002	-0.02	7.56	0.02	0.6	A0 V		230 s		
76262	117263		8 55 10.5	+2 37 43	+0.001	-0.02	7.7	1.1	0.2	K0 III		320 s		
76653	236405		8 55 11.9	-54 57 56	+0.004	-0.08	5.71	0.48	4.0	dF8	-2	22 s		
76376	154745		8 55 12.4	-18 14 29	+0.001	0.00	5.75	1.32		K0				m
76377	154746		8 55 14.6	-18 15 27	-0.002	-0.02	7.1	1.1	-0.1	K2 III		280 s		
76352	154743		8 55 17.2	-11 22 18	0.000	-0.03	6.80	0.1	0.6	A0 V		170 s		m
76514	199789		8 55 18.4	-38 19 38	0.000	+0.04	7.3	1.9	-0.1	K2 III		110 s		
76566	220664		8 55 19.2	-45 02 29	0.000	+0.01	6.26	-0.16	-1.7	B3 V		390 s		m
76468	199787		8 55 21.6	-31 01 12	0.000	+0.01	7.9	1.3	-0.1	K2 III		340 s		
76334	136405		8 55 22.6	-3 08 52	+0.001	-0.04	7.1	1.1	-0.1	K2 III		280 s		
76221	98230		8 55 22.8	+17 13 53	0.000	0.00	6.64	3.36		N7.7	-1			X Cnc, v
76294	117264	16 ζ Hya	8 55 23.6	+5 56 44	-0.007	+0.01	3.11	1.00	0.2	K0 III	+23	38 s		
76118	61135		8 55 25.2	+35 07 11	-0.004	-0.04	7.8	1.1	0.2	K0 III		310 mx		
76484	199790		8 55 25.2	-33 24 06	-0.001	0.00	7.1	0.1	2.6	F0 V		78 s		
76311	117265		8 55 26.0	+5 33 58	-0.002	0.00	7.7	1.1	-0.1	K2 III		360 s		
76533	199793		8 55 27.1	-36 48 43	-0.003	+0.02	7.9	1.7	0.2	K0 III		140 s		
76192	80508		8 55 27.3	+26 24 36	-0.002	-0.02	7.2	0.1	0.6	A0 V		210 s		
76390	154752		8 55 27.7	-14 24 19	0.000	+0.02	8.0	1.4	-0.3	K5 III		450 s		
76369	136408	17 Hya	8 55 29.4	-7 58 13	0.000	-0.04	6.08	0.22	1.7	A3 V	-20	64 s	7093	m
76483	176697	δ Pyx	8 55 31.5	-27 40 55	+0.006	-0.11	4.89	0.11	1.7	A3 V	+5	42 s	7095	m
74776	1412		8 55 33.2	+80 01 49	-0.010	-0.01	7.5	0.1	0.6	A0 V		240 s		
76825	250377		8 55 34.1	-61 12 19	+0.003	+0.01	8.0	0.1	0.6	A0 V		240 s		
76501	199794		8 55 35.1	-31 17 19	-0.001	0.00	7.9	1.1	3.2	G5 IV		56 s		
76500	199795		8 55 36.0	-31 11 23	-0.001	-0.02	7.9	1.5	0.2	K0 III		180 s		
76293	98232		8 55 39.1	+14 10 52	+0.001	-0.03	7.9	0.6	4.4	G0 V		50 s		
76219	80511	58 ρ² Cnc	8 55 39.6	+27 55 39	-0.001	-0.04	5.22	1.00	-0.9	G8 II-III	+17	170 s		
76400	136414		8 55 39.7	-9 08 29	-0.002	0.00	7.2	1.6		M5 III	-3			T Hya, v
76366	136413		8 55 40.5	-2 47 26	-0.001	+0.03	7.0	1.1	0.2	K0 III	+30	230 s		
76564	199799		8 55 47.8	-34 16 09	-0.001	-0.01	7.4	0.8	-1.0	B5.5 V		140 s		
76726	236410		8 55 49.8	-51 35 11	-0.003	+0.01	7.6	0.3	1.4	A2 V		120 s		
76649	220671		8 55 50.4	-46 20 32	-0.001	-0.02	8.0	0.4	0.0	B8.5 V		200 s		
76079	42600		8 55 53.0	+49 05 14	0.000	-0.04	7.8	0.8	3.2	G5 IV		83 s		
76478	154760		8 55 54.1	-15 26 20	-0.002	-0.02	6.6	1.1	0.2	K0 III		190 s		
76351	98235	60 Cnc	8 55 55.5	+11 37 34	-0.001	-0.01	5.41	1.46	0.2	K0 III	+24	58 s		
76218	61138		8 55 55.6	+36 11 44	-0.002	-0.05	7.9	0.9	3.2	G5 IV		88 s		
76512	176711		8 55 55.9	-23 49 03	-0.002	+0.07	6.39	0.16		A3				
76399	117275		8 55 56.5	+0 53 10	-0.001	-0.01	7.8	1.1	0.2	K0 III		330 s		
76551	176714		8 55 57.0	-26 38 50	+0.001	-0.01	8.0	0.8	0.2	K0 III		360 s		
76240	61140		8 56 06.8	+35 23 52	+0.002	-0.11	8.0	1.1	0.2	K0 III		100 mx		
76261	61139		8 56 07.1	+36 26 05	-0.008	-0.28	7.8	0.8	3.2	G5 IV		34 mx		
75973	14702		8 56 07.9	+60 20 11	-0.003	-0.01	7.7	1.1	0.2	K0 III		310 s		
76824	236416		8 56 11.1	-55 31 42	-0.002	+0.01	7.70	1.1	0.2	K0 III		310 s		m
76510	154766		8 56 12.1	-13 54 17	0.000	0.00	8.00	-0.18	-3.5	B1 V	+22	1900 s		
76333	80519		8 56 13.4	+24 26 51	+0.007	-0.08	6.7	0.4	3.0	F2 V		56 s		
76918	250382		8 56 17.3	-60 28 19	-0.001	0.00	7.4	1.7	-0.3	K5 III		280 s		
76607	176722		8 56 18.8	-29 27 08	-0.001	0.00	8.0	1.6	0.2	K0 III		160 s		
76805	236417		8 56 19.3	-52 43 25	0.000	+0.01	4.69	-0.12	-1.1	B5 V	+22	140 s		m
76668	199803		8 56 19.6	-37 07 21	-0.016	-0.01	6.85	0.62		G0		16 mn		m
76260	61141		8 56 22.4	+38 48 42	-0.001	-0.04	7.34	1.34	0.2	K0 III		140 s		
73835	1404		8 56 23.5	+83 53 14	-0.092	-0.12	7.5	0.5	4.0	F8 V		50 s		
76637	176726		8 56 28.8	-29 20 42	+0.002	0.00	7.5	0.6	1.4	A2 V		75 s		
77513	—		8 56 29.5	-76 23 21			7.9	0.8	3.2	G5 IV		86 s		
76292	42611		8 56 30.4	+40 12 04	-0.007	-0.05	5.9	0.4	3.0	F2 V	+26	38 s		
76238	42610		8 56 33.8	+45 45 50	-0.002	-0.01	7.1	0.5	2.6	F0 V	-8	74 s		
76579	154773		8 56 34.0	-16 42 34	-0.001	-0.05	5.96	1.54		K0				
76494	117287		8 56 36.8	+4 14 12	-0.001	0.00	6.14	1.00		G5	-12			
75958	14703	6 UMa	8 56 37.5	+64 36 14	-0.004	-0.08	5.58	0.86	0.3	gG5	+3	110 s		
76349	61142		8 56 39.4	+33 15 28	0.000	-0.03	7.3	0.6	4.4	G0 V		38 s		
77034	250391		8 56 40.4	-63 32 43	-0.004	+0.01	7.7	0.4	2.6	F0 V		100 s		
79837	258515	ζ Oct	8 56 42.2	-85 39 47	-0.095	+0.03	5.42	0.31	0.6	F0 III	-3	92 s		
77033	250390		8 56 42.2	-63 02 55	-0.004	0.00	7.6	0.0	0.4	B9.5 V		270 s		
76445	98241		8 56 46.0	+17 28 52	+0.003	-0.02	7.8	0.8	3.2	G5 IV		84 s		

HD	SAO	Star Name	α 2000	δ 2000	μ(α)	μ(δ)	V	B-V	M_V	Spec	RV	d(pc)	ADS	Notes
			h m s	° ′ ″	s	″								
76788	220683		8 56 46.8	-43 39 24	-0.002	0.00	7.51	1.35	-0.3	K5 III		370 s		
76545	136425		8 56 46.9	-0 24 52	-0.001	0.00	7.86	1.34	-0.1	K2 III		300 s		
76635	154781		8 56 49.9	-17 26 02	-0.001	-0.01	7.30	0.4	3.0	F2 V		72 s	7103	m
76291	42612		8 56 49.9	+45 37 55	-0.012	-0.04	5.74	1.09		K1 IV	+59	94 mx		
--	--		8 56 50.6	+23 50 49			8.00	1.6		R6				
75972	14706		8 56 54.6	+65 31 33	+0.001	-0.02	7.4	0.9	3.2	G5 IV	-34	70 s		
77440	256569		8 56 56.3	-73 55 19	+0.001	+0.02	7.3	1.1	-0.1	K2 III		300 s		
76398	61146	59 σ² Cnc	8 56 56.6	+32 54 37	-0.004	-0.07	5.45	0.12		A3	+5	28 mn		
76169	27052		8 56 56.9	+56 53 20	+0.003	-0.01	7.2	1.1	-0.1	K2 III		290 s		
76875	236429		8 56 57.0	-50 21 24	-0.003	+0.01	7.72	1.11	-0.1	K2 III		370 s		
77002	236436		8 56 58.3	-59 13 45	-0.002	+0.01	4.92	-0.19	-2.3	B3 IV	+25	280 s		
76860	220691		8 56 58.5	-49 40 52	0.000	0.00	7.13	1.71	0.2	K0 III		78 s		
77348	--		8 56 58.6	-72 24 17			7.9	0.4	3.0	F2 V		96 s		
--	220686		8 57 03.5	-43 15 46	-0.004	0.00	7.60	0.00	0.0	B8.5 V		240 mx		m
76397	61148		8 57 04.9	+35 01 24	-0.001	-0.02	7.2	1.1	0.2	K0 III		250 s		
76475	80529		8 57 05.1	+21 51 37	0.000	-0.01	7.0	0.8	3.2	G5 IV		58 s		
--	--		8 57 06.7	-43 15 06			7.60							m
76217	27055		8 57 07.1	+54 40 51	+0.001	-0.02	8.0	1.1	0.2	K0 III		360 s		
76348	42613		8 57 07.4	+44 24 33	+0.004	+0.01	6.9	0.9	3.2	G5 IV		55 s		
76838	220689		8 57 07.6	-43 15 23	0.000	-0.01	7.31	0.00	-1.7	B3 V		550 s		m
76508	98245		8 57 08.2	+17 08 38	-0.003	-0.03	6.17	1.00	0.0	K1 III	+19	170 s		
77324	256568		8 57 08.5	-71 01 00	-0.001	+0.01	6.8	1.1	-0.1	K2 III		240 s		
--	--		8 57 11.3	+36 57 03	-0.001	-0.03	7.7	0.9	3.2	G5 IV		81 s		
76290	27058		8 57 11.7	+50 44 29	-0.003	-0.05	7.2	0.1	0.6	A0 V		73 mx		
77076	250392		8 57 12.5	-61 33 48	+0.001	-0.01	8.0	0.3	0.6	A0 V		180 s		
76583	117292		8 57 13.7	+2 31 29	-0.001	-0.02	6.9	0.4	3.0	F2 V		59 s		
76600	117293		8 57 13.8	+0 32 07	0.000	0.00	7.67	0.93	3.2	G5 IV		60 s		
76543	98247	62 o Cnc	8 57 14.9	+15 19 21	+0.004	+0.01	5.20	0.15		A3	0			
77032	236445		8 57 15.4	-58 35 09	0.000	0.00	8.0	0.2	-1.0	B5.5 V		390 s		
76542	98248		8 57 18.5	+16 16 57	0.000	-0.01	7.9	0.7	4.4	G0 V		51 s		
76461	61152		8 57 18.7	+32 26 43	-0.005	-0.01	6.9	0.2	2.1	A5 V		90 s		
76331	42614		8 57 20.1	+48 28 41	+0.004	-0.03	8.0	0.9	3.2	G5 IV		93 s		
77187	250396		8 57 20.7	-66 12 14	-0.010	+0.04	7.30	1.1	0.2	K0 III		260 s		m
77289	250401		8 57 22.3	-69 38 30	-0.005	+0.01	7.6	0.1	1.4	A2 V		170 s		
76272	27059		8 57 22.4	+54 50 46	-0.008	-0.01	7.8	0.5	4.0	F8 V		58 s		
76216	27056		8 57 25.2	+58 13 08	-0.006	+0.01	6.9	0.1	1.4	A2 V	-27	90 mx		
76898	220696		8 57 28.3	-44 15 56	-0.001	0.00	7.40	-0.16	-1.1	B5 V n		480 s		
76968	236442		8 57 28.9	-50 44 58	0.000	-0.01	7.08	0.13	-6.2	O9 I k		3500 s		
76708	136442		8 57 30.3	-10 05 32	-0.001	-0.04	7.8	1.1	0.2	K0 III		340 s		
76800	176746		8 57 32.6	-27 42 58	-0.001	-0.03	7.5	0.9	1.4	A2 V		51 s		
76573	98249		8 57 33.9	+16 14 01	-0.002	-0.05	7.8	1.1	0.2	K0 III		330 s		
76872	220697		8 57 34.9	-41 21 05	-0.002	-0.03	8.0	0.1	0.6	A0 V		250 s		
76582	98250	63 Cnc	8 57 35.1	+15 34 53	+0.004	+0.02	5.67	0.20		A5	-4			
76799	176747		8 57 40.4	-24 50 29	+0.006	-0.05	7.3	0.2	0.2	K0 III		270 s		
76629	117301		8 57 41.9	+9 23 16	-0.001	0.00	6.19	0.98	0.3	G8 III	-14	140 s		
76897	199829		8 57 43.7	-38 44 34	0.000	0.00	7.8	-0.5	0.4	B9.5 V		300 s		
76493	61155		8 57 46.6	+38 15 34	-0.009	-0.07	6.9	0.5	3.4	F5 V		50 s		
76784	154794		8 57 50.9	-18 53 21	-0.002	-0.04	7.10	0.1	0.6	A0 V		200 s	7112	m
76954	220699		8 57 53.3	-42 31 30	-0.001	-0.01	7.89	-0.12	0.4	B9.5 V		320 s		
77020	220703		8 57 55.7	-48 34 23	0.000	-0.01	5.87	1.07	0.2	gK0		120 s		
76735	136450		8 57 58.5	-4 51 34	0.000	-0.04	6.6	0.4	3.0	F2 V		52 s		
76572	61157	61 Cnc	8 57 58.7	+30 14 01	+0.004	+0.02	6.29	0.43	3.4	F5 V	+8	38 s	7107	m
76757	117311		8 58 08.2	+1 32 29	-0.002	-0.01	6.59	0.06		A0	+26			m
76913	199834		8 58 13.0	-30 57 59	-0.003	+0.08	8.0	0.8	3.4	F5 V		49 s		
76892	176754		8 58 14.7	-26 21 00	+0.002	-0.01	7.9	1.2	0.6	A0 V		56 s		
76952	199836		8 58 14.7	-34 24 32	-0.003	0.00	7.8	0.8	0.4	B9.5 V		100 s		
76023	6699		8 58 17.5	+70 39 17	-0.007	-0.03	7.07	1.18	-0.1	K2 III		270 s		
76507	42623		8 58 18.0	+46 45 05	0.000	-0.04	7.9	1.1	0.2	K0 III		340 s		
76706	98265		8 58 18.2	+18 18 29	-0.002	-0.03	6.6	0.1	0.6	A0 V		160 s		
77185	236458		8 58 18.2	-59 11 39	0.000	-0.01	7.83	-0.06	0.4	B9.5 V		310 s		
76950	199838		8 58 20.4	-32 08 31	-0.004	-0.04	8.0	0.2	3.4	F5 V		82 s		
75544	6692		8 58 22.7	+78 08 44	-0.006	-0.01	7.4	0.4	2.6	F0 V		93 s		
76522	42625		8 58 27.3	+48 02 56	-0.003	-0.03	8.0	1.1	0.2	K0 III		360 s		
76595	61161		8 58 27.4	+35 48 08	+0.001	-0.01	6.50	-0.01	1.2	A1 V	-15	120 s		
76428	27064		8 58 29.0	+55 51 41	-0.003	-0.01	7.8	0.8	3.2	G5 IV		84 s		
76756	98267	65 α Cnc	8 58 29.2	+11 51 28	+0.002	-0.03	4.25	0.14		A m	-14	23 mn	7115	Acubens, m
77119	236455		8 58 30.5	-50 14 59	-0.004	+0.02	7.25	1.07	0.2	K0 III		240 s		
77120	236456		8 58 30.7	-50 48 24	-0.003	+0.04	7.57	1.22	-0.3	K5 III		380 s		
77244	236460		8 58 34.5	-59 44 16	-0.002	+0.02	7.64	-0.09	0.0	B8.5 V		340 s		
76911	154802		8 58 36.3	-17 21 14	0.000	-0.01	8.0	1.1	0.2	K0 III		360 s		
76932	154804		8 58 43.8	-16 07 57	+0.016	+0.21	5.86	0.53	4.2	F9 V	+122	22 s		
76733	80547		8 58 45.3	+23 57 59	-0.001	0.00	7.6	0.1	1.4	A2 V		170 s		
77243	236462		8 58 45.3	-57 28 21	0.000	0.00	7.9	1.3	-0.1	K2 III		350 s		
77017	199846		8 58 45.8	-31 14 41	-0.001	-0.02	7.5	0.4	0.6	A0 V		130 s		
77373	250407		8 58 46.1	-64 31 51	0.000	-0.01	6.5	1.1	0.2	K0 III		180 s		

HD	SAO	Star Name	α 2000	δ 2000	μ(α)	μ(δ)	V	B-V	M_V	Spec	RV	d(pc)	ADS	Notes
			h m s	° ′ ″	s	″								
76947	154806		8 58 46.5	−16 44 57	+0.002	−0.03	7.1	1.1	0.2	K0 III		240 s		
76271	14718		8 58 46.9	+66 00 06	−0.002	−0.03	7.3	0.4	2.6	F0 V		85 s		
77018	199847		8 58 46.9	−31 57 58	0.000	−0.02	7.7	0.5	0.6	A0 V		130 s		
75089	1421		8 58 48.2	+81 17 06	−0.012	−0.01	7.8	0.4	2.6	F0 V		110 s		
76752	80548		8 58 49.8	+25 24 18	−0.005	−0.12	7.50	0.65	4.7	G2 V		34 s		
77140	220717		8 58 52.2	−47 14 05	−0.009	+0.05	5.18	0.25		A m	+20	22 mn		m
76962	154808		8 58 55.3	−15 59 05	−0.002	0.00	6.9	1.1	−0.1	K2 III		250 s		
77226	236463		8 58 55.3	−54 38 01	−0.006	+0.03	8.0	0.0	0.4	B9.5 V		140 mx		
76780	80552		8 58 55.6	+21 09 58	−0.006	−0.05	7.6	0.9	3.2	G5 IV		74 s		
77493	250412		8 58 55.9	−68 08 20	0.000	+0.01	6.92	−0.06	0.4	B9.5 V		200 s		
76766	80550		8 58 56.2	+25 55 33	−0.005	+0.01	7.94	0.51	4.0	F8 V		61 s		
76868	117318		8 58 59.3	+3 39 21	0.000	−0.02	8.00	0.20		B5 e	−15			
76362	14723		8 59 00.6	+63 25 43	−0.003	−0.02	7.20	0.9	3.2	G5 IV		63 s	7106	m
77233	236464		8 59 01.7	−53 20 39	−0.004	−0.01	7.83	1.39	−0.1	K2 III		260 s		
76384	14727		8 59 05.7	+63 21 08	−0.004	−0.05	7.0	0.4	2.6	F0 V		75 s		
76830	98276		8 59 10.7	+18 08 05	−0.003	−0.07	6.38	1.55	−0.5	M4 III	+21	170 mx		
76644	42630	9 ι UMa	8 59 12.4	+48 02 29	−0.044	−0.24	3.14	0.19	2.4	A7 V	+12	15 ts	7114	Talitha, m
77015	154810		8 59 14.1	−16 30 11	0.000	−0.02	7.60	1.1	0.0	K1 III	+56	330 s		
77087	176789		8 59 15.6	−28 48 22	−0.005	+0.01	6.25	1.00		G5				
76884	117321		8 59 19.4	+8 58 56	−0.003	−0.01	7.5	0.0	0.4	B9.5 V		270 s		
76993	136466		8 59 19.8	−8 59 24	−0.002	0.00	7.8	1.1	0.2	K0 III		330 s		
77455	250413		8 59 20.8	−63 28 02	−0.003	0.00	7.0	0.1	0.6	A0 V		180 s		
77224	220724		8 59 22.2	−48 17 30	−0.001	−0.03	7.33	1.14	0.2	K0 III		210 s		
76656	42633		8 59 22.7	+48 16 18	−0.008	+0.03	8.0	0.5	3.4	F5 V		82 s		
77370	236475		8 59 24.1	−59 05 01	−0.023	+0.28	5.16	0.42	2.8	dF1	+11	24 ts		m
76704	42634		8 59 24.2	+45 44 19	−0.003	−0.03	6.6	0.1	0.6	A0 V	+3	160 s		
76908	98278		8 59 27.0	+13 04 27	0.000	−0.02	6.7	1.4	−0.3	K5 III		250 s		
77369	236477		8 59 28.1	−58 44 27	0.000	0.00	7.9	1.1	0.2	K0 III		320 s		
77085	176797		8 59 30.5	−23 10 02	−0.002	+0.02	8.00	1.1	−0.1	K2 III		420 s		m
76813	61177	64 σ³ Cnc	8 59 32.6	+32 25 07	−0.004	−0.04	5.20	0.93	3.2	G5 IV	+23	20 s		m
76977	117328		8 59 35.5	+1 24 42	+0.001	−0.02	7.7	0.9	3.2	G5 IV		80 s		
77616	—		8 59 36.9	−68 44 29			8.0	1.1	0.2	K0 III		360 s		
77084	154815		8 59 39.8	−19 12 28	−0.004	−0.10	6.18	0.46		F8				
77071	154814		8 59 41.4	−15 00 03	−0.001	−0.01	8.0	1.4	−0.3	K5 III		450 s		
77137	176805		8 59 42.5	−27 48 56	−0.005	−0.03	6.9	0.9	4.4	G0 V		20 s		
77347	236480		8 59 44.4	−54 14 14	0.000	−0.03	7.0	1.2	2.6	F0 V		21 s		
77305	236476		8 59 45.2	−51 30 14	−0.005	+0.03	7.3	0.4	1.4	A2 V		90 s		
76829	61181		8 59 46.3	+34 49 29	−0.003	−0.07	7.5	0.7	4.4	G0 V		42 s		
77127	176806		8 59 53.6	−22 41 31	−0.006	+0.01	7.2	−0.1	2.6	F0 V		84 s		
77656	250416		8 59 54.5	−68 05 34	−0.005	+0.02	7.2	0.0	0.4	B9.5 V		230 s		
77069	136478		8 59 57.6	−2 32 52	0.000	−0.01	7.9	1.1	−0.1	K2 III		400 s		
76924	80562		8 59 57.8	+26 26 04	−0.004	+0.02	6.60	0.1	1.4	A2 V		110 s	7123	m
77321	220735		9 00 01.0	−49 33 29	−0.002	0.00	6.88	1.32	0.2	K0 III		110 s		m
77287	220731		9 00 02.1	−44 24 47	−0.002	−0.04	7.9	1.0	−0.3	K5 III		430 s		
74924	1422		9 00 04.5	+82 43 10	0.000	−0.01	7.5	1.1	0.2	K0 III		280 s		
77258	220730		9 00 05.4	−41 15 14	−0.004	+0.04	4.45	0.65	2.4	F8 IV	−7	20 s		
77981	256573		9 00 06.2	−74 23 59	−0.007	+0.02	6.7	0.0	0.4	B9.5 V		180 s		
77554	250414		9 00 07.2	−62 59 39	0.000	0.00	8.0	0.0	0.0	B8.5 V		390 s		
76592	14735		9 00 16.2	+61 56 41	−0.008	−0.05	7.8	1.1	0.2	K0 III		240 mx		
76749	27078		9 00 18.7	+53 19 52	+0.003	0.00	8.0	0.5	3.4	F5 V		82 s		
77151	154823		9 00 20.0	−12 14 24	−0.001	+0.03	7.8	1.4	−0.3	K5 III		410 s		
77134	136486		9 00 21.5	−9 07 25	+0.004	−0.18	7.8	0.5	3.4	F5 V		63 mx		
77320	220738		9 00 22.1	−43 10 24	−0.002	+0.02	6.07	−0.16	−2.1	B2.5 V n		420 s		
76944	61191		9 00 30.8	+37 36 15	−0.001	0.00	6.5	1.4	−0.3	K5 III	−17	230 s		
77832	—		9 00 31.6	−70 09 36			7.8	0.5	3.4	F5 V		75 s		
77177	154826		9 00 32.7	−12 08 29	−0.005	−0.07	7.4	0.4	2.6	F0 V		92 s		
77200	154829		9 00 33.2	−16 09 13	−0.004	+0.01	7.9	0.2	2.1	A5 V		150 s		
77464	236495		9 00 37.9	−51 33 20	−0.001	0.00	6.70	−0.16	−2.5	B2 V		660 s		CV Vel, m,v
77232	176820		9 00 37.9	−21 12 56	−0.004	+0.04	7.4	0.3	2.6	F0 V		91 s		
76943	42642		9 00 38.3	+41 46 57	−0.039	−0.25	3.97	0.44	3.4	F5 V	+26	14 ts		m
77490	236500		9 00 41.5	−54 21 45	+0.001	0.00	7.69	0.02	0.4	B9.5 V		260 s		
77366	199881		9 00 42.1	−38 25 09	−0.001	+0.01	7.28	−0.12	−1.0	B5.5 V		450 s		
77615	250417		9 00 45.7	−60 57 50	−0.001	0.00	5.79	1.21		K0				
77196	136489		9 00 45.8	−9 11 29	+0.003	−0.02	7.30	0.1	0.6	A0 V		220 s	7136	m
77123	98290		9 00 50.5	+12 34 23	+0.001	−0.02	8.0	0.1	0.6	A0 V		300 s		
77299	176824		9 00 51.9	−28 16 24	−0.002	−0.02	8.0	1.8	3.2	G5 IV		22 s		
77285	176822		9 00 53.3	−24 09 13	−0.002	0.00	7.1	0.4	0.6	A0 V		110 s		
77489	236503		9 00 53.5	−50 26 10	−0.008	+0.01	7.95	1.01	0.2	K0 III		240 mx		
78009	—		9 00 57.0	−72 26 42			7.7	0.1	0.6	A0 V		260 s		
77298	176825		9 00 57.6	−24 58 11	−0.004	+0.02	7.4	0.7	2.6	F0 V		52 s		
77231	136494		9 01 02.0	−9 13 11	0.000	−0.02	7.3	0.1	0.6	A0 V		220 s		
77339	176829		9 01 06.1	−27 30 59	−0.002	−0.04	7.5	1.6	−0.5	M2 III		380 s		
77193	117345		9 01 06.2	+1 16 48	−0.003	−0.01	7.5	0.1	0.6	A0 V		250 s		
77511	220759		9 01 08.3	−47 15 05	−0.004	+0.02	7.74	0.23		A m				
77887	250421		9 01 08.5	−68 41 01	+0.003	0.00	5.88	1.63	−0.3	gK6		150 s		

HD	SAO	Star Name	α 2000	δ 2000	μ(α)	μ(δ)	V	B-V	M$_v$	Spec	RV	d(pc)	ADS	Notes
77431	199888		9^h 01^m 09.4^s	-38° 55′ 03″	-0.003^s	+0.03″	7.9	0.7	-0.1	K2 III		390 s		
77283	154840		9 01 09.5	-19 20 33	-0.001	-0.03	7.5	0.1	1.4	A2 V		170 s		
77361	176833		9 01 11.3	-26 39 50	+0.002	-0.06	6.20	1.13		K0				
77236	136495		9 01 14.7	-2 33 41	-0.010	-0.05	7.54	1.15	-0.1	K2 III	+142	340 s		
77382	176836		9 01 15.0	-28 50 01	-0.001	+0.03	8.0	1.1	-0.1	K2 III		410 s		
77796	--		9 01 18.1	-64 55 37			7.9	1.1	-0.1	K2 III		400 s		
77475	220760		9 01 20.7	-41 51 51	-0.003	0.00	5.55	-0.14		B7	+22			
77397	176839		9 01 21.5	-28 19 29	-0.002	-0.01	8.0	0.7	0.6	A0 V		100 s		
77266	136498		9 01 23.0	-3 22 07	-0.001	-0.03	7.6	0.2	2.1	A5 V		130 s		
77282	136501		9 01 23.0	-7 55 26	0.000	+0.02	7.9	1.1	0.2	K0 III		340 s		
77104	61202	66 σ⁴ Cnc	9 01 24.0	+32 15 08	0.000	0.00	5.80	0.00	1.7	A3 V	-13	66 s	7137	m
77415	199889		9 01 24.4	-30 42 25	-0.001	-0.01	7.5	2.2	-0.1	K2 III		80 s		
77316	154842		9 01 26.2	-11 14 18	0.000	-0.01	8.0	1.1	-0.1	K2 III		410 s		
77211	98301		9 01 29.8	+15 11 34	-0.003	+0.02	8.0	0.4	3.0	F2 V		98 s		
77250	117351		9 01 31.3	+5 38 27	-0.002	-0.01	6.07	1.12	-1.1	K1 II-III	+33	270 s		m
77296	136502		9 01 31.3	-4 27 22	-0.003	+0.01	6.7	1.4	-0.3	K5 III		250 s		
77566	220763		9 01 31.4	-49 34 44	-0.002	0.00	7.45	-0.12		B8.5 V		300 s		
77092	42647		9 01 32.3	+40 56 24	-0.008	-0.06	7.8	0.5	3.4	F5 V		74 s		
77281	136505		9 01 37.8	-1 28 37	-0.002	-0.03	7.5	0.1	1.4	A2 V		170 s		
77280	136504		9 01 38.4	-0 34 54	0.000	-0.01	7.5	0.1	1.4	A2 V		170 s		
77093	61203		9 01 40.5	+39 42 48	-0.005	-0.08	6.36	0.30	2.6	F0 V	-8	57 s		
77122	61205		9 01 43.3	+35 51 32	-0.009	-0.07	8.0	0.5	3.4	F5 V		82 s		
77450	176844		9 01 43.3	-25 25 57	-0.002	+0.03	7.7	0.6	0.2	K0 III		320 s		
77230	98302		9 01 43.5	+17 04 52	-0.002	-0.01	7.1	0.0	0.4	B9.5 V	+7	220 s		
77653	236518		9 01 44.6	-52 11 18	-0.001	+0.02	5.23	-0.12		A pe	+32			m
77190	80585	67 Cnc	9 01 48.8	+27 54 10	-0.004	-0.08	6.10	0.2	2.1	A5 V	+12	42 mx		m
77462	176847		9 01 49.0	-26 33 49	-0.018	+0.28	7.28	0.51		F8		13 mn		
77293	117361		9 01 49.1	+5 39 07	-0.003	-0.02	7.19	0.30	1.9	F3 IV		110 s		
83745	--		9 01 49.4	-88 17 15			6.7	0.8	3.2	G5 IV		51 s		
76921	27089		9 01 53.6	+59 24 58	-0.001	-0.02	7.86	1.06	0.2	K0 III		300 s		
77354	136509		9 01 56.8	-1 07 30	-0.002	-0.04	7.8	0.5	4.0	F8 V		57 s		
77381	136513		9 01 57.2	-8 22 26	-0.002	-0.05	7.3	1.1	0.2	K0 III		260 s		
77353	136511		9 01 57.9	-0 28 58	-0.003	+0.07	5.67	1.15	0.2	K0 III	+73	97 s		
77248	98303		9 01 58.5	+17 19 36	-0.003	-0.01	8.0	0.9	3.2	G5 IV		93 s		
77314	117363		9 01 58.7	+2 40 16	0.000	-0.02	7.10	0.1	1.4	A2 V		140 s	7152	m
78312	--		9 02 01.3	-74 11 51			7.9	0.1	0.6	A0 V		290 s		
77536	199898		9 02 01.7	-35 41 17	-0.003	+0.03	7.5	0.2	0.4	B9.5 V		180 s		
77506	176851		9 02 04.9	-27 48 50	-0.003	-0.04	8.0	0.7	3.4	F5 V		61 s		
77507	176852		9 02 05.8	-28 09 09	-0.005	-0.05	7.5	1.2	1.4	A2 V		37 s		
77580	199902		9 02 06.3	-39 24 09	-0.004	+0.02	6.27	1.00		K1 III-IV		400 mx		
77581	220767		9 02 06.7	-40 33 17	-0.001	0.00	6.89	0.47		B0.5 Ib				
77189	61213		9 02 19.0	+38 39 22	+0.002	-0.05	7.30	1.4	-0.3	K5 III	+8	330 s		
77447	136521		9 02 24.0	-2 21 56	-0.003	-0.02	7.6	0.5	3.4	F5 V		71 s		
77795	236528		9 02 24.5	-55 01 51	-0.006	-0.01	7.56	0.05	0.6	A0 V		150 mx		
77592	199906		9 02 25.7	-34 14 44	-0.001	0.00	8.0	0.2	1.4	A2 V		160 s		
78045	250422	α Vol	9 02 26.9	-66 23 46	+0.001	-0.10	4.00	0.14	2.1	A5 V	+5	24 s		
76827	14742	8 ρ UMa	9 02 32.7	+67 37 47	-0.004	+0.02	4.76	1.53		Ma III	+5	23 mn		
77313	80592		9 02 32.9	+25 51 20	-0.003	-0.03	7.2	1.1	0.0	K1 III		280 s		
77591	199908		9 02 34.6	-31 01 02	-0.001	+0.01	7.5	0.7	1.4	A2 V		73 s		
77684	220772		9 02 35.5	-42 42 27	0.000	-0.02	7.29	0.04	0.6	A0 V		200 s		
75370	1428		9 02 42.8	+82 44 40	+0.005	+0.01	7.1	0.4	2.6	F0 V		79 s		
77350	80595	69 υ Cnc	9 02 44.2	+24 27 10	0.000	-0.01	5.45	-0.04		B9 p	-15			
77445	117369		9 02 44.7	+7 17 53	-0.001	-0.01	5.85	1.11	-0.2	gK3	+27	160 s		
77377	80596		9 02 45.1	+21 31 10	0.000	+0.01	7.7	0.9	3.2	G5 IV		80 s		
77628	176873		9 02 50.3	-28 11 49	-0.005	+0.03	7.3	1.1	-0.5	M2 III		260 mx		
77484	117371		9 02 50.5	+0 24 30	-0.001	0.00	7.80	0.00	0.4	B9.5 V		300 s	7159	m
77391	80597		9 02 50.7	+22 15 31	-0.003	-0.02	7.80	0.8	3.2	G6 IV	+43	83 s		
77645	199917		9 02 51.7	-32 26 24	-0.002	0.00	6.9	0.5	0.4	B9.5 V		100 s		
77227	27096		9 03 02.8	+49 56 54	-0.001	-0.01	6.8	0.1	1.4	A2 V		120 s		
77907	236542		9 03 05.2	-53 32 59	-0.003	+0.03	6.40	-0.09		B9	-7			
77609	176878		9 03 06.5	-20 32 49	-0.002	-0.04	7.9	0.2	0.4	B9.5 V		240 s		
77390	61220		9 03 07.2	+30 06 09	+0.002	-0.01	7.9	0.1	1.4	A2 V		200 s		
77590	154862		9 03 07.7	-14 00 12	-0.002	+0.01	7.5	1.4	-0.3	K5 III		370 s		
77573	136533		9 03 08.2	-8 59 24	-0.001	-0.01	8.00	0.1	1.4	A2 V		210 s		m
77665	176881		9 03 08.7	-25 30 15	-0.001	-0.01	6.74	-0.03		B8				
77753	199922		9 03 09.4	-38 36 35	0.000	-0.03	7.4	1.8	-0.5	M2 III		310 s		
77865	220785		9 03 10.3	-49 41 47	-0.003	+0.02	7.04	1.26	0.2	K0 III		150 s		
78084	250425		9 03 11.8	-62 21 07	-0.006	+0.05	7.70	0.4	2.6	F0 V		100 s		m
77767	199926		9 03 12.5	-38 38 51	0.000	0.00	7.4	1.3	0.2	K0 III		190 s		
77518	117374		9 03 14.6	+6 48 58	0.000	+0.02	6.9	1.1	0.2	K0 III		220 s		
77408	61221		9 03 15.7	+32 52 54	-0.033	0.00	7.03	0.50	3.8	F7 V	+71	44 s		
77737	199924		9 03 16.1	-33 36 04	-0.001	-0.02	6.70	0.1	0.6	A0 V		160 s		m
78360	256580		9 03 17.8	-70 36 25	-0.009	-0.02	6.9	0.0	2.6	F0 V		73 s		
77501	98318		9 03 18.6	+14 11 00	0.000	-0.04	7.80	1.1	-0.2	K3 III		400 s		
77407	61224		9 03 27.0	+37 50 27	-0.007	-0.17	7.1	0.6	4.4	G0 V		35 s		

HD	SAO	Star Name	α 2000	δ 2000	μ(α)	μ(δ)	V	B-V	M$_v$	Spec	RV	d(pc)	ADS	Notes
			^{h} ^{m} ^{s}	$^{\circ}$ $'$ $"$	^{s}	$"$								
76539	6717		9 03 29.9	+76 23 52	-0.017	-0.18	7.9	0.6	4.4	G0 V	+15	50 s		
77247	27101		9 03 32.4	+53 06 28	+0.002	-0.02	7.0	1.1	0.2	K0 III		230 s		
77677	154870		9 03 36.1	-14 26 14	-0.003	+0.02	7.9	1.1	-0.1	K2 III		390 s		
77327	42661	12 κ UMa	9 03 37.5	+47 09 23	-0.003	-0.06	3.60	0.00		A0 n	+4	28 mn	7158	m
77788	199933		9 03 38.6	-33 41 38	+0.002	+0.02	7.9	0.3	1.4	A2 V		130 s		
77640	136542		9 03 43.3	-5 10 17	-0.002	0.00	6.80	-0.06		A0				
77735	154871		9 03 45.0	-17 29 49	0.000	-0.03	6.8	1.1	0.2	K0 III		210 s		
77443	61226		9 03 47.1	+38 44 32	-0.001	-0.03	6.6	1.5	-0.5	M4 III		260 s		
78232	250428		9 03 49.4	-64 41 08	-0.004	-0.01	7.40	0.00	0.4	B9.5 V		250 s		m
77902	220796		9 03 51.1	-41 33 19	-0.003	-0.01	7.2	1.0	3.2	G5 IV		63 s		
78026	236550		9 03 53.0	-52 53 57	-0.003	-0.01	8.0	1.5	-0.3	K5 III		450 s		
76905	6729		9 03 58.2	+71 18 19	-0.005	-0.03	7.2	0.4	2.6	F0 V		81 s		
77881	199942		9 03 58.5	-36 44 21	-0.008	+0.03	7.1	0.6	3.4	F5 V		44 s		
77309	27105		9 04 00.3	+54 17 02	-0.001	0.00	5.7	0.1	1.4	A2 V	-2	72 s		
78005	220802		9 04 05.7	-47 26 29	-0.002	0.00	6.45	-0.15	-2.9	B3 III		690 s		
78004	220803		9 04 09.2	-47 05 52	-0.005	-0.02	3.75	1.20	-0.1	K2 III	+24	57 s		
77557	80609	70 Cnc	9 04 09.8	+27 53 53	0.000	-0.01	6.30	0.00	0.6	A0 V	-24	130 s		
77841	176903		9 04 12.3	-24 14 24	-0.002	+0.02	7.6	0.4	1.7	A3 V		96 s		
77921	199947		9 04 12.8	-35 02 38	-0.005	+0.02	8.0	0.7	0.6	A0 V		100 s		
78231	250429		9 04 18.3	-60 33 56	-0.004	+0.03	7.9	1.2	-0.1	K2 III		380 s		
77586	80610		9 04 22.5	+29 16 05	-0.002	-0.02	7.60	1.61	-0.5	M3 III		420 s		
77497	42674		9 04 27.1	+43 26 32	0.000	-0.01	8.0	1.1	-0.1	K2 III		410 s		
77005	6732		9 04 27.8	+71 42 03	+0.003	-0.03	7.8	0.0	0.4	B9.5 V		300 s		
78135	236561		9 04 29.6	-52 30 49	+0.005	-0.16	7.44	0.58		G0				
77938	199952		9 04 29.6	-32 26 56	0.000	-0.04	7.85	1.60		gM5	-2			
---	236565		9 04 32.0	-56 20 34	+0.006	-0.05	7.50							m
78190	236568		9 04 32.8	-56 20 30	-0.002	-0.02	6.89	0.04	0.4	B9.5 V		200 s		m
77710	98338		9 04 35.8	+10 51 00	-0.002	-0.01	7.6	1.1	-0.1	K2 III		340 s		
77760	117399		9 04 39.2	+1 54 36	0.000	-0.01	7.7	1.1	0.2	K0 III		310 s		
77777	117398		9 04 39.6	+3 32 27	-0.003	-0.02	7.5	0.1	0.6	A0 V		250 s		
77778	117400		9 04 39.7	+0 55 06	-0.001	-0.02	7.4	0.4	2.6	F0 V		93 s		
77879	154884		9 04 41.5	-17 14 13	-0.003	-0.03	7.6	0.4	3.0	F2 V		85 s		
77292	14752		9 04 42.1	+60 59 27	-0.002	-0.03	7.7	0.9	3.2	G5 IV		81 s		
78097	220815		9 04 46.8	-44 48 37	-0.002	0.00	7.62	1.67	-0.1	K2 III		350 s		
78293	236578		9 04 48.0	-57 51 08	-0.003	+0.02	6.44	0.25		A3				m
78133	220819		9 04 49.8	-46 48 59	-0.005	+0.02	7.0	0.4	1.4	A2 V		86 s		
77660	61242		9 04 55.1	+32 22 37	-0.003	-0.06	6.4	0.2	2.1	A5 V	+16	54 mx		
77915	136561		9 04 57.2	-10 07 20	-0.005	+0.02	7.3	0.2	2.1	A5 V		110 s		
77776	98343		9 04 58.1	+13 20 38	0.000	-0.02	7.6	0.9	0.3	G8 III		290 s		
77730	80615		9 04 59.9	+22 49 56	-0.009	+0.12	7.2	0.2		A m				
78446	250435		9 05 02.4	-64 17 22	-0.002	-0.01	8.0	1.1	0.2	K0 III		350 s		
77999	176923		9 05 04.7	-24 11 23	0.000	-0.02	8.0	0.4	2.6	F0 V		100 s		
77729	80616		9 05 04.8	+26 09 53	0.000	-0.04	7.64	1.38		K2 IV				
77894	136563		9 05 05.2	-3 46 50	-0.002	+0.01	7.4	1.1	-0.1	K2 III	+16	310 s		
78035	176924		9 05 06.6	-29 01 08	-0.002	-0.04	6.7	1.6	0.2	K0 III		85 s		
78791	256582		9 05 09.3	-72 36 10	+0.003	-0.01	4.48	0.61	-0.6	F6 II-III	+22	92 s		
77874	117411		9 05 15.3	+2 24 51	-0.003	-0.01	7.2	0.1	1.4	A2 V		150 s		
77570	27111		9 05 18.6	+50 49 25	-0.012	-0.07	6.8	0.4	3.0	F2 V	+17	56 s		
78445	250436		9 05 19.3	-61 49 35	+0.008	+0.04	7.5	1.2	-0.3	K5 III		150 mx		
78602	250439		9 05 20.3	-68 06 41	-0.015	-0.02	7.6	1.4	-0.3	K5 III		150 mx		
78165	220824		9 05 23.2	-42 10 00	-0.004	+0.03	7.6	0.3	1.7	A3 V		160 s		
78033	176928		9 05 23.5	-22 58 15	+0.001	+0.02	8.0	2.6	-0.3	K5 III		95 s		
77601	42682		9 05 24.0	+48 31 49	-0.001	-0.02	5.95	0.45	3.4	F5 V	-6	30 s		
76559	6726		9 05 28.4	+79 20 46	-0.004	-0.02	7.9	0.1	0.6	A0 V		290 s		
77836	98353		9 05 29.5	+19 26 09	+0.004	-0.04	8.0	0.5	3.4	F5 V		84 s		
78290	236588		9 05 30.5	-50 17 15	0.000	-0.02	7.27	-0.12	0.0	B8.5 V		280 s		
77745	61245		9 05 33.4	+36 10 48	-0.008	-0.04	6.9	0.4	2.6	F0 V		73 s		
77428	14758		9 05 37.0	+64 03 14	-0.006	-0.05	7.5	1.1	0.2	K0 III		250 mx		
78764	256583		9 05 38.4	-70 32 20	-0.001	-0.01	4.71	-0.15	-2.5	B2 V	+35	280 s		
77892	98358		9 05 45.5	+17 23 25	-0.002	-0.01	8.0	0.1	1.4	A2 V		210 s		
77599	27113		9 05 45.8	+55 31 45	-0.026	+0.01	8.0	0.6	4.4	G0 V		52 s		
78162	199965		9 05 48.8	-31 15 30	-0.001	+0.01	7.9	1.1	-0.3	K5 III		430 s		
78201	199967		9 05 52.5	-34 32 03	0.000	+0.02	8.0	0.4	0.6	A0 V		160 s		
77619	27116		9 05 56.4	+55 23 20	-0.002	-0.06	7.7	0.4	2.6	F0 V		100 s		
78224	199972		9 05 57.0	-36 05 46	0.000	-0.02	7.2	1.5	-0.1	K2 III		190 s		
78130	176946		9 05 57.6	-21 31 56	-0.002	-0.01	8.0	1.3	3.2	G5 IV		44 s		
78355	220846		9 05 57.8	-49 14 02	-0.003	0.00	7.94	1.50	0.2	K0 III		180 s		
77996	117420	18 ω Hya	9 05 58.3	+5 05 32	-0.001	-0.01	4.97	1.22	-1.1	K2 II-III	+25	170 s		
78092	154903		9 06 00.3	-11 20 43	-0.001	-0.06	7.9	1.1	0.2	K0 III		250 mx		
78845	---		9 06 06.9	-70 14 10			8.0	1.1	0.2	K0 III		360 s		
78632	250441		9 06 07.6	-64 29 59	+0.003	+0.03	6.37	1.35		K0				
79175	256587		9 06 10.3	-76 33 59	-0.004	+0.02	7.3	0.0	0.4	B9.5 V		240 s		
77986	98361		9 06 12.5	+15 51 44	-0.001	-0.01	7.3	0.0	0.4	B9.5 V	+2	240 s		
77985	98362		9 06 15.6	+17 06 42	0.000	-0.07	7.6	0.9	3.2	G5 IV	-5	75 s		
78011	98363		9 06 21.5	+15 16 28	-0.001	-0.01	7.9	1.6	-0.5	M2 III	+20	490 s		

HD	SAO	Star Name	α 2000	δ 2000	μ(α)	μ(δ)	V	B-V	M_V	Spec	RV	d(pc)	ADS	Notes
			$h\ m\ s$	$^\circ\ '\ ''$	s	$''$								
77426	14761		9 06 22.3	+68 26 41	+0.003	-0.01	7.0	0.2	2.1	A5 V		94 s		
77770	42690		9 06 22.5	+49 36 41	0.000	0.00	7.53	-0.20	-1.0	B5.5 V		510 s		
78568	236610		9 06 22.5	-59 12 22	0.000	0.00	7.82	-0.06	0.4	B9.5 V		310 s		
78143	136580		9 06 29.7	-2 36 00	+0.002	+0.02	7.6	0.7	4.4	G0 V		44 s		
78257	176961		9 06 30.7	-24 09 28	-0.003	+0.01	8.00	0.4	3.0	F2 V		100 s	7185	m
77912	61254		9 06 31.7	+38 27 08	-0.002	-0.02	4.56	1.04	-3.3	G8 Ib-II	+17	370 s		
78029	80631		9 06 32.3	+20 07 10	-0.001	0.00	7.9	1.1	0.2	K0 III		340 s		
78505	236606		9 06 33.2	-53 02 15	-0.002	-0.01	8.0	1.4	-0.1	K2 III		310 s		
78177	136584		9 06 33.4	-9 45 33	0.000	+0.01	7.8	1.1	0.2	K0 III		340 s		
78548	236611		9 06 34.0	-55 48 12	-0.002	-0.02	6.11	-0.15	-1.1	B5 V n		270 s		
78459	220860		9 06 34.4	-47 55 11	-0.002	0.00	7.9	-0.1	0.4	B9.5 V		320 s		
78126	117428		9 06 38.1	+2 49 06	-0.004	+0.03	8.0	0.5	3.4	F5 V		82 s	7182	m
77969	61259		9 06 38.8	+33 22 13	-0.002	-0.03	8.0	0.1	1.4	A2 V		210 s		
78429	220859		9 06 38.8	-43 29 32	+0.004	+0.17	7.31	0.66		G0				
78050	80634		9 06 42.7	+20 30 36	-0.007	-0.18	7.69	0.80	4.4	G0 V		49 mn		
78529	236612		9 06 42.9	-52 31 45	-0.005	+0.01	7.43	1.02	0.2	K0 III		270 s		
77692	27121		9 06 43.0	+59 20 39	-0.003	-0.02	6.45	0.04		A0	+4			
78351	199983		9 06 43.9	-33 34 06	+0.005	-0.03	8.0	1.0	3.2	G5 IV		67 s		
78725	250444		9 06 45.1	-62 50 58	-0.002	+0.01	7.53	-0.03	0.4	B9.5 V		270 s		
77819	27125		9 06 47.1	+52 58 52	-0.001	-0.03	7.6	0.9	3.2	G5 IV		75 s		
78350	176976		9 06 49.3	-29 17 31	-0.005	+0.01	7.8	0.9	3.4	F5 V		39 s		
77967	61261		9 06 51.1	+38 16 43	+0.020	-0.01	6.6	0.4	2.6	F0 V		53 mx		
78690	250443		9 06 51.4	-60 15 42	0.000	-0.01	7.4	0.1	0.4	B9.5 V		210 s		
78302	154918		9 06 54.8	-18 12 05	-0.001	-0.02	7.10	1.1	0.2	K0 III		240 s		m
78196	117432		9 06 59.9	+1 27 45	-0.001	-0.03	6.17	1.65		Ma	+3			
79027	256585		9 07 00.6	-70 58 18	+0.007	-0.07	7.5	1.1	0.2	K0 III		290 s		
77726	14765		9 07 04.3	+60 31 39	-0.002	-0.03	7.7	1.1	0.2	K0 III		310 s		
78469	199992		9 07 04.9	-39 29 55	-0.002	-0.01	8.0	1.2	0.2	K0 III		290 s		
78195	117433		9 07 10.4	+9 40 17	-0.003	0.00	7.5	0.9	0.2	G9 III	+3	290 s		
78441	199993		9 07 11.9	-34 31 12	-0.002	0.00	7.9	0.0	0.6	A0 V		270 s		
77890	27128		9 07 13.4	+51 37 47	-0.002	-0.02	7.3	1.1	0.2	K0 III		260 s		
78318	154920		9 07 14.6	-13 18 26	-0.001	+0.01	7.0	0.4	2.6	F0 V		75 s		
78599	236618		9 07 14.8	-51 12 43	+0.001	0.00	6.73	1.62		K5				
78255	117434		9 07 16.6	+2 34 52	0.000	0.00	6.6	0.4	3.0	F2 V		52 s		
78484	199999		9 07 17.9	-38 49 04	0.000	0.00	7.3	0.9	-0.1	K2 III		310 s		
78724	236623		9 07 18.6	-57 27 49	-0.003	+0.02	6.6	1.2	3.2	G5 IV		25 s		
78282	117436		9 07 22.1	+0 36 01	-0.002	0.00	7.16	0.03	0.6	A0 V		190 s		
78761	236628		9 07 23.8	-59 09 00	-0.004	0.00	7.8	0.2	1.4	A2 V		150 s		
78254	117435		9 07 24.2	+6 50 20	-0.003	+0.01	8.0	0.1	0.6	A0 V		300 s		
77818	27127		9 07 26.0	+58 52 29	+0.001	-0.11	7.65	1.01		K1 IV	-39	86 mx		
78175	80643		9 07 26.8	+22 58 51	-0.012	0.00	6.90	0.4	3.1	F3 V	+29	58 s	7187	m
78467	176986		9 07 28.5	-30 05 00	-0.003	+0.02	7.6	0.5	3.2	G5 IV		77 s		
78564	200003		9 07 32.8	-39 49 55	-0.001	+0.02	7.90	0.1	1.4	A2 V		200 s		m
78317	117440		9 07 33.0	+0 03 48	-0.007	+0.03	7.8	0.5	4.0	F8 V		57 s		
78139	61274		9 07 37.8	+31 12 00	0.000	-0.05	7.0	1.1	-0.1	K2 III		270 s		
79041	250455		9 07 39.6	-68 45 18	-0.002	0.00	7.54	-0.08	-0.6	B7 V		430 s		
78802	236634		9 07 40.3	-59 11 05	+0.002	0.00	7.5	0.4	1.7	A3 V		91 s		
78616	220873		9 07 42.4	-44 37 58	-0.001	-0.01	6.79	-0.01		B1				m
78424	154927		9 07 43.7	-12 01 16	-0.001	0.00	7.2	1.4	-0.3	K5 III		320 s		
78392	154926		9 07 44.4	-10 29 27	-0.001	-0.01	7.80	0.1	0.6	A0 V		280 s	7197	m
78543	200005		9 07 44.6	-34 07 22	-0.004	0.00	7.9	0.6	3.0	F2 V		64 s		
78316	98378	76 κ Cnc	9 07 44.8	+10 40 06	-0.001	-0.01	5.24	-0.11		B8 p	+24			
80258	258518		9 07 47.6	-82 19 28	+0.019	-0.01	7.9	0.3	3.4	F5 V		79 s		
78423	136595		9 07 47.9	-9 51 11	0.000	+0.03	7.8	0.1	0.6	A0 V		280 s		
78500	176996		9 07 48.6	-28 11 54	0.000	-0.02	7.6	1.0	3.2	G5 IV		53 s		
78520	176997		9 07 48.7	-28 35 37	-0.001	0.00	7.7	1.3	-0.1	K2 III		310 s		
78522	200007		9 07 50.8	-32 02 15	+0.002	+0.01	8.0	1.4	-0.3	K5 III		440 s		m
78251	80649		9 07 53.1	+25 37 34	-0.003	-0.01	6.8	0.2	2.1	A5 V		88 s		
78252	80651		9 07 53.6	+22 27 23	-0.004	+0.02	8.0	0.4	3.0	F2 V		100 s		
78747	236632		9 07 56.9	-50 28 56	-0.003	-0.13	7.74	0.56	4.0	F8 V		54 s		
78422	136598		9 07 58.1	-7 48 18	-0.003	0.00	8.0	0.4	2.6	F0 V		120 s		
76702	1442		9 07 58.2	+81 02 27	-0.020	-0.05	8.00	0.4	3.0	F2 V		100 s	7163	m
78647	220878	λ Vel	9 07 59.7	-43 25 57	-0.002	+0.01	2.21	1.66	-4.4	K5 Ib	+18	150 s		
78235	80650	72 τ Cnc	9 08 00.0	+29 39 15	-0.002	0.00	5.43	0.89	0.3	G8 III	-13	110 s		
78421	136597		9 08 01.5	-2 28 34	0.000	-0.03	7.1	0.9	3.2	G5 IV		61 s		
78277	80653		9 08 02.6	+27 33 32	-0.008	-0.05	7.90	0.7	3.0	G2 IV		96 s		
78541	177002	κ Pyx	9 08 02.8	-25 51 30	+0.003	0.00	4.58	1.59	-0.4	gM0	-45	97 s	7202	m
78542	177004		9 08 02.8	-27 18 19	+0.001	+0.01	7.8	2.2	-0.5	M2 III		210 s		
78234	61276		9 08 04.1	+32 32 26	-0.007	-0.02	6.3	0.4	3.0	F2 V	+41	47 s		
78561	177006		9 08 04.9	-27 45 11	-0.004	0.00	7.8	1.7	0.6	A0 V		26 s		
78883	236646		9 08 05.6	-58 07 46	-0.002	+0.08	6.9	0.7	3.2	G5 IV		55 s		
78108	42708		9 08 06.3	+47 25 24	-0.001	+0.01	7.8	1.1	0.2	K0 III		330 s		
78348	98380		9 08 06.4	+14 02 33	0.000	-0.01	7.6	1.4	-0.3	K5 III		390 s		
78626	200014		9 08 07.1	-37 21 32	-0.002	0.00	7.7	-1.2	0.6	A0 V		260 s		
78593	200011		9 08 07.8	-34 15 48	-0.002	-0.02	8.0	1.8	-0.3	K5 III		290 s		

HD	SAO	Star Name	α 2000	δ 2000	μ(α)	μ(δ)	V	B-V	M_v	Spec	RV	d(pc)	ADS	Notes
			h m s	° ′ ″	s	″								
78347	98381		9 08 08.4	+14 42 40	−0.001	−0.02	7.4	1.1	0.2	K0 III		280 s		
78376	98382		9 08 12.5	+11 56 53	−0.006	−0.09	7.40	1.1	0.2	K0 III		130 mx	7200	m
78494	154933		9 08 16.7	−11 56 31	−0.002	−0.03	8.0	1.1	−0.1	K2 III		410 s		
78705	220885		9 08 17.0	−41 25 35	+0.002	−0.05	7.2	0.3	0.2	K0 III		250 s		
77800	14769	11 σ¹ UMa	9 08 23.5	+66 52 24	−0.003	−0.04	5.14	1.51	−0.3	K5 III	+15	120 s		
78577	177015		9 08 24.8	−23 03 43	0.000	+0.04	8.0	2.2	0.2	K0 III		70 s		
78558	154939		9 08 25.2	−15 08 42	−0.037	−0.21	7.31	0.61	4.7	G2 V	+60	33 s		
78275	61280		9 08 28.8	+36 18 42	−0.003	−0.02	7.5	0.4	3.0	F2 V		79 s		
78419	98384		9 08 34.1	+16 42 05	−0.001	−0.02	7.9	1.1	0.2	K0 III		340 s		
78643	177018		9 08 36.5	−25 50 23	−0.025	−0.02	6.77	0.57	4.5	G1 V		28 s		m
78741	200023		9 08 36.8	−38 47 03	−0.006	0.00	7.4	0.2	1.4	A2 V		120 s		
78931	236657		9 08 40.0	−54 22 16	−0.002	−0.02	7.42	−0.11	0.0	B8.5 V		310 s		
78557	136603		9 08 40.0	−9 23 03	−0.007	+0.04	7.9	0.9	3.2	G5 IV		88 s		
78331	61284		9 08 42.0	+35 21 42	−0.002	−0.03	8.0	1.1	−0.1	K2 III		410 s		
78556	136604	19 Hya	9 08 42.1	−8 35 23	−0.002	−0.01	5.60	−0.06		B8	+23			m
78676	177022		9 08 43.5	−26 46 03	−0.003	+0.01	6.15	0.17		A2				
78614	154944		9 08 44.5	−16 16 36	+0.001	−0.05	6.5	1.1	0.2	K0 III		180 s		
78538	136606		9 08 46.9	−5 07 00	−0.004	+0.05	8.0	0.9	3.2	G5 IV		93 s		
78418	80659	75 Cnc	9 08 47.3	+26 37 45	−0.009	−0.37	5.98	0.66	4.9	dG3	+13	21 ts		m
78612	154945		9 08 49.6	−10 45 29	−0.026	−0.13	7.15	0.61		G0				
78756	200028		9 08 49.8	−35 21 00	0.000	−0.04	8.0	0.8	−0.3	K5 III		460 s		
78366	61288		9 08 51.0	+33 52 55	−0.015	−0.12	5.93	0.60	4.0	F8 V	+27	22 s		
78209	27136	15 UMa	9 08 52.2	+51 36 16	−0.015	−0.04	4.48	0.27		A m	0			
78949	236660		9 08 53.5	−53 35 36	+0.001	+0.01	6.9	1.0	1.4	A2 V		37 s		
78250	42716		9 08 57.7	+49 17 55	+0.001	+0.03	7.7	1.1	0.2	K0 III		320 s		
78642	154946		9 09 01.2	−14 11 05	−0.004	−0.01	7.90	1.1	0.2	K0 III		280 mx	7210	m
78752	177031		9 09 01.7	−28 57 24	−0.008	+0.05	7.7	1.0	4.4	G0 V		25 s		
78535	117455		9 09 01.8	+7 30 47	−0.001	−0.02	7.0	1.1	0.2	K0 III	+72	230 s		
78554	117456		9 09 02.1	+5 36 26	−0.001	−0.01	7.9	0.9	3.2	G5 IV		88 s		
78479	98389		9 09 02.2	+17 28 11	−0.003	−0.04	7.19	1.21	−0.2	K3 III	+77	300 s		
77833	14775		9 09 05.0	+69 12 38	+0.001	−0.01	7.7	1.6	−0.5	M2 III		430 s		
78718	154955		9 09 07.0	−19 25 16	0.000	+0.01	8.0	1.1	0.2	K0 III		360 s		
78610	136612		9 09 08.4	−1 35 19	−0.003	−0.07	7.80	1.4	−0.3	K5 III		200 mx		m
77724	6759		9 09 08.8	+72 18 16	−0.005	−0.07	7.8	0.5	3.4	F5 V		74 s		
78735	177034		9 09 09.5	−24 50 39	+0.001	−0.01	7.8	2.0	−0.5	M4 III		270 s		
78734	177030		9 09 09.5	−20 55 39	−0.001	−0.02	7.4	0.9	0.2	K0 III		270 s		
78668	154953		9 09 11.4	−12 21 28	+0.001	0.00	5.77	0.94	0.3	G6 III	−9	120 s		
79075	236670		9 09 14.6	−59 16 13	−0.008	+0.02	7.6	0.3	1.7	A3 V		110 s		
78515	80666	77 ξ Cnc	9 09 21.5	+22 02 43	0.000	0.00	5.14	0.97	0.2	K0 III	−7	97 s		m
78088	14779		9 09 27.8	+63 16 00	−0.006	−0.04	7.8	1.1	0.2	K0 III		330 s		
78637	––		9 09 31.1	+2 56 28	0.000	−0.04	7.90	0.4	2.6	F0 V		120 s	7215	m
78878	177048		9 09 34.5	−29 48 59	+0.001	+0.03	6.8	1.1	0.2	K0 III		190 s		
78732	136622	20 Hya	9 09 35.5	−8 47 16	−0.002	−0.02	5.46	1.01	0.3	gG6	+26	92 s		
78513	61296		9 09 38.2	+30 42 46	−0.001	−0.05	7.8	0.2	2.1	A5 V		140 s		
78806	177047		9 09 38.4	−22 36 10	−0.002	−0.01	7.9	1.1	0.2	K0 III		300 s		
78751	136625		9 09 40.6	−9 56 18	−0.002	−0.03	7.6	1.1	−0.1	K2 III		350 s		
78780	154961		9 09 41.1	−13 51 27	+0.003	−0.01	7.5	1.1	−0.1	K2 III		330 s		
79092	236672		9 09 43.3	−54 24 53	−0.001	−0.03	7.9	1.7	−0.3	K5 III		320 s		
79025	220910		9 09 45.0	−49 25 29	+0.001	−0.04	6.48	0.17	1.7	A3 V n		90 s		
79173	250462		9 09 45.4	−60 29 55	−0.003	+0.08	7.8	1.0	−0.1	K2 III		240 mx		
78876	177055		9 09 45.7	−25 48 10	−0.001	+0.01	7.40	0.1	0.6	A0 V		230 s	7220	m
78661	98400		9 09 46.4	+11 33 52	−0.003	−0.07	6.5	0.4	2.6	F0 V	−16	60 s		
78477	42727		9 09 48.9	+40 09 23	−0.002	0.00	7.9	1.4	−0.3	K5 III		430 s		
78249	27140		9 09 49.3	+59 09 13	−0.022	−0.03	7.10	0.98		K1 IV	+47	86 mx		
78388	27145		9 09 52.3	+49 49 56	−0.003	−0.03	7.7	0.4	3.0	F2 V		86 s		
79039	220912		9 09 52.7	−47 28 35	−0.002	−0.01	6.84	−0.13		B8				
78922	200047	ε Pyx	9 09 56.4	−30 21 55	0.000	−0.05	5.59	0.19		A pe	−10			
79091	236673		9 09 57.0	−52 04 59	−0.010	+0.04	6.22	1.02	0.2	K0 III		160 s		
78622	80668		9 09 58.1	+22 40 39	+0.001	−0.01	8.0	1.4	−0.3	K5 III		450 s		
79071	220915		9 10 05.4	−48 12 24	−0.002	0.00	7.89	0.88	3.2	G5 IV		71 s		
79072	220916		9 10 05.9	−49 17 23	−0.003	0.00	7.03	−0.11	0.0	B8.5 V		260 s		
79337	250466		9 10 07.7	−65 29 22	−0.005	+0.06	7.5	0.4	3.0	F2 V		74 s		
78891	154967		9 10 10.3	−16 51 44	−0.002	+0.01	7.10	0.4	2.6	F0 V		79 s	7222	m
79019	200054		9 10 10.4	−38 44 54	+0.001	−0.01	7.5	0.8	−0.1	K2 III		330 s		
79550	––		9 10 10.6	−71 58 36			7.9	0.5	3.4	F5 V		79 s		
78413	27149		9 10 13.2	+54 58 43	−0.020	−0.04	8.0	0.5	3.4	F5 V		82 s		
78296	14793		9 10 17.7	+62 26 47	−0.010	−0.06	7.9	0.2	2.1	A5 V		64 mx		
78715	80674	79 Cnc	9 10 20.8	+21 59 47	0.000	0.00	6.01	0.90	3.2	G5 IV	−7	31 s		m
78851	136636		9 10 20.8	−6 09 36	−0.001	−0.02	7.5	1.1	0.2	K0 III		290 s		
78955	177066		9 10 23.0	−23 10 37	−0.002	−0.01	6.53	0.00		A0				
78154	14788	13 σ² UMa	9 10 23.1	+67 08 04	−0.003	−0.08	4.80	0.49	3.1	F7 IV-V	−2	22 ts	7203	m
78920	154973		9 10 24.7	−14 54 00	−0.002	0.00	7.7	0.1	1.4	A2 V		180 s		
78985	177072		9 10 28.6	−26 17 16	0.000	−0.04	7.48	−0.11	−1.1	B5 V		520 s		
78889	136639		9 10 31.9	−3 25 37	−0.003	−0.03	7.6	0.1	1.4	A2 V		180 s		
79060	200062		9 10 33.1	−36 50 38	−0.001	+0.03	7.50	0.92	3.2	G8 IV		69 s		

HD	SAO	Star Name	α 2000	δ 2000	μ(α)	μ(δ)	V	B−V	M_v	Spec	RV	d(pc)	ADS	Notes
			$h\ m\ s$	$\circ\ '\ ''$	s	$''$								
78712	61306		9 10 38.6	+30 57 47	−0.002	−0.04	5.30	1.6		M6 Ib−II	+14			RS Cnc, q
78999	177076		9 10 41.1	−23 06 15	−0.002	+0.03	8.0	1.4	−0.1	K2 III		310 s		
81103	258521		9 10 41.4	−83 44 24	+0.006	+0.05	7.8	1.3	−0.3	K5 III		300 mx		
79206	236685		9 10 41.8	−53 02 45	−0.005	+0.01	7.73	−0.09	0.4	B9.5 V		190 mx		
78698	61304		9 10 45.6	+38 20 42	−0.001	−0.03	7.9	0.9	3.2	G5 IV		88 s		
80257	256593		9 10 48.1	−79 21 31	−0.009	+0.01	7.9	0.1	0.6	A0 V		290 s		
78998	154976		9 10 48.2	−17 48 21	−0.001	0.00	8.01	0.17		A2				
78697	61307		9 10 51.0	+39 47 16	−0.001	−0.04	7.9	0.4	2.6	F0 V		120 s		
79335	236692		9 10 51.7	−59 52 16	+0.001	−0.01	7.8	−0.1	1.4	A2 V		190 s		
78362	14796	14 τ UMa	9 10 55.0	+63 30 49	+0.015	−0.07	4.67	0.35		A m	−9		7211	m
78997	154978		9 10 55.0	−15 27 33	−0.001	+0.01	7.6	0.1	0.6	A0 V		250 s		
79154	220926		9 10 57.4	−43 16 05	−0.002	−0.05	7.3	−0.2	1.4	A2 V		140 mx		
79351	236693		9 10 57.9	−58 58 01	−0.003	+0.01	3.44	−0.19	−3.0	B2 IV	+23	190 s		a Car
78103	6770		9 11 02.9	+71 21 51	+0.011	+0.01	7.1	1.1	0.2	K0 III		240 s		
79186	220928		9 11 04.3	−44 52 04	−0.001	0.00	5.00	0.23	−6.8	B3 Ia	+35	1400 s		
77246	1453		9 11 12.5	+80 49 44	−0.010	−0.02	6.5	0.4	3.0	F2 V		51 s		
79184	200079		9 11 15.9	−39 36 09	−0.002	+0.05	8.0	0.9	0.2	K0 III		360 s		
79447	250471		9 11 16.7	−62 19 02	−0.006	+0.01	3.97	−0.18	−2.3	B3 IV	+17	180 s		
78769	42734		9 11 18.5	+40 18 00	−0.001	−0.01	7.7	1.1	−0.1	K2 III		360 s		
79629	250477		9 11 20.1	−69 15 09	−0.014	+0.08	7.15	0.38	3.0	F2 V		65 s		
79183	200081		9 11 26.1	−35 44 53	−0.001	−0.01	7.4	0.5	1.4	A2 V		89 s		
78969	117482		9 11 26.8	+8 58 33	−0.001	−0.03	7.1	1.1	−0.1	K2 III	−4	270 s		
79011	117483		9 11 29.2	+0 17 27	+0.001	−0.04	6.78	1.32	0.2	K0 III		120 s	7227	m
79149	177091		9 11 31.1	−28 25 39	−0.001	+0.01	7.8	0.4	0.4	B9.5 V		160 s		
79226	220936		9 11 31.7	−41 36 17	−0.001	−0.02	7.4	−0.2	1.4	A2 V		160 s		
79420	236706		9 11 32.4	−57 37 07	−0.001	+0.01	7.55	−0.08	0.4	B9.5 V		270 s		
79421	236707		9 11 32.8	−57 58 06	−0.001	0.00	6.59	−0.13	−2.6	B2.5 IV		610 s		m
79275	220937		9 11 33.2	−46 35 01	−0.002	+0.01	5.79	−0.21	−3.0	B2 IV	+7	570 s		
79387	236704		9 11 37.5	−54 02 14	−0.005	0.00	6.72	−0.05	0.4	B9.5 V		180 s		
79241	200084		9 11 40.9	−39 15 32	−0.001	0.00	6.00	−0.11		B8				
79350	236703		9 11 41.8	−50 55 01	−0.004	+0.04	7.76	1.08	0.2	K0 III		290 s		
78803	42735		9 11 46.3	+44 49 27	−0.005	−0.02	7.5	1.1	0.2	K0 III		290 s		
79743	250480		9 11 47.4	−69 12 11	−0.007	+0.01	7.2	0.5	3.4	F5 V		56 s		
79986	—		9 11 49.2	−74 36 49			7.9	0.1	0.6	A0 V		290 s		
79202	177098		9 11 50.7	−28 18 55	−0.007	+0.02	7.1	0.9	0.2	K0 III		210 mx		
79097	136651		9 11 51.3	−6 58 46	0.000	0.00	7.61	1.63	−0.5	gM2	+7	400 s		
79332	220943		9 11 52.8	−45 14 06	−0.001	−0.01	7.97	−0.09	0.0	B8.5 V		340 s		
79490	250476		9 11 54.0	−60 18 09	−0.002	0.00	7.8	−0.1	0.6	A0 V		270 s		
78917	61324		9 11 55.1	+35 28 54	0.000	−0.04	7.9	1.4	−0.3	K5 III		430 s		
79066	117487		9 11 55.6	+5 28 06	−0.007	−0.03	6.35	0.32		F0	+3			
78792	27165		9 11 56.2	+50 26 23	−0.001	−0.01	6.6	0.1	0.6	A0 V		160 s		
79009	98425		9 11 56.9	+18 02 39	−0.003	−0.02	6.89	0.06		A0	+10			m
79719	250481		9 11 57.2	−67 43 18	−0.002	0.00	7.0	0.4	2.6	F0 V		75 s		
79181	154989		9 11 58.7	−19 44 52	−0.004	+0.04	5.73	0.98	0.2	G9 III	−1	130 s		
79129	136653		9 12 00.8	−4 33 18	−0.002	+0.01	7.5	0.1	1.7	A3 V		140 s		
79446	236712		9 12 03.3	−53 57 44	−0.002	0.00	7.73	−0.08	0.4	B9.5 V		290 s		m
79780	—		9 12 08.2	−68 44 44			7.7	1.1	−0.1	K2 III		360 s		
79108	117492		9 12 12.9	+3 52 02	−0.003	0.00	6.14	−0.01		A0	+20			
80194	256594		9 12 13.7	−76 39 45	+0.024	−0.03	6.14	1.09	0.2	gK0		130 s		
79403	220948		9 12 13.8	−45 50 57	−0.006	−0.04	6.67	0.43		F5				
79250	177105		9 12 16.8	−26 28 42	+0.002	−0.04	8.0	1.6	0.2	K0 III		160 s		
79096	98427	81 Cnc	9 12 17.5	+14 59 46	−0.036	+0.24	6.51	0.73	5.9	K0 V	+45	15 ts		m
79194	154993		9 12 18.5	−11 04 45	−0.001	−0.01	7.9	0.1	0.6	A0 V		290 s		
79290	177110		9 12 21.8	−28 56 52	−0.001	+0.01	6.8	0.6	0.0	B8.5 V		91 s		
79179	136661		9 12 25.3	−6 55 50	0.000	−0.01	7.60	1.1	0.2	K0 III		300 s	7240	m
79268	177111		9 12 25.5	−25 36 04	−0.001	0.00	7.4	0.8	1.4	A2 V		57 s		
80913	258522		9 12 25.6	−81 46 08	−0.036	+0.01	7.7	0.5	3.4	F5 V		65 s		
79193	136662	21 Hya	9 12 26.0	−7 06 35	−0.001	+0.02	6.11	0.22		A2				
78768	27168		9 12 27.9	+58 42 41	+0.002	−0.01	7.8	1.1	0.2	K0 III		330 s		
78899	42742		9 12 28.3	+49 12 24	−0.004	−0.17	7.7	0.9	3.2	G5 IV		56 mx		
79416	220952		9 12 30.5	−43 36 49	−0.002	0.00	5.57	−0.11	−0.2	B8 V	+4	140 s		m
79362	200099		9 12 31.3	−35 41 05	0.000	−0.04	7.9	0.9	0.2	K0 III		340 s		
80020	—		9 12 36.5	−72 47 51			7.5	1.1	−0.1	K2 III		330 s		
—	250486		9 12 40.5	−65 43 46	−0.002	−0.05	7.8							
79218	136665		9 12 41.2	−3 02 00	+0.001	−0.04	7.2	1.1	0.2	K0 III	0	250 s		
78767	14808		9 12 45.7	+61 40 34	0.000	+0.02	7.70	0.1	0.6	A0 V		260 s	7226	m
78766	14809		9 12 47.4	+61 40 54	0.000	+0.01	7.8	0.1	0.6	A0 V		270 s		
79145	98433		9 12 47.5	+16 37 48	−0.002	−0.01	7.9	1.1	0.2	K0 III		340 s		
78865	27170		9 12 50.2	+56 56 30	−0.004	−0.02	7.02	1.11	0.0	K1 III		240 s		
79699	250485		9 12 50.6	−60 54 59	−0.002	+0.03	7.20	0.1	0.6	A0 V		210 s		m
79044	42748		9 12 55.2	+40 23 06	−0.002	−0.04	8.0	0.5	3.4	F5 V		82 s		
79698	236723		9 12 55.6	−59 24 52	0.000	0.00	5.54	0.85		G5	+16			
79413	200105		9 12 56.1	−31 03 56	−0.001	+0.01	7.2	0.9	0.2	K0 III		250 s		
79191	98437		9 12 56.4	+10 40 09	0.000	0.00	7.90	0.5	3.4	F5 V		79 s	7242	m
79308	155003		9 12 57.2	−15 25 10	−0.003	0.00	7.4	0.4	3.0	F2 V		76 s		

HD	SAO	Star Name	α 2000	δ 2000	μ(α)	μ(δ)	V	B-V	M_v	Spec	RV	d(pc)	ADS	Notes
78937	27171		9 12 59.3	+55 22 42	0.000	-0.01	7.9	1.1	0.2	K0 III		340 s		
79625	236721		9 13 08.7	-53 30 18	0.000	-0.01	7.25	-0.06	0.6	A0 V		210 s		
80007	250495	β Car	9 13 12.2	-69 43 02	-0.029	+0.10	1.68	0.00	-0.6	A0 III	-5	26 s		Miaplacidus
79508	200116		9 13 16.4	-39 51 20	-0.004	-0.01	8.0	0.3	1.7	A3 V		130 s		
79524	220962		9 13 18.6	-42 16 25	-0.002	+0.05	6.29	1.25		K0				
79670	236726		9 13 18.7	-53 58 58	0.000	+0.02	7.99	-0.10	0.4	B9.5 V		330 s		
79433	177126		9 13 20.7	-24 37 35	0.000	-0.02	7.7	1.0	-0.1	K2 III		370 s		
78765	14810		9 13 21.0	+65 43 12	-0.009	-0.07	7.9	0.4	2.6	F0 V		110 s		
79456	177127		9 13 21.5	-29 39 51	+0.002	+0.02	6.4	2.1	-0.3	K5 III		99 s		
79283	117506		9 13 22.7	+5 48 57	-0.002	0.00	7.9	0.0	0.4	B9.5 V		320 s		
79264	98440		9 13 23.9	+10 18 22	+0.001	-0.04	7.7	1.1	0.2	K0 III		310 s		
79320	117510		9 13 24.8	+3 49 29	-0.001	0.00	7.6	0.1	1.4	A2 V		180 s		
79144	61336		9 13 25.4	+35 06 22	+0.001	0.00	8.0	0.1	0.6	A0 V		300 s		
79523	200119		9 13 25.8	-38 36 59	0.000	-0.02	6.31	0.00		A0				
79379	155010		9 13 26.3	-11 13 43	-0.001	-0.01	7.9	1.1	-0.1	K2 III		390 s		
79622	220964		9 13 31.2	-49 05 40	+0.004	-0.03	6.61	1.52	-0.3	K5 III		240 s		
79214	80699		9 13 33.0	+24 17 35	-0.002	-0.01	7.80	1.1	0.2	K0 III		330 s		
79621	220965		9 13 34.4	-47 20 19	-0.002	0.00	5.92	-0.05		B9				
79248	80702		9 13 37.2	+21 17 00	-0.001	-0.01	6.10	0.02	1.4	A2 V	+9	87 s		
80364	256596		9 13 38.1	-75 35 24	+0.006	+0.01	7.9	-0.3	0.6	A0 V		250 mx		
79107	42755		9 13 38.8	+46 59 26	+0.003	+0.01	7.80	0.5	4.0	F8 V	-20	58 s	7243	m
78656	6782		9 13 41.1	+69 56 16	-0.016	-0.05	7.7	0.9	3.2	G5 IV		79 s		
78820	14811		9 13 41.9	+65 47 48	-0.008	+0.01	7.9	0.4	3.0	F2 V		94 s		
79505	177138		9 13 42.2	-28 18 20	-0.003	-0.02	7.5	0.5	0.2	K0 III		290 s		
79601	220966		9 13 44.5	-42 18 37	-0.014	-0.06	8.03	0.57		G0				
80160	--		9 13 44.7	-71 36 56			7.2	0.1	0.6	A0 V		210 s		
79484	177137		9 13 47.4	-22 52 08	-0.001	+0.07	7.6	0.7	3.2	G5 IV		75 s		
78374	6778		9 13 47.7	+75 16 47	-0.014	+0.04	7.7	0.9	3.2	G5 IV		79 s		
79158	42759	36 Lyn	9 13 48.1	+43 13 04	-0.003	-0.04	5.32	-0.14		B8	+21			
79522	--		9 13 49.9	-28 05 41			7.6	0.1	0.4	B9.5 V		220 s		
79864	250494		9 13 51.8	-61 16 46	-0.002	+0.01	6.99	-0.04	0.0	B8.5 V		250 s		
79410	117519		9 13 52.5	-0 11 14	-0.005	-0.01	7.9	1.1	0.2	K0 III		220 mx		
79395	117518		9 13 54.5	+2 51 46	-0.001	-0.02	7.9	1.1	-0.1	K2 III		390 s		
79481	155021		9 14 03.1	-14 41 40	0.000	-0.01	6.6	1.1	0.2	K0 III		190 s		
78633	6784		9 14 03.9	+71 39 20	+0.002	-0.05	6.4	0.9	1.8	G8 III-IV	+6	85 s		
79737	220975		9 14 05.1	-49 28 52	-0.001	-0.01	7.51	1.30	0.2	K0 III		230 s		
79905	250496		9 14 07.0	-60 32 08	-0.001	-0.02	7.6	0.0	0.4	B9.5 V		260 s		
79569	177149		9 14 07.8	-28 07 05	-0.004	-0.04	7.8	0.6	0.6	A0 V		110 s		
79668	220970		9 14 07.9	-42 28 04	-0.002	-0.02	8.03	0.41	3.0	F2 V		100 s		
79694	220972		9 14 08.0	-44 08 46	-0.004	0.00	5.85	-0.12	-1.3	B6 IV		220 mx		
80128	250507		9 14 08.2	-68 55 07	-0.001	-0.04	7.4	1.1	0.2	K0 III		270 s		
79985	250501		9 14 10.4	-63 41 36	-0.017	+0.08	7.5	0.8	3.2	G5 IV		71 s		
79846	236749		9 14 18.0	-55 34 11	-0.004	+0.02	5.27	0.99		G5	+9	19 mn		
79028	14819	16 UMa	9 14 20.6	+61 25 24	0.000	-0.03	5.13	0.58	4.2	dF9	-14	20 ts		m
79613	200135		9 14 21.5	-32 33 24	+0.001	-0.03	6.50	1.1	0.2	K0 III		180 s		m
79469	117527	22 θ Hya	9 14 21.8	+2 18 51	+0.009	-0.31	3.88	-0.06	0.6	A0 V p	-8	45 s	7253	m
79210	27178		9 14 22.5	+52 41 11	-0.174	-0.59	7.64	1.43	8.7	M0 V	+9	6.0 t	7251	m
80060	250505		9 14 22.7	-65 03 14	-0.005	0.00	7.06	1.07	0.2	K0 III		210 s		
79735	220978		9 14 24.4	-43 13 39	-0.002	+0.01	5.25	-0.14		B5 n	+32			m
79211	27179		9 14 24.7	+52 41 09	-0.171	-0.68	7.71	1.36	8.8	M0 V	+11	6.0 t		
79581	155023		9 14 26.1	-19 59 57	-0.005	+0.07	8.0	2.0	3.2	G5 IV		16 s		
79643	177157		9 14 27.1	-29 48 23	+0.003	-0.04	6.79	0.56	4.5	G1 V		29 s		
79373	80709		9 14 27.8	+25 00 41	-0.001	-0.06	7.0	1.1	-0.2	K3 III		190 mx		
79664	200140		9 14 28.3	-33 25 26	-0.003	-0.02	8.0	0.2	1.4	A2 V		160 s		
79758	220983		9 14 33.8	-41 53 32	-0.001	0.00	7.8	-0.3	2.1	A5 V		140 s		
82025	258525		9 14 35.0	-84 37 21	-0.052	+0.07	7.8	1.4	-0.1	K2 III		190 mx		
79580	155025		9 14 40.3	-14 56 48	-0.001	0.00	8.0	1.1	-0.1	K2 III		410 s		
79810	220987		9 14 43.8	-43 27 55	-0.002	+0.02	6.82	1.14	0.2	K0 III		170 s		
79567	136686		9 14 44.2	-4 32 36	-0.001	-0.10	8.0	1.1	-0.1	K2 III		150 mx		
79711	177166		9 14 45.1	-30 09 36	-0.001	+0.03	7.5	1.5	-0.5	M4 III		400 s		
79755	200148		9 14 46.2	-35 57 43	-0.003	0.00	7.2	0.2	0.6	A0 V		150 s		
79946	236758		9 14 47.4	-55 38 25	-0.002	+0.02	7.17	-0.11		B8				
79190	27181		9 14 51.6	+55 37 12	-0.002	-0.03	7.5	0.1	0.6	A0 V		240 s		
79566	136687		9 14 51.9	-1 35 14	0.000	-0.01	7.0	0.1	1.4	A2 V		130 s		
79468	80715		9 14 52.7	+22 42 43	0.000	-0.03	7.9	1.1	0.2	K0 III		340 s		
79597	136690		9 14 53.1	-8 45 29	+0.001	-0.01	7.40	0.1	0.6	A0 V		230 s	7258	m
79555	117534		9 14 53.5	+4 26 35	-0.008	+0.03	7.97	1.02	0.2	K0 III		170 mx		
79807	200152		9 14 57.1	-37 36 09	0.000	-0.02	5.86	0.83		G0				
79769	200149		9 14 57.8	-33 04 03	-0.003	-0.01	7.5	0.9	0.2	K0 III		290 s		
79392	61353		9 14 57.9	+38 36 34	-0.011	+0.08	6.72	0.37	3.0	F2 V		56 s		
79768	200151		9 15 03.6	-30 22 03	-0.001	-0.01	7.90	0.1	0.6	A0 V		290 s		m
79709	--		9 15 05.1	-20 06 47			7.50	0.1	0.6	A0 V		240 s	7263	m
79982	236766		9 15 07.8	-53 19 11	-0.002	-0.01	7.80	-0.04	0.6	A0 V		280 s		
79424	61358		9 15 09.1	+37 03 35	+0.001	-0.02	7.86	1.31		K0				
79578	117537		9 15 09.1	+2 05 01	0.000	0.00	7.6	1.1	0.2	K0 III		300 s		

HD	SAO	Star Name	α 2000	δ 2000	μ(α)	μ(δ)	V	B–V	M$_V$	Spec	RV	d(pc)	ADS	Notes
—	61356		9 15 09.2	+38 20 04	−0.002	−0.02	7.9	1.6						m
79498	80717		9 15 09.3	+23 22 31	−0.009	−0.17	8.00	0.9	3.2	G5 IV		56 mx		m
80351	—		9 15 13.1	−70 07 03			8.0	1.1	0.2	K0 III		360 s		
79554	98456	82 π Cnc	9 15 13.8	+14 56 29	−0.003	−0.01	5.34	1.32	0.2	gK0	+26	62 s		
79753	155030		9 15 14.0	−19 35 46	−0.001	0.00	7.4	0.6	0.4	B9.5 V		110 s		
79452	61361		9 15 14.3	+34 38 01	−0.012	+0.05	5.97	0.86	0.3	G6 III	+56	140 s		m
79900	220998		9 15 14.6	−45 33 20	−0.001	−0.01	6.25	−0.08		A0				m
79055	14822		9 15 17.4	+67 08 20	−0.010	−0.04	8.0	0.4	2.6	F0 V		120 s		
80094	236772		9 15 17.5	−58 23 18	−0.003	−0.01	6.02	−0.10	−1.0	B7 IV	+7	250 s		
76990	1461		9 15 21.3	+84 10 51	+0.019	+0.01	6.3	0.4	2.6	F0 V	−6	54 s		
79707	136698		9 15 22.2	−8 28 58	0.000	−0.01	6.9	0.4	2.6	F0 V		73 s		
79752	155032		9 15 24.8	−15 01 29	−0.002	−0.02	6.35	0.02		A0 n	+32			
79553	80722		9 15 26.7	+22 48 29	0.000	−0.02	7.90	0.4	3.0	F2 V		96 s	7261	m
80093	236776		9 15 31.4	−56 15 41	−0.002	0.00	7.9	1.2	−0.3	K5 III		430 s		
79729	136704		9 15 33.0	−8 44 35	0.000	−0.02	7.9	0.1	1.4	A2 V		200 s		
79857	200157		9 15 34.7	−33 19 27	+0.001	−0.03	7.0	1.1	−0.1	K2 III		260 s		
80126	236777		9 15 34.9	−57 34 41	0.000	−0.01	6.32	1.04		K0				
79497	61366		9 15 36.0	+35 58 52	−0.003	−0.02	7.9	1.1	−0.1	K2 III		390 s		
79917	200159		9 15 36.7	−38 34 11	−0.006	−0.01	4.94	1.11	−0.1	gK2	+2	100 s		
79815	177175		9 15 37.1	−23 29 35	−0.004	+0.01	7.8	−0.1	0.6	A0 V		270 s		
79423	42769		9 15 39.5	+48 13 46	+0.001	+0.01	8.0	0.1	0.6	A0 V		300 s		
79804	155039		9 15 40.9	−13 07 53	−0.001	−0.04	7.5	1.1	−0.1	K2 III		330 s		
79940	200163		9 15 45.1	−37 24 47	+0.002	−0.01	4.62	0.45	2.5	F3 IV–V	+12	22 ts		m
79682	98462		9 15 49.4	+10 22 25	−0.003	−0.02	7.8	0.1	1.4	A2 V		190 s		
79354	27185	17 UMa	9 15 49.8	+56 44 28	−0.002	−0.03	5.27	1.56	−0.3	K5 III	−30	120 s		
79961	200168		9 15 51.2	−37 37 44	−0.005	+0.01	6.9	0.6	0.2	K0 III		220 s		
78935	6792		9 15 52.6	+72 56 46	−0.019	−0.07	5.96	0.18		A2	+2	15 mn		
80077	221009		9 15 54.7	−49 58 23	+0.001	+0.01	7.65	1.30	−6.8	B2 Ia pe		7800 s		
79496	42775		9 15 55.3	+42 25 58	−0.001	−0.02	7.8	1.1	0.2	K0 III		330 s		
79894	177186		9 15 58.7	−27 11 44	+0.001	0.00	7.80	1.1	0.2	K0 III		330 s		m
80057	221010		9 16 04.1	−44 53 54	+0.001	+0.01	6.04	0.28	−5.1	A1 Ib		1300 s		m
79825	—		9 16 06.8	−8 20 52			7.60	0.2	2.1	A5 V		130 s	7270	m
79935	177193		9 16 07.4	−28 53 11	−0.003	0.00	7.3	0.3	0.6	A0 V		140 s		
79915	177191		9 16 09.9	−21 11 51	−0.009	0.00	7.9	0.5	3.4	F5 V		72 s		
79679	80731		9 16 11.2	+23 24 07	−0.005	−0.09	7.50	1.1	0.2	K0 III		130 mx	7268	m
79439	27191	18 UMa	9 16 11.3	+54 01 18	+0.006	+0.06	4.83	0.19	2.1	A5 V	−15	34 s		
80230	236787		9 16 12.2	−57 32 28	−0.002	−0.01	4.34	1.63	−0.3	gK5	−5	70 s		
79853	136715		9 16 17.7	−4 32 00	−0.006	−0.02	8.0	1.1	0.2	K0 III		190 mx		
80108	221014		9 16 23.0	−44 15 57	0.000	0.00	5.12	1.67	−5.9	cK	−3	1100 s		
79933	155050		9 16 25.4	−19 27 31	−0.002	−0.01	7.3	0.4	3.0	F2 V		72 s		
79876	136720		9 16 26.0	−4 24 17	−0.008	−0.03	8.0	1.1	−0.1	K2 III		130 mx		
79765	98468		9 16 27.1	+18 48 37	−0.011	−0.02	6.9	0.1	1.7	A3 V	+31	58 mx		
79914	155048		9 16 27.4	−13 50 00	−0.005	+0.04	6.8	0.1	1.4	A2 V		86 mx		
80250	236796		9 16 37.2	−53 52 43	−0.004	0.00	8.0	1.5	0.2	K0 III		190 s		
79873	117553		9 16 39.9	+0 43 45	−0.007	−0.02	6.50	0.5	3.4	F5 V		42 s	7276	m
80268	236801		9 16 40.1	−54 39 22	−0.002	−0.01	7.93	0.01	0.4	B9.5 V		300 s		
80210	221022		9 16 40.2	−49 34 42	+0.001	+0.01	6.98	0.70	4.0	F8 V		28 s		
79931	136728	24 Hya	9 16 41.3	−8 44 41	−0.002	0.00	5.47	−0.09		B9 n	+10			
79910	136725	23 Hya	9 16 41.6	−6 21 11	+0.001	0.00	5.24	1.17	−0.1	K2 III	−8	120 s		m
80308	236804		9 16 43.3	−56 44 59	−0.003	+0.03	8.0	0.3	3.4	F5 V		82 s		
82370	258527		9 16 46.6	−84 39 38	−0.021	0.00	7.1	1.8	−0.1	K2 III		110 s		
80173	221023		9 16 54.0	−41 57 19	−0.004	+0.01	8.0	1.2	0.2	K0 III		360 s		
80459	250521		9 16 55.7	−63 46 31	0.000	0.00	7.39	−0.01	0.4	B9.5 V		250 s		
80171	200184		9 16 56.0	−40 07 00	−0.003	+0.02	7.1	0.8	0.2	K0 III		240 s		
79994	155056		9 16 57.0	−11 06 09	−0.001	−0.05	6.5	1.1	0.2	K0 III		180 s		
80170	200185		9 16 57.1	−39 24 05	−0.001	−0.03	5.33	1.17		K5 III–IV	0	16 mn		
80205	221026		9 16 57.9	−45 00 40	−0.001	+0.02	6.74	−0.01		A0				
79993	136731		9 16 59.7	−9 37 20	−0.002	−0.06	7.6	1.1	0.2	K0 III		250 mx		
80282	236806		9 17 00.1	−51 26 19	−0.004	0.00	7.62	−0.13	0.6	A0 V		250 s		
78846	6794		9 17 00.3	+77 04 23	−0.011	−0.03	6.9	0.1	0.6	A0 V		120 mx		
79800	61378		9 17 01.0	+30 22 29	0.000	+0.01	7.9	0.1	0.6	A0 V		280 s		
80188	221025		9 17 01.1	−41 19 12	−0.006	−0.01	7.5	0.7	0.6	A0 V		130 mx		
80030	155059		9 17 02.4	−15 30 26	−0.001	+0.03	8.0	1.1	−0.1	K2 III		430 s		
80404	236808	ι Car	9 17 05.4	−59 16 31	−0.002	0.00	2.25	0.18	−4.7	F0 Ib	+13	250 s		Aspidiske
80050	155060		9 17 07.4	−14 34 25	−0.004	+0.01	5.84	1.06	0.2	gK0	−37	120 s		
80281	221027		9 17 10.8	−47 44 07	0.000	+0.01	7.75	1.29	0.2	K0 III		200 s		
79971	117563		9 17 11.9	+2 40 43	−0.005	0.00	8.0	0.4	2.6	F0 V		120 s		
79972	117564		9 17 12.2	+1 55 45	−0.003	−0.01	6.9	0.4	2.6	F0 V		71 s		
80603	250524		9 17 13.8	−67 40 34	+0.007	0.00	7.4	1.1	0.2	K0 III		270 s		
80671	250526		9 17 17.2	−68 41 22	−0.019	−0.03	5.39	0.42	3.0	dF2	+32	24 ts		m
79838	61379		9 17 17.5	+33 30 04	−0.006	−0.07	6.8	0.9	3.2	G5 IV		53 s		
79872	80738		9 17 19.1	+23 39 09	−0.004	−0.07	7.2	0.5	3.4	F5 V	+13	59 s	7281	m
80187	200189		9 17 23.1	−32 38 03	−0.001	+0.03	7.6	1.1	3.2	G5 IV		48 s		
80138	177216		9 17 23.3	−21 20 07	+0.011	+0.01	8.0	0.8	−0.3	K5 III		74 mx		
80457	236816		9 17 23.5	−58 43 48	0.000	−0.01	8.0	0.4	3.4	F5 V		82 s		

HD	SAO	Star Name	α 2000	δ 2000	μ(α)	μ(δ)	V	B-V	M$_v$	Spec	RV	d(pc)	ADS	Notes
			h m s	° ′ ″	s	″								
80951	256599		9 17 25.5	-74 53 39	-0.003	+0.03	5.29	0.02		A0	+11			m
80866	256598		9 17 26.6	-73 02 10	-0.006	+0.01	8.0	0.1	1.7	A3 V		180 s		
80347	236809		9 17 27.0	-50 34 35	-0.001	-0.05	7.33	0.98	0.2	K0 III		270 s		
79851	61382		9 17 27.2	+31 18 04	+0.001	-0.01	7.9	1.1	0.2	K0 III		340 s		
80950	256600		9 17 27.5	-74 44 05	-0.009	+0.03	5.87	-0.02		A0				
80105	155064		9 17 29.1	-11 57 43	-0.003	-0.05	7.07	1.03	0.2	gK0	+10	230 s		
80009	117566		9 17 30.6	+8 22 34	-0.001	-0.01	7.7	1.1	0.2	K0 III		310 s		
79763	42790		9 17 31.2	+46 49 02	+0.002	+0.01	5.7	0.1	0.6	A0 V	-12	100 s		
79315	6804		9 17 31.3	+69 52 24	+0.002	-0.01	7.1	1.1	0.2	K0 III		240 s		
80419	236817		9 17 31.7	-55 56 08	-0.003	+0.01	7.36	1.08		K0				
80046	117569		9 17 32.8	+0 33 18	0.000	-0.03	7.00	0.1	0.6	A0 V		190 s		m
80245	200193		9 17 32.9	-36 22 20	-0.001	+0.03	8.0	1.7	0.2	K0 III		140 s		
79929	80741		9 17 35.7	+27 25 18	0.000	-0.02	6.5	0.5	3.4	F5 V	+12	43 s		
79992	98474		9 17 38.1	+16 42 18	0.000	-0.01	8.0	0.1	0.6	A0 V		300 s		
79850	61383		9 17 41.4	+39 11 50	+0.001	-0.04	8.0	0.5	3.4	F5 V		84 s		
80435	236819		9 17 42.3	-54 29 42	-0.002	-0.01	6.33	1.41		K0				
79990	80744		9 17 45.6	+23 04 43	-0.003	-0.02	7.1	1.1	0.2	K0 III		240 s		
80275	200198		9 17 46.6	-35 41 24	-0.002	+0.01	7.8	1.3	-0.1	K2 III		300 s		
79675	27201		9 17 48.8	+57 28 50	-0.007	0.00	7.62	1.12	0.0	K1 III		290 mx		
80065	117573		9 17 50.7	+7 56 48	-0.002	+0.02	7.2	0.8	3.2	G5 IV		63 s		
80064	98476		9 17 51.3	+11 30 04	-0.001	-0.01	6.30	0.07	1.7	A3 V	-3	83 s	7285	m
80710	250529		9 17 51.8	-67 03 03	0.000	+0.01	6.11	1.26		K0				
79969	80745		9 17 53.5	+28 33 39	+0.005	-0.51	7.21	0.98	6.9	K3 V	-18	17 ts	7284	m
80332	200204		9 17 56.1	-37 23 53	-0.014	+0.02	7.0	0.0	4.4	G0 V		33 s		
80807	—		9 17 56.5	-69 47 18			7.80	0.1	0.6	A0 V		280 s		m
80380	221041		9 17 59.4	-45 38 11	-0.003	-0.01	7.21	-0.15	0.0	B8.5 V		280 s		
80300	200202		9 18 00.7	-31 24 13	+0.001	+0.02	8.0	1.4	-0.1	K2 III		310 s		
79702	27203		9 18 01.2	+57 54 20	-0.007	-0.06	7.6	0.9	3.2	G5 IV		77 s		
80456	236824		9 18 05.8	-51 03 03	-0.003	+0.01	5.26	-0.07	-1.2	B8 III	+66	190 s		
80117	117578		9 18 07.5	+8 14 21	-0.002	-0.02	7.9	0.9	3.2	G5 IV		88 s		
80240	155076		9 18 10.3	-18 14 53	+0.001	-0.01	7.9	0.1	1.4	A2 V		200 s		
80163	117583		9 18 16.6	+0 42 35	0.000	-0.04	7.9	1.1	0.2	K0 III		340 s		
80742	—		9 18 17.3	-65 25 17			7.6	1.2	0.2	K0 III		220 s		
80628	236836		9 18 21.7	-59 38 47	0.000	0.00	7.8	0.2	0.6	A0 V		190 s		
80577	236834		9 18 25.0	-57 16 22	-0.001	+0.01	6.9	1.3	-0.1	K2 III		220 s		
80024	61387		9 18 25.9	+35 21 50	-0.003	-0.03	6.40	0.2	2.1	A5 V	+22	72 s	7286	m
80694	250531		9 18 30.2	-61 03 04	0.000	-0.01	8.0	1.6	-0.1	K2 III		240 s		
80432	200219		9 18 32.0	-39 02 38	-0.002	+0.01	6.7	0.8	0.6	A0 V		54 s		
80558	236837		9 18 42.2	-51 33 38	-0.002	0.00	5.87	0.54	-6.4	B7 Iab		1400 s		
80431	200222		9 18 43.9	-34 05 44	+0.002	+0.01	7.4	1.2	-0.5	M4 III		380 s		
80528	221052		9 18 46.6	-47 12 17	-0.002	0.00	7.7	-0.3	0.4	B9.5 V		290 s		
80598	236839		9 18 47.3	-53 36 42	+0.001	-0.01	8.0	0.3	0.0	B8.5 V		230 s		
80081	61391	38 Lyn	9 18 50.6	+36 48 09	-0.002	-0.13	3.82	0.06	1.7	A3 V	+2	27 s	7292	m
80527	221053		9 18 51.5	-45 00 29	-0.001	-0.02	7.20	1.11		K0				
80501	200228		9 18 53.0	-39 26 01	0.000	-0.01	7.80	1.1	0.2	K0 III		320 s		m
80667	236843		9 18 55.6	-56 03 32	-0.001	0.00	7.87	1.18		K2				
80369	155084		9 18 58.0	-16 48 04	+0.001	-0.02	6.9	0.1	0.6	A0 V		190 s		
80218	98488		9 18 58.8	+17 42 19	-0.009	-0.13	6.6	0.5	3.4	F5 V	-15	44 s		
80236	98491		9 18 59.0	+10 57 00	-0.001	-0.01	7.90	0.1	1.4	A2 V		200 s	7296	m
80116	61395		9 18 59.7	+35 40 57	-0.001	-0.01	7.8	1.4	-0.3	K5 III		410 s		
80572	221060		9 19 00.1	-47 58 05	-0.021	+0.18	7.04	1.01		G5		22 mn		
81143	256603		9 19 03.8	-73 19 18	+0.002	-0.01	7.5	0.1	0.6	A0 V		240 s		
80542	221055		9 19 04.1	-41 28 05	-0.005	-0.01	7.6	0.6	3.2	G5 IV		77 s		
80340	136750		9 19 07.5	-2 27 30	-0.002	-0.03	7.8	0.1	1.7	A3 V		170 s		
80429	177268		9 19 07.8	-22 33 12	-0.002	0.00	7.2	0.5	0.6	A0 V		100 s		
80726	236847		9 19 09.4	-56 21 54	-0.005	+0.04	6.82	0.16		A2				
80294	117585		9 19 11.9	+9 04 43	+0.003	-0.14	7.8	1.1	0.2	K0 III		88 mx		
80395	155085		9 19 13.7	-11 09 27	0.000	-0.03	7.7	1.1	0.2	K0 III		310 s		
80217	80757		9 19 15.2	+26 15 06	-0.002	-0.03	6.5	1.4	-0.3	K4 III		230 s		
80079	42803		9 19 15.3	+46 56 38	-0.003	-0.02	7.0	0.1	0.6	A0 V		190 s		
80790	236850		9 19 20.0	-58 55 45	+0.004	0.00	7.2	0.5	4.0	F8 V		43 s		
80465	177274		9 19 20.3	-23 24 16	+0.001	-0.03	8.0	1.2	0.6	A0 V		59 s		
79887	14842		9 19 23.4	+61 35 19	-0.001	+0.02	7.7	1.1	0.2	K0 III		310 s		
80980	250537		9 19 29.3	-67 25 53	+0.002	-0.02	7.6	1.4	-0.3	K5 III		380 s		
80339	98495		9 19 30.4	+9 47 25	-0.002	+0.01	7.3	1.1	0.2	K0 III		260 s		
80781	236851		9 19 32.5	-55 11 12	-0.001	0.00	6.28	-0.10	-1.0	B7 IV	+17	290 s		
80479	155091		9 19 33.1	-15 50 04	+0.003	-0.05	5.78	1.29		K0	-30		7302	m
80447	155090		9 19 33.6	-11 18 51	-0.006	+0.03	6.62	0.08		A2				
80541	200236		9 19 34.7	-32 55 56	-0.002	+0.02	8.0	1.2	-0.3	K5 III		450 s		
80780	236852		9 19 39.0	-53 43 48	-0.004	0.00	8.0	0.6	1.7	A3 V		83 s		
80328	80764		9 19 39.4	+20 32 47	+0.001	-0.02	7.9	1.1	-0.1	K2 III		390 s		
80499	155096	26 Hya	9 19 46.3	-11 58 30	-0.002	+0.01	4.79	0.93	0.3	G8 III	-2	79 s		m
80590	200241		9 19 47.9	-34 06 12	-0.001	-0.01	6.39	-0.10		B8				
80464	136761		9 19 48.1	-5 59 32	-0.001	-0.02	8.0	1.4	-0.3	K5 III		450 s		
80327	80765		9 19 49.5	+24 24 59	+0.004	-0.08	7.90	0.52	4.0	F8 V		60 s		

HD	SAO	Star Name	α 2000	δ 2000	μ(α)	μ(δ)	V	B-V	M_v	Spec	RV	d(pc)	ADS	Notes
			h m s	+° ′ ″	s	″								
80425	117590		9 19 51.2	+5 12 57	-0.002	-0.02	6.5	0.2	2.1	A5 V		76 s		
81178	256604		9 19 51.3	-71 00 42	+0.009	-0.12	7.2	0.4	3.0	F2 V		69 s		
79262	6811		9 19 52.3	+77 58 06	-0.001	+0.02	8.0	0.9	3.2	G5 IV		93 s		
80521	155098		9 19 55.2	-13 18 23	-0.002	-0.02	7.2	1.1	0.2	K0 III		250 s		
79517	6816		9 19 55.7	+74 00 59	-0.009	-0.07	6.5	0.9	0.3	G8 III	+56	170 s		
80641	200245		9 19 57.4	-35 29 52	-0.003	+0.06	7.0	0.5	0.2	K0 III		230 mx		
80338	80766		9 19 58.1	+28 17 33	+0.004	0.00	8.0	1.1	-0.1	K2 III		240 mx		
80621	177289		9 20 00.4	-27 35 33	-0.002	+0.03	8.0	2.8	-0.5	M2 III		99 s		
80761	221073		9 20 01.2	-46 56 59	0.000	+0.03	7.28	-0.10		B8				
80587	177290		9 20 03.9	-25 14 54	-0.003	+0.02	8.0	1.0	0.2	K0 III		360 s		
80926	236862		9 20 07.1	-58 56 03	-0.002	0.00	7.9	1.3	0.2	K0 III		240 s		
80313	61405		9 20 07.2	+37 45 19	+0.003	-0.03	7.9	1.1	0.2	K0 III		340 s		
80759	221074		9 20 11.1	-43 23 55	-0.002	-0.01	7.55	1.20	0.0	K1 III		320 s		
80618	--		9 20 12.0	-23 28 17			7.80	0.4	2.6	F0 V		110 s	7312	m
80721	200251		9 20 13.0	-36 57 15	0.000	-0.01	7.8	1.1	-0.1	K2 III		380 s		
79968	14845		9 20 14.2	+65 00 42	-0.022	-0.30	7.75	0.74	5.0	dG4	+22	30 s		
80777	221077		9 20 15.1	-45 10 22	+0.002	+0.01	6.70	1.22		K0				m
80042	14846		9 20 16.2	+62 29 56	+0.007	-0.08	8.0	0.5	4.0	F8 V		62 s		
81372	256607		9 20 18.7	-73 36 18	-0.002	0.00	7.1	0.1	0.6	A0 V		200 s		
80550	136767		9 20 20.9	-9 36 37	-0.002	-0.03	6.95	0.39	3.0	F2 V	+16	60 s		
80755	200255		9 20 24.6	-37 20 21	-0.001	-0.01	8.0	1.8	-0.3	K5 III		290 s		
81504	256608		9 20 25.9	-75 42 49	+0.004	+0.01	7.41	0.06	0.4	B9.5 V		230 s		
80537	117603		9 20 26.3	+0 17 02	+0.002	-0.01	7.7	0.9	3.2	G5 IV		78 s		
80705	177304		9 20 26.4	-29 07 39	+0.002	-0.05	7.3	0.7	0.2	K0 III		260 s		
80586	136768	27 Hya	9 20 28.9	-9 33 21	-0.001	-0.03	4.80	0.93	1.8	G8 III-IV	+25	39 s	7311	m
80130	27213		9 20 29.4	+59 46 51	+0.005	-0.02	7.4	1.1	0.2	K0 III	-33	280 s		
80233	27214		9 20 29.5	+52 27 04	+0.002	+0.01	6.6	0.1	0.6	A0 V		160 s		
80774	200257		9 20 29.6	-37 34 53	-0.002	-0.02	6.05	1.39		K0				
81141	250542		9 20 30.3	-66 29 15	0.000	+0.05	7.5	0.0	0.4	B9.5 V		260 s		
81038	250540		9 20 31.9	-61 59 36	-0.001	0.00	6.90	-0.09	-1.1	B5 V n		390 s		
80817	221080		9 20 33.0	-44 16 59	-0.003	+0.01	8.03	0.09		A2				
80859	221084		9 20 34.8	-47 33 59	0.000	+0.02	7.5	0.3	0.0	B8.5 V		180 s		
80567	117605		9 20 36.6	+0 10 53	0.000	-0.03	6.7	1.6	-0.5	M4 III	+3	280 s		
80495	98510		9 20 37.8	+19 05 27	-0.002	-0.01	7.3	0.9	3.2	G5 IV		66 s		
80510	98511		9 20 38.2	+17 07 39	-0.001	-0.01	7.6	0.4	3.0	F2 V		85 s		
81961	258529		9 20 38.8	-80 23 23	-0.055	+0.17	7.6	1.1	0.2	K0 III		92 mx		
81206	250546		9 20 39.5	-68 21 34	-0.002	-0.02	7.8	1.1	0.2	K0 III		330 s		
80290	27215		9 20 43.7	+51 15 58	-0.004	+0.15	6.10	0.4	3.0	F2 V	-8	34 ts	7303	m
80773	200258		9 20 44.1	-31 45 38	-0.003	-0.01	6.82	0.00		A0				
80936	236867		9 20 44.3	-53 15 20	-0.005	0.00	7.04	1.06	3.2	G5 IV		36 s		
80633	136773		9 20 44.3	-7 43 49	-0.009	-0.03	7.7	0.5	4.0	F8 V		54 s		
80771	177312		9 20 47.1	-28 48 11	-0.005	+0.03	7.5	0.8	-0.1	K2 III		250 mx		
80735	177311		9 20 50.3	-24 05 52	-0.003	0.00	8.0	0.8	0.2	K0 III		360 s		
80999	236871		9 20 53.9	-55 48 44	-0.001	-0.01	7.94	0.39		A0				
80719	155114		9 20 55.4	-15 37 04	+0.004	-0.11	6.33	0.46	3.7	dF6	-1	34 s		
81101	250544		9 20 57.1	-62 24 17	+0.001	-0.01	4.81	0.94	0.3	gG4	+51	68 s		
80441	61411		9 20 59.2	+38 11 18	-0.004	-0.02	6.60	0.4	3.1	F3 V	+1	50 s	7307	m
80493	61414	40 α Lyn	9 21 03.2	+34 23 33	-0.018	+0.01	3.13	1.55	-0.4	M0 III	+38	51 s		
80922	221089		9 21 03.6	-45 02 30	-0.003	+0.02	7.64	0.09		A2				
81320	--		9 21 06.0	-69 17 36			7.6	0.1	0.6	A0 V		250 s		
81022	236874		9 21 07.5	-54 38 45	+0.002	-0.02	8.0	0.9	0.2	K0 III		360 s		
80678	117611		9 21 08.9	+2 56 25	-0.005	-0.03	7.1	0.4	3.0	F2 V		65 s		
80853	200268		9 21 10.2	-35 21 27	-0.005	0.00	7.50	0.4	3.0	F2 V		79 s		m
80536	80776		9 21 10.5	+25 09 47	+0.004	-0.15	7.29	0.63	4.5	dG1	-38	34 s		
80509	61418		9 21 11.8	+34 56 52	-0.016	0.00	8.0	0.5	3.4	F5 V		82 s		
80701	117613		9 21 12.9	+3 25 14	-0.007	+0.03	8.0	0.4	2.6	F0 V		120 s		
80492	61417		9 21 15.3	+39 39 59	-0.005	-0.02	7.1	1.1	0.2	K0 III		240 s		
80613	98517		9 21 15.3	+15 22 16	-0.002	-0.02	6.53	-0.01	0.6	A0 V	+18	150 s		
80934	221096		9 21 18.1	-42 49 29	-0.004	-0.01	7.29	1.00	0.2	K0 III		260 s		
80654	98520		9 21 18.4	+13 06 41	-0.003	-0.08	6.6	0.5	3.4	F5 V	-8	44 s		
80611	80779		9 21 21.2	+22 53 26	0.000	-0.01	8.0	1.1	0.2	K0 III		360 s		
80461	42820		9 21 23.6	+45 22 12	-0.001	-0.03	6.6	1.1	0.2	K0 III		190 s		
81078	236879		9 21 25.2	-55 24 06	-0.003	-0.01	7.88	1.47		K5				
80508	42822		9 21 25.4	+40 17 00	+0.003	-0.03	7.0	0.1	1.4	A2 V		130 s		
80652	98521		9 21 25.4	+16 35 54	-0.002	-0.04	6.8	0.2	2.1	A5 V	+5	87 s		
80894	200272		9 21 25.9	-30 48 22	-0.002	+0.01	8.0	0.3	1.4	A2 V		150 s		
80546	61424		9 21 27.1	+32 54 06	+0.001	-0.04	6.16	1.09	0.2	gK0	+28	130 s		
80874	177322	θ Pyx	9 21 29.5	-25 57 55	-0.001	-0.01	4.72	1.63	-0.5	M1 III	+20	100 s		
80908	200275		9 21 30.4	-34 20 10	-0.005	+0.03	7.7	0.6	0.2	K0 III		260 mx		
80945	200276		9 21 32.3	-39 41 45	-0.002	0.00	7.6	1.6	-0.1	K2 III		200 s		
81175	250551		9 21 34.3	-60 49 50	-0.002	-0.01	7.50	0.1	0.6	A0 V		240 s		m
80651	80782		9 21 38.3	+22 45 59	-0.001	-0.05	7.9	1.1	-0.1	K2 III		390 s		
80580	61425		9 21 39.3	+32 15 52	-0.004	-0.03	6.80	0.13		A0	+9			
81138	236885		9 21 40.2	-56 54 14	-0.003	+0.01	7.23	0.99		K0				
81035	221100		9 21 40.8	-47 19 12	-0.004	+0.01	7.47	0.13		A2				m

HD	SAO	Star Name	α 2000	δ 2000	μ(α)	μ(δ)	V	B–V	M_v	Spec	RV	d(pc)	ADS	Notes
80649	80781		9^{h}21^{m}43.1^s	+28°30'01"	0.000^s	+0.02"	8.0	0.4	3.0	F2 V		98 s		
80390	27219		9 21 43.2	+56 41 57	−0.001	−0.01	5.47	1.61	−0.5	M4 III	+21	160 s		
80796	136784		9 21 44.6	−2 47 36	−0.002	−0.02	7.1	0.0	0.4	B9.5 V		220 s		
80699	98523		9 21 47.0	+18 44 47	−0.005	−0.11	7.8	0.6	4.4	G0 V		47 s		
66368	1401		9 21 48.5	+88 34 13	−0.023	+0.01	7.10	0.1	0.6	A0 V	−8	200 s		m
81157	236887		9 21 49.9	−55 30 54	−0.008	+0.06	5.63	0.19		A3 p	+59			
81034	221103		9 21 50.8	−42 11 42	+0.004	−0.06	5.58	1.63		M1				
80851	155126		9 21 51.2	−10 25 39	−0.001	−0.02	8.0	0.4	2.6	F0 V		120 s		
80873	155127		9 21 52.0	−13 32 14	−0.001	0.00	7.7	1.1	−0.1	K2 III		370 s		
81137	236888		9 21 59.1	−52 33 52	−0.001	−0.01	7.40	1.6	−4.8	M3 Ib pe		2300 s		WY Vel, q
81077	221105		9 22 00.9	−46 45 32	+0.001	+0.02	7.68	0.99		G5				m
79870	6825		9 22 01.6	+74 54 44	−0.006	−0.02	7.4	0.1	1.4	A2 V		160 s		
81188	236891	κ Vel	9 22 06.8	−55 00 38	−0.001	+0.01	2.50	−0.18	−3.0	B2 IV	+22	120 s		
81639	—		9 22 07.5	−73 31 45			8.00	0.9	3.2	G5 IV		91 s		m
80823	117624		9 22 08.2	+5 34 43	−0.003	−0.03	8.0	0.5	4.0	F8 V		62 s		
80871	136789		9 22 10.1	−0 25 35	−0.002	−0.02	7.9	0.9	3.2	G5 IV		88 s		
81222	236893		9 22 16.3	−55 57 40	+0.001	−0.03	7.5	0.8	3.2	G5 IV		73 s		V Vel, v
81074	200289		9 22 20.8	−39 22 48	0.000	0.00	7.3	1.7	−0.1	K2 III		150 s		
81045	200287		9 22 21.0	−34 00 10	−0.003	+0.04	7.7	0.8	−0.1	K2 III		360 s		
81046	200288		9 22 21.9	−34 58 36	0.000	+0.01	7.6	1.2	0.2	K0 III		240 s		
80609	42824		9 22 22.9	+48 24 37	−0.006	−0.04	7.4	0.4	3.0	F2 V		76 s		
80917	136792		9 22 22.9	−6 03 39	−0.003	0.00	7.1	0.1	1.4	A2 V		140 s		
81136	221109		9 22 23.9	−46 02 52	0.000	+0.01	5.75	0.92		G5				
80715	42826		9 22 25.8	+40 12 04	−0.031	−0.36	7.63	0.99	6.3	K2 V	−43	25 ts		
81238	236895		9 22 28.3	−53 44 12	0.000	−0.01	7.95	0.49	3.4	F5 V		74 s		m
81202	236894		9 22 28.5	−50 14 07	0.000	0.00	7.76	−0.09	0.4	B9.5 V		300 s		
80971	155129		9 22 29.4	−17 53 38	−0.001	−0.03	6.80	0.4	3.0	F2 V		58 s	7331	m
81112	200296		9 22 30.1	−39 18 58	+0.003	−0.01	7.8	1.1	0.2	K0 III		300 s		
80916	136793		9 22 31.9	−1 04 34	−0.001	−0.03	8.00	0.1	0.6	A0 V		300 s	7329	m
80608	42825		9 22 32.1	+49 32 42	−0.003	0.00	7.10	0.00	0.4	B9.5 V		220 s	7324	m
80970	155130		9 22 32.1	−15 14 08	0.000	−0.04	6.6	1.1	0.2	K0 III		190 s		
81134	200299		9 22 36.6	−39 46 29	−0.006	+0.07	6.54	1.13	0.0	K1 III		170 mx		
81091	200300		9 22 47.1	−30 31 33	−0.002	−0.04	7.9	0.3	1.7	A3 V		120 s		
80888	98534		9 22 48.2	+18 07 55	+0.002	−0.01	7.7	1.1	0.2	K0 III		310 s		
81133	200301		9 22 49.5	−35 31 25	−0.008	−0.12	7.2	0.6	4.0	F8 V		40 s		
80968	136797		9 22 50.6	−5 02 22	−0.001	0.00	7.4	1.1	0.2	K0 III		280 s		
81009	136799		9 22 50.8	−9 50 20	−0.002	−0.03	6.53	0.22		A2			7334	m
81435	236920		9 23 03.9	−59 33 23	−0.003	+0.01	8.0	0.3	0.6	A0 V		180 s		
81353	236913		9 23 06.0	−53 49 14	−0.001	−0.01	6.81	1.05	3.2	G5 IV		33 s		m
81351	236912		9 23 06.8	−52 53 09	−0.007	−0.01	8.0	0.1	3.0	F2 V		98 s		
80870	61438		9 23 10.5	+32 55 15	+0.014	−0.03	7.8	0.9	3.2	G5 IV		64 mx		
81276	221120		9 23 10.9	−44 57 54	−0.003	+0.01	7.95	0.04		A0				
81169	177374	λ Pyx	9 23 12.1	−28 50 02	−0.011	+0.02	4.69	0.92	0.3	gG7	+10	76 s		
—	61440		9 23 14.4	+33 05 12	−0.001	−0.04	7.8	1.6						
81109	177371		9 23 14.6	−21 02 16	−0.001	0.00	7.10	1.6	−0.5	M3 III	+17	330 s		
81029	117641		9 23 15.6	+3 30 05	−0.005	+0.01	7.31	0.34		F0			7342	m
80365	14860		9 23 16.9	+69 21 11	−0.010	−0.06	7.3	1.4	−0.3	K5 III		210 mx		
80915	61442		9 23 23.3	+33 27 51	0.000	+0.01	7.8	0.0	0.4	B9.5 V		300 s		
82114	256614		9 23 25.0	−77 53 24	−0.080	+0.37	7.06	0.62		G0		31 mn		m
81347	221126		9 23 25.4	−48 17 15	0.000	−0.02	6.28	−0.14	−1.1	B5 V k		290 s		
81502	250561		9 23 27.4	−60 18 09	−0.001	−0.01	6.30	1.48	0.2	gK0		86 s		
81028	117646		9 23 29.4	+7 42 50	−0.001	−0.04	7.20	1.6	−0.5	M4 III	+58	350 s		
81148	155143		9 23 30.3	−16 39 57	−0.001	+0.02	7.3	0.0	0.4	B9.5 V		240 s		
80956	80797		9 23 31.7	+25 10 58	−0.009	0.00	6.4	0.8	3.2	G5 IV	−1	45 s		
81069	117649		9 23 38.7	+2 24 35	+0.002	−0.02	7.50	1.1	0.2	K0 III	−18	290 s		
81369	221131		9 23 39.6	−46 54 32	−0.002	0.00	6.22	−0.10		B9				
81254	200314		9 23 39.7	−33 12 00	0.000	+0.02	6.8	0.6	0.6	A0 V		73 s		
81067	117648		9 23 40.2	+5 13 11	+0.001	−0.01	6.8	0.1	1.4	A2 V		120 s		
81433	—		9 23 41.7	−52 25 42			7.9	0.0	2.6	F0 V		110 s		
81309	200320		9 23 44.7	−37 45 26	−0.007	+0.01	6.48	0.18		A2				
81040	80800		9 23 47.0	+20 21 53	−0.011	+0.04	7.8	0.7	4.4	G0 V		48 s		
81308	200321		9 23 47.9	−36 39 54	−0.002	+0.02	8.0	2.0	−0.1	K2 III		130 s		
81230	—		9 23 48.4	−23 38 38			7.80	0.4	3.0	F2 V		91 s	7350	m
80195	6837		9 23 49.4	+74 58 33	−0.009	+0.01	7.9	1.1	−0.1	K2 III		390 mx		
81307	200322		9 23 49.8	−35 14 51	+0.001	−0.01	6.7	0.2	0.6	A0 V		120 s		
80792	27235		9 23 50.3	+55 11 37	−0.003	+0.05	7.9	1.1	0.2	K0 III		340 s		
80966	61446		9 23 50.3	+34 32 54	+0.003	−0.06	7.2	1.1	0.2	K0 III		190 mx		
81125	117654		9 23 54.6	+1 25 33	+0.001	+0.01	7.8	0.4	2.6	F0 V		110 s		
81166	136806		9 23 54.6	−5 21 40	−0.002	+0.01	7.7	1.1	−0.1	K2 III		370 s		
81471	236930		9 23 59.4	−51 44 14	+0.001	0.00	6.08	0.57	−6.8	A7 Iab		2100 s		
81183	136809		9 24 00.7	−5 27 57	−0.001	+0.02	7.2	1.4	−0.3	K5 III	−23	310 s		
81613	250563		9 24 05.7	−61 38 56	+0.002	−0.05	5.99	1.06		G5				
80532	14865		9 24 06.5	+68 51 40	+0.002	−0.05	7.9	1.1	−0.1	K2 III		390 s		
81058	80801		9 24 08.0	+25 55 06	+0.001	−0.05	6.8	1.1	−0.1	K2 III	−16	240 s		
82554	258530	ι Cha	9 24 09.3	−80 47 13	−0.059	+0.13	5.36	0.45	0.6	gF2 pe	+7	90 s		

HD	SAO	Star Name	α 2000	δ 2000	μ(α)	μ(δ)	V	B-V	M_V	Spec	RV	d(pc)	ADS	Notes
80783	14868		9 24 11.8	+60 26 51	-0.002	+0.01	7.38	1.41		K2				
81411	200330		9 24 16.2	-39 25 33	+0.001	-0.04	6.06	0.20		A2				m
81107	80805		9 24 18.6	+23 34 42	+0.003	-0.04	7.2	1.1	0.2	K0 III		250 s		
81057	61457		9 24 21.1	+34 20 36	+0.005	-0.08	7.9	1.1	0.2	K0 III		150 mx		
81039	61456		9 24 22.3	+36 35 13	-0.007	-0.03	6.5	0.2	2.1	A5 V	+15	74 s		
81378	177403		9 24 22.8	-29 10 51	0.000	-0.02	7.6	1.2	-0.3	K5 III		380 s		
81212	117661		9 24 28.4	+6 21 00	-0.012	-0.04	7.50	0.5	3.4	F5 V	+45	66 s	7352	m
81267	136816		9 24 29.9	-6 53 13	0.000	-0.04	8.0	1.1	-0.1	K2 III		410 s		
81163	98552		9 24 31.1	+18 08 28	-0.003	-0.04	7.30	0.6	4.4	G0 V		38 s	7341	m
82188	256617		9 24 32.1	-76 43 29	-0.006	+0.09	7.9	1.1	0.2	K0 III		220 mx		
81449	200337		9 24 36.2	-36 57 09	0.000	-0.01	8.0	1.2	0.4	B9.5 V		63 s		
81896	250572		9 24 37.4	-70 04 34	-0.006	-0.02	6.5	1.7	0.2	K0 III		69 s		
81910	256612		9 24 38.4	-70 24 35	-0.004	+0.01	7.80	-0.04	0.4	B9.5 V		300 s		m
81146	80807	1 κ Leo	9 24 39.2	+26 10 56	-0.002	-0.05	4.46	1.23	-0.1	K2 III	+28	73 s	7351	m
81654	236941		9 24 39.5	-58 41 21	-0.001	-0.02	7.93	0.00	-1.6	B3.5 V		670 s		
82423	256620		9 24 40.4	-79 04 28	-0.011	+0.02	7.9	-0.1	0.4	B9.5 V		180 mx		
81193	98554		9 24 41.8	+17 42 36	-0.001	-0.04	7.1	1.1	-0.1	K2 III	+37	280 s		
81192	80809		9 24 45.3	+19 47 11	-0.006	-0.12	6.56	0.94	0.3	G8 III	+135	180 s		
80983	27243		9 24 46.2	+50 16 41	0.000	-0.02	7.1	0.8	3.2	G5 IV		59 s		
81410	177412		9 24 48.8	-23 49 35	-0.004	-0.04	7.90	1.07		K2				
81377	155158		9 24 51.8	-15 38 19	0.000	-0.02	7.5	1.1	0.2	K0 III		290 s		
81394	155159		9 24 52.6	-18 05 26	-0.005	+0.01	8.0	1.1	0.2	K0 III		290 mx		
81342	136821		9 24 53.1	-4 17 00	-0.002	-0.06	6.9	0.5	3.4	F5 V		50 s		
81376	155157		9 24 54.7	-10 56 05	-0.001	-0.02	7.8	0.1	0.6	A0 V		270 s		
81025	27246		9 24 55.6	+51 34 26	+0.004	-0.01	6.31	0.75	4.4	G0 V	-16	19 s	7348	m
81624	236940		9 24 56.2	-50 17 11	+0.001	-0.02	7.83	1.21	0.2	K0 III		340 s		
81806	250570		9 24 56.5	-64 42 30	0.000	+0.02	7.6	0.1	1.4	A2 V		170 s		
81881	250573		9 24 57.5	-67 55 20	-0.008	+0.08	7.9	0.5	3.4	F5 V		79 s		
81409	155163		9 24 59.9	-17 13 07	-0.001	0.00	7.3	0.4	3.0	F2 V		74 s		
81515	200341		9 25 00.2	-35 54 22	-0.004	-0.03	7.6	0.9	1.7	A3 V		53 s		
82303	256618		9 25 00.9	-77 15 08	-0.009	+0.02	7.7	1.4	-0.3	K5 III		400 s		
81576	221144		9 25 01.2	-45 22 35	-0.002	-0.02	7.91	1.66	-0.5	M4 III		330 s		
81543	—		9 25 07.6	-38 04 47			8.0	-0.2	0.4	B9.5 V		330 s		
81575	221146		9 25 08.0	-43 58 37	-0.003	+0.03	6.51	1.57		M5 III				
81933	—		9 25 15.0	-68 20 57			8.0	1.1	0.2	K0 III		360 s		
81144	42848		9 25 15.6	+44 37 31	-0.001	-0.02	7.8	1.1	0.2	K0 III		330 s		
81621	221153		9 25 15.6	-46 56 36	-0.004	+0.01	7.12	1.36	0.2	K0 III		130 s		
81870	250576		9 25 16.8	-65 16 13	-0.001	+0.04	6.9	0.0	2.6	F0 V		71 s		
81145	42849		9 25 17.3	+44 01 01	-0.001	-0.03	8.0	1.1	0.2	K0 III		360 s		
81529	177422		9 25 18.4	-28 58 44	-0.003	0.00	7.6	1.7	-0.1	K2 III		170 s		
81390	136826		9 25 18.5	-0 50 14	-0.004	-0.01	8.0	0.1	1.4	A2 V		150 mx		
81720	236952		9 25 19.4	-54 27 50	+0.002	-0.02	6.7	1.3	-0.1	K2 III		200 s		
81389	117679		9 25 22.4	+4 17 08	-0.001	0.00	7.9	0.1	0.6	A0 V		280 s		
81373	117678		9 25 22.7	+6 15 30	+0.003	-0.02	7.5	0.9	0.2	G9 III	-2	290 s		
81420	136832	28 Hya	9 25 23.9	-5 07 03	-0.001	-0.01	5.59	1.52	-0.3	K5 III	+5	150 s		
81421	136833		9 25 27.0	-6 24 16	-0.003	-0.01	7.9	0.1	1.7	A3 V		170 s		
81830	250575		9 25 27.1	-61 57 01	-0.018	+0.06	5.77	0.15	1.4	A2 V n		58 mx		m
81265	61475		9 25 29.4	+30 29 37	+0.006	-0.18	7.80	0.8	3.2	G7 IV	-1	55 mx		
81361	98561		9 25 32.4	+16 35 08	-0.006	-0.02	6.29	0.97	0.2	gK0	+12	170 s		
81768	236958		9 25 35.1	-55 00 23	-0.002	+0.01	7.9	0.5	0.2	K0 III		340 s		
81734	236956		9 25 36.4	-53 15 09	-0.002	-0.11	7.05	0.51	4.0	F8 V		41 s		m
81512	155172		9 25 36.7	-13 19 53	-0.003	+0.02	8.0	1.1	0.2	K0 III		360 s		
81264	61478		9 25 36.9	+35 09 08	-0.003	+0.01	8.0	0.5	3.4	F5 V		82 s		
81542	177427		9 25 37.0	-25 20 11	-0.002	+0.02	7.00	0.00	0.4	B9.5 V		210 s		m
81104	27249		9 25 37.9	+54 00 57	-0.003	-0.03	7.80	0.1	1.4	A2 V	+19	190 s	7354	m
81552	177428		9 25 38.4	-27 27 05	-0.001	0.00	8.0	3.1	-0.5	M2 III		65 s		
81246	61477		9 25 38.8	+37 53 20	-0.008	-0.05	8.0	0.5	3.4	F5 V		82 s		
81634	221158		9 25 39.2	-40 18 50	-0.004	+0.01	7.3	0.9	-0.1	K2 III		300 s		
81604	200354		9 25 41.3	-34 19 22	-0.004	-0.01	7.5	1.3	0.2	K0 III		200 s		
80953	14875		9 25 44.1	+63 56 27	-0.001	-0.04	6.28	1.46		K2	+7			
81603	200356		9 25 44.4	-33 10 26	-0.001	+0.02	7.7	1.3	-0.3	K5 III		390 s		
81299	61481		9 25 44.9	+32 45 44	0.000	0.00	7.9	0.9	3.2	G5 IV	+17	88 s		
81583	177437		9 25 55.7	-20 45 42	-0.001	-0.02	7.3	0.7	1.7	A3 V		59 s		
81568	155181		9 25 56.4	-17 56 52	-0.006	-0.01	7.1	0.9	3.2	G5 IV		60 s		
81803	236964		9 25 57.4	-53 50 23	-0.003	-0.01	7.09	0.67	3.4	F5 V		36 s		
81630	200364		9 25 58.9	-31 22 10	-0.001	-0.02	7.6	0.5	0.6	A0 V		120 s		
81850	236965		9 25 59.1	-56 05 22	-0.001	+0.02	7.31	-0.02		B8				m
81490	117686		9 26 00.2	+3 01 48	-0.001	-0.02	7.7	0.9	3.2	G5 IV		80 s		
81663	200366		9 26 02.6	-35 58 09	-0.001	-0.02	7.9	0.4	2.1	A5 V		110 s		
81524	136840		9 26 03.9	-1 14 06	0.000	-0.01	6.7	1.1	0.2	K0 III		200 s		
81893	236973		9 26 04.2	-59 44 38	0.000	+0.01	7.4	1.7	0.2	K0 III		110 s		
81783	221165		9 26 08.1	-47 45 48	-0.005	-0.10	7.72	1.27	-0.2	K3 III		110 mx		
81662	177446		9 26 10.9	-30 02 05	-0.002	0.00	7.6	-0.1	1.4	A2 V		180 s		
80842	6854		9 26 11.7	+70 20 56	+0.008	-0.03	7.4	0.9	3.2	G5 IV		69 s		
81848	236972		9 26 18.0	-53 22 45	-0.001	0.00	5.11	-0.11	-1.1	B5 V n	+22	170 s		

HD	SAO	Star Name	α 2000	δ 2000	μ(α)	μ(δ)	V	B−V	M_v	Spec	RV	d(pc)	ADS	Notes
81567	136844		9^{h}26^{m}22.2^s	−1°27'50"	0.000	−0.01	6.01	1.32	−0.2	gK3	−15	170 s		
81782	221166		9 26 24.2	−42 15 06	−0.003	−0.03	7.50	0.1	0.6	A0 V		240 s		m
81802	221168		9 26 25.3	−46 07 32	0.000	−0.01	8.00	0.18		A2				
81628	155186		9 26 25.7	−12 47 59	−0.001	−0.03	8.0	1.1	0.2	K0 III		360 s		
81891	236979		9 26 27.8	−54 52 27	−0.001	+0.01	7.17	−0.13	−1.6	B3.5 V		530 s		
81780	221167		9 26 28.4	−40 30 06	+0.002	0.00	6.20	0.25		B3				
80388	6845		9 26 28.7	+78 26 15	+0.011	+0.02	8.00	0.6	4.5	G1 V	−12	50 s		
81506	98567		9 26 34.4	+19 03 35	−0.001	0.00	7.6	0.1	1.7	A3 V		150 s		
81755	200374		9 26 35.4	−33 53 52	−0.001	−0.06	6.9	1.0	3.2	G5 IV		38 s		
81566	117695		9 26 35.5	+6 31 50	−0.001	+0.01	7.90	0.5	3.4	F5 V		79 s	7371	m
81540	98568		9 26 37.8	+16 41 54	0.000	−0.01	7.8	1.6	−0.5	M2 III	+67	470 s		
81825	221172		9 26 39.4	−44 14 39	−0.003	+0.01	7.49	0.02		A0				
81694	177454		9 26 39.5	−24 10 03	−0.004	+0.01	6.8	−0.1	0.6	A0 V		140 mx		
81673	155190		9 26 41.0	−16 26 39	0.000	+0.02	7.6	0.9	3.2	G5 IV		77 s		
81440	61493		9 26 41.1	+36 27 27	+0.002	−0.03	6.9	0.5	3.4	F5 V		51 s		
81659	155191		9 26 42.7	−14 29 26	+0.002	−0.12	7.8	0.8	3.2	G5 IV		84 s		
81922	236983		9 26 43.9	−53 30 53	−0.001	+0.01	7.1	1.6	−0.5	M4 III		330 s		
82068	250583		9 26 44.2	−64 55 47	−0.008	+0.06	6.05	0.15		A3				
81753	177461		9 26 44.7	−28 47 15	−0.002	0.00	6.10	−0.10		B8			7379	m
81990	236990		9 26 46.9	−59 34 43	−0.003	+0.01	6.52	−0.09	0.4	B9.5 V		170 s		
81581	98571		9 26 48.4	+14 29 37	0.000	−0.05	7.75	0.31		A3			7373	m
81712	155193		9 26 49.0	−19 09 17	0.000	+0.01	6.9	0.1	1.4	A2 V		120 s		
81439	42863		9 26 49.1	+41 01 25	0.000	−0.07	7.7	1.1	−0.1	K2 III		160 mx		
81907	236984		9 26 49.4	−51 46 11	−0.002	0.00	8.04	0.00	0.4	B9.5 V		340 s		
81949	236987		9 26 53.2	−54 21 07	−0.002	−0.01	7.1	0.3	3.2	G5 IV		60 s		
81383	42861		9 26 55.7	+49 41 33	0.000	−0.01	8.0	1.6	−0.5	M2 III		510 s		
81595	98574		9 26 56.7	+14 18 10	−0.001	−0.06	7.0	1.1	−0.1	K2 III	+54	120 mx		
81921	221177		9 26 58.8	−48 41 40	−0.003	−0.01	6.83	−0.05	0.4	B9.5 V		190 s		
81731	155196		9 27 00.7	−16 00 49	−0.005	−0.02	7.1	1.1	−0.1	K2 III		210 mx		
81405	42864		9 27 01.2	+47 46 24	−0.001	0.00	7.6	0.1	1.4	A2 V		170 s		
81729	155197		9 27 03.7	−12 54 33	−0.002	−0.01	7.7	0.1	1.7	A3 V		160 s		
82350	256623		9 27 06.5	−71 36 08	−0.020	+0.07	5.47	1.08		K3	+3	13 mn		
81766	155205		9 27 11.5	−13 54 06	−0.003	−0.04	7.8	1.1	−0.1	K2 III		380 s		
81841	200382		9 27 14.1	−33 47 31	−0.003	−0.07	7.1	0.6	0.2	K0 III		170 mx		
81728	136861	29 Hya	9 27 14.5	−9 13 25	−0.002	0.00	6.54	0.05		A0			7382	m
81946	221181		9 27 14.5	−48 50 46	−0.001	0.00	7.78	−0.11	0.0	B8.5 V		360 s		
82187	−−		9 27 16.1	−66 35 07			7.6	0.0	0.0	B8.5 V		320 s		
81670	117704		9 27 16.5	+6 13 58	−0.012	−0.16	7.50	0.5	3.4	F5 V		58 mx	7380	m
81799	177469		9 27 18.3	−22 20 37	+0.013	−0.16	4.69	1.14	−0.2	gK3	+29	95 s		
81889	200388		9 27 22.8	−36 52 49	−0.001	−0.04	7.7	0.6	0.6	A0 V		120 s		
81357	27259		9 27 23.9	+58 08 33	+0.001	−0.01	7.9	0.0	0.0	B8.5 V		380 s		
81539	42867		9 27 26.4	+41 36 00	+0.003	−0.01	7.9	0.4	3.0	F2 V		94 s		
81538	42869		9 27 31.5	+42 45 53	−0.004	−0.01	7.4	0.4	2.6	F0 V		92 s		
82082	237002		9 27 31.6	−58 05 40	−0.019	0.00	7.4	0.2	4.4	G0 V		40 s		
81764	136867		9 27 32.4	−3 47 52	0.000	+0.01	7.7	0.6	4.4	G0 V		47 s		
81888	200390		9 27 33.1	−31 30 18	+0.001	+0.01	7.7	2.0	−0.3	K5 III		190 s		
81797	136871	30 α Hya	9 27 35.2	−8 39 31	−0.001	+0.03	1.98	1.44	−0.2	K3 III	−4	26 mn		Alphard, m
81707	98580		9 27 36.3	+12 24 00	+0.003	−0.08	7.7	1.1	0.2	K0 III		150 mx		
81919	200392		9 27 38.3	−35 00 28	−0.003	−0.03	6.65	0.20		A3				m
81579	42871		9 27 41.6	+41 11 50	−0.002	−0.04	7.5	0.4	3.0	F2 V		78 s		
81987	221183		9 27 46.2	−42 34 40	−0.002	+0.02	7.20	0.1	1.4	A2 V		140 s		m
81809	136872		9 27 46.7	−6 04 16	−0.015	−0.08	5.38	0.64	4.7	G2 V	+54	15 mn		m
82458	256625		9 27 49.2	−71 46 52	−0.015	+0.04	6.6	1.3	3.0	F2 V		15 s		
81837	136875		9 27 50.4	−9 47 49	−0.002	−0.02	8.0	1.1	0.2	K0 III		360 s		
80930	6858		9 27 51.2	+75 05 55	−0.007	+0.03	6.3	0.1	1.4	A2 V	+1	95 s		
81903	177488		9 27 52.9	−22 31 43	−0.003	+0.02	7.3	0.7	0.2	K0 III		270 s		
81863	155213		9 27 53.8	−14 52 06	0.000	−0.07	7.5	0.5	4.0	F8 V		51 s		
82125	237009		9 27 55.2	−57 19 47	−0.002	−0.03	7.50	1.6	−0.5	M2 III		390 s		m
82111	237007		9 27 57.2	−55 20 05	−0.002	0.00	7.80	−0.10	−0.9	B6 V		550 s		
82101	237005		9 27 57.3	−53 31 02	−0.003	−0.02	7.2	0.2	2.6	F0 V		85 s		
81904	177490		9 27 58.1	−23 40 45	+0.001	−0.02	7.9	0.9	0.2	K0 III		340 s		
81874	136878		9 28 06.1	−10 01 28	−0.004	0.00	7.1	1.1	0.2	K0 III		240 s		
81884	136880		9 28 13.2	−7 43 13	0.000	+0.01	7.0	1.6	−0.5	M2 III		310 s		
81834	117713		9 28 14.5	+5 54 17	−0.001	−0.02	7.9	1.1	0.2	K0 III		340 s		
81985	200398		9 28 14.5	−34 54 15	−0.001	−0.02	7.1	0.4	0.4	B9.5 V		130 s		
82032	200399		9 28 19.0	−38 13 01	−0.003	−0.03	7.6	0.6	0.6	A0 V		100 s		
81902	136883		9 28 20.4	−9 59 14	0.000	−0.02	6.90	1.1	0.2	K0 III		220 s	7393	m
82124	221196		9 28 20.9	−49 39 18	0.000	−0.04	7.0	0.7	3.0	F2 V		38 s		
81487	14887		9 28 24.9	+61 43 53	−0.006	0.00	7.9	1.4	−0.3	K5 III		410 mx		
81915	136884		9 28 25.3	−9 11 48	0.000	−0.02	8.0	1.1	0.2	K0 III		360 s		
82048	200401		9 28 25.5	−38 05 35	−0.003	−0.02	8.0	1.5	−0.1	K2 III		270 s		
81856	98591		9 28 26.7	+12 40 28	−0.009	−0.15	7.6	0.7	4.4	G0 V		44 s		
81858	117717	2 ω Leo	9 28 27.4	+9 03 24	+0.004	−0.01	5.41	0.60	4.0	dF8	−6	17 mn	7390	m
82227	237018		9 28 28.4	−59 09 00	−0.005	0.00	7.9	1.1	0.2	K0 III		340 s		
82567	−−		9 28 28.6	−72 25 18			7.6	0.0	0.4	B9.5 V		280 s		

HD	SAO	Star Name	α 2000	δ 2000	μ(α)	μ(δ)	V	B−V	M_v	Spec	RV	d(pc)	ADS	Notes
			^{h} ^{m} ^{s}	$^\circ$ $'$ $''$	^{s}	$''$								
81873	117718	3 Leo	9 28 29.1	+8 11 18	−0.002	−0.04	5.71	1.04	0.2	gK0	+22	120 s	7391	m
82406	250591		9 28 30.6	−66 42 06	−0.005	+0.04	5.91	0.01		A0				
82244	237019		9 28 31.1	−59 29 14	−0.010	+0.04	8.0	1.5	3.4	F5 V		18 s		
81872	98594		9 28 33.5	+12 23 07	−0.005	−0.01	6.9	0.8	3.2	G5 IV		55 s		
82109	221199		9 28 33.9	−45 29 51	−0.003	0.00	7.14	−0.13		B5				m
82017	177507		9 28 38.5	−25 25 58	−0.003	0.00	7.2	0.4	1.7	A3 V		80 s		
82121	221200		9 28 39.7	−45 30 00	−0.002	+0.01	7.63	−0.14		B8				m
81688	42876		9 28 39.9	+45 36 05	−0.001	−0.13	5.41	0.98	3.2	G5 IV	+39	21 s		m
81956	155220		9 28 42.1	−10 16 15	−0.001	+0.01	7.8	0.1	1.4	A2 V		190 s		
82065	200407		9 28 42.4	−35 36 47	−0.005	+0.01	8.0	0.5	3.4	F5 V		81 s		
82347	250590		9 28 47.0	−62 16 22	−0.011	+0.03	5.92	1.10		K0				
81900	98595		9 28 52.4	+12 35 01	+0.002	−0.07	7.7	0.9	3.2	G5 IV		81 s		
81981	136892		9 28 55.4	−8 28 06	−0.003	−0.01	8.0	1.1	0.2	K0 III		360 s		
81703	27267		9 28 59.1	+50 01 52	−0.002	−0.01	7.7	1.6	−0.5	M2 III		440 s		
82045	177513		9 28 59.4	−20 26 06	−0.004	−0.01	8.00	0.4	2.6	F0 V		120 s		m
81980	136894		9 29 02.2	−1 15 25	−0.005	0.00	6.27	0.27		A5			7396	m
82405	250595		9 29 03.5	−63 26 22	0.000	+0.02	8.0	0.0	0.4	B9.5 V		330 s		
81997	136895	31 τ¹ Hya	9 29 08.8	−2 46 08	+0.008	−0.02	4.60	0.46	3.7	F6 V	+98	15 ts		m
82151	200415		9 29 08.8	−39 00 32	−0.003	0.00	7.4	1.6	0.2	K0 III		130 s		
82077	177521		9 29 12.6	−20 44 55	−0.002	+0.02	5.66	1.60	−0.5	gM1	−8	160 s		
85742	258541		9 29 13.9	−87 14 14	−0.009	0.00	8.0	0.1	0.6	A0 V		300 s		
82150	200416	ε Ant	9 29 14.7	−35 57 05	−0.002	−0.01	4.51	1.44	−0.4	M0 III	+22	96 s		
82278	237029		9 29 16.1	−52 25 29	−0.003	−0.01	7.25	−0.06	0.0	B8.5 V		280 s		
82165	200419		9 29 16.2	−38 24 14	−0.008	−0.02	6.19	0.22	1.4	A2 V n		90 s		
82058	155229		9 29 17.2	−13 44 12	+0.001	0.00	7.2	1.4	−0.3	K5 III		320 s		
82148	200417		9 29 17.9	−33 40 18	−0.004	+0.03	7.50	1.6	−0.5	M4 III		400 s		m
82030	136897		9 29 18.3	−3 09 54	−0.001	+0.01	6.9	0.1	0.4	A0 V		180 s		
82207	221213		9 29 18.5	−44 32 24	−0.011	−0.02	7.00	0.54	4.4	G0 V		33 s		m
81702	27268		9 29 21.3	+56 14 26	−0.016	−0.13	7.0	0.4	3.0	F2 V	−2	62 s		
82224	221214		9 29 22.5	−43 11 24	0.000	−0.01	6.60	0.48		F5				
81954	80855		9 29 22.7	+20 54 54	−0.003	+0.01	7.9	0.9	3.2	G5 IV		88 s		
82043	136899		9 29 24.4	−2 12 19	−0.003	−0.01	6.14	0.22		A3				
82093	155233		9 29 26.7	−17 30 18	+0.001	+0.01	7.1	0.1	1.4	A2 V		140 s		
82241	221216		9 29 28.3	−44 31 57	−0.011	0.00	6.98	0.51	0.6	F8 III		89 mx		
82164	200421		9 29 28.6	−34 01 45	0.000	−0.01	7.9	0.2	4.0	F8 V		60 s		
82388	237036		9 29 28.7	−56 32 14	−0.003	+0.02	8.0	−0.2	1.4	A2 V		210 s		
82074	136900		9 29 32.3	−4 14 50	−0.002	−0.08	6.26	0.84		G5				
81977	98600		9 29 32.9	+17 39 05	−0.001	−0.02	7.4	1.1	0.2	K0 III		280 s		
82468	250597		9 29 32.9	−62 21 40	−0.011	+0.01	6.56	0.09	0.6	A0 V		120 mx		
82297	221219		9 29 33.0	−48 42 52	−0.002	−0.01	7.49	0.89	3.2	G5 IV		59 s		
82073	136903		9 29 36.3	−3 07 21	−0.001	−0.01	7.9	0.1	1.7	A3 V		170 s	7400	m
82421	237039		9 29 36.8	−58 17 54	−0.004	+0.04	7.4	0.4	0.2	K0 III		270 s		
82346	237034		9 29 37.4	−50 36 13	−0.001	+0.01	7.4	0.5	2.6	F0 V		70 s		
83563	258536		9 29 40.3	−82 15 46	−0.016	+0.03	7.8	−0.2	0.6	A0 V		160 mx		
82161	177539		9 29 40.3	−26 35 39	−0.007	−0.03	6.8	1.0	3.2	G5 IV		38 s		
82239	200428		9 29 40.5	−39 58 22	−0.001	0.00	8.03	0.94		G5				
82181	177540		9 29 45.6	−25 31 41	−0.002	−0.03	7.5	1.1	0.2	K0 III		240 s		
81790	27271		9 29 47.5	+55 44 43	−0.016	−0.01	6.5	0.4	3.0	F2 V	+10	50 s		
82180	177541		9 29 49.8	−23 20 43	0.000	−0.01	6.24	1.57		K0				
82320	221223		9 29 50.3	−43 57 31	−0.006	+0.01	7.9	−0.4	1.4	A2 V		140 mx		
82089	117736		9 29 52.3	+2 11 55	−0.002	−0.02	8.0	0.5	3.4	F5 V		82 s		
82321	221225		9 29 53.3	−44 54 33	0.000	−0.04	8.0	0.1	3.0	F2 V		64 s		
82205	177546		9 29 54.4	−26 35 23	−0.002	0.00	5.48	1.36	−0.2	gK3	+12	120 s	7405	m
82106	117737		9 29 54.7	+5 39 17	−0.035	+0.09	7.22	1.00	6.9	K3 V	+27	12 ts		
81964	61518		9 29 55.5	+33 33 06	0.000	0.00	7.8	1.1	0.2	K0 III	−58	330 s		
82220	177552		9 29 58.3	−28 14 38	+0.001	−0.04	7.4	0.0	0.4	B9.5 V		250 s		
−−	237048		9 30 01.9	−57 30 32	+0.003	−0.03	8.00	0.1	1.4	A2 V		210 s		m
82204	177551		9 30 02.3	−22 08 31	+0.002	0.00	8.0	0.5	0.6	A0 V		140 s		
81996	61523		9 30 02.7	+32 02 28	−0.019	−0.03	7.8	0.7	4.4	G0 V		49 s		
82012	80859		9 30 03.2	+28 43 02	−0.001	−0.01	8.0	1.1	0.2	K0 III		360 s		
82236	177556		9 30 05.0	−28 35 39	+0.001	−0.03	7.5	1.2	0.2	K0 III		220 s		
82419	237042		9 30 05.2	−51 31 01	−0.001	+0.01	5.45	−0.10		B5	+10			
82684	250604		9 30 09.0	−67 12 03	−0.004	+0.07	8.0	1.1	0.2	K0 III		300 mx		
82275	177562		9 30 09.9	−30 08 39	+0.001	0.00	7.5	1.2	−0.3	K5 III		370 s		
82010	61525		9 30 11.7	+31 19 37	−0.001	−0.02	7.6	0.0	0.0	B8.5 V	+32	330 s		
81474	6877		9 30 12.0	+72 18 57	−0.004	−0.06	8.0	0.4	2.6	F0 V		120 s		
82234	177561		9 30 13.9	−24 16 26	+0.003	+0.02	6.8	1.3	−0.1	K2 III		210 s		
82201	155243		9 30 15.7	−15 37 23	+0.003	−0.01	8.0	1.1	0.2	K0 III		360 s		
82366	221226		9 30 17.4	−40 20 48	+0.002	−0.02	8.0	1.9	−0.3	K5 III		260 s		
82457	237051		9 30 18.8	−52 18 19	+0.001	+0.01	7.99	−0.08	0.0	B8.5 V		400 s		
82386	221228		9 30 19.1	−44 51 36	−0.002	−0.01	7.58	0.48		G0				
82315	200435		9 30 20.0	−33 24 14	+0.002	−0.01	6.9	0.7	0.2	K0 III		220 s		
82314	200436		9 30 21.9	−31 33 02	−0.005	0.00	7.2	0.5	3.0	F2 V		55 s		
82232	155246		9 30 22.6	−15 34 39	−0.005	−0.07	5.85	1.20	−0.2	gK3	+24	160 s		
82536	237056		9 30 23.4	−58 21 42	−0.004	+0.03	5.88	1.70		Ma				m

HD	SAO	Star Name	α 2000	δ 2000	μ(α)	μ(δ)	V	B–V	M_v	Spec	RV	d(pc)	ADS	Notes
			$9^h 30^m 23.5^s$	$+21°48'46''$	-0.004^s	$-0.02''$								
82105	80864		9 30 23.5	+21 48 46	-0.004	-0.02	6.8	0.1	1.7	A3 V		110 s		
82518	237054		9 30 27.6	-55 13 54	-0.001	0.00	7.31	1.64	-2.4	M3 II		880 s		
82436	221233		9 30 33.0	-45 34 13	-0.002	+0.03	6.63	1.19		K0				
82194	117739		9 30 35.0	+1 15 27	-0.001	+0.01	7.8	0.1	0.6	A0 V		270 s		
81995	42899		9 30 38.1	+44 46 09	-0.007	-0.03	7.1	0.2	2.1	A5 V	+16	100 s		
82363	177582		9 30 40.2	-28 46 00	-0.003	-0.05	6.7	1.1	0.2	K0 III		170 s		
82341	177577		9 30 40.7	-25 36 59	-0.001	-0.03	7.70	0.1	1.7	A3 V		160 s		m
82229	136910		9 30 41.2	-4 03 56	-0.001	-0.01	7.2	1.1	-0.1	K2 III	+28	290 s		
82176	98617		9 30 41.7	+13 29 45	-0.001	-0.05	7.6	1.1	0.2	K0 III		300 s		
82434	221234	ψ Vel	9 30 41.9	-40 28 00	-0.017	+0.07	3.60	0.36	1.9	F2 IV	+12	19 ts		m
82087	61529	7 LMi	9 30 43.1	+33 39 21	-0.001	-0.05	5.85	1.05	0.2	K0 III	+2	130 s		m
82384	200444	ζ¹ Ant	9 30 45.4	-31 53 29	-0.001	-0.03	5.77	0.05	0.6	A0 V		98 s		
82340	177584		9 30 48.3	-23 30 35	-0.003	+0.01	8.0	0.8	0.2	K0 III		360 s		
82104	61533		9 30 50.9	+34 12 40	-0.002	-0.02	7.6	1.4	-0.3	K5 III		380 s		
82517	221239		9 30 51.5	-49 25 00	-0.003	+0.02	7.9	0.3	1.4	A2 V		200 s		
82268	136913		9 30 55.1	-5 39 41	0.000	+0.04	6.8	1.1	0.2	K0 III		210 s		
82040	42901		9 30 56.4	+44 41 02	-0.002	0.00	7.90	1.6	-0.5	M4 III		480 s	7403	m
82683	250606		9 31 01.0	-60 27 11	0.000	0.00	7.9	1.1	0.2	K0 III		280 s		
82215	98621		9 31 01.2	+10 09 19	-0.003	-0.02	7.6	0.2	2.1	A5 V		130 s		
82267	117742		9 31 06.1	+3 04 47	-0.006	0.00	7.3	0.5	4.0	F8 V		45 s		
81787	14903		9 31 09.6	+67 32 28	+0.002	-0.02	7.50	0.1	1.4	A2 V		170 s	7399	m
82552	221242		9 31 12.8	-46 00 27	-0.001	-0.02	7.99	0.29		F0				
81953	27277		9 31 13.1	+58 45 21	-0.001	-0.04	7.6	1.1	-0.1	K2 III		350 s		
82668	237067		9 31 13.3	-57 02 04	-0.004	0.00	3.13	1.55	-0.3	K5 III	-14	45 s		N Vel, q
82191	80877		9 31 17.3	+27 23 14	-0.003	-0.03	6.6	0.1	0.6	A0 V	0	160 s		
82356	136922		9 31 18.8	-3 19 20	-0.002	-0.02	7.5	1.1	0.2	K0 III		290 s		
82578	221246		9 31 18.8	-47 57 09	-0.004	0.00	6.54	0.26		F0				
82432	177599		9 31 20.0	-24 20 37	0.000	-0.01	7.4	1.1	-0.1	K2 III		310 s		
82265	98624		9 31 20.1	+12 51 30	+0.001	-0.04	6.9	1.1	0.2	K0 III		220 s		
82644	237068		9 31 22.1	-53 47 47	-0.002	+0.02	7.7	0.0	0.4	B9.5 V		260 s		
82333	117745		9 31 23.6	+2 16 42	-0.001	-0.01	7.7	0.9	0.3	G5 III	+13	310 s		
83096	256633		9 31 25.6	-73 44 48	-0.002	+0.04	8.00	0.4	2.6	F0 V		120 s		m
82412	155253		9 31 28.6	-12 03 44	-0.002	-0.04	7.1	1.1	0.2	K0 III		240 s		
82450	177602		9 31 29.3	-23 30 22	-0.001	-0.03	7.7	0.0	1.4	A2 V		180 s		
82531	200461		9 31 30.7	-36 45 00	-0.001	+0.03	7.2	1.2	0.2	K0 III		180 s		
82355	117747		9 31 31.4	+1 27 54	+0.001	-0.04	7.40	0.5	4.0	F8 V		48 s	7412	m
81937	14908	23 UMa	9 31 31.7	+63 03 42	+0.016	+0.03	3.67	0.33	1.7	F0 IV	-10	25 ts	7402	m
82513	200459	ζ² Ant	9 31 32.2	-31 52 18	-0.003	0.00	5.93	0.26		A m				
82198	61540	8 LMi	9 31 32.3	+35 06 11	-0.005	-0.11	5.37	1.53	-0.3	K5 III	+38	96 mx		
82514	200462		9 31 32.8	-35 42 54	+0.011	-0.18	5.87	1.29	-0.3	K4 III		100 mx		m
82858	250611		9 31 32.9	-66 43 09	-0.004	-0.01	6.27	1.35		K0				
82264	80882		9 31 34.9	+20 16 15	-0.003	0.00	7.1	1.1	0.2	K0 III		240 s		
82664	237070		9 31 35.0	-51 15 09	-0.002	-0.02	8.03	0.25	2.1	A5 V		140 s		
83095	256634		9 31 36.5	-73 04 52	-0.004	-0.01	5.47	1.56		K2				
82428	155257		9 31 38.9	-10 33 08	+0.001	-0.02	6.14	0.24	0.5	gA8	-18	130 s		
82332	98625		9 31 38.9	+15 46 12	-0.002	-0.02	7.8	1.1	0.2	K0 III		320 s		
82212	61543		9 31 42.2	+36 25 44	-0.004	-0.02	7.0	1.1	-0.1	K2 III		270 s		
82835	250610		9 31 42.8	-64 14 15	-0.003	+0.03	7.7	0.4	2.6	F0 V		100 s		
82308	80885	4 λ Leo	9 31 43.1	+22 58 04	-0.002	-0.04	4.31	1.54	-0.3	K5 III	+27	79 s		Alterf
82309	80886		9 31 43.9	+20 00 18	-0.003	-0.13	7.46	1.28	-0.2	K3 III	-11	99 mx		
83019	256631		9 31 45.9	-70 28 10	+0.001	+0.01	7.03	-0.04	0.0	B8.5 V		260 s		
82425	136929		9 31 46.6	-3 23 04	-0.005	0.00	7.5	0.1	0.6	A0 V		110 mx		
82426	136930		9 31 47.5	-7 29 46	-0.003	0.00	7.8	0.2	2.1	A5 V		140 s		
83093	—		9 31 50.5	-72 28 36			7.74	-0.02	-2.5	B2 V		940 s		m
81759	6884		9 31 53.0	+72 06 40	-0.010	-0.04	7.6	0.9	3.2	G5 IV		74 s		
82764	237074		9 31 53.3	-56 59 20	-0.005	0.00	7.12	0.00		B8.5 V		230 mx		
82511	177607		9 31 55.3	-23 19 53	+0.001	+0.02	8.0	1.2	-0.1	K2 III		410 s		
82477	155262		9 31 55.7	-10 22 14	-0.006	+0.01	6.13	1.19		K0				
82395	98627	5 ξ Leo	9 31 56.7	+11 17 59	-0.006	-0.09	4.97	1.05	0.2	K0 III	+29	84 s		
82381	117751	6 Leo	9 31 57.5	+9 42 57	0.000	-0.02	5.07	1.37	-0.2	K3 III	+19	99 s	7416	m
82446	136932	32 τ² Hya	9 31 58.9	-1 11 06	-0.001	-0.02	4.57	0.10	0.0	A3 III	+6	82 s		
82593	200469		9 32 00.0	-32 14 43	-0.005	0.00	8.0	0.4	1.4	A2 V		120 s		
82627	200474		9 32 00.5	-38 56 14	-0.003	+0.02	7.5	0.1	0.4	B9.5 V		220 s		
82695	221260		9 32 01.3	-48 11 40	-0.005	+0.01	7.64	1.12	0.2	K0 III		250 s		
82834	250612		9 32 03.0	-61 14 05	-0.002	0.00	7.14	0.06	0.6	A0 V		190 s		m
82331	80889		9 32 04.0	+26 49 41	+0.001	-0.02	7.91	1.03	0.0	K1 III		380 s		
82737	237075		9 32 04.2	-53 32 31	-0.002	0.00	7.68	-0.08	0.0	B8.5 V		340 s		
82410	98628		9 32 04.8	+11 01 27	-0.008	-0.05	7.9	0.7	4.4	G0 V		51 s		
82712	221261		9 32 05.2	-49 24 08	0.000	+0.01	7.72	1.01	3.2	G5 IV		80 s		
82490	136935		9 32 11.7	-3 30 55	0.000	-0.03	7.8	1.1	0.2	K0 III		330 s		
82508	155267		9 32 11.7	-10 50 37	-0.001	0.00	7.90	0.7	4.4	G0 V		50 s	7423	m
82394	80892		9 32 13.2	+21 51 24	0.000	0.00	7.45	1.00	-2.1	G7 II	-4	810 s	7419	m
82246	42907		9 32 14.4	+43 44 19	-0.005	-0.13	7.7	0.6	4.4	G0 V		45 s		
82901	250614		9 32 14.7	-62 47 19	-0.005	+0.02	3.8	1.6		M5 III	+28	520 mx		R Car, m,v
82641	200475		9 32 16.5	-36 01 08	-0.001	-0.02	8.0	1.2	0.2	K0 III		290 s		

HD	SAO	Star Name	α 2000	δ 2000	μ(α)	μ(δ)	V	B−V	M_v	Spec	RV	d(pc)	ADS	Notes
			h m s	° ′ ″	s	″								
82729	221265		9 32 16.8	−47 08 46	−0.006	+0.01	7.9	−0.1	2.6	F0 V		110 s		
82610	177619		9 32 18.4	−28 37 40	−0.006	+0.04	6.4	0.4	2.6	F0 V	−5	58 s		S Ant, m,v
82694	221262		9 32 19.2	−40 38 58	+0.001	−0.01	5.35	0.90		G5	−1	16 mn		m
82444	98629		9 32 20.2	+10 46 35	−0.001	+0.01	7.8	0.2	2.1	A5 V		140 s		
82573	155273		9 32 20.5	−19 24 02	−0.001	0.00	5.74	0.14		A2				
82558	155272		9 32 25.4	−11 11 05	−0.018	+0.03	7.80	0.92		K0				
82711	221266		9 32 26.8	−41 10 48	−0.003	0.00	7.9	0.3	3.2	G5 IV		88 s		
82800	237080		9 32 27.8	−54 15 21	−0.002	−0.02	8.0	1.5	−0.3	K5 III		450 s		
82920	250615		9 32 31.9	−60 31 22	−0.003	+0.01	7.4	0.6	4.4	G0 V		38 s		
82504	117755		9 32 32.3	+7 03 49	−0.001	0.00	7.5	1.1	−0.1	K2 III		330 s		
83017	250617		9 32 32.5	−65 22 42	+0.001	−0.03	7.9	1.5	−0.3	K5 III		430 s		
82876	237085		9 32 33.1	−57 31 17	−0.003	+0.02	7.2	0.1	0.6	A0 V		170 s		
82261	42909		9 32 33.6	+48 19 52	−0.003	−0.05	7.7	1.1	0.2	K0 III		310 s		
82287	42911		9 32 38.6	+48 10 00	+0.002	−0.01	7.9	0.4	2.6	F0 V	−14	120 s		
82678	177626		9 32 41.3	−27 24 53	−0.001	−0.02	7.5	1.8	−0.5	M2 III		320 s		
82543	117757		9 32 41.4	+1 51 51	−0.001	−0.04	6.11	0.62		F5	+28			m
82856	237086		9 32 41.4	−54 49 52	+0.001	0.00	7.2	0.2	1.4	A2 V		110 s		
82443	80897		9 32 43.7	+26 59 19	−0.011	−0.24	7.01	0.77	5.7	dG9	+14	20 ts		
82919	237090		9 32 46.7	−57 05 39	−0.004	+0.01	7.11	−0.04	−1.1	B5 V		160 mx		m
83254	−−		9 32 48.9	−72 40 41			8.0	0.1	0.6	A0 V		300 s		
82328	27289	25 θ UMa	9 32 51.3	+51 40 38	−0.103	−0.54	3.17	0.46	2.2	F6 IV	+15	14 mx	7420	m
82661	155280		9 32 54.6	−14 00 01	−0.002	−0.01	7.20	0.9	3.2	G5 IV		63 s	7427	m
82709	177634		9 32 55.3	−27 40 37	0.000	−0.07	7.7	0.4	2.1	A5 V		71 mx		
82660	155281		9 32 55.8	−13 31 02	0.000	−0.02	5.94	1.50		K5				
82392	42915		9 33 01.8	+43 18 38	−0.002	−0.02	7.9	1.1	0.2	K0 III		340 s		
82474	61555		9 33 02.0	+30 34 09	+0.001	−0.05	7.9	1.1	0.2	K0 III		340 s		
82638	136945		9 33 02.1	−8 30 19	−0.002	−0.04	6.12	0.98		K0				
85300	258540		9 33 04.1	−86 00 34	−0.038	−0.02	7.40	0.4	3.0	F2 V		76 s		m
82758	200486		9 33 05.6	−33 04 03	−0.004	+0.04	8.0	1.4	−0.3	K5 III		250 mx		
82380	42914		9 33 07.1	+49 26 18	−0.003	+0.01	6.5	0.1	1.7	A3 V	−10	92 s		
82785	200492		9 33 07.7	−39 07 43	−0.003	+0.05	6.43	0.33		F2				m
82774	200490		9 33 10.0	−36 24 15	+0.002	+0.02	7.60	0.1	0.6	A0 V		250 s		m
82408	42916		9 33 11.3	+45 30 51	0.000	−0.02	6.5	1.1	0.2	K0 III		190 s		
82734	177642		9 33 12.4	−21 06 57	−0.002	+0.01	5.01	1.02	3.2	K0 IV	+13	21 s		
82850	221280		9 33 13.1	−45 30 55	−0.002	−0.01	8.0	1.6	−0.5	M4 III		490 s		U Vel, q
82829	221279		9 33 13.2	−45 12 31	−0.004	+0.02	8.00	0.2	2.1	A5 V e		150 s		S Vel, v
82987	237096		9 33 14.7	−57 52 57	0.000	0.00	7.2	1.5	−0.5	M2 III		350 s		
82932	237093		9 33 16.6	−52 49 45	−0.001	+0.06	8.00	0.01	0.6	A0 V		130 mx		
82798	200493		9 33 16.9	−38 38 16	−0.015	+0.01	7.1	0.9	4.4	G0 V		22 s		
82988	237097		9 33 17.5	−57 57 46	0.000	+0.01	7.80	0.00	0.4	B9.5 V		290 s		m
82724	177647		9 33 17.6	−21 23 30	−0.001	−0.01	6.7	0.0	0.6	A0 V		160 s		
82523	80900		9 33 18.2	+28 22 04	−0.004	−0.04	6.53	0.12	0.0	A3 III	+26	120 mx	7426	m
82674	136951		9 33 20.0	−7 11 24	−0.003	−0.02	6.24	1.17		K0				
82747	177649		9 33 26.1	−22 51 49	−0.004	+0.06	5.91	0.02	0.6	A0 V n		89 mx		
82522	61561	9 LMi	9 33 30.2	+36 29 12	+0.003	−0.04	6.18	1.27	−0.1	K2 III	−17	160 s		
82965	221287		9 33 38.4	−49 45 29	−0.002	−0.04	7.37	1.03	3.2	G5 IV		44 s		m
82624	98640		9 33 38.5	+18 44 11	−0.002	0.00	7.9	1.1	0.2	K0 III		340 s		
83111	250621		9 33 44.0	−60 37 38	−0.003	+0.03	7.0	1.4	0.2	K0 III		130 s		
82984	221288		9 33 44.4	−49 00 18	+0.002	+0.01	5.12	−0.12	−1.4	B4 V n	+27	200 s		m
83138	250623		9 33 44.5	−63 13 53	−0.003	0.00	8.04	1.60	−0.3	K5 III		410 s		
82986	221289		9 33 46.4	−49 43 31	+0.004	−0.01	7.56	0.95	3.2	G5 IV		56 s		
83030	237102		9 33 46.7	−54 02 23	−0.001	+0.01	7.8	1.6	−0.5	M2 III		470 s		
82442	27291		9 33 48.9	+52 26 38	−0.004	−0.02	7.6	0.1	1.7	A3 V		150 s		
83979	258538	ζ Cha	9 33 53.4	−80 56 29	−0.017	+0.01	5.11	−0.14	−1.6	B5 IV	−52	220 s		
82555	42921		9 33 57.8	+45 17 11	−0.005	−0.03	7.9	0.1	1.7	A3 V		110 mx		
82670	80909		9 33 59.0	+23 27 13	−0.004	−0.09	6.25	1.46	−0.3	K5 III	−6	130 mx		
83058	237107		9 34 08.7	−51 15 21	−0.002	−0.02	5.01	−0.18		B3 n	+35			
82635	61570	10 LMi	9 34 13.3	+36 23 51	+0.001	−0.03	4.55	0.92	0.3	G8 III	−12	71 s		
82636	61571		9 34 14.7	+34 48 41	−0.001	0.00	7.9	0.9	3.2	G5 IV		87 s		
83000	221293		9 34 17.6	−41 03 24	−0.002	−0.02	7.9	1.9	−0.5	M2 III		490 s		
82502	27296		9 34 18.7	+54 02 48	−0.003	−0.01	7.1	1.1	0.2	K0 III		240 s		
82285	14923		9 34 19.3	+66 47 42	−0.005	−0.04	7.4	0.5	3.4	F5 V		63 s	7425	m
82582	42924		9 34 19.5	+46 54 08	−0.006	−0.01	6.52	0.22		A5	+11			
82982	200514		9 34 21.1	−36 49 44	−0.002	−0.01	7.6	1.1	−0.1	K2 III		350 s		
82171	6895		9 34 23.8	+71 05 15	−0.004	−0.04	7.2	1.1	0.2	K0 III		260 s		
83183	237117		9 34 26.6	−59 13 46	−0.001	+0.01	4.08	0.01	−3.7	B5 II	+22	330 s		
82928	177680		9 34 27.6	−24 32 12	−0.004	0.00	7.6	−0.2	1.7	A3 V		160 s		
83155	237113		9 34 28.0	−55 22 09	−0.003	+0.03	7.6	1.7	0.2	K0 III		120 s		
82210	6897	24 UMa	9 34 28.8	+69 49 49	−0.012	+0.08	4.56	0.77	3.0	G2 IV	−27	21 ts		
83029	221297		9 34 29.8	−40 27 58	−0.006	0.00	7.9	0.4	4.0	F8 V		61 s		
83312	250628		9 34 30.0	−66 07 19	+0.001	+0.01	7.2	0.0	0.0	B8.5 V		270 s		
82581	42926		9 34 31.8	+48 45 15	+0.003	−0.02	7.6	1.1	0.2	K0 III		310 s		
82870	136964	33 Hya	9 34 32.6	−5 54 54	0.000	−0.06	5.56	1.16	0.0	gK1	+13	120 s		
82842	136962		9 34 32.6	−3 29 17	−0.002	+0.01	7.2	0.8	3.2	G5 IV		63 s		
83297	250627		9 34 35.3	−64 01 49	+0.001	+0.02	6.9	0.0	0.0	B8.5 V		240 s		

HD	SAO	Star Name	α 2000	δ 2000	μ(α)	μ(δ)	V	B−V	M$_v$	Spec	RV	d(pc)	ADS	Notes
			h m s	° ′ ″	s	″								
82927	155310		9 34 37.5	−16 22 32	−0.001	+0.02	7.7	1.6	−0.5	M2 III		440 s		
82819	117777		9 34 41.2	+8 11 15	−0.001	0.00	8.0	1.6	−0.5	M2 III	+46	510 s		
83090	221300		9 34 42.4	−44 48 59	−0.002	−0.01	7.1	0.4	0.4	B9.5 V		170 s		
83170	237118		9 34 45.2	−52 01 41	−0.005	+0.04	7.33	0.41	2.6	F0 V		80 s		
83153	237119		9 34 49.1	−50 33 45	−0.002	0.00	8.00	0.04	−1.0	B5.5 V		520 s		
82621	27298	26 UMa	9 34 49.4	+52 03 05	−0.007	−0.04	4.50	0.01	1.4	A2 V	+23	42 s		
82943	155312		9 34 50.5	−12 07 48	−0.001	−0.19	6.7	0.7	4.4	G0 V		29 s		
82189	6898	22 UMa	9 34 53.5	+72 12 20	+0.017	−0.08	5.72	0.48	3.4	F5 V	−38	28 s		
83359	250629		9 34 56.3	−64 59 58	−0.006	+0.04	7.6	0.5	3.4	F5 V		68 s		
83078	200525		9 34 57.1	−34 22 03	0.000	−0.01	7.2	0.0	0.4	B9.5 V		220 s		
82865	98652		9 34 59.7	+12 39 16	0.000	−0.03	6.82	−0.11	−1.6	B5 IV		480 s		
83169	221304		9 35 00.3	−48 53 23	−0.004	−0.02	7.8	0.6	2.6	F0 V		69 s		
82957	136974		9 35 02.4	−4 53 21	−0.001	−0.01	7.4	0.9	0.3	G8 III	0	270 s		
82741	61578		9 35 03.8	+39 37 17	−0.002	+0.01	4.81	0.99	0.2	K0 III	−12	84 s		
82958	136975		9 35 07.0	−5 22 09	−0.001	−0.05	7.1	1.1	0.2	K0 III		240 s		
82817	80912		9 35 08.8	+26 11 33	−0.001	−0.03	7.66	0.05		B9				
82472	14930		9 35 11.2	+65 24 33	−0.001	−0.02	7.9	1.1	0.2	K0 III		350 s		
83108	200531		9 35 11.8	−35 49 25	−0.006	−0.01	6.49	0.42		F5				
82569	14931		9 35 14.2	+60 53 30	−0.004	−0.07	7.20	0.4	2.6	F0 V		83 s	7432	m
82780	42932		9 35 22.9	+39 57 26	−0.007	0.00	6.48	0.37	3.1	F2.5 V	−42	48 s	7438	m
83151	200534		9 35 24.0	−38 54 06	−0.005	+0.02	7.90	1.1	−0.1	K2 III		290 mx		m
83087	177707		9 35 24.6	−22 41 10	−0.006	−0.04	7.0	1.2	0.2	K0 III		160 mx		
82994	117792		9 35 25.2	+3 54 18	−0.003	−0.09	8.00	0.6	4.4	G0 V		53 s	7444	m
82864	61585		9 35 28.7	+31 04 54	−0.006	−0.03	8.0	0.5	3.4	F5 V		84 s		
83048	155319		9 35 29.9	−14 41 33	−0.007	−0.06	7.9	1.6		M5 III	+42	140 mx		X Hya, v
83104	155323		9 35 33.7	−19 35 01	−0.003	−0.05	6.26	0.08		A1				m
82885	61586	11 LMi	9 35 39.5	+35 48 36	−0.058	−0.25	5.41	0.77	5.6	G8 IV−V	+13	9.2 t	7441	m
83247	221318		9 35 41.2	−44 11 18	−0.005	0.00	7.6	0.2	0.6	A0 V		180 s		
83531	250634		9 35 41.5	−68 12 38	−0.004	+0.01	7.40	0.1	0.6	A0 V		230 s		m
83086	155324		9 35 44.9	−11 56 46	−0.001	−0.01	7.8	1.1	0.2	K0 III		330 s		
82886	61587		9 35 45.2	+34 46 51	+0.002	−0.03	7.7	0.6	4.4	G0 V		45 s		
83335	237130		9 35 47.3	−54 10 06	−0.002	+0.02	7.96	−0.09	0.0	B8.5 V		390 s		
83046	117796		9 35 48.9	+0 12 00	0.000	−0.03	7.7	1.1	0.2	K0 III		310 s		
82792	42933		9 35 50.7	+48 29 29	−0.005	+0.02	6.9	0.5	3.4	F5 V		51 s		
83023	98662	7 Leo	9 35 52.8	+14 22 46	−0.003	−0.01	6.30	0.04	1.2	A1 V	+24	110 s	7448	m
83477	250633		9 35 53.8	−64 33 10	−0.002	+0.03	6.7	0.1	0.6	A0 V		170 s		
83005	80918		9 36 02.0	+20 02 44	0.000	0.00	7.1	1.1	0.2	K0 III		250 s		
83196	177722		9 36 02.8	−27 31 17	−0.004	+0.02	7.60	0.1	0.6	A0 V		130 mx	7452	m
83085	117798		9 36 04.4	+4 30 53	−0.001	−0.02	7.7	1.1	−0.1	K2 III		360 s		
83004	80917		9 36 04.5	+23 11 29	+0.004	−0.09	7.5	1.1	0.2	K0 III		140 mx		
83523	250636		9 36 05.1	−64 57 01	−0.006	+0.03	6.56	0.10		A2				
82327	6900		9 36 06.8	+74 19 03	−0.004	−0.07	6.4	−0.1		B9				
83432	237139		9 36 11.4	−58 01 30	−0.009	+0.05	7.6	1.1	0.2	K0 III		200 mx		
83388	237135		9 36 14.7	−52 32 39	+0.001	0.00	7.8	1.2	−0.5	M2 III		470 s		
83213	177725		9 36 16.2	−21 59 51	−0.001	−0.03	7.4	0.4	1.7	A3 V		88 s		
83387	237136		9 36 18.4	−51 30 49	−0.001	0.00	7.4	0.6	2.6	F0 V		59 s		
83488	250637		9 36 19.7	−61 07 57	+0.001	0.00	8.0	0.1	0.4	B9.5 V		260 s		
83212	177727		9 36 19.8	−20 53 16	−0.003	−0.05	8.0	1.1	0.2	K0 III		330 s		
82719	14936		9 36 21.2	+62 47 25	+0.002	0.00	7.3	0.4	2.6	F0 V		88 s		
83084	98669		9 36 21.8	+11 25 57	0.000	−0.02	7.9	0.9	3.2	G5 IV		88 s		
83368	221339		9 36 25.2	−48 45 05	−0.001	−0.03	6.17	0.27		A5 p				m
83293	200543		9 36 25.8	−36 18 08	+0.016	−0.09	7.5	0.7	3.2	G5 IV		72 s		
83161	136993		9 36 27.6	−2 46 41	0.000	−0.02	6.96	1.02		K0				
83118	98671		9 36 29.1	+10 50 47	−0.001	−0.02	7.6	0.1	0.6	A0 V		260 s		
83209	155335		9 36 31.0	−11 16 27	−0.001	0.00	7.50	0.1	0.6	A0 V		240 s		m
83277	177740		9 36 31.7	−27 57 49	−0.004	−0.03	7.8	0.3	1.4	A2 V		130 s		
83261	177738		9 36 33.6	−24 42 10	−0.009	+0.06	6.53	0.38		F2				m
82862	27310		9 36 33.7	+54 07 20	−0.002	+0.06	7.9	0.5	4.0	F8 V		60 s		
82620	14934		9 36 41.1	+69 38 12	−0.009	−0.04	7.2	0.4	2.6	F0 V		81 s		
82861	27311		9 36 42.8	+56 58 12	+0.001	+0.01	6.9	0.1	1.4	A2 V		130 s		
83069	61594		9 36 42.8	+31 09 42	0.000	−0.04	5.56	1.59	−0.5	M2 III	−20	160 s		
83465	237144		9 36 46.3	−52 56 38	−0.014	+0.08	6.19	1.05	0.3	G5 III		100 mx		
83098	80925		9 36 48.9	+27 45 29	−0.001	−0.03	7.0	1.1	−0.1	K2 III		270 s		
83446	221344		9 36 49.7	−49 21 18	−0.011	+0.03	4.35	0.17	2.1	dA5	+21	28 s		m
83146	80929		9 36 52.4	+22 14 39	+0.001	−0.02	7.9	1.1	0.2	K0 III		340 s		
83317	177746		9 36 52.8	−26 51 29	+0.001	−0.01	7.8	1.3	−0.1	K2 III		320 s		
83067	61596		9 36 53.6	+35 35 57	−0.004	−0.06	7.2	0.5	3.4	F5 V		59 s		
83145	80931		9 36 57.0	+23 01 52	0.000	0.00	7.1	0.1	0.6	A0 V		200 s		
81437	1547		9 36 57.2	+82 55 31	−0.010	−0.01	7.2	0.5	3.4	F5 V		56 s		
83643	—		9 36 57.6	−66 24 58			7.6	0.0	0.4	B9.5 V		270 s		
83332	177748		9 37 00.1	−25 17 48	−0.005	+0.04	5.70	1.12	0.0	gK1	+30	130 s		
83189	98673	8 Leo	9 37 02.5	+16 26 16	−0.001	−0.01	5.69	1.25	0.2	K0 III	+6	89 s		
81817	1551		9 37 05.2	+81 19 35	−0.008	−0.02	4.29	1.48	−0.2	K3 III	−5	57 s		
82602	6908		9 37 09.1	+72 04 49	−0.029	−0.10	7.3	0.5	3.4	F5 V		60 s		
83380	200561		9 37 09.9	−32 10 43	+0.003	−0.03	5.63	1.02	0.2	gK0		120 s		

HD	SAO	Star Name	α 2000	δ 2000	μ(α)	μ(δ)	V	B-V	M_v	Spec	RV	d(pc)	ADS	Notes
83520	237149		9^{h}37^{m}12^s.3	-53°40'07"	-0^s.006	-0".02	5.45	0.15	-6.7	A2 I	-13	140 mx		m
83240	117807	10 Leo	9 37 12.6	+6 50 09	-0.004	0.00	5.00	1.05	0.0	K1 III	+20	100 s		
83379	177754		9 37 14.1	-30 12 40	0.000	-0.02	7.5	0.5	1.4	A2 V		95 s		
83399	200565		9 37 16.9	-30 58 12	0.000	-0.01	7.0	1.4	0.2	K0 III		140 s		
83225	98676		9 37 17.2	+15 15 08	-0.008	-0.01	8.0	0.4	3.0	F2 V		100 s		
83290	117812		9 37 25.1	+1 52 52	-0.007	+0.06	7.9	0.7	4.4	G0 V		51 s		
83441	200573		9 37 28.2	-36 05 45	-0.003	0.00	5.98	1.12		K0				
83529	221353		9 37 28.7	-49 59 27	-0.002	+0.22	6.99	0.59		G0		16 mn		m
83414	177764		9 37 28.9	-29 48 09	-0.001	-0.03	7.1	1.2	0.2	K0 III		170 s		
83614	237160		9 37 31.3	-59 53 22	+0.001	+0.03	7.2	0.2	1.7	A3 V		110 s		
82969	14941		9 37 37.3	+60 12 48	-0.002	-0.05	6.5	0.9	3.2	G5 IV		47 s		
83329	117816		9 37 40.5	+1 41 45	-0.008	+0.02	7.00	0.9	3.2	G5 IV		58 s	7462	m
83375	155350		9 37 42.0	-15 41 59	-0.001	-0.01	7.6	1.1	0.2	K0 III		300 s		
83352	137007		9 37 45.8	-3 10 15	-0.002	-0.03	6.5	1.1	0.2	K0 III		180 s		
83459	200576		9 37 46.3	-30 40 42	-0.003	+0.02	7.8	1.1	0.2	K0 III		270 s		
83273	80935		9 37 49.8	+24 40 13	-0.007	-0.02	6.6	0.5	4.0	F8 V	+31	34 s		
83373	137011	34 Hya	9 37 51.4	-9 25 29	-0.004	0.00	6.40	-0.04		A0				
83363	137009		9 37 54.5	-0 19 12	-0.001	-0.01	8.0	0.1	1.4	A2 V		210 s	7464	m
83517	200581		9 37 55.6	-36 55 33	-0.009	+0.07	7.8	0.7	3.2	G5 IV		81 s		
82685	6915		9 37 56.2	+73 04 49	-0.010	-0.03	7.20	0.4	2.6	F0 V	0	83 s	7446	m
82619	6913		9 37 56.4	+74 20 32	-0.004	-0.05	7.8	0.2	2.1	A5 V		75 mx		
83327	98681		9 37 58.6	+16 50 02	-0.001	-0.03	7.9	0.9	3.2	G5 IV		88 s		
83343	98683		9 38 01.2	+14 20 50	-0.005	-0.09	6.64	0.44		F2	+23			
83548	221355		9 38 01.4	-43 11 27	+0.002	-0.04	5.50	1.00	-0.9	K0 II	+3	200 s		
83173	42954		9 38 03.0	+47 33 40	-0.002	-0.02	7.9	0.5	4.0	F8 V		60 s		
84451	—		9 38 03.2	-80 21 20			7.8	0.0	0.4	B9.5 V		310 s		
82625	237163		9 38 03.2	-54 13 08	-0.001	+0.01	7.1	-0.4		A0 p				
83326	98684		9 38 04.6	+17 21 19	0.000	-0.06	7.8	1.1	0.2	K0 III		200 mx		
83657	237167		9 38 06.1	-56 19 17	-0.001	+0.01	7.7	1.0	0.2	K0 III		310 s		
83596	221356		9 38 06.3	-48 46 23	-0.005	-0.02	7.30	0.4	2.6	F0 V		87 s		m
83426	137015		9 38 12.0	-5 10 00	-0.001	-0.02	8.0	1.1	-0.1	K2 III		410 s		
83342	80939		9 38 13.3	+23 32 48	-0.005	-0.04	8.0	0.5	3.4	F5 V		82 s		
83655	237166		9 38 13.4	-52 41 35	0.000	0.00	7.9	0.2	0.0	B8.5 V		250 s		
83802	250645		9 38 15.0	-63 47 15	-0.001	-0.01	8.0	0.1	0.6	A0 V		290 s		
83362	98687		9 38 16.1	+12 44 10	-0.001	-0.01	6.8	0.9	3.2	G5 IV		52 s		
83408	117818		9 38 16.8	+1 50 07	-0.013	-0.05	7.75	0.53		F8				
82839	6918		9 38 17.0	+70 16 00	+0.004	+0.02	6.9	0.4	3.0	F2 V		59 s		
84698	258543		9 38 20.8	-81 42 32	-0.030	+0.02	6.7	1.2	0.2	K0 III		150 s		
83287	42958	42 Lyn	9 38 21.7	+40 14 22	-0.002	0.00	5.25	0.22		A5 n	-3			
83834	250647		9 38 21.9	-64 24 09	-0.001	0.00	6.69	-0.01	-0.2	B8 V		220 s		
83453	137019		9 38 22.1	-3 51 17	-0.002	+0.01	7.1	1.1	0.2	K0 III		240 s		
83406	117819		9 38 22.4	+7 42 31	-0.004	-0.04	7.60	0.1	1.4	A2 V		78 mx	7467	m
83742	237177		9 38 26.4	-57 31 50	0.000	+0.02	7.17	0.11	0.6	A0 V		170 s		m
83425	117821		9 38 27.2	+4 38 57	-0.011	-0.06	4.68	1.32	-0.2	K3 III	+45	90 s		
83833	250646		9 38 27.3	-62 33 21	-0.002	-0.01	7.59	-0.04	0.4	B9.5 V		270 s		
83340	80941		9 38 27.6	+28 00 22	0.000	+0.01	7.78	0.64	2.8	G0 IV		93 s		
83763	237181		9 38 31.6	-59 10 31	-0.003	0.00	7.7	0.0	0.4	B9.5 V		260 s		
83396	80943		9 38 35.3	+23 02 54	-0.006	-0.12	8.0	0.5	3.4	F5 V		73 mx		
83514	155367		9 38 40.2	-11 47 50	0.000	0.00	7.8	1.1	0.2	K0 III		330 s		
83610	200590		9 38 40.6	-39 36 50	-0.004	-0.07	6.70	0.49		F5				m
83405	80946		9 38 41.2	+21 32 22	-0.002	+0.01	7.7	1.1	0.2	K0 III		310 s		
82701	6917		9 38 41.7	+75 12 24	-0.002	+0.04	6.6	1.1	0.2	K0 III		190 s		
83948	250651		9 38 45.0	-66 51 33	-0.008	+0.03	7.7	0.0	2.6	F0 V		100 s		
83452	98690		9 38 45.7	+10 46 39	-0.004	-0.07	7.60	0.1	1.7	A3 V		48 mx	7471	m
83720	237180		9 38 48.0	-50 30 27	-0.004	-0.01	8.00	0.1	1.4	A2 V		200 mx		m
83395	61614		9 38 48.7	+30 09 11	-0.002	0.00	8.00	0.9	0.3	G6 III	+13	350 s		
83919	250649		9 38 49.8	-64 29 22	+0.003	+0.02	7.2	1.1	0.2	K0 III		240 s		
83434	80950		9 38 54.6	+20 17 52	-0.002	-0.01	6.8	0.0	0.4	B9.5 V	+31	190 s		
82992	14947		9 38 56.4	+69 17 19	-0.004	-0.01	8.00	0.4	2.6	F0 V		120 s	7459	m
83536	137025		9 38 58.0	-4 00 56	+0.003	-0.09	7.8	1.1	-0.1	K2 III		170 mx		
83719	221370		9 38 58.1	-49 03 40	-0.004	+0.01	7.65	0.06	0.4	B9.5 V		220 mx		
83572	155370		9 39 00.0	-17 37 27	-0.005	+0.01	6.9	0.4	2.6	F0 V		72 s		
83205	27325		9 39 00.5	+58 32 52	-0.003	-0.01	7.4	1.6	-0.5	M2 III	+20	390 s		
83706	221367		9 39 03.2	-42 51 56	0.000	-0.02	7.7	0.1	0.6	A0 V		250 s		
83535	137026		9 39 04.0	-2 50 21	0.000	0.00	7.2	0.1	0.6	A0 V		210 s		
83866	237190		9 39 05.4	-58 16 51	0.000	0.00	7.66	-0.09	0.0	B8.5 V		340 s		
83608	177795		9 39 06.0	-23 05 42	0.000	0.00	7.2	0.4	0.4	B9.5 V		130 s		
83674	200597		9 39 07.0	-39 31 45	+0.005	-0.02	7.5	1.0	0.2	K0 III		280 s		
84033	250655		9 39 09.0	-67 34 20	-0.008	0.00	7.7	1.1	0.2	K0 III		310 s		
83945	250652		9 39 10.2	-62 33 16	-0.002	+0.01	7.63	-0.01	0.0	B8.5 V		310 s		
83371	42962		9 39 10.7	+43 08 39	-0.004	-0.07	6.6	1.1	0.2	K0 III		160 mx		
83469	80954		9 39 12.7	+25 22 03	0.000	+0.02	7.21	0.32	2.6	F0 V		81 s		
83470	80956		9 39 12.9	+21 11 50	-0.003	-0.03	7.9	0.9	3.2	G5 IV		86 s		
83509	98696		9 39 17.3	+13 18 44	-0.004	0.00	7.04	0.47	3.8	F7 V		45 s		
83944	250653		9 39 20.9	-61 19 40	-0.006	+0.02	4.52	-0.07	0.2	B9 V	+24	73 s		

HD	SAO	Star Name	α 2000	δ 2000	μ(α)	μ(δ)	V	B-V	M_v	Spec	RV	d(pc)	ADS	Notes
			$9^h39^m23.3^s$	$-52°56'36''$	-0.001	0.00								
83853	237191		9 39 23.3	-52 56 36	-0.001	0.00	7.44	1.62	-0.1	K2 III		160 s		
83865	237193		9 39 26.7	-54 45 12	0.000	0.00	6.84	-0.12	-1.1	B5 V		390 s		
83864	237192		9 39 27.0	-54 15 54	-0.001	-0.02	8.0	1.0	-2.1	K0 II		1000 s		
84001	250656		9 39 27.4	-63 15 17	-0.005	+0.02	7.3	0.0	0.4	B9.5 V		200 mx		
83451	61618		9 39 27.9	+35 14 34	-0.006	-0.01	7.2	0.5	3.4	F5 V		59 s		
83126	14949		9 39 28.1	+67 16 19	-0.001	-0.04	5.94	1.52	-0.3	K5 III	+19	170 s		
82951	6924		9 39 28.6	+72 44 51	-0.011	+0.06	7.0	0.1	1.7	A3 V		110 mx		
83705	177808		9 39 36.7	-29 52 16	0.000	+0.02	7.8	1.2	0.2	K0 III		250 s		
83581	117831		9 39 39.5	+3 11 43	-0.001	-0.01	7.6	0.9	3.2	G5 IV		77 s		
83852	221385		9 39 40.5	-48 14 15	-0.007	+0.01	7.37	0.40	3.4	F5 V		62 s		
83881	237199		9 39 40.8	-53 15 26	-0.001	+0.03	7.55	-0.10	-0.2	B8 V		360 s		
83590	117832		9 39 41.9	+4 17 04	-0.001	+0.02	8.0	1.1	0.2	K0 III		370 s		
84046	250657		9 39 44.9	-62 56 36	-0.005	0.00	6.43	-0.09	0.0	B8.5 V		190 s		
83879	221389		9 39 45.9	-50 08 18	-0.002	-0.01	7.90	0.4	2.6	F0 V		110 s		m
83650	155377	37 Hya	9 39 47.3	-10 34 13	-0.002	0.00	6.31	-0.03		B9				
---	221382		9 39 47.4	-40 49 24	-0.005	+0.02	8.0	0.0	0.2	K0 III		290 mx		
83735	177814		9 39 49.2	-28 54 45	0.000	0.00	7.9	1.6	-0.3	K5 III		380 s		
84000	237211		9 39 50.9	-59 17 04	-0.006	+0.01	7.50	0.2	2.1	A5 V		120 s		m
83618	137035	35 ι Hya	9 39 51.3	-1 08 34	+0.003	-0.07	3.91	1.32	-0.2	K3 III	+23	63 s		
83525	61620		9 39 57.8	+35 20 10	-0.005	0.00	7.0	0.5	3.4	F5 V		52 s		
83394	27331		9 39 59.0	+53 29 50	-0.001	-0.12	7.4	1.1	0.2	K0 III		94 mx		
83712	155382		9 40 02.5	-12 59 49	-0.009	0.00	7.1	0.5	3.4	F5 V		55 s		
83702	137042		9 40 07.1	-9 05 43	+0.001	-0.03	7.5	0.4	2.6	F0 V		97 s		
83998	237214		9 40 07.7	-55 42 40	-0.002	+0.01	7.95	0.14	1.4	A2 V		180 s		
83647	117836		9 40 10.2	+8 16 41	-0.003	0.00	7.7	0.9	3.2	G5 IV		80 s		
83826	200614		9 40 16.7	-30 55 19	-0.002	0.00	7.4	1.5	-0.3	K5 III		340 s		
83754	155388	38 κ Hya	9 40 18.3	-14 19 56	-0.002	-0.02	5.06	-0.15	-1.1	B5 V	+18	170 s		
83731	155387		9 40 20.0	-10 46 09	-0.001	-0.01	6.37	0.08		A2				
83186	6929		9 40 34.9	+71 45 16	-0.047	-0.10	7.48	0.52	4.2	F9 V	-27	45 s		
83683	98704		9 40 35.3	+13 03 13	+0.003	-0.12	6.97	0.47		F8	-22			
83844	177828		9 40 37.9	-21 34 53	-0.006	+0.01	7.6	1.5	0.2	K0 III		160 s		
83508	27334		9 40 40.6	+53 21 53	+0.002	0.00	7.7	1.4	-0.3	K5 III		400 s		
83792	137050		9 40 41.7	-6 53 23	+0.002	-0.06	7.9	0.4	2.6	F0 V		110 s		
84121	237221		9 40 42.5	-57 59 00	-0.004	0.00	5.32	0.20		A2	+7	12 mn		m
83927	200624		9 40 43.4	-37 09 35	0.000	-0.03	7.80	0.5	4.0	F8 V		58 s		m
83630	61624		9 40 45.6	+32 59 59	-0.003	+0.01	7.9	0.4	3.0	F2 V	+25	94 s		
83035	6928		9 40 51.5	+75 25 50	-0.009	-0.06	7.5	1.1	0.2	K0 III		220 mx		
83601	42975		9 40 55.7	+43 59 33	-0.001	0.00	7.8	0.4	2.6	F0 V		110 s		
84101	237222		9 40 56.1	-54 59 00	-0.001	0.00	6.93	-0.12	0.0	B8.5 V		240 s		
83491	27335		9 40 58.8	+58 35 20	-0.001	-0.03	7.8	1.1	0.2	K0 III		330 s		
84153	237225		9 41 00.0	-59 02 26	+0.001	+0.01	7.4	1.1	0.2	K0 III		250 s		
84044	221399		9 41 01.3	-46 09 58	+0.002	-0.03	7.84	1.37	-0.1	K2 III		270 s		
84152	237224		9 41 02.2	-57 15 34	+0.010	-0.04	5.80	1.09	0.2	gK0		120 s		
84261	250664		9 41 06.6	-66 04 49	-0.010	+0.04	6.85	0.86	1.8	G7 III-IV		110 s		
83808	98709	14 o Leo	9 41 09.0	+9 53 32	-0.010	-0.04	3.52	0.49	2.1	A5 V	+27	17 mn	7480	m
83994	200631		9 41 10.1	-36 01 44	-0.010	+0.05	7.30	0.5	4.0	F8 V		46 s		m
83923	155399		9 41 12.3	-17 22 59	-0.002	+0.03	8.0	1.4	-0.3	K5 III		450 s		
84070	221402		9 41 13.2	-46 22 54	-0.007	+0.03	7.91	1.06	0.2	K0 III		220 mx		
83822	117851		9 41 14.5	+8 59 42	-0.002	-0.06	6.9	0.4	2.6	F0 V	+33	71 s		
84137	237227		9 41 15.8	-51 37 57	-0.001	-0.01	7.9	0.6	1.4	A2 V		86 s		
83564	27340		9 41 16.4	+55 51 58	-0.006	-0.05	6.50	1.13		K2 III-IV		210 mx		
83953	177840		9 41 16.9	-23 35 30	-0.002	-0.01	4.77	-0.12	-1.1	B5 V	+26	150 s		m
83698	61629		9 41 21.9	+38 57 00	+0.007	-0.14	7.30	0.8	3.2	G5 IV	+5	66 s	7477	m
84068	221404		9 41 22.6	-41 58 47	-0.004	-0.09	7.6	0.4	0.2	K0 III		130 mx		
84042	200635		9 41 24.6	-33 23 48	-0.002	-0.02	7.0	0.3	0.6	A0 V		120 s		
83937	155401		9 41 27.3	-10 30 08	-0.004	0.00	7.4	1.1	-0.1	K2 III		310 mx		
84174	237231		9 41 30.7	-52 32 17	-0.003	+0.01	8.0	0.4	1.4	A2 V		120 s		
83973	155402		9 41 31.4	-18 29 01	+0.008	+0.04	7.60	0.5	4.0	F8 V		53 s	7488	m
83787	61633		9 41 35.1	+31 16 41	+0.002	0.00	5.89	1.60	-0.3	K5 III	-13	150 s		
83936	137054		9 41 35.9	-4 16 47	-0.002	-0.01	8.0	1.4	-0.3	K5 III		450 s		
84375	250668		9 41 37.4	-68 30 17	-0.001	+0.01	7.0	0.0	0.0	B8.5 V		250 s		
83821	80970	13 Leo	9 41 38.4	+25 54 46	-0.001	-0.04	6.24	1.25	0.2	K0 III	-26	110 s		
82371	1560		9 41 44.6	+82 21 54	-0.002	+0.01	7.5	0.1	0.6	A0 V		240 s		
83986	155404		9 41 44.8	-12 36 53	+0.002	-0.06	7.6	0.4	2.6	F0 V		100 s		
83891	98717		9 41 46.0	+15 45 17	-0.002	-0.02	7.9	1.1	0.2	K0 III		340 s		
83786	61634		9 41 46.5	+37 15 52	-0.005	-0.10	7.9	0.6	4.4	G0 V		50 s		
84228	237237		9 41 47.8	-55 12 50	-0.002	-0.01	6.00	-0.13	-1.4	B4 V		300 s		
83747	42983		9 41 49.0	+43 38 16	+0.003	+0.01	7.3	1.1	0.2	K0 III		260 s		
84227	237238		9 41 51.4	-54 13 02	-0.005	-0.01	8.0	0.4	2.6	F0 V		100 s		
84085	200644		9 41 52.2	-30 18 14	-0.003	-0.01	6.7	1.6	0.2	K0 III		86 s		
83805	61636	43 Lyn	9 42 00.2	+39 45 29	-0.005	-0.04	5.62	0.95	0.3	G8 III	+30	120 s		
84171	221411		9 42 01.4	-42 16 58	-0.004	+0.01	7.2	0.5	0.6	A0 V		98 s		
84134	200646		9 42 01.7	-37 05 00	-0.001	-0.05	7.9	0.4	0.6	A0 V		150 s		
84133	200647		9 42 06.6	-32 37 15	-0.002	+0.02	8.0	2.0	-0.3	K5 III		220 s		
84083	177863		9 42 08.1	-23 42 32	-0.001	-0.03	8.00	0.4	2.6	F0 V		120 s		m

HD	SAO	Star Name	α 2000	δ 2000	μ(α)	μ(δ)	V	B-V	M_v	Spec	RV	d(pc)	ADS	Notes
84416	250672		9 42 13.1	−66 54 53	−0.001	0.00	6.60	0.1	0.6	A0 V		160 s		m
84117	177866		9 42 14.4	−23 54 56	−0.029	+0.25	4.94	0.53	4.4	G0 V	+34	13 ts		
83489	14966		9 42 14.8	+69 14 15	−0.013	−0.07	5.69	1.14	0.2	K0 III	−9	100 s		
84021	117863		9 42 17.0	+2 18 46	−0.002	−0.02	7.90	1.1	−0.1	K2 III		400 s	7493	m
84147	177869		9 42 19.1	−28 14 07	−0.003	+0.01	7.8	1.9	−0.1	K2 III		150 s		
83003	6931		9 42 20.8	+78 49 40	0.000	−0.04	8.00	0.2	2.1	A5 V		150 s		m
83935	80978		9 42 21.2	+25 35 24	0.000	−0.01	7.26	1.10	0.0	K1 III		280 s		
—	—		9 42 22.7	+77 51 05			7.80							Y Dra, v
84050	137065		9 42 23.2	−0 16 46	−0.002	+0.02	6.7	1.1	0.2	K0 III		200 s		
83804	42986		9 42 25.2	+48 46 34	−0.007	−0.20	7.3	1.1	0.2	K0 III		57 mx		
84244	221419		9 42 27.8	−46 51 56	−0.002	−0.01	7.0	1.8	−0.5	M2 III		320 s		
84201	200650		9 42 29.2	−35 34 41	−0.002	0.00	7.9	0.3	0.6	A0 V		190 s		
84129	155423		9 42 33.6	−13 58 42	−0.003	+0.03	7.6	0.1	0.6	A0 V		260 s		
84405	250674		9 42 34.2	−62 44 55	−0.001	−0.01	7.8	0.4	2.6	F0 V		110 s		
83934	61647		9 42 34.4	+32 59 42	−0.001	+0.02	7.1	0.9	3.2	G5 IV		59 s		
84631	256647		9 42 39.2	−72 11 29	−0.004	−0.01	6.9	0.2	0.4	B9.5 V		150 s		
84330	237248		9 42 40.8	−55 49 57	−0.018	+0.03	7.4	0.4	3.2	G5 IV		68 s		m
84224	200652		9 42 41.3	−35 30 07	−0.002	−0.01	6.41	−0.06		B9				
84078	117866		9 42 41.8	+1 47 45	−0.004	−0.01	8.0	0.4	2.6	F0 V		120 s		
83951	61648	13 LMi	9 42 42.7	+35 05 36	−0.001	−0.05	6.14	0.36	3.0	F2 V	−8	42 s		
83869	42990		9 42 43.1	+48 25 52	−0.003	−0.02	6.4	0.1	0.6	A0 V	−12	140 s		
83839	27355		9 42 43.8	+51 16 07	0.000	+0.01	7.30	1.6	−0.5	M2 III	−11	360 s		
84005	61649		9 42 44.5	+30 06 42	−0.002	0.00	6.7	0.2	2.1	A5 V	+1	83 s		
84199	177879		9 42 50.7	−22 57 55	−0.009	+0.02	7.9	0.6	3.0	F2 V		64 s		m
84234	200654		9 42 51.9	−33 24 08	+0.002	−0.04	7.4	0.7	1.4	A2 V		64 s		
84004	61651		9 42 55.0	+32 15 51	−0.001	−0.04	7.2	0.4	3.0	F2 V	+5	70 s		
84359	237254		9 42 56.2	−54 45 57	0.000	0.00	7.97	−0.06	0.0	B8.5 V		390 s		
83506	6936	27 UMa	9 42 57.1	+72 15 09	−0.006	−0.03	5.3	1.1	0.2	K0 III	−17	100 s		
84250	200657		9 42 58.9	−36 57 57	+0.001	−0.01	7.4	1.8	−0.5	M2 III		310 s		
85128	256651		9 43 02.0	−79 35 27	−0.004	+0.06	7.6	1.1	0.2	K0 III		290 mx		
84003	61652		9 43 04.2	+37 07 42	−0.001	0.00	7.4	1.1	−0.1	K2 III		320 s		
83886	27359		9 43 06.8	+54 21 49	−0.005	−0.03	6.4	0.1	1.4	A2 V	+21	98 s		
84110	98726		9 43 11.4	+17 46 00	−0.002	−0.03	7.9	0.0	0.4	B9.5 V		320 s		
84185	137077		9 43 12.6	−6 37 52	−0.004	−0.01	7.9	1.1	−0.1	K2 III		320 mx		
84232	177889		9 43 12.6	−22 19 20	−0.004	−0.01	7.9	0.6	3.0	F2 V		64 s		
—	237258		9 43 12.6	−54 04 33	−0.003	+0.01	7.9	0.1	0.6	A0 V		280 s		
83765	14973		9 43 15.6	+62 55 35	+0.001	−0.03	7.9	0.1	1.7	A3 V		180 s		
84464	237263		9 43 16.6	−60 01 44	−0.002	+0.01	7.49	−0.06	0.0	B8.5 V		320 s		m
84273	177891		9 43 20.2	−29 48 13	−0.007	+0.05	8.0	0.9	3.2	G5 IV		78 s		
84124	80988		9 43 22.4	+20 11 35	+0.001	−0.01	6.9	1.1	0.2	K0 III		220 s		
84184	117871		9 43 27.1	+2 37 39	+0.002	0.00	7.40	0.5	4.0	F8 V	−3	48 s	7500	m
84400	237260		9 43 27.4	−51 13 42	−0.003	−0.01	6.15	−0.10		B8				m
84242	155433		9 43 30.4	−13 41 45	+0.001	−0.05	8.0	0.9	3.2	G5 IV		93 s		
84257	177893		9 43 30.6	−20 21 58	−0.001	−0.02	6.6	1.2	0.2	K0 III		150 s		
84462	237267		9 43 32.6	−56 57 11	+0.001	0.00	7.73	−0.11	0.0	B8.5 V		350 s		
84107	61656	15 Leo	9 43 33.2	+29 58 28	−0.001	−0.11	5.64	0.12	1.7	A3 V	+16	30 mx		
84183	98731		9 43 35.2	+10 31 08	+0.002	−0.12	6.9	0.5	3.4	F5 V		49 s		
84182	98730		9 43 36.3	+13 26 30	+0.004	−0.06	7.1	1.1	0.2	K0 III	−29	240 s		
84354	221427		9 43 36.4	−41 10 21	−0.002	−0.02	8.0	0.9	3.2	G5 IV		93 s		
84461	237268		9 43 42.2	−53 53 29	−0.008	+0.02	5.56	−0.04	0.6	A0 V	+6	98 s		
84106	61657		9 43 42.6	+34 05 50	−0.001	0.00	7.3	1.1	0.2	K0 III		260 s		
84194	98733	16 ψ Leo	9 43 43.8	+14 01 18	0.000	−0.01	5.35	1.63	−0.5	M2 III	+8	140 s		m
84369	221428		9 43 44.4	−41 14 29	−0.001	−0.07	7.6	1.1	−0.1	K2 III		230 mx		
84059	43000		9 43 44.8	+44 42 05	+0.002	0.00	7.6	0.4	3.0	F2 V		84 s		
83950	27364		9 43 45.5	+55 57 09	+0.003	−0.03	7.90	0.66	4.0	F8 V	−46	51 s	7494	W UMa, m,v
83422	6937		9 43 46.6	+77 13 17	−0.008	−0.04	7.9	1.1	0.2	K0 III		340 mx		
84398	221431		9 43 48.0	−42 51 05	0.000	−0.01	7.61	0.01	3.4	F5 V		70 s		
84734	—		9 43 58.3	−68 35 34			8.0	1.1	0.2	K0 III		350 s		
84493	237272		9 44 02.5	−51 22 01	−0.002	+0.02	7.72	−0.11	−1.6	B3.5 V		670 s		
84588	237282		9 44 05.9	−59 52 15	−0.001	+0.03	7.7	0.1	0.4	B9.5 V		230 s		
84123	43003		9 44 06.8	+42 03 02	−0.002	−0.10	6.8	0.4		F0 p	+16			
84523	237278		9 44 07.1	−53 44 43	0.000	0.00	7.95	0.06	−1.6	B3.5 V		600 s		
84367	177908	θ Ant	9 44 12.0	−27 46 10	−0.004	+0.03	4.79	0.51	3.8	F7 V	+24	14 s		m
84181	61661		9 44 15.5	+34 43 14	+0.001	−0.03	7.9	1.6	−0.5	M2 III		470 s		
84447	200681		9 44 15.7	−39 34 16	−0.003	0.00	6.82	0.30		A5				
84299	117881		9 44 23.6	+1 58 04	0.000	−0.02	7.9	1.1	0.2	K0 III		340 s		
83599	6945		9 44 26.4	+74 35 16	0.000	−0.021	7.1	0.4	2.6	F0 V		66 s		
84252	98742		9 44 29.9	+18 51 48	0.000	−0.06	6.60	1.1	0.2	K0 III	−1	190 s		m
84337	137092		9 44 30.4	−4 39 27	−0.003	0.00	7.7	0.1	1.4	A2 V		180 s		
84963	256653		9 44 31.1	−73 54 01	−0.004	+0.01	7.6	0.5	4.0	F8 V		52 s		
84427	177915		9 44 36.0	−23 21 42	−0.007	−0.03	7.5	1.2	0.2	K0 III		160 mx		
84412	177914		9 44 36.3	−22 45 07	−0.004	+0.01	6.8	1.3	3.2	G5 IV		24 s		
83962	14976		9 44 36.6	+64 59 02	−0.008	+0.01	6.2	0.4	3.0	F2 V	−28	43 s		
84552	221446		9 44 37.2	−48 32 55	−0.003	+0.01	6.69	−0.10	0.0	B8.5 V		220 s		
84321	117885		9 44 38.0	+3 21 12	−0.002	+0.02	7.2	1.1	−0.1	K2 III		290 s		

HD	SAO	Star Name	α 2000	δ 2000	μ(α)	μ(δ)	V	B-V	M$_v$	Spec	RV	d(pc)	ADS	Notes
84428	177917		9^{h}44^{m}39^s.3	−23°34′21″	−0^s.001	+0″.01	7.8	0.6	3.2	G5 IV		85 s		
83838	6951		9 44 41.7	+70 17 00	+0.003	+0.02	7.9	1.1	−0.1	K2 III		390 s		
84392	155445		9 44 44.6	−11 39 52	−0.002	+0.01	8.00	1.1	0.2	K0 III		360 s		m
84458	177923		9 44 47.6	−27 37 49	−0.001	−0.01	7.00	0.4	2.6	F0 V		76 s		m
84424	155449		9 44 48.6	−19 27 08	−0.002	0.00	7.8	1.1	0.2	K0 III		330 s		
84423	155450		9 44 51.8	−17 42 23	+0.002	0.00	7.0	1.1	−0.1	K2 III		270 s		
84265	80996		9 44 52.8	+28 42 31	−0.002	−0.01	7.7	0.2	2.1	A5 V		130 s		
84598	221451		9 44 53.0	−50 00 38	−0.002	−0.02	7.47	0.99	3.2	G5 IV		35 s		m
84583	221448		9 44 53.0	−47 28 52	−0.004	−0.01	8.0	−0.1	3.0	F2 V		98 s		
84391	137098		9 44 53.4	−6 42 14	−0.001	−0.02	8.0	0.0	0.4	B9.5 V		330 s		
84676	237297		9 44 53.9	−57 34 09	+0.005	−0.04	7.00	0.9	3.2	G5 IV		57 s		m
84910	250686		9 45 04.8	−69 45 42	0.000	+0.03	7.20	0.1	0.6	A0 V		210 s		m
84320	81000		9 45 10.5	+27 44 29	−0.002	−0.04	7.4	0.4	2.6	F0 V		93 s		
84420	137099		9 45 11.4	−0 27 54	+0.001	−0.03	7.5	1.1	0.2	K0 III		290 s		
84759	237307		9 45 12.6	−59 28 37	−0.003	−0.05	7.50	0.9	3.2	G5 IV		72 s		m
84347	81002		9 45 14.7	+23 28 29	−0.002	−0.02	6.7	1.1	0.2	K0 III		200 s		
84810	250683		9 45 14.8	−62 30 28	−0.002	0.00	4.1	0.6	4.4	G0 V	+4	16 mn		1 Car, v
84866	250685		9 45 15.4	−66 48 32	0.000	0.00	8.0	0.0	0.0	B9.5 V		330 s		
84407	117890		9 45 16.8	+8 52 55	−0.004	−0.05	7.90	0.5	3.4	F5 V		79 s	7508	m
84688	237305		9 45 17.5	−54 14 51	−0.002	+0.01	7.2	0.0	1.4	A2 V		150 s		
84567	177939		9 45 21.8	−30 12 10	−0.003	0.00	6.45	−0.13	−2.5	B2 V var		620 s		
84674	221464		9 45 21.8	−49 48 45	−0.005	−0.03	8.0	0.4	4.0	F8 V		51 s		
84580	200699		9 45 24.1	−32 40 59	−0.011	−0.04	7.0	0.6	3.4	F5 V		44 s		
84487	137103		9 45 25.2	−8 28 58	−0.001	−0.05	7.3	1.1	0.2	K0 III	+14	260 s		
84851	−−		9 45 26.7	−64 29 39			8.0	0.0	0.0	B8.5 V		390 s		
84418	98747		9 45 28.6	+10 13 18	0.000	−0.02	7.4	1.4	−0.3	K5 III		350 s		
−−	200704		9 45 29.0	−37 44 00	−0.005	+0.02	7.8	1.0	0.2	K0 III		300 mx		
84579	177942		9 45 29.7	−29 32 22	+0.001	+0.08	7.6	1.5	0.2	K0 III		140 mx		
83550	6948		9 45 30.8	+78 08 05	+0.002	0.00	6.23	1.35	−0.1	K2 III	−27	82 s		
84346	61669		9 45 34.1	+34 30 44	−0.001	+0.01	6.2	1.6		M5 III	+10			R LMi, v
84624	200705		9 45 37.1	−37 48 44	−0.006	0.00	7.6	0.9	4.0	F8 V		32 s		
85012	250689		9 45 40.5	−70 07 04	0.000	+0.01	7.14	−0.02	0.0	B8.5 V		260 s		m
84809	237314		9 45 40.5	−57 11 08	−0.004	+0.02	6.46	−0.11		B8				
84529	155459		9 45 41.4	−11 21 24	+0.003	−0.09	7.6	1.1	−0.1	K2 III		210 mx		
84757	237312		9 45 42.7	−52 17 05	+0.001	0.00	7.6	0.2	3.0	F2 V		85 s		
84673	−−		9 45 43.1	−41 39 35			7.80	0.00	0.4	B9.5 V		300 s		m
84578	155463		9 45 47.7	−18 42 00	−0.002	+0.01	8.0	0.9	3.2	G5 IV		93 s		
84515	137107		9 45 47.9	−1 54 49	−0.001	−0.09	8.0	0.5	3.4	F5 V		82 s		
84441	81004	17 ε Leo	9 45 51.0	+23 46 27	−0.003	−0.02	2.98	0.80	−2.0	G0 II	+5	95 s		
84455	98750		9 45 51.9	+18 41 01	+0.002	−0.02	6.9	1.1	0.2	K0 III		220 s		
84440	81005		9 45 54.2	+27 02 45	0.000	−0.03	7.90	1.1	0.0	K1 III		380 s		
84850	237321		9 45 55.3	−58 47 39	−0.016	+0.06	6.22	0.46	2.9	F6 IV-V		45 s		
84179	14980	28 UMa	9 45 55.4	+63 39 12	−0.002	−0.05	6.34	0.31	3.0	F2 V	−27	47 s		m
84729	221471		9 45 55.4	−43 40 43	−0.004	+0.01	7.8	0.8	0.2	K0 III		330 s		
85125	256656		9 45 58.2	−71 47 15	−0.010	−0.01	8.00	1.1	−0.1	K2 III		280 mx		m
84687	200710		9 45 59.7	−35 51 51	−0.001	−0.10	6.8	1.1	3.4	F5 V		19 s		
84789	237319		9 45 59.9	−50 52 11	−0.008	+0.01	8.02	1.07	0.2	K0 III		280 mx		
85195	256657		9 46 00.4	−74 06 55	−0.002	0.00	7.3	0.1	1.4	A2 V		150 s		
84864	237324		9 46 00.8	−59 02 46	−0.009	+0.06	8.0	0.4	0.2	K0 III		200 mx		
84485	98751		9 46 01.0	+17 53 21	−0.002	−0.03	7.9	0.9	3.2	G5 IV		88 s		
84165	14981		9 46 02.7	+65 37 37	+0.002	−0.03	7.20	1.6	−0.5	M1 III	−34	350 s		
84563	137110		9 46 04.1	−2 35 29	0.000	+0.03	7.9	1.1	0.2	K0 III		360 s		
84388	43023		9 46 05.4	+39 50 16	0.000	+0.02	7.2	0.4	3.0	F2 V		69 s		
−−	221474		9 46 05.4	−40 38 06	+0.001	−0.02	8.0	1.6						
84620	155466		9 46 06.7	−19 43 40	−0.003	+0.01	7.0	0.8	0.6	A0 V		63 s		
84774	221480		9 46 08.9	−45 55 05	−0.002	−0.01	6.67	0.94	3.2	G5 IV		37 s		m
84542	117898		9 46 10.0	+6 42 30	0.000	−0.03	5.79	1.64	−0.5	gM1	+3	170 s		
84497	81009		9 46 13.2	+20 29 21	−0.001	0.00	7.5	0.9	3.2	G5 IV		71 s		
84636	155467		9 46 14.4	−14 35 24	−0.003	0.00	7.30	1.33	−0.1	gK2	+22	240 s		
84727	200714		9 46 16.0	−33 30 29	−0.002	0.00	7.0	0.4	0.6	A0 V		110 s		
84669	177960		9 46 17.4	−22 39 11	−0.003	−0.02	7.4	1.1	0.2	K0 III		250 s		
85396	256658	ν Cha	9 46 21.2	−76 46 33	+0.030	−0.05	5.45	0.89		G4	+11	17 mn		
84709	177962		9 46 21.6	−27 16 33	−0.004	−0.06	7.0	0.9	3.0	F2 V		29 s		
84561	98755	18 Leo	9 46 23.3	+11 48 36	−0.001	+0.01	5.63	1.49	−0.3	K5 III	+30	150 s		
84607	117901		9 46 23.5	+1 47 08	−0.004	−0.05	5.65	0.34	0.6	gF0	+15	98 s		
84929	237332		9 46 26.4	−59 21 10	−0.002	0.00	7.6	−0.7	0.6	A0 V		260 s		
84816	221484		9 46 30.3	−44 45 18	−0.001	0.00	5.55	−0.18	−2.5	B2 V n	−8	390 s		
84752	200720		9 46 30.4	−31 16 30	−0.005	+0.01	7.7	0.9	0.2	K0 III		300 mx		
84335	27377		9 46 31.6	+57 07 40	0.000	+0.03	5.2	1.6	−0.5	M2 III	+8	140 s		
84616	117902		9 46 32.3	+1 43 08	−0.011	−0.01	7.5	0.5	4.0	F8 V		50 s		
84805	221482		9 46 34.6	−40 40 46	−0.006	+0.02	7.8	0.3	1.7	A3 V		130 s		
84860	237329		9 46 36.2	−50 34 34	−0.001	0.00	7.6	0.6	1.4	A2 V		79 s		
84453	43030		9 46 42.4	+45 06 51	+0.005	−0.13	6.82	0.95	3.2	K0 IV	−44	52 s		
85026	250692		9 46 42.5	−62 42 33	−0.002	+0.01	7.9	0.1	1.7	A3 V		170 s		
84664	137117		9 46 43.0	−3 12 43	−0.003	+0.03	7.4	0.4	3.0	F2 V		77 s		

HD	SAO	Star Name	α 2000	δ 2000	μ(α)	μ(δ)	V	B-V	M_V	Spec	RV	d(pc)	ADS	Notes
			h m s	° ' "	s	"								
84606	98761		9 46 47.1	+18 06 59	-0.005	-0.04	7.9	0.7	2.8	G0 IV	+19	110 s		
84706	155471		9 46 47.3	-11 56 21	0.000	-0.04	7.8	1.1	0.2	K0 III		330 s		
84725	155472		9 46 49.9	-16 32 47	0.000	0.00	7.9	0.5	4.0	F8 V		60 s		
84705	137123		9 46 53.8	-4 17 53	-0.003	-0.04	8.0	1.1	0.2	K0 III		360 s		
84681	117904		9 47 02.1	+7 05 28	+0.002	-0.03	7.9	1.1	-0.1	K2 III		390 s		
84824	200729		9 47 03.8	-31 16 06	-0.004	0.00	6.8	0.6	0.6	A0 V		79 s		
84751	155473		9 47 05.0	-10 31 23	-0.004	0.00	7.1	0.4	3.0	F2 V		66 s		
85123	250695	υ Car	9 47 06.1	-65 04 18	-0.001	+0.01	2.97	0.27	-2.0	A7 II	+14	99 s		m
84803	177984		9 47 07.7	-21 42 01	-0.005	+0.02	7.7	0.8	4.0	F8 V		39 s		
84768	155476		9 47 10.1	-10 44 37	-0.001	-0.01	7.1	0.5	3.4	F5 V		56 s		
84802	155478		9 47 14.1	-17 12 19	-0.001	+0.02	7.0	1.4	-0.3	K5 III		280 s		
92239	--		9 47 16.4	-89 21 27			7.7	1.1	0.2	K0 III		320 s		
84701	117908		9 47 16.5	+8 34 12	-0.005	-0.06	6.8	0.4	2.6	F0 V		68 s		
83727	6956		9 47 18.1	+79 08 11	-0.009	-0.03	6.1	0.4	2.6	F0 V	-7	51 s		
84903	221490		9 47 19.2	-41 27 05	-0.001	-0.02	8.0	0.9	3.2	G5 IV		93 s		
84615	61684		9 47 20.2	+32 46 56	+0.002	-0.03	7.8	1.6	-0.5	M2 III		460 s		
84680	81015		9 47 22.2	+23 38 51	-0.001	-0.01	6.7	1.1	0.2	K0 III		200 s		
--	237342		9 47 24.8	-56 45 44	+0.001	+0.02	7.7	0.0	0.4	B9.5 V		290 s		
84722	98767	19 Leo	9 47 25.9	+11 34 06	-0.004	0.00	6.45	0.43		F0	-4			
84902	200735		9 47 28.4	-37 43 28	-0.001	-0.02	7.70	0.5	4.0	F8 V		55 s		m
84740	98768		9 47 29.3	+10 50 39	+0.001	+0.01	7.8	1.1	0.2	K0 III		330 s		
84406	14987		9 47 30.5	+63 14 51	-0.017	-0.14	6.94	0.94	0.2	gK0	+8	220 s		
84748	98769		9 47 33.4	+11 25 43	0.000	-0.05	4.9	1.6		M5 III	+13	19 mn		R Leo, v
84926	200739		9 47 34.9	-38 11 24	0.000	-0.01	6.70	1.1	-0.1	K2 III		230 s		m
84888	178002		9 47 35.7	-29 41 22	0.000	-0.01	7.89	1.01	0.2	G9 III		330 s		
84900	200736		9 47 36.4	-33 14 38	+0.003	-0.06	6.8	0.9	3.2	G5 IV		42 s		
84887	178005		9 47 40.3	-28 22 46	-0.006	0.00	7.4	0.7	4.0	F8 V		36 s		
84764	98771		9 47 41.2	+10 23 05	-0.002	0.00	7.6	0.2	2.1	A5 V		130 s		
84739	81017		9 47 42.8	+20 36 17	+0.006	-0.01	7.70	0.4	2.6	F0 V	-15	110 s		m
85037	221496		9 47 43.9	-49 56 38	-0.002	-0.03	6.9	0.9	0.6	A0 V		53 s		
84781	98772		9 47 45.7	+10 04 30	-0.004	-0.02	7.60	0.8	3.2	G5 IV		76 s	7517	m
85286	250703		9 47 52.8	-66 51 46	-0.005	-0.01	8.0	1.4	-0.3	K5 III		440 s		
84991	200749		9 48 00.2	-38 21 51	-0.009	+0.12	7.3	0.8	4.4	G0 V		26 s		
84874	137139		9 48 03.9	-8 10 08	+0.002	-0.04	8.0	1.6	-0.5	M2 III		510 s		
84881	137140		9 48 06.6	-5 54 14	+0.001	-0.01	7.9	1.1	-0.1	K2 III		390 s		
84882	137142		9 48 12.6	-7 14 45	-0.002	-0.02	7.1	0.0	0.4	B9.5 V		220 s		
85154	237355		9 48 13.0	-54 23 47	-0.003	-0.01	7.30	1.1	0.2	K0 III		260 s		m
84974	178015		9 48 17.5	-25 12 29	-0.001	+0.02	8.00	0.1	1.4	A2 V		210 s		m
84801	81019		9 48 20.1	+28 31 21	0.000	0.00	7.8	0.1	0.6	A0 V		270 s		
85643	256660		9 48 21.3	-75 46 39	-0.037	+0.11	6.9	0.5	3.4	F5 V		46 s		
85103	221503		9 48 24.0	-44 01 15	-0.008	-0.02	8.0	0.2	1.4	A2 V		100 mx		
85044	200755		9 48 28.1	-37 38 10	0.000	-0.02	7.00	1.1	-0.1	K2 III		260 s		m
85036	200754		9 48 32.5	-30 19 11	-0.003	+0.06	7.9	1.3	3.2	G5 IV		42 s		
84659	27384		9 48 34.5	+54 15 53	-0.006	-0.02	7.3	0.4	2.6	F0 V		87 s		
84737	43046		9 48 35.3	+46 01 16	+0.022	-0.10	5.09	0.62	4.7	G2 V	+5	13 ts		
84763	43047		9 48 35.3	+43 32 11	-0.003	-0.01	7.9	0.9	3.2	G5 IV		88 s		
85230	237369		9 48 40.0	-55 29 42	0.000	-0.01	7.7	1.9	-0.5	M2 III		300 s		
85250	237370		9 48 40.0	-56 24 42	-0.007	+0.01	6.05	0.93		K0				
85043	178027		9 48 42.7	-26 24 51	-0.012	+0.03	6.80	0.4	2.6	F0 V		69 s	7524	m
84984	137151		9 48 43.5	-9 55 08	-0.001	-0.01	7.0	0.1	0.6	A0 V		190 s		
85313	237374		9 48 43.8	-60 07 56	-0.006	+0.03	7.6	-0.6	0.2	K0 III		300 s		
84800	43050		9 48 44.3	+43 39 58	-0.005	0.00	7.8	0.1	1.4	A2 V		170 mx		
85100	200758		9 48 45.3	-35 01 14	+0.002	-0.01	7.33	0.26	1.7	A3 V		110 s		m
85228	237371		9 48 46.5	-52 36 55	-0.029	+0.24	7.93	0.89	6.1	K1 V		22 s		m
84558	14991		9 48 46.8	+64 53 39	+0.001	+0.03	7.1	0.5	3.4	F5 V		55 s		
85035	155505		9 48 47.0	-19 18 46	+0.001	0.00	7.1	1.1	0.2	K0 III		240 s		
85042	178029		9 48 49.7	-22 01 07	+0.002	-0.01	7.3	0.2	0.6	A0 V		160 s		
84633	14993		9 48 50.1	+60 05 51	-0.010	-0.03	7.2	1.1	0.2	K0 III		250 s		
84952	117923		9 48 52.7	+8 18 08	+0.002	0.00	8.0	0.5	3.4	F5 V		82 s		
83339	1579		9 48 55.6	+83 19 46	-0.017	-0.01	7.9	0.4	2.6	F0 V		120 s		
85185	221509		9 48 56.2	-47 54 45	-0.005	+0.01	7.9	0.3	0.6	A0 V		160 mx		
84870	61697		9 49 02.8	+34 05 07	-0.002	-0.06	7.10	0.1	1.7	A3 V		60 mx	7520	m
84950	98783		9 49 05.4	+18 03 28	-0.002	-0.01	7.8	1.1	0.2	K0 III		320 s		
85209	221512		9 49 08.9	-43 29 00	-0.001	0.00	6.54	1.28	0.0	K0 III		120 s		m
85892	256664		9 49 09.3	-78 16 53	-0.001	+0.02	7.8	0.0	0.0	B8.5 V		370 s		
85793	256663		9 49 09.3	-77 00 51	-0.001	-0.01	8.00	1.1	0.2	K0 III		360 s		m
85543	250711		9 49 13.9	-70 04 29	+0.003	-0.02	7.8	0.5	3.4	F5 V		75 s		
85114	178042		9 49 14.2	-24 38 46	+0.003	-0.04	7.5	1.2	0.2	K0 III		230 s		
85424	250707		9 49 15.2	-63 09 06	-0.009	+0.03	8.0	0.5	3.4	F5 V		82 s		
85055	137157		9 49 17.1	-8 50 08	-0.003	-0.02	7.0	1.1	-0.1	K2 III	+17	260 s		
85183	200773		9 49 27.2	-33 24 31	0.000	-0.05	8.0	1.0	-0.1	K2 III		410 s		
85206	200777		9 49 28.0	-37 11 11	-0.008	+0.02	5.97	1.24		K0				
84914	61703		9 49 28.4	+36 44 51	+0.003	-0.02	6.8	1.4	-0.3	K5 III		270 s		
84779	27391		9 49 28.6	+57 13 15	0.000	-0.04	7.67	1.13	0.2	K0 III		240 s		
85017	81032		9 49 33.0	+20 47 38	-0.004	-0.08	7.9	0.9	3.2	G5 IV		88 s		

HD	SAO	Star Name	α 2000	δ 2000	μ(α)	μ(δ)	V	B-V	M$_V$	Spec	RV	d(pc)	ADS	Notes
			h m s	° ′ ″	s	″								
85129	155515		9 49 33.0	−11 47 05	−0.004	+0.01	6.9	1.1	−0.1	K2 III		260 s		
85341	237381		9 49 33.5	−52 07 38	−0.002	0.00	7.59	−0.11	0.0	B8.5 V		330 s		
85146	155517		9 49 35.7	−16 38 34	−0.002	+0.03	8.0	1.1	−0.1	K2 III		410 s		
84868	27393		9 49 45.6	+52 53 27	−0.005	−0.05	7.2	1.1	0.2	K0 III		230 mx		
85016	61705		9 49 46.6	+31 04 37	+0.004	−0.04	6.9	0.2	2.1	A5 V		91 s		
85091	98794		9 49 48.5	+11 06 23	−0.021	−0.07	7.5	0.5	4.0	F8 V	+40	48 mx		
85040	81035	20 Leo	9 49 50.0	+21 10 46	−0.003	−0.02	6.09	0.25	2.6	F0 V	+26	50 s		m
85296	200784		9 49 51.2	−36 16 06	−0.002	0.00	6.37	1.01		K0				
85070	98792		9 49 52.8	+16 50 18	+0.002	−0.04	8.00	1.1	−0.1	K2 III		420 s		m
85469	237389		9 49 53.7	−58 48 34	−0.001	0.00	8.0	−0.2	0.6	A0 V		300 s		
85612	250714		9 49 54.8	−68 25 11	−0.016	+0.02	6.80	0.8	3.2	G5 IV		53 s		m
85355	221523		9 49 57.0	−45 43 58	−0.003	0.00	5.08	−0.10		B7	+12			m
85356	221525		9 49 58.7	−47 56 53	−0.002	−0.02	7.8	0.4		B0.5 II				
85030	61706		9 50 00.5	+33 36 07	−0.002	−0.01	8.0	0.5	3.4	F5 V		82 s		
85323	200787		9 50 01.4	−40 10 08	−0.004	+0.01	7.5	0.9	0.6	A0 V		71 s		
85277	178066		9 50 02.6	−28 53 09	−0.005	−0.06	7.6	1.2	0.2	K0 III		160 mx		
85411	237387		9 50 04.6	−50 37 34	+0.011	−0.07	7.67	0.48	4.0	F8 V		54 s		
85087	81040		9 50 10.3	+24 33 44	−0.004	+0.04	7.0	0.5	3.4	F5 V		52 s		
85409	221531		9 50 10.4	−49 37 14	−0.001	−0.06	7.72	0.48	4.0	F8 V		56 s		m
85496	237393		9 50 11.5	−58 12 56	0.000	0.00	7.98	−0.02	0.0	B8.5 V		380 s		m
85180	117933		9 50 12.1	−0 13 43	−0.003	−0.02	7.3	1.1	0.2	K0 III		270 s		
85179	117934		9 50 13.0	+0 06 13	−0.003	−0.01	7.5	0.5	3.4	F5 V		66 s		
85029	61707		9 50 13.8	+39 37 54	0.000	0.00	6.7	1.4	−0.3	K5 III	+9	250 s		
85495	237396		9 50 14.8	−58 02 45	+0.001	0.00	7.6	0.2	−1.0	B5.5 V		320 s		
85827	—		9 50 16.7	−74 12 08			7.8	0.1	0.6	A0 V		270 s		
85436	237390		9 50 19.6	−50 26 08	−0.001	−0.01	7.5	1.2	−0.5	M2 III		400 s		
85052	61708		9 50 21.5	+35 52 23	−0.002	−0.04	7.2	1.1	0.2	K0 III		260 s		
84812	15000		9 50 23.6	+65 35 36	−0.009	−0.03	6.31	0.28	2.6	F0 V	−7	55 s		
85219	137173		9 50 23.8	−1 25 21	−0.002	−0.02	7.9	1.1	0.2	K0 III		340 s		
85217	117937	4 Sex	9 50 30.0	+4 20 37	−0.010	−0.06	6.24	0.48	3.7	F6 V	+17	32 s		
85218	117938		9 50 30.8	+1 49 39	−0.003	−0.03	8.03	1.20		K0				
85387	200795		9 50 31.3	−36 15 42	−0.005	+0.03	7.6	0.6	1.4	A2 V		78 s		
85596	250718		9 50 38.2	−60 31 39	−0.001	+0.02	7.6	−0.2	0.6	A0 V		250 s		
85294	137179		9 50 41.0	−7 22 56	0.000	−0.04	7.1	0.4	3.0	F2 V		65 s		
85483	221538		9 50 41.9	−46 56 03	−0.004	+0.01	5.73	1.08	0.2	gK0		110 s		
85198	98802		9 50 42.2	+17 43 59	−0.002	−0.06	8.0	0.5	3.7	F6 V	+24	71 s		
85352	178081		9 50 44.6	−20 46 12	−0.001	−0.01	7.9	0.4	0.6	A0 V		160 s		
85162	61710		9 50 45.0	+31 23 34	+0.001	+0.01	7.30	1.6	−0.5	M2 III	−34	360 s		
85451	200801		9 50 48.1	−38 50 30	−0.039	−0.02	7.6	0.5	0.6	A0 V		21 mx		
85259	98806		9 50 49.2	+11 50 32	−0.001	−0.01	6.70	0.1	0.6	A0 V		170 s	7539	m
85335	155525		9 50 53.5	−11 44 14	−0.006	−0.07	8.0	1.1	−0.1	K2 III		140 mx		
85270	117942		9 50 55.1	+8 07 39	0.000	0.00	7.9	0.9	3.2	G5 IV		88 s		
85656	250721		9 50 55.7	−62 44 42	−0.001	0.00	5.57	1.32	0.2	gK0	+12	70 s		
85108	43066		9 50 57.6	+45 32 54	−0.006	−0.02	8.0	0.5	3.4	F5 V		82 s		
84999	27401	29 υ UMa	9 50 59.3	+59 02 19	−0.038	−0.16	3.80	0.29	1.9	F2 IV	+31	26 ts	7534	m
85382	155529		9 50 59.4	−15 52 48	−0.002	−0.01	7.4	1.1	0.2	K0 III		280 s		
86320	256666		9 51 00.7	−80 03 39	−0.005	+0.02	6.6	0.2	0.6	A0 V		120 s		
85268	98809	23 Leo	9 51 02.0	+13 03 58	+0.002	−0.01	6.6	1.4	−0.3	K5 III	−9	240 s		
85405	178088		9 51 03.5	−23 01 02	−0.001	0.00	6.50	2.0		N7.7	+3			Y Hya, v
85404	—		9 51 03.7	−21 04 58			7.9	0.3	4.4	G0 V		51 s		
85449	200804		9 51 03.9	−31 30 34	+0.002	+0.02	7.8	0.3	2.6	F0 V		110 s		
85512	221544		9 51 06.8	−43 30 08	+0.040	−0.45	7.66	1.16	7.6	K5 V		10 t		
85402	155533		9 51 07.9	−18 39 32	−0.007	+0.02	7.3	0.5	3.4	F5 V		61 s		
85655	237418		9 51 12.0	−59 25 32	+0.003	−0.04	5.79	1.36	−0.1	gK2		120 s		
85364	137183	6 Sex	9 51 13.9	−4 14 36	−0.001	−0.03	6.01	0.17	0.3	A5 III	−10	140 s		
85420	155536		9 51 15.7	−13 43 38	0.000	0.00	7.9	1.1	0.2	K0 III		350 s		
85379	137184		9 51 19.5	−1 51 30	+0.006	−0.11	7.3	1.1	−0.1	K2 III		150 mx		
85563	221547		9 51 19.8	−46 11 38	−0.004	+0.03	5.62	1.17		K0				
85380	137185		9 51 21.5	−6 10 54	−0.009	+0.05	6.42	0.58		G0				
85431	155538		9 51 24.7	−16 33 28	−0.005	+0.02	7.96	0.50	3.7	F6 V	+20	68 s		m
85444	155542	39 υ¹ Hya	9 51 28.6	−14 50 48	+0.001	−0.03	4.12	0.92	0.3	G8 III	−15	58 s		
85377	117949		9 51 28.7	+3 22 06	−0.001	−0.02	7.6	0.1	0.6	A0 V		260 s		
85463	155544		9 51 31.2	−19 39 53	0.000	+0.02	8.0	0.8	0.2	K0 III		360 s		
85552	—		9 51 35.0	−38 11 03			7.5	1.2	3.2	G5 IV		40 s		
85604	221551		9 51 38.6	−46 30 44	−0.004	+0.01	7.5	−0.2	0.0	B8.5 V		200 mx		
85622	221553		9 51 40.7	−46 32 52	−0.001	0.00	4.58	1.20	0.3	gG6	+11	46 s		
85461	155545		9 51 41.2	−11 20 22	+0.003	−0.01	6.7	1.6	−0.5	M2 III	+13	280 s		
85741	250728		9 51 41.3	−60 35 37	+0.002	−0.02	7.26	0.95	1.7	K0 III-IV		130 s		
85683	237423		9 51 41.4	−54 39 35	−0.001	−0.13	7.2	0.5	4.4	G0 V		37 s		
85291	61718		9 51 46.3	+39 37 19	−0.001	−0.02	7.7	0.1	1.7	A3 V		160 s		
85621	221554		9 51 46.9	−45 09 28	−0.001	0.00	7.9	0.3	0.4	B9.5 V		320 s		
85510	155549		9 51 49.1	−20 02 29	−0.003	+0.02	8.0	1.1	−0.1	K2 III		410 s		
85489	155548		9 51 49.2	−14 04 44	0.000	−0.03	7.8	1.1	0.2	K0 III		330 s		
85376	81054	22 Leo	9 51 53.0	+24 23 43	+0.001	−0.18	5.32	0.23	2.1	A5 V	−2	23 mx		
85238	43073		9 51 55.4	+49 37 22	+0.005	−0.04	7.8	0.7	4.4	G0 V		48 s		

HD	SAO	Star Name	α 2000	δ 2000	μ(α)	μ(δ)	V	B−V	M_v	Spec	RV	d(pc)	ADS	Notes
85577	200820		9 51 55.6	−32 58 48	0.000	−0.01	7.2	0.5	3.2	G5 IV		63 s		
85519	155553		9 51 59.5	−16 32 05	+0.001	−0.07	6.08	1.04		K0				
85507	137195		9 52 03.1	−9 50 50	−0.001	+0.01	8.0	1.1	−0.1	K2 III		410 s		
85235	27408	30 ø UMa	9 52 06.3	+54 03 51	−0.001	+0.02	4.59	0.03		A3 s	−12	13 mn	7545	m
85780	237433		9 52 09.1	−58 50 33	−0.001	0.00	8.0	1.5	−0.3	K5 III		450 s		
85505	117960		9 52 11.8	+0 04 32	−0.003	−0.03	6.35	0.94	0.2	G9 III	+19	170 s		
85504	117959	7 Sex	9 52 12.1	+2 27 14	−0.012	+0.09	6.02	−0.04	1.2	A1 V	+97	92 s		
85428	81058		9 52 14.6	+25 06 30	−0.004	−0.02	7.80	1.1	−0.1	K2 III		330 mx		
85346	43080		9 52 15.5	+40 58 12	+0.001	−0.02	8.00	0.1	1.7	A3 V		180 s	7546	m
85374	61724		9 52 16.3	+36 17 59	−0.005	−0.13	7.8	0.5	3.4	F5 V		66 mx		
85301	43078		9 52 16.7	+49 11 28	−0.022	−0.06	7.7	0.9	3.2	G5 IV		79 s		
85758	237432		9 52 19.8	−52 38 57	−0.001	0.00	7.31	1.26	0.2	K0 III		170 s		
85373	61725		9 52 21.6	+37 54 52	−0.005	−0.02	6.8	0.4	2.6	F0 V	+13	68 s		
86055	256665		9 52 26.6	−71 56 37	0.000	0.00	6.7	0.0	0.0	B8.5 V		220 s		
85559	137198		9 52 27.1	−9 13 54	+0.001	−0.01	7.8	1.1	−0.1	K2 III		380 s		
85440	81061		9 52 28.6	+27 46 27	0.000	−0.01	7.70	0.9	0.3	G8 III		300 s		
85372	43082		9 52 28.7	+43 27 58	+0.001	−0.05	7.9	1.1	0.2	K0 III		340 s		
85427	61730		9 52 29.0	+31 32 32	+0.010	−0.01	7.8	0.8	3.2	G5 IV		84 s		
85416	61728		9 52 30.1	+35 31 30	−0.001	+0.01	7.2	1.4	−0.3	K5 III		320 s		
85558	137199	8 γ Sex	9 52 30.4	−8 06 18	−0.004	−0.04	5.05	0.04		A0 n	+12	16 mn	7555	m
85397	43083		9 52 35.2	+40 22 36	−0.001	−0.01	7.9	1.1	0.2	K0 III		340 s		
85777	237436		9 52 36.4	−51 42 49	0.000	−0.03	7.49	−0.12	−1.0	B5.5 V		500 s		
85503	81064	24 μ Leo	9 52 45.8	+26 00 25	−0.016	−0.06	3.88	1.22	−0.1	K2 III	+14	55 s		Rasalas
86089	256667		9 52 49.1	−71 27 16	+0.020	−0.03	7.24	1.15		K2				
85693	178129		9 52 56.2	−26 45 18	−0.001	+0.05	7.9	0.7	2.1	A5 V		73 s		
85725	178130		9 52 57.9	−27 19 56	−0.022	+0.09	6.30	0.62	4.5	dG1	+23	22 s		m
85871	237448		9 53 00.0	−55 22 23	−0.002	0.00	6.48	−0.14	−2.5	B2 V nn		590 s		
85838	237446		9 53 00.1	−52 06 34	−0.002	−0.01	7.91	0.81	3.2	G5 IV		84 s		
85557	81066		9 53 04.3	+22 07 24	0.000	0.00	8.01	1.72		K5				
85837	221574		9 53 10.2	−48 55 47	−0.004	0.00	7.84	1.26	0.2	K0 III		340 s		
85736	178138		9 53 14.5	−23 41 01	+0.001	−0.03	7.70	1.1	0.2	K0 III		320 s		m
85821	221575		9 53 16.1	−44 35 06	−0.006	+0.01	6.66	0.99	3.4	F5 V		18 s		
85586	81067		9 53 17.0	+19 56 16	+0.001	−0.03	7.5	0.9	3.2	G5 IV		72 s		
85439	27416		9 53 17.3	+50 37 15	+0.002	−0.04	7.50	0.9	3.2	G5 IV		72 s	7554	m
83615	1588		9 53 18.8	+84 28 56	+0.004	−0.02	8.0	0.8	3.2	G5 IV		90 s		
85690	137206		9 53 21.9	−9 54 14	0.000	−0.01	7.1	0.1	0.6	A0 V		200 s		
85804	200841		9 53 22.9	−38 24 31	0.000	+0.03	7.5	0.6	4.0	F8 V		44 s		
85723	155572		9 53 23.4	−17 55 35	+0.001	−0.02	7.7	1.6	−0.5	M2 III		440 s		
85752	—		9 53 30.5	−19 29 10			7.80	0.8	3.2	G5 IV		83 s	7567	m
85615	81069		9 53 36.8	+25 40 04	+0.001	−0.03	7.0	1.1	−0.1	K2 III		260 s		
85585	61742		9 53 36.9	+34 59 01	−0.002	−0.02	7.0	1.1	−0.1	K2 III		260 s		
85734	137209		9 53 42.6	−3 51 42	−0.001	+0.01	7.3	0.0	0.4	B9.5 V		240 s		
85709	117975		9 53 42.8	+5 57 30	−0.001	−0.01	5.95	1.66	−0.5	gM2	−1	180 s		
85660	98831		9 53 43.7	+18 56 41	+0.001	0.00	8.0	0.1	0.6	A0 V		300 s		
85888	221582		9 53 44.1	−41 50 50	−0.005	+0.01	7.4	0.4	2.6	F0 V		79 s		
86216	250742		9 53 48.0	−69 53 44	−0.003	0.00	7.0	0.2	0.0	B8.5 V		180 s		
85689	98832		9 53 48.3	+14 44 19	0.000	+0.03	7.9	0.6	4.4	G0 V		51 s		
85953	237464		9 53 50.1	−51 08 49	−0.001	−0.01	5.93	−0.15	−3.6	B2 III	+8	740 s		
86405	256668		9 53 50.4	−75 22 56	−0.005	−0.06	7.8	0.0	0.4	B9.5 V		110 mx		
85360	15013		9 53 54.0	+64 47 22	−0.005	−0.01	7.50	1.1	0.2	K0 III		290 s	7556	m
85861	200846		9 53 54.9	−33 14 05	−0.007	+0.02	7.6	0.0	2.6	F0 V		100 s		
86000	237468		9 53 57.9	−54 50 51	−0.001	+0.02	7.7	0.1	0.0	B8.5 V		260 s		
85472	27421		9 53 58.0	+57 41 01	+0.005	−0.09	7.5	0.8	3.2	G5 IV		74 s		
85720	98835		9 53 58.2	+10 15 32	−0.003	+0.01	7.8	1.6	−0.5	M4 III		470 s		
86076	250740		9 53 58.3	−62 07 13	−0.003	−0.02	7.4	0.4	1.4	A2 V		94 s		
85672	81072		9 53 59.2	+27 41 43	−0.002	−0.02	7.6	0.1	0.6	A0 V		250 s		
85213	6990		9 54 02.6	+71 40 52	−0.002	−0.03	7.6	0.1	0.6	A0 V		250 s		
85860	178157		9 54 05.1	−27 59 56	−0.001	−0.01	7.00	0.00	0.4	B9.5 V		210 s	7570	m
85748	117979		9 54 06.6	+8 04 28	−0.001	−0.02	7.1	0.8	3.2	G5 IV		60 s		
85762	117980		9 54 06.6	+4 56 44	−0.002	+0.01	7.00	1.4	−0.3	K5 III	+27	290 s		m
86256	250743		9 54 08.4	−69 37 24	+0.005	−0.03	8.0	1.6	−0.5	M2 III		470 s		
85859	178158		9 54 12.2	−25 55 56	−0.014	+0.06	4.88	1.23	−0.2	gK3	+51	100 s		
85501	15015		9 54 17.4	+59 55 12	+0.004	−0.03	7.1	1.1	−0.1	K2 III		270 s		
85980	221592		9 54 17.5	−45 17 02	−0.002	−0.01	5.71	−0.11	−1.7	B3 V	+26	280 s		
85966	221590		9 54 18.0	−42 55 01	−0.004	+0.02	7.31	0.97	3.2	G5 IV		47 s		
85997	221593		9 54 22.7	−46 38 14	−0.005	−0.01	6.67	0.98	0.2	K0 III		200 s		
85964	200854		9 54 28.9	−37 17 24	−0.001	−0.05	7.5	0.7	3.0	F2 V		48 s		
85905	178164		9 54 31.6	−22 29 18	−0.003	−0.04	6.24	0.04		A2				
85963	200855		9 54 32.2	−34 54 27	−0.004	−0.02	7.50	0.1	1.4	A2 V		170 s		m
86118	237480		9 54 34.0	−58 25 15	+0.001	+0.01	6.64	−0.17	−1.0	B5.5 V		340 s		
85833	117983		9 54 34.3	+0 56 39	−0.002	−0.06	7.8	0.5	4.0	F8 V		56 s		
85883	155582		9 54 34.5	−12 57 06	+0.003	−0.24	7.1	0.5	4.0	F8 V		41 s		
86005	221596		9 54 39.0	−43 19 18	−0.001	−0.02	7.18	1.30	0.2	K0 III		200 s		
86034	221598		9 54 48.6	−43 32 51	0.000	−0.02	7.90	1.25	−0.1	K2 III		400 s		
86087	237483		9 54 51.2	−50 14 37	−0.003	+0.01	5.72	−0.01		A0				

HD	SAO	Star Name	α 2000	δ 2000	μ(α)	μ(δ)	V	B-V	M$_v$	Spec	RV	d(pc)	ADS	Notes
85951	155588		9^{h}54^{m}52.1^s	-19°00′34″	-0.004^s	-0.04″	4.94	1.57	-0.5	M1 III	+50	120 s		
88705	258556		9 54 53.4	-86 54 37	-0.034	+0.01	7.2	2.0	-0.3	K5 III		170 s		
86215	250747		9 54 57.7	-60 38 27	-0.003	0.00	7.8	0.9	0.2	K0 III		330 s		
85949	155592		9 55 00.0	-16 11 37	-0.003	+0.03	6.7	0.6	4.4	G0 V		29 s	7576	m
85904	117989		9 55 00.4	-0 10 57	0.000	0.00	7.97	1.60	-0.5	M2 III	+30	490 s		
85948	155591		9 55 01.9	-12 28 02	-0.001	+0.05	7.7	0.4	2.6	F0 V		110 s		
85583	15023		9 55 03.3	+61 06 58	+0.001	0.00	6.27	1.05	0.2	K0 III	-11	150 s		m
85992	178173		9 55 03.5	-23 29 47	-0.001	0.00	7.9	1.6	-0.1	K2 III		230 s		
86099	221602		9 55 05.2	-45 44 29	-0.003	-0.01	7.7	0.9	0.4	B9.5 V		81 s		
86388	250750		9 55 05.5	-69 11 19	-0.013	+0.03	7.00	0.00	0.4	B9.5 V		100 mx		m
85843	81082		9 55 05.9	+24 38 13	+0.001	-0.13	7.2	0.5	4.0	F8 V		44 s		
86199	237497		9 55 05.9	-57 23 01	-0.003	-0.02	6.8	-0.3	0.4	B9.5 V		190 s		
86049	200866		9 55 10.5	-33 41 29	-0.005	-0.02	7.5	0.7	3.2	G5 IV		73 s		
86048	200868		9 55 11.8	-33 21 48	-0.004	-0.02	7.5	0.6	3.2	G5 IV		73 s		
86016	178179		9 55 12.1	-28 32 14	-0.001	-0.02	7.7	1.5	0.2	K0 III		160 s		
85961	155593		9 55 13.5	-11 02 03	-0.002	+0.01	7.5	1.1	-0.1	K2 III		340 s		
86404	250751		9 55 14.7	-68 49 00	+0.005	-0.11	7.7	0.6	4.4	G0 V		46 s		
86032	178181		9 55 17.5	-29 16 21	-0.001	+0.03	8.0	1.9	-0.3	K5 III		260 s		
--	--		9 55 17.6	-16 16 44			6.70							
86110	221607		9 55 30.0	-40 26 05	-0.004	0.00	7.8	0.4	0.6	A0 V		160 s		
86289	237503		9 55 33.3	-58 44 26	-0.001	+0.01	7.64	-0.08	0.6	A0 V		260 s		
85932	98846		9 55 34.2	+18 49 16	+0.001	-0.06	8.0	0.4	2.6	F0 V		120 s		
85990	137229		9 55 35.0	-1 07 35	+0.001	-0.08	7.98	1.12	0.2	gK0	+1	290 s		
85783	43102		9 55 35.4	+48 28 16	-0.001	+0.01	8.0	0.1	1.4	A2 V		210 s		
86030	155595		9 55 35.4	-17 28 09	0.000	0.00	7.0	1.6	-0.5	M2 III		320 s		
86003	137230		9 55 37.0	-4 58 34	+0.001	-0.03	7.2	1.1	-0.1	K2 III		290 s		
86774	--		9 55 39.8	-77 17 36			7.68	1.4		K5		15 mn		
86319	237506		9 55 42.0	-59 06 24	-0.002	0.00	7.23	-0.07	0.0	B8.5 V		280 s		
85795	27430	31 UMa	9 55 42.9	+49 49 11	-0.001	+0.02	5.23	0.08		A2	-6			SY UMa, q
86066	155597		9 55 43.5	-19 40 38	-0.002	0.00	7.4	-0.2	0.6	A0 V		230 s		
86029	137234		9 55 44.9	-8 50 12	-0.001	-0.01	6.6	0.1	1.7	A3 V		94 s		
86083	178190		9 55 47.7	-20 44 51	+0.012	-0.12	7.71	0.59		G0				
86288	237504		9 55 52.0	-52 58 45	-0.001	-0.01	7.9	0.1	-1.0	B5.5 V		430 s		
86002	117997		9 55 55.0	+7 40 38	0.000	-0.03	6.7	1.1	0.2	K0 III		200 s		
86195	200878		9 55 58.9	-39 33 13	-0.002	-0.07	6.9	1.2	0.2	K0 III		160 s		
86211	221617		9 56 05.4	-40 49 30	-0.001	-0.04	6.41	1.61		Ma				
85829	27432		9 56 05.8	+52 16 06	-0.005	-0.05	6.7	1.1	0.2	K0 III		200 s		
86082	137237		9 56 07.9	-7 38 43	-0.002	-0.01	6.72	1.43	-0.3	K4 III	-29	250 s		
86606	256672		9 56 09.8	-71 23 21	-0.003	0.00	6.35	-0.08		B0	-30			
86137	155604		9 56 13.5	-19 42 47	-0.002	-0.02	8.0	1.3	0.2	K0 III		250 s		
86353	237511		9 56 13.8	-53 37 11	-0.001	-0.01	6.81	-0.09	0.0	B8.5 V		230 s		
86151	178202		9 56 15.0	-25 04 50	+0.001	-0.02	8.0	0.8	-0.1	K2 III		410 s		
86193	200880		9 56 15.5	-31 05 29	-0.001	-0.01	7.1	0.3	0.6	A0 V		130 s		
86230	200883		9 56 15.9	-38 30 28	-0.004	+0.01	7.7	1.6	-0.3	K5 III		360 s		
86303	221623		9 56 18.2	-45 15 24	-0.006	0.00	7.72	1.02	0.2	K0 III		320 s		m
85974	61769		9 56 21.0	+35 19 35	-0.007	-0.09	7.7	1.1	0.2	K0 III		120 mx		
86332	221625		9 56 21.8	-47 45 15	-0.001	-0.01	7.08	0.93	3.2	G5 IV		46 s		
86352	237514		9 56 22.0	-51 20 11	0.000	-0.01	6.37	-0.17		B3	+9			
86229	200885		9 56 25.0	-32 14 35	-0.002	0.00	8.0	2.0	-0.5	M2 III		310 s		
86080	118001		9 56 25.9	+8 55 59	-0.006	+0.01	5.85	1.13	-0.1	gK2	+9	160 s		
85876	27434		9 56 26.1	+54 14 37	-0.005	-0.02	6.7	1.6	-0.5	M2 III	-32	270 s		
86350	221628		9 56 26.6	-48 24 38	-0.003	-0.03	7.9	1.9	-0.3	K5 III		240 s		
86226	178205		9 56 29.9	-24 05 59	-0.011	+0.03	7.9	0.3	4.4	G0 V		51 s		
86173	155605		9 56 31.0	-13 27 51	-0.001	-0.02	7.0	1.1	-0.1	K2 III		260 s		
86012	61771		9 56 31.2	+32 23 04	-0.004	+0.01	6.6	0.4	3.0	F2 V	+8	53 s		
86267	200889		9 56 35.4	-33 25 06	+0.002	+0.02	5.84	1.20	0.2	gK0		100 s		
86247	178211		9 56 35.9	-30 07 21	-0.002	+0.02	7.9	2.1	0.2	K0 III		77 s		
86403	237516		9 56 37.0	-54 25 09	-0.001	-0.01	7.96	0.01	0.4	B9.5 V		300 s		
86091	98854		9 56 37.3	+14 43 26	0.000	-0.14	7.6	0.9	3.2	G5 IV		73 mx		
86385	237515		9 56 38.8	-51 49 54	-0.001	-0.01	7.92	-0.10	0.6	A0 V		290 s		
86135	137239		9 56 39.0	-0 27 43	-0.001	-0.02	7.83	1.48	-0.3	gK5	-7	420 s		
86427	237519		9 56 43.5	-55 05 00	-0.002	0.00	7.89	0.18	1.7	A3 V		150 s		m
86441	237521		9 56 45.2	-57 39 17	-0.001	0.00	7.4	0.7	0.4	B9.5 V		96 s		
86266	178214		9 56 46.5	-26 33 01	-0.009	0.00	6.28	0.22		A4			7591	m
86283	178215		9 56 47.1	-29 15 59	-0.003	+0.02	7.6	0.5	0.6	A0 V		120 s		
86147	118006		9 56 48.5	+4 14 32	-0.013	-0.05	6.8	0.5	3.4	F5 V		48 s		
85345	6998		9 56 49.4	+77 56 16	-0.014	-0.03	7.4	1.1	0.2	K0 III		280 s		
86384	221631		9 56 51.6	-47 40 37	+0.001	-0.01	7.41	1.16	0.2	K0 III		220 s		
86440	237522	ø Vel	9 56 51.7	-54 34 03	-0.001	0.00	3.54	-0.08	-6.0	B5 I-II	+14	770 s		m
86301	178216		9 56 53.9	-27 28 30	-0.006	+0.03	6.32	0.17		A5				
86347	200893		9 56 58.6	-34 49 33	-0.003	+0.01	7.50	0.5	3.4	F5 V		66 s		m
86659	250761		9 56 59.7	-69 06 07	-0.004	-0.01	6.20	-0.10	-2.0	B4 IV	+20	350 pc mx		
86242	137246		9 57 01.3	-9 35 12	0.000	0.00	7.8	0.8	3.2	G5 IV		84 s		
86133	81101		9 57 02.1	+19 45 44	-0.016	-0.02	7.60	0.56	4.0	dF8	+28	50 s	7589	m
86410	221632		9 57 07.8	-46 40 42	+0.004	-0.04	7.9	1.6	-0.3	K5 III		370 mx		

HD	SAO	Star Name	α 2000	δ 2000	μ(α)	μ(δ)	V	B-V	M_v	Spec	RV	d(pc)	ADS	Notes
86466	237526		9^{h}57^{m}10.8	−52°38'20"	−0.001	0.00	6.12	−0.13	−1.7	B3 V	+17	350 s		
86439	221634		9 57 12.5	−49 52 13	−0.004	0.00	7.24	−0.01	0.0	B8.5 V		250 s		m
85945	27438		9 57 13.5	+57 25 05	+0.004	−0.06	5.93	0.89	3.2	G5 IV	−44	30 s		
86131	81104		9 57 14.7	+28 33 41	−0.002	+0.01	7.40	1.1	−0.1	K2 III		320 s		
86634	250760		9 57 15.3	−64 29 23	−0.019	+0.07	6.58	1.13		K0				
86376	200898		9 57 16.3	−32 51 29	−0.002	−0.01	7.1	0.6	1.4	A2 V		68 s		
86453	221640		9 57 21.8	−49 33 21	−0.002	+0.01	7.18	0.79	4.4	G0 V		25 s		
86188	98861		9 57 23.4	+19 17 22	+0.001	−0.01	8.0	1.1	0.2	K0 III		360 s		
85915	15030		9 57 29.6	+63 22 34	−0.008	−0.02	7.7	0.9	3.2	G5 IV		79 s		
86238	98862		9 57 33.7	+16 27 32	−0.001	−0.01	7.3	1.1	0.2	K0 III		270 s		
86572	237533		9 57 37.6	−56 51 51	+0.001	0.00	7.6	0.9	1.4	A2 V		51 s		
86371	155622		9 57 39.7	−16 31 20	+0.001	−0.03	6.7	0.4	2.6	F0 V		66 s		
86146	43115	19 LMi	9 57 41.0	+41 03 20	−0.011	−0.03	5.14	0.46	3.4	F5 V	−10	23 ts		
86523	221644		9 57 42.5	−48 24 52	−0.002	−0.02	6.05	−0.14	−1.7	B3 V nn		350 s		m
86341	137256		9 57 43.7	−1 56 31	0.000	0.00	6.70	0.9	3.2	G5 IV		50 s	7596	m
86391	178235		9 57 46.9	−21 10 56	0.000	−0.04	7.80	0.8	0.3	G7 III	0	320 s		
86675	250762		9 57 49.3	−61 55 54	−0.004	+0.02	7.06	−0.02	0.6	A0 V		200 s		
86202	61779		9 57 49.4	+35 07 29	−0.003	+0.01	7.4	1.4	−0.3	K5 III		340 s		
85582	7003		9 57 51.8	+76 36 44	−0.006	−0.03	7.9	0.8	3.2	G5 IV		87 s		
86166	43117		9 57 56.8	+45 24 51	0.000	−0.03	6.30	1.11		K0	+5			
86144	43118		9 58 01.9	+48 13 12	−0.004	0.00	7.8	0.8	3.2	G5 IV		84 s		
86313	81115		9 58 03.6	+20 58 56	−0.003	−0.05	7.70	0.5	3.4	F5 V		72 s		m
86655	237549		9 58 04.8	−58 51 40	−0.002	+0.01	6.6	1.6	−0.5	M4 III		250 s		RR Car, v
86312	81116		9 58 07.1	+22 39 51	−0.002	−0.01	7.9	0.1	1.4	A2 V		200 s		
86369	118023		9 58 07.5	+8 18 50	+0.001	−0.03	6.04	1.36	6.9	dK3	−19			
86703	250766		9 58 07.5	−62 35 07	−0.001	+0.01	7.2	0.9	3.2	G5 IV		53 s		
86421	155628		9 58 07.9	−13 07 44	−0.002	+0.01	7.4	1.4	−0.3	K5 III		350 s		
86910	256673		9 58 11.7	−72 03 59	−0.001	−0.06	7.50	0.00	0.0	B8.5 V		140 mx		m
86359	98874		9 58 11.7	+15 13 17	−0.003	−0.02	7.6	0.9	3.2	G5 IV	+17	77 s		
86360	98876	27 ν Leo	9 58 13.3	+12 26 41	−0.002	−0.02	5.26	−0.04	0.4	B9.5 V	+19	94 s		
86463	178247		9 58 14.0	−25 07 52	−0.001	0.00	6.7	1.1	1.4	A2 V		27 s		
86553	221652		9 58 14.8	−42 11 44	−0.001	−0.01	8.0	1.3	−0.1	K2 III		360 s		
86810	250768		9 58 14.9	−67 00 48	−0.003	−0.02	8.00	0.1	0.6	A0 V		290 s		m
85841	7009		9 58 22.7	+72 52 46	−0.017	−0.04	5.83	1.14	0.2	K0 III	+4	110 s		
86275	43123		9 58 25.4	+40 44 33	−0.001	−0.04	7.8	1.1	0.2	K0 III		330 s		
86358	81120		9 58 26.0	+27 45 32	−0.009	−0.04	6.4	0.4	2.6	F0 V	+36	58 s		
86481	155630		9 58 30.5	−13 56 29	+0.001	−0.03	7.4	1.4	−0.3	K5 III		350 s		
86129	27445		9 58 34.1	+58 57 06	−0.001	−0.05	6.8	1.1	−0.1	K2 III		240 s		
86540	178259		9 58 34.6	−24 20 11	−0.002	+0.01	7.9	0.3	3.4	F5 V		81 s		
86419	98880		9 58 39.4	+10 57 38	+0.002	+0.03	7.40	0.4	3.0	F2 V		76 s		m
86539	178260		9 58 40.2	−22 19 18	−0.002	0.00	7.8	1.4	0.2	K0 III		200 s		
86564	178261		9 58 43.6	−29 18 17	−0.001	−0.01	7.3	0.4	0.2	K0 III	−22	260 s		
86274	27450		9 58 44.3	+50 07 31	−0.002	−0.01	6.6	0.1	0.6	A0 V		160 s		
86444	118028		9 58 44.8	+6 06 52	+0.002	−0.03	7.9	1.1	0.2	K0 III		340 s		
86630	221656		9 58 45.0	−41 21 35	−0.002	0.00	7.4	1.2	3.2	G5 IV		70 s		
86435	98881		9 58 45.6	+10 27 32	0.000	+0.04	7.3	1.1	0.2	K0 III		260 s		
86476	118029		9 58 48.7	+4 48 31	+0.001	−0.04	7.30	1.6	−0.5	M2 III	−29	360 s		
86538	155633		9 58 51.1	−14 08 36	−0.003	0.00	7.9	0.1	1.4	A2 V		200 s		
86629	200926	η Ant	9 58 52.1	−35 53 28	−0.008	−0.02	5.23	0.31	2.6	F0 V	+30	33 s		m
86699	221658		9 58 55.4	−47 29 23	−0.005	0.00	7.55	1.45	−0.1	K2 III		340 s		
86836	250770		9 58 57.8	−62 18 59	−0.005	+0.03	7.80	0.2	2.1	A5 V		140 s		m
86715	221659		9 58 57.9	−48 29 27	−0.001	−0.04	8.01	1.10	0.2	K0 III		370 s		
86535	137267		9 59 01.9	−3 01 18	+0.002	−0.03	7.4	0.4	3.0	F2 V		77 s		
86893	250772		9 59 05.1	−65 59 01	−0.001	−0.01	7.8	0.0	0.0	B8.5 V		360 s		
86868	250771		9 59 05.6	−63 20 43	+0.004	−0.05	7.6	0.9	3.2	G5 IV		74 s		
86612	178271		9 59 06.0	−23 57 01	−0.004	+0.02	6.21	−0.10		B5				
86593	155634		9 59 08.4	−16 00 41	−0.004	+0.01	7.3	0.1	0.6	A0 V		130 mx		
86739	221665		9 59 12.3	−47 05 33	−0.002	−0.01	7.26	1.50	−0.1	K2 III		300 s		
86823	237564		9 59 13.1	−58 07 43	−0.003	−0.01	7.4	0.4	0.4	B9.5 V		130 s		
86592	155635		9 59 14.1	−12 45 13	−0.004	+0.01	7.7	0.1	0.6	A0 V		130 mx		
86460	81124		9 59 15.8	+27 31 23	−0.025	−0.07	7.77	0.58	2.8	G0 IV		54 mx		
86579	137269		9 59 18.5	−3 04 31	+0.002	−0.04	7.5	0.4	2.6	F0 V		95 s		
86626	178276		9 59 18.7	−20 21 26	−0.001	−0.03	6.9	1.5	−0.1	K2 III		150 s		
86754	221668		9 59 25.9	−44 57 21	−0.004	−0.02	6.90	0.1	0.6	A0 V		180 s		m
86516	81128		9 59 28.1	+21 19 17	−0.003	+0.03	6.6	0.1	1.4	A2 V		110 s		
86623	137272		9 59 33.4	−7 51 24	0.000	0.00	7.8	0.8	3.2	G5 IV		84 s		
86335	27462		9 59 35.3	+56 28 29	−0.001	+0.01	7.3	1.1	0.2	K0 III	+13	260 s		
86513	81129		9 59 36.2	+29 38 43	−0.007	−0.04	5.73	1.06	0.2	K0 III	−1	120 s		
86611	118041	12 Sex	9 59 43.0	+3 23 05	−0.005	+0.02	6.70	0.27	2.6	F0 V	−4	66 s		
86684	155644		9 59 44.1	−18 39 54	−0.001	−0.01	7.9	0.1	1.4	A2 V		200 s		
86712	178285		9 59 46.4	−28 20 10	−0.004	+0.02	7.9	1.9	−0.3	K5 III		240 s		
86378	27464		9 59 51.6	+56 48 42	−0.004	−0.03	5.48	1.46	−0.3	K5 III	−13	140 s		
86735	178286		9 59 52.0	−26 23 51	−0.002	+0.02	7.5	1.3	−0.3	K5 III		370 s		
86765	200944		9 59 56.5	−31 20 59	−0.002	0.00	7.3	1.4	−0.1	K2 III		210 s		
86982	250775		9 59 58.7	−60 29 47	−0.004	−0.03	7.6	0.8	2.6	F0 V		53 s		

HD	SAO	Star Name	α 2000	δ 2000	μ(α)	μ(δ)	V	B-V	M_v	Spec	RV	d(pc)	ADS	Notes
86511	43134		10ʰ00ᵐ03ˢ.3	+42°18′59″	0ˢ.000	−0″.05	7.5	1.1	−0.1	K2 III		210 mx		
86694	137276		10 00 03.9	−5 00 46	0.000	0.00	7.6	1.1	−0.1	K2 III		340 s		
87072	−−		10 00 05.8	−66 16 47			7.8	1.1	0.2	K0 III		320 s		
86819	200949		10 00 06.0	−36 02 35	−0.017	+0.05	7.5	0.3	4.4	G0 V		42 s		
86832	221681		10 00 06.4	−42 28 14	−0.002	0.00	8.0	1.1	−0.1	K2 III		410 s		
86663	118044	29 π Leo	10 00 12.7	+8 02 39	−0.002	−0.03	4.70	1.60	−0.5	M2 III	+23	110 s		
86683	118045		10 00 13.1	+6 15 02	−0.007	+0.01	7.69	0.47		F5			7604	m
86818	200951		10 00 13.7	−35 18 50	−0.001	−0.01	7.5	0.7	0.6	A0 V		85 s		
86846	200952		10 00 17.9	−36 44 25	−0.002	−0.05	8.0	1.5	3.2	G5 IV		33 s		
86997	237580		10 00 18.7	−59 00 45	+0.001	−0.13	7.96	0.52	3.8	F7 V		65 s		
87013	237585		10 00 26.4	−58 46 42	−0.001	0.00	7.85	1.57	−0.1	K2 III		210 s		
86888	221691		10 00 27.1	−42 53 13	−0.003	−0.02	7.9	0.2	1.4	A2 V		200 s		
86706	98892		10 00 31.5	+15 51 50	−0.001	−0.06	7.7	1.1	0.2	K0 III		210 mx		
87089	250776		10 00 33.7	−61 45 31	−0.014	+0.12	8.0	1.0	0.2	K0 III		120 mx		
87030	237586		10 00 34.4	−56 56 47	−0.008	+0.01	6.52	0.98		G5				m
86056	7019		10 00 36.3	+76 02 51	−0.010	+0.02	7.9	1.1	−0.1	K2 III		360 mx		
86680	81137		10 00 39.8	+28 10 26	−0.003	−0.13	7.99	0.61	4.4	G0 V		49 s		
86731	98894		10 00 41.9	+17 34 05	−0.003	+0.06	8.0	0.5	3.4	F5 V		82 s		
87971	258554	μ Cha	10 00 43.7	−82 12 52	−0.016	+0.03	5.52	0.03		A0	+16			
87109	237596		10 00 54.3	−60 13 36	−0.001	−0.02	7.63	1.26		K0				
86560	27473		10 00 55.2	+53 07 28	+0.017	−0.05	7.6	0.5	4.0	F8 V		53 s		
87285	−−		10 00 56.4	−69 47 52			8.0	0.1	0.6	A0 V		290 s		
87026	237592		10 00 58.5	−50 28 40	−0.001	0.00	6.88	−0.15	−1.1	B5 V		400 s		
86829	137290		10 00 59.1	−3 11 21	−0.002	0.00	7.2	1.4	−0.3	K5 III		320 s		
86728	61808	20 LMi	10 01 00.6	+31 55 25	−0.041	−0.43	5.36	0.66	4.7	G2 V	+56	14 ts		m
−−	250777		10 01 03.5	−60 17 36	−0.001	0.00	7.7	1.3	3.2	G5 IV		41 s		
87367	−−		10 01 06.1	−72 18 55			7.7	0.5	4.0	F8 V		55 s		
86917	178305		10 01 10.7	−23 08 48	−0.003	+0.01	7.6	0.6	3.4	F5 V		53 s		
−−	−−		10 01 11.2	−56 10 58			6.50							m
87122	237603		10 01 11.9	−56 05 47	−0.004	0.00	6.41	0.01	0.0	B8.5 V		170 s		m
86992	200965		10 01 14.6	−39 27 04	−0.003	0.00	7.1	1.0	0.4	B9.5 V		54 s		
86916	178307		10 01 15.8	−22 45 53	−0.001	+0.03	7.60	0.1	0.6	A0 V		250 s	7610	m
86778	81143		10 01 18.6	+28 47 04	+0.004	+0.01	6.9	1.1	−0.1	K2 III		260 s		
87351	256676		10 01 19.1	−70 43 40	+0.003	+0.02	7.30	0.4	3.0	F2 V		72 s		m
86826	98897		10 01 20.7	+15 40 13	0.000	−0.01	7.9	1.1	0.2	K0 III		340 s		
86913	155661		10 01 22.0	−15 27 13	−0.003	+0.01	7.7	0.1	0.6	A0 V		260 s		
86777	61813		10 01 24.7	+30 35 14	−0.006	+0.06	7.9	0.1	1.4	A2 V		97 mx		
86458	15046		10 01 32.1	+68 43 05	−0.007	−0.03	7.90	0.4	2.6	F0 V		120 s	7603	m
86776	61814		10 01 32.8	+36 30 38	−0.002	−0.07	7.9	0.7	4.4	G0 V		51 s		
86902	137294		10 01 35.3	−1 01 06	0.000	−0.01	7.2	0.1	0.6	A0 V		210 s		
86950	155667		10 01 37.5	−17 19 58	−0.002	0.00	8.0	1.1	0.2	K0 III		360 s		
86661	27476		10 01 38.0	+55 35 05	−0.022	−0.46	7.96	0.73	4.4	G8 IV-V	+23	23 mx		m
87152	237612		10 01 40.4	−53 21 52	−0.004	−0.02	6.20	−0.13		B5 n	+12			
87241	237617		10 01 42.0	−60 06 45	+0.001	+0.01	7.85	−0.01		B9				
87019	178316		10 01 44.2	−23 48 24	−0.003	−0.01	6.9	1.5	3.2	G5 IV		19 s		
87102	221706		10 01 48.8	−42 29 10	−0.002	−0.03	7.4	1.3	0.2	K0 III		270 s		
86989	155668		10 01 51.6	−12 53 56	−0.001	−0.01	7.9	1.1	0.2	K0 III		350 s		
86746	27479		10 01 51.9	+50 28 50	+0.004	−0.01	7.9	0.4	3.0	F2 V		95 s		
86898	98906		10 01 55.8	+12 14 45	−0.002	−0.01	7.80	1.04		K0				
87221	237619		10 01 55.9	−55 00 12	−0.003	0.00	7.91	−0.04	0.0	B8.5 V		370 s		
87017	155670		10 01 58.0	−14 23 33	−0.001	0.00	7.8	1.4	−0.3	K5 III		420 s		
87238	237621		10 01 58.0	−57 20 58	−0.005	+0.03	6.20	1.12		K0				
86322	7031		10 01 59.2	+74 45 32	−0.014	−0.04	6.90	1.05	0.0	K1 III	+6	240 s		
87283	250782		10 02 00.0	−60 25 15	−0.002	0.00	5.93	0.27		F0				
87099	200982		10 02 00.2	−34 10 25	−0.006	+0.02	7.2	0.7	2.1	A5 V		49 s		
87018	155671		10 02 00.5	−17 59 32	+0.001	+0.01	8.0	1.1	−0.1	K2 III		410 s		
87036	155672		10 02 05.0	−15 01 44	−0.001	−0.02	7.60	0.1	0.6	A0 V		250 s	7615	m
87254	237622		10 02 05.8	−54 58 55	+0.007	−0.03	7.78	1.17	0.2	K0 III		240 s		m
86873	61820		10 02 06.0	+31 31 52	+0.003	−0.05	7.5	0.8	3.2	G5 IV		73 s		
87016	155673		10 02 10.1	−10 25 51	−0.002	−0.02	7.5	1.1	0.2	K0 III		290 s		
87408	250785		10 02 16.7	−65 55 05	0.000	0.00	7.4	0.1	0.0	B8.5 V		240 s		
86791	27487		10 02 27.6	+57 17 00	+0.002	−0.04	8.0	0.5	3.4	F5 V		82 s		
86986	98914		10 02 29.5	+14 33 24	+0.010	−0.23	7.99	0.12	1.2	A1 V	+13	200 s		
87265	221720		10 02 32.3	−48 51 44	−0.004	−0.01	7.8	−0.1	−1.0	B5.5 V		160 mx		
87096	155680		10 02 33.1	−13 17 47	−0.010	+0.02	7.0	0.5	4.0	F8 V	+16	40 s		
87295	237627		10 02 33.6	−53 08 54	−0.002	0.00	7.69	−0.05	−1.6	B3.5 V		610 s		
87323	237629		10 02 34.9	−56 15 25	−0.001	−0.01	7.7	0.5	3.2	G5 IV		80 s		
87130	178336		10 02 37.0	−20 25 25	+0.002	−0.01	6.9	0.4	0.6	A0 V		110 s		
87163	178338		10 02 47.6	−22 06 01	−0.001	0.00	7.8	1.9	−0.3	K5 III		230 s		
87015	81154		10 02 48.9	+21 56 57	−0.001	−0.01	5.66	−0.19	−2.5	B2 V	+3	430 s		
87095	137312		10 02 49.2	−1 04 04	0.000	−0.09	7.0	1.1	0.2	K0 III		160 mx		
87199	200994		10 02 49.2	−30 34 39	−0.002	+0.02	6.54	1.19		K0				
87438	250789		10 02 49.4	−62 09 23	−0.001	+0.01	6.42	1.72		K5				
86871	27490		10 02 49.6	+49 52 42	+0.001	−0.04	7.50	0.9	0.3	G5 III	−11	280 s		m
86966	61825		10 02 55.9	+37 53 35	−0.001	−0.01	8.0	0.5	3.4	F5 V		82 s		

HD	SAO	Star Name	α 2000	δ 2000	μ(α)	μ(δ)	V	B-V	M_v	Spec	RV	d(pc)	ADS	Notes
87144	137315	.	10^{h}02^{m}56^s.3	−5°37′04″	−0^s.002	−0″.01	7.4	1.4	−0.3	K5 III		350 s		
86677	15054		10 02 56.4	+68 47 11	−0.005	−0.01	8.00	0.5	3.4	F5 V		83 s	7611	m
87436	237640		10 02 59.9	−60 10 43	−0.001	0.00	6.18	0.17		A5				
87293	201002		10 03 05.4	−39 25 27	−0.004	−0.01	7.4	0.7	0.2	K0 III		280 s		
86944	43156		10 03 05.5	+46 11 46	+0.001	−0.04	7.8	1.1	0.2	K0 III		340 s		
87161	137318		10 03 05.8	−8 53 12	−0.002	−0.01	8.04	0.67	3.0	G2 IV	+1	99 s		
87458	237645		10 03 08.3	−59 50 27	0.000	+0.01	7.74	−0.03		A0				
87364	237638		10 03 10.6	−52 02 46	−0.003	+0.03	7.36	0.19	1.4	A2 V		130 s		m
86884	27491		10 03 11.8	+58 05 51	−0.022	−0.15	7.4	0.5	4.0	F8 V		49 s		
87230	178348		10 03 12.4	−20 55 52	−0.006	−0.04	8.0	0.4	0.2	K0 III		180 mx		
86942	27494		10 03 13.7	+50 06 33	−0.002	−0.01	7.00	1.1	−0.1	K2 III		260 s		m
87334	221727		10 03 17.2	−43 50 26	−0.005	+0.04	7.49	1.18	0.2	K0 III		210 s		
87363	221728		10 03 20.4	−46 38 10	−0.003	−0.01	6.12	0.02		A1				
87214	155691		10 03 21.0	−10 43 39	−0.002	+0.01	7.2	1.1	0.2	K0 III		250 s		
87479	237650		10 03 21.1	−59 56 53	−0.001	0.00	7.91	1.17		K0				
88369	258560		10 03 22.3	−82 30 28	−0.040	+0.04	7.3	1.6	0.2	K0 III		120 s		
86894	27493		10 03 22.4	+56 40 03	−0.002	−0.03	8.0	0.5	3.4	F5 V		82 s		
87303	201005		10 03 23.4	−33 14 19	−0.003	−0.02	6.7	1.1	0.2	K0 III		170 s		
87034	43162		10 03 24.5	+42 00 32	−0.001	+0.01	7.3	0.5	4.0	F8 V		45 s		
87046	43164		10 03 27.7	+41 18 18	+0.004	0.00	7.7	1.6	−0.5	M2 III		350 mx		
87178	98924		10 03 33.1	+9 53 56	−0.002	−0.02	7.2	0.4	3.0	F2 V		68 s		
87543	250795		10 03 34.3	−61 53 02	0.000	0.00	6.14	−0.04		B8				m
87246	137325		10 03 38.4	−3 28 51	−0.003	0.00	7.8	1.1	0.2	K0 III		330 s		
87274	155695		10 03 38.7	−12 52 38	+0.001	−0.04	7.1	1.1	−0.1	K2 III		280 s		
87213	118081		10 03 38.7	+3 58 18	−0.001	+0.02	7.4	1.1	0.2	K0 III	+6	280 s		
87526	237655		10 03 38.9	−60 12 00	−0.002	−0.01	7.31	0.89		G5				
87319	178356		10 03 39.6	−26 54 34	−0.004	+0.01	7.3	0.8	2.6	F0 V		43 s		
87262	137326		10 03 41.0	−9 34 26	0.000	−0.02	6.12	1.66		K0				
87318	178357		10 03 41.4	−25 18 59	−0.003	+0.01	6.70	0.1		A0			7625	m
87195	98927		10 03 44.4	+18 13 12	−0.002	−0.05	8.0	0.5	4.0	F8 V		62 s		
87398	201009		10 03 44.9	−39 16 44	−0.002	+0.01	6.9	1.3	−0.3	K5 III		280 s		
87211	98929		10 03 53.0	+18 57 04	−0.011	−0.05	7.9	0.9	3.2	G5 IV		88 s		
87127	61837		10 03 53.5	+38 01 14	−0.008	−0.12	6.80	0.52	3.8	dF7	+31	38 s	7621	m
87396	201012		10 03 55.0	−34 52 23	−0.001	−0.01	8.0	1.6	0.2	K0 III		160 s		
87361	178361		10 03 56.2	−28 29 55	−0.002	+0.04	7.8	0.6	4.0	F8 V		54 s		
87158	61840		10 03 58.9	+32 38 48	−0.004	0.00	7.80	0.5	4.0	F8 V		58 s	7624	m
87541	237657		10 04 01.2	−55 54 30	−0.001	−0.01	8.01	−0.02	0.0	B8.5 V		370 s		
87344	155704		10 04 02.8	−18 06 05	−0.001	−0.01	5.86	−0.06		A0			7627	m
88351	258561		10 04 07.8	−81 33 56	−0.020	+0.05	6.60	0.92		G5				
87271	98931		10 04 08.3	+11 37 43	0.000	0.00	7.1	0.1	0.6	A0 V		200 s		
87301	118086	13 Sex	10 04 08.4	+3 12 04	−0.005	−0.10	6.45	0.40	3.1	dF3	0	46 s		
87113	43166		10 04 08.7	+48 35 30	−0.007	+0.02	7.7	0.4	3.0	F2 V		88 s		
87210	61841		10 04 16.0	+32 27 09	−0.002	−0.03	7.3	1.1	0.2	K0 III		270 s		
87416	178366		10 04 17.8	−28 22 43	−0.003	−0.03	7.28	0.47	4.0	F8 V		45 s	7629	m
87394	178365		10 04 18.8	−22 33 32	0.000	+0.01	7.8	1.6	−0.3	K5 III		360 s		
87540	221749		10 04 20.6	−48 26 44	−0.006	+0.03	6.84	1.02	3.2	G5 IV		35 s		
87427	178367		10 04 20.9	−24 17 08	−0.008	+0.02	5.70	0.30	2.4	A8 V	+4	42 s		
87477	201020		10 04 23.3	−39 58 33	−0.004	0.00	6.43	1.30		K0				
87359	137336		10 04 26.8	−7 13 06	+0.010	−0.21	7.5	0.9	3.2	G5 IV		72 s		
87580	237665		10 04 27.8	−51 48 03	−0.001	0.00	7.50							
87286	81165		10 04 33.5	+27 50 50	+0.001	0.00	7.9	0.2	2.1	A5 V		150 s		
87713	250802		10 04 34.2	−64 38 54	−0.001	−0.01	6.7	0.1	0.6	A0 V		170 s		
87425	155710		10 04 35.9	−17 56 23	+0.003	−0.04	8.0	0.1	1.4	A2 V		210 s		
87141	27503		10 04 36.3	+53 53 30	−0.003	−0.01	5.74	0.48	3.4	F5 V	−16	28 s		
87298	81168		10 04 36.7	+22 34 11	−0.002	−0.01	7.9	1.1	0.2	K0 III		340 s		
87424	155709		10 04 37.4	−11 43 48	−0.015	−0.04	7.8	1.1	0.2	K0 III		82 mx		
87537	221753		10 04 41.3	−42 13 30	−0.003	+0.01	7.9	1.0	−0.3	K5 III		430 s		
87653	237673		10 04 45.7	−54 57 01	−0.003	+0.01	7.9	0.1	0.0	B8.5 V		290 s		
87488	178378		10 04 48.9	−28 11 20	−0.003	0.00	6.9	1.1	0.4	B9.5 V		41 s		
87642	237674		10 04 53.6	−52 01 31	0.000	0.00	7.4	1.3	0.2	K0 III		190 s		
87559	201027		10 04 55.7	−35 09 05	+0.001	−0.03	7.7	2.0	−0.3	K5 III		190 s		
87641	221757		10 04 59.6	−46 56 11	+0.002	+0.02	7.88	0.91	0.2	K0 III		340 s		
87652	237675		10 05 01.7	−51 18 49	−0.004	+0.01	7.20	0.00	0.4	B9.5 V		230 s		m
87700	237677		10 05 01.8	−56 53 53	0.000	+0.01	6.9	0.8	2.6	F0 V		39 s		
87390	98936		10 05 03.4	+15 08 02	0.000	−0.05	7.7	1.1	0.2	K0 III		230 mx		
87423	118095		10 05 04.3	+7 59 28	+0.001	+0.01	7.3	0.5	3.4	F5 V		60 s		
87953	256678		10 05 06.4	−72 04 30	−0.012	−0.04	7.8	0.0	0.4	B9.5 V		140 mx		
87640	221759		10 05 06.6	−45 53 43	+0.003	−0.02	7.31	1.08	0.2	K0 III		230 s		m
87504	155713	40 υ² Hya	10 05 07.4	−13 03 53	−0.003	+0.01	4.60	−0.09	−1.2	B8 III	+28	140 s		
87588	201032		10 05 07.9	−36 40 11	−0.002	−0.04	8.0	0.4	1.4	A2 V		120 s		
87761	250805		10 05 08.2	−61 40 53	+0.001	+0.02	7.60	1.4	−0.3	K5 III		370 s		m
87243	27512		10 05 10.4	+52 22 14	−0.001	−0.03	6.2	0.1	1.4	A2 V	−25	89 s		
87556	178385		10 05 11.9	−28 11 50	−0.003	0.00	7.20	0.1	0.6	A0 V		210 s	7635	m
87486	137343		10 05 12.1	−3 31 03	−0.003	0.00	7.5	1.4	−0.3	K5 III		360 s		
87627	221760		10 05 12.7	−42 10 20	+0.005	0.00	6.5	1.5	0.2	K0 III		89 s		

HD	SAO	Star Name	α 2000	δ 2000	μ(α)	μ(δ)	V	B-V	M_v	Spec	RV	d(pc)	ADS	Notes
			h m s	° ′ ″	s	″								
87043	15067		10 05 12.9	+67 16 45	+0.002	-0.04	7.1	1.1	0.2	K0 III		240 s		
87258	27513		10 05 13.2	+50 33 04	-0.001	-0.02	7.4	1.1	-0.1	K2 III		310 s		
88278	256681		10 05 14.2	-79 03 43	-0.024	+0.07	7.8	0.1	1.4	A2 V		53 mx		
87606	201037		10 05 15.1	-36 23 02	-0.001	0.00	6.27	1.11	0.2	K0 III		130 s		
87802	250806		10 05 19.3	-62 51 07	-0.002	0.00	7.8	1.1	0.2	K0 III		320 s		
87502	137344		10 05 20.8	-3 22 28	-0.002	+0.02	7.7	1.1	-0.1	K2 III		370 s		
87639	201039		10 05 26.3	-34 52 58	-0.002	-0.01	6.6	0.9	0.2	K0 III		190 s		
87638	201040		10 05 32.8	-33 23 35	-0.007	-0.05	6.98	0.31	1.9	F3 IV		84 mx		
87312	27516		10 05 35.5	+51 23 27	-0.001	-0.02	7.9	0.1	1.4	A2 V		200 s		
87422	61852		10 05 36.5	+34 41 24	-0.002	-0.02	8.0	1.1	0.2	K0 III		360 s		
88948	258565		10 05 36.9	-84 05 19	-0.081	+0.01	7.80	0.8	3.2	G5 IV		83 s		m
87500	98944		10 05 40.9	+15 45 27	-0.005	-0.02	6.37	0.37	2.6	F0 V	+12	50 s		
87411	43178		10 05 43.5	+41 02 43	-0.002	0.00	8.00	0.4	2.6	F0 V		120 s	7631	m
87846	250808		10 05 44.2	-61 10 20	-0.002	+0.01	7.06	1.44	-0.3	K5 III		300 s		
87660	201045		10 05 45.2	-30 53 30	-0.003	-0.02	6.60	1.1	0.2	K0 III		190 s		m
87357	27517		10 05 47.0	+50 55 10	-0.005	+0.01	7.8	0.5	3.4	F5 V		77 s		
87462	61854		10 05 48.9	+35 00 08	-0.001	-0.04	7.4	1.1	0.2	K0 III		270 s		
87207	15074		10 05 50.8	+64 24 48	-0.010	-0.02	7.7	0.4	3.0	F2 V		87 s		
87969	250813		10 05 51.7	-68 17 54	-0.009	+0.01	6.9	1.1	0.2	K0 III		210 s		
87659	178397		10 05 52.3	-27 16 31	-0.003	+0.01	7.7	1.3	-0.1	K2 III		310 s		
87896	250811		10 05 53.0	-63 50 06	0.000	-0.02	7.2	0.8	3.2	G5 IV		62 s		
87442	61855		10 05 55.8	+39 34 56	-0.006	-0.02	7.30	0.25		A5			7633	m
87800	237687		10 05 56.3	-54 35 33	-0.004	0.00	7.85	-0.07	-1.0	B5.5 V		140 mx		
87749	221771		10 06 04.0	-42 31 30	-0.003	-0.01	7.8	-0.1	3.0	F2 V		89 s		
87685	178405		10 06 05.4	-29 07 09	-0.002	+0.01	7.3	1.3	0.2	K0 III		180 s		
87816	237690		10 06 07.0	-52 11 16	-0.008	+0.01	6.52	0.98	0.2	K0 III		180 s		R Vel, q
87672	178403		10 06 08.7	-20 56 56	0.000	-0.03	7.4	1.8	-0.3	K5 III		220 s		
87550	81181		10 06 10.7	+21 30 52	-0.001	+0.03	7.0	1.1	0.2	K0 III		230 s		
87783	221773		10 06 11.3	-47 22 12	+0.001	-0.06	5.08	0.88	3.2	K0 IV	+20	24 s		m
87726	178414		10 06 25.5	-26 35 52	-0.001	0.00	7.8	0.8	0.2	K0 III		330 s		
87421	27518		10 06 28.5	+55 46 10	-0.001	+0.01	7.3	1.1	0.2	K0 III		260 s		
87531	61863		10 06 32.3	+39 30 15	-0.001	0.00	7.6	1.1	0.2	K0 III		300 s		
87843	221776		10 06 34.7	-45 22 36	-0.009	+0.04	7.67	0.97	3.2	G5 IV		78 s		
87647	98949		10 06 37.2	+12 46 54	0.000	-0.03	7.4	0.2	2.1	A5 V		120 s		
87756	--		10 06 40.0	-23 08 09			7.70	0.5	3.4	F5 V		72 s	7642	m
87646	98950		10 06 40.8	+17 53 42	-0.005	-0.03	7.9	0.7	2.9	G1 IV	+21	100 s		
87682	118111	14 Sex	10 06 47.3	+5 36 41	-0.002	-0.02	6.21	0.94	0.3	gG6	+17	140 s		
87793	178425		10 06 50.5	-24 42 56	-0.002	-0.04	7.90	0.5	4.0	F8 V		60 s	7644	m
87909	221780		10 06 51.4	-49 57 59	-0.002	+0.01	7.9	1.2	-0.3	K5 III		430 s		
87910	237700		10 06 52.0	-51 04 40	-0.002	+0.01	7.1	1.1	0.6	A0 V		46 s		
87768	155734		10 06 53.7	-17 26 38	-0.003	+0.01	6.6	0.4	2.6	F0 V		65 s		
87742	137355		10 06 58.8	-3 17 03	-0.002	-0.01	7.6	0.1	0.6	A0 V		260 s		
88189	256683		10 06 59.8	-71 28 42	-0.009	-0.04	6.6	0.5	4.0	F8 V		32 s		
88031	250818		10 07 03.3	-61 12 52	+0.004	0.00	7.3	0.1	0.6	A0 V		180 s		
87355	15082		10 07 03.9	+66 18 55	-0.004	0.00	7.9	1.4	-0.3	K5 III		430 s		
87764	137357		10 07 05.7	-5 34 00	-0.004	-0.02	7.9	1.1	0.2	K0 III		310 mx		
87828	178432		10 07 06.6	-28 52 02	-0.001	+0.02	7.61	1.54		K3		17 mn		
87810	178431		10 07 07.6	-21 15 21	+0.005	0.00	6.8	0.5	3.0	F2 V		47 s		
87600	43197		10 07 07.8	+45 03 36	-0.002	0.00	7.5	0.4	3.0	F2 V		81 s		
87986	237707		10 07 08.0	-54 33 49	-0.006	+0.01	7.60	0.08	0.6	A0 V		180 mx		m
87667	81186		10 07 08.3	+29 30 49	-0.003	-0.01	7.4	0.5	3.4	F5 V		63 s		
87966	237705		10 07 08.6	-52 32 03	-0.016	+0.03	7.90	0.4	3.0	F2 V		95 s		m
87808	155739		10 07 09.5	-17 08 29	+0.002	-0.05	5.60	1.49	-0.3	gK5	+11	150 s		
87918	221784		10 07 12.0	-43 10 38	-0.007	+0.02	7.9	1.8	0.2	K0 III		210 s		
87481	15086		10 07 14.4	+60 55 13	+0.003	+0.02	7.40	1.4	-0.3	K5 III	+33	350 s		
87680	81189		10 07 14.8	+29 14 17	+0.001	+0.04	7.96	0.67	4.7	G2 V		42 s		
87827	155743		10 07 19.8	-15 27 18	-0.002	+0.01	7.9	0.4	2.6	F0 V		120 s		
87737	98955	30 η Leo	10 07 19.9	+16 45 45	0.000	-0.01	3.52	-0.03	-5.2	A0 Ib	+3	560 s		m
87666	61873		10 07 21.6	+38 17 34	-0.014	-0.10	8.0	1.1	0.2	K0 III		91 mx		
87840	155744		10 07 22.2	-19 42 31	-0.002	+0.02	7.80	0.1	0.6	A0 V		280 s	7647	m
87807	137360		10 07 23.8	-7 43 29	-0.002	+0.01	7.5	0.1	1.4	A2 V		170 s		
87696	61874	21 LMi	10 07 25.7	+35 14 41	+0.004	0.00	4.48	0.18	2.4	A7 V	-18	26 s		
87297	7049		10 07 27.2	+72 36 39	-0.008	-0.06	7.9	1.1	0.2	K0 III		210 mx		
87946	221788		10 07 28.1	-42 57 35	0.000	-0.01	7.77	1.50	-0.3	K5 III		410 s		
87870	178437		10 07 32.0	-22 29 23	-0.001	-0.01	7.02	1.52	-0.5	gM4	+30	320 s		
87806	118120		10 07 33.3	+0 55 05	-0.001	-0.02	6.9	1.6	-0.5	M2 III	-30	300 s		
87838	137363		10 07 33.7	-6 26 21	-0.025	+0.01	7.6	0.7	4.4	G0 V		40 mx		
88015	221792		10 07 35.7	-48 15 37	-0.001	+0.01	6.38	-0.16	-2.9	B3 III		710 s		
87776	98960		10 07 39.2	+15 09 28	-0.004	-0.09	7.2	0.5	4.0	F8 V		44 s		
88092	237718		10 07 44.1	-58 22 21	-0.002	0.00	7.61	0.82		G0				
87855	137364		10 07 45.8	-7 37 53	-0.002	-0.04	6.62	1.61	-0.5	gM2	+32	260 s		
87805	98962		10 07 50.1	+18 30 47	-0.001	-0.02	7.8	1.1	0.2	K0 III		330 s		
88308	--		10 07 51.7	-71 16 22			8.00	1.1	-0.1	K2 III		420 s		m
87509	15088		10 07 54.1	+66 50 16	-0.021	-0.04	7.9	0.7	4.4	G0 V		51 s		
87837	98964	31 Leo	10 07 54.2	+9 59 51	-0.006	-0.07	4.37	1.45	-0.3	K4 III	+41	81 s	7649	m

HD	SAO	Star Name	α 2000	δ 2000	μ(α)	μ(δ)	V	B−V	M$_v$	Spec	RV	d(pc)	ADS	Notes
87887	137366	15 α Sex	10^h 07^m 56.2^s	−0° 22′ 18″	−0.001^s	−0.01″	4.49	−0.04	−1.1	B5 V	+7	100 s		
88158	250826		10 07 56.4	−62 13 17	−0.004	0.00	6.45	−0.10	0.0	B8.5 V		200 s		
88013	201081		10 08 01.7	−37 20 01	−0.001	0.00	6.36	0.98		K0				
87999	201080		10 08 01.9	−32 03 32	−0.001	+0.01	7.6	0.3	1.4	A2 V		120 s		
87441	7054		10 08 06.2	+69 56 00	0.000	+0.01	8.0	0.4	2.6	F0 V		120 s		
87570	15090		10 08 12.1	+66 01 40	−0.002	0.00	7.8	1.4	−0.3	K5 III		410 s		
88210	250828		10 08 14.0	−61 45 39	−0.003	+0.01	7.96	1.04	0.2	K0 III		340 s		
87822	61882		10 08 15.8	+31 36 15	−0.007	−0.09	6.24	0.45		F5	−8		7651	m
88176	237722		10 08 18.2	−58 11 55	0.000	−0.02	7.0	1.0	4.4	G0 V		18 s		
87998	155757		10 08 21.5	−19 45 19	−0.009	−0.33	7.28	0.61	4.7	G2 V	+12	33 s	7655	m
87901	98967	32 α Leo	10 08 22.2	+11 58 02	−0.017	0.00	1.35	−0.11	−0.6	B7 V	+4	26 ts	7654	Regulus, m
88050	201090		10 08 22.8	−35 00 49	−0.004	0.00	7.2	0.9	1.4	A2 V		48 s		
88174	237726		10 08 26.3	−56 26 12	−0.005	−0.01	7.2	1.2	0.2	K0 III		200 s		
87866	81202		10 08 27.8	+24 18 31	−0.004	−0.01	7.7	0.4	2.6	F0 V		110 s		
87836	61885		10 08 29.7	+33 30 51	+0.011	−0.18	7.5	0.8	3.2	G5 IV		62 mx		
87975	137372		10 08 31.3	−3 28 42	−0.001	−0.01	7.8	1.1	0.2	K0 III		320 s		
86321	1637		10 08 34.3	+83 55 06	−0.002	+0.01	6.4	1.1	0.2	K0 III	−12	180 s		
88025	155763		10 08 35.4	−15 36 42	−0.002	+0.01	6.26	0.01		A0				
88066	201093		10 08 37.2	−31 06 10	−0.001	−0.01	7.9	0.9	4.4	G0 V		33 s		
88141	221803		10 08 39.2	−47 18 36	−0.001	−0.02	7.9	0.8	−0.1	K2 III		390 s		
87974	118131		10 08 41.2	+1 09 38	0.000	+0.01	6.6	0.4	2.6	F0 V		62 s		
88322	250833		10 08 42.0	−65 30 41	0.000	−0.01	7.4	−0.2	0.0	B8.5 V		310 s		
88226	237733		10 08 42.5	−56 25 15	+0.001	−0.04	7.9	0.8	0.2	K0 III		340 s		
88323	250836		10 08 42.7	−65 48 55	−0.011	+0.04	5.28	0.98	0.3	gG7	0	91 s		m
87883	61890		10 08 43.1	+34 14 32	−0.005	−0.06	7.6	1.1	0.2	K0 III		170 mx		
87914	81204		10 08 44.3	+23 28 50	+0.001	+0.02	7.7	0.4	2.6	F0 V		100 s		
88024	155765		10 08 45.6	−10 53 05	0.000	−0.01	6.53	0.02		A0				
88088	178467		10 08 52.7	−29 45 01	−0.005	−0.01	7.9	0.9	3.4	F5 V		41 s		
88293	250835		10 08 55.4	−61 11 32	+0.001	+0.01	7.49	1.08	−0.3	K5 III		360 s		
88206	237736		10 08 56.2	−51 48 40	−0.001	0.00	4.86	−0.12	−1.6	B5 IV	+23	200 s		
88205	221805		10 08 56.8	−48 51 21	−0.007	+0.01	7.9	0.3	3.0	F2 V		94 s		
88224	221808		10 08 59.9	−49 46 59	−0.015	+0.02	7.8	0.6	3.4	F5 V		74 s		
88076	155768		10 09 02.6	−13 11 46	+0.005	−0.08	7.9	1.1	0.2	K0 III		200 mx		
88223	221809		10 09 05.6	−48 18 24	−0.007	+0.04	6.9	0.3	2.1	A5 V		71 s		
88185	221807		10 09 06.6	−43 41 43	−0.001	0.00	7.27	1.09	3.2	G5 IV		37 s		
88292	237742		10 09 07.0	−58 02 26	−0.001	0.00	7.50	−0.06	0.0	B8.5 V		320 s		
87734	15098		10 09 07.7	+63 57 09	+0.002	0.00	6.7	1.4	−0.3	K5 III		250 s		
88084	155769		10 09 08.1	−15 29 39	−0.007	−0.20	7.50	0.9	3.2	G5 IV		63 mx		m
87852	27536		10 09 08.2	+50 49 16	0.000	−0.01	7.7	0.1	1.4	A2 V	−11	180 s		
88009	98969		10 09 08.4	+18 31 50	+0.004	−0.07	7.1	0.9	3.2	G5 IV		61 s		
87582	7058		10 09 10.0	+71 41 17	+0.003	−0.02	7.6	1.1	0.2	K0 III		310 s		
87994	81206		10 09 10.3	+21 47 25	−0.003	−0.01	8.00	0.1	0.6	A0 V		300 s		m
88048	118135		10 09 14.8	+6 10 15	−0.001	−0.01	6.8	1.1	0.2	K0 III	+26	210 s		
88109	155771		10 09 15.5	−18 08 20	−0.001	−0.03	7.2	0.4	3.0	F2 V		68 s		
88263	237741		10 09 18.0	−50 38 22	−0.001	+0.01	7.89	0.09	−1.6	B3.5 V		580 s		
88204	201104		10 09 19.0	−39 50 08	−0.004	0.00	8.0	1.0	0.2	K0 III		360 s		
87955	61900		10 09 19.1	+38 25 50	−0.002	−0.03	7.75	1.62	−0.5	M2 III	−32	430 s		
88022	81208		10 09 19.6	+20 19 56	−0.002	−0.03	6.7	0.1	1.4	A2 V	+10	110 s	7662	m
88366	250840		10 09 21.9	−61 32 56	−0.013	+0.07	5.3	1.6		M5 III	+289			S Car, v
88108	155772		10 09 22.6	−13 21 45	−0.004	−0.01	7.30	1.65	−0.5	M2 III	−12	270 mx		
88083	137376		10 09 22.7	−1 40 06	+0.001	−0.01	7.8	1.4	−0.3	K5 III		410 s		
88072	118136		10 09 23.3	+2 22 15	−0.014	−0.02	7.6	0.7	4.4	G0 V		44 s		
88459	250843		10 09 27.6	−68 18 17	+0.002	−0.02	8.0	1.1	−0.1	K2 III		400 s		
88473	250844		10 09 30.5	−68 41 00	−0.001	+0.01	5.81	0.02		A0				m
87703	15100		10 09 31.1	+68 39 52	0.000	−0.02	7.2	1.1	0.2	K0 III		250 s		
88071	118138		10 09 31.3	+9 35 34	−0.003	0.00	7.4	1.6	−0.5	M2 III	+6	390 s		
88201	201107		10 09 31.6	−32 50 48	−0.017	+0.03	7.44	0.56	2.8	G0 IV		75 mx		
88218	201109		10 09 31.7	−35 51 25	−0.036	0.00	6.13	0.60	4.4	dG0	+41	24 ts		m
85013	1628		10 09 32.9	+86 34 35	−0.032	−0.04	8.0	0.9	3.2	G5 IV		91 s		
88135	155773		10 09 37.9	−11 53 43	+0.003	−0.04	7.5	1.1	−0.1	K2 III		330 s		
88321	237747		10 09 41.9	−52 52 33	−0.003	−0.01	7.6	0.8	0.0	B8.5 V		110 s		
88166	155774		10 09 43.1	−14 45 45	−0.002	−0.01	7.8	1.6	−0.5	M2 III		470 s		
88262	201114		10 09 45.4	−37 43 58	−0.001	−0.04	8.0	0.1	0.6	A0 V		300 s		R Ant, q
88304	221816		10 09 48.1	−47 26 47	−0.003	0.00	7.2	1.1	0.2	K0 III		230 s		
88384	237749		10 09 49.4	−56 40 13	−0.001	−0.01	7.91	0.97	3.2	G5 IV		63 s		
88303	221817		10 09 51.5	−45 11 41	0.000	0.00	7.36	1.43	−0.1	K2 III		200 s		
87834	15103		10 09 55.9	+63 45 17	−0.004	+0.04	7.4	1.1	0.2	K0 III		270 s		
88182	155777		10 09 56.4	−12 05 45	−0.001	−0.04	6.24	0.18		A2				
88510	—		10 10 03.0	−66 35 29			7.3	1.1	0.2	K0 III		260 s		
88215	155780		10 10 05.8	−12 48 58	−0.009	−0.12	5.31	0.36	3.4	F5 V	+23	24 s		
88195	137385	17 Sex	10 10 07.5	−8 24 30	−0.002	−0.01	5.91	0.02		A0				
88349	221822		10 10 08.6	−45 47 11	−0.001	−0.01	8.00	0.4	3.0	F2 V		99 s		m
88287	201122		10 10 12.7	−30 53 53	+0.002	0.00	7.8	1.2	−0.1	K2 III		360 s		
88382	221826		10 10 14.3	−48 00 51	−0.004	−0.03	6.7	1.1	0.2	K0 III		190 s		
88046	43216		10 10 16.8	+49 30 26	−0.009	−0.05	7.3	0.4	3.0	F2 V		71 s		

HD	SAO	Star Name	α 2000	δ 2000	μ(α)	μ(δ)	V	B-V	M_v	Spec	RV	d(pc)	ADS	Notes
			$10^h 10^m 21^s.2$	$-67°15'48''$	$-0''.001$	$+0''.01$								
88572	250851		10 10 21.2	-67 15 48	-0.001	+0.01	6.7	0.0	0.4	B9.5 V		180 s		
87925	15105		10 10 22.3	+62 54 55	-0.005	-0.18	7.1	1.1	0.2	K0 III		65 mx		
88162	81215		10 10 22.6	+20 56 43	-0.001	0.00	7.3	0.4	2.6	F0 V		88 s		
88528	250849		10 10 24.3	-64 23 03	-0.002	+0.02	6.91	-0.02	0.6	A0 V		180 s		
88591	250853		10 10 30.1	-67 50 28	-0.002	0.00	7.0	0.0	0.0	B8.5 V		250 s		
88400	221831		10 10 30.8	-45 18 27	-0.006	+0.02	7.3	0.2	1.4	A2 V		120 mx		
88300	155786		10 10 34.4	-18 36 14	0.000	-0.03	7.1	0.4	2.6	F0 V		80 s		
88284	155785	41 λ Hya	10 10 35.2	-12 21 15	-0.014	-0.09	3.61	1.01	0.2	K0 III	+19	46 s	7671	m
88179	81219		10 10 36.6	+23 52 53	0.000	-0.02	7.90	0.9	3.2	G5 IV		87 s	7669	m
88399	221832		10 10 37.8	-41 42 54	+0.004	-0.12	5.98	1.24		K0				
88398	201129		10 10 41.5	-38 54 41	-0.006	-0.01	6.8	0.8	2.6	F0 V		34 s		
88233	98984		10 10 42.9	+15 42 15	-0.007	-0.07	7.8	0.4	2.6	F0 V		110 mx		
88375	178505		10 10 52.0	-25 26 18	+0.002	-0.04	7.6	1.2	-0.5	M2 III		420 s		
88346	155791		10 10 52.2	-15 42 33	0.000	0.00	7.1	1.1	-0.1	K2 III		270 s		
88333	137395	18 Sex	10 10 55.7	-8 25 06	-0.001	-0.05	5.65	1.30	-0.1	gK2	0	110 s		
88542	237764		10 10 55.8	-59 09 15	-0.002	+0.01	7.44	0.06	0.6	A0 V		220 s		
88191	61911		10 10 55.9	+36 28 05	0.000	0.00	7.4	1.1	0.2	K0 III		270 s		
88359	155793		10 10 57.1	-16 55 52	-0.003	-0.02	7.2	1.1	0.2	K0 III		250 s		
88161	43220		10 10 58.8	+40 39 41	-0.001	-0.01	6.32	1.25	0.2	K0 III	+14	120 s		
88332	137396		10 11 01.3	-2 24 54	0.000	0.00	7.7	0.9	3.2	G5 IV		80 s		
87897	7067		10 11 05.9	+70 50 21	-0.007	-0.05	7.6	1.1	0.2	K0 III		230 mx		
88315	118150		10 11 06.1	+5 15 08	-0.003	+0.01	7.8	1.1	0.2	K0 III		330 s		
88270	81220		10 11 08.5	+20 41 47	+0.009	-0.13	6.6	0.4	2.6	F0 V		65 s		
88331	118151		10 11 11.2	+6 10 03	+0.001	-0.01	7.9	1.1	0.2	K0 III		340 s		
88721	--		10 11 11.4	-69 39 35			8.0	1.1	-0.1	K2 III		400 s		
88231	61914		10 11 12.6	+37 24 07	-0.003	-0.03	5.85	1.29	0.2	K0 III	+9	90 s		
88704	250858		10 11 12.7	-67 53 32	-0.006	0.00	7.2	1.1	0.2	K0 III		240 s		
88624	250857		10 11 14.5	-61 46 06	-0.001	-0.01	7.47	1.07	0.2	K0 III		260 s		
88407	155799		10 11 17.4	-17 38 09	-0.002	-0.01	7.7	0.9	3.2	G5 IV		80 s		
88372	137400		10 11 17.7	-7 18 59	0.000	-0.01	6.25	0.01		A0 n	+13			
88419	155800		10 11 19.1	-18 57 26	+0.003	-0.09	6.9	1.6	-0.5	M2 III	+37	160 mx		
88230	43223		10 11 22.0	+49 27 15	-0.140	-0.51	6.61	1.37	8.3	K7 V	-27	4.5 t		Groombridge 1618, m
88269	61918		10 11 22.2	+32 11 30	-0.004	-0.08	8.0	1.1	0.2	K0 III		130 mx		
88556	237769		10 11 23.9	-50 29 27	+0.001	-0.02	7.88	-0.10	-1.6	B3.5 V		560 s		
87850	7069		10 11 30.7	+73 53 33	+0.009	0.00	7.7	0.1	1.4	A2 V		180 s		
88247	43226		10 11 32.5	+41 43 55	+0.001	-0.03	7.8	1.1	-0.1	K2 III		380 s		
88647	237773		10 11 35.2	-58 49 40	-0.006	+0.02	6.40	1.66		Mb				
88356	98990		10 11 35.3	+12 02 10	0.000	-0.02	8.0	1.4	-0.3	K5 III		470 s		
88355	98991	34 Leo	10 11 38.2	+13 21 18	+0.003	-0.04	6.44	0.46	3.7	F6 V	-16	35 s	7674	m
88661	237776		10 11 46.3	-58 03 38	-0.003	0.00	5.72	-0.08	-3.0	B2 IV pne		530 s		
88603	221851		10 11 47.9	-50 09 23	-0.002	+0.01	7.94	-0.09	-1.2	B8 III		670 s		
88718	250864		10 11 48.2	-63 49 22	-0.001	-0.02	8.0	0.1	1.4	A2 V		200 s		
88524	201152		10 11 48.8	-34 20 02	+0.006	-0.04	6.8	0.6	3.0	F2 V		41 s		
88688	237777		10 11 49.2	-59 09 17	-0.001	0.00	7.6	0.6	1.4	A2 V		85 s		
88695	250863		10 11 49.8	-61 32 30	-0.003	0.00	7.17	-0.08	0.0	B8.5 V		270 s		
88569	221849		10 11 51.6	-40 52 29	-0.002	0.00	6.9	1.2	0.2	K0 III		180 s		
88539	201156		10 11 53.8	-35 19 28	0.000	0.00	6.64	2.27		N7.7	+4			
88590	221852		10 11 54.7	-44 43 44	-0.007	+0.04	7.10	0.91	3.2	G5 IV		60 s		
88568	201158		10 11 56.9	-39 59 39	-0.003	-0.01	7.4	0.9	0.2	K0 III		280 s		
88660	237778		10 11 58.7	-54 59 06	+0.002	-0.01	7.7	0.2	0.4	B9.5 V		200 s		
88403	81225		10 12 02.9	+20 07 08	+0.004	0.00	8.00	0.9	3.2	G5 IV		91 s		m
88522	178526		10 12 02.9	-28 36 22	-0.002	-0.02	6.28	0.01		A0			7681	m
88565	201161		10 12 09.6	-32 43 51	-0.002	-0.02	8.0	2.3	-0.5	M2 III		200 s		
88716	237783		10 12 10.1	-59 36 11	-0.003	0.00	7.9	0.7	4.4	G0 V		42 s		
88784	250865		10 12 11.3	-66 10 54	-0.001	+0.01	7.7	1.4	-0.3	K5 III		390 s		
88519	155808		10 12 13.7	-16 10 43	0.000	0.00	8.0	1.6	-0.5	M2 III		510 s		
88446	98999		10 12 19.0	+17 17 57	-0.012	-0.24	7.85	0.55		F8				
88693	237784		10 12 22.9	-52 09 47	-0.005	+0.04	6.16	1.17	0.2	gK0		120 s		
88549	155813		10 12 23.7	-15 04 04	-0.003	0.00	7.0	0.1	0.6	A0 V		190 s		
88827	250867		10 12 25.8	-66 17 13	-0.002	0.00	7.9	0.2	2.1	A5 V		140 s		
88619	201167		10 12 26.0	-32 41 23	-0.005	-0.01	7.3	1.2	0.2	K0 III		200 s		
89081	256689		10 12 29.3	-76 04 37	+0.010	0.00	7.8	0.4	2.6	F0 V		110 s		
88846	--		10 12 29.3	-66 40 39			7.9	1.4	-0.3	K5 III		420 s		
88595	155820		10 12 37.8	-19 09 13	-0.017	-0.12	6.44	0.50	4.0	dF8	+34	31 s		
88279	15116		10 12 38.5	+60 32 00	-0.001	+0.05	7.90	0.5	4.0	F8 V		60 s	7676	m
88476	81233		10 12 42.4	+28 14 31	-0.002	-0.02	6.9	0.9	0.3	G8 III	+5	210 s		
88864	250869		10 12 43.9	-64 23 41	-0.022	+0.03	7.89	0.64	4.5	G1 V		45 s		
88692	221860		10 12 46.0	-44 00 32	-0.004	+0.01	7.82	1.38	-0.1	K2 III		260 s		
88643	178539		10 12 47.6	-29 29 56	-0.005	+0.03	7.4	1.1	3.2	G5 IV		45 s		
88547	118164	19 Sex	10 12 48.3	+4 36 53	-0.004	-0.01	5.77	1.18	0.2	gK0	+32	96 s		
88773	237793		10 12 48.3	-56 09 03	0.000	0.00	7.96	0.01	0.0	B8.5 V		330 s		
88670	201178		10 12 50.5	-32 30 55	0.000	+0.01	7.1	1.3	0.2	K0 III		150 s		
88756	237794		10 12 52.4	-54 34 25	0.000	0.00	7.90	1.33	-0.1	K2 III		330 s		
87386	1649		10 12 55.1	+82 24 04	+0.001	-0.02	7.60	1.1	0.2	K0 III	-35	300 s		
88402	27558		10 12 56.0	+51 28 32	+0.003	-0.06	7.6	0.7	4.4	G0 V		44 s		

HD	SAO	Star Name	α 2000	δ 2000	μ(α)	μ(δ)	V	B-V	M_V	Spec	RV	d(pc)	ADS	Notes
88894	250872		10^h 12^m 57^s.7	-65° 10' 00"	0^s.000	-0".01	7.10	0.00	0.4	B9.5 V		220 s		m
88655	155826		10 13 00.8	-19 41 21	-0.002	+0.01	7.0	1.5	-0.3	K5 III		290 s		
88825	237799		10 13 01.3	-59 55 05	-0.001	0.00	6.10	-0.08	-2.2	B5 III		420 s		
88755	221869		10 13 04.7	-47 55 11	0.000	-0.02	7.0	1.7	-0.3	K5 III		220 s		
88611	118169		10 13 13.2	+3 04 21	-0.006	-0.04	7.3	0.4	2.6	F0 V		89 s		
88654	137417		10 13 16.3	-8 26 45	+0.001	-0.04	7.7	0.9	3.2	G5 IV		81 s		
88793	237800		10 13 17.0	-50 36 40	-0.002	-0.01	7.92	1.19	-0.1	K2 III		400 s		m
88699	178544		10 13 19.4	-27 01 44	-0.004	+0.03	6.25	0.31		F0				
88907	250875		10 13 21.2	-61 39 31	-0.002	0.00	6.41	-0.11	-1.7	B3 V	+11	400 s		
88862	237808		10 13 22.4	-56 35 13	-0.002	-0.01	6.7	1.3	-0.1	K2 III		200 s		
88824	237804		10 13 22.8	-51 13 59	-0.005	-0.03	5.28	0.25		A5	+48			
89050	256690		10 13 23.1	-71 03 27	-0.001	-0.03	8.0	0.1	0.6	A0 V		300 s		
88742	201186		10 13 24.7	-33 01 54	-0.029	+0.06	6.38	0.59	4.5	G1 V	+41	23 ts		
88729	201185		10 13 27.5	-30 25 48	+0.001	+0.01	7.7	1.7	0.2	K0 III		120 s		
88842	237807		10 13 27.9	-51 45 21	-0.005	0.00	5.78	0.14		A3				m
88424	27564		10 13 28.8	+57 59 37	+0.001	0.00	6.9	0.1	1.4	A2 V		130 s		
89067	256691		10 13 28.9	-71 40 09	-0.013	-0.01	7.6	0.4	2.6	F0 V		100 s		
88640	118171		10 13 29.9	+9 11 10	0.000	-0.01	7.5	0.8	3.2	G5 IV		74 s		
88981	250880		10 13 30.6	-66 22 22	-0.006	0.00	5.16	0.21		A m	-15	18 mn		
88682	137421		10 13 31.3	-5 05 14	-0.005	-0.04	7.5	0.4	3.0	F2 V		81 s		
88841	237810		10 13 32.5	-50 27 47	+0.001	-0.02	7.94	0.92	3.2	G8 IV		89 s		
88934	250878		10 13 40.4	-60 15 48	-0.003	+0.01	8.0	-0.2	0.6	A0 V		300 s		
88512	27566		10 13 43.8	+50 29 46	+0.002	0.00	6.6	0.1	1.7	A3 V	-22	98 s		
89080	250885	ω Car	10 13 44.3	-70 02 16	-0.006	0.00	3.32	-0.08	-1.0	B7 IV	+4	70 s		
88697	137424		10 13 44.3	-7 23 03	-0.013	+0.02	7.22	0.49	3.7	dF6	+15	48 s		
88877	237817		10 13 45.3	-53 14 28	-0.001	-0.01	7.71	-0.02	0.6	A0 V		260 s		
88809	221877		10 13 45.8	-40 20 45	-0.007	+0.01	5.90	1.21	0.0	gK1		130 s		m
88709	137425		10 13 46.4	-5 13 12	-0.001	-0.02	7.4	0.1	1.4	A2 V		160 s		
89049	250884		10 13 48.9	-68 03 36	-0.005	-0.02	6.9	0.0	0.0	B8.5 V		240 s		
88639	81243		10 13 49.7	+27 08 09	-0.001	0.00	6.04	0.85	3.2	G5 IV	+10	33 s		
--	27568		10 13 50.5	+50 17 47	-0.003	-0.01	7.9	1.5	0.2	K0 III		180 s		
88767	178549		10 13 52.1	-23 04 48	0.000	+0.03	7.7	0.5	0.2	K0 III		320 s		
88726	137427		10 13 54.5	-4 07 28	0.000	-0.03	7.9	1.1	0.2	K0 III		340 s		
88836	221883		10 13 56.5	-40 18 38	-0.003	-0.01	6.35	0.94		K0				m
89403	256694		10 13 57.2	-79 00 13	-0.003	+0.01	7.71	-0.02	-2.5	B2 V		880 s		
88875	221888		10 14 00.1	-46 43 56	-0.003	0.00	8.0	0.9	2.6	F0 V		120 s		
88874	221887		10 14 02.9	-44 05 32	-0.001	-0.01	7.6	0.0	0.6	A0 V		240 s		
88945	237819		10 14 06.0	-54 59 38	-0.002	-0.01	7.48	-0.09	-1.6	B3.5 V		580 s		
88917	221889		10 14 07.6	-48 43 29	-0.001	-0.05	7.9	0.8	4.4	G0 V		51 s		
88725	118176		10 14 08.3	+3 09 04	+0.016	-0.41	7.76	0.60	4.5	G1 V	-24	32 ts		
88764	137430		10 14 08.3	-7 59 36	-0.001	-0.02	6.97	1.03	0.3	gG7	+9	190 s		
88806	178558		10 14 09.0	-23 48 49	-0.001	+0.02	6.34	1.63	-0.5	gM2	-6	220 s		
88630	43243		10 14 12.6	+45 05 14	0.000	+0.01	7.90	1.16	-0.1	K2 III		400 s		
88873	201195		10 14 12.9	-39 20 40	-0.001	-0.03	7.4	1.0	-0.3	K5 III		350 s		
88872	201196		10 14 13.6	-39 06 13	0.000	-0.02	6.5	1.2	0.2	K0 III		150 s		
88545	27572		10 14 19.0	+56 11 44	+0.001	0.00	8.0	0.1	0.6	A0 V		300 s		
88905	221890		10 14 21.1	-42 48 32	-0.002	-0.01	7.6	0.3	0.4	B9.5 V		160 s		
88530	27571		10 14 24.6	+59 23 59	0.000	-0.02	7.8	1.6	-0.5	M4 III		460 s		
89079	--		10 14 26.3	-64 31 46			8.0	0.1	1.4	A2 V		200 s		
88978	237822		10 14 28.1	-51 48 51	-0.003	-0.01	7.40	-0.14	-1.0	B5.5 V		480 s		
88737	81250		10 14 29.7	+21 10 04	-0.010	-0.09	6.02	0.56	3.4	F5 V	+17	28 s		
88748	99019		10 14 31.7	+16 08 19	0.000	-0.01	7.3	1.1	0.2	K0 III		260 s		
89007	237826		10 14 34.0	-55 17 36	-0.001	0.00	8.0	0.2	3.2	G5 IV		92 s		
88871	178562		10 14 37.4	-21 43 40	-0.003	-0.02	8.0	0.2	1.4	A2 V		160 s		
88832	137435		10 14 42.5	-2 52 34	-0.002	-0.02	7.9	0.2	2.1	A5 V		150 s		
88916	201203		10 14 42.6	-30 49 01	-0.002	+0.02	7.9	0.3	1.4	A2 V		150 s		
88955	221895		10 14 44.1	-42 07 19	-0.013	+0.04	3.85	0.05	1.4	A2 V	+8	31 s		
89029	237829		10 14 47.7	-56 14 52	+0.001	0.00	7.4	1.6	0.2	K0 III		120 s		
88976	221898		10 14 49.9	-41 43 34	-0.013	+0.02	6.57	0.15	0.0	A0 IV		93 mx		
88966	221899		10 14 53.2	-42 22 59	-0.007	0.00	7.7	-0.1	0.4	B9.5 V		130 mx		
88881	155844		10 14 55.8	-12 00 48	+0.002	-0.02	7.2	0.5	3.4	F5 V		58 s		
88830	118183		10 14 57.4	+9 12 40	+0.001	-0.05	7.9	0.5	3.4	F5 V		80 s		
88995	221904		10 15 00.6	-44 28 04	-0.003	0.00	6.59	1.32	0.2	K0 III		110 s		
88786	61953	22 LMi	10 15 06.3	+31 28 05	-0.003	-0.01	6.5	0.9	3.2	G5 IV	+15	47 s		
89205	250894		10 15 06.4	-67 17 10	+0.005	0.00	7.90	0.9	3.2	G5 IV		86 s		m
88651	15129		10 15 07.6	+59 59 07	+0.002	0.00	6.25	1.60	-0.4	M0 III	-21	200 s		U UMa, q
89077	237832		10 15 07.6	-55 32 57	-0.006	0.00	6.70	1.1	0.2	K0 III		200 s		m
88853	99025		10 15 09.5	+11 40 27	-0.003	-0.01	7.3	0.8	3.2	G5 IV		67 s		
89095	237833		10 15 10.3	-55 39 26	+0.002	0.00	7.3	2.2	-0.5	M4 III		160 s		
88926	155849		10 15 12.6	-11 47 24	-0.004	0.00	7.4	0.1	0.6	A0 V		230 s		
89266	250897		10 15 15.6	-69 38 12	+0.011	+0.03	8.0	0.4	3.0	F2 V		99 s		
89104	237834		10 15 16.6	-54 58 27	-0.002	0.00	6.16	-0.17	-1.7	B3 V	+8	370 s		
89105	237835		10 15 17.9	-55 41 23	0.000	-0.01	7.90	-0.02	0.4	B9.5 V		320 s		
89015	201211		10 15 20.8	-36 31 05	-0.003	-0.01	6.19	1.06		K0				
89062	221910		10 15 31.4	-43 06 45	0.000	-0.07	5.60	1.52		K2				

HD	SAO	Star Name	α 2000	δ 2000	μ(α)	μ(δ)	V	B−V	M$_v$	Spec	RV	d(pc)	ADS	Notes
			h m s	° ′ ″	s	″								
89203	250896		10 15 31.9	−62 32 13	+0.001	−0.01	7.06	−0.04	0.4	B9.5 V		220 s		
88775	43248		10 15 33.0	+47 22 10	+0.001	−0.10	7.7	0.6	4.4	G0 V		46 s		
89041	201214		10 15 34.3	−35 28 47	+0.001	+0.02	7.80	1.1	−0.1	K2 III		380 s		m
89103	221912		10 15 34.3	−49 13 27	−0.003	0.00	7.8	0.2	0.4	B9.5 V		300 s		
89137	237842		10 15 39.8	−51 15 26	−0.003	−0.02	7.97	−0.02	−4.4	O9.5 V		2300 s		
89175	237844		10 15 46.5	−52 38 39	−0.002	−0.01	7.74	1.11	−0.1	K2 III		370 s		
89157	221918		10 15 48.4	−50 10 39	+0.001	−0.02	7.3	0.8	0.2	K0 III		260 s		
89176	237845		10 15 48.5	−52 39 09	−0.002	+0.03	7.6	1.1	4.4	G0 V		21 s		
89060	201225		10 15 50.9	−32 10 34	+0.001	+0.01	7.7	2.0	−0.5	M2 III		270 s		
89174	237847		10 15 50.9	−52 12 14	+0.001	+0.01	7.96	0.13	−4.4	B1 III		1900 s		
89201	237850		10 15 52.7	−57 22 29	+0.001	0.00	7.84	0.69	−6.2	B1 I		2000 s		
89192	237851		10 15 56.8	−55 31 10	−0.004	+0.02	6.83	0.06	0.6	A0 V		160 s		
88759	15131		10 16 03.0	+60 09 02	−0.004	−0.03	6.8	0.4	3.0	F2 V		57 s		
89263	237853		10 16 03.1	−59 54 12	−0.006	+0.01	6.22	0.20	1.4	A2 V		75 s		m
89090	178597		10 16 03.9	−28 36 51	−0.018	+0.05	7.21	0.54	4.5	G1 V		35 s		
89033	155855		10 16 09.0	−11 12 12	−0.001	0.00	6.08	1.10		K0				
88924	61959		10 16 09.3	+35 09 51	−0.002	−0.03	7.4	0.1	1.4	A2 V		160 s		
88960	81258	23 LMi	10 16 14.3	+29 18 37	−0.006	−0.03	5.35	0.01	0.6	A0 V	+16	77 mx		
88987	99032		10 16 16.0	+17 44 24	0.000	0.00	7.30	0.4	2.6	F0 V	−8	87 s	7704	m
89088	155858		10 16 16.2	−16 08 30	−0.004	+0.02	7.2	1.1	0.2	K0 III		250 s		
89057	137451		10 16 16.8	−5 49 45	−0.002	0.00	7.6	0.0	0.4	B9.5 V		280 s		
88295	7088		10 16 17.2	+78 56 48	+0.005	0.00	6.7	0.1	0.6	A0 V		170 s		
89117	178601		10 16 19.1	−28 58 42	+0.002	−0.01	7.72	0.40	3.4	F5 V		73 s	7706	m
89171	201229		10 16 21.8	−36 54 29	−0.004	+0.06	7.5	1.1	3.2	G5 IV		43 s		
88923	43259		10 16 25.9	+45 25 59	−0.004	−0.02	7.8	0.4	3.0	F2 V		91 s		
88986	81259	24 LMi	10 16 28.1	+28 40 57	−0.004	−0.09	6.49	0.61		G0	+30			
88935	43261		10 16 31.3	+45 02 36	0.000	+0.02	7.4	0.5	3.4	F5 V		64 s		
89010	81260	35 Leo	10 16 32.2	+23 30 11	−0.015	+0.03	5.97	0.67	4.4	G0 V	−33	18 s		
89132	155862		10 16 32.8	−19 18 31	−0.002	−0.01	6.6	0.1	1.4	A2 V		110 s		
88959	43264		10 16 38.3	+41 16 33	+0.006	−0.02	7.4	0.5	3.4	F5 V		62 s		
89273	237858		10 16 40.2	−51 12 18	−0.003	0.00	6.30	1.54		gM5				
89056	99034	37 Leo	10 16 40.7	+13 43 42	−0.002	−0.02	5.41	1.61	−0.5	M2 III	+3	150 s		
89025	81265	36 ζ Leo	10 16 41.4	+23 25 02	+0.001	−0.01	3.44	0.31	0.6	F0 III	−15	36 s		Adhafera, m
89024	81264		10 16 41.8	+25 22 17	−0.008	+0.03	5.84	1.20	0.2	K0 III	+34	93 s		
89169	178610		10 16 45.5	−20 40 13	−0.009	−0.04	6.57	0.48	3.4	F5 V		41 s	7711	m
89114	137455		10 16 46.6	−2 47 51	−0.006	−0.02	7.7	1.1	0.2	K0 III		230 mx		
89149	155866		10 16 46.7	−13 05 58	−0.001	−0.01	7.2	0.8	3.2	G5 IV		62 s		
89113	118195		10 16 51.5	+1 47 51	−0.003	−0.01	7.9	1.1	−0.1	K2 III		390 s		
89213	201237		10 16 51.9	−34 07 39	0.000	−0.01	7.3	0.3	0.6	A0 V		140 s		
89055	81267		10 16 56.7	+25 51 39	+0.013	−0.29	7.62	0.60	4.4	G0 V		39 mx		m
89086	81268		10 17 00.3	+21 54 47	+0.003	−0.02	7.5	0.5	3.4	F5 V		66 s		
88984	43267		10 17 00.8	+47 18 30	+0.001	−0.13	7.3	0.5	3.4	F5 V		61 s		
89211	178618		10 17 01.0	−29 40 12	−0.001	−0.02	8.0	0.4	3.4	F5 V		82 s		
89388	250905		10 17 04.9	−61 19 56	−0.004	0.00	3.40	1.54	−4.4	K5 Ib	+9	280 mx		q Car, m
89021	43268	33 λ UMa	10 17 05.7	+42 54 52	−0.015	−0.04	3.45	0.03	0.6	A2 IV	+18	37 s		Tania Borealis
89125	81270	39 Leo	10 17 14.5	+23 06 22	−0.030	−0.10	5.82	0.50	3.4	dF5 e	+38	17 ts	7712	m
89053	43270		10 17 17.6	+41 28 02	−0.008	−0.03	6.8	1.6	−0.5	M2 III	−33	120 mx		
89328	221934		10 17 20.1	−46 50 08	−0.002	−0.02	6.7	0.4	2.6	F0 V		54 s		
−−	81273		10 17 25.0	+23 14 00	−0.001	−0.01	7.9	1.4	−0.3	K5 III		430 s		
88999	27591		10 17 30.3	+55 53 55	+0.002	−0.04	7.3	1.1	−0.1	K2 III		310 s		
89518	250907		10 17 31.7	−65 25 48	+0.001	−0.03	6.9	0.5	0.6	A0 V		87 s		
89282	178627		10 17 33.5	−23 52 35	−0.001	−0.01	7.3	0.8	0.2	K0 III		260 s		
89145	61969		10 17 35.5	+30 53 14	+0.001	0.00	7.8	1.6	−0.5	M2 III		460 s		
89254	137469	22 ε Sex	10 17 37.7	−8 04 08	−0.011	0.00	5.26	0.31	0.6	F1 III	+15	85 s		
89371	221937		10 17 37.9	−48 25 04	−0.004	+0.01	7.4	0.9	−0.3	K5 III		350 s		
89280	178630		10 17 40.0	−21 35 20	−0.010	+0.04	7.3	0.4	1.4	A2 V		51 mx		
89270	155877		10 17 43.7	−12 04 45	−0.002	−0.02	8.0	0.1	1.4	A2 V		210 s		
88849	7100		10 17 51.3	+71 03 22	−0.006	−0.05	6.66	0.32	0.5	gA8	+11	160 s	7705	m
89429	237884		10 17 52.8	−54 37 25	−0.002	−0.03	7.89	0.01	0.2	B9 V		320 s		
89312	178638		10 17 53.3	−21 01 35	+0.001	−0.03	7.40	1.4	−1.3	K5 II-III	−4	550 s		
89110	43273		10 17 54.5	+49 34 35	−0.006	+0.10	7.7	0.5	4.0	F8 V		56 s		
89224	81276		10 17 58.8	+21 58 07	−0.001	−0.01	8.0	0.9	3.2	G5 IV		93 s		
88815	7098		10 18 01.1	+73 04 24	−0.013	−0.08	6.40	0.23	2.6	F0 V	+16	58 s		
88983	15135	32 UMa	10 18 02.0	+65 06 30	−0.014	−0.01	5.7	0.1	1.7	A3 V	−6	64 s		
89353	178644		10 18 07.5	−28 59 31	−0.001	+0.01	5.34	0.24		B9	−39			
89239	81278		10 18 10.2	+27 24 55	−0.004	0.00	6.52	−0.02		B9.5 pe	+7			
90317	258569		10 18 12.0	−82 54 38	+0.032	−0.01	6.7	1.7	0.2	K0 III		77 s		
89494	237892		10 18 21.2	−52 19 52	+0.003	+0.01	7.8	0.5	−0.1	K2 III		380 s		
89307	99049		10 18 21.3	+12 37 17	−0.018	−0.04	7.1	0.7	4.4	G0 V		34 s		
89206	43275		10 18 23.3	+43 45 32	−0.007	−0.03	7.9	0.4	2.6	F0 V		110 s		
89351	155882		10 18 23.8	−13 18 16	+0.004	−0.03	7.0	0.8	3.2	G5 IV		58 s		
89391	178654		10 18 25.1	−26 30 00	+0.003	−0.04	7.94	0.94	3.2	K0 IV		89 s		
89461	221948		10 18 28.1	−41 40 06	−0.003	0.00	5.96	−0.06		A0				
89333	118212		10 18 30.6	+1 34 11	−0.002	−0.04	7.9	1.1	−0.1	K2 III		390 s		
89221	43276		10 18 32.8	+43 02 54	−0.011	−0.08	6.6	0.8	3.2	G5 IV		47 s		

HD	SAO	Star Name	α 2000	δ 2000	μ(α)	μ(δ)	V	B-V	M_v	Spec	RV	d(pc)	ADS	Notes
			h m s	° ' "	s	"								
89532	237897		10 18 33.7	-51 44 46	-0.004	-0.01	7.65	0.95	0.2	K0 III		310 s		
89441	201264		10 18 34.8	-35 45 57	-0.002	-0.13	7.7	1.0	3.2	G5 IV		60 s		
89569	237902		10 18 37.5	-56 06 36	-0.030	+0.12	5.81	0.48	4.0	F8 V		23 s		
89442	201266		10 18 37.7	-36 48 17	-0.004	0.00	6.30	1.28		K0				
89422	201265		10 18 38.0	-32 32 18	-0.002	+0.01	7.7	1.3	-0.3	K5 III		400 s		
89531	221951		10 18 38.6	-49 54 46	-0.002	0.00	8.00	1.4	-0.3	K5 III		460 s		m
--	61980		10 18 46.3	+39 18 41	0.000	-0.02	7.8	2.1						
89588	237903		10 18 47.9	-56 02 06	-0.003	-0.09	8.0	0.9	0.2	K0 III		140 mx		
89703	250915		10 18 48.3	-66 57 23	-0.015	-0.06	7.9	1.4	-0.3	K5 III		120 mx		
89420	178665		10 18 50.4	-23 10 53	+0.003	-0.01	7.7	0.7	3.2	G5 IV		81 s		
89269	43279		10 18 51.8	+44 02 54	+0.005	-0.30	6.65	0.66	5.2	dG5	-8	20 s		m
89647	--		10 18 52.1	-61 22 02			7.9	1.9	-0.5	M4 III		320 s		
89363	99056		10 18 57.4	+17 42 20	-0.003	-0.03	6.6	0.1	0.6	A0 V	+17	160 s		
89268	43281		10 18 58.9	+46 45 39	-0.002	-0.03	6.50	1.1	0.0	K1 III	-21	200 s		m
89251	27600		10 18 59.6	+51 24 20	+0.002	0.00	6.8	0.7	4.4	G0 V		30 s		
89418	155887		10 19 00.0	-11 11 07	-0.001	-0.14	7.7	0.7	4.4	G0 V		46 s		
89344	81285		10 19 00.7	+24 42 42	-0.003	-0.01	6.5	1.1	0.2	K0 III	0	190 s		
89613	237907		10 19 02.0	-56 01 15	-0.008	+0.01	7.80	0.1	1.4	A2 V		130 mx		m
89587	237905		10 19 02.7	-50 42 59	0.000	-0.01	6.88	-0.15	-2.2	B5 III		660 s		
89715	250917		10 19 04.8	-64 40 34	-0.006	0.00	5.67	0.05		A0				m
89528	201275		10 19 10.6	-34 36 58	-0.005	+0.04	6.9	-0.1	0.6	A0 V		100 mx		
88998	7107		10 19 11.7	+72 26 12	-0.022	-0.08	7.7	1.1	0.2	K0 III	+18	160 mx		
89361	81287		10 19 13.1	+24 21 50	-0.001	-0.04	7.80	1.1	-0.1	K2 III		380 s		
89551	201277		10 19 13.8	-37 37 56	-0.003	+0.01	8.0	0.3	1.4	A2 V		140 s		
89683	250919		10 19 15.0	-61 06 28	0.000	0.00	8.00	0.00	0.0	B8.5 V		380 s		m
89267	27601		10 19 16.1	+54 43 32	-0.001	-0.01	7.9	1.1	0.2	K0 III		350 s		
89455	155894		10 19 16.8	-12 31 41	-0.001	-0.02	6.00	0.26		F0				m
89417	118218		10 19 22.0	+9 12 46	-0.004	-0.01	7.9	0.5	3.4	F5 V		79 s		
89319	43285		10 19 26.8	+48 23 47	-0.010	-0.12	6.00	1.02	0.2	K0 III	-6	80 mx		
89432	99061		10 19 28.0	+9 55 13	+0.002	-0.02	7.9	1.1	-0.1	K2 III		390 s		
89611	221959		10 19 29.1	-45 38 47	-0.002	+0.01	7.8	0.1	0.6	A0 V		220 s		
89406	99060		10 19 29.3	+16 57 14	-0.001	-0.01	8.0	0.9	3.2	G5 IV		93 s		
89490	137490		10 19 32.1	-5 06 21	-0.004	-0.07	6.37	0.90		K0				
89741	250920		10 19 33.3	-62 11 43	-0.006	0.00	7.4	1.6	2.1	A5 V		16 s		
89507	137493		10 19 35.6	-5 13 39	-0.003	-0.04	7.6	0.4	3.0	F2 V		83 s		
89682	237916		10 19 36.8	-55 01 46	-0.001	-0.01	4.57	1.62	-5.9	cK	+13	910 s		
89488	118222		10 19 39.0	-0 12 36	+0.001	-0.02	8.0	0.4	2.6	F0 V		120 s		
89585	201283		10 19 39.2	-30 59 45	-0.002	+0.01	7.46	0.89	0.3	G5 III		270 s		
89740	237922		10 19 40.9	-59 08 45	+0.002	0.00	6.93	-0.11	-1.7	B3 V n		510 s		
89805	250921		10 19 42.2	-65 08 13	0.000	+0.01	6.5	1.8	-0.1	K2 III		88 s		
89449	99065	40 Leo	10 19 44.1	+19 28 15	-0.017	-0.22	4.79	0.45	2.2	F6 IV	+7	29 mx		
89566	178687		10 19 48.7	-22 31 23	0.000	0.00	7.9	0.8	0.2	K0 III		340 s		
89581	178688		10 19 48.9	-22 58 15	0.000	0.00	7.4	1.3	-0.5	M2 III		380 s		
89681	221961		10 19 51.7	-47 57 40	+0.001	-0.02	6.9	0.6	1.4	A2 V		63 s		
89713	237921		10 19 52.2	-51 34 06	-0.004	-0.01	7.00	0.4	3.0	F2 V		63 s		m
89484	81298	41 γ¹ Leo	10 19 58.3	+19 50 30	+0.022	-0.15	2.28	1.08	0.2	K0 III	-37		7724	Algieba, m
89485	81299	41 γ² Leo	10 19 58.6	+19 50 25	+0.022	-0.17	3.53			G7 III	-36		7724	m
89565	137495		10 19 59.3	-9 03 32	-0.005	-0.07	6.32	0.33		F2				
89891	250923		10 20 03.6	-67 10 06	-0.001	-0.01	7.50	0.00	0.4	B9.5 V		260 s		m
89677	201291		10 20 05.3	-39 42 09	0.000	0.00	7.31	0.01	0.4	B9.5 V		230 s		
89712	221967		10 20 07.0	-46 06 48	-0.002	+0.01	7.9	1.5	-0.3	K5 III		430 s		
89803	237934		10 20 11.7	-59 39 22	0.000	-0.02	7.2	0.9	2.6	F0 V		35 s		
89767	237930		10 20 11.8	-52 36 05	0.000	0.00	7.21	0.16		B0.5 II				
89659	201292		10 20 12.0	-31 34 59	-0.002	+0.01	8.0	0.2	0.2	K0 III		360 s		
89638	178700		10 20 13.1	-23 06 09	-0.001	+0.01	7.5	0.6	3.2	G5 IV		72 s		
89672	201293		10 20 13.5	-33 07 45	-0.003	-0.01	7.60	0.1	1.4	A2 V		170 s		m
91542	258572		10 20 13.8	-86 33 25	-0.054	+0.02	7.8	0.6	2.6	F0 V		76 s		
90438	258570		10 20 14.3	-81 38 51	-0.030	+0.06	7.3	0.0	2.1	A5 V		60 mx		
89389	27606		10 20 14.7	+53 46 45	-0.010	+0.04	6.45	0.54	4.0	F8 V	-21	30 s		
89524	81302		10 20 15.3	+22 00 41	0.000	-0.04	7.4	1.1	0.2	K0 III		280 s		
89736	221970		10 20 16.6	-47 41 57	-0.001	-0.01	5.65	1.67		K0	+16			
89903	250925		10 20 19.1	-66 03 49	0.000	+0.01	7.7	0.1	0.6	A0 V		250 s		
89876	250924		10 20 23.6	-63 20 13	-0.002	-0.01	7.96	0.00	0.6	A0 V		300 s		m
89752	221972		10 20 26.3	-43 14 25	0.000	-0.03	8.0	0.3	4.0	F8 V		62 s		
90038	256701		10 20 27.9	-72 07 52	+0.006	0.00	7.3	1.6	-0.5	M4 III		360 s		
89788	237937		10 20 28.7	-51 27 00	-0.001	+0.01	7.4	1.2	-0.5	M2 III		390 s		
89414	27609		10 20 31.0	+54 13 00	-0.005	-0.01	6.00	1.13	0.2	K0 III	+9	120 s		
89874	237947		10 20 32.1	-59 38 12	+0.001	-0.01	7.4	1.0	3.2	G5 IV		50 s		
89619	118241		10 20 32.4	+6 25 46	+0.001	-0.06	7.34	0.50	0.6	gF7	+2	220 s	7730	m
89709	201297		10 20 33.9	-31 48 49	-0.013	+0.04	6.8	0.4	4.0	F8 V		36 s		
89694	--		10 20 36.4	-23 39 05			7.50	1.1	0.2	K0 III		290 s	7735	m
89844	237948		10 20 37.0	-57 10 12	-0.002	-0.02	7.74	-0.10	0.0	B8.5 V		350 s		
89670	137503		10 20 38.7	-6 11 34	-0.003	-0.01	7.4	0.2	2.1	A5 V		110 s		
89557	81304		10 20 39.0	+28 56 34	0.000	-0.01	7.5	0.9	0.3	G8 III		270 s		
89289	15144		10 20 40.3	+69 01 14	-0.005	-0.01	7.8	1.1	0.2	K0 III		330 s		

HD	SAO	Star Name	α 2000	δ 2000	μ(α)	μ(δ)	V	B-V	M$_v$	Spec	RV	d(pc)	ADS	Notes
			h m s	° ′ ″	s	″								
89839	237945		10 20 40.5	-53 39 50	0.000	+0.02	7.70	0.5	3.4	F5 V		72 s		m
89976	—		10 20 42.0	-68 04 11			7.4	0.0	0.0	B8.5 V		300 s		
89669	137505		10 20 44.1	-5 22 55	-0.004	-0.01	7.0	0.1	1.4	A2 V		130 s		
89708	155918		10 20 46.0	-19 03 01	-0.011	+0.08	7.8	0.5	4.0	F8 V		57 s		
89785	221976		10 20 48.2	-42 52 21	-0.004	-0.02	7.30	0.00	0.4	B9.5 V		200 mx		m
89800	221977		10 20 48.3	-47 33 56	+0.002	0.00	6.8	1.2	-0.3	K5 III		270 s		
89707	155919		10 20 49.9	-15 28 49	-0.016	+0.28	7.18	0.55	4.0	F8 V	+79	43 s		
89595	61991		10 20 51.2	+30 39 50	-0.002	+0.03	7.2	1.1	0.2	K0 III		250 s		
89890	237960		10 20 55.4	-56 02 36	-0.004	+0.01	4.50	-0.12	-2.3	B3 IV	+10	210 s		
89688	118248	23 Sex	10 21 01.9	+2 17 23	-0.001	-0.01	6.66	-0.08	-2.3	B3 IV	+5	550 s		RS Sex, q
89343	15147		10 21 03.4	+68 44 51	-0.009	-0.04	5.8	0.4	2.6	F0 V	+4	44 s		
89572	43296		10 21 03.6	+41 50 57	0.000	0.00	6.7	0.1	0.6	A0 V	-2	170 s		
89720	137509		10 21 07.0	-5 24 55	-0.002	+0.02	6.9	1.1	-0.1	K2 III		250 s		
89747	155920		10 21 07.7	-17 59 06	-0.004	-0.03	6.51	0.40		F5				
89925	237968		10 21 08.2	-58 29 02	-0.003	-0.01	7.29	0.87	-6.3	G0 Iab		5100 s		
89746	155922		10 21 14.6	-13 47 09	-0.002	+0.01	6.6	0.1	1.4	A2 V		110 s		
89594	43298		10 21 16.1	+46 14 47	-0.003	+0.03	7.8	1.1	-0.1	K2 III		380 s		
89854	221981		10 21 17.1	-44 32 24	0.000	-0.02	8.0	0.0	0.6	A0 V		280 s		
90020	250930		10 21 17.8	-65 11 45	-0.001	0.00	7.3	0.1	0.6	A0 V		210 s		
89653	61997		10 21 19.4	+33 28 21	-0.005	-0.02	8.0	1.1	0.2	K0 III		330 mx		
89915	237969		10 21 19.9	-51 44 29	0.000	0.00	7.90	-0.03		B9				
89796	178717		10 21 21.4	-25 22 22	-0.002	0.00	6.8	1.5	0.2	K0 III		110 s		
89734	118251		10 21 22.1	-0 13 14	-0.003	0.00	8.0	0.5	3.4	F5 V		82 s		
89816	178721		10 21 28.6	-23 42 38	-0.004	+0.02	6.50	0.20		A3				
90103	—		10 21 34.1	-69 19 14			7.8	0.9	3.2	G5 IV		82 s		
89921	221984		10 21 34.1	-49 24 05	0.000	-0.03	8.0	1.6	0.2	K0 III		250 s		
89828	178723		10 21 35.9	-22 31 42	-0.002	0.00	6.51	0.07		A0			7739	m
89652	43303		10 21 41.7	+48 02 30	-0.023	0.00	7.7	0.6	4.4	G0 V		45 s		
89546	15153		10 21 47.1	+60 54 46	-0.014	-0.02	7.4	1.1	0.2	K0 III		210 mx		
89774	99080	42 Leo	10 21 50.2	+14 58 32	-0.002	-0.02	6.12	0.02	1.2	A1 V	+9	96 s		
89886	178730		10 21 50.5	-24 26 08	-0.002	+0.02	7.9	0.0	4.0	F8 V		60 s		
89782	99081		10 21 52.8	+10 41 59	-0.005	-0.03	7.77	0.45	4.0	F8 V		57 s		
89884	155931		10 21 59.3	-18 02 06	-0.001	-0.03	7.13	-0.11	-0.9	B6 V	+22	400 s		
89686	43306		10 22 00.3	+43 54 20	-0.005	-0.01	7.1	0.5	3.4	F5 V		54 s	7737	m
89935	201318		10 22 00.3	-36 13 26	-0.002	0.00	7.8	0.9	2.6	F0 V		45 s		
89885	178733		10 22 01.3	-20 34 33	-0.003	-0.01	7.11	1.17	0.0	K1 III	+19	240 s		
89824	137517		10 22 03.3	-0 45 06	-0.002	-0.01	7.5	0.9	3.2	G5 IV		74 s		
90001	221992		10 22 05.7	-49 31 54	-0.001	-0.04	7.9	0.5	2.6	F0 V		110 s		
90082	250932		10 22 07.2	-61 51 02	-0.002	-0.05	7.1	1.9	-0.5	M2 III		230 s		
89481	15152		10 22 08.4	+68 55 34	-0.003	0.00	7.5	1.1	0.2	K0 III		290 s		
90000	221995		10 22 08.8	-49 17 37	-0.001	0.00	7.6	0.8	-1.6	B3.5 V		200 s		
89913	178737		10 22 09.3	-21 16 34	-0.002	0.00	8.0	0.4	0.6	A0 V		160 s		
89813	99084		10 22 09.4	+11 18 36	+0.001	-0.33	7.86	0.76	3.2	G5 IV	-17	40 mx		
89744	43309		10 22 10.4	+41 13 46	-0.011	-0.14	5.76	0.54	3.4	F5 V	-7	19 ts		
90074	237990		10 22 10.4	-58 15 56	-0.007	+0.02	6.5	1.0	0.2	K0 III		170 s		
89897	155934		10 22 10.8	-13 24 36	-0.001	-0.03	6.8	0.0	0.4	B9.5 V		190 s		
89911	155935		10 22 12.9	-19 52 01	-0.002	-0.01	6.12	0.03		A0				
89848	118260		10 22 14.2	+8 57 51	-0.001	-0.01	6.9	1.4	-0.3	K5 III		270 s		
89998	221998		10 22 19.5	-41 39 00	-0.003	+0.05	4.83	1.12	0.0	K1 III	+21	88 s		
89758	43310	34 μ UMa	10 22 19.7	+41 29 58	-0.007	+0.03	3.05	1.59	-0.4	M0 III	-21	48 s		Tania Australis
90087	237995		10 22 20.6	-59 45 20	-0.003	-0.01	7.80	0.00	-4.4	O9.5 V		2000 s		
89948	178741		10 22 21.8	-29 33 19	-0.002	+0.08	7.7	0.2	3.2	G5 IV		80 s		
90034	222001		10 22 22.1	-47 58 16	-0.002	+0.01	7.40	0.1	0.6	A0 V		230 s		m
89847	99086		10 22 28.4	+17 30 49	-0.004	-0.06	7.8	0.4	2.6	F0 V		110 s		
89781	43311		10 22 34.2	+42 36 58	-0.006	-0.05	7.8	0.4	2.6	F0 V		110 s		
89866	99088		10 22 34.8	+16 44 28	+0.001	-0.04	7.8	0.4	2.6	F0 V		110 s		
90151	250936		10 22 37.3	-62 04 23	-0.002	0.00	7.66	-0.02	0.6	A0 V		260 s		
89865	81318		10 22 38.4	+24 35 47	-0.003	0.00	7.9	0.5	3.4	F5 V		78 s		
90086	237997		10 22 41.0	-51 36 51	-0.003	-0.02	7.97	-0.09		B9				
89945	137524		10 22 41.7	-9 23 41	-0.001	+0.01	7.20	1.6	-0.5	M3 III	0	350 s		
89906	99091		10 22 43.7	+15 20 38	-0.018	-0.12	7.50	0.9	3.2	G5 IV	+19	50 mx	7744	m
90028	178752		10 22 51.2	-28 00 50	-0.008	+0.01	7.7	0.9	4.4	G0 V		29 s		
90264	250940		10 22 58.1	-66 54 06	-0.004	0.00	4.99	-0.13		B8	+12			
89930	99092		10 22 59.1	+19 07 52	-0.001	+0.01	7.1	0.9	3.2	G5 IV		60 s		
90027	178755		10 22 59.3	-24 20 01	+0.002	-0.03	6.7	0.1	0.4	B9.5 V		150 s		
89962	118269	43 Leo	10 23 00.4	+6 32 33	-0.001	-0.10	6.07	1.12	-0.2	gK3	-24	180 s		
89905	62008		10 23 01.2	+30 50 14	-0.004	-0.02	7.5	1.1	0.2	K0 III	+35	290 s		
89892	62009		10 23 05.2	+35 13 04	-0.003	-0.01	7.1	0.1	1.4	A2 V		140 s		
90219	250939		10 23 05.3	-61 15 56	0.000	0.00	6.8	1.8	0.2	K0 III		72 s		
89904	62010	27 LMi	10 23 06.3	+33 54 29	-0.001	-0.01	5.8	0.1	1.7	A3 V	-16	67 s		
90134	238004		10 23 07.3	-50 31 38	-0.001	+0.02	7.2	0.9	2.6	F0 V		35 s		
90071	178759		10 23 13.1	-30 09 44	0.000	-0.02	6.27	0.31		F0				
89995	118271		10 23 14.5	+5 41 39	-0.016	-0.08	6.54	0.46		F2	+5	25 mn		m
90012	137531		10 23 16.7	-7 46 23	-0.001	+0.02	8.0	1.1	0.2	K0 III		360 s		
90115	222010		10 23 17.1	-40 41 45	+0.002	-0.02	7.3	1.6	-0.1	K2 III		160 s		

HD	SAO	Star Name	α 2000	δ 2000	μ(α)	μ(δ)	V	B-V	M_v	Spec	RV	d(pc)	ADS	Notes
90045	155947		10^{h}23^{m}17.5	-13°22'38"	-0.009	-0.04	6.60	0.5	4.0	F8 V		33 s	7749	m
89342	7123		10 23 20.6	+78 20 33	-0.010	-0.02	7.6	1.1	0.2	K0 III		300 s		
90175	238008		10 23 24.1	-50 20 55	-0.018	+0.05	8.00	0.5	3.4	F5 V		83 s		m
90044	137533	25 Sex	10 23 26.4	-4 04 27	-0.004	0.00	5.97	-0.10		B9 pe	+23			
90174	222015		10 23 27.0	-49 36 56	-0.001	0.00	7.9	1.7	-0.1	K2 III		220 s		
90043	137532		10 23 28.2	-0 54 09	+0.004	-0.04	6.6	0.8	3.2	G5 IV	0	48 s		
90132	201346		10 23 29.2	-38 00 35	-0.014	-0.06	5.33	0.25		A3	+17	19 mn		
90094	178761		10 23 29.8	-26 03 23	-0.007	+0.03	7.1	0.6	3.0	F2 V		48 s		
90355	--		10 23 32.1	-68 17 22			8.0	1.1	0.2	K0 III		350 s		
90057	137534		10 23 32.6	-3 38 34	-0.001	0.00	6.6	1.4	-0.3	K5 III		240 s		
90081	155950		10 23 34.5	-12 29 12	-0.012	+0.04	7.6	0.9	3.2	G5 IV		77 s		
90111	201347		10 23 35.0	-30 55 35	-0.002	-0.02	7.8	0.5	1.4	A2 V		110 s		
90093	178766		10 23 36.7	-23 28 02	-0.008	-0.01	8.0	0.4	4.0	F8 V		62 s		
90170	222016		10 23 40.4	-41 57 12	-0.012	+0.01	6.27	0.88	3.2	K0 IV		41 s		
89993	81328		10 23 41.7	+29 36 57	-0.001	-0.01	6.39	1.09		K0	-13			
90274	238018		10 23 42.4	-59 12 22	-0.007	+0.03	6.7	1.7	0.2	K0 III		79 s		
90201	222019		10 23 47.2	-44 14 34	-0.002	-0.02	7.70	0.1	0.6	A0 V		260 s		m
90080	137537		10 23 48.8	-2 16 34	-0.003	+0.01	7.7	0.9	3.2	G5 IV		80 s		
90146	201354		10 23 49.2	-31 33 23	-0.004	+0.02	8.0	1.5	0.2	K0 III		190 s		
90009	81329		10 23 49.4	+25 34 04	-0.001	-0.06	6.8	1.1	-0.1	K2 III	-1	180 mx		
90289	238021		10 23 50.9	-57 57 13	0.000	-0.01	6.35	1.52	-5.9	cK		2800 s		
90156	178771		10 23 55.2	-29 38 44	-0.003	+0.10	6.95	0.65	3.2	G5 IV		56 s		
90008	62015		10 24 04.2	+39 02 09	-0.002	-0.01	7.8	0.5	3.4	F5 V		77 s		
90024	62018		10 24 05.9	+34 11 34	-0.001	-0.03	7.2	1.1	0.2	K0 III		260 s		
90127	137539		10 24 06.3	-4 55 38	+0.001	0.00	7.3	1.1	0.2	K0 III		270 s		
89822	15163		10 24 07.8	+65 33 59	-0.001	-0.03	4.97	-0.06		A0 p	0	11 mn		
90040	62019	28 LMi	10 24 08.5	+33 43 06	-0.002	-0.01	5.50	1.18	0.2	K0 III	-22	90 s		
90399	--		10 24 11.3	-64 49 23			7.8	1.1	-0.1	K2 III		370 s		
90258	222028		10 24 12.1	-44 02 17	-0.002	+0.01	7.4	0.3	2.1	A5 V		100 s		
90125	118278		10 24 13.1	+2 22 05	+0.002	-0.03	6.32	1.00		K0	-14		7755	m
90197	178775		10 24 13.7	-24 36 22	-0.025	+0.08	7.11	0.67	4.4	dG0	+58	30 s		
90478	250947		10 24 14.2	-69 55 07	-0.003	-0.03	7.5	1.4	-0.3	K5 III		350 s		
90312	238029		10 24 15.8	-53 43 40	-0.002	-0.02	7.84	0.98	3.2	G8 IV		71 s		
90123	99098		10 24 19.6	+10 35 16	-0.001	0.00	6.8	1.1	0.2	K0 III		210 s		
90068	62021		10 24 22.0	+34 10 34	-0.001	-0.03	7.30	1.6		M6 III	+2			m
90398	250945		10 24 22.7	-61 33 07	-0.003	0.00	7.61	-0.05	0.6	A0 V		250 s		m
89589	256710		10 24 23.7	-74 01 54	-0.004	-0.03	4.00	0.35	0.6	F2 III	-4	17 ts		
90185	137542		10 24 24.8	-7 35 18	-0.002	-0.04	7.9	1.4	-0.3	K5 III		430 s		
90155	118281		10 24 25.1	+2 23 34	-0.004	-0.03	6.7	1.1	0.2	K0 III		200 s		
90382	238038		10 24 25.2	-60 11 32	-0.002	-0.03	7.8	1.8	-0.5	M4 III		350 s		CK Car, v
90227	178776		10 24 26.1	-25 33 58	-0.005	-0.01	7.0	1.4	0.2	K0 III		130 s		
90352	238037		10 24 34.1	-54 00 51	-0.010	-0.01	8.0	0.3	3.4	F5 V		82 s		
90565	--		10 24 37.5	-71 37 26			7.7	1.1	-0.1	K2 III		350 s		
90270	201363		10 24 38.9	-33 29 17	-0.005	+0.05	7.7	1.4	-0.3	K5 III		210 mx		
90371	238040		10 24 39.6	-54 19 18	0.000	0.00	7.0	1.2	-0.5	M4 III		320 s		
89108	1681		10 24 42.1	+82 48 05	+0.005	-0.01	6.9	1.1	0.2	K0 III		220 s		
89717	7133		10 24 42.9	+74 46 25	-0.004	-0.03	8.0	0.6	4.4	G0 V		53 s		
90630	256711		10 24 44.4	-73 58 17	-0.006	+0.01	6.19	0.07		A2				
90212	137545		10 24 45.8	-0 47 25	-0.003	-0.02	7.9	0.1	0.6	A0 V		290 s		
90255	155965		10 24 46.0	-18 38 42	0.000	-0.06	7.0				+5	10 mn		
89704	7131		10 24 48.6	+75 40 35	-0.003	+0.02	7.8	1.1	0.2	K0 III		330 s		
90283	178789		10 24 50.0	-24 58 19	+0.001	0.00	8.0	1.8	-0.3	K5 III		290 s		
90211	137547		10 24 51.0	-0 24 08	-0.003	-0.04	8.0	0.5	4.0	F8 V		62 s		
90396	238043		10 24 53.2	-52 33 05	-0.001	0.00	8.0	0.7	1.4	A2 V		79 s		
90335	201369		10 24 56.9	-38 18 12	-0.001	-0.02	6.8	0.4	0.6	A0 V		93 s		
90208	99104		10 24 58.5	+12 43 48	-0.004	-0.01	8.0	0.1	1.4	A2 V		170 mx		
90308	178794		10 24 59.4	-27 37 13	+0.001	+0.07	7.8	1.6	-0.1	K2 III		200 mx		
90454	238046		10 24 59.4	-58 34 34	-0.010	+0.01	5.95	0.32	2.6	dF0		46 s		
89970	15168		10 25 00.9	+64 09 27	-0.001	-0.04	7.7	1.1	-0.1	K2 III		370 s		
90164	62028		10 25 01.6	+30 22 16	+0.008	-0.13	7.91	0.55	4.0	F8 V	-28	59 s		
90307	178795		10 25 04.7	-26 21 29	-0.003	0.00	7.3	1.2	0.2	K0 III		210 s		
89977	15169		10 25 06.5	+65 54 33	-0.004	-0.02	7.9	1.1	0.2	K0 III		340 s		
90412	222040		10 25 06.5	-50 06 25	+0.006	-0.04	6.9	0.9	2.6	F0 V		34 s		
90452	238048		10 25 10.3	-57 03 50	-0.001	-0.01	7.9	1.3	0.2	K0 III		240 s		
90154	--		10 25 14.8	+42 43 11	0.000	+0.01	7.9	1.1	0.2	K0 III		340 s		
90254	118286	44 Leo	10 25 15.1	+8 47 06	+0.001	-0.04	5.61	1.62		Ma	-20			
90490	238052		10 25 16.5	-58 52 20	0.000	+0.01	6.98	-0.08	0.4	B9.5 V		210 s		
90451	238049		10 25 16.8	-52 33 54	0.000	0.00	7.8	0.6	0.6	A0 V		110 s		
90393	222041		10 25 17.1	-42 28 04	-0.001	-0.04	6.18	1.00	0.2	gK0		160 s		
90366	178801		10 25 23.7	-29 11 37	-0.004	+0.01	6.7	1.7	0.2	K0 III		79 s		
90278	99111		10 25 26.4	+11 09 24	-0.001	+0.01	7.9	1.1	0.2	K0 III		340 s		
90320	155970		10 25 27.9	-12 45 46	0.000	-0.01	7.3	0.9	3.2	G5 IV		65 s		
90206	62031		10 25 30.6	+36 12 15	-0.001	-0.04	6.6	1.1	0.2	K0 III		190 s		
90303	118292		10 25 35.0	+8 46 35	-0.002	-0.01	7.90	0.4	2.6	F0 V	-18	120 s	7764	m
90365	178802		10 25 35.1	-21 57 12	+0.002	-0.02	7.9	0.9	0.2	K0 III		340 s		

HD	SAO	Star Name	α 2000	δ 2000	μ(α)	μ(δ)	V	B-V	M_v	Spec	RV	d(pc)	ADS	Notes
			$10^h25^m35.9^s$	$-19°35'59''$	$-0^s.005$	$-0''.01$								
90364	155973		10 25 35.9	-19 35 59	-0.005	-0.01	7.1	0.1	0.6	A0 V		120 mx		
90362	137557		10 25 44.1	-7 03 36	-0.010	+0.12	5.57	1.52	-0.4	M0 III	+32	87 mx		
90250	62037		10 25 44.8	+35 25 31	-0.009	-0.07	6.48	1.09	0.0	K1 III	+11	130 mx		
90579	250950		10 25 45.7	-60 49 57	+0.001	0.00	8.0	0.1	0.4	B9.5 V		260 s		
90533	238054		10 25 46.3	-54 59 19	-0.005	+0.05	7.1	1.1	-0.5	M2 III		280 mx		
90390	178806		10 25 51.1	-20 19 33	0.000	-0.01	7.3	0.8	0.2	K0 III		260 s		
90387	155977		10 25 51.7	-13 07 58	-0.005	+0.02	7.4	0.4	3.0	F2 V		75 s		
90277	62038	30 LMi	10 25 54.8	+33 47 45	-0.006	-0.07	4.74	0.25	2.6	F0 V	+13	27 s		
90374	137559		10 25 56.7	-7 51 31	-0.001	0.00	7.9	1.1	-0.1	K2 III		410 s		
90361	118296		10 25 58.6	+2 55 44	-0.002	-0.01	7.60	0.1	0.6	A0 V		250 s	7769	m
90204	27639		10 25 59.0	+52 37 17	-0.013	-0.05	7.90	0.5	3.4	F5 V	+15	79 s	7762	m
90276	62039		10 26 02.6	+39 16 00	+0.001	-0.04	7.3	1.4	-0.3	K5 III		330 s		
90432	155980	42 μ Hya	10 26 05.3	-16 50 11	-0.009	-0.08	3.81	1.48	-0.3	K4 III	+40	59 s		
90386	118299		10 26 09.1	+3 55 55	0.000	-0.02	6.6	0.1	1.4	A2 V		110 s	7773	m
90518	222047		10 26 09.4	-42 44 20	-0.012	-0.05	6.13	1.13	0.2	K1 III		110 mx		
90520	222050		10 26 10.5	-45 33 46	-0.009	+0.06	7.52	0.65	4.9	G3 V		31 s		
90519	222049		10 26 10.9	-45 31 54	-0.001	-0.02	7.82	1.36	0.0	K1 III		260 s		
90661	250956		10 26 13.4	-63 17 56	-0.005	-0.01	7.08	-0.01	0.6	A0 V		200 s		
91620	258577		10 26 14.0	-84 20 52	-0.016	+0.03	7.1	1.6	-0.5	M4 III		330 s		X Oct, v
90346	81339		10 26 15.3	+24 42 55	+0.001	-0.01	7.3	1.1	0.0	K1 III		280 s		
90586	238059		10 26 15.6	-53 53 30	-0.001	-0.01	7.30	1.4	-0.3	K5 III		320 s		m
90430	137563		10 26 18.3	-6 25 38	0.000	-0.01	6.9	0.1	0.6	A0 V		190 s		
90107	15176		10 26 18.4	+67 41 09	-0.012	+0.01	7.6	1.1	0.2	K0 III		270 mx		
90249	27642		10 26 22.6	+54 24 26	-0.007	-0.01	7.1	0.9	3.2	G5 IV		61 s		
90549	201391		10 26 25.5	-39 21 42	-0.003	0.00	6.8	0.9	0.2	K0 III		210 s		
90420	99120		10 26 26.4	+10 30 11	+0.003	-0.04	7.6	1.1	0.2	K0 III		300 s		
90222	27641		10 26 30.1	+59 09 11	0.000	-0.01	7.6	1.1	0.2	K0 III		310 s		
--	222056		10 26 34.9	-49 53 07	-0.002	-0.01	8.0	1.9	-0.3	K5 III		250 s		
90473	137565		10 26 36.8	-0 59 18	-0.003	0.00	6.31	1.46	-0.2	gK3	+4	160 s		
90428	99122		10 26 39.4	+12 29 20	-0.002	+0.01	8.0	0.9	3.2	G5 IV		93 s		
90967	256715		10 26 39.4	-75 08 59	-0.021	+0.02	7.82	0.30		A5				
90611	222057		10 26 43.9	-47 39 36	+0.001	-0.07	6.70	0.2	2.1	A5 V		83 s		m
90731	250958		10 26 44.5	-64 21 06	-0.001	0.00	7.4	0.1	0.6	A0 V		220 s		
90445	99123		10 26 45.2	+13 37 35	-0.002	-0.01	7.4	1.1	0.2	K0 III		270 s		
90485	137567		10 26 46.8	-4 23 18	-0.003	0.00	6.54	0.96	0.3	gG7	+6	170 s		
90677	238067		10 26 48.9	-54 52 39	-0.001	-0.01	5.58	1.56		K0				
90914	256714		10 26 49.1	-73 10 05	-0.009	+0.01	6.6	1.6	-0.5	M2 III		270 s		
90444	99127		10 26 53.0	+17 13 10	-0.002	-0.07	7.90	0.5	4.0	F8 V		60 s	7775	m
90460	99125		10 26 54.2	+19 30 46	0.000	-0.02	8.00	0.5	4.0	F8 V		63 s		m
90706	238072		10 26 55.4	-57 36 26	-0.004	-0.01	7.06	0.47	-6.3	B3 I k		3400 s		
90443	81346		10 26 56.4	+24 56 52	-0.001	-0.01	7.90	1.1	0.0	K1 III		380 s		
90471	81349		10 26 58.4	+20 26 02	0.000	-0.03	7.9	0.9	3.2	G5 IV		88 s		
90441	81347		10 26 59.5	+29 40 30	-0.005	-0.03	7.79	0.35	3.0	F2 V	+13	91 s		m
90472	99128		10 27 00.4	+19 21 52	-0.004	-0.01	6.15	1.14	0.2	K0 III	+32	130 s		
90856	--		10 27 05.2	-68 30 34			7.90	0.00	0.4	B9.5 V		310 s		m
90574	155992		10 27 08.1	-18 15 22	-0.002	+0.02	7.8	1.1	0.2	K0 III		330 s		
90610	201405	α Ant	10 27 09.1	-31 04 04	-0.006	+0.01	4.25	1.45	-0.4	M0 III	+13	85 s		
90654	222060		10 27 09.5	-42 03 10	-0.006	-0.01	7.2	0.4	1.4	A2 V		86 s		
90774	250961		10 27 09.9	-62 01 09	-0.015	-0.10	7.56	0.56		F8				
90624	201406		10 27 10.0	-37 12 48	-0.002	0.00	8.0	1.3	-0.5	M4 III		510 s		
90512	99132		10 27 12.2	+11 19 03	-0.002	+0.02	6.66	0.85	0.3	G5 III		190 s		
90524	99133		10 27 16.8	+12 26 43	+0.001	0.00	8.0	1.1	0.2	K0 III		360 s		
90772	238077		10 27 24.4	-57 38 19	-0.002	0.00	4.68	0.51	-8.5	F0 Ia	-1	3600 s		
90606	155994		10 27 25.2	-18 17 11	-0.002	0.00	7.1	0.1	0.6	A0 V		200 s		
90874	250966		10 27 25.3	-65 42 17	-0.013	+0.02	6.01	0.09		A0				
90470	43344		10 27 27.9	+41 36 03	-0.005	-0.08	6.02	0.17		A2	+7			
90572	118314		10 27 30.3	+3 33 49	-0.007	-0.03	7.23	0.97	0.2	gK0	+45	260 s	7779	m
90493	62051		10 27 33.5	+39 19 57	-0.009	-0.03	7.6	0.5	3.4	F5 V		68 s		
90855	250967		10 27 34.8	-63 01 21	-0.003	+0.01	7.5	1.1	-0.1	K2 III		320 s		
90785	238080		10 27 38.1	-53 54 13	-0.002	0.00	8.01	1.08	0.2	K0 III		320 s		
90569	99136	45 Leo	10 27 38.9	+9 45 45	0.000	0.00	6.04	-0.06		A2 p	-7		7781	m
90740	222071		10 27 41.1	-44 20 32	-0.002	-0.01	7.11	0.90	0.3	G5 III		220 s		
90509	43348		10 27 43.5	+41 14 10	-0.002	0.00	7.1	1.1	0.2	K0 III		240 s		
90400	27648		10 27 45.6	+59 35 49	-0.008	-0.03	7.0	1.1	0.2	K0 III		230 s		
90752	222074		10 27 46.6	-45 05 56	-0.002	-0.03	7.4	0.9	0.2	K0 III		250 s		
90725	201415		10 27 47.5	-38 42 14	-0.006	-0.04	7.60	0.5	4.0	F8 V		53 s		m
90712	201414		10 27 47.6	-34 23 58	-0.011	-0.04	7.6	0.3	4.4	G0 V		44 s		
90667	156000		10 27 49.9	-15 55 06	-0.006	-0.02	8.0	0.9	3.2	G5 IV		90 s		
90853	238085		10 27 52.7	-58 44 22	-0.002	-0.01	3.82	0.31	-2.0	F0 II	+9	140 s		
90537	62053	31 β LMi	10 27 52.9	+36 42 26	-0.010	-0.11	4.21	0.90	1.8	G8 III-IV	+6	31 s	7780	m
90816	238084		10 27 53.0	-56 12 23	+0.001	-0.01	7.8	0.5	1.7	A3 V		87 s		
90783	222079		10 27 55.8	-48 01 16	-0.002	-0.01	7.5	0.0	3.4	F5 V		68 s		
90797	222080		10 27 58.5	-49 16 11	-0.002	0.00	8.0	0.7	-1.6	B3.5 V		270 s		
--	222076		10 27 58.9	-42 00 36	-0.002	-0.01	8.0	1.6	-0.3	K5 III		380 s		
90798	222081		10 28 01.9	-49 24 19	0.000	-0.04	6.10	1.51		K2				

HD	SAO	Star Name	α 2000	δ 2000	μ(α)	μ(δ)	V	B−V	M_v	Spec	RV	d(pc)	ADS	Notes
			h m s	° ′ ″	s	″								
90872	238086		10 28 02.7	−57 11 52	−0.003	+0.02	6.85	−0.15	0.4	B9.5 V		200 s		
90507	43350		10 28 03.3	+48 57 49	0.000	−0.01	6.8	0.9	3.2	G5 IV		53 s		
90508	43351		10 28 03.8	+48 47 05	+0.008	−0.89	6.44	0.60	4.5	G1 V	−7	14 mx		m
90651	118319		10 28 05.4	+3 18 55	−0.001	−0.03	7.8	0.1	0.6	A0 V		270 s		
90711	137585		10 28 12.1	−6 36 03	−0.024	−0.30	7.90	0.81	5.9	K0 V	+28	25 s		
90781	201422		10 28 14.4	−36 13 12	−0.001	−0.02	7.5	0.5	3.0	F2 V		62 s		
90566	43353		10 28 15.7	+46 10 39	0.000	−0.04	7.6	0.5	4.0	F8 V		53 s		
90966	250969		10 28 18.2	−63 09 53	−0.002	−0.01	6.45	−0.06	−1.6	B3.5 V		370 s		
90849	222083		10 28 25.0	−46 33 40	−0.001	−0.02	7.1	0.4	0.6	A0 V		110 s		
90684	99144		10 28 25.3	+14 45 12	−0.002	−0.03	7.6	1.1	0.2	K0 III		300 s		
89881	1703		10 28 25.7	+81 11 12	+0.006	−0.01	7.7	0.6	4.4	G0 V		46 s		
90683	99145		10 28 26.3	+15 45 20	−0.009	−0.01	7.2	0.5	4.0	F8 V		44 s		
90736	137588		10 28 28.5	−9 52 52	−0.004	+0.01	7.7	1.1	0.2	K0 III		320 s		
90700	99146		10 28 28.6	+11 29 24	−0.001	−0.01	7.9	0.9	3.2	G5 IV		88 s		
91509	258578		10 28 32.0	−81 03 21	−0.018	+0.04	7.5	1.3	0.2	K0 III		180 s		
90750	137589		10 28 33.8	−6 18 13	−0.002	−0.01	7.8	0.1	0.6	A0 V		270 s		
90602	43356		10 28 36.4	+45 12 43	−0.002	−0.03	6.35	1.32		K0	−4			
90776	156005		10 28 36.9	−19 31 31	+0.001	−0.05	6.8	1.1	0.2	K0 III		210 s		
90868	201429		10 28 39.7	−38 53 31	−0.007	+0.02	7.6	1.4	−0.1	K2 III		240 mx		
90898	222084		10 28 42.3	−43 50 19	−0.004	+0.01	6.9	0.3	0.6	A0 V		120 s		
90719	99149		10 28 42.4	+13 17 19	−0.004	−0.01	7.8	0.1	1.4	A2 V		190 s		
90709	99148		10 28 43.4	+17 08 00	−0.002	0.00	7.1	0.9	3.2	G5 IV		60 s		
90763	137591		10 28 43.9	−3 44 33	0.000	−0.03	6.05	0.05		A0				
90682	81370		10 28 44.8	+26 55 24	+0.002	−0.01	8.00	1.1	−0.2	K3 III		440 s		
90980	238102		10 28 47.9	−57 14 00	−0.003	+0.01	6.74	1.02	0.2	K0 III		200 s		
90718	99150		10 28 48.4	+14 20 37	−0.005	−0.01	7.1	1.1	0.2	K0 III	+38	230 mx		
91024	250973		10 28 49.3	−60 53 41	−0.001	−0.01	7.61	0.27	−6.5	B8 Iab		4600 s		
91056	250979		10 28 52.7	−64 10 20	−0.001	0.00	5.29	1.86		M1	−3	17 mn		
90775	118326		10 28 58.6	+2 38 47	+0.001	−0.07	7.6	0.4	3.0	F2 V		85 s		
90717	81371		10 29 01.6	+29 43 39	+0.001	−0.04	6.61	1.18	0.2	K0 III	+2	140 s		
91094	250981		10 29 03.6	−65 10 34	+0.007	+0.03	6.41	1.67	−0.5	M1 III		180 mx		
91054	238109		10 29 06.6	−59 36 40	+0.003	+0.01	7.67	0.63	−4.0	A0 Ib-II		880 s		
90917	201435		10 29 06.9	−38 03 34	0.000	−0.02	7.2	0.7	3.2	G5 IV		62 s		
90844	156010		10 29 07.6	−12 26 08	−0.001	−0.03	7.9	1.1	0.2	K0 III		340 s		
90791	99153		10 29 11.3	+10 09 21	0.000	−0.03	7.80	0.5	4.0	F8 V		58 s	7792	m
90842	118331		10 29 21.0	+6 15 24	−0.003	+0.01	7.80	1.1	−0.1	K2 III		380 s	7794	m
90974	222093		10 29 24.5	−40 57 45	0.000	−0.01	7.4	0.1	0.6	A0 V		230 s		
90823	99157		10 29 25.5	+12 11 13	−0.005	−0.01	7.8	0.4	3.0	F2 V		89 s		
90882	137600	29 δ Sex	10 29 28.6	−2 44 21	−0.003	−0.02	5.21	−0.06		B9 n	+19			
90958	201440		10 29 28.9	−33 24 32	−0.003	−0.02	8.0	0.3	1.4	A2 V		140 s		
90957	178888		10 29 28.9	−29 39 49	−0.005	+0.01	5.58	1.42	−0.3	gK5	−5	150 s		
90973	201441		10 29 33.6	−32 56 30	−0.004	−0.03	7.4	1.1	−0.1	K2 III		270 mx		
90972	201442	δ Ant	10 29 35.3	−30 36 26	−0.003	0.00	5.56	−0.04		A0 n	+19			m
91128	250982		10 29 36.2	−62 06 33	0.000	0.00	7.54	0.57	−4.8	A5 Ib		1600 s		
91034	222095		10 29 38.1	−43 03 03	−0.003	0.00	6.9	1.2	−0.3	K5 III		280 s		
90955	178890		10 29 38.8	−22 14 49	0.000	−0.01	7.3	0.1	0.6	A0 V		190 s		
89571	1701		10 29 41.4	+84 15 07	−0.088	−0.04	5.6	0.1	1.7	A3 V	+3	61 s		
91298	—		10 29 42.2	−72 37 44			8.0	1.1	−0.1	K2 III		420 s		
90905	118333		10 29 42.2	+1 29 28	−0.010	−0.13	6.9	0.5	3.4	F5 V	+1	50 s		
90954	178895		10 29 42.3	−22 02 39	−0.001	+0.03	7.1	0.7	2.6	F0 V		46 s		
90715	27660		10 29 47.7	+54 49 33	−0.002	+0.01	7.3	1.1	0.2	K0 III		270 s		
90861	81381		10 29 53.6	+28 34 52	+0.001	0.00	6.90	1.12	−0.1	K2 III	+39	250 s		m
90633	15196	35 UMa	10 29 54.3	+65 37 34	0.000	−0.03	6.32	1.14	0.2	K0 III	−25	140 s		
91125	238120		10 29 54.9	−54 58 42	+0.001	−0.02	8.0	0.0	1.4	A2 V		210 s		
90790	43361		10 29 56.4	+45 51 09	−0.007	−0.01	7.9	0.4	2.6	F0 V		110 s		
90878	81383		10 29 58.3	+27 49 34	−0.005	−0.01	7.80	0.48	4.0	F8 V		58 s		
91273	—		10 30 02.8	−68 33 44			8.0	0.0	0.4	B9.5 V		330 s		
91089	222099		10 30 02.9	−43 07 47	−0.004	0.00	7.6	−0.1	0.4	B9.5 V		220 mx		
90860	43362		10 30 04.3	+40 28 12	+0.002	−0.02	7.0	0.5	4.0	F8 V		40 s		
90840	62076	32 LMi	10 30 06.4	+38 55 30	−0.002	−0.01	5.77	0.08	0.1	A4 III	+3	140 s		
91140	238122		10 30 07.1	−53 20 36	−0.001	0.00	8.0	0.3	0.4	B9.5 V		200 s		
91326	256720		10 30 08.1	−70 53 18	−0.004	+0.01	7.0	1.9	−0.5	M4 III		230 s		
91272	250989		10 30 08.8	−66 59 06	−0.003	−0.01	6.19	0.01		B5	−9			
90806	27664		10 30 11.3	+50 59 34	−0.006	−0.04	7.70	0.4	2.6	F0 V		110 s	7796	m
90807	27665		10 30 15.2	+50 34 08	+0.001	−0.04	6.7	0.4	3.0	F2 V		56 s		
90714	15199		10 30 15.9	+63 20 55	+0.002	0.00	7.40	1.1	0.2	K0 III		280 s	7793	m
90994	137608	30 β Sex	10 30 17.4	−0 38 13	−0.003	−0.03	5.09	−0.14	−0.9	B6 V	+12	160 s		
91375	256722		10 30 20.0	−71 59 34	+0.004	−0.03	4.74	0.04		A2	+8	18 mn		
91188	238130		10 30 22.5	−57 04 39	+0.001	−0.01	6.64	−0.10	0.0	B8.5 V		210 s		
90745	15200		10 30 26.4	+64 15 28	−0.009	−0.06	6.0	0.1	1.7	A3 V	−12	45 mx		
91010	118340		10 30 27.3	+7 03 33	−0.004	0.00	7.4	0.9	3.2	G5 IV		69 s		
91124	201456		10 30 30.8	−39 03 51	−0.001	−0.02	7.6	−0.1	3.2	G5 IV		75 s		
91011	118341		10 30 30.9	+2 09 00	+0.004	−0.05	6.99	1.03	0.2	gK0	−1	220 s		
91062	156027		10 30 32.7	−10 48 29	−0.002	+0.01	7.6	1.4	−0.3	K5 III		370 s		
90931	62084		10 30 35.9	+36 16 22	−0.004	−0.09	6.8	0.4	2.6	F0 V		69 s		

HD	SAO	Star Name	α 2000	δ 2000	μ(α)	μ(δ)	V	B–V	M_v	Spec	RV	d(pc)	ADS	Notes
			h m s	° ′ ″	s	″								
90839	27670	36 UMa	10 30 37.5	+55 58 50	−0.021	−0.03	4.84	0.52	4.0	F8 V	+9	13 ts		
91270	250992		10 30 39.2	−61 21 22	0.000	+0.02	6.40	1.6	−0.5	M2 III		230 s		
90859	27671		10 30 41.7	+56 34 09	−0.003	+0.01	7.9	1.1	−0.1	K2 III		390 s		
91242	238136		10 30 44.6	−54 28 53	+0.003	−0.01	7.80	1.4	−0.3	K5 III		400 s		m
92683	258586		10 30 49.6	−86 05 25	−0.002	0.00	6.8	0.2	0.6	A0 V		130 s		
91135	178917		10 30 51.3	−26 29 01	−0.005	+0.01	6.51	0.54		F5				
91214	222109		10 30 57.3	−44 22 02	0.000	+0.02	6.90	0.9	3.2	G5 IV		55 s		m
91307	250994		10 30 58.5	−61 36 47	+0.004	+0.01	7.8	0.3	0.6	A0 V		170 s		
91106	137614		10 30 58.6	−7 38 15	−0.003	0.00	6.20	1.38	−0.3	K5 III	+7	200 s	7808	m
91120	156029		10 30 59.7	−13 35 18	−0.003	0.00	5.58	−0.04		B9 pne	+13			
91496	256723		10 31 02.0	−73 13 18	−0.003	−0.01	4.93	1.68	−0.5	gM1	+11	110 s		m
90990	62091		10 31 02.9	+39 01 13	+0.002	−0.06	7.5	1.1	−0.1	K2 III		230 mx		
90089	1714		10 31 04.4	+82 33 31	−0.044	+0.03	5.26	0.37	2.1	F5 IV	+7	43 s		
91173	178922		10 31 04.8	−24 10 48	0.000	−0.03	7.8	0.3	1.7	A3 V		120 s	7809	m
91223	222112		10 31 06.1	−42 13 16	−0.003	−0.02	8.04	0.10	1.4	A2 V		200 s		m
91407	250998		10 31 07.4	−67 42 10	−0.002	−0.01	7.8	0.0	0.4	B9.5 V		300 s		
91423	250999		10 31 07.7	−68 53 50	−0.001	−0.01	7.90	0.1	0.6	A0 V		280 s		m
91239	222113		10 31 13.2	−42 13 46	−0.002	−0.01	7.37	−0.08	0.4	B9.5 V		250 s		m
91441	--		10 31 13.6	−68 58 48			7.6	0.0	0.4	B9.5 V		270 s		
90838	15207		10 31 19.0	+66 18 40	−0.001	0.00	7.3	0.4	3.0	F2 V		73 s		
90820	15205		10 31 21.4	+68 19 30	+0.007	0.00	8.0	1.1	0.2	K0 III		360 s		
91324	238146		10 31 21.8	−53 42 56	−0.047	+0.20	4.89	0.50	3.7	dF6 s	+20	21 ts		
90696	7154		10 31 22.2	+73 19 50	+0.014	−0.02	7.1	1.1	0.2	K0 III		240 s		
91249	201471		10 31 30.0	−35 07 43	−0.003	−0.08	7.2	1.5	0.2	K0 III		130 s		
91029	43369		10 31 31.9	+46 52 29	−0.001	0.00	7.6	1.1	−0.1	K2 III		350 s		
91208	156036		10 31 33.0	−17 04 37	−0.003	−0.03	7.9	1.1	0.2	K0 III		340 s		
91165	118347		10 31 36.2	+2 50 42	−0.003	−0.02	6.5	1.1	0.2	K0 III		180 s		
91292	222118		10 31 37.7	−40 48 40	−0.002	−0.02	7.3	1.7	−0.1	K2 III		300 s		
91207	156037		10 31 37.8	−14 53 09	−0.002	−0.02	6.9	0.1	0.6	A0 V		180 s		
91323	222122		10 31 43.7	−44 29 06	−0.002	−0.01	7.22	−0.12	−2.2	B5 III		750 s		m
91206	137621		10 31 44.4	−8 57 09	−0.001	−0.04	7.9	1.4	−0.3	K5 III		430 s		
91148	81396		10 31 45.6	+24 04 56	+0.003	−0.04	7.89	0.69	5.6	G8 V		29 s		
91280	178938		10 31 48.6	−28 14 15	−0.007	−0.02	6.05	0.51	4.0	F8 V		26 s		
91452	251005		10 31 50.8	−63 56 25	+0.001	0.00	7.50	0.23	−6.2	B0 I		3000 s		
92029	258581		10 31 51.2	−81 55 15	−0.014	+0.01	7.07	−0.08		B8				m
91130	62101	33 LMi	10 31 51.3	+32 22 46	+0.001	0.00	5.80	0.11	0.0	A0 IV	−12	120 s	7813	m
91370	238152		10 31 51.9	−52 13 54	0.000	−0.01	7.50	0.00	0.0	B8.5 V		310 s		m
91356	222125		10 31 56.5	−45 04 10	−0.002	0.00	6.5	0.0	0.0	B8.5 V		200 s		
91355	222126		10 31 57.5	−45 04 00	0.000	−0.01	5.16	−0.15	0.0	B8.5 V		110 s		m
91164	81398		10 31 57.6	+24 43 23	−0.002	−0.01	7.90	1.1	0.2	K0 III		350 s		
90645	7155		10 32 01.0	+77 29 53	−0.009	−0.01	7.9	0.9	3.2	G5 IV		86 s		
91248	156040		10 32 01.3	−11 25 25	+0.003	−0.03	7.5	0.4	3.0	F2 V		78 s		
91163	--		10 32 01.4	+29 43 56	+0.006	−0.02	7.86	0.61	4.4	G0 V	−20	46 s		
91465	251006		10 32 01.4	−61 41 07	−0.003	+0.01	3.32	−0.09	−1.7	B3 V	+26	96 s		p Car, v
91204	99171		10 32 05.4	+17 59 23	+0.001	−0.02	7.9	0.6	4.4	G0 V		51 s		
91232	99172	46 Leo	10 32 11.7	+14 08 14	−0.003	+0.02	5.46	1.68	−0.5	M2 III	+34	140 s		
91245	99174		10 32 16.8	+12 55 09	−0.006	0.00	8.0	0.4	3.0	F2 V		98 s		
91220	81403		10 32 17.3	+24 26 33	−0.004	+0.01	7.2	0.4	2.6	F0 V		83 s		
91256	118354		10 32 18.4	+4 38 43	−0.001	+0.03	7.2	1.1	0.2	K0 III	+7	250 s		
91477	238159		10 32 20.9	−58 32 45	−0.004	0.00	7.35	−0.11	0.0	B8.5 V		290 mx		
91352	178948		10 32 24.3	−23 36 08	−0.005	−0.01	7.5	0.3	0.6	A0 V		110 mx		
91388	201491		10 32 27.1	−38 44 48	+0.004	−0.08	7.5	0.3	0.2	K0 III		260 mx		
91181	43375		10 32 28.4	+44 10 54	−0.008	−0.02	7.4	0.2	2.1	A5 V	−1	110 mx		
91318	137628		10 32 28.5	−6 04 30	−0.001	−0.04	6.90	1.1	0.2	K0 III		220 s	7822	m
92091	258583		10 32 29.4	−81 36 57	−0.010	+0.01	7.5	0.9	0.2	K0 III		280 s		
91437	222129		10 32 33.5	−44 37 07	−0.002	−0.03	5.91	0.92		K0				
91600	--		10 32 37.9	−66 10 52			8.0	1.6	−0.5	M4 III		480 s		
91418	201497		10 32 40.3	−36 27 23	−0.008	+0.04	7.9	1.1	0.2	K0 III		180 mx		
91369	156047		10 32 41.1	−16 57 31	−0.001	−0.09	7.7	0.8	3.2	G5 IV	+12	78 s		
91491	238164		10 32 42.4	−53 08 25	−0.001	−0.01	7.78	1.12	0.2	K0 III		280 s		
91400	201498		10 32 43.6	−32 33 35	−0.002	−0.04	7.7	0.6	0.2	K0 III		320 s		
91461	222132		10 32 46.4	−43 37 41	−0.003	−0.02	7.8	0.8	0.2	K0 III		330 s		
91533	238168		10 32 47.6	−58 40 00	−0.003	0.00	6.00	0.32	−6.7	A2 Iab		2500 s		
91316	118355	47 ρ Leo	10 32 48.6	+9 18 24	−0.001	−0.01	3.85	−0.14	−5.7	B1 Ib	+42	770 s		m
91416	178955		10 32 51.1	−26 36 33	−0.001	+0.01	7.6	1.5	−0.3	K5 III		380 s		
92106	258584		10 32 52.5	−81 09 44	−0.012	+0.02	8.0	−0.3	0.6	A0 V		160 mx		
91398	178957		10 32 54.6	−22 31 38	−0.002	+0.01	6.9	0.1	0.4	B9.5 V		160 s		
91504	222136		10 32 56.9	−47 00 12	−0.001	0.00	5.02	1.04	−0.3	K4 III	+4	120 s		m
91573	238171		10 32 57.6	−60 06 47	−0.004	+0.01	7.2	0.2	1.7	A3 V		110 s		
91873	256728		10 33 04.6	−76 02 19	+0.001	−0.02	7.9	1.1	0.2	K0 III		350 s		
91286	62117		10 33 08.9	+39 13 19	+0.002	0.00	7.3	1.1	−0.1	K2 III		300 s		
91434	156057		10 33 12.4	−13 24 15	−0.003	−0.01	6.8	1.1	0.2	K0 III		210 s		
91312	43379		10 33 13.8	+40 25 31	−0.012	−0.01	4.75	0.23	1.5	A7 IV	+14	45 s	7826	m
91412	137634		10 33 14.7	−5 21 46	+0.007	−0.17	7.1	1.1	0.2	K0 III		120 mx		
--	238177		10 33 15.2	−55 23 12	+0.001	0.00	6.60	0.8	3.2	G5 IV		48 s		m

HD	SAO	Star Name	α 2000	δ 2000	μ(α)	μ(δ)	V	B-V	M$_v$	Spec	RV	d(pc)	ADS	Notes
91366	81409		10^{h}33^{m}16.8^s	+25°07'29"	0^s.000	-0".02	7.70	1.1	0.0	K1 III		350 s		
91619	238182		10 33 25.3	-58 11 25	-0.002	-0.01	6.14	0.35	-7.0	B5 Ia	+7	2400 s		
91629	238184		10 33 27.4	-59 25 10	-0.002	-0.02	7.3	1.0	3.2	G5 IV		48 s		
91698	251013		10 33 28.8	-65 22 07	+0.001	-0.01	8.02	0.01		B8				
91519	201510		10 33 29.3	-36 03 58	-0.002	-0.01	7.6	0.9	1.7	A3 V		50 s		
91365	62121	34 LMi	10 33 30.8	+34 59 19	-0.003	-0.01	5.60	0.02	1.4	A2 V	+12	69 s		
91538	201512		10 33 31.1	-40 14 19	+0.002	-0.04	6.68	0.92	0.3	G8 III		190 s		
91590	222145		10 33 32.4	-46 58 35	-0.001	-0.02	7.20	0.00	0.4	B9.5 V		230 s		m
91330	43382		10 33 36.6	+45 54 53	0.000	0.00	7.9	1.1	-0.1	K2 III		390 s		
91311	27682		10 33 43.5	+53 29 50	-0.005	-0.03	6.5	0.1	0.6	A0 V	+2	98 mx		
91427	81415		10 33 47.3	+23 21 01	-0.003	-0.02	7.30	0.1	1.4	A2 V		150 s	7833	m
91347	43383		10 33 50.3	+49 11 10	+0.027	+0.13	7.6	0.5	4.0	F8 V	-25	53 s		
91455	81416		10 33 50.4	+20 18 16	-0.007	-0.01	7.3	0.5	3.4	F5 V		62 s		
91551	178978		10 33 51.3	-27 20 57	-0.002	-0.01	7.6	1.1	-0.1	K2 III		350 s		
92210	258585		10 33 55.9	-80 30 58	-0.005	+0.07	7.1	0.4	2.6	F0 V		68 s		
91550	178979	44 Hya	10 34 00.8	-23 44 42	-0.001	+0.02	5.08	1.60	-0.3	K4 III	-4	93 s	7834	m
91498	99185		10 34 07.3	+12 22 28	-0.001	-0.01	7.70	0.1	1.7	A3 V		160 s		
91674	222153		10 34 09.7	-45 44 31	-0.013	+0.02	8.0	0.8	3.4	F5 V		49 s		
91767	251014		10 34 12.9	-60 59 15	+0.001	-0.01	6.23	1.40		K2				
91114	7161		10 34 13.6	+73 50 02	-0.002	-0.04	7.60	0.1	1.7	A3 V		150 s	7824	m
91645	201523		10 34 16.8	-37 23 12	-0.003	0.00	6.90	0.03	0.4	B9.5 V		180 s		m
91566	137648		10 34 18.0	-9 53 57	-0.004	0.00	7.7	0.5	3.4	F5 V		74 s		
91426	43387		10 34 20.8	+49 06 27	-0.002	-0.02	7.1	1.4	-0.3	K5 III		300 s		
91527	81420		10 34 24.0	+21 35 38	0.000	0.00	7.3	1.1	0.2	K0 III		260 s	7836	m
91691	201525		10 34 32.5	-38 07 44	-0.001	-0.01	7.9	1.1	0.2	K0 III		320 s		
91580	118375		10 34 32.7	+1 45 40	-0.007	-0.10	7.8	0.5	3.4	F5 V		74 s		
91394	27688		10 34 32.9	+56 09 50	+0.001	-0.02	8.0	0.5	3.4	F5 V		82 s		
91660	178990		10 34 42.0	-22 12 11	-0.001	0.00	7.70	1.1	0.2	K0 III		320 s		m
91826	251018		10 34 42.4	-60 16 19	0.000	0.00	7.86	-0.05	-2.6	B2.5 IV		1000 s		
91546	81421		10 34 43.7	+26 09 32	+0.001	-0.07	7.8	0.5	3.4	F5 V		76 s		
91612	118376	48 Leo	10 34 47.9	+6 57 14	-0.007	+0.06	5.08	0.94	-0.9	G8 II-III	+5	49 mx		
91707	201527		10 34 48.9	-31 20 46	+0.005	-0.08	7.37	0.44		F2				
91103	7163		10 34 49.1	+76 43 55	-0.002	+0.01	7.54	1.16		K0				
91329	15224		10 34 50.0	+68 42 23	-0.007	-0.02	7.7	1.1	-0.1	K2 III		370 s		
91545	81423		10 34 50.1	+27 57 43	+0.013	-0.08	6.84	1.07	-0.1	K2 III	-30	120 mx		
91638	137653		10 34 50.3	-3 53 59	+0.001	-0.16	6.7	0.5	4.0	F8 V		34 s		
91714	201528		10 34 53.0	-35 05 08	-0.005	+0.01	7.7	1.7	0.2	K0 III		120 s		
91775	238205		10 34 53.2	-50 26 33	0.000	-0.02	8.0	1.0	0.4	B9.5 V		330 s		
91906	251022		10 34 56.9	-64 08 02	+0.001	+0.01	7.40	0.04		A0				m
91706	178993		10 34 57.6	-23 10 34	-0.007	+0.03	6.10	0.50	3.8	dF7	+12	28 s		
91713	178998		10 35 01.9	-26 22 54	-0.004	-0.01	7.6	1.1	0.2	K0 III		260 s		
91636	118380	49 Leo	10 35 02.1	+8 39 01	-0.004	-0.01	5.67	0.05	1.4	A2 V	+17	71 s	7837	TX Leo, m,v
91795	222168		10 35 04.2	-45 53 43	-0.004	+0.04	7.7	0.4	3.0	F2 V		78 s		
91869	238217		10 35 04.5	-57 40 39	-0.004	-0.02	7.10	1.1	0.2	K0 III		230 s		m
91848	238214		10 35 04.6	-55 24 36	-0.007	-0.06	7.20	0.81		G0				
91190	7164		10 35 05.5	+75 42 46	-0.008	-0.01	4.84	0.96	0.2	K0 III	+17	85 s		
91657	118381		10 35 06.0	+2 12 17	-0.002	0.00	6.7	1.1	-0.1	K2 III		230 s		
91985	251025		10 35 06.2	-69 21 06	0.000	-0.03	7.2	1.6	-0.5	M2 III		340 s		
92009	---		10 35 07.7	-70 37 01			8.0	0.1	0.6	A0 V		290 s		
91564	43390		10 35 08.5	+41 54 32	+0.002	-0.02	7.0	1.4	-0.3	K5 III		280 s		
91480	27695	37 UMa	10 35 09.6	+57 04 57	+0.008	+0.04	5.16	0.34	2.8	F1 V	-12	30 s		
91805	222170		10 35 10.5	-43 39 53	-0.003	+0.02	6.08	0.94		G5				m
90343	1722		10 35 11.5	+84 23 58	+0.015	+0.05	7.30	0.82	0.2	K0 III		260 s		
91576	43392		10 35 12.2	+40 52 33	-0.002	-0.08	7.8	1.1	0.2	K0 III		140 mx		
91793	201533		10 35 12.8	-39 33 45	-0.003	-0.01	5.38	2.88		N7.7	+37			U Ant, v
91667	118382		10 35 15.2	+4 49 22	-0.001	-0.01	8.0	1.1	0.2	K0 III		360 s		
91656	81427		10 35 17.3	+21 22 44	-0.007	-0.06	7.7	0.4	3.0	F2 V		87 s		
91728	156084		10 35 22.1	-12 51 32	-0.005	-0.06	6.78	0.34		A5				
92209	256730		10 35 24.6	-76 18 32	-0.006	+0.01	6.30	1.20		K0				
91847	222173		10 35 24.8	-50 10 46	+0.003	-0.03	7.8	1.5	-0.1	K2 III		330 s		
91725	137660		10 35 25.3	-5 21 43	-0.004	0.00	7.3	0.1	1.4	A2 V		140 mx		
91804	201538		10 35 27.3	-37 21 41	-0.004	0.00	7.0	0.9	1.7	A3 V		36 s		
92305	256731	γ Cha	10 35 28.1	-78 36 27	-0.013	+0.02	4.11	1.58	-0.4	M0 III	-22	77 s		
92047	251029		10 35 29.8	-69 49 03	-0.011	0.00	7.9	0.0	0.4	B9.5 V		170 mx		
91603	43396		10 35 31.2	+45 39 13	+0.002	-0.01	7.20	1.1	0.2	K0 III		250 s	7838	m
91942	238222		10 35 35.2	-57 33 27	-0.003	0.00	4.45	1.62	-0.4	gM0	+10	92 s		
91790	156087		10 35 38.8	-18 34 08	0.000	0.00	6.49	0.20		A0				
---	---		10 35 41.8	-58 13 04			7.70	0.1	-5.1	B1 II		2800 s		
91943	238225		10 35 42.0	-58 11 34	0.000	0.00	6.71	0.07		B0.5 Ib				
---	238228		10 35 43.2	-58 14 45	-0.007	-0.02	7.0							
91789	156088		10 35 43.3	-13 57 24	-0.009	-0.05	7.8	0.5	3.4	F5 V		74 s		
91892	222180		10 35 44.2	-46 30 59	-0.001	-0.02	7.9	1.5	0.2	K0 III		180 s		
91969	238230		10 35 49.3	-58 13 29	0.000	-0.02	6.51	0.01	-5.8	B0 Ib		2200 s		
91927	238226		10 35 50.7	-51 25 44	-0.003	-0.01	7.8	1.9	-0.3	K5 III		230 s		
91787	137665		10 35 51.4	-8 32 05	+0.001	-0.06	7.7	1.1	-0.1	K2 III		280 mx		

HD	SAO	Star Name	α 2000	δ 2000	μ(α)	μ(δ)	V	B–V	M_v	Spec	RV	d(pc)	ADS	Notes
			h m s	° ′ ″	s	″								
91904	222183		10 35 51.5	−49 13 53	−0.002	−0.01	7.1	1.0	0.4	B9.5 V		55 s		
92253	256732		10 35 54.1	−75 47 32	−0.016	+0.05	7.4	0.7	0.2	K0 III		270 mx		
91738	99194		10 35 55.1	+18 02 43	+0.004	−0.06	7.9	1.1	0.2	K0 III		260 mx		
91802	118388		10 36 01.4	+2 44 50	−0.004	0.00	7.7	1.1	0.2	K0 III		310 s		
91075	1735		10 36 01.8	+80 29 40	−0.009	−0.01	6.5	0.9	3.2	G5 IV	−11	47 s		
91816	156090		10 36 02.1	−11 54 47	+0.009	−0.25	8.03	0.86	6.9	dK3	+4	17 s		
91881	179014		10 36 04.5	−26 40 30	0.000	−0.07	6.29	0.48	3.7	F6 V	−21	33 s	7846	m
91497	15233		10 36 06.3	+68 56 23	+0.001	−0.01	7.9	1.1	0.2	K0 III		340 s		
91785	99198		10 36 06.7	+11 36 53	−0.001	+0.02	7.90	1.1	−0.1	K2 III		400 s		m
91701	43400		10 36 08.9	+47 45 14	−0.007	−0.02	7.9	1.1	0.2	K0 III		310 mx		
92025	238237		10 36 10.5	−60 10 56	+0.002	−0.03	7.9	0.0	−1.6	B3.5 V		610 s		
91800	99202		10 36 13.6	+13 48 46	−0.002	0.00	7.7	0.1	1.4	A2 V		180 s		
91880	156093		10 36 16.6	−16 20 40	−0.002	−0.01	6.03	1.65	−0.5	gM1	+16	190 s	7847	m
91858	156092		10 36 17.3	−10 35 00	+0.001	−0.04	6.57	0.29		A5				
91981	222188		10 36 17.6	−47 51 00	−0.010	+0.01	7.29	0.58	2.8	G0 IV		79 s		m
91683	27703		10 36 18.0	+52 06 33	−0.001	−0.01	7.4	0.1	1.4	A2 V		160 s		
92219	—		10 36 18.1	−72 31 06			8.0	0.1	1.4	A2 V		210 s		
92063	238242		10 36 20.3	−59 33 53	−0.007	−0.06	5.08	1.18	0.2	gK0	−12	74 s		
91752	62147	35 LMi	10 36 21.4	+36 19 36	+0.003	−0.04	6.28	0.39	3.0	F2 V	−24	44 s		
91811	99203		10 36 21.7	+14 47 35	0.000	−0.03	8.0	1.6	−0.5	M2 III		510 s		
91993	222189		10 36 21.9	−47 54 38	0.000	−0.01	7.7	0.4	0.4	B9.5 V		150 s		
91889	156095		10 36 32.3	−12 13 49	+0.018	−0.68	5.70	0.52	4.0	F8 V	−9	23 ts		m
92072	238247		10 36 32.7	−59 11 25	−0.003	−0.01	7.03	−0.10	−1.0	B5.5 V		400 s		
91955	179025		10 36 35.3	−28 46 18	−0.003	0.00	7.50	0.1	0.6	A0 V		240 s	7852	m
91965	201556		10 36 35.9	−33 16 21	−0.002	−0.03	7.13	0.08	0.6	A0 V		180 s		
92087	238250		10 36 39.7	−59 10 02	−0.004	+0.02	7.9	0.0	−1.0	B5.5 V		490 s		
91964	179027		10 36 42.6	−27 39 23	−0.001	0.00	6.8	1.7	−0.1	K2 III		120 s		
92040	222191		10 36 43.9	−47 15 20	−0.001	−0.03	7.6	0.3	3.2	G5 IV		77 s		
91975	179029		10 36 51.3	−22 39 55	+0.004	−0.02	7.2	1.6	−0.3	K5 III		290 s		
92057	222194		10 36 58.1	−41 20 28	+0.001	−0.01	7.1	0.7	0.2	K0 III		240 s		
92084	238254		10 36 58.6	−50 52 39	0.000	+0.01	7.6	1.0	3.2	G5 IV		56 s		
91962	137678		10 36 59.9	−8 50 24	−0.007	−0.06	7.03	0.62		G0		14 mn	7854	m
91910	99210		10 37 01.7	+12 51 59	0.000	−0.02	7.6	0.1	0.6	A0 V		250 s		
92304	256733		10 37 01.8	−71 41 32	−0.008	0.00	7.5	1.1	0.2	K0 III		280 s		
91840	43410		10 37 10.7	+46 32 52	+0.005	−0.04	7.8	1.1	0.2	K0 III		290 mx		
91992	156105		10 37 11.5	−11 44 55	−0.004	−0.04	6.52	0.29		F0				
92102	222197		10 37 12.3	−45 28 52	−0.002	0.00	7.8	1.0	0.2	K0 III		330 s		
92056	201566		10 37 12.6	−31 45 46	−0.003	−0.01	6.7	1.5	0.2	K0 III		110 s		
92036	179041		10 37 13.7	−27 24 46	−0.008	+0.01	4.89	1.62	−0.3	gK5	+17	93 s		
92138	222198		10 37 15.7	−47 53 18	−0.001	−0.01	7.9	0.2	0.4	B9.5 V		220 s		
91932	81438		10 37 16.0	+23 27 00	−0.002	−0.01	8.0	0.5	3.4	F5 V		82 s		
92155	238260		10 37 16.1	−53 51 18	−0.001	0.00	6.40	−0.15	−1.7	B3 V n		400 s		
92139	222199		10 37 18.0	−48 13 32	−0.016	−0.02	3.84	0.30	2.0	F4 IV var	+19	23 s		
91810	27707		10 37 20.4	+56 25 52	−0.006	−0.01	6.55	1.17	0.2	K0 III		130 s		
92137	222200		10 37 23.4	−43 31 23	−0.005	+0.02	7.1	0.6	3.0	F2 V		46 s		
92207	238271		10 37 26.8	−58 44 00	−0.004	0.00	5.45	0.50	−7.1	A0 Ia	−12	180 mx		
92034	156109		10 37 27.0	−12 23 36	−0.001	+0.05	6.9	1.1	0.2	K0 III		220 s		
91425	7169		10 37 28.3	+78 05 19	0.000	0.00	8.0	0.5	3.4	F5 V		82 s		
92154	222202		10 37 28.3	−46 19 45	+0.006	−0.04	7.9	1.5	0.2	K0 III		180 s		
92055	156110		10 37 33.1	−13 23 04	+0.002	−0.04	4.82	2.68		N2	−25			U Hya, v
92098	179046		10 37 35.9	−23 22 54	−0.002	+0.01	7.3	1.0	0.2	K0 III		270 s		
92263	—		10 37 39.0	−61 38 07			7.9	2.0	−0.3	K5 III		210 s		
92136	201576		10 37 42.1	−35 43 13	0.000	−0.02	7.01	0.01	0.4	B9.5 V		190 s		
92013	99219		10 37 46.0	+17 16 44	−0.009	−0.03	7.6	0.5	3.4	F5 V		71 s		
92204	222207		10 37 50.4	−46 16 05	−0.003	+0.04	8.0	1.2	−0.3	K5 III		450 s		
92000	62167		10 37 52.2	+34 04 43	−0.002	0.00	6.6	1.1	0.2	K0 III	+13	190 s		
93310	—		10 37 59.9	−85 11 24			8.0	0.0	0.4	B9.5 V		340 s		
92287	238278		10 38 02.5	−57 15 22	−0.003	0.00	5.91	−0.14	−2.9	B3 III	+20	530 s		
92385	251042		10 38 17.7	−65 02 32	−0.001	−0.01	6.74	−0.07	−0.2	B8 V		240 s		
91948	15243		10 38 19.8	+60 07 27	−0.003	−0.21	7.00	0.5	4.0	F8 V		40 s	7855	m
94009	258591		10 38 20.8	−86 53 49	−0.043	+0.03	7.51	0.77	5.2	G5 V		26 s		
92762	256740		10 38 25.5	−79 30 33	−0.020	+0.04	7.7	0.2	2.1	A5 V		73 mx		
92151	118407		10 38 27.5	+5 54 49	−0.001	−0.02	7.6	0.1	1.7	A3 V		160 s		
92468	—		10 38 27.6	−69 16 11			7.9	0.1	0.6	A0 V		280 s		
91328	1739		10 38 28.4	+81 34 55	−0.024	−0.06	7.6	1.1	0.2	K0 III		200 mx		
92479	256738		10 38 30.8	−70 18 37	−0.001	−0.01	7.4	0.0	0.0	B8.5 V		300 s		
92214	156122	φ Hya	10 38 34.9	−16 52 36	−0.007	+0.02	4.91	0.92	0.2	K0 III	+18	88 s		
92048	43414		10 38 38.3	+49 12 30	−0.002	+0.02	7.9	1.1	0.2	K0 III		350 s		
92330	222221		10 38 42.9	−49 44 33	0.000	−0.01	8.0	1.7	−0.3	K5 III		250 s		
92125	62173	37 LMi	10 38 43.1	+31 58 34	0.000	0.00	4.71	0.81	−2.1	G2 II	−7	230 s		
92184	118410		10 38 43.1	+5 44 02	0.000	−0.01	8.00	0.4	3.0	F2 V		100 s	7864	m
92346	238290		10 38 43.6	−52 45 51	−0.001	0.00	7.4	0.6	0.6	A0 V		93 s		
92397	238295		10 38 45.0	−59 10 58	−0.001	0.00	4.66	1.48	−5.9	cK	+11	1300 s		
92399	238296		10 38 45.1	−59 15 44	−0.004	0.00	6.48	−0.13	0.6	A0 V		150 s		
92298	222219		10 38 48.1	−40 40 20	+0.001	−0.04	7.8	0.3	1.7	A3 V		120 s		

HD	SAO	Star Name	α 2000	δ 2000	μ(α)	μ(δ)	V	B-V	M$_v$	Spec	RV	d(pc)	ADS	Notes
			h m s	° ′ ″	s	″								
92246	156125		10 38 50.1	-15 43 46	-0.003	+0.02	7.7	1.1	0.2	K0 III		310 s		
92328	222222		10 38 50.2	-42 45 13	-0.003	-0.02	6.11	0.66	0.7	F5 III		84 s		m
92245	156124		10 38 50.3	-12 26 37	-0.004	0.00	6.04	0.00		A0				
---	222223		10 38 50.7	-42 42 54	-0.003	+0.01	8.0	1.7						
92467	251043		10 38 51.4	-64 29 54	-0.002	0.00	6.98	0.03	-0.3	B9 IV		260 s		m
92150	62174		10 38 52.0	+36 00 54	-0.001	-0.02	7.8	0.8	3.2	G5 IV		83 s		
92379	238294		10 38 52.1	-53 36 09	-0.005	+0.02	8.0	0.3	0.0	B8.5 V		170 mx		
92196	99229		10 38 54.5	+16 07 39	+0.004	-0.02	6.6	0.4	3.0	F2 V	-13	53 s		
92421	238303		10 38 57.1	-59 16 08	-0.003	-0.01	7.72	-0.08	0.4	B9.5 V		290 s		
92436	238304		10 38 59.4	-58 49 00	-0.011	+0.01	5.87	1.43	-0.5	M3 III		190 s		m
92405	238301		10 39 00.6	-55 06 50	-0.002	0.00	7.2	0.1	0.0	B8.5 V		210 s		
92478	251046		10 39 01.8	-64 58 28	-0.009	+0.02	7.56	0.05		A0				
92665	256739		10 39 02.0	-75 09 26	-0.007	+0.01	6.7	0.1	1.4	A2 V		120 s		
92095	27724		10 39 05.5	+53 40 06	-0.011	-0.08	5.52	1.27	-0.2	K3 III	+45	130 mx		
92466	251045		10 39 06.0	-61 47 34	-0.001	-0.01	7.3	1.9	0.2	K0 III		76 s		
92168	62178	38 LMi	10 39 07.6	+37 54 36	-0.019	-0.04	5.85	0.57	5.2	dG5 pe	+7	21 ts		
92310	179075		10 39 09.5	-25 51 05	+0.002	-0.06	7.4	0.5	3.4	F5 V		55 s		
92435	238306		10 39 09.5	-54 51 39	-0.004	+0.02	7.5	1.7	-0.3	K5 III		280 s		
92223	99232		10 39 10.0	+14 43 57	0.000	-0.01	7.9	1.1	0.2	K0 III		340 s		
92645	256741		10 39 13.0	-73 33 35	+0.003	+0.01	7.9	0.1	0.6	A0 V		290 s		
92363	201598		10 39 16.1	-36 40 38	-0.005	0.00	6.8	1.1	3.2	G5 IV		35 s		
92682	256742		10 39 16.7	-74 29 37	+0.001	-0.01	6.07	1.71		K5				
92449	238309		10 39 18.3	-55 36 12	-0.003	0.00	4.28	1.04	-2.1	G2 II	+20	150 s		m
92556	---		10 39 19.4	-66 40 13			8.0	0.1	0.6	A0 V		290 s		
92374	201600		10 39 19.6	-37 26 27	+0.002	-0.03	7.7	0.3	0.6	A0 V		160 s		
92536	251050		10 39 22.7	-64 06 43	-0.003	0.00	6.32	-0.07	-0.3	B9 IV		210 s		
92464	238312		10 39 22.9	-55 59 23	-0.001	0.00	7.11	-0.09	-1.1	B5 V n		420 s		
92571	---		10 39 22.9	-67 10 14			8.00	1.1	0.2	K0 III		350 s		m
92463	238313		10 39 24.2	-55 36 25	-0.003	0.00	6.24	-0.08	-0.2	B8 V		190 s		m
92506	251049		10 39 25.7	-61 05 06	-0.001	-0.01	7.9	1.5	3.2	G5 IV		32 s		
92393	201602		10 39 27.4	-32 48 51	+0.001	-0.04	7.6	1.9	0.2	K0 III		88 s		
---	201604		10 39 29.1	-38 48 40	-0.002	+0.01	8.0	0.8	3.2	G5 IV		92 s		
92505	251051		10 39 29.2	-60 59 09	-0.002	0.00	7.02	-0.10	-2.2	B5 III		670 s		
92403	201606		10 39 30.0	-37 48 41	0.000	0.00	8.0	0.9	3.2	G5 IV		75 s		
92340	137711		10 39 30.2	-7 57 48	-0.001	0.00	7.7	1.4	-0.3	K5 III		390 s		
92357	156137		10 39 37.1	-12 59 40	-0.004	0.00	7.3	0.1	1.4	A2 V		150 s		
92323	118418		10 39 42.0	+8 50 34	-0.008	0.00	7.51	0.44	3.4	F5 V	+14	65 s	7871	m
92432	201610		10 39 42.2	-37 26 01	-0.002	-0.03	7.6	1.4	-0.1	K2 III		250 s		
92391	179088		10 39 42.3	-22 31 51	-0.002	-0.02	7.3	1.3	0.2	K0 III		180 s		
92461	222233		10 39 46.0	-42 24 08	+0.001	-0.02	8.0	0.9	0.2	K0 III		250 s		
92030	7181		10 39 46.9	+70 08 29	-0.008	+0.01	7.9	0.5	4.0	F8 V		60 s		
92265	62185		10 39 49.3	+39 24 17	0.000	0.00	7.8	1.1	0.2	K0 III		340 s		
92266	62186		10 39 50.5	+37 50 20	+0.001	-0.05	8.0	1.1	0.2	K0 III		360 s		
92430	201615		10 39 52.8	-33 36 45	+0.002	0.00	7.4	1.2	0.2	K0 III		220 s		
92552	238323		10 39 55.7	-56 16 41	-0.002	-0.01	7.9	0.7	0.0	B8.5 V		140 s		
92518	238321		10 39 57.6	-52 17 04	-0.006	-0.02	7.0	1.0	0.6	A0 V		50 s		
92501	222239		10 39 58.3	-49 16 47	0.000	+0.02	6.9	2.1	-0.3	K5 III		120 s		
92338	99241		10 39 58.7	+18 03 46	-0.001	-0.02	7.5	0.4	2.6	F0 V		98 s		
92278	43423		10 40 07.6	+46 50 25	-0.004	-0.05	7.30	0.17		A2	-7			
92440	156142		10 40 09.8	-16 22 19	+0.012	-0.08	7.75	0.61		G5				m
92664	251059		10 40 11.4	-65 06 01	-0.004	0.00	5.51	-0.17		A0	+30			
92321	62190		10 40 12.4	+38 24 12	-0.002	-0.01	8.00	1.4	-0.3	K4 III	+25	460 s	7873	m
92530	222243		10 40 14.3	-45 03 18	0.000	-0.01	8.0	0.2	1.4	A2 V		190 s		
92550	222245		10 40 14.9	-47 09 37	-0.003	0.00	7.3	0.2	0.0	B8.5 V		190 s		
92426	137717		10 40 15.0	-9 02 39	-0.002	-0.01	7.7	1.1	-0.1	K2 III		360 s		
92371	81463		10 40 21.1	+27 31 31	0.000	-0.01	7.0	0.1	1.4	A2 V	0	130 s		
92663	251060		10 40 21.5	-64 00 30	-0.001	+0.01	7.80	1.53		K0				
92457	99245		10 40 37.3	+12 04 45	0.000	+0.02	7.90	1.1	-0.1	K2 III		400 s		m
92715	251062		10 40 42.0	-64 39 13	-0.003	-0.01	6.82	-0.03	0.2	B9 V		210 s		
92388	43429		10 40 46.2	+41 31 35	+0.002	-0.02	7.90	1.1	0.2	K0 III		350 s	7880	m
92456	81467		10 40 49.9	+25 42 01	0.000	+0.02	7.70	1.1	0.0	K1 III		350 s		
92626	222254		10 40 49.9	-48 01 31	-0.005	-0.01	7.08	1.36		K2 pe				
92589	201631		10 40 51.5	-35 44 29	-0.002	+0.02	6.37	0.92		G5				m
92693	238341		10 40 52.8	-57 56 07	-0.003	0.00	6.99	1.07	-7.5	A2 Ia		2100 s		
92784	251065		10 41 02.5	-66 05 47	-0.005	0.00	7.8	0.5	4.0	F8 V		57 s		
92148	7186		10 41 03.2	+73 46 17	-0.032	-0.08	7.7	0.5	3.4	F5 V		73 s		
92400	43431		10 41 03.5	+48 34 22	+0.001	0.00	7.8	0.1	1.4	A2 V		190 s		
92712	238345		10 41 05.0	-57 15 25	+0.001	-0.01	7.87	0.04	-1.6	B3.5 V		650 s		
92577	156160		10 41 06.0	-14 21 30	0.000	-0.01	8.0	0.4	3.0	F2 V		98 s		
92783	251066		10 41 06.4	-64 28 29	-0.003	0.00	6.73	-0.05	0.2	B9 V n		200 s		
92914	---		10 41 07.4	-73 17 21			7.7	1.6	-0.5	M2 III		430 s		
92511	99248		10 41 09.1	+16 52 26	+0.001	-0.01	7.9	0.1	1.4	A2 V		200 s		
92559	137725		10 41 10.5	-0 16 34	-0.002	-0.03	7.4	0.4	2.6	F0 V		93 s		
92741	238351		10 41 12.3	-59 58 25	-0.001	-0.01	7.25	-0.01	-5.1	B1 II k		2300 s		m
92740	238353		10 41 17.6	-59 40 37	-0.001	0.00	6.42	0.08		WN7	+33			

HD	SAO	Star Name	α 2000	δ 2000	μ(α)	μ(δ)	V	B-V	M$_v$	Spec	RV	d(pc)	ADS	Notes
92408	27735		10^{h}41^{m}17.7^s	+52°58'14"	-0^s.006	-0".01	7.1	0.4	2.6	F0 V		81 s		
92576	118431		10 41 23.4	+8 33 45	-0.002	-0.01	8.0	1.6	-0.5	M2 III		510 s		
92588	137728	33 Sex	10 41 24.1	-1 44 29	-0.009	-0.12	6.26	0.88		K1 IV	+43	13 mn		
92558	81472		10 41 25.1	+20 33 12	-0.001	-0.01	7.9	0.1	1.4	A2 V		200 s		
92757	238355		10 41 26.2	-56 04 06	-0.004	+0.01	6.80	-0.01	0.6	A0 V		170 s		
92678	201641		10 41 30.2	-35 43 52	-0.003	-0.01	6.82	0.04	0.6	A0 V		160 s		
92879	—		10 41 34.9	-67 42 21			8.0	1.6	-0.5	M2 III		470 s		
92837	251070		10 41 35.1	-64 06 24	-0.003	-0.01	7.17	0.00	0.2	B9 V		240 s		
92587	99249		10 41 37.1	+13 58 42	-0.001	0.00	7.8	1.6	-0.5	M2 III	-11	470 s		
92736	222260		10 41 40.7	-45 46 07	-0.012	+0.04	7.5	0.7	3.4	F5 V		45 s		
92604	99251		10 41 41.5	+10 44 23	+0.001	0.00	8.00	0.1	1.4	A2 V		210 s	7888	m
92354	15260		10 41 48.2	+68 26 36	-0.005	-0.03	5.75	1.30	0.2	K0 III	+5	85 s		
93237	256745		10 41 51.4	-79 46 59	-0.010	+0.01	5.97	-0.07	-1.1	B5 V e	0	160 mx		
92675	137734		10 41 52.2	-8 03 23	0.000	-0.03	7.4	1.1	-0.1	K2 III		320 s		
92698	179128		10 41 52.4	-26 02 47	-0.001	-0.01	7.3	0.4	1.4	A2 V		90 s		
92674	137735		10 41 53.6	-6 34 30	0.000	-0.03	7.7	0.5	4.0	F8 V		54 s		
92755	222261		10 41 55.9	-40 41 28	-0.010	+0.01	7.7	0.4	4.0	F8 V		55 s		
92424	15261	38 UMa	10 41 56.5	+65 42 58	-0.027	-0.08	5.12	1.20	-0.1	K2 III	-11	97 mx		
92896	251075		10 41 59.7	-63 30 27	-0.003	-0.01	7.31	0.22	1.2	A5 IV		160 s		m
92773	201651		10 42 09.3	-33 42 35	-0.001	+0.01	7.9	2.2	-0.1	K2 III		97 s		
92620	62206		10 42 11.3	+31 41 49	+0.001	-0.03	6.02	1.62	-0.5	M2 III	+16	200 s		m
92719	156167		10 42 13.1	-13 47 15	+0.015	-0.17	6.8	0.7	4.4	G0 V		30 s		
92938	251078		10 42 14.0	-64 27 58	-0.004	+0.01	4.80	-0.14	-1.7	B3 V	+24	190 s		
92686	99257		10 42 16.6	+10 21 20	-0.003	-0.02	7.2	1.4	-0.3	K5 III		320 s		
92706	118440		10 42 21.8	+0 51 44	-0.001	+0.01	7.6	1.1	-0.1	K2 III	+18	350 s		
92847	222268		10 42 21.8	-47 13 14	-0.002	-0.03	7.10	1.1	0.2	K0 III		240 s		m
92771	179133		10 42 21.9	-20 36 11	-0.004	0.00	7.7	0.7	1.4	A2 V		72 s		
92966	251079		10 42 25.0	-64 23 55	-0.004	-0.01	7.27	0.00		B9				
92804	201660		10 42 26.5	-33 39 20	-0.006	+0.01	7.70	0.36	3.0	F2 V		85 s		
92753	137739		10 42 27.3	-8 43 38	+0.001	0.00	6.9	1.1	0.2	K0 III		220 s		
92770	156170		10 42 31.2	-13 58 30	-0.002	-0.01	6.24	1.54		K2				
92696	99258		10 42 31.7	+18 13 12	0.000	-0.02	7.1	1.1	0.2	K0 III		240 s		
92908	238375		10 42 34.7	-54 31 55	-0.003	+0.01	8.0	0.1	0.6	A0 V		240 s		
92846	222271		10 42 35.8	-41 54 15	0.000	-0.02	8.0	1.8	-0.3	K5 III		140 s		
93165	256747		10 42 36.5	-75 27 43	0.000	+0.01	6.8	0.2	0.4	B9.5 V		140 s		
92749	118443		10 42 37.4	+3 34 59	-0.006	+0.02	6.70	0.5	3.4	F5 V	+19	46 s	7896	m
92989	—		10 42 38.4	-64 41 23			7.59	0.04		A0				
92964	238379		10 42 40.6	-59 12 57	-0.001	0.00	5.38	0.26	-6.8	B3 Ia	-2	1600 s		
92936	238378		10 42 42.0	-56 52 41	-0.001	-0.02	7.05	-0.04	-2.6	B2.5 IV		740 s		
92845	201665		10 42 43.1	-32 42 57	-0.002	0.00	5.64	0.00		A0	+4			m
92788	137743		10 42 48.4	-2 11 01	-0.002	-0.22	7.2	0.8	3.2	G5 IV		52 mx		
92668	27744		10 42 52.1	+50 47 56	-0.002	-0.01	7.40	0.1	1.7	A3 V	-2	140 s	7894	m
92844	179147		10 42 52.2	-23 32 55	0.000	+0.01	6.8	1.1	0.2	K0 III	-12	190 s		
92748	99262		10 42 52.6	+18 22 55	+0.001	-0.01	7.5	0.4	3.0	F2 V		80 s		
92635	27742		10 42 54.2	+55 53 27	+0.001	-0.01	7.6	1.4	-0.3	K5 III		380 s		
92860	179148		10 42 54.2	-22 30 31	0.000	-0.05	7.92	0.43	3.1	F3 V		86 s		
93030	251083	θ Car	10 42 57.4	-64 23 39	-0.003	+0.01	2.76	-0.23	-4.1	B0 V p	+24	230 s		
92949	222274		10 43 01.0	-49 03 25	-0.001	+0.03	7.0	2.0	-0.3	K5 III		140 s		
92769	81485	40 LMi	10 43 01.8	+26 19 32	-0.008	-0.07	5.51	0.17	2.1	A5 V	+16	34 mx	7899	m
93003	251081		10 43 03.6	-60 59 14	-0.003	-0.01	7.12	-0.02	-1.7	B3 V		480 s		
92523	15269		10 43 04.0	+69 04 34	0.000	-0.02	5.00	1.38	-0.2	K3 III	0	96 s		
92948	222275		10 43 04.7	-48 37 11	-0.001	-0.01	7.9	0.2	0.0	B8.5 V		250 s		
93164	256748		10 43 04.7	-72 11 46	-0.012	+0.02	7.5	0.5	4.0	F8 V		50 s		
92319	7191		10 43 08.0	+77 24 40	-0.016	-0.02	7.6	0.0	0.4	B9.5 V		120 mx		
92906	201669		10 43 09.2	-35 00 16	0.000	-0.06	7.50	0.98	0.2	K0 III		250 mx		
93010	251085		10 43 09.4	-61 10 07	-0.002	0.00	6.63	0.00	-2.9	B3 III		610 s		
92768	81486		10 43 11.2	+29 23 06	-0.004	+0.04	7.9	0.4	3.0	F2 V		94 s		
92934	201672		10 43 17.8	-31 45 36	0.000	-0.05	6.6	1.0	0.2	K0 III		180 s		
92633	15270		10 43 20.2	+62 12 44	+0.001	-0.01	7.9	1.1	-0.1	K2 III		400 s		
—	118448		10 43 20.4	+4 44 46	+0.001	-0.06	6.30						7902	m
92841	118449	35 Sex	10 43 20.9	+4 44 52	+0.001	-0.03	5.79	1.17	-0.2	K3 III	-6	160 s	7902	m
92946	201674		10 43 22.4	-32 08 14	0.000	0.00	7.16	0.00	0.6	A0 V		200 s		
92857	118451		10 43 24.1	+4 18 06	-0.002	-0.03	7.7	0.1	1.4	A2 V		180 s		
92825	81490	41 LMi	10 43 24.9	+23 11 18	-0.008	0.00	5.08	0.04	1.4	A2 V	+19	55 s		
92945	179168		10 43 28.1	-29 03 54	-0.017	-0.09	7.76	0.89	0.2	K0 III		73 mx		
93098	251092		10 43 29.2	-64 04 08	-0.005	-0.02	7.60	0.04		A0				
92192	1758		10 43 29.5	+80 25 38	+0.007	+0.01	6.7	0.1	0.6	A0 V		170 s		
93070	251090		10 43 32.1	-60 33 59	-0.004	0.00	4.57	1.71	-0.3	K5 III	+9	75 s		
92787	43444		10 43 32.8	+46 12 14	-0.026	-0.07	5.18	0.33		F0	+4	22 mn		m
92786	43443		10 43 33.6	+48 12 51	-0.035	+0.18	7.9	0.9	3.2	G5 IV		44 mx		
92987	201684		10 43 36.2	-39 03 31	+0.001	0.00	7.1	1.0	4.4	G0 V		18 s		
93115	251093		10 43 37.3	-63 38 46	+0.004	0.00	7.85	1.52		K5				
92973	201681		10 43 37.7	-30 52 01	-0.002	+0.01	7.3	1.6	-0.3	K5 III		280 s		
92728	27748	39 UMa	10 43 43.3	+57 11 57	+0.002	-0.06	5.7	0.0		B9				
92822	43445		10 43 48.0	+42 22 28	-0.002	0.00	7.8	1.1	0.2	K0 III		330 s		

HD	SAO	Star Name	α 2000	δ 2000	μ(α)	μ(δ)	V	B−V	M_v	Spec	RV	d(pc)	ADS	Notes
			h m s	° ′ ″	s	″								
92884	99266		10 43 49.0	+18 18 45	+0.001	−0.06	6.6	1.1	0.2	K0 III		190 s		
93163	251095		10 43 51.2	−64 14 56	−0.002	0.00	5.77	0.00	−1.7	B3 V	+8	250 s		
93131	238394		10 43 52.1	−60 07 04	−0.003	0.00	6.49	−0.03		WN7				
93037	222278		10 43 54.3	−43 44 49	−0.001	0.00	8.0	1.6	−0.3	K5 III		390 s		
92811	27751		10 43 56.3	+49 48 02	−0.002	0.00	7.03	1.16	0.2	K0 III		170 s		
93129	--		10 43 57.8	−59 33 26			7.30	0.1		B1				m
92855	43449		10 44 00.5	+46 12 24	−0.027	−0.06	7.32	0.55	4.2	F9 V	+7	42 s		m
92955	118459		10 44 06.0	+3 13 33	−0.002	+0.01	7.9	1.4	−0.3	K5 III		430 s		
93194	251096		10 44 06.9	−63 57 40	−0.002	+0.01	4.82	−0.14	−1.7	B3 V	+26	200 s		
93093	222282		10 44 07.1	−48 53 48	−0.005	0.00	7.30	0.5	3.4	F5 V		60 s		
93161	238406		10 44 08.8	−59 34 34	−0.003	+0.01	7.0	0.8		O7				m
93127	238402		10 44 08.9	−55 47 12	−0.005	0.00	7.4	1.4	0.2	K0 III		160 s		
93094	238399		10 44 09.6	−51 55 26	−0.004	−0.01	7.1	1.6	0.2	K0 III		110 s		
93325	--		10 44 10.8	−72 03 29			7.9	1.6	−0.5	M2 III		450 s		
92941	81496		10 44 14.5	+19 45 31	−0.008	−0.03	6.27	0.17	1.7	A3 V	+8	55 mx		
--	--		10 44 19.1	−70 51 47			6.30							m
93344	256750		10 44 19.4	−70 51 35	−0.014	−0.01	6.26	0.20		A9				m
93036	156191		10 44 21.0	−19 08 32	−0.002	−0.03	7.9	0.1	0.6	A0 V		280 s		
93206	238414		10 44 23.0	−59 59 36	−0.001	0.00	6.24	0.13	−6.2	O9.5 I		1900 s		m
92940	62224		10 44 25.2	+32 37 11	−0.002	−0.02	7.7	0.2	2.1	A5 V	−5	130 s		
92992	99272		10 44 25.9	+10 22 30	+0.002	−0.03	7.6	1.1	−0.1	K2 III		340 s		
93372	256752		10 44 26.6	−72 26 38	−0.037	+0.03	6.27	0.49		F8				
93064	179180		10 44 28.1	−23 59 06	−0.001	+0.02	6.8	1.7	0.2	K0 III		81 s		
93359	256751		10 44 32.2	−70 51 18	−0.011	0.00	6.46	0.23		A3				
92917	43455		10 44 32.7	+49 26 21	−0.001	−0.03	7.6	1.1	0.2	K0 III		300 s		
--	--		10 44 33.4	−57 33 55			7.58B							VY Car, v
93205	238418		10 44 33.9	−59 44 15	+0.002	0.00	7.75	0.05		O3 V				m
92763	15273		10 44 38.3	+68 46 32	−0.008	−0.03	5.8	1.6		M5 III	+34	340 mx		R UMa, v
93105	201698		10 44 39.3	−36 55 50	+0.007	−0.02	6.8	0.5	4.0	F8 V		37 s		
92954	43456		10 44 43.7	+41 42 51	−0.003	−0.02	7.7	1.4	−0.3	K5 III		390 s		
93199	238420		10 44 48.0	−50 53 06	−0.004	0.00	7.9	0.7	3.2	G5 IV		88 s		
92953	43457		10 44 54.3	+48 32 46	−0.002	+0.01	8.0	1.1	0.2	K0 III		360 s		
93081	118471		10 45 02.2	+3 18 24	+0.006	−0.03	7.1	0.5	3.4	F5 V		54 s		
93308	238429	η Car	10 45 03.6	−59 41 03	0.000	+0.01	6.20	0.61			−25			m, v
93171	179193		10 45 03.8	−29 41 00	−0.005	+0.02	7.8	1.4	−0.1	K2 III		270 mx		
92839	15274		10 45 03.9	+67 24 41	+0.001	0.00	6.00	2.41		N1	−5			VY UMa, v
93307	238428		10 45 04.1	−58 08 55	−0.012	0.00	8.0	0.5	3.2	G5 IV		93 s		
--	238431		10 45 06.1	−59 40 04	+0.001	+0.02	7.90							
93077	99275		10 45 06.7	+16 20 16	−0.001	−0.04	7.8	0.5	3.4	F5 V		74 s		
93186	201709		10 45 07.1	−31 20 14	0.000	+0.01	8.0	1.4	−0.3	K5 III		450 s		
93102	118473	36 Sex	10 45 09.4	+2 29 16	−0.004	−0.03	6.28	1.21	−0.3	gK4	+11	210 s		
93306	238432		10 45 10.8	−57 44 01	−0.001	−0.02	7.1	1.8	0.2	K0 III		80 s		
93101	118474		10 45 13.4	+3 43 48	−0.001	−0.01	8.0	1.1	−0.1	K2 III		410 s		
93779	258592	δ¹ Cha	10 45 15.8	−80 28 10	−0.013	−0.03	5.47	0.95	0.2	gK0	+11	110 s		m
93033	43460		10 45 20.5	+45 33 58	+0.001	0.00	6.9	0.0	0.0	B8.5 V	−13	240 s		TX UMa, m, v
93044	43461		10 45 23.2	+41 18 26	−0.004	−0.06	7.1	0.4	2.6	F0 V		81 s		
93228	201716		10 45 26.6	−37 23 33	−0.002	−0.01	7.7	0.2	0.4	B9.5 V		200 s		
93340	--		10 45 27.0	−55 47 30			7.1	0.2	2.1	A5 V		100 s		
93153	118478		10 45 29.2	+1 00 40	0.000	0.00	7.9	1.1	0.2	K0 III		340 s		
93339	238438		10 45 29.4	−55 47 28	−0.002	+0.01	7.1	2.1	−0.1	K2 III		76 s		
93580	256754		10 45 32.7	−75 00 27	+0.001	−0.01	7.7	0.1	0.6	A0 V		260 s		
93227	179198		10 45 32.9	−25 01 34	−0.001	+0.01	7.80	1.1	0.2	K0 III		330 s	7918	m
93169	118479		10 45 33.6	+7 30 51	−0.001	−0.02	7.9	1.1	0.2	K0 III		340 s		
93336	222303		10 45 41.1	−47 27 36	0.000	−0.01	6.9	1.6	−0.5	M2 III		300 s		
93298	222302		10 45 42.5	−41 29 50	+0.001	0.00	7.6	0.5	1.4	A2 V		180 s		
93216	156201		10 45 43.1	−10 51 58	−0.010	0.00	8.0	0.9	3.2	G5 IV		93 s		
93357	238441		10 45 43.7	−53 37 41	−0.003	+0.02	7.5	1.9	−0.5	M2 III		280 s		
93403	238445		10 45 44.2	−59 24 28	0.000	0.00	7.26	0.21		O5.8				
93135	62234		10 45 46.4	+32 35 50	−0.008	−0.01	7.6	0.4	2.6	F0 V		98 s		
93845	258593	δ² Cha	10 45 46.7	−80 32 24	−0.021	0.00	4.45	−0.19	−1.7	B3 V	+22	170 s		
93226	156202		10 45 48.1	−10 42 48	−0.002	0.00	7.30	0.1	0.6	A0 V		220 s	7917	m
93297	201720		10 45 49.5	−36 56 40	−0.003	−0.03	7.7	1.0	0.2	K0 III		300 s		
93420	238447		10 45 50.4	−59 29 20	−0.003	−0.01	7.90	1.6	−0.5	M2 III		460 s		BO Car, m, q
93316	222305		10 45 50.8	−40 58 26	+0.001	−0.01	6.8	1.5	0.2	K0 III		110 s		
92880	7199		10 45 51.4	+72 17 15	−0.005	−0.02	6.92	1.63		K0				
93152	62236	42 LMi	10 45 51.8	+30 40 56	−0.002	−0.04	5.24	−0.06	0.2	B9 V	+14	100 s		m
93075	27758		10 45 59.7	+56 55 15	+0.001	+0.01	7.1	0.4	2.6	F0 V	−18	78 s		
93244	118483		10 46 05.6	+6 22 24	−0.001	−0.04	6.37	1.12		K0	−9			
93332	179205		10 46 08.6	−23 25 57	−0.001	0.00	7.7	−0.1	1.7	A3 V		160 s		
93484	251111		10 46 12.5	−61 56 32	−0.004	−0.02	7.32	−0.06	−2.1	B2.5 V		640 s		
93469	238457		10 46 12.6	−59 19 03	−0.008	+0.03	7.88	0.27	2.1	A5 V		110 mx		m
93517	--		10 46 12.8	−64 35 33			7.85	0.10		A				
93385	222310		10 46 15.2	−41 27 52	−0.004	−0.06	7.5	0.6	3.2	G5 IV		73 s		
93540	251115		10 46 16.5	−64 30 54	−0.003	−0.01	5.34	−0.10	−0.6	B7 V	+32	150 s		
93502	251113		10 46 16.8	−60 36 12	−0.008	−0.01	6.25	0.04		A0				m

HD	SAO	Star Name	α 2000	δ 2000	μ(α)	μ(δ)	V	B-V	M_V	Spec	RV	d(pc)	ADS	Notes
93347	179209		10 46 17.2	-26 43 35	-0.002	-0.01	7.4	0.5	0.2	K0 III		280 s		
93273	99280		10 46 19.3	+12 44 52	0.000	-0.03	6.8	0.1	1.4	A2 V		120 s		
99685	--		10 46 20.2	-89 47 45			7.6	0.1	0.6	A0 V		250 s		
93132	27760	41 UMa	10 46 22.4	+57 21 56	-0.006	-0.07	6.34	1.56	-0.5	M1 III	-2	160 mx		
93257	99281	51 Leo	10 46 24.5	+18 53 29	+0.007	-0.04	5.49	1.12	0.2	K0 III	-6	97 s		
93331	156211		10 46 24.5	-13 27 35	-0.004	0.00	7.2	0.1	0.6	A0 V		150 mx		
93291	99282	52 Leo	10 46 25.2	+14 11 41	-0.009	-0.07	5.48	0.91	0.2	K0 III	+35	89 mx		
93539	251116		10 46 25.8	-63 18 00	-0.004	-0.07	6.7	1.1	0.2	K0 III		160 mx		
93346	156212		10 46 26.6	-15 02 04	-0.002	-0.01	7.3	0.1	0.6	A0 V		220 s		
93412	222314		10 46 27.3	-42 38 37	-0.007	0.00	7.1	0.7	3.0	F2 V		40 s		
93353	156213		10 46 28.7	-19 45 38	-0.003	0.00	7.1	1.5	-0.1	K2 III		170 s		
93549	251117		10 46 29.7	-64 15 47	-0.001	+0.01	5.23	-0.08		B8	+21			m
93213	43467		10 46 29.9	+44 06 10	-0.013	-0.04	8.0	0.5	4.0	F8 V	+25	65 s		
93365	156216		10 46 32.9	-18 52 28	+0.003	-0.04	7.5	1.1	-0.1	K2 III		330 s		
93330	118488		10 46 33.1	+4 38 43	-0.015	-0.16	7.9	1.1	0.2	K0 III		58 mx		
93453	222318		10 46 37.0	-43 11 34	0.000	-0.06	6.6	0.9	1.7	A3 V		32 s		
93452	222319		10 46 37.5	-42 36 39	+0.001	-0.03	7.3	0.9	3.2	G5 IV		55 s		
93477	238460		10 46 37.8	-51 03 08	-0.001	-0.03	7.9	0.8	0.6	A0 V		85 s		
93328	99286		10 46 41.6	+16 32 33	-0.001	-0.02	7.7	1.1	0.2	K0 III		320 s		
93410	179217		10 46 42.7	-26 02 53	-0.012	+0.05	6.71	1.03		K0		17 mn		
93411	179218		10 46 45.9	-26 23 50	-0.003	0.00	7.8	0.8	-0.3	K5 III		410 s		
93497	222321	μ Vel	10 46 46.1	-49 25 12	+0.007	-0.05	2.69	0.90	0.3	G5 III	+7	30 s		m
93409	179219		10 46 46.6	-25 32 02	-0.003	-0.02	7.5	1.6	-0.3	K5 III		330 s		
93649	251121		10 46 47.9	-69 12 35	-0.004	0.00	6.6	0.8	1.4	A2 V		39 s		
93396	137780		10 46 49.7	-9 23 56	-0.006	-0.07	7.9	1.1	0.2	K0 III		160 mx		
93607	251120		10 46 51.2	-64 22 59	-0.003	+0.01	4.85	-0.15	-2.0	B4 IV	+16	240 s		
93397	156221		10 46 51.9	-17 17 48	-0.002	-0.02	5.42	0.11		A3 m				
93271	43471		10 46 54.3	+43 01 37	-0.001	+0.01	7.4	1.1	-0.1	K2 III		320 s		
93494	222327		10 46 54.9	-44 16 41	0.000	-0.01	7.9	0.6	0.6	A0 V		110 s		
93563	238468		10 46 57.5	-56 45 26	-0.002	-0.01	5.23	-0.08		B8	+31			
93408	156223		10 46 58.4	-15 43 54	-0.003	-0.01	7.0	0.1	0.6	A0 V		190 s		
93466	201741		10 46 59.7	-36 05 49	0.000	-0.02	7.9	1.8	-0.3	K5 III		280 s		
93474	179224		10 47 08.6	-24 32 54	-0.002	+0.02	7.8	0.7	2.1	A5 V		70 s		
93648	251123		10 47 09.1	-64 15 52	-0.003	+0.01	7.84	0.13	0.6	A0 V		230 s		
93350	43472		10 47 18.6	+40 44 17	0.000	-0.04	7.8	1.1	0.2	K0 III		330 s		
93619	238476		10 47 19.0	-57 19 30	-0.001	+0.01	6.94	0.15		B0.5 Ib				
93431	118493		10 47 19.1	+6 20 47	-0.003	-0.03	7.0	0.1	1.4	A2 V	+22	130 s		
93684	251124		10 47 21.2	-65 55 06	+0.002	0.00	7.61	0.11	-2.5	B2 V		680 s		
93391	81523		10 47 22.8	+26 54 38	0.000	-0.01	7.30	1.4	-0.3	K5 III	+9	330 s		
93545	201748		10 47 26.4	-37 31 26	-0.002	0.00	7.90	0.01	0.4	B9.5 V		290 s		
93286	15290		10 47 30.2	+60 07 27	-0.009	-0.03	7.3	0.4	2.6	F0 V	-7	85 s		
93739	251127		10 47 30.3	-69 26 19	-0.004	-0.04	6.7	-0.4	0.0	B8.5 V		160 mx		
93554	201750		10 47 33.2	-31 36 12	-0.003	-0.01	7.1	1.5	-0.5	M2 III		330 s		
93555	201752		10 47 35.3	-36 07 36	-0.002	-0.02	7.7	1.3	0.2	K0 III		220 s		
93527	156234		10 47 36.6	-15 37 07	-0.010	-0.06	7.66	0.50	3.4	dF5	+31	66 s		m
93238	15287		10 47 36.7	+65 37 01	-0.001	-0.03	7.20	1.6	-0.5	M4 III	-18	350 s		
93526	156235		10 47 37.9	-15 15 44	0.000	-0.02	6.67	-0.01		A2	+22		7930	m
93662	238480		10 47 38.7	-57 28 03	+0.001	-0.01	6.36	1.62		K5				
93683	251125		10 47 38.9	-60 37 04	0.000	0.00	7.89	0.13	-2.6	B2.5 IV		900 s		
93714	251126		10 47 39.7	-64 32 50	-0.005	0.00	6.54	0.02	-2.9	B3 III		480 mx		
93525	156236		10 47 39.8	-15 14 41	0.000	-0.02	6.9	0.5	3.4	F5 V		51 s		
93459	99294		10 47 41.2	+18 57 46	-0.001	0.00	8.0	0.9	3.2	G5 IV		93 s		
93524	156238		10 47 42.8	-15 06 36	0.000	0.00	7.0	1.1	-0.1	K2 III		260 s		
93270	15292		10 47 43.8	+65 27 36	-0.015	-0.01	7.68	0.55	3.4	dF5	-1	62 s	7925	m
93695	238483		10 47 44.2	-59 52 31	-0.003	-0.01	6.47	-0.12	-1.1	B5 V		330 s		
93509	99296		10 47 46.5	+10 26 37	-0.002	+0.01	7.9	0.1	1.7	A3 V		170 s		
93738	251128		10 47 53.8	-64 15 45	0.000	+0.02	6.46	0.02		A0				
93668	238484		10 47 55.5	-52 14 47	-0.002	+0.02	7.0	1.0	0.6	A0 V		46 s		
93680	238486		10 47 55.8	-55 58 37	-0.004	+0.02	8.0	1.2	0.2	K0 III		290 s		
93458	43474		10 48 00.8	+39 44 33	+0.002	-0.03	7.0	1.1	0.2	K0 III		230 s		
93457	43475		10 48 02.3	+41 06 34	-0.003	-0.02	7.40	0.1	1.7	A3 V	-5	140 s	7929	m
93737	238493		10 48 05.3	-59 55 09	-0.002	0.00	6.00	0.27	-6.6	A0 I		2500 s		
93626	179236		10 48 05.6	-26 23 53	-0.011	0.00	7.91	0.53	4.4	G0 V		50 s		
93471	43476		10 48 07.4	+44 48 06	-0.004	-0.02	7.8	1.1	0.2	K0 III		330 s		
93657	201766		10 48 14.1	-31 41 17	-0.002	-0.02	5.88	0.03		A1				
93521	62257		10 48 23.4	+37 34 13	0.000	0.00	7.06	-0.27	-4.8	O9 V p	-16	2400 s		
93809	251131		10 48 23.4	-62 33 49	-0.003	+0.01	7.4	1.0	2.1	A5 V		38 s		
93721	222342		10 48 29.2	-44 22 42	-0.001	0.00	7.9	0.3	0.4	B9.5 V		190 s		
93751	222343		10 48 34.9	-48 21 02	-0.016	+0.03	7.05	0.42		F2		16 mn		
93470	27771		10 48 36.2	+57 18 43	-0.003	-0.07	7.8	0.7	4.4	G0 V		49 s		
93843	238504		10 48 37.7	-60 13 26	-0.001	-0.01	7.33	-0.05		O6				
93807	238502		10 48 38.6	-57 40 24	0.000	-0.02	7.09	1.33	-0.2	K3 III		270 s		
93655	137800		10 48 40.5	-1 57 32	-0.001	0.00	5.93	1.60	-0.5	gM2	+3	190 s		
93773	222347		10 48 41.4	-47 45 00	-0.007	+0.02	6.6	1.2	3.2	G5 IV		26 s		
93750	222345		10 48 41.5	-45 18 13	-0.003	-0.01	8.0	0.5	1.7	A3 V		180 s		

HD	SAO	Star Name	α 2000	δ 2000	μ(α)	μ(δ)	V	B-V	M_v	Spec	RV	d(pc)	ADS	Notes
93913	251137		10^h 48^m 42^s.2	−67° 18′ 21″	0^s.000	−0″.03	7.81	0.15	−1.6	B3.5 V		490 s		
93427	15298		10 48 49.9	+65 07 56	+0.001	0.00	6.2	0.0		B9				
93821	238505		10 48 50.1	−55 21 33	−0.002	+0.01	7.4	0.6	0.4	B9.5 V		110 s		
93873	238507		10 48 55.4	−59 26 51	+0.001	−0.03	7.81	0.46		B0.5 Iab				m
93636	81533	43 LMi	10 48 57.1	+29 24 57	−0.007	−0.05	6.15	1.15	0.0	gK1	+10	160 s		
93582	43480		10 48 57.3	+43 55 39	−0.003	−0.05	7.7	0.5	3.4	F5 V		71 s		
93731	179248		10 48 57.6	−27 55 03	−0.004	+0.02	6.9	0.4	1.4	A2 V		77 s		
93704	137805		10 48 57.9	−9 06 00	−0.001	−0.03	7.1	1.1	0.2	K0 III	−12	240 s		
93745	201782		10 49 00.3	−31 03 33	+0.012	−0.09	7.50	0.60	4.0	F8 V		44 s		
93687	99302		10 49 03.2	+10 09 30	−0.003	0.00	7.5	0.1	0.6	A0 V		250 s		
93770	—		10 49 06.4	−30 44 39			7.3	1.2	0.2	K0 III		200 s		
93840	222355		10 49 08.7	−46 46 42	0.000	0.00	7.76	−0.05	−6.2	B1 I k		4700 s		
93719	137807		10 49 09.1	−5 24 27	−0.001	−0.03	7.8	1.1	0.2	K0 III		340 s		
93898	238510		10 49 11.3	−57 58 11	−0.004	−0.01	8.02	−0.05	0.4	B9.5 V		310 mx		
93990	—		10 49 11.5	−68 21 42			7.8	1.1	0.2	K0 III		320 s		
93702	99305	53 Leo	10 49 15.4	+10 32 43	0.000	−0.03	5.25	0.01	1.4	A2 V	−6	59 s		
93870	238509		10 49 15.6	−50 16 17	+0.004	−0.06	8.0	0.9	3.2	G5 IV		93 s		
93785	—		10 49 16.2	−26 48 40			7.90	0.4	3.0	F2 V		96 s	7939	m
93742	137808	40 Sex	10 49 17.2	−4 01 27	−0.004	−0.02	6.61	0.22		A2	+14		7936	m
93729	137809		10 49 18.7	−2 46 36	+0.005	−0.13	7.9	1.1	0.2	K0 III		160 mx		
93837	201794		10 49 21.0	−39 10 14	−0.003	0.00	7.1	0.4	0.6	A0 V		110 s		
93728	118518		10 49 23.4	+3 26 24	−0.001	−0.02	8.0	0.1	1.4	A2 V		210 s		
93943	238514		10 49 24.1	−59 19 25	−0.005	0.00	5.85	0.01		A0				m
93664	43482		10 49 25.7	+41 23 24	−0.006	−0.01	7.80	0.6	4.4	G0 V		48 s		m
93850	222360		10 49 25.7	−42 22 04	−0.006	0.00	7.2	0.7	−0.1	K2 III		280 s		
93717	99308		10 49 26.8	+17 08 47	0.000	−0.01	7.20	1.1	0.2	K0 III		250 s	7935	m
93868	222362		10 49 27.6	−44 17 05	0.000	0.00	8.0	0.2	0.6	A0 V		210 s		
93551	15300		10 49 28.9	+63 48 33	0.000	−0.02	6.5	1.1	0.2	K0 III		180 s		
93663	43483		10 49 30.4	+43 04 04	−0.001	+0.01	7.8	1.1	0.2	K0 III		330 s		
93849	179261		10 49 35.2	−29 59 39	−0.009	−0.02	7.9	0.4	4.4	G0 V		49 s		
93813	156256	ν Hya	10 49 37.4	−16 11 37	+0.006	+0.20	3.11	1.25	−0.1	K2 III	−1	39 s		
94131	256759		10 49 42.3	−71 39 15	−0.011	+0.01	7.6	0.4	2.6	F0 V		99 s		
93833	137815		10 49 43.4	−9 51 10	0.000	−0.04	5.86	1.07	0.3	gG8	+40	100 s		
93765	81542	44 LMi	10 49 53.7	+27 58 26	0.000	+0.03	6.04	0.37	3.4	F5 V	+3	34 s		
93906	201804		10 49 55.2	−35 48 17	0.000	−0.02	7.34	0.09	0.6	A0 V		190 s		
93905	201805		10 49 56.9	−34 03 29	−0.004	+0.01	5.61	0.04		A0				
93864	137819		10 49 58.6	−7 28 57	−0.009	−0.04	7.9	1.1	0.2	K0 III		150 mx		
94066	—		10 49 59.3	−64 32 44			7.87	0.10		B5				
93936	222367		10 50 02.1	−40 57 17	−0.006	−0.07	8.00	0.6	4.4	G0 V		53 s		m
93797	62275		10 50 07.9	+36 05 51	−0.003	−0.03	7.8	0.1	1.7	A3 V		170 s		
94097	251145		10 50 12.9	−62 38 07	−0.004	+0.01	7.20	0.08	0.0	B8.5 V		250 s		
93671	15307		10 50 13.0	+60 36 33	0.000	−0.01	7.9	0.9	3.2	G5 IV		88 s		
93904	137821		10 50 13.4	−8 59 19	−0.002	+0.03	7.9	1.1	0.2	K0 III		340 s		
94019	222369		10 50 17.2	−49 06 43	−0.003	+0.02	7.9	1.1	0.2	K0 III		320 s		
93903	137823	41 Sex	10 50 18.0	−8 53 52	−0.001	−0.02	5.79	0.16		A2			7942	m
93763	27783		10 50 24.4	+52 16 03	−0.001	−0.01	7.9	0.4	2.6	F0 V		110 s		
93932	156264		10 50 25.3	−15 06 16	−0.020	−0.10	7.53	0.62	4.4	dG0	+36	39 s		
93983	201813		10 50 26.1	−33 50 23	−0.007	−0.02	7.6	1.2	0.2	K0 III		210 s		
94096	238523		10 50 26.3	−59 58 58	−0.001	−0.01	7.6	1.9	−0.5	M2 III		290 s		IX Car, v
93876	81546		10 50 29.5	+25 38 26	+0.002	−0.05	8.0	0.5	3.4	F5 V		82 s		
93860	62278		10 50 33.8	+32 01 39	+0.005	−0.01	7.5	0.5	3.4	F5 V		67 s		
93901	81547		10 50 36.5	+23 24 24	−0.004	+0.03	6.6	0.2	2.1	A5 V		80 s		
94144	251148		10 50 39.9	−61 16 09	−0.003	+0.01	6.84	−0.01		B5				
94129	238526		10 50 41.4	−59 12 45	−0.001	+0.01	7.9	0.8	0.4	B9.5 V		110 s		
94174	—		10 50 42.3	−64 28 46			7.75	0.11		A0				
94034	201818		10 50 44.1	−34 29 14	−0.001	−0.08	7.8	0.5	2.6	F0 V		79 s		
94108	238527		10 50 49.7	−52 28 17	0.000	−0.01	7.9	−0.5	0.0	B8.5 V		380 s		
93967	118529		10 50 53.1	+5 35 09	−0.001	−0.02	8.0	0.5	4.0	F8 V		62 s		
94016	156269		10 50 54.1	−14 53 29	−0.010	+0.02	7.6	0.4	3.0	F2 V		82 s		
93982	118530		10 50 57.0	+3 35 26	0.000	0.00	7.1	0.1	1.4	A2 V		140 s		
93740	15310		10 50 58.2	+66 32 17	+0.001	−0.01	8.0	0.1	0.6	A0 V		300 s		
94173	238529		10 50 58.7	−59 57 27	−0.001	−0.02	6.79	−0.05		A0				m
94140	238528		10 50 59.6	−51 25 32	−0.002	+0.02	8.0	1.1	0.2	K0 III		330 s		
93966	99319		10 51 02.1	+16 03 14	+0.001	−0.06	7.80	0.5	4.0	F8 V		58 s	7944	m
94014	137834		10 51 05.4	−3 05 33	−0.003	−0.01	5.95	1.48		K2				
93847	27788		10 51 05.9	+56 14 16	−0.003	−0.02	7.3	0.0	0.4	B9.5 V	−6	240 s		
94046	156271		10 51 05.9	−18 19 56	−0.004	0.00	6.6	0.1	0.6	A0 V		160 s		
93993	99321		10 51 08.4	+11 34 45	−0.006	−0.03	6.8	1.1	0.2	K0 III		210 s		
93859	27791	43 UMa	10 51 11.0	+56 34 56	−0.007	0.00	5.67	1.12	3.2	G5 IV	+15	19 s		
94105	201827		10 51 11.7	−36 57 26	−0.005	−0.01	7.7	1.0	0.2	K0 III		310 s		
94031	118536		10 51 12.2	+2 42 28	−0.006	−0.12	7.9	1.1	−0.1	K2 III		93 mx		
94012	118535		10 51 13.7	+9 13 27	−0.003	−0.22	7.85	0.48	4.0	F8 V		47 mx		
93915	43492		10 51 14.4	+46 47 46	−0.011	−0.03	7.8	0.8	3.2	G5 IV		84 s		
94073	179272		10 51 14.4	−21 31 27	−0.002	0.00	8.00	1.4	−0.3	K5 III		460 s	7947	m
94058	137839		10 51 18.4	−4 44 12	−0.001	−0.02	7.3	1.1	−0.1	K2 III		310 s		

HD	SAO	Star Name	α 2000	δ 2000	μ(α)	μ(δ)	V	B–V	M_v	Spec	RV	d(pc)	ADS	Notes
			^{h} ^{m} ^{s}	′ ″	^{s}	″								
94030	118539		10 51 19.1	+9 08 56	+0.001	−0.04	7.7	1.1	−0.1	K2 III		360 s		
94230	238532		10 51 20.7	−59 44 39	−0.001	−0.02	7.80	0.37		B0.5 Ib-II				
94290	251153		10 51 20.9	−66 48 55	−0.004	−0.02	7.47	0.10	0.4	B9.5 V		210 mx		
93635	7218		10 51 22.6	+75 59 42	−0.022	−0.03	7.2	0.1	1.7	A3 V		69 mx		
93875	27793	42 UMa	10 51 23.6	+59 19 12	−0.004	−0.06	5.58	1.14	0.2	gK0	−17	91 s		
94261	251152		10 51 24.0	−62 21 48	+0.006	−0.01	7.9	1.7		K0 III		140 s		
94454	256765		10 51 30.0	−75 52 57	−0.007	0.00	6.8	0.0	0.4	B9.5 V		170 s		
94056	99325		10 51 31.9	+13 39 59	+0.001	−0.02	8.0	0.1	1.4	A2 V		210 s		
94245	238536		10 51 32.3	−59 09 05	+0.003	−0.01	8.0	0.3	2.6	F0 V		120 s		
−−	179278		10 51 37.3	−21 15 02	0.000	−0.03	6.70	2.0		N7.7				
94151	179280		10 51 39.2	−22 04 14	−0.020	+0.06	7.82	0.72		G5				
94289	251155		10 51 40.6	−62 54 21	−0.002	−0.02	7.81	0.04	0.4	B9.5 V		300 s		
94089	99327		10 51 45.0	+13 33 58	0.000	−0.01	7.9	1.1	0.2	K0 III		340 s		
94181	179284		10 51 46.3	−29 39 07	−0.007	−0.02	7.9	0.5	4.4	G0 V		50 s		
94275	238540		10 51 46.9	−57 16 20	−0.005	0.00	6.29	0.23	2.1	A5 V		62 s		
94256	238538		10 51 49.8	−52 23 05	+0.002	0.00	7.9	0.5	0.2	K0 III		340 s		
94243	222385		10 51 53.2	−49 09 27	+0.007	−0.06	6.7	0.3	3.4	F5 V		47 s		
94304	238544		10 51 56.3	−58 24 54	0.000	+0.02	6.86	0.50	−6.3	B5 Iab		1900 s		
94286	238543		10 51 58.4	−55 08 19	−0.008	0.00	6.9	1.1	−0.3	K5 III		280 s		
94069	62286		10 51 58.5	+31 22 06	−0.003	−0.01	7.8	0.7	4.4	G0 V		49 s		
94196	179286		10 51 59.5	−24 22 15	−0.006	−0.01	8.0	1.6	−0.5	M2 III		270 mx		
94134	118545		10 52 00.0	+9 03 41	+0.001	−0.03	8.00	0.5	3.4	F5 V		83 s		m
94209	179287		10 52 00.9	−27 34 18	−0.001	+0.04	7.8	2.2	−0.3	K5 III		150 s		
93991	27802		10 52 01.2	+55 21 18	0.000	0.00	8.0	0.4	2.6	F0 V		120 s		
94346	251156		10 52 02.6	−61 25 45	+0.002	−0.01	7.47	0.05	0.4	B9.5 V		250 s		
94087	62289		10 52 04.0	+32 37 42	0.000	−0.03	7.7	1.4	−0.3	K5 III		400 s		
94040	27805		10 52 10.1	+50 15 55	+0.001	−0.07	6.9	1.1	−0.1	K2 III		170 mx		
94180	118550		10 52 13.6	+1 01 31	0.000	0.00	6.38	0.08		A2	−9			m
94206	156285		10 52 17.6	−11 06 38	−0.003	−0.02	7.6	0.4	2.6	F0 V		98 s		
94327	238552		10 52 18.2	−51 48 02	−0.005	−0.01	7.6	0.2	0.4	B9.5 V		180 mx		
94370	238555		10 52 23.2	−58 44 48	−0.002	−0.01	7.94	0.09	−1.6	B3.5 V		640 s		m
94369	238556		10 52 24.5	−58 15 09	−0.001	−0.02	7.36	0.25	−6.2	B1 I		2500 s		
94717	256768		10 52 27.7	−79 33 33	−0.013	0.00	6.33	1.46		K2				
94283	201849		10 52 28.6	−33 39 20	0.000	−0.03	7.9	1.1	−0.3	K5 III		430 s		
94282	179294		10 52 30.0	−30 10 58	0.000	−0.03	8.0	1.5	0.2	K0 III		190 s		
94083	27809		10 52 30.7	+52 33 55	−0.002	−0.03	6.7	1.1	0.2	K0 III	−8	200 s		
94367	238557		10 52 30.8	−57 14 25	−0.001	0.00	5.25	0.16	−7.1	A0 Ia	−23	2800 s		
94084	27810		10 52 31.8	+52 30 13	−0.008	−0.06	6.44	1.11	−0.1	K2 III	−3	150 mx		
94118	43501		10 52 32.5	+45 46 40	−0.002	+0.01	7.1	0.1	1.4	A2 V	+6	140 s		
94326	222392		10 52 33.4	−46 13 02	−0.001	0.00	7.9	0.4	1.4	A2 V		120 s		
94162	62290		10 52 35.0	+38 20 03	−0.001	−0.06	7.66	0.46	3.4	F5 V		71 s		
94237	118555		10 52 36.0	−0 12 05	−0.001	−0.02	6.31	1.50		K4	+9			
94394	238559		10 52 36.6	−59 10 55	−0.001	+0.02	7.9	0.7	0.4	B9.5 V		120 s		
94270	−−		10 52 37.2	−17 00 48			7.9	0.7	4.4	G0 V	+21	51 s		
94324	−−		10 52 39.2	−40 29 49			7.9	1.4	−0.3	K5 III		430 s		
94117	27811		10 52 39.4	+52 04 18	−0.004	0.00	7.0	0.4	3.0	F2 V		62 s		
94269	156290		10 52 40.4	−16 35 18	−0.001	0.00	7.6	1.4	−0.3	K5 III		380 s		
94178	62291		10 52 41.0	+32 59 18	0.000	−0.07	7.6	0.8	3.2	G5 IV	+11	76 s		
94177	43503		10 52 42.2	+40 10 13	−0.009	−0.08	7.3	1.1	0.2	K0 III		100 mx		
94268	156292		10 52 44.4	−11 25 25	−0.017	+0.04	7.5	0.6	4.4	G0 V		42 s		
94252	118558		10 52 45.3	+2 06 45	+0.001	−0.01	7.8	1.6	−0.5	M2 III		470 s		
94495	251161		10 52 46.3	−68 12 38	−0.008	0.00	7.3	1.6	−0.5	M2 III		360 s		
−−	62294		10 52 48.2	+30 23 03	−0.001	−0.01	8.0	1.3						
94218	81561		10 52 48.7	+27 51 54	−0.002	+0.02	7.5	0.5	3.4	F5 V		66 s		
94280	137858		10 52 49.0	−6 49 20	−0.013	−0.20	7.3	0.5	4.0	F8 V		45 s		
94311	156294		10 52 53.7	−19 33 39	−0.001	+0.04	8.0	1.1	−0.1	K2 III		410 s		
94522	251163		10 52 54.8	−69 07 17	−0.007	0.00	7.7	1.1	−0.1	K2 III		350 s		
94310	156295		10 52 56.5	−17 16 40	−0.012	+0.04	7.2	0.6	4.4	G0 V		36 s		
94391	222398		10 52 57.9	−47 23 02	−0.001	−0.01	7.9	1.3	0.2	K0 III		340 s		
94366	201856		10 53 01.0	−35 29 20	−0.003	0.00	6.77	−0.06	0.0	B8.5 V		230 s		
94292	118562		10 53 02.2	+4 57 45	−0.004	+0.03	7.8	0.8	3.2	G5 IV		82 s		
94340	179301		10 53 04.4	−20 37 41	−0.012	−0.27	7.08	0.64	4.9	dG3	−13	26 s		
94376	222400		10 53 05.1	−40 25 03	−0.003	−0.02	8.0	0.2	0.6	A0 V		220 s		
94448	238565		10 53 08.2	−57 49 20	−0.002	−0.01	7.2	1.2	0.2	K0 III		200 s		
94375	179309		10 53 14.8	−30 05 47	−0.002	−0.04	7.8	1.1	4.0	F8 V		27 s		
94493	251164		10 53 15.2	−60 48 53	+0.001	−0.01	7.26	0.01	−6.2	B1 I k		3500 s		
94264	62297	46 LMi	10 53 18.6	+34 12 53	+0.007	−0.28	3.83	1.04	1.7	K0 III-IV	+16	23 s		
94491	238570		10 53 21.1	−58 53 36	−0.002	+0.01	6.27	−0.11	−1.1	B5 V		300 s		
94318	99343		10 53 22.3	+13 07 08	−0.004	−0.02	7.7	1.1	−0.1	K2 III		300 mx		
94363	137863		10 53 24.8	−2 15 19	−0.009	−0.09	6.12	0.90	0.2	K0 III		110 mx	7967	m
94388	156301		10 53 29.4	−20 08 20	+0.005	−0.24	5.24	0.47	3.7	F6 V	−5	20 s		m
94510	238574		10 53 29.6	−58 51 12	+0.009	+0.03	3.78	0.95	1.7	K0 III-IV	+9	26 s		u Car, m
94132	7229		10 53 30.7	+69 51 13	−0.077	−0.08	5.93	0.99	5.7	dG9	+15	17 mn		m
94263	43506		10 53 31.3	+49 13 54	+0.001	+0.02	7.8	0.1	1.4	A2 V		190 s		
94386	156302		10 53 32.8	−15 26 44	+0.004	−0.05	6.38	1.16	3.2	K0 IV		33 s		

HD	SAO	Star Name	α 2000	δ 2000	μ(α)	μ(δ)	V	B-V	M_v	Spec	RV	d(pc)	ADS	Notes
94508	238575		10^{h}53^{m}33^s.4	-56°25'14"	-0^s.002	0".00	6.8	1.1	-0.1	K2 III		230 s		
94486	238573		10 53 33.8	-52 20 28	-0.002	-0.01	7.8	0.1	0.2	K0 III		330 s		
94247	27815	44 UMa	10 53 34.3	+54 35 06	-0.008	-0.01	5.10	1.36	-0.2	K3 III	+1	100 s		
94336	81568		10 53 34.7	+26 12 29	-0.002	-0.01	7.30	1.6		M III	-4			
94443	201864		10 53 35.9	-39 07 25	+0.002	-0.06	7.7	0.8	0.2	K0 III		320 s		
94506	238576		10 53 40.6	-51 29 56	-0.008	0.00	7.3	1.2	0.2	K0 III		210 s		
94651	--		10 53 41.0	-70 42 54			6.1	0.0	0.0	B8.5 V		140 s		
94559	251167		10 53 41.5	-61 39 56	-0.001	-0.01	7.73	0.02	-5.4	B1 Ib-II		3100 s		
94430	179311		10 53 41.7	-24 49 09	-0.004	-0.01	8.0	1.5	-0.3	K5 III		450 s		
94650	256770		10 53 42.0	-70 43 13	-0.005	-0.01	5.99	-0.02	0.0	B8.5 V		160 s		m
94402	137871		10 53 43.6	-2 07 45	-0.006	+0.01	5.45	0.96	0.3	gG6	+15	97 s		m
94461	201867		10 53 48.0	-32 19 36	+0.001	-0.01	7.8	0.8	3.0	F2 V		47 s		
94485	222405		10 53 48.4	-43 07 04	-0.001	-0.03	7.4	1.2	0.2	K0 III		220 s		
94317	43511		10 53 55.4	+47 40 17	-0.003	+0.01	7.6	0.4	2.6	F0 V		99 s		
94504	201873		10 53 56.4	-36 27 26	-0.013	-0.01	6.9	0.0	3.4	F5 V		49 s		
94427	156309		10 53 56.8	-12 26 06	-0.002	-0.03	7.3	0.2	2.1	A5 V		110 s		
94483	201872		10 53 56.9	-34 30 59	-0.002	-0.06	7.8	1.1	3.0	F2 V		30 s		
94473	179318		10 53 58.5	-26 44 46	-0.002	+0.01	7.30	0.00	0.4	B9.5 V		240 s		m
94334	43512	45 ω UMa	10 53 58.7	+43 11 24	+0.004	-0.03	4.71	-0.05	1.2	A1 V	-17	50 s		
94726	256771		10 53 59.5	-72 09 21	-0.008	-0.02	7.8	0.1	0.6	A0 V		200 mx		
94598	251168		10 54 05.3	-60 21 10	-0.006	+0.02	7.86	0.51		F5				
94613	251169		10 54 06.2	-62 02 34	-0.002	-0.02	7.71	2.18	-5.6	M2 Iab		2000 s		BZ Car, v
94556	222408		10 54 06.4	-49 09 56	-0.001	-0.02	7.9	1.0	0.2	K0 III		240 s		
94460	99350		10 54 17.5	+12 22 17	-0.001	-0.01	7.4	1.1	0.2	K0 III		280 s		
94481	156310		10 54 17.7	-13 45 28	0.000	+0.01	5.66	0.83	0.3	gG4	+5	120 s		
94426	62306		10 54 25.4	+30 31 55	+0.004	-0.01	7.5	0.5	3.4	F5 V		65 s		
94683	251173		10 54 29.6	-61 49 35	+0.001	-0.01	5.93	1.75		K5				m
93979	7230		10 54 32.8	+79 20 43	+0.002	0.00	7.8	1.1	0.2	K0 III		340 s		
94425	43518		10 54 33.6	+41 20 19	-0.009	+0.04	7.6	1.1	0.2	K0 III		190 mx		
94500	137875		10 54 34.0	-1 31 07	-0.003	-0.02	8.03	0.39	2.0	F4 IV	+16	160 s		
94457	81574		10 54 35.4	+29 37 48	-0.007	-0.05	6.6	0.5	4.0	F8 V		33 s		
94565	201886		10 54 35.4	-38 45 16	-0.002	0.00	7.02	-0.04	0.4	B9.5 V		210 s		m
94645	238589		10 54 37.0	-55 37 08	+0.002	-0.01	6.6	1.4	-0.1	K2 III		150 s		
94644	238591		10 54 41.7	-54 33 37	-0.002	-0.01	7.6	-0.1	0.0	B8.5 V		330 s		
94480	81576	48 LMi	10 54 42.1	+25 29 27	-0.004	0.00	6.20	0.28		A m	+10			
94607	222417		10 54 43.2	-42 01 05	-0.006	-0.02	7.1	1.4	0.2	K0 III		240 s		
94456	62308		10 54 44.6	+36 45 30	-0.005	-0.05	7.3	0.4	2.6	F0 V		88 s		
94515	118571		10 54 45.5	+5 50 53	-0.001	+0.01	7.9	1.1	0.2	K0 III		340 s		
94589	201890		10 54 50.4	-30 20 06	-0.005	+0.02	7.8	0.8	1.4	A2 V		64 s		
94642	222420		10 54 52.2	-46 34 59	-0.003	0.00	7.9	1.7	-0.3	K5 III		430 s		
94497	62310		10 54 58.1	+34 02 05	-0.005	-0.05	5.72	1.01	0.2	K0 III	-28	130 s		
94660	222422		10 55 00.9	-42 15 04	-0.004	0.00	6.11	-0.08		A0 p				
94619	179334		10 55 11.4	-20 39 55	-0.002	-0.02	6.44	1.10		K0				
94190	7237		10 55 13.4	+77 05 14	-0.008	-0.01	6.87	1.59	-0.5	M3 III	-90	300 s		
94604	156325		10 55 14.5	-12 07 17	+0.001	+0.03	7.7	1.1	0.2	K0 III		310 s		
94776	251178		10 55 17.2	-60 31 01	-0.004	+0.08	5.92	1.08	0.2	gK0	-26	130 s		T Car, q
94731	238596		10 55 18.1	-52 39 13	+0.001	-0.02	7.70	0.4	2.6	F0 V		100 s		m
94964	256773		10 55 20.3	-75 11 34	-0.014	-0.10	7.3	0.6	4.4	G0 V		38 s		
94758	238598		10 55 25.3	-54 04 46	-0.005	0.00	7.2	0.0	0.4	B9.5 V		190 mx		
94724	222423		10 55 27.0	-43 01 13	-0.006	+0.01	6.7	0.8	0.6	A0 V		54 s		
94692	179340		10 55 28.4	-26 58 46	+0.002	-0.03	7.1	1.2	0.2	K0 III		170 s		
94587	81582		10 55 29.8	+23 09 50	-0.002	-0.03	7.5	0.8	3.2	G5 IV		73 s		
94757	222426		10 55 34.2	-49 15 01	0.000	-0.01	7.8	1.0	0.2	K0 III		330 s		
94601	81584	54 Leo	10 55 37.2	+24 44 55	-0.005	-0.03	4.32	0.02	1.2	A1 V	+4	42 s	7979	m
94672	118574	55 Leo	10 55 42.4	+0 44 13	+0.007	0.00	5.91	0.42	3.1	dF3	-2	35 s	7982	m
94924	251183		10 55 44.1	-68 47 28	-0.005	-0.02	7.9	0.1	0.6	A0 V		270 mx		
94600	62314	46 UMa	10 55 44.3	+33 30 25	-0.009	-0.03	5.03	1.10	0.0	K1 III	-22	100 s		
94741	201908		10 55 51.0	-31 47 10	-0.004	0.00	7.2	0.5	0.6	A0 V		110 s		
94740	179347		10 55 51.6	-27 49 15	-0.001	-0.03	7.8	2.1	-0.1	K2 III		110 s		
94771	201911		10 55 53.8	-35 06 54	-0.003	-0.08	7.4	0.7	4.4	G0 V		34 s		
94671	99358		10 55 56.0	+18 09 10	-0.004	+0.01	7.3	0.9	3.2	G5 IV	-10	67 s		
94705	118576	56 Leo	10 56 01.4	+6 11 07	-0.002	-0.01	5.81	1.45		M5 III	-13	22 mn		VY Leo, q
94798	222431		10 56 03.2	-41 12 01	-0.001	-0.01	7.9	1.4	-0.3	K5 III	-4	430 s		
94703	99359		10 56 04.8	+16 08 59	-0.004	-0.04	7.8	0.7	4.4	G0 V		49 s		
94738	118577		10 56 10.4	+0 25 58	-0.001	-0.01	6.8	1.1	0.2	K0 III	-28	210 s		
94840	222436		10 56 11.5	-48 05 28	-0.004	-0.01	7.5	0.3	0.2	K0 III		280 s		
94910	251185		10 56 11.6	-60 27 14	0.000	-0.02	6.96	0.61						AG Car, v
94549	15339		10 56 12.0	+64 32 18	-0.016	+0.09	7.3	1.1	0.2	K0 III	+19	160 mx		
94669	43535		10 56 14.4	+42 00 30	+0.001	-0.10	6.03	1.13	-0.1	K2 III	-54	120 mx		
94633	43534		10 56 15.0	+45 46 13	-0.004	+0.01	7.9	1.1	0.2	K0 III		340 s		
94898	238611		10 56 16.2	-58 03 11	-0.001	-0.03	7.13	1.58	-0.5	M4 III		340 s		
94720	81589		10 56 16.8	+22 21 06	-0.002	0.00	6.14	1.55		K2	+25			
94781	156338		10 56 19.8	-10 37 29	0.000	-0.02	7.4	1.1	-0.1	K2 III		310 s		
94719	81590		10 56 21.4	+23 15 02	+0.002	-0.04	7.2	0.8	3.2	G5 IV		64 s		
94909	238613		10 56 24.5	-57 33 05	0.000	0.00	7.34	0.48	-6.2	B0 I		2000 s		m

HD	SAO	Star Name	α 2000	δ 2000	μ(α)	μ(δ)	V	B-V	M_v	Spec	RV	d(pc)	ADS	Notes
94631	27831		10^h 56^m 26^s.8	+57° 30' 10"	-0^s.004	-0".03	6.8	0.9	3.2	G5 IV	+10	52 s		
94765	118578		10 56 30.7	+7 23 16	-0.017	-0.10	7.34	0.92		K0		13 mn		
94766	118579		10 56 30.9	+5 16 24	-0.002	-0.04	7.8	0.1	1.4	A2 V		190 s		
94747	81591	50 LMi	10 56 34.4	+25 30 00	-0.002	-0.02	6.35	1.03	0.2	K0 III	+30	160 s		
94963	251187		10 56 35.7	-61 42 33	-0.001	-0.02	7.17	-0.08		O8.8				
94839	179362		10 56 37.0	-21 16 31	-0.003	+0.01	8.0	0.4	1.4	A2 V		120 s		
94808	137898		10 56 40.4	-1 10 08	-0.002	-0.02	7.91	0.31	0.3	gA5	-10	280 s		
94920	238616		10 56 40.5	-50 36 30	+0.011	-0.07	7.5	0.0	3.0	F2 V		79 s		
94890	201927	ι Ant	10 56 43.0	-37 08 16	+0.006	-0.13	4.60	1.03	0.3	G5 III	0	56 s		
94867	179367		10 56 51.5	-22 02 00	-0.004	+0.01	7.6	1.6	-0.5	M2 III		310 mx		
94906	201929		10 56 51.9	-31 12 17	-0.006	-0.06	7.44	0.35	3.0	F2 V		77 s		
95086	251193		10 57 03.2	-68 40 02	-0.006	+0.01	7.6	0.1	1.4	A2 V		170 s		
94792	43541		10 57 06.5	+41 21 03	-0.009	-0.07	7.4	0.6	4.4	G0 V		40 s		
94985	238622		10 57 07.8	-50 45 55	-0.004	-0.01	5.91	0.18	1.7	A3 V n		69 s		
94864	137904		10 57 07.9	-0 18 43	-0.007	-0.03	6.88	0.42	3.4	F5 V	+3	50 s		
94835	81595		10 57 09.4	+21 48 18	-0.012	-0.22	7.96	0.63	4.4	G0 V		39 mx		
94973	222444		10 57 11.3	-44 48 40	-0.003	-0.05	7.1	0.1	2.6	F0 V		81 s		
94834	81597		10 57 15.1	+24 08 34	-0.005	+0.03	7.70	1.1		K1 IV		250 mx		
95208	256775		10 57 15.6	-75 05 59	-0.004	0.00	6.13	1.53		K2				
95122	251195		10 57 19.9	-69 02 19	-0.007	-0.02	7.00	0.00	0.0	B8.5 V		190 mx		m
94956	179378		10 57 23.8	-29 16 51	-0.002	+0.01	7.8	1.7	3.2	G5 IV		23 s		
94573	7244		10 57 27.5	+74 20 10	+0.001	-0.03	8.0	1.6	-0.5	M4 III		500 s		
94995	201943		10 57 29.3	-37 50 33	-0.003	0.00	7.8	0.7	1.4	A2 V		71 s		
95009	222452		10 57 37.4	-41 07 58	+0.007	-0.06	7.5	1.0	3.0	F2 V		81 s		
94927	99369		10 57 37.8	+15 45 40	-0.007	-0.07	7.8	0.4	3.0	F2 V		92 s		
95027	222454		10 57 37.9	-45 52 32	-0.004	+0.01	6.80	1.1	0.2	K0 III		210 s		m
94832	43544		10 57 39.8	+47 45 29	-0.003	-0.02	7.9	0.5	4.0	F8 V		61 s		
95026	222453		10 57 40.8	-41 18 57	-0.001	-0.03	8.0	1.2	0.2	K0 III		360 s		
94926	99370		10 57 42.0	+17 56 34	-0.002	-0.01	7.9	0.2	2.1	A5 V		150 s		
95083	238634		10 57 45.9	-52 40 54	-0.002	-0.02	7.2	1.4	0.2	K0 III		150 s		
95207	256776		10 57 47.5	-71 06 34	0.000	-0.07	7.6	0.0	0.4	B9.5 V		110 mx		
95109	238635		10 57 48.3	-59 43 54	+0.001	+0.01	6.11	1.10	-6.1	G7 Iab		1000 s		U Car, v
95025	179383		10 57 54.8	-22 01 15	-0.003	0.00	7.6	0.7	3.4	F5 V		50 s		
94862	27849		10 57 57.4	+55 25 48	-0.005	+0.02	7.8	0.8	3.2	G5 IV		84 s		
94882	27850		10 57 58.7	+51 27 28	-0.001	-0.01	7.9	1.1	-0.1	K2 III		390 s		
95119	238639		10 58 04.8	-55 47 01	0.000	-0.02	6.8	1.5	0.2	K0 III		100 s		
94966	81601		10 58 09.1	+24 22 32	0.000	-0.01	7.90	1.1	0.0	K1 III		380 s		
95064	179385		10 58 10.5	-29 14 22	-0.001	0.00	7.8	1.2	-0.1	K2 III		380 s		
95004	118593		10 58 11.8	+9 42 09	-0.005	+0.04	7.8	1.1	-0.1	K2 III		220 mx		
95267	--		10 58 16.2	-69 56 06			7.9	1.4	-0.3	K5 III		420 s		
95709	258598		10 58 17.2	-82 34 49	-0.009	+0.01	7.5	0.0	0.4	B9.5 V		210 mx		
95137	222460		10 58 23.5	-44 56 04	-0.002	+0.01	7.4	1.1	0.2	K0 III		250 s		
95202	238644		10 58 33.5	-57 59 32	-0.004	+0.01	7.4	-0.1	1.4	A2 V		160 s		
95103	179389		10 58 34.1	-24 12 02	-0.001	-0.01	8.0	1.0	-0.1	K2 III		410 s		
95034	81603		10 58 36.7	+21 52 09	-0.001	+0.01	8.0	0.9	3.2	G5 IV		93 s		
95047	99384		10 58 37.4	+19 37 20	+0.001	-0.02	6.9	1.1	0.2	K0 III		220 s		
95149	201962		10 58 37.4	-38 12 21	+0.003	-0.08	7.5	0.9	0.2	K0 III		260 mx		
95046	81605		10 58 41.1	+21 30 32	-0.004	-0.01	7.7	0.2	2.1	A5 V		130 s		
94847	7248		10 58 43.5	+69 48 43	-0.006	-0.03	8.00	0.4	2.6	F0 V		120 s	7993	m
95160	222463		10 58 43.5	-41 34 13	+0.001	-0.02	7.5	0.2	1.4	A2 V		170 s		
95255	251202		10 58 46.2	-60 51 46	0.000	-0.01	8.0	1.3	-0.1	K2 III		360 s		
95089	118598		10 58 47.6	+1 43 45	-0.003	-0.06	7.9	1.1	0.2	K0 III		190 mx		
95223	238649		10 58 51.4	-50 16 28	+0.002	-0.02	7.9	1.2	0.2	K0 III		320 s		
95145	156365		10 58 56.6	-19 36 18	+0.003	-0.08	6.9	1.1	0.2	K0 III		220 s		
95045	43552		10 59 00.5	+45 11 51	-0.005	-0.04	6.9	1.1	0.2	K0 III		210 mx		
95100	99385		10 59 00.7	+17 43 19	0.000	-0.02	7.9	1.1	-0.1	K2 III		390 s		
94791	7247		10 59 01.8	+75 43 15	-0.040	-0.04	8.00	0.5	3.4	F5 V	+15	83 s		m
94902	7250		10 59 01.9	+69 59 20	+0.006	-0.02	7.3	1.6	-0.5	M2 III		360 s		VW UMa, v
95198	201972		10 59 04.1	-34 52 34	-0.003	0.00	7.80	0.00	0.4	B9.5 V		300 s		m
95263	238652		10 59 04.3	-53 32 23	-0.012	0.00	7.2	0.8	4.0	F8 V		31 s		
95182	179397		10 59 05.0	-23 15 16	-0.007	-0.02	8.0	0.3	2.6	F0 V		120 s		
95290	238654		10 59 05.6	-59 10 07	-0.002	-0.01	7.68	-0.08		A0				
95197	201973		10 59 06.3	-34 22 02	-0.002	+0.04	7.8	1.0	0.2	K0 III		330 s		
95112	99387		10 59 08.2	+17 49 36	0.000	-0.03	7.3	1.1	0.2	K0 III		270 s		
95222	201975		10 59 09.5	-36 30 12	-0.006	0.00	7.8	0.5	1.7	A3 V		87 s		
95788	258599		10 59 13.1	-81 33 21	-0.069	+0.07	6.71	0.55		F5				m
95221	201976		10 59 13.8	-33 44 14	+0.001	-0.05	5.71	0.37	2.6	dF0		38 s		m
95324	251205		10 59 14.0	-61 19 12	-0.004	0.00	6.16	-0.06		B9				m
96124	258600	η Oct	10 59 14.1	-84 35 37	-0.044	-0.01	6.19	0.11		A0				
95180	156368		10 59 16.2	-17 37 24	-0.002	-0.02	8.0	0.1	1.4	A2 V		210 s		
95178	156367		10 59 17.8	-10 19 29	-0.002	+0.02	7.1	0.1	0.6	A0 V		200 s		
95057	27858		10 59 17.9	+51 52 57	-0.001	0.00	6.17	1.38		K0	-7			
95156	99388		10 59 21.0	+15 32 25	0.000	0.00	7.6	1.1	0.2	K0 III		300 s		
94686	7246		10 59 21.1	+79 40 35	+0.021	+0.07	7.32	0.55	4.0	F8 V		44 s	7992	m
95274	222475		10 59 24.7	-46 36 20	-0.002	0.00	7.5	1.2	0.2	K0 III		230 s		

HD	SAO	Star Name	α 2000	δ 2000	μ(α)	μ(δ)	V	B-V	M$_v$	Spec	RV	d(pc)	ADS	Notes
95128	43557	47 UMa	10^{h}59^{m}27.9	+40°25'49"	-0^s.028	+0".05	5.05	0.61	4.4	G0 V	+13	13 ts		
95234	156372		10 59 30.8	-16 21 14	-0.003	-0.02	5.89	1.63	-0.5	gM2	-33	180 s		
95129	62345		10 59 32.7	+36 05 35	+0.006	-0.05	6.00	1.59	-0.5	M2 III	-26	200 s		
95190	99391		10 59 32.9	+9 55 51	-0.004	-0.01	7.00	0.1	1.4	A2 V	-3	130 s	8003	m
95260	201983		10 59 33.6	-33 04 29	-0.001	+0.03	8.0	0.2	0.2	K0 III		360 s		
95285	222478		10 59 34.3	-43 30 06	+0.003	-0.01	7.5	0.7	0.2	K0 III		280 s		
95393	251206		10 59 37.4	-62 29 00	0.000	+0.02	7.4	1.5	0.2	K0 III		140 s		
95216	99392		10 59 41.0	+11 42 21	-0.016	+0.04	6.4	0.5	0.7	F5 III	+20	66 mx		
95355	238665		10 59 43.1	-56 12 19	+0.001	+0.02	7.3	0.0	0.6	A0 V		210 s		
95376	238666		10 59 46.0	-53 41 19	0.000	-0.01	7.6	0.1	1.4	A2 V		160 s		
95272	156375	7 α Crt	10 59 46.4	-18 17 56	-0.033	+0.12	4.08	1.09	0.2	K0 III	+47	37 mx		Alkes
95300	201991		10 59 46.8	-35 08 35	-0.005	0.00	7.4	0.2	2.6	F0 V		93 s		
95351	222481		10 59 47.2	-49 41 59	-0.002	-0.02	8.0	1.1	-0.1	K2 III		410 s		
95125	27863		10 59 47.3	+50 54 36	-0.003	-0.01	7.7	0.9	3.2	G5 IV		80 s		
95215	---		10 59 48.2	+16 23 14	-0.002	-0.01	7.9	1.1	0.2	K0 III		340 s		
95098	27861		10 59 49.4	+58 54 22	-0.001	-0.03	7.12	1.19		K2			8001	m
94860	7255		10 59 56.7	+77 46 12	-0.023	-0.03	6.20	0.97	3.2	G5 IV	-50	30 s		
95347	222485		10 59 59.4	-43 48 25	-0.006	+0.01	5.81	-0.08		B9				
95280	137933		11 00 01.9	-3 28 18	+0.004	+0.03	7.50	0.5	4.0	F8 V		50 s	8007	m
95242	81615		11 00 02.5	+22 01 35	-0.002	-0.01	7.2	1.1	0.2	K0 III		250 s		
95558	251210		11 00 08.4	-69 47 59	-0.004	-0.05	8.0	0.0	0.4	B9.5 V		120 mx		
95429	238674		11 00 08.5	-51 49 03	-0.004	0.00	6.15	0.18	1.7	dA3		69 s		m
95543	251209		11 00 09.2	-68 50 28	+0.006	-0.01	7.9	0.0	0.4	B9.5 V		310 s		
95370	222487		11 00 09.2	-42 13 33	+0.002	0.00	4.39	0.11	0.6	A2 IV	-5	52 s		
95314	156382		11 00 11.5	-14 05 00	-0.002	-0.03	5.88	1.50	-0.3	gK5	-6	170 s		
95212	43562		11 00 14.6	+45 31 34	+0.001	0.00	5.47	1.47	-0.1	K2 III	+9	86 s		
95408	222490		11 00 15.4	-44 54 37	-0.001	+0.01	7.30	1.4	-0.3	K5 III		330 s		m
95409	222491		11 00 15.6	-46 36 05	0.000	-0.01	7.9	0.0	0.6	A0 V		270 s		
95407	222489		11 00 15.9	-43 15 06	+0.001	-0.03	7.9	0.2	3.4	F5 V		78 s		
95459	238679		11 00 16.0	-57 33 37	-0.004	0.00	7.3	0.9	-0.5	M2 III		370 s		
95241	43564		11 00 20.5	+42 54 41	-0.010	-0.14	6.02	0.57	4.0	F8 V	-6	24 s		m
95346	137935		11 00 22.3	-9 29 34	-0.001	-0.01	7.5	0.1	1.4	A2 V		170 s		
95233	27868		11 00 25.5	+51 30 07	-0.003	-0.02	6.5	0.9	0.2	G9 III	0	180 s		
95492	238684		11 00 33.5	-51 56 51	+0.001	-0.01	7.9	-0.1	0.6	A0 V		280 s		
95345	118610	58 Leo	11 00 33.6	+3 37 03	+0.001	-0.02	4.84	1.16	0.0	K1 III	+6	83 s		
95509	238687		11 00 36.1	-53 06 43	-0.011	0.00	6.5	1.4	0.2	K0 III		110 s		
95590	---		11 00 36.2	-66 02 11			7.9	1.4	-0.3	K5 III		420 s		
95540	251211		11 00 37.2	-60 18 15	0.000	-0.02	7.91	0.38	-5.2	A0 Ib		2500 s		
95344	118612		11 00 37.4	+3 44 22	-0.002	-0.02	8.0	0.1	0.6	A0 V		300 s		
95403	156390		11 00 37.9	-19 20 36	-0.013	+0.02	7.0	0.8	3.2	G5 IV		59 s		
95456	201998		11 00 40.7	-31 50 21	-0.006	+0.10	6.07	0.52	4.4	G0 V		22 s		
95402	156391		11 00 41.6	-12 38 15	-0.002	+0.04	7.8	0.4	3.0	F2 V		89 s		
95296	43565		11 00 42.0	+42 43 53	-0.001	-0.01	6.70	1.1	0.2	K0 III		200 s	8015	m
95382	118615	59 Leo	11 00 44.7	+6 06 05	-0.003	-0.03	4.99	0.16	2.1	A5 V	-12	38 s	8019	m
95381	99398		11 00 47.9	+11 58 13	-0.002	-0.04	8.0	0.1	1.7	A3 V		180 s		
95295	43566		11 00 48.3	+43 54 42	-0.001	-0.01	7.7	1.1	0.2	K0 III		320 s		
95534	238692		11 00 49.6	-52 04 29	-0.002	0.00	7.1	0.5	0.6	A0 V		92 s		
95521	222501		11 00 50.1	-49 01 12	-0.010	-0.07	8.0	0.1	3.2	G5 IV		93 s		
95310	62354	49 UMa	11 00 50.3	+39 12 43	-0.006	-0.02	5.08	0.24		A m	+3			
95506	222499		11 00 51.1	-40 29 52	-0.021	+0.01	6.79	0.53	4.0	F8 V		35 s		m
95423	156394		11 00 51.4	-12 33 30	-0.004	+0.02	7.5	0.1	0.6	A0 V		110 mx		
95441	156396		11 00 57.1	-15 47 34	+0.004	-0.04	6.34	1.18		K0				
95363	81623		11 00 57.5	+27 07 57	-0.002	+0.02	7.98	0.49	3.8	F7 V		69 s		
95330	43568		11 01 03.4	+47 40 57	-0.005	0.00	7.58	1.31		K2				
95453	118620		11 01 03.6	+0 02 48	-0.003	-0.01	7.90	0.1	1.4	A2 V		200 s		m
95329	43569		11 01 04.5	+47 51 04	0.000	-0.02	7.9	1.1	-0.1	K2 III		390 s		
95438	99401		11 01 05.5	+12 02 47	+0.009	-0.10	7.7	1.1	0.2	K0 III		140 mx		
95256	15360		11 01 05.8	+63 25 16	-0.006	-0.05	6.4	0.1	0.6	A0 V	+11	82 mx		
95635	251218		11 01 10.0	-61 13 18	+0.001	0.00	7.95	0.09	-1.6	B3.5 V		570 s		
95501	137941		11 01 15.0	-9 46 19	0.000	-0.01	7.7	1.6	-0.5	M2 III		430 s		
95379	43572		11 01 20.6	+46 33 48	-0.001	0.00	7.0	1.1	0.2	K0 III		230 s		
95486	99403		11 01 25.8	+15 00 36	0.000	-0.32	7.9	1.1	0.2	K0 III	-56	21 mx		
95532	156406		11 01 29.0	-16 56 47	-0.004	-0.08	8.0	0.5	2.3	F7 IV	-25	50 mx		
95883	256781		11 01 29.9	-75 39 51	-0.001	+0.03	7.3	-0.1	1.4	A2 V		150 s		
95033	7265		11 01 31.1	+78 52 59	+0.015	-0.04	7.6	0.9	3.2	G5 IV		75 s		
95397	27875		11 01 35.6	+52 42 08	-0.001	-0.04	7.2	1.1	0.2	K0 III		250 s		
95294	15363		11 01 35.7	+66 27 01	-0.002	0.00	7.90	1.1	0.2	K0 III		350 s	8020	m
95687	251221		11 01 35.7	-61 02 56	0.000	0.00	7.3	0.6	-0.5	M2 III		370 s		
95707	251223		11 01 39.2	-61 33 44	-0.001	-0.01	7.56	0.24	-5.6	B8 Ib		3100 s		
95466	62359		11 01 41.2	+38 00 31	-0.001	+0.02	8.0	1.1	0.2	K0 III		360 s		
95518	99405		11 01 43.6	+17 21 35	0.000	-0.02	8.0	0.5	4.0	F8 V		62 s		
95485	62360		11 01 45.5	+36 40 41	-0.003	+0.05	7.3	0.4	2.6	F0 V		89 s		
95515	62361		11 01 49.2	+29 52 17	-0.005	+0.01	7.2	1.1	0.2	K0 III		250 s		
95578	137947	61 Leo	11 01 49.6	-2 29 04	+0.001	-0.04	4.74	1.62	-0.3	K5 III	-14	86 s		
95752	251226		11 01 50.0	-64 18 15	+0.002	-0.02	7.0	0.5	2.6	F0 V		60 s		

HD	SAO	Star Name	α 2000	δ 2000	μ(α)	μ(δ)	V	B-V	M_v	Spec	RV	d(pc)	ADS	Notes
95418	27876	48 β UMa	11^{h}01^{m}50^s.4	+56°22'56"	+0^s.010	+0".03	2.37	-0.02	1.2	A1 V	-12	19 ts		Merak
95786	—		11 01 56.0	-66 00 14			7.51	0.07	0.4	B9.5 V		220 s		
95881	256784		11 01 57.6	-71 30 51	-0.002	-0.03	7.9	0.0	0.4	B9.5 V		310 s		
95667	202021		11 02 00.9	-37 50 00	+0.001	0.00	7.8	1.5	-0.3	K5 III		420 s		
95915	256786		11 02 01.9	-73 29 23	+0.008	-0.05	7.3	0.5	3.4	F5 V		59 s		
95894	256785		11 02 02.3	-71 44 58	-0.014	-0.08	7.6	0.1	0.6	A0 V		67 mx		
95624	137949		11 02 03.7	-9 51 37	-0.002	-0.08	7.9	1.1	0.2	K0 III		180 mx		
95716	222508		11 02 13.7	-41 06 51	+0.001	-0.04	6.5	1.1	-0.5	M4 III		260 s		
95717	222510		11 02 17.4	-42 48 18	0.000	-0.03	8.0	1.2	-0.3	K5 III		450 s		
95499	27877		11 02 17.6	+58 39 59	-0.004	-0.03	6.8	1.1	-0.1	K2 III		240 s		
95608	81637	60 Leo	11 02 19.7	+20 10 47	-0.001	+0.03	4.42	0.05		A m	-10	22 mn		
95698	179456		11 02 24.4	-26 49 53	+0.005	-0.12	6.23	0.31	1.7	F0 IV		80 s	8028	m
95620	99415		11 02 24.8	+16 52 47	0.000	0.00	7.9	0.9	3.2	G5 IV		88 s		
95680	179457		11 02 26.6	-21 24 40	-0.004	0.00	7.4	1.1	0.3	G6 III	+8	210 s		
95547	27879		11 02 30.6	+55 01 42	-0.010	+0.09	7.4	0.5	3.4	F5 V		64 s		
95678	137951		11 02 31.2	-9 59 42	-0.002	-0.02	7.5	1.4	-0.3	K5 III		360 s		
95651	118628		11 02 32.6	+9 10 22	-0.005	-0.02	7.1	0.1	0.6	A0 V	-9	130 mx		
95695	137952		11 02 36.7	-3 30 46	-0.002	-0.02	6.80	0.95		G5				m
95618	62370		11 02 38.1	+36 20 11	-0.003	-0.01	7.9	0.9	3.2	G5 IV		88 s		
95713	137954		11 02 45.9	-6 23 38	+0.002	-0.01	7.7	1.1	-0.1	K2 III		360 s		
95880	238715		11 02 46.4	-59 44 27	+0.002	0.00	6.95	0.34	-5.7	B5 Ib		1900 s		m
95773	179465		11 02 52.1	-28 43 30	+0.002	-0.02	7.7	3.0	-0.1	K2 III		28 s		
95758	179467		11 02 55.2	-23 59 14	-0.005	+0.02	8.0	1.2	-0.1	K2 III		280 mx		
96044	256788		11 02 55.2	-72 14 32	-0.002	-0.01	7.1	0.0	0.0	B8.5 V		260 s		
95837	222521		11 02 56.0	-45 26 43	-0.001	-0.01	7.6	0.9	0.2	K0 III		300 s		
95740	137957		11 02 57.9	-2 38 08	-0.002	-0.02	7.3	0.9	3.2	G5 IV		67 s		
95995	—		11 02 59.2	-67 07 16			7.8	1.6	-0.5	M4 III		430 s		KV Car, v
95933	251234		11 03 04.5	-60 49 29	-0.002	0.00	7.52	0.58		F5				
95793	179469		11 03 05.0	-25 34 30	-0.004	+0.01	6.8	1.1	1.4	A2 V		30 s		
95950	251235		11 03 06.2	-60 54 38	0.000	0.00	6.6	2.4	-0.5	M2 III		85 s		
95994	—		11 03 07.8	-64 41 16			7.7	0.0	0.4	B9.5 V		280 s		
95725	81642		11 03 11.4	+28 55 47	+0.001	-0.01	7.07	1.05	-2.2	K1 II		720 s		
95908	238717		11 03 12.8	-50 59 32	+0.003	-0.02	7.80	0.8	3.2	G5 IV		83 s		m
95771	137963		11 03 14.5	-0 45 09	-0.001	-0.12	6.14	0.26		A3				
95808	156421		11 03 14.9	-11 18 13	-0.005	-0.11	5.50	0.94	0.3	gG6	-8	110 s	8037	m
95857	202039		11 03 16.1	-31 57 39	-0.003	-0.03	6.46	1.62		Ma				m
95855	179474		11 03 16.6	-27 31 04	-0.004	0.00	7.80	0.1	0.6	A0 V		150 mx	8038	m
95735	62377		11 03 20.2	+35 58 13	-0.047	-4.74	7.49	1.51	10.5	M2 V	-87	2.5 t		Lalande 21185
95638	15379		11 03 21.8	+61 39 15	-0.017	-0.07	7.15	0.51	3.8	F7 V		46 s		
95572	7272		11 03 27.2	+70 01 51	+0.001	-0.03	6.6	1.1	0.2	K0 III		190 s		
95463	7271		11 03 27.7	+75 26 29	-0.010	-0.02	7.2	1.1	0.2	K0 III		250 s		
96143	256791		11 03 31.7	-71 37 46	+0.031	-0.10	7.7	0.4	3.0	F2 V		86 s		
95870	156427		11 03 36.5	-13 26 05	0.000	0.00	6.34	0.88		G5				
95849	118634	62 Leo	11 03 36.6	-0 00 03	-0.004	0.00	5.95	1.22	-0.2	gK3	-8	170 s		
95768	43594		11 03 39.1	+44 19 46	-0.016	-0.09	7.21	0.78	1.5	G8 III-IV				m
96008	238720		11 03 42.5	-51 21 10	-0.023	+0.01	6.75	0.35		A5				
95689	15384	50 α UMa	11 03 43.6	+61 45 03	-0.017	-0.07	1.79	1.07	0.2	K0 III	-9	23 ts	8035	Dubhe, m
99828	258607		11 03 44.1	-89 14 20	-0.301	+0.02	7.7	-0.2	2.6	F0 V		100 s		
95804	81648		11 03 44.9	+25 46 29	-0.008	0.00	6.9	0.2	2.1	A5 V		91 s		
95900	137970		11 03 50.1	-7 41 19	+0.002	-0.09	7.4	1.1	0.2	K0 III		210 mx		
96021	238724		11 03 50.4	-51 01 47	-0.006	-0.01	7.4	1.4	0.2	K0 III		160 s		
95963	202048		11 03 53.1	-33 26 40	-0.002	-0.03	7.2	1.6	0.2	K0 III		110 s		
95939	156431		11 03 54.1	-19 39 01	-0.006	-0.02	6.6	0.2	2.1	A5 V		78 s		
95938	137971		11 03 57.1	-9 04 52	-0.002	-0.01	8.0	0.4	3.0	F2 V		98 s		
95899	118638		11 03 58.8	+3 38 19	-0.002	0.00	7.80	0.4	2.6	F0 V		110 s	8043	m
96088	238730		11 04 00.0	-57 57 19	-0.004	-0.01	6.17	-0.17	-2.9	B3 III		310 mx		
96216	—		11 04 01.4	-71 00 19			7.7	0.0	0.4	B9.5 V		280 s		
96068	238729		11 04 02.9	-54 11 53	-0.001	-0.02	6.5	1.5	0.2	K0 III		90 s		
96290	256792		11 04 04.4	-74 14 50	-0.005	+0.03	7.3	0.6	4.4	G0 V		38 s		
96033	222533		11 04 07.4	-40 18 32	-0.004	-0.02	7.4	0.3	2.6	F0 V		86 s		
96054	202055		11 04 16.2	-36 43 18	-0.007	0.00	7.2	0.7	1.4	A2 V		59 s		
95884	62385		11 04 18.0	+38 52 06	-0.004	0.00	7.3	0.1	1.7	A3 V		130 s		
95960	118644		11 04 18.9	+5 13 30	-0.003	+0.01	8.0	1.6	-0.5	M2 III		510 s		
95981	137974		11 04 19.2	-1 16 43	+0.001	-0.02	6.8	0.1	0.6	A0 V		180 s		
95959	99427		11 04 21.0	+14 43 29	-0.002	0.00	7.90	1.1	0.2	K0 III		350 s		
96268	251248		11 04 21.6	-69 55 49	-0.012	-0.04	7.7	0.4	2.6	F0 V		100 s		
96158	—		11 04 22.1	-59 48 18			7.63	0.12	-0.4	B V		360 s		
96067	222536		11 04 23.8	-42 40 39	-0.001	-0.01	7.3	0.4	0.2	K0 III		260 s		
96159	238735		11 04 25.6	-59 52 06	-0.001	0.00	7.81	0.12		B				
96082	202057		11 04 25.8	-39 49 13	-0.004	+0.03	7.8	1.2	-0.3	K5 III		420 s		
96135	238734		11 04 28.5	-53 06 31	-0.007	-0.01	7.3	1.2	1.7	A3 V		31 s		
96193	251246		11 04 29.1	-61 29 18	-0.002	-0.02	7.5	1.1	-0.3	K5 III		370 s		
95934	62387	51 UMa	11 04 31.1	+38 14 29	-0.006	0.00	6.00	0.16	1.7	A3 V	+7	67 s	8046	m
96113	222538		11 04 31.2	-47 40 43	-0.011	+0.04	5.67	0.24		A5	-16			
96081	202058		11 04 32.8	-30 50 30	0.000	+0.01	7.8	1.0	-0.1	K2 III		390 s		

HD	SAO	Star Name	α 2000	δ 2000	μ(α)	μ(δ)	V	B-V	M_v	Spec	RV	d(pc)	ADS	Notes
96003	99428		11^h 04^m 33^s.2	+12° 40′ 01″	−0^s.001	−0″.01	6.70	0.1		A3 p				m
95978	81652		11 04 36.2	+29 10 21	+0.005	−0.01	7.86	1.23	−0.1	K2 III		360 s		
95977	62390		11 04 38.9	+31 26 12	−0.002	+0.02	7.3	0.9	3.2	G5 IV		67 s		
96265	251249		11 04 39.5	−65 34 11	+0.003	+0.10	7.9	1.7	−0.3	K5 III		100 mx		
96064	137978		11 04 41.3	−4 13 17	−0.013	−0.12	7.64	0.76		G5			8048	m
96133	222541		11 04 43.1	−46 05 58	−0.004	−0.02	7.70	0.1	1.4	A2 V		150 mx		m
95976	62392		11 04 43.7	+38 14 49	−0.007	+0.01	7.4	0.4	3.0	F2 V	+5	77 s		
96451	256794		11 04 48.5	−75 09 20	−0.019	−0.03	7.1	−0.5	0.6	A0 V		120 mx		
96287	251252		11 04 50.1	−64 36 56	−0.002	+0.01	7.24	0.00	0.4	B9.5 V		220 s		
96146	202067		11 04 54.0	−35 48 17	−0.002	0.00	5.43	0.03		A0	+11			
96264	251253		11 04 55.6	−61 03 05	+0.001	0.00	7.63	−0.06	−4.8	O9 V		2400 s		m
96016	62395		11 04 57.1	+38 24 38	−0.005	−0.03	7.83	0.48	3.0	F2 V		80 s	8047	m
96248	238743		11 04 57.3	−59 51 31	−0.004	+0.01	6.56	0.19	−6.2	B1 Iab		2100 s		
96205	222546		11 04 58.3	−48 11 52	−0.002	−0.02	7.80	0.1	0.6	A0 V		270 s		m
96097	118648	63 χ Leo	11 05 01.0	+7 20 10	−0.023	−0.05	4.63	0.33	1.3	F2 III-IV	+5	47 s		m
96261	238749		11 05 03.4	−59 42 49	+0.001	−0.01	7.78	0.14		B0.5 III				m
96224	222548		11 05 04.1	−49 23 32	−0.003	0.00	6.13	−0.02		A0				
96187	202073		11 05 12.2	−30 26 09	−0.003	0.00	7.4	1.2	−0.1	K2 III		310 s		
96094	81657		11 05 15.4	+25 12 06	−0.030	−0.07	7.64	0.58	4.4	dG0	−8	45 s		
96075	62397		11 05 16.8	+31 56 03	0.000	−0.01	8.0	1.1	0.2	K0 III		360 s		
96238	222551		11 05 17.8	−43 32 33	−0.003	+0.01	7.9	0.2	3.0	F2 V		94 s		
96202	179514	χ¹ Hya	11 05 19.8	−27 17 37	−0.014	−0.01	4.94	0.36	3.3	F4 V	+17	21 s		m
96220	156448		11 05 33.9	−11 05 20	+0.001	−0.09	6.09	0.30		A3				
96001	15393		11 05 36.3	+64 48 15	+0.005	+0.02	7.3	0.4	2.6	F0 V		85 s		
96300	222558		11 05 40.4	−43 07 07	−0.005	+0.02	7.53	0.27	3.0	F2 V		81 s		
96525	256796		11 05 42.4	−71 05 06	+0.001	−0.02	7.7	0.0	0.4	B9.5 V		280 s		
96338	222560		11 05 45.6	−47 26 32	−0.004	−0.01	7.1	0.6	0.6	A0 V		80 s		
96105	27897		11 05 45.9	+50 10 21	−0.011	−0.01	7.1	0.5	3.4	F5 V		55 s		
96127	43608		11 05 46.0	+44 18 06	+0.002	0.00	7.4	1.1	−0.1	K2 III		320 s		
96317	222559		11 05 46.0	−45 13 17	0.000	−0.01	7.9	1.1	3.0	F2 V		94 s		
96161	62401		11 05 48.2	+38 23 38	−0.007	−0.04	7.5	0.8	0.3	G5 III	+13	150 mx		
96570	—		11 05 49.0	−72 40 23			7.7	0.4	3.0	F2 V		86 s		
96218	99433		11 05 49.7	+12 33 57	−0.003	−0.07	7.8	0.5	3.4	F5 V		74 s		
96348	222563		11 05 52.3	−41 54 27	−0.002	−0.01	7.6	0.4	1.4	A2 V		100 s		
96126	27901		11 05 53.2	+52 19 36	−0.005	0.00	7.26	1.22	0.2	K0 III		170 s		
96125	27900		11 05 53.7	+53 51 03	−0.008	−0.01	7.7	1.6	−0.5	M2 III		310 mx		
96314	179522	χ² Hya	11 05 57.5	−27 17 16	+0.002	−0.02	5.71	−0.06		B9	+53			
96675	256798		11 05 57.9	−76 07 49	−0.006	−0.01	8.0	0.0	0.4	B9.5 V		340 s		
96274	118655		11 06 02.1	+1 12 38	−0.001	−0.03	7.8	1.6	−0.5	M2 III		470 s		
96311	137987		11 06 05.7	−7 00 31	0.000	−0.04	7.7	1.1	0.2	K0 III		310 s		
96407	238763		11 06 05.7	−51 12 45	0.000	−0.07	6.30	0.95	3.2	G6 IV		34 s		
96446	238766		11 06 05.7	−59 56 59	−0.003	0.00	6.68	−0.15	−2.5	B2 V		690 s		
96474	238765		11 06 06.4	−59 14 29	−0.001	0.00	7.99	−0.09	0.0	B8.5 V		400 s		
96272	118658		11 06 09.5	+7 08 23	−0.003	+0.04	7.26	0.43		F5				
96074	15401		11 06 09.8	+65 52 31	+0.002	−0.02	7.7	0.9	3.2	G5 IV	−11	79 s		
96569	251265		11 06 11.5	−68 46 55	−0.002	−0.01	7.30	1.68		K5				
96489	238771		11 06 13.0	−58 50 38	−0.004	+0.01	8.0	0.0	0.0	B8.5 V		210 mx		
97437	258602		11 06 13.6	−85 44 56	−0.015	0.00	7.20	0.1	0.6	A0 V		210 s		m
96548	251264		11 06 17.3	−65 30 35	+0.002	−0.01	7.70	0.10		WN8				
96442	238770		11 06 19.6	−50 26 48	−0.001	+0.03	6.8	1.7	−0.3	K5 III		210 s		
96423	222569		11 06 19.8	−44 22 24	+0.009	−0.09	7.6	0.1	3.2	G5 IV		77 s		
96568	251267		11 06 24.3	−64 50 22	−0.007	0.00	6.41	0.12		A2				
96364	156459		11 06 25.9	−16 31 52	0.000	−0.03	7.7	1.1	−0.1	K2 III		360 s		
96484	238775		11 06 27.3	−50 57 24	−0.007	0.00	6.32	1.16	−2.2	K2 II		100 mx		
96544	238779		11 06 29.3	−58 40 30	−0.002	0.00	6.02	1.24	−1.0	K2 II-III		1200 s		m
96566	251269		11 06 32.3	−62 25 26	−0.006	+0.01	4.61	1.03	−6.2	G5 I	−2	160 mx		
96403	179537		11 06 32.8	−21 10 01	−0.003	−0.01	7.3	0.1	0.6	A0 V		180 s		
96441	179542		11 06 38.7	−28 43 40	−0.004	−0.04	6.77	0.04		A0				
96375	118664		11 06 40.4	+5 25 48	−0.005	−0.01	8.0	0.4	2.6	F0 V		120 s		
96233	27907		11 06 40.9	+56 05 30	+0.003	−0.02	7.4	1.1	−0.1	K2 III		320 s		
96400	156466		11 06 41.3	−13 00 13	−0.002	−0.11	6.5	0.4	3.0	F2 V		51 s		
96538	238781		11 06 42.6	−52 57 03	0.000	−0.01	7.2	1.6	0.2	K0 III		110 s		
96373	99445		11 06 43.8	+15 10 57	+0.002	−0.01	7.60	1.6	−0.5	M2 III	−1	420 s		
96372	99444		11 06 43.9	+17 44 14	+0.002	−0.03	6.5	1.4	−0.3	K5 III		230 s		
96327	43612		11 06 46.0	+44 06 55	+0.002	0.00	7.5	0.2	2.1	A5 V		120 s		
96705	256799		11 06 48.2	−70 49 49	−0.002	−0.02	7.90	0.4	3.0	F2 V		95 s		m
96706	256800		11 06 49.9	−70 52 40	−0.005	−0.01	5.57	−0.05	−2.9	B3 III		410 s		m
96483	179545		11 06 51.4	−23 33 20	−0.001	0.00	7.9	0.4	2.1	A5 V		110 s		
96504	202096		11 06 52.2	−30 33 15	−0.003	−0.04	6.5	0.9	0.2	K0 III		190 s		
96436	118668	65 Leo	11 06 54.1	+1 57 20	−0.026	−0.09	5.52	0.97	0.3	gG7	+55	34 ts	8060	m
96395	81672		11 06 55.2	+20 29 01	−0.003	0.00	7.5	1.1	0.2	K0 III		290 s		
96503	179547		11 06 57.6	−24 09 09	+0.002	−0.06	7.80	1.1	0.2	K0 III		330 mx		m
96655	—		11 06 58.7	−63 06 24			8.0	1.6	−0.5	M2 III		470 s		
96418	81674		11 07 04.7	+25 32 13	−0.005	−0.02	6.8	0.5	3.4	F5 V	−9	47 s		
96557	202100		11 07 08.3	−32 35 14	−0.007	−0.06	6.59	0.34	1.9	F2 IV		87 s		

HD	SAO	Star Name	α 2000	δ 2000	μ(α)	μ(δ)	V	B–V	M_v	Spec	RV	d(pc)	ADS	Notes
			h m s	° ′ ″	s	″								
96580	202102		11 07 12.6	−33 33 16	−0.002	+0.01	7.8	1.3	−0.3	K5 III		410 s		
96479	99455		11 07 13.1	+10 12 44	0.000	−0.02	7.3	0.5	4.0	F8 V		47 s		
96670	238793		11 07 13.6	−59 52 23	−0.005	0.00	7.46	0.10		O8				
96616	222581		11 07 16.5	−42 38 19	−0.009	+0.04	5.15	0.03		A pe	+2	24 mn		m
96514	99457		11 07 26.4	+13 00 48	+0.001	−0.02	7.60	1.1	−0.1	K2 III		350 s		
96599	179553		11 07 27.7	−28 29 28	−0.001	0.00	7.5	1.2	0.2	K0 III		220 s		
96497	81677		11 07 28.6	+22 03 10	−0.015	−0.01	7.9	0.7	4.5	G1 V	+3	49 s		
96512	81678		11 07 30.5	+21 09 03	+0.001	−0.02	7.20	0.1	1.4	A2 V		150 s		m
96660	222583		11 07 34.0	−48 38 23	−0.002	+0.02	6.4	1.7	0.2	K0 III		70 s		
96574	99459		11 07 36.1	+13 51 31	−0.026	+0.09	7.32	0.55	4.2	F9 V		39 mx		
96496	62416		11 07 36.5	+36 47 07	−0.004	−0.03	8.0	1.1	−0.1	K2 III		280 mx		
96550	99460		11 07 39.3	+13 29 56	+0.001	−0.08	8.0	0.5	3.4	F5 V		85 s		
96528	81681	64 Leo	11 07 39.6	+23 19 25	−0.001	0.00	6.46	0.16	2.1	A5 V	−2	75 s		
96756	238801		11 07 42.9	−59 40 16	−0.002	0.00	7.49	1.36		K0				
97436	258603		11 07 45.6	−84 26 12	−0.017	+0.02	7.9	1.6	0.2	K0 III		150 s		
96701	202115		11 07 51.8	−38 24 26	0.000	0.00	7.4	1.3	−0.1	K2 III		250 s		
96700	179558		11 07 54.3	−30 10 29	−0.040	−0.14	6.54	0.60	4.7	G2 V	+11	24 ts		
96789	238804		11 07 57.3	−58 17 27	−0.001	0.00	7.9	1.0	0.2	K0 III		340 s		
96527	27918		11 08 00.0	+52 49 17	−0.007	−0.01	7.4	0.5	3.4	F5 V	−37	63 s	8065	m
96749	222590		11 08 02.7	−42 23 29	−0.010	0.00	8.0	0.8	0.2	K0 III		180 mx		
96829	251278		11 08 06.5	−60 49 32	−0.003	0.00	7.32	0.23	−2.9	B3 III		670 s		m
96696	179565		11 08 09.9	−21 31 06	−0.006	+0.02	7.70	0.9	0.3	G7 III	+5	170 mx		
96805	238805		11 08 13.5	−52 24 07	−0.004	+0.12	6.8	1.0	3.0	F2 V		23 s		
96681	118679		11 08 13.7	+3 31 59	−0.001	−0.01	7.6	0.1	1.4	A2 V		180 s		
96723	179568		11 08 15.7	−29 58 22	0.000	−0.02	6.49	0.03		A0				
96694	138001		11 08 16.9	−1 54 12	0.000	−0.02	6.8	1.1	0.2	K0 III		210 s		
96712	156481		11 08 17.4	−15 57 41	+0.004	−0.06	7.90	0.5	3.4	F5 V		79 s	8069	m
96680	118681		11 08 20.2	+4 27 31	0.000	−0.03	8.0	0.5	3.4	F5 V		82 s		
96391	7287		11 08 21.6	+71 57 33	−0.008	−0.01	6.9	0.4	2.6	F0 V		72 s		
96711	138004		11 08 23.5	−0 50 28	−0.003	−0.03	7.7	1.6	−0.5	M2 III		430 s		
96880	238809		11 08 24.6	−59 24 45	0.000	−0.01	7.56	0.48	−5.7	B1 Ib		1800 s		
96920	251284		11 08 26.6	−65 25 41	−0.004	+0.02	7.0	1.8	0.2	K0 III		80 s		
97015	256803		11 08 26.8	−73 58 41	+0.009	−0.06	7.9	1.1	0.2	K0 III		300 mx		
96693	118682		11 08 27.3	+7 37 51	−0.002	0.00	8.0	1.1	0.2	K0 III		360 s		
96883	251281		11 08 27.8	−61 37 34	0.000	+0.01	7.87	0.39	−5.6	B8 Ib		2900 s		
96919	251286		11 08 33.8	−61 56 49	−0.003	+0.01	5.13	0.22	−7.1	B9 Ia	−22	2300 s		
96918	238813		11 08 35.3	−58 58 30	−0.001	0.00	3.91	1.23	−8.0	G0 Ia	+7	1400 s		
96742	138009		11 08 35.4	−0 33 48	−0.001	−0.02	8.0	1.1	−0.1	K2 III		410 s		
96677	62421		11 08 35.7	+36 00 40	−0.001	+0.01	7.3	1.1	0.2	K0 III		260 s		
96844	202125		11 08 39.4	−39 28 19	−0.002	+0.01	7.2	0.2	0.6	A0 V		150 s		
96802	179575		11 08 39.7	−29 44 50	−0.001	−0.01	7.1	1.1	0.2	K0 III		210 s		
96719	99464		11 08 40.7	+17 12 04	−0.002	−0.16	6.9	0.9	3.2	G5 IV		54 s		
97903	258605		11 08 41.6	−86 13 51	−0.028	0.00	6.5	2.7	−0.5	M2 III		53 s		
96917	238815		11 08 42.2	−57 03 57	−0.004	0.00	7.07	0.08		O9 II		200 mx		
96916	238814		11 08 43.2	−55 21 15	−0.002	0.00	7.9	1.7	−0.3	K5 III		320 s		
96819	179577		11 08 43.9	−28 04 50	−0.006	−0.02	5.44	0.07		A2	+16	23 mn		
96738	81692	67 Leo	11 08 49.0	+24 39 30	0.000	0.00	5.68	0.06	0.0	A3 III	−6	140 s	8071	m
96780	99467		11 08 56.4	+15 12 41	−0.005	−0.01	7.9	0.7	0.6	G0 III		280 mx		
96838	156493		11 08 56.9	−19 24 59	0.000	−0.01	6.7	0.1	0.6	A0 V		170 s		
96656	27923		11 08 57.5	+59 12 55	−0.003	+0.01	7.16	1.01		G5	+7			
96778	62426		11 09 02.4	+30 02 25	−0.003	−0.02	7.2	0.8	3.2	G5 IV		63 s		
96874	179584		11 09 05.1	−20 30 48	−0.003	+0.01	7.7	0.6	0.6	A0 V		110 s		
96855	138015		11 09 13.6	−1 20 00	−0.005	−0.01	6.8	0.1	1.4	A2 V		120 mx		
96708	27928		11 09 15.7	+57 52 34	0.000	0.00	7.2	0.9	3.2	G5 IV		64 s		
96734	27930		11 09 16.5	+51 22 44	−0.002	−0.01	7.0	1.6	−0.5	M2 III	−20	320 s		
96813	62427		11 09 19.0	+36 18 33	−0.004	−0.03	5.74	1.51	−0.5	M4 III	+22	180 s		
96906	156498		11 09 19.0	−11 40 40	−0.002	−0.01	7.7	1.4	−0.3	K5 III		390 s		
96834	43627		11 09 38.4	+43 12 27	−0.006	−0.02	5.89	1.57	−0.5	M2 III	+18	190 s		
96833	43629	52 ψ UMa	11 09 39.7	+44 29 54	−0.006	−0.03	3.01	1.14	0.0	K1 III	−4	37 s		
96707	15414		11 09 39.8	+67 12 37	−0.015	−0.03	6.06	0.22		A p	+5			
96937	118693		11 09 40.2	+2 27 23	−0.019	+0.04	7.9	0.9	3.2	G5 IV		54 mx		
97063	238824		11 09 41.0	−54 42 53	−0.002	−0.01	7.4	0.1	0.4	B9.5 V		200 s		
97082	238825		11 09 41.3	−58 50 17	0.000	−0.02	6.5	0.5	4.0	F8 V		31 s		ER Car, v
97025	222601		11 09 43.5	−43 19 57	−0.006	−0.02	7.8	0.1	2.1	A5 V		110 mx		
96758	15415		11 09 44.9	+63 20 06	−0.005	0.00	7.90	0.5	3.4	F5 V		79 s	8073	m
96832	27932		11 09 47.0	+54 06 24	−0.002	0.00	7.2	0.6	4.4	G0 V		36 s		
96759	15416		11 09 47.2	+62 59 09	−0.007	0.00	8.0	0.9	3.2	G5 IV		93 s		
96793	15419		11 09 52.1	+60 46 57	−0.001	+0.01	7.8	0.8	3.2	G5 IV		83 s		
99357			11 09 52.7	−88 57 51			7.8	1.1	0.2	K0 III		330 s		
97023	202149		11 09 53.3	−32 22 03	+0.002	−0.03	5.81	0.04		A1				
96921	62432		11 09 55.5	+32 31 51	−0.007	−0.02	7.2	1.1	0.2	K0 III		250 s		
97022	179604		11 09 58.7	−29 47 40	−0.003	−0.02	6.9	1.5	0.2	K0 III		120 s		
97120	238830		11 10 01.3	−55 14 04	0.000	0.00	8.0	1.3	0.2	K0 III		250 s		
96572	7295		11 10 01.7	+77 47 13	+0.013	−0.01	7.3	1.6	−0.5	M2 III	−26	360 s		
97151	238831		11 10 02.2	−60 05 42	−0.003	0.00	7.74	−0.09	−2.5	B2 V e		1000 s		

HD	SAO	Star Name	α 2000	δ 2000	μ(α)	μ(δ)	V	B-V	M_v	Spec	RV	d(pc)	ADS	Notes
			h m s	° ′ ″	s	″								
96975	99476		11 10 03.8	+15 54 44	-0.003	-0.04	7.4	0.4	3.0	F2 V		77 s		
97044	179609		11 10 04.8	-25 36 25	-0.002	-0.01	7.3	1.4	0.2	K0 III		160 s		
97166	238832		11 10 06.1	-60 14 57	+0.001	-0.01	7.90	0.06		O8				
97056	179610		11 10 07.7	-27 31 56	-0.003	-0.02	7.8	1.7	-0.3	K5 III		310 s		
98795	--		11 10 08.5	-88 20 46			8.0	1.1	0.2	K0 III		360 s		
96951	62433		11 10 12.8	+35 20 06	0.000	0.00	7.86	0.04	0.6	A0 V		260 s		
97037	138023		11 10 13.3	-7 23 23	-0.016	-0.18	6.8	0.7	4.4	G0 V		30 s		
97072	179612		11 10 18.0	-28 16 15	0.000	-0.05	7.7	0.8	3.2	G5 IV		78 s		
97005	81703		11 10 20.9	+22 42 06	+0.001	-0.02	7.50	0.34		F0				
97113	202161		11 10 21.2	-36 05 45	-0.004	-0.01	7.6	1.6	-0.5	M2 III		420 mx		
96973	43635		11 10 27.2	+47 47 53	-0.006	+0.05	7.3	0.5	3.4	F5 V		60 s		
97089	138026		11 10 27.8	-9 01 41	-0.011	-0.13	7.8	0.5	4.0	F8 V		57 s		
97053	118702		11 10 30.9	+7 53 30	+0.001	-0.01	7.9	1.4	-0.3	K5 III		430 s		
97111	179616		11 10 31.7	-25 59 47	+0.003	-0.08	7.4	0.7	2.1	A5 V		54 s		
97067	99479		11 10 33.3	+11 38 37	-0.004	-0.06	7.9	1.4	-0.3	K5 III		160 mx		
97145	202165		11 10 34.5	-32 33 51	-0.005	-0.02	7.30	0.5	3.4	F5 V		60 s		m
97068	99480		11 10 36.6	+11 18 05	-0.002	-0.02	7.4	1.6	-0.5	M4 III	0	380 s		
97185	222614		11 10 39.1	-49 39 13	0.000	-0.02	7.51	-0.09	-1.4	B4 V		600 s		
97253	251302		11 10 42.0	-60 23 04	0.000	0.00	7.11	0.16		O6				
97144	156513		11 10 49.5	-13 06 50	-0.001	0.00	7.6	1.1	0.2	K0 III		310 s		
97201	222619		11 10 49.9	-45 35 06	-0.001	-0.04	7.0	1.0	0.2	K0 III		230 s		
97103	81707		11 10 51.7	+21 20 07	-0.004	-0.01	7.2	0.8	3.2	G5 IV		63 s		
96511	1829		11 10 54.6	+81 43 53	-0.048	-0.19	7.1	0.7	4.4	G0 V	-46	19 mx		
97271	238839		11 10 54.6	-58 27 19	-0.004	0.00	6.88	-0.08		B8	+17			
97217	202172		11 11 01.7	-36 13 33	-0.003	-0.03	7.9	0.3	3.4	F5 V		78 s		
97182	156516		11 11 08.4	-17 07 56	-0.002	0.00	7.8	0.1	1.4	A2 V		190 s		
97197	138039		11 11 19.3	-1 23 41	-0.003	0.00	7.9	1.1	-0.1	K2 III		390 s		
97292	222627		11 11 27.6	-43 27 35	0.000	-0.01	8.0	-0.1	1.7	A3 V		180 s		
97472	256808		11 11 29.6	-71 26 10	-0.003	0.00	6.35	1.37		K0				
97125	27940		11 11 31.2	+54 08 57	-0.006	+0.03	6.6	0.2	2.1	A5 V		80 s		
97310	222631		11 11 35.2	-41 39 54	+0.002	-0.03	7.6	1.1	-0.1	K2 III		340 s		
97400	251307		11 11 39.2	-60 26 36	0.000	0.00	7.80	0.09		B1				m
97277	179624	11 β Crt	11 11 39.4	-22 49 33	0.000	-0.10	4.48	0.03	0.2	A2 III-IV	+6	72 s		
97398	251308		11 11 40.5	-60 18 38	-0.001	0.00	6.71	-0.04	0.4	B9.5 V		180 s		
97396	238850		11 11 41.7	-59 23 29	-0.001	0.00	8.04	0.01	0.6	A0 V		300 s		
97211	81713		11 11 41.9	+26 49 56	-0.001	+0.03	8.0	0.5	3.4	F5 V		82 s		
97140	27941		11 11 43.3	+58 53 59	-0.009	+0.02	7.35	0.61		G0	-25			
97244	99492		11 11 43.6	+14 24 01	-0.005	-0.01	6.30	0.21		A5	+6			
97194	43641		11 11 48.8	+42 49 54	-0.012	-0.25	7.30	0.5	4.0	F8 V		34 mx		m
97309	156522		11 11 53.6	-20 08 52	-0.005	+0.01	8.01	0.22		A0				
97344	179628		11 11 57.8	-26 48 22	-0.002	+0.01	6.5	1.3	3.2	G5 IV		23 s		
97343	179630		11 12 01.2	-26 08 10	+0.020	-0.05	7.06	0.77		G0		14 mn		
97451	238855		11 12 03.7	-58 25 25	-0.004	-0.02	7.1	1.2	-0.5	M4 III		330 s		
97535	256809		11 12 05.8	-71 13 03	-0.002	0.00	7.02	0.01	0.4	B9.5 V		210 s		m
97413	222635		11 12 10.1	-46 16 00	-0.006	0.00	6.5	0.2	0.6	A0 V		110 s		
97138	15431		11 12 10.9	+68 16 18	+0.006	+0.01	6.4	0.1	1.4	A2 V	-18	100 s		
97358	138053		11 12 11.6	-9 29 07	-0.009	-0.02	7.6	0.4	3.0	F2 V		82 s		
97287	62446		11 12 14.6	+36 53 34	-0.006	0.00	7.6	0.9	3.2	G5 IV		75 s		
97393	202197		11 12 14.8	-32 26 02	+0.002	0.00	6.41	1.69		Ma				
97522	251315		11 12 16.3	-65 13 09	0.000	-0.01	7.73	0.31	-5.4	B1 Ib-II		2100 s		
97412	179635		11 12 18.7	-29 46 58	-0.002	-0.02	7.1	1.8	-0.3	K5 III		200 s		
97468	238858		11 12 20.0	-51 32 22	-0.003	+0.01	7.9	1.2	0.2	K0 III		270 s		
97322	62448		11 12 22.6	+34 47 06	-0.008	-0.05	7.1	1.1	0.2	K0 III		130 mx		
97642	256811		11 12 23.4	-74 12 38	-0.003	-0.04	7.7	0.3	3.4	F5 V		71 s		
97618	256810		11 12 30.1	-71 44 56	-0.002	0.00	7.5	0.0	0.4	B9.5 V		260 s		
97411	156528		11 12 30.3	-18 30 00	-0.002	-0.03	5.84	0.05		A0			8086	m
97446	202203		11 12 31.0	-31 08 09	+0.003	-0.07	7.9	1.0	3.4	F5 V		35 s		
97334	62451		11 12 32.1	+35 48 49	-0.022	-0.17	6.41	0.61	4.4	G0 V	-3	24 s		m
97495	222639		11 12 33.0	-49 06 03	-0.010	+0.03	5.36	0.18		A2	-28			
97428	179638		11 12 34.5	-21 44 57	-0.001	+0.01	6.40	1.38		K0				
97600	251318		11 12 34.9	-68 00 24	-0.003	-0.01	8.0	0.0	-1.0	B5.5 V		590 s		
97482	222640		11 12 35.1	-46 58 00	-0.008	0.00	6.9	1.5	4.0	F8 V		38 s		
97534	251316		11 12 36.0	-60 19 03	0.000	0.00	4.60	0.55	-8.5	F0 Ia	-8	3300 s		m
97465	202204		11 12 36.2	-35 24 08	+0.001	0.00	7.9	0.2	1.4	A2 V		150 s		
97556	238863		11 12 44.1	-56 56 09	0.000	-0.01	8.0	0.3	0.6	A0 V		180 s		
97371	62453		11 12 44.2	+35 49 46	-0.005	-0.04	7.21	1.01	0.2	K0 III		200 mx		m
97557	238864		11 12 44.3	-59 40 39	0.000	-0.01	7.23	0.00	-2.9	B3 III		850 s		
97302	27952		11 12 44.4	+54 53 39	0.000	-0.01	6.5	0.1	1.4	A2 V	-6	100 s		
97583	251320		11 12 45.3	-64 10 12	-0.006	-0.01	5.23	-0.06	0.2	B9 V	+21	100 s		m
97443	138063		11 12 46.3	-5 07 01	-0.001	0.00	7.9	1.1	0.2	K0 III		340 s		
97617	251321		11 12 48.7	-67 06 14	-0.005	-0.14	7.76	0.07	0.4	B9.5 V		49 mx		
97321	27953		11 12 49.6	+56 24 09	-0.006	-0.03	7.9	0.4	3.0	F2 V		94 s		
97461	156536		11 12 50.3	-13 56 01	+0.005	-0.06	7.1	0.5	4.0	F8 V		42 s		
97494	156537		11 12 56.5	-12 47 06	-0.003	-0.02	7.3	0.4	2.6	F0 V		88 s		
97550	222643		11 12 56.8	-49 44 11	-0.001	-0.01	6.11	1.06		K0				

HD	SAO	Star Name	α 2000	δ 2000	μ(α)	μ(δ)	V	B-V	M_V	Spec	RV	d(pc)	ADS	Notes
			$11^h 13^m 05^s.1$	$-47°03'19''$	$+0^s.003$	$+0''.01$								
97547	222645		11 13 05.1	-47 03 19	+0.003	+0.01	7.60	0.5	4.0	F8 V		53 s		m
97440	81720		11 13 06.3	+27 03 05	-0.006	-0.04	7.95	0.33		F0				
97595	238870		11 13 06.8	-56 24 36	+0.002	0.00	7.6	1.9	-0.1	K2 III		130 s		
97301	15438		11 13 07.6	+67 19 26	-0.004	+0.05	6.7	0.8	3.2	G5 IV		51 s		
97594	238871		11 13 10.6	-52 49 35	-0.004	0.00	7.7	1.1	0.2	K0 III		290 s		
97576	222647		11 13 14.6	-44 22 20	0.000	0.00	5.80	1.66	-0.3	gK5		130 s		
97592	222649		11 13 17.4	-50 07 22	-0.003	-0.01	7.50	0.1	1.4	A2 V		160 s		m
97502	99505		11 13 18.0	+10 26 50	+0.001	-0.03	7.4	1.1	-0.1	K2 III		310 s		
97476	81722		11 13 18.2	+27 10 30	0.000	-0.03	7.80	1.4	-0.3	K4 III		420 s		
97488	62456		11 13 19.7	+33 26 38	-0.012	-0.06	6.80	1.1	0.2	K0 III		110 mx	8091	m
97543	179651		11 13 20.1	-22 02 13	-0.001	+0.01	7.6	1.9	-0.5	M2 III		290 s		
97670	238877		11 13 30.8	-59 37 09	-0.001	-0.01	5.74	-0.10	-2.9	B3 III	+17	480 s		
97654	238874		11 13 31.0	-57 03 05	0.000	0.00	7.54	0.03	0.6	A0 V		230 s		
97609	222650		11 13 31.9	-41 39 27	-0.005	+0.01	7.9	0.8	2.6	F0 V		110 s		
97455	27957		11 13 34.6	+55 24 56	-0.005	0.00	7.90	0.4	3.0	F2 V		96 s	8092	m
97456	27958		11 13 35.0	+52 51 01	-0.005	-0.03	7.9	1.1	0.2	K0 III		260 mx		
97651	238878		11 13 39.2	-53 13 54	-0.003	+0.03	5.76	1.32		K2				
97501	43649		11 13 40.1	+41 05 19	-0.001	+0.01	6.33	1.15	0.2	K0 III	+12	130 s	8093	m
97561	81725		11 13 40.9	+20 07 43	-0.027	-0.13	6.93	0.75	3.2	G7 IV	+45	27 mx	8094	m
97590	179661		11 13 41.0	-22 59 55	-0.004	0.00	7.3	0.2	1.4	A2 V		130 s		
97585	118731	69 Leo	11 13 45.5	-0 04 11	-0.003	-0.01	5.42	-0.03		A0	+5			
97513	27960		11 13 47.7	+49 51 23	-0.007	+0.01	7.4	1.1	0.2	K0 III		270 s		
97688	238880		11 13 48.5	-51 48 04	0.000	+0.01	7.80	1.1	0.2	K0 III		320 s		m
97901	256813		11 13 49.0	-75 47 26	0.000	-0.01	7.4	0.6	1.4	A2 V		79 s		
97689	238882		11 13 50.4	-52 51 22	-0.001	-0.07	6.82	0.25		A m				
97606	118734		11 13 53.0	+2 16 07	-0.006	-0.04	6.8	0.4	3.0	F2 V		58 s		
97605	118735		11 14 01.7	+8 03 39	+0.003	-0.11	5.79	1.12	-0.2	gK3	+17	160 s		m
97702	222657		11 14 03.4	-43 41 22	-0.002	0.00	7.8	0.2	0.2	K0 III		330 s		
97486	15444		11 14 04.5	+62 16 54	+0.002	-0.03	7.80	0.8	0.3	G5 III	-30	320 s		m
97723	222659		11 14 05.4	-48 36 01	-0.001	+0.01	7.2	1.2	-0.1	K2 III		260 s		
97602	81726		11 14 06.0	+23 50 34	-0.001	-0.01	8.0	0.0	0.4	B9.5 V		330 s		
97603	81727	68 δ Leo	11 14 06.4	+20 31 25	+0.010	-0.14	2.56	0.12	1.9	A4 V	-21	16 ts		Zosma, m
97663	156554		11 14 10.5	-16 53 09	-0.002	+0.05	7.2	0.8	3.2	G5 IV		64 s		
97662	138076		11 14 12.8	-8 19 41	0.000	-0.01	7.3	1.6	-0.5	M2 III		360 s		
97633	99512	70 θ Leo	11 14 14.3	+15 25 46	-0.004	-0.08	3.34	-0.01	1.4	A2 V	+8	24 s		Chertan
97792	238887		11 14 20.2	-56 02 51	+0.002	-0.01	8.04	-0.05	0.6	B8.5 V		410 s		
97808	251327		11 14 21.8	-61 35 14	-0.003	-0.01	7.4	1.0	0.6	A0 V		61 s		
97643	81728		11 14 24.4	+20 01 52	+0.001	0.00	7.9	0.9	3.2	G5 IV		88 s		
97805	238889		11 14 24.6	-57 48 04	0.000	0.00	7.78	-0.03	0.6	A0 V		270 s		
97900	251330		11 14 24.7	-69 31 57	-0.004	-0.05	7.9	1.4	-0.3	K5 III		220 mx		
96571	1834		11 14 29.8	+85 38 24	-0.040	0.00	7.2	0.1	1.4	A2 V	-5	150 s		
97789	222663		11 14 32.3	-42 51 09	-0.002	-0.01	7.6	0.3	2.6	F0 V		99 s		
97698	118742		11 14 32.5	+5 59 34	-0.003	0.00	7.9	1.1	0.2	K0 III		340 s		
97658	81730		11 14 33.1	+25 42 37	-0.008	+0.04	7.76	0.84	5.5	G7 V	+2	24 s		
97601	27961		11 14 33.3	+52 56 49	+0.003	-0.03	7.4	0.8	3.2	G5 IV		68 s		
97716	138079		11 14 37.7	-1 16 15	+0.002	-0.03	6.7	1.1	0.2	K0 III		200 s		
97657	43655		11 14 45.4	+43 19 45	+0.007	-0.12	7.9	0.5	4.0	F8 V		59 s		
97986	256816		11 14 49.3	-70 21 38	-0.004	0.00	7.8	-0.4	0.0	B8.5 V		350 mx		
97866	222671		11 14 53.9	-43 44 03	-0.003	-0.02	6.21	1.61		K5				
97971	251332		11 14 54.7	-68 19 40	-0.004	0.00	7.2	1.4	-0.3	K5 III		310 s		
97537	7318		11 14 58.7	+72 00 30	+0.003	+0.01	7.3	0.0	0.4	B9.5 V	-8	240 s		
97840	202240		11 15 02.3	-33 19 04	-0.012	+0.05	6.99	0.36	2.1	F5 IV		95 s		
97927	238902		11 15 06.4	-58 15 33	-0.002	-0.01	7.71	-0.05	0.6	A0 V		260 s		
97711	43657		11 15 07.1	+45 03 40	-0.004	-0.06	7.6	0.6	4.4	G0 V		43 s		
97881	202243		11 15 08.6	-39 28 36	+0.001	0.00	7.80	0.1	0.6	A0 V		280 s		m
97778	81736	72 Leo	11 15 12.2	+23 05 43	-0.002	-0.01	4.63	1.66	-0.5	gM2	+16	100 s		
97969	238906		11 15 14.5	-60 09 53	-0.001	-0.02	7.72	-0.02	-3.5	B1 V n		1500 s		
97864	156564		11 15 17.5	-19 38 16	-0.004	-0.01	6.8	0.1	0.6	A0 V		130 mx		
97799	81740		11 15 20.9	+27 34 14	-0.002	-0.01	7.70	0.2	2.1	A5 V		130 s	8105	m
97876	156565		11 15 23.5	-11 35 17	+0.001	-0.02	7.30	1.6	-0.5	M4 III	-27	360 s		
97833	99521		11 15 25.5	+15 06 46	-0.001	-0.03	7.3	1.1	0.2	K0 III		270 s		
97584	7321		11 15 25.7	+73 28 16	0.000	-0.01	7.63	1.03	7.4	dK4	+8	13 ts	8100	m
97945	222675		11 15 28.2	-41 30 36	-0.002	0.00	7.9	0.3	0.6	A0 V		150 s		
97944	222676		11 15 29.1	-41 03 26	-0.001	0.00	7.2	0.9	-0.1	K2 III		280 s		
97959	222677		11 15 29.4	-44 23 43	-0.006	+0.01	7.9	0.5	0.6	A0 V		120 mx		
97957	222678		11 15 30.4	-42 36 48	-0.018	+0.08	7.46	0.50		F5		16 mn		
97943	202250		11 15 33.8	-38 15 31	-0.003	-0.03	8.0	0.1	0.2	K0 III		360 s		
97891	118753		11 15 38.9	+4 28 39	0.000	-0.12	8.0	0.5	3.4	F5 V		82 s		
97918	156568		11 15 39.6	-12 35 35	-0.001	-0.03	6.6	1.6	-0.5	M2 III	-1	260 s		
97940	179701		11 15 42.7	-22 54 55	-0.004	+0.04	7.2	0.1	1.4	A2 V		110 mx		
98025	238914		11 15 42.7	-57 21 21	-0.002	-0.01	6.43	0.10	1.4	A2 V		93 s		
98026	238915		11 15 43.5	-57 59 23	-0.002	+0.02	8.0	1.5	-0.1	K2 III		270 s		
97998	202254		11 15 49.1	-39 19 24	-0.014	+0.18	7.37	0.63	3.2	G5 IV		64 mx		
97907	99525	73 Leo	11 15 51.8	+13 18 27	-0.001	-0.01	5.32	1.20	-0.2	K3 III	+15	130 s		
98022	222682		11 15 53.6	-47 55 13	-0.002	0.00	6.90	0.1	0.6	A0 V		180 s		m

HD	SAO	Star Name	α 2000	δ 2000	μ(α)	μ(δ)	V	B-V	M_V	Spec	RV	d(pc)	ADS	Notes
97953	156573		11ʰ 15ᵐ 55.6ˢ	-10° 36′ 59″	-0.002ˢ	-0.02″	7.6	1.1	-0.1	K2 III		350 s		
97811	27967		11 15 55.9	+59 07 15	-0.003	0.00	7.94	0.21		A0				
97938	99526		11 15 56.2	+12 36 49	-0.002	-0.02	6.82	-0.02	0.6	A0 V	+7	180 s		
97997	202258		11 15 56.6	-36 18 50	0.000	-0.02	7.7	2.0	-0.5	M2 III		270 s		
97937	99527		11 15 57.7	+12 50 41	-0.002	-0.06	6.67	0.27	2.6	F0 V	-20	65 s		
98052	238919		11 15 59.3	-55 12 31	-0.003	0.00	7.3	1.0	2.1	A5 V		36 s		
98037	222683		11 16 00.0	-47 18 22	-0.001	-0.05	7.5	0.7	0.2	K0 III		280 s		
97771	15451		11 16 01.5	+64 54 27	+0.002	-0.01	7.0	1.1	0.2	K0 III		230 s		
97976	138094		11 16 02.3	-8 54 00	-0.003	-0.01	7.4	1.1	0.2	K0 III		280 s		
97855	27970		11 16 04.0	+52 46 22	+0.018	+0.05	6.50	0.43	3.0	dF2	-41	48 s	8108	m
98036	222684		11 16 07.3	-41 01 18	-0.003	-0.02	7.3	0.1	0.6	A0 V		180 s		
98064	238921		11 16 07.6	-54 57 07	-0.002	-0.03	8.0	0.9	0.2	K0 III		360 s		
98134	251336		11 16 08.1	-68 11 53	-0.008	+0.01	7.7	1.4	-0.3	K5 III		390 s		
97619	7322		11 16 09.1	+78 18 32	-0.020	-0.02	7.0	1.1	0.2	K0 III		230 s		
97991	138096		11 16 11.6	-3 28 18	+0.001	+0.02	7.42	-0.23	-2.5	B2 V	+25	960 s		
98019	179713		11 16 15.3	-20 41 24	-0.002	-0.01	7.8	0.7	0.3	G7 III	+33	320 s		
97889	15454		11 16 18.7	+59 56 33	-0.005	-0.02	6.7	0.1	1.7	A3 V	-5	99 s		
98048	202266		11 16 19.6	-34 40 04	-0.005	0.00	6.8	0.1	1.4	A2 V		110 s		
98096	222687		11 16 27.6	-45 52 48	-0.013	+0.06	6.31	0.40		F2				m
98195	251339		11 16 30.5	-69 54 28	-0.005	-0.04	7.2	0.1	0.6	A0 V		160 mx		
98062	179722		11 16 32.4	-28 07 45	-0.002	+0.03	7.4	1.1	3.4	F5 V		27 s		
98132	238924		11 16 35.3	-55 33 19	-0.004	+0.01	7.80	0.8	3.2	G5 IV		83 s		m
98046	138101		11 16 36.3	-3 58 00	+0.001	-0.01	6.8	0.1	1.7	A3 V		110 s		
98058	138102	74 φ Leo	11 16 39.6	-3 39 06	-0.007	-0.04	4.47	0.21	1.5	A7 IV	-3	39 s		m
97989	43675		11 16 41.8	+49 28 34	-0.009	-0.01	5.88	1.08	0.2	K0 III	0	120 s		
98090	156579		11 16 48.3	-14 30 29	-0.004	0.00	7.9	1.1	-0.1	K2 III		310 mx		
98137	238926		11 16 49.7	-50 55 35	-0.006	+0.02	8.0	0.5	3.0	F2 V		81 s		
98244	--		11 16 51.5	-69 29 43			7.6	0.0	0.0	B8.5 V		320 s		
98092	179732		11 16 51.9	-27 14 23	0.000	-0.02	7.2	1.1	3.2	G5 IV		43 s		
98126	202276		11 16 53.0	-38 23 58	-0.002	0.00	6.8	0.6	0.2	K0 III		210 s		
98029	43677		11 16 53.4	+41 43 32	0.000	0.00	7.8	1.1	0.2	K0 III		330 s		
98125	202278		11 16 57.5	-34 55 40	-0.002	-0.02	6.9	0.8	0.2	K0 III		220 s		
98088	138106		11 16 58.1	-7 08 06	-0.001	-0.01	6.14	0.20		A2 p	-55		8115	m
98180	238930		11 17 00.1	-54 46 38	+0.007	-0.07	7.0	0.4	3.0	F2 V		62 s		
98192	238932		11 17 02.8	-55 37 21	-0.004	0.00	8.00	0.1	0.6	A0 V		220 mx		m
98179	238931		11 17 04.7	-52 39 26	-0.003	-0.01	7.9	0.6	4.0	F8 V		57 s		
98177	222693		11 17 09.2	-43 49 09	-0.009	-0.02	8.0	0.5	1.4	A2 V		70 mx		
97853	7327		11 17 09.9	+75 21 09	+0.012	-0.06	7.7	1.1	0.2	K0 III		180 s		
98175	222694		11 17 10.5	-41 23 25	-0.011	+0.02	7.1	-0.1	1.4	A2 V		67 mx		
98161	202283		11 17 11.8	-38 00 52	-0.008	+0.01	6.27	0.10	0.6	A0 V n		88 mx		
98190	238933		11 17 13.6	-53 06 22	+0.005	+0.01	7.4	0.8	0.2	K0 III		270 s		
98176	222695		11 17 14.2	-41 56 02	-0.005	0.00	6.9	-0.6	0.6	A0 V		120 mx		
98240	238936		11 17 14.5	-59 46 11	-0.007	-0.01	7.6	0.1	1.7	A3 V		150 s		
98118	118764	75 Leo	11 17 17.3	+2 00 37	+0.004	-0.15	5.18	1.52	-0.4	M0 III	-59	130 s		
98075	43679		11 17 19.0	+47 56 10	-0.002	-0.03	7.8	1.1	0.2	K0 III		330 s		
98292	251341		11 17 19.2	-67 49 25	+0.003	-0.02	6.06	1.76		M1				m
97904	7329		11 17 21.6	+74 20 15	+0.015	-0.03	7.6	0.9	0.3	G7 III	-23	260 mx		
98330	--		11 17 22.2	-69 29 45			7.9	0.0	0.4	B9.5 V		310 s		
98054	15462		11 17 26.8	+60 16 18	-0.001	-0.01	6.7	1.1	0.2	K0 III		200 s		
98278	238939		11 17 27.8	-59 06 19	-0.001	0.00	6.70	-0.01	0.6	A0 V		170 s		m
98314	251342		11 17 31.1	-63 28 35	0.000	0.00	7.70	0.1	0.6	A0 V		260 s		m
98234	222700		11 17 33.6	-44 26 42	-0.003	+0.01	7.8	1.5	0.2	K0 III		170 s		
98341	251347		11 17 36.7	-68 07 06	-0.002	-0.01	7.6	0.9	3.2	G5 IV		74 s		
98221	202289		11 17 39.1	-34 44 14	-0.001	0.00	6.45	0.40	2.6	dF0		51 s		
98220	202288		11 17 39.2	-33 32 48	+0.007	-0.33	6.85	0.51	3.6	G0 IV-V		45 s		
98154	81755		11 17 40.8	+25 27 37	-0.004	-0.04	7.39	0.14	1.4	A2 V		73 mx		
98172	99540		11 17 42.7	+12 23 59	-0.001	-0.03	7.9	1.1	-0.1	K2 III		390 s		
98233	202291		11 17 42.9	-36 32 04	-0.007	+0.02	6.68	0.98		K0				
98387	256821		11 17 43.8	-70 58 42	+0.005	-0.03	7.9	1.4	-0.3	K5 III		420 s		
98153	81756		11 17 44.3	+27 21 44	+0.001	-0.01	6.94	0.20		A2				
98155	81757		11 17 45.6	+25 03 05	+0.001	-0.01	7.80	1.1	0.2	K0 III		330 s		
98218	179746		11 17 47.0	-22 08 45	-0.002	-0.03	7.7	1.2	-0.5	M4 III		440 s		
98329	238946		11 17 47.3	-59 12 28	-0.007	0.00	7.03	-0.08	0.4	B9.5 V		130 mx		
98286	222702		11 17 51.3	-45 30 20	-0.002	-0.01	7.2	0.8	3.2	G5 IV		27 s		
98340	238948		11 17 53.5	-59 14 11	-0.005	0.00	7.2	0.1	0.6	A0 V		170 s		
98363	251348		11 17 58.1	-64 02 33	-0.004	-0.01	7.6	0.1	1.4	A2 V		170 s		
98217	81760		11 18 01.3	+22 10 48	-0.002	-0.01	7.0	0.9	3.2	G5 IV		57 s		
98302	202299		11 18 08.9	-31 45 08	-0.003	-0.03	7.4	1.2	0.2	K0 III		220 s		
98386	251350		11 18 09.4	-62 58 31	0.000	0.00	7.9	1.1	0.2	K0 III		330 s		
98231	62484	53 ξ UMa	11 18 10.9	+31 31 45	-0.034	-0.59	3.79	0.59	4.9	G0 V	-16	7.7 t	8119	Alula Aus., m
98486	--		11 18 14.9	-74 09 46			7.9	0.0	0.4	B9.5 V		320 s		
98359	222707		11 18 19.6	-48 47 15	-0.005	0.00	7.3	0.1	0.4	B9.5 V		150 mx		
98280	99544		11 18 20.9	+11 59 05	0.000	-0.04	6.50	0.07	1.4	A2 V	-35	100 s		
98281	138119		11 18 21.9	-5 04 02	+0.053	-0.15	7.31	0.73	5.6	G8 V	+10	22 ts		
98247	43689		11 18 28.6	+42 18 59	+0.006	-0.03	7.0	0.8	3.2	G5 IV		59 s		

HD	SAO	Star Name	α 2000	δ 2000	μ(α)	μ(δ)	V	B−V	M$_v$	Spec	RV	d(pc)	ADS	Notes
98262	62486	54 ν UMa	11^{h}18^{m}28.7^s	+33°05'39"	−0.002^s	+0.02"	3.48	1.40	−0.2	K3 III	−9	46 s	8123	Alula Bor., m
98214	27980		11 18 31.8	+56 11 41	−0.002	−0.01	7.38	1.08	0.2	K0 III		230 s		
98617	256823		11 18 34.4	−79 40 08	+0.014	−0.04	6.35	0.26		A3				
98346	179760		11 18 36.6	−24 20 36	0.000	0.00	6.7	0.6	0.2	K0 III		200 s		
98537	256822		11 18 41.6	−74 11 25	+0.008	−0.01	7.1	0.5	4.0	F8 V		41 s		
98434	238956		11 18 43.6	−58 11 12	−0.006	−0.01	7.1	1.4	−0.5	M4 III		330 s		
98196	15470		11 18 46.7	+66 41 03	−0.008	−0.01	7.20	0.4	2.6	F0 V		83 s	8122	m
98317	62488		11 18 48.2	+35 29 21	0.000	0.00	6.9	1.1	0.2	K0 III		220 s		
98484	251354		11 18 53.2	−61 38 46	+0.001	0.00	7.6	0.8	0.6	A0 V		77 s		
98366	118778	76 Leo	11 18 54.9	+1 39 01	−0.003	−0.06	5.91	1.04	0.2	gK0	+5	130 s		
98354	99551		11 18 59.8	+14 16 06	+0.003	−0.17	6.90	0.7	4.4	G0 V	+24	32 s	8128	m
98403	179763		11 19 00.7	−25 13 35	−0.004	−0.01	8.0	1.3	2.6	F0 V		28 s		
98261	15472		11 19 03.5	+62 10 58	−0.007	+0.03	7.0	1.1	0.2	K0 III		230 s		
98627	256824		11 19 03.9	−75 49 30	−0.001	−0.02	7.0	2.4	−0.3	K5 III		79 s		
98463	238959		11 19 04.0	−55 39 17	−0.002	+0.02	7.6	1.8	−0.1	K2 III		140 s		
98388	99552		11 19 06.6	+13 23 24	−0.009	−0.02	7.1	0.5	4.0	F8 V	+6	42 s		
98353	62491	55 UMa	11 19 07.8	+38 11 07	−0.005	−0.07	4.78	0.12	1.4	A2 V	−3	42 mx		
98315	27983		11 19 08.6	+58 53 14	−0.002	+0.02	7.96	1.36	0.2	K0 III		190 s		
98440	222710		11 19 10.6	−40 30 00	+0.003	0.00	7.2	0.9	2.1	A5 V		39 s		
98399	118783		11 19 15.4	+0 43 24	−0.005	0.00	7.7	1.1	0.2	K0 III		280 mx		
98560	251357		11 19 16.6	−64 34 57	−0.046	+0.04	5.99	0.46	3.4	F5 V		32 s		
98430	156605	12 δ Crt	11 19 20.4	−14 46 43	−0.009	+0.20	3.56	1.12	1.8	G8 III-IV	−5	22 mn		
98495	222716		11 19 21.1	−45 05 36	0.000	−0.01	7.7	1.5	−0.5	M4 III		350 s		
98427	138130		11 19 22.5	−1 39 18	−0.015	−0.16	7.09	0.52	3.7	dF6	+2	44 s		m
98457	202320		11 19 25.9	−30 19 21	−0.001	0.00	8.0	−0.3	0.6	A0 V		300 s		
98456	179768		11 19 26.3	−28 28 33	−0.002	−0.03	7.9	0.8	0.2	K0 III		340 s		
98183	7335		11 19 27.2	+75 05 12	+0.022	−0.02	8.0	1.1	0.2	K0 III		240 mx		
98397	62494		11 19 31.7	+32 49 31	−0.011	+0.03	8.0	1.1	0.2	K0 III		140 mx		
98517	222718		11 19 32.3	−47 23 48	+0.001	0.00	7.4	1.0	0.2	K0 III		270 s		
98425	81770		11 19 34.8	+20 43 30	−0.001	0.00	7.5	1.1	−0.1	K2 III		340 s		
98672	256826		11 19 36.4	−75 08 32	−0.008	0.00	6.27	−0.03		A0				
98423	62496		11 19 41.5	+38 06 03	−0.004	−0.07	7.2	0.4	3.0	F2 V		69 s		
98542	238964		11 19 42.0	−52 29 31	−0.003	+0.01	6.9	1.0	0.2	K0 III		220 s		
98471	118792		11 19 44.4	+3 37 15	−0.001	−0.03	8.0	0.5	4.0	F8 V		62 s		
98516	179776		11 19 47.5	−28 11 18	−0.004	−0.03	7.8	0.9	0.2	K0 III		300 mx		
98558	238967		11 19 49.8	−53 28 34	−0.001	−0.01	7.90	0.1	1.4	A2 V		200 s		m
98469	99557		11 19 50.4	+19 04 57	0.000	−0.01	8.0	1.1	0.2	K0 III		370 s		
98671	256827		11 19 51.1	−72 57 31	−0.006	−0.03	6.6	0.4	0.6	A0 V		92 s		
98584	238971		11 19 55.0	−58 09 17	0.000	−0.01	7.59	−0.09	0.0	B8.5 V		330 s		
98596	238975		11 20 00.6	−59 11 16	−0.002	0.00	7.31	0.01	0.6	A0 V		220 s		
98695	256828		11 20 04.0	−71 59 39	−0.007	−0.01	6.41	0.05		B3				
98502	81777		11 20 05.9	+21 54 50	0.000	−0.04	7.9	1.1	0.2	K0 III		340 s		
98659	251361		11 20 06.8	−67 03 04	+0.002	−0.07	7.50	0.11	0.4	B9.5 V		120 mx		m
98528	99562		11 20 09.4	+17 06 53	+0.001	−0.05	7.7	1.1	−0.1	K2 III		360 s		
98501	81778		11 20 10.0	+28 18 37	+0.001	+0.01	7.9	1.1	−0.1	K2 III		400 s		
98500	62498		11 20 10.9	+30 07 13	0.000	−0.02	7.40	1.6	−0.4	M0 III	+31	360 s		
98553	156614		11 20 11.6	−19 34 38	+0.006	−0.05	7.5	0.7	4.4	G0 V		42 s		
98551	156615		11 20 12.7	−13 53 05	−0.003	−0.01	7.9	1.1	0.2	K0 III		340 s		
98539	118797		11 20 15.8	+0 16 26	+0.003	−0.02	7.8	0.8	3.2	G5 IV		84 s		
98622	238978		11 20 17.4	−53 38 30	−0.002	+0.02	7.9	−0.2	0.2	K0 III		340 s		
98579	179783		11 20 18.8	−28 19 56	−0.019	+0.01	6.70	1.13	0.2	K0 III		79 mx		m
98669	251362		11 20 21.3	−61 05 50	−0.005	0.00	7.70	0.23	2.6	F0 V		110 s		
98547	99564		11 20 26.4	+17 18 39	+0.001	−0.03	7.15	0.12	0.6	A0 V	−7	170 s		
--	--		11 20 28.7	−61 52 28			7.80							RS Cen, v
98635	222730		11 20 30.0	−45 43 44	+0.003	−0.03	7.0	0.9	2.6	F0 V		33 s		
98591	156618		11 20 31.2	−10 17 41	−0.007	−0.01	6.6	1.1	0.2	K0 III		190 mx		
98526	43702		11 20 31.9	+44 59 53	−0.006	−0.04	6.7	0.4	2.6	F0 V		66 s		
98651	202343		11 20 42.5	−35 43 12	+0.003	−0.07	7.6	0.6	3.0	F2 V		60 s		
98722	238984		11 20 51.8	−58 29 11	−0.005	−0.01	8.02	0.39	1.8	F1 IV		160 s		
98649	179793		11 20 52.0	−23 13 02	−0.011	−0.18	8.03	0.66	4.4	G0 V		45 s		
98499	15478		11 20 53.8	+67 06 03	+0.009	−0.05	6.21	1.01	0.2	K0 III	−56	160 s		
98732	238985		11 20 55.0	−58 42 10	−0.039	+0.07	7.02	1.00		K0		21 mn		
98632	118801		11 20 55.9	+2 25 21	−0.005	−0.04	8.0	0.4	3.0	F2 V		98 s		
98572	27995		11 20 56.6	+51 45 40	−0.019	−0.11	7.2	0.4	2.6	F0 V		53 mx		
98733	238987		11 20 57.3	−60 09 42	0.000	−0.02	7.95	0.16	−5.7	B1 Ib		3300 s		
98629	81786		11 20 59.0	+20 51 29	−0.002	−0.01	7.9	0.4	3.0	F2 V		94 s		
98718	238986	π Cen	11 21 00.4	−54 29 27	−0.003	−0.01	3.89	−0.15	−1.1	B5 V n	+16	99 s		m
98717	222735		11 21 06.6	−45 53 10	0.000	0.00	7.4	1.2	0.2	K0 III		200 s		
98664	118804	77 σ Leo	11 21 08.1	+6 01 45	−0.006	−0.02	4.05	−0.06	0.2	B9 V	−5	59 s		
98836	256829		11 21 19.8	−70 48 12	−0.008	−0.04	8.0	0.3	3.4	F5 V		82 s		
98729	179803		11 21 26.4	−23 59 40	−0.003	−0.01	7.80	1.1	0.2	K0 III		330 s		m
98697	118806		11 21 26.7	+6 38 05	−0.020	−0.01	6.6	0.5	4.0	F8 V		33 s		
98618	27996		11 21 28.8	+58 29 03	+0.002	+0.03	7.66	0.64	3.2	G5 IV		78 s		
98727	156625		11 21 30.9	−12 41 54	−0.011	+0.05	7.8	0.5	4.0	F8 V		57 s		
98818	251368		11 21 33.5	−61 13 33	−0.014	+0.04	7.36	1.12	3.2	K0 IV		51 s		

HD	SAO	Star Name	α 2000	δ 2000	μ(α)	μ(δ)	V	B–V	M_v	Spec	RV	d(pc)	ADS	Notes
			h m s	° ' "	s	"								
98711	118808		11 21 33.8	+4 12 29	-0.006	-0.02	7.9	0.5	3.4	F5 V		78 s		
98753	202356		11 21 35.6	-31 31 20	+0.004	-0.02	8.0	0.4	3.4	F5 V		82 s		
98681	43709		11 21 35.8	+40 25 51	+0.004	-0.07	7.2	0.5	4.0	F8 V		43 s		
98779	222738		11 21 37.6	-43 33 28	-0.004	-0.02	7.20	1.1	0.2	K0 III		250 s		m
98792	222741		11 21 40.5	-48 46 22	-0.003	-0.02	7.6	-0.4	1.4	A2 V		180 s		
98802	222743		11 21 44.9	-48 29 44	-0.001	-0.01	8.0	0.3	0.2	K0 III		360 s		
98763	179807		11 21 45.8	-22 06 21	-0.003	-0.02	7.36	-0.05	-1.0	B7 IV		440 s		
98750	138156		11 21 46.7	-5 43 08	-0.004	-0.03	7.9	0.9	3.2	G5 IV		88 s		
96870	1848		11 21 47.9	+87 38 19	+0.010	+0.01	7.4	0.0	0.0	B8.5 V	-23	310 s		
98673	27999		11 21 49.3	+57 04 30	-0.006	+0.02	6.3	0.1	1.4	A2 V	-20	97 s		
98736	99573		11 21 49.3	+18 11 23	-0.010	-0.10	7.94	0.89	3.2	G5 IV	-4	12 ts	8140	m
98764	179808		11 21 49.8	-24 11 24	+0.003	-0.05	8.00	0.9	3.2	G5 IV		91 s		m
98747	118813		11 21 50.2	+9 10 07	-0.005	-0.02	6.7	0.1	1.4	A2 V		80 mx		
98696	43712		11 21 50.6	+45 19 56	-0.007	-0.01	7.90	0.1	1.4	A2 V		100 mx	8139	m
98791	202362		11 21 51.6	-33 18 05	0.000	+0.01	7.8	1.0	-0.1	K2 III		380 s		
99015	256832		11 21 57.3	-77 36 30	-0.023	-0.01	6.43	0.20		A2				
98735	43714		11 22 01.6	+41 01 39	-0.004	+0.01	7.8	0.8	3.2	G5 IV		84 s		
98746	62513		11 22 05.4	+35 20 13	-0.003	+0.02	8.0	1.1	0.2	K0 III		360 s		
98830	202368		11 22 09.2	-33 46 53	-0.002	-0.04	7.7	1.3	-0.1	K2 III		310 s		
98896	239002		11 22 14.7	-56 36 10	-0.002	+0.01	7.6	1.6	-0.1	K2 III		200 s		
98897	239003		11 22 15.6	-58 23 11	+0.002	-0.07	7.1	1.2	0.2	K0 III		190 s		
98828	179816		11 22 16.2	-25 11 42	+0.001	+0.02	7.8	1.5	0.2	K0 III		170 s		
98867	202373		11 22 21.3	-39 06 19	-0.004	-0.01	7.3	0.4	0.4	B9.5 V		140 s		
98892	222751		11 22 23.0	-44 38 45	-0.005	-0.04	6.12	0.92	3.2	G5 IV		29 s		
98853	138166		11 22 28.6	-8 50 34	-0.004	0.00	6.8	0.1	1.4	A2 V		120 s		
98824	99577		11 22 28.9	+17 26 13	-0.008	-0.02	7.03	1.06	0.0	K1 III	+5	180 mx		
98922	239007		11 22 31.8	-53 22 12	+0.001	-0.02	7.00	0.00	0.4	B9.5 V		210 s		m
98812	62519		11 22 40.1	+35 01 35	0.000	+0.01	7.6	1.1	0.2	K0 III		300 s		
98823	43717		11 22 41.6	+40 10 31	-0.007	-0.01	6.6	0.5	3.4	F5 V		45 s		
98938	222758		11 22 44.2	-42 56 54	-0.001	-0.01	7.4	1.4	-0.3	K5 III		350 s		
98983	239010		11 22 47.4	-58 18 54	-0.007	0.00	7.4	0.9	1.4	A2 V		48 s		
98937	202378		11 22 48.2	-36 14 35	-0.006	-0.03	7.8	-0.3	2.1	A5 V		84 mx		
98839	43719	56 UMa	11 22 49.5	+43 28 58	-0.003	-0.01	4.99	0.99	-2.1	G8 II	+3	260 s		
98661	7348		11 22 49.8	+77 22 25	-0.026	+0.01	7.6	0.5	3.4	F5 V		69 s		
98851	62524		11 22 51.1	+31 49 43	-0.003	+0.01	8.0	0.4	3.0	F2 V		98 s		
98772	15486		11 22 51.2	+64 19 49	-0.001	+0.03	6.0	0.1	0.6	A0 V	+2	120 s		
98982	239013		11 22 52.0	-57 38 13	0.000	-0.01	8.0	0.0	0.4	B9.5 V		300 s		
98980	239012		11 22 54.6	-50 47 11	-0.003	0.00	6.7	1.1	0.2	K0 III		170 s		
98915	179826		11 22 56.8	-20 37 34	-0.005	-0.01	7.6	-0.2	1.4	A2 V		94 mx		
98964	222759		11 22 57.4	-44 01 17	-0.001	-0.02	8.0	1.3	-0.1	K2 III		360 s		
98949	179828		11 23 05.3	-28 53 47	+0.002	+0.01	7.4	0.5	0.6	A0 V		120 s		
99022	239017		11 23 08.0	-56 46 46	-0.005	0.00	5.79	0.01		A0				
98993	202391		11 23 12.5	-36 09 54	-0.003	-0.02	5.00	1.46	-0.3	gK6	-5	120 s		m
98947	118823		11 23 15.0	+6 35 09	-0.001	-0.01	7.0	0.8	3.2	G5 IV		59 s		
99048	239020		11 23 17.6	-58 47 14	-0.001	+0.01	7.1	1.6	0.2	K0 III		110 s		
98960	118825		11 23 17.9	+0 07 54	-0.003	-0.02	6.05	1.46		K3	+22			
98975	179831		11 23 18.4	-23 42 23	-0.003	+0.02	7.2	0.7	0.2	K0 III		250 s		
99104	251383		11 23 21.8	-64 57 18	+0.001	-0.01	5.11	-0.08	-1.0	B5.5 V	+19	150 s		
98991	156646	13 λ Crt	11 23 21.8	-18 46 48	-0.022	-0.04	5.09	0.42	2.1	F5 IV	+12	40 s		
98990	156645		11 23 23.0	-12 05 59	-0.001	-0.01	8.0	1.1	-0.1	K2 III		410 s		
99004	99585		11 23 38.0	+17 08 30	-0.003	-0.01	7.1	0.1	1.4	A2 V	-1	140 s		
99101	239027		11 23 38.5	-58 15 57	-0.003	+0.01	7.2	1.5	0.2	K0 III		130 s		
99148	251384		11 23 39.2	-65 20 33	+0.002	-0.01	7.9	1.1	-0.1	K2 III		380 s		
99062	202399		11 23 47.4	-36 05 23	-0.003	-0.03	6.9	0.6	0.6	A0 V		74 s		
98967	28007		11 23 47.6	+52 02 26	-0.010	-0.01	7.96	0.40		F0				
99060	179839		11 23 52.9	-26 57 29	-0.005	+0.02	7.5	1.4	0.2	K0 III		160 s		
99002	62533		11 23 53.2	+37 14 06	-0.005	-0.01	6.9	0.4	2.6	F0 V	-12	73 s		
99028	99587	78 ι Leo	11 23 55.4	+10 31 45	+0.011	-0.08	3.94	0.41	1.9	F2 IV	-10	24 ts	8148	m
99145	239033		11 23 56.6	-57 52 49	+0.002	-0.02	6.4	1.9	-0.3	K5 III		120 s		
99055	118831	79 Leo	11 24 02.2	+1 24 28	-0.001	0.00	5.39	0.94	0.3	gG7	-10	100 s		
99264	256834		11 24 11.0	-72 15 24	-0.007	-0.01	5.59	0.06	-2.9	B3 III		330 mx		
99126	138184		11 24 21.7	-5 54 36	-0.006	-0.10	7.1	0.5	3.4	F5 V		54 s		
99171	222773		11 24 22.0	-42 40 09	-0.001	0.00	6.12	-0.18	-5.0	B0 III		1500 s		
99138	179845		11 24 24.2	-22 05 14	-0.001	+0.01	8.0	0.8	-0.1	K2 III		420 s		
99219	239042		11 24 25.5	-58 10 51	0.000	0.00	7.2	0.7	0.6	A0 V		74 s		
99108	99592		11 24 25.9	+17 20 37	-0.002	0.00	7.5	0.8	3.2	G5 IV		73 s		
99122	99593		11 24 26.5	+15 53 02	-0.002	-0.02	8.0	0.5	4.0	F8 V		62 s		
99169	202407		11 24 31.0	-34 17 12	-0.002	-0.03	7.5	0.0	0.6	A0 V		240 s		
99073	43732		11 24 34.3	+48 36 21	-0.001	-0.01	7.0	1.4	-0.3	K5 III		290 s		
99167	156658	14 ε Crt	11 24 36.5	-10 51 33	-0.002	+0.02	4.83	1.56	-0.3	K5 III	+3	96 s		
99185	179852		11 24 38.8	-29 10 46	-0.002	+0.05	7.6	1.7	-0.5	M2 III		370 s		
99279	251393		11 24 40.3	-61 38 52	-0.068	+0.06	7.22	1.26	7.3	K7 V		7.3 t		m
99071	--		11 24 44.8	+59 39 49	-0.001	-0.02	7.7	0.9	3.2	G5 IV		81 s		
99166	118833		11 24 45.1	+2 05 47	0.000	-0.01	7.9	0.9	3.2	G5 IV		88 s		
99317	--		11 24 45.6	-67 06 56			7.2	0.0	0.4	B9.5 V		230 s		

HD	SAO	Star Name	α 2000	δ 2000	μ(α)	μ(δ)	V	B−V	M$_v$	Spec	RV	d(pc)	ADS	Notes
99199	179854		11^h 24^m 47.4^s	−22° 49′ 58″	0.000^s	0.00″	7.2	0.8	1.7	A3 V		50 s		
99211	156661	15 γ Crt	11 24 52.8	−17 41 02	−0.007	0.00	4.08	0.21	2.1	A5 V	+1	24 s	8153	m
99316	251397		11 24 55.1	−63 25 54	−0.001	0.00	7.40	0.33	−5.5	B9 Ib		2500 s		
99164	62541		11 24 55.4	+31 49 31	−0.001	0.00	7.69	0.94	0.2	K0 III		320 s		
99372	—		11 24 55.7	−71 47 56			7.7	0.0	0.4	B9.5 V		280 s		
99275	222777		11 24 57.3	−49 02 22	−0.003	−0.03	8.0	1.1	0.2	K0 III		360 s		
99210	138190		11 24 58.2	−2 12 45	−0.004	−0.04	6.7	0.2	2.1	A5 V		64 mx		
99196	99598		11 24 58.8	+11 25 49	−0.007	−0.01	5.80	1.38	−0.3	K4 III	+38	170 s		
99312	239048		11 25 00.2	−59 39 06	−0.004	−0.01	8.00	0.9	3.2	G5 IV		90 s		m
99195	81810		11 25 05.0	+26 08 54	−0.003	−0.01	6.00	1.4	−0.3	K4 III		180 s		
99208	81812		11 25 06.1	+20 25 10	−0.001	−0.02	8.0	0.9	3.2	G5 IV		93 s		
99417	—		11 25 10.0	−72 51 57			8.0	1.1	0.2	K0 III		350 s		
99225	118838		11 25 11.2	+5 44 22	−0.001	−0.02	8.0	1.1	−0.1	K2 III		410 s		
99441	256837		11 25 11.6	−73 37 57	−0.020	+0.04	6.7	0.4	3.0	F2 V		54 s		
99256	156664		11 25 14.9	−18 36 27	−0.003	−0.03	7.1	0.2	2.1	A5 V		100 s		
99207	62545		11 25 16.6	+29 45 32	+0.004	−0.04	7.6	0.4	2.6	F0 V		99 s		
99827	258611		11 25 17.3	−84 57 16	−0.041	+0.01	7.80	0.5	3.4	F5 V		76 s		m
99335	222783		11 25 25.8	−43 26 52	−0.001	−0.03	7.6	0.0	2.6	F0 V		100 s		
99369	—		11 25 26.9	−57 49 57			7.9	2.0	−0.3	K5 III		210 s		
99322	202428		11 25 29.3	−36 03 47	−0.010	+0.02	5.22	0.99	0.3	gG5	+4	83 s		
99334	222784		11 25 31.4	−41 23 53	−0.004	−0.02	7.6	0.7	2.6	F0 V		100 s		
99333	202430		11 25 33.0	−37 44 52	−0.004	−0.02	5.89	1.54	−0.5	M2 III		190 s		m
99285	99601	81 Leo	11 25 36.3	+16 27 23	−0.010	−0.01	5.57	0.36	3.0	F2 V	+18	32 s		m
99267	62550		11 25 38.1	+29 59 13	−0.004	0.00	6.9	0.4	2.6	F0 V	−4	73 s		
99305	118847		11 25 39.4	+3 18 03	−0.002	−0.05	6.7	0.1	1.4	A2 V	+8	54 mx		
99332	179870		11 25 41.7	−27 32 54	−0.005	+0.03	7.93	0.32	1.7	F0 IV		170 s		
99385	222789		11 25 41.8	−45 52 49	−0.004	−0.02	7.0	0.7	0.2	K0 III		230 s		
99331	156670		11 25 42.3	−14 32 14	−0.003	0.00	7.40	1.4	−0.3	K5 III	+25	350 s		
99453	251402		11 25 43.2	−63 58 22	−0.047	−0.08	5.17	0.50	3.1	dF3	−5	22 s		
99437	251401		11 25 45.1	−63 48 32	−0.001	0.00	7.9	0.4	3.0	F2 V		95 s		
99302	81819		11 25 45.7	+26 44 50	−0.003	0.00	7.31	0.23	1.4	A2 V	+8	120 s		
99284	43738		11 25 47.5	+40 56 13	−0.002	−0.06	7.9	1.1	−0.1	K2 III		180 mx		
99329	118851	80 Leo	11 25 50.0	+3 51 36	−0.005	−0.04	6.37	0.33	2.4	dA7 n	−3	62 s		
99363	156672		11 25 53.7	−13 45 03	−0.004	+0.02	6.9	1.6	−0.5	M2 III	+6	310 s		
99266	28016		11 25 55.0	+52 07 55	−0.004	−0.02	7.20	0.4	2.6	F0 V		83 s	8158	m
99283	28017		11 25 57.0	+55 51 01	−0.008	+0.04	5.8	0.9	3.2	G5 IV	−6	33 s		
99409	—		11 26 01.2	−40 14 57			7.8	0.9	0.2	K0 III		330 s		
99467	239060		11 26 01.2	−59 21 11	−0.003	−0.01	7.51	−0.01	0.6	A0 V		240 s		
99380	179872		11 26 01.5	−20 34 57	−0.008	−0.04	6.7	0.7	2.1	A5 V		41 s		
99231	15502		11 26 03.6	+66 45 14	−0.001	−0.03	7.7	1.1	0.2	K0 III		320 s		
99359	81823		11 26 08.7	+22 42 27	−0.007	−0.01	7.3	0.4	2.6	F0 V		85 s		
99327	43741		11 26 09.0	+42 43 56	−0.001	−0.01	7.9	1.1	0.2	K0 III		340 s		
99405	138206		11 26 16.9	−9 52 53	0.000	−0.09	7.7	1.1	−0.1	K2 III		200 mx		
99393	118859		11 26 17.9	+8 39 35	0.000	−0.02	6.8	1.1	0.2	K0 III		210 s		
99480	222798		11 26 22.9	−43 33 49	−0.004	0.00	6.9	0.8	2.6	F0 V		35 s		
99513	239066		11 26 24.3	−59 16 56	−0.003	−0.02	8.0	1.3	0.2	K0 III		250 s		
99373	62551		11 26 25.4	+33 27 01	−0.003	+0.01	6.32	0.43	3.4	F5 V	−25	38 s		
99419	81825		11 26 27.1	+20 31 05	−0.008	−0.07	7.7	0.8	3.2	G5 IV		79 s		
99495	202452		11 26 32.4	−36 53 00	+0.001	−0.02	7.7	1.4	−0.5	M2 III		440 s		
99459	179887		11 26 33.9	−21 21 24	−0.003	−0.01	6.8	1.0	0.6	A0 V		45 s		
99556	251406		11 26 35.3	−61 06 54	−0.003	0.00	5.30	−0.08	−1.6	B5 IV	+9	230 s		m
99508	202453		11 26 42.4	−38 09 29	−0.001	0.00	7.90	0.06	0.6	A0 V		260 s		
99491	118864	83 Leo	11 26 45.3	+3 00 47	−0.048	+0.18	6.16	0.85	3.2	K0 IV	−3	21 ts	8162	m
99492	118865		11 26 46.2	+3 00 21	−0.048	+0.17	7.57	1.01	6.3	K2 V	+2	14 s		
99574	239074		11 26 47.2	−53 09 36	−0.005	+0.01	5.81	0.52	4.4	G0 V		19 s		
99507	138211		11 26 52.0	−9 56 27	−0.001	+0.01	8.0	0.5	3.4	F5 V		82 s		
99473	81831		11 26 57.5	+28 24 26	0.000	−0.01	7.5	0.2	2.2	A6 V		120 s		
99644	251409		11 26 58.7	−69 02 50	−0.006	−0.01	7.80	0.1	1.4	A2 V		190 s		m
99619	251408		11 26 59.4	−61 22 09	−0.003	0.00	6.6	2.6	−0.1	K2 III		31 s		
99586	239078		11 27 00.9	−56 16 49	−0.003	0.00	6.6	1.4	−0.1	K2 III		170 s		
99534	156684		11 27 02.3	−19 12 03	+0.001	−0.06	7.3	1.1	−0.1	K2 III		310 s		
99505	81832		11 27 03.5	+21 51 07	+0.001	−0.18	7.6	0.9	3.2	G5 IV		60 mx		
99602	222804		11 27 04.9	−47 45 42	+0.002	−0.03	7.8	0.7	0.2	K0 III		330 s		
99518	81833		11 27 09.0	+25 01 40	−0.004	−0.01	7.6	0.4	2.6	F0 V		100 s		
99564	156685	16 κ Crt	11 27 09.4	−12 21 24	−0.007	+0.02	5.94	0.49	3.3	dF4	+6	31 s		m
99565	156686		11 27 10.6	−15 38 54	−0.001	−0.18	7.66	0.75		K0			8166	m
99504	62559		11 27 12.0	+33 07 27	+0.002	−0.08	7.23	0.99	0.2	K0 III		170 mx		
99806	256842		11 27 16.4	−78 31 16	−0.011	+0.03	7.80	1.1	0.2	K0 III		330 s		m
99585	179898		11 27 20.9	−24 10 13	−0.001	−0.03	7.4	1.5	−0.3	K5 III		330 s		
99561	118868		11 27 22.7	+1 22 25	+0.003	−0.11	7.7	0.9	3.2	G5 IV		80 s		
99598	156689		11 27 24.0	−18 31 12	−0.001	−0.02	7.5	1.1	0.2	K0 III		290 s		
99611	179901		11 27 29.1	−24 05 08	+0.001	+0.02	7.8	0.8	0.2	K0 III		330 s		
99627	202462		11 27 29.3	−39 52 35	−0.003	−0.01	7.40	0.1	0.6	A0 V		230 s		m
99610	156690		11 27 30.2	−12 26 09	−0.005	−0.05	7.7	0.9	3.2	G5 IV		80 s		
99595	118870		11 27 33.3	+1 18 05	−0.004	0.00	7.7	1.1	0.2	K0 III		310 s		

HD	SAO	Star Name	α 2000	δ 2000	μ(α)	μ(δ)	V	B-V	M_v	Spec	RV	d(pc)	ADS	Notes
			$11^h27^m33.9^s$	$+0°57'23''$	$0^s.000$	$-0''.09$								
99596	118871		11 27 33.9	+0 57 23	0.000	-0.09	7.6	1.1	0.2	K0 III		160 mx		
99680	222806		11 27 35.9	-49 33 05	-0.003	0.00	7.8	1.0	0.2	K0 III		330 s		
99625	179904		11 27 37.6	-25 51 38	-0.005	+0.04	6.8	1.1	0.2	K0 III	-15	170 s		
99559	43744		11 27 39.4	+46 17 34	+0.002	-0.08	7.90	0.5	4.0	F8 V		60 s	8168	m
99517	15514		11 27 40.4	+64 12 15	+0.002	-0.01	8.0	0.9	3.2	G5 IV		93 s		
99736	251412		11 27 43.0	-60 29 45	-0.002	0.00	6.8	1.5	-0.1	K2 III		140 s		
99579	62564		11 27 44.8	+37 56 18	-0.006	-0.09	7.3	0.5	4.0	F8 V		45 s		
99592	43748		11 27 50.4	+45 11 04	0.000	-0.05	6.50	1.6	-0.5	M4 III	-17	250 s		ST UMa, v
99734	239091		11 27 51.0	-58 40 49	-0.002	0.00	7.65	-0.12	0.4	B9.5 V		280 s		
--	--		11 27 52.2	+2 52 58			5.20							m
99651	138216		11 27 53.6	-1 42 00	-0.003	-0.01	6.25	1.04	-0.1	gK2	-10	190 s		m
99607	43750		11 27 56.0	+44 33 58	-0.012	+0.01	6.90	0.4	2.6	F0 V	+17	72 s	8171	m
99695	202477		11 27 56.1	-35 32 58	-0.002	0.00	7.7	0.8	0.2	K0 III		320 s		
99648	118875	84 τ Leo	11 27 56.2	+2 51 22	+0.001	-0.02	4.95	1.00	-0.9	G8 II-III	-9	150 s		m
99606	43751		11 27 57.1	+44 52 39	+0.003	-0.05	6.8	1.1	0.2	K0 III		210 s		
99785	--		11 27 57.8	-66 11 00			7.70	0.11	0.4	B9.5 V		280 s		
99712	202478		11 27 58.4	-35 19 43	-0.001	+0.01	6.4	1.4	-0.1	K2 III		150 s		
99647	99616		11 28 02.0	+11 58 23	-0.001	-0.02	6.6	1.1	-0.1	K2 III		220 s		
99665	138220		11 28 03.5	-0 53 52	0.000	-0.03	7.1	0.1	0.6	A0 V		200 s		
99743	202486		11 28 08.2	-37 05 08	-0.003	0.00	7.5	0.2	0.4	B9.5 V		180 s		
99620	28026		11 28 11.4	+55 40 05	+0.001	-0.01	8.00	0.1	0.6	A0 V		300 s	8173	m
99722	156698		11 28 13.0	-15 53 08	-0.004	+0.03	7.2	0.1	0.6	A0 V		210 s		
99783	239095		11 28 14.2	-52 03 06	-0.007	-0.02	7.5	1.2	0.2	K0 III		230 s		
99872	256843		11 28 18.2	-72 28 28	-0.008	-0.03	6.09	0.16		B3				m
99766	222811		11 28 20.6	-43 27 30	-0.003	-0.01	8.0	1.5	0.2	K0 III		190 s		
99871	256844		11 28 21.3	-71 07 05	-0.012	-0.02	7.4	0.4	3.0	F2 V		75 s		
99754	179915		11 28 23.3	-23 49 38	-0.002	-0.06	7.5	0.8	-0.1	K2 III		240 mx		
99857	251417		11 28 26.8	-66 29 21	-0.004	0.00	7.53	0.15		B0.5 III		590 s		
99781	202488		11 28 29.6	-31 51 24	-0.004	0.00	7.9	-0.1	1.4	A2 V		160 mx		
99706	43758		11 28 30.1	+43 57 58	+0.004	-0.11	7.6	1.1	0.2	K0 III		140 mx		
99823	239097		11 28 34.4	-58 08 31	-0.002	0.00	7.2	0.4	0.6	A0 V		110 s		
99803	222813		11 28 35.0	-42 40 27	-0.004	0.00	5.08	-0.03		B9	+3			m
99720	62571		11 28 35.7	+39 18 31	-0.003	+0.03	7.90	1.1	0.2	K0 III		210 s	8175	m
99804	222814		11 28 37.2	-45 08 28	-0.004	-0.04	8.00	0.61	4.4	G0 V		53 s		m
99845	239099		11 28 38.2	-54 53 01	-0.001	-0.03	7.10	0.1	1.7	A3 V		120 s		m
99819	202492		11 28 42.5	-39 49 14	-0.003	-0.02	7.8	0.4	3.2	G5 IV		84 s		
99789	118884		11 28 47.0	+1 12 13	-0.002	-0.01	7.9	1.1	0.2	K0 III		340 s		
99799	179921		11 28 48.9	-22 23 45	-0.001	-0.03	7.66	1.39		K0				
99761	43764		11 28 51.0	+42 47 34	-0.003	-0.07	7.3	1.1	0.2	K0 III		150 mx		
99814	156703		11 28 52.1	-16 35 41	-0.002	-0.01	7.5	1.6	-0.5	M2 III		390 s		
99748	43763		11 28 52.4	+48 13 51	-0.004	+0.05	8.0	0.9	3.2	G5 IV		93 s		
99854	222815		11 28 55.6	-43 28 48	-0.001	-0.03	7.1	1.1	0.2	K0 III		220 s		
99775	43765		11 28 58.6	+41 10 42	+0.001	+0.04	7.7	0.8	3.2	G5 IV		80 s		
99719	15518		11 29 01.5	+63 34 24	+0.013	+0.02	7.7	0.5	3.4	F5 V		71 s		
99893	239105		11 29 03.9	-58 47 32	-0.001	-0.02	7.60	0.1	0.6	A0 V		250 s		m
99787	62572	57 UMa	11 29 04.1	+39 20 13	-0.005	+0.01	5.31	0.01	1.2	A1 V	-11	210 s	8175	m
99747	15520		11 29 04.4	+61 46 42	-0.017	+0.24	5.83	0.36	2.6	F0 V	-8	43 s		
99737	15519		11 29 09.6	+67 27 28	-0.007	-0.02	7.2	1.1	0.2	K0 III		260 s		
99865	138230		11 29 14.3	-4 27 06	+0.001	-0.10	7.90	0.5	4.0	F8 V		60 s	8178	m
99953	251422		11 29 15.4	-63 33 14	+0.001	-0.01	6.49	0.30	-6.8	B2 Ia		2400 s		
99832	62574		11 29 18.7	+30 25 29	+0.001	-0.14	7.20	0.47	3.4	F5 V	-19	56 s	8177	m
99864	138231		11 29 19.7	-0 20 34	-0.001	0.00	7.9	1.4	-0.3	K5 III		430 s		
99939	239111		11 29 21.4	-57 49 57	+0.001	0.00	7.23	0.04	-6.3	B2 I k		3500 s		
99873	138233		11 29 24.2	-0 50 55	-0.003	+0.02	7.40	1.4	-0.3	K4 III	-15	350 s		
99760	15523		11 29 24.5	+66 55 30	-0.012	-0.03	7.5	0.4	3.0	F2 V		81 s		
99905	118891		11 29 35.8	+3 46 46	-0.003	-0.01	7.9	1.1	0.2	K0 III		340 s		
99923	179934		11 29 37.5	-28 01 50	-0.001	-0.01	6.7	0.9	0.2	K0 III		200 s		
99918	138234		11 29 37.6	-10 03 46	-0.004	-0.02	7.2	0.1	1.7	A3 V		99 mx		
99904	118892		11 29 39.9	+7 35 58	-0.001	-0.03	6.7	0.5	3.4	F5 V	+2	47 s		
99902	99629	85 Leo	11 29 41.8	+15 24 48	-0.002	-0.05	5.74	1.32	0.2	K0 III	-29	82 s		
99952	202508		11 29 42.2	-38 27 25	0.000	-0.01	6.8	0.7	4.4	G0 V		25 s		
99859	28035		11 29 43.5	+56 44 15	-0.011	-0.04	6.28	0.15		A2	+9			
99934	179938		11 29 45.3	-25 00 06	0.000	0.00	7.4	0.7	4.4	G0 V		33 s		
100027	251425		11 29 47.9	-67 04 48	+0.003	-0.09	7.9	0.0	0.4	B9.5 V		82 mx		
100004	239117		11 29 48.1	-51 39 46	-0.037	+0.08	7.38	0.41	3.7	F6 V		47 mx		
99951	179939		11 29 49.8	-24 29 03	-0.002	+0.02	5.80	0.4	3.0	F2 V		36 s	8183	m
99915	81853		11 29 51.8	+24 56 50	-0.008	0.00	7.9	0.5	3.4	F5 V		81 s		
100015	239118		11 29 54.8	-55 19 05	-0.002	+0.02	7.70	1.1	-0.1	K2 III		350 s		m
99969	156713		11 29 57.6	-17 46 11	-0.001	-0.04	7.77	1.61		Ma				
99970	156714		11 29 59.3	-17 53 10	-0.006	-0.04	7.7	1.1	0.2	K0 III		210 mx		
100003	222821		11 30 00.4	-44 41 27	-0.005	0.00	7.3	1.8	-0.1	K2 III		130 s		
99947	81854		11 30 01.7	+24 53 04	-0.003	+0.01	7.72	1.01	0.2	K0 III		320 s		
99946	62579		11 30 04.1	+29 57 52	-0.008	-0.20	6.92	0.35		F0 n	-7			m
99957	81855		11 30 07.3	+25 18 13	-0.004	-0.02	7.71	1.37	-0.2	K3 III		280 mx		
99913	28038		11 30 12.8	+54 21 41	+0.001	-0.06	6.5	0.8	3.2	G5 IV	-22	45 s		

HD	SAO	Star Name	α 2000	δ 2000	μ(α)	μ(δ)	V	B-V	M_v	Spec	RV	d(pc)	ADS	Notes
			$11^h 30^m 16.4^s$	$+37°57'13''$	$+0.006^s$	$-0.03''$								
99956	62582		11 30 16.4	+37 57 13	+0.006	-0.03	7.2	1.1	0.2	K0 III		260 s		
99998	138238	87 Leo	11 30 18.8	-3 00 13	+0.001	-0.02	4.77	1.54	-0.3	K4 III	+19	85 s		
99954	43781		11 30 19.7	+47 14 39	+0.003	-0.03	7.46	0.96	0.2	K0 III		280 s		
100012	179943		11 30 20.6	-25 47 54	-0.003	+0.01	6.6	0.5	0.2	K0 III		190 s		
99967	43784		11 30 24.9	+46 39 27	+0.001	+0.03	6.35	1.27	5.9	K0 V	+27	12 s		
100078	239125		11 30 25.5	-59 37 39	0.000	-0.02	7.9	1.2	-0.1	K2 III		390 s		
99995	43786		11 30 28.4	+43 34 42	-0.025	+0.07	6.7	1.1	0.2	K0 III		69 mx		
100006	99637	86 Leo	11 30 29.0	+18 24 35	-0.006	+0.01	5.52	1.05	0.2	K0 III	+27	110 s		
99966	43785		11 30 29.7	+48 56 07	+0.001	-0.02	7.1	0.0	0.4	B9.5 V		220 s		
99984	43787	58 UMa	11 30 31.0	+43 10 23	-0.005	+0.08	5.94	0.49	4.0	F8 V	-30	24 s		
100043	156722		11 30 38.7	-13 03 01	+0.002	-0.06	7.0	0.4	2.6	F0 V		77 s		
100126	251431		11 30 40.6	-62 47 11	+0.003	+0.01	7.8	0.0	0.4	B9.5 V		300 s		
99983	28041		11 30 42.9	+56 44 47	+0.002	+0.02	7.0	0.4	2.6	F0 V	-3	76 s		
100074	202521		11 30 43.9	-39 14 06	+0.001	+0.01	7.1	1.1	3.2	G5 IV		39 s		
100018	43789		11 30 49.8	+41 17 13	+0.008	-0.09	6.95	0.47	2.8	dF1	-2	59 s	8189	m
100070	138243		11 30 50.9	-6 43 09	+0.002	-0.01	7.70	0.5	4.0	F8 V	+4	55 s	8190	m
100135	251433		11 30 51.4	-60 19 03	-0.002	+0.01	7.27	0.00	0.4	B9.5 V		240 s		
100133	239132		11 30 52.7	-50 56 34	-0.004	-0.02	8.00	0.9	3.2	G5 IV		90 s		m
100030	43790		11 30 52.8	+47 55 44	-0.023	-0.08	6.42	0.88	3.2	G8 IV	+38	44 s		
100041	81863		11 30 54.8	+28 27 05	-0.007	0.00	6.72	1.51		M III	+85			
100119	222832		11 30 56.1	-41 55 36	-0.004	-0.03	7.2	1.0	0.0	B8.5 V		66 s		
100147	239134		11 30 57.8	-52 36 27	-0.001	-0.02	7.8	0.9	0.6	A0 V		71 s		
100067	62584		11 31 00.9	+32 19 01	-0.026	+0.21	7.3	0.5	3.4	F5 V		40 mx		
100118	202524		11 31 02.6	-37 44 58	-0.001	-0.03	7.8	1.4	-0.1	K2 III		290 s		
100132	222833		11 31 04.6	-43 33 11	-0.004	+0.01	8.00	1.1	-0.1	K2 III		360 s		m
--	15530		11 31 07.3	+66 17 17	0.000	+0.01	7.9	2.3						
100145	222835		11 31 09.4	-44 58 05	-0.002	-0.01	7.1	0.6	3.2	G5 IV		60 s		
100055	43793		11 31 10.2	+48 47 21	-0.003	-0.04	6.56	0.93	0.2	G9 III	+6	190 s		
100198	251437		11 31 15.0	-61 16 42	-0.001	0.00	6.38	0.52	-6.6	A0 I		2100 s		
100054	28043		11 31 18.4	+59 42 03	0.000	0.00	7.20	0.2	2.1	A5 V	-15	110 s	8191	m
100128	81865		11 31 21.2	+22 49 03	-0.003	-0.04	7.81	0.21	1.7	A3 V		140 s		
100157	202527		11 31 22.1	-33 43 03	-0.001	+0.03	8.0	2.5	-0.3	K5 III		110 s		
100129	118902		11 31 23.2	+3 03 45	+0.001	0.00	8.00	0.4	2.6	F0 V		120 s	8193	m
100029	15532	1 λ Dra	11 31 24.2	+69 19 52	-0.007	-0.02	3.84	1.62	-0.4	M0 III	+7	65 s		Giausar
100190	222838		11 31 24.6	-41 32 43	-0.009	+0.05	8.0	0.9	0.2	K0 III		160 mx		
100191	222839		11 31 24.7	-44 37 42	-0.002	-0.01	7.1	0.7	1.4	A2 V		53 s		
100082	28044		11 31 27.2	+57 08 05	0.000	-0.01	7.8	1.4	-0.3	K5 III		420 s		
100150	99645		11 31 31.7	+17 44 59	0.000	-0.04	7.12	0.00	0.6	A0 V		200 s		
100209	222844		11 31 31.7	-46 05 37	-0.002	+0.01	7.9	0.6	0.6	A0 V		110 s		
100282	--		11 31 35.3	-67 44 05			7.9	0.4	2.6	F0 V		110 s		
100228	222845		11 31 35.5	-44 44 18	-0.002	+0.01	6.8	1.7	-0.3	K5 III		200 s		
100149	62586		11 31 39.4	+30 58 22	-0.002	0.00	8.03	0.89	5.2	G5 V		25 s		
100168	118905		11 31 40.0	+7 51 48	-0.002	-0.02	7.90	1.1	0.2	K0 III		350 s	8195	m
100180	99647	88 Leo	11 31 44.4	+14 22 07	-0.021	-0.17	6.20	0.57	3.8	dF7	-4	31 ts	8196	m
100261	239145	o¹ Cen	11 31 46.1	-59 26 32	0.000	-0.01	5.13	1.08	-8.0	G0 Ia	-20	3000 s		m
100219	179964		11 31 47.6	-20 46 34	-0.009	+0.04	6.24	0.54	3.4	F5 V		32 s		
100359	256849		11 31 48.0	-73 54 09	0.000	0.00	6.88	0.26		B9				m
100276	251443		11 31 48.1	-60 36 23	0.000	-0.01	7.25	0.04		B0.5 I k				
100262	239146	o² Cen	11 31 48.6	-59 30 56	-0.002	0.00	5.15	0.49	-7.5	A2 Ia	-17	2100 s		
100277	251444		11 31 48.9	-60 41 37	+0.001	-0.01	7.90	0.05	0.6	A0 V		280 s		
99945	1884		11 31 50.3	+81 07 38	-0.066	+0.03	6.1	0.2		A m	+3	15 mn		
100278	251445		11 31 50.3	-61 46 24	-0.002	-0.01	7.69	0.74		G0				
100179	81868		11 31 51.5	+24 18 41	-0.004	-0.01	7.13	1.36	-0.3	K4 III		310 s		
100167	43797		11 31 53.8	+41 26 23	-0.005	+0.10	7.4	0.5	4.0	F8 V		48 s		
100238	138251		11 31 56.6	-6 28 14	-0.002	-0.09	6.8	1.1	0.2	K0 III	+4	150 mx		
100237	138252		11 31 59.6	-1 46 57	-0.006	0.00	7.6	0.1	0.6	A0 V		99 mx		
100204	62588		11 32 00.0	+30 14 19	0.000	-0.01	7.74	1.15		K1 IV				
100236	99650		11 32 03.6	+18 09 18	+0.001	-0.03	7.9	1.6	-0.5	M2 III		490 s		
100235	62590		11 32 12.6	+36 14 50	0.000	-0.03	7.40	1.1	0.2	K0 III		280 s	8198	m
100270	156739		11 32 15.3	-12 23 48	-0.006	0.00	7.5	1.1	-0.1	K2 III		240 mx		
100287	179967		11 32 16.0	-29 15 48	-0.001	+0.13	5.8	0.3	4.4	G0 V	+4	19 s	8202	m
100335	251449		11 32 16.8	-60 51 10	-0.001	+0.02	7.86	-0.05	-1.9	B6 III		790 s		
100214	28050		11 32 19.7	+56 05 42	-0.031	+0.07	8.00	0.5	4.0	F8 V	+13	54 mx	8199	m
100382	251451		11 32 20.0	-66 57 43	0.000	-0.01	5.90	1.13	0.2	gK0		110 s		
100203	15542		11 32 20.8	+61 04 57	0.000	-0.07	5.48	0.50	3.7	F6 V	-46	22 ts	8197	m
100307	179969		11 32 23.2	-26 44 47	-0.006	+0.03	6.16	1.66		Ma				
100255	81872		11 32 25.5	+29 02 43	-0.001	-0.09	7.9	0.5	3.4	F5 V	+14	78 s		
100317	179972		11 32 31.5	-29 08 31	-0.002	-0.01	7.5	1.7	-0.5	M2 III		360 s		
100380	239153		11 32 32.4	-59 43 04	-0.004	-0.01	7.0	0.0	1.4	A2 V		130 s		
100412	239157		11 32 44.4	-60 01 45	-0.005	+0.01	8.0	1.1	0.6	A0 V		67 s		
100343	138265		11 32 47.4	-7 49 39	-0.001	0.00	5.95	1.38	-0.3	gK4	-1	180 s		
100378	222856		11 32 48.0	-40 26 10	-0.006	+0.06	5.64	1.58		Ma				
100339	99655		11 32 49.7	+18 00 07	-0.010	0.00	7.3	0.5	3.4	F5 V		60 s		
100311	62596		11 32 50.1	+38 51 51	-0.001	-0.01	8.0	0.1	1.7	A3 V		180 s		
100395	202553		11 32 51.1	-36 12 33	-0.017	-0.04	6.69	0.60	4.4	G0 V		27 s		m

HD	SAO	Star Name	α 2000	δ 2000	μ(α)	μ(δ)	V	B-V	M_v	Spec	RV	d(pc)	ADS	Notes
100393	202554		11^{h}32^{m}54.0^s	-31°05'14"	-0.003^s	0.00"	5.04	1.58	-0.5	gM2	+19	130 s		
99884	--		11 32 57.9	+84 42 24	+0.005	+0.02	7.3	1.1	0.2	K0 III		260 s		
100375	179979		11 32 57.9	-23 26 37	-0.004	+0.01	7.30	0.1	1.4	A2 V		150 s	8212	m
100407	202558	ξ Hya	11 33 00.1	-31 51 27	-0.016	-0.04	3.54	0.94	0.3	G7 III	-5	44 s		m
100338	62598		11 33 02.0	+34 03 05	-0.001	-0.01	7.20	1.32	-0.1	K2 III		220 s		
100298	15549		11 33 03.5	+62 05 36	-0.004	-0.02	8.0	1.1	0.2	K0 III		360 s		
100441	222860		11 33 07.5	-43 44 22	-0.002	-0.02	8.0	0.5	1.4	A2 V		100 s		
100372	62602		11 33 11.9	+33 07 58	-0.002	-0.06	8.0	0.5	4.0	F8 V		62 s		
100418	156750		11 33 14.6	-16 16 50	0.000	-0.05	6.05	0.60	0.6	gG0	-4	120 s		
100508	--		11 33 14.9	-67 03 07			7.74	0.83		K0				
100416	118917		11 33 19.7	+3 21 45	-0.001	-0.01	7.9	1.1	0.2	K0 III		340 s		
100546	251457		11 33 25.1	-70 11 43	-0.010	-0.02	6.6	-0.3	0.4	B9.5 V		180 s		
100447	99662		11 33 29.4	+19 40 47	-0.001	-0.02	7.01	1.28	-0.1	K2 III		230 s		
100456	118923		11 33 36.2	+2 29 56	+0.001	-0.03	6.7	1.4	-0.3	K5 III		250 s		
100493	222863		11 33 37.2	-40 35 12	-0.007	+0.02	5.39	0.12		A2	+9	15 mn		m
100715	256852		11 33 42.0	-80 15 07	-0.012	-0.05	8.0	0.1	1.4	A2 V		120 mx		
100470	62604		11 33 56.2	+36 48 55	-0.011	-0.06	6.40	1.05	0.2	K0 III	+18	110 mx		
100486	99666		11 33 56.8	+13 33 07	0.000	-0.03	8.0	1.1	-0.1	K2 III		410 s		
100537	156754		11 34 04.0	-16 02 44	+0.003	0.00	7.2	1.1	-0.1	K2 III		280 s		
100613	239177		11 34 06.9	-59 40 54	-0.006	0.00	8.0	-0.5	3.4	F5 V		82 s		
100446	15554		11 34 07.1	+65 14 35	-0.007	-0.20	7.3	0.5	4.0	F8 V	-31	37 mx		
100518	99668		11 34 09.9	+11 01 25	+0.003	-0.04	6.55	0.18	2.1	A5 V	-5	76 s		
100638	251460		11 34 12.8	-65 24 35	-0.001	0.00	7.17	0.11	0.6	A0 V		210 s		
100569	202578		11 34 13.3	-39 56 22	0.000	+0.02	7.8	2.1	-0.3	K5 III		180 s		
100587	222874		11 34 16.0	-41 09 17	+0.001	-0.02	8.00	0.1	0.6	A0 V		81 s		m
100637	239180		11 34 20.4	-57 37 38	-0.001	-0.01	8.0	0.8	0.2	K0 III		360 s		
100565	138275		11 34 21.4	-5 31 37	-0.006	+0.01	6.7	0.1	1.4	A2 V		86 mx		
100586	202580		11 34 21.5	-36 12 18	0.000	-0.03	7.8	0.8	-0.1	K2 III		380 s		
100563	118929	89 Leo	11 34 21.9	+3 03 36	-0.012	-0.11	5.77	0.46	3.4	dF5	+3	29 s		
100623	202583		11 34 29.4	-32 49 53	-0.054	+0.82	5.98	0.81	6.1	K0 V	-23	9.5 t		
100604	179998		11 34 31.4	-28 17 19	-0.006	-0.04	8.0	0.0	3.4	F5 V		82 s		
100675	239184		11 34 38.0	-57 09 09	-0.002	0.00	7.4	1.6	0.2	K0 III		120 s		
100600	99673	90 Leo	11 34 42.4	+16 47 49	-0.001	0.00	5.95	-0.16	-1.7	B3 V	+19	340 s	8220	m
100673	239189		11 34 45.6	-54 15 50	-0.007	+0.01	4.62	-0.08	0.0	B8.5 V	+4	82 s		
100629	138280		11 34 47.4	-2 29 29	-0.002	+0.02	7.7	1.1	0.2	K0 III		310 s		
100713	251463		11 34 51.2	-60 47 09	-0.001	+0.02	7.4	1.7	-0.5	M4 III		350 s		
100671	222880		11 34 52.2	-46 27 15	-0.002	-0.02	7.7	1.7	-0.1	K2 III		360 s		
100645	138283		11 34 52.6	-2 13 50	-0.003	-0.01	8.0	1.6	-0.5	M2 III		510 s		
100660	156765		11 34 54.8	-12 05 14	-0.004	+0.03	7.4	1.1	0.2	K0 III		250 mx		
100708	222883		11 34 56.8	-49 08 11	-0.018	+0.17	5.50	1.04	0.2	gK0	-1	110 s		
100659	138284		11 34 58.8	-4 21 41	-0.003	-0.05	6.5	1.1	0.2	K0 III		180 s		
100670	180006		11 35 01.4	-30 01 56	-0.008	-0.01	7.2	1.0	0.2	K0 III		210 mx		
100643	62611		11 35 01.6	+30 30 32	-0.001	-0.02	7.39	1.07	3.2	K0 IV		53 s		
100686	202595		11 35 02.9	-30 43 28	-0.007	0.00	7.3	0.7	2.6	F0 V		53 s		
100724	239196		11 35 03.1	-53 14 32	+0.001	+0.01	7.4	0.9	3.2	G5 IV		58 s		
100735	239198		11 35 03.5	-56 07 20	+0.001	-0.01	6.97	0.92	1.8	G5 III-IV		97 s		
100655	81886		11 35 03.7	+20 26 29	-0.004	-0.01	6.45	1.01	0.2	G9 III	-7	170 s		
100615	28064		11 35 04.8	+54 47 07	+0.001	0.00	5.63	1.02	3.2	G5 IV	+18	22 s		
100733	222887		11 35 13.2	-47 22 21	-0.009	-0.01	5.71	1.66	-0.5	M3 III	+18	160 s		m
100773	251466		11 35 13.9	-60 53 41	-0.023	-0.03	6.72	0.31	1.9	F2 IV		90 mx		
100721	202599		11 35 14.0	-35 36 44	-0.003	-0.01	7.0	0.3	1.4	A2 V		100 s		
100667	28065		11 35 25.2	+53 34 26	-0.013	+0.08	7.9	0.9	3.2	G5 IV		88 s		
100772	239201		11 35 27.2	-51 17 39	-0.001	0.00	7.6	1.4	-0.1	K2 III		260 s		
100717	99681		11 35 31.8	+11 11 19	+0.004	-0.03	6.6	1.1	0.2	K0 III		190 s		
100742	156771		11 35 32.9	-12 24 40	-0.002	-0.13	7.0	0.5	3.4	F5 V		53 s		
100678	28066		11 35 34.5	+58 12 10	0.000	-0.02	8.0	1.1	-0.1	K2 III		410 s		
100640	15566		11 35 40.0	+69 34 23	-0.007	0.00	7.5	0.8	3.2	G5 IV		73 s		
100726	99682		11 35 40.8	+17 52 29	-0.001	+0.02	7.27	1.18	0.2	K0 III		180 s		
100826	251469		11 35 42.2	-61 17 16	-0.001	-0.01	6.70	0.1	-6.6	A0 I		3200 s		m
100740	99683		11 35 43.3	+10 54 40	+0.002	-0.02	6.40	0.13	0.1	A4 III	-5	180 s		
--	251470		11 35 45.0	-61 34 40	0.000	0.00	7.0	1.9						
100901	256853		11 35 45.3	-72 50 33	-0.034	+0.03	6.53	1.16		K1 IV		130 mx		
100841	251472	λ Cen	11 35 46.8	-63 01 11	-0.006	-0.01	3.13	-0.04	-0.8	B9 III	+8	57 s		m
100818	239206		11 35 47.7	-52 57 54	-0.001	-0.01	7.9	0.4	3.2	G5 IV		88 s		
100782	202609		11 35 48.2	-34 47 11	-0.002	0.00	8.0	1.5	-0.1	K2 III		270 s		
100825	222895		11 35 55.5	-47 38 30	+0.003	-0.05	5.25	0.25		F2 s	+5	17 mn		
100838	239207		11 35 58.6	-50 44 32	-0.003	+0.01	7.8	0.8	2.6	F0 V		53 s		
100696	15567	2 Dra	11 36 02.8	+69 19 22	+0.021	-0.13	5.3	1.1	0.2	K0 III	-2	87 mx		
100854	222897		11 36 05.9	-48 13 44	-0.003	+0.01	8.0	0.9	0.2	K0 III		360 s		
100850	202615		11 36 11.3	-37 03 52	-0.004	-0.17	7.8	1.1	4.4	G0 V		23 s		
100900	239212		11 36 12.0	-60 00 39	-0.004	+0.06	7.2	1.4	3.2	G5 IV		26 s		
100822	118942		11 36 14.6	+3 18 04	-0.002	-0.04	7.20	0.9	3.2	G5 IV		63 s	8233	m
100808	81893		11 36 17.9	+27 46 51	+0.002	0.00	5.80	0.23	2.4	A8 V	+8	48 s	8231	m
100929	251479		11 36 22.2	-61 03 08	-0.002	0.00	5.83	-0.09	-2.3	B3 IV	+9	380 s		
100930	--		11 36 27.1	-61 19 10			8.0	1.6	-0.5	M2 III		470 s		

HD	SAO	Star Name	α 2000	δ 2000	μ(α)	μ(δ)	V	B-V	M_V	Spec	RV	d(pc)	ADS	Notes
			h m s	° ′ ″	s	″								
100943	251480		11 36 28.3	-61 39 55	-0.001	-0.01	7.14	0.11	-6.3	B5 I		3500 s		
100843	81896		11 36 28.6	+25 01 47	+0.003	-0.01	7.01	0.18	1.7	A3 V		100 s		
100873	156783		11 36 28.6	-16 51 01	-0.012	-0.10	7.0	0.6	4.4	G0 V		33 s		
100872	118946		11 36 34.1	+6 06 24	-0.007	-0.10	7.0	1.1	0.2	K0 III		92 mx		
---	251483		11 36 34.6	-61 36 34	-0.004	+0.01	7.5	1.6	-0.5	M2 III		380 s		
101011	---		11 36 34.9	-73 21 10			8.0	0.1	1.7	A3 V		180 s		
100893	202622		11 36 35.0	-33 34 12	+0.003	-0.04	5.74	1.02		K0				m
100831	28071		11 36 35.3	+56 08 07	-0.022	-0.10	7.90	0.9	3.2	G5 IV	-15	54 mx	8236	m
100894	202623		11 36 35.4	-36 07 18	+0.002	-0.02	6.9	0.5	2.6	F0 V		52 s		
100910	202625		11 36 37.1	-34 38 48	-0.021	+0.11	7.58	0.49		F5		16 mn		
100888	138295		11 36 40.4	-9 18 45	-0.003	+0.01	7.4	0.1	0.6	A0 V		230 s		
100911	202627		11 36 40.7	-37 14 16	-0.001	-0.02	6.31	0.06		A0				
100889	138296	21 θ Crt	11 36 40.8	-9 48 08	-0.005	0.00	4.70	-0.08	0.2	B9 V	+1	79 s		
100908	180033		11 36 45.5	-24 26 21	-0.005	-0.03	7.20	1.1	-0.1	K2 III		250 s	8240	m
100934	180034		11 36 55.8	-20 58 06	+0.001	-0.09	7.9	0.4		K0 III		200 mx		
100920	138298	91 υ Leo	11 36 56.9	-0 49 26	0.000	+0.04	4.30	1.00	0.2	G9 III	+1	64 s		
100954	202633		11 36 57.8	-38 57 34	-0.003	-0.01	6.90	0.1	0.6	A0 V		180 s		m
101007	---		11 36 57.8	-61 10 11			7.8	1.2	-0.5	M2 III		470 s		
100949	180036		11 36 59.8	-22 56 54	-0.003	+0.04	6.6	1.1	0.2	K0 III	+22	170 s		
101021	251486		11 37 00.6	-61 17 00	-0.030	0.00	5.14	1.12	0.0	K1 III	+3	92 mx		
100953	202634		11 37 01.1	-32 59 17	-0.001	-0.08	6.29	0.46	2.8	F5 IV-V		48 s		
100965	202635		11 37 04.4	-38 59 42	+0.004	-0.02	7.3	1.5	-0.1	K2 III		190 s		
101035	251487		11 37 04.5	-62 29 07	+0.003	0.00	7.7	2.3	-0.3	K5 III		130 s		
100917	28076		11 37 08.1	+51 09 15	-0.001	-0.04	7.8	0.8	3.2	G5 IV		84 s		
100432	1902		11 37 08.9	+85 37 02	+0.008	+0.04	7.4	0.4	2.6	F0 V		92 s		
100947	81903		11 37 13.8	+27 46 26	+0.001	-0.01	7.65	1.07	0.0	K1 III		340 s		
101088	251489		11 37 14.3	-69 40 28	-0.011	-0.02	6.7	0.2	3.4	F5 V		47 s		
101132	256857	π Cha	11 37 15.6	-75 53 47	-0.035	0.00	5.65	0.35	1.3	F2 III-IV		76 s		
100975	138302		11 37 16.7	-0 34 42	-0.001	-0.01	7.9	1.1	0.2	K0 III		340 s		
100974	118952		11 37 17.8	+6 16 13	-0.001	+0.01	7.1	0.1	1.4	A2 V		140 s		
100998	156788		11 37 20.0	-12 20 50	+0.003	-0.11	7.1	1.1	0.2	K0 III		230 mx		
101069	251488		11 37 20.6	-60 43 41	+0.001	-0.01	8.0	0.9	0.2	K0 III		370 s		
100933	15571		11 37 24.7	+62 11 48	+0.002	+0.02	7.4	1.6	-0.5	M2 III	-28	380 s		
101053	239231		11 37 25.0	-53 44 19	+0.001	-0.04	7.9	1.1	0.2	K0 III		320 s		
101029	202641		11 37 29.1	-36 43 40	-0.001	-0.01	7.6	0.9	4.0	F8 V		29 s		
101105	251494		11 37 32.1	-61 29 06	-0.004	-0.01	7.14	0.01	-2.5	B2 V n		780 s		m
101048	202642		11 37 32.5	-39 21 32	+0.001	-0.02	6.8	1.9	-0.5	M2 III		200 s		
101067	222917		11 37 33.9	-47 44 49	-0.007	+0.02	5.44	1.24	-0.1	gK2	-1	120 s		
101104	251497		11 37 33.9	-60 54 13	-0.010	-0.01	6.6	1.7	-0.5	M4 III		230 s		
101103	251496		11 37 34.1	-60 21 53	+0.001	0.00	7.3	0.0	0.0	B8.5 V		250 s		
101438	258619		11 37 35.0	-85 29 10	+0.017	+0.02	7.5	0.1	2.6	F0 V		96 s		
100972	43831		11 37 36.7	+44 42 59	-0.002	-0.01	6.85	0.04		B9	+17			
101065	222918		11 37 36.9	-46 42 36	-0.005	+0.02	8.01	0.78		B5				
100960	15573		11 37 37.2	+62 42 47	-0.013	+0.02	7.4	0.4	3.0	F2 V		75 s		
101119	251498		11 37 37.7	-60 59 01	-0.003	-0.02	7.34	0.01	0.4	B9.5 V		230 s		
100858	7395		11 37 42.3	+77 35 45	+0.001	+0.01	6.6	1.4	-0.3	K5 III		240 s		
101080	202646		11 37 46.6	-31 13 06	-0.006	-0.05	7.1	0.4	3.2	G5 IV		61 s		
101131	251500		11 37 48.3	-63 19 23	-0.001	0.00	7.13	0.04		O7 k				
101162	251501		11 37 48.4	-67 37 13	-0.014	-0.02	5.96	1.02	0.3	gG8		120 s		
---	---		11 37 51.2	+28 29 07	-0.006	-0.04	8.0	0.6	4.4	G0 V		51 s		
101013	28081		11 37 52.9	+50 37 04	-0.006	-0.04	6.14	1.07		K0	-4			
101059	81912		11 37 54.0	+21 44 43	+0.001	+0.01	7.8	0.1	1.7	A3 V		170 s		
101095	180049		11 37 55.5	-28 15 23	-0.002	+0.01	7.9	1.8	-0.5	M2 III		360 s		
101093	138313		11 37 59.0	-1 36 00	-0.021	+0.11	7.62	0.55	4.2	F9 V		44 mx		
101078	99705		11 38 00.3	+12 57 27	-0.005	-0.07	7.60	1.4	-0.3	K5 III		140 mx		m
101174	251504		11 38 04.1	-65 39 22	+0.001	-0.01	7.42	0.11		B8.5 V		260 s		
101122	156793		11 38 04.7	-18 11 32	-0.007	0.00	7.2	0.1	0.6	A0 V		83 mx		
101091	62632		11 38 05.5	+31 52 51	-0.003	-0.02	7.2	0.4	3.0	F2 V	-13	68 s		
101189	251505		11 38 07.2	-61 49 35	-0.009	+0.01	5.14	-0.02	-0.3	B9 IV	+4	120 s		
101112	118961		11 38 09.7	+8 53 03	-0.004	+0.01	6.17	1.08	0.0	K1 III	+11	170 s		
101190	251506		11 38 10.1	-63 11 50	+0.001	-0.02	7.32	0.06		O7				m
101158	222924		11 38 10.9	-45 44 59	-0.003	-0.01	7.3	0.3	2.6	F0 V		83 s		
101141	180054		11 38 12.0	-28 17 26	-0.009	-0.02	7.6	0.8	0.2	K0 III		190 mx		
101205	251511		11 38 20.4	-63 22 23	+0.001	-0.01	6.48	0.07		O8				m
101107	43837	59 UMa	11 38 20.5	+43 37 31	-0.014	-0.04	5.59	0.33	2.6	F0 V	+2	39 s		
101169	202656		11 38 20.9	-33 36 22	-0.003	-0.01	7.8	-0.4	0.6	A0 V		270 s		
101222	251508		11 38 21.0	-60 39 14	0.000	0.00	7.7	1.6	0.0	K1 III		170 s		
101075	15578		11 38 22.4	+64 10 36	+0.002	0.00	7.9	0.1	0.6	A0 V		290 s		
101154	138314		11 38 24.0	-2 26 10	-0.002	0.00	6.22	1.12	0.0	gK1	-15	170 s	8247	m
101153	118965	1 ω Vir	11 38 27.5	+8 08 03	-0.001	0.00	5.36	1.57		gM6	+4	30 mn		
101151	62635		11 38 32.2	+33 37 32	-0.002	-0.02	6.27	1.32	-0.1	K2 III	-6	140 s		
101133	43839	60 UMa	11 38 33.4	+46 50 03	-0.004	-0.03	6.10	0.38	3.0	F2 V	-24	38 s		
101288	256859		11 38 38.0	-72 22 09	+0.001	-0.06	7.2	0.1	1.4	A2 V		120 mx		
101198	156802	24 ι Crt	11 38 40.0	-13 12 07	+0.006	+0.11	5.48	0.52	3.4	dF5	-24	23 s		m
101215	202660		11 38 40.5	-39 14 51	0.000	0.00	7.8	0.5	0.6	A0 V		120 s		

HD	SAO	Star Name	α 2000	δ 2000	μ(α)	μ(δ)	V	B-V	M_V	Spec	RV	d(pc)	ADS	Notes
			$h \quad m \quad s$	$\circ \quad ' \quad ''$	s	''								
101179	81919		11 38 44.1	+23 19 50	+0.001	+0.02	7.55	0.96	0.2	K0 III		300 s		
101177	43841		11 38 44.8	+45 06 31	−0.057	+0.02	6.27	0.61	6.9	K3 V	−18	11 ts	8250	m
101233	222931		11 38 47.1	−40 18 11	−0.004	−0.02	6.9	1.3	0.2	K0 III		140 s		
101247	222932		11 38 47.5	−47 51 06	−0.006	−0.03	7.5	0.8	0.2	K0 III		270 mx		
101150	15580		11 38 49.1	+64 20 49	+0.002	0.00	6.80	0.1	1.4	A2 V	−22	120 s	8249	m
101209	138316		11 38 49.3	−7 36 00	−0.004	+0.01	8.0	0.4	3.0	F2 V		98 s		
101178	62640		11 38 49.8	+39 10 21	−0.001	+0.01	7.28	1.61	−0.5	M1 III	−36	340 s		m
101245	180064		11 38 53.9	−21 10 27	−0.005	−0.02	7.8	0.3	1.4	A2 V		83 mx		
101265	202664		11 38 56.3	−39 51 37	−0.006	+0.01	7.8	1.2	−0.1	K2 III		270 mx		
101194	43843		11 38 56.4	+40 15 08	−0.007	+0.01	7.6	0.5	3.4	F5 V		71 s		
101259	180065		11 39 00.3	−24 43 15	+0.001	−0.24	6.42	0.82		G5				
101242	118971		11 39 01.0	+6 03 30	−0.022	−0.11	7.61	0.69	3.2	G5 IV		44 mx		
101271	156810		11 39 05.6	−16 36 44	0.000	+0.02	8.0	1.6	−0.5	M2 III		510 s		
101314	239258		11 39 09.1	−58 35 51	−0.001	0.00	7.60	0.8	3.2	G5 IV		75 s		m
101330	251516		11 39 10.0	−62 25 18	+0.001	+0.01	7.31	−0.02	0.6	A0 V		220 s		
101241	99714		11 39 10.2	+17 50 18	0.000	−0.02	7.6	0.5	3.4	F5 V		71 s		
101312	239261		11 39 10.5	−57 44 22	−0.003	−0.01	7.70	0.1	0.6	A0 V		260 s		m
101332	251519		11 39 16.7	−62 56 00	+0.002	−0.02	7.4	0.1		B0.5 II				
101307	180071		11 39 22.3	−24 53 46	+0.001	+0.01	7.5	1.2	0.2	K0 III		230 s		
101328	239262		11 39 22.9	−51 25 28	0.000	0.00	7.4	1.6	−0.1	K2 III		180 s		
101289	81924		11 39 28.4	+25 18 05	−0.007	−0.04	7.74	0.58	4.4	G0 V		45 s	8255	m
101379	251522		11 39 29.4	−65 23 52	−0.006	−0.01	5.17	0.80	0.3	G8 III p	+4	33 s		
101321	156815		11 39 30.0	−13 11 03	+0.005	−0.15	6.7	1.1	0.2	K0 III		160 mx		
101302	99718		11 39 36.4	+18 59 48	−0.004	−0.01	7.00	0.5	3.4	F5 V		53 s	8257	m
101348	202677		11 39 36.4	−35 14 00	−0.020	0.00	7.82	0.72	3.2	G5 IV		67 mx		
101341	138326		11 39 39.5	−9 27 59	−0.001	−0.02	7.4	1.1	0.2	K0 III		280 s		
101358	202681		11 39 45.3	−34 58 56	0.000	−0.01	7.8	1.5	−0.3	K5 III		410 s		
101388	222937		11 39 46.3	−45 58 40	−0.005	0.00	7.9	0.3	2.1	A5 V		110 s		
101319	43853		11 39 49.8	+45 09 24	+0.002	−0.01	8.0	0.4	3.0	F2 V			8250	m
101436	251525		11 39 49.9	−63 28 42	−0.001	+0.01	7.70	0.12		B3				m
101370	156819		11 39 50.3	−16 37 13	−0.001	−0.01	6.4	1.6		Ma	+26			
101369	156820		11 39 51.0	−14 28 07	−0.003	−0.02	6.21	0.00		A0			8259	m
101408	222938		11 39 52.9	−45 48 15	−0.003	−0.10	7.25	1.04	3.2	G8 IV		52 s		
101410	239272		11 39 53.6	−50 29 10	−0.005	0.00	6.80	0.1	0.6	A0 V		150 mx		m
101387	202686		11 39 57.0	−33 27 01	−0.003	−0.05	6.90	1.1	0.2	K0 III		220 s		m
101406	202689		11 39 58.4	−38 06 30	−0.005	−0.02	6.6	0.8	3.2	G5 IV		46 s		
101404	202690		11 40 01.2	−30 28 48	0.000	−0.01	7.90	1.1	0.2	K0 III		350 s		m
101465	239273		11 40 04.4	−59 00 06	−0.007	0.00	7.6	0.4	3.2	G5 IV		77 s		
101366	43854		11 40 04.8	+42 01 32	−0.002	−0.02	7.57	1.54	−0.4	M0 III		390 s		
101466	251526		11 40 06.9	−61 11 08	0.000	−0.01	7.4	−0.1	0.4	B9.5 V		250 s		
101397	81930		11 40 09.3	+23 42 30	−0.005	0.00	6.8	0.1	1.7	A3 V		110 s		
101431	202695	o Hya	11 40 12.9	−34 44 40	0.000	0.00	4.70	−0.07		B9	+6			
101453	222943		11 40 13.4	−43 44 46	−0.003	0.00	7.9	2.3	−0.3	K5 III		140 s		
101452	202696		11 40 13.7	−39 08 48	−0.001	−0.03	7.6	0.7	1.4	A2 V		73 s		
101498	251528		11 40 16.1	−62 25 56	0.000	0.00	8.0	0.4	0.2	K0 III		360 s		
101430	180083		11 40 16.8	−28 29 47	−0.001	−0.01	7.2	1.6	−0.5	M2 III		330 s		
101365	15589		11 40 18.5	+62 23 47	−0.016	−0.04	7.1	0.5	3.4	F5 V		54 s		
101531	251529		11 40 20.5	−69 40 19	−0.002	−0.03	7.4	1.1	0.2	K0 III		270 s		
101445	118981		11 40 24.0	+0 57 08	0.000	−0.01	6.80	0.1	1.7	A3 V		110 s	8261	m
101391	28093		11 40 27.3	+57 58 14	−0.002	+0.02	6.37	−0.12		A p	+4			
101443	118983		11 40 29.2	+5 08 24	+0.001	−0.01	7.9	0.5	3.4	F5 V		78 s		
101527	239279		11 40 32.7	−56 19 11	0.000	+0.01	7.4	1.6	0.2	K0 III		120 s		
101472	138340		11 40 36.5	−8 24 20	−0.001	−0.01	7.5	0.6	4.4	G0 V		41 s		
101545	251533		11 40 36.9	−62 34 09	0.000	−0.04	6.38	0.2	−5.9	O9.5 Ib		2300 s		m
101523	222951		11 40 38.3	−49 30 34	−0.006	−0.03	8.0	0.4	0.6	A0 V		110 mx		
101541	239284		11 40 42.6	−53 58 06	0.000	−0.02	5.96	1.66	−0.4	gM0		160 s		
101484	81941	92 Leo	11 40 47.0	+21 21 10	−0.004	−0.05	5.26	0.98	0.2	K0 III	+9	100 s		
101569	251534		11 40 49.6	−60 58 50	0.000	−0.01	7.1	1.9	−0.3	K5 III		170 s		
101439	15594		11 40 50.5	+60 06 22	+0.003	−0.01	7.8	0.4	2.6	F0 V		110 s		
101517	138343		11 40 50.9	−9 55 02	+0.001	0.00	7.4	1.1	0.2	K0 III		270 s		
101568	239287		11 40 52.9	−55 05 53	−0.002	+0.01	7.7	1.4	0.2	K0 III		190 s		
101570	251535		11 40 53.6	−62 05 23	−0.002	0.00	4.94	1.15	−6.2	cG6	+14	1700 s		
101514	138346		11 40 55.8	−0 40 27	+0.001	−0.02	7.1	1.1	−0.1	K2 III		270 s		
101584	239288		11 40 58.8	−55 34 27	−0.001	−0.02	7.01	0.39	−8.5	F0 Ia pe		6400 s		
101582	239289		11 41 00.3	−54 32 56	−0.001	−0.01	7.8	−0.3	0.6	A0 V		270 s		
101581	222956		11 41 02.3	−44 24 21	−0.063	+0.22	7.77	1.07	8.0	K5 V		11 ts		
101601	251539		11 41 02.6	−60 55 24	−0.007	−0.01	7.4	1.3	0.2	K0 III		190 s		
101782	258621		11 41 02.6	−83 05 59	−0.022	+0.01	6.33	1.08	0.3	gG8		130 s		m
101501	62655	61 UMa	11 41 02.9	+34 12 05	−0.001	−0.39	5.33	0.72	5.5	G8 V	−5	9.1 t		m
101566	222955		11 41 05.7	−40 53 06	+0.002	−0.02	8.0	1.0	−0.3	K5 III		470 s		
101483	15600		11 41 05.8	+60 50 34	−0.007	−0.02	7.2	0.8	3.2	G5 IV		62 s		
101638	—		11 41 06.3	−71 09 14			7.9	0.1	1.4	A2 V		190 s		
101563	180098		11 41 08.3	−29 11 47	−0.025	+0.21	6.44	0.66	4.4	dG0	−20	22 s		
101560	156832		11 41 09.5	−17 42 20	+0.002	−0.06	7.0	1.1	0.2	K0 III		230 s		
101629	251540		11 41 10.4	−61 08 26	−0.005	−0.02	7.30	1.1	0.2	K0 III		260 s		m

HD	SAO	Star Name	α 2000	δ 2000	μ(α)	μ(δ)	V	B-V	M$_v$	Spec	RV	d(pc)	ADS	Notes
			h m s	° ′ ″	s	″								
101575	156833		11 41 10.8	-16 41 00	+0.001	-0.03	7.7	1.4	-0.3	K5 III		390 s		
101615	222960		11 41 19.7	-43 05 45	-0.008	0.00	5.7	0.1	0.6	A0 V	+8	66 mx		
101549	43861		11 41 20.2	+44 00 28	-0.003	+0.01	7.8	0.1	1.4	A2 V	-9	190 s		
101613	202714		11 41 22.3	-35 36 14	-0.006	-0.02	7.1	0.4	1.7	A3 V		83 s		
101612	180103		11 41 22.4	-26 40 03	-0.016	+0.02	7.54	0.47	4.0	F8 V		51 s		
101614	222962		11 41 26.4	-41 01 06	+0.017	-0.13	6.87	0.58	4.4	G0 V		31 s		
101727	256862		11 41 31.9	-77 03 16	0.000	-0.01	6.93	0.42		F2				
101606	62658	62 UMa	11 41 34.2	+31 44 45	-0.027	+0.02	5.73	0.43	3.3	F4 V	+32	34 ts		m
101713	—		11 41 36.4	-71 30 15			7.9	0.0	0.4	B9.5 V		310 s		
101684	251542		11 41 38.1	-63 49 45	0.000	0.00	7.23	0.94	3.2	G5 IV		47 s		
101585	43863		11 41 40.1	+44 11 42	0.000	-0.01	7.93	1.58	-0.4	M0 III	0	450 s		
101604	28101		11 41 43.5	+55 10 20	-0.002	+0.02	6.27	1.49	-0.3	K5 III	-7	210 s		
101666	202717		11 41 43.9	-32 29 59	0.000	-0.05	5.22	1.48	-0.3	K5 III	+34	130 s		m
101645	138352		11 41 44.6	-9 04 36	-0.014	+0.04	7.3	0.9	3.2	G5 IV		67 s		
101682	239298		11 41 45.2	-50 37 32	-0.008	+0.04	7.5	0.7	2.6	F0 V		56 s		
101620	43864		11 41 46.2	+41 14 14	+0.003	-0.05	6.9	0.5	3.4	F5 V	-8	49 s		
101665	202718		11 41 47.4	-32 29 10	-0.002	-0.04	5.30	0.6	4.4	G0 V		15 s		m
101712	251544		11 41 49.3	-63 24 52	-0.002	+0.01	7.86	1.78	-4.8	M2 Ib p		2600 s		
101711	239302		11 41 53.5	-60 13 51	-0.003	-0.01	7.2	0.2	2.6	F0 V		85 s		
101697	222965		11 41 54.9	-44 06 13	-0.005	0.00	8.0	0.3	2.6	F0 V		120 s		
101724	251545		11 41 55.2	-63 47 44	0.000	-0.01	8.03	0.01	0.4	B9.5 V		340 s		
101696	180116		11 41 57.8	-24 23 08	-0.002	-0.02	6.9	0.6	2.6	F0 V		51 s		
101708	222966		11 41 58.9	-46 21 02	-0.004	-0.01	7.9	1.0	0.2	K0 III		270 s		
101694	156839		11 42 02.6	-17 00 59	-0.005	+0.03	7.6	0.5	3.4	F5 V		68 s		
101695	180117		11 42 03.4	-20 17 38	-0.001	-0.05	6.22	0.95		K0				
101676	99743		11 42 03.9	+12 17 06	+0.001	-0.13	7.2	0.5	3.4	F5 V		56 s		
101688	81949		11 42 05.2	+22 12 39	-0.007	-0.07	6.6	0.4	3.0	F2 V	-23	53 s		
101690	119000		11 42 07.2	+4 44 51	-0.015	+0.04	7.6	0.7	4.4	G0 V		44 s		
101805	256864		11 42 14.9	-75 13 37	+0.011	0.00	6.47	0.53	4.5	G1 V		25 s		
101763	239306		11 42 16.7	-57 33 53	-0.006	+0.01	7.3	0.1	0.6	A0 V		140 mx		
101779	251546		11 42 18.9	-60 57 42	0.000	0.00	7.3	2.3	-0.1	K2 III		68 s		
101730	119003		11 42 25.5	+2 21 44	-0.003	-0.04	7.0	0.5	3.4	F5 V	+5	52 s		
101795	251547		11 42 26.2	-63 08 33	0.000	-0.02	7.33	0.06	0.0	B8.5 V		290 s		
101673	15606	3 Dra	11 42 28.3	+66 44 41	-0.008	+0.03	5.30	1.28	-0.2	K3 III	+3	130 s		
101758	180127		11 42 35.3	-29 38 22	+0.005	-0.05	7.6	1.5	0.2	K0 III		160 s		
101814	251548		11 42 37.7	-60 25 14	+0.003	0.00	7.6	1.8	-0.3	K5 III		250 s		
101753	99749		11 42 43.9	+18 14 31	-0.001	-0.01	7.3	0.0	0.4	B9.5 V		240 s		
101740	28106		11 42 48.6	+49 57 32	-0.002	-0.02	8.00	0.9	0.2	G9 III		360 s		
101784	138365		11 42 49.8	-3 32 44	-0.006	-0.07	7.5	0.1	0.6	A0 V		44 mx		
101917	256865		11 42 55.5	-79 18 22	+0.048	-0.01	6.39	0.90		K0				
101808	138366		11 42 56.8	-9 07 13	-0.009	-0.01	7.4	0.9	3.2	G5 IV		70 s		
101949	—		11 43 01.1	-78 13 16			8.0	1.1	0.2	K0 III		360 s		
101849	239316		11 43 05.7	-52 26 36	-0.002	-0.03	8.00	1.1	0.2	K0 III		350 s		m
101915	256866		11 43 07.1	-72 26 43	-0.008	-0.01	7.1	0.0	0.4	B9.5 V		200 mx		
101846	119012		11 43 19.8	+0 11 07	-0.002	-0.02	7.8	0.1	1.4	A2 V		190 s		
101841	81955		11 43 22.8	+28 03 19	0.000	+0.02	7.4	0.4	3.0	F2 V		77 s		
101875	156850		11 43 22.9	-19 49 24	-0.001	-0.02	7.1	1.8	-0.3	K5 III		200 s		
101884	222978		11 43 25.5	-41 32 40	-0.001	-0.03	8.0	1.5	-0.1	K2 III		370 s		
101883	202744		11 43 27.2	-37 11 25	0.000	-0.03	5.98	1.46		K2				
101881	202746		11 43 29.6	-33 53 11	0.000	-0.05	7.7	2.1	-0.1	K2 III		100 s		
101897	156853		11 43 30.5	-15 02 33	-0.004	+0.01	6.9	0.1	0.6	A0 V		140 mx		
101855	62671		11 43 31.2	+31 45 44	-0.001	-0.03	6.77	1.09	0.2	K0 III		180 s		
101947	251555		11 43 31.2	-62 29 21	-0.001	0.00	5.05	0.80	-8.0	G0 Ia	+10	4100 s		
101927	239323		11 43 33.1	-52 37 12	-0.005	-0.01	7.2	0.7	0.2	K0 III		250 s		
101966	251556		11 43 36.6	-68 28 43	+0.002	0.00	7.2	0.6	2.6	F0 V		52 s		
102065	258622		11 43 37.3	-80 28 59	-0.010	0.00	6.6	-0.2	0.4	B9.5 V		170 s		
101906	81958		11 43 47.0	+24 00 37	-0.001	-0.01	7.36	0.90	4.7	G2 V		22 s	8282	m
101936	180138		11 43 47.0	-21 02 18	-0.008	+0.01	7.3	0.7	2.6	F0 V		47 s		
101941	202749		11 43 47.4	-38 55 10	+0.010	-0.06	7.8	0.3	4.0	F8 V		57 s		
101871	28113		11 43 47.5	+52 23 42	0.000	0.00	7.3	0.1	1.4	A2 V		150 s		
101907	99759		11 43 47.8	+16 19 23	-0.009	+0.03	7.7	0.6	4.4	G0 V		47 s		
101892	28115		11 43 50.8	+58 24 40	-0.002	-0.01	8.0	0.4	2.6	F0 V		120 s		
101995	251559		11 43 52.9	-62 52 41	+0.003	-0.02	6.10	0.06		A0				
101933	138375		11 43 55.1	-6 40 38	+0.004	-0.05	6.07	0.96	0.3	gG8	-3	140 s		
101956	138376		11 43 55.1	-9 07 57	+0.001	-0.04	6.8	1.1	0.2	K0 III		210 s		
101959	180141		11 43 56.5	-29 44 51	-0.021	+0.04	6.98	0.55	4.4	G0 V		33 s		
101975	222983		11 43 57.8	-45 14 59	-0.003	0.00	7.9	1.3	0.2	K0 III		240 s		
101968	62677		11 44 12.9	+29 53 44	-0.001	+0.01	7.87	0.04		A2				
101980	81960		11 44 13.1	+25 13 07	-0.001	+0.01	6.02	1.53	-0.3	K5 III	-3	180 s	8285	m
102040	239336		11 44 14.5	-58 26 29	-0.002	0.00	7.6	1.7	-0.1	K2 III		170 s		
101967	43879		11 44 16.4	+44 29 25	-0.020	0.00	7.8	0.5	3.4	F5 V	+15	73 mx		
101978	62680		11 44 20.2	+38 31 37	-0.004	-0.04	7.84	1.48		K2 IV		200 mx		
102036	180152		11 44 23.7	-20 41 43	-0.003	0.00	6.8	0.9	0.6	A0 V		52 s		
101998	43880		11 44 29.0	+48 30 59	-0.003	+0.01	7.14	1.44	7.8	K2 V		10 s		
101828	1926		11 44 34.6	+82 19 16	+0.002	-0.01	7.80	0.8	-2.1	G5 II	-16	960 s		

HD	SAO	Star Name	α 2000	δ 2000	μ(α)	μ(δ)	V	B−V	M_v	Spec	RV	d(pc)	ADS	Notes
102076	222991		11ʰ44ᵐ38.5ˢ	−49°08′40″	+0.ˢ001	−0.″03	7.3	0.8	0.2	K0 III		260 s		
102088	251565		11 44 39.0	−60 51 02	+0.001	−0.01	7.8	1.5	−0.3	K5 III		400 s		
102101	251566		11 44 41.6	−60 24 03	−0.002	+0.01	7.63	0.08	−1.0	B5.5 V		440 s		
101950	7416		11 44 42.4	+73 08 49	−0.003	+0.01	7.2	1.1	−0.1	K2 III		290 s		
102056	81968		11 44 44.0	+28 40 13	0.000	+0.02	7.01	0.01		A m	−11			
102100	239342		11 44 45.4	−55 49 11	+0.001	−0.02	7.8	1.3	−0.1	K2 III		330 s		
102070	156869	27 ζ Crt	11 44 45.7	−18 21 03	+0.002	−0.04	4.73	0.97	0.3	G8 III	−5	74 s		
102030	15622		11 44 45.9	+60 02 48	−0.001	+0.01	7.90	1.1	−0.1	K2 III		400 s		m
102156	256867		11 44 47.7	−73 56 34	−0.007	+0.02	7.52	1.76		K5				
102117	239348		11 44 50.3	−58 42 11	−0.011	−0.05	7.3	0.7	3.2	G5 IV		66 s		
102113	239346		11 44 50.8	−55 27 40	−0.008	+0.01	7.80	0.1	0.6	A0 V		97 mx		m
102155	256868		11 44 52.6	−73 09 02	−0.002	0.00	6.93	1.38		K0				
102098	202765		11 44 52.8	−36 24 37	−0.004	−0.03	7.8	0.5	4.4	G0 V		49 s		
102112	239350		11 44 57.5	−51 34 16	−0.001	−0.01	7.4	0.7	0.6	A0 V		81 s		
102111	222995		11 44 57.8	−44 36 35	−0.005	−0.01	7.8	1.3	3.2	G5 IV		40 s		
102045	15623		11 44 59.9	+66 51 51	−0.001	+0.02	8.0	1.1	0.2	K0 III		360 s		
102044	7418		11 45 00.6	+69 55 53	−0.003	−0.01	7.7	0.1	1.4	A2 V		180 s		
102096	138388		11 45 01.2	−0 31 04	−0.004	−0.10	7.3	1.1	0.2	K0 III		120 mx		
102105	119026		11 45 05.5	+1 34 43	0.000	0.00	7.9	1.4	−0.3	K5 III		430 s		
102106	119027		11 45 07.1	+0 54 33	+0.001	−0.02	7.7	1.1	−0.1	K2 III		360 s		
102103	99774		11 45 11.3	+14 15 48	−0.004	−0.01	6.49	1.15	0.2	K0 III		150 s		
102150	222997		11 45 12.5	−49 04 11	−0.003	+0.01	6.26	1.17	0.2	gK0		130 s		
102124	119029	2 ξ Vir	11 45 17.0	+8 15 30	+0.004	−0.02	4.85	0.18		A3 n	−1	34 t		
102149	202771		11 45 19.2	−36 24 15	+0.003	−0.08	7.86	1.13	0.0	K1 III		220 mx		
102122	81972		11 45 20.9	+19 53 20	0.000	−0.05	7.52	1.05	0.2	K0 III		270 s		
102143	119030		11 45 24.1	+7 01 48	−0.002	−0.03	7.8	1.1	0.2	K0 III		330 s		
102174	222998		11 45 25.0	−40 31 00	−0.002	0.00	7.1	0.9	0.2	K0 III		240 s		
102238	—		11 45 25.1	−71 14 18			7.7	0.8	3.2	G5 IV		78 s		
102142	81973		11 45 25.3	+27 13 04	−0.001	−0.02	7.22	0.73	5.2	G5 V	+9	23 s		
102184	202775		11 45 27.3	−37 37 20	−0.017	−0.01	7.89	0.43	3.0	F2 V		89 mx		
102165	180164		11 45 28.0	−21 28 26	−0.001	−0.03	7.60	0.5	2.3	F7 IV	+19	120 s		
102159	62693		11 45 34.8	+35 53 39	−0.005	+0.02	7.10	1.50		M5 III	+59	240 mx		TV UMa, v
102249	251575	λ Mus	11 45 36.4	−66 43 43	−0.016	+0.03	3.64	0.16	2.1	A5 V	+16	16 mx		m
102198	202779		11 45 37.2	−35 36 18	−0.003	+0.01	7.5	1.3	0.2	K0 III		180 s		
102219	223006		11 45 40.4	−49 50 59	−0.007	−0.01	7.9	0.2	3.4	F5 V		78 s		
102232	223009		11 45 43.8	−45 41 24	−0.005	+0.01	5.29	−0.12	−1.2	B8 III	−7	200 s		
102235	239354		11 45 44.5	−52 41 27	−0.006	−0.03	8.0	0.7	3.0	F2 V		59 s		
102293	256871		11 45 45.7	−76 51 37	0.000	−0.02	7.89	0.08		A0				
102248	251577		11 45 45.9	−61 27 52	+0.002	0.00	7.92	0.00		B1				
102212	119035	3 ν Vir	11 45 51.5	+6 31 45	−0.001	−0.19	4.03	1.51	−0.5	M1 III	+51	51 mx		
102194	28129		11 45 52.9	+58 01 38	−0.020	−0.01	8.0	0.9	3.2	G5 IV		93 s		
102226	62696		11 45 58.2	+39 23 42	−0.001	−0.02	8.04	1.21	−0.1	K2 III		400 s		
102257	223013		11 45 58.7	−43 35 03	0.000	−0.04	7.37	1.12	0.2	K0 III		220 s		
102224	43886	63 χ UMa	11 46 03.0	+47 46 45	−0.014	+0.02	3.71	1.18	0.2	K0 III	−9	37 s		
102325	—		11 46 03.9	−73 36 18			7.4	0.0	0.4	B9.5 V		250 s		
102253	119038		11 46 06.3	+7 10 26	+0.001	−0.05	7.0	1.6	−0.5	M2 III	−19	320 s		
102223	28130		11 46 08.4	+50 33 48	−0.006	−0.01	7.7	0.4	2.6	F0 V		110 s		
102243	81978		11 46 09.6	+20 19 09	+0.002	0.00	7.65	0.93	3.2	G5 IV		59 s		
102276	180169		11 46 09.9	−24 52 23	0.000	−0.02	7.6	1.8	−0.5	M4 III		320 s		
102274	138395		11 46 19.5	−3 00 09	−0.004	−0.01	7.6	1.1	0.2	K0 III		310 mx		
102301	180171		11 46 21.2	−27 57 48	−0.004	0.00	7.6	0.8	4.0	F8 V		37 s		
102288	202790		11 46 22.1	−34 45 12	+0.004	−0.04	7.5	0.0	0.6	A0 V		180 mx		
102262	28136		11 46 24.9	+51 55 57	−0.001	0.00	7.6	0.8	3.2	G5 IV		78 s		
102340	239366		11 46 27.2	−58 01 51	−0.002	0.00	8.00	0.00	0.0	B8.5 V		380 s		m
102300	180174		11 46 27.5	−24 58 34	+0.015	−0.03	7.69	0.57	4.4	G0 V		46 s		
102310	180176		11 46 29.2	−24 29 03	−0.001	0.00	7.4	1.7	−0.5	M2 III		330 s		
102298	180177		11 46 29.9	−23 04 21	+0.001	0.00	7.9	1.8	−0.5	K0 III		120 s		
102350	251579		11 46 30.7	−61 10 42	−0.004	−0.02	4.11	0.90	−2.0	G0 II	−4	140 s		m
102365	223020		11 46 31.0	−40 30 01	−0.134	+0.40	4.91	0.66	5.2	G5 V	+15	10 ts		
102368	251580		11 46 35.7	−62 04 48	+0.001	−0.02	7.60	0.02	−2.6	B2.5 IV		910 s		
102370	251582		11 46 36.4	−64 45 56	−0.003	+0.01	6.88	0.09	0.0	B8.5 V		200 s		
102386	251583		11 46 40.4	−62 17 44	−0.007	−0.02	7.7	1.6	−0.1	K2 III		210 s		
102331	138399		11 46 40.6	−3 44 34	+0.001	−0.06	7.5	0.4	3.0	F2 V		81 s		
102361	180180		11 46 48.9	−21 28 29	−0.008	0.00	7.8	1.3	4.0	F8 V		20 s		
102343	99787		11 46 50.5	+15 00 07	0.000	+0.01	8.00	0.2	2.1	A5 V		150 s	8302	m
102399	251585		11 46 52.5	−61 02 30	−0.001	−0.02	8.02	−0.04	0.0	B8.5 V		400 s		
102328	28142		11 46 55.5	+55 37 41	+0.001	−0.04	5.27	1.27	−0.2	K3 III	+2	120 s		
102357	81988		11 47 00.8	+23 43 13	−0.006	0.00	6.9	0.5	3.4	F5 V	+8	50 s		
102327	15630		11 47 04.1	+64 23 34	0.000	−0.02	7.5	0.1	0.6	A0 V		240 s		
102397	202805		11 47 06.9	−35 54 25	+0.003	−0.04	6.17	0.96	0.3	gG8		150 s		
102355	15631		11 47 07.7	+61 24 06	−0.005	−0.04	6.7	0.4	2.6	F0 V		65 s		
102413	202807		11 47 10.7	−35 23 47	−0.003	+0.02	7.6	0.8	2.6	F0 V		50 s		
102506	256874		11 47 13.7	−76 37 03	−0.003	0.00	7.59	1.67		Mb				
102406	99790		11 47 15.3	+12 39 52	−0.002	−0.03	7.9	0.2	2.1	A5 V		150 s		
102438	202811		11 47 15.6	−30 17 12	−0.021	−0.24	6.48	0.68	5.2	G5 V		19 ts		

HD	SAO	Star Name	α 2000	δ 2000	μ(α)	μ(δ)	V	B-V	M$_v$	Spec	RV	d(pc)	ADS	Notes
102461	239373		11 47 19.0	-57 41 47	-0.004	+0.01	5.41	1.67	-0.5	M2 III	-52	140 s		
102404	81991		11 47 19.1	+24 25 27	-0.001	-0.02	7.85	1.25	0.2	K0 III		220 s		
102437	180190		11 47 19.3	-29 29 32	-0.002	-0.02	8.0	1.0	4.0	F8 V		33 s		
102435	--		11 47 20.8	-25 57 19			7.4	0.6	1.7	A3 V		70 s		
102433	156888		11 47 22.7	-18 22 49	+0.003	-0.05	6.8	1.1	0.2	K0 III		210 s		
102456	202816		11 47 29.7	-33 54 01	0.000	-0.07	8.0	1.8	-0.1	K2 III		180 s		
102534	251593		11 47 41.6	-67 41 32	+0.002	-0.01	7.4	1.1	0.2	K0 III		270 s		
102481	28148		11 47 49.7	+49 49 22	-0.005	+0.01	7.00	0.1	0.6	A0 V		180 mx	8307	m
102528	180197		11 47 53.6	-28 19 22	+0.002	+0.01	7.70	0.1	1.4	A2 V		180 s		m
102493	62706		11 47 54.7	+32 29 28	-0.001	0.00	7.84	1.25	-0.1	K2 III		350 s		
102510	119058	4 Vir	11 47 54.8	+8 14 45	-0.004	+0.01	5.32	0.02		A0	-1	26 mn		m
102494	81996		11 47 56.4	+27 20 25	-0.002	-0.01	7.50	0.87	0.3	G8 III		280 s		
102509	81998	93 Leo	11 47 59.0	+20 13 08	-0.011	-0.01	4.53	0.55		A	0	22 mn		m
102541	223032		11 48 00.0	-40 17 32	+0.002	-0.06	7.9	0.8	1.4	A2 V		65 s		
102540	202824		11 48 03.7	-35 13 30	-0.013	-0.07	7.09	0.77		G5				
102585	--		11 48 09.0	-68 31 20			8.0	0.5	3.4	F5 V		82 s		
102546	156895		11 48 09.3	-16 32 55	+0.001	-0.03	7.8	1.1	-0.1	K2 III		390 s		
102578	239383		11 48 11.1	-58 13 16	+0.002	-0.03	7.6	0.2	4.4	G0 V		44 s		
102561	223033		11 48 13.7	-49 59 39	-0.001	-0.02	7.5	0.6	1.4	A2 V		100 s		
102584	251597	μ Mus	11 48 14.4	-66 48 53	+0.003	-0.02	4.72	1.54	-0.5	gM2	+37	110 s		
102556	138416		11 48 17.2	-8 52 27	-0.002	+0.01	7.8	1.1	0.2	K0 III		330 s		
102574	156896		11 48 23.4	-10 18 48	-0.007	-0.12	6.26	0.58		G0				m
102555	82000		11 48 25.1	+28 25 01	-0.007	-0.02	7.2	0.4	3.0	F2 V	+18	71 s		
102572	138418		11 48 26.0	-3 14 19	-0.002	+0.01	7.8	0.4	2.6	F0 V		110 s		
102582	180204		11 48 26.2	-23 05 27	-0.023	+0.01	7.44	0.77	3.2	G5 IV		51 mx		
102613	239387		11 48 29.3	-55 58 02	-0.001	-0.03	7.3	1.4	2.1	A5 V		20 s		
102593	180205		11 48 30.3	-25 05 08	-0.001	+0.01	7.9	0.6	1.7	A3 V		79 s		
102580	138419		11 48 30.8	-9 22 33	-0.004	+0.02	7.9	1.1	-0.1	K2 III		300 mx		
102569	28152		11 48 35.1	+55 32 24	-0.001	+0.03	7.94	1.12	0.2	K0 III		280 s		
102610	202827		11 48 36.3	-38 44 26	0.000	+0.01	8.0	1.9	-0.3	K5 III		260 s		
102589	82001		11 48 36.7	+28 47 59	-0.001	-0.02	7.06	0.06		A2	-6			
102590	99800		11 48 38.6	+14 17 03	-0.007	0.00	5.88	0.29	2.5	A9 V	+9	46 s	8311	m
102608	202831		11 48 39.2	-35 59 12	0.000	0.00	6.9	1.4	-0.5	M4 III		300 s		
102604	156901		11 48 41.5	-14 20 40	-0.001	-0.06	6.9	1.1	0.2	K0 III		220 s		
102620	180208		11 48 45.0	-26 44 59	-0.003	+0.01	5.11	1.60	-0.5	M4 III	+7	130 s		
102618	99802		11 48 51.2	+18 49 24	-0.006	0.00	7.9	0.5	4.0	F8 V		61 s		
102628	156905		11 48 52.5	-16 35 52	-0.001	+0.01	7.6	1.4	-0.3	K5 III		390 s		
102685	--		11 48 55.9	-70 27 20			8.0	0.5	3.4	F5 V		82 s		
102657	239389		11 48 59.1	-51 24 32	-0.002	-0.02	7.74	-0.02	-1.7	B3 V		640 s		
102634	138420		11 49 01.1	-0 19 07	-0.015	+0.01	6.15	0.52		F8	0			
102627	99806		11 49 01.7	+16 07 57	-0.002	+0.01	6.85	1.06	0.2	K0 III		200 s		
102651	138421		11 49 02.9	-9 07 04	0.000	0.00	7.5	0.8	0.3	G6 III	+14	280 s		
102647	99809	94 β Leo	11 49 03.5	+14 34 19	-0.034	-0.12	2.14	0.09	1.7	A3 V	0	12 mx	8314	Denebola, m
102649	119061		11 49 07.0	+5 11 00	-0.006	-0.15	6.6	1.1	0.2	K0 III		72 mx		
102616	15640		11 49 10.5	+67 19 43	+0.003	-0.01	7.60	0.5	4.0	F8 V		53 s	8313	m
102661	138424		11 49 10.7	-7 21 39	-0.001	-0.01	7.1	1.4	-0.3	K5 III		300 s		
102646	82004		11 49 12.4	+28 07 17	+0.002	0.00	7.39	0.93	0.2	G9 III		270 s		
102693	239392		11 49 12.7	-57 41 30	-0.003	0.00	7.10	0.00	0.4	B9.5 V		210 s		m
102660	99812		11 49 14.8	+16 14 34	+0.004	-0.07	6.04	0.27		A m	-23			
102676	138425		11 49 15.2	-6 53 43	-0.001	0.00	7.3	1.1	-0.1	K2 III		310 s		
102728	251600		11 49 24.3	-60 25 31	-0.003	+0.01	7.2	2.1	-0.3	K5 III		140 s		
102703	223042		11 49 27.9	-46 04 04	-0.004	-0.02	7.18	0.28		A3				m
102686	--		11 49 35.1	+29 29 55	0.000	-0.01	7.59	0.93	0.3	G6 III p	0	290 s		
102715	99816		11 49 38.7	+18 56 07	-0.002	-0.06	7.84	1.26	0.2	K0 III		190 mx		
102776	251602		11 49 41.0	-63 47 18	-0.003	0.00	4.32	-0.15	-1.1	B5 V	+37	120 s		
102713	62718		11 49 41.6	+34 55 54	-0.010	0.00	5.70	0.46	2.1	F5 IV	-7	49 s		
102732	119066		11 49 44.4	+6 31 24	-0.006	-0.11	8.0	0.9	3.2	G5 IV		75 mx		
102753	239404		11 49 44.7	-50 43 14	-0.002	0.00	7.1	0.9	0.2	K0 III		240 s		
102749	156918		11 49 46.6	-13 39 08	+0.003	-0.03	7.4	1.1	-0.1	K2 III		320 s		
102769	223044		11 49 48.2	-46 00 55	-0.003	-0.04	7.61	1.30	0.0	K1 III		250 s		
102839	251604		11 49 56.3	-70 13 32	-0.004	0.00	4.97	1.40	-5.9	cK	+18	1500 s		
102791	202851		11 49 59.5	-40 16 33	-0.003	+0.02	7.3	1.1	0.2	K0 III		240 s		
102817	239413		11 50 03.5	-57 02 49	-0.003	0.00	7.6	2.0	-0.1	K2 III		110 s		
102815	239412		11 50 04.2	-51 42 03	-0.004	+0.01	7.4	0.2	0.6	A0 V		160 s		
102777	28157		11 50 14.0	+54 15 09	+0.004	+0.01	6.8	1.1	0.2	K0 III		210 s		
102828	156924		11 50 15.8	-19 36 19	-0.005	+0.01	8.0	0.4	2.6	F0 V		120 s		
102845	156926		11 50 19.4	-15 51 50	-0.001	0.00	6.13	0.95		K0				
102865	223052		11 50 25.7	-46 28 09	-0.003	-0.06	7.67	0.40		F2				
102844	138437		11 50 26.4	-2 25 09	-0.001	-0.05	7.90	0.7	4.4	G0 V		50 s	8318	m
102878	251605		11 50 27.2	-62 38 58	-0.001	-0.01	5.70	0.26	-6.7	A2 I		2300 s		
102889	202863		11 50 35.6	-37 40 51	-0.001	+0.01	7.7	0.9	0.2	K0 III		310 s		
102857	43909		11 50 36.0	+43 39 31	+0.001	-0.03	7.92	0.78	0.3	G6 III		330 s		
102888	180232		11 50 37.1	-27 16 40	-0.007	+0.01	6.48	0.98		K0				
102885	156930		11 50 40.0	-12 52 00	+0.001	+0.02	7.30	1.4	-0.3	K5 III	-25	330 s		
102870	119076	5 β Vir	11 50 41.6	+1 45 53	+0.049	-0.27	3.61	0.55	3.6	F8 V	+5	10 t		Zavijava, m

HD	SAO	Star Name	α 2000	δ 2000	μ(α)	μ(δ)	V	B-V	M_v	Spec	RV	d(pc)	ADS	Notes
102902	202865		11^{h}50^{m}41.9^s	-33°08'29"	-0.012^s	0.00"	7.34	0.74	4.1	G4 IV-V		43 s		
102868	82012		11 50 43.7	+28 05 47	-0.005	-0.01	8.0	0.5	3.4	F5 V		82 s		
102883	138441		11 50 44.0	-9 23 51	-0.004	0.00	7.7	1.1	0.2	K0 III		320 s		
102882	28160		11 50 50.4	+51 24 45	-0.003	0.00	7.6	1.6	-0.5	M2 III		420 s		
102935	—		11 50 55.3	-39 15 21			7.8	1.7	-0.1	K2 III		190 s		
102910	99827		11 50 55.4	+12 16 43	-0.009	+0.01	6.35	0.27		A3	+7		8320	m
102855	7434		11 50 55.5	+75 21 21	-0.015	-0.04	7.9	0.4	2.6	F0 V		110 s		
102909	62725		11 50 58.0	+33 36 15	-0.002	+0.01	7.16	1.03	0.2	K0 III		240 s		
102945	138446		11 51 02.0	-9 16 24	-0.005	0.00	7.6	0.4	2.6	F0 V		98 s		
102928	138445		11 51 02.2	-5 20 00	0.000	-0.01	5.64	1.06	3.2	K0 IV	+12	25 s		
102948	156935		11 51 02.6	-13 19 25	-0.001	-0.02	7.2	0.1	0.6	A0 V		210 s		
102968	239427		11 51 04.6	-60 10 34	+0.008	+0.04	7.1	1.4	4.0	F8 V		13 s		
102962	180240		11 51 04.8	-27 04 48	-0.006	-0.01	7.2	1.2	0.2	K0 III		180 s		
102985	—		11 51 05.4	-62 38 21			7.9	2.0	-0.3	K5 III		210 s		
102941	62729		11 51 07.2	+36 50 06	+0.003	-0.01	7.56	1.10	0.2	K0 III		250 s		
102964	223062		11 51 08.5	-45 10 25	-0.008	-0.02	4.46	1.30	-0.3	K4 III	+2	90 s		
102942	62731		11 51 09.3	+33 22 29	-0.002	+0.01	6.27	0.32	1.4	A2 V	+2	69 s		
102960	138447		11 51 11.0	-7 59 29	+0.002	-0.02	6.9	1.1	0.2	K0 III		220 s		
102980	202873		11 51 11.2	-36 56 09	-0.001	-0.01	7.90	0.1	0.6	A0 V		290 s		m
102997	251611		11 51 13.0	-61 50 46	-0.001	0.00	6.53	0.30	-7.0	B5 Ia		3100 s		
102981	223064		11 51 13.1	-43 55 59	-0.002	-0.02	6.61	-0.01		A0				
102996	239431		11 51 19.7	-52 06 25	-0.001	-0.01	7.4	0.8	-0.1	K2 III		310 s		
102925	15643		11 51 19.9	+68 50 01	-0.009	-0.04	7.1	0.1	1.4	A2 V	+13	97 mx		
102994	223065		11 51 21.1	-43 04 48	-0.006	0.00	6.8	1.4	0.2	K0 III		130 s		
103021	251614		11 51 21.1	-68 51 39	-0.001	+0.01	7.3	0.5	4.0	F8 V		45 s		
102990	156940		11 51 22.0	-12 11 19	-0.015	+0.01	6.35	0.40		F0				m
102956	28161		11 51 22.4	+57 38 27	-0.002	-0.02	8.0	0.9	3.2	G5 IV		93 s		
102988	99829		11 51 24.4	+11 48 19	-0.018	0.00	7.0	0.4	3.0	F2 V		63 s		
103002	180242		11 51 25.9	-23 50 49	-0.011	-0.04	7.9	1.0	3.4	F5 V		35 s		
103017	223069		11 51 31.8	-40 41 17	-0.003	-0.03	7.9	0.5	3.0	F2 V		62 s		
103014	180244		11 51 32.8	-21 38 08	-0.002	+0.01	7.5	0.4	3.4	F5 V		67 s		
103030	202879		11 51 34.0	-39 48 01	-0.003	+0.01	8.0	1.6	-0.1	K2 III		240 s		
103041	223073		11 51 41.0	-40 18 28	-0.005	-0.03	7.6	1.3	0.2	K0 III		200 s		
103026	202883		11 51 41.5	-30 50 05	-0.001	-0.29	5.85	0.56	3.4	F5 V	+33	26 mx		
103052	251615		11 51 42.9	-61 09 31	0.000	-0.01	7.6	2.2	-0.5	M2 III		190 s		
103051	223075		11 51 47.0	-41 14 52	0.000	+0.01	7.5	0.7	3.4	F5 V		68 s		
103066	251616		11 51 49.1	-64 35 45	-0.001	-0.02	7.19	0.03	0.0	B8.5 V		260 s		m
103079	251617		11 51 51.2	-65 12 22	-0.005	-0.02	4.90	-0.11	-2.0	B4 IV	+26	130 mx		m
103046	119083		11 51 56.5	+8 49 21	+0.001	-0.06	7.93	0.49	3.7	F6 V		70 s		
103047	119084		11 51 57.4	+8 49 47	-0.001	-0.06	7.40	0.97	1.7	K0 III-IV		140 s	8327	m
103077	223078		11 52 04.2	-49 24 05	-0.001	-0.01	7.1	0.0	0.0	B8.5 V		230 s		
103045	15647		11 52 07.3	+61 42 33	-0.006	-0.02	8.0	0.4	3.0	F2 V		99 s		
103101	239443		11 52 10.1	-56 59 16	-0.014	+0.03	5.57	0.07		A2				m
103086	156948		11 52 12.2	-19 12 22	-0.004	+0.01	7.2	0.2	2.1	A5 V		110 s		
103069	28172		11 52 14.2	+49 55 54	-0.005	+0.02	7.10	1.30	0.2	K0 III		140 s		
103125	239445		11 52 18.1	-51 18 53	0.000	-0.03	6.8	1.3	0.2	K0 III		150 s		
103112	99835		11 52 21.0	+9 56 52	-0.022	+0.09	7.8	1.1	0.2	K0 III	+11	52 mx		
103130	156952		11 52 27.0	-14 07 38	-0.014	+0.05	7.8	0.5	4.0	F8 V		57 s		
103111	99838		11 52 30.1	+15 50 44	+0.001	-0.13	7.9	0.9	3.2	G5 IV		88 s		
103110	82020		11 52 30.9	+20 24 36	-0.007	+0.03	7.3	0.5	3.4	F5 V		61 s		
103168	251620		11 52 32.2	-62 26 56	+0.004	-0.05	7.9	1.5	-0.3	K5 III		430 s		
103170	251621		11 52 32.2	-63 10 12	-0.014	0.00	7.70	1.19	0.0	K1 III		220 mx		
103182	251623		11 52 39.0	-62 14 05	+0.002	+0.01	7.22	-0.02	0.0	B8.5 V		280 s		
103174	202899		11 52 42.2	-37 59 13	-0.002	-0.01	7.9	0.1	3.0	F2 V		94 s		
103177	239451		11 52 42.4	-52 29 50	-0.006	-0.03	7.5	1.3	0.2	K0 III		200 s		
103152	99840		11 52 46.2	+15 26 11	+0.002	-0.07	6.85	0.26	1.4	A2 V	-9	95 s		m
103194	—		11 52 50.4	-38 20 22			7.5	1.3	-0.1	K2 III		280 s		
103192	202901	β Hya	11 52 54.5	-33 54 28	-0.004	0.00	4.28	-0.10	-0.3	B9 IV	-1	82 s		m
103188	99843		11 52 55.9	+13 25 16	-0.003	-0.02	7.51	1.00	0.2	K0 III		290 s		
103095	62738		11 52 58.7	+37 43 07	+0.337	-5.81	6.45	0.75	6.7	G8 V p	-98	8.8 t		Groombridge 1830, m
103149	15651		11 52 59.1	+62 55 25	0.000	+0.01	8.0	0.9	3.2	G5 IV		93 s		
103257	202903		11 53 16.7	-36 34 38	+0.001	-0.04	6.7	0.8	2.6	F0 V		34 s		
103270	251625		11 53 20.3	-65 24 16	-0.001	-0.01	7.33	0.03	0.0	B8.5 V		290 s		
103203	7443		11 53 21.7	+73 17 28	+0.001	-0.02	7.6	1.1	0.2	K0 III		300 s		
103253	156964		11 53 22.8	-16 35 34	-0.005	-0.01	7.8	0.4	2.6	F0 V		110 s		
103266	202910		11 53 26.7	-35 04 00	-0.007	-0.03	6.17	0.08		A2				
103276	156966		11 53 33.7	-16 06 52	+0.003	-0.01	7.50	1.1	0.2	K0 III		290 s	8339	m
103281	223089		11 53 35.9	-46 44 31	+0.007	+0.01	7.23	1.03		G5				
103272	43938		11 53 38.1	+42 54 57	+0.002	-0.03	8.00	0.5	3.4	F5 V		83 s	8338	m
103246	7445		11 53 42.8	+73 45 21	-0.013	-0.12	6.80	0.56		F8	-35	14 mn	8337	m
103338	251628		11 53 44.1	-65 17 19	-0.002	+0.03	7.49	0.13	-1.0	B5.5 V		390 s		
103288	62744		11 53 47.4	+33 36 53	-0.001	-0.04	7.03	0.31	2.6	F0 V		73 s		
103287	28179	64 γ UMa	11 53 49.7	+53 41 41	+0.011	+0.01	2.44	0.00	0.6	A0 V	-13	23 s		Phecda
103313	119100		11 53 50.2	+0 33 06	-0.003	-0.01	6.30	0.20		A5	+10			
103309	43939		11 53 50.4	+40 54 47	-0.003	-0.06	6.74	1.06	0.0	K1 III		180 mx		

HD	SAO	Star Name	α 2000	δ 2000	μ(α)	μ(δ)	V	B-V	M_V	Spec	RV	d(pc)	ADS	Notes
			h m s	° ' "	s	"								
103333	—		11 53 50.9	-38 25 22			7.8	1.1	-0.1	K2 III		380 s		
103327	138464		11 53 51.9	-3 46 37	-0.002	-0.05	7.3	0.9	3.2	G5 IV	+26	67 s		
103321	7446		11 54 04.2	+71 55 27	-0.016	-0.02	7.58	0.46	3.4	dF5	-7	67 s	8344	m
103340	119101		11 54 04.7	+4 52 44	+0.002	+0.01	7.9	1.4	-0.3	K5 III		430 s		
103354	99853		11 54 07.6	+18 10 11	-0.002	-0.02	7.8	0.5	4.0	F8 V		58 s		
103356	99854		11 54 07.8	+9 53 35	-0.003	-0.06	7.9	1.1	0.2	K0 III		170 mx		
103391	202922		11 54 11.3	-30 54 24	-0.001	+0.02	7.31	1.62	-0.5	M4 III		370 s		
103400	239475		11 54 11.5	-57 24 36	-0.001	0.00	6.06	0.05		A1				
103389	99855		11 54 22.7	+10 02 57	-0.002	+0.01	7.6	0.5	3.4	F5 V		71 s		
103423	223098		11 54 23.4	-46 29 31	-0.001	-0.04	7.98	0.97		G5				
103437	202926		11 54 25.8	-37 44 56	-0.026	+0.06	6.46	0.52	4.0	dF8		31 s		m
103426	239480		11 54 26.4	-56 58 10	-0.003	0.00	6.9	1.9	-0.5	M2 III		190 s		
103438	223099		11 54 29.9	-42 02 41	-0.001	+0.01	8.0	1.0	-0.3	K5 III		450 s		
103457	251634		11 54 32.7	-63 42 01	-0.003	0.00	7.76	-0.01	0.6	A0 V		270 s		
103420	99859		11 54 33.4	+14 09 04	-0.005	-0.04	7.80	0.44	3.4	F5 V		76 s		
103451	223101		11 54 37.0	-43 39 30	-0.002	-0.02	8.0	1.9	-0.3	K5 III		280 s		
103430	43944		11 54 39.7	+48 55 59	-0.016	-0.09	7.34	0.95	0.2	G9 III		63 mx		
103462	180288		11 54 42.5	-25 42 49	+0.004	+0.08	5.30	0.88	0.3	gG4	-11	93 s		
103482	251637		11 54 44.6	-66 22 34	+0.002	-0.07	6.5	0.6	3.0	F2 V		36 s		
103461	180290		11 54 48.2	-23 32 59	-0.002	+0.02	8.0	0.2	3.4	F5 V		82 s		
103481	223105		11 54 50.0	-43 36 40	+0.001	-0.01	7.43	1.28		K0				
103459	138476		11 54 51.9	-1 27 06	+0.009	-0.32	7.9	0.9	3.2	G5 IV	+18	43 mx		
103516	251640		11 54 59.8	-63 16 44	-0.003	+0.01	5.91	0.19	-6.7	A2 I		2700 s		
103515	251641		11 55 01.2	-62 34 38	-0.004	+0.01	7.80	1.6	-0.5	M2 III		430 s		m
103493	239487		11 55 01.3	-56 05 47	+0.024	-0.21	6.69	0.64	5.0	G4 V		22 s		
—	—		11 55 02.1	-59 15 12			7.70							W Cen, v
103484	119111	6 Vir	11 55 03.1	+8 26 38	-0.002	+0.01	5.58	0.94	0.2	gK0	-10	120 s		
103483	43945	65 UMa	11 55 05.6	+46 28 37	0.000	0.00	6.54	0.10	1.7	A3 V	-8	92 s	8347	m
103498	43946	65 UMa	11 55 11.1	+46 28 11	-0.001	0.00	7.03	0.01		A p	-7		8347	m
103500	62754		11 55 14.0	+36 45 22	-0.005	-0.05	6.49	1.58	-0.4	M0 III	+19	170 mx		
103574	251643		11 55 21.6	-63 42 12	-0.001	0.00	7.98	-0.02	0.0	B8.5 V		400 s		
103520	62756		11 55 22.5	+38 45 25	+0.005	-0.02	7.03	0.99	0.2	K0 III		230 s		
103567	223115		11 55 23.6	-41 54 14	+0.003	-0.02	7.90	0.1	1.4	A2 V		200 s		m
103548	119114		11 55 23.8	+1 05 44	-0.004	-0.05	7.2	0.4	3.0	F2 V		69 s		
103543	82051		11 55 24.4	+25 31 21	-0.005	0.00	6.98	1.22	0.2	K1 III	+8	200 s		
103550	138485		11 55 26.3	-5 08 05	-0.002	-0.02	7.1	0.1	1.4	A2 V		140 s		
103563	202946		11 55 28.0	-33 27 04	-0.001	-0.01	8.0	1.1	-0.1	K2 III		410 s		
103569	239491		11 55 28.3	-53 14 24	-0.005	-0.03	7.7	0.3	0.2	K0 III		280 mx		
103581	180301		11 55 30.3	-24 51 26	-0.005	+0.05	6.8	1.6	0.2	K0 III		95 s		
103580	180304		11 55 32.0	-23 27 02	-0.001	+0.01	7.8	0.8	-0.3	K5 III		420 s		
103586	239494		11 55 37.7	-51 45 39	0.000	-0.02	8.0	1.5	-0.3	K5 III		450 s		
103611	—		11 55 37.7	-68 32 23			7.6	1.6	-0.5	M2 III		390 s		
103596	180307		11 55 40.1	-28 28 37	0.000	-0.03	5.93	1.50	-0.3	gK5	+11	180 s		
103578	99869	95 Leo	11 55 40.5	+15 38 48	+0.001	0.00	5.53	0.11	1.7	A3 V	-21	56 s		m
103539	15658		11 55 40.6	+67 15 50	+0.007	-0.05	7.5	0.7	4.4	G0 V		42 s		
103616	—		11 55 53.6	-21 39 23			7.7	0.3	3.2	G5 IV		80 s		
103637	202950		11 55 54.6	-39 41 21	+0.004	-0.01	6.13	1.02		K0				
103655	—		11 55 58.0	-64 37 23			7.82	1.18	0.2	K0 III		260 s		
103605	28191	66 UMa	11 55 58.3	+56 35 55	+0.001	0.00	5.84	1.10	0.2	K0 III	+13	120 s		
103634	—		11 55 59.5	-22 10 23			7.80	1.1	-0.1	K2 III		380 s	8353	m
103613	62759		11 56 00.1	+35 20 25	-0.003	-0.02	6.9	0.5	3.4	F5 V		49 s		
103632	156988	30 η Crt	11 56 00.8	-17 09 03	-0.004	-0.01	5.18	-0.02	0.6	A0 V	+15	51 mx		
103612	43949		11 56 03.4	+40 38 52	-0.001	-0.02	6.92	1.11	0.2	K0 III		180 s		
103644	62762		11 56 12.1	+36 14 54	-0.007	-0.02	7.90	1.11	0.0	K1 III		210 s		
103661	99874		11 56 12.4	+15 43 44	+0.001	-0.04	7.94	1.48	-0.5	M4 III	-29	490 s		WX Leo, v
103660	82061		11 56 15.1	+28 51 29	-0.003	-0.01	7.30	1.05	0.2	K0 III	-12	240 s		
103659	62763		11 56 17.1	+35 26 52	-0.008	0.00	6.80	0.4	3.0	F2 V		58 s	8355	m
103676	82062		11 56 20.4	+26 40 44	-0.012	-0.01	6.9	0.4	3.0	F2 V	+10	60 s		
103683	43951		11 56 23.0	+42 00 50	-0.002	0.00	7.04	0.98	0.2	K0 III		230 s		
103685	99876		11 56 26.7	+14 11 10	-0.001	-0.02	6.72	1.46	0.2	K0 III		110 s		
103684	62765		11 56 29.5	+34 54 12	-0.001	-0.03	7.33	1.11	0.0	K1 III		290 s		
103681	28194		11 56 29.9	+57 52 16	-0.004	-0.02	6.60	1.6		M5 III	-53	300 mx		Z UMa, v
103690	15663		11 56 37.7	+65 14 34	+0.010	-0.02	6.7	0.9	3.2	G5 IV		50 s		
103742	202965		11 56 42.0	-32 16 06	-0.015	-0.02	7.64	0.64	4.4	G0 V		40 s		
103743	202966		11 56 43.6	-32 16 01	-0.016	0.00	7.81	0.67	4.4	G0 V		41 s		m
103746	223127		11 56 43.8	-47 04 21	-0.013	+0.01	6.26	0.40	2.5	F3 IV-V		54 s		
103705	15665		11 56 44.4	+62 53 13	-0.006	+0.01	7.9	1.1	0.2	K0 III		340 s		
103766	251651		11 56 45.0	-70 09 21	-0.006	+0.01	7.8	0.0	0.4	B9.5 V		180 mx		
103721	157001		11 56 45.5	-10 43 29	+0.005	-0.09	7.5	0.4	3.0	F2 V		78 s		
103719	62767		11 56 46.4	+32 12 10	-0.002	-0.01	7.98	1.16	0.0	K1 III		340 s		
103740	119125		11 56 48.2	+5 20 42	-0.002	-0.02	7.5	1.1	0.2	K0 III		280 s		
103760	202968		11 56 51.1	-38 50 56	+0.003	-0.02	7.8	1.0	3.2	G5 IV		60 s		
103736	15666		11 56 53.2	+61 32 57	-0.005	-0.04	6.22	0.96	0.3	G8 III	+17	150 s		
103774	157002		11 56 55.6	-12 06 28	-0.008	-0.03	7.2	0.5	3.4	F5 V		57 s		
103753	157003		11 56 56.0	-13 45 17	+0.001	-0.16	6.6	0.5	4.0	F8 V		32 s		

HD	SAO	Star Name	α 2000	δ 2000	μ(α)	μ(δ)	V	B-V	M_v	Spec	RV	d(pc)	ADS	Notes
103779	251654		11ʰ56ᵐ57ˢ.5	-63°14'57"	-0ˢ.002	-0".01	7.21	0.02		B0.5 Ib				
103773	138496		11 57 00.2	-4 46 57	-0.011	+0.02	6.9	0.5	3.4	F5 V		50 s		
103789	202971		11 57 03.6	-33 18 55	-0.005	0.00	6.21	-0.04		A0				
103805	239520		11 57 04.9	-52 06 00	+0.010	-0.02	6.7	1.3	0.2	K0 III		130 s		
103803	202972		11 57 05.9	-34 53 57	+0.001	+0.03	7.8	1.4	0.2	K0 III		200 s		
103770	43961		11 57 06.5	+40 17 20	-0.004	0.00	7.14	1.04	0.2	K0 III		230 s		
103807	251655		11 57 06.7	-63 40 36	-0.002	0.00	8.0	0.5	4.0	F8 V		62 s		
103799	43963		11 57 14.4	+40 20 37	-0.015	-0.07	6.62	0.46	3.4	F5 V	+26	43 s		
103822	202977		11 57 15.0	-35 11 35	-0.004	0.00	7.6	1.7	-0.3	K5 III		300 s		
103817	––		11 57 17.8	-22 32 23			8.00	0.5	3.4	F5 V		83 s	8361	m
103875	258628		11 57 18.3	-80 29 50	+0.013	-0.09	7.60	0.1	0.6	A0 V		83 mx		m
103856	223130		11 57 20.3	-49 21 37	-0.017	-0.03	6.67	0.36		A3				m
103813	82071		11 57 25.7	+26 45 44	+0.001	-0.15	7.46	0.95	3.2	G9 IV	+34	67 mx		
103859	251657		11 57 25.9	-61 07 24	-0.004	0.00	7.3	2.0	-0.1	K2 III		96 s		
103847	82073		11 57 28.7	+19 59 01	-0.029	+0.04	8.0	0.9	3.2	G5 IV		37 mx		
103870	202984		11 57 30.5	-38 29 46	-0.001	+0.01	7.2	0.4	0.6	A0 V		110 s		
103868	180329		11 57 31.0	-27 03 22	-0.004	-0.07	7.0	0.6	3.4	F5 V		42 s		
103869	202985		11 57 35.3	-32 57 28	-0.006	+0.01	7.7	1.6	-0.1	K2 III		200 s		
103884	251659		11 57 40.0	-62 26 55	-0.003	-0.01	5.57	-0.15	-1.7	B3 V	+16	280 s		
103877	99886		11 57 42.2	+17 28 04	-0.007	-0.01	6.9	0.4	3.0	F2 V		61 s		
103880	119134		11 57 45.1	+6 58 28	-0.002	-0.07	7.9	1.1	0.2	K0 III		170 mx		
103891	138501		11 57 45.3	-8 32 57	-0.007	0.00	6.6	0.7	4.4	G0 V		27 s		
103901	251660		11 57 45.9	-60 27 04	-0.006	+0.01	6.8	1.2	0.2	K0 III		160 s		
103890	138502		11 57 46.2	-8 16 30	-0.002	-0.07	7.7	0.5	4.0	F8 V		54 s		
103889	119136		11 57 48.3	+8 59 23	-0.008	+0.01	7.9	1.1	0.2	K0 III		160 mx		
103887	15670		11 57 48.6	+60 27 42	-0.005	0.00	8.0	1.1	-0.1	K2 III		420 s		
103910	223136		11 57 50.1	-43 43 13	0.000	-0.02	7.98	0.23		A0				m
103907	157015		11 57 52.8	-18 58 54	+0.003	-0.03	7.7	1.1	0.2	K0 III		310 s		
103932	180337		11 57 56.2	-27 42 26	-0.081	-0.63	6.97	1.16	7.0	K5 V	+52	9.8 t		
103903	15672		11 58 00.1	+67 48 42	-0.012	-0.04	7.0	1.1	-0.1	K2 III		190 mx		
103922	239529		11 58 00.3	-56 51 55	-0.002	-0.02	7.6	1.2	0.2	K0 III		240 s		
103933	202991		11 58 03.5	-31 39 03	-0.008	-0.04	7.3	0.5	3.4	F5 V		58 s		
103938	239531		11 58 05.8	-59 35 33	-0.005	-0.01	7.6	0.1	0.0	B8.5 V		180 mx		
103929	138508		11 58 06.6	-4 22 12	-0.010	+0.08	7.1	0.4	3.0	F2 V		66 s		
103928	62774		11 58 07.1	+32 16 26	-0.009	-0.06	6.42	0.29	2.5	A9 V	+2	36 mx	8368	m
103946	119138		11 58 12.9	+0 52 09	-0.002	-0.05	7.4	1.1	0.2	K0 III		270 s		
103945	119139		11 58 13.9	+3 28 55	+0.001	-0.01	6.8	1.6	-0.5	M2 III	-22	290 s		
103961	239533		11 58 15.2	-56 19 02	-0.002	-0.01	5.44	-0.08		B8	-23			
103975	223138		11 58 15.6	-47 58 39	-0.013	-0.10	6.77	0.53	4.4	G0 V		30 s		
103974	223139		11 58 20.2	-40 56 50	-0.005	+0.02	6.79	0.97		K0				m
103953	15673		11 58 20.5	+61 27 53	+0.006	0.00	6.6	1.1	0.2	K0 III	-26	190 s		
103962	239538		11 58 20.9	-59 45 19	-0.004	-0.01	7.4	-0.3	0.6	A0 V		220 mx		
103965	62779		11 58 28.3	+31 55 14	+0.003	-0.03	7.90	1.06	0.2	G9 III		290 s		
104015	256891		11 58 29.8	-70 44 24	-0.005	-0.01	7.00	0.00		B5				
103984	43980		11 58 31.8	+47 45 54	-0.013	-0.01	7.1	0.4	3.0	F2 V		66 s		
104003	157023		11 58 33.6	-15 06 59	+0.002	-0.02	7.5	0.4	3.0	F2 V		78 s		
104007	223143		11 58 34.9	-43 26 16	-0.005	-0.01	7.2	1.3	3.4	F5 V		28 s		
104036	256892		11 58 35.6	-77 49 34	-0.010	-0.04	6.73	0.22		A2				
104021	239542		11 58 38.9	-52 46 08	-0.003	-0.03	7.6	0.3	1.4	A2 V		120 s		
104017	62781		11 58 45.1	+37 52 38	-0.001	+0.03	7.83	1.04	0.0	K1 III		370 s		
104035	251664		11 58 47.7	-64 20 21	+0.001	+0.01	5.61	0.18	-4.8	A3 Ib		1100 s		
104039	180349		11 58 54.3	-25 54 32	-0.002	-0.02	6.43	0.03		A0			8371	m
104057	138516		11 59 00.9	-6 39 12	-0.001	+0.03	8.0	0.4	3.0	F2 V		98 s		
104072	256893		11 59 01.9	-74 00 33	-0.004	-0.03	7.71	1.40		K0				
104058	180351		11 59 02.3	-23 24 50	+0.001	-0.03	7.7	0.5	3.2	G5 IV		79 s		
104055	119147		11 59 03.3	+0 31 50	-0.004	+0.02	6.17	1.26	0.2	gK0	+12	95 s	8372	m
104078	157028		11 59 09.2	-10 28 35	-0.001	+0.01	6.7	1.1	-0.1	K2 III		230 s		
104067	180353		11 59 09.9	-20 21 14	+0.010	-0.43	7.93	0.99	-0.1	K2 III		43 mx		
104080	223151		11 59 10.7	-45 49 56	-0.003	-0.01	6.36	-0.07		A0				
104081	239548		11 59 10.8	-51 41 48	-0.001	-0.01	6.05	1.28		K2				
104077	138520		11 59 14.5	-3 19 22	+0.001	-0.02	7.6	1.6	-0.5	M2 III		420 s		
104087	15676		11 59 15.7	+61 20 03	-0.004	-0.02	7.6	0.1	1.7	A3 V		120 mx		
104075	62784		11 59 17.5	+33 10 02	0.000	0.00	5.96	1.15	0.2	K0 III	-1	110 s	8374	m
104103	180356		11 59 22.2	-30 03 38	-0.002	0.00	6.9	1.6	-0.3	K5 III		230 s		
104125	239550		11 59 23.3	-57 10 03	-0.008	0.00	6.9	0.5	1.4	A2 V		64 s		
104101	119150		11 59 23.9	+1 49 36	-0.001	-0.04	7.0	1.1	0.2	K0 III		230 s		
104111	251670		11 59 25.6	-62 49 52	-0.001	-0.01	6.6	0.2	2.1	A5 V		79 s		
104122	223152		11 59 28.2	-49 08 25	-0.001	-0.01	7.3	0.6	0.6	A0 V		89 s		
104136	203011		11 59 31.7	-35 18 30	-0.003	+0.01	7.2	1.0	0.2	K0 III		260 s		
104130	138521		11 59 33.9	-1 55 07	-0.003	-0.03	7.1	1.1	0.2	K0 III		240 s		
104138	223155		11 59 35.2	-46 37 58	+0.004	-0.01	6.66	0.56		F8				
104174	256894	ε Cha	11 59 37.4	-78 13 18	-0.016	0.00	4.91	-0.06	0.2	B9 V n	+22	88 s		m
104173	––		11 59 43.6	-68 19 24			7.5	0.0	0.4	B9.5 V		260 s		
104126	7473		11 59 47.3	+79 35 41	-0.013	0.00	7.8	0.5	4.0	F8 V		57 s		
104172	251673		11 59 49.0	-62 49 53	+0.001	0.00	7.9	1.4	-0.3	K5 III		420 s		

HD	SAO	Star Name	α 2000	δ 2000	μ(α)	μ(δ)	V	B−V	M$_v$	Spec	RV	d(pc)	ADS	Notes
			h m s	° ′ ″	s	″								
--	223157		11 59 52.9	−48 18 25	−0.001	0.00	8.0	1.3	−0.1	K2 III		360 s		
104200	239561		11 59 56.2	−56 17 34	−0.004	+0.01	7.9	−0.2		B0.5 V		140 mx		
104181	119156	7 Vir	11 59 56.8	+3 39 18	−0.001	−0.01	5.37	0.00		A0	−3	26 mn		
104179	62786		11 59 57.1	+34 02 06	−0.005	+0.04	6.50	0.22	0.6	A9 III	−8	150 s		
104207	99901		12 00 04.6	+19 25 10	−0.007	−0.03	6.94	1.58	−0.5	gM4	+35	310 s		
104237	256895		12 00 04.7	−78 11 32	−0.016	+0.01	6.5	0.1	0.6	A0 V		99 mx		
--	7476		12 00 05.8	+70 39 35	−0.001	−0.01	8.00						8379	m
104215	251675		12 00 06.7	−62 33 39	−0.003	−0.01	7.3	2.0	−0.1	K2 III		96 s		
104202	7477		12 00 08.1	+70 39 05	−0.009	−0.01	7.90	0.1	0.6	A0 V		120 mx	8379	m
104226	223163		12 00 08.8	−44 51 40	−0.007	−0.03	7.85	0.55		F8				
104221	157033		12 00 09.0	−16 35 07	−0.001	−0.02	8.0	0.4	3.0	F2 V		98 s		
104204	62789		12 00 10.0	+36 43 47	+0.001	+0.01	7.48	0.11		A0				m
104250	239568		12 00 13.5	−52 51 14	−0.002	0.00	8.0	0.3	0.2	K0 III		360 s		
104216	1967		12 00 18.4	+80 51 12	−0.030	−0.04	6.17	1.61	−0.5	M2 III	+32	210 s		
104271	223165		12 00 19.5	−47 39 22	−0.001	−0.03	7.4	0.3	0.2	K0 III		270 s		
104270	203029		12 00 20.4	−36 33 46	−0.004	−0.02	7.2	1.0	1.7	A3 V		38 s		
104241	43991		12 00 24.3	+44 37 48	0.000	−0.01	7.55	0.06		A m				
104288	15686		12 00 37.2	+69 11 07	−0.006	−0.01	7.80	0.4	2.6	F0 V		110 s	8387	m
104290	99904		12 00 38.0	+18 20 04	+0.001	−0.03	7.84	1.03	0.2	K0 III		320 s		
104289	28219		12 00 41.0	+59 21 11	+0.001	+0.01	8.0	0.5	4.0	F8 V		64 s		
104307	180374		12 00 42.4	−21 50 14	+0.002	0.00	6.28	1.22		K0			8389	m
104304	157041		12 00 44.4	−10 26 46	+0.008	−0.48	5.55	0.77	3.2	K0 IV	0	13 ts		
104329	223167		12 00 48.2	−49 34 07	0.000	−0.01	8.0	1.3	0.2	K0 III		250 s		
104346	239581		12 00 49.9	−58 04 14	−0.001	−0.01	7.9	−0.1	0.4	B9.5 V		320 s		
104337	157042		12 00 51.1	−19 39 32	−0.001	+0.01	5.2	0.1	−3.0	B1.5 V	+2	440 s		
104321	119164	8 π Vir	12 00 52.3	+6 36 51	0.000	−0.03	4.66	0.13	1.7	A3 V	−23	37 s		
104316	7480		12 00 52.5	+70 14 16	−0.003	+0.01	6.7	0.1	0.6	A0 V		170 s		
104361	--		12 00 59.0	−56 05 24			6.80	0.1	−5.1	B1 II		1900 s		
104356	138533		12 01 01.7	−1 46 05	−0.001	−0.07	6.31	1.21	0.3	gG8	+36	100 s		
104367	119169		12 01 10.0	+3 37 56	−0.004	−0.03	8.0	0.5	3.4	F5 V		82 s		
104379	62799		12 01 13.1	+30 38 19	−0.008	−0.04	7.9	0.5	4.0	F8 V		59 s		
104381	99910		12 01 13.7	+12 22 40	−0.003	0.00	6.92	0.10	0.6	A0 V		160 s		
104414	180388		12 01 23.2	−20 39 50	−0.002	−0.04	7.4	0.9	0.2	K0 III		280 s		
104390	15690		12 01 23.5	+66 07 22	−0.003	−0.01	7.2	1.1	−0.1	K2 III		290 s		
104417	203037		12 01 24.9	−34 03 02	−0.001	−0.03	7.0	1.4	0.2	K0 III		140 s		
104430	239589		12 01 28.9	−57 30 13	−0.009	−0.02	6.16	0.00	0.2	B9 V		89 mx		
104436	15691		12 01 37.3	+64 56 19	−0.007	−0.05	7.3	0.1	1.7	A3 V	−5	60 mx		
104435	7485		12 01 37.9	+70 50 59	−0.002	+0.01	7.3	1.1	0.2	K0 III	+3	270 s	8395	m
104438	62802		12 01 39.4	+36 02 31	−0.008	−0.09	5.59	1.01	0.2	K0 III	+30	97 mx		
104452	82106		12 01 44.2	+22 05 39	−0.003	−0.02	6.6	0.5	4.0	F8 V	+11	33 s		
104479	251688		12 01 44.6	−69 11 54	+0.002	−0.03	7.3	1.1	0.2	K0 III		260 s		
104471	203046		12 01 46.0	−34 39 01	−0.016	+0.01	6.91	0.59		G0		10 mn		m
104481	256896		12 01 50.7	−73 17 54	−0.010	−0.01	6.87	0.05		A0				
104484	119174		12 01 56.8	+0 06 08	−0.004	−0.02	7.79	0.30	1.4	A2 V		89 mx		m
104497	180395		12 02 02.4	−20 48 31	−0.003	−0.01	7.7	0.3	0.6	A0 V		170 s		
--	--		12 02 06.5	+43 02 48			6.80							m
104513	44002	67 UMa	12 02 06.7	+43 02 44	−0.030	+0.07	5.21	0.26		A p	+6	21 mn		m
104523	--		12 02 11.4	−64 47 24			7.94	0.09	0.0	B8.5 V		320 s		
104555	258632		12 02 20.5	−85 37 53	−0.049	+0.01	6.05	1.29		K2				m
104568	--		12 02 29.3	−69 23 24			6.96	1.27	0.6	A0 V		32 s		cs?
104575	119180		12 02 35.0	+8 04 10	−0.001	−0.04	7.7	1.6	−0.5	M2 III		440 s		
104600	251699		12 02 37.5	−69 11 32	−0.009	−0.02	5.89	−0.08		B8				
104590	82113		12 02 42.3	+24 26 46	−0.002	−0.13	7.75	1.13	0.2	K0 III		91 mx		
104612	203061		12 02 45.6	−36 59 25	+0.003	+0.04	7.8	1.4	0.2	K0 III		200 s		
104592	138549		12 02 45.8	−6 27 58	0.000	−0.03	8.0	0.4	2.6	F0 V		120 s		
104625	138551		12 02 51.5	−7 41 01	−0.003	+0.01	6.22	1.49		K5				
104622	99924		12 02 53.3	+12 48 27	−0.005	−0.02	7.5	0.4	2.6	F0 V		94 s		
104634	--		12 02 54.0	−68 34 24			7.4	0.5	4.0	F8 V		47 s		
104631	251702		12 02 56.2	−62 10 31	−0.003	−0.01	6.77	0.08	−5.1	B1 II		1600 s		
104649	251703		12 02 58.5	−62 40 18	0.000	0.00	7.90	0.01	−4.4	O9.5 V		2000 s		
104638	157061		12 02 59.8	−10 45 09	+0.003	−0.03	7.6	0.5	3.4	F5 V		68 s		
104641	180409		12 03 00.0	−22 09 08	0.000	+0.01	7.9	1.4	−0.3	K5 III		440 s		
104646	239612		12 03 00.9	−57 15 35	−0.003	0.00	7.2	1.0	0.2	K0 III		250 s		
104644	203066		12 03 01.3	−31 41 30	−0.003	−0.02	7.5	1.1	0.2	K0 III		270 s		
104671	251705	θ¹ Cru	12 03 01.6	−63 18 46	−0.021	0.00	4.33	0.27		A p	−2	17 mn		m
104666	223189		12 03 03.9	−49 39 15	0.000	−0.02	7.3	0.7	0.6	A0 V		77 s		
104664	223191		12 03 06.5	−48 11 42	0.000	−0.02	6.62	0.24		A2				
104668	--		12 03 06.8	−55 11 24			7.5	2.2	−0.5	M2 III		180 s		
104662	203070		12 03 07.4	−38 50 38	−0.003	−0.02	7.8	1.5	0.2	K0 III		170 s		
104677	138553		12 03 13.1	−3 23 44	−0.005	−0.08	7.7	1.1	−0.1	K2 III		130 mx		
104683	251707		12 03 13.7	−64 21 05	0.000	−0.01	7.96	−0.01	−3.5	B1 V		1500 s		
104698	1973		12 03 20.8	+81 41 18	−0.002	0.00	7.6	0.9	3.2	G5 IV		77 s		
104704	239616		12 03 22.3	−55 19 16	+0.004	+0.05	6.9	1.5	0.2	K0 III		120 s		
104705	251709		12 03 23.9	−62 41 44	0.000	+0.01	7.79	−0.15		B0.5 III				
104724	--		12 03 30.8	−72 11 24			7.99	0.80	5.2	G5 V		31 s		

302

HD	SAO	Star Name	α 2000	δ 2000	μ(α)	μ(δ)	V	B-V	M_v	Spec	RV	d(pc)	ADS	Notes
			^{h} ^{m} ^{s}	^{o} $'$ $''$	^{s}	$''$								
104722	251711		12 03 32.7	-61 05 54	0.000	-0.01	7.7	-0.3	0.0	B8.5 V		350 s		
104710	—		12 03 33.1	+29 40 42	0.000	-0.05	7.40	1.56	-0.5	M2 III	-4	380 s		
104728	138556		12 03 35.4	-5 28 47	-0.004	-0.02	7.2	0.1	0.6	A0 V		89 mx		
104725	15696		12 03 35.7	+69 01 14	-0.001	+0.02	7.1	1.1	-0.1	K2 III		270 s		
104747	203084		12 03 36.3	-39 00 32	-0.032	-0.04	6.52	0.51	4.4	G0 V		27 s		m
104731	223193		12 03 39.5	-42 26 03	+0.029	-0.12	5.15	0.41	3.3	F4 V	+37	23 s		
104739	15697		12 03 44.2	+68 56 03	+0.011	-0.05	8.0	1.1	-0.1	K2 III		230 mx		
104755	119189		12 03 44.4	+5 33 28	-0.010	-0.09	6.5	0.5	3.4	F5 V	+6	42 s		
104752	256898		12 03 44.4	-74 12 50	-0.005	0.00	6.44	1.23		K0				
104764	239624		12 03 47.7	-54 42 52	-0.001	-0.06	7.60	0.4	3.0	F2 V		82 s		m
104781	7494		12 03 52.2	+76 04 00	-0.019	-0.01	8.0	0.1	1.7	A3 V		73 mx		
104776	239625		12 03 52.2	-53 29 54	-0.001	-0.02	7.2	1.2	3.2	G5 IV		36 s		
104772	203089		12 03 52.7	-34 48 30	+0.003	+0.01	7.8	1.6	-0.3	K5 III		360 s		
104782	15699		12 03 58.8	+61 24 13	+0.003	+0.01	7.9	1.1	0.2	K0 III		350 s		
104788	157069		12 04 00.3	-10 17 46	0.000	+0.01	6.6	0.1	1.4	A2 V		110 s		
104785	99927		12 04 00.9	+18 48 53	-0.002	-0.09	7.7	1.1	-0.1	K2 III		120 mx		
104810	—		12 04 07.7	-64 31 24			7.36	0.03	0.0	B8.5 V		270 s		
104811	251716		12 04 09.0	-69 31 20	-0.002	-0.02	7.1	0.8	3.2	G5 IV		59 s		
104819	180425		12 04 10.8	-22 22 15	-0.014	-0.01	7.7	1.4	-0.3	K5 III		100 mx		
104817	119195		12 04 12.8	+1 27 42	-0.008	-0.03	7.68	0.21		A m				
104822	239631		12 04 15.6	-57 21 53	-0.003	-0.02	8.0	-0.3	0.6	A0 V		300 s		
104827	82123	2 Com	12 04 16.5	+21 27 33	+0.003	-0.01	5.87	0.22	2.1	F0 IV-V	+5	56 s	8406	m
104839	239632		12 04 16.7	-51 28 20	-0.002	-0.01	6.8	0.3	0.6	A0 V		110 s		
104835	180426		12 04 16.9	-21 02 22	-0.002	-0.03	7.90	0.1	1.4	A2 V		200 s		m
104831	119200		12 04 18.7	+4 55 54	0.000	-0.03	7.5	1.6	-0.5	M2 III		400 s		
104841	251717	θ² Cru	12 04 19.2	-63 09 56	-0.002	0.00	4.72	-0.08	-3.0	B2 IV	+16	290 s		
104843	—		12 04 20.3	-73 36 24			8.00	1.19		K0				
104904	1975		12 04 28.0	+85 35 13	-0.051	+0.09	6.27	0.59	3.4	F5 V	+8	31 s		
104860	15703		12 04 33.7	+66 20 10	-0.009	+0.04	7.9	0.5	4.0	F8 V		62 s		
104872	239635		12 04 37.7	-55 25 50	-0.003	0.00	7.8	0.0	0.6	A0 V		250 s		
104878	251720		12 04 38.7	-68 19 44	-0.008	-0.02	5.35	-0.01	0.6	A0 V	+23	89 s		
104879	—		12 04 38.7	-71 12 24			7.7	1.1	0.2	K0 III		310 s		
104862	62826		12 04 40.4	+35 34 01	-0.001	-0.03	7.40	1.05	0.2	K0 III		250 s		
104883	99928		12 04 42.2	+16 16 10	0.000	-0.01	7.6	0.5	4.0	F8 V		52 s		
104884	119207		12 04 42.9	+3 34 24	-0.001	-0.03	7.2	1.1	0.2	K0 III		250 s		
104887	138565		12 04 44.1	-7 58 36	0.000	-0.01	7.8	0.1	0.6	A0 V		270 s		
104900	239644		12 04 45.1	-59 15 11	-0.005	-0.01	6.20	-0.09	0.4	B9.5 V		150 s		
104902	256899	κ Cha	12 04 46.5	-76 31 08	-0.021	+0.04	5.04	1.49	-0.3	gK5	-2	120 s		
104901	251721		12 04 46.8	-61 59 48	-0.001	0.00	7.43	0.18	0.0	B8.5 V		220 s		m
—	—		12 04 47.6	-61 59 42			7.70							m
104914	203106		12 04 49.9	-32 48 10	-0.004	-0.02	7.6	0.8	3.2	G5 IV		71 s		
104920	256900		12 04 51.8	-72 27 31	-0.003	-0.06	7.8	0.0	0.4	B9.5 V		120 mx		
104933	251723		12 04 57.0	-60 58 08	-0.005	-0.04	5.96	1.67		Ma				
104938	—		12 04 57.8	-75 40 24			7.74	1.63		K2				
104954	7497		12 04 59.4	+76 45 44	-0.030	-0.08	7.4	0.5	3.4	F5 V		64 s		
104936	251724		12 05 00.0	-70 00 06	+0.005	0.00	8.0	1.4	-0.3	K5 III		440 s		
104950	239648		12 05 03.4	-51 11 27	-0.003	+0.02	7.3	1.5	-0.1	K2 III		200 s		
104956	62831		12 05 07.1	+33 44 59	-0.008	-0.07	7.4	0.5	4.0	F8 V		49 s		
104971	251727		12 05 08.4	-61 10 47	-0.005	-0.03	6.64	0.96	0.2	K0 III		190 s		m
104977	99933		12 05 09.7	+13 28 22	0.000	-0.01	7.62	1.47	-0.3	K5 III		380 s		
104983	203112		12 05 11.6	-34 12 33	-0.004	-0.01	7.4	1.9	-0.5	M2 III		240 s		
104986	7499		12 05 12.1	+73 27 15	-0.010	+0.01	7.60	0.9	0.2	G9 III	-54	300 s		
104979	119213	9 o Vir	12 05 12.5	+8 43 59	-0.015	+0.04	4.12	0.98	0.3	G8 III	-30	34 ts		
104982	180442		12 05 13.3	-28 43 00	+0.016	-0.39	7.82	0.65	5.0	G4 V		32 ts		
104985	7500		12 05 15.1	+76 54 20	+0.044	-0.10	5.80	1.01	1.7	K0 III-IV	-20	61 s		
104991	223206		12 05 20.0	-50 16 13	0.000	0.00	6.7	1.4	0.2	K0 III		120 s		
104990	223207		12 05 20.0	-49 00 25	-0.005	-0.01	7.5	1.3	0.2	K0 III		200 s		
105016	251730		12 05 22.8	-62 58 34	-0.013	-0.01	6.69	0.29	1.4	A2 V		72 mx		
104997	62834		12 05 23.8	+38 39 24	+0.001	-0.03	8.00	0.5	3.4	F5 V		83 s		m
104999	99936		12 05 25.1	+17 17 22	-0.004	-0.05	7.7	0.4	3.0	F2 V		86 s		
105000	99937		12 05 27.5	+10 57 38	+0.002	-0.01	7.8	0.4	3.0	F2 V		89 s		
105007	203114		12 05 27.9	-37 33 09	-0.002	-0.03	7.8	1.5	-0.3	K5 III		410 s		
105029	15708		12 05 32.8	+68 45 33	-0.010	-0.02	7.74	1.59	-0.5	M2 III	+12	360 mx		
105028	15709		12 05 33.1	+68 47 38	-0.005	-0.04	7.37	1.04	0.2	K0 III	-24	210 mx	8413	m
105031	28241		12 05 34.2	+51 55 53	+0.001	0.00	7.0	0.9	0.3	G8 III	-18	220 s	8414	m
105036	138579		12 05 35.1	-5 50 46	-0.001	-0.01	6.90	1.6	-0.5	M2 III	-19	300 s		m
105043	15710		12 05 39.6	+62 55 58	-0.008	-0.08	6.13	1.17	-0.1	K2 III	-26	110 mx	8417	m
105061	138582		12 05 48.8	-5 51 30	-0.001	-0.07	7.6	1.1	0.2	K0 III		210 mx		
105056	251733		12 05 49.9	-69 34 24	0.000	-0.02	7.41	0.07	-6.2	B0 I pe		3700 s		
105059	119216		12 05 50.9	+5 21 26	0.000	-0.02	7.9	0.2	2.1	A5 V		150 s		
105071	251734		12 05 53.3	-65 32 50	-0.004	-0.02	6.33	0.22	-6.8	B8 Ia-Iab		300 mx		
105078	203119		12 05 56.6	-35 41 38	-0.003	0.00	6.23	-0.08		B9				m
105086	99938		12 05 59.7	+16 46 21	+0.004	-0.10	7.7	1.1	0.2	K0 III		200 mx		
105089	138585		12 05 59.7	-3 07 54	-0.002	-0.02	6.37	1.00	0.3	gG8	+17	150 s		
105095	203121		12 06 00.2	-39 13 13	-0.006	-0.03	7.6	0.7	1.4	A2 V		67 s		

HD	SAO	Star Name	α 2000	δ 2000	μ(α)	μ(δ)	V	B-V	M_v	Spec	RV	d(pc)	ADS	Notes
			h m s	° ′ ″	s	″								
105085	82136		12 06 00.8	+23 12 14	−0.001	−0.03	7.49	0.33	3.4	F5 V		66 s		
105096	239662		12 06 01.2	−54 15 28	+0.005	+0.01	7.2	1.4	0.2	K0 III		150 s		
105122	15713		12 06 01.5	+68 41 56	−0.002	−0.02	7.10	0.46		F5	−12		8419	m
105113	203123		12 06 05.1	−32 57 41	−0.006	−0.20	6.70	0.7	4.4	G0 V		29 s		m
105123	44031		12 06 06.0	+49 09 31	−0.012	+0.01	7.8	0.5	3.4	F5 V		77 s		
105119	251735		12 06 08.3	−69 07 31	−0.008	+0.05	7.3	0.6	4.4	G0 V		38 s		
105121	256902		12 06 11.1	−73 33 23	−0.004	+0.01	6.50	1.83		Ma				
105115	203126		12 06 12.8	−38 40 32	−0.003	−0.02	7.0	0.5	−0.1	K2 III		260 s		
105134	223218		12 06 14.9	−44 30 44	−0.001	−0.01	8.0	0.7	−0.1	K2 III		410 s		
105144	180465		12 06 16.3	−22 47 46	−0.002	+0.01	7.3	0.7	0.2	K0 III		260 s		
105139	251736		12 06 18.0	−69 47 55	−0.005	−0.02	7.57	0.02	−2.9	B3 III		250 mx		
105140	44032		12 06 19.2	+46 17 03	−0.005	−0.02	7.54	1.50		K4 IV		260 mx		
105138	251737		12 06 19.8	−68 39 03	−0.001	+0.01	6.23	1.24		K0				
105150	239665		12 06 19.9	−59 08 42	−0.002	−0.02	7.66	0.13	1.4	A2 V		160 s		
105151	251738		12 06 23.1	−65 42 33	−0.007	0.00	5.93	0.60	2.3	F7 IV		42 s		m
105173	203137		12 06 37.7	−37 51 36	−0.004	0.00	6.70	1.1	0.2	K0 III		200 s		m
105181	62839		12 06 39.4	+33 34 00	0.000	−0.02	7.92	1.51	−0.3	K4 III		390 s		
105197	28250		12 06 44.1	+57 36 03	−0.013	−0.01	7.2	0.4	2.6	F0 V		85 s		
105205	157086		12 06 47.2	−12 14 27	−0.008	−0.01	6.7	0.4	2.6	F0 V		67 s		
105211	251742	η Cru	12 06 52.8	−64 36 49	+0.005	−0.04	4.15	0.34	0.6	F0 III	+9	33 ts		m
105215	44033		12 06 55.7	+47 52 30	0.000	+0.01	8.00	1.1	−0.1	K2 III		420 s		
105241	180479		12 07 01.4	−23 45 59	−0.007	+0.03	7.0	0.9	4.0	F8 V		25 s		
105234	256904		12 07 05.2	−78 44 26	−0.017	0.00	7.2	0.1	1.4	A2 V		110 mx		
105259	44038		12 07 07.9	+47 17 54	−0.003	+0.01	8.00	1.4	−0.3	K4 III		460 s		
105263	119224		12 07 10.7	+9 39 46	−0.005	−0.01	7.3	0.1	1.7	A3 V		100 mx		
105262	99943		12 07 10.9	+12 59 08	+0.003	−0.04	7.08	0.00	0.4	B9.5 V	+41	210 s		
105264	138591		12 07 11.3	−0 37 48	−0.002	−0.01	7.7	1.1	0.2	K0 III		320 s		
105250	180482		12 07 11.7	−22 22 14	−0.001	−0.01	8.0	1.6	−0.5	M2 III		510 s		
105265	138592		12 07 12.9	−6 05 02	−0.006	+0.04	7.60	1.1	−0.1	K2 III		190 mx	8429	m
105266	138594		12 07 14.9	−6 45 56	−0.001	+0.01	7.00	1.6		M5 III				RW Vir, v
105274	251749		12 07 16.0	−64 40 56	−0.024	−0.01	6.84	0.45	3.4	F5 V		49 s		
105279	99946		12 07 16.1	+10 31 53	−0.001	−0.08	8.0	0.9	3.2	G5 IV		93 s		
105295	203147		12 07 22.8	−32 39 44	−0.002	−0.03	7.5	0.1	0.6	A0 V		200 s		
105283	223224		12 07 22.8	−43 14 48	−0.002	−0.01	7.21	0.42	0.2	G9 III		250 s		
105292	157091		12 07 27.7	−18 11 25	−0.001	+0.03	7.1	1.1	0.2	K0 III		240 s		
105303	82143		12 07 28.0	+23 02 33	−0.006	+0.01	7.61	0.28	3.0	F2 V		84 s		
105313	223226		12 07 32.6	−45 34 52	−0.003	−0.01	7.31	0.02		A0				
105307	119227		12 07 32.7	+4 30 46	−0.002	−0.02	7.9	1.1	0.2	K0 III		340 s		
105330	203150		12 07 35.3	−31 24 39	−0.008	−0.12	6.9	0.5	4.0	F8 V		37 s		
105319	44040		12 07 35.5	+42 04 13	0.000	−0.04	7.63	1.58	−0.3	K4 III		310 s		
105321	99953		12 07 35.9	+13 31 00	−0.001	−0.01	7.65	1.01	0.2	K0 III		300 s		
105329	180487		12 07 36.7	−29 03 46	−0.002	0.00	7.8	0.8	1.4	A2 V		62 s		
105328	180488		12 07 38.9	−23 58 32	+0.008	−0.24	6.7	1.2	4.4	G0 V		11 s		
105343	138597		12 07 47.1	−8 29 11	−0.008	−0.01	7.5	0.5	3.4	F5 V		66 s		
105340	256905		12 07 50.0	−75 22 00	−0.023	+0.02	5.18	1.30	−0.4	gM0	−45	130 s		
105353	239683		12 07 51.3	−59 17 34	−0.002	+0.02	8.02	−0.05	−0.6	B7 V		510 s		
105374	119234		12 08 00.5	+0 37 16	0.000	−0.03	7.2	0.4	3.0	F2 V		68 s		
105383	239686		12 08 04.7	−50 45 47	−0.004	0.00	6.37	−0.05		B9 n	+15			m
105382	239687		12 08 05.1	−50 39 40	−0.004	−0.01	4.47	−0.15	−1.6	B5 IV	+17	160 s		m
105388	62851		12 08 05.6	+31 03 00	−0.002	−0.02	7.45	−0.02		A0	−7			
105421	28253		12 08 06.7	+55 27 51	−0.022	−0.02	7.76	0.53	4.0	dF8	+4	57 s	8434	m
105405	62852		12 08 07.1	+39 39 19	−0.011	−0.01	7.4	0.5	4.0	F8 V		47 s		
105401	251754		12 08 08.3	−60 46 51	−0.013	+0.04	7.6	0.4	3.2	G5 IV		77 s		
105409	138602		12 08 12.5	−4 17 13	+0.001	+0.01	7.6	1.1	0.2	K0 III		300 s		
105416	223235		12 08 14.6	−48 41 33	−0.003	−0.02	5.34	−0.01	1.2	A1 V	+6	67 s		
105424	62856		12 08 15.4	+30 16 31	0.000	−0.01	7.60	1.53		K3 IV				
105440	28255		12 08 18.0	+58 33 24	−0.005	+0.02	7.6	0.9	3.2	G5 IV		77 s		
105434	203162		12 08 19.3	−34 40 31	−0.005	−0.01	6.7	1.6	0.2	K0 III		92 s		
105435	239689	δ Cen	12 08 21.5	−50 43 20	−0.003	−0.01	2.60	−0.12	−2.5	B2 V e	+9	100 s		m
105437	251757		12 08 24.6	−60 50 50	−0.001	0.00	6.22	1.74		K2				
105452	180505	1 α Crv	12 08 24.7	−24 43 44	+0.006	−0.04	4.02	0.32	1.9	F2 IV	+4	21 ts		Alchiba
105458	44048		12 08 25.9	+48 58 06	0.000	−0.01	7.9	0.2	2.1	A5 V		140 s		
105485	15723		12 08 33.5	+65 20 58	+0.005	−0.01	7.9	0.5	3.4	F5 V		78 s		
105475	82152		12 08 37.2	+26 29 38	+0.005	−0.03	6.97	1.02	0.2	K0 III	+2	220 s		
105482	239695		12 08 39.0	−51 32 10	−0.002	−0.01	7.8	0.7	0.2	K0 III		330 s		
105491	157103		12 08 43.0	−18 26 06	−0.006	+0.01	7.1	0.4	3.0	F2 V		67 s		
105525	44052		12 08 53.7	+49 11 08	0.000	0.00	7.45	1.03	0.2	K0 III		270 s		
105509	223241		12 08 53.7	−44 19 34	−0.005	−0.05	5.75	0.24	0.0	A3 III		120 s		
105521	223242		12 08 54.4	−41 13 53	−0.002	−0.01	5.48	−0.07		B3	0			
105529	203172		12 09 05.2	−33 05 09	−0.002	−0.04	8.0	2.1	−0.3	K5 III		190 s		
105515	256910		12 09 07.3	−78 46 53	−0.019	−0.02	6.6	0.1	1.7	A3 V		97 s		
105548	99962		12 09 12.9	+17 11 10	+0.001	−0.03	7.23	1.62	−0.5	M1 III	+36	330 s		
105567	99963		12 09 14.1	+14 49 48	−0.003	−0.04	7.8	0.5	4.0	F8 V		57 s		
105545	251763		12 09 16.5	−66 06 30	−0.004	−0.02	7.6	0.1	0.6	A0 V		240 s		
105568	99964		12 09 16.8	+11 17 30	−0.003	−0.04	8.00	0.5	3.4	F5 V		83 s	8438	m

HD	SAO	Star Name	α 2000	δ 2000	μ(α)	μ(δ)	V	B−V	M$_v$	Spec	RV	d(pc)	ADS	Notes
			h m s	° ′ ″	s	″								
105577	223244		12 09 20.3	−42 50 18	−0.004	−0.01	7.93	0.53	4.0	F8 V		47 s		
105584	28262		12 09 20.4	+53 23 40	+0.007	−0.11	7.6	0.4	3.0	F2 V		83 s		
105586	−−		12 09 21.5	+29 27 37	−0.003	−0.02	8.03	1.00	0.2	G9 III		350 s		
105571	157109		12 09 22.1	−12 56 24	+0.001	−0.02	7.7	1.1	0.2	K0 III		310 s		
105580	239702		12 09 23.4	−59 46 10	−0.003	−0.01	7.16	−0.06	−0.9	B6 V		400 s		
105589	138613		12 09 25.7	−0 45 33	−0.003	−0.07	7.8	0.4	3.0	F2 V		93 s		
105601	62866		12 09 27.6	+38 37 54	−0.004	−0.07	7.38	0.29		A2				
105590	157111		12 09 28.4	−11 51 26	+0.021	−0.17	6.56	0.66	4.7	dG2	+5	25 ts	8440	m
105605	180521		12 09 31.9	−21 03 13	0.000	−0.03	8.0	0.0	1.4	A2 V		210 s		
105610	239706		12 09 36.5	−52 08 40	−0.002	−0.02	7.4	0.7	0.4	B9.5 V		96 s		
105631	44058		12 09 37.2	+40 15 07	−0.028	−0.06	7.47	0.79	5.5	G7 V	−3	22 s		
105615	251768		12 09 38.2	−67 27 15	−0.006	−0.01	7.2	0.0	0.4	B9.5 V		210 mx		
105613	239707		12 09 38.7	−58 20 58	−0.005	−0.01	7.6	0.0	1.4	A2 V		160 mx		
105632	62868		12 09 40.3	+33 05 13	−0.003	−0.02	7.57	1.12	0.0	K1 III		300 s		
105639	119245	10 Vir	12 09 41.2	+1 53 52	+0.003	−0.18	5.95	1.12	−0.2	gK3	+3	170 s		m
105646	239710		12 09 46.4	−51 47 02	−0.001	0.00	7.00	0.5	4.0	F8 V		40 s		m
105678	7512		12 09 47.2	+74 39 41	0.000	0.00	6.35	0.50	3.4	F5 V	−19	36 s		
105654	138617		12 09 47.6	−5 13 34	−0.004	0.00	7.2	0.4	3.0	F2 V		71 s		
105643	203180		12 09 47.8	−33 29 15	−0.002	−0.01	7.7	0.3	0.6	A0 V		160 s		
105717	7513		12 09 48.8	+77 23 26	−0.037	+0.05	6.9	0.9	3.2	G5 IV		56 s		
105679	44059		12 09 55.2	+42 18 35	−0.008	+0.02	8.00	0.9	5.5	G7 V	+4	32 s		
105686	203183		12 10 02.4	−34 42 18	−0.004	−0.02	6.17	0.03		A0				m
105702	119249	11 Vir	12 10 03.3	+5 48 25	−0.011	+0.02	5.72	0.35		A m	−9			
105704	157120		12 10 05.7	−17 32 16	−0.006	−0.05	6.7	0.5	3.4	F5 V		45 s		
105690	223254		12 10 06.1	−49 10 49	−0.018	−0.04	7.8	1.1	3.2	G5 IV		53 s		
105707	180531	2 ε Crv	12 10 07.4	−22 37 11	−0.005	+0.01	3.00	1.33	−0.1	K2 III	+5	32 s		
105699	99969		12 10 07.8	+16 25 37	0.000	−0.02	7.06	1.33	−0.1	K2 III		220 s		
105730	−−		12 10 14.7	−19 46 23			7.70	1.4	−0.3	K5 III	−6	400 s		
105733	203189		12 10 16.7	−35 03 23	−0.001	−0.03	7.9	0.7	3.0	F2 V		56 s		
105748	203191		12 10 20.6	−35 14 07	−0.008	−0.01	7.8	0.6	4.0	F8 V		54 s		
105759	138625		12 10 27.1	−7 46 26	−0.001	+0.02	6.6	0.1	0.6	A0 V		160 s		
105771	82165		12 10 27.8	+29 04 06	−0.003	0.00	7.50	0.99	0.2	K0 III	−5	290 s		
105778	99973	3 Com	12 10 31.5	+16 48 33	−0.001	−0.01	6.39	0.06	1.4	A2 V	−11	97 s		
105776	203195		12 10 33.7	−37 52 12	+0.003	−0.03	6.06	0.20		A2				
105785	239716		12 10 42.0	−56 26 33	−0.005	−0.01	7.3	0.4	1.4	A2 V		93 s		
105788	239719		12 10 45.8	−59 16 39	0.000	−0.01	7.3	1.1	0.2	K0 III		240 s		
105805	82166		12 10 46.0	+27 16 53	−0.001	−0.02	6.01	0.11	1.9	A4 V	−9	66 s		
105824	44064		12 10 47.3	+39 53 29	−0.003	−0.02	6.85	0.28	1.7	A3 V		88 s	8446	m
105811	138630		12 10 48.2	−9 23 51	−0.003	−0.06	7.4	1.1	0.2	K0 III		200 mx		
105842	28279		12 10 51.0	+51 47 46	−0.001	−0.02	7.5	1.1	0.2	K0 III		290 s		
105814	180541		12 10 52.2	−28 04 03	−0.001	−0.03	7.7	1.1	0.2	K0 III		290 s		
105815	203202		12 10 52.6	−33 28 07	+0.003	−0.07	7.6	0.8	0.2	K0 III		210 mx		
105817	239722		12 10 54.6	−56 25 33	−0.003	−0.02	7.2	1.5	0.6	A0 V		27 s		
105837	223263		12 10 57.7	−46 19 18	+0.008	+0.17	7.53	0.56		G0				
105943	1991		12 10 59.9	+81 42 36	−0.014	0.00	6.00	1.62	0.2	K0 III	−27	61 s		m
105834	223265		12 11 00.2	−40 57 08	−0.005	−0.01	7.4	1.1	0.2	K0 III		140 s		
105822	251776		12 11 01.2	−68 15 40	+0.001	−0.03	7.4	1.1	0.2	K0 III		270 s		
105846	119256		12 11 01.5	+8 43 03	−0.004	+0.03	7.6	1.1	0.2	K0 III		270 mx		
105865	99975		12 11 02.1	+18 37 29	−0.003	−0.03	8.03	1.14	−0.1	K2 III		420 s		
105852	223266		12 11 02.8	−45 25 21	−0.002	+0.01	6.61	1.08		K0				m
105850	180546	3 Crv	12 11 03.8	−23 36 08	−0.005	−0.02	5.46	0.06		A2	+11	16 mn		
105841	251778		12 11 04.9	−61 16 39	−0.018	−0.02	6.08	0.39	0.6	F0 III		110 mx		
105857	239727		12 11 05.8	−56 24 04	−0.004	−0.01	7.3	0.6	1.4	A2 V		68 s		
105879	28282		12 11 06.6	+54 21 32	−0.001	−0.01	7.9	1.1	0.2	K0 III		350 s		
105770	258636		12 11 08.8	−83 46 39	−0.014	+0.01	7.4	−0.5	0.4	B9.5 V		190 mx		
105881	62883		12 11 09.5	+39 07 49	−0.001	+0.03	8.0	0.5	3.4	F5 V		82 s		
105874	239730		12 11 14.8	−52 13 02	−0.003	0.00	8.00	0.1	1.4	A2 V		200 s		m
105898	82172		12 11 15.1	+24 45 12	−0.008	−0.03	7.54	0.86	4.7	G2 V		26 s		
105912	138638		12 11 21.6	−3 46 44	−0.008	−0.06	6.9	0.5	3.4	F5 V		51 s		
105910	82173		12 11 21.7	+22 35 37	0.000	−0.01	7.50	0.26	2.1	A5 V		110 s		
105900	138637		12 11 22.0	−2 41 57	+0.001	−0.08	7.1	1.1	0.2	K0 III		220 mx		
105913	157133		12 11 22.9	−16 47 28	−0.010	−0.06	7.20	0.9	3.2	G5 IV		63 s	8444	m
105911	119259		12 11 25.4	+0 11 35	−0.001	−0.02	7.7	1.1	0.2	K0 III		320 s		
105963	28286		12 11 26.4	+53 25 07	−0.023	−0.14	8.03	0.88	0.2	K0 III	−9	52 mx	8450	m
105919	223269		12 11 28.5	−44 16 59	+0.012	−0.05	6.59	0.46		F5				
105920	239735		12 11 31.3	−51 21 33	−0.022	−0.07	6.23	0.82		K0				
105946	62886		12 11 32.0	+34 15 30	+0.004	−0.11	7.2	0.4	3.0	F2 V		68 s		
105934	180549		12 11 33.3	−27 04 15	−0.002	0.00	8.0	1.3	−0.5	M2 III		510 s		
105923	256911		12 11 37.7	−71 10 36	−0.011	−0.01	7.9	1.1	0.2	K0 III		340 s		
105938	239736		12 11 38.3	−54 25 02	+0.009	+0.01	7.0	0.4	4.4	G0 V		33 s		
105937	239737	ρ Cen	12 11 39.1	−52 22 07	−0.004	−0.02	3.96	−0.15	−1.4	B4 V	+21	120 s		
105967	119262		12 11 40.1	+4 03 21	0.000	0.00	6.93	0.15	0.6	A0 V	−5	150 s		
105968	157136		12 11 40.2	−10 34 15	−0.001	−0.02	7.2	0.8	3.2	G5 IV		62 s		
105956	223270		12 11 40.3	−47 14 24	−0.002	0.00	7.82	1.13	0.0	K1 III		370 s		
105953	203219		12 11 41.6	−30 36 12	−0.005	0.00	7.60	1.1	0.2	K0 III		300 s		m

HD	SAO	Star Name	α 2000	δ 2000	μ(α)	μ(δ)	V	B-V	M_v	Spec	RV	d(pc)	ADS	Notes
105939	251782		12^{h}11^{m}44.1^s	-64°56′59″	0.000^s	-0.09″	7.4	1.1	0.2	K0 III		130 mx		
106002	28291	68 UMa	12 11 44.8	+57 03 16	+0.001	-0.02	6.4	1.4	-0.3	K5 III	+35	220 s		
105982	82177		12 11 46.0	+23 19 20	+0.002	-0.03	6.71	1.04	-0.1	K2 III		230 s		
105973	239739		12 11 47.2	-50 57 23	-0.003	+0.01	7.7	0.4	0.6	A0 V		160 s		
106053	7521		12 11 48.3	+77 26 27	+0.001	+0.01	6.73	0.03		A0	-15			
105974	239740		12 11 48.3	-53 05 36	-0.002	-0.01	7.9	1.1	-0.1	K2 III		390 s		
105989	157137		12 11 49.2	-19 55 45	-0.007	-0.01	7.3	0.6	4.0	F8 V		42 s		
105981	82178	4 Com	12 11 51.0	+25 52 13	-0.003	-0.03	5.7	1.1	6.3	K2 V	+22	7.7 s		d?
106007	180556		12 11 59.0	-25 56 29	+0.001	-0.01	7.4	1.2	-0.1	K2 III		310 s		
105999	251784		12 11 59.9	-63 22 54	-0.003	-0.06	7.7	0.4	3.0	F2 V		86 s		
106000	251785		12 12 00.5	-64 30 38	-0.003	-0.02	6.55	0.26		A2				
106022	82181		12 12 01.1	+28 32 10	+0.007	-0.07	6.49	0.36	3.0	F2 V	-15	49 s		
--	--		12 12 02.0	+28 32 19			7.00							m
106056	44075		12 12 02.6	+46 36 04	-0.004	+0.02	7.9	0.5	3.4	F5 V		78 s		
106013	223275		12 12 02.7	-49 49 17	-0.002	-0.02	8.0	1.0	0.2	K0 III		360 s		
105941	258637		12 12 02.9	-81 03 57	-0.007	-0.03	8.0	1.1	0.2	K0 III		360 s		
106072	15736		12 12 05.3	+64 27 49	-0.005	-0.02	7.7	1.1	-0.1	K2 III		350 mx		
106057	82182	5 Com	12 12 09.2	+20 32 31	-0.001	-0.03	5.57	0.95	-0.9	K0 II-III	-25	200 s		
106036	251788		12 12 10.3	-63 27 15	-0.006	-0.02	7.01	0.09	0.6	A0 V		170 s		
106112	7522		12 12 11.8	+77 36 58	+0.003	+0.02	5.14	0.33		A m	0	40 t		
106059	138644		12 12 14.2	-4 24 08	+0.001	-0.02	7.9	1.1	-0.1	K2 III		390 s		
106076	119270		12 12 17.7	+5 52 21	0.000	-0.05	7.9	1.1	0.2	K0 III		340 s		
106102	28298		12 12 17.7	+55 42 22	-0.017	-0.01	7.21	1.20		K2 III-IV		140 mx		
106078	157145		12 12 18.5	-19 52 22	-0.003	-0.01	7.3	0.7	3.2	G5 IV		67 s		
106068	251790		12 12 22.0	-62 57 03	0.000	0.00	5.92	0.29	-7.1	B9 Ia		2900 s		
106086	239751		12 12 25.0	-60 04 03	-0.013	+0.01	7.1	0.3	2.6	F0 V		79 s		
106083	223278		12 12 27.1	-43 59 58	-0.003	0.00	7.93	1.16		K0				
106116	138647		12 12 28.8	-3 05 06	-0.040	+0.40	7.45	0.70	5.0	G4 V	+11	27 ts		m
106098	203233		12 12 28.9	-33 24 19	-0.002	-0.02	7.0	0.6	1.4	A2 V		60 s		
106105	119272		12 12 29.9	+3 24 14	-0.005	+0.02	8.0	0.5	3.4	F5 V		82 s		
106101	239753		12 12 35.8	-52 46 40	-0.003	-0.03	7.6	0.4	0.6	A0 V		140 s		
106121	203236		12 12 39.2	-35 05 56	0.000	-0.06	7.8	0.8	0.6	A0 V		81 s		
106127	138648		12 12 42.0	-2 28 21	-0.001	-0.04	7.40	1.4	-0.3	K5 III	+8	350 s		
106150	28303		12 12 45.6	+51 56 24	-0.002	-0.01	7.25	0.91		K0				
106111	251791		12 12 46.9	-70 09 07	-0.003	-0.02	6.2	0.5	4.0	F8 V	0	28 s		S Mus, v
106130	203238		12 12 47.9	-37 31 38	+0.001	-0.01	7.6	1.0	-0.1	K2 III		340 s		
106152	62897		12 12 51.9	+31 47 46	+0.002	-0.03	7.77	0.54		F5				
106132	239757		12 12 52.1	-59 39 18	-0.001	0.00	7.13	1.38	0.2	K0 III		130 s		
106158	157151		12 12 56.7	-12 47 27	-0.003	-0.02	8.0	0.5	3.4	F5 V		82 s		
106156	99991		12 12 57.5	+10 02 14	+0.014	-0.38	7.92	0.79	5.6	G8 V	-9	27 s		
106208	--		12 13 01.5	+59 43 12	-0.004	-0.02	7.9	0.9	3.2	G5 IV		88 s		
106146	251793		12 13 01.9	-62 18 46	+0.003	0.00	7.80	0.29	0.0	B8.5 V		230 s		
106184	82189		12 13 03.0	+28 38 06	0.000	0.00	7.70	1.4	-0.3	K4 III		400 s		
106179	203243		12 13 04.5	-31 31 11	-0.001	-0.06	7.5	1.2	0.2	K0 III		220 s		
106197	203246		12 13 12.8	-33 20 00	-0.004	-0.02	7.3	1.1	0.2	K0 III		230 s		
106198	203245		12 13 12.9	-34 07 32	-0.003	-0.01	6.5	1.9		Mb				
106210	99995		12 13 13.2	+10 49 17	+0.002	-0.60	7.57	0.67	4.9	G3 V	-30	29 mx		
106200	223285		12 13 13.2	-41 48 10	-0.002	-0.01	7.9	1.3	3.4	F5 V		51 s		
106209	99996		12 13 13.9	+11 30 58	-0.001	-0.01	7.52	1.56	-0.3	K5 III		330 s		
106223	62899		12 13 16.8	+30 17 00	-0.005	+0.03	7.43	0.28		A2 pe	-17			
106224	82191		12 13 17.5	+22 01 35	-0.003	+0.01	7.52	0.11		A2	-26			
106231	203250		12 13 25.2	-38 55 44	-0.001	-0.01	5.76	-0.13		B5	-47			
106251	99997	12 Vir	12 13 25.8	+10 15 44	-0.007	-0.02	5.85	0.27		A m	+2			
106252	99998		12 13 29.3	+10 02 31	+0.001	-0.27	7.36	0.64	4.4	G0 V		36 s		
106250	99999		12 13 29.3	+17 10 52	-0.005	-0.04	7.8	1.1	-0.1	K2 III		170 mx		
106268	138654		12 13 36.1	-1 14 16	-0.007	-0.05	8.0	0.4	3.0	F2 V		98 s		
106257	203252		12 13 36.6	-33 47 34	-0.001	-0.02	8.0	0.08		B9				m
106270	138655		12 13 37.2	-9 30 47	-0.006	-0.03	7.9	0.9	3.2	G5 IV		88 s		
106284	180586		12 13 44.7	-20 19 36	-0.011	+0.02	7.5	0.5	3.0	F2 V		63 s		
106261	251800		12 13 45.5	-62 55 52	-0.002	+0.02	7.96	0.13	-3.0	B2 IV		1000 s		
106295	119277		12 13 50.2	+8 58 20	-0.001	-0.03	8.0	0.1	1.4	A2 V		210 s		
106316	138658		12 13 52.7	-0 38 22	-0.003	-0.01	7.9	1.1	-0.1	K2 III		390 s		
106314	119278		12 13 55.7	+2 15 36	-0.006	-0.03	7.0	0.8	3.2	G5 IV		58 s		
106248	256915		12 13 55.8	-78 34 24	-0.009	-0.01	6.35	1.24		K2				
106308	239770		12 13 57.2	-57 56 39	-0.002	-0.01	7.90	0.12	0.6	A0 V		240 s		
106306	239771		12 13 57.7	-53 08 36	-0.003	-0.01	7.8	0.3	0.6	A0 V		170 s		
106336	203259		12 14 00.9	-32 23 13	-0.014	+0.02	7.5	0.6	3.2	G5 IV		72 s		
106309	239772		12 14 01.6	-59 23 48	-0.003	+0.01	7.83	-0.03	-1.1	B5 V e		570 s		
106307	239773		12 14 02.1	-55 28 57	+0.001	-0.03	7.2	1.9	0.2	K0 III		71 s		
106321	223297		12 14 02.5	-45 43 26	-0.004	0.00	5.31	1.43		K5	+7	22 mn		m
106349	100002		12 14 03.9	+12 08 51	-0.002	+0.01	7.93	1.16	0.2	K0 III		260 s		m
106381	15747		12 14 05.5	+66 06 32	-0.006	-0.01	6.7	1.1	0.2	K0 III		200 s		
106338	203261		12 14 06.0	-36 43 46	-0.005	-0.04	7.60	0.1	1.4	A2 V		84 mx		m
106365	62904		12 14 06.5	+32 47 03	-0.009	-0.01	6.88	1.13	-0.1	K2 III	-10	170 mx	8470	m
106337	203262		12 14 06.9	-36 33 23	-0.001	-0.01	7.86	-0.09	-0.9	B6 V		560 s		m

HD	SAO	Star Name	α 2000	δ 2000	μ(α)	μ(δ)	V	B-V	M_v	Spec	RV	d(pc)	ADS	Notes
			h m s	° ′ ″	s	″								
106328	—		12 14 13.3	-72 07 22			7.8	0.0	0.4	B9.5 V		300 s		
106383	62909		12 14 14.0	+33 27 00	-0.005	+0.01	8.04	0.86	0.3	G7 III		230 mx		
106360	239780		12 14 14.3	-53 55 13	-0.007	-0.01	7.2	0.7	0.6	A0 V		74 s		
106384	138664		12 14 15.4	-5 42 59	-0.006	+0.11	6.58	0.25	3.1	dF3	+9	50 s	8471	m
106343	251803		12 14 16.8	-64 24 30	-0.001	0.00	6.22	0.12	-6.8	B2 Ia	-7	2800 s		
106344	251802		12 14 16.8	-66 32 57	-0.004	-0.02	7.14	-0.01	-1.1	B5 V		210 mx		
106398	82199		12 14 18.0	+26 30 17	+0.001	-0.02	7.4	0.9	0.3	G8 III		270 s		
106374	203263		12 14 18.1	-33 46 43	-0.003	+0.01	7.3	0.6	1.4	A2 V		69 s		
106362	251804		12 14 26.4	-66 31 28	+0.001	0.00	7.46	0.00	-6.2	B1 I		4200 s		
106423	119282		12 14 27.3	+8 46 56	+0.004	-0.14	7.60	0.6	4.4	G0 V		44 s	8473	m
106421	82201		12 14 27.4	+23 39 39	-0.003	-0.15	7.4	0.9	3.2	G5 IV		63 mx		
106449	62911		12 14 36.9	+39 20 32	0.000	-0.02	6.98	1.59		K5 IV				
106453	180598		12 14 42.0	-24 46 35	-0.022	-0.08	7.46	0.72		K0		11 mn	8474	m
106478	28309		12 14 43.3	+53 26 04	-0.002	-0.02	6.16	1.04	0.2	K0 III	0	150 s		
106440	223300		12 14 44.4	-44 20 15	+0.001	-0.04	8.0	0.4	-0.1	K2 III		410 s		
106456	223302		12 14 51.6	-44 06 42	-0.003	0.00	7.55	1.11		K0				
106363	258639		12 14 52.3	-81 11 42	+0.002	-0.04	7.6	0.1	1.7	A3 V		150 s		
106489	223305		12 14 57.2	-41 08 23	-0.029	-0.09	7.48	0.65	5.2	G5 V		29 s		
106488	203275		12 14 59.4	-36 13 15	-0.005	-0.01	8.00	0.1	1.4	A2 V		91 mx		m
106485	180602		12 14 59.5	-20 50 39	0.000	0.00	5.83	1.05	0.3	gG7	+16	110 s		
106474	239789		12 14 59.6	-54 49 14	-0.002	-0.03	7.9	1.5	-0.5	M4 III		490 s		V369 Cen, v
106498	138670		12 14 59.9	-1 19 37	-0.003	-0.03	7.70	1.1	0.2	K0 III		320 s		m
106475	239790		12 15 00.0	-56 06 13	-0.004	+0.01	7.4	1.5	0.2	K0 III		140 s		
106515	138673		12 15 06.1	-7 15 27	-0.015	-0.08	7.35	0.82	5.2	dG5	+22	21 s	8477	m
106500	180604		12 15 06.3	-29 14 14	-0.002	-0.03	6.4	1.2	0.2	K0 III		140 s		
106574	7532		12 15 08.4	+70 12 00	-0.005	-0.02	5.71	1.19	0.2	K0 III	-14	97 s		
106490	239791	δ Cru	12 15 08.6	-58 44 55	-0.005	-0.01	2.80	-0.23	-3.0	B2 IV	+26	79 mx		
106516	157168		12 15 10.5	-10 18 45	+0.002	-1.02	6.11	0.46	3.7	F6 V	+6	30 s		m
106542	100009		12 15 13.4	+16 54 27	-0.001	+0.01	6.80	1.22		K2				
106529	119285		12 15 13.8	+4 31 33	-0.009	+0.03	7.9	1.1	0.2	K0 III		130 mx		
106556	44092		12 15 15.5	+47 06 58	0.000	+0.01	7.32	1.01	0.2	K0 III		270 s		
106505	251809		12 15 17.2	-62 53 14	-0.001	+0.02	7.9	0.8	3.2	G5 IV		85 s		
106506	251810		12 15 18.6	-63 25 30	-0.004	-0.02	8.0	0.8	3.2	G5 IV		89 s		
107192	2010		12 15 19.5	+87 42 00	-0.059	+0.05	6.28	0.35	2.6	F0 V	-4	53 s		
106591	28315	69 δ UMa	12 15 25.5	+57 01 57	+0.013	0.00	3.31	0.08	1.7	A3 V	-13	20 ts		Megrez, m
106593	62922		12 15 29.0	+38 39 35	-0.004	0.00	7.69	0.29	2.6	F0 V		100 s		
106572	223309		12 15 30.4	-41 54 47	-0.030	-0.18	6.26	1.01		K0		14 mn		
106581	157172		12 15 30.7	-14 04 20	+0.001	-0.02	7.5	1.1	0.2	K0 III		290 s		
106586	203285		12 15 34.2	-31 41 27	-0.001	-0.04	7.6	1.2	-0.5	M2 III		420 s		
106582	180614		12 15 36.5	-20 31 45	-0.003	+0.03	6.9	0.7	0.2	K0 III		220 s		
106677	7533		12 15 41.3	+72 33 03	-0.004	-0.04	6.29	1.14	0.2	K0 III	-48	140 s		
106590	251816		12 15 45.3	-61 07 47	0.000	-0.01	7.95	0.24	-5.7	B1 Ib		3000 s		
106612	180619		12 15 46.9	-23 21 13	+0.003	-0.04	6.54	0.45	3.4	F5 V		42 s	8481	m
106625	157176	4 γ Crv	12 15 48.3	-17 32 31	-0.011	+0.02	2.59	-0.11	-1.2	B8 III	-4	57 s		Gienah
106624	157177		12 15 49.8	-13 49 00	-0.001	0.00	7.5	1.4	-0.3	K5 III		360 s		
106614	239796		12 15 55.5	-54 03 25	-0.003	-0.02	6.7	1.6	-0.3	K5 III		220 s		
106661	100012	6 Com	12 16 00.1	+14 53 57	-0.006	-0.03	5.10	0.06	1.4	A2 V	+10	48 mx		
106616	251818		12 16 01.5	-65 11 33	-0.005	0.00	8.02	-0.01	0.0	B8.5 V		190 mx		
106690	44097	2 CVn	12 16 07.5	+40 39 36	+0.001	-0.04	5.66	1.55	-0.5	M1 III	-15	170 s	8489	m
106798	2007		12 16 07.9	+80 07 18	+0.005	-0.01	7.8	0.4	2.6	F0 V	-10	110 s		
106668	223322		12 16 11.5	-45 14 18	-0.002	-0.01	8.0	1.4	-0.3	K5 III		450 s		
106799	2009		12 16 11.7	+80 07 31	+0.007	0.00	7.30	0.4	2.6	F0 V		87 s	8494	m
106715	119299		12 16 15.3	+6 13 23	-0.002	0.00	7.9	1.1	0.2	K0 III		340 s		
106694	138685		12 16 16.6	-3 14 10	+0.001	-0.09	8.0	0.4	3.0	F2 V		98 s		
106714	82211	7 Com	12 16 20.4	+23 56 43	-0.002	-0.01	4.94	0.97	0.2	K0 III	-28	89 s		
106717	138688		12 16 22.4	-5 50 44	+0.001	-0.01	7.5	1.1	-0.1	K2 III		340 s		
106676	256919		12 16 23.5	-72 36 53	-0.010	-0.03	6.22	-0.01		A0				
106749	119300		12 16 26.6	+7 21 19	-0.006	+0.01	8.0	0.1	1.7	A3 V		130 mx		
106760	62928		12 16 30.1	+33 03 41	-0.004	-0.12	5.00	1.14	0.0	K1 III	-42	87 mx		
106731	—		12 16 35.1	-65 30 21			7.9	0.1	0.6	A0 V		280 s		
106742	251825		12 16 38.2	-63 11 52	-0.022	0.00	6.96	0.58		G0				
106784	62930		12 16 42.0	+39 35 33	-0.002	0.00	7.22	0.17		A2	+4		8495	m
106813	28323		12 16 49.9	+56 16 53	-0.010	-0.02	7.2	1.1	-0.1	K2 III		200 mx		
106788	180642		12 16 51.7	-26 35 55	+0.001	-0.04	7.2	1.0	0.2	K0 III		260 s		
107113	2012		12 16 51.7	+86 26 09	+0.226	-0.01	6.33	0.43	3.0	F2 V	-6	44 s		
106817	119304		12 16 58.3	+0 21 06	-0.002	-0.02	8.0	1.1	0.2	K0 III		360 s		
106759	256920		12 17 00.0	-74 08 37	-0.008	0.00	7.84	0.35		A0				
106851	44105		12 17 03.2	+48 25 57	+0.001	-0.03	8.04	1.22	0.0	K1 III		330 s		
106819	157184		12 17 03.2	-16 41 37	-0.004	0.00	6.05	0.10		A2				
106797	251826		12 17 06.0	-65 41 33	-0.008	-0.01	6.06	0.03		A0				
106877	119308		12 17 27.0	+6 36 03	+0.002	-0.05	7.91	0.39	3.0	F2 V		91 s	8500	m
106884	28327		12 17 29.4	+53 11 28	+0.003	-0.05	5.81	1.30	-0.1	K2 III	-41	130 s		
106887	82219		12 17 30.4	+28 56 14	-0.003	+0.03	5.68	0.15	1.9	A4 V	-7	55 s	8501	m
106869	223335		12 17 31.6	-48 55 33	-0.029	-0.06	6.82	0.58	4.5	G1 V		29 s		
106849	251830	ε Mus	12 17 34.2	-67 57 38	-0.042	-0.03	4.11	1.58		gM6	+7	12 mn		

HD	SAO	Star Name	α 2000	δ 2000	μ(α)	μ(δ)	V	B-V	M_v	Spec	RV	d(pc)	ADS	Notes
			$12^h 17^m 34.6^s$	$-40°00'14''$	$-0\overset{s}{.}001$	$+0\overset{"}{.}02$								
106866	203312		12 17 34.6	-40 00 14	-0.001	+0.02	7.2	1.3	-0.1	K2 III		250 s		
106926	100023		12 17 44.3	+15 08 38	+0.003	-0.07	6.34	1.37	0.2	K0 III	-42	100 s		
106873	251832		12 17 45.3	-63 37 01	-0.009	-0.06	7.9	1.4	-0.3	K5 III		190 mx		AO Cru, v
106923	203317		12 17 46.0	-36 20 03	-0.007	0.00	8.0	1.1	-0.1	K2 III		240 mx		
106920	180657		12 17 47.0	-28 16 17	0.000	-0.05	7.8	0.5	0.2	K0 III		330 s		
106922	203319		12 17 47.2	-36 05 38	-0.003	-0.01	6.15	0.01		A0				m
106902	223338		12 17 47.6	-45 23 34	+0.001	0.00	7.96	-0.08		B9				
106946	82222		12 17 50.8	+25 34 17	-0.001	-0.01	7.89	0.36	3.0	F2 V	-1	92 s		
106906	239819		12 17 53.1	-55 58 34	-0.005	-0.05	7.78	0.46	3.4	F5 V		73 s		
106952	138701		12 17 53.3	-6 25 52	-0.006	-0.05	7.9	0.5	4.0	F8 V		61 s		
106932	157194		12 17 54.7	-16 57 12	0.000	-0.07	8.0	1.1	0.2	K0 III		270 mx		
106908	239820		12 17 55.2	-57 51 02	-0.002	-0.02	6.7	1.8	0.2	K0 III		69 s		
106955	180662		12 17 55.3	-24 00 48	-0.004	0.00	7.00	0.00	0.4	B9.5 V		210 s	8503	m
106965	119313		12 17 57.4	+1 34 30	-0.003	-0.02	8.0	0.1	1.4	A2 V		210 s		
106910	—		12 17 59.0	-69 56 20			7.9	1.1	-0.1	K2 III		380 s		
107028	15761		12 18 00.3	+68 47 51	-0.022	+0.04	7.80	0.8	3.2	G5 IV	-29	83 s		
106972	100028		12 18 04.2	+18 26 04	0.000	-0.03	7.50	0.5	3.4	F5 V	-27	66 s		m
106943	251837		12 18 06.7	-61 28 12	-0.001	-0.02	7.51	-0.02		B9				
106975	138703		12 18 09.0	-3 57 15	-0.001	+0.01	5.99	0.36	3.1	F3 V	+1	38 s		
106976	138704		12 18 09.5	-3 56 55	-0.001	+0.02	6.54	0.33	2.6	F0 V	-1	60 s	8505	m
106970	239822		12 18 11.0	-52 18 22	-0.001	-0.01	7.70	0.00	0.4	B9.5 V		280 s		m
106962	—		12 18 11.0	-62 38 20			7.6	1.8	-0.5	M2 III		340 s		
106999	82227		12 18 11.2	+27 18 03	-0.003	+0.02	7.46	0.15	1.4	A2 V	-8	150 s		
106991	—		12 18 16.5	-20 50 20			7.4	0.6	4.4	G0 V		37 s		
106979	239825		12 18 20.1	-51 14 41	0.000	0.00	7.7	1.1	-0.1	K2 III		360 s		
106911	256924	β Cha	12 18 20.7	-79 18 43	-0.015	+0.01	4.26	-0.12	-0.9	B6 V	+23	110 s		
106994	223343		12 18 21.2	-42 57 06	-0.002	-0.02	8.0	0.6	1.7	A3 V		100 s		
107001	157200		12 18 22.0	-15 35 39	-0.003	-0.01	7.9	0.1	1.7	A3 V		180 s		
106993	203329		12 18 22.4	-39 16 16	-0.002	+0.01	7.8	0.8	0.2	K0 III		330 s		
107052	15765		12 18 22.9	+60 21 22	-0.002	-0.02	7.9	0.9	3.2	G5 IV		88 s		
107006	203331		12 18 24.4	-31 29 45	-0.017	-0.10	7.5	0.7	4.0	F8 V		40 s		
106981	239828		12 18 26.1	-58 10 46	-0.003	0.00	8.0	1.2	0.2	K0 III		290 s		
106983	251841	ζ Cru	12 18 26.1	-64 00 11	-0.007	-0.02	4.04	-0.17	-2.3	B3 IV	+19	150 mx		m
107003	180666		12 18 26.5	-20 48 35	-0.001	-0.01	7.7	1.2	-0.5	M2 III		430 s		
107035	119318		12 18 27.5	+8 03 54	-0.003	0.00	8.0	1.1	0.2	K0 III		360 s		
107053	62942		12 18 29.0	+32 44 57	-0.001	0.00	6.72	0.22	2.1	A5 V	-9	77 s		
107055	62943		12 18 31.6	+30 14 56	+0.007	-0.13	6.1	0.1	1.4	A2 V	-18	57 mx		
107038	138706		12 18 32.1	-8 54 15	-0.001	-0.08	7.5	0.7	4.4	G0 V		41 s		
107040	157202		12 18 35.4	-14 24 23	-0.003	+0.03	7.6	0.4	2.6	F0 V		100 s		
107045	203332		12 18 37.0	-39 25 05	-0.002	+0.04	7.6	0.3	3.2	G5 IV		75 s		
107070	138710	13 Vir	12 18 40.2	-0 47 14	+0.002	-0.02	5.90	0.17		A3 n	-14			
107086	82229		12 18 40.8	+26 11 25	-0.003	+0.05	7.5	0.5	3.4	F5 V		68 s		
107013	—		12 18 44.7	-71 52 20			7.3	1.1	0.2	K0 III		260 s		
106461	258640		12 18 45.8	-88 24 54	-0.080	0.00	6.6	-0.4	0.6	A0 V		160 s		
107087	100033		12 18 46.2	+19 10 04	-0.006	-0.03	8.0	0.5	4.0	F8 V		64 s		
107193	7540		12 18 50.0	+75 09 38	-0.009	0.00	5.38	-0.02		A2	-4	22 mn		
107106	203338		12 18 59.4	-33 17 33	-0.004	-0.01	7.70	1.1	0.2	K0 III		320 s		m
107082	239836		12 18 59.4	-60 12 09	+0.001	+0.01	7.9	0.8	0.2	K0 III		340 s		
107079	239838		12 18 59.7	-55 08 35	-0.008	-0.02	5.00	1.59	-0.5	gM3	-7	130 s		m
107102	157205		12 19 00.7	-13 32 31	+0.002	-0.01	7.8	0.1	1.4	A2 V		190 s		
107131	82237		12 19 02.0	+26 00 28	0.000	-0.03	6.46	0.18		A m	0			
107122	203344		12 19 05.4	-33 44 42	-0.003	0.00	7.8	1.0	2.6	F0 V		39 s		
107158	44121		12 19 06.5	+40 18 45	0.000	0.00	7.18	0.99	0.2	K0 III		250 s		
107146	100038		12 19 06.5	+16 32 54	-0.012	-0.16	7.07	0.62	4.9	dG3	+6	27 s		
107097	251844		12 19 11.5	-61 08 10	-0.002	-0.01	7.77	0.04		A0				
107159	82238		12 19 15.8	+25 03 37	+0.001	-0.02	7.75	0.29	2.6	F0 V	+10	110 s		
107143	223351		12 19 16.4	-47 49 47	-0.003	0.00	7.9	0.0	1.4	A2 V		200 s		
107149	157207		12 19 18.9	-16 16 26	-0.001	+0.01	7.8	1.1	0.2	K0 III	+14	330 s		
107168	82239	8 Com	12 19 19.1	+23 02 04	-0.002	-0.02	6.26	0.17		A m	+1			
107161	138716		12 19 20.1	-8 54 53	0.000	-0.03	7.00	1.1	0.2	K0 III	-12	230 s	8509	m
107170	100039		12 19 20.2	+14 32 29	-0.002	-0.06	6.53	1.12		K0	+25			
107213	82244	9 Com	12 19 29.5	+28 09 24	-0.015	-0.13	6.3	0.5	3.4	F5 V	-8	38 s		
107224	44124		12 19 31.4	+40 56 21	-0.003	-0.06	7.86	0.49	3.4	F5 V		73 s		
107183	203351		12 19 32.1	-32 35 19	-0.001	+0.01	7.4	1.1	1.4	A2 V		35 s		
107199	157211		12 19 37.9	-19 15 20	0.000	0.00	5.8	1.6		M5 III	-22			R Crv, v
107273	15772		12 19 39.8	+62 21 31	-0.006	-0.02	7.8	0.1	1.7	A3 V		81 mx		
107206	223353		12 19 40.1	-49 30 09	-0.007	-0.03	7.1	1.1	0.2	K0 III		220 s		
107145	256927		12 19 45.2	-76 48 03	-0.048	-0.05	6.84	0.45	4.0	F8 V		37 s		
107238	119333		12 19 47.2	+6 02 27	-0.001	-0.02	7.7	1.1	0.2	K0 III		310 s		
107274	44127	3 CVn	12 19 48.6	+48 59 03	-0.001	0.00	5.29	1.66	7.4	K4 V	+8			
—	100043		12 19 49.1	+16 48 29	+0.001	-0.15	7.9	1.1	0.0	K1 III		95 mx		
107276	82246		12 19 50.5	+28 27 51	-0.001	-0.01	6.67	0.16		A m	-1			
107209	251848		12 19 53.7	-62 51 14	0.000	0.00	6.81	0.24	-6.6	A0 I		3800 s		
107259	138721	15 η Vir	12 19 54.3	-0 40 00	-0.004	-0.02	3.89	0.02	1.4	A2 V	+2	32 s		Zaniah
107246	180689		12 19 55.2	-26 44 12	-0.004	-0.02	6.80	1.4	-0.3	K5 III		260 s		m

HD	SAO	Star Name	α 2000	δ 2000	μ(α)	μ(δ)	V	B-V	M_v	Spec	RV	d(pc)	ADS	Notes
107258	119334		12^{h}19^{m}55.3^s	+5°40'18"	-0.006	+0.02	7.7	0.2	2.1	A5 V		120 mx		
107233	223356		12 19 55.5	-48 19 00	+0.004	-0.03	7.5	-0.1	1.7	A3 V		150 s		
107287	62951		12 19 58.1	+30 29 10	-0.005	-0.02	8.01	1.22	0.0	K1 III		220 mx		
107249	223357		12 19 58.3	-41 21 35	+0.003	-0.02	7.3	0.3	0.2	K0 III		260 s		
107286	44129		12 19 58.5	+43 36 33	-0.002	+0.01	7.69	0.84	0.3	G8 III		300 s		
107261	180691		12 19 59.5	-28 31 32	-0.004	-0.03	7.9	0.3	4.0	F8 V		60 s		
107263	203356		12 20 04.5	-37 48 14	-0.014	-0.15	7.7	0.5	4.0	F8 V		53 s		
107288	100048		12 20 04.8	+13 51 19	-0.004	-0.01	6.94	1.08	0.2	K0 III	+9	200 s	8514	m
107303	100049		12 20 06.0	+15 42 52	-0.005	-0.08	8.0	0.5	4.0	F8 V		62 s		
107250	239851		12 20 06.3	-56 43 03	-0.001	-0.01	7.87	0.01	0.4	B9.5 V		300 s		
107305	100050		12 20 08.0	+9 53 33	-0.005	+0.02	7.57	1.35	0.2	K0 III		150 s		
107295	180695		12 20 10.6	-22 10 32	-0.008	-0.03	5.97	0.82	4.7	dG2	-1	13 s	8515	m
107341	62953		12 20 13.4	+37 54 08	-0.005	0.00	6.73	1.00	0.2	K0 III	+4	200 s	8516	m
107284	239854		12 20 15.1	-53 23 18	-0.002	-0.01	7.9	0.8	-0.1	K2 III		390 s		
107270	251852		12 20 16.1	-64 38 52	-0.002	-0.01	7.7	0.8	3.2	G5 IV		78 s		
107309	138724		12 20 16.8	-4 30 37	0.000	-0.01	7.3	1.1	0.2	K0 III		260 s		
107326	82249		12 20 17.6	+26 00 07	-0.011	+0.02	6.13	0.30	2.6	F0 V	+8	23 ts		
107298	239855		12 20 19.5	-51 13 07	-0.002	-0.02	7.5	0.4	1.4	A2 V		99 s		
107325	82250		12 20 19.6	+26 37 10	-0.005	-0.11	5.56	1.09	0.0	K1 III	-10	90 mx		
107328	119341	16 Vir	12 20 20.9	+3 18 45	-0.020	-0.07	4.96	1.16	0.2	K0 III	+35	67 s		m
107285	—		12 20 22.5	-66 16 19			8.0	1.1	0.2	K0 III		350 s		
107379	15774		12 20 23.0	+66 23 27	+0.002	-0.02	7.2	0.1	0.6	A0 V		210 s		
107314	239857		12 20 26.9	-51 58 26	-0.005	-0.01	7.6	0.7	1.4	A2 V		68 s		
107397	15775		12 20 27.1	+61 18 34	-0.006	-0.02	7.00	1.6	-0.5	M2 III var	-12	320 s		RY UMa, v
107301	251854		12 20 28.1	-65 50 34	-0.008	-0.02	6.21	-0.04		B9 n	-8			
107348	180700	5 ζ Crv	12 20 33.6	-22 12 57	-0.007	-0.03	5.21	-0.10		B8 pne	+2		8517	m
107414	15776		12 20 35.8	+63 40 13	-0.001	0.00	7.4	0.9	3.2	G5 IV		69 s		
107398	82254		12 20 41.3	+27 03 17	0.000	-0.12	6.3	0.4	3.0	F2 V	-15	46 s	8519	m
107383	100053	11 Com	12 20 42.9	+17 47 34	-0.008	+0.08	4.74	1.01	0.3	G8 III	+42	69 s	8521	m
107415	100054		12 20 48.6	+15 32 28	-0.004	0.00	6.46	1.03		K0	-22			
107465	28346	70 UMa	12 20 50.7	+57 51 50	+0.005	-0.08	5.55	1.43	-0.1	K2 III	-43	94 s		
107403	157224		12 20 50.9	-12 51 11	-0.002	+0.02	8.0	1.6	-0.5	M2 III		510 s		
107392	223364		12 20 54.1	-47 27 17	-0.005	0.00	7.12	0.25		A3				
107418	157226		12 20 55.7	-13 33 56	0.000	+0.01	5.14	1.05	0.0	K1 III	+13	110 s		m
107410	223365		12 20 58.0	-46 38 46	-0.003	-0.01	8.0	1.4	-0.3	K5 III		450 s		
107467	44135		12 21 02.5	+45 30 55	-0.006	+0.01	7.39	0.96	0.3	G8 III		230 mx		
107469	82260		12 21 03.3	+25 01 54	-0.017	+0.14	7.31	0.88	0.2	G9 III	+24	51 mx		
107485	62961		12 21 04.1	+38 01 26	-0.001	+0.01	7.49	0.97	0.2	K0 III		290 s		
107422	223366		12 21 06.5	-42 33 44	0.000	-0.02	6.83	0.04		A0				m
107434	203374		12 21 09.5	-38 18 09	-0.005	-0.02	7.9	0.3	4.0	F8 V		60 s		
107484	44136		12 21 10.1	+41 29 00	-0.002	-0.02	7.72	1.16	-0.1	K2 III		370 s		
107486	62964		12 21 10.8	+34 41 12	-0.003	-0.03	7.1	1.1	0.0	K1 III	-19	270 s		
107424	239866		12 21 11.0	-56 46 26	-0.001	-0.01	7.76	-0.01	0.6	A0 V		270 s		
107474	138730		12 21 13.3	-9 32 57	-0.003	0.00	7.7	1.1	0.2	K0 III		310 s		
107475	180711		12 21 16.1	-26 43 57	-0.003	-0.01	7.7	1.3	-0.1	K2 III		310 s		
107447	251861		12 21 21.2	-62 16 52	-0.001	+0.01	6.6	0.7	4.4	G0 V		28 s		T Cru, v
107446	251862	ε Cru	12 21 21.5	-60 24 04	-0.023	+0.08	3.59	1.42		K2	-5	18 mn		
107340	256930		12 21 21.6	-80 03 41	-0.023	0.00	7.6	0.1	0.6	A0 V		95 mx		
107513	82262		12 21 26.6	+24 59 48	-0.001	-0.02	7.42	0.28		A m	0			
107504	203380		12 21 29.3	-34 21 03	-0.008	0.00	8.0	1.2	0.2	K0 III		200 mx		
107502	203381		12 21 29.9	-30 29 04	0.000	+0.02	8.0	2.4	-0.5	M2 III		170 s		
107492	223376		12 21 32.5	-44 16 06	-0.001	0.00	7.6	0.8	3.2	G5 IV		77 s		
107463	251863		12 21 35.5	-69 06 39	0.000	-0.02	7.5	0.6	4.4	G0 V		42 s		
107538	180719		12 21 41.5	-22 53 18	+0.001	+0.02	7.8	1.1	-0.1	K2 III		380 s		
107509	251865		12 21 44.3	-61 32 19	-0.036	-0.06	7.91	0.57		G5				
107413	256932		12 21 44.9	-79 27 57	-0.007	+0.01	7.4	0.1	0.6	A0 V		210 mx		
107569	82267		12 21 45.0	+21 52 22	0.000	+0.01	7.4	0.5	3.4	F5 V		64 s		
107524	223379		12 21 46.7	-49 56 55	-0.002	-0.01	6.4	1.3	-0.1	K2 III		180 s		
107597	44140		12 21 49.9	+40 42 33	0.000	0.00	7.99	0.97	0.2	G9 III		360 s		
107572	138735		12 21 51.0	-4 23 11	-0.001	-0.04	7.7	1.4	-0.3	K5 III		390 s		
107610	44141		12 21 56.0	+47 10 54	-0.008	-0.04	6.35	1.11	0.0	K1 III		160 mx		
107543	239880		12 21 57.5	-56 22 29	-0.001	-0.02	5.8	2.2	-0.5	M2 III		76 s		
107634	62969		12 22 01.4	+33 05 25	-0.004	-0.01	7.48	1.04	0.2	K0 III		270 s		
107561	239884		12 22 01.4	-51 45 21	0.000	-0.01	8.0	0.4	0.2	K0 III		360 s		
107612	100062		12 22 01.8	+16 44 39	-0.005	-0.05	6.68	0.02		A2 p	+3			
107670	15783		12 22 04.1	+61 35 29	+0.004	0.00	6.9	0.9	3.2	G5 IV		54 s		
107566	251866	ζ² Mus	12 22 07.4	-67 31 20	-0.004	-0.01	5.15	0.19		A m	-17	24 mn		m
107603	203398		12 22 08.7	-35 30 44	-0.001	-0.01	6.9	1.4	0.2	K0 III		130 s		
107655	82271		12 22 10.7	+24 46 25	-0.005	-0.01	6.19	-0.02	0.6	A0 V	-3	93 mx		
107760	7552		12 22 11.5	+73 14 52	-0.108	+0.16	8.00	0.73	5.5	G7 V	-98	32 s		AS Dra, v
107567	251868	ζ¹ Mus	12 22 11.9	-68 18 27	-0.002	-0.06	5.74	1.04	0.2	gK0		120 s		
107621	203399		12 22 14.1	-39 31 09	+0.001	0.00	7.8	0.5	0.2	K0 III		330 s		
107620	203402		12 22 14.7	-38 27 06	+0.001	-0.03	7.4	1.1	0.2	K0 III		230 s		
107657	119357		12 22 15.0	+4 51 17	+0.001	-0.02	8.0	1.6	-0.5	M4 III		510 s		
107547	256934		12 22 15.4	-73 30 15	-0.008	-0.04	6.77	0.16		A2				m

HD	SAO	Star Name	α 2000	δ 2000	μ(α)	μ(δ)	V	B-V	M$_v$	Spec	RV	d(pc)	ADS	Notes
			h m s	° ' "	s	"								
107592	—		12 22 17.3	-64 40 18			7.7	1.1	0.2	K0 III		300 s		
107642	157240		12 22 18.3	-15 33 17	+0.002	0.00	6.6	1.1	0.2	K0 III	+10	190 s		
107593	—		12 22 18.3	-65 36 18			7.76	0.04	0.0	B8.5 V		360 s		
107646	223386		12 22 21.8	-43 27 48	-0.005	+0.02	7.8	0.4	3.2	G5 IV		84 s		
107627	239889		12 22 24.5	-50 41 15	0.000	-0.01	7.8	0.4	0.2	K0 III		330 s		
107661	180733		12 22 26.5	-24 14 11	-0.003	-0.01	7.10	1.1	0.2	K0 III		240 s		m
107700	82273	12 Com	12 22 30.2	+25 50 46	-0.001	-0.01	4.79	0.49		F8 p	+1	27 mn	8530	m
107705	119360	17 Vir	12 22 31.9	+5 18 19	-0.011	-0.06	6.40	0.60	3.8	dF7	+5	29 s	8531	m
107740	44145		12 22 33.0	+48 21 11	-0.002	+0.02	7.2	0.1	1.7	A3 V		130 s		
107725	82275		12 22 34.3	+26 37 16	-0.003	-0.04	8.00	1.1	-0.1	K2 III		420 s		
107663	223389		12 22 36.6	-49 19 33	-0.001	0.00	8.0	-0.1	0.6	A0 V		300 s		
107743	82277		12 22 42.0	+20 08 56	-0.001	-0.02	7.89	1.25	-0.1	K2 III		350 s		
107692	203408		12 22 44.7	-39 10 40	+0.015	-0.15	6.7	0.4	4.4	G0 V		29 s		
107695	239900		12 22 46.3	-52 17 36	0.000	-0.03	6.8	0.5	0.2	K0 III		210 s		
107696	239901		12 22 49.3	-57 40 34	-0.005	-0.01	5.39	-0.10	-0.2	B8 V p	+1	130 s		
107762	44147		12 22 49.6	+43 48 31	+0.001	0.00	7.91	0.94	0.2	K0 III		350 s		
107792	15785		12 22 49.7	+60 06 17	+0.001	+0.01	7.8	1.1	0.2	K0 III		330 s		
107716	223393		12 22 50.1	-45 39 31	-0.006	-0.01	7.6	0.6	0.6	A0 V		100 mx		
107733	203411		12 22 51.5	-33 59 39	-0.001	+0.01	7.7	1.6	3.2	G5 IV		25 s		
107746	157248		12 22 54.8	-14 39 46	-0.002	-0.02	7.3	0.4	3.0	F2 V		72 s		
107720	251876		12 22 59.8	-64 43 50	+0.001	-0.02	7.5	1.1	0.2	K0 III		280 s		
107756	180742		12 23 00.0	-24 52 18	-0.004	+0.02	6.9	1.5	0.2	K0 III		120 s		
107780	138747		12 23 09.5	-7 17 57	-0.002	0.00	7.1	1.1	0.2	K0 III		240 s		
107758	239905		12 23 09.8	-52 25 14	-0.001	-0.02	7.8	0.3	0.2	K0 III		330 s		
107773	251877		12 23 13.9	-67 37 54	-0.130	+0.25	6.36	0.89	4.6	K0 IV-V		25 mn		
107783	203418		12 23 14.2	-37 29 20	-0.005	+0.03	7.50	0.4	2.6	F0 V		96 s		m
107794	138750		12 23 15.2	-4 58 29	+0.001	-0.04	6.6	1.1	0.2	K0 III		190 s		
107795	138751		12 23 16.2	-9 46 17	-0.004	-0.07	7.5	1.1	0.2	K0 III		160 mx		
107814	157253		12 23 18.7	-11 48 44	-0.005	-0.02	6.45	1.59	-0.5	gM3	+4	250 s		
107815	180747	6 Crv	12 23 21.5	-24 50 25	-0.001	-0.02	5.68	1.16	0.0	gK1	-2	130 s		
107830	138754		12 23 27.5	-6 07 09	-0.003	+0.03	7.3	0.5	4.0	F8 V		45 s		
107803	239909		12 23 28.4	-51 15 54	-0.002	0.00	8.0	1.1	0.2	K0 III		330 s		
107854	82282		12 23 29.9	+24 35 37	+0.002	0.00	7.40	1.05	0.2	K0 III	+5	250 s		
107832	203420		12 23 35.3	-35 24 46	-0.003	-0.01	5.32	-0.08	-0.8	B9 III	-10	170 s		
107903	15792		12 23 36.2	+60 54 12	-0.003	0.00	7.7	0.8	3.2	G5 IV		80 s		
107833	203421		12 23 36.8	-39 18 10	-0.001	-0.02	6.40	0.29		A5				
107805	251878		12 23 37.9	-61 37 45	+0.002	-0.02	6.6	0.5	3.4	F5 V		44 s		R Cru, v
107844	157258		12 23 39.6	-20 05 16	+0.005	-0.05	7.7	0.7	0.2	K0 III		270 mx		
107821	251879		12 23 42.1	-63 52 11	-0.006	0.00	7.42	0.18	0.6	A0 V		140 mx		
107860	203424		12 23 44.9	-38 54 40	-0.003	-0.01	5.79	-0.08		B9 n	-8			
107904	44155	4 CVn	12 23 47.0	+42 32 33	-0.007	+0.01	6.06	0.33	0.6	F0 III	-10	120 s		
107869	203427		12 23 47.7	-30 20 07	+0.001	-0.01	6.60	1.1	-0.1	K2 III	+69	220 s		m
107879	157260		12 23 52.6	-19 13 30	-0.004	+0.02	7.5	1.1	0.2	K0 III		290 s		
107950	28366	5 CVn	12 24 01.4	+51 33 44	+0.001	+0.01	4.80	0.87	0.3	G7 III	-13	38 ts		
107892	157261		12 24 02.2	-19 34 17	-0.004	-0.04	7.4	0.5	4.0	F8 V		49 s		
107935	82288		12 24 03.4	+25 51 04	-0.001	-0.01	6.72	0.24		A m	0			
107937	119372		12 24 11.5	+5 58 17	+0.001	-0.02	8.00	1.6	-0.5	M4 III	+26	500 s		
107914	203431		12 24 13.1	-38 54 52	+0.002	-0.02	7.00	0.2	2.1	A5 V		96 s		m
107966	82291	13 Com	12 24 18.4	+26 05 55	-0.001	-0.01	5.17	0.08		A4 p	+1	31 mn		
107931	223411		12 24 21.3	-47 22 22	-0.003	-0.01	6.7	0.7	0.4	B9.5 V		64 s		
108005	44161		12 24 24.5	+43 05 15	+0.001	0.00	7.90	0.33		F0			8540	m
108007	82293		12 24 26.7	+25 34 57	-0.001	-0.01	6.40	0.27	2.6	F0 V	-8	58 s	8539	m
107863	256938		12 24 28.4	-77 12 51	-0.004	+0.01	6.95	1.54		K2				
108020	44162		12 24 38.0	+41 19 56	-0.001	-0.02	7.22	1.31		K2 IV				
107960	239930		12 24 40.2	-52 34 15	0.000	-0.05	7.6	0.3	4.4	G0 V		44 s		
108023	100086		12 24 40.3	+17 51 09	+0.001	-0.11	7.9	0.9	3.2	G5 IV		88 s		
108026	119381		12 24 40.6	+1 23 02	-0.002	+0.03	7.6	1.6	-0.5	M2 III		420 s		
107979	251884		12 24 42.3	-60 54 14	-0.003	-0.01	7.12	1.25	-0.1	K2 III		110 s		
107998	223417		12 24 44.7	-41 23 03	-0.006	-0.08	6.25	1.18		K0				m
108076	62989		12 24 45.8	+38 19 06	-0.050	+0.05	8.02	0.56	4.4	G0 V	-1	27 mx		
108078	62988		12 24 49.0	+31 02 02	-0.004	0.00	7.56	1.14	0.0	K1 III		290 s		
107976	239932		12 24 49.4	-58 07 11	+0.001	-0.02	7.60	0.6	4.4	G0 V		43 s		m
107964	251886		12 24 50.8	-67 37 56	+0.008	+0.02	7.5	1.1	0.2	K0 III		280 s		
108077	62990		12 24 51.9	+31 49 22	+0.002	-0.06	4.02	0.96	0.2	G9 III		58 s		
107947	256939		12 24 52.1	-72 36 13	-0.008	0.00	6.8	-0.3	0.6	A0 V		160 mx		
108002	251889		12 24 56.3	-65 12 43	+0.003	-0.04	6.94	0.11	-5.7	B1 Ib		2200 s		
108100	44164		12 24 57.1	+42 51 17	-0.007	+0.02	7.14	0.37		F2				
108135	28375	71 UMa	12 25 03.1	+56 46 39	-0.002	-0.02	5.81	1.62	-0.5	M2 III	-17	180 s		
108134	15800		12 25 05.1	+60 41 51	+0.019	0.00	7.4	0.6		G0 p	-46			
108150	15801		12 25 06.2	+63 48 09	-0.003	0.00	6.32	0.91	3.2	G5 IV	-4	35 s		m
108063	223426		12 25 08.3	-42 30 52	-0.014	-0.03	6.11	0.65	2.8	G0 IV		42 s		
108068	223425		12 25 08.9	-48 18 32	+0.001	0.00	7.3	1.8	-0.3	K5 III		220 s		
108107	157272		12 25 11.6	-11 36 37	-0.005	-0.03	5.95	0.03		A0				
108095	180783		12 25 13.9	-25 59 10	-0.015	+0.05	7.0	0.4	3.0	F2 V		57 s		
108123	82297		12 25 15.0	+23 55 34	+0.004	-0.04	6.04	1.09	0.2	K0 III	-5	130 s		

HD	SAO	Star Name	α 2000	δ 2000	μ(α)	μ(δ)	V	B-V	M$_v$	Spec	RV	d(pc)	ADS	Notes
108054	251893		12^{h}25^{m}17.3^s	-65°46'13"	-0.012^s	-0.13"	6.30	0.96	0.2	gK0		170 s		
108110	180786		12 25 18.3	-27 44 57	0.000	-0.01	6.09	1.27		K0				
108105	--		12 25 18.7	+0 47 42	-0.012	-0.14	7.20	1.8		C6.3 e	+2			SS Vir, v
108137	100090		12 25 20.7	+11 54 35	-0.001	-0.03	7.7	0.6	4.4	G0 V		47 s		
108114	203450		12 25 21.7	-35 11 11	-0.003	-0.01	5.70	-0.06	-0.8	B9 III	-11	200 s		
108115	203452		12 25 24.2	-36 43 00	-0.001	-0.06	7.8	1.8	-0.1	K2 III		160 s		
108129	180787		12 25 25.8	-30 01 33	+0.002	-0.03	7.8	1.9	0.2	K0 III		97 s		
108174	62998		12 25 26.6	+38 20 51	+0.001	-0.04	7.66	1.11	0.0	K1 III		320 s		
108186	44169		12 25 27.9	+48 22 17	-0.004	-0.03	7.78	1.03	0.2	G9 III		200 mx		
108141	157275		12 25 28.6	-19 21 31	-0.006	-0.02	7.9	0.9	3.2	G5 IV		86 s		
108073	251894		12 25 30.2	-69 28 37	-0.005	-0.02	7.51	-0.02	0.0	B8.5 V		290 mx		m
108163	180789		12 25 35.9	-26 48 41	-0.011	-0.06	7.8	0.8	3.4	F5 V		45 s		
107739	258644		12 25 37.7	-86 09 01	-0.013	-0.01	6.33	1.08	-0.1	gK2		190 s		
108201	82302		12 25 39.4	+23 21 04	-0.002	+0.01	7.6	1.1	0.2	K0 III		300 s		
108189	138774		12 25 40.4	-0 16 55	-0.005	+0.03	7.7	0.9	3.2	G5 IV		79 s		
108147	251899		12 25 46.2	-64 01 21	-0.028	-0.08	6.99	0.53		G0				
108225	63000	6 CVn	12 25 50.8	+39 01 07	-0.007	-0.04	5.02	0.96	1.8	G8 III-IV	-4	41 s		
108213	119391		12 25 54.7	+4 11 59	-0.002	0.00	7.5	1.1	0.2	K0 III		290 s		
108215	157283		12 25 59.5	-14 56 53	-0.009	-0.04	7.30	0.5	3.4	F5 V		60 s	8547	m
108218	203461		12 26 00.5	-31 46 18	-0.009	-0.04	8.0	0.9	0.2	K0 III		150 mx		
108196	223436		12 26 00.7	-49 35 18	-0.010	0.00	7.00	0.1	0.6	A0 V		63 mx		m
108228	119392		12 26 00.8	+2 02 32	-0.003	+0.01	7.6	0.1	0.6	A0 V		260 s		
108239	100095		12 26 03.2	+15 51 51	0.000	+0.02	6.73	1.11	0.2	K0 III		170 s		
108316	28386		12 26 20.1	+54 35 20	-0.004	+0.01	7.8	1.1	-0.1	K2 III		390 s		
108399	7563		12 26 23.6	+71 55 48	-0.034	-0.02	6.4	1.1	0.2	K0 III	+6	130 mx		
108283	82310	14 Com	12 26 24.0	+27 16 05	-0.001	-0.01	4.93	0.27		F0 sh	-4	24 mn		
108256	223438		12 26 26.3	-43 15 38	-0.002	-0.01	7.8	2.3	-0.5	M2 III		210 s		
108301	100097		12 26 29.8	+14 34 19	-0.002	-0.01	8.04	1.04	0.2	K0 III		350 s		
108250	251903		12 26 30.8	-63 07 21	-0.005	-0.03	4.86	-0.12	-2.0	B4 IV	+27	89 mx		m
108276	157286		12 26 31.2	-19 18 58	-0.003	-0.02	8.0	0.4	3.0	F2 V		98 s		
108267	223439		12 26 31.5	-44 44 56	-0.001	0.00	7.5	0.4	2.1	A5 V		120 s		
108257	239948		12 26 31.6	-51 27 03	-0.005	-0.02	4.82	-0.14	-1.1	B5 V n	+24	150 s		m
108346	28388		12 26 32.5	+55 09 34	+0.001	+0.01	7.1	0.1	0.6	A0 V		200 s		
108302	119397		12 26 33.3	+7 53 37	0.000	-0.01	7.9	1.1	-0.1	K2 III		390 s		
108248	251904	α¹ Cru	12 26 35.9	-63 05 56	-0.004	-0.02	1.41	0.1	-3.9	B1 IV	-11	110 s		Acrux, m
108249	--	α² Cru	12 26 36.5	-62 05 58	-0.005	-0.02	1.88			B3 n	-1	110 s		m
108318	119399		12 26 45.9	-0 11 02	-0.002	-0.03	7.7	0.2	2.1	A5 V		130 s	8550	m
108320	138781		12 26 47.6	-5 35 32	0.000	-0.03	8.00	0.4	3.0	F2 V		100 s	8551	m
108309	223443		12 26 48.1	-48 54 47	-0.064	-0.08	6.26	0.68	4.2	G5 IV-V		27 mx		
108380	44180		12 26 48.5	+48 39 41	-0.004	-0.01	7.95	1.07	0.2	K0 III		320 s		
108400	28389		12 26 49.9	+51 12 23	-0.007	+0.02	7.4	1.1	-0.1	K2 III		280 mx		
108323	203477		12 26 51.6	-32 49 48	-0.001	-0.03	5.55	0.01		A0				
108381	82313	15 γ Com	12 26 56.2	+28 16 06	-0.006	-0.09	4.35	1.13		K1 III-IV	+4	31 mn		
108311	239952		12 26 58.7	-52 42 10	-0.010	-0.01	7.1	1.1	0.2	K0 III		220 s		
108350	138786		12 26 58.8	-8 49 46	-0.005	+0.02	8.0	0.4	3.0	F2 V		98 s		
108382	82314	16 Com	12 26 59.3	+26 49 32	-0.001	-0.01	4.99	0.08		A4 p	+2	21 mn		
108351	180806		12 27 02.5	-22 23 20	-0.008	+0.05	7.8	1.1	4.0	F8 V		27 s		
108313	251908		12 27 02.5	-61 23 06	-0.006	+0.01	7.7	0.8	0.2	K0 III		310 s		
108408	63007		12 27 12.2	+36 22 29	+0.002	-0.02	7.67	0.15		A2				
108374	239954		12 27 16.2	-55 44 04	-0.004	+0.01	7.14	0.00	0.4	B9.5 V		220 s		m
108355	251911		12 27 24.6	-63 47 21	-0.004	-0.02	6.00	0.07		B9 n	+42			
108395	239961		12 27 28.4	-58 19 00	-0.004	0.00	6.9	0.8	-0.3	K5 III		280 s		
108396	239960		12 27 28.7	-58 59 30	-0.004	0.00	5.3	1.6	-0.5	M4 III	+71	140 s		
108429	180812		12 27 30.2	-21 13 36	-0.007	0.00	7.8	0.9	0.2	K0 III		200 mx		
108464	44184		12 27 32.9	+41 21 18	-0.002	-0.01	6.9	0.5	3.4	F5 V	-6	50 s		
108468	100104		12 27 33.2	+17 50 03	-0.001	-0.03	7.44	0.96	0.3	G5 III	-24	240 s		
108451	100105		12 27 34.6	+12 05 56	-0.005	-0.03	7.9	0.7	4.4	G0 V		51 s		
108502	28394	73 UMa	12 27 35.0	+55 42 45	-0.003	-0.01	5.70	1.55	-0.5	M2 III	+17	170 s		
108518	15806		12 27 37.0	+61 09 10	+0.001	0.00	7.9	0.1	1.4	A2 V		200 s		
108343	256943		12 27 37.9	-75 28 22	-0.006	-0.01	8.03	0.23		A3				
108486	82321		12 27 38.3	+25 54 42	-0.001	-0.03	6.73	0.16		A m	-1			
108442	180814		12 27 40.1	-24 43 39	-0.001	-0.04	7.4	1.1	0.2	K0 III		230 s		
108471	119413		12 27 42.0	+8 36 37	+0.001	-0.01	6.37	0.93	0.3	G8 III	-6	160 s		
108519	82322		12 27 46.2	+27 25 22	-0.003	-0.01	7.72	0.27	2.5	A9 V	-7	110 s		
108474	138796		12 27 46.3	-2 23 07	-0.005	-0.17	7.9	0.9	3.2	G5 IV		62 mx		
108477	157299		12 27 49.3	-16 37 55	-0.001	0.00	6.35	0.84	0.3	gG5	-8	160 s		
108506	138798		12 27 51.5	-4 36 55	-0.006	-0.01	6.22	0.44	2.6	dF0 n	-12	53 s		
108510	138799		12 27 55.4	-8 40 42	-0.015	-0.05	6.7	0.6	4.4	G0 V		29 s		
108490	157302		12 27 56.7	-18 36 36	-0.004	0.00	7.5	1.1	0.2	K0 III		290 s		
108481	203497		12 27 57.0	-36 42 26	-0.001	-0.03	7.5	1.1	0.4	B9.5 V		55 s		
108522	157303		12 28 01.6	-14 27 01	-0.002	-0.01	6.9	1.1	-0.1	K2 III		250 s		
108483	223454	σ Cen	12 28 02.4	-50 13 51	-0.003	-0.02	3.91	-0.19	-1.7	B3 V	+12	130 s		
108495	203498		12 28 02.6	-39 29 00	-0.006	-0.03	7.80	1.4	-0.3	K5 III		250 mx		m
108435	251918		12 28 02.7	-68 43 58	+0.001	-0.17	7.1	0.5	3.4	F5 V		54 s		
108574	44187		12 28 04.3	+44 47 38	-0.018	-0.02	7.50	0.58	4.0	dF8	0	46 s	8561	m

HD	SAO	Star Name	α 2000	δ 2000	μ(α)	μ(δ)	V	B−V	M$_V$	Spec	RV	d(pc)	ADS	Notes
			h m s	° ′ ″	s	″								
108546	82324		12 28 05.7	+20 21 15	−0.001	−0.02	7.80	1.28	0.2	K0 III		200 s		
108498	239970		12 28 07.9	−52 04 31	−0.003	−0.03	7.7	0.2	4.0	F8 V		54 s		
108499	239972		12 28 13.1	−53 34 37	−0.003	−0.03	7.4	0.4	0.2	K0 III		270 s		
108560	100114		12 28 14.6	+11 49 36	−0.006	+0.01	7.84	1.36	0.2	K0 III		170 s		
108447	256945		12 28 14.6	−70 32 47	−0.007	−0.04	7.65	1.44		K2				m
108500	251919		12 28 16.4	−61 45 57	+0.007	−0.17	7.00	0.6	4.4	G0 V		33 s		m
108561	119420		12 28 18.3	+4 23 47	−0.004	−0.03	6.80	0.1	0.6	A0 V	+16	49 mx	8563	m
108501	251920		12 28 18.9	−64 20 29	−0.009	−0.03	6.04	0.02	0.6	A0 V n		100 mx		
108541	203508		12 28 22.4	−39 02 29	−0.002	−0.02	5.44	−0.08		B8	+5			
108530	251924		12 28 25.5	−61 47 42	−0.004	−0.02	6.22	1.26		K0				
108570	239977		12 28 33.3	−56 24 27	−0.028	−0.22	6.15	0.92	5.6	dG8		16 mn		m
108554	239975		12 28 33.6	−56 27 27	−0.002	0.00	8.0	0.2	0.4	B9.5 V		230 s		
108614	100118		12 28 34.3	+12 07 02	0.000	−0.04	7.95	1.02	0.2	K0 III		340 s		
108642	82326		12 28 38.1	+26 13 36	−0.002	−0.02	6.53	0.18		A m	+3			
108599	157315		12 28 38.7	−16 00 21	0.000	−0.03	7.9	1.4	−0.3	K5 III		430 s		
108660	44196		12 28 38.9	+49 38 07	−0.008	+0.01	7.2	0.1	1.4	A2 V		100 mx		
108651	82328		12 28 44.5	+25 53 57	−0.001	−0.02	6.63	0.22		A m	−2		8568	m
108643	138811		12 28 51.6	−6 31 47	0.000	+0.03	7.9	1.1	−0.1	K2 III		390 s		
108633	203515		12 28 53.4	−37 58 01	−0.001	−0.02	7.80	1.1	0.2	K0 III		330 s		m
108662	82330	17 Com	12 28 54.6	+25 54 46	−0.002	−0.02	5.29	−0.05		A0 p	−3	22 mn	8568	m
108610	251928		12 28 54.8	−61 52 15	−0.002	−0.01	6.93	−0.04	−1.0	B5.5 V		360 s		m
108625	239987		12 29 00.0	−50 38 54	−0.005	−0.02	7.9	0.0	0.6	A0 V		120 mx		
108344	258646		12 29 00.9	−83 48 05	−0.006	+0.05	6.6	−0.5	0.4	B9.5 V		120 mx		
108693	63020		12 29 07.6	+31 23 25	+0.011	−0.05	7.94	0.60	4.4	G0 V	−47	49 s	8569	m
108558	---		12 29 07.6	−75 37 12			8.01	1.52		K5				
108656	203517		12 29 08.4	−34 50 03	+0.004	−0.03	7.2	1.0	−0.1	K2 III		290 s		
108639	251931		12 29 09.5	−60 48 17	−0.001	0.00	7.81	0.08	−5.0	B0 III p		2600 s		
108680	138813		12 29 09.6	−2 25 46	−0.002	−0.01	7.5	1.6	−0.5	M2 III	−35	400 s		
108259	258645		12 29 13.9	−85 34 51	+0.026	−0.02	8.0	0.5	4.0	F8 V		63 s		
108713	63022		12 29 14.8	+34 42 12	−0.011	+0.02	7.56	0.56	4.0	F8 V		49 s		
108714	100122		12 29 20.4	+17 19 18	0.000	−0.06	7.73	0.09		A0	+2			
108774	15811		12 29 22.7	+66 53 37	+0.001	−0.01	7.0	0.9	3.2	G5 IV		58 s		
108722	82333	18 Com	12 29 26.9	+24 06 32	−0.002	−0.01	5.48	0.43	0.7	F5 III	−25	90 s		
108723	119428		12 29 28.1	+4 50 19	−0.009	+0.05	7.1	0.4	3.0	F2 V		67 s		
108691	239992		12 29 30.6	−60 13 50	−0.002	0.00	7.95	1.11		K1 IV				
108659	251932		12 29 31.0	−66 51 42	−0.002	−0.05	7.31	0.32	−5.7	B5 Ib		2300 s		
108689	239993		12 29 33.7	−59 05 08	+0.001	−0.01	8.0	0.0	3.0	F2 V		98 s		
108765	82336	20 Com	12 29 43.1	+20 53 46	+0.002	−0.04	5.69	0.07	1.7	A3 V	−6	63 s		
108775	100127		12 29 46.8	+14 39 01	−0.003	+0.01	7.26	1.25	0.2	K0 III		180 s		
108767	157323	7 δ Crv	12 29 51.8	−16 30 55	−0.015	−0.14	2.95	−0.05	0.2	B9 V	+9	36 s	8572	Algorab, m
108732	239999		12 29 53.9	−56 31 29	−0.002	+0.01	5.80	1.56		gMa				
108807	82337		12 29 55.5	+24 20 18	−0.004	−0.05	7.84	0.43	3.4	F5 V	+2	77 s		
108844	28405	74 UMa	12 29 57.2	+58 24 21	−0.008	+0.09	5.32	0.20		A5	+7			
108759	223471		12 29 57.8	−41 44 10	0.000	−0.03	6.02	1.52		Mb				
108760	223472		12 30 00.2	−42 55 47	−0.005	−0.01	6.7	1.6	−0.5	M2 III		280 s		
108796	138816		12 30 01.8	−6 01 25	−0.008	−0.07	7.10	0.4	3.0	F2 V		66 s		m
108845	28407	7 CVn	12 30 02.8	+51 32 07	−0.031	+0.02	6.21	0.51	4.0	F8 V	+19	29 ts		m
108781	203534		12 30 03.2	−32 03 36	−0.003	−0.02	7.8	1.3	0.2	K0 III		230 s		
108861	28408	75 UMa	12 30 04.2	+58 46 03	+0.004	−0.03	6.08	0.98	1.8	G8 III−IV	−17	67 s		
108799	157326		12 30 04.7	−13 23 36	−0.017	−0.05	6.35	0.56	4.0	dF8	0	28 s	8573	m
108782	203537		12 30 06.2	−33 19 42	−0.001	−0.02	7.8	1.3	−0.1	K2 III		300 s		
108907	15816	4 Dra	12 30 06.6	+69 12 04	−0.012	−0.06	4.95	1.62	−0.5	gM4	−13	120 s		
108815	100132		12 30 07.4	+17 53 44	0.000	0.00	7.45	1.58	−0.5	M4 III		390 s		
108771	240002		12 30 09.8	−53 35 38	+0.001	+0.01	8.0	−0.4	0.6	A0 V		300 s		
108816	119432		12 30 11.2	+5 52 38	−0.001	0.00	8.0	0.5	4.0	F8 V		62 s		
108846	63027		12 30 12.4	+39 14 48	−0.011	−0.12	7.7	0.5	4.0	F8 V		55 mx		
108788	240004		12 30 14.9	−52 04 48	−0.003	+0.01	7.8	1.0	−0.5	M2 III		470 s		
108821	180850		12 30 17.4	−23 41 47	−0.002	−0.01	5.63	1.67	−0.4	gM0	−11	140 s		
108735	256947		12 30 18.0	−72 59 09	−0.006	−0.03	7.07	0.28		A3				
108773	251934		12 30 19.3	−60 59 23	−0.004	−0.01	6.68	0.45		F0				
108863	82342		12 30 19.8	+21 56 52	−0.006	−0.05	7.71	0.99	0.2	K0 III		140 mx		
108822	180852		12 30 23.7	−27 05 09	+0.001	0.00	7.6	2.3	−0.5	M2 III		170 s		
108877	119437		12 30 34.4	+3 30 28	−0.002	−0.02	7.4	0.9	0.3	G8 III	−5	260 s	8576	m
108840	223478		12 30 36.6	−44 22 02	−0.001	0.00	7.9	0.6	0.6	A0 V		110 s		
108851	203542		12 30 37.9	−36 14 24	+0.001	−0.02	8.0	1.7	−0.3	K5 III		340 s		
108879	138822		12 30 38.5	−9 49 28	0.000	0.00	7.9	1.1	0.2	K0 III		340 s		
108791	256950		12 30 42.6	−73 05 20	−0.002	−0.01	6.93	1.15		K0				
108954	28413		12 30 50.0	+53 04 35	+0.001	+0.18	6.22	0.54	4.0	F8 V	−21	27 s		
108910	138824		12 30 50.2	−4 03 39	−0.005	0.00	6.88	1.43		K3	+84			
109051	7575		12 30 53.2	+71 18 00	−0.013	−0.01	7.0	1.1	0.2	K0 III		230 s		
109094	7576		12 30 55.1	+76 41 12	−0.015	+0.01	7.5	1.1	0.2	K0 III		280 s		
108792	256951		12 30 56.3	−75 24 05	+0.001	−0.03	7.51	0.13		A0				
108858	251939		12 30 57.3	−61 29 36	+0.007	−0.02	7.9	0.8	3.2	G5 IV		85 s		
109002	15817		12 30 58.0	+65 53 51	−0.007	−0.01	6.7	0.1	1.7	A3 V		100 mx		
108945	82346	21 Com	12 31 00.5	+24 34 02	−0.001	−0.01	5.45	0.05		A3 p	0			UU Com, v

HD	SAO	Star Name	α 2000	δ 2000	μ(α)	μ(δ)	V	B-V	M$_v$	Spec	RV	d(pc)	ADS	Notes
108944	63037		12^{h}31^{m}00.7	+31°25'26"	+0.002	+0.02	7.34	0.50	3.4	F5 V		57 s		
108956	82347		12 31 03.1	+23 46 23	−0.005	+0.04	7.1	0.6	4.4	G0 V	−42	35 s		
108973	63040		12 31 04.2	+39 34 55	0.000	−0.01	6.80	1.04	0.2	K0 III		190 s		
108957	100135		12 31 06.4	+16 36 50	+0.001	−0.05	7.45	1.13		K2	+11			
108932	157333		12 31 09.6	−19 35 39	−0.001	−0.02	7.3	1.1	0.2	K0 III		260 s		
108903	240019	γ Cru	12 31 09.9	−57 06 47	+0.003	−0.27	1.63	1.59	−0.5	M3 III	+21	27 s		[Gacrux], m
108959	119447		12 31 14.5	+1 19 37	−0.003	−0.02	7.80	0.2	2.1	A5 V		140 s	8582	m
108924	240023		12 31 15.5	−52 47 32	0.000	−0.01	7.2	1.3	0.2	K0 III		170 s		
108925	240022	γ Cru	12 31 16.7	−57 04 51	+0.001	0.00	6.42	0.16		A2				m
108985	119453		12 31 21.3	+7 36 15	−0.002	0.00	6.05	1.52		K5	−17			
108986	119454		12 31 22.4	+5 03 10	+0.001	+0.02	8.0	0.4	3.0	F2 V		98 s		
108979	138830		12 31 25.2	−9 10 56	0.000	−0.04	7.2	1.1	0.2	K0 III		250 s		
109029	63046		12 31 30.7	+33 00 53	+0.001	−0.10	7.45	0.38	2.6	F0 V		84 s		
109030	82354		12 31 33.6	+25 52 14	−0.002	−0.01	7.91	0.02		A0 p	+7			
108992	203556		12 31 35.1	−33 40 12	−0.002	−0.01	7.8	1.1	1.4	A2 V		47 s		
109005	157339		12 31 36.4	−11 04 21	−0.003	−0.01	7.97	0.22		A4	−14		8585	
109068	—		12 31 38.3	+45 13 32	−0.003	−0.01	7.66	0.20	2.4	A7 V		110 s		
108991	203557		12 31 38.5	−30 58 55	+0.007	−0.01	6.8	1.3	0.2	K0 III		140 s		
109014	138832		12 31 38.6	−5 03 09	−0.003	+0.03	6.19	1.04	0.2	gG9	+2	140 s		
109032	100140		12 31 39.2	+12 07 39	−0.005	−0.01	7.9	0.4	2.6	F0 V		110 s		
108968	240027		12 31 40.2	−59 25 26	−0.002	−0.01	5.48	0.63	−4.6	F5 Ib	−20	960 s		
109070	82358		12 31 45.9	+22 58 10	+0.001	+0.01	8.03	1.06	0.2	K0 III		320 s		
109034	157341		12 31 49.2	−15 42 55	+0.004	−0.05	7.5	0.4	3.0	F2 V		80 s		
109069	82360		12 31 50.4	+29 18 51	−0.002	−0.01	7.55	0.30	2.6	F0 V		98 s		
109071	119455		12 31 51.0	+5 42 36	−0.003	0.00	7.9	1.1	−0.1	K2 III		390 s		
109083	82361		12 31 51.9	+23 16 08	+0.001	−0.05	7.7	0.4	3.0	F2 V	−20	85 s		
109009	240032		12 31 52.4	−58 17 26	−0.023	0.00	7.2	0.4	3.0	F2 V		64 s		
109035	180867		12 31 53.2	−20 58 52	−0.002	+0.06	7.3	1.2	0.2	K0 III	+4	190 s		
109178	7580		12 31 53.8	+72 10 43	−0.003	+0.01	7.9	0.1	1.4	A2 V		200 s		
109154	15822		12 31 54.5	+68 30 56	+0.008	0.00	7.5	0.5	3.4	F5 V		65 s		
109000	251946		12 31 55.8	−63 30 22	−0.009	0.00	5.95	0.27		A5				
109038	203563		12 31 56.4	−32 40 22	0.000	−0.01	7.8	2.0	−0.3	K5 III		200 s		
109040	223494		12 31 59.1	−41 03 18	+0.002	−0.02	7.4	0.7	−0.1	K2 III		310 s		
109155	15823		12 32 00.3	+63 45 29	−0.004	0.00	7.9	0.9	3.2	G5 IV		88 s		
109044	240035		12 32 01.7	−53 18 56	0.000	0.00	7.8	−0.4	0.6	A0 V		270 s		
109024	240034		12 32 02.0	−56 17 18	0.000	−0.01	7.44	0.05	0.4	B9.5 V		250 s		
109213	7584		12 32 03.0	+74 48 40	+0.002	−0.02	7.39	1.06	−0.9	K0 II-III	−30	470 s	8591	m
109085	157345	8 η Crv	12 32 04.1	−16 11 46	−0.030	−0.07	4.31	0.38	1.7	F0 IV	−4	29 ts		
109098	138836		12 32 04.4	−1 46 20	−0.001	+0.04	7.6	0.7	4.4	G0 V		44 s		
109074	203567		12 32 04.4	−32 32 01	−0.001	−0.01	6.46	0.20		A3				
109061	223495		12 32 05.4	−43 32 10	−0.001	−0.02	7.7	0.9	3.2	G5 IV		62 s		
109046	240036		12 32 08.0	−57 20 25	+0.001	−0.04	7.2	1.7	0.2	K0 III		96 s		
109045	240037		12 32 08.6	−56 07 45	−0.005	+0.01	6.56	1.68	−0.1	K2 III		130 s		
109086	180870		12 32 09.9	−27 33 58	+0.001	+0.01	7.2	1.6	0.2	K0 III		110 s		
108970	256954		12 32 10.1	−73 00 05	+0.010	−0.03	5.88	1.11		K0				
108913	256952		12 32 18.1	−79 47 01	−0.016	+0.01	7.2	0.1	0.6	A0 V		110 mx		
108927	256953		12 32 19.7	−78 11 38	−0.007	0.00	7.6	0.0	0.4	B9.5 V		270 mx		
109106	180876		12 32 25.1	−22 10 15	+0.001	0.00	7.9	0.2	2.1	A5 V		55 s		
109026	256955	γ Mus	12 32 28.1	−72 07 58	−0.010	−0.01	3.87	−0.15	−1.1	B5 V	+14	59 mx		
109132	180879		12 32 32.5	−21 12 44	−0.011	−0.06	6.5	1.5	−0.1	K2 III		100 mx		
109134	203574		12 32 35.0	−32 27 52	−0.002	−0.01	7.5	0.6	0.6	A0 V		100 s		
109081	—		12 32 35.4	−68 06 09			7.0	1.4	−0.3	K5 III		280 s		
109141	157350		12 32 35.9	−13 51 32	−0.010	−0.06	5.74	0.40	2.6	F0 V	−1	39 s		
109142	180881		12 32 37.4	−24 12 04	+0.003	−0.03	7.8	0.2	1.4	A2 V		150 s		
109048	256957		12 32 39.9	−73 39 23	−0.001	−0.02	7.5	0.6	4.4	G0 V		42 s		
109159	157351		12 32 41.0	−17 13 14	−0.009	+0.02	7.9	0.9	3.2	G5 IV		88 s		
109135	223500		12 32 43.6	−45 27 33	+0.004	−0.03	7.4	0.8	3.0	F2 V		43 s		
109123	240043		12 32 43.8	−56 18 30	−0.003	0.00	7.3	2.2	−0.1	K2 III		72 s		
109185	82369		12 32 44.7	+23 25 09	0.000	0.00	7.4	0.4	2.6	F0 V	−8	90 s		
109145	203575		12 32 46.3	−38 04 35	0.000	0.00	7.9	1.7	−0.3	K5 III		320 s		
109160	180884		12 32 50.2	−30 02 42	0.000	−0.01	6.8	1.6	3.2	G5 IV		16 s		
109067	256958		12 32 50.6	−75 26 04	−0.014	−0.02	7.79	0.40		F2				
109204	119461		12 32 58.9	−0 16 25	−0.004	+0.05	8.0	0.4	2.6	F0 V		120 s		
109217	100146	20 Vir	12 33 02.8	+10 17 44	−0.004	0.00	6.26	0.95	0.2	K0 III	+1	160 s		
109173	223503		12 33 06.5	−47 02 45	−0.001	−0.02	7.60	1.1	0.2	K0 III		300 s		m
109164	251952		12 33 10.4	−60 56 49	0.000	0.00	7.84	0.05		B5				m
109195	240050		12 33 12.2	−52 04 58	−0.002	−0.02	6.8	0.1	0.6	A0 V		140 s		
109194	240051		12 33 16.1	−50 39 01	+0.008	+0.03	7.5	0.6	4.0	F8 V		42 s		
109196	240053		12 33 18.4	−53 21 02	+0.001	+0.01	7.9	0.5	0.2	K0 III		340 s		
109197	240056		12 33 19.5	−54 58 51	−0.006	0.00	7.62	0.08	1.4	A2 V		110 mx		
109198	240054		12 33 20.7	−57 42 59	+0.001	−0.02	7.75	−0.01	0.6	A0 V		270 s		
109238	157359		12 33 22.3	−19 47 31	−0.001	−0.01	6.26	0.29		A5				
109270	119465		12 33 24.5	+7 56 45	−0.004	−0.07	7.58	1.10	−0.3	K5 III		130 mx		
109283	82376		12 33 27.5	+20 26 05	−0.002	−0.03	8.0	0.9	3.2	G5 IV		93 s		
109241	223511		12 33 28.6	−41 24 51	−0.004	−0.03	6.8	0.6	2.6	F0 V		43 s		

HD	SAO	Star Name	α 2000	δ 2000	μ(α)	μ(δ)	V	B-V	M_v	Spec	RV	d(pc)	ADS	Notes
			h m s	° ′ ″	s	″								
109387	7593	5 κ Dra	12 33 28.9	+69 47 17	-0.012	+0.01	3.87	-0.13		B7 p	-11	22 mn		
109305	63066		12 33 29.7	+38 04 08	-0.007	-0.01	6.53	1.04	0.2	K0 III		170 s		
--	--		12 33 31.4	-54 39 35			7.20							U Cen, v
109200	251953		12 33 31.7	-68 45 19	-0.097	-0.30	7.13	0.85	5.9	K0 V		15 ts		
109282	82377		12 33 32.1	+24 26 55	-0.001	-0.01	7.4	1.6	-0.5	M2 III	-11	380 s		
109285	119466		12 33 32.6	+7 40 45	-0.006	+0.04	6.91	1.04	0.2	K0 III		170 mx		
109199	--		12 33 33.0	-66 07 08			7.9	0.0	0.0	B8.5 V		370 s		
109413	7594		12 33 33.8	+72 28 44	+0.017	-0.05	7.5	0.5	4.0	F8 V		49 s		
109307	82378	22 Com	12 33 34.1	+24 16 59	-0.001	-0.01	6.29	0.10		A m	+1			
109272	157361		12 33 34.2	-12 49 49	-0.002	+0.05	5.58	0.86	0.3	G8 III	-16	110 s		
109287	119470		12 33 36.9	+4 13 13	-0.003	-0.01	7.8	1.1	0.2	K0 III		340 s		
109317	63070		12 33 38.8	+33 14 51	+0.001	-0.04	5.42	1.00	0.2	K0 III	-20	110 s		
109344	44229		12 33 38.9	+44 05 47	-0.015	-0.05	7.6	0.5	4.0	F8 V		52 s		
109251	223512		12 33 39.0	-44 11 03	-0.007	0.00	7.2	1.6	-0.3	K5 III		260 mx		
109460	7596		12 33 42.7	+75 12 43	-0.012	-0.01	7.1	1.1	0.2	K0 III		240 s		
109358	44230	8 β CVn	12 33 44.4	+41 21 26	-0.063	+0.29	4.26	0.59	4.5	G0 V	+7	9.2 t		Chara
109309	138845	21 Vir	12 33 46.7	-9 27 07	-0.006	0.00	5.48	-0.03		B9 n	-11	17 mn		
109345	63072		12 33 47.3	+33 23 05	0.000	-0.01	6.24	1.05	0.2	K0 III	-43	22 ts		
109310	180901		12 33 48.9	-23 30 58	+0.007	-0.13	7.9	0.6	3.2	G5 IV		88 s		
109331	138847		12 33 55.2	-6 46 40	-0.009	+0.04	8.0	0.5	4.0	F8 V		62 s		
109276	240061		12 33 58.0	-56 34 13	-0.006	-0.01	7.3	0.3	1.4	A2 V		100 s		
109312	223516		12 33 59.1	-49 54 34	-0.017	-0.04	6.38	0.46	2.4	F2 IV-V		53 s		m
109298	240063		12 34 00.0	-56 07 40	-0.007	0.00	8.0	1.5	0.2	K0 III		190 s		
109400	44233		12 34 03.3	+46 44 52	-0.006	-0.03	7.31	1.07	0.2	G9 III		160 mx		
109334	180908		12 34 04.7	-23 32 46	-0.002	+0.02	7.4	0.7	0.2	K0 III		280 s		
109313	240065		12 34 06.1	-53 46 26	-0.001	-0.03	8.0	0.1	0.6	A0 V		240 s		
109324	223518		12 34 08.8	-50 03 37	-0.005	-0.01	7.1	0.8	1.4	A2 V		46 s		
109349	157369		12 34 09.6	-18 11 42	+0.004	-0.05	7.30	0.6	4.4	G0 V		38 s	8597	m
109352	203596		12 34 11.7	-32 05 34	-0.006	0.00	7.10	0.2	2.1	A5 V		93 mx		m
109340	223519		12 34 13.7	-40 49 44	-0.004	-0.03	8.0	0.4	2.1	A5 V		120 mx		
109646	2054		12 34 16.4	+80 15 02	-0.040	-0.01	7.4	0.4	3.0	F2 V	+35	76 s		
109401	100149		12 34 20.1	+10 55 16	-0.005	-0.07	7.4	0.2	2.1	A5 V		42 mx		
109403	138853		12 34 23.1	-1 24 28	-0.002	+0.01	7.1	1.1	-0.1	K2 III		280 s		
109379	180915	9 β Crv	12 34 23.2	-23 23 48	0.000	-0.06	2.65	0.89	-2.1	G5 II	-8	89 s		
109355	240070		12 34 29.8	-58 19 52	-0.002	+0.01	7.9	0.4	4.4	G0 V		51 s		
109417	119479		12 34 30.7	+7 44 05	+0.004	-0.04	6.70	1.10	-0.1	K2 III		230 s		
109421	157372		12 34 36.8	-16 21 17	-0.006	-0.01	8.0	0.4	2.6	F0 V		120 s		
109409	223527		12 34 42.3	-44 40 24	-0.008	-0.22	5.77	0.70	3.2	G5 IV		33 s		m
109496	44236		12 34 42.4	+47 00 28	-0.001	+0.01	7.90	1.1		K3 IV				
109551	7600	6 Dra	12 34 43.9	+70 01 18	-0.007	0.00	4.94	1.31	-0.1	gK2	+5	84 s		
109463	82386		12 34 45.6	+24 13 27	-0.002	0.00	7.80	1.4	-0.3	K5 III		420 s		
109485	82390	23 Com	12 34 51.0	+22 37 45	-0.005	+0.02	4.81	0.00	-0.6	A0 III	-16	110 s		m
110093	2057		12 34 51.7	+85 43 55	-0.028	+0.01	7.1	0.4	2.6	F0 V	-12	80 s		
109486	100156		12 34 51.9	+10 50 10	-0.001	-0.03	7.88	0.28	2.6	F0 V		110 s		
109529	28432		12 34 54.1	+51 21 37	-0.010	0.00	7.9	1.1	0.2	K0 III		230 mx		
109372	251957		12 34 54.2	-67 45 25	-0.004	-0.01	6.7	1.6	-0.5	M4 III		260 s		BO Mus, v
109511	100160	24 Com	12 35 07.7	+18 22 37	0.000	+0.02	5.02	1.15	-0.1	K2 III	+4	110 s	8600	m
109499	119484		12 35 08.0	+7 26 33	+0.011	-0.10	7.90	0.5	4.0	F8 V		60 s	8601	m
109519	82394		12 35 08.0	+21 52 53	+0.001	-0.02	5.85	1.22	0.0	K1 III	-14	130 s		
109520	100162		12 35 10.6	+11 32 52	-0.002	0.00	7.50	1.04	0.2	K0 III		270 s		
109530	63080		12 35 10.8	+36 25 31	-0.002	-0.01	7.32	0.41	0.7	F4 III		210 s		
109399	--		12 35 15.3	-72 43 06			7.64	0.01	-5.7	B1 Ib		3600 s		
109521	119485		12 35 15.9	+2 15 40	-0.007	+0.05	7.1	0.5	3.4	F5 V		54 s		
109531	138861		12 35 27.5	-3 32 46	-0.002	-0.01	8.0	1.4	-0.3	K5 III		460 s		
109492	251962		12 35 29.0	-61 50 30	-0.042	-0.09	6.22	0.73	3.2	G5 IV		40 s		
109475	251961		12 35 29.8	-61 44 26	-0.002	-0.02	6.61	0.47		F5				
109448	251959		12 35 31.8	-68 21 02	-0.004	0.00	7.3	0.1	0.6	A0 V		210 s		
109545	157379		12 35 33.1	-12 01 16	+0.001	-0.03	6.8	0.9	3.2	G5 IV		52 s		
109524	203614		12 35 33.6	-34 52 50	-0.017	-0.08	7.80	1.1	0.2	K0 III		80 mx		m
109525	223540		12 35 36.9	-49 16 58	0.000	-0.02	7.8	1.0	-0.1	K2 III		380 s		
109547	157381		12 35 39.3	-17 09 02	-0.004	-0.05	7.1	0.4	3.0	F2 V		65 s		
109557	157382		12 35 42.6	-16 49 34	-0.006	+0.01	6.60	0.4	3.0	F2 V		53 s	8603	m
109536	223542		12 35 45.4	-41 01 19	-0.010	-0.01	5.13	0.22		A m	-11	16 mn		
109505	251966		12 35 46.4	-61 34 22	-0.001	0.00	8.01	0.22	-1.6	B3.5 V		540 s		
109615	63089		12 35 47.3	+39 41 01	-0.001	-0.02	7.30	-0.05	1.2	A1 V		170 s		
109701	15835		12 35 51.7	+69 00 48	-0.003	-0.03	7.4	0.9	3.2	G5 IV		70 s		
109584	157385		12 35 52.6	-12 04 11	-0.002	0.00	6.6	0.4	2.6	F0 V		63 s		
109655	44240		12 35 53.0	+45 46 51	-0.001	+0.01	7.30	1.60	8.2	K4 V	-1	6.1 s		
109244	--		12 35 55.6	-84 16 07			8.0	0.0	0.4	B9.5 V		340 s		
109625	63093		12 35 57.9	+31 43 57	-0.001	-0.01	7.97	0.55	4.4	G0 V		52 s		
109585	180937		12 35 58.6	-20 31 38	+0.001	-0.04	6.20	0.33	2.6	dF0 n	-2	53 s		
109573	203621		12 36 01.1	-39 52 11	-0.003	-0.04	5.80	0.01		A0				m
109649	63095		12 36 01.2	+32 00 14	-0.002	-0.02	7.37	1.35	-0.1	K2 III	+10	240 s		
109627	82398		12 36 01.8	+25 25 24	+0.001	-0.05	8.00	1.1	0.2	K0 III		360 s		
109590	223545		12 36 05.3	-42 18 34	-0.010	+0.03	7.97	0.44		F2 VI				

HD	SAO	Star Name	α 2000	δ 2000	μ(α)	μ(δ)	V	B-V	M$_v$	Spec	RV	d(pc)	ADS	Notes
			h m s	° ' "	s	"								
109591	223546		12 36 06.5	-42 51 08	-0.012	-0.06	7.8	0.4	4.4	G0 V		49 s		
109604	203624		12 36 08.1	-33 44 47	-0.003	-0.04	7.9	1.8	-0.3	K5 III		290 s		
109702	28441		12 36 10.1	+56 49 31	+0.001	-0.04	7.23	1.18	-0.1	K2 III		280 s	8607	m
109681	44243		12 36 12.2	+40 35 11	-0.005	+0.02	7.61	1.10	0.2	K0 III		250 s		
109593	240086		12 36 12.3	-52 25 04	+0.001	-0.02	6.7	1.1	-0.1	K2 III		230 s		
109729	28444		12 36 23.2	+59 29 12	-0.004	-0.02	5.4	1.6		M5 III	-91			T UMa, v
109638	223549		12 36 25.8	-43 29 17	-0.004	-0.01	7.6	1.3	0.2	K0 III		210 s		
109563	251969		12 36 25.8	-67 10 15	-0.010	0.00	7.6	0.1	0.6	A0 V		92 mx		
109660	203629		12 36 38.0	-39 19 34	-0.003	0.00	6.8	1.0	0.2	K0 III		200 s		
109695	180951		12 36 40.4	-21 18 23	-0.001	+0.01	7.9	0.6	0.2	G9 III	+9	350 s		
109675	240089		12 36 46.5	-50 20 08	-0.005	-0.03	6.7	0.5	0.6	A0 V		79 s		
109704	138873	25 Vir	12 36 47.3	-5 49 55	-0.002	-0.02	5.87	0.07		A0	-6			
109741	82401		12 36 48.6	+20 14 15	-0.001	+0.02	7.90	0.5	3.4	F5 V		79 s		
109709	157397		12 36 50.7	-15 34 18	-0.004	+0.05	7.6	1.1	-0.1	K2 III		240 mx		
109822	7611		12 36 52.6	+70 11 05	-0.001	0.00	6.7	1.1	-0.1	K2 III		230 s		
109698	203632		12 36 52.6	-35 25 47	+0.003	0.00	7.7	1.0	-0.1	K2 III		360 s		
109742	100176	25 Com	12 36 58.2	+17 05 22	-0.003	-0.02	5.68	1.41	-0.3	K5 III	-8	160 s		
109756	138876		12 37 06.7	-2 19 19	+0.010	-0.14	7.0	0.5	3.4	F5 V		53 s		
109764	119497		12 37 07.5	+8 47 49	-0.004	0.00	6.59	0.25	1.4	A2 V		83 s		
109668	251974	α Mus	12 37 11.0	-69 08 07	-0.007	-0.02	2.69	-0.20	-2.3	B3 IV	+18	100 s		m
109771	100179		12 37 12.4	+11 49 05	+0.001	+0.01	7.98	1.00	0.2	K0 III		360 s		
110010	7614		12 37 19.2	+79 12 55	-0.045	+0.01	7.1	0.7	4.4	G0 V	-19	34 s		
109772	138878		12 37 21.1	-8 18 00	-0.003	-0.01	6.9	0.1	0.6	A0 V		180 s		
109838	44256		12 37 33.4	+45 15 12	+0.002	0.00	8.04	0.31	3.0	F2 V		100 s		
109752	240095		12 37 33.8	-58 05 19	-0.002	-0.01	7.98	0.01	0.6	A0 V		300 s		m
109815	100181		12 37 38.3	+10 35 02	+0.003	-0.01	7.58	1.34	-0.1	K2 III		250 s		
109839	63108		12 37 39.0	+38 41 03	0.000	-0.08	7.34	0.48	3.4	F5 V		59 s		
109787	223560	τ Cen	12 37 42.1	-48 32 28	-0.019	-0.01	3.86	0.05	1.4	A2 V	+5	30 mx		
109799	180965		12 37 42.2	-27 08 20	+0.006	-0.10	5.45	0.32	3.0	F2 V	-1	31 s	8612	m
109824	119501		12 37 48.0	+4 59 07	-0.004	-0.03	7.9	1.1	0.2	K0 III		230 mx		
109776	240097		12 37 50.5	-57 21 01	+0.001	-0.03	7.9	1.6	0.2	K0 III		160 s		
109777	240098		12 37 50.8	-57 51 55	-0.004	+0.01	7.56	0.03	0.4	B9.5 V		170 mx		
109840	119502		12 37 51.7	+4 17 20	-0.002	-0.03	7.0	0.1	1.4	A2 V		130 s		
109894	28458		12 37 56.2	+54 51 12	-0.005	0.00	7.1	0.8	3.2	G5 IV		60 s		
109761	251977		12 38 01.3	-69 36 11	-0.005	-0.02	7.40	0.93		G5				
109860	119503		12 38 04.3	+3 16 57	-0.002	-0.01	6.33	0.01		A0	0			
109808	240104		12 38 07.2	-55 55 51	-0.004	0.00	6.95	0.15	1.4	A2 V		120 s		m
109896	119508		12 38 22.3	+1 51 16	-0.005	-0.02	5.71	1.60		Ma	-16	12 mn		
109929	100186		12 38 25.4	+14 52 20	-0.003	-0.01	7.6	0.7	4.4	G0 V		44 s		
109864	203656		12 38 25.7	-38 47 52	+0.001	-0.03	7.4	0.4	0.2	K0 III		270 s		
109914	119509		12 38 29.9	+6 59 18	-0.002	-0.01	7.08	1.44		M4.6 e	-25	17 mn		R Vir, v
109880	223565		12 38 31.9	-47 56 34	0.000	0.00	7.8	0.1	0.6	A0 V		270 s		
109979	44263		12 38 32.6	+45 13 02	-0.014	-0.03	7.1	0.4	3.0	F2 V	+8	67 s		
109942	100187		12 38 35.0	+13 48 20	-0.001	-0.03	7.18	1.17	0.2	K0 III		200 s		
109901	203657		12 38 39.0	-35 09 49	-0.011	+0.02	7.8	0.9	-0.1	K2 III		150 mx		
109902	203659		12 38 41.5	-36 43 52	0.000	-0.03	7.6	1.3	0.2	K0 III		210 s		
109944	138885		12 38 43.2	-4 22 25	-0.003	-0.01	7.00	1.4	-0.3	K5 III	+10	290 s		m
109931	157415		12 38 44.5	-18 15 01	-0.008	+0.01	6.00	0.29	2.2	dA6 n	-13	58 s		
109981	63115		12 38 44.6	+34 34 35	-0.002	0.00	5.69	1.15	0.2	K1 III		120 s		
109907	223568		12 38 44.7	-45 14 45	+0.001	0.00	7.2	2.0	-0.3	K5 III		130 s		
109980	44265	9 CVn	12 38 46.2	+40 52 28	-0.002	-0.02	6.37	0.18	2.1	A5 V	-16	71 s		
109995	63116		12 38 47.4	+39 18 31	-0.012	-0.14	7.62	0.05	0.6	A0 V		19 mx		
109908	240112		12 38 48.9	-51 12 57	-0.008	-0.01	7.8	0.7	4.4	G0 V		40 s		
109866	251981		12 38 50.0	-62 01 53	+0.005	+0.06	7.2	1.6	0.2	K0 III		110 s		
109867	251980		12 38 52.5	-67 11 34	0.000	-0.01	6.25	0.06		B0.5 I k				m
109813	256965		12 38 54.8	-76 03 57	-0.006	-0.01	7.7	1.1	0.2	K0 III		320 s		
109969	138889		12 38 56.2	-0 51 17	-0.006	-0.03	7.2	0.1	0.6	A0 V		64 mx		
110084	15853		12 38 57.8	+61 01 44	+0.004	0.00	7.3	1.1	0.2	K0 III		270 s		
109996	82420		12 39 02.1	+22 39 34	-0.004	-0.02	6.38	1.10	0.0	K1 III	-27	190 s		m
109960	203666		12 39 03.4	-30 25 21	-0.002	-0.02	5.89	1.21		K0				
110024	82421	26 Com	12 39 07.2	+21 03 45	-0.006	-0.01	5.46	0.96	0.2	G9 III	-21	110 s		
110025	100194		12 39 09.1	+16 28 47	-0.007	0.00	7.5	0.4	3.0	F2 V		81 s		
11027	100195		12 39 10.0	+14 19 37	-0.001	-0.02	8.0	0.2		A m			8616	m
110533	2067		12 39 10.3	+83 38 39	-0.129	+0.01	7.3	0.6	4.4	G0 V	-19	38 s		
109963	223573		12 39 12.2	-49 03 59	-0.004	0.00	7.5	1.3	-0.1	K2 III		240 s		
110105	28464		12 39 12.6	+56 00 14	-0.009	+0.01	7.9	0.4	3.0	F2 V		97 s		
109857	256967		12 39 14.6	-75 22 10	-0.006	+0.03	6.49	0.08		B9				m
110014	138892	26 χ Vir	12 39 14.7	-7 59 45	-0.005	-0.03	4.66	1.23	-0.1	K2 III	-20	78 s		
110066	63118		12 39 16.8	+35 57 06	-0.004	-0.01	6.45	0.06		A4 p	-15			
110003	223576		12 39 22.8	-45 27 26	-0.004	+0.03	8.0	1.7	0.2	K0 III		100 s		
110015	180989		12 39 24.2	-29 32 12	-0.002	0.00	7.5	0.6	2.6	F0 V		64 s		
110234	7621		12 39 26.7	+73 00 07	+0.001	-0.01	7.4	1.1	0.2	K0 III		280 s		
110069	138897		12 39 30.7	-6 06 03	+0.001	-0.01	6.8	1.1	-0.1	K2 III		240 s		
110165	15859		12 39 34.5	+60 53 04	-0.010	-0.03	7.1	1.1	0.2	K0 III		170 mx		
110052	180991		12 39 35.2	-23 48 38	-0.002	-0.02	7.8	0.2	2.6	F0 V		110 s		

HD	SAO	Star Name	α 2000	δ 2000	μ(α)	μ(δ)	V	B-V	M$_v$	Spec	RV	d(pc)	ADS	Notes
			h m s	° ′ ″	s	″								
110037	223579		12 39 38.9	-44 55 36	-0.001	+0.03	7.8	1.6	-0.5	M2 III		230 s		
110058	223581		12 39 46.0	-49 11 54	-0.004	-0.01	8.0	0.0	0.6	A0 V		140 mx		
110089	180995		12 39 48.0	-29 39 22	-0.003	-0.01	7.7	1.7	-0.1	K2 III		180 s		
110182	28467		12 39 50.8	+51 43 29	+0.001	-0.01	7.63	0.05		A0				
110073	203681		12 39 52.4	-39 59 15	-0.004	-0.03	4.64	-0.08		B8 p	+15			
110123	138898		12 39 53.5	-3 04 00	-0.003	-0.04	7.9	1.1	0.2	K0 III		340 s		
110020	251987		12 39 55.6	-66 30 41	-0.009	-0.03	6.26	-0.05		B9				
110313	15861		12 40 03.2	+68 48 09	-0.084	+0.04	7.87	0.61	4.5	dG1	-5	46 s		
110194	63130		12 40 08.6	+34 09 57	-0.003	0.00	7.5	1.4	6.3	K2 V	-44	17 s		
110009	—		12 40 09.8	-72 08 00			8.0	1.1	0.2	K0 III		350 s		
110128	203688		12 40 16.0	-37 51 26	-0.002	-0.02	7.6	0.8	1.4	A2 V		57 s		
110259	28471		12 40 21.2	+55 50 47	0.000	0.00	7.70	1.6		M7 II-III				Y UMa, v
110143	223586		12 40 21.8	-49 24 01	-0.004	-0.11	7.1	0.6	3.2	G5 IV		60 s		
110275	—		12 40 23.8	+59 31 32	-0.001	0.00	7.9	0.1	1.7	A3 V		180 s		
110080	256971		12 40 31.6	-70 32 47	-0.012	+0.01	7.40	0.28		A0				
110422	7629		12 40 33.3	+75 54 22	+0.001	0.00	7.2	1.1	0.2	K0 III		260 s		
110248	63132		12 40 34.9	+30 22 38	-0.001	-0.07	7.65	0.31		A m				
110221	119526		12 40 35.1	+7 41 58	0.000	-0.02	8.00	0.89	3.2	G5 IV		74 s		
110198	157440		12 40 36.0	-14 05 58	-0.001	0.00	7.0	1.1	0.2	K0 III		230 s		
110223	138905		12 40 40.7	-8 26 55	-0.003	-0.16	7.3	0.5	4.0	F8 V		46 s		
110249	100203		12 40 43.8	+12 42 56	-0.002	-0.01	7.90	1.53	-0.1	K2 III		210 s		
110177	240141		12 40 46.5	-57 37 52	-0.009	-0.01	7.7	1.5	0.2	K0 III		160 s		
110345	15866		12 40 48.5	+63 46 17	-0.017	+0.02	7.5	0.5	3.4	F5 V		68 s		
110261	119529		12 40 49.9	+3 52 09	-0.004	-0.05	7.8	0.4	2.6	F0 V		110 s		
110225	181011		12 40 51.5	-27 42 49	-0.001	-0.02	7.30	0.00	0.4	B9.5 V		240 s	8624	m
110296	63134		12 40 52.7	+34 06 16	+0.002	-0.01	7.78	1.55	8.2	K4 V	+1	7.8 s		
110280	119530		12 40 54.0	+8 49 44	-0.001	-0.04	7.60	0.4	3.0	F2 V		83 s	8625	m
110297	82439		12 40 55.3	+27 08 20	-0.003	-0.02	7.82	0.44		F5				m
110238	203705		12 40 59.6	-31 44 15	-0.011	-0.03	7.8	1.2	-0.1	K2 III		130 mx		
110315	100205		12 41 06.3	+15 22 34	+0.007	-0.42	7.95	1.14	-0.1	K2 III		40 mx		
110326	63135		12 41 07.6	+30 26 14	-0.002	+0.02	6.94	0.27		A m	-9			
110299	138911		12 41 08.9	-9 14 43	-0.001	+0.01	7.5	1.6	-0.5	M2 III		400 s		
110253	223595		12 41 09.6	-44 06 04	-0.008	0.00	6.72	1.27	-0.2	K3 III		240 mx		
110284	181014		12 41 13.8	-27 54 29	-0.001	0.00	6.8	1.4	-0.5	M2 III		290 s		
110317	157448		12 41 16.1	-13 00 54	-0.009	0.00	5.27	0.42	3.4	F5 V	-14	23 s	8627	m
110285	203711		12 41 18.1	-33 26 41	+0.001	-0.03	7.8	0.8	3.4	F5 V		45 s		
110287	223601		12 41 22.9	-46 08 44	-0.007	+0.05	5.84	1.52	-2.3	K3 II		63 mx		
110350	100206		12 41 24.4	+19 04 29	+0.001	-0.05	7.7	1.1	0.2	K0 III		320 s		
110258	240150		12 41 26.0	-59 47 38	0.000	+0.01	7.90	0.62	3.4	F5 V		57 s		AG Cru, v
110392	44285		12 41 27.0	+40 34 45	-0.001	-0.02	7.63	0.92	0.3	G8 III		290 s		
110409	44286		12 41 28.6	+49 17 28	-0.007	+0.05	6.91	1.30	-0.1	K2 III		210 s		
110273	240152		12 41 28.9	-56 18 16	0.000	+0.02	7.8	0.8	4.0	F8 V		41 s		
110338	181022		12 41 30.5	-20 37 00	-0.007	-0.01	7.8	0.2	2.6	F0 V		110 s		
110289	240156		12 41 30.8	-52 50 58	-0.003	-0.02	7.4	1.4	0.2	K0 III		160 s		
110304	223603	γ Cen	12 41 30.9	-48 57 34	-0.019	-0.01	2.17	-0.01	-0.6	A0 III	-8	34 s		m
110462	15871	76 UMa	12 41 33.8	+62 42 46	-0.005	-0.02	5.9	0.1	0.6	A0 V	-4	100 mx		
110377	100207	27 Vir	12 41 34.3	+10 25 35	-0.007	0.00	6.19	0.19	2.4	A7 V	+18	48 mx		m
110379	138917	29 γ Vir	12 41 39.5	-1 26 57	-0.038	+0.01	2.75	0.36	2.6	F0 V	-20	11 ts	8630	Porrima, m
110412	100210		12 41 48.3	+9 53 10	-0.004	+0.02	7.70	0.4	2.6	F0 V		110 s	8631	m
110385	157451		12 41 49.1	-19 45 31	-0.015	+0.02	6.03	0.39	3.0	dF2 n	-3	40 s		
110310	251995		12 41 52.8	-64 43 43	-0.001	0.00	7.6	0.4	2.6	F0 V		99 s		
110411	100211	30 ρ Vir	12 41 53.0	+10 14 08	+0.006	-0.09	4.88	0.09	0.6	A0 V	+2	63 s		
110611	7635		12 41 53.1	+75 21 21	-0.028	+0.02	8.0	1.1	0.2	K0 III	-34	150 mx		
110356	223608		12 41 55.7	-47 36 21	0.000	0.00	7.5	1.6	-0.1	K2 III		190 s		
110335	240161		12 41 56.6	-59 41 08	-0.003	-0.01	4.93	-0.04	-1.0	B7 IV	+12	140 s		
110423	119538	31 Vir	12 41 57.0	+6 48 24	-0.005	-0.02	5.59	0.00		A0 n	+4	18 mn	8633	m
110418	138920		12 41 57.6	-7 30 01	+0.001	-0.05	7.1	1.4	-0.3	K5 III	-2	310 s		
110440	119541		12 42 02.9	+0 29 42	-0.007	-0.03	8.0	0.4	3.0	F2 V		98 s		
110311	251996		12 42 05.2	-69 24 27	+0.001	-0.01	5.9	0.9	3.2	G5 IV		34 s		R Mus, v
110372	240165		12 42 06.2	-54 45 38	-0.005	+0.02	7.20	1.4	-0.3	K5 III		300 s		m
110500	44292		12 42 14.8	+45 52 37	-0.005	-0.01	7.04	0.24		A m	-8			
110467	119543		12 42 15.8	+8 02 31	-0.001	+0.03	7.8	0.5	4.0	F8 V		59 s		
110468	119542		12 42 16.2	+4 17 33	-0.001	0.00	7.9	1.1	-0.1	K2 III		400 s		
110469	138923		12 42 17.6	-0 18 29	+0.002	+0.01	8.01	0.50	3.4	F5 V		78 s		
110501	63146		12 42 17.6	+33 41 18	-0.018	-0.11	6.6	1.1	0.2	K0 III		53 mx		
110442	157454		12 42 17.9	-15 04 59	+0.006	-0.04	7.8	1.1	0.2	K0 III		230 mx		
110453	157455		12 42 24.5	-15 14 17	-0.002	-0.02	7.7	1.1	-0.1	K2 III		360 s		
110390	252000		12 42 27.1	-61 01 06	-0.005	-0.02	7.3	0.0	1.4	A2 V		140 mx		
110570	28484		12 42 31.5	+56 59 57	0.000	-0.01	7.4	1.1	-0.1	K2 III		320 s		
110456	203731		12 42 31.8	-37 54 10	-0.008	-0.01	7.9	1.0	2.6	F0 V		46 s		
110458	223614		12 42 35.3	-48 48 47	-0.013	-0.03	4.66	1.09	0.0	gK1	-12	86 s		
110524	82449		12 42 39.0	+28 21 36	-0.001	-0.05	7.51	0.38		F2				
110445	240173		12 42 39.0	-54 32 18	-0.004	0.00	7.0	0.2	0.6	A0 V		140 s		
110491	203738		12 42 47.4	-30 56 45	-0.002	+0.03	8.0	0.9	-0.1	K2 III		430 s		
110447	240175		12 42 47.9	-58 28 43	-0.003	0.00	7.9	0.3	0.4	B9.5 V		190 s		

HD	SAO	Star Name	α 2000	δ 2000	μ(α)	μ(δ)	V	B-V	M$_v$	Spec	RV	d(pc)	ADS	Notes
110483	223618		12^{h}42^{m}48^s.7	-47°55'36"	-0^s.005	0".00	8.0	1.4	0.2	K0 III		220 s		
110461	240176		12 42 49.6	-55 56 49	-0.005	-0.02	6.08	-0.03	0.2	B9 V	+37	150 s		
110432	252002		12 42 50.2	-63 03 32	-0.002	-0.02	5.31	0.27		B1 pe				
110434	--		12 42 51.5	-66 26 56			7.3	0.1	0.6	A0 V		210 s		
110527	157462		12 42 55.6	-13 37 53	+0.001	-0.02	7.9	1.1	0.2	K0 III		340 s		
110504	203741		12 42 57.5	-34 17 45	+0.001	-0.04	8.0	1.8	-0.1	K2 III		180 s		
110537	138928		12 42 59.2	-4 02 58	-0.015	-0.19	7.8	0.6	4.4	G0 V		38 mx		
110477	252005		12 43 03.6	-61 08 55	-0.022	-0.04	7.77	0.44	2.2	F6 IV		80 mx		
110678	15877		12 43 04.1	+61 09 20	-0.007	+0.02	6.38	1.26	0.2	K0 III	-6	120 s		m
110582	82453		12 43 04.5	+24 17 59	-0.003	-0.01	8.00	1.01	0.2	K0 III		360 s		
110540	181043		12 43 04.7	-23 50 31	-0.003	-0.01	8.0	0.9	-0.1	K2 III		410 s		
110506	240179		12 43 09.0	-56 10 34	-0.005	-0.01	6.00	-0.08		B8 n	+10			
110560	181047		12 43 13.0	-24 59 25	0.000	-0.01	6.8	1.6	-0.5	M2 III		280 s		
110612	100222		12 43 15.8	+10 06 02	-0.004	-0.06	7.38	1.56	-0.5	gM3	-14	380 s		
110741	15879		12 43 17.7	+65 56 44	-0.011	+0.01	7.5	0.4	3.0	F2 V		80 s		
110628	82456		12 43 18.3	+26 07 35	+0.001	-0.01	6.70	0.36	0.6	F2 III n	-12	170 s		
110575	203746		12 43 26.2	-40 10 40	+0.002	-0.02	6.44	0.25		A m				
110532	240183		12 43 28.1	-58 54 11	-0.010	-0.01	6.40	1.09		K0				m
110593	223625		12 43 30.2	-41 19 20	-0.008	0.00	7.14	1.59	-0.5	M4 III		230 mx		
110602	181053		12 43 31.7	-26 18 05	-0.004	-0.01	7.4	1.3	0.2	K0 III		190 s		
110616	157470		12 43 32.8	-13 51 34	-0.003	-0.03	7.1	0.1	0.6	A0 V		200 s		
110687	44300		12 43 33.7	+41 15 40	-0.001	+0.01	7.90	1.63	-0.5	M3 III		470 s		
110680	82461		12 43 37.8	+25 40 48	+0.001	0.00	7.6	0.1	1.7	A3 V		150 s		
110646	138933		12 43 38.0	-1 34 37	+0.003	-0.08	5.93	0.86	0.3	G8 III p	+1	130 s		
110606	203749		12 43 40.1	-35 24 54	-0.003	-0.07	7.9	0.5	2.6	F0 V		90 s		
110619	203750		12 43 42.6	-37 42 28	-0.053	-0.21	7.52	0.67	5.2	G5 V	-30	29 s		
110662	157473		12 43 47.7	-12 00 49	+0.003	-0.01	6.80	1.1	0.2	K0 III		210 s	8645	m
110661	138935		12 43 48.5	-9 46 07	-0.004	-0.02	7.6	0.4	2.6	F0 V		100 s		
110683	157474		12 43 52.5	-18 17 05	-0.004	0.00	7.6	1.1	0.2	K0 III		300 s		
110682	157476		12 43 55.9	-13 08 31	+0.003	-0.02	7.0	0.2	2.1	A5 V		96 s		
110668	203757		12 43 58.1	-35 07 22	-0.008	-0.12	8.0	1.1	3.2	G5 IV		58 s		
110653	203756		12 43 58.5	-36 20 56	-0.001	0.00	6.39	-0.06		A0				
110721	82462		12 43 59.7	+21 10 23	-0.003	+0.02	7.98	0.25	2.6	F0 V		120 s	8649	m
110666	181063		12 44 00.4	-28 19 26	-0.002	-0.04	5.48	1.34	-0.3	K4 III	+7	140 s		
110787	63165		12 44 08.8	+35 46 08	+0.001	+0.02	7.11	0.30		A m			8651	m
110746	138942		12 44 11.1	-2 50 33	-0.005	-0.01	6.7	0.1	1.4	A2 V		85 mx		
110833	28499		12 44 14.4	+51 45 32	-0.043	-0.19	7.04	0.94	5.9	dK0 e	+9	16 ts		
110789	82466		12 44 22.4	+21 59 36	-0.011	-0.20	7.7	0.5	4.0	F8 V		39 mx		
110778	119565		12 44 25.3	+0 32 12	-0.001	+0.01	8.0	1.6	-0.5	M2 III		510 s		
111112	2080		12 44 25.8	+80 37 15	+0.011	-0.05	6.3	0.1	0.6	A0 V	-26	11 mx		
110834	44307		12 44 27.0	+44 06 10	-0.003	0.00	6.33	0.43	0.7	F5 III	-16	130 s		
110835	44308		12 44 33.6	+43 07 35	-0.001	-0.01	6.95	1.33		K2 IV				
110698	240203		12 44 34.4	-57 17 13	-0.009	-0.02	6.64	0.08	0.6	A0 V		80 mx		m
110757	223639		12 44 37.7	-50 02 00	-0.005	-0.02	7.9	1.2	0.2	K0 III		270 s		
110733	240204		12 44 40.5	-54 05 07	-0.003	0.00	7.2	2.0	0.2	K0 III		63 s		
110822	100232		12 44 42.6	+17 42 02	0.000	0.00	8.0	0.5	4.0	F8 V		62 s		
110805	157484		12 44 42.6	-17 46 43	-0.002	-0.05	7.3	1.1	-0.1	K2 III		300 s		
110770	240206		12 44 51.1	-52 45 21	-0.001	-0.01	7.80	0.4	3.0	F2 V		90 s		m
110871	63176		12 44 54.0	+33 04 58	-0.003	0.00	7.78	0.40	4.0	F8 V		57 s		
111178	2083		12 44 59.2	+79 55 24	-0.009	-0.03	7.7	0.9	3.2	G5 IV		80 s		
110897	63177	10 CVn	12 44 59.4	+39 16 44	-0.031	+0.13	5.95	0.55	4.4	G0 V	+81	17 ts		
110836	138948		12 44 59.7	-8 31 56	+0.002	-0.01	7.80	1.1	0.2	K0 III		330 s		m
110717	--		12 45 01.3	-69 01 53			7.35	0.13		A2				
110716	252012		12 45 01.7	-68 49 52	-0.004	-0.02	6.16	0.69		G0				
110883	82471		12 45 03.2	+27 23 36	-0.001	-0.02	7.46	1.07	0.3	G8 III		210 s		m
110872	82470		12 45 03.7	+23 35 37	-0.002	0.00	7.75	1.63	-0.5	M2 III		410 s		
110772	240208		12 45 05.6	-60 04 33	-0.009	-0.01	6.7	1.4	-0.1	K2 III		180 s		
110914	44317		12 45 07.8	+45 26 25	0.000	+0.01	4.99	2.54		N7.7	+12	21 mn		Y CVn, v
110808	240213		12 45 10.3	-51 50 17	-0.002	+0.01	7.9	1.2	-0.1	K2 III		390 s		
110786	252014		12 45 14.1	-62 13 03	0.000	-0.01	7.68	1.25	-4.8	A4 Ib		1200 s		m
110848	181075		12 45 16.2	-28 45 58	-0.005	-0.02	6.8	0.6	1.4	A2 V		58 s		
110886	138952		12 45 17.3	-3 53 18	-0.002	+0.01	6.80	0.1	1.7	A3 V		110 s	8657	m
110838	223644		12 45 23.4	-48 10 13	0.000	+0.01	6.65	1.14	0.0	K1 III		200 s		
110720	256978		12 45 23.8	-73 47 51	-0.006	-0.03	7.9	0.1	0.6	A0 V		260 mx		
110932	100236		12 45 26.1	+14 22 24	-0.003	-0.01	7.23	-0.11		A0	+4	23 mn	8659	m
110760	256979		12 45 26.1	-71 42 48	-0.009	-0.08	7.8	1.6	-0.5	M4 III		150 mx		
--	100235		12 45 26.4	+14 21 49	-0.005	-0.05	7.75	0.41		A0				
110875	203777		12 45 27.7	-37 26 58	-0.006	-0.02	7.60	0.6	4.4	G0 V		44 s		m
110918	138955		12 45 31.9	-4 48 39	-0.001	+0.01	7.4	1.1	-0.1	K2 III		310 s		
110902	157492		12 45 32.7	-20 08 43	-0.003	0.00	7.57	0.02	0.6	A0 V		240 s		
110951	119574	32 Vir	12 45 37.0	+7 40 24	-0.007	0.00	5.22	0.33		A m	-9	22 mn		
110829	252016	z Cru	12 45 37.8	-60 58 52	+0.014	-0.07	4.69	1.05	0.0	K1 III	+9	87 s		m
110987	63181		12 45 41.0	+38 21 28	-0.003	-0.07	8.0	1.1	0.2	K0 III		140 mx		
110919	181082		12 45 42.0	-24 32 09	-0.003	+0.01	7.8	1.7	-0.1	K2 III		190 s		
110892	223649		12 45 42.1	-50 10 49	-0.030	-0.11	7.5	0.0	3.4	F5 V		48 mx		

HD	SAO	Star Name	α 2000	δ 2000	μ(α)	μ(δ)	V	B−V	M_v	Spec	RV	d(pc)	ADS	Notes
			h m s	° ′ ″	s	″								
110988	63182		12 45 43.9	+33 32 28	+0.001	−0.06	7.53	0.97	0.3	G8 III		47 s		
110907	223653		12 45 50.7	−45 50 44	−0.004	−0.02	7.7	0.5	2.6	F0 V		55 s		
110952	181086		12 45 52.5	−21 46 05	0.000	−0.01	7.1	1.1	0.2	K0 III		190 s		
110953	181087		12 45 53.3	−22 37 48	−0.004	0.00	7.6	0.6	1.4	A2 V		78 s		
110989	119575		12 46 02.3	+2 35 34	+0.002	−0.10	8.0	0.5	4.0	F8 V		62 s		
110924	240229		12 46 05.3	−54 36 44	−0.004	0.00	6.7	1.5	0.2	K0 III		94 s		
110967	223659		12 46 14.0	−43 59 55	+0.001	−0.01	7.6	2.2	−0.5	M2 III		190 s		
111005	119576		12 46 14.8	+2 27 51	−0.003	−0.02	7.70	0.2	2.1	A5 V		130 s	8662	m
110879	252019	β Mus	12 46 16.9	−68 06 29	−0.006	−0.02	3.05	−0.18	−1.7	B3 V	+42	89 s		m
111028	119580	33 Vir	12 46 22.4	+9 32 24	+0.019	−0.45	5.67	0.99		K1 IV	+52	23 mn		m
110956	240235		12 46 22.6	−56 29 20	−0.005	−0.03	4.65	−0.16	−1.7	B3 V	+17	35 mx		m
111006	157498		12 46 22.8	−18 13 11	−0.009	−0.02	7.3	1.1	0.2	K0 III		150 mx		
111043	119582		12 46 26.6	+9 17 08	−0.002	−0.02	7.7	0.4	2.6	F0 V		110 s		
111016	157501		12 46 30.2	−15 26 26	0.000	0.00	8.0	1.6	−0.5	M2 III		510 s		
111031	157502		12 46 30.7	−11 48 44	−0.019	+0.05	6.87	0.70		G5		18 mn		
111066	82490		12 46 32.6	+24 08 42	−0.009	−0.21	6.88	0.55	3.4	F5 V		36 mx		m
111129	44324		12 46 37.4	+47 22 20	−0.002	0.00	8.00	1.6	−0.5	M2 III		500 s		
111067	100252	27 Com	12 46 38.6	+16 34 39	+0.001	0.00	5.12	1.35	−0.2	K3 III	+53	100 s		
111020	223664		12 46 42.2	−47 16 49	−0.004	−0.01	8.0	0.8	0.2	K0 III		360 s		
111032	203797		12 46 46.0	−33 18 56	−0.001	−0.03	5.86	1.34		K0				m
111153	44326		12 46 48.2	+43 09 06	−0.001	−0.04	8.0	0.5	4.0	F8 V		62 s		
111179	28513		12 46 48.3	+51 38 11	−0.002	−0.02	8.0	0.5	3.4	F5 V		82 s		
111096	138963		12 46 58.8	−7 47 59	−0.005	−0.05	7.9	0.9	3.2	G5 IV		88 s		
111166	—		12 47 01.9	+4 09 26	+0.002	−0.02	8.00	1.8		C7.7 e	+2			
111133	119585		12 47 02.2	+5 57 03	+0.002	−0.05	6.34	−0.05		A4 p	+16			
111132	119586		12 47 02.2	+9 03 50	0.000	−0.02	6.86	1.34	0.2	K0 III		130 s		
111097	181096		12 47 03.5	−22 07 45	+0.006	+0.02	7.8	0.1	3.4	F5 V		74 s		
111073	223668		12 47 03.6	−43 15 52	−0.001	0.00	7.6	1.0	0.2	K0 III		300 s		
111155	100258		12 47 04.1	+13 37 02	−0.001	+0.02	7.8	0.5	3.4	F5 V		76 s		
111114	138964		12 47 06.3	−9 12 52	−0.001	−0.01	7.7	1.1	−0.1	K2 III		360 s		
111180	63190		12 47 07.7	+32 34 04	+0.003	−0.03	7.73	1.49	8.0	K3 V		11 s		
111163	100259		12 47 10.5	+15 35 29	0.000	−0.03	6.64	0.89	0.2	K0 III		190 s		
111084	223671		12 47 11.4	−42 55 43	0.000	+0.01	7.9	0.9	3.2	G5 IV		69 s		
111164	100260	34 Vir	12 47 13.6	+11 57 29	+0.003	−0.02	6.07	0.12	1.7	A3 V	−1	72 s		m
111137	157508		12 47 17.8	−12 16 44	−0.001	0.00	7.9	1.1	−0.1	K2 III		390 s		
111270	15901		12 47 18.8	+62 46 51	+0.002	0.00	5.8	0.2	2.1	A5 V	−14	55 s		
111252	28515		12 47 19.7	+52 51 43	−0.005	+0.01	7.7	1.1	0.2	K0 III		310 s		
111156	157509		12 47 20.8	−19 00 14	−0.007	+0.01	7.2	0.8	3.2	G5 IV		64 s		
111199	138967		12 47 33.3	−6 18 06	0.000	−0.05	6.26	0.55	3.4	dF5	+13	32 s		m
111335	15902	7 Dra	12 47 34.2	+66 47 25	0.000	−0.01	5.43	1.56	−0.3	K5 III	+8	130 s		
111105	240253		12 47 35.8	−60 04 52	+0.002	−0.02	7.25	0.20	1.4	A2 V		120 s		
111102	240257		12 47 39.3	−58 17 52	−0.003	−0.03	6.96	0.24		A5				
111185	181103		12 47 41.7	−29 47 42	−0.004	−0.02	7.9	2.1	−0.5	M4 III		240 s		
111213	157511		12 47 43.1	−12 34 48	−0.002	−0.07	7.5	0.5	3.4	F5 V		67 s		
111123	240259	β Cru	12 47 43.2	−59 41 19	−0.005	−0.02	1.25	−0.23	−5.0	B0 III	+20	130 mx		m,v
111306	28519		12 47 48.2	+50 09 23	−0.011	−0.02	6.8	0.4	2.6	F0 V		69 s		
111271	82498		12 47 48.6	+29 31 54	−0.007	−0.01	7.25	0.36		F0				
111238	100263		12 47 48.8	+19 01 27	−0.006	−0.01	7.7	0.5	4.0	F8 V		55 s		
111255	82497		12 47 51.0	+21 38 11	−0.001	−0.01	8.02	1.38	0.2	K0 III		190 s		
111239	119596	35 Vir	12 47 51.3	+3 34 22	−0.001	−0.01	6.41	1.60	−0.5	gM4	+8	240 s		
111214	181104		12 47 53.1	−25 00 54	−0.003	−0.02	6.8	1.6	0.2	K0 III		89 s		
111226	181105		12 47 53.6	−24 51 05	−0.003	+0.04	6.44	−0.06		B9				
111217	203818		12 47 54.9	−34 03 38	−0.005	−0.01	8.0	0.3	2.6	F0 V		120 s		
111285	82500		12 47 59.2	+24 05 44	−0.003	+0.01	7.18	0.96	0.3	G8 III		230 s		
111420	7663		12 47 59.6	+70 56 47	+0.009	−0.06	7.24	1.27	−1.3	K3 II-III	−39	500 s		
111272	100268		12 48 00.3	+18 50 11	+0.002	−0.04	6.91	0.94	0.2	K0 III	−14	220 s		
111160	240265		12 48 02.7	−59 35 37	+0.001	−0.01	7.9	−1.1	0.0	B8.5 V		380 s		m
111275	138973		12 48 06.3	−0 21 34	−0.006	−0.01	7.98	0.55		F8				
111193	240267		12 48 10.3	−60 12 35	−0.001	+0.01	7.96	0.21		B0.5 II				
111381	28521		12 48 13.4	+52 27 12	−0.002	0.00	7.0	0.8	3.2	G5 IV		58 s		
111308	100269	28 Com	12 48 14.3	+13 33 11	−0.003	−0.03	6.56	0.01	0.6	A0 V	0	150 s		
111318	63197		12 48 15.2	+30 23 47	−0.005	0.00	7.66	1.12	0.2	K0 III		260 s		
111235	223677		12 48 16.4	−44 43 08	−0.002	−0.01	6.9	0.4	0.6	A0 V		97 s		
111307	100270		12 48 17.7	+19 19 20	−0.001	−0.01	7.60	1.61	−0.5	M2 III	−7	420 s		
111262	203823		12 48 17.7	−32 19 29	−0.004	−0.01	8.0	0.9	−0.1	K2 III		410 s		
111236	—		12 48 18.0	−48 39 47			8.0	2.3	0.2	K0 III		61 s		
111309	119602		12 48 19.3	+9 06 43	−0.005	+0.02	8.0	0.5	3.4	F5 V		84 s		
111347	63199		12 48 22.0	+30 46 03	−0.005	+0.04	8.03	0.44	4.0	F8 V		64 s		
111295	181114		12 48 26.3	−27 35 51	−0.011	−0.07	5.66	0.95		G5				
111443	15906		12 48 28.4	+60 49 12	+0.005	−0.01	7.6	1.1	−0.1	K2 III		340 s		
111249	240272		12 48 32.4	−52 37 27	−0.001	0.00	7.6	0.7	0.2	K0 III		300 s		
111456	15907		12 48 39.3	+60 19 11	+0.014	0.00	5.85	0.46	3.7	F6 V	−12	27 s		
111421	44332	11 CVn	12 48 41.7	+48 28 00	−0.007	+0.01	6.27	0.18		A m	−2			
111232	—		12 48 46.7	−68 25 47			7.59	0.70	5.2	G5 V		29 s		
111395	82511		12 48 46.9	+24 50 24	−0.025	−0.11	6.31	0.70	5.5	G7 V	−8	16 mn		

318

HD	SAO	Star Name	α 2000	δ 2000	μ(α)	μ(δ)	V	B−V	M_V	Spec	RV	d(pc)	ADS	Notes
			^{h} ^{m} ^{s}	$^{\circ}$ ′ ″	^{s}	″								
111513	—		12 48 50.8	+61 22 44	−0.043	+0.11	7.35	0.62	4.5	G1 V		34 s		
111340	203830		12 48 51.9	−37 02 33	−0.006	−0.05	7.9	0.7	3.4	F5 V		54 s		
111398	100279		12 48 52.4	+12 05 47	+0.016	−0.14	7.07	0.66	3.2	G5 IV		59 s		m
111397	100283	29 Com	12 48 54.2	+14 07 21	+0.002	−0.03	5.70	0.02	1.4	A2 V	−7	72 s		
111781	2097		12 48 54.5	+80 24 22	−0.002	−0.02	7.5	0.8	3.2	G5 IV		73 s		
111384	138978		12 49 00.6	−9 13 13	+0.001	−0.03	7.60	1.1	−0.1	K2 III	+16	350 s		
111371	181123		12 49 01.6	−23 14 18	−0.003	+0.02	7.8	1.1	−0.1	K2 III		390 s		
111283	252038		12 49 02.5	−65 35 35	−0.002	0.00	7.35	−0.01	−1.1	B5 V		420 s		m
111251	252037		12 49 03.5	−69 23 35	−0.019	−0.03	7.4	0.5	4.0	F8 V		47 s		
111341	223683		12 49 03.8	−43 56 57	0.000	0.00	7.9	1.3	1.4	A2 V		37 s		
112014	2101		12 49 06.5	+83 25 04	−0.016	+0.02	5.83	−0.06		A0	+1			
111354	223684		12 49 06.8	−42 38 06	−0.007	−0.02	7.6	0.4	4.4	G0 V		44 s		
—	44337		12 49 07.6	+42 13 24	0.000	−0.04	7.70							m
111373	203835		12 49 09.9	−39 43 03	0.000	−0.02	8.0	−0.1	0.4	B9.5 V		330 s		
112028	2102		12 49 13.4	+83 24 46	−0.018	+0.02	5.28	−0.03		A2	+3	21 mn	8682	m
111469	82515	30 Com	12 49 17.4	+27 33 08	−0.007	+0.02	5.78	0.03	1.4	A2 V	+1	75 s	8674	m
111403	223688		12 49 22.3	−40 46 31	−0.003	+0.01	6.9	0.3	0.2	K0 III		220 s		
111375	240285		12 49 24.0	−52 34 07	0.000	−0.01	7.7	0.9	0.2	K0 III		310 s		
111470	119609		12 49 26.9	+5 10 20	+0.007	−0.07	7.5	0.5	4.0	F8 V		50 s		
111290	—		12 49 30.0	−71 43 46			7.77	0.02	−2.6	B2.5 IV		950 s		
111111	258650		12 49 31.8	−80 42 08	0.000	0.00	7.2	0.0	1.4	A2 V		150 s		
111433	203842		12 49 32.4	−34 47 28	−0.001	0.00	7.6	0.7	3.4	F5 V		50 s		
111498	119613		12 49 34.9	+9 22 24	+0.001	+0.01	7.65	0.90	0.2	K0 III		310 s		
111572	44343		12 49 38.1	+48 44 24	−0.001	−0.01	6.51	1.14	0.2	K0 III		140 s		
111406	240288		12 49 40.3	−53 21 42	−0.002	−0.01	7.0	0.5	3.2	G5 IV		58 s		
111488	138984		12 49 42.6	−6 52 50	−0.002	−0.03	7.7	0.1	0.6	A0 V		260 s		
111315	256983		12 49 44.9	−71 59 11	+0.000	−0.01	5.55	1.17		K0				
111500	157535		12 49 45.7	−15 52 57	0.000	−0.02	7.0	1.1	−0.1	K2 III		260 s		
111499	157536		12 49 47.0	−15 04 42	−0.001	+0.01	6.9	1.6	−0.5	M4 III	−14	310 s		
111541	82519		12 49 48.7	+26 25 54	+0.004	−0.01	6.89	1.07	3.2	G9 IV		41 s		
111436	240292		12 49 48.7	−51 23 34	−0.004	−0.01	7.5	0.2	1.4	A2 V		130 s		
111452	240294		12 49 51.3	−50 19 26	−0.001	−0.02	7.1	0.7	−0.1	K2 III		270 s		
111461	223694		12 49 54.4	−49 52 08	−0.002	−0.04	7.9	0.7	0.2	K0 III		340 s		
111409	—		12 49 55.8	−64 36 45			7.61	0.19	0.6	A0 V		200 s		
111627	28531		12 50 00.5	+54 45 22	−0.001	−0.03	7.1	1.1	−0.1	K2 III		280 s		
111544	119616		12 50 02.4	+0 39 56	−0.003	−0.03	7.9	1.1	−0.1	K2 III		390 s		
111545	138986		12 50 05.1	−7 37 56	−0.016	+0.01	6.9	0.5	3.4	F5 V		50 s		
111532	203852		12 50 10.3	−37 28 55	−0.006	−0.05	7.8	1.8	−0.3	K5 III		200 mx		
111604	63217		12 50 10.6	+37 31 01	−0.008	+0.02	5.89	0.15	1.9	A4 V	−11	62 s		
111463	252047		12 50 11.9	−60 24 03	0.000	−0.02	6.66	0.35		A2				
111464	252046		12 50 14.0	−62 38 28	−0.003	+0.02	6.63	1.41		K0				
111591	82523		12 50 17.3	+22 51 48	+0.008	−0.07	6.43	1.00	0.2	K0 III	+6	170 mx		
111548	203855		12 50 19.5	−33 36 34	0.000	+0.02	8.0	0.7	−0.1	K2 III		410 s		
111519	223698		12 50 19.5	−48 27 35	−0.004	0.00	6.24	0.05		A0				
111478	240300		12 50 19.7	−58 03 14	−0.001	−0.01	7.79	0.46		F2				
111563	181143		12 50 19.9	−24 13 06	+0.001	−0.12	8.0	0.9	3.2	G5 IV		78 s		
111564	203856		12 50 20.1	−30 34 39	−0.033	+0.01	7.61	0.61		G0		21 mn		
111592	119619		12 50 20.5	+6 13 38	+0.002	−0.02	7.3	0.8	3.2	G5 IV		67 s		
111441	—		12 50 22.7	−69 31 44			7.6	1.1	0.2	K0 III		290 s		
111535	223699		12 50 25.3	−47 13 20	−0.006	−0.03	7.98	0.48	2.2	F6 IV		140 s		
111581	157540		12 50 28.4	−13 02 03	−0.001	+0.01	7.5	1.1	−0.1	K2 III		340 s		
111536	240305		12 50 33.2	−52 14 03	−0.002	−0.02	7.4	0.3	1.4	A2 V		110 s		
111584	203860		12 50 36.4	−37 56 15	+0.001	−0.04	8.0	1.6	−0.3	K5 III		400 s		
111538	240307		12 50 38.9	−58 09 58	−0.008	+0.04	7.07	1.53		K5				
111597	203863		12 50 41.1	−33 59 57	−0.002	−0.02	4.91	−0.04	0.0	A0 IV	+18	96 s		m
111596	181148		12 50 41.4	−27 51 02	−0.003	−0.02	7.8	0.6	3.2	G5 IV		84 s		
111794	15919		12 50 48.8	+62 59 10	−0.003	+0.02	7.2	0.1	1.7	A3 V		130 s		
111588	240314		12 50 57.8	−52 47 15	−0.003	−0.02	5.73	0.13		A3				
111717	63221		12 50 59.6	+31 28 35	−0.003	−0.03	7.88	0.37	2.6	F0 V		100 s		
111619	223705		12 51 02.9	−44 01 00	+0.002	−0.03	7.8	1.9	0.2	K0 III		97 s		
111599	240315		12 51 05.4	−55 24 30	−0.002	0.00	7.9	0.3	0.4	B9.5 V		190 s		
111691	119627		12 51 05.7	+5 33 13	0.000	+0.01	7.6	1.6		M5 III	−46			U Vir, v
111651	181155		12 51 06.2	−29 10 39	−0.005	0.00	7.9	0.2	1.4	A2 V		100 mx		
111889	15921		12 51 08.3	+69 17 39	−0.007	+0.02	8.0	0.1	0.6	A0 V		200 mx		
111718	100300		12 51 12.0	+19 09 41	−0.010	+0.02	7.4	0.5	3.4	F5 V		63 s		
111558	—		12 51 12.3	−69 38 43			7.26	0.13	−7.1	B8 Ia		6300 s		
111613	252054		12 51 17.7	−60 19 46	−0.003	0.00	5.74	0.38	−7.3	A1 Ia	−22	2600 s		
111811	44357		12 51 18.3	+49 16 09	−0.003	−0.01	7.80	1.4	−0.3	K5 III		420 s		
111733	119631		12 51 20.2	+8 12 36	−0.002	−0.01	7.63	1.09	3.2	G5 IV		43 s		
111720	157550		12 51 22.8	−10 20 18	−0.001	−0.01	6.41	1.02		K0	−17		8684	m
111721	157552		12 51 25.2	−13 29 28	−0.018	−0.32	7.97	0.81		G0	+25	20 mn		
111850	28539		12 51 26.8	+57 22 24	+0.002	−0.01	7.9	1.1	0.2	K0 III		340 s		
111654	240323		12 51 35.8	−58 29 05	−0.002	−0.01	7.37	1.65		K2				
111765	119633	37 Vir	12 51 36.8	+3 03 24	−0.002	+0.02	6.02	1.29		K0	+3			
111812	82537	31 Com	12 51 41.8	+27 32 26	−0.001	−0.01	4.94	0.67	0.6	G0 III	−1	74 s		

HD	SAO	Star Name	α 2000	δ 2000	μ(α)	μ(δ)	V	B-V	M_v	Spec	RV	d(pc)	ADS	Notes
111737	203877		12 51 44.7	-31 05 00	-0.002	+0.01	7.30	1.6	-0.5	M2 III		360 s		m
111748	203879		12 51 50.5	-31 12 08	-0.002	-0.02	7.0	0.8	0.4	B9.5 V		69 s		
111816	119638		12 51 52.0	+0 05 04	+0.005	-0.06	7.9	0.5	4.0	F8 V		61 s		
111842	82539		12 51 54.0	+25 40 30	0.000	0.00	7.55	1.46		K3 IV				m
111844	100307		12 51 54.7	+19 10 20	-0.006	0.00	7.34	0.32		F2	-15		8690	m
111832	119639		12 51 55.3	+7 54 15	+0.003	-0.05	7.78	1.01	3.2	G5 IV		54 s		
111774	203881		12 51 56.7	-39 40 50	-0.003	-0.03	5.98	-0.10	-0.6	B7 V	+5	210 s		
111771	203884		12 51 56.8	-35 04 57	-0.001	-0.01	7.1	1.2	0.2	K0 III		190 s		
111786	181169		12 51 57.8	-26 44 17	-0.008	+0.04	6.15	0.23		A0				
111601	256985		12 52 01.7	-73 55 16	-0.005	-0.02	7.9	0.1	1.4	A2 V		190 s		
111775	223720		12 52 05.2	-48 05 38	-0.006	+0.01	6.33	0.03	-2.8	A0 II		110 mx		
111755	240331		12 52 09.4	-51 56 40	-0.002	0.00	7.6	1.4	-0.1	K2 III		260 s		
111853	139010		12 52 12.0	-0 55 45	-0.001	-0.02	8.0	1.4	-0.3	K5 III		460 s		
111862	100309	32 Com	12 52 12.1	+17 04 25	0.000	-0.02	6.32	1.59	-0.3	K5 III	-1	190 s		m
111808	223723		12 52 22.7	-49 35 55	-0.003	0.00	7.88	-0.03		B9				
111892	100311		12 52 22.8	+17 06 30	+0.002	-0.04	7.0	0.5	4.0	F8 V	+9	39 s		
111790	240338		12 52 24.5	-53 49 46	-0.002	+0.01	6.30	1.1	0.2	K0 III		160 s		m
111791	240337		12 52 25.0	-58 44 48	-0.001	-0.02	8.0	-0.1	0.4	B9.5 V		330 s		
111893	100312		12 52 27.5	+16 07 20	-0.003	-0.02	6.30	0.16	2.1	A5 V	-28	69 s		
111822	240340		12 52 30.6	-52 40 02	0.000	0.00	7.87	-0.03		B0.5 III k				
111865	181179		12 52 33.7	-23 08 07	-0.004	+0.03	7.4	0.6	3.2	G5 IV		70 s		
111909	119642		12 52 34.1	+1 56 37	+0.003	0.00	7.9	0.9	3.2	G5 IV		88 s		
111849	240344		12 52 48.3	-52 31 51	-0.002	-0.01	7.5	0.4	1.4	A2 V		99 s		
111946	157566		12 52 58.6	-14 37 13	-0.004	0.00	7.8	1.6	-0.5	M2 III		350 mx		
112001	82547		12 53 01.6	+26 47 39	-0.017	-0.10	7.68	0.63	2.8	G0 IV		47 mx		
111997	100317		12 53 03.6	+19 36 17	-0.009	-0.03	7.7	0.5	4.0	F8 V		54 s		
111884	240347		12 53 03.8	-54 57 08	-0.013	+0.01	5.93	1.31		K0				
111858	252066		12 53 06.5	-62 40 56	-0.009	+0.02	7.9	1.1	0.2	K0 III		330 s		
111915	223731		12 53 06.8	-48 56 35	-0.009	-0.02	4.33	1.37	-0.1	gK2	-2	54 s		
111962	157569		12 53 09.3	-14 57 55	-0.002	+0.03	7.7	0.4	3.0	F2 V		85 s		
111828	--		12 53 10.1	-69 52 39			7.9	0.8	3.2	G5 IV		85 s		
112002	100318		12 53 10.4	+12 27 57	-0.001	-0.01	7.89	0.15	1.4	A2 V		170 s		
111998	139022	38 Vir	12 53 11.1	-3 33 11	-0.017	-0.01	6.11	0.50	3.7	dF6	-7	29 s		
111900	240349		12 53 12.1	-53 13 32	-0.001	-0.03	7.3	0.3	0.6	A0 V		140 s		
111930	203904		12 53 13.3	-39 23 25	+0.001	-0.01	6.80	1.1	-0.1	K2 III		240 s		m
112031	82548		12 53 13.9	+28 59 24	-0.001	-0.02	8.02	0.34		F0				
111902	240350		12 53 16.0	-56 11 37	-0.005	-0.01	7.2	2.0	0.2	K0 III		63 s		
111948	181193		12 53 17.0	-30 00 54	-0.002	0.00	7.7	1.2	0.2	K0 III		250 s		
111931	223733		12 53 17.9	-45 08 31	-0.003	0.00	7.3	1.6	0.2	K0 III		120 s		
112033	82551	35 Com	12 53 19.2	+21 14 25	-0.004	-0.02	4.90	0.90	0.3	G8 III	-6	83 s	8695	m
112082	44376		12 53 20.0	+46 39 23	+0.001	0.00	7.45	1.56	-0.5	M3 III	-26	390 s		
111904	252069		12 53 21.7	-60 19 42	-0.001	0.00	5.77	0.34	-7.1	B9 Ia	-15	2500 s		
111968	203907		12 53 26.1	-40 10 44	+0.006	-0.03	4.27	0.21	0.5	A7 III	-3	57 s		
112060	100321		12 53 31.9	+19 28 51	-0.012	-0.19	6.44	0.78	3.2	G5 IV		38 mx		
111984	223737		12 53 33.3	-43 04 31	-0.007	-0.01	7.20	0.2	2.1	A5 V		79 mx		m
112036	139025		12 53 36.8	-9 03 53	+0.001	-0.06	7.8	1.1	-0.1	K2 III		330 mx		
112048	139027		12 53 38.0	-4 13 26	0.000	-0.05	6.44	1.10		K0				
111934	252071		12 53 39.3	-60 21 13	+0.002	-0.01	6.80	0.24	-5.7	B3 Ib		1800 s		
111876	256986		12 53 39.9	-71 16 25	-0.007	-0.02	7.1	0.1	0.6	A0 V		190 s		
112061	119649		12 53 40.0	+4 14 16	-0.001	-0.04	7.7	1.1	0.2	K0 III		320 s		
--	252073		12 53 41.2	-60 20 57	-0.001	-0.01	7.9	1.6	-0.5	M2 III		450 s	--	
112336	7693		12 53 46.6	+76 56 06	-0.008	-0.04	7.9	1.1	0.2	K0 III		310 mx		
112084	100323		12 53 47.9	+19 03 26	-0.004	+0.02	7.06	1.14	0.2	K0 III	0	190 s		
111973	252077	κ Cru	12 53 49.0	-60 22 36	+0.001	-0.03	5.95	0.24	-6.3	B3 Iab	-1	1700 s		
112097	100322	41 Vir	12 53 49.6	+12 25 07	+0.003	-0.03	6.25	0.27		A7 p	-10			
112052	--		12 53 53.3	-29 18 37			7.70	0.9	3.2	G5 IV		79 s	8698	m
112127	82554		12 53 55.6	+26 46 48	-0.001	-0.01	6.91	1.26	0.2	K0 III	+2	130 s		
114282	2125		12 53 56.0	+87 38 54	-0.021	+0.01	7.47	1.42		K2	-55			
111990	252080		12 53 59.7	-60 20 05	+0.001	+0.02	6.78	0.26	-6.3	B2 I k		2900 s		m
112074	157580		12 54 00.2	-18 02 15	-0.002	0.00	6.80	0.28		A0			8699	m
112185	28553	77 ε UMa	12 54 01.7	+55 57 35	+0.013	-0.01	1.77	-0.02		A0 p	-9	19 mn		Alioth, v
112086	--		12 54 10.5	-27 57 36			7.40	0.7	4.4	G0 V		40 s	8700	m
112054	240360		12 54 11.4	-51 21 56	-0.008	+0.01	7.8	0.6	3.0	F2 V		61 s		
112171	63244		12 54 13.0	+33 32 03	-0.008	+0.02	6.26	0.20	2.1	A5 V	+5	66 s		
112651	2112		12 54 13.0	+82 31 03	+0.016	-0.01	7.30	1.1	0.2	K0 III		260 s	8375	m
112371	7695		12 54 14.7	+75 24 32	-0.001	-0.01	7.0	1.1	0.2	K0 III		230 s		
111953	252081		12 54 17.9	-70 02 10	+0.001	+0.01	7.0	1.1	0.2	K0 III		220 s		
112131	157584		12 54 18.6	-11 38 55	-0.010	+0.01	6.00	0.08		A1				
112142	139033	40 ψ Vir	12 54 21.1	-9 32 20	-0.002	-0.02	4.80	1.59	-0.5	M3 III	+18	120 s		
112044	240362		12 54 21.9	-58 25 50	-0.001	-0.01	6.4	0.7	4.4	G0 V		25 s		S Cru, v
111993	252083		12 54 22.8	-68 46 31	0.000	-0.06	7.9	0.0	0.0	B8.5 V		120 mx		
112220	44382		12 54 28.0	+46 46 41	-0.003	0.00	7.49	1.04	0.0	K1 III		320 s		
112144	157586		12 54 30.4	-20 05 51	+0.001	+0.01	6.9	1.0	-0.1	K2 III		250 s		
112092	240366	μ¹ Cru	12 54 35.6	-57 10 40	-0.004	-0.01	4.03	-0.17	-2.3	B3 IV	+12	190 s		m
112091	240367	μ² Cru	12 54 36.8	-57 10 05	-0.004	-0.01	5.18	-0.14		B5 V ne	+19	190 s		m

HD	SAO	Star Name	α 2000	δ 2000	μ(α)	μ(δ)	V	B-V	M_v	Spec	RV	d(pc)	ADS	Notes
112078	240368	λ Cru	12^{h}54^{m}39.1^s	-59°08'47"	-0.004	-0.01	4.62	-0.15	-1.1	B5 V n	+16	54 mx		
112196	82559		12 54 39.9	+22 06 27	+0.003	-0.05	7.0	0.5	4.0	F8 V		40 s		
112264	44383		12 54 56.5	+47 11 48	-0.002	-0.01	5.84	1.55		M5 III	-17			TU CVn, v
112164	223753		12 54 58.4	-44 09 07	-0.021	-0.23	5.89	0.64	3.0	G2 IV		38 s		
111482	258654	ι Oct	12 54 58.7	-85 07 24	+0.051	+0.02	5.46	1.02	0.2	K0 III	+53	110 s		m
112123	252089		12 55 02.8	-62 33 31	-0.001	0.00	7.74	0.08	0.6	A0 V		230 s		
112109	252088		12 55 03.8	-63 38 27	-0.007	-0.02	7.9	0.4	2.6	F0 V		110 s		
112234	100332		12 55 05.1	+19 36 53	-0.006	-0.07	6.68	1.15	0.2	K0 III		110 mx		
112223	157591		12 55 07.0	-13 27 09	0.000	0.00	7.0	1.1	0.2	K0 III		230 s		
112192	223756		12 55 07.0	-42 17 28	0.000	-0.02	6.83	-0.12	-1.1	B5 V n		390 s		
112257	82565		12 55 08.2	+27 45 59	0.000	-0.07	7.84	0.66	4.7	G2 V		39 s		
112179	223754		12 55 08.4	-48 05 29	-0.004	0.00	7.5	1.6	-0.1	K2 III		280 s		
112165	240376		12 55 08.7	-51 48 30	+0.001	0.00	7.8	1.2	0.4	B9.5 V		58 s		
112207	181222		12 55 10.8	-24 57 20	+0.001	-0.01	7.2	1.2	-0.1	K2 III		270 s		
112235	119665		12 55 11.1	+3 02 57	0.000	+0.02	7.9	0.9	3.2	G5 IV		88 s		
112249	119668		12 55 15.1	+5 18 00	+0.001	-0.01	7.7	0.4	2.6	F0 V		110 s		
112599	7703		12 55 15.4	+78 29 58	+0.001	-0.04	7.6	1.1	0.2	K0 III		310 s		
112213	223760		12 55 19.3	-42 54 57	-0.003	-0.02	5.47	1.68	-0.4	M0 III		130 s		
112180	240380		12 55 21.0	-53 31 12	0.000	-0.02	7.9	0.4	3.2	G5 IV		88 s		
112250	139039		12 55 21.7	-4 30 17	-0.013	+0.05	7.2	0.9	3.2	G5 IV		65 s		
112079	256989		12 55 26.5	-73 58 09	+0.005	-0.03	7.5	0.1	0.6	A0 V		210 mx		
112429	15941	8 Dra	12 55 28.4	+65 26 18	-0.001	-0.03	5.24	0.28		F0	+9	37 t		
112277	100336		12 55 29.4	+12 42 08	-0.001	0.00	7.98	0.96	0.2	K0 III		360 s		
112278	100337		12 55 30.3	+11 29 47	-0.001	0.00	6.92	1.53	-0.5	M4 III	-5	310 s		m
112241	181231		12 55 32.3	-30 04 11	+0.001	0.00	6.7	0.6	1.4	A2 V		55 s		
112150	252090		12 55 33.4	-68 59 53	-0.002	-0.03	7.5	0.1	1.4	A2 V		160 s		
112300	119674	43 δ Vir	12 55 36.1	+3 23 51	-0.031	-0.06	3.38	1.58	-0.5	M3 III	-18	45 mx		m
112281	119673		12 55 38.4	+0 03 19	-0.001	-0.01	6.8	1.1	0.2	K0 III		210 s		
112396	28567		12 55 38.9	+54 57 40	-0.002	0.00	8.0	0.1	0.6	A0 V		300 s		
112353	63255		12 55 44.7	+32 00 03	-0.002	-0.02	6.84	1.09	0.2	K0 III		190 s		
112124	256990		12 55 51.3	-73 01 57	+0.004	-0.01	7.4	1.1	0.2	K0 III		270 s		
112242	240384		12 55 52.8	-51 56 04	-0.003	-0.02	7.8	1.8	-0.3	K5 III		270 s		
112304	157599		12 55 53.2	-15 19 38	-0.001	-0.01	6.1	0.1	0.6	A0 V		130 s		
112244	240385		12 55 56.9	-56 50 09	-0.002	-0.01	5.32	0.01	-6.1	O9 Ib	+22	1600 s		m
112412	63256	12 α¹ CVn	12 56 00.3	+38 18 53	-0.020	+0.06	5.52	0.34		F0 V	-3		8706	m
112305	203947		12 56 01.5	-31 28 53	-0.003	+0.01	7.1	0.3	0.6	A0 V		130 s		
112413	63257	12 α² CVn	12 56 01.6	+38 19 06	-0.020	+0.05	2.90	-0.12		A0 p	-3	20 mn	8706	Cor Caroli, m,v
112287	223766		12 56 01.8	-40 59 47	0.000	+0.01	7.8	1.3	0.2	K0 III		330 s		
112326	157601		12 56 04.2	-20 12 50	-0.004	+0.01	7.7	0.9	0.2	K0 III		300 mx		
112288	223767		12 56 07.2	-44 46 52	0.000	-0.02	7.6	2.1	-0.1	K2 III		150 s		
112372	139049		12 56 14.9	-4 51 51	-0.004	-0.01	7.00	0.1	0.6	A0 V		100 mx	8707	m
112252	252096		12 56 15.5	-61 45 17	0.000	0.00	7.9	0.3	0.4	B9.5 V		190 s		
112357	157608		12 56 17.2	-19 37 53	-0.007	+0.01	7.0	0.8	3.2	G5 IV		58 s		
112486	28572		12 56 17.5	+54 05 57	-0.009	0.00	5.82	0.19		A m	0		8710	m
112268	240388		12 56 18.6	-55 33 38	-0.002	+0.01	6.7	1.9	-0.1	K2 III		81 s		
112559	15945		12 56 25.6	+65 59 37	-0.001	-0.03	6.35	3.26		N7.7	-20			RY Dra, v
112385	157609		12 56 25.6	-14 56 59	-0.008	-0.06	7.1	0.4	2.6	F0 V		80 s		
112398	139053		12 56 26.5	-0 57 18	+0.002	-0.10	7.20	0.5	3.4	F5 V		58 s	8708	m
112374	181244		12 56 30.0	-26 27 36	+0.001	-0.02	6.62	0.68	-6.4	cF6	-22	3700 s		
113591	2122		12 56 30.3	+85 53 03	-0.011	0.00	7.8	0.5	4.0	F8 V		58 s		
112219	256992		12 56 31.6	-72 11 07	-0.004	-0.02	5.93	1.13		K0				
112272	252099		12 56 33.7	-64 21 37	-0.002	0.00	7.46	0.83		B0.5 Ia				
112446	119678		12 56 35.3	+9 19 00	-0.003	+0.02	7.58	1.06	0.2	K0 III		270 s		
112501	44395		12 56 38.7	+43 33 07	0.000	0.00	6.97	0.11		A m	-9		8713	m
112361	223774		12 56 39.3	-47 41 09	-0.007	0.00	6.75	0.45		F5				m
112609	15947		12 56 40.3	+68 36 59	-0.001	-0.02	7.4	0.8	3.2	G5 IV		68 s		
112433	157613		12 56 42.7	-11 56 37	-0.002	-0.02	7.40	1.4	-0.3	K5 III		350 s		m
112401	181250		12 56 42.7	-24 27 03	0.000	-0.01	7.9	0.7	-0.1	K2 III		390 s		
112388	203960		12 56 43.0	-40 15 31	-0.001	-0.01	8.00	1.1	0.2	K0 III		360 s		m
112403	181251		12 56 44.8	-27 22 56	-0.006	0.00	7.8	0.7	0.2	K0 III		260 mx		
112405	223779		12 56 54.7	-45 07 51	0.000	-0.01	7.7	1.5	0.2	K0 III		160 s		
112640	15950		12 56 57.8	+67 14 34	-0.005	+0.01	6.6	1.1	0.2	K0 III		190 s		
112381	240405		12 56 58.4	-54 35 15	-0.002	-0.02	6.50	-0.07		A0 p				
112364	240403		12 56 59.7	-59 44 34	0.000	-0.02	7.38	0.21	-6.2	B1 I		3000 s		
112409	240407		12 57 04.3	-51 11 55	-0.003	-0.02	5.16	-0.06	-0.2	B8 V	+25	120 s		
112570	44398		12 57 07.6	+46 10 36	-0.002	-0.05	6.12	1.01	0.2	G9 III	+7	140 s		
112453	203967		12 57 08.3	-39 47 07	-0.002	+0.03	7.5	0.8	0.2	K0 III		290 s		
112478	181259		12 57 10.3	-22 10 20	-0.003	-0.03	7.3	0.3	0.6	A0 V		140 s		
112366	252103		12 57 11.0	-63 27 51	-0.001	-0.02	7.61	0.74	-6.8	B2 Ia		2100 s		
112495	157618		12 57 12.7	-12 04 01	0.000	-0.02	6.4	1.4	-0.3	K5 III		220 s		
112503	119683		12 57 13.6	+8 17 40	-0.005	+0.03	7.90	0.5	3.4	F5 V		79 s		m
112504	139060		12 57 18.9	-8 54 37	0.000	+0.01	6.8	0.0	0.4	B9.5 V		190 s		
112489	181262		12 57 19.8	-28 52 03	-0.005	0.00	7.3	0.5	1.7	A3 V		75 s		
112440	240411		12 57 22.1	-56 18 26	-0.010	-0.03	7.4	1.6	-0.1	K2 III		180 s		
112542	119684		12 57 26.2	+1 25 40	-0.008	-0.02	7.0	0.4	3.0	F2 V		64 s		

HD	SAO	Star Name	α 2000	δ 2000	μ(α)	μ(δ)	V	B−V	M$_v$	Spec	RV	d(pc)	ADS	Notes
			h m s	° ′ ″	s	″								
112383	252105		12 57 26.2	−67 57 40	−0.006	−0.04	6.79	0.06		A0				
112369	252104		12 57 27.9	−69 20 43	−0.008	−0.05	7.8	0.4	2.6	F0 V		110 s		
112505	203969		12 57 28.1	−33 09 16	−0.004	+0.02	8.0	0.9	0.2	K0 III		340 mx		
112517	139061		12 57 28.5	−9 45 40	+0.003	−0.11	7.10	0.5	3.4	F5 V		55 s	8715	m
112410	252106		12 57 32.0	−65 38 48	−0.009	+0.02	6.9	1.5	0.2	K0 III		120 s		
112519	181265		12 57 33.1	−22 45 14	−0.004	−0.03	6.31	1.07		G5				
112612	119686		12 57 51.5	+9 20 20	−0.003	−0.02	7.16	0.24	1.7	A3 V		100 s		
112509	240422		12 57 57.5	−52 36 54	−0.005	−0.02	7.5	1.0	2.6	F0 V		35 s		
112563	203979		12 58 00.9	−38 55 02	−0.002	0.00	6.90	0.1	0.6	A0 V		180 s		m
112564	203981		12 58 03.1	−39 16 02	−0.006	0.00	7.3	0.3	2.1	A5 V		86 mx		
112613	157627		12 58 03.3	−14 59 07	−0.013	−0.01	7.2	0.8	3.2	G5 IV		63 s		
112550	223791		12 58 03.9	−46 25 39	+0.004	−0.04	7.5	1.3	−0.1	K2 III		240 s		
112603	181274		12 58 09.7	−23 03 18	−0.009	−0.02	6.8	0.4	3.0	F2 V		55 s		
112532	240424		12 58 10.1	−54 11 08	−0.002	−0.01	7.17	0.15	0.6	A0 V		170 s		m
112565	223793		12 58 10.6	−47 03 53	−0.003	−0.01	7.0	0.4	0.6	A0 V		100 s		
112844	15954		12 58 11.1	+68 42 21	+0.002	−0.01	7.4	0.9	3.2	G5 IV		70 s		
112826	15955		12 58 18.8	+64 55 28	−0.004	−0.03	6.5	1.1	0.2	K0 III		180 s		
112581	240433		12 58 25.8	−51 30 31	−0.006	+0.01	7.80	0.8	3.2	G5 IV		82 s		m
112827	15956		12 58 30.6	+61 42 58	−0.005	−0.02	7.1	1.1	0.2	K0 III		240 s		
112654	157634		12 58 30.8	−12 30 47	−0.001	0.00	7.9	1.6	−0.5	M2 III		490 s		
112583	240431		12 58 31.8	−58 15 22	+0.002	−0.01	7.8	0.5	2.6	F0 V		79 s		
112734	82589		12 58 34.0	+28 19 10	−0.006	+0.01	6.97	0.23		A5	−6			
112679	139071		12 58 35.8	−6 56 54	−0.002	+0.01	7.7	1.1	0.2	K0 III		310 s		
112696	139072		12 58 39.9	−2 54 14	+0.001	−0.02	6.64	1.45		K2				
112657	203996		12 58 40.7	−36 16 30	−0.002	−0.01	7.1	0.1	1.4	A2 V		130 s		
112735	82591		12 58 40.8	+22 02 34	+0.005	−0.25	7.22	0.57	4.4	G0 V	−41	37 s		
112753	82592		12 58 40.8	+27 28 27	−0.010	−0.11	7.97	0.65	4.4	G0 V		45 s	8721	m
113049	7714		12 58 47.1	+75 28 21	+0.001	+0.01	6.01	0.99		K0	−15			
112680	204001		12 58 52.5	−39 06 21	−0.005	−0.03	6.6	0.7	0.4	B9.5 V		66 s		
112769	100357	36 Com	12 58 55.4	+17 24 33	−0.002	+0.02	4.78	1.56	−0.5	M1 III	−2	110 s		
112814	44408		12 58 56.6	+39 50 40	−0.003	+0.01	6.87	0.95	0.2	G9 III		220 s		
112710	181283		12 58 57.1	−29 59 56	−0.001	−0.02	7.3	1.7	−0.5	M2 III		300 s		
112709	181284		12 58 57.8	−21 32 48	+0.003	−0.03	6.7	1.5	0.2	K0 III		100 s		
112742	139078		12 58 59.5	−8 44 21	−0.006	−0.07	7.7	1.1	0.2	K0 III		120 mx		
112741	139079		12 59 00.0	−6 05 27	−0.002	−0.01	7.3	0.1	1.4	A2 V		150 s		
112607	252114		12 59 01.2	−63 38 26	−0.002	−0.01	8.04	0.19	0.4	B9.5 V		280 s		
112758	139081		12 59 01.5	−9 50 04	−0.055	+0.19	7.56	0.78	5.9	K0 V	−4	18 ts		m
112859	44410		12 59 03.8	+47 09 06	−0.005	+0.02	7.98	0.93	5.3	G6 V		24 s		
112684	223803		12 59 05.4	−44 44 36	+0.001	−0.03	7.8	1.9	−0.1	K2 III		140 s		
112987	7715		12 59 05.8	+70 35 46	+0.005	−0.02	7.1	1.1	−0.1	K2 III		270 s		
112685	223804		12 59 07.4	−45 57 43	−0.004	+0.01	7.86	0.32	1.9	F3 IV		160 s		
112699	223806		12 59 08.3	−43 42 18	−0.001	−0.04	7.1	2.3	0.2	K0 III		39 s		
112802	119692		12 59 09.1	+7 54 03	−0.005	0.00	7.87	1.08	0.2	K0 III		280 mx	8723	m
112815	119696		12 59 20.1	+6 30 18	−0.001	−0.05	7.3	0.9	3.2	G5 IV		67 s	8724	m
112792	181291		12 59 26.5	−22 07 40	+0.001	0.00	8.0	0.8	−0.1	K2 III		410 s		
112887	82595		12 59 32.8	+28 03 56	+0.001	−0.01	7.19	0.42		F5	−8			
112703	252117		12 59 36.8	−64 22 18	−0.058	−0.08	7.5	0.5	4.0	F8 V		42 mx		
112778	223818		12 59 37.0	−43 52 17	−0.007	0.00	8.0	1.7	0.2	K0 III		140 s		
112846	139086	44 Vir	12 59 39.4	−3 48 43	−0.003	0.00	5.79	0.18		A0			8727	m
112929	44418		12 59 43.9	+40 28 30	−0.002	0.00	8.0	0.5	3.4	F5 V		83 s		
112888	119702		12 59 46.0	+9 31 32	−0.002	−0.02	6.71	1.05	0.2	K0 III		190 s		
112764	240455		12 59 48.0	−55 54 42	−0.004	+0.01	7.80	0.04	0.4	B9.5 V		170 mx		m
112848	181297		12 59 48.3	−20 17 30	−0.008	−0.03	7.5	0.5	4.0	F8 V		49 s		
112765	240456		12 59 50.5	−57 24 39	−0.002	−0.02	7.2	1.7	0.2	K0 III		96 s		
112780	240458		12 59 53.3	−54 22 45	+0.003	−0.04	7.97	1.01		G5		22 mn		
113092	15960	9 Dra	12 59 55.0	+66 35 50	−0.024	−0.02	5.32	1.29	0.3	G8 III	−30	56 s		
112820	223821		12 59 59.4	−46 06 43	−0.001	0.00	7.8	1.3	−0.1	K2 III		330 s		
113435	2128		13 00 01.9	+80 52 28	−0.033	−0.02	7.3	0.6	4.4	G0 V		39 s		
112865	204016		13 00 03.0	−36 10 56	+0.001	−0.03	6.8	0.8	−0.1	K2 III		240 s		
112915	119706		13 00 04.5	+0 18 26	−0.001	−0.03	8.00	1.6	−0.5	M2 III		500 s		m
112973	63286		13 00 07.3	+34 55 10	−0.007	0.00	7.7	0.9	3.2	G5 IV		81 s		
112988	63287		13 00 10.3	+34 59 54	−0.005	0.00	7.9	0.6	4.4	G0 V		51 s		
112890	181301		13 00 13.3	−23 54 52	+0.001	−0.04	7.90	0.1	1.4	A2 V		200 s	8728	m
112974	82602		13 00 15.4	+21 33 15	+0.001	+0.02	7.6	0.5	4.0	F8 V		53 s		
112989	63288	37 Com	13 00 16.4	+30 47 06	−0.002	−0.01	4.90	1.17		K1 p	−13	27 mn	8731	m
112851	223827		13 00 16.6	−48 36 12	+0.003	0.00	7.10	1.1	0.2	K0 III		240 s		m
112881	223828		13 00 17.7	−42 30 03	−0.004	0.00	7.4	0.5	3.0	F2 V		38 s		
112961	100364		13 00 22.1	+13 50 40	+0.003	−0.05	7.9	0.9	3.2	G5 IV		86 s		
112894	223830		13 00 24.0	−41 39 22	−0.001	−0.03	7.6	1.3	3.2	G5 IV		36 s		
112945	157650		13 00 25.1	−13 53 32	+0.001	+0.03	7.4	0.4	3.0	F2 V		75 s		
113021	63290		13 00 25.6	+31 46 45	+0.001	−0.01	6.67	0.92	3.2	G5 IV	+3	39 s		
112907	223832		13 00 29.0	−43 05 19	+0.001	−0.02	6.8	1.5	0.2	K0 III		110 s		
112964	139094		13 00 29.4	−9 06 01	−0.003	−0.04	7.1	0.6	4.4	G0 V		35 s		
112975	119712		13 00 30.1	+3 36 15	−0.002	+0.03	7.9	1.4	−0.3	K5 III		430 s		
112842	252122		13 00 31.6	−60 22 33	0.000	−0.02	7.05	0.21	−6.3	B5 I		3100 s		

HD	SAO	Star Name	α 2000	δ 2000	μ(α)	μ(δ)	V	B−V	M_V	Spec	RV	d(pc)	ADS	Notes
			$h\ \ m\ \ s$	$°\ \ '\ \ ''$	s	$''$								
112935	204029		13 00 32.6	−33 30 19	−0.006	−0.08	6.02	0.38		F2				
112934	204030		13 00 33.4	−33 03 04	−0.010	−0.05	6.6	1.2	2.6	F0 V		18 s		
112992	139096	46 Vir	13 00 35.8	−3 22 07	−0.002	+0.05	5.99	1.12	−0.1	gK2	+23	170 s	8732	m
113022	100366		13 00 38.7	+18 22 22	−0.016	+0.06	6.20	0.42	3.4	F5 V	+1	36 s	8735	m
113139	28601	78 UMa	13 00 43.7	+56 21 59	+0.013	−0.01	4.93	0.36	3.0	F2 V	−10	24 s	8739	m
112855	252125		13 00 52.9	−66 28 46	−0.005	+0.07	8.0	1.1	−0.1	K2 III		140 mx		
113003	181313		13 00 57.5	−21 25 33	−0.006	+0.01	8.0	0.4	2.6	F0 V		100 s		
112979	204036		13 00 57.6	−36 34 04	−0.001	−0.04	7.8	1.2	0.2	K0 III		260 s		
113167	28602		13 00 59.4	+52 32 12	−0.009	0.00	7.7	0.4	3.0	F2 V		86 s		
113005	204040		13 01 03.3	−33 37 22	+0.001	0.00	6.7	0.5	3.2	G5 IV		51 s		
113094	82607		13 01 03.6	+24 18 59	+0.002	−0.03	7.81	1.13	0.0	K1 III		350 s		
112951	240476		13 01 04.3	−53 08 10	−0.002	−0.05	7.4	0.6	2.6	F0 V		59 s		
113079	119717		13 01 08.0	+4 21 31	−0.002	0.00	7.24	0.25	2.1	A5 V		97 s		
113095	100374	38 Com	13 01 09.6	+17 07 23	0.000	−0.03	5.96	0.96	0.2	K0 III	−6	140 s		
112911	252126		13 01 10.3	−63 01 23	+0.005	−0.01	7.7	1.1	−0.1	K2 III		350 s		
113051	157653		13 01 11.1	−18 07 17	−0.001	+0.03	7.6	1.4	−0.3	K5 III		380 s		
113052	181319		13 01 13.5	−20 46 39	−0.002	−0.02	7.0	1.1	−0.1	K2 III		260 s		
113009	223842		13 01 15.0	−44 45 08	−0.001	−0.04	7.68	−0.07	−0.9	B6 V		500 s		
113168	63305		13 01 20.2	+38 02 55	+0.001	+0.02	7.87	−0.03		A2				
113082	157657		13 01 22.4	−17 52 57	0.000	−0.03	7.9	1.1	−0.1	K2 III		390 s		
113083	181322		13 01 26.1	−27 22 31	−0.037	−0.24	8.03	0.55	4.2	F9 V	+227	58 s		
113126	119722		13 01 30.4	+1 31 12	+0.001	−0.02	7.8	1.6	−0.5	M2 III	+9	470 s		
113084	204047		13 01 33.1	−32 33 33	−0.009	0.00	7.9	0.7	2.6	F0 V		63 s		
112999	252129		13 01 34.9	−60 40 17	−0.001	−0.02	7.4	1.3	0.0	B8.5 V		46 s		
113170	82613		13 01 35.0	+21 16 08	−0.002	−0.03	7.07	1.46	−0.3	K5 III	−24	300 s		
113103	181325		13 01 36.4	−24 39 58	+0.001	−0.02	7.00	0.4	2.6	F0 V		76 s		m
113253	28604		13 01 38.0	+56 56 58	0.000	−0.01	7.5	1.1	0.2	K0 III		290 s		
113104	204049		13 01 46.0	−39 37 00	−0.016	+0.01	6.6	0.7	0.2	K0 III		110 mx		
113337	15962		13 01 46.7	+63 36 37	−0.028	+0.03	6.00	0.41	3.4	F5 V	−11	32 ts		m
113129	204050		13 01 47.9	−30 49 58	−0.002	−0.01	7.40	1.1	0.2	K0 III		280 s		m
113158	157664		13 01 53.4	−19 46 29	+0.001	−0.05	7.9	0.2	2.4	A7 V	+4	130 s		UY Vir, v
113088	240488		13 01 58.5	−51 04 10	+0.002	−0.01	7.9	0.8	0.2	K0 III		340 s		
112984	252131		13 01 59.0	−69 13 45	−0.001	−0.02	7.6	0.2	2.1	A5 V		120 s		
113226	100384	47 ε Vir	13 02 10.5	+10 57 33	−0.019	+0.02	2.83	0.94	0.2	G9 III	−14	32 ts		Vindemiatrix, m
113216	157668		13 02 12.7	−12 06 32	−0.007	+0.03	7.3	1.1	0.2	K0 III		170 mx		
112985	257000	δ Mus	13 02 16.3	−71 32 56	+0.056	−0.03	3.62	1.18	−0.1	K2 III	+37	54 s		
113137	240495		13 02 18.8	−56 36 05	−0.009	+0.02	7.6	1.4	0.2	K0 III		180 s		
113284	63310		13 02 20.8	+30 21 27	−0.007	+0.02	7.94	0.35	2.6	F0 V	+7	110 s		
113302	63312		13 02 22.3	+37 20 42	+0.001	−0.07	8.0	0.5	3.4	F5 V		84 s		
113203	181334		13 02 22.7	−28 17 11	−0.003	0.00	6.8	1.3	0.2	K0 III		140 s		
113392	28609		13 02 23.9	+59 14 45	−0.003	+0.02	7.9	0.5	3.4	F5 V		80 s		
113204	181335		13 02 25.3	−29 15 57	−0.003	−0.02	7.10	0.1	0.6	A0 V		200 s	8745	m
113303	82618		13 02 30.5	+23 29 57	−0.003	+0.02	7.70	0.4	3.0	F2 V	+6	87 s	8749	m
113319	63314		13 02 33.5	+32 26 00	−0.014	+0.05	7.55	0.65	4.4	G0 V		38 s		
112857	256999		13 02 34.0	−79 21 55	−0.025	−0.02	8.0	0.5	4.0	F8 V		63 s		
113436	28613		13 02 40.3	+59 42 58	−0.004	−0.01	6.3	0.1	0.6	A0 V	−36	140 s		
113379	44439		13 02 44.2	+45 23 04	−0.001	−0.01	7.63	1.64	−0.5	M2 III		390 s		
113256	204070		13 02 47.1	−30 20 18	−0.003	+0.01	7.8	1.4	0.2	K0 III		200 s		
113163	252135		13 02 47.7	−60 44 35	+0.002	−0.02	7.83	0.16	−1.6	B5 IV		560 s		
113290	157674		13 02 51.9	−18 26 18	0.000	−0.01	7.9	1.1	−0.1	K2 III		410 s		
113289	157677		13 02 57.5	−16 52 52	−0.001	+0.01	7.1	1.1	0.2	K0 III		250 s		
113592	7731		13 02 59.0	+71 43 11	−0.011	−0.03	8.0	0.4	2.6	F0 V		120 s		
113365	82622		13 02 59.4	+22 38 15	−0.003	−0.02	6.97	0.08		A0	−12			
113153	—		13 03 02.0	−68 47 20			7.8	0.7	4.4	G0 V		48 s		
113292	204075		13 03 02.1	−34 17 36	+0.001	−0.04	7.7	0.6	1.4	A2 V		83 s		
113380	82624		13 03 02.4	+23 12 07	−0.003	−0.01	7.46	1.12	0.2	K0 III		240 s		
113330	157680		13 03 02.5	−19 04 39	−0.003	−0.04	8.0	0.4	3.0	F2 V		98 s		
113120	257003		13 03 05.3	−71 28 34	−0.003	−0.02	6.03	0.05	−3.5	B1 V ne	−35	750 s		
113406	82627		13 03 10.3	+23 49 35	−0.003	−0.01	6.98	1.59	−0.5	M1 III	+3	310 s		
113237	240503		13 03 11.1	−56 06 38	−0.001	−0.02	8.04	−0.02	0.4	B9.5 V		340 s		m
113293	223866		13 03 17.6	−44 33 47	−0.003	−0.03	8.0	0.8	1.4	A2 V		68 s		
113354	181349		13 03 22.2	−29 39 49	−0.009	+0.01	7.4	0.2	4.0	F8 V		48 s		
113314	223870	ξ¹ Cen	13 03 33.1	−49 31 38	−0.006	−0.02	4.85	0.02	0.6	A0 V	−10	67 s		
113357	204080		13 03 36.4	−38 58 08	−0.007	−0.05	7.7	0.3	2.6	F0 V		100 s		
113297	240511		13 03 36.8	−56 38 43	−0.003	−0.02	7.2	1.4	0.2	K0 III		150 s		
113398	204086		13 03 45.8	−34 15 01	−0.001	−0.01	7.10	0.8	3.2	G5 IV		60 s		m
113415	181357		13 03 46.0	−20 34 59	+0.010	+0.01	5.58	0.56	4.0	dF8	+34	19 s	8757	m
113388	223874		13 03 47.2	−40 51 06	−0.001	−0.04	7.00	1.1	0.2	K0 III		230 s		m
113493	63318		13 03 49.0	+31 00 05	+0.002	−0.03	7.35	1.06	0.2	K0 III		240 s		
113449	139129		13 03 49.5	−5 09 43	−0.014	−0.23	7.5	0.8	5.2	G5 V	0	29 s		
113416	181359		13 03 51.0	−25 48 08	−0.009	−0.01	7.9	0.6	3.0	F2 V		73 s		
113448	139130		13 03 53.0	−3 13 03	−0.002	−0.01	7.4	0.4	2.6	F0 V		90 s		
113545	44446		13 03 53.4	+43 00 27	−0.001	−0.01	6.81	1.63	−0.5	M2 III	+14	270 s		
113459	139131	48 Vir	13 03 54.3	−3 39 48	−0.003	−0.04	6.59	0.29	2.4	dA7 n	+3	69 s	8759	m
113428	181361		13 03 55.7	−24 31 02	−0.003	+0.02	7.6	0.5	0.2	K0 III		300 s		

323

HD	SAO	Star Name	α 2000	δ 2000	μ(α)	μ(δ)	V	B-V	M$_v$	Spec	RV	d(pc)	ADS	Notes
113496	100398		13^{h}03^{m}59.7	+11°13'52"	0.000	-0".01	7.39	1.51	-0.5	M4 III	-3	380 s		
113516	82634		13 04 05.9	+21 53 41	-0.002	+0.01	7.8	0.5	3.8	F7 V	-10	64 s		
113418	240515		13 04 13.1	-52 53 45	-0.001	-0.03	7.8	1.0	-0.5	M4 III		470 s		
113621	28622		13 04 18.9	+50 04 05	+0.006	-0.08	8.00	0.5	4.2	F9 V	-5	58 s		
113502	157694		13 04 19.1	-17 14 14	-0.003	+0.02	8.0	0.4	2.6	F0 V		120 s		
113637	28624		13 04 19.8	+53 57 15	-0.003	-0.01	7.50	1.1	-0.2	K3 III	-32	350 s		
113481	181365		13 04 24.2	-28 24 24	+0.003	-0.07	6.9	1.0	0.6	A0 V		48 s		
113480	181366		13 04 25.8	-26 28 54	0.000	-0.05	7.5	1.4	-0.1	K2 III		240 s		
113048	258656		13 04 29.4	-81 08 59	-0.005	0.00	7.20	1.03	0.2	K0 III		240 s		
113451	223877		13 04 30.0	-48 07 56	-0.002	+0.01	7.64	-0.03		B9				
113564	139139		13 04 38.4	-0 43 44	0.000	-0.05	7.3	1.1	-0.1	K2 III		310 s		
113649	63329		13 04 39.1	+39 19 23	-0.007	-0.02	7.72	0.96	0.2	K0 III		170 mx		
113362	—		13 04 39.8	-69 59 16			7.9	0.0	0.0	B8.5 V		370 s		
113696	28626		13 04 39.9	+53 51 28	-0.002	+0.02	7.6	1.1	-0.1	K2 III		340 s		
113595	119756		13 04 44.4	+0 17 42	-0.006	-0.11	7.1	1.1	0.2	K0 III		87 mx		
113504	223879		13 04 45.2	-40 40 58	-0.008	-0.02	7.9	0.2	3.4	F5 V		81 s		
113565	139140		13 04 46.9	-8 34 15	-0.002	-0.01	7.9	1.1	0.2	K0 III		340 s		
113523	223881		13 04 48.0	-41 11 48	-0.003	-0.04	6.26	1.68		M4				
113889	7741		13 04 49.6	+73 01 31	-0.005	+0.02	6.50	0.2	2.1	A5 V	-15	76 s	8772	m
113566	181370		13 04 52.5	-21 31 11	-0.002	0.00	7.2	0.5	1.7	A3 V		70 s		
113465	—		13 04 58.2	-61 26 15			8.0	2.5	-0.3	K5 III		110 s		
113672	63332		13 04 58.5	+37 16 20	+0.006	-0.06	8.0	1.1	0.2	K0 III		240 mx		
113506	240529		13 05 00.2	-51 02 28	-0.005	-0.03	7.8	1.0	3.2	G5 IV		61 s		
113548	204109		13 05 00.3	-37 31 30	-0.006	-0.03	7.8	1.3	4.4	G0 V		17 s		
113457	252144		13 05 01.9	-64 26 30	-0.007	-0.03	6.64	0.00	0.6	A0 V		130 mx		
113537	223885		13 05 06.8	-47 07 00	-0.003	-0.03	6.44	0.42	0.7	F5 III		140 s		
113433	—		13 05 08.6	-69 36 15			7.7	1.4	-0.3	K5 III		380 s		
113612	181375		13 05 10.4	-28 31 04	-0.012	-0.03	7.69	0.86		G5				
113553	240531		13 05 16.7	-50 51 23	-0.014	-0.01	8.0	0.9	3.2	G5 IV		78 s		
113770	44460		13 05 27.1	+41 40 52	+0.006	-0.02	7.90	0.5	3.4	F5 V		79 s	8769	m
113713	100407		13 05 28.4	+14 43 30	-0.009	-0.03	7.80	0.5	3.4	F5 V		76 s		m
113602	240535		13 05 30.7	-52 06 54	-0.005	+0.03	6.43	1.70		Ma				
113601	223891		13 05 32.6	-49 51 04	-0.001	0.00	7.8	1.6	-0.3	K5 III		350 s		
113731	100409		13 05 34.2	+13 13 36	-0.006	+0.02	7.46	0.97	0.2	K0 III		190 mx		
113771	82642		13 05 40.8	+26 35 08	-0.001	-0.01	7.50	0.93	0.2	K0 III		290 s		
113797	63338	14 CVn	13 05 44.3	+35 47 56	-0.003	+0.02	5.25	-0.08	0.2	B9 V	-13	100 s		
113719	157713		13 05 45.7	-18 15 35	-0.003	-0.01	7.1	0.4	3.0	F2 V		66 s		
113828	44464		13 05 46.0	+47 46 41	+0.004	-0.07	7.3	1.1	0.2	K0 III		270 s		
113811	63339		13 05 46.7	+39 36 17	-0.004	+0.04	7.31	1.51	-0.3	K5 III		280 mx		
113733	157715		13 05 51.7	-14 06 42	-0.001	-0.02	7.2	1.4	-0.3	K5 III		310 s		
113847	44465		13 05 52.2	+45 16 07	-0.001	+0.02	5.63	1.13	0.0	K1 III	-20	130 s	8775	m
113865	82648		13 06 10.1	+29 01 46	-0.005	0.00	6.54	0.05	1.7	A3 V	+3	89 mx	8777	m
113892	44470		13 06 13.0	+40 55 22	-0.004	+0.01	7.25	1.65	-0.5	M1 III	-33	320 s		m
113686	240544		13 06 15.5	-55 36 42	0.000	-0.01	6.8	1.4	-0.1	K2 III		180 s		
113703	223900		13 06 16.6	-48 27 48	-0.003	-0.03	4.71	-0.14	-2.0	B4 IV	+9	220 s		m
113848	82650	39 Com	13 06 21.1	+21 09 12	-0.005	-0.05	5.99	0.39	3.4	F5 V	+1	33 s		m
113559	—		13 06 22.4	-72 48 13			7.6	0.5	4.0	F8 V		52 s		
113866	82651	40 Com	13 06 22.5	+22 36 58	+0.002	-0.05	5.60	1.59		Mb III	-5	10 mn		
113994	15985		13 06 22.6	+62 02 31	0.000	-0.04	6.14	0.99	0.2	gK0	+15	150 s		
113867	82652		13 06 22.6	+22 16 48	0.000	+0.03	6.82	0.29		F0	-5			
113983	28635		13 06 25.0	+57 01 31	+0.003	-0.01	7.5	0.7	4.4	G0 V		42 s		
113817	157723		13 06 26.9	-14 54 57	-0.003	+0.01	7.2	1.1	0.2	K0 III	-12	250 s		
113659	—		13 06 28.5	-65 05 12			7.8	0.2	-5.4	O9 IV		3000 s		
113749	240552		13 06 32.4	-51 02 57	-0.001	0.00	7.5	1.4	-0.5	M2 III		400 s		
113192	258657		13 06 32.8	-82 56 59	-0.005	+0.03	7.7	0.0	1.4	A2 V		170 mx		
113778	223905		13 06 35.0	-41 35 19	+0.004	-0.03	5.59	1.05		K0				
113820	204128		13 06 37.7	-31 36 42	-0.002	-0.03	6.8	0.8	0.2	K0 III		210 s		
113752	240554		13 06 37.7	-54 30 19	-0.002	+0.02	6.8	1.4	0.2	K0 III		130 s		
113740	240556		13 06 47.6	-59 26 39	-0.001	-0.04	7.92	0.28	1.7	A3 V		150 s		
113513	—		13 06 51.6	-77 52 13			7.3	1.1	0.2	K0 III		260 s		
113852	204132		13 06 54.2	-35 51 42	+0.003	-0.08	5.65	0.03		A0	+16			
113791	223909	ξ² Cen	13 06 54.5	-49 54 22	-0.003	-0.01	4.27	-0.19	-3.0	B2 IV	+14	280 s		m
113837	204131		13 06 54.7	-37 33 01	-0.001	-0.03	7.9	1.5	-0.5	M4 III		490 s		
113870	204134		13 06 56.3	-34 07 18	-0.017	-0.09	7.3	0.6	3.4	F5 V		48 s		
113782	240559		13 07 02.2	-58 48 21	-0.003	-0.02	7.1	2.1	0.2	K0 III		54 s		
113996	82659	41 Com	13 07 10.6	+27 37 29	+0.002	-0.08	4.80	1.48	-0.3	K5 III	-16	110 s		
114215	7751		13 07 18.6	+71 23 40	-0.001	+0.03	7.6	1.1	0.2	K0 III		300 s		
113984	119773		13 07 19.5	+0 35 13	-0.008	-0.11	7.05	0.49	3.4	F5 V	-92	52 s	8786	m
113823	240566		13 07 24.1	-59 51 38	-0.003	-0.02	5.99	0.48		B9 n	+4			m
114037	82661		13 07 24.7	+26 31 26	-0.008	+0.02	7.61	1.09	0.0	K1 III		190 mx		m
113887	223910		13 07 24.9	-48 03 02	-0.005	-0.02	7.5	1.2	3.0	F2 V		81 s		
113873	240568		13 07 26.2	-52 45 04	-0.007	-0.03	7.6	1.1	2.1	A5 V		36 s		
113842	—		13 07 27.7	-60 16 09			7.2	3.6	-0.4	M0 III		20 s		
113709	257010		13 07 34.6	-73 35 07	-0.005	0.00	7.60	0.1	0.6	A0 V		250 s		m
114018	119778		13 07 36.6	+1 28 27	+0.002	-0.03	7.6	0.5	3.4	F5 V		71 s		

HD	SAO	Star Name	α 2000	δ 2000	μ(α)	μ(δ)	V	B−V	M_v	Spec	RV	d(pc)	ADS	Notes
			h m s	° ′ ″	s	″								
113902	240573		13 07 38.2	−53 27 35	−0.004	−0.03	5.71	−0.07	−0.2	B8 V		120 mx		
113971	204145		13 07 38.9	−30 26 05	+0.002	−0.01	7.8	0.1	0.6	A0 V		220 s		
114001	157735		13 07 38.9	−14 11 18	−0.011	−0.02	8.0	0.5	4.0	F8 V		62 s		
113807	252157		13 07 38.9	−65 52 47	+0.001	−0.08	8.0	0.1	0.6	A0 V		82 mx		
114019	157736		13 07 49.7	−18 00 17	+0.001	−0.02	7.90	0.9	3.2	G5 IV		87 s	8787	m
114020	157737		13 07 51.9	−19 02 07	−0.002	+0.01	7.80	1.1	−0.1	K2 III		380 s		m
114124	44481		13 07 52.2	+42 22 25	−0.002	−0.04	7.98	1.25	−0.1	K2 III		350 s		
114092	82665		13 07 53.5	+27 33 21	−0.003	−0.07	6.19	1.36	−0.3	K4 III	−9	140 mx		
114038	157739	49 Vir	13 07 53.7	−10 44 25	+0.001	−0.01	5.19	1.14	0.0	K1 III	−9	100 s		
113953	240584		13 08 01.1	−56 23 05	−0.001	−0.01	7.29	−0.03	0.4	B9.5 V		240 s		
113954	240583		13 08 01.2	−56 41 03	−0.004	−0.02	7.27	0.11	0.6	A0 V		150 mx		
114374	7755		13 08 02.1	+73 20 33	−0.016	−0.05	8.0	0.1	0.6	A0 V		45 s		
114093	82666		13 08 02.2	+24 49 54	−0.005	+0.05	6.83	0.91	0.3	G8 III		160 mx		
114062	157742		13 08 06.2	−13 07 05	+0.001	−0.01	8.0	0.1	0.6	A0 V		300 s		
113904	252162	θ Mus	13 08 07.0	−65 18 22	−0.001	−0.01	5.51	−0.02		O9.8	−28			m
114253	15991		13 08 07.2	+62 41 24	+0.002	−0.03	7.9	1.1	0.2	K0 III		340 s		
114146	63362		13 08 13.4	+38 44 21	−0.010	−0.09	7.30	0.8		G5 p			8795	m
114159	44482		13 08 15.0	+43 10 49	−0.003	−0.03	7.64	1.58	−0.5	M2 III	−18	430 s		
113694	257011		13 08 20.3	−78 26 44	−0.014	−0.01	6.60	0.1	0.6	A0 V		130 mx		
114050	204156		13 08 21.5	−36 13 25	−0.012	+0.02	6.8	0.8	2.1	A5 V		37 s		
114065	204157		13 08 23.2	−34 07 08	−0.005	−0.03	7.8	0.9	1.7	A3 V		50 s		
114131	100432		13 08 26.4	+15 29 29	+0.001	−0.07	7.86	0.93		K1 IV		220 mx	8796	m
113919	252163		13 08 27.7	−67 47 50	−0.002	−0.03	7.2	1.6	−0.5	M2 III		340 s		
114114	157745		13 08 28.0	−11 45 45	0.000	−0.08	7.3	1.1	−0.1	K2 III		220 mx		
114125	139174		13 08 29.9	−2 40 43	−0.001	+0.04	8.00	0.4	3.0	F2 V		100 s		m
114113	139175		13 08 32.4	−8 59 03	−0.002	−0.07	5.55	1.18	−0.2	gK3	+16	140 s		
114075	204162		13 08 36.7	−38 05 36	−0.004	+0.01	7.8	0.9	0.2	K0 III		330 s		
114042	240590		13 08 38.3	−54 36 49	−0.002	0.00	7.3	0.7	1.4	A2 V		58 s		
114098	181408		13 08 39.3	−23 06 22	−0.008	−0.05	6.7	1.3	0.2	K0 III		130 mx		
114116	181409		13 08 39.8	−23 49 18	−0.009	+0.01	7.8	0.6	3.4	F5 V		59 s		
114397	15994		13 08 50.6	+65 34 20	−0.004	−0.02	8.0	1.1	0.2	K0 III		360 s		
114174	119789		13 08 50.9	+5 12 25	+0.006	−0.69	6.80	0.67	3.2	G5 IV	+22	17 mx		
114134	204167		13 09 00.1	−34 20 53	0.000	−0.02	7.9	0.5	1.4	A2 V		110 s		
114133	204168		13 09 00.7	−33 30 06	0.000	−0.02	7.2	0.8	0.2	K0 III		250 s		
114149	181410	45 ψ Hya	13 09 03.2	−23 07 05	−0.002	−0.04	4.95	1.05	0.2	K0 III	−19	81 s		
114241	100435		13 09 05.5	+18 37 31	−0.001	+0.01	6.91	1.49	−0.1	K2 III	−40	160 s		
114102	223931		13 09 05.7	−49 29 42	−0.005	−0.01	7.8	0.2	0.6	A0 V		110 mx		
114222	119792		13 09 07.4	+3 08 40	−0.003	0.00	7.6	0.9	3.2	G5 IV		77 s		
—	—		13 09 10.1	+45 22 28	−0.003	+0.02	6.2	0.6	4.4	G0 V		23 s		
114256	100436		13 09 12.3	+10 01 20	+0.001	−0.01	5.78	1.00	0.2	K0 III	0	130 s		
114203	139183		13 09 14.3	−9 32 18	−0.002	−0.01	6.32	1.02		K0				
114164	223935		13 09 19.4	−41 01 09	−0.006	−0.02	7.6	1.1	0.2	K0 III		260 mx		
114155	223937		13 09 27.3	−48 31 27	−0.003	+0.01	7.9	−0.6	3.4	F5 V		78 s		
114243	157757		13 09 30.2	−20 07 24	−0.001	0.00	7.2	0.1	0.6	A0 V		190 s		
114083	—		13 09 31.4	−66 32 05			7.8	1.1	0.2	K0 III		320 s		
114446	28655		13 09 33.1	+56 49 58	−0.011	+0.04	7.01	0.60		F8	−34			
114357	63372		13 09 38.6	+37 25 23	−0.009	0.00	6.02	1.15	−0.2	K3 III	−19	180 s		
114196	223940		13 09 41.3	−50 08 59	−0.003	−0.01	7.8	1.1	0.2	K0 III		330 s		
114376	63374	15 CVn	13 09 41.9	+38 32 01	−0.002	0.00	6.28	−0.12	−1.6	B7 III		380 s		
114268	157759		13 09 42.2	−16 30 55	−0.003	−0.01	6.7	0.1	1.4	A2 V		120 s		
114260	181420		13 09 42.5	−22 11 34	+0.011	−0.35	7.37	0.73	5.5	dG7	−7	17 ts		
114245	204181		13 09 43.4	−36 09 36	−0.005	+0.02	7.7	0.6	0.2	K0 III		320 s		
114287	157760	50 Vir	13 09 45.2	−10 19 46	−0.001	−0.02	5.94	1.49	−0.3	gK5	−7	180 s		
114286	139187		13 09 46.3	−7 39 20	−0.002	−0.02	7.33	1.12		K0				
114326	100439		13 09 47.8	+16 50 55	−0.005	−0.02	5.91	1.45	0.2	K0 III	−17	63 s		
114246	204182		13 09 48.6	−36 31 57	−0.003	−0.02	7.6	1.0	3.2	G5 IV		57 s		
114504	15999		13 09 50.1	+62 13 44	−0.005	−0.01	6.60	0.1	0.6	A0 V	−17	74 mx		m
114231	223945		13 09 56.2	−47 33 22	−0.003	−0.03	7.80	1.1	−0.1	K2 III		380 s		m
114330	139189	51 θ Vir	13 09 56.9	−5 32 20	−0.002	−0.04	4.38	−0.01	1.2	A1 V	−3	43 s	8801	m
114378	100443	42 α Com	13 09 59.2	+17 31 45	−0.030	+0.13	4.32	0.45	3.4	F5 V	−18	18 ts	8804	m
114380	100442		13 10 00.4	+16 08 51	−0.003	−0.03	7.9	0.4	3.0	F2 V		94 s		
114427	63377		13 10 01.2	+38 43 25	−0.003	−0.01	7.18	0.18	1.7	A3 V		110 s		
114447	63380	17 CVn	13 10 03.1	+38 29 56	−0.006	+0.04	5.91	0.29	1.1	A9 III−IV	0	91 s	8805	m
—	—		13 10 05.9	+38 30 03			6.20						8805	m
114312	181423		13 10 09.3	−23 24 54	−0.002	−0.02	7.9	0.2	1.4	A2 V		200 s		
114333	181424		13 10 11.2	−25 32 15	−0.005	−0.02	7.6	1.4	3.2	G5 IV		33 s		
114334	204185		13 10 14.7	−32 05 40	−0.002	0.00	7.4	1.0	0.2	K0 III		280 s		
114345	181426		13 10 21.1	−23 37 30	0.000	−0.02	7.5	−0.2	0.6	A0 V		240 s		
114250	240614		13 10 22.3	−57 44 42	−0.008	−0.04	7.5	0.6	4.0	F8 V		47 s		
114348	204190		13 10 26.9	−32 33 26	+0.001	−0.03	7.9	0.7	0.2	K0 III		350 s		
114430	157768		13 10 31.6	−10 48 55	−0.004	−0.05	7.6	1.1	0.2	K0 III		180 mx		
114449	139195		13 10 42.3	−4 56 30	−0.005	−0.01	7.61	0.40		F2			8807	m
114537	63383		13 10 45.9	+34 18 00	−0.006	−0.06	6.9	0.5	3.4	F5 V		49 s		
114432	181436		13 10 46.5	−24 10 46	−0.023	−0.05	7.70	0.9	3.2	G5 IV		48 mx	8809	m
114493	100446		13 10 47.0	+13 18 27	−0.005	+0.02	7.10	1.11	0.2	K0 III	−18	210 s	8810	m

325

HD	SAO	Star Name	α 2000	δ 2000	μ(α)	μ(δ)	V	B−V	M_v	Spec	RV	d(pc)	ADS	Notes
			$h \quad m \quad s$	$°\quad '\quad ''$	s	$''$								
114351	240623		13 10 49.8	−52 54 26	−0.001	+0.01	6.7	1.5	0.2	K0 III		110 s		
114633	28669		13 10 51.6	+57 33 38	−0.003	+0.03	8.0	0.9	3.2	G5 IV		93 s		
114409	204201		13 10 51.7	−39 29 35	−0.007	−0.07	7.9	0.2	3.0	F2 V		95 s		
114520	82692		13 10 52.3	+21 14 04	+0.001	−0.03	7.00	0.4	3.0	F2 V	−5	63 s		m
114307	240622		13 10 54.0	−60 03 01	−0.002	−0.01	8.0	1.9	−0.1	K2 III		160 s		
114365	240627		13 10 58.3	−52 34 00	−0.004	−0.02	6.06	−0.09		A0 p				
114340	240626		13 11 03.7	−59 44 45	−0.002	−0.02	8.04	0.55	−6.6	B1 Ia		3100 s		
114482	157772		13 11 05.1	−17 05 22	−0.003	−0.02	7.1	0.4	2.6	F0 V		78 s		
114538	100448		13 11 07.2	+13 54 35	−0.002	−0.04	7.93	0.39	1.9	F2 IV		150 s		
114239	257015		13 11 07.6	−71 48 55	−0.007	−0.05	7.8	0.4	3.0	F2 V		90 s		
114435	223960		13 11 08.7	−42 13 58	−0.007	−0.03	5.79	0.52	3.2	F8 IV-V		33 s		
114523	157775		13 11 14.2	−14 14 32	−0.004	−0.01	7.7	1.1	0.2	K0 III		320 s		
114622	63388		13 11 14.7	+33 27 58	+0.003	−0.02	8.0	0.4	2.6	F0 V		120 s		
114473	223965		13 11 15.8	−41 25 11	−0.002	−0.01	7.6	1.4	0.2	K0 III		300 s		
114557	157779		13 11 20.7	−10 30 50	−0.004	+0.02	8.0	1.6	−0.5	M4 III		290 mx		
114545	157778		13 11 21.8	−13 57 49	−0.004	−0.08	7.0	1.1	0.2	K0 III		140 mx		
114776	16008		13 11 22.6	+63 08 46	−0.007	−0.01	7.5	0.5	3.4	F5 V		67 s		
114474	223966		13 11 23.1	−43 22 07	−0.011	−0.03	5.25	1.05	0.3	gG8	−9	85 s		
114674	44499		13 11 28.6	+40 47 32	−0.003	−0.01	7.16	0.92	0.2	K0 III	−2	250 s		
114441	240636		13 11 29.4	−55 21 24	−0.001	−0.02	8.03	0.13	−3.0	B2 IV pe		1300 s		
114509	204212		13 11 30.0	−35 07 48	−0.005	0.00	6.60	1.1	0.2	K0 III		190 s		m
114608	119807		13 11 30.8	+6 47 18	−0.003	−0.02	7.8	0.1	1.7	A3 V		160 s		
114637	82699		13 11 32.2	+21 55 06	−0.002	−0.04	6.82	0.88	3.2	G5 V	+23	46 s		m
114458	240637		13 11 36.2	−56 54 33	0.000	−0.02	6.6	1.9	−0.5	M2 III		180 s		
114576	181446		13 11 39.2	−26 33 06	−0.005	−0.01	6.50	0.19	1.7	A3 V n		91 s		m
114611	157785		13 11 40.6	−11 13 19	−0.003	−0.02	7.0	1.1	0.2	K0 III		230 s		
114371	252185		13 11 51.6	−69 56 31	+0.009	0.00	5.91	0.42		F2				m
114512	240641		13 11 52.2	−52 48 44	−0.003	0.00	7.0	1.9	−0.5	M2 III		220 s		
114710	82706	43 β Com	13 11 52.3	+27 52 41	−0.060	+0.88	4.26	0.57	4.7	G0 V	+6	8.3 t		m
114461	252187		13 11 53.1	−63 18 10	0.000	0.00	6.33	0.44		F0				
114597	181450		13 11 53.9	−24 34 09	0.000	−0.01	7.97	1.34		K0				
114598	204225		13 11 54.4	−32 32 28	−0.009	−0.03	7.8	1.5	0.2	K0 III		160 mx		
114489	252189		13 11 55.4	−61 07 03	−0.002	−0.01	6.74	0.46		F0				
114723	63396		13 12 02.0	+32 05 06	+0.003	−0.02	7.20	0.5	4.0	F8 V	−13	44 s	8814	m
114613	204227		13 12 03.1	−37 48 11	−0.032	+0.04	4.85	0.70	5.2	dG5	−15	13 ts		
114642	157788	53 Vir	13 12 03.5	−16 11 55	+0.007	−0.29	5.04	0.46	1.4	F6 III-IV	−14	23 mx		m
114584	223972		13 12 06.9	−50 04 01	−0.002	−0.03	7.6	1.3	−0.1	K2 III		300 s		
114724	82708		13 12 08.3	+24 15 30	−0.001	−0.03	6.33	0.98	0.0	K1 III	−24	190 s		
114662	181452		13 12 09.7	−20 53 04	−0.003	−0.01	8.0	0.6	−2.1	K0 II	+5	1000 s		
114725	100456		13 12 13.9	+16 07 53	−0.002	+0.07	7.40	0.5	3.4	F5 V		63 s	8816	m
114529	240645		13 12 17.4	−59 55 15	−0.006	−0.03	4.60	−0.08	−0.2	B8 V	+12	63 mx		m
114762	100458		13 12 19.7	+17 31 00	−0.040	−0.02	7.31	0.54	4.2	F9 V	+50	24 mx		m
115179	7777		13 12 23.2	+77 11 37	+0.002	−0.04	7.9	0.1	1.4	A2 V		200 s		
115337	2164		13 12 25.3	+80 28 16	−0.004	+0.01	6.25	0.94	3.2	G5 IV	−11	32 s		m
115299	7780		13 12 27.6	+79 39 08	0.000	−0.01	7.5	1.1	0.2	K0 III		290 s		
114692	204235		13 12 31.9	−34 44 53	−0.019	−0.28	7.76	0.53	3.8	dF7	+1	59 s		m
114780	100460		13 12 32.8	+11 33 23	−0.004	−0.03	5.77	1.51	−0.4	M0 III	+25	170 s		
114683	204236		13 12 35.6	−39 41 34	−0.001	−0.01	7.70	1.1	0.2	K0 III		320 s		m
114793	100461		13 12 35.9	+18 45 05	−0.005	−0.01	6.53	0.88	0.3	G8 III	−20	180 s		
114781	119815		13 12 40.0	+4 04 06	0.000	−0.01	7.4	0.4	2.6	F0 V		93 s		
114783	139218		13 12 43.6	−2 15 54	−0.010	+0.01	7.9	1.1	0.2	K0 III		130 mx		
114729	204237		13 12 44.2	−31 52 24	−0.015	−0.32	6.69	0.62	4.9	G3 V	+64	23 s		
114782	139219		13 12 45.6	−1 45 30	−0.003	−0.01	7.3	1.6	−0.5	M4 III		360 s		
114904	28674		13 12 48.5	+51 54 00	−0.005	+0.02	7.2	1.1	0.2	K0 III		250 s		
114570	252196		13 12 48.9	−66 13 37	+0.005	0.00	5.90	0.06		A0				
114877	44509		13 12 49.5	+42 37 36	−0.001	+0.03	6.78	0.93	3.2	G5 IV		42 s		
114707	223980		13 12 50.8	−42 41 59	−0.002	0.00	6.22	1.06		K0				
114819	82718		13 12 53.6	+19 55 05	−0.001	−0.03	7.37	1.25	0.2	K0 III	+2	180 s		
114630	240653		13 12 55.9	−59 49 00	+0.002	−0.11	6.16	0.60		F5				m
114769	181460		13 12 58.8	−29 05 59	+0.001	0.00	8.00	0.1	0.6	A0 V		300 s	8818	m
115061	16014		13 13 00.1	+67 18 34	−0.028	0.00	6.98	1.15	−0.1	K2 III	+4	130 mx		m
114905	63408		13 13 03.5	+38 16 36	+0.013	−0.12	6.9	0.5	3.4	F5 V		50 s		
114865	100465		13 13 07.4	+15 19 01	−0.006	−0.02	7.65	1.12	0.2	K0 III		170 mx		
114531	—		13 13 08.3	−72 32 57			7.7	0.0	0.0	B8.5 V		340 s		
114889	100467		13 13 12.3	+18 43 37	−0.015	−0.05	6.11	1.20	0.3	G8 III	−24	58 mx		
114842	139227		13 13 13.6	−2 33 23	−0.009	−0.05	8.00	0.5	4.0	F8 V		63 s	8823	m
114883	100469		13 13 18.2	+13 23 29	−0.001	−0.04	7.41	1.22	0.2	K0 III		200 s		
114845	157797		13 13 21.0	−13 28 16	−0.010	−0.06	7.6	0.5	3.4	F5 V		68 s		
114866	139229		13 13 21.1	−3 29 40	−0.002	0.00	7.3	1.1	−0.1	K2 III		300 s		
114772	240655		13 13 23.4	−50 42 00	−0.002	−0.03	6.70	0.1	0.6	A0 V		170 s		m
114846	157798	54 Vir	13 13 26.8	−18 49 36	−0.002	−0.02	6.28	0.09	0.6	A0 V	−41	120 s	8824	m
114655	—		13 13 27.1	−68 29 55			7.9	0.1	1.4	A2 V		190 s		
115136	16018		13 13 27.9	+67 17 14	−0.028	−0.02	6.54	1.14	−0.1	K2 III	+4	110 mx		m
115227	7782		13 13 32.0	+72 47 55	+0.004	−0.03	6.4	0.1	0.6	A0 V	+2	150 s		
114975	63414		13 13 35.9	+36 53 14	+0.001	−0.02	6.48	1.60	−0.5	M2 III	+1	250 s		

HD	SAO	Star Name	α 2000	δ 2000	μ(α)	μ(δ)	V	B−V	M_v	Spec	RV	d(pc)	ADS	Notes
			h m s	° ′ ″	s	″								
115043	28679		13 13 36.9	+56 42 28	+0.013	−0.03	6.83	0.60	4.5	G1 V	−9	29 s		m
114976	63416		13 13 41.9	+29 49 06	0.000	0.00	7.28	1.10	0.2	G9 III		200 s	8826	m
115004	44519		13 13 42.9	+40 09 10	−0.004	+0.01	4.92	1.06	0.2	K0 III	−21	79 s		
114988	63419		13 13 44.6	+32 31 51	−0.001	0.00	6.67	0.78	4.4	G0 V	+2	21 s		
114737	252213		13 13 45.4	−63 35 11	−0.001	0.00	8.01	0.17	−4.8	O9 V		2600 s		
114738	252214		13 13 46.7	−64 18 48	−0.005	−0.03	7.9	0.1	0.6	A0 V		160 mx		
114871	204257		13 13 49.2	−34 40 29	+0.001	−0.02	7.5	0.5	2.6	F0 V		74 s		
114853	223987		13 13 52.1	−45 11 09	−0.010	−0.12	7.1	0.9	4.4	G0 V		22 s		
114989	82730		13 13 52.8	+25 41 54	0.000	−0.04	7.4	0.5	3.4	F5 V		63 s		
114873	223989		13 13 57.4	−43 08 20	−0.016	+0.02	6.16	1.38	−2.3	K5 II		40 mx		
114960	119822		13 13 57.5	+1 27 23	−0.003	−0.06	6.59	1.40	−0.3	K5 III	+7	150 mx		
114917	181471		13 14 01.3	−22 21 44	−0.001	−0.03	7.8	0.6	2.6	F0 V		69 s		
114792	252218		13 14 02.0	−62 39 15	0.000	−0.02	6.80	0.90	−4.6	F5.5 Ib				m
114808	240660		13 14 03.3	−58 19 02	−0.002	−0.03	7.4	0.7	0.6	A0 V		81 s		
114961	139236		13 14 04.2	−2 48 24	−0.004	0.00	7.8	1.6		M6 III	−15	310 mx		SW Vir, v
114962	157807		13 14 09.2	−14 16 56	−0.008	−0.01	7.7	0.6	4.4	G0 V		46 s		
114946	157806	55 Vir	13 14 10.8	−19 55 51	−0.009	+0.16	5.6	0.8	3.2	G5 IV	−45	30 s		
114800	252219		13 14 11.4	−63 22 25	−0.001	−0.01	8.01	0.09	−2.5	B2 V pe		1100 s		
114835	240663		13 14 11.7	−58 41 02	−0.010	−0.02	5.89	1.08		K2	−2			
114837	240666		13 14 14.7	−59 06 12	−0.034	−0.16	4.92	0.48	4.0	F8 V	−65	17 ts		m
114921	223992		13 14 16.1	−41 12 43	−0.004	−0.02	7.60	1.03	3.2	G8 IV		43 s		
114533	257019		13 14 17.5	−78 26 51	−0.001	−0.01	5.85	1.07	0.3	gG5		100 s		
114922	223993		13 14 19.9	−42 51 09	−0.009	−0.05	7.1	0.2	0.2	K0 III		150 mx		
114947	204269		13 14 24.9	−35 26 24	−0.007	−0.04	7.8	0.2	3.4	F5 V		74 s		
114534	—		13 14 24.9	−79 11 56			7.4	1.4	−0.3	K5 III		350 s		
114899	240669		13 14 26.2	−54 57 43	−0.003	−0.01	8.0	1.1	0.2	K0 III		330 s		
114993	181476		13 14 27.5	−24 17 03	−0.005	+0.01	7.30	0.1	1.4	A2 V		100 mx	8831	m
115046	100473		13 14 31.2	+11 19 54	+0.005	−0.06	5.67	1.50	−0.4	M0 III	+12	160 s	8832	m
114555	257020		13 14 39.2	−79 17 36	−0.048	−0.12	7.7	0.8	3.2	G5 IV		78 s		
114981	204273		13 14 40.7	−38 39 06	0.000	0.00	7.12	−0.05	−1.1	B5 V		420 s		
114971	224000		13 14 43.0	−48 57 24	−0.013	−0.08	5.89	1.06		K0				
114886	252220		13 14 44.5	−63 34 51	+0.001	0.00	6.89	0.09	−4.8	O9 V		1700 s		m
115062	157811		13 14 45.0	−10 22 14	−0.002	−0.05	7.1	1.4	−0.3	K5 III	+27	310 s		
115026	204279		13 14 54.4	−34 37 53	0.000	−0.02	7.2	1.2	0.2	K0 III		180 s		
114982	240673		13 14 55.5	−51 48 18	−0.008	−0.06	7.5	0.8	3.4	F5 V		41 s		
115080	157815		13 14 55.7	−11 22 10	−0.014	−0.32	7.04	0.69	4.9	dG3	+8	25 s		m
115079	157816		13 15 00.8	−11 20 54	0.000	−0.01	7.80	1.1	0.2	K0 III	+9	330 s		m
115104	119829		13 15 02.8	+4 31 03	−0.003	−0.01	7.4	1.6	−0.5	M4 III		390 s		
115676	7786		13 15 02.9	+78 42 05	−0.015	+0.02	7.7	1.1	0.2	K0 III		310 s		
115066	181483		13 15 04.1	−30 10 54	−0.005	−0.04	7.9	1.2	0.2	K0 III		200 mx		
114983	240674		13 15 07.0	−54 58 11	−0.002	0.00	7.9	0.7	0.2	K0 III		340 s		
115197	63429		13 15 08.4	+35 26 43	−0.005	−0.01	6.85	0.15		A2				
115050	204282		13 15 09.6	−36 22 16	−0.001	−0.02	6.19	0.97		K0				
115166	100476		13 15 10.2	+18 55 06	−0.001	−0.01	6.57	1.47	0.2	K0 III		98 s		
114035	—		13 15 10.4	−85 34 02			8.0	0.1	0.6	A0 V		300 s		
116459	2186		13 15 11.3	+84 45 08	−0.098	+0.02	7.4	0.6	4.4	G0 V	+11	40 s		
115053	224007		13 15 14.3	−40 23 20	−0.009	−0.03	7.9	0.0	4.4	G0 V		49 s		
114911	252224	η Mus	13 15 14.9	−67 53 41	−0.006	−0.02	4.80	−0.08	−0.2	B8 V	+5	100 s		m
115081	204287		13 15 15.3	−34 09 12	−0.002	−0.02	7.60	0.1	0.6	A0 V		250 s		m
115183	100478		13 15 18.5	+16 11 41	−0.003	+0.02	7.49	1.49	−0.1	K2 III		180 s		
114887	257022		13 15 18.6	−70 28 43	+0.001	−0.03	7.5	−0.5	0.4	B9.5 V		270 s		
114998	240678		13 15 19.1	−57 15 32	0.000	−0.02	7.6	0.6	−1.0	B5.5 V		210 s		
115068	224009		13 15 22.7	−43 03 44	+0.001	+0.01	7.8	0.3	3.2	G5 IV		84 s		
114912	252225		13 15 25.7	−69 40 47	−0.017	−0.06	6.37	1.21		K0				
113283	258659		13 15 28.0	−87 33 37	−0.332	−0.15	7.12	0.71	4.2	G5 IV-V		38 s		
115153	139246		13 15 30.6	−8 03 19	+0.003	+0.05	7.9	0.9	3.2	G5 IV		88 s		
115271	44531	19 CVn	13 15 31.9	+40 51 18	−0.010	+0.01	5.79	0.19	2.4	A7 V	−18	48 s		
115199	119834		13 15 32.1	+3 02 56	−0.005	−0.05	7.6	0.9	3.2	G5 IV		77 s		
115256	82740		13 15 36.4	+28 44 31	−0.002	0.00	7.3	1.1	−0.1	K2 III		310 s		
115111	224014		13 15 37.2	−41 08 29	−0.002	0.00	7.7	0.4	3.2	G5 IV		80 s		
115272	63433		13 15 38.7	+34 10 58	−0.002	−0.02	8.0	0.5	4.0	F8 V		63 s		
115348	28695		13 15 40.3	+54 24 10	−0.001	+0.01	7.6	1.1	0.2	K0 III		310 s		
115245	119837		13 15 46.9	+7 57 10	−0.001	0.00	6.8	1.1	0.2	K0 III		210 s		
115141	204299		13 15 49.4	−40 03 16	−0.001	−0.01	7.7	0.4	0.2	K0 III		310 s		
115217	157822		13 15 54.0	−12 21 04	+0.002	−0.06	7.3	1.1	−0.1	K2 III		280 mx		
115301	82742		13 15 57.6	+21 22 55	−0.003	−0.02	7.43	0.02		B9	0			
115202	157823	57 Vir	13 15 58.7	−19 56 35	+0.022	−0.12	5.22	1.03		K1 IV	+34	27 t		
115001	—		13 16 00.0	−69 32 49			7.7	1.4	−0.3	K5 III		380 s		
115247	139251		13 16 03.0	−5 40 07	−0.006	−0.03	7.8	0.5	4.0	F8 V		58 s		
115274	119841		13 16 03.9	+3 57 51	−0.003	+0.01	7.9	0.7	4.4	G0 V		51 s		
115500	16028		13 16 04.1	+65 41 32	+0.013	−0.05	7.4	0.9	3.2	G5 IV		70 s		
115113	240693		13 16 06.3	−57 03 40	−0.003	+0.01	7.70	0.4	2.6	F0 V		100 s		m
115159	240699		13 16 12.5	−52 24 51	0.000	−0.01	6.8	1.7	0.2	K0 III		83 s		
115277	157826		13 16 13.3	−12 37 16	−0.002	−0.02	8.0	1.1	−0.1	K2 III		410 s		
115319	100484		13 16 14.2	+19 03 05	−0.007	0.00	6.45	0.98	3.2	G5 IV	−45	33 s		

HD	SAO	Star Name	α 2000	δ 2000	μ(α)	μ(δ)	V	B-V	M$_v$	Spec	RV	d(pc)	ADS	Notes
			h m s	° ′ ″	s	″								
115322	119843		13 16 23.9	+6 30 17	-0.001	-0.05	7.20	1.6		M6 III	-23			
115308	139254		13 16 25.2	-1 23 24	-0.004	-0.02	6.68	0.31		F0				
115612	16033		13 16 28.6	+68 24 28	-0.003	+0.01	6.1	-0.1		B9	-23			
115352	100488		13 16 29.6	+17 18 00	-0.001	-0.01	7.7	0.6	4.4	G0 V		45 s		
115365	82751		13 16 32.1	+19 47 07	-0.008	+0.02	6.45	0.25	1.7	A3 V n	-33	77 mx		m
115381	100489		13 16 36.5	+18 54 12	-0.005	0.00	6.73	0.97	0.2	K0 III		200 s		
115403	82753		13 16 44.4	+19 48 54	-0.008	+0.02	7.61	0.26		A2				
115149	252236		13 16 44.8	-65 08 18	-0.015	-0.06	6.07	0.44		F5				
115279	224025		13 16 46.1	-41 16 59	-0.002	-0.01	7.30	0.8	3.2	G5 IV		66 s		m
115383	119847	59 Vir	13 16 46.4	+9 25 27	-0.023	+0.19	5.22	0.59	4.4	G0 V	-26	14 ts		m
--	--		13 16 46.5	-41 16 48			7.30							m
115382	100490		13 16 48.3	+12 24 55	+0.006	-0.22	8.0	0.9	3.2	G5 IV		79 mx		
115404	100491		13 16 51.0	+17 01 01	+0.044	-0.27	6.52	0.94	6.3	K2 V	+6	12 ts	8841	m
115386	119849		13 16 51.7	+1 14 24	-0.003	-0.01	7.4	0.5	3.4	F5 V		64 s		
115407	119851		13 16 52.6	+5 09 41	+0.002	-0.01	7.6	1.1	0.2	K0 III		300 s		
115324	204313		13 16 52.7	-30 35 30	-0.003	+0.01	7.6	0.7	3.4	F5 V		45 s		
115310	204312		13 16 53.1	-31 30 22	+0.003	-0.06	5.10	0.96	0.0	K1 III	+13	110 s		
115368	157831		13 16 58.4	-13 09 32	-0.002	-0.02	7.80	1.1	-0.1	K2 III		380 s		m
115333	224031		13 17 11.4	-46 28 35	-0.001	0.00	8.0	1.5	-0.1	K2 III		200 s		
115211	252240		13 17 13.0	-66 47 01	-0.002	-0.02	4.87	1.50	-0.3	gK6	-10	110 s		
115331	224032		13 17 13.8	-43 58 45	0.000	-0.01	5.84	0.20		A m				
115478	100497		13 17 15.5	+13 40 32	0.000	+0.03	5.33	1.31	-0.2	K3 III	-26	120 s		
115519	63448		13 17 19.8	+32 29 16	+0.009	-0.09	7.8	0.5	4.0	F8 V		56 s		
115446	157836		13 17 21.9	-11 29 00	-0.005	+0.02	6.7	0.1	1.4	A2 V		110 mx		
115283	252241		13 17 25.0	-61 35 02	-0.001	-0.01	7.8	2.2	-0.5	M4 III		200 s		V396 Cen, q
115466	157837		13 17 27.3	-10 32 47	-0.006	+0.01	7.3	0.4	2.6	F0 V	+7	85 s		
115488	139264		13 17 29.8	-0 40 36	-0.004	-0.03	6.37	0.26		F0				m
115677	28707		13 17 31.6	+57 49 18	+0.001	-0.03	7.81	1.49		K0				
115467	157838		13 17 31.6	-15 32 53	-0.003	-0.07	6.7	1.1	0.2	K0 III	+24	130 mx		
115604	44549	20 CVn	13 17 32.5	+40 34 21	-0.011	+0.02	4.73	0.30	-0.7	F0 II-III	+8	52 mx		
115538	100498		13 17 33.2	+17 17 33	-0.001	+0.03	7.6	0.9	3.2	G5 IV		77 s		
115537	100499		13 17 34.5	+19 12 54	-0.002	-0.02	7.94	1.44	-0.1	K2 III		250 s		
115627	28706		13 17 35.3	+49 45 37	-0.002	+0.02	7.9	1.1	-0.1	K2 III		390 s		
115396	224040		13 17 35.4	-43 17 56	+0.001	0.00	7.6	-0.4	0.6	A0 V		260 s		
115521	119855	60 σ Vir	13 17 36.2	+5 28 12	0.000	+0.01	4.80	1.67	-0.5	gM2	-27	100 s		
115539	100500		13 17 37.8	+13 45 48	-0.007	0.00	7.22	0.95	1.8	G8 III-IV	-8	120 s		
115431	204325		13 17 38.3	-37 01 03	-0.001	-0.04	7.40	0.1	1.4	A2 V		160 s		m
115503	139268		13 17 44.0	-8 44 00	+0.001	-0.06	7.1	0.5	4.0	F8 V		42 s		
115720	28710		13 17 44.2	+58 45 01	-0.001	-0.01	8.04	1.05		G5			8851	m
115415	224043		13 17 44.6	-47 35 26	-0.003	-0.03	6.66	0.00		B9				
115286	252245		13 17 55.2	-68 29 46	-0.005	-0.03	7.20	0.00	0.4	B9.5 V		210 mx		m
115214	257027		13 17 59.3	-73 20 41	-0.012	-0.06	7.0	0.1	1.4	A2 V		110 mx		
115470	224047		13 18 04.9	-44 03 20	-0.004	-0.03	7.4	-0.2	0.6	A0 V		110 mx		
115399	252251		13 18 05.4	-60 23 49	-0.003	+0.02	7.6	2.1	-0.1	K2 III		94 s		
115363	252250		13 18 06.7	-63 41 13	-0.004	-0.01	7.76	0.64	-6.6	B1 Ia		290 mx		
116736	2194		13 18 07.5	+83 53 55	-0.062	-0.01	7.7	0.4	3.0	F2 V		87 s		
115434	240717		13 18 08.4	-55 15 07	-0.003	+0.02	7.9	1.3	-0.1	K2 III		350 s		
115866	16037		13 18 11.5	+67 41 18	-0.006	+0.01	7.3	1.1	0.2	K0 III		260 s		
115655	63458		13 18 11.9	+31 37 49	-0.001	-0.01	7.8	0.4	3.0	F2 V		93 s		
115735	44556	21 CVn	13 18 14.5	+49 40 55	-0.003	+0.02	5.15	-0.07		A0	-3	21 mn		
115418	252253		13 18 17.7	-61 19 40	0.000	+0.01	7.1	2.5	-0.5	M4 III		110 s		
115436	240718		13 18 18.0	-57 17 59	-0.002	+0.01	7.0	0.0	0.0	B8.5 V		210 s		
115400	252252		13 18 19.7	-63 26 58	-0.002	-0.02	6.81	0.74	-4.6	F2 Ib				
115577	181536		13 18 23.3	-28 19 56	-0.005	-0.08	6.80	0.98	3.2	G8 IV		46 s		
115116	257025		13 18 24.1	-78 49 23	-0.027	-0.02	7.2	-0.4	1.7	A3 V		67 mx		
115617	157844	61 Vir	13 18 24.2	-18 18 41	-0.075	-1.07	4.74	0.71	5.1	G6 V	-9	8.4 t		m
115527	224051		13 18 24.9	-45 45 52	-0.002	-0.02	7.0	0.1	0.6	A0 V		160 s		
115723	63462		13 18 27.7	+34 05 53	+0.002	0.00	5.82	1.35	0.2	K0 III	-20	82 s		
115640	119863		13 18 28.7	+1 14 47	+0.002	-0.10	7.9	1.1	0.2	K0 III		190 mx		
115631	139272		13 18 32.9	-6 15 56	-0.004	-0.01	7.7	1.1	0.2	K0 III		280 mx		
115529	240721		13 18 34.5	-51 17 09	-0.005	-0.03	6.19	0.01	0.6	A0 V		100 mx		
115708	82769		13 18 37.1	+26 21 56	0.000	0.00	7.83	0.25		A2 p				
115564	224055		13 18 43.1	-49 52 43	-0.001	-0.02	7.34	0.01		B9				
115642	157846		13 18 45.2	-17 08 01	+0.001	0.00	7.4	0.6	4.4	G0 V		40 s		
--	100508		13 18 47.3	+11 49 58	-0.002	-0.03	7.98	1.55						
115088	257026		13 18 48.4	-79 58 32	-0.005	-0.02	6.5	-0.4	0.6	A0 V		150 s		
115679	157848		13 18 49.8	-12 00 55	-0.003	-0.01	7.5	0.1	0.6	A0 V		240 s		
115709	119867		13 18 51.0	+3 41 16	-0.003	-0.01	6.62	0.06	0.3	A1 IV	-1	170 s		
115659	181543	46 γ Hya	13 18 55.2	-23 10 17	+0.005	-0.05	3.00	0.92	0.3	G5 III	-5	32 s		m
115661	181544		13 18 57.9	-28 47 08	-0.005	-0.05	7.8	1.4	-0.3	K5 III		180 mx		
115693	157849		13 18 58.4	-12 38 52	-0.008	0.00	7.2	0.4	2.6	F0 V		84 s		
115162	--		13 19 00.1	-79 44 45			8.0	1.1	0.2	K0 III		360 s		
115782	82777		13 19 02.1	+28 34 04	-0.002	+0.12	7.1	0.5	4.0	F8 V		42 s		
115810	63468		13 19 04.2	+35 07 40	-0.003	+0.01	6.02	0.24	0.6	A9 III	-2	120 s		m
115753	119872		13 19 06.5	+3 43 22	-0.001	-0.01	7.0	1.1	-0.1	K2 III		260 s		

HD	SAO	Star Name	α 2000	δ 2000	μ(α)	μ(δ)	V	B-V	M_v	Spec	RV	d(pc)	ADS	Notes
			$13^h 19^m 18^s.6$	$-72°02'09''$	$-0^s.004$	$-0''.07$								
115439	257028		13 19 18.6	-72 02 09	-0.004	-0.07	6.04	1.35		K2				
115619	240740		13 19 21.0	-55 41 07	-0.003	-0.04	7.8	1.0	-0.1	K2 III		370 s		
115665	---		13 19 25.2	-48 27 38			7.9	2.4	-0.1	K2 III		73 s		
115624	240743		13 19 31.1	-59 14 34	+0.001	-0.02	7.7	1.4	-0.1	K2 III		270 s		
115602	252261		13 19 33.0	-63 45 19	+0.003	-0.03	8.00	0.5	3.4	F5 V		82 s		m
115636	252263		13 19 38.3	-60 38 06	-0.002	0.00	7.0	2.6	-4.4	K5 Ib		410 s		
115830	119881		13 19 38.8	+9 41 34	+0.001	-0.08	7.90	0.4	2.6	F0 V		120 s	8860	m
115928	63474		13 19 42.6	+36 04 55	-0.004	-0.03	7.18	0.97	3.2	G5 IV		47 s		
115770	204359		13 19 43.0	-30 19 58	+0.001	-0.02	8.0	0.3	0.0	B8.5 V		230 s		
115583	252262		13 19 43.5	-67 21 52	-0.002	-0.03	7.2	0.1	0.6	A0 V		200 s		
115813	157860		13 19 44.0	-11 40 23	-0.003	-0.03	7.10	0.1	1.4	A2 V		140 s	8858	m
115927	63476		13 19 47.3	+36 22 25	-0.002	+0.02	7.62	1.04	3.2	G5 IV		53 s		
115742	224064		13 19 47.7	-42 09 17	-0.002	-0.07	8.0	2.0	-0.1	K2 III		73 s		
115652	252265		13 19 50.4	-61 58 07	-0.002	+0.01	7.97	0.04	0.4	B9.5 V		320 s		
115585	257030		13 19 50.5	-70 51 17	-0.079	-0.03	7.42	0.75	4.3	G6 IV-V		38 mx		
115885	119882		13 19 56.9	+8 29 14	+0.001	-0.03	7.08	1.13	0.2	K0 III		190 s		
115773	224066		13 19 58.3	-41 11 28	-0.029	0.00	6.75	0.52		F8		17 mn		
115968	63483		13 20 01.5	+38 09 32	-0.034	-0.08	7.9	1.1	0.2	K0 III	+1	38 mx		
115669	252268		13 20 02.0	-62 31 35	-0.001	0.00	6.6	2.6	0.2	K0 III		20 s		
116091	16046		13 20 07.6	+59 51 42	-0.008	+0.05	8.01	0.63		F8				
115942	100516		13 20 09.2	+15 33 50	-0.008	-0.02	7.25	1.08	3.2	G5 IV		36 s		m
115570	---		13 20 09.8	-72 45 39			5.90	0.9	0.2	G9 III		140 s		
115930	119885		13 20 14.6	+3 17 15	0.000	-0.01	7.0	1.1	-0.1	K2 III		260 s		
115955	100517		13 20 15.6	+17 45 57	-0.002	-0.08	7.80	0.5	3.4	F5 V		76 s	8863	m
116010	44570	23 CVn	13 20 18.9	+40 09 01	-0.005	-0.03	5.60	1.20	0.0	K1 III	-21	120 s		
115903	157869		13 20 19.9	-11 18 15	-0.008	-0.01	6.73	1.00		K0	+50			
115422	257029		13 20 20.2	-77 57 30	-0.004	+0.03	7.6	1.1	-0.1	K2 III		320 mx		
115981	100519		13 20 25.7	+17 34 48	-0.007	+0.02	7.6	0.4	3.0	F2 V		85 s		
115852	224072		13 20 26.4	-41 34 34	-0.001	-0.01	7.4	1.4	-0.1	K2 III		310 s		
115890	181564		13 20 29.7	-27 24 36	-0.003	-0.02	7.7	1.6	-0.3	K5 III		340 s		
115873	204369		13 20 29.9	-33 39 28	-0.001	0.00	7.8	1.4	-0.3	K5 III		410 s		
115906	181567		13 20 30.4	-20 52 52	+0.005	-0.07	7.8	0.5	3.4	F5 V		72 s		
115982	119887		13 20 32.9	+8 14 42	0.000	-0.02	8.0	1.1	-0.1	K2 III		410 s		
115983	119886		13 20 33.7	+4 49 38	-0.003	-0.01	6.9	0.1	1.7	A3 V	-19	110 s		
115778	240756		13 20 34.8	-59 46 23	-0.005	-0.03	6.18	0.43		F2				m
115892	204371	z Cen	13 20 35.8	-36 42 44	-0.028	-0.09	2.75	0.04	1.4	A2 V	0	16 mx		
115823	240762		13 20 37.7	-52 44 52	-0.003	-0.02	5.48	-0.13	-2.2	B5 III	+6	330 s		
115671	252271		13 20 38.7	-69 40 51	-0.003	-0.02	7.4	0.1	1.4	A2 V		150 s		
116029	82792		13 20 39.6	+24 38 55	+0.001	-0.06	7.86	1.00	0.0	K1 III		240 mx		
115995	119889		13 20 41.5	+2 56 30	-0.004	-0.03	6.26	0.10		A0	+4		8864	m
116285	16055		13 20 45.2	+69 06 07	0.000	-0.04	7.2	0.9	3.2	G5 IV		65 s		
115947	181571		13 20 45.2	-23 45 47	-0.006	-0.01	7.2	1.7	-0.1	K2 III		130 s		
115842	240765		13 20 48.2	-55 48 02	-0.002	0.00	6.02	0.29	-6.2	B0 I k		2100 s		
116127	44573		13 20 50.9	+43 59 24	-0.014	+0.02	6.6	0.4	3.0	F2 V		52 s		
115912	224080		13 20 57.6	-46 52 49	-0.007	0.00	5.77	1.12		K0				
115985	181576		13 21 03.1	-22 22 59	+0.002	-0.02	7.27	0.96		G5				
115998	181577		13 21 04.9	-24 28 13	0.000	0.00	7.1	1.8	-0.1	K2 III		120 s		
116172	44577		13 21 07.3	+47 16 12	+0.003	+0.01	7.0	1.1	0.2	K0 III		230 s		
115915	240771		13 21 10.1	-51 40 17	-0.003	-0.01	7.53	-0.05		B9				
116156	63491		13 21 10.8	+37 51 26	+0.002	0.00	6.9	0.5	4.0	F8 V	-10	38 s		
116045	139293		13 21 12.3	-9 59 57	-0.007	+0.04	7.0	0.8	3.2	G5 IV		58 s		
115918	---		13 21 22.3	-57 55 33			7.9	2.2	-0.3	K5 III		160 s		
116110	100527		13 21 22.4	+14 08 57	+0.001	-0.03	7.28	1.49	-0.1	K2 III		190 s		
115987	204378		13 21 22.8	-39 59 25	0.000	-0.03	7.49	0.00	-0.2	B8 V		290 s		
116046	181584		13 21 24.0	-22 47 38	+0.004	+0.03	7.20	0.5	3.4	F5 V		58 s	8867	m
115988	224083		13 21 26.1	-43 38 02	-0.001	-0.01	6.70	-0.02		B9				
115846	252276		13 21 28.7	-67 32 16	-0.002	-0.05	7.05	-0.03	-1.4	B4 V		420 s		
116061	157879		13 21 29.8	-19 29 20	-0.005	0.00	6.21	0.09		A0				
116342	16058		13 21 30.5	+63 57 01	-0.016	+0.02	7.4	1.1	-0.1	K2 III		190 mx		
116204	63494		13 21 32.1	+38 52 51	-0.006	-0.01	7.16	1.15	-0.1	K2 III		250 mx		
115828	---		13 21 34.4	-69 56 34			7.7	1.4	-0.3	K5 III		380 s		
116139	119898		13 21 36.8	+2 51 14	-0.001	-0.04	7.9	1.4	-0.3	K5 III		430 s		
116173	100531		13 21 40.6	+16 04 53	-0.003	-0.05	7.10	1.36	0.2	K0 III		120 s		
116160	119899		13 21 41.5	+2 05 14	-0.004	-0.06	5.69	0.06		A0	-5			
116033	224086		13 21 44.2	-42 07 54	0.000	-0.02	7.8	2.2	-0.1	K2 III		220 s		
115637	---		13 21 44.7	-78 10 36			7.6	1.6	-0.5	M4 III		410 s		
116114	157882		13 21 46.2	-18 44 30	-0.004	0.00	7.1	0.4	2.6	F0 V		78 s		
116513	7802		13 21 49.4	+71 52 17	-0.003	-0.01	7.5	1.1	-0.1	K2 III		320 s		
116078	204390		13 21 49.6	-36 06 43	-0.006	-0.02	7.4	0.1	2.6	F0 V		91 s		
116232	82799		13 21 56.7	+25 59 54	-0.001	+0.01	7.61	0.97	0.3	G8 III		280 s		
116080	204393		13 21 56.9	-38 51 19	-0.006	-0.09	7.1	0.7	4.4	G0 V		30 s		
116038	240778		13 21 57.0	-51 16 57	-0.004	-0.03	7.4	0.5	1.4	A2 V		84 s		
116233	82800		13 21 59.7	+24 52 56	-0.005	-0.02	7.07	0.22	2.1	A5 V		64 mx		
116207	119902		13 22 01.7	+2 45 26	0.000	-0.03	7.8	1.6	-0.5	M2 III		470 s		
116081	224089		13 22 02.7	-43 03 01	0.000	-0.02	7.1	0.1	0.6	A0 V		160 s		

HD	SAO	Star Name	α 2000	δ 2000	μ(α)	μ(δ)	V	B-V	M_v	Spec	RV	d(pc)	ADS	Notes
			h m s	° ′ ″	s	″								
116303	44582		13 22 03.7	+43 54 11	-0.007	+0.01	6.35	0.24	1.7	A3 V	-1	73 s		
116177	157885		13 22 06.3	-14 25 13	0.000	-0.06	7.0	1.1	-0.1	K2 III		260 s		
115881	257032		13 22 06.6	-70 32 48	0.000	-0.01	7.4	-0.4	0.4	B9.5 V		250 s		
116003	240776		13 22 07.4	-60 11 07	0.000	-0.02	6.95	0.01	-1.0	B5.5 V		380 s		
116208	119903		13 22 07.6	-0 04 05	-0.003	-0.01	8.0	1.1	0.2	K0 III		360 s		
116175	157886		13 22 08.1	-12 34 47	0.000	-0.01	6.88	1.64	-0.5	gM1	-33	270 s		
116235	119905	64 Vir	13 22 09.6	+5 09 17	-0.005	-0.04	5.87	0.12		A5	-10			
116287	63499		13 22 15.0	+30 17 31	+0.002	-0.03	6.77	1.34	0.2	K0 III		130 s		
116084	240782		13 22 16.1	-52 10 58	-0.001	0.00	5.83	0.12		B0.5 II	-15			
116180	204404		13 22 23.8	-33 57 06	-0.004	-0.02	7.8	1.9	0.2	K0 III		97 s		
116288	82806		13 22 25.8	+20 41 48	-0.002	+0.02	7.28	1.37	0.2	K0 III		130 s		
116211	181598		13 22 27.3	-25 50 28	-0.010	-0.01	7.1	1.0	3.4	F5 V		26 s		
116258	139303		13 22 28.7	-0 34 45	-0.003	-0.03	8.0	1.1	0.2	K0 III		360 s		
116249	139304		13 22 30.8	-6 12 07	-0.006	-0.14	6.7	1.1	0.2	K0 III		81 mx		
116316	82809		13 22 34.1	+26 06 56	-0.012	+0.01	7.50	0.5	4.0	F8 V		50 s	8879	m
116087	252284		13 22 37.8	-60 59 18	-0.005	-0.02	4.53	-0.13	-1.1	B5 V	+26	130 s		m
116275	157895		13 22 45.5	-13 11 12	-0.003	-0.01	7.77	0.15		A2	-22		8878	m
116197	224095		13 22 52.6	-47 56 35	+0.003	0.00	6.90	0.1	1.4	A2 V		130 s		m
115967	257033		13 22 52.6	-72 08 48	-0.003	-0.02	6.05	0.09	-0.9	B6 V var		240 s		m
116119	252285		13 22 55.6	-62 00 44	0.000	-0.01	7.87	0.71	-6.5	B9 I		3100 s		
116493	28730		13 22 57.1	+51 39 10	-0.002	-0.02	7.8	1.6	-0.5	M2 III		470 s		
116364	100541		13 22 59.8	+15 15 01	-0.004	-0.01	7.59	1.24	0.2	K0 III		190 s		
116292	157899	63 Vir	13 23 01.0	-17 44 07	-0.003	-0.03	5.37	0.99	0.2	K0 III	-27	110 s		
116226	224096		13 23 02.5	-48 33 46	-0.001	-0.01	6.38	-0.07	-1.0	B7 IV		290 s		
116379	100542		13 23 03.0	+18 16 42	-0.002	+0.01	7.92	0.16		A2				
116475	44590		13 23 05.5	+47 00 06	-0.001	+0.01	6.8	1.6	-0.5	M4 III	-9	290 s		
116278	204420		13 23 08.6	-33 11 24	-0.001	-0.01	6.2	2.6	-0.5	M2 III		57 s		
116321	181609		13 23 09.7	-24 44 03	-0.010	0.00	7.8	0.9	4.0	F8 V		36 s		
116186	240791		13 23 15.7	-60 13 36	-0.005	0.00	7.9	1.6	3.2	G5 IV		27 s		
116365	139308	65 Vir	13 23 18.8	-4 55 28	-0.001	-0.02	5.89	1.43	-0.2	gK3	+10	130 s		
116333	181611		13 23 19.9	-24 39 21	-0.002	-0.02	7.6	1.7	-0.5	M4 III		370 s		
116531	44591		13 23 20.5	+46 07 22	+0.001	0.00	7.8	1.1	0.2	K0 III		320 s		
116293	204423		13 23 23.1	-39 37 37	-0.002	+0.02	7.4	-0.6	1.4	A2 V		160 s		
116494	63510		13 23 24.5	+39 25 47	-0.005	+0.02	7.5	0.8	3.2	G5 IV		74 s		
116441	100544		13 23 25.9	+17 17 46	0.000	-0.01	7.53	1.16	0.2	K0 III		210 s		
116655	16062		13 23 27.3	+62 48 05	-0.029	+0.13	8.0	0.6	4.4	G0 V		53 s		
116324	224102		13 23 36.1	-42 31 55	-0.002	-0.04	7.9	0.6	0.6	A0 V		280 s		
116442	119909		13 23 39.0	+2 43 24	0.000	+0.20	7.06	0.77	0.2	K0 III		98 mx	8883	m
116443	119910		13 23 40.7	+2 43 30	0.000	+0.19	7.36	0.82	-0.1	K2 III		100 mx		
116336	224103		13 23 42.1	-45 28 33	-0.003	-0.03	7.7	0.7	-0.1	K2 III		370 s		
116497	100549		13 23 44.6	+19 05 27	+0.001	-0.03	7.9	0.7	4.4	G0 V		51 s		
116153	--		13 23 46.5	-70 23 28			7.9	1.4	-0.3	K5 III		420 s		
116515	82819		13 23 46.7	+25 32 52	-0.003	0.00	7.38	1.03	0.2	K0 III	-8	260 s		
116338	224104		13 23 52.2	-49 49 23	-0.007	-0.05	6.48	0.96		K0				
116337	224106		13 23 52.8	-47 53 26	-0.003	-0.03	6.9	1.3	3.2	G5 IV		28 s		
116581	63514		13 23 53.9	+37 02 02	+0.002	-0.01	6.07	1.68	-0.5	M3 III	0	190 s		
116656	28737	79 ζ UMa	13 23 55.5	+54 55 31	+0.014	-0.02	2.27	0.02	1.4	A2 V	-9	18 ts	8891	Mizar, m
116657	28738	79 ζ UMa	13 23 56.7	+54 55 18	+0.013	-0.03	3.95	0.13		A m	-9		8891	m
116429	181615		13 23 57.0	-20 55 27	-0.002	-0.01	6.60	1.1	0.2	K0 III		190 s	8885	m
116243	252293		13 24 00.5	-64 32 09	+0.005	-0.03	4.53	0.85	1.8	G5 III-IV	+12	35 s		
116356	240802		13 24 08.4	-53 47 35	-0.004	-0.02	7.2	-0.1	0.6	A0 V		140 mx		
116255	252294		13 24 13.5	-67 18 57	-0.002	-0.03	7.5	0.0	0.0	B8.5 V		310 s		
116481	181622		13 24 14.9	-25 07 44	-0.003	+0.02	7.8	0.3	0.2	K0 III		330 s		
116542	119913		13 24 18.3	+1 23 57	-0.002	-0.03	7.60	0.1	1.4	A2 V	-15	170 s	8890	m
116448	204431		13 24 22.9	-38 01 59	0.000	0.00	7.14	1.47	-0.3	K4 III		290 s		
116634	63516		13 24 23.3	+32 05 06	+0.001	+0.01	7.74	0.06		A2				
116398	240807		13 24 25.7	-51 52 30	-0.003	+0.01	6.8	0.9	0.2	K0 III		210 s		
116545	139322		13 24 26.0	-4 18 31	-0.005	-0.13	7.0	1.1	0.2	K0 III		85 mx		
116594	100553		13 24 30.4	+12 25 55	-0.001	+0.04	6.44	1.06	0.2	K0 III	-6	160 s		
--	--		13 24 31.2	-84 13 28			7.10							U Oct, v
116568	139324	66 Vir	13 24 33.1	-5 09 50	+0.010	-0.03	5.75	0.42	3.1	dF3	+14	33 s		
116416	240810		13 24 35.8	-53 54 50	-0.005	+0.02	6.9	1.5	0.2	K0 III		120 s		
116434	240811		13 24 36.0	-51 30 16	-0.004	-0.03	7.90	0.1	1.4	A2 V		120 mx		m
117655	2214		13 24 38.6	+83 18 32	+0.012	+0.10	7.3	0.9	3.2	G5 IV	-25	65 s		
116284	257040		13 24 38.8	-70 25 49	-0.051	-0.13	7.2	0.8	0.2	K0 III		72 mx		
116582	139325		13 24 39.0	-4 55 01	+0.001	-0.05	7.6	1.1	-0.1	K2 III		340 s		
116855	16065		13 24 39.9	+66 09 54	0.000	0.00	8.0	1.1	0.2	K0 III		360 s		
116535	204437		13 24 42.2	-32 40 10	-0.001	-0.01	7.9	1.1	0.2	K0 III		320 s		
116621	119916		13 24 42.5	+4 24 08	+0.001	+0.03	7.8	1.4	-0.3	K5 III		410 s		
116549	204439		13 24 48.1	-34 17 25	-0.001	-0.01	7.6	0.5	1.7	A3 V		84 s		
116737	63525		13 24 58.7	+37 26 05	-0.005	-0.01	7.9	0.5	3.4	F5 V		78 s		
116622	157920		13 25 03.6	-15 37 55	-0.004	0.00	7.60	0.1	1.4	A2 V		120 mx	8893	m
116706	82825		13 25 06.6	+23 51 15	-0.001	-0.01	5.78	0.06	1.7	A3 V	-1	66 s		
116244	257041	ι¹ Mus	13 25 07.2	-74 53 15	-0.026	-0.13	5.05	1.11	0.3	gG7	+29	65 s		
116089	257038		13 25 09.2	-78 57 05	-0.052	-0.05	7.3	0.4	3.0	F2 V		72 s		

HD	SAO	Star Name	α 2000	δ 2000	μ(α)	μ(δ)	V	B-V	M$_v$	Spec	RV	d(pc)	ADS	Notes
116658	157923	67 α Vir	13^h 25^m 11^s.5	-11° 09′ 41″	-0^s.003	-0″.03	0.98	-0.23	-3.5	B1 V	+1	79 s		Spica, m,v
116538	240823		13 25 11.6	-51 50 28	-0.005	+0.02	7.92	-0.05	-3.0	B2 IV n		140 mx		
116842	28751	80 UMa	13 25 13.4	+54 59 17	+0.013	-0.02	4.01	0.16	2.1	A5 V	-8	25 ts		Alcor
116457	252304		13 25 13.8	-64 29 07	-0.024	-0.03	5.31	0.40		F5	-2	24 mn		m
116682	139330		13 25 14.8	-8 47 09	+0.001	0.00	7.7	0.1	1.4	A2 V		180 s		
116154	257039		13 25 15.1	-78 03 16	-0.001	-0.02	7.3	1.1	0.2	K0 III		260 s		
116695	119918		13 25 16.0	+0 50 56	0.000	-0.10	8.00	1.1	0.2	K0 III		150 mx	8895	m
116560	240828		13 25 20.7	-51 53 22	-0.003	-0.01	7.16	0.03	1.4	A2 V		140 s		
116424	252305		13 25 31.4	-68 31 40	0.000	0.00	6.9	1.1	0.2	K0 III		210 s		
116784	100561		13 25 39.9	+15 33 50	-0.005	-0.01	6.87	1.20	0.2	K0 III		140 s		
116604	240830		13 25 42.6	-54 39 39	+0.006	-0.05	7.9	1.2	0.2	K0 III		230 mx		
117113	7812		13 25 43.7	+72 16 22	+0.007	-0.03	7.0	1.1	0.2	K0 III		230 s		
116458	257042		13 25 50.2	-70 37 39	-0.012	-0.02	5.67	-0.03		A0 p				
116651	240834		13 25 55.9	-51 10 36	-0.003	-0.01	8.0	0.1	1.4	A2 V		180 s		
116711	204462		13 25 58.0	-35 04 33	0.000	-0.01	7.0	0.7	-0.3	K5 III		290 s		
117043	16071		13 25 59.8	+63 15 40	-0.059	+0.21	6.5	0.8	3.2	G5 IV	-31	37 mx		
116664	240835		13 26 01.8	-51 28 46	-0.007	-0.01	7.2	0.4	1.7	A3 V		79 s		
116768	157932		13 26 06.2	-16 51 31	+0.002	+0.05	6.6	1.1	-0.1	K2 III		220 s		
116771	181646		13 26 06.6	-26 36 23	-0.003	-0.04	7.7	0.7	3.2	G5 IV		80 s		
116879	63536		13 26 06.9	+30 42 09	-0.004	+0.04	8.0	0.5	3.4	F5 V		82 s		
116713	204465		13 26 07.7	-39 45 19	+0.016	-0.06	5.09	1.20		K0	+68	19 mn		
117187	7814		13 26 07.9	+72 23 29	+0.004	-0.01	5.79	1.63	-0.3	K5 III	-48	140 s		m
116743	204469		13 26 07.9	-32 32 50	-0.005	-0.02	7.10	0.4	2.6	F0 V		79 s		m
116817	139336		13 26 08.6	-3 39 40	0.000	0.00	7.2	1.1	0.2	K0 III		260 s		
116831	139337		13 26 11.3	-1 11 33	-0.008	0.00	5.97	0.19		A3				
116957	44611		13 26 16.5	+46 01 40	+0.002	-0.03	5.88	0.97	0.2	K0 III	+4	140 s		
116717	224138		13 26 16.6	-48 47 12	-0.005	-0.03	6.8	0.6	0.6	A0 V		79 s		
116868	119924		13 26 25.5	+5 42 31	-0.007	+0.01	7.9	1.1	0.2	K0 III		200 mx		
116788	204475		13 26 26.4	-34 00 38	-0.002	-0.20	7.0	1.4	3.2	G5 IV		23 s		
116772	224142		13 26 30.4	-41 51 01	-0.003	-0.03	7.6	1.5	-0.1	K2 III		340 s		
116630	---		13 26 31.1	-65 50 19			7.9	0.6	4.4	G0 V		50 s		
116870	157938	68 Vir	13 26 43.1	-12 42 28	-0.009	-0.02	5.25	1.52	-0.4	M0 III	-29	140 s		
116941	100566		13 26 44.4	+16 56 55	0.000	+0.06	7.8	0.5	4.0	F8 V		57 s		
116914	139342		13 26 53.4	-5 55 50	-0.001	-0.01	8.0	0.1	0.6	A0 V		300 s		
117156	16076		13 26 54.7	+63 00 29	+0.002	-0.03	8.0	1.1	0.2	K0 III		360 s		
116835	224148		13 26 56.0	-41 29 53	-0.001	-0.02	5.69	1.47		K2				
116776	240850		13 26 56.6	-53 47 12	-0.001	-0.02	7.3	0.6	0.6	A0 V		89 s		
117566	7821		13 26 56.9	+78 38 37	-0.048	+0.03	5.77	0.77	3.2	G5 IV	+15	33 s		
116960	100569		13 26 58.8	+11 54 31	0.000	-0.01	7.99	-0.02	0.6	A0 V		300 s		
117200	16078		13 27 04.5	+64 44 07	-0.011	+0.02	6.66	0.37	3.0	F2 V	-13	53 s		m
116836	224149		13 27 06.2	-49 08 37	-0.002	-0.01	7.20	0.1	0.6	A0 V		210 s		m
117201	16079		13 27 10.6	+64 43 09	-0.010	+0.03	7.04	0.39	3.1	F3 V	-15	61 s		m
116928	157942		13 27 10.6	-17 55 44	-0.005	-0.04	7.0	1.1	0.2	K0 III		180 mx		
117317	7817		13 27 11.8	+70 18 59	-0.004	-0.04	7.5	0.4	2.6	F0 V	-40	96 s		
116873	204483		13 27 14.6	-40 09 47	-0.001	-0.01	6.30	1.1	0.2	K0 III		170 s		
116961	157944		13 27 15.7	-13 18 37	-0.001	+0.02	7.7	0.9	0.3	G5 III	+2	310 s		
116579	257047	z² Mus	13 27 18.3	-74 41 30	-0.010	-0.02	6.63	-0.06		B9				
116862	224151		13 27 20.7	-49 22 51	+0.001	+0.01	6.28	-0.12		B3 n	-10			
116749	---		13 27 24.5	-66 45 16			7.9	0.1	0.6	A0 V		270 s		
116781	252312		13 27 25.1	-62 38 55	0.000	0.00	7.90							m
117099	63548		13 27 27.0	+35 43 34	-0.025	-0.33	8.0	1.1	0.2	K0 III		25 mx		
116976	157946	69 Vir	13 27 27.1	-15 58 25	-0.008	+0.02	4.76	1.09	0.0	K1 III	-14	32 ts		
116816	252313		13 27 31.7	-63 49 01	0.000	-0.01	7.9	0.0	0.4	B9.5 V		310 s		
117045	139350		13 27 48.1	-3 10 22	+0.001	0.00	7.9	1.6		M5 III	+33			V Vir, v
116905	240859		13 27 55.2	-57 01 47	-0.026	-0.05	7.2	1.7	0.2	K0 III		82 mx		
117242	28763		13 27 59.4	+52 44 45	-0.013	-0.01	6.34	0.23	2.6	F0 V	-7	56 s		
116875	252317		13 28 01.2	-61 03 47	-0.003	-0.03	7.33	0.04	0.6	A0 V		220 s		
117033	181674		13 28 01.7	-26 24 09	-0.001	-0.01	6.7	1.1	-0.1	K2 III		230 s		
117448	7823		13 28 03.6	+70 07 35	-0.008	-0.01	7.5	1.1	0.2	K0 III	-28	280 mx		
116992	224156		13 28 05.1	-45 04 08	-0.001	-0.01	7.9	0.9	-0.1	K2 III		390 s		
117299	28767		13 28 05.3	+58 12 34	-0.002	0.00	7.9	0.4	2.6	F0 V		110 s		
116991	224157		13 28 05.7	-43 46 09	-0.001	-0.03	7.21	1.08	0.2	K0 III		230 s		m
117115	119940		13 28 08.9	+8 21 29	-0.004	+0.02	8.0	0.4	3.0	F2 V		98 s		
117114	119941		13 28 09.7	+9 27 37	-0.003	-0.03	7.61	1.17	0.2	K0 III		220 s	8912	m
117281	28766		13 28 11.6	+50 35 14	-0.012	+0.04	6.8	0.1	1.7	A3 V	-16	56 mx		
117083	181678		13 28 13.4	-22 24 02	-0.002	-0.05	7.43	1.24	0.2	K0 III		180 s		
117125	119942		13 28 16.4	+2 14 07	-0.003	-0.03	7.60	0.9	0.3	G8 III	-1	290 s	8913	m
117126	139353		13 28 18.6	-0 50 25	+0.014	-0.41	7.5	0.9	3.2	G5 IV	0	48 mx		
117203	63556		13 28 18.7	+31 08 57	0.000	0.00	6.92	1.38	-0.1	K2 III		190 s		
116948	240866		13 28 21.4	-59 31 56	0.000	-0.04	7.07	0.96	3.2	G5 IV		44 s		
116982	240868		13 28 23.2	-55 46 47	+0.001	-0.03	7.3	0.6	2.6	F0 V		57 s		
116865	252318		13 28 23.7	-67 52 11	-0.006	-0.03	7.30	0.1	1.7	A3 V		130 s		m
117176	100582	70 Vir	13 28 25.7	+13 46 43	-0.016	-0.58	4.98	0.71	5.2	G5 V	+4	10 ts		m
117068	224161		13 28 25.8	-41 58 27	-0.002	-0.02	7.8	1.9	-0.1	K2 III		120 s		
117261	44621		13 28 26.1	+40 43 47	+0.001	-0.06	6.47	0.92	0.2	K0 III	-58	180 s		

HD	SAO	Star Name	α 2000	δ 2000	μ(α)	μ(δ)	V	B-V	M$_v$	Spec	RV	d(pc)	ADS	Notes
117104	181681		13^h 28^m 26$.^s$5	-25° 12′ 46″	-0$.^s$004	0$.″$00	7.4	0.7	3.4	F5 V	-34	46 s		
117376	16080		13 28 27.0	+59 56 45	-0.011	+0.04	5.40	-0.01		A0	-7	21 mn	8919	m
117105	181682		13 28 30.2	-27 23 32	-0.020	+0.20	7.20	0.58		F8		14 mn		
117018	240874		13 28 38.4	-53 12 26	-0.004	-0.01	8.0	0.1	1.4	A2 V		150 mx		
117361	28771		13 28 45.6	+50 43 06	+0.003	-0.09	6.43	0.37	2.6	F0 V	-7	54 s		
116890	252321		13 28 46.4	-69 37 41	-0.007	-0.05	6.20	0.02	-0.2	B8 V		140 mx		
117000	252326		13 28 55.1	-62 06 08	+0.002	-0.01	6.62	1.08	4.4	G0 V		14 s		
117163	204511		13 28 58.2	-31 02 38	0.000	-0.01	8.0	1.5	-0.3	K5 III		450 s		
117377	44625		13 29 01.5	+47 50 50	-0.005	0.00	7.1	0.5	3.4	F5 V		56 s		
116591	257050		13 29 02.0	-80 15 29	-0.005	-0.04	7.5	0.4	3.0	F2 V		79 s		
117231	139358		13 29 02.6	-5 57 20	-0.004	0.00	6.9	1.1	0.2	K0 III		220 s		
117378	44624		13 29 03.1	+42 14 18	-0.013	-0.02	7.64	0.56	4.4	G0 V		45 s		
117417	28774		13 29 03.2	+56 14 00	-0.002	+0.02	7.70	0.9	3.2	G5 IV		79 s	8921	m
117024	252328		13 29 05.4	-63 52 27	0.000	-0.02	7.09	0.05		B2				
117165	204512		13 29 06.3	-35 46 29	-0.004	0.00	7.8	0.7	0.2	K0 III		330 s		
117025	252329		13 29 07.6	-64 40 33	-0.012	-0.01	6.11	0.11		A2 pe				
117266	119950		13 29 12.4	+0 07 08	-0.012	-0.02	8.0	0.5	4.0	F8 V		62 s		
117304	100592	71 Vir	13 29 12.9	+10 49 06	-0.004	-0.04	5.65	1.05	0.2	K0 III	-1	120 s		
117094	240880		13 29 14.6	-54 55 05	+0.002	-0.02	7.80	0.8	3.2	G5 IV		82 s		m
117267	139359		13 29 14.8	-1 21 52	-0.003	-0.07	6.43	1.11	0.2	gK0	+39	140 s		
117145	224175		13 29 16.3	-45 32 14	+0.001	+0.01	6.62	-0.03		B9				
117449	28775		13 29 17.8	+53 10 39	-0.002	+0.01	8.0	1.1	0.2	K0 III		360 s		
117207	204517		13 29 21.0	-35 34 14	-0.017	-0.07	7.2	0.4	3.2	G5 IV		62 mx		
117150	240883		13 29 25.1	-51 09 55	0.000	-0.02	5.06	0.07		A2	-2	22 mn		
117246	157962		13 29 29.1	-18 43 44	+0.003	-0.02	6.89	1.38	-0.3	K4 III	+19	270 s		
117247	181690		13 29 30.3	-27 23 21	-0.005	-0.02	7.7	0.2	2.1	A5 V		75 mx		
117193	224179		13 29 35.9	-47 52 32	-0.005	-0.01	7.0	0.4	0.6	A0 V		100 s		
117169	240885		13 29 38.2	-52 45 27	-0.004	-0.01	6.8	1.6	0.2	K0 III		89 s		
117224	224185		13 29 38.6	-41 52 55	-0.002	-0.02	8.0	2.1	-0.3	K5 III		340 s		
117814	7828		13 29 39.3	+74 52 51	-0.008	0.00	7.8	1.1	0.2	K0 III		330 s		
117226	224184		13 29 41.9	-46 11 02	-0.002	-0.02	7.5	1.7	-0.3	K5 III		230 s		
117287	181695		13 29 42.7	-23 16 52	-0.004	+0.01	4.97	1.61		Md	-10		8920	R Hya, m,v
117391	100597		13 29 43.3	+16 54 12	-0.001	0.00	7.4	0.5	3.4	F5 V		63 s		
117170	240889		13 29 44.0	-54 00 12	-0.001	-0.02	7.65	-0.01	-1.6	B3.5 V		600 s		
117362	119959		13 29 47.0	+1 05 43	-0.004	-0.10	6.70	0.1	1.4	A2 V		35 mx		m
117171	240888		13 29 48.5	-57 33 51	+0.008	-0.04	7.3	1.9	0.2	K0 III		76 s		
117111	--		13 29 51.4	-65 30 09			7.6	0.2	-3.5	B1 V pe		1400 s		
117623	16092		13 29 52.8	+64 10 50	-0.006	0.00	7.1	0.6	4.4	G0 V		34 s		
117476	63568		13 29 56.0	+34 31 27	-0.005	+0.01	7.8	0.1	1.7	A3 V		140 mx		
117405	119961		13 29 57.5	+6 00 48	-0.001	+0.05	6.51	0.96		G6	-19			
117587	16091		13 29 59.1	+60 21 14	+0.005	-0.01	8.00	0.9	3.2	G5 IV		91 s		m
117404	119962		13 30 00.0	+7 10 44	0.000	0.00	6.17	1.47		K5	-3			
117434	100600		13 30 01.2	+19 03 42	+0.001	+0.10	7.3	1.1	0.2	K0 III	-5	170 mx		
117543	44630		13 30 12.4	+45 44 07	-0.006	0.00	7.5	1.6	-0.5	M2 III		330 mx		
117419	139367		13 30 14.7	-1 53 43	0.000	-0.03	7.7	0.1	0.6	A0 V		260 s		
117253	240897		13 30 25.0	-58 39 52	-0.005	-0.01	6.77	1.01	0.2	K0 III		200 s		
117436	139370	72 Vir	13 30 25.6	-6 28 13	+0.003	+0.01	6.09	0.33		A5	-9		8924	m
117408	181708		13 30 29.5	-23 39 00	0.000	-0.04	7.89	0.00	0.6	A0 V	+10	290 s	8923	SS Hya, m,q
117382	204540		13 30 38.1	-38 52 59	-0.004	-0.06	7.9	0.3	0.6	A0 V		67 mx		
117498	119971		13 30 39.0	+9 19 00	-0.001	-0.02	7.10	1.1	0.2	K0 III		240 s	8928	m
117452	181713		13 30 50.0	-27 36 41	-0.003	-0.01	7.8	1.3	0.2	K0 III		230 s		
117515	139376		13 30 51.3	-3 02 59	-0.001	+0.03	7.1	1.1	0.2	K0 III		240 s		
117567	82868		13 30 52.1	+24 14 16	0.000	+0.02	7.50	0.4	3.0	F2 V		79 s	8929	m
117501	139377		13 30 54.0	-7 51 48	0.000	0.00	8.0	1.4	-0.3	K5 III		460 s		
117425	204541		13 30 54.7	-39 33 53	-0.003	-0.02	7.7	2.4	-0.3	K5 III		110 s		
117479	157974		13 30 55.1	-19 33 35	-0.005	-0.02	7.1	0.9	3.2	G5 IV		61 s		
117589	82869		13 31 00.8	+26 23 24	+0.002	0.00	7.39	0.29	2.1	A5 V		100 s		
117440	204545		13 31 02.6	-39 24 27	-0.001	-0.02	3.88	1.17	0.3	G8 III	-3	36 s		m
117505	181716		13 31 07.2	-25 39 25	-0.001	-0.03	7.4	-0.1	1.7	A3 V		140 s		
117544	157978		13 31 13.3	-13 26 55	-0.006	+0.01	6.9	0.4	2.6	F0 V		73 s		
117710	44637		13 31 15.7	+42 06 21	-0.009	+0.02	6.08	1.05	-0.1	K2 III	-20	170 s		
117673	63581		13 31 17.1	+36 28 49	-0.001	-0.01	7.54	1.56	-0.5	M2 III	-2	410 s		
117483	224214		13 31 22.7	-42 27 53	+0.004	-0.02	6.70	1.1	0.2	K0 III		200 s		m
117580	139381		13 31 23.2	-9 06 24	+0.001	+0.02	7.9	0.1	1.4	A2 V		200 s		
117962	7833		13 31 28.2	+70 25 22	+0.003	-0.01	7.9	0.2	2.1	A5 V		140 s		
117521	224217		13 31 29.3	-41 35 05	-0.003	+0.01	7.6	1.2	0.2	K0 III		240 s		
117558	181723		13 31 33.2	-28 06 46	-0.005	0.00	6.47	0.10	2.1	A0 IV		170 s		
117399	252342		13 31 33.5	-61 34 55	+0.001	0.00	6.49	0.70	4.0	F8 V		23 s		
117548	204552		13 31 34.1	-32 08 28	-0.007	-0.04	7.6	-0.1	2.6	F0 V		99 s		
118018	7836		13 31 34.7	+71 47 37	-0.007	-0.05	7.9	1.1	-0.1	K2 III		230 mx		
117485	224218		13 31 35.7	-48 09 11	-0.001	-0.02	7.5	1.3	0.2	K0 III		190 s		
117728	63583		13 31 35.9	+37 06 49	-0.005	+0.05	7.6	1.1	0.2	K0 III		220 mx		
117042	--		13 31 36.1	-78 57 08			8.0	1.1	0.2	K0 III		360 s		
117635	139384		13 31 39.9	-2 19 04	-0.056	+0.26	7.36	0.78	5.7	G9 V	-54	22 s		
117696	100606		13 31 42.0	+18 57 59	-0.003	-0.04	7.87	1.12	0.2	K0 III		270 s		

HD	SAO	Star Name	α 2000	δ 2000	μ(α)	μ(δ)	V	B-V	M_v	Spec	RV	d(pc)	ADS	Notes
117582	204558		13^{h}31^{m}48.2^s	-31°03'23"	+0.005^s	-0.07"	7.6	0.7	0.2	K0 III		170 mx		
117730	—		13 31 49.4	+26 36 04	-0.003	-0.01	7.68	1.55	-0.3	K5 III		370 s		
117711	100607		13 31 50.3	+17 07 53	-0.002	-0.01	8.0	0.5	3.4	F5 V		82 s		
117815	44642		13 31 56.0	+47 14 04	+0.003	-0.01	6.9	0.2	2.1	A5 V		92 s		
117432	252345		13 31 57.4	-65 19 19	-0.012	-0.02	7.40	0.1	1.7	A3 V		67 mx		m
117675	139390	74 Vir	13 31 57.8	-6 15 21	-0.007	-0.05	4.69	1.62	-0.5	gM3	+18	110 s		
117697	119985		13 31 58.2	+8 58 38	0.000	+0.02	7.32	0.80	4.0	F8 V	+9	31 s		
117535	240925		13 32 02.2	-51 50 18	-0.002	-0.01	7.4	1.4	-0.1	K2 III		240 s		
117661	157987	73 Vir	13 32 02.7	-18 43 44	-0.007	-0.02	6.01	0.18	1.7	A3 V		59 mx		m
117460	252347		13 32 03.7	-63 02 15	0.000	-0.01	7.70	0.1		B1				
117597	204561		13 32 05.2	-38 23 57	+0.003	-0.05	6.16	1.04	0.2	gK0		150 s		
117598	224224		13 32 08.9	-42 42 11	-0.002	-0.01	7.7	0.4	0.4	B9.5 V		150 s		
117076	257055		13 32 11.0	-79 39 37	-0.008	+0.01	6.7	1.4	-0.1	K2 III		160 s		
117846	63593		13 32 21.8	+36 49 07	+0.001	-0.01	6.81	0.75	0.3	G8 III	-25	200 s	8934	m
117733	157992		13 32 24.6	-12 39 48	-0.004	-0.01	7.90	0.4	2.6	F0 V		120 s		m
117680	204569		13 32 24.7	-32 03 29	+0.001	-0.03	7.7	1.4	-0.1	K2 III		270 s		
117618	224228		13 32 25.4	-47 16 18	+0.002	-0.14	7.2	0.9	4.4	G0 V		24 s		
117444	—		13 32 26.9	-69 10 01			7.7	0.0	0.4	B9.5 V		280 s		
117933	28792		13 32 30.1	+53 57 43	+0.001	-0.11	8.0	1.1	0.2	K0 III		160 mx		
117858	63594		13 32 30.9	+36 02 06	+0.007	-0.29	7.9	0.7	4.4	G0 V		51 s		
117540	252356		13 32 31.9	-62 20 39	-0.002	-0.01	7.20	0.00	0.4	B9.5 V		220 s		m
117445	252353		13 32 32.5	-69 14 01	-0.001	-0.07	7.20	0.4	3.0	F2 V		69 s		m
117718	181735		13 32 34.4	-29 33 55	-0.005	0.00	6.45	0.44	2.8	F5 IV-V		54 s		
117716	181737		13 32 35.8	-28 41 34	-0.008	-0.03	5.69	0.04	1.2	A1 V n		52 mx		
117817	100611		13 32 36.2	+14 23 38	-0.006	+0.01	7.1	0.4	2.6	F0 V		80 s		
117665	224230		13 32 38.9	-44 27 01	-0.005	-0.03	7.3	0.4	0.6	A0 V		94 mx		
117155	257057		13 32 40.8	-79 04 29	-0.009	0.00	7.4	1.4	-0.3	K5 III		350 s		
118327	7843		13 32 47.3	+76 03 44	+0.005	-0.03	7.3	0.9	3.2	G5 IV		65 s		
117893	63597		13 32 47.9	+30 44 55	+0.002	+0.02	7.5	1.1	0.2	K0 III		290 s		
117876	82875		13 32 48.0	+24 20 47	+0.004	-0.21	6.11	0.96	0.2	K0 III	+6	83 mx	8937	m
117735	204574		13 32 50.1	-36 51 14	+0.001	+0.03	7.0	1.3	-0.1	K2 III		230 s		
117902	63599		13 32 50.8	+34 54 25	-0.005	-0.02	6.78	0.20		A3	-25		8939	m
117789	157998	75 Vir	13 32 51.6	-15 21 47	-0.005	-0.01	5.55	1.23	0.2	K0 III	-40	79 s		m
117781	181742		13 32 53.9	-26 35 15	-0.003	-0.01	6.9	0.4	1.7	A3 V		76 s		
117832	139402		13 32 57.0	-2 25 28	-0.003	-0.02	7.90	1.1	0.2	K0 III		350 s	8938	m
117818	139401	76 Vir	13 32 58.0	-10 09 54	-0.002	-0.04	5.21	0.96	0.2	K0 III	-1	100 s		
117833	139403		13 33 00.6	-7 11 42	+0.003	-0.01	5.9	1.6		M5 III	+10			S Vir, v
117736	204576		13 33 00.6	-39 58 20	-0.006	-0.03	7.9	0.0	2.6	F0 V		110 s		
117609	252358		13 33 06.2	-63 39 49	+0.005	+0.04	7.0	0.6	4.4	G0 V		33 s		
117877	119993		13 33 09.2	+5 51 12	+0.003	+0.03	6.8	0.9	3.2	G5 IV		53 s		
117819	158003		13 33 11.2	-19 59 24	-0.003	-0.02	7.0	0.9	3.2	G5 IV		49 s		
117860	139405		13 33 11.2	-8 26 36	-0.017	+0.05	7.2	0.6	4.4	G0 V		37 s		
118019	44653		13 33 13.0	+47 45 33	+0.001	-0.02	8.0	1.1	0.2	K0 III		360 s		
117360	257060		13 33 15.0	-77 34 05	-0.102	-0.11	6.40	0.51		F5	-41	12 mn		S Cha (invar), m
117981	63605		13 33 22.5	+33 39 27	-0.003	-0.02	6.7	1.1	0.2	K0 III		200 s		
117806	204583		13 33 24.1	-38 52 17	-0.002	0.00	7.4	0.4	0.6	A0 V		120 s		
117878	139407		13 33 24.4	-7 37 21	-0.005	0.00	7.1	0.4	2.6	F0 V	-17	80 s		
118020	44656		13 33 26.4	+39 56 01	-0.005	+0.02	7.80	0.4	2.6	F0 V		110 s		m
117905	139408		13 33 27.4	-8 26 07	+0.001	0.00	7.7	0.8	3.2	G5 IV		79 s		
117807	224243		13 33 29.0	-43 25 23	-0.012	-0.03	6.7	0.4	3.4	F5 V		46 s		
117936	119997		13 33 32.4	+8 35 13	-0.033	+0.10	7.99	1.00	0.2	K0 III		39 mx		
117670	—		13 33 34.2	-64 45 57			7.7	0.5	4.0	F8 V		54 s		
117651	252361		13 33 35.7	-65 37 57	-0.007	-0.02	6.37	-0.02		A0				
118178	16108		13 33 39.0	+60 47 15	0.000	+0.02	8.0	0.9	3.2	G5 IV		93 s		
117852	224245		13 33 41.0	-41 19 06	-0.003	0.00	7.50	0.1	1.4	A2 V		170 s		m
118202	16109		13 33 45.4	+60 15 45	-0.008	-0.01	6.9	1.1	0.2	K0 III		220 s		
117851	204597		13 33 47.6	-39 56 45	-0.005	-0.01	6.8	0.6	0.6	A0 V		72 s		
117774	240943		13 33 48.1	-58 13 43	-0.001	-0.02	7.8	1.0	4.0	F8 V		31 s		
117653	—		13 33 48.6	-68 30 57			7.9	0.8	3.2	G5 IV		85 s		
118051	63609		13 33 49.7	+32 20 13	0.000	-0.14	7.80	0.69		G0				
117694	252366		13 33 49.9	-66 04 43	-0.002	-0.02	6.7	0.0	0.4	B9.5 V		180 s		
118095	44660		13 33 51.6	+45 36 00	+0.007	+0.05	7.6	0.4	2.6	F0 V		99 s		
117565	—		13 34 04.0	-74 33 58			7.9	1.1	0.2	K0 III		340 s		
118214	28803	81 UMa	13 34 07.2	+55 20 54	-0.003	-0.01	5.60	-0.03		A0 p	-9	27 mn		
118022	120004	78 Vir	13 34 07.8	+3 39 32	+0.003	-0.03	4.94	0.03		A2 p	-12	26 mn		
118555	7846		13 34 09.5	+75 47 47	-0.001	-0.04	7.7	0.1	0.6	A0 V		260 s		
118066	100623		13 34 11.9	+18 24 57	-0.002	0.00	8.0	0.4	2.6	F0 V		120 s		
117927	204603		13 34 15.6	-34 23 16	+0.004	-0.02	7.7	0.8	0.2	K0 III		310 s		
118036	139416		13 34 16.0	-0 18 50	-0.016	+0.02	7.36	0.91	0.2	K0 III	-2	87 mx	8949	m
118067	100624		13 34 17.4	+14 13 56	-0.004	-0.04	7.9	1.1	0.2	K0 III		180 mx		
118024	139415		13 34 17.6	-8 37 07	-0.006	-0.02	7.90	0.5	4.0	F8 V		60 s	8950	m
118008	158011		13 34 17.8	-11 32 15	-0.005	-0.06	7.90	0.4	3.0	F2 V		96 s	8947	m
118068	100625		13 34 20.8	+14 06 39	+0.002	-0.03	7.20	1.11	0.2	K0 III		210 s		
118156	63616		13 34 21.7	+38 47 21	+0.001	+0.02	6.37	0.21	2.4	A8 V	-9	62 s	8956	m
—	44667		13 34 24.7	+46 24 28	-0.001	+0.01	8.0	1.5						

HD	SAO	Star Name	α 2000	δ 2000	μ(α)	μ(δ)	V	B-V	M_V	Spec	RV	d(pc)	ADS	Notes
117854	240951		13^{h}34^{m}25.3^s	−59°30′45″	−0$.^s$036	−0$.^{\prime\prime}$20	7.57	0.64		G5		15 mn		
118232	44668	24 CVn	13 34 27.2	+49 00 57	−0.013	+0.02	4.70	0.12	1.9	A4 V	−12	36 ts		
118026	158014		13 34 28.1	−13 20 48	−0.001	0.00	7.4	0.1	1.7	A3 V		140 s		
117919	224254		13 34 28.8	−48 16 20	−0.003	−0.02	6.33	−0.05		A0				m
117939	204606		13 34 32.5	−38 54 26	+0.038	−0.40	7.28	0.67	5.0	G4 V	+87	28 s		
117929	224256		13 34 37.2	−49 52 01	−0.005	−0.06	7.6	1.2	4.0	F8 V		29 s		
117973	224258		13 34 38.5	−42 17 06	−0.001	0.00	7.3	1.6	−0.3	K5 III		210 s		
118054	158021		13 34 40.4	−13 12 52	−0.003	−0.02	5.91	0.02		A1	−20		8954	m
118098	139420	79 ζ Vir	13 34 41.5	−0 35 46	−0.019	+0.04	3.37	0.11	1.7	A3 V	−13	23 ts		
118082	139418		13 34 41.9	−9 06 25	0.000	−0.02	7.9	1.1	0.2	K0 III		340 s		
118686	7848		13 34 42.6	+76 32 47	−0.008	−0.01	6.6	1.4	−0.3	K5 III	−14	240 s		
118010	204612		13 34 43.2	−33 18 38	−0.007	−0.01	6.4	1.5	−0.1	K2 III		130 s		
117856	252371		13 34 43.2	−63 20 06	−0.003	0.00	7.35	0.22	−5.8	B0 Ib		2300 s		
118128	120015		13 34 44.3	+4 54 40	−0.004	−0.03	7.9	1.1	−0.1	K2 III		180 mx		
118216	63623		13 34 47.7	+37 10 56	+0.007	−0.01	4.98	0.40	1.9	F2 IV	+7	39 s		
117941	240961		13 34 49.7	−50 47 34	−0.003	−0.04	7.4	1.0	0.2	K0 III		270 s		
118388	16114		13 34 50.3	+65 15 18	+0.006	−0.05	7.70	0.4	3.0	F2 V		87 s		m
118524	7847		13 34 56.3	+70 07 07	−0.006	−0.01	7.3	0.1	0.6	A0 V		100 mx		
118180	100627		13 35 01.6	+12 30 51	−0.002	0.00	6.56	1.02	0.2	K0 III		180 s		
118234	82886		13 35 08.0	+20 46 54	−0.009	−0.05	7.57	1.08	0.2	K0 III		100 mx		
117923	252375		13 35 11.1	−62 37 51	0.000	−0.03	6.6	1.9	0.2	K0 III		55 s		
118244	82888		13 35 11.4	+22 29 58	−0.019	+0.11	7.0	0.5	3.4	F5 V		53 s		
118084	204623		13 35 12.3	−32 12 44	−0.004	−0.03	6.8	0.7	2.6	F0 V		40 s		
118295	44675		13 35 14.0	+44 11 49	−0.002	+0.01	6.84	0.20	2.1	A5 V	−26	81 s		
117947	240970		13 35 14.4	−60 09 28	+0.002	−0.03	7.9	1.3	2.1	A5 V		31 s		
118245	100629		13 35 21.1	+14 53 33	−0.002	−0.04	7.6	0.4	3.0	F2 V		82 s		
118264	82890		13 35 24.4	+22 48 19	−0.003	−0.01	7.97	1.02	3.2	G5 IV		59 s		
118145	181774		13 35 25.0	−22 37 07	−0.002	+0.01	7.31	1.66	−0.5	M2 III		340 s		
118296	63630		13 35 28.4	+34 58 40	+0.001	−0.02	7.8	1.1	0.2	K0 III		330 s		
118131	204630		13 35 28.7	−34 01 06	−0.006	−0.05	7.7	1.2	0.2	K0 III		180 mx		
118015	240976		13 35 29.2	−57 17 29	+0.002	−0.03	7.3	1.6	−0.1	K2 III		170 s		
118219	139428	80 Vir	13 35 31.2	−5 23 46	+0.001	+0.08	5.73	0.95	0.3	G6 III	−8	120 s		
116632	258671		13 35 31.4	−85 49 29	−0.015	−0.02	7.2	−0.1	0.4	B9.5 V		220 s		
118187	181777		13 35 31.7	−22 01 28	−0.009	−0.04	7.11	0.48	4.0	F8 V		42 s		
118266	100630		13 35 33.2	+10 12 17	+0.005	−0.06	6.49	1.03	0.0	K1 III	+33	200 s		m
118163	204632		13 35 38.2	−32 48 00	0.000	+0.02	8.0	2.1	−0.3	K5 III		190 s		
117960	252377		13 35 38.5	−65 09 08	−0.011	−0.02	7.8	0.5	3.4	F5 V		74 s		
118016	252378		13 35 40.3	−61 52 02	−0.001	0.00	8.0	1.0	−1.6	B3.5 V		180 s		
117979	−−		13 35 45.4	−65 19 50			7.7	0.0	0.0	B8.5 V		330 s		
118190	204633		13 35 46.3	−32 22 25	+0.001	−0.02	7.5	1.3	−0.1	K2 III		290 s		
118106	224268		13 35 46.6	−49 11 39	−0.002	−0.02	7.5	1.7	−0.3	K5 III		280 s		
118063	−−		13 35 50.7	−59 11 48			8.0	2.4	−0.1	K2 III		76 s		
118289	120026		13 35 51.9	+8 17 35	−0.003	+0.01	7.0	1.6	−0.5	M4 III	+23	320 s		
118017	−−		13 35 55.5	−64 54 49			7.76	0.96	0.2	K0 III		330 s		
117949	−−		13 35 58.4	−69 33 50			8.0	0.0	0.4	B9.5 V		320 s		
117872	257066		13 35 58.6	−73 19 39	−0.007	−0.01	6.6	0.0	0.4	B9.5 V		170 s		
118135	240986		13 36 05.5	−54 56 21	0.000	−0.01	7.3	1.9	−0.3	K5 III		190 s		
118150	240988		13 36 12.6	−55 15 28	−0.001	−0.03	8.0	1.4	−0.1	K2 III		310 s		
118330	139437		13 36 14.5	−0 55 52	−0.015	−0.06	7.06	0.53	4.0	F8 V	+17	41 s		
118252	224271		13 36 25.9	−42 24 57	0.000	−0.02	7.1	0.1	0.6	A0 V		160 s		
118137	252382		13 36 33.2	−62 20 41	−0.003	0.00	7.3	1.9	4.4	G0 V		38 s		
118575	28813		13 36 33.5	+52 41 14	0.000	−0.04	7.0	1.4	−0.3	K5 III		290 s		
118253	224272		13 36 35.1	−47 17 54	−0.016	−0.03	7.7	0.8	0.2	K0 III		110 mx		
118557	28814		13 36 36.1	+51 17 39	−0.003	−0.01	7.4	1.1	0.2	K0 III		280 s		
118196	240993		13 36 39.1	−58 29 56	0.000	−0.04	7.8	0.3	4.0	F8 V		57 s		
118536	44682		13 36 39.7	+49 29 12	−0.001	−0.02	6.5	1.1	0.0	K1 III	−10	200 s		
118936	7852		13 36 40.3	+75 54 28	+0.066	−0.16	8.0	0.9	3.2	G5 IV	−42	57 mx		
119197	7856		13 36 46.5	+79 21 12	−0.007	−0.01	7.5	1.4	−0.3	K5 III		360 s		
118349	181790		13 36 48.3	−26 29 42	−0.007	+0.02	5.40	0.23	0.5	A7 III	−10	93 s	8966	m
118225	240995		13 36 49.0	−57 47 26	−0.003	−0.02	7.4	1.7	−0.1	K2 III		150 s		
118319	204653		13 36 50.4	−34 28 03	−0.001	−0.03	6.50	1.03	3.2	G5 IV		29 s		
118258	240999		13 36 54.8	−56 09 24	−0.011	−0.05	7.97	0.85	5.3	G6 V		28 s		m
117910	257067		13 36 58.3	−76 30 29	−0.031	+0.03	7.8	0.9	4.0	F8 V		33 s		
118508	82905		13 36 59.0	+24 36 48	−0.002	−0.01	5.74	1.55	−0.5	M2 III	−31	180 s		
118335	204656		13 37 00.3	−37 06 01	0.000	−0.04	7.4	0.5	0.6	A0 V		110 s		
118788	16124		13 37 02.7	+67 02 09	+0.010	−0.06	6.8	0.9	3.2	G5 IV	−1	52 s		
118411	158043		13 37 03.0	−16 26 50	−0.002	−0.03	8.00	0.4	3.0	F2 V		100 s		m
118377	181795		13 37 03.1	−21 51 14	−0.006	0.00	8.0	0.8	2.6	F0 V		57 s		
118338	224275		13 37 05.9	−44 08 35	−0.005	−0.02	5.98	0.94		K0				
118337	224278		13 37 08.7	−42 46 53	0.000	−0.02	7.2	1.7	−0.3	K5 III		240 s		
118412	181797		13 37 10.6	−23 36 58	−0.003	−0.06	7.14	0.27	1.7	A3 V		73 mx		TV Hya, q
118904	7854		13 37 10.8	+71 14 32	−0.009	−0.01	5.51	1.20	0.2	K0 III	+15	87 s		
118261	252387		13 37 12.2	−61 41 30	+0.020	−0.12	5.63	0.50		F5	+40			m
118668	28818		13 37 12.9	+52 35 34	−0.003	−0.02	6.8	1.4	−0.3	K5 III		270 s		
118415	181799		13 37 15.6	−28 41 32	−0.002	+0.01	7.7	1.1	0.2	K0 III		290 s		

HD	SAO	Star Name	α 2000	δ 2000	μ(α)	μ(δ)	V	B-V	M$_v$	Spec	RV	d(pc)	ADS	Notes
118379	224281		13^{h}37^{m}17.8^s	-40°53'53"	-0.001^s	-0.04"	7.6	-0.2	1.4	A2 V		180 s		
118354	224283		13 37 23.6	-46 25 42	-0.001	-0.04	5.90	-0.12		B8				
118242	252388		13 37 25.4	-64 56 11	-0.002	-0.03	7.55	-0.04	-0.2	B8 V		350 s		m
118623	63648	25 CVn	13 37 27.6	+36 17 41	-0.008	+0.02	4.82	0.23	0.5	A7 III	-6	73 s	8974	m
118322	––		13 37 31.6	-56 28 42			7.00	2.0		N3				RV Cen, v
118418	204665		13 37 32.0	-36 34 21	+0.002	-0.02	7.3	0.4	0.6	A0 V		120 s		
118434	204667		13 37 34.1	-35 02 56	-0.005	-0.02	7.26	0.09	0.6	A0 V		81 mx		
118511	139447		13 37 35.2	-7 52 17	-0.001	0.00	7.90	1.1	0.2	K0 III	-5	350 s	8972	m
118643	63649		13 37 38.2	+33 44 17	-0.001	-0.02	7.19	1.35	-2.3	K3 II	-4	790 s		
118669	44687		13 37 38.3	+42 12 03	-0.004	-0.01	7.59	1.52	-0.5	M2 III		350 mx		
118741	28819		13 37 42.9	+50 42 52	-0.002	0.00	6.80	1.6	-1.4	M2 II-III	-48	450 s	8979	m
118578	120042		13 37 43.9	+2 22 55	-0.001	-0.03	6.7	1.1	0.2	K0 III		200 s	8975	m
118559	139449		13 37 46.9	-3 14 10	0.000	-0.03	6.7	1.1	0.2	K0 III		200 s		
118465	204673		13 37 47.9	-35 03 51	-0.006	-0.01	7.20	0.69	4.4	G0 V		30 s		m
118230	––		13 37 49.3	-70 24 44			8.0	0.0	0.4	B9.5 V		320 s		
118468	224289		13 37 51.8	-42 30 53	-0.004	-0.02	7.8	-0.1	0.6	A0 V		100 mx		
118467	204675		13 37 52.2	-38 53 49	-0.002	-0.02	7.5	0.2	2.6	F0 V		95 s		
118421	241012		13 37 56.3	-50 36 06	-0.002	-0.06	7.1	1.8	-0.5	M2 III		230 mx		
118742	63654		13 37 57.2	+39 10 29	-0.019	-0.16	7.78	0.67	5.2	G5 V	-21	33 s	8981	m
118670	82915		13 38 00.8	+22 31 38	-0.010	-0.09	7.02	0.94	3.2	G5 IV		46 s		
118582	158054		13 38 01.6	-14 35 32	+0.003	+0.01	7.9	1.4	-0.3	K5 III		430 s		
118483	224292		13 38 03.0	-42 45 54	-0.003	-0.01	7.0	1.6	-0.3	K5 III		250 s		
118937	16129		13 38 05.4	+66 36 20	+0.002	-0.02	7.8	0.1	1.4	A2 V		190 s		
118384	241013		13 38 07.6	-58 24 54	0.000	-0.04	6.42	1.12	0.2	gK0		150 s		m
118660	100644		13 38 07.8	+14 18 06	+0.002	-0.01	6.52	0.25	2.6	F0 V	-2	61 s		
118450	241018		13 38 08.9	-50 20 57	-0.007	-0.01	6.65	-0.08		B8				
118600	181821		13 38 23.4	-29 50 19	-0.001	+0.02	7.0	1.2	2.1	A5 V		23 s		
118385	252392		13 38 28.3	-64 03 19	-0.002	+0.02	7.39	0.14	0.4	B9.5 V		200 s		
118646	181825		13 38 41.9	-29 33 39	-0.006	-0.07	5.83	0.40	2.9	F6 IV-V		38 s		
118588	224303		13 38 42.8	-44 31 00	-0.002	-0.04	7.6	0.0	0.6	A0 V		240 s		
118744	100646		13 38 43.2	+9 58 52	0.000	+0.01	7.9	1.1	0.2	K0 III		340 s		
118590	224302		13 38 45.0	-46 58 24	0.000	-0.02	7.9	1.0	-0.1	K2 III		390 s		
118344	257070		13 38 45.4	-70 26 42	-0.015	-0.05	6.10	1.42		K2				
118520	241026		13 38 49.0	-57 37 23	0.000	-0.02	6.01	1.14		K0				m
118566	224304		13 38 49.3	-49 29 44	-0.004	-0.04	7.6	0.9	-0.1	K2 III		310 mx		
118648	204692		13 38 50.2	-33 06 41	-0.002	-0.02	6.99	-0.04	0.4	B9.5 V		210 s		
118548	241032		13 38 54.1	-55 01 29	-0.003	-0.04	7.2	0.6	0.6	A0 V		85 s		
118760	139465		13 38 54.3	-2 31 50	-0.002	+0.02	7.9	1.4	-0.3	K5 III		430 s		
118824	82924		13 38 57.0	+24 07 53	+0.001	-0.03	7.8	0.4	2.6	F0 V		110 s		
118792	120052		13 38 57.8	+5 33 46	-0.004	-0.01	8.0	0.1	0.6	A0 V		89 mx		
118473	––		13 38 57.9	-64 42 38			7.5	0.0	0.4	B9.5 V		260 s		
118603	241033		13 38 58.8	-52 12 27	-0.005	-0.02	7.7	0.6	2.1	A5 V		76 s		
118709	181831		13 39 01.5	-24 51 45	-0.003	-0.02	7.6	0.4	0.2	K0 III		300 s		
118839	100650		13 39 02.2	+18 15 55	-0.003	-0.02	6.48	1.20	0.2	K0 III	-11	120 s		
118285	257069		13 39 12.0	-75 41 01	-0.004	-0.01	6.34	0.01		A0				
118954	44695		13 39 12.5	+47 53 46	-0.003	-0.01	7.9	0.2	2.1	A5 V	-5	150 s		
118840	100651		13 39 13.8	+10 30 36	+0.001	-0.04	6.7	1.4	-0.3	K5 III	-7	250 s		
118651	224311		13 39 14.1	-48 59 51	-0.012	-0.01	7.90	1.1	0.2	K0 III		160 mx		m
118970	28831		13 39 14.7	+51 48 14	-0.003	0.00	6.5	1.1	-0.1	K2 III		210 s		
118796	158069		13 39 17.8	-12 05 32	+0.001	-0.06	7.8	0.8	3.2	G5 IV		85 s		
118778	158070		13 39 19.8	-15 12 30	-0.003	-0.01	7.5	0.8	3.2	G5 IV		74 s		
118475	252395		13 39 20.1	-67 40 17	+0.049	-0.21	6.97	0.63		G0		13 mn		
118905	82928		13 39 25.2	+26 41 03	0.000	-0.06	7.16	1.03	0.0	K1 III		230 mx		
118728	204703		13 39 25.7	-36 14 29	0.000	-0.03	7.6	0.5	0.6	A0 V		130 s		
119024	28832	82 UMa	13 39 30.3	+52 55 16	-0.016	+0.06	5.46	0.10		A2	-18	25 mn		
118826	158072		13 39 33.7	-15 36 58	-0.009	-0.04	7.9	0.5	4.0	F8 V		60 s		
118865	120055		13 39 34.2	+1 05 17	-0.007	-0.07	8.0	0.5	4.0	F8 V		62 s		
118841	139469		13 39 34.4	-10 06 48	0.000	+0.02	7.9	1.1	0.2	K0 III		340 s		
118889	100654		13 39 34.5	+10 44 46	-0.008	-0.02	5.57	0.33	2.2	dA6	-18	41 s	8987	m
118781	204708		13 39 40.7	-39 44 53	-0.002	-0.03	6.3	1.6	-0.5	M4 III		220 s		
118780	204710		13 39 44.8	-37 46 01	-0.003	-0.02	8.0	1.3	-0.1	K2 III		360 s		
118697	241045		13 39 45.6	-54 08 52	-0.004	-0.01	7.35	0.00	0.6	A0 V		220 s		
118915	100655		13 39 46.3	+12 46 40	-0.006	-0.01	8.0	0.5	3.4	F5 V		82 s		
118715	241046		13 39 47.9	-52 41 16	0.000	-0.01	8.00	1.1	-0.1	K2 III		390 s		m
118799	204712		13 39 48.5	-40 03 06	-0.004	-0.07	5.6	1.7	0.2	K0 III		49 s		
118753	224315		13 39 49.6	-48 26 02	-0.005	-0.04	7.4	1.5	0.2	K0 III		270 s		m
118716	241047	ε Cen	13 39 53.2	-53 27 58	-0.002	-0.02	2.30	-0.22	-3.5	B1 V	+6	150 s		
117695	258676		13 39 57.9	-83 54 30	+0.068	+0.01	7.4	-0.1	3.0	F2 V		77 s		
118876	158076		13 39 58.4	-16 26 48	0.000	-0.02	6.8	1.1	-0.1	K2 III		240 s		
118767	224317		13 39 59.6	-49 57 00	-0.012	+0.01	5.5	1.6	-0.5	M4 III		160 s		
118971	82934		13 40 00.5	+25 55 36	0.000	+0.02	7.83	0.92	0.3	G8 III	+31	320 s		
118522	257074		13 40 00.6	-70 47 18	-0.003	-0.03	6.59	1.30	0.2	gK0		120 s		
118877	181843		13 40 08.7	-27 34 40	-0.003	-0.01	7.0	1.1	0.2	K0 III		200 s		
118666	252400		13 40 10.9	-64 34 37	-0.008	-0.02	5.79	0.39	0.6	F0 III		100 s		
119146	28835		13 40 11.5	+53 06 37	-0.004	-0.01	7.9	0.1	1.7	A3 V		99 mx		

HD	SAO	Star Name	α 2000	δ 2000	μ(α)	μ(δ)	V	B-V	M_v	Spec	RV	d(pc)	ADS	Notes
119035	63676		13^{h}40^{m}15.5^s	+31°00′42″	-0.006	+0.08	6.21	0.96	-2.1	G5 II	-18	460 s		
--	63677		13 40 16.2	+37 00 09	+0.004	-0.06	8.0	2.3						
118801	241054		13 40 16.2	-50 32 20	0.000	-0.01	7.5	1.4	-0.1	K2 III		320 s		
118769	241049		13 40 18.5	-57 36 49	-0.001	-0.02	5.7	0.9	3.2	G5 IV		32 s		XX Cen, v
119111	44704		13 40 20.4	+44 59 12	0.000	-0.03	8.0	1.1	-0.1	K2 III		410 s		
119007	82940		13 40 21.1	+22 13 56	+0.001	-0.01	7.68	1.45	-0.1	K2 III		210 s		
119213	28838		13 40 21.2	+57 12 27	-0.007	+0.02	6.29	0.10		A2	0			
119124	28836		13 40 23.1	+50 31 09	-0.014	+0.06	6.40	0.5	4.0	F8 V	-10	24 ts	8992	m
118770	241053		13 40 23.7	-57 55 32	-0.001	0.00	7.6	0.2	2.1	A5 V		130 s		
119169	28837		13 40 24.5	+53 11 29	-0.004	-0.02	7.1	1.1	-0.1	K2 III		270 s		
116877	258672		13 40 26.2	-86 43 29	-0.024	0.00	8.0	1.1	0.2	K0 III		320 s		
118816	241055		13 40 26.6	-53 51 37	-0.004	-0.01	7.82	-0.01	0.4	B9.5 V		130 mx		
118771	252405		13 40 27.3	-60 16 10	-0.023	+0.03	7.4	1.8	0.2	K0 III		92 s		
119054	63680		13 40 29.5	+32 50 08	+0.001	-0.02	7.9	0.5	3.4	F5 V	0	78 s		
118894	204720		13 40 30.7	-34 21 30	0.000	-0.01	7.37	0.78	4.4	G0 V		27 s		
119170	44707		13 40 32.9	+46 16 52	-0.008	+0.02	7.4	0.5	4.0	F8 V		49 s		
118878	224322		13 40 37.6	-44 19 49	-0.002	-0.02	6.8	0.4	0.6	A0 V		100 s		
119081	82944		13 40 39.0	+28 03 55	-0.005	+0.01	6.23	1.28	-0.2	K3 III	-63	190 s		m
119055	82942	1 Boo	13 40 40.4	+19 57 20	-0.003	+0.02	5.75	0.01	1.2	A1 V	-25	81 s	8991	m
119056	100658		13 40 40.6	+19 12 00	-0.008	-0.08	7.6	0.5	4.0	F8 V		53 s		
119082	82943		13 40 40.8	+20 00 48	-0.003	+0.01	7.39	0.24		A				
119702	7867		13 40 40.9	+76 50 37	-0.013	0.00	6.70	0.2	2.1	A5 V		83 s	8997	m
118897	224325		13 40 41.9	-44 15 40	-0.005	+0.01	8.0	0.8	0.2	K0 III		360 s		
119214	28842		13 40 43.2	+52 46 19	+0.001	-0.01	6.9	1.1	0.2	K0 III		220 s		
119228	28843	83 UMa	13 40 44.1	+54 40 54	-0.003	-0.01	4.66	1.64	-0.5	M2 III	-17	100 s		
119058	100659		13 40 45.9	+14 08 06	0.000	-0.02	7.7	0.5	3.4	F5 V		74 s		
118984	158084		13 40 50.6	-17 47 33	-0.003	-0.01	7.8	0.5	3.4	F5 V		74 s		
118758	252407		13 40 52.2	-66 37 03	+0.001	-0.04	7.7	1.1	0.2	K0 III		300 s		
117374	258674	κ Oct	13 40 56.0	-85 47 09	-0.071	-0.02	5.58	0.18		A2	-9			
118945	204724		13 40 57.1	-39 43 03	0.000	0.00	7.8	1.2	-0.3	K5 III		410 s		
119009	158085		13 40 59.7	-12 46 58	+0.001	-0.02	7.2	1.1	0.2	K0 III		250 s		
118986	181854		13 41 01.6	-29 43 51	0.000	0.00	7.7	1.4	3.2	G5 IV		33 s		
119126	82946	2 Boo	13 41 02.2	+22 29 45	-0.001	-0.03	5.62	1.01	0.2	G9 III	+5	120 s		
118972	204725		13 41 04.1	-34 27 50	+0.016	-0.17	6.93	0.86	3.2	G5 V		48 s		
118846	252410		13 41 06.9	-62 29 56	-0.002	0.00	7.3	1.9	-0.1	K2 III		110 s		
118870	241067		13 41 09.6	-59 13 39	-0.004	0.00	8.0	-0.2	0.6	A0 V		180 mx		
118932	224331		13 41 11.7	-48 19 07	-0.001	-0.01	7.7	1.7	-0.5	M2 III		440 s		
118962	224333		13 41 13.4	-46 01 44	-0.003	-0.05	7.3	0.6	0.2	K0 III		260 s		
119095	158088		13 41 22.5	-10 27 08	-0.001	+0.03	8.0	1.6	-0.5	M2 III		510 s		
119062	181860		13 41 27.1	-26 31 53	0.000	-0.01	7.3	1.4	-0.1	K2 III		230 s		
119063	181861		13 41 29.2	-27 13 57	-0.003	-0.01	7.8	0.9	3.2	G5 IV		70 s		
119476	16142		13 41 29.7	+64 49 20	+0.008	-0.02	5.7	0.1	0.6	A0 V	-5	100 s		
119086	181863		13 41 30.8	-23 26 59	-0.001	0.00	6.59	0.06		A0			8994	m
118998	224338		13 41 36.3	-45 29 37	+0.001	-0.03	6.9	0.7	-0.1	K2 III		250 s		
119149	139490	82 Vir	13 41 36.7	-8 42 11	-0.007	+0.03	5.01	1.63	-0.5	M2 III	-37	120 s		
118667	257078		13 41 39.1	-75 06 53	-0.006	+0.02	6.70	0.1	0.6	A0 V		170 s		m
120103	2262		13 41 39.5	+80 12 13	-0.014	+0.01	7.1	1.4	-0.3	K5 III		310 s		
119066	204736		13 41 43.8	-39 00 54	-0.001	-0.02	8.0	1.4	-0.1	K2 III		290 s		
119133	158092		13 41 43.9	-19 08 10	-0.003	0.00	7.9	1.6	-0.5	M4 III		490 s		RY Vir, v
118991	241076		13 41 44.6	-54 33 36	-0.005	-0.05	5.01	-0.05	-0.2	B8 V	+2	110 s		
119090	204739		13 41 45.6	-33 35 48	-0.001	+0.02	5.5	1.6		M5 III	+28			T Cen, v
119370	44717		13 41 51.6	+47 24 50	-0.003	-0.03	7.9	1.1	0.2	K0 III		340 s		
119152	158096		13 41 52.7	-18 59 05	-0.004	-0.01	7.8	0.4	2.6	F0 V		110 s		
119820	7872		13 41 58.4	+74 37 14	0.000	0.00	7.8	1.1	0.2	K0 III		340 s		
118978	241080		13 42 01.0	-58 47 13	-0.004	-0.01	5.38	-0.03	-0.3	B9 IV	-30	140 s		
119103	224349		13 42 01.1	-43 08 48	-0.003	-0.01	7.16	-0.06	-1.2	B8 III		410 s		
--	63692		13 42 05.2	+37 13 28	+0.003	-0.02	8.0	1.9						
119581	16143		13 42 06.3	+65 16 24	+0.013	+0.01	6.6	0.1	1.4	A2 V		110 s		
119288	120075		13 42 12.6	+8 23 18	-0.026	-0.09	6.16	0.42	3.3	dF4	-11	37 s		
119191	204750		13 42 20.7	-33 58 45	-0.002	+0.06	6.67	0.42	3.4	F5 V		45 s		m
119290	120076		13 42 21.3	+0 59 49	0.000	-0.02	7.8	1.6	-0.5	M2 III	-8	470 s		
119046	241086		13 42 22.2	-57 19 25	-0.003	-0.02	8.00	0.1	1.4	A2 V		200 s		m
120565	2266		13 42 23.0	+82 45 08	+0.017	-0.05	5.98	1.01	3.2	G5 IV	-50	26 s		
118913	252417		13 42 26.8	-69 14 48	-0.008	-0.01	7.4	0.1	0.6	A0 V		120 mx		
119445	44720		13 42 28.7	+41 40 26	-0.008	0.00	6.30	0.86	0.3	G6 III	-33	130 mx		
119549	28858		13 42 30.8	+55 43 40	-0.004	+0.01	6.9	1.1	0.2	K0 III		220 s		
118925	--		13 42 32.2	-70 25 28			7.8	1.1	0.2	K0 III		320 s		
119350	100672		13 42 32.7	+15 08 50	+0.001	-0.02	7.25	1.30	-0.1	K2 III		250 s		
119073	241091		13 42 33.1	-57 51 09	+0.002	-0.01	8.0	0.4	-0.1	K2 III		410 s		
119392	82957		13 42 34.2	+23 19 08	+0.003	-0.05	7.33	1.55	-0.5	M2 III	-36	370 s		
119292	158105		13 42 35.7	-12 05 12	-0.005	-0.01	7.4	0.5	3.4	F5 V		64 s		
120084	7876		13 42 38.9	+78 03 51	-0.023	+0.04	5.91	1.01	0.2	K0 III	-7	140 s		
--	252421		13 42 40.4	-61 44 16	-0.002	-0.01	6.80	0.27	-6.5	B8 I		3200 s		
119458	63701		13 42 43.4	+34 59 20	+0.001	+0.01	5.98	0.85	0.3	G5 III	-15	140 s		
--	--		13 42 43.5	-61 45 19			6.90	0.00	-6.5	B8 I		3100 s		

HD	SAO	Star Name	α 2000	δ 2000	μ(α)	μ(δ)	V	B-V	M_v	Spec	RV	d(pc)	ADS	Notes
119221	224355		13ʰ42ᵐ43ˢ.8	-43°11'09"	-0ˢ.001	-0".04	7.6	-0.2	1.4	A2 V		180 s		
119446	63700		13 42 43.9	+33 50 41	0.000	-0.01	7.86	1.21	0.2	K0 III		240 s		
119076	252422		13 42 44.7	-61 53 32	+0.001	-0.02	6.80	1.1	0.2	K0 III		200 s		m
119222	224356		13 42 49.7	-44 11 27	-0.002	-0.01	7.01	-0.11		B8				
119193	241098		13 42 54.5	-50 47 26	-0.001	-0.03	6.2	1.4	0.2	K0 III		94 s		
119250	224359		13 42 54.9	-41 24 04	-0.005	-0.05	5.98	1.02		K0				
119223	224357		13 42 55.7	-46 59 25	-0.002	-0.01	7.6	1.8	-0.1	K2 III		260 s		
119159	241096		13 42 55.9	-56 46 04	-0.002	-0.01	6.3	0.3	-3.0	B2 IV	-48	350 s		
119251	224360		13 42 58.1	-42 10 55	-0.002	-0.04	7.6	0.6	4.4	G0 V		42 s		
119583	28860		13 42 58.4	+50 01 36	+0.001	-0.02	7.60	1.1	0.2	K0 III		300 s		m
119307	181891		13 43 00.8	-30 10 58	-0.004	-0.01	7.00	0.1	1.4	A2 V		110 mx		m
119425	120082	84 Vir	13 43 03.6	+3 32 16	-0.020	-0.07	5.36	1.11	-0.1	K2 III	-42	56 mx	9000	m
119022	252423		13 43 08.5	-69 07 38	-0.006	-0.02	7.2	0.8	3.2	G5 IV		61 s		
119308	204765		13 43 12.7	-35 17 25	-0.002	-0.01	7.7	-0.2	0.4	B9.5 V		280 s		
119164	252426		13 43 14.2	-61 42 32	-0.002	0.00	7.19	1.29		K0				
119637	44727		13 43 25.9	+45 31 13	0.000	+0.01	6.6	1.1	0.2	K0 III		190 s		
92081	224364		13 43 27.4	-42 40 56	-0.002	0.00	8.0	1.0	0.2	K0 III		360 s		
119461	139507		13 43 30.2	-4 16 28	-0.003	-0.03	7.10	1.29	-0.1	K2 III	+5	230 s	9002	m
119550	100676		13 43 35.6	+14 21 55	-0.022	-0.02	6.96	0.64	4.4	G0 V		29 s		
119256	241111		13 43 36.3	-57 35 24	+0.001	-0.01	7.30	1.1	-0.1	K2 III		290 s		m
119616	63711		13 43 38.4	+35 42 53	0.000	-0.02	7.96	0.98	3.2	G5 IV		60 s		
119361	224365		13 43 39.9	-42 04 03	-0.002	-0.01	5.98	-0.08		B8				m
119584	82969		13 43 45.1	+22 42 01	+0.004	-0.04	6.13	1.42	-0.3	K4 III	+9	190 s		
119585	100680		13 43 48.6	+17 24 27	-0.005	-0.03	7.7	1.1	0.2	K0 III		160 mx		
119415	204777		13 43 49.2	-40 10 40	-0.007	-0.01	7.60	0.2	2.1	A5 V		73 mx		m
119378	224367		13 43 51.5	-45 26 14	-0.001	-0.03	7.5	0.4	2.6	F0 V		82 s		
119842	16153		13 43 52.1	+60 08 43	-0.007	-0.01	7.1	0.1	0.6	A0 V	-11	64 mx		
119341	241118		13 43 52.7	-52 05 59	-0.005	-0.02	7.8	0.8	-0.1	K2 III		320 mx		
119537	139516		13 43 54.1	-5 29 56	-0.004	-0.02	6.51	0.05		A0	-22			
119765	28866		13 43 54.6	+52 03 51	-0.003	-0.01	6.02	0.00	0.6	A0 V	-12	120 s		
119283	241114		13 43 57.3	-59 14 10	-0.004	-0.02	6.70	0.00	0.4	B9.5 V		160 mx		m
119430	224377		13 44 08.9	-43 26 27	+0.002	0.00	7.10	-0.02		B9				
119686	82975		13 44 10.2	+27 35 10	-0.002	0.00	7.12	0.26	1.7	A3 V		100 s		
119864	28874		13 44 11.8	+59 21 30	-0.005	0.00	7.90	1.1	0.2	K0 III		350 s	9009	m
119419	241120		13 44 15.9	-51 00 47	-0.004	-0.04	6.47	-0.13		A0 p				
119590	158128		13 44 21.9	-14 45 38	0.000	-0.02	7.5	1.1	-0.1	K2 III		340 s		
119109	257085		13 44 21.9	-73 38 11	+0.001	0.00	7.47	0.01	-0.6	B7 V		380 s		
119650	120092		13 44 27.0	+0 42 06	-0.002	-0.01	7.6	1.1	-0.1	K2 III		340 s		
119605	158131	83 Vir	13 44 29.7	-16 10 45	+0.001	-0.01	5.60	0.81	-2.0	G0 II	+1	320 s		
119608	158132		13 44 31.2	-17 56 13	-0.001	0.00	7.57	-0.07	-5.7	B1 Ib	+23	3700 s		
119510	224387		13 44 38.2	-47 13 02	+0.001	+0.01	7.0	0.8	3.2	G5 IV		58 s		
119638	158135		13 44 44.4	-14 13 30	-0.001	-0.17	6.91	0.54		G0				
119623	181921		13 44 45.6	-25 30 03	-0.004	-0.02	6.2	1.1	0.2	K0 III		130 s		
119706	139527		13 44 56.8	-7 38 03	-0.003	0.00	7.1	1.1	0.2	K0 III	-32	240 s		
119705	139528		13 44 56.9	-6 00 08	-0.003	+0.01	7.2	1.1	0.2	K0 III		250 s		
119823	100687		13 45 06.6	+18 21 35	-0.001	+0.03	7.80	0.5	3.4	F5 V		76 s		m
119720	158141		13 45 08.6	-14 46 02	-0.002	-0.01	7.7	1.6	-0.5	M2 III		440 s		
119992	28878		13 45 13.1	+55 52 45	+0.012	-0.37	6.50	0.47	3.4	F5 V	-4	41 s		
119784	100689		13 45 13.3	+10 19 31	-0.005	-0.05	6.5	1.1	-0.1	K2 III		150 mx		
118934	257084		13 45 15.2	-79 46 11	-0.008	-0.08	8.0	1.4	-0.3	K5 III		180 mx		
119423	252431		13 45 18.4	-66 45 18	-0.001	-0.02	7.60	0.00	0.0	B8.5 V		320 s		m
120162	16166		13 45 19.0	+69 00 00	+0.020	+0.08	8.0	0.5	4.0	F8 V		62 s		
119629	224394		13 45 27.2	-48 47 30	-0.015	+0.02	6.77	0.54	3.9	F V		35 s		
119693	204806		13 45 28.3	-35 55 32	-0.008	-0.06	7.2	0.6	3.2	G5 IV		62 s		
121623	2280		13 45 29.7	+84 30 48	-0.022	-0.01	7.9	1.1	0.2	K0 III		350 s		
120134	16165		13 45 29.9	+64 37 51	+0.002	+0.01	7.9	0.1	1.4	A2 V		200 s		
119712	204805		13 45 30.3	-38 30 42	-0.003	-0.05	7.5	1.1	0.2	K0 III		240 s		
119713	204807		13 45 31.8	-39 21 20	-0.005	-0.01	7.8	0.2	2.6	F0 V		110 s		
119843	120098		13 45 32.1	+3 30 29	-0.002	0.00	8.00	0.5	4.0	F8 V		63 s	9014	m
119914	63722		13 45 34.2	+34 38 51	+0.001	0.00	7.5	0.8	3.2	G5 IV		74 s		
119786	158147	85 Vir	13 45 35.0	-15 46 03	-0.003	-0.03	6.19	0.05		A0 n	-41			m
119752	181931		13 45 36.8	-26 06 58	-0.005	-0.01	5.81	0.02	0.6	A0 V n		80 mx		
119600	241142		13 45 39.4	-58 14 36	-0.004	-0.04	6.9	1.6	-0.1	K2 III		150 s		
119756	204812	1 Cen	13 45 43.0	-33 02 30	0.000	0.00	4.9	0.38	0.6	F2 III	-22	28 ts		
119944	82987		13 45 52.1	+27 13 35	-0.001	+0.02	8.00	1.24	-0.1	K2 III		400 s		
119853	158152	86 Vir	13 45 56.2	-12 25 35	-0.001	0.00	5.80	1.1	0.2	K0 III	-11	130 s	9018	m
120004	63730		13 46 00.3	+37 31 30	-0.016	-0.06	7.6	0.5	4.0	F8 V		53 s		
119993	63731		13 46 05.7	+34 03 35	-0.009	-0.01	8.0	0.5	3.4	F5 V		84 s		
119931	120102		13 46 06.8	+5 06 55	-0.006	-0.05	7.16	0.55	4.4	G0 V		36 s	9019	m
119726	241150		13 46 07.7	-52 30 51	-0.003	0.00	8.0	-0.3	1.7	A3 V		180 s		
120047	44742		13 46 13.4	+41 05 19	-0.010	-0.05	5.87	0.21	1.7	A3 V	-12	28 mx		
120005	63733		13 46 13.6	+30 53 50	-0.021	-0.10	6.6	0.5	3.4	F5 V	-8	36 mx		
119727	241152		13 46 14.3	-54 41 01	-0.005	-0.03	6.7	0.3	0.6	A0 V		110 mx		
119780	224409		13 46 16.1	-46 21 58	+0.001	0.00	7.6	0.1	1.4	A2 V		180 s		
119781	224410		13 46 17.4	-47 09 22	-0.002	-0.01	7.7	1.7	-0.5	M2 III		440 s		

HD	SAO	Star Name	α 2000	δ 2000	μ(α)	μ(δ)	V	B-V	M_v	Spec	RV	d(pc)	ADS	Notes
			$13^h 46^m 17.6^s$	$-62°27'06''$	$-0\overset{s}{.}004$	$-0\overset{''}{.}01$								
119646	252439		13 46 17.6	-62 27 06	-0.004	-0.01	6.59	0.11	-6.3	B2 I		230 mx		
120048	63735		13 46 18.9	+38 30 14	-0.005	-0.01	5.94	0.94	0.2	G9 III	-14	140 s		
119871	204826		13 46 28.6	-33 46 27	-0.004	-0.02	8.0	0.4	2.1	A5 V		94 mx		
119661	—		13 46 30.0	-64 41 12			7.9	1.6	-0.5	M2 III		440 s		
120198	28885	84 UMa	13 46 35.5	+54 25 58	-0.003	0.00	5.70	-0.08		A2 p	-5			m
119699	252442		13 46 36.7	-62 54 36	+0.001	-0.01	6.6	1.1	0.6	A0 V		33 s		
119872	224415		13 46 39.1	-42 15 44	-0.003	-0.03	8.0	1.2	0.2	K0 III		250 s		
119834	241157		13 46 39.3	-51 25 58	+0.001	-0.04	4.65	0.96	0.2	gK0	-6	78 s		m
120064	82993	3 Boo	13 46 43.2	+25 42 08	-0.002	-0.07	5.95	0.49	2.9	F6 IV-V	+8	38 s		
119949	181947		13 46 45.4	-20 51 09	-0.010	+0.11	7.9	0.5	4.0	F8 V		60 s		
119921	204835		13 46 56.3	-36 15 08	-0.001	-0.02	5.15	-0.02		A0 n	-10	19 mn		m
120066	120108		13 46 57.0	+6 21 01	-0.034	-0.11	6.33	0.63	4.4	G0 V	-31	22 s		
120164	63739		13 46 59.7	+38 32 34	-0.012	-0.02	5.50	1.03	0.2	K0 III	-10	110 mx		m
119857	241164		13 47 03.4	-53 16 55	-0.002	+0.03	7.3	0.6	-0.1	K2 III		300 s		
--	--		13 47 06.5	+6 21 01			7.56	0.82	4.7	dG2		26 s		
119884	241165		13 47 10.3	-52 16 09	-0.005	-0.03	6.95	0.02	0.6	A0 V		100 mx		
120231	44750		13 47 10.7	+47 43 44	-0.005	-0.02	7.9	1.6	-0.5	M2 III		290 mx		
119796	252448		13 47 10.7	-62 35 24	-0.002	-0.02	6.51	1.98	-8.0	G8 Ia		2900 s		m
--	63743		13 47 12.9	+37 36 43	+0.006	-0.10	7.9	1.3						
120025	158161		13 47 13.2	-19 15 19	-0.003	-0.01	6.74	0.32	1.4	A2 V		83 s		
120033	139544		13 47 13.3	-9 42 33	0.000	-0.04	6.05	1.42	-0.3	gK5	+7	190 s		m
120136	100706	4 τ Boo	13 47 15.6	+17 27 24	-0.034	+0.03	4.50	0.48	3.8	F7 V	-16	16 ts	9025	m
120086	139546		13 47 19.1	-2 26 37	-0.001	-0.01	7.90	-0.17	-2.9	B3 III		1400 s		
120088	139548		13 47 25.2	-6 42 19	-0.005	-0.01	6.9	0.1	1.7	A3 V		75 s		
120052	158165	87 Vir	13 47 25.3	-17 51 36	+0.004	-0.04	5.43	1.62	-0.5	M1 III	+64	140 s		
--	241172		13 47 27.4	-50 14 56	+0.005	+0.02	6.1	0.4	1.7	A3 V		50 s		
119938	241173		13 47 27.6	-50 15 00	+0.007	-0.01	5.91	0.29		A3				
120245	63747		13 47 31.6	+37 53 35	+0.002	-0.01	6.95	1.13	-0.1	K2 III		260 s		
120315	44752	85 η UMa	13 47 32.3	+49 18 48	-0.013	-0.01	1.86	-0.19	-1.7	B3 V	-11	33 mx		Alkaid
119700	257089		13 47 34.2	-71 13 52	-0.005	-0.02	6.8	0.9	0.4	B9.5 V		53 s		
119971	241177		13 47 38.4	-50 19 16	-0.016	-0.03	5.45	1.36	-0.3	gK5	+30	140 s		
119970	224425		13 47 40.5	-50 07 31	-0.001	-0.02	6.8	1.4	0.6	A0 V		26 s		
119999	224432		13 47 41.0	-43 36 17	-0.003	-0.03	7.6	1.0	0.2	K0 III		300 s		
119926	241175		13 47 50.4	-59 52 35	-0.002	-0.01	7.9	-0.3	-1.0	B5.5 V		610 s		
120121	158171		13 47 55.5	-18 34 21	-0.001	0.00	7.6	1.6	-0.5	M2 III		420 s		
120405	28889		13 47 55.6	+50 14 22	-0.002	+0.03	7.4	0.2	2.1	A5 V		120 s		
120232	100713		13 47 57.3	+18 56 39	-0.002	-0.03	7.69	1.66	-0.3	K5 III		320 s		
119974	241181		13 47 57.7	-55 57 37	-0.001	0.00	6.9	0.8	0.4	B9.5 V		66 s		
120151	158172		13 47 58.2	-16 03 54	-0.004	-0.01	7.9	1.1	-0.1	K2 III		290 mx		
120186	139556		13 48 02.0	-8 01 29	-0.007	-0.10	7.71	0.55		G0				
120247	100714		13 48 04.7	+13 20 12	-0.002	-0.01	7.55	1.10	0.2	K0 III		250 s		
120348	44754		13 48 05.1	+42 02 50	+0.002	-0.06	6.59	1.09	0.0	K1 III	-1	210 s		
119959	252457		13 48 08.6	-60 48 32	+0.001	-0.01	7.3	2.2	-0.1	K2 III		72 s		
120123	204854		13 48 14.9	-38 15 44	-0.002	-0.02	7.8	1.3	0.2	K2 III		320 s		
120235	139559		13 48 17.9	-6 50 15	-0.003	-0.02	6.5	1.1	0.2	K0 III	-1	180 s		
120349	63756		13 48 19.9	+31 24 05	-0.003	+0.01	7.54	0.28	2.6	F0 V		97 s		
120110	224444		13 48 23.6	-43 29 14	-0.002	-0.03	7.5	1.4	-0.1	K2 III		240 s		
120334	83000		13 48 24.5	+23 26 57	0.000	-0.01	7.97	1.16	0.2	K0 III		260 s		
120098	224445		13 48 28.7	-48 01 15	+0.001	-0.02	7.8	1.0	0.2	K0 III		260 s		
120219	181973		13 48 31.2	-22 09 20	-0.003	+0.02	7.7	1.4	3.2	G5 IV		33 s		
120364	63759		13 48 32.3	+29 51 41	0.000	0.00	7.70	1.01	0.2	K0 III		310 s		
120435	44757		13 48 32.7	+40 14 50	-0.004	+0.02	7.76	1.53	-0.5	M2 III		350 mx		
120317	120121		13 48 34.6	+7 57 36	+0.003	-0.01	6.7	0.4	3.0	F2 V		54 s		
120475	44759		13 48 36.1	+48 21 23	+0.004	-0.02	7.50	0.41	2.6	F0 V		96 s	9030	m
120420	63760		13 48 38.6	+31 11 25	-0.001	+0.04	5.62	1.01	0.2	K0 III	+11	120 s		
120297	139564		13 48 38.8	-2 50 25	+0.001	-0.01	6.8	1.1	-0.1	K2 III		240 s		
120144	224449		13 48 39.4	-46 58 26	-0.007	-0.02	7.4	0.6	0.2	K0 III		230 mx		
120207	204864		13 48 41.8	-37 07 38	+0.003	-0.01	6.8	0.0	2.6	F0 V		69 s		
120419	63762		13 48 42.6	+32 58 35	-0.005	-0.06	7.7	0.5	3.4	F5 V		74 s		
120463	44760		13 48 42.8	+42 20 15	-0.005	-0.05	7.10	1.09	-0.1	K2 III		150 mx		
120421	83007		13 48 45.0	+27 52 51	-0.003	-0.03	7.15	1.07	0.0	K1 III		270 s		
120129	241194		13 48 48.3	-52 49 48	-0.001	+0.02	8.00	0.5	4.0	F8 V		63 s		m
120365	120122		13 48 51.0	+9 42 48	-0.005	-0.03	7.9	1.4	-0.3	K5 III		170 mx		
120177	224450		13 48 52.4	-46 45 41	-0.004	-0.01	7.7	0.8	0.2	K0 III		310 s		
120250	181978		13 48 53.1	-29 39 39	+0.003	0.00	8.0	0.9	3.4	F5 V		43 s		
120237	204867		13 48 55.1	-35 42 14	-0.042	-0.18	6.53	0.57	3.9	G3 IV-V	+6	21 mx		
120302	181980		13 48 57.0	-20 44 55	-0.004	-0.01	7.0	1.1	0.2	K0 III		190 s		
120499	63763		13 48 57.1	+39 32 33	+0.002	-0.01	6.9	1.6		M5 III	-6	29 mn		R CVn, v
120270	204870		13 49 00.2	-30 42 41	+0.001	-0.03	7.7	1.3	-0.1	K2 III		310 s		
120285	181981		13 49 02.0	-28 22 03	-0.003	-0.06	7.0	1.6		M5 III	+42	140 mx		W Hya, m,v
120476	83011		13 49 03.9	+26 58 46	-0.033	-0.09	7.04	1.12	8.2	dK6	-21	13 ts	9031	m
121366	2282		13 49 06.3	+79 55 18	-0.013	+0.02	7.4	0.9	3.2	G5 IV		70 s		
120132	252464		13 49 15.0	-60 45 10	-0.001	-0.02	7.6	-0.1	0.4	B9.5 V		280 s		
120113	252462		13 49 15.5	-62 05 59	+0.001	+0.02	7.60	0.00	0.0	B8.5 V		320 s		
122882	2298		13 49 15.5	+85 44 51	-0.041	+0.02	7.5	0.4	2.6	F0 V		95 s		

HD	SAO	Star Name	α 2000	δ 2000	μ(α)	μ(δ)	V	B−V	M$_v$	Spec	RV	d(pc)	ADS	Notes
120272	224458		13^{h}49^{m}16.3	−40°31'00"	−0.009^s	−0.01"	7.10	0.4	2.6	F0 V		79 s		m
120287	224463		13 49 22.2	−40 30 54	0.000	+0.01	7.1	1.5	−0.3	K5 III		300 s		
120323	204875	2 Cen	13 49 26.6	−34 27 03	−0.004	−0.06	4.19	1.50	−0.5	M1 III	+41	87 s		
120477	100725	5 υ Boo	13 49 28.6	+15 47 52	−0.007	+0.04	4.06	1.52	−0.3	K5 III	−6	72 s		
120566	63770		13 49 29.0	+37 17 18	−0.013	+0.03	8.0	1.1	0.2	K0 III		130 mx		
120448	120127		13 49 29.4	+6 20 37	−0.001	−0.04	6.8	0.1	0.6	A0 V	+1	170 s		
120531	63768		13 49 29.6	+30 43 14	−0.002	0.00	7.8	1.1	0.2	K0 III		330 s		
120307	224469	ν Cen	13 49 30.2	−41 41 16	−0.002	−0.03	3.41	−0.22	−2.5	B2 V	+9	150 s		
121128	7888		13 49 32.0	+75 34 00	+0.002	−0.01	7.4	0.4	2.6	F0 V		92 s		
120275	224467		13 49 33.3	−47 22 13	+0.004	−0.01	8.00	0.5	4.0	F8 V		39 s		m
120510	100726		13 49 36.1	+13 00 37	−0.009	+0.02	6.7	0.5	3.4	F5 V		46 s		
120324	224471	μ Cen	13 49 36.9	−42 28 25	−0.002	−0.03	3.04	−0.17	−1.7	B3 V e	+13	89 s		m, v
120158	252467		13 49 39.1	−63 03 51	−0.004	−0.04	8.0	0.1	0.6	A0 V		160 mx		
120369	204881		13 49 41.8	−36 36 26	−0.003	−0.05	7.6	1.3	−0.5	M2 III		420 s		
120539	83015	6 Boo	13 49 42.7	+21 15 50	+0.001	+0.01	4.91	1.43	−0.3	K4 III	−3	110 s		
120500	120130		13 49 44.1	+8 24 31	−0.001	+0.01	6.59	0.13		A0				
120451	158184		13 49 44.8	−12 33 06	−0.002	−0.02	7.70	1.6	−0.5	M2 III		440 s		m
120600	63772		13 49 44.9	+36 37 58	−0.006	+0.02	6.4	0.1	1.7	A3 V	−12	86 s		
120787	16186		13 49 45.4	+61 29 21	+0.009	−0.10	5.96	0.96	0.2	K0 III	−11	140 s		
120816	16188		13 49 46.7	+62 48 29	−0.004	−0.04	8.0	1.1	−0.1	K2 III		410 s		
120540	100730		13 49 50.7	+19 20 49	−0.003	0.00	7.64	1.28	0.2	K0 III		180 s		
120452	158186	89 Vir	13 49 52.2	−18 08 03	−0.007	−0.04	4.97	1.06	0.0	K1 III	−40	99 s		
120541	100728		13 49 52.3	+13 11 31	−0.003	+0.01	6.66	0.19	1.4	A2 V		94 s		
120426	181996		13 49 54.3	−28 11 12	+0.001	−0.04	7.8	0.6	1.7	A3 V		83 s		
120650	63774		13 49 58.6	+32 31 18	0.000	−0.01	7.46	1.37		K2				
120427	204884		13 49 59.4	−32 39 57	−0.005	−0.03	7.6	0.7	0.2	K0 III		220 mx		
121457	7895		13 50 01.5	+78 59 45	−0.018	0.00	6.6	0.8	3.2	G5 IV	−4	48 s		
120601	100734		13 50 06.1	+18 37 44	−0.004	−0.03	6.7	0.4	2.6	F0 V		67 s		
120455	181999		13 50 06.4	−29 04 52	−0.004	−0.03	6.1	0.5		B9				
120828	16190		13 50 07.6	+60 31 12	+0.001	−0.02	7.8	0.1	0.6	A0 V		270 s		
120702	44772		13 50 08.0	+42 33 27	0.000	0.00	6.9	0.4	2.6	F0 V	−18	73 s		
120516	182004		13 50 14.7	−20 52 10	0.000	−0.01	7.0	1.8	−0.1	K2 III		110 s		
120457	204888		13 50 19.3	−39 54 03	0.000	−0.02	6.44	0.99		K0				
120636	83021		13 50 20.5	+21 15 21	−0.003	+0.02	7.32	0.88	4.4	G0 V	−22	25 s		
120543	158190		13 50 20.7	−13 06 34	−0.003	−0.04	7.5	1.6	−0.5	M2 III		410 s		
120483	204892		13 50 21.1	−31 04 05	−0.003	−0.04	7.2	0.7	2.6	F0 V		45 s		
120651	83022		13 50 23.4	+21 16 36	−0.003	+0.01	6.84	0.93	4.4	G0 V	−19	19 s		m
120602	120132		13 50 24.7	+5 29 49	+0.002	−0.01	6.01	0.90		G5	−7			
120485	204894		13 50 27.1	−33 10 41	−0.002	−0.01	8.0	1.4	0.2	K0 III		220 s		
120874	28901		13 50 27.6	+58 32 22	−0.004	+0.01	6.4	0.1	0.6	A0 V	−40	140 s		
120486	204893		13 50 27.9	−39 05 54	−0.005	−0.08	7.9	0.1	0.6	A0 V		58 mx		
120584	139581		13 50 32.4	−7 35 55	−0.001	−0.04	7.3	1.1	0.2	K0 III		270 s		
120544	158192		13 50 34.4	−19 53 50	−0.003	+0.03	6.3	1.1	4.0	F8 V		14 s		
120534	204900		13 50 40.2	−31 12 22	−0.002	−0.04	7.1	0.6	2.1	A5 V		59 s		
120546	182011		13 50 43.7	−29 52 30	0.000	−0.02	7.70	0.9	3.2	G5 IV		79 s	9033	m
120400	241219		13 50 43.9	−57 34 49	−0.004	+0.01	6.9	0.9	3.2	G5 IV		56 s		V381 Cen, v
120572	182015		13 50 47.2	−24 58 47	+0.001	+0.01	7.6	1.3	−0.1	K2 III		300 s		
120622	158194		13 50 49.0	−11 21 32	+0.001	−0.02	7.8	1.6	−0.5	M2 III		450 s		
120548	204901		13 50 49.7	−38 08 10	+0.001	0.00	7.9	0.4	3.0	F2 V		85 s		
120637	139584		13 50 50.8	−7 47 05	0.000	−0.03	7.0	1.1	0.2	K0 III		230 s		
120817	44774		13 50 51.8	+42 09 19	+0.002	+0.02	7.67	0.16	1.4	A2 V	−9	160 s		
121129	16196		13 50 52.8	+69 09 07	+0.004	−0.08	7.7	1.1	0.2	K0 III		220 mx		
120623	182020		13 50 56.8	−20 59 05	−0.004	−0.01	7.5	1.3	2.6	F0 V		23 s		
121146	16197		13 50 58.8	+68 18 54	−0.035	−0.07	6.40	1.17	0.2	K0 III	−45	36 mx		m
120915	28903		13 50 59.6	+53 17 28	−0.011	+0.07	6.8	0.5	3.4	F5 V		48 s		
121778	2289		13 51 00.6	+80 46 04	+0.019	−0.01	6.7	1.1	0.2	K0 III		200 s		
120818	63779		13 51 04.4	+34 46 20	+0.002	−0.02	6.65	0.12	1.9	A4 V	−12	89 s		
120655	158197		13 51 05.8	−13 40 45	−0.003	−0.02	7.1	0.4	2.6	F0 V		78 s		
120459	241221		13 51 06.6	−57 28 24	−0.007	−0.01	7.98	0.21		A2		16 mn		
120819	63781		13 51 09.1	+34 39 52	0.000	−0.06	5.87	1.62		Ma III	−40	17 mn		
120803	83035		13 51 16.2	+24 41 43	−0.001	−0.07	7.61	1.16	0.0	K1 III		190 mx		
120690	182026		13 51 20.3	−24 23 27	−0.042	−0.31	6.45	0.69	4.4	dG0	+2	19 ts		
121130	16199	10 Dra	13 51 25.8	+64 43 23	−0.001	−0.01	4.66	1.57	−0.5	gM3	−11	110 s	9039	m
120638	204909		13 51 26.4	−37 23 35	0.000	−0.04	7.8	0.8	0.2	K0 III		320 s		
120230	257092		13 51 26.8	−73 08 37	+0.010	−0.01	7.1	−0.3	2.1	A5 V		100 s		
120592	224485		13 51 32.1	−48 17 34	−0.013	−0.02	7.37	0.80	5.2	G5 V		23 s		m
120593	224486		13 51 34.3	−48 17 55	−0.014	−0.02	7.47	0.49	3.7	F6 V		56 s		
120756	158203		13 51 35.0	−13 00 54	−0.009	0.00	7.0	0.9	3.2	G5 IV		58 s		
120672	204911		13 51 36.5	−36 26 00	−0.008	−0.12	6.35	0.48	4.4	dG0		25 s		
120691	204913		13 51 38.5	−31 19 08	−0.025	−0.06	7.16	0.52		F5		16 mn		
120559	241230		13 51 40.3	−57 26 09	−0.045	−0.43	7.97	0.66	5.2	G5 V		36 s		
120792	139592		13 51 44.0	−7 10 22	−0.005	0.00	7.7	0.9	3.2	G5 IV		80 s		
120950	63794		13 51 46.9	+39 40 10	−0.001	−0.02	7.40	1.56	−0.5	M4 III	+33	380 s		
120640	224489		13 51 47.1	−46 53 56	−0.002	−0.03	5.77	−0.16	−2.6	B4 III	−6	470 s		
120933	63793		13 51 47.4	+34 26 39	−0.002	−0.03	4.74	1.66		gMa	−44	19 mn		

HD	SAO	Star Name	α 2000	δ 2000	μ(α)	μ(δ)	V	B-V	M_v	Spec	RV	d(pc)	ADS	Notes
			^{h} ^{m} ^{s}	^{o} $'$ $''$	^{s}	$''$								
120404	252481		13 51 47.4	−69 24 05	+0.002	−0.01	5.75	1.73		K2				
120732	204918		13 51 48.6	−30 20 03	−0.005	−0.02	8.0	1.4	−0.1	K2 III		240 mx		
120709	204917	3 Cen	13 51 50.0	−32 59 41	−0.004	−0.03	4.32	−0.13	−1.6	B5 IV	+14	91 mx		m
121046	28909		13 51 50.5	+52 50 27	−0.001	−0.05	7.3	1.1	0.2	K0 III		270 s		
120865	100742		13 51 51.0	+11 56 14	−0.010	+0.02	6.8	0.5	3.4	F5 V		48 s		
120848	100741		13 51 51.7	+10 08 10	−0.003	−0.02	7.00	1.4	−0.3	K5 III		290 s		m
120917	83040		13 51 53.9	+25 02 12	−0.008	+0.06	8.0	0.5	4.0	F8 V		63 s		
120759	204922		13 52 00.8	−31 37 10	−0.004	−0.06	6.12	0.48	4.4	G0 V		22 s		m
120642	241238		13 52 02.7	−52 48 35	−0.006	−0.03	5.25	−0.09	0.2	B9 V n	+27	100 s		m
120675	224497		13 52 02.8	−47 51 58	−0.001	−0.04	7.10	0.6	4.4	G0 V		35 s		m
120850	158207		13 52 15.5	−14 40 34	+0.002	+0.02	7.3	0.4	2.6	F0 V		88 s		
120497	—		13 52 17.2	−69 05 51			7.9	1.1	0.2	K0 III		330 s		
120807	204927		13 52 17.9	−30 47 07	−0.001	−0.01	7.80	0.1	1.7	A3 V		170 s		m
120934	100745		13 52 18.3	+12 09 54	+0.002	−0.01	6.04	0.04	1.4	A2 V	−16	85 s		
120578	252486		13 52 21.8	−62 46 29	−0.006	+0.01	7.9	0.0	0.0	B8.5 V		140 mx		
120736	241246		13 52 28.3	−51 12 37	−0.011	0.00	7.6	1.0	0.2	K0 III		180 mx		
120997	100747		13 52 32.6	+16 43 44	−0.001	−0.03	6.65	1.52	−0.3	K5 III		240 s		
120780	241249		13 52 35.6	−50 55 18	−0.064	−0.05	7.39	0.90	6.1	K1 V		15 ts		m
120967	120146		13 52 38.2	+0 49 29	−0.001	−0.01	7.6	1.1	0.2	K0 III		300 s		
120902	158210		13 52 38.7	−18 42 30	−0.003	+0.02	7.1	0.1	1.4	A2 V	−38	140 s		
120697	241247		13 52 44.6	−59 05 46	−0.003	−0.02	7.9	−0.2	0.4	B9.5 V		320 s		
120834	224505		13 52 47.8	−42 29 01	−0.007	−0.08	7.3	1.0	0.2	K0 III		140 mx		
121063	83048		13 52 49.5	+23 20 02	−0.003	−0.02	8.01	1.52	−0.1	K2 III		220 s		
120764	241250		13 52 49.8	−54 34 45	−0.029	−0.12	6.9	0.7	3.4	F5 V		35 s		
120969	158212		13 52 53.7	−10 40 56	−0.004	−0.02	7.8	1.1	0.2	K0 III		270 mx		
120678	252495		13 52 56.3	−62 43 14	0.000	−0.02	7.62	0.25		O pe				
120564	252491		13 52 59.1	−70 02 51	−0.005	−0.02	7.0	0.1	0.6	A0 V		180 s		
120739	252497		13 53 02.6	−61 21 19	0.000	0.00	7.86	0.36	0.0	B8.5 V		240 s		
120508	257098		13 53 05.9	−73 16 09	−0.009	−0.05	7.00	0.1	0.6	A0 V		120 mx		m
120954	182048		13 53 07.0	−23 30 47	+0.006	−0.01	8.0	1.4	4.0	F8 V		19 s		
121197	—		13 53 08.4	+40 20 16	0.000	−0.02	6.53	1.55	−0.3	K5 III		220 s		
120837	241257		13 53 09.5	−51 09 53	−0.008	−0.02	7.6	0.7	0.2	K0 III		220 mx		
121164	83055		13 53 10.2	+28 38 53	−0.009	+0.02	5.90	0.20	2.1	A5 V	−12	55 s		
120999	158214		13 53 11.7	−15 59 54	−0.015	−0.04	8.0	0.5	4.0	F8 V		62 mx		
121032	139603		13 53 12.0	−3 32 32	0.000	−0.04	7.60	0.5	4.0	F8 V		53 s		
120955	204944	4 Cen	13 53 12.4	−31 55 39	−0.001	−0.02	4.73	−0.14	−1.6	B5 IV	+5	190 s		
121107	100751	7 Boo	13 53 12.8	+17 55 58	−0.003	0.00	5.70	0.84	0.3	G5 III	−10	120 s		
121109	100752		13 53 16.0	+12 44 30	−0.003	−0.03	7.19	0.28		A3	−18			
120294	—		13 53 17.9	−78 37 54			7.9	1.1	−0.1	K2 III		400 s		
120680	252498		13 53 18.0	−66 30 45	−0.001	−0.02	7.09	0.10	−2.5	B2 V k		750 s		
121297	28922		13 53 20.2	+52 19 22	−0.001	−0.01	7.0	1.6	−0.5	M4 III	−19	310 s		
121247	44787		13 53 20.7	+42 11 05	−0.003	0.00	6.88	0.05	0.6	A0 V		160 s	9044	m
121184	83060		13 53 22.8	+24 09 34	0.000	−0.04	7.94	1.47	−0.2	K3 III		310 s		
121212	63811		13 53 23.1	+33 47 11	−0.003	−0.02	6.90	1.48	0.2	K0 III		110 s		
120958	204949		13 53 28.2	−39 03 26	0.000	−0.01	7.61	−0.09	−1.7	B3 V ne		730 s		
120160	258682		13 53 30.9	−80 16 26	−0.001	−0.04	7.80	0.4	2.6	F0 V		110 s		m
121490	16204		13 53 32.0	+64 14 42	−0.006	−0.03	8.0	0.9	3.2	G5 IV		93 s		
121343	28924		13 53 32.5	+52 06 39	+0.003	−0.07	7.4	0.5	3.4	F5 V		63 s		
120987	204955		13 53 32.7	−35 39 51	−0.007	−0.02	5.54	0.44	3.0	F2 V	−8	29 s		m
120974	204954		13 53 33.0	−38 16 01	+0.001	−0.07	7.7	−0.2	0.6	A0 V		110 mx		
—	204956		13 53 33.3	−35 38 43	0.000	−0.01	6.30							m
120859	241259		13 53 37.3	−58 26 23	−0.002	−0.01	7.8	0.2	0.6	A0 V		190 s		
121111	139605		13 53 38.0	−6 11 17	+0.003	−0.08	7.69	0.55		G0				
121052	182053		13 53 42.0	−26 42 28	0.000	−0.04	7.7	1.3	−0.1	K2 III		310 s		
120908	241262		13 53 43.0	−53 22 25	−0.002	−0.03	5.89	0.01	−1.1	B5 V	+8	210 s		
121409	28928	86 UMa	13 53 50.9	+53 43 43	−0.004	−0.01	5.70	−0.05	0.6	A0 V	−21	64 mx		
121136	158220		13 53 51.7	−14 39 50	+0.003	−0.01	7.90	0.4	2.6	F0 V		120 s		m
121056	204963		13 53 52.2	−35 18 52	−0.022	−0.08	6.19	1.02	4.6	dK0	+12	21 s		
121388	44790		13 53 55.8	+47 52 21	−0.008	0.00	7.5	0.4	3.0	F2 V		81 s		
120991	224514		13 53 56.8	−47 07 40	−0.004	0.00	6.09	−0.07		B3	−21			m
121248	100762		13 54 01.2	+13 50 03	−0.002	−0.03	7.76	0.38	2.1	A5 V		100 s		
121138	182059		13 54 02.6	−23 32 35	+0.001	−0.01	7.7	0.8	0.2	K0 III		310 s		
120786	252505		13 54 03.7	−68 30 19	−0.003	−0.01	7.7	0.0	0.4	B9.5 V		280 s		
121185	158223		13 54 07.8	−11 41 43	−0.001	+0.01	7.5	1.1	−0.1	K2 III		330 s		
121319	83066		13 54 09.2	+28 19 40	−0.002	−0.02	7.73	1.01	0.2	K0 III		320 s		
121087	204967		13 54 10.0	−36 51 51	+0.001	−0.02	7.2	2.0	−0.3	K5 III		160 s		
121157	182064		13 54 15.8	−28 44 34	−0.004	−0.01	6.7	1.2	0.2	K0 III		160 s		
121156	182065		13 54 16.5	−28 34 11	−0.013	−0.07	4.55	0.76	0.2	K0 III		74 s		
121139	204972		13 54 19.7	−32 26 48	−0.012	−0.08	6.9	0.8	2.6	F0 V		35 s		
121057	224519		13 54 19.8	−48 41 34	−0.007	−0.02	7.4	0.1	1.4	A2 V		76 mx		
121811	7903		13 54 21.0	+73 18 50	−0.003	+0.05	7.8	1.1	0.2	K0 III		320 s		
121221	158225		13 54 21.3	−17 10 48	−0.002	0.00	6.8	1.1	0.2	K0 III		210 s		
121009	241271		13 54 23.9	−54 24 42	−0.001	0.00	6.7	0.9	0.2	K0 III		200 s		
121369	63822		13 54 25.3	+29 54 56	−0.006	0.00	7.47	0.29	3.0	F2 V		78 s	9051	m
121059	241275		13 54 29.5	−51 31 26	−0.002	−0.02	8.00	1.1	0.2	K0 III		340 s		m

HD	SAO	Star Name	α 2000	δ 2000	μ(α)	μ(δ)	V	B-V	M_v	Spec	RV	d(pc)	ADS	Notes
121223	182070		13 54 31.6	-22 14 44	+0.003	-0.11	6.70	1.1	0.2	K0 III		120 mx	9050	m
120863	252507		13 54 33.6	-68 14 15	-0.008	-0.04	8.0	0.1	1.7	A3 V		100 mx		
121039	241276		13 54 35.3	-55 04 16	-0.007	0.00	7.2	1.4	0.2	K0 III		150 s		
121187	204977		13 54 38.3	-34 35 50	-0.009	-0.07	6.90	0.4	3.0	F2 V		60 s		m
121040	241277		13 54 38.4	-55 44 18	-0.003	-0.03	7.5	0.2	1.7	A3 V		120 s		
121458	63828		13 54 38.6	+37 03 54	+0.003	-0.15	7.5	0.5	3.4	F5 V		67 s		
122188	7908		13 54 40.4	+78 58 28	+0.008	-0.05	7.8	0.9	3.2	G5 IV		83 s		
121093	241281		13 54 40.7	-50 41 26	-0.001	-0.03	7.9	1.0	3.4	F5 V		35 s		
121370	100766	8 η Boo	13 54 41.0	+18 23 51	-0.004	-0.36	2.68	0.58	2.7	G0 IV	0	9.8 t		Muphrid, m
121141	224526		13 54 41.7	-48 08 06	-0.011	-0.03	7.18	0.36	3.0	F2 V		68 s		
121299	139613	90 Vir	13 54 42.1	-1 30 12	-0.005	-0.03	5.15	1.08	-0.1	K2 III	-7	110 s		
121284	139612		13 54 42.5	-4 27 29	-0.001	-0.03	8.0	0.5	4.0	F8 V		62 s		
121094	241282		13 54 43.3	-51 37 22	-0.004	-0.02	7.9	0.4	1.4	A2 V		120 s		
120993	252515		13 54 43.3	-61 20 04	-0.002	-0.01	7.7	0.0	0.4	B9.5 V		260 s		
121371	100767		13 54 44.3	+15 16 39	+0.002	-0.04	7.93	0.97	0.2	K0 III		350 s		
121646	16215		13 54 44.6	+62 22 38	+0.006	-0.05	7.2	1.1	0.2	K0 III		250 s		
120913	252511		13 54 48.9	-67 39 09	-0.004	-0.03	5.71	1.49	0.2	gK0		55 s		
121325	139618		13 54 58.1	-8 03 32	-0.012	-0.03	6.19	0.53	3.8	dF7	-19	29 s	9053	m
121122	241287		13 54 59.3	-55 01 17	-0.001	0.00	7.1	2.8	-0.4	M0 III		51 s		
121303	158229		13 55 01.7	-19 13 00	0.000	0.00	7.1	0.8	3.2	G5 IV		59 s		
120948	252516		13 55 10.5	-68 52 54	+0.005	+0.01	7.8	0.0	0.0	B8.5 V		350 s		
120616	—		13 55 10.9	-78 08 46			7.9	0.1	1.4	A2 V		200 s		
121190	241294		13 55 12.1	-52 09 39	-0.003	-0.03	5.71	-0.08	-0.2	B8 V	+8	150 s		
121226	224534		13 55 13.1	-47 05 32	-0.005	-0.03	7.4	0.1	0.6	A0 V		87 mx		
121476	100771		13 55 16.2	+18 41 24	-0.003	-0.03	7.90	1.49	-0.1	K2 III		220 s		
121287	204988		13 55 16.2	-32 05 57	-0.003	-0.02	7.10	0.8	3.2	G5 IV		60 s		m
121160	241292		13 55 19.1	-57 11 06	-0.001	-0.02	7.4	0.2	0.0	B8.5 V		200 s		
121331	182081		13 55 19.2	-26 25 57	-0.004	-0.08	7.7	1.5	-0.5	M4 III		160 mx		
121492	100772		13 55 21.6	+19 23 14	-0.002	0.00	7.88	1.42	0.2	K0 III		160 s		
121920	7906		13 55 27.4	+69 56 05	-0.001	-0.02	7.3	1.1	0.2	K0 III		270 s		
121240	241299		13 55 30.0	-51 58 37	-0.002	-0.01	7.7	1.0	-0.3	K5 III		390 s		
121422	158236		13 55 31.4	-12 13 21	-0.001	+0.03	7.90	0.9	3.2	G5 IV		87 s		m
121274	224542		13 55 31.6	-44 35 12	-0.003	0.00	7.6	1.1	4.4	G0 V		21 s		
121263	224538	ζ Cen	13 55 32.3	-47 17 17	-0.006	-0.04	2.55	-0.22	-3.0	B2 IV	+7	110 mx		
121288	224543		13 55 35.2	-42 37 30	0.000	-0.01	7.70	1.22	6.3	K2 V		19 s		
121444	—		13 55 35.4	-9 31 31			7.60	0.4	2.6	F0 V		100 s	9056	m
120213	258683		13 55 39.2	-82 39 58	-0.009	-0.02	5.95	1.46		K2				
121192	241298		13 55 40.8	-60 03 10	-0.006	-0.02	7.6	0.6	2.6	F0 V		66 s		
121291	224547		13 55 41.6	-44 38 59	-0.002	-0.03	8.0	0.8	1.4	A2 V		110 s		
121333	224548		13 55 44.1	-41 59 44	0.000	-0.03	6.6	2.0	-0.3	K5 III		120 s		
121397	204994		13 55 44.4	-31 17 06	-0.004	+0.01	6.4	1.5	0.2	K0 III		86 s		
121824	16220		13 55 45.3	+63 08 58	-0.011	+0.05	8.0	0.4	3.0	F2 V		98 s		
121447	158240		13 55 46.9	-18 14 56	-0.003	+0.01	7.81	1.76		M II p				
121481	139626		13 55 47.8	-9 33 35	-0.002	-0.01	6.8	1.4	-0.3	K5 III		270 s		
121559	83076		13 55 48.1	+22 40 49	-0.008	-0.01	7.57	1.49	-0.1	K2 III		160 mx		
121513	120177		13 55 48.3	+1 31 11	0.000	-0.02	8.0	0.1	0.6	A0 V		300 s		
121560	100776		13 55 49.9	+14 03 23	-0.020	0.00	6.16	0.50	3.7	F6 V	-13	30 s		
121101	252521		13 55 50.3	-66 23 13	-0.010	-0.06	8.0	0.8	3.2	G5 IV		89 s		
121293	241305		13 55 52.1	-50 39 42	-0.006	-0.05	7.0	1.0	0.2	K0 III		220 mx		
121228	241302		13 55 52.3	-59 22 16	+0.001	-0.02	7.81	0.19	-2.6	B2.5 IV		760 s		
121647	63836		13 55 52.8	+39 27 48	+0.001	-0.01	7.36	1.22	-0.1	K2 III		290 s		
121496	139627		13 55 53.1	-9 45 19	+0.001	0.00	6.85	0.47		F5				
121414	205000		13 55 54.2	-30 34 41	0.000	-0.01	7.1	0.8	3.4	F5 V		34 s		
121399	204997		13 55 58.5	-39 24 34	0.000	+0.01	7.3	0.7	0.6	A0 V		81 s		
121429	205003		13 55 58.9	-32 45 21	-0.003	-0.05	7.8	0.9	0.2	K0 III		320 s		
121523	139628		13 56 00.4	-3 39 41	0.000	0.00	6.9	1.1	-0.1	K2 III		250 s		
121280	241306		13 56 04.4	-55 57 09	-0.004	-0.01	6.8	1.1	0.2	K0 III		200 s		
121626	83079		13 56 04.8	+28 40 21	-0.001	+0.01	6.99	0.05	0.6	A0 V	-9	170 s		
121209	252525		13 56 05.7	-63 53 30	+0.003	-0.02	7.2	0.1	0.6	A0 V		200 s		
121359	224553		13 56 06.7	-48 53 28	+0.005	-0.02	7.6	0.3	0.2	K0 III		300 s		
121648	83080		13 56 09.3	+25 55 07	-0.009	-0.01	6.7	0.4	2.6	F0 V	-27	66 s		ZZ Boo, m,v
121682	63837		13 56 10.3	+32 01 57	-0.010	+0.05	6.32	0.37	2.6	F4 IV-V	-22	54 s		
121763	44808		13 56 10.3	+48 21 42	-0.019	-0.02	7.9	1.1	0.2	K0 III		80 mx		
121452	205007		13 56 15.3	-38 39 47	-0.001	-0.02	7.60	0.4	3.0	F2 V		83 s		m
121103	257102		13 56 19.1	-70 48 21	-0.023	-0.05	7.9	0.5	3.4	F5 V		78 s		
121416	224555		13 56 19.4	-46 35 33	-0.016	-0.08	5.83	1.14	3.2	K0 IV		26 s		
121336	241309		13 56 19.8	-54 07 55	-0.004	-0.02	6.14	0.07		A2				m
121696	63841		13 56 20.4	+31 52 54	+0.001	0.00	7.84	0.22	2.6	F0 V		110 s		
121953	16225		13 56 25.2	+65 21 02	-0.003	-0.29	7.6	0.5	3.4	F5 V	-28	34 mx		
121780	44810		13 56 26.1	+49 00 44	-0.002	-0.03	7.0	1.1	0.2	K0 III		230 s		
121402	241313		13 56 26.2	-52 02 08	-0.005	-0.01	7.0	0.9	1.7	A3 V		41 s		
121607	120185	92 Vir	13 56 27.8	+1 03 02	-0.002	+0.01	5.91	0.20		A3 n	-27			
121606	120187		13 56 28.9	+2 14 39	0.000	0.00	7.8	1.1	0.2	K0 III		330 s		
121524	182102		13 56 31.2	-27 38 17	+0.002	-0.01	6.8	1.2	3.2	G5 IV		27 s		
121384	241315		13 56 32.9	-54 42 16	-0.004	-0.22	6.00	0.78	4.4	G0 V		16 s		m

HD	SAO	Star Name	α 2000	δ 2000	μ(α)	μ(δ)	V	B-V	M_v	Spec	RV	d(pc)	ADS	Notes
121710	83084	9 Boo	13 56 34.1	+27 29 31	+0.002	-0.05	5.01	1.42	-0.2	K3 III	-40	85 s		
121608	139640		13 56 36.2	-10 02 03	-0.017	-0.04	7.68	0.54		F8				
121683	100780		13 56 36.9	+15 53 29	-0.003	+0.06	6.83	0.94	0.2	K0 III		210 s		
121665	120189		13 56 40.5	+2 59 11	-0.001	-0.02	6.90	0.2	2.1	A5 V		91 s		m
121527	205016		13 56 43.0	-38 19 09	+0.002	-0.03	7.0	0.9	0.2	K0 III		230 s		
121483	224560		13 56 47.8	-46 23 10	-0.002	-0.01	6.95	-0.13	-2.5	B2 V		740 s		
121539	205017		13 56 50.1	-38 23 50	-0.001	0.00	7.8	1.4	-0.1	K2 III		290 s		
121579	182105		13 56 52.6	-27 39 41	-0.006	-0.06	7.81	0.49	3.4	F5 V		71 s	9062	m
121825	44813		13 56 55.0	+44 16 56	-0.002	-0.01	7.7	0.6	4.4	G0 V	+23	45 s		
121611	182110		13 56 59.2	-23 06 12	0.000	-0.01	7.6	-0.1	0.4	B9.5 V		270 s		
121922	28942		13 56 59.9	+54 35 08	+0.003	-0.01	7.0	1.4	-0.3	K5 III		290 s		
122020	16229		13 57 03.8	+62 47 32	-0.011	+0.02	7.4	0.5	3.4	F5 V		62 s		
121631	182112		13 57 05.4	-24 24 50	0.000	-0.02	7.8	0.2	0.6	A0 V		190 s		
121764	83089		13 57 06.2	+20 57 14	+0.002	-0.03	6.71	1.55	0.2	K0 III		94 s		
121596	205020		13 57 08.4	-34 54 15	-0.002	-0.03	7.7	1.0	-0.1	K2 III		360 s		
121653	182117		13 57 16.9	-28 38 07	-0.004	-0.01	7.2	1.4	-0.1	K2 III		200 s		
121504	241321		13 57 16.9	-56 02 24	-0.031	-0.08	7.60	0.8	3.2	G5 IV		50 mx		m
121454	252530		13 57 21.4	-62 29 20	-0.016	-0.07	7.50	0.6	4.4	G0 V		41 s		m
121635	205026		13 57 22.2	-37 09 56	+0.002	-0.03	7.8	1.4	-0.1	K2 III		290 s		
121699	182123		13 57 27.6	-23 01 22	-0.003	0.00	6.2	1.6	0.2	K0 III		69 s		
121518	241324		13 57 28.1	-57 42 40	0.000	0.00	7.5	1.6	-0.5	M4 III		400 s		V412 Cen, m,v
121844	83096		13 57 28.7	+24 59 55	-0.005	-0.04	7.89	1.11	0.0	K1 III		160 mx		
121365	252529		13 57 29.9	-68 00 23	-0.001	-0.03	7.8	1.4	-0.3	K5 III		400 s		
122064	16230		13 57 31.9	+61 29 33	-0.005	+0.21	6.3	1.4	-0.3	K5 III	-25	77 mx		
121829	100792		13 57 35.1	+18 15 38	-0.003	-0.05	7.70	0.9	1.8	G6 III-IV	-5	160 s		
121474	252531		13 57 38.9	-63 41 11	-0.005	-0.03	4.71	1.11	-0.3	K4 III	+22	100 s		
121617	224570		13 57 41.1	-47 00 34	-0.002	-0.03	7.5	0.3	0.6	A0 V		150 s		
122007	28947		13 57 42.6	+53 54 36	0.000	-0.02	6.9	0.1	0.6	A0 V		180 s		
121845	100793		13 57 45.4	+11 57 44	-0.003	-0.01	7.5	0.2	2.1	A5 V		120 s		
121691	205036		13 57 46.3	-39 20 02	-0.001	-0.02	7.8	1.2	0.2	K0 III		260 s		
121638	224572		13 57 48.6	-49 01 07	+0.006	-0.01	7.3	0.6	0.2	K0 III		260 s		
121880	100795		13 57 52.1	+16 12 06	+0.001	-0.06	7.60	0.05	0.6	A0 V		110 mx		
121730	205042		13 57 55.1	-30 57 54	-0.004	-0.06	7.58	1.07	0.2	K0 III		190 mx		
121620	241330		13 57 56.4	-53 42 16	-0.005	-0.02	7.08	0.96	0.2	K0 III		240 s		
121862	120196		13 57 56.8	+5 07 04	0.000	-0.02	7.7	0.9	3.2	G5 IV		80 s		
121934	63857		13 57 56.9	+32 36 03	0.000	0.00	7.20	0.94	0.2	K0 III		250 s		
121758	182126		13 57 58.7	-25 59 50	+0.001	+0.03	6.5	1.6	0.2	K0 III		82 s		
121733	205041		13 58 01.2	-38 10 08	+0.002	-0.03	7.5	0.0	1.4	A2 V		170 s		
121860	120197		13 58 01.7	+7 27 48	-0.002	0.00	7.8	1.6	-0.5	M2 III		470 s		
121907	100796		13 58 06.2	+16 24 11	+0.003	-0.02	7.40	0.19	0.6	A0 V	+23	170 s		
121881	120198		13 58 07.2	+2 47 01	0.000	-0.01	7.6	1.1	0.2	K0 III		300 s		
121760	205046		13 58 07.7	-34 23 42	-0.003	-0.02	8.0	0.5	0.2	K0 III		360 s		
121769	182128		13 58 09.8	-29 44 39	-0.002	-0.04	7.2	1.3	0.2	K0 III		160 s		
121743	224577	ø Cen	13 58 16.2	-42 06 02	-0.002	-0.02	3.83	-0.21	-2.5	B2 V	+7	190 s		
121315	257104		13 58 18.2	-73 38 24	-0.005	-0.05	7.3	0.0	0.0	B8.5 V		150 mx		
121865	158267		13 58 26.8	-12 03 33	+0.009	-0.15	7.05	0.98		G5				
121746	224581		13 58 27.7	-48 27 54	-0.010	-0.07	7.17	0.48	2.1	F5 IV		89 mx		
122251	16234		13 58 28.7	+64 53 37	0.000	0.00	6.9	1.1	0.2	K0 III		220 s		
121883	158269		13 58 29.8	-14 07 19	+0.001	0.00	6.8	0.8	3.2	G5 IV		52 s		
121847	182134	47 Hya	13 58 31.1	-24 58 20	-0.004	-0.03	5.15	-0.10		B8	+5			
121557	252534		13 58 31.2	-65 48 02	-0.004	-0.04	6.20	1.05	0.2	gK0		150 s		m
122149	28953		13 58 35.6	+53 34 45	-0.007	-0.01	7.90	0.7	3.0	G2 IV	-2	96 s		
121833	205050		13 58 36.6	-34 23 49	-0.006	-0.02	7.80	1.4	-0.3	K5 III		220 mx		m
122298	16235		13 58 37.8	+66 06 02	-0.002	+0.01	7.5	0.5	3.4	F5 V		65 s		
121996	83103	10 Boo	13 58 38.8	+21 41 46	-0.001	-0.05	5.76	-0.03	0.6	A0 V	+6	110 s		
121980	100801		13 58 39.8	+14 38 57	-0.004	-0.06	6.00	1.44	0.2	K0 III	-41	65 s		
121790	224585	υ¹ Cen	13 58 40.7	-44 48 13	-0.002	-0.02	3.87	-0.20	-2.3	B3 IV	+7	170 s		
121661	252536		13 58 42.4	-62 43 08	-0.002	-0.03	7.9	0.1	0.6	A0 V		270 s		
122132	44824		13 58 45.8	+46 35 45	0.000	-0.02	7.20	1.6	-0.5	M2 III	-59	350 s		
122021	100804		13 58 50.6	+14 33 46	-0.005	+0.01	7.9	1.1	0.2	K0 III		270 mx		
121804	224587		13 58 52.1	-47 48 27	-0.001	-0.03	7.7	1.3	-0.5	M2 III		440 s		
122052	83105		13 58 54.1	+24 41 31	-0.003	-0.06	7.29	1.20	0.6	G0 III		100 s		
122200	28955		13 58 55.4	+53 06 24	+0.002	0.00	6.80	0.1	1.4	A2 V		120 s	9077	m
121981	139660		13 58 59.8	-6 55 30	+0.001	-0.06	6.96	1.00		G5				
121957	158275		13 59 03.6	-18 37 21	+0.003	+0.02	6.9	1.1	0.2	K0 III		220 s		
122080	83108		13 59 05.2	+25 48 59	+0.001	0.00	7.11	0.18	2.1	A5 V		98 s	9076	m
121896	205062		13 59 08.9	-37 35 45	-0.001	-0.02	7.7	1.1	-0.1	K2 III		370 s		
121852	224592		13 59 09.6	-45 28 09	-0.005	-0.09	7.38	0.52	3.8	F7 V		50 s		
122363	16239		13 59 10.4	+64 23 06	-0.009	+0.02	7.2	1.1	-0.1	K2 III		280 s		
121853	241352		13 59 17.2	-50 22 12	+0.001	-0.04	6.0	1.5	0.2	K0 III		74 s		
121662	252538		13 59 21.2	-67 50 26	-0.001	-0.01	7.9	0.0	0.0	B8.5 V		370 s		
121961	205066		13 59 22.7	-31 44 38	-0.002	-0.03	7.8	0.9	2.6	F0 V		45 s		
122040	139663		13 59 27.6	-5 25 27	+0.004	-0.21	7.4	0.5	4.0	F8 V		47 s		
121914	224596		13 59 28.2	-43 42 12	-0.003	-0.02	7.8	1.3	-0.2	K0 III		230 s		
122236	44831		13 59 32.9	+48 32 57	-0.005	+0.01	7.9	1.1	-0.1	K2 III		400 s		

HD	SAO	Star Name	α 2000	δ 2000	μ(α)	μ(δ)	V	B-V	M_V	Spec	RV	d(pc)	ADS	Notes
			^{h} ^{m} ^{s}	$^\circ$ $'$ $''$	^{s}	$''$								
121809	252544		13 59 38.2	−61 43 24	+0.001	−0.01	7.9	1.4	0.2	K0 III		210 s		
122253	44833		13 59 41.4	+42 03 00	+0.001	+0.03	7.9	1.1	0.2	K0 III		340 s		
122104	120211		13 59 44.6	+0 02 57	+0.001	−0.03	7.8	0.8	3.2	G5 IV		85 s		
121972	205068		13 59 45.1	−39 55 26	−0.002	0.00	8.0	1.2	−0.1	K2 III		410 s		
121930	224599		13 59 45.9	−50 13 43	+0.001	−0.11	7.5	1.4	0.2	K0 III		110 mx		
122106	139666		13 59 49.2	−3 32 59	−0.002	−0.06	6.40	0.49	3.4	F5 V	−8	37 s		
122727	7924		13 59 50.8	+73 23 53	−0.001	−0.01	7.7	0.9	3.2	G5 IV		79 s		
121796	252546		13 59 58.5	−65 32 32	−0.006	−0.02	7.7	0.4	0.6	A0 V		130 mx		
122066	182152	48 Hya	14 00 00.0	−25 00 37	−0.015	−0.10	5.8	1.7	3.1	F3 V	−17	14 mn		
122135	139669		14 00 04.6	−8 09 42	−0.002	−0.05	6.6	1.1	0.2	K0 III	−3	190 s		
121857	252549		14 00 05.3	−62 46 54	−0.003	+0.01	7.7	0.0	0.4	B9.5 V		280 s		
122014	224605		14 00 13.7	−44 13 19	−0.001	−0.01	8.0	1.2	0.2	K0 III		290 s		
121901	252552		14 00 17.3	−61 28 52	−0.009	−0.04	6.49	0.33	3.0	F2 V		50 s		
122015	224606		14 00 18.7	−46 07 39	−0.003	−0.04	6.8	1.0	1.4	A2 V		33 s		
122203	120218		14 00 20.1	+2 40 27	−0.002	−0.04	7.4	1.1	0.2	K0 III		270 s		
122070	205079		14 00 22.2	−37 23 55	+0.004	−0.04	7.9	1.3	0.2	K0 III		220 s		
122326	63881		14 00 23.4	+39 01 47	+0.002	−0.01	6.6	1.1	−0.1	K2 III		220 s		
122171	158285		14 00 28.3	−14 57 07	−0.001	−0.02	7.9	1.1	0.2	K0 III		340 s		
121439	257107		14 00 33.0	−78 35 25	−0.003	−0.02	6.09	0.03		A0				
122156	182161		14 00 36.0	−26 15 59	−0.002	−0.16	6.7	1.3	4.0	F8 V		12 s		
122112	224614		14 00 41.6	−41 52 33	−0.002	−0.01	7.5	1.3	−0.5	M2 III		400 s		
121949	252555		14 00 43.1	−62 38 00	−0.004	−0.01	8.0	1.1	0.2	K0 III		350 s		
122093	224615		14 00 51.7	−49 31 35	−0.004	−0.03	8.0	0.4	2.6	F0 V		100 s		
121932	252554		14 00 52.2	−66 16 07	−0.021	−0.02	5.97	0.35	2.6	F0 V		44 s		m
121545	−−		14 00 55.8	−78 12 24			7.6	0.5	4.0	F8 V		52 s		
122422	63887		14 01 03.4	+32 29 21	−0.004	−0.05	7.74	0.22	2.1	A5 V		130 s		
122364	83127		14 01 03.5	+21 58 36	+0.002	−0.03	7.11	0.91	3.2	G5 IV		50 s		
122327	100813		14 01 04.7	+13 06 40	+0.002	−0.04	7.7	0.9	3.2	G5 IV		80 s		
122158	224618		14 01 08.5	−44 11 51	−0.005	−0.04	7.0	0.2	1.7	A3 V		80 mx		
122195	205093		14 01 08.6	−37 28 13	+0.009	−0.05	6.8	1.8	0.2	K0 III		67 s		
122405	83130	11 Boo	14 01 10.4	+27 23 11	−0.006	+0.01	6.23	0.17	0.5	A7 III	−23	130 mx		
122208	−−		14 01 18.0	−39 15 09			7.9	0.4	1.4	A2 V		120 s		
122096	241371		14 01 18.1	−57 13 44	−0.003	−0.01	6.9	1.9	−0.1	K2 III		85 s		
122210	205096		14 01 18.9	−40 13 20	−0.003	−0.01	6.13	1.25		K0				m
122386	100815		14 01 19.6	+13 43 58	−0.002	+0.04	7.1	0.9	3.2	G5 IV		61 s		
122116	241372		14 01 19.8	−56 13 26	0.000	−0.01	7.7	0.5	0.0	B8.5 V		150 s		
122365	120228		14 01 20.3	+8 53 43	+0.002	+0.01	5.99	0.09		A2 n	−14			
122222	205099		14 01 21.6	−37 19 52	−0.004	0.00	7.4	1.7	−0.5	M4 III		340 s		
122456	63889		14 01 22.3	+31 33 49	−0.001	−0.04	6.88	1.42	0.2	K0 III		120 s		
122387	120232		14 01 26.2	+9 17 55	0.000	−0.04	6.8	1.1	0.2	K0 III		210 s		
122062	252556		14 01 29.0	−62 57 02	0.000	+0.01	7.60	1.1	0.2	K0 III		290 s		m
122517	63894		14 01 34.2	+36 13 27	−0.014	−0.04	7.53	0.84	0.2	K0 III		78 mx		
122406	120235		14 01 34.4	+4 15 10	−0.002	0.00	7.1	0.8	3.2	G5 IV		59 s		
122443	100819		14 01 36.7	+17 40 06	−0.004	−0.09	6.36	1.22	3.2	G5 IV		23 s		
122457	100820		14 01 37.2	+16 45 27	−0.008	+0.01	7.4	0.5	4.0	F8 V		47 s		
122408	120238	93 τ Vir	14 01 38.7	+1 32 40	+0.001	−0.02	4.26	0.10	1.7	A3 V	−2	32 s	9085	m
122097	252557		14 01 38.7	−61 50 07	+0.002	−0.01	6.8	2.1	−0.1	K2 III		69 s		
122223	224621	υ² Cen	14 01 43.3	−45 36 12	0.000	−0.02	4.34	0.60	−2.0	F5 II	−1	160 s		
122259	205105		14 01 45.6	−39 26 19	−0.003	−0.03	7.3	1.0	1.4	A2 V		42 s		
122574	44851		14 01 49.3	+40 25 11	−0.002	−0.02	6.81	1.17	−0.1	K2 III		240 s		
122909	16254		14 01 50.5	+68 40 43	−0.007	0.00	6.34	1.40	−0.3	K5 III	−22	210 s		
122141	252566		14 01 53.9	−60 15 26	−0.005	−0.01	7.3	0.6	2.6	F0 V		57 s		
122142	252565		14 01 55.2	−60 47 35	−0.001	−0.01	7.9	0.1	0.0	B8.5 V		290 s		
122143	252563		14 01 56.6	−62 10 22	−0.011	−0.05	8.0	1.0	0.2	K0 III		190 mx		
122518	83134		14 01 57.5	+21 33 21	−0.008	0.00	7.94	0.59	4.4	G0 V		50 s		
122740	28984		14 02 00.5	+57 13 28	0.000	−0.02	7.76	0.47		F5			9089	m
123011	7930		14 02 00.7	+70 20 10	+0.001	+0.01	7.5	0.9	0.3	G8 III	−38	280 s		
122651	44856		14 02 07.0	+43 33 32	−0.004	−0.01	7.8	0.9	3.2	G5 IV		82 s		
122675	44858		14 02 12.0	+45 45 13	+0.001	−0.07	6.27	1.32		K5	−49			
122637	63903		14 02 20.2	+36 06 41	−0.004	−0.10	7.1	0.5	3.4	F5 V		56 s		
122562	83135		14 02 21.1	+20 52 52	+0.008	−0.08	7.69	0.95	3.2	G5 IV		62 s		
122394	205112		14 02 22.5	−36 10 46	+0.001	−0.01	7.05	0.04	0.6	A0 V		180 s		
122430	182182		14 02 22.7	−27 25 48	−0.002	−0.01	5.7	1.6	0.2	K0 III	0	50 s		
122118	252570		14 02 23.2	−67 32 09	−0.004	−0.04	8.0	1.1	0.2	K0 III		350 s		
122548	100823		14 02 23.7	+10 19 38	−0.002	+0.02	7.2	1.1	0.2	K0 III		250 s		
122692	44860		14 02 25.4	+41 21 42	−0.001	−0.01	8.0	0.9	3.2	G5 IV		93 s		
122652	63905		14 02 31.6	+31 39 37	−0.008	−0.01	7.2	0.5	4.0	F8 V		43 s		
122563	120251		14 02 31.7	+9 41 11	−0.013	−0.07	6.20	0.90	2.8	G0 IV	−22	32 s		
122491	158300		14 02 32.4	−19 48 35	−0.003	−0.01	7.2	0.8	4.0	F8 V		30 s		
122432	205117		14 02 33.3	−35 47 46	−0.008	−0.06	8.0	1.4	3.2	G5 IV		38 s		
122476	182187		14 02 38.4	−26 50 52	−0.001	−0.02	7.3	1.0	3.2	G5 IV		46 s		
122944	16258		14 02 46.7	+61 40 44	−0.001	0.00	7.3	0.1	1.4	A2 V		150 s		
122603	120252		14 02 47.6	+3 32 43	−0.012	−0.10	7.7	0.6	4.4	G0 V		47 s		
122865	28988		14 02 49.7	+54 39 58	−0.004	−0.01	7.9	0.8	3.2	G5 IV		89 s		
122653	100828		14 02 50.3	+17 59 16	+0.001	0.00	8.04	1.50	0.2	K0 III		150 s		

HD	SAO	Star Name	α 2000	δ 2000	μ(α)	μ(δ)	V	B-V	M_v	Spec	RV	d(pc)	ADS	Notes
			h m s	° ′ ″	s	″								
122676	100829		14 02 56.8	+14 58 30	−0.004	−0.01	7.2	0.9	3.2	G5 IV		62 s		
122866	28989		14 02 59.6	+50 58 18	−0.003	−0.01	6.15	0.01		A0	−8			
122510	205123		14 03 01.6	−31 41 01	+0.002	+0.08	6.18	0.48	4.0	F8 V		27 s		m
122577	158306		14 03 04.0	−17 22 00	−0.013	0.00	6.5	1.1	0.2	K0 III		100 mx		
121317	258686		14 03 06.0	−82 55 12	−0.034	+0.02	7.1	1.8	−0.1	K2 III		120 s		
122523	205124		14 03 06.6	−34 32 25	+0.004	−0.01	7.9	0.4	2.6	F0 V		96 s		
122784	63911		14 03 11.9	+32 48 47	0.000	−0.04	7.22	1.49	−0.1	K2 III		160 s		
122767	83143		14 03 15.6	+24 35 50	0.000	−0.03	7.96	1.33	−0.2	K3 III		410 s		
122657	139700		14 03 16.1	−7 08 23	−0.012	−0.01	7.8	0.5	3.4	F5 V		76 s		
122554	205127		14 03 18.7	−33 26 43	−0.011	−0.05	7.8	1.4	0.2	K0 III		100 mx		
123262	7934		14 03 20.1	+70 17 19	−0.004	+0.07	7.6	1.4	−0.3	K5 III		290 mx		
122768	83145		14 03 21.9	+22 29 45	−0.001	−0.02	6.85	0.31	2.6	F0 V	+6	68 s		
122415	241401		14 03 23.4	−57 20 58	0.000	−0.04	7.6	1.6	0.2	K0 III		140 s		
122699	139702		14 03 24.6	−0 37 41	+0.002	−0.02	7.9	0.4	2.6	F0 V		120 s		
122438	241403		14 03 26.2	−56 12 49	−0.009	−0.03	5.92	1.22	0.2	gK0		100 s		
122249	−−		14 03 26.8	−68 30 06			6.9	0.0	0.0	B8.5 V		230 s		
122658	158312		14 03 27.0	−10 58 02	0.000	0.00	7.50	0.1	1.4	A2 V		170 s		m
122796	83148		14 03 27.1	+27 30 35	−0.006	−0.02	7.14	1.06	0.0	K1 III		180 mx		
122532	224641		14 03 27.3	−41 25 24	−0.003	−0.02	6.11	−0.11	0.6	A0 V		130 s		
122479	241404		14 03 30.7	−52 03 34	−0.002	−0.01	7.36	−0.06		B8				
122679	158313		14 03 31.2	−10 43 51	+0.001	+0.01	7.7	1.4	−0.3	K5 III		390 s		
122742	100832		14 03 32.2	+10 47 12	+0.005	−0.31	6.30	0.74	5.6	G8 V	−17	16 ts		
121951	257108		14 03 33.7	−77 45 28	−0.014	−0.02	7.3	0.8	3.2	G5 IV		65 s		
122744	120261		14 03 36.7	+7 32 47	−0.003	−0.02	6.26	0.94	0.2	G9 III	−20	160 s		
122769	120262		14 03 43.2	+8 29 14	+0.002	−0.01	7.58	0.45	3.4	dF5	−4	68 s	9094	m
122451	252582	β Cen	14 03 49.4	−60 22 22	−0.003	−0.02	0.61	−0.24	−5.1	B1 II		140 s		m
122703	182204		14 03 53.0	−22 25 19	0.000	−0.01	6.30	0.45		F2				
122833	100833		14 03 55.1	+12 43 35	0.000	−0.01	7.1	1.1	0.2	K0 III		240 s		
122797	120265		14 03 55.6	+4 54 04	−0.001	0.00	6.24	0.39		F2				
122834	100834		14 03 56.9	+11 16 57	−0.006	−0.02	6.8	1.1	0.2	K0 III		160 mx		
122771	139704		14 03 57.6	−9 44 42	−0.001	+0.02	7.2	1.1	0.2	K0 III		250 s		
122749	158320		14 04 00.9	−12 45 09	+0.002	0.00	8.0	0.1	0.6	A0 V		300 s		
122641	205136		14 04 01.5	−38 27 34	−0.003	0.00	7.8	1.1	0.2	K0 III		300 s		
122683	205139		14 04 07.0	−37 15 51	−0.002	−0.03	7.20	0.5	4.0	F8 V		44 s		m
122968	63919		14 04 11.3	+37 22 09	+0.002	−0.05	8.0	1.1	0.2	K0 III		360 s		
122800	158322		14 04 12.0	−12 07 47	−0.003	−0.01	7.9	0.4	3.0	F2 V		95 s		
122666	224651		14 04 12.2	−41 48 46	−0.002	−0.08	7.2	0.5	3.4	F5 V		53 s		
122815	139711		14 04 14.5	−5 22 53	−0.001	−0.01	6.39	1.32		K0				
122867	120267		14 04 19.2	+2 36 26	−0.003	−0.01	7.8	1.1	0.2	K0 III		330 s		
122816	139713		14 04 21.4	−9 15 27	−0.003	−0.01	6.6	0.1	0.6	A0 V		160 s		
123299	16273	11 α Dra	14 04 23.2	+64 22 33	−0.009	+0.01	3.65	−0.05	−0.6	A0 III	−16	71 s		Thuban
122755	182213		14 04 23.6	−29 53 59	−0.002	−0.01	8.0	1.1	−0.5	M2 III		510 s		
122559	241413		14 04 24.9	−58 12 03	−0.005	−0.03	8.0	0.0	0.6	A0 V		110 mx		
122775	182215		14 04 26.3	−29 03 44	−0.005	+0.02	7.1	0.9	3.2	G5 IV		48 s		
122837	158325		14 04 26.9	−14 58 18	−0.003	−0.02	6.28	1.08	0.3	G6 III	−15	120 s		
122992	83151		14 04 27.6	+29 08 31	+0.002	−0.07	7.91	1.56	−0.5	M2 III	−17	330 mx		
123133	28997		14 04 28.2	+52 24 57	−0.006	−0.02	7.9	1.1	0.2	K0 III		230 mx		
122946	100838		14 04 32.3	+19 10 39	−0.005	−0.10	7.5	0.5	3.4	F5 V		68 s		
122776	205143		14 04 37.2	−35 38 42	+0.002	0.00	7.90	0.4	2.6	F0 V		120 s		m
122910	120269		14 04 37.4	+2 17 51	−0.002	−0.01	6.28	1.02		K0	−29			
122341	257111		14 04 40.2	−72 04 16	+0.001	−0.07	8.0	0.7	4.4	G0 V		52 s		
122928	120270		14 04 41.5	+6 59 54	−0.002	0.00	8.0	0.4	2.6	F0 V		120 s		
122705	224654		14 04 42.0	−50 04 19	−0.003	−0.05	7.8	0.3	1.4	A2 V		190 s		
122419	252586		14 04 45.4	−69 52 22	−0.002	−0.03	7.3	0.4	3.0	F2 V		71 s		
123033	83152		14 04 45.8	+25 49 04	0.000	−0.11	6.95	0.45	3.4	F5 V	−17	51 s		m
122911	139718		14 04 48.7	−6 33 10	−0.003	−0.02	8.00	0.4	2.6	F0 V		120 s	9100	m
123300	29003		14 04 51.5	+58 52 03	−0.023	+0.01	7.6	0.9	3.2	G5 IV		75 s		
122806	205149		14 04 55.7	−38 16 46	−0.002	0.00	7.8	1.9	−0.3	K5 III		230 s		
122900	182226		14 05 05.7	−22 37 20	−0.004	−0.01	7.2	1.5	0.2	K0 III		130 s		
122993	139726		14 05 12.4	−7 16 35	−0.005	−0.02	8.0	1.1	0.2	K0 III		200 mx		
122956	158333		14 05 12.9	−14 51 25	−0.007	−0.05	7.3	0.9	3.2	G5 IV		66 s		
122958	158331		14 05 13.9	−16 20 09	0.000	0.00	6.5	0.1	1.4	A2 V		100 s		
123253	44877		14 05 15.8	+46 35 32	−0.003	+0.05	8.0	0.4	2.6	F0 V		120 s		
123338	29004		14 05 17.6	+56 31 13	−0.005	0.00	7.2	1.1	0.2	K0 III		250 s		
122250	257112	θ Aps	14 05 19.8	−76 47 48	−0.025	−0.03	6.0	1.6	−0.5	M4 III		200 s		v
123802	7947		14 05 30.3	+74 42 56	+0.012	−0.12	7.48	1.16	−0.2	gK3	−69	340 s		
123279	44880		14 05 35.7	+44 20 27	+0.001	+0.02	7.9	1.1	0.2	K0 III		350 s		
123190	83159		14 05 41.2	+22 43 51	−0.004	−0.08	7.67	1.16	0.2	K0 III		110 mx		
122844	241430		14 05 46.2	−54 40 09	−0.008	−0.03	6.17	0.23		A3				
122763	252593		14 05 51.1	−61 55 58	0.000	0.00	7.1	2.4	−0.5	M2 III		110 s		
123003	205173		14 05 56.9	−37 43 45	0.000	−0.01	7.7	0.3	0.6	A0 V		160 s		
123103	158344		14 05 57.6	−13 04 25	+0.002	0.00	8.0	0.1	1.7	A3 V		180 s	9104	m
122980	224673	χ Cen	14 06 02.7	−41 10 47	−0.002	−0.02	4.36	−0.19	−2.5	B2 V	+12	240 s		
122982	224674		14 06 07.7	−44 41 22	−0.007	−0.04	7.5	−0.1	0.6	A0 V		65 mx		
123004	224676		14 06 10.8	−43 05 30	−0.001	−0.04	6.20	0.98	0.2	gK0		160 s		

HD	SAO	Star Name	α 2000	δ 2000	μ(α)	μ(δ)	V	B-V	M_v	Spec	RV	d(pc)	ADS	Notes
123106	182240		14^{h}06^{m}11.3^s	-24°44'27"	+0.001^s	-0.03"	7.6	1.5	-0.3	K5 III		380 s		
--	44886		14 06 11.9	+49 39 04	0.000	+0.01	7.6	2.0	0.2	K0 III		80 s		
123232	100849		14 06 12.9	+10 49 35	-0.001	+0.01	8.0	1.4	-0.3	K5 III		450 s		
123282	83163		14 06 13.5	+22 10 03	-0.001	-0.02	7.41	1.05	3.2	G5 IV		47 s		
123071	205179		14 06 15.0	-36 09 19	+0.001	0.00	7.8	1.1	-0.1	K2 III		380 s		
123177	139732		14 06 17.7	-8 53 30	0.000	+0.02	6.53	0.02		A0				
123422	44888		14 06 20.3	+48 52 52	-0.003	+0.03	7.6	1.4	-0.3	K5 III		380 s		
123123	182244	49 π Hya	14 06 22.2	-26 40 56	+0.003	-0.14	3.27	1.12	-0.1	K2 III	+27	47 s		
123254	120289		14 06 24.9	+7 00 17	+0.002	-0.02	7.3	1.1	0.2	K0 III		270 s		
122879	241439		14 06 25.0	-59 42 55	-0.001	+0.01	6.42	0.12	-6.2	B0 I		2100 s		
123351	63932		14 06 26.0	+30 50 47	+0.002	-0.14	7.5	1.0	0.2	K0 III		130 mx		
123303	100852		14 06 29.5	+16 58 12	-0.001	+0.01	6.7	1.6	-0.5	M2 III	+14	270 s		
123110	205184		14 06 29.8	-34 53 54	-0.003	0.00	7.0	1.8	-0.5	M2 III		230 s		
123323	100853		14 06 33.3	+17 54 43	-0.004	-0.01	7.70	1.06	0.2	K0 III		240 mx		
122926	241443		14 06 38.6	-59 45 19	+0.001	-0.03	7.4	0.3	0.4	B9.5 V		150 s		
123139	205188	5 θ Cen	14 06 40.9	-36 22 12	-0.043	-0.52	2.06	1.01	1.7	K0 III-IV	+1	14 ts		[Menkent], m
123518	29009		14 06 41.2	+53 18 47	0.000	+0.07	7.0	1.1	0.2	K0 III		230 s		
123408	63935		14 06 41.5	+34 46 42	+0.001	-0.01	7.00	1.00	0.2	K0 III	-3	230 s	9112	m
123214	158357		14 06 41.5	-14 12 18	-0.001	-0.02	6.6	1.6	-0.5	M4 III	+8	260 s		
123237	158358		14 06 42.2	-11 30 09	-0.002	-0.05	7.9	1.1	0.2	K0 III		340 s		
123255	139736	95 Vir	14 06 42.7	-9 18 48	-0.010	+0.01	5.46	0.34	2.4	dA8 n	-36	41 s		
122905	252598		14 06 47.3	-62 22 26	-0.002	0.00	7.6	1.6	-0.5	M2 III		420 s		
123324	120293		14 06 53.8	+0 57 10	-0.003	-0.06	8.0	1.1	0.2	K0 III		170 mx		
123409	83172		14 06 55.7	+28 26 16	-0.006	-0.01	6.90	0.99		K0	-55			
124063	7953	3 UMi	14 06 56.1	+74 35 37	-0.016	+0.01	6.4	1.1	1.7	A3 V	-4	85 s		
123112	224687		14 06 58.0	-47 35 21	-0.003	-0.02	6.9	0.1	0.6	A0 V		150 s		
122938	252599		14 07 05.1	-63 26 43	+0.001	-0.01	6.70							m
123244	205197		14 07 07.1	-31 24 00	-0.008	+0.01	7.6	0.5	3.0	F2 V		66 s		
123222	205196		14 07 07.7	-35 29 30	-0.004	0.00	7.19	0.96	5.3	G6 V		17 s		
123573	44895		14 07 11.5	+46 50 12	-0.009	+0.05	8.0	0.5	4.0	F8 V		62 s		
123307	158363		14 07 13.7	-16 11 25	-0.003	-0.01	7.0	0.0	0.4	B9.5 V		210 s		
123532	44896		14 07 21.0	+42 05 55	+0.001	-0.02	7.2	1.1	0.2	K0 III		250 s		
123056	252605		14 07 25.6	-60 28 15	0.000	-0.02	8.0	0.0	-4.4	O9.5 V		2100 s		
123058	252606		14 07 29.4	-61 33 44	-0.008	-0.03	7.6	0.6	3.0	F2 V		58 s		
123148	241462		14 07 39.5	-56 53 26	0.000	0.00	7.94	0.02	0.4	B9.5 V		320 s		
123247	224703		14 07 40.6	-48 42 16	-0.005	-0.05	6.8	0.4	0.6	A0 V		82 mx		
123227	224702		14 07 41.2	-49 52 03	-0.004	+0.11	7.20	0.7	4.4	G0 V		36 s		m
--	--		14 07 45.0	-53 41 14			7.60							m
122455	257114		14 07 46.7	-79 15 08	-0.015	-0.03	7.4	1.1	0.2	K0 III		270 s		
123344	205209		14 07 47.3	-34 56 57	-0.002	0.00	7.2	1.1	0.6	A0 V		47 s		
123472	120297		14 07 50.0	-0 03 22	-0.001	-0.01	7.3	1.4	-0.3	K5 III		340 s		
123519	100860		14 07 52.5	+16 52 00	-0.001	-0.02	7.9	1.1	-0.1	K2 III		390 s		
123657	44901		14 07 55.7	+43 51 16	+0.001	-0.03	5.27	1.59		Mb III	-36	18 mn		
123691	44904		14 07 56.5	+48 12 38	-0.008	-0.02	6.8	0.4	3.0	F2 V		57 s		
123169	252613		14 08 00.3	-60 46 31	-0.001	+0.01	8.0	-0.1	0.0	B8.5 V		410 s		
123453	158372		14 08 04.0	-12 55 42	+0.007	-0.11	7.63	0.58		G0	+14		9115	m
122342	258690		14 08 04.2	-80 45 48	-0.010	+0.02	7.8	-0.6	2.6	F0 V		110 s		
123151	252616		14 08 14.0	-63 12 29	+0.001	+0.03	6.40	1.02		K0				
123612	83190		14 08 15.8	+24 18 54	-0.001	0.00	6.6	1.4	-0.3	K5 III		240 s		
123782	44905	13 Boo	14 08 17.2	+49 27 29	-0.007	+0.06	5.25	1.65		gMa	-13	19 mn		m
123428	205222		14 08 21.5	-33 03 10	-0.004	-0.02	7.26	1.02	1.7	K0 III-IV		120 s		
123614	100864		14 08 23.1	+17 06 28	-0.003	+0.02	7.7	1.1	0.2	K0 III		310 s		
123188	252617		14 08 23.8	-62 51 48	+0.001	0.00	8.0	0.0	0.4	B9.5 V		320 s		
122862	257116		14 08 27.1	-74 51 01	-0.061	+0.17	6.02	0.58	4.5	G1 V		20 s		
123523	158379		14 08 28.9	-11 49 46	-0.001	-0.01	6.8	0.1	1.4	A2 V		120 s		
123673	100870		14 08 43.8	+17 38 19	+0.001	0.00	7.80	0.8	0.3	G6 III	-8	320 s		
123977	29019		14 08 45.9	+59 20 15	-0.017	-0.03	6.46	1.02	0.2	K0 III	+11	130 mx		
124547	7958	4 UMi	14 08 50.8	+77 32 51	-0.011	+0.03	4.82	1.36	-0.2	K3 III	+11	90 s		
123445	224721		14 08 51.8	-43 28 16	-0.002	-0.03	6.17	-0.06		B9				m
123432	224720		14 08 55.6	-48 23 03	0.000	-0.03	7.8	0.4	0.2	K0 III		330 s		
123171	252620		14 08 55.7	-68 08 26	-0.013	-0.03	8.0	1.1	0.2	K0 III		220 mx		
123335	241478		14 08 56.3	-59 16 35	-0.001	-0.02	6.34	0.05	-1.6	B5 IV	+3	320 s		
123598	158383		14 09 00.2	-19 14 44	0.000	-0.06	7.20	1.6	-0.5	M3 III	+58	190 mx		
123630	158385	96 Vir	14 09 00.5	-10 20 04	-0.001	+0.02	6.47	1.00	0.3	gG7	-20	160 s		
123618	158384		14 09 01.2	-18 30 03	-0.004	+0.02	7.7	0.9	3.2	G5 IV		80 s		
123712	120311		14 09 11.1	+7 23 06	-0.002	-0.01	7.20	1.1	0.2	K0 III		250 s	9127	m
124169	16290		14 09 19.8	+65 33 30	-0.002	+0.01	8.0	1.1	-0.1	K2 III		410 s		
123913	44914		14 09 25.0	+40 02 20	-0.003	-0.02	7.9	1.1	0.2	K0 III		350 s		
123603	205242		14 09 25.0	-34 52 58	-0.004	-0.05	7.7	0.5	4.0	F8 V		55 s		
123760	100878		14 09 26.4	+10 14 38	-0.007	-0.15	7.90	0.9	5.2	G5 V	-1	35 s		
124290	16291		14 09 26.4	+69 35 25	-0.006	+0.01	7.1	0.0	0.4	B9.5 V		150 mx		
123583	205241		14 09 26.9	-38 25 58	-0.001	-0.03	7.9	-0.4	1.4	A2 V		200 s		
123464	241487		14 09 27.6	-55 21 41	-0.006	+0.02	7.8	0.8	3.4	F5 V		45 s		
123739	120316		14 09 28.1	+2 47 51	-0.002	0.00	6.9	1.1	0.2	K0 III		220 s		
123435	241485		14 09 31.0	-58 34 24	-0.002	-0.03	7.9	1.5	3.2	G5 IV		32 s		

HD	SAO	Star Name	α 2000	δ 2000	μ(α)	μ(δ)	V	B-V	M_v	Spec	RV	d(pc)	ADS	Notes
123844	83196		14 09 31.4	+20 28 17	-0.011	-0.04	8.0	0.4	3.0	F2 V		67 mx		
123515	241491		14 09 34.8	-51 30 18	-0.005	-0.02	6.00	-0.06		B9				m
123845	100882		14 09 37.2	+15 37 17	+0.004	-0.07	6.7	0.5	3.4	F5 V	-2	46 s		
123804	100881		14 09 38.6	+10 15 31	-0.004	-0.01	8.0	0.4	2.6	F0 V		120 s		
123877	83199		14 09 42.0	+25 50 45	+0.001	0.00	7.93	1.49	-0.3	K5 III		440 s		
123944	63962		14 09 44.4	+37 19 47	-0.004	0.00	7.6	0.4	2.6	F0 V		100 s		
123929	63961		14 09 44.9	+31 15 23	+0.001	+0.09	7.26	0.85	0.2	K0 III		200 mx		
123860	100884		14 09 45.1	+18 19 36	-0.001	-0.03	8.0	1.1	0.2	K0 III		360 s		
123879	100885		14 09 46.4	+19 44 24	-0.003	0.00	7.75	1.29	0.2	K0 III		190 s		
123825	139765		14 09 53.7	-0 39 06	-0.002	-0.03	7.4	1.4	-0.3	K5 III		340 s		
123569	241496		14 09 54.7	-53 26 20	-0.016	-0.10	4.75	0.94	0.3	G5 III	-17	69 s		m
123635	224735		14 09 55.7	-44 17 02	+0.002	-0.05	7.70	0.00	0.4	B9.5 V		290 s		m
123880	100886		14 09 56.8	+11 22 24	+0.001	-0.03	8.0	0.1	1.4	A2 V		210 s		
123449	252630		14 09 59.4	-62 47 56	-0.006	-0.02	8.0	1.1	-0.1	K2 III		350 mx		
124018	44919		14 09 59.5	+40 46 39	-0.002	-0.03	7.1	1.1	-0.1	K2 III		280 s		
124101	29024		14 09 59.9	+52 15 11	-0.004	+0.03	7.9	1.1	0.2	K0 III		340 s		
124369	16297		14 10 08.7	+68 51 47	-0.001	-0.01	6.5	1.1	0.2	K0 III		180 s		
123664	224737		14 10 11.0	-45 54 51	0.000	-0.02	7.7	0.6	0.6	A0 V		110 s		
123767	182302		14 10 12.2	-25 24 02	-0.006	0.00	7.7	-0.2	3.4	F5 V		71 s		
123746	205253		14 10 13.0	-30 37 14	-0.003	-0.02	6.8	0.7	1.4	A2 V		51 s		
123769	182305		14 10 23.4	-30 05 16	+0.001	0.00	7.00	0.1	1.4	A2 V		130 s	9135	m
123999	83203	12 Boo	14 10 23.9	+25 05 30	-0.002	-0.06	4.83	0.54	2.4	F8 IV	+11	29 ts		d Boo
124085	63968		14 10 25.7	+39 31 34	0.000	-0.06	7.9	1.1	0.2	K0 III		270 mx		
123919	139770		14 10 28.6	-6 33 35	-0.003	-0.01	7.7	1.1	0.2	K0 III		320 s		
123377	257119		14 10 30.6	-70 18 18	-0.003	0.00	6.05	1.74		K0				m
123814	205259		14 10 34.4	-33 14 05	0.000	+0.04	7.8	1.2	-0.1	K2 III		380 s		
123732	224743		14 10 41.4	-47 46 08	-0.007	-0.02	7.0	0.1	4.0	F8 V		54 s		
123790	205260		14 10 42.1	-39 42 54	0.000	+0.02	8.0	0.6	3.2	G5 IV		93 s		
124034	100894		14 10 43.3	+15 17 27	-0.001	0.00	7.0	1.1	-0.1	K2 III		260 s		
123590	252638		14 10 43.9	-62 28 43	+0.001	+0.01	7.62	0.14	-1.0	B5.5 V		450 s		
123979	120327		14 10 44.8	+5 48 18	-0.002	+0.02	8.0	0.4	2.6	F0 V		120 s		
123980	120326		14 10 47.7	+0 47 58	-0.003	-0.04	6.7	0.1	1.4	A2 V		120 s		
123718	241505		14 10 47.7	-51 21 02	-0.003	-0.03	7.5	1.2	0.2	K0 III		230 s		
123591	252639		14 10 50.3	-63 02 32	-0.005	+0.04	8.0	1.1	0.2	K0 III		190 mx		
123934	158401		14 10 50.4	-16 18 07	0.000	-0.01	4.91	1.72	-0.5	gM3	+18	97 s		
123888	205270		14 10 51.9	-32 04 33	0.000	-0.01	6.6	1.6	0.2	K0 III		78 s		
123905	182316		14 10 54.5	-25 19 17	-0.002	-0.01	7.7	1.4	0.2	K0 III		190 s		
123982	139773		14 10 56.9	-5 58 26	-0.004	-0.01	7.4	0.5	4.0	F8 V		47 s		
124138	63975		14 10 57.1	+31 52 25	-0.011	-0.02	8.0	0.4	3.0	F2 V		83 mx		
123906	182317		14 10 58.1	-28 21 07	-0.001	-0.01	7.4	0.8	3.2	G5 IV		64 s		
124002	139774		14 10 58.5	-2 40 27	0.000	+0.02	7.80	1.1	0.2	K0 III		330 s	9140	m
124319	29033		14 11 01.3	+58 32 56	+0.003	-0.06	6.9	1.1	-0.1	K2 III		250 s		
123492	252636		14 11 01.9	-69 43 11	-0.005	-0.03	6.06	0.18		A3				
123797	224747		14 11 02.3	-48 46 57	-0.008	-0.08	6.62	0.92	3.2	G5 IV		39 s		
123983	158405		14 11 07.7	-11 57 10	-0.002	-0.05	7.5	1.1	0.2	K0 III		280 s		
123774	241513		14 11 11.4	-52 12 44	-0.002	-0.02	7.9	1.4	-0.3	K5 III		430 s		
124186	63979		14 11 15.0	+32 17 43	-0.002	+0.02	6.11	1.26	-0.1	K2 III	-22	150 s		
124330	29034		14 11 17.1	+54 24 32	-0.018	-0.04	7.80	0.8	3.1	G4 IV	-30	50 mx		
123951	182319		14 11 19.3	-29 47 06	0.000	-0.03	7.4	1.5	0.2	K0 III		130 s		
123969	182323		14 11 20.8	-26 20 38	0.000	+0.02	7.8	1.4	-0.3	K5 III		410 s		
123775	241515		14 11 23.1	-56 22 59	-0.005	-0.06	7.2	1.1	0.2	K0 III		230 s		
123816	241519		14 11 23.8	-52 17 44	-0.002	-0.01	7.5	1.1	0.2	K0 III		260 s		
123869	224751		14 11 28.8	-49 26 06	-0.002	-0.01	7.9	0.4	3.2	G5 IV		88 s		
123990	205284		14 11 29.3	-32 10 06	+0.002	0.00	7.3	0.6	3.0	F2 V		52 s		
124115	120334		14 11 31.1	+1 21 44	-0.008	+0.02	6.43	0.48	3.4	dF5	-18	38 s		
124026	182330		14 11 36.0	-25 10 53	0.000	-0.01	7.7	1.6	0.2	K0 III		140 s		
124291	44929		14 11 36.9	+42 20 27	-0.005	+0.01	6.7	1.1	0.2	K0 III		200 s		
124117	139778		14 11 41.5	-5 51 27	-0.004	-0.01	7.2	1.1	0.2	K0 III		250 s		
124087	158414		14 11 45.9	-19 01 02	+0.001	0.00	7.90	1.1	0.2	K0 III		350 s		m
124106	158416		14 11 46.1	-12 36 41	-0.018	-0.17	7.92	0.87	5.9	dK0	+8	25 s		
123779	252644		14 11 47.7	-62 01 36	-0.003	-0.02	7.6	0.1	0.4	B9.5 V		220 s		
124159	139780		14 11 52.1	-6 07 35	-0.004	+0.02	7.4	0.2	2.1	A5 V		120 s		
124730	16305		14 12 04.0	+69 25 57	-0.005	-0.05	5.2	1.6	-0.5	M2 III	-23	140 s		
124270	83215		14 12 10.6	+20 38 33	-0.002	-0.02	7.6	0.2	2.1	A5 V		130 s		
124224	120339		14 12 15.7	+2 24 34	-0.003	-0.03	5.01	-0.12		B9 p	+3	34 mn	9152	CU Vir, m,v
124162	182343		14 12 24.4	-24 21 50	+0.001	-0.03	6.4	1.4	0.2	K0 III		96 s		
124346	83219		14 12 26.5	+28 43 01	-0.004	0.00	7.80	0.39		F2			9158	m
124248	139785		14 12 33.4	-9 54 01	+0.003	-0.02	7.3	0.4	2.6	F0 V	0	88 s		
123029	--		14 12 34.6	-81 32 41			8.0	1.4	-0.3	K5 III		460 s		
124013	241530		14 12 35.3	-55 28 43	-0.002	-0.03	8.00	1.1	0.2	K0 III		340 s		m
124347	83220		14 12 37.6	+22 22 30	+0.002	-0.04	7.97	0.43		F0				
123994	241531		14 12 44.1	-58 19 00	+0.004	+0.04	8.0	-0.5	4.0	F8 V		62 s		
124292	139790		14 12 45.1	-3 19 12	-0.011	-0.32	7.9	0.7	4.4	G0 V		29 mx		
124206	182349	50 Hya	14 12 45.9	-27 15 40	-0.001	-0.04	5.1	1.1	0.2	K0 III	+27	98 s		
124321	120344		14 12 47.0	+8 00 28	+0.001	-0.04	7.9	1.1	0.2	K0 III		340 s		

HD	SAO	Star Name	α 2000	δ 2000	μ(α)	μ(δ)	V	B−V	M_v	Spec	RV	d(pc)	ADS	Notes
			$14^h12^m53^s.5$	$-41°38'34''$	$-0^s.004$	$-0''.01$								
124163	224766		14 12 53.5	−41 38 34	−0.004	−0.01	7.1	0.3	−0.3	K5 III		310 s		
124294	158427	98 κ Vir	14 12 53.6	−10 16 25	0.000	+0.14	4.19	1.33	−0.2	K3 III	−4	71 s		
124176	224767		14 12 55.1	−40 50 00	−0.001	−0.01	7.0	0.3	0.6	A0 V		130 s		
124226	182350		14 12 56.1	−25 30 13	−0.002	+0.02	7.40	0.9	3.2	G5 IV		69 s		
124568	29041		14 12 56.2	+50 42 46	+0.001	0.00	7.7	0.1	0.6	A0 V		270 s		
124459	63990		14 12 57.5	+35 36 02	−0.001	−0.01	8.04	0.48	4.0	F8 V		64 s		
124304	158431		14 13 09.6	−13 51 36	−0.003	−0.02	7.20	1.6		M5 II	−45			
124281	182354		14 13 13.1	−26 36 44	+0.001	−0.02	6.2	1.2	0.2	K0 III	−10	120 s		
124264	205315		14 13 13.9	−33 34 09	+0.003	−0.03	7.9	0.7	3.0	F2 V		56 s		
124147	241543		14 13 16.3	−53 39 57	−0.001	−0.02	5.4	1.6	0.2	K0 III	−3	47 s		
124353	139794		14 13 18.5	−8 26 37	−0.001	+0.03	7.0	1.1	0.2	K0 III		230 s		
124818	16310		14 13 21.8	+62 42 15	−0.005	0.00	7.7	0.4	2.6	F0 V		110 s		
124283	205320		14 13 23.8	−33 31 06	−0.002	−0.03	7.8	0.2	2.6	F0 V		110 s		
124675	29045	17 κ Boo	14 13 27.6	+51 47 15	+0.005	−0.02	4.40	0.20	1.5	A7 IV	−16	38 s	9173	m
−−	100920		14 13 30.8	+18 09 44	0.000	−0.13	7.00							m
122470	258691		14 13 33.5	−84 32 38	−0.034	+0.02	6.9	0.8	0.2	K0 III		220 s		
124254	224774		14 13 38.5	−45 59 09	+0.003	+0.02	7.4	0.0	1.7	A3 V		140 s		
124548	83231		14 13 39.3	+24 46 42	−0.004	+0.03	8.03	0.05		A0				
124195	241552		14 13 39.8	−54 37 32	−0.002	−0.02	6.11	0.05		B9				
124425	139798		14 13 40.7	−0 50 44	+0.014	−0.14	5.91	0.47	2.2	F6 IV	+18	54 s		
124793	29050		14 13 40.7	+58 29 48	+0.001	−0.01	7.9	1.1	0.2	K0 III		340 s		
124401	158439		14 13 42.9	−11 50 19	+0.001	−0.03	6.98	1.02		G5				
124324	205325		14 13 46.0	−36 37 18	−0.006	−0.06	7.2	0.7	2.6	F0 V		48 s		
124586	63999		14 13 46.1	+31 11 43	−0.003	−0.01	7.52	−0.08	0.6	A0 V	−12	240 s		
124795	29051		14 13 46.9	+56 19 06	−0.003	+0.01	8.0	1.1	0.2	K0 III		360 s		
124641	44943		14 13 47.3	+41 11 16	−0.005	+0.02	8.00	1.1	−0.1	K2 III		310 mx	9175	m
124410	158440		14 13 47.4	−13 20 27	0.000	0.00	7.9	1.4	−0.3	K5 III		440 s		
124517	100922		14 13 49.4	+11 59 52	−0.001	−0.03	6.70	1.1	0.2	K0 III		200 s	9168	m
124694	44946		14 13 51.2	+46 19 32	−0.015	+0.03	7.19	0.52	4.4	G0 V		36 s		
124493	120354		14 13 54.1	+7 07 17	+0.001	−0.05	7.7	1.1	0.2	K0 III		310 s		
124587	83235		14 13 54.6	+29 06 18	+0.003	−0.02	6.8	0.4	2.6	F0 V	−8	69 s	9174	m
124642	64005		14 13 56.9	+30 13 00	−0.032	+0.16	8.0	1.4	−0.3	K5 III		42 mx		
124731	44949		14 13 58.8	+45 25 27	0.000	−0.04	7.3	1.1	0.2	K0 III		240 s		
124167	252656		14 14 02.4	−62 13 53	−0.006	+0.01	8.0	1.2	0.2	K0 III		290 s		
124570	100925	14 Boo	14 14 05.1	+12 57 34	−0.018	−0.06	5.54	0.54	4.0	F8 V	−39	24 mn		
124819	29053		14 14 05.2	+54 38 15	0.000	0.00	7.7	0.5	3.4	F5 V		73 s		
124051	252652		14 14 05.9	−68 11 25	−0.001	−0.01	7.6	0.1	0.6	A0 V		240 s		
124549	120357		14 14 08.7	+7 52 52	−0.001	0.00	7.9	1.4	−0.3	K5 III		430 s		
124309	224777		14 14 10.0	−48 34 38	−0.001	0.00	7.9	0.7	0.2	K0 III		340 s		
124297	241558		14 14 11.1	−51 09 45	−0.001	−0.02	7.1	0.7	3.2	G5 IV		60 s		
124696	64012		14 14 11.4	+35 36 13	−0.004	−0.01	6.90	1.60	−0.1	K2 III		140 s		
124605	100926		14 14 12.1	+18 05 07	−0.006	−0.13	7.5	0.7	4.4	G0 V		42 s		
124553	139806		14 14 21.2	−5 56 51	−0.021	+0.08	6.36	0.60	4.0	dF8	−33	26 s		
124755	44952		14 14 23.4	+41 31 07	−0.002	−0.11	6.24	1.04	−0.1	K2 III	−10	120 mx		
123842	257128		14 14 29.6	−75 45 48	0.000	−0.10	8.00	0.1	1.4	A2 V		77 mx		m
123961	−−		14 14 30.7	−73 30 21			7.8	1.1	0.2	K0 III		320 s		
124796	64019		14 14 33.4	+39 16 37	+0.001	+0.03	7.3	0.9	3.2	G5 IV		67 s		
124182	252660		14 14 35.1	−66 09 15	+0.001	−0.01	6.94	−0.01	−1.3	B6 IV		420 s		m
124678	100929		14 14 35.5	+13 58 32	−0.007	+0.08	6.6	1.1	−0.1	K2 III		140 mx		
124197	252661		14 14 39.2	−65 42 07	−0.005	−0.03	6.71	−0.02	−0.9	B6 V		130 mx		m
124713	83242		14 14 40.9	+21 52 24	+0.003	−0.01	6.39	0.18	2.4	A8 V	−4	63 s		
124502	182368		14 14 41.6	−27 45 42	−0.010	−0.01	7.7	0.3	4.0	F8 V		55 s		
124433	224791		14 14 42.5	−41 50 15	−0.012	−0.02	5.61	0.93	0.3	gG5		110 s		
124155	−−		14 14 42.7	−68 43 17			7.7	0.4	2.6	F0 V		100 s		
124503	205343		14 14 43.1	−31 03 06	−0.001	−0.07	7.1	0.8	2.6	F0 V		40 s		
124679	100934	15 Boo	14 14 50.8	+10 06 03	−0.002	−0.16	5.29	1.00	0.2	K0 III	+17	82 mx		m
124575	182371		14 14 51.7	−21 03 53	−0.004	0.00	7.70	1.4	−1.3	K5 II−III	−21	78 mx		
−−	7973		14 14 52.2	+75 37 02	+0.002	−0.01	7.9	1.8						
124681	120364		14 14 52.9	+3 20 10	−0.003	−0.02	6.45	1.60	−0.5	gM4	−48	250 s		
124367	241563		14 14 56.9	−57 05 09	−0.004	−0.01	5.07	−0.08	−1.7	B3 V e	+19	36 mx		m
124576	182374		14 15 01.2	−29 16 55	−0.002	−0.02	6.0	1.1	0.6	A0 V		25 s		
124483	224794		14 15 01.4	−40 54 02	−0.001	0.00	6.9	0.9	−0.1	K2 III		250 s		
124797	83251		14 15 01.5	+23 41 10	−0.002	−0.03	6.7	0.4	2.6	F0 V	+6	67 s		
124314	252665		14 15 01.6	−61 42 25	0.000	−0.01	6.64	0.21		O8 nn				m
124328	252668		14 15 08.4	−61 55 03	+0.003	−0.01	7.0	1.9	−0.3	K5 III		170 s		
124239	252663		14 15 14.1	−67 39 35	−0.003	−0.08	7.60	0.6	4.4	G0 V		43 s		m
124539	205351		14 15 15.3	−40 06 02	−0.010	−0.03	7.7	0.4	2.6	F0 V		90 s		
125019	29059		14 15 16.7	+52 32 09	−0.004	−0.01	6.58	0.10		A2	−15			
124757	120370		14 15 19.3	+3 07 51	−0.013	+0.03	7.05	0.54	3.8	dF7	−45	35 ts	9182	m
124454	241569		14 15 21.1	−53 30 35	−0.003	0.00	6.39	1.57		K2				
124703	158449		14 15 22.9	−12 14 34	−0.003	−0.03	7.9	0.2	2.1	A5 V		150 s		
124683	158448		14 15 24.0	−18 12 03	−0.003	−0.02	5.43	−0.03		B9	−19			
124883	83254		14 15 26.4	+27 44 00	+0.002	−0.01	7.29	0.25	1.4	A2 V	−4	120 s		
125020	29061		14 15 27.6	+50 26 24	+0.003	−0.04	8.00	0.5	3.4	F5 V		83 s	9187	m
124968	64028		14 15 35.2	+39 40 35	+0.001	−0.01	7.6	0.9	3.2	G5 IV		74 s		

HD	SAO	Star Name	α 2000	δ 2000	μ(α)	μ(δ)	V	B-V	M$_v$	Spec	RV	d(pc)	ADS	Notes
			h m s	° ′ ″	s	″								
124385	252672		14 15 38.2	-63 06 50	+0.001	0.00	7.5	1.1	0.2	K0 III		280 s		
124580	224798		14 15 38.6	-45 00 03	+0.013	-0.15	6.31	0.60	5.0	G4 V		18 s		
124897	100944	16 α Boo	14 15 39.6	+19 10 57	-0.077	-2.00	-0.04	1.23	-0.2	K2 III p	-5	11 t		Arcturus
124801	139818		14 15 42.1	-7 39 55	+0.001	0.00	7.4	1.1	0.2	K0 III		280 s		
124316	—		14 15 48.4	-68 32 11			7.6	0.1	0.6	A0 V		240 s		
124759	158453		14 15 49.9	-16 04 52	-0.005	+0.01	8.0	1.1	0.2	K0 III		330 mx		
124436	252674		14 15 57.2	-62 29 18	-0.004	+0.03	7.9	1.0	0.2	K0 III		270 mx		
125306	16321		14 15 59.5	+64 51 09	0.000	-0.02	7.2	0.1	0.6	A0 V		210 s		
125193	29070		14 15 59.8	+56 41 20	0.000	-0.10	6.60	0.5	4.0	F8 V	-29	56 s	9197	m
124850	139824	99 ι Vir	14 16 00.8	-6 00 02	-0.001	-0.43	4.08	0.52	0.7	F6 III	+12	22 mx		Syrma
125076	44963		14 16 02.2	+45 33 58	-0.004	+0.06	6.53	1.03	0.2	K0 III		180 s		
124953	100949		14 16 04.1	+18 54 42	+0.003	-0.03	5.98	0.26		A m	+4			
125141	29068		14 16 05.2	+50 10 09	-0.032	+0.19	7.7	0.8	3.2	G5 IV		41 mx		
125229	29074		14 16 08.3	+56 42 44	-0.006	-0.01	7.10	0.4	2.6	F0 V		56 s	9197	m
125161	29071	21 ι Boo	14 16 09.8	+51 22 02	-0.017	+0.09	4.75	0.20	2.4	A7 V	-17	28 ts	9198	m
124986	83257		14 16 11.1	+21 14 35	+0.002	-0.01	7.7	0.5	4.0	F8 V		55 s		
124780	205371		14 16 18.2	-33 14 29	+0.002	0.00	6.6	1.2	2.6	F0 V		19 s		
124811	205372		14 16 18.5	-32 03 30	-0.004	-0.03	7.1	0.8	1.4	A2 V		47 s		
124915	139828		14 16 21.4	-6 37 18	-0.002	-0.02	6.44	0.28		A3				
124649	241582		14 16 21.8	-53 19 37	-0.007	-0.04	7.3	1.6	-0.3	K5 III		210 mx		
125162	44965	19 λ Boo	14 16 22.9	+46 05 18	-0.018	+0.16	4.18	0.08		A0 p	-8	29 t		
125111	64040		14 16 24.1	+39 44 41	-0.013	0.00	6.38	0.36	3.0	F2 V	-24	47 s		
125112	64038		14 16 24.3	+37 07 47	-0.002	-0.02	8.0	0.9	3.2	G5 IV		93 s		
125272	29078		14 16 26.1	+58 04 59	-0.019	+0.12	8.00	0.5	4.2	F9 V	-22	58 s		
124931	139830		14 16 30.0	-3 11 47	-0.002	-0.04	6.15	-0.01		A0	+2			
125260	29077		14 16 30.6	+55 17 54	+0.005	-0.04	6.90	1.13	-0.1	K2 III		250 s		
124620	241583		14 16 31.7	-57 17 51	-0.004	-0.02	7.18	0.20		A0				m
125040	83259		14 16 32.7	+20 07 17	-0.011	-0.11	6.25	0.49	3.4	F5 V	-8	35 s	9192	m
125113	64041		14 16 33.8	+32 12 31	+0.001	+0.01	6.73	0.93	3.2	G5 IV		41 s		
124601	241580		14 16 34.2	-59 54 50	-0.002	-0.02	5.2	1.6		M5 III	-20			R Cen, m,v
124471	252678		14 16 38.6	-66 35 16	-0.001	-0.01	5.75	-0.06	-5.7	B2 Ib	-20	1700 s		m
124622	241585		14 16 41.2	-58 19 31	+0.002	-0.02	7.2	1.1	0.2	K0 III		230 s		
124856	205380		14 16 42.5	-35 00 30	-0.002	-0.01	7.5	0.4	-0.1	K2 III		330 s		
125273	29079		14 16 44.6	+53 15 50	-0.003	+0.02	7.6	0.2	2.1	A5 V		130 s		
124602	252684		14 16 47.0	-61 11 03	+0.001	-0.01	8.0	1.3	0.2	K0 III		250 s		
124973	139834		14 16 48.6	-8 53 05	-0.001	-0.03	6.5	1.1	0.2	K0 III		180 s		
124988	139835		14 16 49.2	-5 38 39	0.000	-0.02	6.9	0.4	2.6	F0 V		73 s		
124099	257131		14 16 54.9	-77 39 50	-0.002	+0.01	6.47	1.42		K0				
124689	241587		14 16 57.0	-57 51 17	-0.008	-0.04	7.2	0.4	3.0	F2 V		70 s		RR Cen, m,v
125056	120379		14 17 00.5	+3 05 17	+0.010	+0.03	7.9	0.9	3.2	G5 IV		88 s		
124136	257132		14 17 03.5	-77 40 52	-0.011	-0.01	7.7	1.1	0.0	K1 III		350 s		
124990	158462		14 17 03.7	-18 35 07	0.000	0.00	6.4	0.9	3.2	G5 IV		43 s		
125333	29083		14 17 03.7	+54 32 34	-0.019	-0.06	7.10	0.8	3.2	G5 IV		60 s	9208	m
125043	139840		14 17 05.5	-5 44 21	-0.007	+0.02	7.7	1.1	0.2	K0 III		210 mx		
124531	—		14 17 06.7	-68 24 05			8.0	0.0	0.0	B8.5 V		390 s		
124584	—		14 17 07.7	-66 19 04			7.4	0.6	4.4	G0 V		40 s		
124857	224815		14 17 09.3	-46 54 39	-0.001	-0.01	7.1	1.8	-0.3	K5 III		190 s		
—	—		14 17 19.7	+66 47 39	-0.002	+0.01	7.40							U UMi, v
125349	29086		14 17 21.0	+51 18 26	-0.003	-0.01	6.20	0.04	0.6	A0 V	-11	120 s		
125059	139847		14 17 22.4	-9 01 22	-0.002	-0.01	6.6	1.1	0.2	K0 III		190 s		
124812	241595		14 17 25.0	-55 45 55	-0.004	-0.01	6.63	0.24		B9				
125048	158468		14 17 27.7	-19 57 49	-0.005	-0.02	6.9	0.3	1.7	A3 V		71 mx		
125180	100956		14 17 28.4	+15 15 48	+0.001	+0.01	5.80	1.71	-0.5	M2 III	-10	150 s		
125027	182414		14 17 33.3	-26 57 30	-0.009	+0.01	7.4	0.2	3.4	F5 V		62 s		
125194	100958		14 17 36.9	+11 20 07	-0.002	-0.01	7.6	1.1	0.2	K0 III		300 s		
124961	224822		14 17 39.3	-40 51 42	0.000	-0.03	7.9	-0.2	0.6	A0 V		290 s		
125081	182422		14 17 41.7	-21 49 48	-0.005	-0.05	7.3	0.7	2.6	F0 V		47 s		
124317	257133		14 17 42.8	-76 05 15	-0.003	+0.02	7.2	2.5	-0.5	M2 III		99 s		
125672	16329		14 17 45.4	+68 45 06	-0.010	+0.02	7.4	0.4	3.0	F2 V		75 s		
125435	29089		14 17 46.1	+51 51 30	-0.004	+0.10	7.1	1.1	-0.1	K2 III		190 mx		
125308	64051		14 17 46.5	+33 15 18	-0.007	+0.03	7.6	0.4	3.0	F2 V		83 s		
125469	29090		14 17 47.7	+55 25 31	+0.001	-0.01	6.7	0.1	1.4	A2 V		120 s		
125406	44982		14 17 49.1	+48 00 06	-0.002	-0.05	6.32	0.44	3.4	F5 V	-17	38 s		
124672	252686		14 17 51.4	-68 02 50	-0.019	-0.12	7.7	0.5	4.0	F8 V		54 s		
124749	252689		14 17 51.8	-63 24 47	-0.002	0.00	7.4	0.0	0.4	B9.5 V		240 s		
125007	224823		14 17 52.9	-42 35 29	-0.002	-0.02	7.02	-0.04		B9				
125183	139853		14 17 56.9	-5 58 18	-0.002	0.00	7.8	1.1	-0.1	K2 III		380 s		
125374	44980		14 17 57.8	+39 54 46	-0.005	+0.03	7.4	0.4	3.0	F2 V		75 s		
125185	139854		14 17 58.5	-7 58 02	0.000	-0.05	7.1	1.1	0.2	K0 III		250 s		
125375	64054		14 17 59.1	+37 39 30	0.000	0.00	8.0	0.4	2.6	F0 V		120 s		
125351	64053		14 17 59.7	+35 30 34	0.000	+0.01	4.81	1.06	0.0	K1 III	-26	92 s		
125184	139856		14 18 00.5	-7 32 33	+0.017	-0.25	6.47	0.73	4.4	G0 V		19 s		m
125068	205404		14 18 09.9	-38 53 16	-0.002	-0.21	7.9	1.3	-0.1	K2 III		69 mx		
123998	258693	η Aps	14 18 13.6	-81 00 27	-0.011	-0.06	4.91	0.25		A2 pe	-9			
125352	83283		14 18 21.3	+21 18 15	-0.003	-0.04	7.0	1.1	0.2	K0 III		230 s		

HD	SAO	Star Name	α 2000	δ 2000	μ(α)	μ(δ)	V	B-V	M_v	Spec	RV	d(pc)	ADS	Notes
125150	205412		14^{h}18^{m}23^s.7	-33°13'14"	-0^s.006	-0".05	6.54	0.27	2.6	F0 V		61 s		
125309	120388		14 18 29.9	+3 40 27	+0.001	-0.04	7.1	1.1	0.2	K0 III		240 s		
125011	241615		14 18 30.3	-54 07 45	-0.001	-0.03	7.7	1.1	0.2	K0 III		290 s		
125376	83285		14 18 31.1	+21 28 23	-0.004	-0.02	8.0	1.1	0.2	K0 III		220 mx		
125557	29094		14 18 31.2	+52 01 59	-0.001	-0.01	6.8	0.1	1.4	A2 V		120 s		
125408	83286		14 18 33.7	+28 26 21	-0.005	+0.07	7.6	0.5	3.4	F5 V		69 s		
125335	100968		14 18 35.8	+10 30 37	-0.003	-0.01	7.23	0.33		A m	-28			
126125	7991		14 18 36.6	+75 40 08	-0.003	-0.02	8.0	1.1	0.2	K0 III		360 s		
125248	158481		14 18 38.2	-18 42 58	-0.004	-0.05	5.90	0.00		A p	-9			CS Vir, v
125310	139859		14 18 41.1	-7 00 48	-0.001	0.00	7.3	0.1	1.4	A2 V		150 s		
125071	241622		14 18 42.0	-52 14 14	-0.006	-0.01	7.5	0.1	2.6	F0 V		98 s		
125391	100970		14 18 43.4	+17 21 12	-0.003	0.00	7.8	1.1	0.2	K0 III		330 s		
125294	158485		14 18 49.2	-12 03 47	0.000	0.00	7.9	1.1	-0.1	K2 III		390 s		
125015	241620		14 18 50.6	-58 41 11	-0.004	-0.02	7.28	0.01	0.4	B9.5 V		150 mx		m
125200	205419		14 18 50.8	-37 19 50	+0.001	-0.04	7.8	0.5	2.1	A5 V		92 s		
125558	44991		14 18 52.4	+48 35 32	+0.005	-0.06	7.2	0.2	2.1	A5 V		110 s		
125038	241621		14 18 53.8	-58 58 21	+0.002	-0.04	7.77	0.36	2.6	F0 V		99 s		
125538	64064		14 18 55.6	+38 46 03	0.000	+0.03	6.86	1.06	3.2	G5 IV	-10	33 s		
125632	29098		14 18 55.7	+54 51 50	-0.004	+0.01	6.6	0.1	1.7	A3 V	-3	94 s		
125547	44990		14 18 57.1	+42 00 18	+0.003	-0.07	7.2	1.1	0.2	K0 III		250 s		
125276	182433		14 19 00.7	-25 48 56	-0.027	+0.35	5.87	0.50	3.3	dF4	-21	27 ts	9212	m
125450	100973		14 19 04.1	+19 03 55	0.000	0.00	6.8	1.1	0.2	K0 III		210 s		
125072	241627		14 19 04.4	-59 22 46	-0.061	-0.83	6.66	1.03	6.7	K3 V		9.6 t		
125299	182437		14 19 05.7	-22 57 13	-0.001	-0.02	7.9	-0.1	0.6	A0 V		280 s		
125337	158489	100 λ Vir	14 19 06.5	-13 22 16	-0.001	+0.02	4.52	0.13		A m	-11	20 mn		
125279	182435		14 19 07.1	-27 08 26	-0.001	+0.01	6.5	1.8	-0.1	K2 III		84 s		
125104	241629		14 19 08.8	-55 52 56	-0.003	-0.01	7.33	0.10	0.4	B9.5 V		240 s		
125088	241630		14 19 13.3	-57 25 21	-0.001	-0.04	7.4	2.3	-0.1	K2 III		66 s		
125357	158493		14 19 14.7	-14 24 28	-0.001	-0.02	8.0	1.6	-0.5	M2 III		510 s		
125451	100975	18 Boo	14 19 16.2	+13 00 15	+0.007	-0.03	5.41	0.38	2.1	F5 IV	-2	46 s		m
125283	205430		14 19 23.8	-37 00 12	-0.004	-0.06	5.94	0.08	0.6	A0 V n		65 mx		
125238	224833	ι Lup	14 19 24.1	-46 03 28	-0.001	0.00	3.55	-0.18	-1.7	B3 V	+22	110 s		
125436	139864		14 19 24.4	-2 05 56	-0.006	-0.01	7.8	0.5	3.4	F5 V		74 s		
125379	158495		14 19 25.8	-18 31 24	-0.003	0.00	7.90	0.1	1.4	A2 V		200 s	9218	m
124654	257136		14 19 27.6	-75 10 50	+0.006	-0.06	7.71	0.07	0.4	B9.5 V		100 mx		
125253	224835		14 19 30.5	-46 05 59	-0.004	-0.01	8.0	0.0	1.4	A2 V		120 mx		
125454	139866	102 υ Vir	14 19 32.4	-2 15 55	-0.008	-0.07	5.14	1.02	0.3	G8 III	-27	80 mx		
125117	241633		14 19 33.6	-59 45 09	0.000	-0.02	7.96	0.09	0.6	A0 V		280 s		
125455	139867		14 19 34.6	-5 09 06	-0.044	-0.14	7.58	0.84	5.9	K0 V	-8	21 ts		m
125489	120400		14 19 40.8	+0 23 03	-0.003	-0.02	6.19	0.20		A3 n	-13			
125343	205436		14 19 44.3	-36 51 32	+0.001	-0.05	6.90	1.6	-0.5	M4 III		300 s		m
125560	100980	20 Boo	14 19 45.1	+16 18 24	-0.010	+0.06	4.86	1.23	-0.2	K3 III	-8	100 s		
125642	64072		14 19 47.6	+38 47 38	+0.002	-0.02	6.33	0.05	1.4	A2 V	-11	97 s		
125158	252703		14 19 51.4	-61 16 23	-0.023	-0.10	5.23	0.29		A m	+21	15 mn		
125726	44997		14 19 52.3	+49 17 57	-0.006	+0.04	8.0	0.5	4.0	F8 V		63 s		
125490	139868		14 19 53.1	-6 44 46	-0.006	-0.03	6.5	0.8	3.2	G5 IV		47 s		
126028	16342		14 19 54.9	+67 46 57	-0.006	+0.01	6.80	0.1	0.6	A0 V		140 mx	9231	m
124834	257140		14 19 57.6	-73 58 05	+0.001	-0.02	6.6	-0.1	0.0	B8.5 V		210 s		
125438	182447		14 20 08.4	-28 02 56	-0.001	-0.02	7.1	0.4	0.6	A0 V		120 s		
125658	64074		14 20 08.5	+30 25 45	-0.001	0.00	6.30	0.15		A m	+1			
125206	252705		14 20 09.1	-61 04 54	+0.002	-0.01	7.92	0.26	-4.4	O9.5 V		1500 s		
125383	224838		14 20 09.6	-43 03 32	-0.001	+0.01	5.56	0.92	3.2	G5 IV		23 s		m
125287	241640		14 20 14.0	-55 47 27	-0.006	+0.04	7.11	0.96		G5				
125440	205451		14 20 17.2	-35 23 23	-0.001	0.00	8.0	1.1	-0.1	K2 III		410 s		
125796	45000		14 20 17.7	+48 30 25	-0.006	-0.01	7.39	0.54	4.0	dF8	-18	46 s	9229	
125288	241641		14 20 19.4	-56 23 12	-0.002	-0.02	4.33	0.12	-3.7	B5 II	+5	310 s		
125608	120406		14 20 20.8	+8 34 56	+0.001	-0.03	7.90	0.9	3.2	G5 IV		87 s	9227	
125550	158498		14 20 27.5	-17 31 41	-0.004	0.00	7.7	1.1	0.2	K0 III		310 s		
125609	120407		14 20 28.9	+0 10 53	-0.006	-0.09	6.9	1.1	0.2	K0 III		92 mx		
125473	205453	ψ Cen	14 20 33.3	-37 53 07	-0.006	-0.02	4.05	-0.03	0.0	A0 IV	-4	65 s		m
125317	241645		14 20 34.7	-58 30 58	-0.001	-0.02	7.90	0.00	0.0	B8.5 V		360 s		m
125540	182453		14 20 35.9	-24 30 22	0.000	+0.02	7.7	0.7	0.2	K0 III		320 s		
125509	205456		14 20 37.4	-34 40 40	-0.001	-0.01	7.70	-0.03	0.4	B9.5 V		280 s		
125400	241651		14 20 42.3	-52 19 34	+0.005	-0.05	7.2	0.6	3.4	F5 V		47 s		
125442	224843		14 20 42.4	-45 11 14	+0.003	-0.08	4.77	0.31	2.6	dF0	0	27 s		
125960	29112		14 20 42.7	+57 10 52	-0.003	-0.06	8.0	0.9	3.2	G5 IV		93 s		
125728	83298		14 20 45.1	+26 04 31	+0.001	-0.01	6.78	0.91	-2.1	G8 II	+23	600 s		
125798	64085		14 20 50.5	+36 23 31	-0.005	-0.01	7.36	0.18	0.6	A0 V	-5	75 mx		
125494	224851		14 20 50.6	-42 25 29	+0.002	-0.02	7.60	0.1	1.7	A3 V		150 s		m
125444	224848		14 20 54.8	-49 22 01	-0.005	-0.02	7.6	-0.2	1.4	A2 V		98 mx		
125751	100985		14 21 02.7	+19 43 49	+0.003	-0.03	7.5	0.4	2.6	F0 V		95 s		
125858	64090		14 21 04.8	+38 59 15	0.000	+0.03	8.0	0.5	3.4	F5 V		84 s		
126228	16352		14 21 07.7	+67 48 10	-0.006	+0.01	7.10	0.1	1.4	A2 V		130 mx		m
126168	16351		14 21 08.8	+65 21 49	-0.007	0.00	7.0	0.5	3.4	F5 V		53 s		
125771	100986		14 21 09.2	+19 10 48	-0.001	0.00	7.0	1.4	-0.3	K5 III		290 s		

HD	SAO	Star Name	α 2000	δ 2000	μ(α)	μ(δ)	V	B-V	M$_v$	Spec	RV	d(pc)	ADS	Notes
			h m s	° ′ ″	s	″								
125676	158507		14 21 16.0	-11 23 56	-0.002	-0.03	8.0	0.1	1.7	A3 V		180 s		
125752	100987		14 21 18.3	+10 50 15	+0.002	0.00	7.8	0.5	3.4	F5 V		74 s		
125332	252712		14 21 18.6	-64 14 29	+0.001	0.00	7.8	1.4	-0.3	K5 III		390 s		V418 Cen, v
126565	8000		14 21 21.8	+75 03 39	-0.015	0.00	7.9	1.1	0.2	K0 III		270 mx		
125785	100989		14 21 28.0	+11 38 32	0.000	-0.09	8.0	0.5	4.0	F8 V		62 s		
125531	241662		14 21 29.0	-52 00 55	0.000	-0.01	7.5	0.4	0.2	K0 III		280 s		
125729	158512		14 21 32.7	-10 22 17	-0.002	-0.04	7.4	0.1	1.4	A2 V		160 s		
125666	182462		14 21 40.3	-28 45 03	+0.001	0.00	8.0	1.0	3.2	G5 IV		67 s		
125681	182463		14 21 40.6	-27 19 21	-0.006	-0.04	7.8	0.5	2.6	F0 V		79 s		
125647	205476		14 21 48.2	-38 40 49	-0.003	-0.01	7.8	-0.1	0.6	A0 V		280 s		
125774	158515		14 21 49.2	-10 40 00	+0.001	-0.01	8.0	1.1	0.2	K0 III		360 s		
125466	252718		14 21 55.7	-63 09 30	+0.002	+0.01	7.8	0.0	0.4	B9.5 V		290 s		
125817	139890		14 21 58.4	-1 59 18	-0.005	-0.01	6.8	0.1	1.4	A2 V		77 mx		
126138	29117		14 21 59.7	+53 31 16	-0.002	+0.01	7.4	0.0	0.4	B9.5 V		260 s		
125920	83308		14 22 01.4	+19 48 19	-0.001	-0.03	8.00	1.1	0.2	K0 III		360 s		Y Boo, q
125734	182468		14 22 01.6	-28 18 32	-0.006	-0.02	6.9	1.3	3.2	G5 IV		27 s		
125818	139889		14 22 02.2	-8 05 29	-0.004	-0.04	7.30	1.1	-0.1	K2 III		170 mx	9233	m
125500	---		14 22 02.3	-62 00 39			8.0	2.3	-0.5	M4 III		190 s		
125545	241669		14 22 02.9	-58 17 28	-0.002	-0.01	7.40	0.15	-1.6	B3.5 V		450 s		
125649	224867		14 22 03.6	-44 39 33	+0.001	-0.05	7.74	1.22		G5				
126009	83312		14 22 14.0	+29 22 11	-0.003	-0.03	6.5	1.6		M III	-18			
125389	252717		14 22 16.3	-69 05 13	-0.022	-0.10	7.9	0.5	3.4	F5 V		78 s		
125745	205485		14 22 19.6	-34 47 11	-0.002	+0.01	5.56	-0.09		B8 n	-37			
124771	257142	ε Aps	14 22 22.7	-80 06 31	-0.006	-0.01	5.06	-0.10	-2.0	B4 IV	+5	240 s		
126323	16355		14 22 28.8	+60 57 57	-0.002	-0.08	7.3	0.6	4.4	G0 V		37 s		
125862	182474		14 22 34.8	-20 45 22	-0.001	+0.07	6.8	1.2	0.2	K0 III		160 s		
124845	257143		14 22 36.7	-79 42 59	-0.004	-0.02	7.2	1.6	-0.5	M2 III		350 s		
125628	241673		14 22 36.9	-58 27 33	-0.006	+0.01	4.79	0.80	0.3	G8 III	+15	79 s		m
126324	16356		14 22 37.8	+59 46 45	-0.003	-0.01	7.6	1.1	0.2	K0 III		300 s		
126081	64105		14 22 38.1	+32 30 13	-0.006	-0.02	7.10	1.1	0.2	K0 III		160 mx	9243	m
125906	139897		14 22 38.5	-7 46 05	0.000	-0.13	6.8	0.6	4.4	G0 V	-35	30 s	9237	m
125721	224870		14 22 38.6	-48 19 12	-0.001	-0.02	6.09	-0.13	-1.7	B3 V	-18	360 s		m
125669	241677		14 22 43.2	-55 25 51	-0.001	0.00	7.21	0.11	0.2	B9 V		210 s		
125979	120420		14 22 43.3	+2 16 29	+0.002	+0.02	7.8	0.4	3.0	F2 V		89 s		
125981	139902		14 22 46.3	-0 38 14	-0.004	-0.01	6.7	0.1	0.6	A0 V		80 mx		
126439	16360		14 22 46.6	+64 16 34	0.000	-0.03	7.7	1.1	-0.1	K2 III		370 s		
126051	100998		14 22 49.3	+17 27 12	0.000	-0.01	7.9	1.1	0.2	K0 III		340 s		
126031	100997		14 22 49.6	+14 56 20	-0.002	-0.01	7.55	0.33	1.4	A2 V		120 s		
125805	205495		14 22 51.6	-39 37 29	-0.005	-0.04	7.0	1.0	0.2	K0 III		200 mx		
125687	241679		14 22 52.0	-55 57 43	-0.001	0.00	7.8	1.7	-0.5	M4 III		410 s		
126289	29125		14 22 52.9	+53 48 35	+0.001	-0.03	7.8	1.6		M5 III	-17			S Boo, v
125963	158523		14 22 56.1	-10 45 03	0.000	-0.01	8.0	1.1	-0.1	K2 III		420 s		
125823	205497		14 23 02.1	-39 30 44	-0.002	-0.04	4.42	-0.18	-1.9	B6 III	+8	180 s		
125932	182483	51 Hya	14 23 05.7	-27 45 14	-0.015	-0.12	4.77	1.31	-0.3	K5 III	+20	48 mx		
126141	83321		14 23 06.7	+25 20 18	-0.012	+0.07	6.22	0.37	3.0	F2 V	-10	44 s		
125809	224881		14 23 15.0	-47 25 00	-0.003	0.00	6.5	1.0	0.2	K0 III		180 s		
126053	120424		14 23 15.2	+1 14 30	+0.015	-0.48	6.27	0.63	4.5	G1 V	-18	17 ts		
125866	224884		14 23 16.2	-42 15 12	-0.003	-0.02	7.08	-0.06		B9				
125810	241686		14 23 20.1	-50 46 20	-0.001	-0.01	6.02	1.36	0.2	gK0		89 s		m
126128	120426		14 23 22.6	+8 26 48	-0.005	-0.01	4.87	0.05	2.5	A9 V	-18	33 mn	9247	m
125968	182489		14 23 23.3	-27 49 21	-0.019	-0.26	7.76	0.66	4.2	G5 IV-V		31 mx		
126035	158528	2 Lib	14 23 25.5	-11 42 51	-0.001	-0.06	6.21	0.99	0.3	G7 III	-1	140 s		
125829	224883		14 23 26.9	-49 51 05	-0.001	-0.02	7.9	0.3	0.4	B9.5 V		290 s		
126266	64115		14 23 27.3	+37 12 22	-0.007	+0.04	7.4	0.4	3.0	F2 V		77 s		
126265	64118		14 23 31.2	+39 19 39	+0.002	-0.16	7.2	0.5	4.0	F8 V		44 s		
125630	252725		14 23 31.8	-66 38 41	-0.002	-0.02	6.7	0.5		A2 p				
126304	45026		14 23 33.7	+43 27 31	-0.002	-0.01	7.5	1.1	0.2	K0 III		300 s		
126378	45030		14 23 36.4	+49 24 19	-0.001	-0.01	7.4	1.1	0.2	K0 III		280 s		
126581	16367		14 23 39.9	+64 44 43	-0.001	-0.05	7.3	1.1	-0.1	K2 III		300 s		
126326	45028		14 23 41.8	+40 50 37	-0.008	0.00	6.7	0.5	3.4	F5 V		45 s		
125767	241687		14 23 43.2	-60 12 45	-0.003	-0.01	7.4	1.9	-0.1	K2 III		120 s		
125869	241691		14 23 48.3	-53 10 35	-0.008	-0.03	6.00	1.10		K0				
126379	45031		14 23 50.4	+42 42 35	0.000	+0.03	8.0	1.1	-0.1	K2 III		410 s		
126057	182498		14 23 55.6	-29 40 33	+0.002	0.00	7.0	0.9	1.4	A2 V		43 s		
126086	182502		14 23 57.2	-21 42 30	-0.001	0.00	8.0	1.0	-0.5	M2 III		510 s		
126131	158539		14 23 57.7	-16 06 06	-0.004	0.00	6.7	0.1	0.6	A0 V		120 mx		
126105	158538		14 24 00.2	-19 48 03	+0.002	-0.05	7.70	1.1	0.2	K0 III		320 s		m
126201	120432		14 24 00.6	+5 49 23	-0.003	+0.02	7.8	1.1	0.2	K0 III	-40	330 s		
126200	120433		14 24 00.8	+8 14 37	0.000	-0.03	5.95	0.06		A0	-4			
126307	83329		14 24 01.9	+27 24 47	0.000	-0.01	6.38	1.61	-0.3	K4 III	+31	170 s		
126452	45038		14 24 05.3	+49 31 59	-0.009	+0.04	7.3	0.4	3.0	F2 V		71 s		
126246	101009		14 24 05.5	+11 14 48	+0.004	-0.01	6.7	0.5	3.4	F5 V		46 s	9251	m
126394	45033		14 24 06.9	+41 02 22	+0.001	-0.03	7.6	0.5	3.4	F5 V		70 s		
126453	45036		14 24 07.4	+46 46 16	-0.018	+0.07	8.0	0.5	4.0	F8 V		62 s		
126269	101011		14 24 11.2	+16 16 25	-0.002	0.00	6.8	0.5	3.4	F5 V	-20	48 s		

HD	SAO	Star Name	α 2000	δ 2000	μ(α)	μ(δ)	V	B-V	M_v	Spec	RV	d(pc)	ADS	Notes
126248	120434		14ʰ 24ᵐ 11.3ˢ	+5° 49' 12"	-0.005ˢ	0".00	5.10	0.12		A3 n	-5	15 mn		
126089	205523		14 24 14.7	-33 05 13	-0.001	-0.01	6.9	1.0	0.2	K0 III		200 s		
126271	120436		14 24 18.2	+8 05 06	-0.008	-0.10	6.19	1.20	-0.3	gK4	-31	200 s		
126146	182510		14 24 22.5	-26 45 33	0.000	-0.04	8.0	0.5	0.6	A0 V		140 s		
124639	258697		14 24 23.2	-82 50 55	+0.004	-0.02	6.42	0.02		B8 n	+27			
125880	252732		14 24 32.4	-61 23 21	+0.001	0.00	7.68	0.25		B9				
124862	--		14 24 33.2	-82 13 50			8.0	1.1	0.2	K0 III		360 s		
126273	139916		14 24 35.7	-2 20 36	-0.002	-0.01	7.18	1.62		Ma	-28			
126213	158546		14 24 35.7	-14 04 52	-0.006	0.00	7.7	0.9	3.2	G5 IV		80 s		
126062	224901		14 24 36.9	-47 10 41	-0.002	-0.04	8.0	-0.5	0.6	A0 V		300 s		
126531	45045		14 24 38.9	+47 49 48	-0.005	-0.05	7.4	0.5	3.4	F5 V		63 s	9259	m
126093	224903		14 24 40.6	-44 18 40	-0.005	-0.05	6.6	0.4	0.2	K0 III		190 s		
126251	158550		14 24 40.8	-11 40 11	-0.005	-0.03	6.49	0.42	2.8	dF1	-36	49 s	9254	m
126135	224905		14 24 43.7	-40 45 18	-0.003	-0.02	6.96	-0.06		B9				
125881	252736		14 24 45.9	-63 42 22	-0.039	-0.13	7.25	0.60		G0		19 mn		
126410	83336		14 24 48.4	+21 21 28	0.000	-0.04	7.4	1.1	-0.1	K2 III		320 s		
126218	182517		14 24 48.5	-24 48 23	-0.004	-0.02	5.32	0.96	0.3	G8 III	-22	100 s		
126750	16372		14 24 53.8	+62 56 11	+0.005	-0.02	7.2	0.5	4.0	F8 V		43 s		
126111	224906		14 24 54.6	-46 31 43	-0.002	+0.01	7.1	0.6	-0.3	K5 III		310 s		
126194	205531		14 25 00.9	-37 26 47	-0.003	-0.02	6.70	0.11	1.4	A2 V		110 s		
126233	205533		14 25 02.2	-30 48 01	+0.001	-0.05	7.9	1.3	0.2	K0 III		240 s		
126381	120442		14 25 03.4	+5 36 56	+0.001	-0.01	7.6	0.9	0.3	G4 III	+7	290 s		
126440	101017		14 25 05.3	+18 15 15	0.000	-0.03	7.9	1.1	-0.1	K2 III		390 s		
125835	252735		14 25 06.2	-68 11 43	-0.002	-0.01	5.61	0.49	-7.5	A2 Ia	-34	2500 s		
125854	--		14 25 10.1	-68 22 27			7.9	1.1	0.2	K0 III		330 s		
126660	29137	23 θ Boo	14 25 11.7	+51 51 02	-0.026	-0.40	4.05	0.50	3.8	F7 V	-11	13 ts		m
126279	182524		14 25 13.1	-28 07 51	+0.001	-0.01	7.70	1.1	-0.1	K2 III		360 s	9256	m
126675	29138		14 25 15.7	+53 35 06	+0.009	-0.07	7.3	0.5	4.0	F8 V		46 s		
126363	158554		14 25 17.5	-13 21 11	-0.004	+0.01	6.60	1.1	0.2	K0 III		190 s		m
126115	241702		14 25 22.5	-55 01 06	-0.003	-0.04	6.9	0.8	3.2	G5 IV		53 s		
126366	158556		14 25 27.4	-19 57 57	-0.003	-0.01	6.98	0.12	1.9	A4 V		120 s	9258	m
126334	182527		14 25 27.6	-28 53 43	+0.004	-0.06	6.9	0.9	0.2	K0 III		170 mx		
126597	64137		14 25 28.9	+38 23 35	-0.001	-0.02	6.27	1.23	0.2	K0 III	+25	110 s		
--	--		14 25 29.3	-19 57 08			6.62	0.13	1.4	A2 V				
126367	158558		14 25 29.7	-19 58 11	-0.003	-0.01	6.61	0.14	0.6	A0 V		120 s	9258	m
126238	224910		14 25 29.8	-43 38 38	+0.002	-0.12	8.00	0.6	4.4	G0 V		46 s		m
126512	83342		14 25 30.0	+20 35 25	+0.009	-0.58	7.30	0.59	4.2	F9 V	-53	40 s		
126313	205541		14 25 31.1	-32 58 19	+0.001	+0.01	7.5	1.0	0.2	K0 III		300 s		
126912	16377		14 25 33.1	+65 59 12	-0.005	-0.03	7.2	0.4	3.0	F2 V		68 s		
126351	205543		14 25 35.4	-30 51 53	-0.001	-0.06	7.9	1.1	4.0	F8 V		28 s		
125990	252737		14 25 39.5	-66 10 24	-0.008	-0.03	6.36	0.13	1.2	A1 V		88 mx		
126198	241708		14 25 43.8	-53 27 14	0.000	0.00	7.6	0.4	0.4	B9.5 V		140 s		
126399	182536		14 25 46.0	-21 25 53	-0.003	-0.03	7.7	1.4	0.2	K0 III		190 s		
126400	182535		14 25 47.6	-25 51 09	-0.001	-0.07	6.48	0.94	3.2	G7 IV		40 s		
125517	--		14 25 47.7	-78 27 34			7.7	1.6	-0.5	M4 III		430 s		
--	16376		14 25 50.9	+60 45 03	-0.005	+0.02	7.8	2.0						
126182	241709		14 25 51.2	-56 55 21	-0.001	+0.01	7.6	2.0	-0.1	K2 III		110 s		
126598	83349		14 25 54.4	+26 15 54	-0.005	-0.04	7.40	1.31	-0.3	K4 III	+3	150 mx		
126514	120453		14 25 54.7	+4 51 27	-0.003	0.00	7.8	0.1	1.7	A3 V		160 s		
126515	120452		14 25 55.8	+0 59 33	-0.002	-0.05	7.12	0.01		A2 p				
126415	205549		14 26 00.0	-30 17 11	-0.006	-0.03	8.0	0.9	3.2	G5 IV		78 s		
126077	--		14 26 00.9	-64 56 21			8.0	1.1	0.2	K0 III		340 s		
126341	224919	τ¹ Lup	14 26 08.1	-45 13 17	-0.001	-0.02	4.56	-0.15	-3.0	B2 IV	-18	310 s		m,v
126354	224920	τ² Lup	14 26 10.7	-45 22 45	+0.001	-0.01	4.35	0.43	4.0	dF8	-1	28 mn		m
126120	252742		14 26 11.9	-64 28 35	-0.001	-0.02	7.5	0.1	0.6	A0 V		230 s		
126386	224923		14 26 13.2	-42 19 08	-0.011	-0.07	6.32	1.19	-0.2	K3 III		100 mx		
126284	241713		14 26 22.2	-55 15 33	+0.001	-0.01	8.0	2.0	-0.3	K5 III		220 s		
126661	101025	22 Boo	14 26 27.3	+19 13 37	-0.005	+0.02	5.39	0.23	2.1	A5 V	-28	40 s		
126459	205557		14 26 27.5	-33 15 00	0.000	0.00	7.1	1.1	3.4	F5 V		22 s		
126447	205559		14 26 37.5	-39 31 25	-0.002	-0.02	7.2	0.3	4.4	G0 V		36 s		
126500	182546		14 26 38.4	-26 51 26	-0.001	-0.01	8.0	1.6	-0.1	K2 III		240 s		
126647	120464		14 26 42.1	+5 56 06	0.000	-0.09	7.2	1.1	0.2	K0 III		160 mx		
127806	8023		14 26 43.4	+78 29 55	-0.002	+0.08	7.20	0.5	3.4	F5 V		58 s		m
126676	120465		14 26 47.3	+8 22 55	0.000	+0.01	7.9	0.1	0.6	A0 V		280 s		
126475	205561		14 26 49.8	-39 52 26	-0.002	-0.02	6.35	-0.08		B9				
126285	252747		14 26 54.3	-60 45 11	-0.002	-0.01	7.84	1.10		K0				
124882	258698	δ Oct	14 26 55.0	-83 40 04	-0.058	-0.01	4.32	1.31	-0.1	gK2	+5	60 s		
125782	257154		14 26 56.4	-76 31 10	+0.016	0.00	7.3	0.5	3.4	F5 V		59 s		
126343	241714		14 26 58.5	-59 14 18	-0.003	-0.03	7.2	1.8	0.2	K0 III		84 s		
126502	205568		14 26 58.7	-38 51 04	0.000	-0.01	7.7	1.2	-0.1	K2 III		350 s		
126375	241720		14 27 00.3	-57 17 19	-0.011	-0.09	7.7	0.7	4.0	F8 V		45 s		
126721	101028		14 27 01.3	+9 58 52	+0.002	-0.06	7.1	1.1	0.2	K0 III		240 s		
126697	120469		14 27 01.5	+5 57 44	+0.001	-0.05	7.7	1.1	0.2	K0 III		310 s		
126699	120468		14 27 01.9	+1 42 54	-0.001	0.00	7.6	0.1	0.6	A0 V		260 s		
126241	252745		14 27 07.0	-65 49 18	-0.003	-0.02	5.85	1.50		K5				

HD	SAO	Star Name	α 2000	δ 2000	μ(α)	μ(δ)	V	B-V	M$_v$	Spec	RV	d(pc)	ADS	Notes
126226	252746		14^{h}27^{m}11.0^s	−66°34'07"	−0.003^s	−0.03"	6.6	0.4	1.4	A2 V		69 s		
126504	224929		14 27 12.1	−46 08 04	−0.015	−0.08	5.83	0.31		A m	−26			m
126764	101030		14 27 14.1	+12 22 56	+0.001	−0.04	8.0	0.9	3.2	G5 IV		93 s		
125690	257152		14 27 16.1	−78 37 12	−0.038	−0.04	6.6	1.1	0.2	K0 III		140 mx		
126487	224930		14 27 17.2	−48 38 12	−0.007	−0.08	7.7	1.1	−0.1	K2 III		150 mx		
127029	29153		14 27 18.5	+53 18 40	−0.010	+0.14	8.0	0.6	4.4	G0 V		52 s		
126679	158573		14 27 21.3	−14 50 20	−0.012	−0.06	7.3	0.5	3.4	F5 V		60 s		
126722	139942	104 Vir	14 27 24.3	−6 07 13	−0.005	−0.06	6.17	0.09		A1 n	−15			
126701	158575		14 27 25.4	−15 48 47	−0.001	+0.01	7.9	1.1	0.2	K0 III		340 s		
126943	45058		14 27 27.2	+41 01 30	−0.005	−0.02	6.63	0.37	2.6	F0 V	−17	61 s		
126391	252750		14 27 29.5	−61 59 47	−0.002	−0.03	7.8	0.1	3.0	F2 V		89 s		
126812	101032		14 27 30.1	+14 45 17	−0.001	+0.02	7.3	1.1	0.2	K0 III		260 s		
126548	224935		14 27 30.7	−47 05 56	0.000	−0.02	7.3	0.7	0.4	B9.5 V		110 s		
126524	224932		14 27 31.0	−48 23 02	0.000	−0.02	7.6	1.3	0.2	K0 III		270 s		
126620	205574		14 27 31.3	−35 26 50	+0.002	−0.05	6.9	0.9	4.0	F8 V		22 s		
127700	8024	5 UMi	14 27 31.4	+75 41 45	+0.001	+0.02	4.25	1.44	−0.3	K4 III	+10	79 s	9286	m
126525	241727		14 27 33.0	−51 56 00	−0.032	+0.02	7.82	0.70	5.2	G5 V		33 s		
126682	182562		14 27 33.3	−21 59 32	−0.003	−0.01	7.2	1.2	0.2	K0 III		190 s		
126561	224936		14 27 33.4	−46 12 47	−0.003	−0.01	7.3	0.3	0.6	A0 V		140 s		
127164	29158		14 27 36.2	+59 43 21	−0.003	−0.02	8.0	0.5	3.4	F5 V		84 s		
126575	224939		14 27 39.1	−46 12 05	−0.003	0.00	7.9	0.1	0.6	A0 V		230 s		
127147	29160		14 27 42.6	+58 20 04	+0.001	+0.02	7.6	0.1	1.4	A2 V		170 s		
127064	45063		14 27 43.7	+49 23 47	−0.001	+0.01	7.3	0.9	3.2	G5 IV		65 s		
126766	158577		14 27 45.1	−13 21 33	−0.005	−0.03	6.65	0.42		F5	−18			
126606	224941		14 27 45.3	−44 19 49	−0.004	−0.03	7.1	0.4	1.4	A2 V		82 s		
126827	120474		14 27 46.7	+3 37 17	−0.002	−0.01	7.9	1.1	0.2	K0 III		340 s		
126668	205582		14 27 52.7	−36 27 09	−0.001	0.00	8.0	−0.1	−0.1	K2 III		410 s		
127455	16392		14 27 58.0	+69 15 13	+0.001	+0.03	7.8	0.7	4.4	G0 V		49 s		
126990	64153		14 27 58.1	+35 34 38	+0.006	−0.09	7.4	1.1	0.2	K0 III		190 mx		
126970	83369		14 27 58.7	+29 15 50	−0.002	0.00	7.6	0.8	3.2	G5 IV		77 s		
127370	16389		14 28 01.3	+66 20 25	−0.002	0.00	7.9	1.1	−0.1	K2 III		390 s		
−−	64155		14 28 03.3	+31 47 16	−0.005	−0.02	8.0	1.7						
126846	139948		14 28 04.6	−3 00 14	−0.001	0.00	7.5	0.1	0.6	A0 V		250 s		
127030	64157		14 28 09.5	+37 33 23	0.000	0.00	8.0	0.9	3.2	G5 IV		93 s		
126769	182570	52 Hya	14 28 10.3	−29 29 30	−0.002	−0.03	4.97	−0.07	−0.6	B8 IV	+6	130 s	9270	m
126868	139951	105 ∅ Vir	14 28 12.0	−2 13 40	−0.009	0.00	4.81	0.70	0.4	G2 III	−10	29 ts	9273	m
126945	101037		14 28 15.1	+16 07 08	+0.006	−0.18	7.6	0.9	3.2	G5 IV		77 s		
126991	83371		14 28 16.0	+24 31 04	−0.017	+0.13	7.90	0.80	4.7	G2 V		33 s		
127065	64161		14 28 16.3	+36 11 48	−0.003	−0.01	6.22	1.16	0.2	K0 III	−18	130 s		
126818	182573		14 28 17.7	−20 33 52	−0.003	−0.02	7.6	0.3	0.6	A0 V		170 s		
126772	205592		14 28 20.9	−33 51 38	−0.003	0.00	8.0	1.8	0.2	K0 III		120 s		
126819	182574		14 28 23.3	−23 00 45	−0.005	−0.05	8.00	0.5	4.0	F8 V		63 s	9274	m
126757	205593		14 28 26.3	−38 33 23	−0.003	+0.01	7.8	0.7	0.2	K0 III		330 s		
126947	120480		14 28 28.5	+5 40 51	−0.002	−0.01	7.4	1.6	−0.5	M2 III	+18	390 s		
127066	64162		14 28 29.6	+30 25 53	−0.001	0.00	7.9	0.9	3.2	G5 IV		88 s		
126961	120481		14 28 31.0	+2 47 20	−0.013	+0.05	7.2	0.5	4.0	F8 V		43 s		
127043	83373		14 28 31.4	+28 17 21	+0.002	−0.02	7.62	−0.01	0.6	A0 V	−15	250 s		
126896	139953		14 28 31.8	−10 00 16	−0.004	−0.03	6.7	1.1	0.2	K0 III		190 mx		
126957	101040		14 28 32.1	+10 20 33	−0.003	−0.05	7.9	1.1	0.2	K0 III		340 s		
127067	83374		14 28 33.2	+28 17 27	+0.001	0.00	7.12	−0.03	0.6	A0 V	−13	200 s	9277	m
127243	29165	24 Boo	14 28 37.7	+49 50 41	−0.032	−0.05	5.59	0.85	4.4	G0 V	−6	18 mn		
126927	139957	106 Vir	14 28 41.6	−6 54 02	−0.001	−0.06	5.42	1.49	−0.3	gK5	−49	140 s		
126655	241739		14 28 41.9	−55 13 17	+0.004	−0.02	7.6	0.4	1.4	A2 V		100 s		
126610	241737		14 28 43.4	−59 11 52	−0.001	−0.02	6.45	0.14		A0				
127093	83375		14 28 46.0	+25 51 13	+0.001	0.00	6.69	1.59		M III	0			
126962	139959		14 28 48.0	−5 13 06	−0.002	+0.04	6.7	1.1	0.2	K0 III		200 s		
126759	224950		14 28 51.9	−47 59 30	−0.001	−0.01	6.40	−0.11		B9				
126992	120485		14 28 52.8	+1 37 58	−0.001	−0.02	8.0	0.1	0.6	A0 V		300 s		
127185	64167		14 28 55.0	+39 37 07	−0.003	+0.01	7.6	1.1	0.2	K0 III		310 s		
127263	45069		14 28 55.6	+47 09 05	−0.004	+0.08	7.9	0.1	0.6	A0 V		86 mx		
127032	120489		14 28 56.4	+3 43 10	+0.001	+0.01	7.0	1.4	−0.3	K5 III		290 s		
127106	83376		14 28 57.4	+20 46 03	−0.004	+0.04	7.4	1.4	−0.3	K5 III		260 mx		
125873	−−		14 28 57.4	−79 45 22			7.7	0.8	3.2	G5 IV		78 s		
127411	16394		14 28 57.8	+60 23 11	−0.003	0.00	7.4	0.1	1.4	A2 V		160 s		
126692	241742		14 29 00.3	−56 08 12	−0.002	0.00	6.91	0.52		A5				
127069	101045		14 29 02.7	+11 02 34	0.000	−0.02	7.7	1.1	0.2	K0 III		310 s		
−−	64169		14 29 09.2	+38 16 38	+0.016	−0.27	7.4	2.7						
126933	182587		14 29 09.8	−29 06 50	+0.001	−0.01	7.1	0.4	4.0	F8 V		41 s		
125838	−−		14 29 10.3	−80 21 22			7.5	0.1	0.6	A0 V		240 s		
127319	45073		14 29 12.3	+47 59 25	−0.010	+0.07	7.8	1.1	0.2	K0 III		150 mx		
127285	45071		14 29 14.9	+44 15 42	−0.004	0.00	6.6	1.1	0.2	K0 III		190 s		
127207	64170		14 29 18.9	+31 00 35	−0.004	−0.01	8.0	0.5	3.4	F5 V		82 s		
128642	2392		14 29 21.6	+80 48 35	−0.040	−0.13	7.1	0.9	3.2	G5 IV		60 s		
126823	224959		14 29 23.3	−50 06 34	0.000	−0.02	7.2	2.0	−0.1	K2 III		91 s		
125856	257158		14 29 25.5	−80 06 09	−0.019	+0.03	6.9	0.5	4.0	F8 V		37 s		m

HD	SAO	Star Name	α 2000	δ 2000	μ(α)	μ(δ)	V	B−V	M_v	Spec	RV	d(pc)	ADS	Notes
127245	64172		14^{h}29^{m}27.8^s	+30°40'46"	−0.007^s	+0.04"	7.7	0.4	2.6	F0 V		110 s		
127072	158598		14 29 31.2	−10 34 00	−0.003	0.00	7.9	1.1	0.2	K0 III		350 s		
126973	182593		14 29 31.2	−26 00 48	+0.001	0.00	7.0	1.3	−0.1	K2 III		210 s		
126903	224960		14 29 31.8	−43 46 50	−0.003	−0.07	8.0	1.1	−0.5	M4 III		230 mx		
126935	205609		14 29 32.5	−37 02 23	−0.010	−0.05	7.90	0.7	4.4	G0 V		50 s		m
127334	45075		14 29 36.7	+41 47 45	+0.014	−0.22	6.5	0.6	4.4	G0 V	−1	31 ts		
126209	257163		14 29 36.8	−76 43 44	−0.009	−0.03	6.07	1.18		K0				
127434	45077		14 29 43.4	+48 53 45	−0.002	+0.02	7.7	0.9	3.2	G5 IV		79 s		
127335	64180		14 29 45.3	+38 51 39	+0.002	−0.03	6.3	1.6		M5 III	−38			V Boo, v
127304	64178		14 29 49.6	+31 47 28	−0.002	0.00	6.06	−0.03		B9.5 pe	−9		9288	m
127037	182599		14 29 49.8	−25 32 34	−0.001	−0.03	7.70	1.1	0.2	K0 III		320 s	9280	m
127167	120499		14 29 50.3	+0 49 44	0.000	0.00	5.94	0.16		A3 n	−9			
127227	101053		14 29 53.6	+16 12 32	−0.004	−0.02	7.48	1.53	−0.3	K5 III	−41	210 mx		
126154	257162		14 29 54.1	−78 09 17	+0.005	+0.05	7.9	1.4	−0.3	K5 III		250 mx		
126859	241751		14 29 58.1	−56 07 53	−0.005	−0.03	6.97	0.23		A5				
127077	182603		14 29 58.7	−29 38 01	−0.004	+0.01	7.9	0.7	2.6	F0 V		63 s		
127168	139969		14 30 00.1	−4 14 50	−0.006	−0.02	7.02	0.40	2.8	dF1	−28	65 s	9284	m
127403	64183		14 30 07.3	+37 08 39	−0.009	+0.01	7.9	0.5	3.4	F5 V		81 s		
127247	120501		14 30 08.3	+8 21 12	−0.003	−0.02	7.6	0.4	3.0	F2 V		85 s		
126981	224969		14 30 08.5	−45 19 18	−0.004	−0.04	5.50	−0.08	−1.3	B6 IV	+10	100 mx		m
126997	224970		14 30 10.1	−43 51 48	−0.001	−0.01	7.2	0.2	1.4	A2 V		110 s		
126793	252763		14 30 12.7	−62 51 44	−0.001	−0.07	7.7	0.6	4.4	G0 V		46 s		
127170	158603		14 30 13.9	−15 14 54	−0.004	+0.07	7.1	0.5	3.4	F5 V		56 s		
127097	205621		14 30 15.2	−33 37 08	0.000	−0.02	7.8	1.1	0.2	K0 III		300 s		
126983	224972		14 30 20.8	−49 31 09	−0.005	−0.05	5.37	0.05	0.6	A0 V	+4	61 mx		m
127132	182606		14 30 26.1	−28 00 41	0.000	−0.04	8.0	1.0	0.2	K0 III		360 s		
127082	224977		14 30 26.2	−40 28 33	−0.002	+0.01	7.2	1.6	−0.5	M2 III		340 s		
127265	139974		14 30 30.5	−4 03 54	−0.002	−0.01	7.3	0.5	3.4	F5 V		61 s		
127174	182608		14 30 36.4	−29 18 37	−0.002	0.00	7.7	1.6	−0.1	K2 III		210 s		
127208	182611		14 30 40.3	−22 27 38	−0.002	0.00	6.9	0.5	0.0	B8.5 V		110 s		
127337	120504		14 30 45.3	+4 46 20	0.000	−0.02	6.02	1.42	−0.1	gK2	+6	110 s		m
127821	16406		14 30 45.9	+63 11 08	−0.027	0.00	6.09	0.41	3.4	F5 V	−3	35 s		
127250	182620		14 30 51.8	−20 43 05	+0.002	−0.02	7.0	1.1	0.2	K0 III		210 s		
127945	16410		14 30 53.5	+68 04 56	−0.008	+0.01	7.2	0.1	1.7	A3 V		110 mx		
127152	224982		14 30 56.4	−40 50 42	−0.003	0.00	6.39	1.44		K2				
127233	182618		14 30 58.7	−30 05 52	−0.001	−0.03	6.1	1.6		M6 III				Y Cen, q
127352	139981		14 31 00.5	−5 48 10	−0.014	−0.03	7.69	0.78		G5				m
127269	182624		14 31 00.6	−25 18 55	−0.002	−0.02	7.8	0.1	1.4	A2 V		190 s		
127193	205637		14 31 10.8	−38 52 11	+0.001	+0.03	5.97	1.06	0.2	gK0		130 s		
127214	224987		14 31 11.1	−40 23 10	−0.004	−0.01	6.7	0.9	0.2	K0 III		200 s		
127254	205639		14 31 12.8	−34 40 54	−0.002	−0.03	7.5	0.2	0.6	A0 V		170 s		
127215	224988		14 31 14.3	−40 39 22	−0.002	−0.04	7.6	−0.2	1.4	A2 V		180 s		
127662	45088		14 31 14.4	+45 35 44	−0.003	+0.03	7.9	1.4	−0.3	K5 III		430 s		
127615	45086		14 31 15.4	+39 54 44	−0.001	+0.01	8.0	0.9	3.2	G5 IV		93 s		
126862	252767		14 31 16.4	−67 43 02	+0.008	−0.07	5.83	1.00	0.2	gK0		130 s		m
127154	224986		14 31 18.0	−47 04 20	−0.002	−0.02	7.30	−0.03		B9				
127356	158612		14 31 19.6	−15 38 19	+0.015	−0.37	8.00	0.72	5.2	G5 V	+29	35 s	9291	m
127663	45090		14 31 20.1	+43 22 57	−0.004	+0.01	6.8	0.1	1.4	A2 V		110 mx		
127195	224989		14 31 29.0	−47 43 45	−0.007	+0.08	7.5	0.4	0.2	K0 III		97 mx		
127539	101063		14 31 33.7	+17 38 35	−0.002	−0.03	7.2	0.5	3.4	F5 V	−22	57 s		
126673	257168		14 31 39.4	−74 23 32	+0.006	−0.04	7.9	1.1	−0.1	K2 III		260 mx		
127540	101065		14 31 40.7	+12 51 23	−0.008	+0.08	7.5	0.5	3.4	F5 V		65 s		
127294	224996		14 31 41.3	−43 05 19	−0.007	−0.05	7.86	0.50		F8				
127929	16411		14 31 42.7	+60 13 32	−0.007	+0.02	6.2	0.4	2.6	F0 V	−19	52 s		
127435	158620		14 31 46.9	−13 11 32	−0.001	−0.01	7.9	0.9	3.2	G5 IV		88 s		
127273	224997		14 31 46.9	−44 48 24	−0.001	−0.01	7.0	1.5	0.2	K0 III		120 s		
127391	182635		14 31 47.1	−24 33 33	+0.001	−0.02	7.7	2.3	−0.3	K5 III		130 s		
127324	205646		14 31 48.1	−38 33 33	+0.002	−0.08	8.0	1.8	−0.5	M4 III		150 mx		
127665	64202	25 ρ Boo	14 31 49.7	+30 22 17	−0.008	+0.12	3.58	1.30	−0.2	K3 III	−14	56 s	9296	m
126595	257167		14 31 50.4	−76 16 26	+0.003	−0.01	7.00	1.1	0.2	K0 III		230 s		m
126954	252770		14 31 55.3	−68 48 36	−0.030	−0.20	7.9	0.5	4.0	F8 V		60 s		
127481	158622		14 31 57.1	−16 21 45	−0.002	−0.01	7.8	1.1	0.2	K0 III		330 s		
127407	205656		14 31 59.9	−32 11 10	−0.003	−0.02	7.50	1.1	−0.1	K2 III		330 s		m
127276	224999		14 32 01.8	−48 50 57	−0.001	0.00	7.7	1.0	3.2	G5 IV		58 s		
127456	182639		14 32 02.0	−20 56 51	0.000	−0.05	8.0	0.8	0.2	K0 III		360 s		
127667	101068		14 32 04.5	+18 50 08	−0.016	−0.01	7.6	0.5	4.0	F8 V		52 s		
127762	64203	27 γ Boo	14 32 04.6	+38 18 30	−0.010	+0.15	3.03	0.19	0.5	A7 III	−36	32 s	9300	Seginus, m,q
127419	182637		14 32 05.0	−27 42 30	−0.009	−0.07	7.48	0.29	2.4	A7 V		34 mx		
127457	182641		14 32 07.8	−23 26 47	+0.003	−0.01	7.0	0.5	2.6	F0 V		59 s		
127459	182642		14 32 12.2	−26 47 05	0.000	−0.05	7.4	0.6	0.2	K0 III		280 s		
127618	120516		14 32 12.9	+4 08 33	−0.001	0.00	7.3	1.6	−0.5	M2 III	0	370 s		
127314	241775		14 32 16.9	−51 20 18	−0.002	−0.01	7.5	1.6	0.2	K0 III		130 s		
127726	83394		14 32 20.1	+26 40 38	−0.005	−0.03	6.01	0.22	1.9	A7 IV-V	−5	65 s	9301	m
127346	241780		14 32 28.9	−51 58 14	−0.001	−0.01	7.2	1.0	0.6	A0 V		55 s		
127930	45099		14 32 30.7	+49 11 02	−0.003	0.00	7.80	0.5	3.4	F5 V	−4	76 s	9306	m

HD	SAO	Star Name	α 2000	δ 2000	μ(α)	μ(δ)	V	B-V	M_v	Spec	RV	d(pc)	ADS	Notes
127296	241778		14ʰ32ᵐ30.ˢ7	-55°27'49"	-0.ˢ004	-0."01	7.39	0.29		F0		160 s		
128000	29191		14 32 30.8	+55 23 52	0.000	-0.02	5.76	1.50	-0.3	K5 III	+3	160 s		
127739	83395	26 Boo	14 32 32.4	+22 15 36	-0.009	+0.03	5.92	0.35	2.6	F0 V	-12	43 s		
127297	241777		14 32 32.8	-56 53 16	-0.003	-0.01	6.3	0.5	3.4	F5 V		38 s		V Cen, v
127381	241781	σ Lup	14 32 36.8	-50 27 25	-0.004	-0.01	4.42	-0.19	-2.5	B2 V	-2	130 mx		
127740	101073		14 32 49.5	+14 03 39	+0.002	-0.04	7.1	0.5	3.4	F5 V		54 s		
127584	182651		14 32 52.7	-24 25 31	+0.001	-0.01	8.0	1.1	0.2	K0 III		330 s		
127743	120520		14 32 57.6	+3 39 34	0.000	+0.01	8.0	0.4	3.0	F2 V		98 s		
127741	120521		14 32 58.1	+6 55 59	-0.001	-0.05	7.9	1.1	0.2	K0 III		340 s		
127742	120522		14 33 02.9	+5 19 25	0.000	-0.08	6.9	0.9	3.2	G5 IV		55 s		
127643	182654		14 33 09.4	-24 01 08	-0.003	0.00	7.2	0.6	3.2	G5 IV		64 s		
127624	205681		14 33 09.5	-30 42 51	+0.003	-0.02	6.09	1.03	0.2	K0 III		150 s		m
127202	252784		14 33 14.6	-67 11 18	-0.002	-0.06	8.0	0.4	2.6	F0 V		120 s		
131616	2433		14 33 17.0	+85 56 18	-0.023	-0.01	7.10	1.1	0.2	K0 III		240 s	9358	m
127427	241788		14 33 18.6	-56 12 57	-0.002	0.00	7.4	2.0	0.2	K0 III		69 s		
127986	64212		14 33 20.2	+36 57 34	-0.001	-0.06	6.5	0.5	3.4	F5 V	+2	41 s		
127764	139994		14 33 21.1	-6 56 09	+0.001	0.00	7.9	1.1	0.2	K0 III		340 s		
127672	205690		14 33 27.4	-33 18 58	0.000	0.00	7.00	1.1	0.2	K0 III		230 s		m
128165	29196		14 33 28.6	+52 54 31	-0.023	+0.24	7.23	0.99	6.9	K3 V	+14	11 s		
127501	241793		14 33 29.8	-52 40 48	0.000	-0.03	5.87	1.09	0.2	gK0		120 s		m
127486	241792		14 33 32.3	-54 59 54	-0.011	0.00	5.87	0.48	4.0	F8 V		24 s		
128385	16431		14 33 35.5	+65 23 43	+0.009	+0.04	6.6	0.5	3.4	F5 V		45 s		
128041	64213		14 33 36.2	+35 35 08	-0.016	+0.06	8.00	0.9	3.2	G5 IV		78 mx	9312	m
127449	241791		14 33 36.8	-58 49 15	0.000	-0.01	7.0	0.9	-1.6	B3.5 V		130 s		
129245	8054		14 33 38.3	+79 39 37	-0.038	+0.08	6.26	1.30	-0.2	K3 III	-23	160 mx		
127502	241796		14 33 42.0	-55 57 39	-0.002	+0.01	8.0	1.7	-0.3	K5 III		340 s		
128092	45113		14 33 44.4	+43 00 28	-0.005	+0.04	6.8	0.1	1.4	A2 V		110 mx		
126596	257170		14 33 46.0	-79 57 19	-0.007	-0.01	6.8	1.1	0.2	K0 III		210 s		
127317	252787		14 33 50.0	-67 37 04	+0.002	-0.02	6.93	-0.05	-1.7	B3 V n		500 s		
127629	225022		14 33 51.6	-46 27 54	-0.003	-0.04	7.50	0.9	3.2	G5 IV		72 s		m
127470	252789		14 33 53.3	-60 48 05	-0.002	-0.02	7.9	0.4	1.4	A2 V		120 s		
127844	158642		14 34 03.5	-16 49 11	-0.002	-0.02	7.2	0.1	0.6	A0 V		210 s		
128184	45117		14 34 04.9	+46 47 11	-0.004	0.00	6.6	0.1	0.6	A0 V	+6	150 mx		
127774	205704		14 34 05.0	-31 57 27	-0.004	-0.03	7.7	1.3	0.2	K0 III		210 s		
127987	101082		14 34 06.4	+10 31 20	+0.002	-0.01	8.0	0.1	1.4	A2 V		210 s		
127716	225027		14 34 07.9	-42 05 59	-0.002	-0.03	6.60	0.06		A2				
128093	64221		14 34 11.7	+32 32 04	+0.009	0.00	6.33	0.40	3.0	F2 V	-8	45 s		m
128332	29202		14 34 15.9	+57 03 54	+0.026	-0.24	6.48	0.49	3.4	F5 V	-22	39 s		
--	83414		14 34 25.8	+24 23 59	+0.001	-0.01	7.9	1.0	4.4	G0 V		30 s		
127874	182669		14 34 26.7	-22 23 29	0.000	-0.03	7.4	0.4	3.2	G5 IV		70 s		
128422	16432		14 34 29.6	+59 51 29	-0.002	-0.01	7.6	1.4	-0.3	K5 III		370 s		
128577	16435		14 34 30.2	+66 47 39	-0.003	-0.02	7.6	0.5	4.0	F8 V		53 s		
128492	16433		14 34 30.3	+63 12 56	-0.010	-0.04	7.3	1.1	-0.1	K2 III		160 mx		
128198	64227		14 34 38.4	+36 37 33	-0.003	-0.06	6.03	1.38	0.2	K0 III	-12	87 s		
128333	45121		14 34 39.5	+49 22 06	-0.005	+0.05	5.74	1.4	-0.3	K5 III	-20	150 s		
127596	252799		14 34 39.8	-61 44 54	0.000	-0.02	7.9	1.4	0.0	K1 III		260 s		
127610	241803		14 34 40.1	-59 47 04	+0.002	-0.01	8.0	0.6	2.6	F0 V		76 s		
128167	83416	28 σ Boo	14 34 40.7	+29 44 42	+0.014	+0.13	4.46	0.36	3.0	F2 V	0	17 ts		m
128608	16437		14 34 43.3	+65 58 17	-0.009	+0.02	7.9	0.4	2.6	F0 V		120 s		
127861	205715		14 34 43.9	-35 27 48	-0.007	-0.05	7.3	0.5	3.4	F5 V		52 s		
128185	83418		14 34 44.0	+28 23 38	-0.002	-0.06	7.89	0.56	4.0	F8 V		57 s		
128027	140010		14 34 49.1	-4 39 06	0.000	-0.03	7.7	0.5	4.0	F8 V		54 s		
127964	182676		14 34 50.6	-20 26 21	+0.001	0.00	6.5	0.7	0.6	A0 V		58 s		
128128	101087		14 34 55.2	+9 53 12	+0.002	+0.02	7.9	0.5	4.0	F8 V		60 s		
127894	205718		14 34 56.5	-37 12 14	0.000	+0.02	7.3	0.6	3.2	G5 IV		65 s		
127453	252795		14 34 59.7	-69 27 17	-0.002	-0.06	7.2	0.0	0.0	B8.5 V		140 mx		
128168	101090		14 35 00.7	+13 05 44	-0.004	-0.04	7.2	0.1	1.7	A3 V		56 mx		
128219	83420		14 35 05.9	+23 30 55	-0.005	-0.07	8.0	0.9	3.2	G5 IV		90 mx		
127753	241812		14 35 08.6	-56 33 47	-0.002	-0.03	7.0	1.9		K2				
127864	225039		14 35 10.5	-46 27 43	-0.003	-0.01	6.89	-0.02		B9				
128097	140015		14 35 15.5	-9 05 44	-0.003	-0.02	7.9	1.1	-0.1	K2 III		390 s		
127724	241811		14 35 17.1	-60 00 57	-0.002	-0.03	6.40	1.26		K0				
127952	205723		14 35 18.1	-38 30 18	-0.001	0.00	7.7	1.4	-0.1	K2 III		270 s		
110994	258660		14 35 19.5	-89 46 18	-0.176	-0.01	6.60	1.70	-0.5	M4 III		240 s		
128143	140018		14 35 27.0	-5 49 58	-0.001	-0.02	7.8	0.5	3.4	F5 V		76 s		
127369	257174		14 35 29.6	-73 41 39	-0.002	-0.02	6.7	1.9	-0.3	K5 III		150 s		
127755	--		14 35 29.9	-60 38 30			7.5	2.7	-0.3	K5 III		70 s		
127972	225044	η Cen	14 35 30.3	-42 09 28	-0.003	-0.04	2.31	-0.19	-2.9	B3 III	0	110 s		
127971	225046		14 35 31.4	-41 31 02	-0.002	-0.02	5.80	0.00	0.0	B8.5 V		150 s		m
127756	252809		14 35 33.2	-61 00 29	0.000	-0.01	7.58	0.13	-1.1	B5 V e		500 s		
127575	252805		14 35 34.7	-68 44 21	-0.004	-0.01	7.8	0.0	0.4	B9.5 V		230 mx		
128221	120548		14 35 36.0	+3 28 04	-0.004	-0.02	7.6	0.4	3.0	F2 V		85 s		
127897	241826		14 35 41.7	-52 28 25	0.000	-0.02	7.6	1.2	-0.1	K2 III		340 s		
128200	140021		14 35 43.6	-5 15 59	-0.003	0.00	7.50	1.1	0.2	K0 III	+2	290 s		
127614	--		14 35 47.1	-69 01 33			7.8	0.0	0.0	B8.5 V		350 s		

HD	SAO	Star Name	α 2000	δ 2000	μ(α)	μ(δ)	V	B-V	M_v	Spec	RV	d(pc)	ADS	Notes
127866	241825		14^h 35^m 47^s.6	-56° 37' 05"	-0^s.001	0".00	8.0	0.3	0.0	B8.5 V		230 s		
128202	140022		14 35 48.4	-8 34 30	+0.001	-0.01	7.1	0.1	0.6	A0 V		200 s		
128296	101097		14 35 49.2	+11 04 06	-0.002	-0.02	7.5	0.1	1.7	A3 V		140 s		
127976	225049		14 35 50.1	-46 48 57	-0.001	-0.02	6.78	0.96		K0				
128369	83424		14 35 52.5	+24 56 20	+0.008	-0.08	8.0	0.5	3.4	F5 V		83 s		
127204	257173		14 35 54.4	-77 06 45	-0.008	-0.04	7.2	0.0	0.4	B9.5 V		150 mx		
127597	252806		14 35 54.7	-69 58 15	0.000	-0.03	7.80	1.6	-0.5	M2 III		430 s		m
128272	120553		14 35 58.9	+0 13 09	-0.003	-0.01	7.70	0.98		G5				
128311	120554		14 36 00.4	+9 44 46	+0.013	-0.27	7.51	0.99	0.2	K0 III		63 mx		
127785	252813		14 36 03.1	-64 00 58	0.000	-0.04	7.1	0.4	2.6	F0 V		78 s		
128120	205738		14 36 06.6	-34 54 22	0.000	+0.01	7.6	1.3	-0.5	M2 III		420 s		
128402	83427		14 36 06.8	+23 15 01	-0.001	+0.02	6.4	1.1	0.2	K0 III	+7	180 s		
126124	258701		14 36 07.1	-84 08 52	+0.012	+0.02	7.1	1.9	-0.1	K2 III		100 s		
128189	182687		14 36 07.6	-22 10 39	-0.003	0.00	7.1	0.4	2.6	F0 V		73 s		
128425	83429		14 36 15.5	+23 59 02	-0.006	+0.04	7.7	1.1	0.2	K0 III		210 mx		
128815	16441		14 36 17.6	+63 27 57	-0.004	+0.02	7.9	0.1	1.7	A3 V		170 s		
128275	140027		14 36 18.0	-9 36 45	-0.003	-0.03	7.6	1.1	0.2	K0 III		300 s		
128068	225054		14 36 18.9	-46 14 43	-0.004	+0.02	5.55	1.48	-0.1	K2 III	-60	79 s		m
128494	83431		14 36 24.0	+27 29 07	-0.002	0.00	7.0	0.4	3.0	F2 V		63 s		
128152	205744		14 36 24.1	-39 35 50	-0.003	-0.02	6.1	0.9	0.2	K0 III		150 s		
128643	45132		14 36 28.6	+49 31 35	-0.001	+0.01	7.8	1.4	-0.3	K5 III		410 s		
128237	182692		14 36 30.7	-28 15 25	+0.001	-0.04	8.0	1.7	-0.3	K5 III		340 s		
128480	83432		14 36 32.3	+23 37 16	+0.001	-0.03	8.0	1.4	-0.3	K5 III		450 s		
128481	101105		14 36 43.6	+12 52 39	-0.001	-0.03	6.9	0.1	0.6	A0 V	-7	180 s		
128207	205751		14 36 44.1	-40 12 42	-0.001	-0.03	5.74	-0.12		B8 n	+14			
128224	205754		14 36 46.7	-36 39 55	+0.001	0.00	7.80	0.00	0.0	B8.5 V		360 s		m
128279	182701		14 36 48.3	-29 06 45	+0.004	-0.32	7.8	0.2	4.4	G0 V		49 s		
128428	140035		14 36 53.6	-4 16 44	-0.023	+0.01	7.77	0.76	4.9	dG3	-43	31 s		
128547	83436		14 36 53.9	+24 27 46	+0.002	-0.01	7.6	0.1	0.6	A0 V		250 s		
128781	29219		14 36 53.9	+54 51 18	-0.006	0.00	7.5	1.1	0.2	K0 III		290 s		
128660	45133		14 36 55.7	+42 49 52	-0.010	-0.06	6.7	0.5	4.0	F8 V	-2	35 s		
128718	45134		14 36 56.1	+48 13 18	-0.002	+0.06	7.72	0.45		F2			9324	m
128429	158677		14 36 59.7	-12 18 19	-0.059	+0.36	6.20	0.46	3.4	F5 V	-70	34 ts		
128527	101107		14 37 03.3	+11 12 30	-0.001	-0.05	7.5	1.1	-0.1	K2 III		330 s		
128299	205764		14 37 03.4	-33 10 48	-0.003	-0.03	7.8	1.1	2.6	F0 V		34 s		
128356	182707		14 37 04.8	-25 48 08	-0.002	-0.13	8.0	1.3	0.2	K0 III		120 mx		
128625	64238		14 37 07.1	+33 40 41	-0.001	-0.04	7.4	1.1	0.2	K0 III		280 s		
128661	64240		14 37 08.5	+35 55 46	-0.004	-0.02	7.0	0.1	0.6	A0 V	+32	190 s		
128461	140036		14 37 08.6	-3 53 32	+0.002	-0.03	7.1	0.6	4.4	G0 V		34 s		
128284	205766		14 37 09.3	-37 31 56	-0.002	-0.02	7.90	0.1	0.6	A0 V		290 s		m
128609	83440		14 37 11.6	+26 44 11	-0.001	+0.01	5.8	1.6		M5 III	-58			R Boo, v
128037	252822		14 37 12.0	-62 00 39	+0.001	-0.01	6.8	1.1	3.2	G5 IV		33 s		
128154	241848		14 37 12.3	-53 03 39	-0.002	-0.02	6.8	1.0	0.2	K0 III		210 s		
128358	182709		14 37 13.1	-30 10 38	+0.004	-0.05	7.7	1.8	0.2	K0 III		110 s		
128266	225062		14 37 20.0	-46 08 02	-0.001	-0.02	5.41	1.01	0.2	K0 III	-16	110 s		m
128267	225063		14 37 20.6	-46 07 46	-0.003	-0.05	5.40	0.01	1.2	A1 V		69 s		m
128563	120569		14 37 28.4	+2 16 38	-0.002	-0.07	6.60	0.5	4.0	F8 V	-2	33 s	9323	m
128343	225070		14 37 36.5	-43 06 41	0.000	0.00	6.9	0.7	0.2	K0 III		220 s		
128288	225067		14 37 38.3	-50 05 08	-0.004	-0.05	7.2	1.2	3.2	G5 IV		55 s		
128409	205774		14 37 43.1	-35 31 13	-0.011	-0.04	7.6	0.8	0.2	K0 III		110 mx		
128020	252824		14 37 45.9	-67 55 55	-0.063	-0.28	6.04	0.50	4.0	F8 V		25 ts		
128901	29222		14 37 46.2	+52 01 57	-0.006	+0.02	8.0	0.4	2.6	F0 V		120 s		
128594	140041		14 37 52.8	-5 32 55	-0.002	-0.01	7.3	0.1	0.6	A0 V		220 s		
128345	225071	ρ Lup	14 37 53.1	-49 25 32	-0.003	-0.02	4.05	-0.15	-1.1	B5 V	+14	110 s		
128344	225073		14 37 54.8	-48 01 26	-0.001	-0.02	6.65	-0.02		B9				
128529	182722		14 37 59.5	-23 09 50	-0.001	+0.01	7.6	1.7	-0.3	K5 III		280 s		
128941	29224		14 38 00.7	+51 34 42	-0.006	+0.01	7.40	0.4	3.0	F2 V	-24	76 s	9329	m
128595	140042		14 38 04.1	-9 44 16	-0.002	0.00	7.9	1.1	-0.1	K2 III		390 s		
128022	252826		14 38 05.3	-68 57 06	-0.003	-0.03	7.7	0.0	0.4	B9.5 V		280 s		
128596	158686		14 38 07.0	-12 54 36	-0.016	-0.03	7.48	0.65		G0				
128468	205782		14 38 08.8	-37 00 29	+0.002	-0.02	7.7	0.7	0.2	K0 III		320 s		
128902	45145		14 38 12.4	+43 38 31	-0.011	+0.03	5.70	1.48	-0.3	K4 III	-49	110 mx		
128413	225079		14 38 13.3	-45 52 20	-0.004	-0.06	6.84	1.18		K0				
128750	101121		14 38 13.9	+18 17 54	-0.002	-0.08	5.91	1.10	-0.1	gK2	-14	160 s		
128998	29227		14 38 15.1	+54 01 24	+0.001	-0.02	5.5	0.1	0.6	A0 V	-1	96 s		
128415	225078		14 38 17.3	-49 54 04	-0.005	-0.02	7.40	0.1	0.6	A0 V		100 mx		m
128488	205786		14 38 19.5	-38 47 39	+0.006	-0.03	6.1	1.0	0.2	K0 III		140 s		
128532	205789		14 38 24.9	-35 16 35	-0.003	-0.04	6.80	0.1	1.4	A2 V		120 s		m
128684	140046		14 38 28.9	-3 36 40	0.000	-0.02	7.40	1.6	-0.5	M2 III	-7	380 s		
128380	241868		14 38 31.2	-55 32 57	+0.001	-0.01	8.0	0.3	-1.0	B5.5 V		340 s		
127284	258704		14 38 34.8	-80 44 09	-0.012	0.00	7.8	1.1	-0.1	K2 III		380 s		
128707	140049		14 38 35.6	-7 09 46	-0.002	-0.08	7.9	0.9	3.2	G5 IV		88 s		
129079	29232		14 38 37.3	+54 51 12	-0.002	0.00	7.7	0.4	3.0	F2 V		85 s		
128885	64252		14 38 40.0	+30 00 54	-0.002	0.00	7.9	1.1	0.2	K0 III		340 s		
129307	16450		14 38 42.7	+65 51 57	-0.004	-0.04	7.9	1.1	-0.1	K2 III		310 mx		

HD	SAO	Star Name	α 2000	δ 2000	μ(α)	μ(δ)	V	B-V	M$_V$	Spec	RV	d(pc)	ADS	Notes
			h m s	° ′ ″	s	″								
128555	225088		14 38 48.4	-43 00 04	-0.003	-0.03	6.7	1.8	3.2	G5 IV		12 s		
129002	45153	33 Boo	14 38 50.1	+44 24 16	-0.007	-0.02	5.39	0.00		A0	-10	25 mn		
128614	205800		14 38 55.7	-35 36 00	-0.004	-0.19	7.4	0.4	3.4	F5 V		62 s		
128711	182732		14 38 56.0	-23 18 40	-0.002	+0.02	7.7	0.7	3.2	G5 IV		78 s		
130235	2431		14 38 56.1	+79 47 16	+0.003	+0.01	7.4	1.1	0.2	K0 III		280 s		
129599	8061		14 38 57.6	+72 17 23	0.000	-0.04	8.0	0.5	3.4	F5 V		84 s		
128398	241873		14 38 59.4	-59 08 28	-0.012	-0.11	6.7	0.7	4.0	F8 V		28 s		
128783	140050		14 38 59.9	-5 47 14	-0.002	-0.03	7.7	1.1	0.2	K0 III		320 s		
129333	16453		14 38 59.9	+64 17 32	-0.024	-0.01	7.5	0.5	4.0	F8 V	-31	26 ts		
128752	158696		14 39 00.3	-10 33 17	-0.001	+0.01	6.73	1.00		G5				
128476	241875		14 39 04.6	-56 29 50	0.000	+0.03	7.43	1.05		K0				
128582	225089		14 39 10.9	-46 35 03	-0.018	-0.21	6.07	0.51	3.2	F8 IV-V	-10	38 s		
128691	205806		14 39 16.2	-33 22 44	-0.001	-0.02	7.4	0.9	3.4	F5 V		36 s		
128669	205805		14 39 16.2	-36 13 52	-0.001	-0.02	7.8	1.7	-0.1	K2 III		190 s		
129390	16455		14 39 22.3	+64 09 14	-0.013	+0.02	7.5	0.4	3.0	F2 V		81 s		
129207	29235		14 39 24.1	+53 26 23	-0.006	-0.04	7.2	0.9	3.2	G5 IV		64 s		
128617	225091		14 39 24.6	-49 03 19	-0.017	-0.13	6.39	0.44	2.9	F6 IV-V		47 mx		
128293	252835		14 39 31.7	-68 12 13	0.000	-0.03	6.76	-0.05	-1.7	B3 V ne		470 s		m
128621	--	α² Cen	14 39 35.4	-60 50 13	-0.494	+0.69	1.39			K1 V	-21	1.3 t		m
128620	252838	α¹ Cen	14 39 36.7	-60 50 02	-0.494	+0.69	0.00	0.68	4.4	G2 V	-25	1.3 t		Rigil Kentaurus, m
128725	205808		14 39 36.8	-39 33 53	-0.001	-0.01	7.6	0.7	0.2	K0 III		300 s		
128418	252836		14 39 38.1	-64 42 54	-0.008	-0.08	8.00	0.4	3.0	F2 V		99 s		m
128456	252837		14 39 39.2	-63 32 42	+0.005	+0.02	7.8	1.1	0.2	K0 III		310 s		
129499	16458		14 39 39.6	+66 20 48	-0.005	-0.03	7.4	0.9	3.2	G5 IV		71 s		
128787	182739		14 39 40.9	-26 43 25	+0.003	-0.03	6.99	0.45	3.4	F5 V		52 s	9330	m
--	182743		14 39 46.9	-26 41 53	+0.003	-0.04	7.00						9330	m
128853	182746		14 39 59.1	-27 08 04	+0.006	-0.10	7.3	1.0	3.2	G5 IV		50 s		
129170	64261		14 39 59.4	+38 06 38	-0.002	+0.01	7.7	1.1	0.2	K0 III		310 s		
128673	241892		14 40 02.0	-52 03 58	-0.002	-0.03	7.5	1.4	0.2	K0 III		170 s		
128924	158715		14 40 08.4	-16 12 08	+0.001	-0.08	7.6	1.1	-0.1	K2 III		180 mx		
129425	29239		14 40 10.5	+57 42 47	-0.012	+0.01	7.9	0.5	4.0	F8 V		59 s		
128618	241890		14 40 16.1	-59 45 04	-0.002	+0.01	8.0	1.6	-0.1	K2 III		240 s		
128819	225106		14 40 17.4	-40 50 32	-0.003	-0.04	6.65	-0.08		B9				m
129171	64262		14 40 18.2	+30 26 38	+0.007	-0.04	7.7	0.6	4.4	G0 V		47 s		
135294	2460		14 40 18.4	+87 12 53	-0.017	+0.02	7.1	1.1	0.2	K0 III	-27	240 s		
128775	225104		14 40 19.4	-45 47 39	0.000	-0.04	6.62	-0.11		A0 p				m
129132	83458		14 40 21.8	+21 58 32	-0.001	+0.03	6.2	0.5	3.4	F5 V	+1	36 s		
128907	205818		14 40 24.8	-30 32 10	0.000	-0.01	8.0	0.8	-0.1	K2 III		420 s		
128928	182752		14 40 25.0	-26 15 19	-0.001	+0.01	7.20	0.1	1.4	A2 V		150 s	9333	m
128931	182751		14 40 25.1	-29 18 05	-0.007	-0.04	7.82	0.46	3.1	F3 V		79 s		
128674	241894		14 40 28.1	-57 01 46	+0.047	-0.31	7.38	0.68	5.2	G5 V		27 s		m
129209	64264		14 40 28.2	+30 31 13	+0.006	-0.06	7.90	0.7	3.0	G2 IV	-7	96 s		
128777	241900		14 40 29.8	-50 44 48	-0.002	-0.02	8.0	1.7	0.2	K0 III		140 s		
128855	225111		14 40 30.9	-40 36 29	+0.001	-0.01	7.6	0.1	0.6	A0 V		210 s		
128987	158720		14 40 31.0	-16 12 34	-0.008	-0.08	7.5	0.8	3.2	G5 IV		72 s		
128986	158722		14 40 31.9	-14 02 48	-0.001	-0.01	6.98	1.63		K5				
128713	241898		14 40 32.4	-56 26 25	-0.004	0.00	6.30	1.18		K0				
128571	--		14 40 37.1	-65 26 05			8.0	0.3	4.0	F8 V		62 s		
129134	120597		14 40 38.2	+8 58 42	-0.001	-0.01	7.9	1.1	0.2	K0 III		340 s		
129154	101136		14 40 41.0	+12 16 43	-0.002	0.00	8.0	0.1	1.7	A3 V		180 s		
128875	225114		14 40 42.1	-42 12 44	-0.001	-0.01	7.5	1.3	-0.5	M2 III		170 s		
129153	101137		14 40 42.3	+13 32 02	+0.004	-0.03	5.91	0.21	2.1	A5 V	-8	54 s		
129174	101138	29 π¹ Boo	14 40 43.5	+16 25 05	+0.001	+0.01	4.93	-0.03		B9 p	-1	40 mn	9338	m
129175	101139	29 π² Boo	14 40 43.9	+16 25 03	0.000	0.00	5.85	0.24		A m	-6		9338	m
129008	182759		14 40 51.1	-24 03 27	-0.001	-0.01	7.2	0.6	3.2	G5 IV		64 s		
128973	205822		14 40 52.0	-32 19 32	0.000	-0.03	7.8	1.4	0.2	K0 III		180 s		
--	--		14 40 54.7	-32 19 49			8.00							m
129010	182760		14 40 55.7	-26 25 44	-0.003	-0.01	7.4	0.8	4.0	F8 V		34 s		
129354	64272		14 40 58.1	+38 09 03	-0.008	+0.18	7.4	0.5	3.4	F5 V		63 s		
129580	29244		14 41 00.8	+57 57 25	+0.016	-0.19	7.42	0.93	5.9	dK0	-9	17 s	9346	m
128974	205823		14 41 01.3	-36 08 06	-0.002	-0.01	5.67	-0.08		A0 p				
129500	29243		14 41 02.2	+52 14 11	-0.004	-0.04	7.2	1.1	0.2	K0 III		210 mx		
129135	140068		14 41 02.3	-2 25 26	0.000	-0.16	7.9	0.7	4.4	G0 V		51 s		
129426	45171		14 41 05.8	+45 24 49	-0.013	-0.17	7.7	0.9	3.2	G5 IV		44 mx		
129065	182764		14 41 06.6	-22 37 08	-0.003	-0.02	7.90	0.1	1.7	A3 V		170 t		m
129248	101144		14 41 07.2	+12 06 35	0.000	-0.02	8.0	0.9	3.2	G5 IV		93 s		
129246	101145	30 ζ Boo	14 41 08.8	+13 43 42	+0.004	-0.02	3.78	0.05	-0.2	A2 III	-5	63 s	9343	m
129229	120599		14 41 18.7	+3 26 55	0.000	-0.04	7.9	0.9	4.4	G0 V		51 s		
129013	205828		14 41 19.2	-36 53 06	+0.003	-0.08	7.9	0.4	3.4	F5 V		78 s		
129357	83469		14 41 22.5	+29 03 33	+0.003	-0.16	7.81	0.63	4.7	G2 V		40 s		
129138	158731		14 41 25.2	-19 55 39	-0.003	-0.01	7.2	0.9	0.2	K0 III		250 s		
129231	140074		14 41 30.1	-2 02 28	-0.003	-0.02	7.60	0.1	1.4	A2 V		170 s		m
129358	101150		14 41 35.2	+18 29 02	-0.002	0.00	8.0	0.9	3.2	G5 IV		93 s		
129600	29246		14 41 35.3	+51 23 50	-0.003	-0.05	8.00	0.4	2.6	F0 V		120 s	9350	m
129249	140076		14 41 35.8	-1 23 19	0.000	-0.04	7.7	1.4	-0.3	K5 III		390 s		

HD	SAO	Star Name	α 2000	δ 2000	μ(α)	μ(δ)	V	B-V	M_v	Spec	RV	d(pc)	ADS	Notes
			h m s	° ′ ″	s	″								
129312	120601	31 Boo	14 41 38.7	+8 09 42	0.000	0.00	4.86	1.00	0.3	G8 III	−22	75 s		
129336	101152	32 Boo	14 41 43.4	+11 39 38	−0.011	−0.12	5.56	0.94	0.3	G8 III	−23	48 mx		
129412	83472		14 41 45.8	+24 32 17	−0.003	−0.02	7.65	0.49	3.8	F7 V		59 s		
129160	182772		14 41 45.8	−29 41 51	−0.001	−0.02	7.80	0.4	2.6	F0 V		110 s	9344	m
129017	225126		14 41 50.3	−47 16 30	+0.006	−0.03	7.1	0.7	−0.5	M2 III		240 mx		
129161	205841		14 41 51.0	−30 56 00	−0.002	−0.03	6.90	0.00	0.4	B9.5 V		200 s		m
129391	101156		14 41 51.4	+18 29 44	−0.001	−0.06	7.6	0.9	0.3	G7 III	−15	190 mx		
128400	257183		14 41 52.5	−75 08 20	+0.034	−0.01	6.6	1.5	4.4	G0 V		28 s		
129430	83474		14 41 54.2	+21 07 25	−0.001	−0.06	6.4	0.8	3.2	G5 IV	−11	44 s		
128917	241912		14 41 55.6	−58 36 58	+0.003	−0.05	6.22	0.45	4.0	dF8		28 s		
129056	225128	α Lup	14 41 55.7	−47 23 17	−0.002	−0.02	2.30	−0.20	−4.4	B1 III	+7	210 s		m
129195	182775		14 41 57.2	−28 03 56	+0.001	+0.01	7.3	1.2	0.2	K0 III		210 s		
129338	120605		14 41 57.4	+1 37 33	0.000	0.00	7.9	1.1	0.2	K0 III		340 s		
129116	205839		14 41 57.5	−37 47 37	−0.002	−0.04	4.00	−0.17	−1.7	B3 V	+8	140 s		
128294	257182		14 41 59.7	−77 00 41	−0.005	−0.02	6.6	0.0	0.0	B8.5 V		210 s		
129674	29248		14 42 00.6	+52 00 21	+0.001	+0.02	7.5	0.4	2.6	F0 V		96 s		
129293	158742		14 42 02.2	−12 14 06	+0.001	0.00	7.1	1.1	0.2	K0 III		240 s		
129070	225132		14 42 02.2	−46 17 53	−0.002	−0.01	7.36	−0.02		A0				
129798	16466		14 42 03.1	+61 15 43	+0.009	−0.03	6.25	0.41	0.6	F2 III	−6	130 s	9357	m
129179	205845		14 42 05.2	−34 18 57	−0.005	0.00	7.7	1.6	0.2	K0 III		140 s		
129118	225133		14 42 09.2	−42 47 20	0.000	−0.03	7.4	0.5	3.2	G5 IV		70 s		
--	225134		14 42 09.2	−42 47 23	0.000	−0.05	8.0	−0.1	3.2	G5 IV		93 s		
129216	205846		14 42 13.0	−32 46 13	−0.011	−0.13	6.65	0.92		G5		19 mn		
129217	205847		14 42 16.3	−33 37 58	+0.001	+0.04	7.4	0.9	0.2	K0 III		280 s		
129447	101159		14 42 19.2	+11 04 33	+0.002	−0.02	7.9	1.1	0.2	K0 III		340 s		
--	64285		14 42 20.2	+36 36 44	0.000	−0.04	8.0	1.7						
129779	29251		14 42 23.1	+54 48 13	+0.001	−0.01	7.62	1.61		Mb				
129272	182782		14 42 23.2	−23 25 31	−0.003	+0.01	7.2	0.6	0.2	K0 III		250 s		
129275	182780		14 42 25.6	−27 47 16	−0.002	−0.04	7.7	1.4	−0.3	K5 III		400 s		
128898	252852	α Cir	14 42 28.0	−64 58 43	−0.032	−0.27	3.19	0.24	2.6	F0 V p	+7	14 ts		m
129449	120611		14 42 31.2	+2 17 42	0.000	−0.02	8.0	0.9	3.2	G5 IV		93 s		
129088	241923		14 42 31.4	−54 35 55	0.000	0.00	7.6	−0.1	0.6	A0 V		260 s		
129470	120614		14 42 34.8	+4 29 12	0.000	−0.04	7.8	1.1	0.2	K0 III		340 s		
129653	64289		14 42 39.5	+36 45 24	+0.003	−0.01	7.3	0.1	1.4	A2 V		150 s		
129537	101163		14 42 42.3	+14 42 30	−0.002	−0.03	6.6	0.4	3.0	F2 V	−20	53 s		
129379	158752		14 42 45.0	−19 18 41	−0.004	+0.03	6.9	1.1	0.2	K0 III		220 s		
129517	120616		14 42 47.5	+6 35 27	−0.012	+0.07	7.21	0.63	4.0	F8 V		38 s	9353	m
128544	--		14 42 48.8	−75 50 03			7.4	1.4	0.2	K0 III		170 s		
129090	241927		14 42 51.9	−58 05 11	0.000	−0.01	7.7	−0.4	0.0	B8.5 V		350 s		
129073	241925		14 42 54.7	−60 04 48	+0.001	+0.01	8.0	0.0	0.4	B9.5 V		300 s		
129538	120618		14 42 55.1	+8 04 35	0.000	+0.01	8.00	0.4	3.0	F2 V		100 s	9355	m
128982	--		14 43 02.0	−65 14 52			7.5	0.0	0.4	B9.5 V		250 s		X Cir, v
129502	140090	107 μ Vir	14 43 03.5	−5 39 30	+0.007	−0.32	3.88	0.38	1.9	F3 IV	+5	26 ts		
129041	252857		14 43 04.2	−62 12 26	+0.003	−0.03	7.2	0.2	1.4	A2 V		110 s		
129125	241933		14 43 09.3	−58 15 59	+0.001	0.00	7.6	1.1	0.2	K0 III		270 s		
129584	120622		14 43 12.9	+1 49 29	−0.003	−0.02	7.6	0.1	0.6	A0 V		260 s		
129433	182795	4 Lib	14 43 13.5	−24 59 51	−0.001	−0.01	5.7	0.5		B9	−4			
129281	225145		14 43 15.3	−47 34 14	0.000	−0.01	6.92	−0.05		B8				
129919	29256		14 43 21.3	+53 47 40	−0.003	+0.03	7.2	0.9	3.2	G5 IV		65 s		
129265	241941		14 43 22.7	−51 33 30	+0.001	−0.02	8.0	1.2	0.2	K0 III		290 s		
129712	83488	34 Boo	14 43 25.3	+26 31 40	−0.001	−0.02	4.81	1.66		gMa	+6	30 mn		W Boo, q
129242	241940		14 43 25.3	−54 51 07	+0.004	−0.02	7.8	0.0	1.7	A3 V		160 s		
129566	140094		14 43 25.6	−9 41 58	0.000	−0.02	6.6	0.8	3.2	G5 IV		48 s		
129454	182799		14 43 27.4	−30 12 27	−0.004	−0.03	7.9	0.4	3.4	F5 V		79 s		
129092	252860		14 43 27.8	−62 58 01	−0.003	−0.01	6.41	−0.08	−1.7	B3 V		380 s		m
129436	205867		14 43 37.6	−39 27 51	−0.001	−0.02	7.4	0.5	3.2	G5 IV		69 s		
129437	205868		14 43 38.0	−39 38 59	−0.006	−0.05	7.90	1.1	0.2	K0 III		170 mx		m
129456	205871		14 43 39.3	−35 10 25	−0.005	−0.19	4.05	1.35	−0.3	K5 III	−39	71 mx		
129505	182805		14 43 44.3	−29 59 47	−0.001	−0.06	7.2	0.3	3.2	G5 IV		63 s		
129846	45190		14 43 44.4	+40 27 34	−0.001	+0.03	5.73	1.39	0.2	K0 III	+13	75 s		
129655	140097		14 43 46.4	−2 30 20	−0.002	−0.03	7.2	0.1	1.4	A2 V		150 s		
130173	16478		14 44 03.5	+61 05 53	+0.004	−0.04	6.6	0.4	3.0	F2 V		53 s	9371	m
128862	--		14 44 07.1	−73 58 54			7.8	0.4	2.6	F0 V		110 s		
129527	205878		14 44 09.2	−37 18 50	−0.003	−0.02	8.0	1.9	0.2	K0 III		110 s		
129814	101175		14 44 11.6	+18 27 44	−0.005	−0.16	7.5	0.9	3.2	G5 IV		55 mx		
129698	140100		14 44 13.6	−8 15 07	−0.014	+0.07	6.6	0.5	3.4	F5 V		44 s		
129060	252862		14 44 14.3	−69 40 28	−0.014	−0.11	6.9	0.5	4.0	F8 V		37 s		
129635	182822		14 44 17.5	−24 07 43	−0.003	+0.03	7.0	0.6	0.2	K0 III		230 s		
129713	140103		14 44 21.2	−9 29 22	−0.002	−0.03	8.0	0.4	3.0	F2 V		98 s		
129742	140104		14 44 23.9	−6 23 19	+0.001	0.00	7.60	1.1	0.2	K0 III		300 s		m
129439	241962		14 44 25.1	−53 16 20	−0.001	−0.02	7.9	1.2	0.2	K0 III		270 s		
130236	16479		14 44 26.0	+60 46 20	−0.004	+0.03	7.8	1.1	0.2	K0 III		330 s		
130044	45197		14 44 28.1	+45 11 10	+0.006	−0.01	6.9	0.4	2.6	F0 V	−7	71 s		
129755	140107		14 44 30.5	−3 21 48	−0.005	−0.01	7.5	0.4	3.0	F2 V		80 s		
130251	16481		14 44 34.1	+60 03 13	−0.003	−0.02	7.4	1.1	0.2	K0 III		280 s		

HD	SAO	Star Name	α 2000	δ 2000	μ(α)	μ(δ)	V	B-V	M_V	Spec	RV	d(pc)	ADS	Notes
129847	101179		14 44 34.5	+10 45 50	-0.001	+0.01	8.0	0.5	3.4	F5 V		82 s		
128679	257186		14 44 35.3	-77 44 03	-0.007	+0.02	7.9	1.1	-0.1	K2 III		380 mx		
129784	140109		14 44 44.9	-9 42 14	-0.007	-0.05	6.9	0.4	3.0	F2 V		59 s		
129147	--		14 44 45.9	-70 04 46			8.00	0.6	4.4	G0 V		52 s		m
129462	241966		14 44 55.4	-58 28 40	-0.010	-0.06	6.11	1.02	0.2	gK0		150 s		
129887	120636		14 44 55.5	+3 29 03	0.000	-0.01	7.7	0.9	3.2	G5 IV		80 s		
130002	64303		14 44 55.6	+30 17 41	-0.001	-0.02	7.6	1.1	0.2	K0 III		310 s		
129577	241969		14 44 56.8	-50 31 50	-0.002	-0.02	8.0	0.9	-0.1	K2 III		410 s		
129685	205899		14 44 59.0	-35 11 31	0.000	-0.01	4.92	0.01		A0 n	-5	17 mn		
129988	83500	36 ε Boo	14 44 59.1	+27 04 27	-0.004	+0.02	2.37	0.97	-0.9	K0 II-III	-17	46 s	9372	Izar, m
129405	252868		14 44 59.1	-61 31 17	+0.001	0.00	7.3	1.9	-0.1	K2 III		110 s		
129623	225159		14 44 59.7	-44 39 55	+0.001	-0.02	7.53	0.96		G5				
--	225161		14 45 09.5	-49 54 58	-0.070	-0.38	8.0	1.1	0.2	K0 III		19 mx		
129512	241968		14 45 09.9	-57 32 11	-0.001	-0.01	7.3	0.7	3.2	G5 IV		66 s		
129557	241971		14 45 10.8	-55 36 07	-0.001	-0.02	6.10	-0.06	-3.0	B2 IV	-4	550 s		m
129902	140116		14 45 11.6	-1 25 04	-0.004	-0.01	6.07	1.61	-0.5	gM1	-47	200 s		
130083	64307		14 45 11.9	+34 22 34	0.000	0.00	7.8	1.6	-0.5	M2 III	-25	450 s		
130084	64306		14 45 13.6	+32 47 17	+0.003	-0.08	6.28	1.58	-0.5	M2 III	+30	230 s		
129972	101184	35 o Boo	14 45 14.4	+16 57 51	-0.004	-0.05	4.60	0.98	0.2	K0 III	-9	29 ts		
130460	16485		14 45 16.5	+65 19 49	-0.019	+0.04	7.8	0.5	3.4	F5 V		74 s		
129422	252869		14 45 17.2	-62 52 33	+0.010	-0.09	5.36	0.29	2.4	A7 V n	+7	39 s		m
129578	241972		14 45 18.8	-55 36 23	+0.001	0.00	7.6	1.1	0.2	K0 III		280 s		
129624	241976		14 45 19.0	-51 18 52	0.000	-0.01	6.8	1.9	0.2	K0 III		63 s		
129903	140120		14 45 20.2	-6 44 04	-0.006	+0.04	7.9	0.7	4.4	G0 V		51 s		
129732	205904		14 45 20.5	-36 08 52	-0.002	-0.01	8.00	0.1	1.4	A2 V		210 s		m
130025	101187		14 45 20.7	+18 53 05	+0.002	+0.01	6.13	0.83	0.2	K0 III	-4	150 s		
129974	101185		14 45 21.7	+10 35 55	+0.001	-0.02	7.1	0.8	3.2	G5 IV		59 s		
130004	101188		14 45 24.3	+13 50 45	-0.014	-0.25	7.7	1.1	0.2	K0 III		35 mx		
130188	45208		14 45 29.5	+42 22 57	-0.007	+0.05	7.34	0.45	3.4	dF5	-21	60 s	9378	m
129956	120642	108 Vir	14 45 30.1	+0 43 02	-0.003	-0.01	5.69	-0.03	0.2	B9 V	-17	120 s		
129764	205906		14 45 30.2	-37 57 23	-0.013	-0.05	7.8	1.3	4.4	G0 V		17 s		
130045	101191		14 45 33.4	+17 53 05	+0.003	+0.09	7.9	0.9	3.2	G5 IV		88 s		
130005	120643		14 45 36.1	+8 09 35	+0.003	-0.05	7.0	1.1	0.2	K0 III		230 s		
129330	--		14 45 41.6	-69 37 41			7.5	1.1	0.2	K0 III		280 s		
129908	182852		14 45 44.2	-22 24 37	-0.003	-0.02	8.00	0.1	0.6	A0 V		300 s		m
130834	8081		14 45 50.9	+71 57 58	-0.008	+0.02	7.4	0.2	2.1	A5 V		120 s		
130086	101197		14 45 51.8	+13 06 06	-0.001	-0.07	6.6	1.1	0.2	K0 III		180 mx		
129942	158784		14 45 52.9	-17 41 54	+0.004	-0.05	7.4	1.4	-0.3	K5 III		210 mx		
130108	101199		14 45 53.9	+16 47 53	0.000	0.00	7.35	1.20	-0.1	K2 III		300 s		
130087	101198		14 45 56.7	+10 03 07	-0.006	-0.03	7.5	0.5	3.4	F5 V		67 s		
129805	225175		14 45 57.0	-43 18 20	0.000	0.00	7.9	1.1	-0.3	K5 III		430 s		
129791	225174		14 45 57.6	-44 52 03	-0.002	-0.04	6.92	0.04		B9				m
129978	158788	5 Lib	14 45 57.7	-15 27 35	-0.002	-0.01	6.60	1.1	0.2	K0 III	-40	190 s	9376	m
129926	182855	54 Hya	14 46 00.0	-25 26 35	-0.011	-0.11	4.94	0.35	0.6	F0 III	-13	51 mx	9375	m
130600	16489		14 46 01.3	+64 37 32	-0.001	-0.05	7.9	0.1	0.6	A0 V		290 s		
130215	83509		14 46 03.0	+27 30 46	+0.001	-0.02	7.98	0.84	6.3	K2 V		22 s		
129594	252873		14 46 03.5	-62 32 08	+0.004	0.00	7.5	0.5	4.0	F8 V		49 s		
130144	101200		14 46 05.9	+15 07 54	-0.006	+0.01	5.63	1.57		gM5	-22			
129877	205916		14 46 06.4	-38 36 15	-0.004	-0.05	7.8	0.9	0.2	K0 III		210 mx		
129944	182857		14 46 06.6	-23 09 10	+0.002	-0.06	5.8	1.3	0.2	K0 III	+7	92 s		
128813	--		14 46 08.5	-79 07 52			7.8	1.4	-0.3	K5 III		420 s		
129980	182858		14 46 10.8	-21 10 34	-0.004	-0.11	6.40	0.58	4.2	dF9	0	26 s		m
130048	140127		14 46 13.5	-7 47 49	0.000	-0.02	7.2	0.9	3.2	G5 IV		64 s		
130109	120648	109 Vir	14 46 14.9	+1 53 34	-0.008	-0.03	3.72	-0.01	0.6	A0 V	-6	38 ts		
130145	120651		14 46 15.5	+9 38 48	+0.005	-0.26	7.26	0.62	4.7	G2 V	+25	32 s	9380	m
130125	120649		14 46 16.2	+3 30 14	+0.001	-0.06	7.5	0.7	4.4	G0 V		42 s		
130110	140129		14 46 17.8	-0 50 01	0.000	+0.03	7.1	1.1	0.2	K0 III		240 s		
130155	101201		14 46 18.2	+14 30 37	0.000	-0.02	7.30	1.6	-0.5	M1 III	+6	360 s		
129821	225177		14 46 18.2	-48 39 19	+0.002	-0.04	7.8	1.3	0.2	K0 III		230 s		
129946	205923		14 46 22.3	-33 45 58	+0.007	-0.02	8.0	0.9	3.2	G5 IV		78 s		
130499	29274		14 46 23.3	+56 36 59	-0.003	0.00	6.5	1.1	0.2	K0 III		190 s		
--	182861		14 46 27.9	-22 24 14	-0.007	-0.05	7.5	0.4	3.0	F2 V		80 s		
129858	225178		14 46 29.0	-47 26 28	-0.002	-0.02	5.74	0.07		A2				
130217	101204		14 46 29.4	+14 51 18	-0.001	-0.01	7.8	1.1	0.2	K0 III		330 s		
130445	29273		14 46 29.8	+50 23 52	-0.007	+0.05	7.5	0.5	3.4	F5 V		66 s		
129981	205930		14 46 33.5	-32 10 16	0.000	-0.02	7.80	0.8	-2.1	G5 II p		960 s		V553 Cen, v
129708	252879		14 46 42.0	-61 27 42	0.000	+0.01	7.8	0.5	3.0	F2 V		71 s		
130218	120655		14 46 44.1	+7 16 53	-0.002	-0.01	7.8	1.6	-0.5	M4 III		470 s		
129774	252880		14 46 57.9	-60 51 55	+0.002	+0.01	8.0	0.7	2.6	F0 V		66 s		
130317	83516		14 46 59.6	+19 47 58	+0.002	-0.02	7.6	1.1	0.2	K0 III		300 s		
129893	241992		14 47 01.2	-52 23 01	-0.002	-0.09	5.21	0.98	0.3	G6 III	-21	88 s		m
130684	29278		14 47 02.4	+59 42 39	+0.005	+0.01	7.2	0.1	1.4	A2 V		150 s		
129894	241993		14 47 04.4	-53 46 29	-0.005	-0.03	7.1	0.7	0.2	K0 III		240 s		
130055	205937		14 47 05.0	-38 17 27	+0.004	-0.09	5.94	1.33		K0				
130256	120657		14 47 05.3	+0 58 14	-0.001	-0.04	6.70	0.1	0.6	A0 V	+6	170 s	9383	m

HD	SAO	Star Name	α 2000	δ 2000	μ(α)	μ(δ)	V	B-V	M_v	Spec	RV	d(pc)	ADS	Notes
			$14^h 47^m 09\overset{s}{.}8$	$+76° 02' 30''$	$-0\overset{s}{.}011$	$-0\overset{''}{.}02$								
131358	8087		14 47 09.8	+76 02 30	-0.011	-0.02	7.3	0.1	1.4	A2 V		82 mx		
129932	241995		14 47 12.2	-52 12 20	-0.002	-0.02	6.30	0.1	0.6	A0 V		140 s		m
130157	182873		14 47 13.6	-21 19 29	-0.001	0.00	6.0	2.0	-0.1	K2 III	-24	55 s		
130307	120663		14 47 16.0	+2 42 11	-0.020	-0.09	7.80	0.91	3.2	G5 IV		36 mx		
130461	64330		14 47 21.1	+35 34 03	-0.002	-0.02	7.7	0.1	1.4	A2 V		180 s		
130158	182875	55 Hya	14 47 22.5	-25 37 27	-0.001	-0.02	5.7	0.1		A0 p	-18			
129739	--		14 47 23.2	-65 50 29			7.9	0.1	0.6	A0 V		280 s		
130582	45217		14 47 23.3	+47 54 33	-0.003	+0.01	7.9	0.5	4.0	F8 V		61 s		
130321	120667		14 47 25.6	+2 02 09	-0.001	-0.04	7.60	0.1	1.4	A2 V		170 s	9384	m
129740	252882		14 47 28.2	-66 10 47	-0.004	-0.03	7.35	-0.07	-2.2	B5 III		260 mx		
130396	101216		14 47 31.8	+19 03 00	+0.001	0.00	7.4	0.5	3.4	F5 V		63 s		
130073	225183		14 47 32.0	-43 33 26	-0.001	-0.03	6.30	1.08	0.0	K1 III		180 s		
129915	--		14 47 36.8	-59 09 24			7.5	2.8	-0.3	K5 III		59 s		
130446	83524		14 47 39.1	+24 05 50	-0.003	+0.02	7.60	1.1	0.2	K0 III		300 s	9386	m
130914	16503		14 47 40.3	+65 15 06	-0.005	+0.02	7.2	0.2	2.1	A5 V		110 s		
130259	182882	56 Hya	14 47 44.7	-26 05 15	+0.003	-0.01	5.24	0.94	0.3	G5 III	-1	89 s		
131207	8085		14 47 44.8	+71 36 04	+0.009	-0.01	7.8	0.4	3.0	F2 V		92 s		
130163	205948		14 47 46.6	-39 55 35	-0.006	-0.01	7.0	0.6	0.6	A0 V		77 s		
129795	252884		14 47 48.4	-66 19 54	+0.004	-0.01	7.1	0.0	0.0	B8.5 V		260 s		
130555	64331		14 47 49.8	+35 39 15	-0.004	+0.02	7.6	0.5	3.4	F5 V		70 s		
129078	257193	α Aps	14 47 51.6	-79 02 41	-0.001	-0.02	3.83	1.43	-0.3	K5 III	0	67 s		
130223	205953		14 47 53.4	-32 14 50	-0.002	-0.02	7.6	1.5	-0.5	M2 III		420 s		
130325	158808		14 47 54.8	-12 50 24	+0.003	-0.09	6.35	1.10	3.2	G5 IV		24 s		
129750	252883		14 47 56.8	-68 31 11	-0.004	-0.02	7.60	0.1	0.6	A0 V		190 mx		m
--	225193		14 47 57.2	-41 37 52	-0.002	-0.02	8.0	0.5	0.2	K0 III		360 s		
130274	182883	57 Hya	14 47 57.4	-26 38 46	-0.001	-0.01	5.60	-0.02	0.2	B9 V	+6	120 s		
129842	252887		14 48 00.0	-65 44 25	-0.007	-0.04	7.9	0.0	0.0	B8.5 V		94 mx		
130386	140145		14 48 02.2	-5 30 24	-0.001	-0.02	7.7	1.1	-0.1	K2 III		360 s		
130225	225198		14 48 09.0	-41 01 02	-0.001	-0.06	7.7	0.6	4.0	F8 V		54 s		
129996	242003		14 48 09.0	-58 34 19	-0.002	-0.03	8.0	1.3	0.2	K0 III		250 s		
130060	242006		14 48 12.6	-56 00 32	-0.004	-0.04	7.2	0.5	0.6	A0 V		98 s		
129862	252891		14 48 13.3	-66 26 22	+0.003	-0.03	7.8	0.0	0.4	B9.5 V		290 s		
--	16509		14 48 21.1	+63 43 30	-0.006	-0.01	7.1	1.7						
130603	83535		14 48 23.2	+24 21 59	-0.008	+0.04	6.14	0.48	3.4	F5 V	-31	33 s	9389	m
130277	205961		14 48 23.7	-39 01 42	-0.001	+0.04	7.9	0.5	1.7	A3 V		92 s		
130583	83534		14 48 25.4	+22 23 54	+0.001	-0.03	7.9	1.4	-0.3	K5 III		430 s		
130360	182888		14 48 29.0	-25 29 12	-0.009	+0.17	7.0	1.0	4.0	F8 V		20 s		
129918	252893		14 48 30.5	-66 31 36	-0.003	-0.03	7.9	0.1	1.4	A2 V		190 s		
130412	158813		14 48 31.6	-17 20 27	+0.003	-0.08	7.40	0.7	4.4	G0 V		40 s	9387	m
130311	205964		14 48 32.7	-35 50 29	+0.003	-0.01	7.20	1.1	0.2	K0 III		250 s		m
130328	205966		14 48 37.9	-36 38 05	-0.001	-0.05	6.0	1.4	-0.5	M4 III		200 s		
130485	140149		14 48 40.6	-7 06 27	-0.003	-0.03	8.0	0.4	2.6	F0 V		120 s		
129811	257197		14 48 42.5	-70 35 57	0.000	-0.02	7.4	1.7	0.2	K0 III		110 s		
129954	252894		14 48 44.3	-66 35 38	-0.002	-0.03	5.91	-0.07	-2.5	B2 V		420 s		m
130226	242021		14 48 45.0	-51 23 02	0.000	-0.01	7.9	0.9	0.2	K0 III		340 s		
130463	158815		14 48 48.9	-13 07 10	0.000	-0.02	7.9	1.1	0.2	K0 III		340 s		
130652	101226		14 48 51.5	+18 11 59	+0.002	0.00	7.4	1.1	-0.1	K2 III	-28	310 s		
130604	120673		14 48 53.1	+5 57 15	-0.001	-0.09	6.84	0.44	3.4	dF5	-1	49 s	9392	m
130373	205971		14 48 53.4	-37 37 58	-0.003	0.00	7.8	1.6	3.2	G5 IV		26 s		
130557	140152		14 48 54.0	-0 50 51	-0.001	+0.02	6.14	-0.04		A0	-16			
130363	225205		14 48 58.5	-40 25 18	-0.001	-0.03	7.5	1.6	-0.1	K2 III		190 s		
130169	242019		14 48 58.5	-58 07 12	0.000	-0.02	7.3	0.0	4.0	F8 V		45 s		
130388	205974		14 49 01.6	-35 32 59	+0.001	-0.02	7.7	0.3	1.4	A2 V		140 s		
130817	64344		14 49 06.6	+37 48 40	-0.022	+0.11	6.16	0.36	2.6	F0 V	-35	50 s		
131005	29293		14 49 06.7	+54 13 53	-0.004	-0.01	7.3	1.1	-0.1	K2 III		300 s		
130227	242026		14 49 06.8	-56 40 04	-0.013	-0.12	6.23	1.13	-0.1	K2 III		99 mx		
130558	158817		14 49 10.9	-10 49 37	+0.004	+0.01	7.80	0.5	4.0	F8 V		58 s	9395	m
130206	242025		14 49 14.8	-59 24 28	-0.001	-0.02	7.30	0.1	1.7	A3 V		130 s		m
130766	83546		14 49 14.9	+25 09 02	0.000	-0.03	6.7	1.1	-2.3	K3 II	-12	640 s		
130449	205978		14 49 17.2	-34 01 10	+0.001	+0.01	7.7	1.3	-0.1	K2 III		310 s		
130945	45226	38 Boo	14 49 18.6	+46 06 58	-0.001	-0.08	5.74	0.48	3.4	F5 V	-5	28 s		
130529	182898		14 49 18.6	-24 15 06	-0.002	-0.01	5.80	1.1	0.2	K0 III	-26	130 s	9394	m
130559	158821	7 μ Lib	14 49 19.0	-14 08 56	-0.004	-0.02	5.40	0.07		A0 p	-4	22 mn	9396	m
130264	242033		14 49 23.5	-57 26 16	-0.001	-0.04	7.76	0.61		F5				
130042	252899		14 49 23.5	-67 14 11	-0.021	-0.35	7.50	1.1	0.2	K0 III		46 mx		m
130080	252901		14 49 24.3	-65 40 56	-0.006	-0.03	7.1	0.5	3.4	F5 V		53 s		
130620	140159		14 49 25.0	-8 30 13	-0.001	-0.01	7.1	0.5	3.4	F5 V		55 s		
130705	101232		14 49 26.1	+10 02 39	-0.003	-0.08	6.64	1.26	0.2	K0 III		120 s		
130563	158822		14 49 27.5	-19 54 13	0.000	0.00	7.1	1.0	1.7	A3 V		34 s		
130767	101233		14 49 28.7	+19 30 37	-0.003	-0.01	6.9	0.0	0.4	B9.5 V	-14	200 s		
131040	29296		14 49 32.2	+51 22 29	+0.001	0.00	6.50	0.4	3.0	F2 V	-5	50 s	9405	m
130987	45228		14 49 35.0	+46 34 28	+0.008	-0.15	7.60	0.5	4.0	F8 V		53 s		
130021	252900		14 49 37.1	-68 56 08	-0.001	0.00	6.50	-0.10	-2.9	B3 III		690 s		
130707	120679		14 49 38.4	+1 30 17	-0.002	0.00	8.0	1.1	0.2	K0 III		360 s		
130768	101235		14 49 41.1	+10 12 03	0.000	-0.02	7.5	0.9	0.2	G9 III	+16	290 s		

HD	SAO	Star Name	α 2000	δ 2000	μ(α)	μ(δ)	V	B-V	M_v	Spec	RV	d(pc)	ADS	Notes
131041	45231	39 Boo	14ʰ49ᵐ41ˢ.2	+48°43'14"	-0ˢ.008	+0".09	5.69	0.47	3.4	F5 V	-32	28 s	9406	m
130726	120683		14 49 42.0	+7 59 11	-0.002	-0.01	7.70	0.5	3.4	F5 V	-31	72 s	9400	m
130818	83548		14 49 42.9	+23 01 56	+0.004	-0.01	7.14	0.42	2.2	F6 IV	+2	43 ts		RY Boo, q
130744	120682		14 49 44.2	+3 52 15	-0.003	-0.03	7.7	1.1	0.2	K0 III		310 s		
130672	140161		14 49 46.0	-9 12 10	-0.005	-0.02	7.6	0.5	4.0	F8 V		52 s		
130418	225215		14 49 46.2	-48 38 09	-0.002	-0.02	7.7	1.3	-0.3	K5 III		390 s		
130377	242042		14 49 47.0	-52 44 40	-0.001	-0.01	7.2	0.9	0.6	A0 V		64 s		
130788	101237		14 49 48.0	+10 29 35	-0.001	-0.02	7.0	0.1	1.4	A2 V		130 s		
130539	205988		14 49 49.2	-32 25 41	+0.005	-0.09	7.0	1.0	3.2	G5 IV		42 s		
130770	120684		14 49 52.4	+3 55 30	+0.002	-0.04	8.0	0.1	1.7	A3 V		180 s		
130335	242039		14 49 52.7	-57 20 48	0.000	-0.02	7.6	0.1	1.4	A2 V		160 s		
130870	83550		14 49 56.3	+24 43 11	-0.001	-0.02	6.5	0.0	0.4	B9.5 V		170 s		
130626	182905		14 49 57.3	-29 02 04	-0.006	+0.01	7.30	1.1	0.2	K0 III		220 mx		m
130917	83551		14 49 58.3	+28 36 57	+0.002	0.00	5.80	0.05	0.0	A3 III	-3	150 s		m
130708	158825		14 49 59.0	-13 30 31	-0.004	-0.02	7.9	0.4	2.6	F0 V		120 s		
130838	101239		14 50 00.7	+12 20 08	+0.003	-0.09	7.7	0.6	4.4	G0 V		47 s		
130286	252907		14 50 01.0	-62 19 49	0.000	0.00	7.40	0.00	0.0	B8.5 V		290 s		m
130771	120687		14 50 03.2	-0 00 21	+0.002	-0.03	7.9	0.1	0.6	A0 V		280 s		
130589	205991		14 50 05.4	-37 23 56	-0.007	-0.12	7.80	0.5	4.0	F8 V		58 s		m
130709	158828		14 50 06.4	-15 31 33	0.000	-0.01	7.2	1.1	0.2	K0 III		250 s		
131194	29303		14 50 13.5	+52 24 53	-0.004	+0.10	7.1	0.5	3.4	F5 V		56 s		
130948	83553		14 50 15.8	+23 54 42	+0.011	+0.03	5.85	0.56	4.4	G0 V	-1	20 s		
130380	242045		14 50 16.0	-58 58 35	-0.001	+0.01	7.3	1.3	3.2	G5 IV		31 s		
130694	182911	58 Hya	14 50 17.2	-27 57 37	-0.018	-0.06	4.41	1.40	-0.3	gK4	-10	88 s		
130454	242050		14 50 18.1	-52 23 07	0.000	-0.01	7.8	0.0	0.4	B9.5 V		270 s		
133002	2459		14 50 20.2	+82 30 42	+0.088	-0.23	5.64	0.68	4.4	G0 V	-43	27 mn		
131444	16524		14 50 21.9	+65 38 44	-0.004	0.00	7.2	1.4	-0.3	K5 III	-28	310 s		
132610	2458		14 50 29.1	+80 44 59	-0.020	+0.02	7.2	0.1	0.6	A0 V		94 mx		
131111	64355		14 50 29.5	+37 16 19	-0.018	+0.09	5.48	1.02	1.7	K0 III-IV	-66	53 s		
130287	252909		14 50 34.2	-66 15 35	+0.001	-0.03	7.8	0.0	0.0	B8.5 V		350 s		
130950	101241		14 50 35.0	+12 47 02	0.000	-0.04	7.5	1.1	-0.1	K2 III		330 s		
130801	158835		14 50 36.3	-12 01 03	-0.001	+0.01	7.2	0.4	3.0	F2 V		68 s		
130714	206000		14 50 37.9	-32 27 59	-0.008	-0.15	7.5	0.2	4.4	G0 V		42 s		
130819	158836	8 α¹ Lib	14 50 41.1	-15 59 50	-0.007	-0.07	5.15	0.41	2.1	F5 IV	-23			m
131873	8102	7 β UMi	14 50 42.2	+74 09 19	-0.009	+0.01	2.08	1.47	-0.3	K4 III	+17	29 ts		Kochab, m
130518	242057		14 50 43.9	-55 00 34	+0.001	-0.01	6.9	1.8	-0.1	K2 III		110 s		
130572	242061		14 50 50.8	-53 22 07	0.000	-0.05	6.8	0.6	0.6	A0 V		78 s		
130898	140174		14 50 51.6	-4 08 00	-0.006	+0.03	8.0	0.9	3.2	G5 IV		93 s		
130841	158840	9 α² Lib	14 50 52.6	-16 02 30	-0.007	-0.07	2.75	0.15		A m	-10	22 t		Zubenelgenubi, m
130790	182924		14 50 53.9	-28 19 28	-0.005	-0.03	7.9	0.0	3.0	F2 V		94 s		
131494	16527		14 50 54.4	+62 46 48	-0.005	-0.02	7.8	1.1	0.2	K0 III		310 mx		
130697	225234		14 50 58.5	-42 49 21	-0.004	-0.03	6.85	0.13		A2				
130970	140177		14 51 00.0	-0 15 26	-0.002	+0.01	6.15	1.39		K5	-20			
130952	140176	11 Lib	14 51 00.9	-2 17 57	+0.005	-0.13	4.94	0.98	1.8	G8 III-IV	+83	39 s		
131023	120697		14 51 02.2	+9 43 24	-0.015	+0.05	7.60	1.1	0.2	K0 III	-36	81 mx	9410	m
131043	101245		14 51 02.8	+12 09 51	0.000	-0.01	7.8	1.6	-0.5	M2 III		470 s		
130900	158842		14 51 03.4	-15 01 38	-0.004	+0.01	7.19	0.58	3.4	F5 V		47 s		
127163	258706		14 51 03.6	-86 23 39	+0.007	+0.01	7.8	1.7	-0.1	K2 III		190 s		
130659	225233		14 51 07.9	-49 49 07	-0.008	-0.08	7.4	1.1	0.2	K0 III		130 mx		
130953	158845		14 51 14.3	-13 36 25	-0.001	+0.01	7.8	1.1	-0.1	K2 III		380 s		
130855	182931		14 51 15.3	-23 51 35	+0.002	-0.05	7.6	1.0	3.2	G5 IV		56 s		
131132	101247		14 51 16.0	+18 38 59	0.000	-0.11	7.9	1.1	0.2	K0 III		130 mx		
130717	225238		14 51 16.2	-47 24 29	-0.002	-0.03	8.00	1.1	0.2	K0 III		360 s		m
130576	242063		14 51 21.3	-59 30 09	-0.002	+0.01	7.2	2.1	-0.1	K2 III		82 s		
131156	101250	37 ξ Boo	14 51 23.2	+19 06 04	+0.010	-0.10	4.55	0.76	5.5	G8 V	+4	6.8 t	9413	m
131507	29315		14 51 26.3	+59 17 38	-0.017	+0.14	5.46	1.36	-0.3	K4 III	+11	92 mx		
130081	257204		14 51 30.2	-74 55 59	-0.001	-0.01	6.80	0.00	0.4	B9.5 V		190 s		m
130761	225245		14 51 30.5	-47 05 00	-0.005	-0.04	8.00	0.44		F2				
129899	257202		14 51 30.5	-77 10 32	+0.001	0.00	6.5	0.2	0.6	A0 V		110 s		
130551	252918		14 51 31.6	-60 55 51	+0.007	+0.11	7.18	0.44	4.0	F8 V		43 s		
131158	101251		14 51 32.3	+13 28 39	-0.006	-0.09	7.9	1.1	0.2	K0 III		90 mx		
131157	101252		14 51 33.9	+14 45 51	+0.006	-0.11	7.9	1.1	0.2	K0 III		150 mx		
130989	158855		14 51 34.8	-17 47 25	-0.001	-0.13	6.50	0.46	3.4	dF5	+26	41 s		
131301	64365		14 51 38.1	+39 06 13	0.000	0.00	8.0	0.5	3.4	F5 V		84 s		
130807	225248	o Lup	14 51 38.3	-43 34 31	-0.002	-0.03	4.32	-0.15	-1.9	B6 III	+7	180 s		m
133872	2465		14 51 40.2	+83 56 15	-0.010	0.00	7.1	1.1	0.2	K0 III	-12	240 s		
130992	182935		14 51 40.5	-24 18 14	-0.068	-0.42	7.83	1.00	8.0	dK5	-65	15 ts		
130991	182937		14 51 44.8	-20 36 51	-0.002	+0.01	7.50	0.9	0.3	G7 III	-5	280 s		
131333	64367		14 51 45.9	+38 24 14	-0.004	-0.06	7.8	1.1	0.2	K0 III		150 mx		
130903	225253		14 51 57.4	-40 48 21	-0.001	-0.01	7.6	0.3	0.4	B9.5 V		280 s		
130663	252923		14 51 58.5	-60 20 05	+0.001	-0.02	8.0	0.1	0.6	A0 V		240 s		
131446	45246		14 52 03.1	+46 28 37	0.000	-0.05	7.1	1.1	0.2	K0 III		240 s		
131525	29316		14 52 05.5	+52 35 21	-0.006	+0.05	7.6	0.4	2.6	F0 V		99 s		
131265	83560		14 52 07.3	+20 17 25	0.000	-0.02	7.00	0.1	0.6	A0 V	-2	190 s	9420	m
130904	225256		14 52 15.5	-44 55 39	-0.001	-0.08	7.20	0.54		G0				

HD	SAO	Star Name	α 2000	δ 2000	μ(α)	μ(δ)	V	B−V	M_V	Spec	RV	d(pc)	ADS	Notes
			$h \quad m \quad s$	$\circ \quad ' \quad ''$	s	$''$								
131302	101262		14 52 20.4	+17 46 52	−0.004	+0.01	8.0	1.1	−0.1	K2 III		340 mx		
131526	45252		14 52 20.8	+48 40 16	−0.012	−0.01	7.6	0.6	4.4	G0 V		44 s		
130679	252927		14 52 23.4	−62 34 18	+0.001	+0.02	8.0	0.5	4.0	F8 V		62 s		
130932	225260		14 52 23.9	−46 45 37	+0.001	−0.03	6.87	1.69		Ma III				
131347	83562		14 52 29.1	+23 48 05	−0.002	+0.02	6.80	1.1	0.2	K0 III		210 s		m
130933	225261		14 52 29.2	−48 14 50	+0.001	−0.01	7.4	1.0	0.2	K0 III		270 s		
131617	29317		14 52 31.0	+55 44 42	−0.002	0.00	7.1	0.2	2.1	A5 V		100 s		
131196	158870		14 52 31.7	−15 23 15	−0.002	0.00	7.8	0.4	2.6	F0 V		110 s		
131117	206035		14 52 33.0	−30 34 36	−0.025	−0.03	6.29	0.60	4.5	dG1	−27	23 s		
130907	242081		14 52 34.3	−50 34 57	−0.002	−0.02	7.8	1.0	−0.5	M2 III		450 s		
130701	252928		14 52 35.0	−63 48 36	−0.003	−0.01	5.8	0.8	−2.1	G3 II		370 s		
129955	—		14 52 36.3	−78 51 16			7.8	1.1	−0.1	K2 III		380 s		
130737	252929		14 52 36.5	−61 34 07	0.000	−0.02	7.8	0.3	0.6	A0 V		170 s		
131183	182950		14 52 37.0	−25 27 17	−0.014	−0.12	8.0	1.0	3.2	G5 IV		49 mx		
131162	182951		14 52 38.0	−24 07 03	+0.001	−0.01	7.7	−0.7	0.0	B8.5 V		350 s		
131451	64373		14 52 43.7	+31 55 29	0.000	−0.03	7.1	1.1	0.2	K0 III		240 s		
131120	206037		14 52 50.9	−37 48 11	−0.003	−0.02	5.03	−0.16	−0.9	B6 V	+5	150 s		
131140	206041		14 52 51.1	−33 37 39	−0.003	−0.01	7.8	0.4	2.6	F0 V		91 s		
131252	158874		14 52 52.9	−12 23 01	−0.003	−0.02	7.8	0.4	3.0	F2 V		89 s		
131495	64377		14 52 53.4	+32 41 28	+0.003	+0.01	6.9	0.4	3.0	F2 V		60 s		
131253	158875		14 52 54.7	−12 38 40	−0.002	−0.05	8.0	0.5	4.0	F8 V		62 s		
131383	101271		14 52 56.4	+15 07 56	−0.003	−0.02	6.8	0.1	1.4	A2 V		120 s		
130764	252932		14 53 00.7	−64 19 31	+0.002	−0.03	7.8	0.0	0.0	B8.5 V		350 s		
130831	252935		14 53 01.9	−60 50 40	−0.004	−0.02	7.1	1.8	−0.1	K2 III		120 s		
132698	8111		14 53 03.1	+78 10 36	−0.010	0.00	6.50	1.1	0.2	K0 III		180 s	9445	m
131165	206048		14 53 06.6	−38 15 24	−0.001	−0.02	6.6	0.5	0.6	A0 V		76 s		
131509	83568		14 53 11.8	+28 30 29	−0.007	+0.08	7.96	0.85	5.9	K0 V		25 s		
130458	257206		14 53 13.7	−73 11 24	+0.007	+0.03	5.60	0.82		G5	+38			m
131599	45260		14 53 15.6	+40 49 00	−0.001	+0.02	8.0	0.5	3.4	F5 V		84 s		
131168	225273		14 53 22.1	−45 51 21	0.000	−0.01	7.05	−0.09	−1.7	B3 V e		510 s		m
131496	101274		14 53 22.9	+18 14 07	+0.003	−0.04	7.9	1.1	0.2	K0 III		350 s		
131473	101273		14 53 23.3	+15 42 16	−0.002	0.00	6.90	0.7	4.4	G0 V	+21	32 s	9425	m
131511	101276		14 53 23.6	+19 09 10	−0.032	+0.21	6.01	0.83	6.1	K1 V	−34	11 ts		
130721	—		14 53 24.1	−68 22 56			7.8	1.1	0.2	K0 III		320 s		
131528	83569		14 53 31.3	+19 53 17	+0.002	−0.07	7.0	1.1	0.2	K0 III		230 s		
131104	242097		14 53 33.6	−53 14 14	+0.001	−0.01	7.9	1.0	0.2	K0 III		340 s		
132188	16544		14 53 35.1	+69 45 46	0.000	−0.05	7.70	1.4	−0.3	K5 III		400 s		m
131477	120717		14 53 38.9	+2 14 14	−0.005	−0.03	7.1	1.4	−0.3	K5 III		170 mx		
131476	120719		14 53 39.7	+6 14 30	−0.002	0.00	6.6	1.1	0.2	K0 III		190 s		
131337	158880		14 53 40.0	−19 58 21	+0.001	−0.02	7.8	0.1	3.0	F2 V		90 s		
131338	182971		14 53 46.6	−28 24 20	−0.001	0.00	7.5	0.8	3.2	G5 IV		69 s		
131225	225276		14 53 46.8	−43 09 43	−0.001	−0.01	7.9	1.4	−0.1	K2 III		260 s		
131257	206057		14 53 46.9	−39 44 52	+0.003	−0.01	7.1	0.5	3.4	F5 V		49 s		
131295	206060		14 53 47.1	−33 41 29	−0.002	+0.01	7.8	0.4	1.4	A2 V		120 s		
131745	45265		14 53 49.2	+45 41 00	−0.001	+0.01	8.0	1.1	−0.1	K2 III		410 s		
131479	140209		14 53 49.9	−0 24 22	+0.003	−0.08	8.00	1.1	0.2	K0 III		190 mx		m
131429	140205		14 53 51.4	−9 05 11	+0.002	−0.02	7.3	0.1	0.6	A0 V		220 s		
131306	206061		14 53 53.0	−33 38 30	−0.002	0.00	7.7	1.5	−0.3	K5 III		390 s		
131368	182974		14 53 53.2	−23 58 27	−0.005	0.00	7.2	−0.1	1.4	A2 V		100 mx		
131122	242100		14 53 53.9	−55 02 54	−0.006	−0.05	7.3	1.9	3.2	G5 IV		13 s		
131498	140210		14 54 00.7	−5 34 18	+0.002	−0.04	7.80	1.1	0.2	K0 III		330 s		
133085	8121		14 54 04.5	+79 31 50	−0.016	−0.01	7.8	0.1	1.4	A2 V		82 mx		
130798	252941		14 54 05.0	−69 22 25	−0.004	−0.04	7.8	0.0	0.4	B9.5 V		150 mx		
131969	29331		14 54 06.3	+59 30 53	−0.004	−0.01	7.3	1.1	−0.1	K2 III		300 s		
130940	252945		14 54 11.1	−66 25 14	−0.044	−0.20	6.98	0.58		G0		18 mn		
131724	64388		14 54 18.0	+32 01 00	−0.007	+0.03	7.0	1.1	0.2	K0 III		230 s		
131430	182983	12 Lib	14 54 20.0	−24 38 32	−0.001	−0.03	5.3	1.2	0.2	K0 III	+9	77 s		
131415	182982		14 54 20.6	−28 20 55	−0.004	−0.02	6.6	0.5	0.2	K0 III		190 s		
131530	158887	13 ξ¹ Lib	14 54 22.8	−11 53 55	−0.004	−0.02	5.8	1.1	0.2	K0 III	−24	130 s		
131861	45269		14 54 24.5	+45 17 55	−0.006	+0.05	7.9	0.5	3.4	F5 V		80 s		
131399	206071		14 54 25.2	−34 08 34	−0.002	−0.04	7.00	0.1	0.6	A0 V		190 s		m
132166	16546		14 54 25.3	+64 38 47	+0.002	−0.03	7.9	1.1	0.2	K0 III		340 s		
131541	140214		14 54 28.8	−8 23 18	+0.001	0.00	7.4	1.1	0.2	K0 III		280 s		
131762	64395		14 54 33.4	+34 24 40	+0.002	0.00	7.9	0.5	3.4	F5 V		80 s		
131883	45270		14 54 33.6	+46 35 38	+0.005	−0.03	7.4	1.1	0.2	K0 III		280 s		
131764	64393		14 54 35.4	+30 03 50	−0.004	+0.03	6.9	0.4	3.0	F2 V	−34	59 s		
—	—		14 54 36.6	−33 18 09			6.00							m
131403	206075		14 54 37.2	−39 18 33	−0.002	−0.02	7.8	1.5	0.2	K0 III		170 s		
131432	206079		14 54 37.9	−33 18 01	0.000	0.00	6.00	1.1	0.2	K0 III		150 s		m
131389	225286		14 54 39.8	−42 02 44	−0.001	−0.04	7.6	1.4	0.2	K0 III		300 s		
130580	257207		14 54 40.3	−75 36 07	−0.004	−0.02	7.3	1.4	−0.1	K2 III		220 s		
131058	252951	ζ Cir	14 54 42.4	−65 59 29	−0.002	−0.03	6.09	−0.06	−1.4	B4 V	−21	290 s		
135280	2477		14 54 46.7	+85 07 15	−0.063	−0.03	7.4	1.1	0.2	K0 III		180 mx		
131714	101281		14 54 49.9	+15 19 31	−0.001	+0.02	7.1	0.1	1.7	A3 V		120 s		
131766	83577		14 54 53.9	+23 35 34	+0.001	−0.02	8.0	1.4	−0.3	K5 III		450 s		

HD	SAO	Star Name	α 2000	δ 2000	μ(α)	μ(δ)	V	B–V	M_v	Spec	RV	d(pc)	ADS	Notes
			h m s	° ′ ″	s	″								
130942	252948		14 54 54.0	−69 51 39	−0.001	−0.02	6.6	−0.5	0.4	B9.5 V		170 s		
131461	206083		14 54 54.4	−36 25 49	−0.002	−0.03	7.20	0.1	0.6	A0 V		210 s		m
131435	225289		14 54 54.8	−41 21 51	−0.011	−0.08	7.2	0.7	3.4	F5 V		40 s		
131487	206084		14 54 56.1	−36 30 55	−0.004	−0.01	8.0	1.4	−0.1	K2 III		290 s		
131217	252955		14 54 56.9	−61 04 32	−0.001	0.00	7.2	2.1	−0.5	M2 III		180 s		
131374	225287		14 54 57.2	−48 41 47	−0.001	−0.01	8.0	−0.5	0.6	A0 V		300 s		
131634	140219		14 54 57.3	−3 37 49	+0.001	+0.03	7.8	0.4	3.0	F2 V		89 s		
132770	8120		14 55 00.5	+74 52 52	0.000	0.00	6.9	1.6	−0.5	M2 III	+30	300 s		
131747	101282		14 55 00.9	+14 58 45	+0.001	+0.01	7.0	1.1	0.2	K0 III		230 s		
131437	—		14 55 02.1	−43 29 37			8.0	2.4	−0.5	M2 III		170 s		
131586	158890		14 55 05.5	−20 00 48	−0.001	−0.04	7.4	0.5	0.2	K0 III		270 s		
131229	252957		14 55 05.5	−61 52 25	−0.006	−0.03	7.2	0.0	0.4	B9.5 V		91 mx		
131783	101283		14 55 06.7	+17 42 05	−0.002	−0.05	7.8	1.1	0.2	K0 III		340 s		
129652	258708		14 55 17.4	−82 55 01	+0.004	+0.07	8.0	0.8	3.2	G5 IV		90 s		
131893	64400		14 55 18.8	+32 49 27	−0.001	−0.01	7.3	1.1	0.2	K0 III		260 s		
131464	225291		14 55 18.8	−46 37 53	+0.002	−0.03	7.31	0.98		G5				
131786	120739		14 55 22.7	+6 47 04	0.000	−0.02	6.90	1.4	−0.3	K5 III		280 s		m
131503	225292		14 55 25.8	−44 20 02	−0.001	0.00	7.99	0.24		A2				
131785	120740		14 55 26.4	+7 45 44	−0.002	−0.01	7.5	1.4	−0.3	K5 III		370 s		
131589	206094		14 55 26.8	−31 25 08	+0.001	−0.02	7.7	1.9	0.2	K0 III		92 s		
131826	101286		14 55 28.6	+12 25 51	−0.002	−0.05	6.8	0.1	1.4	A2 V		120 s		
131971	64403		14 55 29.0	+39 39 09	0.000	−0.01	7.5	1.1	0.2	K0 III		290 s		
131574	206092		14 55 29.7	−37 09 42	−0.002	−0.02	7.5	0.5	3.0	F2 V		62 s		
131342	242111		14 55 34.4	−60 06 50	−0.017	−0.11	5.20	1.16	0.0	K1 III	−15	100 s		
132167	29340		14 55 37.3	+52 34 03	−0.028	0.00	7.8	0.8	3.2	G5 IV		48 mx		
130425	257208		14 55 39.1	−78 52 57	−0.004	−0.01	7.2	0.1	0.6	A0 V		210 s		
131625	206099		14 55 44.6	−33 51 21	+0.002	0.00	5.30	0.1	0.6	A0 V	−9	87 s		m
130943	257210		14 55 46.3	−73 10 02	−0.018	−0.06	7.7	0.5	3.4	F5 V		71 s		
132027	64407		14 55 46.4	+39 38 53	−0.001	+0.01	8.03	0.40		F0			9441	m
131698	183005		14 55 48.6	−24 16 48	−0.001	+0.02	8.0	0.0	0.0	B8.5 V		360 s		
131544	225297		14 55 48.6	−44 41 51	+0.004	−0.02	7.94	1.54		K2				
131789	158901		14 55 53.1	−13 54 03	−0.001	−0.02	7.59	0.32	2.6	F0 V		97 s		
132047	64409		14 55 53.5	+36 25 49	−0.009	−0.03	7.8	1.1	0.2	K0 III		120 mx		
131376	252960		14 55 57.9	−60 54 22	−0.005	−0.02	6.80	0.8	3.2	G5 IV		52 s		m
132029	64408		14 55 58.6	+32 18 01	−0.004	0.00	6.12	0.10	1.7	A3 V	−12	76 s	9442	m
132089	45280		14 55 59.8	+41 08 12	−0.001	+0.04	7.2	0.1	1.4	A2 V		150 s		
131972	83586		14 56 00.4	+24 23 09	+0.003	−0.03	6.9	1.1	−0.1	K2 III		250 s		
132131	45282		14 56 01.0	+42 47 32	−0.005	−0.02	7.9	0.9	3.2	G5 IV		88 s		
131790	158903		14 56 02.3	−15 39 22	−0.001	+0.03	8.0	0.5	4.0	F8 V		63 s		
132130	45284		14 56 07.9	+43 19 03	−0.007	+0.08	7.9	0.7	4.4	G0 V		51 s		
131951	101293		14 56 13.1	+14 26 46	−0.001	0.00	5.77	−0.06	0.4	B9.5 V	−11	120 s		
131627	225299		14 56 14.8	−44 05 31	+0.004	0.00	7.89	0.16		A2				
131424	252961		14 56 15.5	−61 16 03	−0.010	−0.06	7.71	0.41	3.0	F2 V		81 s		
131917	120743		14 56 16.7	+3 24 58	−0.003	−0.05	7.1	0.4	3.0	F2 V		67 s		
131562	242120		14 56 17.2	−52 48 35	+0.004	0.00	5.38	0.13	1.4	A2 V	+7	56 s		
131811	158906		14 56 19.4	−18 55 41	+0.005	−0.01	7.8	0.5	3.4	F5 V		76 s		
131827	158907		14 56 19.6	−16 48 04	−0.002	−0.05	7.1	0.4	3.0	F2 V		67 s		
131773	183011		14 56 21.7	−29 57 35	0.000	+0.04	7.8	1.2	2.6	F0 V		29 s		
132254	45288		14 56 22.9	+49 37 43	+0.011	−0.23	5.63	0.50	3.4	F5 V	−15	26 s		
131637	225304		14 56 24.7	−44 42 17	+0.002	−0.03	6.77	0.04		A0				
131721	206107		14 56 25.5	−39 37 43	−0.003	−0.03	7.2	0.7	2.6	F0 V		48 s		
132031	101295		14 56 29.8	+17 06 12	−0.003	−0.04	7.7	0.6	4.4	G0 V		47 s		
131751	206111		14 56 30.6	−34 37 55	−0.007	−0.04	7.35	0.53		F8				m
131774	206112		14 56 30.7	−32 38 12	−0.001	−0.03	6.1	1.8	0.2	K0 III		51 s		
131830	183017		14 56 31.6	−24 15 29	−0.004	−0.01	7.5	0.8	0.2	K0 III		260 mx		
131657	225306		14 56 31.8	−47 52 45	−0.004	−0.02	5.64	−0.04	0.2	B9 V	+8	78 mx		
—	—		14 56 32.5	−34 38 04			7.40							m
131678	225308		14 56 34.2	−46 57 02	0.000	0.00	7.10	0.58		A2				
131752	206110		14 56 35.7	−39 24 58	−0.002	−0.03	6.5	0.5	0.6	A0 V		72 s		
131703	225310		14 56 36.2	−43 27 23	+0.001	−0.03	7.5	1.4	0.2	K0 III		230 s		
131491	252964		14 56 38.2	−62 21 54	−0.002	−0.03	6.38	−0.04	−1.6	B3.5 V		320 s		m
131989	120749		14 56 42.1	+5 23 56	−0.002	−0.02	8.0	0.1	1.4	A2 V		210 s		
132032	101296		14 56 43.9	+13 08 55	0.000	−0.12	7.8	0.8	3.2	G5 IV		84 s		
131492	252965	θ Cir	14 56 44.0	−62 46 51	−0.001	0.00	5.11	0.00	−1.7	B3 V ne	−4	220 s		m
131659	242125		14 56 44.4	−50 35 18	−0.002	−0.02	7.6	0.8	0.6	A0 V		77 s		
131918	158915	15 ξ² Lib	14 56 46.0	−11 24 35	0.000	0.00	5.46	1.49	−0.3	K4 III	+15	130 s		
133086	8127		14 56 48.1	+74 54 04	+0.001	+0.01	6.84	1.00		K0				
131470	252966		14 56 56.6	−65 06 01	0.000	−0.09	7.7							
131885	183025		14 56 57.3	−26 17 06	−0.002	−0.02	6.8	−0.2	0.6	A0 V		180 s		
131733	225314		14 57 01.2	−49 04 56	−0.004	−0.02	7.6	0.9	0.2	K0 III		300 s		
131705	242128		14 57 01.3	−51 26 49	−0.008	−0.01	6.4	1.5	−0.5	M2 III		240 s		
132145	83596		14 57 03.5	+21 33 19	−0.001	−0.03	6.49	−0.01	1.2	A1 V	−11	110 s		
132071	120756		14 57 04.3	+5 25 51	−0.002	−0.01	7.9	1.1	0.2	K0 III		340 s		
131643	242127		14 57 06.3	−55 54 44	+0.005	−0.03	7.44	0.39	2.6	F0 V		87 s		
131442	—		14 57 06.3	−66 52 34			8.00	1.1	0.2	K0 III		350 s		m

HD	SAO	Star Name	α 2000	δ 2000	μ(α)	μ(δ)	V	B-V	M$_v$	Spec	RV	d(pc)	ADS	Notes
			h m s	° ′ ″	s	″								
132276	64418		14 57 06.8	+35 29 24	0.000	-0.05	7.30	0.4	3.0	F2 V		72 s	9449	m
132369	45297		14 57 08.2	+48 23 43	-0.002	+0.12	7.9	0.6	4.4	G0 V		49 s		
132052	140240	16 Lib	14 57 10.9	-4 20 47	-0.007	-0.16	4.49	0.32	1.7	F0 IV	+22	31 ts		
132146	101299		14 57 11.7	+16 23 18	0.000	0.00	5.71	0.94	0.2	K0 III	-16	130 s		
131837	225318		14 57 12.3	-42 08 17	-0.010	-0.04	7.20	1.1	0.2	K0 III		120 mx		m
131919	183030		14 57 13.5	-29 09 28	-0.002	-0.04	6.1	0.5		B9				
131901	206126		14 57 19.7	-32 50 03	-0.003	-0.04	7.3	0.3	0.6	A0 V		140 s		
132560	29354		14 57 20.5	+57 38 58	-0.005	-0.02	7.2	0.4	3.0	F2 V	-9	68 s		
132256	83597		14 57 23.2	+25 19 25	-0.004	+0.02	7.31	0.66	3.0	G2 IV	-3	73 s		
131991	183041		14 57 24.8	-22 09 12	-0.009	-0.09	7.8	0.6	4.0	F8 V		54 s		
131902	206127		14 57 26.4	-37 07 49	-0.002	-0.01	7.9	1.9	-0.5	M2 III		340 s		
131977	183040		14 57 27.9	-21 24 56	+0.074	-1.74	5.64	1.14	7.1	K4 V	+20	5.6 t	9446	33 G. Librae, m
--	--		14 57 30.7	-0 05 00			5.80							m
131992	183042		14 57 31.8	-25 26 29	+0.002	+0.01	7.2	0.2	1.4	A2 V		120 s		
132132	120758		14 57 33.2	-0 10 03	+0.004	-0.03	5.53	1.13	0.0	K1 III	+20	120 s		m
131921	206128		14 57 33.4	-35 23 03	0.000	-0.02	7.10	1.1	0.2	K0 III		240 s		m
132813	16558		14 57 34.8	+65 55 56	-0.014	+0.03	4.60	1.59		M5 III	+7	30 mn		RR UMi, q
132072	158927		14 57 36.5	-12 38 20	-0.002	-0.05	7.7	0.1	0.6	A0 V		260 s		
131980	206133		14 57 40.6	-32 49 51	-0.003	-0.02	7.8	0.4	1.7	A3 V		110 s		
132304	83599		14 57 41.5	+24 40 26	-0.004	+0.01	7.2	1.1	-0.2	K3 III		300 s		
131855	225322		14 57 41.5	-49 52 17	-0.003	-0.01	8.0	0.5	0.6	A0 V		190 s		
132445	45303		14 57 42.8	+44 27 56	+0.001	-0.03	7.3	0.1	1.4	A2 V	-11	150 s		
132133	140243		14 57 43.0	-4 01 54	-0.002	-0.05	7.6	0.5	3.4	F5 V		71 s		
132296	101303		14 57 44.9	+18 59 13	-0.010	+0.01	6.7	1.1	0.2	K0 III		130 mx		
132112	158929		14 57 46.6	-12 26 15	+0.002	-0.01	7.5	1.6	-0.5	M4 III	+4	400 s		
132277	101302		14 57 47.5	+15 03 31	+0.002	-0.03	7.6	1.1	-0.1	K2 III		340 s		
131109	257212		14 57 52.8	-76 39 45	-0.020	-0.02	5.3	2.4	-0.1	K2 III	-31	23 s		R Aps, q
132150	158931		14 58 05.1	-17 21 51	-0.004	0.00	6.7	1.1	0.2	K0 III		200 s		
131923	225328		14 58 08.7	-48 51 45	-0.002	-0.31	6.35	0.71	5.2	G5 V	+44	16 s		
132408	83601		14 58 12.8	+26 49 00	+0.003	0.00	7.6	0.5	4.0	F8 V		52 s		
132230	158935	17 Lib	14 58 13.3	-11 09 18	-0.001	-0.02	6.4	0.1	0.6	A0 V	-17	150 s		
132041	206143		14 58 14.0	-36 03 54	0.000	-0.04	7.81	-0.06	-2.2	B5 III		840 s		
132343	101306		14 58 16.1	+14 02 15	-0.002	+0.01	6.7	1.1	-0.2	K3 III		240 s		
132094	206149		14 58 24.1	-37 21 45	-0.002	-0.03	7.3	-0.4	0.6	A0 V		220 s		
132173	183057		14 58 30.6	-28 42 38	-0.006	-0.15	7.7	0.4	4.0	F8 V		55 s		
132374	120768		14 58 31.0	+9 29 52	+0.003	-0.03	7.7	0.4	3.0	F2 V		88 s		
131999	225333		14 58 31.4	-47 26 08	0.000	-0.01	7.30	1.1	0.2	K0 III		250 s		m
132058	225335	β Lup	14 58 31.8	-43 08 02	-0.003	-0.04	2.68	-0.22	-2.5	B2 V	0	110 s		
132095	206153		14 58 34.3	-38 56 14	-0.001	0.00	7.9	1.6	0.2	K0 III		150 s		
132298	140248		14 58 34.9	-8 43 03	-0.003	0.00	7.2	0.8	3.2	G5 IV		64 s		
132096	206154		14 58 36.6	-39 54 24	-0.004	-0.07	6.2	1.4	-0.1	K2 III		130 s		
132890	16561		14 58 39.0	+61 40 00	-0.004	+0.01	7.1	0.1	1.4	A2 V	+1	140 s		
132219	183058	59 Hya	14 58 39.1	-27 39 26	-0.003	-0.01	6.30	0.2	2.1	A5 V	-16	69 s	9453	m
132234	183060		14 58 40.6	-22 24 05	-0.007	-0.04	7.70	0.5	3.4	F5 V		72 s		m
132309	158942		14 58 42.7	-12 46 05	+0.004	+0.01	7.9	1.1	-0.1	K2 III		330 mx		
131782	252979		14 58 44.9	-65 00 14	-0.004	-0.03	8.0	0.1	1.7	A3 V		180 s		
132504	83603		14 58 48.4	+24 10 10	-0.001	+0.01	7.2	0.4	2.6	F0 V		83 s		
--	64437		14 58 48.8	+37 14 02	-0.010	+0.05	7.8	2.1						
132524	83604		14 58 49.6	+25 02 52	-0.001	-0.01	7.4	1.1	0.2	K0 III		280 s		
133541	8137		14 58 49.6	+76 31 40	+0.018	0.00	7.9	1.1	0.2	K0 III		340 s		
132446	101311		14 58 52.7	+12 04 47	+0.002	0.00	7.7	0.2	2.1	A5 V		130 s		
132375	140256		14 58 52.7	-4 59 21	-0.024	-0.10	6.09	0.50	3.7	dF6	-29	32 mn	9457	m
132345	158946	18 Lib	14 58 53.5	-11 08 39	-0.007	-0.07	5.84	1.25		K3 III-IV p	-12	100 mx	9456	m
132505	101312		14 58 58.1	+17 19 38	-0.002	+0.01	7.9	0.9	3.2	G5 IV		88 s		
132098	225339		14 59 01.7	-48 14 06	+0.001	-0.03	7.6	0.7	0.0	B8.5 V		220 s		
132121	225340		14 59 05.2	-47 57 28	-0.002	-0.03	7.4	1.5	0.2	K0 III		91 s		
132200	225344	κ Cen	14 59 09.6	-42 06 15	-0.002	-0.03	3.13	-0.20	-2.5	B2 V	+9	130 s		m
132122	225341		14 59 10.8	-48 49 33	+0.001	-0.06	7.5	0.2	4.4	G0 V		42 s		
132063	242142		14 59 11.3	-53 29 25	-0.002	-0.02	8.0	1.4	0.2	K0 III		220 s		
132238	206167		14 59 13.8	-37 52 53	-0.002	-0.03	6.5	0.2	0.0	B8.5 V		140 s		
132101	242146		14 59 19.0	-51 34 34	-0.003	-0.02	6.9	0.2	0.0	B8.5 V		160 s		
132202	225349		14 59 19.2	-43 28 22	-0.005	-0.06	7.04	0.36		F0				
132525	120774		14 59 23.0	+4 34 04	0.000	-0.01	5.93	1.60	-0.5	gM1	-12	190 s		
132736	64447		14 59 24.8	+39 38 42	-0.001	+0.04	6.8	0.4	3.0	F2 V		58 s		
132242	225351		14 59 27.0	-43 09 36	-0.002	-0.02	6.10	0.60		F8				
132347	206173		14 59 28.8	-30 42 41	-0.004	-0.03	7.20	0.1	1.4	A2 V		76 mx		m
132909	29370		14 59 32.9	+53 52 16	0.000	0.00	7.64	0.29		F0				
132910	29372		14 59 34.3	+53 51 37	+0.001	0.00	6.84	0.31		F0			9474	m
132772	64449	40 Boo	14 59 36.9	+39 15 55	-0.003	+0.03	5.64	0.31	3.0	F2 V	+12	34 s		
132348	206176		14 59 42.8	-36 53 28	0.000	-0.01	7.6	0.7	0.2	K0 III		310 s		
132753	64450		14 59 43.4	+35 06 03	0.000	-0.02	7.90	0.9	3.2	G5 IV		87 s		m
132301	225355		14 59 45.0	-43 48 40	-0.018	-0.13	6.58	0.48	3.4	F5 V		41 s		
131844	--		14 59 46.6	-69 12 20			8.0	1.1	0.2	K0 III		350 s		
132285	225354		14 59 49.5	-46 09 37	-0.003	-0.04	7.4	0.4	1.4	A2 V		94 s		
131615	257220		14 59 50.4	-73 47 21	+0.003	-0.01	7.9	1.1	0.2	K0 III		350 s		

HD	SAO	Star Name	α 2000	δ 2000	μ(α)	μ(δ)	V	B−V	M_v	Spec	RV	d(pc)	ADS	Notes
			^{h} ^{m} ^{s}	$^\circ$ ' "	^{s}	"								
132507	158959		14 59 53.5	−10 55 51	+0.003	−0.05	7.1	0.4	3.0	F2 V		67 s		
132302	225357		14 59 54.7	−46 14 51	−0.001	−0.01	7.15	−0.01	0.4	B9.5 V		220 s		
131551	257219		14 59 56.0	−75 01 57	+0.001	−0.02	6.20	−0.04		B9				m
132679	101320		14 59 57.6	+14 49 59	0.000	−0.02	7.1	1.1	−0.1	K2 III		280 s		
132396	206185		15 00 00.2	−36 01 51	0.000	−0.05	7.2	1.0	0.2	K0 III		250 s		
132414	206187		15 00 00.8	−32 49 58	−0.001	+0.03	7.8	0.6	1.4	A2 V		92 s		
132718	101321		15 00 02.4	+17 59 34	−0.004	−0.03	7.9	1.1	0.2	K0 III		180 mx		
133229	16570		15 00 02.5	+65 28 35	−0.001	−0.01	6.9	1.1	0.2	K0 III		220 s		
132892	45324		15 00 03.4	+44 35 10	−0.007	−0.01	6.9	1.1	0.2	K0 III		190 mx		
132381	206184		15 00 04.7	−39 47 51	+0.001	0.00	7.3	1.2	0.2	K0 III		200 s		
133328	16572		15 00 07.9	+67 05 23	−0.005	−0.02	7.5	1.1	0.2	K0 III		290 s		
133482	8138		15 00 10.3	+70 36 51	0.000	+0.07	7.7	0.5	4.0	F8 V		55 s		
131425	257218		15 00 11.8	−77 09 36	+0.002	0.00	5.93	1.05		K0				
132223	242161		15 00 13.7	−57 07 45	0.000	−0.01	7.2	1.8	3.2	G5 IV		15 s		
132208	242160		15 00 16.2	−58 18 22	−0.001	−0.01	7.2	2.4	−0.5	M2 III		120 s		
132701	120785		15 00 18.7	+7 38 55	+0.002	0.00	7.0	1.1	−0.1	K2 III		270 s		
132571	158962		15 00 22.6	−18 37 36	+0.003	+0.01	7.9	1.1	0.2	K0 III		340 s		
133109	29377		15 00 23.8	+55 37 05	+0.005	+0.02	7.2	0.4	2.6	F0 V		81 s		
132864	64460		15 00 26.1	+31 36 55	−0.015	+0.10	7.7	0.5	4.0	F8 V		54 s		
133621	8140		15 00 26.8	+71 45 57	−0.087	+0.09	6.66	0.61	4.4	G0 V	−45	27 s		
132572	183079		15 00 30.6	−22 50 57	+0.003	+0.01	8.0	1.2	−0.5	M2 III		510 s		
132614	158964		15 00 31.7	−11 52 28	−0.001	−0.01	8.0	1.6	−0.5	M2 III		510 s		
132756	120789		15 00 35.0	+8 36 01	−0.006	−0.31	7.31	0.70	4.4	G0 V		31 s		
133050	29378		15 00 35.2	+50 34 41	−0.002	+0.01	7.7	0.1	1.4	A2 V		180 s		
133029	45326		15 00 38.6	+47 16 39	−0.001	+0.02	6.37	−0.14			−14		9477	m
132249	242164		15 00 44.8	−59 48 43	+0.004	−0.06	7.1	0.7	4.0	F8 V		34 s		
132879	83616		15 00 52.3	+22 02 44	+0.001	0.00	6.4	1.1	0.2	K0 III	−26	170 s		
132742	140270	19 δ Lib	15 00 58.2	−8 31 08	−0.004	−0.01	4.92	0.00	0.6	A0 V	−39	73 s		v
133030	45328		15 01 00.6	+40 05 05	0.000	−0.03	7.1	0.8	3.2	G5 IV		60 s		
132832	120794		15 01 01.3	+2 53 52	−0.006	−0.07	6.8	0.5	3.4	F5 V	−9	49 s		
132209	252990		15 01 02.3	−64 34 34	−0.001	−0.07	6.9	−0.7	2.1	A5 V		89 s		
132602	206206		15 01 02.7	−33 50 22	−0.007	−0.03	7.5	0.8	3.4	F5 V		38 s		
132604	206208		15 01 12.9	−38 03 30	+0.001	−0.04	5.89	1.24		K2				
132127	252988		15 01 15.2	−67 59 00	−0.001	−0.01	6.99	−0.06		B9				m
132833	140276		15 01 19.8	−2 45 18	+0.002	−0.02	5.52	1.68		K5	−15			
131911	−−		15 01 20.4	−73 20 15			7.9	1.1	0.2	K0 III		350 s		
132004	257225		15 01 22.6	−71 29 29	−0.002	−0.04	6.8	−0.2	0.0	B8.5 V		230 s		
132224	252992		15 01 25.9	−65 54 49	+0.003	0.00	7.8	0.0	0.4	B9.5 V		290 s		
133207	45335		15 01 26.2	+49 11 40	0.000	−0.04	7.4	1.1	0.2	K0 III		270 s		
133388	16574		15 01 26.9	+60 12 16	−0.004	+0.01	5.9	0.1	1.4	A2 V	−9	79 s		
132778	158974		15 01 30.9	−18 48 11	0.000	0.00	8.0	1.6	−0.5	M2 III		510 s		
132883	140278		15 01 35.3	−3 09 51	+0.001	−0.09	6.65	1.17		K1 IV		180 mx	9479	m
132322	252993		15 01 35.8	−63 55 36	−0.004	−0.02	7.3	0.2		A6 p				
132481	242174		15 01 38.5	−56 15 38	+0.002	−0.03	6.87	−0.07	−2.5	B2 V n		670 s		
132761	206222		15 01 48.7	−31 32 39	−0.001	−0.04	7.7	0.7	1.4	A2 V		71 s		
132933	120798		15 01 48.8	−0 08 26	0.000	−0.03	5.71	1.52		K0	−34		9480	m
130650	258713	π¹ Oct	15 01 50.1	−83 13 39	+0.006	+0.05	5.65	0.95		K0				
132648	225381		15 01 53.0	−47 00 27	−0.004	−0.08	7.8	0.8	3.2	G5 IV		65 s		
133208	45337	42 β Boo	15 01 56.6	+40 23 26	−0.004	−0.03	3.50	0.97	0.3	G8 III	−20	42 s		Nekkar
132667	225383		15 01 56.7	−47 09 12	−0.002	−0.02	7.8	0.7	3.2	G5 IV		42 s		
132606	242180		15 01 57.0	−51 55 05	−0.002	−0.02	7.10	1.1	0.2	K0 III		230 s		m
132763	206223		15 01 58.0	−34 21 31	−0.002	−0.02	6.4	0.4	1.7	A3 V		60 s		
132935	140280		15 02 04.2	−8 20 40	+0.001	+0.03	6.8	0.8	3.2	G5 IV		54 s		
132851	183099	60 Hya	15 02 06.4	−28 03 39	+0.007	−0.04	5.85	0.16		A5				
133252	45342		15 02 06.4	+42 03 09	0.000	+0.01	8.0	1.1	0.2	K0 III		360 s		
133124	83624	41 ω Boo	15 02 06.4	+25 00 29	0.000	−0.05	4.81	1.50	−0.3	K4 III	+13	91 s		
131968	−−		15 02 07.4	−74 43 13			8.0	1.4	−0.3	K5 III		460 s		
132953	140281		15 02 08.5	−7 34 32	−0.001	−0.01	6.40	0.18		A3				
133704	16580		15 02 20.4	+65 57 53	+0.001	−0.01	7.7	1.1	0.2	K0 III		320 s		
133253	64472		15 02 29.1	+33 16 28	+0.001	−0.01	7.3	1.1	0.2	K0 III		260 s		
133008	140283		15 02 29.9	−7 50 24	−0.004	0.00	6.61	0.17		A2				
133254	64471		15 02 30.6	+31 41 01	−0.002	+0.02	6.7	1.6	−0.5	M4 III	−28	280 s		
133445	29388		15 02 31.2	+51 37 10	−0.001	−0.04	7.2	1.4	−0.3	K5 III		310 s		
133161	101345		15 02 33.0	+16 03 17	−0.015	+0.08	7.0	0.6	4.4	G0 V	−34	33 s		
133112	140286		15 02 44.8	−3 01 53	−0.003	−0.02	6.61	0.20		A2				
133164	120808		15 02 49.1	+7 54 43	−0.001	−0.03	8.0	0.4	3.0	F2 V		98 s		
133073	158990		15 02 51.4	−10 23 33	+0.002	−0.03	7.7	1.1	0.2	K0 III		310 s		
133165	120809	110 Vir	15 02 53.9	+2 05 28	−0.004	+0.01	4.40	1.04	0.2	K0 III	−16	65 s		
132840	225399		15 02 57.1	−47 56 32	−0.004	−0.02	7.3	1.6	0.2	K0 III		120 s		
133826	16585		15 02 58.7	+65 46 41	+0.010	−0.13	7.3	0.6	4.4	G0 V		38 s		
132955	206239		15 02 59.1	−32 38 36	−0.002	−0.03	5.44	−0.12	−2.0	B4 IV	+6	300 s		m
133392	64476		15 03 06.0	+35 12 21	−0.004	+0.01	5.51	1.02	0.2	K0 III	−27	110 s		
133093	158992		15 03 06.3	−17 37 59	+0.003	−0.04	7.0	1.1	0.2	K0 III		230 s		
133330	83632		15 03 06.4	+28 16 02	+0.002	0.00	7.03	0.12		A0	+9			
133484	45348		15 03 06.5	+44 38 39	−0.009	+0.01	6.65	0.46	3.4	F5 V	−20	43 s		

HD	SAO	Star Name	α 2000	δ 2000	μ(α)	μ(δ)	V	B-V	M_v	Spec	RV	d(pc)	ADS	Notes
			h m s	° ′ ″	s	″								
132731	--		15 03 06.5	-58 13 53			7.8	2.7	-0.5	M4 III		110 s		
133130	158994		15 03 08.0	-12 51 28	0.000	-0.01	7.6	0.1	0.6	A0 V		250 s		
133312	101353		15 03 13.0	+17 58 32	+0.001	0.00	7.2	1.1	0.2	K0 III		250 s		
132366	257229		15 03 13.1	-70 47 15	+0.006	-0.02	7.6	0.0	1.4	A2 V		180 s		
132996	206243		15 03 17.9	-36 55 29	-0.000	-0.40	7.77	0.61		G0		21 mn		
133233	120814		15 03 20.3	+4 50 03	0.000	0.00	7.7	0.5	3.4	F5 V		73 s		
132960	225404		15 03 20.7	-41 16 17	-0.001	0.00	7.40	-0.15	-3.9	B1 IV		1800 s		
133666	29393		15 03 21.5	+56 02 06	-0.002	-0.08	6.8	1.1	-0.1	K2 III		230 mx		
133234	120813		15 03 22.3	+0 53 22	-0.001	-0.04	7.9	1.1	0.2	K0 III		340 s		
133768	16586		15 03 26.5	+60 01 32	-0.004	0.00	6.7	1.4	-0.3	K5 III		260 s		
133173	158995		15 03 33.6	-16 35 32	+0.001	-0.03	7.9	1.1	-0.1	K2 III		390 s		
133077	206247		15 03 34.5	-35 56 34	-0.001	+0.01	6.8	0.7	2.1	A5 V		40 s		
133351	101355		15 03 34.8	+15 25 18	-0.004	+0.02	8.0	0.5	3.4	F5 V		82 s		
133485	64484		15 03 36.5	+34 33 57	0.000	-0.02	6.59	1.02	1.8	G8 III-IV	-25	80 s		
133640	45357	44 Boo	15 03 47.2	+47 39 15	-0.041	+0.03	4.76	0.65	4.4	G0 V	-25	12 ts	9494	i Boo, m,v
133147	206256		15 03 48.2	-30 28 24	+0.004	+0.01	7.7	1.3	0.2	K0 III		220 s		
133460	83637		15 03 49.1	+26 02 19	-0.001	-0.02	8.0	0.5	3.8	F7 V	-9	68 s		
133459	83638		15 03 49.2	+27 04 52	-0.001	-0.01	6.8	1.4	-0.3	K4 III		270 s		
133994	16587		15 03 57.6	+65 55 11	+0.003	-0.01	6.1	0.1	0.6	A0 V	-5	130 s		
133194	183138		15 04 02.0	-28 03 40	+0.003	0.00	7.8	0.1	3.0	F2 V		90 s		
133432	101359		15 04 03.7	+10 44 06	-0.002	-0.01	7.9	1.4	-0.3	K5 III		430 s		
133216	183139	20 σ Lib	15 04 04.1	-25 16 55	-0.005	-0.05	3.29	1.70	-0.5	M4 III	-4	51 s		
--	120821		15 04 06.1	+5 29 43	-0.001	-0.04	7.36	0.31		F0 n			9493	m
133408	120822		15 04 06.3	+5 29 33	-0.001	-0.04	6.50	0.31		F0	-8		9493	m
133352	140297		15 04 06.8	-6 53 13	-0.015	+0.01	7.80	0.47	3.2	G5 IV		69 mx	9492	m
132273	--		15 04 07.6	-75 52 03			7.9	1.1	0.2	K0 III		350 s		
133294	159000		15 04 08.9	-15 27 45	0.000	-0.06	8.0	0.0	0.0	B8.5 V		110 mx		
133544	83643		15 04 16.0	+29 04 43	+0.001	0.00	7.80	0.1	1.4	A2 V	-18	190 s		m
133909	29397		15 04 17.4	+59 32 03	-0.004	-0.03	7.3	0.1	1.4	A2 V	-9	150 s		
133487	101362		15 04 19.9	+13 38 50	-0.002	0.00	7.90	1.1	-0.1	K2 III		400 s		
131062	--		15 04 20.9	-83 38 33			7.8	1.1	0.2	K0 III		330 s		
133024	242204		15 04 22.9	-53 37 22	+0.001	-0.05	7.7	1.6	3.2	G5 IV		25 s		
132984	242202		15 04 23.4	-56 55 01	-0.002	-0.01	7.42	-0.02	-1.6	B3.5 V		560 s		
133409	140303		15 04 24.4	-0 54 22	+0.001	-0.02	8.0	0.1	0.6	A0 V		300 s		
133217	206261		15 04 26.0	-36 17 05	-0.003	-0.01	7.9	0.3	1.7	A3 V		130 s		
133582	83645	43 ψ Boo	15 04 26.6	+26 56 51	-0.013	-0.01	4.54	1.24	-0.1	K2 III	-26	75 s		
133395	140301		15 04 27.9	-8 02 49	-0.002	-0.02	7.7	1.1	-0.1	K2 III		360 s		
133528	83644		15 04 29.3	+19 50 29	-0.001	+0.01	7.80	1.1	0.2	K0 III		330 s	9495	m
132985	242203		15 04 29.6	-57 41 55	-0.002	-0.02	7.40	0.08	0.6	A0 V		230 s		
133295	183145		15 04 32.8	-28 17 59	+0.002	-0.04	7.2	0.4	4.4	G0 V		36 s		
132888	253004		15 04 33.7	-62 53 23	-0.008	-0.05	8.0	0.8	3.2	G5 IV		89 s		
133220	225422		15 04 42.8	-40 51 41	-0.001	-0.04	6.41	1.46		Mb				
133725	64491		15 04 43.5	+38 36 17	-0.001	+0.01	7.49	0.48	4.0	F8 V		50 s		
131246	258714	π² Oct	15 04 46.6	-83 02 17	-0.005	-0.01	5.65	1.30		K0	-21			
133275	206264		15 04 47.4	-37 30 28	-0.001	-0.03	7.8	1.7	0.2	K0 III		130 s		
132905	253005	η Cir	15 04 48.1	-64 01 53	+0.016	0.00	5.2	0.9	3.2	G5 IV	+45	21 s		
134350	8154		15 04 52.7	+70 45 55	+0.007	-0.06	7.9	1.1	0.2	K0 III		310 mx		
132924	253009		15 04 53.0	-63 19 33	-0.001	-0.02	7.4	0.1	1.4	A2 V		150 s		
129289	258710		15 04 55.3	-86 28 17	+0.021	+0.02	7.8	1.5	-0.1	K2 III		240 s		
133817	45364		15 05 00.1	+41 06 31	0.000	-0.04	7.5	0.8	3.2	G5 IV		72 s		
133300	206273		15 05 02.1	-39 59 07	-0.004	-0.01	7.4	1.2	-0.1	K2 III		310 s		
134584	8160		15 05 06.9	+73 53 31	-0.004	+0.04	7.2	0.1	0.6	A0 V	-16	210 s		
133242	225426	π Lup	15 05 07.1	-47 03 04	-0.002	-0.02	3.89	-0.14	-1.6	B5 IV	+17	130 s		m
133321	225432		15 05 09.8	-40 34 12	-0.003	-0.02	8.0	0.7	-0.1	K0 III		410 s		
132675	257234		15 05 10.5	-72 11 23	-0.006	-0.03	7.4	0.9	3.2	G5 IV		67 s		
133466	183155		15 05 16.7	-23 00 47	-0.001	0.00	7.3	0.7	0.4	B9.5 V		86 s		
133434	183153		15 05 18.2	-29 59 24	-0.001	-0.01	7.20	1.1	0.2	K0 III		250 s		m
--	--		15 05 18.6	+30 55 39	+0.006	-0.03	7.9	1.6	-0.5	M2 III		220 mx		
133340	225435		15 05 19.0	-41 04 02	+0.002	-0.01	5.30	1.1	0.2	K0 III	-3	110 s		m
132252	257231		15 05 22.5	-78 10 03	-0.003	-0.03	7.2	0.4	2.6	F0 V		83 s		
134023	29401		15 05 23.4	+55 40 36	-0.007	+0.01	7.6	0.5	4.0	F8 V		53 s		
133962	45370	47 Boo	15 05 25.7	+48 09 03	-0.007	+0.03	5.57	0.00	0.6	A0 V	-13	70 mx	9500	m
133686	101367		15 05 29.6	+11 51 26	-0.001	-0.01	7.7	1.4	-0.3	K5 III		390 s		
133584	--		15 05 30.6	-7 01 27			8.00	0.6	4.4	G0 V		53 s	9497	m
133469	206284		15 05 33.7	-30 33 02	0.000	+0.09	6.7	0.7	3.4	F5 V		33 s		
133645	140318		15 05 42.7	-1 59 10	-0.001	-0.04	7.7	1.1	0.2	K0 III		310 s		
134585	8162		15 05 45.4	+71 52 58	-0.023	+0.03	7.44	1.14	0.0	gK1	-18	270 s		
133529	183163		15 05 47.6	-25 47 23	0.000	-0.02	6.67	-0.01		B8				
133873	64497		15 05 49.1	+31 03 53	0.000	-0.02	7.9	1.4	-0.3	K5 III		430 s		
133745	120835		15 05 58.1	+4 14 00	-0.004	-0.01	7.3	0.4	2.6	F0 V		87 s		
133399	225439		15 06 02.4	-48 52 58	+0.001	0.00	6.50	-0.13	-1.6	B3.5 V		400 s		
133603	183168		15 06 03.0	-22 24 14	+0.001	-0.01	7.6	1.1	0.2	K0 III		270 s		
--	83659		15 06 04.6	+28 59 59	-0.001	0.00	8.0	1.6	0.2	K0 III		170 s		
133947	64498		15 06 07.9	+34 45 11	0.000	-0.04	7.8	1.1	-0.1	K2 III		380 s		
133627	183170		15 06 10.5	-24 07 53	0.000	-0.08	6.9	0.9	3.2	G5 IV		46 s		

HD	SAO	Star Name	α 2000	δ 2000	μ(α)	μ(δ)	V	B-V	M_v	Spec	RV	d(pc)	ADS	Notes
132907	257236		15 06 10.9	-71 07 58	-0.001	-0.05	7.69	-0.04	-2.2	B5 III		790 s		
133604	183171		15 06 11.4	-23 19 23	-0.003	-0.03	7.1	1.3	3.0	F2 V	-28	19 s		
133550	206292		15 06 13.8	-36 15 51	-0.001	0.00	6.27	1.65		K5				
133649	159021		15 06 14.8	-19 22 39	+0.002	-0.04	7.8	1.1	0.2	K0 III		340 s		
134190	29407		15 06 16.6	+54 33 22	+0.005	+0.01	5.25	0.96	0.3	G8 III	+16	98 s		
133818	120837		15 06 18.0	+5 33 01	-0.001	-0.01	7.2	0.4	2.6	F0 V		82 s		
134301	16594		15 06 18.1	+59 49 31	+0.003	-0.04	7.6	0.4	3.0	F2 V		82 s		
133651	183172		15 06 21.5	-26 06 28	+0.004	-0.05	7.9	1.5	-0.1	K2 III		170 mx		
133367	242223		15 06 22.5	-55 49 24	-0.004	-0.02	8.0	1.7	0.2	K0 III		140 s		
132874	257237		15 06 22.6	-72 10 13	-0.003	-0.03	7.50	0.1	0.6	A0 V		230 s		m
133670	183176		15 06 27.0	-22 01 55	+0.004	-0.06	6.0	1.8	0.2	K0 III	+5	50 s		
133841	120839		15 06 27.8	+3 00 32	0.000	+0.01	7.6	0.4	2.6	F0 V		100 s		
133652	206300		15 06 33.2	-30 55 07	-0.001	-0.04	6.0	1.1		A0 p				
133772	159027		15 06 34.1	-12 54 26	+0.001	-0.03	7.4	0.1	0.6	A0 V		230 s		
134044	64503		15 06 35.0	+36 27 20	-0.005	+0.02	6.35	0.52	3.4	F5 V	-5	35 s		
133774	159028	21 v Lib	15 06 37.5	-16 15 24	-0.003	-0.03	5.2	1.4	-0.3	K5 III	-15	120 s		
133307	253019		15 06 43.0	-62 24 04	0.000	-0.01	7.3	2.0	-0.3	K5 III		170 s		
133800	159030		15 06 49.0	-16 29 03	-0.004	-0.02	6.39	0.16		A0				m
133749	183183		15 06 51.0	-26 49 32	-0.001	-0.04	8.0	1.2	-0.1	K2 III		410 s		
133518	242229		15 06 55.9	-52 01 48	+0.001	-0.03	6.39	-0.10	-2.9	B3 III		660 s		
132544	--		15 06 59.1	-78 20 52			8.0	1.1	0.2	K0 III		360 s		
133778	183188		15 07 06.1	-28 49 20	-0.004	+0.01	7.8	1.0	3.4	F5 V		34 s		
133716	206306		15 07 06.2	-37 56 44	+0.002	-0.02	7.18	0.06	1.4	A2 V		140 s		
133948	120845		15 07 06.8	+2 21 44	0.000	-0.03	6.7	1.1	-0.1	K2 III		230 s		
133928	140332		15 07 08.6	-2 16 53	-0.004	-0.02	7.66	0.57	4.0	F8 V		50 s	9504	m
133049	257238		15 07 08.6	-71 54 19	+0.007	-0.02	6.52	1.59	-0.3	K4 III		180 s		
134191	45383		15 07 09.2	+40 06 28	-0.007	+0.06	8.0	0.5	3.4	F5 V		82 s		
133995	101376		15 07 13.1	+10 17 45	0.000	0.00	8.0	0.4	2.6	F0 V		120 s		
133750	206315		15 07 13.5	-32 54 37	-0.001	-0.01	7.4	0.2	0.6	A0 V		170 s		
133507	242230		15 07 14.5	-57 26 24	-0.001	-0.04	7.20	1.31	0.2	K0 III		160 s		
134063	83668		15 07 15.3	+22 33 51	0.000	-0.04	7.82	0.92	0.3	G8 III	-11	320 s		
133843	183193		15 07 16.4	-24 11 38	0.000	+0.01	7.4	1.0	3.2	G5 IV		53 s		
134062	83670		15 07 17.2	+22 54 28	-0.001	-0.03	7.2	1.1	0.2	K0 III		250 s		
134083	83671	45 Boo	15 07 18.0	+24 52 08	+0.014	-0.17	4.93	0.43	3.4	F5 V	-7	19 ts		m
134012	120847		15 07 19.2	+8 57 30	-0.001	-0.03	7.0	1.4	-0.3	K5 III		280 s		
134064	101379		15 07 20.3	+18 26 30	+0.003	-0.06	6.02	0.06	1.4	A2 V	-5	83 s	9505	m
134351	29410		15 07 20.5	+52 44 35	0.000	-0.04	7.2	1.1	0.2	K0 III		250 s		
134027	101377		15 07 21.1	+10 31 45	-0.011	-0.11	7.7	1.1	0.2	K0 III		58 mx		
133912	159035		15 07 25.2	-14 40 44	-0.002	-0.01	8.0	1.6	-0.5	M2 III		510 s		
133631	225456		15 07 25.8	-49 05 18	+0.002	+0.02	5.77	0.92		K0				
133913	159034		15 07 25.9	-15 25 01	+0.003	-0.04	7.7	0.4	3.0	F2 V		85 s		
133656	225457		15 07 27.3	-48 17 54	0.000	-0.01	7.50	0.37	2.1	A5 V		120 s		
134645	16603		15 07 31.0	+64 43 10	-0.010	+0.02	7.3	1.1	0.2	K0 III		260 s		
134066	120851		15 07 32.9	+9 13 34	-0.012	+0.03	6.7	0.9	3.2	G5 IV	-35	50 s	9507	m
134404	29415		15 07 34.5	+52 59 37	+0.001	+0.01	7.5	1.1	0.2	K0 III		290 s		
134047	120852		15 07 40.2	+5 29 53	-0.001	-0.02	6.16	0.94	0.3	gG6	+3	140 s		
134225	64512		15 07 45.1	+32 15 33	+0.001	-0.01	7.9	0.0		B9.5 V		320 s		
134646	16606		15 07 50.0	+63 07 02	-0.001	0.00	6.80	0.5	0.7	F4 III	0	170 s	9520	m
133557	253025		15 07 55.7	-61 07 44	-0.003	-0.02	6.7	0.0	3.4	F5 V		46 s		
133456	253024		15 07 56.8	-65 16 32	+0.002	-0.01	6.17	1.47		K2				
133822	225468		15 07 57.6	-45 34 45	-0.011	-0.12	7.74	0.73	3.2	G5 IV		73 mx		
134013	159040		15 07 58.4	-14 35 35	-0.001	-0.02	8.0	0.1	0.6	A0 V		300 s		
134246	83678		15 08 00.3	+28 31 06	-0.002	-0.01	7.3	0.9	0.3	G8 III		250 s		
133820	225469		15 08 01.1	-43 42 43	-0.001	-0.02	7.02	0.04	0.6	A0 V		180 s		
134068	140338		15 08 05.7	-3 45 23	-0.002	+0.03	7.7	1.1	-0.1	K2 III		370 s		
134048	159042		15 08 06.4	-12 03 06	-0.009	-0.07	7.6	0.9	3.2	G5 IV		72 mx		
133880	225474		15 08 12.0	-40 35 03	-0.003	-0.05	5.79	-0.14		A0 p				
134088	140339		15 08 12.5	-7 54 48	-0.011	-0.45	8.00	0.59	4.4	G0 V	-59	53 s		
134282	83680		15 08 14.0	+26 42 40	+0.002	-0.01	8.00	0.9	-2.1	G8 II		1000 s		
134607	29421		15 08 15.0	+57 39 25	+0.001	-0.03	7.8	1.4	-0.3	K5 III		410 s		
134169	120858		15 08 17.9	+3 55 50	0.000	-0.02	8.0	0.5	3.4	F5 V		82 s		
134352	64515		15 08 18.4	+33 42 23	-0.002	0.00	6.8	1.1	0.2	K0 III		210 s		
134807	16608		15 08 19.0	+65 47 18	-0.006	+0.01	6.7	1.6	-0.5	M2 III	-27	280 s		
134493	29420		15 08 19.4	+50 03 18	-0.001	-0.02	6.39	1.03	0.2	K0 III	-29	170 s		
134228	101394		15 08 20.5	+10 06 50	-0.005	-0.02	8.0	0.5	4.0	F8 V	+7	62 s		
134320	83682	46 Boo	15 08 23.7	+26 18 04	0.000	-0.02	5.67	1.24	0.2	gK0 pe	+21	120 s		
132501	258719		15 08 29.7	-80 18 48	-0.008	+0.01	6.9	-0.2	0.0	B8.5 V		240 mx		
133525	253026		15 08 29.8	-66 28 31	-0.005	0.00	7.6	0.0	0.4	B9.5 V		190 mx		
135143	8174		15 08 30.0	+72 22 09	+0.014	-0.02	7.9	0.7	4.4	G0 V		51 s		
134586	29423		15 08 33.1	+53 15 44	-0.008	+0.03	7.3	0.4	2.6	F0 V		87 s		
134335	83685		15 08 35.4	+25 06 31	-0.001	0.00	5.81	1.24	0.2	gK0	-16	95 s		m
134284	101399		15 08 38.8	+10 05 58	0.000	-0.02	7.9	1.1	-0.1	K2 III		390 s		
133937	225479		15 08 39.0	-42 52 04	-0.003	-0.02	5.85	-0.11	-0.9	B6 V	+2	220 s		
134001	206330		15 08 41.6	-37 35 21	0.000	-0.03	7.6	1.2	-0.5	M2 III		420 s		
134052	183206		15 08 42.8	-29 21 54	-0.001	-0.01	8.0	0.2	3.4	F5 V		82 s		

HD	SAO	Star Name	α 2000	δ 2000	μ(α)	μ(δ)	V	B–V	M_v	Spec	RV	d(pc)	ADS	Notes
134851	16614		15^{h}08^{m}43.4^s	+64°32'53"	−0.003	+0.07	7.2	1.1	0.2	K0 III		250 s		
134305	101402		15 08 45.0	+12 29 19	0.000	−0.01	7.2	0.2		A7 p	−33			
133699	253028		15 08 45.5	−61 38 03	+0.003	−0.02	7.1	−0.8	0.0	B8.5 V		260 s		
134136	183209		15 08 47.3	−20 41 41	0.000	−0.02	7.9	1.0	−0.1	K2 III		400 s		
--	--		15 08 47.8	+28 13 52	−0.004	−0.25	8.0 B							
133426	257242		15 08 48.9	−70 42 39	−0.003	−0.05	7.0	0.2	0.4	B9.5 V		150 s		
133955	225483	λ Lup	15 08 50.5	−45 16 47	−0.001	−0.02	4.05	−0.18	−2.3	B3 IV	+18	190 s		m
134353	101404		15 08 51.9	+19 02 19	−0.011	−0.04	7.9	1.1	0.2	K0 III		84 mx		
134138	183210		15 08 52.8	−23 03 59	−0.002	−0.02	7.9	2.0	−0.5	M2 III		290 s		
134323	101403		15 08 53.4	+13 14 06	−0.004	+0.06	6.10	0.96	5.9	dK0	−49	11 s		d?
133790	242251		15 08 53.9	−59 08 32	0.000	−0.02	6.9	1.3	0.2	K0 III		140 s		
134193	159049		15 08 56.6	−11 47 27	+0.002	−0.02	8.0	0.4	3.0	F2 V		98 s		
134454	64519		15 08 57.1	+33 15 57	−0.004	+0.06	7.2	1.1	−0.1	K2 III		220 mx		
134285	120863		15 08 57.3	+1 41 21	+0.005	−0.01	7.80	0.4	2.6	F0 V	+11	110 s	9517	m
134140	183211		15 08 57.6	−26 29 50	+0.002	0.00	6.95	1.72	−0.5	M4 III		280 s		
133385	257241		15 09 01.0	−72 18 28	+0.002	−0.02	6.8	−0.1	−2.5	B2 V k		730 s		
134214	159050		15 09 02.3	−13 59 58	−0.003	+0.02	7.5	0.4	2.6	F0 V		94 s		
134055	206335		15 09 02.3	−38 32 03	−0.002	−0.03	7.1	0.2	2.1	A5 V		92 s		
133738	253030		15 09 03.0	−61 53 15	−0.003	−0.01	6.8	−0.4		B5 pne				
134141	183215		15 09 04.6	−28 21 44	+0.003	0.00	7.63	0.31		A3		16 mn		
135119	16622		15 09 05.8	+69 39 14	−0.006	+0.05	7.1	0.4	3.0	F2 V		65 s		
133958	242264		15 09 08.6	−50 51 37	−0.006	−0.02	8.00	0.9	3.2	G5 IV		90 s		m
134142	206342		15 09 14.6	−32 50 11	−0.002	−0.04	7.9	1.7	−0.5	M2 III		450 s		
134788	29427		15 09 22.4	+57 06 45	+0.003	−0.09	7.2	0.5	4.0	F8 V		45 s		
133792	253034		15 09 25.4	−63 38 34	−0.001	−0.02	6.28	0.06		A0 p				
134370	120870		15 09 28.0	+3 51 58	+0.002	−0.01	7.1	0.4	3.0	F2 V		65 s		
134495	83692		15 09 29.5	+24 40 34	+0.002	+0.03	7.3	0.5	3.4	F5 V		60 s		
133683	253031		15 09 29.9	−67 05 03	+0.001	−0.01	5.76	0.69	−6.4	cF		2600 s		
134358	140354		15 09 34.6	−5 23 42	−0.011	−0.06	7.1	1.1	0.2	K0 III		73 mx		
133638	253029		15 09 36.4	−68 43 18	−0.005	−0.04	6.85	−0.02		A0				T TrA, m,q
--	--		15 09 38.1	−68 42 55			7.00							m
134197	206345		15 09 40.9	−35 51 57	−0.002	0.00	6.9	1.1	0.2	K0 III		200 s		
134438	120873		15 09 47.4	+3 42 30	−0.002	+0.01	7.7	1.1	0.2	K0 III		310 s		
134328	183232		15 09 48.5	−20 31 15	−0.002	−0.06	7.2	0.6	2.6	F0 V		52 s		
134329	183231		15 09 51.2	−23 59 09	−0.002	−0.02	6.80	1.4	−0.3	K5 III	−14	260 s		m
134679	64528		15 09 54.0	+38 58 33	−0.008	+0.02	8.00	0.5	4.0	F8 V	−12	63 s	9527	m
134456	140360		15 09 58.2	−0 52 48	−0.004	−0.01	7.7	0.4	2.6	F0 V		110 s		
134159	225497		15 09 59.8	−45 59 49	−0.001	0.00	7.7	1.1	−0.1	K2 III		360 s		
134218	225499		15 10 00.5	−41 10 45	−0.001	0.00	7.49	−0.04	0.4	B9.5 V		260 s		
134255	206352		15 10 07.4	−38 47 33	−0.001	−0.03	6.0	1.2	3.2	G5 IV		20 s		
134712	64530		15 10 07.5	+37 41 25	+0.002	−0.05	8.00	0.1	1.4	A2 V		210 s		m
133369	257243		15 10 09.1	−75 27 43	−0.005	−0.03	6.9	1.3	0.2	K0 III		140 s		
134388	159063		15 10 11.3	−18 03 25	−0.001	−0.01	8.0	1.1	−0.1	K2 III		420 s		
134373	183237		15 10 18.5	−26 19 58	−0.002	−0.01	5.9	1.7	0.2	K0 III		48 s		
134102	242273		15 10 28.1	−57 07 57	−0.005	−0.05	7.5	1.6	0.2	K0 III		130 s		
134627	101412		15 10 31.4	+11 40 26	0.000	+0.03	7.0	1.6	−0.5	M2 III	−20	320 s		
133684	257244		15 10 31.6	−71 39 52	−0.005	−0.05	7.1	1.4	0.2	K0 III		140 s		
134772	64534		15 10 38.7	+33 04 44	−0.001	0.00	7.1	1.1	0.2	K0 III		240 s		
134331	225508		15 10 41.6	−43 43 48	+0.002	−0.10	7.01	0.62	5.2	G5 V		23 s		m
134330	225510		15 10 43.1	−43 42 58	+0.001	−0.08	7.61	0.72	4.6	K0 IV-V		41 s		
135384	16630		15 10 44.0	+67 46 52	−0.001	0.00	6.2	0.1	1.4	A2 V	−8	89 s		
134060	253043		15 10 44.6	−61 25 21	−0.026	−0.02	6.30	0.63	3.0	G3 IV		46 s		
134792	83705		15 10 53.3	+29 13 45	0.000	−0.03	7.1	0.5	3.4	F5 V	+15	56 s		
131596	258717	ω Oct	15 11 08.2	−84 47 15	−0.001	+0.01	5.91	−0.06		A0				
134313	242288		15 11 09.1	−52 32 11	−0.002	−0.02	7.2	1.0	0.2	K0 III		250 s		
134590	159074		15 11 11.6	−16 57 51	+0.001	+0.01	8.0	0.1	0.6	A0 V		300 s		
134630	159075		15 11 13.4	−13 03 18	−0.002	−0.02	7.6	0.9	0.3	G7 III	−37	280 s		
134426	225514		15 11 13.6	−40 42 57	+0.004	−0.02	7.9	0.5	3.0	F2 V		62 s		
134270	242287		15 11 15.8	−55 20 46	−0.001	−0.01	5.43	1.12	0.3	gG5	−4	76 s		m
--	101420		15 11 16.3	+18 36 28	0.000	−0.01	8.0							
134649	159077		15 11 20.6	−11 28 17	0.000	−0.02	7.9	1.1	0.2	K0 III		340 s		
135244	29439		15 11 24.0	+59 03 14	−0.002	+0.01	7.6	1.4	−0.3	K5 III	−11	370 s		
134259	242289		15 11 24.6	−57 55 15	−0.001	−0.01	8.0	−0.3	0.0	B8.5 V		400 s		
135420	16634		15 11 26.3	+65 38 21	−0.001	+0.03	7.7	0.1	1.4	A2 V		180 s		
134754	120884		15 11 29.9	+2 49 42	−0.003	−0.02	7.9	0.2	2.1	A5 V		150 s		
134443	225516		15 11 31.7	−45 16 46	+0.002	−0.06	7.39	1.08	0.2	K0 III		200 mx		m
134852	83708		15 11 34.2	+20 02 36	+0.003	−0.02	7.1	0.4	3.0	F2 V		67 s		
134793	120885		15 11 34.2	+8 31 03	+0.001	+0.01	7.57	0.14		A3 p				
134444	225517		15 11 34.8	−45 16 39	−0.001	+0.01	6.44	1.04	0.2	K0 III		170 s		m
135120	29436		15 11 34.9	+49 54 11	−0.001	0.00	7.5	1.6	−0.5	M2 III		410 s		
134737	140368		15 11 35.1	−2 15 37	−0.001	−0.03	7.1	0.4	2.6	F0 V		78 s		
134395	242296		15 11 36.0	−51 03 46	+0.001	−0.02	8.0	0.5	1.7	A3 V		96 s		
134869	83709		15 11 37.4	+21 31 34	−0.003	+0.01	7.7	0.1	1.4	A2 V		180 s		
134038	253046		15 11 38.4	−67 42 38	−0.004	−0.04	7.46	0.04		A0				
135437	16637		15 11 44.2	+63 37 17	−0.005	−0.02	7.9	0.1	1.4	A2 V		77 mx		

HD	SAO	Star Name	α 2000	δ 2000	μ(α)	μ(δ)	V	B-V	M_v	Spec	RV	d(pc)	ADS	Notes
			h m s	° ′ ″	s	″								
134574	206385		15 11 45.9	-33 38 20	+0.005	-0.02	7.3	0.7	3.2	G5 IV		66 s		
134854	101423		15 11 47.6	+10 13 00	-0.002	+0.02	6.8	0.1	0.6	A0 V	-13	170 s		
134853	101424		15 11 49.8	+11 28 31	-0.003	+0.04	7.8	0.5	3.4	F5 V		74 s		
134700	159085		15 11 49.8	-16 09 34	-0.001	-0.02	6.7	1.1	0.2	K0 III		200 s		
135421	16635		15 11 49.9	+61 51 40	-0.025	+0.09	7.4	0.7	4.4	G0 V		39 s	9537	m
134701	159086		15 11 51.4	-16 36 41	0.000	-0.02	7.8	0.5	4.0	F8 V		57 s		
134756	140373		15 11 51.6	-8 50 24	-0.001	0.00	7.9	1.4	-0.3	K5 III		430 s		
135466	16638		15 11 52.7	+63 39 10	-0.006	-0.01	7.6	0.1	0.6	A0 V		71 mx		
134871	120886		15 11 54.0	+9 46 37	-0.011	-0.04	7.4	1.1	0.2	K0 III		82 mx		
134481	225525	κ Lup	15 11 56.0	-48 44 16	-0.010	-0.05	3.72	-0.04	0.2	B9 V	+3	39 mx		m
134482	225526		15 11 57.5	-48 44 37	-0.010	-0.04	5.69	0.14	0.6	A0 V	0	43 mx		
135078	64549		15 11 57.6	+36 45 16	-0.017	+0.02	7.21	0.79		G5				
134963	83714		15 12 03.8	+22 18 49	-0.003	-0.02	6.7	1.6	-0.5	M4 III	-28	280 s		
134557	225528		15 12 03.9	-45 00 42	-0.003	-0.01	7.89	0.04	0.4	B9.5 V		290 s		m
134943	101429		15 12 04.2	+18 58 34	-0.001	+0.01	5.89	1.53	-0.5	M4 III	-35	190 s		m
134483	242300		15 12 09.7	-52 06 22	-0.011	-0.06	6.69	0.50	4.0	F8 V		35 s		
134664	206400		15 12 10.2	-30 53 11	+0.007	-0.11	7.7	0.0	3.2	G5 IV		76 mx		
134758	159089		15 12 11.8	-19 06 22	-0.004	-0.01	6.7	1.1	0.2	K0 III		200 s		
134759	159090	24 ι Lib	15 12 13.2	-19 47 30	-0.003	-0.04	4.54	-0.08	-0.3	B9 IV	-12	93 s	9532	m
135200	45430		15 12 15.7	+43 58 51	+0.001	-0.04	8.0	0.4	2.6	F0 V		120 s		
134505	242304	ζ Lup	15 12 17.0	-52 05 57	-0.012	-0.07	3.41	0.92	0.3	G8 III	-10	42 s		m
133429	--		15 12 18.6	-79 01 22			8.0	1.1	0.2	K0 III		360 s		
134276	253051		15 12 19.1	-64 32 27	-0.001	-0.01	7.4	0.1	1.4	A2 V		150 s		
134685	206406		15 12 19.3	-34 58 04	-0.002	-0.03	7.8	0.2	0.6	A0 V		190 s		
134796	159093		15 12 21.0	-16 24 41	+0.001	-0.02	7.3	0.0	0.4	B9.5 V		240 s		
135162	64557		15 12 23.0	+38 59 16	-0.001	+0.02	7.7	1.1	0.2	K0 III		320 s		
135339	29443		15 12 26.1	+52 55 49	+0.002	+0.03	7.4	1.1	0.2	K0 III		280 s		
134812	159094		15 12 26.1	-18 05 25	+0.003	-0.01	7.0	0.9	3.2	G5 IV		58 s		
134830	159096		15 12 26.2	-15 42 53	-0.001	+0.03	7.7	1.1	-0.1	K2 III		370 s		
134345	253052		15 12 29.6	-63 02 04	-0.004	-0.07	6.84	1.41	-0.2	K3 III		210 s		
134597	225533		15 12 31.2	-48 13 07	-0.001	-0.03	6.33	1.12	0.2	gK0		140 s		
133981	257247		15 12 33.8	-72 46 13	-0.003	-0.01	6.01	0.00		A0				
134945	140379		15 12 37.5	-0 22 54	+0.002	-0.02	7.5	1.4	-0.3	K5 III		370 s		
133386	--		15 12 39.0	-79 43 22			7.8	1.4	-0.3	K5 III		420 s		
134449	242305		15 12 40.7	-60 05 29	+0.009	-0.02	7.2	1.6	3.2	G5 IV		20 s		
134885	159098		15 12 41.8	-14 41 39	-0.005	-0.02	8.00	0.4	3.0	F2 V		100 s		
135101	101438		15 12 43.6	+19 17 32	-0.043	+0.28	6.27	0.70	5.2	G5 V	-37	21 mn	9535	m
135364	45436		15 12 43.7	+48 34 51	-0.002	+0.04	7.40	1.1	0.2	K0 III		280 s	9539	m
135004	120893		15 12 45.0	+3 32 41	-0.006	+0.06	7.6	0.5	3.4	F5 V		71 s		
135080	101436		15 12 45.6	+19 05 44	0.000	0.00	7.3	1.1	-0.1	K2 III		310 s		
134450	253056		15 12 45.6	-60 23 37	+0.001	-0.04	7.50	0.5	4.0	F8 V		50 s		m
134742	206412		15 12 46.1	-37 31 21	-0.001	0.00	7.5	1.2	-0.5	M2 III		410 s		
135061	101435		15 12 47.1	+10 42 09	-0.002	-0.06	6.6	1.1	-0.1	K2 III		160 mx		
--	--		15 12 49.2	+19 16 48			7.53	0.73	5.2	dG5		26 s	9535	m
134687	225539		15 12 49.3	-44 30 02	-0.003	-0.03	4.82	-0.17	-2.9	B3 III	+11	340 s		
134704	225542		15 12 54.7	-44 57 32	+0.001	-0.01	8.0	0.3	3.2	G5 IV		93 s		
135501	29446		15 12 57.8	+56 02 46	-0.003	-0.01	7.40	0.1	1.4	A2 V		160 s	9546	m
134642	242318		15 12 59.0	-51 58 21	0.000	-0.03	7.8	1.3	-0.1	K2 III		330 s		
135401	45438		15 13 00.3	+49 46 31	-0.001	+0.02	7.2	1.1	0.2	K0 III		250 s		
134468	253059		15 13 00.9	-61 44 37	-0.001	-0.01	6.0	2.8	-4.4	K4 Ib		120 s		
134799	--		15 13 03.0	-37 14 45			7.33	0.23	1.7	A3 V		110 s		m
134837	206418		15 13 07.3	-36 05 29	-0.004	-0.02	6.10	-0.08		B9				
135027	159107		15 13 14.8	-11 00 24	-0.003	-0.03	6.6	0.1	1.4	A2 V		110 s		
134946	183269		15 13 17.3	-24 00 30	-0.001	-0.02	6.47	-0.04		B9				
134967	159105	25 Lib	15 13 19.1	-19 38 51	-0.004	-0.04	6.08	0.12	0.6	A0 V n	+1	59 mx		
134987	183275	23 Lib	15 13 28.6	-25 18 33	-0.029	-0.07	6.45	0.70	5.0	dG4	+3	19 s		m
134672	242323		15 13 28.9	-55 43 54	-0.004	-0.03	7.6	0.5	4.4	G0 V		43 s		
135263	83723		15 13 31.8	+22 59 00	+0.004	+0.09	6.20	0.06	1.4	A2 V	-5	81 mx		
135166	120902		15 13 31.9	+5 02 39	-0.001	-0.04	7.9	1.1	-0.1	K2 III		390 s		
134527	253061		15 13 34.7	-62 56 37	0.000	-0.05	8.0	0.5	4.0	F8 V		62 s		
134528	253060		15 13 35.3	-63 53 32	-0.002	0.00	7.7	1.6	-0.5	M4 III		410 s		
135402	64572		15 13 35.5	+38 15 53	-0.001	-0.04	6.20	1.20	0.2	K0 III	-62	120 s		
134874	225551		15 13 39.1	-41 29 38	-0.002	-0.01	7.67	-0.05	0.4	B9.5 V		280 s		
134909	206427		15 13 41.3	-37 32 43	0.000	-0.02	8.0	1.3	-0.5	M2 III		510 s		
135107	159111		15 13 47.0	-14 01 22	-0.002	0.00	7.4	0.9	3.2	G5 IV		71 s		
135204	140392		15 13 50.7	-1 21 04	-0.086	-0.49	6.60	0.77	5.6	G8 V	-70	10 mx	9544	m
134082	257250		15 13 51.4	-74 54 41	+0.003	-0.03	7.7	1.4	-0.3	K5 III		400 s		
135051	183285		15 13 53.2	-26 11 37	-0.001	0.00	6.10	1.1	0.2	K0 III	-28	150 s	9538	m
135066	183286		15 13 53.9	-24 22 43	+0.003	-0.03	7.9	0.4	3.2	G5 IV		88 s		
135544	45442		15 13 54.1	+47 30 06	-0.002	+0.01	8.0	1.1	0.2	K0 III		360 s		
134913	225552		15 13 54.2	-43 21 23	-0.004	-0.01	8.00	0.9	3.2	G5 IV		91 s		m
135151	140390		15 13 57.2	-9 29 37	+0.005	-0.06	8.02	0.51	4.4	G0 V		53 s		
135558	45444		15 13 58.8	+47 58 08	+0.002	-0.01	7.5	0.1	1.7	A3 V		150 s		
134657	253064		15 13 59.0	-61 20 33	-0.001	-0.01	6.36	-0.05	-1.1	B5 V		280 s		m
135517	45440		15 13 59.1	+43 02 54	-0.007	+0.05	6.64	0.18		A5				

HD	SAO	Star Name	α 2000	δ 2000	μ(α)	μ(δ)	V	B-V	M_V	Spec	RV	d(pc)	ADS	Notes
134930	225554		15^{h}13^{m}59^s.2	-43°47'37"	-0^s.004	0".00	8.00	0.1	1.4	A2 V		170 mx		m
135205	140398		15 14 01.3	-1 53 11	-0.003	-0.01	7.9	1.6	-0.5	M2 III	-59	490 s		Y Ser, v
135438	64574		15 14 06.0	+31 47 17	+0.003	-0.02	5.99	1.52	-0.3	K5 III	+4	180 s		m
134990	206436		15 14 09.9	-38 29 32	-0.001	-0.03	7.70	0.1	1.4	A2 V		180 s		m
135530	45445		15 14 10.2	+42 10 17	+0.001	-0.02	6.13	1.66	-0.5	M2 III	-7	190 s		
134403	257254		15 14 12.6	-70 33 33	+0.002	-0.01	7.8	0.6	1.7	A3 V		86 s		
135207	159117		15 14 18.9	-14 12 34	+0.001	-0.02	7.0	1.6	-0.5	M2 III	+6	310 s		
135168	159116		15 14 20.0	-18 57 12	0.000	0.00	7.4	0.2	2.1	A5 V		110 s		
135297	120910		15 14 20.9	+0 22 11	-0.001	0.00	8.03	-0.01		A0 p				
134977	225564		15 14 27.6	-47 02 15	-0.004	-0.01	7.8	-0.3	1.7	A3 V		140 mx		
135208	159118		15 14 28.0	-18 25 43	-0.008	-0.02	6.74	0.44	3.1	dF3	-24	49 s		m
135298	140403		15 14 28.5	-3 21 52	-0.002	-0.01	7.2	0.1	0.6	A0 V		210 s		
135502	83729	48 χ Boo	15 14 29.1	+29 09 51	-0.005	+0.03	5.26	0.03	1.4	A2 V	-16	59 s		
135034	225567		15 14 30.5	-43 23 13	-0.005	-0.02	7.10	0.1	1.4	A2 V		83 mx		m
134785	242334		15 14 30.9	-59 44 06	+0.001	-0.03	6.9	2.0	-0.3	K5 III		140 s		
135230	159122	26 Lib	15 14 33.6	-17 46 07	-0.002	-0.02	6.17	-0.05		B9	-26			
134708	253066		15 14 34.1	-63 37 48	+0.005	+0.02	7.3	1.1	0.2	K0 III		250 s		
135153	206445	1 Lup	15 14 37.2	-31 31 09	0.000	0.00	4.91	0.37	-6.6	F0 I	-23	1800 s		
136064	16660		15 14 38.2	+67 20 48	+0.037	-0.39	5.13	0.53	4.0	F8 V	-47	19 ts		
134890	242339		15 14 38.3	-54 24 24	-0.001	0.00	7.7	1.4	0.2	K0 III		190 s		
134189	257253		15 14 43.8	-75 32 58	+0.003	-0.01	7.5	1.6	-0.1	K2 III		190 s		
135088	225574		15 14 49.4	-42 19 11	-0.003	0.00	7.4	0.6	0.6	A0 V		230 s		
135367	140408		15 14 50.5	-5 30 10	-0.001	0.00	6.28	1.47		K2				
136174	16662		15 14 51.7	+68 56 43	+0.003	-0.01	6.5	0.1	0.6	A0 V	-11	150 s		
135631	64585		15 14 56.5	+38 18 04	-0.003	+0.04	7.20	0.4	2.6	F0 V	-34	83 s	9553	m
135386	140410		15 14 57.6	-2 24 53	0.000	-0.04	7.9	1.6	-0.5	M4 III		480 s		
134709	--		15 14 57.9	-66 09 44			7.87	0.16		A0				
135482	120916	3 Ser	15 15 11.3	+4 56 21	-0.001	0.00	5.33	1.09	0.2	gK0	-34	93 s		
135302	183308		15 15 14.7	-27 51 24	-0.001	0.00	7.4	0.3	0.6	A0 V		160 s		
134606	257257		15 15 15.1	-70 31 11	-0.033	-0.17	6.9	1.3	0.2	K0 III		72 mx		
135174	225582		15 15 19.5	-44 08 58	-0.002	-0.02	6.67	-0.04	0.4	B9.5 V		180 s		m
133921	257252		15 15 27.3	-79 08 09	-0.012	-0.02	6.6	0.9	0.2	K0 III		190 s		
135722	64589	49 δ Boo	15 15 30.1	+33 18 53	+0.007	-0.12	3.47	0.95	0.3	G8 III	-12	43 s	9559	m
134894	253073		15 15 30.7	-63 19 02	-0.013	-0.04	8.0	0.4	2.6	F0 V		120 s		
134567	--		15 15 34.3	-72 24 47			8.0	1.6	-0.5	M2 III		490 s		
135576	101461		15 15 35.3	+10 07 35	+0.004	-0.03	7.2	0.4	3.0	F2 V		68 s		
135253	225588		15 15 36.7	-40 57 02	-0.001	0.00	7.6	1.8	0.2	K0 III		130 s		
--	64591		15 15 38.1	+33 19 13	+0.005	-0.13	7.82	0.58						
135390	183316		15 15 39.1	-27 35 45	-0.004	+0.03	7.10	0.4	3.0	F2 V		66 s	9552	m
135679	83733		15 15 41.4	+25 38 33	0.000	0.00	6.97	-0.09	0.6	A0 V	-4	190 s		
135344	206462		15 15 48.3	-37 09 16	-0.002	-0.03	7.90	0.1	1.4	A2 V		200 s		
135559	120920	4 Ser	15 15 49.0	+0 22 20	-0.007	+0.01	5.63	0.18		A1 n	-8	21 mn		
135944	29464		15 15 50.6	+50 56 18	+0.002	-0.05	6.70	0.9	3.2	G5 IV		50 s		m
135343	206465		15 15 52.6	-36 59 55	-0.003	-0.05	7.90	0.4	3.0	F2 V		96 s		m
135615	120923		15 15 53.5	+6 27 55	0.000	0.00	6.6	1.1	0.2	K0 III		190 s		
135235	225590		15 15 53.7	-48 04 27	+0.002	-0.03	5.95	0.21	1.4	A2 V		67 s		m
135255	225592		15 15 56.0	-47 09 29	-0.001	-0.02	8.0	-0.2	1.4	A2 V		210 s		
134735	257258		15 15 58.9	-70 24 48	+0.002	-0.04	6.7	1.0	2.6	F0 V		27 s		
135599	120922		15 15 59.1	+0 47 47	+0.012	-0.14	7.1	1.1	0.2	K0 III		66 mx		
135578	140421		15 16 01.5	-4 53 52	+0.001	-0.02	7.20	0.51		F8			9557	m
135345	225600		15 16 03.9	-41 29 28	0.000	-0.01	5.1	1.3	-8.0	G5 Ia	-27	3000 s		
135697	101464		15 16 07.2	+13 28 12	+0.002	-0.03	7.6	0.9	3.2	G5 IV		74 s		
135724	101466		15 16 08.3	+15 32 29	-0.001	-0.09	7.9	0.9	3.2	G5 IV		88 s		
135348	225602		15 16 10.4	-43 29 05	-0.001	0.00	6.04	-0.14	-1.0	B5.5 V	-21	260 s		
135310	225601		15 16 16.9	-48 02 39	-0.001	0.00	7.2	-0.3	1.4	A2 V		150 s		
135058	253079		15 16 21.2	-62 21 48	0.000	-0.01	7.4	-0.2	0.4	B9.5 V		250 s		
135891	64602		15 16 21.3	+37 04 09	-0.023	+0.01	7.1	0.5	4.0	F8 V		42 s		
135534	183328		15 16 22.9	-22 23 58	-0.003	0.00	5.6	2.1	-0.1	K2 III	-5	39 s		
134333	257255		15 16 23.9	-77 21 12	-0.001	+0.02	7.4	1.1	0.2	K0 III		270 s		
135792	101471		15 16 25.6	+16 47 39	0.000	-0.18	7.90	0.7	4.4	G0 V		50 s	9562	m
135415	225605		15 16 28.3	-43 57 01	-0.001	+0.01	7.90	-0.10	0.0	B8.5 V		380 s		
135416	225604		15 16 28.3	-44 30 15	-0.008	-0.05	7.1	0.3	-0.1	K2 III		150 mx		
135637	159140		15 16 31.2	-13 13 01	-0.003	0.00	8.01	0.28		A2				
135452	206475		15 16 32.1	-39 33 39	0.000	-0.01	7.0	1.0	0.2	K0 III		210 s		
135375	225603		15 16 34.6	-47 41 50	+0.004	-0.02	7.4	1.3	0.2	K0 III		190 s		
135775	120927		15 16 35.6	+9 42 46	+0.001	-0.01	6.6	0.6	4.4	G0 V	-14	28 s		
135160	253082		15 16 36.4	-60 54 14	-0.003	0.00	5.73	-0.08	-3.5	B1 V		610 s		m
135662	140426		15 16 36.5	-9 12 20	-0.006	-0.06	8.0	0.4	3.0	F2 V		88 mx		
135454	225612		15 16 37.0	-42 22 12	-0.002	-0.02	6.76	-0.03	0.4	B9.5 V		190 s		
135429	225611		15 16 39.2	-44 21 30	+0.003	-0.02	7.8	-0.1	4.0	F8 V		57 s		
136244	16666		15 16 39.7	+60 05 57	-0.011	0.00	8.0	1.4	-0.3	K5 III		180 mx		
134899	--		15 16 43.7	-69 52 36			7.9	0.4	3.0	F2 V		95 s		
135138	253081		15 16 44.7	-63 20 55	+0.001	-0.03	7.4	0.0	0.0	B8.5 V		290 s		
135617	183333		15 16 47.8	-24 00 34	-0.002	+0.01	7.2	0.1	1.4	A2 V		140 s		
135681	159146		15 16 48.4	-13 02 21	+0.001	0.00	7.1	0.1	1.4	A2 V		140 s		

HD	SAO	Star Name	α 2000	δ 2000	μ(α)	μ(δ)	V	B-V	M_v	Spec	RV	d(pc)	ADS	Notes
			^{h} ^{m} ^{s}	$^\circ$ $'$ $''$	^{s}	$''$								
135394	242379		15 16 49.7	−51 02 52	+0.001	−0.01	7.9	1.8	−0.1	K2 III		170 s		
135430	225613		15 16 51.7	−47 51 50	−0.006	−0.01	6.5	1.2	0.2	K0 III		150 s		
135725	140428		15 16 53.0	−8 17 09	−0.006	−0.24	7.87	0.74	3.2	G5 IV	−35	43 mx	9564	m
135468	225614		15 16 54.7	−46 12 08	−0.015	−0.03	6.6	0.8	3.4	F5 V		25 s		
135240	253084	δ Cir	15 16 56.7	−60 57 27	−0.002	−0.01	5.09	−0.06	−4.8	O9 V	+88	830 s		m
135549	206480		15 16 59.1	−38 26 53	−0.007	−0.03	7.0	0.6	2.6	F0 V		52 s		
135742	140430	27 β Lib	15 17 00.3	−9 22 58	−0.007	−0.02	2.61	−0.11	−0.2	B8 V	−35	37 s		Zubeneschamali
136726	8207	11 UMi	15 17 05.8	+71 49 26	0.000	+0.01	5.02	1.37	−0.3	K4 III	−16	120 s		
--	101475		15 17 06.1	+17 04 00	−0.001	0.00	8.0							
135377	242381		15 17 07.7	−56 23 02	−0.006	−0.05	7.4	1.0	0.2	K0 III		210 mx		
135619	206484		15 17 10.7	−34 34 35	−0.001	0.00	8.00	0.1	1.4	A2 V		210 s		m
135793	140431		15 17 13.5	−7 51 25	0.000	0.00	7.5	1.1	0.2	K0 III		290 s		
135139	253083		15 17 13.7	−66 05 32	+0.003	−0.01	6.84	0.01		B9				
136727	8208		15 17 16.6	+71 12 40	−0.001	−0.02	7.30	0.5	4.0	F8 V		46 s		m
135564	225619		15 17 17.4	−41 55 37	0.000	−0.01	7.7	1.6	−0.5	M2 III		260 s		
136402	16672		15 17 17.6	+62 46 16	−0.002	0.00	6.7	1.1	0.2	K0 III		200 s		
136115	45469		15 17 18.6	+42 33 58	−0.007	+0.03	7.4	0.2	2.1	A5 V		68 mx		
135355	242382		15 17 21.3	−58 22 04	−0.002	−0.02	7.71	1.64		K2				
136919	8211		15 17 22.6	+74 02 43	−0.004	+0.04	6.6	1.1	0.2	K0 III		190 s		
135991	83749		15 17 27.4	+22 17 58	−0.009	+0.05	7.9	0.6	4.4	G0 V		50 s		
136009	83750		15 17 28.5	+27 00 59	0.000	0.00	8.0	1.1	0.2	K0 III		360 s		
135379	242384	β Cir	15 17 30.8	−58 48 04	−0.012	−0.14	4.07	0.09	1.7	A3 V	+9	22 mx		
135825	159155		15 17 32.0	−10 30 02	−0.004	−0.02	7.3	0.4	2.6	F0 V		87 s		
136691	16677		15 17 37.8	+69 09 16	−0.010	+0.03	7.3	1.1	0.2	K0 III		260 s		
136230	45473		15 17 38.1	+47 44 23	−0.003	−0.09	7.7	1.1	0.2	K0 III		160 mx		
135291	253088	ε Cir	15 17 38.8	−63 36 37	+0.001	+0.01	4.86	1.25	−0.3	gK4	−5	110 s		
135894	140437		15 17 40.3	−0 59 05	−0.002	+0.02	7.3	0.8	3.2	G5 IV		66 s		
135872	140435		15 17 41.1	−5 12 09	−0.005	−0.02	7.1	1.1	0.2	K0 III		180 mx		
135895	140436		15 17 42.1	−4 29 39	0.000	+0.02	7.5	0.1	0.6	A0 V		240 s		
135667	206492		15 17 43.1	−39 48 42	−0.002	0.00	7.33	1.25		K0				
135624	225623		15 17 47.7	−45 42 53	−0.002	0.00	7.4	1.2	−0.3	K5 III		340 s		
135758	183346	2 Lup	15 17 49.7	−30 08 55	−0.001	−0.01	4.34	1.10	0.2	gK0	−4	55 s		
135796	183349		15 17 51.9	−26 59 46	−0.001	−0.01	7.80	1.1	0.2	K2 III		380 s	9569	m
136137	64616		15 17 53.0	+33 35 40	+0.001	0.00	6.7	0.9	3.2	G5 IV		50 s		
136159	64617		15 17 59.5	+30 49 58	0.000	−0.08	6.8	1.1	0.2	K0 III	−16	190 mx		
135730	225631		15 18 09.2	−41 03 39	+0.002	0.00	6.28	0.18		A m				
135669	225627		15 18 09.6	−47 54 00	0.000	+0.01	7.1	0.7	3.4	F5 V		39 s		
136175	64619		15 18 11.1	+31 38 50	−0.002	−0.01	7.80	0.00	−0.6	B7 V	−8	480 s		U CrB, m,v
137527	8217		15 18 12.6	+78 23 39	+0.015	−0.01	7.8	1.1	−0.1	K2 III		380 s		
136290	45475		15 18 14.8	+45 37 12	−0.002	+0.01	6.7	1.4	−0.3	K5 III		260 s		
136494	29475		15 18 15.8	+58 47 38	−0.001	+0.01	7.3	1.1	0.2	K0 III		260 s		
135760	225633		15 18 17.1	−41 25 13	−0.012	−0.11	7.6	0.4	0.2	K0 III		77 mx		
136176	83756		15 18 20.1	+26 50 25	+0.007	+0.08	6.58	0.56	4.0	dF8	−20	31 s	9578	m
136027	120938		15 18 22.5	+0 56 21	+0.002	−0.02	6.70	1.1	0.2	K0 III		200 s	9574	m
136138	83755		15 18 24.4	+20 34 22	−0.001	−0.02	5.70	0.97	3.2	G5 IV	−8	24 s		
136028	140444		15 18 26.1	−0 27 41	0.000	−0.01	5.89	1.51	−0.3	gK5	−13	170 s		
135734	225638	μ Lup	15 18 31.9	−47 52 30	−0.002	−0.04	4.27	−0.08	−0.2	B8 V n	+15	77 s		m
135748	225639		15 18 33.7	−47 52 44	−0.002	−0.03	5.10	0.1	0.6	A0 V		78 s		m
134583	257260		15 18 35.1	−78 28 18	+0.008	+0.01	7.60	0.4	2.6	F0 V		100 s		m
136010	140446		15 18 35.2	−8 23 50	−0.002	−0.01	7.60	1.1	−1.1	K1 II-III	+1	550 s		
135951	183358		15 18 37.6	−25 00 01	+0.002	−0.04	8.0	0.0	1.4	A2 V		210 s		
136362	45480		15 18 39.7	+45 01 01	−0.002	+0.01	7.9	1.1	0.2	K0 III		340 s		
135896	206509		15 18 41.2	−31 12 34	0.000	−0.01	6.10	1.1	0.2	K0 III		150 s		m
136160	101481		15 18 41.8	+10 25 28	−0.008	+0.01	6.75	0.52			−46		9580	m
136577	29479		15 18 45.5	+58 30 12	+0.001	+0.01	6.8	1.1	0.2	K0 III		210 s		
135689	242399		15 18 46.0	−54 21 49	0.000	+0.01	7.00	1.75	−0.5	M3 III		260 s		
135591	253101		15 18 48.8	−60 29 47	−0.003	−0.01	5.46	−0.10	−6.1	O9 Ib	−6	1600 s		m
137292	8215		15 18 50.1	+74 51 27	−0.011	−0.01	7.8	1.1	−0.1	K2 III		340 mx		
135382	253097	γ TrA	15 18 54.5	−68 40 46	−0.011	−0.03	2.89	0.00	0.6	A0 V	0	28 s		
136118	140452		15 18 55.4	−1 35 33	−0.008	+0.01	7.1	0.5	4.0	F8 V		41 s		
135970	183366		15 18 55.6	−29 13 17	−0.004	−0.02	7.9	2.0	−0.1	K2 III		130 s		
135876	225647		15 18 56.3	−40 47 18	−0.001	−0.03	5.59	−0.10	−4.1	B0 V	+16	620 s		
135672	242401		15 18 57.8	−57 32 28	−0.002	−0.01	8.0	1.9	−0.5	M2 III		350 s		
136274	83761		15 18 58.9	+25 41 29	−0.042	−0.13	7.96	0.74	5.6	G8 V	−29	22 mx		
136418	45483		15 19 06.2	+41 44 04	−0.001	−0.13	7.9	0.9	3.2	G5 IV		88 s		
135878	225650		15 19 06.9	−42 26 14	−0.002	−0.01	7.9	0.5	0.6	A0 V		130 s		
136032	183368		15 19 09.2	−24 15 57	−0.005	−0.04	7.50	0.1	1.4	A2 V		61 mx	9579	m
135674	242402		15 19 09.3	−60 08 02	−0.001	−0.02	7.4	0.2	0.4	B9.5 V		190 s		
134624	257261		15 19 14.5	−79 00 48	−0.001	−0.01	8.0	1.1	0.2	K0 III		360 s		
136749	16685		15 19 18.5	+61 22 30	−0.007	−0.01	7.4	0.7	4.4	G0 V		39 s		
136202	120946	5 Ser	15 19 18.7	+1 45 55	+0.025	−0.52	5.05	0.55	3.2	F8 IV-V	+54	25 ts	9584	m
136033	183369		15 19 20.0	−28 58 28	+0.003	−0.02	7.7	1.5	3.2	G5 IV		29 s		
136140	140456		15 19 21.7	−9 08 49	+0.001	−0.06	7.19	1.42	−0.5	gM4	+10	350 s		
135692	253109		15 19 22.0	−60 22 31	0.000	−0.02	7.7	−0.2	0.0	B8.5 V		350 s		
136013	206522		15 19 22.2	−34 01 58	−0.001	−0.03	7.7	0.3	0.6	A0 V		160 s		

HD	SAO	Star Name	α 2000	δ 2000	μ(α)	μ(δ)	V	B-V	M_V	Spec	RV	d(pc)	ADS	Notes
136562	29482		15^{h}19^{m}24.1^s	+50°12'52"	-0.001^s	+0.01"	7.4	0.1	0.6	A0 V	-10	230 s		
136102	159169		15 19 26.6	-19 10 09	-0.002	-0.03	7.9	0.1	1.4	A2 V		200 s		
136101	159171		15 19 26.9	-18 17 17	0.000	+0.01	8.0	0.1	1.4	A2 V		210 s		
135338	257264		15 19 28.9	-71 47 33	-0.003	-0.03	6.4	2.4	-0.5	M2 III		74 s		
136403	64630		15 19 30.1	+32 30 55	-0.001	+0.01	6.10	0.24		A m	-25			
136014	206523		15 19 31.5	-37 05 48	-0.008	-0.13	6.20	0.96		G5				
136121	183377		15 19 40.3	-24 16 13	-0.002	-0.04	7.90	0.4	3.0	F2 V		96 s	9586	m
135592	253107		15 19 45.6	-66 29 47	0.000	-0.02	6.6	0.5	3.4	F5 V	-19	44 s		R TrA, v
136205	159175		15 19 50.1	-19 33 01	-0.005	+0.01	7.0	0.3	3.4	F5 V		54 s		
136881	16693		15 19 52.2	+62 28 27	-0.010	+0.04	6.6	1.1	-0.1	K2 III		220 s		
135161	257263		15 19 52.4	-75 19 47	-0.015	-0.11	7.8	0.9	3.4	F5 V		39 s		
135786	253113		15 19 54.1	-60 33 40	+0.002	-0.01	8.0	-0.3	-1.0	B5.5 V		620 s		
136848	16692		15 19 59.3	+60 20 22	-0.001	+0.01	7.7	0.4	3.0	F2 V		86 s		m
136257	140461		15 20 00.0	-8 39 42	+0.006	-0.20	7.55	0.54	4.2	F9 V	+29	47 s		
136580	45491		15 20 02.2	+40 59 03	-0.003	+0.19	6.8	0.5	3.4	F5 V		48 s		
136124	206528		15 20 02.6	-32 51 56	+0.002	-0.06	7.8	0.4	3.0	F2 V		81 s		
136729	29487		15 20 05.0	+51 57 30	+0.001	+0.01	5.5	0.1	1.7	A3 V	+8	58 s		
136882	16694		15 20 05.0	+60 22 48	-0.005	+0.03	7.5	0.4	2.6	F0 V		95 s		m
136404	101491		15 20 06.1	+14 33 38	-0.001	-0.01	7.7	1.6	-0.5	M2 III		430 s		
--	--		15 20 07.0	+60 14 15			7.60							
--	--		15 20 07.0	+60 14 15			7.40							
136512	83768	1 o CrB	15 20 08.4	+29 36 58	-0.009	-0.05	5.51	1.02	0.2	K0 III	-53	62 mx		m
136440	101493		15 20 08.5	+16 11 07	-0.003	-0.04	7.5	0.7	4.4	G0 V		42 s		
135902	242421		15 20 12.9	-57 23 53	-0.005	-0.04	7.4	1.0	-0.5	M4 III		290 mx		
135219	--		15 20 13.1	-75 34 23			6.90	1.1	0.2	K0 III		220 s		m
136164	206531		15 20 13.4	-34 55 31	0.000	-0.03	7.8	0.7	1.7	A3 V		76 s		
136527	83772		15 20 19.5	+29 44 36	-0.005	0.00	7.8	1.1	0.2	K0 III		230 mx		
136107	225667		15 20 22.6	-43 40 31	-0.001	-0.02	7.9	0.9	-0.1	K2 III		390 s		
136477	101495		15 20 23.3	+16 33 19	-0.005	+0.04	7.9	0.9	3.2	G5 IV		88 s		
136246	183389		15 20 31.3	-28 17 14	-0.002	-0.03	7.1	0.5	0.6	A0 V		100 s		
138366	2535		15 20 31.6	+81 02 33	-0.008	+0.04	7.3	0.9	3.2	G5 IV		66 s		
135917	242425		15 20 32.5	-59 32 38	+0.001	-0.01	8.0	0.0	-1.0	B5.5 V		570 s		
136380	140471		15 20 36.5	-5 07 11	+0.003	-0.01	6.6	1.1	0.2	K0 III		190 s		
136344	159185		15 20 37.1	-14 44 29	-0.002	-0.03	7.6	1.1	0.2	K0 III		300 s		
136579	45498		15 20 37.4	+46 55 31	0.000	-0.03	7.7	1.1	0.2	K0 III		310 s		
135737	253115		15 20 40.7	-67 28 53	+0.001	-0.01	6.28	-0.09	-1.7	B3 V		360 s		m
136295	183392		15 20 41.3	-25 59 25	-0.003	-0.15	7.1	0.7	0.2	K0 III		96 mx		
136751	45497		15 20 41.4	+44 26 03	+0.002	-0.11	6.19	0.39	2.6	F0 V	0	47 s		
136003	242427		15 20 42.4	-56 07 58	-0.004	-0.02	6.81	0.19	-3.6	B2 III		730 s		
136497	120953		15 20 42.9	+8 54 49	-0.001	-0.03	7.9	0.5	3.4	F5 V		78 s		
137422	8220	13 γ UMi	15 20 43.6	+71 50 02	-0.005	+0.02	3.05	0.05	-1.1	A3 II-III	-4	69 s		Pherkad
136442	140473		15 20 47.0	-2 24 49	-0.017	-0.18	6.35	1.06	6.3	dK2	-41	18 mn		
135772	--		15 20 47.5	-67 29 08			8.0	0.0	0.0	B8.5 V		380 s		
136654	64647		15 20 50.0	+31 28 49	-0.014	+0.15	6.9	0.5	3.4	F5 V	-30	50 s		
136795	45500		15 20 52.8	+45 39 52	+0.004	-0.04	6.7	1.1	0.2	K0 III		200 s		
136366	159187	28 Lib	15 20 53.5	-18 09 31	-0.001	-0.06	6.17	1.02	0.3	gG8	+3	140 s		
136406	159188		15 20 57.0	-15 22 25	-0.001	-0.02	7.5	1.1	0.2	K0 III	-20	280 s		
136407	159191	29 o Lib	15 21 01.2	-15 32 54	+0.002	-0.02	6.30	0.41	3.4	F5 V	+5	38 s		m
136514	120955	6 Ser	15 21 01.9	+0 42 55	-0.003	-0.11	5.35	1.19	-0.2	K3 III	+9	100 mx	9596	m
136921	29493		15 21 03.6	+53 55 43	-0.001	+0.02	7.6	0.1	1.4	A2 V		180 s		
136187	225685		15 21 04.8	-47 45 37	+0.009	-0.02	7.3	0.3	3.4	F5 V		61 s		
136643	83778		15 21 06.8	+24 57 28	-0.002	-0.03	6.39	1.23	0.2	K0 III	-2	130 s		
136479	140474		15 21 07.5	-5 49 30	-0.004	-0.02	5.54	1.04	6.1	dK1	-33			m
136480	140475		15 21 09.2	-6 36 46	+0.001	+0.01	7.4	1.1	-0.1	K2 III	+20	320 s		
135739	253116		15 21 10.1	-69 59 14	+0.009	+0.04	8.0	1.1	0.2	K0 III		350 s		
136113	242436		15 21 12.4	-54 17 30	-0.011	-0.03	7.4	0.3	3.4	F5 V		64 s		
136618	101502		15 21 19.6	+13 31 01	-0.003	-0.01	7.9	0.6	4.4	G0 V		50 s		
136298	225691	δ Lup	15 21 22.2	-40 38 51	-0.001	-0.03	3.22	-0.22	-3.0	B2 IV	+2	180 s		
136458	183401		15 21 23.6	-20 23 16	-0.004	+0.01	7.9	1.6		M5 III	+294			S Lib, v
136753	64652		15 21 23.8	+31 22 03	-0.002	-0.01	5.9	1.6		M5 III	-1			S CrB, v
137000	29494		15 21 26.9	+52 20 40	-0.003	0.00	7.5	0.4	3.0	F2 V		80 s		
136922	45504		15 21 29.3	+46 21 25	0.000	0.00	7.9	1.1	0.2	K0 III		340 s		
137686	8227		15 21 29.9	+73 28 35	-0.030	+0.07	7.4	0.5	3.4	F5 V		63 s		
136347	206543		15 21 30.0	-38 13 09	-0.001	-0.04	6.48	-0.06		A0 p				
136544	140480		15 21 32.5	-6 49 35	-0.013	-0.03	7.6	0.5	4.0	F8 V		53 s		
136300	225692		15 21 34.3	-44 56 19	-0.001	-0.03	7.2	0.1	0.4	B9.5 V		190 s		
136754	83785		15 21 34.6	+24 20 36	-0.002	-0.01	7.2	0.1	0.6	A0 V	-15	210 s		
136334	225695		15 21 35.2	-40 42 59	-0.001	-0.04	6.20	0.07		A2				
136695	101503		15 21 39.5	+14 18 53	0.000	0.00	7.5	1.6		M5 III	+12			S Ser, v
136711	101504		15 21 43.7	+18 26 22	-0.002	+0.02	7.75	1.19	-1.3	K3 II-III	-76	630 s		
136352	225697	ν^2 Lup	15 21 48.1	-48 19 03	-0.161	-0.27	5.65	0.65	4.7	G2 V	-69	16 mx		
136422	206552	φ^1 Lup	15 21 48.3	-36 15 41	-0.007	-0.09	3.56	1.54	-0.3	K5 III	-29	56 s		m
136849	64656	50 Boo	15 21 48.5	+32 56 01	-0.004	+0.01	5.37	-0.07	0.2	B9 V	+22	110 s		
136336	225698		15 21 51.0	-45 44 19	-0.005	-0.05	7.9	1.4	-0.1	K2 III		200 mx		
136755	101506		15 21 54.8	+16 30 08	+0.001	0.00	7.60	0.5	3.4	F5 V		69 s	9609	m

HD	SAO	Star Name	α 2000	δ 2000	μ(α)	μ(δ)	V	B−V	M_V	Spec	RV	d(pc)	ADS	Notes
			$15^h 21^m 59\overset{s}{.}6$	$-57°08'43''$	$-0\overset{s}{.}004$	$-0\overset{''}{.}01$								
136225	242445		15 21 59.6	−57 08 43	−0.004	−0.01	7.0	1.1	−0.1	K2 III		260 s		
136696	120960		15 22 00.0	+5 35 48	−0.002	+0.01	7.8	0.5	3.4	F5 V		74 s		
137180	29500		15 22 07.4	+55 41 25	−0.002	0.00	7.9	1.4	−0.3	K5 III		440 s		
136351	225703	ν¹ Lup	15 22 08.2	−47 55 40	−0.013	−0.14	5.00	0.50	3.4	dF5	−11	20 mn		
136501	206560		15 22 08.5	−32 11 22	−0.005	−0.01	7.4	1.3	−0.5	M4 III		300 mx		
136482	206559		15 22 11.1	−37 38 08	−0.002	−0.03	6.8	−0.1	0.4	B9.5 V		190 s		
136863	83792		15 22 13.3	+25 14 39	0.000	−0.01	7.9	0.1	0.6	A0 V		280 s		
136301	242452		15 22 14.0	−53 36 17	−0.010	−0.09	7.3	0.4	3.4	F5 V		61 s		
136316	242453		15 22 17.8	−53 14 15	−0.003	−0.07	7.62	1.21	0.4	G2 III		140 s		
136239	242448		15 22 20.1	−59 08 50	+0.001	−0.01	7.80	0.90	−6.8	B2 Ia		2000 s		
136603	183408		15 22 21.6	−25 27 37	−0.002	−0.04	7.6	1.5	−0.1	K2 III		230 s		
136831	101508	7 Ser	15 22 23.2	+12 34 02	0.000	−0.01	6.28	−0.02	0.4	B9.5 V	+8	140 s		
136550	206563		15 22 24.8	−34 42 31	0.000	−0.03	7.8	1.4	−0.1	K2 III		280 s		
136901	83795		15 22 25.4	+25 37 27	+0.003	−0.01	7.21	1.25	0.0	K1 III	−12	230 s		m
136061	253125		15 22 25.4	−66 35 57	−0.010	−0.04	7.92	0.70		G5				
136130	253127		15 22 26.3	−64 07 40	−0.002	+0.03	7.9	0.8	3.2	G5 IV		85 s		
136549	206570		15 22 28.4	−34 09 29	−0.001	−0.01	7.70	0.1	0.6	A0 V		260 s		m
135719	257268		15 22 29.2	−74 23 58	−0.003	−0.02	6.6	0.0	0.6	A0 V		150 s		
137002	64661		15 22 32.0	+34 13 45	−0.010	+0.06	7.3	1.1	0.2	K0 III		150 mx		
136093	253126		15 22 32.0	−66 12 43	−0.007	−0.04	7.06	0.12		A0				
136486	225711		15 22 33.0	−44 17 33	−0.001	−0.01	6.75	0.04	1.4	A2 V		120 s		
136466	225709		15 22 36.4	−47 55 17	−0.036	−0.27	7.69	0.70	5.2	G5 V		20 ts		m
136285	242454		15 22 36.5	−59 10 02	0.000	−0.08	7.60	0.4	2.6	F0 V		98 s		m
136713	159207		15 22 36.6	−10 39 40	−0.004	−0.20	7.9	1.1	0.2	K0 III		62 mx		
137389	16712		15 22 37.1	+62 02 50	+0.001	−0.04	5.98	−0.03		B9	−24			
137071	64667		15 22 37.2	+39 34 53	0.000	−0.02	5.50	1.59	−0.3	K5 III	−12	130 s		
137443	16713		15 22 38.2	+63 20 28	−0.004	−0.10	5.79	1.27	−0.1	K2 III	−46	100 mx		
136504	225712	ε Lup	15 22 40.7	−44 41 21	−0.002	−0.01	3.37	−0.18	−2.3	B3 IV	+4	140 s		m
136646	183417		15 22 45.2	−29 20 29	+0.003	+0.01	6.5	0.5	0.2	K0 III		190 s		
137003	83797		15 22 45.9	+28 03 25	0.000	0.00	7.38	0.99	0.3	G8 III	−12	250 s		
136923	101515		15 22 46.8	+18 55 07	−0.016	+0.06	7.20	0.76	0.2	K0 III	−11	84 mx		
136703	183423		15 22 57.0	−26 41 20	+0.002	0.00	6.52	1.12		K0				
137491	16717		15 23 00.6	+63 07 48	−0.009	+0.01	7.4	1.1	0.2	K0 III		270 s		
136888	120969		15 23 01.3	+0 27 51	−0.003	0.00	7.2	1.1	0.2	K0 III		250 s		
136801	159212		15 23 01.7	−15 08 03	0.000	+0.01	6.6	1.6	−0.4	M0 III	+14	260 s		
136664	206580	φ² Lup	15 23 09.2	−36 51 30	−0.002	−0.02	4.54	−0.15	−2.3	B3 IV	−1	220 s		
136505	242466		15 23 09.4	−51 44 10	−0.001	−0.02	6.9	0.9	0.2	K0 III		220 s		
136359	253136		15 23 10.3	−60 39 25	−0.011	−0.02	5.67	0.48	3.4	F5 V		27 s		
137107	64673	2 η CrB	15 23 12.2	+30 17 16	+0.010	−0.19	4.98	0.58	4.4	G0 V	−7	14 ts	9617	m
137147	64675		15 23 21.2	+31 48 43	−0.002	+0.02	7.94	−0.29		F0				
136415	242463	γ Cir	15 23 22.6	−59 19 14	−0.001	−0.04	4.51	0.19	−1.0	B5.5 V	−17	84 s		
136607	225721		15 23 24.7	−46 08 26	−0.005	−0.02	7.0	0.6	3.2	G5 IV		58 s		
136905	140499		15 23 25.8	−6 36 37	−0.001	−0.12	7.4	1.1	0.2	K0 III		120 mx		
136866	159218		15 23 30.7	−16 33 52	−0.001	−0.05	7.60	1.4	−1.3	K4 II-III	+35	600 s		
136470	—		15 23 31.0	−58 08 44			7.7	2.6	−0.3	K5 III		82 s		
136890	159219		15 23 31.4	−14 21 07	+0.012	−0.09	7.6	1.1	0.2	K0 III		71 mx		
137125	101522		15 23 41.5	+17 52 00	−0.001	−0.03	7.7	0.4	2.6	F0 V		110 s		
137006	140502	8 Ser	15 23 43.6	−1 01 21	+0.005	−0.03	6.12	0.25	2.6	F0 V	−2	51 s		
136537	242471		15 23 51.1	−57 19 50	+0.002	−0.01	6.8	1.3	0.2	K0 III		150 s		
136956	159227		15 23 52.1	−12 22 11	−0.003	−0.04	5.7	1.1	0.2	K0 III	−26	130 mx		
137127	120977		15 24 00.1	+8 54 23	+0.001	+0.03	7.5	1.4	−0.3	K5 III		370 s		
136894	183433		15 24 02.5	−27 18 18	+0.001	−0.04	7.50	0.9	3.2	G5 IV		72 s	9616	m
136981	159230		15 24 04.8	−14 18 36	−0.001	−0.02	7.6	1.1	0.2	K0 III		310 s		
137390	45521		15 24 04.9	+45 16 15	−0.003	0.00	6.01	1.20		K2	−10			
137052	159234	31 ε Lib	15 24 11.8	−10 19 20	−0.005	−0.16	4.94	0.44	3.4	F5 V	−10	23 ts		
—	64684		15 24 12.7	+38 09 04	0.000	−0.07	8.0	2.3						
137182	101527		15 24 14.5	+10 33 18	0.000	0.00	7.2	0.1	0.6	A0 V	−19	210 s		
137687	16722		15 24 17.8	+60 32 57	−0.050	+0.17	7.5	0.8	3.2	G5 IV		43 mx		
137368	64685		15 24 18.6	+38 11 16	−0.010	−0.11	7.6	1.1	0.2	K0 III		71 mx		
137184	120983		15 24 19.8	+8 28 54	+0.001	−0.01	8.0	0.5	4.0	F8 V		62 s		
136707	242482		15 24 20.0	−52 42 56	0.000	−0.01	7.78	1.58	0.2	K0 III		170 s		
137128	140510		15 24 21.0	−0 32 39	−0.002	+0.01	7.1	0.1	1.4	A2 V		140 s		
130339	258721		15 24 21.1	−87 42 44	−0.016	−0.05	7.7	1.1	0.2	K0 III		320 s		
137391	64686	51 μ¹ Boo	15 24 29.3	+37 22 38	−0.012	+0.08	4.31	0.31	2.6	F0 V	−10	18 s	9626	Alkalurops, m
137054	159237		15 24 29.9	−19 57 49	0.000	0.00	7.7	1.2	−0.1	K2 III		350 s		
137392	64687	51 μ² Boo	15 24 30.8	+37 20 51	−0.012	+0.09	6.50	0.59	5.9	dK0	−8	18 s	9626	m
136857	225745		15 24 31.1	−43 55 38	−0.002	0.00	7.9	−0.2	0.4	B9.5 V		320 s		
137588	29514		15 24 35.1	+54 12 48	+0.003	−0.04	7.50	1.1	0.2	K0 III		290 s	9628	m
—	45524		15 24 36.4	+43 18 40	−0.002	+0.02	7.9	1.0						
136739	242485		15 24 38.2	−52 51 13	−0.001	0.00	7.81	1.26	3.2	G5 IV		40 s		
137035	183442		15 24 39.1	−25 40 02	−0.004	−0.03	7.5	1.1	−0.1	K2 III		230 mx		
136961	206599		15 24 41.6	−35 55 01	+0.001	0.00	6.8	0.4	1.4	A2 V		74 s		
136933	206597	υ Lup	15 24 44.8	−39 42 36	−0.003	−0.05	5.40	0.1		A0 p	−8	24 mn		m
136289	257272		15 24 48.3	−71 19 26	−0.016	−0.08	7.4	1.1	0.2	K0 III		150 mx		
137294	101533		15 24 49.1	+10 59 34	+0.003	−0.13	7.0	0.4	3.0	F2 V		64 s		

HD	SAO	Star Name	α 2000	δ 2000	μ(α)	μ(δ)	V	B-V	M_v	Spec	RV	d(pc)	ADS	Notes
			h m s	° ' "	s	"								
--	45526		15 24 51.9	+44 21 54	+0.004	0.00	8.0	1.4						
137759	29520	12 ι Dra	15 24 55.6	+58 57 58	-0.002	+0.01	3.29	1.16	-0.1	K2 III	-11	48 s		Edasich, m
137295	120989		15 24 56.3	+7 47 29	0.000	-0.02	7.9	0.9	3.2	G5 IV		88 s		
136272	257273		15 24 57.4	-72 02 46	+0.003	-0.01	7.2	0.3	0.6	A0 V		140 s		
137115	183447		15 24 57.5	-22 02 36	-0.001	-0.01	7.7	1.3	-0.1	K2 III		320 s		
135961	257271		15 25 00.0	-76 31 28	-0.002	-0.01	7.3	0.0	0.4	B9.5 V		240 s		
137717	29519		15 25 02.6	+55 36 39	-0.002	+0.02	7.76	1.14	-0.1	K2 III		370 s		
137055	206608		15 25 02.6	-32 09 22	-0.012	-0.02	7.4	0.7	2.6	F0 V		52 s		
137241	140519		15 25 04.4	-6 14 45	-0.003	0.00	7.5	1.1	0.2	K0 III		280 s		
137015	206607		15 25 06.5	-38 10 10	-0.001	-0.03	7.03	0.09		A2				m
137240	140520		15 25 06.8	-5 55 08	+0.001	-0.01	7.6	1.6	-0.5	M2 III		420 s		
137629	45529		15 25 08.0	+47 03 39	-0.006	-0.03	6.79	0.60		G5	-17			
137131	183448		15 25 09.2	-26 45 10	+0.001	0.00	8.0	1.6	-0.5	M2 III		500 s		m
136937	225754		15 25 11.0	-47 13 03	-0.001	0.00	6.7	1.3	3.2	G5 IV		24 s		
137425	83811		15 25 13.4	+24 49 00	0.000	-0.10	8.0	0.4	3.0	F2 V		100 s		
137058	206616		15 25 20.1	-38 44 01	-0.004	-0.02	4.60	0.00	0.0	A0 IV	-3	80 s		m
137210	--		15 25 24.2	-21 55 21			7.80	1.1	0.2	K0 III		330 s	9625	m
136966	225757		15 25 24.4	-49 04 00	-0.003	0.00	8.0	1.1	0.2	K0 III		290 s		
136508	253148		15 25 26.0	-68 41 36	+0.005	-0.03	8.0	1.1	0.2	K0 III		330 mx		
137099	206621		15 25 29.3	-38 45 48	-0.004	-0.08	4.70	0.4	2.6	F0 V		26 s		m
137664	45533		15 25 29.7	+45 56 51	-0.003	+0.07	7.2	0.8	3.2	G5 IV		63 s		
137119	206625		15 25 29.9	-36 11 58	-0.003	-0.03	7.6	0.1	0.6	A0 V		220 s		
137718	29521		15 25 32.0	+50 35 37	-0.003	-0.02	7.7	1.4	-0.3	K5 III		390 s		
136945	242503		15 25 35.6	-51 20 28	-0.004	0.00	7.2	1.2	3.2	G5 IV		35 s		
137039	225763		15 25 37.9	-47 05 59	-0.007	-0.06	8.0	0.9	3.2	G5 IV		78 s		
137212	183456		15 25 40.5	-28 05 43	-0.002	0.00	7.9	1.1	3.2	G5 IV		56 s		
137373	140526		15 25 44.3	-1 15 17	-0.006	-0.03	7.1	0.4	3.0	F2 V		65 s		
136947	242504		15 25 44.8	-53 03 27	+0.004	+0.01	6.71	0.29	2.6	F0 V		65 s		
136899	242500		15 25 44.9	-57 00 49	-0.002	+0.01	7.4	0.3	0.0	B8.5 V		170 s		
137471	101545	9 τ¹ Ser	15 25 47.3	+15 25 40	-0.001	-0.01	5.17	1.66	-0.5	M1 III	-20	120 s		
137318	159253		15 25 52.0	-18 30 58	-0.001	-0.01	7.6	0.1	1.4	A2 V		180 s		
137510	101548		15 25 53.2	+19 28 51	-0.004	-0.01	6.27	0.60	4.4	G0 V	-3	23 s		
136969	242505		15 25 53.2	-54 19 15	0.000	-0.02	7.3	0.9	0.2	K0 III		260 s		
137719	45538		15 25 58.9	+44 18 08	+0.001	0.00	7.40	1.4	-0.3	K5 III	-14	350 s		
137630	64699		15 25 59.6	+32 28 28	-0.001	-0.01	7.0	1.1	0.2	K0 III		230 s		
137230	206638		15 26 03.7	-36 28 23	-0.001	0.00	7.6	1.2	0.2	K0 III		230 s		
137339	183466		15 26 05.3	-21 22 49	-0.003	+0.01	7.1	0.9	0.2	K0 III		240 s		
137193	206637		15 26 06.7	-39 53 20	-0.001	-0.01	7.3	0.4	0.6	A0 V		120 s		
137531	101553		15 26 12.7	+12 56 27	-0.001	0.00	8.0	0.9	3.2	G5 IV		93 s		
136672	253155		15 26 14.7	-68 18 33	+0.023	+0.01	5.89	1.02	0.2	K0 III		140 s		m
137157	225770		15 26 16.2	-48 38 19	-0.006	-0.03	8.0	0.7	3.4	F5 V		82 s		
137261	206647		15 26 17.1	-36 35 53	0.000	-0.04	6.8	0.9	0.2	K0 III		210 s		
137704	64701		15 26 17.3	+34 20 09	-0.009	+0.05	5.46	1.40	-0.3	K4 III	-48	140 s		
137569	101555		15 26 20.7	+14 41 37	+0.001	-0.01	7.86	0.24	-2.2	B5 III	-48	610 s		
137805	45541		15 26 26.7	+44 00 11	+0.011	-0.09	7.60	0.8	3.2	G5 IV		76 s	9639	m
137340	206652		15 26 27.0	-31 19 15	-0.001	-0.05	7.5	0.8	3.4	F5 V		38 s		
137895	29526		15 26 27.9	+52 03 11	-0.004	+0.04	7.0	1.1	-0.1	K2 III		260 s		
--	8253		15 26 29.0	+76 54 15	-0.016	+0.10	8.0	2.1						
137688	83821		15 26 30.0	+28 07 39	+0.001	-0.03	7.45	1.35	-0.2	K3 III		310 s		
137570	101556		15 26 30.2	+10 02 09	-0.001	-0.02	7.0	1.6	-0.5	M2 III	-53	310 s		
137928	29527		15 26 32.4	+54 01 12	+0.005	-0.02	6.2	0.1	1.4	A2 V	-5	91 s		
137247	225782		15 26 41.6	-45 30 47	+0.002	-0.02	7.4	0.7	0.2	K0 III		280 s		
137870	45546		15 26 43.5	+45 36 15	-0.003	+0.06	7.2	0.5	4.0	F8 V		44 s		
137235	225781		15 26 43.6	-47 25 17	-0.001	-0.02	7.7	1.2	3.2	G5 IV		40 s		
137591	121003		15 26 45.4	+5 14 12	-0.003	-0.03	8.0	1.1	0.2	K0 III		360 s		
137350	206656		15 26 46.7	-35 46 51	-0.002	-0.03	7.6	0.1	0.6	A0 V		210 s		
136996	253165		15 26 46.8	-61 42 12	+0.003	-0.01	8.0	1.3	-0.1	K2 III		360 s		
137410	183473		15 26 49.2	-27 58 19	+0.001	+0.04	7.6	1.0	4.0	F8 V		27 s		
137065	242528		15 26 50.2	-59 57 01	+0.002	+0.01	8.0	-0.3	0.4	B9.5 V		330 s		
137463	159263		15 26 50.5	-20 00 19	+0.003	-0.01	7.1	0.8	4.0	F8 V		27 s		
137160	242529		15 26 51.5	-55 01 05	+0.001	-0.03	7.9	0.2	0.6	A0 V		200 s		
136999	253168		15 26 55.1	-62 45 57	0.000	-0.02	7.9	0.1	0.6	A0 V		280 s		
137323	225793		15 26 57.4	-44 06 27	0.000	-0.01	8.0	-0.2	1.4	A2 V		210 s		
--	64708		15 27 00.2	+36 33 49	-0.001	-0.05	7.8	3.0						
137312	225790		15 27 04.1	-48 27 21	0.000	-0.01	8.0	1.5	-0.5	M4 III		230 s		
137896	45549		15 27 08.9	+41 33 13	+0.001	+0.01	8.0	0.9	3.2	G5 IV		93 s		
141155	2566		15 27 09.7	+84 49 30	-0.001	-0.06	6.9	1.1	0.2	K0 III		220 s		
137722	101561		15 27 10.4	+13 22 46	-0.003	0.00	7.7	0.4	2.6	F0 V		110 s		
137313	225795		15 27 13.4	-49 27 19	-0.005	-0.06	7.4	0.5	1.4	A2 V		67 mx		
138003	29531		15 27 14.8	+50 37 08	-0.002	-0.02	7.7	0.0	0.4	B9.5 V		290 s		
--	83828		15 27 17.9	+23 27 12	-0.001	-0.03	7.9	1.9	-0.3	K7 III		270 s		
137723	121010		15 27 18.0	+9 41 59	+0.004	-0.11	7.99	0.55		G0			9643	m
137432	206660		15 27 18.1	-36 46 04	-0.001	-0.04	5.45	-0.15	-1.4	B4 V	+7	230 s		m
137343	225799		15 27 25.4	-49 47 03	-0.003	-0.02	7.5	0.6	0.2	K0 III		290 s		
136977	253172		15 27 27.3	-66 07 20	+0.003	-0.05	6.92	-0.01		B9				

HD	SAO	Star Name	α 2000	δ 2000	μ(α)	μ(δ)	V	B-V	M_v	Spec	RV	d(pc)	ADS	Notes
			$15^h27^m32.8^s$	$-64°31'54''$	-0.004^s	$-0.03''$								
137066	253174		15 27 32.8	-64 31 54	-0.004	-0.03	5.71	1.64	-0.3	gK5		130 s		
137666	140545		15 27 38.3	-8 56 55	-0.007	-0.01	7.7	1.1	0.2	K0 III		150 mx		
137853	83830		15 27 38.7	+25 06 05	0.000	-0.02	6.02	1.62	-0.3	K5 III	-7	160 s		
138004	45551		15 27 40.3	+42 52 53	-0.005	-0.25	7.60	0.8	3.2	G5 IV		48 mx		m
138074	29534		15 27 40.5	+50 18 54	+0.002	+0.01	7.8	1.1	-0.1	K2 III		390 s		
138245	16740		15 27 40.7	+62 16 32	+0.004	-0.03	6.4	0.1	0.6	A0 V	-5	140 s		
137650	159275		15 27 41.6	-14 57 15	-0.003	-0.01	7.3	1.1	-0.1	K2 III		310 s		
137164	253175		15 27 45.5	-63 01 16	-0.010	-0.06	8.0	0.8	3.2	G5 IV		89 s		
136135	257274		15 27 47.1	-79 18 20	-0.005	+0.05	7.2	1.1	-0.1	K2 III		220 mx		
137613	183485		15 27 48.2	-25 10 11	-0.001	-0.03	7.50	1.18		R	+55			
137909	83831	3 β CrB	15 27 49.7	+29 06 21	-0.014	+0.08	3.68	0.28		F0 p	-19	18 mn		Nusakan, m
138265	16741		15 27 51.3	+60 40 12	-0.004	0.00	5.90	1.44	-0.3	gK5	-47	170 s		
137476	225811		15 27 51.6	-40 39 34	-0.001	0.00	6.7	1.5	-0.5	M2 III		270 s		
137364	242553		15 27 53.0	-53 42 06	+0.001	-0.01	7.4	0.1	3.4	F5 V		64 s		
137595	206669		15 28 02.2	-33 32 42	-0.002	-0.03	7.50	0.04	-1.7	B3 V n		690 s		
137854	101569		15 28 04.6	+11 51 44	0.000	+0.04	7.0	0.4	2.6	F0 V		76 s		
137763	140550		15 28 09.5	-9 20 53	+0.005	-0.36	6.98	0.81	5.8	dK1	+2	16 t		m
137778	140552		15 28 12.2	-9 21 29	+0.006	-0.37	7.58	0.86	6.6	dK5	+7	16 t		m
137597	206673		15 28 13.2	-37 21 31	+0.003	-0.03	7.03	1.63	-0.5	M4 III		320 s		m
137779	159281		15 28 13.6	-11 26 56	+0.001	-0.01	7.7	0.1	0.6	A0 V		270 s		
137795	140553		15 28 13.7	-6 26 16	0.000	-0.05	7.9	0.9	3.2	G5 IV		88 s		
137744	159280	32 Lib	15 28 15.3	-16 42 59	+0.001	-0.04	5.8	1.4	-0.3	K5 III	-21	170 s		
137518	225814		15 28 17.6	-45 08 05	+0.003	0.00	7.81	0.12		B0.5 II				
137725	183495		15 28 18.6	-21 12 36	+0.003	-0.02	7.7	0.0	3.2	G5 IV		80 s		
129723	258720		15 28 19.0	-88 07 59	-0.176	-0.08	6.48	0.30	0.6	F0 III		110 mx		
137617	206677		15 28 21.7	-37 37 48	+0.003	-0.02	7.4	0.3	0.2	K0 III		280 s		
137831	140555		15 28 23.0	-6 00 12	0.000	-0.01	7.9	1.1	-0.1	K2 III		390 s		
137465	242569		15 28 27.1	-51 35 51	0.000	0.00	6.10	1.09		K0				m
137946	101571		15 28 28.8	+18 00 49	-0.004	-0.04	7.7	1.1	0.2	K0 III		160 mx		
138100	64721		15 28 31.2	+38 43 25	+0.003	0.00	6.67	0.36	2.6	F0 V	-17	62 s		
137884	121017		15 28 32.9	+0 36 26	-0.003	+0.02	7.0	0.1	1.7	A3 V		120 s		
137898	121020	10 Ser	15 28 38.1	+1 50 31	-0.006	-0.04	5.17	0.23		A5 n	-10	21 mn		
137812	159285		15 28 38.1	-15 42 08	-0.005	-0.02	6.8	0.9	3.2	G5 IV		52 s		
137845	140559		15 28 38.5	-8 44 51	-0.003	0.00	7.9	1.1	-0.1	K2 III		390 s		
138406	16743		15 28 39.0	+61 43 37	-0.003	+0.02	6.8	0.1	0.6	A0 V	-1	180 s		
137747	183500		15 28 41.2	-29 34 06	-0.002	-0.02	7.6	1.1	1.7	A3 V		41 s		
138057	83844		15 28 43.0	+27 08 18	-0.005	+0.04	7.6	0.5	4.0	F8 V		52 s		
138213	45562		15 28 44.3	+47 12 04	-0.002	-0.01	6.0	0.1	0.6	A0 V	-16	120 s		
137728	206688		15 28 45.8	-31 28 33	-0.002	-0.02	7.70	0.1	1.4	A2 V		180 s		m
138367	29542		15 28 51.7	+57 26 41	-0.034	+0.15	6.87	0.49	2.9	F6 IV-V	-32	26 mx	9672	m
138338	29540		15 28 56.7	+55 11 41	-0.002	+0.03	6.3	0.1	1.4	A2 V	-7	96 s		
137913	140562		15 28 58.0	-6 16 09	0.000	+0.01	8.0	0.1	0.6	A0 V		300 s		
137522	242583		15 28 59.4	-55 37 38	+0.001	0.00	7.9	0.1	0.0	B8.5 V		300 s		
138039	101577		15 29 08.0	+13 01 57	-0.001	0.00	7.1	0.9	3.2	G5 IV		60 s		
137859	183513		15 29 09.9	-24 27 05	0.000	-0.01	7.6	1.1	0.2	K0 III		270 s		
139777	2556		15 29 10.8	+80 26 55	-0.094	+0.11	6.13	0.71	3.6	G0 IV-V	-14	28 ts	9696	m
138085	101578		15 29 12.0	+16 23 44	-0.003	+0.06	6.9	0.9	1.8	G8 III-IV		89 s		
137676	225824		15 29 18.8	-49 57 11	-0.024	-0.09	7.69	0.76	5.2	G5 V		32 s		
137785	206695		15 29 19.1	-38 38 07	-0.004	-0.09	6.6	0.7	2.1	A5 V		40 s		
138524	16746		15 29 21.0	+62 05 59	-0.004	0.00	6.4	1.4	-0.3	K5 III	-40	220 s		
138302	45565		15 29 23.3	+47 42 46	+0.001	-0.02	6.80	0.1	0.6	A0 V		170 s	9673	m
139813	2558		15 29 23.8	+80 27 00	-0.086	+0.10	7.31	0.80	5.2	dG5	-29	22 s		
137709	225825		15 29 24.1	-46 43 58	-0.001	-0.01	5.24	1.74	-4.4	K5 Ib	-18	660 s		
137678	242603		15 29 28.8	-50 44 04	-0.002	-0.01	7.76	0.31	1.7	A3 V		120 s		
137679	242602		15 29 30.2	-51 23 19	-0.002	-0.01	7.80	1.1	-0.1	K2 III		360 s		
137799	225834		15 29 31.4	-40 31 11	-0.003	-0.02	7.40	0.9	3.2	G5 IV		69 s		m
137730	225830		15 29 31.4	-46 54 24	-0.002	-0.03	7.9	1.3	-0.3	K5 III		430 s		
137583	242593		15 29 32.9	-58 21 06	-0.002	-0.06	7.20	0.1	1.7	A3 V		100 mx		m
137582	242596		15 29 33.0	-57 14 50	-0.003	-0.02	7.9	0.3	1.7	A3 V		120 s		
--	64726		15 29 33.6	+36 04 37	-0.001	0.00	7.7	2.5						
137949	159292		15 29 34.6	-17 26 27	-0.005	+0.01	6.69	0.38		F0 p	-31			
--	--		15 29 36.1	+78 38 02			8.00							S UMi, v
138157	101580		15 29 44.4	+16 11 31	-0.002	+0.01	7.3	1.1	0.2	K0 III		270 s		
138174	101581		15 29 50.7	+18 30 00	-0.001	+0.01	7.1	1.1	0.2	K0 III		240 s		
--	64729		15 29 50.8	+34 44 09	-0.001	-0.03	8.0	1.4						
140363	2564		15 29 51.0	+82 15 49	+0.032	-0.02	6.9	0.4	2.6	F0 V		74 s		
137384	253183		15 29 51.5	-67 29 07	-0.001	-0.03	7.36	-0.09		B8				
139951	2562		15 29 53.3	+80 46 16	0.000	+0.02	6.9	1.1	0.2	K0 III		220 s		
138214	83850		15 29 54.0	+23 49 00	-0.002	-0.03	7.6	0.1	1.4	A2 V		170 s		
137787	225835		15 29 55.6	-49 20 24	0.000	-0.01	8.0	1.6	-0.5	M2 III		310 s		
137753	242614		15 29 58.0	-52 18 10	-0.003	-0.01	7.0	0.0	0.4	B9.5 V		200 s		
138232	83852		15 29 58.4	+25 30 29	0.000	-0.01	7.90	1.4	-0.3	K4 III		440 s	9675	m
138061	159298		15 29 59.7	-12 46 35	-0.002	+0.01	7.9	0.9	3.2	G5 IV		88 s		
138104	140569		15 30 04.8	-6 50 43	-0.001	-0.01	7.4	1.1	-0.1	gK2 III		320 s		
138339	45569		15 30 05.4	+39 54 32	+0.002	+0.01	7.7	0.5	4.0	F8 V		54 s		

HD	SAO	Star Name	α 2000	δ 2000	μ(α)	μ(δ)	V	B-V	M_v	Spec	RV	d(pc)	ADS	Notes
			h m s	° ′ ″	s	″								
137901	206703		15 30 07.0	-38 35 11	-0.004	-0.05	7.4	0.4	1.4	A2 V		65 mx		
138354	45570		15 30 07.1	+40 23 06	-0.004	+0.04	7.9	0.1	1.4	A2 V		200 s		
138421	45575		15 30 09.4	+46 23 09	-0.001	-0.01	7.7	1.1	-0.1	K2 III		370 s		
138266	83854		15 30 10.4	+23 45 20	-0.001	-0.05	7.5	0.1	0.6	A0 V		240 s		
137919	225846		15 30 21.0	-41 55 09	-0.003	-0.04	6.6	0.2	0.6	A0 V		120 s		
138341	64736		15 30 22.6	+31 17 10	-0.002	-0.01	6.30	0.19	2.1	A5 V	-4	66 s		
138216	101585		15 30 23.7	+12 00 01	-0.001	+0.01	7.2	1.1	0.2	K0 III		260 s		
138383	64741		15 30 27.8	+36 48 15	-0.005	+0.02	6.5	1.1	0.2	K0 III	+2	180 s		
137819	242628		15 30 29.7	-52 59 01	-0.005	-0.05	7.5	0.5	2.6	F0 V		72 s		
137979	206706		15 30 35.8	-39 54 15	-0.002	-0.01	8.0	1.4	0.2	K0 III		220 s		
138105	183533		15 30 36.2	-20 43 42	+0.001	-0.02	6.1	0.4	1.4	A2 V		51 s		
137990	206709		15 30 37.0	-37 41 28	+0.001	-0.08	8.0	-0.1	2.1	A5 V		94 mx		
138137	159307	34 Lib	15 30 40.3	-16 36 34	+0.001	0.00	5.8	1.1	0.2	K0 III	-2	130 s		
138123	183534		15 30 42.9	-21 52 43	-0.001	+0.01	7.9	0.2	4.4	G0 V		51 s		
--	225852		15 30 45.1	-41 39 09	+0.002	+0.04	7.8	0.7	4.0	F8 V		43 s		
137992	--		15 30 45.5	-41 39 49			7.9	0.6	4.0	F8 V		57 s		
138422	64745		15 30 46.1	+34 27 57	-0.001	0.00	6.7	0.1	0.6	A0 V		170 s		
137626	--		15 30 47.6	-65 36 00			8.0	0.3	3.2	G5 IV		90 s		
137866	242636		15 30 48.1	-53 22 27	-0.002	-0.02	6.8	0.3	0.2	K0 III		210 s		
137957	225851		15 30 48.3	-45 25 28	-0.001	-0.02	7.8	-0.1	0.6	A0 V		270 s		
137366	257284		15 30 49.2	-71 39 15	0.000	-0.02	6.7	-0.2	0.0	B8.5 V		220 s		
139152	8266		15 30 49.6	+72 56 14	-0.002	+0.01	7.80	1.1	0.2	K0 III		330 s		m
137993	225854		15 30 51.8	-43 51 54	-0.015	+0.05	7.6	0.4	3.2	G5 IV		64 mx		
138290	121038		15 30 55.3	+8 34 45	+0.002	0.00	6.56	0.37		F2	-1			
138852	16754		15 30 55.6	+64 12 31	-0.018	+0.08	5.79	0.96	3.2	G5 IV	+10	35 mn		
--	45581		15 30 55.7	+42 54 39	+0.002	-0.03	7.3	1.6						
138481	45580	52 ν¹ Boo	15 30 55.7	+40 49 59	+0.001	-0.01	5.02	1.59	-0.3	K5 III	-9	100 s		
137958	225853		15 30 58.9	-47 55 13	+0.001	-0.02	7.4	0.3	4.4	G0 V		40 s		
137838	242638		15 31 00.9	-57 29 09	-0.002	+0.01	8.0	0.7	0.0	B8.5 V		150 s		
137903	242643		15 31 03.3	-52 38 54	+0.001	-0.06	7.5	0.5	1.4	A2 V		86 s		
138089	206718		15 31 03.3	-35 38 23	-0.005	-0.03	7.9	0.0	1.7	A3 V		76 mx		
138187	159310		15 31 05.0	-15 50 43	0.000	0.00	7.7	0.2	2.1	A5 V		130 s		
138175	183537		15 31 06.8	-23 39 26	-0.005	-0.08	7.3	1.1	0.2	K0 III		120 mx		
136813	257280		15 31 10.6	-78 45 07	-0.001	-0.04	7.6	0.4	3.0	F2 V		83 s		
137700	--		15 31 11.2	-65 31 58			7.9	-0.4	0.4	B9.5 V		310 s		
137683	253192		15 31 16.8	-65 56 36	-0.002	-0.01	6.7	0.0	0.4	B9.5 V		180 s		
138138	206720		15 31 17.1	-33 49 10	-0.002	-0.02	7.70	0.1	1.4	A2 V		180 s		m
138253	--		15 31 19.1	-12 59 39			7.70	0.5	4.0	F8 V		55 s	9680	m
138525	64754		15 31 22.3	+36 36 59	0.000	-0.03	6.38	0.50	3.4	F5 V	-50	36 s		
139669	8274	15 θ UMi	15 31 24.7	+77 20 57	-0.016	+0.01	5.2	1.4	-0.3	K5 III	-25	130 s		
137509	257290		15 31 26.6	-71 03 45	-0.007	-0.03	7.0	-0.3	0.0	B8.5 V		110 mx		
137387	257289	κ¹ Aps	15 31 30.7	-73 23 23	+0.001	-0.02	5.49	-0.12	-2.3	B3 IV	+96	350 s		m
138441	101594		15 31 39.3	+14 15 33	-0.002	-0.03	7.9	0.9	3.2	G5 IV		88 s		
139115	16762		15 31 40.0	+69 09 33	-0.001	+0.03	8.0	1.1	-0.1	K2 III		410 s		
138268	159316		15 31 42.5	-20 09 54	-0.004	-0.05	6.22	0.22	2.6	F0 V	-40	53 s	9681	m
138369	121042		15 31 42.8	+0 53 00	-0.001	-0.09	7.7	0.6	4.4	G0 V		47 s		
138236	183540		15 31 43.7	-28 24 26	-0.002	-0.04	7.6	1.5	-0.1	K2 III		230 s		
138594	64758		15 31 44.2	+39 40 44	+0.004	+0.06	7.8	1.1	0.2	K0 III		270 mx		
138482	83866		15 31 45.3	+20 22 38	+0.001	-0.02	7.9	0.4	3.0	F2 V		94 s		
138629	45590	53 ν² Boo	15 31 46.9	+40 53 58	-0.002	-0.01	5.02	0.07		A2 n	-16	21 mn	9688	m
138483	101597		15 31 48.4	+16 14 03	-0.001	-0.01	7.6	0.1	0.6	A0 V		260 s		
--	29560		15 31 49.0	+54 28 17	+0.002	-0.01	7.5	2.6	-0.5	M2 III		95 s		
138221	206734		15 31 50.1	-32 52 53	-0.001	-0.04	6.46	0.09		B8				
138093	242659		15 31 56.3	-50 56 34	0.000	-0.01	7.9	1.1	0.2	K0 III		320 s		
137714	--		15 32 01.3	-69 09 56			7.9	0.1	0.6	A0 V		280 s		
138541	83869		15 32 02.9	+22 34 13	+0.005	-0.18	7.7	1.1	0.2	K0 III		140 mx		
138204	206736		15 32 04.3	-38 37 23	-0.002	-0.09	6.25	0.20		A3				
138295	183546		15 32 07.6	-28 10 00	+0.001	-0.02	7.9	0.5	0.6	A0 V		140 s		
138527	101600	12 τ² Ser	15 32 09.6	+16 03 22	0.000	0.00	6.10	-0.05	-0.2	B8 V	-8	170 s		
138343	183547		15 32 10.2	-21 58 00	0.000	-0.02	7.0	0.8	0.4	B9.5 V		63 s		
138191	225876		15 32 13.8	-45 04 03	+0.001	-0.01	6.82	-0.02	0.0	B8.5 V		210 s		
138344	183548		15 32 14.9	-23 52 48	0.000	0.00	6.89	1.61		M5 III				
138190	225878		15 32 15.3	-44 24 26	0.000	-0.02	7.5	1.2	-0.1	K2 III		320 s		
138140	242663		15 32 16.5	-50 52 16	0.000	-0.02	7.4	1.0	-0.5	M4 III		390 s		
138425	159327		15 32 17.8	-10 26 21	+0.001	-0.07	6.7	0.6	4.4	G0 V		29 s		
138542	101601		15 32 19.4	+15 11 19	-0.002	+0.02	7.9	0.2	2.1	A5 V		150 s		
138557	101602		15 32 23.6	+15 37 25	-0.001	-0.03	8.0	0.9	3.2	G5 IV		93 s		
139797	8279		15 32 29.6	+76 46 27	+0.003	+0.09	8.00	0.6	4.4	G0 V		53 s	9712	m
138285	206742		15 32 30.8	-38 04 18	-0.001	-0.04	7.6	-0.4	0.6	A0 V		250 s		
138413	159330		15 32 36.6	-19 40 14	-0.002	-0.04	5.52	0.18		A m	-33	18 mn		
138109	242668		15 32 42.9	-57 24 27	+0.001	-0.01	7.70	0.4	2.6	F0 V		100 s		m
138573	101603		15 32 43.6	+10 58 04	0.000	+0.14	7.2	0.6	4.4	G0 V	-33	36 s		
138485	159335	35 ζ Lib	15 32 55.1	-16 51 10	-0.001	-0.02	5.50	-0.14	-2.5	B2 V nn	+11	400 s		
138749	64769	4 θ CrB	15 32 55.7	+31 21 32	-0.002	-0.01	4.14	-0.13	-1.1	B5 V	-25	110 s		m
138562	140596	11 Ser	15 32 57.9	-1 11 10	-0.001	-0.04	5.51	1.09	0.2	gG9	-16	91 s		

HD	SAO	Star Name	α 2000	δ 2000	μ(α)	μ(δ)	V	B-V	M_v	Spec	RV	d(pc)	ADS	Notes
138685	101610		15^{h}33^{m}04^s.7	+16°00′42″	+0^s.002	-0″.02	6.6	1.1	-0.1	K2 III		220 s		
138472	183563		15 33 07.5	-25 48 03	-0.003	-0.04	7.5	1.5	-0.1	K2 III		220 s		
138488	183565		15 33 09.0	-24 29 22	0.000	-0.04	7.50	0.1	1.7	A3 V		150 s	9689	m
138546	159341		15 33 15.1	-13 00 45	-0.001	+0.01	7.60	0.8	3.2	G5 IV		76 s		m
137701	257292		15 33 16.3	-73 32 39	+0.008	-0.03	7.9	1.1	0.2	K0 III		290 mx		
138701	101612		15 33 17.6	+13 05 48	-0.001	+0.02	7.9	0.9	3.2	G5 IV		88 s		
138644	121058		15 33 17.9	+3 19 21	-0.001	-0.04	7.6	1.1	0.2	K0 III		300 s		
138395	225898		15 33 24.2	-40 29 27	+0.004	-0.06	6.5	1.3	0.2	K0 III		110 s		
138153	253207		15 33 26.9	-61 16 23	-0.007	-0.03	6.8	0.1	1.4	A2 V		79 mx		
--	64777		15 33 29.0	+35 55 41	-0.005	-0.05	7.0	2.8						
138347	225891		15 33 29.2	-49 00 03	+0.001	-0.02	7.6	1.1	0.2	K0 III		270 s		
138362	225895		15 33 33.3	-47 32 16	-0.002	-0.07	7.10	0.4	2.6	F0 V		78 s		m
138670	140605		15 33 38.8	-3 18 17	-0.001	-0.02	8.0	1.1	-0.1	K2 III		410 s		
138750	121061		15 33 40.1	+9 21 12	-0.012	-0.08	7.9	0.5	4.0	F8 V		53 mx		
138820	83883		15 33 41.9	+23 26 38	-0.004	-0.01	7.8	1.1	0.2	K0 III		230 mx		
138492	206758		15 33 46.9	-37 38 24	-0.001	-0.03	7.9	1.2	-0.5	M2 III		470 s		
138803	101618		15 33 52.7	+17 08 15	-0.003	-0.04	6.5	0.4	2.6	F0 V	-21	59 s		
138549	206764		15 33 53.3	-31 01 08	-0.016	-0.02	7.97	0.71	5.2	G5 V		33 s		
138600	183572		15 33 54.6	-25 06 39	-0.004	-0.02	7.63	0.59	4.0	F8 V		48 s		
138505	206763		15 34 01.4	-40 03 58	-0.005	-0.03	5.8	1.5	-0.5	M2 III		180 s		
138686	159351		15 34 03.4	-14 13 48	+0.002	-0.03	7.2	0.5	4.0	F8 V		43 s		
138575	206768		15 34 10.0	-33 10 15	-0.003	-0.02	7.1	0.6	0.6	A0 V		90 s		
138716	140609	37 Lib	15 34 10.6	-10 03 53	+0.021	-0.24	4.62	1.01	0.0	K1 III	+48	57 mx		
138276	253215		15 34 14.3	-60 48 29	-0.003	-0.05	7.4	0.0	0.4	B9.5 V		250 s		
138450	242706		15 34 19.4	-50 51 34	-0.001	-0.02	7.8	1.6	-0.5	M2 III		470 s		
138115	--		15 34 19.9	-68 36 39			7.9	0.5	3.4	F5 V		78 s		
138763	140613		15 34 20.7	-5 41 41	-0.006	+0.01	6.51	0.58		G0				
138564	206769		15 34 20.8	-39 20 58	-0.002	-0.05	6.36	-0.03	0.4	B9.5 V		150 s		
138934	83890		15 34 21.2	+23 12 35	0.000	-0.03	7.7	1.1	-0.1	K2 III		360 s		
138363	242701		15 34 22.4	-57 04 22	-0.003	-0.02	7.30	1.1	0.2	K0 III		250 s		m
140084	8283		15 34 23.2	+76 27 01	+0.001	-0.01	7.6	0.1	1.4	A2 V	+21	170 s		
138506	--		15 34 24.6	-47 13 26			7.80	1.1	0.2	K0 III		310 s		m
138764	140614		15 34 26.5	-9 11 00	-0.001	-0.03	5.17	-0.09	-1.3	B6 IV	-5	190 s		
--	225913		15 34 30.0	-47 13 47	-0.003	-0.01	7.9	1.1	0.2	K0 III		320 s		
138808	140617		15 34 31.2	-2 56 30	-0.002	-0.01	8.0	0.5	4.0	F8 V		62 s		
138833	121067		15 34 31.6	+4 43 53	-0.003	+0.02	7.1	0.1	1.4	A2 V		140 s		
139586	16773		15 34 36.0	+67 48 20	+0.003	-0.15	6.9	0.8	3.2	G5 IV	-34	55 s		
138688	183580	36 Lib	15 34 37.2	-28 02 49	+0.001	-0.04	5.1	1.4	0.2	K0 III	+12	55 s		
138935	101622		15 34 37.6	+18 14 29	-0.002	0.00	7.9	1.4	-0.3	K5 III		430 s		
139006	83893	5 α CrB	15 34 41.2	+26 42 53	+0.009	-0.09	2.23	-0.02	0.6	A0 V	+2	24 ts		Alphecca, v
139133	45620		15 34 47.4	+44 04 02	0.000	+0.06	7.2	1.1	0.2	K0 III		250 s		
138917	101624	13 δ Ser	15 34 48.0	+10 32 21	-0.005	+0.01	3.80	0.26	1.7	F0 IV	-38	27 mn	9701	m
138870	121069		15 34 49.5	-0 08 47	0.000	-0.01	7.6	1.1	0.2	K0 III		310 s		
138170	253214		15 34 50.3	-68 53 21	0.000	-0.02	7.3	1.6	-0.5	M2 III		360 s		
138507	242715		15 34 51.3	-53 01 30	-0.002	-0.01	8.0	0.7	0.2	K0 III		360 s		
138810	159358		15 34 59.4	-17 08 19	-0.001	-0.02	7.0	1.1	0.2	K0 III		230 s		
138936	121071		15 35 04.5	+1 40 08	-0.005	-0.04	6.56	0.28		A0	-20			
138690	225938	γ Lup	15 35 08.4	-41 10 00	-0.001	-0.03	2.78	-0.20	-1.7	B3 V	+6	79 s		m
139193	45623		15 35 09.8	+43 10 03	-0.005	+0.04	6.8	0.9	3.2	G5 IV		53 s		
139283	29579		15 35 12.0	+49 59 41	+0.004	-0.07	7.0	1.1	-0.1	K2 III		270 s		
138779	183592		15 35 13.3	-28 28 26	-0.002	-0.05	8.0	1.0	4.0	F8 V		34 s		
139607	16778		15 35 14.0	+65 16 43	-0.001	+0.06	7.5	0.1	1.4	A2 V		130 mx		
139307	45625		15 35 14.7	+49 41 54	0.000	-0.01	7.5	1.4	-0.3	K5 III	-11	360 s		
139153	64790	6 μ CrB	15 35 14.9	+39 00 35	+0.002	0.00	5.11	1.64	-0.5	M2 III	-19	120 s		
138813	183594		15 35 16.0	-25 44 02	-0.002	-0.02	7.3	0.3	0.6	A0 V		150 s		
139357	29583		15 35 16.1	+53 55 18	-0.002	-0.01	5.97	1.18	0.2	K0 III	-10	110 s		
138319	--		15 35 16.2	-66 15 31			7.8	0.0	0.0	B8.5 V		350 s		
138886	159363		15 35 18.9	-12 30 26	+0.003	-0.04	7.8	0.5	4.0	F8 V		57 s		
138622	225934		15 35 19.3	-49 40 11	+0.001	0.00	6.9	1.1	-1.0	B5.5 V		74 s		
139455	29585		15 35 20.2	+59 40 44	-0.006	+0.03	7.9	1.4	-0.3	K5 III		340 mx		
138791	183596		15 35 20.4	-29 03 01	-0.003	-0.03	7.6	0.5	0.6	A0 V		130 s		
138519	242724		15 35 26.4	-57 25 00	-0.003	+0.01	7.5	0.6	0.0	B8.5 V		130 s		
138741	225943		15 35 28.1	-41 30 22	+0.004	-0.11	7.6	0.6	3.4	F5 V		57 s		
--	29587		15 35 31.1	+59 05 51	-0.003	+0.02	8.0	2.3	-0.3	K7 III		150 s		
138905	159370	38 γ Lib	15 35 31.5	-14 47 23	+0.004	0.00	3.91	1.01	1.8	G8 III-IV	-28	23 s	9704	m
139074	101631	15 τ³ Ser	15 35 33.1	+17 39 20	-0.005	-0.02	6.12	0.94	0.2	K0 III	-22	150 s		
--	64795		15 35 38.2	+35 55 50	0.000	-0.05	8.0	2.2						
138452	253217		15 35 38.5	-63 03 46	-0.002	-0.04	7.9	1.1	0.2	K0 III		330 s		
139044	121081		15 35 48.8	+1 13 39	-0.004	+0.01	6.7	0.4	2.6	F0 V		67 s		
139284	64797		15 35 49.2	+38 22 26	+0.002	-0.01	6.4	1.1	-0.1	K2 III	+3	200 s		
139086	101633		15 35 51.1	+12 54 59	-0.003	+0.02	8.00	0.5	4.0	F8 V		63 s	9708	m
139087	101634		15 35 53.1	+11 15 56	-0.003	-0.01	6.07	1.09	3.2	G5 IV	-26	24 s		
138769	225950		15 35 53.1	-44 57 31	-0.002	-0.03	4.54	-0.18	-2.3	B3 IV	+8	230 s		m
138742	225949		15 35 56.4	-47 23 18	0.000	0.00	7.7	0.8	-0.3	K5 III		390 s		
139323	64799		15 35 56.4	+39 49 52	-0.039	+0.06	7.90	1.1	0.2	K0 III	-72		9716	m

HD	SAO	Star Name	α 2000	δ 2000	μ(α)	μ(δ)	V	B-V	M_v	Spec	RV	d(pc)	ADS	Notes
139493	29588		15^h 35^m 57^s.0	+54° 37' 49"	−0^s.005	−0".01	5.7	0.1	0.6	A0 V	−19	35 mx		
137607	257293		15 35 57.3	−79 05 07	−0.001	+0.02	7.1	0.1	0.6	A0 V		200 s		
138743	225948		15 35 57.5	−49 30 29	+0.002	−0.01	6.3	1.6	−0.5	M4 III		220 s		R Nor, m,v
139057	121082		15 36 00.7	+0 04 01	+0.002	−0.02	8.0	0.5	3.4	F5 V		82 s		
139341	64800		15 36 02.2	+39 48 08	−0.039	+0.04	6.77	0.91	7.4	dK4	−70	15 ts	9716	m
139088	121084		15 36 03.4	+6 11 01	−0.003	0.00	7.4	0.1	1.4	A2 V		160 s		
139033	140630		15 36 03.6	−7 37 56	−0.002	−0.01	7.7	1.1	0.2	K0 III		310 s		
139478	29589		15 36 04.0	+52 04 11	−0.003	+0.08	6.74	0.32	2.6	F0 V	−16	66 s		
138263	257299		15 36 04.8	−71 13 05	−0.010	−0.06	8.0	1.1	−0.1	K2 III		210 mx		
138973	183608		15 36 08.3	−21 44 47	+0.004	−0.09	7.8	1.2	3.2	G5 IV		47 s		
138923	206795		15 36 11.3	−33 05 33	−0.001	−0.02	7.00	0.00	0.4	B9.5 V		210 s		m
138497	253221		15 36 11.9	−65 06 52	−0.006	−0.05	7.20	0.00	0.4	B9.5 V		84 mx		m
138816	225957		15 36 12.0	−44 23 48	−0.004	−0.05	5.43	1.50	−0.4	M0 III	−19	150 s		
139324	64802		15 36 12.5	+35 42 19	−0.015	−0.08	7.6	0.8	3.2	G5 IV		49 mx		
138196	257296		15 36 12.5	−72 50 21	−0.006	−0.01	7.6	0.0	0.4	B9.5 V		170 mx		
138785	225955		15 36 12.6	−48 42 48	0.000	0.00	7.9	1.2	−0.3	K5 III		430 s		
139059	140631		15 36 12.7	−6 23 19	−0.007	−0.18	7.9	0.9	3.2	G5 IV		47 mx		
138938	183606		15 36 12.9	−28 59 56	0.000	−0.01	7.0	1.9	−0.3	K5 III		160 s		
138498	253222		15 36 17.4	−65 36 48	−0.013	−0.07	6.51	0.35	0.6	F0 III		93 mx		m
138860	225962		15 36 26.7	−44 00 37	0.000	−0.01	7.5	−0.4	0.4	B9.5 V		270 s		
138694	242749		15 36 26.9	−57 00 53	−0.001	−0.03	7.7	0.0	0.4	B9.5 V		260 s		
139216	101641	17 τ⁴ Ser	15 36 28.1	+15 06 05	0.000	+0.01	6.7	1.6	−0.5	M4 III	−26	280 s		v
138499	253223		15 36 28.3	−66 32 46	+0.005	−0.01	6.7	2.6	−0.1	K2 III		34 s		
139225	101642	18 τ⁵ Ser	15 36 29.2	+16 07 08	+0.005	−0.01	5.93	0.29	2.6	F0 V	−2	46 s		
139195	101640	16 Ser	15 36 29.5	+10 00 36	+0.003	−0.13	5.26	0.95		K0 p	+8	17 mn		
139137	140643	14 Ser	15 36 33.6	−0 33 42	−0.002	−0.02	6.51	0.72	3.4	dF5	−23	26 s		
139061	159379		15 36 35.7	−17 00 54	−0.005	−0.06	7.7	0.5	3.4	F5 V		73 s		
139268	101648		15 36 42.3	+14 55 31	−0.004	−0.01	6.9	0.1	0.6	A0 V	−8	70 mx		
138538	253226	ε TrA	15 36 43.1	−66 19 02	+0.005	−0.07	4.11	1.17	0.2	K0 III	−16	44 s		m
138940	225972		15 36 43.9	−41 03 36	−0.001	0.00	7.6	−0.2	0.6	A0 V		260 s		
139245	101646		15 36 44.7	+11 36 04	0.000	−0.06	8.0	1.1	0.2	K0 III		290 mx		
141218	2576		15 36 48.3	+80 36 52	−0.019	+0.05	7.0	0.9	3.2	G5 IV		57 s		
139178	140648		15 36 52.7	−6 01 36	−0.001	−0.02	7.20	1.1	−0.1	K2 III		290 s		m
138826	242757		15 36 52.9	−53 43 08	+0.002	−0.01	8.00	0.2	2.1	A5 V		150 s		m
139389	64808		15 36 53.3	+29 59 28	+0.007	−0.06	6.5	0.5	3.4	F5 V	−13	42 s		
138994	206803		15 36 53.4	−38 10 49	−0.001	−0.01	7.9	−0.1	3.0	F2 V		94 s		
139118	159382		15 36 54.1	−15 30 51	−0.002	−0.04	8.0	1.1	−0.1	K2 III		430 s		
138908	225973		15 36 55.7	−46 09 03	0.000	−0.01	7.8	0.5	2.6	F0 V		79 s		
139063	183619	39 υ Lib	15 37 01.4	−28 08 06	−0.001	0.00	3.58	1.38	−0.3	K5 III	−25	39 ts	9705	m
138996	225982		15 37 01.6	−40 38 33	+0.001	−0.02	7.7	0.8	0.2	K0 III		320 s		
139094	183622		15 37 06.8	−26 29 33	−0.001	−0.01	7.38	0.08	−0.6	B8 IV		320 s		
137852	257295		15 37 10.5	−78 52 39	−0.022	−0.04	7.7	0.4	3.0	F2 V		87 s		
—	225989		15 37 10.8	−41 01 04	+0.001	+0.01	7.3	1.1	0.2	K0 III		250 s		
138975	225984		15 37 10.9	−44 21 51	−0.001	0.00	7.9	−0.2	1.4	A2 V		200 s		
139095	206810		15 37 16.7	−32 03 27	−0.005	−0.04	7.9	0.1	2.6	F0 V		120 s		
138878	242763		15 37 18.1	−54 28 08	−0.002	0.00	7.96	0.04	0.0	B8.5 V		340 s		
137316	—		15 37 22.8	−82 08 59			8.0	1.6	−0.5	M2 III		490 s		
139375	101651		15 37 23.0	+12 18 04	−0.009	+0.01	7.1	0.4	2.6	F0 V		78 s		
138895	242767		15 37 25.8	−54 21 53	−0.005	−0.06	7.79	0.91	3.2	G5 IV		68 s		m
139080	206811		15 37 28.3	−37 02 33	−0.002	−0.01	8.0	0.5	2.6	F0 V		91 s		
139160	183631		15 37 28.4	−26 16 48	−0.001	−0.03	6.19	0.01	−0.2	B8 V		180 s		
139308	140659		15 37 28.8	−0 53 05	−0.001	0.00	7.78	1.28	−0.1	K2 III	−25	320 s		
139202	183632		15 37 31.2	−22 06 57	+0.001	+0.03	7.0	0.3	0.6	A0 V		120 s		
139778	29597		15 37 31.9	+54 30 32	−0.005	−0.02	5.87	1.07	0.2	K0 III	−23	120 s		
139422	101654		15 37 33.2	+16 31 45	−0.003	−0.05	7.7	1.1	0.2	K0 III		310 s		
139391	101653		15 37 34.0	+11 09 04	−0.003	−0.02	7.6	0.9	3.2	G5 IV		76 s		
138773	253236		15 37 34.3	−62 40 53	0.000	−0.01	7.6	0.1	0.6	A0 V		240 s		
139081	206812		15 37 34.6	−39 05 24	−0.001	−0.03	7.4	0.8	0.2	K0 III		280 s		
138172	—		15 37 36.2	−76 41 32			8.0	1.1	−0.1	K2 III		420 s		
140227	16794		15 37 39.0	+69 17 00	−0.010	+0.05	5.62	1.35	0.2	K0 III	−29	75 s		
—	64817		15 37 41.3	+36 17 30	−0.007	−0.05	8.0	2.3						
140342	8300		15 37 45.5	+71 09 13	−0.007	+0.02	7.1	1.1	0.2	K0 III		240 s		
139106	206816		15 37 47.3	−39 16 02	−0.005	−0.05	7.9	0.9	0.2	K0 III		160 mx		
139254	183637		15 37 48.0	−23 08 30	−0.002	−0.08	5.7	1.2	0.2	K0 III	+7	100 s		
139641	45643	54 φ Boo	15 37 49.5	+40 21 12	+0.005	+0.06	5.24	0.88	3.2	G8 IV	−10	26 s		
139184	206821		15 37 51.0	−31 15 08	−0.007	−0.02	8.00	0.5	3.4	F5 V		83 s		m
139512	83917		15 37 53.7	+21 26 26	0.000	−0.04	7.9	0.4	2.6	F0 V		110 s		
139779	29598		15 37 55.1	+50 04 55	−0.003	−0.26	7.8	0.6	4.4	G0 V		48 s		
139125	206818		15 37 58.3	−40 06 32	+0.001	−0.01	7.7	1.6	−0.5	M4 III		440 s		
139378	140664		15 37 58.6	−5 25 19	−0.002	−0.03	8.0	1.4	−0.3	K5 III		450 s		
139457	101658		15 37 59.2	+10 14 23	+0.010	−0.37	7.10	0.50	3.7	F6 V	+38	47 s		
139327	159398		15 37 59.9	−14 31 49	−0.003	−0.02	7.3	0.4	2.6	F0 V		89 s		
138380	257302		15 37 59.9	−74 25 14	+0.002	0.00	8.0	1.1	0.2	K0 III		360 s		
139000	242780		15 38 01.9	−54 31 20	−0.003	−0.03	6.99	0.14	1.4	A2 V		120 s		
139144	206820		15 38 02.4	−39 51 32	+0.001	−0.01	8.0	1.4	0.2	K0 III		220 s		

HD	SAO	Star Name	α 2000	δ 2000	μ(α)	μ(δ)	V	B−V	M$_v$	Spec	RV	d(pc)	ADS	Notes
			h m s	° ′ ″	s	″								
139127	226004	ω Lup	15 38 03.1	−42 34 02	−0.013	+0.06	4.33	1.42	−0.4	M0 III	−7	69 mx		m
138910	242777		15 38 04.1	−59 52 35	0.000	0.00	8.0	0.0	−1.6	B3.5 V		640 s		
140023	16791		15 38 05.6	+61 25 13	−0.023	+0.03	7.9	0.7	4.4	G0 V		51 s		
139690	64820		15 38 09.0	+38 01 41	0.000	+0.01	7.8	1.1	−0.1	K2 III		380 s		
139231	206824		15 38 09.3	−34 16 59	−0.004	+0.01	8.0	2.1	−0.3	K5 III		190 s		
139691	64821		15 38 12.8	+36 14 47	−0.006	+0.03	7.0	0.5	3.4	F5 V	−22	52 s	9731	m
139290	183641		15 38 15.6	−28 12 24	0.000	−0.03	6.3	1.1	−0.1	K2 III		190 s		
139798	45650		15 38 16.1	+46 47 51	+0.008	−0.13	5.75	0.36	3.0	F2 V	−2	36 s		
139329	183646		15 38 16.2	−21 00 58	+0.005	−0.06	5.9	1.3	0.2	K0 III		85 s		
139608	83921		15 38 16.5	+24 31 18	−0.001	−0.02	7.10	1.6		M III	−24			
139291	183642		15 38 21.5	−29 35 47	−0.001	−0.04	8.0	0.5	2.6	F0 V		87 s		
139780	45651		15 38 22.7	+43 36 15	−0.001	0.00	6.8	0.1	1.4	A2 V	−4	120 s		
139621	83922		15 38 24.6	+22 40 29	−0.003	0.00	7.60	1.1	0.2	K0 III		300 s		
139068	242785		15 38 25.2	−53 18 16	+0.001	+0.01	7.4	0.2	0.4	B9.5 V		170 s		
139364	159402		15 38 25.2	−19 54 47	−0.015	−0.10	6.8	0.3	3.0	F2 V		44 mx		
139408	159406		15 38 29.0	−14 31 02	−0.004	−0.09	6.8	1.1	0.2	K0 III		110 mx		
139233	206826		15 38 32.4	−39 09 38	−0.003	−0.02	6.57	−0.07	0.4	B9.5 V		170 s		
140250	16799		15 38 33.2	+65 47 26	+0.004	−0.08	7.0	1.1	0.2	K0 III		230 s		
139906	29600		15 38 34.2	+50 25 23	0.000	−0.04	5.84	0.83	3.2	G5 IV	−14	31 s		
139234	226011		15 38 37.3	−43 08 44	0.000	−0.02	7.6	1.5	0.2	K0 III		170 s		
139992	29602		15 38 38.7	+55 53 57	−0.003	+0.02	7.2	0.0	0.4	B9.5 V		230 s		
139365	183649	40 τ Lib	15 38 39.3	−29 46 40	−0.001	−0.03	3.66	−0.17	−1.4	B4 V	+1	100 s		
139552	121105		15 38 39.7	+6 14 24	−0.002	+0.02	7.9	1.1	0.2	K0 III		340 s		
139460	140671		15 38 39.9	−8 47 40	+0.001	−0.02	5.79	0.50	3.8	F7 V	+4	25 s		
139461	140672		15 38 40.0	−8 47 28	+0.002	−0.02	6.48	0.52	3.8	F7 V	0	34 s	9728	m
139409	159409		15 38 41.2	−17 39 53	−0.004	−0.03	7.2	1.1	0.2	K0 III		200 mx		
139514	140675		15 38 41.7	−1 47 14	+0.001	−0.05	7.9	0.9	3.2	G5 IV		88 s		
139206	226010		15 38 41.8	−45 42 31	+0.002	−0.02	6.97	−0.08	0.0	B8.5 V		250 s		
139271	206827		15 38 42.2	−39 07 41	+0.002	−0.01	6.04	0.21		A3				m
139724	64824		15 38 42.5	+31 33 06	0.000	−0.01	7.8	0.1	1.7	A3 V		160 s		
139609	101663		15 38 43.9	+12 15 13	−0.002	0.00	7.1	0.0	0.4	B9.5 V	+6	220 s		
139411	183654		15 38 44.4	−23 02 57	−0.002	0.00	7.2	1.7	−0.3	K5 III		250 s		
139761	64825		15 38 48.8	+34 40 30	0.000	−0.02	6.11	1.02	0.2	K0 III	+4	150 s		
139129	242793		15 38 49.4	−52 22 22	−0.004	−0.04	5.44	0.00	0.2	B9 V	−12	110 s		m
139610	101664		15 38 49.5	+10 36 57	+0.002	0.00	7.1	0.8	3.2	G5 IV		60 s		
139446	159411	41 Lib	15 38 54.4	−19 18 07	+0.006	−0.08	5.38	0.86	0.3	G8 III	+47	100 s		
140064	29607		15 38 54.5	+57 27 41	−0.003	−0.02	7.60	1.6	−0.5	M4 III	+1	420 s		m
139084	242791		15 38 57.5	−57 42 27	−0.006	−0.11	7.4	1.3	3.2	G5 IV		33 s		
139920	45661		15 38 58.7	+46 55 35	−0.001	−0.03	6.8	1.4	−0.3	K5 III		270 s		
143802	2603		15 38 58.8	+85 17 10	−0.058	+0.04	7.1	0.2	2.1	A5 V	−10	46 mx		
139622	121107		15 39 01.0	+3 28 04	−0.003	−0.04	7.6	1.1	−0.1	K2 III		340 s		
139590	140680		15 39 01.0	−0 18 42	−0.013	−0.10	7.51	0.55	4.0	F8 V	−28	44 mx		
−−	183657		15 39 01.3	−22 28 24	0.000	+0.01	7.7	1.4	−0.3	K5 III		390 s		
138212	257301		15 39 01.4	−78 14 07	−0.008	−0.06	7.1	0.1	1.4	A2 V		110 mx		
139486	−−		15 39 03.5	−19 43 47			7.64	0.03	0.4	B9.5 V		260 s		
139261	226015		15 39 06.2	−48 36 50	−0.001	−0.01	7.7	1.9	−0.5	M2 III		350 s		
138682	257304		15 39 07.0	−72 17 52	0.000	−0.02	7.60	0.1	0.6	A0 V		250 s		m
140117	29609		15 39 09.4	+57 55 28	−0.001	+0.01	6.45	1.09	0.2	K0 III	−8	150 s		
138289	257303		15 39 18.6	−77 55 04	−0.022	−0.13	6.18	1.20	−2.2	K2 II		42 mx		
139518	183665		15 39 21.2	−23 09 01	−0.003	−0.03	6.2	0.4	0.6	A0 V		71 s		
139891	64833	7 ζ¹ CrB	15 39 22.1	+36 38 12	−0.001	−0.02	6.0	−0.1		B6 V	−19	130 s	9737	m
139892	64834	7 ζ² CrB	15 39 22.6	+36 38 09	−0.001	−0.01	5.0	0.0	−0.6	B7 V	−24	130 s	9737	m
139236	242799		15 39 34.8	−55 25 37	−0.001	−0.01	7.1	1.5	0.2	K0 III		120 s		
−−	8311		15 39 35.9	+75 16 21	+0.001	−0.03	8.0	1.5						
141652	2588		15 39 38.4	+79 59 00	−0.015	+0.04	7.30	0.4	3.0	F2 V	−32	72 s	9769	
139430	226030		15 39 38.8	−41 35 19	−0.001	−0.02	7.4	1.5	0.2	K0 III		210 s		
139799	101670		15 39 39.9	+16 18 50	−0.003	+0.04	6.9	0.1	1.4	A2 V		130 s		
139431	226033		15 39 45.6	−42 46 03	+0.001	−0.01	7.40	0.03	−1.1	B5 V e		460 s		
139521	206843	3 ψ¹ Lup	15 39 45.9	−34 24 42	+0.001	0.00	4.67	0.99	0.3	gG5	−23	63 s		
140453	16806		15 39 50.4	+64 39 43	+0.001	+0.01	7.3	1.1	−0.1	K2 III		300 s		
139503	206844		15 39 52.1	−36 26 07	0.000	+0.01	7.8	0.6	4.4	G0 V		46 s		
139432	226036		15 39 52.4	−43 37 06	+0.001	−0.02	7.59	0.06	−1.0	B5.5 V		150 s		
139628	159421		15 39 54.4	−19 46 06	+0.003	0.00	7.60	0.4	3.0	F2 V		83 s	9735	m
139211	242803		15 39 56.3	−59 54 30	−0.016	−0.22	5.95	0.48	4.0	F8 V		25 s		
139465	226041		15 40 09.1	−45 01 18	0.000	−0.09	7.40	1.26	−0.3	K4 III		180 mx		
139862	101673		15 40 10.3	+12 03 10	−0.001	−0.01	6.25	0.94	−0.9	G8 II−III	−21	270 s	9740	m
139312	242810		15 40 10.9	−56 54 52	−0.001	−0.03	7.1	1.1	0.2	K0 III		220 s		
138965	257308		15 40 11.6	−70 13 39	−0.006	−0.05	6.44	0.08	1.4	A2 V		94 mx		
139613	206858		15 40 15.3	−31 12 48	−0.005	−0.02	6.4	2.0	−0.1	K2 III		64 s		
139663	183686	42 Lib	15 40 16.8	−23 49 05	−0.001	−0.02	4.96	1.33	−0.3	K4 III	−22	110 s		
140138	45671		15 40 17.5	+44 49 51	−0.003	0.00	7.5	0.4	3.0	F2 V		80 s		
140208	45673		15 40 19.6	+48 03 17	−0.010	+0.08	7.8	0.5	3.4	F5 V		74 s		
138800	257307	κ² Aps	15 40 21.2	−73 26 48	−0.003	−0.03	5.65	−0.04		B8 s	−19			m
139298	242811		15 40 21.4	−59 08 11	−0.001	−0.01	8.0	1.4	0.2	K0 III		220 s		
140139	45672		15 40 21.5	+43 50 46	−0.001	+0.06	7.2	0.4	2.6	F0 V	−23	81 s		

HD	SAO	Star Name	α 2000	δ 2000	μ(α)	μ(δ)	V	B-V	M_v	Spec	RV	d(pc)	ADS	Notes
			h m s	° ′ ″	s	″								
139524	226048		15 40 23.4	-44 14 48	-0.001	-0.01	8.0	0.1	0.6	A0 V		240 s		
139525	226049		15 40 24.6	-44 17 19	-0.001	-0.01	6.98	-0.07	0.0	B8.5 V		250 s		
140086	64840		15 40 25.8	+37 31 00	-0.001	-0.03	7.0	1.4	-0.3	K5 III		290 s		
139212	253248		15 40 28.5	-63 50 16	-0.002	-0.01	7.70	0.1	0.6	A0 V		250 s		m
140101	64841		15 40 30.1	+37 01 01	-0.001	-0.01	7.0	0.1	0.6	A0 V	-13	190 s		
140512	16808		15 40 32.7	+61 52 50	-0.005	+0.02	7.8	0.2	2.1	A5 V		110 mx		
139597	226054		15 40 33.8	-41 56 39	-0.001	-0.02	7.6	0.5	0.6	A0 V		260 s		
139817	140692		15 40 35.9	-9 06 32	-0.010	-0.11	7.9	1.1	0.2	K0 III		63 mx		
140155	64844		15 40 45.2	+38 43 07	+0.001	0.00	7.8	1.4	-0.3	K5 III		410 s		SW CrB, q
140087	64842		15 40 45.2	+29 58 21	-0.003	+0.04	7.6	1.1	-0.1	K2 III		340 s		
139676	206863		15 40 47.1	-35 25 37	-0.007	-0.06	7.6	0.3	2.1	A5 V		43 mx		
139452	242822		15 40 49.3	-55 05 25	-0.002	-0.01	8.0	0.7	0.0	B8.5 V		150 s		
139598	226058		15 40 51.2	-44 37 24	-0.004	-0.07	7.8	-0.1	1.7	A3 V		59 mx		
139300	253249		15 40 56.3	-63 11 39	-0.001	0.00	6.9	0.0	0.0	B8.5 V		230 s		
139599	226059		15 40 58.2	-47 44 07	-0.003	-0.03	6.30	1.4	-0.3	K5 III		200 s		m
140027	101678	19 τ⁶ Ser	15 40 59.1	+16 01 29	+0.002	-0.02	6.01	0.90	0.3	G5 III	+3	140 s		
139004	257309		15 40 59.3	-71 44 43	-0.001	-0.03	7.6	0.0	0.4	B9.5 V		280 s		
140006	121126		15 41 04.2	+9 18 08	-0.003	-0.02	8.0	0.9	3.2	G5 IV		93 s		
139677	206867		15 41 06.8	-39 58 56	+0.002	-0.04	7.00	0.5	3.4	F5 V		53 s		m
139664	226064		15 41 11.2	-44 39 40	-0.017	-0.26	4.64	0.40	2.8	F5 IV-V	-7	22 ts		
139909	159438		15 41 15.8	-13 58 13	0.000	-0.01	6.8	0.0	0.4	B9.5 V		190 s		
139993	140700		15 41 24.6	-2 38 05	-0.002	-0.01	8.00	0.9	3.2	G5 IV		91 s		m
139339	253254		15 41 25.5	-65 04 05	-0.001	-0.03	6.9	0.4	2.6	F0 V		71 s		
140297	64848		15 41 26.1	+38 33 26	+0.001	-0.04	7.1	1.6	-0.5	M4 III	-50	330 s		RR CrB, v
140157	83944		15 41 28.7	+22 19 26	+0.001	-0.02	7.9	1.1	-0.1	K2 III		390 s		
139616	242835		15 41 30.7	-53 23 29	+0.001	-0.02	7.0	0.6	0.4	B9.5 V		92 s		
139867	183705		15 41 32.8	-29 17 58	-0.001	-0.01	7.28	1.72	-0.5	M4 III		320 s		
140159	101682	21 ι Ser	15 41 33.0	+19 40 13	-0.004	-0.05	4.52	0.04	1.2	A1 V	-17	36 mx	9744	m
--	64851		15 41 41.7	+35 59 51	-0.003	-0.01	8.0	2.0						
140029	140702		15 41 45.2	-6 26 39	-0.001	0.00	7.4	1.1	0.2	K0 III	-4	270 s		
140160	101683	20 χ Ser	15 41 47.3	+12 50 51	+0.003	-0.02	5.33	0.04		A0 p	+2	17 mn		v
139681	242841		15 41 47.6	-51 28 17	+0.002	-0.01	7.4	1.3	-0.1	K2 III		270 s		
140364	64853		15 41 48.5	+37 02 54	-0.002	+0.01	8.0	0.1	1.4	A2 V		210 s		
--	--		15 41 53.4	-30 08 03			7.70							m
139911	183710		15 41 54.2	-30 08 44	-0.003	+0.03	7.70	1.1	0.2	K0 III		320 s		m
140232	101686	22 τ⁷ Ser	15 41 54.6	+18 27 50	-0.005	+0.05	5.8	0.1	1.7	A3 V	-30	66 s		
138867	257310		15 41 54.8	-76 04 54	0.000	-0.03	5.95	-0.04		A0				
139997	159442	43 κ Lib	15 41 56.7	-19 40 44	-0.003	-0.11	4.74	1.57	-0.3	K5 III	-4	91 s		m
139977	183713		15 41 57.8	-24 18 02	-0.002	-0.05	7.5	0.6	2.1	A5 V		69 s		
140187	101685		15 41 59.2	+10 34 13	0.000	-0.01	7.9	1.1	0.2	K0 III		340 s		
139472	253258		15 41 59.6	-64 17 57	0.000	+0.02	7.8	0.0	0.4	B9.5 V		290 s		
140122	121129		15 42 01.4	+0 27 26	+0.002	-0.01	7.30	0.20		A p	+3		9747	m
139471	253259		15 42 01.7	-64 10 50	+0.001	-0.02	7.0	1.1	-0.1	K2 III		250 s		
140233	121131		15 42 13.0	+7 49 23	-0.004	+0.06	7.3	0.6	4.4	G0 V		38 s		
140254	121133		15 42 15.8	+9 39 03	-0.002	-0.02	7.7	0.5	3.4	F5 V		71 s		
139828	226085		15 42 25.5	-49 13 03	-0.002	-0.01	7.0	1.8	-0.1	K2 III		120 s		
140320	101692		15 42 26.8	+18 12 58	-0.002	+0.04	7.9	0.5	3.4	F5 V		80 s		
139979	206886		15 42 29.4	-35 45 39	-0.005	-0.06	7.9	0.9	2.6	F0 V		47 s		
139870	226088		15 42 33.7	-48 41 31	0.000	-0.02	8.0	1.3	-0.1	K2 III		270 s		
139871	226089		15 42 37.0	-49 29 23	-0.002	-0.03	6.0	1.7	0.2	K0 III		53 s		
140612	45692		15 42 37.9	+45 45 41	-0.004	+0.03	6.9	0.4	2.6	F0 V	-27	73 s		
139980	206887		15 42 38.2	-37 25 30	-0.004	-0.02	5.2	1.6	0.2	K0 III	-16	45 s		
140037	206892		15 42 40.0	-32 11 09	-0.001	0.00	7.4	0.4	0.0	B8.5 V		160 s		
140472	64862		15 42 40.6	+31 42 33	+0.001	+0.01	7.0	1.1	-0.1	K2 III	-38	260 s		
140008	206889	4 ψ² Lup	15 42 40.9	-34 42 37	-0.002	-0.03	4.75	-0.15	-0.9	B6 V	+8	140 s		
140436	83958	8 γ CrB	15 42 44.5	+26 17 44	-0.008	+0.04	3.84	0.00	-0.3	A0 III-IV	-11	64 s	9757	m
140164	159453		15 42 46.2	-16 00 48	-0.009	-0.04	7.30	0.5	2.3	F7 IV		68 mx	9751	m
140728	29628		15 42 50.7	+52 21 39	-0.008	+0.03	5.51	-0.07		A0 p	-16	30 mn		
140346	121144		15 42 51.6	+6 53 02	-0.003	-0.01	7.1	1.1	0.2	K0 III		240 s		
139960	226096		15 42 52.2	-44 14 11	+0.003	+0.02	7.4	1.1	-0.1	K2 III		310 s		
139793	242853		15 42 58.2	-58 06 53	+0.001	0.00	7.3	1.8	1.7	A3 V		13 s		
140167	183725		15 42 59.4	-24 15 55	0.000	-0.01	7.9	2.3	-0.3	K5 III		140 s		
139901	242860		15 42 59.9	-51 19 18	-0.001	0.00	8.0	0.0	0.0	B8.5 V		350 s		
142297	2597		15 43 01.1	+80 07 22	-0.009	+0.03	7.9	0.4	2.6	F0 V		110 s		
140283	159459		15 43 02.8	-10 56 02	-0.078	-0.32	7.20	0.49	2.4	A8 V	-171	75 s		
139984	226104		15 43 07.1	-46 04 40	-0.002	-0.03	7.3	1.1	-0.5	M2 III		370 s		
140269	159458		15 43 08.7	-13 03 22	+0.005	-0.09	6.8	0.7	4.4	G0 V		30 s		
140438	101699		15 43 10.5	+13 40 04	-0.001	-0.03	6.90	0.9	1.8	G8 III-IV	-10	110 s	9758	m
140514	83962		15 43 10.5	+21 42 04	-0.002	-0.03	7.90	0.7	3.0	G2 IV	+14	96 s		
140192	183731		15 43 11.4	-25 24 56	+0.001	-0.01	7.30	0.4	2.6	F0 V		87 s	9754	m
140571	64865		15 43 14.2	+31 23 03	+0.004	-0.14	7.4	0.5	4.0	F8 V		47 s		
139947	242866		15 43 15.6	-51 06 49	-0.001	-0.01	7.9	0.0	0.4	B9.5 V		290 s		
137333	258731	ρ Oct	15 43 16.8	-84 27 55	+0.095	+0.09	5.57	0.11		A2	-11			
140456	101700		15 43 22.1	+11 42 42	0.000	+0.01	7.1	1.1	-0.1	K2 III		280 s		
140301	159461		15 43 24.7	-15 02 36	-0.001	-0.10	6.4	1.1	0.2	K0 III	+21	140 mx		

HD	SAO	Star Name	α 2000	δ 2000	μ(α)	μ(δ)	V	B–V	M_v	Spec	RV	d(pc)	ADS	Notes
			$15^h 43^m 25.0^s$	$-33°24'21''$	-0.007^s	$-0.02''$								
140172	206904		15 43 25.0	-33 24 21	-0.007	-0.02	7.5	0.9	0.2	K0 III		170 mx		
140194	206906		15 43 26.4	-30 32 06	+0.001	-0.01	7.4	0.4	0.6	A0 V		140 s		
143173	2606		15 43 26.8	+82 56 50	+0.002	0.00	7.4	0.1	1.4	A2 V	-8	160 s		
139508	257312		15 43 31.6	-70 16 00	+0.001	-0.04	7.5	0.1	0.6	A0 V		200 s		
140370	140718		15 43 32.6	-10 07 24	-0.004	+0.08	7.8	0.5	3.4	F5 V		75 s		
139912	242868		15 43 32.6	-56 09 41	0.000	-0.01	7.35	1.14	0.2	K0 III		230 s		
140572	83964		15 43 33.0	+22 55 11	+0.003	-0.22	7.3	0.5	4.0	F8 V		45 s		
140002	242880		15 43 37.9	-51 37 47	+0.001	-0.02	6.70	1.1	-0.1	K2 III		220 s		m
139913	242871		15 43 40.1	-57 27 06	-0.001	0.00	7.4	0.7	1.4	A2 V		62 s		
140399	140722		15 43 44.9	-9 18 55	-0.001	-0.01	7.5	0.4	2.6	F0 V		96 s		
140489	121150		15 43 46.2	+2 26 24	0.000	-0.01	7.6	0.9	0.3	G8 III	-12	290 s		
141323	16830		15 43 46.7	+68 20 52	+0.002	-0.04	8.0	1.1	0.2	K0 III		360 s		
140197	206910		15 43 49.9	-39 27 41	-0.001	-0.03	7.70	0.1	0.6	A0 V		260 s		m
139915	253269		15 43 55.0	-60 17 14	-0.001	-0.01	6.49	1.05	0.3	gG5		140 s		
140387	159465		15 43 57.0	-16 52 09	0.000	+0.01	7.4	1.1	0.2	K0 III		220 s		
140716	64870	9 π CrB	15 43 59.2	+32 30 57	-0.003	-0.01	5.56	1.06	0.2	K0 III	-4	110 s		
140538	121152	23 ψ Ser	15 44 01.6	+2 30 54	-0.006	-0.15	5.88	0.68	5.2	dG5	+14	19 ts	9763	m
140772	64873		15 44 01.7	+39 34 26	-0.002	0.00	7.5	1.6	-0.5	M2 III		400 s		
142105	8328	16 ζ UMi	15 44 03.3	+77 47 40	+0.004	0.00	4.32	0.04	1.7	A3 V	-16	33 s		
140079	242888		15 44 03.4	-51 54 17	-0.001	-0.01	7.23	-0.02		B9				
140041	242884		15 44 03.7	-54 59 13	-0.001	-0.02	6.2	1.6		M5 III	-31			T Nor, v
140329	206919		15 44 04.1	-31 36 11	+0.001	-0.08	6.8	1.6	0.2	K0 III		86 s		
140417	159466	44 η Lib	15 44 04.3	-15 40 22	-0.002	-0.07	5.41	0.23		A m				
140965	29636		15 44 05.6	+52 54 21	-0.007	+0.06	7.7	1.1	-0.1	K2 III		200 mx		
140664	83971		15 44 05.9	+24 27 37	-0.002	0.00	7.2	0.1	1.4	A2 V		150 s		
140328	206920		15 44 06.9	-30 42 37	-0.002	+0.04	7.4	1.1	-0.1	K2 III		320 s		
140537	121154		15 44 09.1	+4 40 53	0.000	+0.01	8.0	0.1	1.4	A2 V		210 s		
140130	242892		15 44 10.2	-50 12 55	-0.002	-0.01	6.82	0.36	2.6	F0 V		65 s		m
140220	226126		15 44 10.8	-44 06 48	-0.001	+0.02	7.6	0.6	0.6	A0 V		79 s		
139903	253270		15 44 11.1	-62 09 35	+0.002	0.00	8.0	1.0	-0.1	K2 III		410 s		
140573	121157	24 α Ser	15 44 16.0	+6 25 32	+0.009	+0.04	2.65	1.17	-0.1	K2 III	+3	26 ts	9765	Unukalhai, m
140042	242889		15 44 16.1	-56 09 55	-0.001	-0.02	7.06	0.01	0.4	B9.5 V		220 s		
140043	242887		15 44 17.0	-57 06 45	-0.001	-0.02	7.7	0.7	0.0	B8.5 V		130 s		
140419	183745		15 44 17.5	-24 23 43	-0.004	0.00	7.8	-0.2	0.6	A0 V		110 mx		
141039	29638		15 44 21.6	+52 59 01	-0.028	+0.06	7.3	0.9	3.2	G5 IV	-34	34 mx		
140285	226132		15 44 22.5	-41 49 10	-0.003	-0.05	5.94	0.00		A0				
139658	257313		15 44 23.1	-70 17 37	+0.014	+0.03	8.0	0.7	4.4	G0 V		46 s		
141285	16831		15 44 27.0	+64 28 45	-0.006	+0.02	8.0	1.1	0.2	K0 III		360 s		
140700	101710		15 44 29.4	+16 30 37	+0.001	-0.01	7.40	1.4	-1.3	K5 II-III	-26	550 s		
140614	121159		15 44 30.3	+3 22 11	-0.004	-0.06	7.2	0.5	3.4	F5 V		59 s		
140667	101709		15 44 31.8	+11 15 59	-0.008	-0.04	7.5	0.5	4.0	F8 V		50 s		
142795	2604		15 44 36.1	+80 55 54	-0.002	-0.01	6.9	1.1	-0.1	K2 III		260 s		
140286	226136		15 44 36.2	-44 02 55	-0.002	0.00	7.6	1.5	0.2	K0 III		170 s		
140666	101711		15 44 36.5	+11 25 21	-0.002	-0.01	7.5	1.1	0.2	K0 III		280 s		
142123	8332		15 44 38.7	+77 09 00	-0.011	+0.02	7.98	0.99		G5				
140729	101712	26 τ⁸ Ser	15 44 42.0	+17 15 50	-0.002	0.00	5.90	0.00	0.6	A0 V	-6	110 s		
141173	29643		15 44 42.0	+58 26 01	-0.010	+0.05	7.4	0.4	2.6	F0 V		90 s		
141506	16839		15 44 45.5	+68 40 48	-0.009	+0.06	7.2	0.1	1.4	A2 V		95 mx		
140442	206928		15 44 46.8	-30 41 11	+0.001	-0.03	7.3	0.9	0.6	A0 V		63 s		
140176	242902		15 44 47.9	-55 06 04	-0.004	-0.05	7.7	0.4	2.6	F0 V		86 s		
140730	101713		15 44 48.1	+16 44 15	-0.005	-0.01	7.0	1.1	-0.1	K2 III		210 mx		
140813	83979		15 44 55.8	+20 52 51	+0.003	-0.03	7.6	0.1	1.4	A2 V		180 s		
140812	83980		15 44 56.0	+21 26 38	+0.005	+0.12	7.5	0.4	3.0	F2 V		79 s		
140833	83981		15 44 57.2	+24 42 30	-0.002	-0.03	7.5	0.1	1.7	A3 V		140 s		
140274	242915		15 44 59.3	-50 47 05	-0.007	0.00	6.80	0.1	1.4	A2 V		97 mx		m
--	--		15 44 59.7	-59 06 54			7.60							m
140111	242901		15 44 59.9	-59 07 24	-0.004	-0.01	7.60	0.1	1.4	A2 V		140 mx		m
140097	253279		15 45 05.1	-61 17 52	-0.015	-0.09	7.09	0.40		F2				
140544	183754		15 45 12.2	-28 31 54	-0.006	+0.02	8.00	1.1	0.2	K0 III		160 mx		m
140177	242911		15 45 12.5	-58 40 58	+0.003	0.00	7.70	0.1	1.4	A2 V		180 s		m
141174	29645		15 45 18.1	+52 40 38	-0.002	+0.01	7.9	1.1	0.2	K0 III		340 s		
140775	121170		15 45 23.4	+5 26 49	+0.002	-0.02	5.58	0.04	1.4	A2 V	-10	69 s		
140652	159480		15 45 23.7	-19 06 26	-0.001	-0.02	7.8	0.1	1.7	A3 V		140 s		
140018	--		15 45 24.6	-65 45 19			7.8	1.1	0.2	K0 III		320 s		
140354	242925		15 45 26.4	-52 34 00	0.000	+0.01	8.0	1.7	-0.3	K5 III		340 s		
140732	140734		15 45 27.5	-6 08 43	-0.002	-0.01	7.3	1.1	0.2	K0 III	-35	270 s		
142005	8331		15 45 28.8	+73 50 11	-0.003	+0.04	7.9	1.1	0.2	K0 III		340 s		
141243	29648		15 45 29.4	+55 15 04	+0.001	+0.02	7.9	1.4	-0.3	K5 III		430 s		
141068	45711		15 45 33.7	+40 10 38	+0.002	-0.02	7.6	0.1	1.4	A2 V		170 s		
140355	242928		15 45 36.4	-53 26 16	0.000	0.00	8.00	1.1	0.2	K0 III		340 s		m
139604	257315		15 45 36.4	-74 51 21	-0.013	-0.01	7.9	0.8	3.2	G5 IV		86 s		
140815	121173		15 45 39.6	+0 53 29	+0.003	+0.01	6.33	1.19		K0	+14			
139757	257317		15 45 42.7	-72 18 23	+0.005	-0.03	7.5	1.1	0.2	K0 III		290 s		
141336	29650		15 45 42.8	+57 15 46	-0.007	+0.06	7.9	0.5	3.4	F5 V		81 s		
140671	183765		15 45 42.9	-22 45 13	-0.001	-0.03	7.60	1.1	0.2	K0 III		300 s		m

HD	SAO	Star Name	α 2000	δ 2000	μ(α)	μ(δ)	V	B-V	M_v	Spec	RV	d(pc)	ADS	Notes
			$h\ m\ s$	$^{\circ}\ '\ ''$	s	$''$								
140636	183763		15 45 46.2	-30 02 29	-0.005	-0.03	7.0	1.1	0.2	K0 III		180 mx		
140672	183766		15 45 46.3	-25 29 43	-0.001	+0.01	7.9	1.1	-0.3	K5 III		430 s		
140520	226159		15 45 46.7	-43 47 23	0.000	0.00	8.0	0.1	0.4	B9.5 V		330 s		
141398	29651		15 45 47.7	+59 18 44	-0.001	-0.05	7.6	0.4	3.0	F2 V		83 s		
140687	183767		15 45 49.7	-24 43 00	-0.006	-0.06	7.5	0.7	0.2	K0 III	+2	120 mx		
140721	159481		15 45 50.4	-16 57 04	-0.002	-0.04	7.6	0.5	3.4	F5 V		65 s		
--	64885		15 45 51.8	+36 25 05	-0.005	0.00	7.8	2.0						
140873	140740	25 Ser	15 46 05.5	-1 48 16	-0.002	-0.04	5.40	-0.03		B8	-12			
--	183771		15 46 10.9	-28 04 25	-0.001	-0.05	6.40	0.7	4.4	G0 V		25 s		m
141003	101725	28 β Ser	15 46 11.2	+15 25 18	+0.005	-0.05	3.67	0.06	0.6	A2 IV	-1	37 ts	9778	m
140722	183772		15 46 12.7	-28 03 41	-0.005	-0.03	6.51	0.32	1.2	A5 IV		91 s	9775	m
141204	45718		15 46 13.6	+42 28 07	-0.002	+0.02	7.40	1.1	0.2	K0 III		280 s	9782	m
141244	45719		15 46 16.2	+45 01 17	-0.001	-0.05	7.9	1.1	0.2	K0 III		310 mx		
140406	242940		15 46 18.7	-57 48 45	-0.001	0.00	6.9	1.7	0.2	K0 III		87 s		
141024	101728		15 46 22.4	+13 43 15	-0.006	0.00	8.0	0.9	3.2	G5 IV		93 s		
141069	83994		15 46 23.2	+19 54 27	-0.004	+0.03	7.8	0.7	4.4	G0 V		49 s		
141040	101729		15 46 23.9	+15 31 35	-0.001	-0.01	6.8	0.1	1.7	A3 V	-48	110 s		
141186	64893		15 46 24.3	+36 26 46	-0.009	+0.06	8.00	0.5	3.4	F5 V		83 s		m
141004	121186	27 λ Ser	15 46 26.5	+7 21 12	-0.015	-0.07	4.43	0.60	4.4	G0 V	-66	11 ts		
141472	29655		15 46 34.7	+55 28 29	-0.015	+0.07	5.92	1.39	-0.1	K2 III	-4	120 s		
140019	257319		15 46 39.4	-70 42 03	0.000	-0.07	8.0	1.1	0.0	K1 III		220 mx		
141653	16848		15 46 39.8	+62 35 58	+0.005	-0.06	5.19	0.04		A2	-6	26 mn		
140619	226174		15 46 40.4	-48 44 13	0.000	-0.03	7.0	0.2	0.0	B8.5 V		160 s		
140657	226176		15 46 40.8	-46 42 00	-0.001	-0.03	8.0	0.4	3.2	G5 IV		93 s		
141070	121188		15 46 43.6	+9 47 10	+0.002	0.00	7.0	0.1	1.4	A2 V	-28	130 s		
140784	206962		15 46 44.1	-34 40 57	-0.002	-0.03	5.61	-0.11	-0.9	B6 V n	-5	200 s		m
140604	242952		15 46 44.7	-50 13 25	-0.004	-0.01	7.9	0.3	0.6	A0 V		140 mx		
140986	140751		15 46 45.4	-6 07 11	0.000	+0.02	6.24	1.17		K0				
140605	242951		15 46 47.4	-52 08 48	+0.001	-0.03	7.07	-0.04	-1.1	B5 V nn		400 s		
142006	8335		15 46 48.7	+70 22 27	-0.007	-0.03	7.8	0.5	4.0	F8 V		57 s		
140504	242947		15 46 50.9	-58 07 30	-0.002	-0.04	7.60	0.1	1.4	A2 V		170 s		m
141374	45723		15 46 51.5	+46 17 48	-0.006	+0.02	7.8	1.1	-0.1	K2 III		290 mx		
140919	159489		15 46 52.0	-17 06 33	-0.001	-0.03	7.7	1.1	0.2	K0 III		260 s		
141399	45726		15 46 53.7	+46 59 12	-0.011	+0.02	7.21	0.77		K0				
140360	--		15 46 53.7	-64 25 08			7.8	0.0	0.4	B9.5 V		290 s		
140704	226186		15 46 55.9	-44 14 39	+0.001	-0.01	8.00	0.1	0.6	A0 V		120 s		m
141987	16854		15 46 59.5	+69 28 44	0.000	+0.09	7.5	0.1	1.4	A2 V		96 mx		
140408	253290		15 47 00.2	-62 47 48	+0.014	+0.08	7.11	0.49		F8				
141127	101736		15 47 04.1	+11 31 40	-0.002	0.00	7.6	0.1	1.4	A2 V		170 s		
140817	206968		15 47 04.4	-35 30 37	-0.001	-0.03	7.20	0.00	0.4	B9.5 V		230 s		m
139919	--		15 47 04.7	-74 03 20			7.7	0.0	0.4	B9.5 V		290 s		
140840	206969		15 47 06.1	-35 31 04	-0.001	-0.02	6.80	0.00	0.4	B9.5 V		190 s		m
141268	64902		15 47 09.6	+30 33 56	-0.002	-0.03	7.7	0.1	0.6	A0 V		260 s		
140785	226198		15 47 13.0	-42 12 17	+0.002	-0.14	7.6	0.5	4.4	G0 V		44 s		
--	64905		15 47 14.5	+38 59 51	-0.001	0.00	8.0	2.3						
141042	140755		15 47 15.3	-8 27 09	-0.002	-0.01	7.7	1.1	0.2	K0 III		310 s		
141103	140761		15 47 17.0	-0 16 12	-0.017	-0.03	7.3	0.5	3.4	F5 V		46 mx		
141187	101739	31 υ Ser	15 47 17.2	+14 06 55	-0.004	+0.04	5.70	0.09	1.4	A2 V	-34	69 s		
140990	183790		15 47 20.1	-20 28 11	-0.007	-0.13	7.8	0.4	4.4	G0 V		47 s		
140841	226202		15 47 21.5	-41 01 35	-0.004	0.00	7.6	0.6	0.2	K0 III		300 s		
141454	45729		15 47 22.7	+44 35 52	+0.001	-0.02	7.4	1.1	0.2	K0 III		270 s		
140968	183789		15 47 23.3	-25 52 00	-0.001	+0.02	8.0	0.3	1.4	A2 V		140 s		
140861	226206		15 47 25.3	-40 11 38	-0.015	-0.04	6.42	0.88	1.8	G5 III-IV		71 mx		
--	101741		15 47 25.8	+18 50 57	+0.001	-0.03	8.00							m
141144	121191		15 47 26.8	+1 32 46	+0.001	0.00	6.6	1.1	0.2	K0 III		190 s		
140901	206976		15 47 28.8	-37 54 59	-0.036	-0.22	6.01	0.72	5.3	G6 V	-4	15 ts		m
141029	183793		15 47 31.3	-21 14 01	+0.001	-0.01	8.00	0.9	3.2	G5 IV		91 s	9784	m
141729	29671		15 47 31.7	+59 34 07	-0.006	-0.02	6.8	1.1	0.2	K0 III		210 s		
141675	29668		15 47 37.8	+55 22 36	+0.001	+0.01	5.80	0.2		A m	-2		9793	m
141128	140764		15 47 38.0	-9 28 49	-0.005	+0.02	6.9	0.5	3.4	F5 V		51 s		
141352	84005		15 47 40.7	+28 28 10	-0.003	+0.05	7.5	0.4	3.0	F2 V		78 s		
141525	45732		15 47 46.2	+45 44 00	-0.004	+0.02	6.8	1.1	-0.1	K2 III		240 s		
141207	121193		15 47 48.2	+4 17 53	-0.002	-0.03	7.2	1.1	0.2	K0 III	+2	250 s		
141208	121194		15 47 50.2	+3 32 34	-0.001	0.00	8.0	1.1	0.2	K0 III		360 s		
141008	206988		15 47 51.0	-33 08 00	-0.007	-0.01	7.5	1.1	0.2	K0 III		190 mx		
140483	253297		15 47 53.5	-65 26 33	+0.001	-0.04	5.53	0.23	2.1	A5 V		45 s		m
141063	183797		15 47 56.3	-25 59 13	+0.003	-0.01	6.9	1.1	0.6	A0 V		39 s		
140973	206991		15 48 01.5	-37 40 33	+0.004	-0.05	7.3	0.5	2.6	F0 V		65 s		
141456	64915		15 48 01.9	+31 44 09	+0.002	-0.04	6.5	1.4	-0.3	K5 III	-20	230 s		
141091	183802		15 48 03.3	-25 12 57	0.000	-0.03	7.3	0.3	0.6	A0 V		150 s		
141676	29675		15 48 05.7	+50 00 15	0.000	-0.02	8.0	1.1	-0.1	K2 III		420 s		
140842	242990		15 48 06.3	-50 35 24	-0.001	-0.02	6.7	1.1	2.1	A5 V		22 s		
141272	121196		15 48 09.4	+1 34 19	-0.011	-0.16	7.9	0.9	3.2	G5 IV	-28	38 mx		
141247	140766		15 48 11.7	-4 47 10	-0.007	-0.04	8.0	0.5	4.2	F9 V	-7	57 s		
140882	226218		15 48 11.9	-49 44 33	0.000	0.00	7.45	1.58		K5				

HD	SAO	Star Name	α 2000	δ 2000	μ(α)	μ(δ)	V	B-V	M$_v$	Spec	RV	d(pc)	ADS	Notes
			h m s	° ′ ″	s	″								
141353	101744		15 48 13.2	+13 47 19	0.000	−0.12	6.00	1.25	0.2	K0 III	−54	94 mx		
141712	29676		15 48 13.3	+50 57 16	−0.002	−0.02	8.0	1.1	0.2	K0 III		360 s		
140977	226223		15 48 19.2	−44 21 05	0.000	0.00	6.7	1.1	1.4	A2 V		29 s		
140925	226220		15 48 20.3	−49 45 49	−0.010	−0.06	7.4	0.7	4.0	F8 V		39 s		
141107	183810		15 48 22.5	−28 47 20	−0.003	−0.02	7.6	0.6	3.0	F2 V		58 s		
141221	159506		15 48 24.9	−13 29 57	0.000	0.00	7.0	1.1	0.2	K0 III		230 s		
141164	183813		15 48 26.5	−23 50 03	−0.004	−0.03	6.7	−0.1	0.6	A0 V		77 mx		
141476	84011		15 48 28.7	+22 24 56	−0.002	−0.04	7.0	1.1	0.2	K0 III		230 s		
141527	84015		15 48 34.3	+28 09 24	0.000	−0.02	5.8	0.6	4.4	G0 V	+25	26 mn		R CrB, v
144463	2625		15 48 41.9	+83 37 12	−0.007	+0.01	7.60	0.1	1.7	A3 V		150 s	9853	m
141477	101752	35 κ Ser	15 48 44.3	+18 08 29	−0.004	−0.09	4.09	1.62	−0.5	M1 III	−39	78 s		
140926	242997		15 48 47.8	−54 23 46	−0.001	−0.02	7.8	0.3	0.0	B8.5 V		210 s		
140979	243001		15 48 50.2	−52 26 16	0.000	0.00	6.07	1.38		K0				m
141458	101753		15 48 50.5	+12 43 26	0.000	0.00	6.81	0.03	0.6	A0 V	−5	170 s		
141377	140773		15 48 52.7	−1 00 05	−0.003	−0.01	6.9	1.6	−0.5	M2 III		310 s		
141529	84016		15 48 53.0	+23 14 31	−0.003	−0.03	7.8	0.5	4.0	F8 V		57 s		
141274	183822		15 48 54.8	−20 26 57	−0.001	−0.03	7.9	1.5	−0.5	M2 III		470 s		
141134	207003		15 48 56.0	−38 36 46	−0.001	−0.02	8.0	2.2	−0.3	K5 III		170 s		
140826	253302		15 48 56.5	−60 22 03	−0.001	+0.07	6.65	0.42		F0				
141378	140775		15 48 56.7	−3 49 07	−0.002	+0.01	5.53	0.12		A3	−16			
141053	226233		15 49 01.7	−48 45 53	−0.007	−0.06	7.7	0.0	0.2	K0 III		150 mx		
139991	—		15 49 05.1	−77 12 14			7.6	1.1	−0.1	K2 III		350 s		
141438	140781		15 49 11.7	−1 59 19	−0.002	−0.04	7.8	0.1	0.6	A0 V		270 s		
141054	243008		15 49 13.4	−52 12 53	−0.001	+0.03	7.1	1.2	−0.5	M2 III		330 s		
141166	226246		15 49 14.8	−41 47 31	0.000	−0.04	7.39	−0.01	0.4	B9.5 V		250 s		
141989	29680		15 49 16.7	+56 14 10	+0.002	−0.06	7.3	1.1	−0.1	K2 III		300 s		
141359	183829		15 49 21.2	−20 17 04	−0.007	0.00	7.9	0.9	0.2	K0 III		170 mx		
141067	243011		15 49 26.5	−52 53 27	−0.001	−0.03	8.00	0.12	0.6	A0 V		250 s		
141312	207015		15 49 30.0	−31 04 23	0.000	−0.04	7.6	1.8	−0.1	K2 III		150 s		
141589	101758		15 49 33.9	+12 33 31	0.000	+0.02	6.7	1.1	0.2	K0 III	−6	200 s		
141714	84019	10 δ CrB	15 49 35.6	+26 04 06	−0.006	−0.07	4.63	0.80	1.8	G5 III-IV	−19	38 s		
141513	140787	32 μ Ser	15 49 37.1	−3 25 49	−0.006	−0.03	3.54	−0.04	0.6	A0 V	−9	44 mn		
141691	84020		15 49 38.9	+23 40 15	+0.001	−0.03	7.8	0.9	3.2	G5 IV		82 s		
141404	183833		15 49 40.0	−20 46 40	−0.003	−0.01	7.70	0.13	0.2	B9 V		250 s		
141327	207020		15 49 43.1	−32 48 29	0.000	−0.02	7.4	0.4	0.4	B9.5 V		140 s		
141930	45752		15 49 49.8	+44 31 19	−0.003	−0.01	7.90	0.1	0.6	A0 V	−16	290 s	9802	m
141610	121210		15 49 51.5	+2 30 03	0.000	−0.01	7.1	0.1	1.4	A2 V		140 s		
141465	159518		15 49 52.2	−17 54 07	+0.002	−0.01	6.7	0.4	3.0	F2 V		53 s		
141692	101761		15 49 53.0	+14 03 56	−0.006	0.00	7.7	0.5	4.0	F8 V		54 s		
141609	121211		15 49 53.9	+2 47 43	0.000	−0.01	7.9	1.1	−0.1	K2 III		390 s		
141466	183837		15 49 56.4	−21 29 27	0.000	−0.02	7.7	1.0	0.2	K0 III		310 s		
141194	226254		15 49 57.4	−48 54 43	−0.003	0.00	5.84	0.07		A2				
141569	140789		15 49 57.6	−3 55 16	−0.002	−0.02	6.90	0.00	0.4	B9.5 V		200 s		m
141295	226257		15 49 59.8	−42 38 59	+0.002	−0.01	7.9	0.8	2.1	A5 V		85 s		
141515	159520		15 50 03.1	−18 08 33	0.000	−0.03	7.8	0.4	2.6	F0 V		99 s		
141168	243022		15 50 06.8	−53 12 34	−0.005	−0.03	5.77	−0.08	−0.2	B8 V	+21	32 mx		m
142342	16870		15 50 07.7	+64 47 59	−0.003	+0.01	7.2	0.1	0.6	A0 V		210 s		
141693	121214		15 50 11.8	+5 57 18	−0.001	−0.01	8.0	0.1	0.6	A0 V		300 s		
141296	226263		15 50 16.2	−45 24 06	+0.004	−0.04	6.12	0.29	2.1	A5 V		53 s		m
141297	226262		15 50 16.2	−47 01 25	−0.001	−0.02	7.9	0.9	0.2	K0 III		340 s		
141257	226260		15 50 17.3	−49 02 24	−0.002	−0.07	7.2	1.0	0.2	K0 III		220 mx		
141680	121215	34 ω Ser	15 50 17.4	+2 11 47	+0.002	−0.05	5.23	1.02	0.3	G8 III	−12	85 s		
141120	243020		15 50 21.3	−58 30 06	0.000	0.00	7.3	1.5	0.2	K0 III		130 s		
141849	84028		15 50 31.1	+20 20 25	+0.002	−0.04	7.8	1.1	0.2	K0 III		330 s		
141554	183847		15 50 34.7	−26 17 21	−0.007	−0.01	6.6	1.3	0.2	K0 III		130 s		
141597	183850		15 50 39.6	−23 15 29	−0.002	−0.04	7.5	0.5	4.0	F8 V		49 s		
141850	101771		15 50 41.6	+15 08 00	0.000	−0.04	5.5	1.6		M5 III	+24			R Ser, v
142143	45758		15 50 46.6	+48 29 00	0.000	−0.01	6.7	1.6		M6 III	−29			ST Her, v
141795	121218	37 ε Ser	15 50 48.9	+4 28 40	+0.008	+0.06	3.71	0.15		A m	−9	33 t		
—	8357		15 50 51.4	+77 51 31	+0.004	+0.04	8.0	1.3						
142035	64944		15 50 51.6	+37 33 13	−0.001	+0.04	7.7	1.1	0.2	K0 III		320 s		
142108	45757		15 50 57.0	+42 33 56	−0.002	+0.04	7.3	0.5	3.4	F5 V		59 s		
141556	207040	5 χ Lup	15 50 57.4	−33 37 38	−0.001	−0.03	3.95	−0.04	−0.3	B9 IV	−18	68 s		
141637	183854	1 Sco	15 50 58.6	−25 45 05	−0.001	−0.03	4.64	−0.05	−1.7	B3 V	−10	170 s		
141972	84033		15 51 04.6	+22 19 02	−0.001	−0.05	7.87	0.44		F0				
141683	183858		15 51 05.2	−22 37 35	−0.005	−0.02	7.6	0.5	3.4	F5 V		65 s		
141318	243044		15 51 06.8	−55 03 21	0.000	−0.02	5.73	0.06	−3.6	B2 III	−3	530 s		m
142282	29691		15 51 09.9	+52 54 25	−0.002	0.00	6.60	0.1	0.6	A0 V	−9	160 s	9816	m
142091	64948	11 κ CrB	15 51 13.8	+35 39 26	−0.001	−0.35	4.82	1.00	1.7	K0 III-IV	−24	33 ts		m
141851	140801	36 Ser	15 51 15.5	−3 05 26	−0.006	−0.03	5.11	0.12		A0 n	−8	29 mn		
141992	84037	38 ρ Ser	15 51 15.8	+20 58 40	−0.004	+0.02	4.76	1.54	−0.3	K5 III	−62	98 s		
142393	29695		15 51 28.2	+57 08 32	−0.001	−0.02	7.5	1.4	−0.3	K5 III		370 s		
141171	253315		15 51 28.6	−64 10 55	+0.008	−0.02	7.75	0.45		F8				
142053	84040		15 51 29.3	+25 18 15	+0.001	−0.02	7.47	1.09	−1.1	K1 II-III		520 s		
141036	—		15 51 29.6	−68 39 40			7.8	1.4	−0.3	K5 III		400 s		

382

HD	SAO	Star Name	α 2000	δ 2000	μ(α)	μ(δ)	V	B-V	M$_V$	Spec	RV	d(pc)	ADS	Notes
			h m s	o ' "	s	"								
141774	183864		15 51 29.8	-20 35 14	+0.001	0.00	7.70	0.09	0.2	B9 V		270 s		
141544	226295		15 51 31.4	-47 03 38	-0.011	-0.03	6.01	1.16		K1 IV		130 mx		
141698	207050		15 51 32.0	-32 41 09	-0.001	+0.02	7.1	1.0	0.2	K0 III		230 s		
141993	101779		15 51 32.8	+13 54 20	-0.004	+0.03	7.6	1.1	-0.1	K2 III		260 mx		
141697	207052		15 51 32.9	-30 42 12	0.000	-0.02	7.9	0.6	4.4	G0 V		48 s		
--	226293		15 51 34.0	-48 36 24	+0.001	-0.04	8.0	0.8	3.2	G5 IV		93 s		
141643	207048		15 51 35.0	-37 26 14	-0.002	+0.03	7.3	0.8	0.2	K0 III		260 s		
141812	159540		15 51 37.9	-18 56 22	-0.002	-0.05	7.3	1.1	0.2	K0 III		220 s		
141853	159544		15 51 38.3	-14 08 01	-0.002	-0.01	6.2	0.7	4.4	G0 V	-22	23 s		
142924	8354		15 51 41.4	+71 20 56	-0.002	-0.01	7.9	1.4	-0.3	K5 III		430 s		
143077	8355		15 51 47.9	+73 19 04	-0.015	+0.06	8.0	0.9	3.2	G5 IV		93 s		
141897	140805		15 51 49.9	-8 02 36	+0.002	+0.01	7.6	1.1	-0.1	K2 III		350 s		
141545	243057		15 51 53.8	-51 37 09	-0.001	0.00	7.5	0.2	0.0	B8.5 V		210 s		
142093	101782		15 52 00.4	+15 14 09	-0.005	-0.14	7.3	0.6	4.4	G0 V		38 s		
141645	226306		15 52 01.2	-45 30 18	-0.004	-0.01	7.7	0.6	0.4	B9.5 V		140 mx		
142176	64956		15 52 02.2	+29 52 53	-0.001	-0.01	7.40	1.4	-0.3	K5 III	-53	350 s		
141037	--		15 52 04.1	-70 49 39			7.9	0.0	0.0	B8.5 V		380 s		
141831	183872		15 52 04.3	-26 31 23	-0.002	-0.01	7.1	1.0	2.6	F0 V		31 s		
141562	243058		15 52 06.4	-51 08 22	-0.001	-0.01	7.8	-0.1	0.0	B8.5 V		360 s		
141832	183873		15 52 12.7	-29 53 12	-0.010	-0.09	6.40	0.98	0.2	K0 III		78 mx	9813	m
141687	226308		15 52 13.2	-45 17 02	0.000	+0.01	6.7	0.9	0.2	K0 III		200 s		
141815	207065		15 52 14.0	-34 07 19	-0.006	-0.10	7.6	0.4	3.4	F5 V		70 s		
142531	29698		15 52 16.5	+55 49 36	-0.003	+0.05	5.81	0.96	0.2	K0 III	-30	130 s		
141937	159551		15 52 17.4	-18 26 10	+0.007	+0.02	7.4	0.7	4.4	G0 V		38 s		
141779	207062		15 52 18.7	-39 30 54	-0.002	+0.01	7.9	0.3	1.4	A2 V		130 s		
142243	84053		15 52 30.8	+28 54 46	-0.001	-0.02	7.80	1.1	-0.2	K3 III		400 s		
142072	140812		15 52 33.6	-1 54 23	-0.008	0.00	7.9	0.9	3.2	G5 IV		88 s		
142070	140814		15 52 34.9	-1 01 53	-0.002	-0.03	8.0	0.1	0.6	A0 V		300 s		
141938	183883		15 52 35.3	-25 00 56	+0.001	-0.03	8.0	1.2	-0.5	M2 III		510 s		
141261	253319		15 52 35.5	-67 38 23	-0.001	-0.02	7.7	0.1	1.7	A3 V		150 s		
141646	243068		15 52 38.3	-53 26 14	0.000	0.00	7.7	1.3	0.2	K0 III		220 s		
142374	45771		15 52 38.5	+42 13 04	-0.001	-0.01	8.0	0.9	3.2	G5 IV		93 s		
142373	45772	1 χ Her	15 52 40.4	+42 27 05	+0.040	+0.63	4.62	0.56	4.2	F9 V	-55	15 ts		
140316	257326		15 52 40.7	-79 24 13	+0.001	-0.01	7.2	0.1	0.6	A0 V		210 s		
141647	243069		15 52 42.5	-54 14 28	-0.002	+0.01	7.4	1.7	-0.3	K5 III		270 s		
141000	--		15 52 42.5	-73 20 38			8.0	0.2	2.1	A5 V		150 s		
141743	226318		15 52 43.8	-48 49 43	+0.002	-0.02	7.6	1.4	-0.1	K2 III		340 s		
141724	243078		15 52 51.4	-50 36 55	+0.002	+0.01	6.4	1.4	-0.1	K2 III		150 s		
141885	207076		15 52 52.0	-37 23 56	+0.001	-0.05	7.7	0.2	4.4	G0 V		45 s		
141583	243067		15 52 54.7	-59 03 42	-0.012	-0.11	7.4	0.7	4.0	F8 V		39 s		
142244	101789		15 52 56.1	+17 24 12	-0.003	0.00	6.4	1.1	0.2	K0 III	-12	170 s		
142245	101788		15 52 56.3	+15 25 51	-0.003	-0.02	7.6	1.1	0.2	K0 III		300 s		
141413	253323		15 52 56.7	-65 09 09	+0.007	0.00	6.54	0.19		A3				
141905	207077		15 53 02.5	-39 53 38	-0.003	-0.04	7.9	0.6	0.6	A0 V		110 s		
142453	45777		15 53 04.2	+41 17 43	0.000	-0.11	7.9	1.1	0.2	K0 III		190 mx		
142532	45778		15 53 08.4	+46 02 23	-0.001	+0.02	7.3	0.9	3.2	G5 IV		67 s		
142267	101792	39 Ser	15 53 12.0	+13 11 48	-0.010	-0.56	6.10	0.60	4.7	G2 V	+36	19 s		m
142016	207084		15 53 13.1	-30 46 43	0.000	-0.03	7.2	0.6	1.7	A3 V		62 s		
142268	101793		15 53 14.4	+12 21 12	+0.002	+0.01	7.5	1.1	-0.1	K2 III		330 s		
142395	64966		15 53 14.8	+31 14 12	-0.005	-0.08	7.9	1.4	-0.3	K5 III		110 mx		
142112	159555		15 53 16.9	-17 58 06	-0.004	-0.02	8.0	0.4	3.0	F2 V		91 s		
142131	159559		15 53 17.6	-14 42 59	0.000	-0.07	8.0	0.5	4.0	F8 V		59 s		
142096	183895	45 λ Lib	15 53 20.0	-20 10 02	-0.001	-0.03	5.03	-0.01	-1.7	B3 V	-4	180 s		
142197	140822		15 53 22.6	-3 01 46	-0.005	-0.04	7.4	0.5	3.4	F5 V		63 s		
141585	253328		15 53 22.8	-62 36 25	-0.001	-0.02	6.19	1.47		K0				
141943	226339		15 53 27.1	-42 16 01	-0.005	-0.08	7.6	0.8	3.2	G5 IV		77 s		
142961	16888		15 53 28.5	+65 17 17	-0.007	+0.03	6.9	0.8	3.2	G5 IV		55 s		
142357	101796		15 53 34.8	+16 04 30	+0.002	-0.02	6.10	0.4	3.0	F2 V	+2	42 s	9828	m
143105	16890		15 53 36.0	+68 43 14	-0.005	-0.01	6.8	0.5	3.4	F5 V		49 s		
142114	183896	2 Sco	15 53 36.6	-25 19 38	-0.001	-0.03	4.59	-0.07	-2.1	B2.5 V n	-12	220 s	9823	m
142376	101799		15 53 47.2	+17 37 00	-0.001	+0.02	8.0	0.1	1.4	A2 V		210 s		
142198	159563	46 θ Lib	15 53 49.4	-16 43 46	+0.007	+0.13	4.15	1.02	1.8	G8 III-IV	+3	25 s		
140535	257329		15 53 51.6	-78 09 31	-0.014	0.00	7.5	0.0	0.4	B9.5 V		110 mx		
142165	183900		15 53 53.7	-24 31 59	-0.002	-0.03	5.39	-0.02	-0.9	B6 V	+13	160 s		
142184	183901		15 53 55.7	-23 58 41	-0.001	-0.03	5.42	-0.04	-2.5	B2 V nn	-27	380 s		
142215	159564		15 53 56.1	-18 57 03	-0.002	0.00	7.7	1.6	-0.5	M2 III		340 s		
142185	183903		15 54 00.2	-25 14 20	-0.004	+0.04	4.90	0.1	-2.5	B2 V nn		50 mx		
141303	257332		15 54 02.3	-71 40 33	+0.018	0.00	7.9	0.1	1.4	A2 V		140 mx		
141963	243099		15 54 08.6	-50 29 24	-0.002	+0.01	7.8	0.1	-1.0	B5.5 V		420 s		
142313	159571		15 54 13.5	-11 05 26	+0.001	-0.03	7.1	1.1	0.2	K0 III		240 s		
141945	243097		15 54 14.6	-53 35 00	+0.005	-0.03	7.1	0.3	3.0	F2 V		67 s		
142132	226351		15 54 24.3	-41 10 23	-0.001	+0.01	7.6	0.9	3.2	G5 IV		77 s		
142399	121246		15 54 24.4	+3 17 24	+0.002	-0.10	7.7	1.4	-0.3	K5 III		160 mx		
142443	121248		15 54 25.4	+5 17 38	+0.003	+0.04	7.0	1.1	0.2	K0 III		230 s		
142217	207103		15 54 26.8	-30 25 31	-0.003	-0.08	8.0	0.5	3.4	F5 V		76 s		

HD	SAO	Star Name	α 2000	δ 2000	μ(α)	μ(δ)	V	B−V	M$_v$	Spec	RV	d(pc)	ADS	Notes
142250	183907		15^h 54^m 29^s.9	−27° 20′ 19″	−0^s.002	−0″.03	6.14	−0.07	−0.6	B7 V		220 s		
142019	243104		15 54 31.2	−51 14 15	−0.003	−0.01	7.5	−0.2	0.4	B9.5 V		260 s		
142574	84070		15 54 34.5	+20 18 40	−0.006	+0.04	5.44	1.59	−0.3	K4 III	−61	110 s		
142780	45788	2 Her	15 54 37.7	+43 08 19	−0.003	+0.06	5.37	1.65		Ma III	−10	13 mn		
142301	183914	3 Sco	15 54 39.4	−25 14 37	−0.001	−0.02	5.87	−0.06	−1.0	B7 IV		240 s		
142500	121254	40 Ser	15 54 40.2	+8 34 49	0.000	0.00	6.29	0.18		A2 n	−24			
142478	121251		15 54 41.8	+5 12 34	+0.002	−0.03	7.6	1.1	0.2	K0 III		310 s		
142315	183916		15 54 41.8	−22 45 57	+0.003	−0.01	6.86	0.04	0.2	B9 V		210 s		
141396	257333		15 54 44.9	−72 37 44	−0.001	−0.01	7.9	0.0	0.4	B9.5 V		320 s		
142553	101809		15 54 49.0	+11 30 54	0.000	+0.05	7.8	0.1	1.4	A2 V		190 s		
142080	243111		15 54 50.9	−50 20 05	−0.001	−0.04	6.8	0.2	1.4	A2 V		96 s		
141913	253344		15 54 52.5	−60 44 36	0.000	0.00	6.15	0.10		B8	−5			m
142064	243109		15 54 56.0	−53 28 59	−0.002	+0.01	7.3	0.6	0.4	B9.5 V		110 s		
142117	226353		15 54 56.7	−49 24 10	0.000	−0.03	7.8	0.2	3.4	F5 V		74 s		
142742	64970		15 54 56.8	+34 21 44	−0.001	0.00	7.16	0.16		A3	−11		9838	m
142378	159572	47 Lib	15 55 00.2	−19 22 59	−0.001	−0.02	5.94	−0.01	−1.1	B5 V	−6	230 s	9834	m
143803	8376		15 55 00.7	+75 34 27	+0.001	−0.02	6.9	0.9	3.2	G5 IV	−19	56 s		
141891	253346	β TrA	15 55 08.4	−63 25 50	−0.028	−0.40	2.85	0.29	3.0	F2 V	0	10 ts		m
142254	226361		15 55 15.6	−42 36 19	−0.009	−0.04	6.69	0.40	2.6	F0 V		61 s		
142277	226364		15 55 19.7	−42 31 14	0.000	−0.01	7.60	1.08		K0				
142255	226362		15 55 20.0	−43 23 31	−0.001	0.00	7.2	0.0	1.7	A3 V		120 s		
142639	121262		15 55 21.4	+9 13 25	−0.001	−0.01	6.9	1.1	0.2	K0 III	+8	220 s		
142256	226363		15 55 22.8	−44 31 39	−0.001	−0.07	7.2	0.5	0.0	B8.5 V		82 mx		
142190	226358		15 55 24.5	−49 58 40	−0.004	−0.05	7.2	0.8	−0.1	K2 III		220 mx		
142796	84078		15 55 25.8	+29 32 12	0.000	0.00	7.80	0.1	0.6	A0 V	−14	280 s	9844	m
142279	226365		15 55 25.8	−43 58 25	0.000	0.00	7.9	−0.1	0.0	B8.5 V		380 s		
141059	257331		15 55 28.5	−77 43 54	−0.003	−0.01	6.8	1.1	0.2	K0 III		210 s		
141767	253342	κ TrA	15 55 29.5	−68 36 11	−0.001	−0.01	5.09	1.13	−6.2	G6 I	+5	1700 s		
142445	183931	4 Sco	15 55 29.9	−26 15 57	−0.003	−0.02	5.62	0.15	1.4	A2 V		60 s		
142407	207119		15 55 30.3	−31 05 01	+0.001	+0.02	6.3	1.4	0.2	K0 III		89 s		
142926	45790	4 Her	15 55 30.5	+42 33 58	−0.003	+0.01	5.75	−0.11	−0.2	B8 V	−17	160 s		
142049	253349		15 55 32.3	−60 10 40	−0.006	−0.08	5.77	0.35		A m				m
142763	101821		15 55 39.7	+18 37 14	−0.001	−0.02	6.20	−0.11	−1.6	B7 III		350 s		
143641	8375		15 55 40.3	+72 23 31	−0.014	+0.01	7.4	1.1	0.2	K0 III		250 mx		
142456	183933		15 55 41.8	−26 44 50	−0.002	−0.08	7.32	0.49	4.0	F8 V		46 s	9836	m
142596	140838		15 55 42.2	−8 38 54	0.000	−0.08	7.2	0.8	3.2	G5 IV		63 s		
142304	226369		15 55 43.1	−44 31 59	+0.001	−0.04	7.0	0.6	0.0	B8.5 V		97 s		
142218	243131		15 55 44.2	−51 07 32	0.000	0.00	7.9	−0.1	0.0	B8.5 V		380 s		
142908	64974	12 λ CrB	15 55 47.6	+37 56 49	+0.003	+0.08	5.45	0.33		F2	−12	25 t		m
143187	29723		15 55 49.6	+58 54 41	−0.004	+0.02	6.1	0.0		B9	−6			
142429	207123		15 55 50.7	−35 40 31	0.000	+0.03	7.9	0.7	3.2	G5 IV		85 s		
142661	140842		15 55 54.5	−2 09 52	−0.006	−0.07	7.00	0.6	4.4	G0 V	−38	33 s	9842	m
142384	226374		15 55 56.4	−40 47 15	0.000	−0.08	7.2	1.1	0.2	K0 III		220 mx		
−−	64976		15 55 56.7	+36 43 21	−0.016	−0.01	7.7	1.4						
142898	84084		15 56 03.5	+27 03 05	+0.003	0.00	8.00	1.1		K1 IV				
142431	226375		15 56 04.9	−40 42 18	−0.004	−0.03	7.6	−0.4	1.7	A3 V		85 mx		
142781	121270		15 56 05.9	+7 56 10	+0.001	−0.01	7.9	1.1	−0.1	K2 III		390 s		
142139	253353		15 56 05.9	−60 28 57	−0.004	−0.08	5.76	0.06	1.2	A1 V		66 mx		
142448	207128		15 56 06.4	−39 51 51	−0.003	−0.02	6.03	0.15		B9				m
142978	64978		15 56 07.1	+39 25 46	+0.001	+0.03	8.0	0.4	3.0	F2 V		99 s		
142798	101823		15 56 12.0	+10 17 51	+0.001	−0.01	7.7	1.1	0.2	K0 III		310 s		
142542	207134		15 56 13.8	−31 47 09	+0.003	0.00	6.5	0.8	3.4	F5 V		27 s		
142640	159584		15 56 14.3	−14 23 58	+0.002	−0.08	6.37	0.48	3.4	F5 V	−6	36 s		
142942	64977		15 56 15.3	+31 13 18	−0.001	−0.01	8.0	1.6	−0.5	M2 III		510 s		
141949	−−		15 56 21.9	−68 15 04			8.0	0.1	0.6	A0 V		290 s		
142860	101826	41 γ Ser	15 56 27.1	+15 39 42	+0.021	−1.29	3.85	0.48	3.7	F6 V	+7	12 ts		m
142703	159587		15 56 33.3	−14 49 45	+0.005	−0.03	6.13	0.23	−0.6	A0 III		97 mx		
142862	121273		15 56 41.3	+5 54 39	+0.001	−0.06	7.9	1.1	0.2	K0 III		280 mx		
142348	243142		15 56 43.9	−53 47 20	+0.003	+0.01	7.5	0.6	0.2	K0 III		280 s		
142910	101829		15 56 45.3	+12 28 41	−0.005	0.00	7.40	0.4	3.0	F2 V		76 s	9850	m
142669	183957	5 ρ Sco	15 56 53.0	−29 12 50	−0.001	−0.02	3.88	−0.20	−2.5	B2 V	+3	190 s	9846	m
142629	207144	ξ¹ Lup	15 56 53.4	−33 57 58	+0.002	−0.04	5.15			A1	−10	17 mn		m
142630	207145	ξ² Lup	15 56 54.1	−33 57 51	+0.002	−0.04	5.57	0.11		A1 V	−12			m
142881	140851		15 57 00.8	−0 44 43	0.000	−0.01	7.9	1.4	−0.3	K5 III		430 s		
142529	226392		15 57 03.7	−48 09 43	−0.011	−0.10	6.31	0.38	3.0	F2 V		45 s		
142643	207146		15 57 04.8	−35 55 39	−0.002	−0.02	7.1	−0.2	1.4	A2 V		140 s		
142804	159598		15 57 05.1	−16 02 04	+0.001	−0.03	6.56	1.78	−0.5	gM1	−11	190 s	9848	m
142865	140852		15 57 07.5	−6 49 42	−0.001	−0.01	7.9	0.2	2.1	A5 V		150 s		
142850	159600		15 57 07.8	−10 53 17	−0.001	−0.02	7.7	1.1	0.2	K0 III		310 s		
142864	140853		15 57 08.7	−6 17 47	+0.002	+0.01	7.0	0.1	0.6	A0 V	−25	190 s		
142882	140854		15 57 09.3	−5 23 54	0.000	−0.01	7.8	0.5	3.4	F5 V		74 s		
142805	183969		15 57 12.5	−21 29 10	−0.002	−0.03	7.14	0.16	0.2	B9 V		190 s		
142749	183965		15 57 13.6	−27 38 34	−0.002	−0.07	7.2	0.9	3.0	F2 V		34 s		
−−	121276		15 57 14.3	+3 24 27	−0.002	0.00	7.20						9855	m
142980	101834	φ Ser	15 57 14.5	+14 24 52	−0.008	+0.09	5.54	1.14		K1 IV	−68			

HD	SAO	Star Name	α 2000	δ 2000	μ(α)	μ(δ)	V	B-V	M_v	Spec	RV	d(pc)	ADS	Notes
			$^h \ ^m \ ^s$	$^{\circ} \ ' \ ''$	^{s}	$''$								
142930	121277		15 57 14.8	+3 24 19	-0.002	+0.01	7.0	0.1	0.6	A0 V	-14	190 s	9855	m
142691	207152		15 57 21.1	-36 11 06	-0.002	0.00	5.9	1.0	4.4	G0 V		12 s		
141416	---		15 57 23.5	-77 49 17			7.9	0.5	3.4	F5 V		79 s		
142468	243156		15 57 24.5	-54 19 55	0.000	0.00	7.89	0.59		B0.5 I				
142546	243160		15 57 28.3	-50 16 12	-0.003	-0.05	8.0	-0.1	1.4	A2 V		210 s		
144344	8384		15 57 29.6	+76 05 23	-0.045	+0.06	8.0	0.9	3.2	G5 IV		78 mx		
143209	64988		15 57 29.7	+39 41 43	-0.006	+0.06	6.31	1.07	0.2	K0 III	-14	150 s		
143107	84098	13 ε CrB	15 57 35.2	+26 52 40	-0.006	-0.06	4.15	1.23	-0.2	K3 III	-31	74 s	9859	m
142883	183972		15 57 40.3	-20 58 58	-0.001	-0.02	5.85	0.02	-1.7	B3 V		250 s		
142415	253358		15 57 40.7	-60 12 00	-0.015	-0.10	7.33	0.63		G0				
142172	---		15 57 42.3	-68 20 54			7.7	1.1	-0.1	K2 III		350 s		
143466	29727		15 57 47.3	+54 44 59	-0.018	+0.11	4.95	0.26	1.7	F0 IV	-8	45 s		
142884	183973		15 57 48.7	-23 31 37	-0.001	-0.01	6.79	0.01		B9				
---	84102		15 57 51.3	+24 23 06	+0.001	-0.01	7.9	1.5	-0.5	M2 III		490 s		
142490	243163		15 57 54.1	-58 17 42	-0.003	+0.01	7.5	0.4	0.0	B8.5 V		160 s		
142943	159605		15 57 57.6	-17 05 23	-0.005	-0.09	8.0	0.5	3.4	F5 V		75 mx		
142851	207168		15 57 59.4	-31 43 44	0.000	-0.03	7.1	0.4	0.4	B9.5 V		120 s		
142567	243171		15 58 05.3	-57 08 51	0.000	-0.03	7.5	0.4	0.6	A0 V		130 s		
142584	243172		15 58 06.0	-56 23 37	-0.001	-0.01	6.9	1.4	0.2	K0 III		130 s		
142512	243169		15 58 09.0	-59 01 28	-0.001	-0.01	7.7	1.3	0.2	K0 III		220 s		
142983	159607	48 Lib	15 58 11.3	-14 16 46	-0.001	-0.02	4.88	-0.10		B sh	-6			
144061	8382		15 58 21.0	+70 53 41	-0.013	+0.26	7.29	0.66	3.2	G5 IV	-9	57 mx	9878	m
142808	226408		15 58 23.9	-44 04 51	-0.004	-0.03	7.0	0.4	1.4	A2 V		83 s		
143328	64994		15 58 24.3	+37 49 35	+0.003	-0.04	7.3	0.4	3.0	F2 V		72 s		
143162	101844		15 58 24.3	+12 50 06	-0.001	-0.03	7.9	1.1	0.2	K0 III		340 s		
142809	226410		15 58 26.2	-44 31 34	+0.003	-0.02	6.9	1.3	-0.1	K2 III		220 s		
142051	257340		15 58 26.4	-73 08 55	+0.001	-0.03	7.7	0.0	0.0	B8.5 V		350 s		
143033	159612		15 58 26.6	-13 26 34	-0.001	-0.04	7.2	1.1	-0.1	K2 III		240 s		
143129	121287		15 58 28.7	+4 45 17	0.000	-0.01	7.0	1.1	0.2	K0 III		230 s		
142988	183983		15 58 30.3	-20 53 38	0.000	-0.06	7.3	0.0	1.4	A2 V		94 mx		
142889	207178		15 58 30.6	-37 30 11	-0.002	0.00	6.4	0.7	3.2	G5 IV		44 s		
142852	226415		15 58 31.3	-42 44 32	-0.002	-0.01	7.5	0.8	1.7	A3 V		52 s		
143291	84112		15 58 32.0	+27 44 25	-0.057	+0.31	8.01	0.77	5.9	K0 V	-70	17 mx		
142822	226413		15 58 32.3	-44 37 06	0.000	-0.03	7.5	1.1	-0.1	K2 III		330 s		
142990	183982		15 58 34.8	-24 49 53	-0.001	-0.02	5.43	-0.09	-1.7	B3 V	-11	250 s		
143080	140870		15 58 35.3	-6 42 14	+0.001	0.00	7.9	1.1	0.2	K0 III		340 s		
142964	207185		15 58 40.4	-30 43 27	-0.002	-0.02	7.9	1.0	-0.5	M2 III		490 s		
142648	243183		15 58 43.9	-57 45 55	-0.002	0.00	7.6	2.2	-0.5	M2 III		190 s		
143349	64997		15 58 44.7	+30 00 17	0.000	+0.01	7.6	1.4	-0.3	K5 III		370 s		
143257	101846		15 58 49.1	+16 12 40	0.000	-0.03	8.0	1.1	-0.1	K2 III		410 s		
143018	183987	6 π Sco	15 58 51.0	-26 06 50	-0.001	-0.03	2.89	-0.19	-3.5	B1 V	-3	190 s	9862	m
143435	65001		15 58 57.6	+36 38 37	+0.002	+0.03	5.62	1.49	-0.3	K5 III	+11	150 s		
142514	253362		15 58 58.1	-65 02 16	-0.001	-0.02	5.75	-0.06		B8				m
143468	65002		15 58 59.7	+36 56 30	-0.003	-0.03	7.7	1.1	0.2	K0 III		320 s		
143292	101852		15 59 01.8	+18 19 35	-0.003	+0.02	7.7	0.5	4.0	F8 V		54 s		
143584	29737		15 59 04.1	+49 52 52	0.000	-0.06	6.05	0.29	1.7	F0 IV	+4	74 s		
143393	84124		15 59 05.7	+29 25 54	+0.002	-0.05	7.10	1.13	-0.1	K2 III	+18	280 s		
143213	121294		15 59 05.7	+0 35 45	+0.002	-0.01	7.5	0.1	0.6	A0 V		240 s		
143595	29739		15 59 07.4	+50 38 20	-0.005	+0.02	8.0	1.1	-0.1	K2 III		360 mx		
142994	207192		15 59 10.8	-38 44 55	-0.002	-0.04	7.4	-0.3	2.6	F0 V		90 s		
143051	207195		15 59 20.5	-33 00 43	+0.003	-0.02	7.1	0.4	0.4	B9.5 V		110 s		
142758	243196		15 59 23.9	-58 43 35	+0.001	-0.02	7.05	0.30	-6.2	B1 I k		2700 s		
142325	257342		15 59 26.7	-71 06 18	-0.008	-0.02	7.80	0.1	1.4	A2 V		110 mx		m
143112	183995		15 59 27.7	-27 00 58	+0.001	0.00	7.01	1.11		G5				
143455	84128		15 59 28.8	+25 26 07	-0.003	+0.03	7.9	1.1	-0.1	K2 III		410 s		
143009	226425		15 59 30.2	-41 44 39	-0.003	-0.02	4.99	1.00	0.3	G8 III	-27	81 s		
142341	257343		15 59 31.7	-71 06 46	-0.009	-0.05	7.80	0.8	3.2	G5 IV		83 s		m
143146	184001		15 59 32.4	-24 11 48	-0.001	-0.06	7.50	0.6	4.4	G0 V		42 s		m
144012	16911		15 59 33.0	+64 53 35	-0.004	-0.01	7.7	1.1	-0.1	K2 III		370 s		
141432	---		15 59 33.9	-80 11 13			8.0	0.4	2.6	F0 V		120 s		
142699	253367		15 59 36.0	-62 48 53	+0.001	0.00	7.03	1.57		K2				
142840	243207		15 59 37.5	-56 46 12	+0.002	-0.01	7.9	0.2	0.6	A0 V		200 s		
143114	183999		15 59 38.2	-29 37 58	-0.005	-0.05	7.2	0.5	4.4	G0 V		37 s		
143115	184000		15 59 39.7	-30 05 01	0.000	-0.03	7.2	0.8	1.4	A2 V		50 s		
143259	140882		15 59 41.2	-7 18 07	-0.001	-0.01	6.7	0.1	0.6	A0 V		160 s		
143296	140885		15 59 46.7	-6 07 38	-0.001	-0.01	7.4	1.1	0.2	K0 III	-6	270 s		
143098	207200		15 59 47.8	-35 50 44	+0.010	-0.07	7.6	0.6	4.4	G0 V		44 s		
143148	207203		15 59 49.7	-31 50 05	-0.002	-0.01	7.5	0.1	2.6	F0 V		95 s		
142676	253368		15 59 50.2	-64 51 16	+0.003	-0.02	7.9	1.1	-0.1	K2 III		380 s		
143297	140886		15 59 51.6	-7 01 36	-0.002	0.00	7.7	1.1	-0.1	K2 III		360 s		
143147	207206		15 59 52.8	-31 10 04	+0.001	-0.02	6.7	0.3	3.2	G5 IV		49 s		
142919	243219		15 59 53.9	-54 01 16	-0.002	-0.02	6.10	0.00		B9 n	-38			
142892	243216		15 59 54.4	-56 48 40	-0.004	-0.03	7.4	0.4	0.6	A0 V		120 s		
141846	257341		15 59 54.5	-78 01 36	-0.005	-0.05	7.10	0.5	3.4	F5 V		55 s		m
143023	226429		15 59 54.6	-46 17 57	-0.002	-0.04	7.5	0.3	0.4	B9.5 V		160 s		

HD	SAO	Star Name	α 2000	δ 2000	μ(α)	μ(δ)	V	B-V	M_v	Spec	RV	d(pc)	ADS	Notes
143149	207205		15ʰ59ᵐ55ˢ.1	−33°23′06″	+0ˢ.001	−0″.02	7.0	0.6	0.6	A0 V		82 s		
142876	243215		15 59 56.0	−58 02 51	−0.002	0.00	7.80	0.00	0.4	B9.5 V		290 s		m
143514	84132		15 59 57.9	+21 37 18	+0.001	−0.03	7.6	1.1	−0.1	K2 III		350 s		
143084	226431		15 59 57.9	−40 39 09	+0.002	+0.01	6.2	1.3	0.2	K0 III		110 s		
143099	207202		15 59 58.0	−38 24 33	−0.003	−0.06	3.60	0.5	3.4	F5 V		11 s		m
143513	84133		16 00 00.9	+22 59 45	+0.002	0.00	7.7	0.0	0.4	B9.5 V		280 s		
143395	121301		16 00 02.1	+0 37 25	0.000	+0.01	7.5	1.4	−0.3	K5 III		360 s		
143118	207208	η Lup	16 00 07.1	−38 23 48	−0.002	−0.03	3.41	−0.22	−2.5	B2 V	+7	150 s		
143722	45829		16 00 11.7	+44 16 46	+0.002	−0.05	6.9	1.1	0.2	K0 III		220 s		
144652	8392		16 00 15.3	+73 56 09	−0.009	+0.05	8.00	1.1	0.2	K0 III		300 mx		m
143333	159625	49 Lib	16 00 19.4	−16 32 00	−0.044	−0.40	5.47	0.52	4.0	F8 V	−25	20 s		m
143275	184014	7 δ Sco	16 00 19.9	−22 37 18	−0.001	−0.03	2.32	−0.12	−4.1	B0 V	−14	170 s		
144282	16925		16 00 20.8	+67 37 38	−0.001	+0.01	6.8	0.4	3.0	F2 V		59 s		
143914	29747		16 00 20.9	+54 31 21	−0.004	+0.03	7.90	0.1	0.6	A0 V		140 mx	9889	m
143119	226435		16 00 21.8	−43 49 20	0.000	+0.01	7.60	1.1	0.2	K0 III		280 s		
143181	207215		16 00 22.1	−38 59 37	−0.002	−0.02	7.2	0.2	0.6	A0 V		170 s		
143120	226434		16 00 22.9	−45 27 09	−0.014	−0.12	7.53	0.74	3.2	G5 IV		56 mx		
143915	29748		16 00 25.0	+53 58 04	+0.001	+0.03	7.6	1.1	0.2	K0 III		310 s		
143215	207220		16 00 31.4	−36 05 17	+0.001	−0.04	7.9	0.6	4.0	F8 V		57 s		
144108	16922		16 00 35.8	+62 05 14	−0.007	+0.02	7.8	1.4	−0.3	K5 III		320 mx		
145742	2652		16 00 36.3	+80 37 39	−0.011	+0.02	7.57	1.06	0.2	K0 III		270 s		
142921	253374		16 00 37.1	−61 18 12	−0.037	−0.20	7.8	0.6	4.4	G0 V		35 mx		
143687	65019		16 00 42.4	+31 34 06	+0.005	−0.05	6.6	1.1	0.2	K0 III		190 s		
143704	65020		16 00 43.6	+32 02 25	+0.001	+0.03	7.3	1.1	0.2	K0 III		260 s		
142497	257347		16 00 46.8	−72 27 48	−0.004	−0.06	6.5	1.7	−0.1	K2 III		110 s		
143232	207224		16 00 46.9	−39 05 18	+0.002	−0.04	6.7	0.4	2.6	F0 V		57 s		
143459	140897	50 Lib	16 00 47.5	−8 24 41	−0.001	−0.02	5.55	0.05		A1	−19			
142721	—		16 00 49.7	−68 41 32			7.6	1.4	−0.3	K5 III		360 s		
−−	226439		16 00 49.9	−48 34 38	0.000	−0.01	8.0	1.0	0.4	B9.5 V		84 s		
143553	121315		16 00 51.0	+4 25 39	−0.003	+0.07	5.83	1.00	0.2	gK0	−4	130 s		
143248	226442		16 00 53.6	−40 26 07	+0.001	+0.03	6.21	0.01		A0				m
143597	101874		16 00 54.1	+13 16 17	−0.001	−0.03	7.50	0.5	3.4	F5 V		66 s	9880	m
143806	65025		16 00 54.9	+39 10 39	+0.002	+0.01	6.70	0.1	0.6	A0 V		170 s	9887	m
143705	84143		16 00 56.9	+28 56 54	+0.005	0.00	7.97	0.61	4.4	G0 V		49 s		
144283	16927		16 00 57.4	+64 56 59	−0.001	−0.01	7.1	0.1	1.4	A2 V		140 s		
143483	159633		16 00 58.3	−12 20 14	−0.003	−0.04	7.8	0.1	0.6	A0 V		220 s		
143470	159634		16 01 02.1	−14 25 24	−0.004	−0.01	7.8	0.5	3.4	F5 V		70 s		
143761	65024	15 ρ CrB	16 01 02.6	+33 18 13	−0.016	−0.77	5.41	0.60	4.7	G2 V	+18	16 ts		m
143707	84144		16 01 03.7	+26 10 19	−0.006	+0.02	8.00	0.4	3.0	F2 V		100 s		m
143438	184029		16 01 05.9	−21 09 21	+0.001	−0.02	7.1	0.9	0.2	K0 III		250 s		
143101	243246		16 01 06.3	−54 34 40	−0.005	−0.04	6.13	0.25	0.0	A3 III		140 s		
143440	184027		16 01 08.1	−24 16 00	+0.002	−0.01	7.3	1.0	−0.1	K2 III		310 s		
142941	253377		16 01 10.6	−63 46 35	0.000	−0.01	6.4	0.8	3.2	G5 IV	+2	43 s		S TrA, v
142308	257345		16 01 11.1	−75 37 09	+0.001	0.00	7.6	0.3	0.6	A0 V		150 s		
143666	101879	5 Her	16 01 14.2	+17 49 06	−0.004	+0.15	5.12	0.99	0.2	K0 III	−19	96 s		
143404	207234		16 01 19.4	−31 53 21	+0.001	−0.01	6.3	1.4	0.2	K0 III		100 s		
142894	253376		16 01 25.1	−66 50 56	−0.004	−0.03	7.9	1.6	−0.5	M4 III		440 s		
144109	29760		16 01 25.3	+55 31 17	−0.001	+0.01	8.0	1.4	−0.3	K5 III		450 s		
143807	84152	14 ι CrB	16 01 26.5	+29 51 04	−0.002	−0.01	4.99	−0.07	−0.6	A0 III	−19	130 s		
143472	184032		16 01 26.6	−25 11 55	+0.001	−0.03	7.9	0.2	1.4	A2 V		150 s		
144002	45850		16 01 32.3	+47 08 34	−0.002	+0.02	7.4	1.1	0.2	K0 III		270 s		
143708	101882		16 01 37.1	+12 03 04	+0.001	0.00	7.7	0.9	3.2	G5 IV		80 s		
142033	257344		16 01 37.2	−78 42 40	−0.014	−0.08	7.4	0.9	3.2	G5 IV		68 s		
143615	140903		16 01 41.1	−3 45 49	−0.005	+0.02	6.8	0.4	3.0	F2 V		58 s		
143893	65028		16 01 45.4	+31 52 33	−0.001	0.00	7.9	0.7	4.4	G0 V		51 s		
143462	207243		16 01 47.2	−34 37 36	0.000	0.00	7.6	1.8	−0.3	K5 III		240 s		
144302	29766		16 01 51.5	+59 37 51	−0.017	+0.06	7.8	0.9	3.2	G5 IV		83 s		
143028	—		16 01 51.6	−64 24 19			7.5	0.0	0.0	B8.5 V		310 s		
143762	101883		16 01 52.5	+13 28 20	−0.002	0.00	7.5	0.2	2.1	A5 V		120 s		
144284	29765	13 θ Dra	16 01 53.2	+58 33 55	−0.042	+0.34	4.01	0.52	3.2	F8 IV-V	−9	16 ts		
143567	184043		16 01 55.3	−21 58 49	−0.001	−0.01	7.19	0.08	0.2	B9 V		220 s		
143473	207245		16 01 58.9	−37 32 06	0.000	−0.05	7.8	−0.5	0.4	B9.5 V		300 s		
144015	45855		16 02 01.9	+40 01 22	−0.005	+0.03	7.0	1.1	0.2	K0 III		230 s		
143321	243262		16 02 04.4	−51 07 12	−0.001	0.00	6.80	0.0	0.0	B8.5 V		220 s		m
143488	207246		16 02 04.6	−36 44 38	−0.003	−0.03	7.2	−0.2	0.6	A0 V		210 s		
144204	29763		16 02 05.4	+52 54 57	−0.001	−0.04	5.93	1.48	−0.1	K2 III	−7	110 s		
143723	140907		16 02 06.6	−0 49 17	−0.001	−0.03	7.4	0.1	1.4	A2 V		160 s		
143463	226457		16 02 10.3	−42 42 00	0.000	−0.17	6.92	0.49	3.7	F6 V		43 s		
143600	184045		16 02 13.5	−22 41 15	0.000	−0.03	7.33	0.10	0.2	B9 V		220 s		
142678	257350		16 02 15.9	−73 30 54	−0.003	−0.06	6.8	0.2	1.7	A3 V		91 s		
143894	84155	44 π Ser	16 02 17.7	+22 48 16	+0.001	+0.02	4.83	0.07	1.7	A3 V	−26	42 s		
142309	257348		16 02 23.4	−77 28 37	0.000	+0.03	7.5	1.1	0.2	K0 III		290 s		
143306	243265		16 02 28.2	−56 51 15	−0.018	−0.12	6.8	0.8	4.0	F8 V		27 s		
143617	184047		16 02 29.0	−26 08 52	+0.003	+0.01	7.6	1.2	−0.5	M4 III		420 s		
143936	101890		16 02 37.3	+18 06 50	0.000	+0.01	7.4	0.1	0.6	A0 V		230 s		

HD	SAO	Star Name	α 2000	δ 2000	μ(α)	μ(δ)	V	B-V	M_v	Spec	RV	d(pc)	ADS	Notes
			$16^h 02^m 38.3^s$	$+26°36'44''$	-0.002^s	$-0.06''$								
144004	84160		16 02 38.3	+26 36 44	-0.002	-0.06	7.7	0.5	3.4	F5 V		74 s		
144205	45863		16 02 39.2	+47 14 25	-0.005	+0.06	5.7	1.6		M6 III	-92			X Her, v
143619	184049		16 02 39.3	-29 08 09	+0.003	-0.02	6.20	1.1	0.2	K0 III		160 s	9896	m
143445	243273		16 02 42.3	-50 15 25	+0.001	-0.02	7.9	0.4	0.6	A0 V		210 s		
143838	140913		16 02 43.1	-0 24 46	-0.002	-0.01	7.6	0.8	3.2	G5 IV		·76 s		
143991	84161		16 02 44.5	+22 48 21	+0.001	+0.02	7.8	0.1	1.7	A3 V		160 s		
143561	226463		16 02 45.0	-42 30 26	-0.003	-0.08	8.0	1.3	0.2	K0 III		140 s		
143692	184055		16 02 45.3	-23 10 51	+0.001	+0.01	8.0	-0.6	1.4	A2 V		210 s		
144247	45867		16 02 45.5	+48 21 05	-0.001	+0.03	8.0	1.1	0.2	K0 III		360 s		
143992	84162		16 02 46.1	+22 14 21	-0.001	0.00	7.0	0.1	1.4	A2 V	-35	130 s		
143593	207262		16 02 46.2	-36 11 02	-0.001	-0.01	7.7	1.1	0.2	K0 III		290 s		
144206	45865	6 υ Her	16 02 47.8	+46 02 12	+0.005	-0.06	4.76	-0.11		B9 p	+3			
143839	140915		16 02 48.1	-2 28 13	-0.002	-0.02	8.00	1.1	-0.1	K2 III		420 s		m
143649	207266		16 02 51.7	-30 56 40	-0.008	-0.09	7.8	0.2	3.4	F5 V		73 mx		
143715	184058		16 02 52.0	-25 00 53	0.000	-0.04	7.40	0.1	0.6	A0 V		230 s	9899	m
143221	253389		16 02 52.0	-63 07 50	0.000	0.00	7.25	1.62	-0.3	K5 III		270 s		
143238	253390		16 02 52.3	-62 32 29	-0.005	-0.02	6.25	-0.04		A0				
143287	253393		16 02 53.1	-60 20 10	-0.001	-0.02	6.8	2.0	0.0	B8.5 V		13 s		
142814	--		16 02 53.5	-73 36 25			7.4	0.1	0.6	A0 V		230 s		
143572	226464		16 02 55.9	-43 26 29	0.000	-0.02	7.6	0.0	0.6	A0 V		230 s		
143898	121335		16 02 57.0	+1 41 54	+0.002	-0.02	8.0	0.1	0.6	A0 V		300 s		
143727	184059		16 02 57.3	-24 10 27	-0.001	-0.04	7.6	1.8	-0.1	K2 III		150 s		
143876	140917		16 02 57.9	-1 38 59	-0.002	-0.02	7.9	1.1	-0.1	K2 III		390 s		
143841	140916		16 02 58.7	-8 02 54	0.000	-0.02	7.20	0.4	3.0	F2 V		69 s		m
143857	140918		16 03 03.4	-5 50 05	0.000	-0.03	7.0	1.1	-0.1	K2 III	-38	260 s		
142610	--		16 03 05.8	-76 20 31			7.9	1.4	-0.3	K5 III		440 s		
143785	159655		16 03 06.9	-19 50 28	-0.001	-0.02	7.2	0.4	1.4	A2 V		87 s		
144542	29777		16 03 09.1	+59 24 39	-0.004	-0.02	6.19	1.58	-0.5	M2 III	-5	220 s		
--	65046		16 03 12.1	+37 07 09	-0.002	-0.01	8.0	1.8						
143546	226466	η Nor	16 03 12.6	-49 13 47	+0.003	0.00	4.65	0.92	0.3	gG4	0	64 s		
143766	184064		16 03 16.2	-26 42 59	+0.001	-0.16	6.9	0.8	4.0	F8 V		27 s		
144208	65049		16 03 19.2	+36 37 54	+0.001	-0.02	5.83	0.56	3.4	F5 V	-1	26 s		
143787	184068		16 03 20.4	-25 51 55	-0.005	-0.04	5.0	1.1	0.2	K0 III	-39	91 s		
143289	253397		16 03 22.4	-63 22 00	+0.009	-0.08	7.84	0.53		F8				m
143698	207278		16 03 22.5	-37 07 58	-0.001	0.00	8.0	0.8	0.2	K0 III		360 s		
143699	207276		16 03 24.0	-38 36 09	-0.003	-0.03	4.89	-0.14	-1.7	B3 V	0	200 s		
144653	16939		16 03 28.2	+61 20 36	+0.001	-0.01	7.80	0.1	1.7	A3 V		170 s	9916	m
145622	8417		16 03 31.1	+76 47 37	-0.008	+0.02	5.6	0.1	0.6	A0 V	-25	99 s		
143474	243279	ι¹ Nor	16 03 31.9	-57 46 31	-0.015	-0.08	4.63	0.24	2.1	A5 V	-18	30 s		m
143878	159659		16 03 33.7	-16 24 07	+0.004	0.00	7.9	1.1	0.2	K0 III		270 s		
143790	207282		16 03 34.2	-32 00 02	-0.003	-0.03	6.1	0.9	3.4	F5 V		18 s		
144064	101898		16 03 39.6	+11 26 00	-0.003	-0.04	7.30	1.1	-0.1	K2 III		300 s	9908	m
144654	29780		16 03 41.2	+59 05 22	-0.005	-0.01	7.0	1.1	0.2	K0 III		230 s		
143448	253406		16 03 44.2	-60 29 54	-0.001	0.00	6.9	1.0	-2.3	B3 IV		150 s		
143449	253407		16 03 44.6	-60 30 46	+0.002	+0.01	6.9	1.8		B1				
144046	121339	43 Ser	16 03 45.5	+4 59 12	-0.003	+0.01	6.08	0.96	0.2	G9 III	-44	150 s		m
143324	253401		16 03 47.6	-64 27 37	+0.053	+0.12	7.7	1.1	0.2	K0 III		55 mx		
143823	207285		16 03 48.1	-33 04 17	-0.009	-0.07	7.80	0.4	2.6	F0 V		78 mx		m
144149	101900		16 03 49.7	+17 48 09	-0.002	+0.01	6.9	1.1	0.2	K0 III	-45	220 s		
143735	226468		16 03 50.2	-42 29 35	-0.004	-0.08	8.0	0.7	3.4	F5 V		80 s		
143625	243284		16 03 52.9	-50 50 24	+0.002	-0.01	7.9	0.0	0.6	A0 V		270 s		
143900	184075		16 03 54.6	-24 43 35	0.000	+0.01	6.4	1.7	0.2	K0 III		68 s		
145368	8412		16 03 59.8	+73 08 53	-0.016	+0.09	7.0	0.5	3.4	F5 V	-14	52 s		
144287	84179		16 04 03.6	+25 15 16	-0.039	+0.67	7.10	0.77	5.6	G8 V	-41	20 ts		
143548	243283		16 04 06.0	-59 10 34	-0.001	0.00	6.6	1.4	-0.3	K5 III		240 s		
143845	207289		16 04 08.6	-35 26 58	-0.002	-0.06	7.2	1.0	0.2	K0 III		210 mx		
143846	207288		16 04 09.4	-37 18 27	-0.028	-0.02	7.86	0.60		G0		16 mn		
144047	140925		16 04 09.7	-9 55 09	0.000	0.00	7.4	1.1	0.2	K0 III	-40	270 s		
144151	121344		16 04 10.0	+6 03 12	-0.001	-0.04	8.0	1.1	-0.1	K2 III		420 s		
144578	29781		16 04 13.3	+50 29 54	0.000	-0.04	7.80	1.8		C7.2 e	-37			RR Her, v
144359	65057		16 04 13.4	+34 10 43	-0.002	0.00	6.73	0.06		A0	-8			
--	8419		16 04 15.0	+75 47 10	-0.007	-0.05	7.8	1.5						
143824	226476		16 04 15.5	-40 50 42	-0.001	-0.04	7.9	-0.1	1.4	A2 V		200 s		
143902	207292		16 04 17.7	-33 12 51	-0.002	-0.06	6.10	0.34	2.5	A9 V		48 s		
144171	121349		16 04 19.6	+6 00 41	+0.002	-0.03	7.1	1.1	0.2	K0 III		240 s		
143658	243296		16 04 21.5	-53 42 37	+0.003	-0.01	6.7	-0.2	1.4	A0 V		170 s		
144069	159665	ξ Sco	16 04 22.0	-11 22 24	-0.004	-0.03	4.16	0.45	2.2	F6 IV	-29	26 ts	9909	m
144071	159667		16 04 25.2	-11 51 24	-0.003	+0.02	7.51	1.82	-0.3	K5 III		230 s		
144087	159668		16 04 25.7	-11 26 58	-0.005	-0.03	6.96	0.77	5.6	G8 V	-34	21 ts	9910	m
144088	159670		16 04 26.5	-11 27 00	-0.005	-0.02	8.03	0.84		K0				
144172	121350		16 04 29.5	+0 40 21	0.000	-0.08	7.1	0.5	4.0	F8 V		42 s		
143976	184083		16 04 30.4	-28 55 51	0.000	0.00	7.7	1.4	-0.3	K5 III		390 s		
143928	207297		16 04 36.6	-37 51 47	-0.012	-0.12	5.90	0.40	2.1	F0 IV-V		47 mx		m
144211	140934		16 04 41.8	-1 09 37	0.000	-0.03	7.8	1.1	-0.1	K2 III		380 s		
143927	207300		16 04 42.0	-37 48 57	-0.002	-0.02	7.0	0.1	0.4	B9.5 V		170 s		

HD	SAO	Star Name	α 2000	δ 2000	μ(α)	μ(δ)	V	B-V	M_v	Spec	RV	d(pc)	ADS	Notes
			^{h} ^{m} ^{s}	$^{\circ}$ $'$ $''$	^{s}	$''$								
143939	207299		16 04 44.3	−39 26 07	−0.001	−0.05	7.20	0.00		B9 p				m
144903	16945		16 04 48.7	+60 02 40	−0.002	−0.01	7.60	1.1	0.2	K0 III		300 s		m
145309	8415		16 04 49.2	+70 15 43	+0.004	+0.03	6.70	0.1	0.6	A0 V		170 s		m
144304	121356		16 04 49.3	+9 38 41	0.000	−0.01	7.8	0.2	2.1	A5 V		140 s		
143494	−−		16 04 51.5	−65 50 58			7.8	0.0	0.0	B8.5 V		350 s		
143826	226482		16 04 56.1	−50 00 51	+0.002	−0.01	7.3	0.7	0.2	K0 III		260 s		
144579	65065		16 04 56.7	+39 09 23	−0.049	+0.05	6.66	0.73	5.6	G8 V	−60	13 ts		m
143996	207304		16 05 01.0	−39 12 59	−0.002	−0.03	7.9	1.4	−0.5	M4 III		470 s		
144073	184092		16 05 01.2	−26 56 53	+0.001	−0.02	7.80	0.8	3.2	G5 IV		83 s		m
144134	159677		16 05 02.8	−18 32 23	−0.004	−0.04	6.8	0.5	3.4	F5 V		45 s		
144514	65062		16 05 04.2	+33 09 08	+0.002	0.00	8.0	1.1	0.2	K0 III		360 s		
144271	140938		16 05 08.4	−3 31 38	−0.001	+0.02	6.9	0.1	0.6	A0 V	−29	180 s		
144704	45883		16 05 08.9	+47 33 25	−0.001	+0.01	8.0	1.1	0.2	K0 III		360 s		
144562	65066		16 05 11.3	+34 14 11	+0.001	+0.01	7.7	1.1	0.2	K0 III		320 s		
145068	16949		16 05 15.1	+62 19 31	−0.003	0.00	7.9	1.1	0.2	K0 III		340 s		
143796	243305		16 05 15.5	−56 20 24	+0.001	−0.01	7.34	1.89		M5 III				
143817	243308		16 05 18.3	−55 48 41	+0.001	−0.02	8.0	0.5	3.2	G5 IV		93 s		
144308	140941		16 05 21.1	−3 20 55	−0.004	−0.04	7.8	1.1	0.2	K0 III		160 mx		
142878	257354		16 05 22.0	−77 14 12	−0.004	−0.01	7.9	1.1	−0.1	K2 III		400 s		
144115	207315		16 05 23.7	−31 27 24	−0.003	−0.03	7.1	0.6	1.4	A2 V		62 s		
144217	159682	8 β¹ Sco	16 05 26.1	−19 48 19	0.000	−0.02	2.64	−0.07		B0.5 V	−7		9913	Graffias, m
144218	159683	8 β² Sco	16 05 26.4	−19 48 07	−0.001	−0.02	4.92	−0.02	−2.5	B2 V	−5	250 s	9913	m
144722	45885		16 05 27.2	+43 43 09	−0.002	0.00	7.7	1.1	0.2	K0 III		320 s		
144600	65069		16 05 30.2	+31 34 59	−0.001	0.00	7.8	0.1	1.4	A2 V		190 s		
144099	207316		16 05 31.2	−36 10 02	+0.001	+0.02	7.7	0.7	0.2	K0 III		320 s		
144222	184104		16 05 34.8	−24 13 26	−0.006	−0.08	7.4	0.9	0.2	K0 III		110 mx		
144426	121361		16 05 37.7	+8 05 46	−0.001	0.00	6.29	0.08		A2	−21			
144544	84189		16 05 37.8	+21 53 34	−0.004	−0.10	7.2	0.5	3.4	F5 V		59 s		
144179	207321		16 05 39.1	−32 51 45	−0.028	−0.23	7.85	0.82		G5		11 mn		m
144253	184105		16 05 40.5	−20 26 59	+0.023	−0.34	7.40	1.04	6.3	dK2	+34	19 ts		
144254	184106		16 05 43.3	−21 50 20	0.000	−0.03	7.9	−0.3	0.6	A0 V		280 s		
143549	253414		16 05 44.0	−68 26 08	0.000	−0.01	7.70	−0.10	0.4	B9.5 V		290 s		
144362	140945		16 05 44.3	−6 17 29	+0.002	−0.01	6.35	0.41		F5	−11		9918	m
144682	65073		16 05 45.0	+32 14 52	−0.003	−0.02	6.8	1.4	−0.3	K5 III		270 s		
144723	65075		16 05 48.6	+36 31 39	−0.003	−0.02	7.4	0.8	3.2	G5 IV		68 s		
144516	101918		16 05 52.1	+9 56 09	−0.001	+0.01	7.1	0.1	1.4	A2 V		140 s		
143346	257357		16 05 55.8	−72 24 03	−0.005	+0.07	5.70	1.17	0.2	gK0		89 s		
144390	140947		16 05 59.6	−6 08 22	−0.002	−0.01	6.41	1.03		K0				m
144563	101921		16 06 01.0	+13 19 44	−0.002	−0.02	6.6	1.1	0.2	K0 III		190 s		
144564	101922		16 06 02.7	+13 19 16	+0.001	−0.02	6.90	1.1	0.2	K0 III		220 s	9922	m
145310	16958		16 06 05.5	+64 41 14	−0.009	+0.04	7.3	1.1	0.2	K0 III		260 s		
144334	184113		16 06 06.2	−23 36 23	−0.001	−0.03	5.92	−0.08	−0.8	B9 III		220 s		
144622	101927		16 06 11.8	+18 22 12	+0.002	−0.01	7.9	1.6		M5 III	−30			R Her, v
144490	140951		16 06 13.7	−0 22 01	−0.002	−0.03	7.4	0.9	3.2	G5 IV		71 s		
144517	121364		16 06 14.5	+1 42 40	0.000	−0.01	7.5	0.8	3.2	G5 IV		72 s		
145454	16962		16 06 19.5	+67 48 36	−0.007	+0.06	5.44	−0.02		A0	−18	27 mn		
144582	121367		16 06 23.0	+5 24 45	−0.003	+0.02	8.0	0.4	2.6	F0 V		120 s		
144779	65078		16 06 25.9	+29 54 15	−0.005	−0.02	7.8	1.1	−0.1	K2 III		170 mx		
143832	253423		16 06 27.3	−63 20 27	−0.004	−0.02	7.51	0.09		A0				
144197	226500	δ Nor	16 06 29.3	−45 10 24	0.000	+0.03	4.72	0.23		A m	−16	18 mn		
144259	226503		16 06 30.7	−40 26 59	−0.001	0.00	7.80	0.6	4.4	G0 V		47 s		m
144450	159696		16 06 31.3	−14 18 45	−0.011	−0.13	8.0	0.9	3.2	G5 IV		47 mx		
144493	159698		16 06 35.4	−11 54 18	−0.002	+0.02	7.56	0.30	2.6	F0 V		97 s		
144294	207332	θ Lup	16 06 35.4	−36 48 08	−0.002	−0.03	4.23	−0.17	−2.3	B3 IV	+15	200 s		
143853	−−		16 06 44.2	−64 33 43			7.7	1.1	0.2	K0 III		300 s		
144227	226504		16 06 45.3	−46 22 44	+0.004	−0.02	8.0	−0.1	4.0	F8 V		62 s		
144470	184123	9 ω¹ Sco	16 06 48.3	−20 40 09	−0.001	−0.02	3.96	−0.04	−3.5	B1 V	−4	250 s		
144724	101931		16 06 50.4	+16 21 08	+0.002	−0.05	8.0	0.4	2.6	F0 V		120 s		
145082	45895		16 06 59.5	+47 30 18	−0.001	+0.01	6.61	−0.01		A0	−8			
144636	140966		16 07 00.9	−3 52 40	−0.002	−0.02	7.6	1.6	−0.5	M2 III		420 s		
143967	253428		16 07 01.4	−62 45 44	−0.003	0.00	7.70	−0.04		B9				
144585	159706		16 07 03.2	−14 04 15	−0.018	+0.02	6.32	0.66	4.1	G4 IV-V		28 s		
144569	159704		16 07 04.6	−16 56 36	0.000	−0.03	7.6	0.1	0.6	A0 V		210 s		
145145	45900		16 07 08.1	+49 05 01	−0.001	+0.03	7.7	0.1	1.4	A2 V		180 s		
144026	253431		16 07 08.5	−60 21 26	−0.004	−0.05	7.9	1.2	3.0	F2 V		32 s		
144433	207340		16 07 10.2	−33 08 53	+0.001	−0.04	7.6	0.0	2.1	A5 V		120 s		
143265	257358		16 07 11.9	−76 15 22	−0.007	−0.01	7.8	1.1	−0.1	K2 III		380 s		
144338	226513		16 07 13.2	−45 20 48	+0.001	−0.02	7.3	1.1	0.2	K0 III		210 s		
144350	226517		16 07 13.6	−42 04 08	+0.002	0.00	7.3	1.1	−0.1	K2 III		300 s		
144351	226515		16 07 14.0	−43 17 20	−0.001	−0.03	7.2	0.9	1.4	A2 V		89 s		
144586	159707		16 07 14.8	−17 56 09	0.000	−0.01	7.5	0.1	0.6	A0 V		200 s		
144415	207341		16 07 16.2	−36 45 19	+0.006	−0.05	5.73	0.29	2.6	F0 V		42 s		m
144455	207342		16 07 17.2	−33 42 35	+0.001	−0.05	8.00	0.1	0.6	A0 V		300 s		m
143999	253430		16 07 19.3	−62 54 39	+0.004	−0.02	7.6	0.5	3.4	F5 V		68 s		U TrA, v
145049	65093		16 07 20.9	+38 49 46	0.000	−0.01	7.9	1.1	0.2	K0 III		340 s		

HD	SAO	Star Name	α 2000	δ 2000	μ(α)	μ(δ)	V	B-V	M$_v$	Spec	RV	d(pc)	ADS	Notes
144889	84202		16^{h}07^{m}22^s.1	+21°49'21"	-0^s.001	-0".04	6.14	1.37	-0.3	K4 III	+57	190 s		
144183	243339		16 07 23.9	-56 11 28	0.000	-0.01	6.3	1.3	2.6	F0 V		14 s		
144608	184135	10 ω² Sco	16 07 24.2	-20 52 07	+0.003	-0.04	4.32	0.84	0.4	gG2	-5	53 s		
144316	226514		16 07 25.1	-48 25 21	+0.004	-0.02	6.9	0.2	0.6	A0 V		140 s		
144839	101938		16 07 25.6	+13 20 12	-0.006	+0.03	7.2	0.4	3.0	F2 V	-30	70 s		
144762	121375		16 07 29.6	+3 24 33	0.000	-0.02	8.0	1.1	0.2	K0 III		360 s		
144475	207348		16 07 34.7	-36 43 10	+0.004	+0.02	6.8	1.0	0.2	K0 III		190 s		
144708	159715	11 Sco	16 07 36.3	-12 44 44	-0.003	-0.03	5.60	0.1	0.6	A0 V	-25	93 s	9924	m
144874	101939	45 Ser	16 07 37.4	+9 53 30	-0.001	-0.01	5.63	0.20	2.1	A5 V	-28	48 s		
144999	84205		16 07 38.1	+28 59 43	-0.002	-0.01	7.80	0.1	1.7	A3 V	-23	170 s	9930	m
144570	207356		16 07 38.7	-30 33 47	-0.002	-0.01	7.9	0.7	0.2	K0 III		350 s		
144742	140974		16 07 43.2	-7 57 24	-0.002	0.00	8.0	1.1	0.2	K0 III		360 s		
144399	226521		16 07 43.9	-46 31 33	-0.002	-0.06	7.4	-0.4	0.6	A0 V		82 mx		
144904	101941		16 07 45.2	+15 42 38	+0.001	-0.02	7.9	1.1	-0.1	K2 III		400 s		
144383	226520		16 07 45.4	-47 58 01	0.000	-0.04	7.9	1.9	-0.3	K5 III		180 s		
144922	101943		16 07 45.5	+16 44 03	-0.001	-0.02	7.9	0.4	3.0	F2 V		94 s		
144905	101942		16 07 48.9	+12 22 24	+0.001	-0.03	7.4	0.9	3.2	G5 IV		70 s		
144661	184142		16 07 51.8	-24 27 43	-0.001	-0.02	6.33	-0.06	-1.0	B7 IV		290 s		
143720	257363		16 07 57.6	-71 56 38	-0.004	-0.01	7.1	0.3	2.6	F0 V		79 s		
145201	45903		16 08 01.0	+40 37 39	0.000	-0.01	7.7	1.1	0.2	K0 III		320 s		
144478	226524		16 08 03.9	-43 49 43	+0.001	0.00	8.0	-0.2	0.4	B9.5 V		260 s		
144591	207362		16 08 04.3	-36 13 54	0.000	-0.02	6.9	-0.3	0.4	B9.5 V		200 s		
145001	101951	7 κ Her	16 08 04.4	+17 02 49	-0.002	-0.01	5.00	0.95	0.3	G8 III	-9	87 s	9933	m
145000	101952	7 κ Her	16 08 04.8	+17 03 16	-0.002	-0.03	6.25	1.14	-0.1	K2 III	+38		9933	m
144263	243345		16 08 04.9	-58 58 14	-0.001	-0.02	6.9	0.7	1.4	A2 V		49 s		
144937	101947		16 08 05.6	+10 04 50	-0.002	-0.03	6.7	0.1	1.7	A3 V	-22	100 s		
144766	159724		16 08 07.0	-18 14 37	+0.002	-0.15	7.0	0.6	4.4	G0 V		32 s		
144663	184143		16 08 07.1	-27 43 50	-0.001	-0.01	7.17	1.58		K5				
144690	184144		16 08 07.5	-26 19 36	+0.008	-0.01	5.38	1.65	-0.5	gM2	-18	140 s		
144386	243355		16 08 10.7	-53 04 44	0.000	0.00	7.0	1.0	-0.3	K5 III		290 s		
143326	257359		16 08 10.8	-77 17 21	-0.008	-0.02	7.2	0.0	0.4	B9.5 V		140 mx		
144821	159728		16 08 16.3	-13 46 11	-0.010	-0.11	7.4	0.7	4.4	G0 V		38 s		
144417	243357		16 08 16.4	-51 38 10	-0.007	-0.01	7.4	1.4	-0.1	K2 III		240 s		
143389	257360		16 08 27.0	-77 14 19	-0.002	0.00	7.0	1.1	0.2	K0 III		230 s		
144892	140981		16 08 27.1	-10 06 08	+0.002	-0.11	6.70	0.50	3.8	F7 V		38 s	9932	m
145002	121383	47 Ser	16 08 27.9	+8 32 03	-0.001	-0.01	5.73	1.61	-0.5	gM3	-22	180 s		
144231	253437		16 08 33.0	-62 58 54	+0.001	-0.01	6.89	-0.08		B9				
144668	207367		16 08 34.1	-39 06 19	-0.002	-0.03	6.65	-0.07	0.6	A0 V		160 s		m
144667	207368		16 08 34.3	-39 05 35	-0.002	-0.04	7.05	0.36	0.6	A0 V		120 s		m
145050	121385		16 08 36.4	+8 36 48	-0.002	-0.03	6.6	1.6	-0.5	M4 III	+52	270 s		
145051	121386		16 08 40.7	+5 24 22	0.000	+0.02	7.9	1.1	0.2	K0 III		340 s		
144844	184164		16 08 43.6	-23 41 07	-0.001	-0.03	5.88	0.02	0.2	B9 V		140 s		
145389	45911	11 φ Her	16 08 46.0	+44 56 05	-0.003	+0.04	4.26	-0.07		B9 p	-16	23 mn		
145122	101962	8 Her	16 08 46.5	+17 12 21	-0.002	-0.03	6.10	0.00	0.6	A0 V	-16	130 s		
145286	65106		16 08 47.8	+35 33 01	-0.001	+0.03	7.7	0.2	2.1	A5 V		130 s		
144323	253442		16 08 50.7	-61 48 01	0.000	0.00	7.7	1.9	-0.1	K2 III		130 s		
145147	101963		16 08 51.6	+15 47 46	0.000	-0.01	7.6	1.4	-0.3	K5 III		380 s		
144479	243362		16 08 53.8	-54 31 54	0.000	-0.01	7.95	0.04	0.0	B8.5 V		370 s		
145328	65108	16 τ CrB	16 08 58.2	+36 29 27	-0.005	+0.33	4.76	1.01	0.2	K0 III	-18	52 mx	9939	m
145085	121390		16 08 58.8	+3 27 16	-0.002	+0.01	5.91	1.47	-0.3	gK5	+9	180 s		
145004	140990		16 09 01.3	-8 13 07	0.000	0.00	7.6	1.1	0.2	K0 III		300 s		
145005	140989		16 09 02.0	-9 38 54	-0.002	-0.04	7.3	0.2	2.1	A5 V		100 s		
144925	159731		16 09 02.5	-18 59 44	0.000	-0.03	7.6	0.1	0.6	A0 V		210 s		
145674	29823		16 09 02.7	+57 56 15	-0.004	+0.02	6.30	0.1	0.6	A0 V	-6	140 s	9944	m
145202	101964		16 09 05.7	+17 22 00	0.000	+0.02	7.7	0.1	1.4	A2 V		180 s		
144534	243365		16 09 08.9	-56 03 27	0.000	0.00	6.50	1.1	0.2	K0 III		180 s		
145532	29819		16 09 09.1	+50 10 51	-0.002	+0.02	6.9	1.1	0.2	K0 III		220 s		
144880	207388		16 09 11.0	-32 06 00	0.000	-0.30	7.45	0.52		G0		22 mn		
145148	121392		16 09 11.1	+6 22 44	+0.016	-0.73	5.97	1.00	3.2	K0 IV	-4	26 mx		
144556	243369		16 09 12.1	-55 10 46	+0.001	0.00	7.27	1.78		K5				
145184	101965		16 09 12.4	+12 55 55	-0.001	+0.07	7.8	0.4	2.6	F0 V		110 s		
144009	257366		16 09 13.4	-71 04 10	-0.040	-0.39	7.25	0.72	5.6	G8 V		25 ts		
144106	--		16 09 16.1	-69 35 30			7.8	1.1	0.2	K0 III		330 s		
145435	45913		16 09 17.3	+41 05 44	-0.019	+0.07	6.9	0.5	3.4	F5 V		50 s		
144480	243368	z² Nor	16 09 18.4	-57 56 03	-0.001	-0.05	5.57	-0.04	0.2	B9 V	0	100 mx		
145313	84217		16 09 19.3	+27 22 31	+0.001	-0.01	8.0	0.4	2.6	F0 V		120 s		
145056	159741		16 09 22.4	-12 22 34	-0.002	0.00	7.60	0.05	0.6	A0 V		240 s		
145057	159740		16 09 22.9	-13 03 11	-0.002	-0.02	7.13	0.22	1.4	A2 V		110 s		
145870	16979		16 09 23.4	+63 24 27	0.000	-0.01	6.7	0.1	1.7	A3 V		100 s		
145694	29825		16 09 25.9	+55 49 44	-0.005	+0.02	6.5	1.1	0.2	K0 III	-15	180 s		
145229	101968		16 09 26.6	+11 34 28	-0.006	+0.10	7.6	0.7	4.4	G0 V		44 s		
144926	207395		16 09 26.9	-31 03 09	-0.004	-0.05	7.80	0.4	2.6	F0 V		110 s		m
145228	101970		16 09 28.0	+11 44 52	+0.001	0.00	7.0	0.4	3.0	F2 V	-54	61 s		
145059	159743		16 09 30.0	-17 11 16	-0.002	-0.08	8.00	0.6	4.5	G1 V	+45	48 s		
144927	207397		16 09 32.2	-32 38 56	-0.003	-0.04	6.19	0.79	3.2	G5 IV		38 s		m

HD	SAO	Star Name	α 2000	δ 2000	μ(α)	μ(δ)	V	B-V	M_v	Spec	RV	d(pc)	ADS	Notes
144793	226540		16ʰ 09ᵐ 32.5ˢ	-43° 39′ 14″	0.000ˢ	+0.01″	7.8	1.9	0.2	K0 III		130 s		
144696	243383		16 09 38.5	-52 06 55	+0.001	-0.03	7.6	0.2	0.6	A0 V		180 s		
144985	184175		16 09 40.6	-27 53 42	-0.002	-0.07	7.90	1.1	-0.1	K2 III		170 mx		m
145204	121396		16 09 41.3	+0 49 11	0.000	-0.03	6.7	1.1	-0.1	K2 III		230 s		
144628	243380		16 09 42.7	-56 26 43	-0.015	+0.32	7.12	0.86	6.9	K3 V		13 ts		
145206	141001		16 09 50.4	-3 28 00	-0.002	-0.01	5.37	1.45	-0.3	K4 III	-46	140 s		
144987	207403		16 09 52.4	-33 32 46	-0.002	-0.06	5.5	0.2	0.0	B8.5 V	-45	91 s		
145153	159748		16 09 53.3	-12 53 10	+0.002	-0.01	7.47	1.18	0.3	G8 III	-15	180 s		
145100	159745		16 09 55.0	-18 20 27	-0.006	-0.08	6.47	0.44	3.4	dF5		41 s		
145886	16980		16 09 56.0	+59 52 17	-0.001	-0.02	6.7	0.1	0.6	A0 V		170 s		
145101	184181		16 09 57.9	-22 24 35	-0.002	-0.03	7.4	0.6	1.7	A3 V		66 s		
145247	141003		16 09 58.8	-1 36 57	+0.002	-0.04	7.8	0.4	3.0	F2 V		89 s		
144481	253444		16 09 59.2	-62 58 00	-0.001	-0.05	6.49	0.23		A2				
144988	207405		16 10 00.4	-35 38 32	+0.002	-0.06	7.2	0.9	4.4	G0 V		23 s		
144814	243389		16 10 02.9	-50 45 35	+0.002	0.00	7.2	1.1	-0.3	K5 III		320 s		
145457	84223		16 10 03.8	+26 44 34	-0.002	+0.04	6.6	0.9	0.2	G9 III	-5	190 s		
145458	84224		16 10 06.3	+25 29 15	-0.001	+0.01	7.38	0.94	-0.9	G8 II-III		450 s		
145127	184182		16 10 09.7	-24 34 57		-0.03	6.6	0.1	0.6	A0 V		140 s		
144965	207407		16 10 10.5	-40 07 45	0.000	-0.03	7.06	0.15	-1.0	B5.5 V		300 s		m
145316	121403		16 10 12.1	+1 36 04	+0.001	-0.06	6.5	1.1	0.2	K0 III		180 s		
145605	45929		16 10 15.6	+40 33 19	-0.007	+0.08	7.6	0.5	3.4	F5 V		69 s		
145727	45935		16 10 15.8	+48 36 42	-0.004	0.00	7.5	0.8	3.2	G5 IV		73 s		
145102	184184		16 10 16.0	-26 54 32	+0.001	-0.03	6.59	0.07	0.2	B9 V p		190 s		
142022	258738		16 10 16.0	-84 13 52	-0.216	-0.02	7.69	0.78	5.9	K0 V		23 s		m
144947	226556		16 10 18.8	-43 39 24	+0.001	-0.02	7.9	1.4	3.2	G5 IV		38 s		
145188	184188		16 10 19.7	-22 09 27	-0.001	-0.04	6.9	0.1		A0 V		160 s		
145103	184186		16 10 23.3	-29 17 01	+0.001	-0.03	7.4	2.2	-0.5	M2 III		170 s		
145675	45933		16 10 24.1	+43 49 03	+0.011	-0.30	6.67	0.90	6.1	dK1	-6	16 ts		
145189	184189		16 10 25.3	-23 06 24	-0.001	-0.03	7.5	0.0	1.4	A2 V		170 s		
144596	253446		16 10 27.2	-62 55 22	+0.003	-0.01	7.45	-0.03		B9				
144816	243394		16 10 31.1	-55 20 47	-0.002	-0.01	7.62	0.08		B3				m
145768	45940		16 10 31.7	+47 48 23	-0.001	-0.01	7.70	1.1	0.2	K0 III		320 s	9950	m
145832	29830		16 10 32.2	+53 02 33	+0.002	0.00	8.0	1.1	-0.1	K2 III		410 s		
146585	8442		16 10 37.0	+72 23 36	-0.004	+0.01	7.8	0.1	0.6	A0 V		200 mx		
145406	121404		16 10 37.4	+2 37 16	-0.001	0.00	7.9	1.1	0.2	K0 III		340 s		
145436	101985		16 10 42.3	+10 02 16	-0.002	+0.01	7.4	0.5	3.4	F5 V		64 s		
144858	243403		16 10 46.8	-56 04 17	-0.002	0.00	7.16	0.01	-1.0	B5.5 V		360 s		
145348	141014		16 10 48.2	-8 17 59	+0.002	-0.02	7.9	0.1	1.7	A3 V		170 s		
146926	8446	19 UMi	16 10 49.3	+75 52 39	-0.002	+0.01	5.5	-0.1		B8	-1			
145645	65123		16 10 54.0	+30 59 08	-0.006	-0.17	7.8	0.8	3.2	G5 IV		54 mx		
145644	65125		16 10 55.5	+32 00 10	-0.001	-0.02	8.0	0.1	1.4	A2 V		210 s		
145801	45943		16 10 59.2	+43 03 16	-0.003	0.00	7.9	1.1	0.2	K0 III		340 s		
145277	184198		16 11 01.7	-25 07 04	0.000	+0.01	8.0	1.3	0.2	K0 III		250 s		
145250	184197		16 11 01.9	-29 24 58	-0.007	-0.09	5.1	1.1	0.2	K0 III	-27	94 s		
145676	84233		16 11 07.8	+26 53 32	+0.002	-0.08	7.4	0.5	3.4	F5 V		63 s		
145549	101990		16 11 08.1	+16 18 31	+0.003	-0.13	8.0	0.5	3.4	F5 V		82 s		
145588	101991		16 11 09.1	+17 55 37	0.000	+0.01	8.0	0.1	0.6	A0 V		300 s		
145173	207423		16 11 09.5	-40 07 43	-0.005	-0.11	7.9	0.2	3.0	F2 V		79 mx		
145873	45953		16 11 11.8	+47 33 36	-0.001	+0.04	7.80	0.4	2.6	F0 V		110 s	9956	m
145061	243413		16 11 14.6	-50 32 36	+0.001	-0.01	7.2	0.8	1.4	A2 V		52 ts		
144483	---		16 11 16.0	-69 29 15			8.0	1.1	0.2	K0 III		360 s		
145191	226564		16 11 17.4	-41 07 11	-0.009	-0.13	5.86	0.27		F0				
145158	226562		16 11 18.7	-45 19 59	-0.009	-0.05	6.64	0.48	4.0	F8 V		34 s		
145568	121412		16 11 23.0	+7 27 57	+0.001	-0.01	8.0	1.1	-0.1	K2 III		410 s		
145729	84239		16 11 23.7	+28 26 19	-0.007	-0.07	7.6	0.5	4.0	F8 V		53 s		
145646	101992		16 11 25.9	+16 58 38	-0.004	-0.01	7.9	1.4	-0.3	K5 III		250 mx		
145547	101994		16 11 28.6	+16 39 56	0.000	0.00	6.08	0.02	0.6	A0 V	-15	120 s		
145589	121414		16 11 29.6	+9 42 45	+0.001	-0.01	6.53	0.24		A3	-27			m
145353	184205		16 11 33.5	-27 09 02	-0.001	-0.02	6.93	0.12	0.2	B9 V		180 s		
145713	84240	10 Her	16 11 37.9	+23 29 41	-0.002	-0.01	5.8	1.6	-0.5	M4 III	-25	190 s		LQ Her, q
145802	65129		16 11 39.6	+33 20 33	+0.001	+0.01	6.29	1.20	0.2	K0 III	0	130 s	9958	m
145931	45957		16 11 47.5	+42 22 28	-0.001	+0.03	5.87	1.45	-2.3	K4 II	-21	430 s	9962	m
145849	65132		16 11 47.9	+36 25 29	-0.001	-0.04	5.63	1.34	-0.2	K3 III	-31	140 s		m
145428	184209		16 11 51.0	-25 52 58	-0.011	-0.03	7.7	0.9	3.2	G5 IV		67 s		
145110	243422		16 11 57.6	-54 21 12	-0.002	-0.03	6.53	0.02	0.4	B9.5 V		160 s		
145501	159763	14 ν Sco	16 11 58.5	-19 26 59	-0.001	-0.01	6.27	0.12	0.0	A0 IV		170 s	9951	m
145502	159764	14 ν Sco	16 11 59.6	-19 27 38	-0.001	-0.03	4.00	0.05	-2.8	B2 IV-V	-7	170 s	9951	
145570	141022	15 ψ Sco	16 11 59.9	-10 03 51	-0.001	-0.02	4.94	0.09		A m	-6	26 mn		
145518	159765		16 12 02.2	-18 13 57	+0.001	-0.03	7.5	0.7	4.4	G0 V		40 s		
145519	159767		16 12 05.8	-19 03 44	-0.002	-0.02	7.98	0.25		B9				
145957	65139		16 12 06.4	+39 03 19	0.000	-0.02	6.7	1.1	0.2	K0 III		200 s		
145607	141024	16 Sco	16 12 07.1	-8 32 51	+0.003	0.00	5.43	0.12		A1 n	+5	47 mn		
145505	184216		16 12 09.8	-23 55 19	-0.001	-0.04	8.00	0.1	0.6	A0 V		300 s		m
146079	45964		16 12 14.9	+46 13 33	-0.015	+0.22	7.9	0.5	4.0	F8 V		54 mx		
145483	184217	12 Sco	16 12 15.8	-28 25 02	-0.002	-0.05	5.68	0.01	0.2	B9 V		120 s	9953	m

HD	SAO	Star Name	α 2000	δ 2000	μ(α)	μ(δ)	V	B-V	M_v	Spec	RV	d(pc)	ADS	Notes
			h m s	° ′ ″	s	″								
145357	226568		16 12 16.7	−42 22 31	−0.001	−0.03	8.00	0.4	2.6	F0 V		120 s		m
144702	253453		16 12 17.1	−69 37 16	−0.004	−0.06	7.9	0.0	0.0	B8.5 V		94 mx		
145911	65140		16 12 17.3	+32 39 18	−0.008	+0.02	7.9	1.1	0.2	K0 III		160 mx		
145482	184221	13 Sco	16 12 18.1	−27 55 35	−0.001	−0.03	4.58	−0.16	−2.1	B2.5 V n	+10	220 s		
144888	253458		16 12 18.9	−65 21 53	−0.009	−0.06	8.0	1.4	0.2	K0 III		170 mx		
145554	159770		16 12 21.7	−19 34 44	−0.001	−0.02	7.65	0.14	0.2	B9 V		250 s		
145020	253459		16 12 22.3	−61 55 44	−0.001	−0.03	7.3	0.2	0.6	A0 V		160 s		
146603	16993		16 12 25.0	+67 08 39	−0.005	−0.04	6.21	0.99	0.3	G8 III	−10	140 s		
144300	257368		16 12 27.0	−75 17 35	−0.009	−0.03	7.90	0.25		A3				
147142	8452	20 UMi	16 12 32.0	+75 12 38	−0.012	+0.03	6.4	1.1	−0.1	K2 III	−26	200 s		
145445	207447		16 12 39.3	−39 44 31	−0.001	−0.01	7.8	0.6	3.0	F2 V		61 s		
146154	45968		16 12 42.8	+45 53 39	−0.003	+0.01	7.6	1.1	0.2	K0 III		310 s		
144633	257370		16 12 42.9	−71 34 30	+0.001	−0.01	7.9	0.0	0.4	B9.5 V		320 s		
146080	45965		16 12 43.2	+40 46 54	−0.001	0.00	7.80	0.1	0.6	A0 V		280 s	9968	m
145631	159777		16 12 44.0	−19 30 11	0.000	−0.04	7.58	0.15	0.4	B9.5 V		220 s		
145976	84247		16 12 45.2	+26 40 15	+0.004	−0.03	6.50	0.39	3.0	F2 V	−8	49 s	9966	m
145774	141028		16 12 45.5	−1 43 20	−0.001	−0.02	7.5	0.0	0.0	B8.5 V	+30	310 s		
146025	65147		16 12 50.9	+32 35 58	0.000	+0.01	7.9	1.1	−0.1	K2 III	−7	390 s		
145912	102014		16 12 51.3	+19 06 04	−0.003	−0.03	7.4	1.1	−0.1	K2 III		310 s		
145788	141031		16 12 56.4	−4 13 15	−0.002	−0.02	6.25	0.13		A0	−16			
145891	102012		16 12 58.0	+12 48 02	+0.003	+0.02	7.0	0.1	1.7	A3 V	−24	110 s		
145324	243442		16 13 02.7	−54 21 38	0.000	−0.01	7.8	0.5	1.7	A3 V		87 s		
146010	84250		16 13 08.0	+21 33 58	−0.001	+0.01	6.6	0.1	1.4	A2 V	−23	110 s		
145934	102017		16 13 09.7	+13 14 21	−0.001	−0.01	8.0	1.1	0.2	K0 III		360 s		
145748	159784		16 13 10.9	−15 07 00	−0.001	−0.01	7.40	1.6	−0.4	M0 III	−34	290 s		
145852	141033		16 13 11.1	−0 31 09	−0.004	−0.03	7.1	1.1	0.2	K0 III		180 mx		
145412	226576		16 13 11.8	−49 53 05	−0.001	−0.02	6.85	0.17	0.6	A0 V		140 s		
145596	207462		16 13 12.3	−34 18 26	0.000	−0.02	7.5	1.8	−0.5	M2 III		290 s		
146135	65153		16 13 12.8	+38 04 20	−0.002	+0.02	7.1	0.4	3.0	F2 V		65 s		
145853	141034		16 13 14.5	−1 24 08	−0.003	−0.08	8.0	0.5	4.0	F8 V		62 s		
145510	226580		16 13 14.8	−44 04 27	+0.004	−0.03	7.2	0.9	2.6	F0 V		35 s		
148589	2686		16 13 15.1	+82 37 55	−0.019	+0.05	7.9	0.1	1.7	A3 V		90 mx		
145892	121431	9 Her	16 13 15.4	+5 01 16	+0.003	−0.01	5.48	1.47	−0.3	gK5	−2	140 s		
145384	243448		16 13 16.7	−53 40 17	−0.002	−0.02	5.9	2.1	−0.5	M2 III		93 s		
145958	102018		16 13 18.2	+13 31 40	+0.012	−0.41	6.68	0.76	5.9	dK0	+18	20 ts	9969	m
145913	121432		16 13 18.7	+6 02 17	0.000	+0.01	7.9	0.2	2.1	A5 V		150 s		
145361	243449		16 13 22.4	−55 32 27	−0.012	−0.04	5.81	0.34	1.7	F0 IV		63 s		
145342	243447		16 13 24.3	−56 24 53	−0.006	−0.03	7.90	0.5	4.0	F8 V		60 s		m
145876	141036		16 13 25.6	−4 03 06	−0.002	−0.02	6.9	0.2	2.1	A5 V		89 s		
146211	65159		16 13 27.4	+39 49 22	−0.001	−0.04	7.2	1.1	0.2	K0 III		260 s		
145114	253460		16 13 27.7	−64 59 08	−0.003	−0.01	7.8	0.0	0.4	B9.5 V		290 s		
145397	243454	κ Nor	16 13 28.6	−54 37 50	0.000	−0.03	4.94	1.04	0.3	G4 III	−14	61 s		m
145854	159789		16 13 37.6	−13 02 05	+0.001	+0.01	7.9	0.1	0.6	A0 V		230 s		
145736	184234		16 13 38.2	−29 03 25	−0.004	−0.03	7.5	0.2	4.0	F8 V		49 s		
145936	141040		16 13 38.4	−1 28 32	−0.004	−0.01	6.5	1.1	0.2	K0 III		180 s		
145809	184240		16 13 40.2	−21 24 00	−0.009	+0.01	6.7	0.4	4.4	G0 V	+17	29 s		
145894	141039		16 13 45.2	−8 07 03	−0.003	+0.01	6.84	1.11		K0				
145792	184241		16 13 45.5	−24 25 19	+0.001	−0.02	6.41	0.04	−1.0	B7 IV		260 s	9967	m
146168	84258		16 13 47.6	+28 43 54	−0.003	−0.02	7.70	0.4	3.0	F2 V		87 s	9973	m
145417	243462		16 13 48.3	−57 34 12	−0.108	−1.40	7.53	0.81	6.6	K0 V	+11	15 t		
146026	121437		16 13 50.5	+6 59 38	0.000	+0.01	7.4	0.1	1.7	A3 V		140 s		
145897	159793	17 χ Sco	16 13 50.7	−11 50 15	−0.001	−0.01	5.4	1.1	−0.2	K3 III	−25	120 s		
145779	207476		16 13 50.8	−30 12 43	0.000	−0.03	7.9	0.2	4.0	F8 V		59 s		
146278	45974		16 13 51.3	+40 20 55	−0.003	−0.11	7.8	1.1	0.2	K0 III		130 mx		
146081	102021		16 13 52.3	+16 26 02	0.000	0.00	7.9	1.6	−0.5	M2 III		470 s		
146101	102024		16 13 53.7	+18 29 09	−0.001	0.00	8.0	0.4	2.6	F0 V		120 s		
144869	—		16 13 59.2	−72 19 59			8.0	0.5	4.0	F8 V		63 s		
145836	184246		16 14 02.0	−23 46 27	+0.003	−0.01	8.0	0.3	3.2	G5 IV		93 s		
145542	243472		16 14 02.0	−51 54 23	+0.001	−0.01	7.9	1.0	−0.3	K5 III		430 s		
146047	121440		16 14 04.8	+4 01 38	−0.001	−0.02	7.9	0.9	3.2	G5 IV		88 s		
145449	243468		16 14 05.3	−58 02 22	+0.002	+0.01	7.7	0.6	0.2	K0 III		310 s		
145597	226588		16 14 07.8	−48 55 58	+0.002	−0.02	7.3	0.2	−0.1	K2 III		300 s		
145523	243474		16 14 11.1	−54 10 16	+0.001	−0.06	7.82	0.58	4.4	G0 V		48 s		
145825	207479		16 14 11.9	−31 39 49	−0.005	−0.26	6.6	1.0	4.4	G0 V		16 s		
146084	121443		16 14 13.4	+5 54 08	+0.002	−0.01	6.31	1.15	−0.2	gK3	−21	200 s		
145942	184251		16 14 20.4	−22 08 09	0.000	−0.01	7.6	0.3	1.4	A2 V		120 s		
146051	141052	1 δ Oph	16 14 20.6	−3 41 39	−0.003	−0.15	2.74	1.58	−0.5	M1 III	−20	43 s		Yed Prior, m
146102	121445		16 14 22.0	+2 38 49	0.000	−0.07	7.1	0.5	3.4	F5 V	+3	55 s		
145838	207480		16 14 22.2	−33 00 41	−0.001	−0.04	6.0	1.0	0.2	K0 III		140 s		
145964	184253		16 14 28.7	−21 06 27	−0.002	−0.03	6.3	0.2	0.6	A0 V		99 s		
147321	8462		16 14 33.1	+73 23 41	−0.005	+0.03	5.98	0.08		A0	−15			
146116	141056		16 14 38.3	−0 23 56	+0.001	−0.02	7.8	1.1	0.2	K0 III		330 s		
145997	159807		16 14 39.0	−18 32 07	−0.008	−0.11	6.3	1.1	0.2	K0 III		75 mx		
146361	65165	17 σ CrB	16 14 40.6	+33 51 30	−0.022	−0.08	5.22	0.58	4.4	G0 V	−11	21 ts	9979	m
146169	121451		16 14 42.1	+7 51 30	−0.001	−0.01	6.8	1.1	0.2	K0 III	−20	210 s		

HD	SAO	Star Name	α 2000	δ 2000	μ(α)	μ(δ)	V	B−V	M$_v$	Spec	RV	d(pc)	ADS	Notes
			16^{h}14^{m}47^s.9	+18°13′00″	0^s.000	−0″.01								
146264	102033		16 14 47.9	+18 13 00	0.000	−0.01	7.5	0.9	0.3	G8 III	−21	270 s		
147231	8459		16 14 49.8	+70 55 46	−0.005	−0.29	7.8	0.8	3.2	G5 IV	−18	69 mx		
146029	184259		16 14 53.3	−22 22 48	0.000	−0.01	7.38	0.07	0.2	B9 V		240 s		
146001	184258		16 14 53.5	−25 28 37	+0.001	−0.02	6.05	0.04	−0.6	B8 IV		190 s		
145566	243485		16 14 55.1	−59 41 32	−0.006	−0.04	7.5	1.5	3.2	G5 IV		26 s		
145812	226596		16 14 56.2	−45 59 30	−0.003	−0.03	8.0	0.2	3.4	F5 V		82 s		
145880	207487		16 14 57.5	−39 37 40	−0.002	−0.04	7.1	0.3	0.6	A0 V		130 s		
146279	102034		16 15 01.2	+11 29 24	0.000	−0.08	7.5	0.4	2.6	F0 V		94 s		
146193	141061		16 15 04.1	−1 46 30	−0.002	−0.02	7.0	0.1	0.6	A0 V		190 s		
145666	243495		16 15 04.1	−56 22 27	−0.023	−0.22	7.2	1.1	4.4	G0 V		17 s		
147023	17011		16 15 06.4	+65 04 21	−0.007	+0.06	8.0	1.1	0.2	K0 III		240 mx		
146638	45993		16 15 11.8	+47 02 25	−0.002	+0.03	7.9	1.1	0.2	K0 III		350 s		
146489	65174		16 15 14.7	+34 42 46	−0.003	−0.05	7.3	1.1	0.2	K0 III		260 s		
145842	226600	θ Nor	16 15 15.2	−47 22 20	−0.003	−0.05	5.3	0.2	0.0	B8.5 V	+2	80 s		
146070	184262		16 15 19.0	−27 12 37	−0.005	−0.14	7.5	0.5	4.4	G0 V		42 s		
145921	226603		16 15 23.8	−42 53 58	−0.002	−0.01	6.14	1.11	0.2	gK0		130 s		
146213	159814		16 15 24.6	−11 02 49	−0.011	−0.02	7.6	0.7	4.4	G0 V		43 s		
145544	253474	δ TrA	16 15 26.2	−63 41 08	+0.001	−0.01	3.85	1.11	−2.1	G2 II	−5	110 s		m
146388	102040	16 Her	16 15 28.6	+18 48 30	−0.005	−0.08	5.69	1.12	0.2	gK0	−18	100 s		
146214	159816		16 15 33.2	−12 40 48	+0.001	0.00	7.3	0.1	0.6	A0 V		190 s		
146233	141066	18 Sco	16 15 37.1	−8 22 11	+0.015	−0.50	5.50	0.65	4.5	G1 V	+11	17 ts		m
145815	243513		16 15 42.8	−54 02 50	0.000	−0.03	7.97	0.46	3.4	F5 V		80 s		
146870	29858		16 15 47.1	+53 14 12	+0.001	−0.06	7.1	0.4	2.6	F0 V		78 s		
146537	84269		16 15 47.2	+27 25 20	−0.002	−0.04	6.14	1.31	−0.1	K2 III	−11	150 s		
146234	159819		16 15 47.4	−15 35 38	−0.002	0.00	7.9	0.6	4.4	G0 V		48 s		
145782	243509		16 15 49.6	−57 54 44	−0.002	−0.05	5.63	0.14	0.0	A0 IV		110 s		
146216	159818		16 15 50.4	−20 04 36	−0.002	−0.04	7.7	0.1	2.6	F0 V		100 s		
146639	65183		16 15 50.6	+36 33 14	−0.001	−0.03	7.0	1.4	−0.3	K5 III		290 s		
146254	159821		16 15 51.4	−14 50 55	+0.001	+0.01	6.09	0.06		A0	−11			
146255	159822		16 15 53.6	−15 53 15	0.000	−0.05	8.0	1.1	0.2	K0 III		290 s		
146658	65184		16 15 56.4	+35 52 13	−0.001	−0.03	8.0	1.4	−0.3	K5 III		450 s		
146452	102046		16 15 59.7	+11 25 31	−0.003	+0.04	7.60	1.1	0.2	K0 III	−26	300 s		m
146177	207494		16 16 03.8	−30 37 20	0.000	−0.04	7.5	0.5	2.6	F0 V		72 s		m
146057	226612		16 16 08.9	−43 07 30	+0.001	0.00	7.1	0.6	0.2	K0 III		240 s		
147093	17016		16 16 11.4	+60 53 24	−0.006	0.00	8.0	1.1	0.2	K0 III		360 s		
146588	102051		16 16 14.4	+19 31 29	+0.002	+0.27	7.7	0.6	4.4	G0 V		47 s		
146604	84274		16 16 16.6	+23 07 22	−0.001	−0.01	6.5	1.1	0.2	K0 III	+14	180 s		
146433	141069		16 16 19.5	−1 38 55	−0.008	+0.01	7.00	0.5	4.0	F8 V		40 s	9984	m
146512	121465		16 16 23.9	+8 02 09	0.000	−0.02	7.9	0.9	3.2	G5 IV		88 s		
146285	184277		16 16 25.0	−24 59 20	−0.001	−0.03	7.94	0.23	−0.2	B8 V		280 s		
146284	184278		16 16 26.6	−24 16 56	0.000	−0.01	6.70	0.16	−0.2	B8 V		180 s		m
145987	243524		16 16 31.4	−54 10 42	0.000	−0.01	8.0	0.9	−0.1	K2 III		410 s		
146828	65195		16 16 42.4	+38 38 41	+0.001	−0.01	8.0	1.1	−0.1	K2 III	−34	420 s		
146003	243526		16 16 43.0	−53 48 39	0.000	0.00	5.44	1.73	−0.5	M2 III	−28	130 s		
146738	84281	18 υ CrB	16 16 44.6	+29 09 00	+0.001	−0.02	5.78	0.07	0.0	A3 III	+2	140 s	9990	m
146871	46006		16 16 46.3	+40 02 19	−0.001	+0.03	6.8	1.1	−0.1	K2 III		240 s		
145421	—		16 16 46.7	−71 54 37			7.9	0.1	0.6	A0 V		290 s		
146643	102056		16 16 48.1	+13 21 00	−0.002	−0.02	8.0	1.1	−0.1	K2 III		410 s		
146093	226617		16 16 48.1	−49 27 45	0.000	−0.02	7.4	0.5	1.4	A2 V		160 s		
146490	141074		16 16 48.3	−5 29 49	0.000	−0.02	7.2	0.1	1.4	A2 V		150 s		
146332	184280		16 16 52.5	−29 44 39	+0.001	−0.02	7.60	0.19	−3.7	B5 II		1300 s	9983	m
147187	17022		16 16 52.9	+60 32 06	+0.013	+0.01	8.0	0.9	3.2	G5 IV		93 s		
146124	226618		16 16 52.9	−49 51 23	−0.002	−0.03	7.69	0.77	0.2	K0 III		320 s		
146514	141075		16 16 55.1	−3 57 12	+0.003	+0.01	6.18	0.31	2.2	dA6 n	−8	63 s		
146698	102059		16 16 57.4	+18 50 47	+0.001	+0.01	7.5	1.1	−0.1	K2 III	+9	340 s		
146416	184285		16 16 58.7	−21 18 14	−0.001	−0.02	6.61	0.02	0.4	B9.5 V		170 s		
146059	243533		16 16 58.8	−53 41 46	−0.007	−0.10	6.41	0.78	5.2	G5 V		15 s		m
146436	159829		16 16 59.0	−20 06 14	−0.001	0.00	6.5	1.1	0.2	K0 III	−33	170 s		
146143	226619	γ¹ Nor	16 17 00.7	−50 04 05	0.000	0.00	4.99	0.80	−6.3	F8 Iab	−19	1700 s		
144266	258740		16 17 00.7	−80 47 58	+0.015	0.00	7.7	1.1	0.2	K0 III		320 s		
146183	226621		16 17 03.3	−47 15 13	−0.002	−0.09	7.9	0.2	2.1	A5 V		60 mx		
145689	253481		16 17 05.4	−67 56 29	−0.007	−0.10	5.75	0.15	1.4	A2 V		54 mx		
146829	84287		16 17 09.2	+29 25 42	−0.001	+0.02	7.6	0.4	2.6	F0 V		98 s		
147160	29870		16 17 09.7	+56 55 06	−0.003	+0.01	8.0	0.4	3.0	F2 V		98 s		
145547	257376		16 17 13.2	−70 59 29	0.000	0.00	7.3	0.1	1.4	A2 V		150 s		
147232	29874		16 17 15.2	+59 45 18	0.000	+0.02	5.40	1.63	−0.5	M4 III	−36	150 s		AT Dra, v
146830	84288		16 17 19.0	+27 38 41	−0.002	0.00	7.9	0.1	0.6	A0 V		280 s		
146145	243544		16 17 20.7	−53 05 11	−0.006	−0.06	6.33	0.27	2.1	A5 V		57 mx		m
146543	159831		16 17 20.9	−15 19 40	+0.003	0.00	7.5	0.4	2.6	F0 V	+6	87 s		
148048	8470	21 η UMi	16 17 30.2	+75 45 19	−0.023	+0.25	4.95	0.37	2.6	dF0	−10	28 ts		m
147291	17030		16 17 31.3	+61 22 07	+0.015	−0.07	8.0	0.4	3.0	F2 V		98 s		
145585	—		16 17 36.6	−71 40 30			7.6	0.0	0.4	B9.5 V		280 s		
146760	121479		16 17 36.9	+7 44 02	−0.001	0.00	7.7	1.1	0.2	K0 III		310 s		
146740	121478		16 17 43.3	+1 29 48	0.000	0.00	6.5	1.1	0.2	K0 III		180 s		
146418	207519		16 17 45.2	−37 45 15	−0.004	−0.09	7.5	0.3	3.0	F2 V		80 s		

HD	SAO	Star Name	α 2000	δ 2000	μ(α)	μ(δ)	V	B−V	M_V	Spec	RV	d(pc)	ADS	Notes
			$16^h\,17^m\,45\overset{s}{.}8$	$+31°\,48'\,14''$	$+0\overset{s}{.}012$	$+0\overset{''}{.}29$								
146946	65206		16 17 45.8	+31 48 14	+0.012	+0.29	6.9	0.6	4.4	G0 V	−32	32 s		
146224	243553		16 17 48.5	−53 33 10	+0.001	+0.03	7.53	−0.04	−1.6	B3.5 V		560 s		
146662	159834		16 17 49.8	−13 22 20	−0.003	−0.03	7.3	1.1	0.2	K0 III		220 s		
147005	65209		16 17 51.2	+35 41 57	0.000	−0.02	7.5	0.1	0.6	A0 V		240 s		
146915	84290		16 17 51.4	+23 36 26	+0.002	−0.03	7.1	1.1	−0.1	K2 III		280 s		
146872	102067		16 17 51.6	+17 46 36	−0.001	−0.01	8.0	0.1	0.6	A0 V		300 s		
146677	159836		16 17 59.5	−13 26 42	−0.003	−0.05	7.3	0.4	2.6	F0 V		81 s		
146815	121480		16 18 01.1	+6 04 43	+0.001	−0.03	7.6	0.9	−2.1	G7 II	+31	880 s		
147044	65213		16 18 05.7	+34 28 58	−0.006	−0.06	7.7	0.6	4.4	G0 V		45 s		
147407	17032		16 18 06.0	+62 25 24	+0.002	−0.04	7.5	1.1	0.2	K0 III		280 s		
147662	17036		16 18 09.2	+68 33 16	−0.009	+0.05	6.4	1.1	0.2	K0 III	−11	170 s		
146545	207525		16 18 14.2	−34 54 53	0.000	0.00	7.4	1.5	−0.3	K5 III		340 s		
147113	65216		16 18 16.1	+38 16 31	0.000	+0.01	8.0	0.1	1.4	A2 V		210 s		
146606	184300		16 18 16.1	−28 02 31	0.000	−0.03	6.9	0.8	0.6	A0 V		57 s		
146422	226636		16 18 17.3	−46 27 00	−0.001	−0.02	7.3	0.2	−0.5	M4 III		370 s		
146624	184301		16 18 17.8	−28 36 51	−0.002	−0.11	4.78	0.02	1.4	A2 V	−12	44 mx		
146791	141086	2 ε Oph	16 18 19.1	−4 41 33	+0.005	+0.04	3.24	0.96	0.3	G8 III	−10	32 ts		Yed Posterior, m
146546	207528		16 18 21.0	−36 33 56	+0.002	+0.04	7.80	0.5	4.0	F8 V		57 s		m
145619	257379		16 18 23.2	−73 02 35	−0.002	−0.01	6.9	−0.6	0.6	A0 V		190 s		
147025	84295		16 18 23.5	+25 53 48	−0.001	0.00	6.6	0.8	3.2	G5 IV	−9	48 s		
146247	243565		16 18 26.1	−58 23 33	−0.006	−0.08	7.1	1.5	0.2	K0 III		120 s		
146741	159839		16 18 27.0	−17 23 16	+0.002	−0.06	7.2	0.4	3.0	F2 V		66 s		
146706	184305		16 18 28.0	−23 16 28	−0.002	−0.03	7.55	0.14	0.2	B9 V		240 s		
145673	—		16 18 29.2	−72 57 26			7.8	1.1	0.2	K0 III		330 s		
148386	8481		16 18 32.5	+77 32 55	+0.006	−0.05	7.5	1.1	0.2	K0 III		290 s		
146444	226640		16 18 38.8	−49 24 50	−0.004	−0.01	7.71	0.05	−1.4	B4 V e		100 mx		
147006	102078		16 18 44.5	+16 36 54	+0.002	−0.03	8.0	1.1	0.2	K0 III		360 s		
146324	243583		16 18 48.8	−57 55 50	0.000	+0.01	7.93	0.03		B8				
146323	243586		16 18 51.8	−57 53 59	+0.001	0.00	6.6	0.6	4.4	G0 V	−3	27 s		S Nor, v
145308	257377		16 18 52.0	−77 08 38	−0.002	+0.02	8.0	0.1	0.6	A0 V		300 s		
146722	207546		16 18 55.4	−30 57 26	+0.001	−0.02	7.9	2.0	0.2	K0 III		88 s		
146850	159846		16 19 00.2	−14 52 21	−0.002	+0.01	5.94	1.52	0.2	gK0	−42	54 s		
147216	65223		16 19 00.5	+37 45 56	−0.001	−0.05	7.2	1.1	−0.1	K2 III		300 s		
146501	226641		16 19 02.6	−49 24 23	+0.002	−0.02	7.28	0.06	0.6	A0 V		220 s		
146834	184309		16 19 07.5	−20 13 04	+0.001	−0.01	6.29	1.08	−0.3	K5 III	+8	210 s		
146775	184308		16 19 08.4	−28 17 38	+0.005	−0.26	7.69	0.61	4.4	G0 V		42 s		
146949	141096		16 19 09.4	−6 52 35	+0.001	−0.09	6.9	0.8	3.2	G5 IV		53 s		
147352	46025		16 19 11.1	+49 02 17	−0.003	+0.03	5.91	1.37	0.2	K0 III	−32	83 s		
147275	46022		16 19 14.3	+41 39 33	−0.003	+0.01	7.9	0.2	2.1	A5 V		150 s	10006	m
146745	207551		16 19 16.9	−35 29 28	−0.004	−0.05	7.40	0.4	3.0	F2 V		74 s		m
146404	243603		16 19 17.0	−58 05 29	+0.001	0.00	7.67	1.75		Ma				
146667	226650	λ Nor	16 19 17.5	−42 40 26	+0.001	−0.01	5.45	0.10	1.7	A3 V n	−22	56 s		m
146525	243619		16 19 17.6	−52 42 19	−0.003	−0.01	8.0	0.9	0.2	K0 III		360 s		
146481	243616		16 19 20.5	−54 30 48	+0.014	−0.09	7.4	0.4	4.4	G0 V		40 s		
147062	121502		16 19 25.9	+5 32 17	−0.001	−0.03	7.7	0.6	4.4	G0 V		47 s		
146835	207557		16 19 31.3	−30 54 06	+0.005	+0.02	7.09	0.57	4.2	F9 V		38 s		
146836	207558		16 19 32.6	−30 54 24	+0.007	+0.02	5.49	0.47	0.7	F5 III	−8	91 s		m
148004	8475		16 19 32.6	+70 22 05	+0.011	0.00	7.5	0.5	4.0	F8 V		51 s		
146952	159849		16 19 35.0	−18 49 40	−0.001	+0.01	6.9	0.1	0.6	A0 V		160 s		
146596	243637		16 19 42.3	−52 46 18	−0.002	+0.01	7.98	0.08	−1.0	B5.5 V		490 s		
146630	243641		16 19 42.6	−51 02 05	−0.001	−0.02	7.4	0.3	−0.1	K2 III		310 s		
147394	46028	22 τ Her	16 19 44.3	+46 18 48	−0.002	+0.04	3.89	−0.15	−1.6	B5 IV	−14	130 s	10010	m
147204	102091		16 19 45.1	+19 50 01	0.000	+0.01	7.6	1.1	0.2	K0 III		310 s		
146711	226660		16 19 48.8	−47 13 07	−0.001	−0.02	7.90	0.4	2.6	F0 V		120 s		m
148432	8485		16 19 49.0	+76 08 34	−0.004	0.00	6.9	0.1	1.7	A3 V	−5	110 s		
146686	243643	γ² Nor	16 19 50.3	−50 09 20	−0.016	−0.06	4.02	1.08	0.3	G8 III	−29	40 s		
150075	2706		16 19 52.3	+83 41 49	−0.008	+0.01	7.1	0.1	1.4	A2 V		140 s		
147365	65233		16 19 55.1	+39 42 31	−0.011	0.00	5.50	0.4	3.0	F2 V	−29	29 ts		m
146817	226665		16 19 59.5	−40 49 07	+0.004	−0.14	7.3	0.7	4.4	G0 V		32 s		
146349	253500		16 20 00.3	−64 38 57	−0.002	−0.05	6.71	0.00		A0				m
147266	84306		16 20 04.2	+21 07 57	−0.001	−0.05	6.05	0.94	3.2	G5 IV	−25	29 s		
146380	253506		16 20 04.6	−63 56 25	+0.001	0.00	7.9	1.1	−0.1	K2 III		380 s		
147010	159860		16 20 05.4	−20 03 22	−0.001	−0.02	7.40	0.16		A2 p	−9			m
147308	84308		16 20 09.2	+27 19 59	−0.001	+0.01	7.5	0.2	2.1	A5 V		120 s		
146779	226664		16 20 11.9	−47 14 58	−0.001	−0.02	7.9	1.5	−0.1	K2 III		260 s		
146905	207566		16 20 12.0	−39 30 00	−0.003	−0.01	7.2	−0.1	0.6	A0 V		210 s		
146799	226666		16 20 13.0	−46 20 58	+0.001	+0.01	7.8	0.1	0.6	A0 V		200 s		
146906	207570		16 20 14.6	−39 37 48	+0.002	+0.01	7.1	0.2	3.2	G5 IV		60 s		
147428	46031		16 20 17.0	+40 15 33	+0.001	−0.02	8.0	0.4	3.0	F2 V		98 s		
145869	257384		16 20 17.4	−74 18 20	+0.009	+0.01	7.2	1.1	0.2	K0 III		250 s		
147395	65237		16 20 18.1	+36 58 31	−0.001	−0.03	6.7	1.6	−0.5	M2 III	−15	270 s		
147764	29901		16 20 19.3	+59 46 14	−0.004	−0.03	7.6	0.9	3.2	G5 IV		75 s		
145366	257380	δ¹ Aps	16 20 20.7	−78 41 44	−0.002	−0.04	4.68	1.69	−0.5	M4 III	−12	110 s		m
147608	29897		16 20 24.3	+52 02 20	0.000	−0.01	6.7	0.1	1.4	A2 V		110 s		
146690	243654		16 20 25.1	−55 08 23	+0.003	−0.02	5.8	1.3	0.2	K0 III		92 s		

HD	SAO	Star Name	α 2000	δ 2000	μ(α)	μ(δ)	V	B−V	M$_V$	Spec	RV	d(pc)	ADS	Notes
			h m s	° ′ ″	s	″								
147252	102093		16 20 26.3	+11 56 17	0.000	−0.04	7.9	1.1	0.2	K0 III		340 s		
145388	257381	δ² Aps	16 20 26.7	−78 40 02	0.000	−0.02	5.27	1.41	−0.3	K5 III	−10	130 s		
147135	159867		16 20 28.1	−12 54 39	−0.004	+0.03	6.6	0.4	2.6	F0 V		59 s		
147103	159865		16 20 30.2	−20 06 53	−0.003	−0.03	7.56	0.50	0.6	A0 V	+18	130 s	10005	m
146954	207574		16 20 32.5	−39 25 51	−0.001	−0.07	6.12	−0.07		A0				m
147046	184325		16 20 33.8	−29 30 54	+0.001	−0.02	7.70	0.1	0.6	A0 V		260 s	10004	m
146295	253505		16 20 33.9	−68 32 12	−0.002	−0.02	7.4	0.0	0.4	B9.5 V		250 s		
147544	46037		16 20 34.6	+45 34 43	−0.005	+0.04	7.5	1.1	0.2	K0 III		300 s		
147368	84310		16 20 36.4	+25 53 12	−0.001	−0.02	8.0	0.4	3.0	F2 V		98 s		
147084	184329	19 o Sco	16 20 38.0	−24 10 10	0.000	−0.03	4.55	0.84	−2.1	A5 II	−9	92 s		
147119	184331		16 20 43.2	−21 50 26	+0.001	0.00	7.0	1.6	−0.3	K5 III		260 s		
147733	29905		16 20 51.1	+54 23 11	−0.001	+0.04	7.8	0.1	1.7	A3 V		160 s		
146807	243670		16 20 58.5	−54 54 52	−0.001	−0.05	7.80	0.1	1.7	A3 V		160 s		m
147430	84315		16 21 10.6	+22 14 23	0.000	−0.01	7.0	1.1	−0.1	K2 III		260 s		
147165	184336	20 σ Sco	16 21 11.2	−25 35 34	−0.001	−0.02	2.89	0.13	−4.4	B1 III	0	180 s	10009	m,v
147639	46044		16 21 11.9	+43 49 09	+0.003	−0.03	7.9	0.4	2.6	F0 V		110 s		
146911	243680		16 21 12.4	−52 28 13	−0.002	−0.05	7.5	−0.4	0.2	K0 III		280 s		
147442	84317		16 21 14.6	+22 59 18	0.000	−0.04	8.00	0.5	4.0	F8 V		63 s	10017	m
147196	184337		16 21 19.0	−23 42 28	−0.002	−0.02	7.04	0.17	−1.1	B5 V		270 s		
146920	243681		16 21 22.3	−53 06 28	−0.003	−0.02	8.0	0.2	0.4	B9.5 V		230 s		
147664	46046		16 21 23.4	+41 26 05	0.000	−0.05	7.91	1.05	0.2	K0 III		330 s		
147149	207586		16 21 25.9	−33 18 05	−0.001	−0.03	7.30	0.5	3.4	F5 V		59 s		m
146980	226677		16 21 26.2	−48 32 23	0.000	0.00	7.9	1.8	−0.5	M2 III		160 s		
147001	226678		16 21 26.9	−48 11 19	−0.001	−0.03	6.7	0.4	0.0	B8.5 V		120 s		
147167	207587		16 21 34.1	−33 14 07	+0.001	−0.03	7.4	0.6	−0.1	K2 III		320 s		
147066	226681		16 21 34.4	−44 56 41	0.000	0.00	7.7	0.1	0.2	K0 III		310 s		
147370	121529		16 21 34.7	+0 04 36	−0.002	−0.04	7.0	1.1	0.2	K0 III	+38	230 s		
147923	29914		16 21 36.1	+56 52 37	−0.001	+0.01	7.6	1.6	−0.5	M2 III		410 s		
146981	243687		16 21 36.7	−51 45 03	−0.004	−0.07	6.9	0.9	0.2	K0 III		170 mx		
146310	257389		16 21 37.2	−71 52 27	−0.007	−0.04	7.3	−0.2	0.6	A0 V		92 mx		
148344	8490		16 21 38.8	+70 29 51	+0.005	−0.04	7.2	1.1	0.2	K0 III		260 s		
147410	121532		16 21 42.0	+5 48 25	−0.003	−0.02	8.0	1.1	0.2	K0 III		360 s		
147123	226686		16 21 45.6	−42 00 15	−0.001	−0.04	7.6	0.2	1.7	A3 V		140 s		
148293	17062		16 21 48.5	+69 06 34	−0.005	−0.01	5.3	1.1	−0.1	K2 III	−8	120 s		
147546	84327		16 21 51.8	+21 48 43	−0.001	+0.01	7.1	1.1	−0.1	K2 III		270 s		
147508	102104		16 21 54.4	+13 27 19	−0.003	−0.07	7.3	1.1	−0.1	K2 III		140 mx		
146964	243688		16 21 54.4	−55 33 06	−0.002	−0.04	7.1	0.6	1.4	A2 V		70 s		
147547	102107	20 γ Her	16 21 55.1	+19 09 11	−0.003	+0.04	3.75	0.27	0.6	A9 III	−35	42 s	10022	m
147449	121540	50 σ Ser	16 22 04.2	+1 01 44	−0.010	+0.05	4.82	0.34	2.6	F0 V	−46	28 ts		
147677	65254	19 ξ CrB	16 22 05.7	+30 53 32	−0.008	+0.11	4.85	0.97	0.2	K0 III	−29	85 s		m
147049	243699		16 22 09.5	−52 43 35	−0.001	−0.01	7.70	0.35	−2.6	B2.5 IV		590 s		m
145327	——		16 22 11.5	−80 45 29			7.8	0.5	3.4	F5 V		75 s		
147470	121542		16 22 12.8	+0 29 53	−0.001	−0.03	7.7	0.1	0.6	A0 V		260 s		
147181	226694		16 22 14.4	−43 12 00	+0.002	−0.01	7.2	0.6	0.0	B8.5 V		180 s		
147510	121543		16 22 15.0	+2 52 31	+0.001	−0.01	7.3	1.1	−0.1	K2 III	+9	310 s		
147431	141122		16 22 17.3	−8 44 31	−0.002	−0.02	7.2	0.1	1.4	A2 V		130 s		
147531	121547		16 22 19.2	+3 05 06	−0.002	−0.02	7.4	0.5	3.4	F5 V		62 s		
147749	65257	20 ν¹ CrB	16 22 21.3	+33 47 56	0.000	−0.04	5.30	1.6	−0.5	M2 III	−13	150 s		m
147040	243702		16 22 25.2	−56 25 23	+0.002	+0.01	8.0	1.4	−0.3	K5 III		450 s		
147152	226693		16 22 27.9	−49 34 20	−0.001	−0.02	5.33	−0.04	−1.3	B6 IV	−9	180 s		
147075	243708		16 22 28.0	−54 38 04	−0.003	−0.03	7.90	1.20	−0.1	K2 III		380 s		
147225	226696		16 22 28.7	−43 54 43	−0.002	−0.02	5.85	1.17	3.2	G5 IV		18 s		m
147767	65259	21 ν² CrB	16 22 29.1	+33 42 13	−0.001	+0.06	5.39	1.53	−0.3	K5 III	−39	130 s		m
148855	8506		16 22 33.6	+76 26 42	−0.007	+0.02	7.5	1.4	−0.3	K5 III		370 s		
147550	141129		16 22 38.7	−2 04 47	−0.001	0.00	6.23	0.07		B9				
147924	46056		16 22 45.8	+44 05 03	+0.001	+0.01	7.5	0.1	0.6	A0 V		240 s		
146958	253524		16 22 46.9	−62 34 09	+0.002	0.00	7.8	0.9	3.2	G5 IV		81 s		
148204	17065		16 22 50.8	+61 26 07	−0.001	−0.02	8.0	1.1	−0.1	K2 III		410 s		
147432	184350		16 22 51.8	−23 07 05	+0.001	+0.01	7.70	0.1	1.4	A2 V		180 s	10024	m
147473	159888		16 22 53.5	−17 01 09	+0.001	−0.01	6.7	0.4	2.6	F0 V		62 s		
146827	253522		16 22 54.4	−66 53 15	−0.004	−0.03	6.9	0.2	0.4	B9.5 V		130 mx		
146190	257390		16 22 54.4	−75 58 26	0.000	0.00	6.9	1.4	−0.3	K5 III		280 s		
147156	243716		16 22 55.5	−55 02 16	−0.003	−0.01	8.0	−0.1	0.4	B9.5 V		340 s		
147835	65262		16 22 56.4	+32 19 58	+0.001	−0.01	6.40	0.08	1.7	A3 V	−3	87 s	10031	m
146584	——		16 22 58.3	−71 55 49			8.0	1.1	−0.1	K2 III		420 s		
147575	141133		16 22 59.3	−7 12 33	−0.002	−0.01	7.6	1.6	−0.5	M4 III		330 s		
147018	253526		16 23 00.2	−61 41 19	−0.022	−0.33	8.0	1.1	0.2	K0 III		41 mx		
147851	65265		16 23 01.8	+32 37 34	−0.003	−0.03	7.9	1.4	−0.3	K5 III	−1	430 s		
147769	102118		16 23 04.4	+19 30 21	−0.001	+0.01	7.3	0.1	0.6	A0 V		220 s		
147735	102116		16 23 05.5	+13 50 24	−0.002	+0.02	8.00	0.4	3.0	F2 V	−42	100 s	10030	m
147645	141136		16 23 05.9	−0 51 22	0.000	0.00	7.30	0.2	2.1	A5 V		110 s	10028	m
148570	8503		16 23 06.0	+70 47 50	−0.006	+0.02	7.6	1.4	−0.3	K5 III		380 s		
148049	29922		16 23 07.0	+50 20 40	0.000	0.00	7.4	0.5	4.0	F8 V		48 s		
147371	207611		16 23 07.6	−37 25 28	−0.004	−0.05	6.9	0.5	0.2	K0 III		180 mx		
149502	2707		16 23 08.5	+80 21 06	−0.018	+0.05	7.8	0.8	3.2	G5 IV		84 s		

HD	SAO	Star Name	α 2000	δ 2000	μ(α)	μ(δ)	V	B-V	M_v	Spec	RV	d(pc)	ADS	Notes
			$16^h 23^m 09^s.3$	$-0°42'29''$	$-0^s.007$	$-0''.20$								
147644	141138		16 23 09.3	-0 42 29	-0.007	-0.20	8.0	0.5	4.2	F9 V	+10	46 mx		
147387	207614		16 23 11.3	-34 55 31	-0.002	-0.07	7.3	-0.6	3.0	F2 V		73 s		
147925	65270		16 23 17.6	+35 36 43	+0.002	-0.03	7.4	1.1	0.2	K0 III		270 s		
147417	226714		16 23 39.9	-41 04 26	+0.001	-0.01	7.3	1.3	0.2	K0 III		180 s		
147274	243735		16 23 39.9	-53 02 37	0.000	+0.01	8.0	-0.1	0.0	B8.5 V		400 s		
148374	17073		16 23 46.8	+61 41 47	-0.006	+0.03	5.67	0.96	3.2	G5 IV	-24	24 s	10052	m
147836	102128		16 23 47.1	+11 35 27	0.000	-0.03	7.7	1.1	0.2	K0 III		310 s		
148085	46063		16 23 47.6	+42 06 40	-0.003	0.00	8.0	0.4	2.6	F0 V		120 s		
147579	207623		16 23 51.1	-30 10 51	-0.001	+0.02	7.8	1.7	-0.1	K2 III		180 s		
--	--		16 23 55.4	-33 11 54			7.10							m
147980	84351		16 23 55.7	+28 23 02	+0.001	-0.01	7.50	1.1	-1.1	K1 II-III		530 s		
147553	207625		16 23 56.5	-33 11 58	-0.001	-0.03	6.6	0.5	0.6	A0 V		75 s		m
148069	65275		16 23 57.4	+38 38 31	-0.002	+0.01	7.9	1.1	0.2	K0 III		340 s		
148387	17074	14 η Dra	16 23 59.3	+61 30 51	-0.003	+0.06	2.74	0.91	0.3	G8 III	-14	26 ts	10058	m
147952	102134		16 24 00.4	+19 38 04	-0.001	-0.01	8.0	0.8	3.2	G5 IV		90 s		
148005	84354		16 24 00.9	+29 36 57	-0.001	-0.02	7.9	0.4	2.1	A5 V		140 s		
147242	243736		16 24 01.0	-58 36 20	-0.004	-0.07	7.2	0.5	1.4	A2 V		67 mx		
147513	207622		16 24 01.2	-39 11 36	+0.007	-0.01	5.40	0.62	5.2	dG5	+10	15 ts		
147302	243741		16 24 01.2	-55 27 14	0.000	-0.02	7.74	0.01	-1.7	B3 V ne		680 s		
146893	257394		16 24 04.4	-70 08 41	-0.002	-0.04	6.9	1.4	0.2	K0 III		120 s		
147700	159892	4 ψ Oph	16 24 06.0	-20 02 15	-0.002	-0.05	4.50	1.01	0.2	K0 III	0	72 s		
147886	102133		16 24 08.2	+11 25 29	+0.002	-0.04	7.9	1.4	-0.3	K5 III		430 s		
147869	121568	21 Her	16 24 10.7	+6 56 54	0.000	+0.01	5.85	-0.01		A0	-33			
147775	159896		16 24 11.8	-11 53 52	-0.001	-0.03	7.5	0.4	3.0	F2 V		75 s		
147347	243748		16 24 16.0	-55 13 10	-0.001	+0.01	7.9	-0.4	0.0	B8.5 V		380 s		
148329	29930		16 24 17.4	+55 19 51	-0.001	-0.35	7.0	0.5	4.0	F8 V		40 s		
147928	102139		16 24 24.0	+10 15 15	+0.001	-0.01	7.1	1.4	-0.3	K5 III		310 s		
147742	184366		16 24 25.0	-22 39 16	+0.003	+0.01	7.9	0.9	4.0	F8 V		37 s		
148330	29931		16 24 25.2	+55 12 18	+0.001	+0.02	5.74	0.00	1.4	A2 V	-4	74 s		
148281	29929		16 24 26.2	+52 17 22	-0.002	+0.01	6.7	0.1	0.6	A0 V		170 s		
147703	184365		16 24 29.9	-27 09 04	-0.005	-0.03	7.47	0.18	0.2	B9 V n		59 mx		
147628	207637		16 24 31.6	-37 33 57	-0.002	-0.03	5.42	-0.11	-0.6	B8 IV	+8	160 s		m
147652	207639		16 24 31.9	-35 10 29	+0.005	0.00	7.6	0.2	3.4	F5 V		69 s		
147537	226725		16 24 35.5	-45 51 04	+0.003	-0.04	7.9	1.0	-0.3	K5 III		430 s		
147907	141154		16 24 39.3	-2 29 16	-0.003	-0.01	7.0	0.4	3.0	F2 V		64 s		
147722	184368		16 24 39.5	-29 42 12	+0.005	-0.11	5.41	0.60	2.8	G0 IV		33 s	10035	m
147556	226726		16 24 40.4	-45 10 49	-0.001	-0.01	7.9	0.0	0.6	A0 V		280 s		
147683	207641		16 24 43.8	-34 53 38	+0.002	-0.02	7.1	0.2	0.0	B8.5 V		190 s		
146847	257395		16 24 44.4	-72 35 14	-0.008	-0.05	8.0	1.1	0.2	K0 III		250 mx		
148052	84360		16 24 45.1	+22 22 57	-0.002	-0.02	7.9	0.1	1.4	A2 V		200 s		
148253	46073		16 24 47.9	+44 41 24	0.000	+0.01	7.2	0.5	3.4	F5 V		58 s		
147375	243755		16 24 48.2	-57 00 22	-0.003	-0.04	7.4	1.3	-0.3	K5 III		350 s		
147614	226730		16 24 54.0	-45 20 56	+0.002	0.00	6.50	0.1	1.4	A2 V		100 s		m
147685	207643		16 24 54.2	-38 26 59	-0.001	-0.09	8.0	0.6	3.4	F5 V		65 s		
147422	243763		16 24 57.5	-56 19 08	+0.004	-0.02	8.0	0.7	0.2	K0 III		300 mx		
148365	29933		16 24 59.7	+50 27 06	-0.003	+0.01	7.80	0.8	0.3	G6 III	-43	320 s		
148364	29934		16 25 00.1	+51 43 00	-0.001	+0.02	7.3	0.1	1.7	A3 V		130 s		
147559	243772		16 25 01.9	-50 44 54	-0.001	-0.01	7.88	0.07	0.4	B9.5 V		280 s		
147724	207646		16 25 03.4	-36 08 39	0.000	-0.04	8.0	0.6	1.7	A3 V		83 s		
147779	207648		16 25 04.6	-31 25 19	0.000	-0.03	7.1	0.1	0.6	A0 V		160 s		
148127	84364		16 25 06.7	+24 03 23	+0.001	-0.01	7.8	1.4	-0.3	K5 III		410 s		
147338	253538		16 25 11.9	-62 12 42	+0.002	0.00	8.0	1.6	-0.1	K2 III		240 s		
148128	102152		16 25 15.8	+19 14 21	+0.002	-0.05	7.2	1.4	-0.3	K5 III		320 s		
147633	226738		16 25 17.7	-49 08 53	-0.008	-0.09	8.00	0.9	3.2	G5 IV		91 mx		
147479	243770		16 25 17.8	-56 30 18	+0.001	+0.01	7.3	1.6	-0.1	K2 III		170 s		
148346	46080		16 25 19.1	+45 22 43	-0.005	+0.02	7.4	0.9	3.2	G5 IV		69 s		
148727	17088		16 25 20.3	+67 04 20	+0.001	0.00	7.4	1.1	0.2	K0 III		270 s		
147349	253539		16 25 21.9	-63 07 29	-0.003	+0.02	6.15	0.03		A2				
147657	226743		16 25 22.7	-47 31 42	-0.004	-0.10	8.0	0.2	3.4	F5 V		82 s		
148331	46077		16 25 23.0	+42 31 11	0.000	-0.06	7.7	0.2	2.1	A5 V		94 mx		
148283	65290	25 Her	16 25 24.0	+37 23 38	0.000	-0.01	5.5	0.1	1.7	A3 V	-1	58 s		
147888	184377		16 25 24.1	-23 27 38	-0.001	-0.04	6.74	0.31	-1.7	B3 V	-10	230 s	10049	m
147889	184376		16 25 24.1	-24 27 57	-0.001	-0.03	7.90	0.84	-2.5	B2 V	-3	320 s		m
148112	102153	24 ω Her	16 25 24.8	+14 02 00	+0.003	-0.06	4.57	0.00		A0 p	-7	33 t	10054	m
148147	102154		16 25 26.8	+17 18 13	-0.003	-0.05	8.00	0.6	4.4	G0 V	-31	53 s	10057	m
148941	8513		16 25 28.8	+71 23 04	+0.001	-0.01	7.2	0.5	4.0	F8 V		44 s		
148433	29937		16 25 31.8	+51 08 19	-0.006	-0.02	7.3	0.4	2.6	F0 V	-21	88 s		
148038	141164		16 25 33.0	-3 51 59	-0.003	+0.01	7.9	0.4	2.6	F0 V		110 s		
148205	84368		16 25 33.1	+25 13 18	0.000	-0.04	7.9	1.1	0.2	K0 III		340 s		
147933	184381	5 ρ Oph	16 25 34.9	-23 26 46	-0.001	-0.02	4.59	0.24	-2.5	B2 V	-10	230 s	10049	m
147932	184383		16 25 35.1	-23 24 19	0.000	-0.03	7.27	0.32	-1.1	B5 V	-19	230 s	10049	m
147857	207654		16 25 36.6	-32 04 37	0.000	-0.06	7.6	0.5	3.0	F2 V		67 s		
148039	141165		16 25 38.5	-5 53 52	-0.001	-0.01	8.0	1.1	-0.1	K2 III		320 s		
147890	184380		16 25 39.1	-29 24 00	0.000	-0.01	7.65	0.23		A0 p			10048	m
147707	226748		16 25 39.6	-49 25 29	-0.001	-0.03	7.1	0.9	0.2	K0 III		240 s		

HD	SAO	Star Name	α 2000	δ 2000	μ(α)	μ(δ)	V	B-V	M_v	Spec	RV	d(pc)	ADS	Notes
			$16^h 25^m 40.6^s$	$+9°22'15''$	$+0.002^s$	$-0.01''$								
148148	121588		16 25 40.6	+9 22 15	+0.002	-0.01	6.8	1.1	-0.1	K2 III		250 s		
149681	8527		16 25 42.9	+78 57 50	-0.041	+0.11	5.5	0.1	1.7	A3 V	-12	22 mx		
147873	207656		16 25 45.0	-33 34 03	-0.001	-0.03	8.00	0.6	4.4	G0 V		52 s		m
148206	102160		16 25 47.6	+18 53 33	+0.002	-0.01	6.6	1.6		M5 III	-28	17 mn		U Her, v
148228	102165		16 26 11.4	+11 24 27	-0.002	+0.01	6.11	1.03	0.2	K0 III	-21	150 s		
148229	121595		16 26 12.4	+8 30 46	-0.006	-0.09	7.6	1.1	0.2	K0 III		84 mx		
147841	226760		16 26 19.7	-45 47 26	+0.001	+0.02	7.7	1.3	-0.3	K5 III		390 s		
148207	121596		16 26 21.2	+2 30 28	-0.001	+0.02	6.6	0.1	0.6	A0 V		160 s		
148434	46085		16 26 21.4	+40 48 34	+0.002	-0.05	6.91	1.11	0.2	K0 III		180 s		
147746	243799		16 26 22.5	-53 25 56	0.000	-0.01	8.0	1.7	-0.5	M2 III		470 s		
147670	243796		16 26 23.8	-56 53 04	+0.001	0.00	7.1	0.9	0.0	B8.5 V		73 s		
148286	102168		16 26 25.0	+15 05 40	-0.002	+0.02	7.4	1.1	0.2	K0 III		280 s		
148543	46090		16 26 31.9	+47 57 27	-0.007	+0.12	6.5	1.1	0.2	K0 III		120 mx		
147860	226764		16 26 36.5	-48 25 13	0.000	-0.02	8.0	0.0	1.4	A2 V		210 s		
148317	102174		16 26 39.2	+15 58 17	+0.001	-0.04	6.8	0.6	4.4	G0 V	-37	30 s		
148297	121602		16 26 41.0	+9 16 26	-0.001	+0.02	6.8	1.1	0.2	K0 III		210 s		
148296	102172		16 26 43.2	+10 59 27	-0.002	-0.03	7.0	1.6	-0.5	M2 III	-30	310 s		
148182	159916		16 26 43.7	-12 25 38	+0.001	-0.01	7.00	2.0		N7.7 e	-37			V Oph, v
147247	257398		16 26 49.2	-71 46 05	+0.006	0.00	7.8	1.1	0.2	K0 III		330 s		
148347	102178		16 26 49.4	+16 10 16	+0.002	-0.05	7.6	1.1	0.2	K0 III		300 s		
147894	226767		16 26 49.6	-48 02 38	-0.001	0.00	7.50	0.00	0.0	B8.5 V		300 s		m
148287	121604		16 26 49.9	+2 20 51	+0.001	-0.03	6.07	0.92		G5	+4			
148551	46091		16 26 51.1	+44 57 13	-0.005	-0.03	7.3	0.1	0.6	A0 V		45 mx		
147985	226769		16 26 56.5	-43 47 58	-0.001	-0.02	7.9	-0.2	0.0	B8.5 V		380 s		
145621	258741		16 26 59.9	-82 33 34	-0.011	-0.01	6.9	0.9	0.2	K0 III		220 s		
146987	--		16 27 01.2	-75 55 29			7.9	1.4	-0.3	K5 III		440 s		
148184	159918	7 χ Oph	16 27 01.3	-18 27 23	-0.001	-0.03	4.42	0.28	-2.5	B2 V	-5	150 s		v
148198	159919		16 27 01.6	-17 46 08	+0.002	-0.04	7.2	0.5	4.0	F8 V		42 s		
147845	243813		16 27 06.5	-54 57 52	-0.001	0.00	7.5	0.3	2.6	F0 V		94 s		
148152	184392		16 27 07.5	-26 15 28	+0.001	-0.03	7.6	1.7	0.2	K0 III		120 s		
147799	243809		16 27 08.7	-58 07 18	-0.003	-0.05	7.8	0.2	0.2	K0 III		240 mx		
147970	226772		16 27 09.7	-47 32 56	-0.003	-0.02	7.5	0.1	0.6	A0 V		230 s		
147971	226773	ε Nor	16 27 10.9	-47 33 18	0.000	-0.03	4.47	-0.07	-1.7	B3 V	-12	150 s		m
147958	226771		16 27 12.0	-48 47 17	-0.002	-0.05	7.50	1.4	-0.3	K5 III		360 s		m
148153	184393		16 27 12.3	-27 11 21	-0.001	-0.01	7.80	0.4	3.0	F2 V		91 s	10062	m
148211	184396		16 27 12.7	-22 07 37	-0.021	-0.31	7.68	0.54	4.0	dF8	-19	35 ts		
148199	184398		16 27 29.5	-29 17 17	-0.001	-0.01	7.01	0.09		B9				
147465	253549		16 27 29.7	-69 19 35	+0.002	-0.04	6.6	0.1	0.4	B9.5 V		150 s		
148390	121614		16 27 32.1	+2 52 15	0.000	-0.01	6.6	1.4	-0.3	K5 III		240 s		
148681	46100		16 27 33.7	+47 20 30	-0.001	-0.02	7.4	1.1	0.2	K0 III		280 s		
147974	243829		16 27 35.0	-52 46 08	-0.005	-0.06	7.8	0.8	0.2	K0 III		170 mx		
148349	141186		16 27 43.4	-7 35 53	+0.001	-0.16	5.23	1.72	-0.5	M2 III	+100	110 s		
145990	258742		16 27 45.6	-81 57 40	-0.021	-0.04	7.3	1.2	-0.1	K2 III		270 mx		
148367	141187	3 υ Oph	16 27 48.1	-8 22 18	-0.005	+0.01	4.63	0.17		A m	-31	21 mn		
148122	226782		16 27 49.8	-44 01 54	0.000	-0.01	7.8	0.1	-1.0	B5.5 V		580 s		
147500	253552		16 27 52.9	-69 51 41	-0.002	-0.01	7.4	0.2	0.4	B9.5 V		180 s		
148554	84393		16 27 54.6	+25 59 02	0.000	0.00	7.10	0.1	1.7	A3 V		120 s	10070	m
147848	253557		16 27 56.0	-61 37 32	0.000	0.00	7.7	1.4	-0.1	K2 III		270 s		
147787	253555	ι TrA	16 27 57.2	-64 03 28	+0.008	+0.02	5.25	0.36	2.6	F0 V	-5	32 s		m
148616	65320		16 27 57.6	+32 41 58	-0.002	-0.01	6.9	1.1	0.2	K0 III	-15	220 s		
147350	257400		16 27 58.3	-73 01 17	-0.003	+0.01	7.9	0.8	3.2	G5 IV		86 s		
148103	226784		16 27 58.7	-46 15 04	-0.004	-0.04	7.1	1.2	-0.3	K5 III		200 mx		
149212	17107	15 Dra	16 27 58.8	+68 46 05	-0.006	+0.03	5.00	-0.06	-0.3	B9 IV	-7	110 s		
147883	253559		16 28 00.2	-60 44 21	-0.004	-0.02	7.8	1.8	-0.1	K2 III		160 s		
148801	29953		16 28 01.1	+51 35 31	-0.003	+0.01	7.50	1.1	0.2	K0 III		290 s	10076	m
148494	102188		16 28 01.5	+15 44 11	+0.001	-0.02	7.6	0.1	0.6	A0 V		260 s		
148511	102189		16 28 05.0	+15 21 05	-0.001	+0.02	7.6	0.5	4.0	F8 V		52 s		
148247	207688		16 28 14.5	-37 10 46	+0.003	-0.01	5.8	1.4	0.2	K0 III		74 s		
147977	243836		16 28 14.9	-58 35 59	-0.004	-0.04	5.69	0.00	-0.8	B9 III		160 mx		
148321	184403		16 28 15.0	-25 27 14	+0.001	+0.01	6.97	0.19		A m				
148394	159929		16 28 15.7	-16 12 42	-0.003	-0.01	7.60	0.5	3.4	F5 V		65 s		m
148013	243840		16 28 16.3	-57 12 41	+0.001	-0.02	7.4	0.1	0.6	A0 V		190 s		
148156	226791		16 28 17.2	-46 19 03	+0.005	+0.03	7.8	0.3	4.0	F8 V		57 s		
149198	17108		16 28 21.0	+67 02 36	-0.004	-0.03	6.57	1.58	-0.5	gM3	-81	260 s		
148556	102191		16 28 24.0	+15 25 51	-0.001	-0.02	7.4	1.1	0.2	K0 III		280 s		
148352	184405		16 28 25.0	-24 45 00	-0.004	-0.06	7.51	0.41	3.1	F3 V		74 s		
148427	159932		16 28 27.8	-13 23 59	-0.004	0.00	6.8	0.8	3.2	G5 IV		51 s		
147584	253554	ζ TrA	16 28 28.1	-70 05 04	+0.040	+0.10	4.91	0.55	4.4	G0 V	+9	12 ts		
148978	29960		16 28 29.5	+58 14 18	0.000	-0.06	8.0	1.1	0.2	K0 III		360 s		
148667	84402		16 28 31.7	+29 04 32	-0.001	+0.01	7.1	0.1	1.4	A2 V		140 s		
148513	121623		16 28 33.8	+0 39 54	0.000	-0.07	5.39	1.46	-0.3	K4 III p	+7	140 s		
148619	102194		16 28 37.7	+19 01 05	+0.001	-0.02	7.2	0.8	3.2	G5 IV		63 s		
148783	46108	30 Her	16 28 38.4	+41 52 54	+0.002	0.00	5.04	1.52		M6 III	+3	23 mn		g Her, v
148531	121627		16 28 42.3	+0 03 18	-0.004	-0.09	6.6	1.4	-0.3	K5 III		110 mx		
148893	29958		16 28 42.9	+51 45 03	-0.001	-0.01	7.7	1.1	0.2	K0 III		320 s		

HD	SAO	Star Name	α 2000	δ 2000	μ(α)	μ(δ)	V	B-V	M_v	Spec	RV	d(pc)	ADS	Notes
148880	29956		16^{h}28^{m}43^s.2	+51°24'27"	+0^s.002	0".00	6.29	1.05	0.2	K0 III	−16	150 s	10079	m
148621	102196		16 28 45.5	+13 36 45	0.000	+0.01	7.7	0.4	2.6	F0 V		100 s		
148438	159935		16 28 45.5	−17 59 05	+0.003	0.00	7.1	0.1	0.6	A0 V		180 s		
148260	226799		16 28 48.2	−45 02 03	+0.002	−0.02	7.89	0.44	−2.6	B2.5 IV		570 s		
148515	141195		16 28 48.8	−8 07 44	−0.004	−0.08	6.48	0.40	3.1	dF3	0	46 s	10072	m
148259	226801		16 28 49.8	−44 48 45	+0.001	0.00	7.20	0.14	−2.5	B2 V e		740 s		
148653	102200		16 28 52.7	+18 24 52	−0.023	+0.40	7.02	0.84	6.3	dK2	−36	18 ts	10075	m
148428	184412		16 28 58.9	−24 32 17	−0.005	−0.20	7.8	0.5	3.2	G5 IV		48 mx		
149222	17111		16 29 00.9	+64 47 16	−0.023	+0.15	7.8	0.7	4.4	G0 V	−34	48 s		
148963	29961		16 29 06.9	+50 45 58	+0.004	−0.02	8.0	0.5	3.4	F5 V		82 s		
147694	257403		16 29 14.9	−71 07 25	−0.007	−0.03	7.1	1.1	3.4	F5 V		22 s		
148261	243873		16 29 22.4	−51 34 05	−0.002	−0.01	7.2	0.7	0.2	K0 III		250 s		
148895	46117		16 29 23.1	+43 00 56	−0.002	0.00	7.6	0.4	2.6	F0 V		100 s		
148478	184415	21 α Sco	16 29 24.3	−26 25 55	0.000	−0.02	0.96	1.83	−4.7	M1 Ib	−3	100 s	10074	Antares, m,v
148683	102204		16 29 24.9	+10 35 31	−0.001	−0.01	7.60	0.93	0.2	G5 III	−14	270 s	10077	m
148046	253572		16 29 25.0	−62 53 53	+0.004	+0.03	8.0	0.5	3.4	F5 V		82 s		
148560	159942		16 29 25.6	−13 34 11	+0.001	+0.04	7.1	1.1	−0.1	K2 III		230 s		
148441	207708		16 29 27.6	−33 19 14	−0.002	+0.04	7.5	0.3	3.4	F5 V		65 s		
148668	121641		16 29 34.5	+4 34 33	−0.002	−0.02	7.9	0.9	3.2	G5 IV		88 s		
148418	226814		16 29 40.0	−41 06 40	0.000	−0.03	7.3	0.1	0.6	A0 V		180 s		
148379	226813		16 29 42.1	−46 14 36	−0.001	−0.01	5.35	0.56	−6.8	B2 Ia	−19	1000 s		
148218	243874		16 29 44.9	−57 45 21	−0.001	+0.01	5.9	1.6	0.2	K0 III		61 s		
148882	65339		16 29 45.0	+35 12 07	−0.003	−0.01	8.0	1.1	0.2	K0 III		360 s		
148710	121645		16 29 46.5	+7 44 55	0.000	−0.03	7.2	1.1	0.2	K0 III		250 s		
148604	159948		16 29 46.8	−14 33 03	+0.002	+0.01	5.7	0.6	0.4	G2 III	−31	110 s		
148632	159949		16 29 48.4	−11 07 44	−0.001	+0.01	6.7	0.0	0.4	B9.5 V		160 s		
148711	121646		16 29 50.9	+5 58 15	0.000	+0.01	7.0	0.5	3.4	F5 V		52 s		
148562	184424		16 29 54.6	−24 58 45	0.000	−0.02	7.81	0.18	1.7	A3 V		150 s		
147482	257402		16 29 54.8	−75 12 21	−0.008	−0.02	7.2	1.8	−0.1	K2 III		120 s		
148579	184425		16 29 59.0	−25 08 52	−0.002	−0.03	7.33	0.27	0.2	B9 V		180 s		
148928	65342		16 30 05.4	+34 47 09	0.000	−0.03	7.9	1.1	−0.1	K2 III		410 s		
149081	46128	34 Her	16 30 05.9	+48 57 39	−0.006	−0.06	6.2	0.1	0.6	A0 V	−8	47 mx		
148832	102212		16 30 10.1	+15 07 42	+0.002	+0.01	7.5	0.1	1.7	A3 V		150 s		
148605	184429	22 Sco	16 30 12.4	−25 06 54	−0.001	−0.02	4.79	−0.11	−2.5	B2 V	−4	260 s		
148856	84411	27 β Her	16 30 13.1	+21 29 22	−0.007	−0.02	2.77	0.94	0.3	G8 III	−26	31 s		Kornephoros, m
149105	46129		16 30 14.8	+47 57 07	−0.010	−0.29	7.00	0.57	4.4	G0 V	−46	33 s		
148382	243890		16 30 15.4	−52 26 27	−0.001	0.00	7.3	0.2	0.4	B9.5 V		160 s		
148594	184428		16 30 15.5	−27 54 58	−0.001	−0.01	6.89	0.10	0.2	B9 V		200 s		
149025	46127		16 30 22.6	+40 44 11	−0.006	+0.04	7.9	0.6	4.4	G0 V		50 s		
148546	207720		16 30 23.2	−37 58 21	0.000	−0.01	7.71	0.30	−6.2	O9.5 I		2900 s		
148504	226830		16 30 24.9	−44 06 53	+0.001	+0.03	7.5	2.2	−0.5	M2 III		180 s		
148729	159955		16 30 25.5	−10 26 29	−0.009	−0.08	7.3	0.6	4.4	G0 V		37 s		
149411	17118		16 30 28.1	+63 34 14	−0.005	+0.02	8.0	0.4	2.6	F0 V		120 s		
148816	121653		16 30 28.3	+4 10 42	−0.030	−1.39	7.27	0.54	4.2	F9 V	−52	41 s		
148995	65348		16 30 29.5	+33 45 01	−0.002	+0.02	7.0	1.1	0.2	K0 III		230 s		
148743	141206		16 30 29.7	−7 30 54	−0.001	−0.01	6.50	0.37	−4.8	A7 Ib	+2	1300 s		
148897	84416		16 30 33.5	+20 28 45	−0.006	−0.07	5.25	1.29		G8 p	+18	33 mn		
148643	184430		16 30 34.9	−27 10 37	−0.001	−0.02	7.8	1.1	−0.1	K2 III		380 s		
150275	8548		16 30 38.6	+77 26 47	−0.032	+0.28	6.34	1.00	0.0	K1 III	−32	69 mx		
148911	102215		16 30 41.0	+18 38 03	−0.001	+0.01	7.93	0.41	3.0	F2 V		94 s		
148081	253579		16 30 41.3	−68 04 27	0.000	−0.09	7.8	0.1	0.6	A0 V		83 mx		
148655	184432		16 30 42.0	−29 28 59	−0.001	−0.03	7.6	0.8	1.4	A2 V		66 s		
148291	253582		16 30 49.3	−61 38 00	0.000	−0.01	5.20	1.23	−0.1	gK2	+4	110 s		
148484	243900		16 30 53.6	−50 58 16	−0.007	−0.04	7.42	0.38	2.6	F0 V		87 s		
148857	121658	10 λ Oph	16 30 54.7	+1 59 02	−0.002	−0.08	3.82	0.01	1.2	A1 V	−15	33 s	10087	Marfik, m
149044	65354		16 30 55.2	+30 56 27	−0.002	−0.01	7.3	0.5	3.4	F5 V		61 s		
148567	226839		16 31 02.6	−46 28 46	0.000	−0.02	7.7	0.0	−1.6	B3.5 V		800 s		m
149084	65356		16 31 02.8	+35 13 29	0.000	−0.03	6.4	1.4	−0.3	K5 III	+24	220 s		
148833	141218		16 31 05.4	−7 32 28	0.000	−0.03	7.88	0.14		A0				
148506	243904		16 31 07.7	−51 39 52	0.000	−0.02	8.0	0.5	0.0	B8.5 V		170 s		
148786	159963	8 φ Oph	16 31 08.2	−16 36 46	−0.004	−0.04	4.28	0.92	0.3	G8 III	−34	63 s	10086	m
148672	207731		16 31 10.6	−34 19 58	0.000	0.00	7.0	0.5	3.2	G5 IV		59 s		
149009	84423		16 31 13.3	+22 11 43	−0.001	0.00	5.76	1.61	−0.3	K5 III	−26	140 s		
148930	121663		16 31 16.0	+6 53 31	−0.002	+0.02	8.0	0.5	3.4	F5 V		82 s		
150706	8557		16 31 17.1	+79 47 24	+0.031	−0.08	7.06	0.62	4.9	dG3	−14	21 ts		
148859	141219		16 31 18.2	−7 55 15	0.000	−0.03	7.9	0.4	3.0	F2 V		87 s		
149412	29975		16 31 20.0	+57 50 20	−0.001	0.00	7.72	1.40		K5				
148912	121661		16 31 21.0	+1 18 32	−0.001	+0.03	6.79	1.39	0.2	K0 III		110 s		
148760	184437		16 31 22.7	−26 32 16	−0.002	−0.04	6.10	1.08		K0				
148703	207732		16 31 22.8	−34 42 15	−0.001	−0.02	4.23	−0.16	−2.5	B2 V	0	220 s		
149012	102219		16 31 23.0	+16 25 11	−0.003	−0.08	7.5	1.1	−0.1	K2 III		120 mx		
148473	243905		16 31 23.1	−55 56 25	+0.001	−0.03	7.8	0.5	0.4	B9.5 V		140 s		
149011	102221		16 31 24.3	+17 29 59	0.000	−0.02	7.7	1.1	0.2	K0 III		310 s		
149067	84425		16 31 24.8	+25 50 27	+0.002	−0.08	8.00	0.9	−2.1	G8 II		73 mx		
150010	8544		16 31 27.9	+72 36 43	−0.011	+0.04	6.30	1.32		K0	−33	17 mn		

HD	SAO	Star Name	α 2000	δ 2000	μ(α)	μ(δ)	V	B–V	M_v	Spec	RV	d(pc)	ADS	Notes
			h m s	° ′ ″	s	″								
148704	207733		16 31 30.0	-39 00 44	-0.036	-0.33	7.24	0.86	6.1	dK1	-59	20 ts		
149141	65360		16 31 30.4	+33 30 48	-0.002	-0.02	6.70	0.1	0.6	A0 V	-30	170 s	10100	m
148979	121665		16 31 30.5	+8 17 39	+0.002	+0.08	6.98	0.91		G5				m
149223	46144		16 31 31.2	+42 09 45	+0.002	-0.02	7.6	1.1	0.2	K0 III		300 s		
148834	184440		16 31 37.0	-22 48 04	-0.003	0.00	8.0	1.1	3.2	G5 IV		58 s		
148980	121667		16 31 38.4	+5 26 00	-0.002	-0.04	7.6	0.1	1.7	A3 V	-14	160 s	10094	m
148860	159964		16 31 38.5	-17 42 49	-0.002	-0.02	8.04	0.15		B9				
148689	226851		16 31 40.0	-43 13 07	0.000	-0.05	7.5	0.3	1.7	A3 V		140 s		
--	--		16 31 40.7	-41 48 54			5.50							m
148688	226855		16 31 41.5	-41 49 01	-0.001	0.00	5.33	0.33	-6.6	B1 Ia	-14	1300 s		m
148359	253586		16 31 44.7	-64 28 25	+0.003	-0.02	6.6	-0.1	0.0	B8.5 V		200 s		
149303	46147		16 31 47.1	+45 35 54	-0.001	+0.04	5.61	0.12	1.2	dA1	-16	66 s	10105	m
148931	141223		16 31 47.8	-7 01 25	-0.001	-0.03	7.41	0.18	0.6	A0 V		180 s	10092	m
149142	84432		16 31 54.1	+26 01 54	-0.001	-0.07	7.5	0.9	0.3	G8 III		150 mx		
149304	46148		16 31 55.9	+44 15 52	-0.004	+0.02	7.6	0.5	3.4	F5 V		70 s		
147978	--		16 31 56.1	-74 01 44			7.9	1.1	-0.1	K2 III		400 s		
149059	121671		16 31 56.3	+9 24 53	-0.001	-0.03	6.7	0.4	3.0	F2 V	-28	55 s		
149575	17129		16 31 59.3	+60 28 48	-0.003	+0.04	8.0	1.1	0.2	K0 III		360 s		
148708	226858		16 31 59.4	-44 53 16	+0.002	-0.05	7.4	1.7	-0.5	M2 III		350 s		
148967	141226		16 32 00.7	-9 41 45	-0.002	+0.03	7.3	0.5	3.4	F5 V		57 s		
149029	121672		16 32 03.7	+3 21 43	-0.002	+0.01	8.0	0.1	1.4	A2 V		210 s		
148968	159966		16 32 07.9	-12 25 56	-0.002	-0.02	7.1	0.1	0.6	A0 V		170 s		
148898	184450	9 ω Oph	16 32 08.0	-21 27 59	+0.001	+0.03	4.45	0.13		A7 p	+3	19 mn		
148691	243923		16 32 12.9	-50 14 55	-0.001	-0.01	7.7	0.6	0.2	K0 III		310 s		
149213	84437		16 32 19.9	+24 50 55	-0.001	-0.03	7.9	0.1	0.6	A0 V		290 s		
149251	65370		16 32 21.6	+30 07 25	+0.004	-0.05	7.3	1.1	0.2	K0 III		230 mx		
148899	184451		16 32 24.4	-29 02 36	-0.001	-0.02	7.5	0.2	1.4	A2 V		130 s		
149650	17130		16 32 25.5	+60 49 24	+0.002	-0.01	5.8	0.1	0.6	A0 V	-14	110 s		
149241	84443		16 32 26.8	+27 42 32	+0.001	0.00	7.90	1.4	-0.3	K5 III		440 s		
149178	102231		16 32 29.2	+16 50 51	-0.003	+0.01	7.5	0.2	2.1	A5 V		120 s		
149160	102230		16 32 31.0	+13 09 23	+0.004	-0.11	8.0	0.5	4.0	F8 V		62 s		
149379	46152		16 32 33.0	+40 06 48	-0.002	+0.01	7.70	0.9	3.2	G5 IV		79 s	10111	m
146164	258743		16 32 33.0	-83 35 30	+0.008	-0.03	7.91	0.39	2.8	F1 V		97 s		
148825	226870		16 32 35.3	-43 44 02	-0.002	-0.02	7.60	0.1	1.4	A2 V		170 s		m
149121	121676	28 Her	16 32 35.6	+5 31 16	+0.001	0.00	5.63	-0.06		B8 sh	-27			
149161	102234	29 Her	16 32 36.2	+11 29 17	-0.012	-0.08	4.84	1.49	-0.3	K5 III	+3	70 mx		
149013	159971		16 32 38.0	-15 59 15	-0.004	-0.13	7.0	0.5	4.0	F8 V		38 s		
149748	17133		16 32 41.8	+62 51 18	-0.006	+0.03	7.36	0.33		A m	-21			
148848	226872		16 32 42.0	-41 02 08	-0.001	-0.06	7.5	0.4	0.6	A0 V		84 mx		
148720	243929		16 32 44.2	-52 22 18	-0.002	-0.01	7.7	0.2	-1.0	B5.5 V		340 s		
149108	159972		16 32 55.6	-10 33 44	+0.001	+0.02	6.80	0.2	2.1	A5 V		81 s	10104	m
149522	29983		16 32 56.6	+50 08 31	-0.004	-0.03	7.0	0.9	3.2	G5 IV		57 s		
148597	253595		16 32 59.4	-61 51 30	+0.001	-0.01	8.0	0.3	0.4	B9.5 V		200 s		
148950	207759		16 33 00.3	-33 31 58	+0.001	-0.04	7.70	0.1	0.6	A0 V		250 s		m
149199	121685		16 33 00.6	+6 36 42	+0.003	-0.02	7.9	0.9	3.2	G5 IV		88 s		
148984	207761		16 33 01.4	-31 33 16	-0.003	-0.04	7.9	1.8	-0.3	K5 III		310 s		
148587	253594		16 33 05.7	-63 50 32	-0.065	-0.19	7.39	0.58	4.4	G0 V		26 mx		
147979	257408		16 33 08.6	-75 56 59	+0.010	+0.03	7.7	0.1	0.6	A0 V		230 mx		
149123	159973		16 33 12.1	-12 47 56	0.000	-0.02	7.6	0.8	3.2	G5 IV		71 s		
149394	65376		16 33 12.3	+30 54 28	0.000	+0.02	8.00	0.4	3.0	F2 V	-12	100 s	10113	m
148837	226873		16 33 12.3	-49 35 21	0.000	-0.02	7.81	1.82	-0.3	K5 III		300 s		
149380	84451		16 33 21.0	+25 41 24	+0.017	-0.07	7.9	0.5	4.0	F8 V		60 s		
149200	141237		16 33 23.4	-4 15 42	-0.003	-0.04	7.2	0.5	3.4	F5 V		58 s		
149002	207764		16 33 26.8	-37 14 41	+0.002	0.00	8.0	0.0	2.6	F0 V		120 s		
147675	257407	γ Aps	16 33 27.1	-78 53 49	-0.040	-0.07	3.89	0.91	3.2	K0 IV	+5	14 s		
149420	65380		16 33 29.0	+30 29 56	-0.003	-0.01	6.70	0.4	2.6	F0 V	-15	66 s	10116	m
149346	102243		16 33 30.0	+17 49 43	-0.002	-0.04	8.0	0.5	3.4	F5 V		82 s		
149165	159975		16 33 30.9	-12 33 36	-0.002	+0.01	7.37	1.97		K5				
149305	102242		16 33 33.9	+10 22 09	+0.001	-0.02	6.7	0.1	1.7	A3 V	+4	100 s		
148679	253601		16 33 34.8	-62 07 17	+0.003	-0.03	6.7	2.0	0.2	K0 III		52 s		
148852	243945		16 33 35.8	-53 12 00	+0.001	-0.04	7.0	0.5	3.2	G5 IV		58 s		
148985	226890		16 33 36.5	-40 46 11	+0.002	+0.02	7.5	0.4	3.0	F2 V		74 s		
149327	121694		16 33 38.1	+8 50 24	+0.001	-0.02	8.0	0.1	1.4	A2 V		210 s		
149036	207767		16 33 40.8	-37 22 57	-0.002	-0.03	7.5	0.3	1.7	A3 V		110 s		
149504	65390		16 33 41.8	+38 05 28	-0.003	+0.08	6.6	0.5	3.4	F5 V	-46	44 s		
148922	226886		16 33 45.2	-49 10 21	0.000	-0.03	7.69	0.66	-0.9	A5 II-III		490 s		
149061	207771		16 33 48.0	-34 26 26	-0.005	-0.02	7.6	0.5	2.6	F0 V		74 s		
149843	29993		16 33 48.2	+59 40 38	-0.002	-0.01	7.8	1.1	0.2	K0 III		340 s		
148878	243947		16 33 48.4	-53 12 34	-0.001	+0.01	7.70	0.39		B0.5 II				
149381	102245		16 33 50.3	+13 23 41	+0.001	-0.05	7.0	1.1	0.2	K0 III		230 s		
148952	226892		16 33 50.9	-46 33 32	+0.001	-0.02	7.7	1.6	-0.3	K5 III		280 s		
149216	159978		16 33 50.9	-14 28 00	-0.004	-0.02	7.8	1.1	0.2	K0 III		210 mx		
149062	207772		16 33 52.0	-35 32 33	0.000	0.00	7.71	1.00		G5				
148937	226891		16 33 52.4	-48 06 40	+0.001	0.00	6.72	0.36		O6.8				m
150142	17150		16 33 54.7	+68 00 44	-0.005	-0.03	7.2	1.1	0.2	K0 III		260 s		

HD	SAO	Star Name	α 2000	δ 2000	μ(α)	μ(δ)	V	B-V	M$_v$	Spec	RV	d(pc)	ADS	Notes
			h m s	° ′ ″	s	″								
149718	29989		16 33 59.0	+52 14 16	0.000	0.00	6.7	1.1	0.2	K0 III		200 s		
149348	121697		16 34 01.4	+1 25 54	−0.002	−0.03	7.71	0.26	1.4	A2 V		150 s		
149474	84456		16 34 04.2	+25 28 35	−0.001	−0.03	7.80	1.1	−0.2	K3 III		400 s		
149038	226900	μ Nor	16 34 04.8	−44 02 43	−0.001	−0.01	4.90	0.10	−6.2	B0 Ia	+5	1100 s		
149630	46161	35 σ Her	16 34 06.0	+42 26 13	−0.001	+0.04	4.20	−0.01	0.2	B9 V	−11	61 s		
148974	226897		16 34 06.1	−48 27 43	+0.002	−0.02	7.0	−0.1	2.1	A5 V		96 s		
149719	29992		16 34 09.0	+50 58 08	+0.002	−0.02	8.0	0.8	3.2	G5 IV		90 s		
149090	207777		16 34 10.3	−38 23 25	0.000	−0.04	7.6	0.3	3.0	F2 V		84 s		
148938	243953		16 34 11.7	−52 37 19	−0.002	−0.03	8.0	1.5	−0.1	K2 III		270 s		
148488	257409		16 34 19.0	−70 59 18	−0.005	−0.03	5.50	1.22	0.2	gK0	−3	77 s		
148778	253608		16 34 21.3	−62 16 27	+0.003	−0.02	7.2	0.3	3.0	F2 V		70 s		
149019	226901		16 34 22.7	−49 46 12	−0.002	−0.02	7.40	0.89	−8.1	A Ia		5300 s		
148414	—		16 34 23.8	−72 55 22			7.8	1.1	−0.1	K2 III		380 s		
--	65399		16 34 27.6	+38 54 54	0.000	−0.03	7.8	2.1						
149433	121703		16 34 27.7	+8 40 25	0.000	−0.02	6.8	1.1	0.2	K0 III		210 s	10118	m
149363	141251		16 34 28.2	−6 08 09	+0.001	0.00	7.80	0.02		B0.5 III	+115			
149383	141252		16 34 33.6	−8 08 57	−0.001	−0.01	7.8	1.1	−0.1	K2 III		300 s		
150076	17152		16 34 35.7	+63 15 33	−0.019	+0.09	7.9	1.1	−0.1	K2 III		140 mx		
149054	226907		16 34 35.9	−49 19 23	0.000	−0.03	7.91	1.04	3.2	G5 IV		88 s		
149076	226912		16 34 38.4	−47 00 15	+0.001	−0.01	7.38	0.48	−7.1	B6 Ia		6700 s		
150099	17154		16 34 41.0	+63 56 06	−0.001	0.00	7.9	1.1	0.2	K0 III		340 s		
148740	253607		16 34 41.8	−66 00 54	+0.003	−0.04	7.35	−0.08	−1.9	B6 III		660 s		
149077	226914		16 34 45.5	−49 23 45	0.000	−0.02	7.43	0.44	0.0	B8.5 V		150 s		
149098	226920		16 34 52.3	−47 59 23	−0.001	0.00	7.9	−0.1	0.4	B9.5 V		320 s		
148725	—		16 34 52.7	−67 18 08			7.9	0.5	3.4	F5 V		79 s		
150077	17155		16 35 00.7	+60 28 06	−0.008	+0.02	6.7	1.6	−0.5	M4 III	+52	48 mx		TX Dra, v
149174	226926		16 35 07.1	−45 14 41	−0.002	−0.07	6.4	0.8	0.2	K0 III		170 s		
149021	243963		16 35 11.1	−57 21 09	0.000	−0.02	7.2	1.0	0.2	K0 III		250 s		
149749	65409		16 35 12.0	+37 20 42	−0.002	0.00	7.3	1.6		M5 III	−51		10121	W Her, m,v
149274	207794		16 35 13.8	−35 43 27	0.000	−0.02	6.8	0.1	0.6	A0 V		150 s		
149100	243965		16 35 19.8	−53 38 52	−0.001	−0.02	7.21	−0.08	−1.7	B3 V k		540 s		
151043	8571		16 35 21.0	+77 45 17	+0.002	−0.02	7.0	0.5	4.0	F8 V		39 s		
149386	184476		16 35 25.0	−29 31 51	+0.001	0.00	7.8	1.8	0.2	K0 III		110 s		
148879	—		16 35 25.6	−64 41 00			7.8	1.1	0.2	K0 III		320 s		
149631	102258		16 35 25.9	+17 06 01	−0.002	−0.01	7.3	0.2	2.1	A5 V		110 s		
149632	102259		16 35 26.2	+17 03 26	0.000	0.00	6.30	0.05	1.4	A2 V	−9	96 s		m
--	102260		16 35 26.2	+18 57 22	−0.001	−0.02	8.0	1.1	−0.1	K2 III		420 s		
150116	30011		16 35 41.1	+57 46 51	+0.002	−0.05	7.4	0.1	1.7	A3 V		140 s		
148890	253614	θ TrA	16 35 44.6	−65 29 44	+0.006	−0.04	5.52	0.93	0.3	gG5	+10	99 s		
148650	257413		16 35 49.6	−71 53 37	+0.001	−0.02	6.8	−0.1	0.6	A0 V		170 s		
149139	243972		16 35 51.6	−56 07 37	−0.009	−0.08	7.8	0.9	3.2	G5 IV		82 s		
149438	184481	23 τ Sco	16 35 52.8	−28 12 58	−0.001	−0.02	2.82	−0.25	−4.1	B0 V	−1	240 s		
149340	226945		16 35 53.3	−43 12 45	−0.001	−0.01	8.0	0.0	0.6	A0 V		280 s		
151340	8580		16 35 55.6	+79 12 17	+0.012	−0.05	7.8	0.4	3.0	F2 V		90 s		
149652	121722		16 36 02.9	+1 22 51	−0.010	−0.10	7.25	0.52	3.4	F5 V		49 mx		
149249	243980		16 36 05.1	−51 14 24	+0.001	−0.01	7.50	0.00	0.2	B8.5 V		300 s		m
149754	102267		16 36 07.7	+17 28 18	−0.001	0.00	7.5	0.4	3.0	F2 V		81 s		
150030	46184		16 36 11.1	+46 36 49	−0.002	+0.01	5.79	1.04	3.2	G5 IV	−15	23 s		
150100	30012	16 Dra	16 36 11.3	+52 54 00	−0.002	+0.03	5.53	−0.07	0.4	B9.5 V	−9	100 s	10129	m
150117	30013	17 Dra	16 36 13.6	+52 55 28	−0.002	+0.03	5.08	−0.04	0.2	B9 V	−11	100 s	10129	m
149439	207811		16 36 15.5	−34 12 00	0.000	−0.03	7.8	0.2	1.4	A2 V		170 s		
149258	243982		16 36 19.6	−52 03 50	0.000	−0.02	8.0	0.8	0.2	K0 III		360 s		
149661	141269	12 Oph	16 36 21.3	−2 19 29	+0.030	−0.31	5.75	0.82	5.9	K0 V	−15	11 ts		m
149930	65427		16 36 22.0	+33 49 17	−0.007	−0.01	7.20	1.1	0.2	K0 III		170 mx	10127	m
149447	207814		16 36 22.4	−35 15 21	+0.002	0.00	4.16	1.57		Ma	−2	17 mn		
149404	226953		16 36 22.4	−42 51 32	−0.001	−0.01	5.47	0.40	−6.2	O9 I	−48	1500 s		
149594	159995		16 36 22.5	−14 38 30	−0.002	−0.05	8.0	1.1	0.2	K0 III		290 s		
149890	65426		16 36 25.8	+30 56 29	−0.001	−0.48	7.12	0.54	4.0	F8 V	−8	33 mx		
149425	226957		16 36 28.5	−40 18 10	−0.002	−0.02	7.1	0.5	0.4	B9.5 V		110 s		
149056	253619		16 36 30.7	−64 13 55	−0.002	+0.02	7.9	0.5	4.0	F8 V		60 s		
149662	141270		16 36 31.4	−8 51 17	−0.002	−0.02	6.8	0.4	3.0	F2 V		55 s		
149956	65430		16 36 31.8	+36 02 23	+0.001	−0.03	7.14	1.58	−0.5	gM3	−52	340 s		
148976	253618		16 36 31.9	−67 00 17	−0.004	−0.02	7.0	0.7	0.6	A0 V		70 s		
149511	—		16 36 34.0	−31 14 32			7.80	1.9		S4.7				ST Sco, v
149406	226955		16 36 36.0	−44 59 41	+0.002	−0.04	8.0	1.8	−0.1	K2 III		180 s		
148992	—		16 36 37.2	−66 58 53			7.6	1.1	0.2	K0 III		300 s		
150631	17167		16 36 38.9	+69 47 37	−0.006	+0.06	8.04	0.43		F5			10140	m
148385	257411		16 36 39.6	−77 20 44	−0.004	−0.07	7.1	0.2	2.1	A5 V		93 mx		
149530	207826		16 36 40.7	−32 06 35	−0.002	−0.03	7.9	1.2	2.6	F0 V		31 s		
149481	207820		16 36 41.4	−35 55 00	+0.004	0.00	7.0	0.2	3.0	F2 V		63 s		
149416	226958		16 36 41.7	−45 09 30	0.000	+0.02	7.7	0.2	2.1	A5 V		130 s		
149316	243984		16 36 41.9	−53 55 16	+0.001	−0.01	8.0	1.6	0.2	K0 III		160 s		
149822	102271		16 36 42.9	+15 29 53	−0.001	−0.01	6.30	−0.11		A0 p	0			
150188	30017		16 36 45.3	+52 15 23	−0.002	+0.04	8.0	1.1	0.2	K0 III		360 s		
149804	102270		16 36 45.6	+10 43 13	0.000	−0.01	8.0	0.4	2.6	F0 V		120 s		

HD	SAO	Star Name	α 2000	δ 2000	μ(α)	μ(δ)	V	B–V	M$_v$	Spec	RV	d(pc)	ADS	Notes
			h m s	° ′ ″	s	″								
149724	141271		16 36 48.4	−6 17 39	−0.002	0.00	8.0	0.9	3.2	G5 IV		93 s		
149931	84485		16 36 48.9	+26 31 54	0.000	−0.09	7.80	0.5	3.4	F5 V		76 s		m
149805	121728		16 36 51.5	+7 06 26	−0.001	0.00	6.9	0.0	0.4	B9.5 V		200 s		
149907	84484		16 36 53.8	+22 52 04	+0.007	−0.14	6.8	1.1	0.2	K0 III		93 mx		
150429	17165		16 36 54.8	+63 04 22	−0.001	−0.09	6.16	1.53	−0.3	K5 III	−42	190 s		
149079	—		16 36 55.1	−65 34 49			7.7	0.6	4.4	G0 V		46 s		
149342	243988		16 36 57.1	−54 59 03	0.000	+0.03	7.19	1.48		K2		24 mn		
149881	102273		16 36 58.0	+14 28 31	0.000	0.00	7.05	−0.18		B0.5 III	+13			
150693	17169		16 36 59.0	+69 49 18	−0.008	+0.04	8.00	0.4	2.6	F0 V	0	120 s		m
149533	207834		16 37 02.4	−38 40 33	−0.003	−0.01	7.4	0.1	0.4	B9.5 V		210 s		
149401	243994		16 37 08.2	−52 22 56	−0.001	−0.05	6.90	1.6	−0.5	M2 III		280 s		m
149806	121731		16 37 08.3	+0 15 17	+0.006	+0.08	7.08	0.87	0.2	K0 III	+6	130 mx		m
149757	160006	13 ζ Oph	16 37 09.4	−10 34 02	+0.001	+0.02	2.56	0.02	−4.4	O9.5 V	−19	170 s		
149932	102274		16 37 15.8	+15 42 40	−0.004	−0.01	7.8	1.1	−0.1	K2 III		230 mx		
149845	141281		16 37 21.0	−0 24 50	−0.005	−0.09	7.97	1.31	−0.1	K2 III		96 mx		
150119	65437		16 37 26.4	+37 29 24	0.000	−0.04	7.7	1.4	−0.3	K5 III		400 s		
149807	160011		16 37 29.8	−11 52 11	−0.001	−0.02	7.9	0.4	3.0	F2 V		87 s		
150203	46196		16 37 30.1	+43 33 54	+0.001	+0.04	7.2	0.1	1.4	A2 V	−16	140 s		
149908	121738		16 37 33.6	+5 16 38	0.000	−0.02	6.6	1.1	0.2	K0 III		190 s		
150086	84491		16 37 36.8	+28 50 09	0.000	−0.06	7.70	0.9	0.3	G8 III	0	280 mx		
150047	84489		16 37 37.5	+22 26 41	−0.001	−0.01	8.0	1.6	−0.5	M2 III		510 s		
150048	102279		16 37 43.9	+19 33 23	−0.001	−0.02	8.00	1.1	0.2	K0 III		360 s		m
149237	253631		16 37 46.3	−64 14 56	+0.007	−0.06	7.54	0.97		K1 IV		170 mx		
150012	102278		16 37 47.9	+13 41 14	−0.002	−0.06	6.2	0.4	3.0	F2 V	−21	44 s		
150102	84493		16 37 48.8	+27 02 37	0.000	−0.04	7.10	1.6	−0.5	M2 III	+1	330 s		
149455	244003		16 37 49.1	−54 51 32	−0.001	−0.03	7.8	0.2	0.0	B8.5 V		240 s		
151623	8592		16 37 52.6	+78 55 06	−0.009	+0.04	6.32	1.14	0.2	K0 III	−20	130 s		
149418	244002		16 37 58.7	−58 37 51	0.000	−0.02	7.7	0.2	0.0	B8.5 V		230 s		
149609	226978		16 37 59.6	−46 09 17	0.000	0.00	6.5	1.7	−0.1	K2 III		100 s		
150825	17175		16 37 59.6	+69 09 45	0.000	−0.02	8.0	0.4	2.6	F0 V		120 s		
150449	30026		16 38 00.3	+56 00 56	−0.001	+0.07	5.29	1.08	0.0	K1 III	−19	110 s		
149911	141284		16 38 01.5	−6 32 16	0.000	−0.01	6.09	0.16		A p				
—	102281		16 38 02.5	+17 45 05	0.000	−0.01	7.9			K0 IV				
149640	226981		16 38 02.5	−44 18 01	+0.001	−0.03	7.97	1.15	3.2	K0 III		69 s		
150255	65444		16 38 03.8	+39 34 38	−0.003	−0.06	6.9	1.1	0.2	K0 III		180 mx		
150578	17170		16 38 09.4	+61 10 22	−0.005	+0.01	6.9	1.4	−0.3	K5 III		280 s		
150205	84499		16 38 15.9	+29 40 20	−0.005	−0.20	7.51	0.73	5.2	G5 V		27 s		
149866	184508		16 38 19.1	−22 53 35	−0.003		7.6	0.2	3.0	F2 V		85 s		
150121	102282		16 38 22.0	+14 47 37	+0.001	−0.01	8.0	0.1	1.4	A2 V		210 s		
149976	141288		16 38 25.3	−7 09 08	0.000	+0.02	7.6	1.1	−0.1	K2 III		280 s		
149711	226989		16 38 26.1	−43 23 55	−0.002	−0.03	5.83	−0.02	−2.3	B3 IV	+2	340 s		m
150122	102283		16 38 29.9	+12 39 44	−0.002	+0.04	7.8	0.7	4.4	G0 V		48 s		
149764	207863		16 38 30.6	−39 09 09	−0.003	−0.04	7.0	0.5	0.6	A0 V		93 s		
150409	46209		16 38 32.3	+48 51 44	−0.001	−0.02	6.6	1.6	−0.5	M2 III	−37	270 s		
150050	141290		16 38 33.2	−1 13 55	−0.003	−0.03	6.7	1.1	0.2	K0 III		200 s		
149934	160018		16 38 34.5	−18 49 31	−0.002	0.00	7.0	1.1	0.2	K0 III		230 s		
149239	—		16 38 40.3	−68 43 40			6.9	0.0	0.4	B9.5 V		200 s		
150450	46210	42 Her	16 38 44.7	+48 55 42	−0.005	+0.03	4.90	1.55	−0.5	gM2	−55	120 s	10144	m
150052	141291		16 38 47.7	−8 37 07	+0.001	−0.01	6.7	1.4	−0.3	K5 III		210 s		
149485	253638		16 38 52.6	−60 59 24	0.000	−0.02	6.18	−0.08	−0.2	B8 V		190 s		
149779	226997		16 38 53.9	−44 09 28	+0.002	−0.03	7.5	−0.4	−1.6	B3.5 V		650 s		
150068	160025		16 38 58.8	−10 15 51	−0.004	−0.01	7.9	1.1	0.2	K0 III		250 mx		
150509	30032		16 39 01.6	+50 49 47	−0.004	+0.03	7.0	1.1	0.2	K0 III		230 s		
150256	84505		16 39 01.7	+21 41 16	0.000	−0.04	7.8	0.7	4.4	G0 V		49 s		
149678	244024		16 39 01.8	−52 28 13	+0.001	−0.01	7.1	0.8	0.6	A0 V		61 s		
149612	244021		16 39 03.8	−58 15 29	−0.031	−0.29	7.03	0.61	4.9	G3 V		27 s		
149886	207878		16 39 05.0	−37 13 01	−0.001	−0.03	5.90	0.00		A0				m
150295	84509		16 39 05.8	+24 41 45	−0.003	−0.11	8.0	0.5	3.4	F5 V		83 mx		
149812	227002		16 39 08.8	−44 21 28	0.000	−0.03	7.7	2.2	−0.3	K5 III		150 s		
150391	65461		16 39 08.9	+38 20 42	0.000	0.00	7.6	0.1	1.4	A2 V		170 s		
150144	141295		16 39 15.4	−1 45 29	+0.001	+0.02	7.6	1.1	0.2	K0 III		300 s		
149980	184523		16 39 16.2	−29 55 27	0.000	0.00	7.2	0.4	1.7	A3 V		88 s		
150296	84513		16 39 17.3	+22 13 16	−0.001	+0.01	7.8	1.1	0.2	K0 III		330 s		
150257	102295		16 39 20.1	+13 02 18	+0.002	−0.02	7.9	1.1	0.2	K0 III		340 s		
149981	207889		16 39 20.6	−30 27 52	−0.001	+0.02	7.6	0.4	4.4	G0 V		44 s		
150304	84517		16 39 21.8	+22 00 21	−0.004	−0.10	6.8	1.1	0.2	K0 III	+59	130 mx		
150361	84519		16 39 25.9	+29 13 04	−0.001	+0.02	7.2	0.1	0.6	A0 V	−25	210 s		
149765	244035		16 39 26.2	−51 49 34	0.000	−0.02	7.6	1.2	−0.1	K2 III		340 s		
150340	84522		16 39 37.4	+23 00 05	+0.002	−0.03	7.0	0.4	2.6	F0 V		77 s	10146	m
150177	141298		16 39 38.9	−9 33 17	0.000	−0.15	6.35	0.48		F5				
149835	244045		16 39 43.4	−50 12 44	+0.004	−0.01	6.9	1.3	0.2	K0 III		150 s		
149730	244037		16 39 44.5	−56 59 38	−0.001	−0.01	6.8	0.0	0.4	B9.5 V		190 s		R Ara, m,v
150463	65468		16 39 47.8	+34 37 31	0.000	0.00	7.9	1.1	0.2	K0 III		360 s		
150004	207894		16 39 57.7	−39 47 19	−0.002	−0.02	7.8	0.3	0.6	A0 V		170 s		
149902	227021		16 39 58.1	−48 02 58	−0.002	0.00	7.93	0.95	3.2	G5 IV		88 s		

HD	SAO	Star Name	α 2000	δ 2000	μ(α)	μ(δ)	V	B-V	M_v	Spec	RV	d(pc)	ADS	Notes
			$16^h39^m59^s.7$	$-72°13'36''$	^{s}	$''$								
149264	—		$16^h39^m59^s.7$	$-72°13'36''$			8.0	0.1	0.6	A0 V		300 s		
150258	141302		16 40 01.7	-6 04 52	-0.002	-0.10	6.8	0.5	4.0	F8 V		37 s		
—	—		16 40 05.7	+29 00 41	0.000	+0.01	6.7	0.6	4.4	G0 V		29 s		
149784	244046		16 40 06.0	-57 18 18	-0.002	-0.01	7.2	2.4	-0.3	K5 III		90 s		
150126	184532		16 40 10.6	-28 56 25	-0.002	-0.03	6.9	1.3	-0.3	K5 III		280 s		
150393	102305		16 40 12.1	+17 42 26	+0.001	0.00	7.4	1.1	-0.1	K2 III		310 s		
150090	207903		16 40 12.8	-33 44 35	0.000	+0.01	6.5	1.2	0.2	K0 III		150 s		
150146	184535		16 40 13.1	-26 19 20	+0.002	-0.02	8.0	1.0	0.2	K0 III		360 s		
150566	65475		16 40 15.1	+37 21 19	-0.003	-0.02	7.8	1.1	-0.1	K2 III		380 s		
149967	227026		16 40 15.8	-48 18 28	0.000	-0.01	8.0	1.0	-0.5	M4 III		510 s		
150070	207902		16 40 23.0	-39 55 23	-0.002	-0.05	7.70	0.4	3.0	F2 V		85 s		m
150038	227035		16 40 26.3	-44 34 27	+0.002	-0.02	7.1	1.2	-0.3	K5 III		310 s		
150537	84532		16 40 28.1	+29 51 31	0.000	+0.03	7.9	0.1	0.6	A0 V		290 s		
150039	227036		16 40 31.3	-45 57 27	-0.001	-0.04	7.9	0.3	0.2	K0 III		340 s		
150259	184541		16 40 34.4	-20 24 31	+0.001	+0.03	6.4	1.4	0.2	K0 III		97 s		
150379	121774	36 Her	16 40 35.0	+4 12 26	0.000	-0.01	6.93	0.13	2.1	A5 V	-31	100 s	10149	m
149922	244053		16 40 35.9	-54 08 13	-0.002	-0.02	7.7	0.3	-1.0	B5.5 V		290 s		
150871	30042		16 40 36.9	+56 10 01	-0.002	-0.03	8.0	1.1	0.2	K0 III		360 s		
150151	207909		16 40 37.7	-35 41 13	0.000	0.00	6.7	0.6	0.0	B8.5 V		91 s		
—	—		16 40 38.3	-32 22 35			7.50	2.0		N7.7				
150378	121776	37 Her	16 40 38.5	+4 13 11	0.000	-0.01	5.77	-0.02	0.6	A0 V	-34	100 s	10149	m
150567	84533		16 40 42.2	+28 54 52	-0.003	-0.04	8.00	1.1	-0.2	K3 III		440 s		
150093	227043		16 40 42.2	-41 07 38	-0.002	-0.02	7.3	0.7	0.0	B8.5 V		110 s		
151481	8598		16 40 42.5	+72 40 18	+0.001	+0.01	6.8	1.6	-0.5	M2 III	-62	300 s		
150208	207913		16 40 43.2	-30 49 58	+0.001	-0.04	8.0	1.3	0.2	K0 III		250 s		
149989	244058		16 40 44.3	-51 28 42	+0.002	+0.13	6.5	1.1	2.1	A5 V		22 s		
—	227039		16 40 44.5	-48 45 23	0.000	-0.02	7.3	-0.7	-1.0	B5.5 V		450 s		
150041	227038		16 40 44.6	-48 45 22	+0.001	-0.01	7.07	0.09	-5.6	B0 II k		2100 s		
150164	207912		16 40 46.3	-37 09 13	-0.001	0.00	7.2	0.3	0.6	A0 V		130 s		
149392	257421		16 40 47.0	-72 18 01	+0.001	+0.01	6.70	1.1	0.2	K0 III		200 s		m
149669	—		16 40 49.3	-65 55 18			7.6	1.1	0.2	K0 III		300 s		
149837	253651		16 40 50.3	-60 26 46	+0.008	-0.07	6.18	0.48	4.0	dF8		27 s		m
150483	102314		16 40 51.3	+12 23 42	-0.002	-0.01	6.00	0.05	1.4	A2 V	-27	83 s		
151101	17188	18 Dra	16 40 55.0	+64 35 20	-0.001	-0.02	4.83	1.22		K1 p	0	33 mn		
150381	141307		16 40 56.3	-8 18 32	-0.001	+0.02	6.5	1.1	0.2	K0 III		160 s		
150554	84534		16 40 56.3	+21 56 53	-0.006	0.00	7.90	0.5	4.0	F8 V		60 s	10155	m
150709	46225		16 40 56.5	+42 27 44	-0.002	-0.01	7.5	1.1	-0.1	K2 III		340 s		
151186	17190		16 40 59.7	+66 57 31	-0.003	-0.04	8.0	1.1	-0.1	K2 III		420 s		
150580	84536		16 41 00.5	+24 51 32	-0.002	0.00	6.06	1.31	-0.1	K2 III	-68	140 s		
150679	65481		16 41 04.5	+36 12 01	+0.003	-0.05	7.2	0.1	1.4	A2 V		150 s		
150083	227044		16 41 05.4	-47 44 44	0.000	+0.01	7.25	0.01	-1.9	B6 III		530 s		m
151067	17187		16 41 06.2	+62 18 29	-0.002	0.00	7.2	0.0	0.0	B8.5 V		280 s		
150382	160043		16 41 07.7	-11 50 19	0.000	-0.02	7.0	1.4	-0.3	K5 III		230 s		
150433	141309		16 41 08.1	-2 51 25	-0.002	-0.44	7.24	0.64	4.7	dG2	-42	25 ts		
150451	141310		16 41 11.4	-1 00 02	0.000	-0.02	6.24	0.30		A5				
150493	121783		16 41 16.6	+1 14 43	-0.001	-0.02	6.41	1.04	0.2	K0 III		170 s		
150071	244066		16 41 17.1	-51 23 47	-0.001	0.00	8.0	0.0	0.6	A0 V		280 s		
150680	65485	40 ζ Her	16 41 17.1	+31 36 10	-0.037	+0.39	2.81	0.65	3.0	G0 IV	-70	9.6 t	10157	m
149839	253654		16 41 18.2	-62 53 10	+0.002	-0.01	8.0	1.4	-0.3	K5 III		440 s		
150365	160044		16 41 18.6	-18 03 35	+0.003	-0.04	6.6	0.2	2.1	A5 V	-2	81 s		
150600	84539		16 41 19.0	+22 26 11	0.000	-0.09	7.2	1.1	0.2	K0 III		190 mx		
150136	227049		16 41 20.2	-48 45 46	-0.001	0.00	5.62	0.16		O7	+23			m
150681	65486		16 41 21.4	+30 16 22	-0.005	-0.04	7.5	0.4	2.6	F0 V		96 s		
150568	102318		16 41 22.6	+16 06 23	+0.002	-0.04	7.7	1.1	-0.1	K2 III		360 s		
149671	253649	η TrA	16 41 23.2	-68 17 46	+0.001	-0.01	5.91	-0.08	-1.0	B7 IV		240 s		
150167	227056		16 41 24.0	-46 23 05	+0.002	+0.01	7.9	0.3	0.4	B9.5 V		190 s		
150214	227059		16 41 24.7	-42 03 37	0.000	0.00	7.7	1.9	-0.3	K5 III		190 s		
150525	121788		16 41 30.2	+4 52 19	+0.001	-0.02	6.8	0.1	0.6	A0 V	+2	170 s		
150632	102321		16 41 33.0	+18 55 17	0.000	+0.02	7.4	1.1	-0.1	K2 III		310 s		
150665	84542		16 41 33.8	+26 04 47	+0.001	-0.06	7.60	1.1	0.2	K0 III		300 s		
150416	160046		16 41 34.2	-17 44 32	-0.002	0.00	4.96	1.11	-2.1	G8 II	-25	240 s		
150710	65487		16 41 34.6	+30 52 33	-0.004	0.00	7.0	0.5	3.4	F5 V		53 s		
150366	184549		16 41 36.0	-24 28 05	-0.004	-0.01	6.09	0.20	2.1	A5 V		59 s		
150682	84543	39 Her	16 41 36.6	+26 55 00	0.000	-0.04	5.92	0.40	3.0	F2 V	-12	38 s		
150367	184547		16 41 37.4	-27 48 34	+0.001	-0.04	6.5	1.5	0.2	K0 III		85 s		
150749	65488		16 41 39.3	+34 35 21	-0.005	-0.04	7.7	0.5	3.4	F5 V		73 s		
150615	102325		16 41 39.6	+15 57 05	-0.002	-0.02	7.8	1.6	-0.5	M2 III		470 s		
150168	227058		16 41 40.2	-49 39 04	+0.001	+0.01	5.65	-0.03		B0.5 I k				
150557	121790	14 Oph	16 41 42.3	+1 10 52	-0.007	+0.05	5.74	0.32	3.0	dF2	-45	35 s		
150323	207929		16 41 43.2	-32 49 17	-0.001	-0.02	7.61	0.07	-1.0	B5.5 V		390 s		
150331	207930		16 41 45.3	-33 08 47	-0.005	-0.08	5.87	0.65	0.6	G0 III		99 mx		
150095	244070		16 41 48.9	-54 53 23	0.000	-0.01	7.5	1.3	-0.1	K2 III		290 s		
150248	227066		16 41 49.6	-45 22 08	+0.006	-0.11	7.02	0.68	4.9	G3 V		25 s		
150633	102327		16 41 50.0	+13 08 56	0.000	-0.25	7.7	1.1	0.2	K0 III		61 mx		
150111	244071		16 41 50.4	-54 44 59	0.000	-0.16	7.7	1.5	3.2	G5 IV		29 s		

HD	SAO	Star Name	α 2000	δ 2000	μ(α)	μ(δ)	V	B-V	M_v	Spec	RV	d(pc)	ADS	Notes
			h m s	° ′ ″	s	″								
150683	102330		16 41 53.2	+18 37 43	0.000	+0.01	7.8	0.4	2.6	F0 V		110 s		
150453	160052		16 41 53.5	−19 55 28	+0.002	+0.04	5.60	0.5	3.4	F5 V	+5	28 s		m
151746	8602		16 41 56.8	+73 53 14	−0.007	+0.03	7.30	0.1	1.4	A2 V	−10	120 mx		m
150813	65492		16 42 02.0	+33 06 06	−0.002	+0.04	7.9	0.5	3.4	F5 V		78 s		
150417	207939		16 42 05.5	−30 31 46	+0.002	−0.03	7.6	1.1	0.2	K0 III		270 s		
150618	121792		16 42 07.1	+1 14 28	0.000	−0.05	7.60	0.41	3.0	F2 V		80 s		
150437	184553		16 42 07.2	−29 07 30	+0.004	−0.10	7.86	0.68	4.7	G2 V		37 s		
−−	8604		16 42 20.3	+74 28 31	−0.001	+0.01	8.0	1.6						
150216	244078		16 42 21.7	−54 10 24	−0.005	0.00	7.4	1.9	−0.1	K2 III		120 s		
150781	84550		16 42 26.1	+21 35 33	+0.001	−0.01	7.30	0.1	0.6	A0 V	−24	220 s	10167	m
151044	30055		16 42 27.6	+49 56 11	+0.012	−0.11	6.47	0.54	4.0	F8 V		31 ts		
150185	244076		16 42 27.9	−56 48 00	0.000	−0.02	7.3	2.1	−0.3	K5 III		150 s		
150420	207943		16 42 28.8	−37 04 47	−0.003	−0.11	7.00	0.9	3.2	G5 IV		57 s		m
150352	227077		16 42 29.2	−43 44 56	−0.004	−0.03	7.9	0.4	1.7	A3 V		85 mx		
149197	257422		16 42 30.3	−77 32 46	−0.011	+0.05	7.7	0.4	3.0	F2 V		87 s		
150620	160057		16 42 35.4	−11 10 28	−0.001	−0.02	7.2	0.1	1.4	A2 V		130 s		
151541	17200		16 42 38.3	+68 06 08	−0.052	+0.43	7.56	0.76	6.1	K1 V	+6	23 ts		
150085	253669		16 42 39.0	−61 47 04	+0.002	+0.01	6.59	0.79	0.4	G3 III		170 s		
149771	257426		16 42 40.3	−70 34 52	−0.001	−0.05	7.5	−0.5	0.6	A0 V		240 s		
150474	207949		16 42 46.5	−37 20 33	−0.001	−0.09	7.1	1.1	3.2	G5 IV		41 s		
150096	253671		16 42 46.7	−62 22 14	+0.005	−0.02	7.3	1.1	0.2	K0 III		260 s		
150543	184563		16 42 46.9	−27 11 16	−0.001	−0.04	7.9	1.5	−0.1	K2 III		260 s		
150097	253672		16 42 48.4	−62 33 14	+0.002	−0.01	8.00	1.1	−0.1	K2 III		400 s		m
149057	257420		16 42 52.1	−79 01 25	−0.002	+0.01	7.4	1.1	−0.1	K2 III		320 s		
150997	65504	44 η Her	16 42 53.7	+38 55 20	+0.003	−0.08	3.53	0.92	1.8	G8 III-IV	+8	21 ts		m
150889	84555		16 42 55.1	+25 51 17	0.000	−0.03	7.40	1.1	−0.1	K2 III		320 s		
150830	102339		16 42 57.0	+15 39 15	+0.001	−0.02	7.6	1.1	−0.1	K2 III		360 s		
150800	102338		16 42 57.7	+10 56 30	+0.001	0.00	7.9	1.4	−0.3	K5 III		430 s		
151199	30062		16 42 58.3	+55 41 25	+0.005	+0.08	6.16	0.07		A5 p	−46			
150486	207953		16 43 00.4	−38 40 23	−0.001	−0.08	7.7	−0.3	0.4	B9.5 V		67 mx		
150487	207952		16 43 01.5	−39 30 33	−0.006	+0.01	7.7	1.1	3.2	G5 IV		51 s		
150764	121807		16 43 02.8	+3 27 15	−0.001	−0.01	7.80	0.2	2.1	A5 V		140 s		m
150421	227092		16 43 03.3	−46 04 14	−0.001	−0.02	6.23	0.85	−4.6	F2 Ib		780 s		
150998	65508		16 43 04.1	+36 30 34	0.000	0.00	6.89	1.50	−0.1	K2 III		160 s		
149324	257424	β Aps	16 43 04.5	−77 31 02	−0.086	−0.35	4.24	1.06	0.2	K0 III	−31	42 mx		m
150250	244084		16 43 04.9	−58 28 15	−0.002	0.00	8.0	−0.2	0.0	B8.5 V		400 s		
150530	207954		16 43 05.5	−36 38 03	+0.001	−0.13	7.9	1.1	0.2	K0 III		130 mx		
152303	8612		16 43 05.8	+77 30 50	+0.014	+0.21	6.10	0.4	3.0	F2 V	+7	41 ts	10214	m
150918	84559		16 43 06.6	+26 03 24	+0.002	0.00	8.0	1.6	−0.5	M2 III		510 s		
150477	227099		16 43 11.3	−43 39 11	+0.001	−0.03	8.0	0.0	0.6	A0 V		280 s		
150765	141328		16 43 12.3	−4 09 01	0.000	−0.02	7.8	1.4	−0.3	K5 III		420 s		
150857	102342		16 43 14.5	+12 52 42	−0.001	−0.03	7.6	1.4	−0.3	K5 III		380 s		
150251	244087		16 43 17.6	−59 12 26	−0.001	−0.01	7.8	1.1	0.6	A0 V		61 s		
150026	253673		16 43 22.0	−67 25 57	+0.004	−0.04	6.03	0.02	0.6	A0 V n		120 s		
150519	227105		16 43 29.4	−44 20 20	−0.004	−0.01	7.8	1.9	−0.3	K5 III		230 s		
150933	84562		16 43 30.0	+20 42 57	+0.002	−0.04	7.10	0.5	4.0	F8 V		42 s		m
150905	102346		16 43 33.3	+12 25 06	0.000	0.00	7.2	1.1	−0.1	K2 III		290 s		
150637	207963		16 43 34.0	−31 13 33	+0.002	+0.02	7.8	0.0	1.7	A3 V		170 s		
150713	184579		16 43 37.6	−20 41 46	0.000	−0.01	7.7	0.7	0.6	A0 V		93 s		
151058	65510		16 43 38.3	+32 41 30	−0.002	−0.04	7.8	1.1	0.2	K0 III		330 s		
150638	207967		16 43 38.5	−32 06 21	−0.001	−0.02	6.46	−0.08		B9				
150891	121816		16 43 38.8	+6 37 06	0.000	−0.01	7.70	1.1	0.2	K0 III		320 s		m
150714	184578		16 43 39.4	−22 44 10	−0.003	−0.01	7.5	0.3	0.6	A0 V		160 s		
150859	141330		16 43 41.7	−0 47 03	−0.001	0.00	7.7	1.1	0.2	K0 III		320 s		
150500	227109		16 43 44.2	−47 06 20	0.000	−0.02	7.04	−0.03		B9 pe				
150573	227118		16 43 45.3	−41 07 08	−0.004	−0.02	6.20	0.14	1.7	A3 V		73 s		
150608	207966		16 43 47.5	−38 09 23	−0.001	−0.05	6.05	−0.06		A0				m
150698	184581		16 43 51.2	−26 48 32	−0.003	−0.10	6.9	0.7	4.4	G0 V		28 s		
151087	65512		16 43 51.6	+34 02 19	−0.006	+0.05	5.99	0.29	3.0	F2 V	−10	40 s		
150591	227123		16 43 53.8	−41 06 47	−0.002	−0.01	6.12	−0.07	0.0	B8.5 V	−2	170 s		m
150641	207970		16 43 56.3	−38 20 02	−0.002	−0.03	7.9	−0.4	0.6	A0 V		290 s		
150719	207975		16 43 59.8	−30 15 36	0.000	−0.03	8.0	1.5	−0.3	K5 III		450 s		
150965	121821		16 44 00.2	+9 01 18	−0.001	0.00	8.0	1.1	0.2	K0 III		360 s		
150359	244093		16 44 04.5	−59 28 25	−0.003	+0.03	7.8	0.5	0.6	A0 V		120 s		
150738	207976		16 44 10.1	−31 00 27	−0.001	+0.01	7.9	2.1	−0.5	M2 III		250 s		
151070	84572		16 44 10.6	+23 31 00	0.000	−0.01	7.40	0.4	3.0	F2 V	−38	76 s	10184	m
150768	184591		16 44 17.2	−27 27 22	−0.001	−0.01	6.60	0.1	0.6	A0 V		150 s	10173	m
151554	17204		16 44 17.2	+60 58 15	−0.005	+0.02	7.8	0.1	0.6	A0 V		92 mx		
150814	184593		16 44 17.5	−22 31 26	−0.002	−0.04	7.6	0.4	0.6	A0 V		140 s		
150935	141335		16 44 17.6	−0 33 27	−0.001	−0.03	7.50	0.1	0.6	A0 V		240 s		m
150689	−−		16 44 18.1	−38 56 32			7.7	0.6	0.2	K0 III		310 s		
150563	227128		16 44 18.5	−49 46 20	+0.002	−0.01	7.9	1.5	−0.1	K2 III		260 s		
151287	46256		16 44 20.6	+46 15 10	−0.004	+0.02	8.0	0.0	0.4	B9.5 V		82 mx		
150645	227135		16 44 21.1	−44 22 44	+0.001	−0.03	7.8	0.3	1.4	A2 V		130 s		
151071	102354		16 44 26.0	+18 34 07	+0.001	−0.05	8.0	1.1	0.2	K0 III		360 s		

HD	SAO	Star Name	α 2000	δ 2000	μ(α)	μ(δ)	V	B-V	M_v	Spec	RV	d(pc)	ADS	Notes
151072	102355		16h 44m 33.8s	+13° 36′ 40″	0.000	0.00	7.90	0.1	0.6	A0 V		290 s	10186	m
151387	46258		16 44 34.4	+49 51 47	+0.001	-0.05	7.4	0.4	2.6	F0 V		90 s		
151352	46257		16 44 36.4	+48 15 54	0.000	+0.04	7.7	1.4	-0.3	K5 III		400 s		
150576	244106		16 44 39.6	-53 09 07	-0.001	0.00	6.00	1.1	0.2	K0 III		140 s		m
150922	160076		16 44 41.9	-14 04 50	-0.002	-0.02	8.0	1.6	-0.5	M2 III		400 s		
150756	207979		16 44 42.3	-39 23 36	-0.002	-0.05	7.2	0.4	0.2	K0 III		250 s		
150742	227146		16 44 42.3	-40 50 23	-0.002	-0.03	5.71	-0.12	-2.1	B2.5 V	+12	330 s		m
151029	121826		16 44 42.4	+2 19 55	-0.001	-0.07	8.00	0.1	1.4	A2 V		50 mx	10185	m
151088	102359		16 44 51.6	+9 57 06	+0.001	-0.02	7.90	1.4	-0.3	K5 III		440 s		m
150743	227149		16 44 52.4	-41 55 28	-0.001	-0.02	8.0	0.4	0.6	A0 V		300 s		
150675	227144		16 44 52.7	-47 38 41	0.000	-0.01	7.07	1.20	-4.4	K1 Ib		2000 s		
151046	141343		16 44 59.7	-0 36 53	0.000	-0.01	7.9	1.4	-0.3	K5 III		430 s		
151090	121831		16 44 59.9	+6 05 17	-0.014	-0.27	6.58	0.88	5.3	dG6	-14	20 ts		m
150894	184602		16 45 00.1	-28 30 35	-0.002	0.00	6.00	0.1	1.4	A2 V		82 s		m
151202	84574		16 45 02.2	+23 43 03	-0.012	-0.03	7.3	1.1	0.2	K0 III		83 mx		
150577	244108		16 45 02.6	-55 43 08	+0.003	-0.01	7.3	1.2	0.0	B8.5 V		52 s		
151237	84577		16 45 05.1	+28 21 28	-0.001	+0.03	7.30	0.5	3.4	F5 V	-46	60 s	10194	m
150878	207990		16 45 05.2	-31 22 49	-0.003	-0.02	8.0	1.5	-0.3	K5 III		450 s		
151353	46260		16 45 05.7	+40 38 19	0.000	+0.03	8.0	0.1	0.6	A0 V		300 s		
150937	184607		16 45 08.5	-23 11 02	-0.002	-0.02	6.9	0.8	3.4	F5 V	-36	30 s		
151061	141344		16 45 11.2	-3 05 06	-0.001	-0.01	7.20	1.6		M6 III	-8			
151388	46262		16 45 11.6	+43 13 02	-0.002	-0.05	6.05	1.40	-0.1	K2 III	-13	120 s		
151613	30076		16 45 17.6	+56 46 54	+0.001	+0.06	4.85	0.38	3.0	F2 V	0	24 ts		
150924	207995		16 45 19.0	-30 48 23	0.000	+0.02	6.9	1.0	-0.1	K2 III		260 s		
151876	17209		16 45 19.0	+66 53 17	-0.001	-0.01	8.0	1.1	-0.1	K2 III		410 s		
151203	102365		16 45 22.4	+15 44 43	+0.002	-0.05	5.56	1.66	-0.5	M4 III	-19	160 s		
150139	257429		16 45 26.8	-71 50 23	+0.010	+0.06	7.70	0.7	4.4	G0 V		46 s		m
150548	253686		16 45 27.8	-60 33 49	+0.002	-0.01	7.6	0.2	0.0	B8.5 V		220 s		
151133	121834	16 Oph	16 45 29.6	+1 01 13	0.000	+0.01	6.03	-0.02		B9 n	-14			
151011	160080		16 45 30.2	-19 08 13	+0.001	0.00	6.8	1.4	-0.3	K5 III		270 s		
150970	184612		16 45 34.2	-28 50 27	0.000	-0.01	7.4	-0.2	1.4	A2 V		160 s		
150926	207999		16 45 35.5	-33 51 01	+0.004	-0.07	7.3	0.5	4.0	F8 V		46 s		
151669	30081		16 45 37.9	+55 18 49	-0.003	-0.04	7.1	0.4	3.0	F2 V		66 s		
151428	65537		16 45 48.1	+35 37 53	+0.004	+0.05	7.40	0.7	4.4	G0 V	+3	40 s	10203	m
151012	184617		16 45 48.3	-26 38 57	-0.002	-0.02	6.9	0.5	0.6	A0 V		89 s		
151217	121843	43 Her	16 45 49.8	+8 34 57	0.000	+0.01	5.15	1.53	-0.3	K5 III	-21	120 s		m
151543	46271		16 45 51.6	+46 02 43	-0.001	+0.02	7.4	1.1	0.2	K0 III		270 s		
148527	258748		16 45 53.4	-83 14 20	+0.009	+0.02	6.3	2.6	-0.3	K5 III		47 s		
151035	184619		16 45 55.3	-29 12 23	+0.002	-0.06	7.8	0.9	0.2	K0 III		200 mx		
150596	253687		16 45 56.7	-61 39 48	-0.005	-0.04	7.00	0.1	0.6	A0 V		87 mx		m
150408	253685		16 45 57.0	-67 45 01	+0.006	-0.03	7.0	0.1	1.4	A2 V		130 s		
151219	121845		16 45 57.8	+4 02 30	0.000	-0.02	7.1	0.4	3.0	F2 V		67 s		
153751	2770	22 ε UMi	16 45 57.8	+82 02 14	+0.005	0.00	4.23	0.89	0.3	G5 III	-11	61 s	10242	m,v
150884	227171		16 46 03.8	-46 29 07	+0.001	+0.03	7.02	0.86	-8.0	F8 Ia		8800 s		
150897	227172		16 46 06.2	-46 31 56	-0.001	-0.04	6.47	0.10	1.2	A1 V		100 s		
--	--		16 46 11.0	+33 38 03	-0.005	+0.02	7.0	1.1	0.0	K1 III		240 mx		
151269	121847		16 46 17.8	+3 50 56	0.000	-0.09	7.9	0.9	3.2	G5 IV		88 s		
150745	244122		16 46 21.0	-58 30 13	0.000	-0.02	5.9	0.0	-1.6	B3.5 V	-16	300 s		
150597	--		16 46 22.1	-64 27 31			7.8	1.1	-0.1	K2 III		380 s		
151651	46278		16 46 23.8	+47 32 37	+0.004	+0.03	8.0	0.5	3.4	F5 V		84 s		
150958	227184		16 46 27.4	-45 23 51	+0.001	0.00	7.30	0.35		O6				m
151003	227190		16 46 33.9	-41 36 37	-0.001	0.00	7.09	0.17		O9 II				
151307	141357		16 46 38.2	-3 11 42	0.000	+0.01	8.00	0.4	2.6	F0 V		120 s	10204	m
150549	253688		16 46 39.9	-67 06 35	-0.001	-0.02	5.13	-0.08		A0 p	-2	23 mn		m
150775	244127		16 46 40.2	-59 02 34	-0.005	-0.02	7.7	2.0	-0.1	K2 III		120 s		
151238	160091		16 46 47.0	-13 20 18	-0.002	+0.02	8.00	0.1	1.4	A2 V		180 s		m
151078	208020		16 46 47.7	-39 22 38	-0.003	-0.03	5.4	2.0	0.2	K0 III		29 s		
151051	227197		16 46 50.2	-41 28 53	0.000	0.00	7.35	1.78	-0.5	M4 III		370 s		
151179	184630	25 Sco	16 46 51.2	-25 31 43	0.000	-0.02	6.5	1.5	0.2	K0 III	+2	90 s		
151837	30095		16 46 51.6	+55 24 36	+0.003	0.00	6.9	1.1	-0.1	K2 III	-7	260 s		
151154	208026		16 46 52.7	-31 27 10	-0.001	0.00	7.6	0.6	0.6	A0 V		120 s		
151372	121857		16 46 52.8	+2 14 32	-0.004	+0.02	6.70	0.4	2.6	F0 V		66 s		m
152048	17220		16 46 54.2	+63 23 09	0.000	+0.02	7.9	1.4	-0.3	K5 III		430 s		
150550	253689		16 46 55.5	-68 06 07	+0.004	+0.03	6.6	0.3	0.6	A0 V		97 s		
151222	184634		16 46 57.0	-20 06 14	+0.002	-0.15	7.6	1.0	0.2	K0 III		100 mx		
151171	208028		16 46 58.2	-30 59 51	-0.002	-0.03	7.9	2.2	-0.3	K5 III		160 s		
151109	208024		16 47 01.7	-39 32 04	0.000	-0.05	7.00	-0.06	-0.2	B8 V		270 s		
152222	17226		16 47 04.5	+67 16 03	+0.002	-0.01	7.2	1.1	-0.1	K2 III		290 s		
151195	184632		16 47 04.6	-29 35 37	+0.003	-0.02	8.0	0.4	0.2	K0 III		360 s		
151431	121859	19 Oph	16 47 09.6	+2 03 53	-0.001	-0.01	6.10	0.14		A2 n	-6		10207	m
151493	102384		16 47 09.7	+13 01 54	0.000	+0.02	8.0	1.1	-0.1	K2 III		420 s		
152124	17223		16 47 15.6	+63 31 53	-0.012	+0.06	7.0	1.4	-0.3	K5 III		210 mx		
150898	244133		16 47 19.4	-58 20 29	-0.002	-0.02	5.58	-0.08	-6.2	B0 Iab	-51	1900 s		
151732	46288		16 47 19.6	+42 14 20	0.000	-0.02	5.87	1.61	-0.5	M4 III	-7	190 s		
150915	244136		16 47 19.8	-57 27 38	-0.001	-0.01	7.0	1.1	0.4	B9.5 V		45 s		

HD	SAO	Star Name	α 2000	δ 2000	μ(α)	μ(δ)	V	B-V	M_v	Spec	RV	d(pc)	ADS	Notes
			h m s	° ' "	s	"								
153448	2769		16 47 20.4	+80 07 05	−0.015	+0.05	8.0	0.8	3.2	G5 IV		89 s		
152375	17228		16 47 22.8	+69 05 53	−0.001	−0.01	7.4	0.1	0.6	A0 V		230 s		
151139	227215		16 47 23.6	−41 48 33	+0.001	−0.01	7.60	0.29	−4.0	B1.5 III		1100 s		
150886	244134		16 47 25.3	−59 40 12	+0.001	−0.05	7.3	0.9	0.4	B9.5 V		69 s		
151082	227205		16 47 28.6	−48 19 09	+0.007	0.00	7.30	1.4	−0.3	K5 III		230 mx		m
151097	227209		16 47 30.8	−47 35 25	0.000	−0.01	7.24	1.13	4.4	G0 V		37 s		
151748	65560		16 47 35.0	+39 26 44	0.000	−0.01	7.9	1.1	−0.1	K2 III		400 s		
151240	208034		16 47 36.0	−33 42 00	−0.002	−0.06	7.9	0.0	3.4	F5 V		79 s		
151749	65559		16 47 37.6	+37 33 36	−0.001	−0.01	7.0	0.0	0.4	B9.5 V		210 s		
153372	8634		16 47 38.8	+79 30 13	−0.001	+0.03	6.8	1.1	0.2	K0 III	−29	210 s		
152376	17232		16 47 46.0	+67 50 30	−0.005	+0.02	7.6	1.1	0.2	K0 III		310 s		
151525	121865	45 Her	16 47 46.3	+5 14 48	−0.001	−0.04	5.24	−0.02		A p	−16	32 mn		m
151346	184646		16 47 46.4	−23 58 26	−0.001	0.00	7.90	0.41		B8				
151626	84610		16 47 47.4	+20 12 19	−0.001	+0.03	7.9	1.1	0.2	K0 III		340 s		
151173	227225		16 47 49.5	−45 23 23	+0.002	+0.01	7.4	0.0	0.0	B8.5 V		280 s		
151348	184647		16 47 55.9	−28 57 25	0.000	+0.01	7.9	0.2	4.0	F8 V		60 s		
151241	208036		16 47 58.0	−39 57 55	−0.002	−0.01	7.88	0.18	0.0	A0 IV		280 s		
152084	30104		16 48 00.9	+57 29 21	+0.002	−0.01	7.0	0.8	3.2	G5 IV		58 s		
151450	160101		16 48 01.8	−15 57 51	+0.006	−0.11	7.3	0.7	4.4	G0 V		39 s		
151196	227233		16 48 05.2	−45 56 41	−0.002	−0.02	6.67	0.37	1.5	A7 IV		94 s		
151616	102391		16 48 05.9	+11 07 57	0.000	+0.04	7.4	1.1	0.2	K0 III		270 s		
151627	102393		16 48 08.7	+13 35 25	+0.001	−0.02	6.35	0.87	3.2	G5 IV	+1	37 s		
151315	208043		16 48 11.1	−36 52 59	−0.002	+0.03	7.40	0.2	2.1	A5 V		110 s		m
151591	121871		16 48 13.7	+1 05 26	−0.001	−0.02	8.0	0.1	1.7	A3 V		180 s		m
151415	184652		16 48 13.7	−24 31 35	0.000	−0.01	7.10	1.69	−0.4	gM0	−75	260 s		m
151227	227236		16 48 14.7	−44 56 10	−0.003	0.00	8.0	1.5	−0.1	K2 III		270 s		
151495	160103		16 48 14.8	−15 21 39	−0.002	−0.01	8.0	1.1	0.2	K0 III		370 s		
151213	227235		16 48 18.6	−47 16 51	0.000	−0.01	7.64	0.29	−4.6	B0 IV		2400 s		
151701	102396		16 48 23.9	+17 07 57	−0.002	−0.01	7.3	0.1	1.4	A2 V		150 s		
151395	208050		16 48 24.6	−31 12 12	0.000	−0.01	6.8	0.0	0.0	B8.5 V		220 s		
152105	30106		16 48 25.2	+55 23 25	+0.001	−0.02	7.9	1.1	0.2	K0 III		340 s		
147661	—		16 48 25.6	−85 33 23			8.0	1.1	0.2	K0 III		360 s		
151527	160104		16 48 26.8	−14 54 33	−0.002	0.00	6.1	0.1	0.6	A0 V		130 s		
151528	160105		16 48 28.2	−16 20 05	−0.006	−0.28	8.0	0.9	3.2	G5 IV		34 mx		
151005	253703		16 48 30.6	−61 29 21	+0.001	−0.01	7.2	1.6	0.2	K0 III		110 s		
151780	84619		16 48 32.0	+26 35 58	+0.002	−0.01	7.90	1.1	0.0	K1 III		380 s		
151379	208051		16 48 37.6	−37 15 06	−0.002	−0.10	7.4	−0.3	3.4	F5 V		62 s		
150869	—		16 48 38.9	−67 05 16			6.87	1.56	−0.3	K4 III		220 s		
150798	253700	α TrA	16 48 39.8	−69 01 39	+0.005	−0.03	1.92	1.44	−0.1	K2 III	−4	17 s		[Atria]
151878	65569		16 48 41.3	+35 55 19	−0.002	0.00	7.30	0.4	3.0	F2 V		72 s	10224	m
151766	102402		16 48 42.6	+18 19 07	−0.001	+0.02	7.7	0.5	4.0	F8 V		54 s		
151396	208054		16 48 43.0	−38 04 27	−0.003	−0.06	7.9	1.1	0.2	K0 III		180 mx		
151655	141374		16 48 51.3	−4 30 58	0.000	−0.10	7.3	0.9	3.2	G5 IV		65 s		
151231	244158		16 48 53.5	−52 56 27	−0.002	−0.01	7.0	0.1	0.0	B8.5 V		200 s		
151066	253705		16 48 55.8	−61 23 17	−0.002	−0.01	7.8	0.0	0.6	A0 V		250 s		
151337	227251		16 48 59.0	−47 43 07	−0.011	−0.03	7.39	0.91	5.9	K0 V		18 s		
151632	141377		16 48 59.6	−8 55 49	−0.003	−0.04	7.8	0.5	3.4	F5 V		72 s		
151782	102405		16 49 00.1	+12 52 55	−0.002	0.00	7.4	1.4	−0.3	K5 III		350 s		
151936	65573		16 49 00.9	+34 04 18	−0.001	+0.04	7.8	0.5	3.4	F5 V		76 s		
152031	46300		16 49 01.9	+41 39 49	−0.001	+0.04	7.5	0.1	1.7	A3 V		150 s		
151935	65574		16 49 04.2	+34 57 26	0.000	+0.07	6.7	1.1	0.2	K0 III		200 s		
152106	46304		16 49 05.2	+48 14 11	−0.002	+0.07	7.5	0.2	2.1	A5 V		120 s		
151796	102406		16 49 09.3	+12 54 33	−0.001	+0.01	6.6	1.1	0.2	K0 III		190 s		
152107	46305	52 Her	16 49 14.1	+45 59 00	+0.002	−0.06	4.82	0.09		A2 p	−1	33 mn	10227	m
151937	65577		16 49 15.1	+29 57 53	−0.005	+0.08	6.58	1.25	−1.1	K1 II-III	−43	310 s		
151511	208065		16 49 19.1	−34 01 01	−0.001	−0.05	7.10	0.1	1.4	A2 V		130 s		m
152223	30112		16 49 23.2	+52 55 01	−0.003	0.00	7.1	1.1	0.2	K0 III		240 s		
151783	121882		16 49 25.5	+0 53 27	+0.001	−0.04	7.3	0.4	2.6	F0 V		86 s		
151439	227261		16 49 25.7	−44 59 29	0.000	0.00	7.9	0.9	0.2	K0 III		350 s		
151363	227256		16 49 26.4	−49 59 09	0.000	−0.02	7.7	0.0	0.4	B9.5 V		290 s		
152644	17238		16 49 26.9	+67 35 45	−0.001	−0.04	8.0	1.1	0.2	K0 III		360 s		
151676	160116		16 49 27.6	−15 40 03	+0.001	+0.03	6.1	0.1	1.7	A3 V		76 s		
151473	227264		16 49 30.6	−43 56 41	−0.001	−0.02	7.47	0.19	0.6	A0 V		200 s		m
153143	8636		16 49 31.1	+75 22 45	−0.013	+0.04	7.6	0.4	3.0	F2 V	−28	84 s		
151579	208069		16 49 33.1	−31 39 04	−0.001	−0.01	6.70	1.1	0.2	K0 III		190 s		m
151862	102410		16 49 34.5	+13 15 40	−0.002	−0.02	6.00	0.03	1.2	A1 V	−23	91 s	10225	m
151658	184681		16 49 34.9	−21 51 09	−0.001	−0.02	7.60	1.6	−0.5	M2 III	−102	420 s		
151635	184678		16 49 35.0	−25 04 20	+0.002	−0.01	7.4	0.9	0.2	K0 III		280 s		
152153	46311		16 49 40.4	+43 25 49	−0.002	−0.01	6.13	1.26		K0	−20			
152013	84628		16 49 43.9	+28 23 06	−0.002	+0.02	8.0	0.5	4.0	F8 V		62 s		
151659	184683		16 49 44.3	−24 38 26	0.000	−0.02	7.1	0.1	1.4	A2 V		120 s		
151581	208070		16 49 44.5	−36 24 22	+0.004	0.00	7.9	0.9	0.2	K0 III		340 mx		
151249	244168	η Ara	16 49 47.0	−59 02 29	+0.005	−0.03	3.76	1.57	−0.3	K5 III	+9	58 s		m
151515	227272		16 49 47.7	−42 00 07	−0.003	−0.01	7.16	0.16	−3.9	B3 II		1200 s		
151475	227266		16 49 48.6	−47 07 44	0.000	0.00	8.01	0.17	−2.5	B2 V		1300 s		

HD	SAO	Star Name	α 2000	δ 2000	μ(α)	μ(δ)	V	B-V	M$_v$	Spec	RV	d(pc)	ADS	Notes
151769	160118	20 Oph	16ʰ49ᵐ49.9ˢ	-10°46'59"	+0.006ˢ	-0.10"	4.65	0.47	0.7	F6 III	-1	56 mx		
151550	227274		16 49 55.9	-43 06 26	+0.001	-0.06	7.0	0.3	2.6	F0 V		75 s		
151939	102415		16 49 57.1	+15 22 36	-0.003	-0.03	7.0	0.5	3.4	F5 V		53 s		
151879	121893		16 49 58.2	+9 25 18	+0.001	+0.02	6.91	0.91		G5	-26			
152238	46315		16 49 59.1	+46 38 58	-0.004	+0.05	6.9	1.1	-0.1	K2 III		250 s		
151476	227268		16 50 00.0	-49 56 53	+0.001	+0.01	7.8	1.5	-0.3	K5 III		340 s		
151564	227278		16 50 01.4	-41 37 17	0.000	-0.01	7.96	0.11		B2				
152032	84631		16 50 04.0	+26 12 31	+0.002	-0.01	7.25	0.94	-0.9	G8 II-III		430 s		
151798	160122		16 50 04.9	-12 23 15	-0.006	-0.11	7.9	0.6	4.4	G0 V		50 s		
151972	102419		16 50 06.2	+16 29 45	-0.001	+0.01	7.6	0.1	0.6	A0 V		260 s		
151099	—		16 50 08.8	-67 02 04			7.8	0.1	0.6	A0 V		270 s		
151680	208078	26 ε Sco	16 50 09.7	-34 17 36	-0.049	-0.25	2.29	1.15	-0.1	K2 III	-3	20 mx		
151954	102418		16 50 10.2	+11 04 32	+0.001	+0.01	8.0	1.4	-0.3	K5 III		450 s		
151721	184689		16 50 10.6	-26 44 34	0.000	-0.04	7.4	0.3	0.6	A0 V		150 s		
151956	121895	47 Her	16 50 19.3	+7 14 51	+0.003	-0.01	5.49	0.10		A0	-4			
151900	141393		16 50 22.2	-2 39 14	0.000	-0.03	6.32	0.42		F2				
151517	244184		16 50 29.6	-51 32 51	+0.002	-0.02	7.8	0.8	0.0	B8.5 V		120 s		
152207	65592		16 50 32.7	+33 47 50	0.000	-0.02	7.7	0.1	1.4	A2 V		180 s		
151566	227281		16 50 35.7	-50 02 44	-0.002	-0.02	6.47	0.31		A5				m
152262	46317		16 50 35.9	+41 53 48	-0.007	+0.07	6.29	1.08	0.2	K0 III	-37	150 s		
152173	84641	50 Her	16 50 38.8	+29 48 23	0.000	0.00	5.72	1.60	-0.3	K5 III	-10	140 s		
151759	208085		16 50 39.5	-31 38 12	-0.002	-0.03	8.0	0.7	3.2	G5 IV		93 s		
152224	65595		16 50 43.0	+32 33 13	+0.001	+0.04	6.13	1.01	0.2	K0 III	-30	150 s		
151814	184700		16 50 43.2	-25 25 40	-0.002	-0.01	7.9	1.7	-0.3	K5 III		320 s		
151726	208083		16 50 45.4	-38 15 24	-0.001	-0.04	7.2	-0.2	0.4	B9.5 V		230 s		
152414	46322		16 50 51.3	+48 20 40	+0.004	+0.02	7.7	1.1	0.2	K0 III		310 s		
151707	227296		16 50 54.3	-41 49 49	+0.001	-0.01	7.7	0.3	0.2	K0 III		310 s		
151884	160127		16 50 55.4	-16 32 49	0.000	-0.02	6.9	0.0	0.0	B8.5 V	-13	250 s		
152067	121902		16 50 58.8	+5 29 19	-0.002	-0.04	7.8	0.1	1.4	A2 V		190 s		
151771	208089		16 50 59.7	-37 30 52	-0.002	-0.01	6.11	0.12		B9 p				m
152112	121905		16 51 02.9	+9 52 43	-0.001	0.00	7.30	1.6	-0.5	M3 III	-45	360 s		
151856	184705		16 51 05.4	-27 58 50	-0.001	-0.03	7.88	1.30	-1.3	K3 II-III		670 s		
152113	121908		16 51 07.3	+9 24 14	+0.001	-0.13	6.66	0.49	4.0	F8 V	-35	34 s	10229	m
152174	102428		16 51 07.9	+18 04 38	+0.001	+0.01	7.6	1.1	0.2	K0 III		300 s		
152377	46324		16 51 08.5	+42 04 48	-0.001	+0.04	7.6	0.7	4.4	G0 V		44 s		
152155	102429		16 51 12.0	+15 47 40	+0.001	-0.01	7.3	0.1	0.6	A0 V	-23	220 s		
152264	84644		16 51 12.8	+29 34 16	+0.008	-0.08	7.78	0.55	4.4	G0 V		47 s		
151500	244190		16 51 18.7	-59 31 02	-0.001	-0.01	8.0	1.4	0.2	K0 III		220 s		
152225	102430		16 51 21.5	+19 18 57	-0.001	+0.01	8.0	0.1	0.6	A0 V		300 s		
152551	30125		16 51 23.1	+50 46 23	-0.005	+0.04	7.9	1.1	0.2	K0 III		310 mx		
151904	184714		16 51 24.7	-28 01 16	+0.002	+0.03	8.00	0.4	3.0	F2 V		98 s	10228	m
152127	121911	21 Oph	16 51 24.8	+1 12 58	-0.002	-0.01	5.51	0.05		A0	-26	23 mn	10230	m
152306	84648		16 51 27.2	+28 07 19	0.000	+0.03	7.06	0.91	0.3	G8 III		230 s		
152275	84647		16 51 31.3	+23 37 32	-0.004	+0.03	7.5	1.1	0.2	K0 III		230 mx		
151804	227313		16 51 33.6	-41 13 50	0.000	0.00	5.22	0.07		O8.8	-63			
150995	257437		16 51 45.1	-73 43 20	-0.001	-0.01	7.0	-0.5	0.6	A0 V		190 s		
152326	84651	51 Her	16 51 45.2	+24 39 23	+0.001	+0.01	5.04	1.25	-1.1	K2 II-III	-16	170 s		
152307	84650		16 51 45.3	+20 34 25	-0.002	+0.02	7.1	1.1	0.2	K0 III		240 s		
152342	84654		16 51 46.4	+25 24 01	-0.004	0.00	6.9	0.4	3.0	F2 V	-29	61 s		
152380	84655		16 51 49.9	+28 39 58	-0.001	+0.03	6.70	0.5	3.4	F5 V	-24	46 s	10235	m
152674	30132		16 51 51.7	+54 19 53	-0.003	+0.02	7.3	1.1	0.2	K0 III		260 s		
151772	227314		16 51 51.7	-49 06 55	-0.002	-0.04	7.4	0.1	3.4	F5 V		64 s		
151890	208102	μ¹ Sco	16 51 52.1	-38 02 51	-0.001	-0.03	3.04	-0.21	-3.0	B1.5 V	-25	160 s		m,v
152276	102434		16 51 53.7	+14 56 29	0.000	-0.01	5.8	1.6		M5 III	-10			S Her, v
151942	208109		16 51 53.7	-32 10 56	0.000	+0.01	8.0	1.5	0.2	K0 III		190 s		
151441	253717		16 51 53.7	-65 22 31	0.000	-0.01	6.13	-0.02		B8 n	-10			
153058	17253		16 51 54.8	+67 54 53	-0.004	0.00	8.0	0.1	0.6	A0 V		300 s		
151943	208108		16 51 58.8	-35 27 31	-0.001	+0.02	7.50	1.1	0.2	K0 III		270 s		m
152144	141408		16 51 59.2	-8 51 50	+0.002	-0.01	8.0	1.1	0.2	K0 III		360 s		
151908	208106		16 52 02.6	-39 14 41	-0.001	+0.03	7.8	1.3	0.6	A0 V		46 s		
152128	160133		16 52 04.5	-14 48 11	+0.004	-0.03	7.9	0.5	3.4	F5 V		78 s		
152308	102435	49 Her	16 52 04.7	+14 58 27	+0.001	0.00	6.40	-0.05		A0 p	-23			
151891	227323		16 52 05.9	-40 40 39	+0.002	-0.06	7.40	1.1	0.2	K0 III		230 mx		m
152210	141410		16 52 07.7	-2 47 53	+0.002	+0.01	7.0	0.1	0.6	A0 V	+2	190 s		
151688	244208		16 52 08.7	-57 34 50	-0.001	-0.03	7.3	0.3	0.4	B9.5 V		160 s		
152875	17248		16 52 10.5	+61 39 55	-0.005	0.00	7.8	1.1	0.2	K0 III		330 s		
152495	65618		16 52 11.3	+35 29 24	+0.001	+0.02	7.7	1.4	-0.3	K5 III		390 s		
151892	227324		16 52 13.8	-43 32 35	-0.001	-0.02	8.0	0.9	3.4	F5 V		82 s		
152212	141413		16 52 15.3	-5 47 19	+0.005	-0.02	8.0	1.1	0.2	K0 III		170 mx		
152071	184735		16 52 15.8	-25 36 02	0.000	-0.01	6.90	0.1	0.6	A0 V		180 s	10232	m
151404	253718		16 52 17.3	-67 40 55	-0.011	-0.07	6.32	1.28	0.2	gK0		96 s		
151932	227328		16 52 19.0	-41 51 15	-0.002	0.00	6.49	0.23		WR7	+25			
152328	102438		16 52 19.6	+11 14 30	-0.006	-0.03	7.0	1.4	-0.3	K5 III		140 mx		
151985	208116	μ² Sco	16 52 20.0	-38 01 03	-0.001	-0.03	3.57	-0.21	-3.0	B2 IV	+2	210 s		m
151021	—		16 52 24.1	-74 50 02			8.0	0.5	3.4	F5 V		82 s		

HD	SAO	Star Name	α 2000	δ 2000	μ(α)	μ(δ)	V	B-V	M_V	Spec	RV	d(pc)	ADS	Notes
			$16^h 52^m 27.2^s$	$-40°43'22''$	-0.001^s	$-0.01''$								
151965	227332		16 52 27.2	-40 43 22	-0.001	-0.01	6.35	-0.14	0.4	B9.5 V		160 s		
152255	141415		16 52 30.1	-5 08 56	-0.002	+0.04	7.8	1.1	0.2	K0 III		340 s		
151966	227331		16 52 31.4	-42 59 33	-0.005	-0.05	6.8	0.1	0.6	A0 V		63 mx		
152791	30136		16 52 31.7	+54 34 46	-0.001	-0.03	7.8	1.1	0.2	K0 III		330 s		
152002	208121		16 52 32.1	-38 26 12	-0.001	-0.05	7.4	0.1	0.0	B8.5 V		250 s		
151745	244218		16 52 34.6	-57 50 28	-0.002	-0.03	7.4	1.4	-0.1	K2 III		240 s		
152040	208125		16 52 35.8	-37 35 57	-0.001	-0.03	6.8	0.6	3.2	G5 IV		53 s		
151602	--		16 52 39.8	-64 12 40			7.9	1.4	-0.3	K5 III		440 s		
152445	84660		16 52 40.2	+20 58 24	-0.002	-0.02	7.9	1.4	-0.3	K5 III		430 s		
152446	102445		16 52 40.9	+18 03 42	+0.001	-0.04	6.9	0.5	3.4	F5 V		50 s		
153845	8654		16 52 43.7	+76 51 08	0.000	+0.01	7.2	0.4	2.6	F0 V	-2	84 s		
--	65624		16 52 46.5	+37 20 13	0.000	-0.03	7.9	1.7						
152003	227336		16 52 47.1	-41 47 08	-0.001	0.00	7.02	0.36	-6.2	B0 I	-28	2000 s		
152075	208130		16 52 48.6	-37 00 45	-0.004	-0.09	7.4	1.1	0.2	K0 III		120 mx		
153720	8651		16 52 54.9	+75 23 33	+0.007	-0.02	6.9	0.4	2.6	F0 V	-8	71 s		
152447	102448		16 52 57.3	+11 18 04	0.000	+0.01	8.0	0.1	0.6	A0 V		300 s		
152598	65627	53 Her	16 52 57.9	+31 42 06	-0.008	-0.02	5.32	0.29	2.6	F0 V	-22	35 s		m
152391	121921		16 52 58.6	-0 01 37	-0.048	-1.50	6.64	0.76	5.6	G6 V	+41	16 ts		
151988	227338		16 53 03.6	-47 14 24	0.000	-0.01	7.5	0.1	0.4	B9.5 V		260 s		
152180	208136		16 53 06.4	-31 52 49	0.000	-0.02	7.4	0.1	0.0	B8.5 V		230 s		
152192	208138		16 53 07.1	-30 25 46	+0.001	0.00	7.04	0.16		A m				
152158	208135		16 53 07.9	-33 16 56	+0.001	+0.02	7.66	0.10		A0				
--	65631		16 53 09.7	+34 41 47	0.000	+0.01	7.9	1.5						
152482	102450		16 53 13.4	+10 52 28	+0.001	-0.01	6.8	1.4	-0.3	K5 III		270 s		
152812	46349		16 53 17.5	+47 25 01	-0.004	+0.10	6.00	1.32	-0.1	K2 III	-63	130 s		
152531	102453		16 53 21.1	+15 24 30	+0.001	+0.02	7.5	0.1	0.6	A0 V		240 s		
152311	184754		16 53 25.0	-20 24 56	-0.004	-0.03	5.9	1.1	3.2	G5 IV	-17	22 s		
151948	244238		16 53 25.8	-54 16 38	+0.005	-0.05	7.9	0.5	0.6	A0 V		99 mx		
151479	--		16 53 28.3	-70 36 43			7.8	0.1	0.6	A0 V		270 s		
152147	227356		16 53 28.6	-42 07 17	+0.001	-0.01	7.25	0.36	-6.2	B0 I	-28	2300 s		
152449	141421		16 53 29.0	-1 07 50	-0.003	-0.02	7.8	0.5	4.0	F8 V		56 s		
156401	2799		16 53 29.0	+84 45 52	-0.021	+0.05	7.8	0.1	0.6	A0 V		88 mx		
151763	253726		16 53 30.4	-63 14 06	+0.006	-0.04	7.7	0.1	0.6	A0 V		130 mx		
152792	46350		16 53 32.2	+42 49 30	+0.010	-0.33	6.81	0.65	4.4	G0 V	+7	32 ts		
152813	46354		16 53 37.2	+43 47 01	-0.004	-0.07	7.6	1.1	0.2	K0 III		200 mx		
153752	8655		16 53 37.8	+74 17 32	+0.001	+0.02	7.7	0.4	2.6	F0 V	-21	100 s		
152161	227359		16 53 42.3	-43 03 03	-0.002	-0.03	5.5	1.9	-0.5	M4 III		120 s		
152581	102455		16 53 43.5	+11 58 27	+0.001	0.00	8.0	1.1	0.2	K0 III		360 s		
152484	141423		16 53 45.2	-4 19 17	+0.002	-0.11	7.70	0.9	0.2	G9 III	+45	180 mx		
151442	--		16 53 48.3	-72 06 43			8.0	1.1	-0.1	K2 III		420 s		
151540	257440		16 53 49.7	-70 10 13	+0.001	+0.01	7.4	-0.5	0.0	B8.5 V		310 s		
152025	244247		16 53 49.9	-54 21 48	-0.003	-0.03	8.0	0.8	0.2	K0 III		360 s		
152219	227370		16 53 55.4	-41 52 52	0.000	-0.01	7.65	0.18	-5.0	O9.5 IV		3400 s		
152006	244250		16 53 55.4	-55 04 21	-0.002	-0.04	8.0	0.1	3.0	F2 V		99 s		
152418	160151		16 53 56.8	-17 58 37	0.000	-0.05	7.7	1.1	0.2	K0 III		310 s		
152366	184762		16 53 57.6	-25 04 10	+0.001	-0.02	7.9	0.1	0.0	B8.5 V		290 s		
152149	227362		16 53 57.7	-46 55 04	+0.002	-0.03	7.92	0.12	0.0	B8.5 V		380 s		m
152235	227374		16 53 58.7	-41 59 41	0.000	-0.02	6.32	0.50		B0.5 Ia	-36			
152236	227375		16 53 59.6	-42 21 44	0.000	-0.01	4.73	0.49	-6.7	B1.5 Ia	-26	780 s		
151967	244245		16 54 00.3	-57 54 34	-0.003	-0.13	5.94	1.59	-0.4	M0 III		130 mx		
152614	102458	25 ι Oph	16 54 00.4	+10 09 55	-0.004	-0.04	4.38	-0.08	-0.6	B8 IV	-21	99 s		
153166	17263		16 54 01.7	+60 21 44	+0.002	-0.01	7.1	1.1	0.2	K0 III	-42	240 s		
152234	227377		16 54 01.7	-41 48 23	-0.001	-0.01	5.48	0.22		B0.5 Ia	-6			m
152046	244253		16 54 02.5	-54 26 18	0.000	-0.01	7.0	0.6	0.0	B8.5 V		110 s		
152233	227379		16 54 03.0	-41 47 26	+0.002	-0.01	7.70	0.20		O6	-16			m
153265	17266		16 54 03.2	+63 52 13	+0.001	-0.02	8.0	1.1	-0.1	K2 III		410 s		
152246	227381		16 54 05.4	+04 04 47	+0.002	-0.01	7.30	0.17	-6.0	O9 III		2400 s		
152258	--		16 54 07.3	-40 27 12			8.0	2.2	-0.5	M2 III		230 s		
152555	141426		16 54 08.0	-4 20 23	-0.003	-0.09	7.6	0.7	4.4	G0 V		44 s		
152382	184765		16 54 09.7	-29 23 14	+0.001	+0.01	8.0	1.1	0.2	K0 III		330 s		
152248	227382		16 54 09.8	-41 49 31	-0.001	-0.02	6.14	0.16		O8	-35			m
152654	102461		16 54 10.0	+15 37 25	0.000	-0.01	7.3	0.5	3.4	F5 V	-32	59 s		
152569	141427		16 54 10.5	-1 36 44	+0.001	-0.07	6.25	0.28		F0 n	-20			m
152247	227384		16 54 11.3	-41 38 32	0.000	-0.01	7.17	0.18		O9	-17			
152249	227383		16 54 11.6	-41 51 00	+0.001	-0.04	6.50	0.20	-6.2	O9 I	-24	2100 s		
152951	46362		16 54 18.7	+46 32 21	0.000	-0.02	6.7	0.1	1.4	A2 V	-2	120 s		
152905	46360		16 54 18.8	+43 24 23	-0.002	0.00	7.1	1.1	0.2	K0 III		240 s		
152270	227390		16 54 19.5	-41 49 12	-0.001	-0.01	6.64	0.25		WR7	-44			m
152429	184775		16 54 20.5	-25 49 47	-0.006	-0.05	7.2	0.5	4.0	F8 V		43 s		
152250	227387		16 54 23.2	-44 53 34	+0.001	-0.06	7.42	0.39	2.6	F0 V n		90 s		
--	102465		16 54 24.6	+18 44 36	-0.002	-0.01	8.0							
--	--		16 54 24.8	-41 49 50			8.00							m
--	227393		16 54 25.7	-41 49 56	-0.001	0.00	8.0	0.2		B1				
152384	208169		16 54 26.5	-33 28 30	+0.001	-0.02	7.03	0.04		A0				
152600	141430		16 54 26.8	-4 09 50	0.000	+0.01	7.7	0.8	3.2	G5 IV		78 s		

HD	SAO	Star Name	α 2000	δ 2000	μ(α)	μ(δ)	V	B−V	M$_v$	Spec	RV	d(pc)	ADS	Notes
			16^h 54^m 26^s.8	−42° 28′ 44″	0^s.000	−0″.01								
152293	227392		16 54 26.8	−42 28 44	0.000	−0.01	5.88	0.64	0.7	gF5		83 s		
152422	208170		16 54 26.8	−30 43 13	−0.001	−0.03	8.00	1.1	0.2	K0 III		340 s		m
153596	8657		16 54 28.3	+70 27 49	−0.003	−0.05	7.0	0.1	1.4	A2 V		130 s		
−−	65644		16 54 28.5	+36 40 47	+0.003	−0.04	8.0	1.8						
152220	227386		16 54 31.3	−49 42 39	0.000	−0.01	6.5	1.4	0.2	K0 III		110 s		
152314	227400		16 54 31.7	−41 48 20	−0.001	−0.01	7.90	0.20	−5.1	O9 IV−V	−34	4000 s		
151917	253728		16 54 33.2	−63 27 20	0.000	−0.03	7.3	1.4	−0.3	K5 III		330 s		
−−	65645		16 54 33.8	+34 51 39	0.000	+0.02	8.0	1.9						
152334	227402	ζ Sco	16 54 34.9	−42 21 41	−0.011	−0.24	3.62	1.37	−0.3	K5 III	−19	50 mx		
152601	141431	23 Oph	16 54 35.5	−6 09 14	−0.002	−0.02	5.25	1.08	−0.1	K2 III	−17	120 s		
152333	227403		16 54 35.7	−41 25 24	0.000	−0.01	8.02	0.22		B3				
152431	208176		16 54 35.9	−30 35 14	+0.003	0.00	6.35	0.21		A m				
152516	184782		16 54 38.2	−21 52 47	−0.001	−0.01	8.04	0.08	−1.0	B5.5 V	−30	520 s		
148940	−−		16 54 38.8	−85 02 19			8.0	1.1	−0.1	K2 III		420 s		
152585	160159		16 54 40.1	−11 47 33	+0.001	−0.02	6.5	0.1	0.6	A0 V		150 s		
152433	208177		16 54 40.2	−32 11 24	+0.003	−0.03	7.6	0.3	4.0	F8 V		52 s		
152335	227404		16 54 41.7	−43 33 36	+0.001	+0.01	8.00	0.5	4.0	F8 V		63 s		m
152952	46363		16 54 42.6	+40 42 31	+0.001	−0.01	7.7	1.1	0.2	K0 III		310 s		
152295	−−		16 54 43.0	−46 06 09			7.9	2.2	0.2	K0 III		67 s		
153344	17271		16 54 45.8	+62 06 00	−0.048	−0.05	7.07	0.68	3.2	G5 IV	−82	59 s		
152473	208179		16 54 46.2	−30 58 48	−0.001	−0.05	7.50	1.15	0.2	K0 III		220 s		
152501	184783		16 54 49.4	−27 44 53	−0.001	+0.03	7.6	1.0	3.2	G5 IV		56 s		
152500	184784		16 54 49.5	−26 24 47	−0.006	−0.01	7.5	1.0	2.6	F0 V		35 s		
152534	184786		16 54 50.6	−23 30 44	−0.001	−0.03	7.0	0.9	0.2	K0 III	−27	230 s		
152815	84687		16 54 55.1	+20 57 31	+0.004	0.00	5.40	0.98	0.3	G8 III	−3	100 s		
152405	227424		16 54 55.1	−40 31 29	−0.002	0.00	7.17	0.14	−6.2	O9.5 I k	+19	2500 s		
152877	84688		16 54 55.2	+28 08 14	+0.001	+0.03	7.2	0.4	2.6	F0 V	−37	81 s		
152298	227408		16 54 57.9	−48 47 19	+0.002	−0.01	7.6	0.7	0.4	B9.5 V		250 s		
152408	227425		16 54 58.2	−41 09 04	−0.002	−0.01	5.77	0.15		O7.5 p	−138			m
152272	244271		16 54 58.3	−50 11 19	−0.008	−0.01	7.2	1.0	0.2	K0 III		170 mx		
152406	227428		16 54 58.5	−40 34 50	+0.001	−0.01	7.70	1.4	−0.3	K5 III		360 s		m
152204	244268		16 55 00.8	−56 02 54	−0.001	−0.06	7.1	1.7	0.2	K0 III		92 s		
152896	84694		16 55 01.2	+29 02 20	+0.001	−0.01	7.3	0.2	2.1	A5 V	+2	110 s		
152863	84692	56 Her	16 55 02.0	+25 43 50	+0.001	−0.02	6.08	0.92	0.3	G5 III	+1	140 s	10259	m
152424	227430		16 55 03.1	−42 05 27	−0.001	0.00	6.28	0.38	−6.2	O9 I	−18	1900 s		m
−−	−−		16 55 04.2	−42 00 17			6.60							m
152456	208182		16 55 06.1	−38 24 48	+0.001	−0.02	7.2	0.1	0.0	B8.5 V		230 s		
152603	160160		16 55 06.2	−18 53 03	−0.001	−0.01	7.9	0.1	1.4	A2 V		200 s		
152677	141437		16 55 07.7	−6 39 04	0.000	−0.02	6.9	1.1	0.2	K0 III		220 s		
152521	208187		16 55 08.5	−31 24 09	−0.001	−0.05	6.73	0.06	0.6	A0 V		150 s		m
152680	160162		16 55 12.5	−11 18 20	−0.001	−0.04	7.8	0.5	3.4	F5 V		74 s		
152830	102474		16 55 15.8	+13 37 11	+0.002	−0.04	6.2	0.4	3.0	F2 V	−5	43 s		
152879	102476	54 Her	16 55 22.1	+18 26 00	−0.008	+0.01	5.35	1.41	−0.3	K4 III	+12	90 mx	10262	m
152878	84695		16 55 22.5	+21 29 32	−0.002	−0.01	8.0	0.5	4.0	F8 V		62 s		
151836	253729		16 55 22.5	−69 16 45	−0.002	0.00	7.1	0.0	0.4	B9.5 V		220 s		
152082	253734		16 55 24.6	−63 16 11	+0.002	−0.02	6.10	0.1	0.6	A0 V		130 s		m
152574	208192		16 55 27.9	−32 30 12	+0.002	0.00	7.18	0.58		F5				
152490	227443		16 55 29.4	−40 58 46	−0.002	−0.05	7.8	−0.3	1.4	A2 V		190 s		
152618	184798		16 55 32.2	−29 16 34	−0.001	+0.02	7.7	1.0	0.2	K0 III		320 s		
152655	184804		16 55 32.4	−21 34 11	−0.001	−0.02	7.50	0.00	0.0	B8.5 V		320 s	10257	m
152186	253736		16 55 34.4	−60 40 38	−0.002	−0.02	7.1	2.3	−0.1	K2 III		57 s		
152491	227444		16 55 35.0	−43 18 57	−0.003	−0.01	6.77	0.03	0.6	A0 V		160 s		
152524	208191		16 55 37.5	−39 30 19	−0.002	−0.01	7.20	1.4	−0.3	K5 III		290 s		m
152476	227442		16 55 37.9	−45 06 10	+0.005	−0.03	6.4	1.6		M5 III	+7	200 mx		RS Sco, v
152237	244272		16 55 42.5	−59 25 02	−0.004	−0.03	7.3	0.5	3.4	F5 V		56 s		
−−	17283		16 55 42.9	+67 53 13	−0.005	+0.06	7.9	1.4						
152657	184805		16 55 44.3	−25 32 00	0.000	+0.01	7.4	0.4	0.0	B8.5 V		150 s		
153680	17282		16 55 46.0	+67 28 42	−0.004	0.00	6.7	0.1	0.6	A0 V		130 mx		
152796	141443		16 55 49.1	−6 07 24	+0.003	−0.04	7.5	1.1	0.2	K0 III	−54	280 s		
154527	8673		16 55 50.9	+77 57 36	−0.005	+0.02	7.4	1.1	0.2	K0 III		270 s		
152635	208204		16 55 52.2	−31 28 13	+0.001	−0.02	7.69	−0.01	0.0	B8.5 V		340 s		
152636	208205		16 55 57.7	−33 30 26	+0.002	0.00	6.37	1.71		K2	−92			
153597	17281	19 Dra	16 56 01.5	+65 08 05	+0.037	+0.05	4.89	0.48	3.7	F6 V	−23	17 ts		
152781	160171		16 56 01.7	−16 48 22	+0.005	+0.04	6.50	1.1	3.2	K0 IV	−3	46 s		m
153286	46378		16 56 04.5	+47 22 15	−0.003	+0.05	6.9	0.2		A m	−18			
152898	121946		16 56 05.4	+1 25 15	−0.001	−0.02	7.9	1.1	0.2	K0 III		340 s		
153299	30161		16 56 06.2	+50 02 20	−0.003	−0.01	6.6	1.6	−0.5	M2 III	−31	260 s		
152478	244280		16 56 08.7	−50 40 30	0.000	−0.02	6.33	−0.02		B3 pne	+28			
152637	208206		16 56 09.1	−37 11 16	−0.003	−0.03	7.40	0.9	3.2	G5 IV		68 s		m
154528	8674		16 56 09.5	+77 39 10	−0.008	+0.02	6.7	0.1	0.6	A0 V	−4	160 s		
152972	102482		16 56 09.6	+15 23 01	0.000	0.00	6.6	1.1	0.2	K0 III		190 s		
152700	208213		16 56 13.6	−31 31 08	−0.001	−0.02	8.00	0.1	1.7	A3 V		170 s		m
152541	227453		16 56 14.4	−46 50 55	+0.001	+0.01	7.45	0.06	0.0	B8.5 V		310 s		m
152544	227451		16 56 14.6	−47 43 26	+0.002	−0.03	7.5	0.3	1.7	A3 V		110 s		
152623	227464		16 56 14.8	−40 39 36	−0.001	−0.01	6.68	0.09	−5.2	O8 V		1600 s		

HD	SAO	Star Name	α 2000	δ 2000	μ(α)	μ(δ)	V	B-V	M_v	Spec	RV	d(pc)	ADS	Notes
152273	253740		16 56 16.5	-61 38 26	-0.001	-0.01	6.9	0.2	0.6	A0 V		130 s		
154099	8667		16 56 16.6	+73 07 40	-0.002	-0.02	6.2	0.2	2.1	A5 V	-7	67 s		
152720	208214		16 56 16.7	-31 18 28	-0.004	-0.05	7.30	1.1	0.2	K0 III		170 mx		m
153345	30164		16 56 18.2	+52 41 54	+0.001	-0.01	7.1	0.1	0.6	A0 V		200 s		AI Dra, v
154181	8670		16 56 18.8	+74 16 57	+0.015	-0.09	7.20	0.5	3.4	F5 V	-9	58 s	10299	m
153062	84703		16 56 20.1	+22 06 24	0.000	-0.01	8.0	0.1	1.4	A2 V		210 s		
153598	17284		16 56 21.6	+62 22 11	-0.004	+0.01	6.8	0.1	0.6	A0 V		100 mx		
153015	102484		16 56 22.0	+16 17 05	-0.001	-0.02	7.1	0.2	2.1	A5 V		94 s		
155153	2793		16 56 24.2	+80 51 34	+0.002	+0.01	6.7	0.8	3.2	G5 IV		51 s		
153697	17285	20 Dra	16 56 25.1	+65 02 20	-0.008	+0.03	6.90	0.4	2.6	F0 V	-21	72 s	10279	m
152527	244285		16 56 28.5	-52 17 02	-0.005	-0.05	5.94	-0.08	-0.6	A0 III		120 mx		
152667	227473		16 56 35.9	-40 49 25	0.000	-0.01	6.15	0.26	-6.2	B0 Ia p	+51	1800 s		
152783	208221		16 56 37.9	-30 34 49	+0.001	-0.02	5.4	1.6		M5 III	-36			RR Sco, v
153063	102489		16 56 39.1	+15 38 34	0.000	-0.03	7.7	0.5	3.4	F5 V		71 s		
153064	102488		16 56 39.4	+14 08 25	-0.002	+0.03	7.00	1.1	0.2	K0 III		230 s	10270	m
152685	227477		16 56 45.2	-41 09 16	0.000	0.00	7.43	0.17	-6.3	B2 I		5600 s		
152168	253739		16 56 46.8	-67 09 55	+0.010	-0.05	7.6	1.1	0.2	K0 III		160 mx		
152849	184822	24 Oph	16 56 47.9	-23 09 00	0.000	0.00	6.20	0.1	0.6	A0 V		130 s	10265	m
153648	17286		16 56 48.3	+61 28 30	-0.005	+0.01	7.3	1.1	0.2	K0 III		260 s		
153033	121954		16 56 50.9	+6 30 05	-0.003	0.00	7.50	1.4	-0.3	K5 III	-42	360 s		
152723	227479		16 56 54.4	-40 30 42	-0.001	+0.01	7.31	0.10		O6 k	+14			m
--	--		16 56 56.3	-70 36 44			6.60							m
152805	208228		16 56 57.7	-32 20 17	0.000	-0.06	7.50	0.1	1.4	A2 V		82 mx		m
152010	257446		16 57 01.3	-71 06 41	+0.002	-0.03	6.7	0.0	1.4	A2 V		110 s		m
153146	102491		16 57 01.6	+18 14 01	0.000	-0.01	7.6	1.1	0.2	K0 III		300 s		
152909	160180		16 57 03.8	-19 32 24	-0.001	-0.02	6.1	0.9	0.0	B8.5 V		42 s	10266	m
152322	253746		16 57 06.1	-64 11 18	+0.004	-0.01	7.7	0.6	4.4	G0 V		46 s		
153129	102492		16 57 06.6	+14 40 25	0.000	-0.01	7.6	1.4	-0.3	K5 III		390 s		
153496	30172		16 57 10.0	+51 53 56	-0.001	+0.01	7.9	1.1	0.2	K0 III		340 s		
--	65670		16 57 10.6	+35 17 11	+0.001	-0.01	7.8	1.6						
152820	208232	27 Sco	16 57 10.9	-33 15 33	-0.001	-0.01	5.48	1.59		K5				
153113	121959		16 57 20.3	+6 12 40	-0.001	0.00	6.9	1.1	0.2	K0 III		220 s		
153399	46393		16 57 21.4	+43 41 02	+0.003	+0.01	7.6	0.9	3.2	G5 IV		75 s		
153021	160186		16 57 26.0	-10 57 48	+0.001	-0.08	6.19	1.00	1.8	G5 III-IV		61 s		
152511	253748		16 57 29.6	-60 36 00	-0.002	-0.01	7.20	0.00	0.0	B8.5 V		270 s		m
153287	84720	57 Her	16 57 30.9	+25 21 10	0.000	+0.01	6.6	1.1	0.2	K0 III	+9	190 s		
153225	--		16 57 31.0	+14 01 01	-0.001	-0.01	7.6	1.1	-0.1	K2 III		340 s		
153226	102496		16 57 31.8	+13 53 03	-0.006	+0.07	6.37	0.94	3.2	K0 IV	-31	43 s		
153116	141457		16 57 38.3	-3 00 59	-0.001	0.00	7.4	0.1	1.4	A2 V		160 s		
153210	121962	27 κ Oph	16 57 40.0	+9 22 30	-0.020	+0.01	3.20	1.15	-0.1	K2 III	-56	36 mx		q
153312	84726		16 57 42.2	+24 22 52	0.000	-0.03	6.32	1.10	0.2	K0 III	-22	150 s		
153472	46401		16 57 50.0	+42 30 44	-0.001	-0.05	6.34	1.28	-0.2	K3 III	+28	200 s		
152885	208237		16 57 50.0	-39 06 56	-0.001	+0.01	6.5	1.2	-0.5	M2 III		260 s		
152901	208238		16 57 52.2	-37 59 47	-0.001	0.00	7.30	0.00	-1.0	B5.5 V		410 s		m
152989	184838		16 57 53.8	-27 36 38	-0.001	-0.01	7.60	0.1	0.6	A0 V		240 s	10271	m
153134	160193		16 58 02.2	-12 05 04	+0.001	-0.07	7.7	0.5	3.4	F5 V		73 s		
152902	227504		16 58 06.5	-40 21 08	0.000	+0.03	7.3	1.4	-0.1	K2 III		210 s		
153271	121964		16 58 06.6	+9 01 57	0.000	0.00	8.0	0.4	2.6	F0 V		120 s		
152668	244304		16 58 07.1	-57 46 09	0.000	-0.01	7.3	0.8	0.4	B9.5 V		80 s		
152853	227500		16 58 07.8	-45 58 56	+0.001	0.00	7.94	0.11	-1.6	B3.5 V		600 s		
152642	244303		16 58 07.9	-58 59 34	+0.001	-0.02	7.7	0.4	0.4	B9.5 V		150 s		
153135	160195		16 58 09.3	-14 22 19	-0.001	+0.01	7.2	1.1	-0.1	K2 III		290 s		
152691	244306		16 58 10.3	-57 47 15	-0.002	+0.03	7.3	0.0	0.4	B9.5 V		220 s		
151233	257443		16 58 10.6	-80 04 17	-0.003	+0.02	8.0	-0.1	0.6	A0 V		300 s		
152493	253753		16 58 15.5	-65 12 14	-0.001	-0.04	6.6	0.8	1.7	A3 V		39 s		
153069	184849		16 58 16.7	-26 03 32	+0.002	0.00	7.5	1.4	-0.1	K2 III		250 s		
152824	244313		16 58 17.8	-50 38 29	-0.001	-0.05	5.7	1.0	0.4	B9.5 V	-44	28 s		
153004	208246		16 58 19.6	-33 36 34	0.000	-0.02	6.2	0.5	3.4	F5 V	-22	37 s		RV Sco, m,v
152443	253751		16 58 24.3	-67 31 24	+0.006	-0.13	7.90	0.1	0.6	A0 V		69 mx		m
153374	84730		16 58 24.7	+20 08 42	-0.002	-0.04	7.7	1.1	0.2	K0 III		310 s		
--	65691		16 58 29.3	+37 35 01	-0.002	-0.02	8.0	1.8						
152917	227508		16 58 32.8	-45 55 10	-0.001	+0.01	7.62	0.26	2.1	A5 V		120 s		
153085	208255		16 58 35.3	-30 50 36	+0.003	+0.06	7.8	0.9	3.2	G5 IV		70 s		
152786	244315	ζ Ara	16 58 37.1	-55 59 24	-0.002	-0.04	3.13	1.60	-0.3	K5 III	-6	42 s		
153376	102505		16 58 37.8	+15 27 17	-0.002	+0.15	7.0	0.9	5.2	G5 V		23 s		
154273	17302		16 58 40.0	+69 38 07	-0.001	+0.04	7.9	1.1	0.2	K0 III		350 s		
153229	160205		16 58 41.4	-14 52 11	0.000	-0.03	6.59	0.40	3.0	F2 V		49 s		
152671	253755		16 58 41.6	-61 44 27	+0.007	-0.06	7.0	0.7	3.4	F5 V		37 s		
153072	208259		16 58 52.3	-37 37 16	+0.001	-0.06	6.09	0.19	1.7	A3 V		67 s		m
153050	208260		16 58 55.1	-39 10 06	+0.001	0.00	7.3	0.7		K0 III		260 s		
153455	102507		16 58 58.3	+17 32 12	-0.002	+0.01	8.0	0.5	4.0	F8 V		62 s		
152946	227514		16 59 00.3	-49 44 38	0.000	-0.01	7.8	0.1	0.0	B8.5 V		370 s		
153136	208265		16 59 02.3	-33 31 36	+0.001	+0.01	7.81	0.58		A5				
154319	17305		16 59 02.4	+69 11 10	-0.001	-0.04	6.5	1.1	0.2	K0 III	-27	180 s		
153454	102508		16 59 04.8	+17 41 18	+0.001	0.00	7.6	0.1	0.6	A0 V		250 s		

HD	SAO	Star Name	α 2000	δ 2000	μ(α)	μ(δ)	V	B-V	M_v	Spec	RV	d(pc)	ADS	Notes
			h m s	° ′ ″	s	″								
152966	227518		16 59 05.3	-48 52 29	0.000	+0.02	7.0	0.2	2.1	A5 V		86 s		
153415	102506		16 59 05.5	+11 30 46	+0.001	-0.03	7.9	1.6	-0.5	M2 III		490 s		
153102	208262		16 59 06.9	-38 33 58	-0.001	-0.01	7.59	0.04	-1.1	B5 V nn		480 s		
153473	102509		16 59 11.7	+14 04 46	-0.002	-0.05	7.4	0.4	1.9	F2 IV		130 s		
154320	17308		16 59 17.5	+68 40 41	+0.001	-0.04	7.4	0.2	2.1	A5 V		110 s		
152921	244322		16 59 20.2	-54 52 29	-0.002	-0.03	7.8	-0.1	0.2	K0 III		330 s		
153377	141470		16 59 21.4	-1 41 15	+0.002	+0.01	7.5	0.4	3.0	F2 V		80 s		
153956	30190		16 59 21.4	+56 41 19	-0.007	+0.01	6.03	1.16	0.2	K0 III	-15	120 s		
153316	160214		16 59 23.3	-16 03 52	+0.003	-0.02	7.9	1.4	-0.3	K5 III		450 s		
153475	121980		16 59 29.4	+9 42 10	0.000	-0.01	7.90	0.4	2.6	F0 V		120 s	10295	m
153560	102513		16 59 30.1	+18 33 21	-0.001	+0.01	7.9	1.4	-0.3	K5 III		430 s		
152564	253756		16 59 33.7	-69 16 06	-0.002	-0.02	5.79	-0.10		A0				
152652	253758		16 59 34.2	-67 12 06	-0.013	-0.08	7.7	0.6	4.4	G0 V		46 s		
153650	84742		16 59 34.7	+29 32 57	0.000	-0.02	7.8	0.1	0.6	A0 V	+3	280 s		
152980	244331	ε¹ Ara	16 59 34.9	-53 09 38	+0.001	+0.02	4.06	1.45	-0.5	M1 III	+23	82 s		
153651	84741		16 59 40.0	+24 12 16	-0.001	-0.08	7.8	1.1	0.2	K0 III		190 mx		
153106	227530		16 59 40.3	-46 28 56	+0.001	+0.02	7.76	-0.01		B0.5 V				
153361	160216		16 59 44.2	-18 14 55	-0.002	-0.15	6.5	0.7	4.4	G0 V		27 s		
153540	102514		16 59 47.8	+10 54 55	+0.002	+0.01	7.30	1.4	-0.3	K4 III	-18	330 s		
153437	141473		16 59 49.3	-4 13 16	+0.001	+0.01	7.6	1.1	-0.1	K2 III		360 s		
154081	30196		16 59 49.9	+58 27 44	-0.002	0.00	6.7	0.1	1.7	A3 V		100 s		
153294	208281		16 59 50.8	-31 42 35	-0.001	-0.01	8.0	0.1	0.0	B8.5 V		300 s		
153601	102519		16 59 51.2	+17 48 22	0.000	+0.01	7.3	0.4	2.6	F0 V		89 s		
153698	84745		16 59 52.7	+27 18 56	-0.003	-0.01	7.30	1.6		M III	-22			
153140	227533		16 59 54.4	-46 19 00	0.000	+0.01	7.50	0.35	-1.6	B3.5 V		640 s		
152923	244329		16 59 54.9	-59 19 47	+0.005	-0.10	7.17	0.48	3.7	F6 V		49 s		m
152339	257448		16 59 54.9	-73 25 20	+0.003	-0.01	7.20	0.00	0.4	B9.5 V		230 s		m
153336	184892		16 59 57.5	-25 05 31	0.000	-0.01	5.86	1.62	10.6	M3 V	-32			
153562	121987		17 00 00.5	+9 47 53	0.000	-0.01	7.4	0.1	0.6	A0 V		230 s		
153669	102521		17 00 01.3	+19 35 36	+0.005	-0.02	8.0	0.9	3.2	G5 IV		93 s		
154009	30195		17 00 03.3	+53 02 41	-0.001	+0.04	7.7	1.1	0.2	K0 III		310 s		
155246	8695		17 00 03.6	+78 05 58	-0.011	+0.03	7.2	1.1	-0.1	K2 III		290 s		
153832	65717		17 00 05.3	+39 06 02	-0.005	+0.12	7.2	1.1	0.2	K0 III		140 mx		
153053	244338		17 00 06.1	-54 35 48	-0.001	-0.06	5.65	0.19	1.4	A2 V		58 s		m
153363	184897	26 Oph	17 00 09.3	-24 59 21	+0.005	-0.06	5.8	0.3	2.6	F0 V	+19	41 s		
153807	65715		17 00 09.8	+34 20 19	-0.002	+0.02	7.1	0.9	3.2	G5 IV	-16	59 s		
153234	227541		17 00 14.1	-44 59 18	-0.001	-0.02	6.51	0.38	2.1	A5 V		63 s		
153627	102522		17 00 16.3	+10 49 22	+0.007	-0.11	7.6	0.5	4.0	F8 V		52 s		
153028	244340		17 00 16.8	-56 36 24	0.000	+0.02	7.3	0.0	2.1	A5 V		110 s		
153808	65716	58 ε Her	17 00 17.2	+30 55 35	-0.004	+0.03	3.92	-0.01	0.6	A0 V	-25	26 mx		
153723	102525		17 00 23.3	+17 54 59	+0.001	0.00	7.5	1.4	-0.3	K5 III		370 s		
153330	208287		17 00 25.4	-37 56 12	-0.004	-0.01	7.7	0.6	4.4	G0 V		42 s		
153221	227542		17 00 27.0	-48 38 52	0.000	-0.08	6.00	0.88	3.2	G5 IV		31 s		m
154955	8690		17 00 27.5	+75 13 30	+0.011	-0.05	6.9	0.4	3.0	F2 V		61 s		
154790	8686		17 00 27.6	+73 11 40	-0.003	+0.01	7.4	1.1	-0.1	K2 III		320 s		
153701	102526		17 00 28.6	+15 09 34	+0.004	+0.13	7.80	0.8	5.2	G5 V		33 s		
153393	184901		17 00 28.7	-28 38 04	0.000	0.00	8.0	-0.4	0.6	A0 V		300 s		
153970	46435		17 00 28.8	+45 46 42	-0.013	+0.06	7.6	0.5	4.0	F8 V		53 s		
153653	121995		17 00 29.2	+6 35 02	+0.003	-0.04	6.59	0.23		A5	-10			
153438	184906		17 00 29.5	-21 27 43	-0.002	-0.09	7.4	1.0	0.2	K0 III		130 mx		
153075	244344		17 00 31.4	-57 17 48	-0.021	-0.29	7.00	0.60	4.4	G0 V		33 s		
153896	65719		17 00 31.8	+38 54 30	+0.001	-0.05	7.33	1.02	0.2	K0 III		270 s		
153258	227547		17 00 32.1	-45 27 06	0.000	0.00	6.65	1.80		K2				
153368	208293		17 00 36.7	-35 56 03	-0.001	-0.07	6.0	0.8	0.2	K0 III		140 s		
--	17311		17 00 38.4	+61 01 26	-0.001	-0.02	8.0	0.7						
--	65720		17 00 39.4	+35 09 25	-0.001	+0.03	8.0	2.6						
152260	257449		17 00 54.5	-76 13 07	-0.009	-0.16	7.1	0.8	4.0	F8 V		29 s		
153162	244351		17 00 55.5	-56 28 48	-0.001	-0.03	7.6	1.5	-0.5	M4 III		420 s		
153834	84758		17 00 58.0	+22 37 56	-0.001	-0.02	5.65	1.33	0.2	K0 III	+11	78 s		
148542	258751		17 00 58.1	-86 21 51	+0.010	0.00	6.04	0.05	0.9	A3 IV		110 s		
154082	46438		17 01 00.5	+46 16 29	0.000	+0.02	8.00	1.4	-0.3	K5 III		460 s	10311	m
152893	--		17 01 00.5	-66 31 33			7.8	1.1	-0.1	K2 III		380 s		
153484	184912		17 01 01.2	-29 40 19	-0.002	-0.05	7.4	1.4	-0.5	M4 III		370 s		
153809	102532		17 01 02.6	+16 35 36	+0.002	-0.02	7.20	0.1	0.6	A0 V	0	210 s	10308	m
153503	184914		17 01 03.2	-28 33 12	0.000	-0.03	7.8	0.9	0.2	K0 III		330 s		
153687	141483	30 Oph	17 01 03.4	-4 13 21	-0.003	-0.08	4.82	1.48	-0.3	K4 III	-7	93 s		m
153739	122002		17 01 05.5	+2 33 08	-0.002	-0.03	7.8	0.4	3.0	F2 V		89 s		
153382	227552		17 01 05.8	-42 04 20	-0.001	-0.03	7.33	0.12	-1.2	B8 III		380 s		m
153201	244356		17 01 06.8	-56 33 18	-0.001	-0.02	6.60	0.00	0.4	B9.5 V		170 s		m
153123	244353		17 01 06.9	-58 51 00	0.000	0.00	7.80	0.00	0.0	B8.5 V		350 s		m
153847	84759		17 01 07.1	+21 29 50	+0.001	-0.05	7.1	0.4	2.6	F0 V		78 s		
153796	102531		17 01 07.2	+11 20 28	0.000	+0.02	7.7	0.4	2.6	F0 V		100 s		
153897	84765		17 01 09.4	+27 11 47	-0.001	-0.06	6.55	0.41	3.4	F5 V	-31	43 s		
153631	160227		17 01 10.6	-13 34 02	-0.002	-0.32	7.14	0.60		G3		21 t		
153608	160226		17 01 11.3	-17 20 45	0.000	-0.04	7.8	0.0	0.4	B9.5 V		300 s		

HD	SAO	Star Name	α 2000	δ 2000	μ(α)	μ(δ)	V	B-V	M$_V$	Spec	RV	d(pc)	ADS	Notes
			h m s	° ′ ″	s	″								
153515	184918		17 01 12.5	-29 41 04	+0.001	0.00	7.90	0.9	3.2	G5 IV		85 s	10304	m
153426	208304		17 01 12.8	-38 12 14	-0.001	-0.03	7.47	0.14		O8.5				
153688	141484		17 01 13.3	-7 01 42	-0.002	0.00	7.5	1.1	-0.1	K2 III	-69	320 s		
154199	30200		17 01 14.2	+52 36 17	-0.001	+0.02	6.82	0.09	0.6	A0 V		150 s		
153609	184922		17 01 14.7	-20 26 14	-0.001	-0.03	7.20	0.1	0.6	A0 V		210 s		m
154391	17312		17 01 16.6	+60 38 57	-0.008	+0.06	6.13	1.00	0.0	K1 III	-17	170 s		
153798	122008		17 01 17.8	+7 25 46	-0.001	+0.01	6.8	1.1	0.2	K0 III		210 s		
154928	8693		17 01 17.8	+73 18 45	-0.002	+0.01	7.7	1.1	0.2	K2 III	-5	360 s		
153571	184921		17 01 19.2	-27 15 05	+0.002	-0.04	7.6	0.4	1.4	A2 V		100 s		
153835	102533		17 01 27.0	+11 09 39	+0.002	+0.02	8.0	0.1	0.6	A0 V		300 s		
154355	30209		17 01 30.9	+58 33 23	-0.009	-0.04	7.2	0.5	3.4	F5 V		57 s		
153882	102536		17 01 32.9	+14 56 58	0.000	-0.01	6.31	0.04		A4 p	-32		10310	m
154115	46444		17 01 33.3	+42 43 38	-0.002	+0.02	7.30	0.8	3.2	G5 IV		66 s	10316	m
153759	141489		17 01 34.7	-6 44 11	0.000	-0.01	7.2	0.0	0.4	B9.5 V		230 s		
153504	208312		17 01 35.2	-38 07 15	-0.002	-0.06	7.5	0.6	0.2	K0 III		200 mx		
154029	65736	59 Her	17 01 36.2	+33 34 06	0.000	0.00	5.25	0.02	0.0	A3 III	-13	110 s		
155154	8697		17 01 39.9	+75 17 50	+0.002	-0.08	6.3	0.4	2.6	F0 V	+1	56 s		
153573	208318		17 01 43.4	-32 26 10	-0.001	-0.03	7.7	1.4	3.2	G5 IV		33 s		
153574	208317		17 01 43.5	-33 22 02	+0.002	-0.02	7.18	0.28		A2				m
153370	244370		17 01 46.0	-51 07 51	-0.003	-0.04	6.45	0.27	1.7	A3 V		70 s		m
153261	244362		17 01 47.3	-58 57 30	0.000	-0.01	6.11	-0.03	-2.5	B2 V nne		510 s		
153612	208322		17 01 48.7	-30 46 44	-0.002	-0.01	7.5	0.4	1.4	A2 V		110 s		
152904	253767		17 01 49.8	-69 24 37	+0.002	-0.02	7.7	1.1	0.2	K0 III		320 s		
153727	160231	29 Oph	17 01 51.1	-18 53 08	-0.003	-0.02	6.3	1.1	0.2	K0 III	+43	170 s		
153613	208324		17 01 52.6	-32 08 37	0.000	-0.05	5.03	-0.10	-0.2	B8 V	+15	110 s		m
153262	244363		17 01 53.7	-59 02 54	-0.001	-0.04	7.8	-0.5	0.4	B9.5 V		300 s		
153182	253772		17 01 54.1	-62 08 24	-0.002	-0.05	7.3	0.5	3.0	F2 V		58 s		
153575	208319		17 01 55.1	-37 18 05	-0.001	-0.01	7.8	0.2	0.0	B8.5 V		240 s		
153672	184936		17 01 56.4	-27 56 21	+0.002	-0.02	7.3	0.4	1.7	A3 V		84 s		
153914	122023		17 01 59.0	+8 27 02	+0.003	0.00	6.33	0.09		A0	-1		10312	m
153993	102539		17 02 01.2	+19 03 29	-0.001	-0.01	8.0	0.1	0.6	A0 V		300 s		
153741	184944		17 02 02.0	-20 35 45	-0.001	-0.02	7.60	0.8	-2.1	G6 II	-27	870 s		
154049	84774		17 02 02.2	+25 02 16	+0.001	+0.02	7.90	1.1	-0.2	K3 III		420 s		
153371	244373		17 02 06.8	-54 09 06	+0.005	+0.01	8.0	0.2	0.2	K0 III		360 s		
154344	30211		17 02 08.2	+51 56 32	-0.001	-0.06	7.83	0.24		A2				
154633	17324		17 02 15.1	+64 36 01	-0.009	+0.03	6.10	0.96	0.2	K0 III	-25	150 s		
154323	46450		17 02 16.0	+49 35 38	+0.001	-0.01	7.1	1.1	0.2	K0 III		240 s		
154322	30212		17 02 16.2	+50 18 30	-0.002	+0.01	7.4	1.1	0.2	K0 III		270 s		
153431	244379		17 02 16.3	-51 41 50	+0.001	+0.01	8.0	1.2	0.2	K0 III		290 s		
154126	65745		17 02 17.1	+31 53 04	+0.003	+0.04	6.36	1.13	0.2	K0 III	-13	140 s		
154084	84776		17 02 18.5	+25 30 19	+0.004	+0.09	5.75	1.02	0.2	K0 III	-50	130 s		
154225	46445		17 02 20.9	+40 04 23	-0.001	-0.04	7.9	0.2	2.1	A5 V		150 s		
153870	160239		17 02 22.5	-10 57 07	-0.002	+0.02	8.0	1.1	0.2	K0 III		360 s		
152827	257456		17 02 26.7	-72 36 31	-0.003	-0.02	7.2	1.1	0.2	K0 III		250 s		
153731	208333		17 02 27.4	-32 15 35	0.000	-0.05	7.4	0.0	1.4	A2 V		160 s		
153386	244377		17 02 28.1	-56 40 45	-0.007	-0.09	7.3	-0.1	4.0	F8 V		45 s		
154127	84782		17 02 30.9	+29 17 48	+0.001	0.00	7.6	0.1	1.4	A2 V	-27	170 s		
154085	102549		17 02 31.4	+19 23 11	-0.001	-0.01	7.2	0.1	1.4	A2 V		140 s		
154015	122030		17 02 33.1	+9 49 10	+0.003	-0.07	7.0	0.1	1.4	A2 V	-9	100 mx		
154345	46452		17 02 36.2	+47 04 54	+0.012	+0.85	6.77	0.73	5.6	G8 V	-46	17 ts		
154100	84781		17 02 38.4	+20 43 32	+0.001	0.00	7.0	1.6	-0.5	M2 III	-13	320 s		
154413	30213		17 02 39.5	+50 56 33	-0.002	0.00	7.7	1.1	0.2	K0 III		320 s		
153466	244381		17 02 44.1	-54 33 40	-0.002	-0.05	7.5	-0.4	1.4	A2 V		170 s		
153639	227572		17 02 47.9	-44 51 14	0.000	0.00	7.02	1.35	0.2	K0 III		140 s		
153579	244386		17 02 51.9	-50 09 37	-0.005	-0.02	7.40	0.1	0.6	A0 V		85 mx		m
155057	8699		17 02 52.5	+71 03 40	-0.001	-0.02	7.6	1.4	-0.3	K5 III		380 s		
153747	208337		17 02 53.7	-38 27 36	+0.001	+0.01	7.4	0.1	0.4	B9.5 V		210 s		
153766	208338		17 02 54.3	-36 51 40	0.000	+0.02	7.95	0.12	-3.9	B1 IV		1500 s		m
153974	141499		17 02 57.4	-6 21 08	-0.003	+0.02	7.4	0.1	1.7	A3 V		140 s		
153660	227577		17 02 59.2	-46 11 14	0.000	0.00	7.7	1.2	0.2	K0 III		290 s		
153767	208340		17 03 00.1	-37 43 11	-0.002	-0.02	7.5	0.0	0.6	A0 V		240 s		
154227	84794		17 03 02.6	+29 28 45	0.000	-0.01	7.72	1.25	-0.1	K2 III	-3	320 s		
153873	184955		17 03 04.5	-27 05 52	+0.002	0.00	7.4	0.9	0.2	K0 III		280 s		
154184	84790		17 03 05.0	+21 33 43	0.000	-0.01	7.4	1.1	0.2	K0 III		270 s		
154274	65755		17 03 05.1	+34 32 24	+0.001	-0.02	7.7	0.4	2.6	F0 V		110 s		
154143	102553		17 03 07.7	+14 05 30	+0.001	-0.07	4.98	1.60	-0.5	M3 III	+43	120 s		
153580	244388	ε² Ara	17 03 08.6	-53 14 12	0.000	-0.14	5.40	0.5	4.0	F8 V	+7	19 s		m
154160	102554		17 03 10.3	+14 30 40	-0.012	-0.19	6.5	0.9	3.2	G5 IV	-56	38 ts		
153678	227579		17 03 13.2	-48 20 10	+0.002	+0.04	7.3	0.7	0.2	K0 III		260 s		
154375	46457		17 03 13.5	+40 57 14	0.000	+0.03	7.9	0.5	4.0	F8 V		59 s		
153855	208347		17 03 17.9	-31 36 53	0.000	-0.01	7.00	-0.07	-2.6	B2.5 IV		770 s		
153662	244391		17 03 19.2	-50 24 08	-0.001	+0.02	7.7	0.4	3.0	F2 V		79 s		
153932	184962		17 03 21.4	-23 09 05	-0.001	0.00	7.6	1.7	-0.3	K5 III		280 s		
153805	227587		17 03 22.2	-40 05 18	0.000	-0.05	7.4	0.4	2.1	A5 V		81 s		
154346	65760		17 03 26.0	+34 22 59	+0.001	-0.01	6.7	1.1	-0.1	K2 III		230 s		

HD	SAO	Star Name	α 2000	δ 2000	μ(α)	μ(δ)	V	B-V	M$_v$	Spec	RV	d(pc)	ADS	Notes
153663	244393		17^{h}03^{m}26.2	−51°12'10"	+0.001	+0.01	7.90	0.1	1.7	A3 V		170 s		m
153942	184964		17 03 30.0	−24 14 28	0.000	−0.01	7.2	1.3	−0.1	K2 III		240 s		
154356	65761	61 Her	17 03 30.1	+35 24 51	+0.002	−0.04	6.7	1.6	−0.5	M4 III	−12	270 s		
152463	257453		17 03 31.7	−78 04 43	−0.008	+0.01	7.3	0.1	0.6	A0 V		210 mx		
−−	−−		17 03 35.9	+13 36 51			6.10							m
154507	46463		17 03 36.8	+46 01 26	+0.002	−0.04	8.0	1.1	0.2	K0 III		370 s		
154228	102564		17 03 39.1	+13 36 19	−0.002	−0.04	5.90	0.00	1.2	A1 V	−32	87 s		m
154035	160248		17 03 39.8	−15 51 58	−0.001	−0.01	8.0	0.1	1.4	A2 V		210 s		
154145	141503		17 03 41.3	−0 08 47	−0.002	−0.03	6.8	0.1	1.4	A2 V		120 s		
153791	227588		17 03 41.6	−47 09 35	+0.001	+0.03	6.30	0.1	1.4	A2 V		93 s		m
154275	102566		17 03 45.6	+18 17 56	−0.001	+0.01	7.7	1.1	−0.1	K2 III		360 s		
153890	208355		17 03 50.7	−38 09 08	+0.006	−0.03	5.92	0.37	2.6	dF0		42 s		
154300	102570		17 03 50.7	+19 48 02	+0.001	+0.03	7.80	0.20	3.2	G5 IV		83 s		
154301	102571		17 03 52.6	+19 41 26	−0.001	+0.01	6.35	1.52		K4 p	−39		10323	m
154431	65766		17 03 53.3	+34 47 25	−0.006	0.00	6.0	0.1	1.7	A3 V	−17	31 mx		
154277	102568		17 03 54.4	+16 01 21	−0.001	−0.03	7.87	1.11	0.2	K0 III		280 s		
154542	46465		17 03 54.9	+45 26 57	−0.001	0.00	8.0	1.1	0.2	K0 III		360 s		
153840	227590		17 03 55.7	−44 08 58	+0.003	−0.01	8.0	−0.1	0.0	B8.5 V		400 s		
153919	208356		17 03 56.5	−37 50 38	0.000	+0.01	6.55	0.26		O7				
154278	102569		17 03 57.9	+13 34 03	+0.001	−0.13	6.08	1.03	0.0	K1 III	+46	160 s		m
154002	184978		17 03 59.3	−28 15 25	+0.003	−0.03	8.0	−0.1	0.4	B9.5 V		330 s		
154021	184982		17 04 00.6	−25 41 53	−0.001	−0.03	6.7	0.5	0.4	B9.5 V		90 s		
154186	141504		17 04 01.0	−2 35 00	−0.001	+0.01	8.0	1.4	−0.3	K5 III		450 s		
154038	184981		17 04 02.0	−27 12 56	+0.002	−0.05	7.9	0.4	4.0	F8 V		60 s		
153920	208359		17 04 05.0	−39 04 19	0.000	−0.03	7.6	3.2	−0.5	M2 III		46 s		
154859	17337		17 04 09.1	+61 11 03	−0.001	+0.03	7.9	1.1	0.2	K0 III		340 s		
154508	65773		17 04 15.4	+36 45 10	0.000	−0.02	7.9	1.1	−0.1	K2 III		390 s		
153892	227595		17 04 15.8	−45 14 26	+0.004	+0.01	7.63	1.21	0.0	K1 III		290 s		
153716	244401		17 04 24.6	−57 42 44	−0.001	−0.03	5.73	−0.10	−1.4	B4 V	+6	250 s		
154104	184991		17 04 27.6	−24 14 24	+0.003	+0.01	7.4	0.8	0.2	K0 III		270 s		
154117	184992		17 04 27.6	−23 23 28	−0.001	−0.04	7.3	0.9	0.2	K0 III		260 s		
154088	184990		17 04 27.7	−28 34 57	+0.007	−0.26	6.59	0.83	4.4	G8 IV−V	+17	19 ts		
154147	160257		17 04 29.3	−17 29 32	+0.002	−0.11	7.2	0.9	3.2	G5 IV		62 s		
154202	160261		17 04 29.4	−12 40 33	−0.003	−0.06	7.00	1.1	0.2	K0 III		150 mx	10324	m
153950	227597		17 04 30.8	−43 18 38	+0.011	−0.17	7.42	0.56	3.9	G2 IV−V		52 s		
154509	65775		17 04 32.1	+31 24 40	+0.001	−0.02	7.7	1.1	−0.1	K2 III		360 s		
154212	160264		17 04 33.2	−11 05 18	−0.001	−0.01	7.0	1.1	0.2	K0 III	−24	230 s		
153389	253783		17 04 33.9	−68 16 44	+0.003	−0.03	6.8	1.6	−0.5	M4 III		280 s		
154441	102579		17 04 41.1	+19 35 56	+0.001	0.00	6.10	−0.01	0.4	B9.5 V	−25	130 s	10326	m
154132	184995		17 04 43.3	−25 38 42	−0.001	−0.08	6.8	0.2	2.1	A5 V		52 mx		
154204	184999		17 04 45.2	−20 29 41	0.000	−0.03	6.1	−0.1	−1.6	B3.5 V	−11	340 s		
154510	84810		17 04 45.6	+28 05 29	0.000	+0.02	7.30	1.1	0.0	K1 III	+2	290 s	10332	m
154107	208379		17 04 46.0	−31 21 48	−0.001	−0.01	7.4	1.1	4.4	G0 V		19 s		
153908	244411		17 04 46.5	−50 04 17	+0.001	−0.01	7.9	−0.3	0.0	B8.5 V		390 s		
−−	84813		17 04 47.4	+29 40 29	+0.001	−0.02	7.99	1.35	−0.3	K5 III		460 s		
−−	−−		17 04 47.4	+29 39 45			7.95	1.40		Ma				
154214	160265		17 04 48.1	−17 57 26	+0.001	−0.01	7.9	0.1	1.4	A2 V		200 s		
154090	208377		17 04 49.3	−34 07 22	+0.001	0.00	4.87	0.26	−6.2	B1 Iab	+8	890 s		m
154732	46476		17 04 49.6	+48 48 14	+0.003	−0.07	6.09	1.09	0.2	K0 III	+12	130 s		
153980	227599		17 05 02.1	−46 58 17	−0.001	−0.03	7.3	0.1	0.4	B9.5 V		190 s		
154363	141508		17 05 03.2	−5 03 58	−0.062	−1.13	7.73	1.16	7.5	K5 V	+29	11 t		m
154904	30236		17 05 03.5	+56 32 15	−0.001	0.00	8.0	1.1	0.2	K0 III		360 s		
154713	46478		17 05 04.8	+43 48 44	0.000	0.00	6.43	0.09		A0	−9			
154025	227601		17 05 05.1	−45 30 06	−0.003	−0.01	6.28	0.07	1.4	A2 V		91 s		
154417	122056		17 05 16.7	+0 42 09	−0.001	−0.34	6.01	0.58	4.0	dF8	−18	23 ts		
154043	227604		17 05 18.9	−47 04 07	+0.002	+0.01	7.10	0.63	−6.2	B1 I k		2500 s		
154906	30239	21 μ Dra	17 05 19.5	+54 28 13	−0.009	+0.08	4.92	0.48		F5	−15	26 t	10345	Alrakis, m
154494	102584	60 Her	17 05 22.6	+12 44 27	+0.003	−0.01	4.91	0.12	0.9	A3 IV	−4	62 s	10334	m
153860	244414		17 05 23.1	−58 04 36	−0.003	−0.01	8.00	1.6	−0.5	M2 III		480 s		m
154443	122061		17 05 23.2	+3 26 18	0.000	−0.01	7.3	0.4	2.6	F0 V		85 s		
155058	17342		17 05 24.3	+60 35 03	−0.005	+0.01	7.8	0.8	3.2	G5 IV		84 s		
154815	46482		17 05 25.7	+46 21 22	−0.001	+0.02	7.4	1.1	0.2	K0 III		280 s		
154293	185012		17 05 30.2	−22 04 25	0.000	−0.05	7.2	0.2	0.0	B8.5 V		190 s		
154445	141513		17 05 32.1	−0 53 32	0.000	0.00	5.64	0.16	−3.5	B1 V	+15	410 s		
154516	102589		17 05 32.8	+10 59 42	+0.001	−0.01	7.9	1.1	0.2	K0 III		340 s		
154478	122066		17 05 34.5	+3 46 02	0.000	−0.02	7.5	1.6	−0.5	M2 III		400 s		
154651	84827		17 05 37.6	+29 37 21	0.000	−0.02	7.6	0.1	1.4	A2 V	−10	170 s		
154986	30244		17 05 40.5	+56 08 07	−0.009	+0.06	7.2	1.1	0.2	K0 III		200 mx		
154349	185015		17 05 42.2	−21 16 54	−0.003	−0.06	7.5	0.9	0.2	K0 III		170 mx		
154218	208390		17 05 42.9	−36 44 25	+0.001	−0.01	7.7	0.1	−1.0	B5.5 V		410 s		
154383	160270		17 05 48.2	−18 06 56	0.000	+0.02	7.6	0.0	0.0	B8.5 V		340 s		
154153	227615		17 05 48.4	−44 06 18	+0.003	+0.02	6.19	0.28		A3				
154333	185017		17 05 56.5	−28 52 22	0.000	0.00	7.4	0.4	0.6	A0 V		120 s		
154543	122069		17 05 57.5	+5 02 17	+0.001	−0.01	7.9	1.1	−0.1	K2 III		390 s		
155859	8714		17 05 59.6	+75 06 13	−0.031	+0.01	7.2	0.7	4.4	G0 V		37 s		

HD	SAO	Star Name	α 2000	δ 2000	μ(α)	μ(δ)	V	B-V	M_v	Spec	RV	d(pc)	ADS	Notes
			$^h \quad ^m \quad ^s$	$^\circ \quad ' \quad ''$	^{s}	$''$								
154581	122072		17 06 00.2	+7 51 18	-0.001	-0.01	7.6	0.1	1.7	A3 V		160 s		
154196	227617		17 06 00.6	-44 22 22	+0.002	-0.01	8.0	-0.3	0.6	A0 V		300 s		
154691	84830		17 06 03.3	+23 19 20	-0.003	0.00	7.8	1.1	0.2	K0 III		330 s		
154365	185020		17 06 05.3	-26 34 49	0.000	+0.01	6.7	0.8	3.2	G5 IV	-32	51 s		BF Oph, v
154611	122073		17 06 06.7	+8 37 21	-0.004	-0.03	7.0	1.1	-0.1	K2 III	-3	190 mx		
154610	122075		17 06 09.6	+9 44 00	+0.001	-0.01	6.37	1.45		K5	-6			
154652	102593		17 06 11.2	+16 16 14	+0.001	0.00	7.7	1.1	0.2	K0 III		310 s		
154418	185024		17 06 11.6	-21 33 53	-0.001	-0.08	6.30	0.13		A0				
154598	122074		17 06 11.8	+5 10 05	+0.002	-0.02	7.9	0.9	3.2	G5 IV		88 s		
153937	253792		17 06 11.9	-60 25 14	+0.002	0.00	7.1	1.6	-0.1	K2 III		160 s		
154619	102592		17 06 13.0	+10 27 15	+0.003	+0.02	6.4	0.9	1.8	G8 III-IV	-24	86 s		
154653	102595		17 06 15.8	+15 13 46	-0.007	-0.08	7.3	1.1	5.9	K0 V		19 s		
154733	84835		17 06 17.8	+22 05 02	-0.007	-0.04	5.56	1.30	-0.3	K4 III	-96	150 s	10343	m
154206	227620		17 06 20.0	-46 03 10	+0.002	-0.03	7.7	1.3	0.2	K0 III		190 s		
154310	208406		17 06 20.1	-37 13 39	+0.001	-0.02	5.98	0.07	1.4	A2 V		79 s		m
155207	17348		17 06 21.2	+60 38 09	-0.003	-0.01	8.0	1.1	0.2	K0 III		370 s		
154519	160279		17 06 21.9	-12 24 39	0.000	+0.03	7.9	0.4	3.0	F2 V		95 s		
154715	102596		17 06 27.4	+17 52 36	-0.001	+0.02	7.9	1.1	0.2	K0 III		340 s		
154368	208410		17 06 28.3	-35 27 04	+0.001	-0.01	6.13	0.50	-6.2	O9.5 Iab	+13	1100 s		m
154620	122076		17 06 28.5	+3 38 03	0.000	+0.01	7.9	1.1	-0.1	K2 III		390 s		
154404	208415		17 06 29.7	-30 08 55	0.000	+0.01	7.5	0.2	1.7	A3 V		130 s		
154791	84844		17 06 34.4	+23 58 17	-0.001	-0.02	8.0	1.6	-0.5	M2 III		510 s		
154250	227625		17 06 40.8	-48 00 44	-0.002	-0.07	7.7	1.1	3.2	G5 IV		40 s		
154385	208417		17 06 40.9	-36 04 42	+0.001	-0.01	7.37	0.35		B0.5 I k				
154888	65794		17 06 41.4	+35 19 24	-0.001	-0.02	7.2	0.1	0.6	A0 V	-19	210 s		
154660	141522		17 06 52.8	-1 39 23	+0.002	-0.04	6.31	0.23		A2 n	+7		10347	m
154481	185033		17 06 53.1	-26 30 47	0.000	-0.01	6.29	-0.04	-0.6	A0 III		240 s		
155038	46493		17 06 54.7	+47 40 16	0.000	0.00	7.8	1.1	-0.1	K2 III		380 s		
156109	8716		17 06 56.9	+75 52 09	-0.020	+0.06	7.9	1.1	-0.1	K2 III		190 mx		
154408	208421		17 06 57.0	-38 25 35	-0.001	+0.01	8.0	0.3	-1.0	B5.5 V		340 s		
155059	46496		17 07 07.7	+44 15 19	-0.008	-0.04	8.0	0.9	3.2	G5 IV		93 s		
154762	122095		17 07 08.5	+9 00 33	+0.001	+0.01	7.9	0.0	0.4	B9.5 V		320 s		
155305	30257		17 07 10.2	+58 16 07	+0.005	-0.06	7.2	1.1	0.2	K0 III		260 s		
154796	102600		17 07 10.8	+12 07 35	+0.005	+0.01	8.0	0.5	3.4	F5 V		82 s		
154697	141524		17 07 14.2	-2 51 59	+0.007	-0.07	7.8	0.8	3.2	G5 IV		85 s		
153298	257460		17 07 18.7	-76 13 24	-0.001	-0.06	7.0	0.8	0.6	A0 V		63 s		
154373	227636		17 07 19.6	-46 53 28	+0.002	-0.04	7.3	1.3	0.2	K0 III		180 s		
154410	227637		17 07 21.4	-44 26 34	+0.001	-0.02	7.20	0.1	0.6	A0 V		200 s		m
155136	46501		17 07 22.3	+48 23 56	-0.003	+0.03	6.8	1.1	0.2	K0 III		210 s		
154290	244447		17 07 26.4	-53 45 21	-0.001	-0.08	8.00	0.1	1.4	A2 V		67 mx		m
153985	253798		17 07 26.8	-65 44 51	-0.002	-0.06	7.1	0.4	3.0	F2 V		66 s		
154503	208434		17 07 28.6	-37 18 28	-0.001	-0.02	8.0	1.3	-0.1	K2 III		360 s		
154175	244443		17 07 30.3	-59 40 02	0.000	-0.02	6.8	1.6	-0.1	K2 III		140 s		
154797	122102		17 07 30.5	+3 58 24	-0.003	+0.03	7.8	0.4	3.0	F2 V		89 s		
155479	17359		17 07 33.9	+63 26 25	-0.001	+0.04	7.8	0.5	3.4	F5 V		76 s		
154426	227642		17 07 35.1	-46 40 47	-0.009	+0.01	6.89	0.26		A5				
154613	185040		17 07 36.4	-27 43 52	-0.001	-0.01	7.7	1.2	3.2	G5 IV		44 s		
154485	227646		17 07 42.8	-42 45 11	+0.001	-0.01	7.95	0.05	0.0	B8.5 V		340 s		
154604	208446		17 07 44.1	-31 04 38	+0.002	-0.03	7.0	0.6	0.2	K0 III		230 s		
154721	160299		17 07 45.6	-16 05 33	-0.001	-0.01	5.9	1.6		M5 III	-47			R Oph, v
155102	46502		17 07 46.5	+40 30 58	-0.004	-0.03	6.3	0.1	1.4	A2 V	-7	73 mx		
154112	253804		17 07 47.2	-63 50 11	-0.002	-0.01	7.7	0.1	0.6	A0 V		260 s		
155343	30260		17 07 48.2	+55 46 04	-0.001	+0.03	6.9	1.1	0.2	K0 III		220 s		
153987	253801		17 07 50.1	-67 11 23	0.000	-0.08	7.0	0.1	0.6	A0 V		82 mx		
155060	65809		17 07 55.7	+32 06 18	-0.013	-0.06	7.3	0.5	4.0	F8 V		45 s		
154679	185049		17 07 55.9	-24 59 53	-0.001	0.00	7.2	0.8	3.2	G5 IV		63 s		
155025	84864		17 07 59.0	+27 06 05	-0.001	+0.01	7.9	0.4	3.0	F2 V		94 s		
155061	65810		17 08 00.5	+31 12 23	0.000	-0.01	6.60	1.1	-0.1	K2 III	-10	220 s	10356	m
155208	46509		17 08 00.7	+44 10 09	0.000	-0.02	8.0	1.1	-0.1	K2 III		410 s		
154850	141526		17 08 01.8	-4 19 40	-0.001	+0.01	8.0	1.1	0.2	K0 III		360 s		
155103	65812		17 08 01.9	+35 56 07	-0.002	-0.01	5.39	0.31		A5	-30	21 mn	10360	m
154569	227651		17 08 05.3	-41 36 53	-0.002	-0.05	7.70	0.1	1.7	A3 V		150 s		m
155513	17360		17 08 07.3	+61 09 34	-0.004	+0.06	6.7	0.5	3.4	F5 V	-7	46 s		
154486	227648		17 08 08.2	-48 53 01	0.000	-0.02	6.8	1.6	-0.3	K5 III		240 s		
154895	141528		17 08 13.4	-1 04 46	-0.001	-0.03	6.06	0.08		A0 n	-21		10355	m
154643	208452		17 08 13.6	-35 00 14	-0.001	+0.01	7.15	0.28	-4.4	O9.5 V		1000 s		
154779	160305		17 08 14.7	-17 36 32	-0.001	-0.03	6.1	1.1	0.2	K0 III	-14	150 s		
154974	102611		17 08 15.5	+16 05 36	-0.004	-0.03	6.7	0.5	4.0	F8 V	-26	35 s		
155328	30262		17 08 16.9	+50 50 32	-0.001	+0.03	6.30	0.00		B9	-15		10369	m
154703	185055		17 08 17.1	-30 03 07	-0.001	-0.03	7.2	0.8	0.2	K0 III		250 s		
154589	227660		17 08 17.2	-41 01 29	0.000	-0.01	7.1	0.2	3.4	F5 V		54 s		
154930	122113		17 08 20.7	+5 23 50	0.000	-0.01	7.9	1.1	-0.1	K2 III		390 s		
155092	84868		17 08 21.1	+28 14 02	-0.001	-0.13	7.0	0.4	3.0	F2 V	+4	64 s		
154931	122112		17 08 21.3	+4 25 28	-0.004	-0.20	7.3	0.6	4.4	G0 V	-18	38 s		
154865	160310		17 08 24.9	-12 02 00	-0.004	-0.01	7.7	0.5	4.0	F8 V		54 s		

HD	SAO	Star Name	α 2000	δ 2000	μ(α)	μ(δ)	V	B-V	M_V	Spec	RV	d(pc)	ADS	Notes
			^{h} ^{m} ^{s}	$^\circ$ $'$ $''$	^{s}	$''$								
156011	8720		17 08 29.3	+72 08 01	-0.002	+0.06	7.5	1.1	-0.1	K2 III		310 mx		
154753	185061		17 08 30.8	-29 59 07	-0.001	+0.01	7.8	1.0	0.2	K0 III		310 s		
155179	65821		17 08 31.6	+35 02 01	-0.003	+0.08	6.7	1.1	0.2	K0 III		200 s		
154831	185067		17 08 32.2	-20 13 03	0.000	-0.01	7.50	0.1	0.6	A0 V		220 s		m
154682	208457		17 08 32.6	-36 38 34	+0.004	-0.07	7.9	-0.2	4.4	G0 V		51 s		
155104	84873		17 08 33.3	+24 29 11	0.000	-0.05	6.8	0.1	0.6	A0 V	+21	180 s		
154685	208461		17 08 42.7	-39 20 13	-0.001	+0.02	7.6	0.2	0.6	A0 V		180 s		
155763	17365	22 ζ Dra	17 08 47.0	+65 42 53	-0.004	+0.02	3.17	-0.12	-1.9	B6 III	-14	97 s		
154783	208467		17 08 47.6	-30 24 14	+0.001	-0.07	5.97	0.25	0.6	F0 III		120 s		
154896	160312		17 08 47.9	-13 07 35	-0.002	+0.03	7.6	1.1	-0.1	K2 III		350 s		
154821	185071		17 08 49.4	-28 13 02	+0.002	-0.02	7.9	1.8	3.2	G5 IV		21 s		
155227	65824		17 08 53.9	+33 18 02	+0.001	-0.01	7.6	0.1	0.6	A0 V		260 s		
154962	141535		17 08 54.3	-3 52 58	-0.004	-0.16	6.36	0.69	4.2	G5 IV-V		27 s		
154866	185074		17 08 57.6	-23 59 38	-0.001	-0.04	8.0	0.0	3.4	F5 V		82 s		
154909	160313		17 09 01.1	-16 54 12	-0.001	-0.07	7.7	0.4	2.6	F0 V		110 s		
155093	102615		17 09 01.8	+14 57 40	+0.001	-0.02	7.1	1.1	0.2	K0 III		240 s		
155118	102617		17 09 04.7	+16 27 44	-0.001	0.00	8.0	0.4	2.6	F0 V		120 s		
154140	253810		17 09 10.0	-68 50 49	0.000	-0.04	6.8	-0.1	0.6	A0 V		170 s		
154459	253815		17 09 24.7	-60 37 06	+0.001	-0.03	7.5	1.6	3.2	G5 IV		23 s		
153793	257464		17 09 26.5	-75 03 48	+0.008	+0.03	7.4	1.4	0.2	K0 III		150 s		
155228	84877		17 09 26.9	+22 05 29	+0.003	-0.06	6.9	0.5	3.4	F5 V		51 s		
154671	244468		17 09 29.3	-51 53 12	+0.003	-0.01	7.7	0.0	2.1	A5 V		130 s		
154744	227672		17 09 29.9	-47 15 39	+0.002	-0.01	7.74	0.11	0.0	B8.5 V		350 s		
155410	46524		17 09 32.9	+40 46 37	-0.005	+0.01	5.08	1.28	-0.2	K3 III	-56	92 mx		
155121	122134		17 09 33.7	+3 55 56	-0.001	-0.06	7.9	0.9	3.2	G5 IV		88 s		
155358	65834		17 09 34.4	+33 21 23	-0.019	-0.19	7.6	0.6	4.4	G0 V		33 mx		
155394	65838		17 09 35.5	+38 17 36	-0.001	+0.03	7.9	0.5	3.4	F5 V		78 s		
155193	102623		17 09 41.0	+10 02 18	-0.005	-0.15	7.1	0.5	3.4	F5 V		53 mx		
—	65837		17 09 42.1	+33 04 09	0.000	-0.01	8.0	1.3						
155078	160324		17 09 47.7	-10 31 24	+0.004	-0.11	5.56	0.52	3.4	F5 V	-3	25 s		
152565	258760		17 09 51.6	-82 19 07	+0.004	-0.02	7.20	0.1	0.6	A0 V		210 s		m
154811	227679		17 09 52.8	-47 01 52	0.000	+0.01	6.93	0.40	-5.9	O9.5 Ib		1500 s		
155344	84885		17 09 56.6	+26 27 19	-0.001	+0.01	6.9	1.1	-0.1	K2 III	+3	260 s		
155048	185095		17 09 57.1	-20 40 49	0.000	-0.02	7.54	1.26	0.2	K0 III		200 s		
155105	141540		17 10 00.0	-8 31 26	0.000	-0.06	8.00	0.6	4.9	G3 V	-33	42 s		
154978	208487		17 10 05.2	-30 50 16	-0.003	-0.03	7.8	0.2	1.4	A2 V		170 s		
154555	253818		17 10 06.0	-61 40 31	-0.001	-0.01	6.5	0.1	0.4	B9.5 V		130 s		
155229	122145		17 10 08.2	+5 23 22	0.000	-0.03	8.0	0.4	2.6	F0 V		120 s		
155010	185097		17 10 08.8	-27 46 25	-0.001	-0.05	7.70	0.4	3.0	F2 V		85 s	10368	m
155195	141541		17 10 10.1	-0 45 37	-0.001	-0.01	8.00	0.1	0.6	A0 V		300 s	10376	m
155480	65843		17 10 10.2	+38 44 10	-0.001	0.00	7.80	1.24	-0.1	K2 III		350 s		
154577	253819		17 10 10.3	-60 43 42	+0.013	+0.60	7.41	0.89	5.9	K0 V		19 ts		
155095	160326		17 10 14.7	-19 26 09	0.000	+0.01	7.03	0.08	0.0	B8.5 V		210 s		
155231	122146		17 10 15.3	+0 28 52	-0.001	-0.01	6.8	0.8	3.2	G5 IV		54 s		
153880	257465		17 10 16.2	-75 22 34	+0.001	+0.03	7.60	0.4	3.0	F2 V		83 s		m
155523	46531		17 10 18.5	+40 41 23	+0.003	+0.02	7.7	1.6	-0.5	M4 III		440 s		
159251	2841		17 10 20.2	+84 35 56	+0.015	-0.02	7.2	0.8	3.2	G5 IV		63 s		
154873	227683		17 10 20.8	-46 44 25	0.000	0.00	6.69	0.29	-2.6	B2.5 IV		390 s		
155125	160332	35 η Oph	17 10 22.5	-15 43 30	+0.003	+0.09	2.43	0.06	1.4	A2 V	-1	18 ts	10374	Sabik, m
155014	208493		17 10 25.5	-32 37 18	+0.003	-0.04	8.0	0.5	0.4	B9.5 V		150 s		
155395	84891		17 10 30.5	+21 12 56	0.000	+0.02	8.0	0.5	3.4	F5 V		82 s		
155711	30277		17 10 30.5	+52 24 31	-0.002	-0.01	6.1	0.0		B9	-42			
155559	65845		17 10 32.1	+39 51 08	+0.002	+0.04	7.5	0.5	3.4	F5 V		65 s		
155656	46542		17 10 40.1	+46 33 21	+0.002	-0.09	7.6	1.1	0.2	K0 III		240 mx		
154948	227688		17 10 42.1	-44 33 27	-0.003	-0.06	5.08	0.86	0.4	gG2	-7	72 s		m
155213	160337		17 10 43.1	-12 42 05	-0.003	-0.06	6.7	0.1	1.7	A3 V		50 mx		
155375	102632		17 10 45.6	+12 28 02	+0.002	-0.01	6.57	0.08		A m	+5			
155696	46545		17 10 45.7	+49 20 48	-0.006	+0.01	7.5	1.4	-0.3	K5 III		250 mx		
155524	65846		17 10 48.0	+32 10 44	+0.001	-0.09	7.21	0.98		K2				
156677	8732		17 10 49.2	+75 05 51	0.000	+0.01	7.9	0.1	0.6	A0 V		290 s		
155525	65848		17 10 56.4	+31 10 26	0.000	-0.01	7.5	1.1	0.2	K0 III		290 s		
—	65851		17 10 58.3	+39 04 42	-0.001	-0.01	8.0	1.6						
154208	257470		17 11 01.2	-72 51 12	-0.005	-0.07	7.7	0.5	3.4	F5 V		71 s		
155249	160340		17 11 01.4	-16 38 06	-0.002	-0.01	7.4	1.1	0.2	K0 III		170 mx		
155031	227693		17 11 01.7	-41 26 23	-0.001	-0.02	7.80	0.11	0.6	A0 V		230 s		
155514	84896	63 Her	17 11 03.0	+24 14 16	-0.001	+0.03	6.2	0.1	1.7	A3 V	-2	79 s		
155233	185116		17 11 04.2	-20 39 16	-0.001	-0.13	6.81	1.04	0.2	K0 III		100 mx		
155019	227692		17 11 05.1	-43 29 21	+0.003	+0.02	7.7	0.9	-0.1	K2 III		270 s		
155234	185117		17 11 05.5	-20 49 20	-0.001	-0.02	7.99	1.15	0.2	K0 III		310 s		
155051	227694		17 11 05.8	-41 41 54	-0.001	-0.02	8.0	-0.1		B0.5 II				
155902	30285		17 11 08.4	+56 39 32	+0.001	-0.08	7.1	0.8	3.2	G5 IV		59 s		
154813	244489		17 11 09.4	-58 01 32	+0.002	-0.01	6.8	0.4	0.2	K0 III		210 s		
154857	244491		17 11 15.7	-56 40 50	+0.013	-0.05	7.4	0.3	4.4	G0 V		40 s		
155543	84900		17 11 16.8	+24 15 09	0.000	+0.04	7.1	0.4	3.0	F2 V		65 s		
155838	30283		17 11 17.5	+51 45 14	+0.001	-0.01	7.1	0.1	0.6	A0 V		200 s		

HD	SAO	Star Name	α 2000	δ 2000	μ(α)	μ(δ)	V	B−V	M_v	Spec	RV	d(pc)	ADS	Notes
			h m s	° ′ ″	s	″								
154473	—		17 11 19.2	−69 21 12			8.0	1.1	0.2	K0 III		360 s		
155291	185122		17 11 21.8	−20 25 30	−0.001	−0.04	7.52	0.03	0.6	A0 V		220 s		
155423	122157		17 11 22.5	+4 41 25	−0.003	0.00	6.8	0.5	4.0	F8 V		37 s		
154526	—		17 11 24.8	−68 38 10			8.00	0.1	1.4	A2 V		210 s		m
154983	244502		17 11 25.5	−50 07 09	−0.001	−0.01	7.5	−0.1	0.6	A0 V		240 s		
155526	102639		17 11 33.5	+16 24 39	0.000	−0.01	8.0	1.1	0.2	K0 III		360 s		
155164	208509		17 11 37.0	−37 51 12	−0.001	−0.01	7.8	1.5	−0.1	K2 III		250 s		
155035	227699		17 11 38.7	−48 52 27	+0.003	−0.03	5.8	1.8	−0.5	M2 III		140 s		
155816	46556		17 11 39.6	+45 19 26	−0.001	−0.03	6.8	1.1	−0.1	K2 III		240 s		
155860	46561		17 11 40.0	+49 44 48	+0.001	+0.03	6.00	0.1	1.4	A2 V	−11	83 s	10397	m
154886	244497		17 11 42.6	−59 10 03	0.000	+0.02	7.2	1.7	−0.3	K5 III		240 s		
155424	141550		17 11 43.1	−4 38 24	−0.001	+0.04	7.70	0.1	1.4	A2 V		180 s		m
154901	244498		17 11 44.1	−58 48 53	−0.002	−0.01	7.1	0.0	0.4	B9.5 V		200 s		
155500	122164		17 11 45.0	+7 53 41	+0.002	+0.01	6.33	1.04	0.2	K0 III	−6	160 s		
155784	46558		17 11 49.9	+41 43 48	−0.001	0.00	6.8	1.1	−0.1	K2 III		240 s		
155114	227703		17 11 51.5	−45 52 28	0.000	−0.11	7.5	0.7	4.4	G0 V		35 s		
154970	244505		17 11 51.5	−55 37 48	+0.001	−0.01	8.0	0.2	0.0	B8.5 V		260 s		
156051	30291		17 11 54.4	+57 58 03	−0.006	+0.06	7.3	0.5	3.4	F5 V		60 s		
155413	160351		17 11 58.5	−14 37 15	−0.001	−0.06	7.3	0.6	4.4	G0 V		38 s		
155581	102643		17 11 59.3	+14 29 15	−0.001	−0.01	7.40	1.4	−0.3	K5 III	+13	350 s		
155642	84910		17 12 05.9	+21 13 44	+0.002	0.00	7.1	1.1	0.2	K0 III	−49	240 s	10394	m
154440	—		17 12 08.0	−72 16 11			7.7	1.4	−0.3	K5 III		400 s		
155203	227707	η Sco	17 12 09.0	−43 14 21	+0.002	−0.28	3.33	0.41	0.6	F2 III	−28	21 ts		
155363	185137		17 12 10.8	−27 02 33	−0.003	−0.09	6.90	0.9	3.2	G5 IV		54 s	10388	m
155379	185138		17 12 13.3	−25 15 17	−0.001	−0.04	6.3	0.6	0.6	A0 V		57 s		
155066	244511		17 12 14.5	−53 22 33	0.000	0.00	7.7	0.6	−0.5	M2 III		440 s		
155427	160354		17 12 15.2	−17 14 48	0.000	−0.03	7.9	0.1	0.6	A0 V		280 s		
155259	208521		17 12 16.0	−39 30 25	−0.001	−0.07	5.67	0.04		A0	+12			
155276	208522		17 12 16.3	−38 49 20	+0.002	−0.04	6.4	0.6	0.2	K0 III		170 s		
155275	208523		17 12 16.6	−38 24 26	0.000	−0.02	8.0	2.3	−0.3	K5 III		150 s		
154556	257472		17 12 19.7	−70 43 16	+0.010	−0.08	6.22	1.06		K0				
155546	141555		17 12 22.3	−0 46 26	0.000	−0.06	7.5	0.4	2.6	F0 V		98 s		
155921	46565		17 12 22.7	+43 44 14	+0.002	+0.18	7.3	0.9	3.2	G5 IV		66 s		
156279	17390		17 12 23.2	+63 21 09	+0.002	+0.18	8.0	1.1	0.2	K0 III		98 mx		
155414	185143		17 12 23.2	−22 55 30	−0.002	0.00	7.9	0.1	1.7	A3 V		160 s		
155401	185142		17 12 24.8	−27 45 43	−0.001	−0.04	6.1	−0.1	0.4	B9.5 V		140 s		
154775	253824		17 12 24.8	−66 36 36	−0.002	−0.03	8.0	1.4	−0.3	K5 III		460 s		
155644	102646	37 Oph	17 12 27.7	+10 35 07	+0.001	−0.03	5.60	1.4	−0.3	K5 III	+26	150 s		m
155415	185145		17 12 31.0	−25 13 37	0.000	−0.05	7.8	1.1	3.4	F5 V		32 s		
155515	160364		17 12 31.2	−10 16 43	+0.006	−0.02	7.8	0.6	4.4	G0 V		47 s		
156295	17391		17 12 32.3	+62 52 28	+0.001	+0.05	5.56	0.21		A3	−3	26 mn		
154902	253826		17 12 34.4	−63 41 07	+0.001	−0.01	7.0	0.0	0.4	B9.5 V		210 s		
155503	160362		17 12 34.8	−15 09 49	−0.001	−0.01	7.9	0.0	0.4	B9.5 V		320 s		
155658	102648		17 12 36.6	+11 35 45	+0.001	−0.03	7.9	1.4	−0.3	K5 III		430 s		
155469	185150		17 12 39.5	−21 36 22	+0.003	0.00	6.9	1.0	2.6	F0 V		28 s		
155402	208535		17 12 47.0	−33 21 44	+0.002	−0.06	7.80	0.00	−1.0	B5.5 V		89 mx		m
155646	122182		17 12 54.2	+0 21 06	+0.001	−0.08	6.65	0.50		F5	+58			
153014	258763		17 12 54.9	−82 23 16	+0.014	0.00	7.5	−0.1	2.6	F0 V		93 s		
155922	65872		17 12 57.5	+35 09 54	−0.001	+0.03	7.7	1.4	−0.3	K5 III		400 s		
156161	30297		17 12 58.1	+54 07 29	−0.003	+0.11	7.20	0.9	3.2	G5 IV		70 s	10410	m
155450	208539		17 12 58.5	−32 26 19	+0.001	−0.01	6.01	0.07	−4.4	B1 III	+7	840 s		m
155099	244518		17 12 58.9	−58 35 49	−0.008	−0.10	6.85	0.38	2.6	F4 IV-V		69 s		m
155714	122186		17 12 59.2	+7 44 59	0.000	+0.04	7.00	0.4	2.6	F0 V	−42	76 s	10398	m
155169	244522		17 13 03.0	−54 51 06	+0.001	−0.04	7.2	0.5	0.2	K0 III		250 s		
156162	30299		17 13 05.8	+54 08 25	−0.003	+0.10	7.00	0.4	2.6	F0 V	−18	70 s	10410	m
155839	84926		17 13 08.2	+24 59 07	+0.001	−0.04	7.0	1.4	−0.3	K5 III		290 s		
155416	208541		17 13 09.1	−37 57 09	0.000	−0.01	6.7	0.4	0.0	B8.5 V		120 s		
155463	208545		17 13 10.0	−32 37 23	−0.002	−0.09	7.9	0.4	2.6	F0 V		96 s		
155612	160372		17 13 13.8	−12 43 49	−0.002	−0.05	8.0	0.9	3.2	G5 IV		93 s		
154903	253827		17 13 17.4	−67 11 48	−0.029	−0.09	5.89	1.06	1.7	K0 III-IV		59 s		m
155978	65877		17 13 22.6	+35 25 51	0.000	0.00	8.0	0.1	1.4	A2 V		210 s		
156012	65878		17 13 27.1	+38 59 06	+0.001	−0.01	6.98	1.18	0.2	K0 III		170 s		
156110	46575		17 13 27.4	+45 22 20	−0.002	−0.01	7.58	−0.16	−1.7	B3 V n	−43	720 s		
155506	208553		17 13 27.8	−33 20 23	+0.002	+0.01	7.75	−0.06		B5				
155904	84933		17 13 30.0	+22 44 45	+0.002	−0.02	8.0	0.1	1.4	A2 V		210 s		
156280	30302		17 13 30.0	+55 28 31	−0.007	+0.06	7.5	0.5	3.4	F5 V		68 s		
155117	253830		17 13 30.1	−60 30 41	0.000	−0.01	7.7	1.3	4.4	G0 V		16 s		
156074	46574		17 13 31.0	+42 06 21	−0.001	−0.06	7.59	1.15		C1.2	−16			
155389	227721		17 13 31.9	−47 19 09	−0.001	−0.02	7.11	0.03	0.6	A0 V		190 s		
155663	160380		17 13 36.6	−15 33 37	−0.001	−0.02	6.6	0.1	1.7	A3 V		98 s		
155718	141562		17 13 37.1	−9 17 01	−0.001	0.00	7.60	0.1	0.6	A0 V		250 s	10400	m
155923	84934		17 13 39.2	+21 25 52	+0.001	−0.01	6.9	1.1	0.2	K0 III		220 s		
155408	227724		17 13 40.8	−47 56 52	−0.002	+0.04	8.00	0.9	3.2	G5 IV		91 s		m
155550	208558		17 13 44.7	−32 51 12	−0.001	−0.03	8.01	0.06		B9				FV Sco, v
155299	244536		17 13 52.8	−56 48 03	−0.004	−0.04	7.3	−0.2	3.4	F5 V		61 s		

HD	SAO	Star Name	α 2000	δ 2000	μ(α)	μ(δ)	V	B−V	M$_V$	Spec	RV	d(pc)	ADS	Notes
			$17^h 13^m 53^s.0$	$-41°41'11''$	$+0^s.002$	$-0''.03$								
155508	227726		17 13 53.0	−41 41 11	+0.002	−0.03	7.8	0.1	2.1	A5 V		140 s		
155536	208556		17 13 53.6	−38 17 39	+0.003	+0.01	6.8	−0.6	3.4	F5 V		48 s		
155600	208562		17 13 57.2	−32 14 29	+0.001	0.00	7.93	0.15		B8				
155650	185175		17 14 01.8	−27 47 47	+0.001	0.00	6.6	0.7	3.0	F2 V		32 s		
156389	30306		17 14 06.7	+56 08 01	−0.001	+0.01	7.77	0.45	3.3	dF4	−3	77 s	10425	m
155068	253831		17 14 07.3	−65 57 47	+0.002	−0.10	7.2	1.5	−0.1	K2 III		180 s		
155802	141567		17 14 07.9	−8 24 13	−0.007	+0.08	8.00	1.1	0.2	K0 III		87 mx	10404	m
155979	102674		17 14 09.8	+19 10 45	−0.003	+0.02	8.0	0.5	4.0	F8 V		62 s		
155938	102669		17 14 10.3	+12 02 22	−0.003	−0.03	7.1	1.1	0.2	K0 III		240 s		
155341	244539		17 14 13.0	−56 53 18	−0.001	0.00	6.0	1.8	−0.3	K5 III		120 s		
155685	185178		17 14 14.0	−26 59 04	0.000	−0.07	6.9	0.9	2.6	F0 V		34 s		
155190	253841		17 14 16.1	−62 44 05	0.000	−0.03	7.3	0.0	0.0	B8.5 V		290 s		
155633	208568		17 14 17.0	−35 07 45	−0.002	−0.07	8.0	0.3	3.0	F2 V		99 s		
155980	102676		17 14 17.5	+17 05 04	0.000	−0.03	7.9	0.1	0.6	A0 V		290 s		
155967	102675		17 14 19.8	+14 33 08	0.000	0.00	7.43	0.44	3.7	F6 V	−16	56 s		
155151	253839		17 14 22.0	−64 37 53	−0.002	−0.04	7.6	1.1	0.2	K0 III		300 s		
155864	141569		17 14 23.0	−4 09 42	−0.001	−0.01	7.8	0.1	0.6	A0 V		270 s		
157602	8745		17 14 24.5	+77 21 04	−0.006	+0.01	7.9	0.9	3.2	G5 IV		89 s		
155552	227732		17 14 25.4	−43 38 13	+0.001	−0.02	7.9	1.1	0.2	K0 III		180 s		
155603	208569		17 14 27.5	−39 46 01	0.000	−0.01	6.60	2.21	−8.0	G5 Ia		1900 s		m
156281	46581		17 14 29.4	+45 11 13	−0.001	−0.01	7.9	1.6	−0.5	M4 III		480 s		
156093	84948		17 14 33.1	+26 03 07	−0.001	−0.03	6.9	1.1	−0.2	K3 III		260 s		
156649	17402		17 14 33.8	+63 21 10	−0.002	−0.03	7.40	1.4	−0.3	K5 III	+16	350 s		
156014	102680	64 α¹ Her	17 14 38.8	+14 23 25	−0.001	+0.03	3.19	1.44		M5 II	−33	67 mn	10418	Rasalgethi, m,v
156015	102681	64 α² Her	17 14 39.1	+14 23 23	−0.001	+0.04	5.41			G5 III	−37		10418	m
155243	253845		17 14 42.0	−63 51 54	−0.002	+0.05	7.9	1.1	−0.1	K2 III		400 s		
156282	46582		17 14 43.9	+42 12 13	+0.002	−0.08	8.0	0.5	4.0	F8 V	−1	62 s		
155846	185190		17 14 53.1	−20 58 11	+0.001	−0.02	8.0	1.0	−0.1	K2 III		410 s		
155589	227740		17 14 56.8	−48 33 37	0.000	−0.01	8.0	1.0	0.2	K0 III		360 s		
156111	102688		17 14 56.9	+19 41 00	−0.006	−0.16	7.4	0.9	3.2	G5 IV		54 mx		
155705	208577		17 14 59.2	−38 56 48	+0.002	−0.04	7.6	1.0	3.2	G5 IV		56 s		
155754	208580		17 14 59.5	−33 44 44	+0.002	0.00	7.92	−0.02		B5				
156164	84951	65 δ Her	17 15 01.8	+24 50 21	−0.002	−0.16	3.14	0.08	0.9	A3 IV	−41	28 s	10424	m
156283	65890	67 π Her	17 15 02.6	+36 48 33	−0.002	0.00	3.16	1.44	−2.3	K3 II	−26	120 s		
155634	227743		17 15 05.5	−45 52 43	−0.002	−0.05	7.8	−0.2	3.0	F2 V		89 s		
156129	102689		17 15 07.7	+17 48 05	−0.001	0.00	7.4	1.4	−0.3	K5 III		350 s		
155849	185195		17 15 13.3	−26 31 50	+0.004	+0.03	5.30	1.1	0.2	K0 III		100 s		
156946	17408		17 15 14.2	+67 23 18	+0.001	−0.03	7.5	1.4	−0.3	K5 III		360 s		
155823	208587		17 15 15.1	−30 32 11	−0.002	+0.03	7.8	2.2	−0.5	M4 III		200 s		
155723	227748		17 15 15.7	−41 00 13	−0.002	−0.04	7.9	0.5	0.2	K0 III		340 s		
154343	257474		17 15 18.0	−78 23 55	+0.008	+0.01	7.2	0.4	3.0	F2 V		69 s		
155806	208585		17 15 19.1	−33 32 54	+0.001	0.00	5.53	−0.01		O8 e	+5			
155970	160402		17 15 20.1	−14 35 02	−0.001	0.00	6.20	1.1	0.0	K1 III		170 s	10419	m
155886	185198	36 Oph	17 15 20.7	−26 36 04	−0.037	−1.13	4.31	0.86	6.4	K0 V	−1	5.4 t	10417	m
155775	208582		17 15 22.1	−38 12 47	0.000	−0.01	6.72	−0.01		O9.5 k				
156296	65895		17 15 22.1	+32 34 01	−0.002	0.00	8.0	1.4	−0.3	K5 III		450 s		
155852	208593		17 15 31.0	−33 06 20	−0.002	−0.06	7.9	1.7	−0.5	M2 III		200 mx		V727 Sco, v
156206	102694		17 15 34.2	+17 49 54	+0.002	+0.01	7.6	1.1	−0.1	K2 III		340 s		
156023	160405		17 15 34.5	−10 17 51	0.000	+0.03	7.31	0.46	3.4	F5 V		59 s	10421	m
155826	208591		17 15 35.7	−38 35 38	−0.016	−0.41	5.96	0.58	3.0	G3 IV	−51	27 mx		m
155981	160404		17 15 37.4	−17 13 38	+0.002	+0.02	7.8	1.1	−0.1	K2 III		380 s		
156361	65900		17 15 37.8	+34 42 27	−0.001	−0.01	6.9	1.1	−0.1	K2 III		250 s		
155854	208598		17 15 39.8	−34 21 08	+0.003	0.00	7.8	−0.4	2.6	F0 V		110 s		
156034	141578		17 15 40.9	−9 48 32	+0.001	−0.02	6.93	0.83	3.4	F5 V		30 s	10423	m
156284	84955		17 15 41.5	+23 44 34	−0.002	+0.02	5.96	1.31	−0.1	K2 III	−42	130 s		
156166	122217		17 15 43.2	+6 22 49	0.000	+0.03	8.0	0.1	1.4	A2 V		210 s		
155478	253850		17 15 43.4	−60 47 55	0.000	−0.01	7.8	0.4	1.4	A2 V		110 s		
155670	244562		17 15 44.4	−51 04 59	0.000	−0.01	7.4	0.1	0.0	B8.5 V		230 s		
156452	46589		17 15 49.5	+41 38 55	−0.002	+0.01	7.7	1.4	−0.3	K5 III		390 s		
155889	208604		17 15 50.5	−33 44 13	0.000	−0.01	6.55	−0.01	−4.8	O9 V		1300 s		m
155940	208606		17 15 51.3	−30 12 37	0.000	−0.04	6.2	−0.1	0.6	A0 V		130 s		
155778	227756		17 15 54.3	−46 20 13	+0.001	−0.01	7.9	0.0	0.0	B8.5 V		330 s		
148451	258754		17 15 57.8	−87 33 58	−0.198	−0.14	6.57	0.91	0.3	G5 III		69 mx		
156362	84958		17 15 58.6	+27 08 02	−0.004	−0.09	6.7	1.1	−0.1	K2 III		120 mx		
155708	244568		17 15 58.7	−52 18 30	0.000	−0.04	7.5	0.8	0.2	K0 III		280 s		
155672	244566		17 15 58.9	−53 56 32	0.000	0.00	8.0	0.5	−0.3	K5 III		450 s		
156058	160410		17 15 59.4	−16 11 46	−0.003	−0.04	7.6	0.5	3.4	F5 V		69 s		
156890	17410		17 16 04.7	+60 42 49	0.000	+0.03	6.80	0.4	2.6	F0 V	−22	69 s	10448	m
155912	208607		17 16 10.5	−38 20 11	−0.003	0.00	8.0	0.3	0.4	B9.5 V		200 s		
157370	8748		17 16 12.9	+71 47 33	−0.001	0.00	6.8	1.1	−0.1	K2 III	−4	240 s		
156026	185213		17 16 13.2	−26 32 46	−0.036	−1.12	6.34	1.16	7.7	K5 V	−6	5.4 t		
156208	122224		17 16 14.0	+2 11 10	0.000	−0.02	6.17	0.22		A0	−7			
156115	160413		17 16 15.6	−15 13 26	−0.001	0.00	6.59	1.82	−0.4	gM0	−8	150 s		
155088	257479		17 16 17.2	−72 33 19	0.000	+0.01	8.00	1.1	−0.1	K2 III		420 s		m
155896	227768		17 16 17.3	−42 20 20	0.000	0.00	6.75	0.13	−0.6	B7 V e		280 s		m

HD	SAO	Star Name	α 2000	δ 2000	μ(α)	μ(δ)	V	B-V	M$_v$	Spec	RV	d(pc)	ADS	Notes
			h m s	° ′ ″	s	″								
155811	227760		17 16 20.1	−49 33 43	−0.001	−0.08	7.5	0.5	2.6	F0 V		71 s		
156341	102701		17 16 20.2	+16 39 39	−0.001	0.00	7.5	0.1	0.6	A0 V	−14	250 s		
155974	208610		17 16 21.3	−35 44 58	−0.010	−0.32	6.12	0.48	3.6	G0 IV-V		30 mx		m
156079	185220		17 16 23.6	−21 51 28	−0.006	−0.17	7.5	0.8	4.4	G0 V		30 s		
156061	185219		17 16 27.5	−25 18 19	+0.001	−0.01	7.0	0.9	0.2	K0 III		230 s		
156286	122228		17 16 28.8	+5 07 41	+0.001	0.00	8.0	0.1	1.4	A2 V		210 s		
156947	17414		17 16 29.2	+60 40 14	−0.007	+0.01	6.32	1.09	0.2	K0 III	+17	150 s		VW Dra, v
156062	185218		17 16 29.6	−27 33 57	−0.005	−0.14	7.5	0.7	4.4	G0 V		35 s		
156377	102705		17 16 31.0	+18 01 09	0.000	−0.03	6.9	0.0	0.4	B9.5 V	+2	200 s		
156247	122226		17 16 31.6	+1 12 38	0.000	−0.01	5.90	0.06	−1.1	B5 V	−11	200 s	10428	U Oph, m,v
156004	208612		17 16 33.3	−32 20 20	−0.004	−0.02	7.81	0.01	−2.6	B4 III		160 mx		
155914	227771		17 16 34.5	−44 45 46	0.000	−0.04	7.9	1.0	0.2	K0 III		340 s		
154972	257478		17 16 35.6	−74 31 59	−0.003	−0.06	6.25	−0.01	0.6	A0 V		92 mx		
156772	30322		17 16 35.9	+52 47 04	−0.001	−0.02	7.5	1.1	0.2	K0 III		290 s		
156266	141586	41 Oph	17 16 36.5	−0 26 43	−0.002	−0.07	4.73	1.14	−0.1	K2 III	−2	93 s	10429	m
156728	30320		17 16 39.0	+50 36 23	−0.002	+0.11	8.0	0.9	3.2	G5 IV		93 s		
156227	141585		17 16 42.6	−6 14 44	−0.001	−0.03	6.09	1.10		K0				
156169	160418		17 16 44.0	−17 54 54	0.000	−0.04	7.4	0.9	3.2	G5 IV		68 s		
156753	46605		17 16 48.4	+49 41 28	+0.002	0.00	7.6	1.1	−0.1	K2 III		340 s		
156632	65910		17 16 53.8	+39 28 06	−0.004	+0.14	7.4	0.5	3.4	F5 V		63 s		
156096	208621		17 16 54.3	−30 21 03	+0.001	+0.01	7.2	0.4	2.6	F0 V		73 s		
183030	3020	λ UMi	17 16 56.0	+89 02 15	−0.104	−0.01	6.38	1.57	−0.5	M1 III	+2	240 s		m
156430	102708		17 16 58.5	+15 49 15	−0.001	−0.01	7.4	1.1	0.2	K0 III		280 s		
156228	160424		17 16 59.9	−13 06 03	−0.001	+0.01	7.5	0.8	3.2	G5 IV		72 s		
156098	208626		17 17 03.4	−32 39 46	−0.008	−0.05	5.5	−0.1	3.4	F5 V	−36	27 s		
156029	208619		17 17 04.9	−39 25 44	+0.001	0.00	7.2	1.1	3.2	G5 IV		42 s		
155985	227778		17 17 05.3	−44 46 43	0.000	−0.01	6.47	0.25		B0.5 ne				
−−	17419		17 17 05.4	+62 37 13	−0.014	−0.03	7.8	1.0						
156679	46603		17 17 07.7	+39 58 07	0.000	+0.02	7.5	1.1	0.2	K0 III		290 s		
156456	102711		17 17 10.0	+13 08 02	0.000	−0.09	6.7	1.1	−0.1	K2 III		180 mx		
156547	84977		17 17 10.8	+25 52 26	+0.003	+0.02	7.1	0.0	0.4	B9.5 V	0	210 s		
156796	46611		17 17 12.6	+46 48 46	0.000	+0.02	8.0	0.9	3.2	G5 IV		93 s		
156484	102713		17 17 15.5	+13 52 08	−0.001	−0.01	7.9	0.4	3.0	F2 V		94 s		
156152	208633		17 17 19.1	−32 03 58	+0.009	−0.11	7.9	0.7	4.4	G0 V		42 s		
156633	65913	68 Her	17 17 19.4	+33 06 00	−0.001	0.00	4.82	−0.17	−2.9	B3 III	−21	350 s	10449	u Her, m,v
156184	208636		17 17 20.4	−30 09 53	−0.001	−0.01	6.95	0.82	3.2	G5 IV		53 s		m
155555	253856		17 17 25.5	−66 57 03	−0.002	−0.13	6.67	0.80		G5				m
156651	65915		17 17 26.3	+31 31 06	0.000	−0.01	7.2	0.0	−1.0	B5.5 V	−15	440 s		
156212	185228		17 17 27.3	−27 46 03	−0.001	−0.02	7.92	0.53	−5.5	O9.5 III		1600 s		
156984	30326		17 17 29.2	+55 03 22	−0.005	−0.04	7.6	1.1	0.2	K0 III		300 mx		
156485	122245		17 17 33.4	+6 45 43	+0.001	+0.02	7.9	1.1	0.2	K0 III		340 s		
156652	84983		17 17 34.5	+28 54 47	0.000	+0.01	7.10	1.6	−0.5	M2 III	−38	330 s	10451	m
156593	84982		17 17 35.7	+23 05 27	0.000	0.00	6.5	1.1	−0.1	K2 III	−15	210 s		
156458	122244		17 17 36.8	+1 44 33	−0.002	−0.01	6.7	0.4	2.6	F0 V		67 s		
157222	17422		17 17 37.0	+64 02 01	−0.008	+0.06	7.2	0.4	3.0	F2 V		70 s		
156252	185233		17 17 39.3	−26 37 46	−0.001	−0.04	7.10	0.1	0.6	A0 V		190 s	10436	m
156729	65921	69 Her	17 17 40.1	+37 17 29	−0.003	+0.06	4.65	0.05	1.4	A2 V	−10	45 s		
156201	208643		17 17 45.2	−35 13 24	−0.001	+0.03	7.90	0.65		B0				
156757	65925		17 17 48.6	+36 05 38	0.000	0.00	7.6	0.2	2.1	A5 V		130 s		
156985	30328		17 17 50.3	+52 26 49	+0.003	−0.20	8.0	1.1	−0.1	K2 III		110 mx		
156696	84990		17 17 55.2	+28 54 22	−0.002	−0.03	7.3	0.4	2.6	F0 V		89 s		
155797	253861		17 17 55.5	−62 42 10	+0.001	−0.05	7.6	1.4	−0.3	K5 III		380 s		
156614	102722		17 17 58.3	+17 07 27	+0.001	−0.02	7.1	1.4	−0.3	K5 III		300 s		
156172	227800		17 17 59.5	−42 03 37	−0.001	0.00	8.0	0.1		B0.5 II				
156070	227793		17 17 59.7	−49 27 09	+0.004	−0.02	7.54	−0.01	−3.6	B2 III		180 mx		
156349	185238	39 o Oph	17 18 00.5	−24 17 13	−0.004	−0.01	5.20	1.10	0.0	K1 III	−29	110 s	10442	m
156157	227799		17 18 01.2	−43 34 40	−0.001	−0.03	7.02	−0.04	0.4	B9.5 V		210 s		
157009	30330		17 18 01.7	+52 20 14	0.000	−0.04	8.0	1.1	0.2	K0 III		360 s		
156653	102724		17 18 04.8	+17 19 04	+0.001	−0.01	5.90	0.01	1.2	A1 V	−2	87 s		
156539	122251		17 18 04.8	+3 08 50	−0.001	+0.05	6.8	0.4	3.0	F2 V	+3	58 s		
156189	227801		17 18 05.3	−42 53 55	0.000	−0.05	8.0	−0.6	1.4	A2 V		210 s		
156232	208650		17 18 05.4	−38 09 50	0.000	−0.01	6.7	0.3	0.6	A0 V		100 s		
156365	185239		17 18 06.9	−24 04 22	+0.007	−0.07	6.7	1.1	3.2	G5 IV	−15	31 s		
156216	208651		17 18 11.7	−39 41 27	−0.001	−0.02	7.1	1.3	−0.1	K2 III		240 s		
156549	141599		17 18 14.6	−0 45 55	−0.002	−0.05	8.0	0.5	4.0	F8 V		62 s		
156461	160439		17 18 17.8	−15 47 55	+0.002	−0.05	7.1	1.1	0.2	K0 III	+19	240 s		
156462	160440		17 18 19.0	−16 18 43	0.000	+0.01	6.43	1.66		K5				
156325	208657		17 18 20.2	−32 33 13	−0.001	−0.02	6.36	0.15	−1.3	B6 IV	−14	250 s		m
156891	65930		17 18 23.1	+38 48 41	−0.002	+0.08	5.94	1.00	0.2	K0 III	−38	140 s		
156123	244608		17 18 23.7	−50 12 37	−0.001	−0.01	7.1	0.2	0.6	A0 V		140 s		
156774	84995		17 18 24.4	+26 56 12	−0.001	−0.03	7.51	1.26	−0.1	K2 III		290 s		
156086	244604		17 18 25.5	−52 59 50	−0.006	−0.05	8.0	0.5	−0.1	K2 III		180 mx		
156775	84996		17 18 29.6	+25 48 27	−0.002	+0.05	6.7	1.1	0.0	K1 III		220 s		
155355	−−		17 18 30.3	−73 54 22			7.9	0.1	0.6	A0 V		290 s		
156290	208659		17 18 32.9	−39 32 31	−0.002	−0.01	7.9	2.0	−0.5	M2 III		290 s		

416

HD	SAO	Star Name	α 2000	δ 2000	μ(α)	μ(δ)	V	B−V	M_v	Spec	RV	d(pc)	ADS	Notes
			$17^h 18^m 36\fs8$	$+10°51'53''$	$0\fs000$	$-0\farcs09$								
156681	102725		17 18 36.8	+10 51 53	0.000	−0.09	5.03	1.55	−1.3	K4 II−III	+40	160 s		
156519	160445		17 18 39.0	−16 01 46	+0.001	+0.01	7.4	0.1	1.4	A2 V		160 s		
156567	160448		17 18 40.2	−10 13 09	0.000	−0.01	7.9	0.1	0.6	A0 V		280 s		
156236	227807		17 18 41.8	−46 47 57	0.000	−0.01	7.22	0.62		F2				
156382	208666		17 18 43.8	−33 26 30	−0.001	−0.04	7.7	0.6	2.6	F0 V		67 s		
156049	244605		17 18 45.1	−57 26 20	+0.001	−0.02	8.0	−0.1	0.4	B9.5 V		330 s		
156292	227814		17 18 45.5	−42 53 32	0.000	−0.02	7.51	0.28	−4.1	B0 V		1100 s		
156293	227813		17 18 47.8	−44 07 47	0.000	−0.02	5.76	−0.05		B9				
156910	65932		17 18 48.2	+32 51 54	−0.001	−0.09	7.9	1.1	−0.1	K2 III		200 mx		
156874	85001		17 18 48.3	+28 49 23	+0.003	0.00	5.65	0.98	0.2	K0 III	−14	120 s		
156635	141602		17 18 50.2	−2 48 37	−0.009	−0.06	6.7	0.5	4.0	F8 V		34 s		
157409	17426		17 18 51.2	+62 33 00	−0.001	+0.01	7.3	1.1	−0.1	K2 III		310 s		
156219	227809		17 18 51.6	−49 20 19	0.000	0.00	7.50	1.1	−0.1	K2 III		330 s		m
155951	253867		17 18 51.6	−62 52 28	−0.007	−0.02	6.9	1.1	0.2	K0 III		220 s		
160034	2869		17 18 52.1	+83 20 13	+0.034	+0.11	7.5	0.5	4.0	F8 V		51 s		
156697	122270		17 18 52.6	+6 05 07	+0.001	+0.01	6.51	0.39		F0 n	−25			
157462	17427		17 18 54.3	+63 43 23	−0.003	+0.03	7.3	0.1	0.6	A0 V		210 s		
156384	208670		17 18 57.0	−34 59 24	+0.095	−0.17	5.91	1.04	7.0	K2 V comp	0	7.1 t		m
157774	8763		17 18 57.0	+70 47 15	−0.005	+0.02	7.0	0.1	0.6	A0 V		130 mx		
156586	160451		17 18 57.6	−15 13 13	+0.001	−0.10	7.8	0.5	4.0	F8 V		58 s		
156425	208673		17 18 58.0	−32 57 55	0.000	0.00	8.0	1.7	−0.3	K5 III		340 s		
156465	208676		17 19 00.3	−31 06 39	+0.005	+0.03	8.0	−0.3	0.0	B8.5 V		130 mx		
155453	257482		17 19 01.0	−73 31 18	+0.014	−0.06	8.0	1.1	−0.1	K2 III		150 mx		
156396	208671		17 19 01.8	−37 27 49	+0.027	−0.01	7.9	1.6	−0.5	M2 III		43 mx		
156275	227815		17 19 02.0	−48 41 49	−0.001	−0.03	7.87	0.01	0.4	B9.5 V		310 s		
156274	227816		17 19 03.0	−46 38 02	+0.095	+0.22	5.48	0.80	6.1	G8 V	+19	7.6 t		m
156636	160456		17 19 09.4	−11 59 09	−0.001	−0.02	7.9	0.1	0.6	A0 V		290 s		
156091	244613		17 19 12.1	−59 41 40	−0.001	−0.01	6.00	1.1	−0.1	K2 III		170 s		m
158731	8782		17 19 14.2	+79 18 21	−0.004	−0.02	7.3	0.2	2.1	A5 V		110 s		
156442	208680		17 19 15.4	−37 21 33	0.000	−0.02	7.9	1.6	−0.3	K5 III		380 s		
156506	208685		17 19 16.4	−31 21 42	−0.001	−0.04	7.0	−0.1	1.4	A2 V		130 s		
156822	102738		17 19 17.0	+11 10 21	0.000	−0.01	7.7	0.1	0.6	A0 V		270 s		
156508	—		17 19 18.1	−32 23 36			6.14	0.23	0.6	A0 V		91 s		
—	17434		17 19 19.1	+67 16 53	−0.005	−0.04	7.9	1.1						
156715	141606		17 19 21.1	−2 44 56	0.000	0.00	7.3	1.1	0.2	K0 III		260 s		
156987	85006		17 19 23.7	+28 01 34	−0.001	0.00	7.30	0.1	0.6	A0 V	−11	220 s		m
156966	85005		17 19 23.8	+27 17 00	0.000	−0.03	6.70	1.65	−0.5	M1 III	+59	250 s		
156468	208681		17 19 23.9	−38 00 14	+0.001	−0.01	7.76	0.20	−3.5	B1 V e		1300 s		
156398	227821		17 19 24.3	−44 13 23	−0.001	−0.01	6.65	0.20		A0				m
156637	160458		17 19 29.5	−18 57 18	−0.002	−0.02	7.9	1.4	−0.3	K5 III		370 s		
156385	227822		17 19 29.7	−45 38 24	+0.001	−0.01	6.92	0.05		WC7 p				
156331	244631		17 19 30.2	−50 03 47	−0.001	−0.01	7.20	0.4	2.6	F0 V		83 s		m
156241	244623		17 19 30.3	−55 07 00	0.000	−0.03	7.5	−0.1	1.4	A2 V		160 s		
156911	102749		17 19 33.8	+19 21 46	−0.001	−0.04	7.7	1.4	−0.3	K5 III		390 s		
158996	2859		17 19 36.7	+80 08 11	+0.005	0.00	5.72	1.50	−0.1	K2 III	−7	93 s		
156276	244629		17 19 38.5	−53 49 03	+0.001	+0.01	7.8	0.1	0.4	B9.5 V		240 s		
156572	208691		17 19 43.1	−31 41 28	−0.004	−0.03	7.9	1.6	0.2	K0 III		160 s		
156860	122279		17 19 46.2	+2 08 21	+0.001	+0.02	6.8	1.6	−0.5	M4 III	−26	290 s		
156493	227827		17 19 46.7	−42 01 51	0.000	−0.04	7.2	1.5	−0.1	K2 III		190 s		
155497	257483		17 19 47.0	−74 36 22	+0.003	−0.04	7.2	0.4	2.6	F0 V		83 s		
156825	141607		17 19 49.9	−4 02 58	0.000	−0.03	7.8	0.5	3.4	F5 V		76 s		
157224	46638		17 19 50.8	+45 18 31	−0.002	+0.09	6.7	0.4	2.6	F0 V		65 s		
156411	227825		17 19 51.4	−48 32 57	−0.001	−0.21	6.67	0.62		G0		16 mn		
156717	160462		17 19 53.1	−17 45 23	−0.001	−0.02	6.30	0.1	0.6	A0 V		130 s	10465	m
156826	141611		17 19 59.3	−5 55 02	+0.002	−0.18	6.32	0.85	5.2	G5 V		13 s		
156926	122284		17 19 59.9	+7 58 09	0.000	−0.01	8.0	0.1	0.6	A0 V		300 s		
156802	141608		17 19 59.9	−8 01 23	−0.001	−0.22	7.97	0.68	4.7	G2 V	−89	42 s		m
156989	102751		17 20 00.4	+15 53 04	−0.001	−0.01	7.4	1.1	0.2	K0 III		280 s		
156720	160464		17 20 01.5	−19 58 51	+0.003	−0.02	8.00	0.4	3.0	F2 V		96 s	10466	m
157891	17443		17 20 02.7	+69 45 13	−0.001	−0.01	7.1	1.1	−0.1	K2 III		280 s		
157440	30345		17 20 05.3	+55 56 08	−0.004	+0.02	7.9	0.1	1.4	A2 V		100 mx		
157087	85016		17 20 09.7	+25 32 15	+0.001	−0.02	5.38	0.03	0.0	A3 III	−8	120 s		
156968	122289		17 20 11.4	+9 27 39	−0.002	−0.31	7.97	0.60	4.4	G0 V	−12	37 mx		
155875	253870		17 20 12.6	−70 02 43	−0.007	−0.20	6.53	0.60	3.9	G2 IV−V		34 s		m
157049	102757		17 20 18.7	+18 03 26	0.000	−0.06	5.00	1.62	−0.5	gM2	−46	120 s		
157325	46645	74 Her	17 20 20.9	+46 14 27	−0.004	+0.04	5.59	1.57	−0.4	M0 III	−57	160 s		
156333	244636		17 20 23.0	−57 30 39	−0.002	−0.07	7.5	0.3	3.0	F2 V		64 s		
157050	102758		17 20 24.8	+16 05 57	0.000	−0.01	7.9	1.1	0.2	K0 III		340 s		
156357	244637		17 20 26.6	−57 14 52	+0.002	+0.01	7.2	1.8	−0.3	K5 III		210 s		
156721	185278		17 20 27.5	−29 21 54	0.000	−0.03	6.8	0.9	1.4	A2 V		39 s		
157012	122291		17 20 27.5	+9 11 30	−0.001	−0.03	7.9	1.1	−0.1	K2 III		390 s		
157213	65960		17 20 27.7	+35 22 16	0.000	0.00	7.9	0.1	0.6	A0 V		280 s		
157067	102759		17 20 28.5	+15 59 52	−0.002	0.00	7.9	0.9	3.2	G5 IV		88 s		
157326	46647		17 20 29.8	+44 11 31	−0.003	+0.04	6.9	1.1	0.2	K0 III		220 s		
156780	185282		17 20 30.6	−26 32 59	−0.001	−0.01	7.0	0.9	0.2	K0 III		230 s		

HD	SAO	Star Name	α 2000	δ 2000	μ(α)	μ(δ)	V	B−V	M_v	Spec	RV	d(pc)	ADS	Notes
157411	46652		17ʰ20ᵐ31ˢ.8	+49°00′45″	−0ˢ.002	−0″.03	7.7	1.1	−0.1	K2 III		370 s		
157373	46651		17 20 33.4	+48 11 18	+0.018	−0.02	6.43	0.43	3.0	F2 V	+31	46 s		
156846	160474		17 20 34.0	−19 19 57	−0.011	−0.10	6.52	0.58	2.8	G0 IV		54 mx	10476	m
157068	102760		17 20 36.0	+15 09 34	+0.001	−0.03	7.3	0.5	3.4	F5 V		60 s		
156335	244638		17 20 37.0	−59 26 26	0.000	−0.02	7.80	0.1	0.6	A0 V		280 s		m
156575	227837		17 20 37.3	−46 02 47	+0.001	−0.01	7.33	0.21	−4.0	B1.5 III		880 s		
156641	227842		17 20 38.0	−40 34 01	0.000	−0.03	8.0	1.4	0.2	K0 III		90 s		
157051	122294		17 20 39.3	+9 25 26	0.000	+0.04	7.9	1.1	0.2	K0 III		340 s		
157214	65963	72 Her	17 20 39.4	+32 28 04	+0.010	−1.04	5.39	0.62	4.4	G0 V	−78	14 ts	10488	m
157804	17444		17 20 41.8	+65 38 21	−0.001	−0.03	7.7	0.4	3.0	F2 V		86 s		
156688	208703		17 20 42.8	−38 01 13	+0.001	+0.01	7.19	0.09	−3.6	B2 III k		1100 s		
157255	65965		17 20 45.4	+32 40 29	+0.001	+0.02	7.03	0.03		A0	−28			
157121	102762		17 20 45.7	+17 02 24	+0.001	−0.01	7.7	1.1	−0.1	K2 III		360 s		
157151	85020		17 20 46.3	+21 31 11	0.000	+0.01	6.9	0.0	0.4	B9.5 V	−8	200 s		
156928	160479	53 ν Ser	17 20 49.4	−12 50 48	+0.003	0.00	4.33	0.03	1.2	A1 V	+5	42 s	10481	m
156623	227843		17 20 50.4	−45 25 13	−0.001	−0.02	7.26	0.09		A0				
156953	160481		17 20 52.3	−12 25 22	−0.001	+0.02	8.0	1.1	−0.1	K2 III		410 s		
156971	160482		17 20 52.6	−10 41 46	+0.004	−0.01	6.46	0.33		F0				m
157198	85021	70 Her	17 20 54.0	+24 29 58	−0.002	0.00	5.12	−0.03	1.2	A1 V	−17	61 s		m
157892	17446		17 20 56.6	+66 33 45	−0.003	+0.06	7.0	1.1	−0.1	K2 III		260 s		
156782	208711		17 20 57.1	−32 49 42	+0.002	−0.03	7.4	1.8	−0.3	K5 III		230 s		
156897	185296	40 ξ Oph	17 21 00.0	−21 06 46	+0.017	−0.21	4.39	0.39	3.0	F2 V	−9	19 ts		m
156705	227851		17 21 01.9	−41 52 53	−0.001	−0.07	7.9	1.4	−0.5	M2 III		200 mx		
157088	122300		17 21 02.4	+4 37 08	−0.002	−0.03	7.3	1.1	0.2	K0 III		260 s		
156882	185295		17 21 05.2	−27 25 06	−0.002	−0.03	7.5	0.9	3.2	G5 IV		59 s		
156662	227847		17 21 05.8	−45 58 57	0.000	−0.02	7.83	0.17	−1.0	B5.5 V		450 s		
157089	122301		17 21 06.9	+1 26 35	−0.010	+0.27	6.95	0.60	4.2	F9 V	−162	33 ts		
158063	17451		17 21 10.2	+69 48 02	−0.004	+0.03	7.5	0.5	4.0	F8 V		49 s		
156999	160489		17 21 11.4	−12 03 27	−0.004	−0.11	7.6	0.5	4.0	F8 V		51 s		
156415	244645		17 21 13.2	−60 00 57	0.000	+0.01	7.8	0.1	0.6	A0 V		220 s		
157481	46660		17 21 13.2	+46 09 01	−0.007	0.00	7.6	1.1	0.2	K0 III		200 mx		
156310	253878		17 21 13.4	−63 34 55	0.000	−0.01	7.0	1.1	−0.1	K2 III		260 s		
157105	122303		17 21 14.1	+1 33 23	0.000	0.00	8.04	0.53		F8			10486	m
157329	65971		17 21 14.5	+31 15 26	−0.002	+0.02	6.9	0.5	4.0	F8 V		38 s		
157199	102765		17 21 20.8	+13 37 49	0.000	+0.01	7.9	1.1	−0.1	K2 III		390 s		
157029	160491		17 21 21.6	−11 56 47	−0.002	+0.01	8.0	1.1	0.2	K0 III		360 s		
156955	185305		17 21 25.6	−26 30 05	−0.002	−0.03	8.0	−0.4	0.6	A0 V		300 s		
156883	208721		17 21 25.7	−31 34 57	+0.002	−0.05	7.5	0.8	1.4	A2 V		61 s		
157463	46661		17 21 29.2	+42 16 10	0.000	+0.02	7.6	0.9	0.3	G5 III	+8	290 s		
157358	85028		17 21 31.0	+28 45 29	0.000	+0.01	6.35	0.70	4.0	F8 V	−6	23 s		m
157257	102770		17 21 33.2	+16 43 51	−0.001	−0.03	6.5	1.6	−0.5	M2 III	+39	250 s		
155245	257484		17 21 35.4	−79 03 52	+0.009	+0.03	7.7	0.2	2.1	A5 V		130 s		
156675	244660		17 21 40.1	−51 58 33	−0.001	0.00	7.9	0.0	0.4	B9.5 V		290 s		
156992	185313		17 21 41.3	−24 54 22	−0.002	−0.04	6.6	0.9	0.2	K0 III	−12	190 s		
−−	46663		17 21 41.4	+40 05 31	+0.001	0.00	8.0	1.6						
157330	85031		17 21 42.2	+22 55 16	−0.001	0.00	7.4	1.6		M5 III	−41			RS Her, v
157482	46664		17 21 43.4	+39 58 28	0.000	−0.07	5.51	0.68	4.0	F8 V	+3	22 mn		
157681	30354		17 21 45.2	+53 25 13	+0.002	0.00	5.67	1.47	−0.3	K5 III	−8	160 s		
157359	85035		17 21 49.5	+23 41 10	0.000	0.00	7.9	0.1	0.6	A0 V		280 s		
157258	122312		17 21 50.3	+9 44 10	0.000	−0.01	7.4	1.1	0.2	K0 III		280 s		
157072	185318		17 21 51.4	−20 34 11	−0.001	−0.01	7.10	1.4		K4 IV	+1			
156747	227863		17 21 52.9	−49 46 07	−0.003	0.00	8.0	−0.3	2.6	F0 V		120 s		
157297	102773		17 21 56.9	+11 40 21	0.000	0.00	7.1	0.1	0.6	A0 V	+10	200 s		
156884	227871		17 21 57.0	−42 10 53	0.000	+0.02	7.6	0.0	1.4	A2 V		180 s		
156277	253882	ζ Aps	17 21 59.3	−67 46 13	−0.005	0.00	4.78	1.21	−0.3	gK5	+13	100 s		
157056	185320	42 θ Oph	17 22 00.4	−24 59 58	0.000	−0.02	3.27	−0.22	−3.0	B2 IV	−4	180 s		
157375	85036		17 22 01.0	+21 09 17	−0.003	+0.04	7.0	1.1	−0.1	K2 III		270 s		
157261	122316		17 22 04.0	+5 00 03	+0.001	−0.07	6.70	0.9	3.2	G5 IV		50 s	10498	m
156190	257491	ι Aps	17 22 05.8	−70 07 24	+0.001	−0.01	5.41	−0.04		B9	−4			m
157073	185322		17 22 06.0	−23 34 25	−0.002	0.00	7.8	1.3	0.2	K0 III		230 s		
157016	208730		17 22 09.0	−30 30 06	+0.001	−0.02	7.4	0.1	0.4	B9.5 V		200 s		
157279	122319		17 22 09.2	+5 12 50	0.000	−0.01	8.0	0.9	3.2	G5 IV		93 s		
155918	257489		17 22 12.2	−75 20 49	−0.253	−0.20	7.01	0.59	4.7	G2 V		18 mx		
156936	227874		17 22 12.9	−41 07 17	+0.001	−0.03	7.1	1.0	−0.3	K5 III		310 s		
157035	208734		17 22 16.3	−31 35 07	+0.005	−0.01	7.6	0.1	1.4	A2 V		130 mx		
157094	185328		17 22 22.2	−26 12 56	−0.002	−0.07	7.5	2.2	−0.1	K2 III		80 s		
157170	160500		17 22 22.5	−17 20 24	+0.001	0.00	7.95	0.14	0.0	A0 IV		310 s		
156853	227872		17 22 23.5	−49 45 44	0.000	−0.02	7.6	0.0	0.0	B8.5 V		330 s		
157075	208738		17 22 24.4	−30 12 18	+0.003	−0.13	8.00	0.5	3.4	F5 V		82 s		m
157393	102777		17 22 24.8	+15 08 14	−0.004	+0.04	7.8	0.5	4.0	F8 V		57 s		
157616	46675		17 22 25.0	+40 41 09	−0.001	+0.03	7.8	1.1	−0.1	K2 III		380 s		
157360	102775		17 22 27.5	+10 11 47	0.000	+0.03	7.2	1.1	−0.1	K2 III		290 s		
157466	85045		17 22 27.6	+24 52 45	+0.006	−0.17	6.8	0.5	3.4	F5 V	+28	48 s		
156766	244672		17 22 29.7	−54 17 08	+0.008	−0.10	7.0	0.2	3.4	F5 V		53 s		
156244	257492		17 22 30.7	−70 26 32	−0.014	−0.05	7.2	1.6	0.2	K0 III		110 s		

HD	SAO	Star Name	α 2000	δ 2000	μ(α)	μ(δ)	V	B-V	M_v	Spec	RV	d(pc)	ADS	Notes
157060	208741		17 22 37.8	-35 54 36	+0.007	+0.11	6.47	0.54	4.2	dF9		28 s		
157038	208740		17 22 39.2	-37 48 18	+0.001	-0.01	6.41	0.64	-6.9	B4 Ia		1700 s		
158692	8788		17 22 40.3	+74 39 46	-0.004	+0.05	7.5	0.1	1.7	A3 V		120 mx		
156904	227878		17 22 45.5	-49 52 25	0.000	-0.02	8.0	0.5	0.2	K0 III		360 s		
156979	227880		17 22 46.3	-45 36 51	0.000	0.00	6.9	0.6	3.2	G5 IV		55 s		V636 Sco, v
156751	244675		17 22 47.3	-58 28 21	-0.010	-0.06	6.76	0.24	2.1	A5 V		45 mx		m
157565	85050		17 22 47.9	+27 51 34	-0.004	+0.04	8.0	0.5	3.4	F5 V		82 s		
157347	141642		17 22 51.1	-2 23 18	+0.003	-0.11	6.29	0.68	3.2	G5 IV		42 s		
156869	244683		17 22 52.2	-52 58 43	+0.001	-0.06	7.9	-0.2	0.6	A0 V		120 mx		
157097	208747		17 22 54.7	-37 13 14	+0.003	-0.03	5.93	1.08		K0				m
156768	244678		17 22 54.9	-58 00 36	-0.002	-0.01	5.88	1.07		K0				m
157443	122334		17 22 55.9	+9 21 53	0.000	-0.01	7.8	1.1	0.2	K0 III		330 s		
157263	185343		17 22 56.5	-20 12 56	-0.001	-0.02	8.0	1.0	0.2	K0 III		360 s		
157061	227885		17 22 57.1	-42 05 17	+0.001	-0.01	7.6	0.1	0.6	A0 V		250 s		
157415	122332		17 23 01.9	+0 50 27	0.000	+0.03	7.9	0.9	3.2	G5 IV		84 s		
157348	141643		17 23 04.2	-7 06 15	+0.002	-0.04	7.0	0.6	4.4	G0 V		33 s		
156854	244685		17 23 06.7	-56 31 31	0.000	+0.01	5.80	1.00	0.2	gK0		130 s		
157234	185346		17 23 09.6	-25 05 51	0.000	+0.01	7.4	0.0	2.1	A5 V		120 s		
157062	227889		17 23 12.8	-44 43 00	+0.001	-0.04	7.87	1.19		K0				
157282	185349		17 23 12.9	-23 00 32	-0.001	+0.01	7.4	0.4	0.4	B9.5 V		130 s		
158537	8787		17 23 14.9	+71 52 10	-0.006	0.00	7.0	1.6	-0.5	M4 III	-22	310 s		
157042	227886	ı Ara	17 23 15.9	-47 28 05	-0.001	-0.02	5.25	-0.11		B3 pne	-19			m
157495	122338		17 23 19.7	+9 28 00	+0.002	-0.01	7.1	0.1	1.4	A2 V	-5	140 s		
157236	185350	43 Oph	17 23 21.4	-28 08 35	0.000	-0.03	5.3	1.1	-0.1	K2 III	-14	120 s		
158013	30366		17 23 22.5	+57 00 42	0.000	+0.01	6.6	0.1	1.4	A2 V	-7	110 s		
157467	141645		17 23 26.7	-3 07 19	-0.003	-0.01	7.6	0.7	4.4	G0 V		44 s		
157192	208758		17 23 29.5	-37 04 40	-0.003	-0.03	8.0	0.2	0.4	B9.5 V		230 s		
157582	102789		17 23 30.0	+16 54 10	0.000	-0.01	7.70	0.1	0.6	A0 V	-21	260 s	10516	m
157264	208761		17 23 32.3	-31 42 00	0.000	-0.01	7.7	0.3	3.0	F2 V		86 s		
159048	8795		17 23 33.8	+76 03 18	-0.011	-0.03	7.1	1.1	0.2	K0 III		240 s		
157379	160515		17 23 36.1	-16 02 18	+0.002	-0.01	6.6	0.5	3.4	F5 V		43 s		
158147	17458		17 23 37.1	+60 59 56	-0.006	+0.04	7.2	1.1	0.2	K0 III		250 s		
157606	102790		17 23 37.7	+13 23 50	0.000	-0.05	7.40	1.4	-0.3	K4 III	-8	350 s	10517	m
157778	66001	75 ρ Her	17 23 40.8	+37 08 45	-0.003	+0.01	4.17	-0.01	0.6	A0 V	-19	52 s	10526	m
156905	244693		17 23 44.0	-58 39 55	-0.001	-0.01	7.0	-0.1	0.0	B8.5 V		250 s		
157906	46691		17 23 44.4	+47 16 14	-0.001	+0.03	8.00	0.5	4.0	F8 V	-27	63 s	10530	m
157117	227900		17 23 47.1	-48 42 26	0.000	0.00	7.56	1.03	3.2	G5 IV		75 s		
157335	--		17 23 48.5	-28 29 56			8.0	1.6	-0.5	M2 III		460 s		
157907	46692		17 23 48.6	+47 01 17	+0.002	+0.01	7.9	1.1	0.2	K0 III		340 s		
157637	102794		17 23 52.8	+13 05 58	-0.002	+0.08	7.9	0.9	3.2	G5 IV		88 s		
156817	253901		17 23 53.6	-63 10 03	+0.001	+0.05	8.0	0.8	3.2	G5 IV		90 s		
157336	185354		17 23 53.7	-29 49 15	-0.002	-0.03	7.4	1.5	0.2	K0 III		140 s		
157617	122346		17 23 57.5	+8 51 09	+0.001	0.00	5.77	1.25	0.0	gK1	+16	110 s		
157498	141647		17 23 57.6	-9 21 29	-0.002	-0.01	7.83	0.62	4.4	G0 V	-38	45 s		m
156709	253899		17 23 57.6	-65 42 08	+0.002	-0.02	6.8	0.0	0.0	B8.5 V		190 s		
157618	122347		17 23 58.1	+8 36 16	0.000	-0.05	7.3	1.1	0.2	K0 III		260 s		
157702	102795		17 24 00.0	+18 51 04	0.000	0.00	7.5	0.1	0.6	A0 V		240 s		
156838	253903		17 24 00.9	-62 51 51	0.000	-0.01	5.70	-0.14	-2.5	B2 V	-1	420 s		
157853	66006		17 24 02.1	+38 34 58	-0.001	+0.04	6.49	0.73	4.0	F8 V	-24	23 s	10531	m
157398	185357		17 24 03.4	-23 50 39	+0.001	-0.02	6.7	1.4	0.2	K0 III		120 s		
157728	85062	73 Her	17 24 06.5	+22 57 36	-0.003	-0.04	5.7	0.1	1.7	A3 V	-20	63 s		
157338	208769		17 24 08.6	-34 47 55	+0.001	-0.20	6.9	0.5	4.4	G0 V		32 s		
157879	66008		17 24 11.0	+36 55 23	-0.001	+0.02	7.9	1.1	0.2	K0 III		340 s		
157243	227911		17 24 12.8	-44 09 45	-0.001	-0.03	5.12	-0.06	-0.9	B6 V	+8	150 s		
157382	208773		17 24 17.1	-31 59 06	-0.001	0.00	7.9	1.7	0.2	K0 III		140 s		
156942	253908		17 24 18.5	-60 40 25	0.000	0.00	5.77	-0.08		B8	-10			
157416	185358		17 24 19.5	-29 52 15	0.000	-0.02	7.8	0.2	0.4	B9.5 V		210 s		
157383	208774		17 24 22.4	-33 13 40	0.000	-0.02	7.7	0.3	0.6	A0 V		160 s		
157434	185360		17 24 23.2	-28 39 10	+0.001	+0.01	7.5	0.7	0.4	B9.5 V		100 s		
157686	122353		17 24 25.2	+6 31 14	-0.003	+0.04	7.9	0.9	3.2	G5 IV		85 s		
158064	30371		17 24 25.9	+50 25 30	-0.002	-0.06	7.9	0.4	3.0	F2 V		97 s		
157910	66014		17 24 26.9	+36 57 07	-0.002	+0.05	6.24	0.86	0.3	G5 III	-16	160 s	10535	m
157740	102805		17 24 31.3	+16 18 04	+0.001	-0.03	5.71	0.07	0.0	A3 III	+11	140 s		
157741	102806		17 24 33.6	+15 36 22	+0.001	+0.01	6.20	-0.02	-0.1	B9.5 IV	-25	180 s	10528	
157546	160523		17 24 36.8	-18 26 45	+0.001	-0.01	6.21	0.03	0.6	A0 V		130 s		
158096	30372		17 24 36.9	+50 41 10	-0.010	-0.09	7.6	0.8	3.2	G5 IV		76 s		
157911	66018		17 24 39.0	+32 40 12	-0.001	0.00	6.8	1.1	0.2	K0 III		210 s		
157527	185367		17 24 41.9	-21 26 30	-0.001	-0.03	6.00	1.1	0.2	K0 III	-56	140 s	10522	m
157585	160527		17 24 42.3	-13 17 40	0.000	0.00	7.8	0.2	2.1	A5 V		140 s		
157418	208780		17 24 42.9	-36 27 35	0.000	-0.02	7.7	1.8	-0.5	M2 III		350 s		
157316	227920		17 24 42.9	-45 00 28	-0.001	+0.04	6.66	0.38		F0				
157081	244705		17 24 45.6	-59 12 49	-0.003	-0.01	8.00	0.1	0.6	A0 V		300 s		m
157179	244717		17 24 47.7	-54 33 43	-0.002	-0.03	7.2	-0.1	1.4	A2 V		150 s		
157387	227925		17 24 50.5	-41 30 01	+0.002	+0.04	7.6	0.7	3.4	F5 V		49 s		
157317	227923		17 24 52.1	-46 12 05	+0.004	0.00	6.76	-0.06	-1.0	B5.5 V		150 mx		

HD	SAO	Star Name	α 2000	δ 2000	μ(α)	μ(δ)	V	B−V	M$_V$	Spec	RV	d(pc)	ADS	Notes
			h m s	° ′ ″	s	″								
157822	102808		17 24 57.5	+16 56 05	+0.001	−0.01	7.6	1.1	−0.1	K2 III		340 s		
157354	227924		17 24 59.1	−45 53 38	+0.001	0.00	8.02	1.1		K0				
158633	17474		17 24 59.9	+67 18 23	−0.092	0.00	6.43	0.76	5.9	K0 V	−40	13 ts		
157687	141655		17 25 01.6	−9 12 42	−0.001	−0.12	8.0	0.9	3.2	G5 IV		85 mx		
157486	208786		17 25 02.5	−34 41 46	0.000	−0.03	6.3	0.2	0.6	A0 V		110 s		
157304	244727		17 25 05.6	−50 26 56	0.000	−0.07	7.7	0.2	4.0	F8 V		54 s		
157588	185374		17 25 06.0	−24 14 37	+0.001	+0.01	6.19	1.10	0.0	gK1	+20	170 s		
158222	30377		17 25 08.1	+53 07 57	−0.002	+0.08	7.8	0.7	4.4	G0 V		48 s		
157516	208787		17 25 08.5	−34 33 42	+0.001	0.00	7.50	0.1	0.6	A0 V		230 s		m
157515	208789		17 25 13.2	−34 25 49	−0.001	−0.07	6.8	0.7	0.2	K0 III		200 mx		
157244	244725	β Ara	17 25 17.9	−55 31 47	0.000	−0.02	2.85	1.46	−4.4	K3 Ib	0	240 s		
157823	122366		17 25 18.5	+8 55 27	−0.004	−0.07	6.8	1.4	−0.3	K5 III		120 mx		
157528	208790		17 25 18.8	−34 14 56	+0.002	0.00	7.2	0.7	2.1	A5 V		46 s		
157246	244726	γ Ara	17 25 23.5	−56 22 39	0.000	−0.01	3.34	−0.13	−4.4	B1 III	−4	330 s		m
158259	30380		17 25 23.8	+52 47 26	−0.010	−0.05	6.6	0.6	4.4	G0 V		27 s		
157895	102813		17 25 25.1	+15 32 56	0.000	0.00	7.9	1.4	−0.3	K5 III		430 s		
157744	141656		17 25 26.1	−8 49 44	−0.002	0.00	7.1	0.1	1.4	A2 V		140 s		
156960	253913		17 25 28.4	−66 23 24	+0.005	−0.07	7.4	1.1	0.2	K0 III		220 mx		
158084	66035		17 25 29.4	+37 47 11	0.000	−0.01	7.9	1.1	−0.1	K2 III		390 s		
157611	185381		17 25 29.5	−29 40 16	0.000	−0.04	6.8	0.6	1.4	A2 V		55 s		
157824	122368		17 25 29.6	+3 18 40	0.000	0.00	7.4	0.1	1.4	A2 V		160 s		
157519	227943		17 25 37.5	−41 58 41	0.000	0.00	7.84	0.05	0.6	A0 V		280 s		
157896	122372		17 25 39.2	+7 30 06	−0.002	0.00	7.9	0.7	4.4	G0 V		51 s		
157935	102816		17 25 40.7	+16 23 02	−0.001	+0.03	6.7	0.4	3.0	F2 V	−52	55 s		
157487	227940		17 25 40.7	−44 46 46	0.000	0.00	7.63	1.26		K0				
158460	17472		17 25 41.1	+60 02 54	−0.002	+0.03	5.65	0.03		A2	+7			
157881	122374		17 25 45.0	+2 06 40	−0.039	−1.19	7.54	1.36	8.2	K7 V	−28	7.5 t		
157826	141658		17 25 50.9	−8 40 20	+0.001	−0.02	7.8	1.1	0.2	K0 III		330 s		
157967	102819		17 25 54.2	+16 55 02	−0.001	+0.01	5.98	1.62	−0.5	M4 III	−10	200 s		m
158098	66039		17 25 54.3	+33 59 52	−0.001	+0.02	8.0	0.1	0.6	A0 V		300 s		
157856	141661		17 25 57.7	−1 39 06	+0.004	+0.05	6.44	0.46		F5	−24			
157455	227944		17 25 58.6	−49 44 06	0.000	−0.01	7.2	0.3	0.0	B8.5 V		150 s		
157841	141660		17 25 59.2	−6 34 53	−0.001	−0.01	6.8	0.0	0.4	B9.5 V		190 s		
157457	244734	κ Ara	17 25 59.8	−50 38 01	+0.001	0.00	5.23	1.06	0.0	K1 III	+18	110 s		m
157594	227951		17 25 59.9	−40 45 30	0.000	−0.01	7.6	1.4	−0.1	K2 III		220 s		
158067	85080		17 26 00.7	+26 52 43	0.000	+0.02	6.40	0.2	2.1	A5 V	−27	72 s		m
157747	185397		17 26 00.9	−23 10 17	−0.007	−0.04	7.4	0.9	3.4	F5 V		34 s		
158485	30387		17 26 04.7	+58 39 07	−0.002	+0.02	6.5	0.1	1.4	A2 V	−30	110 s		
157736	185395		17 26 05.8	−27 35 57	0.000	−0.03	7.40	0.00	0.4	B9.5 V		240 s	10538	m
157719	185394		17 26 06.4	−28 32 35	−0.013	−0.09	7.4	1.0	3.2	G5 IV		50 s		
158296	46715		17 26 09.8	+46 51 32	−0.001	+0.03	8.0	1.1	0.2	K0 III		360 s		
158116	85089		17 26 13.6	+29 27 21	0.000	+0.01	7.68	0.30		A m	−25		10553	m
157749	185400		17 26 15.4	−28 25 00	−0.001	−0.04	7.3	0.6	3.4	F5 V		50 s		
157857	160545		17 26 17.1	−10 59 33	−0.001	+0.02	7.79	0.18		O7.8	+59			
157978	122381		17 26 18.8	+7 35 44	0.000	−0.01	6.06	0.58	4.4	G0 V	−4	20 s		
157792	185401	44 Oph	17 26 22.1	−24 10 31	0.000	−0.12	4.17	0.28	2.5	A9 V	−37	25 ts		
157555	227954		17 26 22.2	−48 36 53	+0.004	−0.01	7.04	0.48	4.0	F8 V		41 s		
159218	8803		17 26 26.9	+72 48 22	−0.004	+0.03	7.5	1.1	0.2	K0 III		290 s		
157707	208814		17 26 29.6	−37 06 42	+0.003	0.00	7.4	1.8	−0.3	K5 III		220 s		
157999	122387	49 σ Oph	17 26 30.7	+4 08 25	0.000	+0.01	4.34	1.50	−2.3	K3 II	−27	190 s		
157983	122382		17 26 31.4	+0 49 20	0.000	0.00	6.9	1.1	0.2	K0 III		210 s		
157624	227962		17 26 31.9	−44 37 45	+0.001	−0.01	7.88	−0.05		B8				
157708	208815		17 26 34.3	−38 19 45	−0.001	−0.04	7.5	1.1	0.2	K0 III		240 s		
157750	208819		17 26 34.8	−32 58 10	−0.001	−0.12	8.04	0.67	4.7	G2 V		43 s		
158166	85093		17 26 35.5	+27 13 52	0.000	+0.03	7.8	0.2	2.1	A5 V		140 s		
157290	253921		17 26 35.8	−62 55 23	−0.002	−0.03	7.1	1.1	−0.1	K2 III		270 s		
157950	141665		17 26 37.7	−5 05 12	−0.006	−0.04	4.54	0.39	3.1	F3 V	0	21 ts		
157984	141667		17 26 39.8	−1 00 08	+0.001	−0.04	7.6	0.1	0.6	A0 V		250 s		
157781	208824		17 26 40.8	−32 01 21	−0.002	−0.05	7.5	−0.1	0.6	A0 V		240 s		
157751	208821		17 26 41.0	−34 05 31	+0.001	0.00	7.5	0.1	0.4	B9.5 V		210 s		
157862	185402		17 26 42.4	−21 28 11	+0.001	−0.03	7.6	1.1	3.2	G5 IV		51 s		
158414	46723	77 Her	17 26 44.1	+48 15 36	0.000	−0.01	5.8	0.1	1.4	A2 V	−9	76 s		
157649	227966		17 26 44.3	−45 49 24	−0.001	−0.01	6.00	0.1	0.6	A0 V		120 s		m
158225	66052		17 26 45.8	+31 13 09	+0.001	−0.13	7.1	0.5	3.4	F5 V	0	54 s		
158261	66054		17 26 45.9	+34 41 44	−0.003	+0.04	5.90	−0.01		A1 p	−22			
158148	85095		17 26 48.9	+20 04 51	0.000	+0.02	5.54	−0.13	−0.9	B6 V	−30	190 s		
157865	185404		17 26 50.6	−26 19 56	0.000	−0.01	7.50	0.00	0.4	B9.5 V		250 s	10547	m
157661	227971		17 26 51.3	−45 50 36	−0.004	−0.04	5.29	−0.07		B9 n	−9			m
157864	185406		17 26 55.0	−25 56 36	−0.002	−0.03	6.3	−0.4	0.6	A0 V		140 s		
157599	244749		17 26 56.1	−51 56 56	+0.001	−0.02	6.4	0.7	0.0	B8.5 V		67 s		
157968	160553		17 27 01.9	−12 30 45	+0.002	−0.07	6.3	0.5	4.0	F8 V	−40	29 s		
157698	227974		17 27 03.4	−47 08 11	0.000	0.00	7.13	−0.05	−1.6	B5 IV		500 s		
156961	—		17 27 06.3	−71 56 03			7.9	1.1	0.2	K0 III		350 s		
157969	160554		17 27 09.6	−15 51 12	−0.001	−0.01	6.6	1.1	0.2	K0 III		170 s		
157653	244753		17 27 11.0	−51 40 07	−0.003	+0.03	8.0	0.1	3.4	F5 V		82 s		

HD	SAO	Star Name	α 2000	δ 2000	μ(α)	μ(δ)	V	B-V	M_v	Spec	RV	d(pc)	ADS	Notes
157662	244755		17 27 12.2	-50 37 49	-0.001	0.00	6.10	0.00	0.4	B9.5 V	+11	140 s		m
157918	185410		17 27 16.3	-28 46 11	0.000	-0.02	7.6	-0.1	0.6	A0 V		260 s		
158211	102827		17 27 17.0	+17 53 51	+0.002	0.00	7.5	0.9	0.2	G9 III	-1	290 s		
157919	185412	45 Oph	17 27 21.1	-29 52 01	+0.002	-0.14	4.29	0.40	2.1	F5 IV	+38	27 s		
157830	208832		17 27 21.6	-38 03 42	-0.016	-0.03	7.8	0.4	3.2	G5 IV		58 mx		
158227	102830		17 27 23.1	+18 07 49	+0.002	+0.02	7.8	1.4	-0.3	K5 III		410 s		
158404	66067		17 27 29.2	+35 43 13	+0.001	+0.03	7.8	0.1	1.7	A3 V		160 s		
158251	102832		17 27 32.6	+16 27 24	0.000	+0.01	7.30	0.4	2.6	F0 V	-11	87 s	10560	m
157795	227984		17 27 32.7	-43 32 37	+0.002	-0.13	7.25	0.44		F2				
157676	244757		17 27 33.2	-53 13 20	-0.001	+0.01	6.6	1.4	-0.5	M2 III		260 s		
157955	185417		17 27 37.3	-29 43 28	-0.001	-0.03	5.9	0.2	0.4	B9.5 V		86 s		
157902	208838		17 27 41.2	-37 11 05	0.000	+0.02	6.6	1.0	0.2	K0 III		190 s		
157480	253927		17 27 43.4	-63 15 48	0.000	+0.01	8.0	1.4	-0.3	K5 III		460 s		
158228	122405		17 27 43.9	+8 26 30	0.000	-0.05	6.5	1.6	-0.5	M2 III	+4	260 s		
157988	185420		17 27 45.5	-28 21 30	0.000	+0.02	7.8	0.1	0.6	A0 V		220 s		
157928	208842		17 27 47.4	-36 31 33	+0.001	-0.01	7.5	1.5	0.2	K0 III		150 s		
158101	160561		17 27 47.6	-12 11 12	-0.001	0.00	8.00	0.4	3.0	F2 V		100 s	10559	m
158263	102835		17 27 52.0	+11 23 22	-0.002	-0.03	7.12	0.13		A5	-26		10562	m
157832	227990		17 27 54.6	-47 01 35	0.000	-0.01	6.65	0.02	-1.1	B5 V nne		350 s		m
157887	227994		17 27 54.9	-42 24 32	+0.003	-0.09	7.9	0.7	3.2	G5 IV		88 s		
157753	244763		17 27 57.4	-52 17 49	0.000	-0.05	5.80	1.1	0.2	K0 III		130 s		m
157461	—		17 28 00.1	-64 57 43			7.2	1.8	-0.3	K5 III		210 s		
158170	141679		17 28 02.1	-8 12 30	-0.006	-0.13	6.37	0.58	2.4	F8 IV		46 mx		
157957	208846		17 28 03.4	-36 52 56	+0.001	0.00	7.78	0.18	0.0	B8.5 V		330 s		
157524	253928		17 28 07.6	-63 02 11	-0.002	-0.03	6.3	0.2	0.4	B9.5 V	-3	110 s		
158020	208852		17 28 11.6	-31 12 24	0.000	+0.01	7.80	1.1	-0.1	K2 III		350 s		m
157560	253931		17 28 13.2	-62 56 42	-0.001	0.00	7.9	0.0	0.4	B9.5 V		320 s		
157274	—		17 28 14.9	-70 04 49			7.9	1.3	3.2	G5 IV		43 s		
158122	185428		17 28 16.0	-20 57 49	-0.001	+0.04	7.96	0.45	3.4	dF5	-10	82 s	10561	m
157817	244769		17 28 20.8	-52 38 09	-0.002	-0.08	8.0	0.0	0.4	B9.5 V		60 mx		
157799	244768		17 28 22.0	-53 30 01	-0.002	-0.04	7.3	1.3	-0.3	K5 III		340 s		
157786	244767		17 28 23.3	-54 01 16	0.000	+0.01	6.9	0.4	1.4	A2 V		74 s		
157973	228002		17 28 24.2	-40 20 23	+0.001	-0.02	8.0	0.4		B0.5 II				
158088	185429		17 28 27.4	-29 03 31	0.000	-0.01	7.5	1.3	-0.3	K5 III		360 s		
158933	30409		17 28 31.0	+57 06 37	-0.003	-0.04	8.0	0.9	3.2	G5 IV		93 s		
158821	30406		17 28 35.6	+51 21 40	-0.001	-0.05	7.7	0.5	3.4	F5 V		72 s		
157819	244770		17 28 38.5	-55 10 11	-0.001	-0.01	6.00	1.1	0.2	K0 III		140 s		m
158140	185433		17 28 38.6	-25 30 36	+0.001	-0.01	7.70	0.4	2.6	F0 V		100 s	10564	m
158103	208857		17 28 40.7	-31 23 00	-0.002	+0.01	7.8	0.6	2.6	F0 V		69 s		
158172	185437		17 28 41.4	-22 34 54	-0.001	-0.02	8.00	1.1	-0.1	K2 III		370 s		m
157715	253934		17 28 42.5	-60 24 24	-0.004	-0.06	7.7	2.5	-0.3	K5 III		95 s		
155454	258768		17 28 44.0	-82 46 46	+0.005	0.00	6.80	0.00	0.0	B8.5 V		230 s		m
158123	208859		17 28 45.5	-30 55 56	+0.001	-0.03	7.8	-0.1	0.6	A0 V		270 s		
158418	102846		17 28 47.6	+10 32 03	-0.001	0.00	8.0	0.1	1.4	A2 V		210 s		
158352	122418		17 28 49.5	+0 19 49	-0.004	+0.01	5.44	0.22		A5 n	-36	32 mn		
158173	185438		17 28 49.9	-26 43 47	+0.002	-0.05	7.9	1.2	4.4	G0 V		21 s		
158693	46742		17 28 50.0	+41 23 58	-0.001	0.00	8.0	1.4	-0.3	K5 III		460 s		
157342	257502		17 28 51.0	-70 38 02	-0.002	-0.04	7.3	1.1	0.2	K0 III		260 s		
158105	208860		17 28 55.9	-36 46 42	0.000	-0.01	6.0	1.2	0.2	K0 III		100 s		
158868	30413		17 28 57.9	+50 52 12	-0.002	+0.01	7.70	0.4	2.6	F0 V		110 s	10597	m
157678	253935		17 28 59.4	-63 25 26	-0.001	-0.03	7.4	0.0	0.4	B9.5 V		250 s		
158042	228010		17 29 00.6	-43 58 27	-0.001	-0.02	6.19	-0.02		B9				m
158896	30414		17 29 02.0	+51 29 50	-0.001	-0.01	7.9	1.4	-0.3	K5 III		440 s		
157801	244775		17 29 02.1	-59 01 56	0.000	-0.01	7.0	1.8	-0.5	M4 III		240 s		
158654	66078		17 29 03.1	+33 45 24	+0.001	0.00	7.9	0.4	3.0	F2 V		94 s		
158869	30415		17 29 06.4	+50 05 14	-0.001	-0.02	7.8	1.4	-0.3	K5 III		410 s		
158302	160572		17 29 07.1	-14 04 51	+0.001	-0.02	7.9	0.5	4.0	F8 V		58 s		
158449	122426		17 29 09.3	+6 08 54	-0.002	+0.02	7.9	1.1	0.2	K0 III		340 s		
158186	208869		17 29 12.5	-31 32 06	-0.002	-0.02	7.00	0.03	-4.1	B0 V		1200 s		
158490	102850		17 29 13.1	+10 53 57	-0.001	-0.02	7.5	0.1	0.6	A0 V		250 s		
158822	46743		17 29 14.5	+43 28 11	-0.002	-0.01	8.0	0.9	3.2	G5 IV		93 s		
157873	244780		17 29 16.6	-57 24 42	0.000	-0.02	7.3	1.0	0.2	K0 III		260 s		
158492	122429		17 29 18.7	+7 41 21	0.000	0.00	7.9	1.1	0.2	K0 III		340 s		
158107	—		17 29 19.2	-42 20 14			7.8	1.0	-0.1	K2 III		380 s		
158078	228015		17 29 23.3	-46 41 47	-0.002	-0.01	7.62	-0.06	-1.1	B5 V		520 s		
158156	208870		17 29 25.4	-38 30 59	+0.003	+0.02	6.39	0.09	1.4	A2 V		94 s		m
158756	66085		17 29 27.9	+34 56 18	-0.002	0.00	7.50	1.1	0.2	K0 III		290 s	10594	m
157802	253938		17 29 28.7	-61 45 48	-0.001	+0.01	6.90	0.4	2.6	F0 V		72 s		m
158027	244787		17 29 29.3	-50 23 55	0.000	-0.02	7.9	0.4	0.2	K0 III		350 s		
158374	160578		17 29 29.5	-16 02 03	+0.001	+0.03	7.7	1.1	0.2	K0 III		280 s		
158462	141689		17 29 32.4	-0 10 11	-0.001	-0.04	7.9	0.4	2.6	F0 V		110 s		
158509	122433		17 29 36.8	+1 33 19	-0.001	0.00	7.9	0.0	0.4	B9.5 V		300 s		
159329	17504		17 29 44.1	+63 51 08	-0.002	-0.20	7.63	0.59	4.2	F9 V	-29	47 s		
158463	141691		17 29 47.2	-5 55 11	-0.002	-0.07	6.37	0.93		G5	+4		10583	m
158656	102862		17 29 51.3	+15 57 51	0.000	-0.02	7.60	0.5	3.4	F5 V		69 s	10592	m

HD	SAO	Star Name	α 2000	δ 2000	μ(α)	μ(δ)	V	B-V	M$_v$	Spec	RV	d(pc)	ADS	Notes
158175	228029		17^{h}29^{m}51.8^s	-44°44'29"	-0.001	-0.01	7.54	-0.05		B8				
158420	160584		17 29 58.5	-17 48 53	+0.002	-0.10	7.8	0.7	4.4	G0 V		48 s		
159266	30431		17 30 01.4	+59 41 18	+0.004	-0.05	8.01	1.00		K0	-15			
158320	208881		17 30 05.3	-33 43 00	-0.001	-0.01	6.67	0.13	-2.0	B4 IV		420 s		m
--	17510		17 30 06.9	+66 56 23	-0.003	+0.02	8.0	0.5						
--	208883		17 30 07.3	-33 42 06	-0.004	+0.01	6.80	0.00	0.0	B8.5 V		88 mx		m
158358	208889		17 30 15.5	-32 07 18	+0.001	0.00	7.4	1.6	-0.3	K5 III		310 s		
159062	46762		17 30 16.0	+47 24 10	+0.016	+0.10	7.3	0.8	3.2	G5 IV		66 s		
158789	85153		17 30 18.5	+22 02 08	-0.001	+0.02	7.8	1.1	0.2	K0 III		330 s		
159428	17509		17 30 19.4	+63 46 40	-0.002	+0.04	7.2	1.1	-0.1	K2 III		290 s		
158576	141701		17 30 19.8	-4 22 09	0.000	0.00	6.60	0.4	2.6	F0 V		63 s		m
157772	--		17 30 21.3	-66 53 25			7.0	0.0	0.4	B9.5 V		210 s		
158717	102868		17 30 22.0	+10 46 06	+0.001	-0.10	7.1	1.1	0.2	K0 III		170 mx		
158716	102869		17 30 22.2	+11 55 30	-0.002	+0.05	6.40	0.06	1.4	A2 V	-25	100 s		
158614	141702		17 30 23.6	-1 03 45	-0.008	-0.17	5.31	0.72	4.4	G8 IV-V	-77	19 ts	10598	m
158111	244795		17 30 25.4	-54 14 24	-0.005	-0.02	7.78	-0.01	-1.9	B6 III		180 mx		
159181	30429	23 β Dra	17 30 25.8	+52 18 05	-0.002	+0.01	2.79	0.98	-2.1	G2 II	-20	82 s	10611	Rastaban, m
158824	85157		17 30 26.7	+22 52 22	0.000	-0.01	7.7	1.1	0.2	K0 III		310 s		
158525	160591		17 30 28.2	-15 38 14	0.000	-0.06	7.5	1.1	0.2	K0 III		250 s		
158289	228046		17 30 28.8	-43 45 14	0.000	-0.03	7.8	0.7	3.4	F5 V		52 s		
158395	208893		17 30 30.4	-34 32 20	-0.001	0.00	7.8	1.2	2.1	A5 V		34 s		
159712	17512		17 30 31.3	+68 44 53	-0.001	-0.03	8.0	1.4	-0.3	K5 III		460 s		
158306	228048		17 30 33.7	-43 13 32	+0.004	-0.01	6.90	0.5	3.4	F5 V		50 s		m
158806	102873		17 30 35.0	+17 31 13	-0.007	+0.12	6.9	0.5	3.4	F5 V		50 s		
158422	208898		17 30 40.0	-32 03 43	+0.002	-0.01	7.5	-0.2	0.6	A0 V		250 s		
159026	66102		17 30 40.1	+38 52 56	+0.001	+0.01	6.43	0.49	3.0	F2 V	-27	38 s		
159330	30434		17 30 43.6	+57 52 35	+0.003	-0.03	6.4	1.1	-0.1	K2 III	-14	200 s		
158899	85163	76 λ Her	17 30 44.1	+26 06 39	+0.001	+0.02	4.41	1.44	-0.3	K4 III	-26	85 s		
158408	208896	34 υ Sco	17 30 45.6	-37 17 45	0.000	-0.03	2.69	-0.22	-5.7	B3 Ib	+18	480 s		Lesath
166926	2940	24 UMi	17 30 47.6	+86 58 04	+0.069	0.00	5.79	0.25		A m	+1			
158409	208897		17 30 48.0	-37 26 12	+0.002	-0.03	7.60	0.1	0.6	A0 V		240 s		m
158443	208899		17 30 48.0	-33 36 32	-0.002	+0.04	8.0	0.8	3.4	F5 V		48 s		V482 Sco, v
158527	185466		17 30 49.4	-23 50 30	-0.002	-0.01	7.4	0.5	2.6	F0 V		73 s		
158737	122453		17 30 51.3	+1 07 11	-0.001	-0.01	7.1	0.4	2.6	F0 V		78 s		
158410	228059		17 30 54.5	-40 39 36	0.000	+0.01	8.0	0.4	0.2	K0 III		360 s		
158974	66103		17 30 55.2	+31 09 30	+0.001	+0.02	5.61	0.95	0.3	G8 III	-26	120 s		
158738	141713		17 31 02.0	-2 49 35	-0.001	0.00	7.2	1.1	0.2	K0 III		240 s		
158094	253945	δ Ara	17 31 05.8	-60 41 01	-0.007	-0.09	3.62	-0.10	-0.2	B8 V	+12	29 mx		m
158873	102880		17 31 12.2	+13 06 18	0.000	-0.01	8.0	0.4	2.6	F0 V		120 s		
159206	46771		17 31 12.8	+43 33 32	-0.004	+0.02	8.0	1.1	0.2	K0 III		280 mx		
159906	17517		17 31 17.4	+69 36 00	-0.005	-0.02	7.3	1.1	0.2	K0 III	+6	260 s		
158719	160610		17 31 17.6	-10 58 17	0.000	-0.06	7.80	0.5	3.4	F5 V		76 s	10605	m
158955	102883		17 31 18.4	+19 31 17	-0.003	-0.07	7.0	1.4	-0.3	K5 III		150 mx		
158468	208903		17 31 19.3	-39 01 06	0.000	0.00	7.50	0.5	4.0	F8 V		50 s		
158469	208904		17 31 19.5	-39 09 55	-0.003	+0.01	7.8	-0.1	4.0	F8 V		58 s		
158837	122465		17 31 21.1	+2 43 28	-0.001	+0.02	5.59	0.84	0.4	gG3	-29	99 s	10607	m
158220	244808		17 31 22.8	-56 55 14	-0.002	+0.01	5.95	-0.08		B8	-3			
158280	244812		17 31 23.8	-54 41 38	-0.005	-0.22	7.5	0.9	3.2	G5 IV		53 mx		
158643	185470	51 Oph	17 31 24.8	-23 57 46	0.000	-0.03	4.81	0.00		A0 n	-12	13 mn		
159387	30438		17 31 26.7	+54 21 21	-0.002	+0.01	7.1	0.1	0.6	A0 V		200 s		
156513	258769		17 31 27.1	-80 51 32	-0.003	-0.04	5.88	1.67		Mb				
160077	8830		17 31 27.6	+71 14 53	-0.002	0.00	7.7	1.1	-0.1	K2 III		370 s		
158855	122468		17 31 30.6	+1 40 19	-0.002	-0.03	7.19	1.35	-0.2	K3 III	-16	280 s		
159349	30437		17 31 33.7	+50 40 43	-0.003	-0.05	7.7	0.9	3.2	G5 V		80 s		
158530	208910		17 31 34.2	-38 38 51	0.000	-0.03	7.6	0.2	0.6	A0 V		180 s		
158618	208916		17 31 35.1	-30 14 45	-0.004	0.00	8.0	0.3	-1.0	B5.5 V		65 mx		
159118	66109		17 31 42.3	+30 19 04	0.000	-0.13	7.20	1.1	0.2	K0 III		150 mx		
158531	228078		17 31 42.5	-41 02 24	0.000	-0.02	7.19	0.06	0.4	B9.5 V		210 s		m
158426	228066		17 31 43.8	-48 45 01	-0.001	+0.05	7.70	0.4	2.6	F0 V		100 s		m
158704	185474		17 31 44.2	-26 16 11	0.000	-0.03	6.05	-0.06	0.4	B9.5 V		140 s		
158619	208921		17 31 47.3	-33 42 10	0.000	-0.02	6.4	1.1	0.2	K0 III		160 s		
158476	228075		17 31 48.9	-46 02 11	0.000	+0.01	6.30	0.6	4.4	G0 V		24 s		m
159005	102888		17 31 49.2	+11 55 54	+0.001	+0.02	8.0	0.1	0.6	A0 V		300 s		
159139	85182	78 Her	17 31 49.4	+28 24 27	0.000	+0.03	5.60	0.00	0.2	B9.5 V	-26	100 s		
158427	228069	α Ara	17 31 50.3	-49 52 34	-0.002	-0.07	2.95	-0.17	-1.7	B3 V	-2	58 mx		m
158043	253947		17 31 50.9	-65 55 04	+0.009	-0.04	8.0	0.8	3.2	G5 IV		90 s		
159063	102891		17 31 53.8	+16 49 27	-0.003	-0.01	6.9	0.5	4.0	F8 V	-7	37 s		
158499	228079		17 31 56.7	-46 05 33	+0.002	-0.02	7.1	1.7	-0.1	K2 III		130 s		
158921	122477		17 31 57.4	+0 02 34	0.000	0.00	7.0	1.1	0.2	K0 III		220 s		
159966	17526	27 Dra	17 31 57.6	+68 08 06	-0.003	+0.13	5.05	1.08	0.2	K0 III	-73	83 s		m
159222	66118		17 32 00.8	+34 16 15	-0.019	+0.05	6.56	0.65	5.2	G5 V	-52	20 ts		
158682	208927		17 32 01.3	-31 49 38	0.000	0.00	7.9	0.1	0.0	B8.5 V		290 s		
158976	122481		17 32 02.8	+2 49 24	+0.001	-0.03	7.90	0.00	0.0	B8.5 V		350 s	10614	m
159050	102894		17 32 05.3	+10 23 32	0.000	0.00	8.0	1.1	-0.1	K2 III		410 s		
158705	208928		17 32 06.9	-31 32 55	0.000	+0.02	7.8	0.8		B0.5 II				

HD	SAO	Star Name	α 2000	δ 2000	μ(α)	μ(δ)	V	B–V	M_v	Spec	RV	d(pc)	ADS	Notes
			^{h} ^{m} ^{s}	$^\circ$ ' "	^{s}	"								
159541	30447	24 ν¹ Dra	17 32 10.3	+55 11 03	+0.016	+0.06	4.88	0.26		A m	−15		10628	m
166205	2937	23 δ UMi	17 32 12.7	+86 35 11	+0.011	+0.05	4.36	0.02	1.2	A1 V	−8	44 mn		
159410	46784		17 32 13.4	+46 19 49	−0.002	−0.02	7.50	1.1	−0.2	K3 III	−52	350 s		
159082	102897		17 32 14.9	+11 55 49	+0.002	+0.02	6.20	−0.01	−0.6	A0 III	−12	230 s		
159560	30450	25 ν² Dra	17 32 15.8	+55 10 22	+0.017	+0.05	4.87	0.28		A m	−16	19 mn	10628	m
159153	102898		17 32 18.3	+18 23 10	0.000	−0.03	7.6	1.1	−0.1	K2 III		350 s		
159029	122484		17 32 21.0	+3 16 41	0.000	0.00	8.0	1.1	−0.1	K2 III		370 s		
159052	122485		17 32 22.6	+2 03 21	+0.001	−0.02	7.9	1.4	−0.3	K5 III		390 s		
159119	102899		17 32 23.9	+14 23 01	0.000	−0.05	7.40	1.4	−0.3	K5 III	−20	350 s		
158741	208935		17 32 24.4	−34 16 47	−0.001	−0.04	6.3	0.6	3.0	F2 V		31 s		
159223	85188		17 32 24.9	+26 26 21	0.000	+0.03	6.9	0.2	2.1	A5 V	−29	90 s		
159051	122486		17 32 25.3	+2 24 43	0.000	−0.01	7.9	1.1	0.2	K0 III		320 s		
159303	66124		17 32 26.7	+35 46 24	+0.001	0.00	7.9	0.0	0.4	B9.5 V		320 s		
160538	8842		17 32 40.8	+74 13 39	−0.019	+0.04	6.55	1.05	−0.1	K2 III	−9	180 mx		
158742	208936		17 32 42.9	−39 12 58	0.000	−0.01	8.0	1.6	0.2	K0 III		160 s		
159411	46787		17 32 43.0	+40 39 34	−0.004	−0.01	7.8	0.8	3.2	G5 IV		84 s		
158311	253955		17 32 44.0	−62 14 09	+0.015	0.00	7.53	0.90	3.2	G8 IV		71 s		
160052	17528		17 32 44.8	+66 58 56	−0.003	+0.02	7.9	0.4	3.0	F2 V		94 s		
159225	102905		17 32 48.8	+19 22 37	0.000	−0.05	7.9	0.5	4.0	F8 V		61 s		
159543	46791		17 32 49.5	+47 53 20	+0.001	+0.03	7.70	0.1	0.6	A0 V		260 s	10630	m
159241	102906		17 32 50.0	+19 05 11	0.000	−0.01	7.9	1.4	−0.3	K5 III		430 s		
159226	102904		17 32 50.3	+16 31 31	−0.001	+0.01	8.0	0.1	1.4	A2 V		210 s		
159008	160627		17 32 50.5	−13 28 56	−0.002	−0.01	7.1	0.1	0.6	A0 V		180 s		
158312	253956		17 32 52.7	−63 20 31	0.000	+0.02	7.7	0.1	1.7	A3 V		160 s		
158605	244833		17 32 53.4	−50 40 30	+0.001	0.00	8.0	1.7	−0.3	K5 III		280 s		
159011	160629		17 32 58.9	−17 29 45	+0.001	+0.02	6.8	1.1	0.2	K0 III		190 s		
158883	208943		17 32 59.0	−30 22 40	−0.008	−0.04	7.46	0.97	3.2	K0 IV		70 s		
158628	244835		17 33 01.2	−50 59 52	+0.001	−0.02	7.9	1.4	−0.1	K2 III		260 s		
159607	46793		17 33 02.0	+49 20 43	−0.002	0.00	7.3	0.1	0.6	A0 V		220 s		
158902	185490		17 33 03.2	−29 39 03	0.000	0.00	7.0	0.6	0.0	B8.5 V		100 s		
158859	208945		17 33 05.9	−33 03 29	+0.001	−0.02	7.27	0.17	−1.1	B5 V		350 s		
159501	46792		17 33 07.0	+41 14 36	−0.007	−0.06	5.74	1.09	0.0	gK1	−29	140 s		
158799	228110		17 33 07.2	−41 10 25	0.000	−0.02	5.84	0.06	0.4	B9.5 V		110 s		
159267	102908		17 33 09.0	+16 46 08	0.000	−0.01	7.0	1.4	−0.3	K5 III		280 s		
159227	122501		17 33 10.3	+7 43 08	+0.001	+0.01	7.7	1.1	0.2	K0 III		310 s		
158550	244832		17 33 11.4	−56 17 02	+0.002	+0.03	7.8	0.1	0.6	A0 V		220 s		
159186	122499		17 33 13.6	+2 26 20	−0.002	−0.04	8.0	1.4	−0.3	K5 III		400 s		
158747	228107		17 33 14.0	−45 30 42	+0.002	−0.02	7.08	0.01	0.4	B9.5 V		210 s		m
158764	228109		17 33 19.2	−46 45 53	0.000	−0.02	8.0	0.9	0.2	K0 III		360 s		
157976	—		17 33 20.5	−72 11 10			8.0	1.1	0.2	K0 III		360 s		
159332	102912		17 33 22.6	+19 15 24	−0.002	−0.09	5.64	0.48	3.4	F5 V	−59	27 s		
158846	228115		17 33 28.9	−42 23 58	+0.001	0.00	7.34	−0.04	−1.6	B5 IV		550 s		m
159170	141730		17 33 29.6	−5 44 42	−0.003	−0.10	5.62	0.18		A0 n	−26			
159870	30464		17 33 31.5	+57 33 32	+0.001	+0.01	6.17	0.59	3.0	F2 V	−1	29 s		
158926	208954	35 λ Sco	17 33 36.4	−37 06 14	0.000	−0.03	1.63	−0.22	−3.0	B2 IV	0	84 s		Shaula, m
159353	102917		17 33 39.2	+16 19 03	−0.001	−0.06	5.69	1.01	0.2	gK0	−22	120 s		
159354	102918		17 33 42.6	+14 50 30	+0.001	−0.07	6.6	1.6	−0.5	M4 III	+30	260 s		
158781	244846		17 33 46.6	−50 43 51	+0.001	0.00	7.9	0.5	0.0	B8.5 V		380 s		
159189	160641		17 33 50.8	−13 30 16	−0.001	+0.01	7.8	1.1	0.2	K0 III		280 s		
158748	244844		17 33 50.9	−52 45 21	+0.001	−0.01	7.3	1.2	0.2	K0 III		210 s		
159333	122512		17 33 52.7	+8 06 14	−0.001	−0.06	7.6	0.7	4.4	G0 V		44 s		
159442	85210		17 33 54.8	+21 33 54	0.000	−0.02	7.9	1.4	−0.3	K5 III		430 s		
158551	253962		17 33 58.5	−61 31 54	+0.004	−0.02	7.3	1.9	−0.3	K5 III		190 s		
159172	160642		17 33 58.9	−17 50 10	0.000	0.00	7.9	0.4	2.6	F0 V		110 s		
159090	208965		17 34 01.3	−30 24 40	−0.003	0.00	7.41	0.16	−5.0	B0 III		1800 s		
159160	185512		17 34 02.1	−23 01 55	−0.002	−0.09	7.2	0.5	2.6	F0 V		68 s		
158906	228123		17 34 04.2	−46 36 01	+0.003	0.00	7.66	−0.13	0.0	B8.5 V		340 s		
158967	228128		17 34 04.7	−41 38 53	+0.001	−0.04	7.6	1.4	−0.1	K2 III		260 s		
158630	244841		17 34 05.0	−59 46 31	−0.007	−0.38	7.63	0.60		G0		16 mn		
159307	141736		17 34 08.2	−3 03 20	−0.003	−0.03	7.2	0.5	4.0	F8 V		44 s		
160078	17536		17 34 08.7	+60 05 22	+0.002	−0.13	7.32	0.48		F5				
158928	228127		17 34 10.4	−45 00 22	0.000	+0.01	7.00	−0.02	0.4	B9.5 V		210 s		
158890	228124		17 34 11.5	−48 28 04	−0.001	+0.01	7.9	1.3	−0.5	M2 III		490 s		
158783	244852		17 34 11.8	−54 53 43	+0.006	−0.28	6.8	0.9	3.2	G5 IV		42 s		
159091	208967		17 34 14.1	−35 44 01	+0.001	−0.01	7.60	0.00	0.0	B8.5 V		310 s		m
159173	185514		17 34 14.6	−24 52 34	+0.002	−0.01	7.9	2.5	−0.3	K5 III		100 s		
159036	228130		17 34 15.7	−41 01 34	−0.001	−0.02	7.6	0.1	1.4	A2 V		160 s		
158907	228126		17 34 16.8	−48 31 35	−0.003	+0.04	6.6	1.1	3.2	G5 IV		30 s		
159466	102925		17 34 21.8	+13 09 39	0.000	−0.03	6.70	0.9	3.2	G5 IV	−60	50 s	10633	m
159035	228135		17 34 22.5	−40 31 52	+0.001	−0.02	7.13	0.03	0.0	B8.5 V		230 s		
159520	102927		17 34 22.5	+19 14 31	−0.001	+0.01	7.3	0.2	2.1	A5 V		110 s		
159413	122519		17 34 22.8	+6 31 18	+0.001	−0.01	7.9	1.1	0.2	K0 III		340 s		
159503	102928		17 34 27.0	+16 30 14	0.000	−0.02	6.4	0.2	2.1	A5 V	−41	73 s		
159357	141739		17 34 29.2	−1 45 41	+0.007	−0.19	7.8	0.5	4.0	F8 V		62 s		
158829	244855		17 34 30.7	−55 10 47	−0.007	−0.02	7.8	1.0	0.2	K0 III		240 mx		

HD	SAO	Star Name	α 2000	δ 2000	μ(α)	μ(δ)	V	B-V	M_V	Spec	RV	d(pc)	ADS	Notes
			h m s	° ' "	s	"								
159626	85218		17 34 32.7	+28 12 31	0.000	0.00	8.0	0.1	0.6	A0 V		300 s		
159273	160649		17 34 33.3	−18 13 14	0.000	0.00	7.2	0.4	2.6	F0 V		79 s		
--	122525		17 34 36.0	+9 34 32	0.000	0.00	7.81	0.29						
159480	122526	53 Oph	17 34 36.5	+9 35 12	0.000	−0.01	5.49	0.02	0.6	A2 IV	−14	95 s	10635	m
159445	122522		17 34 38.2	+4 05 56	−0.002	−0.01	7.9	1.1	0.2	K0 III		310 s		
159174	208975		17 34 38.9	−31 52 20	+0.002	−0.01	7.1	0.9	0.2	K0 III		240 s		
159110	228143		17 34 40.5	−41 19 25	+0.003	+0.01	7.4	0.3	−1.0	B5.5 V		250 s		
159796	46806		17 34 41.1	+40 54 33	0.000	+0.02	7.2	0.1	0.6	A0 V		210 s		
159833	46809		17 34 41.3	+43 23 51	+0.004	−0.04	7.3	1.1	−0.1	K2 III		260 mx		
159176	208977		17 34 42.3	−32 34 54	+0.001	−0.01	5.70	0.2		O7	−4			m
157008	258770		17 34 42.3	−81 04 04	0.000	+0.02	7.8	0.1	1.4	A2 V		170 s		
159482	122527		17 34 42.9	+6 00 52	−0.031	+0.39	7.41	0.52	4.4	G0 V	−144	40 s		
159695	66147		17 34 44.1	+32 25 01	+0.001	0.00	7.9	0.1	1.4	A2 V		200 s		
159733	66150		17 34 45.2	+34 45 05	−0.001	+0.02	6.8	0.4	2.6	F0 V	−19	71 s		
159358	160653		17 34 46.1	−11 14 31	0.000	0.00	5.55	0.02	−0.2	B8 V		120 s		m
159583	102931		17 34 46.8	+19 08 14	−0.001	−0.02	8.0	0.5	3.4	F5 V		82 s		
159193	208978		17 34 48.3	−34 54 30	−0.002	+0.03	8.0	−0.4	0.4	B9.5 V		330 s		
159481	122529		17 34 48.4	+6 01 26	+0.001	−0.06	7.39	0.52	4.0	dF8	+3	46 s	10638	m
158910	244863		17 34 49.3	−53 54 42	−0.001	−0.03	7.9	0.3	0.6	A0 V		170 s		
159447	141741		17 34 52.5	−1 36 19	0.000	−0.01	8.0	0.1	0.6	A0 V		280 s		
159609	102934		17 34 53.0	+17 48 32	0.000	0.00	8.0	0.1	0.6	A0 V		300 s		
159584	102933		17 34 53.9	+15 18 59	0.000	−0.01	8.0	1.1	0.2	K0 III		360 s	10641	m
159111	228149		17 34 54.4	−42 55 01	+0.001	−0.01	8.04	−0.01	0.4	B9.5 V		340 s		
159561	102932	55 α Oph	17 34 55.9	+12 33 36	+0.008	−0.23	2.08	0.15	0.3	A5 III	+13	19 ts		Rasalhague
159147	228153		17 34 56.0	−40 04 15	+0.002	−0.02	8.0	0.2	1.4	A2 V		170 s		
158852	244862		17 34 56.9	−56 54 52	+0.001	−0.01	7.3	0.0	0.6	A0 V		210 s		
160269	17546	26 Dra	17 34 59.2	+61 52 29	+0.035	−0.51	5.23	0.61	4.5	G1 V	−13	15 ts	10660	m
159644	102937		17 35 01.6	+18 13 22	−0.002	+0.01	7.9	0.9	3.2	G5 IV		88 s		
159610	102935		17 35 02.1	+13 08 16	+0.001	0.00	7.3	0.1	0.6	A0 V		220 s		
158730	253964		17 35 02.9	−62 06 39	−0.001	−0.03	7.40	0.2	2.1	A5 V		120 s		m
159194	208980		17 35 06.1	−38 47 57	−0.001	+0.04	6.6	1.1	0.2	K0 III		160 s		
159415	160660		17 35 07.2	−13 38 13	−0.002	0.00	6.8	0.1	1.4	A2 V		110 s		
159309	185524		17 35 10.1	−28 12 12	−0.001	0.00	7.9	1.3	0.2	K0 III		240 s		
160361	17549		17 35 10.4	+62 27 42	−0.005	0.00	7.1	1.1	−0.1	K2 III		280 s		
159714	85226		17 35 10.6	+24 20 56	+0.001	−0.04	7.40	1.4	−0.3	K4 III	−20	350 s		
159376	185526	52 Oph	17 35 18.3	−22 02 38	0.000	−0.01	6.57	0.01		A p	−12			
159018	244870		17 35 19.8	−53 21 10	0.000	−0.01	6.3	0.3	0.6	A0 V		91 s		
159673	102943		17 35 26.5	+13 21 42	−0.002	−0.02	7.80	0.1	1.7	A3 V		170 s	10648	m
159525	141746		17 35 29.9	−5 55 55	0.000	−0.02	7.80	0.1	0.6	A0 V		280 s		m
159279	208993		17 35 29.9	−37 50 51	−0.005	−0.03	7.7	0.2	3.4	F5 V		73 s		
159545	141747		17 35 31.1	−2 53 29	+0.001	+0.01	7.4	0.0	0.4	B9.5 V		250 s		
159674	102944		17 35 32.4	+11 22 02	−0.001	−0.04	7.7	1.4	−0.3	K5 III		390 s		
159397	185532		17 35 33.9	−24 37 41	−0.004	−0.07	8.0	0.0	3.4	F5 V		82 s		
158895	244866		17 35 34.7	−59 50 46	0.000	−0.01	6.28	−0.08		A0				
--	46824		17 35 36.5	+48 49 31	+0.001	−0.03	8.0	2.3						
159217	228162	σ Ara	17 35 39.4	−46 30 20	−0.003	−0.03	4.59	−0.03		A0 n	+4	47 mn		
159925	66165		17 35 42.2	+37 18 04	+0.001	−0.01	6.10	0.98	0.2	G9 III	+4	150 s		
159564	141748		17 35 42.3	−6 07 15	−0.001	+0.03	6.7	1.1	0.2	K0 III		200 s		
159312	209001		17 35 42.8	−37 26 24	−0.001	−0.04	6.48	0.01		A0				
159527	160669		17 35 43.5	−14 38 37	0.000	−0.02	7.9	1.4	−0.3	K5 III		350 s		
159797	102952		17 35 47.5	+19 50 37	−0.002	−0.03	6.7	0.9	3.2	G5 IV	−11	49 s		
158785	253967		17 35 50.1	−65 02 44	−0.003	−0.04	7.4	1.4	3.2	G5 IV		30 s		
159736	102948		17 35 50.7	+12 02 49	0.000	+0.01	7.0	1.4	−0.3	K5 III		290 s		
159993	66170		17 35 53.5	+38 00 43	−0.001	+0.01	7.9	1.1	0.2	K0 III		340 s		
159341	209003		17 35 57.3	−38 28 24	+0.002	−0.02	8.0	0.7	−0.1	K2 III		410 s		
159834	85232		17 35 59.5	+20 59 46	+0.001	−0.02	6.10	0.18	0.5	A7 III	−17	130 s	10655	m
159340	209004		17 36 00.5	−38 06 08	−0.001	−0.05	7.1	0.2		A m				
159778	102953		17 36 04.0	+11 18 45	0.000	0.00	7.6	1.1	−0.1	K2 III		340 s		
158752	253969		17 36 06.1	−66 51 53	−0.002	−0.02	7.3	1.1	0.2	K0 III		260 s		
159926	85236		17 36 07.7	+28 11 05	−0.003	+0.02	6.4	1.4	−0.3	K5 III	−34	220 s		
159058	244874		17 36 10.0	−58 54 53	+0.001	0.00	8.0	−0.1	0.4	B9.5 V		330 s		
161128	8858		17 36 10.0	+74 00 33	+0.002	+0.01	7.7	0.5	3.4	F5 V		71 s		
159285	228170		17 36 12.4	−47 21 22	0.000	−0.01	8.0	0.9	0.2	K0 III		360 s		
159589	160678		17 36 13.2	−17 51 49	−0.002	−0.08	7.5	0.6	4.4	G0 V		41 s		
159798	122556		17 36 13.8	+9 42 55	−0.006	−0.11	7.3	1.1	0.2	K0 III		82 mx		
158932	253973		17 36 15.4	−63 11 36	0.000	−0.01	7.9	1.1	0.2	K0 III		350 s		
160229	46831		17 36 17.3	+49 26 43	0.000	−0.03	7.5	0.1	1.4	A2 V		160 s		
158819	253970		17 36 17.3	−65 59 51	+0.003	−0.02	7.6	−0.2	0.6	A0 V		250 s		
159758	122552		17 36 20.3	+1 14 48	−0.001	+0.02	7.8	1.1	0.2	K0 III		300 s		
159968	85239		17 36 21.5	+27 34 00	+0.002	−0.07	6.42	1.59	−2.4	M1 II	−36	580 s		
159948	85240		17 36 24.5	+25 36 58	0.000	−0.05	7.37	1.21	−0.1	K2 III		300 s		
160832	17569		17 36 29.7	+68 29 20	−0.002	−0.01	7.9	1.4	−0.3	K5 III		430 s		
159433	209019		17 36 32.6	−38 38 07	−0.001	−0.20	4.29	1.09	0.2	gK0	−49	58 s		
159889	102957		17 36 33.1	+13 15 33	0.000	−0.05	7.1	1.1	0.2	K0 III		240 s		
160291	46836		17 36 33.8	+48 27 27	+0.009	−0.15	7.6	0.5	4.0	F8 V		54 s		

HD	SAO	Star Name	α 2000	δ 2000	μ(α)	μ(δ)	V	B-V	M_v	Spec	RV	d(pc)	ADS	Notes
			h m s	° ′ ″	s	″								
159907	102959		17 36 35.8	+16 23 52	0.000	-0.02	7.7	0.1	0.6	A0 V		260 s		
160054	66175		17 36 36.6	+30 47 06	+0.002	-0.01	5.80	0.16	2.4	A7 V	-17	48 s		
159908	102958		17 36 36.6	+13 48 49	0.000	+0.02	6.8	0.1	1.7	A3 V		110 s		
160290	46838	82 Her	17 36 37.5	+48 35 08	+0.003	+0.06	5.37	1.15	0.0	gK1	+29	110 s		
160933	17572		17 36 39.5	+69 34 15	-0.012	-0.21	6.42	0.58	4.0	F8 V	-53	28 s		
159434	228188		17 36 40.7	-41 55 40	+0.001	+0.01	8.00	1.1	-0.1	K2 III		160 s		m
159384	228184		17 36 41.5	-44 52 43	+0.001	+0.01	7.36	1.43	-0.3	K4 III		340 s		
160362	30480		17 36 46.1	+50 59 03	+0.001	-0.02	8.0	0.1	1.7	A3 V		180 s		
159530	209022		17 36 47.6	-36 31 57	-0.003	+0.03	8.0	1.3	0.2	K0 III		250 s		
161095	8860		17 36 48.1	+72 03 16	-0.004	-0.01	7.3	0.1	0.6	A0 V		200 mx		
159909	102962		17 36 50.0	+10 34 28	-0.010	-0.01	7.90	0.9	3.2	G5 IV		79 mx		
159927	102963		17 36 54.8	+10 02 23	0.000	-0.01	7.9	0.1	0.6	A0 V		290 s		
159286	244885		17 36 56.3	-55 08 17	-0.001	-0.03	7.8	0.4	0.4	B9.5 V		170 s		
160922	17576	28 ω Dra	17 36 56.9	+68 45 28	0.000	+0.32	4.80	0.43	3.4	F5 V	-14	22 ts		m
159513	209024		17 36 57.2	-39 40 06	-0.001	0.00	8.0	1.1	0.2	K0 III		330 s		
--	85248		17 36 58.2	+24 31 58	-0.001	+0.02	8.0	1.3	3.2	G5 IV		46 s		
159595	209028		17 36 58.5	-32 07 43	+0.001	-0.03	6.9	0.6	1.4	A2 V		57 s		V449 Sco, q
159743	160691		17 37 01.3	-18 59 31	-0.008	-0.06	7.4	0.8	3.2	G5 IV		66 s		
159994	102969		17 37 02.1	+14 51 06	0.000	+0.02	7.6	1.1	0.2	K0 III		300 s		
160056	85249		17 37 04.2	+20 15 31	0.000	+0.02	7.9	0.2	2.1	A5 V		150 s		
161178	8862		17 37 08.6	+72 27 21	+0.003	+0.02	5.86	1.01	0.2	G9 III	+8	130 s		
160657	17567		17 37 09.6	+61 42 16	-0.005	+0.02	8.0	1.1	0.2	K0 III		360 s		
159800	160696		17 37 13.0	-14 00 07	+0.002	-0.05	7.8	0.4	3.0	F2 V		85 s		
160520	30488		17 37 14.1	+55 44 26	-0.007	+0.02	7.1	1.1	0.2	K0 III		230 mx		
159439	228194		17 37 16.6	-49 14 41	+0.001	0.00	7.60	0.00	0.4	B9.5 V		270 s		m
159532	228201	θ Sco	17 37 19.0	-42 59 52	+0.001	0.00	1.87	0.40	-5.6	F0 I-II	+1	280 s		
159574	228204		17 37 19.5	-40 19 13	0.000	-0.01	7.20	0.00	0.0	B8.5 V		280 s		m
159891	141766		17 37 20.6	-2 52 26	+0.003	+0.07	7.5	0.9	3.2	G5 IV		70 s		
160245	66189		17 37 20.8	+35 47 56	+0.001	+0.03	7.4	0.7	4.4	G0 V		40 s		
159652	209035		17 37 22.4	-35 43 35	-0.001	-0.02	8.0	0.0	0.0	B8.5 V		350 s		
159874	141765		17 37 23.7	-8 02 14	-0.001	-0.03	7.6	0.1	0.6	A0 V		250 s		
159046	253979		17 37 25.3	-66 31 17	-0.003	+0.01	7.6	1.1	0.2	K0 III		300 s		
159633	209034		17 37 26.6	-38 03 56	-0.001	+0.03	6.26	1.25		K0				
159463	244894		17 37 27.0	-50 03 36	-0.001	-0.10	5.93	1.10	1.7	K0 III-IV		59 s		
160137	85261		17 37 29.5	+20 35 46	0.000	-0.02	7.5	0.4	2.6	F0 V		97 s		
159684	209039		17 37 30.8	-35 20 01	+0.001	-0.03	7.6	0.5	-2.6	B2.5 IV		420 s		
160181	85264	79 Her	17 37 30.9	+24 18 36	-0.001	0.00	5.70	0.09	1.4	A2 V	-3	70 s		
159876	160700	55 ξ Ser	17 37 35.0	-15 23 55	-0.003	-0.06	3.54	0.26	1.7	F0 IV	-43	23 s		m
159877	160701		17 37 35.9	-15 34 16	-0.001	0.00	5.9	0.2	2.1	A5 V		56 s		
162092	8878		17 37 35.9	+79 12 49	-0.001	+0.05	7.2	0.4	3.0	F2 V		70 s		
159702	209044		17 37 37.4	-34 21 05	+0.004	0.00	6.9	0.2	2.6	F0 V		73 s		
159654	228211		17 37 37.6	-40 48 49	+0.002	-0.01	7.6	0.5	3.4	F5 V		65 s		
160419	46850		17 37 40.9	+42 50 02	+0.001	+0.02	8.0	0.1	0.6	A0 V		310 s		
160779	17575		17 37 42.5	+61 57 13	-0.005	+0.01	8.0	0.1	0.6	A0 V		88 mx		
159782	185562		17 37 44.6	-28 26 15	+0.002	0.00	8.0	-0.2	0.6	A0 V		300 s		
159975	141772	57 μ Oph	17 37 50.5	-8 07 08	-0.001	-0.03	4.62	0.11	-0.2	B8 V	-19	73 s		
159784	209051		17 37 50.7	-30 54 34	-0.002	-0.01	8.0	0.6	4.0	F8 V		60 s		
159704	209047		17 37 50.9	-37 51 41	-0.002	-0.12	6.68	0.77	5.6	G8 V		16 s		m
159656	228216		17 37 53.7	-42 34 02	+0.016	-0.35	7.16	0.65	5.2	G5 V		26 ts		
159328	244892		17 37 54.3	-59 25 58	+0.001	-0.01	6.7	2.2	-0.3	K5 III		94 s		
160017	141775		17 38 00.9	-9 01 49	-0.003	-0.06	7.8	0.7	4.4	G0 V		46 s		
160486	46854		17 38 03.1	+43 35 55	-0.001	-0.02	7.2	0.1	1.4	A2 V	+14	140 s		
160506	46855		17 38 03.6	+45 33 24	+0.001	+0.02	7.7	0.7	4.4	G0 V		45 s		
159894	185572		17 38 03.7	-22 34 54	0.000	-0.01	7.8	-0.1	2.6	F0 V		110 s		
159492	244896	π Ara	17 38 05.5	-54 30 01	-0.004	-0.15	5.25	0.20	2.4	A7 V	-4	31 ts		
159441	244895		17 38 05.5	-56 49 15	-0.007	-0.03	7.3	0.2	1.7	A3 V		72 mx		
160246	85267		17 38 06.4	+20 55 17	-0.001	-0.01	7.9	1.1	0.2	K0 III		340 s		
159707	228220		17 38 08.3	-42 52 50	0.000	-0.04	6.10	-0.06	-0.2	B8 V		180 s		
160018	160708		17 38 09.3	-10 55 35	-0.001	-0.01	5.8	1.1	0.2	K0 III	-33	120 s		
160402	66201		17 38 10.4	+34 58 31	0.000	0.00	7.9	1.4	-0.3	K5 III		430 s		
159881	185573		17 38 11.9	-28 02 47	+0.001	-0.01	6.84	1.88		K5				
159846	209059		17 38 16.1	-32 52 25	-0.002	-0.02	7.8	-0.3	0.4	B9.5 V		300 s		
160451	66203		17 38 18.9	+36 44 51	+0.004	+0.01	6.8	1.1	0.2	K0 III		210 s		
160570	46858		17 38 20.5	+47 08 45	-0.001	-0.05	7.8	0.5	4.0	F8 V		57 s		
160230	102984		17 38 22.2	+11 29 10	0.000	0.00	7.6	0.4	3.0	F2 V		85 s		
160061	160709		17 38 23.5	-14 06 15	0.000	0.00	7.9	1.6	-0.5	M2 III		390 s		
158606	257513		17 38 23.6	-74 54 41	0.000	0.00	7.5	-0.5	0.6	A0 V		240 s		
160247	102987		17 38 25.2	+12 44 14	0.000	0.00	7.6	1.1	0.2	K0 III		310 s		
159116	253984		17 38 25.8	-68 31 21	+0.004	-0.02	7.9	1.1	-0.1	K2 III		400 s		
159749	228221		17 38 30.2	-44 52 34	-0.002	-0.03	7.87	1.52		K2				
159807	228228		17 38 36.4	-42 44 00	-0.002	-0.01	7.33	0.03	0.6	A0 V		210 s		
160780	30494		17 38 38.2	+55 45 35	0.000	0.00	7.80	0.5	3.4	F5 V		76 s	10699	m
161052	17584		17 38 38.8	+64 50 37	+0.001	-0.03	8.0	1.1	0.2	K0 III		370 s		
160042	185584		17 38 44.7	-21 54 46	-0.002	-0.01	6.7	0.1	4.4	G0 V	0	29 s		
160507	66214		17 38 49.5	+32 44 22	+0.001	-0.01	6.4	1.1	0.2	K0 III	-15	170 s		

HD	SAO	Star Name	α 2000	δ 2000	μ(α)	μ(δ)	V	B-V	M_v	Spec	RV	d(pc)	ADS	Notes
			h m s	° ′ ″	s	″								
159809	228233		17 38 52.8	−45 45 35	−0.008	−0.15	7.45	1.01		K1 IV		69 mx		m
160365	102999		17 38 57.6	+13 19 45	−0.002	+0.04	6.12	0.56	3.4	F5 V	0	30 s		
160295	122600		17 38 57.7	+2 39 31	0.000	+0.02	7.8	0.4	2.6	F0 V		100 s		
159981	209072		17 38 58.8	−34 29 04	−0.004	−0.09	7.3	0.4	3.4	F5 V		60 s		
159868	228234		17 38 59.5	−43 08 43	−0.019	−0.16	7.24	0.72	5.2	G5 V		24 s		
160043	185588		17 39 00.3	−28 24 44	−0.007	−0.07	7.71	0.42	3.7	F6 V		63 s		
160314	122605		17 39 04.3	+2 03 14	+0.002	−0.04	7.74	0.41	2.0	F4 IV		140 s		m
159098	257517		17 39 04.9	−70 41 38	+0.002	+0.01	7.3	0.1	1.4	A2 V		150 s		
160487	85286		17 39 05.0	+27 03 02	0.000	−0.11	7.4	0.5	3.4	F5 V		63 s		
159937	209073		17 39 05.4	−37 46 34	−0.001	−0.02	7.1	0.6	0.2	K0 III		240 s		
160089	185589		17 39 05.7	−26 56 15	−0.002	−0.07	7.4	0.3	4.4	G0 V		40 s		
160315	122607		17 39 08.2	+2 01 41	+0.002	−0.02	6.26	1.03	0.2	K0 III	0	160 s		m
160541	66218		17 39 09.3	+31 53 08	−0.003	−0.03	7.8	1.4	−0.3	K5 III		410 s		
--	46867		17 39 13.2	+46 09 18	+0.002	0.00	7.9	1.7						
160637	46865		17 39 13.5	+41 02 31	0.000	−0.01	7.9	1.1	0.2	K0 III		340 s		
160542	66220		17 39 15.8	+30 15 12	0.000	−0.05	7.1	0.4	3.0	F2 V		67 s		
159752	244915		17 39 16.6	−53 12 34	−0.002	−0.04	7.8	0.4	1.7	A3 V		100 s		
160346	122610		17 39 16.7	+3 33 19	−0.012	−0.10	6.52	0.96	5.9	dK0	+19	12 ts		
160108	185591		17 39 20.2	−29 31 52	0.000	−0.02	6.7	0.9	0.2	K0 III		200 s		
159941	228239		17 39 20.6	−43 09 53	0.000	−0.01	7.9	1.5	−0.1	K2 III		190 s		
160470	103004		17 39 22.5	+18 00 25	0.000	+0.03	7.9	1.1	−0.1	K2 III		390 s		
160109	185594		17 39 26.7	−29 57 34	+0.001	0.00	7.5	0.6	0.0	B8.5 V		130 s		
160762	46872	85 ι Her	17 39 27.7	+46 00 23	−0.001	0.00	3.80	−0.18	−1.7	B3 V	−20	130 s		m
160276	160725		17 39 28.0	−11 16 02	0.000	−0.01	7.9	0.4	2.6	F0 V		110 s		
160001	209078		17 39 29.3	−39 20 41	+0.007	−0.03	7.2	0.8	4.4	G0 V		25 s		
160254	160724		17 39 29.9	−13 38 59	−0.001	−0.01	7.50	0.4	2.6	F0 V		90 s	10685	m
160740	46871		17 39 30.2	+44 00 06	−0.003	+0.04	7.2	0.1	1.4	A2 V	−30	150 s		
160385	122615		17 39 31.0	+3 23 33	0.000	−0.04	6.80	1.1	0.2	K0 III		200 s	10688	m
160002	209080		17 39 31.5	−39 46 50	+0.002	−0.03	7.6	0.5	3.4	F5 V		65 s		
159958	228243		17 39 33.4	−43 51 35	+0.001	−0.03	7.77	−0.04	0.0	B8.5 V		360 s		
160124	209092		17 39 37.5	−32 19 12	+0.001	0.00	7.0	0.3	−2.3	B3 IV		390 s		
160488	103009		17 39 40.0	+13 17 03	0.000	−0.01	7.6	0.5	3.4	F5 V		69 s		
159810	244923		17 39 46.0	−54 21 24	0.000	−0.02	8.0	0.4	−0.1	K2 III		410 s		
160096	209091		17 39 46.9	−38 22 08	−0.002	−0.04	8.0	0.4	0.2	K0 III		360 s		
160388	141793		17 39 47.2	−4 58 04	+0.001	+0.01	7.76	0.47		F2			10693	m
160883	46879		17 39 47.8	+49 46 48	−0.001	+0.01	6.6	0.1	0.6	A0 V		160 s		
160069	228255		17 39 51.2	−40 16 43	+0.001	−0.01	8.0	0.3	0.0	B8.5 V		400 s		
160509	122627		17 39 55.8	+7 48 04	−0.002	−0.01	7.1	1.1	0.2	K0 III		220 s		
161053	30509		17 39 56.8	+56 52 13	−0.010	−0.08	7.4	1.1	−0.1	K2 III		160 mx		
160677	66231		17 39 57.3	+31 12 09	−0.001	+0.01	6.03	1.58	−0.5	M2 III	−9	200 s		
160438	141795		17 39 57.5	−0 38 21	0.000	−0.01	7.50	0.1	0.6	A0 V		230 s	10696	m
160189	209102		17 39 59.3	−32 26 35	+0.002	+0.01	8.0	0.2	0.4	B9.5 V		230 s		
160202	209105		17 40 01.0	−32 12 03	+0.001	0.00	6.8	0.0	0.0	B8.5 V		220 s		
160678	85301		17 40 03.6	+29 14 14	0.000	−0.06	7.60	1.1	0.2	K0 III		300 s	10703	m
160127	209099		17 40 09.3	−39 57 31	−0.001	−0.02	7.9	0.2	1.4	A2 V		180 s		
160147	209100		17 40 09.6	−39 12 19	+0.001	−0.02	7.5	0.3	0.0	B8.5 V		200 s		
160454	141797		17 40 09.6	−3 32 10	−0.001	0.00	7.6	0.4	2.6	F0 V		99 s		
160221	209111		17 40 11.1	−32 15 22	+0.002	+0.01	7.1	0.4	0.0	B8.5 V		140 s		
160471	141798		17 40 11.6	−2 09 09	−0.002	−0.02	6.19	1.67	−0.5	gM4	−49	200 s		
161538	17601		17 40 17.0	+69 09 00	+0.012	+0.11	7.1	0.4	3.0	F2 V		66 s		
161285	17599		17 40 18.0	+63 40 30	+0.002	+0.13	7.10	0.5	3.4	F5 V		55 s	10728	m
160558	122630		17 40 19.8	+6 33 53	0.000	−0.02	7.3	1.1	−0.1	K2 III		270 s		
160032	228257	λ Ara	17 40 23.4	−49 24 56	+0.008	−0.17	4.77	0.40	3.4	dF5	+4	19 s		
160319	185604		17 40 23.9	−28 55 22	+0.002	+0.01	7.0	0.4	0.0	B8.5 V		140 s		
160071	228259		17 40 33.0	−49 42 17	+0.002	−0.03	7.5	1.3	0.2	K0 III		260 s		
161162	30515		17 40 35.9	+57 18 38	−0.002	+0.03	6.8	1.1	0.2	K0 III	−14	210 s		
160282	209119		17 40 36.2	−35 38 53	−0.001	−0.05	7.90	0.5	3.4	F5 V		78 s		
160950	46883		17 40 37.4	+43 28 15	+0.005	+0.06	6.6	1.1	0.2	K0 III	−29	190 s		
160172	228268		17 40 37.6	−44 40 22	+0.002	−0.02	7.8	1.0	0.2	K0 III		330 s		
160822	66243		17 40 40.9	+31 17 15	−0.005	−0.07	6.28	1.05	0.2	K0 III	−6	150 mx		m
160370	185607		17 40 43.0	−27 53 35	+0.002	−0.02	7.7	0.9	0.4	B9.5 V		84 s		
160335	209125		17 40 43.4	−32 09 18	+0.001	+0.01	8.0	−0.4	0.0	B8.5 V		400 s		
159558	253990		17 40 44.6	−67 51 15	−0.005	−0.02	6.6	0.7	3.2	G5 IV		47 s		
160609	122634		17 40 44.7	+4 09 02	−0.001	−0.03	7.7	0.0	0.0	B8.5 V		320 s		
160430	185611		17 40 49.7	−23 50 10	0.000	0.00	7.90	0.36	0.0	B8.5 V		260 s		
160113	228263		17 40 49.9	−49 57 28	+0.012	−0.18	7.4	1.0	3.2	G5 IV		53 s		
160072	244940		17 40 50.7	−52 37 36	+0.002	−0.06	7.7	1.4	0.2	K0 III		190 s		
160408	185610		17 40 52.1	−27 23 45	−0.003	−0.01	7.8	1.7	−0.5	M2 III		430 s		V551 Oph, v
160371	209132		17 40 58.2	−32 12 51	−0.001	0.00	5.9	1.1	0.2	K0 III		130 s		BM Sco, v
160337	209130		17 41 03.7	−37 39 15	−0.003	−0.03	7.8	0.6	−2.6	B2.5 IV		400 s		
160835	85310		17 41 05.3	+24 30 48	−0.001	+0.06	6.30	1.14	0.0	K1 III	−32	170 s	10715	m
160322	228282		17 41 06.0	−40 05 39	−0.002	−0.07	7.4	0.3	4.4	G0 V		40 s		
160765	103020		17 41 10.8	+15 10 41	−0.001	−0.02	6.34	0.03	1.4	A2 V	−18	97 s		
160951	66253		17 41 13.6	+35 44 18	+0.002	+0.01	8.0	1.1	0.2	K0 III		360 s		
161016	46884		17 41 13.6	+41 39 18	−0.001	+0.02	7.60	0.1	1.4	A2 V	−38	170 s	10722	m

HD	SAO	Star Name	α 2000	δ 2000	μ(α)	μ(δ)	V	B-V	M$_V$	Spec	RV	d(pc)	ADS	Notes
160411	209138		17^{h}41^{m}15^s.2	-34°58'08"	-0^s.007	-0".03	7.8	0.3	3.2	G5 IV		84 s		
160263	228279		17 41 16.0	-46 55 18	+0.001	-0.01	5.79	-0.01		A0				
160725	122641		17 41 16.4	+5 55 57	0.000	+0.01	7.9	0.1	1.7	A3 V		160 s		
161193	30522		17 41 21.5	+51 49 04	-0.004	-0.02	5.99	1.05	0.2	K0 III	-9	130 s		
161617	17608		17 41 23.1	+67 09 06	-0.002	0.00	6.8	1.4	-0.3	K5 III		260 s		
160613	160747	56 o Ser	17 41 24.7	-12 52 31	-0.005	-0.06	4.26	0.08	1.4	A2 V	-30	36 s		
160978	66256		17 41 27.0	+34 16 45	-0.001	+0.04	7.9	0.9	3.2	G5 IV		88 s		
161001	66259		17 41 27.7	+37 51 43	-0.001	+0.01	7.8	0.7	4.4	G0 V		48 s		
160196	244945		17 41 30.3	-53 28 18	-0.004	-0.03	7.90	0.1	1.7	A3 V		170 s		m
160461	209144		17 41 30.7	-34 04 28	+0.003	-0.06	7.9	-0.3	0.6	A0 V		110 mx		
160781	122646		17 41 32.1	+6 18 46	0.000	-0.01	5.95	1.26	0.3	gG7	-31	78 s		
160865	103025		17 41 32.1	+19 08 50	+0.001	0.00	7.9	0.1	0.6	A0 V		290 s		
160953	85319		17 41 35.6	+28 12 46	+0.001	-0.02	8.0	1.1	0.2	K0 III		360 s		
160413	228294		17 41 35.9	-40 07 11	-0.001	-0.02	7.5	1.7	-0.1	K2 III		190 s		
160491	209147		17 41 37.7	-32 06 45	0.000	-0.4	8.0	-0.4	0.4	B9.5 V		330 s		
160866	103029		17 41 40.1	+16 59 19	0.000	-0.08	8.0	0.5	3.4	F5 V		82 s		
160923	85323		17 41 43.0	+24 34 21	-0.001	-0.03	7.4	1.1	0.2	K0 III		270 s		
161238	30523		17 41 47.1	+50 29 11	+0.001	+0.03	7.8	1.1	0.2	K0 III		340 s		
160935	85324		17 41 47.6	+21 30 05	+0.002	-0.13	6.90	0.5	3.4	F5 V	-27	50 s		m
160965	85328		17 41 49.6	+23 45 12	-0.001	-0.01	7.8	0.0	0.0	B8.5 V		370 s		
160355	228292		17 41 49.9	-47 30 46	-0.002	+0.01	7.6	-0.1	3.4	F5 V		71 s		
--	103031		17 41 50.0	+15 24 56	+0.001	-0.01	7.8							
160414	228296		17 41 53.9	-44 47 02	+0.001	-0.02	7.8	0.5	0.6	A0 V		170 s		
161129	46892		17 41 55.0	+41 01 29	+0.001	-0.02	7.9	0.1	0.6	A0 V		290 s		
160823	122652		17 41 55.5	+4 22 00	0.000	-0.01	6.9	0.6	4.4	G0 V		32 s		
161368	30525		17 41 55.8	+56 03 58	0.000	-0.01	7.9	1.1	0.2	K0 III		340 s		
162003	8890	31 ψ Dra	17 41 56.1	+72 08 56	+0.003	-0.27	4.58	0.42	2.8	F5 IV-V	-10	23 ts	10759	m
162004	8891	31 ψ Dra	17 41 57.7	+72 09 24	+0.004	-0.28	5.79	0.53	4.0	F8 V	-10	23 s	10759	m
160910	103033		17 41 58.5	+15 57 07	+0.001	+0.10	5.6	0.5	3.4	F5 V	-44	27 s	10723	m
160529	209151		17 41 58.8	-33 30 12	0.000	+0.01	6.67	1.28	-7.5	A2 Ia	-34	1500 s		
--	--		17 41 59.5	+15 57 06			5.60							
161019	85334		17 42 01.5	+27 38 43	+0.001	+0.04	6.5	0.1	1.7	A3 V	-18	93 s		
162130	8893		17 42 03.6	+73 27 40	0.000	-0.01	7.7	0.1	0.6	A0 V		260 s		
160342	244954		17 42 03.7	-50 30 38	+0.003	-0.01	6.1	2.2	-0.5	M2 III		85 s		
160627	185630		17 42 05.1	-26 50 45	0.000	-0.03	8.0	0.6	2.6	F0 V		76 s		
163240	2928		17 42 11.3	+80 16 57	-0.008	0.00	7.0	1.6	-0.5	M2 III	-28	320 s		
161163	66273		17 42 13.1	+38 04 14	-0.008	+0.11	7.3	0.6	4.4	G0 V		38 s		
159776	253999		17 42 13.2	-69 01 10	+0.006	-0.02	7.60	0.1	1.4	A2 V		170 s		m
160415	228301		17 42 14.4	-48 38 33	+0.002	0.00	8.00	0.4	3.0	F2 V		99 s		m
160966	103038		17 42 15.0	+17 13 42	-0.001	-0.01	7.1	0.1	0.6	A0 V		200 s		
160531	228309		17 42 15.3	-40 27 45	+0.002	-0.02	7.9	1.2	-0.3	K5 III		430 s		
160589	209160		17 42 16.7	-32 31 24	+0.002	+0.01	7.9	-0.6	2.6	F0 V		110 s		V703 Sco, v
160416	228302		17 42 17.1	-49 16 16	-0.001	-0.03	8.0	0.8	3.4	F5 V		39 s		
160575	209158		17 42 17.8	-35 48 03	-0.001	+0.01	7.8	0.2	0.0	B8.5 V		240 s		
160703	185634		17 42 20.1	-23 20 56	-0.006	+0.06	7.7	0.9	3.2	G5 IV		67 s		
161020	103041		17 42 20.6	+19 37 10	-0.003	+0.03	7.7	1.1	0.2	K0 III		310 s		
160480	228307		17 42 20.9	-46 17 57	+0.001	-0.02	7.6	0.3	0.6	A0 V		160 s		
160936	122659		17 42 22.3	+9 26 57	0.000	0.00	7.4	0.1	0.6	A0 V		220 s		
160516	228311		17 42 23.6	-43 34 25	+0.003	-0.03	7.3	1.5	-0.1	K2 III		150 s		
160980	103039		17 42 26.0	+13 43 16	0.000	-0.01	7.7	1.1	0.2	K0 III		310 s		
159495	257521		17 42 26.7	-73 23 34	+0.001	-0.03	6.8	0.1	0.6	A0 V		180 s		
160665	185632		17 42 26.9	-29 26 19	+0.001	-0.01	7.9	0.7	2.6	F0 V		63 s		
161074	85344	83 Her	17 42 28.2	+24 33 50	-0.004	-0.10	5.52	1.46	-0.3	K4 III	-27	89 mx		m
161073	85345		17 42 28.9	+25 48 04	+0.001	+0.03	7.9	1.1	-0.1	K2 III		390 s		
160578	209163	κ Sco	17 42 29.0	-39 01 48	-0.001	-0.03	2.41	-0.22	-3.0	B2 IV	-10	120 s		
160868	141820		17 42 29.3	-2 25 26	0.000	-0.02	8.0	0.9	3.2	G5 IV		91 s		
160869	141821		17 42 35.0	-4 50 58	+0.001	0.00	6.7	1.6	-0.5	M2 III	+39	270 s		
161130	85347		17 42 36.3	+29 25 04	-0.004	-0.03	6.6	0.2	2.1	A5 V	-34	78 s		
162465	8896		17 42 37.6	+75 32 20	-0.004	-0.01	7.9	1.1	-0.1	K2 III		400 s		
161112	85346		17 42 38.2	+26 33 07	+0.002	+0.02	7.51	0.99	0.2	K0 III		290 s	10731	m
160825	160769		17 42 40.0	-15 33 35	+0.001	-0.02	7.4	0.1	0.6	A0 V		210 s		
161501	30532		17 42 44.3	+55 37 43	0.000	0.00	8.0	1.1	0.2	K0 III		360 s		
160482	228313		17 42 46.8	-50 00 03	-0.007	-0.14	7.75	0.51	3.7	F6 V		53 s		
166798	2958		17 42 49.1	+85 40 12	-0.004	-0.02	7.5	0.4	2.6	F0 V		98 s		
160668	209172		17 42 50.9	-36 56 45	0.000	-0.03	5.5	1.8	-0.1	K2 III	-4	54 s		
160399	244960		17 42 53.2	-55 37 39	-0.002	-0.02	7.2	0.7	0.2	K0 III		250 s		
161113	103048		17 43 01.9	+17 40 45	+0.001	-0.01	7.8	1.6	-0.5	M2 III		470 s		
161287	66281		17 43 03.3	+37 34 10	+0.001	-0.01	7.8	0.4	3.0	F2 V		89 s		
161369	46906		17 43 05.4	+44 05 04	-0.004	+0.04	6.5	1.1	-0.1	K2 III	-60	95 mx		
160748	209176		17 43 06.7	-33 03 04	+0.001	0.00	6.4	1.7	-0.5	M2 III		210 s		
161197	85356		17 43 10.5	+24 47 10	-0.003	-0.04	7.92	0.73	3.0	G2 IV		86 s	10737	m
160982	141825		17 43 11.0	-2 17 53	+0.001	-0.02	8.0	1.1	0.2	K0 III		340 s		
161198	85357		17 43 15.4	+21 36 32	-0.010	-0.64	7.50	0.77	5.9	K0 V	+20	22 ts		
161288	66282		17 43 15.6	+32 59 03	-0.002	+0.01	7.6	0.1	1.4	A2 V		170 s		
161164	103050		17 43 15.8	+17 41 23	-0.002	+0.01	8.00	0.1	0.6	A0 V		300 s	10738	m

HD	SAO	Star Name	α 2000	δ 2000	μ(α)	μ(δ)	V	B−V	M$_v$	Spec	RV	d(pc)	ADS	Notes
			h m s	° ′ ″	s	″								
160839	185655		17 43 17.6	−27 53 03	+0.001	−0.01	6.4	0.9	2.1	A5 V		24 s		
161180	103053		17 43 18.3	+18 49 27	0.000	+0.02	8.0	1.1	−0.1	K2 III		410 s		
160435	244964		17 43 18.8	−57 43 30	0.000	−0.05	7.17	3.70		N7.7				V Pav, m,v
161239	85360	84 Her	17 43 21.3	+24 19 40	−0.008	+0.07	5.71	0.65	4.4	G0 V	−26	26 mn		
161149	103052		17 43 21.8	+14 17 43	−0.001	+0.03	6.24	0.42	3.4	F5 V	−42	35 s		
161299	66284		17 43 22.8	+33 56 41	0.000	−0.01	8.0	0.5	3.4	F5 V		82 s		
160648	228325		17 43 24.0	−46 35 17	0.000	−0.01	7.56	−0.04	0.0	B8.5 V		330 s		
160915	185660	58 Oph	17 43 25.6	−21 41 00	−0.007	−0.05	4.87	0.47	3.4	F5 V	+11	20 ts		
160810	209180		17 43 25.7	−35 17 55	−0.001	−0.03	6.8	1.7	−0.1	K2 III		120 s		
159047	257520		17 43 25.8	−78 41 33	−0.008	+0.02	7.4	1.1	−0.1	K2 III		320 s		
161096	122671	60 β Oph	17 43 28.2	+4 34 02	−0.003	+0.16	2.77	1.16	−0.1	K2 III	−12	37 s		Cebalrai
160852	209187		17 43 29.7	−30 10 37	−0.001	−0.01	7.6	0.2	2.1	A5 V		120 s		
160483	244967		17 43 32.8	−57 01 20	−0.001	−0.01	6.64	−0.02	0.4	B9.5 V		170 s		m
160840	209188		17 43 35.0	−32 39 45	+0.001	0.00	6.7	1.2	−0.5	M4 III		280 s		
160841	209186		17 43 35.7	−34 04 53	−0.003	−0.02	7.9	−0.3	0.6	A0 V		280 s		
160715	228335		17 43 36.9	−45 58 14	0.000	+0.02	6.93	−0.03		B9				
160794	209182		17 43 40.8	−39 48 23	0.000	−0.04	8.0	1.1	0.2	K0 III		330 s		
161098	141833		17 43 44.0	−3 55 05	−0.009	−0.25	7.7	0.8	3.2	G5 IV		33 mx		
161183	122674		17 43 44.6	+8 36 54	−0.002	+0.01	7.9	1.1	0.2	K0 III		310 s		
161056	141832		17 43 46.8	−7 04 46	0.000	−0.01	6.30	0.38	−1.7	B3 V n	−26	310 s		
161023	160784		17 43 48.3	−13 30 31	−0.004	−0.11	6.39	0.37	3.0	F2 V		48 s		
160197	—		17 43 50.0	−67 24 30			8.0	1.4	−0.3	K5 III		460 s		
161184	122676		17 43 51.6	+5 50 40	+0.001	−0.02	8.03	0.06	0.6	A0 V		300 s	10741	m
162195	17624		17 43 55.5	+69 04 27	−0.012	−0.03	8.0	0.5	4.0	F8 V		62 s		
160972	185665		17 43 56.6	−26 18 25	+0.001	−0.04	7.9	0.8	2.1	A5 V		63 s		
161186	122679		17 43 56.6	+4 20 00	0.000	+0.02	8.0	1.4	−0.3	K5 III		400 s		
161693	30538		17 43 59.1	+53 48 06	+0.002	−0.02	5.7	0.1	0.6	A0 V	−3	100 s		
161151	141839		17 44 00.6	−1 48 25	0.000	0.00	7.6	0.1	0.6	A0 V		250 s		
161131	141836		17 44 00.7	−5 55 46	0.000	−0.01	7.3	1.1	0.2	K0 III	+1	260 s		
161223	122683		17 44 03.4	+6 03 43	−0.001	+0.01	7.43	0.33		A2				
160754	228345		17 44 04.0	−48 18 42	+0.001	0.00	7.9	1.1	3.2	G5 IV		88 s		
161569	46917		17 44 04.2	+45 02 03	0.000	−0.01	6.6	0.0	0.4	B9.5 V	−9	170 s		
160874	209197		17 44 05.4	−38 48 21	−0.001	−0.05	6.8	0.6	3.2	G5 IV		52 s		
160691	244981	μ Ara	17 44 08.5	−51 50 02	−0.001	−0.19	5.15	0.70	5.2	G5 V	−12	12 ts		
161058	160785		17 44 08.7	−17 44 39	+0.001	−0.04	7.6	0.5	3.4	F5 V		66 s		
160568	244974		17 44 09.2	−58 00 35	0.000	0.00	7.9	1.1	−0.5	M2 III		490 s		
160876	228351		17 44 12.3	−41 39 40	−0.001	+0.01	7.4	0.2	0.0	B8.5 V		300 s		
161242	122686		17 44 12.9	+5 15 02	+0.001	0.00	7.81	1.28		K2				
160940	209204		17 44 14.8	−34 46 14	0.000	+0.03	8.0	0.1	3.4	F5 V		82 s		
161101	160788		17 44 16.6	−17 26 12	−0.001	0.00	7.4	0.8	3.2	G5 IV		65 s		
161464	66290		17 44 16.8	+33 13 32	−0.001	−0.02	7.0	0.0	0.0	B8.5 V	−12	260 s		
161321	103069		17 44 17.1	+14 24 37	0.000	+0.02	6.19	0.21		A m	−31		10749	m
161038	185674		17 44 17.5	−23 40 46	+0.002	−0.03	7.1	0.5	2.6	F0 V		59 s		
161322	103070		17 44 18.3	+13 47 04	0.000	+0.04	6.8	0.1	1.7	A3 V	−48	110 s		
161100	160789		17 44 19.4	−16 51 55	−0.001	−0.06	7.4	0.5	4.0	F8 V		47 s		
159964	257525		17 44 19.6	−72 13 15	+0.002	+0.11	6.49	0.48	4.0	F8 V		32 s		m
162483	8899		17 44 20.6	+72 37 30	−0.003	+0.02	7.4	1.4	−0.3	K5 III		340 s		
161549	46920		17 44 21.8	+40 26 50	0.000	−0.06	7.90	1.4	−0.3	K5 III		390 mx	10760	m
161083	185678		17 44 23.8	−22 11 42	+0.002	−0.01	6.7	0.4	2.6	F0 V	+8	58 s		
160798	228349		17 44 28.6	−49 56 57	+0.001	0.00	7.80	1.1	0.2	K0 III		330 s		m
160877	228357		17 44 30.4	−43 54 18	+0.001	0.00	7.4	1.1	0.2	K0 III		270 s		
161270	122690	61 Oph	17 44 33.9	+2 34 46	0.000	+0.01	5.59	0.07	0.6	A0 V	−31	91 s	10750	m
160602	244980		17 44 35.1	−59 31 02	+0.003	−0.08	8.0	0.8	3.2	G5 IV		89 s		
161289	122691		17 44 35.2	+2 34 44	0.000	+0.01	6.7	0.1	0.6	A0 V	−30	160 s		
160928	228363		17 44 41.8	−42 43 44	0.000	+0.01	5.87	0.16	1.4	A2 V		69 s		
160917	228360		17 44 43.7	−46 02 32	+0.002	0.00	6.72	−0.01	0.4	B9.5 V		180 s		
161303	122696		17 44 44.1	+2 26 50	0.000	+0.02	7.8	0.4	3.0	F2 V		87 s		
160900	228362		17 44 48.3	−47 42 53	+0.003	0.00	7.4	0.0	0.6	A0 V		220 s		
161583	66299		17 44 49.7	+35 13 03	−0.001	+0.01	6.8	1.1	0.2	K0 III		210 s		
161550	66298		17 44 50.7	+31 07 54	−0.001	+0.01	8.00	0.5	4.0	F8 V	−37	63 s	10765	m
161619	66301		17 44 51.8	+38 19 25	0.000	+0.01	7.6	0.1	1.4	A2 V		170 s		
161136	185688		17 44 54.4	−24 41 03	+0.005	−0.02	7.9	1.3	0.2	K0 III		180 mx		
161796	30548		17 44 55.3	+50 02 39	0.000	−0.01	7.3	0.4	−4.6	F3 Ib		2400 s		
160720	244989		17 44 55.7	−57 32 42	+0.002	0.00	6.1	1.2	3.2	G5 IV		22 s		
—	17627		17 44 56.3	+67 23 58	−0.005	−0.04	7.9	1.5						m
160943	—		17 44 57.0	−45 11 00			8.0	0.2	0.6	A0 V		220 s		
160818	—		17 44 59.0	−54 08 05			8.0	0.1	0.6	A0 V		300 s		m
161424	103076		17 45 00.3	+13 01 29	0.000	+0.01	7.8	0.8	3.2	G5 IV		84 s		
161086	209214		17 45 01.9	−35 31 54	0.000	−0.03	7.0	0.5	0.4	B9.5 V		110 s		
160976	228369		17 45 02.4	−44 07 29	+0.002	−0.03	7.2	1.4	0.2	K0 III		250 s		
161479	103077		17 45 02.7	+19 17 25	+0.010	+0.04	7.9	1.1	0.2	K0 III		97 mx		
161674	66307		17 45 09.2	+38 59 27	+0.001	−0.04	8.0	1.1	−0.1	K2 III		410 s		
161355	141853		17 45 13.3	−2 45 45	0.000	−0.02	7.3	0.1	0.6	A0 V		220 s		
160993	228373		17 45 17.5	−45 38 13	0.000	0.00	7.74	0.03	−6.2	B1 I		4600 s		
161502	103079		17 45 27.0	+11 08 40	+0.002	0.00	7.1	0.8	3.2	G5 IV	−21	58 s		

HD	SAO	Star Name	α 2000	δ 2000	μ(α)	μ(δ)	V	B-V	M_v	Spec	RV	d(pc)	ADS	Notes
161712	66309		17 45 28.4	+35 53 35	+0.002	-0.01	7.8	0.2	2.1	A5 V		140 s		
161540	103081		17 45 31.8	+13 01 09	0.000	-0.02	7.7	1.1	0.2	K0 III		310 s		
161480	122709		17 45 33.5	+5 42 58	0.000	0.00	7.68	0.02	-0.2	B8 V		360 s		
161428	141855		17 45 34.0	-3 30 15	+0.001	+0.02	7.70	0.1	0.6	A0 V		260 s	10763	m
160996	245004		17 45 38.7	-50 46 57	0.000	-0.12	7.90	0.5	4.0	F8 V		60 s		m
161695	66311		17 45 40.0	+31 30 17	0.000	0.00	6.23	0.00	-5.2	A0 Ib	+2	1900 s		
161065	228382		17 45 43.2	-46 55 08	+0.004	-0.01	8.00	0.1	0.6	A0 V		300 s		m
160635	254020	η Pav	17 45 43.8	-64 43 25	-0.001	-0.05	3.62	1.19	0.0	K1 III	-8	45 s		
160468	254013		17 45 44.2	-68 38 54	+0.004	0.00	7.3	0.4	3.0	F2 V		72 s		
161372	160804		17 45 48.8	-14 01 21	+0.002	0.00	7.91	1.88	-0.5	M4 III		350 s		
160675	254021		17 45 49.2	-64 19 31	-0.001	-0.02	7.1	0.0	1.7	A3 V		120 s		
161404	160806		17 45 50.5	-10 21 15	0.000	-0.01	7.8	0.1	1.4	A2 V		170 s		
161815	66315		17 45 53.6	+38 52 53	0.000	-0.03	6.52	1.00	0.2	K0 III	-12	180 s		
161542	122715		17 45 54.8	+5 54 27	-0.003	0.00	7.51	0.12		A0	-30			
161068	228385		17 45 56.6	-48 56 03	0.000	0.00	6.7	1.4	-0.3	K5 III		250 s		
161572	122716		17 45 56.8	+5 41 39	-0.001	-0.02	7.59	0.00	-0.2	B8 V	-19	360 s		
161832	66317		17 45 58.3	+39 19 21	+0.001	+0.01	6.70	1.1	-0.2	K3 III	-32	240 s	10782	m
161327	185716		17 46 01.4	-26 58 20	0.000	0.00	7.4	1.0	0.4	B9.5 V		62 s		
161573	122723		17 46 07.2	+5 31 47	+0.001	-0.03	6.86	0.00	-0.2	B8 V	-7	260 s		
162035	30559		17 46 08.1	+52 28 46	-0.001	0.00	8.0	0.1	0.6	A0 V		310 s		
161603	122725		17 46 10.6	+5 39 28	-0.001	-0.02	7.35	0.01	0.2	B9 V	-8	270 s		
160326	—		17 46 22.2	-72 49 19			7.9	1.6	-0.5	M2 III		470 s		
161587	122729		17 46 24.6	+1 02 38	-0.001	-0.01	6.8	1.1	-0.1	K2 III		230 s		
160505	—		17 46 24.6	-70 20 13			8.00	0.9	3.2	G5 IV		91 s		m
161405	185725		17 46 25.0	-24 08 29	-0.002	-0.03	8.0	1.5	0.2	K0 III		190 s		
161797	85397	86 μ Her	17 46 27.3	+27 43 15	-0.023	-0.75	3.42	0.75	3.9	G5 IV	-16	8.1 t	10786	m
161728	103093		17 46 27.7	+17 34 09	-0.002	-0.17	7.8	0.7	4.4	G0 V		49 s		
161330	209237		17 46 28.2	-34 53 10	0.000	-0.04	7.5	0.0	0.0	B8.5 V		300 s		
161622	122731		17 46 28.3	+5 23 48	+0.003	-0.05	7.93	0.45		F0				
161638	122732		17 46 28.4	+6 14 09	0.000	-0.02	7.88	1.03		K0				
162131	30567		17 46 30.0	+53 34 44	0.000	-0.03	7.4	0.1	1.4	A2 V		160 s		
161277	209236		17 46 30.6	-39 15 52	+0.001	0.00	6.9	0.4	0.4	B9.5 V		110 s		
161555	141859		17 46 31.6	-7 59 08	-0.001	-0.15	7.3	0.9	3.2	G5 IV		64 s		
161605	141861		17 46 32.4	-2 51 41	-0.003	-0.05	7.6	1.1	-0.1	K2 III		320 s		
160859	254025		17 46 35.6	-62 45 06	-0.009	-0.38	7.6	0.7	4.4	G0 V		31 mx		
161660	122734		17 46 36.3	+6 07 13	+0.001	-0.02	7.75	0.00		B9				
161388	185729		17 46 37.2	-29 39 57	0.000	-0.01	8.0	1.4	-0.3	K5 III		450 s		
161408	185731		17 46 39.6	-29 23 34	-0.003	0.00	7.9	1.5	-0.3	K5 III		430 s		
161817	85402		17 46 40.3	+25 45 00	-0.005	-0.01	6.97	0.16		A2 VI	-363			
161677	122735		17 46 40.7	+5 46 27	-0.001	-0.01	7.12	0.02	-0.2	B8 V	-26	290 s		
161345	228407		17 46 51.1	-40 23 29	+0.005	0.00	7.4	0.2	2.6	F0 V		93 s		
161346	228408		17 46 52.9	-41 16 54	+0.001	0.00	7.5	1.6	-0.3	K5 III		230 s		
161958	66326		17 46 53.0	+36 05 15	+0.001	-0.02	6.5	1.4	-0.3	K5 III		230 s		
161662	141871		17 46 53.5	-1 43 09	-0.002	+0.02	7.4	1.1	-0.1	K2 III		300 s		
161312	—		17 46 55.5	-43 29 42			7.64	-0.06	0.0	B8.5 V		340 s		m
161434	209247		17 46 57.3	-32 24 50	+0.003	+0.02	8.0	0.2	0.4	B9.5 V		230 s		
—	228409		17 46 57.8	-43 29 37			8.0	-0.3	0.0	B8.5 V		400 s		
162363	30576		17 47 00.8	+59 15 13	-0.004	+0.08	7.8	1.1	0.2	K0 III		210 mx		
161733	122742		17 47 01.8	+5 41 30	-0.002	-0.01	7.99	0.07		B8				m
161378	228410		17 47 05.7	-40 51 36	-0.001	-0.03	8.0	-0.4	0.4	B9.5 V		330 s		
161390	209246		17 47 07.1	-38 06 41	0.000	-0.01	6.43	-0.02		B9				m
161032	245015		17 47 07.2	-59 59 38	-0.001	-0.03	6.7	0.4	2.1	A5 V		64 s		
161833	103106		17 47 07.9	+17 41 50	+0.001	-0.01	5.60	0.01	0.6	A0 V	+2	97 s	10795	m
162132	46954		17 47 07.9	+47 36 44	-0.001	0.00	6.43	0.11	0.6	A0 V	-26	120 s		
162554	17635		17 47 11.6	+63 07 21	-0.003	+0.02	7.8	0.1	1.4	A2 V		190 s		
161469	209250		17 47 13.2	-34 51 01	-0.001	-0.01	7.8	0.9	-0.1	K2 III		380 s		
161959	85413		17 47 13.4	+28 54 37	-0.002	0.00	7.4	0.1	0.6	A0 V	-39	230 s		
161049	254029		17 47 13.9	-60 15 27	-0.002	-0.10	7.3	0.4	2.6	F0 V		80 s		
161920	85415		17 47 24.9	+24 14 47	-0.001	+0.02	7.6	1.4	-0.3	K5 III		380 s		
161643	160819		17 47 27.1	-18 06 26	+0.002	-0.01	7.50	0.4	2.6	F0 V		91 s	10788	m
161511	209256		17 47 27.9	-35 42 07	-0.002	-0.03	7.95	2.85		N7.7	-36			SX Sco, v
161644	185760		17 47 31.2	-20 50 15	0.000	-0.01	7.1	0.3	1.7	A3 V		96 s		
161921	103110		17 47 31.8	+18 53 20	+0.001	-0.01	6.7	0.1	0.6	A0 V	-23	160 s		
163127	8924		17 47 32.4	+72 38 43	-0.007	+0.01	8.0	1.1	-0.1	K2 III		410 s		
161592	185755	3 Sgr	17 47 33.4	-27 49 51	0.000	-0.01	4.56	0.78	-2.0	F7 II	-14	200 s		X Sgr, v
161471	228420	ι¹ Sco	17 47 34.9	-40 07 37	0.000	-0.01	3.03	0.51	-8.4	F2 Ia	-28	1700 s		m
161394	228417		17 47 35.1	-45 29 43	+0.003	-0.03	7.5	-0.1	1.4	A2 V		170 s		
161701	160822		17 47 36.7	-14 43 33	-0.001	-0.02	6.0	0.1		B9				
161900	103111		17 47 36.9	+15 02 05	0.000	0.00	7.8	0.1	0.6	A0 V		270 s		
161148	245021		17 47 36.9	-59 49 34	+0.001	-0.01	7.2	2.3	-0.3	K5 III		100 s		
161191	245024		17 47 37.5	-57 24 14	+0.001	+0.01	8.0	1.1	-0.1	K2 III		410 s		
163183	8927		17 47 37.8	+73 07 57	+0.009	-0.07	7.6	0.7	4.4	G0 V		43 s		
161192	245026		17 47 41.9	-57 26 55	+0.001	+0.02	7.5	0.3	2.6	F0 V		95 s		
161664	185765		17 47 45.4	-22 28 41	0.000	-0.01	6.2	1.3	0.2	K0 III		98 s		
162093	66334		17 47 45.5	+34 55 23	+0.003	-0.01	7.9	0.1	1.4	A2 V		200 s		

HD	SAO	Star Name	α 2000	δ 2000	μ(α)	μ(δ)	V	B-V	M_v	Spec	RV	d(pc)	ADS	Notes
			h m s	° ′ ″	s	″								
161050	254033		17 47 45.9	−63 33 45	−0.001	−0.07	6.9	0.6	4.4	G0 V		31 s		
161575	209263		17 47 47.5	−34 18 40	+0.002	−0.03	6.9	0.4	0.6	A0 V		97 s		
162053	66333		17 47 48.8	+30 21 11	0.000	−0.01	7.8	1.6	−0.5	M2 III		470 s		
162094	66336		17 47 51.8	+34 16 39	−0.001	−0.02	6.60	0.00	−1.6	B3.5 V	−36	430 s	10807	m
161885	122757		17 47 52.5	+6 24 10	−0.002	−0.04	7.54	1.62		Ma				
161868	122754	62 γ Oph	17 47 53.4	+2 42 26	−0.002	−0.07	3.75	0.04	0.6	A0 V	−5	35 ts		
162133	66338		17 47 54.1	+36 00 05	−0.001	−0.03	7.3	0.1	1.4	A2 V		150 s		
162159	66341		17 47 56.4	+36 33 17	+0.001	−0.02	6.6	1.6	−0.5	M2 III	−15	270 s		
161612	209265		17 47 57.2	−34 01 07	−0.018	−0.55	7.20	0.71	5.6	G8 V		21 mx		
162208	66344		17 47 58.3	+39 58 50	−0.002	+0.13	7.6	0.1	0.6	A0 V		52 mx		
162299	46964		17 47 59.2	+47 35 03	−0.001	−0.02	7.0	1.1	0.2	K0 III		230 s		
161922	122761		17 47 59.8	+9 46 57	−0.001	+0.01	7.1	0.1	1.4	A2 V		140 s		
162261	46963		17 48 00.9	+44 06 24	−0.006	0.00	7.8	0.8	3.2	G5 IV		84 s		
161013	254034		17 48 06.5	−66 12 15	+0.004	−0.01	7.5	0.0	0.4	B9.5 V		270 s		
162230	46961		17 48 06.9	+40 35 31	−0.006	−0.02	7.8	0.1	1.7	A3 V		30 mx		
160860	254028		17 48 10.5	−69 33 41	+0.019	−0.08	8.0	1.1	0.2	K0 III		100 mx		
161649	209269		17 48 11.1	−34 34 58	+0.002	−0.05	7.5	0.3	0.0	B8.5 V		200 s		
162209	66346		17 48 13.1	+38 13 56	+0.007	−0.09	7.8	0.6	4.4	G0 V		47 s		
161051	254037		17 48 13.5	−65 34 15	−0.007	−0.02	7.1	0.1	0.6	A0 V		110 mx		
161439	245039		17 48 14.3	−51 26 49	+0.003	+0.02	8.0	1.0	3.4	F5 V		37 s		
161628	209270		17 48 15.1	−35 53 43	+0.001	−0.03	6.9	0.5	0.4	B9.5 V		94 s		
161902	141884		17 48 15.8	−1 01 01	−0.001	−0.02	7.7	1.4	−0.3	K5 III		360 s		
161665	209277		17 48 16.9	−32 40 16	0.000	−0.04	7.2	0.5	1.4	A2 V		74 s		
161516	228428		17 48 17.3	−47 15 18	+0.002	+0.01	7.4	0.1	0.6	A0 V		190 s		
161903	141886		17 48 19.0	−1 48 30	0.000	0.00	7.7	0.1	1.4	A2 V		170 s		
161941	122766		17 48 20.0	+3 48 15	0.000	0.00	6.22	0.16		A0	−44			
−−	103124		17 48 20.2	+19 04 10	0.000	+0.03	8.0							
161773	185780		17 48 22.5	−23 13 58	+0.003	−0.01	8.0	0.0	3.0	F2 V		98 s		
162076	85429		17 48 24.6	+20 33 55	+0.002	+0.01	5.69	0.94	0.3	gG5	−26	110 s		
161718	209283		17 48 26.0	−30 17 56	−0.004	+0.02	7.8	1.0	4.4	G0 V		26 s		
161756	185779		17 48 27.7	−26 58 30	0.000	−0.01	6.35	0.12	−2.3	B3 IV		350 s		
161653	209278		17 48 28.3	−38 08 00	+0.003	0.00	7.20	0.00		B0.5 I k				
161961	141889		17 48 36.8	−2 11 46	+0.002	+0.01	7.81	0.23		B0.5 III	−11			
161420	245044		17 48 37.7	−55 24 07	−0.004	+0.03	6.3	0.7	2.6	F0 V		34 s		
160921	254035		17 48 38.1	−69 55 51	+0.001	−0.01	7.5	−0.2	0.4	B9.5 V		270 s		
162897	17642		17 48 40.1	+64 46 02	−0.004	+0.06	7.6	0.1	0.6	A0 V		95 mx		
161667	209282		17 48 42.6	−39 53 56	+0.001	−0.02	6.99	−0.03	0.4	B9.5 V		210 s		
162028	122776		17 48 43.2	+5 42 04	0.000	−0.01	7.49	0.03	0.2	B9 V	−6	290 s		
162500	46977		17 48 46.0	+48 50 08	−0.002	0.00	7.5	1.1	−0.1	K2 III		340 s		
161741	209291		17 48 47.5	−35 03 25	+0.001	0.00	7.5	0.0	1.4	A2 V		160 s		V393 Sco, m,v
162161	103131		17 48 47.7	+19 15 19	−0.001	+0.02	6.00	0.02	0.0	A0 IV	−22	150 s		
162211	85437	87 Her	17 48 49.0	+25 37 22	0.000	−0.04	5.12	1.16	−0.1	K2 III	−26	110 s		
161719	209289		17 48 49.4	−37 24 45	−0.005	−0.04	8.0	0.1	0.2	K0 III		190 mx		
164372	8943		17 48 51.3	+79 19 58	−0.015	+0.07	7.4	0.4	3.0	F2 V		77 s		
161743	209292		17 48 57.9	−38 07 07	+0.002	−0.01	7.6	−0.4	0.6	A0 V		260 s		
162338	66352		17 48 58.3	+37 03 42	+0.006	−0.09	7.1	0.6	4.4	G0 V		35 s		
163148	17648		17 49 01.8	+68 20 29	−0.001	−0.02	8.0	0.1	1.4	A2 V		210 s		
162643	30593		17 49 02.2	+54 11 40	+0.002	+0.04	6.6	0.1	1.4	A2 V		110 s		
162898	17645		17 49 03.8	+62 47 53	−0.009	+0.04	8.00	0.4	3.0	F2 V		100 s		m
161981	160835		17 49 04.1	−11 20 29	−0.004	−0.02	8.00	1.1	0.2	K0 III		220 mx		m
162579	30591	30 Dra	17 49 04.1	+50 46 52	−0.006	+0.21	5.02	0.02		A2	−55	42 mn		
161721	228447		17 49 05.9	−41 27 42	+0.002	−0.01	7.4	0.5	2.6	F0 V		93 s		
−−	17646		17 49 05.9	+63 03 00	−0.002	+0.03	8.0	0.4						
160931	257530		17 49 07.0	−71 05 43	+0.002	−0.01	7.6	0.1	0.6	A0 V		250 s		
161840	209303		17 49 10.3	−31 42 12	+0.001	−0.02	4.83	−0.04	−0.2	B8 V	−13	98 s		
162231	85441		17 49 11.6	+20 11 51	−0.002	+0.05	7.7	1.1	0.2	K0 III		310 s		
161853	209306		17 49 16.4	−31 15 17	0.000	+0.01	7.94	0.23		O7.5				
161852	209307		17 49 17.2	−30 35 51	+0.006	−0.04	6.80	0.4	2.6	F0 V		68 s		m
162113	122787		17 49 18.8	+1 57 40	−0.002	+0.06	6.47	1.24		K2	−58			
161945	185799		17 49 20.6	−23 22 07	+0.001	0.00	8.0	−0.3	2.6	F0 V		120 s		
161807	209304		17 49 24.6	−38 58 59	0.000	+0.02	7.01	−0.08	−1.7	B3 V nn		520 s		
161908	185797		17 49 26.4	−29 26 20	+0.005	−0.04	8.0	0.7	2.6	F0 V		66 s		
163989	8939	35 Dra	17 49 26.8	+76 57 46	+0.011	+0.25	5.04	0.49	2.2	F6 IV	−23	35 s		
163214	17651		17 49 27.7	+68 41 20	−0.003	+0.07	6.9	0.1	1.4	A2 V		100 mx		
161855	209309		17 49 28.8	−35 22 39	0.000	+0.01	7.44	0.06		A0				
162163	122793		17 49 33.7	+0 54 22	−0.001	0.00	7.9	1.4	−0.3	K5 III		390 s		
162825	30596		17 49 33.7	+56 49 37	−0.001	0.00	7.6	1.4	−0.3	K5 III		380 s		
162300	85445		17 49 34.1	+20 38 13	−0.001	−0.03	7.6	1.4	−0.3	K5 III		380 s		
162178	122796		17 49 35.6	+4 22 36	−0.002	−0.02	7.6	0.2	2.1	A5 V		120 s		
162010	−−		17 49 37.7	−20 00 09			7.70	0.1	1.4	A2 V		170 s		m
162525	66361		17 49 40.5	+39 36 27	+0.002	−0.03	7.9	1.1	0.2	K0 III		360 s		
162484	66358		17 49 40.8	+36 24 58	−0.001	+0.01	7.9	0.2	2.1	A5 V		150 s		
162318	85448		17 49 42.2	+20 55 22	0.000	0.00	7.7	1.1	−0.1	K2 III		360 s		
161841	228459		17 49 46.8	−41 26 31	+0.001	0.00	7.5	0.4	0.4	B9.5 V		260 s		
160580	257527		17 49 47.7	−76 12 06	+0.009	−0.04	7.9	1.1	0.2	K0 III		290 mx		

HD	SAO	Star Name	α 2000	δ 2000	μ(α)	μ(δ)	V	B-V	M_v	Spec	RV	d(pc)	ADS	Notes
161892	209318		17^{h}49^{m}51.3^s	−37°02'36"	+0^s.005	+0".03	3.21	1.17	−0.1	K2 III	+25	46 s		G Sco, m
162467	66359		17 49 52.3	+31 02 13	0.000	−0.02	7.9	0.4	2.6	F0 V		110 s		
161475	254048		17 49 52.6	−61 42 52	0.000	−0.02	6.6	0.0	0.4	B9.5 V		160 s		
162118	160845		17 49 55.2	−12 36 35	−0.002	−0.02	8.0	1.1	0.2	K0 III		290 s		
162385	85452		17 49 56.7	+20 52 11	0.000	−0.03	7.6	1.4	−0.3	K5 III		380 s		
162030	185815		17 49 57.3	−24 12 24	−0.004	−0.03	7.1	1.3	0.2	K0 III		160 s		
162622	46991		17 49 58.4	+41 57 17	0.000	−0.05	8.00	0.4	3.0	F2 V		100 s	10829	m
161877	228462		17 49 58.7	−41 22 45	+0.001	+0.01	7.93	−0.06	0.4	B9.5 V		320 s		
162428	85456		17 50 01.0	+24 28 05	0.000	0.00	7.0	0.1	0.6	A0 V	−14	190 s		
162732	46997	88 Her	17 50 03.1	+48 23 39	0.000	+0.01	6.68	−0.11		A pe	−16			
161708	---		17 50 05.5	−54 08 20			8.0	0.9	−0.1	K2 III		410 s		
161441	254047		17 50 07.1	−64 18 16	−0.001	+0.03	6.6	0.4	2.6	F0 V		56 s		
162365	103141		17 50 07.5	+15 29 43	+0.001	+0.01	8.04	−0.12	−2.5	B2 V	−9	1200 s		
161985	209324		17 50 07.6	−35 09 08	−0.001	−0.04	7.80	1.1	−0.1	K2 III		350 s		m
162015	209328		17 50 08.3	−30 57 38	−0.005	−0.03	7.8	1.2	−0.1	K2 III		210 mx		
162249	141900		17 50 09.0	−0 40 45	−0.001	−0.02	7.9	0.0	0.0	B8.5 V		350 s		
162319	122806		17 50 09.0	+9 50 54	−0.002	−0.04	6.75	1.39		K5				
162970	30604		17 50 09.5	+59 02 23	0.000	+0.01	7.6	0.4	2.6	F0 V		98 s		
164428	8946		17 50 10.3	+78 18 24	+0.004	+0.02	6.24	1.44	−0.3	K5 III	−7	200 s		
161912	228466	ι² Sco	17 50 10.9	−40 05 26	+0.001	−0.01	4.81	0.26		A2	−18	31 mn		m
161498	254049		17 50 13.7	−62 40 49	+0.004	−0.01	7.4	1.1	0.2	K0 III		270 s		
162485	85459		17 50 14.8	+25 17 27	0.000	+0.01	6.80	0.1	1.4	A2 V	−35	120 s	10827	m
161931	228467		17 50 15.0	−41 26 29	+0.001	+0.01	8.0	0.4	0.4	B9.5 V		320 s		
161858	228464		17 50 18.8	−47 13 48	+0.003	+0.01	7.5	0.2	0.4	B9.5 V		180 s		
162083	185826		17 50 21.0	−27 03 42	−0.001	−0.04	6.8	0.8	1.7	A3 V		42 s		
162555	85464		17 50 22.7	+29 19 20	+0.002	+0.05	5.50	1.05	0.0	K1 III	−15	130 s		
161566	254050		17 50 25.4	−61 55 27	−0.007	+0.04	6.8	0.7	3.4	F5 V		33 s		
161969	228473		17 50 26.6	−41 38 55	0.000	−0.02	7.20	1.4	−0.3	K5 III		310 s		m
161783	245065		17 50 28.1	−53 36 44	−0.001	0.00	6.00	0.1	−1.7	B3 V	−8	350 s		m
161880	228465		17 50 28.5	−48 17 06	−0.001	0.00	6.80	1.1	0.2	K0 III		210 s		m
161970	228474		17 50 28.5	−42 18 04	−0.002	0.00	8.0	−0.4	0.6	A0 V		300 s		
162751	47003		17 50 28.7	+44 29 20	+0.005	−0.02	7.7	0.1	1.4	A2 V	−1	130 mx		
162119	185830		17 50 29.7	−24 51 11	+0.004	−0.01	7.9	1.6	−0.3	K5 III		220 mx		
161784	245063		17 50 29.9	−54 43 52	−0.003	−0.03	7.8	0.6	3.2	G5 IV		83 s		
163010	30607		17 50 31.1	+57 26 35	−0.002	−0.01	7.9	0.0	0.4	B9.5 V		320 s		
162667	66373		17 50 31.2	+36 50 46	−0.001	+0.01	7.0	0.6	4.4	G0 V		33 s		
161859	228468		17 50 34.4	−49 29 42	−0.001	+0.03	7.20	0.2	2.1	A5 V		100 s		m
162404	122820		17 50 37.0	+9 32 32	−0.001	−0.01	7.5	0.1	0.6	A0 V		230 s		
162900	30605		17 50 37.7	+51 14 32	−0.002	+0.04	7.9	1.1	0.2	K0 III		350 s		
162120	185832		17 50 41.1	−29 16 44	+0.001	−0.03	8.00	0.1	1.4	A2 V		200 s	10820	m
162468	103145		17 50 43.3	+11 56 48	−0.002	−0.02	6.17	1.25		K1 III−IV	−49			
162085	209337		17 50 43.4	−35 31 58	−0.001	−0.04	7.6	0.2	0.4	B9.5 V		190 s		
162047	209336		17 50 48.1	−38 28 30	−0.003	+0.01	8.0	−0.1	0.0	B8.5 V		400 s		
162570	85468		17 50 48.2	+22 18 59	+0.001	−0.02	5.9	0.1	1.7	A3 V	+4	69 s		
161973	228478		17 50 49.1	−46 21 49	+0.002	0.00	7.6	1.3	0.2	K0 III		300 s		
162143	209343		17 50 50.2	−30 57 21	−0.001	−0.05	7.9	1.4	0.2	K0 III		210 s		
162021	228481		17 50 51.7	−42 19 47	−0.001	−0.05	6.67	1.04	0.2	K0 III		190 s		
162102	209340		17 50 52.3	−33 42 17	+0.002	+0.03	7.5	0.8	3.2	G5 IV	−18	72 s		RY Sco, m,v
160819	257531		17 50 55.0	−76 03 28	+0.004	−0.06	7.2	1.1	0.2	K0 III		250 s		
162668	66378		17 50 57.1	+29 59 45	0.000	0.00	6.7	0.1	1.4	A2 V	−21	120 s		
161915	245070		17 50 58.2	−50 17 25	0.000	+0.01	7.4	1.9	−0.5	M2 III		250 s		
162571	103149		17 50 58.7	+18 21 00	+0.001	0.00	7.9	1.1	0.2	K0 III		340 s		
162144	209345		17 51 02.0	−35 04 15	+0.001	−0.02	7.5	0.5	0.4	B9.5 V		120 s		
162501	122829		17 51 04.0	+8 31 49	+0.002	+0.01	6.8	0.4	2.6	F0 V		69 s		
162185	209353		17 51 04.5	−30 58 12	−0.001	0.00	7.8	0.4	0.4	B9.5 V		160 s		
162486	122827		17 51 04.8	+5 13 43	+0.001	+0.01	7.1	1.4	−0.3	K5 III		280 s		
162145	209346		17 51 05.8	−35 20 50	−0.001	−0.01	6.9	0.5	0.4	B9.5 V		99 s		
162255	185842		17 51 07.5	−22 55 15	+0.003	−0.07	7.1	0.5	4.4	G0 V		35 s		
162807	66386		17 51 07.7	+38 26 33	+0.002	+0.01	7.7	0.8	3.2	G5 IV		79 s		
162880	47012		17 51 10.1	+44 54 29	+0.007	+0.02	8.00	0.4	2.6	F0 V	0	120 s	10849	m
161917	245072		17 51 10.7	−53 07 50	+0.001	−0.01	6.09	0.00	0.6	A0 V		120 s		
162220	209357		17 51 12.3	−30 33 26	0.000	−0.01	6.47	0.04	0.6	A0 V n		140 s		
162826	47009		17 51 13.9	+40 04 21	−0.002	+0.01	6.5	0.6	4.4	G0 V	+2	27 s		
164056	8945		17 51 24.7	+74 34 08	0.000	−0.01	6.9	0.1	1.7	A3 V		110 s		
161935	245075		17 51 25.6	−53 50 18	0.000	−0.05	7.0	0.1	1.7	A3 V		110 s		
162189	228489		17 51 32.5	−40 46 21	−0.003	−0.05	6.00	1.6	−0.5	M2 III		200 s		m
161747	254054		17 51 33.9	−62 38 07	0.000	−0.01	7.7	1.1	−0.1	K2 III		360 s		
161814	254057		17 51 35.3	−60 09 52	0.000	−0.03	5.80	1.1	0.2	K0 III		130 s		m
162272	209368		17 51 35.6	−31 24 45	0.000	+0.03	7.5	1.4	0.2	K0 III		170 s		
161568	---		17 51 36.6	−67 37 22			7.9	1.1	−0.1	K2 III		400 s		
162343	185850		17 51 38.9	−23 40 40	+0.001	−0.02	7.2	0.0	4.0	F8 V		44 s		
162240	209365		17 51 41.4	−36 28 02	−0.006	+0.01	7.5	0.3	0.2	K0 III		140 mx		
162123	228490		17 51 44.4	−45 36 02	+0.001	0.00	6.2	1.5	3.2	G5 IV		14 s		
163203	30616		17 51 44.6	+56 07 11	+0.002	−0.01	7.1	0.1	1.7	A3 V		120 s		
162705	103156		17 51 47.9	+15 00 23	+0.002	−0.02	7.6	0.4	2.6	F0 V		100 s		

HD	SAO	Star Name	α 2000	δ 2000	μ(α)	μ(δ)	V	B-V	M_v	Spec	RV	d(pc)	ADS	Notes
			h m s	° ′ ″	s	″								
162434	160868		17 51 48.3	−19 31 31	0.000	−0.07	6.9	0.2	2.6	F0 V		73 s		
158651	258778		17 51 50.2	−84 06 37	−0.003	+0.02	7.8	0.7	3.2	G5 IV		83 s		
162287	209371		17 51 51.7	−35 21 51	+0.002	+0.03	7.2	0.2	0.4	B9.5 V		160 s		V720 Sco, v
162690	103157		17 51 52.0	+12 00 28	0.000	−0.02	7.8	1.1	0.2	K0 III		310 s		
162628	122841		17 51 52.4	+2 53 59	0.000	−0.02	8.00	0.1	0.6	A0 V		280 s	10844	m
162305	209373		17 51 52.9	−34 47 00	+0.004	−0.04	7.9	−0.3	0.6	A0 V		100 mx		
162389	185854		17 51 55.0	−25 46 23	+0.002	−0.02	7.8	1.1	0.2	K0 III		300 s		
162734	103161		17 51 58.4	+15 19 33	0.000	+0.03	6.80	1.1	0.2	K0 III	−43	210 s	10850	m
162596	141913		17 51 59.2	−1 14 13	−0.001	0.00	6.35	1.12		K0				
162648	122846		17 51 59.9	+4 29 14	0.000	0.00	7.0	1.4	−0.3	K5 III	−49	260 s		
162651	122843		17 52 00.5	+1 05 58	−0.001	+0.02	7.2	0.1	0.6	A0 V	−29	190 s		
163075	47023		17 52 00.7	+46 38 36	+0.003	−0.12	6.5	1.1	0.2	K0 III	−28	180 s		
162691	122850		17 52 00.7	+6 43 48	0.000	+0.04	8.0	0.4	3.0	F2 V		95 s		
162828	85485		17 52 02.3	+24 18 15	−0.001	−0.02	7.0	1.1	−0.1	K2 III		260 s		
162772	103162		17 52 02.3	+16 37 57	−0.001	0.00	7.4	0.1	1.4	A2 V		160 s		
162915	66396		17 52 03.0	+33 53 19	0.000	0.00	7.2	1.1	0.2	K0 III		250 s		
162989	66402		17 52 04.6	+39 58 55	−0.002	+0.05	6.04	1.33	−0.3	gK4	−66	190 s		
161918	254059		17 52 04.9	−60 23 38	+0.006	−0.07	7.50	0.4	2.6	F0 V		96 s		m
162652	122847		17 52 05.1	+1 06 40	0.000	0.00	6.80	1.1	−0.1	K2 III		220 s		m
162864	85486		17 52 05.7	+26 17 26	+0.001	−0.04	8.0	0.1	0.6	A0 V		300 s		
162708	122854		17 52 07.0	+8 49 23	0.000	0.00	7.8	0.1	1.4	A2 V		180 s		
162243	228497		17 52 07.6	−43 12 19	+0.001	−0.02	7.9	0.5	0.4	B9.5 V		320 s		
163076	47022		17 52 08.9	+42 51 26	−0.001	−0.01	7.7	0.0	0.4	B9.5 V		280 s		
162510	185874		17 52 09.5	−20 04 44	−0.002	−0.02	8.0	−0.3	0.6	A0 V		300 s		
––	47029		17 52 12.8	+46 28 30	−0.002	−0.07	7.4	1.7						
162936	66399		17 52 13.1	+32 00 50	−0.001	−0.01	7.1	0.1	0.6	A0 V	−21	200 s		
162374	209383		17 52 13.5	−34 47 57	0.000	0.00	5.90	−0.12	−0.2	B8 V	−14	170 s		
162438	––		17 52 13.6	−28 29 49			8.00	0.4	3.2	G5 IV		89 s	10838	m
162415	209391		17 52 16.7	−31 19 42	−0.002	−0.01	6.9	1.7	−0.3	K5 III		210 s		
162949	66403		17 52 16.8	+34 12 03	+0.001	−0.06	8.0	0.4	3.1	F3 V	−11	94 s		
162151	245082		17 52 18.4	−51 42 50	+0.004	+0.03	7.8	0.6	3.4	F5 V		59 s		
164445	8952		17 52 18.6	+76 00 02	−0.006	+0.01	7.2	0.4	2.6	F0 V		81 s		
162492	185877		17 52 18.7	−22 40 09	−0.001	+0.02	7.3	0.3	1.4	A2 V		110 s		
162561	160877		17 52 19.1	−16 58 10	−0.005	−0.04	7.9	1.1	0.2	K0 III		170 mx		
162563	160875		17 52 19.2	−19 07 18	−0.001	−0.03	7.3	0.2	0.6	A0 V		160 s		
162391	209390		17 52 19.6	−34 24 59	0.000	0.00	5.8	1.5	0.2	K0 III		61 s		
162692	141921		17 52 22.0	−1 25 49	−0.002	−0.02	7.8	0.1	1.4	A2 V		180 s		
162562	160879		17 52 23.3	−18 05 24	0.000	+0.01	7.5	0.2	2.1	A5 V		110 s		
163466	17664		17 52 25.3	+60 23 47	0.000	+0.04	6.8	0.1	1.4	A2 V	−16	120 s		
162808	103164		17 52 25.4	+11 59 29	0.000	+0.05	8.0	0.9	3.2	G5 IV		91 s		
162736	122857		17 52 26.5	+2 59 38	+0.002	−0.01	7.9	0.1	1.4	A2 V		180 s		
162865	103168		17 52 31.1	+16 54 01	+0.003	−0.03	6.6	0.5	3.4	F5 V	−4	44 s		
162950	85497		17 52 32.0	+27 11 38	+0.002	−0.02	7.90	0.2		A m				m
162754	122859		17 52 33.8	+1 42 34	+0.001	0.00	7.7	0.9	3.2	G5 IV		78 s		
––	66412		17 52 34.6	+35 45 51	0.000	+0.01	7.9	1.8						
162774	122861		17 52 35.2	+1 18 18	−0.003	−0.02	5.95	1.58		K5	−65			
162713	141925		17 52 36.9	−5 15 42	0.000	0.00	6.9	1.1	0.2	K0 III		190 s		
162714	141926		17 52 38.6	−6 08 38	0.000	−0.01	6.21	1.40	−4.5	G0 Ib var	−5	890 s		Y Oph, v
162829	122864		17 52 38.8	+8 38 52	+0.001	−0.04	7.6	0.4	2.6	F0 V		99 s		
162418	––		17 52 41.9	−38 38 49			7.6	−0.1	0.0	B8.5 V		330 s		
163113	66421		17 52 42.1	+38 49 13	−0.001	−0.01	7.00	1.59	0.2	K0 III		82 s		
162775	141932		17 52 47.5	−3 33 49	0.000	+0.03	7.5	1.1	−0.1	K2 III		270 s		
162496	209397		17 52 49.0	−34 06 51	0.000	+0.02	6.0	1.3	0.2	K0 III		100 s		
162916	103172		17 52 49.1	+17 50 00	+0.004	−0.03	8.0	0.4	3.0	F2 V		98 s		
162154	245086		17 52 51.8	−56 09 17	+0.001	−0.02	6.7	0.8	0.2	K0 III		200 s		
162755	141933		17 52 52.1	−5 21 04	+0.002	−0.02	7.8	1.1	0.2	K0 III		270 s		
162396	228510		17 52 52.4	−41 59 48	+0.014	−0.20	6.20	0.54	4.0	F8 V		27 s		
162515	209401		17 52 55.7	−35 01 07	+0.001	+0.02	6.5	0.1	0.6	A0 V		150 s		
163788	17671		17 52 55.8	+66 25 06	0.000	−0.04	7.6	1.1	0.2	K0 III		310 s		
163204	47034		17 52 56.5	+42 12 45	−0.003	0.00	7.9	0.1	1.4	A2 V		200 s		
162716	160884		17 52 57.6	−12 24 12	0.000	0.00	8.0	1.1	0.2	K0 III		290 s		
162517	209404		17 52 57.7	−35 37 27	+0.004	−0.05	6.03	0.34	1.9	F2 IV		67 s		
162756	141935		17 52 58.6	−7 55 11	−0.003	−0.26	7.64	0.62	3.6	G0 IV-V	−124	63 s	10858	m
162812	141939		17 53 03.0	−2 34 46	−0.004	−0.05	7.47	1.63		Mc				V533 Oph, v
162757	160885		17 53 03.5	−10 53 59	+0.003	−0.04	6.3	0.9	3.2	G5 IV	−35	41 s		
162739	––		17 53 03.6	−13 38 38			7.50	0.5	3.4	F5 V		63 s	10856	m
163184	66428		17 53 05.4	+39 41 14	+0.002	−0.01	7.9	1.4	−0.3	K5 III		430 s		
162155	245088		17 53 10.9	−58 51 52	+0.001	−0.01	7.3	0.5	3.0	F2 V		58 s		
162796	160889		17 53 11.2	−10 22 18	−0.004	−0.04	7.9	0.5	3.4	F5 V		74 s		
162585	209410		17 53 12.3	−32 01 56	0.000	0.00	6.7	1.5	0.2	K0 III		110 s		
162917	122880		17 53 14.0	+6 06 05	−0.008	+0.07	5.77	0.42	3.3	dF4	−33	29 ts		
162565	209407		17 53 15.0	−37 04 49	−0.001	−0.03	7.5	0.6	0.2	K0 III		280 s		
162834	141942		17 53 15.0	−5 55 43	−0.001	−0.02	6.90	1.1	0.2	K0 III		190 s	10861	m
162576	209409		17 53 16.0	−34 37 14	+0.002	0.00	6.8	0.6	0.4	B9.5 V		79 s		
162655	185892		17 53 16.7	−27 37 44	0.000	−0.01	7.8	0.2	1.4	A2 V		150 s		

HD	SAO	Star Name	α 2000	δ 2000	μ(α)	μ(δ)	V	B-V	M$_v$	Spec	RV	d(pc)	ADS	Notes
163217	47037	90 Her	17^{h}53^{m}17.8	+40°00'28"	+0^s.001	+0".05	5.16	1.18	-0.2	K3 III	-35	120 s	10875	m
161955	254063		17 53 18.2	-65 29 20	-0.001	-0.08	6.49	1.08	0.2	gK0		160 s		
163077	85508		17 53 18.2	+24 59 29	-0.007	-0.05	7.92	0.74	6.1	K1 V	+6	23 s	10871	m
162904	122881		17 53 18.9	+2 40 04	0.000	-0.01	7.3	1.1	0.2	K0 III		250 s		
162586	209411		17 53 19.4	-34 43 50	+0.001	0.00	6.14	-0.06	0.2	B9 V		150 s		m
162588	209413		17 53 23.0	-35 00 59	-0.002	+0.01	7.3	0.5	0.4	B9.5 V		120 s		
162587	209416		17 53 23.1	-34 53 43	-0.001	-0.01	6.40	1.1	0.2	K0 III		170 s		m
163731	17672		17 53 24.1	+63 04 14	-0.006	+0.02	8.0	0.4	3.0	F2 V		100 s		
163186	66431		17 53 27.3	+32 22 05	-0.004	-0.02	7.9	0.5	3.4	F5 V		78 s		
163588	30631	32 ξ Dra	17 53 31.5	+56 52 21	+0.011	+0.08	3.75	1.18	-0.1	K2 III	-26	58 s		Grumium, m
163218	66438		17 53 31.7	+35 40 21	0.000	-0.02	7.3	1.1	-0.1	K2 III		300 s		
162719	185900		17 53 33.9	-27 16 56	+0.001	+0.02	6.8	0.4	1.4	A2 V		71 s		
162883	141946		17 53 34.7	-8 42 30	+0.001	0.00	7.9	0.4	2.6	F0 V		110 s		
162954	122886		17 53 35.6	+4 51 33	0.000	0.00	7.4	0.0	0.0	B8.5 V		270 s		
163714	17673		17 53 36.2	+61 02 38	-0.001	+0.04	7.80	0.5	4.0	F8 V		58 s	10887	m
162577	209421		17 53 39.3	-39 52 43	+0.002	-0.01	8.00	1.1	0.2	K0 III		350 s		m
162545	228519		17 53 39.7	-43 43 08	+0.003	-0.02	7.4	1.2	0.2	K0 III		220 s		
162075	254064		17 53 39.7	-63 57 34	+0.002	0.00	7.5	0.1	0.6	A0 V		240 s		
162361	245099		17 53 40.7	-53 25 30	+0.010	-0.01	7.8	0.7	0.2	K0 III		150 mx		
162678	209425		17 53 45.3	-34 47 09	+0.001	0.00	6.4	0.1	0.6	A0 V		140 s		
162679	—		17 53 46.0	-34 47 38			7.3	0.5	0.6	A0 V		110 s		
162680	209426		17 53 46.8	-34 50 45	+0.001	+0.01	7.9	0.2	0.6	A0 V		230 s		
162836	160891		17 53 46.9	-19 53 14	+0.001	0.00	7.2	0.8	0.2	K0 III		260 s		
163219	66443		17 53 49.5	+30 22 14	-0.001	+0.01	7.5	0.1	1.7	A3 V	-38	150 s		
162761	—		17 53 50.4	-28 42 35			8.0	0.9	0.2	K0 III		360 s		
162616	228525		17 53 51.0	-41 10 33	+0.001	0.00	7.6	1.5	-0.1	K2 III		300 s		
162724	209428		17 53 54.7	-34 45 09	+0.002	0.00	6.50	0.1	1.2	A1 V		110 s		m
162297	245098		17 53 55.6	-58 16 20	+0.006	+0.01	7.8	0.6	0.6	A0 V		110 s		
163014	122893		17 53 56.4	+0 35 52	-0.001	0.00	7.9	1.1	-0.1	K2 III		360 s		
163079	122900		17 53 56.7	+7 55 23	+0.003	-0.02	7.8	0.5	3.4	F5 V		74 s		
162725	209430		17 53 58.0	-34 49 53	+0.001	-0.02	6.42	0.01		B9 p				
163308	66447		17 53 59.9	+34 40 01	-0.003	-0.03	7.6	1.1	0.2	K0 III		300 s		
164462	8958		17 54 00.9	+73 07 16	-0.009	+0.03	8.0	1.1	-0.1	K2 III		320 mx		
163418	47043		17 54 01.2	+42 38 49	+0.001	0.00	7.6	0.8	0.3	G6 III	-1	290 s		
163015	141952		17 54 03.5	-1 37 08	-0.001	-0.01	6.5	0.1	0.6	A0 V		140 s		
162745	209431		17 54 03.7	-35 12 22	+0.003	-0.09	8.0	0.1	0.2	K0 III		160 mx		
163265	85526		17 54 05.0	+27 20 33	-0.003	-0.11	8.0	0.5	4.0	F8 V		62 s		
165309	8965		17 54 05.6	+78 40 47	+0.003	+0.03	7.5	0.4	3.0	F2 V		79 s		
162956	160896		17 54 07.3	-11 20 16	0.000	-0.01	6.60	1.1	0.2	K0 III		170 s	10872	m
162781	209435		17 54 12.8	-34 49 24	+0.001	+0.01	7.7	0.1	0.6	A0 V		220 s		
163151	103194		17 54 13.9	+11 07 49	-0.005	-0.17	6.38	0.45	3.4	F5 V	-41	26 mx		m
162780	209436		17 54 14.3	-34 43 39	+0.001	0.00	6.9	0.5	0.6	A0 V		94 s		
163964	17676		17 54 19.9	+64 04 59	-0.008	+0.02	8.0	1.1	-0.1	K2 III		230 mx		
162804	209443		17 54 22.0	-34 52 41	+0.002	+0.01	7.2	-0.1	0.6	A0 V		210 s		
162763	209439		17 54 24.0	-38 37 31	0.000	-0.08	7.1	0.4	3.0	F2 V		61 s		
162000	254065		17 54 24.8	-68 39 26	-0.003	0.00	8.0	0.8	3.2	G5 IV		90 s		
164780	8962		17 54 26.3	+75 10 15	-0.005	+0.02	6.36	0.98	0.2	K0 III	-18	170 s		
162817	209446		17 54 27.0	-34 27 59	+0.001	+0.01	6.1	0.1	1.2	A1 V		91 s		
162783	209444		17 54 27.2	-37 26 59	+0.002	+0.02	7.8	0.9	0.2	K0 III		330 s		
162381	245103		17 54 27.2	-58 54 42	0.000	-0.03	7.1	0.2	0.4	B9.5 V		150 s		
163132	122909		17 54 28.1	+3 43 56	0.000	-0.02	6.6	1.1	-0.1	K2 III		200 s		
163102	141954		17 54 29.6	-3 27 46	0.000	-0.16	7.2	0.5	3.4	F5 V		56 s		
162156	—		17 54 32.2	-66 06 54			7.9	1.1	0.2	K0 III		350 s		
162874	209449		17 54 38.0	-33 57 21	0.000	+0.02	8.0	-0.1	0.4	B9.5 V		330 s		
162785	228534		17 54 38.5	-40 15 11	0.000	0.00	8.0	1.4	-0.1	K2 III		320 s		
163285	103203		17 54 40.5	+18 11 09	+0.001	-0.01	8.0	0.1	0.6	A0 V		300 s		
162873	209450		17 54 41.6	-33 54 18	+0.004	+0.02	7.6	-0.1	0.6	A0 V		180 mx		
162520	245108		17 54 43.9	-55 53 54	-0.001	0.00	7.3	0.0	0.4	B9.5 V		220 s		
163608	47057		17 54 44.0	+45 13 00	-0.001	0.00	8.0	0.1	0.6	A0 V	-26	310 s		
162888	209451		17 54 48.1	-34 32 29	+0.004	+0.01	6.6	0.5	0.6	A0 V		77 s		
162314	254071		17 54 51.6	-63 16 31	+0.001	-0.04	7.5	1.1	-0.1	K2 III		330 s		
162978	185928		17 54 53.8	-24 53 14	0.000	0.00	6.20	0.04		O8	-11			
163117	160899		17 54 55.6	-11 37 55	+0.001	+0.04	7.00	0.4	3.0	F2 V		60 s	10878	m
163153	141956		17 54 57.5	-7 44 02	-0.003	-0.05	6.93	0.76		G5				
162822	228538		17 54 57.8	-43 09 23	0.000	0.00	8.0	0.8	0.2	K0 III		330 s		
162926	209458		17 55 07.7	-36 28 32	-0.001	-0.01	6.1	0.9	1.4	A2 V		27 s		
164613	8961	34 Dra	17 55 10.9	+72 00 19	0.000	0.00	5.5	0.4	3.0	F2 V	-2	32 s		
164446	17694		17 55 11.6	+69 36 56	+0.001	-0.02	6.9	1.1	0.2	K0 III	-8	220 s		
163363	103215		17 55 15.3	+15 16 49	+0.002	-0.10	8.0	0.5	4.0	F8 V		62 s		
163311	122916		17 55 16.7	+5 42 14	0.000	-0.04	7.4	1.1	-0.1	K2 III		290 s		
163440	103218		17 55 20.1	+19 27 28	+0.001	-0.01	7.9	1.1	0.2	K0 III		340 s		
163929	30645		17 55 23.4	+55 58 16	+0.003	+0.12	6.10	0.32	0.6	gF1 p	-27	130 s		
163621	66472		17 55 24.5	+36 11 18	-0.012	-0.04	8.0	0.9	3.2	G5 IV		83 mx		
163506	85545	89 Her	17 55 25.0	+26 03 00	0.000	+0.01	5.46	0.34	-8.4	F2 Ia	-29	5900 s		V441 Her, v
163590	66469		17 55 25.2	+32 26 18	0.000	0.00	7.2	0.1	0.6	A0 V	-13	210 s		

HD	SAO	Star Name	α 2000	δ 2000	μ(α)	μ(δ)	V	B-V	M_v	Spec	RV	d(pc)	ADS	Notes
			h m s	° ′ ″	s	″								
--	85544		17 55 25.6	+25 14 40	+0.002	-0.02	8.0	1.5	3.2	G5 IV		36 s		
163385	122921		17 55 27.7	+8 33 33	+0.001	-0.09	8.0	0.5	4.0	F8 V		61 s		
163064	209471		17 55 28.7	-30 01 18	0.000	0.00	7.6	0.9	0.2	K0 III		300 s		
163042	209473		17 55 34.1	-31 57 24	-0.001	-0.02	8.0	1.7	-0.3	K5 III		340 s		
163529	85548		17 55 36.0	+25 07 54	0.000	0.00	8.00	0.1	1.7	A3 V		180 s		m
163346	122922		17 55 37.3	+2 04 30	0.000	+0.01	6.8	0.1	1.7	A3 V		99 s		
163003	209465		17 55 38.5	-38 08 56	+0.001	-0.03	7.5	1.7	-0.1	K2 III		160 s		
161462	257536		17 55 40.1	-77 33 33	+0.001	-0.03	8.0	0.8	3.2	G5 IV		90 s		
164135	17687		17 55 42.6	+61 23 44	-0.003	+0.01	7.2	0.1	1.7	A3 V		130 s		
163106	209480		17 55 44.5	-30 34 15	-0.003	+0.03	7.70	0.1	0.6	A0 V		250 s		m
163005	209476		17 55 49.0	-39 20 51	+0.001	-0.03	8.0	1.1	0.2	K0 III		330 s		
162048	--		17 55 49.2	-71 54 55			8.0	1.1	0.2	K0 III		360 s		
163547	85552		17 55 50.7	+22 27 51	0.000	0.00	5.58	1.24	-0.1	K2 III	-44	120 s		
163245	160909		17 55 54.8	-18 48 08	+0.001	-0.02	6.4	0.1	0.6	A0 V	+4	140 s		
163069	209481		17 55 55.7	-37 03 30	-0.002	0.00	8.0	0.3	0.2	K0 III		360 s		
163402	122924		17 55 56.5	+3 44 21	+0.001	-0.01	8.0	1.4	-0.3	K5 III		400 s		
162913	228548		17 55 59.3	-48 56 23	+0.004	-0.01	7.2	0.8	3.2	G5 IV		61 s		
163470	122931		17 55 59.9	+9 45 17	0.000	0.00	7.8	1.6	-0.5	M2 III		420 s		
163442	122929		17 56 01.1	+5 09 42	0.000	0.00	7.9	0.0	0.0	B8.5 V		340 s		
163025	228557		17 56 04.2	-42 19 21	0.000	-0.01	7.4	0.3	2.1	A5 V		100 s		
162399	254077		17 56 06.2	-67 15 03	-0.002	-0.04	7.6	1.1	0.2	K0 III		300 s		
163139	209486		17 56 07.6	-35 02 59	0.000	-0.03	6.7	0.5	0.4	B9.5 V		91 s		
163246	185961		17 56 10.5	-22 27 37	0.000	-0.05	8.0	-0.5	3.2	G5 IV		93 s		
163489	122932		17 56 11.7	+4 50 00	0.000	0.00	7.8	1.1	-0.1	K2 III		340 s		
163789	66487		17 56 11.9	+39 26 21	+0.007	-0.04	7.90	0.7	4.4	G0 V		50 s	10911	m
163007	228556		17 56 13.3	-46 42 06	+0.001	+0.01	7.51	-0.03	-1.1	B5 V e		510 s		
163422	141968		17 56 14.7	-4 34 35	-0.001	-0.02	7.7	0.0	0.4	B9.5 V		240 s		
163227	185960		17 56 14.8	-25 40 21	-0.002	+0.01	7.9	-0.3	0.4	B9.5 V		320 s		
163770	66485	91 θ Her	17 56 15.1	+37 15 02	0.000	+0.01	3.86	1.35	-2.2	K1 II	-27	130 s		
163181	209489		17 56 15.9	-32 28 30	+0.001	0.00	6.60	0.2		O9.5	-41			V453 Sco, m,v
163472	122935		17 56 18.2	+0 40 13	-0.001	0.00	5.82	0.09	-2.5	B2 V	-18	330 s		
163336	160915		17 56 18.8	-15 48 45	-0.001	-0.07	6.10	0.1	0.6	A0 V		54 mx	10891	m
164330	17695		17 56 19.9	+62 36 35	-0.008	+0.09	7.20	1.1	0.2	K0 III		200 mx		m
163296	185966		17 56 21.2	-21 57 21	0.000	-0.03	6.85	0.08	0.6	A0 V	-3	160 s		
162895	245127		17 56 23.8	-54 22 41	-0.002	-0.01	7.3	0.2	0.4	B9.5 V		160 s		
162521	254079		17 56 24.0	-65 43 20	+0.012	-0.33	6.36	0.45	4.0	F8 V		30 s		
163272	185965		17 56 24.2	-26 46 22	-0.001	-0.06	7.4	0.7	4.4	G0 V		33 s		
163640	103227		17 56 24.3	+18 19 37	+0.001	-0.01	6.6	0.1	-0.6	A0 III		280 s	10905	m
163639	103229		17 56 24.4	+19 53 20	0.000	-0.02	7.4	0.1	1.7	A3 V		140 s		
--	--		17 56 24.4	+58 13 10			7.50	2.0		N7.7 e				T Dra, v
163230	209495		17 56 25.9	-30 11 54	0.000	-0.06	7.32	1.60	-0.3	K5 III		240 mx		
164329	17696		17 56 26.0	+62 37 10	0.000	-0.02	7.80	0.1	1.4	A2 V		190 s		m
163317	185972		17 56 29.6	-23 23 23	+0.002	+0.02	8.0	0.4	2.6	F0 V		100 s		
164058	30653	33 γ Dra	17 56 36.2	+51 29 20	-0.001	-0.02	2.23	1.52	-0.3	K5 III	-28	31 s	10923	Eltanin, m
163675	103234		17 56 37.7	+18 36 45	+0.015	+0.01	6.6	1.1	0.2	K0 III		62 mx		
163610	122943		17 56 40.3	+6 50 26	-0.001	+0.01	7.7	1.4	-0.3	K5 III		350 s		
163318	185975		17 56 41.7	-28 03 55	+0.003	-0.02	5.8	0.7	1.7	A3 V		28 s		
163966	47077		17 56 42.7	+44 59 14	-0.001	-0.01	6.8	0.0	0.4	B9.5 V	-30	190 s		
163251	209500		17 56 46.0	-35 12 08	-0.003	-0.03	7.8	-0.4	0.4	B9.5 V		310 s		
163592	122944		17 56 47.0	+4 22 36	-0.001	-0.01	7.9	0.0	0.4	B9.5 V		290 s		
163145	228562		17 56 47.1	-44 20 32	-0.001	-0.01	4.86	1.21	-0.3	gK5	+45	110 s		
163532	141979		17 56 47.6	-4 04 55	-0.001	-0.01	5.47	1.16	0.2	gG9	-39	82 s		
163990	47080		17 56 48.2	+45 21 03	0.000	-0.03	6.02	1.64		M6 s	+13			OP Her, v
--	66496		17 56 48.8	+34 57 18	+0.001	-0.04	8.0	1.1						
163274	209503		17 56 49.9	-32 41 20	+0.001	0.00	6.6	0.8	0.4	B9.5 V		57 s		
163196	209497		17 56 51.0	-39 55 54	+0.004	+0.03	7.4	0.6	1.4	A2 V		73 s		
163338	209507		17 56 52.0	-30 17 10	-0.009	+0.02	7.9	1.0	0.0	B8.5 V		40 mx		
163611	122946		17 56 52.4	+4 59 16	+0.005	+0.08	7.5	0.5	3.3	F4 V		68 s		V566 Oph, m,v
163234	228564		17 56 55.6	-40 18 20	+0.001	+0.01	6.4	1.6	-0.3	K5 III		180 s		
163641	122949		17 56 55.8	+6 29 16	+0.002	0.00	6.29	0.00		A0	-14			
162861	254086		17 56 59.0	-60 19 26	+0.003	+0.01	7.9	1.3	0.2	K0 III		210 s		
163624	122950		17 57 04.2	+0 03 59	+0.002	-0.02	5.97	0.10		A2	-11		10912	m
163732	103239		17 57 04.5	+13 58 29	0.000	-0.01	7.8	0.1	1.4	A2 V		190 s		
164059	47084		17 57 04.9	+45 51 22	0.000	+0.11	6.90	0.5	3.4	F5 V	-28	50 s	10934	m
162702	254082		17 57 04.9	-65 06 52	-0.003	-0.06	7.31	1.21	6.3	K2 V		11 s		
163428	185981		17 57 07.3	-23 56 20	+0.001	+0.01	6.65	2.06	9.2	M0 V	-12			
163968	66502		17 57 08.7	+36 51 38	+0.003	+0.07	7.4	0.4	2.6	F0 V		91 s		
163967	66505		17 57 08.8	+37 48 10	+0.003	+0.07	6.9	1.1	0.2	K0 III		220 s		
163568	141983		17 57 09.9	-8 04 07	+0.003	0.00	7.8	0.7	4.4	G0 V		47 s		
163254	228569		17 57 10.2	-41 58 42	0.000	0.00	6.73	-0.08	-1.1	B5 V k		350 s		
164077	47089		17 57 10.4	+47 13 34	-0.001	-0.01	7.8	1.1	0.2	K0 III		330 s		
163028	245131		17 57 12.8	-55 22 53	-0.002	-0.03	7.00	0.5	4.0	F8 V		40 s		m
163840	85575		17 57 14.1	+23 59 45	-0.002	+0.08	6.30	0.63	4.4	G0 V	-34	22 s		
163304	209510		17 57 17.0	-38 46 57	-0.001	0.00	7.80	0.00	0.4	B9.5 V		290 s		m
162401	257544		17 57 17.7	-71 05 41	-0.001	-0.13	7.7	1.1	0.2	K0 III		120 mx		

HD	SAO	Star Name	α 2000	δ 2000	μ(α)	μ(δ)	V	B-V	M_v	Spec	RV	d(pc)	ADS	Notes
163750	103240		17ʰ57ᵐ18.5ˢ	+12°37'36"	+0.004ˢ	-0.08"	7.45	0.51		F8				
163991	66507		17 57 20.8	+35 40 30	0.000	+0.01	7.80	1.1	0.2	K0 III		330 s	10932	m
163948	66506		17 57 24.2	+33 24 04	+0.002	-0.01	6.7	1.1	0.2	K0 III	-2	200 s		
163772	103243		17 57 26.7	+11 02 39	-0.001	-0.02	6.36	0.11	1.4	A2 V	-16	91 s		
164181	47095		17 57 30.2	+48 17 45	0.000	+0.01	7.9	1.1	-0.1	K2 III		400 s		
164345	30659		17 57 31.4	+54 39 55	+0.001	-0.02	7.00	1.4		K5 p				UW Dra, v
163992	66510		17 57 31.6	+33 50 49	+0.001	0.00	7.9	1.1	0.2	K0 III	-25	340 s		
163374	209517		17 57 34.7	-38 05 25	+0.002	-0.05	7.4	0.7	0.2	K0 III		270 s		
163625	160941		17 57 35.4	-11 32 02	-0.010	-0.04	7.8	0.7	4.4	G0 V		47 s		
163375	209519		17 57 40.6	-39 06 06	-0.002	-0.02	6.7	0.8	0.6	A0 V		58 s		
161988	257542		17 57 41.6	-76 10 38	+0.003	+0.02	6.07	1.20		K2				m
163071	245133		17 57 42.6	-56 53 47	0.000	-0.02	6.26	-0.05	0.0	B8.5 V		170 s		
160288	258780		17 57 43.1	-83 14 08	-0.012	-0.09	7.3	-0.2	3.0	F2 V		74 s		
163993	85590	92 ξ Her	17 57 45.7	+29 14 52	+0.006	-0.02	3.70	0.94	0.2	K0 III	-2	50 s		
163908	85583		17 57 45.9	+20 20 46	-0.001	+0.02	7.6	1.1	0.2	K0 III		300 s		
163376	228578		17 57 47.5	-41 42 58	-0.001	-0.01	4.88	1.65	-0.5	gM1	+4	110 s		
--	17706		17 57 47.8	+64 11 16	-0.004	-0.03	7.5	1.7						
163792	122964		17 57 50.7	+2 15 12	-0.001	-0.01	6.7	0.1	0.6	A0 V		160 s		
164212	47096		17 57 56.3	+43 25 02	-0.001	+0.01	6.9	0.0	0.4	B9.5 V	-30	200 s		
163482	209525		17 57 56.4	-36 00 31	-0.001	-0.03	7.00	0.1	0.6	A0 V		190 s		m
163433	209524		17 57 57.6	-39 08 13	-0.001	-0.04	6.60	0.1	0.6	A0 V		160 s		m
163413	228582		17 57 59.0	-41 56 34	0.000	+0.01	7.4	0.5	3.2	G5 IV		70 s		
163555	209529		17 58 00.3	-30 03 54	+0.002	+0.01	7.7	0.4	0.0	B8.5 V		190 s		
164182	47097		17 58 01.1	+42 00 05	-0.002	-0.03	7.9	0.9	3.2	G5 IV		88 s		
163909	103252		17 58 02.0	+15 24 24	0.000	-0.04	7.9	1.6	-0.5	M2 III		490 s		
163826	122965		17 58 03.1	+2 43 12	+0.001	-0.01	7.1	1.4	-0.3	K5 III		280 s		
163483	209526		17 58 03.4	-38 14 39	-0.003	+0.01	8.0	1.8	-0.3	K5 III		290 s		
163378	228581		17 58 04.9	-45 27 21	+0.001	0.00	7.70	0.1	1.7	A3 V		160 s		m
164078	66518		17 58 04.9	+32 38 54	-0.002	-0.01	6.6	0.5	3.4	F5 V	+3	43 s		
164394	30665		17 58 05.2	+52 13 06	0.000	-0.02	7.52	0.27		A4	-1		10953	m
163930	103254		17 58 06.8	+15 08 22	-0.002	+0.07	7.30	0.4		F2	-46			Z Her, m,v
161071	258782		17 58 08.7	-81 27 09	-0.008	-0.08	7.3	0.1	3.0	F2 V		74 s		
164042	85596		17 58 10.4	+27 23 50	+0.003	+0.01	8.00	1.1	-0.1	K2 III		420 s	10942	m
163973	103258		17 58 20.4	+14 30 38	+0.001	-0.02	7.3	0.1	0.6	A0 V	0	220 s		
164213	66525		17 58 25.6	+36 52 29	-0.007	+0.06	7.5	1.1	0.2	K0 III		130 mx		
164136	66524	94 v Her	17 58 30.0	+30 11 22	0.000	+0.01	4.41	0.39	-2.0	F2 II	-22	190 s		
163578	209537		17 58 36.3	-37 58 25	0.000	0.00	8.0	0.0	0.2	K0 III		360 s		
164395	47103		17 58 37.8	+44 56 13	-0.002	-0.01	7.9	0.6	4.4	G0 V		51 s		
163685	186025		17 58 38.9	-28 45 33	0.000	-0.01	6.01	-0.08	-2.3	B3 IV		390 s		
164251	66530		17 58 40.3	+35 37 54	+0.002	0.00	7.0	1.1	0.2	K0 III		230 s		
155392	258773		17 58 40.3	-87 21 52	-0.003	-0.03	7.7	1.8	-0.1	K2 III		160 s		
163797	160954		17 58 41.9	-16 51 28	+0.001	-0.03	7.1	0.2	2.1	A5 V		93 s		
164280	66531		17 58 42.1	+36 17 15	0.000	-0.06	6.00	0.94	0.2	K0 III	+10	150 s		
163884	142000		17 58 45.6	-5 25 31	-0.002	-0.04	7.6	1.1	-0.1	K2 III		280 s		
163239	245148		17 58 46.2	-58 39 22	+0.001	-0.06	7.3	2.0	0.2	K0 III		66 s		
164043	103265		17 58 46.8	+14 50 35	-0.003	-0.12	7.20	0.53		F8	-8			
164429	47106		17 58 52.1	+45 28 34	0.000	+0.03	6.48	-0.08		B9	-19			
163652	209545		17 58 55.5	-36 51 30	+0.001	+0.01	5.80	0.8	3.2	G5 IV		33 s		
163800	186037		17 58 56.9	-22 31 01	-0.002	+0.01	7.02	0.30		O8	+5			
163917	142004	64 v Oph	17 59 01.4	-9 46 25	-0.001	-0.12	3.34	0.99	0.2	K0 III	+12	42 s		
164964	17715		17 59 02.5	+63 57 27	-0.004	-0.01	7.9	1.4	-0.3	K5 III		430 s		
164214	85606		17 59 03.1	+23 49 41	0.000	-0.03	8.0	1.4	-0.3	K5 III		450 s		
164252	66535		17 59 03.4	+30 02 57	0.000	+0.02	6.91	0.68	-0.9	G8 II-III	-21	360 s	10955	m
163848	186042		17 59 03.5	-20 03 47	+0.001	-0.05	8.0	0.3	0.0	B8.5 V		250 s		
164045	122981		17 59 04.0	+4 57 17	0.000	-0.02	8.0	0.1	1.4	A2 V		200 s		
163653	209547		17 59 04.0	-38 49 36	-0.001	-0.03	7.70	0.1	0.6	A0 V		260 s		m
163976	142007		17 59 04.7	-5 02 54	+0.003	-0.04	7.6	0.5	4.0	F8 V		51 s		
163755	209553		17 59 05.3	-30 15 12	-0.001	0.00	4.99	1.62	-3.6	M2 Ib-II	-20	460 s		m
163707	209551		17 59 08.9	-36 02 03	-0.004	-0.05	7.7	0.6	3.2	G5 IV		79 s		
163691	209550		17 59 10.3	-38 34 23	0.000	-0.01	6.9	1.1	-0.3	K5 III		280 s		
164137	103271		17 59 11.1	+13 54 18	0.000	-0.01	7.8	0.1	0.6	A0 V		270 s		
164807	30674		17 59 12.5	+58 45 42	+0.002	+0.01	8.0	1.1	-0.1	K2 III		410 s		
163708	209552		17 59 13.5	-36 56 19	+0.002	-0.01	6.8	0.3	0.6	A0 V		110 s		V1647 Sgr, m,v
164984	17717		17 59 13.6	+64 08 33	+0.002	+0.01	7.5	0.4	3.0	F2 V		80 s	10985	m
163802	186041		17 59 14.9	-29 23 26	+0.002	-0.03	7.1	0.9	0.2	K0 III		240 s		
163813	186047		17 59 16.1	-25 49 09	+0.001	-0.02	7.99	0.05	0.4	B9.5 V		330 s		
164138	103273		17 59 18.5	+12 05 09	+0.001	+0.01	7.7	0.5	4.0	F8 V		54 s		
162049	257545		17 59 20.3	-77 49 26	+0.001	+0.01	6.7	0.8	0.2	K0 III		200 s		
163892	186053		17 59 26.2	-22 28 00	+0.001	0.00	7.44	0.11	-5.4	O9 IV	-14	2200 s		
164115	122988		17 59 26.9	+6 25 08	+0.001	-0.01	7.1	0.4	2.6	F0 V		78 s		
163758	209560		17 59 28.2	-36 01 16	+0.001	-0.01	7.31	0.03		O8				
164503	47111		17 59 31.4	+41 43 56	-0.003	-0.02	7.9	0.5	3.4	F5 V		78 s		
164064	142012		17 59 36.5	-4 49 17	-0.001	-0.09	5.87	1.56	-0.3	gK5	-32	160 s		
163894	186055		17 59 36.6	-27 04 13	+0.002	-0.03	7.8	2.2	-0.3	K5 III		150 s		
164575	47113		17 59 36.7	+43 14 07	-0.001	+0.08	6.9	1.1	0.2	K0 III		220 s		

HD	SAO	Star Name	α 2000	δ 2000	μ(α)	μ(δ)	V	B–V	M$_v$	Spec	RV	d(pc)	ADS	Notes
163486	245156		17^{h}59^{m}37$.^s$8	−56°15′19″	+0$.^s$001	0$.^{''}$00	7.2	0.9	−0.5	M2 III		350 s		
163745	228606		17 59 40.8	−41 29 16	+0.001	−0.03	7.2	−0.2	0.0	B8.5 V		280 s		
164026	160971		17 59 41.8	−12 20 55	−0.003	+0.03	8.0	0.9	3.2	G5 IV		86 s		
163655	228603		17 59 42.4	−47 45 42	+0.001	0.00	7.6	0.1	0.6	A0 V		210 s		
164235	103278		17 59 42.4	+14 27 49	0.000	−0.01	7.8	0.1	0.6	A0 V		270 s		
163955	186061	4 Sgr	17 59 47.4	−23 48 58	0.000	−0.04	4.76	−0.04		A0 n	−22	21 mn		
164646	47121		17 59 55.9	+45 30 05	−0.001	−0.03	5.67	1.57	−0.4	gM0	−10	160 s		
163868	209569		17 59 56.4	−33 24 30	+0.002	0.00	7.38	0.00	−1.1	B5 V e		450 s		
163617	245159		17 59 57.3	−52 28 24	+0.001	+0.01	7.4	1.1	−0.1	K2 III		310 s		
165028	17719		17 59 57.5	+60 26 43	−0.004	+0.03	8.0	0.1	1.7	A3 V		98 mx		
163819	209564		17 59 57.9	−39 03 46	0.000	−0.01	7.20	1.1	0.2	K0 III		240 s		V394 CrA, m,v
164028	186070		17 59 59.9	−20 20 21	0.000	−0.01	6.4	0.9	−0.9	K0 II−III		300 s		
−−	186069		18 00 00.7	−22 32 51	+0.001	+0.04	7.5	0.0	−1.0	B5.5 V		410 s		
166865	8994	40 Dra	18 00 03.1	+80 00 03	+0.016	+0.13	6.04	0.51	3.7	F6 V	+4	29 s		
164349	103285	93 Her	18 00 03.2	+16 45 03	0.000	−0.01	4.67	1.26	−0.9	K0 II−III	−23	100 s		
163636	245162		18 00 04.4	−53 05 56	−0.003	−0.06	8.0	−0.2	0.2	K0 III		200 mx		
164396	85621		18 00 06.0	+20 44 29	−0.002	−0.07	8.0	0.5	4.0	F8 V		62 s		
164257	123001		18 00 07.2	+6 33 15	+0.001	+0.01	6.7	0.1	0.6	A0 V		160 s		
166866	8996	41 Dra	18 00 09.0	+80 00 14	+0.016	+0.12	5.68	0.50	6.3	K2 V	+10	23 mn	11061	m
164031	186074		18 00 11.4	−24 17 02	+0.002	−0.03	6.7	0.9	0.2	K0 III	−26	200 s		
164309	123007		18 00 12.4	+8 51 21	−0.002	−0.02	7.10	0.8	3.2	G5 IV		59 s		m
164104	160978		18 00 12.5	−16 03 59	−0.003	+0.01	7.6	0.9	3.2	G5 IV		71 s		
163900	209576		18 00 15.1	−37 24 27	+0.002	+0.01	7.0	0.6	0.0	B8.5 V		100 s		
164258	123004		18 00 15.3	+0 37 46	−0.001	0.00	6.37	0.15		A2 p	−34			
164284	123005	66 Oph	18 00 15.5	+4 22 07	0.000	−0.01	4.64	−0.03	−2.5	B2 V e	−11	250 s		
164506	85625		18 00 16.0	+28 43 11	+0.001	0.00	7.4	0.1	1.4	A2 V	+2	160 s		
163761	228612		18 00 16.1	−47 33 13	+0.002	−0.02	6.9	0.2	0.4	B9.5 V		150 s		
164310	123008		18 00 16.5	+8 51 40	−0.002	+0.02	7.1	1.1	0.2	K0 III		220 s		
163874	209575		18 00 18.2	−39 14 56	−0.004	−0.09	7.3	0.9	4.0	F8 V		27 s		
164285	123006		18 00 19.9	+1 36 37	0.000	−0.01	7.9	1.1	0.2	K0 III		310 s		
164144	160981		18 00 20.4	−14 12 29	+0.001	−0.02	7.99	1.69	0.2	K0 III		140 s		
163729	228610		18 00 21.4	−49 54 59	+0.001	+0.01	7.8	0.3	4.0	F8 V		58 s		
164124	160980		18 00 24.5	−19 13 52	0.000	0.00	8.0	1.3	−0.3	K5 III		450 s		
164165	160984		18 00 26.9	−12 59 19	+0.003	−0.02	7.4	0.0	0.0	B8.5 V		250 s		
164447	103291		18 00 27.5	+19 30 21	+0.001	0.00	6.40	−0.06		B7 ne	−29			
164032	186079		18 00 27.7	−29 49 27	+0.001	0.00	7.48	0.13	−5.7	B1 Ib		2800 s		
164259	142025	57 ζ Ser	18 00 28.7	−3 41 25	+0.010	−0.05	4.62	0.38	3.1	F3 V	−43	22 ts		
163940	209581		18 00 31.1	−38 21 26	+0.002	−0.01	7.8	0.9	0.2	K0 III		340 s		
163618	245163		18 00 31.3	−56 56 35	+0.002	−0.02	7.3	0.9	0.2	K0 III		260 s		
164448	103292		18 00 31.3	+17 06 12	0.000	0.00	7.5	1.6	−0.5	M2 III		390 s		
163958	−−		18 00 32.4	−37 41 40			7.5	1.1	0.2	K0 III		280 s		
163925	209580		18 00 32.8	−39 40 01	+0.002	−0.02	8.0	0.8	0.2	K0 III		360 s		
164614	66551		18 00 36.3	+33 12 50	0.000	−0.02	5.99	1.51	−0.3	K5 III	−16	180 s		
164353	123013	67 Oph	18 00 38.5	+2 55 53	0.000	−0.01	3.97	0.02	−5.7	B5 Ib	−4	740 s	10966	m
164595	85632		18 00 38.9	+29 34 18	−0.009	+0.17	7.08	0.64	4.5	dG1	+7	29 ts		
164004	209585		18 00 39.1	−36 58 01	0.000	−0.02	7.9	0.4	0.2	K0 III		340 s		
164481	103295		18 00 42.2	+17 10 00	+0.001	+0.01	7.6	0.1	1.7	A3 V		160 s		
164430	123016		18 00 43.8	+9 33 50	−0.002	+0.01	7.1	1.1	0.2	K0 III		230 s		
164754	47131		18 00 48.5	+41 28 46	−0.001	−0.02	8.0	1.1	0.2	K0 III		360 s		
164986	30696		18 00 49.6	+53 37 44	+0.001	+0.02	8.0	1.1	−0.1	K2 III		410 s		
164432	123017		18 00 52.7	+6 16 06	0.000	0.00	6.34	−0.08	−3.0	B2 IV	−22	640 s		
164377	142032		18 00 57.0	−2 34 44	0.000	−0.04	7.6	1.1	0.2	K0 III		250 s		
164507	103299		18 00 57.0	+15 05 36	−0.004	−0.11	6.26	0.69		G5	+4			
164088	209592		18 00 57.8	−34 00 20	0.000	+0.05	8.0	0.7	0.2	K0 III		360 s		
165587	17728		18 01 00.2	+68 35 07	−0.002	0.00	7.8	1.4	−0.3	K5 III		410 s		
164940	47141		18 01 03.2	+48 27 56	−0.003	+0.01	7.0	1.1	−0.1	K2 III		270 s		
164921	47140		18 01 04.9	+47 24 13	+0.003	+0.02	7.2	1.1	0.2	K0 III		250 s		
164898	47139		18 01 07.0	+45 21 03	0.000	+0.02	7.56	0.09		A0	−13			m
162298	−−		18 01 07.9	−78 23 34			7.9	1.4	−0.3	K5 III		440 s		
164050	228629		18 01 11.2	−40 20 08	0.000	−0.02	7.6	0.0	0.0	B8.5 V		290 s		
163986	228625		18 01 11.9	−45 49 33	−0.004	−0.01	7.70	0.9	3.2	G5 IV		79 s		m
164198	186099		18 01 12.3	−29 20 32	+0.001	−0.04	8.0	0.3	3.4	F5 V		82 s		
−−	−−		18 01 12.3	−45 49 33			7.70							m
164175	186100		18 01 12.8	−29 26 39	+0.006	−0.05	7.7	1.0	0.2	K0 III		140 mx		
165169	30703		18 01 14.3	+56 58 03	0.000	+0.02	7.3	1.1	0.2	K0 III		270 s		
164987	47147		18 01 21.9	+47 53 25	0.000	+0.05	8.0	1.1	0.2	K0 III		360 s		
164358	160998	6 Sgr	18 01 22.9	−17 09 25	0.000	−0.01	6.2	1.1	−0.2	K3 III	−22	180 s		
165522	17729		18 01 23.0	+65 56 55	−0.001	+0.04	7.80	0.5	3.4	F5 V	−20	76 s	11016	m
164755	66558		18 01 24.5	+30 38 43	+0.001	−0.02	7.0	1.4	−0.3	K5 III	−29	290 s		
164696	85646		18 01 25.3	+22 46 27	0.000	+0.02	7.5	1.1	−0.1	K2 III		340 s		
163822	245174		18 01 25.8	−55 58 29	−0.003	−0.06	8.0	0.2	0.2	K0 III		180 mx		
164244	186109		18 01 26.5	−29 35 15	−0.001	+0.02	7.4	0.5	0.6	A0 V		110 s		
164730	85649		18 01 29.1	+24 14 58	−0.001	−0.02	7.2	1.1	0.2	K0 III	+14	250 s		
164669	85648	95 Her	18 01 30.3	+21 35 44	+0.001	+0.03	4.27	0.39	0.5	A7 III	−30	48 mn	10993	m
164129	209602		18 01 31.9	−38 04 56	+0.004	−0.01	6.6	0.6	0.4	B9.5 V		75 s		

HD	SAO	Star Name	α 2000	δ 2000	μ(α)	μ(δ)	V	B−V	M$_v$	Spec	RV	d(pc)	ADS	Notes
			h m s	° ′ ″	s	″								
164615	103308		18 01 32.8	+11 17 09	−0.001	0.00	7.0	0.4	3.0	F2 V		62 s		
159517	258779		18 01 33.0	−85 12 53	−0.013	−0.14	6.45	0.44	3.3	F4 V		41 s		
164728	85651		18 01 34.2	+25 29 23	+0.001	0.00	7.59	−0.02	0.6	A0 V		250 s		
164824	66562		18 01 35.7	+33 18 41	+0.002	+0.03	6.15	1.55	−0.3	K5 III	−10	180 s	10998	m
164359	186126		18 01 38.8	−22 07 52	−0.002	+0.02	7.54	0.00	−5.0	B0 III	−14	2300 s		
164670	103309		18 01 40.2	+16 43 05	+0.001	0.00	8.0	0.1	0.6	A0 V		300 s		
164577	123035	68 Oph	18 01 45.0	+1 18 19	+0.001	−0.01	4.45	0.02	1.2	A1 V	+4	45 s	10990	m
164245	209608		18 01 48.1	−36 22 40	−0.001	−0.02	6.3	1.0	0.4	B9.5 V		34 s		
164385	186132		18 01 50.3	−24 09 26	+0.002	−0.01	8.04	0.00	0.0	B8.5 V		410 s		
164438	161004		18 01 52.2	−19 06 22	+0.001	0.00	7.48	0.34	−5.4	O9 IV	−27	2000 s		
164402	186135		18 01 54.1	−22 46 50	0.000	−0.01	5.84	0.01	−5.8	B0 Ib	−13	1600 s	10983	m
165109	47157		18 01 57.5	+47 53 05	+0.003	−0.07	7.4	1.1	0.2	K0 III		270 mx		
164714	103314		18 01 58.1	+12 45 27	−0.001	−0.02	7.8	0.1	0.6	A0 V		270 s		
164073	228630		18 02 00.5	−48 48 37	+0.001	0.00	8.04	0.04	−1.0	B5.5 V		430 s		
164320	209614		18 02 05.1	−35 40 10	+0.002	−0.01	7.6	−0.1	0.0	B8.5 V		330 s		
164321	209616		18 02 09.5	−36 21 33	−0.002	0.00	7.3	0.2	0.0	B8.5 V		190 s		
164039	245185		18 02 11.3	−52 57 28	−0.002	−0.05	7.7	1.7	−0.1	K2 III		180 s		
165240	30709		18 02 16.7	+51 38 23	+0.002	+0.01	7.3	0.4	2.6	F0 V		86 s		
164651	123047		18 02 17.5	+0 06 15	+0.010	+0.04	8.0	0.9	3.2	G5 IV		86 s		
164579	161017		18 02 22.0	−12 19 06	−0.001	0.00	7.67	1.96	−0.5	M4 III		280 s		
163878	254118		18 02 22.6	−60 08 12	+0.001	0.00	6.9	1.5	0.2	K0 III		110 s		
164492	186143		18 02 22.7	−23 02 01	−0.001	−0.01	7.63	0.00			+4		10991	m
164852	85672	96 Her	18 02 22.9	+20 50 01	0.000	−0.01	5.28	−0.09	−1.7	B3 V	−15	230 s		
164300	228643		18 02 26.1	−42 54 17	+0.001	−0.04	7.4	1.4	−0.1	K2 III		240 s		
161352	258783		18 02 27.5	−82 44 06	−0.008	0.00	7.9	1.1	−0.1	K2 III		400 s		
164900	85676	97 Her	18 02 30.0	+22 55 23	0.000	−0.01	6.10	−0.10	−1.7	B3 V	−36	340 s		
165144	47162		18 02 30.3	+43 16 07	+0.001	+0.02	7.4	1.1	0.2	K0 III		280 s		
165459	30711		18 02 30.6	+58 37 38	−0.002	+0.02	6.7	0.1	1.4	A2 V		110 s		
164922	85678		18 02 30.7	+26 18 48	+0.029	−0.59	6.99	0.80	5.9	K0 V	+23	19 ts	11003	m
165170	47163		18 02 30.7	+44 14 03	−0.004	−0.06	7.40	0.4	3.0	F2 V	−19	76 s	11010	m
164514	186149		18 02 31.6	−22 54 20	+0.001	0.00	7.42	1.09	−7.7	A5 Ia	−1	2700 s		
165008	66573		18 02 34.5	+30 33 19	−0.002	+0.09	6.8	0.4	3.0	F2 V	−37	57 s		
164581	186158		18 02 37.1	−20 44 15	0.000	0.00	6.81	0.11	−3.5	B1 V	−6	750 s		
164177	228642		18 02 37.2	−49 37 11	+0.001	0.00	7.9	0.8	0.0	B8.5 V		380 s		
164536	186152		18 02 38.4	−24 15 20	0.000	−0.01	7.11	−0.03		B3	−11			m
––	186154		18 02 39.5	−24 14 49	−0.001	−0.02	6.90							m
––	66577		18 02 40.0	+34 57 47	+0.001	−0.03	7.9	2.0						
164455	209631		18 02 42.6	−33 53 22	−0.001	−0.02	7.43	−0.08	−3.3	B2.5 III		1100 s		
164901	103323		18 02 42.8	+16 48 54	−0.001	+0.11	7.5	0.6	4.4	G0 V		42 s		
164132	245190		18 02 44.1	−53 42 07	−0.002	−0.01	7.5	0.7	1.4	A2 V		65 s		
164716	142045		18 02 46.1	−5 21 31	+0.001	−0.04	6.76	0.16		A0				
164496	209635		18 02 46.3	−30 58 10	0.000	+0.01	7.9	0.8	0.4	B9.5 V		110 s		
164760	142046		18 02 47.2	−1 20 12	−0.001	−0.04	7.97	0.20	0.4	B9.5 V		240 s		
166655	8999		18 02 50.6	+75 47 25	−0.004	+0.04	6.90	0.1	0.6	A0 V		120 mx	11072	m
164584	186163	7 Sgr	18 02 50.9	−24 16 56	0.000	−0.01	5.35	0.52	−2.0	F5 II	−12	270 s		
165502	30715		18 02 52.5	+56 25 41	−0.004	+0.05	7.1	0.2	2.1	A5 V		100 s	11035	m
163696	––		18 02 53.3	−67 50 44			7.7	0.5	4.0	F8 V		55 s		
164562	186161		18 02 54.2	−27 49 36	+0.001	0.00	6.70	1.1	0.2	K0 III		190 s		m
164942	103328		18 02 57.9	+16 05 00	0.000	0.00	7.9	1.1	0.2	K0 III		340 s		
165298	47171		18 02 59.6	+46 14 09	−0.002	+0.05	7.9	0.5	4.0	F8 V		60 s		
164924	103329		18 03 00.3	+15 00 08	+0.001	0.00	7.9	1.6	−0.5	M2 III	−30	490 s		
164409	228653		18 03 02.4	−43 27 47	0.000	−0.02	7.50	1.1	0.2	K0 III		280 s		m
164249	245201		18 03 03.3	−51 38 58	+0.001	−0.11	7.4	0.3	3.4	F5 V		64 s		
164789	142049		18 03 04.1	−8 20 24	−0.001	+0.03	7.7	0.5	3.4	F5 V		68 s		
164764	142050	69 τ Oph	18 03 04.8	−8 10 49	+0.002	−0.04	4.79	0.38	2.6	F0 V	−40	22 ts	11005	m
165171	66586		18 03 05.6	+36 07 14	+0.001	−0.01	8.0	0.4	2.6	F0 V		120 s		
164700	161024		18 03 05.9	−17 24 50	+0.001	0.00	8.0	0.0	−1.0	B5.5 V	+2	480 s		
––	103331		18 03 07.9	+19 41 13	+0.001	+0.02	7.9							
165358	47173		18 03 08.8	+48 27 51	+0.002	+0.01	6.21	0.03		A0	−13		11028	m
165029	103334		18 03 14.6	+19 36 46	+0.002	−0.01	6.50	0.01	0.6	A0 V	−32	150 s		
164704	186179		18 03 18.7	−22 53 03	0.000	+0.01	7.7	0.4	−1.0	B5.5 V	−5	270 s		
165042	103336		18 03 20.5	+19 33 21	+0.001	−0.01	7.20	1.6		M5 II−III	−22			
164967	123069		18 03 20.9	+8 25 09	0.000	−0.01	7.0	0.1	0.6	A0 V	−12	180 s		
165607	30720		18 03 20.9	+56 56 01	−0.003	−0.01	7.6	1.4	−0.3	K5 III		380 s		
164738	161028		18 03 21.7	−17 36 36	0.000	0.00	7.1	0.0	−1.0	B5.5 V	+6	340 s		
164681	186180		18 03 23.6	−26 19 09	0.000	+0.01	7.2	0.7	0.6	A0 V		76 s		
164929	123067		18 03 26.2	+2 30 52	0.000	−0.02	7.90	0.1	0.6	A0 V		250 s	11008	m
164988	103335		18 03 29.6	+10 17 48	−0.002	−0.01	8.0	1.1	−0.1	K2 III		390 s		
165043	103337		18 03 31.5	+15 25 34	+0.001	−0.04	8.0	1.1	0.2	K0 III		360 s		
165906	17739		18 03 31.5	+64 51 48	−0.003	−0.05	7.6	1.1	−0.1	K2 III		340 s		
163880	254122		18 03 33.6	−66 32 15	+0.003	−0.01	7.2	1.1	0.2	K0 III		250 s		
165074	103338		18 03 34.9	+14 05 34	+0.002	0.00	7.7	0.4	2.6	F0 V		100 s		
164685	186186		18 03 36.2	−29 51 44	0.000	−0.05	7.5	0.7	3.2	G5 IV		72 s		
164210	245202		18 03 36.5	−58 18 10	−0.002	−0.01	8.0	1.2	3.2	G5 IV		51 s		
164589	209649		18 03 38.7	−39 24 23	0.000	−0.05	7.9	1.4	−0.5	M2 III		490 s		

HD	SAO	Star Name	α 2000	δ 2000	μ(α)	μ(δ)	V	B-V	M_v	Spec	RV	d(pc)	ADS	Notes
			h m s	° ′ ″	s	″								
164719	186189		18 03 40.0	−26 52 08	−0.001	−0.02	7.99	0.18	−1.0	B5.5 V		430 s		
164211	245205		18 03 43.0	−58 37 44	−0.001	−0.01	8.0	1.0	0.2	K0 III		360 s		
164303	245206		18 03 43.7	−56 16 10	+0.002	−0.01	7.3	2.0	−0.5	M2 III		220 s		
−−	47181		18 03 50.9	+41 21 52	+0.001	−0.01	7.9	1.7						
165372	47180		18 03 51.0	+40 49 50	−0.003	−0.02	7.8	1.4	−0.3	K5 III		410 s		
164794	186204	9 Sgr	18 03 52.3	−24 21 38	0.000	0.00	5.98	0.02		O5	+9			
164477	228660		18 03 52.7	−48 55 52	+0.003	0.00	7.4	1.3	0.2	K0 III		270 s		
−−	66600		18 03 53.1	+36 23 41	−0.002	−0.02	8.0	1.4						
165281	66596		18 03 53.2	+30 22 39	−0.004	−0.27	6.7	0.5	3.4	F5 V	+2	45 s		
165339	66598		18 03 53.9	+35 58 13	+0.004	+0.03	7.6	0.8	3.2	G5 IV		76 s		
165241	85696		18 03 54.0	+26 39 04	+0.002	−0.06	7.00	0.4	3.0	F2 V		63 s	11031	m
164546	228662		18 03 55.6	−45 10 42	+0.002	0.00	7.9	0.2	0.4	B9.5 V		260 s		
164833	186208		18 03 55.8	−22 50 15	+0.001	0.00	7.15	−0.01	−5.0	B0 III	−23	2000 s		m
164816	186207		18 03 56.8	−24 18 45	+0.001	−0.01	7.07	0.1	−4.1	B0 V		1100 s		
165700	30724		18 04 00.3	+55 16 22	−0.007	−0.13	7.7	0.5	4.0	F8 V		54 s		
164844	186211		18 04 03.0	−22 33 57	+0.003	0.00	8.00	0.1	−3.5	B1 V	−9	1200 s		
165523	47187		18 04 04.9	+46 26 31	+0.004	+0.02	8.0	0.9	3.2	G5 IV		92 s		
165434	66603		18 04 06.6	+39 28 50	−0.001	−0.03	7.5	0.1	0.6	A0 V		240 s		
164774	209659		18 04 09.3	−34 03 10	+0.003	+0.01	6.9	1.1	−0.1	K2 III		250 s		
165172	103349		18 04 09.8	+14 46 50	+0.001	−0.01	8.00	0.1	0.6	A0 V		300 s	11032	m
164883	186216		18 04 12.4	−22 30 04	0.000	0.00	7.28	0.01	−4.1	B0 V	−10	1400 s		m
164865	−−		18 04 13.3	−24 11 03			7.62	0.84	−6.5	B9 Iab		2300 s		
165503	47188		18 04 16.7	+42 05 44	−0.001	+0.01	7.9	0.4	2.6	F0 V		110 s		
165011	161038		18 04 20.3	−10 54 37	0.000	+0.01	7.6	0.7	4.4	G0 V		43 s		
164945	161037		18 04 21.3	−17 21 12	+0.001	0.00	7.3	0.0	0.4	B9.5 V		210 s		
164867	186218		18 04 22.4	−27 23 16	0.000	+0.01	7.79	0.03	0.4	B9.5 V		280 s		
164906	186220		18 04 25.7	−24 23 10	+0.001	−0.02	7.46	0.21	−3.9	B1 IV pe	+18	1200 s		
164371	245211		18 04 26.5	−59 10 40	+0.002	+0.01	7.0	2.0	−0.1	K2 III		86 s		
165566	47191		18 04 29.8	+42 51 38	+0.001	0.00	7.5	1.1	0.2	K0 III	−14	280 s		
164909	186222		18 04 30.0	−26 50 57	0.000	+0.01	7.5	0.8	3.4	F5 V		41 s		
165146	123088		18 04 32.1	+1 14 15	+0.001	0.00	7.6	0.4	2.6	F0 V		96 s		
166091	17748		18 04 33.2	+63 47 49	−0.007	+0.06	7.40	1.4	−1.3	K5 II-III	−69	550 s		
165174	123090		18 04 37.1	+1 55 08	0.000	0.00	6.14	0.00		B0.5 III	+17			V986 Oph, q
164776	228675		18 04 37.7	−40 38 16	+0.002	−0.02	7.4	0.0	0.4	B9.5 V		230 s		
164777	228673		18 04 38.7	−41 58 12	−0.001	−0.03	7.6	0.2	0.6	A0 V		260 s		
165398	85707		18 04 39.0	+27 06 55	+0.003	0.00	7.20	0.1	1.4	A2 V	+6	150 s	11040	m
165195	123093		18 04 39.9	+3 46 45	−0.002	−0.08	7.9	0.9	3.2	G5 IV		86 s		
165373	85706		18 04 40.0	+23 56 32	+0.001	−0.06	6.34	0.30	2.6	F0 V	−33	56 s		
164427	245217		18 04 42.2	−59 12 35	−0.026	−0.06	6.89	0.61	3.1	G4 IV		53 mx		m
165567	47193		18 04 43.1	+40 05 03	+0.002	+0.02	6.52	0.46	3.4	F5 V	−1	41 s		
163766	257558		18 04 47.7	−72 22 00	−0.004	−0.04	7.9	1.4	−0.3	K5 III		340 mx		
164972	186235		18 04 48.3	−25 36 20	+0.001	0.00	7.0	0.3	2.6	F0 V		76 s		
165623	47194		18 04 48.6	+42 57 27	0.000	+0.01	7.2	0.1	0.6	A0 V	−21	210 s		
164724	228676		18 04 49.0	−45 35 00	+0.001	+0.02	7.8	1.7	−0.3	K5 III		410 s		
165540	66613		18 04 49.0	+37 50 02	−0.002	0.00	7.8	0.8	3.2	G5 IV		84 s		
164870	209671		18 04 50.1	−35 54 06	−0.002	−0.04	5.80	1.1	0.2	K0 III		130 s		m
164662	245229		18 04 51.4	−50 26 26	+0.001	0.00	8.0	1.3	0.2	K0 III		190 s		
165504	66611		18 04 52.3	+33 16 21	+0.001	−0.19	7.60	0.6	4.4	G0 V	−7	44 s	11047	m
165717	47200		18 04 56.9	+48 08 09	+0.001	0.00	8.00	0.1	1.4	A2 V		210 s	11058	m
165098	161048		18 04 57.3	−15 17 20	+0.002	−0.05	7.9	0.6	4.4	G0 V		48 s		
165016	186240		18 04 58.0	−24 40 49	0.000	+0.02	7.33	−0.04	−5.0	B0 III		2300 s		
165473	85713		18 04 58.9	+29 04 50	0.000	−0.03	7.60	0.9	0.2	G9 III		300 s		
165374	103362		18 05 00.0	+16 55 36	−0.002	+0.01	7.20	1.6	−0.5	M2 III	−13	350 s		
165645	47195		18 05 00.6	+41 56 47	−0.002	+0.11	6.34	0.26	2.6	F0 V	−20	56 s	11054	m
164975	186237		18 05 01.1	−29 34 48	+0.001	0.00	5.09	0.95		F8	−29		11029	W Sgr, m,v
165435	85712		18 05 01.6	+22 54 50	0.000	−0.01	7.3	0.4	3.0	F2 V		74 s		
165505	85715		18 05 09.8	+27 33 52	+0.001	+0.02	8.0	0.1	0.6	A0 V		300 s		
165052	186247		18 05 10.4	−24 23 55	0.000	−0.01	6.87	0.09		O7	+3			
166379	17756		18 05 11.0	+66 56 44	+0.003	+0.02	6.9	0.5	3.4	F5 V		50 s		
164936	209679		18 05 13.8	−37 43 09	−0.002	0.00	7.3	0.4	0.2	K0 III		270 s		
166330	17754		18 05 14.6	+65 43 23	−0.002	−0.02	7.8	0.1	0.6	A0 V		270 s		
165417	103365		18 05 18.4	+15 38 50	0.000	0.00	7.7	1.4	−0.3	K5 III		390 s		
168518	2993		18 05 19.5	+81 29 08	+0.004	+0.03	7.70	0.2	2.1	A5 V		130 s	11156	m
165988	30740		18 05 20.6	+55 06 19	−0.002	−0.03	8.0	0.5	4.0	F8 V		64 s		
165081	186250		18 05 20.7	−27 50 04	0.000	−0.01	8.0	1.2	−0.3	K5 III		450 s		
162337	258787		18 05 26.0	−81 29 11	+0.007	−0.04	6.35	1.50		K2				
165589	85719		18 05 26.6	+28 41 08	+0.002	0.00	7.80	1.1	0.0	K1 III		360 s		
165845	47209		18 05 27.0	+49 20 38	−0.003	+0.01	7.3	1.1	−0.1	K2 III		310 s		
165341	123107	70 Oph	18 05 27.2	+2 29 58	+0.017	−1.09	4.03	0.86	5.7	K0 V	−7	5.1 t	11046	m
165524	85718		18 05 30.0	+21 38 49	+0.001	+0.01	6.15	1.23	0.2	K0 III	−35	110 s		
165907	30739		18 05 30.7	+51 56 50	+0.003	+0.07	6.9	1.1	0.2	K0 III		220 s		
165244	161057		18 05 31.1	−11 01 09	+0.001	+0.02	7.3	0.2	2.1	A5 V		100 s		
166356	17759		18 05 31.8	+65 04 52	−0.011	+0.10	7.4	1.1	0.2	K0 III	−2	160 mx		
165084	209688		18 05 32.5	−30 50 21	−0.003	−0.02	8.0	1.2	0.2	K0 III		290 s		
164999	209684		18 05 32.8	−37 28 17	−0.002	−0.02	6.7	0.7	0.2	K0 III		200 s		

HD	SAO	Star Name	α 2000	δ 2000	μ(α)	μ(δ)	V	B-V	M$_v$	Spec	RV	d(pc)	ADS	Notes
165202	161056		18^h 05^m 33.1^s	-19° 45' 15"	-0.001^s	+0.01"	6.8	0.1	0.4	B9.5 V		170 s		
165020	209685		18 05 35.4	-36 45 17	-0.003	-0.04	7.6	-0.2	2.1	A5 V		130 s		
165401	123112		18 05 37.3	+4 39 27	-0.001	-0.31	6.79	0.63	4.4	G0 V	-124	25 ts		
165118	186256		18 05 40.0	-29 16 39	0.000	-0.04	7.1	0.6	1.4	A2 V		62 s		
165569	85720		18 05 41.3	+21 25 34	-0.001	0.00	7.8	0.4	2.6	F0 V		110 s		
165475	103373		18 05 43.0	+12 00 14	-0.001	0.00	7.04	0.31	1.2	A5 IV	+13	120 s	11056	m
165063	209691		18 05 44.1	-36 34 46	-0.002	0.00	7.70	0.00	0.0	B8.5 V		330 s		m
165474	--		18 05 44.9	+12 00 20			7.45	0.31		A7 p	+13			
165419	123115		18 05 47.3	+1 59 09	-0.003	-0.05	7.7	1.1	-0.1	K2 III		300 s		
165135	209696	10 γ Sgr	18 05 48.3	-30 25 26	-0.004	-0.19	2.99	1.00	0.2	K0 III	+22	36 s		Alnasl
165683	66626		18 05 49.4	+32 13 50	+0.001	-0.03	5.71	1.16		K0	+1			
165590	85723		18 05 49.5	+21 26 44	-0.002	-0.05	7.07	0.66	4.5	dG1	+4	30 s	11060	m
165476	123116		18 05 52.1	+7 05 26	-0.014	-0.10	7.7	0.9	3.2	G5 IV		46 mx		
165136	209700		18 05 54.5	-30 39 39	+0.003	-0.06	7.9	1.1	0.2	K0 III		200 mx		
165319	161061		18 05 58.6	-14 11 52	-0.001	+0.01	7.94	0.59	-6.2	B0 Ia	+30	2300 s		
170489	3005		18 05 59.6	+84 39 21	+0.002	+0.02	7.4	0.4	3.0	F2 V		77 s		
166155	30748		18 05 59.8	+57 21 32	+0.001	+0.05	7.0	1.4	-0.3	K5 III		290 s		
165360	142081		18 06 01.2	-8 06 31	+0.003	+0.08	7.00	0.5	4.0	F8 V		39 s	11052	m
167407	9020		18 06 01.5	+76 07 54	-0.024	+0.02	7.7	0.5	4.0	F8 V		55 s		
165625	85725	98 Her	18 06 01.7	+22 13 08	-0.001	-0.01	5.2	1.6	-0.5	M2 III	-20	140 s		
164955	228691		18 06 03.6	-47 41 22	-0.002	-0.03	7.3	1.4	0.2	K0 III		160 s		
165121	209698		18 06 04.4	-37 25 59	-0.001	-0.01	7.6	0.9	0.2	K0 III		300 s		
165246	186268		18 06 04.5	-24 11 45	+0.001	-0.02	7.71	0.10	0.4	B9.5 V		290 s	11049	m
163858	257559		18 06 04.6	-74 00 13	+0.001	+0.03	7.2	0.1	0.6	A0 V		210 s		
164749	245235		18 06 05.2	-56 49 18	+0.001	+0.01	8.00	1.1	-0.1	K2 III		420 s		m
166011	47217		18 06 07.1	+49 27 59	+0.003	+0.04	7.8	0.4	2.6	F0 V	-14	110 s		
165402	142083		18 06 07.2	-8 19 26	0.000	-0.02	5.85	0.21	-0.6	B8 IV	-27	140 s		
165462	142084		18 06 07.2	-0 26 48	-0.001	-0.03	6.34	1.06		G5				
165183	209706		18 06 08.8	-31 33 16	+0.004	-0.05	7.8	1.4	-0.1	K2 III		270 s		
--	47218		18 06 08.8	+49 31 14	0.000	-0.01	7.5	2.0						
164896	245238		18 06 09.0	-51 44 39	0.000	-0.02	6.7	0.9	1.4	A2 V		38 s		
165801	66631		18 06 09.0	+37 46 33	0.000	-0.02	7.9	0.9	3.2	G5 IV		86 s		
165137	209702		18 06 12.8	-37 29 58	+0.001	-0.12	7.9	-0.4	3.0	F2 V		90 mx		
165511	123121		18 06 12.9	+1 34 57	0.000	-0.03	8.0	0.1	0.6	A0 V		250 s		
165269	186271		18 06 13.6	-27 29 35	+0.002	-0.06	7.2	0.8	4.4	G0 V		26 s		
165438	142085		18 06 15.0	-4 45 05	+0.009	-0.03	5.77	0.96	0.0	gK1	-19	140 s		
165022	228693		18 06 15.5	-46 28 32	0.000	-0.01	7.1	2.0	-0.5	M2 III		200 s		
165684	85732		18 06 16.9	+24 24 47	0.000	-0.02	6.9	1.1	0.2	K0 III		220 s		
166044	47222		18 06 17.9	+48 50 54	+0.001	+0.02	8.0	0.1	1.4	A2 V		210 s		
165185	209710		18 06 23.5	-36 01 11	+0.009	+0.01	5.95	0.62	5.2	G5 V		14 s		
165627	103384		18 06 25.7	+12 20 54	+0.002	+0.03	7.7	1.1	0.2	K0 III		310 s		
165139	228698		18 06 26.8	-41 45 26	+0.002	+0.01	7.70	0.1	0.6	A0 V		260 s		m
--	47223		18 06 27.4	+48 35 47	-0.004	0.00	7.8	2.3						
164806	245237		18 06 29.5	-58 34 16	+0.001	-0.01	6.84	-0.09	-2.2	B5 III		590 s		
165941	47220		18 06 30.3	+40 21 41	-0.001	+0.01	7.40	0.1	0.6	A0 V		230 s	11074	m
165442	161074		18 06 34.2	-14 58 22	+0.005	-0.02	7.9	0.4	2.6	F0 V		110 s		
169839	3003		18 06 34.3	+83 40 58	-0.005	-0.04	7.8	0.8	3.2	G5 IV		85 s		
165024	245242	θ Ara	18 06 37.6	-50 05 30	-0.001	-0.02	3.66	-0.08	-5.1	B1 II	+3	480 s		
166067	47224		18 06 38.7	+46 16 23	0.000	-0.02	7.5	0.4	3.0	F2 V		80 s		
166578	17768		18 06 40.3	+64 13 15	-0.007	-0.01	6.8	1.1	0.2	K0 III		210 s		
165421	161077		18 06 42.7	-18 59 02	+0.001	+0.01	7.80	0.1	0.6	A0 V		240 s		m
165365	186286		18 06 45.6	-28 21 49	0.000	0.00	8.0	-0.7	0.6	A0 V		300 s		
166068	47225		18 06 47.2	+43 27 03	-0.004	-0.07	8.0	1.6	-0.5	M4 III		260 mx		
165672	123130		18 06 48.6	+6 24 38	-0.001	-0.04	7.9	0.9	3.2	G5 IV		86 s		
165423	186291		18 06 49.4	-22 02 46	0.000	+0.02	7.72	1.79	-0.3	K5 III		270 s		
165189	228708		18 06 49.6	-43 25 29	0.000	-0.10	4.95	0.22	2.1	A5 V		34 s		m
165802	85738		18 06 50.6	+24 18 39	+0.003	-0.01	7.7	1.1	0.2	K0 III		310 s		
165670	123133		18 06 50.7	+8 52 33	+0.004	-0.15	7.80	0.5	3.4	F5 V	+32	75 s		m
165963	66649		18 06 51.8	+37 47 54	0.000	-0.02	8.0	1.1	0.2	K0 III		360 s		
165003	245243		18 06 52.8	-54 10 49	+0.003	-0.07	7.3	0.9	3.4	F5 V		32 s		
166207	30751		18 06 53.3	+50 49 23	-0.001	+0.10	6.29	1.04	0.2	K0 III	-57	150 mx		
165383	186288		18 06 55.2	-29 04 30	0.000	-0.02	7.9	0.0	0.4	B9.5 V		290 s		
165610	142092		18 06 55.5	-3 14 00	+0.002	+0.08	7.0	0.5	4.0	F8 V		38 s		
165141	228707		18 07 00.2	-48 14 46	+0.001	+0.05	6.8	1.2	0.2	K0 III		170 s		
166728	17772		18 07 00.6	+65 53 14	-0.004	-0.02	7.8	0.1	1.4	A2 V		180 mx		
165908	66648	99 Her	18 07 01.3	+30 33 43	-0.007	+0.07	5.04	0.52	3.8	F7 V	+1	17 ts	11077	m
165825	85740		18 07 01.9	+21 46 35	0.000	+0.01	8.0	1.1	0.2	K0 III		360 s		
165847	85743		18 07 03.6	+23 24 13	0.000	+0.01	7.2	0.1	1.7	A3 V		120 s		
166012	66650		18 07 05.7	+36 25 14	-0.001	0.00	8.0	0.1	1.4	A2 V		210 s		
166228	47233		18 07 06.0	+49 42 37	-0.001	+0.02	6.30	0.1	0.6	A0 V	-26	140 s	11090	m
165191	228711		18 07 06.5	-47 22 02	+0.002	0.00	7.5	1.0	0.2	K0 III		280 s		
165329	--		18 07 08.8	-37 45 42			7.9	2.4	-0.3	K5 III		120 s		
165516	186302		18 07 11.1	-21 26 38	-0.001	0.00	6.28	0.12		B0.5 Ib	-11			
166013	66652		18 07 11.3	+34 49 31	0.000	0.00	7.3	1.6	-0.5	M2 III		370 s		
165741	123139		18 07 16.9	+6 42 05	0.000	-0.06	7.1	1.1	0.2	K0 III		220 mx		

HD	SAO	Star Name	α 2000	δ 2000	μ(α)	μ(δ)	V	B−V	M_v	Spec	RV	d(pc)	ADS	Notes
			h m s	° ′ ″	s	″								
165760	123140	71 Oph	18 07 18.2	+8 44 02	+0.001	+0.03	4.64	0.96	1.8	G8 III-IV	−3	34 s		
165826	103397		18 07 20.5	+15 14 05	−0.001	0.00	7.4	1.6	−0.5	M2 III		370 s		
165742	123138		18 07 20.8	+2 28 54	−0.001	+0.03	6.6	1.4	−0.3	K5 III		210 s		
165777	123142	72 Oph	18 07 20.8	+9 33 50	−0.004	+0.08	3.73	0.12	1.9	A4 V	−24	28 ts	11076	
165848	103398		18 07 21.0	+15 54 59	−0.001	−0.15	6.71	1.12	0.0	gK1	+18	210 s		
165761	123141		18 07 22.4	+6 32 34	+0.001	−0.01	7.8	1.6	−0.5	M4 III		410 s		
167963	9030		18 07 24.8	+77 35 34	−0.003	−0.01	7.4	1.1	0.2	K0 III		280 s		
165271	228716		18 07 26.0	−46 53 55	−0.003	−0.10	7.64	0.65	3.2	G5 IV		77 s		
166381	30759		18 07 26.3	+53 11 01	+0.002	+0.01	7.9	1.1	0.2	K0 III		340 s		
164726	254139		18 07 26.6	−65 31 20	+0.009	−0.12	7.8	1.1	−0.1	K2 III		130 mx		
166208	47237		18 07 28.6	+43 27 42	0.000	−0.06	5.00	0.91		K0 p	−16	25 mn		
165886	103403		18 07 29.4	+19 40 00	0.000	+0.02	7.70	0.1	1.4	A2 V		180 s	11080	m
165989	85748		18 07 29.9	+26 24 53	0.000	+0.01	7.3	0.9	0.3	G8 III		250 s		
165804	123145		18 07 30.9	+9 29 45	0.000	+0.02	7.7	1.1	0.2	K0 III		290 s		
166014	85750	103 o Her	18 07 32.4	+28 45 45	0.000	+0.01	3.83	−0.03	0.2	B9 V	−30	52 s		q
165686	142100		18 07 33.7	−9 10 41	+0.002	0.00	7.8	0.0	0.4	B9.5 V		240 s		
165164	245250		18 07 35.8	−54 09 00	0.000	−0.02	7.4	0.4	1.4	A2 V		94 s		
165351	228719		18 07 39.9	−43 23 33	−0.001	−0.04	7.6	1.1	0.2	K0 III		270 s		
165413	209736		18 07 41.7	−39 55 35	−0.001	−0.06	7.8	1.3	−0.5	M2 III		230 mx		
—	103407		18 07 45.3	+18 52 40	+0.001	−0.01	7.9	1.1	0.2	K0 III		350 s		
166434	30762		18 07 45.3	+52 01 40	−0.001	+0.03	7.8	1.1	−0.1	K2 III		380 s		
165864	123150		18 07 47.7	+9 50 55	+0.005	0.00	6.6	0.1	0.6	A0 V		91 mx		
166974	17779		18 07 47.7	+67 24 58	+0.001	+0.01	7.7	1.1	0.2	K0 III		320 s		
165687	161093		18 07 48.1	−17 09 15	−0.007	+0.06	5.52	1.11	0.0	K1 III	−32	84 mx		
164871	254141		18 07 48.2	−64 33 00	−0.001	−0.05	6.41	1.26		K2				
165910	103406		18 07 48.2	+13 04 16	+0.001	+0.01	6.63	0.05	1.4	A2 V	−17	110 s	11086	m
166045	85752	100 Her	18 07 49.4	+26 05 51	0.000	+0.03	5.86	0.12		A3	−15		11089	m
166046	85753	100 Her	18 07 49.4	+26 06 05	0.000	+0.03	5.90	0.14		A3	−17		11089	m
166253	47242		18 07 50.8	+41 43 06	+0.001	−0.04	7.63	1.66	−0.5	gM4	−17	400 s		
166093	85760		18 07 52.1	+29 49 03	+0.002	−0.02	7.1	1.1	−2.3	K3 II		760 s		
165470	209745		18 07 54.2	−38 33 57	0.000	−0.01	7.9	−0.5	0.0	B8.5 V		380 s		
166276	66669		18 07 59.2	+39 55 24	−0.006	+0.05	7.7	0.1	1.4	A2 V		57 mx		
165273	245256		18 08 01.8	−52 14 17	−0.001	−0.01	7.3	1.4	0.2	K0 III		160 s		
166229	66666		18 08 02.0	+36 24 05	−0.008	−0.18	5.48	1.17	−0.1	K2 III	−7	95 mx		
165598	209751		18 08 02.6	−31 00 06	0.000	+0.01	7.9	0.7	3.4	F5 V		54 s		
165634	186328		18 08 04.8	−28 27 25	+0.002	−0.03	4.57	0.94		G pe	−5	16 mn		
166300	47246		18 08 04.9	+40 54 51	−0.002	−0.01	7.4	1.1	0.2	K0 III		270 s		
165887	123156		18 08 07.1	+2 13 07	+0.001	−0.01	6.50	0.1	1.4	A2 V		98 s	11088	m
165807	142105		18 08 08.5	−7 07 12	−0.001	0.00	7.50	1.1	0.2	K0 III		240 s		m
166180	66668		18 08 12.2	+30 59 56	+0.001	0.00	7.4	0.1	0.6	A0 V	−30	230 s		
165256	245258		18 08 15.8	−55 41 28	−0.002	0.00	7.7	1.0	−0.1	K2 III		360 s		
166181	85767		18 08 16.0	+29 41 27	+0.010	−0.03	7.66	0.72	5.2	G5 V		31 s		
165830	161101		18 08 17.2	−10 33 01	+0.001	0.00	7.81	0.32		A m				
166409	47250		18 08 21.6	+44 06 45	+0.002	+0.03	6.7	0.5	3.4	F5 V	−17	45 s		
165782	—		18 08 28.2	−18 33 24			7.68	2.14	−8.0	G8 Ia		3900 s		AX Sgr, v
165763	186341		18 08 28.3	−21 15 10	−0.001	0.00	7.78	−0.25		WC6				
165808	161102		18 08 28.6	−16 24 58	0.000	+0.02	7.67	0.20	−1.6	B3.5 V	−16	430 s		
165493	228734		18 08 29.8	−45 46 02	−0.001	−0.02	6.15	−0.08		B8	−35			m
166095	103414		18 08 33.6	+14 17 03	−0.001	−0.01	6.37	0.17		A m	−9			
164250	257563		18 08 34.6	−75 06 18	+0.004	−0.02	6.9	2.4	−0.3	K5 III		79 s		
165040	254147	π Pav	18 08 34.6	−63 40 06	+0.003	−0.19	4.35	0.22		A m	−16	17 mn		
—	66674		18 08 34.7	+38 11 41	0.000	−0.02	8.0	1.2						
165723	—		18 08 36.0	−27 52 26			7.50	0.9	3.2	G5 IV		71 s	11084	m
165583	228736		18 08 36.2	−41 56 19	−0.001	+0.01	7.3	0.6	0.6	A0 V		89 s		
165784	186343		18 08 38.3	−21 27 00	−0.001	−0.01	6.54	0.85	−7.5	A2 Ia	−16	2300 s		
166780	30770		18 08 38.4	+57 58 47	−0.006	0.00	7.33	1.48	−0.3	K5 III	−43	150 mx		
165965	142115		18 08 39.0	−2 54 27	+0.001	0.00	6.8	0.8	3.2	G5 IV		52 s		
—	—		18 08 40.5	−86 47 46			7.40							S Oct, v
165471	228735		18 08 43.7	−49 20 44	−0.002	−0.03	7.4	0.7	3.4	F5 V		54 s		
165812	186348		18 08 44.9	−22 09 36	+0.001	+0.02	7.93	0.01	−1.0	B5.5 V	−24	580 s		
166182	85769	102 Her	18 08 45.4	+20 48 52	0.000	−0.01	4.36	−0.16	−2.5	B2 V	−15	240 s	11102	m
165991	142117		18 08 45.9	−3 59 29	+0.001	−0.02	8.0	0.1	0.6	A0 V		250 s		
164750	—		18 08 46.0	−70 04 57			8.0	1.1	0.2	K0 III		360 s		
166073	123172		18 08 52.3	+1 59 45	0.000	+0.04	6.9	0.4	3.0	F2 V		59 s		
166072	123173		18 08 52.3	+2 14 34	0.000	−0.02	7.3	0.1	1.4	A2 V		140 s		
167534	9028		18 08 52.5	+72 09 12	−0.002	+0.03	8.0	0.5	3.4	F5 V		82 s		
166230	85770	101 Her	18 08 52.7	+20 02 43	+0.001	−0.02	5.10	0.15		A4	−16	25 mn		
166280	85773		18 08 52.7	+24 57 59	−0.001	−0.02	6.8	0.1	0.6	A0 V		170 s		
165662	209760		18 08 52.9	−39 29 41	0.000	−0.04	7.7	1.0	−0.5	M2 III		440 s		
165617	228742		18 08 53.4	−42 39 26	+0.001	−0.01	7.5	−0.3	0.0	B8.5 V		320 s		
165814	186350		18 08 53.9	−25 28 21	0.000	+0.02	6.61	0.02	0.0	B8.5 V		200 s	11069	m
165787	186349		18 08 54.0	−27 47 02	+0.001	0.00	7.4	1.3	0.2	K0 III		190 s		
165767	209769		18 08 54.5	−29 59 33	+0.002	+0.01	7.0	1.2	3.2	G5 IV		31 s		
167274	17792		18 08 55.4	+67 59 16	−0.001	0.00	8.0	0.1	0.6	A0 V		310 s		
166301	85776		18 08 58.5	+24 09 29	+0.007	+0.07	7.6	0.7	4.4	G0 V		44 s		

HD	SAO	Star Name	α 2000	δ 2000	μ(α)	μ(δ)	V	B−V	M_v	Spec	RV	d(pc)	ADS	Notes
			h m s	° ′ ″	s	″								
165770	209771		18 08 58.9	−31 09 37	+0.001	−0.03	7.7	0.2	0.6	A0 V		190 s		
165967	161113		18 08 59.8	−11 44 39	+0.001	−0.03	7.8	0.1	0.6	A0 V		230 s		
166382	66680		18 09 06.1	+31 01 16	+0.001	+0.01	6.8	1.6		M5 III	−122			T Her, m,v
166411	66682		18 09 10.0	+30 28 09	+0.005	+0.13	6.38	1.19	0.0	gK1	−80	160 s		
166255	103428		18 09 11.2	+14 01 34	+0.001	−0.01	7.9	0.1	0.6	A0 V		280 s		
165395	245266		18 09 14.2	−57 47 11	+0.001	−0.02	7.8	1.5	−0.1	K2 III		250 s		
165896	186362		18 09 14.9	−26 06 58	−0.003	−0.32	7.59	0.68		G0		15 mn	11096	m
165873	186361		18 09 16.2	−27 44 05	0.000	−0.01	7.3	1.4	−0.3	K5 III		330 s		
165750	209773		18 09 17.0	−38 36 39	0.000	−0.02	7.9	0.7	3.2	G5 IV		88 s		
165921	186366		18 09 17.7	−23 59 18	+0.003	+0.01	7.28	0.12		O7.5				
166435	85784		18 09 21.2	+29 57 08	+0.006	+0.08	6.84	0.62	5.2	dG5	−14	21 s		
165793	209779		18 09 22.2	−36 40 21	−0.001	0.00	6.58	−0.03	−5.7	B1 Ib k		2100 s		
165819	209780		18 09 24.8	−35 42 27	−0.001	−0.02	7.9	0.8	−0.1	K2 III		390 s		
166282	103431		18 09 25.0	+12 24 37	0.000	−0.01	8.0	1.1	−0.1	K2 III		400 s		
165836	209782		18 09 26.4	−33 14 03	+0.001	0.00	7.6	1.3	−0.1	K2 III		290 s		
165696	228747		18 09 27.3	−45 56 18	+0.003	−0.13	7.36	0.50	4.0	F8 V		47 s		
166183	123181		18 09 30.8	+0 23 28	+0.003	−0.04	8.0	0.5	4.0	F8 V		60 s		
165666	228744		18 09 31.8	−48 40 48	0.000	0.00	8.0	0.5	2.6	F0 V		120 s		
−−	66697		18 09 31.9	+38 56 39	0.000	−0.03	7.9	1.9						
166233	123187	73 Oph	18 09 33.7	+3 59 36	+0.002	−0.01	5.73	0.37	2.6	dF0 n	−17	41 s	11111	m
−−	17791		18 09 34.6	+61 59 25	−0.002	+0.01	8.0	1.9						
165999	186375		18 09 36.2	−23 34 05	+0.002	−0.01	7.64	0.27	1.4	A2 V		130 s		
166160	142125		18 09 36.3	−5 39 14	+0.001	−0.08	7.4	0.7	4.4	G0 V		38 s		
166620	66700		18 09 37.2	+38 27 27	−0.026	−0.47	6.40	0.87	6.3	K2 V	−19	11 ts		
166234	123192		18 09 38.7	+3 40 48	+0.002	−0.02	7.8	1.1	−0.1	K2 III		320 s		
166820	30776		18 09 40.5	+50 24 08	+0.001	+0.02	7.50	0.1	1.4	A2 V		170 s	11128	m
166161	142126		18 09 40.5	−8 46 47	−0.007	−0.21	8.0	0.9	3.2	G5 IV		39 mx		
166052	−−		18 09 41.0	−18 51 14			7.5	0.0	0.0	B8.5 V		270 s		
166303	123196		18 09 42.9	+6 12 24	−0.003	+0.03	7.1	1.1	0.2	K0 III	−38	220 s		
166105	161124		18 09 43.0	−15 16 58	+0.001	−0.01	7.90	2.24	−0.5	M2 III		200 s		
166103	161125		18 09 43.2	−13 56 04	+0.001	−0.01	6.4	1.1	0.2	K0 III		160 s		
165796	228759		18 09 44.9	−42 24 12	−0.001	−0.01	7.4	1.2	−0.1	K2 III		310 s		
166144	161127		18 09 45.0	−11 43 30	−0.001	−0.03	7.33	1.79	−0.1	K2 III		130 s		
166284	123197		18 09 47.4	+3 12 03	+0.001	−0.02	7.38	1.23	−0.1	K2 III	−68	280 s		
165753	228758		18 09 48.3	−44 56 45	+0.002	−0.03	7.04	1.10	3.2	K0 IV		45 s		
167159	17794		18 09 53.0	+61 52 52	−0.010	+0.04	7.1	0.5	4.0	F8 V		42 s		
166285	123198		18 09 53.8	+3 07 11	+0.001	−0.19	5.69	0.47	3.3	dF4	−14	29 ts	11113	m
164872	257568		18 09 54.6	−71 38 45	+0.003	−0.06	8.0	0.4	3.0	F2 V		100 s		
165715	228756		18 09 56.2	−49 06 52	−0.003	−0.03	7.6	1.7	−0.1	K2 III		260 s		
165497	245273		18 09 57.4	−59 02 24	−0.002	0.00	6.40	1.4	−0.3	K5 III		220 s		m
166640	66703		18 09 58.8	+36 27 59	+0.001	+0.01	5.58	0.91	3.2	G5 IV	−26	25 s		
165978	209794		18 09 59.8	−32 43 11	0.000	−0.15	6.43	1.02	3.2	G5 IV		31 s		
166003	209799		18 10 03.1	−32 08 38	+0.004	−0.01	7.6	0.4	0.6	A0 V		140 mx		
167027	30783		18 10 03.1	+56 16 02	+0.002	+0.04	7.5	1.1	−0.1	K2 III	−52	330 s		
165822	228760		18 10 05.5	−45 54 14	0.000	−0.01	7.7	0.2	0.4	B9.5 V		200 s		
166023	209803		18 10 05.6	−30 43 43	0.000	−0.03	5.53	0.97		K0	0			
−−	47275		18 10 06.2	+47 05 21	−0.002	+0.01	8.0	0.8						
166107	186389		18 10 06.8	−23 46 24	+0.002	−0.01	7.96	0.07	−1.1	B5 V		570 s		m
166479	103443		18 10 08.5	+16 28 36	0.000	−0.01	6.1	0.4	3.0	F2 V	−13	43 s	11123	m
170151	3007		18 10 09.2	+82 56 24	−0.012	+0.04	7.3	0.0	0.4	B9.5 V		98 mx		
166757	66711		18 10 17.1	+39 04 32	−0.001	−0.01	8.0	0.1	1.4	A2 V		210 s		
166058	209805		18 10 19.8	−31 31 57	−0.001	0.00	7.7	1.5	0.2	K0 III		160 s		
166709	66710		18 10 23.5	+34 33 30	0.000	+0.02	8.0	1.1	0.2	K0 III		360 s		
166384	123205		18 10 23.8	+0 32 36	0.000	−0.01	8.00	0.1	0.6	A0 V		260 s	11119	m
165499	254157	ι Pav	18 10 26.0	−62 00 08	−0.011	+0.22	5.49	0.58	3.9	G3 IV−V	+29	23 ts		
166682	66709		18 10 26.1	+32 21 09	+0.001	−0.01	8.00	0.1	1.4	A2 V		210 s	11131	m
166729	66713		18 10 27.1	+35 10 18	−0.001	−0.02	7.0	0.5	4.0	F8 V		41 s		
167042	30784		18 10 31.5	+54 17 12	+0.013	+0.25	5.95	0.94	0.0	K1 III	−16	65 mx		
166194	186403		18 10 33.5	−26 01 59	+0.002	−0.01	7.3	0.8	1.7	A3 V		52 s		
164644	257567		18 10 34.1	−75 11 28	+0.001	+0.01	7.7	−0.7	0.6	A0 V		260 s		
166580	103453		18 10 36.4	+18 17 19	0.000	+0.02	7.6	0.1	1.7	A3 V		160 s		
166498	123215		18 10 37.4	+7 38 07	0.000	+0.01	7.4	1.1	−0.1	K2 III		300 s		
166287	161142		18 10 37.8	−16 49 29	+0.001	−0.01	7.90	0.21		B0.5 III	−17			
166460	123212		18 10 40.2	+3 19 28	+0.001	0.00	5.51	1.19	−0.1	gK2	+10	130 s		
166133	209811		18 10 41.6	−33 55 08	0.000	−0.03	7.3	1.0	0.2	K0 III		260 s		
166581	103455		18 10 43.1	+16 03 22	+0.001	−0.02	7.7	1.1	−0.1	K2 III		360 s		
166621	103456		18 10 43.7	+18 30 15	+0.001	0.00	7.2	0.1	1.4	A2 V		150 s		
166263	186410		18 10 43.9	−22 14 22	0.000	+0.01	7.78	1.23	3.2	G5 IV		44 s		
167103	30788		18 10 45.0	+53 30 01	−0.001	+0.02	8.0	0.1	1.4	A2 V		210 s		
166361	161147		18 10 45.1	−11 01 20	+0.001	−0.02	7.7	1.1	−0.1	K2 III		290 s		
166499	123217		18 10 49.7	+4 26 02	−0.001	−0.03	7.6	1.1	0.2	K0 III		260 s		
166955	47278		18 10 49.7	+44 35 54	−0.001	−0.02	6.8	1.1	0.2	K0 III		210 s		
166197	209817		18 10 55.1	−33 47 59	−0.001	+0.01	6.16	−0.14	−3.6	B2 III k	−32	780 s		
167104	30792		18 11 01.1	+51 44 24	−0.002	+0.02	7.2	1.1	0.2	K0 III		250 s		
165984	228771		18 11 03.3	−48 18 14	+0.001	−0.03	7.9	1.5	−0.3	K5 III		430 s		

HD	SAO	Star Name	α 2000	δ 2000	μ(α)	μ(δ)	V	B-V	M_v	Spec	RV	d(pc)	ADS	Notes
			$^h \quad ^m \quad ^s$	$^{\circ} \quad ' \quad ''$	^{s}	$''$								
166781	85804		18 11 03.4	+26 40 15	+0.001	+0.02	7.7	0.6	-2.1	G3 II	-39	930 s		
166006	228774		18 11 04.2	-47 30 47	+0.001	-0.03	6.07	1.20		K1 III-IV				m
166295	186420		18 11 04.9	-25 45 44	0.000	+0.02	6.8	1.3	0.2	K0 III		140 s		
166114	228778		18 11 05.4	-41 21 32	+0.003	-0.03	5.86	0.29	2.1	A5 V n	-32	56 s		
167387	17803		18 11 06.8	+60 24 35	-0.002	+0.01	6.3	0.1	0.6	A0 V	-22	140 s		
--	47286		18 11 08.6	+48 01 42	-0.001	-0.02	8.0	1.5						
166583	123223		18 11 12.0	+1 58 55	-0.001	-0.04	8.0	1.1	-0.1	K2 III		340 s		
166956	66730		18 11 12.7	+39 06 03	-0.001	+0.01	6.9	0.1	0.6	A0 V		180 s		
166063	228777	ε Tel	18 11 13.6	-45 57 15	-0.001	-0.03	4.53	1.01	0.3	G5 III	-26	57 s		m
166393	161153		18 11 14.6	-19 50 32	0.000	-0.03	6.90	0.1	1.4	A2 V		120 s	11127	m
164712	257569		18 11 15.4	-75 53 29	+0.003	-0.29	5.86	1.24	-0.2	K3 III		62 mx		m
166309	186422		18 11 15.7	-29 38 22	+0.001	-0.01	8.00	0.9	3.2	G5 IV		90 s		m
166562	142139		18 11 16.2	-0 39 33	-0.001	-0.05	7.9	1.1	-0.1	K2 III		320 s		
166198	228782		18 11 20.3	-40 18 26	-0.001	-0.01	7.8	-0.2	0.0	B8.5 V		320 s		
166867	85808		18 11 23.5	+29 54 32	0.000	-0.03	7.3	1.1	3.2	K0 IV		67 s		
166297	209827		18 11 23.6	-33 19 37	+0.003	-0.06	7.5	0.7	3.4	F5 V		46 s		
166868	85810		18 11 24.4	+29 40 12	+0.002	-0.02	7.5	0.1	1.4	A2 V	-6	170 s		
166563	142140		18 11 26.7	-5 12 18	-0.001	0.00	6.60	0.00	0.4	B9.5 V		160 s	11135	m
166521	161157		18 11 27.7	-11 18 32	-0.002	+0.02	7.5	0.9	3.2	G5 IV		68 s		
166842	85811		18 11 31.5	+25 33 41	-0.004	+0.02	6.77	1.12	0.0	K1 III		220 s		
--	47291		18 11 33.9	+47 39 35	-0.002	+0.02	7.9	1.8						
167190	47293		18 11 34.9	+48 17 22	-0.002	-0.05	6.6	1.1	0.2	K0 III		190 s		
166326	209832		18 11 40.5	-35 01 32	0.000	-0.02	7.0	0.5	0.0	B8.5 V		120 s		
166642	142145		18 11 40.8	-2 43 35	-0.001	-0.04	7.2	1.1	0.2	K0 III	-50	220 s		
166298	228791		18 11 41.9	-40 12 40	-0.008	-0.07	7.5	1.2	0.2	K0 III		110 mx		
165986	245295		18 11 42.8	-55 21 44	-0.002	-0.04	6.2	2.5	-0.3	K5 III		48 s		
166464	186437	1 Sgr	18 11 43.2	-23 42 04	+0.001	-0.02	4.98	1.05	0.2	K0 III	+4	83 s	11133	m
166447	186434		18 11 44.2	-28 01 55	-0.002	-0.02	7.9	0.9	-0.3	K5 III		430 s		
166988	66733		18 11 44.9	+33 26 49	+0.001	+0.01	5.80	0.01	-0.2	A2 III	-32	160 s	11149	m
167043	66739		18 11 46.6	+37 53 14	+0.001	-0.02	7.7	0.6	4.4	G0 V		47 s		
166246	228789		18 11 46.8	-43 04 57	-0.001	-0.03	8.0	-0.2	0.6	A0 V		300 s		
167779	17813		18 11 49.7	+66 08 09	-0.001	+0.03	7.3	1.1	0.2	K0 III	-15	260 s		
165987	245300		18 11 51.0	-56 24 04	0.000	0.00	7.5	0.5	1.4	A2 V		95 s		
167006	66737	104 Her	18 11 54.0	+31 24 19	-0.001	+0.02	4.97	1.65	-0.5	gM3	0	120 s		
166662	142146		18 11 55.2	-5 37 26	-0.001	+0.01	7.0	0.1	1.4	A2 V		120 s		
165937	245297		18 11 55.5	-58 11 26	+0.004	-0.04	7.3	1.9	3.2	G5 IV		13 s		
167348	30803		18 11 55.3	+52 19 00	0.000	+0.05	7.5	1.1	-0.1	K2 III		330 s		
166546	186450		18 11 56.9	-20 25 25	+0.001	-0.01	7.25	0.05	-5.5	O9.5 III	+1	2300 s		
166469	186444		18 11 58.0	-28 54 05	+0.001	-0.01	6.4	-0.1		A0 p				
166140	245307		18 11 59.9	-51 39 15	-0.002	-0.04	7.9	0.1	0.4	B9.5 V		250 s		
166470	186445		18 12 00.6	-29 34 17	+0.001	-0.02	7.9	0.3	0.4	B9.5 V		190 s		
166664	142147		18 12 01.2	-8 43 50	0.000	-0.01	7.1	0.2	2.1	A5 V		91 s		
166026	245304		18 12 02.8	-56 25 40	+0.002	+0.02	7.60	0.1	0.6	A0 V		250 s		m
167260	47298		18 12 04.8	+46 44 11	-0.007	-0.05	8.0	0.9	3.2	G5 IV		93 s		
168497	9050		18 12 05.5	+74 20 04	-0.004	+0.02	7.8	1.4	-0.3	K5 III		420 s		
166450	209841		18 12 06.1	-34 04 05	+0.001	+0.02	7.1	0.1	0.0	B8.5 V		200 s		
166844	123239		18 12 10.0	+8 13 37	-0.002	-0.05	7.2	1.1	0.2	K0 III		240 s		
167063	66743		18 12 10.8	+33 17 31	-0.001	0.00	7.1	1.4	-0.3	K5 III	-1	300 s		
165239	--		18 12 11.4	-72 57 35			7.7	1.1	0.2	K0 III		320 s		
165776	254167		18 12 13.2	-63 59 10	-0.001	+0.04	7.6	1.1	-0.1	K2 III		360 s		
166402	209842		18 12 13.9	-38 11 06	+0.003	-0.03	7.5	0.1	2.6	F0 V		98 s		
166628	161172		18 12 14.5	-19 25 59	0.000	-0.01	7.17	0.59	-6.8	B3 Ia	+3	2300 s		
167426	30806		18 12 14.8	+52 43 10	0.000	+0.02	7.8	1.4	-0.3	K5 III		410 s		
167327	47302		18 12 15.0	+48 07 13	+0.001	-0.01	7.6	1.1	0.2	K0 III		300 s		
166761	142150		18 12 23.8	-7 17 49	-0.002	-0.06	7.60	0.6	4.4	G0 V		42 s	11165	m
--	66749		18 12 26.6	+37 39 32	-0.001	-0.03	7.6	2.0						
166473	--		18 12 26.7	-37 44 56			7.5	0.3	2.1	A5 V		110 s		
166689	161180		18 12 27.3	-16 22 53	-0.002	-0.03	7.52	0.15	-5.4	B1 Ib-II	-5	2400 s		
166454	--		18 12 27.7	-39 34 57			8.0	0.0	0.6	A0 V		280 s		
166716	161181		18 12 28.3	-15 22 24	0.000	0.00	8.00	0.1	-5.3	B0 II-III	-6	2100 s	11146	m
166453	209847		18 12 29.2	-39 20 34	0.000	-0.02	6.9	0.2	0.0	B8.5 V		170 s		
167388	47306		18 12 31.1	+49 05 56	-0.001	-0.02	7.9	0.1	1.4	A2 V		200 s		
167044	85823		18 12 32.1	+24 27 05	+0.001	0.00	7.0	1.1	0.2	K0 III	-29	230 s		
167008	103482		18 12 32.1	+18 01 38	+0.001	0.00	7.8	1.4	-0.3	K5 III		410 s		
165259	257571		18 12 34.3	-73 40 18	-0.011	-0.23	5.85	0.46	3.2	F8 IV-V		34 s		m
166976	103481		18 12 35.5	+12 23 36	+0.001	-0.02	7.3	0.4	2.6	F0 V	-32	87 s		
166613	186461		18 12 37.3	-28 14 11			7.4	0.7	0.6	A0 V		88 s		
166917	123250		18 12 40.1	+2 48 46	-0.001	+0.01	6.70	0.05	-0.6	B7 V		270 s		
167304	47305		18 12 42.3	+41 08 48	-0.002	-0.05	6.36	1.03	0.2	K0 III	-48	160 s		
170047	3010		18 12 45.9	+81 28 35	-0.008	+0.03	7.5	1.1	0.2	K0 III		280 s		
166765	161187		18 12 47.9	-16 34 34	-0.002	-0.03	7.73	1.80	-0.1	K2 III		150 s		
167106	85826		18 12 51.8	+22 49 31	0.000	-0.02	7.0	1.6	-0.5	M2 III	+17	320 s		
167132	85828		18 12 52.8	+25 38 50	+0.001	-0.03	8.00	1.1	0.0	K1 III		400 s		
167369	47310		18 12 52.9	+42 48 11	0.000	+0.01	8.0	1.1	0.2	K0 III		360 s		
165961	254172		18 12 53.1	-63 36 57	0.000	+0.02	7.0	1.6		M5 III				R Pav, v

HD	SAO	Star Name	α 2000	δ 2000	μ(α)	μ(δ)	V	B−V	M_v	Spec	RV	d(pc)	ADS	Notes
166376	245320		18^{h}12^{m}56.1	−50°33′32″	0^s.000	−0″.08	6.80	0.4	2.6	F0 V		69 s		m
165260	—		18 12 56.5	−74 17 32			7.6	0.1	0.6	A0 V		250 s		
167028	103489		18 12 57.0	+11 27 06	−0.001	−0.01	7.6	0.1	0.6	A0 V		250 s		
166553	209855		18 12 59.0	−39 09 42	+0.003	−0.14	7.30	0.6	4.4	G0 V		38 s		
166767	186478		18 13 02.2	−23 07 02	0.000	−0.01	6.2	0.5	3.4	F5 V	−18	36 s		AP Sgr, v
166991	123257		18 13 03.7	+2 58 50	0.000	−0.01	6.6	0.1	1.4	A2 V		100 s		
167370	66765		18 13 04.5	+38 46 25	−0.001	+0.01	6.04	−0.08	−0.2	B8 V	−9	170 s		
167192	85833		18 13 05.7	+25 17 22	+0.001	+0.01	8.0	0.9	3.2	G5 IV		93 s		
167389	47313		18 13 07.1	+41 28 32	+0.005	−0.12	7.41	0.65		F8				
166154	254174		18 13 07.9	−60 08 34	+0.001	+0.02	7.1	0.2	3.0	F2 V		67 s		
166960	142159		18 13 09.8	−4 00 40	0.000	+0.04	6.59	0.27		A m				
166596	228815		18 13 12.4	−41 20 10	−0.001	0.00	5.47	−0.17		B3	−15			
167134	103495		18 13 15.3	+16 16 22	−0.003	−0.05	6.7	0.5	3.4	F5 V	−21	47 s		
166992	142161		18 13 15.8	−1 43 11	+0.001	−0.01	7.3	0.1	0.6	A0 V		190 s		
167349	66768		18 13 16.1	+35 11 50	0.000	+0.02	7.8	0.2	2.1	A5 V		140 s		
167193	85836		18 13 16.3	+21 52 49	+0.004	+0.05	6.12	1.47	−0.3	K4 III	−66	180 s		
166617	209860		18 13 16.3	−39 15 14	0.000	0.00	7.5	0.1	0.4	B9.5 V		210 s		
167065	123264		18 13 18.7	+9 05 51	+0.002	0.00	8.0	0.7	4.4	G0 V		52 s		
167082	123265		18 13 20.6	+7 47 21	+0.002	0.00	8.0	0.1	0.6	A0 V		260 s		
166807	186484		18 13 22.3	−25 09 04	+0.001	−0.03	7.6	0.2	1.4	A2 V		140 s		
167135	103497		18 13 24.8	+11 53 05	−0.002	+0.01	6.8	0.1	0.6	A0 V		170 s		
167275	85841		18 13 26.9	+26 14 37	+0.002	−0.06	7.26	1.22	0.0	K1 III		240 s		
167513	47318		18 13 28.9	+43 19 54	−0.003	−0.03	7.7	0.9	3.2	G5 IV		80 s		
166921	—		18 13 33.7	−18 23 40			7.81	0.04		B8				
167195	103499		18 13 34.2	+13 38 34	−0.001	+0.01	7.6	1.1	−0.1	K2 III		340 s		
167587	47323		18 13 34.9	+45 39 58	0.000	−0.01	7.4	1.1	0.2	K0 III		280 s		
167161	123268		18 13 36.7	+9 46 24	0.000	−0.04	8.0	1.1	0.2	K0 III		340 s		
166790	209873		18 13 37.4	−31 58 02	+0.002	−0.01	7.60	0.00	0.0	B8.5 V		230 s		m
167160	103500		18 13 38.0	+10 49 33	−0.001	0.00	7.1	1.4	−0.3	K5 III		300 s		
166962	—		18 13 38.6	−15 36 39			7.30	0.5	3.4	F5 V		58 s	11166	m
166701	228821		18 13 40.8	−41 07 38	+0.006	−0.06	7.3	0.8	0.2	K0 III		150 mx		
165277	257572		18 13 42.7	−75 32 59	0.000	+0.01	6.8	0.1	0.6	A0 V		180 s		
167108	123267		18 13 43.1	+0 41 06	−0.002	0.00	8.0	1.1	0.2	K0 III		300 s		
166937	186497	13 μ Sgr	18 13 45.6	−21 03 32	0.000	0.00	3.86	0.23	−7.1	B8 Ia	−6	1200 s	11169	m, v
166636	228819		18 13 47.4	−46 18 44	+0.002	−0.04	7.3	0.2	0.6	A0 V		160 s		
168266	17829		18 13 48.3	+66 28 22	−0.004	+0.02	8.0	0.4	2.6	F0 V		120 s		
166675	228823		18 13 48.5	−43 12 15	−0.001	−0.01	6.9	1.1	3.2	G5 IV		34 s		
167162	123271		18 13 51.9	+2 23 37	−0.001	0.00	6.5	1.1	−0.1	K2 III		190 s		
168151	17828	36 Dra	18 13 53.5	+64 23 50	+0.053	+0.03	5.03	0.38	3.4	F5 V	−35	22 ts		
167277	103507		18 13 55.2	+16 24 57	−0.003	−0.03	7.8	1.1	−0.1	K2 III		380 s		
167350	85845		18 13 58.8	+20 16 36	−0.001	−0.01	7.50	1.1	0.2	K0 III		290 s		m
166677	228827		18 13 59.5	−44 39 05	+0.002	−0.02	7.5	1.1	0.2	K0 III		240 s		
167241	123276		18 14 01.0	+9 26 33	+0.001	+0.02	7.6	1.1	0.2	K0 III		280 s		
166861	209885		18 14 02.4	−34 17 21	+0.003	0.00	7.9	−0.1	0.0	B8.5 V		380 s		
166886	209889		18 14 05.6	−34 35 50	−0.001	−0.01	7.5	0.7	0.2	K0 III		280 s		
166810	209880		18 14 06.9	−38 24 44	+0.001	−0.03	7.2	−0.2	0.6	A0 V		210 s		
167472	85848		18 14 07.5	+28 12 59	−0.001	−0.04	6.80	1.09	−2.2	K1 II	−2	630 s		
167218	123277		18 14 08.3	+4 13 41	−0.001	−0.01	7.9	1.4	−0.3	K5 III		350 s		
166833	209886		18 14 09.8	−37 37 17	+0.002	−0.03	7.70	0.1	0.6	A0 V		250 s		m
166968	186505		18 14 10.6	−27 30 11	0.000	−0.03	7.16	−0.02	0.4	B9.5 V		230 s	11173	m
168267	17831		18 14 11.3	+64 45 08	−0.009	+0.02	7.30	1.1	−0.1	K2 III		200 mx		m
166811	209884		18 14 12.6	−39 57 04	+0.002	−0.07	7.9	0.0	0.6	A0 V		73 mx		
168600	9054		18 14 14.1	+70 50 06	+0.001	+0.02	7.1	0.1	1.7	A3 V		120 s		
167036	186509	14 Sgr	18 14 15.7	−21 42 48	−0.001	−0.02	5.44	1.52	0.2	gK0	−59	46 s		
166251	254179		18 14 16.1	−63 41 22	−0.004	−0.05	6.3	2.0	−0.1	K2 III		62 s		
166793	228832		18 14 19.4	−43 09 16	+0.001	−0.06	7.4	0.9	3.0	F2 V		35 s		
166834	228834		18 14 20.8	−41 54 33	+0.001	−0.01	7.9	−0.3	0.4	B9.5 V		320 s		
165861	257575		18 14 23.6	−70 45 05	−0.003	−0.02	6.73	−0.03		B9				
166751	228831		18 14 26.3	−47 57 16	−0.001	+0.02	7.5	1.5	0.2	K0 III		280 s		
167244	142173		18 14 29.7	−5 34 34	0.000	−0.04	8.0	0.1	0.6	A0 V		250 s		
167016	186512		18 14 33.1	−28 50 18	0.000	+0.02	7.95	−0.02	0.0	B8.5 V		370 s		
167278	123283		18 14 33.4	+0 10 34	+0.001	−0.02	7.59	0.44	3.0	F2 V	−8	73 s	11186	m
166252	254181		18 14 33.9	−65 11 55	+0.002	−0.05	8.0	0.2	2.6	F0 V		120 s		
167279	142176		18 14 36.5	−2 36 09	0.000	+0.02	7.47	2.08	−0.3	K5 III		160 s		
167940	30833		18 14 38.7	+52 33 27	0.000	−0.01	8.0	0.1	1.4	A2 V		210 s		
168092	30836		18 14 40.8	+56 35 18	−0.002	+0.03	6.37	0.34	2.6	F0 V	−8	55 s	11213	m
166679	245333		18 14 41.0	−53 04 43	+0.001	−0.03	8.0	0.9	0.2	K0 III		360 s		
167493	103518		18 14 43.1	+19 02 55	−0.002	+0.01	7.7	1.1	−0.1	K2 III		360 s		
166776	228836		18 14 43.3	−48 48 46	+0.005	0.00	7.8	0.9	−0.1	K2 III		310 mx		
167588	85856		18 14 43.8	+29 12 26	+0.001	−0.25	6.56	0.54	4.2	dF9	+3	30 s		
166835	228838		18 14 44.0	−45 35 42	0.000	0.00	7.0	1.8	−0.3	K5 III		190 s		
168129	30838		18 14 48.8	+58 00 08	−0.002	−0.02	7.8	0.0	0.6	A0 V		250 s		
166226	254183		18 14 49.2	−66 48 49	−0.006	+0.04	7.60	0.00	0.0	B8.5 V		100 mx		m
—	66793		18 14 51.0	+36 22 44	−0.001	−0.04	7.7	1.2						
167022	209902		18 14 52.0	−35 04 51	0.000	−0.02	7.1	1.2	0.2	K0 III		170 s		

HD	SAO	Star Name	α 2000	δ 2000	μ(α)	μ(δ)	V	B-V	M_v	Spec	RV	d(pc)	ADS	Notes
			$^h \quad ^m \quad ^s$	$^{\circ} \quad ' \quad ''$	^{s}	$''$								
--	66798		18 14 52.5	+39 43 57	-0.001	-0.03	7.5	1.7						
167941	47336		18 14 55.9	+49 09 30	-0.001	+0.08	6.8	1.1	-0.1	K2 III		210 mx		
167538	103524		18 14 57.4	+18 28 28	-0.001	-0.02	7.8	0.4	2.6	F0 V		110 s		
167246	161212		18 14 57.5	-15 23 06	0.000	-0.01	7.3	1.1	0.2	K0 III	-18	220 s		
166839	245344		18 14 59.8	-50 18 28	-0.003	-0.02	8.00	0.1	0.6	A0 V		300 s		m
167121	186531		18 15 04.8	-29 49 21	-0.001	0.00	6.7	1.5	-0.5	M2 III		280 s		
166653	245335		18 15 08.2	-57 51 16	+0.007	-0.12	7.3	1.2	3.2	G5 IV		36 s		m
167204	186537		18 15 10.4	-26 18 45	+0.001	-0.02	8.0	1.6	-0.5	M2 III		510 s		
167264	186543	15 Sgr	18 15 12.7	-20 43 42	0.000	0.00	5.38	0.07	-6.2	B0 Ia	-6	1400 s		
167263	186544	16 Sgr	18 15 12.7	-20 23 16	0.000	0.00	5.98	0.04		O9 II	-5		11191	m
165736	--		18 15 13.0	-74 03 12			8.0	0.4	3.0	F2 V		100 s		
167177	186536		18 15 13.1	-28 36 49	-0.004	-0.01	7.9	-0.1	1.7	A3 V		170 s		
167026	228847		18 15 14.7	-40 52 36	-0.002	0.00	7.9	1.6	0.2	K0 III		240 s		
168130	30841		18 15 15.1	+54 08 28	-0.010	-0.08	6.7	1.1	-0.1	K2 III		160 mx		
164134	--		18 15 16.3	-81 51 02			8.0	0.1	0.6	A0 V		300 s		
168653	17837	37 Dra	18 15 16.9	+68 45 21	+0.002	-0.06	5.95	1.06	0.2	K0 III	-10	130 s		
168093	30839		18 15 19.7	+50 58 17	+0.001	-0.04	7.8	0.5	4.0	F8 V		57 s		
167335	161222		18 15 20.1	-16 49 52	+0.002	0.00	7.4	0.0	0.6	B8.5 V		260 s		
167146	209915		18 15 21.4	-31 44 22	+0.003	0.00	8.0	0.0	2.6	F0 V		120 s		
167314	--		18 15 23.2	-18 56 24			7.82	1.99	-0.3	K5 III		220 s		
--	--		18 15 25.9	-48 51 01			6.80							m
166949	228845		18 15 26.6	-48 51 07	+0.001	+0.01	6.80	1.1	0.2	K0 III		210 s		m
167148	209914		18 15 26.8	-35 02 17	+0.002	-0.01	7.2	1.2	0.2	K0 III		200 s		
167147	209916		18 15 27.3	-34 35 41	+0.001	-0.01	6.8	0.0	0.0	B8.5 V		210 s		
174878	3059		18 15 28.7	+86 39 28	-0.004	+0.03	6.7	1.6	-0.5	M2 III	-36	280 s		
167653	85865		18 15 28.7	+20 15 50	0.000	-0.01	8.0	1.1	0.2	K0 III		360 s		
167356	161227		18 15 30.7	-18 39 41	+0.001	0.00	6.10	0.1	-7.1	A0 Ia	-1	2200 s	11196	m
168268	30849		18 15 31.3	+56 53 12	-0.002	-0.04	7.6	0.1	0.6	A0 V		250 s		
168009	47343		18 15 32.4	+45 12 34	-0.007	-0.11	6.29	0.62	4.4	G0 V	-64	23 ts		
169925	9066		18 15 33.8	+79 03 54	-0.010	0.00	7.8	0.7	4.4	G0 V		49 s		
168066	47345		18 15 33.8	+47 33 40	-0.002	-0.01	7.6	0.5	3.4	F5 V		68 s		
167450	142182		18 15 34.0	-5 28 31	-0.001	-0.02	7.2	0.9	3.2	G5 IV		61 s		
165416	257574		18 15 35.6	-77 04 25	+0.008	0.00	7.5	0.1	0.6	A0 V		240 s		
167208	209918		18 15 37.0	-32 20 35	+0.001	-0.04	8.0	0.0	2.6	F0 V		120 s		
167965	47342		18 15 38.6	+42 09 33	0.000	0.00	5.59	-0.10	-0.9	B6 V	-21	200 s		
167715	85867		18 15 38.9	+22 03 02	+0.001	-0.04	8.0	0.9	3.2	G5 IV		93 s		
168152	30844		18 15 39.9	+51 37 46	-0.003	+0.14	7.8	1.1	0.2	K0 III		130 mx		
167230	209919		18 15 40.2	-33 05 29	+0.003	-0.01	6.9	0.5	0.4	B9.5 V		95 s		
166599	254189		18 15 40.7	-63 03 20	-0.002	-0.03	5.60	1.1	0.2	K0 III	-7	120 s		m
167517	142184		18 15 42.3	-3 21 28	0.000	+0.01	7.30	1.1	0.2	K0 III		220 s	11205	m
166891	245351		18 15 48.5	-55 21 23	-0.001	-0.01	7.5	1.0	3.0	F2 V		32 s		
167231	209923		18 15 49.8	-35 38 16	+0.001	-0.04	7.3	0.2	0.6	A0 V		170 s		
167095	228852		18 15 49.9	-43 10 06	-0.003	+0.01	7.3	0.1	0.6	A0 V		190 s		
167233	209922		18 15 50.6	-36 34 26	0.000	-0.01	7.01	-0.10	-1.6	B3.5 V		480 s		
167411	--		18 15 51.3	-18 15 20			7.71	0.27	-5.6	B0 II	-8	2300 s		
167096	228851		18 15 52.5	-44 12 23	+0.007	+0.01	5.46	0.96		K0				
--	228854		18 15 53.2	-44 12 24	+0.006	0.00	5.4	0.7	0.2	K0 III		110 s		
167856	66804		18 15 53.2	+30 24 00	-0.002	+0.02	6.9	1.1	0.2	K0 III	-58	220 s		
167496	161236		18 15 55.7	-11 01 03	-0.002	-0.04	8.0	0.5	4.0	F8 V		59 s		
167433	161233		18 15 57.7	-17 36 28	0.000	0.00	7.51	0.07		B8				
167564	142185		18 15 57.8	-3 37 03	+0.001	+0.02	6.36	0.20		A3				
168293	30855		18 16 00.7	+53 17 49	-0.004	+0.05	7.2	1.1	0.2	K0 III		260 s		
167319	209928		18 16 01.8	-33 45 23	0.000	+0.05	7.9	0.4	0.2	K0 III		340 s		
167498	161237		18 16 02.0	-14 55 26	+0.001	+0.01	7.9	0.1	0.6	A0 V		240 s		
168221	30853		18 16 03.0	+50 31 56	+0.001	-0.04	7.4	0.1	0.6	A0 V		230 s		
168037	66810		18 16 03.3	+39 23 03	-0.002	-0.03	7.8	1.1	0.2	K0 III		330 s		
167518	161242		18 16 04.4	-11 10 48	-0.003	-0.05	7.6	0.5	3.4	F5 V		64 s		
167654	123308		18 16 05.4	+2 22 40	0.000	-0.02	6.01	1.59	-0.5	gM4	+22	200 s		
166912	245354		18 16 06.6	-56 39 03	-0.003	-0.03	6.90	1.1	0.2	K0 III		220 s		m
--	103539		18 16 07.1	+16 40 37	0.000	+0.02	8.0							
167828	85874		18 16 08.8	+21 02 23	0.000	+0.01	7.6	0.0	0.4	B9.5 V		280 s		
167829	103544		18 16 10.4	+18 55 10	0.000	0.00	7.9	0.2	2.1	A5 V		150 s		
167363	209933		18 16 10.6	-31 09 33	+0.001	-0.01	7.65	0.04	-1.0	B5.5 V		420 s		m
167153	228857		18 16 11.2	-46 01 30	+0.001	0.00	7.50	0.1	1.4	A2 V		170 s		m
167342	209932		18 16 12.8	-35 14 25	-0.002	+0.02	7.1	1.3	-0.1	K2 III		220 s		
167591	142186		18 16 15.2	-9 48 36	+0.004	-0.02	7.9	0.1	1.4	A2 V		100 mx		
167382	209937		18 16 19.2	-31 19 17	+0.006	-0.03	7.2	0.7	2.6	F0 V		46 s		
167785	103546		18 16 22.1	+10 48 52	0.000	-0.02	7.9	0.1	-2.5	B2 V	-13	1100 s		
168269	47352		18 16 24.4	+48 22 08	-0.004	+0.01	7.6	1.4	-0.3	K5 III		310 mx		
168343	30860		18 16 27.4	+50 37 00	0.000	+0.01	7.7	1.1	0.2	K0 III		310 s		
167212	228863		18 16 28.2	-46 33 17	0.000	-0.01	7.8	-0.3	0.4	B9.5 V		300 s		
166972	245359		18 16 28.5	-57 26 50	+0.003	-0.03	7.2	0.6	3.4	F5 V		45 s		
167441	209941		18 16 30.6	-30 24 17	-0.001	+0.02	7.78	0.07	0.0	B8.5 V		300 s		
167156	228862		18 16 30.9	-49 12 48	-0.002	-0.07	7.9	0.9	2.1	A5 V		64 mx		
167943	85884		18 16 31.3	+23 23 54	+0.001	+0.01	7.60	1.1	0.2	K0 III		300 s		m

HD	SAO	Star Name	α 2000	δ 2000	μ(α)	μ(δ)	V	B–V	M_v	Spec	RV	d(pc)	ADS	Notes
			h m s	° ′ ″	s	″								
167808	123317		18 16 33.4	+8 21 41	+0.001	0.00	7.8	0.1	1.4	A2 V		180 s		
167503	186573		18 16 34.1	−23 54 09	+0.001	−0.05	7.4	0.1	2.6	F0 V		93 s		
167570	186575		18 16 35.3	−20 32 41	+0.001	−0.01	7.1	0.7	3.2	G5 IV	−16	60 s		
168320	47355		18 16 37.0	+48 06 18	−0.001	−0.04	7.0	1.1	−0.1	K2 III		270 s		
167568	161247		18 16 37.6	−18 48 22	0.000	0.00	7.77	1.21	0.2	K0 III		240 s		
166953	245361		18 16 46.1	−59 33 47	+0.001	−0.02	7.8	−0.4	0.6	A0 V		270 s		
168428	30861		18 16 49.2	+51 13 44	−0.001	−0.01	7.7	1.4	−0.3	K5 III		390 s		
167677	161256		18 16 49.3	−11 24 19	+0.005	−0.19	7.9	0.7	4.4	G0 V		49 s		
167768	142189		18 16 52.9	−3 00 26	+0.001	−0.27	6.00	0.89	0.3	G8 III	+2	48 mx		
168382	47360		18 16 53.9	+49 05 41	0.000	+0.02	7.4	1.1	−0.1	K2 III		320 s		
167576	186582		18 16 57.3	−27 42 57	0.000	−0.11	6.68	1.26	−0.2	K3 III		110 mx		
169027	17847		18 16 58.3	+68 44 29	−0.006	−0.09	6.7	0.1	0.6	A0 V	−26	120 mx		
167659	161255		18 16 58.4	−18 58 05	0.000	0.00	7.39	0.21		O8				
167944	103557		18 16 59.4	+12 04 04	0.000	0.00	7.2	0.5	3.4	F5 V	+2	57 s		
167257	245372		18 17 00.7	−51 04 07	0.000	−0.02	6.2	0.5	0.4	B9.5 V		71 s		
167858	123320		18 17 04.6	+1 00 21	−0.001	−0.03	6.63	0.31		F0				
168178	66827		18 17 06.2	+33 32 07	−0.002	+0.04	7.0	1.1	−0.1	K2 III		270 s		
168322	47359		18 17 06.7	+40 56 12	−0.015	+0.07	6.11	0.99	0.2	K0 III	−73	150 s		
167128	245369		18 17 07.2	−56 01 24	−0.001	−0.01	5.33	−0.05	−1.7	B3 V	+12	230 s		
167720	161260		18 17 11.4	−17 22 27	0.000	−0.02	5.9	1.4	−1.3	K4 II-III	−7	240 s		
167946	123325		18 17 14.2	+5 45 13	0.000	−0.02	8.0	0.1	0.6	A0 V		260 s		
165844	257576		18 17 15.5	−76 35 01	+0.008	−0.06	7.7	1.1	0.2	K0 III		300 mx		
167599	209953		18 17 16.1	−31 17 50	+0.002	0.00	7.7	0.0	0.0	B8.5 V		300 s		
167578	209952		18 17 16.6	−32 56 19	0.000	+0.03	7.9	0.0	0.0	B8.5 V		330 s		
167506	209949		18 17 17.6	−38 10 39	+0.001	0.00	6.7	0.9	0.2	K0 III		200 s		
167665	186593		18 17 23.5	−28 17 21	+0.010	−0.16	6.40	0.54	4.4	G0 V		25 s		
167666	186594		18 17 23.9	−28 39 09	0.000	−0.03	6.0	0.4	1.7	A3 V		48 s		
167833	142192		18 17 23.9	−9 45 31	0.000	−0.06	6.31	0.38	0.3	A5 III		110 mx		
167812	161268		18 17 26.0	−14 50 56	+0.001	0.00	8.0	0.0	0.0	B8.5 V		320 s		
167772	161265		18 17 26.8	−19 29 19	+0.001	−0.04	7.9	0.2	3.4	F5 V		81 s		
167771	161267		18 17 28.3	−18 27 48	0.000	0.00	6.54	0.11		O8	+9			m
167810	161271		18 17 28.7	−12 43 26	+0.003	0.00	7.74	1.76	−0.3	K5 III		280 s		
167702	186598		18 17 35.2	−27 51 59	+0.001	−0.07	8.00	0.00	0.4	B9.5 V		63 mx	11224	m
167647	209959		18 17 36.0	−34 06 26	−0.001	−0.01	6.1	0.0	−1.0	B5.5 V	+10	250 s		RS Sgr, m,v
167838	161275		18 17 37.5	−15 25 49	0.000	+0.02	6.73	0.46	−7.0	B5 Ia	−6	2700 s		
167815	161273		18 17 40.0	−19 40 19	+0.002	+0.01	7.56	0.17		B2	−8		11228	m
167406	245379		18 17 40.4	−50 43 17	0.000	+0.01	7.9	1.7	−0.5	M2 III		330 s		
167618	209957	η Sgr	18 17 37.5	−36 45 42	−0.011	−0.17	3.11	1.56	−2.4	M3 II	+1	130 s		m
167968	142199		18 17 42.7	−1 27 53	0.000	−0.02	7.5	0.1	1.4	A2 V		150 s		
167385	245380		18 17 46.8	−52 02 25	+0.001	+0.03	7.3	0.4	3.4	F5 V		61 s		
167795	186608		18 17 47.5	−23 16 38	−0.002	−0.02	7.67	0.05	0.0	B8.5 V		300 s		
166817	—		18 17 47.9	−67 52 34			7.7	1.1	0.2	K0 III		320 s		
168439	47364		18 17 48.0	+40 18 08	−0.003	−0.06	7.2	0.1	0.6	A0 V		39 mx		
167488	228881		18 17 48.6	−46 52 30	+0.001	0.00	7.7	1.7	−0.3	K5 III		390 s		
167270	245377		18 17 50.3	−56 28 40	−0.003	+0.02	7.3	0.9	2.6	F0 V		37 s		
167863	161278		18 17 50.9	−18 47 53	0.000	+0.01	6.71	0.04		B5	−15		11232	m
167686	209966		18 17 51.0	−33 23 46	+0.002	0.00	7.0	0.5	0.0	B8.5 V		110 s		
166864	254195		18 17 51.2	−67 21 22	+0.002	−0.02	7.6	0.4	2.6	F0 V		100 s		
167774	186606		18 17 51.5	−27 21 19	+0.001	−0.03	7.2	1.6	−0.5	M2 III		350 s		
169282	17848		18 17 51.7	+70 00 44	−0.002	+0.03	7.7	0.5	3.4	F5 V		74 s		
168131	103572		18 17 51.9	+11 52 41	0.000	0.00	7.0	0.0	0.0	B8.5 V	−3	250 s		
168198	103576		18 17 54.7	+17 58 52	+0.001	+0.04	7.4	1.6	−0.5	M4 III		390 s		IQ Her, v
—	103575		18 17 55.0	+15 50 24	0.000	0.00	7.8							
168760	30878		18 17 55.5	+57 53 27	−0.003	−0.01	7.5	1.1	0.2	K0 III		290 s		
167842	186613		18 17 56.8	−22 20 36	0.000	−0.01	8.0	0.4	3.2	G5 IV		91 s		
167926	161289		18 17 58.0	−12 52 12	−0.001	−0.01	7.4	0.1	0.6	A0 V		200 s		
166841	254196		18 18 00.8	−68 13 44	+0.001	−0.01	6.33	−0.04		A0				m
168179	103577		18 18 02.4	+12 11 55	+0.002	0.00	7.6	0.0	0.4	B9.5 V		280 s		
168199	103578		18 18 02.7	+13 46 36	−0.001	−0.02	6.30	−0.03	−1.1	B5 V	−21	270 s		
167818	186612		18 18 03.0	−27 02 33	+0.001	+0.01	4.65	1.66	−0.3	gK5	−17	84 s		
168073	142206		18 18 05.3	−1 19 49	0.000	0.00	7.5	0.1	0.6	A0 V		210 s		
167971	161292		18 18 05.7	−12 14 32	−0.001	+0.01	7.52	0.76		O8.8	+1			
168323	85906		18 18 07.5	+23 17 48	0.000	−0.02	6.70	1.4	−0.3	K5 III	+3	250 s		m
168270	103581		18 18 07.5	+18 07 53	−0.001	−0.01	6.00	0.03	−0.7	B9.5 III	−24	190 s		
168014	142203		18 18 08.4	−8 38 01	+0.001	0.00	7.8	1.1	0.2	K0 III		260 s		
168201	103580		18 18 11.8	+11 23 46	−0.001	+0.01	7.3	1.4	−0.3	K5 III		330 s		
168564	47369		18 18 11.9	+45 09 24	0.000	0.00	7.9	0.1	0.6	A0 V		290 s		
168619	47376		18 18 17.8	+46 24 25	+0.004	−0.03	7.0	0.8	3.2	G5 IV		59 s		
167259	254199		18 18 18.8	−61 12 31	+0.001	−0.03	6.8	1.4	3.2	G5 IV		21 s		
167465	245386		18 18 20.2	−54 19 55	+0.001	0.00	7.3	1.3	−0.1	K2 III		260 s		
167976	161294		18 18 22.5	−18 28 32	+0.001	+0.01	7.39	2.01	−0.3	K5 III		170 s		
168000	161295		18 18 23.9	−17 32 03	−0.001	0.00	7.9	1.1	0.2	K0 III		280 s		
168271	103584		18 18 24.3	+12 58 50	0.000	0.00	6.5	0.0	0.4	B9.5 V	−12	170 s		
168202	123343		18 18 28.0	+5 08 06	−0.001	−0.03	7.4	0.0	0.4	B9.5 V		230 s		
168384	85910		18 18 28.9	+24 26 20	−0.001	−0.06	8.0	1.1	−0.1	K2 III		250 mx		

HD	SAO	Star Name	α 2000	δ 2000	μ(α)	μ(δ)	V	B-V	M$_V$	Spec	RV	d(pc)	ADS	Notes
167846	209978		18^{h}18^{m}30^s.3	-34°41'11"	-0^s.001	-0".01	6.8	0.1	0.0	B8.5 V		190 s		
168363	103587		18 18 32.3	+18 22 50	0.000	-0.09	8.0	0.5	3.4	F5 V		82 s		
168386	103591		18 18 33.5	+19 27 02	0.000	-0.01	7.9	0.0	0.4	B9.5 V		320 s		
168889	30887		18 18 33.6	+58 00 29	-0.002	-0.02	7.7	1.1	0.2	K0 III		310 s		
168364	103590		18 18 36.4	+17 00 10	+0.001	-0.02	7.5	0.7	4.4	G0 V		42 s		
168344	103588		18 18 37.4	+14 22 44	0.000	-0.02	7.7	1.1	-0.1	K2 III		360 s		
167820	209979		18 18 39.0	-37 13 27	-0.001	-0.01	7.5	0.9	0.2	K0 III		280 s		
167756	228895		18 18 39.8	-42 17 18	-0.001	0.00	6.30	-0.11		B0.5 II	-25			
167979	186629		18 18 41.5	-25 36 17	0.000	-0.03	6.51	1.34	0.2	K0 III		100 s		
168021	161304		18 18 43.1	-18 37 10	+0.001	0.00	6.84	0.31	-5.8	B0 Ib	-1	1700 s	11240	m
167558	245392		18 18 50.5	-53 44 16	+0.004	-0.03	7.4	1.1	3.2	G5 IV		44 s		
168549	66844		18 18 51.3	+33 02 42	+0.001	-0.02	7.7	1.4	-0.3	K5 III		390 s		
167849	228901		18 18 52.8	-40 26 30	-0.001	-0.03	7.5	1.1	3.2	G5 IV		73 s		
168136	161315		18 18 55.0	-13 32 28	+0.001	+0.01	7.94	1.76	-0.3	K5 III		310 s		
168116	--		18 18 58.0	-18 31 53			7.6	1.1	0.2	K0 III		260 s		
168638	66852		18 18 58.9	+39 16 00	+0.001	0.00	7.9	1.1	-0.1	K2 III		400 s		
168245	142218		18 18 59.1	-4 06 19	+0.001	0.00	7.6	0.9	-2.1	G7 II	+41	610 s		
168774	47385		18 18 59.3	+48 34 14	+0.001	+0.13	7.6	0.5	3.4	F5 V		71 s		
167301	254204		18 19 01.1	-63 51 12	0.000	+0.01	7.7	0.1	0.6	A0 V		260 s		
168654	47382		18 19 04.1	+40 55 38	+0.001	+0.03	8.0	0.6	4.4	G0 V		53 s		
167674	245395		18 19 04.9	-51 57 52	+0.003	+0.02	7.2	0.6	0.2	K0 III		250 s		
168431	103602		18 19 05.8	+12 11 38	-0.001	+0.01	6.9	0.0	0.0	B8.5 V	-7	240 s		
169666	9072		18 19 07.9	+71 31 03	-0.014	+0.03	7.9	0.5	3.4	F5 V	-40	40 mx		
168603	66851		18 19 08.7	+33 13 54	+0.001	+0.16	8.0	1.1	0.2	K0 III		110 mx		
168387	123353		18 19 09.3	+7 15 35	-0.003	0.00	5.39	1.07	-0.1	gK2	-8	130 s	11254	m
168532	85921	105 Her	18 19 10.5	+24 26 45	+0.001	0.00	5.27	1.53	-2.3	K4 II	-14	320 s		
168440	103605		18 19 14.0	+12 32 01	+0.001	-0.01	7.3	0.0	0.0	B8.5 V	-5	290 s		
168481	103606		18 19 16.1	+15 49 34	+0.001	0.00	7.0	0.1	1.7	A3 V	+1	120 s		
168024	209986		18 19 17.8	-34 44 27	-0.004	+0.02	7.5	0.9	3.2	G5 IV		61 s		
168912	30889		18 19 18.0	+52 39 09	-0.002	+0.01	7.4	1.4	-0.3	K5 III		350 s		
167188	254202		18 19 19.4	-67 19 42	-0.002	-0.03	7.9	1.4	-0.3	K5 III		440 s		
--	47390		18 19 22.5	+48 17 56	+0.001	-0.02	7.9	2.1						
165278	258791		18 19 23.4	-81 08 38	-0.006	+0.02	8.0	1.1	0.2	K0 III		360 s		
168413	123355		18 19 25.5	+0 50 30	0.000	-0.02	7.93	1.95	-0.3	K5 III		240 s		
167825	228904		18 19 33.8	-49 20 51	-0.002	-0.05	7.4	0.3	3.0	F2 V		77 s		
168498	103611		18 19 34.8	+11 52 13	0.000	-0.01	7.4	0.0	0.6	A0 V		210 s		
167425	254209		18 19 40.1	-63 53 13	+0.007	-0.29	6.18	0.58	3.0	G3 IV		43 s		m
169508	17857		18 19 40.4	+67 26 07	0.000	+0.03	7.0	1.6	-0.5	M2 III	-23	320 s		
167954	228906		18 19 41.3	-45 41 54	-0.005	-0.11	6.85	0.53		F8		12 mn		
168794	66868		18 19 45.8	+39 05 29	0.000	-0.01	8.0	0.1	0.6	A0 V		310 s		
167325	254206		18 19 46.9	-66 39 22	-0.002	-0.07	7.3	1.1	0.2	K0 III		210 mx		
168827	47394		18 19 48.9	+42 00 05	-0.002	+0.03	7.8	0.8	0.2	K0 III		330 s		
168329	161338		18 19 49.8	-17 45 03	-0.002	0.00	7.0	1.1	0.2	K0 III		200 s		
168605	103618		18 19 50.2	+19 10 17	0.000	-0.01	7.8	0.1		A0 p				
168775	66869	1 κ Lyr	18 19 51.5	+36 03 52	-0.002	+0.04	4.33	1.17	-0.1	K2 III	-22	77 s		
168393	161345		18 19 51.9	-11 17 14	-0.001	-0.01	7.41	0.55	-6.6	F0 I		4900 s		
168694	85933		18 19 51.9	+29 39 58	0.000	0.00	5.99	1.29	0.2	K0 III	-36	96 s		
167985	228909		18 19 53.6	-46 13 55	-0.001	-0.02	7.9	0.4	0.6	B9.5 V		320 s		
168281	186660		18 19 54.5	-23 18 54	-0.002	-0.01	7.8	1.6	-0.5	M2 III		470 s		
169028	30897		18 19 55.9	+51 20 52	-0.004	-0.05	6.30	1.10	0.0	gK1	-10	180 s		
167916	228907		18 19 56.6	-49 25 36	-0.001	-0.04	7.9	0.2	0.6	A0 V		280 s		
167826	245401		18 19 58.7	-54 41 20	-0.002	-0.02	7.5	2.1	-0.1	K2 III		92 s		
167956	228910		18 20 01.5	-48 45 54	-0.001	-0.01	6.8	0.5	1.4	A2 V		69 s		
168443	142228		18 20 03.7	-9 35 47	-0.007	-0.25	7.0	0.9	3.2	G5 IV		34 mx		
168459	142229		18 20 05.0	-7 58 52	+0.002	-0.01	6.60	0.5	3.4	F5 V	-49	42 s	11262	m
168415	161348		18 20 08.6	-15 49 54	+0.002	-0.04	5.39	1.47	-0.3	K4 III	+31	130 s		
168212	210002		18 20 08.7	-34 46 29	-0.001	-0.01	7.6	0.2	0.4	B9.5 V		210 s		
168060	228917		18 20 11.5	-45 55 17	+0.004	-0.13	7.32	0.76		G5		13 mn		
168579	123370		18 20 11.6	+7 04 10	0.000	+0.01	7.9	1.4	-0.3	K5 III		360 s		
168236	210005		18 20 16.7	-35 25 43	+0.001	-0.01	6.7	0.3	0.4	B9.5 V		110 s		
168720	85941	106 Her	18 20 17.7	+21 57 41	+0.001	-0.06	4.95	1.59	-0.4	M0 III	-33	110 s		
168696	85939		18 20 17.9	+21 04 25	-0.001	0.00	7.6	0.0	0.4	B9.5 V		280 s		
169486	17859		18 20 19.2	+64 12 11	-0.003	+0.02	7.4	1.1	0.2	K0 III		270 s		
168062	228921		18 20 19.4	-47 22 24	+0.002	0.00	7.9	0.1	0.0	B8.5 V		290 s		
168355	186671		18 20 21.0	-27 24 29	+0.002	-0.03	7.2	1.1	0.2	K0 III		230 s		
168007	228918		18 20 21.0	-49 07 39	0.000	-0.04	7.60	0.1	1.4	A2 V		120 s		m
167852	245403		18 20 21.2	-56 12 55	+0.003	0.00	6.8	1.4	0.2	K0 III		130 s		
165737	258793		18 20 22.2	-80 14 20	+0.003	0.00	7.30	1.1	-0.1	K2 III		300 s		m
164920	258790		18 20 23.3	-82 30 15	-0.005	+0.02	7.7	1.8	-0.3	K5 III		260 s		
168812	85947		18 20 25.1	+28 59 01	-0.001	0.00	6.5	0.0	0.0	B8.5 V	-7	200 s		
168356	186672		18 20 25.4	-29 03 59	+0.002	0.00	7.2	0.7	3.2	G5 IV		65 s		
168310	210012		18 20 26.5	-31 18 58	+0.001	+0.03	7.4	0.2	2.1	A5 V		110 s		
168214	228931		18 20 26.6	-40 16 25	-0.003	-0.06	7.5	1.0	3.2	G5 IV		73 s		
168147	228928		18 20 28.5	-44 04 41	+0.004	-0.05	7.18	1.62	-0.5	M4 III		210 mx		
168334	210015		18 20 35.1	-32 21 07	+0.003	-0.05	8.0	0.5	1.4	A2 V		100 s		

HD	SAO	Star Name	α 2000	δ 2000	μ(α)	μ(δ)	V	B-V	M$_V$	Spec	RV	d(pc)	ADS	Notes
			^{h} ^{m} ^{s}	$^\circ$ $'$ $''$	^{s}	$''$								
168655	103627		18 20 36.1	+10 06 13	0.000	+0.01	7.9	1.4	-0.3	K5 III		380 s		
168795	85950		18 20 38.5	+22 47 49	0.000	-0.02	6.80	0.00	0.4	B9.5 V		190 s	11273	m
168313	210014		18 20 39.5	-36 46 30	-0.001	0.00	7.50	1.6	-0.5	M2 III		370 s		m
167806	245405		18 20 41.9	-59 50 24	+0.003	0.00	7.2	-1.0	0.0	B8.5 V		280 s		
--	103630		18 20 44.1	+17 23 14	-0.001	0.00	7.8							
170000	9084	43 φ Dra	18 20 45.2	+71 20 16	-0.002	+0.04	4.22	-0.10		A0 p	-17	33 mn	11311	m
167918	245411		18 20 47.4	-56 55 57	-0.001	+0.01	7.8	0.4	0.4	B9.5 V		160 s		
169905	9083		18 20 48.7	+70 04 22	+0.002	0.00	7.9	0.6	4.4	G0 V		51 s		
168874	85952		18 20 49.1	+27 31 50	+0.004	+0.10	7.04	0.61	5.0	dG4	-19	26 s	11275	m
167736	254212		18 20 51.9	-62 56 48	+0.006	-0.11	7.6	0.5	4.0	F8 V		52 s		
168656	123377	74 Oph	18 20 51.9	+3 22 38	0.000	+0.01	4.86	0.91	0.3	G8 III	+5	82 s	11271	m
168357	210017		18 20 55.1	-37 29 15	+0.001	-0.01	6.4	1.0	0.2	K0 III		180 s		
168453	186680		18 20 55.4	-28 07 09	-0.003	-0.01	8.0	0.1	0.6	A0 V		240 s		
168400	210019		18 20 55.6	-33 19 58	+0.001	+0.01	7.0	0.2	0.6	A0 V		130 s		
168913	85956	108 Her	18 20 56.9	+29 51 32	+0.001	+0.06	5.63	0.21	2.2	A6 V	-20	49 s		
167761	254213		18 20 58.4	-63 15 01	+0.002	-0.10	7.8	1.1	0.2	K0 III		160 mx		
168454	186681	19 δ Sgr	18 20 59.5	-29 49 42	+0.003	-0.03	2.70	1.38	-0.1	K2 III	-20	25 s	11264	Kaus Media, m
168761	123381		18 21 00.6	+8 37 43	+0.001	+0.01	7.9	0.1	0.6	A0 V		260 s		
168914	85957	107 Her	18 21 00.8	+28 52 12	0.000	+0.05	5.12	0.20		A5	-29	21 mn		
168287	228939		18 21 01.4	-43 46 32	+0.001	-0.01	7.4	-0.1	1.7	A3 V		140 s		
168567	161370		18 21 02.2	-15 13 50	+0.001	+0.01	6.93	1.76	-0.1	K2 III		110 s		
168796	103632		18 21 02.4	+13 49 01	-0.002	0.00	7.8	0.1	0.6	A0 V		270 s		
170153	9087	44 χ Dra	18 21 03.0	+72 43 58	+0.117	-0.36	3.57	0.49	4.1	F7 V	+33	7.8 t		m
168852	85955		18 21 06.1	+20 56 05	0.000	0.00	7.4	0.0	0.4	B9.5 V	-19	250 s		
169221	47411		18 21 06.9	+49 43 31	-0.001	+0.02	6.40	1.07	0.0	K1 III	-17	190 s		
168571	161372		18 21 08.8	-17 22 53	0.000	+0.01	7.79	0.53	-5.4	B1 Ib-II	-5	1600 s		
166973	257582		18 21 09.4	-74 40 26	+0.004	-0.02	8.0	0.1	0.6	A0 V		300 s		
167854	254214		18 21 10.6	-62 11 34	-0.006	-0.14	7.8	0.5	4.0	F8 V		57 s		
169029	66890		18 21 11.6	+32 52 40	-0.001	-0.08	7.4	0.5	3.4	F5 V		64 s		
168525	186687		18 21 12.2	-26 05 06	+0.001	+0.02	6.7	0.5	1.7	A3 V		61 s		
168403	210022		18 21 14.2	-39 01 20	0.000	-0.03	6.7	0.1	0.4	B9.5 V		140 s		
--	47412		18 21 16.0	+49 35 25	-0.004	0.00	8.0	1.0						
168493	210026		18 21 16.4	-30 56 29	+0.001	-0.01	7.0	0.3	0.4	B9.5 V		140 s		
168723	142241	58 η Ser	18 21 18.4	-2 53 56	-0.037	-0.70	3.26	0.94	1.7	K0 III-IV	+9	16 mx		m
169060	66893		18 21 19.5	+32 56 22	+0.001	0.00	7.2	1.1	-0.1	K2 III		290 s		
168494	210027		18 21 19.9	-32 10 52	+0.002	0.00	7.1	0.5	1.7	A3 V		71 s		
168608	161376		18 21 22.9	-18 51 36	+0.001	-0.01	5.4	0.5	-6.3	F8 I	-3	1200 s		Y Sgr, v
167062	257583		18 21 24.6	-74 22 31	-0.002	-0.12	7.9	0.5	3.4	F5 V		79 s		
168957	85964		18 21 27.7	+25 03 23	+0.001	-0.01	7.01	-0.10		B3 e	-41			
168797	123385		18 21 28.3	+5 26 09	+0.001	0.00	6.13	-0.04	-1.7	B3 V e	-9	320 s		
169355	30917		18 21 29.1	+51 18 04	-0.002	-0.02	6.6	1.4	-0.3	K5 III		240 s		
168574	186699		18 21 31.2	-24 54 54	+0.001	+0.01	6.25	1.84		gM5	+3			
169305	47417		18 21 32.4	+49 07 18	-0.003	+0.05	5.05	1.66	-0.5	gM2	+14	120 s		
169412	30919		18 21 33.3	+52 54 09	0.000	+0.02	7.7	0.1	0.6	A0 V		270 s		
168509	210033		18 21 34.2	-33 31 45	0.000	+0.05	7.9	1.6	-0.3	K5 III		370 s		
--	66905		18 21 40.0	+38 20 17	0.000	-0.03	8.0	1.3						
169128	66903		18 21 48.0	+32 14 51	0.000	+0.02	7.7	1.4	-0.3	K5 III		390 s		
169031	85969		18 21 48.3	+21 30 28	-0.001	0.00	6.80	0.00	0.0	B8.5 V		230 s	11287	m
169509	30926		18 21 48.3	+55 07 48	+0.001	+0.01	7.8	1.4	-0.3	K5 III		410 s		
168701	161385		18 21 48.8	-16 19 28	+0.001	0.00	7.9	1.1	0.2	K0 III	+22	280 s		
168959	103641		18 21 49.1	+16 08 22	+0.002	0.00	7.5	1.1	0.2	K0 III		280 s		
168746	161386		18 21 49.6	-11 55 21	-0.002	-0.05	7.9	0.9	3.2	G5 IV		82 s		
168960	103642		18 21 50.2	+15 37 35	-0.003	-0.14	7.6	0.9	3.2	G5 IV		73 mx		
169243	47418		18 21 51.8	+41 12 47	-0.002	-0.04	7.7	1.1	0.2	K0 III		320 s		
169129	66906		18 21 59.2	+30 02 47	+0.002	+0.03	7.9	1.1	-0.1	K2 III		410 s		
168646	186704		18 22 00.0	-28 25 48	+0.001	0.00	6.16	0.26	1.4	A2 V		66 s		
168708	186717		18 22 03.0	-22 55 18	+0.002	-0.01	7.1	0.0	0.0	B8.5 V		240 s		
168707	186719		18 22 04.3	-22 17 51	+0.001	-0.01	8.0	-0.4	0.6	A0 V		300 s		
170709	9101		18 22 04.5	+75 51 43	+0.001	+0.09	8.0	0.5	4.0	F8 V		63 s		
167991	254218		18 22 05.7	-63 22 54	0.000	+0.01	7.5	0.8	3.2	G5 IV		71 s		
168890	123392		18 22 08.2	+0 09 30	-0.001	+0.01	7.69	1.72	-0.3	K5 III		290 s		
169110	85975		18 22 08.5	+23 17 06	+0.001	+0.08	5.41	1.60	-0.3	K5 III	-58	120 s		
169431	47429		18 22 08.5	+48 21 29	-0.005	+0.05	8.00	0.9	3.2	G5 IV		91 s	11301	m
169244	66911		18 22 08.7	+36 18 06	+0.004	-0.05	7.6	0.5	4.0	F8 V		53 s		
168731	186721		18 22 10.0	-22 14 19	+0.001	0.00	7.9	0.5	0.2	K0 III		340 s		
168856	142250		18 22 10.5	-7 29 59	-0.001	-0.04	6.9	0.0	0.4	B9.5 V		180 s		
169487	30932		18 22 11.7	+51 32 30	+0.002	-0.01	6.7	0.1	1.4	A2 V		110 s		
168815	161390		18 22 13.6	-15 05 20	-0.001	-0.02	7.90	0.53	3.4	F5 V	-29	75 s	11282	m
168814	161391		18 22 16.1	-14 23 20	+0.003	0.00	7.3	0.1		A0 p	-15			
168496	228951		18 22 17.4	-47 10 34	+0.001	+0.03	8.0	1.4	0.2	K0 III		360 s		
169006	123399		18 22 17.9	+7 32 57	-0.006	-0.04	7.1	0.6	4.4	G0 V		35 s		
168592	210048		18 22 18.4	-38 39 24	-0.003	-0.03	5.0	1.3	0.2	K0 III	+18	58 s		
168591	210049		18 22 18.5	-38 32 43	+0.001	-0.03	7.9	0.8	0.2	K0 III		340 s		
169356	47425		18 22 22.7	+40 22 31	-0.002	-0.01	7.8	0.1	1.4	A2 V		190 s		
168932	142253		18 22 23.6	-1 12 09	0.000	-0.02	7.1	0.0	0.4	B9.5 V		190 s		

HD	SAO	Star Name	α 2000	δ 2000	μ(α)	μ(δ)	V	B-V	M$_v$	Spec	RV	d(pc)	ADS	Notes
169306	66913		18^{h}22^{m}24.6^s	+36°54'03"	0.000^s	-0.04"	7.9	0.5	4.0	F8 V		59 s		
169451	47433		18 22 32.9	+43 26 55	+0.002	-0.02	8.0	1.4	-0.3	K5 III		460 s		
169111	103648		18 22 35.1	+12 01 44	+0.001	-0.01	5.90	0.04	1.4	A2 V	-55	79 s		
172668	3049		18 22 39.3	+83 22 07	-0.004	+0.03	7.3	0.9	3.2	G5 IV		67 s		
169413	66920		18 22 39.9	+39 15 29	0.000	-0.02	7.9	1.1	0.2	K0 III		350 s		
169471	47437		18 22 42.6	+43 01 13	-0.003	+0.03	7.8	1.1	0.2	K0 III		330 s		
--	103652		18 22 42.8	+15 53 26	-0.001	0.00	7.7							
169112	123408		18 22 43.7	+9 46 44	+0.001	+0.01	7.9	1.1	-0.1	K2 III		350 s		
169154	103653		18 22 44.6	+15 47 16	-0.005	-0.05	7.5	0.8	3.2	G5 IV		73 s		
169191	103655		18 22 48.8	+17 49 36	+0.005	+0.02	5.25	1.27	-0.2	K3 III	-19	120 s		
169113	123409		18 22 48.9	+7 12 24	0.000	0.00	7.04	1.36	-0.1	K2 III	-32	210 s		
169169	103654		18 22 49.4	+15 00 06	-0.001	-0.02	7.4	0.1	1.4	A2 V	-12	160 s		
168733	210061		18 22 52.9	-36 40 09	0.000	-0.02	5.3	0.3	0.0	B8.5 V	-12	73 s		
169246	85988		18 22 56.4	+20 57 49	-0.001	0.00	7.3	1.1	0.2	K0 III		260 s		
169571	47448		18 22 59.3	+46 21 46	+0.001	+0.08	7.2	0.4	2.6	F0 V		82 s		
169009	161412		18 23 02.0	-10 13 08	+0.001	-0.01	6.3	0.1	0.6	A0 V		130 s		
169488	66923		18 23 02.2	+39 30 58	+0.001	-0.03	7.8	1.1	0.2	K0 III		330 s		
169223	103660		18 23 02.7	+16 41 16	-0.001	-0.02	6.22	1.20	0.2	K0 III	+15	120 s		
168176	254221		18 23 03.8	-64 22 47	-0.004	-0.05	8.0	0.1	0.6	A0 V		77 mx		
169572	47451		18 23 04.0	+45 13 14	0.000	+0.03	8.0	0.1	0.6	A0 V		310 s		
168338	254225		18 23 04.3	-60 44 56	-0.001	-0.01	7.00	0.1	1.7	A3 V		120 s		m
167586	--		18 23 07.1	-73 01 59			7.8	0.5	4.0	F8 V		57 s		
168987	161411		18 23 08.8	-16 35 37	0.000	-0.01	8.01	0.86		B1				
169745	30938		18 23 09.3	+54 25 25	-0.002	+0.02	7.7	0.1	1.4	A2 V		180 s		
169033	161415		18 23 12.0	-12 00 54	-0.001	-0.03	5.7	0.0		B8				
168339	254226	ξ Pav	18 23 13.3	-61 29 38	0.000	0.00	4.36	1.48	-0.5	M1 III	+12	94 s		m
168651	228963		18 23 16.9	-48 21 39	+0.001	-0.05	7.4	0.4	2.1	A5 V		88 s		
169064	161419		18 23 18.9	-12 18 01	-0.001	0.00	7.5	0.1	0.6	A0 V		200 s		
169247	103665		18 23 19.1	+14 42 00	-0.002	-0.01	6.7	0.0	0.4	B9.5 V	-16	180 s		
168988	186750		18 23 19.7	-20 39 13	-0.001	0.00	7.63	2.10	-0.5	M2 III		210 s		
169532	66928		18 23 23.3	+38 18 08	0.000	-0.03	7.7	1.4	-0.3	K5 III		390 s		
170092	17870		18 23 23.6	+62 46 49	-0.005	+0.03	8.0	0.1	1.4	A2 V		88 mx		
168942	186747		18 23 26.1	-27 30 09	+0.002	0.00	8.0	0.0	0.6	A0 V		280 s		
168838	210075		18 23 28.7	-36 14 18	+0.001	-0.01	5.5	1.4	0.2	K0 III		69 s		
168786	228970		18 23 29.9	-42 19 24	-0.001	-0.06	7.6	0.0	0.2	K0 III		200 mx		
169224	123422		18 23 32.5	+5 19 54	0.000	-0.01	7.50	-0.01	0.4	B9.5 V		260 s		
169155	142266		18 23 34.8	-6 15 02	-0.001	-0.01	8.0	0.5	4.0	F8 V		60 s		
167468	257584	φ Oct	18 23 36.0	-75 02 38	+0.001	+0.03	5.47	0.02	0.6	A0 V	+1	89 s		
168839	210076		18 23 36.3	-38 51 13	-0.002	0.00	7.5	0.1	0.2	K0 III		280 s		
169156	142267	ζ Sct	18 23 39.3	-8 56 03	+0.003	+0.04	4.68	0.95	0.2	K0 III	-6	79 s		
169414	86003	109 Her	18 23 41.7	+21 46 11	+0.014	-0.24	3.84	1.18	-0.1	K2 III	-58	33 mx		m
169617	66933		18 23 43.9	+38 20 41	0.000	0.00	6.80	0.1	0.6	A0 V	-13	170 s	11317	m
169472	86006		18 23 46.6	+23 40 58	-0.001	+0.06	7.7	1.1	0.2	K0 III		270 mx		
169359	103670		18 23 46.9	+14 54 28	+0.008	-0.06	7.9	0.6	4.4	G0 V		50 s		
169885	30943		18 23 47.6	+53 18 03	0.000	-0.01	6.2	0.1	1.4	A2 V	-4	92 s		
169225	142270		18 23 49.3	-1 08 48	-0.001	-0.01	7.5	0.0	0.4	B9.5 V		230 s		
168409	254227		18 23 52.8	-63 23 09	+0.001	-0.03	7.8	1.1	0.2	K0 III		330 s		
170073	30949	39 Dra	18 23 54.4	+58 48 02	-0.006	+0.06	4.98	0.08	1.2	A1 V	-13	53 s	11336	m
169550	66932		18 23 55.7	+31 21 23	-0.001	-0.04	8.0	0.9	3.2	G5 IV		93 s		
--	66931		18 23 56.8	+30 26 30	-0.003	-0.01	8.0	1.1						
169646	66936		18 23 57.2	+38 44 20	+0.002	+0.01	6.40	1.1	-0.1	K2 III	-40	200 s	11320	m
169267	142273		18 23 59.6	-2 28 20	+0.002	-0.04	7.7	1.1	0.2	K0 III		260 s		
169746	47462		18 23 59.8	+43 54 28	+0.002	0.00	7.00	1.6	-0.5	M2 III	-46	320 s		
169268	142274		18 24 03.3	-3 35 00	-0.002	-0.05	6.38	0.34		F2				
169159	161424		18 24 03.5	-19 23 38	+0.001	0.00	7.69	1.98	-0.3	K5 III		200 s		
169021	210088		18 24 03.8	-34 20 01	+0.001	0.00	6.8	0.3	0.4	B9.5 V		130 s		
168904	228981		18 24 04.2	-41 36 08	+0.001	-0.03	7.9	0.7	0.2	K0 III		340 s		
169490	86007		18 24 05.4	+20 27 06	-0.002	+0.03	6.60	0.4	2.6	F0 V	-43	63 s		m
168993	210089		18 24 05.5	-35 00 17	+0.002	0.00	7.6	1.0	-0.1	K2 III		360 s		
171606	9130		18 24 07.7	+79 13 23	-0.011	+0.08	6.5	1.1	0.2	K0 III		190 s		
172864	3056		18 24 08.8	+83 10 31	+0.005	-0.03	6.2	0.1	1.4	A2 V	-11	89 s		
169416	103677		18 24 08.8	+10 57 27	-0.001	-0.01	7.8	0.1	0.6	A0 V		260 s		
169647	66938		18 24 09.9	+34 29 40	+0.003	0.00	7.7	1.1	0.2	K0 III		310 s		
169118	186770		18 24 09.9	-26 26 50	0.000	+0.01	7.57	1.50		K2				
169022	210091	20 ε Sgr	18 24 10.2	-34 23 05	-0.003	-0.13	1.85	-0.03	-0.3	B9 IV	-11	26 s		Kaus Australis, m
171299	9123		18 24 10.3	+77 33 40	+0.010	+0.03	7.8	1.1	0.2	K0 III	-7	330 s		
169778	47465		18 24 10.8	+42 28 00	+0.001	-0.02	6.7	0.1	1.4	A2 V		120 s		
169702	66943	2 μ Lyr	18 24 13.6	+39 30 26	-0.002	-0.01	5.12	0.03	0.0	A3 III	-25	110 s		
169686	66942		18 24 16.2	+37 42 15	0.000	-0.02	8.0	1.1	0.2	K0 III		360 s		
169310	142277		18 24 17.1	-4 05 19	-0.001	-0.01	7.6	0.1	1.7	A3 V		140 s		
168905	228982		18 24 18.2	-44 06 37	+0.001	-0.02	5.25	-0.19	-1.7	B3 V n	+14	250 s		m
168980	228986		18 24 18.2	-40 43 55	-0.001	0.00	8.0	-0.4	0.0	B8.5 V		400 s		
169491	103679		18 24 22.4	+15 39 20	0.000	0.00	7.4	0.0	0.4	B9.5 V	-19	250 s		
--	103680		18 24 22.9	+16 14 49	0.000	-0.01	8.0							
168596	245441		18 24 23.2	-59 36 59	+0.001	-0.02	7.6	-0.6	0.4	B9.5 V		270 s		

HD	SAO	Star Name	α 2000	δ 2000	μ(α)	μ(δ)	V	B-V	M_V	Spec	RV	d(pc)	ADS	Notes
			^{h} ^{m} ^{s}	$^\circ$ $'$ $''$	^{s}	$''$								
169367	142282		18 24 23.3	−1 03 50	−0.002	−0.10	7.5	0.7	4.4	G0 V		41 s		
169119	210101		18 24 24.1	−30 41 19	0.000	+0.01	8.00	0.1	0.6	A0 V		290 s		m
169668	66941		18 24 25.4	+32 07 51	−0.001	−0.01	7.9	0.1	0.6	A0 V		280 s		
168805	245445		18 24 27.9	−51 49 42	−0.001	−0.01	7.6	0.0	2.1	A5 V		130 s		
169142	186777		18 24 29.6	−29 46 51	0.000	−0.05	8.00	0.1	1.4	A2 V		200 s		m
169332	142281		18 24 29.8	−8 10 30	+0.002	−0.04	7.3	0.4	2.6	F0 V		82 s		
169074	210098		18 24 30.7	−37 13 36	0.000	+0.01	7.5	0.2	1.7	A3 V		120 s		
169533	103682		18 24 31.3	+13 42 23	0.000	+0.01	7.8	1.1	−0.1	K2 III		380 s		
168871	228983		18 24 33.0	−49 39 10	+0.004	−0.14	6.47	0.58	4.0	F8 V		29 s		
168825	245447		18 24 35.1	−52 06 28	+0.001	0.00	7.7	0.7	0.2	K0 III		310 s		
169143	210107		18 24 36.7	−32 17 27	0.000	+0.01	7.2	1.2	0.2	K0 III		180 s		
169274	161436		18 24 38.0	−19 43 59	−0.001	−0.01	7.51	2.08	−0.3	K5 III		160 s		
168791	245446		18 24 38.6	−54 57 26	0.000	−0.01	7.4	1.8	−0.1	K2 III		130 s		
169511	103686		18 24 40.1	+10 10 58	0.000	−0.02	7.9	1.4	−0.3	K5 III		390 s		
169370	142288		18 24 41.9	−7 04 31	+0.009	−0.01	6.31	1.16		K0				
170421	17875		18 24 42.4	+65 15 07	0.000	+0.02	8.0	1.1	0.0	K1 III		400 s		
169392	142289		18 24 42.7	−6 36 20	−0.004	−0.04	7.50	0.6	2.8	G0 IV		82 s	11318	m
169436	142292		18 24 43.2	−1 52 17	+0.002	−0.01	7.7	0.4	3.0	F2 V		81 s		
169512	123443		18 24 45.1	+7 13 53	+0.001	+0.01	7.80	0.00	0.4	B9.5 V		270 s	11323	m
169334	161447		18 24 52.8	−16 16 36	−0.001	0.00	8.01	1.99	−0.3	K5 III		230 s		
169648	86016		18 24 55.0	+21 11 52	+0.002	0.00	7.7	1.1	0.2	K0 III		310 s		
169493	142294		18 24 56.9	−1 34 47	−0.001	−0.01	6.15	0.37	3.0	dF2 n	−10	42 s	11324	m
169123	210112		18 24 57.5	−38 34 30	0.000	+0.01	7.8	1.3	−0.5	M2 III		470 s		
169576	123448		18 24 57.9	+9 44 09	+0.001	0.00	7.9	1.4	−0.3	K5 III	−17	370 s		
−−	103692		18 24 59.4	+16 41 54	+0.001	−0.02	7.9							
169335	161449		18 24 59.7	−18 32 20	0.000	−0.02	7.27	1.74	−0.5	M2 III		290 s		
169233	210116	18 Sgr	18 25 01.3	−30 45 23	−0.010	−0.07	5.60	1.14	1.7	K0 III−IV		48 s		
169628	103693		18 25 02.7	+14 08 42	−0.001	−0.01	7.7	0.0	0.4	B9.5 V		290 s		
167961	−−		18 25 03.9	−73 33 44			8.0	0.1	1.7	A3 V		180 s		
169554	123449		18 25 06.4	+3 30 04	0.000	0.00	8.0	1.1	0.2	K0 III		310 s		
169578	123453		18 25 08.6	+5 05 05	0.000	+0.01	6.74	0.02		B9				
170527	17878		18 25 09.6	+64 50 21	−0.007	+0.09	7.2	1.1	0.2	K0 III		200 mx		
170108	47481		18 25 11.4	+49 51 51	−0.003	−0.06	7.1	1.1	0.2	K0 III		240 s		
168690	254235		18 25 13.3	−62 30 04	+0.002	+0.01	8.0	0.4	3.0	F2 V		100 s		
169454	161457		18 25 15.1	−13 58 41	+0.001	+0.02	6.61	0.94	−6.6	B1 Ia	−25	950 s		
168849	245452		18 25 16.7	−57 05 45	+0.005	−0.02	7.0	0.7	3.4	F5 V		37 s		
170074	47479		18 25 17.1	+46 53 08	−0.004	+0.01	7.80	0.1	0.6	A0 V		63 mx	11343	m
170109	47482		18 25 17.9	+48 45 42	+0.001	+0.01	7.90	0.9	3.2	G5 IV		87 s	11344	m
169818	86027		18 25 20.1	+28 04 58	−0.001	−0.01	8.00	0.00	0.4	B9.5 V		330 s	11338	m
169797	86025		18 25 20.4	+26 05 02	−0.001	−0.01	7.65	0.98	0.3	G8 III		290 s	11337	m
169420	186794	21 Sgr	18 25 20.8	−20 32 30	0.000	−0.03	4.81	1.31	0.2	K0 III	−12	57 s	11325	m
169236	210119		18 25 21.5	−35 59 30	−0.003	+0.01	6.2	1.0	3.2	G5 IV		31 s		
170110	47485		18 25 21.8	+48 17 48	+0.002	−0.01	7.7	1.1	−0.1	K2 III		360 s		
169952	66955		18 25 23.0	+38 26 27	+0.001	−0.01	7.2	0.1		A0 p	−9			
169798	86026		18 25 27.6	+22 42 24	−0.001	−0.01	6.7	0.0	−1.6	B3.5 V	−17	450 s		
169337	210128		18 25 29.7	−30 15 12	−0.001	+0.01	7.40	0.00	0.0	B8.5 V		290 s		m
169257	210122		18 25 31.1	−37 08 57	−0.001	0.00	7.9	−0.3	0.4	B9.5 V		320 s		
168740	254237		18 25 31.3	−63 01 17	0.000	−0.10	6.14	0.20	1.4	A2 V		50 mx		
169720	103700		18 25 32.5	+14 45 04	0.000	+0.01	7.5	1.1	0.2	K0 III		280 s		
169516	161459		18 25 37.4	−15 09 10	0.000	0.00	7.5	1.1	0.2	K0 III		240 s		
169689	123462		18 25 38.6	+8 01 55	0.000	−0.01	5.65	0.92	4.4	G0 V	−8	10 s		
170027	66961		18 25 38.9	+39 07 20	0.000	−0.02	7.8	1.1	−0.1	K2 III		380 s		
169106	229003		18 25 40.5	−49 42 24	+0.002	0.00	8.0	0.7	0.2	K0 III		360 s		
169476	186800		18 25 43.7	−23 04 53	0.000	−0.02	7.8	1.3	0.2	K0 III		230 s		
169953	66959		18 25 46.2	+31 54 53	+0.001	0.00	8.0	1.1	0.2	K0 III		360 s		
170003	66963		18 25 47.0	+35 20 34	+0.001	−0.02	7.4	0.9	3.2	G5 IV		69 s		
169058	245461		18 25 48.8	−53 38 31	−0.001	+0.01	7.30	0.4	2.6	F0 V		87 s		m
169721	123465		18 25 50.0	+7 06 07	0.000	+0.04	8.0	0.5	3.4	F5 V		79 s		
171044	9124		18 25 54.1	+71 43 54	0.000	+0.05	7.4	0.5	3.4	F5 V		64 s		
169398	210135		18 25 54.4	−33 56 43	0.000	+0.01	6.4	0.5	0.0	B8.5 V	−6	88 s		
169425	210138		18 25 54.6	−31 45 17	+0.001	−0.01	7.38	0.03	0.0	B8.5 V		280 s		
169820	103709		18 25 55.2	+14 58 00	+0.001	−0.01	6.40	0.01	0.2	B9 V	−25	170 s		
169151	245464		18 25 57.0	−51 23 19	+0.002	−0.01	7.8	−0.1	0.4	B9.5 V		300 s		
169981	86043		18 25 58.6	+29 49 44	+0.002	−0.02	5.83	0.06	1.4	A2 V	+9	75 s		
170693	17888	42 Dra	18 25 58.8	+65 33 49	+0.016	−0.03	4.82	1.19	−0.1	K2 III	+32	96 s		
169840	103710		18 26 00.8	+15 45 30	0.000	−0.05	7.4	0.4	3.0	F2 V		77 s		
169725	123471		18 26 03.2	+0 46 50	+0.001	+0.01	7.60	0.1	1.7	A3 V		140 s	11339	m
169650	142310		18 26 07.3	−8 09 08	+0.001	−0.02	8.0	0.1	0.6	A0 V		240 s		
169186	245466		18 26 08.2	−50 52 44	+0.001	+0.02	7.8	0.5	3.2	G5 IV		84 s		
168264	−−		18 26 08.5	−73 05 33			7.8	1.1	0.2	K0 III		330 s		
172099	9139		18 26 09.1	+79 37 44	−0.002	+0.05	7.1	0.2	2.1	A5 V		98 s		
169822	123474		18 26 10.0	+8 46 38	−0.013	−0.48	7.83	0.69	5.5	G7 V	−22	23 mx		m
169928	103715		18 26 15.4	+18 52 09	0.000	−0.01	7.7	0.0	0.4	B9.5 V		290 s		
167100	257587		18 26 16.3	−79 55 39	−0.002	−0.01	7.3	2.5	−0.5	M2 III		95 s		
170093	66973		18 26 17.6	+31 36 08	0.000	0.00	8.0	1.1	−0.1	K2 III		410 s		

449

HD	SAO	Star Name	α 2000	δ 2000	μ(α)	μ(δ)	V	B-V	M$_v$	Spec	RV	d(pc)	ADS	Notes
170028	86048		18^{h}26^{m}19.4^s	+26°13'40"	-0.001	0.00	6.8	0.0	-1.6	B3.5 V	-25	470 s		
169954	86045		18 26 20.6	+20 28 43	0.000	+0.01	8.0	1.1	0.2	K0 III		360 s		
169955	103718		18 26 21.4	+19 10 53	-0.001	+0.01	7.4	1.4	-0.3	K5 III		350 s		
169673	161469		18 26 23.5	-15 37 48	0.000	+0.02	7.2	0.1	-5.1	B1 II	-17	1300 s		
169403	229020		18 26 24.9	-42 32 11	-0.001	+0.01	7.2	1.0	-0.5	M2 III		350 s		
168599	254238		18 26 27.3	-69 30 33	+0.001	-0.02	7.8	1.1	0.2	K0 III		330 s		
169752	142315		18 26 28.2	-8 11 07	0.000	0.00	7.5	0.1	0.6	A0 V		210 s		
170051	86055		18 26 28.9	+26 27 40	-0.001	0.00	7.11	-0.09	-1.6	B3.5 V	-24	490 s		
169753	142317		18 26 33.3	-9 12 06	+0.001	-0.01	7.55	0.72	-2.5	B2 V	-28	310 s		RZ Sct, m,v
170614	30977		18 26 38.5	+59 42 21	0.000	0.00	7.1	1.1	0.2	K0 III		240 s		
170198	66979		18 26 39.0	+35 42 49	0.000	0.00	8.0	1.1	-0.1	K2 III		410 s		
170111	86060		18 26 40.7	+26 26 57	0.000	0.00	6.53	-0.11	-1.7	B3 V	-18	410 s	11356	m
169586	210159		18 26 40.8	-30 23 37	+0.002	-0.06	6.8	0.8	4.4	G0 V		22 s		
170228	66980		18 26 41.2	+36 09 53	+0.001	0.00	7.70	1.1	0.2	K0 III		320 s	11361	m
169657	186822		18 26 41.4	-23 26 09	-0.002	-0.01	7.9	-0.1	0.4	B9.5 V		320 s		
170199	66981		18 26 50.9	+30 58 06	-0.002	+0.02	7.3	1.1	0.2	K0 III		270 s		
170260	66985		18 26 51.9	+37 26 25	-0.002	-0.01	8.0	0.1	0.6	A0 V		300 s		
169931	123489		18 26 52.6	+3 54 48	0.000	-0.02	8.0	1.6	-0.5	M4 III		420 s		V988 Oph, q
169405	229021		18 26 53.7	-48 07 01	-0.001	-0.04	5.4	0.9	3.2	G5 IV	+4	22 s		
169757	161477		18 26 55.8	-17 11 51	0.000	+0.01	7.90	0.1	1.7	A3 V		160 s	11345	m
170261	66987		18 26 56.1	+36 04 27	-0.002	-0.01	8.0	0.9	3.2	G5 IV		93 s		
167448	257590		18 26 56.8	-79 16 08	-0.006	-0.02	7.7	1.1	-0.1	K2 III		360 s		
169467	229023	α Tel	18 26 58.2	-45 58 06	-0.001	-0.05	3.51	-0.17	-2.9	B3 III	-1	180 s		
170558	30978		18 26 59.0	+54 53 08	+0.002	+0.07	7.6	1.1	0.2	K0 III		260 mx		
170008	123495		18 27 01.4	+9 12 35	-0.004	-0.04	7.5	0.8	3.2	G5 IV		71 s		
170009	123494		18 27 01.7	+7 43 57	+0.001	0.00	8.00	0.09	0.0	B8.5 V		310 s		
168741	254242		18 27 03.3	-69 22 54	0.000	-0.04	7.9	0.1	0.6	A0 V		290 s		
169660	210166		18 27 03.6	-31 22 23	0.000	-0.03	7.4	0.7	0.2	K0 III		280 s		
169326	245474		18 27 04.2	-53 06 37	+0.002	+0.01	6.4	1.1	0.2	K0 III		170 s		
170326	66992		18 27 07.8	+36 40 38	+0.001	+0.02	7.9	0.1	0.6	A0 V		280 s		
169913	142326		18 27 08.7	-8 02 26	0.000	+0.01	6.60	1.1	0.2	K0 III		170 s		
170288	66990		18 27 10.6	+31 59 35	0.000	-0.01	7.8	1.1	0.2	K0 III		330 s		
169986	123497	59 Ser	18 27 12.2	+0 11 46	-0.001	0.00	5.21	0.50	0.6	G0 III	-22	84 s	11353	d Ser, m,q
169383	245481		18 27 12.7	-51 53 03	+0.015	-0.10	7.98	0.60	4.4	G0 V		50 s		
170053	123502		18 27 14.0	+7 00 33	-0.001	0.00	7.60	1.1	-2.2	K2 II	-31	680 s		
169303	245475		18 27 17.5	-55 57 15	+0.002	-0.03	7.4	0.5	4.4	G0 V		40 s		
169349	245482		18 27 18.6	-53 38 01	+0.008	-0.03	7.2	1.1	3.2	G5 IV		38 s		
169679	210170		18 27 23.0	-36 00 59	0.000	-0.02	6.8	0.6	0.4	B9.5 V		76 s		
--	103733		18 27 24.4	+18 44 27	+0.001	0.00	7.9							
169680	210171		18 27 25.9	-36 53 11	-0.001	-0.02	8.0	-0.3	0.6	A0 V		300 s		
170541	47510		18 27 27.2	+49 49 08	+0.001	+0.07	7.5	0.1	1.7	A3 V		140 mx		
169915	161490		18 27 31.1	-15 22 26	+0.003	+0.02	6.80	0.5	3.4	F5 V		46 s		m
169706	210176		18 27 31.3	-33 24 39	0.000	0.00	7.06	1.02	-2.1	K0 II		680 s		
170593	30983		18 27 32.8	+51 37 45	+0.002	+0.02	8.0	1.4	-0.3	K5 III		450 s		
169429	245487		18 27 38.1	-54 02 01	-0.001	-0.05	8.00	1.1	0.2	K0 III		360 s		m
170645	30985		18 27 39.9	+53 57 03	+0.001	-0.01	7.3	0.8	3.2	G5 IV		66 s		
170811	30990		18 27 42.1	+59 32 57	+0.006	+0.04	6.4	1.1	3.2	K0 IV	-10	44 s		
169851	186837		18 27 43.6	-26 38 05	0.000	-0.02	6.90	0.2	2.1	A5 V		90 s	11354	m
170314	86083		18 27 45.7	+24 41 51	0.000	+0.02	7.00	0.00	0.4	B9.5 V	-26	210 s	11373	m
170375	66998		18 27 47.7	+31 11 54	-0.001	-0.02	7.6	1.6	-0.5	M4 III		420 s		
169830	186838		18 27 49.3	-29 48 59	0.000	+0.03	5.8	1.2	4.4	G0 V		19 s		d?
170115	123512		18 27 49.7	+0 11 59	0.000	-0.01	7.9	0.1	0.6	A0 V		250 s		
170137	123513		18 27 50.1	+3 44 55	-0.001	0.00	6.07	1.62	-0.2	gK3	-19	110 s		
170559	47514		18 27 50.3	+46 03 59	0.000	+0.01	7.8	1.1	0.2	K0 III		330 s		
--	--		18 27 55.5	-33 19 20			7.20							RV Sgr, v
169990	161493		18 27 56.3	-17 48 00	0.000	+0.01	6.0	0.0		B8	-35			
169165	--		18 27 56.4	-64 19 00			7.8	1.6	-0.5	M2 III		450 s		
169916	186841	22 λ Sgr	18 27 58.1	-25 25 18	-0.003	-0.18	2.81	1.04	-0.1	K2 III	-43	30 ts		Kaus Borealis
170200	123516		18 27 58.5	+6 11 39	0.000	-0.02	5.71	-0.03		B8				
170812	30993		18 27 59.8	+57 23 52	0.000	-0.02	7.7	0.1	0.6	A0 V		260 s		
170694	30988		18 28 01.5	+51 02 12	+0.002	+0.02	7.7	1.1	0.2	K0 III		310 s		
170201	123519		18 28 02.9	+5 45 36	0.000	-0.02	8.0	1.4	-0.3	K5 III		380 s		
169220	254251		18 28 03.2	-63 18 48	+0.003	-0.04	7.8	1.1	0.2	K0 III		330 s		
170849	30994		18 28 04.3	+57 55 20	+0.001	0.00	7.8	0.8	3.2	G5 IV		84 s		
169791	210188		18 28 05.6	-38 21 25	+0.001	-0.02	7.9	-0.7	0.4	B9.5 V		320 s		
169938	186843		18 28 06.0	-26 45 26	0.000	-0.04	6.3	-0.1	1.7	A3 V		83 s		
165338	258795		18 28 06.5	-84 23 12	+0.013	+0.01	6.6	-0.3	0.6	A0 V		160 s		
169966	186848		18 28 07.0	-22 59 57	+0.001	+0.03	7.1	0.4	3.2	G5 IV		60 s		
168951	254249		18 28 07.2	-69 18 50	+0.001	+0.01	8.0	1.1	0.2	K0 III		360 s		
170594	47517		18 28 10.2	+43 55 34	-0.003	-0.01	7.9	0.1	1.7	A3 V	+3	170 s		
170270	123522		18 28 10.6	+7 57 15	0.000	+0.02	7.5	1.4	-0.3	K5 III		330 s		
169872	210194		18 28 10.9	-32 28 13	+0.002	-0.02	8.0	-0.7	0.6	A0 V		300 s		
170616	47518		18 28 14.9	+44 07 52	-0.002	+0.01	7.9	0.1	0.6	A0 V		280 s		
169643	229033		18 28 16.3	-49 31 09	0.000	-0.02	8.0	1.0	0.2	K0 III		290 s		
170615	47519		18 28 18.5	+44 15 23	-0.001	-0.01	7.7	1.1	0.2	K0 III	-30	320 s		

HD	SAO	Star Name	α 2000	δ 2000	μ(α)	μ(δ)	V	B-V	M$_v$	Spec	RV	d(pc)	ADS	Notes
170850	30995		18^h 28^m 19.1^s	+55° 40' 27"	0.000	+0.02	7.7	0.5	3.4	F5 V		72 s		
170247	123525		18 28 19.8	+3 57 55	−0.001	−0.02	7.92	0.12	0.6	A0 V		270 s		
170274	123529		18 28 23.9	+3 46 45	0.000	−0.03	7.89	0.28	2.6	F0 V		110 s		
170232	142337		18 28 24.6	−3 52 35	−0.013	−0.32	7.58	0.70		G0				
169970	210202		18 28 25.5	−31 08 00	−0.002	−0.09	7.7	0.6	0.2	K0 III		120 mx		
169853	210197		18 28 26.9	−38 59 44	0.000	−0.04	5.6	0.6	1.4	A2 V		33 s		
169766	229043		18 28 33.0	−46 44 57	−0.001	−0.01	7.9	0.6	4.4	G0 V		51 s		
170316	123539		18 28 33.4	+4 50 55	0.000	−0.01	7.69	0.02	0.6	A0 V		260 s		m
169877	210200		18 28 34.2	−38 51 48	0.000	+0.03	7.6	0.0	0.4	B9.5 V		250 s		
170294	123536		18 28 34.2	+0 47 49	−0.007	−0.03	7.9	0.9	3.2	G5 IV		84 s		
168241	257593		18 28 34.3	−77 21 55	+0.003	−0.03	6.8	1.1	0.2	K0 III		210 s		
170595	67010		18 28 34.9	+38 46 06	0.000	−0.05	7.7	1.1	0.2	K0 III		310 s		
170710	47524		18 28 35.3	+46 25 25	−0.002	+0.02	7.2	0.4	2.6	F0 V		84 s		
168717	257595		18 28 36.1	−73 32 06	+0.002	−0.02	7.8	1.1	0.2	K0 III		330 s		
169814	229046		18 28 40.1	−46 11 51	+0.001	−0.02	7.7	1.2	−0.1	K2 III		360 s		
168873	—		18 28 40.6	−71 47 07			7.7	1.1	0.2	K0 III		320 s		
169645	245502		18 28 40.7	−53 51 00	0.000	0.00	7.4	1.0	−0.1	K2 III		310 s		
170826	30998		18 28 42.5	+51 56 42	+0.002	−0.03	7.8	0.1	1.4	A2 V		190 s		
169898	229051		18 28 43.8	−41 40 30	0.000	−0.01	7.9	1.3	−0.1	K2 III		340 s		
—	103758		18 28 44.4	+16 46 26	+0.013	−0.03	7.9	0.8	3.2	G5 IV		57 mx		
169942	210203		18 28 46.3	−38 01 41	+0.003	+0.02	7.65	1.58	−0.3	K4 III		310 s		
169972	210205		18 28 46.7	−36 59 47	−0.003	−0.01	7.9	1.0	3.2	G5 IV		64 s		
169505	254259		18 28 49.4	−60 36 34	+0.003	+0.01	8.0	1.4	−0.1	K2 III		310 s		
169767	229047	ζ Tel	18 28 49.7	−49 04 15	+0.015	−0.24	4.13	1.02	0.3	gG8	−31	51 s		
170617	67015		18 28 51.0	+35 47 49	−0.001	0.00	8.0	1.4	−0.3	K5 III		450 s		
170423	103761		18 28 52.5	+14 59 22	0.000	+0.01	7.8	1.1	−0.1	K2 III		340 s		
170141	186863		18 28 57.2	−26 34 53	0.000	+0.01	6.50	0.1	0.6	A0 V		150 s		m
169351	254256		18 29 01.2	−65 11 24	−0.002	−0.08	7.7	0.3	4.4	G0 V		47 s		
170448	103763		18 29 01.8	+11 39 02	−0.001	−0.01	7.6	0.7	4.4	G0 V		44 s		
169881	229054		18 29 03.4	−47 10 56	−0.001	−0.01	7.9	0.6	3.2	G5 IV		88 s		
170778	47529		18 29 03.6	+43 56 21	+0.006	+0.16	7.6	0.9	3.2	G5 IV		77 s		
170038	—		18 29 05.6	−37 00 31			7.9	0.7	4.4	G0 V		42 s		
170039	—		18 29 06.7	−37 18 31			7.9	1.1	0.2	K0 III		320 s		
170469	103765		18 29 10.8	+11 41 44	−0.003	−0.01	8.0	0.9	3.2	G5 IV		90 s		
170619	86109		18 29 11.4	+29 33 13	0.000	+0.04	7.70	0.9	3.2	G8 IV		79 s	11393	m
170296	161520	γ Sct	18 29 11.7	−14 33 57	0.000	−0.01	4.70	0.06	1.4	A2 V	−41	45 s		
170142	186867		18 29 11.8	−29 15 34	+0.001	−0.04	6.80	0.5	3.4	F5 V		48 s		m
169943	229056		18 29 12.6	−43 50 45	0.000	+0.03	6.4	0.7	0.2	K0 III		180 s		
170021	229057		18 29 15.3	−41 57 24	0.000	−0.02	8.00	1.1	0.2	K0 III		360 s		m
170040	210213		18 29 16.5	−38 51 04	0.000	−0.01	6.63	−0.02		B8 n	−28			
170144	210220		18 29 19.0	−31 33 33	0.000	−0.05	7.8	0.7	3.4	F5 V		52 s		
167714	258796		18 29 19.2	−80 13 57	−0.012	−0.06	5.95	1.16	0.0	gK1		130 s		
170562	103769		18 29 20.6	+17 59 03	−0.001	0.00	6.5	0.0	0.4	B9.5 V		170 s		
170235	186873		18 29 21.8	−25 15 23	0.000	0.00	6.59	0.07		B1.5 e	0			
170875	47535		18 29 22.7	+45 58 54	−0.004	0.00	7.4	1.1	0.2	K0 III		270 mx		
172114	9146		18 29 24.7	+76 13 31	−0.006	−0.04	7.4	1.1	0.2	K0 III		270 s		
169977	229058		18 29 27.1	−47 05 07	−0.001	+0.03	7.7	1.0	0.2	K0 III		310 s		
170542	103770		18 29 27.5	+13 51 18	−0.001	−0.07	6.9	0.4	3.0	F2 V	−2	60 s		
170512	123559		18 29 27.5	+7 21 33	−0.002	+0.03	7.70	0.7	4.4	G0 V		45 s	11390	m
170543	103772		18 29 30.6	+12 27 03	0.000	+0.01	7.6	0.1	0.6	A0 V		240 s		
170236	186874		18 29 31.1	−29 18 47	+0.006	−0.01	7.9	0.3	2.1	A5 V		99 mx		
170544	103773		18 29 31.5	+11 54 50	+0.001	−0.01	8.0	0.4	3.0	F2 V		95 s		
170648	86119		18 29 35.2	+26 03 22	0.000	+0.02	8.0	0.1	1.4	A2 V		210 s		
170650	86115		18 29 35.6	+23 51 58	0.000	−0.01	5.89	−0.10	−1.3	B6 IV	−17	270 s		
170597	103778		18 29 36.3	+18 00 47	0.000	−0.01	7.7	0.1	0.6	A0 V		260 s		
170213	210226		18 29 37.9	−33 56 03	+0.002	−0.02	7.5	0.1	0.4	B9.5 V		210 s		
170474	142348	60 Ser	18 29 40.8	−1 59 07	+0.002	−0.03	5.39	0.96	0.3	G8 III	+28	100 s		
172340	9151		18 29 44.6	+77 32 49	−0.002	0.00	5.64	1.18	0.2	K0 III	+1	96 s		
170630	103781		18 29 45.8	+17 35 57	0.000	−0.02	7.6	0.1	1.4	A2 V		180 s		
170397	161528		18 29 46.5	−14 34 54	+0.001	+0.02	5.96	−0.04		A0 pe	−16			
170563	123563		18 29 48.9	+6 45 47	+0.001	0.00	8.0	0.1	0.6	A0 V		270 s		
170364	186885		18 29 51.0	−20 57 10	0.000	−0.04	7.4	0.5	0.6	A0 V		120 s		
170579	123567		18 29 52.6	+9 11 56	−0.006	−0.19	7.5	0.5	3.4	F5 V		41 mx		
170513	142351		18 29 54.7	−3 59 07	−0.001	+0.02	8.0	0.5	3.4	F5 V		77 s		
170069	229064		18 29 55.6	−47 13 13	+0.001	0.00	5.6	1.5	0.2	K0 III		62 s		
169836	245510		18 29 56.4	−57 31 24	+0.006	−0.03	5.80	1.1	0.2	K0 III		130 s		m
170598	103782		18 29 57.1	+10 28 59	+0.002	+0.05	6.8	0.4	2.6	F0 V	−34	68 s		
170651	103784		18 29 59.5	+15 56 40	+0.001	+0.01	7.4	1.1	0.2	K0 III		250 s		
170279	210235		18 30 01.6	−33 02 54	+0.002	−0.02	7.12	0.22		A3				
170280	210234		18 30 01.8	−33 29 44	0.000	−0.02	8.02	0.01	0.4	B9.5 V		330 s		
170876	67034		18 30 01.9	+37 03 17	−0.001	−0.07	8.0	1.1	0.2	K0 III		340 mx		
170303	210236		18 30 02.2	−31 44 49	−0.001	−0.07	7.9	2.0	0.2	K0 III		88 s		
170698	103785		18 30 03.8	+18 35 16	+0.001	0.00	7.6	0.9	0.3	G6 III	0	290 s		
170149	229069		18 30 04.3	−43 50 01	+0.002	−0.15	7.8	−0.7	4.4	G0 V		49 s		
170580	123571		18 30 04.9	+4 03 54	−0.001	−0.02	6.69	0.11	−2.5	B2 V	−22	460 s	11399	m

HD	SAO	Star Name	α 2000	δ 2000	μ(α)	μ(δ)	V	B–V	M_v	Spec	RV	d(pc)	ADS	Notes
170433	161540		18 30 11.6	-18 43 44	+0.003	-0.10	5.7	1.1	0.2	K0 III	-1	120 s		
170896	67040		18 30 11.9	+37 13 12	+0.001	-0.02	7.6	1.1	0.2	K0 III		300 s		
170091	229068		18 30 14.1	-49 29 22	-0.001	-0.02	7.9	0.5	3.4	F5 V		78 s		
170547	142353		18 30 14.2	-5 43 28	0.000	-0.02	6.28	0.95		G5				
170756	86134		18 30 16.1	+21 52 01	-0.001	0.00	7.9	0.5	-4.6	F4 Ib p	-30	3100 s		AC Her, v
170304	210238		18 30 16.6	-36 49 25	0.000	-0.03	7.9	1.4	3.2	G5 IV		36 s		
170105	229071		18 30 19.7	-49 40 03	+0.001	0.00	8.0	0.7	0.2	K0 III		360 s		
169714	254265		18 30 21.1	-63 09 07	+0.001	-0.02	7.8	1.1	0.2	K0 III		330 s		
169569	254262		18 30 23.5	-66 56 06	-0.001	+0.01	7.7	0.1	0.6	A0 V		260 s		
170457	186894		18 30 28.7	-23 15 03	-0.002	-0.01	7.00	1.1	0.2	K0 III		220 s		m
169883	245515		18 30 30.2	-59 17 32	+0.001	-0.02	8.0	1.8	-0.3	K5 III		290 s		
170415	186892		18 30 30.9	-28 47 43	0.000	-0.02	7.6	0.5	1.4	A2 V		88 s		
170931	67044		18 30 31.4	+32 38 31	-0.001	+0.03	7.62	0.14		A0				
171008	47547		18 30 32.7	+40 58 32	0.000	-0.02	6.9	1.1	-0.1	K2 III		250 s		
170566	---		18 30 32.9	-10 53 10			7.8	0.4	3.0	F2 V		83 s		
170496	186900		18 30 32.9	-20 17 17	-0.003	-0.05	7.8	1.6	0.2	K0 III		150 s		
171124	47554		18 30 33.7	+48 17 55	0.000	+0.01	7.9	0.1	0.6	A0 V		280 s		
170970	67045		18 30 34.3	+36 14 56	-0.001	-0.01	7.5	1.6	-0.5	M2 III		410 s		
170399	210245		18 30 34.7	-30 49 38	-0.001	-0.03	8.0	1.4	0.2	K0 III		220 s		
170757	103793		18 30 39.0	+12 36 41	0.000	+0.01	7.30	0.1	0.6	A0 V	-25	200 s		RX Her, m,v
170699	123585		18 30 40.9	+4 30 40	-0.001	+0.01	6.80	0.1	1.4	A2 V	-39	110 s		m
170829	86142		18 30 41.4	+20 48 54	+0.001	-0.26	6.50	0.79	3.2	G8 IV	-59	46 s		
171065	47551		18 30 48.1	+40 29 18	-0.001	-0.07	7.8	0.4	2.6	F0 V		110 s		
170712	123587		18 30 49.3	+3 09 16	-0.001	-0.01	7.83	0.12	0.6	A0 V		240 s		
170499	186903		18 30 53.4	-27 57 06	+0.001	+0.01	7.4	1.8	-0.3	K5 III		230 s		
170416	210251		18 30 54.3	-35 19 18	-0.003	+0.03	7.5	1.1	3.2	G5 IV		47 s		
170780	123591		18 30 54.4	+8 04 42	0.000	-0.02	7.5	1.4	-0.3	K5 III	-23	330 s		
170283	229077		18 30 56.1	-48 00 56	-0.004	-0.05	6.70	0.7	4.4	G0 V		29 s		m
170224	245521		18 30 56.5	-51 31 04	-0.002	-0.04	7.4	1.3	-0.3	K5 III		350 s		
171087	47556		18 30 57.6	+41 06 23	-0.003	+0.03	7.2	0.5	3.4	F5 V		58 s		
170897	86145		18 30 58.8	+21 39 16	0.000	-0.04	7.40	1.07		K0				
170782	123593		18 31 01.3	+5 45 09	0.000	-0.01	7.81	0.14	1.4	A2 V		170 s		
170150	245520		18 31 02.0	-54 38 46	-0.007	-0.04	7.9	0.0	3.4	F5 V		80 s		
170531	186906		18 31 02.2	-24 53 30	+0.003	+0.01	7.5	0.1	0.4	B9.5 V		220 s		
172083	9149		18 31 02.7	+73 05 02	+0.012	+0.11	7.8	0.7	4.4	G0 V		49 s		
169884	254269		18 31 02.7	-63 02 26	+0.004	+0.03	7.9	0.1	0.6	A0 V		240 mx		
170384	229080		18 31 02.8	-41 54 49	0.000	-0.02	6.3	-0.4	2.1	A5 V		68 s		
170310	229078		18 31 03.8	-47 49 30	+0.004	-0.01	7.2	0.8	0.2	K0 III		250 s		
170783	123594		18 31 04.3	+4 37 37	-0.001	-0.01	7.73	0.19	-1.0	B5.5 V	-10	360 s		
170878	103800		18 31 04.3	+16 55 43	-0.003	-0.02	5.70	0.04	1.4	A2 V	-9	72 s		
170479	210257		18 31 04.7	-32 59 21	0.000	-0.04	5.34	0.16		A3	+9	18 mn		m
170986	86151		18 31 05.2	+28 29 45	+0.001	+0.02	7.5	0.0	0.4	B9.5 V		260 s		
171026	67059		18 31 06.9	+32 14 44	+0.001	0.00	7.80	0.1	0.6	A0 V		280 s	11424	m
170714	142359		18 31 08.4	-5 47 23	0.000	-0.01	7.39	0.37	-1.0	B5.5 V	-16	260 s		
170193	245523		18 31 08.4	-53 31 41	-0.002	-0.04	6.8	1.1	0.2	K0 III		200 s		
170461	210256		18 31 11.4	-36 48 33	+0.004	0.00	7.0	0.6	2.6	F0 V		48 s		
--	67066		18 31 11.5	+35 02 42	0.000	-0.02	7.3	2.3						
170385	229084		18 31 13.7	-43 43 47	+0.002	0.00	7.91	-0.13	-1.7	B3 V		810 s		m
170239	245524		18 31 13.9	-52 27 41	-0.003	-0.05	7.9	-0.4	0.6	A0 V		280 s		
171653	17912		18 31 14.5	+65 26 10	-0.004	+0.07	6.30	0.1	1.7	A3 V	-9	83 s		m
171066	67067		18 31 14.8	+33 37 50	-0.003	-0.04	7.3	1.1	0.2	K0 III		260 s		
171549	17911		18 31 18.8	+62 32 16	0.000	+0.05	7.00	0.4	3.0	F2 V		63 s	11443	m
170657	161557		18 31 18.8	-18 54 32	-0.010	-0.20	6.84	0.84	6.9	K3 V	-47	12 ts		
170622	186912		18 31 20.3	-23 18 56	0.000	0.00	7.4	0.9	0.2	K0 III		270 s		
170354	229083		18 31 20.3	-47 18 46	-0.003	+0.01	8.0	0.2	3.2	G5 IV		93 s		
170795	142362		18 31 21.8	-0 28 54	-0.001	0.00	7.0	0.0	0.0	B8.5 V		220 s		
169978	254273	ν Pav	18 31 22.2	-62 16 42	+0.001	-0.04	4.64	-0.11	-1.2	B8 III	+59	150 s		
170535	210263		18 31 22.8	-34 45 01	-0.004	-0.03	8.0	0.7	1.4	A2 V		77 mx		
170501	210262		18 31 23.9	-37 35 34	-0.002	-0.03	7.6	0.4	3.2	G5 IV		76 s		
170682	161562		18 31 24.0	-19 09 29	-0.001	+0.02	7.98	0.30	-1.6	B7 III	+4	470 s		
170740	161569		18 31 25.5	-10 47 45	0.000	-0.02	5.72	0.24	-2.5	B2 V	-17	250 s	11414	m
170899	123603		18 31 26.0	+8 01 36	+0.001	-0.02	7.6	0.6	4.4	G0 V		43 s		
170715	161566		18 31 26.0	-12 01 15	-0.001	-0.18	7.5	0.5	4.0	F8 V		48 s		
170680	161564		18 31 26.2	-18 24 10	0.000	-0.03	5.20	0.1	0.6	A0 V	-37	79 s	11411	m
170818	142364		18 31 28.8	-1 38 06	+0.001	-0.03	7.6	0.4	3.0	F2 V		81 s		
170464	229090		18 31 29.1	-42 05 48	-0.002	-0.03	5.40	1.1	0.2	K0 III		110 s		
170338	245531		18 31 30.0	-50 11 10	0.000	-0.07	7.4	0.4	2.6	F0 V		81 s		
170046	254275		18 31 31.7	-62 17 53	+0.002	-0.01	7.6	0.1	1.4	A2 V		170 s		
--	245526		18 31 33.1	-56 59 00	+0.001	+0.01	7.5	0.1	0.6	A0 V		240 s		
170502	229093		18 31 38.3	-41 39 29	-0.001	-0.02	7.9	-0.3	0.4	B9.5 V		320 s		
170935	123606		18 31 42.9	+7 13 53	0.000	-0.01	7.38	0.15	0.0	B8.5 V		210 s		
171046	86159		18 31 44.8	+21 54 08	0.000	-0.03	7.8	1.1	0.2	K0 III		330 s		
170465	229092	δ¹ Tel	18 31 45.2	-45 54 54	-0.001	-0.03	4.96	-0.11	-1.3	B6 IV	-15	180 s		
171285	47569		18 31 45.2	+45 59 20	+0.002	+0.01	7.2	1.1	0.2	K0 III		260 s		
170857	142368		18 31 46.0	-2 25 07	+0.002	-0.01	7.4	1.1	-0.1	K2 III	-32	260 s		

HD	SAO	Star Name	α 2000	δ 2000	μ(α)	μ(δ)	V	B-V	M$_v$	Spec	RV	d(pc)	ADS	Notes
			h m s	° ′ ″	s	″								
171341	47572		18 31 51.6	+49 53 30	+0.003	+0.04	7.5	0.5	3.4	F5 V		65 s		
170764	161571		18 31 53.0	−19 07 30	−0.001	0.00	6.5	0.5	−8.2	F5 Ia	−2	3900 s	11433	U Sgr, m,v
171566	31034		18 31 54.4	+59 25 11	+0.003	−0.04	7.9	0.4	2.6	F0 V		110 s		
170521	229094		18 31 56.0	−43 30 28	0.000	−0.02	5.6	1.1	0.2	K0 III		110 s		
170920	142372	61 Ser	18 31 56.8	−1 00 11	+0.001	−0.01	5.94	0.16		A2	−27			
170987	123609		18 32 01.5	+6 46 48	−0.003	−0.08	7.70	0.5	3.4	F5 V		70 s	11432	m
170523	229095	δ² Tel	18 32 01.8	−45 45 26	0.000	−0.01	5.07	−0.14	−1.6	B5 IV	−6	220 s		
171125	86164		18 32 02.6	+25 04 00	−0.001	−0.01	7.70	0.2	2.1	A5 V		130 s		m
---	31033		18 32 06.7	+55 51 54	+0.001	+0.01	7.7	2.1	−0.3	K5 III		170 s		
170973	123610		18 32 06.8	+3 39 35	+0.002	+0.02	6.43	−0.04		A0 p				
170444	245536		18 32 07.9	−51 07 29	0.000	−0.02	7.6	−0.2	0.6	A0 V		260 s		
171067	103819		18 32 10.4	+13 44 12	+0.008	+0.13	7.2	0.8	3.2	G5 IV		62 s		
171461	31032		18 32 11.2	+52 06 56	−0.002	+0.01	6.4	−0.1		B9				
170901	142374		18 32 11.6	−9 22 13	0.000	−0.01	7.70	0.1		A0 p			11429	m
170820	161576		18 32 12.9	−19 07 26	−0.001	0.00	7.30	0.8	−2.1	G6 II	−13	530 s		
170640	210276		18 32 13.1	−37 12 44	−0.001	0.00	7.5	0.6	0.4	B9.5 V		120 s		
170641	210275		18 32 13.6	−38 05 55	−0.003	+0.02	8.0	0.0	3.0	F2 V		98 s		
170747	186932		18 32 13.9	−29 11 25	+0.001	0.00	6.9	0.5	4.4	G0 V		32 s		
170770	186937		18 32 15.5	−27 13 18	−0.001	−0.03	7.76	0.12	0.0	B8.5 V		300 s		
171275	67085		18 32 16.9	+36 25 33	+0.002	0.00	7.8	0.4	2.6	F0 V		110 s		
171992	17922		18 32 18.6	+67 46 36	−0.003	−0.01	6.7	1.4	−0.3	K5 III		260 s		
---	67087		18 32 20.0	+36 59 56	+0.001	+0.01	8.02	3.67		C6.5				T Lyr, v
---	---		18 32 20.5	+36 58 27			8.00							
170902	161580		18 32 20.6	−14 38 39	+0.001	−0.02	6.4	0.1	1.4	A2 V		91 s		
171383	47577		18 32 20.7	+45 46 28	−0.001	+0.01	7.0	1.4	−0.3	K5 III		280 s		
170642	210277		18 32 21.2	−39 42 14	+0.003	−0.04	5.16	0.08	1.4	A2 V n	−2	57 s		
170724	210283		18 32 23.4	−32 56 20	0.000	−0.01	7.64	0.25	1.7	A3 V		120 s		
170886	161582		18 32 29.0	−18 58 22	0.000	−0.01	6.95	1.38	−2.1	G3 II	−14	350 s		
171215	86169		18 32 29.3	+23 07 46	+0.001	−0.05	8.0	0.9	3.2	G5 IV		93 s		
172066	17924		18 32 30.0	+68 36 51	−0.004	+0.08	7.6	0.9	3.2	G5 IV		77 s		
169614	---		18 32 33.0	−72 41 36			8.0	0.8	3.2	G5 IV		90 s		
170750	210284		18 32 33.3	−35 20 58	+0.002	−0.01	7.0	1.1	0.2	K0 III		180 s		
171635	31039	45 Dra	18 32 34.2	+57 02 45	−0.001	0.00	4.77	0.61	−4.6	F7 Ib	−12	750 s		
171232	86172		18 32 35.9	+25 29 21	+0.002	0.00	7.5	0.9	3.2	G5 IV		74 s		
171438	47581		18 32 37.6	+46 05 33	−0.001	+0.01	6.7	1.1	0.2	K0 III		200 s		
170938	161585		18 32 37.7	−15 42 07	+0.001	−0.03	7.87	0.85	−6.6	B1 Ia	+27	1900 s		
171089	123619		18 32 38.3	+4 17 46	−0.001	−0.02	7.36	1.84	−0.3	K5 III	−4	210 s		
170975	161587		18 32 43.0	−14 51 56	0.000	0.00	5.50	1.97	−0.1	K2 III	+1	44 s		
171245	86175		18 32 45.9	+23 37 01	+0.001	+0.01	5.84	1.46	−0.3	K5 III	−4	170 s		
171244	86178		18 32 47.7	+24 40 44	+0.002	+0.01	7.8	0.4	2.6	F0 V		110 s		
171439	47582		18 32 48.0	+44 05 11	−0.001	+0.01	7.9	1.1	−0.1	K2 III		390 s		
171567	31038		18 32 48.0	+52 23 45	−0.005	+0.04	7.9	1.1	0.2	K0 III		200 mx		
171301	67090		18 32 49.8	+30 33 15	+0.001	+0.01	5.48	−0.10	−0.6	B8 IV	−10	160 s	11446	m
171384	67097		18 32 51.4	+38 50 00	0.000	−0.08	6.94	0.63	4.0	F8 V		33 s		m
170196	254280		18 32 52.9	−65 15 45	+0.002	−0.05	7.3	0.1	0.6	A0 V		220 s		
171363	67095		18 32 53.0	+37 25 54	+0.001	+0.02	7.9	0.1	1.4	A2 V		200 s		
169570	257601		18 32 55.3	−73 57 58	+0.004	−0.11	5.89	0.98	−0.1	gK2		160 s		
170773	210286		18 33 00.3	−39 53 32	+0.003	−0.09	6.23	0.42	1.7	F0 IV		68 s		
170991	161590		18 33 01.8	−16 33 39	0.000	−0.02	6.94	1.46	0.2	K0 III		120 s		
171109	142380		18 33 06.9	−5 09 46	+0.001	−0.02	7.1	0.9	3.2	G5 IV		58 s		
171170	123625		18 33 07.0	+3 57 22	0.000	−0.01	8.0	1.1	−0.1	K2 III		350 s		
171484	47586		18 33 08.4	+40 31 57	+0.003	+0.01	6.9	1.1	0.2	K0 III		220 s		
171012	161592		18 33 10.0	−18 22 07	0.000	−0.01	6.88	0.46		B0.5 Ia	−18			
170978	186959		18 33 14.1	−24 06 34	0.000	0.00	6.83	0.04	−2.3	B3 IV		490 s		
171286	103839		18 33 14.4	+18 54 54	−0.001	−0.09	6.8	1.1	0.2	K0 III		170 mx		
171485	47587		18 33 14.6	+40 09 35	−0.002	+0.06	7.26	0.51	3.4	F5 V		56 s		m
171219	123633		18 33 17.6	+5 26 43	0.000	−0.01	7.67	0.19	0.0	B8.5 V		250 s		
170865	---		18 33 18.6	−37 19 54			7.9	1.1	4.4	G0 V		24 s		
171128	142382		18 33 20.6	−8 34 00	+0.001	0.00	7.9	0.2	2.1	A5 V		130 s		
171149	142386		18 33 22.6	−5 54 41	+0.001	−0.02	6.36	0.02		A0				
170867	210295	κ² CrA	18 33 22.9	−38 43 34	0.000	−0.03	5.9	0.0	0.0	B8.5 V	−20	150 s		
171406	67102		18 33 22.9	+30 53 32	+0.001	+0.01	6.59	−0.12	−1.4	B4 V	−4	380 s		
170868	210296	κ¹ CrA	18 33 23.1	−38 43 13	+0.002	−0.03	5.90	0.00	0.4	B9.5 V	−16	130 s		
171247	123634		18 33 23.2	+8 16 06	0.000	0.00	6.42	−0.03		B8			11448	m
170525	245547		18 33 29.3	−58 42 33	+0.004	−0.12	6.44	0.68	2.8	G0 IV		47 s		
170993	186964		18 33 29.6	−28 06 09	+0.001	+0.06	8.0	2.3	−0.5	M2 III		200 s		
170845	229111	θ CrA	18 33 29.9	−42 18 45	+0.003	−0.02	4.64	1.01	0.3	G5 III	−2	60 s		
171234	123635		18 33 31.5	+1 54 13	0.000	0.00	7.95	0.20	1.7	A3 V		160 s		
170707	245551		18 33 33.0	−50 12 40	+0.001	−0.01	7.6	1.1	0.2	K0 III		300 s		
171129	161604		18 33 33.0	−12 42 17	+0.001	+0.04	7.8	0.5	3.4	F5 V		70 s		
171263	123641		18 33 35.9	+5 36 23	0.000	−0.02	7.76	0.19		A0 p				
171130	161605		18 33 38.7	−14 51 14	+0.001	−0.01	5.7	0.1	0.6	A0 V		98 s		
170943	210304		18 33 39.6	−36 48 20	+0.001	0.00	8.00	0.00	0.4	B9.5 V		330 s		m
171058	186975		18 33 45.1	−25 26 56	+0.002	−0.04	7.8	0.0	0.4	B9.5 V		280 s		
171365	103853		18 33 45.4	+17 43 54	0.000	−0.02	7.70	0.5	3.4	F5 V	−48	71 s	11454	m

HD	SAO	Star Name	α 2000	δ 2000	μ(α)	μ(δ)	V	B-V	M_v	Spec	RV	d(pc)	ADS	Notes
			$18^h\,33^m\,47.3^s$	$-71°\,46'\,08''$	$+0.001^s$	$0.00''$								
169979	257603		18 33 47.3	-71 46 08	+0.001	0.00	7.2	0.6	2.6	F0 V		52 s		
171654	47600		18 33 47.5	+46 13 09	0.000	+0.01	6.7	0.1	0.6	A0 V	-10	160 s		
171115	186981	24 Sgr	18 33 53.3	-24 01 57	0.000	-0.01	5.6	1.7	-0.1	K2 III	-14	69 s		
171703	47604		18 33 54.6	+47 52 14	-0.001	+0.01	7.7	1.1	-0.1	K2 III		360 s		
172199	17930		18 33 56.0	+66 21 23	+0.001	-0.04	7.2	0.1	0.6	A0 V		210 s		
171779	31051		18 33 56.5	+52 21 13	-0.001	+0.01	5.36	1.09	0.2	K0 III	-24	93 s	11468	m
171034	210312		18 33 57.6	-33 00 59	+0.001	-0.01	5.28	-0.11	-2.3	B3 IV	-17	310 s		
171060	210314		18 34 05.1	-32 57 40	+0.001	-0.01	7.17	0.22	0.4	B9.5 V		170 s		
170981	229115		18 34 05.1	-41 12 58	0.000	-0.01	7.5	0.1	0.4	B9.5 V		240 s		
171761	47608		18 34 06.4	+49 32 15	-0.002	0.00	7.9	0.9	3.2	G5 IV		88 s		
171520	86202		18 34 07.5	+28 17 42	+0.001	-0.01	7.5	0.1	0.6	A0 V		240 s		
171619	67119		18 34 08.7	+38 26 14	0.000	+0.04	7.6	1.1	-0.1	K2 III		340 s		
171568	67117		18 34 08.8	+34 26 52	+0.002	-0.03	7.3	1.1	0.2	K0 III	+3	270 s		
171279	142391		18 34 12.3	-7 42 44	0.000	0.00	7.2	0.0		B9 p				
171134	186988		18 34 12.7	-28 20 45	+0.003	+0.01	8.0	0.7	0.6	A0 V		100 s		
171407	103858		18 34 13.8	+14 28 07	+0.002	0.00	7.9	0.5	4.0	F8 V		58 s		
170407	254286		18 34 14.2	-66 16 46	0.000	-0.01	7.00	0.9	3.2	G5 IV		58 s		m
172611	9169		18 34 16.0	+72 24 40	-0.002	-0.01	7.3	1.1	0.2	K0 III		260 s		
171117	210318		18 34 16.1	-30 52 59	0.000	-0.02	7.1	-0.5	0.4	B9.5 V		220 s		
171000	229117		18 34 17.9	-42 35 27	0.000	+0.01	7.9	0.3	3.2	G5 IV		88 s		
171550	86204		18 34 19.4	+29 44 30	+0.003	-0.09	6.7	1.1	0.2	K0 III		170 mx		
171487	86203		18 34 19.4	+20 27 59	0.000	0.00	6.40	0.13	1.7	A3 V	-9	83 s		
171488	103862		18 34 19.9	+18 41 25	-0.002	-0.04	7.3	0.6	4.4	G0 V		39 s		
171871	31054		18 34 23.7	+51 06 48	0.000	+0.02	7.82	-0.17	-3.3	B2 III-IV	-58	1600 s		
170806	245556		18 34 25.0	-54 05 18	-0.001	-0.04	6.9	0.9	3.4	F5 V		27 s		
171367	123650		18 34 25.0	+3 49 50	-0.006	-0.05	7.68	1.12	0.2	K0 III		140 mx		
171236	161619		18 34 25.3	-19 47 06	+0.003	-0.02	7.3	1.2	0.2	K0 III		190 s		
172022	31058		18 34 26.1	+57 48 05	-0.002	-0.03	7.3	0.2	2.1	A5 V		110 s		
171119	210321		18 34 26.9	-34 48 57	0.000	+0.03	7.60	0.4	2.6	F0 V		98 s		m
171583	86207		18 34 29.0	+28 45 22	0.000	0.00	7.6	0.1	1.4	A2 V		180 s		
171620	67124		18 34 30.4	+34 24 56	+0.015	+0.20	7.9	0.5		F6 p	-32	30 mn		
171911	31055		18 34 30.6	+51 46 56	-0.005	-0.03	6.67	1.58	-0.5	gM4	-87	270 s		
170873	245559		18 34 31.0	-52 53 31	-0.001	-0.05	6.2	1.5	0.2	K0 III		83 s		
171427	123652		18 34 31.1	+6 27 46	0.000	+0.02	7.2	1.1	-0.1	K2 III	-19	260 s		
171533	86205		18 34 31.7	+21 10 40	-0.001	0.00	7.9	1.1	-0.1	K2 III		390 s		
171993	31057		18 34 32.5	+55 56 01	-0.002	-0.07	7.5	0.1	1.7	A3 V		150 s		
171175	210328		18 34 32.6	-31 15 46	0.000	-0.04	8.0	0.5	0.2	K0 III		370 s		
171237	186995	25 Sgr	18 34 32.6	-24 13 21	0.000	0.00	6.51	0.54	2.6	F0 V	+10	40 s		
171678	67126		18 34 34.6	+36 07 48	+0.001	-0.02	7.60	1.1	0.2	K0 III		300 s	11471	m
171388	123653		18 34 35.2	+3 08 09	0.000	0.00	7.5	0.2		A m				
171489	103864		18 34 35.4	+12 32 20	+0.002	+0.03	7.9	0.9	3.2	G5 IV		86 s		
171622	86211		18 34 38.0	+29 59 03	0.000	0.00	7.5	0.1	0.6	A0 V		240 s		
171679	67129		18 34 40.0	+35 48 16	+0.001	0.00	7.6	0.1	0.6	A0 V		260 s		
170946	245561		18 34 41.5	-50 43 21	-0.002	-0.02	6.7	1.3	0.2	K0 III		140 s		
171238	186998		18 34 43.5	-28 04 19	-0.002	-0.09	8.0	1.1	3.2	G5 IV		58 s		
171569	86209		18 34 46.9	+21 29 10	+0.002	0.00	7.2	0.1	0.6	A0 V		210 s		
171505	103867		18 34 47.3	+10 53 30	0.000	-0.01	6.40	0.08	1.2	A1 V	-36	100 s		
170984	229119		18 34 50.8	-49 21 08	+0.001	-0.03	7.3	0.8	0.2	K0 III		260 s		
171534	103870		18 34 51.1	+14 43 58	+0.001	+0.04	7.9	0.5	3.4	F5 V		79 s		
170948	245562		18 34 53.2	-52 23 16	+0.001	-0.07	7.4	0.3	0.2	K0 III		190 mx		
171039	229126		18 34 56.4	-48 09 04	+0.003	0.00	7.9	0.8	0.2	K0 III		340 s		
172041	31060		18 34 59.3	+54 12 33	0.000	+0.05	7.9	1.1	-0.1	K2 III		390 s		
171391	161632		18 35 02.2	-10 58 38	+0.003	0.00	5.14	0.92	0.3	G8 III	+7	93 s		
171827	67133		18 35 02.9	+39 31 49	0.000	-0.01	7.70	-0.01	0.6	A0 V		260 s		
170286	257605		18 35 03.0	-71 19 28	-0.002	-0.07	7.7	1.1	-0.1	K2 III		200 mx		
171347	161631		18 35 07.8	-16 59 09	+0.001	-0.01	7.0	0.1	1.4	A2 V		120 s		
170928	245564		18 35 09.1	-55 34 59	0.000	-0.02	7.1	1.9	-0.3	K5 III		170 s		
171443	142408	α Sct	18 35 12.2	-8 14 39	-0.001	-0.31	3.85	1.33	-0.2	K3 III	+36	55 mx		
171623	103879		18 35 12.4	+18 12 12	+0.001	+0.01	5.70	0.00	-0.1	B9.5 IV	-20	140 s		
171780	67134		18 35 13.3	+34 27 28	0.000	0.00	6.10	-0.11	-1.1	B5 V n	-13	280 s		
171886	47620		18 35 14.3	+41 34 54	0.000	+0.09	7.3	0.5	3.4	F5 V		60 s		
172783	9176		18 35 14.8	+72 16 27	+0.001	0.00	7.4	1.1	0.2	K0 III		270 s		
171206	210339		18 35 16.1	-39 41 39	+0.001	-0.02	7.0	0.2	2.6	F0 V		76 s		
171348	187010		18 35 20.1	-22 05 27	-0.004	0.00	7.95	0.18		B3	-9			m
171413	161636		18 35 20.3	-16 08 28	-0.001	-0.08	7.7	0.5	3.4	F5 V		68 s		
171430	161639		18 35 21.0	-12 11 07	0.000	-0.07	8.0	0.5	4.0	F8 V		59 s		
171369	187012		18 35 21.1	-20 50 27	+0.002	-0.01	6.5	0.9	2.1	A5 V		29 s		
171872	67143		18 35 21.4	+38 53 41	-0.001	0.00	6.96	0.03	0.6	A0 V		180 s		
171394	161635		18 35 23.4	-19 16 06	+0.001	-0.01	7.1	0.5	-0.5	M4 III	-24	330 s		
171296	210344		18 35 26.5	-32 53 26	+0.002	+0.02	6.9	0.4		A0 V		100 s		
171745	86224		18 35 30.2	+23 36 20	0.000	+0.01	5.61	1.00	0.3	gG8	+16	100 s	11479	m
171585	123670		18 35 30.8	+5 46 02	-0.003	-0.03	8.0	1.4	-0.3	K5 III		380 s		
171350	187014		18 35 33.7	-27 20 32	+0.002	-0.03	7.6	1.5	-0.5	M4 III		420 s		
171432	—		18 35 33.8	-18 33 28			7.12	0.21	-6.6	B1 Ia	+13	3200 s		
171586	123673		18 35 36.3	+4 56 09	+0.002	-0.03	6.44	0.08		A2 p	-11		11477	m

HD	SAO	Star Name	α 2000	δ 2000	μ(α)	μ(δ)	V	B-V	M_v	Spec	RV	d(pc)	ADS	Notes
171945	47625		18 35 36.8	+40 19 55	+0.001	+0.03	7.6	0.1	1.4	A2 V		180 s		
171944	47626		18 35 36.9	+40 46 12	0.000	+0.01	7.6	0.2	2.1	A5 V		130 s		
171352	187019		18 35 43.5	-29 56 14	+0.002	+0.01	6.9	1.4	0.2	K0 III		130 s		
172975	9182		18 35 45.6	+73 16 53	-0.006	0.00	7.5	0.8	3.2	G5 IV		71 s		
171746	103886		18 35 53.0	+16 58 32	+0.003	-0.07	6.21	0.53	4.7	G2 V	+10	20 s	11483	m
172922	9181		18 35 54.5	+72 22 48	+0.003	+0.01	6.9	1.1	-0.1	K2 III		250 s		
172043	47632		18 35 54.5	+43 03 20	-0.002	0.00	7.3	1.1	0.2	K0 III		260 s		
171799	86227		18 35 55.5	+22 45 20	+0.001	+0.01	7.6	0.1	0.6	A0 V		250 s		
171143	245575		18 35 59.0	-53 30 26	-0.002	-0.05	7.9	0.8	0.6	A0 V		85 s		
171416	187024		18 35 59.5	-29 41 56	+0.001	-0.01	6.4	1.0	0.2	K0 III		180 s		
171554	161645		18 35 59.9	-14 03 57	-0.002	-0.02	7.2	0.0	0.4	B9.5 V		200 s		
173269	9188		18 36 01.0	+75 30 24	+0.001	0.00	8.0	0.2	2.1	A5 V		150 s		
172803	9178		18 36 03.4	+70 28 02	-0.001	+0.02	7.6	1.1	0.2	K0 III		300 s		
171764	103889		18 36 06.4	+11 51 13	+0.001	-0.02	7.9	0.1	0.6	A0 V		270 s		
172006	67153		18 36 06.5	+36 58 08	0.000	0.00	7.5	1.1	0.2	K0 III		290 s		
171610	142415		18 36 07.7	-6 44 30	0.000	-0.01	7.0	1.1	-0.1	K2 III	-61	220 s		
171765	103891		18 36 11.4	+11 14 34	0.000	-0.03	7.6	0.1	0.6	A0 V		240 s		
172068	47639		18 36 11.9	+41 16 39	+0.001	+0.02	7.70	0.1	0.6	A0 V		260 s	11500	m
172023	67156		18 36 12.9	+38 02 03	+0.006	-0.03	7.4	0.5	3.4	F5 V		63 s		
172569	17947		18 36 13.0	+65 29 19	+0.002	+0.08	6.0	0.1	1.7	A3 V	-16	72 s		
171946	67152		18 36 14.7	+30 35 05	-0.001	+0.03	6.6	0.8	3.2	G5 IV		48 s		
171588	161653		18 36 17.3	-13 44 02	+0.002	-0.01	7.7	0.1	1.4	A2 V		160 s		
171374	210356		18 36 17.6	-38 08 33	-0.001	-0.03	7.30	1.6	-0.5	M2 III		360 s	11474	m
171873	103895		18 36 20.3	+20 00 24	+0.002	0.00	7.9	0.1	0.6	A0 V		290 s		
172669	17955		18 36 23.5	+66 54 45	+0.014	-0.08	7.6	0.7	4.9	G3 V	-6	35 s		
171451	210361		18 36 25.6	-35 26 50	-0.003	-0.04	7.04	1.65		Mb				
172148	47641		18 36 26.2	+43 47 48	-0.001	-0.01	7.7	1.1	-0.1	K2 III		360 s		
171706	142422		18 36 27.0	-4 34 31	+0.002	-0.20	8.0	0.5	4.2	F9 V	+20	56 s		
171802	123690		18 36 27.7	+9 07 21	-0.001	-0.13	5.39	0.37	3.0	dF2	-22	30 s		
171767	123688		18 36 28.9	+4 57 18	+0.001	-0.03	6.80	1.1	-0.1	K2 III		220 s	11494	m
171782	123689		18 36 29.5	+5 17 19	+0.001	-0.01	8.0	0.1		A0 p				
170691	254297		18 36 29.6	-69 04 52	+0.004	-0.02	8.0	0.4	2.6	F0 V		120 s		
172007	67162		18 36 33.6	+30 26 33	+0.001	-0.02	7.5	0.9	3.2	G5 IV		72 s		
172227	47646		18 36 33.6	+48 51 21	+0.001	+0.04	7.8	0.1	1.7	A3 V		160 s		
170825	254302		18 36 36.2	-66 53 19	+0.003	-0.03	7.9	1.1	-0.1	K2 III		400 s		
171874	103897		18 36 36.2	+12 59 12	-0.002	-0.02	7.4	0.5	3.4	F5 V		62 s		
171948	86239		18 36 36.9	+22 06 27	-0.001	-0.01	6.70	0.00	0.4	B9.5 V		180 s	11498	m
172044	67164		18 36 37.0	+33 28 08	-0.002	0.00	5.42	-0.10	-3.4	B8 II p	-27	580 s	11504	m
171834	123693		18 36 38.9	+6 40 19	-0.002	-0.14	5.45	0.37	2.8	dF1	-21	33 s	11496	m
171611	187040		18 36 39.2	-20 18 51	0.000	-0.01	7.40	0.00	-1.0	B5.5 V	-23	450 s	11487	m
170487	257607		18 36 41.1	-72 47 15	-0.014	+0.08	7.2	1.1	-0.1	K2 III		140 mx		
171835	123695		18 36 41.1	+6 02 32	0.000	-0.01	8.0	0.1	1.4	A2 V		190 s		
171212	245584		18 36 41.3	-56 13 35	0.000	-0.06	6.6	1.7	0.2	K0 III		76 s		
171913	103903		18 36 42.0	+15 26 52	+0.001	+0.02	6.9	1.1	0.2	K0 III	-30	200 s		
172187	47644		18 36 45.4	+43 13 18	+0.002	-0.01	6.3	0.2	2.1	A5 V	+2	68 s		
171662	161660		18 36 47.5	-15 56 42	+0.002	+0.02	7.40	1.4	-0.3	K5 III	-18	280 s		
172149	67171		18 36 48.7	+38 31 15	+0.001	0.00	7.9	0.5	3.4	F5 V		78 s		
171974	103905		18 36 48.9	+18 56 48	+0.001	+0.01	7.4	1.4	-0.3	K5 III	+5	330 s		
172391	31073		18 36 52.8	+54 48 51	0.000	+0.02	7.6	1.1	0.2	K0 III		310 s		
171852	123697		18 36 54.0	+3 47 23	0.000	0.00	6.6	1.1	0.2	K0 III		170 s		
171191	245585		18 36 55.0	-58 35 10	+0.002	+0.01	7.5	1.1	0.2	K0 III		260 s		
172824	17957		18 36 55.9	+67 48 18	-0.002	+0.05	7.4	0.1	0.6	A0 V		230 s		
172167	67174	3 α Lyr	18 36 56.2	+38 47 01	+0.017	+0.28	0.03	0.00	0.5	A0 V	-14	8.1 t	11510	Vega, m
171512	210367		18 36 57.0	-37 24 50	0.000	+0.02	8.0	0.5	1.4	A2 V		100 s		
171887	123698		18 36 59.2	+6 24 58	0.000	-0.02	7.2	1.4	-0.3	K5 III	-21	290 s		
171377	229142		18 37 01.8	-48 12 34	+0.001	+0.03	7.2	0.2	4.4	G0 V		37 s		
172150	67175		18 37 02.2	+37 24 41	+0.001	-0.01	8.0	1.1	-0.1	K2 III		410 s		
171836	142432		18 37 02.4	-4 36 56	0.000	+0.01	7.9	0.4	2.6	F0 V		110 s		
171627	187048		18 37 03.2	-28 30 47	+0.002	-0.14	6.79	0.98	6.1	K1 V		12 s		
169506	—		18 37 05.4	-79 38 29			8.0	0.8	3.2	G5 IV		90 s		
171736	161665		18 37 06.9	-19 12 33	0.000	0.00	7.20	1.1	-0.1	K2 III		280 s	11497	m
171994	103913		18 37 08.8	+16 11 54	+0.001	+0.05	6.29	0.90	3.2	G8 IV	-46	40 s		
171770	161668		18 37 09.0	-15 01 34	+0.005	-0.02	7.6	0.4	3.0	F2 V		78 s		
171888	123699		18 37 09.3	+0 57 04	+0.001	+0.02	7.1	0.5	4.0	F8 V		40 s		
172085	86248		18 37 10.2	+24 25 59	-0.005	-0.13	7.5	0.5	4.0	F8 V	-37	38 mx		
171914	123701		18 37 12.0	+2 58 34	-0.001	-0.02	8.0	0.1		A0 p				
171975	103912		18 37 12.4	+11 25 17	+0.002	-0.01	6.40	-0.02	-0.2	B8 V	-27	190 s		
171665	187054		18 37 12.8	-25 40 16	+0.014	-0.27	7.46	0.68	5.2	G5 V		28 s		
171930	123704		18 37 14.3	+5 48 08	+0.001	-0.01	8.0	1.1	0.2	K0 III		320 s		
171951	123709		18 37 15.2	+7 31 41	+0.004	-0.08	7.70	0.7	4.4	G0 V		45 s	11506	m
171545	229150		18 37 15.5	-40 48 12	-0.001	-0.02	6.8	1.7	0.2	K0 III		83 s		
171643	187053		18 37 17.2	-29 28 27	-0.001	-0.01	7.9	-0.3	3.4	F5 V		78 s		
172010	103918		18 37 20.3	+11 21 19	0.000	-0.01	7.0	0.1	0.6	A0 V		180 s		
173831	9199		18 37 24.1	+77 41 12	+0.005	+0.01	7.60	0.4	2.6	F0 V		100 s	11584	m
171953	123711		18 37 24.9	+2 37 46	-0.002	0.00	7.80	0.5	3.4	F5 V		73 s	11507	m

HD	SAO	Star Name	α 2000	δ 2000	μ(α)	μ(δ)	V	B-V	M_v	Spec	RV	d(pc)	ADS	Notes
			^{h} ^{m} ^{s}	$^{\circ}$ $'$ $''$	^{s}	$''$								
171502	229148		18 37 25.7	−47 04 27	+0.001	0.00	7.00	0.4	3.0	F2 V		63 s		m
171952	123712		18 37 26.8	+3 13 12	−0.001	0.00	7.9	1.4	−0.3	K5 III		360 s		
172169	86256		18 37 28.7	+29 35 05	−0.001	+0.09	6.7	1.4	−0.3	K4 III		220 mx		
171917	142439		18 37 29.5	−2 55 12	0.000	0.00	6.6	1.1	0.2	K0 III		170 s		
172243	67180		18 37 31.2	+37 46 50	+0.002	+0.01	7.1	1.4	−0.3	K5 III		310 s		
172045	103925		18 37 32.3	+14 52 27	+0.003	0.00	8.0	1.1	0.2	K0 III		330 s		
171854	161675		18 37 32.7	−12 21 01	−0.003	−0.06	6.9	0.5	4.0	F8 V		38 s		
172728	17958		18 37 33.2	+62 31 35	−0.002	+0.04	5.6	0.1	0.6	A0 V	−11	99 s		
172379	47661		18 37 34.6	+46 09 00	−0.003	+0.04	7.3	0.4	3.0	F2 V		74 s		
172392	47662		18 37 34.9	+47 58 55	−0.003	+0.06	7.9	0.1	0.6	A0 V		130 mx		
171978	142444		18 37 35.7	−0 18 34	+0.001	−0.02	5.75	0.07		A2	+12			
172468	31077		18 37 40.9	+51 56 45	0.000	0.00	7.8	1.1	−0.1	K2 III		380 s		
171918	142443		18 37 42.5	−7 20 13	+0.008	−0.08	7.9	0.7	4.4	G0 V		49 s		
171696	210378		18 37 42.9	−34 32 44	−0.005	−0.02	7.5	0.7	0.0	B8.5 V		79 mx		
171616	229154		18 37 44.0	−41 36 51	−0.001	−0.03	7.3	0.7	2.6	F0 V		49 s		
171954	142446		18 37 45.2	−4 48 24	−0.002	+0.05	6.7	0.4	3.0	F2 V		54 s		
172134	103929		18 37 48.3	+18 26 29	+0.002	0.00	7.5	1.1	−0.1	K2 III		310 s		
172046	123726		18 37 48.5	+5 53 01	+0.002	+0.01	6.6	0.0	0.0	B8.5 V		190 s		
171893	161683		18 37 51.4	−17 13 53	0.000	0.00	6.8	0.4	3.0	F2 V	−5	56 s		
170776	257611		18 37 51.8	−71 58 27	−0.005	+0.01	8.0	1.6	−0.5	M2 III		420 mx		
−−	67185		18 37 53.1	+35 05 42	−0.001	−0.04	8.0	1.1						
171856	187071		18 37 54.2	−21 23 52	0.000	−0.07	5.8	−0.1		A m				
−−	47664		18 37 55.1	+45 03 41	+0.004	−0.02	8.0	1.6						
171920	161686		18 37 55.2	−13 28 21	0.000	+0.01	7.3	0.7	4.4	G0 V		37 s		
172028	142454		18 37 56.6	−0 23 10	+0.002	−0.02	7.83	0.55	−3.9	B3 II	−13	870 s	11515	m
172013	142452		18 37 56.9	−2 35 26	0.000	−0.02	7.0	1.1	0.2	K0 III	−24	200 s		
172269	67184		18 37 57.0	+31 19 00	+0.001	0.00	7.7	0.9	3.2	G5 IV		80 s		
172244	86264		18 37 59.8	+28 37 49	0.000	+0.02	6.6	0.1	0.6	A0 V		160 s		
171957	161687		18 38 04.4	−14 00 17	+0.001	0.00	6.40	0.21		B9			11512	m
172380	67193		18 38 06.3	+39 40 05	0.000	+0.01	6.04	1.65	−0.5	M4 III	−19	200 s		XY Lyr, v
172188	103936		18 38 10.6	+16 30 49	+0.002	+0.01	7.8	0.4	2.6	F0 V		100 s		
171645	229160		18 38 17.5	−45 27 58	−0.005	−0.02	7.5	0.8	4.0	F8 V		35 s		
172103	142460		18 38 18.9	−1 06 48	−0.002	−0.01	6.66	0.42		F2				
171840	210384		18 38 19.7	−32 41 09	+0.002	−0.13	7.20	0.5	3.4	F5 V		57 s		m
172171	123744		18 38 20.8	+8 50 02	−0.001	−0.01	5.90	1.1	0.0	K1 III	−71	140 s	11524	X Oph, m,v
172212	103940		18 38 21.2	+14 22 51	−0.001	−0.01	8.0	0.5	4.0	F8 V		61 s		
172088	142461		18 38 23.4	−3 11 37	−0.001	+0.03	6.49	0.55	4.0	dF8	−21	30 s	11520	m
173084	17964		18 38 23.4	+67 07 37	−0.023	+0.20	7.73	0.63	4.5	dG1	−43	43 s	11568	m
172923	17963		18 38 23.9	+63 31 59	−0.005	+0.02	7.10	1.1	0.2	K0 III		240 s	11559	m
172825	17961		18 38 24.1	+60 42 40	−0.001	+0.06	6.70	0.4	3.0	F2 V		55 s		m
−−	31082		18 38 24.4	+55 26 43	−0.001	+0.01	8.0	1.8	−0.3	K5 III		310 s		
172394	67198		18 38 25.4	+36 03 13	+0.001	−0.03	7.50	0.2	2.1	A5 V		120 s	11534	m
172289	86270		18 38 26.0	+25 16 03	0.000	0.00	7.8	0.4	2.6	F0 V		110 s		
171960	187078		18 38 26.3	−23 06 12	0.000	−0.03	7.4	1.1	−0.1	K2 III		320 s		
171961	187080		18 38 30.5	−23 30 18	0.000	−0.02	5.7	0.0		B9				
171861	210391		18 38 30.8	−33 28 46	+0.001	−0.01	7.7	0.9	−0.3	K5 III		390 s		
172449	67206		18 38 30.9	+39 25 53	+0.002	−0.02	7.9	0.1	0.6	A0 V		280 s		
171648	229162		18 38 31.4	−48 43 16	+0.001	−0.03	6.5	0.9	0.2	K0 III		180 s		
172711	31083		18 38 31.8	+55 14 43	0.000	+0.07	7.6	0.5	2.4	F8 IV	−13	110 s		
172228	103942		18 38 34.3	+15 05 04	+0.001	+0.04	6.90	0.1	0.6	A0 V		170 s		m
172155	123745		18 38 34.5	+0 06 54	0.000	+0.02	8.02	0.46		F5				
171649	229163		18 38 35.6	−49 27 03	+0.003	0.00	7.9	0.5	2.1	A5 V		150 s		
171898	210396		18 38 36.0	−31 21 17	+0.003	+0.01	8.0	−0.1	2.6	F0 V		120 s		
172032	161689		18 38 36.5	−16 18 35	−0.001	+0.01	7.70	0.49	2.6	F0 V		79 s		
172246	103945		18 38 36.6	+16 32 27	−0.002	−0.05	7.77	0.60		G5			11530	m
172190	123756		18 38 39.7	+4 51 17	−0.001	−0.04	6.60	1.1	0.2	K0 III		180 s	11526	m
172381	67202		18 38 39.7	+30 27 07	+0.001	−0.01	7.30	1.6	−0.5	M2 III	−61	360 s		
171936	210399		18 38 42.0	−30 01 39	+0.003	+0.01	6.8	0.5	3.0	F2 V		50 s		
172395	67204		18 38 42.1	+31 26 29	0.000	−0.03	8.0	1.1	−0.1	K2 III		410 s		
171042	257612		18 38 42.8	−70 14 07	+0.007	−0.01	7.2	2.2	−0.3	K5 III		120 s		
171962	187081		18 38 47.5	−29 39 10	0.000	−0.02	7.9	−0.4	4.0	F8 V		60 s		
172713	31086		18 38 51.0	+52 20 39	+0.001	0.00	6.9	0.4	3.0	F2 V	+18	60 s	11558	m
171963	210403		18 38 52.2	−30 31 08	+0.002	0.00	7.9	−0.7	0.0	B8.5 V		380 s		
170506	257609		18 38 52.5	−76 29 56	0.000	−0.03	7.8	0.5	3.4	F5 V		75 s		
172051	187086		18 38 53.2	−21 03 06	−0.006	−0.15	5.9	0.7	3.2	G5 IV	+36	34 s		
172230	123760		18 38 54.9	+6 16 14	+0.002	0.00	7.1	0.2	2.1	A5 V		96 s		
171900	210400		18 38 55.3	−37 52 23	+0.002	−0.01	7.5	0.4	1.4	A2 V		99 s		
171881	210398		18 38 56.3	−39 31 11	+0.003	−0.01	7.9	1.2	3.2	G5 IV		48 s		
171650	245605		18 38 59.0	−52 59 10	−0.001	−0.06	7.7	1.3	0.2	K0 III		200 mx		
172052	187088		18 39 02.0	−23 10 56	0.000	−0.01	6.8	0.3	−4.6	F5 Ib		1900 s		
172137	161694		18 39 02.3	−15 02 20	−0.001	+0.02	7.7	1.4	−0.3	K5 III		310 s		
171902	210402		18 39 06.2	−39 41 52	−0.002	−0.01	7.5	1.5	−0.1	K2 III		210 s		
172312	103950		18 39 06.5	+11 01 36	0.000	−0.01	7.9	0.1	0.6	A0 V		260 s		
171106	257614		18 39 09.4	−70 48 16	+0.002	−0.04	7.12	1.26	0.0	K1 III		210 s		
172233	123764		18 39 10.1	+0 06 21	+0.005	−0.01	8.04	0.61		G0				

HD	SAO	Star Name	α 2000	δ 2000	μ(α)	μ(δ)	V	B-V	M$_v$	Spec	RV	d(pc)	ADS	Notes
			$18^h39^m11.3^s$	$-64°51'47''$	$+0.003^s$	$-0.02''$								
171381	254321		18 39 11.3	-64 51 47	+0.003	-0.02	7.8	1.5	0.2	K0 III		160 s		
172383	103953		18 39 12.4	+18 04 24	+0.001	+0.02	7.9	1.1	-0.1	K2 III		360 s		
172234	142471		18 39 13.3	-1 57 07	-0.002	-0.04	7.1	1.1	0.2	K0 III	-36	220 s		
172054	187089		18 39 13.7	-28 10 52	+0.002	-0.01	7.5	0.5	0.4	B9.5 V		130 s		
171819	229165		18 39 14.1	-47 54 35	+0.003	+0.02	6.0	0.4	2.1	A5 V		42 s		
172397	86287		18 39 14.6	+20 29 51	-0.001	-0.01	7.5	0.0	0.0	B8.5 V		290 s		
171864	229166		18 39 15.5	-46 29 10	+0.001	-0.03	7.9	0.0	0.2	K0 III		340 s		
171582	245603		18 39 16.6	-59 07 20	-0.004	-0.11	7.3	1.2	0.2	K0 III		110 mx		
171938	210404		18 39 16.7	-39 54 41	0.000	-0.07	7.6	1.2	0.2	K0 III		190 mx		
172016	210408		18 39 17.7	-34 10 00	+0.001	-0.03	6.7	1.1	0.6	A0 V		34 s		
172421	86288		18 39 19.0	+20 55 58	0.000	-0.01	7.80	0.00	0.0	B8.5 V		330 s	11546	m
172957	31092		18 39 19.2	+59 31 40	+0.001	-0.01	7.0	1.4	-0.3	K5 III		290 s		
172003	--		18 39 24.1	-37 18 01			7.7	1.5	-0.3	K5 III		400 s		
172057	210410		18 39 25.1	-32 02 02	+0.003	+0.03	7.9	0.8	3.2	G5 IV		85 s		
172203	161698		18 39 26.3	-12 14 51	-0.001	+0.01	7.7	0.1	0.6	A0 V		220 s		
171966	229170		18 39 27.9	-41 26 23	+0.001	-0.01	7.9	0.8	0.4	B9.5 V		310 s		
171882	229169		18 39 28.8	-45 44 54	+0.001	+0.06	7.40	1.1	0.2	K0 III		250 mx		m
172844	31091		18 39 30.0	+52 56 33	-0.002	0.00	7.9	1.1	-0.1	K2 III		390 s		
172059	210412		18 39 30.1	-32 59 42	0.000	-0.03	7.5	0.4	0.6	A0 V		130 s		
172613	67218		18 39 32.8	+37 31 01	+0.001	-0.02	7.7	1.4	-0.3	K5 III		390 s		
172671	47676		18 39 32.9	+40 56 06	+0.002	0.00	6.25	-0.06	0.6	A0 V	-15	140 s		
172499	86293		18 39 34.1	+26 35 15	0.000	0.00	7.8	0.4	2.6	F0 V		110 s		
172343	123773		18 39 34.7	+3 05 54	+0.001	0.00	8.0	1.1	-0.1	K2 III		350 s		
171967	229172		18 39 34.9	-43 11 10	-0.005	-0.05	5.2	1.5	-0.5	M4 III	+29	140 s		
172365	123778		18 39 36.8	+5 15 50	+0.001	-0.01	6.37	0.79	-3.3	F8 Ib-II	-19	800 s		
172422	103961		18 39 37.4	+11 51 39	0.000	-0.01	7.9	1.1	-0.1	K2 III		350 s		
172401	123779		18 39 39.5	+8 43 58	0.000	0.00	6.99	1.07	0.2	K0 III		210 s		
151285	--		18 39 39.8	-88 59 09			8.0	1.1	0.2	K0 III		360 s		
172649	67221		18 39 42.0	+37 59 34	-0.001	+0.04	7.5	0.5	3.4	F5 V		66 s		
172692	47678		18 39 42.4	+40 07 03	0.000	-0.05	8.0	1.1	0.2	K0 III		360 s		
172036	210414		18 39 43.9	-39 23 32	+0.001	+0.04	7.8	1.7	-0.5	M2 III		430 s		
171674	245612		18 39 47.4	-58 52 09	-0.005	0.00	6.9	0.9	0.2	K0 III		220 s		
171675	245611		18 39 48.5	-59 18 07	+0.002	-0.02	7.6	1.6	-0.5	M2 III		410 s		
172471	103966		18 39 49.3	+15 05 23	+0.001	+0.04	7.4	0.1	1.4	A2 V		150 s		
172424	123782		18 39 51.4	+7 21 30	0.000	-0.06	6.28	0.96	0.3	G8 III	-41	120 mx	11555	m
172883	31093		18 39 52.6	+52 11 46	0.000	+0.03	5.8	-0.1		B9				
172226	187108		18 39 53.5	-20 04 16	+0.003	-0.03	6.8	0.7	2.1	A5 V		40 s		
172502	103969		18 39 54.1	+17 36 10	0.000	0.00	7.6	1.1	0.2	K0 III		280 s		
172317	161706		18 39 56.3	-11 06 24	-0.003	-0.06	7.2	0.6	4.4	G0 V		35 s		
172450	123785		18 39 56.9	+9 58 22	+0.002	-0.04	7.9	1.6	-0.5	M4 III		430 s		
172503	103970		18 39 59.7	+14 09 29	0.000	-0.01	7.7	0.2	2.1	A5 V		130 s		
172348	142480		18 40 00.3	-7 47 27	0.000	-0.03	5.84	1.55	-0.3	gK4	-23	140 s	11552	m
172631	67226		18 40 01.7	+30 50 58	-0.001	+0.03	6.4	1.1	0.2	K0 III	-50	180 s		
171548	254324		18 40 02.8	-64 43 34	0.000	-0.05	7.6	0.8	1.4	A2 V		62 s		
172426	123783		18 40 03.9	+1 17 28	-0.003	-0.01	7.9	0.4	3.0	F2 V		91 s		
172472	123787		18 40 04.2	+7 45 46	+0.002	-0.01	7.7	1.1	-0.1	K2 III		320 s		
172672	67230		18 40 07.4	+31 39 30	+0.001	-0.04	8.0	1.1	0.0	K1 III		400 s		
171941	245625		18 40 08.4	-51 39 08	+0.001	-0.02	7.7	0.0	4.0	F8 V		54 s		
172542	103973		18 40 08.7	+16 33 28	+0.002	+0.03	8.0	0.1	1.7	A3 V		170 s		
172741	67233		18 40 12.0	+38 22 02	+0.002	0.00	6.5	0.1	1.7	A3 V	+17	91 s		
172715	67232		18 40 15.9	+34 25 00	-0.001	-0.01	7.7	1.1	0.2	K0 III		310 s		
172196	210430		18 40 18.1	-35 09 32	+0.001	-0.01	7.7	1.3	3.2	G5 IV		37 s		
172543	103975		18 40 18.6	+12 03 14	0.000	-0.02	7.1	0.1	0.6	A0 V		180 s		
171795	245622		18 40 20.6	-58 46 47	-0.002	0.00	6.9	1.2	3.2	G5 IV		30 s		
172742	67234		18 40 20.8	+33 41 04	+0.002	-0.06	7.2	0.5	3.4	F5 V	+6	58 s		
172558	103976		18 40 21.7	+13 22 11	+0.002	-0.03	7.5	0.7	4.4	G0 V		42 s		
172650	86307		18 40 21.7	+26 07 52	+0.001	+0.02	6.7	0.0	0.4	B9.5 V	-12	180 s		
172522	123793		18 40 22.4	+8 52 08	-0.001	+0.01	6.92	0.22	-0.2	A2 III		210 s		
171986	245626		18 40 24.5	-52 12 05	0.000	0.00	7.4	-0.1	1.4	A2 V		160 s		
172506	123794		18 40 31.2	+2 41 03	-0.002	-0.02	7.9	0.4	3.0	F2 V		89 s		
172238	210436		18 40 35.9	-33 51 26	+0.003	+0.01	7.1	0.5	0.2	K0 III		240 s		
172587	103979		18 40 41.8	+10 57 54	+0.001	-0.01	7.6	0.0	0.4	B9.5 V		260 s		
172144	229185		18 40 42.2	-44 10 20	+0.001	+0.02	7.39	0.55	3.0	G2 IV		76 s		
172128	229184		18 40 43.7	-47 01 38	0.000	+0.01	7.0	0.1	0.6	A0 V		160 s		
172297	210441		18 40 44.1	-30 55 09	-0.002	+0.01	8.0	1.2	1.4	A2 V		45 s		
172588	123802		18 40 45.9	+8 47 35	0.000	+0.05	7.23	0.35	-0.7	F0 II-III		370 s		
172508	142491		18 40 47.1	-4 29 48	-0.001	-0.02	7.61	1.81	-0.9	K0 II-III	-12	190 s		
172488	142489		18 40 47.9	-8 43 08	-0.001	-0.02	7.62	0.54		B0.5 V	+50			
171988	245630		18 40 50.7	-56 05 23	-0.006	-0.02	7.3	1.4	0.2	K0 III		160 s		
172475	161716		18 40 54.6	-14 30 20	+0.002	+0.01	7.97	1.60		B1			11564	m
173398	17975		18 40 56.1	+62 44 59	-0.001	+0.06	6.09	0.98	0.2	K0 III	-26	150 s		
172112	245632		18 40 59.3	-51 11 06	+0.004	+0.01	7.8	0.0	0.4	B9.5 V		270 s		
172945	47700		18 40 59.6	+41 55 02	+0.001	+0.02	7.7	0.1	0.6	A0 V		270 s		
172356	210453		18 41 00.5	-30 31 45	+0.002	-0.04	7.7	0.8	-0.1	K2 III		360 s		
172976	47702		18 41 03.2	+44 16 16	-0.001	0.00	7.4	0.4		F0 p	-8			

HD	SAO	Star Name	α 2000	δ 2000	μ(α)	μ(δ)	V	B-V	M_v	Spec	RV	d(pc)	ADS	Notes
169166	258802		18 41 03.6	-82 53 48	-0.031	+0.06	8.0	1.4	-0.3	K5 III		180 mx		
172388	187125		18 41 04.9	-28 15 19	0.000	+0.01	8.0	1.1	0.2	K0 III		320 s		
172281	210448		18 41 06.9	-39 05 18	+0.001	-0.01	7.4	2.1	-0.5	M2 III		200 s		
172693	103986		18 41 07.6	+12 12 37	-0.001	0.00	7.9	0.8	3.2	G5 IV		85 s		
172373	210457		18 41 08.2	-31 04 20	+0.001	-0.02	7.9	0.5	0.0	B8.5 V		160 s		
172530	161723		18 41 10.6	-13 51 48	+0.001	-0.01	8.0	0.1	0.6	A0 V		240 s		
173128	31103		18 41 11.4	+50 42 57	+0.005	+0.08	7.8	1.1	0.2	K0 III		200 mx		
172261	229191		18 41 12.9	-42 10 00	+0.001	-0.01	7.60	0.1	0.6	A0 V		250 s		m
172865	67250		18 41 16.3	+30 17 40	+0.001	-0.04	6.94	0.83		G5			11579	m
172358	210458		18 41 17.0	-36 18 30	+0.001	0.00	7.9	1.7	-0.1	K2 III		200 s		
172431	187128		18 41 18.2	-28 49 32	0.000	+0.01	7.7	0.2	0.0	B8.5 V		230 s		
172511	161724		18 41 19.7	-17 04 27	-0.002	+0.01	8.0	1.6	-0.5	M2 III		480 s		
172675	123808		18 41 20.2	+4 33 28	+0.004	-0.05	7.10	0.5	4.0	F8 V		41 s		m
172181	245637		18 41 21.0	-50 28 20	-0.002	-0.03	6.6	1.2	3.2	G5 IV		26 s		
172718	123811		18 41 21.6	+7 54 14	+0.003	-0.09	8.0	0.5	4.0	F8 V		61 s		
172440	187132		18 41 24.1	-27 57 27	+0.001	+0.01	7.5	2.5	-0.1	K2 III		52 s		
172651	123809		18 41 26.5	+0 33 51	-0.001	-0.03	7.48	1.43	-0.1	K2 III	+13	200 s		
174257	9215		18 41 26.8	+75 18 00	+0.005	+0.09	7.5	0.4	2.6	F0 V		94 s		
173024	47704		18 41 27.3	+40 28 36	-0.001	+0.07	8.0	0.9	3.2	G5 IV		93 s		
172744	103989		18 41 28.3	+12 14 11	0.000	0.00	7.0	0.9	3.2	G5 IV		57 s		
172763	103990		18 41 30.2	+12 58 30	-0.001	-0.01	7.6	0.1	1.4	A2 V		170 s		
172223	229194		18 41 30.4	-48 05 42	-0.002	-0.12	6.49	1.22	-0.2	K3 III		100 mx		
173052	47706		18 41 33.0	+41 23 21	-0.001	-0.02	8.0	0.1	0.6	A0 V		300 s		
173129	47711		18 41 33.2	+45 49 19	-0.001	+0.01	8.0	0.1	0.6	A0 V		300 s		
172654	142508		18 41 34.2	-3 07 07	+0.004	0.00	6.8	1.1	0.2	K0 III		190 s		
173109	47709		18 41 34.7	+43 33 00	-0.001	0.00	8.0	0.4	2.6	F0 V		120 s		
172826	103994		18 41 36.7	+17 41 00	0.000	0.00	7.9	1.4	-0.3	K5 III		390 s		
172977	67258		18 41 39.9	+34 05 55	-0.001	0.00	7.44	-0.16	0.4	B9.5 V		260 s		
173511	17977		18 41 40.5	+61 32 46	-0.002	+0.05	7.40	1.4	-0.3	K5 III	-8	350 s		
173086	47708		18 41 40.7	+40 43 04	+0.001	-0.04	6.7	0.4	2.6	F0 V		66 s		
172958	67256		18 41 41.1	+31 37 03	0.000	+0.01	6.41	-0.04	-0.2	B8 V	-16	200 s		
172594	161730		18 41 42.3	-14 33 51	0.000	0.00	6.5	0.5	3.4	F5 V		41 s		
173287	31111		18 41 44.8	+51 07 33	0.000	0.00	7.90	0.1	1.4	A2 V		200 s	11599	m
172483	210472		18 41 46.4	-31 10 13	-0.001	-0.03	8.0	0.8	3.2	G5 IV		89 s		
172827	103995		18 41 49.4	+12 15 21	0.000	+0.01	7.3	0.1	1.4	A2 V		150 s		
172546	187146	26 Sgr	18 41 51.5	-23 50 00	+0.002	-0.02	6.1	0.5	1.4	A2 V	+1	47 s		
172210	245640		18 41 52.3	-53 08 25	-0.004	-0.05	8.0	1.3	0.2	K0 III		210 mx		
172512	210474		18 41 57.0	-31 28 42	0.000	-0.01	7.4	0.7	0.2	K0 III		270 s		
172959	86332		18 41 59.9	+25 01 51	-0.001	-0.02	7.9	1.4	-0.3	K5 III		430 s		
172513	210475		18 42 02.2	-34 27 58	+0.001	-0.06	7.9	0.8	3.2	G5 IV		85 s		
172867	103998		18 42 04.7	+11 45 39	-0.004	-0.11	7.9	0.7	4.4	G0 V		50 s		
172806	123819		18 42 06.1	+4 02 03	0.000	-0.02	8.0	0.0	0.4	B9.5 V		290 s		
172412	229206		18 42 06.7	-43 55 33	-0.001	-0.03	7.9	0.7	0.2	K0 III		340 s		
173087	67265		18 42 08.0	+34 44 47	+0.001	+0.01	6.47	-0.13	-1.1	B5 V	-19	330 s	11593	m
173326	47722		18 42 08.2	+49 23 05	-0.004	-0.07	7.8	1.4	-0.3	K5 III		300 mx		
172535	210478		18 42 10.0	-32 22 04	+0.002	-0.04	7.79	-0.09	0.0	B8.5 V		360 s		m
172441	229210		18 42 10.1	-41 54 29	-0.008	-0.14	7.2	1.3	3.2	G5 IV		30 s		
172680	161738		18 42 10.2	-17 00 41	+0.002	-0.01	7.3	1.1	0.2	K0 III		260 s		
169904	258804		18 42 13.9	-81 48 28	+0.004	0.00	6.27	-0.13	-0.2	B8 V	-12	200 s		
172623	187156		18 42 14.2	-23 17 34	+0.001	0.00	7.4	1.5	-0.1	K2 III		200 s		
172748	142515	δ Sct	18 42 16.2	-9 03 09	0.000	0.00	4.72	0.35	1.3	F3 III-IV	-45	49 s	11581	m,v
172768	142518		18 42 17.8	-7 20 15	+0.002	0.00	7.0	1.6	-0.5	M2 III		270 s		
173169	67274		18 42 19.8	+36 33 05	0.000	-0.01	7.42	-0.08	0.4	B9.5 V		250 s		
172416	229208		18 42 22.0	-47 45 41	-0.008	-0.04	6.8	0.3	3.4	F5 V		48 s		
172021	254343		18 42 22.3	-64 38 35	0.000	-0.03	6.37	0.15		B3				
173088	86340		18 42 23.9	+28 07 29	-0.001	-0.01	7.5	0.1	0.6	A0 V		240 s		
--	31117		18 42 24.7	+55 37 35	+0.002	+0.02	7.5	1.5	0.2	K0 III		140 s		
172696	--		18 42 26.2	-20 18 30			7.10	0.19		B9				
171722	257619		18 42 30.1	-70 18 10	+0.004	-0.01	7.4	0.1	0.6	A0 V		230 s		
172831	142525		18 42 36.0	-7 04 24	0.000	0.00	6.15	1.00		G5				
173025	104010		18 42 36.0	+17 43 13	+0.001	+0.04	7.9	1.1	-0.1	K2 III		370 s		
173000	104008		18 42 36.9	+15 11 59	+0.001	0.00	6.9	0.1	0.6	A0 V	-6	170 s		
173524	31119	46 Dra	18 42 37.7	+55 32 22	-0.001	+0.02	5.04	-0.09	-6.8	A I pe	-30	2300 s		m
172601	210484		18 42 37.8	-33 37 55	0.000	+0.05	7.9	1.2	-0.1	K2 III		390 s		
172772	161751		18 42 39.6	-15 31 21	+0.001	-0.06	7.0	0.9	3.2	G5 IV		58 s		
173026	104011		18 42 44.6	+14 50 33	+0.001	+0.03	7.3	0.1	0.6	A0 V		210 s		
172961	123833		18 42 45.7	+5 41 14	-0.001	-0.07	7.6	0.5	3.4	F5 V		69 s		
173448	47730		18 42 46.7	+49 26 41	-0.002	-0.01	7.9	1.1	-0.1	K2 III		400 s		
--	86343		18 42 48.3	+20 40 40	0.000	0.00	7.9	0.8	3.2	G5 IV		88 s		
173605	31123		18 42 49.3	+57 51 54	+0.015	+0.01	7.8	0.8	3.2	G5 IV		84 s		
173415	47729		18 42 50.4	+47 34 31	+0.003	+0.04	6.9	0.1	1.4	A2 V	-17	130 s		
173170	86350		18 42 51.8	+28 18 11	-0.003	+0.01	7.5	0.0	0.4	B9.5 V		270 s		
172579	210488		18 42 53.2	-39 17 08	+0.003	0.00	7.80	0.00	0.0	B8.5 V		360 s		m
173054	104013		18 42 54.1	+13 34 00	+0.001	-0.01	7.4	0.1	0.6	A0 V		220 s		
--	86352		18 42 54.8	+28 38 45	0.000	-0.06	8.0	1.4	0.2	K0 III		210 s		

HD	SAO	Star Name	α 2000	δ 2000	μ(α)	μ(δ)	V	B-V	M$_V$	Spec	RV	d(pc)	ADS	Notes
172816	161754		18^{h}42^{m}54.9^s	-19°17'02"	0^s.000	-0".02	6.4	1.5	-0.5	M4 III		240 s		
173399	47727		18 42 55.0	+44 55 31	-0.003	-0.02	7.21	0.90	3.2	G5 IV	-35	53 s	11616	m
172603	--		18 42 55.7	-39 22 31			7.9	1.9	-0.1	K2 III		150 s		
173580	31122		18 42 56.1	+55 27 05	-0.001	+0.04	7.2	1.1	0.2	K0 III		250 s		
172751	187173		18 42 57.6	-23 58 01	+0.001	+0.05	7.7	-0.2	4.4	G0 V		47 s		
173155	86351		18 43 00.9	+22 50 27	0.000	-0.02	7.7	1.1	-0.1	K2 III		330 s		
173239	67279		18 43 01.4	+31 08 08	0.000	-0.01	7.91	-0.08	0.4	B9.5 V		320 s		
172702	210492		18 43 01.7	-32 45 00	+0.001	-0.07	8.0	0.3	2.6	F0 V		120 s		
171759	257620	ζ Pav	18 43 02.0	-71 25 42	+0.002	-0.16	4.01	1.14	-0.1	K2 III	-17	66 s		m
171043	257615		18 43 03.2	-78 03 57	+0.009	+0.02	6.8	0.2	3.4	F5 V		48 s		
173133	104019		18 43 05.0	+18 04 40	0.000	-0.01	7.7	0.0	0.4	B9.5 V		270 s		
173111	104016		18 43 05.2	+16 38 34	+0.004	0.00	7.9	0.5	3.4	F5 V		77 s		
173156	86354		18 43 05.9	+21 33 26	-0.001	-0.02	7.7	1.1	0.2	K0 III		290 s		
172517	229219		18 43 08.7	-48 49 28	0.000	-0.03	7.7	0.0	0.2	K0 III		310 s		
174205	9222		18 43 09.9	+70 47 34	-0.001	0.00	6.5	1.1	-0.1	K2 III	-5	210 s		
174123	17991		18 43 10.0	+69 20 59	-0.016	0.00	8.0	0.8	3.2	G5 IV		90 s		
173512	47733		18 43 13.4	+49 43 02	-0.002	0.00	7.9	1.4	-0.3	K5 III		440 s		
173171	104023		18 43 14.4	+19 27 59	+0.001	0.00	7.10	0.00	0.4	B9.5 V	-25	210 s	11609	m
173383	67287		18 43 16.6	+39 18 01	+0.001	0.00	6.60	1.4	-0.3	K5 III	-34	240 s		m
173003	142542		18 43 17.7	-1 38 46	-0.001	-0.02	7.7	0.0	-1.0	B5.5 V	-11	430 s		
173700	31134		18 43 17.8	+57 49 05	-0.010	-0.12	7.9	0.5	4.0	F8 V		60 s		
173172	104024		18 43 18.5	+17 29 22	+0.001	-0.03	7.2	0.1	0.6	A0 V		200 s		
172794	187180		18 43 19.5	-28 26 58	0.000	-0.01	8.0	1.4	0.2	K0 III		220 s		
172854	187185		18 43 19.9	-22 24 39	-0.002	0.00	7.70	0.00	0.4	B9.5 V		280 s	11595	m
173173	104026		18 43 22.6	+16 11 36	+0.001	0.00	7.7	0.2	2.1	A5 V		130 s		
173005	142544		18 43 23.9	-5 41 41	0.000	+0.02	7.8	1.1	-0.1	K2 III		310 s		
173242	86366		18 43 24.0	+23 06 40	-0.001	-0.02	7.9	0.9	3.2	G5 IV		86 s		
--	--		18 43 25.2	-55 46 00			7.60							m
173073	123846		18 43 26.2	+3 47 09	0.000	-0.03	7.9	0.1	0.6	A0 V		250 s		
172904	161764		18 43 26.4	-16 47 06	+0.001	0.00	8.0	0.1	0.6	A0 V		280 s		
173664	31133		18 43 28.8	+53 52 18	0.000	-0.01	6.1	0.1	1.4	A2 V	0	86 s		
173009	142546	ε Sct	18 43 31.1	-8 16 31	+0.001	+0.01	4.90	1.12	-2.1	G8 II	-11	230 s	11601	m
172982	161766		18 43 31.3	-11 20 13	0.000	0.00	7.1	1.4	-0.3	K5 III		250 s		
172905	187189		18 43 32.5	-20 55 09	0.000	-0.01	7.3	1.2	0.2	K0 III		200 s		
--	67289		18 43 33.0	+34 59 42	-0.003	-0.04	7.7	1.5						
173034	142549		18 43 34.0	-8 22 07	0.000	-0.02	7.0	1.1	0.2	K0 III	+42	200 s		
173416	67292		18 43 35.9	+36 33 24	+0.002	+0.07	6.01	1.04	0.2	K0 III	-61	140 s		
172211	254353		18 43 37.1	-64 33 04	+0.002	-0.03	5.78	0.96		K0				
173074	142553		18 43 37.2	-1 33 33	0.000	0.00	7.18	1.93	-0.5	M2 III		220 s		
172934	187196		18 43 42.7	-22 40 13	-0.002	-0.06	7.80	0.5	3.4	F5 V		75 s		m
172777	210501	λ CrA	18 43 46.7	-38 19 25	0.000	-0.06	5.13	0.09	1.2	A1 V	-26	56 s		m
173216	123860		18 43 51.0	+8 37 31	0.000	-0.02	7.2	0.5	4.0	F8 V	+20	44 s		
173093	142557		18 43 51.1	-6 49 07	+0.003	-0.06	6.31	0.48		F5				
173215	104033		18 43 51.1	+10 54 04	-0.002	0.00	6.8	1.4	-0.3	K5 III		240 s		
173417	67293		18 43 51.4	+31 55 35	-0.003	-0.13	5.70	0.34	2.6	F0 V	-2	40 s		
172984	161772		18 43 51.5	-16 56 30	0.000	-0.02	7.8	0.1	1.4	A2 V		180 s		
173581	47741		18 43 51.5	+45 45 07	-0.007	+0.02	7.8	1.1	-0.1	K2 III		150 mx		
173630	47744		18 43 54.2	+47 37 00	+0.001	-0.02	6.7	1.1	-0.1	K2 III		230 s		
172519	245660		18 43 55.2	-56 22 35	+0.002	-0.06	7.8	0.5	3.0	F2 V		71 s		
173272	104035		18 43 55.8	+12 10 07	-0.001	-0.02	7.2	1.1	-0.1	K2 III		270 s		
173525	67302		18 44 04.9	+38 31 59	-0.002	+0.01	6.7	1.4	-0.3	K5 III		260 s		
172875	210507		18 44 07.5	-36 43 07	0.000	-0.06	6.3	1.2	0.2	K0 III		130 s		
173059	161777		18 44 08.7	-19 19 03	+0.001	+0.03	6.7	0.8	3.0	F2 V		31 s		
173198	142563		18 44 12.6	-1 33 17	0.000	-0.04	7.78	0.31	-3.5	B1 V	-22	890 s		
173368	104044		18 44 13.8	+18 11 50	-0.001	-0.04	7.9	1.1	-0.1	K2 III		360 s		
173435	86384		18 44 14.2	+26 13 51	-0.001	+0.01	7.70	0.9	0.3	G7 III	-12	300 s		
173949	17995		18 44 18.0	+61 02 53	-0.001	+0.02	5.99	0.96	0.2	K0 III	-25	140 s	11661	m
172910	210509		18 44 19.2	-35 38 32	0.000	-0.04	4.87	-0.18	-2.5	B2 V	+3	300 s		m
173582	67309	4 ε¹ Lyr	18 44 20.1	+39 40 15	0.000	+0.06	4.67	0.16	1.7	A3 V	-31	38 s	11635	m
173471	86388		18 44 21.3	+28 06 05	+0.002	-0.01	7.3	0.4	2.6	F0 V		86 s		
173607	67315	5 ε² Lyr	18 44 22.7	+39 36 46	0.000	+0.06	5.1	0.2	2.1	A5 V	-24	38 s	11635	m
173666	47750		18 44 23.3	+44 53 27	+0.001	+0.06	7.9	0.4	3.0	F2 V	-35	94 s		
173451	86389		18 44 29.0	+23 30 46	+0.003	+0.08	7.8	1.1	0.2	K0 III		230 mx		
173493	86391		18 44 29.0	+27 31 54	+0.001	+0.04	8.0	0.9	3.2	G5 IV		93 s		
172689	245666		18 44 30.4	-53 48 06	+0.001	-0.02	7.3	0.6	0.4	B9.5 V		110 s		
173548	67311		18 44 30.5	+33 59 45	0.000	-0.02	8.00	0.5	4.0	F8 V		63 s		m
171425	--		18 44 32.1	-77 17 10			8.0	1.1	0.2	K0 III		360 s		
173219	142567		18 44 33.2	-7 06 38	+0.001	+0.01	7.81	0.22	-3.5	B1 V pne	+6	940 s		
174156	18003		18 44 34.6	+64 48 39	-0.001	+0.11	7.3	1.1	0.2	K0 III	+46	160 mx		
173701	47755		18 44 34.8	+43 50 02	-0.012	+0.03	7.54	0.84	0.2	K0 III		86 mx		
173631	67317		18 44 35.0	+39 04 16	-0.001	+0.01	7.8	0.1	1.4	A2 V		190 s		
172913	229240		18 44 36.7	-41 22 43	+0.001	-0.02	7.9	1.3	0.2	K0 III		200 s		
173329	123876		18 44 37.8	+1 52 28	0.000	-0.02	7.72	1.81	-0.5	M2 III		330 s		
172914	229241		18 44 39.2	-41 26 09	+0.008	-0.09	7.5	0.4	3.4	F5 V		68 s		
172952	210513		18 44 39.9	-38 07 42	0.000	-0.02	8.0	0.3	3.0	F2 V		98 s		

HD	SAO	Star Name	α 2000	δ 2000	μ(α)	μ(δ)	V	B-V	M_V	Spec	RV	d(pc)	ADS	Notes
			h m s	° ′ ″	s	″								
173494	86393		18 44 40.1	+23 35 23	+0.001	−0.09	6.31	0.40	3.0	F2 V	−12	44 s		
173278	142572		18 44 40.7	−6 32 10	0.000	0.00	6.63	1.90	−0.3	K5 III	−4	140 s		
173064	210519		18 44 41.9	−29 58 49	−0.001	0.00	7.50	1.1	−0.1	K2 III		310 s		m
173648	67321	6 ζ¹ Lyr	18 44 46.2	+37 36 18	+0.002	+0.02	4.36	0.19		A m	−26		11639	m
173649	67324	7 ζ² Lyr	18 44 48.0	+37 35 40	+0.002	+0.02	5.73	0.28	1.7	F0 IV	−24	64 s	11639	m
173117	187216		18 44 49.5	−25 00 40	0.000	−0.02	5.83	0.05	−1.1	B5 V		190 s		
173370	123879	4 Aql	18 44 49.7	+2 03 36	0.000	−0.02	5.02	−0.06	0.2	B9 V	−13	92 s		
173549	86395		18 44 49.9	+26 11 56	0.000	+0.01	7.8	0.0	0.4	B9.5 V		300 s		
--	47760		18 44 50.7	+43 33 46	+0.001	0.00	8.0	2.1						
173632	67322		18 44 54.0	+33 11 14	+0.002	−0.03	7.7	1.4	−0.3	K5 III		390 s		
173920	31151		18 44 55.2	+54 53 48	0.000	−0.02	6.23	0.82	3.2	G5 IV	+7	38 s		
173371	142582		18 44 55.7	−0 22 24	0.000	−0.01	6.8	0.0	0.4	B9.5 V	−18	170 s		
173526	86397		18 44 56.5	+22 33 44	0.000	−0.01	7.5	0.8	−2.1	G4 II	+12	700 s		
172991	210518		18 44 56.9	−39 41 11	+0.001	−0.01	5.4	0.8	−2.3	K3 II	−17	340 s		
173760	47764		18 44 57.4	+44 21 33	−0.001	−0.01	7.90	0.5	3.4	F5 V		79 s	11654	m
172993	210520		18 44 59.1	−39 45 01	−0.001	−0.18	6.7	0.5	4.4	G0 V		29 s		
173689	67325		18 45 03.4	+35 00 31	−0.001	+0.01	7.18	−0.02	0.4	B9.5 V		230 s		
173814	47767		18 45 03.6	+46 08 08	+0.001	+0.02	8.00	0.1	0.6	A0 V		300 s	11660	m
173320	161788		18 45 04.2	−11 27 17	+0.001	−0.02	7.0	0.0	0.0	B8.5 V		210 s		
173527	104059		18 45 07.2	+17 57 53	+0.005	+0.03	7.7	1.1	0.2	K0 III		210 mx		
174018	31154		18 45 08.8	+57 06 56	+0.003	0.00	7.6	1.1	0.2	K0 III		300 s		
173179	187225		18 45 08.9	−28 13 12	−0.001	+0.03	8.0	0.1	0.4	B9.5 V		260 s		
173741	67330		18 45 10.7	+38 18 52	−0.002	−0.06	7.60	1.4	−0.3	K5 III	−6	370 mx	11656	m
172630	254359		18 45 11.4	−61 05 42	+0.004	−0.02	6.1	1.6	−0.1	K2 III		92 s		
172916	229244		18 45 14.2	−48 59 22	−0.001	−0.04	7.4	0.3	3.2	G5 IV		70 s		
173564	104062		18 45 18.2	+17 43 39	0.000	+0.01	7.3	0.2	2.1	A5 V		110 s		
173282	187234		18 45 18.5	−21 00 05	+0.002	−0.02	6.4	0.7	3.4	F5 V		29 s		
173122	210532		18 45 18.5	−35 51 25	−0.001	−0.06	6.6	1.8	0.2	K0 III		65 s		
173761	67334		18 45 21.4	+36 34 03	−0.001	+0.02	6.9	0.0	0.4	B9.5 V		200 s		
172781	245681		18 45 23.5	−56 52 53	−0.006	0.00	6.2	2.2	0.2	K0 III		31 s		
172555	254358		18 45 26.5	−64 52 17	+0.003	−0.16	4.79	0.20	2.4	A7 V	+5	29 mx		
173550	104064		18 45 26.9	+11 35 50	−0.001	−0.05	7.9	1.4	−0.3	K5 III		390 s		
173742	67337		18 45 28.1	+34 12 48	+0.003	+0.03	7.9	0.9	3.2	G5 IV		88 s		
173495	123886		18 45 28.2	+5 29 59	+0.001	−0.01	5.83	0.04		A0	−10		11640	m
173633	104071		18 45 31.0	+19 58 12	−0.001	0.00	6.6	1.1	0.2	K0 III		180 s		
172917	245685		18 45 32.7	−51 52 07	+0.001	−0.02	7.9	0.8	0.2	K0₁ III		340 s		
172996	229246		18 45 33.6	−48 27 35	−0.003	−0.03	7.3	0.8	0.2	K0 III		260 s		
173388	161796		18 45 33.8	−12 35 12	+0.001	+0.03	7.7	1.1	0.2	K0 III		250 s		
173650	86405		18 45 35.5	+21 59 06	+0.001	+0.01	6.51	0.02		A0 p	−17			
172995	229248		18 45 36.6	−48 15 50	−0.002	+0.01	7.2	0.1	2.1	A5 V		100 s		
173375	161794		18 45 36.6	−17 32 43	−0.003	−0.03	7.16	0.19	−1.1	B5 V	−12	290 s		
175938	9256		18 45 38.0	+79 56 34	+0.007	+0.07	6.3	0.2	2.1	A5 V	−5	70 s		
173300	187239	27 φ Sgr	18 45 39.2	−26 59 27	+0.004	0.00	3.17	−0.11	−1.2	B8 III	+22	75 s		
173667	86406	110 Her	18 45 39.6	+20 32 47	−0.001	−0.33	4.19	0.46	3.7	F6 V	+24	15 ts	11658	m
173404	161799		18 45 39.8	−13 40 16	+0.001	−0.02	8.00	0.4	3.0	F2 V		93 s	11637	m
173609	104074		18 45 41.4	+15 43 30	+0.003	−0.01	6.6	0.1	1.4	A2 V	−21	110 s		
173610	104072		18 45 43.6	+11 29 48	0.000	−0.01	7.1	0.1	0.6	A0 V		180 s		
174980	9241		18 45 46.5	+74 05 07	0.000	+0.08	5.3	1.1	−0.9	K0 II-III	+3	77 mx		
174619	18014		18 45 47.7	+68 43 39	−0.003	0.00	8.0	0.1	0.6	A0 V		300 s		
173815	67350		18 45 49.6	+34 31 07	0.000	−0.01	7.20	0.2	2.1	A5 V		110 s	11669	m
173457	161805		18 45 51.2	−10 29 34	+0.001	−0.01	7.20	0.4	3.0	F2 V		66 s	11642	m
173165	229253		18 45 51.6	−41 28 12	+0.001	−0.05	7.4	1.5	0.2	K0 III		180 s		
174220	31160		18 45 51.6	+59 33 33	−0.002	+0.01	7.9	1.4	−0.3	K5 III		430 s		
173284	210543		18 45 55.5	−34 34 28	0.000	0.00	7.9	1.4	0.2	K0 III		210 s		
173047	245693		18 45 55.7	−50 52 22	0.000	−0.02	6.6	−0.7	0.0	B8.5 V		210 s		
173182	229259		18 45 57.9	−42 31 53	−0.002	0.00	7.50	1.05	0.3	G8 III		240 s		
173566	123897		18 45 58.2	+1 19 02	+0.001	+0.01	7.6	0.1	0.6	A0 V		230 s		
173425	161803		18 46 01.0	−19 36 23	0.000	0.00	6.4	1.5	−0.5	M4 III	−40	250 s		
173634	123901		18 46 01.0	+9 49 06	−0.001	0.00	7.4	0.5	3.4	F5 V		61 s		
174259	31162		18 46 01.6	+59 41 14	+0.007	+0.05	7.3	1.1	0.2	K0 III		260 s		
173183	229260		18 46 03.3	−42 34 00	+0.003	−0.02	6.92	0.42	1.9	F2 IV		92 s		
173378	187246		18 46 03.5	−27 29 56	+0.001	+0.03	6.7	1.1	3.2	G5 IV		31 s		
173780	86418		18 46 04.3	+26 39 43	+0.001	+0.02	4.83	1.20	−0.2	K3 III	−17	100 s		
173720	104078		18 46 04.3	+18 29 15	0.000	−0.01	6.8	0.1	0.6	A0 V		160 s		
173184	229262		18 46 08.7	−44 13 21	+0.001	−0.02	6.6	1.4	−0.3	K5 III		250 s		
174019	47784		18 46 10.2	+46 37 50	+0.002	0.00	8.0	0.1	0.6	A0 V		300 s		
173978	47781		18 46 12.0	+44 08 10	0.000	+0.03	7.9	0.2	2.1	A5 V		150 s		
173936	47779		18 46 12.8	+41 26 29	−0.001	−0.01	5.8	−0.1		B9				
172266	257626		18 46 15.6	−72 48 49	−0.012	−0.07	7.2	0.4	3.0	F2 V		69 s		
173258	229266		18 46 17.8	−41 55 06	−0.001	−0.03	7.1	1.0	0.2	K0 III		240 s		
173669	123904		18 46 19.5	+5 09 10	+0.001	−0.03	7.9	0.1	0.6	A0 V		260 s		
173460	187255	28 Sgr	18 46 20.4	−22 23 32	+0.002	0.00	5.80	1.1	−0.1	K2 III	−3	150 s	11652	m
175286	9250	50 Dra	18 46 22.0	+75 26 02	−0.005	+0.08	5.35	0.05		A0	−8	37 mn		
173980	47787		18 46 22.7	+41 41 18	−0.001	−0.02	8.0	0.1	1.4	A2 V		210 s		
174020	47790		18 46 27.9	+43 33 38	−0.001	0.00	7.5	1.4	−0.3	K5 III		370 s		

HD	SAO	Star Name	α 2000	δ 2000	μ(α)	μ(δ)	V	B-V	M_v	Spec	RV	d(pc)	ADS	Notes
173654	142607	5 Aql	18ʰ46ᵐ29.1ˢ	-0°57'48"	+0.001ˢ	0.00"	5.78	0.13		A m	+19		11667	m
173410	210559		18 46 30.7	-33 54 31	+0.002	-0.02	7.8	1.2	4.4	G0 V		19 s		
173431	210561		18 46 34.9	-31 54 27	+0.001	-0.01	7.28	1.01	0.2	K0 III		260 s		
173611	142605		18 46 36.5	-8 07 11	+0.001	+0.01	7.70	0.1	0.6	A0 V		220 s		m
173869	86425		18 46 36.5	+25 32 20	0.000	-0.01	7.93	0.17	0.6	A0 V		230 s		
173908	67369		18 46 39.2	+30 35 12	+0.002	0.00	7.7	1.1	-0.1	K2 III		360 s		
173357	210557		18 46 39.5	-39 43 54	+0.001	-0.05	6.7	1.6	-0.5	M2 III		270 s		
173833	104087		18 46 41.2	+18 42 21	+0.002	-0.02	6.17	1.59	-0.3	K5 III	-13	170 s		
174237	31165		18 46 42.9	+52 59 17	+0.001	0.00	5.88	-0.09	-1.7	B3 V	-20	320 s		
174546	18016		18 46 43.0	+63 15 59	+0.002	+0.04	7.2	1.1	0.2	K0 III		260 s		
173638	161817		18 46 43.1	-10 07 30	-0.001	0.00	5.71	0.59	-3.3	F2 Ib-II	+10	460 s	11670	m
174732	18022		18 46 45.8	+67 46 15	+0.001	-0.02	7.15	-0.03		A0				
173614	161816		18 46 46.2	-14 27 54	-0.002	-0.25	7.10	0.5	4.0	F8 V		38 mx		m
173673	142610		18 46 46.4	-6 41 24	+0.001	+0.01	7.65	0.01	0.0	B8.5 V		320 s		
173484	187263		18 46 47.1	-29 37 53	-0.001	-0.03	6.7	1.4	-0.5	M4 III		280 s		
173339	229271		18 46 47.8	-43 47 19	+0.007	-0.06	7.42	0.27	2.4	A7 V		80 mx		
173816	104088		18 46 49.9	+12 33 17	+0.001	+0.01	7.8	0.1	0.6	A0 V		250 s		
173640	161819		18 46 50.5	-14 54 47	0.000	-0.03	8.0	1.1	0.2	K0 III		350 s		
172801	254364		18 46 52.1	-65 14 59	+0.002	0.00	7.1	2.1	-0.3	K5 III		140 s		
173693	142612		18 46 54.5	-7 34 44	+0.001	+0.01	7.2	0.0	0.4	B9.5 V		200 s		
173799	123914		18 46 55.4	+9 25 09	+0.001	-0.01	7.66	1.85	-0.5	M2 III		300 s		
174177	47798		18 46 58.8	+46 18 54	-0.001	-0.01	6.52	0.07		A0	-1			
173263	245702		18 46 58.9	-50 05 39	0.000	-0.03	6.6	0.7	2.6	F0 V		38 s		
173834	104091		18 46 58.9	+12 31 00	-0.001	-0.01	7.9	0.1	0.6	A0 V		260 s		
173411	229276		18 46 59.7	-41 09 45	+0.001	-0.01	7.8	1.3	3.2	G5 IV		36 s		
173880	104093	111 Her	18 47 01.1	+18 10 53	+0.005	+0.11	4.36	0.13	1.7	A3 V	-45	34 ts		m
173881	104095		18 47 05.5	+16 31 33	+0.001	0.00	7.7	1.4	-0.3	K5 III		350 s		
175305	9252		18 47 06.2	+74 43 32	+0.080	+0.08	7.16	0.77	0.3	G5 III	-181	240 s		
174063	67380		18 47 07.6	+35 52 33	-0.001	+0.01	7.3	0.1	0.6	A0 V		220 s		
173744	142616		18 47 08.0	-5 53 55	0.000	0.00	7.05	0.09		A0				
—	31170		18 47 09.7	+53 49 45	0.000	+0.02	7.9	1.7	-0.3	K5 III		320 s		
173764	142618	β Sct	18 47 10.3	-4 44 52	-0.001	-0.02	4.22	1.10	-2.1	G5 II	-22	150 s		
173538	210572		18 47 11.7	-31 44 23	-0.001	+0.04	7.6	0.9	3.4	F5 V		36 s		
174022	67378		18 47 13.0	+31 24 20	0.000	+0.02	7.20	0.9	-2.1	G8 II	-22	720 s	11685	m
174064	67382		18 47 15.5	+34 05 52	-0.001	-0.02	8.0	1.1	0.2	K0 III		360 s		
173505	210570		18 47 18.5	-37 32 55	-0.003	-0.03	6.9	0.4	0.4	B9.5 V		110 s		
173921	104098		18 47 20.0	+16 54 31	+0.001	0.00	6.6	0.0	0.0	B8.5 V	-3	200 s		
174413	31173		18 47 20.9	+55 47 05	0.000	+0.01	8.0	0.1	0.6	A0 V		300 s		
174178	47802		18 47 22.2	+42 02 52	+0.002	-0.03	7.8	1.4	-0.3	K5 III		410 s		
173069	254368		18 47 22.8	-60 22 31	+0.003	-0.02	8.0	-0.6	0.4	B9.5 V		330 s		
173360	229277		18 47 23.9	-49 37 49	-0.002	-0.02	7.10	0.00	0.4	B9.5 V		220 s		m
173819	142620		18 47 28.8	-5 42 18	-0.003	-0.03	5.20	1.47	-5.4	G5 I-II var	+44	770 s		R Sct, v
173922	104103		18 47 29.1	+15 30 45	0.000	0.00	7.2	0.0	0.4	B9.5 V		210 s		
174344	47811		18 47 29.3	+49 25 55	+0.001	-0.01	7.2	0.1	0.6	A0 V	-18	210 s	11698	m
173678	187284		18 47 31.2	-24 19 11	-0.001	+0.01	7.9	1.5	0.2	K0 III		180 s		
173490	229281		18 47 33.5	-43 32 48	0.000	-0.02	8.0	1.0	-0.3	K5 III		450 s		
173657	187286		18 47 37.0	-28 16 47	+0.001	0.00	7.2	0.3	0.4	B9.5 V		140 s		
173621	210576		18 47 38.3	-32 33 42	-0.001	+0.02	7.5	1.0	0.2	K0 III		280 s		
174366	47815		18 47 39.8	+49 04 30	0.000	+0.02	6.4	0.1	0.6	A0 V	-17	150 s		
173952	104106		18 47 41.6	+13 26 31	-0.001	-0.01	7.1	0.0	0.4	B9.5 V	+3	200 s		
173883	142626		18 47 44.1	-0 14 11	-0.003	+0.04	8.0	0.5	4.2	F9 V	-82	55 s		
173540	229285	μ CrA	18 47 44.2	-40 24 22	+0.001	-0.01	5.24	0.78		G2	-18	18 mn		
173803	161833		18 47 44.7	-12 19 47	-0.001	-0.01	7.6	1.4	-0.3	K5 III		300 s		
173467	229283		18 47 46.3	-47 47 09	+0.001	+0.03	7.6	2.2	-0.1	K2 III		84 s		
171990	257625		18 47 49.3	-77 52 02	-0.003	+0.19	6.39	0.60	4.7	G2 V		22 s		
173645	210578		18 47 51.1	-35 29 54	-0.001	+0.02	8.0	1.0	3.4	F5 V		37 s		
173787	187294		18 47 52.2	-20 16 28	+0.001	0.00	6.8	0.2	-1.7	B3 V		300 s		V356 Sgr, m,v
173983	104109		18 47 53.4	+11 09 48	0.000	-0.02	7.3	0.0	0.4	B9.5 V		230 s		
174179	67396		18 47 57.3	+31 45 25	0.000	0.00	6.06	-0.13	-2.3	B3 IV p	-15	470 s		
173822	161835		18 47 59.0	-16 46 37	-0.001	-0.04	7.10	1.1	-0.1	K2 III		270 s		m
173954	123928		18 48 02.5	+4 14 28	0.000	0.00	6.21	1.51		K5				
174260	67399		18 48 02.8	+37 04 26	+0.001	0.00	7.34	-0.09	0.0	B8.5 V		290 s		
173167	254372		18 48 06.2	-62 13 47	+0.003	-0.08	7.2	0.5	4.0	F8 V		43 s		
173853	161840		18 48 07.6	-16 25 10	+0.002	0.00	7.9	1.1	0.2	K0 III		330 s		
173576	229290		18 48 07.9	-43 34 23	+0.002	-0.02	7.7	0.8	-0.3	K5 III		390 s		
173751	210585		18 48 10.8	-30 21 24	+0.003	+0.02	7.4	0.7	3.0	F2 V		46 s		
173885	161843		18 48 11.8	-14 42 27	-0.002	0.00	6.9	0.6	4.4	G0 V		31 s		
174159	86450		18 48 12.9	+23 42 33	-0.001	-0.01	7.8	0.0	0.4	B9.5 V		280 s		
173854	161842		18 48 13.2	-19 12 01	+0.001	-0.04	7.0	1.7	-0.3	K5 III	+7	230 s		
174043	104115		18 48 13.6	+10 38 34	-0.002	+0.01	8.00	0.9	3.2	G5 IV		89 s	11693	m
173732	210583		18 48 14.5	-33 54 31	+0.001	-0.03	7.1	0.5	1.4	A2 V		76 s		
174481	47823		18 48 15.9	+48 46 03	-0.002	+0.05	6.0	0.1	1.7	A3 V	-31	73 s		
174160	86451		18 48 16.2	+23 30 50	+0.001	-0.03	6.15	0.49	-6.4	F7 Iab	0	3200 s		
173660	210581		18 48 17.8	-39 36 11	-0.001	-0.01	7.0	0.5	0.6	A0 V		97 s		
173912	161844		18 48 20.0	-11 23 30	0.000	-0.03	7.3	0.9	3.2	G5 IV		62 s		

HD	SAO	Star Name	α 2000	δ 2000	μ(α)	μ(δ)	V	B-V	M_v	Spec	RV	d(pc)	ADS	Notes
			h m s	° ′ ″	s	″								
174321	67405		18 48 21.3	+37 49 12	-0.001	-0.04	7.8	1.1	0.2	K0 III		330 s		
173712	210584		18 48 21.5	-37 00 49	+0.004	-0.02	6.90	1.1	0.2	K0 III		220 s		m
173770	210588		18 48 22.4	-31 51 55	+0.002	-0.01	7.40	0.00	0.0	B8.5 V		300 s		m
174105	104118		18 48 22.5	+15 23 39	0.000	0.00	6.90	0.00	0.0	B8.5 V		220 s		m
174297	67403		18 48 27.5	+32 46 47	0.000	+0.02	7.5	1.4	-0.3	K5 III		360 s		
174345	67408		18 48 28.0	+36 26 27	0.000	-0.01	7.3	0.0	0.4	B9.5 V		240 s		
174296	67407		18 48 29.0	+33 46 24	-0.002	+0.02	7.8	1.1	0.2	K0 III		330 s		
174080	104119		18 48 29.2	+10 44 43	+0.010	-0.44	7.99	1.06	8.0	dK5	-18	16 ts	11696	m
174504	47824		18 48 36.0	+45 15 41	+0.002	+0.08	6.91	0.34	2.6	F0 V	-14	71 s		
173733	229296		18 48 37.2	-40 53 23	-0.002	-0.05	7.3	1.3	0.2	K0 III		260 s		
173168	254374	θ Pav	18 48 37.6	-65 04 40	-0.006	-0.08	5.73	0.24	2.5	A9 V	0	44 s		
174346	67410		18 48 38.6	+34 26 36	0.000	-0.03	8.0	0.9	3.2	G5 IV		93 s		
174005	142640		18 48 39.2	-6 00 17	0.000	+0.02	6.52	0.25		A2			11695	m
174223	104123		18 48 39.3	+19 48 35	0.000	0.00	7.9	0.0	0.4	B9.5 V		300 s		
174140	123939		18 48 43.8	+7 56 12	-0.002	+0.01	8.0	0.1	1.4	A2 V		190 s		
173928	161848		18 48 45.2	-18 36 05	-0.001	0.00	7.00	0.1	0.6	A0 V		190 s		m
173601	245724		18 48 45.3	-51 07 32	0.000	-0.01	8.00	1.1	0.2	K0 III		360 s		m
--	67418		18 48 46.2	+35 51 54	-0.002	-0.01	8.0	1.5						
173940	161850		18 48 48.5	-18 37 55	-0.002	+0.01	7.1	0.1	1.4	A2 V		130 s		
173697	229298		18 48 50.1	-45 15 46	+0.002	-0.05	7.26	0.99	1.8	G5 III-IV		100 s		
173715	229299	η¹ CrA	18 48 50.3	-43 40 48	+0.002	-0.02	5.49	0.13	1.4	A2 V n	-6	66 s		
173772	229300		18 48 51.0	-41 20 13	0.000	0.00	7.60	0.8	3.2	G5 IV		76 s		m
174262	104129		18 48 53.2	+19 19 43	+0.001	-0.02	5.80	0.03	1.4	A2 V	+6	76 s		
174298	86459		18 48 55.3	+24 03 22	-0.001	0.00	6.56	-0.07	-1.6	B3.5 V	-15	400 s		
174261	86458		18 48 57.7	+21 10 02	+0.001	0.00	7.10	0.00	-1.0	B5.5 V	-15	380 s	11715	m
174069	142648		18 48 58.8	-8 27 32	0.000	-0.01	7.81	0.13	-1.6	B3.5 V	+6	520 s		m
174029	161856		18 48 59.2	-13 54 19	0.000	-0.03	7.8	0.4	2.6	F0 V		110 s		
173875	210597		18 49 03.2	-32 42 38	+0.002	0.00	6.7	0.8	0.2	K0 III		200 s		
173344	254375		18 49 03.5	-63 16 11	+0.003	0.00	8.0	0.1	0.6	A0 V		300 s		
178564	3133		18 49 07.6	+83 54 19	-0.002	-0.02	6.8	0.1	1.4	A2 V		120 s		
174637	47830		18 49 07.7	+47 30 55	-0.002	-0.04	7.7	1.6	-0.5	M4 III		430 s		
229604	104137		18 49 10.7	+16 19 16	0.000	-0.01	7.9	1.1	0.2	K0 III		320 s		
174263	104135		18 49 11.0	+14 39 30	0.000	-0.01	7.1	0.1	0.6	A0 V		190 s		
174414	86464		18 49 13.2	+27 43 13	-0.001	+0.01	6.9	1.1	0.2	K0 III		220 s		
174369	86462		18 49 14.2	+25 02 47	0.000	-0.01	6.6	0.1	0.6	A0 V	-9	150 s		
174162	142655		18 49 15.1	-0 58 54	-0.001	-0.01	8.0	0.1	0.6	A0 V		260 s		
173902	210600		18 49 17.0	-34 44 55	+0.004	-0.10	6.5	1.8	0.2	K0 III		58 s		
174109	142654		18 49 19.6	-8 18 45	-0.001	-0.03	7.2	1.1	0.2	K0 III		220 s		
174621	47831		18 49 20.4	+43 44 03	-0.001	+0.04	6.8	0.8	3.2	G5 IV	-26	52 s		
173944	210603		18 49 21.7	-30 28 11	+0.003	+0.05	7.9	1.5	0.2	K0 III		180 s		
174530	67427		18 49 23.7	+33 42 03	-0.002	+0.03	6.9	0.1	0.6	A0 V		180 s		
173791	229306		18 49 27.0	-45 48 36	+0.006	+0.06	5.81	0.90	3.2	G6 IV		30 s		
174089	161870		18 49 28.3	-16 43 24	-0.002	-0.01	7.6	0.9	3.2	G5 IV	+19	76 s		YZ Sgr, v
174142	161876		18 49 31.9	-10 22 48	-0.001	-0.02	7.9	1.1	0.2	K0 III	-49	280 s		
173861	229307	η² CrA	18 49 34.8	-43 26 02	0.000	-0.02	5.6	0.5	0.4	B9.5 V	-23	50 s		
174115	161871		18 49 35.4	-19 08 32	+0.001	+0.01	6.75	0.20		A0				
174322	104147		18 49 36.3	+11 39 23	-0.001	-0.01	7.9	1.1	-0.1	K2 III		350 s		
174033	187318		18 49 36.6	-27 07 28	+0.001	+0.02	7.5	0.6	0.6	A0 V		98 s		
174240	123947		18 49 36.9	+0 50 09	0.000	-0.02	6.25	0.04		A0	-45			
174601	67444		18 49 37.9	+38 25 39	0.000	0.00	7.8	0.1	0.6	A0 V		280 s		
174116	187324	29 Sgr	18 49 39.9	-20 19 29	0.000	+0.04	5.40	1.1	0.2	K0 III	-18	110 s	11713	m
174208	142661		18 49 40.8	-5 54 46	0.000	-0.01	5.99	1.60		K0			11719	m
174391	104154		18 49 42.3	+15 56 06	+0.001	+0.01	6.69	-0.01	-1.7	B3 V	-7	400 s		
172881	257630		18 49 43.3	-72 59 41	+0.001	+0.03	6.06	0.00		A0				m
174349	104150		18 49 43.7	+10 31 13	+0.001	-0.01	7.50	1.4	-1.3	K4 II-III	-21	500 s		
174567	67438		18 49 43.8	+31 37 45	-0.001	-0.01	6.50	0.02	0.4	B9.5 V		160 s		
174585	67441	8 Lyr	18 49 45.8	+32 48 46	0.000	-0.01	5.91	-0.16	-2.3	B3 IV	-17	430 s	11732	m
173993	210607		18 49 46.7	-34 31 57	-0.004	-0.02	8.0	1.4	-0.5	M2 III		220 mx		
174371	104155		18 49 48.3	+11 10 03	0.000	0.00	7.9	0.1	0.6	A0 V		260 s		
174677	47834		18 49 49.7	+40 44 11	-0.001	+0.01	7.7	1.1	-0.1	K2 III		360 s		
174350	123950		18 49 50.1	+6 51 50	0.000	+0.03	7.89	1.16	-0.1	K2 III		400 s		
174602	67446	9 ν Lyr	18 49 52.7	+32 33 03	-0.001	-0.01	5.25	0.08	1.7	A3 V	+10	51 s	11737	m
174549	86481		18 49 55.6	+26 25 31	0.000	+0.03	6.90	0.1	1.4	A2 V		130 s	11733	m
173794	245738		18 49 57.1	-52 07 19	0.000	-0.01	7.1	0.1	1.4	A2 V		130 s		
174457	104160		18 50 01.9	+15 18 41	+0.008	+0.14	7.6	0.5	4.0	F8 V		52 s		
174586	86484		18 50 03.1	+29 48 54	+0.001	-0.01	7.50	0.00	0.0	B8.5 V		320 s	11741	m
174638	67452	10 β Lyr	18 50 04.6	+33 21 46	0.000	0.00	3.45	0.00	-0.6	B7 V	-19	92 mn	11745	Sheliak, m,v
174664	67453		18 50 06.5	+33 21 07	0.000	0.00	7.23	-0.08	-0.6	B7 V	-12	370 s		v
173972	229308		18 50 08.2	-42 12 41	0.000	-0.03	8.0	1.2	-0.1	K2 III		410 s		
174323	142673		18 50 08.3	-3 37 22	-0.003	-0.08	6.95	1.51		K5	-84			
176667	9273		18 50 11.5	+78 57 58	-0.010	+0.02	7.8	0.4	2.6	F0 V		110 s		
174170	187331		18 50 11.7	-22 50 56	-0.001	-0.02	7.4	1.1	3.2	G5 IV		45 s		
173795	245741		18 50 13.5	-54 21 35	0.000	-0.05	8.0	1.1	0.2	K0 III		240 mx		
174485	--		18 50 14.8	+11 31 22	0.000	+0.01	7.30	0.1	0.6	A0 V	-36	210 s	11735	m
175148	18039		18 50 17.6	+60 50 16	-0.006	+0.02	6.7	1.1	-0.1	K2 III		230 s		

HD	SAO	Star Name	α 2000	δ 2000	μ(α)	μ(δ)	V	B-V	M_v	Spec	RV	d(pc)	ADS	Notes
			^{h} ^{m} ^{s}	$^{\circ}$ $'$ $''$	^{s}	$''$								
174325	142674		18 50 19.8	−7 54 27	0.000	0.00	6.80	3.09		N7.7	0		11726	S Sct, m,v
174981	31196		18 50 21.4	+54 18 05	−0.001	0.00	7.3	0.0	0.4	B9.5 V		240 s		
174171	187332		18 50 22.0	−28 36 28	+0.002	0.00	7.5	1.8	3.2	G5 IV		17 s		
−−	31191		18 50 22.1	+51 08 48	−0.001	+0.04	7.8	1.9	−0.1	K2 III		140 s		
174623	86487		18 50 24.4	+24 06 22	−0.001	−0.01	7.0	1.4	−0.3	K5 III	−72	270 s		
174487	123962		18 50 25.4	+7 27 30	−0.001	−0.04	6.86	1.32		K4 III-IV p	−2			
174120	210619		18 50 30.3	−36 48 54	+0.002	+0.01	7.8	0.7	4.4	G0 V		41 s		
174829	47844		18 50 30.9	+42 54 42	0.000	+0.01	7.4	1.1	0.2	K0 III		280 s		
174328	161893		18 50 30.9	−13 34 20	−0.001	0.00	6.4	1.1	−0.1	K2 III		200 s		
173947	229309		18 50 32.4	−48 22 45	0.000	−0.01	8.0	0.7	0.2	K0 III		360 s		
174229	187336		18 50 37.3	−26 46 14	0.000	−0.03	7.20	0.1	1.4	A2 V		140 s	11724	m
−−	47852		18 50 39.8	+49 46 04	+0.007	+0.01	7.7	1.3						
174809	67463		18 50 39.9	+39 20 24	+0.001	−0.01	7.3	0.5	4.0	F8 V		45 s		
171192	258808		18 50 40.8	−82 44 00	+0.001	0.00	7.9	1.1	0.2	K0 III		350 s		
173994	229311		18 50 40.9	−47 46 45	−0.002	+0.01	7.08	−0.14	−0.2	B8 V k		290 s		
174880	47847		18 50 43.0	+43 57 26	−0.002	+0.02	7.2	1.1	0.2	K0 III		250 s		
174093	229316		18 50 44.9	−40 24 21	−0.001	0.00	7.9	1.3	0.2	K0 III		340 s		
174569	104170		18 50 45.5	+10 58 35	+0.001	+0.01	6.90	1.4	−0.3	K5 III	−24	260 s	11750	m
174196	210625		18 50 46.8	−34 18 17	+0.002	+0.01	7.2	0.4	0.4	B9.5 V		120 s		
174696	86496		18 50 49.9	+23 25 29	+0.005	−0.08	8.0	1.1	0.2	K0 III		120 mx		
174309	187342	30 Sgr	18 50 50.2	−22 09 44	−0.002	−0.03	6.61	0.38	0.5	A7 III	−35	130 s	11731	m
173933	245752		18 50 53.2	−52 18 02	+0.003	−0.05	7.7	−0.1	3.2	G5 IV		80 s		
175783	9265		18 50 55.4	+70 19 47	0.000	+0.03	7.6	0.4	2.6	F0 V		100 s		
174249	210631		18 50 57.7	−30 57 40	0.000	0.00	7.9	0.3	0.6	A0 V		200 s		
174152	229318		18 50 58.0	−41 03 45	−0.003	0.00	6.43	−0.09	−2.2	B5 III		500 s		
174464	142692		18 50 58.3	−9 46 27	0.000	0.00	5.83	0.61	0.6	gF2	−18	76 s		
174383	187349		18 50 59.8	−20 17 45	+0.001	−0.03	7.14	1.11	4.4	G0 V	+8	14 s		BB Sgr, v
174095	229317		18 50 59.9	−45 01 54	+0.002	−0.01	7.8	0.1		A m				
174830	67468		18 51 01.1	+34 23 40	+0.002	+0.03	6.9	0.1	1.4	A2 V		130 s		
174532	142696		18 51 05.2	−3 15 40	−0.003	−0.05	6.9	0.1	1.4	A2 V	−29	120 s		
174217	210630		18 51 05.8	−37 14 53	0.000	−0.01	7.6	1.4	−0.1	K2 III		250 s		
174403	187358		18 51 06.3	−20 18 01	+0.002	−0.01	7.37	0.30	−1.1	B6 IV-V var	+12	460 s		
174512	142695		18 51 07.8	−6 16 44	0.000	+0.01	8.00	0.00	0.0	B8.5 V	+24	320 s	11752	m
174199	210629		18 51 08.5	−39 10 31	+0.001	0.00	7.0	1.2	−0.1	K2 III		250 s		
170848	258807		18 51 09.7	−83 33 43	+0.004	+0.04	7.7	1.5	−0.1	K2 III		240 s		
175306	31219	47 o Dra	18 51 12.0	+59 23 18	+0.010	+0.03	4.66	1.19	−0.9	K0 II-III	−20	67 mx	11779	m
174515	142699		18 51 13.4	−8 00 38	+0.001	−0.02	7.3	1.1	0.2	K0 III	+5	230 s		
174606	123971		18 51 15.8	+0 49 58	+0.001	−0.03	7.9	0.4	3.0	F2 V		89 s		
174153	229322		18 51 19.5	−44 28 33	−0.001	−0.11	7.55	0.53	4.4	G0 V		43 s		
174467	161909		18 51 19.6	−17 09 15	0.000	0.00	6.8	0.1	1.4	A2 V		120 s		
174589	142706	8 Aql	18 51 21.9	−3 19 05	0.000	−0.02	6.10	0.30	2.2	dA6 n	+12	58 s		
174232	229324		18 51 22.7	−41 31 41	+0.002	−0.05	7.3	0.8	3.2	G5 IV		66 s		
174076	245754		18 51 23.4	−51 31 12	−0.003	−0.02	7.2	0.2	3.2	G5 IV		63 s		
174912	67481		18 51 25.0	+38 37 35	+0.027	+0.04	7.2	0.5	4.0	F8 V	−12	44 s		
174405	187363		18 51 26.3	−27 09 51	0.000	−0.03	7.3	1.2	0.2	K0 III		200 s		
174958	67484		18 51 26.7	+39 19 15	+0.001	+0.02	7.27	−0.01	0.6	A0 V		220 s		
175149	31209		18 51 26.8	+51 03 55	0.000	+0.01	8.0	1.4	−0.3	K5 III		450 s		
175055	47863		18 51 29.4	+45 08 04	0.000	+0.02	7.2	0.0	0.4	B9.5 V	−22	230 s		
173813	254387		18 51 31.3	−62 38 48	0.000	−0.03	7.6	0.1	0.6	A0 V		250 s		
175225	31217		18 51 34.7	+52 58 30	−0.005	+0.27	5.51	0.84	5.6	dG8	+2	14 ts		
174881	86512		18 51 35.8	+28 47 01	+0.001	+0.01	6.18	1.18	0.2	K0 III	−22	120 s		
174959	67485		18 51 36.2	+36 32 19	−0.001	−0.02	6.09	−0.11	−1.3	B6 IV	−21	300 s		
174361	210642		18 51 37.0	−35 22 59	−0.003	−0.03	7.80	1.1	0.2	K0 III		330 s		m
174336	210641		18 51 37.7	−36 27 42	−0.003	+0.02	7.4	1.6	−0.3	K5 III		310 s		
174060	245755		18 51 42.4	−55 08 49	+0.001	−0.01	8.0	1.1	0.2	K0 III		330 s		
174121	245759		18 51 45.5	−52 23 42	−0.007	−0.02	7.5	0.1	0.2	K0 III		180 mx		
174447	210646		18 51 48.1	−31 15 49	0.000	−0.05	7.6	0.8	3.2	G5 IV		75 s		
174719	123984		18 51 48.3	+3 01 51	0.000	−0.15	7.70	0.9	5.3	G6 V	−18	30 s		
175080	47869		18 51 48.3	+41 27 54	−0.002	0.00	7.3	1.1	0.2	K0 III		260 s		
175307	31222		18 51 53.2	+53 03 58	−0.001	−0.01	7.9	0.1	1.4	A2 V		200 s		
171161	258810		18 51 56.5	−83 18 56	−0.010	+0.01	7.16	1.26		K2				
175081	67494		18 52 01.3	+37 30 59	+0.001	−0.01	7.34	−0.07	−1.6	B3.5 V	−26	490 s	11777	m
174853	104196		18 52 01.7	+13 57 55	−0.001	−0.02	6.10	0.00	−0.2	B8 V		170 s		
175511	31232		18 52 02.1	+59 40 04	+0.001	+0.04	6.7	0.0	0.4	B9.5 V		180 s		
174963	86520		18 52 06.8	+27 15 38	0.000	+0.05	7.6	0.2	2.1	A5 V		130 s		
175132	47874		18 52 06.9	+41 22 59	−0.002	0.00	6.2	−0.1		B9				
174596	187381		18 52 08.1	−21 55 17	−0.001	−0.04	7.0	−0.4	1.4	A2 V		110 s		
174337	229331		18 52 08.6	−44 32 10	−0.003	−0.01	7.1	1.0	0.2	K0 III		230 s		
174407	229334		18 52 11.7	−41 42 33	−0.001	−0.03	6.50	1.1	−0.1	K2 III		210 s		m
173346	−−		18 52 12.0	−73 36 48			7.9	0.1	1.4	A2 V		200 s		
173948	254393	λ Pav	18 52 12.8	−62 11 16	0.000	−0.02	4.22	−0.14	−3.5	B1 V e	+20	350 s		m
174984	86526		18 52 14.8	+26 01 35	+0.001	0.00	7.8	0.4	3.0	F2 V		87 s		
−−	229337		18 52 15.7	−40 55 05	−0.002	−0.06	7.2	0.1	1.4	A2 V		60 mx		
174472	210651		18 52 16.2	−37 56 47	−0.003	+0.05	7.5	1.7	−0.1	K2 III		160 s		
175226	47879		18 52 16.2	+43 42 35	−0.003	−0.01	7.5	0.5	3.4	F5 V		67 s		

HD	SAO	Star Name	α 2000	δ 2000	μ(α)	μ(δ)	V	B-V	M_v	Spec	RV	d(pc)	ADS	Notes
			h m s	° ′ ″	s	″								
174933	86521	112 Her	18 52 16.3	+21 25 30	-0.001	-0.01	5.48	-0.07	-1.9	B9 II-III	-20	310 s		
174897	104203		18 52 18.3	+14 32 07	-0.001	-0.03	6.54	1.06		K1 III-IV	-15			m
174882	104202		18 52 18.9	+12 28 19	+0.001	-0.05	7.9	1.1	-0.1	K2 III		360 s		
174796	142731		18 52 25.5	-3 43 33	-0.001	-0.02	7.17	1.91	-0.3	K5 III	-42	180 s		
174387	229336		18 52 26.9	-46 35 42	+0.002	0.00	5.4	1.6	-0.5	M2 III	-28	150 s		
174139	245768		18 52 27.0	-57 11 44	+0.004	+0.01	7.8	0.7	0.6	A0 V		94 s		
174630	187388		18 52 28.3	-26 39 01	+0.001	-0.05	6.3	1.4	3.2	G5 IV		18 s		
174541	210655		18 52 32.4	-36 35 04	0.000	-0.12	7.9	1.1	4.4	G0 V		24 s		
174631	187389		18 52 36.8	-29 22 46	0.000	-0.04	6.2	1.6	0.2	K0 III		71 s		
174525	210657		18 52 37.5	-38 35 38	0.000	+0.01	8.0	0.8	3.2	G5 IV		89 s		
174295	245772	κ Tel	18 52 39.4	-52 06 27	+0.004	-0.10	5.2	1.2	0.2	K0 III	-44	74 s		
174834	123998		18 52 40.1	+4 03 12	0.000	0.00	8.0	0.0	0.0	B8.5 V		350 s		
174632	210663		18 52 41.5	-30 44 03	0.000	-0.02	6.6	-0.1	0.0	B8.5 V		210 s		
175172	67507		18 52 42.6	+35 28 47	-0.002	-0.02	7.4	1.1	-0.1	K2 III		320 s		
174913	124000		18 52 45.2	+9 14 23	0.000	-0.02	7.9	1.1	-0.1	K2 III		350 s		
174836	142738		18 52 45.4	-5 50 55	+0.001	+0.02	7.7	0.0	0.4	B9.5 V		240 s		
174633	210664		18 52 53.1	-34 33 18	-0.004	0.00	8.0	1.3	-0.1	K2 III		190 mx		
175085	86536		18 52 55.3	+21 27 23	0.000	-0.02	7.9	0.1	1.4	A2 V		190 s		
174709	187398		18 52 58.3	-28 08 40	+0.001	+0.01	7.7	1.7	0.2	K0 III		120 s		
175290	67517		18 52 59.0	+37 19 04	0.000	+0.03	8.0	0.5	3.4	F5 V		82 s		
174687	210669		18 52 59.1	-30 16 54	0.000	-0.02	8.0	2.0	-0.1	K2 III		130 s		
174500	229343		18 52 59.4	-46 35 09	-0.003	+0.03	6.4	0.6	1.4	A2 V		45 s		
175370	47888		18 52 59.6	+43 42 47	-0.002	-0.04	7.4	1.1	0.2	K0 III		270 s		
174801	161933		18 53 00.2	-16 22 55	0.000	-0.02	6.90	0.00	0.4	B9.5 V		190 s		m
176282	9276		18 53 00.4	+71 47 14	-0.002	+0.02	7.0	0.1	1.4	A2 V		130 s		
174866	142741		18 53 01.8	-9 34 34	+0.003	0.00	6.34	0.20		A3				
175013	124005		18 53 01.9	+8 43 27	0.000	+0.01	8.0	0.0	0.4	B9.5 V		300 s		
174474	229342		18 53 02.3	-48 21 36	0.000	-0.04	6.19	0.14		A0				
176523	9281		18 53 04.3	+74 05 17	-0.006	+0.06	7.9	0.9	3.2	G5 IV		88 s		
175015	124008		18 53 05.5	+7 53 09	0.000	-0.01	8.0	0.1	0.6	A0 V		270 s		
174429	245781		18 53 05.6	-50 10 49	+0.001	-0.08	7.9	0.1	3.2	G5 IV		88 s		
174802	161935		18 53 06.3	-18 38 12	-0.001	-0.02	6.90	1.1	0.2	K0 III		210 s	11776	m
174742	187403		18 53 07.4	-27 45 23	0.000	+0.03	7.70	0.2	2.1	A5 V		130 s	11774	m
174966	124004		18 53 07.7	+1 45 20	+0.001	+0.03	7.8	0.1	1.7	A3 V		150 s		
174916	142749		18 53 11.1	-4 43 55	0.000	+0.01	7.40	0.39		A m				
174430	245783		18 53 11.9	-51 55 52	+0.001	0.00	6.5	0.5	0.0	B8.5 V	-23	92 s		
175535	31241		18 53 13.4	+50 42 30	0.000	-0.02	4.92	0.90	0.3	G8 III	+8	84 s		
173545	257636		18 53 13.6	-73 37 46	+0.004	-0.01	7.1	-0.5	0.0	B8.5 V		270 s		
174409	245782		18 53 13.6	-52 35 19	+0.001	+0.02	7.9	-0.2	3.4	F5 V		78 s		
175331	67527		18 53 13.6	+38 55 51	+0.001	-0.02	7.80	-0.08	0.4	B9.5 V		300 s		
175056	104218		18 53 13.8	+10 10 45	0.000	-0.01	8.0	0.1	0.6	A0 V		280 s		
174886	161940		18 53 14.3	-10 13 13	+0.001	-0.01	7.73	0.17		B3 e	+4			
174887	161939		18 53 14.5	-11 21 04	+0.002	0.00	7.90	0.00	0.4	B9.5 V		300 s	11783	m
--	31243		18 53 16.1	+50 39 21	-0.001	-0.06	7.8	1.4	-0.3	K5 III		400 mx		
175404	47891		18 53 18.2	+40 59 43	-0.003	+0.02	6.5	1.6	-0.5	M4 III	+9	250 s		
175204	86543		18 53 20.1	+25 22 33	0.000	-0.02	7.70	0.9	0.3	G5 III		280 s		m
174476	245786		18 53 23.7	-52 15 18	-0.001	+0.01	7.8	-0.3	2.6	F0 V		110 s		
174689	210676		18 53 24.2	-37 16 26	-0.002	-0.02	7.1	0.3	0.6	A0 V		130 s		
175173	104225		18 53 27.7	+18 01 15	0.000	0.00	8.0	0.1	0.6	A0 V		280 s		
175037	124011		18 53 27.7	+0 52 13	-0.002	0.00	7.8	0.1	0.6	A0 V		240 s		
175466	47895		18 53 28.3	+42 54 09	+0.002	0.00	6.8	1.1	-0.1	K2 III	-11	240 s		
175490	47896		18 53 28.7	+44 07 45	+0.001	+0.06	7.7	0.4	3.0	F2 V		88 s		
175100	124013		18 53 29.8	+9 39 30	0.000	-0.01	7.32	1.87	-0.5	M2 III		250 s		
176795	9286		18 53 33.0	+75 47 15	+0.002	0.00	6.60	0.1	0.6	A0 V	-17	160 s	11870	m
175425	67535		18 53 38.0	+37 59 07	-0.003	-0.08	7.9	0.7	4.4	G0 V		51 s		
174805	210686		18 53 40.9	-31 59 31	+0.003	-0.01	7.0	1.9	-0.3	K5 III		180 s		
173491	257637		18 53 42.2	-74 57 28	+0.001	0.00	7.4	0.6	0.6	A0 V		98 s		
175426	67537	11 δ¹ Lyr	18 53 43.4	+36 58 18	0.000	0.00	5.58	-0.15	-1.6	B3.5 V	-26	270 s		m
175863	31253		18 53 44.5	+60 01 03	-0.001	-0.01	7.06	-0.14	-1.4	B4 V e	-20	490 s		
175823	31252		18 53 46.0	+57 29 12	+0.002	0.00	6.22	1.23	0.2	K0 III	-5	120 s		
174778	210685		18 53 47.0	-36 17 17	+0.001	-0.03	6.9	0.4	1.7	A3 V		68 s		
174919	161947		18 53 48.4	-19 47 49	0.000	0.00	7.2	1.0	0.2	K0 III		240 s		
175442	67540		18 53 52.3	+35 46 21	+0.001	-0.02	7.6	1.4	-0.3	K5 III		380 s		
178738	3146		18 53 53.3	+82 22 12	+0.004	0.00	6.9	0.1	0.6	A0 V		180 s		
175537	47897		18 53 53.8	+40 19 26	-0.003	+0.02	7.8	0.4	2.6	F0 V		110 s		
175058	142764		18 53 55.6	-9 22 15	-0.002	-0.02	6.9	0.1	1.4	A2 V		120 s		
176197	18066		18 53 59.0	+67 07 15	-0.002	0.00	7.5	0.1	0.6	A0 V		240 s		
175104	142767		18 53 59.8	-4 37 36	+0.001	+0.01	8.0	0.0	0.4	B9.5 V		280 s		
174947	187422	33 Sgr	18 53 59.9	-21 21 35	+0.001	-0.01	5.7	1.2	-6.1	K1 I	-4	2300 s		
174870	210694		18 54 00.3	-31 11 55	+0.001	+0.02	7.9	2.1	-0.1	K2 III		110 s		
174691	229354		18 54 01.3	-47 16 27	-0.001	-0.03	7.40	0.5	3.4	F5 V		63 s		m
175292	104236		18 54 07.6	+15 00 23	-0.001	-0.02	7.4	0.4	3.0	F2 V		74 s		
174973	187425		18 54 07.7	-22 34 12	+0.001	+0.02	7.4	0.3	0.0	B8.5 V		180 s		
174713	229356		18 54 08.8	-47 15 48	+0.001	-0.02	7.40	0.1	0.6	A0 V		230 s		m
174974	187426	32 ν¹ Sgr	18 54 10.0	-22 44 42	+0.001	-0.01	4.83	1.41	-6.0	cK2	-12	1400 s	11794	m

HD	SAO	Star Name	α 2000	δ 2000	μ(α)	μ(δ)	V	B−V	M$_v$	Spec	RV	d(pc)	ADS	Notes
175060	161952		18^{h}54^{m}10.7^s	−13°37′36″	+0.001^s	−0.02″	7.9	0.0	0.4	B9.5 V		310 s	11803	m
174675	229355		18 54 12.6	−49 40 10	−0.001	+0.01	8.0	0.3	0.2	K0 III		370 s		
175443	86558		18 54 13.0	+27 54 34	−0.001	−0.07	5.62	1.35	−0.1	K2 III	+15	110 s		
175538	67550		18 54 13.4	+36 44 01	0.000	0.00	7.8	0.0	0.4	B9.5 V		300 s		
175576	47902		18 54 14.1	+41 13 32	+0.003	+0.02	7.3	0.5	3.4	F5 V		61 s		
174994	187429		18 54 15.5	−22 41 14	+0.006	−0.01	7.9	0.4	3.4	F5 V		79 s		
175334	104241		18 54 21.4	+13 23 25	−0.003	0.00	7.21	0.04		B9	−17			
175293	104239		18 54 21.8	+10 48 32	+0.001	+0.03	6.8	1.1	−0.1	K2 III	−49	220 s		
174996	187431		18 54 22.7	−24 46 08	+0.001	0.00	7.2	0.1	0.0	B8.5 V		210 s		
176524	9283	52 υ Dra	18 54 23.6	+71 17 50	+0.009	+0.04	4.82	1.15	0.2	K0 III	−7	64 s		
175138	161960		18 54 26.2	−11 35 55	+0.002	−0.01	8.0	0.4	2.6	F0 V		120 s		
175405	86561		18 54 28.0	+20 21 36	+0.001	+0.01	7.1	0.9	3.2	G5 IV		58 s		
174453	254406		18 54 28.3	−60 37 17	+0.005	0.00	8.0	1.5	−0.3	K5 III		400 mx		
175309	104242		18 54 28.8	+10 37 58	0.000	−0.01	7.65	1.76	−0.5	M4 III		360 s		V913 Aql, v
174953	210700		18 54 28.8	−31 37 51	−0.004	+0.02	7.9	1.2	0.4	B9.5 V		60 s		
174779	229359		18 54 29.0	−45 29 10	+0.002	0.00	7.1	0.3	0.6	A0 V		130 s		
175840	31259		18 54 29.6	+52 42 05	0.000	+0.02	7.7	1.1	−0.1	K2 III		360 s		
175588	67559	12 δ² Lyr	18 54 30.0	+36 53 56	−0.001	+0.01	4.30	1.68	−2.4	M4 II	−27	220 s	11825	m,q
175987	31264		18 54 30.4	+58 44 08	−0.003	−0.08	8.0	0.4	3.0	F2 V		98 s		
175272	124028		18 54 30.5	+1 53 50	0.000	−0.07	7.60	0.5	3.4	F5 V		67 s		m
174730	229358		18 54 32.2	−49 52 43	−0.001	−0.01	6.6	0.6	1.4	A2 V		50 s		
175427	86563		18 54 32.7	+20 36 55	0.000	0.00	7.00	0.1	0.6	A0 V	−20	180 s	11816	m
175025	187432		18 54 35.2	−28 42 14	0.000	−0.02	7.7	2.1	−0.1	K2 III		100 s		
175336	124033		18 54 35.8	+9 41 09	−0.002	−0.04	7.4	1.1	0.2	K0 III		250 s		
175294	124030		18 54 36.2	+1 53 15	0.000	−0.01	7.80	1.1	0.2	K0 III		290 s		m
175046	187438		18 54 37.7	−26 53 23	−0.001	+0.01	7.9	0.2	0.0	B8.5 V		250 s		
174872	229367		18 54 39.7	−40 34 50	−0.001	−0.01	7.5	1.3	3.2	G5 IV		64 s		
175468	104251		18 54 40.7	+19 50 31	−0.001	−0.03	7.74	0.10	0.6	A0 V		230 s		
175156	161964		18 54 42.9	−15 36 11	−0.001	−0.01	5.0	0.1	−2.9	B3 III	−2	350 s		
175492	86567	113 Her	18 54 44.7	+22 38 43	0.000	+0.01	4.5	0.1	1.7	A3 V	−24	36 s	11820	m
164461	258799	χ Oct	18 54 45.5	−87 36 21	−0.052	−0.14	5.28	1.28	−2.3	K3 II	+34	330 s		`
174781	229361		18 54 45.5	−48 56 54	−0.002	−0.02	7.1	0.1	0.6	A0 V		160 s		
175824	47913		18 54 46.8	+48 51 35	−0.007	−0.12	5.77	0.43	3.4	F5 V	−11	30 s	11846	m
175428	104253		18 54 48.1	+15 20 32	0.000	−0.03	7.0	0.0	0.4	B9.5 V	−14	200 s		
175578	86571		18 54 50.8	+30 01 58	+0.002	+0.02	7.7	0.8	0.3	G5 III		300 s		
175634	67565		18 54 51.7	+33 58 51	0.000	−0.01	7.70	−0.06		A				
175740	47909		18 54 52.0	+41 36 10	0.000	0.00	5.44	1.03	0.2	K0 III	−9	110 s	11840	m
175589	67563		18 54 52.1	+31 08 50	−0.001	+0.01	7.3	1.1	−0.1	K2 III		300 s		
175635	67566		18 54 52.4	+33 58 07	−0.001	0.00	6.02	0.91	−0.9	G5 II-III	−16	240 s	11834	m
175900	31262		18 54 53.3	+51 18 30	+0.002	+0.08	8.0	0.9	3.2	G5 IV		93 s		
175337	124039		18 54 54.1	+1 07 17	−0.002	−0.03	7.7	0.5	3.4	F5 V		70 s		
175250	161974		18 54 54.5	−10 30 57	0.000	+0.01	6.8	0.0	0.4	B9.5 V		190 s		
175190	187445	35 ν² Sgr	18 55 06.9	−22 40 17	+0.007	−0.03	4.99	1.33		K3	−110	13 mn		
175114	210714		18 55 12.8	−32 27 12	−0.002	+0.05	8.0	0.8	3.4	F5 V		49 s		
175700	67571		18 55 13.5	+32 36 17	0.000	+0.02	7.5	0.6	4.4	G0 V		42 s		
175701	67572		18 55 14.7	+32 24 01	0.000	+0.01	7.66	−0.08	0.6	A0 V		260 s		
175297	161979		18 55 15.2	−14 41 30	−0.001	+0.01	7.2	0.4	3.0	F2 V		70 s		
175191	187448	34 σ Sgr	18 55 15.7	−26 17 48	+0.001	−0.05	2.02	−0.22	−2.0	B3 IV-V	−11	64 s		Nunki, m
175073	210713		18 55 18.6	−37 29 54	+0.011	−0.35	7.95	0.85		G5		16 mn		
174874	245806		18 55 19.6	−50 03 44	0.000	−0.04	8.0	0.6	2.6	F0 V		120 s		
175717	67575		18 55 19.7	+32 38 37	0.000	+0.02	7.6	0.7	4.4	G0 V		44 s		
175865	47919	13 Lyr	18 55 19.9	+43 56 46	+0.002	+0.08	4.04	1.59		M5 III	−28	40 mn		R Lyr, v
175128	210717		18 55 20.2	−32 32 10	−0.002	−0.24	7.9	1.4	4.4	G0 V		16 s		
175093	210716		18 55 23.5	−37 23 13	+0.001	−0.01	6.5	0.5	0.2	K0 III		180 s		
175613	104262		18 55 23.5	+19 12 54	0.000	0.00	7.8	0.1	1.4	A2 V		180 s		
175841	47918		18 55 25.5	+40 10 37	0.000	0.00	6.8	0.1	1.4	A2 V		120 s		
175193	187452		18 55 26.0	−29 12 47	+0.003	0.00	6.9	1.2	3.2	G5 IV		31 s		
175515	124050		18 55 27.2	+6 36 55	+0.001	−0.09	5.57	1.04	0.3	gG5	+23	85 s		
176251	18069		18 55 27.2	+60 56 46	−0.003	0.00	7.8	1.4	−0.3	K5 III		410 s		
175579	104261		18 55 28.0	+13 13 20	−0.001	0.00	7.04	0.05		A0	−19			
175592	104263		18 55 30.2	+14 23 05	+0.003	−0.03	6.6	0.4	3.0	F2 V		51 s		
175317	161984		18 55 30.8	−16 22 36	−0.002	−0.19	5.79	0.36	3.4	F5 V	−42	30 s		
175445	142803		18 55 33.1	−2 18 24	0.000	−0.01	7.79	0.11		A0				
175955	47924		18 55 33.3	+47 26 24	−0.004	−0.08	7.1	1.1	0.2	K0 III		240 s		
175516	124051		18 55 33.9	+4 26 07	+0.001	−0.03	7.5	0.8	3.2	G5 IV		72 s		
175884	47922		18 55 35.5	+42 03 36	−0.001	+0.03	6.50	1.31		K0				
172942	—		18 55 35.7	−80 11 53			8.0	0.5	4.0	F8 V		63 s		
175718	86589		18 55 41.1	+25 41 42	−0.001	−0.01	7.93	0.06	0.6	A0 V		270 s		
175343	187461		18 55 42.6	−20 00 59	+0.001	0.00	7.8	0.2	2.1	A5 V		140 s		
174785	245805		18 55 43.5	−58 14 33	+0.005	0.00	7.9	0.4	2.6	F0 V		96 s		
175382	161989		18 55 44.6	−16 20 58	+0.003	−0.05	7.1	0.5	3.4	F5 V		55 s		
175580	124060		18 55 44.9	+7 10 33	−0.002	−0.03	6.8	0.7	−6.2	G2 I		2000 s		
174851	245809		18 55 45.4	−56 25 25	+0.001	0.00	6.8	1.3	−0.3	K5 III		260 s		
175544	124055		18 55 46.5	+0 15 54	0.000	−0.01	7.35	0.10	−1.7	B3 V	−4	490 s		
175902	47923		18 55 46.5	+40 12 12	−0.002	+0.02	7.5	1.4	−0.3	K5 III		370 s		

465

HD	SAO	Star Name	α 2000	δ 2000	μ(α)	μ(δ)	V	B-V	M_v	Spec	RV	d(pc)	ADS	Notes
			h m s	° ′ ″	s	″								
175543	124058		18 55 48.4	+3 26 57	0.000	-0.02	7.07	0.16		A2			11842	m
175785	67588		18 55 49.1	+30 18 43	+0.001	-0.01	7.62	-0.01	0.6	A0 V	-24	250 s		
175432	161994		18 55 50.6	-14 48 55	+0.003	0.00	8.0	0.4	3.0	F2 V		97 s		
175545	142812		18 55 51.2	-0 44 22	+0.001	-0.09	7.50	1.1	-0.1	K2 III	-19	140 mx		
175518	142809		18 55 52.8	-5 44 43	-0.014	-0.39	7.46	0.75	4.6	K0 IV-V	-77	38 s		
175498	142807		18 55 53.1	-8 47 41	0.000	-0.01	7.6	1.1	0.2	K0 III		250 s		
175866	67595		18 55 53.2	+35 48 38	-0.003	0.00	6.7	1.1	0.2	K0 III		200 s		
176022	47930		18 55 54.3	+46 30 48	-0.001	+0.01	8.0	0.0	0.4	B9.5 V		330 s		
176003	47928		18 55 56.9	+44 13 42	+0.001	+0.03	7.20	0.1	1.4	A2 V	-13	150 s	11863	m
175260	210726		18 55 58.6	-33 15 28	+0.001	+0.02	7.4	0.7	3.2	G5 IV		70 s		
172882	258815		18 55 59.4	-80 43 40	+0.002	0.00	6.8	-0.2	0.6	A0 V		170 s		
175345	187465		18 56 00.2	-25 02 45	+0.002	+0.04	7.40	0.7	4.4	G0 V		40 s	11832	m
175637	124067		18 56 00.3	+7 07 35	-0.001	+0.01	7.3	0.4	2.6	F0 V		86 s		
175360	187468		18 56 00.4	-23 10 25	0.000	-0.01	5.8	0.0		B8				
175744	104271		18 56 03.7	+17 59 42	+0.001	-0.02	6.4	0.0		B9 pe				
175702	104268		18 56 05.6	+15 21 56	-0.002	-0.03	7.9	0.7	4.4	G0 V		51 s		
175743	104272		18 56 05.9	+18 06 19	-0.003	-0.16	5.69	1.09	0.0	K1 III	+44	130 mx		
175346	187467		18 56 08.6	-29 28 37	+0.001	0.00	7.2	0.6	1.4	A2 V		66 s		
175677	124069		18 56 09.8	+7 56 08	0.000	0.00	8.0	0.0	0.4	B9.5 V		300 s		
175638	124068	63 θ¹ Ser	18 56 13.0	+4 12 13	+0.003	+0.03	4.06	0.17	2.1	A5 V	-45	31 s	11853	Alya, m
175453	162001		18 56 14.2	-18 42 33	0.000	-0.01	6.6	0.8	3.2	G5 IV		48 s		
175639	124070	63 θ² Ser	18 56 14.5	+4 12 07	+0.003	+0.03	4.98	0.20	2.1	A5 V n	-54	31 s	11853	m
175219	229383		18 56 16.6	-42 42 38	-0.003	-0.02	5.36	1.00	1.8	G6 III-IV	-21	44 s		
175825	86595		18 56 17.4	+23 04 13	+0.003	+0.01	7.9	0.2	2.1	A5 V		140 s		
175615	142822		18 56 17.5	-0 48 16	+0.001	-0.01	8.00	0.1	1.4	A2 V		190 s	11854	m
175803	104275		18 56 18.5	+19 50 52	+0.001	+0.01	8.03	0.02	-1.7	B3 V	-30	710 s		
175322	210732		18 56 19.7	-33 16 13	+0.004	-0.06	7.5	0.4	4.4	G0 V		42 s		
176071	47938		18 56 21.8	+45 30 25	0.000	+0.05	8.00	0.5	4.0	F8 V		63 s		m
175640	142825		18 56 22.4	-1 47 59	-0.001	-0.02	6.22	-0.05		A0	-26			
175562	162005		18 56 23.0	-11 19 24	-0.003	-0.03	7.20	1.4	-0.3	K5 III		300 s		m
174808	254412		18 56 23.1	-61 49 09	+0.003	+0.03	7.80	0.4	3.0	F2 V		91 s		m
176598	18079		18 56 25.3	+65 15 30	-0.005	-0.02	5.63	0.95	0.3	gG5	-5	100 s		
175679	124073		18 56 25.4	+2 28 16	0.000	-0.01	6.15	0.97	0.3	G8 III	-15	140 s		m
175263	229389		18 56 26.8	-40 34 57	-0.001	-0.02	7.9	1.4	-0.3	K5 III		420 s		
175390	210733		18 56 27.0	-31 41 19	+0.001	0.00	6.6	1.5	3.2	G5 IV		17 s		
174583	254409		18 56 29.8	-67 40 11	0.000	+0.01	8.0	1.1	0.2	K0 III		360 s		
175145	229384		18 56 31.9	-48 59 31	0.000	-0.03	6.8	0.9	3.2	G5 IV		42 s		
176131	47941		18 56 32.0	+46 45 59	+0.001	-0.03	7.1	0.1	1.4	A2 V	-4	140 s		
175501	187480		18 56 34.0	-23 08 55	+0.001	-0.10	7.9	-0.2	3.0	F2 V		94 s		
175054	245816		18 56 34.2	-54 34 02	0.000	-0.01	7.6	0.3	0.6	A0 V		170 s		
175726	124077		18 56 37.0	+4 15 54	-0.001	-0.09	6.6	0.9	3.2	G5 IV		48 s		
175196	229385		18 56 37.6	-48 04 33	+0.001	0.00	7.6	1.9	-0.1	K2 III		300 s		
175940	86605		18 56 37.9	+28 11 56	+0.001	+0.06	7.1	1.1	-0.1	K2 III		270 s		
175362	210734		18 56 40.3	-37 20 36	+0.001	-0.03	5.38	-0.14	-0.6	B8 IV	+3	160 s		
174731	254411		18 56 40.7	-65 28 42	+0.007	-0.10	7.6	1.1	0.2	K0 III		180 mx		
175886	86604		18 56 41.1	+21 22 45	+0.001	+0.01	8.0	1.1	0.2	K0 III		330 s		
175526	187483		18 56 42.9	-24 37 13	+0.001	+0.01	7.4	1.1	0.2	K0 III		230 s		
175664	142829		18 56 44.1	-8 25 46	-0.001	-0.01	7.60	0.1	0.6	A0 V		240 s		m
175583	162010		18 56 44.4	-17 26 39	0.000	-0.01	7.2	1.1	-0.1	K2 III		280 s		
176408	31284	48 Dra	18 56 44.7	+57 48 53	-0.005	-0.07	5.66	1.15	-0.2	gK3	-34	150 s		
175623	162012		18 56 45.9	-14 51 28	0.000	0.00	7.16	0.22		B9				m
175644	162014		18 56 46.8	-14 01 43	-0.003	-0.05	7.82	0.08		A0				
175348	210736		18 56 47.0	-39 30 27	0.000	0.00	7.9	0.6	4.4	G0 V		48 s		
176841	18085		18 56 49.1	+68 17 32	-0.020	0.00	7.5	0.8	3.2	G5 IV		72 s		
175459	210746		18 56 51.6	-32 21 12	+0.004	-0.01	8.0	1.0	3.2	G5 IV		67 s		
175394	210741		18 56 51.7	-37 51 38	-0.001	0.00	8.0	0.9	1.4	A2 V		59 s		
175786	124082		18 56 52.9	+4 41 02	0.000	-0.01	7.82	1.78	-0.3	K5 III		290 s		
174877	254415		18 56 54.6	-62 48 04	+0.001	+0.03	6.4	1.6	0.2	K0 III		79 s		
176409	31286		18 56 54.9	+56 44 58	+0.002	+0.05	7.10	0.4	3.0	F2 V		66 s	11883	m
175478	210749		18 56 56.2	-33 19 47	0.000	-0.02	7.10	0.00	0.4	B9.5 V		220 s		m
174694	254413	κ Pav	18 56 56.9	-67 14 01	0.000	+0.01	4.1	0.5	3.4	F5 V	+37	23 mn		v
175805	124083		18 56 57.8	+2 27 42	+0.002	-0.01	7.63	0.54		F8				m
176051	67612		18 57 01.4	+32 54 05	+0.013	-0.16	5.22	0.59	4.4	G0 V	-47	17 ts	11871	m
--	--		18 57 01.5	+32 54 58			5.34	0.56	4.4	G0 V				
175956	86613		18 57 02.2	+22 48 31	-0.001	0.00	7.8	0.4	3.0	F2 V		87 s		
176209	47947		18 57 02.9	+45 51 21	+0.001	+0.02	7.30	0.1	0.6	A0 V	-19	220 s		m
175751	142838	η Sct	18 57 03.5	-5 50 46	+0.004	-0.04	4.83	1.08	-0.1	K2 III	-93	36 ts		
175806	124085		18 57 04.3	+1 30 00	+0.001	-0.02	7.6	0.5	3.4	F5 V		68 s		
175437	210747		18 57 04.5	-38 54 52	+0.003	-0.01	7.0	1.0	0.2	K0 III		220 s		
176341	31282		18 57 04.8	+51 43 15	+0.005	-0.01	7.8	0.5	3.4	F5 V		75 s		
176053	67614		18 57 05.0	+32 00 03	+0.002	+0.02	7.1	0.1	1.7	A3 V	-36	120 s		
176132	67620		18 57 06.2	+38 47 54	0.000	-0.02	7.5	1.4	-0.3	K5 III		370 s		
176087	67618		18 57 07.3	+34 51 21	-0.002	-0.08	7.4	0.5	4.0	F8 V		49 s		
176005	86619		18 57 07.7	+26 05 46	+0.001	+0.01	8.00	0.1	0.6	A0 V		280 s	11869	m
175326	229395		18 57 09.5	-47 23 28	+0.003	-0.02	7.30	1.10	0.0	K1 III		290 s		

HD	SAO	Star Name	α 2000	δ 2000	μ(α)	μ(δ)	V	B-V	M_v	Spec	RV	d(pc)	ADS	Notes
176133	67621		18^{h}57^{m}10^s.4	+36°28'11"	0^s.000	+0".01	6.6	1.1	0.0	K1 III		210 s		
175921	104283		18 57 11.5	+14 50 00	+0.002	+0.02	7.20	0.9	3.2	G5 IV	-22	62 s	11867	m
175922	104280		18 57 12.2	+13 22 22	0.000	+0.01	7.21	0.33		A m				
175395	229397		18 57 13.9	-43 55 02	-0.001	-0.03	6.5	0.1	0.6	A0 V		130 s		
175869	124089	64 Ser	18 57 16.4	+2 32 07	0.000	-0.01	5.57	0.00		B8 n	-11			
174787	254414		18 57 16.9	-66 19 56	+0.015	-0.06	8.0	0.4	2.6	F0 V		120 s		
176668	18082		18 57 17.1	+62 23 48	0.000	-0.04	6.39	0.92	3.2	G5 IV	-8	34 s	11901	m
175166	245821		18 57 17.8	-55 49 58	+0.002	-0.03	7.5	0.7	2.6	F0 V		54 s		
175396	229398		18 57 17.8	-44 18 06	+0.001	-0.03	7.7	0.7	3.2	G5 IV		80 s		
230217	104285		18 57 18.4	+18 50 53	-0.001	-0.01	7.9	1.1	0.2	K0 III		320 s		
175687	187498	36 ξ¹ Sgr	18 57 20.3	-20 39 23	0.000	-0.01	5.0	0.1	0.6	A0 V	+2	76 s		
176229	47950		18 57 25.0	+42 06 13	-0.009	-0.13	7.5	0.4	2.6	F0 V		94 s		
175773	162019		18 57 25.5	-13 06 05	+0.001	0.00	7.9	0.0	0.0	B8.5 V		360 s		
176007	104288		18 57 25.8	+18 35 13	0.000	-0.01	8.0	0.0	0.4	B9.5 V		310 s		
176560	31292		18 57 28.2	+58 13 31	+0.001	+0.05	6.80	0.1	1.4	A2 V	+1	120 s	11897	m
175669	187496		18 57 28.9	-27 27 16	+0.002	-0.01	7.9	1.6	-0.3	K5 III		370 s		
175529	210754		18 57 34.5	-39 49 24	+0.003	+0.03	6.4	0.4	1.7	A3 V		57 s		
175754	162021		18 57 35.5	-19 09 10	+0.001	+0.01	7.03	-0.07		O8.8	-11			
175905	142846		18 57 35.7	-0 31 35	-0.003	-0.04	7.50	1.1	0.0	K1 III	+22	280 s		
175775	187504	37 ξ² Sgr	18 57 43.6	-21 06 24	+0.002	-0.01	3.51	1.18	0.0	K1 III	-20	44 s		
--	18093		18 57 55.0	+68 30 20	+0.001	-0.02	7.8	1.6						
175626	--		18 57 55.9	-38 10 22			7.8	0.2	4.4	G0 V		49 s		
175585	229406		18 57 56.2	-42 08 26	0.000	-0.03	7.8	0.4	0.2	K0 III		330 s		
178089	9296		18 57 56.8	+77 03 03	-0.013	-0.06	6.5	0.4	2.6	F0 V	-27	60 s		
176390	47955		18 57 58.5	+44 59 18	-0.003	-0.02	8.0	0.1	1.7	A3 V		180 s		
176318	67642		18 58 01.7	+38 15 58	0.000	0.00	5.80	-0.17	-0.9	B6 V	-28	220 s		
176624	31295		18 58 01.8	+56 24 30	+0.002	+0.02	8.0	0.1	1.4	A2 V		210 s		
176319	--		18 58 04.6	+36 36 05	0.000	-0.02	8.0	0.0	0.4	B9.5 V		330 s		
175732	210765		18 58 05.4	-31 48 46	0.000	-0.05	7.7	1.4	-0.1	K2 III		270 s		
175758	210768		18 58 06.3	-30 49 29	+0.001	-0.01	7.4	0.6	3.4	F5 V		54 s		
176092	104294		18 58 07.7	+12 01 39	0.000	0.00	7.6	0.1	0.6	A0 V		240 s		
172226	258814		18 58 09.6	-83 25 16	+0.010	+0.02	7.40	0.1	0.6	A0 V		230 s		m
175876	187514		18 58 10.7	-20 25 25	+0.001	0.00	6.94	-0.10		O6	+14		11872	m
176212	86634		18 58 12.4	+24 42 31	0.000	0.00	7.5	0.0	0.4	B9.5 V		240 s		
175793	187511		18 58 14.1	-28 45 08	+0.004	+0.01	7.8	0.8	0.4	B9.5 V		100 s		
176155	104296		18 58 14.5	+17 21 39	0.000	-0.01	5.38	0.80	-6.5	F4 I	-22	1600 s	11884	FF Aql, m,v
175714	210766		18 58 14.7	-37 19 02	-0.003	-0.03	7.8	1.5	-0.3	K5 III		410 s		
175778	210771		18 58 17.2	-32 10 35	0.000	-0.08	7.4	1.0	0.2	K0 III		160 mx		
176392	67653		18 58 19.9	+39 38 30	+0.001	0.00	7.5	0.1	0.6	A0 V		240 s		
176112	124111		18 58 20.3	+8 32 19	+0.002	-0.02	7.5	0.4	2.6	F0 V		95 s		
175852	187517		18 58 20.3	-24 52 36	+0.001	0.00	6.6	0.2	0.6	A0 V		110 s		
--	67654		18 58 20.8	+39 17 43	0.000	-0.03	7.8	1.7						
175794	210773		18 58 21.1	-31 02 10	-0.003	-0.05	6.1	2.4	0.2	K0 III		23 s		
176095	124112		18 58 23.6	+6 14 24	0.000	-0.10	6.21	0.46		F5	-9			
175892	187519		18 58 24.4	-22 31 46	-0.002	+0.02	6.0	0.1	1.4	A2 V		79 s		
176074	124109		18 58 25.6	+1 11 53	+0.001	-0.01	6.9	0.1	1.4	A2 V		120 s		
176968	18092		18 58 26.3	+63 26 01	-0.001	+0.07	7.8	0.4	2.6	F0 V		110 s		
176410	67658		18 58 27.1	+39 46 41	0.000	-0.01	7.4	1.1	0.2	K0 III		280 s		
175510	245834	λ Tel	18 58 27.5	-52 56 18	+0.001	0.00	5.0	0.0	-0.8	B9 III	-6	140 s		
175797	210776		18 58 27.7	-34 12 51	+0.001	+0.02	7.2	-0.1	0.4	B9.5 V		230 s		
176252	86645		18 58 27.8	+23 48 08	-0.002	-0.11	7.4	1.1	0.2	K0 III	-36	85 mx		
175147	254422		18 58 28.1	-64 00 06	+0.010	-0.07	7.5	1.1	0.2	K0 III		170 mx		
176214	104300		18 58 28.4	+18 53 43	0.000	-0.01	8.0	0.1	0.6	A0 V		280 s		
176215	104301		18 58 29.5	+18 31 29	0.000	+0.02	7.9	1.1	0.0	K1 III		350 s		
176284	86648		18 58 31.4	+25 14 54	0.000	-0.02	7.90	1.4	-0.3	K5 III		400 s	11895	m
176893	18091		18 58 33.4	+60 40 04	-0.002	+0.05	6.7	0.1	0.6	A0 V		170 s		
175329	254423	ω Pav	18 58 36.2	-60 12 02	-0.016	+0.04	5.14	1.37		K1 III-IV	+180	20 mn		
177807	9293		18 58 37.4	+74 07 58	-0.001	-0.05	7.0	0.4	2.6	F0 V		77 s		
175853	210783		18 58 39.2	-30 47 29	-0.001	+0.02	8.0	1.7	0.2	K0 III		140 s		
175763	229412		18 58 40.6	-41 04 16	+0.001	0.00	7.9	1.8	-0.3	K5 III		230 s		
176254	86651		18 58 41.4	+20 37 23	+0.001	-0.01	6.77	0.05	-1.6	B3.5 V	-7	350 s		
176730	31303		18 58 41.7	+54 52 12	+0.003	+0.03	8.0	0.2	2.1	A5 V		150 s		
175813	210781	ε CrA	18 58 43.2	-37 06 26	-0.011	-0.10	4.8	0.4	2.6	F0 V	+53	28 s		v
175764	229413		18 58 43.9	-41 34 17	-0.001	0.00	7.9	1.0	0.2	K0 III		340 s		
176301	104306		18 58 44.9	+19 47 39	0.000	0.00	6.20	-0.04	-0.9	B6 V	-1	240 s		
176502	47965		18 58 46.4	+40 40 45	0.000	0.00	6.22	-0.16	-2.0	B4 IV	-19	440 s	11910	m
176232	104303	10 Aql	18 58 46.7	+13 54 24	0.000	-0.05	5.89	0.25		A4 p	+15			
176076	142865		18 58 47.2	-5 38 06	+0.002	-0.01	7.3	0.0	0.4	B9.5 V		210 s		
176377	67661		18 58 50.8	+30 10 51	+0.004	+0.20	6.78	0.59	4.4	dG0	-40	27 ts		
175674	229410		18 58 52.5	-48 30 21	-0.003	0.00	6.65	1.30	-0.3	K5 III		250 s		m
177410	18103		18 58 52.7	+69 31 48	+0.003	-0.04	6.4	-0.1		B9				
176118	142868		18 58 52.8	-4 43 36	-0.001	0.00	7.6	0.5	4.0	F8 V		51 s		
176437	67663	14 γ Lyr	18 58 56.4	+32 41 22	0.000	0.00	3.24	-0.05	-0.8	B9 III	-22	59 s	11908	Sulafat, m
176707	31304		18 58 59.5	+50 48 33	+0.002	+0.02	6.30	0.98	3.2	G5 IV	-21	31 s		
176708	47975		18 59 00.8	+49 19 10	+0.001	+0.01	8.0	0.1	1.4	A2 V		210 s		

HD	SAO	Star Name	α 2000	δ 2000	μ(α)	μ(δ)	V	B-V	M_V	Spec	RV	d(pc)	ADS	Notes
			$18^h 59^m 01^s.2$	$-12°35'17''$	$+0^s.001$	$0''.00$								
176078	162043		18 59 01.2	-12 35 17	+0.001	0.00	7.0	0.1	0.6	A0 V		190 s		
176466	67667		18 59 02.6	+33 07 02	+0.002	-0.01	6.9	0.1	1.4	A2 V	-39	130 s		
176626	47971		18 59 05.3	+43 43 16	-0.003	+0.02	6.9	0.1	1.4	A2 V	-26	130 s		
176079	162044		18 59 05.4	-15 51 36	+0.002	0.00	8.0	0.1	1.4	A2 V		200 s		
176303	104308	11 Aql	18 59 05.6	+13 37 21	+0.001	-0.12	5.23	0.53	1.5	F8 III-IV	+16	56 s	11902	m
175979	187535		18 59 07.3	-28 35 30	-0.001	+0.05	7.7	1.3	4.4	G0 V		16 s		
176141	142876		18 59 07.5	-8 58 16	+0.003	-0.05	7.5	0.5	3.4	F5 V		64 s		
175855	210786		18 59 10.9	-39 32 05	+0.001	-0.03	6.50	0.1	0.6	A0 V		150 s		m
176582	67675		18 59 12.1	+39 13 03	0.000	+0.01	6.41	-0.17	-1.6	B5 IV	-14	400 s		
176321	104314		18 59 12.2	+15 07 47	0.000	-0.01	7.6	0.1	0.6	A0 V		240 s		
176158	142879		18 59 12.6	-6 50 32	0.000	-0.01	7.7	0.0	0.4	B9.5 V		250 s		
176304	104313		18 59 17.3	+10 08 27	+0.001	-0.01	6.75	0.25	-2.5	B2 V	-23	390 s		
176645	47977		18 59 21.2	+41 38 12	-0.001	-0.02	8.0	1.6	-0.5	M2 III		510 s		
176162	162052		18 59 23.6	-12 50 26	0.000	-0.02	5.90	0.00	-1.1	B5 V	-13	240 s		m
176124	162049		18 59 23.7	-19 16 42	+0.001	+0.01	6.6	1.4	-0.5	M2 III		260 s		
176669	47980		18 59 25.9	+43 00 50	-0.002	+0.04	7.53	-0.06	0.0	B8.5 V	-21	300 s		
176123	162050		18 59 26.7	-18 34 02	0.000	-0.04	6.3	0.9	3.2	G5 IV		42 s		
174584	257643		18 59 26.8	-75 23 23	+0.010	+0.01	6.6	2.2	-0.1	K2 III		53 s		
175007	257644		18 59 29.3	-70 28 08	+0.003	-0.12	6.8	0.0	2.1	A5 V		40 mx		
176081	187543		18 59 29.4	-23 15 57	+0.001	-0.01	8.0	1.4	0.2	K0 III		220 s		
176438	104322		18 59 36.3	+19 29 11	0.000	0.00	7.51	-0.01	0.4	B9.5 V		260 s		
176411	104318	13 ε Aql	18 59 37.2	+15 04 06	-0.004	-0.07	4.02	1.08	-0.1	K2 III	-48	65 mx		m
176287	142890		18 59 37.3	-2 50 07	-0.001	0.00	7.5	0.1	0.6	A0 V		210 s		
176439	104323		18 59 37.8	+18 25 26	0.000	-0.02	7.6	0.1	0.6	A0 V		240 s		
176186	162059		18 59 38.3	-16 06 31	0.000	-0.01	7.9	1.1	0.2	K0 III		330 s		
175951	229423		18 59 41.2	-40 37 22	0.000	-0.01	7.5	0.0	0.4	B9.5 V		260 s		
176441	104328		18 59 42.3	+16 15 08	-0.007	-0.14	7.0	0.5	3.4	F5 V	+24	51 s		
176599	67679		18 59 42.6	+34 05 42	-0.001	-0.01	8.0	0.4	2.6	F0 V		120 s		
176527	86673		18 59 45.3	+26 13 50	+0.006	-0.01	5.27	1.24	-0.1	K2 III	-24	110 s		
175817	245854		18 59 46.9	-51 16 56	+0.001	-0.01	7.9	0.9	3.2	G5 IV		74 s		
175894	229420		18 59 46.9	-47 19 49	-0.006	-0.02	7.9	0.5	3.2	G5 IV		88 s		
176445	104329		18 59 50.7	+11 26 41	0.000	+0.02	8.0	0.4	2.6	F0 V		110 s		
176399	124135		18 59 51.3	+6 09 25	-0.002	-0.01	7.3	0.9	3.2	G5 IV		65 s		
175675	245849		18 59 51.7	-57 58 39	+0.001	-0.01	7.4	0.9	0.2	K0 III		270 s		
176444	104330		18 59 53.8	+12 18 41	0.000	-0.01	7.5	1.1	-0.1	K2 III		310 s		
176798	47986		18 59 56.9	+43 15 10	0.000	+0.01	7.2	0.1	1.7	A3 V	-28	130 s		
175880	229421		18 59 57.1	-49 16 23	-0.001	-0.02	7.7	1.5	0.2	K0 III		160 s		
176541	86675		18 59 57.9	+22 48 53	-0.002	-0.01	6.29	1.75	-0.5	M2 III	-53	180 s		
175008	257645		18 59 58.5	-71 55 56	+0.007	-0.01	6.9	-0.1	0.4	B9.5 V		200 s		
--	86677		18 59 58.5	+26 21 07	+0.001	-0.02	8.0	1.1	3.2	G5 IV		57 s		
176486	104332		18 59 58.6	+12 53 25	+0.001	0.00	7.1	1.4	-1.3	K4 II-III	-2	420 s	11916	m
176583	86678		18 59 58.8	+26 57 54	-0.001	-0.02	6.7	0.1	0.6	A0 V		160 s		
175933	229424		18 59 59.6	-46 50 41	+0.002	-0.01	7.8	0.5	2.1	A5 V		92 s		
176165	187552		19 00 00.2	-28 03 02	+0.003	0.00	7.8	0.2	1.4	A2 V		150 s		
176670	67682	15 λ Lyr	19 00 00.7	+32 08 44	+0.001	+0.01	4.93	1.47	-2.3	K3 II	-16	260 s		
176002	229428		19 00 01.2	-43 20 52	-0.001	-0.03	7.9	1.5	3.2	G5 IV		32 s		
175401	254426		19 00 03.3	-66 39 12	+0.002	-0.04	6.01	0.97		K0				
176166	187554		19 00 04.2	-28 42 10	+0.004	-0.05	8.0	0.8	3.4	F5 V		49 s		
176128	210800		19 00 04.2	-32 15 21	-0.001	-0.02	8.0	0.1	4.0	F8 V		62 s		
175652	254433		19 00 11.7	-60 12 14	+0.002	-0.01	7.4	0.4	3.4	F5 V		64 s		
177275	18107		19 00 11.8	+60 58 39	-0.003	-0.03	7.9	1.1	-0.1	K2 III		390 s		
177003	31311		19 00 13.7	+50 32 00	+0.002	0.00	5.38	-0.18	-1.7	B3 V	-19	260 s		
176468	124142		19 00 13.7	+2 57 02	0.000	+0.02	7.9	1.1	-0.1	K2 III		340 s		
175838	245856		19 00 17.6	-55 00 55	+0.001	+0.03	7.1	1.4	-0.1	K2 III		210 s		
176148	210805		19 00 18.0	-32 52 27	0.000	0.00	7.7	1.4	0.2	K0 III		190 s		
176844	47993		19 00 18.8	+40 41 03	0.000	-0.01	6.7	1.6	-0.5	M2 III	-5	270 s		
180499	3160		19 00 21.9	+82 40 18	-0.011	-0.01	8.04	0.02	-0.2	B8 V		280 mx		
176799	67691		19 00 22.0	+37 59 44	+0.002	-0.01	7.9	0.1	1.7	A3 V		160 s		
176774	67690		19 00 22.9	+35 49 14	0.000	+0.02	7.9	0.0	0.4	B9.5 V		320 s		
176488	124143		19 00 23.3	+1 44 25	-0.001	-0.01	8.0	1.1	-0.1	K2 III		350 s		
176775	67689		19 00 24.3	+33 44 31	+0.001	0.00	6.8	1.1	0.2	K0 III		210 s		
176246	187562		19 00 24.7	-24 56 32	+0.004	+0.04	6.3	1.4	0.2	K0 III	-25	96 s		
176328	162076		19 00 25.6	-17 29 05	+0.004	-0.02	8.0	0.4	3.0	F2 V		97 s		
176414	142903		19 00 29.5	-7 23 08	-0.002	-0.10	7.7	0.5	4.0	F8 V		54 s		
176894	47996		19 00 31.6	+41 41 03	-0.001	+0.01	7.8	0.4	2.6	F0 V		110 s		
176266	187563		19 00 32.5	-27 38 43	0.000	-0.02	8.0	0.2	0.6	A0 V		210 s		
176869	67695		19 00 34.0	+39 50 42	0.000	0.00	7.86	-0.06	0.0	B9.5 V		310 s		
176584	104344		19 00 34.2	+12 13 34	0.000	-0.01	8.0	0.1	1.4	A2 V		200 s	11926	m
176646	104349		19 00 34.3	+18 27 48	+0.001	-0.01	8.0	0.9	3.2	G5 IV	-24	88 s		
176895	67697		19 00 35.4	+39 51 58	-0.001	0.00	7.65	0.85	3.2	G5 IV		68 s		
176350	162079		19 00 36.0	-19 39 30	+0.003	0.00	7.3	1.0	-0.1	K2 III		310 s		
176037	229432		19 00 37.9	-48 17 04	-0.003	-0.02	6.9	1.3	0.2	K0 III		150 s		
176530	142913		19 00 41.7	-0 13 32	0.000	+0.01	8.0	0.1	1.4	A2 V		190 s		
177249	31323	49 Dra	19 00 43.2	+55 39 30	-0.003	-0.01	5.48	0.86	3.2	G5 IV	+10	25 s		
176531	142915		19 00 45.2	-0 27 19	+0.001	-0.01	7.20	0.9	3.2	G5 IV		61 s	11927	m

HD	SAO	Star Name	α 2000	δ 2000	μ(α)	μ(δ)	V	B-V	M_v	Spec	RV	d(pc)	ADS	Notes
176191	210810		19^{h}00^{m}47^s.8	−39°34′51″	+0^s.002	−0″.06	8.0	0.1	0.6	A0 V		85 mx		
176674	104351		19 00 49.3	+14 32 49	+0.001	0.00	7.9	0.0	0.4	B9.5 V		290 s		
176969	48000		19 00 49.9	+42 15 10	−0.002	−0.13	6.7	0.5	4.0	F8 V		35 s		
176206	210811		19 00 52.3	−39 04 25	+0.001	−0.04	7.5	1.1	0.2	K0 III		260 s		
176896	67699		19 00 55.0	+33 48 08	0.000	0.00	6.01	0.97	0.2	K0 III	−28	150 s		
176293	210817		19 00 59.9	−33 13 52	+0.003	+0.02	7.3	0.5	2.6	F0 V		64 s		
176649	124155		19 00 59.9	+9 25 55	+0.001	+0.01	8.02	0.10		A0				
176760	104360		19 01 01.1	+20 01 24	+0.001	0.00	7.9	0.0	0.4	B9.5 V		300 s		
176269	210815		19 01 03.1	−37 03 39	+0.001	−0.02	6.8	−0.8	0.0	B8.5 V	+10	230 s		
176270	210816		19 01 04.1	−37 03 43	+0.001	−0.04	6.60	0.00	0.0	B8.5 V	−27	210 s		m
176452	162088		19 01 04.7	−19 20 59	0.000	−0.05	7.4	0.7	0.2	K0 III		280 s		
175782	254436		19 01 05.0	−62 51 54	+0.004	+0.01	7.5	1.1	−0.1	K2 III		330 s		
176776	104362		19 01 05.3	+19 18 35	+0.001	0.00	6.4	1.1	0.0	K1 III	−29	190 s		
176367	187573		19 01 06.0	−28 42 51	+0.002	−0.10	8.0	1.1	4.0	F8 V		30 s		
177151	48006		19 01 06.2	+48 02 09	−0.002	−0.02	7.0	1.1	0.2	K0 III		230 s		
176803	86699		19 01 08.9	+20 09 25	+0.002	−0.03	7.34	0.06	0.2	B8.5 V	−17	250 s		m
176271	—		19 01 09.2	−38 36 55			5.51	0.77	−0.1	K2 III		130 s		
176588	142919		19 01 09.4	−4 26 20	+0.001	+0.02	7.1	1.1	−0.1	K2 III	+3	240 s		
176650	124156		19 01 11.2	+2 28 56	0.000	−0.23	7.2	1.1	0.2	K0 III	+25	88 mx		
176651	124157		19 01 13.2	+2 02 01	0.000	+0.02	7.0	1.1	0.2	K0 III	−16	210 s		
176105	245864		19 01 14.9	−50 10 54	−0.002	+0.02	7.9	1.6	0.2	K0 III		340 s		
176871	86707		19 01 17.1	+26 17 29	−0.001	−0.01	5.69	−0.08	−1.1	B5 V	−14	220 s		
176818	86702		19 01 17.6	+21 30 50	0.000	0.00	6.9	0.0	−1.6	B3.5 V	−9	430 s		
176914	86708		19 01 17.7	+28 24 46	+0.001	+0.03	6.8	0.0	−1.0	B5.5 V	−6	330 s		
176938	86709		19 01 19.2	+29 31 07	+0.001	−0.01	6.7	0.1	0.6	A0 V	−17	160 s		
176676	124159		19 01 20.1	+1 44 40	0.000	−0.02	8.0	1.1	−0.1	K2 III		350 s		
177152	48008		19 01 21.5	+44 23 13	0.000	+0.02	7.4	0.0	0.4	B9.5 V	−19	250 s		
176819	86704		19 01 22.5	+20 50 01	0.000	+0.02	6.69	0.02	−2.8	B2 IV-V	−10	600 s		
176473	187581		19 01 23.9	−23 13 41	+0.002	0.00	8.0	0.6	3.2	G5 IV		93 s		
177196	48011	16 Lyr	19 01 26.2	+46 56 05	+0.001	−0.08	5.01	0.19		A7	+8	18 mn	11964	m
176698	124160		19 01 26.7	+1 17 18	−0.001	−0.03	7.7	1.1	0.2	K0 III		280 s		
176630	142927		19 01 27.2	−6 11 33	+0.001	+0.01	7.7	0.0	−2.0	B4 IV	−7	790 s		
176383	210825		19 01 28.2	−34 22 38	−0.002	−0.09	8.0	0.6	3.4	F5 V		65 s		
177006	67716		19 01 29.9	+32 23 34	+0.001	+0.01	7.25	−0.13	−1.0	B5.5 V		450 s		
176737	124163		19 01 33.4	+2 35 19	+0.002	+0.03	7.30	1.4	−1.3	K4 II-III	−46	440 s		
176593	162097		19 01 33.4	−15 16 57	0.000	0.00	6.32	1.00	0.3	gG6	+20	140 s		
176939	86714		19 01 34.7	+25 01 33	0.000	0.00	6.9	1.1	−0.1	K2 III	−21	230 s		
176537	187584		19 01 37.6	−22 41 44	0.000	0.00	6.3	2.2	−0.3	K5 III		80 s		
176386	210828		19 01 38.7	−36 53 28	+0.001	−0.04	7.20	0.1	0.6	A0 V	+7	210 s		TY CrA, m,v
177153	48012		19 01 39.4	+41 29 27	−0.010	+0.06	7.3	0.6	4.4	G0 V		39 s		
176678	142931	12 Aql	19 01 40.7	−5 44 20	−0.001	−0.03	4.02	1.09	0.0	K1 III	−44	64 s		
176514	187583		19 01 41.3	−26 00 18	0.000	+0.03	8.0	0.4	0.6	A0 V		160 s		
176422	210833		19 01 46.2	−35 52 06	0.000	+0.05	8.0	−0.2	0.6	A0 V		300 s		
176848	104378		19 01 47.7	+11 27 19	−0.001	0.00	7.6	1.1	0.2	K0 III		270 s		
177109	67721		19 01 48.2	+33 37 17	0.000	0.00	6.39	−0.12	−1.6	B5 IV	−23	400 s	11965	m
176873	104379		19 01 48.4	+12 32 28	0.000	0.00	6.70	0.00	0.4	B9.5 V		170 s	11952	m
179215	9311		19 01 48.7	+77 40 32	−0.006	−0.01	7.4	1.1	−0.1	K2 III		320 s		
176699	142934		19 01 49.1	−6 22 46	+0.001	0.00	7.8	0.1	1.4	A2 V		180 s		
176971	86716		19 01 49.3	+22 15 49	+0.001	−0.01	6.4	0.1	1.7	A3 V	−38	85 s		
176495	210836		19 01 52.6	−32 35 49	+0.002	+0.04	8.0	0.5	3.2	G5 IV		93 s		
176714	142937		19 01 55.6	−6 10 54	+0.001	0.00	8.0	1.1	0.2	K0 III		350 s		
176973	104384		19 01 57.3	+19 07 28	+0.004	+0.02	6.80	0.8	3.2	G5 IV		52 s	11957	m
176110	245870		19 01 58.7	−56 58 18	−0.004	−0.13	8.0	0.9	3.2	G5 IV		76 mx		
176406	229445		19 02 01.6	−40 39 47	+0.001	−0.02	7.3	−0.5	0.0	B8.5 V		290 s		
176978	104385		19 02 04.6	+14 25 47	0.000	+0.01	7.60	0.1	1.4	A2 V		170 s	11958	m
177696	18121		19 02 04.8	+60 07 43	0.000	+0.03	6.8	1.4	−0.3	K5 III		270 s		
176497	210840		19 02 06.5	−36 21 44	−0.001	−0.05	8.0	0.9	0.6	A0 V		78 s		
177483	31337		19 02 06.8	+52 15 40	−0.001	−0.03	6.31	1.00	−0.1	K2 III	+4	190 s	11979	m
176515	210843		19 02 07.9	−35 00 28	+0.002	−0.07	7.1	1.1	3.2	G5 IV		40 s		
176425	229446		19 02 08.5	−41 54 36	+0.004	0.00	6.23	0.00	0.6	A0 V		130 s		
177391	48025		19 02 10.0	+47 17 15	−0.001	+0.04	7.9	0.7	4.4	G0 V		51 s		
177250	67725		19 02 10.2	+37 48 42	+0.001	0.00	7.8	1.1	0.2	K0 III		330 s		
176942	104386		19 02 12.5	+10 57 53	−0.001	+0.02	7.5	0.2		A m				
176745	162112		19 02 13.2	−13 15 13	0.000	−0.01	7.9	0.0	0.4	B9.5 V		310 s		
172690	258816		19 02 13.3	−83 56 23	−0.008	−0.01	7.7	−0.2	0.6	A0 V		260 s		
176613	187594		19 02 19.6	−29 20 29	+0.002	−0.02	7.8	1.0	2.6	F0 V		39 s		
176981	124184		19 02 21.5	+8 22 27	+0.002	+0.02	6.30	1.62		K2	−9			
176614	210848		19 02 22.9	−30 56 16	+0.005	−0.01	7.8	1.3	0.2	K0 III		220 mx		
176662	187596		19 02 24.0	−25 41 18	+0.002	+0.02	8.0	1.3	0.2	K0 III		250 s		
174565	257647		19 02 25.5	−79 12 49	−0.016	+0.06	7.3	0.5	3.4	F5 V		59 s		
176498	229451		19 02 25.6	−41 16 20	+0.005	−0.08	8.0	0.4	4.0	F8 V		48 s		
176980	124186		19 02 25.6	+9 38 00	0.000	+0.01	7.4	1.1	0.2	K0 III	+11	250 s		
176704	187599		19 02 27.5	−24 50 49	−0.002	−0.18	5.65	1.23	−0.2	K3 III	0	68 mx		
177929	18123		19 02 30.8	+62 42 21	−0.002	+0.02	7.0	1.1	0.2	K0 III		230 s		
176615	210852		19 02 32.4	−34 37 55	+0.001	−0.01	7.2	−0.2	0.4	B9.5 V		230 s		

HD	SAO	Star Name	α 2000	δ 2000	μ(α)	μ(δ)	V	B−V	M_v	Spec	RV	d(pc)	ADS	Notes
			$h \quad m \quad s$	$\circ \quad ' \quad ''$	s	$''$								
176783	162119		19 02 33.1	−16 35 52	+0.001	0.00	7.9	0.0	0.4	B9.5 V		310 s		
176429	229448		19 02 33.9	−47 57 55	0.000	−0.03	7.5	0.8	0.2	K0 III		280 s		
176853	162122		19 02 33.9	−10 43 17	−0.001	+0.01	6.65	0.22	−1.0	B5.5 V	−13	220 s		V599 Aql, m,v
176687	187600	38 ζ Sgr	19 02 36.5	−29 52 49	−0.001	0.00	2.59	0.08	0.6	A2 IV	+22	24 s	11950	Ascella, m
177082	104394		19 02 38.0	+14 34 02	+0.013	+0.01	6.8	0.6	4.7	G2 V	+3	26 s		
177303	67731		19 02 38.1	+34 34 57	−0.001	−0.02	7.7	1.1	0.2	K0 III		310 s		
177412	48028		19 02 39.4	+43 07 03	+0.006	0.00	7.9	0.5	3.4	F5 V		78 s		
176616	210853		19 02 42.6	−36 05 47	+0.002	0.00	7.90	0.1	0.6	A0 V		290 s		m
176752	187602		19 02 42.6	−24 21 55	−0.001	−0.04	7.7	1.4	−0.3	K5 III		390 s		
178304	18132		19 02 43.0	+68 19 31	+0.003	+0.01	7.0	0.1	1.4	A2 V		130 s		
230532	104400		19 02 46.5	+19 44 38	0.000	+0.01	7.9	1.4	−0.3	K5 III		400 s		
177199	104405		19 02 52.4	+19 39 39	0.000	0.00	6.09	1.34	0.0	K1 III	−7	120 s		
177064	124193		19 02 53.5	+5 20 33	+0.001	0.00	7.9	0.1	1.4	A2 V		180 s		
176984	142959	14 Aql	19 02 54.3	−3 41 56	+0.001	+0.01	5.42	0.00		A0	−39			m
178119	18128		19 02 54.5	+64 21 22	−0.001	+0.02	7.9	1.1	0.2	K0 III		340 s		
177778	31348		19 02 56.3	+56 36 54	+0.001	+0.04	7.9	1.1	0.2	K0 III		340 s		
176499	229454		19 02 59.7	−49 20 24	0.000	−0.04	8.0	0.9	2.6	F0 V		120 s		
177280	86739		19 03 01.3	+24 06 59	+0.001	0.00	8.0	0.0	0.4	B9.5 V		310 s		
176721	210856		19 03 02.2	−32 44 45	+0.002	−0.02	8.0	0.2	0.6	A0 V		210 s		
176172	254447		19 03 03.0	−62 54 01	−0.001	+0.01	7.4	0.9	3.2	G5 IV		68 s		
176884	162130		19 03 03.6	−19 14 43	0.000	+0.01	6.00	0.9	3.2	G5 IV	−20	36 s	11972	m
177486	67741		19 03 05.7	+39 57 15	+0.001	+0.01	8.0	1.1	−0.1	K2 III		420 s		
176986	162137		19 03 05.7	−11 02 38	−0.009	−0.23	7.9	1.1	0.2	K0 III		43 mx		
176638	229461	ζ CrA	19 03 06.7	−42 05 43	+0.005	−0.05	4.75	−0.02	0.6	A0 V n	−7	68 s		
176903	162133		19 03 06.8	−19 06 12	0.000	−0.01	6.3	0.7	3.4	F5 V		28 s		
176578	229459		19 03 10.0	−47 03 04	−0.001	−0.12	6.86	0.96	3.2	K0 IV		54 s		
176354	245882		19 03 12.1	−56 59 48	+0.002	−0.15	7.06	0.89		K1 IV		90 mx		
177370	86744		19 03 12.5	+30 01 31	0.000	−0.03	7.9	1.1	−0.1	K2 III		370 s		
177011	162139		19 03 12.8	−11 11 22	+0.001	−0.02	7.1	0.1	0.6	A0 V		190 s		
177067	142963		19 03 17.4	−5 33 35	0.000	−0.02	7.1	0.9	3.2	G5 IV		60 s		
176723	210859		19 03 17.5	−38 15 12	+0.001	+0.01	5.74	0.32	0.6	F0 III n	+16	110 s		
177347	86743		19 03 17.5	+25 49 31	−0.002	−0.03	6.97	−0.03	0.0	B8.5 V		250 s		
177089	142964		19 03 17.8	−2 47 18	−0.001	0.00	7.7	1.1	−0.1	K2 III		300 s		
177115	124198		19 03 19.2	+0 34 35	0.000	0.00	7.0	1.1	−0.1	K2 III	−14	240 s		
176371	245885		19 03 21.5	−57 55 27	+0.001	−0.05	7.4	0.8	0.2	K0 III		270 s		
177012	162143		19 03 21.9	−12 42 34	−0.001	0.00	7.5	0.0	0.4	B9.5 V		260 s		
177177	124202		19 03 25.4	+5 59 39	+0.002	0.00	7.5	0.1	1.4	A2 V		180 s		
177305	104418		19 03 26.4	+17 04 01	−0.008	−0.07	7.5	0.6	4.4	G0 V		41 s		
176557	245894		19 03 27.1	−50 19 00	+0.002	−0.02	7.19	1.46		K2				
178326	18137		19 03 27.7	+65 35 15	0.000	+0.06	7.1	1.1	0.2	K0 III	+20	240 s		
175304	257648		19 03 28.4	−76 06 56	−0.004	−0.10	8.0	1.1	−0.1	K2 III		140 mx		
178156	18133		19 03 28.5	+62 05 41	0.000	0.00	7.0	0.9	3.2	G5 IV		58 s		
178944	9314		19 03 29.2	+73 22 45	0.000	+0.03	7.9	0.4	2.6	F0 V		110 s		
175986	254446		19 03 29.6	−68 45 19	−0.001	0.00	5.88	0.56	4.4	dG0		19 s		m
177256	124206		19 03 31.7	+8 17 38	+0.001	−0.01	7.6	0.6	4.4	G0 V		43 s		
177178	124203		19 03 32.0	+1 49 08	0.000	−0.07	5.83	0.18		A2	−20			
177015	187618		19 03 33.0	−20 07 43	0.000	−0.01	7.80	−0.04		B7 e	+6			
176250	254452		19 03 34.1	−62 37 41	+0.003	−0.10	8.0	0.5	3.4	F5 V		82 s		
177330	104420		19 03 34.2	+17 16 50	−0.001	−0.05	7.1	1.1	−0.1	K2 III	+14	260 s		
177205	124204		19 03 35.2	+2 32 46	+0.001	+0.02	6.9	1.1	0.2	K0 III		200 s		
176770	210862		19 03 36.7	−39 27 23	+0.004	−0.02	8.00	0.9	3.2	G5 IV		91 s		m
176859	210864		19 03 38.5	−34 54 01	+0.001	0.00	7.9	1.2	−0.1	K2 III		380 s		
177592	67757		19 03 39.1	+39 54 40	+0.001	+0.01	7.9	0.1	0.6	A0 V		290 s		
178001	--		19 03 39.2	+57 27 25	−0.001	−0.01	8.0	0.1	0.6	A0 V		300 s	12019	m
177392	86753		19 03 42.4	+21 16 04	+0.001	0.00	6.5	0.4	3.0	F2 V	+5	50 s		
177282	124213		19 03 45.4	+9 22 13	−0.001	0.00	7.4	1.4	−0.3	K5 III		320 s		
177258	124210		19 03 46.4	+2 29 00	0.000	−0.03	8.0	0.1	1.4	A2 V		190 s		
177828	48047		19 03 46.6	+49 37 58	+0.001	0.00	7.9	0.1	1.7	A3 V		170 s		
177673	48042		19 03 48.5	+41 58 29	0.000	+0.01	8.0	0.1	0.6	A0 V		310 s		
176725	229467		19 03 51.6	−46 32 56	+0.001	−0.01	7.9	−0.2	0.6	A0 V		280 s		
177352	104424		19 03 53.8	+10 51 24	0.000	+0.01	8.0	0.5	4.0	F8 V		61 s		
177981	31361		19 03 54.8	+54 23 22	−0.001	−0.01	7.6	1.4	−0.3	K5 III		380 s		
177593	67759		19 03 55.4	+34 09 03	0.000	−0.02	7.30	−0.09	−1.0	B5.5 V	−24	440 s	12003	m
176664	245899		19 03 57.2	−51 01 07	+0.003	−0.14	5.93	1.24		K5				m
177372	104427		19 03 59.8	+11 18 09	−0.001	−0.03	7.2	0.2	2.1	A5 V		100 s		
176522	245896		19 04 00.7	−57 57 36	0.000	0.00	6.5	0.9	3.2	G5 IV		39 s		
--	67764		19 04 00.7	+36 39 57	+0.001	+0.01	7.6	2.5						
178472	18141		19 04 01.0	+66 25 44	+0.002	0.00	8.0	0.1	1.7	A3 V		180 s		
177697	48044		19 04 01.4	+40 07 00	−0.002	0.00	7.9	1.4	−0.3	K5 III		440 s		
177414	104431		19 04 01.6	+16 27 48	−0.002	−0.05	7.0	1.1	−0.1	K2 III	+50	250 s		
177876	48055		19 04 04.2	+48 52 01	−0.001	+0.03	7.6	0.1	1.4	A2 V		180 s		
176959	210872		19 04 05.1	−34 22 21	+0.003	−0.05	7.9	1.6	0.2	K0 III		160 s		
176463	245895		19 04 06.3	−59 44 33	+0.012	−0.06	7.8	0.6	3.2	G5 IV		83 s		
177644	67766		19 04 07.1	+35 47 03	0.000	+0.03	7.9	0.1	1.7	A3 V		170 s		
177433	104433		19 04 08.2	+15 06 03	+0.001	0.00	7.60	1.1	−0.9	K0 II-III	−31	450 s		

HD	SAO	Star Name	α 2000	δ 2000	μ(α)	μ(δ)	V	B-V	M_v	Spec	RV	d(pc)	ADS	Notes
			h m s	+17°33'22"	+0.008	+0".08								
177459	104437		19 04 09.5	+17 33 22	+0.008	+0.08	6.6	0.5	3.4	F5 V	-67	44 s		
177332	124219		19 04 10.5	+3 19 50	0.000	+0.01	6.73	0.13		A2	-13			
177645	67767		19 04 12.0	+33 19 16	0.000	-0.02	7.8	0.4	2.6	F0 V		110 s		
177120	187632		19 04 14.0	-22 53 47	0.000	+0.01	6.90	0.1	0.6	A0 V		180 s	11987	m
178473	18143		19 04 14.4	+65 12 19	0.000	-0.03	7.7	1.1	0.2	K0 III		320 s		
177829	48054		19 04 16.0	+43 52 47	+0.005	-0.01	6.80	0.00	0.4	B9.5 V	-23	120 mx	12016	m
177780	48053		19 04 16.1	+41 00 10	+0.001	-0.03	8.0	0.9	3.2	G5 IV		93 s		
177334	142984		19 04 16.8	-0 17 57	-0.002	-0.04	8.03	1.84	-0.3	K5 III		290 s		
177698	67771		19 04 17.0	+34 05 12	+0.001	+0.01	7.2	1.1	-0.1	K2 III		290 s		
177166	187634		19 04 20.2	-21 31 54	+0.006	-0.05	7.18	0.52	4.4	G0 V		36 s	11989	m
177595	86766		19 04 21.1	+27 19 16	-0.001	+0.02	7.0	0.0	0.4	B9.5 V	-16	200 s		
177781	67774		19 04 23.5	+38 06 27	-0.001	+0.02	7.9	0.9	3.2	G5 IV		88 s		
177336	142985		19 04 24.1	-5 41 06	+0.001	0.00	6.90	4.19		N7.7	+37			V Aql, v
177074	210883		19 04 24.9	-31 02 49	+0.001	0.00	5.50	0.03	0.0	A0 IV	-20	120 s		
177931	48058		19 04 33.0	+45 55 18	+0.003	+0.02	6.8	0.0	0.4	B9.5 V	-10	190 s		
177597	86770		19 04 35.6	+21 28 06	-0.001	0.00	7.8	1.4	-0.3	K5 III		370 s		
177596	86771		19 04 36.5	+22 05 25	0.000	-0.05	7.8	0.5	4.0	F8 V		56 s		
177648	86774		19 04 38.3	+23 19 46	0.000	0.00	7.24	0.12	-1.6	B3.5 V	-14	400 s	12010	m
176862	229472		19 04 39.6	-48 24 14	+0.002	-0.03	6.6	1.3	3.2	G5 IV		22 s		
177241	187643	39 o Sgr	19 04 40.8	-21 44 30	+0.006	-0.06	3.77	1.01	0.3	gG8	+25	43 s	11996	m
177122	210890		19 04 41.1	-31 22 22	-0.003	-0.07	7.9	0.5	4.4	G0 V		51 s		
177076	210888		19 04 44.2	-36 50 40	0.000	-0.01	8.0	-0.1	0.6	A0 V		300 s		
177213	187644		19 04 45.8	-25 13 52	+0.002	-0.02	6.9	0.5	0.6	A0 V		92 s		
178157	31366		19 04 46.4	+52 45 39	+0.001	+0.01	7.9	1.4	-0.3	K5 III		430 s		
177399	142991		19 04 49.7	-8 39 05	+0.006	+0.02	7.4	1.1	0.2	K0 III	-71	270 s		
--	48061		19 04 50.5	+44 41 29	-0.001	0.00	7.9	1.3						
177649	104454		19 04 52.3	+19 08 38	0.000	+0.01	6.5	1.1	0.2	K0 III		180 s		
177599	104452		19 04 53.1	+15 44 11	+0.001	+0.02	6.9	0.1	0.6	A0 V	-12	170 s		
180427	9326		19 04 53.8	+79 38 47	-0.024	-0.06	7.9	1.1	0.2	K0 III	-72	340 s		
177398	142992		19 04 54.3	-8 07 22	-0.001	-0.07	7.4	1.1	0.2	K0 III		160 mx		
177782	86785		19 04 54.7	+28 11 18	0.000	0.00	7.6	0.1	0.6	A0 V		240 s		
178207	31371	51 Dra	19 04 55.0	+53 23 48	-0.001	+0.02	5.38	-0.01		A0	-24	22 mn		
177676	104456		19 04 55.7	+19 16 09	+0.001	+0.01	7.5	1.1	0.2	K0 III		270 s		
177442	142993		19 04 56.1	-4 02 27	0.000	-0.03	7.1	1.1	0.2	K0 III	-59	210 s		
177463	142996	15 Aql	19 04 57.4	-4 01 53	+0.001	-0.03	5.42	1.12	0.0	K1 III	-18	120 s	12007	m
177808	67782		19 04 57.6	+31 44 39	+0.006	-0.07	5.56	1.54	-0.4	M0 III	+6	130 mx		
176706	245907		19 04 58.1	-58 18 05	+0.003	0.00	8.0	0.1	0.4	B9.5 V		260 s		
177809	67781		19 04 58.1	+30 44 00	+0.002	-0.02	6.06	1.55	-0.5	M2 III	-16	210 s		
177699	86782		19 04 58.9	+21 14 25	-0.002	0.00	7.7	1.1	0.2	K0 III		290 s		
177123	210895		19 04 59.8	-37 57 42	+0.002	-0.02	7.2	0.0	0.4	B9.5 V		210 s		
177744	86787		19 05 02.9	+25 50 18	+0.001	+0.02	7.6	0.1	0.6	A0 V		240 s		
177359	187649		19 05 03.7	-20 56 51	0.000	-0.04	7.9	1.2	-0.3	K5 III		430 s		
176622	254456		19 05 04.7	-61 30 27	+0.002	-0.02	7.3	1.1	3.2	G5 IV		42 s		
177550	124242		19 05 05.2	+1 36 31	-0.001	0.00	7.94	2.21	-0.5	M2 III		210 s		
177059	229478		19 05 05.4	-43 51 54	+0.004	-0.01	8.0	0.8	0.4	B9.5 V		110 s		
177624	124246		19 05 08.2	+9 38 32	-0.001	+0.01	6.86	0.19		B5	-10			
178208	48071		19 05 09.7	+49 55 24	-0.001	+0.02	6.60	1.4	-0.3	K5 III	+8	240 s	12034	m
177316	187650		19 05 11.3	-26 08 18	+0.001	+0.02	7.4	1.6	-0.1	K2 III		180 s		
177878	67792		19 05 12.1	+31 26 20	0.000	-0.01	8.0	0.9	3.2	G5 IV		91 s		
177551	143001		19 05 15.5	-0 56 09	0.000	-0.01	7.0	0.1	0.6	A0 V		170 s		
177552	143003		19 05 18.4	-1 30 46	0.000	-0.01	6.53	0.35		F0				
177830	86791		19 05 20.6	+25 55 16	-0.002	-0.03	7.22	1.03	6.3	dK2	-74	11 s		
177724	104461	17 ζ Aql	19 05 24.4	+13 51 48	-0.001	-0.10	2.99	0.01	0.2	B9 V	-26	32 ts	12026	m
178090	48069		19 05 25.3	+42 10 38	0.000	-0.02	7.4	1.4	-0.3	K5 III		350 s		
177725	104462		19 05 29.6	+11 16 12	0.000	0.00	7.61	0.10		B9				
177320	210914		19 05 36.8	-35 22 17	-0.002	-0.02	7.5	2.0	-0.1	K2 III		110 s		
177426	187660		19 05 37.3	-24 40 29	+0.001	+0.01	7.3	1.4	0.2	K0 III		160 s		
177556	162178		19 05 37.9	-12 00 15	0.000	+0.06	7.9	1.1	0.2	K0 III		320 mx		
177319	210915		19 05 38.9	-33 47 01	+0.001	+0.01	7.5	0.7	0.2	K0 III		280 s		
178327	48078		19 05 40.6	+49 18 21	+0.003	+0.02	7.9	0.2	2.1	A5 V		150 s		
177517	162177		19 05 40.9	-15 39 37	0.000	-0.01	5.90	0.1		A0 p	-26			m
177702	124254		19 05 42.9	+4 16 42	+0.001	0.00	6.9	0.4	2.6	F0 V		72 s		
177786	124258		19 05 44.4	+9 46 12	-0.003	-0.03	7.6	1.1	-0.1	K2 III		310 s		
177427	187661		19 05 46.7	-29 04 52	+0.001	-0.01	6.9	0.1	0.0	B8.5 V		180 s		
178003	86796		19 05 46.9	+29 55 18	-0.001	-0.01	6.5	1.4	-0.3	K5 III	-28	220 s		
177749	124257		19 05 47.4	+6 32 49	0.000	-0.07	7.10	0.5	3.4	F5 V	-18	54 s	12029	m
177810	104467		19 05 48.0	+12 12 27	+0.001	0.00	7.50	1.94	0.2	K0 III		79 s		
178030	86799		19 05 52.2	+29 06 48	+0.001	0.00	7.60	0.00	0.6	A0 V		250 s		
178031	86802		19 05 54.5	+28 49 06	0.000	-0.02	7.7	1.4	-0.3	K5 III		360 s		
177606	162182		19 05 55.4	-14 39 05	-0.001	+0.01	8.0	1.1	-0.1	K2 III		390 s		
178685	18148		19 05 55.8	+60 06 00	+0.006	+0.04	7.4	1.1	0.2	K0 III		270 s		
176729	254460		19 05 57.8	-63 44 42	+0.002	-0.03	7.90	0.1	0.6	A0 V		290 s		
176771	254461		19 06 00.0	-63 00 53	+0.001	-0.02	7.8	0.9	3.2	G5 IV		82 s		
177520	187668		19 06 04.9	-26 14 49	+0.001	+0.03	8.0	1.9	-0.5	M2 III		350 s		
177472	210925		19 06 09.0	-31 41 43	-0.002	-0.01	8.0	0.2	0.6	A0 V		230 s		

HD	SAO	Star Name	α 2000	δ 2000	μ(α)	μ(δ)	V	B–V	M_v	Spec	RV	d(pc)	ADS	Notes
			h m s	° ′ ″	s	″								
177101	245919		19 06 09.8	−53 32 32	+0.002	−0.03	8.0	1.0	0.2	K0 III		360 s		
177706	143015		19 06 10.6	−9 37 56	+0.001	0.00	6.6	0.8	3.2	G5 IV		47 s		
177756	143021	16 λ Aql	19 06 14.7	−4 52 57	−0.001	−0.09	3.44	−0.09	0.0	B8.5 V	−14	30 mx		
177405	––		19 06 16.3	−39 04 12			8.0	0.7	2.6	F0 V		66 s		
178329	48084		19 06 16.8	+41 24 49	0.000	−0.01	6.49	−0.15	−1.7	B3 V	−21	430 s		
177708	162189		19 06 18.2	−11 38 57	+0.004	−0.02	7.7	0.1	1.4	A2 V		110 mx		V805 Aql, m,v
177562	187672		19 06 19.0	−27 17 14	0.000	−0.01	7.2	−0.2	0.0	B8.5 V		270 s		
178210	67816		19 06 19.5	+32 18 48	+0.002	+0.01	8.0	1.4	−0.3	K5 III		410 s		
177983	104478		19 06 19.6	+15 51 42	+0.001	+0.03	7.3	0.2		A m	−44			
177171	245921	ρ Tel	19 06 19.8	−52 20 27	+0.004	−0.11	5.16	0.53	3.0	dF2	+2	21 s		
177709	162191		19 06 20.9	−13 17 56	0.000	+0.02	8.0	1.1	0.2	K0 III		340 s		
177940	124266		19 06 22.0	+8 13 48	0.000	−0.07	5.4	1.6		M5 III	+32	340 mx		R Aql, m,v
––	124264		19 06 22.3	+7 09 24	0.000	+0.06	7.20					55 s	12037	m
177904	124265		19 06 22.8	+7 09 22	+0.001	+0.06	6.7	0.4	3.0	F2 V		55 s	12037	m
177474	210928	γ CrA	19 06 24.9	−37 03 48	+0.008	−0.27	4.21	0.52	4.0	F8 V	−52	12 ts		m
177758	162195		19 06 24.9	−11 53 49	−0.014	−0.39	7.26	0.56	4.4	G0 V	−3	37 s		
177710	162194		19 06 25.3	−13 27 22	+0.001	−0.03	7.9	1.1	−0.1	K2 III		370 s		
178307	67821		19 06 28.1	+37 23 18	0.000	+0.02	7.4	1.1	0.2	K0 III		270 s		
177662	187679		19 06 32.3	−23 15 12	+0.002	+0.01	7.4	0.6	0.2	K0 III		270 s		
178121	86812		19 06 33.0	+22 53 29	0.000	−0.02	7.8	1.1	0.2	K0 III		310 s		
177959	124270		19 06 33.5	+6 53 25	+0.002	−0.01	7.2	0.1	1.7	A3 V		120 s		
177880	143029		19 06 35.0	−1 20 46	+0.001	−0.01	6.77	0.10		B8			12038	m
181178	3175		19 06 35.5	+80 43 48	−0.003	+0.02	7.5	1.1	−0.1	K2 III		330 s		
177103	245920		19 06 35.5	−57 06 42	+0.004	+0.02	7.9	0.4	1.7	A3 V		110 s		
177942	124267		19 06 35.7	+1 31 21	0.000	+0.02	7.5	1.4	−0.3	K5 III		320 s		
178233	86819		19 06 37.5	+28 37 43	+0.006	+0.09	5.55	0.29		A5	−19	22 mn		
177300	245923		19 06 38.2	−51 25 03	+0.002	0.00	7.4	0.3	4.0	F8 V		48 s		BL Tel, m,v
178187	86817		19 06 38.2	+24 15 03	+0.004	+0.02	5.7	0.2	2.1	A5 V	−22	52 s		
177455	229494		19 06 43.6	−42 48 53	0.000	−0.04	7.4	0.8	0.2	K0 III		270 s		
177795	162198		19 06 43.6	−15 28 22	+0.001	−0.01	7.90	1.1	0.2	K0 III		330 s		m
––	229495		19 06 45.0	−41 32 24	+0.001	−0.02	7.4	1.6	−0.5	M4 III		380 s		
174341	––		19 06 45.7	−82 30 43			7.6	0.7	4.4	G0 V		44 s		
178352	67828		19 06 48.4	+36 07 41	−0.002	−0.01	7.3	1.1	0.2	K0 III		260 s		
178234	86820		19 06 49.8	+23 49 17	0.000	−0.02	7.7	1.1	0.2	K0 III		290 s		
178612	48099		19 06 51.2	+48 55 34	−0.003	0.00	7.30	1.4	−0.3	K4 III	−9	330 s		
177817	162201		19 06 52.0	−16 13 44	+0.001	0.00	5.90	−0.04		B9			12039	m
177565	210937		19 06 52.3	−37 48 37	−0.016	−0.35	6.16	0.72	3.2	G5 IV		21 ts		
178211	86821		19 06 52.8	+22 10 21	−0.001	−0.07	7.40	0.4	2.6	F0 V	+12	89 s	12050	m
177365	245925		19 06 54.4	−50 19 24	−0.002	−0.03	6.31	−0.09		A0				
177406	229493		19 06 55.4	−48 17 57	+0.002	−0.01	6.1	0.0	0.6	A0 V		120 s		
177716	187683	40 τ Sgr	19 06 56.2	−27 40 13	−0.004	−0.25	3.32	1.19	0.0	K1 III	+45	40 s		
176196	257656		19 06 56.4	−74 45 25	+0.006	0.00	7.4	0.6	0.4	B9.5 V		100 s		
177963	143032		19 06 56.8	−1 44 05	−0.001	−0.03	7.9	0.0	0.4	B9.5 V		280 s		
177799	187686		19 06 58.0	−22 29 50	0.000	−0.01	6.8	0.9	0.2	K0 III		210 s		
177456	229497		19 06 58.1	−45 58 13	+0.001	0.00	7.3	1.6	−0.5	M2 III		370 s		RX Tel, v
178125	104488	18 Aql	19 06 58.4	+11 04 17	0.000	−0.03	5.09	−0.07	−0.6	B7 V	−19	130 s		Y Aql, v
178739	31391		19 07 06.0	+51 39 28	+0.001	+0.02	8.0	0.5	3.4	F5 V		83 s		
179729	9323		19 07 06.3	+72 04 27	+0.002	+0.04	7.60	0.5	4.0	F8 V		53 s	12113	m
179039	18155		19 07 06.5	+61 18 09	−0.003	+0.03	7.9	0.4	3.0	F2 V		94 s		
177801	187691		19 07 07.0	−24 52 58	0.000	−0.01	7.9	1.1	0.2	K0 III		320 s		
178277	86828		19 07 07.9	+22 35 04	+0.001	+0.04	7.48	0.24		F0			12053	m
177522	229499		19 07 08.1	−44 13 34	0.000	−0.01	7.7	2.0	−0.1	K2 III		120 s		
177863	162204		19 07 08.2	−18 44 16	+0.001	0.00	6.3	0.0		B8				
178065	124282		19 07 08.9	+0 38 30	0.000	+0.01	6.56	0.05		B9				
178713	48107		19 07 11.4	+49 36 53	+0.002	−0.01	8.0	0.5	3.4	F5 V		82 s		
177822	187694		19 07 14.6	−24 31 25	+0.001	+0.02	7.7	0.4	0.6	A0 V		150 s		
178129	124288		19 07 17.9	+3 26 35	0.000	0.00	7.41	0.50	−6.8	B3 Ia	+23	2800 s		
178475	67834	18 ι Lyr	19 07 18.0	+36 06 01	0.000	0.00	5.28	−0.11	−1.0	B7 IV	−18	180 s		
177913	162211		19 07 18.3	−17 14 31	−0.001	−0.01	7.3	0.0	0.0	B8.5 V		280 s		
178591	48102		19 07 19.0	+41 03 14	+0.001	0.00	7.17	−0.04	−1.0	B5.5 V	0	350 s		
178796	31395		19 07 19.3	+51 22 46	0.000	−0.01	7.8	1.4	−0.3	K5 III		410 s		
177104	254466		19 07 19.6	−62 08 49	−0.002	+0.03	7.6	0.5	3.4	F5 V		68 s		
178989	31400		19 07 19.6	+58 04 29	0.000	0.00	7.8	1.1	0.2	K0 III		330 s		
178615	48106		19 07 19.7	+43 18 57	−0.002	+0.06	7.3	0.4	3.0	F2 V		74 s		
178165	124293		19 07 20.4	+5 13 08	−0.003	−0.02	7.60	1.1		K3 III-IV	−2			
177523	229500		19 07 22.1	−47 31 58	−0.001	0.00	7.5	0.3	1.7	A3 V		110 s		
178330	86832		19 07 23.7	+20 25 32	−0.002	−0.17	7.4	1.1	0.2	K0 III	−46	40 mx		
178449	67835	17 Lyr	19 07 25.4	+32 30 06	+0.010	+0.02	5.23	0.34	2.6	dF0	+4	32 s	12061	m
178131	143042		19 07 26.2	−1 07 33	−0.001	−0.05	7.6	1.6	−0.5	M2 III		350 s		
178190	124294		19 07 26.4	+6 09 45	0.000	+0.01	7.1	0.1	1.4	A2 V		130 s		
177846	187701		19 07 30.7	−28 38 12	+0.001	−0.01	6.1	1.1	−0.1	K2 III		180 s		
178450	67836		19 07 32.3	+30 15 16	+0.010	+0.11	7.63	0.73	5.6	G8 V	+9	26 s		
178356	104500		19 07 33.0	+18 33 45	0.000	−0.02	7.2	0.1	1.4	A2 V		140 s		
178538	67841		19 07 33.7	+34 23 36	+0.001	+0.01	7.2	0.4	2.6	F0 V		82 s		
178331	104503		19 07 36.8	+16 17 24	0.000	0.00	7.6	1.1	0.2	K0 III		280 s		

472

HD	SAO	Star Name	α 2000	δ 2000	μ(α)	μ(δ)	V	B-V	M_v	Spec	RV	d(pc)	ADS	Notes
178265	124298		19ʰ07ᵐ38ˢ.8	+6°59′00″	+0ˢ.001	-0″.01	7.1	0.4	2.6	F0 V		77 s		
178357	104506		19 07 43.6	+16 04 13	+0.001	+0.02	7.9	1.1	0.2	K0 III		320 s		
181467	3178		19 07 44.6	+80 45 12	-0.002	-0.01	7.5	0.1	1.4	A2 V		170 s		
178240	124300		19 07 46.1	+1 47 06	0.000	+0.01	7.9	0.1	1.4	A2 V		180 s		
178874	48119		19 07 46.7	+49 29 07	-0.001	+0.04	8.0	0.5	3.4	F5 V		82 s		
178661	67846		19 07 47.5	+38 55 42	+0.001	+0.01	7.60	0.26	1.7	A3 V	-29	130 s	12075	m
176865	254465		19 07 47.7	-68 53 01	+0.007	-0.03	6.8	0.1	1.4	A2 V		120 s		
177871	210957		19 07 48.9	-32 34 49	-0.002	-0.06	7.9	0.6	3.2	G5 IV		88 s		
178378	104508		19 07 49.7	+16 14 13	0.000	-0.02	7.8	1.4	-0.3	K5 III		370 s		
177917	210961		19 07 50.6	-31 09 04	-0.002	-0.08	7.5	2.0	-0.3	K5 III		140 mx		
178426	104510		19 07 51.4	+19 51 52	+0.001	+0.03	7.8	0.1	1.7	A3 V		160 s		
177971	187710		19 07 54.5	-25 42 06	+0.003	+0.01	7.5	0.9	3.2	G5 IV		65 s		
178427	104512		19 07 56.2	+17 44 18	+0.002	+0.03	8.0	1.1	0.2	K0 III		330 s		
178428	104511		19 07 57.1	+16 51 12	+0.004	-0.31	6.07	0.70	5.2	dG5	+14	17 ts		m
179141	31410		19 07 59.7	+58 08 40	-0.001	+0.19	7.9	1.1	0.2	K0 III		99 mx		
178821	48120		19 08 00.5	+43 58 02	+0.002	+0.02	8.0	0.5	3.4	F5 V		82 s		
178218	143056		19 08 00.6	-7 50 51	-0.002	-0.02	6.9	1.1	0.2	K0 III		210 s		
178539	86850		19 08 01.5	+26 00 43	-0.001	-0.01	7.2	1.1	0.2	K0 III	-23	240 s		
178540	86847		19 08 01.8	+24 43 34	+0.001	0.00	6.6	0.0	-1.0	B5.5 V	-19	320 s		
177920	210967		19 08 02.4	-34 28 32	-0.001	+0.01	7.80	1.1	-0.1	K2 III		380 s		
178476	86843		19 08 03.4	+21 41 56	+0.004	+0.07	6.23	0.40	3.0	F2 V	-40	42 s		
178910	48124		19 08 04.6	+49 04 56	-0.003	+0.02	8.00	1.1	-0.1	K2 III		420 s	12089	m
178616	67849		19 08 04.8	+30 37 44	-0.001	-0.01	7.6	1.4	-0.3	K5 III		360 s		
179068	31408		19 08 05.3	+55 23 55	0.000	+0.03	7.9	0.1	1.7	A3 V		170 s		
178049	187716		19 08 08.0	-26 50 02	0.000	+0.03	7.00	0.9	3.2	G5 IV		58 s		m
178847	48123		19 08 08.2	+44 02 12	0.000	+0.01	7.7	0.0	0.4	B9.5 V		290 s		
178541	86851		19 08 08.3	+21 46 45	-0.001	-0.05	7.7	1.1	-0.1	K2 III		330 s		
178404	124308		19 08 11.4	+9 34 51	+0.001	+0.02	7.6	0.4	3.0	F2 V		80 s		
178770	67859		19 08 13.0	+39 09 18	0.000	+0.01	7.5	1.6	-0.5	M4 III	-17	390 s		
178431	124309		19 08 13.6	+8 59 11	0.000	+0.03	7.80	1.1	0.2	K0 III		300 s		m
178359	124305		19 08 13.6	+1 17 56	0.000	+0.01	7.8	0.5	-6.4	F5 I	0	36 mn		TT Aql, v
178452	104515		19 08 14.3	+12 15 01	0.000	0.00	7.5	0.8	3.2	G5 IV		71 s		
178075	187718		19 08 14.4	-24 39 25	+0.001	0.00	6.2	0.0		B9				
178593	86853		19 08 16.3	+25 22 29	+0.001	-0.08	7.6	0.5	4.0	F8 V		53 s		
178175	162229		19 08 16.5	-19 17 24	0.000	+0.01	5.54	-0.11	-2.5	B2 V e	-20	410 s		
179142	31417		19 08 20.0	+55 19 51	0.000	+0.03	7.40	0.1	1.7	A3 V		140 s		m
177873	229513	δ CrA	19 08 20.6	-40 29 48	+0.003	-0.02	4.59	1.09	0.2	gK0	+20	63 s		
178512	104520		19 08 24.3	+13 05 51	0.000	-0.03	7.26	0.01		B8	-7			
179332	18170		19 08 25.5	+60 46 43	+0.005	-0.04	7.8	1.1	0.2	K0 III		240 mx		
179094	31413		19 08 25.6	+52 25 32	-0.011	-0.06	5.81	1.09		K1 IV	+4	33 mn		
178268	162232		19 08 26.0	-12 57 22	0.000	-0.02	7.4	1.1	0.2	K0 III		270 s		
--	31411		19 08 31.2	+50 15 24	+0.001	+0.02	7.9	1.5	0.2	K0 III		160 s		
178076	210975		19 08 31.7	-30 58 22	+0.005	-0.07	7.90	0.9	3.2	G5 IV		87 s		m
178077	210976		19 08 33.3	-31 05 42	-0.002	-0.01	8.0	0.9	0.2	K0 III		360 s		
179095	31415		19 08 33.3	+50 21 50	0.000		6.94	-0.04	0.6	A0 V		190 s		
--	--		19 08 33.5	+37 54 36	0.000	-0.02	8.0	1.1	0.2	K0 III		370 s		
178222	187724		19 08 33.8	-20 59 30	-0.001	-0.05	7.9	1.4	0.2	K0 III		210 s		
178568	104522		19 08 35.3	+14 26 07	+0.001	-0.01	6.7	0.0	0.4	B9.5 V	-26	180 s		
178690	86859		19 08 37.5	+24 10 45	0.000	0.00	6.9	1.1	-0.1	K2 III		240 s		
178875	67868		19 08 37.8	+38 30 32	-0.001	+0.01	7.80	0.2	2.1	A5 V		140 s		m
177975	229517		19 08 39.0	-39 57 36	-0.001	0.00	7.9	-0.1	0.4	B9.5 V		320 s		
178619	104524		19 08 40.0	+16 51 05	-0.002	-0.09	6.48	0.52	3.4	dF5	+10	34 s		
177897	229515		19 08 42.5	-45 04 34	+0.001	+0.01	7.8	1.5	0.2	K0 III		170 s		
178078	210979		19 08 43.5	-33 47 40	+0.001	-0.07	8.0	0.3	3.4	F5 V		82 s		
178291	162234		19 08 43.9	-18 57 11	0.000	+0.01	6.6	1.5	0.2	K0 III		100 s		
178250	187729		19 08 44.2	-23 11 25	-0.001	-0.02	6.50	0.9	3.2	G5 IV		46 s		m
178849	67870		19 08 45.1	+34 45 37	0.000	0.00	6.60	0.00	-1.6	B3.5 V	-8	400 s	12093	m
177540	--		19 08 45.8	-59 23 03			8.0	2.3	-0.1	K2 III		88 s		
178224	187728		19 08 46.8	-25 04 45	+0.001	0.00	6.7	-0.2	0.4	B9.5 V		190 s		
177996	229520		19 08 50.1	-42 25 41	+0.001	-0.12	7.88	0.86	6.1	K1 V		23 s		
177693	245937		19 08 51.8	-55 43 12	+0.004	-0.11	6.49	1.10	3.2	K0 IV		38 s		
178798	67869		19 08 51.8	+30 18 04	+0.001	0.00	6.8	1.1	-0.2	K3 III	-15	240 s		
178484	143070		19 08 52.0	-2 17 15	+0.001	+0.01	6.56	1.35		K0				
178692	104528		19 08 54.4	+18 36 39	+0.002	+0.02	8.0	0.5	3.4	F5 V		80 s		
178251	187731		19 08 56.0	-28 44 28	+0.002	-0.08	7.8	1.2	0.2	K0 III		190 mx		
178341	162239		19 08 56.8	-18 40 58	0.000	-0.01	8.0	0.4	2.6	F0 V		120 s		
177345	254471		19 08 57.1	-65 32 01	+0.003	-0.07	7.9	0.4	2.6	F0 V		120 s		
178637	104526		19 08 57.9	+11 17 38	-0.001	-0.01	6.64	1.16	0.2	K0 III	+8	140 s		
178693	104529		19 08 59.4	+14 46 52	+0.001	+0.07	6.7	1.1	0.2	K0 III		190 s		
178596	124318	19 Aql	19 08 59.7	+6 04 24	-0.001	-0.08	5.22	0.35	-6.6	F0 I s	-47	1400 s		
178574	124317		19 09 01.7	+1 21 09	+0.003	0.00	7.80	0.5	3.4	F5 V		73 s		m
178971	67881		19 09 01.8	+38 53 58	+0.001	-0.04	7.9	0.5	3.4	F5 V		78 s		
178459	143071		19 09 01.8	-7 01 02	-0.001	-0.02	7.9	0.9	3.2	G5 IV		87 s		
178911	67879		19 09 04.2	+34 36 02	+0.004	+0.20	6.74	0.64	4.5	dG1	-41	26 s	12101	m
178252	210987		19 09 07.7	-30 37 32	0.000	+0.02	7.6	0.0	0.6	A0 V		240 s		

HD	SAO	Star Name	α 2000	δ 2000	μ(α)	μ(δ)	V	B−V	M_v	Spec	RV	d(pc)	ADS	Notes
			h m s	° ′ ″	s	″								
177776	245940		19 09 09.3	−55 49 20	+0.002	−0.01	7.9	0.3	0.6	A0 V		170 s		
180777	9341	59 Dra	19 09 09.6	+76 33 37	+0.013	−0.12	5.13	0.31	3.0	F2 V	−4	25 ts		
178850	86871		19 09 11.7	+25 19 03	−0.001	−0.09	8.0	0.5	4.0	F8 V		63 s		
178146	229522		19 09 12.9	−41 15 50	−0.001	−0.02	7.8	1.4	−0.5	M2 III		270 s		
178973	67882		19 09 17.9	+32 52 08	0.000	−0.02	8.0	1.1	0.2	K0 III		340 s		
178715	104534		19 09 18.3	+11 37 37	+0.001	−0.01	7.1	0.9	3.2	G5 IV	+11	59 s		
178272	210989		19 09 21.7	−34 29 23	+0.003	−0.03	7.2	1.8	−0.1	K2 III		110 s		
177325	254472		19 09 21.8	−67 32 28	−0.001	−0.04	7.9	1.1	0.2	K0 III		350 s		
178717	104535		19 09 21.8	+10 14 27	0.000	−0.01	7.4	1.4	−0.3	K5 III	+5	320 s		
178947	67883		19 09 22.9	+30 33 59	0.000	+0.01	6.90	−0.04	0.4	B9.5 V	−28	200 s	12110	m
178492	162255		19 09 24.0	−13 27 25	−0.003	−0.07	7.10	1.1	0.2	K0 III		150 mx		m
178464	162251		19 09 24.3	−16 53 10	+0.001	−0.03	7.9	0.9	3.2	G5 IV		87 s		
178253	210990	α CrA	19 09 28.2	−37 54 16	+0.007	−0.10	4.11	0.04		A2 n	−18	14 mn		
178772	104539		19 09 33.9	+11 06 54	+0.001	−0.01	7.2	0.4	3.0	F2 V		67 s		
178439	187748		19 09 34.5	−23 11 25	0.000	+0.02	8.0	−0.2	0.6	A0 V		300 s		
178299	210996		19 09 36.1	−36 09 53	−0.001	−0.01	6.6	0.0	0.4	B9.5 V		170 s		
179557	31434		19 09 37.4	+58 16 30	+0.004	+0.06	7.8	0.4	2.6	F0 V		110 s		
178254	210994		19 09 39.5	−39 49 41	0.000	−0.05	6.4	0.9	0.2	K0 III		170 s		
179556	31438		19 09 42.1	+58 28 07	0.000	−0.01	7.6	1.4	−0.3	K5 III		380 s		
178469	187749		19 09 43.6	−27 06 51	−0.001	−0.02	6.7	1.0	0.2	K0 III		200 s		
179933	18187	55 Dra	19 09 45.6	+65 58 43	−0.001	+0.03	6.2	0.1	1.4	A2 V	−22	91 s		
178524	187756	41 π Sgr	19 09 45.6	−21 01 25	0.000	−0.03	2.89	0.35	−2.0	F2 II	−10	95 s		m
179143	67896		19 09 45.9	+37 47 38	−0.001	−0.04	6.8	0.4	2.6	F0 V		69 s		
179755	18182		19 09 46.3	+61 58 36	+0.001	0.00	7.7	1.4	−0.3	K5 III		400 s		
179505	31435		19 09 47.7	+55 53 18	+0.001	−0.03	7.9	1.1	−0.1	K2 III		380 s		
178555	162260		19 09 48.0	−19 48 13	+0.001	−0.08	6.13	1.16	0.2	K0 III		120 s	12096	m
180372	9338		19 09 49.5	+72 14 36	+0.001	−0.01	7.4	0.4	2.6	F0 V		93 s		
178744	143087		19 09 51.4	−0 25 41	0.000	0.00	6.34	−0.04		B8				
177389	254475		19 09 52.6	−68 25 29	+0.028	−0.05	5.33	0.91	0.2	gG9	−10	110 s		
179145	67899		19 09 56.1	+35 39 40	−0.001	+0.01	6.5	1.1	0.2	K0 III		180 s		
177409	−−		19 09 56.6	−68 18 04			7.2	0.8	3.2	G5 IV		62 s		
178322	229531		19 09 57.5	−41 53 33	+0.001	−0.02	5.88	−0.08	−1.1	B5 V	+6	230 s		
178345	211005	β CrA	19 10 01.5	−39 20 27	0.000	−0.04	4.11	1.20	0.3	gG5	+3	34 s		
−−	245950		19 10 01.7	−49 57 31	+0.002	−0.05	7.9	0.5	2.6	F0 V		83 s		
178746	143088		19 10 02.8	−7 16 41	−0.001	−0.07	7.4	1.1	0.2	K0 III	−58	88 mx		
178857	124337		19 10 07.0	+0 55 57	−0.001	+0.01	7.8	0.1	0.6	A0 V		240 s		
177222	257662		19 10 07.9	−71 32 56	−0.001	0.00	6.7	1.4	0.2	K0 III		120 s		
178880	124339		19 10 09.6	+2 21 26	0.000	−0.01	7.86	0.39		A0				
179166	67903		19 10 10.7	+32 02 48	+0.001	+0.01	7.6	1.1	0.2	K0 III		280 s		
179367	48149		19 10 11.8	+44 33 36	+0.001	0.00	7.2	0.2		A5 p				
179216	67911		19 10 16.0	+35 35 41	+0.001	+0.03	7.5	0.2	2.1	A5 V		120 s		
179394	48153		19 10 26.4	+42 36 03	0.000	−0.01	7.51	−0.10	0.0	B8.5 V		320 s		
179306	67915		19 10 26.8	+38 33 41	−0.001	−0.01	8.0	1.1	0.2	K0 III		360 s		
178606	211011		19 10 27.4	−30 00 25	+0.005	−0.06	6.6	0.6	3.4	F5 V		35 s		
179483	48160		19 10 27.4	+48 26 40	+0.002	−0.01	7.2	0.1	1.4	A2 V		150 s		
176464	257660		19 10 33.5	−78 03 34	+0.019	−0.05	7.9	0.6	4.4	G0 V		50 s		
179395	48156		19 10 33.7	+42 01 20	0.000	+0.01	7.09	0.01	0.4	B9.5 V		200 s		
177999	254482		19 10 37.8	−60 02 47	0.000	−0.03	7.70	0.00	0.4	B9.5 V		290 s		m
178673	187768		19 10 41.6	−29 48 51	−0.009	−0.11	7.8	1.0	4.0	F8 V		28 s		
179185	86894		19 10 42.0	+23 51 05	+0.001	0.00	7.9	0.4	2.6	F0 V		110 s		
176794	257661		19 10 44.4	−76 24 13	−0.024	−0.08	6.9	1.3	0.2	K0 III		120 mx		
179167	86893		19 10 44.7	+21 11 21	0.000	0.00	7.9	0.1	0.6	A0 V		280 s		
178152	245953		19 10 45.1	−56 46 06	0.000	−0.01	7.4	1.4	0.2	K0 III		160 s		
179280	67920		19 10 45.4	+31 38 06	+0.003	−0.03	7.2	0.4	2.6	F0 V	−16	81 s		
179396	67925		19 10 46.7	+38 54 20	+0.001	−0.01	7.8	1.1	−0.1	K2 III		380 s		
179281	67921		19 10 46.7	+31 19 19	0.000	+0.02	7.80	−0.02	0.4	B9.5 V		300 s		
178558	211014		19 10 47.5	−38 54 34	+0.001	−0.03	8.0	−0.5	3.4	F5 V		82 s		
179100	104560		19 10 49.6	+10 20 45	0.000	−0.02	7.4	1.1	0.2	K0 III	−15	260 s		
179421	67927		19 10 50.4	+39 10 18	+0.002	+0.03	7.9	1.4	−0.3	K5 III		430 s		
178881	162279		19 10 50.6	−12 08 34	+0.001	−0.01	7.70	0.1	0.6	A0 V		250 s	12119	m
178882	162278		19 10 51.4	−15 04 41	+0.001	−0.01	7.7	1.6	−0.5	M4 III		410 s		
178674	211018		19 10 52.3	−31 48 27	−0.001	+0.01	7.5	1.1	0.2	K0 III		260 s		
178756	187774		19 10 53.8	−28 05 01	−0.001	+0.08	7.9	1.4	0.2	K0 III		210 s		
178231	245956		19 10 54.1	−56 11 07	+0.001	−0.02	8.0	1.1	−0.1	K2 III		410 s		
179368	67924		19 10 55.4	+33 19 34	−0.001	0.00	7.6	1.1	0.2	K0 III		280 s		
179123	124355		19 10 56.8	+8 07 18	−0.004	−0.01	7.50	0.2	2.1	A5 V		68 mx	12129	m
179459	67931		19 10 56.8	+39 52 23	−0.001	−0.02	7.8	1.4	−0.3	K5 III		420 s		
179506	48165		19 10 57.3	+41 46 34	−0.004	+0.01	7.90	−0.04	0.0	B8.5 V		83 mx	12142	m
178954	143104		19 10 58.0	−6 37 16	+0.001	0.00	6.7	0.1	0.6	A0 V		160 s		
178812	187779		19 10 59.7	−24 11 07	+0.001	−0.02	7.4	0.9	0.2	K0 III		280 s		
179002	143106		19 11 01.1	−7 25 38	−0.002	−0.03	6.76	0.80		G0			12126	m
178628	211017		19 11 01.6	−39 00 18	−0.001	−0.02	6.2	0.3	−0.9	B6 V	−1	150 s		
179104	124356		19 11 05.3	+5 10 48	+0.001	0.00	7.6	1.1	0.2	K0 III		270 s		
178924	162280		19 11 05.4	−15 55 12	+0.001	+0.01	7.9	1.1	0.2	K0 III		330 s		
179124	124358		19 11 06.3	+5 16 29	0.000	+0.01	6.9	0.0	0.4	B9.5 V		190 s		

HD	SAO	Star Name	α 2000	δ 2000	μ(α)	μ(δ)	V	B-V	M_v	Spec	RV	d(pc)	ADS	Notes
179484	67936		19 11 07.7	+38 46 51	-0.021	-0.09	7.57	0.76	5.2	G5 V	+28	36 mn	12145	m
176298	257659		19 11 09.6	-79 29 28	+0.010	-0.01	7.5	1.1	0.2	K0 III		290 s		
179079	143111		19 11 09.6	-2 38 19	-0.010	-0.10	7.7	0.9	3.2	G5 IV		62 mx		
179309	86904		19 11 10.3	+23 28 48	0.000	0.00	6.8	0.0	0.4	B9.5 V	-18	190 s		
179218	104567		19 11 11.0	+15 47 16	+0.001	-0.01	7.1	0.0	0.4	B9.5 V	-3	210 s		
178814	187785		19 11 11.5	-28 32 05	-0.003	-0.03	8.0	1.3	0.2	K0 III		250 s		
178531	229538		19 11 11.8	-47 36 25	-0.001	-0.03	7.7	1.7	-0.1	K2 III		73 s		
179780	31457		19 11 13.5	+52 01 29	+0.003	+0.01	7.9	0.1	0.6	A0 V		290 s		
179187	124362		19 11 13.5	+8 29 56	+0.002	0.00	7.9	1.1	-0.1	K2 III		350 s		
179152	124360		19 11 15.2	+4 11 28	+0.001	-0.01	7.5	0.8	3.2	G5 IV		71 s		
180160	18201		19 11 17.7	+63 12 37	-0.003	-0.06	7.80	0.5	4.0	F8 V		58 s		
178840	187786		19 11 18.7	-29 30 08	+0.003	-0.02	6.2	0.2	0.4	B9.5 V		110 s		
178929	187789		19 11 19.1	-20 20 51	0.000	-0.01	7.9	0.0	0.4	B9.5 V		290 s		
178349	245959		19 11 21.8	-56 18 23	+0.002	+0.02	7.2	1.7	-0.3	K5 III		240 s		
179310	86907		19 11 22.6	+21 15 38	0.000	-0.01	7.90	0.1	1.4	A2 V		190 s	12140	m
179583	48168		19 11 22.9	+40 25 45	0.000	-0.02	6.18	0.09	0.6	A0 V	+6	110 s		
178789	211024		19 11 23.6	-34 54 12	-0.001	+0.02	7.4	1.7	-0.1	K2 III		160 s		
177928	254485		19 11 24.5	-65 52 38	+0.003	-0.01	7.5	0.1	1.4	A2 V		160 s		
179220	124366		19 11 25.6	+7 58 41	0.000	-0.02	7.9	0.9	3.2	G5 IV		86 s		
180024	31467		19 11 30.0	+58 32 02	0.000	+0.03	8.0	0.1	0.6	A0 V		300 s		
179422	86912		19 11 30.7	+26 44 09	+0.002	-0.03	6.36	0.41	3.4	F5 V	-27	39 s		
179485	86914		19 11 31.4	+29 53 25	+0.002	0.00	7.40	1.1	0.2	K0 III	-8	260 s	12150	m
179460	86916		19 11 37.6	+26 16 42	+0.001	-0.01	7.5	0.1	0.6	A0 V		220 s		
179313	104575		19 11 39.0	+11 59 56	0.000	0.00	8.0	0.0	0.0	B8.5 V		360 s		
180006	31468	53 Dra	19 11 40.3	+56 51 33	+0.004	+0.06	5.12	1.01	0.2	G8 III	-16	81 s		
175862	258821		19 11 40.8	-81 24 43	0.000	+0.06	7.8	0.3	2.6	F0 V		110 s		
179398	104580		19 11 43.4	+17 53 00	0.000	-0.01	7.90	0.1	0.6	A0 V		270 s	12146	m
179461	86917		19 11 44.7	+26 14 57	0.000	-0.01	7.4	0.1	1.7	A3 V		140 s	12153	m
179527	67946	19 Lyr	19 11 45.8	+31 17 01	-0.001	0.00	5.98	-0.07		B8 p	-30			
179423	104583		19 11 46.4	+18 05 16	-0.001	-0.01	7.6	0.4	2.6	F0 V		100 s		
179707	67950		19 11 47.7	+39 57 26	-0.002	+0.04	7.2	1.1	-0.1	K2 III		290 s		
179424	104584		19 11 53.2	+14 34 56	0.000	-0.01	8.01	0.00		B9				
179130	162299		19 11 54.8	-14 35 04	+0.003	+0.01	7.3	1.1	0.2	K0 III	-38	260 s		
179817	48183		19 11 56.1	+43 53 27	-0.004	+0.01	7.9	0.0	0.4	B9.5 V		87 mx		
178732	229544		19 11 56.6	-46 08 41	-0.002	-0.01	7.6	2.1	-0.1	K2 III		340 s		
179315	124374		19 11 59.1	+4 21 18	+0.001	0.00	7.7	1.1	-0.1	K2 III		320 s		
178869	211033		19 11 59.4	-39 16 39	+0.001	-0.02	7.5	0.1	0.0	B8.5 V		240 s		
179733	67956		19 12 00.2	+39 25 15	+0.001	+0.01	7.4	0.1	0.6	A0 V		230 s		
179343	124376		19 12 03.1	+2 37 21	0.000	-0.01	6.92	0.11		A0 e	-11		12147	m
178983	211038		19 12 03.3	-32 37 58	+0.001	-0.05	7.7	0.9	0.2	K0 III		240 mx		
179425	104587		19 12 04.4	+11 29 15	+0.002	+0.01	7.7	1.1	0.2	K0 III		290 s		
179958	48192		19 12 04.4	+49 51 15	-0.020	+0.64	6.57	0.65	5.0	G4 V	-38	25 ts	12169	m
179113	187806		19 12 05.8	-21 55 40	+0.003	+0.02	8.00	1.4	-0.3	K5 III		460 s	12139	m
178679	245967		19 12 06.0	-50 14 38	+0.001	-0.04	8.0	1.5	-0.1	K2 III		270 s		
179837	48187		19 12 09.3	+41 50 15	+0.001	-0.01	8.0	0.4	2.6	F0 V		120 s		
178937	211037		19 12 09.7	-37 34 58	+0.001	+0.01	6.57	1.02	0.4	gG2		120 s		
179734	67957		19 12 10.9	+34 22 38	0.000	+0.01	7.9	1.1	0.2	K0 III		320 s		
180161	31474		19 12 11.1	+57 40 18	+0.026	+0.40	7.03	0.79		K0		27 t		
174930	258820		19 12 12.1	-83 29 00	+0.086	-0.12	7.5	-0.1	3.4	F5 V		66 s		
179708	67959		19 12 14.8	+32 15 10	+0.001	+0.01	7.37	0.05	0.6	A0 V		210 s	12161	m
179959	48194		19 12 15.5	+47 22 30	0.000	0.00	6.70	1.1	0.2	K0 III		200 s		m
179756	67965		19 12 15.8	+36 37 47	-0.001	0.00	7.8	1.4	-0.3	K5 III		380 s		
179869	48191		19 12 17.3	+41 14 14	+0.001	-0.04	7.30	1.6	-0.5	M3 III	-11	360 s		
178087	254489		19 12 21.8	-67 55 49	-0.004	-0.05	7.9	1.1	0.2	K0 III		220 mx		
179709	67963		19 12 23.4	+30 20 52	0.000	0.00	7.65	0.01	0.6	A0 V		260 s	12162	m
178734	245970		19 12 23.9	-51 48 20	0.000	-0.02	7.05	1.48		K2				m
179782	67969		19 12 23.9	+36 10 55	-0.002	-0.01	6.93	-0.03	0.6	A0 V	-12	190 s		
179346	143132		19 12 27.2	-5 24 52	-0.001	-0.04	7.6	0.5	4.0	F8 V		52 s		
179558	104594		19 12 27.3	+16 50 59	+0.009	-0.21	7.94	0.76	5.2	G5 V	+37		12160	m
179201	187816		19 12 27.8	-21 39 29	+0.002	-0.01	6.4	1.3	0.2	K0 III	-5	120 s		
179586	104597		19 12 28.4	+18 00 07	+0.001	+0.03	7.3	0.4	2.6	F0 V	-34	84 s		
180711	18222	57 δ Dra	19 12 33.1	+67 39 41	+0.016	+0.09	3.07	1.00	0.2	G9 III	+25	36 s		Altais, m
179588	104602		19 12 34.2	+16 50 47	-0.001	-0.01	6.73	-0.01	-0.2	B8 V	-18	140 s	12160	m
179029	211042		19 12 34.3	-37 31 26	-0.001	-0.01	6.8	0.3	0.0	B8.5 V		140 s		
179648	86930		19 12 36.5	+21 33 16	0.000	0.00	5.90	0.03	1.4	A2 V	-6	79 s		
179491	124388		19 12 36.8	+6 02 22	0.000	-0.01	7.5	0.5	4.0	F8 V		49 s		
179492	124387		19 12 37.4	+5 17 50	+0.001	+0.03	7.8	1.1	0.2	K0 III		310 s		
179406	143134	20 Aql	19 12 40.5	-7 56 23	+0.001	-0.01	5.34	0.13	-2.3	B3 IV	-15	220 s		
179030	211044		19 12 40.9	-38 46 26	0.000	-0.03	7.9	1.8	-0.3	K5 III		280 s		
178938	229551		19 12 41.2	-45 16 13	0.000	-0.01	7.9	0.7	3.2	G5 IV		88 s		
179736	86934		19 12 42.3	+24 34 35	+0.005	-0.02	8.00	0.5	3.4	F5 V		82 s	12166	m
179380	162313		19 12 43.5	-11 33 01	+0.002	+0.01	7.7	1.1	0.2	K0 III		300 s		
179116	211045		19 12 43.7	-33 51 14	+0.004	-0.14	6.79	0.12	0.6	A0 V		32 mx		
179117	211046		19 12 43.9	-33 50 46	0.000	-0.01	7.30	0.1	0.6	A0 V		220 s		m
179564	124393		19 12 44.4	+8 36 39	0.000	+0.01	7.6	0.1	0.6	A0 V		240 s		

HD	SAO	Star Name	α 2000	δ 2000	μ(α)	μ(δ)	V	B-V	M_v	Spec	RV	d(pc)	ADS	Notes
			h m s	° ' "	s	"								
178941	229550		19 12 44.7	-47 09 17	-0.001	-0.01	7.9	1.8	0.2	K0 III		130 s		
179232	187822		19 12 45.5	-24 29 15	+0.004	-0.02	6.9	0.4	0.6	A0 V		100 s		
178845	245976		19 12 45.9	-50 29 11	+0.004	-0.04	6.13	0.95	0.2	gK0		150 s		m
178630	245971		19 12 46.6	-58 15 58	-0.001	-0.07	8.0	1.1	3.2	G5 IV		58 s		
179838	86941		19 12 55.9	+29 13 54	+0.001	+0.01	7.0	0.1	0.6	A0 V	-22	180 s		
--	67990		19 12 56.3	+39 38 49	-0.002	-0.07	8.0	0.9						
--	229553		19 12 59.4	-47 09 40	-0.001	-0.01	8.0	1.9	0.0	B8.5 V		27 s		
179596	124397		19 13 06.8	+4 56 09	-0.001	-0.01	8.0	0.5	3.4	F5 V		79 s		
179160	211052		19 13 07.8	-38 40 51	+0.002	-0.01	6.8	0.6	1.4	A2 V		56 s		
177078	257665		19 13 09.0	-77 51 51	+0.005	+0.01	6.8	1.5	-0.1	K2 III		160 s		
179711	104613		19 13 09.3	+12 01 22	-0.001	-0.01	7.6	0.1	0.6	A0 V		240 s		
179032	229555		19 13 10.4	-46 59 12	+0.004	0.00	7.9	1.0	0.2	K0 III		340 s		
179138	229559		19 13 12.1	-40 44 30	-0.003	-0.03	7.60	1.1	0.2	K0 III		300 s		m
179323	187835		19 13 13.4	-25 54 24	0.000	-0.01	5.80	1.38	-8.0	K0 Ia	+1	5400 s		
179784	104617		19 13 15.1	+15 02 08	-0.001	-0.01	6.69	1.39	-4.5	G5 Ib		1000 s		
179497	162326		19 13 15.3	-12 16 57	+0.001	-0.02	5.5	1.1	0.2	K0 III	-18	120 s		
179984	67994		19 13 16.4	+35 32 34	-0.001	-0.01	7.9	0.1	1.7	A3 V		170 s		
179785	104618		19 13 17.6	+14 56 33	0.000	0.00	7.26	1.51	-1.3	K3 II-III	-30	390 s		
179786	104620		19 13 21.8	+14 37 04	0.000	-0.02	7.57	1.67	-0.3	K5 III	+35	300 s		
--	67997		19 13 22.1	+35 53 44	+0.001	-0.01	8.0	1.2						
179758	104619		19 13 24.1	+11 20 42	+0.002	-0.02	6.9	0.9	3.2	G5 IV		55 s		
179325	211060		19 13 25.4	-31 41 32	+0.002	-0.01	7.8	0.7	3.2	G5 IV		84 s		
179737	124404		19 13 25.5	+9 41 32	+0.001	0.00	7.90	0.9	-0.9	G8 II-III		500 s		
179518	162329		19 13 25.6	-14 26 48	+0.003	-0.04	7.65	0.47		F5			12168	m
179739	124405		19 13 30.5	+6 39 06	+0.001	-0.01	7.8	0.1	1.4	A2 V		180 s		
179912	86948		19 13 33.1	+20 11 56	0.000	-0.02	8.00	1.1	0.2	K0 III		340 s	12187	m
178632	254496		19 13 33.3	-62 55 22	-0.006	-0.08	6.7	1.1	0.2	K0 III		150 mx		
179742	124406		19 13 33.3	+4 15 50	0.000	-0.01	7.6	0.4	2.6	F0 V		99 s		
179390	211064		19 13 40.7	-31 43 15	-0.002	-0.03	8.0	1.3	0.2	K0 III		250 s		
179761	124408	21 Aql	19 13 42.5	+2 17 38	+0.001	0.00	5.15	-0.07		B8 p	-5	28 mn	12182	m
179791	124410		19 13 43.9	+5 30 57	+0.001	-0.01	6.49	0.09		A2	+14			
180163	68010	20 η Lyr	19 13 45.3	+39 08 46	0.000	0.00	4.39	-0.15	-3.0	B2 IV	-8	270 s	12197	m
180138	68013		19 13 52.4	+36 25 28	0.000	+0.01	7.0	0.1	0.6	A0 V		180 s		
180008	86960		19 13 53.1	+25 45 31	-0.001	-0.01	7.00	0.05	0.6	A0 V		190 s		
179870	124415		19 13 53.4	+9 02 01	0.000	+0.01	7.3	1.1	0.2	K0 III	-15	250 s		
180348	48224		19 13 53.5	+48 20 56	0.000	-0.02	7.3	1.1	0.2	K0 III		270 s		
179687	162341		19 13 53.5	-11 12 11	+0.001	+0.01	7.9	1.1	-0.1	K2 III		390 s		
180610	31497	54 Dra	19 13 55.0	+57 42 18	-0.002	-0.07	4.99	1.16	-0.1	K2 III	-27	100 s		
179520	187848		19 13 56.2	-26 52 33	-0.001	-0.08	7.1	0.4	2.6	F0 V		70 s		
180186	68017		19 13 59.0	+38 58 35	+0.001	-0.02	7.7	1.1	-0.1	K2 III		360 s		
179892	124417		19 13 59.6	+7 31 03	-0.001	+0.01	7.6	0.2		A m				
179688	162344		19 13 59.8	-14 02 18	0.000	+0.02	7.2	1.1	0.2	K0 III		250 s		
179359	211067		19 14 00.1	-39 05 29	0.000	-0.03	8.0	0.2	0.6	A0 V		210 s		
180712	31502		19 14 01.3	+59 33 07	+0.006	+0.12	7.98	0.59	4.0	F8 V		57 s		
181043	18234		19 14 03.9	+67 06 55	+0.001	+0.01	6.8	0.1	1.4	A2 V		120 s		
180261	48223		19 14 04.5	+40 34 00	+0.001	0.00	8.0	1.1	0.2	K0 III		360 s		
180842	18230		19 14 05.7	+62 47 40	-0.002	+0.01	7.9	0.1	0.6	A0 V		280 s		
179609	187856		19 14 08.6	-22 03 36	0.000	+0.01	6.9	0.4	0.6	A0 V		100 s		
179939	124419		19 14 09.9	+7 45 52	0.000	+0.01	7.0	0.1	1.7	A3 V		110 s		
178563	254497		19 14 10.9	-67 27 15	0.000	-0.05	7.5	-0.3	1.4	A2 V		170 s		
180025	104632		19 14 11.2	+18 26 54	0.000	-0.01	7.9	0.0	0.0	B8.5 V		350 s		
180214	68019		19 14 13.5	+34 12 54	0.000	0.00	7.68	-0.07	0.4	B9.5 V		290 s	12211	m
180054	104635		19 14 15.6	+19 03 51	+0.003	-0.02	7.90	0.5	4.0	F8 V	-8	59 s	12201	m
179799	143163		19 14 15.8	-8 43 07	+0.002	-0.01	6.50	1.1	0.2	K0 III		180 s	12188	m
180778	31506		19 14 17.0	+59 41 21	+0.001	+0.02	7.67	0.15		A2 p	-29			
179874	143166		19 14 17.3	-3 30 06	+0.002	-0.05	7.9	0.7	4.4	G0 V		51 s		
179872	143168		19 14 18.0	-1 08 28	+0.001	-0.04	7.4	1.1	-0.1	K2 III		310 s		
179987	104631		19 14 18.1	+11 42 46	0.000	-0.01	6.9	1.1	-0.1	K2 III		230 s		
180124	86969		19 14 19.1	+25 00 47	0.000	-0.01	7.2	0.0	0.0	B8.5 V		250 s		
180656	31503		19 14 20.0	+55 56 26	-0.001	+0.02	6.8	1.1	-0.1	K2 III		230 s		
179846	143167		19 14 20.6	-5 41 33	-0.002	-0.01	7.9	0.0	0.0	B8.5 V		370 s		
180187	68020		19 14 20.9	+30 21 03	+0.001	0.00	7.8	0.1	0.6	A0 V		260 s		
179722	162349		19 14 21.9	-17 54 07	-0.001	-0.03	7.9	0.9	3.2	G5 IV		88 s		
180312	68024		19 14 25.3	+39 39 08	-0.001	-0.04	8.0	1.1	0.2	K0 III		360 s		
179847	143169		19 14 26.2	-7 51 51	+0.001	-0.02	8.0	1.1	0.2	K0 III		340 s		
180501	48232		19 14 26.2	+48 43 10	0.000	-0.02	7.44	0.03	0.6	A0 V		220 s		
180286	68022		19 14 28.9	+34 33 53	0.000	+0.09	7.1	0.5	3.4	F5 V		55 s	12215	m
180026	104637		19 14 32.3	+10 33 19	-0.001	+0.01	7.90	0.4	2.6	F0 V		110 s	12203	m
180313	68026		19 14 35.8	+36 09 55	-0.001	-0.01	7.8	0.0	0.4	B9.5 V		290 s		
179669	187862		19 14 36.4	-29 49 52	+0.003	0.00	6.88	1.62		K5				
182126	9365		19 14 38.3	+76 18 33	-0.001	-0.02	7.6	1.4	-0.3	K5 III		380 s		
179433	229573		19 14 39.5	-45 11 36	+0.006	-0.03	5.9	1.8	0.2	K0 III		45 s		
179695	187865		19 14 39.6	-29 14 35	0.000	-0.03	7.2	0.7	1.7	A3 V		51 s		
181142	18237		19 14 43.2	+65 16 22	0.000	+0.01	7.4	0.1	0.6	A0 V		230 s		
180028	124432		19 14 44.6	+6 02 54	+0.001	-0.01	6.97	0.81	-4.6	F6 Ib	-6	1600 s		

HD	SAO	Star Name	α 2000	δ 2000	μ(α)	μ(δ)	V	B-V	M$_v$	Spec	RV	d(pc)	ADS	Notes
179140	245983		19^{h}14^{m}47^s.5	-58°00'25"	+0^s.010	-0".24	7.23	0.64	4.7	G2 V		30 s		m
180083	124433		19 14 48.6	+8 05 47	+0.004	+0.03	7.8	0.5	3.4	F5 V		73 s		
179696	211087		19 14 49.9	-30 50 16	+0.002	-0.01	8.0	0.5	3.2	G5 IV		93 s		
180314	68027		19 14 49.9	+31 51 37	+0.002	+0.02	6.6	1.1	0.2	K0 III		190 s		
179853	162361		19 14 50.6	-17 20 47	+0.002	0.00	7.40	0.1	1.4	A2 V		160 s		m
179551	229575		19 14 54.0	-40 21 21	-0.001	-0.03	7.5	1.2	0.2	K0 III		280 s		
180351	86985		19 14 57.7	+29 31 39	0.000	0.00	8.0	1.1	-0.1	K2 III		380 s		
180125	104642		19 14 57.8	+10 24 33	-0.001	-0.01	7.1	0.0	0.0	B8.5 V		240 s		
180315	86984		19 14 58.4	+28 23 40	-0.001	-0.03	8.0	1.1	-0.1	K2 III		380 s		
180316	86982		19 14 58.5	+27 57 41	-0.001	+0.01	6.7	0.0	0.0	B8.5 V	-10	210 s		
179434	229574		19 15 00.3	-48 57 45	-0.001	+0.01	7.3	0.7	0.2	K0 III		260 s		
181873	9362		19 15 01.2	+73 49 47	+0.007	+0.03	8.0	0.1	0.6	A0 V		230 mx		
180126	124437		19 15 01.3	+9 48 27	+0.001	-0.01	8.03	0.18		B3 p	-9			
180242	86981		19 15 02.4	+20 12 12	-0.001	0.00	6.00	0.88	0.3	gG5	+7	140 s		
184146	3209		19 15 07.5	+83 27 46	+0.003	+0.01	6.30	0.1	1.4	A2 V	-15	96 s		m
180216	104650		19 15 08.7	+16 11 48	+0.001	-0.02	7.1	0.1	1.4	A2 V	-24	130 s		
180843	31517		19 15 11.5	+55 10 15	-0.001	+0.02	8.0	0.1	0.6	A0 V		290 s		
179391	245991		19 15 12.1	-53 22 04	-0.006	+0.01	7.8	0.3	3.4	F5 V		74 s		
180165	104649		19 15 13.5	+10 12 46	+0.001	+0.02	7.0	1.1	0.2	K0 III	-11	210 s		
179522	229577		19 15 13.6	-47 52 34	-0.002	-0.07	7.4	0.8	3.2	G5 IV		70 s		
180243	104653		19 15 13.6	+15 04 56	-0.001	-0.01	7.64	0.08	1.2	A1 V	-27	180 s		
179771	211091		19 15 16.6	-34 17 14	-0.002	-0.02	7.90	1.1	0.2	K0 III		350 s		m
180582	48238		19 15 16.8	+40 06 48	-0.003	-0.01	8.0	0.0	0.4	B9.5 V		330 s		
180317	86987	1 Sge	19 15 17.2	+21 13 56	+0.003	+0.01	5.64	0.11		A3	-23			
180756	48247		19 15 19.1	+50 04 15	0.000	0.00	6.30	0.8	3.2	G5 IV	+6	42 s	12240	m
180262	104655		19 15 19.9	+15 05 01	0.000	-0.01	5.42	0.87		G5 p	-25			m
179699	211089		19 15 20.0	-39 39 47	-0.002	+0.02	7.9	0.6	4.4	G0 V		50 s		
180086	143183		19 15 22.0	-6 03 00	0.000	0.00	6.6	0.4	2.6	F0 V		62 s		
180552	68043		19 15 22.2	+38 42 41	-0.001	+0.08	7.6	0.4	2.6	F0 V		100 s		
--	143184		19 15 23.3	-7 02 50	+0.002	+0.01	8.00	1.9		S7.7 e				W Aql, v
179808	--		19 15 24.3	-34 48 54			7.5	0.0	0.6	A0 V		230 s		
177369	257669		19 15 24.4	-79 01 41	-0.009	-0.04	7.7	0.5	3.4	F5 V		71 s		
180450	68040		19 15 24.6	+30 31 35	+0.002	-0.02	5.85	1.67	-0.5	M2 III	-63	170 s		
180015	162370		19 15 25.8	-16 05 58	-0.001	-0.04	7.5	0.9	0.3	G8 III	+41	270 s		
180681	48245		19 15 25.9	+45 20 09	0.000	0.00	7.4	0.1	0.6	A0 V		230 s		
179997	162368		19 15 26.1	-19 47 12	+0.001	-0.02	7.7	0.4	1.7	A3 V		100 s		
180127	143185		19 15 26.8	-3 06 49	0.000	+0.01	7.9	1.1	-0.1	K2 III		380 s		
180352	104660		19 15 27.6	+19 25 12	+0.001	-0.01	7.19	1.07	0.2	K0 III	-27	230 s		
179950	187882	42 ψ Sgr	19 15 32.2	-25 15 24	+0.003	-0.03	4.85	0.56	3.4	dF5	-34	21 mn	12214	m
181984	9366	60 τ Dra	19 15 32.8	+73 21 20	-0.033	+0.11	4.45	1.25	-0.2	K3 III	-30	70 mx		
179949	187883		19 15 33.0	-24 10 45	+0.009	-0.09	6.25	0.54	4.0	F8 V		27 s		
180377	104662		19 15 33.2	+18 30 58	-0.001	0.00	6.43	1.75	-0.5	M2 III		200 s		m
180263	104659		19 15 34.9	+11 33 17	+0.012	-0.17	7.9	1.1	0.2	K0 III		50 mx		
179831	211097		19 15 35.0	-36 56 57	-0.001	-0.02	6.8	0.6	1.7	A3 V		54 s		
179810	211095		19 15 36.8	-38 50 44	+0.002	-0.01	7.7	1.0	0.2	K0 III		310 s		
179811	--		19 15 39.7	-39 12 53			7.8	0.5	0.0	B8.5 V		160 s		
180110	162376		19 15 42.2	-14 50 14	+0.001	+0.02	7.80	0.02		B9	-9			
180682	48251		19 15 43.5	+40 21 36	-0.006	-0.05	7.0	1.1	0.2	K0 III		230 mx		
179904	211100		19 15 46.6	-33 31 49	+0.001	0.00	7.54	0.01	0.6	A0 V		240 s		
180683	68054		19 15 48.6	+38 22 49	+0.002	+0.23	7.7	0.6	4.4	G0 V		47 s		
180659	68053		19 15 50.0	+37 17 56	0.000	-0.01	7.5	0.1	0.6	A0 V		220 s		
180614	68049		19 15 55.8	+31 02 07	-0.007	-0.10	6.8	1.1	0.2	K0 III	-21	120 mx		
180553	87005		19 15 56.8	+27 27 19	0.000	-0.02	6.30	-0.06	-2.2	B5 III		450 s		m
180613	68051		19 15 57.6	+31 14 26	+0.001	-0.01	6.79	-0.08	0.4	B9.5 V	+9	190 s		
180398	104665		19 15 57.8	+13 06 29	+0.001	-0.01	7.92	0.07		B6 e	-33			
180583	87008		19 15 59.2	+27 55 37	0.000	+0.02	6.16	0.61	-2.0	F6 II	-16	370 s		
180451	104666		19 16 01.7	+16 09 39	+0.002	+0.02	7.30	0.4	2.6	F0 V	-51	85 s	12236	m
180067	187895		19 16 02.6	-24 42 59	+0.001	+0.03	7.6	0.2	1.4	A2 V		140 s		
179303	254509		19 16 05.4	-62 23 35	-0.002	-0.01	7.9	1.1	0.2	K0 III		350 s		
180584	87009		19 16 08.0	+24 46 44	-0.006	-0.18	8.0	0.9	3.2	G5 IV		66 mx		
180615	87012		19 16 08.4	+26 52 07	+0.002	-0.01	7.2	0.5	4.0	F8 V	+7	43 s		
180091	187898		19 16 11.2	-25 39 59	0.000	-0.02	7.3	1.0	0.2	K0 III		270 s		
180554	87010	1 Vul	19 16 12.8	+21 23 26	0.000	0.00	4.77	-0.05	-2.3	B3 IV	-17	210 s	12243	m
180068	211111		19 16 13.3	-30 27 33	+0.002	+0.01	7.2	0.5	1.7	A3 V		76 s		
180196	--		19 16 16.5	-16 58 41			8.00	1.9		S7.7 e	+2			T Sgr, v
180293	143200		19 16 16.5	-8 41 24	+0.001	0.00	7.3	1.1	0.2	K0 III		260 s		
179814	229582		19 16 17.5	-48 27 59	-0.001	-0.03	7.6	0.3	4.0	F8 V		52 s		
179886	229584		19 16 21.5	-45 27 59	0.000	0.00	5.2	2.4	0.2	K0 III	+6	23 mn		
180809	68065	21 θ Lyr	19 16 21.9	+38 08 01	0.000	0.00	4.36	1.26	-2.1	K0 II	-31	170 s		m
178274	257671		19 16 22.1	-75 48 00	+0.001	-0.02	6.8	0.4	1.4	A2 V		78 s		
181020	48270		19 16 22.2	+49 25 28	+0.001	+0.03	7.4	1.1	0.2	K0 III		270 s		
180556	104667		19 16 24.6	+14 05 51	+0.001	-0.06	8.0	0.5	4.0	F8 V		61 s		
180966	48269		19 16 26.0	+46 35 48	-0.002	+0.01	8.0	0.1	0.6	A0 V		300 s		
180555	104668		19 16 26.6	+14 32 41	+0.001	+0.01	5.63	-0.02	0.2	B9 V	-19	120 s	12248	m
180295	162391		19 16 27.3	-11 34 08	+0.002	-0.01	7.6	1.1	-0.1	K2 III		340 s		

HD	SAO	Star Name	α 2000	δ 2000	μ(α)	μ(δ)	V	B-V	M_v	Spec	RV	d(pc)	ADS	Notes
			h m s	° ′ ″	s	″								
180070	211115		19 16 27.6	−33 48 34	−0.001	0.00	7.55	0.12	1.4	A2 V		150 s		
179009	254508	τ Pav	19 16 28.4	−69 11 26	+0.002	−0.02	6.27	0.17		A p				
180482	124455	22 Aql	19 16 30.9	+4 50 05	+0.001	−0.01	5.59	0.08		A2	−22	29 mn		
181021	48271		19 16 31.0	+48 21 24	+0.001	−0.02	7.5	0.9	3.2	G5 IV		71 s		
180093	211117		19 16 32.6	−33 31 18	−0.001	+0.02	6.1	0.6	4.4	G0 V		22 s		RY Sgr, v
180504	124457		19 16 34.4	+6 35 48	−0.001	−0.01	6.5	0.1	1.4	A2 V		100 s		
179775	246003		19 16 36.3	−52 26 17	0.000	−0.04	6.77	1.02		K0				
180133	211119		19 16 40.6	−31 53 07	0.000	−0.03	7.1	0.9	0.2	K0 III		240 s		
180275	162394		19 16 41.6	−19 18 25	+0.001	+0.02	6.6	1.6		M5 III	−45			R Sgr, v
180844	68071		19 16 43.3	+33 07 39	0.000	−0.01	7.22	−0.12	−1.0	B5.5 V	−30	440 s	12263	m
180810	68067		19 16 43.4	+31 08 21	0.000	−0.02	7.9	1.6	−0.5	M2 III		450 s		
180585	104674		19 16 43.5	+11 50 40	+0.001	−0.02	7.90	0.1	1.4	A2 V		190 s		m
179887	246010		19 16 46.7	−49 59 01	+0.005	−0.02	7.94	0.37	2.6	F0 V		110 s		m
180233	187905		19 16 48.6	−28 40 01	+0.001	0.00	7.5	1.4	−0.1	K2 III		250 s		
180684	104678		19 16 48.6	+18 58 33	+0.003	−0.08	7.0	0.6	4.4	G0 V	−3	33 s		
179931	229589		19 16 49.7	−48 38 02	−0.001	−0.01	7.5	0.4	3.0	F2 V		81 s		
181096	48278		19 16 51.1	+46 59 56	−0.002	+0.29	6.00	0.44	3.4	F5 V	−44	33 s		
180409	162401		19 16 52.0	−10 58 18	+0.010	+0.01	7.00	0.6	4.4	G0 V	−59	33 s	12244	m
181566	18257		19 16 52.9	+63 12 28	+0.003	0.00	7.20	0.5	3.4	F5 V	0	58 s	12296	m
179279	254512		19 16 58.2	−67 09 58	+0.004	−0.01	6.5	1.1	0.2	K0 III		180 s		
179419	254514		19 17 00.5	−65 13 40	+0.002	−0.03	6.7	0.3	0.4	B9.5 V		110 s		
181276	31537	1 κ Cyg	19 17 06.0	+53 22 07	+0.006	+0.13	3.77	0.96	0.2	K0 III	−29	52 s		
180590	124461		19 17 06.4	+0 44 32	0.000	0.00	7.7	1.1	0.2	K0 III		310 s		
179616	254517		19 17 08.8	−61 48 50	+0.001	−0.05	7.4	0.4	2.6	F0 V		91 s		
181068	48282		19 17 08.9	+41 15 52	+0.001	−0.01	7.1	0.9	3.2	G5 IV		60 s		
181022	68084		19 17 08.9	+39 01 43	+0.002	+0.03	7.0	1.4	−0.3	K5 III		290 s		
180622	124464		19 17 10.8	+0 29 51	−0.001	−0.01	7.5	1.1	−0.1	K2 III		320 s		
180811	87027		19 17 11.1	+22 25 33	0.000	−0.02	8.0	0.0	0.4	B9.5 V		310 s		
179366	254515		19 17 11.9	−66 39 41	+0.001	−0.02	5.53	0.18		A2 n	+12			m
181252	48295		19 17 18.3	+49 51 00	+0.004	−0.07	7.4	0.4	3.0	F2 V		76 s		
181069	68089		19 17 19.9	+39 07 14	+0.001	−0.04	6.56	1.13	0.2	K0 III		150 s		
182189	9371		19 17 21.7	+71 20 50	−0.001	−0.04	7.1	0.8	3.2	G5 IV		59 s		
180117	229593		19 17 22.9	−45 25 27	0.000	−0.09	7.40	0.2	2.1	A5 V		45 mx		m
180511	162407		19 17 23.5	−16 04 25	−0.002	−0.04	7.7	1.1	0.2	K0 III		310 s		
180938	87032		19 17 30.8	+25 14 18	+0.001	+0.01	7.5	0.1	0.6	A0 V		220 s		
180889	87031		19 17 31.7	+21 48 51	+0.002	+0.01	6.89	0.17		A3	−20			
181278	48299		19 17 32.4	+47 57 08	+0.002	+0.02	7.8	0.3	3.2	G5 IV		84 s		
181328	31540		19 17 33.1	+51 20 47	−0.001	+0.02	7.5	1.6	−0.5	M2 III		400 s		
181118	68093		19 17 34.3	+37 36 36	+0.001	−0.02	7.9	1.6	−0.5	M2 III		450 s		
180890	104688		19 17 37.6	+19 09 40	+0.002	−0.10	7.9	0.9	3.2	G5 IV		86 s		
180540	162413	43 Sgr	19 17 37.9	−18 57 11	−0.001	−0.01	4.9	1.2	−2.1	G8 II	+15	210 s		
180562	162417		19 17 39.7	−15 58 02	+0.002	−0.01	7.00	0.1	1.7	A3 V		120 s	12266	m
180939	87035		19 17 39.9	+22 26 29	+0.001	+0.01	6.90	0.00	−1.1	B5 V	−21	370 s		RS Vul, m,v
180968	87036	2 Vul	19 17 43.5	+23 01 32	0.000	0.00	5.43	0.02		B0.5 IV	+1		12287	ES Vul, m,q
182827	9386		19 17 45.7	+76 37 08	−0.003	+0.01	7.8	0.4	2.6	F0 V		110 s		
180782	124478		19 17 48.0	+2 01 54	0.000	−0.03	6.19	0.02		A0	−27			
180868	104691	25 ω Aql	19 17 48.8	+11 35 43	0.000	+0.02	5.28	0.20		A5	−14	23 mn		
181675	31553		19 17 49.2	+59 50 11	+0.001	+0.03	7.64	0.05		A0				
180940	104692		19 17 49.8	+18 50 55	+0.002	−0.01	7.6	0.7	1.7	G2 III-IV	+4	150 s		
180629	162421		19 17 50.3	−16 55 18	−0.003	−0.02	7.90	0.00	−1.0	B5.5 V	−11	600 s		m
181774	18273		19 17 51.1	+61 54 22	−0.002	+0.03	8.0	1.6	−0.5	M2 III		510 s		
180783	143235		19 17 57.0	−0 56 37	0.000	−0.02	7.20	0.2	2.1	A5 V		100 s	12281	m
181025	87042		19 17 57.7	+20 11 34	0.000	−0.02	8.00	0.2	2.1	A5 V		150 s		m
180236	229602		19 17 58.9	−47 04 28	−0.001	−0.01	7.4	−0.1	0.4	B9.5 V		250 s		
180237	229601		19 18 00.3	−48 06 02	−0.001	−0.02	7.9	0.3	0.0	B8.5 V		380 s		
180727	143234		19 18 00.5	−9 41 10	+0.001	+0.02	8.0	0.4	2.6	F0 V		120 s		
181119	68095		19 18 00.8	+31 01 20	−0.001	+0.02	6.50	0.08	1.7	A3 V	−25	91 s		
180255	229603		19 18 00.7	−45 37 29	−0.001	−0.01	7.9	1.1	−0.3	K5 III		430 s		
180785	143237		19 18 06.9	−5 25 24	−0.001	0.00	7.90	1.1	−0.1	K2 III		380 s	12283	m
180134	246017		19 18 09.2	−53 23 12	−0.001	−0.06	6.38	0.49	3.4	F5 V		37 s		
180742	162436		19 18 10.4	−11 26 54	−0.005	−0.03	7.9	1.1	0.2	K0 III		190 mx		
180699	162432		19 18 10.8	−18 51 50	0.000	−0.01	6.80	0.00	0.4	B9.5 V		190 s		m
181098	87047		19 18 11.7	+24 24 54	+0.006	+0.15	7.20	1.08	−0.1	K2 III	−74	290 s		
181799	18275		19 18 11.8	+60 57 38	+0.003	+0.05	7.13	−0.08		B9	−18			
180445	211142		19 18 12.5	−38 23 05	+0.009	−0.10	7.9	1.0	3.2	G5 IV		64 s		
180700	—		19 18 16.2	−19 21 25			7.9	1.3	−0.5	M4 III		490 s		
180119	246018		19 18 18.5	−55 36 09	0.000	−0.02	8.00	1.1	0.2	K0 III		360 s		m
180022	246016		19 18 20.1	−58 52 27	+0.002	−0.03	7.3	0.8	1.7	A3 V		47 s		
180543	211148		19 18 21.3	−33 09 10	0.000	−0.03	7.66	−0.06	0.0	B8.5 V		340 s		
181048	104698		19 18 22.6	+14 28 27	+0.001	+0.04	7.7	1.1	0.2	K0 III		290 s		
180945	124484		19 18 23.1	+5 07 51	+0.004	+0.01	7.3	0.5	3.4	F5 V		59 s		
180471	211143		19 18 23.7	−39 52 31	0.000	−0.04	7.8	0.6	0.2	K0 III		330 s		
181226	87054		19 18 26.8	+29 57 25	−0.001	0.00	7.86	−0.03	0.6	A0 V		280 s		
181164	87053		19 18 28.0	+26 03 26	0.000	+0.01	7.3	0.0	−1.0	B5.5 V	−8	420 s		
181279	68108		19 18 28.5	+33 56 44	−0.006	0.00	7.7	0.9	3.2	G5 IV		79 s		

HD	SAO	Star Name	α 2000	δ 2000	μ(α)	μ(δ)	V	B-V	M_v	Spec	RV	d(pc)	ADS	Notes
			h m s	° ' "	s	"								
180768	162442		19 18 29.6	-18 41 49	+0.001	+0.01	7.4	0.4	2.6	F0 V		92 s		
181596	31556		19 18 29.9	+50 13 39	0.000	0.00	7.6	1.4	-0.3	K5 III		380 s		
180973	124486		19 18 31.2	+0 25 25	+0.002	+0.04	6.8	0.4	2.6	F0 V		68 s		m
181099	104700		19 18 31.2	+16 41 51	+0.001	-0.03	7.46	0.24		A m	-36			
180972	124487	23 Aql	19 18 32.3	+1 05 07	+0.001	+0.02	5.10	1.15	-1.1	K2 II-III	-24	180 s	12289	m
180896	143245		19 18 33.0	-6 41 17	0.000	0.00	7.9	1.1	0.2	K0 III		330 s		
181597	48315		19 18 37.6	+49 34 11	+0.001	+0.06	6.31	1.12	0.0	K1 III	-14	180 s		
180796	—		19 18 38.6	-18 43 22			8.0	1.1	0.2	K0 III		360 s		
181120	104704		19 18 39.3	+15 40 49	+0.001	-0.03	7.92	0.07		A0	-22			
180879	162452		19 18 40.8	-14 32 21	+0.001	-0.01	7.9	1.1	0.2	K0 III		340 s		
180183	246019		19 18 41.4	-56 08 41	-0.003	-0.01	6.82	-0.16	-2.3	B3 IV		670 s		m
181824	31567		19 18 42.6	+57 00 52	+0.001	0.00	7.8	0.2	2.1	A5 V		140 s		
180546	211153		19 18 43.1	-39 32 52	+0.004	-0.08	7.5	0.5	3.4	F5 V		62 s		
181144	104707		19 18 45.0	+16 29 17	-0.003	-0.02	6.9	0.5	3.4	F5 V	-4	51 s		
180976	143250		19 18 47.0	-5 46 22	0.000	-0.03	7.90	0.1	0.6	A0 V		280 s	12290	m
181182	104711		19 18 48.4	+19 36 37	+0.001	0.00	6.30	0.00	-1.6	B7 III	-17	350 s		U Sge, m,v
181380	68120		19 18 49.5	+37 35 08	-0.002	-0.03	7.7	1.1	-0.1	K2 III		340 s		
181166	104708		19 18 49.7	+16 15 01	0.000	0.00	8.0	0.1	0.6	A0 V		280 s		V889 Aql, v
181053	124492	24 Aql	19 18 50.7	+0 20 20	+0.001	+0.01	6.41	1.05	0.2	gG9	-29	160 s		m
183051	9392		19 18 52.1	+76 47 33	-0.005	+0.01	7.1	1.1	0.2	K0 III		240 s		
181122	124497		19 18 52.5	+9 37 05	0.000	-0.04	6.32	1.05	0.2	G9 III	-12	150 s		
182689	9390		19 18 55.7	+73 33 03	+0.001	+0.06	6.7	0.2	2.1	A5 V		84 s		
181521	48316		19 18 57.9	+40 21 37	0.000	+0.01	6.85	0.02	0.6	A0 V		170 s		
181469	68130		19 18 58.0	+39 16 01	0.000	-0.01	7.80	0.1	1.4	A2 V		180 s	12310	m
180702	211161		19 18 58.9	-33 16 42	-0.001	-0.13	6.94	0.58	4.4	G0 V		31 s		m
180633	229614		19 18 59.2	-41 18 24	0.000	0.00	7.5	0.5	0.6	A0 V		250 s		
180928	162462		19 18 59.9	-15 32 11	-0.007	-0.27	6.06	1.43	-0.2	K3 III	-18	38 mx		m
181470	68129		19 19 01.0	+37 26 43	+0.001	+0.02	6.20	-0.03	-0.6	A0 III	-14	230 s		
181409	68125		19 19 03.6	+33 23 20	-0.001	-0.03	6.60	-0.19	-3.0	B2 IV	+10	830 s		
181330	87061		19 19 05.4	+27 15 39	0.000	-0.02	7.0	1.4	-0.3	K5 III	-23	270 s		
180748	211164		19 19 06.9	-33 09 33	+0.009	-0.03	7.80	0.69	3.0	G3 IV		90 s		m
180548	229613		19 19 08.3	-46 01 34	+0.002	-0.03	7.4	0.8	-0.1	K2 III		310 s		
180953	162465		19 19 09.5	-15 54 30	+0.002	0.00	6.87	2.31		N7.7	-45			V1942 Sgr, v
179392	257676		19 19 10.5	-72 38 57	+0.002	0.00	7.6	0.5	3.4	F5 V		68 s		
181359	87062		19 19 10.5	+26 55 55	-0.001	-0.02	7.4	0.1	1.4	A2 V		150 s		
183076	9393		19 19 15.4	+76 28 50	+0.004	+0.06	8.0	0.5	4.0	F8 V		63 s		
181360	87063		19 19 16.0	+23 18 14	0.000	-0.01	7.62	0.06	-1.6	B3.5 V	-14	500 s		
181253	104717		19 19 16.9	+14 10 39	+0.001	0.00	7.4	0.6	4.4	G0 V	-31	39 s		
181214	124505		19 19 19.4	+7 08 36	0.000	+0.01	7.8	0.5	0.6	F8 III		250 s		
180135	254525		19 19 21.5	-62 52 31	-0.001	-0.01	7.9	1.1	-0.1	K2 III		400 s		
180771	211166		19 19 21.5	-36 53 39	-0.011	-0.03	6.8	0.8	0.2	K0 III		85 mx		
181361	104720		19 19 26.1	+19 43 07	+0.001	+0.01	8.00	0.1	0.6	A0 V		280 s	12309	m
181492	68136		19 19 27.6	+31 57 58	0.000	0.00	6.84	-0.08	-1.0	B5.5 V	-19	360 s		
181776	48336		19 19 27.8	+48 58 07	0.000	-0.03	7.6	1.6	-0.5	M2 III		420 s		
180802	211169		19 19 30.8	-36 39 30	+0.003	-0.09	7.20	0.5	3.4	F5 V		58 s		m
181410	87068		19 19 31.8	+22 34 14	+0.001	-0.01	7.3	1.6	-0.5	M4 III		340 s		
181679	48331		19 19 33.1	+41 16 17	-0.008	0.00	7.8	0.9	3.2	G5 IV		83 s		
181680	48332		19 19 35.7	+40 35 08	-0.003	+0.01	7.9	0.1	0.6	A0 V		290 s		
181960	31574		19 19 36.3	+54 22 34	+0.001	-0.02	6.26	0.03	0.6	A0 V	-6	130 s		
184311	3219		19 19 36.9	+81 47 45	-0.019	-0.02	8.0	1.4	-0.3	K5 III		310 mx		
180575	246028		19 19 37.2	-51 34 21	+0.003	-0.04	6.59	0.04		A0				m
181681	48333		19 19 37.8	+40 16 02	+0.002	0.00	7.40	1.4	-0.3	K4 III	-21	350 s		
181655	68144		19 19 38.7	+37 19 49	-0.006	-0.19	6.31	0.68	5.6	G8 V	+2	14 s		
182400	18290		19 19 39.0	+66 58 24	-0.001	-0.03	8.0	1.6	-0.5	M2 III		510 s		
181333	104722	28 Aql	19 19 39.2	+12 22 29	0.000	+0.02	5.53	0.26	0.6	gF0	+3	97 s		m
180885	211175		19 19 39.8	-35 25 17	0.000	-0.01	5.59	-0.13	-2.0	B4 IV	-10	320 s		
181304	124510		19 19 40.8	+7 47 45	0.000	+0.01	7.8	0.1	1.7	A3 V		150 s		
181362	104724		19 19 44.5	+12 55 16	0.000	-0.01	7.6	0.1	1.4	A2 V		170 s		
181656	68145		19 19 45.7	+36 45 50	0.000	+0.01	7.5	0.5	3.4	F5 V		64 s		
182308	18287		19 19 45.9	+64 23 27	-0.001	+0.01	6.30	-0.04		B9				m
180238	254530		19 19 47.3	-62 53 50	+0.003	-0.10	8.00	0.9	3.2	G5 IV		91 s		m
181778	48341		19 19 49.9	+43 33 24	-0.002	+0.03	7.7	1.1	0.2	K0 III		320 s		
181630	68146		19 19 51.7	+34 46 18	-0.002	-0.03	8.0	1.1	0.2	K0 III		340 s		
181382	104726		19 19 51.9	+12 01 44	+0.004	-0.10	7.4	0.5	3.4	F5 V		62 s		
181657	68150		19 19 52.7	+35 32 16	+0.001	0.00	7.80	1.1	-0.2	K3 III		370 s	12328	m
181383	104728	29 Aql	19 19 52.9	+11 32 06	+0.003	+0.03	6.02	0.08	1.7	A3 V	-22	73 s		
181827	48343		19 19 55.0	+45 01 56	+0.001	+0.04	7.3	1.1	0.2	K0 III		260 s		
181365	124514		19 20 00.9	+5 35 16	-0.001	+0.01	8.00	0.9	3.2	G5 IV		89 s		m
181235	143272		19 20 02.3	-9 24 41	+0.002	0.00	7.5	0.4	2.6	F0 V		95 s		
181255	143275		19 20 02.6	-6 37 41	0.000	-0.01	7.10	1.1	0.2	K0 III		230 s	12314	m
181366	124515		19 20 06.0	+4 28 49	-0.001	+0.01	8.0	1.1	0.2	K0 III		320 s		
181602	87082		19 20 07.3	+26 38 42	+0.002	-0.05	7.40	0.5	3.4	F5 V		62 s		m
181601	87083		19 20 07.4	+26 39 44	0.000	+0.01	7.40	1.1	0.0	K1 III	-20	280 s		m
181219	162482		19 20 08.0	-14 27 09	+0.002	-0.01	7.8	1.1	0.2	K0 III		330 s		
181631	87087		19 20 08.8	+27 54 37	+0.001	+0.02	8.0	0.1	1.7	A3 V		170 s		

HD	SAO	Star Name	α 2000	δ 2000	μ(α)	μ(δ)	V	B-V	M$_v$	Spec	RV	d(pc)	ADS	Notes
			h m s	° ′ ″	s	″								
181386	124516		19 20 10.3	+4 02 46	+0.001	-0.01	7.90	0.9	-2.1	G5 II	+7	770 s	12322	m
181494	104735		19 20 11.5	+14 08 19	0.000	-0.01	7.88	0.41		B9				
181036	211183		19 20 11.6	-33 27 43	0.000	-0.01	8.0	0.0	1.7	A3 V		180 s		
181012	211181		19 20 13.2	-36 13 42	+0.001	-0.03	7.8	1.1	-0.5	M2 III		470 s		
181368	143282		19 20 15.9	-2 35 57	+0.002	+0.09	7.9	1.1	0.2	K0 III		210 mx		
182190	31587		19 20 15.9	+57 38 43	+0.002	+0.01	5.91	1.58	-0.5	M2 III	-21	190 s		
182191	31588		19 20 18.8	+56 53 10	+0.001	+0.01	8.0	0.1	0.6	A0 V		280 s		
181414	124520		19 20 19.3	+4 46 40	-0.002	-0.10	6.9	0.1	1.4	A2 V		23 mx		
181013	211185		19 20 23.9	-38 09 33	+0.001	0.00	8.0	0.7	2.6	F0 V		66 s		
181312	162488		19 20 24.5	-10 33 40	+0.001	-0.03	7.30	1.6		M5 III	-7		12320	m
181750	68159		19 20 25.3	+32 05 56	-0.001	-0.01	6.78	-0.10	0.4	B9.5 V		190 s		
181109	211191		19 20 26.0	-31 49 04	0.000	+0.01	6.58	1.67	-0.5	M3 III		240 s		
181420	143285		19 20 26.8	-1 18 37	+0.005	+0.03	6.6	0.4	3.0	F2 V		52 s		
181878	68166		19 20 27.0	+39 55 34	+0.001	-0.01	7.0	0.9	3.2	G5 IV		57 s		
181800	68161		19 20 29.4	+34 53 58	-0.002	0.00	8.0	1.1	0.2	K0 III		340 s		
181633	104739		19 20 29.7	+19 05 47	+0.001	-0.01	8.0	0.1	1.4	A2 V		200 s		
181524	124524		19 20 29.8	+9 44 26	+0.002	+0.02	7.7	1.1	0.2	K0 III		290 s		
182076	31586		19 20 31.2	+50 20 25	+0.002	+0.04	7.50	1.4	-0.3	K5 III		360 s	12343	m
181391	143286	26 Aq1	19 20 32.7	-5 24 57	+0.008	+0.05	5.01	0.92	1.8	G8 III-IV	-19	42 ts		f Aq1, m,v
181828	68164		19 20 32.8	+35 11 10	+0.001	+0.01	6.30	-0.11	-1.6	B7 III		380 s		m
181087	211190		19 20 34.4	-37 23 48	-0.001	+0.02	7.9	1.7	3.2	G5 IV		24 s		
181039	229629		19 20 35.1	-39 59 22	0.000	-0.02	7.6	0.3	1.4	A2 V		100 s		
181440	143292	27 Aq1	19 20 35.5	-0 53 32	0.000	+0.01	5.49	-0.04		B9	-27	28 mn		
181526	124527		19 20 36.3	+6 38 39	-0.001	-0.07	7.80	0.5	4.0	F8 V		57 s	12332	m
182077	48358		19 20 36.7	+49 27 35	0.000	0.00	7.9	1.4	-0.3	K5 III		430 s		
181421	143291		19 20 37.0	-2 28 32	+0.001	-0.01	7.7	1.1	0.2	K0 III		300 s		
181240	187992		19 20 37.9	-22 24 09	-0.001	+0.04	5.5	1.0	2.1	A5 V		14 s		
182564	18299	58 π Dra	19 20 39.9	+65 42 52	+0.002	+0.04	4.59	0.02	0.6	A2 IV	-29	63 s		
231195	104744		19 20 40.6	+14 25 09	-0.002	0.00	7.73	1.41	-8.2	F5 Ia		4600 s		m
181241	187993		19 20 40.9	-24 12 36	0.000	-0.13	6.9	0.5	0.2	K0 III		94 mx		
183627	9403		19 20 42.3	+78 09 37	+0.004	0.00	8.0	0.4	3.0	F2 V		98 s		
179555	257678		19 20 42.9	-74 39 35	+0.007	-0.01	7.8	0.5	0.6	A0 V		140 s		
181604	104742		19 20 42.9	+10 39 09	0.000	-0.05	8.0	0.5	2.3	F7 IV	-20	130 s		
182486	18297		19 20 45.2	+64 04 35	+0.002	-0.02	7.6	0.2	2.1	A5 V		130 s		
182951	9397		19 20 46.0	+72 06 04	-0.001	-0.01	7.1	1.1	0.2	K0 III	-12	240 s		
181725	87098		19 20 47.1	+21 48 41	+0.001	0.00	8.0	1.1	0.2	K0 III		330 s		
181110	211195		19 20 47.3	-38 44 47	+0.003	-0.02	7.2	1.5	-0.1	K2 III		190 s		
181475	143296		19 20 48.0	-4 30 08	-0.001	+0.01	7.40	1.4	-2.3	K5 II	+3	790 s		
179866	257679		19 20 48.7	-72 37 45	-0.020	-0.08	7.5	0.5	4.0	F8 V		50 s		
181423	162498		19 20 51.6	-11 18 05	+0.001	-0.02	7.3	1.1	0.2	K0 III		260 s		
181199	211200		19 20 53.4	-33 03 08	0.000	-0.11	7.5	0.7	4.4	G0 V		35 s		
181683	104749		19 20 53.7	+13 34 35	+0.001	-0.05	6.90	1.11	0.2	K0 III		190 s		
181751	87102		19 20 54.3	+22 11 55	+0.001	-0.01	6.6	0.0	0.0	B8.5 V		200 s		
181555	124531		19 20 56.2	+0 54 55	+0.002	+0.02	7.7	0.2	2.1	A5 V		130 s		
181752	104753		19 20 59.9	+19 08 43	+0.006	+0.06	7.00	0.5	3.4	F5 V	-47	52 s	12336	m
181041	229635		19 21 00.5	-45 22 26	+0.004	-0.03	7.3	1.1	-0.5	M2 III		260 mx		
182054	48360		19 21 02.5	+42 41 14	-0.003	-0.03	7.1	1.1	0.2	K0 III		240 s		
181636	124533		19 21 03.2	+5 06 41	-0.001	-0.06	7.2	1.4	-0.3	K5 III	-6	220 mx		
181342	188005		19 21 04.1	-23 37 10	-0.003	-0.02	7.9	0.4	0.2	K0 III		340 s		
180909	246040		19 21 05.4	-52 43 22	+0.012	+0.07	7.2	0.5	4.4	G0 V		37 s		
181609	143303		19 21 06.7	-1 10 36	+0.001	-0.01	7.0	1.1	-0.1	K2 III	+8	260 s		
181707	124537		19 21 10.9	+9 48 59	0.000	0.00	7.74	2.09	-0.3	K5 III		180 s		m
181829	87107		19 21 17.2	+20 57 46	0.000	0.00	7.43	0.00	0.6	A0 V		230 s		
181801	104754		19 21 17.4	+14 38 01	0.000	-0.01	7.67	0.00		A0				
--	68192		19 21 22.7	+37 04 16	-0.001	-0.01	8.0	1.2						
182440	31604		19 21 25.2	+57 46 00	+0.002	+0.04	6.4	1.1	-0.1	K2 III	+7	200 s		
182031	68189		19 21 26.8	+32 40 39	+0.002	-0.01	7.4	1.1	0.2	K0 III	-5	260 s		
181963	87111		19 21 27.6	+25 36 14	-0.001	-0.01	7.45	-0.02	-2.5	B2 V	-16	820 s		m
181321	211206		19 21 29.7	-34 59 02	+0.009	-0.10	6.48	0.63	5.2	G5 V		18 s		
180987	246045		19 21 34.7	-55 22 40	+0.001	0.00	7.5	0.9	3.2	G5 IV		61 s		
183556	9404		19 21 35.2	+76 33 35	-0.004	0.00	6.10	2.85		N7.7	+6			UX Dra, v
182353	31603		19 21 35.8	+52 22 35	0.000	+0.03	7.70	0.2	2.1	A5 V		130 s	12366	m
181732	124545		19 21 35.9	+0 26 37	+0.001	-0.01	7.6	0.2	2.1	A5 V		120 s		
181934	104763		19 21 35.9	+19 48 50	+0.001	+0.01	8.0	0.1	0.6	A0 V		280 s		
181558	162511		19 21 36.9	-19 14 03	0.000	0.00	6.40	-0.10		B8				m
182056	68195		19 21 37.9	+30 21 54	+0.001	+0.03	8.00	1.1	-2.2	K2 II		910 s		
181987	87113		19 21 39.0	+25 34 29	0.000	0.00	7.00	0.00	-1.6	B3.5 V	-15	460 s	12352	Z Vul, m,v
181428	188014		19 21 39.1	-29 36 19	+0.009	-0.01	7.10	0.57	2.8	G0 IV		72 s		
184102	9414		19 21 39.9	+79 36 10	+0.003	-0.03	6.0	0.1	1.4	A2 V	-3	83 s		
181577	162512	44 ρ¹ Sgr	19 21 40.2	-17 50 50	-0.002	+0.02	3.93	0.22	1.7	F0 IV	+1	28 s		
182008	87114		19 21 40.3	+26 38 07	0.000	-0.03	7.8	0.5	3.4	F5 V		73 s		
182057	87117		19 21 40.9	+29 54 18	0.000	-0.02	7.2	0.1	1.4	A2 V		140 s		
181576	162516		19 21 41.3	-17 14 43	-0.007	-0.03	7.40	0.5	4.0	F8 V		48 s	12337	m
181615	162518	46 υ Sgr	19 21 43.5	-15 57 18	0.000	-0.01	4.61	0.10	3.0	F2 V	+9	21 s		v
181480	188017		19 21 44.0	-26 09 59	-0.001	-0.03	7.27	1.44		K2				

HD	SAO	Star Name	α 2000	δ 2000	μ(α)	μ(δ)	V	B-V	M_v	Spec	RV	d(pc)	ADS	Notes
181612	162520		19^h 21^m 44^s.7	−14°02′00″	+0^s.002	0″.00	8.0	0.1	1.7	A3 V		180 s		
--	18308		19 21 47.9	+62 48 03	+0.002	0.00	8.0	1.2						
181935	104766		19 21 48.3	+14 32 35	−0.001	−0.01	7.7	1.1	0.2	K0 III		290 s		
181645	162521	45 ρ² Sgr	19 21 50.7	−18 18 30	+0.007	−0.09	5.9	1.1	0.2	K0 III	−13	100 mx		
181882	104767		19 21 53.1	+10 55 15	+0.001	+0.06	7.30	1.1	−0.1	K2 III	−72	280 s	12351	m
182032	87120		19 21 57.1	+22 30 22	0.000	−0.02	7.45	0.00	−1.0	B5.5 V	−18	430 s	12355	m
182237	68209		19 22 01.6	+39 29 14	+0.001	0.00	7.8	1.1	0.2	K0 III		300 s		
181806	143315		19 22 03.4	−4 44 20	0.000	−0.05	7.70	0.5	3.4	F5 V		72 s	12347	m
181544	188025		19 22 03.8	−29 31 24	+0.011	−0.01	7.10	0.58	2.8	G0 IV		72 s		
182010	104771		19 22 03.9	+17 45 12	+0.001	−0.01	7.03	0.07		A0	−28			
181401	229643		19 22 09.3	−42 00 58	−0.005	−0.03	6.34	1.14		K0				
181482	211212		19 22 10.8	−37 13 12	+0.002	+0.01	7.9	1.6	3.2	G5 IV		27 s		
182078	87124		19 22 12.2	+22 32 09	0.000	−0.02	7.82	−0.04	−1.6	B3.5 V		660 s		
182512	31610		19 22 19.0	+52 07 06	−0.001	−0.02	7.7	1.4	−0.3	K5 III		390 s		
181858	143321		19 22 20.6	−8 12 04	−0.001	−0.02	6.67	−0.03	−2.3	B3 IV p	−14	580 s		
181906	124553		19 22 21.1	+0 22 59	+0.005	+0.05	7.9	0.5	4.0	F8 V		59 s		
181907	143324		19 22 21.3	−0 15 09	+0.003	−0.03	5.83	1.09	0.3	gG8	−11	98 s		
181620	211215		19 22 25.5	−31 56 12	+0.001	−0.01	6.91	1.69		Ma				
182100	104773		19 22 29.9	+15 03 30	0.000	0.00	7.6	1.1	−0.1	K2 III		320 s		
182146	87129		19 22 30.6	+22 11 16	+0.001	−0.01	7.9	0.0	0.0	B8.5 V		350 s		
182218	87132		19 22 31.9	+27 09 35	−0.001	−0.03	8.00	1.1	0.0	K1 III		370 s		
182565	31611		19 22 33.0	+53 02 48	−0.001	0.00	7.9	0.1	0.6	A0 V		280 s		
182272	68215		19 22 33.2	+33 31 05	0.000	−0.03	6.06	1.03	0.2	K0 III	−16	140 s		
181939	143328		19 22 33.8	−3 43 06	0.000	−0.02	7.1	0.1	0.6	A0 V		200 s		
181295	246053		19 22 37.3	−51 13 52	+0.005	0.00	6.5	0.4	2.6	F0 V		53 s		
181454	229646	β¹ Sgr	19 22 38.1	−44 27 32	+0.001	−0.02	3.93	−0.08	−0.2	B8 V	−9	67 s		Arkab, m
181809	188043		19 22 40.1	−20 38 34	0.000	−0.09	6.8	1.1	0.2	K0 III		130 mx		
181484	229647		19 22 40.7	−44 27 24	+0.001	−0.01	7.21	0.33		A7				
182354	68225		19 22 40.8	+37 34 52	0.000	+0.01	6.6	1.1	0.2	K0 III		180 s		
--	--		19 22 44.1	+32 43 14	+0.002	+0.02	7.2	1.4	−0.3	K5 III		300 s		
182292	68221		19 22 46.5	+30 16 15	+0.004	+0.01	7.3	1.1	0.2	K0 III	−26	250 s		
182147	104776		19 22 47.0	+12 54 54	0.000	−0.02	7.8	1.1	0.2	K0 III		300 s		
182101	124564		19 22 48.2	+9 54 46	+0.001	+0.10	6.35	0.44	−2.0	F2 II	−20	47 mx		
182255	87136	3 Vul	19 22 50.7	+26 15 45	0.000	−0.01	5.18	−0.12	−1.9	B6 III	−12	250 s		
181296	246055	η Tel	19 22 50.9	−54 25 25	+0.002	−0.07	5.05	0.02		A0 n	+12	18 mn		
181720	211218		19 22 52.7	−32 55 10	+0.008	−0.43	7.85	0.59		G2		22 mn		
182128	124565		19 22 52.7	+8 36 24	0.000	−0.01	7.6	0.4	2.6	F0 V		99 s		
180838	--		19 22 55.2	−66 35 13			7.9	1.1	−0.1	K2 III		400 s		
181788	188045		19 22 56.6	−29 03 52	+0.003	0.00	7.4	1.5	−0.3	K5 III		340 s		
182715	31618		19 22 57.6	+55 04 07	0.000	+0.02	7.3	1.1	0.2	K0 III		260 s		
181327	246056		19 22 58.7	−54 32 16	+0.003	−0.08	7.2	0.6	3.4	F5 V		47 s		
182735	31620		19 23 00.2	+56 13 34	−0.002	+0.09	7.4	0.4	2.6	F0 V		89 s		
182487	48383		19 23 00.3	+42 58 26	+0.001	−0.01	7.01	0.02		A0				
--	68234		19 23 02.5	+38 35 03	−0.001	−0.02	7.6	2.2						
182549	48387		19 23 03.2	+46 17 19	+0.001	−0.03	7.80	0.8	−2.1	G6 II	−25	960 s		
182038	143340		19 23 04.4	−7 24 01	+0.003	−0.01	6.32	1.45		K0				
182379	68232		19 23 04.4	+33 03 43	0.000	−0.05	7.9	0.5	4.0	F8 V		59 s		
181622	229651		19 23 05.7	−43 12 39	−0.002	−0.01	7.7	1.9	−0.3	K5 III		390 s		
182239	104779		19 23 08.0	+14 55 16	−0.002	−0.01	6.6	0.1	1.7	A3 V	+9	92 s		
182040	162551		19 23 09.9	−10 42 10	0.000	+0.02	6.99	1.07		C	−47			
182293	87140		19 23 12.0	+20 16 40	−0.001	+0.10	7.13	1.14	−0.2	K3 III	−110	290 s		
182274	104781		19 23 12.4	+19 22 32	0.000	−0.08	7.81	0.48	3.7	F6 V		66 s		
181623	229654	β² Sgr	19 23 12.9	−44 47 59	+0.009	−0.05	4.29	0.34	0.6	F0 III n	+22	52 mx		
182614	48391		19 23 14.7	+47 11 46	+0.001	−0.01	8.00	0.1	0.6	A0 V		300 s	12397	m
182716	31622		19 23 16.9	+52 34 07	0.000	+0.01	8.0	0.1	0.6	A0 V		290 s		
181018	254547		19 23 22.8	−65 34 45	+0.001	−0.03	7.1	0.0	0.4	B9.5 V		220 s		
182691	31623		19 23 23.6	+50 16 17	0.000	+0.02	6.3	−0.1		B9	−24			
182335	87145		19 23 24.5	+20 34 06	0.000	−0.20	7.93	0.56	4.0	F8 V		56 s		
182380	87146		19 23 29.4	+23 58 32	+0.002	0.00	7.4	0.5	3.4	F5 V		63 s		
182692	48397		19 23 30.7	+48 04 12	0.000	0.00	7.7	1.1	0.2	K0 III		310 s		
180808	254545		19 23 31.7	−69 24 28	−0.002	+0.07	7.80	0.5	4.0	F8 V		58 s		m
182488	68239		19 23 33.8	+33 13 20	+0.007	+0.17	6.37	0.81	5.9	K0 V	−21	16 ts		
183317	18327		19 23 33.8	+69 00 55	−0.005	0.00	7.1	1.6	−0.5	M2 III		320 s		
182277	124578		19 23 33.8	+8 16 56	0.000	+0.01	7.5	0.1	1.4	A2 V		160 s		
181920	211228		19 23 35.3	−30 14 13	−0.001	+0.02	8.0	1.7	−0.3	K5 III		340 s		
181177	254552		19 23 36.4	−63 11 30	−0.011	−0.13	7.3	0.8	3.2	G5 IV		65 s		
181517	246065		19 23 36.6	−53 56 47	+0.001	−0.02	6.6	1.0	0.2	K0 III		190 s		
182090	162560		19 23 36.9	−15 48 45	0.000	−0.02	7.9	0.0	0.4	B9.5 V		310 s		
182296	124580		19 23 38.6	+8 39 37	0.000	+0.01	7.07	1.32	−4.5	G3 Ib	−14	1200 s		
181972	188059		19 23 39.2	−28 40 12	0.000	−0.02	7.8	2.4	−0.1	K2 III		69 s		
181459	246064		19 23 39.5	−55 56 07	+0.011	−0.06	7.2	0.4	3.0	F2 V		64 s		
182615	48396		19 23 44.5	+40 47 15	+0.001	+0.02	8.0	0.0	0.0	B8.5 V		370 s		
181893	211227		19 23 44.6	−37 04 38	0.000	−0.03	7.5	0.8	3.4	F5 V		41 s		
182592	68249		19 23 45.8	+39 49 08	−0.002	−0.02	7.8	0.1	0.6	A0 V		260 s		
182422	87148		19 23 46.7	+20 15 52	−0.001	+0.01	6.40	0.01	−0.2	B8 V	−27	180 s		

HD	SAO	Star Name	α 2000	δ 2000	μ(α)	μ(δ)	V	B-V	M_v	Spec	RV	d(pc)	ADS	Notes
			h m s	° ' "	s	"								
182024	188062		19 23 47.1	-28 40 42	+0.001	-0.01	7.8	1.9	-0.1	K2 III		140 s		
182593	68251		19 23 47.6	+39 35 24	+0.003	+0.01	7.1	1.1	0.2	K0 III	+1	230 s		
182110	162564		19 23 48.7	-17 11 45	0.000	-0.03	6.90	0.00	0.4	B9.5 V		200 s		m
181951	211229		19 23 50.0	-34 58 01	+0.002	0.00	6.9	1.2	0.2	K0 III		160 s		
182381	104788		19 23 50.4	+16 00 53	+0.001	+0.02	7.61	0.15		B9 p	-10			
182566	68247		19 23 51.8	+36 10 49	+0.002	+0.01	7.9	1.1	0.2	K0 III		320 s		
181869	229659	α Sgr	19 23 53.0	-40 36 58	+0.003	-0.12	3.97	-0.10	-0.2	B8 V	0	36 mx		Rukbat
182694	48401		19 23 56.4	+43 23 17	+0.001	-0.02	5.84	0.92	3.2	G5 IV	0	27 s		
182779	48404		19 23 56.4	+48 22 32	+0.001	+0.01	7.8	1.4	-0.3	K5 III		410 s		
182634	68254		19 23 57.2	+38 06 51	0.000	0.00	7.9	0.1	1.7	A3 V		170 s		
182567	68248		19 23 57.9	+32 43 25	+0.004	+0.03	6.8	1.1	0.2	K0 III		200 s		
182357	124584		19 23 58.3	+8 24 41	0.000	0.00	7.8	1.1	0.2	K0 III		310 s		
181998	211234		19 24 01.5	-33 32 30	+0.001	+0.01	7.7	-0.1	2.1	A5 V		130 s		
181975	211232		19 24 02.2	-35 56 28	+0.003	-0.04	7.9	1.4	0.2	K0 III		210 s		
182246	143353		19 24 03.1	-6 58 38	0.000	+0.02	8.0	0.4	2.6	F0 V		120 s		
182736	48403		19 24 03.2	+44 56 00	+0.008	+0.04	7.1	0.6	4.4	G0 V		35 s		
182383	104790		19 24 03.6	+11 26 28	+0.001	0.00	7.6	1.1	0.2	K0 III		280 s		
182281	143354		19 24 04.0	-5 30 04	-0.001	0.00	7.6	0.5	3.4	F5 V		69 s		
182444	104794		19 24 04.3	+18 01 07	0.000	0.00	7.84	0.14		A2				
182754	48406		19 24 05.1	+46 26 18	-0.004	+0.01	7.56	-0.06	0.6	A0 V		250 s	12412	m
181019	254551		19 24 05.3	-68 22 16	-0.003	0.00	6.34	1.23		K0				
182635	68258		19 24 05.9	+36 27 07	+0.001	+0.08	6.4	1.1	0.0	K1 III	-33	180 s		
182568	87159	2 Cyg	19 24 07.4	+29 37 17	+0.001	+0.01	4.97	-0.10	-2.3	B3 IV	-21	260 s		
183098	18325		19 24 07.8	+61 12 12	0.000	-0.02	7.68	1.04	-0.1	K2 III		360 s		m
183077	18324		19 24 09.6	+60 20 56	-0.003	+0.02	7.66	1.07		K0				
182671	68262		19 24 09.8	+39 12 42	-0.004	-0.04	7.46	1.38	-0.1	K2 III		220 s		
181064	254554		19 24 12.3	-68 15 32	+0.003	-0.02	7.5	0.0	0.4	B9.5 V		270 s		
181351	254558		19 24 12.3	-62 59 52	-0.013	+0.04	6.8	0.5	3.4	F5 V		46 s		
182489	104795		19 24 12.6	+18 44 28	0.000	-0.02	7.91	0.01	-0.2	B8 V		380 s		
178737	--		19 24 14.7	-81 33 01			7.8	1.1	0.2	K0 III		330 s		
182616	68260		19 24 17.8	+31 02 51	0.000	0.00	7.2	0.0	0.0	B8.5 V	-16	260 s		
180964	257682		19 24 17.9	-70 06 30	+0.007	-0.18	7.7	0.4	4.0	F8 V		55 s		
181042	254553		19 24 18.6	-69 08 31	+0.003	+0.02	7.5	1.1	0.2	K0 III		290 s		
181925	229660		19 24 21.2	-43 43 23	+0.002	-0.03	6.0	1.2	-0.5	M2 III		200 s		
182617	87166		19 24 21.6	+28 34 48	0.000	-0.01	7.60	1.1	0.2	K0 III		280 s		
182490	104797	2 Sge	19 24 22.0	+16 56 16	0.000	-0.01	6.25	0.08	-0.2	A2 III	+12	190 s		m
182618	87165		19 24 22.2	+28 05 16	-0.002	+0.01	6.53	-0.08	-1.7	B3 V		390 s		
182407	124586		19 24 22.8	+5 33 51	0.000	0.00	7.6	0.6	4.4	G0 V		43 s		
182298	143359		19 24 23.0	-9 20 18	-0.001	-0.03	7.1	1.4	-0.3	K5 III		300 s		
181378	254560		19 24 25.6	-63 21 41	+0.004	-0.01	7.8	1.6	-0.5	M4 III		430 mx		
181703	246070		19 24 26.1	-55 53 20	+0.001	-0.03	7.2	-0.4	0.0	B8.5 V		280 s		
182491	104796		19 24 27.3	+12 16 57	+0.001	0.00	7.1	1.1	-0.1	K2 III	-26	260 s		
182250	162571		19 24 29.1	-18 55 40	-0.003	-0.01	7.9	0.8	3.2	G5 IV		85 s		
182180	188079		19 24 30.0	-27 51 56	0.000	0.00	6.04	-0.13	-1.6	B5 IV	0	340 s		
182737	68272		19 24 30.4	+37 11 16	-0.001	-0.01	7.90	0.15	0.6	A0 V		260 s		
182181	188081		19 24 32.7	-28 36 50	0.000	+0.01	7.8	2.2	-0.3	K5 III		150 s		
182917	31632		19 24 32.9	+50 14 29	-0.001	-0.01	7.10	1.6		M7 III	-54			CH Cyg, v
182156	211241		19 24 34.3	-30 48 09	+0.004	-0.11	7.66	1.15	3.2	K0 IV		60 s		
182342	162577		19 24 41.2	-12 36 55	0.000	+0.02	7.4	0.9	3.2	G5 IV		67 s		
181870	246074		19 24 43.0	-51 47 27	+0.001	-0.02	7.7	1.2	0.2	K0 III		250 s		
182620	104806		19 24 44.3	+19 56 23	+0.003	+0.03	7.16	0.14	1.4	A2 V		130 s		
182410	143368		19 24 44.3	-5 25 04	0.000	+0.01	7.6	0.1	1.4	A2 V		170 s		
182571	104804		19 24 45.2	+16 57 26	+0.003	-0.01	6.82	0.15		A0	-1			
182071	229666		19 24 55.7	-44 11 41	-0.006	-0.01	6.9	0.9	0.2	K0 III		170 mx		
182572	104807	31 Aql	19 24 58.0	+11 56 39	+0.049	+0.64	5.16	0.77	3.2	G8 IV	-100	18 ts		m
181819	246075		19 24 58.8	-55 06 23	+0.001	0.00	7.5	0.1	2.6	F0 V		98 s		
182475	143373		19 25 01.4	-4 53 03	+0.003	+0.02	6.52	0.33		F0				
182286	188093		19 25 04.0	-29 18 33	0.000	-0.05	6.10	1.1	0.2	K0 III		150 s	12400	m
182346	188098		19 25 06.2	-23 02 31	+0.004	-0.02	7.50	0.5	4.0	F8 V		50 s		m
182390	162588		19 25 07.2	-18 33 45	+0.003	+0.01	7.0	0.5	4.0	F8 V		40 s		
182429	162590		19 25 07.9	-13 26 20	+0.001	0.00	7.9	1.6	-0.5	M4 III		470 s		
182842	68282		19 25 08.2	+35 59 58	+0.001	0.00	8.0	0.1	0.6	A0 V		280 s		
181820	246078		19 25 09.3	-56 59 00	+0.001	-0.02	7.9	0.3	3.0	F2 V		94 s		
181980	246085		19 25 09.6	-51 06 00	-0.001	0.00	7.3	0.8	-0.5	M2 III		370 s		
181847	246080		19 25 09.8	-56 10 54	+0.010	-0.03	7.7	1.7	0.2	K0 III		120 s		
182072	229667		19 25 10.1	-47 27 25	+0.006	-0.01	7.2	0.9	3.0	F2 V		53 s		
182369	188101	47 χ¹ Sgr	19 25 16.3	-24 30 31	+0.004	-0.05	5.03	0.23	2.1	dA5	-42	35 s		
--	--		19 25 17.1	-24 30 32			5.80							
182896	68285		19 25 18.0	+37 14 42	-0.001	-0.03	6.9	1.1	0.2	K0 III		210 s		
182453	162592		19 25 18.8	-16 50 14	-0.002	-0.08	7.9	1.1	-0.1	K2 III		130 mx		
182477	162595		19 25 21.6	-13 53 51	+0.005	+0.05	5.7	1.1	0.2	K0 III	-34	130 s		
182391	188102		19 25 21.9	-24 24 44	-0.002	0.00	7.1	0.0	0.4	B9.5 V		220 s		
182761	87186		19 25 22.2	+20 16 17	-0.001	-0.03	6.31	-0.02	1.2	A1 V	-32	110 s		
182158	229669		19 25 23.4	-45 00 44	+0.001	-0.01	7.9	0.9	-0.1	K2 III		390 s		
182807	87190		19 25 25.6	+24 54 46	-0.013	-0.63	6.19	0.51	3.7	F6 V	-4	25 mx		m

HD	SAO	Star Name	α 2000	δ 2000	μ(α)	μ(δ)	V	B-V	M$_v$	Spec	RV	d(pc)	ADS	Notes
182989	48421		19^{h}25^{m}27$.^s$6	+42°47'04"	-0$.^s$011	-0$.''$18	7.2	0.5	3.4	F5 V	-72	59 s		RR Lyr, v
182762	104818	4 Vul	19 25 28.5	+19 47 55	+0.007	-0.06	5.16	0.98	0.2	K0 III	+1	98 s	12425	m
182416	188105	49 χ³ Sgr	19 25 29.5	-23 57 44	-0.001	0.00	5.43	1.43	-0.3	K4 III	+40	140 s		
182640	124603	30 δ Aql	19 25 29.7	+3 06 53	+0.017	+0.08	3.36	0.32	2.1	F0 IV-V	-30	16 ts		m
182897	68290		19 25 30.6	+34 11 38	0.000	-0.01	7.3	0.5	4.0	F8 V		44 s		
182623	143386		19 25 37.9	-2 03 39	+0.001	+0.02	7.7	0.1	0.6	A0 V		260 s		
182865	87195		19 25 38.8	+26 06 14	0.000	0.00	7.3	0.0	0.0	B8.5 V		270 s		
181773	254568		19 25 43.1	-62 11 11	+0.008	-0.14	7.61	0.46	2.1	F5 IV		73 mx		m
182481	188112		19 25 43.4	-22 27 01	+0.005	-0.01	7.0	1.0	0.2	K0 III		200 mx		
182782	104820		19 25 45.7	+14 27 06	+0.001	-0.03	8.0	0.5	4.0	F8 V		61 s		
183339	31652		19 25 46.5	+58 01 39	-0.001	+0.01	6.60	-0.15		B8	-22			
182808	104821		19 25 48.1	+14 36 31	+0.001	-0.01	8.00	0.5	3.4	F5 V		81 s	12429	m
183203	31649		19 25 49.4	+50 51 06	0.000	-0.02	7.36	1.63	-0.3	K5 III		290 s		
183055	48430		19 25 53.5	+41 11 51	-0.002	-0.01	7.6	1.1	-0.1	K2 III		320 s		
182215	229672		19 25 55.9	-49 50 32	-0.024	-0.07	6.6	1.3	0.2	K0 III		48 mx		
183124	48433		19 25 56.4	+44 55 51	-0.004	-0.08	6.7	0.9	3.2	G5 IV		50 s		
182899	87200		19 25 57.6	+25 03 11	-0.001	0.00	7.95	0.16		B9				
182990	68296		19 25 58.1	+36 11 12	+0.001	+0.01	8.0	1.4	-0.3	K5 III		410 s		
181551	254566		19 25 59.3	-68 00 00	-0.003	+0.01	7.8	1.1	-0.1	K2 III		380 s		
183142	48438		19 26 01.0	+45 00 46	0.000	0.00	6.7	0.0	0.0	B8.5 V		220 s		
179778	258831		19 26 02.1	-80 14 19	-0.016	-0.07	7.8	1.1	0.2	K0 III		170 mx		
183318	31654		19 26 03.5	+54 53 12	+0.001	+0.02	7.7	1.1	-0.1	K2 III		360 s		
182740	124619		19 26 05.0	+3 31 09	+0.002	0.00	8.0	0.1	1.4	A2 V		200 s		
183125	48437		19 26 05.3	+42 19 33	+0.002	+0.04	7.8	0.1	1.4	A2 V		180 s		
182843	104827		19 26 06.7	+12 35 54	0.000	-0.01	7.8	0.1	1.4	A2 V		180 s		
183056	68301	4 Cyg	19 26 09.0	+36 19 04	0.000	+0.01	5.15	-0.12		B8 p	-22	51 mn		
182645	162609		19 26 10.9	-15 03 11	+0.001	+0.01	5.72	0.02		B8	-7			
183100	48439		19 26 11.8	+40 37 25	+0.003	-0.03	8.0	1.1	-0.1	K2 III		380 s		
182919	104831	5 Vul	19 26 13.0	+20 05 52	0.000	-0.03	5.63	-0.01	0.6	A0 V	-21	100 s		
182141	246097		19 26 18.3	-55 22 29	+0.002	-0.01	7.4	1.1	0.2	K0 III		230 s		
182629	188121	50 Sgr	19 26 18.9	-21 46 36	+0.002	0.00	5.59	1.22	-0.2	gK3	-20	140 s		
181354	257685		19 26 19.8	-71 27 59	+0.004	-0.04	7.7	0.1	0.6	A0 V		160 mx		
183471	31662		19 26 21.0	+59 25 43	+0.001	0.00	7.99	0.96	0.2	K0 III		360 s		
182160	246098		19 26 21.3	-54 51 47	-0.002	-0.03	7.55	1.06	0.2	K0 III		270 s		
181794	254570		19 26 21.7	-65 11 52	+0.003	-0.06	7.6	0.1	0.6	A0 V		85 mx		
182767	143398		19 26 22.4	-1 56 46	0.000	+0.01	7.9	0.0	0.0	B8.5 V		370 s		
181872	254573		19 26 23.3	-63 13 48	+0.011	0.00	7.5	0.5	4.0	F8 V		50 s		
183256	48447		19 26 23.8	+49 10 46	-0.001	-0.02	7.7	1.4	-0.3	K5 III		390 s		
182900	104832		19 26 23.9	+13 01 26	+0.001	+0.06	5.74	0.47	-4.6	F5 Ib	-34	1200 s		
182678	162615		19 26 24.3	-14 33 04	+0.002	-0.01	6.7	0.1	0.6	A0 V		160 s		
183611	18343		19 26 26.0	+62 33 27	+0.002	+0.05	6.38	1.39		K5	-40			
182648	188123		19 26 27.4	-21 14 45	-0.001	-0.02	7.4	1.2	-0.1	K2 III		320 s		
183257	48450		19 26 28.2	+48 52 17	0.000	+0.03	7.7	0.1	0.6	A0 V		260 s		
183280	48449		19 26 28.4	+47 58 12	+0.001	0.00	7.2	0.1	1.4	A2 V		150 s		
182955	104839		19 26 28.5	+19 53 29	0.000	-0.04	5.81	1.55	-4.4	K5 Ib	-36	810 s	12445	m
182972	87209		19 26 30.4	+20 15 32	0.000	-0.01	6.65	0.00	1.2	A1 V		120 s		
182835	124628	32 ν Aql	19 26 30.9	+0 20 19	0.000	0.00	4.66	0.60	-4.6	F2 Ib	-1	530 s		m
182901	104835		19 26 31.9	+11 51 00	+0.008	-0.01	6.9	0.5	3.4	F5 V	-43	49 s		
183032	87213		19 26 33.6	+27 19 20	+0.008	+0.08	7.44	0.54	4.2	dF9	-11	45 s	12447	m
183081	68312		19 26 33.8	+31 19 53	+0.001	-0.02	7.9	0.7	4.4	G0 V		51 s		
182161	246099		19 26 34.6	-56 55 18	+0.004	0.00	7.4	2.0	-0.5	M2 III		230 s		
182394	229677		19 26 41.3	-46 34 45	-0.001	-0.01	6.5	1.7	-0.3	K5 III		180 s		
182649	188125		19 26 41.9	-26 19 07	-0.001	-0.03	7.50	0.1	1.4	A2 V		170 s	12432	m
183204	68318		19 26 44.1	+39 57 41	0.000	0.00	7.39	-0.01	0.6	A0 V		230 s		
183013	87216		19 26 44.4	+21 39 10	0.000	-0.02	7.38	0.09		B3	-6			
181464	257686		19 26 45.0	-71 27 59	+0.010	-0.01	8.0	0.4	3.0	F2 V		100 s		
183361	31661		19 26 47.6	+50 08 43	+0.005	+0.07	7.30	0.5	3.4	F5 V		60 s	12470	m
183014	87218		19 26 48.2	+21 09 47	+0.001	+0.01	7.34	0.01	-0.6	B7 V		350 s	12451	m
183184	68320		19 26 53.1	+35 54 25	0.000	+0.01	7.72	-0.02	0.6	A0 V		270 s		
182681	188127		19 26 56.4	-29 44 35	+0.001	-0.04	5.60	0.00	0.2	B9 V	+2	120 s		
182991	104842		19 26 56.6	+12 03 43	0.000	-0.02	6.90	0.1	0.6	A0 V	+2	170 s	12452	m
183185	68323		19 26 57.0	+35 10 44	0.000	0.00	7.8	0.1	0.6	A0 V		260 s		
183058	87222		19 26 58.7	+21 06 22	0.000	-0.01	7.12	0.16		B5				m
183205	68327		19 27 03.4	+34 51 39	0.000	0.00	7.7	1.1	-0.1	K2 III		340 s		
181957	—		19 27 05.0	-65 40 36			7.9	1.1	-0.1	K2 III		400 s		
182706	188132		19 27 07.0	-29 18 23	+0.001	-0.01	8.0	1.6	-0.3	K5 III		390 s		
182506	229681		19 27 17.5	-48 30 40	0.000	0.00	8.0	0.8	0.2	K0 III		360 s		
183383	48463		19 27 24.3	+42 13 45	+0.002	-0.02	7.4	0.0	0.4	B9.5 V		240 s		
183534	31673	7 Cyg	19 27 25.8	+52 19 14	-0.002	-0.02	5.75	0.00	0.6	A0 V	+2	110 s		
183143	104860		19 27 26.3	+18 17 45	-0.001	0.00	6.87	1.24	-7.1	B7 Ia	+13	1200 s		
183489	48471		19 27 32.6	+49 15 23	0.000	+0.02	7.8	0.2	2.1	A5 V		140 s		
183144	104862		19 27 33.6	+14 16 56	-0.001	-0.01	6.32	-0.06	-2.6	B4 III	+4	510 s		
183222	87234		19 27 33.8	+24 33 03	+0.001	0.00	7.97	1.71	-0.3	K5 III		340 s		
180709	257684		19 27 35.0	-78 13 16	+0.004	+0.02	7.9	0.6	4.4	G0 V		50 s		
182507	246107		19 27 35.1	-52 35 55	+0.001	-0.06	7.5	0.6	0.6	A0 V		82 mx		

HD	SAO	Star Name	α 2000	δ 2000	μ(α)	μ(δ)	V	B-V	M_v	Spec	RV	d(pc)	ADS	Notes
			h m s	° ′ ″	s	″								
183362	68346		19 27 36.3	+37 56 28	+0.002	−0.01	6.34	−0.14	−1.7	B3 V e	−16	380 s		
182926	162637		19 27 39.9	−18 21 02	0.000	−0.02	7.3	0.5	3.4	F5 V	0	60 s		
183363	68347		19 27 40.4	+36 31 45	0.000	+0.01	7.57	−0.03	0.6	A0 V	−11	250 s	12478	m
183085	124646		19 27 41.0	+4 42 36	+0.002	+0.02	6.7	0.4	2.6	F0 V		65 s		
183145	104863		19 27 42.0	+12 29 15	+0.001	−0.03	7.8	1.1	0.2	K0 III	−1	300 s		
183104	124649		19 27 42.3	+6 11 32	0.000	+0.02	8.0	0.0	0.4	B9.5 V		320 s		
182708	229690		19 27 43.2	−40 05 51	+0.004	−0.12	7.2	0.4	4.4	G0 V		37 s		
183507	48476		19 27 45.4	+47 32 14	−0.001	+0.05	7.6	1.1	0.2	K0 III		300 s		
182509	246110		19 27 47.9	−54 19 30	0.000	+0.02	5.69	1.40	−0.1	K2 III		94 s		m
183162	124653		19 27 51.5	+8 58 11	+0.006	+0.10	7.9	0.7	4.4	G0 V		51 s		
183105	124651		19 27 51.7	+2 10 28	+0.001	0.00	7.80	1.1	−0.1	K2 III		370 s		m
183061	143423		19 27 52.0	−4 49 06	+0.001	−0.01	7.5	1.1	−0.1	K2 III		320 s		
183364	68349		19 27 53.0	+33 22 26	0.000	−0.02	7.4	0.4	2.6	F0 V		90 s		
183261	87240		19 27 53.8	+20 14 51	0.000	0.00	6.92	−0.04	−3.9	B3 II	+7	1300 s		
183241	104867		19 28 02.8	+14 04 20	+0.001	−0.01	7.6	0.1	0.6	A0 V		240 s		
182978	162642		19 28 03.2	−19 20 34	0.000	−0.01	8.0	0.8	3.2	G5 IV		90 s		
182998	162643		19 28 05.1	−18 21 33	+0.001	0.00	6.8	1.1	−0.1	K2 III	−32	240 s		
183262	104871		19 28 05.9	+17 50 48	0.000	−0.04	6.85	0.33		A m	−4			
181958	254578		19 28 06.5	−69 37 54	−0.007	+0.08	6.8	0.5	4.0	F8 V		36 s		
183772	31686		19 28 09.7	+54 46 48	0.000	+0.01	7.50	0.9	3.2	G5 IV		72 s	12509	m
183063	−−		19 28 10.5	−12 09 02			7.56	0.74	4.4	G0 V	−23	32 s	12469	m
181673	257690		19 28 11.2	−73 05 28	+0.004	−0.09	7.5	0.1	1.7	A3 V		53 mx		
183301	104875		19 28 11.8	+18 55 02	−0.001	−0.03	7.4	0.5	3.4	F5 V		63 s		
183612	48488		19 28 12.9	+47 01 34	0.000	+0.02	7.9	0.9	3.2	G5 IV		88 s		
183399	87251		19 28 13.5	+29 27 01	−0.001	−0.02	6.7	1.1	0.2	K0 III	−15	190 s		
183227	124661		19 28 20.6	+2 55 49	0.000	0.00	5.85	0.01		B9				
183786	31688		19 28 22.4	+53 58 13	+0.002	+0.01	8.0	0.1	0.6	A0 V		290 s		
183647	48490		19 28 22.8	+47 46 33	+0.001	0.00	8.0	0.1	1.4	A2 V		210 s		
183263	124664		19 28 24.4	+8 21 29	−0.001	−0.03	7.7	0.6	3.0	G2 IV		87 s		
182750	229696		19 28 25.1	−47 52 56	−0.002	−0.02	7.7	1.4	0.2	K0 III		190 s		
182912	211288		19 28 25.7	−36 59 22	+0.001	−0.03	7.5	1.5	1.7	A3 V		22 s		
183400	87256		19 28 25.8	+24 59 44	−0.002	0.00	7.2	1.1	−0.1	K2 III	−11	270 s		
182964	211290		19 28 27.3	−31 23 14	−0.001	−0.10	7.8	0.7	3.2	G5 IV		84 s		
183385	87253		19 28 27.8	+22 43 03	0.000	0.00	7.8	1.1	0.2	K0 III		310 s		
181465	257689		19 28 28.8	−75 10 26	+0.007	−0.01	7.9	0.4	2.6	F0 V		120 s		
183133	162651		19 28 35.1	−15 06 09	0.000	0.00	6.78	−0.02	−1.1	B5 V	−24	340 s		
183303	124670		19 28 35.7	+8 51 50	+0.001	+0.01	7.4	0.4	0.6	F0 III		210 s		
183132	162652		19 28 36.5	−14 09 36	+0.001	+0.03	7.9	0.9	3.2	G5 IV		88 s		
183559	68371		19 28 37.5	+37 11 18	0.000	−0.02	7.90	0.1	1.7	A3 V		170 s	12504	m
183968	31696		19 28 38.3	+59 45 51	+0.001	+0.02	7.8	1.4	−0.3	K5 III	−25	380 s		
183585	68373		19 28 38.4	+39 56 27	+0.001	0.00	6.9	0.8	3.2	G5 IV		55 s		
183418	87259		19 28 40.4	+21 58 52	0.000	+0.01	7.73	0.97	−2.1	G5 II		930 s		
183508	68367		19 28 41.8	+32 04 30	−0.001	−0.02	7.9	0.5	4.0	F8 V		59 s		
183439	87261	6 α Vul	19 28 42.2	+24 39 54	−0.009	−0.10	4.44	1.50		M0	−86	26 mn		m
182652	246113		19 28 42.5	−56 05 48	+0.003	0.00	7.5	1.2	3.2	G5 IV		40 s		
183341	104879		19 28 46.2	+11 01 36	−0.001	−0.04	7.4	0.8	3.2	G5 IV		68 s		
182985	211295		19 28 46.7	−36 00 01	+0.001	0.00	7.1	0.6	0.4	B9.5 V		90 s		
184273	18362		19 28 46.7	+67 37 36	−0.003	0.00	7.7	0.9	3.2	G5 IV		80 s		
−−	−−		19 28 47.2	+46 02 13			8.00	2.0		N7.7				
183509	68372		19 28 47.2	+31 19 58	0.000	−0.03	7.8	1.1	0.2	K0 III		310 s		
182929	229706		19 28 48.6	−42 06 57	−0.001	+0.02	8.0	1.6	−0.5	M2 III		350 s		
183091	188174		19 28 53.8	−28 58 13	+0.002	−0.01	7.5	1.1	0.6	A0 V		56 s		
177671	258829		19 28 54.1	−84 43 12	+0.051	−0.07	8.0	1.2	3.2	G5 IV		51 s		
−−	18358		19 28 55.5	+63 11 21	−0.001	+0.01	8.0	1.1						
183491	87267	8 Vul	19 28 56.9	+24 46 07	0.000	+0.02	5.81	1.03	0.2	gK0	−27	130 s		
183401	104884		19 28 59.5	+12 28 15	0.000	+0.01	7.94	0.31	0.5	gA8	−31	300 s		m
183421	104885		19 28 59.8	+12 51 30	0.000	−0.01	7.2	0.1	0.6	A0 V	−22	200 s		
231555	104887		19 29 00.5	+14 29 38	+0.003	+0.07	7.9	0.5	4.0	F8 V		59 s		
183324	124675	35 Aql	19 29 00.8	+1 57 01	0.000	−0.03	5.80	0.08		A0 n	+10			
183047	211298		19 29 01.3	−35 48 02	0.000	−0.05	7.5	1.2	0.6	A0 V		48 s		
183460	104891		19 29 12.3	+13 11 08	+0.001	+0.02	8.0	0.4	3.0	F2 V	−31	96 s		
183629	68381		19 29 15.3	+30 32 56	+0.002	0.00	7.4	1.1	0.2	K0 III	−44	260 s		
183387	124681		19 29 17.8	+0 14 46	0.000	−0.01	6.25	1.32		K2	−60			
183369	143456		19 29 18.9	−1 22 55	+0.001	−0.02	7.9	1.1	−0.1	K2 III		380 s		
183537	87269	7 Vul	19 29 20.7	+20 16 47	0.000	−0.02	6.33	−0.10	−1.1	B5 V	−43	300 s		
182586	254586		19 29 21.1	−62 47 17	+0.005	+0.01	7.9	1.1	0.2	K0 III		350 s		
183344	143454		19 29 21.3	−7 02 38	+0.001	0.00	6.61	1.10		F8	−7	26 mn	12503	U Aql, m,v
183650	68384		19 29 21.4	+31 36 31	−0.001	−0.41	6.96	0.71	5.2	dG5	−12	22 ts		
183492	104896		19 29 21.9	+14 35 45	+0.003	−0.02	5.56	1.05	0.2	gK0	−40	110 s		
183007	229712		19 29 23.7	−43 26 45	+0.009	−0.13	5.71	0.22		A0		19 mn		
183511	104897		19 29 25.3	+15 11 06	0.000	−0.01	8.0	1.4	−0.3	K5 III		410 s		
184168	18363		19 29 27.6	+60 40 07	−0.004	−0.10	7.61	1.18	−0.1	K2 III		200 mx		m
185497	9452		19 29 29.6	+78 15 56	+0.006	+0.01	7.1	0.4	2.6	F0 V		79 s	12608	m
183752	68394		19 29 29.8	+36 17 14	+0.001	+0.09	6.6	0.4	2.6	F0 V		63 s		
183442	124685		19 29 29.8	+3 05 24	0.000	+0.01	8.0	0.0	−1.0	B5.5 V	−45	590 s	12508	m

HD	SAO	Star Name	α 2000	δ 2000	μ(α)	μ(δ)	V	B-V	M_v	Spec	RV	d(pc)	ADS	Notes
183306	162669		19^h 29^m 33^s.7	-15° 21' 33"	+0^s.002	-0".01	7.1	0.1	0.6	A0 V		190 s		
183512	104900		19 29 35.6	+12 24 15	-0.001	-0.01	8.00	0.1	1.7	A3 V	-16	170 s	12515	m
183153	211303		19 29 37.6	-35 05 08	+0.002	-0.08	7.6	1.4	0.2	K0 III		160 mx		
183347	162674		19 29 37.9	-12 38 50	+0.010	-0.05	7.40	0.5	4.0	F8 V		48 s	12505	m
183216	211305		19 29 40.4	-30 47 52	+0.001	-0.26	7.14	0.60	4.7	G2 V		31 s		
184006	31702	10 ι Cyg	19 29 42.1	+51 43 47	+0.002	+0.13	3.79	0.14	2.1	A5 V	-20	41 mn		
182485	254584		19 29 45.3	-67 18 30	+0.003	-0.01	7.70	1.1	0.2	K0 III		320 s		m
183197	211307		19 29 49.8	-36 17 46	0.000	-0.04	7.9	1.4	-0.1	K2 III		300 s		
183275	188192		19 29 52.1	-26 59 08	+0.003	-0.04	5.46	1.12	-0.2	K3 III		140 s	12506	m
182893	246125		19 29 52.4	-55 26 29	+0.003	-0.05	6.13	0.98	0.2	K0 III		150 s		
183681	87278		19 29 52.6	+22 42 24	0.000	+0.01	7.31	1.81	-0.4	M0 III	-10	240 s		
183615	104902		19 29 54.0	+13 08 15	0.000	+0.01	7.6	0.2	2.1	A5 V	-35	120 s		
183050	229715		19 29 54.3	-48 27 31	-0.001	-0.01	7.9	2.1	-0.1	K2 III		300 s		
183011	246127		19 29 55.2	-50 16 13	-0.002	+0.01	8.0	0.3	3.4	F5 V		82 s		
183955	48515		19 29 56.5	+44 41 27	-0.001	+0.02	7.9	0.5	3.4	F5 V		78 s		
183969	48516		19 29 58.8	+46 56 44	+0.002	-0.04	6.7	0.1	0.6	A0 V		170 s		
182969	246126		19 30 01.2	-53 37 07	+0.001	-0.02	7.5	-0.2	0.6	A0 V		250 s		
183430	162682		19 30 01.9	-15 58 34	-0.001	-0.03	7.6	0.1	0.6	A0 V		240 s		
183754	87282		19 30 03.8	+25 03 21	-0.003	+0.01	8.0	1.1	0.2	K0 III		340 s		
183518	143469		19 30 04.4	-0 26 43	0.000	-0.02	7.2	0.1	1.7	A3 V		130 s	12518	m
183408	162685		19 30 08.3	-18 14 36	-0.001	-0.01	7.8	0.1	1.7	A3 V		160 s		
183589	124698		19 30 10.2	+2 54 15	0.000	0.00	6.09	1.82		K5	-7		12520	m
184170	31711		19 30 12.1	+55 25 22	+0.002	+0.15	6.90	1.1	0.2	K0 III		120 mx		m
184195	31716		19 30 12.4	+56 38 40	-0.002	-0.02	7.30	0.1	0.6	A0 V		210 s	12552	m
184008	48521		19 30 12.5	+46 08 53	+0.003	+0.01	6.3	1.6	-0.5	M4 III	-15	230 s		AF Cyg, v
183312	211315		19 30 14.5	-32 05 32	+0.007	-0.06	6.60	0.39	2.8	F5 IV-V		59 s		
183886	68402		19 30 17.1	+33 43 43	0.000	-0.02	6.6	1.1	0.2	K0 III		180 s		
183728	104909		19 30 22.0	+16 42 40	0.000	0.00	7.1	0.1	0.6	A0 V	-19	190 s		
183755	104913		19 30 25.0	+17 41 41	0.000	-0.01	7.2	0.1	0.6	A0 V		200 s		
183334	211317		19 30 28.8	-33 54 42	-0.001	-0.02	7.9	1.4	0.2	K0 III		210 s		
183775	104916		19 30 31.7	+16 48 04	+0.002	-0.01	8.0	0.5	3.4	F5 V		81 s		
183756	104914		19 30 31.7	+12 49 02	-0.004	-0.04	6.9	1.1	0.2	K0 III		200 s		
183656	124704		19 30 32.9	+3 26 40	+0.001	+0.01	6.05	0.01	-1.1	B5 V pe	-42	260 s		V923 Aql, v
184240	31724		19 30 34.1	+56 08 10	-0.002	-0.01	6.6	0.1	0.6	A0 V		160 s		
183028	246131		19 30 34.3	-55 06 36	+0.004	-0.01	6.4	0.4	3.4	F5 V		40 s		
--	--		19 30 34.4	+44 57 18	-0.001	-0.02	7.1	1.6	-0.5	M2 III		320 s		
183909	68410		19 30 35.7	+31 58 48	+0.001	0.00	7.5	1.1	-0.1	K2 III		310 s		
183630	143482	36 Aql	19 30 39.7	-2 47 20	+0.001	-0.01	5.03	1.75	-0.5	M1 III	-11	100 s		
182970	246129		19 30 40.3	-59 13 17	+0.003	-0.01	7.5	-0.1	0.4	B9.5 V		260 s		
183499	162699		19 30 40.9	-18 19 25	+0.002	0.00	7.2	1.1	0.2	K0 III		260 s		
--	48531		19 30 40.9	+43 02 15	-0.001	-0.01	7.9	1.3						
183593	143481		19 30 41.7	-8 11 11	0.000	-0.01	7.6	0.0	0.4	B9.5 V		260 s		
182510	257699		19 30 42.1	-70 02 43	+0.002	+0.04	7.9	1.7	-0.3	K5 III		330 s		
183912	87301	6 β Cyg	19 30 43.1	+27 57 35	0.000	0.00	3.08	1.13	-2.3	K3 II	-24	120 s	12540	Albireo, m
183914	87302	6 β Cyg	19 30 45.2	+27 57 55	-0.001	0.00	5.11	-0.10	-0.2	B8 V	-18	120 s	12540	m
183570	162703		19 30 46.6	-16 10 00	+0.001	0.00	7.44		-1.0	B5.5 V	-4	410 s		
183986	68417		19 30 46.7	+36 13 43	0.000	-0.01	6.25	-0.04	-0.7	B9.5 III	+7	230 s	12545	m
183849	104924		19 30 49.4	+18 15 02	-0.001	-0.01	7.3	1.1	0.2	K0 III	-23	250 s		
183658	143485		19 30 52.6	-6 30 53	-0.009	-0.14	7.29	0.64	4.4	G0 V		33 s	12529	m
184056	48534		19 30 53.1	+40 30 48	-0.004	+0.03	8.0	1.1	-0.1	K2 III		380 s		
183545	188219		19 30 53.9	-21 18 44	+0.001	-0.02	6.0	0.6	1.4	A2 V		37 s		
183524	188216		19 30 54.8	-23 57 09	+0.004	-0.03	7.6	0.9	3.2	G5 IV		64 s		
184104	48537		19 30 56.8	+43 50 41	-0.004	-0.02	8.0	1.4	-0.3	K5 III		330 mx		
183452	211325		19 30 57.3	-32 41 57	+0.004	-0.06	7.2	0.1	3.0	F2 V		68 s		
183631	162712		19 30 57.9	-11 29 03	+0.006	0.00	7.0	0.5	3.4	F5 V		51 s		
183632	162710		19 30 58.0	-14 33 45	0.000	+0.03	7.2	1.1	0.2	K0 III		240 s		
183791	124716		19 30 58.4	+6 23 06	0.000	-0.02	7.8	0.7	-2.1	G2 II	+15	890 s		
183736	143489		19 30 59.8	-2 02 05	0.000	0.00	7.7	1.4	-0.3	K5 III		370 s		
183887	104925		19 30 59.8	+19 26 03	+0.001	-0.01	7.55	1.55	-0.1	K2 III		200 s		
184958	9447		19 31 00.1	+70 59 24	-0.002	+0.06	6.07	1.41	-0.1	K2 III	-43	120 s		
184147	48541		19 31 00.9	+46 19 56	0.000	+0.02	6.9	0.1	0.6	A0 V		180 s		
182254	257696		19 31 01.7	-73 59 06	+0.014	-0.02	7.9	0.1	0.6	A0 V		130 mx		
184241	31729		19 31 02.3	+50 11 49	+0.001	+0.01	7.60	1.1	0.2	K0 III		300 s	12563	m
184563	18378		19 31 03.8	+63 18 53	-0.003	-0.02	7.40	1.1	0.2	K0 III		270 s	12586	m
184562	—		19 31 03.8	+63 18 56	-0.002	+0.03	8.00	0.1	0.6	A0 V		270 s	12586	m
183793	124718		19 31 04.9	+2 00 43	-0.001	+0.01	6.80	1.1	0.2	K0 III		200 s		m
184215	48544		19 31 05.6	+48 34 56	0.002	+0.03	7.2	0.0	0.0	B8.5 V		280 s		
183501	188220		19 31 06.4	-29 29 43	0.000	-0.06	7.5	1.2	-0.1	K2 III		200 mx		
184467	31745		19 31 07.3	+58 35 10	-0.067	-0.39	6.59	0.87	6.1	K1 V	+11	15 ts		
182709	254590		19 31 10.8	-68 26 02	+0.003	-0.01	5.96	1.64		K5				
183548	188228		19 31 11.6	-28 20 01	+0.001	0.00	8.0	1.5	0.2	K0 III		190 s		
184398	31741		19 31 13.3	+55 43 55	-0.001	-0.01	6.37	1.16	-1.1	K2 II-III	-6	320 s		
183794	143494		19 31 15.6	-2 06 36	-0.001	0.00	6.80	0.00	0.0	B8.5 V		220 s	12538	V822 Aql, m,v
184057	68429		19 31 16.0	+33 28 13	+0.001	-0.02	6.71	0.08	1.4	A2 V		110 s		
184105	68431		19 31 16.0	+38 24 01	+0.001	-0.02	7.6	0.1	1.4	A2 V		170 s		

485

HD	SAO	Star Name	α 2000	δ 2000	μ(α)	μ(δ)	V	B-V	M$_v$	Spec	RV	d(pc)	ADS	Notes
			h m s	+° ′ ″	s	″								
184242	48548		19 31 16.0	+47 28 54	-0.003	-0.01	7.70	0.1	1.7	A3 V		160 s	12567	m
183575	188231		19 31 16.4	-25 44 14	0.000	-0.06	7.3	1.0	0.2	K0 III		200 mx		
184148	48545		19 31 17.9	+42 59 13	-0.001	+0.01	7.2	1.1	0.2	K0 III		240 s		
184293	31737		19 31 19.2	+50 18 24	-0.003	+0.05	5.53	1.25	0.2	K0 III	-9	82 s		
184677	18383		19 31 20.4	+64 17 13	-0.002	-0.02	7.8	1.1	-0.1	K2 III		390 s		
184010	87314		19 31 21.6	+26 37 02	+0.002	+0.03	5.87	0.93	0.3	gG8	-2	130 s		
183888	104926		19 31 21.8	+11 10 16	0.000	-0.02	7.9	0.0	0.0	B8.5 V		350 s		
183355	229728		19 31 22.1	-46 16 24	+0.003	-0.02	7.0	0.8	3.2	G5 IV		59 s		
183665	188234		19 31 24.4	-21 02 25	-0.001	-0.03	7.9	0.6	-0.3	K5 III		430 s		
183915	104927		19 31 25.3	+11 37 40	0.000	0.00	7.90	1.3		K p				
--	68437		19 31 25.9	+36 42 37	-0.001	-0.02	7.90						12561	m
182860	254591		19 31 26.7	-66 33 16	0.000	-0.05	7.8	1.1	-0.1	K2 III		240 mx		
184423	31746		19 31 28.3	+55 15 53	+0.008	+0.14	6.9	1.1	-0.1	K2 III		110 mx		
184058	87316		19 31 29.8	+28 43 02	+0.001	+0.01	7.50	0.4	2.6	F0 V		94 s		
183989	104936		19 31 34.4	+17 57 30	+0.001	+0.02	7.53	0.11		A0				
184059	87317		19 31 36.2	+26 59 12	0.000	0.00	7.7	0.0	0.0	B8.5 V		330 s		
183919	124728		19 31 38.2	+5 01 16	0.000	-0.01	7.3	0.1	0.6	A0 V		220 s		
183741	162723		19 31 42.4	-19 23 12	+0.001	-0.01	7.2	1.1	-0.1	K2 III		300 s		
183990	104938		19 31 43.0	+13 37 09	+0.001	0.00	7.8	1.1	-0.1	K2 III		340 s		
183936	124732		19 31 43.1	+5 46 07	+0.007	-0.01	7.2	0.4	0.6	F2 III		78 mx		
--	68444		19 31 44.4	+33 48 02	+0.001	+0.03	7.50							m
183960	124735		19 31 44.8	+6 51 33	+0.001	+0.01	7.8	0.1	1.4	A2 V		180 s		
184171	68447	8 Cyg	19 31 46.1	+34 27 11	0.000	0.00	4.74	-0.14	-2.3	B3 IV	-22	240 s		
183605	211331		19 31 47.3	-35 57 31	-0.001	+0.01	8.0	0.4	3.4	F5 V		82 s		
183504	229731		19 31 47.7	-42 25 48	+0.004	0.00	8.0	-0.8	4.0	F8 V		62 s		
184040	104939		19 31 53.5	+14 58 08	0.000	-0.01	7.4	0.1	0.6	A0 V		220 s		
184041	104941		19 31 55.5	+14 18 59	+0.001	0.00	8.0	0.1	0.6	A0 V		280 s		
184150	68449		19 31 55.8	+30 11 16	0.000	-0.01	7.70	1.1	-0.1	K2 III		340 s		
183357	246143		19 31 56.9	-54 09 53	-0.001	+0.01	7.3	0.8	-0.5	M4 III		370 s		
184756	18388		19 31 59.7	+63 07 34	-0.002	-0.01	8.0	1.6	-0.5	M4 III		450 s		
184013	124742		19 32 00.6	+7 29 17	+0.001	-0.03	7.0	1.1	-0.1	K2 III		250 s		
184108	87320		19 32 01.7	+20 55 47	0.000	+0.01	6.96	-0.02	-0.8	B9 III		330 s		
183764	188244		19 32 03.2	-28 12 43	+0.002	+0.01	7.4	0.2	0.4	B9.5 V		190 s		
184151	87323		19 32 05.2	+25 35 49	+0.004	+0.06	6.91	0.42	3.0	F2 V	+12	57 s		
184172	87325		19 32 05.7	+28 16 01	0.000	+0.01	8.00	0.1	0.6	A0 V		280 s	12572	m
183870	162730		19 32 06.4	-11 16 29	+0.015	+0.03	7.4	0.8	3.2	G5 IV		61 mx		
184042	124747		19 32 10.1	+9 49 19	0.000	-0.01	7.8	1.1	0.2	K0 III		310 s		
183237	254592		19 32 11.6	-60 16 11	+0.002	-0.03	7.60	0.1	1.4	A2 V		170 s		m
184061	124749		19 32 12.6	+9 20 01	+0.002	-0.02	6.90	0.1	0.6	A0 V		170 s	12568	m
184026	124745		19 32 13.0	+6 30 12	+0.001	+0.02	7.60	0.1	0.6	A0 V		250 s	12569	m
183577	229734		19 32 13.8	-44 32 47	-0.014	+0.04	6.7	0.1	4.4	G0 V		29 s		
183782	188249		19 32 13.9	-28 22 29	+0.002	-0.01	7.8	1.7	-0.1	K2 III		190 s		
183765	211340		19 32 15.7	-31 04 31	+0.006	0.00	7.6	1.4	-0.1	K2 III		180 mx		
183841	188252		19 32 15.7	-21 31 02	+0.002	-0.01	7.1	1.0	1.7	A3 V		35 s		
184294	68459		19 32 19.1	+35 12 06	0.000	-0.02	7.9	0.1	0.6	A0 V		280 s		
185144	18396	61 σ Dra	19 32 21.5	+69 39 40	+0.114	-1.74	4.68	0.79	5.9	K0 V	+27	5.7 t		m
183578	229736		19 32 21.8	-45 22 05	+0.001	0.00	7.6	0.8	3.2	G5 IV		74 s		
183857	188255		19 32 24.5	-22 44 26	0.000	+0.02	7.6	0.5	3.0	F2 V		67 s		
183921	162739		19 32 25.9	-15 45 06	+0.001	-0.01	8.0	0.1	1.4	A2 V		200 s		
184088	124755		19 32 26.5	+9 35 48	0.000	+0.03	8.0	0.4	2.6	F0 V		110 s		
183799	211345		19 32 28.4	-30 52 09	+0.003	-0.02	6.6	1.8	-0.5	M2 III		190 s		
184110	124761		19 32 37.5	+7 54 41	0.000	+0.01	8.0	0.4	0.6	F2 III		290 s		
183877	188257		19 32 40.1	-28 01 11	+0.005	-0.74	7.15	0.68	3.2	G5 IV	-43	62 s		
182687	257701		19 32 48.6	-73 29 47	+0.007	0.00	7.4	1.1	0.2	K0 III		270 s		
185731	9461		19 32 52.3	+75 19 53	+0.002	+0.01	7.8	0.1	1.4	A2 V		190 s		
183671	229739		19 32 52.5	-46 46 39	+0.006	-0.16	7.50	0.9	3.2	G5 IV		69 mx		m
183924	188263		19 32 52.7	-26 01 42	-0.002	-0.01	7.6	0.4	2.6	F0 V		93 s		
183552	246151		19 32 53.6	-53 11 09	+0.004	-0.01	5.75	0.30		A m				
184381	68472		19 32 54.7	+31 14 11	+0.001	+0.02	6.8	0.5	3.4	F5 V	-29	47 s		
184469	68478		19 32 55.8	+39 39 45	0.000	-0.01	7.75	-0.08	0.4	B9.5 V		300 s		
184275	87338		19 32 56.4	+21 27 22	0.000	0.00	7.83	1.23	0.3	G8 III		220 s		
184296	87340		19 33 00.7	+21 00 09	+0.001	-0.03	7.74	0.07	1.4	A2 V		180 s		
183925	211355		19 33 03.0	-30 21 49	+0.001	-0.04	6.7	1.3	0.2	K0 III		130 s		
184201	124777		19 33 07.6	+5 01 45	+0.001	0.00	6.7	1.6		M5 III	+1			
184602	48581		19 33 08.5	+46 02 22	+0.001	+0.01	7.4	0.1	1.4	A2 V	-23	160 s		
184903	31771		19 33 08.8	+59 23 46	0.000	+0.02	7.81	-0.02		A0 p				
183977	188269		19 33 09.2	-25 27 14	-0.002	0.00	8.0	1.6	-0.3	K5 III		390 s		
184936	18395		19 33 10.0	+60 09 31	0.000	0.00	6.29	1.57	-0.3	K4 III	-19	160 s		m
184200	124779		19 33 10.2	+5 44 34	+0.001	0.00	7.90	1.1	0.2	K0 III		330 s		
184244	124782		19 33 12.0	+7 46 20	+0.001	0.00	7.6	0.1	0.6	A0 V		250 s		
184564	48580		19 33 12.9	+42 48 28	0.000	0.00	7.9	1.1	0.2	K0 III		320 s		
182894	257703		19 33 14.9	-72 06 02	+0.007	0.00	8.0	0.4	2.6	F0 V		120 s		
184075	162747		19 33 15.1	-15 00 27	-0.001	-0.04	8.0	0.5	4.0	F8 V		62 s		
184360	87342		19 33 17.0	+20 24 50	+0.003	+0.05	7.28	0.26		A m	-52	40 mn	12594	m
184077	162748		19 33 20.2	-16 36 36	+0.001	-0.02	8.0	1.6	-0.5	M2 III		500 s		

HD	SAO	Star Name	α 2000	δ 2000	μ(α)	μ(δ)	V	B-V	M$_v$	Spec	RV	d(pc)	ADS	Notes
			^{h} ^{m} ^{s}	° ' "	^{s}	"								
183806	229741		19 33 21.4	−45 16 18	−0.001	−0.03	5.9	0.5		A mp				
183926	211357		19 33 23.6	−37 37 47	−0.001	−0.03	8.0	1.3	3.4	F5 V		24 s		
184385	87351		19 33 25.4	+21 50 26	−0.001	−0.20	6.89	0.70	5.2	G5 V		21 s		
184499	68491		19 33 26.9	+33 12 05	−0.037	+0.21	6.61	0.59	4.4	G0 V	−163	29 ts		
184297	124787		19 33 28.1	+8 13 45	0.000	−0.01	7.70	1.1	0.2	K0 III		310 s		
186195	9472		19 33 29.5	+78 03 34	+0.004	+0.04	7.9	0.9	3.2	G5 IV		88 s		
184522	68494		19 33 29.6	+36 01 18	−0.003	−0.02	8.0	0.1	1.7	A3 V		170 s		
−−	68498		19 33 35.8	+36 02 45	−0.001	−0.03	7.7	1.2						
184603	68499		19 33 36.3	+38 45 44	0.000	+0.03	6.50	0.03	1.7	A3 V	−17	91 s		
184279	124788		19 33 36.9	+3 45 41	+0.001	0.00	6.94	0.03		B0.5 IV	−9			
184069	188275		19 33 41.0	−28 32 04	−0.001	0.00	7.6	0.7	0.6	A0 V		89 s		
184786	48589		19 33 41.5	+49 15 45	0.000	0.00	5.96	1.64	−0.5	M4 III	−10	200 s		
183997	211361		19 33 41.9	−34 11 52	0.000	−0.03	7.0	0.3	0.6	A0 V		120 s		
184826	31773		19 33 42.3	+52 07 22	+0.003	0.00	7.9	0.9	3.2	G5 IV		88 s		
184629	68504		19 33 43.0	+39 43 40	+0.001	−0.01	7.7	0.1	1.4	A2 V		170 s		
184248	143536		19 33 43.1	−4 44 31	0.000	0.00	7.8	1.1	−0.1	K2 III		360 s		
184336	124790		19 33 44.7	+7 49 45	+0.002	−0.01	7.9	0.1	0.6	A0 V		280 s		
184313	124789		19 33 45.7	+5 27 56	−0.001	−0.03	6.6	1.6		M8 V				V450 Aql, v
184362	124792		19 33 46.3	+7 55 06	−0.001	0.00	7.3	0.1	0.6	A0 V		220 s		
185145	18400		19 33 46.4	+62 36 46	0.000	+0.02	7.7	1.1	0.2	K0 III		290 s		
183470	254594		19 33 47.1	−63 03 31	+0.002	−0.05	7.5	0.1	1.7	A3 V		140 s		
184363	124795		19 33 51.9	+6 08 34	−0.001	0.00	8.00	0.9	0.3	G5 III		340 s		
184959	31778		19 33 53.7	+56 27 37	+0.001	+0.05	7.0	1.1	0.2	K0 III		230 s		
184450	104975		19 33 58.0	+16 47 18	0.000	−0.01	8.0	0.1	0.6	A0 V		280 s		
184537	87364		19 34 00.3	+26 03 45	0.000	+0.02	6.9	0.2		A m				
185306	18408		19 34 01.5	+65 31 51	−0.001	−0.01	7.9	0.1	1.7	A3 V		170 s		
184538	87365		19 34 02.0	+25 48 20	0.000	−0.02	7.41	1.32	−0.1	K2 III		250 s		
183316	254593		19 34 03.5	−67 56 37	+0.004	−0.03	7.97	0.74		F5				
184406	124799	38 μ Aql	19 34 05.2	+7 22 44	+0.014	−0.16	4.45	1.17	−0.2	K3 III	−24	39 ts	12607	m
183771	246162		19 34 05.8	−56 06 01	+0.003	−0.03	7.0	1.4	3.2	G5 IV		23 s		
184630	68509		19 34 06.8	+33 00 36	0.000	0.00	8.00	1.1	0.2	K0 III		340 s		m
184035	229746		19 34 08.2	−40 02 05	−0.001	0.00	5.90	0.09	0.0	A3 III		150 s		
183928	229743		19 34 08.8	−48 40 40	0.000	−0.01	7.8	0.4	0.6	A0 V		270 s		
184451	124801		19 34 10.9	+7 47 11	−0.001	+0.01	6.9	0.1	0.6	A0 V		180 s		
183906	246167		19 34 13.4	−50 33 56	+0.002	−0.01	7.7	1.4	−0.1	K2 III		270 s		
184590	87367		19 34 14.0	+25 21 08	0.000	−0.02	7.20	1.6	−1.4	M1 II-III	+19	490 s		
184266	162774		19 34 15.0	−16 18 58	+0.006	−0.18	7.6	0.7	4.4	G0 V		44 s		
185600	9463		19 34 17.5	+70 15 38	0.000	+0.04	7.7	0.1	0.6	A0 V		270 s		
184787	48596		19 34 18.0	+41 55 40	+0.001	+0.01	6.68	−0.01	0.6	A0 V		160 s		
183809	246164		19 34 18.1	−56 25 03	+0.004	−0.01	7.5	0.5	1.7	A3 V		79 s		
184502	104978		19 34 18.5	+16 15 54	0.000	0.00	7.03	−0.01	−1.6	B3.5 V	−23	440 s		
184283	162777		19 34 18.8	−16 22 26	−0.001	+0.01	7.34	3.37		N7.7	+14			AQ Sgr, v
184788	48597		19 34 18.9	+41 25 45	−0.001	+0.01	7.14	−0.06	0.4	B9.5 V		220 s		
184960	31782		19 34 19.6	+51 14 11	+0.003	−0.19	5.73	0.48	4.0	F8 V	+1	22 s		
183846	246166		19 34 21.0	−54 32 03	−0.001	+0.03	7.8	1.2	3.2	G5 IV		46 s		
−−	87373		19 34 23.1	+27 33 57	0.000	−0.02	8.03	1.65	−0.5	M2 III		470 s		
185713	9467		19 34 24.8	+71 36 19	−0.023	−0.07	6.90	0.5	3.4	F5 V	+16	50 s	12690	m
184317	162781		19 34 25.9	−12 55 37	−0.012	−0.10	7.5	0.7	4.4	G0 V		42 s		
184525	104982		19 34 29.0	+11 15 30	−0.005	−0.10	7.9	0.7	4.4	G0 V		51 s		
183675	254597		19 34 32.7	−62 22 56	+0.002	+0.01	7.5	1.1	0.2	K0 III		290 s		
184268	188296		19 34 33.7	−23 51 31	0.000	0.00	6.43	1.68		K5	−12			
184606	104990	9 Vul	19 34 34.7	+19 46 24	0.000	0.00	5.00	−0.09	−0.6	B7 V	+5	130 s	12622	m
183277	−−		19 34 37.3	−70 30 42			8.0	0.2	2.1	A5 V		150 s		
184591	104991		19 34 37.4	+18 07 42	+0.002	+0.03	7.50	0.9	3.2	G5 IV		71 s	12623	m
184541	104987		19 34 38.4	+10 57 09	+0.002	0.00	6.9	1.1	0.2	K0 III		210 s		
184977	48606		19 34 39.7	+48 09 53	−0.002	−0.07	6.7	0.2	2.1	A5 V	−1	84 s		
184875	48601		19 34 41.0	+42 24 46	−0.001	−0.02	5.35	0.05		A2	0	41 mn		
184905	48604		19 34 43.7	+43 56 46	+0.002	−0.03	6.62	−0.03		A0 p	−10			
184607	104993		19 34 45.1	+16 38 30	+0.001	0.00	7.2	1.1	0.2	K0 III	−65	240 s		
184938	48608		19 34 45.4	+46 28 03	+0.002	+0.03	7.5	0.8	3.2	G5 IV		72 s		
185394	18413		19 34 46.2	+63 26 04	+0.002	+0.01	6.6	1.1	−0.1	K2 III	+9	200 s		
184759	87385	9 Cyg	19 34 50.8	+29 27 47	+0.001	+0.03	5.38	0.55	3.4	F5 V	−11	22 mn		
184258	211379		19 34 51.2	−32 47 31	+0.003	0.00	7.9	1.1	0.2	K0 III		320 s		
184142	229749		19 34 51.3	−41 09 18	+0.001	−0.02	8.0	−0.3	1.7	A3 V		180 s		
184592	104996		19 34 55.4	+11 25 26	+0.017	0.00	7.93	0.69	3.2	G5 IV		49 mx		
184827	68529		19 34 55.9	+33 47 44	+0.001	+0.01	6.6	1.6	−0.5	M2 III		250 s		
184761	87388		19 34 58.8	+27 13 31	+0.005	+0.04	6.6	0.1	1.7	A3 V	−26	100 s		
181466	258833		19 34 59.5	−81 45 09	+0.004	+0.02	6.6	1.3	0.2	K0 III		120 s		
185114	31792		19 35 00.7	+52 30 09	−0.004	−0.01	6.5	1.1	0.2	K0 III		180 s		
184571	124817		19 35 01.7	+4 18 41	−0.001	−0.02	7.6	0.4	3.0	F2 V		84 s		
184052	246173		19 35 02.5	−50 07 27	−0.002	+0.01	7.9	−0.1	0.6	A0 V		280 s		
184021	246172		19 35 05.3	−53 31 16	+0.004	−0.01	7.4	0.4	2.6	F0 V		79 s		
184303	211383		19 35 05.3	−31 36 34	+0.002	−0.06	7.4	0.7	1.4	A2 V		64 s		
184492	162792	37 Aql	19 35 07.1	−10 33 38	0.000	0.00	5.12	1.13	0.3	G8 III	−31	70 s		
184722	105005		19 35 07.2	+19 00 17	−0.003	−0.03	6.83	1.65	−4.4	K5 Ib		1600 s		

HD	SAO	Star Name	α 2000	δ 2000	μ(α)	μ(δ)	V	B–V	M_v	Spec	RV	d(pc)	ADS	Notes
			h m s	° ′ ″	s	″								
184740	87390		19 35 07.8	+22 09 31	0.000	0.00	7.39	0.08		B9				
183414	—		19 35 08.7	−69 58 37			7.92	0.66	3.1	G4 IV		92 s		
186063	9477		19 35 10.1	+74 22 39	+0.002	+0.02	7.10	1.1	0.2	K0 III		240 s		m
185413	18416		19 35 10.3	+62 39 33	+0.001	−0.01	7.7	1.1	−0.1	K2 III		330 s		
184127	229751	ι Tel	19 35 12.7	−48 05 57	−0.001	−0.04	4.90	1.09	0.2	G9 III	+22	74 s		
183505	254599		19 35 16.5	−69 05 37	−0.003	−0.18	7.8	0.7	4.4	G0 V		48 s		
184940	68541		19 35 21.7	+34 41 17	0.000	+0.01	7.10	−0.11	0.0	B8.5 V	−14	260 s		
184724	105009		19 35 22.9	+11 56 18	+0.001	−0.01	7.92	1.74	−0.3	K5 III		320 s		m
185799	18428		19 35 23.9	+69 48 21	0.000	+0.04	8.0	1.6	−0.5	M4 III	−11	510 s		
184697	105010		19 35 24.7	+10 23 41	0.000	0.00	7.7	1.1	0.2	K0 III		300 s		
184663	124823		19 35 25.0	+2 54 48	+0.001	+0.05	6.38	0.41		F2	+4			
184741	105014		19 35 27.1	+15 36 36	0.000	−0.01	6.5	1.1	0.2	K0 III		180 s		
184742	105013		19 35 28.4	+13 14 29	0.000	0.00	8.0	1.1	0.2	K0 III		330 s		
184573	143564		19 35 29.6	−7 27 37	+0.002	−0.04	6.34	1.12		K0				
184287	229757		19 35 31.6	−41 07 16	+0.001	0.00	7.3	0.8	0.2	K0 III		260 s		
185263	31799		19 35 31.6	+53 32 48	+0.002	+0.07	7.3	0.4	2.6	F0 V		86 s		
184927	68542		19 35 31.8	+31 16 36	0.000	+0.01	7.46	−0.17	−2.6	B2.5 IV		1000 s		
185148	48627		19 35 33.0	+47 26 11	−0.001	+0.01	7.6	0.1	0.6	A0 V		260 s		
184574	162797		19 35 33.4	−12 15 10	+0.001	−0.01	6.2	1.1	0.2	K0 III		160 s		
184879	87410		19 35 33.5	+27 06 30	+0.001	−0.01	8.0	0.0	0.4	B9.5 V		310 s		
184509	188317		19 35 35.6	−20 46 55	+0.007	−0.16	6.7	0.8	4.4	G0 V		22 s		
184849	87409		19 35 37.7	+22 30 07	+0.001	0.00	7.80	0.06		A0				
184494	188318		19 35 38.6	−23 18 34	+0.001	−0.01	7.90	0.2	2.1	A5 V		150 s	12625	m
184830	105019		19 35 41.5	+17 06 31	+0.001	+0.01	7.6	0.4	3.0	F2 V		83 s		
184880	87411		19 35 42.6	+22 08 25	+0.002	−0.01	7.8	1.1	−0.1	K2 III		350 s		
184438	211392		19 35 44.6	−32 49 17	+0.006	+0.08	7.7	1.7	−0.3	K5 III		160 mx		
183951	254602		19 35 45.9	−61 37 51	+0.004	−0.02	7.3	0.8	3.4	F5 V		37 s		
184419	211391		19 35 46.7	−34 30 35	0.000	+0.03	7.9	1.4	0.2	K0 III		210 s		
184928	87417		19 35 47.0	+25 10 27	0.000	−0.01	8.0	0.0	0.4	B9.5 V		310 s		
184942	87418		19 35 47.7	+25 48 51	−0.001	+0.01	7.5	0.0	−1.0	B5.5 V	−17	450 s		
184881	105023		19 35 48.0	+18 34 27	−0.001	−0.02	7.80	0.8	1.8	G5 III–IV	+3	160 s		
185037	68552	11 Cyg	19 35 48.2	+36 56 40	0.000	0.00	5.90	−0.11	−0.2	B8 V	−15	170 s		
—	48630		19 35 49.7	+42 21 54	0.000	−0.06	7.9	1.7						
184701	143575		19 35 50.6	−2 27 21	−0.005	−0.05	6.9	0.5	3.4	F5 V		50 s		
185414	31813		19 35 55.5	+56 59 01	−0.001	−0.21	6.8	0.7	4.4	G0 V		29 s		
185264	31804		19 35 55.7	+50 14 19	−0.001	+0.04	6.6	0.9	0.2	G9 III	+8	190 s		
184192	246178		19 35 56.9	−53 00 29	−0.002	−0.05	6.9	1.2	−0.5	M4 III		300 s		
184883	105027		19 35 59.2	+13 50 30	0.000	−0.01	8.0	0.1	0.6	A0 V	−27	280 s		
184768	143582		19 36 00.5	+0 05 28	−0.001	−0.38	7.57	0.68	3.2	G5 IV	−14	75 s	12644	m
184767	124830		19 36 00.7	+0 14 58	−0.001	−0.01	7.10	0.1	1.4	A2 V	−18	140 s		m
184552	188326	51 Sgr	19 36 01.5	−24 43 09	+0.001	−0.02	5.65	0.19		A m	−45	19 mn		
184553	—		19 36 01.6	−25 55 02			8.02	0.50	3.3	F4 V		78 s		
184533	188325		19 36 03.0	−28 40 05	0.000	+0.01	7.5	0.1	0.6	A0 V		200 s		
184909	105031		19 36 03.0	+14 30 28	−0.002	−0.01	7.50	1.1	−0.2	K3 III	−21	320 s		
185055	68557		19 36 05.3	+33 44 46	+0.001	−0.04	7.42	1.18	−0.1	K2 III		310 s		
185118	68560		19 36 07.2	+35 12 50	−0.001	0.00	6.9	1.1	−0.1	K2 III		240 s		
184884	105032		19 36 07.8	+11 09 00	0.000	−0.01	6.50	0.1	1.4	A2 V	−5	100 s	12660	m
184961	87426		19 36 08.2	+22 35 07	0.000	−0.02	6.32	−0.07	−0.8	B9 III p	−31	250 s		
184349	229764		19 36 08.6	−46 09 04	+0.002	−0.01	7.1	1.4	−0.3	K5 III		310 s		
184910	105034		19 36 11.3	+12 09 40	−0.002	−0.02	6.7	1.1	−0.1	K2 III		220 s		
184853	124835		19 36 12.4	+6 00 28	0.000	0.00	6.80	0.9	0.3	G8 III	+12	200 s	12661	m
185056	68559		19 36 12.7	+31 31 39	0.000	0.00	8.0	1.4	−0.3	K5 III		410 s		
184944	105036		19 36 15.7	+14 23 30	+0.002	−0.02	6.38	1.04	0.2	K0 III	−42	160 s		
185119	68565		19 36 16.7	+33 27 41	−0.001	0.00	7.27	−0.10	0.0	B8.5 V		280 s		
181983	—		19 36 18.7	−81 15 17			7.8	1.1	0.2	K0 III		330 s		
185016	87435		19 36 19.5	+23 41 22	+0.001	0.00	7.95	0.27		A3				
184554	211401		19 36 19.7	−31 50 00	+0.002	+0.01	7.9	1.4	0.2	K0 III		210 s		
184887	124839		19 36 22.7	+4 11 14	+0.001	0.00	7.9	1.1	−0.1	K2 III		380 s		
184962	105038		19 36 22.8	+15 53 33	+0.003	+0.07	7.30	0.5	4.0	F8 V		45 s		m
184620	188330		19 36 25.5	−28 40 28	0.000	0.00	7.6	1.2	−0.1	K2 III		340 s		
184705	162809		19 36 25.9	−18 51 10	+0.002	0.00	6.1	0.4	2.1	A5 V		44 s		
185395	31815	13 θ Cyg	19 36 26.2	+50 13 16	−0.002	+0.26	4.48	0.38	2.1	F5 IV	−28	19 ts	12695	m
185329	48644		19 36 29.4	+45 47 52	+0.003	+0.05	7.4	0.4	2.6	F0 V		90 s		
185286	48642		19 36 32.2	+43 42 23	0.000	+0.01	6.7	1.4	−0.3	K5 III		240 s		
184858	143593		19 36 33.9	−4 46 19	0.000	−0.01	8.0	0.1	1.4	A2 V		200 s		
184685	188336		19 36 34.4	−24 33 08	+0.002	−0.02	8.0	2.5	−0.3	K5 III		110 s		
184597	211406		19 36 35.9	−32 41 27	+0.001	−0.01	6.9	0.1	0.0	B8.5 V		200 s		
185396	48651		19 36 37.4	+48 30 59	+0.001	−0.01	7.2	1.1	−0.1	K2 III		290 s		
185059	87447		19 36 37.5	+20 19 57	0.000	−0.01	6.5	0.5	−6.4	F5 I var	−11	2300 s		U Vul, v
185351	48649		19 36 37.7	+44 41 42	−0.010	−0.10	5.17	0.93	−0.4	K III	−5	130 s		
184772	162811		19 36 41.6	−17 20 05	−0.001	−0.02	7.9	0.0	0.4	B9.5 V		320 s		
184707	188337	52 Sgr	19 36 42.3	−24 53 01	+0.005	−0.02	4.60	−0.07		B9	−19		12654	m
185017	105042		19 36 42.3	+13 21 45	−0.001	0.00	7.8	0.1	0.6	A0 V		250 s		
185151	87451		19 36 42.4	+27 53 03	+0.001	0.00	7.80	1.1	0.0	K1 III		340 s		
184930	143597	41 ι Aql	19 36 43.1	−1 17 11	0.000	−0.02	4.36	−0.08	−2.2	B5 III	−22	180 s	12663	m

488

HD	SAO	Star Name	α 2000	δ 2000	μ(α)	μ(δ)	V	B-V	M_v	Spec	RV	d(pc)	ADS	Notes
			$19^h 36^m 44.4^s$	$+7° 33' 04''$	0.000^s	$+0.01''$								
184982	124844		19 36 44.4	+7 33 04	0.000	+0.01	7.59	0.17	1.7	A3 V		140 s		
185456	31822		19 36 49.2	+50 11 59	-0.001	-0.01	6.50	1.9		S7.7 e	-25			R Cyg, m,v
184794	--		19 36 49.5	-18 46 53			8.0	0.6	3.4	F5 V		65 s		
185084	105047		19 36 50.0	+19 27 48	-0.001	-0.04	8.0	1.1	0.2	K0 III		330 s		
184889	162815		19 36 50.6	-10 09 28	0.000	-0.02	6.9	1.1	0.2	K0 III		210 s		
185224	68578		19 36 51.9	+30 19 30	0.000	0.00	7.59	-0.08	0.4	B9.5 V		270 s	12688	m
185018	105045		19 36 52.3	+11 16 24	+0.001	+0.01	5.98	0.88	3.2	G5 IV	-1	31 s	12670	m
184915	143600	39 κ Aql	19 36 53.3	-7 01 39	0.000	0.00	4.95	0.00		B0.5 III	-20			
184795	188346		19 36 54.7	-21 54 10	0.000	-0.01	7.9	0.4	3.2	G5 IV		88 s		
185330	68585		19 36 56.4	+38 23 02	0.000	0.00	6.50	-0.15	-2.9	B3 III		690 s		
184890	162820		19 37 02.3	-13 10 23	+0.003	-0.01	7.0	0.0	0.4	B9.5 V		200 s		
184835	162816		19 37 03.2	-18 13 52	+0.001	-0.01	5.8	1.1	0.2	K0 III	-7	130 s		
185061	124853		19 37 03.4	+9 33 33	+0.001	-0.02	8.02	1.26	-2.1	G8 II		810 s		
184376	246183		19 37 05.0	-54 44 27	+0.004	+0.01	8.0	1.6	-0.1	K2 III		220 s		
185193	87459		19 37 06.5	+22 00 21	+0.002	-0.01	6.8	0.1	1.7	A3 V		100 s		
185374	48655		19 37 08.6	+40 12 49	-0.001	-0.05	7.9	0.9	3.2	G5 IV		88 s		
185268	87462		19 37 09.3	+29 20 01	-0.001	0.00	6.43	-0.09	-1.1	B5 V	-20	310 s	12696	m
184421	246184		19 37 10.4	-54 09 14	-0.001	-0.04	7.7	0.7	3.2	G5 IV		80 s		
184918	162822		19 37 10.8	-14 25 54	-0.001	-0.01	7.1	1.1	0.2	K0 III		230 s		
184711	211415		19 37 11.7	-39 44 38	-0.001	-0.06	7.99	1.35	-0.3	K4 III		140 s		
185269	87464		19 37 11.7	+28 30 00	-0.002	-0.07	6.7	0.6	2.8	G0 IV	+1	59 s		
184731	211416		19 37 13.7	-34 17 25	+0.001	+0.02	7.5	1.8	-0.1	K2 III		140 s		
185288	68587		19 37 14.3	+31 53 55	0.000	-0.02	7.80	1.1	0.2	K0 III		310 s		
185194	105061	4 ε Sge	19 37 17.2	+16 27 46	+0.001	+0.02	5.66	1.02	0.3	G8 III	-33	100 s	12693	m
185026	143608		19 37 17.3	-0 29 32	+0.001	0.00	7.7	0.0	0.4	B9.5 V		280 s		
185397	68592		19 37 17.4	+38 35 38	0.000	0.00	6.9	0.2	2.1	A5 V		88 s		
185153	105056		19 37 17.5	+13 06 57	-0.002	-0.01	7.9	1.1	0.2	K0 III		320 s		
185289	87471		19 37 21.6	+26 21 32	0.000	0.00	7.4	0.9	0.3	G8 III		250 s		
185225	105065		19 37 23.2	+18 58 34	+0.001	-0.02	6.7	1.4	-0.3	K5 III		240 s		
184753	211418		19 37 27.0	-35 14 26	-0.002	0.00	8.0	1.0	4.0	F8 V		34 s		
--	48661		19 37 29.4	+43 22 08	-0.003	-0.02	8.0	2.0						
185332	87474		19 37 29.7	+29 36 54	+0.002	+0.01	7.3	0.1	1.4	A2 V		150 s		
185695	31831		19 37 32.1	+54 57 28	-0.001	-0.01	7.5	1.6	-0.5	M2 III		380 s		
185090	143612		19 37 33.4	-0 07 44	+0.001	+0.03	7.5	0.2	2.1	A5 V	+8	120 s		
184985	162827		19 37 34.2	-14 18 06	-0.007	-0.14	5.47	0.50	4.0	F8 V	-21	20 s		
185206	105067		19 37 35.4	+12 41 24	-0.002	-0.01	8.0	0.4	2.6	F0 V		110 s		
185043	143609		19 37 35.6	-9 43 51	+0.002	+0.02	7.9	0.1	1.4	A2 V		190 s		
185976	18440		19 37 35.9	+64 01 01	-0.003	-0.01	7.9	0.0	0.4	B9.5 V		290 s		
185435	68596		19 37 40.7	+35 01 22	0.000	+0.01	6.6	1.4	-0.3	K5 III		230 s		
185068	162829		19 37 42.7	-9 58 29	0.000	0.00	7.30	0.1	1.4	A2 V		150 s	12694	m
185334	87476		19 37 43.3	+24 15 40	0.000	-0.01	7.67	0.14		A0				
185094	143618		19 37 43.5	-5 03 42	+0.001	-0.01	7.5	0.4	2.6	F0 V		95 s		
185124	143621	42 Aql	19 37 47.1	-4 38 51	+0.007	-0.05	5.46	0.43	2.8	dF1	-38	30 s		
185044	162831		19 37 48.3	-13 57 21	0.000	+0.01	6.9	1.1	0.2	K0 III		210 s		
185353	87482		19 37 51.9	+22 47 49	-0.001	-0.03	7.90	0.7	-2.1	G2 II		830 s	12704	m
184954	188360		19 37 52.9	-28 48 57	+0.003	-0.03	7.6	1.1	4.4	G0 V		21 s		
185198	124868		19 37 53.3	+1 29 57	0.000	0.00	7.5	0.0	0.4	B9.5 V		260 s		
184690	229781		19 37 53.5	-48 06 49	0.000	0.00	7.5	0.3	1.4	A2 V		170 s		
185375	87486		19 37 53.8	+24 42 45	+0.001	-0.01	7.73	0.13		A2				
185337	105078		19 37 55.9	+18 35 31	+0.001	-0.02	6.9	0.2	2.1	A5 V		90 s		
185657	48673		19 37 56.5	+49 17 04	+0.003	+0.15	6.60	1.1	0.2	K0 III	-85	190 s		m
185416	87487		19 37 57.2	+26 00 59	+0.001	0.00	7.80	0.09		B9				
184922	211429		19 38 00.3	-32 54 46	+0.003	-0.02	6.7	1.0	0.2	K0 III		200 s		
185526	68603		19 38 01.3	+35 15 31	0.000	-0.04	6.5	1.1	0.2	K0 III		180 s		
184899	211428		19 38 07.1	-38 45 26	+0.004	+0.02	7.7	2.8	-0.5	M2 III		86 s		
185354	105081		19 38 09.6	+17 14 34	0.000	-0.02	6.70	0.1	0.6	A0 V		160 s	12711	m
184356	254609		19 38 10.4	-63 54 56	-0.001	-0.01	7.9	1.1	-0.1	K2 III		400 s		
185310	124875		19 38 11.2	+8 44 54	0.000	0.00	7.79	0.05	-0.6	A0 III		430 s		
185752	31840		19 38 11.5	+51 14 09	+0.002	+0.02	7.2	1.1	0.2	K0 III		260 s		
184840	229785		19 38 11.9	-41 48 17	+0.003	-0.03	8.0	-0.3	1.7	A3 V		180 s		
185247	143637		19 38 12.7	-0 29 22	-0.001	-0.01	7.5	0.1	1.4	A2 V		170 s		
185417	87488		19 38 14.9	+21 53 05	-0.001	-0.02	7.8	1.1	0.2	K0 III		310 s		
185436	87489		19 38 17.3	+20 46 58	-0.004	-0.05	6.4	1.1	0.2	K0 III	+5	170 s		
185527	68606		19 38 18.0	+31 45 32	+0.002	+0.02	7.90	0.9	0.3	G8 III		310 s		
185297	124878		19 38 21.5	+0 20 43	+0.001	+0.02	7.22	0.28	1.4	A2 V	+3	110 s	12708	m
184585	246188		19 38 25.8	-57 58 59	+0.004	-0.05	6.18	0.97	3.2	G5 IV	-2	27 s		
185418	105087		19 38 27.3	+17 15 26	0.000	-0.01	7.45	0.22		B0.5 V			12723	m
184327	254610		19 38 28.9	-65 54 53	0.000	0.00	7.3	0.1	0.6	A0 V		220 s		
185620	68615		19 38 29.1	+37 44 34	+0.001	-0.03	7.6	1.1	0.2	K0 III		280 s		
185528	87493		19 38 30.0	+25 58 43	0.000	0.00	7.92	0.23		A3				
185419	105085		19 38 32.2	+12 21 30	+0.001	-0.01	8.0	1.1	0.2	K0 III		330 s		
185009	211433		19 38 33.3	-34 32 56	-0.006	-0.13	7.2	0.3	3.4	F5 V		59 s		
185529	87494		19 38 38.2	+23 02 32	+0.001	-0.01	7.00	0.1	0.6	A0 V		180 s	12731	m
185603	68614		19 38 38.6	+31 13 08	+0.001	-0.01	8.0	0.1	0.6	A0 V		280 s		
185602	68617		19 38 39.2	+32 24 21	0.000	0.00	7.3	0.0	0.4	B9.5 V		230 s		

HD	SAO	Star Name	α 2000	δ 2000	μ(α)	μ(δ)	V	B-V	M_V	Spec	RV	d(pc)	ADS	Notes
			h m s	° ' "	s	"								
185912	31850		19 38 41.0	+54 58 26	+0.004	+0.17	5.82	0.44	3.4	F5 V	-15	30 ts		
185298	162843		19 38 43.2	-10 09 23	0.000	-0.02	6.80	0.2	2.1	A5 V		86 s	12715	m
185547	87497		19 38 45.8	+23 06 50	0.000	-0.01	7.7	1.1	0.2	K0 III		290 s		
185530	87495		19 38 46.1	+21 54 33	+0.001	+0.04	7.9	0.9	3.2	G5 IV		87 s		
184519	254618		19 38 47.0	-62 53 23	+0.003	+0.02	7.8	0.4	2.6	F0 V		110 s		
185423	124892		19 38 48.8	+3 22 54	0.000	+0.01	6.35	0.04	-2.9	B3 III	-1	510 s		
185636	68624		19 38 49.9	+33 20 48	+0.001	0.00	7.9	0.4	3.0	F2 V		92 s		
185163	211438		19 38 52.4	-30 16 53	+0.005	-0.02	7.5	0.5	0.6	A0 V		110 s		
185183	188386		19 38 56.1	-28 36 26	+0.002	-0.01	6.5	0.1	0.4	B9.5 V		140 s		
185342	162847		19 38 56.4	-10 20 21	+0.001	-0.01	7.89	0.29		A0			12725	m
185660	68626		19 38 57.7	+32 33 43	0.000	-0.01	7.9	0.0	0.4	B9.5 V		300 s		
185978	31858		19 39 05.5	+54 10 53	+0.004	+0.02	7.9	0.5	4.0	F8 V		59 s		
185780	48681		19 39 07.2	+40 37 35	-0.002	-0.03	7.75	-0.07	-2.6	B2.5 IV	-5	1100 s		
185733	68631		19 39 08.2	+34 56 59	+0.001	-0.01	7.64	1.58	-0.3	K5 III		350 s		
186426	18457		19 39 10.0	+68 39 09	-0.006	+0.03	6.9	0.8	3.2	G5 IV		55 s		
185835	48684		19 39 10.0	+43 50 23	+0.001	+0.01	7.97	-0.04	0.0	B8.5 V		370 s		
185507	124903	44 σ Aql	19 39 11.4	+5 23 52	0.000	0.00	5.17	0.03	-1.7	B3 V	-5	190 s	12737	m, v
185344	162853		19 39 12.8	-16 54 27	+0.002	+0.03	6.6	0.1	1.4	A2 V		110 s	12728	m
185753	68639		19 39 13.9	+37 02 37	0.000	0.00	8.0	1.1	0.2	K0 III		340 s		
185605	105099		19 39 13.9	+18 41 00	-0.001	-0.02	7.9	0.0	-1.0	B5.5 V	-9	540 s		
184844	246199		19 39 18.5	-56 18 50	+0.003	-0.02	7.8	-0.1	2.6	F0 V		110 s		
185033	229793		19 39 19.0	-45 25 43	+0.004	-0.05	7.4	0.3	3.0	F2 V		77 s		
185871	48687		19 39 19.3	+44 26 10	+0.003	-0.05	6.9	1.1	0.2	K0 III		210 s		
185734	68637	12 φ Cyg	19 39 22.4	+30 09 12	0.000	+0.04	4.69	0.97	1.8	G8 III-IV	+6	34 s		
184358	254617		19 39 24.5	-69 20 15	+0.009	+0.01	6.95	1.30	-0.2	K3 III		260 s		
185622	105104		19 39 25.2	+16 34 17	+0.001	0.00	6.37	2.09	-5.2	M0 Iab-Ib	-4	850 s	12750	m
185661	87508		19 39 26.1	+22 15 24	0.000	+0.02	7.70	0.4	3.0	F2 V		85 s	12752	m
185872	48691	14 Cyg	19 39 26.3	+42 49 06	+0.002	+0.03	5.40	-0.08		B8	-28			
185137	229796		19 39 28.5	-41 11 08	+0.001	-0.02	7.9	1.1	-0.1	K2 III		390 s		
186174	31872		19 39 29.4	+59 20 47	0.000	-0.01	8.0	1.4	-0.3	K5 III		410 s		
185755	68641		19 39 29.5	+30 24 30	-0.001	0.00	7.11	-0.08	0.4	B9.5 V	-18	220 s		
185662	87513		19 39 30.6	+21 20 36	+0.001	0.00	7.39	1.47	-0.3	gK4	-25	340 s		
185663	105109		19 39 32.4	+19 10 10	-0.001	-0.02	7.70	1.1	-2.2	K2 II		790 s		
184820	246200		19 39 32.8	-58 13 55	+0.006	+0.01	6.8	1.0	2.6	F0 V		25 s		
185955	48697		19 39 34.2	+45 57 29	-0.001	+0.05	6.20	0.90	0.2	K0 III	-10	160 s		
184359	254619		19 39 35.0	-69 36 11	+0.005	-0.03	6.9	1.5	0.2	K0 III		120 s		
185756	87522		19 39 35.0	+29 44 56	-0.001	+0.01	7.39	-0.04	0.6	A0 V		230 s	12759	m
184628	254621		19 39 36.0	-64 32 09	+0.003	-0.06	7.2	1.0	0.2	K0 III		250 s		
185277	211449		19 39 36.7	-32 28 21	+0.001	+0.01	7.7	1.3	0.2	K0 III		200 s		
185697	87518		19 39 37.3	+20 46 58	+0.001	-0.01	8.0	0.0	0.4	B9.5 V		310 s		
185608	124911		19 39 39.3	+7 50 40	+0.002	+0.04	7.89	0.31	0.6	A9 III		280 s		
185511	143658		19 39 40.1	-6 42 31	+0.002	-0.04	8.0	0.1	1.4	A2 V		200 s		
185217	211448		19 39 41.0	-39 44 50	-0.002	-0.08	7.70	0.9	3.2	G5 IV		79 s		m
185139	229800		19 39 41.7	-45 16 41	0.000	+0.02	6.25	0.28		A m				
185837	68654		19 39 44.8	+33 58 45	0.000	+0.02	6.10	0.08	1.7	A3 V	-32	75 s	12765	m
185426	162859		19 39 47.0	-19 13 52	+0.006	-0.04	7.8	0.5	3.4	F5 V		74 s		
184404	188407	53 Sgr	19 39 49.3	-23 25 40	0.000	-0.03	6.90	0.1	0.6	A0 V		180 s	12741	m
184900	246201		19 39 50.2	-58 09 10	-0.003	+0.03	7.4	1.3	3.2	G5 IV		33 s		
185781	87525		19 39 51.1	+24 32 13	-0.001	-0.03	7.0	1.1	0.2	K0 III	-80	220 s		
185487	162862		19 39 52.1	-15 10 03	0.000	-0.02	6.7	0.0	0.0	B8.5 V		220 s		
184586	254622		19 39 52.2	-66 41 08	+0.008	-0.05	6.39	0.02	0.6	A0 V n		140 mx		m
185514	162864		19 39 53.9	-14 50 57	+0.001	0.00	7.30	0.00	0.0	B8.5 V		290 s	12748	m
185257	211451		19 39 55.4	-39 26 00	+0.005	-0.06	6.61	0.23		A2				
185447	188414		19 39 55.5	-22 03 45	0.000	+0.01	7.90	0.1	0.6	A0 V		290 s	12743	m
185737	105116		19 39 55.5	+19 23 58	0.000	-0.04	7.8	0.8	3.2	G5 IV		83 s		
186253	31881		19 39 56.4	+59 50 28	+0.004	+0.03	7.8	0.2	2.1	A5 V		130 s		
--	31873		19 39 56.6	+52 05 41	0.000	0.00	8.0	1.8	-0.1	K2 III		170 s		
185667	124915		19 39 59.1	+7 34 02	0.000	+0.01	7.22	0.97		G5				
185141	246203		19 40 02.2	-49 56 57	-0.003	-0.02	7.30	1.1	-0.1	K2 III		300 s		m
185321	211455		19 40 04.8	-37 24 03	-0.001	+0.03	7.4	1.9	-0.3	K5 III		200 s		
185533	162870		19 40 05.0	-14 02 40	+0.002	-0.02	7.1	0.1	0.6	A0 V		200 s		
185588	143664		19 40 05.2	-7 58 47	+0.005	-0.19	7.6	0.7	4.4	G0 V		44 s		
186196	31880		19 40 05.2	+55 54 28	+0.004	+0.01	7.4	1.1	0.2	K0 III		260 s		
185758	105120	5 α Sge	19 40 05.6	+18 00 50	+0.001	-0.02	4.37	0.78	-2.0	G0 II	+2	190 s	12766	m
185467	188419		19 40 06.9	-23 25 43	+0.002	0.00	6.1	0.9	0.2	K0 III	-28	150 s		
185516	188421		19 40 09.7	-20 32 52	0.000	-0.01	7.9	1.4	-0.3	K5 III		430 s		
--	105121		19 40 10.8	+17 53 57	0.000	0.00	7.6	1.1	0.2	K0 III		280 s		
185896	68666		19 40 11.6	+32 03 20	0.000	0.00	8.02	-0.07	0.4	B9.5 V		330 s		
186340	18461		19 40 12.8	+60 30 26	-0.002	0.00	6.20	0.1	1.4	A2 V	-1	89 s	12789	m
185856	87538		19 40 13.3	+26 59 43	0.000	-0.01	7.2	0.0	0.0	B8.5 V		260 s		
186239	31882		19 40 16.2	+55 47 38	+0.006	+0.05	7.4	0.2	2.1	A5 V		110 s		
185534	188422		19 40 16.4	-21 18 09	+0.001	-0.01	7.84	-0.02	-1.0	B5.5 V	-12	480 s		
185075	246204		19 40 18.1	-54 25 03	+0.005	-0.01	6.26	1.00	0.2	gK0		160 s		
185639	143669		19 40 21.1	-5 26 51	0.000	-0.01	6.7	1.1	-0.1	K2 III		220 s		
185803	105122		19 40 22.4	+16 20 00	0.000	-0.01	7.3	0.0	0.0	B8.5 V		270 s		

HD	SAO	Star Name	α 2000	δ 2000	μ(α)	μ(δ)	V	B−V	M_v	Spec	RV	d(pc)	ADS	Notes
185858	87541		19^{h}40^{m}23^s.7	+23°28'46"	+0^s.003	0".00	6.9	0.1	1.4	A2 V		130 s		
185914	87545		19 40 25.5	+26 18 58	+0.001	+0.02	7.7	0.0	0.0	B8.5 V		330 s		
185859	87542		19 40 28.1	+20 28 36	0.000	−0.02	6.50	0.39		B0.5 Ia	+5			
185874	87546		19 40 32.2	+21 34 34	0.000	−0.01	7.8	0.1	0.6	A0 V		250 s		
186303	31885		19 40 35.7	+56 43 09	+0.001	+0.02	8.0	0.1	0.6	A0 V		280 s		
186120	48715		19 40 38.6	+44 47 20	+0.003	+0.03	7.19	1.12	0.2	gK0	−18	200 s		
185915	87551		19 40 39.5	+23 43 03	+0.001	−0.02	6.64	0.01	−1.3	B6 IV	−20	330 s	12778	m
185999	68679		19 40 41.0	+31 24 24	+0.002	+0.01	7.1	0.9	0.3	G8 III	−9	220 s		
186121	48714		19 40 41.1	+43 04 40	+0.001	0.00	6.16	1.56	−0.5	M2 III	−5	220 s		
185187	246208		19 40 41.8	−54 12 38	0.000	−0.02	7.3	−0.3	1.7	A3 V		130 s		
185644	162883	54 Sgr	19 40 43.1	−16 17 36	+0.005	−0.05	5.30	1.13		K1 IV	−58	14 mn	12767	m
185762	143678	45 Aql	19 40 43.1	−0 37 17	+0.001	+0.02	5.67	0.11		A0	−46	25 mn	12775	m
185739	143677		19 40 44.5	−4 02 00	+0.001	0.00	8.0	0.1	0.6	A0 V		280 s		
185982	87559		19 40 45.0	+27 43 47	−0.001	−0.01	8.00	0.9	0.3	G8 III		330 s		
186221	48720		19 40 45.1	+48 30 45	−0.003	+0.01	8.0	0.4	2.6	F0 V		120 s		
185720	143674		19 40 45.2	−7 30 46	+0.007	−0.07	6.9	0.6	4.4	G0 V		31 s		
186140	48716		19 40 46.3	+44 31 43	+0.004	+0.03	8.0	0.5	4.0	F8 V		62 s		
186176	48717		19 40 46.6	+46 23 08	+0.002	−0.05	7.80	0.8	0.3	G6 III	−1	300 s		
195252	3327		19 40 47.5	+87 55 20	+0.006	−0.01	8.0	1.1	−0.1	K2 III		420 s		
185592	188429		19 40 48.8	−27 41 46	+0.003	−0.05	7.6	1.6	−0.1	K2 III		200 s		
186155	48718		19 40 50.1	+45 31 29	+0.008	+0.11	5.06	0.40	0.6	F2 III	−20	72 s		m
185808	124926		19 40 50.9	+1 40 23	0.000	+0.01	7.9	0.0	0.4	B9.5 V		310 s		
185113	246207		19 40 52.4	−58 16 10	−0.003	−0.02	6.7	1.8	0.2	K0 III		69 s		
185875	105128		19 40 53.3	+11 55 27	+0.001	−0.01	7.6	1.6	−0.5	M2 III		410 s		
186047	68688		19 40 56.8	+32 37 06	0.000	0.00	7.30	2.0		N7.7	−49			TT Cyg, m,v
185984	87562		19 40 57.6	+23 29 18	+0.003	0.00	8.0	0.4	3.0	F2 V		96 s		
185823	124928		19 40 57.7	+0 42 12	−0.001	−0.01	7.9	1.1	−0.1	K2 III	−3	380 s		
185958	105133	6 β Sge	19 41 02.8	+17 28 33	+0.001	−0.03	4.37	1.05	−2.1	G8 II	−22	200 s		
185936	105132		19 41 05.3	+13 48 56	0.000	−0.01	6.01	−0.08	−1.1	B5 V	−14	250 s		QS Aql, m,v
185898	105131		19 41 06.5	+10 40 48	−0.004	−0.07	8.00	0.5	4.0	F8 V		63 s		m
185825	143690		19 41 07.3	−3 55 09	−0.001	−0.01	8.0	1.6	−0.5	M4 III		460 s		
185938	105134		19 41 07.8	+13 10 12	+0.003	−0.01	7.3	0.5	4.0	F8 V		45 s		
185575	211474		19 41 10.3	−36 19 54	0.000	0.00	7.9	0.5	2.1	A5 V		97 s		
185348	246214		19 41 10.4	−51 31 34	−0.007	+0.01	8.0	0.4	4.0	F8 V		62 s		
185842	143695		19 41 11.8	−2 18 48	0.000	+0.01	7.1	0.0	−1.0	B5.5 V	−10	390 s		
185615	211476		19 41 13.7	−32 52 22	−0.002	−0.09	7.9	0.5	3.2	G5 IV		88 s		
186304	31888		19 41 14.0	+50 22 43	+0.001	0.00	7.9	0.9	3.2	G5 IV		88 s		
186021	87569		19 41 14.7	+22 27 10	+0.002	0.00	6.36	1.53		K2	−23			
186222	48725		19 41 14.8	+42 08 29	+0.001	−0.01	8.03	−0.07	0.4	B9.5 V		340 s		
186097	68695		19 41 15.3	+30 43 17	0.000	+0.01	7.2	0.9	3.2	G5 IV		61 s	12786	m
185050	254626		19 41 18.0	−62 49 39	+0.001	0.00	7.5	1.1	0.2	K0 III		290 s		
186075	87571		19 41 21.3	+27 42 19	+0.002	0.00	7.90	0.22	1.7	A3 V		150 s		
186200	68700		19 41 22.9	+38 27 27	−0.002	−0.02	7.52	−0.02	0.4	B9.5 V		270 s		
186305	48734		19 41 27.4	+47 19 41	−0.001	+0.01	7.4	1.6	−0.5	M2 III		350 s		
186355	48736		19 41 35.2	+49 41 43	−0.001	+0.03	7.8	0.8	3.2	G5 IV		84 s		
185495	229816		19 41 36.5	−48 40 00	+0.002	−0.01	8.0	0.8	0.2	K0 III		360 s		
184996	254627		19 41 37.3	−65 51 15	+0.006	−0.06	6.09	1.55		K2				
185349	246216		19 41 38.1	−55 15 55	+0.006	+0.02	7.3	1.3	−0.1	K2 III		260 s		
186177	68705		19 41 38.8	+33 05 00	+0.002	+0.01	7.0	0.2	−4.8	A5 Ib		1600 s		
186255	68708		19 41 39.6	+40 01 21	+0.001	+0.02	6.91	0.16	1.7	A3 V		100 s		
186178	68704		19 41 40.5	+30 48 31	+0.002	+0.02	6.8	1.6	−0.4	M0 III		270 s		
184039	257716		19 41 44.6	−77 34 26	+0.003	+0.03	7.5	1.1	0.2	K0 III		290 s		
185946	143703		19 41 47.5	−4 17 21	+0.001	−0.01	7.9	0.0	0.4	B9.5 V		300 s		
186408	31898	16 Cyg	19 41 48.7	+50 31 31	−0.016	−0.15	5.96	0.64	4.7	G2 V	−26	19 ts	12815	m
185691	211482		19 41 49.2	−36 53 14	+0.001	+0.01	7.1	0.2	0.6	A0 V		150 s		
185770	188453		19 41 50.9	−28 51 06	+0.004	+0.01	7.8	1.3	−0.1	K2 III		330 s		
186427	31899		19 41 51.7	+50 31 03	−0.015	−0.16	6.20	0.66	5.2	G5 V	−28	16 s		
186179	87584		19 41 52.2	+27 22 56	−0.002	0.00	7.20	0.00	0.0	B8.5 V	−24	260 s	12798	m
186257	68713		19 41 57.2	+36 13 32	+0.014	−0.02	7.5	0.5	3.4	F5 V		67 s		
186307	48737		19 41 57.3	+40 15 14	−0.002	+0.03	6.23	0.18	1.7	A3 V	−32	74 s		m
186258	68710		19 42 01.2	+31 29 44	+0.004	−0.01	7.9	0.4	2.6	F0 V		110 s		
185966	143705		19 42 02.8	−9 11 36	+0.003	−0.07	6.5	1.1	−0.1	K2 III		190 mx		
186532	31907		19 42 07.1	+55 28 44	−0.001	+0.01	6.48	1.61	−0.5	M4 III	−28	250 s		
186122	105156	46 Aql	19 42 12.6	+12 11 36	0.000	0.00	6.30	−0.03	−1.2	B8 III	−32	290 s		
185771	211487		19 42 12.8	−36 38 00	0.000	−0.06	6.9	0.2	3.4	F5 V		51 s		
186672	18481		19 42 14.0	+60 33 45	−0.003	−0.02	6.7	0.1	1.4	A2 V		110 s		
185559	246224		19 42 15.4	−52 56 57	0.000	−0.03	7.60	0.1	1.7	A3 V		150 s		m
185472	246221		19 42 19.0	−57 08 26	+0.001	−0.03	7.8	0.7	3.2	G5 IV		84 s		
186201	105163		19 42 19.0	+18 27 56	0.000	0.00	6.8	0.1	0.6	A0 V		170 s		
185596	246225		19 42 20.8	−50 29 32	+0.002	−0.01	7.4	0.7	2.6	F0 V		57 s		
185905	188470		19 42 22.4	−24 51 34	+0.001	−0.02	6.6	0.9	2.1	A5 V		29 s		
185883	188469		19 42 23.4	−26 26 38	+0.001	−0.01	8.0	2.1	−0.1	K2 III		120 s		
186104	124958		19 42 29.5	+1 35 13	+0.010	−0.04	7.7	0.6	4.4	G0 V		46 s		
186005	162915	55 Sgr	19 42 30.9	−16 07 27	+0.004	−0.01	5.06	0.33	0.6	F0 III	−28	76 s		m
186202	105167		19 42 32.3	+12 47 20	0.000	0.00	7.9	1.1	−0.1	K2 III		380 s		

HD	SAO	Star Name	α 2000	δ 2000	μ(α)	μ(δ)	V	B-V	M_V	Spec	RV	d(pc)	ADS	Notes
185634	246226		19^{h}42^{m}33^s.4	-50°40'02"	+0^s.002	+0".01	6.8	1.2	0.2	K0 III		150 s		
185454	246222		19 42 33.6	-59 00 35	+0.025	-0.16	7.49	0.71	5.2	G5 V		27 s		m
186203	105168	47 χ Aql	19 42 33.8	+11 49 36	0.000	-0.01	5.3	0.5	3.4	F5 V	-22	35 mn	12808	m
186410	68731		19 42 34.2	+40 01 34	+0.001	-0.01	7.94	0.00	0.6	A0 V		290 s		m
186143	124960		19 42 34.4	+4 57 15	0.000	+0.01	7.9	0.0	0.0	B8.5 V		370 s		
--	68733		19 42 36.4	+39 59 52	0.000	-0.04	8.0	0.3						
185969	188473		19 42 37.1	-24 22 46	+0.002	-0.03	7.9	0.4	2.6	F0 V		96 s		
186377	68730		19 42 44.3	+32 25 35	-0.001	-0.01	5.90	0.19	0.1	A4 III	-8	130 s		
186226	124969		19 42 45.7	+8 22 58	+0.001	+0.06	6.72	0.49	4.0	F8 V		35 s	12813	m
186514	48759		19 42 47.3	+47 12 31	-0.001	0.00	7.3	0.9	3.2	G5 IV		65 s		
186272	105179		19 42 48.2	+17 58 06	+0.001	0.00	7.9	0.0	-1.6	B3.5 V	-15	670 s		
186310	87607		19 42 48.3	+22 51 07	+0.003	-0.01	6.5	1.1	0.2	K0 III		180 s		
186429	68737		19 42 48.4	+37 41 11	-0.001	-0.01	7.90	1.1	-0.1	K2 III		370 s		m
186357	87612		19 42 48.9	+29 19 54	+0.005	+0.05	6.49	0.33	0.6	F2 III	-25	150 s		
186438	68739		19 42 52.6	+37 40 41	-0.002	-0.01	8.0	0.4	-4.6	F3 Ib		2100 s		
185907	211493		19 42 55.4	-37 26 24	+0.002	-0.02	7.3	0.7	2.6	F0 V		89 s		
186465	48756		19 42 56.2	+40 43 18	+0.001	0.00	6.71	-0.02	0.6	A0 V		170 s	12831	m
185928	211494		19 42 58.9	-35 58 56	-0.004	-0.01	7.5	0.8	0.2	K0 III		230 mx		
186378	87618		19 42 58.9	+27 24 30	-0.002	0.00	7.4	1.1	0.2	K0 III	-43	260 s		
186439	68741		19 43 00.2	+34 25 17	-0.001	-0.03	7.96	-0.01	0.6	A0 V		300 s		
184588	257718		19 43 00.9	-75 23 51	-0.003	+0.09	8.01	0.54		F8				
186158	143722		19 43 03.0	-8 18 26	+0.001	-0.05	7.00	0.5	3.4	F5 V		52 s		m
186505	68748		19 43 04.4	+39 59 49	0.000	+0.02	6.93	0.30	2.1	A5 V		82 s		
186293	124982		19 43 05.1	+9 29 23	-0.001	0.00	7.79	1.17	-4.4	K0 Ib		2700 s		
186379	87619		19 43 06.8	+24 35 52	+0.006	-0.27	6.8	0.5	4.0	F8 V	-8	31 ts		
186159	143724		19 43 06.8	-8 19 47	+0.002	-0.06	7.4	0.5	4.0	F8 V		47 s		
186395	87623		19 43 07.0	+26 55 56	+0.001	+0.05	7.9	0.4	2.6	F0 V		110 s		
186440	68744		19 43 09.3	+30 40 43	-0.001	+0.04	6.05	0.00	1.2	A1 V	-31	93 s		
186506	68751		19 43 10.5	+38 40 19	0.000	+0.01	6.70	1.1	0.2	K0 III		190 s		m
186160	162928		19 43 11.3	-10 40 47	+0.004	-0.14	8.0	0.5	4.0	F8 V		62 s		
185682	246228		19 43 13.5	-56 15 37	0.000	-0.06	7.6	1.2	-0.5	M4 III		230 mx		
186760	31923		19 43 14.4	+58 00 59	+0.017	-0.06	6.22	0.56	4.0	F8 V	-22	24 ts		
187339	9519		19 43 14.4	+72 21 46	-0.003	0.00	8.0	0.1	0.6	A0 V		300 s		
186441	68745		19 43 15.0	+30 13 56	+0.002	0.00	7.23	0.07	0.6	A0 V		190 s		
186311	105185		19 43 15.7	+12 01 06	-0.001	-0.03	7.7	1.1	0.2	K0 III		300 s		
186312	124989		19 43 16.1	+9 29 50	0.000	-0.02	7.79	1.47	-2.3	K3 II		990 s		
186673	31922		19 43 16.7	+52 33 41	-0.001	0.00	7.8	0.1	1.7	A3 V		160 s		
185730	246229		19 43 19.5	-53 48 00	0.000	+0.01	7.8	1.1	-0.1	K2 III		380 s		
186278	124987		19 43 20.9	+4 10 27	0.000	-0.02	8.00	0.2	2.1	A5 V		150 s	12825	m
186031	211502		19 43 21.9	-33 38 46	+0.002	0.00	7.4	1.2	-0.1	K2 III		300 s		
186618	48768		19 43 22.3	+47 14 40	-0.001	-0.02	7.77	-0.18	-2.6	B2.5 IV		1200 s	12849	m
184735	257720		19 43 22.8	-74 46 08	+0.001	-0.01	7.60	1.31		K2				
185850	229830		19 43 23.2	-48 12 16	+0.001	-0.02	8.0	1.1	0.2	K0 III		290 s		
186412	87627		19 43 24.7	+22 29 40	-0.001	-0.02	6.84	-0.07	-1.6	B3.5 V	-41	410 s		
185831	246235		19 43 26.2	-50 10 40	0.000	-0.01	8.0	1.2	-0.1	K2 III		360 s		
185832	246236		19 43 30.5	-50 56 44	+0.001	0.00	7.8	1.9	-0.1	K2 III		140 s		
186185	162931		19 43 33.3	-15 28 12	+0.010	-0.18	5.49	0.46	2.2	F6 IV	+13	31 mx		m
186381	105187		19 43 33.3	+12 12 24	+0.001	-0.01	7.9	1.1	0.2	K0 III		330 s		
185391	254633		19 43 33.9	-66 57 41	-0.001	-0.06	7.8	0.4	2.6	F0 V		110 s		
186032	211504		19 43 36.7	-38 52 32	0.000	-0.04	8.0	0.5	0.6	A0 V		140 s		
186042	211506		19 43 37.2	-37 32 19	-0.001	-0.01	6.1	0.5	0.0	B8.5 V	-29	77 s		
186686	48773		19 43 37.7	+48 46 42	-0.001	+0.02	6.2	1.6		M5 III	-116			RT Cyg, v
186815	31933		19 43 39.4	+57 02 33	+0.001	+0.02	6.27	0.88	3.2	G5 IV	-26	35 s		
186515	68761		19 43 40.0	+32 04 47	+0.002	+0.01	7.7	0.1	1.4	A2 V		180 s		
185993	229835		19 43 42.5	-44 08 02	0.000	+0.01	7.28	1.23	-0.2	K3 III		310 s		
186486	87633	10 Vul	19 43 42.8	+25 46 19	+0.001	+0.02	5.49	0.93	0.3	G8 III	-9	110 s		
188281	9532		19 43 42.9	+79 32 23	0.000	+0.02	7.6	0.1	0.6	A0 V		250 s		
186727	31926		19 43 42.9	+51 04 00	-0.001	-0.01	7.9	1.6	-0.5	M2 III		490 s		
186604	68766		19 43 44.1	+39 44 31	+0.001	+0.02	7.8	0.1	1.4	A2 V		180 s		
186619	48771		19 43 44.9	+41 46 23	+0.001	+0.01	5.84	1.57	-0.3	K5 III	-41	150 s		
186605	68767		19 43 46.8	+38 19 21	0.000	+0.01	7.60	0.00	0.0	B8.5 V		310 s	12851	m
186761	31931		19 43 48.6	+52 36 39	+0.002	+0.02	7.7	0.1	1.4	A2 V		180 s		
186568	68764		19 43 51.2	+34 09 45	0.000	0.00	6.05	-0.01	-2.3	B8 II-III	-11	430 s	12852	m
186085	211509		19 43 52.4	-35 15 14	+0.013	-0.24	7.9	0.7	4.4	G0 V		42 s		
186518	87640		19 43 55.7	+27 08 09	-0.001	+0.01	6.80	1.1	0.2	K0 III	-12	200 s	12850	m
186517	87642		19 43 56.1	+27 26 29	+0.001	+0.04	8.00	1.1	0.0	K1 III		370 s		
186743	48778		19 43 56.3	+49 46 58	-0.003	-0.01	7.6	1.1	0.2	K0 III		300 s		
185523	254636		19 43 59.7	-66 17 51	-0.003	-0.21	7.70	0.7	4.4	G0 V		46 s		m
186636	68771		19 44 00.9	+38 00 24	+0.002	0.00	7.9	0.1	1.4	A2 V		190 s		
186442	125000		19 44 01.4	+9 31 05	-0.001	+0.02	6.52	1.23	-4.5	G5 Ib		1300 s		
186329	162933		19 44 05.8	-11 40 08	-0.001	+0.01	7.8	1.6	-0.5	M2 III		430 s		
186455	105193		19 44 05.8	+12 22 21	-0.001	-0.01	7.80	0.1	1.4	A2 V		190 s	12848	m
186675	68778	15 Cyg	19 44 16.4	+37 21 15	+0.006	+0.04	4.89	0.95	0.3	G8 III	-24	83 s		
184997	257722		19 44 17.2	-74 27 22	-0.002	-0.02	7.62	0.36		F5				
187340	18508		19 44 18.3	+69 20 13	+0.002	-0.02	5.92	0.05	0.6	A0 V	0	100 s		

HD	SAO	Star Name	α 2000	δ 2000	μ(α)	μ(δ)	V	B-V	M$_v$	Spec	RV	d(pc)	ADS	Notes
			h m s	° ′ ″	s	″								
186244	188501		19 44 22.0	−27 38 29	+0.002	−0.04	7.8	0.8	0.2	K0 III		330 s		
186133	229843		19 44 27.5	−41 36 33	0.000	−0.01	6.9	0.7	0.2	K0 III		220 s		
186192	211521		19 44 28.6	−37 50 40	+0.004	0.00	7.9	0.5	1.4	A2 V		99 s		
186369	162937		19 44 30.7	−17 24 01	+0.002	−0.01	8.0	0.1	0.6	A0 V		300 s		
186569	105201		19 44 31.3	+18 35 20	+0.001	+0.01	6.80	1.69	−0.5	M2 III		250 s		
186490	125008		19 44 31.4	+4 58 46	0.000	+0.01	7.80	1.4	−0.3	K5 III		400 s		m
186332	188509		19 44 32.3	−22 51 20	−0.001	+0.01	7.4	−0.1	3.0	F2 V		77 s		
186547	105199	48 ψ Aql	19 44 33.9	+13 18 10	−0.001	0.00	6.10	−0.03	−0.6	B8 IV	−4	200 s		
186548	105200		19 44 35.7	+13 13 48	+0.001	−0.01	7.0	1.4	−0.3	K5 III	0	280 s		
186702	68783		19 44 38.0	+34 24 50	0.000	−0.01	6.7	1.6	−0.5	M2 III	+9	260 s		
186347	188511		19 44 40.3	−23 15 52	−0.001	−0.05	7.4	1.4	−0.1	K2 III		240 s		
186535	125011		19 44 41.2	+8 43 35	+0.004	−0.02	6.41	0.92	0.3	G8 III		170 s		
186534	105202		19 44 41.3	+10 54 46	0.000	+0.01	7.3	0.1	1.7	A3 V		130 s		
186606	105205		19 44 44.7	+18 43 43	+0.009	+0.03	7.8	0.5	4.0	F8 V		56 s		
186688	87659		19 44 48.5	+29 15 53	0.000	0.00	6.80	0.4	0.6	F2 III	−36	170 s		SU Cyg, v
186776	48789		19 44 48.8	+40 43 00	−0.006	−0.02	6.34	1.64	−0.5	M3 III	−97	220 s		
184194	258835		19 44 54.0	−80 05 20		−0.03	7.7	1.6	−0.5	M2 III		430 s		
186461	162945		19 44 54.1	−10 34 23	0.000	−0.06	6.8	1.4		K5 III		190 mx		
186882	48796	18 δ Cyg	19 44 58.4	+45 07 51	+0.005	+0.05	2.87	−0.03	−0.6	A0 III	−21	49 s	12880	m
187053	31956		19 44 59.8	+55 50 52	−0.001	0.00	8.00	1.6	−0.5	M2 III	−6	470 s		
186925	48799		19 45 04.9	+46 37 10	+0.002	+0.01	7.9	0.9	3.2	G5 IV		87 s		
186637	105211		19 45 05.8	+16 17 09	+0.001	0.00	7.5	0.0	0.4	B9.5 V		250 s		
186352	211533		19 45 05.9	−31 47 17	+0.004	−0.02	7.6	1.6	−0.1	K2 III		190 s		
186498	162948		19 45 09.4	−12 43 35	−0.001	−0.01	7.3	1.1	0.2	K0 III		260 s		
186540	143760		19 45 09.4	−3 02 07	+0.001	+0.01	7.3	1.1	0.2	K0 III		250 s		
185560	257725		19 45 11.0	−70 01 00	−0.003	−0.06	7.60	0.5	3.4	F5 V		69 s		m
186589	125020		19 45 12.1	+2 54 44	−0.005	−0.02	8.0	0.9	3.2	G5 IV		92 s		
186401	188524		19 45 14.6	−27 23 48	0.000	−0.02	8.0	1.6	−0.3	K5 III		390 s		
186777	68793		19 45 17.0	+31 25 27	0.000	+0.02	7.41	−0.06		B5	−11			
186657	125025		19 45 22.5	+7 54 48	+0.002	+0.04	8.00	0.47	3.8	F7 V		69 s		
186590	143768		19 45 22.6	−1 29 55	+0.002	+0.02	7.4	1.1	0.2	K0 III		260 s		
186745	87671		19 45 24.1	+23 56 32	−0.002	−0.02	7.03	0.93	−7.1	B8 Ia		2000 s		
186417	211537		19 45 24.2	−30 54 11	+0.001	−0.01	6.8	0.7	0.6	A0 V		63 s		
186746	—		19 45 26.3	+23 55 25			7.0	0.0	0.0	B8.5 V	+1	240 s		
186575	143767		19 45 28.2	−7 01 12	+0.004	−0.09	7.6	0.5	4.0	F8 V		52 s		
186237	246254		19 45 31.5	−49 59 51	0.000	−0.11	8.0	0.8	0.2	K0 III		110 mx		
186858	68799		19 45 33.3	+33 36 08	+0.001	−0.43	7.68	0.99	8.0	dK5	+6	17 ts	12889	m
186994	48806		19 45 38.0	+44 57 51	+0.002	+0.01	7.51	−0.12	−5.0	B0 III		2800 s		
187201	31968		19 45 38.9	+57 53 36	0.000	+0.10	7.9	1.1	0.2	K0 III		190 mx		
185618	257727		19 45 39.0	−70 37 34	+0.015	−0.04	7.2	1.2	0.2	K0 III		180 s		
186901	68805		19 45 39.4	+36 05 28	0.000	+0.02	6.40	0.00	−0.7	B9.5 III	−19	250 s	12893	m
186689	125032	49 υ Aql	19 45 39.8	+7 36 48	+0.003	0.00	5.91	0.18		A2	−30	22 mn		
186902	68806		19 45 40.5	+36 05 19	0.000	+0.01	6.0	0.1	0.6	A0 V		120 s		
186641	143775		19 45 41.1	−0 41 50	+0.003	−0.07	7.3	1.1	0.2	K0 III		190 mx		
186860	68801		19 45 42.7	+30 15 26	0.000	+0.01	7.52	1.70		M III				
186978	48807		19 45 48.6	+40 33 18	0.000	−0.01	7.56	−0.07	0.0	B8.5 V		330 s	12904	m
186927	68810		19 45 51.3	+35 00 46	+0.001	0.00	5.98	0.91	−0.9	K0 II-III	−19	240 s	12900	m
187071	48811		19 45 51.5	+46 29 07	+0.001	−0.01	7.1	1.1	−0.1	K2 III	+31	260 s		
186660	143776		19 45 52.1	−2 53 01	0.000	0.00	6.48	0.05	−2.9	B3 III	−17	540 s		
186841	87685		19 45 53.9	+24 05 46	0.000	−0.01	7.85	0.78	−6.6	B1 Ia		2100 s		
186268	246260		19 45 54.3	−52 31 53	+0.002	−0.09	7.6	−0.2	3.4	F5 V		69 s		
186996	68820		19 45 55.0	+39 53 21	0.000	+0.01	7.48	−0.10	0.4	B9.5 V		260 s	12906	m
186704	125036		19 45 57.2	+4 14 54	+0.005	−0.02	7.00	0.6	4.4	G0 V	−8	33 s	12882	m
186500	211541		19 46 01.0	−31 54 31	0.000	−0.02	5.5	0.0	−0.2	B8 V n	−31	140 s		
187610	18521		19 46 02.1	+67 44 16	0.000	+0.03	7.8	0.4	2.6	F0 V		110 s		
185890	254642		19 46 03.4	−67 02 56	+0.003	−0.07	7.9	0.5	4.0	F8 V		60 s		
187317	31984		19 46 13.9	+58 22 01	0.000	0.00	8.0	0.5	0.7	F6 III	−21	270 s		
186791	105223	50 γ Aql	19 46 15.4	+10 36 48	+0.001	0.00	2.72	1.52	−2.3	K3 II	−2	87 s		Tarazed, m
186980	68823		19 46 15.8	+32 06 57	+0.001	−0.02	7.48	0.08		O7.5	+4			
187120	48820		19 46 17.1	+45 44 11	−0.002	0.00	7.5	1.1	0.2	K0 III		270 s		
188791	9545		19 46 17.2	+79 27 37	−0.013	−0.01	7.4	1.1	−0.1	K2 III		310 s		
187104	48819		19 46 18.8	+44 33 52	0.000	−0.03	7.8	1.1	0.2	K0 III		310 s		
186579	188544		19 46 19.2	−27 15 54	+0.001	+0.06	7.6	0.7	0.2	K0 III		250 mx		
186648	162964	56 Sgr	19 46 21.5	−19 45 40	−0.009	−0.09	5.0	1.1	0.0	K1 III	+20	98 s		
187275	31979		19 46 24.5	+54 24 21	0.000	+0.01	6.7	1.1	0.2	K0 III		200 s		
187013	68827	17 Cyg	19 46 25.4	+33 43 39	+0.001	−0.44	4.99	0.47		F5	+5	23 t	12913	m
186944	87700		19 46 25.4	+24 47 29	−0.001	−0.02	8.0	1.4	−0.3	K5 III		410 s		
186666	188551		19 46 30.3	−21 31 23	+0.005	−0.01	7.50	0.2	2.1	A5 V		120 s	12887	m
186796	125045		19 46 30.6	+1 28 48	0.000	−0.05	7.7	1.1	−0.1	K2 III		350 s		
186526	211543		19 46 31.5	−39 25 33	0.000	0.00	7.5	1.2	0.2	K0 III		230 s		
187217	48828		19 46 32.6	+50 00 38	−0.001	−0.01	6.7	1.1	0.2	K0 III		200 s		
187038	68835		19 46 34.8	+32 53 19	−0.003	0.00	6.18	1.13		K2	−46		12920	m
186595	211544		19 46 35.3	−31 56 06	+0.001	+0.01	7.6	0.1	1.7	A3 V		150 s		
187122	68840		19 46 35.8	+39 30 38	+0.001	−0.03	7.8	0.1	1.4	A2 V		180 s		
186998	87706		19 46 39.3	+25 08 02	+0.005	0.00	6.62	0.27		F0	+13			

HD	SAO	Star Name	α 2000	δ 2000	μ(α)	μ(δ)	V	B-V	M_v	Spec	RV	d(pc)	ADS	Notes
			$19^h 46^m 41^s.0$	$+44°20'54''$	$+0^s.012$	$0''.00$								
187160	48827		19 46 41.0	+44 20 54	+0.012	0.00	7.1	0.7	4.4	G0 V	+4	34 s		
186962	105242		19 46 43.1	+18 49 01	+0.001	+0.02	7.60	1.1	-0.9	K0 II-III	-18	460 s		
187764	18530		19 46 44.5	+68 26 18	0.000	+0.01	6.34	0.28	0.6	F0 III	-12	140 s		
188013	9535		19 46 53.8	+72 27 52	-0.002	+0.03	7.4	1.1	0.2	K0 III		280 s		
187123	68845		19 46 58.0	+34 25 09	+0.012	-0.13	8.0	0.9	3.2	G5 IV		56 mx		
187014	105249		19 46 59.5	+19 10 50	0.000	0.00	8.0	0.0	0.4	B9.5 V		310 s		
186694	188559		19 47 00.2	-26 29 26	+0.002	-0.03	7.7	1.1	-0.1	K2 III		360 s		
186251	254648		19 47 05.3	-62 27 13	-0.002	-0.05	6.8	1.4	-0.3	K5 III		260 s		
186847	143795		19 47 07.9	-8 09 10	+0.002	-0.02	8.04	1.17		K0			12911	m
186780	162980		19 47 08.9	-17 04 39	+0.002	+0.01	7.0	1.8		Ma				
187161	68851		19 47 17.4	+30 50 24	+0.001	0.00	7.62	-0.09		B9				
186803	162982		19 47 17.7	-18 44 46	-0.007	-0.09	7.5	0.6	4.4	G0 V		40 s		
187075	87719		19 47 18.3	+21 46 18	-0.001	+0.02	7.11	1.67	-0.3	K5 III	-43	240 s		
187234	48837		19 47 19.6	+40 59 41	0.000	+0.03	7.9	0.2	2.1	A5 V		140 s		
186908	143798		19 47 20.7	-6 20 36	-0.001	-0.06	7.9	1.1	0.2	K0 III		210 mx		
187076	105259	7 δ Sge	19 47 23.0	+18 32 03	0.000	+0.01	3.82	1.41	-2.4	M2 II	+3	170 s		
186651	229865		19 47 24.8	-43 20 43	-0.009	-0.07	7.12	0.56	4.4	G0 V		35 s		
187372	48842		19 47 26.6	+47 54 27	-0.003	-0.03	6.12	1.64	-0.5	M2 III	+3	190 s		
187545	31999		19 47 27.3	+57 06 01	-0.002	0.00	7.6	0.1	1.4	A2 V		170 s		
187235	68859		19 47 27.7	+38 24 27	+0.001	0.00	5.77	-0.06	-0.2	B8 V		160 s	12944	m
186713	211553		19 47 28.0	-35 10 38	0.000	-0.07	7.5	1.5	-0.1	K2 III		170 mx		
187546	32001		19 47 30.3	+56 54 39	0.000	-0.07	7.1	1.1	-0.1	K2 III		260 s		
187003	125069		19 47 33.1	+1 05 19	-0.002	-0.23	6.80	0.59		G5		18 mn		m
186806	188570		19 47 39.5	-25 37 34	0.000	+0.02	7.6	1.5	-0.1	K2 III		230 s		
187128	105262		19 47 39.6	+15 54 28	+0.001	+0.02	7.4	0.0	0.4	B9.5 V		240 s		
186892	162989		19 47 40.1	-17 04 14	0.000	-0.06	7.3	0.5	3.4	F5 V		60 s		
187342	48843		19 47 41.0	+42 19 08	0.000	+0.01	8.0	1.4	-0.3	K5 III		410 s		
187059	125074		19 47 45.8	+5 46 56	0.000	+0.01	6.7	1.1	0.2	K0 III		200 s		
187792	18536		19 47 46.3	+63 51 35	-0.007	+0.01	7.9	0.6	4.4	G0 V		49 s		
186829	188571		19 47 48.3	-28 29 29	+0.002	-0.01	6.9	0.6	4.4	G0 V		31 s		
187193	87729		19 47 48.4	+25 23 02	+0.006	-0.02	5.95	0.99	0.2	gK0	-18	140 s		
187044	143805		19 47 49.8	-0 03 48	-0.001	+0.02	8.0	0.0	0.0	B8.5 V		370 s		
186853	188572		19 47 50.9	-29 09 35	-0.007	+0.01	7.5	0.9	4.4	G0 V		28 s		
186682	229866		19 47 52.4	-45 43 04	+0.006	-0.03	7.24	0.14	1.7	A3 V		120 s		
187130	125080		19 47 56.9	+10 02 22	+0.001	-0.01	8.00	1.4	-0.3	K5 III		440 s		m
187279	68870		19 47 59.4	+31 30 23	0.000	+0.02	6.83	0.11	0.6	A0 V		150 s		
186938	188580		19 47 59.7	-20 57 27	-0.001	0.00	6.8	0.6	4.4	G0 V		30 s		
187237	87733		19 48 00.7	+27 52 10	-0.001	+0.23	6.88	0.63	5.2	dG5	-36	22 s		
186543	246271	ν Tel	19 48 00.9	-56 21 45	+0.011	-0.14	5.35	0.20		A5 p	-16			
186984	162998		19 48 02.8	-13 42 12	+0.002	-0.01	6.11	0.20	1.7	A3 V		66 s		
187255	87734		19 48 04.3	+27 42 10	+0.002	-0.01	7.56	-0.02		A0	-22			
187182	105274		19 48 08.4	+13 27 16	-0.004	-0.07	7.0	0.5	3.4	F5 V		52 s		
187793	18538		19 48 08.8	+61 24 40	-0.001	-0.01	7.1	1.4	-0.3	K5 III		280 s		
188119	9540	63 ε Dra	19 48 10.2	+70 16 04	+0.015	+0.04	3.83	0.89	0.3	G8 III	+3	51 s	13007	m
187280	87738		19 48 11.2	+28 19 33	-0.001	-0.03	7.93	1.18	-0.1	K2 III		400 s		
187238	87735		19 48 11.7	+22 45 46	0.000	0.00	7.70	1.1	-5.2	K3 Iab-Ib		2300 s		
187374	68880		19 48 11.9	+37 43 31	-0.001	0.00	7.5	1.1	-0.1	K2 III		310 s		
186831	211562		19 48 15.0	-37 50 54	0.000	-0.02	7.8	2.1	-0.5	M2 III		240 s		
187748	32016		19 48 15.3	+59 25 22	+0.003	+0.12	6.7	0.6	4.4	G0 V	-3	28 s		
186563	246274		19 48 17.0	-57 43 33	+0.004	-0.03	7.4	0.2	1.7	A3 V		120 s		
186782	229874		19 48 20.9	-42 04 42	-0.001	-0.02	7.4	0.3	0.2	K0 III		270 s		
187299	87740		19 48 21.4	+25 00 35	0.000	0.00	7.12	1.60	-5.4	G5 Iab-Ib	+1	1400 s		
187220	105277		19 48 27.8	+12 23 16	-0.001	-0.01	7.6	0.1	0.6	A0 V		250 s		
187258	105280		19 48 27.9	+18 39 19	0.000	+0.03	7.6	0.2		A m				
187203	105278		19 48 30.3	+10 41 39	-0.001	0.00	6.44	0.96	-4.5	G0 Ib	-5	1300 s		
186832	229877		19 48 30.6	-40 58 26	+0.004	-0.06	7.4	0.6	3.2	G5 IV		70 s		
187343	87748		19 48 34.1	+24 57 44	0.000	0.00	7.41	0.00	0.4	B9.5 V		250 s	12964	m
187749	32018		19 48 35.3	+56 30 04	-0.001	+0.01	7.4	1.4	-0.3	K5 III		330 s		
187728	32017		19 48 36.0	+55 43 30	+0.008	+0.03	7.6	0.5	4.0	F8 V		53 s		
187168	143810		19 48 38.9	-1 49 35	+0.001	0.00	7.4	0.0	0.4	B9.5 V		230 s		
187111	163006		19 48 39.5	-12 07 20	+0.003	-0.06	7.9	0.9	3.2	G5 IV		88 s		
187283	105284		19 48 39.5	+15 03 32	+0.002	+0.01	7.63	0.44		F5			12961	m
187427	68890		19 48 40.3	+30 49 42	-0.002	-0.08	7.6	0.8	3.2	G5 IV		76 s		
184005	258837		19 48 41.2	-82 55 44	-0.018	-0.06	7.3	1.4	-0.3	K5 III		220 mx		
187399	87754		19 48 41.8	+29 24 07	+0.001	-0.02	7.01	0.18	-7.1	B7 Ia e	-19	3400 s		
187259	105282	52 π Aql	19 48 41.9	+11 48 57	+0.001	0.00	5.7	0.1	1.4	A2 V	+13	72 s	12962	m
187321	105288		19 48 43.5	+18 52 02	0.000	0.00	7.1	0.6	-6.3	G0 I		2600 s		
187458	68893		19 48 43.7	+35 18 41	+0.007	+0.06	6.53	0.44	3.0	dF2	-27	48 s	12972	m
187523	68897		19 48 43.8	+39 55 00	+0.005	+0.03	7.0	0.4	2.6	F0 V		74 s		
187320	105289		19 48 44.2	+19 39 40	0.000	-0.01	7.41	0.15	-1.7	B3 V n	-1	580 s		
187323	105291		19 48 45.8	+18 22 08	+0.001	-0.01	8.0	0.0	-1.0	B5.5 V	-36	560 s		
186772	229875		19 48 47.0	-49 31 30	+0.002	0.00	7.8	1.6	-0.5	M2 III		470 s		
187150	163007		19 48 50.1	-12 19 09	+0.002	-0.01	6.6	1.4	-0.3	K5 III		240 s		
187459	68895		19 48 50.4	+33 26 14	0.000	0.00	6.44	0.20		B0.5 Ib	-10			
187284	105290		19 48 53.2	+11 41 07	0.000	+0.01	6.5	1.1	0.2	K0 III		180 s		

HD	SAO	Star Name	α 2000	δ 2000	μ(α)	μ(δ)	V	B-V	M_v	Spec	RV	d(pc)	ADS	Notes
186756	246277		$19^h48^m54.9$	$-52°53'17''$	$+0.002$	-0.05	6.25	1.13	0.2	K0 III		130 s		
--	143813		19 48 56.3	-4 29 46	0.000	0.00	7.8	0.1	0.6	A0 V		260 s		
186854	229880		19 48 57.9	-46 08 44	$+0.001$	-0.01	8.0	1.1	0.2	K0 III		360 s		
187344	105296		19 48 57.9	$+16$ 37 40	0.000	-0.01	7.6	0.1	0.6	A0 V		240 s		
187362	105298	8 ζ Sge	19 48 58.5	$+19$ 08 32	$+0.001$	$+0.03$	5.00	0.10	1.7	A3 V	-7	46 s	12973	m
187613	48866		19 49 00.7	$+44$ 22 45	0.000	0.00	7.18	-0.03	0.0	B8.5 V	-12	260 s	12986	m
187195	163012		19 49 02.0	-10 52 15	$+0.002$	-0.02	6.2	1.1	0.2	K0 III	-37	160 s		
187503	68902		19 49 03.2	$+32$ 47 57	$+0.001$	0.00	7.6	1.4	-0.3	K5 III		350 s		
187767	32025		19 49 03.8	$+53$ 46 03	$+0.001$	$+0.01$	7.10	-0.11	0.0	B8.5 V		260 s		
187207	163014		19 49 05.9	-9 58 58	$+0.001$	$+0.01$	7.9	1.1	0.2	K0 III		350 s		
186602	254655		19 49 07.1	-61 48 52	$+0.007$	-0.08	7.26	0.49	4.4	G0 V		37 s		m
187462	87766		19 49 08.2	$+27$ 43 53	$+0.005$	-0.22	6.91	0.60	4.4	G0 V	$+3$	30 s		
187347	125096		19 49 09.3	$+8$ 16 37	$+0.001$	-0.02	7.5	0.2	2.1	A5 V		120 s		
185712	257731		19 49 09.6	-76 54 30	$+0.007$	-0.02	7.15	0.01		A0				
187098	188603		19 49 11.4	-28 47 21	$+0.009$	-0.10	6.05	0.40	2.1	F0 IV-V		53 s		
187428	105303		19 49 11.4	$+19$ 47 39	0.000	-0.02	7.92	0.74	-3.3	F8 Ib-II		1800 s		
187876	32031		19 49 12.0	$+57$ 24 34	-0.001	-0.06	7.76	0.59		G0				
187402	105302		19 49 15.8	$+15$ 07 29	$+0.001$	$+0.02$	7.7	0.4	3.0	F2 V		88 s		
187401	105304		19 49 16.8	$+15$ 12 15	0.000	$+0.01$	7.70	0.9	-2.1	G5 II	-16	840 s		
187524	87771		19 49 19.0	$+28$ 34 44	0.000	-0.01	7.39	0.15	0.6	A0 V		180 s		
186117	257735		19 49 23.3	-73 31 24	$+0.007$	-0.02	7.35	0.14		A0				
186219	257736		19 49 25.2	-72 30 12	$+0.004$	$+0.02$	5.41	0.22		A m	0	17 mn		
188053	18547		19 49 25.6	$+63$ 12 15	$+0.003$	$+0.01$	6.8	1.1	0.2	K0 III		190 s		
187548	87774		19 49 26.8	$+28$ 36 37	$+0.007$	$+0.08$	7.92	0.52	4.4	G0 V		51 s		
187638	68909		19 49 27.3	$+38$ 42 36	0.000	-0.01	6.11	0.90	3.2	G5 IV	$+11$	32 s	12992	m
186975	229883		19 49 27.8	-45 44 51	$+0.008$	-0.02	7.27	1.08	3.2	K0 IV		55 s		
186530	254654		19 49 28.5	-65 58 44	$+0.003$	-0.03	7.9	1.1	-0.1	K2 III		400 s		
190224	3299		19 49 30.1	$+82$ 27 17	-0.017	-0.02	7.9	1.1	0.2	K0 III		290 mx		
187565	87776		19 49 31.5	$+29$ 22 29	$+0.004$	$+0.02$	8.01	0.49	4.0	F8 V		63 s		
187085	211579		19 49 33.8	-37 46 50	$+0.001$	-0.10	7.3	0.4	4.0	F8 V		45 s		
187406	125103		19 49 42.7	$+2$ 57 14	$+0.001$	$+0.01$	7.9	0.5	3.4	F5 V		77 s		
187154	211583		19 49 44.0	-32 45 48	$+0.009$	-0.03	7.5	0.8	4.4	G0 V		30 s		
187505	105320		19 49 44.8	$+16$ 22 31	0.000	$+0.01$	7.8	0.7	-4.5	G2 Ib		1900 s		
187306	163020		19 49 46.6	-11 24 33	$+0.003$	-0.07	7.20	0.5	3.4	F5 V		58 s		m
187464	105316		19 49 46.7	$+11$ 13 51	-0.002	-0.03	7.9	1.1	-0.1	K2 III		380 s		
186632	254659		19 49 49.1	-64 54 24	$+0.004$	$+0.09$	8.00	0.6	4.4	G0 V		53 s		m
187353	163022		19 49 50.9	-10 43 31	0.000	-0.01	7.8	0.4	2.6	F0 V		110 s		
186584	254658		19 49 53.2	-66 48 46	0.000	$+0.02$	6.45	1.48		K2				
187488	125107		19 49 54.2	$+8$ 01 43	0.000	0.00	8.0	0.5	3.4	F5 V		81 s		
187640	87786		19 49 54.3	$+28$ 26 25	-0.002	$+0.03$	6.38	-0.06	-2.9	B3 III	-4	570 s		
187614	87785		19 49 55.7	$+27$ 05 06	$+0.002$	0.00	6.44	0.94	0.3	G8 III		170 s		
186725	254662		19 49 55.8	-62 34 00	$+0.001$	-0.01	7.5	0.5	4.0	F8 V		50 s		
187226	188607		19 49 56.3	-26 21 42	$+0.002$	-0.01	8.0	1.4	-0.1	K2 III		310 s		
--	87787		19 50 00.1	$+27$ 09 58	0.000	-0.01	8.0	1.4	0.2	K0 III		220 s		
186483	254656		19 50 02.6	-69 20 19	$+0.004$	0.00	7.3	1.1	-0.1	K2 III		300 s		
186810	246284		19 50 02.7	-59 15 50	0.000	-0.01	7.0	0.5	1.4	A2 V		72 s		
187550	105323		19 50 05.5	$+11$ 46 46	$+0.001$	-0.03	7.9	1.4	-0.3	K5 III		410 s		
187979	32037		19 50 07.3	$+54$ 39 08	0.000	$+0.02$	8.00	1.4	-0.3	K5 III		430 s	13018	m
187028	246291		19 50 08.3	-50 36 08	$+0.004$	$+0.02$	7.7	0.4	2.6	F0 V		94 s		
187410	163026		19 50 09.6	-10 00 24	$+0.001$	$+0.02$	6.90	0.9	3.2	G5 IV		55 s		m
187099	229888		19 50 10.7	-45 22 47	$+0.001$	-0.02	7.20	1.1	0.2	K0 III		250 s		
187919	48889		19 50 12.9	$+48$ 23 29	0.000	-0.02	7.9	1.1	-0.1	K2 III		370 s		
187086	229887		19 50 13.5	-47 33 26	-0.002	-0.01	5.9	1.7	-0.5	M2 III		170 s		
187596	105329		19 50 14.9	$+17$ 42 36	$+0.002$	-0.01	6.8	1.1	0.2	K0 III		200 s		
187333	188612		19 50 16.1	-22 46 48	$+0.001$	0.00	7.6	-0.3	0.4	B9.5 V		280 s		
187567	125116		19 50 17.3	$+7$ 54 09	0.000	0.00	6.51	-0.10	-2.6	B2.5 IV e	-28	630 s		
187795	68935		19 50 19.6	$+36$ 05 52	0.000	-0.01	7.1	0.1	0.6	A0 V		190 s		
186837	254666		19 50 21.5	-61 03 41	0.000	0.00	6.21	-0.14	-1.1	B5 V	-16	290 s		
188166	18556		19 50 21.7	$+61$ 12 34	-0.004	$+0.05$	6.7	0.5	3.4	F5 V		45 s		
187308	188613		19 50 21.8	-26 42 57	$+0.001$	0.00	7.4	1.0	0.2	K0 III		260 s		
187615	105331		19 50 24.9	$+14$ 30 50	$+0.002$	0.00	7.6	0.1	1.7	A3 V		150 s		
187878	48891		19 50 25.1	$+43$ 39 42	0.000	$+0.04$	7.8	1.1	0.2	K0 III		310 s		
187439	163030		19 50 26.3	-13 55 25	-0.002	0.00	7.2	0.0	0.0	B8.5 V		280 s		
187750	68934		19 50 26.4	$+30$ 45 58	-0.001	$+0.01$	7.34	-0.03	0.6	A0 V		220 s		
187309	211591		19 50 29.1	-30 53 06	-0.005	-0.02	7.6	2.0	-0.1	K2 III		110 s		
188054	32041		19 50 29.3	$+55$ 28 27	$+0.002$	$+0.02$	7.9	1.1	-0.1	K2 III		370 s		
187655	105332		19 50 32.0	$+17$ 16 07	$+0.001$	0.00	7.6	0.1	0.6	A0 V		240 s		
186393	257740		19 50 33.1	-72 52 30	$+0.011$	$+0.01$	8.0	1.4	-0.3	K5 III		280 mx		
187796	68943	χ Cyg	19 50 33.7	$+32$ 54 51	-0.002	-0.04	4.23	1.82	0.2	K0 III	-2	25 s		m,v
187849	68947	19 Cyg	19 50 33.8	$+38$ 43 21	$+0.001$	$+0.11$	5.12	1.69	-0.5	M2 III	-39	120 s	13014	m
187879	48892		19 50 37.1	$+40$ 35 59	0.000	-0.01	5.69	-0.04	-3.9	B1 IV	-4	640 s		V380 Cyg, m,v
188056	32042	20 Cyg	19 50 37.6	$+52$ 59 17	-0.002	-0.07	5.03	1.28	-0.2	K3 III	-20	110 s		
188055	32046		19 50 38.6	$+53$ 54 41	$+0.001$	$+0.02$	7.9	0.1	1.4	A2 V		190 s		
187383	188619		19 50 39.5	-25 18 19	0.000	0.00	8.0	0.8	0.2	K0 III		360 s		
187211	229890		19 50 40.2	-41 51 34	$+0.003$	-0.01	7.60	0.1	1.4	A2 V		170 s		m

HD	SAO	Star Name	α 2000	δ 2000	μ(α)	μ(δ)	V	B-V	M_v	Spec	RV	d(pc)	ADS	Notes
			$19^h 50^m 40.8^s$	$+34°57'00''$	$+0.027^s$	$+0.20''$								
226099	68946		19 50 40.8	+34 57 00	+0.027	+0.20	7.2	0.8	3.2	G5 IV		41 mx		
187712	87803		19 50 41.3	+22 06 46	−0.001	−0.02	7.2	0.9	3.2	G5 IV		64 s		
186484	257742		19 50 43.0	−71 46 18	−0.001	−0.01	7.2	1.6		M5 III	+38			T Pav, v
186957	246293		19 50 44.6	−59 11 36	+0.003	+0.01	5.80	0.1	1.4	A2 V	+4	76 s		m
187730	87804		19 50 44.8	+20 12 40	−0.001	0.00	6.7	0.1	1.4	A2 V	−26	110 s		
187356	211596		19 50 45.1	−32 36 40	+0.001	−0.01	7.8	−0.1	0.6	A0 V		270 s		
187368	211597		19 50 45.2	−30 55 40	0.000	0.00	7.6	1.7	0.2	K0 III		120 s		
187532	163036	51 Aql	19 50 46.6	−10 45 49	−0.002	+0.03	5.60	0.4	1.7	F0 IV	+6	60 s	13017	m
187642	125122	53 α Aql	19 50 46.8	+8 52 06	+0.036	+0.39	0.77	0.22	2.2	A7 IV–V	−26	5.1 t	13009	Altair, m
187880	68953		19 50 46.8	+37 49 35	+0.002	+0.01	6.06	1.70		Ma	−16	34 mn		
188014	32044		19 50 47.8	+50 21 22	+0.003	+0.01	8.0	0.4	3.0	F2 V		99 s		
187570	143839		19 50 48.5	−4 41 37	+0.001	+0.01	6.8	0.1	1.4	A2 V		120 s		
187617	125124		19 50 51.2	+4 26 43	+0.001	0.00	7.7	1.1	−0.1	K2 III		350 s		
187751	105339		19 50 52.8	+20 02 23	+0.002	+0.02	7.3	0.2		A m				
187212	229891		19 50 52.9	−46 21 38	0.000	−0.03	7.2	0.3	3.4	F5 V		59 s		
187619	125123		19 50 54.6	+0 50 40	0.000	+0.01	7.8	0.1	0.6	A0 V		260 s		
187797	87810		19 50 54.7	+24 37 36	0.000	−0.01	7.1	0.1	0.6	A0 V		190 s		
188057	48902		19 50 55.1	+49 42 01	−0.001	+0.01	7.4	0.9	3.2	G5 IV		67 s		
187731	105340		19 50 58.8	+16 51 28	0.000	−0.02	7.0	0.0	0.0	B8.5 V	−19	240 s		
186786	254669		19 51 01.1	−65 36 18	+0.018	−0.15	6.05	0.30		A5				
187691	105338	54 o Aql	19 51 01.5	+10 24 56	+0.016	−0.14	5.11	0.55	4.0	F8 V	0	18 ts	13012	m
189589	9562		19 51 03.5	+78 37 27	−0.005	−0.01	7.8	1.4	−0.3	K5 III		410 s		
187811	87813	12 Vul	19 51 03.9	+22 36 36	+0.001	−0.02	4.95	−0.14	−1.7	B3 V	−26	210 s		
187851	87816		19 51 04.5	+27 43 01	0.000	−0.01	7.74	0.13	−2.5	B2 V nn		930 s		
187473	188626		19 51 10.1	−27 28 20	+0.002	−0.01	7.2	0.2	0.4	B9.5 V		150 s		
187660	143853		19 51 10.9	−2 27 39	−0.001	−0.03	6.13	1.57		K5				
188522	18564		19 51 13.3	+65 32 49	−0.006	−0.02	7.9	0.9	3.2	G5 IV		83 s		
186502	257743		19 51 14.0	−72 47 43	+0.005	−0.06	7.31	0.47		F8				m
187516	—		19 51 15.1	−24 54 00			7.4	1.0	0.2	K0 III		270 s		
187414	211604		19 51 17.4	−35 18 37	−0.001	0.00	7.9	0.1	0.4	B9.5 V		250 s		
187753	125139		19 51 17.5	+9 37 49	0.000	−0.01	6.25	0.10		A0	+21			
187388	—		19 51 18.3	−38 17 03			7.9	2.9	−0.1	K2 III		36 s		
188074	48907		19 51 19.2	+47 22 38	+0.001	+0.02	6.20	0.36	3.0	F2 V	−18	43 s		
187732	125138		19 51 20.3	+5 43 48	0.000	+0.01	7.9	1.1	−0.1	K2 III		380 s		
226195	48906		19 51 24.7	+40 44 07	+0.006	+0.02	7.8	0.9	3.2	G5 IV		81 s		
187734	125141		19 51 26.6	+4 05 14	0.000	−0.01	6.70	1.1	0.2	K0 III		200 s	13019	m
187369	229895		19 51 27.1	−42 05 37	+0.003	−0.03	7.85	0.58	3.0	G2 IV		93 s		
226184	68970		19 51 28.1	+35 01 51	0.000	0.00	8.0	0.1	0.6	A0 V		280 s		
187921	87829		19 51 30.7	+27 27 37	0.000	0.00	7.60	0.8	−8.0	G4 Ia var				SV Vul, v
187881	87824		19 51 31.8	+22 16 50	−0.008	−0.12	8.0	0.9	3.2	G5 IV		88 mx		
188168	32058		19 51 33.5	+50 46 26	+0.001	+0.02	7.90	1.1	0.2	K0 III		320 s	13049	m
187663	163048		19 51 35.1	−12 37 15	+0.003	+0.03	7.5	0.9	0.2	G9 III	−11	290 s		
187604	163046		19 51 35.7	−19 12 39	+0.001	+0.02	7.8	1.1	−0.1	K2 III		370 s		
226222	68977		19 51 35.7	+39 32 09	0.000	0.00	7.9	1.1	0.2	K0 III		320 s		
187943	87830		19 51 37.0	+26 13 32	+0.001	0.00	7.3	0.0	0.4	B9.5 V		230 s		
187882	87828		19 51 37.1	+22 09 22	+0.005	−0.06	8.0	0.9	3.2	G5 IV		91 s		
187578	188639		19 51 40.1	−27 04 55	+0.001	0.00	7.6	0.3	0.4	B9.5 V		170 s		
187981	68974		19 51 41.1	+31 08 29	−0.002	0.00	7.06	0.30		A5	+7	16 mn	13038	m
187664	163050		19 51 41.7	−14 46 32	0.000	+0.05	6.8	0.1	1.4	A2 V		120 s		
187294	246305		19 51 42.6	−50 58 43	+0.002	−0.09	6.9	0.7	4.0	F8 V		31 s		
187696	163054		19 51 42.7	−10 57 00	+0.007	0.00	7.2	0.5	3.4	F5 V		57 s		
187697	163055		19 51 44.7	−13 01 59	−0.001	−0.04	7.0	1.1	−0.1	K2 III		260 s		
187957	87832		19 51 45.0	+26 14 15	0.000	+0.02	7.8	0.1	0.6	A0 V		260 s		
187518	211609		19 51 45.9	−37 20 11	−0.002	−0.07	6.6	1.4	0.2	K0 III		100 s		
226228	68981		19 51 46.4	+35 36 41	+0.003	−0.01	8.0	0.7	4.4	G0 V		52 s		
187854	125150		19 51 46.7	+8 04 17	−0.001	−0.02	7.6	0.1	1.4	A2 V		180 s		
187605	188643		19 51 50.3	−28 33 26	0.000	−0.01	7.7	1.3	−0.1	K2 III		290 s		
187474	229903		19 51 50.5	−39 52 28	+0.002	−0.01	5.33	−0.06		A0 p	0	17 mn		
188101	48913		19 51 50.6	+41 20 56	−0.003	+0.02	7.7	0.0	0.4	B9.5 V		270 s		
187958	87835		19 51 57.9	+20 55 03	0.000	−0.01	7.8	0.1	0.6	A0 V		250 s		
187626	188644		19 51 58.0	−28 29 54	−0.002	0.00	8.0	1.3	0.2	K0 III		240 s		
187538	211611		19 51 58.4	−38 24 09	+0.001	−0.02	7.5	1.5	−0.1	K2 III		210 s		
188036	68985		19 51 58.4	+31 36 10	+0.001	−0.01	7.71	0.01	0.6	A0 V		260 s		
188209	48917		19 51 58.9	+47 01 38	−0.001	0.00	5.62	−0.07	−5.5	O9.5 III	−6	1300 s		
188035	68987		19 52 00.0	+31 42 27	+0.001	+0.01	7.84	0.00	0.6	A0 V		280 s	13042	m
187855	125152		19 52 00.3	+1 36 12	−0.002	0.00	8.0	0.1	1.4	A2 V		200 s		
187982	87840		19 52 01.4	+24 59 33	0.000	+0.01	5.57	0.71	−7.5	A2 Ia	−3	1800 s		
187774	163059		19 52 01.5	−10 21 15	−0.002	−0.07	7.50	0.4	3.0	F2 V		87 s	13028	m
187923	105348		19 52 03.3	+11 37 44	−0.023	−0.32	5.78	0.53	4.7	G2 V	−17	16 s		m
188323	32067		19 52 05.8	+51 59 42	+0.001	0.00	8.0	0.1	1.4	A2 V		200 s		
187272	246306		19 52 06.0	−56 43 12	+0.005	−0.07	7.2	1.0	0.2	K0 III		210 mx		
187475	229904		19 52 06.2	−45 39 10	+0.001	−0.01	7.4	1.6	−0.3	K5 III		310 s		
188252	48920		19 52 06.9	+47 55 54	−0.001	−0.01	5.91	−0.18	−3.6	B2 III	−18	790 s		
187775	163064		19 52 07.2	−10 53 31	+0.001	+0.03	6.9	1.1	0.2	K0 III		220 s		
188342	32068		19 52 07.8	+52 16 52	0.000	+0.02	7.2	0.1	0.6	A0 V		210 s		

HD	SAO	Star Name	α 2000	δ 2000	μ(α)	μ(δ)	V	B-V	M$_v$	Spec	RV	d(pc)	ADS	Notes
187897	125154		19^{h}52^{m}09^s.1	+7°27'37"	+0^s.007	+0".08	7.2	0.8	3.2	G5 IV		62 s		
187959	105354		19 52 09.4	+15 19 23	+0.002	+0.01	7.7	0.1	1.4	A2 V		180 s		
187739	163060	57 Sgr	19 52 11.8	−19 02 42	0.000	−0.06	6.0	0.4	3.2	G5 IV	−26	36 s		
187984	105357		19 52 12.5	+19 51 07	0.000	−0.04	7.90	1.1	0.2	K0 III		320 s	13044	m
187961	105355		19 52 15.4	+10 21 05	0.000	−0.01	6.54	−0.01	−0.6	B7 V	−13	240 s	13041	m
188058	87851		19 52 15.4	+28 15 09	−0.004	−0.04	6.9	1.1	−0.1	K2 III	−47	240 s		
188149	69004		19 52 16.2	+36 25 56	0.000	0.00	6.10	1.43	−0.3	K4 III	−21	190 s		
187820	163067		19 52 20.8	−11 21 59	+0.001	+0.02	7.7	0.9	3.2	G5 IV		80 s		
188001	105360	9 Sge	19 52 21.6	+18 40 19	0.000	0.00	6.23	0.01		O8.8	+9			
187960	105358		19 52 23.4	+10 56 25	+0.002	0.00	7.9	1.1	−0.1	K2 III		380 s		
187840	163070		19 52 25.8	−10 57 21	+0.002	+0.02	7.9	1.1	−0.1	K2 III		390 s		
187801	163066		19 52 26.5	−16 53 10	+0.002	−0.01	7.9	1.1	0.2	K0 III		340 s		
187929	125159	55 η Aql	19 52 28.1	+1 00 20	+0.001	−0.01	3.90	0.89	−4.5	G0 Ib var	−15	440 s		v
188002	105361		19 52 29.0	+16 38 14	+0.002	+0.01	7.7	1.1	−0.1	K2 III		350 s		
187973	125164		19 52 29.0	+9 07 51	0.000	0.00	8.0	0.1	0.6	A0 V		290 s		
188016	105364		19 52 31.2	+17 25 53	+0.002	+0.01	7.8	0.1	1.7	A3 V		160 s		
188852	18576		19 52 34.4	+66 42 20	+0.001	−0.01	7.6	0.4	2.6	F0 V		97 s		
185191	258841		19 52 35.8	−82 09 40	−0.010	−0.02	7.9	1.4	−0.3	K5 III		440 s		
187420	246311		19 52 37.4	−54 58 16	+0.001	0.00	5.74	0.92	0.2	K0 III		130 s		m
188062	105371		19 52 38.7	+18 44 28	0.000	+0.01	7.9	0.0	0.0	B8.5 V		350 s		
187421	246312		19 52 38.8	−54 58 36	+0.001	0.00	6.50	0.10	0.6	A0 V n		150 s		
187741	211624		19 52 44.2	−31 58 36	0.000	0.00	7.1	0.1	0.6	A0 V		200 s		
187742	211625		19 52 45.3	−32 28 16	+0.002	+0.03	7.9	1.2	0.2	K0 III		270 s		
187699	211620		19 52 46.7	−37 39 43	0.000	0.00	7.9	1.2	−0.1	K2 III		390 s		
188772	18575		19 52 47.3	+64 10 34	−0.002	−0.02	6.90	0.9	3.2	G5 IV		54 s	13092	m
188210	69014		19 52 47.3	+34 11 12	0.000	0.00	7.8	0.1	0.6	A0 V		260 s		
188170	87863		19 52 49.9	+28 59 33	−0.001	+0.01	7.32	−0.08		A0	−11			
187841	188667		19 52 50.4	−23 09 22	+0.003	−0.02	7.8	−0.4	0.6	A0 V		270 s		
188307	48932		19 52 52.3	+41 04 49	−0.002	+0.05	7.9	0.5	4.0	F8 V		59 s		
188284	69021		19 52 53.2	+38 40 19	+0.002	+0.01	8.0	1.4	−0.3	K5 III		410 s		
188076	105374		19 52 54.4	+16 14 39	−0.001	−0.03	7.8	1.1	0.2	K0 III		320 s		
188103	105376		19 52 56.9	+17 48 54	0.000	−0.01	8.0	0.1	0.6	A0 V		280 s		
186503	—		19 52 59.7	−76 11 30			7.99	0.04		A0				
188439	48940		19 53 01.0	+47 48 27	−0.001	−0.01	6.29	−0.11		B0.5 III p	−65			V819 Cyg, q
188326	69027		19 53 01.4	+38 46 24	−0.003	+0.34	7.56	0.78	3.2	G8 IV	−71	75 s		m
187949	163080		19 53 06.2	−14 36 11	0.000	−0.04	6.4	0.1	1.2	A1 V	−2	110 s		V505 Sgr, m,v
185079	258842		19 53 09.8	−82 37 47	+0.003	+0.04	7.9	1.4	−0.3	K5 III		440 s		
187580	246316		19 53 10.3	−53 05 35	0.000	−0.11	6.9	0.6	0.2	K0 III		110 mx		
188105	125176		19 53 10.8	+8 03 45	0.000	0.00	7.9	1.1	0.2	K0 III		330 s		
188664	32084		19 53 11.5	+58 09 53	−0.002	−0.03	7.5	1.4	−0.3	K5 III		340 s		
226373	69032		19 53 12.5	+39 27 45	0.000	0.00	8.0	1.4	−0.3	K5 III		430 s		
188007	143880		19 53 13.6	−5 32 25	−0.002	−0.04	7.4	0.6	4.4	G0 V		40 s		
188258	87877		19 53 15.9	+28 05 59	+0.002	+0.01	6.94	1.09	−0.1	K2 III		260 s		
188665	32085	23 Cyg	19 53 17.2	+57 31 25	+0.001	+0.02	5.14	−0.13	−1.1	B5 V	−25	180 s		
188041	143883		19 53 18.6	−3 06 52	+0.001	+0.02	5.65	0.20		A5 p	−19	35 mn		
188259	87880		19 53 19.1	+26 29 57	+0.003	+0.02	7.54	1.09	0.0	K1 III		320 s		
186879	—		19 53 19.6	−72 29 16			7.3	1.1	−0.1	K2 III		300 s		
188344	69031		19 53 20.7	+33 57 00	−0.003	+0.03	7.39	1.66	−0.5	M2 III		340 s		V449 Cyg, v
188107	125182		19 53 22.4	+4 24 02	+0.001	+0.01	6.53	0.02		A0	−1			
188211	87874		19 53 22.4	+20 20 14	+0.002	+0.01	7.13	0.13		A0				m
187521	246317		19 53 23.6	−57 05 28	0.000	+0.01	8.0	0.2	0.2	K0 III		360 s		
188150	105385		19 53 23.6	+13 14 47	−0.001	0.00	8.0	1.1	−0.1	K2 III		400 s		
188212	87876		19 53 24.0	+20 19 37	+0.001	+0.01	7.32	0.30		A0				
187886	188677		19 53 24.8	−26 56 35	+0.003	+0.01	7.8	1.0	0.2	K0 III		330 s		
188260	87883	13 Vul	19 53 27.5	+24 04 47	+0.002	+0.04	4.58	−0.06	−0.6	A0 III	−28	110 s		m
188793	32093		19 53 35.3	+59 42 32	+0.005	+0.06	6.06	0.02	0.6	A0 V	−13	99 mx		m
188461	48943		19 53 38.3	+41 21 21	0.000	−0.01	7.00	−0.14	−1.6	B3.5 V	−13	510 s		
188236	105389		19 53 42.0	+14 19 38	+0.002	−0.01	7.9	1.4	−0.3	K5 III		410 s		
187394	254679		19 53 43.8	−63 34 22	+0.004	−0.02	7.9	0.4	2.6	F0 V		120 s		
187117	—		19 53 44.4	−69 57 08			7.9	1.1	−0.1	K2 III		400 s		
188626	32087		19 53 44.5	+51 09 15	+0.001	+0.02	7.9	1.1	−0.1	K2 III		360 s		
188366	87894		19 53 45.1	+27 60 00	0.000	0.00	7.73	0.03		A0				
188262	105392		19 53 45.8	+16 46 40	0.000	0.00	7.6	0.7	4.4	G0 V	−1	45 s		
188537	48949		19 53 45.9	+45 28 13	−0.002	0.00	7.7	1.1	0.2	K0 III	−18	300 s		
188418	69040		19 53 47.1	+34 35 01	0.000	0.00	7.02	−0.13	0.6	A0 V		190 s		
193135	3325		19 53 48.8	+85 45 32	−0.005	−0.01	7.9	1.1	−0.1	K2 III		400 s		
187051	257745		19 53 48.9	−71 24 22	+0.006	−0.02	7.7	0.1	0.6	A0 V		230 s		
188500	48947		19 53 52.1	+41 13 45	0.000	0.00	8.0	0.1	1.4	A2 V		200 s		
188383	87897		19 53 52.4	+26 35 37	0.000	−0.01	6.81	0.10		A0 pe	−6			
187671	246324		19 53 53.1	−54 21 58	+0.010	−0.02	7.3	1.6	0.2	K0 III		120 s		
188216	125200		19 53 58.2	+5 53 06	−0.002	+0.02	7.7	1.4	−0.3	K5 III		370 s		
188649	48959		19 54 00.8	+49 14 36	+0.004	+0.04	8.00	1.1	0.2	K0 III		340 s	13100	m
188328	105396		19 54 01.8	+15 17 31	−0.001	−0.01	7.30	0.5	4.0	F8 V	−5	46 s	13082	m
188154	143894	56 Aql	19 54 08.1	−8 34 27	0.000	−0.02	5.79	1.64		K5	−50			m
187953	211636		19 54 08.3	−35 59 11	−0.004	−0.02	7.9	0.5	1.4	A2 V		100 s		

497

HD	SAO	Star Name	α 2000	δ 2000	μ(α)	μ(δ)	V	B-V	M_v	Spec	RV	d(pc)	ADS	Notes
187936	229920		19 54 09.1	-40 10 49	-0.001	0.00	7.8	1.2	-0.1	K2 III		380 s		
188484	69052		19 54 09.5	+33 46 52	+0.002	+0.03	6.8	1.1	0.2	K0 III		210 s		
188028	188688		19 54 10.1	-26 41 46	+0.001	-0.02	7.7	0.7	3.2	G5 IV		79 s		
188310	125210	59 ξ Aql	19 54 14.7	+8 27 41	+0.007	-0.08	4.71	1.05	0.2	K0 III	-42	73 s		
188088	188692		19 54 17.6	-23 56 28	-0.010	-0.41	6.18	1.02		K0	-7	15 t	13072	m
187316	—		19 54 19.9	-68 09 00			7.9	1.1	0.2	K0 III		350 s		
188196	163097		19 54 27.4	-12 21 39	0.000	+0.01	7.9	0.2	2.1	A5 V		150 s		
188541	69059		19 54 28.7	+33 07 52	0.000	-0.03	7.7	1.1	-0.1	K2 III		340 s		
--	48966		19 54 29.1	+47 54 50	0.000	-0.02	7.84	1.57						
188155	163095		19 54 30.1	-18 56 09	0.000	-0.04	7.9	1.4	-0.3	K5 III		430 s		
188485	87908		19 54 30.9	+24 19 10	+0.002	0.00	5.52	-0.02	-0.6	A0 III	-8	170 s		
188384	125220		19 54 35.0	+9 46 03	-0.001	+0.01	7.2	1.1	0.2	K0 III	-20	250 s		
188667	48963		19 54 35.1	+41 27 52	-0.001	-0.02	7.3	1.1	0.2	K0 III	-2	250 s		
189295	18595		19 54 37.4	+67 30 35	-0.004	-0.03	7.8	1.1	0.2	K0 III		290 s		
188293	143898	57 Aql	19 54 37.5	-8 13 38	0.000	-0.02	5.71	-0.08	-0.6	B7 V n	-6	180 s	13087	m
188294	143899	57 Aql	19 54 37.9	-8 14 14	+0.001	-0.03	6.49	-0.04		B	-5		13087	m
188385	125221		19 54 40.1	+7 08 25	+0.002	0.00	6.15	0.03		A0 n	-16		13093	m
187653	254683		19 54 40.4	-61 10 15	+0.001	+0.01	6.3	0.1	1.7	A3 V		81 s		
188220	163099		19 54 40.7	-15 54 49	0.000	-0.02	8.0	0.1	0.6	A0 V		300 s		
187396	254681		19 54 42.1	-68 15 34	+0.004	-0.02	7.9	0.1	0.6	A0 V		290 s		
189063	18589		19 54 43.8	+60 49 25	0.000	0.00	7.00	1.6	-0.4	M0 III	-22	280 s		
188507	87913		19 54 44.4	+22 25 45	+0.002	-0.01	6.71	1.53	-1.3	K4 II-III		360 s		
188350	125219	58 Aql	19 54 44.6	+0 16 25	+0.003	-0.01	5.61	0.10		A0 n	-42			
188650	69072		19 54 48.1	+36 59 46	+0.001	+0.02	5.76	0.75		G5	-24			
189231	18593		19 54 49.3	+64 43 17	-0.002	-0.01	6.9	0.8	3.2	G5 IV	0	55 s		
189084	18590		19 54 49.7	+60 36 54	-0.002	0.00	7.40	1.6	-0.4	M0 III	-6	330 s		
--	48971		19 54 50.0	+47 45 10	-0.001	0.00	6.70	1.08						
189035	32108		19 54 50.7	+59 36 09	-0.005	+0.04	7.2	1.1	-0.1	K2 III		220 mx		
188542	87915		19 54 50.9	+24 53 18	+0.002	+0.01	7.0	1.1	0.2	K0 III	-37	220 s		
188566	87916		19 54 53.6	+25 19 51	-0.001	-0.03	7.80	1.1	-0.1	K2 III		350 s		
187807	246331		19 54 54.6	-57 34 01	+0.005	-0.05	7.97	1.10		K0				
188403	125225		19 54 55.3	+2 26 23	-0.001	-0.02	7.9	0.0	0.0	B8.5 V		360 s		
188753	48968		19 54 58.0	+41 52 17	-0.006	+0.28	7.43	0.79	5.9	dK0	-24	26 mn	13125	m
187911	246335		19 54 59.3	-51 48 34	-0.001	-0.01	8.00	1.1	0.2	K0 III		360 s		m
188668	69080		19 55 00.1	+34 54 01	+0.002	+0.02	7.1	1.1	0.2	K0 III	-30	230 s		
189344	18601		19 55 01.1	+66 44 54	-0.003	-0.03	7.2	1.1	0.2	K0 III	0	230 s		
188113	211651		19 55 02.4	-36 12 07	+0.003	-0.02	7.1	0.1	1.4	A2 V		130 s		
188158	211653		19 55 05.0	-33 02 47	-0.001	0.00	6.46	1.48		K0				
188669	69078		19 55 05.1	+30 41 16	+0.004	+0.03	7.0	0.9	5.7	G9 V	-70	18 s		
188853	48976		19 55 05.1	+47 32 27	+0.002	+0.01	6.6	1.1	0.2	K0 III		190 s		
188508	105415		19 55 05.8	+13 11 39	0.000	0.00	8.0	1.6	-0.5	M2 III		490 s		
188651	69079		19 55 06.3	+30 11 43	+0.001	+0.02	6.57	-0.08	-1.1	B5 V	-12	320 s	13117	m
189106	32111		19 55 07.5	+59 27 08	-0.002	-0.01	7.90	0.1	1.4	A2 V		190 s		m
188011	229922		19 55 10.4	-46 47 34	+0.003	-0.03	7.69	1.08	1.7	K0 III-IV		140 s		
189207	18598		19 55 11.0	+62 35 51	0.000	-0.02	8.0	0.4	3.0	F2 V		98 s		
188854	48979		19 55 11.8	+46 39 55	-0.002	-0.05	7.5	0.2		A7 p				
188390	143908		19 55 13.0	-7 48 25	+0.001	-0.02	7.9	1.1	-0.1	K2 III		390 s		
187835	246334		19 55 13.5	-59 11 46	0.000	-0.05	6.7	1.6		M6 III	-22			S Pav, v
188114	229927	ι Sgr	19 55 15.5	-41 52 06	+0.001	+0.06	4.13	1.08	0.2	K0 III	+36	55 s		
188512	125235	60 β Aql	19 55 18.5	+6 24 24	+0.003	-0.48	3.71	0.86	3.2	G8 IV	-40	11 mx	13110	Alshain, m
188405	143911		19 55 19.4	-6 44 03	+0.002	-0.04	6.51	0.39		F2			13104	m
188032	229923		19 55 20.3	-48 16 57	+0.001	-0.04	7.8	-0.1	2.6	F0 V		110 s		
187336	257747		19 55 21.8	-71 32 45	+0.007	+0.03	8.0	1.6	-0.5	M4 III		380 mx		
189127	32116		19 55 21.9	+58 15 01	+0.002	-0.07	6.09	1.02	0.2	K0 III	-17	150 s		
188352	163107		19 55 23.7	-19 17 35	+0.003	-0.01	6.9	0.1	2.6	F0 V		72 s		
187977	246337		19 55 28.5	-53 56 50	-0.001	-0.01	7.8	1.2	-0.3	K5 III		410 s		
188754	87938		19 55 34.5	+29 30 53	+0.001	-0.02	7.9	0.2	1.9	A4 V		160 s		
189037	32114	24 ψ Cyg	19 55 37.6	+52 26 20	-0.004	-0.03	4.92	0.12	1.3	A3 IV-V	-11	52 s	13148	m
188875	48982		19 55 40.5	+40 10 22	+0.002	0.00	6.7	1.1	-0.1	K2 III	-18	220 s		
187477	—		19 55 41.7	-70 11 53			7.9	1.1	-0.1	K2 III		400 s		
188891	48983		19 55 44.6	+40 23 32	0.000	+0.03	7.30	-0.01		B3	-24			m
188376	188722	58 ω Sgr	19 55 50.2	-26 17 58	+0.015	+0.08	4.70	0.75	5.2	dG5	-21	11 ts		
188892	69101	22 Cyg	19 55 51.5	+38 29 12	0.000	0.00	4.94	-0.08	-1.9	B6 III	-30	220 s		
188377	188724		19 55 52.0	-26 32 59	-0.002	+0.04	7.6	0.6	3.4	F5 V		55 s		
189013	48993		19 55 53.8	+47 05 26	+0.001	+0.01	6.88	0.14	1.4	A2 V	+8	110 s		
189276	32122		19 55 55.2	+58 50 46	-0.002	-0.02	4.96	1.59	-1.3	K5 II-III	+5	160 s		
188652	105431		19 55 57.2	+10 12 27	+0.011	+0.01	6.9	0.8	3.2	G5 IV		55 s		
188716	105434		19 55 59.9	+15 02 50	+0.002	0.00	7.1	1.4	-0.3	K5 III	-45	290 s		
188727	105436	10 Sge	19 56 01.1	+16 38 05	0.000	0.00	5.36	0.67	-4.5	G0 Ib var	-10	860 s		S Sge, v
186154	258844		19 56 01.3	-81 20 59	+0.007	0.00	6.39	1.40		K0				
188726	105437		19 56 01.9	+17 53 13	+0.001	0.00	7.90	0.00	0.4	B9.5 V		310 s	13140	m
188246	229934		19 56 04.8	-44 00 45	-0.002	-0.02	7.18	-0.07		A0				
188876	69103		19 56 07.6	+33 04 29	0.000	0.00	7.25	-0.06	0.6	A0 V	-17	210 s		
188894	69106		19 56 08.6	+33 29 34	0.000	0.00	7.7	0.0	0.0	B8.5 V		330 s		
188339	211664		19 56 10.6	-38 19 35	+0.003	-0.01	6.70	1.4	-0.3	K5 III		250 s		m

HD	SAO	Star Name	α 2000	δ 2000	μ(α)	μ(δ)	V	B−V	M$_v$	Spec	RV	d(pc)	ADS	Notes
188728	105438	61 φ Aql	19^{h}56^{m}14$.^s$0	+11°25′25″	+0$.^s$002	+0$.^″$01	5.28	−0.01	1.2	A1 V	−27	66 s		
188432	211668		19 56 16.3	−31 36 48	−0.004	−0.16	7.5	0.6	4.4	G0 V		40 s		
188947	69116	21 η Cyg	19 56 18.2	+35 05 00	−0.003	−0.03	3.89	1.02	0.2	K0 III	−27	52 s	13149	m
189296	32128		19 56 18.8	+56 41 13	+0.002	+0.02	6.1	0.1	1.4	A2 V	−29	86 s		
188969	69119		19 56 24.4	+35 47 25	0.000	+0.01	8.0	0.9	3.2	G5 IV		91 s		
188856	105446		19 56 29.7	+19 11 14	+0.001	+0.01	7.6	0.1	0.6	A0 V		240 s		
188777	125255		19 56 33.3	+8 15 22	0.000	−0.01	8.0	0.1	0.6	A0 V		290 s		
188617	163122		19 56 34.1	−15 42 12	+0.001	+0.03	7.6	0.9	3.2	G5 IV		77 s		
189016	69124		19 56 35.3	+36 19 10	+0.001	−0.02	7.80	−0.05		B9				
189427	18610		19 56 36.6	+60 16 44	+0.002	+0.01	7.8	1.1	0.2	K0 III		310 s		
188474	211673		19 56 36.9	−31 20 08	+0.033	+0.02	7.8	1.6	0.2	K0 III		36 mx		
188895	87956		19 56 37.6	+23 53 18	0.000	0.00	7.0	1.1	−0.1	K2 III	−20	250 s		
188579	188737		19 56 39.9	−23 03 59	0.000	−0.05	7.6	0.0	1.4	A2 V		170 s		
−−	69123		19 56 41.5	+32 01 43	0.000	0.00	8.0	1.4						
188993	69122		19 56 41.5	+30 56 51	−0.002	+0.06	6.8	0.6	0.4	G2 III		150 mx		
189066	69129		19 56 44.0	+36 15 03	0.000	+0.01	6.02	−0.15	−1.6	B5 IV	−23	330 s		
188476	211674		19 56 44.4	−37 06 23	+0.002	−0.02	7.5	1.0	0.2	K0 III		280 s		
189253	32130		19 56 45.0	+50 54 09	0.000	+0.01	6.43	−0.01	0.6	A0 V	−19	150 s		
189976	9580		19 56 45.4	+72 18 37	+0.003	+0.03	7.8	1.1	0.2	K0 III		330 s		
188548	211677		19 56 48.9	−30 34 12	+0.004	0.00	6.6	0.7	1.4	A2 V		45 s		
188929	87959		19 56 49.6	+23 46 35	+0.003	+0.03	6.7	0.4	2.6	F0 V		66 s		
189160	49005		19 56 50.0	+44 16 18	+0.001	+0.02	7.86	−0.01		A0 p			13167	m
188858	125269		19 56 54.4	+8 26 52	+0.001	0.00	7.9	1.4	−0.3	K5 III		410 s		
188603	188742	59 Sgr	19 56 56.6	−27 10 12	+0.001	−0.01	4.52	1.46	−0.2	gK3	−16	69 s		
188161	246348		19 56 57.5	−57 55 32	−0.001	−0.03	6.3	1.9	−0.3	K5 III		120 s		
190315	9586		19 56 58.2	+75 42 51	−0.002	−0.01	7.30	1.4	−0.3	K4 III	−27	330 s		
188971	87963		19 57 00.0	+20 59 52	−0.002	−0.02	6.50	0.07	0.0	A3 III	+8	200 s		
188162	246349		19 57 06.1	−58 54 05	+0.003	−0.02	5.3	0.1	0.6	A0 V	−2	88 s		
187654	257750		19 57 08.1	−72 06 50	+0.003	−0.01	8.0	0.7	0.2	K0 III		360 s		
189086	69137		19 57 08.1	+30 46 39	+0.001	0.00	6.93	−0.03		A0	−18			
189107	69142		19 57 10.4	+30 51 23	0.000	+0.01	7.48	−0.06	−0.2	B8 V		330 s		
188806	143939		19 57 11.5	−7 24 18	−0.001	0.00	7.7	0.1	0.6	A0 V		260 s		
189040	87972		19 57 11.6	+25 29 20	0.000	+0.01	7.6	0.1	0.6	A0 V		240 s		
189256	−−		19 57 12.7	+44 15 39	+0.005	−0.02	7.85	3.43		N7.7	−6			m
188859	125272		19 57 13.1	+0 21 06	+0.001	−0.01	6.9	0.1	0.6	A0 V		180 s		
189148	69146		19 57 13.1	+35 01 54	−0.001	0.00	7.26	0.46	3.7	F6 V	−9	52 s		
189087	87976		19 57 13.2	+29 49 27	+0.007	+0.25	7.89	0.79	5.9	K0 V	−30	23 ts		
189178	49011		19 57 13.7	+40 22 04	0.000	0.00	5.45	−0.10	−1.1	B5 V p	−26	200 s		m
188640	211684		19 57 14.2	−30 11 59	+0.002	+0.02	7.8	0.9	3.2	G5 IV		70 s		
189635	18614		19 57 15.6	+62 52 35	0.000	0.00	7.4	0.4	2.6	F0 V		89 s		
188951	125275		19 57 18.5	+10 07 56	+0.001	−0.06	7.9	0.9	3.2	G5 IV		87 s		
189067	87974		19 57 19.5	+24 05 15	−0.009	−0.15	7.2	0.6	4.4	G0 V	−7	35 s		
188551	211682		19 57 20.5	−39 23 05	+0.002	−0.07	7.7	1.3	0.2	K0 III		200 mx		
189017	105460		19 57 20.5	+19 47 51	0.000	−0.01	7.9	1.1	0.2	K0 III		320 s		
189108	87979		19 57 23.4	+28 41 36	−0.001	−0.03	6.8	0.8	5.2	G5 V		21 s		
188844	143942		19 57 24.5	−6 41 54	+0.001	−0.15	6.7	0.9	3.2	G5 IV		50 s		
189235	69155		19 57 28.9	+38 27 29	+0.001	0.00	6.7	0.1	1.4	A2 V		110 s		
189088	87977		19 57 29.0	+21 45 25	0.000	0.00	7.8	1.1	0.2	K0 III		310 s		
188861	163138		19 57 33.7	−10 02 17	0.000	−0.01	7.6	0.1	0.6	A0 V		250 s		
188827	163137		19 57 34.4	−15 27 15	0.000	0.00	7.9	0.9	3.2	G5 IV		88 s		
188641	211685		19 57 37.1	−37 44 42	+0.002	+0.04	7.2	0.2	4.4	G0 V		37 s		
189280	69161		19 57 39.2	+38 24 05	−0.001	+0.01	8.0	1.1	0.2	K0 III		340 s		
188642	211686		19 57 41.0	−38 03 30	+0.007	−0.08	6.55	0.39	3.7	F6 V		37 s		
187978	254692		19 57 41.8	−67 49 38	+0.001	0.00	8.0	0.7	4.4	G0 V		53 s		
188721	211688		19 57 45.1	−31 22 11	−0.002	−0.01	8.0	2.8	−0.1	K2 III		43 s		
189090	105471	11 Sge	19 57 45.3	+16 47 20	+0.001	+0.02	5.33	−0.09	−0.3	B9 IV	−26	130 s		
189179	87983		19 57 47.1	+26 22 24	−0.001	+0.01	7.9	0.0	0.4	B9.5 V		300 s		
189180	87984		19 57 47.8	+26 17 39	+0.002	0.00	7.6	0.1	0.6	A0 V		240 s		
188677	211687		19 57 48.8	−37 39 40	0.000	+0.03	7.9	1.2	−0.1	K2 III		390 s		
189257	69163		19 57 49.3	+34 14 48	+0.001	+0.01	8.0	0.1	0.6	A0 V		280 s		
188303	246352		19 57 50.5	−59 31 35	+0.001	−0.02	7.8	1.2	0.2	K0 III		240 s		
189301	69166		19 57 51.4	+38 07 16	−0.001	0.00	7.25	1.55	−0.2	K3 III		200 s		
188938	143951		19 57 53.6	−9 03 30	+0.001	0.00	7.70	1.1	−0.1	K2 III		360 s	13163	m
189213	87991		19 57 54.3	+28 52 27	+0.002	+0.02	7.24	0.16	2.4	A7 V	−5	93 s		
189092	105473		19 57 55.4	+11 35 24	−0.001	−0.01	7.6	0.1	0.6	A0 V		250 s		
189093	105474		19 57 56.0	+11 24 54	+0.002	+0.01	7.40	0.1	1.7	A3 V		140 s	13168	m
189377	49031		19 57 56.0	+42 15 40	0.000	0.00	6.70	0.1	1.4	A2 V	−6	110 s	13186	m
189724	18621		19 57 56.7	+60 51 19	+0.001	+0.07	7.4	0.8	3.2	G5 IV		67 s		
188899	163141	61 Sgr	19 57 56.8	−15 29 29	+0.001	−0.10	5.02	0.05	0.6	A2 IV	−4	71 mx		
188957	143952		19 57 57.3	−6 26 41	+0.002	+0.01	7.6	0.1	0.6	A0 V		250 s		
189315	69169		19 57 59.2	+36 19 41	+0.002	+0.03	7.31	−0.04		B9				
188321	254695		19 58 01.0	−60 35 23	+0.003	0.00	7.9	1.8	−0.5	M2 III		390 s		
189349	49034		19 58 02.3	+40 55 36	+0.002	+0.01	7.5	0.9	3.2	G5 IV		70 s		
188766	211695		19 58 09.9	−35 11 15	−0.002	0.00	7.5	1.1	3.2	G5 IV		46 s		
188960	163146		19 58 11.6	−15 58 10	−0.001	−0.03	7.6	0.1	0.6	A0 V		250 s		

HD	SAO	Star Name	α 2000	δ 2000	μ(α)	μ(δ)	V	B-V	M_v	Spec	RV	d(pc)	ADS	Notes
			h m s	° ' "	s	"								
188787	211696		19 58 13.0	-36 41 13	+0.004	-0.05	7.1	0.3	2.6	F0 V		78 s		
189335	69174		19 58 13.4	+34 25 21	0.000	-0.01	7.73	-0.12		B9				
189183	105485		19 58 14.1	+16 29 35	0.000	-0.03	6.8	0.0	0.0	B8.5 V		220 s	13182	m
189132	105479		19 58 14.3	+10 43 39	+0.001	-0.02	8.0	0.4	2.6	F0 V		120 s		
188557	246357		19 58 14.9	-51 53 41	+0.004	-0.03	7.50	0.4	2.6	F0 V		96 s		m
189073	--		19 58 18.3	-2 14 01			7.40	0.5	3.4	F5 V		62 s	13178	m
189317	88002		19 58 21.1	+28 36 40	+0.006	+0.04	7.85	0.44	3.1	F3 V		84 s		
189316	88003		19 58 23.2	+28 59 18	0.000	-0.01	7.67	0.00	0.2	B9 V		310 s	13193	m
189350	69180		19 58 24.7	+31 28 38	0.000	0.00	7.89	1.82	-0.5	M2 III		350 s		
189900	18627		19 58 28.5	+63 32 03	-0.001	-0.02	6.0	0.1	0.6	A0 V	-9	110 s		
188864	211699		19 58 30.8	-32 25 34	+0.002	+0.02	7.8	1.2	0.2	K0 III		260 s		
189114	143959		19 58 32.3	-3 33 13	-0.001	+0.01	7.1	1.4	-0.3	K5 III		280 s		
189186	125296		19 58 32.4	+7 55 17	+0.001	+0.04	7.5	1.1	0.2	K0 III		280 s		
189378	69186		19 58 32.5	+33 16 39	+0.004	+0.06	7.24	0.41	-5.6	F2 I-II	-22	61 mx	13196	m
188357	254697		19 58 32.7	-63 07 42	+0.005	-0.03	7.4	0.1	0.6	A0 V		240 mx		
188884	211701		19 58 33.9	-32 37 44	+0.001	+0.01	7.7	0.9	3.2	G5 IV		65 s		
189432	69193		19 58 34.2	+38 06 19	+0.001	0.00	6.32	-0.09	-1.6	B5 IV	-14	370 s	13198	m
189379	88007		19 58 36.4	+29 56 28	0.000	0.00	7.4	0.2	-2.0	A9 II	-6	680 s		
190109	18635		19 58 37.0	+68 24 43	0.000	+0.03	7.5	0.1	1.4	A2 V		160 s		
185545	--		19 58 37.4	-83 47 49			7.8	1.4	-0.3	K5 III		420 s		
189395	69188		19 58 37.8	+30 59 01	+0.002	0.00	5.49	-0.06	-0.8	B9 III	-7	180 s		
188097	254693		19 58 41.3	-69 09 50	+0.016	-0.10	5.75	0.23		A m		16 mn		
190252	9592		19 58 41.8	+70 22 02	+0.008	+0.06	6.4	0.9	0.3	G8 III	-10	170 s		
188813	229959		19 58 42.8	-41 51 04	-0.001	-0.12	6.7	1.6		M5 III	-68			RU Sgr, v
189188	125298		19 58 43.7	+1 40 00	-0.002	-0.02	6.8	1.1	0.2	K0 III		200 s		
189319	105500	12 γ Sge	19 58 45.3	+19 29 32	+0.005	+0.03	3.47	1.57	-0.3	K5 III	-33	51 s		
188980	188775		19 58 49.2	-28 39 11	+0.005	-0.01	7.9	0.8	-0.1	K2 III		410 s		
189474	69199		19 58 51.3	+35 29 51	+0.001	0.00	7.00	0.02	0.6	A0 V	+5	180 s		
188164	254696		19 58 52.9	-68 45 45	+0.008	-0.07	6.39	0.16		A2				
189259	125302		19 58 54.4	+6 35 28	+0.003	0.00	8.0	0.4	3.0	F2 V		97 s		
188981	211708		19 58 56.2	-30 32 17	+0.007	-0.05	6.28	1.05		K0				
189005	188778	60 Sgr	19 58 57.0	-26 11 44	+0.003	+0.03	4.83	0.90	0.3	gG5	-49	80 s		
188815	229960		19 58 58.5	-46 05 20	+0.009	+0.01	7.48	0.48	3.7	F6 V		57 s		
190145	18637		19 58 59.6	+67 28 21	+0.006	+0.02	7.4	0.1		A2 p				
189818	32173		19 59 01.3	+57 48 35	-0.002	0.00	7.30	-0.18	-1.0	B5.5 V	-3	460 s		
190316	9598		19 59 06.4	+70 22 22	+0.007	+0.07	8.0	0.8	3.2	G5 IV		92 s		
189702	32167		19 59 06.8	+50 39 47	-0.002	+0.02	7.40	1.1	0.2	K0 III		260 s	13221	m
189475	88017		19 59 08.9	+27 30 45	+0.001	+0.01	6.8	1.1	-0.1	K2 III		230 s		
189410	88016	14 Vul	19 59 10.3	+23 06 05	-0.005	+0.01	5.7	0.4	2.6	F0 V	-38	41 s		
189117	188789		19 59 10.5	-21 55 26	-0.016	-0.06	8.00	0.6	4.4	G0 V		53 s		m
189337	105505		19 59 11.9	+11 18 21	+0.001	+0.01	6.5	1.1	0.2	K0 III		180 s		
189574	69213		19 59 12.5	+38 52 49	+0.002	+0.01	7.6	0.2		A m				
189775	32170		19 59 15.2	+52 03 21	+0.001	+0.01	6.15	-0.19	-2.2	B5 III	-16	470 s		
193214	3332		19 59 19.6	+84 40 07	-0.007	-0.04	6.6	0.1	1.4	A2 V		110 s		
189684	49058		19 59 20.3	+45 46 20	+0.001	-0.02	5.92	0.16		A2	+6			
187915	257751		19 59 22.3	-74 07 49	-0.001	-0.03	7.52	0.43		F5				
189322	125310		19 59 22.5	+1 22 39	+0.001	+0.06	6.17	1.13		G6	+6			
189264	143970		19 59 24.4	-8 05 22	0.000	-0.01	7.9	1.1	0.2	K0 III		340 s		
189077	211712		19 59 26.8	-32 42 05	+0.003	-0.02	7.9	0.5	0.2	K0 III		340 s		
189193	188795		19 59 33.9	-22 12 40	+0.001	0.00	6.9	0.7	1.4	A2 V		50 s		
189286	163166		19 59 33.9	-11 43 01	-0.001	-0.04	7.6	0.2	2.1	A5 V		130 s		
189596	69219		19 59 35.8	+32 53 42	+0.001	0.00	7.54	-0.11		B9				
190960	9606	69 Dra	19 59 36.4	+76 28 53	-0.009	-0.06	6.20	1.61	-0.5	M3 III	-69	220 s		
189613	69222		19 59 39.0	+31 49 35	-0.001	+0.01	6.80	0.1	0.6	A0 V	-17	170 s	13223	m
190396	18648		19 59 41.2	+69 24 32	-0.004	-0.03	8.0	0.1	0.6	A0 V		260 s		
189359	143974		19 59 41.3	-1 58 13	0.000	-0.03	6.6	0.1	1.4	A2 V		110 s		
189597	69223		19 59 41.8	+30 54 50	+0.002	0.00	8.00	0.00	-3.7	B6 II		1600 s	13224	m
189079	211714		19 59 42.5	-37 22 30	+0.002	-0.01	7.2	0.2	0.6	A0 V		160 s		
189103	211716	θ¹ Sgr	19 59 44.0	-35 16 35	+0.001	-0.02	4.37	-0.15	-2.3	B3 IV	+1	210 s		
188906	246373		19 59 46.9	-50 51 21	-0.001	0.00	7.5	1.6	0.2	K0 III		130 s		
189340	163168		19 59 47.2	-9 57 30	-0.019	-0.40	5.88	0.58	4.4	G0 V	+23	22 ts		m
189118	211717	θ² Sgr	19 59 51.1	-34 41 52	+0.008	-0.07	5.30	0.1	1.7	A3 V	-18	53 s		m
189859	32182		19 59 53.5	+52 09 01	0.000	+0.03	7.0	1.6	-0.5	M4 III		290 s		
191469	9615		19 59 53.6	+79 33 04	+0.004	+0.01	7.9	1.1	0.2	K0 III		340 s		
189687	69231	25 Cyg	19 59 55.0	+37 02 35	0.000	+0.01	5.19	-0.17	-1.7	B3 V	-4	240 s		
188723	254703		19 59 55.3	-60 01 16	+0.004	0.00	7.6	1.3	-0.1	K2 III		300 s		
189688	69230		19 59 55.6	+35 17 47	0.000	0.00	7.67	1.75	-0.3	K5 III		280 s		
189550	105520		19 59 55.7	+19 53 22	+0.001	0.00	7.67	-0.05	-2.5	B2 V	-13	920 s		
189479	125324		19 59 58.0	+6 42 28	-0.002	0.00	8.0	0.0	0.4	B9.5 V		320 s		
189614	88040		20 00 01.8	+25 10 33	0.000	+0.01	7.7	0.1	0.6	A0 V		250 s		
189577	105522	13 Sge	20 00 03.1	+17 30 59	0.000	-0.01	5.37	1.67	-0.5	M4 III	-17	140 s	13230	VZ Sge, m,q
190424	18650		20 00 04.0	+68 26 10	+0.002	-0.02	7.9	0.0	0.4	B9.5 V		280 s		
189638	88043		20 00 06.4	+25 57 01	-0.001	-0.02	7.70	1.1	-0.1	K2 III	-1	340 s		m
189689	69234		20 00 06.9	+32 47 22	0.000	-0.01	7.27	-0.07		B9	+4			
189509	125331		20 00 08.0	+7 45 38	+0.003	0.00	7.8	0.5	3.4	F5 V		74 s		

HD	SAO	Star Name	α 2000	δ 2000	μ(α)	μ(δ)	V	B-V	M$_v$	Spec	RV	d(pc)	ADS	Notes
			^{h} ^{m} ^{s}	$+^{\circ}$ $'$ $''$	^{s}	$''$								
189576	105523		20 00 08.1	+17 36 39	0.000	+0.01	7.10	1.1	0.2	K0 III		230 s		
188986	246379		20 00 08.2	-50 12 58	+0.006	+0.03	6.8	0.6	3.0	F2 V		41 s		
189751	69238		20 00 11.2	+36 24 50	+0.001	0.00	7.01	1.09	0.2	K0 III	-15	190 s	13240	m
189510	125334		20 00 13.4	+7 17 16	+0.001	0.00	7.7	0.0	0.4	B9.5 V		280 s		
189671	88046		20 00 15.1	+26 11 18	0.000	+0.01	6.49	1.15	-2.1	G8 II	-22	450 s	13237	m
189821	49069		20 00 15.1	+42 13 04	-0.002	-0.01	7.3	1.1	0.2	K0 III	+10	250 s		
189690	88049		20 00 15.5	+29 55 15	+0.002	0.00	7.42	0.04	0.6	A0 V	-36	220 s	13236	m
189195	211719		20 00 15.7	-37 42 08	+0.001	-0.02	5.95	1.00	0.2	K0 III		140 s		m
188946	246380		20 00 16.2	-52 41 13	0.000	-0.01	7.8	0.8	-0.1	K2 III		380 s		
189777	69241		20 00 18.8	+35 21 28	0.000	-0.03	7.8	1.1	-0.1	K2 III		360 s		
189706	88051		20 00 19.0	+29 49 16	0.000	-0.01	7.76	-0.08	0.2	B9 V	+1	330 s		
189598	105529		20 00 19.4	+13 06 03	0.000	0.00	7.8	0.4	3.0	F2 V	-45	88 s		
190693	9604		20 00 19.8	+71 54 03	+0.002	+0.03	7.9	1.4	-0.3	K5 III		430 s		
189245	211724		20 00 20.1	-33 42 14	+0.011	-0.30	5.66	0.49	4.0	F8 V	-6	23 ts		
189533	125339		20 00 21.1	+3 19 56	0.000	-0.01	6.8	0.8	3.2	G5 IV	-5	52 s		
189511	125338		20 00 21.8	+1 41 53	0.000	-0.01	7.5	0.8	3.2	G5 IV		72 s		
188584	254702	μ¹ Pav	20 00 22.9	-66 56 58	-0.002	-0.20	5.76	1.04	3.2	K0 IV		28 s		
189080	229973		20 00 25.0	-49 21 05	-0.009	-0.01	6.17	1.06	0.2	K0 III		130 mx		
189140	229977		20 00 26.4	-43 02 36	+0.002	+0.02	6.14	1.64	-0.4	M0 III		180 s		
189437	163173		20 00 29.8	-15 53 12	-0.001	-0.01	7.1	0.5	3.4	F5 V		56 s		
189398	163171		20 00 30.5	-19 51 28	-0.003	-0.02	7.8	1.3	0.2	K0 III		210 s		
189845	69249		20 00 32.7	+39 12 25	-0.002	-0.02	7.22	0.03	0.6	A0 V		200 s		
188228	257757	ε Pav	20 00 35.4	-72 54 38	+0.019	-0.13	3.96	-0.03	0.6	A0 V	0	41 mx		
189365	188808		20 00 35.7	-28 35 10	0.000	-0.01	6.9	0.3	0.2	K0 III		220 s		
189731	88057		20 00 38.5	+24 08 56	0.000	-0.01	7.8	0.8	3.2	G5 IV		83 s		
189328	211729		20 00 38.8	-32 10 34	+0.003	-0.03	7.4	0.1	0.6	A0 V		230 s		
189122	229976		20 00 39.7	-49 36 32	-0.005	0.00	6.8	1.7	-0.5	M4 III		220 mx		
189246	229983		20 00 41.5	-40 11 59	0.000	+0.01	7.7	1.9	-0.5	M2 III		300 s		
189732	88061		20 00 43.2	+23 47 55	0.000	-0.01	7.37	-0.16	0.4	B9.5 V		250 s		
189733	88060		20 00 43.5	+22 42 43	-0.001	-0.22	7.6	0.9	3.2	G5 IV		71 mx		
189864	69252		20 00 44.7	+36 35 24	+0.001	+0.01	6.68	-0.07		B9				m
189796	88066		20 00 46.1	+29 49 19	+0.001	-0.12	7.88	0.61	4.4	G0 V		47 s	13251	m
--	188815		20 00 48.0	-22 38 16	0.000	-0.03	7.9	1.1	0.2	K0 III		350 s		
189198	229981		20 00 48.1	-45 06 47	0.000	0.00	5.81	0.28	0.5	A7 III		110 s		
189582	143985		20 00 48.4	-4 18 41	-0.001	-0.03	7.6	1.1	-0.1	K2 III		340 s		
189439	--		20 00 50.4	-22 37 45			8.0	0.9	0.2	K0 III		360 s		
189662	125353		20 00 50.5	+7 58 54	0.000	+0.01	7.5	0.1	0.6	A0 V		240 s		
189247	229985		20 00 50.6	-43 58 57	+0.002	-0.15	7.66	0.44	2.1	F5 IV		54 mx		
189692	105543		20 00 51.7	+13 22 48	+0.002	0.00	7.1	0.1	0.6	A0 V	-4	200 s		
189754	105545		20 00 51.9	+20 00 18	0.000	+0.02	7.9	1.1	0.2	K0 III		330 s		
189419	188817		20 00 54.6	-26 13 24	-0.001	+0.02	7.8	0.7	1.4	A2 V		83 s		
188229	257758		20 00 56.5	-74 00 44	+0.013	-0.01	6.59	1.04		K0				
189847	69254		20 00 58.6	+31 13 51	+0.001	+0.01	6.86	-0.08	-0.6	B7 V	-16	310 s		
189695	125355		20 00 58.7	+8 33 28	0.000	-0.01	5.91	1.52	-0.3	gK5	-40	170 s		
189901	69258		20 00 59.5	+37 41 57	+0.001	0.00	7.99	-0.02		B5			13262	m
189957	49080		20 00 59.8	+42 00 32	-0.001	+0.02	7.82	0.02	-5.0	B0 III		2600 s		
189558	163181		20 01 00.2	-12 15 20	-0.021	-0.36	7.72	0.55	4.2	F9 V	-15	51 s		m
188769	254704		20 01 00.4	-64 48 30	+0.023	-0.25	6.9	0.1	3.4	F5 V		45 mx		
188987	246383		20 01 01.2	-58 23 16	+0.002	-0.03	7.3	1.2	0.2	K0 III		210 s		
188012	257755		20 01 05.4	-76 13 57	+0.010	-0.01	7.6	1.1	-0.1	K2 III		350 s		
189849	88071	15 Vul	20 01 05.9	+27 45 13	+0.004	+0.01	4.64	0.18		A m	-21	18 mn		
189559	163182		20 01 06.2	-13 56 27	+0.001	-0.01	7.1	1.1	0.2	K0 III		240 s		
189756	105555		20 01 08.0	+14 31 23	0.000	0.00	7.1	0.1	0.6	A0 V	+5	190 s		
189782	105556		20 01 08.7	+14 48 06	0.000	+0.01	7.4	1.1	-0.1	K2 III	-12	300 s		
189958	49084		20 01 09.8	+41 05 31	+0.002	0.00	8.0	1.1	-0.1	K2 III		380 s		
189781	105558		20 01 10.9	+15 12 39	+0.001	0.00	7.9	0.2	2.1	A5 V		150 s		
190025	49087		20 01 12.6	+43 02 39	-0.006	+0.02	7.53	-0.07	-1.0	B5.5 V	-14	45 mx		
189386	211734		20 01 12.7	-38 35 13	+0.004	-0.07	7.70	0.40	3.0	F2 V		81 s		m
190425	18653		20 01 12.7	+62 23 29	-0.002	-0.02	8.0	1.6	-0.5	M2 III		450 s		
189387	211735		20 01 13.7	-38 40 46	+0.002	-0.17	7.5	0.6	3.4	F5 V		53 s		
190043	49089		20 01 14.3	+44 15 51	+0.001	-0.02	7.8	1.1	0.2	K0 III		310 s		
189884	88078		20 01 14.8	+27 11 23	+0.001	+0.01	7.40	1.1	-0.1	K2 III	-21	300 s		
189942	69267		20 01 15.1	+37 05 56	+0.003	+0.05	6.20	1.31	0.2	K0 III	-16	100 s		
189624	163188		20 01 15.3	-10 46 53	+0.001	-0.02	7.7	0.1	0.6	A0 V		260 s		
189712	125360		20 01 15.8	+5 38 54	-0.001	-0.01	7.46	0.46	3.4	F5 V		63 s		
189607	163189		20 01 19.7	-14 44 35	+0.001	-0.01	7.9	0.1	1.4	A2 V		200 s		
189982	69270		20 01 19.8	+37 59 03	+0.001	-0.03	7.79	-0.04		A0				
190147	49098	26 Cyg	20 01 21.4	+50 06 17	+0.002	+0.01	5.05	1.11	-1.1	K1 II-III	+1	170 s	13278	m
190065	49092		20 01 21.4	+43 13 42	-0.004	+0.03	7.68	1.67	-0.5	M2 III		180 mx		
189665	143995		20 01 23.5	-9 29 14	+0.001	+0.01	7.5	0.1	1.4	A2 V		160 s		
189561	188829		20 01 23.8	-22 44 14	+0.001	-0.01	6.01	0.97	3.2	G5 IV	+8	25 s		
189009	254708		20 01 24.2	-60 44 10	+0.002	-0.01	7.5	-0.1	0.6	A0 V		250 pc		
189307	229991		20 01 25.3	-47 23 48	+0.003	+0.02	6.96	0.35		F0				m
189388	229993		20 01 26.2	-40 48 51	-0.001	-0.04	6.29	0.11		A2				
235083	--		20 01 26.2	+53 24 43			8.02	1.48		G5				

HD	SAO	Star Name	α 2000	δ 2000	μ(α)	μ(δ)	V	B−V	M$_v$	Spec	RV	d(pc)	ADS	Notes
189783	105560		20^h 01^m 26.6^s	+10° 44′ 55″	+0.006	+0.04	7.60	0.4	3.0	F2 V	−43	83 s	13256	m
190046	49094		20 01 28.2	+40 17 57	+0.001	+0.01	7.78	0.02	0.6	A0 V		270 s	13271	m
190544	18658	64 Dra	20 01 28.3	+64 49 16	+0.001	−0.01	5.40	1.6	−0.5	M1 III	−34	140 s		m
189943	69272		20 01 31.6	+30 13 06	0.000	−0.01	7.74	0.82	−2.1	G5 II		930 s		
189625	163190		20 01 32.6	−16 52 08	+0.010	−0.06	7.5	0.6	4.4	G0 V		41 s		
189441	211739		20 01 32.9	−36 40 44	−0.002	0.00	7.5	1.2	−0.1	K2 III		300 s		
189500	211744		20 01 38.9	−34 35 06	−0.001	−0.01	8.0	1.0	0.2	K0 III		360 s		
189759	144002		20 01 39.0	−0 11 57	+0.002	0.00	7.50	0.1	0.6	A0 V		230 s	13259	m
188322	257759		20 01 39.4	−74 45 49	+0.001	0.00	7.51	1.67		K2				
190490	18657		20 01 40.0	+62 09 11	+0.001	−0.01	7.8	1.1	0.2	K0 III		300 s		
190001	69280		20 01 43.3	+33 04 00	0.000	−0.01	7.88	−0.08	−0.9	B6 V		550 s		
190112	49101		20 01 43.9	+41 48 54	0.000	−0.06	7.3	0.1	0.6	A0 V		140 mx		
189124	246389		20 01 44.4	−59 22 34	+0.002	−0.03	5.13	1.53	−0.5	M4 III	−10	130 s		
189944	88088		20 01 44.5	+24 48 01	0.000	0.00	5.80	−0.12	−1.6	B5 IV	−15	300 s		
190149	49105		20 01 45.1	+44 07 09	−0.001	−0.01	6.96	1.64	0.2	G9 III	−61	69 s		
190397	32219		20 01 45.2	+57 39 05	−0.001	−0.02	7.68	−0.08	0.6	A0 V		260 s		
190088	69284		20 01 45.3	+38 44 08	+0.001	+0.01	7.85	−0.06	0.6	A0 V		280 s		
190165	49108		20 01 46.5	+45 28 37	+0.002	+0.04	7.6	0.1	1.4	A2 V	−17	170 s		
189811	125371		20 01 46.8	+4 42 55	−0.002	−0.06	7.6	1.1	−0.1	K2 III		220 mx		
189885	105567		20 01 48.5	+16 30 45	+0.002	−0.04	7.5	0.4	2.6	F0 V		93 s		
190130	49106		20 01 51.2	+40 51 32	+0.001	0.00	7.00	0.1	0.6	A0 V		180 s		m
188748	254705		20 01 51.6	−69 11 17	+0.041	−0.14	7.9	0.8	3.2	G5 IV		61 mx		
188887	254707	μ² Pav	20 01 52.2	−66 56 39	+0.007	−0.07	5.31	1.22	0.2	gK0	+42	70 s		
190047	69283		20 01 55.4	+31 07 04	0.000	+0.02	6.53	−0.05	−0.6	B7 V	−11	260 s		
189158	246391		20 01 55.5	−59 26 52	+0.002	−0.03	7.5	1.3	0.2	K0 III		190 s		
189741	163195	63 Sgr	20 01 58.4	−13 38 14	+0.002	+0.02	5.8	0.1	1.4	A2 V		74 s		
189698	163193		20 02 01.0	−19 05 56	+0.001	−0.01	7.6	0.2	3.4	F5 V		68 s		
190004	88098	16 Vul	20 02 01.2	+24 56 17	+0.006	+0.07	5.22	0.36	−2.0	F5 II	−33	280 s	13277	m
190114	69286		20 02 01.7	+35 19 58	+0.001	+0.01	7.41	−0.07	−0.2	B8 V		330 s		
190113	69288		20 02 02.6	+35 38 28	−0.001	−0.01	7.91	1.46	−4.5	G5 Ib		1800 s		
189643	188838		20 02 06.0	−27 59 49	+0.002	−0.01	8.0	0.8	0.2	K0 III		360 s		
189921	105572		20 02 10.2	+10 44 20	0.000	−0.01	6.8	0.0	−1.1	B5 V	−3	370 s		
189502	230000		20 02 10.3	−44 28 00	+0.003	−0.02	7.91	−0.03		A0				
189825	144009		20 02 10.5	−4 59 24	+0.002	−0.01	6.7	0.1	0.6	A0 V		170 s		
189483	229999		20 02 13.0	−45 56 27	+0.001	−0.03	8.0	1.6	0.2	K0 III		160 s		
189985	105575		20 02 19.6	+13 54 35	−0.002	−0.02	6.6	1.1	−0.1	K2 III		210 s		
190713	18669	65 Dra	20 02 20.1	+64 38 04	+0.007	+0.01	6.60	0.8	3.2	G5 IV	+9	47 s		m
190066	88105		20 02 22.0	+22 09 06	0.000	0.00	6.48	0.18	−6.2	B1 Iab	+16	2100 s		
191994	9629		20 02 26.0	+79 40 52	−0.002	0.00	6.6	1.1	0.2	K0 III		190 s		
189826	163199		20 02 27.1	−11 50 01	+0.001	+0.02	7.90	1.1	−0.1	K2 III		400 s		m
227294	49117		20 02 28.9	+41 20 56	−0.001	+0.03	8.0	1.1	0.2	K0 III		340 s		
190067	105581		20 02 34.0	+15 35 30	−0.011	−0.59	7.16	0.71	5.5	G7 V	+12	23 ts		
189563	230002		20 02 35.7	−45 11 46	−0.003	0.00	6.58	1.22		K0				
190028	105580		20 02 36.1	+13 11 19	0.000	+0.01	8.0	1.1	0.2	K0 III		350 s		
189763	188844	62 Sgr	20 02 39.4	−27 42 35	+0.003	+0.02	4.58	1.65	−0.5	M4 III	+10	100 s		
189631	230007		20 02 40.5	−41 25 05	−0.001	0.00	7.56	0.30	2.5	A9 V		99 s		
190050	105583		20 02 41.0	+12 21 30	0.000	−0.02	6.8	1.4	−0.3	K5 III		260 s		
227310	--		20 02 42.3	+38 23 49			6.92	0.28	−1.6	B5 IV		290 s		
190090	105589		20 02 45.8	+14 34 58	0.000	0.00	6.80	1.1	0.2	K0 III		210 s	13290	m
190150	88113		20 02 46.3	+21 51 15	0.000	0.00	8.0	0.0	0.4	B9.5 V		310 s		
190275	69310		20 02 46.4	+37 48 47	0.000	+0.02	7.2	0.2		A m				
190007	125379		20 02 46.8	+3 19 34	−0.006	+0.12	7.46	1.14	−0.1	K2 III	−31	16 ts		
190167	88115		20 02 47.0	+28 30 55	0.000	+0.01	6.8	0.1	1.2	A1 V	−18	130 s		
190227	69307		20 02 48.6	+31 57 30	+0.001	0.00	6.42	1.20	0.2	K0 III	−20	120 s		
190940	18676	67 ρ Dra	20 02 48.9	+67 52 25	+0.002	+0.05	4.51	1.32	−0.2	K3 III	−9	82 s		m
190070	105591		20 02 51.9	+10 54 41	+0.003	−0.01	6.8	0.1	1.4	A2 V		120 s		
190072	125380		20 02 53.9	+8 23 52	+0.001	+0.02	7.90	0.00	0.4	B9.5 V		310 s	13294	m
189893	163208		20 02 54.7	−18 32 47	+0.004	−0.07	7.6	0.6	4.4	G0 V		44 s		
189869	--		20 02 57.8	−20 44 29			7.9	1.5	3.2	G5 IV		32 s		
190625	32238		20 02 57.8	+57 05 05	−0.001	+0.03	7.6	0.1	0.6	A0 V		240 s		
190228	88118		20 03 00.5	+28 18 26	+0.007	−0.06	7.6	0.9	3.2	G5 IV		75 s		
190697	32241		20 03 00.6	+59 39 08	+0.001	+0.02	7.8	1.1	−0.1	K2 III		340 s		
189829	188849		20 03 00.9	−26 02 29	+0.004	+0.01	7.5	0.6	0.6	A0 V		98 s		
189948	144024		20 03 00.9	−9 02 35	−0.002	−0.14	8.0	0.5	4.0	F8 V		62 s		
190359	49122		20 03 02.0	+40 10 23	0.000	−0.01	7.4	0.1	0.6	A0 V		220 s		
190073	125381		20 03 02.3	+5 44 18	0.000	0.00	7.88	0.10		A pe	−1	37 mn		
190151	105603		20 03 02.4	+18 17 45	+0.004	−0.01	6.8	0.6	4.4	G0 V		30 s		
189788	211764		20 03 04.8	−35 19 16	+0.002	−0.03	7.5	1.8	0.2	K0 III		97 s		
189926	163210		20 03 05.6	−17 20 52	0.000	0.00	7.6	1.1	0.2	K0 III		300 s		
190257	88121		20 03 05.6	+26 53 32	0.000	+0.01	7.7	0.0	0.0	B8.5 V		330 s		
227362	49125		20 03 06.8	+41 16 42	−0.003	+0.02	7.7	1.1	0.2	K0 III		300 s		
190401	49126		20 03 08.6	+41 28 27	−0.002	+0.03	7.01	0.36		A m				
--	69313		20 03 11.4	+31 55 09	0.000	−0.01	8.00	0.5	−8.0	F8 Ia				
190095	125384		20 03 12.5	+4 43 47	+0.001	−0.01	6.7	1.4	−0.3	K5 III		240 s		
190547	32236		20 03 12.8	+50 18 32	+0.002	+0.03	7.5	1.4	−0.3	K5 III		330 s		

HD	SAO	Star Name	α 2000	δ 2000	μ(α)	μ(δ)	V	B-V	M$_v$	Spec	RV	d(pc)	ADS	Notes
189927	163211		20^h 03^m 12.8^s	-18°14'32"	-0.001^s	+0.01"	7.70	0.5	3.4	F5 V		72 s		m
190584	32239		20 03 13.1	+52 34 33	+0.006	+0.03	7.7	0.2	2.1	A5 V		130 s		
190546	32237		20 03 13.4	+50 27 56	+0.001	0.00	7.50	0.1	0.6	A0 V		230 s	13317	m
190211	105608		20 03 16.2	+18 30 02	+0.001	-0.03	5.96	1.42	-1.3	K3 II-III	+9	250 s		m
189719	230010		20 03 16.5	-42 55 25	-0.001	-0.01	7.58	0.37		F2				m
189274	254716		20 03 20.6	-65 17 28	+0.011	-0.07	8.0	0.5	3.4	F5 V		82 s		
190118	144032		20 03 26.7	+0 05 19	+0.002	-0.03	8.0	0.5	3.4	F5 V		80 s		
190429	69324		20 03 29.3	+36 01 29	0.000	-0.01	6.63	0.15		O5.8	-16		13312	m
190229	105615		20 03 29.8	+16 01 53	0.000	0.00	5.67	-0.12	-2.3	B8 II-III	-22	390 s		
189830	211769		20 03 31.2	-36 35 44	-0.002	0.00	7.30	0.1	1.4	A2 V		150 s		m
188136	257761		20 03 31.3	-78 49 48	0.000	+0.02	7.8	0.5	4.0	F8 V		57 s		
189831	211767		20 03 33.3	-37 56 27	+0.006	-0.09	4.77	1.41	-0.4	gM0	-38	110 s		
190033	163213		20 03 33.4	-15 24 48	+0.002	0.00	7.2	0.5	3.4	F5 V		57 s		
189832	211768		20 03 34.9	-38 51 04	+0.003	+0.06	6.9	0.5		F0 p				
189854	211770		20 03 35.9	-37 31 46	+0.004	-0.12	6.9	0.7	0.2	K0 III		100 mx		
190360	88133		20 03 37.2	+29 53 48	+0.052	-0.53	5.71	0.73	3.2	G6 IV	-46	12 mx		
190322	88129		20 03 38.6	+22 56 29	-0.001	0.00	6.6	1.1	-0.1	K2 III		210 s		
190466	69335		20 03 39.5	+38 19 35	0.000	-0.04	7.15	1.73	-0.5	M2 III		280 s	13318	m
190430	69330		20 03 40.4	+33 10 34	0.000	-0.01	7.81	1.36	-2.2	K1 II		840 s		
189769	230013		20 03 41.6	-46 59 06	+0.005	-0.05	8.0	0.1	3.4	F5 V		82 s		
189874	211772		20 03 42.1	-36 03 33	+0.001	+0.01	7.1	1.1	0.2	K0 III		220 s		
190337	88131		20 03 42.1	+21 29 47	+0.002	+0.01	6.8	1.6	-0.5	M2 III		280 s		
190403	88137		20 03 43.2	+29 59 23	-0.001	+0.01	6.81	1.15	-3.3	G5 Ib-II	-12	910 s		
190009	188863		20 03 44.1	-22 35 44	-0.003	+0.02	6.45	0.50	3.7	dF6	+6	34 s		
190277	105619		20 03 48.8	+10 27 39	+0.002	0.00	7.9	0.4	3.0	F2 V		94 s		
190323	105624		20 03 49.4	+14 58 59	0.000	0.00	6.83	0.84	-8.0	G0 Ia	+25	9100 s	13310	m
--	--		20 03 51.0	+31 06 44			7.80	0.00	-0.6	B7 V		440 s		
190361	88136		20 03 51.7	+21 03 01	+0.001	0.00	7.4	1.1	0.2	K0 III	-21	270 s		
190404	88138		20 03 51.9	+23 20 26	-0.073	-0.91	7.28	0.82	6.1	K1 V	-3	19 ts		
--	105627		20 03 55.0	+17 05 03	-0.001	-0.02	6.80							m
190384	88139		20 03 55.3	+23 50 49	+0.001	-0.01	7.60	1.1	0.2	K0 III		280 s		
190469	69337		20 03 56.0	+31 07 25	0.000	0.00	7.81	-0.05		A0				
189930	--		20 03 59.1	-36 55 25			7.9	1.1	3.2	G5 IV		56 s		
190338	105629		20 03 59.2	+17 07 22	+0.001	+0.01	6.80	0.1	1.4	A2 V	-28	120 s		m
190339	105628		20 03 59.3	+16 42 33	+0.001	-0.03	7.6	1.1	0.2	K0 III		300 s		
190587	49141		20 03 59.5	+40 39 23	+0.004	0.00	7.6	0.0	0.4	B9.5 V		180 mx		
190173	144036		20 04 00.1	-9 07 37	+0.001	-0.05	7.5	0.1	0.6	A0 V		240 s		
190296	105625		20 04 00.2	+10 40 50	+0.001	-0.03	8.0	0.1	0.6	A0 V		290 s		
190172	144038		20 04 01.0	-7 28 11	+0.001	-0.05	6.72	0.35		A5				
188304	--		20 04 02.1	-78 16 12			7.7	1.1	0.2	K0 III		320 s		
189931	211778		20 04 02.4	-37 52 35	+0.007	-0.39	6.93	0.60		G6		17 mn		
190536	69341		20 04 04.7	+34 18 54	-0.003	-0.04	7.71	1.05	-2.1	G5 II		820 s		m
190406	105635	15 Sge	20 04 06.0	+17 04 12	-0.028	-0.41	5.80	0.61	4.5	G1 V	+4	18 ts		m
190780	32254		20 04 07.1	+54 27 59	+0.007	-0.12	8.00	1.1	0.2	K0 III	-9	110 mx		
190913	32260		20 04 07.7	+59 39 45	-0.002	+0.01	7.9	1.1	0.2	K0 III		320 s		
190327	125403	63 τ Aql	20 04 08.1	+7 16 40	+0.001	+0.01	5.52	1.06	0.2	gK0	-28	110 s		
190470	88147		20 04 09.9	+25 47 25	-0.006	-0.03	7.88	0.91	6.9	K3 V		16 s		
190537	69343		20 04 13.0	+31 14 36	+0.004	+0.03	6.82	0.29		A m	-29			
190405	105638		20 04 14.1	+17 44 05	+0.001	-0.01	6.86	0.48	-6.6	F0 I	-18	3800 s		
190770	32253		20 04 14.5	+51 17 15	-0.002	-0.04	7.2	1.1	-0.1	K2 III	-27	270 s		
--	--		20 04 14.6	-55 43 16			6.50	0.5		F5 pe				RR Tel, v
190102	188873		20 04 18.0	-26 19 47	+0.003	-0.08	7.6	0.8	4.4	G0 V		32 s		
190056	211782		20 04 19.4	-32 03 22	+0.003	0.00	4.99	1.21	0.2	K0 III	-12	64 s		m
190299	144045	62 Aql	20 04 23.0	-0 42 33	0.000	-0.12	5.68	1.30	-0.3	gK4	0	160 s		
190571	69353		20 04 23.3	+33 33 01	+0.001	-0.03	7.49	0.84	5.6	G8 V		21 s		
190104	211784		20 04 24.4	-30 09 59	+0.005	-0.03	7.5	1.9	-0.1	K2 III		120 s		
190103	188875		20 04 25.5	-27 44 50	0.000	-0.03	7.4	1.3	-0.1	K2 III		250 s		
191632	9631		20 04 26.6	+73 54 35	+0.003	+0.03	6.8	1.1	0.2	K0 III		210 s		
190781	49152		20 04 28.6	+48 13 47	+0.001	+0.01	6.0	0.1	0.6	A0 V	-14	120 s		
189951	230024		20 04 29.5	-44 56 48	+0.001	0.00	7.84	0.29		A5				
190434	125415		20 04 34.1	+9 14 23	0.000	+0.02	7.9	0.0	0.4	B9.5 V		310 s	13330	m
190603	69362		20 04 36.0	+32 13 07	-0.001	-0.01	5.64	0.54	-6.7	B1.5 Ia	+21	1100 s	13335	m
190432	105643		20 04 36.6	+11 37 45	0.000	-0.05	7.7	1.1	-0.1	K2 III		350 s		
191009	32268		20 04 39.1	+57 36 37	-0.002	-0.01	7.9	1.1	0.2	K0 III		320 s		
190471	105644		20 04 40.6	+13 00 39	+0.009	+0.05	7.4	1.1	0.2	K0 III		140 mx		
191174	18692		20 04 44.4	+63 53 26	+0.001	+0.05	6.26	0.10		A3	-19		13371	m
190412	125417		20 04 46.4	+1 09 21	+0.001	-0.03	7.5	0.8	3.2	G5 IV		72 s		
190516	105650		20 04 46.7	+15 54 12	+0.002	+0.06	7.30	0.6	4.4	G0 V		38 s		m
190630	69368		20 04 48.5	+30 30 23	-0.001	-0.03	7.90	1.1		K1 IV				
191634	9634		20 04 53.0	+72 46 56	+0.003	+0.02	7.8	0.1	0.6	A0 V		270 s		
191372	18699		20 04 53.1	+68 01 38	-0.004	-0.01	6.28	1.65	-0.5	M2 III	-42	210 s		
190605	88162		20 04 54.9	+26 03 16	+0.003	-0.37	7.72	0.63	4.7	G2 V	+23	38 s		
190283	188889		20 04 57.8	-21 18 51	+0.003	-0.03	7.13	1.03	0.3	G6 III	-20	200 s		
190590	88163		20 04 58.4	+23 12 37	0.000	0.00	6.4	0.1	1.7	A3 V	-22	86 s		
190331	163240		20 05 02.0	-18 46 16	+0.002	+0.01	8.0	0.3	1.7	A3 V		130 s		

HD	SAO	Star Name	α 2000	δ 2000	μ(α)	μ(δ)	V	B-V	M$_v$	Spec	RV	d(pc)	ADS	Notes
			h m s	° ' ''	s	''								
190437	--		20 05 04.4	-4 19 09			8.00	0.5	3.4	F5 V		83 s	13334	m
190390	163245		20 05 05.2	-11 35 58	0.000	-0.01	6.50	0.5	3.4	F5 V	-12	42 s		m
190572	105652		20 05 05.4	+13 45 51	+0.005	-0.01	7.3	1.1	-0.1	K2 III		200 mx		
190592	105654		20 05 06.0	+17 52 43	+0.001	+0.01	8.0	0.1	1.4	A2 V		200 s		
190964	32272		20 05 06.5	+51 50 22	+0.003	+0.04	6.14	1.60	-0.5	M2 III	-56	210 s		
190057	230031		20 05 07.8	-46 05 49	0.000	-0.01	6.97	1.50		K0				
189721	254722		20 05 08.3	-63 03 39	0.000	-0.01	7.90	0.1	1.4	A2 V		200 s		m
190631	88169		20 05 08.5	+23 39 54	+0.001	-0.03	7.9	1.1	-0.1	K2 III		370 s		
190608	105659	16 η Sge	20 05 09.3	+19 59 28	+0.002	+0.08	5.10	1.06	-0.1	K2 III	-40	110 s		
190771	69377		20 05 09.6	+38 28 42	+0.022	+0.11	6.17	0.65	3.2	G5 IV	-24	39 s	13348	m
190498	125423		20 05 09.9	+2 07 07	0.000	-0.11	6.7	0.5	4.0	F8 V	-12	35 s		
190812	49159		20 05 10.1	+40 48 51	0.000	-0.01	7.1	0.2	2.1	A5 V		100 s		
190521	125426		20 05 10.7	+3 25 50	0.000	-0.01	7.8	1.1	0.2	K0 III		310 s		
190285	188892		20 05 11.8	-26 48 47	+0.001	0.00	7.1	0.7	0.6	A0 V		78 s		
190701	88172		20 05 12.1	+27 01 57	+0.001	+0.01	8.0	0.0		B8.5 V		370 s		
190438	144052		20 05 13.8	-8 06 23	-0.002	-0.05	8.0	0.1	1.7	A3 V		52 mx		
190202	211794		20 05 16.2	-36 22 46	+0.002	-0.05	7.5	0.4	1.7	A3 V		91 s		
190286	188894		20 05 17.6	-27 48 42	+0.003	0.00	7.6	1.2	0.2	K0 III		240 s		
190988	32274		20 05 18.7	+50 26 57	-0.001	+0.03	7.4	0.1	0.6	A0 V		210 s		
190305	188895		20 05 19.4	-29 04 47	+0.004	-0.12	7.3	0.8	3.4	F5 V		38 s		
190522	125429		20 05 20.1	+0 27 17	-0.001	0.00	6.90	0.1	0.6	A0 V		180 s	13340	m
191096	32276		20 05 21.2	+56 20 29	-0.001	+0.08	6.2	0.4	2.6	F0 V	-12	52 s		
190268	211796		20 05 25.6	-34 39 59	+0.001	-0.02	7.1	0.1	0.6	A0 V		170 s		
190454	163253	65 Sgr	20 05 26.2	-12 39 55	0.000	-0.05	6.55	0.04		A0				
190658	105663		20 05 26.3	+15 30 01	+0.002	+0.02	6.34	1.64		M2	-112		13344	m
227594	69386		20 05 28.2	+39 21 11	0.000	0.00	8.0	1.1	0.2	K0 III		340 s		
190306	211798		20 05 31.9	-33 00 00	+0.002	-0.02	6.90	0.00	0.0	B8.5 V	-18	240 s		m
191277	18700	66 Dra	20 05 32.6	+61 59 44	+0.017	+0.08	5.39	1.18	-0.2	K3 III	+6	130 s		
189567	254721		20 05 32.7	-67 19 16	+0.147	-0.68	6.07	0.64	4.7	G2 V	-14	20 ts		
227586	--		20 05 36.2	+35 37 22	0.000	-0.02	7.36	0.21	-5.0	B0 III		1700 s		
190916	49173		20 05 38.4	+41 16 47	0.000	-0.01	7.64	0.45	-6.5	B9 Iab		3700 s		
190661	105668		20 05 39.6	+12 09 10	+0.001	+0.01	7.8	1.1	-0.1	K2 III		370 s		
190864	69391		20 05 39.6	+35 36 30	-0.001	+0.02	7.76	0.19		O6	0		13361	m
191022	49176		20 05 40.2	+48 33 59	-0.009	-0.02	7.2	0.7	4.4	G0 V		37 s		
190289	211799		20 05 41.1	-37 42 52	+0.002	+0.02	8.0	0.3	2.6	F0 V		120 s		
190750	88178		20 05 44.1	+22 47 51	+0.001	0.00	6.9	0.1	0.6	A0 V		180 s		
190307	211802		20 05 45.8	-36 29 57	-0.001	-0.05	7.4	0.8	3.4	F5 V		39 s		
190787	88180		20 05 47.9	+28 07 55	0.000	+0.02	7.54	1.72	-0.5	M2 III		340 s		
190788	88179		20 05 50.1	+25 36 03	-0.001	0.00	7.93	2.30	-2.4	M2 II		490 s		m
190917	69401		20 05 55.3	+36 19 16	-0.001	-0.02	7.90	0.9	0.3	G8 III		310 s		
190662	125438		20 05 55.6	+4 46 30	-0.001	-0.02	6.8	0.1	0.6	A0 V		170 s		
190919	69399		20 05 56.0	+35 40 20	0.000	0.00	7.33	0.22	-5.7	B1 Ib	-16	2300 s		
--	--		20 05 57.1	+35 49 00			7.80	0.1	-5.7	B1 Ib		2900 s		
190918	69402		20 05 57.1	+35 47 18	0.000	-0.01	6.75	0.13	-6.2	O9.5 I	-22	2300 s	13374	m
--	69403		20 05 58.4	+35 47 50	-0.001	0.00	8.00						13374	m
190772	105682		20 05 58.5	+19 34 50	+0.003	-0.03	7.59	1.15	3.2	G5 IV		37 s		
227634	69405		20 06 01.3	+35 45 57	0.000	+0.01	7.90	0.24	-5.8	B0 Ib		2900 s	13376	m
190737	125443		20 06 01.8	+9 05 38	0.000	0.00	7.7	1.4	-0.3	K5 III		370 s		
188520	257764		20 06 03.5	-79 25 20	-0.003	+0.02	7.8	0.1	1.4	A2 V		190 s		
190457	211806		20 06 05.8	-30 48 23	+0.001	-0.03	8.0	1.7	-0.1	K2 III		210 s		
190525	188911		20 06 07.4	-21 00 03	-0.001	-0.03	8.0	1.2	0.2	K0 III		290 s		
--	--		20 06 09.0	+24 38 52	+0.003	0.00	7.4	1.1	0.0	K1 III		290 s		
190967	69410		20 06 09.9	+35 23 10	+0.002	0.00	8.04	0.41	-5.4	B1 Ib-II		2200 s		V448 Cyg, v
190309	230037		20 06 11.0	-44 20 33	+0.002	-0.08	7.87	1.11	0.0	K1 III		150 mx		
190664	144062		20 06 12.0	-4 04 42	+0.003	-0.04	6.47	1.16		K0				
191195	32286		20 06 13.7	+53 09 57	+0.024	+0.26	5.85	0.39	3.4	dF5	-41	31 s		m
190576	163261		20 06 13.7	-19 29 25	0.000	-0.01	7.6	0.0	0.6	A0 V		240 s		
190844	88191		20 06 14.5	+20 31 08	+0.003	+0.01	7.8	0.4	2.6	F0 V		110 s		
191026	69413	27 Cyg	20 06 21.6	+35 58 21	-0.019	-0.43	5.36	0.85	3.2	K0 IV	-34	31 ts		m
192419	9649		20 06 23.3	+77 32 17	+0.001	+0.02	8.0	1.6	-0.5	M2 III		510 s		
190205	246433		20 06 23.3	-52 31 17	0.000	+0.03	7.5	0.6	2.6	F0 V		62 s		
190866	105692		20 06 23.6	+16 15 20	0.000	+0.01	7.90	1.1	0.2	K0 III		340 s	13375	m
191045	69417		20 06 24.8	+39 05 55	+0.002	+0.03	6.85	1.53	-0.3	K5 III	+8	260 s		
191804	9642		20 06 27.7	+70 39 35	+0.003	-0.02	7.8	0.4	2.6	F0 V		110 s		
190528	211810		20 06 27.7	-31 34 10	-0.001	-0.14	8.0	0.5	3.2	G5 IV		68 mx		
191046	69416		20 06 28.6	+36 13 38	-0.007	-0.08	7.02	1.15	0.2	K0 III	-100	180 s		
190642	163264		20 06 29.1	-18 18 27	-0.001	+0.01	8.0	0.9	3.2	G5 IV		93 s		
190617	188917		20 06 32.4	-21 40 36	-0.014	-0.09	7.9	0.7	4.4	G0 V		43 s		
190868	105697		20 06 33.9	+12 40 53	+0.003	-0.01	7.60	1.4	-0.3	K5 III		370 s	13377	m
191047	69421		20 06 35.1	+35 07 53	-0.001	-0.03	7.89	1.00	-2.1	G5 II		940 s		
190739	144070		20 06 36.1	-4 25 02	0.000	-0.01	7.2	0.1	0.6	A0 V		210 s		
190850	125455		20 06 38.0	+7 33 29	-0.001	-0.01	7.59	0.17	1.7	A3 V		140 s		
190849	125456		20 06 38.5	+7 34 34	-0.001	-0.01	7.14	0.12	1.2	A1 V		140 s	13379	m
190887	105700		20 06 43.3	+12 56 11	+0.002	-0.01	7.50	0.4	3.0	F2 V	-4	79 s	13384	m
190723	163267		20 06 45.6	-12 55 39	0.000	-0.02	7.60	1.1	0.2	K0 III		300 s	13370	

HD	SAO	Star Name	α 2000	δ 2000	μ(α)	μ(δ)	V	B–V	M$_v$	Spec	RV	d(pc)	ADS	Notes
191564	18716		20^h 06^m 46.1^s	+63° 42' 26"	−0.001^s	0.00"	7.4	0.1	0.6	A0 V		210 s		
190618	188920		20 06 46.4	−26 26 05	−0.002	+0.01	7.6	1.9	−0.1	K2 III		130 s		
190122	254727		20 06 47.5	−60 21 39	+0.001	−0.01	7.7	1.4	0.2	K0 III		190 s		
190920	105703		20 06 49.7	+16 52 15	0.000	+0.03	7.6	1.1	0.2	K0 III		290 s		
190740	163268		20 06 50.1	−12 26 24	0.000	0.00	7.5	0.1	1.7	A3 V		150 s		
191082	69429		20 06 51.6	+32 36 10	0.000	0.00	8.03	−0.02	0.6	A0 V		310 s		
190993	88212	17 Vul	20 06 53.3	+23 36 52	+0.001	0.00	5.07	−0.18	−1.7	B3 V	−5	230 s		
190619	188922		20 06 54.5	−29 24 58	+0.003	−0.07	7.7	1.3	0.2	K0 III		180 mx		
190774	144074		20 06 55.7	−8 54 52	+0.001	−0.01	7.80	0.4	2.6	F0 V		110 s	13380	m
191139	69434		20 06 57.4	+36 23 48	0.000	−0.01	7.96	0.22		B0.5 III				
190313	246439		20 06 58.0	−54 11 17	+0.002	0.00	7.7	1.9	−0.3	K5 III		220 s		
191700	18722		20 07 00.0	+66 18 30	+0.001	−0.01	7.00	0.1	0.6	A0 V		180 s	13424	m
190969	105705		20 07 01.7	+13 19 27	0.000	0.00	7.4	1.4	−0.3	K5 III		330 s		
190795	144078		20 07 03.3	−8 10 54	−0.001	+0.01	7.9	0.4	2.6	F0 V		110 s		
190756	163272		20 07 03.5	−17 11 40	+0.003	+0.02	7.9	0.4	3.0	F2 V		94 s		
190558	—		20 07 04.8	−38 52 02			8.0	2.4	−0.3	K5 III		130 s		
191158	69438		20 07 06.0	+36 49 36	+0.001	0.00	6.9	0.2		A m				
191329	32296		20 07 11.3	+50 13 45	+0.001	−0.02	6.5	0.1	1.4	A2 V	+3	100 s		
189839	254725		20 07 13.6	−69 33 21	+0.002	−0.04	7.4	0.1	1.4	A2 V		140 s		
193532	3347		20 07 15.4	+82 12 58	+0.003	+0.01	7.4	0.5	3.4	F5 V		63 s		
190559	230046		20 07 16.2	−41 48 31	+0.003	−0.02	7.7	1.1	0.2	K0 III		310 s		
190994	105712		20 07 17.8	+11 03 14	−0.002	−0.04	7.0	0.4	2.6	F0 V		75 s		
190647	211821		20 07 19.8	−35 32 19	+0.001	−0.21	7.5	0.7	3.2	G5 IV		47 mx		
191048	105714		20 07 20.5	+16 04 34	−0.001	−0.01	7.5	0.0	0.6	B9.5 V	−22	260 s	13394	m
190421	246443	ξ Tel	20 07 23.0	−52 52 51	−0.001	+0.01	4.94	1.62	−0.5	M2 III	+36	120 s		
190727	188927		20 07 23.6	−29 43 14	0.000	+0.04	6.70	1.1	0.2	K0 III		200 s		m
191201	69446		20 07 23.6	+35 43 05	+0.001	−0.01	7.23	0.13	−5.0	B0 III	−5	1700 s	13405	m
191226	69449		20 07 26.4	+36 34 04	+0.001	0.00	7.31	1.80	−2.2	K2 II	−24	400 s		
191736	18723		20 07 26.6	+65 11 16	+0.003	+0.06	7.7	0.2	2.1	A5 V		130 s		
190599	230050		20 07 28.6	−41 46 16	−0.003	+0.03	7.6	0.7	4.4	G0 V		44 s		
191066	105716		20 07 28.8	+17 01 13	0.000	−0.01	7.7	0.9	3.2	G5 IV		80 s		
191257	—		20 07 30.6	+38 38 18			8.00	0.1	−4.8	A3 Ib		2300 s	13408	m
191084	105717		20 07 31.0	+16 38 51	−0.002	0.00	7.0	1.1	0.2	K0 III	−17	230 s		
191527	32301		20 07 32.7	+56 40 53	0.000	−0.01	8.0	1.1	−0.1	K2 III		380 s		
191331	49205		20 07 34.3	+45 09 12	+0.001	−0.01	7.5	0.8	3.2	G5 IV		71 s		
190422	246444		20 07 34.9	−55 00 59	+0.002	+0.03	6.26	0.53	4.0	dF8		28 s		
190707	211825		20 07 35.2	−35 39 19	+0.001	−0.01	7.6	0.3	1.7	A3 V		110 s		
191014	125474		20 07 36.9	+4 46 16	0.000	−0.03	7.9	1.1	−0.1	K2 III		370 s		
190743	211829		20 07 38.7	−33 20 14	+0.001	−0.10	7.6	1.5	0.2	K0 III		120 mx		
190542	230049		20 07 39.2	−48 39 10	+0.001	−0.10	7.4	1.0	3.0	F2 V		29 s		
190824	188932		20 07 40.2	−23 52 58	+0.001	−0.03	7.5	1.7	−0.3	K5 III		290 s		
191243	69457		20 07 41.2	+34 25 21	−0.001	−0.01	6.11	0.16	−5.7	B5 Ib		1600 s		
191805	18727		20 07 43.2	+65 10 05	0.000	+0.05	7.9	0.2	2.1	A5 V		140 s		
191029	125475		20 07 43.6	+2 26 36	+0.001	+0.01	7.1	0.2	2.1	A5 V		99 s		
191609	32303		20 07 45.3	+57 48 00	−0.001	−0.01	7.18	−0.02	0.6	A0 V		210 s		
191104	125478		20 07 50.2	+9 23 59	+0.003	+0.03	6.43	0.46	3.0	dF2	−27	44 s	13403	m
190730	211830		20 07 53.5	−39 50 34	+0.001	−0.08	7.5	1.2	3.4	F5 V		23 s		
189898	257774		20 07 55.8	−70 48 55	+0.001	0.00	8.00	1.1	0.2	K0 III		360 s		m
191292	69467		20 07 57.7	+32 35 12	0.000	+0.01	7.88	0.02		B9			13410	m
192763	9657		20 07 58.5	+77 49 28	+0.005	−0.02	7.9	0.5	3.4	F5 V		78 s		
191394	49217		20 08 01.0	+42 23 07	0.000	+0.02	7.00	1.1	0.2	K0 III	−20	220 s	13415	m
191067	144095	64 Aql	20 08 01.7	−0 40 42	+0.008	−0.07	5.99	1.02	0.0	gK1 n	−4	160 s		
190507	246451		20 08 02.1	−55 50 30	0.000	−0.04	7.6	1.3	0.2	K0 III		210 s		
191444	49223		20 08 02.8	+45 13 25	+0.001	+0.07	8.0	1.1	−0.1	K2 III		270 mx		
191178	105733		20 08 06.3	+16 39 51	0.000	0.00	6.6	1.6	−0.5	M2 III	+11	250 s		
191423	49221		20 08 06.9	+42 36 20	−0.001	−0.03	8.03	0.16	2.5	A9 V		130 s		
191030	144096		20 08 07.4	−6 45 40	−0.001	−0.01	6.9	0.0	0.0	B8.5 V		240 s		
190580	246453		20 08 09.3	−52 34 39	−0.003	−0.01	7.3	0.6	3.2	G5 IV		66 s		
191379	69476		20 08 10.8	+36 40 16	0.000	−0.01	7.54	1.74	−0.5	M2 III		330 s		
190979	163285		20 08 12.8	−18 48 14	+0.001	−0.01	7.0	0.3	0.0	B8.5 V		140 s		
191397	69481		20 08 14.9	+37 58 11	+0.003	+0.04	7.9	0.5	0.7	F5 III		260 s		
190581	246454		20 08 15.9	−53 33 01	+0.002	−0.05	7.1	1.8	−0.5	M2 III		260 s		
190855	211844		20 08 18.4	−34 37 38	0.000	+0.03	8.0	0.5	0.2	K0 III		360 s		
191312	88234		20 08 19.1	+27 12 21	+0.001	+0.02	8.0	0.0	0.4	B9.5 V		310 s		
191204	105735		20 08 19.5	+11 42 15	0.000	−0.01	8.0	0.1	0.6	A0 V		290 s		
190222	254731		20 08 20.3	−66 21 17	0.000	0.00	6.45	1.58		K5				
191205	105736		20 08 20.7	+11 34 53	−0.002	−0.04	7.7	1.1	0.2	K0 III		300 s		
191260	88230		20 08 20.8	+20 13 11	+0.003	+0.01	8.0	0.4	3.0	F2 V		97 s		
191737	32313		20 08 25.3	+58 47 16	0.000	+0.01	7.8	0.7	0.4	G3 III	−5	290 s		
191262	105740		20 08 26.7	+15 40 30	−0.006	−0.09	7.8	0.8	3.2	G5 IV		83 s		
191508	49229		20 08 27.6	+45 11 48	−0.001	+0.01	7.8	0.8	3.2	G5 IV		82 s		
189450	257770		20 08 28.9	−77 05 30	+0.012	−0.01	7.9	1.1	−0.1	K2 III		350 pc mx		
191049	163286		20 08 29.1	−15 01 55	+0.002	−0.13	6.7	1.1	0.2	K0 III		96 mx		
191146	144105		20 08 29.2	−3 39 41	+0.001	0.00	7.65	−0.03	1.4	A2 V		180 s	13409	m
190745	230060		20 03 29.7	−47 33 34	−0.002	+0.01	7.9	0.5	2.6	F0 V		110 s		

HD	SAO	Star Name	α 2000	δ 2000	μ(α)	μ(δ)	V	B−V	M_v	Spec	RV	d(pc)	ADS	Notes
			h m s	° ′ ″	s	″								
191087	163288		20 08 29.8	−10 18 45	+0.004	+0.04	7.2	0.8	3.2	G5 IV		63 s		
190905	211846		20 08 30.2	−34 30 48	−0.001	−0.02	7.7	0.8	3.2	G5 IV		74 s		
189723	257772		20 08 30.3	−74 45 26	−0.003	−0.04	7.25	1.07	0.2	gK0		230 s		
191110	163290		20 08 31.1	−10 03 46	0.000	−0.04	6.2	0.1	0.6	A0 V	−16	130 s		
191246	125492		20 08 32.9	+10 03 15	0.000	+0.02	7.4	1.1	0.2	K0 III		270 s		
195146	3362		20 08 34.9	+85 07 29	−0.063	0.00	7.2	0.5	4.0	F8 V		43 s		
191068	163289		20 08 35.5	−14 15 13	+0.002	+0.01	7.6	0.9	3.2	G5 IV		75 s		
191033	188948		20 08 35.8	−23 35 27	−0.003	−0.12	8.0	−0.3	4.0	F8 V		62 s		
191456	69491		20 08 36.0	+36 40 20	+0.001	0.00	7.44	0.06		B0.5 III				
191263	105743		20 08 38.1	+10 43 32	0.000	0.00	6.2	0.0	−1.0	B5.5 V	−38	270 s		
191357	88243		20 08 39.7	+24 45 29	−0.003	−0.05	7.9	0.9	3.2	G5 IV		87 s		
191358	88245		20 08 40.9	+24 39 04	+0.001	+0.02	8.0	0.9	3.2	G5 IV		91 s		
191359	88244		20 08 42.7	+22 09 12	−0.003	−0.06	7.1	1.1	0.2	K0 III		230 s		
191649	32310		20 08 42.8	+50 35 10	+0.007	+0.07	7.6	0.7	4.4	G0 V		42 s		
191381	88246		20 08 42.9	+26 16 30	−0.001	+0.01	7.8	0.0	−1.6	B3.5 V		660 s		
190248	254733	δ Pav	20 08 43.3	−66 10 56	+0.199	−1.14	3.56	0.76	4.8	G5 IV	−22	5.7 t		
191493	69496		20 08 46.8	+36 08 53	0.000	0.00	7.32	1.63	−2.3	K5 II		720 s		
191529	69502		20 08 48.4	+39 47 55	+0.002	+0.01	7.3	0.0		B8.5 V		270 s		
191295	105748		20 08 49.5	+12 13 41	0.000	−0.02	7.0	0.0	0.4	B9.5 V	+10	210 s		
191635	49242		20 08 50.3	+48 54 40	+0.002	+0.02	7.2	0.4	2.6	F0 V		83 s		
191231	144110		20 08 50.6	−0 09 15	0.000	+0.01	7.5	0.0	0.4	B9.5 V		260 s		
190907	230068		20 08 51.7	−40 41 34	+0.003	−0.03	8.0	1.3	0.2	K0 III		290 s		
192907	9665	1 κ Cep	20 08 53.1	+77 42 41	+0.003	+0.03	4.39	−0.05	−0.8	B9 III	−23	100 s	13524	m
−−	32314		20 08 54.4	+52 54 49	0.000	+0.01	8.0	1.6	−0.3	K7 III		390 s		
191940	18732		20 08 56.5	+62 04 36	−0.001	−0.01	7.20	0.1	0.6	A0 V		200 s	13449	m
191337	105750		20 09 00.6	+13 00 57	+0.001	−0.01	7.9	0.0	−1.0	B5.5 V	−10	580 s		
191496	69505		20 09 03.1	+33 11 13	+0.003	+0.02	7.48	0.23	−4.8	A9 Ib		2900 s		
190879	230067		20 09 03.7	−47 04 03	+0.004	−0.05	6.45	1.47	−0.3	K5 III		220 s		
189899	257777		20 09 04.4	−74 13 11	−0.006	−0.15	7.60	0.51	3.2	F8 IV-V		56 mx		
191089	188955		20 09 05.1	−26 13 26	+0.004	−0.06	7.7	0.4	4.0	F8 V		54 s		
191382	105756		20 09 07.9	+16 54 26	0.000	−0.05	6.9	1.1	−0.1	K2 III		250 s		
191530	69509		20 09 11.6	+33 55 13	0.000	+0.01	7.90	0.02	−3.1	B9 II		1600 s		
191019	211856		20 09 12.0	−38 18 13	+0.003	−0.01	7.3	1.1	−0.1	K2 III		300 s		
191636	49245		20 09 12.0	+42 13 49	0.000	+0.01	8.0	0.1	0.6	A0 V		280 s		
191446	105760		20 09 13.3	+19 04 23	+0.001	0.00	7.0	1.1	0.2	K0 III	−31	220 s		
191566	69512		20 09 14.2	+35 29 03	+0.001	+0.01	7.34	0.17		B0.5 IV	−37		13429	m
191113	188957		20 09 18.4	−28 26 24	+0.003	−0.03	7.1	1.1	3.2	G5 IV		40 s		
191317	144112		20 09 20.0	−2 36 04	−0.002	−0.02	8.0	0.4	2.6	F0 V		120 s		
192274	18741		20 09 20.2	+68 35 59	−0.004	−0.02	7.0	1.1	0.2	K0 III		210 s		
190936	230070		20 09 22.0	−47 10 15	+0.001	−0.03	7.5	1.8	−0.5	M4 III		320 s		
191892	32322		20 09 22.0	+56 01 36	−0.004	−0.01	6.5	1.1	0.2	K0 III		180 s		
191401	105761		20 09 22.6	+10 47 10	0.000	−0.01	7.5	0.1	0.6	A0 V		240 s		
191053	211860		20 09 22.8	−36 30 19	0.000	−0.16	8.0	0.9	3.2	G5 IV		61 mx		
191315	144115		20 09 23.1	−1 16 27	+0.001	−0.01	7.7	1.1	0.2	K0 III		310 s		
191091	211863		20 09 24.6	−33 52 24	+0.001	+0.03	7.2	0.5	0.4	B9.5 V		110 s		
191610	69518	28 Cyg	20 09 25.4	+36 50 23	0.000	+0.02	4.93	−0.13	−1.7	B3 V	−14	210 s		
191114	211864		20 09 25.7	−32 03 37	+0.003	+0.01	8.0	1.8	−0.3	K5 III		290 s		
191007	230071		20 09 26.5	−43 55 00	−0.002	−0.01	7.85	1.19		K0				
191806	32320		20 09 28.0	+52 16 35	+0.011	+0.10	8.0	1.1	0.2	K0 III		120 mx		
191612	69520		20 09 28.4	+35 44 02	−0.001	0.00	7.78	0.28		O8	−21			
191476	105764		20 09 31.3	+17 25 35	−0.002	−0.03	7.9	1.4	−0.3	K5 III		410 s		
190909	246465		20 09 32.7	−50 26 25	+0.001	−0.02	6.10	0.1	1.7	A3 V		76 s		m
191589	69522		20 09 32.9	+33 40 54	+0.001	+0.01	7.26	1.52	−0.3	K5 III		320 s		
191250	188965		20 09 33.3	−20 35 37	+0.002	−0.07	7.3	0.7	3.0	F2 V	−7	44 s		
191499	105765		20 09 34.1	+16 48 20	0.000	+0.17	7.56	0.82	5.7	dG9	−35	23 s	13434	m
191498	105768		20 09 34.2	+18 46 56	0.000	−0.01	8.0	0.1	1.4	A2 V		200 s		
192017	18737		20 09 35.5	+60 24 17	+0.009	+0.07	8.0	0.4	2.6	F0 V		110 s		
191613	69524		20 09 36.7	+33 24 39	−0.002	−0.01	7.81	0.01	1.4	A2 V		190 s	13441	m
188206	258851		20 09 37.3	−82 48 21	+0.048	−0.08	7.8	0.5	4.0	F8 V		57 s		
228038	49253		20 09 40.4	+40 25 42	+0.001	+0.01	7.7	1.1	0.2	K0 III		300 s		
191569	88269		20 09 40.5	+28 19 26	+0.001	−0.01	7.77	1.73	−0.5	M2 III		370 s		
191287	163301		20 09 41.7	−18 20 47	−0.001	−0.01	8.0	0.0	0.4	B9.5 V		330 s		
191590	88272		20 09 43.3	+29 43 36	+0.002	0.00	8.00	1.1	−0.1	K2 III		390 s		
191614	88273		20 09 47.6	+27 19 53	−0.001	−0.02	7.95	1.66	−0.3	K5 III		360 s		
190954	246466		20 09 47.8	−51 43 07	−0.014	−0.07	7.6	2.0	3.2	G5 IV		14 s		
192018	32329		20 09 48.5	+57 31 48	+0.001	+0.01	7.9	1.1	−0.1	K2 III		360 s		
191703	69530		20 09 50.4	+39 08 13	+0.002	−0.01	7.3	0.1	1.7	A3 V		130 s		
188750	−−		20 09 53.2	−81 26 47			8.0	1.6	−0.5	M4 III		500 s		
191168	211869		20 09 54.8	−35 36 21	+0.001	0.00	7.6	0.6	2.6	F0 V		66 s		
191570	88276		20 09 56.4	+20 54 54	+0.004	+0.10	6.48	0.38	3.0	dF2	−40	36 ts	13442	m
191615	88279		20 09 57.2	+25 32 12	0.000	+0.05	7.80	0.94	3.2	G8 IV		70 s		
191720	69535		20 10 01.4	+36 58 44	0.000	+0.02	7.79	0.03	0.2	B9 V	−15	300 s		
191671	88280		20 10 02.2	+28 16 09	0.000	+0.01	7.6	0.0	−1.0	B5.5 V	−1	480 s		
191117	230077		20 10 05.5	−43 53 37	0.000	0.00	6.94	1.02		K0				
191301	188977		20 10 06.6	−25 17 01	+0.003	+0.01	7.3	0.8	2.6	F0 V		44 s		

HD	SAO	Star Name	α 2000	δ 2000	μ(α)	μ(δ)	V	B-V	M_v	Spec	RV	d(pc)	ADS	Notes
			h m s	° ′ ″	s	″								
191533	125517		20 10 07.6	+8 26 38	-0.005	-0.16	6.60	0.5	4.0	F8 V	+17	33 s	13443	m
191216	211873		20 10 08.5	-36 39 18	-0.002	-0.02	8.0	1.4	-0.1	K2 III		310 s		
191872	49261		20 10 09.0	+45 01 02	-0.001	+0.03	7.79	-0.16	0.0	B8.5 V		360 s		
190733	254737		20 10 09.9	-63 37 22	+0.002	-0.03	7.0	0.1	1.4	A2 V		120 s		
191271	211874		20 10 11.9	-31 10 00	0.000	-0.07	7.8	0.8	2.6	F0 V		52 s		
191854	49262		20 10 13.1	+43 56 43	0.000	+0.08	7.45	0.69	5.0	dG4	-40	29 s	13461	m
192438	18746		20 10 15.0	+68 17 54	+0.006	+0.01	7.9	0.4	3.0	F2 V		90 s		
189487	257775		20 10 15.3	-79 06 38	-0.013	+0.04	7.40	0.9	3.2	G5 IV		69 s		m
192178	18742		20 10 15.8	+61 20 43	+0.002	+0.01	7.8	1.6	-0.5	M2 III		410 s		
191434	163313		20 10 25.3	-19 23 01	+0.002	-0.12	7.3	0.4	4.0	F8 V		46 s		
191548	125522		20 10 25.6	+5 35 03	+0.002	+0.04	7.9	0.4	3.0	F2 V		92 s		
191704	88291		20 10 27.9	+25 02 56	+0.001	0.00	7.30	0.1	0.6	A0 V		210 s		m
191943	49268		20 10 28.6	+44 36 59	-0.003	-0.05	7.5	1.1	0.2	K0 III		270 s		
191500	144133		20 10 29.5	-6 09 49	-0.002	-0.01	7.8	0.6	4.4	G0 V		48 s		
191655	105782		20 10 30.9	+14 33 40	0.000	-0.01	7.7	1.4	-0.3	K5 III		380 s		
191304	211876		20 10 31.4	-33 32 44	-0.004	+0.01	7.8	1.2	-0.1	K2 III		190 mx		
191746	88296		20 10 31.7	+28 26 04	+0.001	+0.01	7.19	0.02	-1.6	B3.5 V	-4	460 s		
191766	69547		20 10 32.0	+30 16 25	+0.001	-0.01	7.8	0.2	2.1	A5 V		130 s		
191745	88297		20 10 32.1	+29 21 50	0.000	-0.01	7.4	0.0	0.4	B9.5 V	-3	240 s		
192034	32334		20 10 32.6	+52 23 06	0.000	+0.02	7.4	1.6	-0.5	M2 III		360 s		
191747	88295	18 Vul	20 10 33.4	+26 54 15	+0.001	+0.02	5.51	0.08	0.0	A3 III	-12	130 s		
191998	49276		20 10 33.5	+48 13 49	0.000	-0.03	6.6	1.1	0.2	K0 III		180 s		
191812	69553		20 10 34.9	+33 38 03	-0.001	-0.01	7.69	0.08	0.2	B9 V		260 s	13463	m
192161	32338		20 10 37.6	+57 23 13	+0.007	+0.03	7.5	0.5	3.4	F5 V		65 s		
191811	69555		20 10 38.2	+33 51 21	+0.002	0.00	7.58	-0.04	-1.7	B3 V		590 s		
191190	230083		20 10 38.9	-46 44 07	+0.004	-0.05	6.81	1.17		K1 IV		270 mx		m
191944	49274		20 10 44.1	+41 14 48	+0.004	-0.01	6.9	0.1	1.4	A2 V		120 s		
191721	88299		20 10 45.0	+20 29 12	+0.001	-0.01	7.4	0.0	0.0	B8.5 V		290 s		
191706	105786		20 10 45.1	+15 12 04	+0.002	0.00	8.0	0.1	0.6	A0 V		290 s		
191897	--		20 10 48.5	+37 02 42			7.34	0.79	-5.4	F9 I-II var		2300 s	13473	m
192035	49282		20 10 49.2	+47 48 49	-0.002	+0.01	7.5	0.0	-1.6	B3.5 V		580 s		
191768	88302		20 10 49.9	+23 31 38	+0.001	+0.01	7.8	0.1	0.6	A0 V		260 s		
191273	230084		20 10 50.0	-43 36 10	+0.002	-0.01	7.84	0.26		A3				
191977	49279		20 10 53.2	+43 05 49	0.000	-0.02	7.79	0.00	0.0	B8.5 V		330 s		
191855	69562		20 10 54.0	+30 47 21	-0.002	-0.02	6.7	0.0	0.4	B9.5 V	-11	180 s		
191407	211883		20 10 54.1	-32 19 25	0.000	-0.03	7.6	0.2	0.6	A0 V		180 s		m
195147	3367		20 10 55.8	+84 32 40	+0.024	+0.04	7.2	0.4	2.6	F0 V		83 s		
192575	18750		20 10 56.7	+68 16 20	-0.002	0.00	6.84	0.16	-1.6	B3.5 V	-38	340 s		
191917	69564		20 10 57.0	+35 57 12	+0.002	0.00	7.78	0.14	-4.4	B1 III	+4	1700 s		
191978	49280		20 10 57.9	+41 21 11	-0.004	+0.01	8.02	0.14		O8				
192002	49283		20 11 02.9	+40 56 42	-0.004	0.00	8.0	1.4	-0.3	K5 III		250 mx		
191814	88309		20 11 03.3	+21 08 05	+0.001	+0.04	6.22	0.90	0.2	K0 III	-7	160 s		
191657	144143		20 11 04.0	-3 34 26	0.000	+0.03	7.8	1.1	0.2	K0 III		330 s		
191875	88313		20 11 04.2	+29 42 22	+0.001	-0.01	8.00	1.1	-0.2	K3 III		400 s		
191305	230087		20 11 04.6	-43 28 29	+0.001	0.00	7.70	0.23		F0				
191918	69566		20 11 04.9	+32 17 13	0.000	+0.01	7.1	0.1	0.6	A0 V	-18	190 s		
191619	144142		20 11 05.7	-6 05 16	0.000	0.00	6.9	0.1	0.6	A0 V		180 s		
191785	105793		20 11 05.9	+16 11 17	-0.029	+0.40	7.33	0.84	6.1	K1 V	-51	19 ts		m
191095	246470		20 11 07.0	-57 31 26	0.000	-0.02	6.37	0.06		A0				m
191707	125533		20 11 09.5	+6 20 55	+0.001	+0.01	7.7	1.6	-0.5	M4 III	+15	410 s		
191639	144144		20 11 09.9	-8 50 33	0.000	0.00	6.49	-0.15	-3.5	B1 V	-7	950 s		
191536	188998		20 11 11.0	-20 15 55	+0.001	-0.02	7.7	1.2	0.2	K0 III		250 s		
191408	211885		20 11 11.8	-36 06 04	+0.037	-1.57	5.32	0.87	6.6	K3 V	-131	5.6 t		
191837	105798		20 11 12.7	+19 21 55	-0.001	-0.02	7.5	0.5	4.0	F8 V		49 s		
191349	230089		20 11 15.5	-43 39 43	0.000	+0.01	6.55	0.88		G5				
191692	144150	65 θ Aql	20 11 18.1	-0 49 17	+0.002	+0.01	3.23	-0.07	-0.8	B9 III	-27	61 s		m
191709	144151		20 11 18.7	-0 07 35	+0.002	-0.02	7.10	0.4	2.6	F0 V	-19	79 s		m
191769	125536		20 11 20.4	+8 43 03	0.000	0.00	7.1	1.6	-0.5	M2 III		320 s		
191877	88315		20 11 20.9	+21 52 32	0.000	+0.01	6.26	-0.02	-5.7	B1 Ib	-18	2000 s		
190983	254741		20 11 23.1	-63 56 24	+0.017	+0.06	8.0	0.5	3.4	F5 V		82 s		
192020	69576		20 11 24.9	+38 24 01	+0.003	+0.12	7.93	0.88	5.6	G8 V	-6	24 s		
192041	69578		20 11 25.7	+38 49 03	+0.002	-0.01	8.00	1.1	-2.2	K2 II		920 s		
191945	88325		20 11 25.7	+29 10 01	0.000	-0.01	7.30	1.6	-0.4	M0 III		320 s	13487	m
191770	125541		20 11 26.5	+6 04 19	-0.002	-0.02	7.7	1.4	-0.3	K5 III		360 s		
191521	188999		20 11 26.6	-28 41 41	0.000	-0.03	7.3	1.8	0.2	K0 III		91 s		
192021	69577		20 11 32.7	+33 58 34	-0.002	+0.04	7.76	0.45	3.7	F6 V		65 s		
192455	18751	68 Dra	20 11 34.7	+62 04 43	+0.017	+0.08	5.75	0.47		F5	-15	39 t		m
191711	144152		20 11 36.6	-7 17 56	-0.001	-0.04	8.0	1.1	-0.1	K2 III		410 s		
191880	105804		20 11 36.8	+14 26 14	0.000	-0.02	8.0	0.1	0.6	A0 V		290 s		
192078	69581		20 11 37.4	+38 52 47	-0.001	0.00	7.70	0.9	-2.1	G5 II		790 s	13495	m
191879	105806		20 11 38.0	+14 38 56	+0.002	-0.01	7.5	0.1	0.6	A0 V	+13	240 s		
192512	18752		20 11 39.4	+63 31 38	+0.004	-0.01	7.9	0.0	0.4	B9.5 V		230 mx		
--	32345		20 11 45.8	+51 40 57	0.000	-0.01	7.9	1.5	-0.1	K2 III		250 s		
192102	69583		20 11 46.3	+38 40 57	+0.001	-0.01	8.0	0.0	-0.6	B8 IV		470 s		
192143	49301		20 11 47.2	+40 19 47	+0.001	0.00	7.05	-0.08	0.4	B9.5 V		210 s		

507

HD	SAO	Star Name	α 2000	δ 2000	μ(α)	μ(δ)	V	B−V	M$_v$	Spec	RV	d(pc)	ADS	Notes
191325	246477		20^{h}11^{m}47.8^s	−52°14'48"	0^s.000	−0".02	7.8	2.1	−0.3	K5 III		180 s		
192004	88330	19 Vul	20 11 47.8	+26 48 32	0.000	−0.01	5.51	1.39	−1.3	K3 II−III	−23	210 s		
191921	105813		20 11 49.4	+15 52 29	−0.001	−0.03	7.1	1.4	−0.3	K5 III		290 s		
191056	254746		20 11 50.2	−63 37 02	−0.001	−0.01	7.5	0.1	0.6	A0 V		240 s		m
192022	88331		20 11 50.3	+26 53 47	−0.002	+0.01	7.1	0.0	0.0	B8.5 V	−11	250 s		
192043	88335		20 11 56.9	+26 44 37	−0.001	−0.01	7.6	0.0	−1.2	B8 III		510 s		
191753	163328	1 Cap	20 11 57.7	−12 23 33	−0.001	−0.02	6.4	1.1	0.2	K0 III	+1	170 s		
192490	32350		20 11 58.8	+59 41 29	+0.005	+0.08	6.9	0.5	3.4	F5 V		51 s		
192044	88339	20 Vul	20 12 00.5	+26 28 44	0.000	−0.01	5.9	0.0	−0.6	B7 V e	−22	190 s		
192124	69588		20 12 01.7	+34 28 42	+0.002	−0.04	7.28	0.22	0.3	A5 III		230 s	13508	m
191924	125559		20 12 01.8	+7 41 05	0.000	+0.01	7.4	1.6	−0.5	M4 III		360 s		
192513	32353		20 12 03.2	+60 08 55	−0.001	0.00	7.86	0.01	0.6	A0 V		280 s		
191965	105820		20 12 03.2	+10 54 59	0.000	0.00	8.0	1.1	0.2	K0 III		340 s		
192276	49314		20 12 03.6	+47 44 14	+0.001	+0.01	6.6	−0.1		B8				
192491	32351		20 12 03.9	+59 37 36	+0.001	+0.04	7.8	0.9	3.2	G5 IV		80 s		
191822	163330		20 12 05.3	−10 42 57	−0.003	0.00	7.5	0.1	0.6	A0 V		250 s		
192163	69592		20 12 06.3	+38 21 17	−0.002	−0.01	7.48	−0.01		WN6	0			
191925	125560		20 12 11.0	+2 18 27	0.000	−0.01	8.0	1.1	−0.1	K2 III		390 s		
192182	69597		20 12 13.0	+38 26 36	0.000	+0.02	7.21	1.02	0.3	G8 III		210 s	13515	m
191775	163331		20 12 14.8	−18 08 44	−0.001	−0.05	7.8	1.1	0.2	K0 III		330 s		
191841	144164		20 12 15.0	−8 05 30	−0.001	−0.01	7.50	0.5	4.0	F8 V		50 s	13493	m
192479	32352		20 12 15.1	+56 48 13	−0.001	+0.03	7.7	1.1	−0.1	K2 III		330 s		
193962	3356		20 12 15.9	+80 28 05	+0.013	+0.08	7.5	0.4	2.6	F0 V		98 s		
192492	32356		20 12 16.8	+57 23 15	0.000	+0.01	7.8	0.1	0.6	A0 V		260 s		
192164	69595		20 12 18.1	+34 51 10	+0.001	−0.01	7.81	1.50	6.3	K2 V		20 s		d?, m
192045	105828		20 12 18.3	+16 13 02	+0.001	+0.01	7.4	0.9	3.2	G5 IV		70 s		
192165	69594		20 12 18.5	+32 04 52	+0.002	−0.01	8.00	0.1	1.4	A2 V		200 s	13513	m
228297	69600		20 12 22.9	+39 56 52	+0.001	+0.01	7.9	1.1	0.2	K0 III		320 s		
191662	211898		20 12 23.4	−35 46 59	0.000	−0.01	7.8	2.1	−0.5	M4 III		230 s		
191584	230100		20 12 23.7	−42 46 50	−0.001	−0.11	6.22	1.23	−0.1	K2 III		99 mx		
191862	163337	2 ξ Cap	20 12 25.7	−12 37 03	+0.013	−0.19	5.90	0.5	4.0	F8 V	+23	26 ts		m
192439	32354		20 12 31.6	+51 27 49	−0.002	−0.01	6.01	1.14	−0.1	K2 III	+13	170 s	13535	m
192281	49319		20 12 33.0	+40 16 06	0.000	+0.01	7.55	0.38		O5.8	−60			
192144	88353		20 12 33.1	+25 39 13	+0.003	+0.12	7.60	0.4	3.0	F2 V		82 s	13518	m
191371	246481		20 12 34.3	−58 06 59	+0.010	−0.05	7.8	0.3	3.2	G5 IV		84 s		
191984	125567		20 12 35.0	+0 52 03	+0.001	−0.01	6.27	0.02		A0	−19		13506	m
192800	18764		20 12 35.1	+66 27 40	−0.004	−0.07	7.9	1.1	0.2	K0 III	−50	300 mx		
191949	144170		20 12 38.5	−6 21 58	0.000	−0.01	6.9	0.1	0.6	A0 V		180 s		
192225	69604		20 12 39.7	+33 29 45	+0.001	+0.01	8.02	−0.04	−1.2	B8 III		640 s		
192260	69608		20 12 43.3	+35 10 40	−0.001	+0.02	7.51	1.09	3.2	K0 IV	−26	55 s		
191953	163349		20 12 44.7	−10 50 24	0.000	0.00	7.6	1.1	0.2	K0 III		300 s		
192007	144173		20 12 47.1	−2 59 45	0.000	0.00	7.00	0.1	0.6	A0 V		190 s	13511	m
192145	105841		20 12 50.7	+16 05 43	−0.003	−0.01	7.6	0.5	4.0	F8 V	+5	53 s		
192167	105842		20 12 53.7	+19 32 38	−0.001	−0.01	7.5	1.1	−0.1	K2 III	+12	320 s		
191986	144174		20 12 54.3	−7 27 29	0.000	0.00	8.0	0.9	3.2	G5 IV		93 s		
192381	49325		20 12 56.5	+40 43 16	−0.003	−0.02	7.95	0.03	0.6	A0 V		270 s		
192261	88360		20 12 58.4	+27 17 05	+0.002	+0.01	7.9	0.0	0.4	B9.5 V		300 s		
192085	125577		20 12 59.9	+3 29 19	−0.001	0.00	7.8	0.2	2.1	A5 V		140 s		
191826	211906		20 13 00.0	−32 44 43	−0.002	−0.08	8.0	1.6	3.2	G5 IV		29 s		
192403	49326		20 13 00.1	+42 50 07	+0.003	+0.02	7.9	0.1	1.4	A2 V		190 s		
192320	69616		20 13 01.8	+34 49 47	+0.001	−0.01	8.03	1.55	−0.1	K2 III		250 s		
192286	69615		20 13 02.4	+30 28 38	+0.002	−0.04	7.90	0.9	0.3	G8 III		310 s		
192168	105843		20 13 03.1	+11 51 02	−0.002	−0.07	6.6	1.1	−0.1	K2 III		200 mx		
191826	144177		20 13 04.1	−0 19 51	−0.002	0.00	6.6	1.4	−0.3	K5 III		240 s		
192494	49329		20 13 07.2	+49 11 18	+0.002	+0.01	7.86	0.02		A0	−23		13545	m
192107	144181	66 Aql	20 13 13.7	−1 00 33	+0.001	−0.03	5.47	1.43	−0.3	K5 III	−28	140 s		
192287	88368		20 13 17.4	+25 14 29	0.000	0.00	7.48	1.64		M III				
192514	49332	30 Cyg	20 13 17.8	+46 48 57	+0.001	0.00	4.83	0.09	0.0	A3 III	−21	92 s		
193138	9683		20 13 18.4	+71 06 31	+0.005	+0.03	8.0	0.4	2.6	F0 V		110 s		
194006	9699		20 13 19.0	+79 39 07	−0.008	−0.01	7.0	1.4	−0.3	K5 III		290 s		
192383	69628		20 13 20.2	+35 36 01	0.000	+0.02	7.12	1.11	0.3	G5 III		170 s		
193265	9684		20 13 20.4	+72 53 50	−0.007	+0.01	7.1	0.4	2.6	F0 V	−2	81 s		
192422	69635		20 13 22.1	+38 45 55	−0.001	0.00	7.09	0.50		B0.5 Ib	−2			
192170	125587		20 13 22.6	+5 18 34	0.000	−0.01	7.7	0.1	1.4	A2 V		180 s		
192443	69636		20 13 23.4	+38 43 44	−0.001	0.00	8.03	3.31		N pe	−50			RS Cyg, m,v
192696	32378	33 Cyg	20 13 23.7	+56 34 04	+0.007	+0.08	4.30	0.11	1.3	A3 IV−V	−26	39 s		
192496	49331		20 13 23.8	+43 48 38	+0.002	−0.03	7.76	0.02	0.6	A0 V		260 s		
192404	69633		20 13 24.9	+34 54 47	+0.002	−0.02	7.70	1.22	−2.1	K0 II		850 s		
192621	32374		20 13 25.2	+51 29 42	+0.001	+0.02	7.8	1.4	−0.3	K5 III		380 s		
192246	105850		20 13 25.6	+14 24 36	+0.001	−0.02	7.4	1.1	0.2	K0 III		270 s		
192677	32377		20 13 27.1	+55 07 53	−0.002	−0.01	6.7	1.1	0.2	K0 III		190 s		
192781	18766		20 13 27.5	+60 38 26	+0.005	+0.06	5.79	1.47	−0.1	K2 III	−1	100 s		
189310	——		20 13 31.3	−82 02 25			8.0	0.8	3.2	G5 IV		90 s		
191732	230106		20 13 32.0	−47 42 49	−0.001	−0.05	6.5	1.4	0.2	K0 III		100 s		
192445	69639		20 13 32.6	+36 19 43	−0.001	+0.01	7.23	−0.08		B0.5 III	+4			

HD	SAO	Star Name	α 2000	δ 2000	μ(α)	μ(δ)	V	B-V	M_v	Spec	RV	d(pc)	ADS	Notes
192247	125589		20 13 33.0	+9 44 03	-0.001	0.00	7.9	1.1	0.2	K0 III		330 s		
191889	211913		20 13 33.6	-38 26 53	+0.001	-0.17	6.9	1.0	3.4	F5 V		24 s		
191796	230108		20 13 34.5	-45 35 25	+0.001	0.00	7.80	-0.02		A0				
192289	105856		20 13 34.5	+16 15 54	0.000	0.00	6.9	1.4	-0.3	K5 III		270 s		
192678	32379		20 13 36.1	+53 39 32	-0.005	-0.05	7.38	-0.03		A4 p				
192577	49337	31 o¹ Cyg	20 13 37.7	+46 44 29	0.000	+0.01	3.79	1.28	-2.2	K2 II	-7	160 s	13554	V695 Cyg, m,v
192579	49338		20 13 38.9	+46 42 42	-0.001	0.00	6.99	-0.15		B9				
192342	88377		20 13 40.3	+24 14 20	+0.004	+0.03	6.50	0.26		A m	-37		13543	m
---	---		20 13 40.4	+46 44 09			5.00						13554	m
192679	32380		20 13 40.6	+53 07 31	+0.005	+0.18	7.10	0.5	3.4	F5 V	-33	54 s	13560	m
191957	211916		20 13 41.4	-34 07 07	+0.002	-0.02	7.00	0.6	4.4	G0 V		33 s		m
192323	105860		20 13 42.1	+18 10 26	-0.001	-0.02	7.9	1.1	-0.1	K2 III		380 s		
191507	254753		20 13 42.3	-61 32 59	+0.001	0.00	7.2	-0.1	0.4	B9.5 V		230 s		
192535	49336		20 13 42.5	+43 22 44	-0.001	+0.01	6.14	1.50	-0.3	K4 III	-24	170 s	13555	m
192680	32381		20 13 47.2	+51 14 24	+0.002	+0.06	7.4	0.1	1.4	A2 V		120 mx		
191849	230110		20 13 53.2	-45 09 48	+0.073	-0.13	7.97	1.46	9.0	M0 V	-30	6.1 t		
192536	69649		20 13 55.3	+39 09 31	+0.004	+0.01	7.11	0.19		A m				
192152	163364		20 13 58.8	-16 40 08	+0.002	-0.04	7.6	0.5	3.4	F5 V		71 s		
192153	163363		20 13 59.4	-17 40 31	+0.006	+0.03	8.0	0.9	3.2	G5 IV		93 s		
192207	144188		20 14 00.3	-9 37 05	+0.002	-0.01	7.7	0.1	0.6	A0 V		260 s		
192386	105873		20 14 03.4	+18 12 58	-0.001	-0.02	7.9	1.4	-0.3	K5 III		410 s		
192424	88384		20 14 04.3	+22 13 21	0.000	+0.01	7.90	0.1	0.6	A0 V		280 s	13553	m
192538	69653		20 14 04.8	+36 36 19	+0.002	0.00	6.45	0.00	-0.6	A0 III	-20	240 s		
192406	105874		20 14 06.0	+17 47 49	0.000	-0.04	7.6	1.1	-0.1	K2 III		330 s		
191560	254755		20 14 06.5	-62 29 23	-0.002	+0.03	7.1	1.1	0.2	K0 III		240 s		
192343	125595		20 14 09.0	+6 34 37	-0.009	-0.06	8.00	0.68	5.0	G4 V	-3	37 s		
192344	125597		20 14 09.6	+6 35 20	-0.009	-0.05	7.71	0.70	3.1	G4 IV	-1	84 s		m
193030	18775		20 14 10.4	+64 45 55	-0.006	+0.04	7.3	0.9	3.2	G5 IV	-67	63 s		
192345	125599		20 14 13.6	+3 48 23	+0.003	-0.06	7.9	1.1	0.2	K0 III		180 mx		
192557	69661		20 14 13.8	+35 21 41	+0.004	-0.09	7.59	0.52	3.7	F6 V		56 s	13562	m
192518	88391	21 Vul	20 14 14.4	+28 41 41	+0.001	-0.02	5.18	0.18		A3	+5	27 mn		
192539	69659		20 14 15.8	+31 59 53	-0.001	+0.01	7.29	0.12	-3.6	B2 III	-23	990 s		
192425	105878	67 ρ Aql	20 14 16.4	+15 11 51	+0.004	+0.06	4.95	0.08	1.4	A2 V	-23	51 s		
192070	211926		20 14 17.0	-33 47 11	+0.001	-0.02	7.6	2.0	0.2	K0 III		76 s		
192558	69664		20 14 17.4	+34 59 25	+0.001	0.00	7.90	1.17	4.7	G2 V		20 s		
192190	189055		20 14 18.5	-22 02 20	+0.003	0.00	8.0	0.7	2.6	F0 V		66 s		
192517	69660		20 14 18.5	+30 10 38	+0.001	-0.01	7.08	-0.14	-1.6	B3.5 V	-11	530 s		
191829	246495		20 14 18.9	-52 26 44	+0.003	-0.04	5.65	1.50		K5		16 mn		
192659	49345		20 14 21.4	+42 06 13	0.000	-0.02	6.71	-0.04		B8			13572	m
192290	144199		20 14 24.1	-6 02 50	+0.002	+0.01	7.30	0.4	2.6	F0 V		87 s	13552	m
192660	49346		20 14 26.0	+40 19 46	+0.001	0.00	7.54	0.63	-6.2	B0 Ia		1700 s		
194375	3364		20 14 26.1	+80 31 55	+0.006	-0.02	6.80	0.1	0.6	A0 V		170 s	13708	m
191603	254756		20 14 26.8	-63 24 57	-0.005	+0.03	6.09	0.31		F0				
192710	49351		20 14 27.5	+44 51 48	+0.003	-0.02	7.9	1.1	0.2	K0 III		320 s		
192232	163370		20 14 27.5	-18 05 37	-0.002	+0.04	7.8	0.8	3.2	G5 IV		84 s		
192090	211929		20 14 28.2	-35 50 11	0.000	-0.01	7.9	2.0	-0.5	M2 III		290 s		
192639	69676		20 14 30.2	+37 21 15	-0.001	+0.01	7.11	0.35		O8.8	-2			
192519	88394		20 14 30.4	+24 50 41	+0.001	-0.01	8.00	0.00	0.4	B9.5 V		320 s	13566	m
192641	69677		20 14 31.5	+36 39 38	-0.001	-0.03	7.91	0.28		WC7	0	26 mn		m
192640	69678	29 Cyg	20 14 31.9	+36 48 23	+0.005	+0.07	4.97	0.14		A2 p	-17	33 t		m
191585	254757		20 14 34.7	-64 25 48	+0.001	-0.03	7.00	0.1	1.4	A2 V		130 s		m
192661	69681		20 14 39.5	+36 45 08	+0.001	0.00	6.60	0.9	0.3	G8 III		180 s		m
192952	32389		20 14 39.7	+58 56 22	0.000	+0.02	7.3	1.1	-0.1	K2 III		280 s		
194422	3365		20 14 41.0	+80 34 33	+0.006	+0.03	7.8	0.2	2.1	A5 V		140 s		
192783	49356		20 14 42.2	+47 42 37	-0.001	+0.05	7.7	1.1	0.2	K0 III		300 s		
192269	163373		20 14 42.8	-19 12 25	+0.002	+0.01	8.0	1.0	-0.3	K5 III		450 s		
---	---		20 14 44.2	-46 57 31			7.60							R Tel, v
192802	49358		20 14 45.6	+47 26 26	+0.002	+0.01	7.24	-0.09	0.4	B9.5 V		230 s		
192561	88398		20 14 46.8	+21 58 32	+0.003	+0.01	7.1	1.4	-0.3	K5 III	-23	300 s		
192348	163378		20 14 48.3	-10 53 25	+0.001	-0.01	7.5	1.1	0.2	K0 III		290 s		
192349	163379		20 14 51.3	-11 38 12	+0.001	-0.01	7.9	0.1	0.6	A0 V		280 s		
192501	105887		20 14 51.6	+10 28 22	+0.001	+0.02	7.0	0.1	0.6	A0 V		180 s		
192460	125608		20 14 53.2	+3 24 24	0.000	0.00	7.7	1.1	0.2	K0 III		300 s		
192684	69685		20 14 53.7	+32 52 08	-0.001	+0.03	6.7	0.0	0.4	B9.5 V	-24	180 s		
192502	125610		20 14 54.9	+9 05 20	-0.001	-0.02	7.5	1.6		M5 III	-46			R Del, v
193533	9696		20 14 56.0	+72 36 21	+0.003	+0.02	7.30	1.6	-0.5	M3 III	-25	310 s		
191869	246499		20 14 56.3	-56 58 31	+0.004	-0.11	8.00	0.5	3.4	F5 V		83 s		m
192786	49363		20 15 00.9	+42 53 34	0.000	-0.01	7.95	-0.02	0.4	B9.5 V		310 s		
192541	105890		20 15 02.0	+13 05 16	+0.004	+0.03	7.8	0.1	0.6	A0 V		100 mx		
192744	69696		20 15 04.5	+37 41 46	+0.003	+0.04	7.46	0.38	2.6	F0 V		89 s		
192173	211933		20 15 05.4	-39 43 12	0.000	-0.05	7.9	0.8	3.0	F2 V		55 s		
192504	125611		20 15 06.4	+2 17 56	0.000	-0.03	7.90	1.1	0.2	K0 III		350 s		m
192389	163383		20 15 06.8	-13 23 01	-0.001	-0.01	7.1	1.1	0.2	K0 III		240 s		
192803	49365		20 15 08.1	+41 17 56	+0.001	0.00	7.87	0.03	0.6	A0 V		280 s	13595	m
192461	144212		20 15 10.4	-3 30 13	-0.003	-0.06	6.92	0.37		F0			13574	m

HD	SAO	Star Name	α 2000	δ 2000	μ(α)	μ(δ)	V	B-V	M$_V$	Spec	RV	d(pc)	ADS	Notes
			^{h} ^{m} ^{s}	$^\circ$ ' "	^{s}	"								
192982	32393		20 15 10.7	+55 39 26	-0.003	+0.01	6.6	1.1	0.2	K0 III		180 s		
192766	69699		20 15 12.3	+38 36 31	0.000	0.00	7.9	0.0	-0.3	B9 IV		410 s		
192410	163384		20 15 13.3	-16 51 04	+0.002	-0.02	7.80	1.4	-0.3	K5 III		420 s		
192967	32392		20 15 13.4	+54 08 45	+0.002	-0.06	7.50	1.6	-0.5	M2 III		350 mx	13610	m
228583	69702		20 15 14.4	+40 05 50	+0.007	+0.05	8.0	0.7	4.4	G0 V		52 s		
192685	88410		20 15 15.7	+25 35 31	0.000	0.00	4.78	-0.18	-1.7	B3 V	-2	200 s	13589	m
192866	49373		20 15 16.9	+45 44 28	-0.001	0.00	7.13	-0.10	1.7	A3 V		120 s		
192310	189065		20 15 17.2	-27 01 58	+0.093	-0.18	5.73	0.88	6.1	K0 V	-55	8.3 t		
192819	49370		20 15 19.5	+40 57 37	0.000	0.00	7.7	1.4	-0.3	K5 III		370 s		
192732	88414		20 15 20.3	+30 01 13	-0.001	0.00	8.00	1.1	0.2	K0 III		340 s		
192663	105900		20 15 21.7	+18 27 58	-0.001	-0.06	7.7	0.5	3.4	F5 V		71 s		
191716	--		20 15 22.6	-64 50 14			7.3	1.1	0.2	K0 III		260 s		
192483	144219		20 15 23.2	-5 32 12	+0.002	0.00	7.0	0.1	1.4	A2 V		130 s		
192868	49377		20 15 23.5	+44 00 03	-0.001	0.00	7.1	0.4	3.0	F2 V		66 s		
192787	69701		20 15 23.5	+33 43 46	-0.004	-0.10	5.66	0.91	0.3	gG5	-10	110 s	13596	m
192609	125614		20 15 24.1	+10 09 13	+0.006	+0.06	7.3	0.5	4.0	F8 V	-37	45 s		
192867	49379		20 15 26.7	+44 08 20	-0.001	-0.02	7.24	1.61	-0.5	M1 III	-9	340 s		
192645	105901		20 15 28.0	+13 38 53	+0.001	0.00	8.0	0.1	0.6	A0 V		290 s		
192909	49385	32 o² Cyg	20 15 28.1	+47 42 51	0.000	+0.01	3.98	1.52	-3.4	K3 Ib-II	-14	280 s		v
192505	144221		20 15 28.2	-7 31 52	0.000	+0.01	6.8	0.1	1.4	A2 V		120 s		
192644	--		20 15 28.6	+13 37 11			7.9	0.7	4.4	G0 V		51 s		
192712	88417		20 15 29.9	+23 44 30	+0.001	-0.02	7.15	1.00		G5				
192713	88416	22 Vul	20 15 30.0	+23 30 31	0.000	-0.01	5.15	1.04	-4.5	G2 Ib	-23	740 s		
192586	125616		20 15 32.4	+2 50 48	+0.004	0.00	7.60	0.5	3.4	F5 V		69 s	13586	m
193053	32402		20 15 34.1	+55 37 10	+0.001	+0.04	7.7	1.1	0.2	K0 III		300 s		
192869	49383		20 15 36.3	+42 21 41	0.000	-0.03	7.83	0.54	2.2	F6 IV	-28	120 s		
192622	125620		20 15 37.0	+7 48 34	+0.002	+0.03	7.20	0.1	1.4	A2 V		140 s	13590	m
193341	18785		20 15 41.2	+65 51 12	-0.001	-0.02	7.7	0.5	3.4	F5 V		70 s		
192804	69709		20 15 42.1	+31 14 33	-0.008	-0.03	7.8	0.5	4.0	F8 V	-25	56 s		
192589	144225		20 15 42.9	-2 33 51	0.000	+0.01	8.0	0.1	0.6	A0 V		300 s		
194191	9712		20 15 43.0	+78 01 24	-0.004	0.00	7.9	1.1	0.2	K0 III		340 s		
192983	32400		20 15 43.2	+50 13 58	+0.002	+0.01	6.3	0.1	0.6	A0 V	-28	140 s		
192715	105905		20 15 45.5	+15 19 03	+0.004	+0.01	6.85	0.26	2.6	F0 V	-32	71 s		
192806	88428	23 Vul	20 15 45.9	+27 48 51	-0.003	+0.01	4.52	1.26	-0.2	K3 III	+3	88 s		
193054	32404		20 15 49.1	+52 30 08	-0.001	-0.01	7.30	1.4	-0.3	K5 III	-56	310 s		
192433	211940		20 15 50.5	-30 00 20	+0.004	+0.01	6.3	1.9	-0.1	K2 III		69 s		
192354	211939		20 15 57.0	-39 29 26	+0.011	-0.22	7.8	1.1	0.2	K0 III		59 mx		
192985	49395		20 16 00.4	+45 34 46	0.000	-0.05	5.91	0.38		F5	-40			
192544	163399		20 16 02.7	-16 17 45	+0.002	-0.05	7.3	0.1	1.4	A2 V		150 s		
192934	69720		20 16 03.3	+38 53 52	0.000	-0.04	6.27	0.01	0.4	B9.5 V p	+4	150 s		
192699	125628		20 16 05.8	+4 34 51	-0.003	-0.05	6.6	0.8	3.2	G5 IV		47 s		
192646	144228		20 16 08.3	-7 43 56	0.000	-0.02	8.00	0.00	0.4	B9.5 V		330 s	13598	m
192547	189085		20 16 11.6	-23 30 37	+0.003	-0.01	7.6	0.4	3.2	G5 IV		77 s		
192968	49399		20 16 13.6	+40 57 49	-0.001	0.00	7.86	-0.02	-2.6	B2.5 IV		1000 s		
192394	230127		20 16 14.5	-41 28 44	0.000	-0.02	7.2	1.1	0.2	K0 III		230 s		
192567	189088		20 16 15.5	-20 01 23	+0.001	-0.04	7.9	0.2	1.7	A3 V		160 s		
193089	32411		20 16 15.8	+51 59 55	-0.001	-0.01	7.3	0.0	0.4	B9.5 V		230 s		
192747	125632		20 16 18.1	+7 04 18	-0.002	0.00	7.9	1.1	-0.1	K2 III		370 s		
192836	88433	18 Sge	20 16 19.5	+21 35 55	0.000	-0.02	6.13	1.04	0.0	K1 III	-4	170 s		
192850	88434		20 16 20.4	+22 17 15	-0.001	+0.02	6.9	1.1	-0.1	K2 III		240 s		
--	105914		20 16 21.2	+17 47 35	-0.001	-0.03	8.0							
193139	32415		20 16 22.2	+52 06 18	-0.001	-0.01	7.7	1.1	0.2	K0 III		290 s		
192892	88438		20 16 22.3	+26 29 25	+0.002	+0.02	7.2	0.9	0.2	G9 III		240 s		
192666	163402	3 Cap	20 16 22.7	-12 20 14	+0.001	-0.01	6.40	0.01		B9			13600	m
192472	211948		20 16 23.5	-36 27 16	+0.001	+0.08	6.10	1.6	-0.5	M4 III		210 s		
192871	88437		20 16 25.4	+22 23 46	-0.001	-0.02	7.2	0.4	-2.0	F3 II		640 s		
192717	144231		20 16 25.7	-1 30 01	-0.001	-0.04	7.5	1.1	0.2	K0 III		290 s		
192486	211950		20 16 26.3	-35 11 52	+0.001	+0.08	6.53	0.37	3.0	F2 V		49 s		
192913	88444		20 16 27.1	+27 46 33	+0.001	-0.01	6.68	-0.07		A0 p	-6			
192987	69725		20 16 28.1	+37 03 23	+0.001	0.00	6.48	-0.09	-1.6	B5 IV	-6	380 s		
190808	257783		20 16 29.9	-79 00 24	+0.003	+0.05	7.8	0.5	3.4	F5 V		75 s		
193062	49403		20 16 30.8	+43 36 12	-0.006	-0.04	7.9	0.7	4.4	G0 V		49 s		
192989	69727		20 16 32.6	+36 16 35	+0.001	+0.02	7.09	1.08	3.2	G5 IV		39 s		
192894	88443		20 16 34.3	+21 24 10	-0.002	-0.01	7.1	0.1	0.6	A0 V		200 s		
193181	32423		20 16 35.0	+52 23 50	-0.001	0.00	8.0	0.1	0.6	A0 V		280 s		
193090	49406		20 16 35.7	+45 20 22	-0.001	+0.03	7.10	1.50	-0.3	K5 III		300 s		
192953	88449		20 16 39.4	+28 31 24	+0.001	+0.01	7.8	0.1	0.6	A0 V		260 s		
192990	69735		20 16 39.6	+35 28 07	+0.001	+0.01	7.10	-0.03	-0.3	B9 IV		280 s		
192700	163408		20 16 45.7	-15 10 52	+0.002	+0.01	7.7	1.1	0.2	K0 III	-21	320 s		
192992	69736		20 16 46.4	+32 06 09	-0.001	-0.02	8.0	0.9	3.2	G5 IV		91 s		
192944	88451	24 Vul	20 16 46.9	+24 40 16	+0.001	-0.01	5.4	0.9	0.3	G8 III	+15	100 s		
193009	69737		20 16 48.1	+32 22 48	0.000	0.00	7.16	0.09	-3.5	B1 V nnpe	-22	1100 s		
193287	32433		20 16 48.8	+56 02 24	+0.001	-0.01	6.9	1.1	-0.1	K2 III		230 s		
192791	144237		20 16 50.0	-3 29 30	0.000	-0.10	7.00	0.5	3.4	F5 V		53 s		m
192509	230131		20 16 50.1	-40 54 22	+0.006	-0.06	7.3	0.3	3.4	F5 V		59 s		

HD	SAO	Star Name	α 2000	δ 2000	μ(α)	μ(δ)	V	B-V	M_v	Spec	RV	d(pc)	ADS	Notes
			h m s	° ' "	s	"								
193115	49413		20 16 51.5	+43 41 00	0.000	0.00	7.7	0.1	0.6	A0 V		250 s		
192969	88452		20 16 51.6	+27 03 17	+0.001	+0.02	8.0	0.1	0.6	A0 V		280 s		
192614	211959		20 16 54.6	-32 36 24	+0.003	+0.07	7.5	0.3	3.2	G5 IV		72 s		m
193010	69743		20 16 54.9	+31 30 23	+0.001	+0.03	6.90	-0.01	0.6	A0 V		180 s	13630	m
193092	49410		20 16 55.0	+40 21 54	0.000	0.00	5.24	1.65	-2.3	K4 II	-20	260 s	13640	m
192217	246511		20 16 55.6	-59 38 34	+0.003	-0.04	7.8	1.0	0.2	K0 III		340 s		
191686	257786		20 16 56.2	-72 11 05	-0.003	0.00	7.7	1.1	0.2	K0 III		320 s		
192938	105922		20 16 56.3	+16 45 57	+0.001	-0.01	7.9	1.1	0.2	K0 III		330 s		
193034	69748		20 16 56.6	+34 08 24	+0.001	+0.01	7.81	-0.07	0.4	B9.5 V		300 s		
192937	105923		20 16 57.6	+18 21 14	+0.001	+0.02	7.8	0.1	0.6	A0 V		270 s		
193011	88454		20 16 57.9	+30 06 45	+0.001	0.00	8.00	1.1	0.0	K1 III		370 s		
193076	69754		20 16 58.8	+37 40 53	0.000	0.00	7.64	0.31		B0.5 II				
192652	211961		20 16 59.9	-31 12 28	-0.003	-0.05	7.9	0.6	0.2	K0 III		340 s		
192736	163412		20 17 00.0	-18 25 52	0.000	+0.02	7.2	0.1	1.4	A2 V		150 s		
193077	69755		20 17 00.0	+37 25 26	+0.001	+0.02	7.97	0.31		WN5	0		13641	m
192873	125646		20 17 00.0	+7 04 18	0.000	0.00	7.4	1.6	-0.5	M4 III		360 s		
192895	125647		20 17 00.1	+8 57 49	-0.001	+0.03	6.6	0.4	3.0	F2 V	-10	52 s		
193064	69753		20 17 00.3	+35 57 27	+0.001	0.00	7.67	-0.03		A0				
192772	163415		20 17 08.5	-16 58 38	0.000	+0.02	7.9	0.5	3.4	F5 V		78 s		
192594	211960		20 17 08.7	-39 00 04	0.000	-0.03	7.2	1.7	0.2	K0 III		90 s		
191993	254763		20 17 09.1	-67 27 18	+0.005	-0.07	6.8	0.4	1.4	A2 V		68 mx		
192737	--		20 17 09.5	-21 19 43			7.41	4.02		N7.7	-30		13616	RT Cap, m,v
192954	105926		20 17 09.5	+15 52 21	-0.001	-0.01	7.49	-0.09	0.4	B9.5 V	+18	260 s		
193159	49421		20 17 12.5	+40 35 27	+0.002	-0.01	7.14	0.00	0.0	B8.5 V		230 s		
192773	189104		20 17 15.6	-19 57 40	0.000	-0.03	7.8	0.3	0.2	K0 III		330 s		
193183	69763		20 17 22.5	+38 14 09	-0.001	0.00	6.98	0.45	-5.7	B1.5 Ib	0	1400 s		
192897	144253		20 17 24.6	-2 03 48	+0.001	-0.02	7.8	0.3	3.2	G5 IV		84 s		
193035	88463		20 17 24.7	+21 45 05	+0.001	+0.01	7.7	1.1	-0.1	K2 III		350 s		
193182	69764		20 17 25.1	+39 35 38	+0.001	+0.01	6.51	-0.09	-3.1	B9 II	-20	840 s		
192823	163418		20 17 26.6	-15 49 39	0.000	-0.04	7.4	0.2	2.1	A5 V		110 s		
193217	49425		20 17 28.9	+42 43 19	+0.001	0.00	6.29	1.64	-2.3	K4 II	-17	420 s		
193664	18796		20 17 31.0	+66 51 13	+0.078	+0.30	5.93	0.58	5.2	G5 V	-5	15 ts		
193118	69761		20 17 31.2	+30 19 23	0.000	+0.03	8.0	1.4	-0.3	K5 III		420 s		
193094	88473		20 17 31.3	+29 08 52	-0.001	+0.03	6.22	1.01	0.2	G9 III	-20	150 s	13648	m
193343	32440		20 17 35.7	+51 09 09	+0.011	+0.03	7.6	1.1	-0.1	K2 III		170 mx		
192752	189105		20 17 36.0	-29 23 36	+0.001	+0.04	8.0	3.2	-0.5	M2 III		56 s		
192738	211971		20 17 36.3	-33 41 01	-0.003	-0.02	7.6	1.3	3.2	G5 IV		36 s		
192919	144256		20 17 37.0	-6 24 15	+0.002	+0.01	8.0	0.9	3.2	G5 IV		93 s		
192876	163422	5 α¹ Cap	20 17 38.6	-12 30 30	+0.001	0.00	4.24	1.07	-4.5	G3 Ib	-26	490 s	13632	m
192530	246517		20 17 43.6	-51 55 03	-0.001	-0.05	6.7	1.2	3.4	F5 V		15 s		
193080	105932		20 17 43.6	+19 14 52	+0.001	-0.01	8.0	0.5	4.0	F8 V		62 s		
--	--		20 17 44.2	-39 06 43			6.00							RT Sgr, v
193237	69773	34 Cyg	20 17 47.0	+38 01 59	-0.001	0.00	4.81	0.42		B1 p	-9	44 mn		P Cyg, v
192724	230134		20 17 49.1	-40 11 06	0.000	+0.01	7.1	0.4	3.4	F5 V		55 s		m
193468	32443		20 17 49.7	+54 19 21	0.000	+0.06	6.8	0.5	3.4	F5 V		47 s		
193218	69775		20 17 54.4	+34 05 26	0.000	+0.01	7.4	1.6	-0.5	M4 III		350 s		
192945	144260		20 17 56.3	-8 03 55	-0.002	-0.01	7.4	1.1	0.2	K0 III		270 s		
193238	69778		20 17 59.8	+33 11 30	+0.001	+0.01	7.56	-0.02	0.4	B9.5 V		270 s	13660	m
192922	163426		20 18 00.2	-18 15 49	-0.002	0.00	7.5	0.1	0.6	A0 V		240 s		
192879	189114	4 Cap	20 18 01.2	-21 48 36	+0.002	-0.03	5.87	1.00	3.2	G8 IV	-18	29 s		
193268	69783		20 18 01.9	+37 09 46	+0.001	-0.06	7.8	0.5	2.1	F5 IV		130 s		
192947	163427	6 α² Cap	20 18 03.1	-12 32 42	+0.004	+0.01	3.57	0.94	0.2	G9 III	0	36 ts	13645	Algedi, m
192956	163428		20 18 03.2	-10 35 51	+0.002	-0.02	7.6	1.1	0.2	K0 III		300 s		
192923	189120		20 18 05.7	-21 59 55	+0.002	-0.04	7.7	0.4	3.4	F5 V		72 s		
193322	49438		20 18 06.8	+40 43 56	0.000	0.00	5.84	0.10		O8	-7	40 mn		
193017	144266		20 18 09.8	-4 43 43	-0.002	-0.02	7.4	0.5	4.0	F8 V		48 s		
192673	230136		20 18 10.4	-48 49 49	-0.003	-0.02	7.9	1.8	0.2	K0 III		200 s		
194031	9720		20 18 10.6	+70 47 12	-0.002	-0.05	7.3	0.2	2.1	A5 V		100 s		
193442	49447		20 18 11.1	+49 02 37	+0.001	+0.01	7.9	0.0	0.0	B8.5 V		360 s		
193630	32456		20 18 13.8	+58 48 35	+0.001	+0.02	7.5	0.1	0.6	A0 V		220 s		
193018	144268		20 18 13.9	-6 18 45	+0.002	0.00	7.8	0.2	2.1	A5 V		140 s		
193161	105940		20 18 14.1	+18 19 42	0.000	+0.02	7.80	0.1	0.6	A0 V		270 s	13662	m
193221	88488		20 18 14.3	+25 30 35	-0.001	-0.02	7.80	1.1	-0.1	K2 III		370 s		
192758	230140		20 18 15.4	-42 51 35	+0.003	-0.04	7.03	0.31		F0				
193512	32451		20 18 16.8	+51 52 01	-0.001	0.00	8.0	0.9	3.2	G5 IV		91 s		
193220	88490		20 18 16.8	+25 38 56	-0.001	+0.03	6.99	-0.13	-1.6	B3.5 V	-5	510 s	13666	m
192674	246522		20 18 19.2	-51 05 13	+0.001	-0.03	7.2	0.4	0.6	A0 V		110 s		
192397	254768		20 18 19.3	-63 09 39	+0.006	-0.03	7.0	0.1	1.7	A3 V		110 s		
193082	125660		20 18 19.7	+3 40 49	+0.006	-0.05	7.9	1.1	0.2	K0 III		140 mx		
192815	230141		20 18 24.0	-40 54 21	0.000	-0.02	8.0	1.3	-0.1	K2 III		370 s		
193292	69794		20 18 24.1	+32 06 37	+0.001	-0.03	7.3	0.2		A m				
193592	32455		20 18 24.6	+55 23 50	-0.001	-0.02	5.76	0.11		A0	+1	37 mn	13692	m
193812	18799		20 18 26.3	+64 52 04	-0.003	-0.01	7.7	1.4	-0.3	K5 III		350 s		
191937	257788		20 18 27.0	-72 58 47	+0.005	-0.05	6.57	1.41		K2				
193344	69799		20 18 27.5	+36 15 15	+0.001	-0.03	7.58	-0.04		B9 p				

HD	SAO	Star Name	α 2000	δ 2000	μ(α)	μ(δ)	V	B–V	M_V	Spec	RV	d(pc)	ADS	Notes
193369	69803	36 Cyg	20 18 28.5	+37 60 00	+0.003	+0.03	5.58	0.06	1.7	A3 V	-9	60 s		
191973	257789		20 18 30.4	-72 48 31	+0.001	0.00	6.94	1.03		K0				
193222	105947		20 18 34.9	+17 21 45	0.000	+0.01	6.9	1.4	-0.3	K5 III		270 s		
193248	88495		20 18 35.7	+21 31 15	-0.001	+0.02	7.5	0.1	1.7	A3 V		140 s		
193122	125665		20 18 37.3	+0 38 29	+0.005	+0.02	6.9	0.1	1.7	A3 V		110 s		
192826	230143		20 18 38.9	-42 37 13	+0.003	-0.12	7.46	0.59		G0				
193370	69806	35 Cyg	20 18 38.9	+34 58 58	0.000	-0.01	5.17	0.65	-4.6	F5 Ib	-14	780 s		
193426	49453		20 18 39.5	+40 13 37	-0.002	-0.01	7.73	1.23	-7.1	B9 Ia		1800 s		
192740	246524		20 18 42.5	-51 30 21	+0.006	+0.24	7.69	0.42	3.0	F2 V		52 mx		
192510	254770		20 18 43.5	-61 54 26	+0.003	-0.06	6.8	0.9	0.6	A0 V		53 s		
193037	163434		20 18 44.5	-17 51 19	+0.006	+0.05	7.9	0.5	4.0	F8 V		60 s		
192975	189129		20 18 47.0	-27 04 30	0.000	+0.03	8.0	1.3	0.2	K0 III		250 s		
193536	49462		20 18 49.5	+46 19 21	-0.001	+0.01	6.45	-0.13	-2.5	B2 V	-9	570 s		
193347	88502		20 18 50.4	+26 59 28	-0.001	-0.02	6.6	1.6	-0.5	M2 III	-36	260 s		
193443	69815		20 18 51.6	+38 16 48	0.000	+0.01	7.24	0.40	-6.0	O9 III		2600 s	13686	m
193314	88500		20 18 51.9	+21 12 35	+0.002	-0.01	7.9	0.1	0.6	A0 V		280 s		
193040	189132		20 18 54.4	-23 53 37	-0.001	+0.03	7.8	0.9	-0.1	K2 III		370 s		
193665	32466		20 18 54.4	+55 32 43	-0.001	+0.04	8.0	0.5	3.4	F5 V		82 s		
193792	18801		20 18 54.6	+60 37 04	-0.002	0.00	7.8	0.0	0.0	B8.5 V		330 s		
192827	230144		20 18 55.7	-47 42 38	+0.001	0.00	6.2	1.8	-0.5	M2 III		170 s		
193146	144278		20 18 56.3	-6 38 20	-0.001	-0.08	7.8	0.8	3.2	G5 IV		84 s		
192844	230146		20 18 56.4	-44 31 28	+0.001	+0.01	7.56	0.95		K0				
193104	163437		20 18 57.0	-15 51 28	+0.002	-0.02	7.9	1.6	-0.5	M4 III		490 s		AE Cap, v
193469	69820		20 18 57.4	+39 00 15	0.000	-0.01	6.33	1.89	-4.4	K5 Ib		740 s	13690	m
193251	125676		20 18 57.7	+10 05 19	-0.001	0.00	7.7	1.1	-0.1	K2 III		340 s		
193225	125673		20 19 00.5	+2 14 00	+0.001	0.00	7.3	0.4	2.6	F0 V		87 s		
193325	88504		20 19 00.8	+20 27 49	0.000	-0.03	7.4	0.0	0.4	B9.5 V	-39	240 s		
193551	49463		20 19 01.9	+44 52 38	0.000	-0.03	7.9	0.9	3.2	G5 IV		87 s		
193102	163440		20 19 02.1	-14 17 26	+0.002	+0.02	7.40	1.1	0.2	K0 III	-48	280 s		m
192531	254772		20 19 02.9	-63 13 52	-0.001	-0.08	6.3	0.7	0.2	K0 III		150 mx		
193147	163442		20 19 04.7	-10 48 00	0.000	+0.01	7.9	0.1	0.6	A0 V		280 s		
193487	69822		20 19 07.0	+36 45 08	0.000	-0.03	7.45	0.39	-2.0	F4 II		780 s	13693	m
193515	69825		20 19 08.3	+38 09 40	+0.003	-0.02	7.70	1.1	-2.2	K1 II		810 s		
193514	69826		20 19 08.5	+39 16 24	+0.002	-0.01	7.41	0.44		O7.8	-20			
193315	105955		20 19 09.0	+14 23 02	0.000	-0.01	7.1	0.9	3.2	G5 IV	+12	59 s		
193410	88512		20 19 10.3	+29 29 57	+0.001	0.00	7.2	0.5	3.4	F5 V		58 s		
193021	211995		20 19 11.1	-32 43 56	+0.003	-0.01	7.3	0.4	0.2	K0 III		270 s		
192886	230150		20 19 17.7	-47 34 49	+0.020	-0.18	6.13	0.46	3.7	F6 V		31 s		
193372	105958		20 19 19.4	+18 49 29	0.000	0.00	7.6	0.1	1.7	A3 V		150 s		
193594	49465		20 19 20.4	+41 23 28	-0.002	-0.01	7.79	0.00		B9				
193350	105957		20 19 21.0	+14 22 15	+0.001	+0.02	6.80				-17		13688	m
193084	189139		20 19 23.3	-29 37 27	+0.001	-0.01	7.5	0.2	0.4	B9.5 V		180 s		
193150	163445	7 σ Cap	20 19 23.4	-19 07 07	0.000	-0.01	5.50	1.1	-2.3	K3 II	-11	360 s	13675	m
193151	189142		20 19 25.5	-20 57 03	-0.001	+0.01	6.6	0.8	2.6	F0 V		31 s		
193631	49470		20 19 26.4	+45 12 56	-0.002	+0.03	7.8	1.1	0.2	K0 III		310 s		
193071	212001		20 19 27.2	-32 17 12	+0.003	-0.04	8.0	1.1	-0.1	K2 III		410 s		
193373	105961		20 19 29.1	+13 13 01	-0.002	-0.01	6.21	1.63	-0.5	M2 III	+23	210 s		
193517	69830		20 19 31.1	+32 02 50	+0.004	-0.01	7.3	0.2	2.1	A5 V		110 s		
193389	105962		20 19 31.3	+15 32 42	-0.001	-0.01	7.7	0.5	4.0	F8 V		54 s		
193576	69833		20 19 32.2	+38 43 54	-0.001	0.00	8.00	0.51		WN5	+3			V444 Cyg, v
193632	49471		20 19 34.3	+43 29 48	+0.002	+0.13	7.5	0.5	4.0	F8 V		49 s		
193964	18807	71 Dra	20 19 36.4	+62 15 27	+0.001	+0.03	5.6	0.0		B9	-25			
193109	212003		20 19 37.5	-31 43 43	+0.001	+0.03	7.2	0.4	3.0	F2 V		71 s		
193700	49479		20 19 41.5	+47 54 20	+0.001	0.00	7.87	0.80	4.4	G0 V		32 s		U Cyg, v
193227	163450		20 19 42.7	-17 29 16	-0.002	+0.01	8.0	1.1	0.2	K0 III		360 s		
193329	144296		20 19 43.1	-1 04 43	+0.003	+0.03	6.06	1.09		K0				
193633	49476		20 19 44.6	+41 08 18	0.000	-0.01	7.14	0.03		B9				m
193193	189151		20 19 45.0	-25 13 43	-0.008	-0.14	7.0	0.7	4.4	G0 V		28 s		
193328	144299		20 19 47.2	-0 54 27	-0.001	-0.04	7.4	0.1	1.4	A2 V		160 s		
193621	69841		20 19 48.1	+37 07 57	+0.001	-0.01	6.57	0.00	-0.4	A1 III	-17	250 s		
193634	69844		20 19 49.1	+38 20 34	+0.001	0.00	7.46	0.30	-2.9	B3 III		620 s		
193701	49481		20 19 49.5	+45 21 49	+0.001	-0.03	6.67	0.45	-4.6	F2 Ib		1600 s	13723	m
193553	88528		20 19 49.5	+29 43 37	-0.001	0.00	6.72	-0.16	0.0	B8.5 V		220 s		
193354	144302		20 19 51.0	-1 18 01	-0.004	0.00	8.0	1.1	-0.1	K2 III		160 mx		
193074	212004		20 19 52.0	-39 40 41	+0.001	0.00	7.4	1.5	0.2	K0 III		140 s		
193353	144306		20 19 52.5	-0 38 43	0.000	+0.01	7.7	1.1	-0.1	K2 III	+4	360 s		
193666	49480		20 19 54.0	+41 36 08	-0.002	-0.01	7.8	0.0	0.0	B8.5 V		340 s		
193111	212009		20 19 55.1	-36 30 21	-0.001	-0.06	7.6	1.3	3.2	G5 IV		36 s		
193722	49482		20 19 55.9	+46 50 15	0.000	+0.01	6.1	-0.1		B9				
193636	69847		20 19 57.6	+36 16 08	0.000	-0.01	7.97	0.28	0.5	A7 III		290 s	13719	m
193174	212012		20 19 58.1	-32 36 58	0.000	+0.01	7.1	0.6	2.6	F0 V		55 s		
193472	105974		20 20 00.1	+13 32 53	+0.001	0.00	5.95	0.31		A5	-8			
193473	105973		20 20 02.3	+10 18 32	+0.001	+0.03	7.9	1.1	0.2	K0 III		330 s		
194258	18817		20 20 05.9	+68 52 50	+0.003	+0.04	5.55	1.59	-0.5	M4 III	-43	160 s		AC Dra, q
193490	105980		20 20 09.6	+11 28 09	-0.001	-0.02	6.8	1.1	-0.1	K2 III		230 s		

HD	SAO	Star Name	α 2000	δ 2000	μ(α)	μ(δ)	V	B-V	M$_v$	Spec	RV	d(pc)	ADS	Notes
			h m s	° ′ ″	s	″								
193194	212015		20 20 14.3	-34 35 06	+0.003	-0.04	7.10	0.6	4.4	G0 V		35 s		m
193244	189160		20 20 14.8	-29 07 53	+0.004	-0.01	7.70	1.1	0.2	K0 III		290 mx		m
193702	69856		20 20 15.0	+39 24 12	0.000	-0.02	6.23	0.06	1.4	A2 V	-1	93 s	13728	m
193555	105984		20 20 15.4	+15 32 35	+0.006	+0.04	6.9	0.5	4.0	F8 V	-27	38 s		
193230	212016		20 20 20.3	-32 57 32	+0.002	+0.01	8.0	1.0	0.2	K0 III		360 s		
193556	105988		20 20 20.3	+14 34 09	-0.001	+0.01	6.13	0.91		G5	+8			
193579	105991		20 20 21.3	+17 47 35	-0.001	-0.03	5.80	1.49	-0.3	K5 III	-33	170 s		
193132	230161		20 20 22.4	-42 54 07	-0.002	-0.01	7.77	1.01		K0				
193539	125694		20 20 25.4	+7 36 53	-0.001	-0.01	7.4	1.1	-0.1	K2 III		300 s		
193557	105990		20 20 25.9	+10 48 39	-0.001	-0.02	8.0	0.1	0.6	A0 V		280 s		
193429	144313		20 20 26.0	-6 21 42	-0.005	-0.08	6.63	1.55		K5				
193793	49491		20 20 27.8	+43 51 16	-0.001	-0.01	6.85	0.40		WC6	+21	26 mn	13736	m
193281	189164		20 20 27.9	-29 11 50	+0.002	+0.01	6.40	0.1	0.6	A0 V		150 s	13702	m
193335	163462		20 20 28.4	-19 38 51	+0.002	-0.04	7.0	1.1	0.2	K0 III		180 s		
193944	32491		20 20 30.2	+53 35 46	-0.002	+0.02	6.18	1.56	-0.3	K5 III	-4	180 s	13743	m
193683	69858		20 20 30.6	+32 01 24	0.000	0.00	7.51	-0.09	-1.0	B5.5 V	+14	500 s		
193002	246535		20 20 32.1	-55 03 05	+0.001	-0.05	6.27	1.59	-0.3	K5 III		180 s		
192949	246534		20 20 38.2	-58 43 46	+0.005	-0.05	7.30	0.1	0.6	A0 V		220 s		m
193597	105995		20 20 38.4	+10 45 15	-0.001	+0.01	8.0	0.1	0.6	A0 V		280 s		
193736	69863		20 20 39.0	+35 15 08	+0.002	-0.01	7.95	-0.07	0.4	B9.5 V		320 s		
193432	163468	8 ν Cap	20 20 39.6	-12 45 33	+0.001	-0.02	4.76	-0.05	-0.3	B9 IV	-2	99 s	13714	m
193668	88544		20 20 41.9	+22 16 35	0.000	-0.01	7.0	0.0	0.4	B9.5 V		200 s		
193559	125698		20 20 43.1	+1 23 19	0.000	0.00	8.0	0.0	0.4	B9.5 V		330 s		
193258	212020		20 20 46.3	-36 51 10	0.000	0.00	7.9	1.6	-0.5	M4 III		490 s		
193452	163471		20 20 46.4	-14 47 06	+0.003	0.00	6.20	-0.02		B9			13717	m
193814	69867		20 20 48.3	+38 15 24	+0.002	+0.01	7.58	-0.07	-0.2	B8 V		360 s		
--	49500		20 20 50.1	+44 32 32	0.000	+0.02	8.0	0.9						
193302	212025		20 20 51.7	-35 40 25	+0.003	+0.02	6.4	2.1	-0.1	K2 III		58 s		
193454	163472		20 20 53.2	-18 09 22	+0.002	-0.03	7.6	0.5	4.0	F8 V		54 s		
192979	246537		20 20 53.3	-59 02 08	0.000	-0.02	7.9	0.4	0.6	A0 V		150 s		
193213	230165		20 20 54.1	-45 32 45	+0.005	-0.08	7.33	0.51		F5				m
193479	163474		20 20 54.9	-16 49 45	+0.002	0.00	7.8	1.1	0.2	K0 III		330 s		
141502	230164		20 20 58.3	-48 18 04	+0.001	-0.01	8.0	0.7	3.2	G5 IV		93 s		
193477	163480		20 20 59.0	-15 25 44	-0.001	-0.03	8.0	1.1	0.2	K0 III		360 s		
193888	49502		20 21 00.1	+43 22 35	-0.004	-0.04	8.0	0.1	0.6	A0 V		120 mx		
193495	163481	9 β Cap	20 21 00.5	-14 46 53	+0.003	0.00	3.08	0.79	4.0	F8 V	-19	32 mn		Dabih, m
193455	163479		20 21 01.1	-18 19 25	+0.002	0.00	7.1	1.1	0.2	K0 III		240 s		
193542	163484		20 21 01.1	-10 58 57	+0.003	+0.01	6.8	1.1	0.2	K0 III		210 s		
193177	246542		20 21 01.9	-50 01 04	+0.002	-0.04	7.6	-0.4	1.4	A2 V		170 s		
193855	69880		20 21 07.5	+39 01 53	+0.001	-0.02	7.80	0.28	-3.6	B2 III		960 s		
193582	144328		20 21 08.6	-6 21 21	0.000	-0.10	7.6	0.9	3.2	G5 IV		75 s		
193361	212029		20 21 10.1	-35 55 06	0.000	-0.05	8.0	1.3	-0.1	K2 III		360 s		
193688	106003		20 21 10.1	+12 50 35	0.000	-0.01	7.6	0.1	0.6	A0 V		250 s		
194298	18826		20 21 11.2	+63 58 50	-0.001	+0.03	5.69	1.56	-0.3	K5 III	+30	150 s	13769	m
193402	212030		20 21 12.0	-32 44 22	+0.001	-0.05	7.6	0.4	3.0	F2 V		73 s		
194056	32499		20 21 13.9	+52 24 37	+0.003	-0.01	7.4	1.4	-0.3	K5 III	-53	330 s		
193926	49510		20 21 14.4	+43 35 21	+0.002	0.00	7.86	-0.04	0.6	A0 V		280 s	13753	m
192418	257795		20 21 14.5	-73 58 40	+0.007	+0.01	7.7	1.1	-0.1	K2 III		360 s		
193707	106007		20 21 15.3	+15 06 15	+0.001	-0.02	6.6	0.1	0.6	A0 V	-22	150 s		
--	106008		20 21 15.7	+17 48 13	-0.001	-0.02	8.0							
194299	18828		20 21 21.6	+63 25 55	+0.006	+0.03	7.30	1.6	-0.4	M0 III	+21	310 s		
193856	69881		20 21 24.1	+30 51 51	-0.001	0.00	7.64	1.71	-0.3	K5 III		290 s		
191122	--		20 21 24.5	-81 37 20			8.0	0.7	4.4	G0 V		53 s		
193723	106009		20 21 25.4	+13 25 16	+0.001	-0.02	7.7	1.1	-0.1	K2 III		340 s		
192887	254775		20 21 28.1	-66 44 58	-0.021	+0.01	6.6	0.3	3.4	F5 V		44 s		
194376	18829		20 21 28.7	+64 41 36	-0.003	+0.01	7.7	1.1	-0.1	K2 III		320 s		
193773	106015		20 21 28.9	+19 39 43	+0.001	-0.01	7.9	0.9	3.2	G5 IV		88 s		
193857	69888		20 21 29.5	+30 35 27	+0.001	+0.04	6.8	0.2		A m				
193624	163495		20 21 29.6	-10 08 39	+0.002	0.00	7.6	0.1	0.6	A0 V		260 s		
193545	189192		20 21 29.7	-23 28 35	0.000	+0.01	7.6	0.4	0.2	K0 III		300 s		
193818	88566		20 21 31.7	+22 50 56	-0.001	0.00	6.7	1.4	-0.3	K5 III	+10	240 s		
193565	189195		20 21 32.5	-21 25 09	-0.011	-0.12	7.5	0.3	3.4	F5 V		61 mx		
193891	69890		20 21 32.9	+32 18 50	+0.002	+0.01	7.9	1.1	0.2	K0 III		320 s		
192865	254776		20 21 35.3	-67 18 46	+0.019	+0.02	6.7	0.5	4.0	F8 V		34 s		
193738	106013		20 21 36.1	+10 59 31	0.000	+0.01	7.9	1.4	-0.3	K5 III		400 s		
193708	125715		20 21 36.1	+1 40 16	-0.001	+0.01	7.5	0.1	0.6	A0 V		240 s		
193910	69891		20 21 36.6	+32 27 14	-0.001	-0.01	7.6	0.1	1.4	A2 V		170 s		
194057	49520		20 21 40.1	+44 48 50	-0.001	-0.02	7.51	0.86	-5.7	B1 Ib		1000 s		
193307	246546		20 21 40.8	-49 59 59	-0.037	-0.25	6.27	0.55	3.9	G2 IV-V		28 mx		
193284	246545		20 21 41.3	-50 46 34	+0.005	-0.04	7.3	2.1	-0.5	M2 III		190 s		
193464	212038		20 21 42.8	-36 36 37	+0.005	-0.05	7.3	0.8	4.0	F8 V		30 s		
191735	257793		20 21 43.0	-79 35 50	+0.006	+0.01	7.5	-0.1	0.4	B9.5 V		270 s		
193777	125720		20 21 47.8	+8 31 09	0.000	-0.01	7.8	0.1	0.6	A0 V		260 s		
193437	230169		20 21 47.9	-40 45 18	0.000	+0.03	7.1	1.9	-0.5	M2 III		230 s		
193860	106025		20 21 53.6	+19 37 39	0.000	+0.01	7.9	1.1	0.2	K0 III		330 s		

HD	SAO	Star Name	α 2000	δ 2000	μ(α)	μ(δ)	V	B-V	M$_v$	Spec	RV	d(pc)	ADS	Notes
			h m s	° ′ ″	s	″								
193819	106024		20 21 55.8	+13 35 32	+0.001	−0.02	7.40	0.1	0.6	A0 V	−10	220 s	13755	m
193799	125726		20 21 56.9	+7 10 18	0.000	+0.02	7.60	0.9	0.2	G9 III	−2	290 s		
194069	49524		20 22 02.8	+41 07 53	−0.001	−0.03	6.39	1.07	−2.1	G2 II	−4	370 s		
193911	88580	25 Vul	20 22 03.2	+24 26 46	0.000	0.00	5.40	−0.06	−1.0	B7 IV e	−13	190 s		
194152	49531		20 22 05.2	+45 47 42	+0.002	+0.05	5.58	1.08	0.2	K0 III	−26	100 s		
193656	189202		20 22 06.7	−21 57 25	+0.002	−0.01	7.5	0.6	1.4	A2 V		74 s		
194204	32506		20 22 12.4	+50 32 28	+0.001	+0.01	7.3	0.1	0.6	A0 V		210 s		
193689	163499		20 22 12.6	−15 37 33	0.000	−0.04	7.7	0.1	0.6	A0 V		270 s		
194093	49528	37 γ Cyg	20 22 13.5	+40 15 24	0.000	0.00	2.20	0.68	−4.6	F8 Ib	−8	230 s	13765	Sadr, m
194095	69913		20 22 13.5	+38 05 46	−0.002	−0.04	7.56	1.14	0.0	K1 III		310 s		
194681	18834		20 22 13.6	+69 30 45	+0.003	−0.01	7.9	0.1	1.7	A3 V		160 s		
193726	163500		20 22 22.0	−15 49 39	+0.002	0.00	8.0	0.5	4.0	F8 V		62 s		
193950	106035		20 22 26.5	+16 55 00	−0.001	−0.01	7.4	1.6	−0.5	M2 III		370 s		
193571	230177	κ¹ Sgr	20 22 27.4	−42 02 59	+0.003	−0.09	5.59	0.00	0.6	A0 V	−17	45 mx		m
194333	32512		20 22 30.7	+54 55 55	+0.001	−0.02	7.5	0.1	0.6	A0 V		230 s		
194096	69916		20 22 32.0	+34 59 25	+0.002	−0.03	6.83	0.04		A0				
194033	88597		20 22 32.4	+26 17 54	+0.001	0.00	7.10	1.1	−0.1	K2 III	−22	270 s		m
193822	144355		20 22 34.9	−8 49 30	+0.003	+0.01	8.0	0.1	1.7	A3 V		180 s		
193507	230176		20 22 36.2	−48 51 53	+0.001	+0.03	7.9	0.5	0.2	K0 III		340 s		
194097	69917		20 22 37.2	+31 15 54	0.000	−0.03	6.09	1.35	−0.1	K2 III	+12	140 s		
194071	88599		20 22 37.4	+28 14 47	+0.001	+0.01	8.00	0.9	0.3	G8 III		340 s		
193407	246550		20 22 38.3	−54 15 48	+0.004	−0.03	8.00	1.1	0.2	K0 III		360 s		m
193862	144357		20 22 39.9	−5 04 52	0.000	+0.01	7.7	1.1	−0.1	K2 III		370 s		
192074	257794		20 22 41.3	−78 57 40	+0.013	−0.04	6.7	0.1	0.4	B9.5 V		130 mx		
193951	106036		20 22 43.8	+11 43 00	+0.001	+0.01	8.0	1.6	−0.5	M2 III		480 s		
193968	106039		20 22 44.8	+12 26 35	+0.005	−0.03	6.8	1.1	−0.1	K2 III		180 mx		
194035	106041		20 22 44.8	+19 51 00	−0.007	−0.03	7.2	0.8	3.2	G5 IV		62 s		
194193	49546		20 22 45.1	+41 01 34	0.000	−0.04	5.93	1.60	−0.3	K5 III	+1	150 s		
195191	9753		20 22 47.0	+76 02 36	−0.003	−0.01	8.0	1.1	−0.1	K2 III	+10	420 s		
194110	69922		20 22 47.0	+31 32 59	+0.007	+0.03	7.1	1.1	0.2	K0 III	−27	200 mx		
194354	32516		20 22 49.3	+53 25 05	+0.001	+0.04	7.70	0.1	0.6	A0 V		250 s	13795	m
193528	246553		20 22 49.4	−50 29 33	+0.001	+0.02	7.9	1.8	−0.1	K2 III		190 s		
194012	106042		20 22 52.1	+14 33 05	+0.005	0.00	6.17	0.51	−4.6	F8 Ib	+2	93 mx		
194220	49550		20 22 55.3	+42 59 00	+0.004	+0.04	6.20	0.95	0.2	K0 III	−20	160 s	13786	m
194206	69929		20 22 57.5	+39 12 40	+0.001	+0.01	6.74	−0.10	−0.2	B8 V		240 s	13783	m
194072	88607		20 22 58.5	+20 19 43	+0.002	0.00	7.9	0.9	3.2	G5 IV		86 s		
193716	212062		20 23 00.1	−35 22 39	+0.003	−0.01	7.1	0.7	3.0	F2 V		39 s		
193896	144361		20 23 00.8	−9 39 18	+0.003	−0.02	6.30	0.91		G5				
194444	32519		20 23 01.0	+54 40 23	+0.002	0.00	6.6	0.1	1.4	A2 V		110 s		
194905	9748		20 23 02.3	+71 04 56	+0.006	+0.03	7.9	1.1	0.2	K0 III		300 s		
193749	212064		20 23 05.3	−33 02 52	+0.004	+0.03	7.60	1.4	−0.3	K5 III		380 s		m
194241	49556		20 23 06.5	+40 48 02	+0.002	+0.01	7.4	1.1	−0.1	K2 III	−20	300 s		
193864	163510		20 23 07.8	−18 20 32	+0.001	−0.02	7.8	0.8	3.2	G5 IV		84 s		
194355	49567		20 23 07.9	+48 48 44	−0.001	−0.01	7.93	1.88	−0.3	K5 III		260 s		
193953	144366		20 23 08.1	−3 48 42	0.000	+0.01	7.8	1.1	0.2	K0 III		330 s		
194013	125747		20 23 10.5	+5 20 35	−0.002	−0.04	5.31	0.97	1.8	G8 III-IV	−12	46 s		
193881	163515		20 23 12.0	−15 47 27	+0.001	−0.03	8.0	1.1	0.2	K0 III		360 s		
194075	125749		20 23 14.9	+9 58 08	−0.001	−0.01	8.0	1.4	−0.3	K5 III		420 s		
194074	106048		20 23 15.5	+11 05 07	+0.001	−0.01	7.8	0.1	1.4	A2 V		180 s		
192905	257798		20 23 17.6	−72 39 07	+0.005	−0.04	7.6	0.9	3.2	G5 IV		75 s		
194279	49563		20 23 18.0	+40 45 32	−0.001	−0.01	7.02	1.02	−6.7	B1.5 Ia	−31	1000 s		
--	88618		20 23 18.7	+28 48 08	−0.001	−0.02	8.04	1.20	0.2	K0 III		280 s		
194014	144369		20 23 18.9	−0 27 10	−0.002	−0.02	7.1	1.6	−0.5	M2 III		320 s		
194115	106051		20 23 19.1	+15 21 41	0.000	−0.03	7.1	0.1	0.6	A0 V	−20	200 s		
193933	163519		20 23 26.0	−14 15 24	+0.001	0.00	6.7	0.0	0.4	B9.5 V		180 s		
194116	106054		20 23 26.8	+10 21 48	+0.003	+0.02	6.7	1.1	−0.1	K2 III		230 s		
193883	189223		20 23 33.3	−26 44 09	−0.001	−0.04	7.30	1.1	0.2	K0 III		260 s	13770	m
194040	144372		20 23 38.1	−5 15 57	+0.011	+0.01	6.9	0.5	3.4	F5 V		49 s		
194282	69944		20 23 40.8	+31 40 37	0.000	−0.02	8.01	−0.07	0.4	B9.5 V		330 s		
194042	163522		20 23 43.6	−10 40 19	+0.002	0.00	7.9	0.1	1.4	A2 V		200 s		
194335	69951		20 23 44.2	+37 28 35	0.000	0.00	5.90	−0.20	−2.5	B2 V p	−31	440 s		
193850	212074		20 23 44.5	−35 07 48	0.000	−0.01	7.4	1.2	0.2	K0 III		220 s		
194357	69952		20 23 48.0	+37 01 33	+0.002	−0.01	6.78	0.12	−2.8	A0 II		810 s		
193694	246558		20 23 48.2	−50 50 16	+0.001	−0.02	7.9	0.9	0.2	K0 III		340 s		
194154	125759		20 23 48.3	+6 37 22	−0.001	−0.06	7.5	0.5	3.4	F5 V		68 s		
193851	212075		20 23 50.5	−37 29 58	+0.003	+0.02	8.0	1.0	3.2	G5 IV		67 s		
194317	69950	39 Cyg	20 23 51.4	+32 11 25	+0.003	+0.04	4.43	1.33	−0.2	K3 III	−15	78 s		
193529	254784		20 23 52.5	−60 36 34	+0.002	−0.01	7.6	1.3	−0.1	K2 III		300 s		
193807	230184	κ² Sgr	20 23 53.0	−42 25 22	0.000	+0.03	5.64	0.20	0.0	A3 III	+2	120 s		m
194078	144377		20 23 54.0	−7 35 47	0.000	0.00	8.0	1.1	−0.1	K2 III		410 s		
192316	257796		20 23 54.2	−79 03 02	+0.017	−0.03	7.9	0.1	0.6	A0 V		170 mx		
194211	106059		20 23 54.2	+16 02 49	−0.001	0.00	6.7	0.0	0.4	B9.5 V	−14	180 s		
193830	230185		20 24 00.9	−41 47 45	+0.003	−0.02	7.3	1.1	0.2	K0 III		240 s		
--	257801		20 24 01.8	−71 04 48	+0.018	−0.01	8.0							
194424	69962		20 24 02.7	+39 32 28	+0.001	−0.02	7.87	−0.04	−0.6	B8 IV		480 s		

HD	SAO	Star Name	α 2000	δ 2000	μ(α)	μ(δ)	V	B-V	M_V	Spec	RV	d(pc)	ADS	Notes
			$20^h\ 24^m\ 05^s.4$	$-21°10'18''$	$-0^s.001$	$0''.00$								
194021	189240		20 24 05.4	-21 10 18	-0.001	0.00	7.9	1.3	1.4	A2 V		37 s		
194479	49582		20 24 05.4	+44 42 04	-0.003	+0.04	7.45	1.08		K1 III-IV				
193677	246560		20 24 08.7	-54 32 36	-0.001	-0.03	7.9	0.2	1.4	A2 V		160 s		
194164	144384		20 24 08.9	-0 34 03	+0.002	-0.02	7.9	1.1	0.2	K0 III		340 s		
194358	88629		20 24 08.9	+29 00 08	+0.002	+0.04	7.20	0.2	2.1	A5 V		100 s	13807	m
193980	212082		20 24 10.7	-30 29 22	+0.004	-0.01	7.9	0.1	2.6	F0 V		120 s		
194121	163532		20 24 13.8	-14 06 51	-0.002	-0.02	6.9	1.1	0.2	K0 III		220 s		
194467	69971		20 24 17.9	+37 34 31	+0.001	-0.01	8.03	-0.01	0.2	B9 V		350 s		
--	69965		20 24 20.2	+31 03 00	0.000	0.00	8.0	1.6						
194318	106068		20 24 22.4	+19 34 31	-0.001	0.00	7.6	0.1	1.4	A2 V		180 s	13808	m
194359	88631		20 24 25.5	+24 16 40	0.000	+0.02	7.40	0.7	4.4	G0 V		40 s		
194480	69973		20 24 25.9	+38 57 53	+0.002	-0.01	8.00	-0.15	1.4	A2 V		210 s		
194319	106070		20 24 29.8	+15 43 37	+0.001	-0.02	8.0	1.1	-0.1	K2 III		390 s		
194668	32534		20 24 32.2	+53 33 07	+0.001	+0.02	6.4	0.0		B9				
194262	125770		20 24 33.2	+5 30 34	+0.001	-0.02	7.20	0.00	0.4	B9.5 V		230 s	13810	m
194614	49590		20 24 35.4	+48 41 34	+0.004	+0.01	7.71	-0.04	0.0	B8.5 V		260 mx	13831	m
194337	106073		20 24 35.5	+16 08 37	0.000	-0.01	7.2	1.4	-0.3	K5 III		300 s		
194102	189249		20 24 36.9	-25 50 02	+0.003	-0.02	6.9	1.7	0.2	K0 III		78 s		
193853	246567		20 24 37.3	-49 53 33	0.000	+0.03	7.8	2.2	-0.5	M4 III		280 s		
194244	125769		20 24 37.3	+1 04 07	0.000	+0.01	6.15	-0.04		A0 n	+4		13811	m
194305	125773		20 24 38.2	+8 35 51	+0.001	0.00	6.6	0.0	0.4	B9.5 V		170 s		
194719	32539		20 24 38.5	+55 26 36	0.000	-0.01	7.3	0.1	0.6	A0 V		210 s		
194263	125772		20 24 41.5	+1 22 07	0.000	-0.02	6.7	1.4	-0.3	K5 III		260 s		
194086	189251		20 24 42.5	-29 04 40	+0.001	-0.05	6.9	1.1	0.2	K0 III		180 s		
193234	257803		20 24 43.7	-71 39 33	+0.008	-0.03	7.8	0.9	3.2	G5 IV		82 s		
194233	144392		20 24 44.5	-8 46 12	+0.001	-0.02	7.40	0.1	1.7	A3 V		140 s		m
194737	32544		20 24 46.0	+55 00 41	+0.011	+0.05	7.4	1.1	-0.9	K0 II-III	-47	430 s		
194451	88641		20 24 46.5	+25 10 46	0.000	-0.01	7.70	0.1	0.6	A0 V		260 s		m
194450	88644		20 24 48.0	+26 07 38	0.000	-0.02	7.1	1.1	-0.1	K2 III	-42	270 s		
194558	69986		20 24 48.6	+40 09 13	+0.001	0.00	6.74	1.27	-0.1	K2 III		200 s	13833	m
194047	212093		20 24 48.9	-36 59 47	-0.002	-0.03	7.6	1.7	0.2	K0 III		120 s		
194682	32542		20 24 50.9	+51 17 23	+0.002	0.00	7.9	0.1	1.4	A2 V		190 s		
191220	258856		20 24 54.2	-83 18 38	+0.011	+0.02	6.17	0.20		A2				
194385	125778		20 24 58.9	+9 11 43	+0.005	+0.02	7.6	0.5	3.4	F5 V		70 s		
194932	18853		20 25 01.5	+61 17 52	+0.001	0.00	8.0	0.1	1.4	A2 V		190 s		
194525	69987		20 25 02.4	+30 34 05	0.000	-0.01	7.94	0.80	0.6	G0 III		240 s		
194406	106084		20 25 02.7	+13 23 11	+0.006	+0.01	7.6	0.1	1.4	A2 V		120 mx		
196768	3404		20 25 03.1	+83 10 59	-0.009	+0.01	6.8	0.9	3.2	G5 IV		53 s		
194541	69990		20 25 03.7	+32 28 07	+0.001	0.00	7.8	1.1	0.2	K0 III		310 s		
194388	125780		20 25 03.9	+7 28 25	-0.002	-0.03	7.7	0.4	2.6	F0 V		100 s		
194213	163545		20 25 04.3	-19 26 00	+0.002	+0.02	6.8	1.4	0.2	K0 III		120 s		
194882	32562		20 25 04.9	+59 36 00	+0.001	0.00	7.20	0.1	0.6	A0 V	-22	200 s	13850	m
194285	163549		20 25 07.9	-11 42 15	+0.001	+0.01	7.4	0.2	2.1	A5 V		120 s		
--	18855		20 25 10.0	+60 26 03	+0.001	0.00	8.0	1.3	0.2	K0 III		250 s		
194495	88648		20 25 10.7	+21 29 19	+0.001	0.00	7.1	0.0	0.4	B9.5 V	+11	210 s		
194511	88653		20 25 14.1	+23 26 42	-0.001	0.00	8.0	1.1	0.2	K0 III		350 s		
--	18856		20 25 14.6	+60 28 04	0.000	+0.02	7.9	2.8	-0.5	M2 III		91 s		
194669	49596		20 25 15.1	+42 10 44	+0.002	-0.03	7.9	0.4	3.0	F2 V		92 s		
194933	32568		20 25 16.4	+60 00 43	0.000	-0.01	7.3	0.1	1.4	A2 V		140 s		
194181	212106		20 25 20.7	-29 59 05	+0.001	0.00	7.4	1.6	-0.5	M4 III		380 s		
194452	125792		20 25 24.5	+7 52 32	0.000	+0.02	8.0	0.1	1.4	A2 V		200 s		
194453	125791		20 25 25.3	+6 38 30	0.000	0.00	6.6	0.1	0.6	A0 V		160 s		
194708	49603		20 25 25.5	+42 36 19	+0.004	+0.04	6.90	0.45	-2.0	F2 II		98 mx		
194684	49602		20 25 26.1	+41 54 37	-0.001	-0.01	6.9	0.1	1.4	A2 V		130 s		
194215	189264		20 25 26.7	-28 39 47	+0.001	+0.01	5.85	1.10	6.9	K3 V		14 mn		
194670	69999		20 25 28.5	+39 47 37	+0.001	+0.01	7.56	-0.09	-0.2	B8 V		360 s		
194685	70001		20 25 31.1	+40 05 52	+0.009	-0.01	7.79	0.59	4.0	F8 V		52 s	13847	m
194883	32566		20 25 32.5	+54 41 03	-0.002	0.00	7.36	-0.04		B3 e	-28			m
196156	--		20 25 33.0	+80 13 14	-0.002	+0.03	7.5	0.1	0.6	A0 V		240 s		
194237	212110		20 25 38.4	-32 32 21	-0.004	-0.07	7.6	2.0	-0.1	K2 III		110 s		
193924	246574	α Pav	20 25 38.7	-56 44 06	+0.002	-0.08	1.94	-0.20	-2.3	B3 IV	+2	71 s		[Peacock], m
194577	88664		20 25 40.3	+21 24 35	+0.001	-0.01	5.66	0.93	0.2	K0 III	-22	120 s		
194454	144412		20 25 42.3	-2 48 01	-0.001	-0.03	6.11	1.19	0.0	gK1	+24	140 s		
195013	18858		20 25 42.6	+60 10 40	+0.004	+0.03	7.40	0.1	1.7	A3 V		130 s	13864	m
194526	125797		20 25 43.7	+10 03 23	0.000	0.00	6.33	1.56	-0.3	K5 III	-77	190 s		
195193	18866		20 25 45.7	+67 08 29	+0.001	+0.02	7.9	1.1	0.2	K0 III		300 s		
194884	32571		20 25 47.4	+53 07 46	-0.001	+0.02	8.0	1.1	0.2	K0 III		330 s		
194184	230197		20 25 47.8	-40 47 47	-0.009	-0.09	6.09	1.36	0.2	K0 III		77 s		
194310	189273		20 25 48.6	-26 33 22	+0.003	0.00	8.0	-0.3	0.4	B9.5 V		330 s		
194779	49609		20 25 54.9	+41 20 13	-0.001	+0.02	7.77	0.23	-3.9	B3 II		1300 s		
194272	212115		20 25 56.2	-33 38 07	0.000	0.00	7.8	0.2	0.6	A0 V		220 s		
194597	106096		20 25 59.0	+14 56 35	0.000	+0.01	7.6	1.4	-0.3	K5 III		370 s		
194738	70011		20 26 00.7	+35 48 40	-0.001	-0.01	7.7	0.5	3.4	F5 V		70 s		
194616	106100		20 26 01.0	+19 51 54	+0.002	0.00	6.41	1.02	0.2	K0 III	-30	170 s		
194631	88671		20 26 01.1	+20 34 55	+0.001	-0.01	7.9	0.1	1.4	A2 V		190 s		

HD	SAO	Star Name	α 2000	δ 2000	μ(α)	μ(δ)	V	B-V	M$_v$	Spec	RV	d(pc)	ADS	Notes
194578	106097		20^{h}26^{m}01.5^s	+13°54'42"	+0.002^s	-0.02"	6.4	1.4	-0.3	K5 III		220 s		
194497	144418		20 26 02.1	-3 51 53	+0.003	0.00	7.6	0.5	3.4	F5 V		71 s		
194789	49612		20 26 03.7	+40 24 07	+0.001	+0.01	6.61	-0.11		B8				
194188	230199		20 26 04.2	-46 39 36	-0.001	-0.02	6.74	0.27		A5				
194458	163564		20 26 04.2	-14 58 47	+0.001	+0.03	7.4	1.1	-0.1	K2 III		320 s		
195352	18870		20 26 08.4	+69 19 36	+0.002	+0.01	7.3	0.9	3.2	G5 IV		64 s		
194790	70020		20 26 10.0	+38 51 43	0.000	+0.02	7.92	0.26	0.9	A3 IV		200 s		
188909	258855		20 26 12.1	-86 06 15	+0.010	+0.03	7.5	1.1	0.2	K0 III		290 s		
194579	125805		20 26 15.2	+2 57 26	+0.002	0.00	6.6	0.9	3.2	G5 IV		49 s		
194839	49618		20 26 21.6	+41 22 46	+0.003	0.00	7.49	0.99		B0.5 Ia	-21			
194473	163568		20 26 21.8	-19 09 06	0.000	+0.02	7.7	0.2	3.4	F5 V		72 s		
194688	106101		20 26 23.0	+17 18 55	+0.001	-0.01	6.22	1.01	0.2	K0 III	-17	160 s		
195066	32590		20 26 23.3	+56 38 20	+0.001	+0.01	6.21	0.01	0.0	A0 IV	-24	160 s	13870	m
194396	212120		20 26 25.8	-32 19 42	+0.004	+0.02	7.3	0.7	0.2	K0 III		260 s		
194515	163575		20 26 31.4	-17 22 38	0.000	+0.01	7.1	0.6	4.4	G0 V		35 s		
194432	189281		20 26 32.5	-29 04 26	0.000	-0.04	7.1	1.0	0.2	K0 III		240 s		
--	70027		20 26 32.6	+35 41 01	-0.002	0.00	7.9	1.6						
195014	32588		20 26 37.4	+51 06 21	-0.004	-0.03	7.5	0.1	1.4	A2 V		160 s		
194461	212125		20 26 39.5	-30 40 33	+0.002	-0.02	7.6	0.9	4.0	F8 V		33 s		
194908	49626		20 26 43.5	+40 20 52	-0.001	+0.02	7.5	0.0	0.4	B9.5 V		250 s		
194885	70033		20 26 43.7	+39 29 46	+0.002	+0.01	7.09	-0.07		A0	-16			
194909	70037		20 26 46.2	+39 27 44	+0.002	0.00	7.76	-0.01	0.6	A0 V		270 s		
196565	3405		20 26 48.3	+81 26 03	+0.019	+0.02	6.67	0.99	0.2	gG9	-4	190 s		
194433	212126		20 26 52.7	-37 24 11	-0.020	-0.11	6.25	0.97		K2 IV-V	+16	15 mn		m
--	--		20 26 53.6	+39 24 41			7.6	0.2	2.1	A5 V		120 s		
194350	230209		20 26 54.1	-45 35 37	+0.001	-0.01	7.8	1.7	-0.1	K2 III		380 s		
194863	70036		20 26 57.9	+30 22 12	0.000	+0.01	7.33	0.00	0.6	A0 V		220 s		
194328	230212		20 27 00.7	-47 15 35	+0.001	-0.01	8.0	0.2	2.6	F0 V		120 s		
194739	125822		20 27 02.0	+9 05 32	+0.001	0.00	7.9	0.0	-1.0	B5.5 V	-10	560 s		
195068	49643	43 Cyg	20 27 02.1	+49 23 01	+0.007	+0.06	5.69	0.26	2.6	F0 V	-20	42 s		
195047	49641		20 27 04.3	+47 55 23	+0.003	-0.02	7.9	0.9	3.2	G5 IV		87 s		
194475	212129		20 27 05.1	-35 45 02	+0.003	-0.06	8.0	1.3	-0.1	K2 III		220 mx		
194692	144440		20 27 06.7	-2 11 12	+0.003	0.00	8.0	1.1	0.2	K0 III		360 s		
194951	70044		20 27 07.5	+34 19 44	-0.001	0.00	6.39	0.48	-6.6	F2 Iab	-14	3300 s		
194654	163588		20 27 08.4	-11 32 50	+0.003	0.00	7.9	0.4	0.4	F0 V		110 s		
194476	--		20 27 12.0	-38 58 34			7.8	1.8	-0.5	M2 III		370 s		
195033	49644		20 27 14.0	+43 08 51	-0.001	+0.01	7.67	-0.06	0.4	B9.5 V		280 s		
194841	88693		20 27 14.1	+20 28 34	+0.001	-0.01	6.6	1.1	0.2	K0 III		190 s		
194822	106120		20 27 15.5	+16 42 39	0.000	-0.03	8.0	0.1	1.4	A2 V		200 s		
194636	163592	10 π Cap	20 27 19.0	-18 12 42	+0.001	-0.01	5.25	-0.07		B8	-13		13860	m
195527	18877		20 27 19.9	+68 46 04	+0.002	+0.01	7.2	1.1	0.2	K0 III	+8	230 s		
--	49646		20 27 20.0	+42 54 58	0.000	+0.01	7.8	0.7						
194911	88696		20 27 20.6	+23 47 25	0.000	+0.02	6.9	0.1	1.4	A2 V		120 s		
195016	70052		20 27 23.3	+39 43 51	0.000	-0.05	7.7	0.6	4.4	G0 V		46 s		
194864	106126		20 27 24.8	+20 04 50	0.000	+0.01	8.0	0.1	0.6	A0 V		290 s		
194764	144448		20 27 25.0	-1 33 31	-0.003	-0.05	7.1	0.7	4.4	G0 V		34 s		
193049	257807		20 27 25.5	-78 45 18	+0.007	+0.02	7.5	0.5	4.0	F8 V		50 s		
194766	144449		20 27 26.6	-2 07 09	-0.005	-0.06	7.50	0.52		F8			13868	m
194765	144450		20 27 27.3	-2 06 10	-0.005	-0.06	6.70	0.52	4.0	dF8	-16	40 mn	13868	m
194587	189302		20 27 28.3	-28 48 50	0.000	0.00	7.2	0.8	3.4	F5 V		34 s		
194823	106123		20 27 29.8	+12 47 55	+0.001	+0.01	7.9	1.1	0.2	K0 III		330 s		
194528	212132		20 27 33.4	-39 31 56	+0.004	-0.02	7.9	0.4	2.1	A5 V		110 s		
193607	257809		20 27 33.8	-74 17 04	+0.012	0.00	7.96	1.18		K0				
195050	70056	40 Cyg	20 27 34.1	+38 26 25	-0.002	-0.07	5.62	0.06	1.7	A3 V	0	61 s		
195100	49652		20 27 34.5	+43 03 24	-0.004	-0.02	7.57	0.89	0.3	G5 III	+2	280 s		
195089	49653		20 27 36.5	+42 02 06	0.000	-0.02	7.31	-0.01		B3	-7			
194934	88700		20 27 36.6	+22 14 15	-0.001	-0.01	7.8	1.4	-0.3	K5 III		390 s		
194437	230218		20 27 41.0	-49 46 00	0.000	+0.01	7.9	2.1	-0.5	M4 III		460 s		
194914	106131		20 27 43.2	+14 27 21	+0.001	0.00	7.8	0.1	0.6	A0 V		260 s		
194640	212140		20 27 44.1	-30 52 05	-0.001	-0.52	6.62	0.73	5.2	G5 V	-3	19 ts		
194794	144454		20 27 45.9	-5 39 15	0.000	-0.01	7.0	1.6	-0.5	M4 III		320 s		
194989	88704		20 27 47.3	+26 08 49	0.000	-0.01	7.9	0.1	1.7	A3 V		170 s		
194935	106132		20 27 49.8	+14 56 39	+0.002	0.00	8.0	0.1	0.6	A0 V		280 s		
194795	144455		20 27 50.6	-6 40 25	+0.002	-0.01	7.9	0.1	1.7	A3 V		170 s		
195503	18879		20 27 51.4	+65 45 21	0.000	-0.03	6.6	0.1	1.4	A2 V		110 s		
194889	125833		20 27 54.1	+6 52 50	-0.001	-0.08	7.9	1.1	0.2	K0 III		190 mx		
194676	189308		20 27 55.1	-28 15 42	+0.001		7.0	1.6		M5 III	+18			T Mic, v
194606	--		20 27 57.9	-38 22 29			8.0	1.6	0.2	K0 III		160 s		
194825	144457		20 27 58.5	-5 04 46	0.000	-0.02	8.0	1.1	0.2	K0 III		360 s		
195391	32611		20 27 59.9	+59 44 04	-0.001	+0.02	7.96	-0.03	0.6	A0 V		300 s		
194711	189309		20 28 01.0	-25 36 22	+0.002	+0.04	6.6	0.6	4.0	F8 V		30 s		
194798	163599		20 28 02.8	-13 59 50	0.000	-0.01	7.9	0.1	1.4	A2 V		200 s		
195102	70063		20 28 03.6	+33 53 17	-0.001	0.00	7.00	-0.06		B9	-9			
194696	189310		20 28 05.6	-29 22 27	-0.002	-0.02	7.3	1.3	0.2	K0 III		180 s		
195150	49660		20 28 05.6	+40 36 39	+0.001	+0.02	7.9	0.1	0.6	A0 V		270 s		

HD	SAO	Star Name	α 2000	δ 2000	μ(α)	μ(δ)	V	B–V	M_v	Spec	RV	d(pc)	ADS	Notes
194937	125841		20ʰ28ᵐ07.3ˢ	+8°26′15″	+0.003ˢ	+0.02″	6.25	1.08	0.2	G9 III	−11	130 s		
195018	106136		20 28 07.6	+19 45 46	+0.001	+0.01	8.0	0.4	2.6	F0 V		120 s		
194768	189316		20 28 08.7	−23 00 47	+0.001	+0.06	7.9	1.7	−0.3	K5 III		260 mx		
195034	88711		20 28 11.6	+22 07 44	−0.002	−0.24	7.2	0.8	3.2	G5 IV		62 s		
196787	3408	75 Dra	20 28 14.1	+81 25 22	+0.012	+0.02	5.46	1.02	0.2	G9 III	−6	100 s		m
195124	70068		20 28 15.9	+35 05 39	−0.001	−0.01	8.0	1.1	0.2	K0 III		340 s		
194953	125843		20 28 16.7	+2 56 13	+0.003	0.00	6.21	0.90	0.3	G8 III	−22	150 s		
195019	106138		20 28 18.4	+18 46 09	+0.024	−0.06	6.80	0.6	3.9	G3 IV-V	−93	37 s	13886	m
195053	106143		20 28 20.8	+20 06 48	+0.002	+0.03	6.8	0.1	0.6	A0 V	−34	170 s		
195052	88715		20 28 23.8	+20 30 45	+0.002	0.00	8.0	0.5	4.0	F8 V		62 s		
194939	144468	68 Aql	20 28 24.6	−3 21 28	+0.001	−0.01	6.13	−0.06		B9				m
195194	70075		20 28 25.7	+39 19 44	+0.002	+0.02	6.97	0.97	0.3	G8 III		210 s		
194916	144465		20 28 26.4	−9 22 15	+0.001	−0.01	7.1	0.9	3.2	G5 IV		60 s		
195020	106139		20 28 26.5	+11 43 15	+0.002	+0.02	8.0	0.1	1.7	A3 V		170 s		
195178	70077		20 28 28.4	+37 47 22	0.000	−0.01	7.56	1.60		Ma				
194810	189321		20 28 29.3	−24 09 47	+0.005	−0.10	7.0	1.1	4.4	G0 V		16 s		
195229	49670		20 28 29.9	+42 00 36	−0.002	0.00	7.66	0.16		B0.5 III				
195254	49672		20 28 30.4	+43 31 23	−0.002	0.00	7.69	0.01		B9				
195307	49675		20 28 30.5	+48 55 08	−0.001	−0.02	6.6	1.1	−0.1	K2 III		210 s		
194727	212153		20 28 32.3	−36 11 40	+0.004	−0.03	7.6	1.4	0.2	K0 III		180 s		
194747	212155		20 28 32.6	−34 24 34	+0.003	0.00	6.8	1.8	0.2	K0 III		73 s		
194917	163610		20 28 33.0	−11 47 47	0.000	−0.02	7.40	1.6	−0.4	M0 III	+17	360 s		
195021	125849		20 28 34.6	+8 15 23	+0.001	−0.06	7.7	1.1	−0.1	K2 III		260 mx		
194990	125847		20 28 35.7	+0 53 01	−0.001	−0.01	7.2	0.0	0.0	B8.5 V		280 s		
194976	144474		20 28 37.0	−4 25 39	0.000	+0.04	7.9	0.5	4.0	F8 V		60 s		
195054	106144		20 28 37.1	+10 59 13	+0.001	+0.01	8.0	0.1	0.6	A0 V		280 s		
195151	88728		20 28 38.4	+28 46 34	+0.002	0.00	7.9	0.1	0.6	A0 V		280 s		
194189	254796		20 28 41.1	−67 45 29	+0.010	+0.01	7.3	0.4	3.0	F2 V		72 s		
195125	88723		20 28 41.2	+21 24 23	0.000	+0.01	8.0	1.1	−0.1	K2 III		400 s		
194918	163612		20 28 43.4	−15 44 30	0.000	−0.01	6.41	0.99		K0				
194848	189330		20 28 44.0	−23 59 01	0.000	−0.04	6.8	1.5	0.2	K0 III		110 s		
195338	49677		20 28 44.1	+47 36 41	−0.001	0.00	7.48	1.17	−2.1	G7 II	−24	660 s		
194783	212160		20 28 46.5	−35 35 45	+0.001	−0.02	6.2	0.1	0.6	A0 V		130 s		
194958	163615		20 28 47.8	−11 43 38	−0.002	−0.06	7.3	0.4	3.0	F2 V		73 s		
195322	49678		20 28 48.4	+45 43 09	−0.002	+0.02	7.40	−0.03	0.4	B9.5 V		250 s		
195353	49680		20 28 49.5	+48 51 41	−0.003	−0.14	7.8	0.7	4.4	G0 V		48 s		
194772	212161		20 28 49.8	−36 50 28	−0.001	0.00	7.8	1.5	0.2	K0 III		170 s		
195005	144480		20 28 49.9	−1 44 05	+0.005	−0.04	6.9	0.7	4.4	G0 V		31 s		
194943	163614	11 ρ Cap	20 28 51.4	−17 48 49	−0.001	−0.02	4.78	0.38	0.6	F2 III	+18	32 ts	13887	m
195373	49682		20 28 52.2	+49 15 51	+0.001	−0.03	7.3	0.8	3.2	G5 IV		65 s		
195038	144482		20 28 55.1	−0 50 01	−0.002	−0.04	7.7	1.1	0.2	K0 III		310 s		
195103	106149		20 28 57.6	+10 19 52	0.000	−0.02	7.9	1.4	−0.3	K5 III		410 s		
194959	163616		20 28 58.8	−17 26 05	+0.001	−0.02	6.8	0.6	4.4	G0 V	−14	30 s		
194944	189335		20 29 00.1	−20 54 05	0.000	+0.01	6.8	0.5	0.6	A0 V		91 s		
194960	163617		20 29 00.5	−17 52 31	+0.003	−0.12	6.55	1.07	0.0	K1 III	+4		13887	m
195216	88738		20 29 00.5	+27 51 06	+0.002	+0.01	7.70	1.62	−0.3	K5 III		340 s		
195215	88741		20 29 01.4	+28 27 43	+0.003	0.00	7.6	0.1	1.4	A2 V		180 s		
197508	3418		20 29 02.8	+83 37 32	+0.014	−0.02	6.2	0.1	1.4	A2 V	+10	90 s		
195154	106151		20 29 03.0	+16 15 15	0.000	0.00	7.3	0.1	0.6	A0 V		210 s		
194679	246604		20 29 03.4	−51 05 05	−0.002	+0.02	8.00	0.4	2.6	F0 V		120 s		m
195271	70092		20 29 05.7	+35 49 52	0.000	−0.02	7.32	1.45		K5	−22		13905	m
195231	88742		20 29 06.8	+28 21 40	+0.007	+0.07	7.6	0.4	3.0	F2 V		84 s		
195134	106152		20 29 08.6	+12 40 52	+0.001	+0.01	6.98	0.03	0.4	B9.5 V		190 s		
195162	106153		20 29 09.5	+16 16 24	+0.003	−0.01	8.0	1.6		M6 III				RS Del, q
195104	125857		20 29 11.6	+4 27 24	+0.003	+0.03	7.8	0.5	4.0	F8 V		57 s		
195323	−−		20 29 13.8	+37 31 04			7.74	0.66		F5			13909	m
195607	18891		20 29 18.5	+60 25 18	−0.002	−0.02	7.4	0.1	0.6	A0 V		210 s		
195324	70096	42 Cyg	20 29 20.2	+36 27 17	0.000	0.00	5.88	0.52	−5.1	A1 Ib	−18	820 s		
195447	32622		20 29 20.7	+50 27 38	+0.001	+0.02	7.4	0.0	0.4	B9.5 V		230 s		
195217	106156		20 29 20.9	+20 05 16	−0.002	0.00	6.40	0.23		A m	+4			
194898	212171		20 29 21.4	−34 06 40	+0.001	+0.01	8.0	1.6	0.2	K0 III		160 s		
195295	70095	41 Cyg	20 29 23.6	+30 22 07	0.000	0.00	4.01	0.40	−2.0	F5 II	−18	160 s		
194753	230235		20 29 26.5	−49 46 58	+0.002	0.00	8.0	0.6	0.6	A0 V		300 s		
195554	32627		20 29 26.9	+56 04 06	0.000	+0.01	6.00	−0.04		B9	−22			m
196925	3413	74 Dra	20 29 27.4	+81 05 29	+0.028	+0.22	5.96	0.92	0.2	K0 III	−14	81 mx		m
195405	49692		20 29 28.2	+42 18 59	0.000	−0.08	7.99	0.63	−6.3	G0 I p		130 mx		
195273	88750		20 29 28.6	+26 56 21	+0.002	+0.01	7.45	1.03	0.0	K1 III		310 s		
195075	163621		20 29 29.8	−12 35 29	0.000	−0.01	7.5	0.8	0.3	G7 III	+28	280 s		
195006	189345		20 29 31.2	−22 23 29	+0.001	−0.02	6.16	1.55	−0.5	gM1	+56	210 s		
195725	18897	2 θ Cep	20 29 34.7	+62 59 39	+0.006	−0.01	4.22	0.20		A m	−8	34 t		
194876	230239		20 29 34.9	−41 23 58	−0.001	+0.01	7.6	0.3	2.6	F0 V		100 s		
195688	18894		20 29 35.4	+60 59 32	−0.007	+0.01	8.0	0.5	4.0	F8 V		61 s		
195041	189348		20 29 37.6	−22 50 43	0.000	−0.01	7.7	0.9	0.2	K0 III		310 s		
195555	32630		20 29 37.8	+54 07 29	+0.001	−0.01	7.9	0.5	3.4	F5 V		76 s		
195135	144495	69 Aql	20 29 38.8	−2 53 08	+0.005	−0.02	4.91	1.15	−0.1	K2 III	−23	100 s		

HD	SAO	Star Name	α 2000	δ 2000	μ(α)	μ(δ)	V	B-V	M_v	Spec	RV	d(pc)	ADS	Notes
			$20^h 29^m 40.7^s$	$-4°10'57''$	0.000^s	$-0.01''$								
195155	144496		20 29 40.7	-4 10 57	0.000	-0.01	8.0	1.4	-0.3	K5 III		450 s		
195198	125863		20 29 44.7	+2 59 46	+0.001	0.00	7.0	0.1	0.6	A0 V		190 s		
195407	70108		20 29 47.6	+36 58 50	0.000	0.00	7.80	0.33	-4.6	B0 IV pe		2000 s		
195093	163625	12 o Cap	20 29 52.3	-18 35 12	+0.001	-0.09	6.74	0.22		A3			13902	m
195094	163626	12 o Cap	20 29 53.7	-18 35 00	+0.002	-0.08	5.94	0.08		A2			13902	m
195590	32632		20 29 55.0	+53 42 16	+0.005	-0.01	8.0	1.1	0.2	K0 III		290 mx		
195274	106164		20 29 55.1	+10 18 07	+0.001	-0.02	6.7	1.1	0.2	K0 III	-13	200 s		
195395	70109		20 29 58.6	+30 33 23	-0.001	-0.03	7.8	0.1	1.7	A3 V		160 s		
195506	49704		20 29 59.7	+45 55 43	+0.007	+0.16	6.41	1.13	-0.1	K2 III	-31	100 mx		
196142	9793		20 30 00.4	+72 31 54	-0.001	-0.02	6.27	1.35	-0.1	gK2	-43	140 s		
195200	144501		20 30 01.5	-4 26 17	+0.003	-0.10	7.8	0.5	4.0	F8 V		58 s		
193549	257810		20 30 02.1	-78 54 23	+0.003	0.00	7.7	0.4	2.6	F0 V		110 s		
195556	49712	45 ω¹ Cyg	20 30 03.4	+48 57 06	+0.001	+0.01	4.95	-0.09	-2.5	B2 V	-22	270 s	13932	m
--	--		20 30 04.3	+44 40 52	0.000	+0.01	6.4	1.6	-0.5	M2 III		230 s		
195340	106170		20 30 04.6	+19 40 03	0.000	0.00	6.7	0.0	0.0	B8.5 V	-7	210 s		
195557	49714		20 30 04.9	+48 38 29	+0.003	+0.01	7.61	0.03	0.6	A0 V		240 s		
195528	49708		20 30 07.0	+45 26 56	+0.009	+0.06	7.9	0.5	4.0	F8 V		59 s		
195341	106171		20 30 07.1	+19 25 49	0.000	+0.01	7.0	0.0	0.4	B9.5 V	-9	200 s		
195408	88771		20 30 13.4	+27 51 58	0.000	0.00	8.0	0.0	0.4	B9.5 V		320 s		
195358	106177		20 30 14.2	+19 25 16	0.000	0.00	6.60	0.00	0.4	B9.5 V	-2	170 s	13921	m
194441	254800		20 30 17.3	-69 04 09	+0.003	-0.04	7.30	0.1	1.4	A2 V		150 s		m
193734	--		20 30 17.6	-77 36 50			8.0	1.1	0.2	K0 III		360 s		
195325	106172	1 Del	20 30 17.8	+10 53 46	+0.001	+0.01	5.94	0.02		A0	-16	27 mn	13920	m
195663	32636		20 30 18.5	+54 40 52	+0.010	+0.11	7.4	0.5	3.4	F5 V		63 s		
195608	49722		20 30 19.7	+49 46 48	+0.002	-0.08	6.6	1.1	0.2	K0 III		180 s		
195326	125878		20 30 20.5	+9 22 36	+0.003	-0.05	7.9	1.4	-0.3	K5 III		430 s		
--	9794		20 30 21.3	+70 36 42	+0.007	+0.02	7.9	0.9						
195529	49715		20 30 23.1	+40 39 52	+0.001	0.00	8.00	-0.12		A0				
195432	88774		20 30 24.9	+27 49 58	-0.001	+0.01	6.9	0.7	-6.2	G2 I	-22	3100 s		
195664	32637		20 30 26.8	+52 17 39	+0.001	+0.01	7.5	0.1	1.4	A2 V		160 s		
195543	49719		20 30 31.1	+40 30 26	+0.002	+0.01	7.5	1.4	-0.3	K5 III		330 s		
195574	49720		20 30 31.1	+43 27 50	-0.001	0.00	7.4	0.1	1.7	A3 V		140 s		
194663	254803		20 30 33.2	-64 11 25	+0.003	-0.01	7.5	0.2	3.4	F5 V		65 s		
195397	106180		20 30 33.3	+13 49 01	+0.001	-0.02	8.0	0.5	4.0	F8 V		62 s		
195140	212190		20 30 33.6	-31 39 37	-0.001	-0.07	7.8	1.5	-0.1	K2 III		150 mx		
195277	144512		20 30 34.1	-5 23 23	0.000	-0.01	7.7	1.1	-0.1	K2 III		360 s		
195787	32645		20 30 34.4	+58 17 14	-0.003	-0.11	7.5	1.1	0.2	K0 III		210 mx		
195592	49724		20 30 34.7	+44 18 54	-0.001	-0.01	7.08	0.87	-6.2	O9.5 Ia	-28	970 s		
195243	163639		20 30 39.6	-16 33 03	+0.003	-0.04	7.9	0.5	3.4	F5 V		78 s		
195222	189370		20 30 40.9	-20 42 35	0.000	-0.01	7.1	1.4	0.2	K0 III		150 s		
195544	70129		20 30 44.8	+36 23 40	+0.001	0.00	7.76	0.38	3.4	F5 V		75 s		
195258	163641		20 30 45.9	-15 46 17	+0.001	+0.01	7.1	0.1	0.6	A0 V		200 s		
194970	246618		20 30 48.6	-52 48 29	+0.004	-0.02	7.5	1.9	-0.1	K2 III		120 s		
195710	49731		20 30 52.6	+49 12 49	0.000	0.00	6.62	0.04		A0	+6			m
195328	163643		20 30 55.6	-10 02 06	+0.002	-0.04	7.8	0.8	3.2	G5 IV		84 s		
195509	88786		20 30 56.4	+26 41 09	-0.001	0.00	7.40	0.99	0.2	K0 III		280 s		
195206	189374		20 30 56.6	-29 06 45	+0.001	+0.01	6.1	0.9	2.1	A5 V		22 s		
195479	88783		20 30 57.9	+20 36 21	+0.007	+0.05	6.00	0.15		A m	-40		13913	m
195593	70135	44 Cyg	20 30 59.1	+36 56 09	0.000	0.00	6.19	1.01	-6.4	F5 Iab	-22	1700 s	13949	m
195480	106192		20 31 02.1	+17 04 34	+0.001	+0.01	7.5	1.4	-0.3	K5 III		350 s		
195629	49729		20 31 02.5	+40 25 37	-0.001	+0.02	7.60	-0.08	0.4	B9.5 V		280 s		
195330	163645		20 31 04.2	-15 03 23	-0.003	-0.05	6.12	0.79	4.4	G0 V	+30	15 s		m
195481	106194		20 31 05.9	+15 48 27	-0.001	-0.01	6.85	0.10		A2			13944	m
195647	49730		20 31 06.6	+40 51 30	0.000	+0.06	7.4	1.1	-0.1	K2 III	-27	270 mx		
195482	106195		20 31 11.7	+11 15 33	-0.001	-0.01	7.12	0.02			+2		13946	m
195483	106196		20 31 12.8	+11 15 39	0.000	0.00	7.11	0.03	0.6	A0 V	-11	200 s	13946	m
193441	258860		20 31 16.9	-80 32 37	+0.010	-0.02	7.6	0.4	2.6	F0 V		91 s		
195774	49741	46 ω² Cyg	20 31 18.6	+49 13 13	+0.001	-0.03	5.44	1.55	-0.5	M2 III	-64	150 s		m
195363	163650		20 31 20.2	-17 08 19	+0.001	0.00	7.7	1.6	-0.5	M2 III		440 s		
195820	32649		20 31 20.9	+52 18 35	+0.003	+0.07	6.18	1.01	0.2	K0 III	-10	160 s		
195141	230251		20 31 21.4	-44 22 42	+0.004	-0.02	7.9	1.4	0.2	K0 III		240 s		
195689	70152		20 31 26.2	+39 56 20	+0.001	+0.02	7.55	0.04		A0				
195490	125894		20 31 28.5	+8 55 17	+0.004	+0.03	7.8	0.1	1.4	A2 V		190 s		
195649	70147		20 31 29.8	+33 05 55	-0.001	-0.02	7.1	1.1	0.2	K0 III	-20	230 s		
196502	9802	73 Dra	20 31 30.2	+74 57 17	+0.001	-0.01	5.20	0.07		A2 p	+9	33 mn		AF Dra, v
195575	88798		20 31 31.7	+21 53 43	-0.001	0.00	6.9	0.1	1.7	A3 V		110 s		
195917	32662		20 31 34.8	+56 21 22	+0.001	+0.01	7.66	-0.13	0.4	B9.5 V		280 s		
195690	70153		20 31 36.1	+34 19 51	0.000	-0.04	6.48	0.37		F2	+14			
195788	49746		20 31 39.5	+45 10 17	+0.003	+0.01	7.93	-0.07	0.6	A0 V		290 s		
195666	70154		20 31 43.3	+30 15 42	-0.001	+0.05	7.3	1.1	-0.1	K2 III	-21	290 s		
195533	125900		20 31 43.4	+4 24 57	+0.002	+0.01	7.1	0.4	2.6	F0 V		79 s		
195366	189392		20 31 46.0	-28 04 23	+0.002	+0.01	7.7	0.6	3.2	G5 IV		80 s		
195964	32667		20 31 46.2	+56 46 48	+0.003	+0.01	6.14	1.43	-0.3	K5 III	-15	190 s		
195612	106204		20 31 52.1	+16 59 30	+0.001	-0.01	7.1	0.0	0.4	B9.5 V	-8	210 s		
196053	32668		20 31 52.8	+60 03 47	+0.003	+0.03	7.6	0.1	0.6	A0 V		240 s		

HD	SAO	Star Name	α 2000	δ 2000	μ(α)	μ(δ)	V	B-V	M_v	Spec	RV	d(pc)	ADS	Notes
			h m s	° ′ ″	s	″								
195534	144534		20 31 54.4	−0 09 14	−0.002	+0.04	7.6	0.9	0.3	G8 III	−49	290 s		
195512	144532		20 31 56.6	−5 36 10	0.000	+0.02	8.0	1.1	0.2	K0 III		360 s		
195692	88808		20 31 58.0	+25 48 17	+0.002	−0.03	6.3	0.7	4.4	G0 V	−18	24 s	13964	m
195189	246622		20 31 58.8	−50 11 05	+0.003	−0.03	7.6	1.3	0.2	K0 III		300 s		
194881	254807		20 32 00.8	−65 48 56	−0.010	−0.05	7.4	0.5	3.4	F5 V		62 s		
196085	18914		20 32 04.3	+60 25 21	+0.001	−0.08	7.1	0.4	3.0	F2 V		66 s		
195535	144537		20 32 05.0	−5 14 33	+0.001	+0.02	6.70	1.1	0.2	K0 III		200 s	13956	m
195713	88812		20 32 05.4	+25 21 07	+0.001	0.00	7.9	0.1	0.6	A0 V		280 s		
195694	88809		20 32 06.1	+22 00 50	−0.001	−0.01	7.9	0.0	0.4	B9.5 V		310 s		
195651	106210		20 32 07.5	+15 11 12	+0.007	+0.11	8.00	0.6	4.4	G0 V		52 s	13965	m
195668	106213		20 32 07.8	+18 37 39	+0.001	−0.01	7.20	1.6	−1.4	M4 II-III	−3	500 s		
195714	88813		20 32 08.9	+22 49 30	0.000	+0.02	6.5	1.1	0.2	K0 III		180 s		
194574	257817		20 32 09.5	−72 58 18	−0.008	−0.01	7.7	1.1	0.2	K0 III		310 mx		
195669	106212		20 32 10.1	+15 52 57	−0.001	−0.01	7.6	0.9	3.2	G5 IV		74 s		
196229	18918		20 32 13.3	+65 23 56	+0.001	+0.02	6.7	1.1	0.2	K0 III		190 s		
195453	189403		20 32 13.3	−22 09 16	+0.002	+0.02	7.80	0.4	2.6	F0 V		110 s		m
194646	257819		20 32 14.0	−71 51 53	−0.004	−0.05	7.2	0.1	0.6	A0 V		81 mx		
195715	88814		20 32 14.6	+20 51 35	0.000	0.00	7.3	0.0	0.0	B8.5 V		280 s		
194593	257818		20 32 16.8	−72 50 00	−0.002	+0.08	7.9	0.8	3.2	G5 IV		86 s		
196017	32670		20 32 16.8	+54 09 23	0.000	+0.01	7.41	−0.10	0.4	B9.5 V		250 s		
195616	125910		20 32 17.4	+2 13 12	−0.001	0.00	7.5	1.1	−0.1	K2 III		330 s		
195470	189406		20 32 18.9	−24 52 13	+0.002	+0.01	7.1	1.2	−0.1	K2 III		260 s		
195617	125911		20 32 20.1	+2 07 57	+0.001	−0.01	6.6	1.1	−0.1	K2 III		220 s		
195564	163665		20 32 23.5	−9 51 13	+0.021	+0.10	5.65	0.69	4.9	G3 V	+9	15 ts	13960	m
196420	9804		20 32 25.0	+70 31 55	−0.001	−0.02	6.7	0.1	0.6	A0 V		160 s		
195416	212212		20 32 25.1	−36 50 19	−0.002	−0.04	8.0	1.9	0.2	K0 III		110 s		
195965	49765		20 32 25.2	+48 12 58	−0.003	0.00	6.98	−0.05	−4.1	B0 V	+10	1300 s		
194238	257816		20 32 31.3	−77 17 39	−0.003	0.00	7.8	1.1	0.2	K0 III		330 s		
195536	163666		20 32 31.5	−16 36 33	+0.005	0.00	7.90	0.4	2.6	F0 V		120 s	13961	m
195349	230263		20 32 36.3	−47 14 35	+0.001	−0.02	8.0	2.0	0.2	K0 III		300 s		
195653	144546		20 32 40.4	−1 27 41	−0.002	−0.01	7.9	1.4	−0.3	K5 III		450 s		
195834	88818		20 32 41.2	+29 03 14	0.000	−0.01	8.00	1.1	−2.3	K3 II		1000 s		
195500	189413		20 32 42.3	−28 35 43	+0.004	0.00	7.1	0.3	0.6	A0 V		130 s		
195486	212221		20 32 42.9	−31 22 56	+0.001	0.00	7.70	0.1	1.4	A2 V		180 s		m
195775	106218		20 32 43.7	+16 46 02	+0.001	0.00	6.7	0.0	0.4	B9.5 V		180 s		
196087	32676		20 32 43.8	+53 01 30	+0.001	+0.01	7.7	0.1	1.4	A2 V		170 s		
195456	212218		20 32 46.0	−36 13 44	+0.002	−0.02	7.8	0.4	0.0	B8.5 V		180 s		
195985	49770		20 32 47.0	+44 58 44	+0.002	0.00	7.66	−0.09	−1.0	B5.5 V	−7	520 s		
195835	88820		20 32 50.9	+27 12 43	−0.001	0.00	7.1	1.1	0.2	K0 III	−16	230 s		
195987	49769		20 32 51.4	+41 53 54	−0.015	+0.45	7.09	0.78	5.7	G9 V	−10	21 ts		
195549	189416		20 32 52.1	−24 56 38	+0.001	−0.04	6.2	0.3	0.6	A0 V		80 s		
195986	49772		20 32 52.2	+43 11 30	0.000	+0.01	6.60	−0.11	−2.6	B4 III	−17	630 s		
195729	125930		20 32 56.8	+2 25 38	+0.001	−0.01	8.0	0.1	1.4	A2 V		210 s		
194972	254811		20 32 57.1	−68 22 18	+0.004	−0.01	7.50	0.1	0.6	A0 V		240 s		m
196089	49782		20 32 59.0	+49 50 28	+0.001	0.00	6.7	0.7	4.4	G0 V		29 s		
195145	254813		20 33 00.4	−62 29 47	+0.004	−0.20	8.0	0.8	3.2	G5 IV		51 mx		
195988	49775		20 33 01.3	+41 28 24	+0.001	+0.03	7.1	0.6	4.4	G0 V		35 s		
--	70179		20 33 02.5	+31 44 07	+0.001	−0.01	8.00						13995	m
195907	70181		20 33 04.9	+31 39 26	0.000	−0.01	7.82	0.03		B2 e				
196143	32681		20 33 06.4	+54 27 26	+0.001	+0.01	7.0	1.6	−0.5	M2 III		290 s		
195518	212225		20 33 08.8	−35 52 28	−0.002	+0.01	8.0	2.0	−0.3	K5 III		220 s		
195874	88826		20 33 09.0	+22 08 57	−0.001	−0.02	7.8	0.0	0.4	B9.5 V		290 s		
195810	106230	2 ε Del	20 33 12.6	+11 18 12	+0.001	−0.02	4.03	−0.13	−1.9	B6 III	−19	150 s		
196090	49786		20 33 14.0	+47 09 58	−0.003	−0.02	7.79	1.42	0.3	G7 III	−42	130 s		
193721	258861		20 33 17.6	−80 57 54	+0.005	−0.01	5.77	1.14		K0				m
196132	49790		20 33 22.0	+50 07 09	+0.003	+0.02	7.8	1.4	−0.3	K5 III		390 s		
195967	88833		20 33 22.5	+29 31 26	−0.001	−0.02	7.50	1.1	−0.1	K2 III	+13	320 s		
195521	230268		20 33 22.8	−41 31 30	−0.006	−0.19	7.0	0.4	4.4	G0 V		33 s		
196195	32688		20 33 23.2	+54 48 08	+0.003	+0.02	7.1	0.4	3.0	F2 V		64 s		
196541	18929		20 33 24.6	+69 40 18	−0.004	−0.01	7.9	1.1	0.2	K0 III		300 s		
195767	144563		20 33 25.3	−6 13 09	+0.001	+0.02	7.30	1.6	−0.5	M3 III	−21	360 s		m
196177	32687		20 33 26.3	+52 48 14	0.000	−0.01	7.6	1.4	−0.3	K5 III		350 s		
196006	70194		20 33 30.2	+32 54 27	0.000	−0.01	7.31	−0.15	−1.6	B3.5 V	−27	590 s		
195680	189425		20 33 30.9	−22 13 54	−0.005	−0.05	7.40	0.5	4.0	F8 V		48 s	13988	m
195824	125947		20 33 36.9	+1 16 09	+0.001	+0.01	8.0	1.1	0.2	K0 III		360 s		
195641	212230		20 33 36.9	−30 11 04	−0.002	−0.04	7.7	1.6	−0.1	K2 III		210 s		
195968	88840		20 33 37.5	+25 52 39	+0.001	+0.02	7.8	0.1	0.6	A0 V		270 s		
195428	246631		20 33 39.1	−53 29 54	+0.006	−0.08	7.9	1.2	0.2	K0 III		160 mx		
196133	49795		20 33 39.3	+45 10 33	+0.001	+0.02	6.69	0.03		A0 p	−8			
196092	70198		20 33 42.6	+38 34 58	0.000	−0.02	8.00	0.03		A0			14007	m
196022	88842		20 33 45.2	+27 52 21	+0.002	+0.03	7.8	0.2	2.1	A5 V		140 s		
--	--		20 33 45.7	−40 33 13			7.90							m
196239	32693		20 33 45.9	+52 49 36	−0.001	−0.01	7.3	0.1	0.6	A0 V		210 s		
195599	230275		20 33 46.5	−40 33 27	+0.001	−0.01	7.1	0.8	0.6	A0 V		61 s		m
196134	49796		20 33 48.2	+41 46 20	−0.002	−0.08	6.49	1.00	0.2	K0 III	+1	180 s		

HD	SAO	Star Name	α 2000	δ 2000	μ(α)	μ(δ)	V	B−V	M$_v$	Spec	RV	d(pc)	ADS	Notes
199135	3444		20^{h}33^{m}48.5^s	+85°39'19"	+0.009^s	+0.01"	7.9	1.1	0.2	K0 III		350 s		
195704	189430		20 33 49.4	−29 17 34	+0.005	−0.01	7.0	0.4	1.7	A3 V		72 s		
195922	125960		20 33 53.4	+10 03 35	−0.001	+0.01	6.56	0.08		A0	−13			
196093	70203	47 Cyg	20 33 54.0	+35 15 03	0.000	0.00	4.61	1.60	−4.4	K2 Ib comp	−4	570 s		
195769	189437		20 33 54.2	−20 32 20	+0.001	−0.03	8.0	0.9	0.2	K0 III		360 s		
196178	49804		20 33 54.7	+46 41 38	+0.001	0.00	5.78	−0.16		B8 p	−22			m
195569	230276	ν Mic	20 33 54.9	−44 30 58	+0.001	−0.04	5.11	1.01	0.2	K0 III	+9	96 s		
195488	246633		20 33 55.8	−52 50 22	+0.002	−0.04	7.9	0.8	4.4	G0 V		37 s		
196407	18928		20 33 56.0	+61 44 36	+0.001	+0.01	7.04	0.07	0.6	A0 V		170 s		
195943	106248	3 η Del	20 33 56.9	+13 01 38	+0.005	+0.03	5.38	0.07	1.4	A2 V	−18	61 s		
196196	49806		20 33 58.2	+46 24 59	+0.001	0.00	7.63	−0.03		B9				
195720	212235		20 33 58.9	−31 12 45	−0.003	−0.02	8.0	2.5	−0.5	M2 III		150 s		
196281	32702		20 33 59.4	+52 35 52	−0.001	−0.03	7.7	1.1	−0.1	K2 III		330 s		
195909	125961		20 33 59.7	+4 53 55	+0.001	0.00	6.41	1.02		K0				
195993	106253		20 34 01.3	+18 11 10	+0.002	+0.03	7.30	1.4	−0.3	K5 III	−3	320 s		
195827	163683		20 34 01.8	−12 20 38	+0.002	−0.01	7.0	0.1	0.6	A0 V		190 s		
196120	70206		20 34 01.8	+34 40 46	0.000	+0.01	6.67	−0.12	0.0	B8.5 V	−28	220 s	14016	m
196023	106256		20 34 07.2	+19 57 03	0.000	−0.01	7.7	1.4	−0.3	K5 III		380 s		
196007	106254		20 34 08.4	+15 49 34	+0.003	+0.05	7.7	0.8	3.2	G5 IV		79 s		
196215	49810		20 34 09.2	+45 07 08	−0.004	−0.07	8.0	1.1	0.2	K0 III		290 mx		
196035	88846		20 34 09.8	+20 59 07	+0.001	0.00	6.48	−0.14	−2.3	B3 IV	+3	540 s		
195838	163686		20 34 11.6	−13 43 16	+0.005	+0.07	6.3	0.5	4.0	F8 V	−43	28 s		
195251	254815		20 34 13.3	−66 44 25	+0.004	−0.03	7.7	1.1	0.2	K0 III		320 s		
196216	49814		20 34 16.0	+43 21 40	+0.002	−0.07	7.28	0.38		F2	−7			
196036	106259		20 34 16.1	+19 31 51	0.000	+0.02	7.5	1.6	−0.5	M4 III		380 s		
196282	49821		20 34 16.3	+49 46 25	+0.001	+0.02	7.30	1.4	−0.3	K4 III	−26	310 s		
195659	230279		20 34 16.4	−43 26 54	+0.006	−0.10	7.4	0.4	0.2	K0 III		130 mx		
196024	106257		20 34 19.8	+10 16 05	0.000	+0.01	7.9	1.1	0.2	K0 III		340 s		
195290	254816		20 34 22.4	−66 43 56	+0.003	−0.06	7.2	0.1	0.6	A0 V		72 mx		
196179	70214		20 34 22.4	+37 51 01	0.000	−0.03	7.4	1.1	0.2	K0 III	+10	260 s		
195881	163695		20 34 26.3	−14 52 46	0.000	−0.02	8.0	1.1	−0.1	K2 III		410 s		
195859	163693		20 34 27.5	−19 23 53	+0.001	−0.02	7.7	0.4	0.6	A0 V		140 s		
196025	125969		20 34 31.7	+6 52 42	+0.001	+0.02	6.99	−0.11	−2.5	B2 V	−4	720 s	14017	m
196240	49822		20 34 32.7	+41 59 48	−0.003	−0.01	7.95	0.19		A0				
196421	32708		20 34 33.2	+57 55 40	−0.001	0.00	8.03	0.12	−1.0	B5.5 V		510 s		
195994	144580		20 34 33.8	−0 40 29	−0.001	+0.01	7.9	1.1	0.2	K0 III		340 s		
196241	49823		20 34 35.9	+41 05 54	0.000	+0.02	6.7	1.4	−0.3	K5 III		240 s		
196330	49827		20 34 40.5	+49 10 38	−0.003	−0.01	7.0	0.0	0.4	B9.5 V		200 s		
195812	212242		20 34 42.2	−34 17 41	−0.001	−0.02	7.9	2.0	−0.5	M4 III		280 s		
196197	70220		20 34 44.3	+32 30 22	0.000	+0.02	6.84	1.12	−1.1	K1 II-III	+4	390 s	14027	m
195861	189449		20 34 45.1	−27 13 52	+0.003	0.00	7.8	0.7	0.2	K0 III		330 s		
197910	3429		20 34 45.5	+82 00 41	+0.009	+0.05	7.2	1.1	0.2	K0 III		250 s		
195830	212245		20 34 46.7	−33 55 24	−0.001	−0.06	7.80	0.5	3.4	F5 V		76 s		m
195843	212249		20 34 47.2	−30 28 25	+0.002	−0.02	6.5	−0.2	0.4	B9.5 V		170 s		
196110	106265		20 34 49.0	+13 27 52	0.000	−0.02	7.4	0.1	1.4	A2 V		150 s		
196379	32709		20 34 50.3	+51 51 15	−0.001	0.00	6.11	0.41	2.6	F0 V	−13	39 s		
195781	230282		20 34 54.2	−41 59 12	0.000	−0.01	7.6	−0.3	1.4	A2 V		180 s		
195814	212246		20 34 55.3	−38 05 23	+0.002	0.00	6.5	1.1	1.4	A2 V		24 s		
196305	49828		20 34 56.5	+41 43 05	+0.002	+0.02	7.4	0.0	0.4	B9.5 V		240 s		
196112	125979		20 34 59.7	+6 32 08	+0.002	−0.02	7.9	1.1	0.2	K0 III		340 s		
195862	212252		20 35 01.5	−34 43 49	+0.003	0.00	7.6	1.8	0.2	K0 III		100 s		
196028	144595		20 35 03.6	−9 34 49	+0.003	−0.01	7.5	0.1	0.6	A0 V		240 s		
195190	257823		20 35 05.7	−71 11 20	−0.010	+0.01	6.5	1.0	0.2	K0 III		170 s		
196243	70232		20 35 07.3	+30 15 10	0.000	0.00	7.4	0.0	−1.0	B5.5 V	−22	460 s		
196295	70238		20 35 10.4	+36 50 50	+0.002	−0.02	7.9	1.1	0.2	K0 III		320 s		
196359	49836		20 35 14.0	+44 24 31	−0.004	−0.06	7.3	0.4	2.6	F0 V	−14	84 s		
196136	125988		20 35 14.6	+6 16 47	+0.002	+0.01	7.8	0.1	1.4	A2 V		190 s		
196360	49835		20 35 17.8	+41 53 24	0.000	+0.02	6.64	0.92	0.2	K0 III		190 s		
196180	106274	4 ζ Del	20 35 18.4	+14 40 27	+0.003	+0.02	4.68	0.11	1.7	A3 V	−25	38 s		
196181	106273		20 35 20.4	+10 45 30	0.000	+0.01	8.0	0.1	0.6	A0 V		300 s		
196113	144603		20 35 22.3	−5 31 43	0.000	−0.11	7.7	0.5	4.0	F8 V		55 s		
195782	246646		20 35 27.5	−50 19 46	+0.001	0.00	7.7	1.9	−0.3	K5 III		280 s		
196848	18941		20 35 28.8	+69 40 48	+0.007	−0.07	6.9	0.9	3.2	G5 IV		53 s		
196182	125996		20 35 29.5	+6 56 50	+0.001	0.00	6.5	0.4	0.4	B9.5 V		170 s		
196078	163712		20 35 32.0	−16 31 33	+0.006	−0.02	6.19	0.20		A5				
196482	32716		20 35 33.5	+50 25 31	+0.006	+0.04	7.9	0.5	3.4	F5 V		79 s		
195627	254823	φ¹ Pav	20 35 34.6	−60 34 54	+0.010	−0.18	4.76	0.28		F0 n	−19	16 mn		
196361	70242		20 35 38.2	+36 28 31	−0.011	−0.20	8.0	0.9	3.2	G5 IV		85 mx		
195980	212263		20 35 39.1	−34 07 40	−0.002	+0.02	8.0	1.1	−0.1	K2 III		410 s		
196147	144608		20 35 39.7	−4 43 17	+0.001	−0.08	8.0	0.9	3.2	G5 IV		93 s		
196218	126000		20 35 42.7	+3 18 12	−0.008	−0.04	8.0	0.5	4.0	F8 V		62 s		
196203	144611		20 35 46.8	−0 00 06	+0.002	−0.04	7.09	0.48	4.0	F8 V		42 s)		
195902	230289		20 35 47.6	−44 20 14	−0.001	+0.03	6.5	1.2	0.2	K0 III		130 s		
196080	189469		20 35 48.8	−24 22 25	+0.001	+0.01	7.4	0.8	0.2	K0 III		280 s		
--	49842		20 35 51.2	+49 19 44	+0.001	0.00	8.0	0.5						

HD	SAO	Star Name	α 2000	δ 2000	μ(α)	μ(δ)	V	B-V	M$_v$	Spec	RV	d(pc)	ADS	Notes
			h m s	° ′ ″	s	″								
195402	254820		20 35 51.6	-69 36 40	+0.007	-0.06	6.11	1.29		K2				
196332	88880		20 35 53.7	+25 28 44	0.000	+0.01	7.5	0.1	0.6	A0 V		240 s		
196081	189471		20 35 54.3	-26 46 24	+0.007	+0.03	7.21	0.42	2.1	F5 IV		100 s		
--	32721		20 35 58.2	+50 43 14	-0.002	0.00	7.8	1.6	0.2	K0 III		140 s		
197637	9827		20 36 00.2	+79 25 49	-0.004	+0.01	6.96	-0.12	-1.6	B3.5 V	-31	500 s		
195762	246649		20 36 04.7	-59 40 43	+0.002	-0.03	7.7	1.2	-0.1	K2 III		360 s		
196362	88884	26 Vul	20 36 08.1	+25 52 57	+0.001	+0.01	6.41	0.21	0.1	A4 III	-19	160 s		
195903	246654		20 36 11.9	-50 53 48	+0.002	0.00	7.96	0.98	3.2	G6 IV		72 s		
195390	257825		20 36 12.1	-71 15 43	+0.003	-0.03	7.6	0.1	0.6	A0 V		250 s		
196345	106287		20 36 15.8	+16 48 52	+0.002	+0.01	6.6	1.1	-0.1	K2 III		210 s		
196707	32729		20 36 16.0	+59 12 49	+0.001	+0.03	8.0	0.1	0.6	A0 V		280 s		
196222	163720		20 36 16.3	-12 22 51	+0.003	+0.03	7.0	1.1	-0.1	K2 III		260 s		
196805	18943		20 36 23.8	+62 00 53	-0.001	+0.01	7.84	-0.08	0.6	A0 V		280 s		
196489	70259		20 36 24.1	+39 11 41	0.000	0.C0	8.0	0.0		B8.5 V		370 s		
196207	189484		20 36 26.3	-20 35 07	+0.001	-0.01	7.2	1.1	0.2	K0 III		230 s		
196503	49847		20 36 27.8	+41 02 37	0.000	-0.05	8.0	0.1	0.6	A0 V		280 s		
196014	230292		20 36 30.7	-45 33 27	+0.006	+0.03	7.80	0.1	1.7	A3 V		140 mx		m
195459	257829		20 36 35.6	-71 04 18	+0.001	-0.03	7.10	0.1	0.6	A0 V		200 s		m
195772	254827		20 36 38.0	-63 07 17	+0.001	-0.09	6.6	0.4	3.0	F2 V		51 s		
196346	126012		20 36 42.5	+2 29 44	+0.002	-0.01	7.60	0.9	0.2	G9 III	-49	300 s		
196670	32731		20 36 42.8	+50 53 24	0.000	+0.01	7.90	0.1	0.6	A0 V		270 s	14072	m
196321	144624	70 Aql	20 36 43.5	-2 32 59	0.000	0.00	4.89	1.60	-2.3	K5 II	-10	240 s		
195573	254825		20 36 48.9	-69 27 03	0.000	-0.02	6.8	1.1	0.2	K0 III		170 s		
196671	49859		20 36 50.3	+49 54 33	-0.001	0.00	7.1	1.1	0.2	K0 III	-38	230 s		
196310	163730		20 36 54.7	-12 44 14	0.000	-0.03	7.65	0.42		F0	-32		14054	m
196151	212280		20 36 54.8	-37 23 13	-0.003	0.00	7.4	1.7	0.2	K0 III		100 s		
196523	70268		20 36 58.6	+32 21 04	+0.002	+0.01	7.3	0.1	1.4	A2 V		150 s		
196449	106306		20 37 03.7	+14 51 12	+0.003	0.00	8.0	0.4	3.0	F2 V		97 s		
196425	126022		20 37 03.8	+8 58 40	+0.003	0.00	7.9	0.2	2.1	A5 V		150 s		
197053	18955		20 37 03.9	+67 30 17	+0.001	+0.04	7.4	0.0	0.4	B9.5 V		220 s		
196504	88903	27 Vul	20 37 04.5	+26 27 43	+0.001	0.00	5.50	-0.07	0.2	B9 V	-10	120 s		
196589	70273		20 37 07.7	+36 35 03	-0.001	-0.02	7.48	0.06		A0				
196903	18948		20 37 07.9	+61 21 52	+0.005	-0.04	7.0	0.4	3.0	F2 V		61 s		
196396	144630		20 37 10.7	-1 18 40	0.000	+0.02	7.7	0.9	3.2	G5 IV		80 s		
195805	254831		20 37 10.9	-65 02 02	+0.005	-0.03	7.1	-0.4	0.4	B9.5 V		170 mx		
194277	--		20 37 12.0	-81 32 22			8.0	1.1	-0.1	K2 III		420 s		
196770	32737		20 37 12.5	+52 58 22	-0.002	-0.02	6.6	1.1	-0.1	K2 III		210 s		
196275	189492		20 37 15.7	-28 42 49	+0.004	-0.04	7.9	1.4	-0.1	K2 III		250 mx		
196426	144632		20 37 18.1	+0 05 49	0.000	-0.01	6.21	-0.08		B8	-23			
196348	163740		20 37 21.0	-15 08 50	+0.004	-0.04	6.76	1.25	-0.2	K3 III	+19	160 mx		
197433	9828		20 37 21.5	+75 36 02	+0.090	+0.56	7.38	0.85	5.9	K0 V	-35	21 ts		VW Cep, m,v
196643	70286		20 37 22.3	+38 05 55	-0.002	+0.01	7.08	1.58	-0.3	K5 III	-14	270 s		
196687	49867		20 37 22.8	+42 59 28	0.000	+0.01	7.27	-0.09	0.6	A0 V	-10	220 s		
196642	70288		20 37 23.4	+38 19 43	+0.001	-0.04	6.20	0.99	0.2	K0 III	-37	160 s		
196672	49866		20 37 24.9	+41 23 35	-0.001	+0.02	7.9	0.5	4.0	F8 V		59 s		
--	70294		20 37 25.5	+39 48 24	+0.002	+0.01	7.9	1.6						
196605	70282		20 37 28.1	+32 03 17	0.000	0.00	7.84	-0.04	0.6	A0 V		280 s		
196880	32743		20 37 28.6	+58 51 37	+0.001	+0.01	7.9	1.4	-0.3	K5 III		390 s		
196608	70283		20 37 31.1	+31 00 09	+0.001	-0.01	7.9	1.6	-0.5	M2 III		470 s		
196606	70287	48 Cyg	20 37 31.5	+31 34 21	+0.001	0.00	6.20	-0.09	-0.6	B8 IV	-19	230 s		m
196629	70289		20 37 32.4	+31 31 19	-0.004	-0.03	6.49	0.34	2.6	F0 V	+1	59 s		
196524	106316	6 β Del	20 37 32.8	+14 35 43	+0.008	-0.03	3.54	0.55	0.7	F5 III	-23	33 s	14073	m
196484	126031		20 37 33.6	+6 30 03	+0.001	-0.01	7.5	0.1	1.4	A2 V		170 s		
196171	230300	α Ind	20 37 33.9	-47 17 29	+0.005	+0.07	3.11	1.00	0.2	K0 III	-1	38 s		m
195961	254835	ρ Pav	20 37 35.1	-61 31 47	+0.008	-0.06	4.88	0.43	3.4	F5 V	+8	24 mn		
196688	70299		20 37 37.9	+37 55 42	0.000	+0.01	7.7	1.1	-0.1	K2 III		340 s		
196689	70301		20 37 42.6	+35 11 59	-0.001	+0.02	7.9	1.1	-0.1	K2 III		370 s		
196384	189502		20 37 43.2	-22 26 35	-0.002	+0.01	7.2	0.7	3.4	F5 V		41 s		
196277	230302		20 37 43.5	-41 13 40	+0.002	-0.06	6.9	0.5	3.4	F5 V		48 s		
196673	70300		20 37 44.3	+33 22 01	+0.002	+0.01	7.40	1.1	0.2	K0 III	-27	260 s	14078	m
--	32741		20 37 46.3	+51 27 56	-0.002	0.00	8.0	0.4	-1.6	B3.5 V		370 s		
196988	18958		20 37 48.1	+60 45 22	+0.003	+0.02	7.20	0.7	4.4	G0 V		36 s	14102	m
196413	163746		20 37 48.7	-17 07 19	+0.001	0.00	7.0	0.1	0.6	A0 V		190 s		
196544	106322	5 z Del	20 37 49.0	+11 22 40	+0.003	0.00	5.43	0.06	1.4	A2 V	-4	64 s		
196655	--		20 37 50.0	+29 43 18	+0.001	-0.01	8.00	0.1	1.4	A2 V		210 s	14079	m
196808	49879		20 37 51.1	+47 56 44	+0.001	+0.02	7.69	0.10	0.6	A0 V		240 s		m
196385	189503		20 37 52.0	-25 06 32	+0.005	0.00	6.36	0.33	2.6	F0 V		54 s		
196341	212298		20 37 53.3	-33 26 38	0.000	-0.04	7.6	0.4	0.6	A0 V		140 s		
196610	106329		20 37 54.4	+18 16 08	+0.001	+0.10	6.25	1.48		M6 III	-66			EU Del, v
196469	163752		20 37 55.8	-11 02 00	+0.001	-0.03	7.0	0.0	0.4	B9.5 V		210 s		
196236	230303		20 37 58.3	-48 32 44	+0.001	-0.01	7.8	0.1	0.4	B9.5 V		300 s		
196571	106327		20 37 58.5	+13 05 32	+0.002	0.00	7.5	0.1	0.6	A0 V		240 s		
196789	49877		20 37 58.7	+42 50 44	+0.008	+0.17	7.04	0.48		F8	+2	30 mn		
197665	9836		20 38 04.5	+76 49 59	+0.002	+0.07	7.1	0.4	3.0	F2 V	-8	66 s		
196430	189509		20 38 04.7	-24 13 43	+0.004	0.00	6.7	0.6	1.7	A3 V		48 s		

HD	SAO	Star Name	α 2000	δ 2000	μ(α)	μ(δ)	V	B-V	M_v	Spec	RV	d(pc)	ADS	Notes
196832	49883		20^h 38^m 05.6^s	+46° 53′ 09″	-0.001	+0.01	7.97	-0.06	0.4	B9.5 V		330 s		
196905	32750		20 38 06.8	+52 00 23	-0.005	-0.01	7.7	1.4	-0.3	K5 III		310 mx		
196723	70311		20 38 08.2	+30 34 40	+0.001	+0.02	7.5	0.1	0.6	A0 V		240 s		
196527	144642		20 38 08.5	-4 22 53	+0.001	0.00	7.2	1.1	0.2	K0 III		250 s		
195320	257830		20 38 08.9	-76 53 34	+0.004	-0.02	7.72	0.96		G5				m
196386	212302		20 38 09.2	-30 33 30	+0.005	-0.01	8.0	0.0	2.6	F0 V		120 s		
196387	212301		20 38 09.2	-30 58 40	0.000	+0.02	7.4	1.3	-0.1	K2 III		250 s		
196237	246670		20 38 11.6	-50 37 22	0.000	-0.04	7.4	1.8	-0.1	K2 III		130 s		
196739	70313		20 38 12.6	+33 21 02	-0.002	-0.01	7.9	1.1	0.2	K0 III		320 s		
196833	49885		20 38 15.6	+44 19 58	+0.001	+0.01	6.65	-0.13		B8	-19			
196573	126040		20 38 16.1	+1 00 59	-0.001	0.00	7.88	1.64		K5				
196865	49886		20 38 16.7	+48 04 12	+0.004	+0.02	6.60	0.8	3.2	G5 IV	-25	47 s	14100	m
196819	49884		20 38 16.8	+42 04 24	-0.002	-0.01	7.50	1.87	-2.3	K3 II		480 s		
196416	212303		20 38 16.9	-30 57 11	-0.002	-0.05	7.6	1.3	-0.1	K2 III		200 mx		
194612	258864		20 38 18.4	-81 17 20	+0.001	-0.02	5.91	1.71		K5				
197035	32757		20 38 18.9	+58 56 52	+0.001	+0.01	7.9	1.1	0.2	K0 III		320 s		
196574	144649	71 Aql	20 38 20.1	-1 06 19	+0.001	-0.02	4.32	0.95	0.3	G8 III	-6	64 s	14081	m
196645	106335		20 38 20.8	+13 19 53	+0.008	+0.10	7.9	1.1	0.2	K0 III		120 mx		
196939	32754		20 38 20.9	+52 22 42	+0.003	+0.02	7.8	0.1	0.6	A0 V		260 s		
196554	144647		20 38 21.1	-4 30 43	+0.001	0.00	7.9	0.6	4.4	G0 V		50 s		
196325	230305		20 38 23.4	-44 31 30	+0.009	-0.03	7.3	0.4	0.2	K0 III		160 mx		
195832	257833		20 38 30.3	-70 04 23	+0.006	-0.03	7.8	1.1	0.2	K0 III		330 s		
196724	88944	29 Vul	20 38 31.1	+21 12 04	+0.005	+0.01	4.82	-0.02	0.6	A0 V	-18	70 s		
196740	88945	28 Vul	20 38 31.7	+24 06 58	+0.001	0.00	5.04	-0.14	-1.1	B5 V	-22	170 s		
196753	88946		20 38 34.9	+23 40 50	+0.001	0.00	5.91	0.98	0.2	K0 III	+9	140 s		
196850	70321		20 38 39.9	+38 38 07	+0.016	-0.19	6.75	0.62	4.4	G0 V	-21	24 ts		m
196773	88951		20 38 43.3	+24 46 40	-0.002	-0.01	7.7	0.0	0.4	B9.5 V		260 s		
196725	106342	8 θ Del	20 38 43.7	+13 18 54	0.000	0.00	5.72	1.53	-4.4	K3 Ib	-14	780 s		
196772	88952		20 38 44.2	+25 33 03	-0.001	-0.01	7.8	0.1	0.6	A0 V		260 s		
196474	212310		20 38 46.2	-33 40 43	0.000	-0.04	7.2	1.2	-0.1	K2 III		280 s		
197442	9833		20 38 53.8	+71 10 10	+0.004	+0.04	7.9	1.4	-0.3	K5 III		360 s		
195849	257834		20 38 55.9	-71 05 40	-0.002	0.00	7.3	1.6	-0.5	M2 III		370 s		
196726	126055		20 38 56.2	+8 31 21	+0.001	+0.01	7.6	0.4	3.0	F2 V		85 s		
196989	49895		20 38 56.6	+47 41 12	+0.001	-0.02	7.60	0.1	0.6	A0 V		240 s	14119	m
196633	144658		20 38 57.2	-8 23 57	+0.003	+0.01	7.7	0.4	3.0	F2 V		88 s		
196852	70323		20 38 59.3	+30 20 04	-0.003	-0.06	5.68	1.09	-0.1	K2 III	+13	140 s		
197101	32763		20 39 00.0	+56 00 18	-0.001	-0.03	6.5	0.4	2.6	F0 V	-1	60 s		
196711	126054		20 39 01.1	+1 11 56	-0.001	+0.01	7.75	1.03	0.2	K0 III		320 s		
196390	230307		20 39 02.1	-49 19 51	-0.017	+0.03	7.6	0.7	4.4	G0 V		39 s		
196531	189523		20 39 02.6	-28 25 34	+0.008	-0.01	7.95	0.54	4.0	F8 V		60 s		
196617	163768		20 39 04.2	-16 33 34	+0.008	-0.02	7.2	0.5	4.0	F8 V		43 s		
196497	212313		20 39 04.3	-36 21 23	+0.001	-0.03	7.6	0.9	0.6	A0 V		67 s		
196775	106347		20 39 04.8	+15 50 18	0.000	-0.02	5.97	-0.14	-1.4	B4 V n	+2	290 s	14106	m
196676	144663		20 39 05.6	-4 55 45	0.000	+0.02	6.6	1.1	0.2	K0 III		190 s		
196811	106349		20 39 07.2	+19 33 38	+0.002	+0.02	7.5	0.1	1.4	A2 V		170 s		
196755	126059	7 κ Del	20 39 07.6	+10 05 10	+0.022	+0.02	5.05	0.72	3.2	G5 IV	-52	23 s	14101	m
196532	212316		20 39 09.5	-33 10 38	+0.004	-0.01	8.0	0.8	0.2	K0 III		300 mx		
196821	88959		20 39 10.4	+21 49 02	+0.001	+0.02	5.90	-0.06	-0.6	A0 III	-37	200 s		
196712	144666		20 39 13.1	-2 24 46	0.000	+0.01	6.22	-0.10		B9 pne	-14			
196662	163771	14 τ Cap	20 39 16.1	-14 57 18	0.000	-0.02	5.22	-0.12	-1.9	B6 III	-5	260 s	14099	m
196906	88963		20 39 19.0	+29 05 56	-0.001	0.00	8.00	0.5	3.4	F5 V		83 s	14117	m
196866	88962		20 39 20.0	+26 04 04	-0.004	-0.10	6.99	1.1	-0.1	K2 III	-78	260 s		
196265	254841		20 39 21.3	-61 32 08	+0.001	0.00	7.1	0.2	0.4	B9.5 V		150 s		
197036	49898		20 39 22.9	+45 40 01	0.000	0.00	6.58	-0.06	-1.6	B5 IV	-15	410 s		
196692	163772		20 39 23.2	-11 45 50	+0.002	+0.02	7.8	1.1	0.2	K0 III		330 s		
196758	126062	1 Aqr	20 39 24.7	+0 29 11	+0.007	-0.01	5.16	1.06	0.0	K1 III	-43	110 s	14108	m
196729	144671		20 39 26.8	-7 43 58	-0.001	-0.02	8.0	0.9	3.2	G5 IV		93 s		
196928	88970		20 39 29.6	+28 05 19	-0.001	-0.02	8.00	1.4	-0.3	K4 III		440 s		
196834	106352		20 39 29.7	+13 20 27	-0.001	-0.02	7.9	1.4	-0.3	K5 III		400 s		
196908	88969		20 39 31.2	+26 41 52	+0.001	+0.02	7.1	0.4	2.6	F0 V		78 s		
197008	70338		20 39 32.5	+38 22 31	-0.001	-0.02	7.90	0.1	1.7	A3 V		170 s		
197037	49900		20 39 32.6	+42 14 55	-0.007	-0.22	6.82	0.50		G0				
197018	49899		20 39 33.1	+40 34 46	0.000	0.00	6.06	-0.16		B8			14126	m
196972	70333		20 39 36.2	+30 48 51	-0.001	-0.01	7.4	1.1	-2.1	K0 II	+15	750 s		
196795	126068		20 39 37.5	+4 58 19	+0.058	+0.09	7.88	1.23	8.0	K5 V	-43	11 ts		m
196867	106357	9 α Del	20 39 38.1	+15 54 43	+0.004	0.00	3.77	-0.06	0.2	B9 V	-6	52 s	14121	m
197118	49908		20 39 44.6	+47 20 43	-0.003	+0.01	7.93	-0.06	0.6	A0 V		290 s	14142	m
196317	254844		20 39 51.2	-62 54 28	+0.003	-0.08	6.3	1.3	0.2	K0 III		110 s		
196885	106360		20 39 51.7	+11 14 59	+0.004	+0.09	6.42	0.55	4.0	F8 V	-28	29 s		m
196884	106362		20 39 52.1	+13 25 12	0.000	0.00	7.2	1.1	-0.1	K2 III		280 s		
196638	212321		20 39 52.4	-36 02 01	+0.004	-0.05	7.2	0.7	3.0	F2 V		40 s		
196910	106363		20 39 52.9	+13 18 21	+0.002	+0.01	7.9	1.1	-0.1	K2 III		370 s		
196796	144675		20 39 53.6	-7 34 23	+0.001	-0.01	8.0	1.1	0.2	K0 III		360 s		
197054	70343		20 39 55.2	+35 23 27	+0.002	+0.05	6.8	1.1	0.2	K0 III		200 s		
196478	246681		20 39 55.5	-53 23 34	-0.001	+0.02	8.0	0.8	1.7	A3 V		63 s		

HD	SAO	Star Name	α 2000	δ 2000	μ(α)	μ(δ)	V	B-V	M_v	Spec	RV	d(pc)	ADS	Notes
196697	189538		20 39 55.6	-27 43 00	+0.002	0.00	7.6	0.7	3.2	G5 IV		78 s		
196912	126081		20 39 59.7	+8 26 49	+0.002	-0.02	7.6	1.1	-0.1	K2 III		340 s		
196378	254846	φ² Pav	20 40 02.4	-60 32 56	+0.042	-0.57	5.12	0.53	4.0	F8 V	-32	21 ts		
196777	163779	15 υ Cap	20 40 02.8	-18 08 19	-0.001	-0.02	5.10	1.66	-0.5	M2 III	-13	120 s		
197139	49912		20 40 03.0	+43 27 31	-0.007	-0.06	5.95	1.19	0.2	gK0	-19	110 s		
196699	212328		20 40 10.0	-34 03 00	+0.004	-0.01	8.0	1.7	-0.1	K2 III		210 s		
196941	106364		20 40 10.0	+10 34 20	0.000	0.00	7.9	0.1	1.4	A2 V		200 s		
196761	189549		20 40 11.5	-23 46 26	+0.036	+0.46	6.37	0.72	5.3	G6 V	-50	15 ts		
196870	144681		20 40 12.3	-2 39 18	-0.001	-0.07	6.6	1.1	0.2	K0 III		180 mx		
196664	212327		20 40 13.1	-37 34 54	-0.003	+0.02	7.6	1.6	3.2	G5 IV		24 s		
197167	49920		20 40 17.3	+45 34 21	-0.002	+0.03	8.0	0.0	-1.2	B8 III		620 s		
196929	126088		20 40 17.5	+3 26 29	+0.003	+0.01	6.90	1.1	0.2	K0 III		220 s		m
197373	18979		20 40 17.7	+60 30 19	+0.001	+0.19	6.01	0.46	3.4	F5 V	-13	33 s		
196701	212332		20 40 18.9	-35 47 58	+0.003	-0.04	7.5	0.7	2.6	F0 V		54 s		
196737	212333		20 40 19.7	-33 25 54	+0.003	+0.04	5.47	1.12	-0.1	K2 III	+14	130 s		
197140	70349		20 40 20.6	+38 46 50	+0.012	+0.05	8.0	0.9	3.2	G5 IV		91 s		
196800	189555		20 40 22.1	-24 07 04	+0.013	-0.28	7.22	0.60		G0		20 mn		
196639	230316		20 40 22.5	-46 49 36	+0.003	-0.01	7.7	1.2	0.2	K0 III		250 s		
196762	189553		20 40 25.3	-29 11 57	+0.001	-0.01	7.6	1.1	0.2	K0 III		280 s		
196857	163783		20 40 32.3	-16 07 27	-0.005	+0.07	5.80	1.00	0.2	K0 III	-4	120 mx		
197308	32791		20 40 32.6	+55 06 25	-0.001	0.00	7.32	-0.02	0.6	A0 V		220 s		
197039	106369		20 40 35.1	+15 38 36	+0.006	+0.05	6.8	0.5	3.4	F5 V	-33	48 s		
196816	189558		20 40 35.1	-27 35 12	-0.001	0.00	7.4	0.4	1.4	A2 V		97 s		
197075	88991		20 40 35.5	+21 55 37	0.000	0.00	7.80	0.00	0.4	B9.5 V		290 s	14148	m
196815	189559		20 40 35.8	-26 38 42	+0.003	-0.01	6.4	1.0	4.4	G0 V		14 s		
197120	88997		20 40 36.0	+29 48 19	+0.001	+0.04	6.10	0.14	1.7	A3 V	-27	70 s	14149	m
197119	70352		20 40 36.3	+31 21 28	+0.001	-0.01	7.88	0.08	1.4	A2 V		190 s		
197102	89000		20 40 42.0	+26 04 45	+0.001	-0.01	6.8	0.1	0.6	A0 V		170 s		
197040	106372		20 40 42.3	+14 31 37	+0.002	+0.01	7.7	0.1	1.4	A2 V	-36	180 s		
197076	106373		20 40 45.0	+19 56 07	+0.009	+0.31	6.45	0.63	4.7	dG2	-37	22 ts		m
196859	189565		20 40 49.4	-25 03 00	0.000	-0.02	7.7	1.6	-0.3	K5 III		350 s		
197176	70358		20 40 49.8	+35 02 23	-0.002	-0.06	7.2	0.4	3.0	F2 V		68 s		
196838	189564		20 40 52.1	-28 32 58	+0.001	-0.01	6.6	1.6	0.2	K0 III		86 s		
197204	49929		20 40 53.3	+41 12 09	+0.003	0.00	7.01	-0.07		B9				
197103	106376		20 40 53.9	+17 34 17	+0.001	+0.03	6.7	0.1	0.6	A0 V		170 s		
196748	230321		20 40 55.2	-42 23 49	0.000	+0.01	6.50	0.9	3.2	G5 IV		46 s		m
196946	163793		20 40 59.9	-14 30 41	-0.001	+0.01	7.70	0.4	2.6	F0 V		110 s	14145	m
197226	70367		20 41 00.2	+39 04 57	0.000	-0.01	6.51	-0.12	-1.3	B6 IV		370 s		m
197205	70365		20 41 01.7	+37 02 01	+0.002	-0.01	7.9	1.1	-0.1	K2 III		370 s		
197177	70362	49 Cyg	20 41 02.3	+32 18 26	0.000	-0.01	5.51	0.88	0.2	K0 III	-29	120 s	14158	m
196623	246689		20 41 06.4	-55 36 20	+0.015	-0.04	7.8	1.0	0.2	K0 III		120 mx		
196995	163795		20 41 12.7	-13 30 01	+0.001	-0.01	7.6	1.1	0.2	K0 III		300 s		
197121	106384	10 Del	20 41 16.0	+14 34 59	0.000	+0.01	5.99	1.24	-0.1	K2 III	-32	150 s		
--	49940		20 41 18.0	+48 08 32	-0.002	+0.03	6.8							V Cyg, v
197206	70370		20 41 18.1	+31 15 00	+0.001	-0.03	7.40	1.1		K1 IV	-26			
196979	163798		20 41 18.9	-17 22 42	-0.001	0.00	6.9	1.1	0.2	K0 III		220 s		
197923	9849		20 41 22.3	+72 58 30	+0.004	+0.02	7.3	1.1	-0.1	K2 III		270 s		
196917	212345		20 41 23.4	-31 35 54	+0.008	-0.05	5.76	1.53				16 mn		
196947	189575		20 41 24.0	-26 00 00	0.000	-0.03	6.3	1.5	0.2	K0 III		79 s		
196738	246692		20 41 24.2	-52 28 46	-0.004	-0.07	7.6	1.3	3.2	G5 IV		36 s		
197058	144702		20 41 24.3	-5 59 52	0.000	0.00	7.6	1.1	0.2	K0 III		300 s		
196829	230323		20 41 24.4	-42 08 01	+0.004	-0.02	6.31	1.57	-2.4	M3 II		160 mx		
197434	32798		20 41 25.1	+54 12 33	-0.001	-0.02	7.98	0.15	0.6	A0 V		270 s		
197345	49941	50 α Cyg	20 41 25.8	+45 16 49	0.000	+0.01	1.25	0.09	-7.5	A2 Ia	-5	560 s	14172	Deneb, m
196919	212347		20 41 28.0	-34 06 07	0.000	-0.02	8.0	1.6	-0.3	K5 III		390 s		
197292	70379		20 41 31.8	+35 54 40	-0.004	-0.05	7.5	1.1	-0.1	K2 III		310 s		
197168	106386		20 41 32.7	+15 23 09	+0.001	+0.01	8.0	0.8	3.2	G5 IV		91 s		
196963	189579		20 41 33.3	-27 59 43	+0.003	-0.02	7.5	-0.1	0.4	B9.5 V		270 s		
197169	106387		20 41 37.5	+13 48 31	+0.001	0.00	6.8	0.1	1.4	A2 V		120 s		
197310	70384		20 41 40.9	+35 32 43	+0.006	+0.04	7.5	0.6	4.4	G0 V		41 s		
197180	106388		20 41 41.3	+11 33 57	0.000	-0.02	7.6	1.1	-0.1	K2 III		340 s		
197245	89016		20 41 41.6	+24 10 57	-0.001	0.00	6.8	0.1	0.6	A0 V	+7	170 s		
197309	70385		20 41 41.7	+36 45 48	0.000	-0.01	7.9	1.4	-0.3	K5 III		400 s		
197375	49944		20 41 42.2	+43 26 35	0.000	0.00	7.86	0.02		A0				
196067	257836	μ² Oct	20 41 43.6	-75 21 02	+0.040	-0.15	6.03	0.62	4.5	G1 V		19 s		m
197089	163803		20 41 45.6	-10 56 44	+0.003	-0.01	7.7	0.6	4.4	G0 V		47 s		
197179	106390		20 41 45.8	+12 31 06	+0.003	-0.04	7.70	0.5	4.0	F8 V		55 s	14168	m
196999	189583		20 41 46.8	-28 45 44	+0.006	+0.02	7.5	0.2	3.0	F2 V		80 s		
197376	49945		20 41 47.0	+42 00 33	0.000	+0.02	7.95	0.59	-2.0	F5 II		830 s		
197246	89020		20 41 49.6	+22 59 00	0.000	0.00	7.80	1.1	0.2	K0 III		320 s	14169	m
197229	106394		20 41 51.1	+19 31 15	+0.002	0.00	8.00	0.1	1.4	A2 V		200 s		m
197228	89019		20 41 51.5	+20 42 49	0.000	-0.06	7.40	0.1	1.7	A3 V		51 mx	14170	m
197195	106391		20 41 53.2	+12 58 49	-0.005	-0.08	8.0	0.9	3.2	G5 IV		93 s		
197510	32806		20 41 53.4	+54 25 09	-0.001	0.00	7.86	0.26	2.6	F0 V		110 s		
197248	89024		20 41 53.7	+20 41 13	+0.002	-0.01	8.0	0.9	3.2	G5 IV		89 s		

HD	SAO	Star Name	α 2000	δ 2000	μ(α)	μ(δ)	V	B-V	M_V	Spec	RV	d(pc)	ADS	Notes
			h m s	° ′ ″	s	″								
197377	70397		20 41 54.2	+39 31 05	+0.002	0.00	7.9	0.1	1.4	A2 V		190 s		
197126	144716		20 41 54.8	-8 46 21	+0.003	+0.04	8.0	1.1	0.2	K0 III		360 s		
197407	49947		20 41 56.0	+42 31 30	+0.004	+0.06	7.4	0.5	3.4	F5 V		64 s		
197392	49946		20 41 56.4	+41 43 01	+0.001	+0.01	5.67	-0.12		B8	-27			
196519	254854	υ Pav	20 41 57.0	-66 45 39	+0.003	-0.02	5.15	-0.06	-0.2	B8 V	+8	120 s		
197249	106396		20 41 58.0	+17 31 17	+0.001	+0.05	6.22	0.94	0.3	G8 III	-2	150 s		
196843	246699		20 42 01.2	-50 15 44	+0.002	-0.02	7.5	1.7	-0.5	M2 III		310 s		
197329	70395		20 42 02.2	+31 18 44	+0.002	+0.03	7.0	0.1	0.6	A0 V		190 s		
196876	230327		20 42 02.8	-48 47 02	+0.002	-0.01	7.9	1.7	-0.3	K5 III		420 s		
196051	257838	μ¹ Oct	20 42 03.0	-76 10 51	+0.058	-0.01	6.00	0.44	2.1	F5 IV n		41 mx		
196898	230328		20 42 03.8	-46 24 01	0.000	0.00	7.7	1.6	-0.5	M2 III		440 s		
197028	212359		20 42 05.5	-32 02 48	+0.001	-0.01	8.0	-0.3	4.0	F8 V		62 s		
197196	126116		20 42 05.7	+6 30 28	+0.001	0.00	6.8	0.0	0.0	B8.5 V		230 s		
197274	106401		20 42 09.5	+19 51 52	+0.001	-0.02	7.4	1.1	0.2	K0 III	+12	270 s		
197275	106402		20 42 10.6	+19 37 42	-0.001	-0.01	7.6	0.1	0.6	A0 V		240 s		
196175	257840		20 42 11.8	-75 03 19	+0.014	+0.02	7.7	1.1	0.2	K0 III		230 mx		
197511	32809	51 Cyg	20 42 12.5	+50 20 24	0.000	+0.01	5.39	-0.10	-2.5	B2 V	-3	330 s	14189	m
197231	126118		20 42 14.9	+7 10 54	-0.001	-0.01	8.0	1.4	-0.3	K5 III		450 s		
196952	230330		20 42 15.4	-44 53 04	-0.001	+0.03	7.1	-0.1	0.6	A0 V		200 s		
197618	32812		20 42 16.6	+57 23 05	-0.002	-0.01	7.30	0.1	1.7	A3 V		130 s	14196	m
197639	32815		20 42 18.7	+57 45 28	-0.001	-0.01	7.8	1.1	0.2	K0 III		300 s		
197276	106404		20 42 19.9	+12 37 58	+0.001	-0.01	7.5	1.4	-0.3	K5 III		360 s		
197488	49960		20 42 20.0	+45 49 24	-0.007	-0.19	7.70	0.7	2.8	G0 IV	+11	94 s	14190	m
197419	70406		20 42 22.0	+35 27 22	-0.002	-0.01	6.66	-0.16	-2.5	B2 V e	-7	600 s		V568 Cyg, q
197311	106406		20 42 22.8	+19 23 47	+0.002	+0.02	7.85	1.03	0.2	K0 III		330 s		
197233	144726		20 42 27.3	-2 19 36	+0.001	+0.01	7.1	0.1	1.4	A2 V		140 s		
197210	144725		20 42 29.2	-5 18 04	-0.005	-0.15	7.8	0.7	4.4	G0 V		49 s		
197312	106409		20 42 33.2	+12 43 43	-0.001	-0.01	8.00	0.5	3.4	F5 V		83 s	14184	m
194149	258866		20 42 33.7	-84 24 26	+0.043	-0.02	7.1	-0.1	0.6	A0 V		150 mx		
197313	106410		20 42 33.7	+12 43 43	-0.002	-0.02	8.00	0.5	3.4	F5 V		83 s	14184	m
199095	3458	76 Dra	20 42 35.0	+82 31 52	+0.014	+0.03	5.7	0.1	0.6	A0 V	-20	25 mx		
197197	163813		20 42 37.0	-13 05 39	-0.010	-0.13	7.8	0.7	4.4	G0 V		49 s		
197185	163812		20 42 38.7	-15 40 15	+0.001	+0.01	8.0	0.9	3.2	G5 IV		93 s		
197809	19000		20 42 39.2	+63 12 48	-0.002	-0.01	7.9	0.1	0.6	A0 V		260 s		
197734	18998		20 42 39.5	+60 36 05	-0.001	-0.01	6.15	0.01	0.6	A0 V	-5	120 s		
197187	163811		20 42 39.6	-18 06 37	0.000	-0.01	7.2	0.4	2.6	F0 V		81 s		
197211	163815		20 42 40.3	-10 58 11	+0.001	0.00	7.7	0.5	3.4	F5 V		72 s		
197620	32819		20 42 46.8	+51 01 31	0.000	+0.02	8.0	1.4	-0.3	K5 III		410 s		
197894	19002		20 42 48.3	+66 19 34	+0.002	-0.02	6.9	0.0	0.4	B9.5 V		180 s		
197093	212369		20 42 52.8	-39 33 31	+0.002	0.00	6.29	1.08		K0				
196480	257843		20 42 54.2	-71 55 40	-0.005	+0.01	6.9	0.4	2.6	F0 V		72 s		
197603	49976		20 42 55.2	+49 03 09	+0.004	-0.02	7.2	0.4	2.6	F0 V		83 s		
197016	246703		20 42 59.0	-50 28 44	+0.004	+0.01	8.0	0.0	2.6	F0 V		120 s		
197666	32824		20 43 01.7	+52 17 43	-0.004	+0.16	7.7	0.6	4.4	G0 V		46 s		
197950	19004	4 Cep	20 43 10.8	+66 39 27	+0.004	+0.04	5.5	0.2	2.1	A5 V	+35	48 s		
197770	32832		20 43 13.4	+57 06 51	-0.002	+0.01	6.32	0.33	-3.0	B2 IV	-15	360 s		
196227	257842		20 43 14.1	-76 32 52	+0.021	-0.16	7.67	0.60	4.5	G1 V		41 s		
197549	70417		20 43 14.3	+33 49 53	+0.001	+0.01	7.48	-0.05	0.6	A0 V		240 s		
197214	189606		20 43 15.8	-29 25 26	-0.002	-0.20	6.95	0.67	5.2	G5 V		22 s		
196845	254856		20 43 16.0	-62 32 15	+0.003	-0.02	7.7	0.4	3.0	F2 V		87 s		
197330	163823		20 43 16.0	-10 44 31	+0.001	+0.02	7.8	1.1	0.2	K0 III		330 s		
197489	89048		20 43 19.8	+25 49 30	-0.001	+0.01	7.04	0.34	-2.0	A7 II		560 s		m
197911	19003		20 43 21.3	+63 12 34	-0.004	+0.01	7.64	0.06	-1.0	B5.5 V		110 mx		
197514	89050		20 43 22.1	+27 15 05	0.000	-0.01	7.7	1.6	-0.5	M4 III		430 s		
197561	70422		20 43 23.5	+33 52 08	+0.002	+0.02	7.24	-0.12	0.6	A0 V		210 s		
197572	70423		20 43 23.9	+35 35 16	-0.001	0.00	6.40	0.7	-4.5	G3 Ib var	+10	1200 s		X Cyg, v
197516	89049		20 43 24.9	+23 38 41	0.000	0.00	8.0	0.1	0.6	A0 V		290 s		
197515	89051		20 43 25.2	+25 36 09	+0.001	+0.01	7.30	1.4	-0.3	K5 III		320 s	14208	m
197284	189613		20 43 26.2	-22 27 15	-0.001	-0.02	7.2	1.5	3.2	G5 IV		24 s		
197461	106425	11 δ Del	20 43 27.3	+15 04 28	-0.001	-0.04	4.43	0.32	0.5	A7 III	+9	56 s		v
197795	32834		20 43 29.3	+55 17 30	+0.001	-0.01	6.97	-0.08	0.4	B9.5 V	+1	210 s		
197473	106427		20 43 29.6	+19 52 53	-0.002	-0.01	7.80	0.8	3.2	G5 IV		82 s	14205	m
197435	126136		20 43 34.3	+6 04 04	0.000	0.00	7.9	0.9	3.2	G5 IV		88 s		
197364	163830		20 43 34.9	-11 38 30	+0.003	-0.02	6.8	0.9	3.2	G5 IV		52 s		
197097	246707		20 43 35.9	-51 48 04	+0.002	-0.04	8.00	0.4	2.6	F0 V		120 s		m
197268	212387		20 43 37.2	-30 13 45	+0.002	-0.01	7.8	0.7	-0.1	K2 III		390 s		
197462	126143		20 43 38.6	+9 06 51	-0.001	-0.05	8.0	0.1	1.7	A3 V		180 s		
197447	126140		20 43 38.7	+6 34 42	-0.001	+0.01	7.3	1.1	0.2	K0 III		260 s		
197320	189617		20 43 39.3	-23 28 13	+0.002	0.00	8.0	0.3	4.0	F8 V		62 s		
197448	126141		20 43 39.9	+5 23 25	+0.002	0.00	6.7	0.1	0.6	A0 V		160 s		
197216	230340		20 43 41.8	-40 03 38	+0.002	-0.01	8.0	1.4	0.2	K0 III		290 s		
197463	126142		20 43 41.8	+4 04 42	-0.001	-0.01	7.9	1.4	-0.3	K5 III		430 s		
197421	144739		20 43 42.9	-2 27 01	+0.004	-0.02	8.0	1.1	0.2	K0 III		230 mx		
197717	50000		20 43 43.9	+46 19 06	-0.001	0.00	7.20	1.1	3.2	K0 IV		62 s	14219	m
197562	89058		20 43 47.5	+23 47 28	+0.001	-0.01	6.8	0.1	0.6	A0 V	-26	170 s		

HD	SAO	Star Name	α 2000	δ 2000	μ(α)	μ(δ)	V	B-V	M_v	Spec	RV	d(pc)	ADS	Notes
197667	70435		$20^h 43^m 49^s.9$	$+39°27'35''$	$-0^s.002$	$-0''.02$	7.80	1.1	0.2	K0 III		310 s		m
197336	189620		20 43 51.6	-25 49 33	+0.005	0.00	7.8	1.3	0.2	K0 III		210 s		
197700	50001		20 43 53.4	+42 13 13	0.000	-0.01	6.99	0.30		F0				
197382	163833		20 43 54.7	-19 20 41	-0.004	-0.07	7.3	1.0	0.2	K0 III		150 mx		
197451	144743		20 43 56.8	-5 35 25	+0.001	-0.01	7.2	0.4	2.6	F0 V		83 s		
197573	89061		20 43 59.8	+21 24 03	0.000	+0.02	7.1	0.1	1.4	A2 V		140 s		
197157	246709	η Ind	20 44 02.2	-51 55 16	+0.017	-0.05	4.51	0.27	2.4	dA7 n	-2	26 s		
197339	212390		20 44 03.9	-31 55 43	0.000	-0.01	7.4	1.2	-0.1	K2 III		310 s		
197491	144748		20 44 12.4	-3 54 57	0.000	-0.01	7.2	1.1	0.2	K0 III		250 s		
197466	144747		20 44 16.2	-9 02 57	+0.002	+0.01	7.8	0.4	2.6	F0 V		110 s		
197735	70444		20 44 21.3	+36 42 50	0.000	0.00	7.9	1.4	-0.3	K5 III		400 s		
197939	32849		20 44 21.8	+56 29 17	0.000	-0.01	5.78	1.67	-0.5	M2 III	-28	160 s		
197702	70442		20 44 26.4	+31 41 45	0.000	0.00	7.90	0.19	-1.6	B3.5 V		530 s		
198236	19018		20 44 32.9	+69 45 08	-0.006	-0.02	6.5	0.9	0.3	G8 III	-9	160 s		
197497	163841		20 44 33.0	-14 11 08	0.000	0.00	6.9	1.6	-0.5	M2 III		310 s		
197426	189630		20 44 35.2	-28 12 17	+0.004	-0.03	7.1	0.5	1.7	A3 V		72 s		
197850	50011		20 44 36.3	+47 17 50	0.000	0.00	6.7	1.1	0.2	K0 III		190 s		
197645	106438		20 44 38.7	+13 33 18	+0.004	-0.02	7.52	0.05	0.2	B9 V		140 mx		
197402	212396		20 44 40.8	-35 49 38	+0.001	+0.02	6.9	1.0	0.2	K0 III		200 s		
197703	89079		20 44 46.7	+20 29 23	0.000	-0.04	7.1	0.2	2.1	A5 V		98 s		
197682	106442		20 44 46.9	+19 42 57	0.000	-0.02	8.00	1.4	-0.3	K5 III		430 s		m
197523	189638		20 44 49.0	-20 53 36	-0.002	-0.05	7.70	1.1	0.2	K0 III		320 s	14218	m
197752	89084	30 Vul	20 44 52.4	+25 16 15	-0.002	-0.17	4.91	1.18	-0.1	K2 III	+31	100 s		
197577	144762		20 44 53.8	-8 00 22	+0.001	+0.01	8.00	0.6	0.4	G2 III	+6	330 s		
198013	32855		20 44 54.1	+56 48 06	+0.001	0.00	8.0	1.1	0.2	K0 III		330 s		
197684	106443		20 44 54.5	+12 18 44	+0.002	-0.01	6.70	0.1	0.6	A0 V	-10	170 s	14233	m
197623	126166		20 44 56.9	+0 17 31	+0.008	-0.08	7.53	0.65	3.2	G5 IV	-71	74 s		
197051	254862	β Pav	20 44 57.4	-66 12 12	-0.006	0.16	3.42	0.16	1.2	A5 IV	+10	28 s		
197683	106446		20 45 00.1	+12 43 37	-0.001	+0.01	8.04	0.44		F	+17		14238	m
198894	9879		20 45 02.5	+79 14 19	-0.008	-0.03	7.9	1.1	0.2	K0 III		340 s		
197705	106448		20 45 06.7	+11 46 56	+0.003	+0.07	8.0	1.1	0.2	K0 III		230 mx		
197341	246717		20 45 08.2	-50 30 24	+0.001	-0.01	7.30	0.8	3.2	G5 IV		66 s		m
197685	126173		20 45 08.6	+6 01 29	+0.002	+0.02	7.8	0.1	0.6	A0 V		270 s		
197482	212401		20 45 11.6	-35 09 56	0.000	+0.01	6.90	0.7	4.4	G0 V		32 s		m
197657	144766		20 45 12.0	-1 02 35	-0.002	-0.01	7.90	0.9	3.2	G5 IV		87 s		m
197540	189641		20 45 13.0	-27 14 50	+0.001	+0.01	6.5	0.9	3.2	G5 IV		38 s		
197839	70458		20 45 13.0	+31 46 26	0.000	-0.02	8.0	1.1	-0.1	K2 III	-13	400 s		
197593	163848		20 45 14.7	-15 47 54	+0.004	0.00	7.0	0.4	2.6	F0 V		75 s		
198149	19019	3 η Cep	20 45 17.2	+61 50 20	+0.013	+0.82	3.43	0.92	3.2	K0 IV	-87	14 ts	14276	m
197961	50027		20 45 20.8	+46 21 18	0.000	-0.02	6.70	-0.03	0.3	A1 IV	-4	190 s		
198084	32862		20 45 20.9	+57 34 47	-0.009	-0.23	4.51	0.54	2.4	F8 IV	-31	26 ts		m
197626	163851		20 45 22.9	-15 02 05	+0.003	+0.02	7.6	0.9	3.2	G5 IV		75 s		
197086	--		20 45 23.6	-66 53 42			7.9	1.6	-0.5	M2 III		470 s		
197541	212410		20 45 23.7	-30 28 39	0.000	+0.01	6.60	0.4	2.6	F0 V		63 s		m
197459	230351		20 45 26.6	-43 45 01	+0.001	0.00	8.0	1.3	0.2	K0 III		360 s		
197737	126179		20 45 27.7	+5 52 53	+0.005	-0.09	7.6	1.1	-0.1	K2 III		130 mx		
197812	106458		20 45 28.0	+18 05 25	0.000	+0.01	6.38	1.68		M5 II-III	-21			U Del, v
198055	32861		20 45 28.7	+54 01 00	+0.001	0.00	8.00	1.1	0.2	K0 III		340 s	14272	m
197851	89097		20 45 29.0	+26 06 23	0.000	+0.01	8.0	0.0	0.4	B9.5 V		320 s		
197165	254864		20 45 30.3	-64 25 41	+0.002	-0.04	7.6	0.1	0.6	A0 V		220 s		
197796	106459		20 45 35.2	+11 38 28	+0.001	-0.05	7.7	1.1	0.2	K0 III		310 s		
--	89099		20 45 37.1	+21 04 11	-0.001	0.00	7.9	2.1		M3				
197912	70467	52 Cyg	20 45 39.6	+30 43 11	-0.001	+0.03	4.22	1.05	0.2	K0 III	-1	59 s		
197868	89103		20 45 40.2	+23 01 02	+0.001	0.00	8.0	0.9	3.2	G5 IV		92 s		
198039	32865		20 45 43.0	+50 40 26	+0.001	+0.01	7.10	0.5	3.4	F5 V		54 s	14273	m
197148	--		20 45 46.5	-66 49 39			8.0	1.1	0.2	K0 III		360 s		
197832	126187		20 45 57.0	+7 22 46	+0.003	+0.03	7.0	0.1	0.6	A0 V		190 s		
197799	144784		20 45 58.9	-0 49 30	-0.001	-0.02	7.7	0.9	3.2	G5 IV		80 s		
197814	144785		20 46 00.9	-0 20 24	-0.001	+0.01	6.74	0.97	3.2	G5 IV		37 s		
197505	230360		20 46 05.2	-48 58 15	+0.001	-0.01	7.5	0.7	3.2	G5 IV		73 s		
197692	189664	16 ψ Cap	20 46 05.5	-25 16 16	-0.004	-0.16	4.14	0.43	3.4	F5 V	+26	12 ts		
197711	189666		20 46 09.7	-21 44 43	+0.009	-0.25	8.0	0.9	3.2	G5 IV		39 mx		
197725	189667	17 Cap	20 46 09.8	-21 30 50	+0.002	-0.01	5.9	0.7	0.6	A0 V		45 s		
197989	70474	53 ε Cyg	20 46 12.5	+33 58 13	+0.028	+0.33	2.46	1.03	0.2	K0 III	-10	25 ts	14274	m
197913	106467		20 46 13.2	+15 54 29	+0.006	+0.04	7.50	1.1	0.2	K0 III	-26	180 mx	14270	m
197726	189669		20 46 16.5	-23 43 27	-0.001	-0.01	7.2	0.1	0.6	A0 V		190 s		
197649	212417		20 46 18.4	-36 07 13	+0.004	-0.05	6.49	0.39	3.0	dF2		47 s		
197673	212419		20 46 18.4	-31 31 22	0.000	-0.01	7.8	0.8	2.6	F0 V		52 s		
197630	212416		20 46 19.8	-39 11 57	+0.004	-0.03	5.5	0.5	0.0	B8.5 V	-49	54 s		
198181	32877		20 46 21.0	+52 59 43	-0.010	-0.10	6.33	1.12	0.2	K0 III	-29	140 s		
198224	32878		20 46 24.4	+56 50 07	+0.007	+0.05	7.5	0.5	3.4	F5 V		65 s		
197940	106472		20 46 25.9	+16 54 13	0.000	-0.01	7.1	1.1	-0.1	K2 III		270 s		
197544	246725		20 46 26.7	-50 48 18	+0.001	+0.01	6.6	0.8	0.2	K0 III		190 s		
197746	189675		20 46 27.1	-26 52 08	-0.001	-0.02	6.9	0.9	3.0	F2 V		28 s	14258	m
197784	189680		20 46 30.0	-22 09 54	+0.001	-0.07	7.2	1.4	0.2	K0 III		140 s		

HD	SAO	Star Name	α 2000	δ 2000	μ(α)	μ(δ)	V	B-V	M$_v$	Spec	RV	d(pc)	ADS	Notes
			h m s	° ′ ″	s	″								
197925	106473		20 46 30.6	+12 37 20	+0.001	+0.01	7.96	0.20	-0.2	A2 III		360 s		
198549	19038		20 46 36.8	+69 56 10	-0.006	+0.01	7.8	0.1	0.6	A0 V		200 mx		
197819	163869		20 46 38.3	-19 26 12	+0.003	+0.05	7.6	0.2	2.6	F0 V		100 s		
198151	50050		20 46 38.3	+46 31 54	-0.003	-0.01	6.30	0.03	1.7	A3 V	-9	83 s		m
197963	106475	12 γ¹ Del	20 46 38.6	+16 07 27	-0.002	-0.20	5.22	0.49		F8 IV-V	-8		14279	m
197964	106476	12 γ² Del	20 46 39.3	+16 07 27	-0.002	-0.19	4.27	1.04	3.2	G5 IV	-7	23 mn	14279	m
197730	212423		20 46 39.6	-33 36 15	+0.003	-0.02	8.0	1.3	3.2	G5 IV		44 s		
197785	189683		20 46 40.1	-27 14 02	+0.001	-0.05	7.9	1.5	-0.5	M4 III		490 s		
198014	89116		20 46 44.4	+24 27 45	+0.002	+0.01	7.1	0.1	1.4	A2 V		140 s		
197359	254867		20 46 46.2	-64 54 34	+0.014	-0.07	6.8	1.4	0.2	K0 III		120 s		
198056	70480		20 46 47.3	+32 25 11	+0.001	-0.01	7.04	-0.04	0.4	B9.5 V		210 s		
198413	19035		20 46 48.8	+63 59 10	+0.001	+0.06	8.0	0.4	2.6	F0 V		110 s		
197942	126200		20 46 49.3	+2 26 15	+0.001	0.00	7.7	1.6		M5 III	-44			V Aqr, v
198182	50054		20 46 52.8	+47 06 42	-0.001	+0.02	8.0	0.1	1.2	A1 V		210 s		
198343	32890		20 47 00.6	+58 25 01	-0.004	+0.01	7.3	0.5	4.0	F8 V		46 s		
197954	144802		20 47 03.4	-2 29 12	-0.001	-0.01	6.27	1.55		K2				
198015	106484		20 47 05.0	+14 43 47	+0.003	-0.01	7.8	0.1	1.4	A2 V		190 s		
198108	70498		20 47 08.3	+32 10 20	+0.002	+0.01	7.42	0.07	0.6	A0 V		200 s		
198058	89124		20 47 09.0	+21 58 41	+0.002	+0.01	7.7	1.1	0.2	K0 III		300 s		
198195	50055		20 47 09.7	+42 24 36	0.000	+0.01	7.18	-0.05	0.4	B9.5 V	-20	230 s	14295	m
198134	70499		20 47 10.6	+34 22 26	+0.004	+0.01	4.92	1.32	-0.2	K3 III	-23	100 s	14290	T Cyg, m,q
198284	32885		20 47 10.7	+52 36 33	+0.002	+0.02	7.7	1.1	0.2	K0 III		290 s		
198285	32886		20 47 13.8	+51 47 41	+0.005	0.00	7.90	0.4	2.6	F0 V		110 s	14302	m
197713	230367		20 47 15.9	-47 10 21	+0.005	+0.02	7.4	0.1	2.6	F0 V		93 s		
197873	189694		20 47 16.5	-27 41 55	-0.002	-0.01	7.7	0.6	3.4	F5 V		59 s		
197889	189697		20 47 20.0	-26 24 55	+0.002	0.00	7.30	0.1	1.4	A2 V		150 s	14280	m
198237	50060		20 47 20.6	+45 34 47	0.000	-0.02	6.40	1.61	-0.2	K3 III	-6	130 s	14298	m
197696	246731		20 47 22.0	-50 32 46	0.000	-0.06	7.5	1.1	0.2	K0 III		200 mx		
197748	230370		20 47 23.5	-45 51 16	+0.001	+0.01	6.8	0.8	-0.1	K2 III		250 s		
198183	70505	54 λ Cyg	20 47 24.3	+36 29 27	+0.001	0.00	4.53	-0.11	-1.1	B5 V	-23	130 s	14296	m
197944	163879		20 47 26.9	-15 36 54	-0.001	-0.01	7.6	1.1	0.2	K0 III		300 s		
--	--		20 47 27.7	+26 07 46	+0.001	+0.01	7.5	1.4	-0.3	K5 III		360 s		
198023	126213		20 47 28.0	+6 48 54	+0.004	+0.04	8.0	0.4	3.0	F2 V		98 s		
--	50061		20 47 28.3	+43 04 34	+0.001	+0.01	8.0	1.7						
198060	126217		20 47 29.8	+9 51 52	+0.002	+0.04	7.5	0.8	3.2	G5 IV		73 s		
198024	126214		20 47 29.9	+5 17 57	0.000	-0.01	7.9	1.4	-0.3	K5 III		440 s		
199437	3467		20 47 33.2	+80 33 08	-0.013	-0.03	5.39	1.12	0.2	K0 III	-26	93 s		
198025	126215		20 47 34.4	+1 49 54	-0.002	-0.01	7.30	1.4	-0.3	K4 III	+19	330 s		
196520	258870		20 47 36.0	-79 51 11	-0.005	+0.03	7.6	1.0	3.2	G5 IV		51 s		
198089	106494		20 47 37.8	+12 58 11	+0.003	+0.05	7.5	0.5	4.0	F8 V		50 s		
198109	106495		20 47 38.2	+16 14 17	+0.008	-0.04	7.5	0.5	4.0	F8 V		51 s		
198061	126219		20 47 38.5	+6 58 34	-0.001	-0.01	7.3	0.5	3.4	F5 V		61 s		
198000	144812		20 47 39.4	-7 06 40	0.000	0.00	8.0	1.1	0.2	K0 III		360 s		
198001	144810	2 ε Aqr	20 47 40.3	-9 29 45	+0.002	-0.03	3.77	0.00	1.2	A1 V	-16	33 s		Albali
198253	50066		20 47 40.5	+42 04 06	-0.001	0.00	7.50	0.05		A0			14306	m
198135	89130		20 47 40.7	+22 41 43	+0.009	-0.02	7.9	0.5	4.0	F8 V		59 s		
198197	70511		20 47 43.6	+31 54 45	0.000	-0.01	7.9	0.1	0.6	A0 V		280 s		
198026	144814	3 Aqr	20 47 44.0	-5 01 40	0.000	-0.04	4.42	1.65	-0.5	M3 III	-22	91 s		
198110	106498		20 47 45.0	+10 27 55	0.000	+0.01	7.5	0.1	0.6	A0 V		240 s		
198137	106501		20 47 46.3	+16 46 11	+0.002	-0.01	7.9	1.6	-0.5	M2 III		440 s		
198198	89133		20 47 47.4	+29 38 16	+0.001	-0.01	8.00	0.9	0.3	G8 III		340 s		
198184	89132		20 47 47.5	+26 10 41	0.000	0.00	7.0	0.0	0.4	B9.5 V	-2	200 s		
198070	126221		20 47 47.6	+3 18 24	+0.002	+0.02	6.40	-0.02		A0	-21			
198213	70512		20 47 47.9	+31 40 19	0.000	+0.01	7.9	0.1	1.4	A2 V		190 s		
198069	126222	13 Del	20 47 48.2	+6 00 30	0.000	0.00	5.58	-0.02		A0	-8	34 mn	14293	m
197969	189703		20 47 49.3	-23 28 31	+0.002	-0.04	7.5	2.0	-0.5	M2 III		220 s		
198345	50073		20 47 49.3	+47 49 54	+0.001	-0.02	5.57	1.46	-0.3	K5 III	-30	150 s	14318	m
198199	89134		20 47 52.3	+27 35 43	+0.001	-0.01	7.0	0.2	2.1	A5 V		93 s		
198387	32897		20 47 52.7	+52 24 26	+0.007	-0.16	6.27	0.89	5.9	dK0	-41	23 mn	14322	m
197956	189701		20 47 52.7	-27 45 03	+0.002	+0.01	7.50	0.9	3.2	G5 IV		72 s	14288	m
198288	70517		20 47 59.4	+39 17 15	-0.001	-0.01	6.95	0.71	-6.9	A5 Iab	-2	2500 s	14314	V367 Cyg, m,v
--	--		20 48 05.6	+27 35 11	+0.003	-0.02	7.3 B							
198028	163886		20 48 07.5	-16 20 54	-0.001	-0.05	7.5	0.4	2.6	F0 V		94 s		
198074	144816		20 48 08.0	-9 44 56	0.000	+0.01	7.3	1.1	0.2	K0 III		260 s		
198124	126224		20 48 09.5	+2 42 55	+0.002	-0.03	6.9	1.1	-0.1	K2 III		250 s		
198254	89140		20 48 10.5	+28 31 46	-0.001	0.00	7.70	1.1	0.0	K1 III		340 s		
198400	50081		20 48 12.0	+49 59 44	0.000	-0.01	8.0	0.1	1.4	A2 V		200 s		
198041	163889		20 48 12.7	-16 31 16	-0.004	-0.05	7.1	0.4	3.0	F2 V		65 s		
196752	257850		20 48 13.7	-78 57 04	+0.018	-0.03	7.8	0.4	3.0	F2 V		91 s		
198031	163887		20 48 13.9	-19 15 38	+0.006	-0.02	6.8	0.5	3.4	F5 V		43 s		
198238	89142		20 48 14.7	+26 23 42	+0.001	+0.01	8.00	1.4	-0.3	K5 III		440 s	14315	m
198075	163891		20 48 15.2	-12 27 14	+0.001	+0.06	8.00	0.6	4.9	G3 V	-16	42 s		
198111	144824		20 48 15.4	-4 38 25	0.000	-0.05	7.1	1.1	-0.1	K2 III		270 s		
198032	189708		20 48 16.7	-22 50 44	+0.001	+0.01	8.0	0.0	1.4	A2 V		210 s		
198265	89145		20 48 17.5	+28 14 57	0.000	+0.02	7.8	1.6	-0.5	M4 III		450 s		

HD	SAO	Star Name	α 2000	δ 2000	μ(α)	μ(δ)	V	B-V	M$_v$	Spec	RV	d(pc)	ADS	Notes
197900	230375		20^{h}48^{m}18.0^s	-44°11'51"	+0.007^s	-0.07"	6.47	1.16		K1 IV		180 mx		
197934	230377		20 48 19.1	-40 00 05	+0.005	-0.06	7.4	1.3	0.2	K0 III		160 s		
198166	126228		20 48 21.4	+7 35 40	+0.003	0.00	7.1	1.1	0.2	K0 III		240 s		
198044	189712		20 48 21.6	-22 44 26	+0.003	-0.16	7.3	0.5	4.0	F8 V	+12	46 s		
198329	70523		20 48 22.4	+36 53 30	-0.010	-0.06	7.9	1.1	-0.1	K2 III		180 mx		
198239	89144		20 48 23.2	+21 13 59	0.000	0.00	7.0	1.1	0.2	K0 III		220 s		
198167	126227		20 48 24.5	+3 38 56	+0.001	0.00	6.6	0.0	0.4	B9.5 V		170 s		
--	163894		20 48 24.7	-18 12 00	+0.001	-0.01	7.7	0.9	3.2	G5 IV		80 s		
198737	19045		20 48 25.7	+68 02 00	+0.010	+0.13	7.0	0.4	2.6	F0 V		72 s		
198063	163895		20 48 25.8	-18 12 05	+0.001	0.00	6.70	0.9	0.2	G9 III	-10	200 s	14299	m
198414	50084		20 48 26.3	+45 27 06	+0.002	-0.01	7.63	-0.10	-1.6	B7 III	-20	640 s		
198153	126226		20 48 27.0	+0 11 45	-0.005	-0.08	7.6	0.1	0.6	A0 V		36 mx		
198168	126230		20 48 27.6	+1 43 22	-0.001	-0.03	7.40	1.1	0.2	K0 III		280 s		
197937	230379	z Mic	20 48 29.0	-43 59 19	+0.017	-0.10	5.11	0.35	2.8	F1 V	-18	28 s		m
198792	19048		20 48 29.5	+69 39 28	+0.003	+0.01	7.5	0.1	0.6	A0 V		210 s		
197790	246733		20 48 30.7	-57 42 21	+0.005	+0.01	7.9	0.3	3.2	G5 IV		88 s		
197901	230376		20 48 30.9	-48 40 12	-0.002	0.00	7.9	0.1	1.7	A3 V		170 s		
198267	89151		20 48 31.9	+23 00 41	0.000	-0.01	7.7	0.0	0.4	B9.5 V		280 s		
198268	89150		20 48 33.1	+21 37 37	0.000	-0.01	7.8	1.1	0.2	K0 III		320 s		
198227	126238		20 48 35.7	+9 28 00	+0.002	-0.01	7.7	0.5	4.0	F8 V		54 s		
197957	230381		20 48 35.9	-43 00 28	0.000	-0.03	6.8	2.1	-0.1	K2 III		69 s		
198330	70526		20 48 37.4	+30 47 34	-0.003	-0.05	7.40	1.4	-0.3	K4 III	-1	340 s		
198513	32908		20 48 42.5	+51 54 38	0.000	+0.01	6.50	-0.07		B9			14336	m
198401	50089		20 48 42.8	+41 45 43	+0.001	-0.05	7.6	0.8	3.2	G5 IV		76 s		
197971	230382		20 48 43.8	-43 59 00	0.000	+0.02	7.70	0.5	3.4	F5 V		72 s		m
197417	257852		20 48 48.4	-72 12 44	+0.008	-0.01	7.5	0.1	0.6	A0 V		240 s		
197827	246736		20 48 50.5	-59 14 04	+0.005	-0.03	7.5	0.5	0.6	A0 V		110 s		
198315	89156		20 48 50.7	+22 59 38	+0.001	-0.01	7.7	1.6	-0.5	M4 III		410 s		
198270	106521		20 48 52.7	+11 51 54	-0.001	-0.01	7.7	1.4	-0.3	K5 III		390 s		
197569	254870		20 48 56.0	-68 47 31	+0.007	-0.09	7.1	0.6	2.1	A5 V		46 mx		
198478	50099	55 Cyg	20 48 56.2	+46 06 51	0.000	0.00	4.84	0.41	-6.8	B3 Ia	-7	1000 s	14337	m
198366	89161		20 48 59.4	+25 40 20	+0.001	0.00	7.8	0.0	0.4	B9.5 V		290 s		
198424	70539		20 49 00.3	+38 51 55	0.000	+0.01	7.50	-0.12		A0				
198436	70541		20 49 00.5	+39 47 30	+0.001	0.00	7.71	-0.05		A0	-21		14334	m
197431	257854		20 49 00.7	-72 31 59	+0.004	-0.02	6.5	0.1	0.6	A0 V		150 s		
198346	89162		20 49 03.1	+22 23 28	+0.001	+0.03	7.8	0.4	2.6	F0 V		110 s		
198271	126248		20 49 04.3	+6 05 02	-0.001	0.00	7.8	0.1	1.4	A2 V		190 s		
198379	89165		20 49 04.8	+25 40 13	+0.001	+0.01	7.69	-0.02	0.4	B9.5 V		290 s		
198654	32917		20 49 06.5	+58 44 50	+0.017	+0.10	7.81	0.91		K0		14 mn		
198480	50102		20 49 07.0	+42 57 15	-0.003	-0.04	7.30	0.00	0.0	B8.5 V	-11	280 s		
198596	32913		20 49 08.1	+54 14 02	-0.001	+0.10	7.2	1.1	0.2	K0 III	0	140 mx		
198481	50101		20 49 08.6	+40 54 57	+0.002	0.00	7.8	0.6	4.4	G0 V		47 s		
198081	212460		20 49 08.6	-36 29 07	-0.005	0.00	8.0	1.1	3.2	G5 IV		58 s		
198456	70546		20 49 09.3	+39 17 24	+0.001	+0.02	7.89	1.06		K0				m
198437	70544		20 49 11.8	+35 33 49	0.000	-0.03	6.6	1.4	-0.3	K5 III		240 s		
197959	246743		20 49 14.0	-52 42 27	+0.004	-0.06	8.0	0.5	3.4	F5 V		75 s		
198010	230388		20 49 14.6	-48 13 29	-0.001	+0.02	7.2	1.2	0.2	K0 III		200 s		
198332	106529		20 49 14.6	+14 36 19	0.000	-0.01	7.1	0.1	0.6	A0 V		200 s		
198739	19049		20 49 14.9	+62 31 47	+0.002	-0.01	7.96	0.04	0.0	B8.5 V		360 s		
198009	230389		20 49 15.9	-46 38 57	+0.005	-0.05	7.81	1.21	1.7	K0 III-IV		120 s		
198495	50104		20 49 16.4	+41 08 35	+0.005	+0.06	8.0	0.5	4.0	F8 V		62 s		
198142	212463		20 49 16.9	-30 11 31	0.000	-0.02	7.26	1.50	-0.3	K5 III		330 s		
198272	144841		20 49 17.0	-0 33 48	-0.002	-0.02	6.40	1.54		M				
198781	19051		20 49 17.2	+64 02 32	-0.002	0.00	6.45	0.07		B0.5 V	-27			
198174	189733		20 49 17.5	-25 46 53	+0.001	-0.02	5.70	-0.07	-0.9	B6 V	-12	200 s		
197635	254871	σ Pav	20 49 17.9	-68 46 36	-0.012	-0.05	5.41	1.12	0.2	gK0	+19	87 s		
198679	32922		20 49 18.8	+58 44 55	-0.001	0.00	6.70	0.00	0.4	B9.5 V		170 s		m
198240	144840		20 49 19.3	-8 47 37	+0.001	0.00	8.0	1.6	-0.5	M2 III		510 s		
198208	163910		20 49 20.4	-18 02 10	0.000	-0.03	6.40	1.1	0.2	K0 III	+44	170 s		m
200545	3475		20 49 22.1	+83 56 31	+0.015	+0.11	7.3	0.4	3.0	F2 V		71 s		
--	19052		20 49 22.7	+63 20 57	-0.002	+0.05	8.0	1.6						
198258	163913		20 49 24.7	-11 27 18	-0.002	+0.01	7.9	0.4	2.6	F0 V		110 s		
198104	230393		20 49 28.7	-41 42 42	-0.001	+0.02	7.4	1.2	3.4	F5 V		64 s		
198048	230391	ζ Ind	20 49 28.8	-46 13 37	+0.005	+0.03	4.89	1.52	-0.5	gM1	-5	120 s		
198611	32919		20 49 32.9	+50 47 00	-0.001	-0.01	7.6	1.4	-0.3	K5 III		350 s		
198482	70549		20 49 36.1	+30 39 00	+0.001	-0.03	7.90	1.1	-0.1	K2 III		380 s		
198638	32921		20 49 36.3	+51 18 53	-0.002	-0.03	6.7	0.4	3.0	F2 V		54 s		
196818	258871		20 49 36.4	-80 07 59	-0.004	+0.03	7.6	1.3	3.2	G5 IV		38 s		
198624	50116		20 49 37.3	+50 07 38	-0.003	+0.02	6.50	1.62	-1.4	M4 II-III		390 s	14345	m
198390	106536	15 Del	20 49 37.6	+12 32 43	+0.004	+0.10	5.98	0.43	3.4	F5 V	+2	33 s		m
198389	106538		20 49 38.9	+13 57 47	+0.007	+0.08	7.1	0.5	4.0	F8 V	-41	41 s		
197325	257853		20 49 39.2	-75 50 01	+0.008	-0.05	7.4	1.1	0.2	K0 III		270 s		
198403	106539		20 49 46.0	+11 29 42	+0.001	0.00	7.4	1.1	-0.1	K2 III		310 s		
198483	89176		20 49 46.3	+25 46 16	+0.007	+0.02	7.70	0.58	4.4	G0 V		44 s		
198178	212466		20 49 47.9	-36 51 18	+0.003	0.00	7.6	1.3	0.2	K0 III		210 s		

HD	SAO	Star Name	α 2000	δ 2000	μ(α)	μ(δ)	V	B-V	M$_v$	Spec	RV	d(pc)	ADS	Notes
			h m s	° ' "	s	"								
198391	126265	14 Del	20 49 48.0	+7 51 51	+0.001	+0.02	6.33	0.02		A0	-30			
198334	144849		20 49 48.7	-6 41 49	+0.002	+0.02	7.9	0.5	4.0	F8 V		61 s		
198371	144852		20 49 49.8	-0 19 28	+0.001	0.00	6.9	1.1	-0.1	K2 III		260 s		
198625	50119		20 49 54.5	+46 39 40	0.000	0.00	6.33	-0.07	1.2	A1 V	-15	110 s	14350	m
198373	144854		20 49 56.2	-5 08 49	-0.001	-0.01	6.7	1.6		M5 III	-39			T Aqr, v
—	70556		20 49 57.2	+40 02 36	+0.002	+0.01	7.7	1.8						
198232	212472	α Mic	20 49 57.8	-33 46 47	+0.001	-0.02	4.90	1.00	0.3	gG6	-15	73 s		m
198526	89181		20 49 57.8	+28 59 34	+0.002	+0.01	8.00	1.1	0.0	K1 III		380 s		
198405	126266		20 49 58.5	+2 26 01	0.000	-0.01	7.5	1.4	-0.3	K5 III		370 s		
198404	126267		20 49 58.9	+5 32 41	+0.003	+0.01	6.21	0.98		K0	-22			m
198426	106546		20 49 59.4	+11 54 29	+0.003	+0.03	7.9	1.1	-0.1	K2 III		390 s		
198484	89178		20 50 00.9	+20 52 48	+0.001	0.00	7.9	0.1	1.7	A3 V		170 s		
198639	50121	56 Cyg	20 50 04.8	+44 03 34	+0.011	+0.14	5.04	0.20		A5 p	-21	21 mn		m
198278	189747		20 50 04.9	-27 22 03	0.000	-0.03	7.00	1.1	0.2	K0 III		230 s	14335	m
198527	89185		20 50 05.2	+27 22 40	0.000	+0.02	7.0	0.0	0.4	B9.5 V	-7	210 s		
198701	32926		20 50 06.0	+52 35 13	+0.003	0.00	8.0	0.5	4.0	F8 V		62 s		
—	19059		20 50 07.9	+61 57 42	-0.001	+0.02	7.8	1.0						
198458	106549		20 50 09.4	+13 17 54	+0.001	-0.03	7.1	1.1	-0.1	K2 III		280 s		
198459	106551		20 50 11.7	+12 05 01	0.000	0.00	7.8	0.1	0.6	A0 V		270 s		
197324	257856		20 50 15.8	-75 02 18	+0.006	-0.06	8.0	0.5	3.4	F5 V		82 s		
198280	212480		20 50 17.1	-30 25 23	+0.003	+0.02	7.8	1.7	3.4	F5 V		13 s		
198793	32929		20 50 17.9	+56 47 55	-0.001	+0.01	7.06	-0.08	0.0	B8.5 V		260 s		
198158	230397		20 50 18.1	-48 34 23	-0.001	0.00	8.0	1.3	-0.3	K5 III		320 s		
198597	70559		20 50 23.5	+31 21 12	+0.001	0.00	7.8	0.1	0.6	A0 V		260 s		
198551	89193		20 50 27.7	+22 53 03	+0.005	-0.07	8.0	1.1	0.2	K0 III		130 mx		
198354	189757		20 50 29.9	-24 58 58	-0.006	-0.10	6.7	0.5	4.0	F8 V		34 s		
—	—		20 50 34.1	+30 54 45			6.90						14355	m
198626	70564		20 50 35.9	+30 54 45	+0.004	+0.04	6.8	0.1	1.4	A2 V	-29	93 mx		
198552	106562		20 50 36.9	+18 03 05	+0.005	+0.02	6.50	0.04	1.2	A1 V	+13	110 s		
198584	89197		20 50 38.4	+22 43 49	+0.001	0.00	7.9	0.1	1.4	A2 V		190 s		
198324	212485		20 50 38.7	-31 43 32	+0.001	-0.06	7.4	1.2	-0.1	K2 III		200 mx		
198690	50125		20 50 39.5	+42 22 54	+0.001	-0.01	7.28	-0.10	0.0	B8.5 V	-24	290 s		
198568	89199		20 50 41.2	+20 28 30	-0.001	-0.03	8.0	1.4	-0.3	K5 III		440 s		
198431	163924		20 50 41.6	-12 32 42	+0.008	-0.07	5.88	1.07	0.2	K0 III	-44	120 s		
198819	32932		20 50 44.0	+55 01 25	+0.004	+0.01	7.6	0.1	1.4	A2 V		170 s		
198356	212487		20 50 46.9	-32 03 16	-0.001	-0.06	6.4	1.9	-0.3	K5 III		130 s		
198444	163926		20 50 48.6	-14 12 02	+0.001	-0.01	8.0	0.9	3.2	G5 IV		93 s		
198420	189763		20 50 53.9	-21 18 45	+0.001	-0.08	7.1	0.7	3.4	F5 V		36 s		
198692	70575		20 50 55.8	+34 45 15	+0.002	+0.01	6.6	1.1	0.2	K0 III		190 s		
198794	50131		20 50 56.8	+48 01 50	+0.002	+0.04	6.95	1.43	-4.4	K3 Ib	-23	1900 s		
198357	212488		20 51 00.5	-37 54 49	-0.001	-0.02	5.52	1.38	-2.1	K0 II	+15	250 s		
198569	126289		20 51 03.5	+6 23 14	-0.001	0.00	7.90	1.1	0.0	K1 III		380 s	14359	m
198833	32934		20 51 03.9	+52 23 45	-0.006	+0.01	7.1	0.9	3.2	G5 IV		60 s		
198834	32933		20 51 05.2	+51 25 03	+0.003	+0.05	7.30	0.4	2.6	F0 V		85 s	14370	m
197343	257857		20 51 09.3	-78 07 55	-0.011	+0.04	7.7	0.5	3.4	F5 V		71 s		
198449	189767		20 51 11.3	-25 34 59	+0.003	-0.02	7.6	1.3	0.2	K0 III		210 s		
198383	212490		20 51 14.0	-38 16 05	+0.001	-0.03	8.0	0.5	2.6	F0 V		87 s		
198783	70584		20 51 23.9	+39 39 12	0.000	+0.01	8.0	0.9	3.2	G5 IV		91 s		
198571	144877	4 Aqr	20 51 25.5	-5 37 35	+0.006	0.00	5.99	0.46	3.1	dF3	-25	34 s	14360	m
198702	89214		20 51 25.5	+27 25 44	-0.001	-0.01	7.8	0.1	1.7	A3 V		160 s		
198797	70588		20 51 27.6	+39 23 32	0.000	-0.02	8.02	0.46	-6.6	F2 I		8300 s		
198726	89216		20 51 28.0	+28 15 02	0.000	0.00	5.77	0.72	-4.6	F5 Ib	-1	950 s		T Vul, v
198308	246762	ι Ind	20 51 29.9	-51 36 30	+0.001	-0.01	5.05	1.13	0.2	K0 III	+21	75 s		
198541	163936		20 51 30.2	-17 30 23	0.000	-0.04	8.0	1.1	0.2	K0 III		360 s		
198858	50145		20 51 30.5	+47 42 26	-0.001	0.00	7.36	1.13	0.0	K1 III		280 s		
198959	32945		20 51 30.6	+56 21 57	+0.006	-0.02	7.9	0.7	4.4	G0 V		51 s		
198784	70590		20 51 30.8	+37 59 23	0.000	+0.01	7.29	0.07	-1.0	B5.5 V	-4	360 s		
198500	189775		20 51 33.3	-27 14 41	-0.001	-0.02	6.7	0.5	4.0	F8 V		34 s		
198877	50146		20 51 34.0	+49 39 17	-0.001	-0.04	7.0	1.1	-0.1	K2 III	-6	250 s		
198501	189777		20 51 36.2	-27 59 43	+0.004	+0.01	6.87	0.26		A m				
197808	257859		20 51 37.7	-73 41 29	-0.006	-0.04	7.5	0.8	3.2	G5 IV		71 s		
198160	254883		20 51 38.2	-62 25 46	+0.012	-0.05	5.66	0.17	1.4	A2 V		62 s		m
198502	189778		20 51 38.8	-29 26 31	-0.007	-0.10	7.1	0.7	3.4	F5 V		40 s		
198503	189780		20 51 40.4	-29 46 50	+0.004	+0.01	7.30	1.1	-0.1	K2 III		300 s		m
198433	230405		20 51 40.5	-40 54 21	-0.001	0.00	7.20	1.1	0.2	K0 III		250 s		m
—	50148		20 51 44.1	+49 41 49	-0.004	-0.02	8.0	1.4						
199476	9899		20 51 44.7	+74 46 48	+0.102	+0.56	7.77	0.71	5.6	G8 V	-30	26 ts		
198727	89219		20 51 45.8	+20 12 01	+0.001	0.00	8.0	0.1	0.6	A0 V		290 s		
198599	163938		20 51 47.2	-10 19 03	0.000	-0.01	7.5	0.7	4.4	G0 V		41 s		
198703	106585		20 51 48.2	+13 55 08	0.000	0.00	7.7	1.1	0.2	K0 III		320 s		
198542	189781	18 ω Cap	20 51 49.1	-26 55 09	0.000	0.00	4.11	1.64	-0.3	K5 III	+9	64 s		
198759	89220		20 51 49.7	+21 52 07	+0.001	0.00	7.7	0.6	4.4	G0 V		46 s		
198758	89221		20 51 49.7	+23 19 52	+0.001	0.00	7.5	1.4	-0.3	K5 III		360 s		
199492	9901		20 51 49.9	+74 38 23	0.000	+0.02	7.4	0.1	1.7	A3 V	-14	130 s		
198704	126306		20 51 55.6	+8 46 23	0.000	0.00	7.1	1.4	-0.3	K5 III		310 s		

HD	SAO	Star Name	α 2000	δ 2000	μ(α)	μ(δ)	V	B-V	M_v	Spec	RV	d(pc)	ADS	Notes
198915	50150		20ʰ51ᵐ57ˢ.0	+46°44'05"	-0ˢ.001	0".00	7.50	-0.07	-1.4	B4 V	-21	550 s		
198529	212499	β Mic	20 51 58.6	-33 10 38	+0.001	+0.02	6.04	0.03	1.4	A2 V n	-7	85 s		
198820	70596		20 52 00.2	+32 50 57	-0.001	+0.01	6.44	-0.15	-2.9	B3 III	-18	680 s		
198647	163940		20 52 00.6	-11 26 23	+0.006	-0.02	7.1	1.1	0.2	K0 III		190 mx		
198705	126309	.	20 52 03.0	+7 34 57	+0.002	-0.01	8.0	0.4	2.6	F0 V		120 s		
198846	70599		20 52 03.5	+34 39 28	+0.002	0.00	7.29	-0.06	-4.6	B0 IV	-57	1900 s		Y Cyg, m,v
198543	212500		20 52 05.0	-33 52 52	+0.002	+0.02	7.80	1.4	-0.3	K5 III		420 s		m
198821	89229		20 52 06.3	+28 51 15	+0.001	0.00	8.00	1.1	-0.1	K2 III		400 s		
198809	89228	31 Vul	20 52 07.5	+27 05 49	-0.005	-0.06	4.59	0.83	0.3	G8 III	+1	38 ts		
198667	144889	5 Aqr	20 52 08.5	-5 30 25	0.000	0.00	5.55	-0.08		B8	-2	84 mn		
198579	212502		20 52 14.0	-30 41 50	+0.003	-0.02	7.5	1.6	-0.1	K2 III		190 s		
198916	70612		20 52 14.0	+39 30 04	+0.001	+0.01	7.3	0.1	1.4	A2 V		140 s		
198707	144891		20 52 15.6	-0 07 08	+0.002	0.00	8.0	0.5	3.4	F5 V		82 s		
198648	163943		20 52 15.8	-19 38 35	0.000	+0.01	7.2	0.3	0.6	A0 V		140 s		
199067	32955		20 52 20.9	+55 03 05	+0.004	+0.02	6.72	0.06	1.4	A2 V		110 s		
198649	189794		20 52 21.3	-21 13 52	-0.001	+0.01	7.3	0.7	0.2	K0 III		260 s		
198729	144892		20 52 21.3	-3 13 15	-0.001	-0.04	7.9	1.6	-0.5	M4 III	-30	490 s		
198811	106598		20 52 21.7	+20 07 43	-0.001	+0.02	7.3	0.1	0.6	A0 V		210 s	14379	m
198222	--		20 52 24.5	-64 46 53			8.0	0.8	3.2	G5 IV		90 s		
199120	32959		20 52 28.7	+58 39 18	-0.004	-0.01	7.6	0.9	-0.9	G7 II-III	-6	170 mx		
198835	106603		20 52 29.7	+18 01 38	0.000	-0.01	6.8	0.8	3.2	G5 IV	+17	51 s		
198694	163946		20 52 30.9	-13 12 10	+0.002	+0.01	6.9	1.1	-0.1	K2 III		250 s		
198974	50162		20 52 34.0	+41 58 46	-0.001	0.00	7.9	0.1	0.6	A0 V		270 s		
198581	212504		20 52 37.3	-38 54 24	+0.001	+0.01	8.0	0.1	1.7	A3 V		170 s		
198743	144895	6 μ Aqr	20 52 39.0	-8 59 00	+0.003	-0.03	4.73	0.32		A m	-9	30 mn		
198282	254888		20 52 40.2	-64 32 14	-0.001	-0.06	7.8	0.9	2.6	F0 V		50 s		
198472	246765		20 52 41.6	-53 16 25	+0.006	-0.06	7.3	0.5	4.0	F8 V		45 s		
198590	212506		20 52 43.0	-39 06 31	+0.001	+0.01	6.9	0.4	0.2	K0 III		220 s		
199136	32961		20 52 43.0	+56 35 05	-0.001	0.00	7.44	-0.13	0.0	B8.5 V		310 s		
198975	70620		20 52 43.2	+38 48 21	+0.003	+0.01	8.0	1.1	-0.1	K2 III		380 s		
198933	89236		20 52 53.4	+24 54 52	0.000	-0.02	6.8	1.4	-0.3	K5 III		260 s		
198946	89238		20 52 56.1	+29 19 55	+0.002	0.00	8.0	0.0	0.4	B9.5 V		320 s		
198822	126327		20 52 58.0	+1 57 51	-0.002	0.00	7.9	1.1	0.2	K0 III		340 s		
198732	189801		20 53 01.0	-23 46 59	+0.007	-0.05	6.33	0.88	3.2	G5 IV	-40	35 s	14380	m
199007	70626		20 53 03.3	+34 45 07	0.000	-0.01	7.9	0.0	0.4	B9.5 V		310 s		
198474	246767		20 53 04.9	-56 13 13	+0.004	-0.01	7.9	1.6	0.2	K0 III		160 s		
198802	163953		20 53 05.5	-11 34 25	+0.003	+0.05	6.4	0.6	4.4	G0 V	-1	25 s		
198976	89241		20 53 07.2	+29 38 58	0.000	-0.04	6.34	1.09	-0.1	K2 III	-10	190 s		
198546	246771		20 53 08.5	-51 42 59	+0.003	-0.05	7.4	-0.2	0.6	A0 V		73 mx		
198592	230414		20 53 08.9	-46 37 03	+0.006	+0.03	7.1	0.4	0.6	A0 V		110 s		
198696	212518		20 53 11.0	-31 30 54	+0.007	-0.06	7.33	0.90	4.4	G0 V		23 s		
198934	106619		20 53 13.4	+16 20 30	+0.002	+0.01	7.5	0.1	1.4	A2 V		170 s		
198688	212515		20 53 14.4	-36 30 51	+0.007	-0.07	7.2	0.9	-0.1	K2 III		170 mx		
199081	50180	57 Cyg	20 53 14.5	+44 23 14	+0.001	0.00	4.78	-0.14	-1.1	B5 V	-20	150 s		
199191	32968		20 53 18.1	+54 31 05	+0.002	+0.18	7.14	0.96	0.2	K0 III	-196	240 s		
199098	50182		20 53 18.4	+45 10 55	+0.001	+0.01	5.45	1.10	0.3	G8 III	-24	83 s		
198978	89243		20 53 24.7	+20 55 53	0.000	0.00	7.8	1.6	-0.5	M2 III		440 s		
198751	212521		20 53 24.8	-30 43 07	+0.003	-0.01	6.4	1.3	0.2	K0 III		110 s		
199154	50185		20 53 25.2	+48 16 04	-0.001	+0.02	7.37	0.18	1.2	A5 IV	-23	160 s		
198398	254891		20 53 25.4	-64 11 51	-0.002	+0.01	7.6	0.1	1.4	A2 V		170 s		
199099	50183		20 53 26.3	+42 24 37	+0.001	+0.01	6.66	-0.04		A0	-7			
198881	126338		20 53 26.9	+0 19 33	0.000	-0.01	7.9	1.1	-0.1	K2 III		390 s		
198920	126341		20 53 29.7	+3 00 07	-0.001	0.00	7.6	0.4	3.0	F2 V		83 s		
198824	163954		20 53 31.9	-19 06 49	0.000	+0.02	7.0	1.0	0.2	K0 III		230 s		
199042	70633		20 53 32.4	+30 57 06	-0.001	+0.02	8.00	0.00	0.4	B9.5 V		320 s		m
199055	70636		20 53 33.1	+31 37 35	+0.002	+0.03	6.9	0.2	2.1	A5 V	-31	91 s		
198717	230421		20 53 34.9	-41 04 29	+0.003	-0.01	8.0	1.0	0.2	K0 III		360 s		
199008	106626		20 53 38.3	+19 35 31	+0.001	-0.01	7.60	0.1	1.7	A3 V		150 s	14402	m
199022	89247		20 53 39.3	+21 15 44	-0.002	-0.03	8.00	0.4	2.6	F0 V		120 s		m
198752	212523		20 53 39.9	-36 11 46	+0.004	-0.05	7.13	1.44	-0.2	K3 III		230 s		
199122	50189		20 53 39.9	+41 02 58	+0.002	0.00	7.56	-0.12	1.4	A2 V		170 s		
198716	212522		20 53 40.0	-39 48 36	+0.004	-0.10	5.35	1.32	-0.3	gK5	+20	140 s		
198979	106625		20 53 40.9	+11 42 18	0.000	-0.02	8.0	1.1	-0.1	K2 III		410 s		
199354	19085		20 53 41.0	+61 46 34	+0.008	+0.14	7.2	0.4	2.6	F0 V		82 s		
199306	32976		20 53 41.4	+59 18 27	+0.006	-0.03	6.8	0.4	3.0	F2 V	+5	58 s	14412	m
199100	70640		20 53 41.5	+36 07 48	-0.005	-0.09	7.75	0.73	4.2	G5 IV-V		49 s		
198698	230422		20 53 44.2	-44 20 24	+0.001	+0.03	7.3	1.9	0.2	K0 III		88 s		
198534	246773		20 53 44.5	-59 42 29	+0.002	0.00	7.6	0.0	0.6	A0 V		250 s		
199308	32978		20 53 52.0	+56 21 47	-0.001	0.00	7.51	0.14	-2.5	B2 V		680 s		
199216	50203		20 53 52.2	+49 32 01	0.000	+0.01	7.03	0.48	-5.1	B1 II	-7	1100 s		
199178	50198		20 53 53.3	+44 23 12	+0.001	+0.01	7.6	0.6	4.7	G2 V	-23	38 s		
199101	70645		20 53 53.8	+33 26 16	-0.001	+0.03	5.47	1.52	-0.1	K2 III	-10	80 s		m
198839	189812		20 53 55.7	-26 19 10	+0.003	-0.09	7.6	0.6	3.4	F5 V		57 s		
199102	89259		20 53 56.5	+29 29 59	0.000	0.00	7.6	0.0	0.4	B9.5 V	-2	260 s		
198949	144915		20 53 58.2	-6 53 23	+0.002	-0.02	6.44	0.37		F0				

HD	SAO	Star Name	α 2000	δ 2000	μ(α)	μ(δ)	V	B-V	M_v	Spec	RV	d(pc)	ADS	Notes
199331	32981		20ʰ53ᵐ58.8ˢ	+57°10′34″	-0.004	+0.01	7.16	0.06	0.6	A0 V		160 mx	14415	m
198867	189814		20 54 01.0	-24 16 49	+0.003	-0.01	7.2	0.7	3.2	G5 IV		64 s		
199217	50206		20 54 01.1	+48 26 16	0.000	+0.02	7.69	-0.03	0.6	A0 V		260 s		
198951	163966		20 54 02.9	-10 55 27	-0.003	-0.01	7.1	1.1	0.2	K0 III		240 s		
198962	144918		20 54 04.8	-4 32 30	+0.005	+0.02	6.7	0.1	1.7	A3 V		100 s		
198902	163965		20 54 05.6	-18 59 46	0.000	0.00	7.5	0.5	4.4	G0 V		42 s		
199206	50205		20 54 05.7	+45 06 37	0.000	+0.01	7.34	-0.03	-3.4	B8 II	-21	1400 s	14411	m
199057	106632		20 54 06.6	+18 01 42	+0.001	0.00	7.7	0.9	3.2	G5 IV		79 s		
198853	189815		20 54 06.7	-27 55 32	+0.002	-0.02	6.4	1.2	-0.5	M2 III		240 s		
199072	106634		20 54 10.4	+19 45 16	+0.003	+0.02	7.4	1.1	0.2	K0 III		260 s		
199276	32983		20 54 16.0	+50 31 10	-0.003	-0.02	8.0	1.1	0.2	K0 III		340 s		
198563	254895		20 54 19.7	-62 50 08	0.000	-0.01	6.8	1.1	0.2	K0 III		210 s		
199218	50209		20 54 22.1	+40 42 11	+0.001	+0.01	6.70	-0.07		B8 e	-22		14413	m
199660	19093		20 54 22.2	+69 56 50	-0.002	0.00	7.80	1.1	0.2	K0 III	+10	290 s	14441	m
199140	89265		20 54 22.2	+28 31 19	0.000	-0.01	6.56	-0.13	-3.6	B2 III	-12	1000 s		BW Vul, v
198924	189818		20 54 23.8	-25 34 36	+0.002	-0.06	7.5	1.3	0.2	K0 III		180 s		
198721	246781		20 54 24.0	-53 35 23	+0.001	-0.01	8.0	1.6	-0.1	K2 III		220 s		
199083	106637		20 54 25.6	+12 04 28	+0.002	-0.01	7.9	1.1	-0.1	K2 III		390 s		
199290	50214		20 54 25.6	+48 55 46	-0.001	-0.02	7.98	0.27		A m				
199169	89272	32 Vul	20 54 33.4	+28 03 27	0.000	0.00	5.01	1.48	-0.3	K4 III	+8	100 s		
198756	246785		20 54 33.8	-52 31 25	-0.002	-0.01	7.6	0.1	3.4	F5 V		71 s		
199103	106639		20 54 34.5	+11 26 34	+0.001	+0.03	7.5	1.1	0.2	K0 III		280 s		
199438	32991		20 54 34.6	+57 40 21	0.000	+0.01	7.7	1.1	-0.1	K2 III		330 s		
198766	246786		20 54 34.9	-50 43 40	0.000	-0.03	6.4	0.7	0.4	B9.5 V	-4	58 s		
198939	189823		20 54 35.7	-27 55 50	+0.001	+0.01	7.6	1.2	-0.1	K2 III		310 s		
198855	230430		20 54 36.9	-39 55 52	-0.004	-0.05	7.1	0.9	0.2	K0 III		230 mx		
199311	50218		20 54 39.8	+46 13 52	+0.002	0.00	6.68	0.06		A m	-40			
199234	70659		20 54 41.9	+37 04 25	+0.002	-0.01	7.15	0.02	0.6	A0 V	-26	200 s	14420	m
199084	126358		20 54 42.8	+3 36 28	-0.001	-0.01	7.7	1.4	-0.3	K5 III		390 s		
198994	189827		20 54 43.3	-20 56 57	-0.001	-0.01	8.0	0.0	1.4	A2 V		210 s		
200039	9911		20 54 44.2	+75 55 32	+0.009	+0.05	6.05	0.93	3.2	G5 IV	-25	30 s		
199312	50219		20 54 45.1	+45 08 10	+0.001	-0.01	7.59	-0.07	-5.2	A0 Ib		3600 s		
199035	144928		20 54 46.2	-8 52 58	+0.002	-0.01	7.0	1.1	0.2	K0 III		230 s		
199014	163973		20 54 46.9	-18 47 34	+0.006	+0.02	6.7	0.6	2.6	F0 V		45 s		
199012	163975	19 Cap	20 54 47.7	-17 55 23	-0.004	-0.02	5.78	1.12	0.2	K0 III	-39	110 s		
198700	246784	β Ind	20 54 48.5	-58 27 15	+0.003	-0.02	3.65	1.25	0.2	K0 III	-5	37 mn		m
198828	230431		20 54 49.0	-46 36 02	+0.008	-0.13	7.39	0.54	4.0	F8 V		46 s		m
199235	70662		20 54 50.4	+34 21 22	0.000	0.00	8.0	0.0	0.4	B9.5 V		320 s		
198609	--		20 54 53.0	-64 12 36			8.0	1.1	0.2	K0 III		360 s		
199085	144935		20 54 54.0	-1 22 33	+0.003	-0.01	7.8	0.5	3.4	F5 V		77 s		
199314	50222		20 54 54.2	+41 34 43	+0.003	+0.07	7.8	1.1	-0.1	K2 III		230 mx		
199047	163976		20 54 55.9	-15 16 58	-0.001	0.00	7.7	0.2	2.1	A5 V		130 s		
199114	126361		20 54 56.3	+3 01 52	+0.001	+0.01	7.9	1.1	0.2	K0 III		340 s		
199251	70663		20 54 57.5	+33 45 50	-0.001	-0.01	7.3	1.6	-0.5	M2 III	-7	350 s		
199180	106651		20 55 02.1	+17 15 34	0.000	-0.01	7.7	0.1	0.6	A0 V		270 s		
199221	89278		20 55 02.2	+28 05 25	+0.007	-0.08	7.85	0.63	4.7	dG2	+5	41 s	14424	m
199086	144937		20 55 03.3	-7 10 43	0.000	-0.18	7.8	0.7	4.4	G0 V		49 s		
199316	70666		20 55 05.2	+39 12 59	+0.001	-0.01	8.0	1.1	-0.1	K2 III		380 s		
199156	126366		20 55 06.9	+7 31 43	+0.002	-0.03	7.4	0.4	3.0	F2 V		77 s		
199124	144941		20 55 07.9	-1 22 24	0.000	+0.02	6.55	0.29		F0				
198829	246790		20 55 08.2	-52 07 12	+0.013	-0.12	7.00	0.9	3.2	G5 IV		58 s		m
199141	126364		20 55 08.4	+1 48 49	+0.003	+0.01	6.8	0.4	3.0	F2 V		56 s		
199394	50225		20 55 09.1	+46 21 00	-0.001	-0.04	7.02	1.03	-2.1	G5 II	0	610 s		
199017	212541		20 55 13.5	-30 08 27	+0.001	-0.04	7.9	0.9	4.4	G0 V		32 s		
199373	50228		20 55 16.2	+43 47 26	-0.004	-0.06	7.7	0.5	3.4	F5 V		71 s		
199194	126368		20 55 17.3	+9 43 16	+0.001	+0.04	7.4	0.4	3.0	F2 V		76 s		
199355	50226		20 55 17.7	+42 30 48	0.000	0.00	7.00	-0.12	0.4	B9.5 V	-23	210 s	14432	m
198842	246792		20 55 20.7	-53 35 20	+0.004	-0.01	7.0	0.7	0.2	K0 III		230 s		
199609	19097		20 55 22.0	+62 25 16	-0.001	+0.05	8.0	1.1	0.2	K0 III		330 s		
199356	50230		20 55 22.9	+40 18 00	0.000	0.00	7.13	0.15		B0 p	-13			
199317	70668		20 55 23.6	+32 41 22	-0.002	+0.02	7.2	1.1	-0.1	K2 III	-57	280 s		
199128	163983		20 55 24.1	-11 18 52	-0.003	+0.02	8.0	1.1	0.2	K0 III		360 s		
199395	50236		20 55 26.3	+43 22 03	-0.001	-0.01	6.70	1.44	-0.3	K4 III		240 s		
199713	19104		20 55 28.9	+65 41 14	-0.002	0.00	7.40	-0.07	0.4	B9.5 V		250 s		
199266	89284		20 55 33.0	+20 24 06	+0.001	+0.03	7.3	1.1	-0.1	K2 III		290 s		
199610	32999		20 55 34.7	+60 01 20	0.000	+0.01	7.6	1.4	-0.3	K5 III		350 s		
199253	106665	17 Del	20 55 36.5	+13 43 17	+0.001	-0.01	5.17	1.12	0.2	K0 III	-10	82 s		
199397	70675		20 55 36.6	+38 54 08	0.000	-0.02	7.8	0.1	0.6	A0 V		260 s		
199062	212543		20 55 37.9	-30 53 46	+0.002	-0.03	7.6	1.5	-0.1	K2 III		210 s		
199254	106666	16 Del	20 55 38.4	+12 34 07	+0.003	+0.02	5.50	0.12	1.9	A4 V	-1	53 s	14429	m
198678	254899		20 55 40.5	-66 09 06	0.000	-0.05	7.9	0.6	4.4	G0 V		50 s		
199223	126373		20 55 40.5	+4 31 58	+0.004	+0.01	6.05	0.82	0.3	gG6	-31	140 s	14430	m
199292	89287		20 55 40.6	+21 42 33	0.000	-0.01	7.8	0.2	2.1	A5 V		140 s		
199357	70673		20 55 42.3	+34 01 36	0.000	+0.02	8.0	1.1	0.2	K0 III		350 s		
199358	70672		20 55 43.1	+32 10 10	0.000	-0.01	7.9	0.1	1.4	A2 V		190 s		

HD	SAO	Star Name	α 2000	δ 2000	μ(α)	μ(δ)	V	B-V	M_v	Spec	RV	d(pc)	ADS	Notes
			h m s	° ' "	s	"								
198725	254900		20 55 45.6	-65 10 32	-0.001	-0.01	7.8	1.1	-0.1	K2 III		380 s		
199000	230442		20 55 46.0	-44 34 33	+0.004	0.00	7.28	1.62		K2				
199255	126376		20 55 46.9	+7 39 59	-0.003	-0.05	7.5	0.9	3.2	G5 IV		71 s		
199143	163989		20 55 47.3	-17 06 49	+0.002	-0.05	7.2	0.6	4.4	G0 V		36 s		
199091	212547		20 55 49.6	-31 33 13	+0.003	-0.03	7.6	1.5	3.2	G5 IV		29 s		
199478	50246		20 55 49.7	+47 25 04	-0.001	0.00	5.67	0.47	-7.1	B8 Ia	-16	2000 s		
199374	70680		20 55 51.9	+32 05 40	+0.001	-0.02	7.70	1.1	0.2	K0 III		310 s		m
199493	50248		20 55 54.2	+47 05 59	+0.002	+0.01	7.21	0.82	3.2	G8 IV		63 s		
199090	212550		20 55 54.3	-31 30 20	+0.004	+0.01	7.7	0.1	0.6	A0 V		200 mx		
199479	50247		20 55 58.7	+44 22 27	-0.001	+0.01	6.80	-0.05	0.2	B9 V	-6	210 s		
199375	89297		20 56 02.0	+27 34 32	+0.001	+0.02	6.80	1.1	-0.1	K2 III		230 s	14440	m
199320	126383		20 56 05.1	+9 15 38	-0.002	-0.03	7.1	1.4	-0.3	K5 III		300 s		
--	106677		20 56 08.1	+19 50 22	-0.002	-0.02	8.0							
199360	89298		20 56 08.3	+21 26 25	+0.001	+0.01	7.8	0.2	2.1	A5 V		140 s		
198967	246798		20 56 08.4	-53 52 58	+0.001	+0.01	8.0	1.5	-0.5	M4 III		500 s		
199197	163996		20 56 09.3	-17 14 34	+0.001	+0.02	7.6	1.1	0.2	K0 III		300 s		
198910	246796		20 56 09.5	-57 34 13	+0.001	0.00	7.6	1.5	3.2	G5 IV		27 s		
--	--		20 56 09.5	+27 47 54	-0.001	-0.02	7.1	0.9	3.2	G5 IV		60 s		
198956	246797		20 56 10.8	-55 13 22	+0.006	-0.05	7.8	1.0	0.2	K0 III		240 mx		
199511	50253		20 56 11.6	+43 25 27	+0.003	+0.01	6.82	-0.12	0.0	B8.5 V	-30	230 s		
200251	9917		20 56 12.0	+75 43 10	+0.023	+0.03	6.8	1.1	-0.1	K2 III		220 s		
199578	33001		20 56 12.7	+51 04 30	+0.001	+0.01	6.3	-0.1		B9				
199398	89301		20 56 13.5	+22 35 04	+0.001	-0.01	7.9	0.1	1.4	A2 V		200 s		
199661	33005		20 56 16.9	+56 53 16	0.000	-0.18	6.23	-0.18	-1.7	B3 V	-19	390 s		
199149	212554		20 56 18.0	-32 46 35	+0.004	-0.02	8.0	2.1	-0.1	K2 III		120 s		
199280	144957		20 56 18.0	-3 33 41	+0.002	+0.02	6.57	-0.10		B9				
199512	50257		20 56 18.7	+42 46 06	-0.003	-0.03	6.7	1.1	0.2	K0 III		190 s		
199440	89303		20 56 23.2	+27 30 15	+0.002	0.00	8.00	1.1	0.0	K1 III		380 s		
199496	70691		20 56 23.6	+36 22 38	0.000	-0.02	7.83	-0.01	0.6	A0 V		280 s		
199547	50260		20 56 24.1	+43 54 00	-0.002	-0.05	7.05	1.13	0.2	K0 III	-8	190 s		
199869	19112		20 56 24.7	+65 49 42	-0.001	+0.03	7.6	1.4	-0.3	K5 III		330 s		
199611	33004		20 56 25.2	+50 43 43	+0.003	-0.02	5.81	0.30		F0	-15		14460	m
199378	106679		20 56 25.2	+14 48 58	+0.002	+0.01	7.6	0.6	2.8	G0 IV	-15	90 s		
199612	50262		20 56 25.7	+49 11 45	+0.001	+0.01	5.90	1.04	-0.9	G8 II-III	-15	210 s		
198943	246799		20 56 26.6	-59 16 26	+0.002	-0.04	8.00	0.6	4.4	G0 V		53 s		m
199514	70695		20 56 30.3	+38 05 28	0.000	+0.01	8.0	1.1	-0.1	K2 III		380 s		
199399	106680		20 56 30.7	+14 41 00	0.000	0.00	7.4	1.4	-0.3	K5 III		340 s		
199579	50263		20 56 34.6	+44 55 30	-0.001	+0.01	5.96	0.05		O6	-6			
195819	--		20 56 44.3	-85 28 02			7.8	1.1	0.2	K0 III		330 s		
199363	144967		20 56 44.4	-3 40 27	+0.001	+0.01	7.3	1.1	0.2	K0 III		260 s		
199260	189856		20 56 47.1	-26 17 47	+0.007	-0.06	5.70	0.50	4.0	F8 V		22 s		
199381	126391		20 56 47.2	+1 42 56	0.000	-0.19	7.0	1.1	0.2	K0 III		72 mx		
199580	50265		20 56 47.9	+42 53 44	+0.021	+0.22	7.21	0.97		K1 IV	-19	31 mn		m
199227	212566		20 56 51.6	-33 59 57	-0.006	-0.01	7.3	0.6	3.4	F5 V		45 s		
199285	189861		20 56 52.6	-22 00 20	+0.001	+0.01	7.5	1.2	0.2	K0 III		210 s		
199400	126394		20 56 53.0	+0 30 10	-0.001	-0.02	7.8	1.1	0.2	K0 III		330 s		
199345	144968	7 Aqr	20 56 53.8	-9 41 51	0.000	-0.01	5.70	1.1	-0.1	K2 III	-33	150 s	14449	m
199662	50272		20 56 56.7	+47 34 44	-0.001	0.00	7.65	0.07	0.0	A0 IV		300 s		
199516	89310		20 57 00.8	+23 40 27	+0.001	0.00	8.00	1.1	-0.1	K2 III		400 s	14461	m
199480	106689		20 57 01.3	+17 05 50	0.000	0.00	6.9	0.1	1.4	A2 V		130 s		
199497	106694		20 57 10.2	+19 38 58	+0.011	+0.07	7.9	0.9	3.2	G5 IV		87 s		
199629	50274	58 v Cyg	20 57 10.2	+41 10 02	+0.001	-0.01	3.94	0.02	0.6	A0 V	-27	45 s		
199442	126396		20 57 10.4	+0 27 49	+0.001	-0.06	6.05	1.22		K2	-26		14457	m
198843	--		20 57 10.5	-67 41 24			7.8	1.1	0.2	K0 III		330 s		
199498	106690		20 57 11.0	+16 15 05	+0.001	0.00	7.6	1.6	-0.5	M4 III		420 s		
200099	19127		20 57 12.7	+69 03 28	-0.013	-0.03	7.0	1.1	-0.1	K2 III	-40	240 s		
199597	70712		20 57 12.8	+35 18 42	0.000	+0.01	6.7	1.1	-0.1	K2 III		230 s		
199850	33019		20 57 12.9	+58 49 01	+0.001	+0.01	8.0	0.1	0.6	A0 V		280 s		
199348	189868		20 57 17.1	-23 50 45	+0.001	0.00	7.8	-0.3	3.4	F5 V		74 s		
200040	19125		20 57 19.9	+66 45 09	+0.005	+0.11	7.90	1.1	0.0	K1 III		180 mx	14496	m
198969	254904		20 57 22.4	-64 56 17	+0.003	-0.02	7.71	0.89	3.2	K0 IV		80 s		
203836	3515		20 57 22.8	+87 01 58	+0.023	+0.02	7.4	0.1	1.7	A3 V	-1	140 s		
199679	50279		20 57 24.5	+41 41 21	-0.003	-0.01	7.34	0.13		A2				
199693	50281		20 57 28.6	+42 26 41	+0.002	+0.03	7.5	1.1	0.2	K0 III	-34	270 s		
199761	50284		20 57 29.1	+47 10 57	0.000	-0.03	7.92	0.45	0.7	F4 III	-16	260 s		
199694	70721		20 57 35.3	+39 39 47	-0.001	+0.02	7.47	-0.11	0.4	B9.5 V		260 s		
199598	89320		20 57 39.4	+26 24 17	+0.019	+0.09	6.94	0.58	4.4	G0 V		31 s		
199288	230458		20 57 39.8	-44 07 44	-0.049	-0.96	6.53	0.58	4.7	G0 V	-16	23 ts		
199469	144978		20 57 39.9	-7 42 48	0.000	+0.01	7.4	0.1	0.6	A0 V		230 s		
199443	164013		20 57 40.5	-16 01 54	+0.004	0.00	5.87	0.18		A m				
198928	254905		20 57 46.5	-68 25 30	+0.009	-0.03	7.5	0.1	1.4	A2 V		160 mx		
199908	33026		20 57 48.4	+55 29 15	+0.002	+0.02	7.3	0.4	2.3	F1 IV-V		100 s		DQ Cep, v
199560	126414		20 57 48.7	+4 11 43	0.000	0.00	6.8	1.1	-0.1	K2 III		240 s		
200149	19129		20 57 49.5	+67 45 50	-0.004	+0.02	7.7	1.4	-0.3	K5 III		330 mx		
--	--		20 57 51.4	+45 32 11	-0.002	-0.02	7.5	1.1	0.2	K0 III		270 s		

HD	SAO	Star Name	α 2000	δ 2000	μ(α)	μ(δ)	V	B-V	M_v	Spec	RV	d(pc)	ADS	Notes
199799	50291		20 57 52.7	+44 47 16	-0.006	-0.02	7.31	1.67	-0.5	M2 III		250 mx		
199762	50288		20 57 55.5	+40 16 21	-0.001	-0.01	7.7	0.1	0.6	A0 V		250 s		
199583	126417		20 57 59.8	+8 03 48	+0.005	+0.04	7.1	1.1	0.2	K0 III		230 mx		
199938	33031		20 58 00.1	+56 07 10	0.000	-0.02	8.0	0.4	2.6	F0 V		110 s		
199523	164018		20 58 01.7	-11 57 18	-0.002	-0.01	7.5	0.9	0.3	G7 III	+13	280 s		
199695	89327		20 58 02.1	+29 23 42	+0.003	+0.02	7.8	0.5	3.4	F5 V		74 s		
199584	126418		20 58 02.8	+6 53 49	-0.001	0.00	7.7	1.4	-0.3	K5 III		390 s		
199716	70730		20 58 02.9	+31 18 00	-0.002	-0.01	8.0	1.1	-0.1	K2 III		400 s		
199647	106708		20 58 06.5	+17 05 01	0.000	-0.02	7.7	1.1	-0.1	K2 III		360 s		
199033	---		20 58 10.4	-66 45 17			7.7	1.1	0.2	K0 III		320 s		
198971	254907		20 58 11.4	-69 08 47	+0.003	-0.03	7.1	-0.2	0.4	B9.5 V		220 s		
199696	89331		20 58 12.8	+23 53 40	0.000	-0.02	8.0	1.4	-0.3	K5 III		430 s		
199890	50299		20 58 15.0	+47 37 05	-0.001	+0.02	7.50	-0.10	-0.6	B8 IV	-22	420 s		
199763	70732		20 58 15.5	+30 23 46	+0.005	+0.03	6.6	0.9	0.2	G9 III		190 s		
199697	89332	33 Vul	20 58 16.2	+22 19 33	0.000	-0.02	5.31	1.40	-0.3	K5 III	-28	130 s		
199631	126420		20 58 17.3	+8 30 54	-0.001	-0.01	8.0	0.4	3.0	F2 V		98 s		
199719	89334		20 58 18.7	+23 02 47	+0.002	0.00	6.6	0.8	3.2	G5 IV		47 s		
199870	50298		20 58 19.3	+44 28 18	+0.010	+0.08	5.55	0.97	0.3	G8 III	-21	100 s		
---	230465		20 58 22.8	-40 59 36	+0.001	-0.05	7.7	0.4	3.0	F2 V		87 s		
199698	106713		20 58 22.8	+17 16 23	+0.001	-0.01	7.9	1.1	-0.1	K2 III		390 s		
199891	50304		20 58 24.9	+46 36 17	+0.005	+0.08	7.42	0.41	3.7	F6 V		56 s		
199835	70740		20 58 25.4	+39 28 39	-0.001	-0.01	7.7	0.2	2.1	A5 V		130 s		
199665	106712	18 Del	20 58 25.7	+10 50 21	-0.004	-0.03	5.48	0.93	0.2	K0 III	0	110 s		m
199871	50301		20 58 27.1	+41 21 27	0.000	+0.01	7.12	1.53	-0.4	M0 III		320 s		m
199740	89339		20 58 28.0	+23 14 50	+0.002	0.00	7.8	0.2	2.1	A5 V		140 s		
199955	33034		20 58 29.9	+50 27 44	+0.001	+0.01	5.61	-0.15		B8	-21	31 mn	14504	m
199892	50303		20 58 30.7	+41 56 25	0.000	+0.02	6.16	-0.08		B9				
199721	106719		20 58 31.9	+16 26 12	0.000	0.00	7.2	0.1	0.6	A0 V		210 s	14490	m
199722	---		20 58 33.4	+16 25 06			7.2	0.0	0.4	B9.5 V		230 s	14490	
199587	164025		20 58 35.6	-13 22 02	+0.009	-0.11	7.6	0.6	4.4	G0 V		43 s		
199851	70746		20 58 38.9	+38 09 00	+0.002	+0.01	8.0	0.9	3.2	G5 IV		92 s		
199723	106720		20 58 39.5	+14 55 56	+0.003	0.00	7.8	1.1	0.2	K0 III		330 s		
199836	70744		20 58 40.1	+34 43 26	-0.003	-0.03	7.8	1.1	0.2	K0 III		320 s		
199603	164027		20 58 41.8	-14 28 58	-0.003	0.00	6.01	0.23		A3				
199939	50310		20 58 43.3	+44 24 51	0.000	-0.03	7.45	1.24	-0.3	K5 III		360 s		
199837	70743		20 58 43.6	+31 38 53	+0.002	+0.04	7.2	0.0	0.4	B9.5 V	-11	220 s		
200150	19133		20 58 44.7	+62 30 13	-0.008	-0.03	7.9	1.1	0.2	K0 III		310 s		
199651	144993		20 58 44.9	-3 50 28	-0.002	+0.03	6.8	0.1	1.7	A3 V		110 s		
199765	106724		20 58 46.3	+16 25 22	0.000	-0.04	8.0	0.5	3.4	F5 V		82 s		
199681	144997		20 58 49.4	-3 19 06	+0.001	-0.03	7.9	0.4	3.0	F2 V		94 s		
199784	106728		20 58 52.4	+17 52 01	0.000	-0.01	7.8	0.2	2.1	A5 V		140 s		
199986	50314		20 58 54.5	+46 15 09	+0.002	-0.03	7.03	0.19	2.1	A5 V	-1	95 s		
199875	70750		20 58 55.0	+31 28 38	+0.001	+0.03	7.9	1.4	-0.3	K5 III		410 s		
199956	50313		20 58 55.5	+44 03 38	+0.002	+0.02	6.64	1.07	0.2	K0 III		170 s		
199838	89347		20 59 03.4	+22 21 02	0.000	+0.02	7.0	1.4	-0.3	K5 III		280 s		
199766	126428	1 Equ	20 59 04.2	+4 17 37	-0.008	-0.14	5.23	0.46	2.1	F5 IV	-10	40 s		m
200018	50316		20 59 04.7	+46 26 22	-0.001	-0.01	7.72	1.11	0.2	K0 III		270 s		
199911	70754		20 59 04.8	+34 29 19	-0.002	+0.07	8.0	0.9	3.2	G5 IV		92 s		
199702	145002		20 59 06.5	-6 54 25	+0.001	-0.02	8.0	0.1	1.7	A3 V		180 s		
200019	50318		20 59 11.5	+44 47 27	+0.004	+0.04	7.14	0.83	3.2	G5 IV		55 s		
199957	70761		20 59 13.8	+37 55 48	0.000	-0.02	7.8	1.4	-0.3	K5 III		390 s		
194314	---		20 59 14.1	-86 47 42			7.4	0.1	0.6	A0 V		230 s		
199969	70764		20 59 20.8	+36 44 50	0.000	0.00	7.9	1.6	-0.5	M4 III		460 s		
199987	70765		20 59 22.9	+36 25 43	0.000	0.00	8.0	0.10	0.4	B9.5 V		300 s	14512	m
199683	189903		20 59 24.2	-20 26 39	+0.002	-0.01	7.8	0.6	4.4	G0 V		46 s		
200030	50319		20 59 24.4	+42 19 28	+0.002	+0.01	6.47	-0.08		B9				
200205	33048		20 59 25.1	+59 26 19	+0.005	+0.01	5.51	1.40	-0.2	gK3	-17	110 s		
199618	212598		20 59 25.2	-30 30 45	+0.003	-0.08	7.2	0.7	3.2	G5 IV		64 s		
199768	145009		20 59 26.8	-7 37 33	0.000	-0.02	8.0	1.1	0.2	K0 III		360 s		
200021	50321		20 59 28.0	+40 57 37	+0.002	0.00	7.0	1.1	0.2	K0 III	-34	220 s		
199620	212597		20 59 28.7	-33 37 02	+0.004	-0.17	7.36	0.84	4.4	G0 V		25 s		
200167	33047		20 59 28.9	+55 56 25	-0.002	-0.01	6.8	1.1	-0.1	K2 III		230 s		
199671	189904		20 59 31.2	-27 01 29	-0.004	-0.24	8.0	0.9	3.2	G5 IV		40 mx		
199789	145011		20 59 31.3	-1 28 06	0.000	-0.05	8.0	1.1	0.2	K0 III		360 s		
200060	50326		20 59 34.0	+44 13 36	-0.004	0.00	7.54	1.18	0.2	K0 III	-23	210 s		
199728	164043	20 Cap	20 59 35.9	-19 02 07	+0.001	-0.01	6.2	0.1		A0 p				
200707	9927		20 59 39.8	+74 17 11	+0.007	+0.05	8.00	1.1	0.2	K0 III		320 s	14557	m
199342	254916		20 59 41.0	-62 52 39	+0.006	0.00	7.1	1.1	-0.1	K2 III		270 s		
200031	70770		20 59 42.7	+38 49 22	0.000	0.00	6.75	0.80	-4.5	G2 Ib		1800 s		
200206	33050		20 59 49.1	+55 13 06	-0.003	+0.01	7.0	1.1	-0.1	K2 III	-20	250 s		
200120	50335	59 Cyg	20 59 49.3	+47 31 16	0.000	0.00	4.74	-0.05	-3.9	B1 IV e	+1	480 s	14526	V832 Cyg, m,v
199941	106738		20 59 50.6	+16 49 27	+0.003	-0.01	6.66	0.38		F2	+2			m
200102	50333		20 59 51.0	+44 59 47	-0.001	+0.01	6.62	1.05	-4.5	G1 Ib	-24	1300 s		
199828	164046		20 59 54.6	-13 03 06	-0.002	0.00	6.6	0.1	1.7	A3 V		96 s		
200077	50331		20 59 55.2	+40 15 32	+0.021	+0.21	6.56	0.55	4.0	F8 V	-36	31 s		m

HD	SAO	Star Name	α 2000	δ 2000	μ(α)	μ(δ)	V	B-V	M_v	Spec	RV	d(pc)	ADS	Notes
			h m s	° ′ ″	s	″								
199639	230472		20 59 57.8	-43 00 55	+0.001	+0.01	7.28	0.16	1.4	A2 V		130 s		
199971	106742		20 59 58.0	+18 37 07	+0.001	-0.01	7.2	1.1	0.2	K0 III		260 s		
200043	70771		20 59 58.6	+32 29 44	-0.001	-0.03	7.20	1.6	-0.5	M3 III	-16	340 s		
--	19148		20 59 58.7	+66 25 06	+0.006	+0.04	7.8	1.4						
199684	212608		20 59 59.5	-36 07 45	+0.008	-0.01	6.11	0.39	2.6	F0 V		45 s		
200103	50336		21 00 00.1	+42 55 54	+0.003	+0.04	7.7	0.1	1.7	A3 V		150 s		
199942	126447		21 00 03.7	+7 30 59	+0.002	+0.02	5.99	0.26		A3 n	-24			m
199190	254915		21 00 05.6	-69 34 46	+0.088	-0.28	6.87	0.62	3.2	G5 IV		36 mx		
200177	50340		21 00 06.4	+48 40 47	+0.001	+0.01	7.34	0.01		B9 p			14531	m
199881	164050		21 00 08.5	-10 37 42	0.000	+0.01	7.4	0.5	3.4	F5 V		63 s		
199999	106745		21 00 10.4	+19 57 40	+0.005	+0.02	7.4	0.5	3.4	F5 V		64 s		
199755	212615		21 00 12.6	-34 14 03	+0.002	0.00	7.1	0.7	0.6	A0 V		74 s		
200104	70777		21 00 13.2	+36 57 46	-0.003	-0.01	7.90	1.4	-0.3	K5 III		420 s	14530	m
200386	19147		21 00 16.9	+61 29 51	-0.010	-0.02	7.50	1.1	0.2	K0 III	-13	220 mx	14544	m
199809	189918		21 00 19.1	-27 20 37	+0.004	-0.07	7.7	0.9	0.2	K0 III		170 mx		
199623	246820		21 00 21.1	-51 15 55	-0.011	+0.13	5.76	0.47	3.7	F6 V		26 s		
200044	106747		21 00 27.5	+19 19 47	-0.001	-0.05	5.65	1.61	-0.5	M2 III	-15	170 s		m
200168	50341		21 00 28.4	+40 19 58	+0.002	+0.02	8.0	1.1	0.2	K0 III		340 s		
197909	258875		21 00 31.4	-83 08 55	+0.001	-0.07	7.2	1.1	0.2	K0 III		200 mx		
199944	145021		21 00 32.4	-5 28 39	+0.001	-0.01	6.6	0.1	0.6	A0 V		160 s		
199960	145022	11 Aqr	21 00 33.6	-4 43 49	+0.003	-0.13	6.21	0.63	4.5	G1 V	-17	20 s		
199810	212617		21 00 40.3	-35 17 31	+0.005	-0.06	6.70	1.1	0.2	K0 III		200 mx		m
200079	106750		21 00 42.8	+16 03 51	-0.001	+0.02	7.9	1.1	0.2	K0 III		340 s		
199642	246824		21 00 43.1	-53 44 21	0.000	-0.02	6.8	0.6	-0.3	K5 III		270 s		
200078	106752		21 00 43.5	+17 26 51	+0.019	-0.01	7.9	0.9	3.2	G5 IV		54 mx		
200047	126462		21 00 44.5	+9 59 46	0.000	+0.01	7.6	0.1	0.6	A0 V		260 s		
200269	50353		21 00 49.7	+46 34 42	0.000	-0.01	7.28	-0.10	-1.1	B5 V		450 s		
199947	164061		21 00 51.6	-17 31 51	-0.002	-0.01	6.5	1.1	0.2	K0 III	+1	180 s		
199845	212623		21 00 53.0	-36 03 15	+0.003	+0.06	6.6	0.8	0.2	K0 III		190 s		
200207	70788		21 00 53.1	+37 39 11	0.000	0.00	8.0	0.9	3.2	G5 IV		92 s		
199861	212627		21 00 54.1	-32 57 38	+0.003	-0.06	8.0	1.2	3.2	G5 IV		51 s		
199917	189930		21 01 03.1	-29 43 44	+0.001	-0.05	7.1	0.6	0.6	A0 V		81 s		
200091	126466		21 01 03.8	+9 45 50	0.000	+0.04	8.0	0.5	3.4	F5 V		82 s		
200063	126464		21 01 03.8	+1 02 05	-0.001	+0.01	7.4	1.4	-0.3	K5 III		350 s		
199862	212630		21 01 04.7	-38 44 23	+0.002	-0.10	8.0	0.8	-0.1	K2 III		120 mx		
200004	164064		21 01 08.2	-13 31 48	0.000	+0.01	6.6	0.8	3.2	G5 IV	+23	48 s		
200065	145030		21 01 08.4	-3 26 28	+0.001	0.00	8.0	1.4	-0.3	K5 III		450 s		
200123	106757		21 01 09.2	+11 57 13	-0.003	-0.01	7.8	0.4	3.0	F2 V		89 s		
200387	33069		21 01 09.4	+52 08 14	-0.002	+0.01	7.8	1.1	-0.1	K2 III		360 s		
200310	50359	60 Cyg	21 01 10.7	+46 09 21	0.000	+0.01	5.37	-0.21	-3.5	B1 V	-10	590 s	14549	m
200253	70794		21 01 12.6	+36 01 35	-0.001	+0.01	5.97	0.98	0.2	K0 III	-10	140 s		
199918	212635		21 01 13.3	-35 11 05	-0.013	-0.08	7.6	0.5	4.4	G0 V		43 s		
200311	50358		21 01 14.0	+43 43 19	-0.002	+0.01	7.70	-0.10		B9 p				
199951	212636	γ Mic	21 01 17.3	-32 15 28	0.000	+0.01	4.67	0.89	0.3	G4 III	+18	70 s		m
200108	126469		21 01 22.0	+1 04 10	+0.001	-0.05	7.2	1.4	-0.3	K5 III		320 s		
199818	230482		21 01 22.2	-49 21 03	+0.006	+0.03	7.8	1.1	3.4	F5 V		30 s		
200024	189940		21 01 24.8	-23 04 43	-0.001	-0.01	7.5	2.0	-0.3	K5 III		180 s		
199952	212638		21 01 25.9	-32 53 43	+0.002	+0.01	6.8	0.3	0.6	A0 V		110 s		
200325	50360		21 01 26.0	+40 37 06	+0.005	+0.01	7.2	0.5	3.4	F5 V		56 s		
199475	254918		21 01 28.0	-68 12 35	0.000	0.00	6.37	0.10		A0				
200007	189941		21 01 32.7	-27 48 11	+0.001	+0.02	7.3	1.2	0.2	K0 III		210 s		
200069	164069		21 01 35.8	-16 26 08	0.000	0.00	8.0	1.1	-0.1	K2 III		410 s		
200775	19158		21 01 36.7	+68 09 49	+0.001	+0.01	7.35	0.38		B3 e	-3			
200678	19155		21 01 37.6	+63 41 08	-0.004	-0.02	7.9	1.4	-0.3	K5 III		380 s		
200406	50366		21 01 38.6	+47 30 07	+0.010	+0.05	7.93	0.52	-2.0	F5 II		57 mx		
200139	145037		21 01 40.3	-4 07 53	0.000	+0.02	7.3	1.6	-0.5	M4 III		370 s		
200270	89378		21 01 41.5	+28 06 05	+0.002	-0.04	7.7	0.5	4.0	F8 V		54 s		
200289	89380		21 01 43.3	+29 37 47	0.000	-0.01	7.90	-0.11		B3				
200052	189942		21 01 45.1	-26 52 52	+0.001	-0.02	5.9	0.5	1.4	A2 V		41 s		
---	---		21 01 47.0	+29 26 41	+0.002	+0.01	6.3	1.4	-0.3	K5 III		200 s		
200407	50368		21 01 47.3	+44 11 14	+0.002	-0.07	6.72	0.30		A m	-8		14560	m
200195	126477		21 01 49.5	+7 31 48	+0.001	-0.01	8.0	0.1	0.6	A0 V		300 s		
200370	70808		21 01 49.9	+39 15 40	-0.001	0.00	8.00	0.4	2.6	F0 V		120 s	14558	m
200008	212643		21 01 50.6	-36 34 25	+0.002	+0.02	7.1	1.2	-0.1	K2 III		250 s		
200254	89379		21 01 50.9	+21 06 19	+0.008	+0.03	7.6	0.9	3.2	G5 IV		75 s		
200593	33076		21 01 54.6	+57 45 58	0.000	+0.04	6.9	0.0	0.4	B9.5 V		190 s		
200575	33075		21 01 54.7	+57 04 19	-0.001	+0.01	6.70	0.00	0.0	B8.5 V		210 s	14574	m
200290	89382		21 01 54.7	+23 39 56	0.000	-0.02	7.50	0.5	4.0	F8 V		50 s		m
200388	50369		21 01 55.0	+40 18 40	+0.003	+0.01	7.8	0.1	0.6	A0 V		260 s		
199983	230490		21 01 55.2	-42 04 45	0.000	-0.01	7.5	2.1	-0.3	K5 III		150 s		
200478	50373		21 01 57.3	+47 01 09	+0.004	-0.01	7.78	0.08	0.3	A1 IV		290 s		
200157	164074		21 01 59.6	-11 41 47	-0.001	-0.03	7.1	0.5	3.4	F5 V		55 s		
200389	70810		21 02 00.5	+36 58 10	-0.001	-0.01	7.9	0.1	1.4	A2 V		190 s		
200477	50376		21 02 05.5	+47 36 57	+0.003	+0.04	8.01	0.96	3.2	G8 IV		82 s		
200614	33078		21 02 08.9	+56 40 12	+0.001	+0.01	6.10	-0.07		B9			14575	m

HD	SAO	Star Name	α 2000	δ 2000	μ(α)	μ(δ)	V	B−V	M_v	Spec	RV	d(pc)	ADS	Notes
			h m s	° ′ ″	s	″								
200256	126482		21 02 12.2	+7 10 47	−0.001	−0.01	7.40	0.5	4.0	F8 V	−5	48 s	14556	m
200011	230492		21 02 12.4	−43 00 07	+0.005	−0.10	6.64	0.68	3.0	G3 IV		54 s		m
200390	70811		21 02 14.6	+30 31 19	+0.001	+0.03	7.9	0.1	1.4	A2 V	−23	190 s		
200409	70813		21 02 16.9	+33 44 20	+0.003	+0.08	7.8	0.5	3.4	F5 V		74 s		
200026	230494		21 02 17.5	−42 59 52	+0.007	−0.12	6.90	0.96	3.2	K0 IV		54 s		
200465	70818		21 02 20.7	+39 30 33	0.000	+0.01	6.50	1.1	−1.3	K3 II−III	−9	330 s	14567	m
200352	89391		21 02 21.6	+23 19 27	0.000	0.00	7.38	0.01		B9				
200527	50381		21 02 23.9	+44 47 28	−0.002	+0.01	6.19	1.69	−3.6	M3 Ib−II	+1	680 s		
200391	89396		21 02 25.7	+27 48 26	+0.006	+0.01	7.2	0.6	0.6	G0 III	−26	200 s		ER Vul, v
200212	164080		21 02 26.0	−12 26 50	+0.002	−0.03	7.3	1.1	−0.1	K2 III		300 s		
200073	212653		21 02 27.1	−38 31 51	+0.014	−0.16	5.94	1.11	3.2	K0 IV		29 s		
200227	145044		21 02 28.7	−9 26 52	+0.002	−0.01	7.7	1.1	−0.1	K2 III		370 s		
199994	246831		21 02 30.6	−50 00 14	+0.004	−0.01	7.6	0.6	1.4	A2 V		78 s		
200372	89395		21 02 31.2	+21 40 51	+0.003	+0.01	7.70	1.1	0.2	K0 III		310 s	14570	m
200172	189957		21 02 31.3	−24 52 31	+0.005	0.00	7.6	1.6	3.2	G5 IV		24 s		
200213	164082		21 02 34.2	−15 28 25	0.000	−0.04	7.8	0.9	3.2	G5 IV		83 s		
200425	89398		21 02 34.7	+26 09 27	+0.001	+0.02	7.80	0.55	4.0	F8 V		55 s		
200576	50387		21 02 34.7	+48 01 51	−0.002	+0.01	6.92	1.69	−4.4	K5 Ib		1500 s		
200228	164086		21 02 35.1	−9 59 45	0.000	+0.01	6.7	1.6	−0.5	M4 III		280 s		
200561	50386		21 02 37.7	+44 36 25	−0.001	−0.03	7.7	1.1	0.2	K0 III		290 s		
200354	106786		21 02 40.1	+11 39 26	0.000	0.00	7.4	0.1	1.7	A3 V		140 s		
200560	50388		21 02 40.3	+45 53 06	+0.035	+0.15	7.68	0.97	−0.1	K2 III	−12	25 ts	14585	m
200451	89400		21 02 41.0	+26 30 30	−0.001	0.00	7.40	1.4	−0.3	K5 III	−25	330 s		
200326	126488		21 02 42.7	+3 21 01	+0.001	−0.01	6.6	1.1	0.2	K0 III		190 s		
200233	164087		21 02 45.1	−19 14 55	+0.003	0.00	7.1	1.3	0.2	K0 III		150 s		
200393	106792		21 02 47.0	+15 57 59	−0.001	−0.02	7.0	1.6	−0.5	M2 III		310 s		
200373	106790		21 02 47.3	+11 16 49	+0.001	+0.01	8.0	0.4	2.6	F0 V		120 s		
200426	89402		21 02 47.9	+21 48 35	0.000	−0.01	7.3	0.0	0.4	B9.5 V		240 s		
200595	50390		21 02 48.3	+45 50 56	−0.001	0.00	6.48	−0.15		B8	−12		14585	m
200468	89404		21 02 50.6	+28 19 24	+0.002	+0.02	7.8	0.1	1.4	A2 V		180 s		
200530	70825		21 02 50.8	+37 09 02	+0.007	+0.02	7.75	0.63	0.2	K0 III		260 mx		
200510	70824		21 02 55.0	+32 20 59	0.000	−0.01	7.1	1.1	−0.1	K2 III	−9	270 s		
200199	189963		21 02 56.2	−29 06 39	0.000	−0.02	6.8	0.3	0.6	A0 V		120 s		
200146	212663		21 02 56.4	−39 05 46	−0.008	−0.16	7.3	0.6	4.4	G0 V		37 s		
200315	145049		21 02 57.2	−7 19 17	+0.004	+0.01	7.9	0.5	4.0	F8 V		60 s		
200163	212666	ζ Mic	21 02 57.8	−38 37 54	−0.002	−0.11	5.3	0.4	2.6	F0 V	+5	35 s		
200340	145050		21 02 59.4	−0 55 29	0.000	0.00	6.50	−0.10		B8				
200428	106797		21 03 00.9	+15 45 36	+0.002	−0.02	7.7	0.8	3.2	G5 IV		80 s		
200430	106796		21 03 01.5	+14 43 48	+0.001	+0.01	6.31	1.67	−0.5	M1 III	−39	200 s		
200375	126491		21 03 02.9	+1 31 55	−0.008	−0.05	6.21	0.48		F5	+7	14 mn	14573	m
200374	126492		21 03 04.0	+1 41 40	0.000	0.00	7.4	1.1	0.2	K0 III		280 s		
200577	70832		21 03 04.7	+38 39 27	+0.001	−0.01	6.07	1.01	0.2	K0 III	−3	150 s		
200059	246836		21 03 06.3	−52 46 06	−0.012	−0.05	7.80	0.8	3.2	G5 IV		83 s		m
200356	145052		21 03 07.4	−1 18 31	+0.001	−0.02	7.4	0.4	3.0	F2 V		75 s		
200245	189966		21 03 10.0	−27 43 55	+0.003	−0.04	6.25	0.95	0.2	K0 III		160 s	14565	m
200512	89407		21 03 10.1	+24 27 06	−0.001	−0.01	7.7	1.6	−0.5	M2 III		410 s		
200299	189968		21 03 17.8	−25 04 29	+0.003	−0.01	7.6	0.6	0.6	A0 V		110 s		
200546	89413		21 03 19.8	+27 19 54	−0.001	−0.02	7.20	1.6	−0.5	M2 III		330 s		
200413	145054		21 03 25.4	−5 22 29	+0.001	−0.01	8.0	1.1	0.2	K0 III		360 s		
200740	33091		21 03 25.8	+50 21 08	+0.006	+0.06	6.37	0.98	0.2	K0 III	−22	170 s		
−−	−−		21 03 25.8	+25 30 28	0.000	+0.03	7.9	1.6		M3				
200435	145057		21 03 29.3	−2 19 20	−0.001	−0.04	7.0	1.1	0.2	K0 III		230 s		
200578	89415		21 03 29.4	+29 05 33	0.000	−0.02	7.00	0.9	0.3	G8 III	−26	210 s	14590	m
200433	145058		21 03 29.5	−1 34 47	+0.002	−0.01	6.8	0.5	3.4	F5 V		48 s		
200563	89414		21 03 29.6	+23 59 46	0.000	−0.01	7.40	1.6	−0.5	M4 III	−9	360 s	14589	DY Vul, m,v
200378	164101		21 03 30.2	−15 09 44	+0.002	−0.04	7.16	1.29	0.2	K0 III		150 s		
200114	246838		21 03 31.0	−52 53 20	−0.004	+0.02	7.43	1.56		Ma				
201998	3504		21 03 35.7	+81 09 39	+0.006	+0.03	7.0	0.4	2.6	F0 V		74 s		
200722	50404		21 03 38.2	+45 22 04	0.000	−0.01	7.86	−0.06	0.6	A0 V		280 s		
200494	126501		21 03 39.1	+2 55 35	−0.007	−0.10	7.46	1.30	−0.1	gK2	+6	280 s		
200631	70845		21 03 39.8	+31 03 44	+0.002	−0.01	7.80	1.1	−0.1	K2 III	−15	370 s	14597	m
201249	9944		21 03 42.8	+71 18 51	+0.002	+0.01	7.9	0.1	0.6	A0 V		250 s		
200753	50408		21 03 43.1	+46 51 43	−0.006	−0.11	6.32	0.25	0.6	F2 V	−15	140 s		
199868	254927		21 03 46.9	−68 04 22	+0.020	−0.03	7.8	0.5	4.0	F8 V		57 s		
199997	254928		21 03 47.0	−63 28 51	0.000	0.00	7.66	0.84	−6.2	G3 I		5900 s		
200828	33099		21 03 47.0	+54 20 40	+0.003	+0.02	7.7	0.4	3.0	F2 V		85 s		
200817	33098		21 03 47.6	+53 17 08	+0.006	+0.02	5.90	0.99	0.2	K0 III	−27	140 s		
200262	230500		21 03 48.7	−40 41 28	−0.001	−0.03	8.0	0.9	3.2	G5 IV		84 s		
200723	50409		21 03 52.0	+41 37 42	0.000	−0.05	6.33	0.38	0.6	gF1	−8	130 s		m
200857	33102		21 03 52.6	+55 13 50	−0.001	+0.01	7.13	0.56	−2.9	B3 III	−14	370 s		
200776	50411		21 03 53.7	+46 19 50	+0.001	+0.01	7.79	0.01	−3.9	B1 IV p		1500 s		
200617	106810		21 03 56.0	+18 44 06	+0.001	−0.02	7.90	1.1	0.2	K0 III		350 s		m
200754	50412		21 03 58.7	+43 35 34	−0.008	−0.04	7.8	1.1	−0.1	K2 III		240 mx		
200598	106812		21 04 03.3	+14 22 24	−0.003	+0.01	8.0	0.1	1.4	A2 V		210 s		
200686	70852		21 04 04.1	+33 14 00	0.000	−0.01	8.0	1.1	−0.1	K2 III		400 s		

HD	SAO	Star Name	α 2000	δ 2000	μ(α)	μ(δ)	V	B-V	M_v	Spec	RV	d(pc)	ADS	Notes
200497	145064	12 Aqr	21^{h}04^{m}04$.^s$4	-5°49'24"	0$.^s$000	+0."02	5.9	0.5	3.4	F5 V	-1	31 s	14592	m
201032	19169		21 04 06.1	+63 22 50	-0.002	+0.03	7.3	0.2	2.1	A5 V	+7	110 s		
200580	126509		21 04 07.2	+2 59 39	-0.019	-0.37	7.32	0.54	4.2	F9 V	-6	42 s		
200248	230501		21 04 07.7	-47 57 40	0.000	-0.01	7.10	1.1	0.2	K0 III		240 s		m
200249	246846		21 04 13.5	-50 26 06	0.000	+0.01	6.4	1.9	-0.1	K2 III		77 s		
200687	89429		21 04 22.5	+23 49 18	+0.002	+0.01	7.0	1.6		M5 III	-12			R Vul, v
200346	230506		21 04 22.8	-43 05 08	+0.004	0.00	7.4	1.0	0.2	K0 III		270 s		
200499	189986	22 η Cap	21 04 24.1	-19 51 18	-0.003	-0.03	4.84	0.17		A4 sh	+24	12 mn		m
201134	19177		21 04 26.4	+65 01 16	-0.006	+0.01	8.0	0.1	0.6	A0 V		140 mx		
200858	50418		21 04 27.7	+46 02 55	-0.002	+0.01	7.92	1.10	0.3	G8 III		270 s		
200334	230507		21 04 31.8	-46 15 19	+0.011	-0.06	7.05	0.58	3.0	G3 IV		65 s		
200203	246845		21 04 34.1	-58 55 59	0.000	-0.02	8.0	-0.5	1.4	A2 V		210 s		
200644	126518	3 Equ	21 04 34.5	+5 30 10	+0.001	0.00	5.61	1.65	-0.3	gK5	-16	120 s		m
200943	33110		21 04 34.8	+52 24 00	+0.003	-0.02	7.59	-0.04		A0			14615	m
201063	19174		21 04 35.5	+60 15 24	-0.001	-0.02	7.5	0.5	3.4	F5 V		64 s		
200302	246850		21 04 37.6	-51 16 26	+0.002	+0.03	7.3	1.4	0.2	K0 III		160 s		
200689	106818		21 04 37.6	+12 01 29	+0.002	-0.02	7.60	1.1	0.2	K0 III		300 s		m
200661	126519		21 04 41.5	+2 56 32	+0.001	0.00	6.42	1.05		K0	-10			
199532	257879	σ Oct	21 04 42.7	-77 01 25	+0.003	-0.36	5.15	0.49	0.7	F4 III	+60	70 s		
200663	126522		21 04 45.2	+2 16 12	+0.006	-0.06	6.33	0.97		G5	-12			
--	--		21 04 47.1	+41 43 50	-0.001	-0.01	6.7	0.1	1.4	A2 V		110 s		
201033	33117		21 04 48.6	+55 35 18	+0.001	+0.04	7.8	0.2		A m				
200582	164119		21 04 51.0	-17 09 48	-0.001	0.00	7.3	0.2	2.1	A5 V		110 s		
200745	106819		21 04 52.4	+12 26 42	-0.002	-0.02	7.60	0.4	3.0	F2 V		83 s	14607	m
200905	50424	62 ξ Cyg	21 04 55.7	+43 55 40	0.000	0.00	3.72	1.65	-4.4	K5 Ib	-20	290 s		
200842	70876		21 04 57.5	+35 25 47	+0.002	+0.03	7.40	0.4	2.6	F0 V		90 s		
200945	50429		21 04 57.5	+47 08 48	+0.001	-0.01	7.73	1.77	-4.4	K2 Ib		1700 s		
--	33119		21 05 00.4	+56 22 09	-0.001	0.00	7.8	0.9	2.6	F0 V		51 s		
--	--		21 05 00.6	+41 24 49	-0.002	-0.01	6.7	0.1	1.4	A2 V		110 s		
200713	126527		21 05 02.8	+2 32 48	-0.001	0.00	7.5	0.0	0.4	B9.5 V		260 s		
200538	212692		21 05 04.8	-32 49 35	-0.002	-0.02	7.5	0.7	4.4	G0 V		35 s		
200788	106825		21 05 05.7	+18 39 21	+0.002	+0.01	7.5	0.1	1.4	A2 V		170 s		
200521	212691		21 05 08.6	-37 13 53	+0.007	-0.09	6.9	1.0	0.2	K0 III		130 mx		
200365	246854	μ Ind	21 05 14.0	-54 43 37	+0.001	-0.03	5.16	1.21	0.2	K0 III	+12	67 s		
200831	89437		21 05 14.4	+23 36 01	+0.001	-0.01	7.5	1.4	-0.3	K5 III		350 s		
200747	126532		21 05 19.0	+0 43 38	-0.001	+0.02	7.8	0.4	2.6	F0 V		110 s		
200502	230513		21 05 23.1	-45 14 31	+0.003	-0.02	8.0	0.7	3.4	F5 V		57 s		
199066	258877		21 05 23.2	-81 23 47	-0.002	+0.03	8.0	0.7	0.2	K0 III		370 s		
200790	126535	4 Equ	21 05 26.5	+5 57 30	-0.006	-0.12	5.94	0.54	3.8	dF7	-22	25 s		m
200928	70888		21 05 26.7	+34 48 43	-0.001	-0.01	8.0	0.1	0.6	A0 V		290 s		
201908	9959		21 05 29.1	+78 07 35	+0.006	+0.03	5.91	-0.07		B9	-16			
200861	89444		21 05 29.7	+21 52 25	0.000	0.00	8.0	1.1	-0.1	K2 III		390 s		
200893	89447		21 05 29.9	+29 05 46	+0.001	-0.01	8.0	1.1	-0.1	K2 III		400 s		
200758	145084		21 05 31.7	-4 21 43	-0.002	0.00	7.2	1.4	-0.3	K5 III		320 s		
200553	230516		21 05 31.8	-43 31 21	-0.004	-0.07	7.20	0.94	3.2	G8 IV		58 s		
201267	19188		21 05 32.2	+62 09 33	0.000	+0.02	7.81	0.09	0.6	A0 V		240 s	14634	m
201065	50438		21 05 35.2	+46 57 48	-0.006	-0.01	7.56	1.79	-4.4	K2 Ib		74 mx		
--	230515		21 05 36.0	-47 49 23	-0.002	+0.01	7.5	2.5	-0.1	K2 III		51 s		
200729	164131		21 05 38.4	-14 54 17	+0.002	+0.01	7.9	1.1	0.2	K0 III		350 s		
200554	230518		21 05 40.7	-44 56 57	-0.003	-0.02	7.1	0.9	0.2	K0 III		240 s		
200539	230517		21 05 41.3	-46 30 25	+0.003	0.00	8.0	0.4	2.6	F0 V		73 s		
201076	50439		21 05 43.8	+47 48 17	0.000	0.00	7.45	-0.05	0.0	A0 IV		310 s	14627	m
200877	106829		21 05 48.5	+15 19 37	+0.008	-0.09	6.6	0.5	0.7	F5 III	-21	65 mx		
200907	106830		21 05 51.0	+20 02 28	0.000	-0.06	8.0	0.5	3.4	F5 V		82 s		
200930	89451		21 05 55.2	+20 56 47	0.000	0.00	6.6	0.0	0.4	B9.5 V	+10	180 s		
201114	50445		21 05 56.1	+48 02 39	0.000	-0.01	7.57	0.05	0.2	B9 V	-16	260 s		
200761	164132	23 θ Cap	21 05 56.6	-17 13 58	+0.006	-0.06	4.07	-0.01	0.6	A0 V	-11	49 s		
200961	89453		21 05 58.3	+24 29 55	-0.001	-0.04	8.0	0.4	2.6	F0 V		120 s		
200670	212704		21 05 58.7	-36 15 35	-0.005	-0.02	7.9	0.2	3.4	F5 V		79 s		
201343	19193		21 06 00.1	+62 55 20	-0.001	0.00	7.0	1.4	-0.3	K5 III	-4	270 s		
200718	212709	δ Mic	21 06 00.9	-30 07 31	+0.003	-0.07	5.68	1.03	0.2	gK0		120 s		
200625	230519		21 06 16.8	-48 31 34	0.000	-0.03	7.5	0.5	4.4	G0 V		41 s		
200931	126542		21 06 17.5	+7 30 26	+0.006	+0.01	8.0	0.5	4.0	F8 V		62 s		
200719	212713		21 06 21.1	-37 35 16	-0.004	-0.01	7.5	1.2	2.6	F0 V		26 s		
201636	9957		21 06 23.0	+71 55 55	-0.011	-0.11	5.87	0.40	3.0	F2 V	+2	36 s		
200673	230521		21 06 23.1	-45 47 46	+0.001	0.00	7.70	1.1	0.0	K0 III		250 s		m
201051	89459		21 06 23.3	+26 55 28	+0.003	-0.01	6.12	1.05	0.0	gK1	-6	170 s		
200750	212715		21 06 23.8	-34 37 44	+0.004	-0.01	6.8	1.1	0.2	K0 III		180 s		
200763	212716		21 06 24.5	-32 20 30	0.000	+0.01	5.2	1.1	0.2	K0 III	+3	98 s		
201344	33139		21 06 25.1	+59 26 22	0.000	-0.01	7.3	0.1	0.6	A0 V	+11	210 s		
200702	230523	η Mic	21 06 25.3	-41 23 10	+0.002	-0.01	5.53	1.35	0.2	K0 III		62 s		m
200285	254936		21 06 29.2	-68 38 46	0.000	0.00	8.0	1.4	-0.3	K5 III		460 s		
201078	70917		21 06 30.0	+31 11 05	0.000	0.00	5.82	0.56	-5.5	F6 I-II var	0	1600 s		DT Cyg, v
200589	246870		21 06 30.5	-54 57 25	+0.003	-0.02	7.5	0.1	3.0	F2 V		81 s		
200932	145101		21 06 33.3	-0 06 18	+0.001	+0.01	6.91	1.51	6.3	K2 V		13 s		

535

HD	SAO	Star Name	α 2000	δ 2000	μ(α)	μ(δ)	V	B-V	M$_v$	Spec	RV	d(pc)	ADS	Notes
			h m s	° ' "	s	"								
201154	70918		21 06 34.1	+38 02 20	-0.002	-0.07	7.6	0.5	3.4	F5 V		69 s		
201293	33137		21 06 35.6	+54 13 03	+0.003	+0.01	8.0	0.5	3.4	F5 V		81 s		
201251	50456	63 Cyg	21 06 35.9	+47 38 54	+0.001	0.00	4.55	1.57	-2.3	K4 II	-26	220 s	14649	m
200733	230526		21 06 36.8	-41 23 06	+0.002	-0.05	5.60	0.5	4.0	F8 V		21 s		m
200898	164138		21 06 37.5	-13 30 13	0.000	+0.04	8.0	1.1	0.2	K0 III		360 s		
200964	126546		21 06 39.7	+3 48 12	+0.007	+0.06	6.6	1.1	0.2	K0 III		160 mx		
201280	33136		21 06 40.5	+51 25 33	0.000	-0.01	7.5	0.0	0.4	B9.5 V		250 s		
200868	190014		21 06 41.3	-20 10 48	-0.003	-0.01	6.6	1.3	0.2	K0 III		130 s		
201269	50460		21 06 43.5	+48 11 20	0.000	0.00	7.63	0.01	0.2	B9 V	-6	290 s		
199391	258879		21 06 44.0	-80 41 53	+0.027	-0.13	7.30	0.2	2.1	A5 V		30 mx		m
201319	33142		21 06 45.2	+53 57 14	+0.003	+0.02	8.0	0.4	2.6	F0 V		110 s		
201156	70925		21 06 46.7	+34 07 56	0.000	-0.04	8.00	0.1	1.4	A2 V		210 s	14645	m
202490	3514		21 06 49.4	+81 01 13	-0.009	0.00	7.1	1.1	-0.1	K2 III		270 s		
200881	190018		21 06 51.1	-23 12 56	+0.003	+0.02	7.5	0.7	3.2	G5 IV		71 s		
200987	145109		21 06 52.1	-0 57 50	-0.001	-0.04	7.2	0.4	3.0	F2 V		70 s		
201270	50461		21 06 52.3	+45 40 32	-0.001	-0.01	7.4	0.6	4.4	G0 V		39 s		
200884	190019		21 06 53.5	-25 33 44	+0.001	+0.01	7.6	0.7	0.6	A0 V		89 s		
201091	70919	61 Cyg	21 06 53.9	+38 44 58	+0.353	+3.20	5.22	1.26	7.6	K5 V	-64	3.4 t	14636	A component, m
---	70930		21 06 54.4	+38 32 21	-0.002	-0.01	7.9	1.7						
201429	33146		21 06 54.7	+59 53 24	+0.002	+0.04	7.6	0.1	0.6	A0 V	-12	230 s		
201092	---	61 Cyg	21 06 55.1	+38 44 32	+0.353	+3.20	6.04	1.37	8.4	K7 V	-64	3.4 t	14636	B component, m
201430	33147		21 06 59.8	+58 59 44	0.000	0.00	7.9	0.5	3.4	F5 V		76 s	14669	m
200913	190024		21 07 03.4	-23 08 56	+0.008	+0.03	7.0	0.4	3.4	F5 V		52 s		
201053	126550		21 07 04.8	+5 26 35	+0.004	+0.02	7.3	1.1	0.2	K0 III		260 s		
201274	70935		21 07 07.4	+38 18 54	-0.001	-0.01	7.8	0.1	0.6	A0 V		260 s		
200914	190025	24 Cap	21 07 07.5	-25 00 21	-0.002	-0.04	4.50	1.61	-0.5	M1 III	+32	95 s	14632	
201194	70931		21 07 08.2	+30 35 47	0.000	+0.01	7.5	0.0	0.0	B8.5 V	-19	310 s		
200970	164145		21 07 09.5	-15 58 23	+0.004	-0.02	7.2	1.1	0.2	K0 III		250 mx		
200968	164147		21 07 10.2	-13 55 21	+0.027	-0.03	7.10	0.90	6.1	K1 V comp	-37	16 ts	14638	m
200993	164149		21 07 11.3	-15 44 49	-0.010	-0.13	7.6	0.5	4.0	F8 V		51 s		
201252	70936		21 07 12.0	+36 57 00	-0.001	0.00	7.90	0.1	0.6	A0 V		280 s		m
201320	50471		21 07 13.0	+47 44 03	-0.001	+0.01	7.33	0.05	0.6	A0 V		200 s	14665	m
201306	50468		21 07 14.7	+44 40 26	-0.001	0.00	7.38	-0.05	0.4	B9.5 V		250 s	14664	m
201415	33151		21 07 16.7	+55 00 13	0.000	-0.03	7.9	1.1	0.2	K0 III		320 s		
201358	50475		21 07 18.1	+49 47 24	+0.002	-0.04	7.56	0.27	1.7	A3 V		130 s		
201097	126554		21 07 18.8	+4 08 30	0.000	-0.01	8.0	0.5	4.0	F8 V		62 s		
201253	70940		21 07 19.9	+32 30 08	+0.001	-0.01	7.7	0.0	0.4	B9.5 V		280 s		
---	---		21 07 21.4	+45 43 11			7.26	0.58	0.4	A6 III		140 s		
201066	145120		21 07 21.7	-0 45 55	0.000	+0.02	6.4	1.1	0.2	K0 III		180 s		
200952	190028		21 07 22.7	-27 17 30	+0.005	-0.03	7.1	-0.3	1.7	A3 V		120 mx		
201359	50476		21 07 23.6	+47 16 40	+0.002	+0.01	7.28	-0.05	-0.2	B8 V		290 s		
200655	254939		21 07 24.5	-59 59 31	-0.006	-0.01	6.78	0.64	3.2	G5 IV		52 s		
201038	145118		21 07 26.1	-8 14 07	-0.001	-0.01	7.50	0.1	0.6	A0 V		240 s	14648	m
201140	126555		21 07 27.0	+9 17 45	0.000	0.00	7.6	1.1	0.2	K0 III		300 s		
201013	164152		21 07 29.7	-19 05 17	+0.002	-0.05	6.8	0.8	3.2	G5 IV		48 s		
200885	212725		21 07 30.9	-38 41 30	+0.001	-0.03	7.4	0.7	2.6	F0 V		56 s		
201098	145121		21 07 32.6	-0 09 50	-0.001	0.00	6.80	1.6	-0.5	M4 III		290 s	14654	m
201582	19208		21 07 32.8	+63 04 14	-0.007	-0.03	8.00	1.1	0.2	K0 III		330 s		m
201196	106853		21 07 33.4	+15 39 31	+0.003	-0.06	6.5	1.1		K2 IV	-34	180 mx		
201431	33153		21 07 36.2	+51 35 02	+0.002	+0.01	7.37	-0.02		A0				
201817	9963		21 07 36.5	+71 40 55	+0.005	+0.02	8.0	0.4	3.0	F2 V		94 s		
200872	230532		21 07 37.7	-44 17 33	+0.005	-0.03	7.7	1.6	-0.1	K2 III		210 s		
201416	50479		21 07 40.0	+48 51 18	0.000	-0.05	7.78	0.96	0.3	G5 III		280 s	14673	m
201731	19214		21 07 41.1	+68 15 15	+0.005	+0.01	6.80	0.12	1.4	A2 V		110 s		
201120	145126		21 07 42.4	-2 03 09	+0.006	0.00	6.8	0.1	1.7	A3 V		100 s		
201099	145122		21 07 43.7	-5 33 57	+0.014	+0.23	7.58	0.54		G0		14 mn		m
200887	230535		21 07 44.3	-45 22 46	+0.003	-0.01	6.7	0.7	1.7	A3 V		45 s		
201057	164156		21 07 44.4	-17 27 19	+0.001	-0.02	6.0	0.1	0.6	A0 V		120 s		
200973	212731		21 07 44.6	-31 03 27	+0.017	+0.08	7.22	0.46		F6 V-VI				
201397	50480		21 07 47.2	+44 57 50	-0.001	-0.02	7.52	0.01		A0				
201159	126556		21 07 49.0	+1 09 14	0.000	-0.04	6.8	1.4	-0.3	K5 III		260 s		
200998	212733		21 07 49.2	-34 11 57	+0.004	-0.06	7.7	1.4	0.2	K0 III		190 s		
200798	246880		21 07 50.8	-54 13 01	+0.001	-0.09	6.9	0.0	1.7	A3 V		45 mx		
200787	246879		21 07 53.4	-56 38 09	+0.004	0.00	7.9	0.8	4.4	G0 V		37 s		
200856	230536		21 07 53.6	-48 56 24	+0.002	-0.04	6.67	0.00		A0				m
201345	70953		21 07 55.2	+33 23 50	0.000	0.00	7.68	-0.12		O9 p	+20			
201058	190040		21 07 56.0	-22 20 07	+0.003	-0.03	7.8	0.4	2.1	A5 V		91 s		
201254	106859		21 07 57.0	+14 40 24	-0.001	-0.01	6.8	0.1	-1.7	B3 V	+8	510 s		
201104	164158		21 07 58.2	-15 37 26	+0.002	-0.02	7.6	1.4	-0.3	K5 III		380 s		
201221	126562		21 08 00.4	+5 09 06	-0.001	-0.01	6.90	0.1	1.4	A2 V		130 s	14666	m
200706	254940		21 08 00.8	-62 56 46	+0.008	-0.02	7.1	1.1	0.2	K0 III		240 s		
200918	230538		21 08 00.9	-46 29 33	+0.003	+0.02	7.3	-0.1	0.6	A0 V		220 s		
201257	126563		21 08 00.9	+9 21 46	-0.001	+0.02	7.6	0.5	3.4	F5 V		71 s		
201294	89476		21 08 01.1	+21 45 52	+0.002	-0.06	7.6	0.4	3.0	F2 V		83 s		
201398	70959		21 08 06.3	+38 31 45	+0.003	0.00	8.0	0.4	3.0	F2 V		97 s		

HD	SAO	Star Name	α 2000	δ 2000	μ(α)	μ(δ)	V	B-V	M_v	Spec	RV	d(pc)	ADS	Notes
201222	145132		21 08 09.2	-0 59 23	+0.004	+0.01	6.8	0.2	2.1	A5 V		87 s		
201297	106864		21 08 15.2	+12 43 40	-0.001	-0.01	7.9	1.1	0.2	K0 III		340 s		
201456	50487		21 08 17.0	+43 44 52	+0.011	+0.06	7.81	0.52	4.0	F8 V		57 s		
201543	33163		21 08 18.1	+53 13 32	+0.007	+0.02	7.3	0.1	0.6	A0 V		130 mx		
201361	89483		21 08 24.3	+24 07 09	-0.001	-0.01	8.0	0.1	1.4	A2 V		200 s		
201469	50488		21 08 26.3	+41 35 11	-0.001	+0.01	8.0	0.0	0.4	B9.5 V		310 s		
201611	33167		21 08 26.5	+56 06 16	+0.001	+0.02	7.9	0.1	0.6	A0 V		270 s		
--	33168		21 08 27.5	+56 02 43	0.000	0.00	7.0	1.3		B1				
201298	126566		21 08 27.9	+6 59 21	-0.001	0.00	6.15	1.66		K6	+20			
201522	50489		21 08 29.4	+47 15 26	+0.001	+0.01	7.88	-0.05	-4.1	B0 V		1900 s		
201185	190049		21 08 31.9	-24 12 20	+0.001	-0.03	7.3	1.3	-0.1	K2 III		260 s		
200751	254941		21 08 32.8	-63 55 44	+0.002	-0.01	5.76	1.18		K0				
201184	190050	25 χ Cap	21 08 33.4	-21 11 37	+0.001	-0.06	5.28	0.01	0.6	A0 V	-7	86 s		m
201203	190052		21 08 36.7	-21 40 58	+0.005	0.00	7.8	0.5	4.4	G0 V		49 s		
201433	70968		21 08 38.7	+30 12 21	+0.002	-0.01	5.59	-0.10		A0	-26	48 mn	14682	m
201108	212743		21 08 41.0	-37 14 34	+0.001	0.00	6.90	0.00	0.0	B8.5 V		240 s		
201348	126570		21 08 42.3	+9 49 10	+0.001	0.00	7.9	1.1	0.2	K0 III		340 s		
201284	145137		21 08 44.4	-6 59 12	-0.001	0.00	7.8	1.1	0.2	K0 III		330 s		
177482	258857	σ Oct	21 08 44.9	-88 57 24	+0.095	0.00	5.47	0.27		A7 n	+12			
201457	70971		21 08 45.1	+33 37 20	0.000	+0.01	8.0	0.1	0.6	A0 V		300 s		
200266	257887		21 08 47.8	-76 12 46	-0.005	-0.02	6.58	1.23		K0				
201332	126571		21 08 51.4	+1 47 16	+0.001	0.00	7.1	0.0	0.4	B9.5 V		210 s		
200475	257889		21 08 52.9	-73 23 49	+0.010	-0.06	8.0	0.1	1.7	A3 V		72 mx		
201311	145144		21 08 55.8	-6 19 45	-0.001	0.00	7.9	0.1	1.7	A3 V		170 s		
201322	145145		21 08 57.2	-5 34 55	-0.004	-0.05	6.7	1.1	0.2	K0 III		200 s		
201599	50498		21 08 57.6	+47 16 16	+0.001	+0.01	6.96	0.38		F2			14693	m
201226	190055		21 08 58.3	-28 28 40	+0.003	-0.13	7.00	0.9	3.2	G5 IV		58 s	14674	m
201376	126577		21 09 00.9	+7 35 23	+0.001	+0.03	7.4	0.1	1.4	A2 V		160 s		
201470	89487		21 09 03.1	+26 59 23	+0.002	+0.07	7.9	0.5	4.0	F8 V		59 s		
201350	145147		21 09 03.9	-3 48 19	-0.001	+0.01	8.0	1.1	-0.1	K2 III		410 s		
201560	50496		21 09 04.9	+41 34 51	0.000	-0.02	8.0	1.1	-0.1	K2 III		380 s		
200800	254944		21 09 05.3	-65 47 59	+0.004	-0.06	7.8	0.1	1.4	A2 V		68 mx		
201262	190058		21 09 07.2	-27 06 26	+0.005	+0.02	7.7	1.2	-0.1	K2 III		310 mx		
202505	9976		21 09 07.4	+78 39 46	+0.001	-0.02	7.4	0.1	1.4	A2 V	-14	160 s		
201561	70977		21 09 08.2	+38 43 46	+0.001	+0.01	7.80	1.1	-0.1	K2 III	-9	370 s		m
201523	70976		21 09 08.5	+35 02 37	+0.003	+0.01	7.8	0.1	1.7	A3 V		160 s		
201490	70975		21 09 10.5	+30 22 37	+0.008	+0.03	8.00	0.51	3.8	F7 V		69 s		
201286	190061		21 09 12.0	-23 37 36	0.000	+0.02	7.0	1.2	0.2	K0 III		180 s		
201110	230547		21 09 12.1	-46 46 04	+0.002	+0.01	7.9	1.9	-0.5	M4 III		250 s		
201732	33173		21 09 12.2	+57 12 09	+0.002	-0.02	8.0	0.5	4.0	F8 V		63 s		
201301	190065		21 09 15.2	-20 11 35	+0.002	0.00	6.8	1.1	0.2	K0 III	-46	190 s		
201377	145152		21 09 15.9	-0 14 06	+0.005	0.00	6.5	0.1	1.7	A3 V		93 s		
201002	246887		21 09 16.5	-57 51 55	0.000	+0.02	7.9	0.4	2.6	F0 V		96 s		
201228	212751		21 09 17.9	-37 56 00	0.000	+0.01	7.0	1.2	3.2	G5 IV		34 s		
201378	145153		21 09 19.0	-3 53 33	-0.002	+0.04	7.8	0.7	4.4	G0 V		49 s		
201435	126581		21 09 21.0	+5 01 42	0.000	-0.02	7.90	1.4	-0.3	K5 III		440 s		
201263	212754		21 09 21.4	-32 01 26	+0.001	-0.01	7.6	1.3	0.2	K0 III		210 s		
199509	258880		21 09 21.5	-82 01 37	+0.142	-0.04	6.9	0.8	4.4	G0 V		22 s		
200525	257890		21 09 22.2	-73 10 23	+0.102	-0.33	5.68	0.59	3.0	G3 IV	-14	29 mx		m
201242	212753		21 09 22.2	-36 42 22	+0.003	-0.12	6.6	0.9	4.4	G0 V		18 s		
201335	164177		21 09 24.9	-16 36 57	+0.002	+0.01	6.8	1.1	-0.1	K2 III		240 s		
201888	19223		21 09 28.7	+63 17 44	+0.002	+0.01	6.54	-0.13		B8	-24			
202012	19229		21 09 31.8	+68 29 25	-0.006	-0.06	7.33	1.49		M5 III e	-12	23 mn		T Cep, v
201352	190069	27 Cap	21 09 32.8	-20 33 23	+0.008	-0.12	6.2	0.6	2.6	F0 V	-43	35 s		
201666	50510		21 09 34.5	+45 44 20	+0.002	+0.02	7.64	-0.02		B5				
201382	164180		21 09 35.0	-14 07 00	+0.002	+0.02	7.5	1.1	-0.1	K2 III		330 s		
202345	9975		21 09 35.4	+75 13 41	-0.001	0.00	7.0	0.5	3.4	F5 V	-10	51 s		
201381	164182	13 ν Aqr	21 09 35.4	-11 22 18	+0.006	-0.01	4.51	0.94	0.3	G8 III	-12	70 s		
201652	50509		21 09 38.6	+42 26 01	-0.003	0.00	7.8	0.1	0.6	A0 V		250 s		
201169	246891		21 09 42.8	-50 54 38	+0.009	-0.05	7.7	0.9	4.4	G0 V		29 s		
201701	50514		21 09 42.8	+46 09 32	+0.001	+0.02	7.62	0.52	4.4	G0 V		44 s		
201245	230552		21 09 43.9	-44 12 35	+0.002	+0.01	6.51	1.16	0.0	K1 III		190 s		
201545	106883		21 09 45.7	+19 12 40	+0.001	+0.04	7.0	0.5	3.4	F5 V	-23	53 s		
201702	50519		21 09 52.2	+44 12 01	-0.007	+0.04	7.5	1.1	0.2	K0 III		110 mx		
201507	126587		21 09 58.1	+2 56 37	+0.003	+0.02	6.45	0.37		F2	-44			
201733	50521		21 09 58.4	+45 30 09	0.000	0.00	6.66	-0.16		B5 e	+9			
201656	70997		21 10 01.9	+32 12 21	+0.001	+0.01	7.4	1.1	0.2	K0 III		270 s		
201390	190079		21 10 05.8	-29 43 16	+0.002	-0.01	6.5	1.8	-0.3	K5 III		150 s		
201667	71002		21 10 06.8	+35 14 52	0.000	-0.01	8.0	1.1	0.2	K0 III		350 s		
202030	19232		21 10 07.2	+65 41 11	0.000	0.00	7.79	-0.03	0.4	B9.5 V		290 s		
201020	254948		21 10 07.4	-64 01 34	0.000	-0.08	7.3	0.1	1.4	A2 V		48 mx		
201462	164187		21 10 07.8	-15 42 12	-0.003	-0.06	7.4	0.4	2.6	F0 V		92 s		
201584	106886		21 10 09.5	+12 33 21	0.000	-0.01	8.0	1.4	-0.3	K5 III		450 s		
201317	230555		21 10 09.8	-43 22 56	+0.001	+0.01	6.57	-0.08	0.4	B9.5 V		170 s		
201703	71006		21 10 10.5	+36 27 46	-0.002	-0.01	8.0	1.1	0.2	K0 III		350 s		

HD	SAO	Star Name	α 2000	δ 2000	μ(α)	μ(δ)	V	B-V	M$_v$	Spec	RV	d(pc)	ADS	Notes
			h m s	° ′ ″	s	″								
201391	212766		21 10 10.9	−31 29 33	+0.005	−0.07	7.6	1.2	3.2	G5 IV		42 s		
201639	89501		21 10 14.2	+20 44 37	−0.004	+0.07	7.8	0.5	4.0	F8 V		59 s		
201834	33185		21 10 15.5	+53 33 48	+0.002	0.00	5.7	−0.1		A pe	−21			
201669	89502		21 10 15.7	+27 17 59	−0.003	−0.01	7.80	0.9	0.3	G8 III		310 s		
201835	33186		21 10 18.5	+52 25 50	+0.010	+0.03	8.0	0.5	3.4	F5 V		80 s		
201601	126593	5 γ Equ	21 10 20.3	+10 07 53	+0.004	−0.15	4.69	0.26		F0 p	−17	21 mn	14702	m
201409	212768		21 10 20.8	−32 50 59	+0.003	+0.01	7.2	0.6	3.2	G5 IV		64 s		
201172	246893		21 10 24.4	−59 24 26	+0.003	+0.01	7.2	0.1	1.7	A3 V		130 s		
201247	246894		21 10 24.9	−54 34 26	+0.012	−0.01	7.80	0.6	4.4	G0 V		48 s		m
201474	—		21 10 27.9	−22 28 45			7.3	1.1	3.2	G5 IV		40 s		
201837	50535		21 10 28.6	+47 39 17	−0.002	−0.02	7.29	1.54		M5 III				
201448	190085		21 10 29.4	−28 29 29	+0.004	−0.01	7.8	1.1	−0.1	K2 III		300 mx		
—	190087		21 10 30.5	−22 28 39	+0.002	+0.02	7.3	0.9	3.2	G5 IV		66 s		
201670	89504		21 10 30.6	+22 27 26	+0.001	0.00	7.67	0.18		A0				
201836	50536		21 10 30.8	+47 41 32	0.000	0.00	6.46	−0.01	−1.3	B6 IV	−9	310 s	14720	m
201616	126597	6 Equ	21 10 31.1	+10 02 56	−0.001	+0.02	6.07	0.02		A2	+7	29 mn		
201671	89505		21 10 31.8	+22 27 17	+0.001	0.00	6.68	0.03		A0	−12		14710	m
201494	164192		21 10 32.4	−18 19 51	+0.002	−0.01	8.0	1.1	0.2	K0 III		360 s		
201750	71015		21 10 32.6	+36 47 45	+0.003	−0.04	7.6	0.4	3.0	F2 V		84 s		
202142	19236		21 10 36.4	+66 09 35	−0.001	+0.04	7.2	1.1	0.2	K0 III		230 s		
201795	71018		21 10 38.4	+38 57 43	+0.001	+0.02	7.53	−0.03	−3.5	B1 V		1300 s		
201777	71019		21 10 41.3	+37 29 52	0.000	−0.02	7.9	1.1	−0.1	K2 III		380 s		
201569	164195		21 10 44.0	−10 12 33	−0.001	0.00	8.0	0.9	3.2	G5 IV		93 s		
201424	230562		21 10 44.6	−39 58 32	+0.001	−0.05	7.5	1.4	3.2	G5 IV		12 s		
201567	145171		21 10 46.7	−9 21 14	+0.008	−0.05	6.27	1.16		K0				
202181	19238		21 10 51.0	+66 42 35	+0.006	−0.01	7.3	1.1	0.2	K0 III		240 s		
202000	33196		21 11 01.5	+55 19 56	−0.001	+0.01	7.8	0.0	0.0	B8.5 V		330 s		
201687	126600		21 11 03.0	+3 55 10	+0.003	−0.02	7.7	1.1	−0.1	K2 III		370 s		
201819	71032		21 11 03.7	+36 17 57	+0.001	−0.01	6.54	−0.14	−3.5	B1 V p	−6	930 s	14724	m
201857	71035		21 11 11.6	+36 39 17	+0.002	−0.03	7.6	0.2	2.1	A5 V		130 s		
201935	50547		21 11 15.2	+45 40 28	−0.002	0.00	6.7	0.1	0.6	A0 V		160 s		
201858	71034		21 11 15.4	+33 08 00	0.000	0.00	7.6	0.9	3.2	G5 IV		75 s		
200924	257893		21 11 20.7	−72 32 39	+0.008	−0.02	6.20	1.08	0.2	gK0		140 s		
201910	50546		21 11 21.8	+41 11 06	−0.001	−0.01	7.40	−0.14	−1.1	B5 V	−12	500 s		
201735	126604		21 11 22.5	+4 15 03	0.000	+0.01	7.7	1.1	0.2	K0 III		310 s		
201427	246896		21 11 22.8	−52 20 23	+0.001	+0.02	7.30	0.5	4.0	F8 V		46 s		m
202107	33201		21 11 23.9	+57 37 13	−0.002	−0.01	7.88	0.15	0.0	B8.5 V		330 s		
201859	89515		21 11 26.3	+27 33 13	+0.003	−0.01	7.73	0.04	0.6	A0 V		260 s		
202900	9993		21 11 27.5	+78 58 31	+0.002	0.00	6.9	0.0	−1.6	B3.5 V	−14	500 s		
202013	—		21 11 28.6	+50 24 08	+0.002	+0.02	7.8	1.1	0.2	K0 III		310 s		
202431	9982		21 11 32.3	+70 49 34	+0.003	+0.02	7.6	0.1	1.7	A3 V		140 s		
201718	145182		21 11 33.3	−2 08 57	0.000	−0.02	7.5	0.1	1.4	A2 V		170 s		
202249	19240		21 11 33.7	+64 11 37	+0.007	+0.01	7.9	1.1	−0.1	K2 III		350 s		
201411	246897		21 11 34.1	−56 31 02	+0.003	−0.02	6.9	0.8	0.2	K0 III		220 s		
201719	145185		21 11 36.2	−3 07 12	+0.001	0.00	7.60	1.1	0.2	K0 III		300 s	14725	m
201937	71048		21 11 36.8	+40 02 13	+0.003	+0.01	8.02	1.60	−0.5	M2 III		500 s		
201753	145186		21 11 38.1	−0 52 16	−0.001	−0.01	7.8	0.1	1.4	A2 V		190 s		
202250	19241		21 11 38.3	+63 46 52	−0.008	+0.02	7.9	1.1	0.2	K0 III		180 mx		
202084	33203		21 11 39.2	+53 44 47	+0.002	0.00	7.05	0.01	0.6	A0 V		190 s		
202432	9984		21 11 40.0	+70 26 30	+0.002	+0.05	7.1	1.1	−0.1	K2 III		240 s		
201707	164204		21 11 41.1	−14 28 20	+0.002	+0.01	6.48	0.28	0.3	gA5	−39	150 s		
202162	33207		21 11 43.7	+58 27 42	+0.002	+0.01	7.8	0.1	1.4	A2 V		180 s		
201841	106904		21 11 46.8	+12 57 06	0.000	0.00	7.4	1.4	−0.3	K5 III		350 s		
202214	33210		21 11 48.1	+59 59 11	0.000	0.00	5.64	0.11	−4.1	B0 V	−16	560 s	14749	m
201912	89520		21 11 48.3	+29 42 44	−0.001	+0.01	6.86	−0.13		B5	−4			m
201890	89519		21 11 54.7	+23 04 49	−0.001	−0.04	7.9	1.1	−0.1	K2 III		380 s		
201977	50556		21 11 54.9	+40 46 26	+0.004	−0.02	7.93	−0.11	0.0	B8.5 V		150 mx		
201872	106906		21 11 56.6	+17 53 42	−0.001	−0.02	7.8	0.1	1.7	A3 V		160 s		
201939	71050		21 11 56.7	+30 37 10	+0.005	+0.04	6.7	0.9	3.2	G5 IV	−20	51 s		
201780	145191		21 11 58.3	−5 48 48	0.000	+0.01	7.90	1.1	0.2	K0 III		350 s	14730	m
201891	106908		21 11 59.0	+17 43 40	−0.008	−0.89	7.38	0.51		F9 VI	−45	27 t		
201889	89522		21 11 59.2	+24 10 07	+0.031	+0.13	8.04	0.58	4.5	G1 V	−103	51 s	14738	m
201968	71052		21 11 59.8	+35 38 42	+0.002	+0.01	7.6	1.1	−0.1	K2 III		330 s		
202182	33212		21 12 00.4	+56 08 26	−0.002	+0.02	7.5	0.1	1.4	A2 V		160 s		
201725	190110		21 12 02.5	−24 07 16	+0.002	+0.02	7.5	0.4	0.6	A0 V		130 s		
202765	9990		21 12 03.7	+76 18 39	+0.002	+0.03	7.00	0.00	0.4	B9.5 V		190 s	14782	m
201821	126606		21 12 04.5	+1 38 00	+0.003	−0.04	7.9	1.1	−0.1	K2 III		390 s		
201875	106910		21 12 04.8	+14 43 05	+0.003	−0.01	7.9	1.4	−0.3	K5 III		430 s		
201768	164211		21 12 05.9	−14 33 30	+0.003	0.00	7.7	0.2	2.1	A5 V		130 s		
202068	50561		21 12 06.7	+45 42 14	+0.001	+0.02	7.90	0.00	0.0	B8.5 V		350 s	14745	m
201695	212786		21 12 07.2	−30 35 24	0.000	−0.05	7.50	0.5	3.4	F5 V		66 s		m
201892	106913		21 12 12.1	+11 46 53	+0.001	−0.01	7.1	1.4	−0.3	K5 III		310 s		
202032	71058		21 12 12.6	+40 09 01	−0.002	−0.02	7.7	1.1	−0.1	K2 III		350 s		
201647	230575		21 12 13.4	−40 16 10	+0.004	−0.22	5.83	0.45	3.8	F7 V	+11	26 s		
202196	33216		21 12 16.9	+53 49 54	+0.004	+0.03	7.7	1.1	0.2	K0 III		290 s		

HD	SAO	Star Name	α 2000	δ 2000	μ(α)	μ(δ)	V	B-V	M_v	Spec	RV	d(pc)	ADS	Notes
201744	190118		21 12 17.3	-27 54 18	+0.001	-0.02	7.7	1.2	-0.1	K2 III		360 s		
201796	--		21 12 20.2	-14 59 33			8.00	0.9	3.2	G5 IV		91 s	14736	m
202124	50567		21 12 28.2	+44 31 56	-0.002	+0.01	7.80	0.22	-5.9	O9.5 Ib		2900 s		
202001	89527		21 12 28.6	+28 57 13	+0.001	0.00	8.0	1.4	-0.3	K5 III		430 s		
202049	71060		21 12 28.9	+33 14 21	-0.001	-0.02	7.5	1.1	0.2	K0 III		280 s		
202088	71065		21 12 29.3	+38 33 59	-0.001	-0.01	7.34	-0.10	0.4	B9.5 V		240 s		m
202069	71066		21 12 35.1	+35 25 08	+0.001	0.00	8.0	1.1	0.2	K0 III		350 s		
201809	190125		21 12 36.5	-21 00 42	-0.002	+0.02	7.6	1.8	0.2	K0 III		100 s		
202215	50576		21 12 44.0	+49 19 49	-0.001	+0.03	8.0	0.1	1.4	A2 V		200 s		
201941	126618		21 12 45.1	+2 38 35	-0.002	-0.01	7.0	0.1	1.4	A2 V		130 s		
204149	3541		21 12 45.5	+84 15 31	+0.014	0.00	7.1	0.8	3.2	G5 IV	+5	59 s		
201771	212791		21 12 46.8	-35 45 45	+0.004	-0.03	7.3	1.2	0.2	K0 III		190 s		
202380	33232		21 12 46.9	+60 05 52	-0.002	-0.01	6.62	2.39	-4.8	M3 Ib	-15	670 s		
201649	246910		21 12 47.2	-51 30 52	+0.003	-0.05	7.2	0.6	3.4	F5 V		47 s		
201980	126621		21 12 53.0	+7 41 33	-0.004	-0.04	7.40	0.43		F2				
202126	71073		21 12 53.1	+35 47 51	-0.003	-0.02	6.7	0.1	1.4	A2 V	-6	110 s		
202050	89530		21 12 55.2	+21 04 07	+0.001	-0.01	7.9	1.1	-0.1	K2 III		400 s		
202109	71070	64 ζ Cyg	21 12 56.0	+30 13 37	0.000	-0.05	3.20	0.99	-2.1	G8 II	+17	120 s		m
202108	71071		21 12 57.6	+30 48 35	-0.001	+0.12	7.3	0.7	4.4	G0 V		39 s		
201982	126622		21 12 59.6	+2 48 09	+0.002	+0.04	7.80	0.5	3.4	F5 V		76 s	14750	m
203398	3532		21 13 00.9	+80 48 19	+0.004	+0.01	7.2	0.1	0.6	A0 V		210 s		
201772	212793		21 13 02.8	-39 25 31	+0.016	-0.12	5.2	0.5	3.4	F5 V	-44	23 s		
--	50584		21 13 07.7	+46 46 54	-0.002	0.00	8.0	0.1						
201983	145209		21 13 07.9	-1 07 46	-0.001	0.00	7.9	1.4	-0.3	K5 III		430 s		
--	89535		21 13 08.8	+30 01 09	-0.001	-0.02	7.9	1.5	0.2	K0 III		170 s		
202253	50583		21 13 09.3	+43 52 28	-0.001	+0.03	7.75	0.14	-3.6	B2 III		1200 s		
202165	71076		21 13 09.4	+35 20 30	-0.002	0.00	8.0	1.1	-0.1	K2 III		400 s		
202184	71079		21 13 13.9	+35 29 24	0.000	+0.01	7.73	-0.04	0.0	B8.5 V		350 s		
202198	71077		21 13 15.2	+33 41 37	+0.001	+0.02	7.1	0.0	0.0	B8.5 V	-6	250 s		
202091	106927		21 13 16.4	+16 55 08	-0.002	0.00	7.40	1.1	-0.1	K2 III		320 s	14758	m
201901	190129		21 13 17.2	-27 37 10	+0.007	-0.12	5.42	1.42	-0.3	K5 III	-42	82 mx		
201852	212800		21 13 18.7	-36 25 26	+0.002	-0.01	5.96	0.98	0.2	K0 III		140 s		
201921	190133		21 13 18.7	-24 50 46	+0.001	-0.02	7.5	1.0	-0.1	K2 III	-8	330 s		
201371	257896	o Pav	21 13 20.3	-70 07 35	+0.009	-0.03	5.02	1.58	-0.5	M2 III	-19	130 s		
203501	3534		21 13 21.3	+81 13 51	-0.002	0.00	6.0	0.1	1.4	A2 V	-1	84 s		
202072	106928		21 13 22.2	+11 12 41	+0.011	-0.01	7.9	0.9	3.2	G5 IV		88 s		
202254	71087		21 13 24.7	+38 24 08	-0.001	0.00	7.8	1.4	-0.3	K5 III		390 s		
201930	190135		21 13 24.9	-22 45 42	-0.001	-0.01	7.5	0.2	4.0	F8 V		51 s		
202240	71086		21 13 26.1	+36 38 01	-0.002	+0.01	6.05	0.21	-2.0	A7 II	-13	410 s		
202073	126625		21 13 26.9	+7 13 06	0.000	0.00	7.25	0.31		F0			14759	m
201945	190137		21 13 27.0	-22 32 46	0.000	-0.03	7.6	0.5	3.4	F5 V		65 s		
201757	246913		21 13 27.6	-49 47 42	-0.003	+0.02	8.0	1.0	3.2	G5 IV		40 s		
202596	19256		21 13 28.3	+66 31 25	0.000	+0.02	7.8	1.1	-0.1	K2 III		320 s		
202128	106930		21 13 28.6	+15 58 57	+0.003	-0.02	6.27	0.24		A3	-30	26 mn	14761	
202093	126627		21 13 28.7	+7 10 05	+0.001	+0.02	7.15	0.33		A2				
201789	230581		21 13 29.4	-47 33 02	+0.002	+0.01	6.9	1.5	-0.5	M2 III		300 s		
202019	145215		21 13 30.0	-5 54 42	+0.001	0.00	7.6	0.8	3.2	G5 IV		76 s		
201711	246912		21 13 32.5	-54 59 23	+0.001	-0.03	7.1	1.6	0.2	K0 III		110 s		
202312	50590		21 13 35.8	+45 10 07	-0.001	-0.01	7.32	0.87	-0.9	G5 II-III		440 s		
202034	145217		21 13 37.1	-6 27 57	+0.002	0.00	7.3	1.1	-0.1	K2 III		310 s		
202347	50592		21 13 41.6	+45 36 40	-0.001	-0.01	7.49	-0.11	-3.5	B1 V	-9	1400 s		
202582	19257		21 13 42.4	+64 24 14	+0.004	-0.10	6.39	0.60		G0	+30		14783	m
202348	50593		21 13 43.7	+45 32 13	+0.001	+0.01	7.8	1.4	-0.3	K5 III		390 s		
202719	19261		21 13 46.8	+70 01 40	-0.002	0.00	6.7	1.1	0.2	K0 III		190 s		
201865	230582		21 13 48.3	-43 07 47	+0.007	-0.03	7.1	0.8	0.2	K0 III		230 mx		
202038	190146		21 13 57.3	-20 05 16	0.000	+0.02	7.8	1.5	0.2	K0 III		170 s		
202443	33246		21 14 00.5	+52 42 43	-0.003	-0.01	7.3	1.1	-0.1	K2 III	-6	280 s		
202025	190147		21 14 01.6	-22 12 44	+0.002	-0.01	6.9	0.2	0.6	A0 V		140 s		
201989	190143		21 14 01.7	-29 39 48	+0.020	-0.03	7.5	0.6	3.2	G5 IV		60 mx		
202519	33251		21 14 04.8	+58 17 51	0.000	-0.01	7.80	0.1	0.6	A0 V	-4	260 s	14784	m
202349	71104		21 14 05.3	+37 46 54	0.000	+0.01	7.39	-0.19		B0.5 V				
202111	164238		21 14 07.2	-12 27 57	0.000	+0.03	7.6	1.1	0.2	K0 III		310 s		
202313	71101		21 14 09.7	+30 57 42	+0.001	+0.01	7.90	0.1	0.6	A0 V	-2	280 s		m
202258	106944		21 14 10.1	+12 59 16	+0.002	-0.01	8.0	1.4	-0.3	K5 III		450 s		
202314	89549		21 14 10.1	+29 54 04	0.000	+0.01	6.17	1.09	-4.5	G2 Ib	-5	1100 s		
202302	89548		21 14 11.2	+26 15 38	-0.001	+0.01	7.9	0.4	2.6	F0 V		120 s		
202507	33250		21 14 13.5	+54 28 55	+0.003	+0.08	7.9	0.6	4.4	G0 V		50 s		
202009	212818		21 14 14.4	-35 22 28	0.000	0.00	7.8	0.3	0.6	A0 V		170 s		
201950	230586		21 14 16.1	-44 29 29	0.000	-0.04	7.8	0.9	0.2	K0 III		330 s		
202149	164240		21 14 16.5	-10 36 19	+0.001	+0.01	6.5	-0.1		B9				
201931	230585		21 14 16.8	-45 47 00	+0.001	-0.09	7.0	0.8	3.2	G5 IV		52 s		
202403	50604		21 14 18.7	+41 08 49	+0.003	+0.03	7.80	0.8	3.2	G5 IV	-12	82 s	14778	m
202242	126641		21 14 19.0	+8 39 35	-0.002	-0.02	7.9	1.1	-0.1	K2 III		390 s		
202386	71109		21 14 19.4	+34 17 41	-0.002	+0.01	7.00	0.00	0.0	B8.5 V		250 s		m
202920	9997		21 14 19.8	+73 10 25	+0.007	+0.03	7.6	0.1	1.4	A2 V		160 mx		

HD	SAO	Star Name	α 2000	δ 2000	μ(α)	μ(δ)	V	B-V	M_v	Spec	RV	d(pc)	ADS	Notes
			h m s	° ' "	s	"								
202529	33252		21 14 22.3	+54 39 59	-0.005	-0.04	7.5	0.4	3.0	F2 V		79 s		
202404	71113		21 14 23.3	+39 14 36	0.000	+0.03	7.2	1.1	-0.1	K2 III	-46	290 s		
202303	106949		21 14 23.4	+17 39 13	+0.001	-0.03	8.0	0.1	1.4	A2 V		210 s		
202385	71111		21 14 23.5	+34 46 24	+0.001	+0.02	7.9	0.4	2.6	F0 V		110 s		
201802	246920		21 14 28.0	-59 42 34	+0.004	-0.01	7.6	1.5	-0.1	K2 III		220 s		
202275	126643	7 δ Equ	21 14 28.7	+10 00 25	+0.003	-0.30	4.49	0.50	4.0	F8 V	-10	15 ts		m
202276	126642		21 14 30.8	+4 41 22	+0.002	-0.02	7.50	1.4	-0.3	K5 III		360 s	14774	m
202077	212821		21 14 31.1	-30 45 25	0.000	0.00	7.9	1.5	3.2	G5 IV		32 s		
202330	89555		21 14 34.1	+22 11 33	+0.001	0.00	7.6	1.4	-0.3	K5 III		380 s		
202583	33254		21 14 36.1	+56 53 01	-0.002	+0.01	7.1	1.4	-0.3	K5 III	+1	270 s		
202259	145229		21 14 36.8	+0 05 33	+0.002	0.00	6.38	1.61		K5				
202304	106952		21 14 37.9	+14 08 32	0.000	-0.01	7.9	1.1	0.2	K0 III		340 s		
202260	145230		21 14 40.8	-0 49 55	+0.001	+0.01	7.46	0.11		A0		35 mn	14775	m
202351	106954		21 14 46.0	+16 28 44	+0.002	-0.02	6.8	0.4	2.6	F0 V	-20	68 s		
202734	19265		21 14 46.5	+64 45 01	+0.002	-0.01	6.9	1.1	0.2	K0 III		200 s		
202444	71121	65 τ Cyg	21 14 47.3	+38 02 44	+0.013	+0.44	3.72	0.39	1.7	F0 IV	-21	21 mx	14787	m
202134	212824		21 14 52.4	-31 11 05	-0.007	-0.09	7.7	1.5	0.2	K0 III		130 mx		
202279	145232		21 14 54.3	-7 05 16	0.000	-0.02	7.5	0.5	4.0	F8 V		50 s		
202735	19266		21 14 55.7	+64 29 56	+0.012	+0.10	8.0	0.5	4.0	F8 V		61 s		
202223	190162		21 14 56.9	-20 04 16	+0.002	-0.01	7.5	1.2	0.2	K0 III		220 s		
201867	246922		21 14 57.0	-59 18 41	+0.006	+0.03	7.7	0.9	0.2	K0 III		320 mx		
202206	190163		21 14 57.4	-20 47 21	-0.005	-0.12	7.9	1.1	3.2	G5 IV		52 s		
202281	145234		21 14 58.0	-8 21 10	+0.003	0.00	7.34	0.42		F5				
202026	230591		21 15 01.9	-49 05 50	-0.002	+0.01	7.00	1.1	0.2	K0 III		230 s		m
202173	212828		21 15 02.0	-30 59 01	+0.001	-0.06	7.4	1.5	3.2	G5 IV		26 s		
202224	190165		21 15 02.9	-21 48 55	+0.002	-0.01	7.2	1.2	0.2	K0 III		180 s		
202316	145237		21 15 03.6	-2 34 19	-0.001	+0.01	7.1	0.1	0.6	A0 V		200 s		
202307	145235		21 15 03.7	-5 33 06	-0.001	-0.03	7.1	1.1	-0.1	K2 III		280 s		
202261	164249		21 15 06.4	-17 20 42	0.000	-0.02	6.04	0.96		G5				
202615	33259		21 15 07.0	+51 24 38	+0.005	0.00	7.8	1.1	-0.1	K2 III		350 s		
202291	164251		21 15 10.0	-15 07 32	+0.002	0.00	7.7	0.1	1.7	A3 V		160 s		
202423	106963		21 15 10.7	+19 51 14	+0.001	-0.02	8.0	0.1	0.6	A0 V		300 s		
202616	33260		21 15 10.7	+51 16 53	-0.001	-0.01	7.10	1.1	0.2	K0 III	0	230 s		m
202135	230596		21 15 14.6	-40 30 23	+0.012	-0.01	6.21	1.14	0.0	K1 III		120 mx		
201933	246924		21 15 17.3	-58 55 47	+0.008	-0.07	6.9	1.3	0.2	K0 III		150 s		
202445	89565		21 15 23.9	+21 46 46	+0.001	+0.02	8.0	1.1	-0.1	K2 III		410 s		
202597	50624		21 15 25.1	+46 25 14	-0.002	+0.02	8.0	1.6	-0.5	M2 III		460 s		
202568	50623		21 15 27.1	+42 01 36	+0.006	+0.14	6.8	0.7	4.4	G0 V		30 s		
202569	71128		21 15 33.8	+37 15 10	-0.001	+0.02	7.75	-0.06	0.0	B8.5 V		360 s	14796	m
202617	50627		21 15 35.2	+44 57 15	-0.002	-0.01	7.44	-0.07		A0				
203516	10010		21 15 35.7	+78 35 54	+0.005	+0.01	7.20	0.1	0.6	A0 V		210 s	14845	m
202654	50631		21 15 36.5	+47 58 25	-0.004	-0.02	6.46	-0.15	-2.0	B4 IV	-26	220 mx		
202552	71127		21 15 37.0	+33 53 07	+0.002	+0.01	8.0	1.1	0.2	K0 III		350 s		
202320	190173	28 φ Cap	21 15 37.7	-20 39 06	+0.001	0.00	5.3	1.1	0.2	K0 III	-5	100 s		
202027	246927		21 15 40.9	-57 37 53	+0.004	-0.01	6.8	-0.3	2.6	F0 V		69 s		
203399	10007		21 15 42.0	+77 00 44	+0.003	+0.02	5.95	1.50	-0.1	K2 III	+15	100 s		
202478	106973		21 15 42.0	+16 43 39	-0.005	-0.08	6.8	1.1	0.2	K0 III		210 s		
202986	19281		21 15 42.5	+68 21 06	+0.003	-0.06	7.85	0.42	1.3	F2 III-IV		120 mx		
202369	164263	29 Cap	21 15 44.7	-15 10 18	+0.002	+0.01	5.28	1.64	-0.5	gM3	-38	140 s		
202103	246929		21 15 45.5	-53 15 48	+0.003	-0.01	5.75	0.19	0.3	A5 III		120 s		m
202463	126661		21 15 45.5	+8 46 38	0.000	0.00	8.0	0.1	0.6	A0 V		300 s		
202287	212843		21 15 46.6	-36 12 38	+0.002	0.00	6.12	1.37		K0				
202397	145246		21 15 48.8	-9 07 15	+0.002	0.00	7.5	1.1	0.2	K0 III		290 s		
202447	126662	8 α Equ	21 15 49.3	+5 14 52	+0.004	-0.08	3.92	0.53	0.6	G0 III	-16	46 s		Kitalpha
202293	212844		21 15 50.9	-35 30 18	+0.001	-0.04	7.7	2.1	-0.3	K5 III		170 s		
202522	106977		21 15 51.9	+19 42 52	+0.001	+0.01	7.4	0.1	1.4	A2 V		160 s		
202492	106975		21 15 51.9	+11 28 44	-0.002	-0.03	7.3	1.1	0.2	K0 III		270 s		
202406	145247		21 15 54.8	-9 23 28	0.000	+0.01	7.8	0.4	3.0	F2 V		92 s		
200307	258884		21 15 55.4	-83 15 55	-0.004	+0.02	7.60	1.1	0.2	K0 III		300 s		m
202573	89570		21 15 57.0	+25 26 02	-0.005	-0.03	6.99	0.89	5.2	G5 V		16 s		
202664	50644		21 15 57.5	+45 43 53	-0.003	0.00	7.83	-0.02	0.4	B9.5 V		310 s		
202407	164265		21 15 59.5	-13 12 11	+0.001	-0.07	6.5	1.1	0.2	K0 III		180 mx		
202508	106979		21 16 01.6	+10 41 09	-0.001	0.00	7.6	1.4	-0.3	K5 III		380 s		
202630	71139		21 16 06.8	+34 37 57	+0.003	0.00	7.9	1.1	0.2	K0 III		330 s		
202523	106980		21 16 07.3	+10 55 49	0.000	0.00	7.8	1.6	-0.5	M2 III		470 s		
202232	246934		21 16 08.6	-50 32 43	+0.003	-0.02	7.6	1.7	0.2	K0 III		120 s		
202358	212849		21 16 08.7	-31 48 59	+0.002	-0.09	7.4	1.2	3.2	G5 IV		40 s		
202618	89573		21 16 16.6	+26 21 02	-0.001	+0.01	7.3	0.4	3.0	F2 V		72 s		
202710	50651		21 16 17.0	+44 14 18	+0.001	+0.02	6.5	1.1	0.2	K0 III		180 s		
202585	89572		21 16 17.4	+21 54 49	-0.001	-0.01	7.7	1.1	-0.1	K2 III		360 s		
202466	145251		21 16 17.6	-9 12 53	-0.001	+0.01	7.60	1.6	-0.5	M2 III	+8	420 s		m
202195	246937		21 16 25.5	-54 51 22	+0.005	+0.02	7.5	0.8	3.2	G5 IV		70 s		
202766	50657		21 16 26.7	+48 47 33	-0.001	+0.01	7.5	0.1	1.4	A2 V		160 s		
202720	50656		21 16 29.4	+42 15 05	0.000	-0.02	6.5	1.1	-0.1	K2 III	+8	200 s		
202494	145256		21 16 29.5	-9 09 28	+0.001	+0.01	7.4	0.1	0.6	A0 V		230 s		

540

HD	SAO	Star Name	α 2000	δ 2000	μ(α)	μ(δ)	V	B−V	M$_v$	Spec	RV	d(pc)	ADS	Notes
			h m s	° ′ ″	s	″								
202575	126671		21 16 32.3	+9 23 37	+0.011	−0.12	7.95	1.12		K8				
202532	145257		21 16 34.3	−0 37 18	0.000	0.00	7.9	1.1	0.2	K0 III		340 s		
202554	145259		21 16 39.4	−1 36 28	+0.002	−0.01	6.48	0.98		K0				
202711	71150		21 16 39.5	+36 19 17	0.000	−0.02	7.4	0.0	0.4	B9.5 V		250 s		
202480	190191		21 16 39.5	−20 10 29	−0.004	−0.06	7.5	0.7	0.2	K0 III		190 mx		
202586	126673		21 16 40.0	+7 30 53	+0.002	−0.03	7.6	0.4	3.0	F2 V		85 s		
202311	246944		21 16 45.6	−49 57 59	−0.004	−0.02	7.5	1.6	0.2	K0 III		62 s		
202587	126675		21 16 45.8	+5 15 08	−0.001	0.00	7.8	0.1	0.6	A0 V		270 s		
202644	106990		21 16 54.6	+13 57 12	+0.001	+0.01	7.4	0.0	0.0	B8.5 V	−14	310 s		
202577	164275		21 16 58.0	−10 08 06	0.000	−0.02	6.9	0.1	0.6	A0 V		180 s		
202923	33281		21 17 01.8	+53 59 51	+0.003	+0.04	6.13	0.05		A0	−8			
204169	3544		21 17 08.0	+81 45 41	−0.026	+0.02	7.7	1.1	0.2	K0 III		210 mx		
201230	258886		21 17 08.5	−80 20 43	+0.009	−0.02	7.90	0.00	0.4	B9.5 V		170 mx		m
202657	126679		21 17 11.9	+3 54 44	0.000	−0.01	7.7	1.1	−0.1	K2 III		360 s		
202606	164279		21 17 13.3	−13 16 44	−0.002	−0.01	6.2	0.1	0.6	A0 V		130 s		
202987	33287		21 17 14.0	+55 47 53	+0.002	+0.02	5.98	1.45	−0.2	K3 III	−19	130 s		
202970	33285		21 17 14.6	+55 03 32	0.000	+0.02	8.0	0.1	1.4	A2 V		190 s		
202560	212866		21 17 15.1	−38 52 05	−0.279	−1.16	6.68	1.41	8.8	M0 V	+23	3.8 t		Lacaille 8760
203025	33288		21 17 18.6	+58 36 42	0.000	+0.01	6.42	0.20	−3.6	B2 III	−17	580 s	14832	m
202779	71157		21 17 21.1	+31 53 58	+0.002	0.00	7.8	0.7	4.4	G0 V		49 s		
202375	246947		21 17 22.4	−54 01 02	+0.002	−0.02	7.4	1.1	0.2	K0 III		250 s		
202811	71159		21 17 22.8	+34 25 18	+0.001	−0.01	7.3	1.1	−0.1	K2 III	+6	300 s		
202862	50671		21 17 23.0	+42 41 00	+0.001	0.00	6.1	−0.1		B8				
202850	71165	67 σ Cyg	21 17 24.7	+39 23 41	0.000	0.00	4.23	0.12	−7.1	B9 Ia	−4	1600 s		
202699	126685		21 17 25.3	+8 14 03	+0.005	+0.02	7.2	1.1	−0.1	K2 III		230 mx		
202540	212867		21 17 25.5	−30 44 46	+0.002	−0.01	6.9	1.3	0.2	K0 III		140 s		
202658	145266		21 17 25.7	−5 21 55	+0.002	+0.01	8.0	0.9	3.2	G5 IV		93 s		
203026	33290		21 17 27.8	+56 45 46	0.000	−0.01	7.53	−0.01	0.4	B9.5 V		270 s		
202769	89584		21 17 32.4	+22 32 29	0.000	−0.04	7.2	0.4	3.0	F2 V		70 s		
−−	71166		21 17 32.5	+35 19 58	−0.001	−0.01	8.0	1.6						
202780	89585		21 17 35.7	+20 50 14	+0.001	+0.01	8.0	1.4	−0.3	K5 III		450 s		
202880	71169		21 17 36.2	+39 12 36	+0.003	−0.05	6.99	0.34	3.4	F5 V		52 s		
202879	71170		21 17 36.7	+39 44 47	+0.003	+0.02	8.00	0.4	2.6	F0 V		120 s	14828	m
202686	145272		21 17 37.9	−5 46 56	+0.003	−0.04	7.5	0.5	3.4	F5 V		67 s		
203231	19292		21 17 39.1	+66 38 18	−0.001	−0.03	7.9	0.9	3.2	G5 IV		84 s		
202863	71168		21 17 39.4	+33 45 30	−0.001	−0.03	7.7	0.2	2.1	A5 V		130 s		
202781	106999		21 17 41.4	+20 00 22	+0.001	0.00	7.6	1.1	0.2	K0 III		300 s		
202799	89587		21 17 42.6	+20 30 19	+0.001	−0.01	8.0	0.1	1.4	A2 V		210 s		
203046	33295		21 17 50.8	+52 28 32	−0.002	−0.04	7.5	0.1	1.7	A3 V		140 s		
202669	164285		21 17 51.4	−15 53 28	−0.001	−0.02	7.9	0.5	3.4	F5 V		78 s		
202501	230614		21 17 52.4	−48 43 06	−0.002	−0.08	6.6	1.0	0.2	K0 III		150 mx		
203028	33294		21 17 52.5	+51 12 26	+0.001	+0.01	8.0	0.1	1.4	A2 V		200 s		
202904	71173	66 υ Cyg	21 17 54.9	+34 53 48	+0.001	0.00	4.43	−0.11	−2.5	B2 V e	+4	240 s	14831	m
202627	212874	ε Mic	21 17 56.1	−32 10 21	+0.004	−0.02	4.71	0.06		A2 p	−1	14 mn		
202815	107005		21 17 56.3	+17 37 42	−0.001	+0.01	7.8	0.1	1.4	A2 V	−1	190 s		
202671	164286	30 Cap	21 17 57.1	−17 59 07	+0.001	0.00	5.43	−0.12		B8	−11			
204337	3551		21 17 57.5	+82 01 07	−0.001	−0.01	7.4	1.1	0.2	K0 III		280 s		
202299	254966		21 18 00.2	−64 40 54	+0.002	−0.04	6.31	−0.07		A0				
199692	258883		21 18 01.1	−85 12 05	+0.060	−0.03	6.9	1.4	0.2	K0 III		130 s		
203013	50681		21 18 02.9	+46 51 26	+0.003	0.00	7.0	0.0	0.4	B9.5 V		200 s		
202752	145276		21 18 04.6	−2 08 30	+0.002	−0.01	7.9	1.1	−0.1	K2 III		390 s		
202660	190209		21 18 05.7	−28 41 38	0.000	−0.01	7.6	1.6	0.2	K0 III		140 s		
202865	89597		21 18 10.4	+22 06 50	−0.001	−0.04	7.2	0.1	1.7	A3 V		120 s		
202753	145278	15 Aqr	21 18 10.9	−4 31 10	+0.001	+0.02	5.82	−0.13		B8		31 mn		
202723	164289	31 Cap	21 18 15.4	−17 27 44	+0.002	+0.01	6.3	0.2	2.1	A5 V		70 s		
201906	257904		21 18 15.8	−75 20 48	+0.004	−0.03	6.63	0.03		A0				
202905	89599		21 18 16.5	+23 55 35	0.000	0.00	8.0	1.4	−0.3	K5 III		450 s		
202835	107009		21 18 16.7	+10 14 54	+0.010	+0.03	7.3	1.1	0.2	K0 III		130 mx		
202649	212881		21 18 17.1	−33 59 55	+0.004	−0.04	7.4	1.5	−0.1	K2 III		210 s		
203232	19294		21 18 17.4	+61 47 15	−0.008	−0.02	7.5	1.1	−0.1	K2 III		300 s		
202592	230619		21 18 23.4	−46 14 53	+0.005	−0.02	7.7	1.6	0.2	K0 III		58 s		
202065	257905		21 18 25.1	−73 01 36	0.000	0.00	6.9	1.1	0.2	K0 III		220 s		
202907	107014		21 18 25.2	+17 43 15	+0.003	0.00	7.5	1.4	−0.3	K5 III	+5	370 s		
202612	230621		21 18 25.3	−45 17 25	+0.002	0.00	7.4	0.7	0.6	A0 V		81 s		
202725	190214		21 18 25.9	−20 20 07	−0.002	−0.02	6.6	0.7	0.2	K0 III		190 s		
203135	33303		21 18 26.7	+54 10 26	−0.002	−0.01	7.40	1.1	−1.3	K3 II-III	−42	480 s		
202628	230622		21 18 26.9	−43 20 04	+0.021	+0.03	6.75	0.63	5.2	G5 V		20 s		
202818	145283		21 18 26.9	−3 40 48	+0.004	+0.02	7.9	0.4	2.6	F0 V		110 s		
203064	50690	68 Cyg	21 18 27.0	+43 56 46	0.000	−0.01	5.00	−0.01		O8	+1			
202593	230620		21 18 27.1	−47 00 07	+0.009	−0.04	7.0	1.4	−0.3	K5 III		200 mx		
204129	3547		21 18 30.4	+80 21 12	+0.060	+0.11	7.32	0.50	3.7	F6 V	+28	52 s	14916	
202926	107016		21 18 31.8	+17 59 16	−0.001	−0.03	7.3	0.5	3.4	F5 V	+9	59 s		
203265	19298		21 18 32.4	+61 11 03	0.000	−0.01	6.7	1.6	−0.5	M4 III		260 s		
202908	107015		21 18 34.5	+11 34 09	0.000	−0.04	7.20	0.7	4.4	G0 V	+6	36 s	14839	m
203280	19302	5 α Cep	21 18 34.6	+62 35 08	+0.022	+0.05	2.44	0.22	1.9	A7 IV-V	−10	14 ts	14858	Alderamin, m

HD	SAO	Star Name	α 2000	δ 2000	μ(α)	μ(δ)	V	B-V	M_V	Spec	RV	d(pc)	ADS	Notes
202884	126700		21 18 37.2	+8 57 35	+0.011	−0.03	7.4	0.5	3.4	F5 V		64 s		
202975	89604		21 18 39.2	+24 39 33	−0.001	−0.02	6.7	1.1	0.2	K0 III	+12	200 s		
203136	50698		21 18 40.4	+50 10 57	+0.002	+0.01	7.76	0.88	0.2	K0 III		330 s		
202457	254970		21 18 44.0	−61 21 10	+0.067	−0.43	6.60	0.69	5.2	G5 V		21 ts		
203137	50701		21 18 45.4	+50 04 10	+0.003	0.00	6.97	1.85	−0.3	K5 III	−9	180 s		
202772	190219		21 18 47.7	−26 36 57	+0.002	−0.04	7.80	0.5	4.0	F8 V		58 s	14834	m
203425	19310		21 18 50.9	+65 59 21	−0.006	−0.01	7.6	1.1	0.2	K0 III		270 s		
202951	107020		21 18 51.9	+11 12 12	+0.002	+0.02	5.96	1.65	−0.3	K5 III	−37	150 s		
202868	145288		21 18 52.0	−6 26 22	+0.002	−0.02	7.8	1.1	0.2	K0 III		330 s		
202773	190220		21 18 54.2	−28 45 56	−0.014	−0.05	6.40	0.97	3.2	G8 IV		39 s		
203096	50699		21 18 55.1	+41 02 27	−0.001	+0.01	6.20	0.2	2.1	A5 V	+7	66 s	14849	m
203065	71192		21 18 58.0	+32 10 50	+0.001	−0.01	7.3	1.1	−0.1	K2 III		290 s		
202887	145289		21 18 59.4	−7 19 47	0.000	0.00	7.8	0.7	4.4	G0 V		49 s		
202886	145290		21 18 59.6	−7 08 21	+0.002	−0.02	8.0	1.1	0.2	K0 III		360 s		
203112	71195		21 19 00.0	+39 44 58	+0.002	+0.01	6.67	−0.03	0.6	A0 V		160 s	14850	m
203048	89613		21 19 01.8	+29 44 35	+0.001	−0.01	7.28	−0.02	0.2	B9 V nn		250 s		
202774	212895		21 19 03.0	−32 21 24	−0.001	−0.05	7.0	1.6	−0.1	K2 III		150 s		
203374	19309		21 19 07.0	+61 51 30	−0.003	−0.01	6.68	0.30	−4.6	B0 IV pe	−7	1200 s	14868	m
203015	107024		21 19 10.9	+17 49 20	−0.002	−0.07	6.6	0.4	3.0	F2 V	+5	54 s		
203050	89614		21 19 12.4	+22 22 40	−0.002	−0.02	7.5	1.1	−0.1	K2 III		330 s		
202676	230625		21 19 12.5	−49 36 40	+0.002	+0.02	7.3	0.6	0.2	K0 III		270 s		
202890	164301		21 19 14.8	−16 10 46	+0.002	−0.03	6.9	1.1	0.2	K0 III	−36	220 s		
203338	33318		21 19 15.6	+58 37 25	0.000	0.00	5.66	1.38	−4.7	M1 Ib pe	−21	910 s	14864	m
202775	212898		21 19 18.0	−39 20 54	+0.001	−0.01	7.4	1.6	−0.3	K5 III		300 s		
203155	71202		21 19 19.6	+38 48 24	0.000	0.00	7.0	0.5	3.4	F5 V		51 s		
202929	164303		21 19 20.4	−13 02 53	+0.003	0.00	6.70	0.1	1.7	A3 V		100 s		
202729	230628		21 19 20.7	−47 03 17	+0.003	−0.03	7.10	0.8	3.2	G5 IV		60 s		m
203467	19313	6 Cep	21 19 22.1	+64 52 18	+0.001	+0.01	5.18	−0.04	−1.7	B3 V	−18	200 s		
203156	71203		21 19 22.1	+38 14 15	+0.001	0.00	5.83	0.50	3.0	F2 V	−7	30 s	14859	m
200204	258885		21 19 23.9	−84 45 25	+0.057	−0.08	7.8	0.4	4.0	F8 V		58 s		
202704	230627		21 19 24.2	−49 23 54	+0.004	0.00	7.8	0.6	0.2	K0 III		320 s		
203283	33315		21 19 24.3	+52 19 27	0.000	−0.01	7.30	−0.01		A0			14865	m
203376	33322		21 19 24.9	+58 19 35	+0.003	+0.02	7.0	1.1	−0.1	K2 III	−11	250 s		
203375	33324		21 19 26.4	+58 53 29	−0.005	−0.01	8.0	0.1	0.6	A0 V		270 s		
203245	50713		21 19 28.5	+49 30 38	+0.001	+0.01	5.76	−0.15	−0.9	B6 V	−23	220 s		
202790	230630		21 19 29.0	−41 03 10	−0.001	−0.01	7.8	0.4	2.6	F0 V		91 s		
203284	50717		21 19 35.6	+49 13 17	−0.009	−0.05	7.70	0.55	3.2	G5 IV		79 s		
203067	126707		21 19 39.2	+9 31 31	+0.002	0.00	7.20	0.1	1.4	A2 V		150 s	14856	m
203170	71208		21 19 40.6	+33 15 15	+0.004	+0.02	8.00	0.6	4.4	G0 V		52 s	14861	m
203320	33323		21 19 40.6	+53 03 29	+0.001	0.00	6.80	1.1	−0.1	K2 III		230 s		m
203378	33329		21 19 42.5	+55 26 59	+0.001	−0.01	7.20	1.6		M6 III	−27			
203377	33330		21 19 44.0	+56 12 00	+0.005	+0.05	7.9	1.1	0.2	K0 III		290 mx		
203356	33327		21 19 44.7	+53 57 05	0.000	0.00	7.7	0.0	0.4	B9.5 V		270 s		
202940	190236		21 19 45.6	−26 21 10	−0.040	−0.35	6.56	0.73	5.2	G5 V	−30	19 ts	14847	m
202990	164310		21 19 48.6	−14 01 02	+0.003	+0.02	7.0	0.1	0.6	A0 V		190 s		
202730	246964	θ Ind	21 19 51.1	−53 26 57	+0.011	−0.07	4.39	0.19	2.1	A5 V	−15	28 s		
202941	190238		21 19 51.2	−27 12 34	+0.001	−0.01	6.9	0.0	0.6	A0 V		180 s		
203139	89625		21 19 51.6	+22 43 30	+0.002	+0.02	7.8	1.1	0.2	K0 III		330 s		
203380	33334		21 19 58.6	+52 58 44	0.000	+0.06	7.68	0.57	3.7	dF6	+34	55 s	14878	m
202846	230632		21 19 58.7	−45 28 44	+0.006	−0.02	7.9	0.7	4.4	G0 V		34 s		
203140	107039		21 20 03.9	+12 57 46	+0.001	+0.01	7.30	1.6	−0.4	M0 III	−17	350 s		
202871	230634		21 20 04.7	−42 40 47	+0.013	−0.05	7.6	0.9	4.4	G0 V		27 s		
202982	--		21 20 05.3	−27 18 49			8.00	0.9	3.2	G5 IV		91 s	14852	m
203054	164315		21 20 09.0	−13 30 08	+0.001	+0.01	7.0	0.0	0.4	B9.5 V		210 s		
202874	230635		21 20 09.2	−45 01 21	0.000	−0.01	6.00	2.33		N7.7	+2			T Ind, v
203534	19323		21 20 13.8	+60 37 51	−0.001	0.00	7.50	1.4	−0.3	K5 III	−31	330 s	14892	m
203206	89628		21 20 13.9	+22 01 35	0.000	+0.01	6.20	−0.08	−1.0	B7 IV	−17	280 s		
203437	33336		21 20 13.9	+54 11 02	0.000	+0.01	7.9	0.0	0.4	B9.5 V		300 s		
202984	212917		21 20 16.1	−32 30 48	0.000	−0.03	7.7	1.2	−0.1	K2 III		360 s		
203468	33338		21 20 16.1	+56 29 43	+0.003	+0.02	7.7	1.1	0.2	K0 III		290 s		
202993	212919		21 20 18.3	−30 26 33	−0.002	−0.05	7.37	1.44	−0.2	K3 III		260 s		
203416	50731		21 20 18.8	+49 21 09	+0.002	+0.01	7.99	1.61					14885	m
203185	107042		21 20 20.0	+16 48 41	0.000	+0.08	7.5	0.5	4.0	F8 V		51 s		
203286	71220		21 20 20.4	+33 29 07	+0.001	−0.01	6.9	0.1	0.6	A0 V		180 s		
202945	230640		21 20 21.7	−40 27 55	+0.002	−0.04	7.8	0.4	3.4	F5 V		74 s		
203551	19329		21 20 25.3	+60 41 14	0.000	−0.01	6.70	0.5	3.4	F5 V	−14	45 s	14898	m
203186	107044		21 20 26.4	+12 08 30	+0.003	−0.03	7.56	0.32		F0				
203004	212922		21 20 31.8	−35 42 15	−0.002	−0.05	7.7	1.1	0.2	K0 III		280 s		
203574	19333		21 20 33.3	+60 45 25	−0.006	0.00	6.11	1.00		K0	−27			
203287	89635		21 20 35.5	+28 49 59	0.000	+0.01	7.7	0.5	4.0	F8 V		54 s		
203207	126715		21 20 37.2	+8 12 43	0.000	0.00	7.8	0.1	1.4	A2 V		190 s		
203247	107054		21 20 37.2	+19 50 31	0.000	−0.02	7.9	1.4	−0.3	K5 III		430 s		
203288	89636		21 20 38.0	+26 15 01	−0.001	−0.03	7.70	1.4	−0.3	K5 III		380 s		
202809	246972		21 20 38.6	−57 50 56	+0.001	+0.01	7.8	0.2	2.1	A5 V		140 s		
203711	19344		21 20 45.0	+67 13 15	+0.001	0.00	7.8	0.1	0.6	A0 V		240 s		

HD	SAO	Star Name	α 2000	δ 2000	μ(α)	μ(δ)	V	B-V	M_V	Spec	RV	d(pc)	ADS	Notes
			h m s	° ′ ″	s	″								
203006	230644	θ¹ Mic	21 20 45.4	−40 48 35	+0.006	0.00	4.82	0.02		A2 p	+2	21 mn		
203162	164322		21 20 46.6	−12 53 11	0.000	−0.02	8.0	1.1	0.2	K0 III		360 s		
203453	50737		21 20 46.9	+46 56 26	+0.001	0.00	7.8	1.1	0.2	K0 III		310 s		
203160	164323		21 20 49.0	−11 20 59	+0.001	−0.01	8.0	1.1	0.2	K0 III		360 s		
203358	71230		21 20 49.9	+32 27 10	+0.004	−0.03	5.68	1.08	4.4	G8 IV-V	−29	11 s	14889	m
203177	145313		21 20 49.9	−9 09 04	−0.001	−0.02	8.0	1.1	−0.1	K2 III		410 s		
202996	230643		21 20 50.7	−45 48 40	−0.009	−0.01	7.6	0.9	4.4	G0 V		27 s		
203086	212930		21 20 52.9	−30 57 20	+0.002	−0.01	7.6	1.3	0.2	K0 III		210 s		
203426	50736		21 20 53.6	+41 26 33	+0.001	−0.01	7.80	0.8	3.2	G5 IV		82 s	14896	m
203085	212932		21 20 56.6	−30 54 24	+0.001	−0.04	8.00	1.1	−0.1	K2 III		420 s		m
203454	50739		21 21 01.3	+40 20 44	−0.002	−0.20	6.40	0.53	4.0	F8 V	+1	30 ts		
203142	190252		21 21 03.0	−20 49 16	+0.002	−0.06	7.1	0.6	3.0	F2 V	−7	44 s		
203222	145317	16 Aqr	21 21 04.1	−4 33 36	−0.001	+0.01	5.87	0.92	0.3	gG7	−6	130 s		
203344	89640		21 21 04.2	+23 51 21	+0.017	−0.12	5.57	1.05	1.7	K0 III-IV	−89	52 s		
203291	126719	9 Equ	21 21 04.6	+7 21 16	+0.003	−0.01	5.82	1.66	−0.5	gM2	−20	170 s		
203482	50743		21 21 04.6	+46 21 11	−0.006	−0.01	7.8	1.1	0.2	K0 III		250 mx		
203591	33344		21 21 05.1	+56 32 30	−0.002	−0.01	7.5	0.1	0.6	A0 V		220 s		
203469	50742		21 21 09.3	+41 58 47	0.000	−0.03	8.0	0.1	0.6	A0 V		290 s		
203010	246983		21 21 16.2	−49 56 16	−0.002	−0.15	6.38	1.32	−0.2	K3 III		79 mx		
203088	230646		21 21 19.5	−41 21 18	−0.001	+0.01	8.0	2.0	−0.5	M2 III		170 s		
203535	50751		21 21 20.3	+46 43 47	+0.004	+0.01	6.7	0.1	1.4	A2 V		110 s		
203345	107061		21 21 21.4	+10 19 57	+0.003	−0.02	6.75	0.53		F7	+11		14893	m
203439	71237		21 21 21.8	+32 36 46	+0.001	−0.01	6.00	0.03	1.4	A2 V	−3	83 s		
203306	126723		21 21 21.9	+2 58 28	+0.001	0.00	7.1	1.1	0.2	K0 III		240 s		
203401	89643		21 21 22.7	+22 53 22	−0.001	−0.01	7.9	0.1	1.7	A3 V		170 s		
203021	246985		21 21 23.0	−51 45 09	−0.016	−0.08	6.8	0.5	4.4	G0 V		30 s		
203427	89645		21 21 27.3	+21 18 59	−0.002	−0.01	7.8	1.1	−0.1	K2 III		380 s		
203577	50755		21 21 29.1	+48 10 01	+0.005	+0.02	7.22	1.55	−0.5	M2 III		340 mx		
203249	164332		21 21 30.2	−11 27 30	+0.005	−0.05	8.0	0.5	4.0	F8 V		62 s		
203105	230647		21 21 30.6	−42 57 06	+0.001	−0.02	8.0	1.3	−0.1	K2 III		270 s		
203419	107065		21 21 31.7	+17 10 39	0.000	0.00	7.9	1.4	−0.3	K5 III		430 s		
203729	19347		21 21 33.4	+60 34 54	+0.003	−0.03	7.7	0.5	3.4	F5 V		70 s		
203837	19352		21 21 43.2	+64 21 53	−0.002	+0.02	7.5	1.1	−0.1	K2 III	+4	300 s		
203382	126729		21 21 44.5	+3 04 41	+0.001	0.00	7.6	0.9	3.2	G5 IV		77 s		
203578	50758		21 21 45.5	+43 08 39	−0.002	0.00	8.00	0.1	0.6	A0 V		280 s	14907	m
203131	230652		21 21 46.4	−43 24 00	0.000	+0.01	7.9	2.0	−0.5	M2 III		280 s		
203383	126730		21 21 47.4	+2 01 39	0.000	0.00	8.0	0.1	1.4	A2 V		210 s	14901	m
203242	190261		21 21 48.9	−29 09 59	+0.001	−0.03	6.6	1.6	0.2	K0 III		87 s		
202876	—		21 21 49.3	−64 24 49			7.5	2.7	−0.5	M2 III		83 s		
203405	126733		21 21 50.0	+1 21 42	0.000	−0.02	6.8	0.4	3.0	F2 V		59 s		
203311	164338		21 21 51.0	−16 16 26	+0.010	−0.18	7.2	0.7	4.4	G0 V		37 s		
203483	89649		21 21 52.1	+24 40 21	+0.001	−0.01	7.9	0.1	1.4	A2 V		200 s		
204087	10032		21 21 57.8	+72 03 33	−0.003	−0.01	7.1	0.1	0.6	A0 V		180 s		
203364	145333		21 21 58.6	−9 19 40	+0.001	−0.04	6.8	1.1	−0.1	K2 III	−52	240 s		
203472	107070		21 21 58.9	+17 12 37	−0.002	−0.03	7.7	1.1	−0.1	K2 III		360 s		
203124	246994		21 20 59.5	−50 11 00	+0.005	−0.03	7.5	0.9	0.2	K0 III		220 s		
203644	50761		21 22 00.2	+49 23 20	+0.003	+0.07	5.69	1.10	0.2	gK0	−2	110 s		
203349	164343		21 22 00.7	−15 09 33	0.000	−0.01	8.0	1.6		M5 III	+42			T Cap, v
203440	126735		21 22 01.5	+4 20 44	+0.002	−0.01	6.9	1.1	0.2	K0 III		220 s		
203125	246995		21 22 04.2	−51 52 32	+0.001	0.00	7.3	0.1	1.7	A3 V		130 s		
203504	107073	1 Peg	21 22 05.0	+19 48 16	+0.007	+0.07	4.08	1.11	0.0	K1 III	−76	63 s	14909	m
203147	246998		21 22 11.2	−51 07 30	+0.011	−0.04	7.4	0.4	3.4	F5 V		64 s		
203457	126739		21 22 12.8	+4 29 18	+0.002	0.00	7.6	0.1	1.4	A2 V		180 s		
201414	258891		21 22 13.6	−83 12 34	+0.019	−0.04	7.4	0.6	2.6	F0 V		62 s		
202046	258892		21 22 14.3	−80 06 50	+0.090	−0.10	7.60	0.8	3.2	G5 IV		71 mx		m
203387	164346	32 ι Cap	21 22 14.6	−16 50 05	+0.002	+0.01	4.28	0.90	0.3	G8 III	+12	63 s		
202731	257915		21 22 16.2	−71 34 17	+0.002	−0.01	7.1	1.6	−0.5	M2 III		340 s		
203473	126740		21 22 18.7	+5 01 25	+0.012	0.00	7.7	0.9	3.2	G5 IV		80 s		
204408	10034		21 22 21.3	+77 05 25	−0.003	−0.03	6.7	0.1	0.6	A0 V		160 s		
203061	254987		21 22 23.2	−59 51 52	+0.011	−0.06	7.80	1.1	−0.1	K2 III		200 mx		m
203611	71254		21 22 24.6	+37 09 54	−0.001	0.00	7.6	0.4	3.0	F2 V		81 s		
203458	145341		21 22 26.5	−3 07 48	−0.003	−0.01	6.7	0.4	3.0	F2 V		54 s		
203506	126741		21 22 28.2	+3 45 13	+0.002	+0.02	7.9	1.1	0.2	K0 III		340 s		
203612	71253		21 22 28.3	+33 51 40	0.000	+0.01	7.8	0.1	1.7	A3 V		160 s		
203746	50768		21 22 28.8	+49 29 29	+0.001	0.00	6.84	0.03		B9				
203973	19359		21 22 34.9	+64 13 42	+0.010	+0.08	7.8	1.1	0.2	K0 III		180 mx		
203522	126744		21 22 35.7	+2 55 07	+0.001	−0.01	6.6	0.5	4.0	F8 V	−13	33 s		
203148	247000		21 22 35.8	−55 47 01	+0.005	−0.02	7.60	0.4	3.0	F2 V		83 s		m
203486	145344		21 22 36.9	−5 37 59	0.000	−0.01	7.1	1.4	−0.3	K5 III		300 s		
203353	212952		21 22 40.4	−34 54 44	+0.001	+0.03	7.4	1.1	0.2	K0 III		250 s		
203392	190276		21 22 40.6	−29 26 07	+0.001	−0.01	7.5	0.3	3.0	F2 V		78 s		
203630	71259		21 22 41.7	+30 18 36	+0.001	+0.01	6.05	1.08	0.0	K1 III	−25	160 s		
204426	10037		21 22 41.7	+76 33 18	−0.037	+0.14	6.9	0.7	4.4	G0 V		32 s		
203819	33371		21 22 42.8	+54 13 51	−0.001	0.00	7.8	0.1		A0 p				
203257	230656		21 22 45.1	−48 27 27	0.000	−0.15	7.1	0.8	0.2	K0 III		77 mx		

HD	SAO	Star Name	α 2000	δ 2000	μ(α)	μ(δ)	V	B-V	M_V	Spec	RV	d(pc)	ADS	Notes
203696	71266		21^{h}22^{m}46^s.7	+38°38'03"	0^s.000	-0".02	6.40	0.01	1.4	A2 V	-15	100 s		
203712	50770		21 22 48.6	+40 55 58	0.000	-0.01	7.30	1.6		M7 III	-50			
203731	50772		21 22 51.1	+40 41 50	+0.001	0.00	7.52	0.15		B2 e	+5			
--	50781		21 22 53.2	+49 22 36	-0.002	-0.05	7.9	1.4						
203562	126749	10 β Equ	21 22 53.5	+6 48 40	+0.004	+0.01	5.16	0.05		A2	-11	21 mn	14920	m
203713	71267		21 22 54.5	+38 42 30	+0.002	0.00	7.3	1.1	0.2	K0 III		250 s		
203525	145351	17 Aqr	21 22 56.0	-9 19 10	-0.002	-0.02	5.99	1.54	-0.4	gM0	+18	190 s		
203475	190285		21 23 00.3	-22 40 08	+0.003	+0.01	5.60	1.63	-0.5	gM1	-7	150 s		
203277	247004		21 23 02.6	-51 56 34	-0.005	-0.05	7.80	0.8	3.2	G5 IV		83 s		m
203448	212961		21 23 06.3	-30 48 42	+0.027	+0.11	7.83	0.57	2.8	G0 IV		49 mx		
200267	258887		21 23 06.9	-85 38 20	-0.001	+0.01	7.3	2.0	-0.3	K5 III		160 s		
203631	107089		21 23 07.6	+16 29 48	+0.001	+0.01	7.6	1.4	-0.3	K5 III	-68	380 s		
203839	50783		21 23 08.2	+48 31 07	+0.012	-0.04	7.9	0.4	2.6	F0 V		110 mx		
203698	89665		21 23 09.0	+29 38 23	+0.001	-0.12	7.8	0.5	4.0	F8 V		56 s		
203581	145355		21 23 10.3	-0 46 23	0.000	-0.02	7.7	1.1	0.2	K0 III		310 s		
204089	19370		21 23 17.1	+65 60 00	0.000	+0.02	7.59	-0.04	0.4	B9.5 V		270 s		
203783	71273		21 23 17.3	+39 20 57	+0.001	-0.02	7.8	0.0	0.4	B9.5 V		290 s		
203733	89667		21 23 18.7	+29 48 49	+0.005	+0.02	8.00	1.1	0.2	K0 III		260 mx		
203632	126755		21 23 19.2	+9 27 13	-0.001	-0.02	7.8	0.1	1.4	A2 V		190 s		
203840	50784		21 23 20.5	+44 31 05	-0.002	-0.01	7.1	0.1	0.6	A0 V		190 s		
203565	164356		21 23 20.6	-13 30 48	0.000	+0.02	7.8	1.1	0.2	K0 III		330 s		
203432	230659		21 23 22.3	-41 33 08	-0.006	-0.12	7.7	1.0	3.2	G5 IV		46 s		
203784	71276		21 23 22.8	+37 24 24	+0.004	+0.03	6.6	0.5	4.0	F8 V	-27	33 s		
203633	126754		21 23 23.6	+2 45 58	+0.001	-0.01	7.5	0.2	2.1	A5 V		120 s		
204100	19371		21 23 24.8	+65 02 07	-0.002	+0.05	7.34	-0.08	0.4	B9.5 V		240 s		
202732	257916		21 23 27.2	-75 29 37	+0.012	+0.12	7.7	0.6	4.4	G0 V		46 s		
203767	71275		21 23 28.3	+31 06 58	+0.001	-0.01	7.6	1.1	-0.1	K2 III		340 s		
203463	230660		21 23 31.7	-40 15 15	+0.002	-0.03	7.5	0.6	2.6	F0 V		39 s		
203492	212965		21 23 34.1	-37 43 43	+0.004	-0.02	7.5	0.5	1.4	A2 V		92 s		
203699	107095		21 23 35.1	+14 03 01	-0.001	0.00	6.86	-0.11		B3 e	-13			
203856	71278		21 23 35.5	+40 01 09	+0.004	+0.03	7.2	0.1	0.6	A0 V		130 mx		
203335	247006		21 23 39.7	-55 32 35	+0.014	-0.13	7.5	0.8	4.0	F8 V		35 s		
204000	33383		21 23 39.9	+55 17 40	-0.002	0.00	7.90	0.1	0.6	A0 V		270 s	14945	m
203493	212968		21 23 40.5	-39 46 44	+0.004	-0.04	7.6	0.4	3.4	F5 V		57 s		
204260	19385		21 23 44.5	+69 32 07	-0.004	+0.02	7.0	0.0	0.0	B8.5 V		170 mx		
203682	126765		21 23 46.5	+1 40 53	0.000	+0.04	7.4	1.1	0.2	K0 III		280 s		
203857	71280		21 23 48.2	+37 21 05	-0.001	0.00	6.60	1.4	-0.3	K5 III	-3	230 s		m
203938	50792		21 23 49.9	+47 09 52	-0.001	-0.01	7.08	0.46		B0.5 IV			14944	m
204067	33389		21 23 50.5	+57 56 26	-0.006	-0.01	7.8	1.1	0.2	K0 III		300 s		
203667	145361		21 23 50.5	-6 35 02	-0.010	0.00	8.00	0.5	4.0	F8 V		63 s		m
203786	89675		21 23 54.5	+21 55 52	-0.001	+0.01	7.6	0.0	0.4	B9.5 V		280 s		
200526	--		21 23 55.5	-85 17 51			7.9	0.1	1.4	A2 V		200 s		
203607	190292		21 23 55.7	-25 12 09	+0.004	-0.02	6.6	1.0	0.2	K0 III		190 s		
203988	--		21 23 57.1	+50 26 28	+0.001	+0.01	7.90	1.58	-0.3	K5 III		390 s		
204001	33386		21 23 57.7	+51 39 25	-0.001	0.00	7.10	-0.11	0.4	B9.5 V		220 s		
204210	19382		21 23 58.4	+66 08 31	-0.003	-0.03	7.9	1.1	-0.1	K2 III		350 s		
203803	89678		21 23 58.7	+24 16 27	+0.010	+0.03	5.71	0.32	2.6	F0 V	-18	40 s		m
203555	230662		21 24 03.5	-40 01 48	-0.003	-0.02	7.8	0.4	3.4	F5 V		74 s		
203989	50798		21 24 05.4	+47 42 24	-0.002	0.00	8.01	0.02		A0				
204002	50799		21 24 05.9	+48 33 23	-0.001	-0.05	7.56	-0.06		A0				
--	--		21 24 06.1	+50 27 00			7.80	0.6	-4.5	G0 Ib		1800 s		
203858	89680		21 24 07.2	+25 18 44	+0.003	0.00	6.20	0.1	1.4	A2 V	-19	91 s	14943	m
203639	190293		21 24 07.7	-22 44 49	-0.002	0.00	6.4	0.9	0.2	K0 III		180 s		
203638	190295	33 Cap	21 24 09.5	-20 51 08	0.000	-0.13	5.41	1.15	-0.1	gK2	+22	130 s		
204023	50802		21 24 09.9	+48 49 11	+0.005	+0.06	7.43	0.12		A2				
203705	164364	18 Aqr	21 24 11.3	-12 52 42	+0.006	+0.01	5.49	0.29	0.6	A9 III		95 s		m
204022	--		21 24 11.3	+50 26 11	-0.001	+0.01	7.7	1.1	0.2	K0 III		300 s		
204150	19380		21 24 13.6	+60 47 50	-0.005	-0.02	7.72	0.02	-2.5	B2 V	-37	890 s		
203651	190298		21 24 14.9	-23 38 02	0.000	0.00	7.4	-0.2	2.1	A5 V		110 s		
203940	71288		21 24 16.0	+37 39 44	+0.012	0.00	7.8	0.8	3.2	G5 IV		83 s		
204211	19386		21 24 16.2	+62 59 38	-0.001	+0.04	7.2	0.1	0.6	A0 V	-11	200 s		
203133	254990		21 24 16.6	-69 44 04	+0.003	-0.02	6.41	2.82		N7.7				Y Pav, v
203706	164366		21 24 20.4	-16 03 44	-0.002	+0.01	8.0	0.1	0.6	A0 V		300 s		
203548	230665		21 24 20.6	-46 36 54	+0.004	-0.02	6.31	0.20		A2				
203652	190299		21 24 20.6	-24 25 23	+0.004	-0.04	6.7	0.1	2.6	F0 V		66 s		
--	--		21 24 22.6	-41 00 05			6.40							m
203924	71287		21 24 22.6	+30 56 11	+0.003	0.00	6.7	0.1	1.4	A2 V	-12	120 s		
203674	190304		21 24 22.7	-23 17 27	+0.002	-0.01	7.4	0.3	3.2	G5 IV		69 s		
203886	89682		21 24 22.9	+24 31 44	+0.002	0.00	6.32	1.04	0.2	K0 III	-24	160 s		
204373	19392		21 24 23.9	+69 59 34	+0.005	-0.01	7.8	0.1	0.6	A0 V		240 mx		
203842	126774		21 24 24.4	+10 10 27	+0.005	+0.03	6.35	0.47	0.7	F5 III	-33	130 s		
203585	230667	θ² Mic	21 24 24.6	-41 00 24	+0.003	0.00	5.77	-0.05		A0 p	+11			m
204624	10043		21 24 24.8	+75 58 17	+0.015	0.00	7.5	1.1	0.2	K0 III		260 s		
203958	71295		21 24 25.2	+36 59 13	+0.001	-0.01	8.0	1.1	0.2	K0 III		350 s		
204212	19387		21 24 26.8	+61 42 06	+0.010	+0.06	7.8	1.1	0.2	K0 III		190 mx		

HD	SAO	Star Name	α 2000	δ 2000	μ(α)	μ(δ)	V	B-V	M_v	Spec	RV	d(pc)	ADS	Notes
--	--		21ʰ24ᵐ29.8ˢ	+55°22′52″	ˢ	″	8.00	0.1	-3.5	B1 V e		1400 s		
204116	33401		21 24 30.1	+55 22 00	-0.002	0.00	7.94	0.49	-3.5	B1 V pe	-23	1300 s		
203959	71297		21 24 32.3	+32 57 26	+0.001	+0.02	7.8	1.4	-0.3	K5 III		390 s		
203925	89685		21 24 33.8	+26 10 28	+0.003	+0.01	5.68	0.31	2.6	F0 V	-3	40 s		
203793	145370		21 24 35.8	-6 42 55	-0.001	-0.05	7.0	0.4	3.0	F2 V		64 s		
203888	107108		21 24 39.4	+14 28 34	-0.001	-0.03	8.0	1.1	-0.1	K2 III		410 s		
203655	212975		21 24 39.6	-34 58 03	+0.001	-0.01	7.00	1.1	0.2	K0 III		230 s		m
203244	254993		21 24 40.7	-68 13 41	+0.028	+0.16	6.97	0.73	5.2	G5 V		21 s		
203826	145375		21 24 42.8	-5 34 52	+0.003	-0.02	7.4	1.1	0.2	K0 III		270 s		
203497	247014		21 24 45.9	-57 15 26	+0.007	-0.05	7.4	0.1	2.1	A5 V		80 mx		
204102	50811		21 24 47.2	+49 23 30	+0.001	-0.03	8.04	0.09		A0				
203977	89687		21 24 48.0	+25 52 50	+0.001	0.00	7.06	0.03	0.6	A0 V		190 s		
205072	3568		21 24 49.4	+80 31 29	+0.017	-0.01	5.97	0.92	0.2	K0 III	+3	140 s		
204091	50810		21 24 51.4	+46 29 19	0.000	-0.04	8.0	0.4	2.6	F0 V		120 s		
203843	145376	20 Aqr	21 24 51.5	-3 23 54	-0.001	-0.05	6.36	0.33	0.6	gA9	-23	130 s		
204427	19398		21 24 52.6	+69 47 39	+0.016	+0.04	7.8	0.7	4.4	G0 V		48 s		
204131	50817		21 24 55.2	+49 19 24	-0.002	0.00	6.58	-0.03		A0	+1		14962	m
203844	145377		21 24 56.4	-8 10 45	+0.007	+0.12	8.00	0.6	4.5	G1 V	-18	50 s		
204007	89692		21 24 59.6	+27 25 06	0.000	+0.01	6.9	0.0	0.0	B8.5 V		230 s		
204231	33415		21 25 03.1	+58 05 06	+0.015	+0.11	7.1	0.5	4.0	F8 V	-41	40 mx		
202919	257917		21 25 04.2	-76 31 33	+0.013	-0.02	7.7	0.4	3.0	F2 V		87 s		
203944	126783		21 25 05.6	+9 23 03	+0.002	+0.02	7.8	0.1	1.4	A2 V		190 s	14954	m
203991	107115		21 25 09.2	+18 27 45	0.000	0.00	7.50	0.1	0.6	A0 V	-2	240 s		
203875	145382	19 Aqr	21 25 12.9	-9 44 55	+0.001	-0.17	5.70	0.20	1.4	A2 V n		21 mx		
203754	212982		21 25 15.5	-34 51 26	+0.001	-0.09	7.0	-0.1	3.4	F5 V		52 s		
204039	89695		21 25 15.9	+24 29 29	0.000	-0.01	8.0	0.1	1.4	A2 V		210 s		
204521	10045		21 25 16.7	+70 28 39	+0.009	+0.04	7.2	0.9	3.2	G5 IV		63 s		
203926	145384	21 Aqr	21 25 16.7	-3 33 25	-0.001	-0.07	5.49	1.46	-0.3	gK4	-25	130 s		
203212	257920		21 25 18.0	-71 47 58	+0.006	-0.01	6.09	1.26		K0				m
204153	50824		21 25 19.4	+46 42 51	+0.019	+0.05	5.60	0.32		F0	+1	28 t		
204051	89697		21 25 19.7	+29 27 43	0.000	+0.01	7.70	0.1	0.6	A0 V		260 s		m
204093	71315		21 25 20.7	+34 44 39	+0.004	+0.02	8.0	0.5	3.4	F5 V		83 s		
204132	50822		21 25 21.6	+42 47 04	+0.002	+0.01	7.5	0.4	2.6	F0 V		95 s		
204194	33417		21 25 22.9	+51 07 57	-0.002	-0.02	7.8	0.1	1.4	A2 V		180 s		
203725	230676		21 25 23.3	-45 09 42	+0.001	-0.01	7.2	0.2	0.6	A0 V		170 s		
203213	257921		21 25 24.2	-72 35 45	-0.002	-0.04	6.4	1.1	0.2	K0 III		170 s		
203893	164378		21 25 24.8	-14 16 37	+0.001	+0.02	6.9	0.1	0.6	A0 V		180 s		
203755	230678		21 25 28.1	-41 42 06	+0.002	+0.02	7.5	1.5	-0.5	M4 III		400 s		
203993	126787		21 25 30.6	+2 02 34	+0.001	-0.02	7.53	0.01		A0			14960	m
204025	107128		21 25 32.2	+11 39 03	+0.006	+0.02	8.0	0.4	3.0	F2 V		98 s		
204321	33422		21 25 33.4	+58 14 04	-0.001	-0.03	7.80	0.1	0.6	A0 V		260 s	14974	m
203978	145391		21 25 34.8	-0 13 54	+0.001	0.00	8.0	1.4	-0.3	K5 III		450 s		
203758	230679		21 25 42.0	-46 43 21	+0.003	-0.03	7.3	0.5	1.7	A3 V		72 s		
203929	164380		21 25 43.0	-18 35 15	+0.004	-0.03	8.0	1.1	0.2	K0 III		290 mx		
204070	107129		21 25 46.3	+15 37 05	-0.001	-0.01	6.9	1.1	-0.1	K2 III		250 s		
204291	33423		21 25 46.9	+52 51 43	+0.002	+0.01	7.5	1.4	-0.3	K5 III		330 s		
204172	71329	69 Cyg	21 25 46.9	+36 40 03	0.000	0.00	5.94	-0.08	-5.8	B0 Ib	+3	1900 s	14969	m
204154	71326		21 25 47.8	+33 28 56	0.000	-0.01	8.00	0.00	0.4	B9.5 V		320 s		
203931	190329		21 25 48.4	-25 14 18	+0.006	0.00	7.6	0.9	0.2	K0 III		230 mx		
203913	190330		21 25 48.4	-23 49 19	+0.002	-0.01	6.6	0.8	0.2	K0 III		190 s		
203849	212992		21 25 49.4	-35 50 16	-0.003	-0.05	7.7	1.2	0.2	K0 III		250 s		
204041	126789		21 25 51.3	+0 32 03	+0.003	+0.02	6.46	0.16		A0	-9			
203397	--		21 26 00.6	-70 43 30			7.8	1.1	-0.1	K2 III		380 s		
204262	50840		21 26 01.2	+44 23 42	0.000	-0.01	8.0	0.1	0.6	A0 V		280 s		
204410	33428		21 26 06.3	+58 42 15	+0.001	-0.03	7.9	1.1	0.2	K0 III		310 s		
203968	190331		21 26 07.0	-24 29 04	-0.004	-0.02	7.1	1.1	3.2	G5 IV		41 s		
204133	107136		21 26 07.3	+18 03 56	+0.001	+0.02	6.87	0.00		A0				
--	10054		21 26 15.0	+79 48 18	+0.001	+0.04	7.9	2.1						
203760	247031	γ Ind	21 26 15.3	-54 39 38	+0.001	+0.04	6.12	0.34	0.6	F0 III		130 s		
204244	71337		21 26 17.4	+35 50 21	-0.001	-0.04	7.7	1.6	-0.5	M2 III		420 s		
204134	126792		21 26 17.9	+10 08 43	+0.001	+0.04	7.7	0.9	3.2	G5 IV		79 s		
204275	50845		21 26 18.8	+41 22 17	-0.003	-0.02	8.0	0.1	1.7	A3 V		180 s		
204029	190337		21 26 21.8	-20 59 56	+0.002	0.00	7.2	0.9	0.2	K0 III		260 s		
203949	212998		21 26 22.7	-37 49 46	+0.015	-0.01	5.63	1.19	-0.2	K3 III		100 mx		
204073	164388		21 26 24.9	-12 05 45	+0.002	-0.21	6.6	1.1	0.2	K0 III		58 mx		
204188	107138		21 26 26.5	+19 22 32	+0.006	+0.02	6.1	0.1	1.7	A3 V	-11	75 s		
203608	254999	γ Pav	21 26 26.6	-65 21 59	+0.015	+0.80	4.22	0.49	4.5	F6 V	-30	8.6 t		
204121	126794		21 26 27.9	+1 06 13	+0.007	-0.15	6.13	0.44	3.4	dF5	+11	24 ts		
204374	50854		21 26 28.0	+49 16 49	-0.002	0.00	7.96	-0.11	0.6	A0 V		300 s	14987	m
203969	212999		21 26 28.1	-36 26 36	+0.003	-0.05	7.8	1.3	-0.1	K2 III		200 mx		
204401	33431		21 26 28.9	+52 44 52	+0.001	-0.01	7.80	0.00	0.4	B9.5 V		280 s	14989	m
204263	71340		21 26 31.6	+33 35 07	+0.002	-0.09	8.0	0.4	2.6	F0 V		120 s		
--	--		21 26 35.2	-46 03 59			7.00							m
204105	145399		21 26 35.3	-6 00 04	+0.001	0.00	8.0	0.9	3.2	G5 IV		93 s		
203915	230688		21 26 35.5	-44 50 48	-0.008	-0.03	7.88	0.45		F8				

HD	SAO	Star Name	α 2000	δ 2000	μ(α)	μ(δ)	V	B-V	M$_v$	Spec	RV	d(pc)	ADS	Notes
			h m s	° ′ ″	s	″								
204323	50851		21 26 36.7	+40 44 19	+0.003	-0.01	7.8	1.1	0.2	K0 III		330 s		
203934	230687		21 26 37.1	-46 04 02	+0.003	-0.11	7.00	0.5	4.0	F8 V		40 s		m
204233	89709		21 26 37.6	+20 42 45	0.000	-0.03	7.8	0.1	1.4	A2 V		190 s		
204075	190341	34 ζ Cap	21 26 39.9	-22 24 41	0.000	+0.03	3.74	1.00	-4.5	G4 Ib pe		450 s	14971	m
204215	107139		21 26 41.0	+13 41 17	0.000	-0.01	7.30	0.1	1.4	A2 V		150 s	14977	m
204030	213002		21 26 41.4	-30 29 23	+0.002	-0.06	7.6	1.9	-0.1	K2 III		130 s		
204044	190340		21 26 42.6	-29 00 46	-0.001	-0.06	7.6	1.6	-0.1	K2 III		180 mx		
--	--		21 26 44.0	-29 51 02			7.80							S Mic, v
204428	33434		21 26 44.8	+52 53 55	+0.001	+0.01	5.9	-0.1		B8		23 mn		
204375	50859		21 26 46.3	+43 48 44	-0.002	-0.01	7.3	0.0	0.4	B9.5 V		230 s		
204245	89711		21 26 50.2	+22 09 50	0.000	0.00	8.0	1.1	-0.1	K2 III		410 s		
204411	50867		21 26 51.5	+48 50 06	+0.006	+0.03	5.31	0.07		A p	-13	30 mn		
--	19414		21 26 54.2	+64 24 38	-0.005	+0.02	7.8	2.0						
204234	107140		21 26 54.4	+10 40 04	+0.003	-0.01	7.1	0.4	3.0	F2 V		66 s		
204598	19413		21 26 55.2	+63 35 08	-0.009	-0.04	7.6	0.1	0.6	A0 V		170 mx		
204412	50869		21 26 59.1	+44 53 01	-0.002	-0.02	7.9	1.1	0.2	K0 III		320 s		
204641	19417		21 26 59.7	+65 07 11	-0.002	-0.01	8.0	1.1	-0.1	K2 III		370 s		
204402	50868		21 27 00.6	+43 13 52	-0.004	0.00	7.8	0.1	0.6	A0 V		150 mx		
203985	230691		21 27 01.1	-44 48 31	+0.023	+0.18	7.48	0.90	5.9	K0 V		18 ts		
204018	230692		21 27 01.4	-42 32 52	-0.005	+0.02	5.51	0.39		A m	+18			m
204125	164395		21 27 03.1	-19 03 28	+0.004	-0.01	7.8	1.2	-0.1	K2 III		270 mx		
204277	107142		21 27 06.5	+16 07 29	-0.005	-0.07	6.8	0.5	4.0	F8 V	+14	37 s		
204376	71355		21 27 08.3	+36 55 33	-0.002	-0.01	8.0	1.1	0.2	K0 III		350 s		
204247	126802		21 27 10.4	+6 54 40	-0.001	0.00	7.7	1.1	0.2	K0 III		310 s		
204139	190349	35 Cap	21 27 14.6	-21 11 46	-0.002	-0.03	5.78	1.44	-0.3	K5 III	+23	160 s		
204198	145406		21 27 15.8	-8 47 24	0.000	-0.05	7.5	0.2	2.1	A5 V		120 s		
204082	213010		21 27 15.9	-34 54 13	+0.004	-0.06	7.7	1.6	-0.1	K2 III		200 mx		
204265	126805		21 27 16.7	+5 23 31	-0.003	-0.02	6.8	0.4	2.6	F0 V		68 s		
204612	19419		21 27 17.5	+62 06 02	-0.005	-0.02	7.5	1.1	0.2	K0 III		270 s		
204403	71358	70 Cyg	21 27 21.2	+37 07 01	0.000	+0.01	5.31	-0.14	-2.3	B3 IV	-20	310 s		
204236	145409		21 27 23.0	-7 00 58	0.000	-0.10	7.6	0.5	4.0	F8 V		52 s		
204220	164400		21 27 23.2	-13 35 15	+0.002	+0.01	6.8	0.0	0.4	B9.5 V		190 s		
204161	190353		21 27 25.1	-23 37 01	+0.004	-0.04	7.9	1.2	-0.1	K2 III		290 mx		
204599	33443		21 27 25.1	+59 45 00	-0.002	-0.01	6.10	1.75	-0.5	M2 III	-16	170 s	14998	m
204413	71362		21 27 27.8	+36 15 13	+0.001	-0.04	7.8	1.1	0.2	K0 III		320 s		
--	--		21 27 28.2	+27 53 05			8.00	1.4	-0.3	K5 III		440 s		
204203	190356		21 27 28.9	-20 12 46	-0.004	-0.08	7.5	1.0	0.2	K0 III		150 mx		
204140	213015		21 27 29.2	-30 50 46	0.000	+0.01	7.6	1.9	-0.1	K2 III		130 s		
204497	50883		21 27 29.3	+47 48 08	-0.002	-0.03	7.52	0.03		A0				
204388	89719		21 27 30.3	+27 52 27	+0.002	+0.01	8.0	1.4	-0.3	K5 III		430 s		
204482	50882		21 27 30.7	+46 25 00	+0.001	+0.01	7.7	1.1	0.2	K0 III		290 s		
204178	190354		21 27 32.9	-27 43 44	+0.004	-0.07	7.5	1.0	0.2	K0 III		160 mx		
204142	213017		21 27 33.4	-32 52 31	+0.005	-0.06	7.9	0.9	0.2	K0 III		180 mx		
204268	164405		21 27 38.3	-11 40 00	+0.002	-0.06	6.9	0.5	3.4	F5 V		49 s		
204179	213021		21 27 39.9	-31 47 47	+0.001	-0.01	7.6	0.9	3.0	F2 V		38 s		
204414	89720	35 Vul	21 27 39.9	+27 36 31	+0.003	+0.02	5.41	0.04	1.2	A1 V	-8	68 s		
204751	19426		21 27 40.0	+66 39 31	+0.001	+0.01	8.0	0.4	3.0	F2 V		95 s		
204752	19427		21 27 43.1	+66 04 29	-0.002	+0.01	8.0	1.6	-0.5	M2 III		440 s		
204280	145411		21 27 44.9	-8 21 36	0.000	-0.01	7.7	1.6	-0.5	M2 III		440 s		
204441	71363		21 27 44.9	+31 47 42	0.000	-0.01	7.9	1.1	0.2	K0 III		330 s		
204164	213019		21 27 45.2	-38 16 15	+0.004	-0.02	8.0	1.5	0.2	K0 III		190 s		
204770	19432	7 Cep	21 27 45.9	+66 48 33	-0.003	-0.01	5.44	-0.11	-0.6	B7 V	+3	160 s		
203970	247035		21 27 47.2	-59 21 50	+0.007	-0.05	7.9	0.2	3.0	F2 V		97 s		
204415	89721		21 27 47.9	+21 44 09	+0.002	+0.03	7.1	0.9	3.2	G5 IV	-17	59 s		
204720	19425		21 27 50.0	+64 00 48	0.000	-0.01	7.9	0.5	3.4	F5 V		76 s		
204484	71368		21 27 53.4	+35 24 32	0.000	0.00	7.1	0.1	1.7	A3 V		120 s		
204536	50890		21 27 54.2	+46 33 42	0.000	+0.01	6.87	-0.07	-1.0	B5.5 V	-15	360 s		
204614	33446		21 27 59.7	+53 17 21	+0.003	-0.01	7.18	0.01	0.6	A0 V		210 s		
203835	255002		21 27 59.8	-66 47 38	+0.002	-0.01	7.6	1.3	0.2	K0 III		190 s		
204442	89725		21 28 00.6	+24 36 49	0.000	-0.01	7.8	1.4	-0.3	K5 III		410 s		
204378	126812		21 28 02.4	+3 10 49	+0.003	-0.04	7.5	0.5	3.4	F5 V		65 s		
204721	19433		21 28 05.4	+62 05 48	+0.002	0.00	6.8	1.1	0.2	K0 III		200 s		
204485	71371		21 28 08.0	+32 13 31	+0.010	+0.08	5.80	0.32	2.6	F0 V	-24	43 s		
204110	247038		21 28 10.1	-52 18 08	+0.003	+0.01	6.6	1.1	0.2	K0 III		180 mx		
204363	164415		21 28 13.8	-11 34 07	0.000	-0.05	6.5	0.5	3.4	F5 V		42 s		
204600	50904		21 28 14.8	+47 57 13	+0.002	-0.01	8.04	-0.08	0.4	B9.5 V		340 s		
204391	145420		21 28 18.7	-2 36 45	+0.003	-0.06	7.8	1.1	-0.1	K2 III		220 mx		
204301	190363		21 28 19.3	-26 32 14	+0.003	+0.01	7.6	1.8	0.2	K0 III		100 s		
204404	145421		21 28 21.6	-2 53 07	+0.001	-0.01	6.8	1.1	0.2	K0 III		210 s		
204626	50910		21 28 24.3	+48 26 03	0.000	-0.03	7.5	0.1	1.4	A2 V		160 s		
204709	33452		21 28 24.4	+56 23 28	-0.003	-0.01	7.9	1.1	0.2	K0 III		310 s		
204445	126818		21 28 24.7	+8 11 45	+0.001	-0.02	6.40	1.64		M2	-6			
204523	71373		21 28 27.5	+30 13 38	-0.002	-0.01	7.7	1.1	-0.1	K2 III		350 s		
205234	10061		21 28 28.0	+76 24 13	-0.001	-0.01	7.7	0.4	2.6	F0 V	-4	100 s		EI Cep, v
204673	--		21 28 31.3	+50 26 42	-0.002	-0.03	7.37	-0.01	0.6	A0 V		230 s		

HD	SAO	Star Name	α 2000	δ 2000	μ(α)	μ(δ)	V	B-V	M_v	Spec	RV	d(pc)	ADS	Notes
--	50913		21ʰ28ᵐ34.ˢ3	+47°05′23″	-0.ˢ001	-0.″01	8.0	1.4		K2 III		240 s		
--	--		21 28 39.3	+25 57 23			6.80	1.1	-0.1	K2 III		240 s		
205021	10057	8 β Cep	21 28 39.4	+70 33 39	+0.002	+0.01	3.23	-0.22	-3.6	B2 III	-8	230 s	15032	Alfirk, m,v
204539	89734		21 28 41.0	+26 24 42	+0.003	0.00	7.58	1.24	-0.2	K3 III		360 s		
204540	89735		21 28 42.2	+25 55 16	+0.002	-0.02	6.55	1.32	-0.1	K2 III		170 s		
204541	89733		21 28 42.2	+24 40 12	+0.003	+0.01	7.9	0.1	0.6	A0 V		280 s		
204381	190374	36 Cap	21 28 43.2	-21 48 26	+0.010	0.00	4.51	0.91	0.3	gG5	-22	67 s		
204559	89736		21 28 43.2	+26 34 12	+0.004	+0.09	7.7	0.5	4.0	F8 V		54 s		
203881	255003		21 28 44.8	-69 30 20	+0.017	-0.05	5.34	1.55	-0.5	M4 III	+43	140 mx		SX Pav, q
205395	10063		21 28 47.1	+77 56 09	-0.002	0.00	7.30	1.1	0.2	K0 III		240 s	15054	m
204699	50922		21 28 48.1	+49 47 43	-0.004	+0.01	6.79	0.29	1.7	A3 V		84 s		
203203	258900		21 28 50.8	-79 55 32	+0.009	-0.01	7.4	1.3	0.2	K0 III		180 s		
204754	33458		21 28 52.5	+55 25 07	+0.001	+0.01	6.0	0.1		B9				
204509	107165		21 28 52.7	+11 05 06	+0.005	+0.03	7.50	0.4	3.0	F2 V		79 s	15007	m
204447	145426		21 28 52.7	-8 59 27	+0.002	-0.04	7.3	1.1	0.2	K0 III		260 s		
204334	213032		21 28 53.6	-37 24 13	+0.004	-0.01	7.5	1.3	-0.1	K2 III		290 s		
204827	33461		21 28 57.4	+58 44 24	-0.002	0.00	7.95	0.80	-4.1	B0 V		610 s		
204585	89737		21 28 59.7	+22 10 46	+0.003	+0.01	5.93	1.53		Mb	-22			m
204560	107168		21 28 59.7	+17 54 21	-0.001	0.00	6.44	1.39		K5	-12			
204228	247043		21 29 00.1	-53 42 21	+0.010	-0.02	6.4	1.1	-0.1	K2 III		200 s		
--	71382		21 29 00.8	+37 34 45	0.000	+0.01	7.9	1.5						
204889	19447		21 29 02.9	+61 25 59	-0.004	0.00	7.1	0.5	3.4	F5 V	-13	55 s		
204394	213034	5 PsA	21 29 03.6	-31 14 19	-0.002	-0.01	6.50	0.03		A0				
204674	50923		21 29 04.4	+40 18 18	+0.003	+0.05	7.9	0.1	1.4	A2 V		190 s		
204710	50925		21 29 06.2	+44 55 23	0.000	0.00	6.95	0.26						m
204571	107170		21 29 11.8	+11 57 10	+0.004	-0.06	7.4	1.1	-0.1	K2 III		170 mx		
204287	247045		21 29 14.1	-50 19 03	+0.005	-0.04	7.4	0.3	4.4	G0 V		40 s		
204722	50930		21 29 14.6	+44 20 17	-0.001	0.00	7.67	-0.02	-2.5	B2 V pne	0	880 s		
204642	89742		21 29 15.3	+28 35 01	-0.001	-0.12	6.74	1.10	-0.1	K2 III	+19	150 mx		
204675	71386		21 29 15.9	+34 58 22	0.000	0.00	7.9	0.1	1.4	A2 V		190 s		
--	--		21 29 18.5	+44 20 16			7.70	0.1	-2.5	B2 V nne		890 s		
205022	19451		21 29 20.3	+67 03 03	+0.003	+0.01	6.90	0.1	1.4	A2 V		120 s	15040	m
204771	50934	71 Cyg	21 29 26.8	+46 32 26	+0.004	+0.11	5.24	0.97	0.2	K0 III	-19	100 s		
204453	213038		21 29 27.3	-31 59 41	+0.001	-0.01	7.9	1.9	-0.1	K2 III		150 s		
204421	213036		21 29 30.7	-38 05 06	+0.003	0.00	7.5	0.8	3.2	G5 IV		70 s		
204603	126834		21 29 34.3	+6 34 52	0.000	-0.01	6.6	1.1	0.2	K0 III		190 s		
204812	50935		21 29 36.7	+46 25 38	+0.002	+0.03	6.8	0.1	1.7	A3 V		100 s		
204602	126836		21 29 37.3	+6 52 46	+0.002	-0.02	7.8	1.1	-0.1	K2 III		380 s		
204813	50936		21 29 39.0	+46 07 56	+0.004	+0.02	7.7	0.4	2.6	F0 V		100 s		
204454	213040		21 29 40.1	-38 20 47	+0.008	-0.01	7.8	1.0	2.6	F0 V		39 s		
204548	164430		21 29 40.1	-14 01 38	+0.002	-0.07	7.1	1.1	-0.1	K2 III		170 mx		
204370	247049		21 29 42.3	-50 20 34	+0.004	-0.03	7.6	0.1	1.4	A2 V		180 s		
204964	19452		21 29 46.0	+60 22 43	-0.003	0.00	7.47	0.17	0.0	B8.5 V	-19	250 s		
204814	50940		21 29 46.6	+45 53 41	+0.040	+0.36	7.90	0.76	5.6	G8 V	-84	29 s		
204475	213044		21 29 49.7	-37 41 33	-0.003	-0.07	7.1	1.0	4.4	G0 V		19 s		
204437	230713		21 29 51.4	-43 30 26	+0.007	-0.04	7.1	1.0	3.2	G5 IV		44 s		
204905	33468		21 29 52.8	+52 56 03	+0.001	+0.02	7.20	0.1	0.6	A0 V	-14	200 s	15035	m
204772	71396		21 29 53.1	+36 10 42	+0.004	+0.04	7.6	1.1	0.2	K0 III		290 s		
204700	89750		21 29 53.7	+20 56 55	+0.001	-0.01	8.0	1.4	-0.3	K5 III		450 s		
204724	89752	2 Peg	21 29 56.8	+23 38 20	+0.002	+0.01	4.57	1.62	-0.5	M1 III	-19	97 s	15027	m
204756	71397		21 29 57.5	+32 48 37	0.000	+0.01	7.6	0.0	0.0	B8.5 V		320 s		
204916	33471		21 29 58.4	+54 39 10	+0.002	+0.03	8.0	1.1	-0.1	K2 III		370 s		
204577	164433		21 29 59.4	-19 08 52	+0.002	-0.04	6.6	0.3	3.0	F2 V	-12	51 s		
204829	50941		21 29 59.6	+41 30 37	-0.002	-0.03	7.3	0.1	0.6	A0 V		210 s		
204860	50942		21 30 00.9	+45 29 40	+0.002	+0.01	6.90	-0.04	-1.0	B5.5 V	-1	350 s		
204352	247051		21 30 04.2	-56 16 39	+0.004	+0.01	8.0	0.7	1.7	A3 V		72 s		
204396	247052		21 30 04.9	-52 08 06	+0.006	-0.03	7.9	0.0	1.4	A2 V		110 mx		
204815	71402		21 30 05.8	+35 28 20	-0.001	-0.01	7.2	0.1	0.6	A0 V		210 s		
204725	107181		21 30 06.7	+18 34 44	-0.001	0.00	7.2	0.2	2.1	A5 V		110 s		
204712	107179		21 30 08.4	+12 16 15	-0.006	-0.13	7.7	0.6	4.4	G0 V	-23	47 s		
204476	230716		21 30 11.9	-44 04 33	-0.001	-0.04	7.6	1.0	3.2	G5 IV		56 s		
204861	50945		21 30 14.1	+42 02 42	0.000	-0.04	7.6	1.1	-0.1	K2 III		340 s		
204677	145443		21 30 15.1	-4 30 07	+0.004	+0.01	7.7	0.1	1.7	A3 V		160 s		
204424	247053		21 30 15.2	-53 08 58	-0.007	-0.04	7.9	0.9	3.4	F5 V		41 s		
204702	126847		21 30 18.3	+2 47 36	+0.003	+0.01	7.8	0.4	2.6	F0 V		110 s		
204917	50948		21 30 18.7	+48 23 28	-0.001	-0.01	7.4	0.1	0.6	A0 V		220 s		
204965	33477		21 30 20.2	+52 57 29	+0.002	+0.02	6.02	0.08	0.6	A0 V	-17	100 s		
204353	255009		21 30 24.9	-60 12 26	+0.006	-0.05	7.60	0.8	3.2	G5 IV		76 s		m
204918	50949		21 30 25.7	+44 52 30	+0.001	0.00	6.8	0.2	2.1	A5 V		85 s		
204609	190394		21 30 25.8	-25 11 36	-0.001	-0.02	7.31	1.54		K5				
204919	50950		21 30 34.8	+42 12 40	-0.001	0.00	7.90	0.1	1.4	A2 V		200 s	15045	m
205235	19467		21 30 35.7	+68 17 36	-0.002	+0.01	7.1	0.0	0.4	B9.5 V		200 s		
205100	19460		21 30 35.8	+61 53 03	+0.001	+0.01	7.9	0.0	0.4	B9.5 V		290 s		
204692	164444		21 30 40.2	-14 17 28	-0.001	-0.03	6.8	1.1	0.2	K0 III	+3	210 s		
--	19463		21 30 40.5	+61 50 14	-0.001	0.00	7.8	0.5						

HD	SAO	Star Name	α 2000	δ 2000	μ(α)	μ(δ)	V	B-V	M$_v$	Spec	RV	d(pc)	ADS	Notes
			h m s	° ' "	s	"								
204592	213054		21 30 42.2	−38 38 16	−0.001	−0.03	7.60	0.5	3.4	F5 V		69 s		m
204668	190396		21 30 44.1	−25 42 36	0.000	+0.01	7.3	1.4	−0.1	K2 III		230 s		
203971	257930		21 30 45.3	−75 12 11	+0.005	−0.01	7.9	0.8	3.2	G5 IV		86 s		
204994	50960		21 30 46.7	+47 40 11	+0.001	−0.02	7.9	1.6	−0.5	M2 III		440 s		
204385	255011		21 30 47.8	−62 10 06	+0.023	−0.10	7.3	0.1	4.4	G0 V		38 s		
205050	33488		21 30 48.2	+55 24 53	+0.002	+0.05	7.6	1.1	−0.1	K2 III		320 s		
204920	71417		21 30 48.4	+37 58 01	0.000	+0.03	7.6	0.1	1.4	A2 V		170 s		
204890	71415		21 30 48.6	+31 57 28	+0.003	+0.03	7.02	0.29		A5				
204818	126856		21 30 53.4	+10 09 24	−0.001	−0.01	7.80	0.92	0.2	K0 III		330 s		
204843	107191		21 30 56.8	+16 34 16	−0.002	−0.03	7.9	0.2	2.1	A5 V		150 s		
205372	10069		21 30 58.9	+70 49 24	+0.002	0.00	7.0	0.1	0.6	A0 V		170 s		
205139	19466		21 30 59.1	+60 27 34	−0.001	0.00	5.53	0.12	−5.1	B1 II	−15	870 s		
205113	33495		21 31 01.3	+59 25 06	+0.019	+0.05	6.8	1.1	0.2	K0 III		130 mx		
204844	107192		21 31 01.8	+11 22 26	0.000	−0.02	7.8	1.6	−0.5	M2 III		470 s		
204923	89769		21 31 07.5	+26 02 48	+0.005	0.00	8.03	1.37	−0.1	K3 III		300 mx		
204921	71419		21 31 08.1	+30 16 45	+0.001	+0.01	7.52	1.32	−0.1	K2 III		270 s		
204862	107195		21 31 09.5	+12 08 15	+0.001	−0.01	5.90	−0.05	0.2	B9 V	−10	140 s		
204478	247057		21 31 09.8	−59 24 25	−0.004	0.00	7.8	1.7	−0.1	K2 III		190 s		
−−	71422		21 31 11.3	+38 45 31	−0.002	−0.01	7.5	1.6						
204978	71423		21 31 14.8	+35 52 25	−0.001	−0.01	7.6	1.1	0.2	K0 III		290 s		
205084	33496		21 31 17.9	+52 11 44	+0.001	+0.06	6.69	0.98	0.2	K0 III		200 s		
204864	126862		21 31 18.8	+3 49 06	+0.006	−0.01	7.0	0.1	1.7	A3 V		120 mx		
204935	89774		21 31 22.1	+22 44 30	−0.001	0.00	7.7	1.1	0.2	K0 III		310 s		
203153	258901		21 31 22.2	−83 02 02	−0.004	+0.02	7.1	−0.1	1.4	A2 V		140 s		
205073	50974		21 31 23.1	+48 21 26	+0.001	+0.02	7.7	0.1	0.6	A0 V		240 s		
205024	71424		21 31 23.2	+40 03 56	0.000	−0.01	7.40	0.5	3.4	F5 V		63 s	15057	m
204779	164449		21 31 25.3	−19 14 16	+0.004	−0.01	7.30	0.4	2.6	F0 V		87 s	15046	m
205114	33497		21 31 27.3	+52 37 12	+0.001	+0.01	6.16	0.90	−4.5	G2 Ib	−23	1400 s		
204800	190404		21 31 32.0	−22 00 40	−0.001	−0.01	7.8	0.3	0.2	K0 III		330 s		
204867	145457	22 β Aqr	21 31 33.3	−5 34 16	+0.001	−0.01	2.91	0.83	−4.5	G0 Ib	+7	300 s	15050	Sadalsuud, m
204848	145455		21 31 34.0	−9 44 42	−0.010	−0.13	7.4	0.6	4.4	G0 V		40 s		
205060	50977		21 31 36.7	+42 42 03	−0.001	+0.01	7.21	−0.01		B6 e	−7			
205196	33506		21 31 38.2	+57 30 10	−0.001	0.00	7.43	0.58	−5.8	B0 Ib	−14	1500 s		
204728	213067		21 31 38.9	−35 04 40	−0.001	−0.03	7.8	0.8	3.0	F2 V		46 s		
204781	190409		21 31 40.0	−25 18 04	+0.005	−0.01	7.9	0.6	3.4	F5 V		67 s		
205051	71428		21 31 41.2	+39 07 38	+0.001	−0.01	7.9	0.1	1.7	A3 V		170 s		
205116	50984		21 31 42.2	+48 35 05	−0.001	−0.01	6.7	0.0	0.4	B9.5 V		180 s		
205509	10075		21 31 42.5	+72 02 19	+0.002	+0.05	7.8	0.5	3.4	F5 V		72 s		
204925	126867		21 31 42.9	+6 45 00	+0.001	−0.01	7.9	0.9	3.2	G5 IV		88 s		
205086	50983		21 31 47.9	+41 44 38	+0.001	0.00	7.8	0.0	0.4	B9.5 V		290 s		
204997	89780		21 31 48.5	+20 59 25	0.000	−0.01	8.0	1.4	−0.3	K5 III		460 s		
204802	213072		21 31 49.3	−31 32 55	+0.001	−0.04	7.9	0.4	3.0	F2 V		87 s		
205011	89782		21 31 50.0	+23 50 43	+0.001	+0.01	6.5	1.1	0.2	K0 III		180 s		
204822	190412		21 31 51.1	−25 24 45	+0.001	0.00	7.2	0.2	1.7	A3 V		120 s		
−−	−−		21 31 52.6	+24 41 16	0.000	−0.01	7.6	0.4		F0				
204968	126868		21 31 54.8	+5 17 36	0.000	−0.01	7.9	1.1	−0.1	K2 III		390 s		
205012	107208		21 31 55.9	+18 24 39	0.000	0.00	8.0	0.4	3.0	F2 V		98 s		
204611	247066		21 32 00.6	−58 48 56	+0.001	−0.03	8.0	1.7	−0.5	M2 III		470 s		
205222	33511		21 32 03.6	+54 00 55	−0.001	0.00	7.8	0.0	0.4	B9.5 V		280 s		
202418	258899		21 32 04.1	−84 48 36	+0.038	−0.03	6.45	1.40		K2				
204783	230726	ξ Gru	21 32 05.6	−41 10 47	+0.002	+0.01	5.29	1.10	−0.3	gK5	−8	130 s		
205197	33509		21 32 05.9	+51 53 35	−0.001	−0.02	7.4	0.1	0.6	A0 V		210 s		
205013	107209		21 32 11.5	+10 54 40	−0.001	+0.01	7.2	0.1	0.6	A0 V		210 s		
204854	213078	6 PsA	21 32 14.5	−33 56 41	0.000	0.00	6.00	0.1	1.4	A2 V		83 s		m
205210	51001		21 32 16.8	+48 26 38	−0.002	−0.02	6.6	0.0	−0.1	B9.5 IV		210 s		
205118	71441		21 32 21.0	+33 10 33	−0.003	+0.06	7.9	1.1	0.2	K0 III		140 mx		
204971	164461		21 32 21.9	−12 16 04	+0.003	−0.02	6.70	1.64		K5				
205087	89786		21 32 26.9	+23 23 40	+0.003	+0.01	6.40	−0.09		A2 p	−16			
205061	107217		21 32 32.1	+11 51 03	+0.001	+0.01	7.7	1.4	−0.3	K5 III		390 s		
205053	126875		21 32 32.3	+4 52 24	+0.001	0.00	7.30	0.4	2.6	F0 V		87 s	15067	m
204943	190423		21 32 33.1	−24 35 24	+0.005	+0.02	6.43	0.20	2.1	A5 V		70 s		
205212	51006		21 32 39.9	+42 17 44	+0.001	−0.05	7.1	1.1	0.2	K0 III	−19	240 s		
204507	255017		21 32 41.1	−69 29 22	0.000	−0.01	7.52	1.22	0.2	K0 III		200 s		
205201	71448		21 32 50.6	+32 46 37	0.000	+0.01	7.45	−0.05		B9	−2			
204652	255019		21 32 52.3	−64 09 06	+0.011	−0.08	6.9	2.3	−0.1	K2 III		50 s		
205237	51009		21 32 53.4	+40 24 27	+0.001	0.00	7.4	0.9	3.2	G5 IV		70 s		
205314	51019		21 32 56.4	+49 58 40	+0.002	+0.02	5.75	−0.03	0.6	A0 V	−33	110 s		
205160	89792		21 32 58.0	+20 42 43	−0.002	−0.03	7.60	0.5	3.4	F5 V	+16	69 s	15076	m
204730	247075		21 32 59.2	−59 41 59	+0.003	0.00	7.7	0.2	0.6	A0 V		190 s		
205127	107224		21 32 59.4	+12 32 24	+0.006	−0.01	7.4	1.4	−0.3	K5 III		190 mx		
204086	257932		21 33 07.5	−78 32 34	+0.006	+0.03	7.2	1.1	0.2	K0 III		250 s		
205286	51015		21 33 08.7	+40 49 25	+0.012	−0.17	7.8	1.1	−0.1	K2 III		57 mx		
205331	51024		21 33 09.7	+48 18 12	−0.002	−0.02	6.9	0.0	−0.3	B9 IV		260 s		
205129	126884		21 33 10.9	+4 33 47	+0.004	0.00	7.6	1.1	0.2	K0 III		300 s		
205397	33525		21 33 12.6	+53 35 26	−0.002	+0.01	7.7	1.1	0.2	K0 III		290 s		

HD	SAO	Star Name	α 2000	δ 2000	μ(α)	μ(δ)	V	B–V	M_v	Spec	RV	d(pc)	ADS	Notes
205758	10084		21 33 16.8	+72 57 38	+0.010	+0.06	7.8	0.9	3.0	F2 V		40 s		
204873	247080		21 33 17.5	−52 44 15	0.000	0.00	6.3	1.2	−0.3	K5 III		210 s		
205349	51027		21 33 17.6	+45 51 15	0.000	+0.01	6.25	1.81	−4.4	K1 Ib	−5	580 s		
203955	258904		21 33 20.4	−80 02 21	+0.006	−0.01	6.47	0.04	0.6	A0 V				m
205270	71457		21 33 20.4	+33 48 07	−0.001	−0.01	7.6	1.1	0.2	K0 III		300 s		
205224	89796		21 33 20.8	+21 35 24	+0.003	0.00	7.2	1.4	−0.3	K5 III		310 s		
204960	230737		21 33 23.3	−44 50 56	−0.002	0.00	5.57	1.04	0.2	K0 III		110 s		
205239	89797		21 33 29.3	+23 02 58	+0.001	0.00	7.8	0.4	3.0	F2 V		92 s		
205067	190437		21 33 30.8	−27 53 25	+0.016	−0.07	7.59	0.66	4.7	G2 V		35 s		
205131	164475		21 33 34.0	−13 27 00	0.000	+0.01	8.0	0.9	3.2	G5 IV		93 s		
205130	145483		21 33 35.2	−9 39 38	0.000	0.00	7.9	0.0	0.4	B9.5 V		310 s		
205333	71461		21 33 35.2	+38 31 38	+0.001	−0.01	8.0	0.0	0.4	B9.5 V		320 s		
205522	19493		21 33 38.1	+60 20 07	0.000	0.00	7.90	0.1	0.6	A0 V		270 s	15094	m
205132	164476		21 33 40.0	−16 12 04	+0.002	−0.11	7.10	0.5	3.4	F5 V	−40	54 mx	15080	m
205482	33528		21 33 45.6	+54 18 58	−0.001	−0.01	7.5	0.1	0.6	A0 V		230 s		
205334	71465		21 33 46.9	+33 12 51	+0.001	−0.01	7.8	0.1	0.6	A0 V		260 s		
203532	258902		21 33 53.7	−82 40 59	+0.007	+0.01	6.38	0.13	−2.3	B3 IV		340 s		
205300	107236		21 33 58.2	+17 26 42	−0.001	−0.02	7.7	1.1	−0.1	K2 III		360 s		
205435	51035	73 ρ Cyg	21 33 58.7	+45 35 30	−0.003	−0.09	4.02	0.89	0.3	G8 III	+7	56 s		
205048	230739		21 34 01.3	−46 36 39	+0.001	−0.04	7.0	1.3	0.2	K0 III		150 s		
205685	19505		21 34 07.2	+65 44 23	−0.001	−0.01	7.8	0.1	0.6	A0 V		250 s		
205244	145493		21 34 07.4	−4 22 06	+0.003	0.00	6.80	0.1	1.4	A2 V		120 s	15085	m
205653	19501		21 34 09.7	+63 45 03	−0.005	+0.01	7.4	0.1	1.4	A2 V		150 s		
205664	19502		21 34 11.1	+64 00 14	−0.003	−0.05	7.80	0.1	0.6	A0 V		250 s	15108	m
205260	145496		21 34 15.0	−3 24 39	0.000	−0.03	6.9	1.1	0.2	K0 III		220 s		
205249	164484		21 34 16.4	−13 29 01	+0.001	+0.01	8.0	0.9	3.2	G5 IV		93 s		
205096	230741		21 34 16.8	−42 55 31	−0.004	−0.02	6.3	1.3	0.2	K0 III		120 s		
205496	51041		21 34 19.9	+45 59 43	−0.003	−0.01	7.99	−0.06	0.4	B9.5 V		330 s		
205354	107241		21 34 23.4	+12 15 26	−0.004	−0.12	7.8	0.7	4.4	G0 V		49 s		
205741	19510		21 34 25.6	+66 43 35	−0.005	−0.03	7.00	1.1	0.0	K1 III	−13	230 s		m
205355	107242		21 34 26.0	+10 26 05	0.000	0.00	8.0	0.5	3.4	F5 V		82 s		
205551	33540		21 34 27.4	+51 41 55	0.000	0.00	5.9	0.0		B9	−22	49 mn		
205356	—		21 34 29.2	+9 29 54	+0.004	−0.08	7.7	0.6	4.4	G0 V		47 s	15092	m
205374	107247		21 34 33.4	+10 55 09	−0.001	+0.01	7.9	1.1	0.2	K0 III		340 s		
205420	89807		21 34 33.8	+22 45 17	+0.001	−0.04	6.47	0.51	4.0	F8 V	+14	31 s		
205511	51048		21 34 37.6	+42 24 16	−0.001	−0.01	7.60	0.18	1.4	A2 V		150 s		
205552	51053		21 34 37.9	+49 31 35	+0.001	0.00	8.0	0.1	0.6	A0 V		280 s		
205318	145501		21 34 39.1	−7 23 32	+0.001	0.00	8.00	0.4	2.6	F0 V		120 s	15093	m
205776	19511		21 34 39.2	+66 46 21	+0.002	0.00	7.2	1.1	−0.1	K2 III	−13	260 s		
205229	190454		21 34 39.3	−28 53 16	−0.003	0.00	7.9	1.4	0.2	K0 III		210 s		
205186	213103		21 34 39.7	−36 38 48	−0.001	0.00	6.9	1.3	−0.1	K2 III		220 s		
205358	126901		21 34 42.6	+1 49 45	+0.001	−0.01	6.7	1.6	−0.5	M4 III		270 s		
205321	145502		21 34 43.1	−9 05 13	0.000	−0.02	8.0	1.1	0.2	K0 III		360 s		
205422	107250		21 34 45.6	+18 19 44	+0.002	+0.01	6.6	1.1	−0.1	K2 III		220 s		
205512	71480	72 Cyg	21 34 46.4	+38 32 03	+0.010	+0.10	4.90	1.08	0.0	K1 III	−66	96 s		
205436	107251		21 34 46.9	+19 56 09	0.000	−0.03	7.1	1.1	0.2	K0 III		240 s		
203853	258905		21 34 47.8	−81 55 38	+0.042	−0.03	7.7	0.1	1.7	A3 V		100 mx		
205289	190461	37 Cap	21 34 50.8	−20 05 04	−0.003	+0.03	5.69	0.40	3.4	F5 V	+6	29 s		
205265	190458		21 34 52.8	−29 41 44	+0.001	0.00	6.5	0.3	0.0	B8.5 V		120 s		
205437	107252		21 34 52.8	+18 55 02	−0.001	−0.03	8.0	0.4	2.6	F0 V		120 s		
205123	247093		21 34 52.8	−51 49 38	−0.023	−0.06	6.8	0.9	3.2	G5 IV		45 s		
205306	190463		21 34 54.8	−20 15 10	0.000	−0.06	7.1	0.0	4.0	F8 V	−13	41 s		
205538	71484		21 34 57.5	+39 26 09	0.000	−0.01	7.6	1.1	−0.1	K2 III		330 s		
205600	51057		21 35 00.2	+47 05 25	0.000	+0.01	7.84	−0.05	0.6	A0 V		280 s		
205514	71483		21 35 01.7	+31 47 01	0.000	−0.01	7.2	0.1	0.6	A0 V		210 s		
205601	51058		21 35 08.4	+43 42 12	−0.001	+0.01	6.76	−0.12	−0.9	B6 V		340 s		
205497	89810		21 35 11.8	+21 24 14	+0.004	−0.01	7.60	0.5	4.0	F8 V		53 s	15109	m
206362	10100		21 35 12.6	+78 37 29	+0.001	+0.01	7.90	2.0		N7.7 e	−34			S Cep, v
205849	19515		21 35 13.4	+65 35 16	0.000	+0.01	8.0	0.1	0.6	A0 V		270 s		
205294	213115		21 35 13.6	−34 19 45	+0.004	−0.09	6.9	0.8	4.4	G0 V		24 s		
205342	190469		21 35 15.7	−23 27 15	+0.006	−0.01	6.3	1.3	0.2	K0 III	−15	120 s		
205343	190470		21 35 17.5	−25 19 04	+0.006	+0.03	7.8	0.3	4.0	F8 V		57 s		
205423	145510		21 35 17.5	−3 59 00	−0.001	0.00	5.77	1.11	0.2	gG9	−2	100 s		
205539	89815		21 35 18.8	+28 11 51	+0.010	−0.04	6.33	0.34	2.6	F0 V	−42	24 mx		
205573	71490		21 35 20.0	+34 50 08	−0.001	−0.01	7.8	0.1	0.6	A0 V		260 s		
205455	126910		21 35 23.0	+0 58 33	−0.006	−0.07	7.4	0.5	4.0	F8 V		49 s		
205938	19521		21 35 25.8	+68 13 10	+0.003	+0.02	7.5	0.0	0.4	B9.5 V		250 s		
205541	89819		21 35 26.8	+24 27 08	+0.001	−0.01	6.60	0.1	1.7	A3 V	−28	96 s	15115	m
205524	107258		21 35 29.1	+18 58 01	+0.001	−0.04	8.0	0.4	3.0	F2 V		98 s		
205553	89820		21 35 34.0	+23 58 39	0.000	+0.01	7.8	1.1	0.2	K0 III		330 s		
—	71495		21 35 39.6	+40 08 06	+0.002	+0.03	7.9	0.4						
205158	247097		21 35 40.3	−59 41 36	−0.003	−0.03	8.0	0.5	4.4	G0 V		52 s		
205097	255030		21 35 42.4	−62 58 09	+0.013	−0.07	7.9	0.8	3.2	G5 IV		86 s		
205617	71494		21 35 45.6	+31 00 36	+0.002	+0.03	8.00	1.1	0.2	K0 III		350 s	15120	m
205602	89828		21 35 52.2	+23 57 53	+0.002	0.00	7.3	1.1	0.2	K0 III		270 s		

HD	SAO	Star Name	α 2000	δ 2000	μ(α)	μ(δ)	V	B-V	M_v	Spec	RV	d(pc)	ADS	Notes
			h m s	° ′ ″	s	″								
205555	126917		21 35 58.8	+6 04 26	+0.001	-0.01	7.8	0.4	2.6	F0 V		110 s		
205730	51079		21 36 02.2	+45 22 29	+0.004	+0.01	5.53	1.58		Mc	-14			W Cyg, v
205603	107262		21 36 04.9	+15 04 56	-0.006	-0.07	6.7	0.9	3.2	G5 IV	0	50 s		
205732	51082		21 36 08.3	+41 55 56	-0.001	-0.01	7.7	0.9	3.2	G5 IV		80 s		
205697	71504		21 36 08.6	+37 52 35	0.000	+0.01	7.6	0.1	0.6	A0 V		250 s		
205947	19525		21 36 09.0	+63 42 03	+0.002	0.00	7.9	0.1	0.6	A0 V		260 s		
205795	33567		21 36 09.5	+50 30 08	+0.002	+0.03	7.17	0.00	0.4	B9.5 V		210 s	15135	m
205471	190478	8 PsA	21 36 10.8	-26 10 17	+0.008	-0.02	5.73	0.22		A m				m
205716	71507		21 36 12.7	+38 46 30	+0.005	-0.01	8.0	1.1	-0.1	K2 III		270 mx		
205808	33568		21 36 13.1	+50 40 46	0.000	+0.01	7.26	-0.10	0.4	B9.5 V		240 s		
205688	89834		21 36 13.7	+30 03 20	-0.005	+0.07	6.50	0.9	1.8	G8 III-IV	-20	80 mx	15126	m
205584	126918		21 36 13.9	+6 08 14	+0.001	-0.01	7.72	1.26		K2				
205619	126920		21 36 16.7	+9 47 05	-0.002	-0.02	8.0	1.1	0.2	K0 III		360 s		
205462	213128		21 36 20.0	-35 10 50	+0.001	-0.05	7.7	2.1	-0.3	K5 III		170 s		
205527	164508		21 36 22.0	-18 23 31	-0.002	+0.01	7.7	0.4	3.0	F2 V		87 s		
205448	230748		21 36 23.0	-40 19 49	-0.001	-0.01	7.1	1.9	-0.3	K5 III		190 s		
205605	145520		21 36 32.4	-5 03 09	0.000	-0.01	7.0	1.1	0.2	K0 III		230 s		
205744	71514		21 36 32.8	+33 14 04	-0.002	-0.01	7.3	1.1	0.2	K0 III		260 s		
205733	71513		21 36 33.6	+32 06 09	+0.002	+0.02	7.40	1.6	-0.5	M4 III e	-7	360 s		AB Cyg, v
205895	33572		21 36 34.5	+54 49 03	+0.001	0.00	7.90	0.2	2.1	A5 V		140 s	15140	m
205745	71515		21 36 35.3	+32 25 42	-0.001	-0.01	7.7	1.1	0.2	K0 III		300 s		
205606	145521		21 36 36.4	-9 02 48	0.000	-0.02	8.0	1.1	0.2	K0 III		360 s		
205391	247108		21 36 38.9	-51 24 01	+0.001	+0.01	7.7	1.8	-0.1	K2 III		150 s		
205390	247109		21 36 41.1	-50 50 46	+0.045	-0.22	7.15	0.88	6.3	K2 V		15 ts		
205690	107275		21 36 43.1	+11 52 11	0.000	+0.01	7.9	1.1	-0.1	K2 III		390 s		
205577	190487		21 36 43.6	-21 30 08	+0.001	-0.01	7.8	1.0	0.2	K0 III		330 s		
205780	71522		21 36 47.6	+32 42 13	+0.003	-0.01	8.0	0.5	3.4	F5 V		81 s		
205529	213136	7 PsA	21 36 48.7	-33 02 53	+0.007	0.00	6.11	0.22	2.4	A7 V n		55 s		
205702	126926		21 36 53.3	+5 48 53	-0.005	-0.11	7.7	0.5	4.0	F8 V		55 s		
205634	164519		21 36 55.0	-10 10 27	+0.004	+0.02	7.4	0.4	2.6	F0 V		90 s		
205835	51101	74 Cyg	21 36 56.8	+40 24 49	0.000	+0.02	5.01	0.18		A5	+7	22 mn		
205949	33576		21 36 57.2	+54 38 17	-0.001	+0.01	7.9	0.1	0.6	A0 V		260 s		
205530	213139		21 36 57.9	-35 53 05	+0.008	-0.03	7.4	0.6	3.0	F2 V		53 s		
205624	164516		21 36 58.6	-18 44 40	+0.006	-0.07	7.6	0.6	-0.1	K2 III		160 mx		
205489	230754		21 36 59.6	-45 43 35	+0.002	-0.02	8.0	1.3	0.2	K0 III		250 s		
205588	190489		21 37 00.0	-28 36 35	0.000	-0.03	7.5	0.9	0.2	K0 III		300 s		
205762	107280		21 37 03.0	+19 46 57	0.000	-0.06	7.3	1.1	0.2	K0 III		260 s		
205637	164520	39 ε Cap	21 37 04.7	-19 27 58	+0.001	+0.01	4.68	-0.17	-2.3	B3 IV p	-24	250 s		m
205984	33579		21 37 05.9	+55 07 04	+0.002	+0.02	8.0	0.1	0.6	A0 V		280 s		
205746	107279		21 37 06.3	+11 43 09	+0.001	+0.01	7.2	0.1	0.6	A0 V	+12	210 s		
205950	33578		21 37 06.4	+53 22 36	-0.001	-0.02	7.6	0.1	0.6	A0 V		230 s		
206078	19528		21 37 10.2	+62 18 16	+0.005	+0.13	7.13	0.97	0.3	G8 III	-75	230 s		
205734	126930		21 37 14.1	+5 51 05	+0.001	-0.02	8.0	0.4	2.6	F0 V		120 s		
205965	33583		21 37 15.3	+51 21 49	-0.001	0.00	7.68	1.58	-0.3	K5 III		350 s		
205966	33586		21 37 17.3	+51 03 53	-0.003	+0.01	7.18	1.87	-0.4	M0 III	-23	240 s		
205693	164525		21 37 19.8	-11 27 38	0.000	+0.01	7.5	1.6	-0.5	M2 III		400 s		
205674	164524		21 37 20.7	-18 26 28	+0.002	-0.10	7.1	0.5	3.4	F5 V		55 s		
205880	71531		21 37 21.4	+35 19 38	-0.001	-0.05	7.2	1.1	0.2	K0 III		250 s		
205881	71532		21 37 23.7	+34 36 11	+0.005	+0.04	7.9	0.7	4.4	G0 V		49 s		
205939	51109		21 37 27.7	+44 41 47	0.000	-0.02	6.1	0.1	1.7	A3 V	+4	76 s		CP Cyg, q
205824	89851		21 37 27.9	+20 22 04	+0.001	0.00	7.9	1.1	0.2	K0 III		340 s		
205782	126933		21 37 29.0	+4 40 57	0.000	+0.01	6.9	0.2	2.1	A5 V		91 s		
--	19542		21 37 30.7	+69 15 55	-0.001	-0.02	8.0	1.4						
205765	145533		21 37 33.6	-0 23 25	-0.001	-0.02	6.21	0.06		A2 n	-8			m
205677	190501		21 37 34.0	-24 27 09	-0.001	-0.02	8.0	0.3	4.0	F8 V		62 s		
205705	164528		21 37 37.5	-19 13 53	+0.001	-0.01	7.3	-0.2	0.4	B9.5 V		240 s		
206080	33602		21 37 38.5	+58 45 30	-0.001	-0.01	7.9	0.1	0.6	A0 V		260 s		
206040	33596		21 37 38.5	+54 02 32	-0.002	0.00	6.15	0.99	0.0	K1 III	+2	170 s		
205837	107284		21 37 41.3	+15 12 59	-0.002	-0.02	7.5	0.9	0.3	G4 III	-29	280 s		
--	126939		21 37 42.9	+6 37 44	+0.003	-0.01	7.67	0.36						
205591	230762		21 37 43.2	-45 35 07	0.000	-0.11	7.3	0.0	3.4	F5 V		61 s		
205811	126940	3 Peg	21 37 43.5	+6 37 06	+0.004	0.00	5.93	0.08	0.6	A0 V	+3	100 s	15147	m
205767	145537	23 ξ Aqr	21 37 44.9	-7 51 15	+0.008	-0.02	4.69	0.17	2.4	A7 V	-18	36 mn		
205852	107288	5 Peg	21 37 45.3	+19 17 02	+0.007	+0.02	5.45	0.30	1.7	F0 IV	-25	55 s		
205853	107285		21 37 46.4	+15 03 28	+0.002	-0.01	8.0	0.4	3.0	F2 V		98 s		
205952	71538		21 37 48.5	+39 19 02	+0.001	0.00	6.75	-0.05	0.4	B9.5 V		190 s		
206165	19541	9 Cep	21 37 55.0	+62 04 55	-0.001	0.00	4.73	0.30	-5.7	B2 Ib	-13	640 s		
205941	71539		21 37 55.5	+33 12 02	+0.008	+0.06	7.39	1.05	5.6	dG8	-30	28 mn		
205769	164534		21 37 56.2	-17 12 35	+0.004	-0.01	7.8	1.1	0.2	K0 III		320 mx		
206041	51123		21 37 59.3	+48 29 15	+0.001	+0.01	7.57	0.07	0.0	B8.5 V		320 s	15158	m
205998	51120		21 38 01.0	+41 04 54	-0.001	-0.05	7.40	1.4	-0.3	K5 III	-37	330 s		
205417	255040		21 38 02.7	-64 49 27	+0.006	-0.01	6.20	0.03		A2				
206243	19545		21 38 03.3	+65 22 30	+0.005	+0.05	6.9	1.1	0.2	K0 III		200 s		
205967	71541		21 38 04.3	+33 07 28	0.000	0.00	7.5	1.1	0.2	K0 III	+9	290 s		
205696	213151		21 38 04.7	-37 09 51	+0.003	+0.01	8.0	1.5	0.2	K0 III		190 s		

HD	SAO	Star Name	α 2000	δ 2000	μ(α)	μ(δ)	V	B-V	M_v	Spec	RV	d(pc)	ADS	Notes
			h m s	° ′ ″	s	″								
205771	190510		21 38 10.5	-25 26 42	+0.002	+0.01	7.5	0.8	3.2	G5 IV		71 s		
205829	164538		21 38 13.8	-14 54 36	+0.002	+0.03	7.1	0.9	3.2	G5 IV		62 s		
205348	255038		21 38 17.0	-69 37 48	+0.009	-0.01	6.75	-0.09	0.4	B9.5 V		190 s		
205921	126953		21 38 23.5	+10 11 05	-0.002	-0.01	7.9	0.2	2.1	A5 V		150 s		
205789	190511		21 38 24.2	-29 38 27	+0.004	+0.02	7.7	1.0	1.4	A2 V		49 s		
206183	33615		21 38 26.0	+56 58 25	-0.002	-0.01	7.41	0.14	-4.1	B0 V	-4	1200 s	15174	m
206066	51127		21 38 27.1	+43 17 27	0.000	+0.01	7.90	1.1	0.2	K0 III		330 s	15167	m
206121	51134		21 38 28.6	+49 47 44	-0.002	0.00	7.01	0.83	-2.1	G5 II		660 s		
206111	51133		21 38 28.8	+48 42 49	-0.002	-0.01	7.7	1.4	-0.3	K5 III		360 s		
205954	107293		21 38 29.6	+13 41 03	-0.001	-0.01	7.6	1.1	-0.1	K2 III		340 s		
--	71550		21 38 29.9	+39 30 39	+0.001	0.00	8.0	1.0						
205818	190513		21 38 31.4	-27 53 39	+0.003	-0.02	8.0	-0.1	3.4	F5 V		82 s		
205924	126956	4 Peg	21 38 31.7	+5 46 18	+0.008	+0.03	5.67	0.25	2.2	dA6 n	-19	38 ts	15157	m
206166	33618		21 38 36.1	+53 34 48	0.000	+0.01	7.1	0.4	2.6	F0 V		77 s		
206311	19550		21 38 40.6	+64 23 09	-0.004	0.00	7.40	1.4	-0.3	K5 III	-8	310 s		
205925	145555		21 38 41.0	-4 09 01	-0.001	-0.03	7.0	1.6	-0.5	M2 III		310 s		
206027	89870		21 38 44.9	+25 29 56	-0.002	+0.01	6.16	1.02	3.2	G5 IV	-14	28 s		
205888	164544		21 38 50.4	-19 37 32	+0.003	-0.01	7.3	0.4	4.0	F8 V		46 s		
205860	213163		21 38 54.5	-31 54 35	+0.005	-0.02	7.8	1.2	3.4	F5 V		25 s		
204904	257942		21 38 56.5	-79 26 36	+0.028	-0.04	6.18	0.46	2.1	F5 IV		62 s		
206267	33626		21 38 57.5	+57 29 21	-0.001	0.00	5.62	0.21		O6	-8		15184	m
205889	190517		21 38 57.7	-24 27 00	+0.001	-0.03	7.6	1.4	-0.1	K2 III		240 s		
206014	126959		21 39 00.7	+9 53 18	-0.001	-0.04	7.7	1.1	0.2	K0 III		310 s		
206043	89871		21 39 01.0	+20 15 55	+0.008	0.00	5.85	0.32	2.6	F0 V	-13	44 s		
205844	213164		21 39 03.3	-39 31 43	+0.002	0.00	7.8	1.4	-0.3	K5 III		410 s		
206030	107302		21 39 03.4	+11 38 59	+0.001	+0.02	8.0	0.9	3.2	G5 IV		93 s		
205872	213168		21 39 05.9	-33 40 45	+0.005	-0.05	6.28	0.92		K0				
205903	190522		21 39 08.5	-21 42 49	+0.003	0.00	7.7	0.9	0.2	K0 III		310 s		
205905	190520		21 39 09.9	-27 18 24	+0.028	-0.08	6.74	0.62	4.1	G4 IV-V		29 ts		
206554	10113		21 39 09.9	+71 18 32	+0.021	-0.08	7.2	0.5	3.4	F5 V	-7	55 s		
206212	51150		21 39 14.7	+46 10 31	0.000	0.00	7.6	0.0	0.4	B9.5 V	-16	260 s		
205972	164548		21 39 15.0	-13 53 43	+0.001	-0.07	7.3	1.1	0.2	K0 III		170 mx		
206259	33631		21 39 17.3	+52 21 45	-0.001	-0.01	7.54	0.05	-2.9	B3 III		890 s		
205929	213175		21 39 19.8	-30 18 18	-0.001	-0.02	7.50	1.6	-0.5	M2 III		400 s		m
206114	89878		21 39 22.2	+24 29 40	0.000	+0.01	7.2	1.1	-0.1	K2 III		280 s		
--	33637		21 39 27.1	+57 29 01	-0.002	0.00	8.00	0.00	-1.6	B3.5 V		670 s		m
206005	164555		21 39 28.0	-10 34 36	+0.002	-0.04	6.08	1.03	0.2	K0 III		140 s		
206224	51153		21 39 28.5	+41 43 34	+0.001	-0.03	7.56	1.21	5.7	dG9	-15	33 mn	15186	m
206058	145566		21 39 31.3	-0 03 04	+0.015	+0.02	6.63	0.52	3.8	dF7	-28	35 s	15176	m
205874	230773		21 39 32.1	-44 34 57	+0.002	0.00	8.0	0.2	1.7	A3 V		160 s		
206067	126965	25 Aqr	21 39 33.1	+2 14 37	-0.002	-0.08	5.10	1.04	0.2	K0 III	-35	89 s		m
206268	51158		21 39 37.7	+47 12 03	0.000	+0.02	8.00	1.1	0.2	K0 III		330 s		m
205847	247123		21 39 40.5	-51 34 22	-0.001	-0.01	8.0	0.3	2.6	F0 V		120 s		
206348	33639		21 39 41.3	+55 22 06	+0.001	+0.01	7.6	1.1	0.2	K0 III		280 s		
205913	230776		21 39 42.3	-42 26 44	+0.001	0.00	7.0	-0.6	1.7	A3 V		110 s		
206101	126967		21 39 45.5	+2 08 20	+0.002	-0.01	7.9	1.1	0.2	K0 III		340 s		
205876	247124		21 39 50.3	-50 29 32	-0.004	-0.06	7.7	0.9	2.1	A5 V		50 s		
205933	230778		21 39 51.7	-43 07 55	-0.003	+0.01	6.8	0.3	0.2	K0 III		210 s		
206261	71574		21 39 53.5	+39 30 54	+0.003	+0.01	6.80	0.1	1.4	A2 V		120 s	15191	m
206280	51161		21 39 53.9	+44 25 55	+0.002	0.00	6.72	-0.06	0.2	B9 V	-13	200 s		
206349	33642		21 39 54.9	+51 28 38	0.000	+0.01	6.73	1.31	-0.1	K2 III		190 s		
206312	51163		21 39 55.5	+49 07 58	+0.001	0.00	7.13	1.22	-2.2	K1 II	-19	740 s	15201	m
205877	247128		21 39 59.5	-52 21 32	-0.002	+0.01	6.21	0.60	3.4	F5 V		27 s		
206383	33646		21 40 00.8	+53 58 08	-0.002	0.00	7.57	0.13	-1.0	B5.5 V		390 s		
206155	126971		21 40 01.8	+9 11 04	+0.005	+0.01	6.8	0.2	2.4	A7 V		76 s		EE Peg, m,v
206088	164560	40 γ Cap	21 40 05.2	-16 39 45	+0.013	-0.02	3.68	0.32		F0 p	-31	18 mn		Nashira
205848	247125		21 40 05.6	-57 26 31	+0.007	-0.05	7.3	1.2	0.2	K0 III		210 mx		
206330	51167	75 Cyg	21 40 10.9	+43 16 26	+0.005	+0.02	5.11	1.60	-0.5	M1 III	-28	130 s	15208	m
206365	51168		21 40 11.3	+49 41 01	-0.001	+0.02	7.18	-0.07	0.4	B9.5 V		230 s		
206294	71579		21 40 12.0	+36 25 52	+0.001	-0.02	8.0	0.9	3.2	G5 IV		92 s		
206482	33652		21 40 21.2	+57 34 54	-0.006	-0.01	7.40	0.46	3.3	dF4	-22	62 s	15214	m
205582	257945		21 40 23.6	-71 30 15	+0.083	-0.08	7.44	0.53		G2 V-VI				
206173	145578		21 40 25.6	-6 58 31	-0.001	-0.02	7.0	1.1	0.2	K0 III		230 s		
206063	213190		21 40 27.0	-34 32 05	-0.005	-0.10	6.7	1.7	0.2	K0 III		77 s		
206038	230784		21 40 29.3	-41 11 02	0.000	+0.02	7.0	0.9	0.2	K0 III		230 s		
205536	257944		21 40 30.4	-74 04 28	-0.025	+0.21	6.6	0.9	3.2	G5 IV		49 s		
206187	145581		21 40 31.8	-2 30 59	+0.004	+0.02	7.8	1.4	-0.3	K5 III		310 mx		
206730	10118		21 40 31.8	+70 47 12	-0.003	+0.04	7.4	1.4	-0.3	K5 III		300 s		
206536	33657		21 40 32.4	+59 45 10	+0.008	+0.03	6.9	1.1	0.2	K0 III		210 s		
206228	107316		21 40 33.2	+10 36 46	-0.001	+0.01	7.1	1.1	0.2	K0 III		240 s		
205935	247132		21 40 33.5	-55 44 15	+0.002	+0.03	6.33	1.06	1.7	K0 III-IV		76 s		
206146	164567		21 40 34.6	-18 53 46	-0.001	-0.01	7.3	1.3	-0.4	M0 III	-2	350 s		
206483	33654		21 40 38.7	+54 19 29	-0.004	0.00	6.8	1.6		M6 III		460 mx	15220	RU Cyg, m,v
206025	230786		21 40 42.7	-48 42 58	+0.005	-0.03	7.6	0.6	4.4	G0 V		42 s		
206332	89899		21 40 42.8	+28 45 21	+0.009	+0.04	7.4	0.7	4.4	G0 V		39 s		

HD	SAO	Star Name	α 2000	δ 2000	μ(α)	μ(δ)	V	B-V	M$_V$	Spec	RV	d(pc)	ADS	Notes
206509	33656		21 40 43.1	+54 52 19	0.000	0.00	6.20	1.16	0.2	K0 III	+4	130 s		
206262	126985		21 40 48.0	+3 53 49	0.000	+0.01	7.1	1.4	-0.3	K5 III		300 s		
206366	71585		21 40 48.9	+32 48 47	-0.001	+0.01	8.0	1.4	-0.3	K5 III		430 s		
206537	33660		21 40 49.3	+55 21 23	+0.001	0.00	8.0	1.1	0.2	K0 III		330 s		
206229	145583		21 40 57.5	-9 08 30	0.000	+0.01	8.0	0.9	3.2	G5 IV		93 s		
206161	213199		21 40 59.3	-31 55 22	+0.003	+0.02	8.0	1.4	-0.1	K2 III		310 s		
206385	71590		21 41 00.2	+30 17 50	0.000	-0.01	7.40	1.4	-0.3	K5 III		330 s		
205727	257949		21 41 00.9	-70 48 26	+0.001	-0.02	8.0	0.5	3.4	F5 V		82 s		
206298	126987		21 41 02.0	+5 01 15	+0.002	0.00	7.1	0.2	2.1	A5 V		100 s		
206367	89903		21 41 02.9	+22 11 42	+0.001	+0.01	7.40	1.6	-0.5	M2 III	-17	380 s		
206401	71595		21 41 03.6	+34 40 30	0.000	0.00	7.2	1.1	-0.1	K2 III		280 s		
205981	247135		21 41 03.8	-57 44 18	+0.006	-0.01	8.00	1.1	0.2	K0 III		360 s		m
206374	89905		21 41 06.1	+26 45 03	+0.026	-0.09	7.46	0.70	5.6	G8 V	-40	24 s		
206523	51182		21 41 09.9	+47 32 42	+0.001	-0.01	7.3	1.1	-0.1	K2 III	+5	290 s		
--	51184		21 41 16.1	+48 20 04	0.000	+0.07	7.9	1.4						
206181	213203		21 41 18.3	-35 35 05	-0.002	+0.02	7.0	1.5	0.2	K0 III		110 s		
206462	71604		21 41 18.6	+39 28 24	-0.001	-0.03	7.90	1.1	0.2	K0 III		330 s	15224	m
--	71603		21 41 21.6	+35 27 05	+0.001	-0.03	8.0	1.6						
206403	89907		21 41 22.6	+21 56 38	-0.001	-0.01	7.0	0.1	0.6	A0 V	-19	190 s		
206288	164579		21 41 27.9	-14 50 29	0.000	+0.02	7.10	0.4	2.6	F0 V		79 s	15216	m
205478	257948	v Oct	21 41 28.6	-77 23 24	+0.018	-0.24	3.76	1.00	0.2	K0 III	+34	32 ts		
206266	190552		21 41 30.4	-21 55 42	+0.007	0.00	7.8	0.6	0.2	K0 III		190 mx		
206274	190553		21 41 32.6	-21 39 49	0.000	-0.03	7.8	1.2	-0.1	K2 III		360 s		
206301	164580	42 Cap	21 41 32.7	-14 02 51	-0.008	-0.30	5.18	0.65	3.0	G2 IV	-1	27 s		
206538	51189	76 Cyg	21 41 34.1	+40 48 19	-0.001	-0.04	6.11	0.07	1.4	A2 V	+3	86 s		m
206430	89911		21 41 38.3	+21 36 07	+0.001	-0.04	7.0	0.1	1.7	A3 V		110 s		
206252	213206		21 41 39.3	-32 31 52	+0.006	-0.06	6.9	1.1	0.2	K0 III		170 mx		
206432	107330		21 41 41.6	+15 02 32	+0.002	-0.05	7.9	1.4	-0.3	K5 III		430 s		
206317	190556		21 41 45.7	-19 48 23	0.000	-0.01	7.0	1.4	0.2	K0 III		130 s		
206601	51200		21 41 45.7	+49 17 56	-0.006	-0.10	7.70	1.1	0.0	K1 III		200 mx		
206291	190555		21 41 45.9	-25 06 07	-0.001	0.00	6.4	1.7	0.2	K0 III		67 s		
206952	10126	11 Cep	21 41 55.1	+71 18 42	+0.024	+0.10	4.56	1.10	0.2	K0 III	-37	60 s		
206404	126992		21 41 56.4	+0 20 45	+0.004	-0.01	7.66	0.48	3.4	F5 V	+17	67 s		
206354	164583		21 41 58.4	-17 13 19	+0.018	+0.05	7.2	1.1	0.2	K0 III		79 mx		
206586	51201		21 41 58.9	+42 34 09	+0.003	+0.01	7.7	0.6	4.4	G0 V		46 s		
206485	107334		21 41 59.7	+18 56 19	-0.002	-0.06	7.70	0.9	0.3	G7 III	-10	300 mx		m
206356	190559	41 Cap	21 42 00.5	-23 15 46	+0.007	-0.09	5.30	1.1	0.2	K0 III	-44	95 mx	15223	m
206570	71613		21 42 00.9	+35 30 36	0.000	0.00	6.07	2.52		C6.3	+10			V460 Cyg, v
206486	107333		21 42 02.4	+13 17 21	0.000	0.00	7.9	1.4	-0.3	K5 III		430 s		
206579	71614		21 42 03.7	+36 52 52	+0.001	+0.01	8.0	1.1	-0.1	K2 III		390 s		
206341	190560		21 42 03.8	-27 41 26	-0.004	-0.11	7.65	1.06	3.2	K0 IV		68 s		
206340	190561		21 42 04.1	-25 51 37	+0.001	-0.01	7.4	1.4	-0.5	M2 III		390 s		
206671	33667		21 42 05.5	+52 17 31	+0.004	+0.01	7.1	1.1	0.2	K0 III		220 s		
206672	33665	80 π¹ Cyg	21 42 05.5	+51 11 23	0.000	0.00	4.67	-0.12	-1.7	B3 V	-8	180 s		
206632	51204		21 42 08.2	+45 45 57	0.000	-0.01	6.17	1.55	-0.5	M4 III	+9	220 s		
206467	126998		21 42 08.6	+4 50 03	-0.002	-0.02	7.4	1.1	-0.1	K2 III		320 s		
206445	126997	26 Aqr	21 42 09.9	+1 17 07	0.000	0.00	5.67	1.44	-0.3	gK4	+10	150 s		
206342	213215		21 42 12.4	-32 30 22	0.000	+0.03	7.1	1.2	-0.1	K2 III		270 s		
206602	71619		21 42 14.8	+39 31 34	+0.004	+0.01	7.4	1.1	-0.1	K2 III		310 s		
206487	127002	7 Peg	21 42 15.3	+5 40 48	+0.001	0.00	5.30	1.64	-0.5	gM2	-4	130 s		
206696	33672		21 42 21.0	+50 51 30	+0.001	+0.01	7.24	-0.05	0.4	B9.5 V		230 s		
206255	247140		21 42 21.9	-50 05 38	+0.010	0.00	7.6	0.6	3.2	G5 IV		77 s		
206644	51207	77 Cyg	21 42 22.7	+41 04 38	+0.002	+0.01	5.69	0.01	0.6	A0 V	-25	100 s		m
206773	33677		21 42 24.2	+57 44 10	+0.002	0.00	6.87	0.23	-4.1	B0 V pe	-22	1100 s		
206897	19593		21 42 26.2	+66 21 42	+0.002	0.00	8.00	0.17	2.1	A5 V		150 s		
206655	51211		21 42 28.2	+41 49 16	+0.003	+0.01	8.0	0.2	2.1	A5 V		150 s		
206343	213217		21 42 28.7	-37 56 00	-0.001	-0.01	7.60	0.4	2.6	F0 V		100 s		m
206540	107340		21 42 32.8	+10 49 29	+0.001	0.00	6.00	-0.11	-1.6	B5 IV	+6	320 s		
206450	164589		21 42 33.0	-15 45 42	0.000	-0.02	8.0	1.1	0.2	K0 III		360 s		
206656	51212		21 42 33.3	+41 02 51	+0.003	+0.01	7.60	0.34		F0			15251	m
206673	51214		21 42 33.9	+42 26 27	0.000	-0.01	7.40	0.1	0.6	A0 V		220 s	15249	m
206557	127009		21 42 36.8	+10 12 28	0.000	-0.01	7.6	1.1	0.2	K0 III		310 s		
206731	51217		21 42 38.7	+49 36 01	0.000	0.00	6.09	1.00	-2.1	G8 II	-2	440 s		
206453	164593	43 κ Cap	21 42 39.3	-18 51 59	+0.010	0.00	4.73	0.88	0.3	G8 III	-3	77 s		
206763	33678		21 42 39.4	+53 15 01	+0.001	-0.01	7.95	0.08	0.0	B8.5 V		330 s		
206913	19595		21 42 40.4	+63 43 19	-0.022	-0.05	7.0	0.5	3.4	F5 V		53 s		
206842	33683		21 42 45.2	+59 16 16	-0.001	+0.01	6.08	1.34	-0.1	K2 III	-2	140 s		
206748	51219		21 42 46.0	+50 10 04	+0.002	+0.03	7.80	0.9	-3.3	G8 Ib-II		1200 s		
207030	19603		21 42 49.0	+69 03 12	+0.001	-0.02	7.9	1.1	-0.1	K2 III		340 s		
206790	33681		21 42 53.1	+53 46 04	+0.002	0.00	7.7	1.1	0.2	K0 III		290 s		
206646	89921		21 42 54.4	+23 19 18	-0.002	-0.01	6.9	1.1	0.2	K0 III	-6	220 s		
206823	33682		21 42 55.6	+54 33 44	0.000	-0.01	7.1	1.1	-0.1	K2 III	-49	260 s		
206699	71636		21 42 58.3	+36 00 39	-0.001	-0.02	8.0	1.1	0.2	K0 III		350 s		
206499	190573		21 42 59.2	-23 10 12	+0.001	+0.03	7.54	1.28	0.0	K1 III		250 s		
206395	230807		21 43 02.0	-43 29 46	+0.019	-0.06	6.66	0.55	2.8	G0 IV		59 s		

HD	SAO	Star Name	α 2000	δ 2000	μ(α)	μ(δ)	V	B−V	M_v	Spec	RV	d(pc)	ADS	Notes
			^{h} ^{m} ^{s}	$^\circ$ ' "	^{s}	"								
207130	10131		21 43 03.8	+72 19 13	−0.010	−0.03	5.17	1.05	0.0	K1 III	−39	110 s		
206561	164600	44 Cap	21 43 04.3	−14 23 59	0.000	+0.03	6.0	0.2		A m				
206923	19598		21 43 05.7	+62 01 56	+0.006	+0.02	8.0	0.1	0.6	A0 V		180 mx		
206749	51221		21 43 06.2	+41 09 18	−0.002	−0.01	5.49	1.59	−0.5	M2 III	−23	160 s		
--	--		21 43 07.3	+62 12 47			8.00						15269	m
206674	89923		21 43 09.1	+25 31 30	+0.002	−0.02	7.7	0.1	1.4	A2 V		180 s		
206546	164601		21 43 13.3	−19 37 15	+0.005	−0.01	6.22	0.27		A m	−25			
206750	71642		21 43 16.1	+38 01 03	0.000	0.00	7.10	2.0		N7.7	+2			RV Cyg, m,v
206857	33688		21 43 17.8	+52 24 30	+0.001	+0.03	7.8	1.1	−0.1	K2 III		340 s		
206548	190577		21 43 20.6	−24 08 29	+0.001	−0.01	7.2	1.2	0.2	K0 III		200 s		
206774	71643	79 Cyg	21 43 25.5	+38 17 02	+0.003	+0.01	5.60	0.00	0.4	B9.5 V	−23	100 s		m
206701	107354		21 43 29.7	+19 38 07	−0.005	−0.08	7.5	0.4	2.6	F0 V		94 s		
206936	33693	μ Cep	21 43 30.2	+58 46 48	0.000	0.00	4.08	2.35	−7.0	M2 Ia	+19	480 s	15271	m,v
206551	213238		21 43 31.7	−31 15 00	+0.002	−0.06	7.4	1.2	0.2	K0 III		190 mx		
206182	255055		21 43 33.5	−68 03 51	+0.002	+0.03	7.0	2.3	−0.5	M2 III		120 s		
206807	71646		21 43 36.5	+38 18 17	+0.002	−0.02	6.90	0.1	0.5	A0 V	−18	180 s		m
206660	127018		21 43 39.3	+0 32 07	+0.002	0.00	7.1	0.8	3.2	G5 IV	−30	59 s		
206689	127021		21 43 39.6	+7 31 43	−0.002	−0.06	6.7	1.1	0.2	K0 III		200 s		
207018	19605		21 43 42.7	+62 38 08	+0.004	+0.02	8.0	0.5	3.4	F5 V		79 s		
206381	247148		21 43 45.8	−57 22 07	+0.004	−0.05	6.80	1.1	0.2	K0 III		170 s		m
206720	107355		21 43 46.2	+11 06 06	−0.001	0.00	7.4	1.4	−0.3	K5 III		350 s		
206427	247150		21 43 48.3	−53 58 39	−0.002	+0.02	6.3	2.4	3.4	F5 V		38 s		
206792	89934		21 43 54.0	+27 50 49	+0.007	0.00	7.45	0.44	3.4	dF5	−57	65 s	15267	m
206443	247152		21 43 54.7	−54 30 00	−0.002	−0.01	7.1	1.7	−0.5	M2 III		290 s		
206793	89935		21 43 58.1	+22 48 55	0.000	−0.03	6.6	1.1	−0.1	K2 III		220 s		
206429	247151		21 43 58.7	−57 19 31	−0.014	−0.05	6.50	0.48	4.4	G0 V		26 s		m
206428	247153		21 44 00.1	−57 16 57	−0.012	−0.03	6.87	0.46	4.4	G0 V		31 s		
206677	164612	45 Cap	21 44 00.8	−14 44 58	−0.002	+0.01	5.9	0.2		A m	−4			
206459	247154		21 44 01.2	−55 03 26	+0.004	−0.05	7.8	1.3	−0.1	K2 III		220 mx		
206847	71650		21 44 02.6	+33 20 06	−0.001	0.00	7.9	0.5	3.4	F5 V		77 s		
206751	127028		21 44 07.8	+7 09 30	−0.001	+0.05	7.90	0.4	3.0	F2 V		96 s		m
206827	89939	78 μ² Cyg	21 44 08.2	+28 44 35	+0.017	−0.22	6.14			dF3	+17		15270	m
206826	89940	78 μ¹ Cyg	21 44 08.4	+28 44 34	+0.022	−0.24	4.78	0.48	3.7	F6 V	+18	17 ts	15270	m
206829	89937		21 44 08.4	+25 10 43	+0.002	0.00	7.9	0.1	1.7	A3 V		180 s		
206778	127029	8 ε Peg	21 44 11.0	+9 52 30	+0.002	0.00	2.38	1.52	−4.4	K2 Ib	+5	160 s	15268	Enif, m
206397	255061		21 44 14.4	−62 06 51	+0.008	−0.02	7.3	0.3	2.6	F0 V		86 s		
206954	51243		21 44 16.8	+49 16 08	−0.001	−0.01	7.54	−0.09	0.6	A0 V		240 s		
207067	19611		21 44 17.4	+60 54 45	+0.004	−0.05	7.50	0.9	3.2	G5 IV		71 s	15288	m
206502	247155		21 44 17.9	−55 28 23	+0.005	−0.02	6.9	1.4	−0.5	M4 III		310 s		
206874	89944		21 44 19.0	+28 46 54	−0.001	−0.05	6.90	0.2	2.1	A5 V	+5	91 s	15275	m
--	--		21 44 19.9	+28 28 10	−0.002	0.00	7.2 B							
207019	33709		21 44 20.0	+55 22 49	+0.002	+0.01	7.8	1.1	−0.1	K2 III		340 s		
206830	107362		21 44 21.5	+17 02 44	+0.002	+0.01	7.8	0.4	2.6	F0 V		110 s		
208306	3636		21 44 22.4	+84 02 18	+0.049	+0.03	7.0	0.2	2.1	A5 V		97 s		
206875	89945		21 44 22.5	+26 17 42	−0.002	−0.04	7.8	1.1	−0.1	K2 III		380 s		
206889	89946		21 44 23.3	+29 15 55	−0.002	−0.09	7.2	1.1	0.0	K1 III		250 mx		
206766	145625		21 44 23.8	−4 43 49	−0.001	+0.02	6.7	0.1	1.4	A2 V		120 s		
206963	51244		21 44 25.3	+46 51 47	0.000	+0.04	6.6	0.5	3.4	F5 V	+10	44 s		
206900	89948		21 44 28.0	+29 11 37	−0.001	−0.02	7.2	0.1	0.6	A0 V		210 s		
206642	213248		21 44 29.3	−38 33 09	+0.007	−0.16	6.30	1.12	0.3	G5 III		57 mx		
206859	107365	9 Peg	21 44 30.5	+17 21 00	+0.001	−0.01	4.34	1.17	−4.5	G5 Ib	−22	490 s		
206860	107364		21 44 31.4	+14 46 21	+0.018	−0.09	5.94	0.59	4.4	G0 V	−19	20 s		
207092	190592		21 44 32.3	−24 54 38	+0.001	−0.07	7.3	0.8	3.0	F2 V		37 s		
--	19622		21 44 33.1	+68 23 49	−0.005	−0.02	8.0	1.4						
206901	89949	10 κ Peg	21 44 38.5	+25 38 42	+0.003	+0.02	4.13	0.43	2.1	F5 IV	−8	27 ts	15281	m
207086	33717		21 44 38.8	+58 13 32	+0.009	+0.02	7.8	1.1	0.2	K0 III		270 mx		
207001	51245		21 44 44.0	+43 25 45	0.000	−0.01	7.8	1.4	−0.3	K5 III		400 s		
206667	230819		21 44 44.7	−42 07 45	0.000	+0.02	7.8	0.4	4.4	G0 V		49 s		
206863	127036		21 44 48.0	+2 59 32	0.000	−0.01	7.70	1.1	−0.1	K2 III		360 s	15278	m
206741	190595		21 44 50.2	−26 18 00	+0.005	−0.03	7.7	1.5	0.2	K0 III		160 s		
206915	89950		21 44 50.8	+24 51 46	0.000	−0.05	7.8	0.2	2.1	A5 V		140 s		
206862	127038		21 44 51.7	+6 25 57	+0.004	−0.01	8.0	0.5	4.0	F8 V		62 s		
207198	19621		21 44 53.1	+62 27 38	−0.001	0.00	5.95	0.31		O9 II	−18			m
206833	145635		21 44 55.1	−9 02 11	+0.003	+0.02	7.0	1.1	0.2	K0 III	−26	230 s		
206742	213258	9 ι PsA	21 44 56.7	−33 01 33	+0.003	−0.09	4.34	−0.05	0.6	A0 V	+2	42 mx		m
206709	213255		21 44 57.3	−38 26 28	+0.002	−0.01	7.4	0.6	1.4	A2 V		77 s		
206834	145637	46 Cap	21 45 00.1	−9 04 57	+0.001	0.00	5.09	1.11	−0.9	G8 II-III	−5	130 s		
206768	190596		21 45 00.4	−26 29 25	+0.002	+0.03	7.6	1.7	−0.3	K5 III		300 s		
206535	255062		21 45 00.6	−61 50 56	0.000	−0.01	8.0	0.2	2.1	A5 V		150 s		
206797	190597		21 45 04.9	−22 33 50	+0.002	−0.03	7.3	0.6	3.0	F2 V		53 s		
206979	89959		21 45 10.5	+29 14 21	+0.003	+0.01	8.00	1.1	−0.1	K2 III		420 s		
206683	247161		21 45 11.9	−50 26 46	−0.003	−0.06	8.0	1.1	3.2	G5 IV		58 s		
207119	33722		21 45 12.1	+52 16 03	−0.001	0.00	6.42	1.91	−4.4	K5 Ib		700 s		
206690	230822		21 45 18.8	−49 29 55	+0.010	+0.06	6.45	1.15		K0				
207021	71670		21 45 20.7	+32 46 50	+0.001	−0.01	7.5	0.2	2.1	A5 V		120 s		

HD	SAO	Star Name	α 2000	δ 2000	μ(α)	μ(δ)	V	B−V	M_v	Spec	RV	d(pc)	ADS	Notes
			h m s	° ′ ″	s	″								
206991	89963		21 45 21.0	+27 00 00	+0.001	+0.03	7.2	0.0	0.4	B9.5 V	+2	230 s		
206893	164627		21 45 21.7	−12 47 00	+0.007	0.00	6.7	0.5	3.4	F5 V		45 s		
207260	19624	10 ν Cep	21 45 26.8	+61 07 15	−0.001	0.00	4.29	0.52	−7.5	A2 Ia	−21	1200 s		
207484	10142		21 45 27.6	+72 09 59	−0.005	+0.01	7.3	0.0	0.4	B9.5 V		220 s		
206399	257955		21 45 28.5	−71 00 31	+0.003	0.00	6.01	−0.10		B8				
206854	190604		21 45 28.8	−28 07 39	+0.002	−0.03	7.5	1.2	−0.3	K5 III		360 s		
206868	190605		21 45 33.4	−28 43 29	+0.009	−0.02	7.64	0.40	3.0	F2 V		79 s		
207308	19629		21 45 41.7	+62 18 30	−0.006	0.00	7.49	0.25		B0.5 V	−23			
207088	71675		21 45 44.3	+35 51 26	+0.007	+0.01	6.5	0.9	0.3	G8 III	−5	170 s		
206993	127045		21 45 46.0	+7 59 16	−0.004	+0.01	7.4	0.8	3.2	G5 IV		68 s		
207803	10153		21 45 53.2	+78 14 00	−0.002	0.00	7.2	0.1	0.6	A0 V		200 s		
207071	89971		21 45 56.3	+25 35 00	+0.001	−0.01	6.56	−0.06	0.0	B8.5 V		200 s		
207232	33731		21 46 02.7	+50 40 27	+0.002	+0.01	7.02	−0.06	0.4	B9.5 V		210 s		
207089	89972	12 Peg	21 46 04.2	+22 56 56	0.000	−0.01	5.29	1.41	−4.4	K0 Ib	−12	690 s		
207183	71680		21 46 12.4	+37 23 23	−0.001	−0.01	8.0	0.0	0.4	B9.5 V		320 s		
206957	190610		21 46 12.9	−24 49 00	−0.001	+0.02	8.0	0.0	1.7	A3 V		180 s		
206855	230832		21 46 14.4	−47 24 09	−0.001	+0.02	7.2	1.8	−0.1	K2 III		120 s		
207005	145648	47 Cap	21 46 16.1	−9 16 33	+0.001	+0.01	6.00	1.66	−0.5	gM3	+21	190 s		
207218	51277		21 46 16.5	+43 03 39	+0.001	+0.02	6.4	0.1	0.6	A0 V	−19	150 s		
207328	33741		21 46 16.5	+58 03 46	0.000	+0.01	7.42	1.95	−0.5	M2 III		230 s		
207121	89974		21 46 16.5	+25 56 05	−0.002	−0.01	8.0	1.4	−0.3	K5 III		450 s		
207006	164636		21 46 17.9	−12 35 55	+0.003	−0.02	7.2	0.8	3.2	G5 IV		64 s		
206691	255068		21 46 19.6	−62 33 01	+0.006	−0.05	6.9	1.1	0.2	K0 III		190 s		
−−	−−		21 46 19.9	+25 33 42			6.50	1.1	−0.2	K3 III		220 s		
207135	89975		21 46 21.4	+22 56 46	+0.001	−0.01	7.7	0.1	1.7	A3 V		160 s		
207134	89976		21 46 23.8	+25 33 48	+0.012	+0.05	6.28	1.21	−0.2	K3 III	−45	100 mx		
206958	213273		21 46 24.0	−32 42 52	+0.003	−0.02	6.8	1.6	0.2	K0 III		86 s		
207007	164637		21 46 27.1	−17 55 25	−0.003	−0.11	7.5	1.1	0.2	K0 III		120 mx		
207147	89978		21 46 28.7	+22 10 27	+0.001	−0.01	7.10	0.1	1.4	A2 V		140 s	15311	m
207090	127053		21 46 31.1	+6 32 06	+0.001	−0.04	7.80	0.5	3.4	F5 V		76 s		m
207076	145652		21 46 31.6	−2 12 46	+0.002	+0.03	7.20	1.6		M8 III	−37			
207052	164639	48 λ Cap	21 46 32.0	−11 21 58	+0.002	−0.01	5.59	−0.01	1.4	A2 V	+1	69 s		
−−	−−		21 46 33.6	+47 23 47	+0.003	+0.01	7.7	1.4	−0.3	K5 III		370 s		
207329	33746		21 46 34.2	+52 07 26	−0.001	−0.01	7.60	0.29	−5.7	B1.5 Ib e	−25	2400 s		
206653	255067		21 46 37.9	−67 35 48	+0.005	−0.03	7.2	0.4	0.4	B9.5 V		110 mx		
207061	164640		21 46 41.8	−11 41 52	+0.002	−0.13	7.1	0.5	4.0	F8 V	+2	41 s		
207164	107393		21 46 43.2	+19 28 38	0.000	+0.01	7.6	0.4	3.0	F2 V		85 s		
207263	71691		21 46 46.1	+37 39 28	+0.002	−0.01	7.6	1.6	−0.5	M2 III		400 s		
207165	107392		21 46 46.2	+13 43 12	+0.003	+0.02	6.6	0.1	1.7	A3 V	−2	95 s		
207330	51293	81 π² Cyg	21 46 47.4	+49 18 35	0.000	0.00	4.23	−0.12	−2.9	B3 III	−12	250 s		
206997	213277		21 46 48.3	−35 42 28	+0.001	−0.02	8.0	0.6	0.6	A0 V		120 s		
207242	71689		21 46 48.7	+32 59 45	−0.001	−0.03	8.0	0.4	3.0	F2 V		98 s		
206948	230836		21 46 51.8	−46 23 20	+0.001	+0.02	7.55	1.15	−0.2	K3 III		360 s		
207201	107394		21 46 52.0	+19 55 25	+0.001	+0.05	7.12	1.43	−0.1	K2 III		180 s		
207166	127057		21 46 58.8	+7 10 32	+0.002	+0.01	7.9	1.4	−0.3	K5 III		430 s		
207350	51296		21 47 00.5	+47 59 01	0.000	−0.03	7.70	0.4	2.6	F0 V		100 s	15329	m
207636	19646		21 47 00.8	+70 09 03	−0.001	−0.02	6.4	0.1	0.6	A0 V	−2	140 s		
207098	164644	49 δ Cap	21 47 02.3	−16 07 38	+0.018	−0.29	2.87	0.29		A m	−6	15 t	15314	Deneb Algedi, m,v
207223	107395		21 47 04.5	+17 11 38	+0.006	−0.01	6.21	0.34	2.6	F0 V	−19	51 s		
−−	71704		21 47 05.6	+37 18 37	0.000	+0.05	8.0	1.6						
206841	255070		21 47 06.2	−62 12 56	+0.001	+0.01	8.0	0.4	3.0	F2 V		100 s		
207150	145664		21 47 07.9	−2 17 33	+0.008	−0.08	7.6	0.5	3.4	F5 V		70 s		
207264	89993		21 47 09.5	+24 08 42	−0.004	+0.01	7.8	1.1	0.2	K0 III		200 mx		
207138	164647		21 47 10.1	−14 30 49	+0.001	+0.02	7.6	0.5	3.4	F5 V		71 s		
207202	127059		21 47 10.1	+4 24 10	0.000	−0.01	7.7	1.1	0.2	K0 III		310 s		
207266	89994		21 47 13.0	+22 01 28	+0.002	+0.03	7.4	0.2	2.1	A5 V		110 s		
207203	127060	11 Peg	21 47 13.8	+2 41 10	+0.001	0.00	5.64	0.00		A0	+17	35 mn		
207204	145668		21 47 24.7	−4 36 31	+0.003	0.00	6.9	0.1	1.4	A2 V		130 s		
207528	19642	12 Cep	21 47 25.2	+60 41 34	−0.001	0.00	5.52	1.52	−0.5	M2 III	−20	160 s		
207174	164651		21 47 28.5	−16 45 47	−0.001	−0.04	8.0	1.1	0.2	K0 III		360 s		
207104	213288		21 47 34.4	−37 57 07	−0.002	−0.01	6.9	0.8	3.2	G5 IV		49 s		
207188	164653		21 47 36.1	−17 17 43	0.000	−0.01	7.30	0.00	0.4	B9.5 V		240 s	15325	m
207224	145670		21 47 36.8	−7 39 58	+0.001	+0.01	8.0	0.9	3.2	G5 IV		93 s		
207235	145671		21 47 37.9	−5 55 02	+0.003	0.00	6.17	0.22		A3				
207538	33763		21 47 39.5	+59 42 03	−0.001	+0.02	7.31	0.33	−4.1	B0 V	−15	910 s		
207190	164654		21 47 40.0	−19 34 40	−0.004	−0.02	7.5	0.6	4.0	F8 V		46 s		
207175	190629		21 47 40.7	−27 45 04	−0.003	−0.03	7.4	0.7	−0.1	K2 III		320 s		
207431	51309		21 47 40.9	+44 00 40	−0.003	−0.03	7.6	0.1	0.6	A0 V	−6	250 s		
207155	213292	10 θ PsA	21 47 44.1	−30 53 54	−0.002	0.00	5.01	0.04	−0.2	A2 III n	+14	110 s		m
207208	164657		21 47 44.7	−18 12 47	+0.003	+0.01	7.6	0.1	1.7	A3 V		160 s		
206692	257958		21 47 49.2	−72 40 55	+0.003	−0.01	7.6	0.4	2.6	F0 V		100 s		
207114	213290		21 47 50.2	−39 30 41	+0.004	+0.01	7.6	0.2	1.4	A2 V		140 s		
207043	247185		21 47 55.2	−52 55 50	−0.013	−0.08	7.4	1.0	3.2	G5 IV		50 s		
207515	33764		21 47 57.6	+51 37 37	−0.002	+0.01	7.8	0.1	0.6	A0 V		250 s		
207300	145675		21 48 00.4	−6 27 48	+0.001	−0.05	7.9	0.1	1.4	A2 V		200 s		

HD	SAO	Star Name	α 2000	δ 2000	μ(α)	μ(δ)	V	B–V	M_v	Spec	RV	d(pc)	ADS	Notes
			h m s	° ′ ″	s	″								
207487	51313		21 48 04.2	+44 20 39	0.000	−0.01	7.9	1.1	0.2	K0 III		330 s		
207446	71716		21 48 08.2	+36 34 50	−0.002	+0.01	6.5	1.4	−0.3	K5 III	−31	230 s		
207486	51317		21 48 08.4	+45 48 15	+0.004	−0.01	7.9	1.1	−0.1	K2 III		390 s		
207238	190635		21 48 09.3	−25 52 41	+0.001	−0.02	7.7	1.1	3.2	G5 IV		51 s		
207488	51316		21 48 10.3	+42 09 38	−0.004	+0.02	8.0	0.9	3.2	G5 IV		92 s		
207419	90012		21 48 10.7	+24 27 51	−0.002	−0.02	7.2	1.4	−0.3	K5 III		320 s		
207543	33768		21 48 10.9	+53 05 23	+0.001	0.00	7.60	−0.01	0.6	A0 V		250 s		
207684	19655		21 48 15.1	+65 10 15	+0.006	+0.01	6.9	1.1	0.2	K0 III		210 s		
207129	230846		21 48 15.6	−47 18 13	+0.016	−0.29	5.58	0.60	4.7	G2 V	−7	15 ts		m
207489	71720		21 48 18.1	+38 57 22	−0.001	0.00	7.2	0.5	−4.6	F5 Ib	−58	1900 s		
207469	71718		21 48 23.3	+32 47 46	0.000	+0.01	6.8	0.1	0.6	A0 V	−8	170 s		
207516	71722		21 48 29.2	+38 38 55	+0.002	0.00	5.80	−0.08	−0.2	B8 V	−20	160 s		
––	––		21 48 32.2	−46 54 49			7.40							R Gru, v
207193	230848		21 48 34.5	−47 46 32	−0.002	+0.01	7.0	0.6	2.1	A5 V		51 s		
207373	145680		21 48 34.5	−6 14 57	+0.001	−0.01	7.90	0.1	1.7	A3 V		170 s	15337	m
207530	71724		21 48 36.3	+39 04 56	0.000	−0.01	7.7	1.4	−0.3	K5 III		370 s		
207015	255076		21 48 43.6	−65 30 13	−0.003	−0.02	7.30	0.4	3.0	F2 V		72 s		m
206309	258913		21 48 45.1	−80 23 11	+0.001	−0.05	7.9	1.4	−0.3	K5 III		440 s		
207490	90020		21 48 47.0	+22 25 40	0.000	−0.04	7.1	0.1	1.4	A2 V		140 s		
207158	247190		21 48 54.3	−56 16 26	+0.003	+0.01	6.7	0.3	0.4	B9.5 V		110 s		
207608	51327		21 48 54.5	+46 54 09	−0.001	−0.02	8.0	0.5	4.0	F8 V		62 s		
207498	90023		21 48 55.2	+25 42 35	+0.003	−0.01	7.94	0.35	2.6	F0 V		110 s		
207228	247192		21 48 58.1	−51 21 59	+0.005	−0.06	7.6	0.1	1.7	A3 V		62 mx		
207517	90024		21 49 00.3	+21 52 11	0.000	0.00	7.9	0.1	1.4	A2 V		200 s		
207435	145685		21 49 01.8	−5 24 14	−0.003	−0.04	6.8	1.1	0.2	K0 III	0	210 s		
207647	51332		21 49 04.2	+49 40 39	0.000	+0.02	7.7	0.9	−4.5	G4 Ib		1700 s		
207609	51329		21 49 06.6	+42 21 31	+0.003	+0.02	7.9	1.4	−0.3	K5 III		420 s		
207826	19665		21 49 08.0	+66 47 33	−0.004	−0.05	6.43	0.40		F2	−14		15366	
207438	164673		21 49 12.5	−15 07 15	+0.001	−0.01	7.7	0.1	1.4	A2 V		190 s		
207661	33780		21 49 13.4	+50 30 44	−0.001	+0.02	7.80	0.1	0.6	A0 V		260 s	15355	m
207532	107413		21 49 15.9	+17 34 21	0.000	−0.02	7.9	1.1	0.2	K0 III		340 s		
207780	19663		21 49 18.8	+61 16 22	0.000	−0.02	6.17	1.67	−1.4	M1 II–III	−19	310 s		
207439	164674		21 49 19.8	−18 23 11	+0.004	+0.01	7.9	0.1	1.4	A2 V		140 mx		
207545	107415		21 49 21.4	+15 45 41	+0.001	+0.02	7.2	1.1	0.2	K0 III		250 s		
207563	90027		21 49 26.6	+20 27 45	−0.001	0.00	6.29	−0.10	−1.7	B3 V	−12	370 s		
207637	71740		21 49 27.8	+37 04 52	−0.001	−0.01	8.0	1.1	−0.1	K2 III		390 s		
207623	71742		21 49 34.9	+33 18 45	−0.003	+0.01	7.1	0.8	3.2	G5 IV		59 s		
206096	258912		21 49 35.9	−82 26 02	+0.038	−0.04	7.1	0.4	3.4	F5 V		54 s		
207649	71744		21 49 36.1	+36 06 05	+0.002	+0.01	8.00	0.5	4.0	F8 V		63 s	15356	m
207673	51344		21 49 39.9	+41 08 56	−0.001	0.00	6.48	0.42	−5.0	A2 Ib	−2	1200 s		
207503	164679		21 49 40.9	−12 43 23	0.000	+0.02	6.1	0.1	0.6	A0 V	0	130 s		
207625	90034		21 49 42.6	+24 05 21	+0.001	+0.01	7.8	1.4	−0.3	K5 III		410 s		
207884	19670		21 49 44.9	+65 14 00	−0.002	−0.01	7.80	1.6	−0.5	M2 III	+4	400 s		m
207376	230854		21 49 45.2	−46 36 40	+0.002	−0.03	7.1	1.3	0.2	K0 III		160 s		
207663	71747		21 49 48.7	+34 55 00	0.000	+0.01	7.8	0.1	0.6	A0 V		260 s		
207650	90040	14 Peg	21 49 50.5	+30 10 27	+0.001	−0.02	5.04	−0.03	0.6	A0 V	−23	77 s		
207581	127100		21 49 51.1	+5 12 46	+0.002	+0.01	7.5	1.1	0.2	K0 III		280 s		
207480	190660		21 49 53.7	−27 24 15	+0.002	−0.01	7.2	−0.2	0.6	A0 V		210 s		
207229	255080		21 50 00.0	−64 42 44	+0.001	−0.02	5.62	1.02	0.2	K0 III	−1	120 s		
207257	255084		21 50 01.5	−62 52 25	+0.003	−0.01	7.6	0.7	4.4	G0 V		44 s		
207400	230858		21 50 01.8	−47 43 30	+0.001	0.00	6.6	1.1	0.2	K0 III		150 s		
207793	33794		21 50 02.2	+52 41 50	−0.001	−0.01	6.60	0.38		B0.5 III	−9			
207626	107423		21 50 04.8	+10 48 26	+0.003	+0.02	7.8	0.4	2.6	F0 V		110 s		
207450	213325		21 50 05.0	−38 36 54	+0.013	−0.18	7.3	0.1	4.0	F8 V		45 s		
207703	71749		21 50 05.1	+31 50 53	+0.002	0.00	7.30	1.1	0.2	K0 III		260 s	15364	m
207702	71750		21 50 05.4	+34 49 29	0.000	+0.01	7.70	1.4	−0.3	K5 III		380 s	15367	m
206729	257961		21 50 06.7	−78 13 18	+0.013	−0.04	7.8	1.1	−0.1	K2 III		310 mx		
207651	107427		21 50 08.1	+19 25 27	+0.004	+0.02	7.1	0.2	2.1	A5 V		100 s		
207652	107425	13 Peg	21 50 08.6	+17 17 08	+0.005	−0.06	5.29	0.37	0.6	F2 III	−4	87 s		m
207872	33799		21 50 09.1	+60 10 48	−0.001	0.00	7.98	0.43	−1.0	B5.5 V		330 s		
207755	51351		21 50 09.5	+42 51 33	−0.003	−0.06	7.9	1.1	0.2	K0 III		330 s		
207754	51353		21 50 10.7	+43 53 18	−0.004	−0.04	7.3	1.1	0.2	K0 III	−18	250 s		
207675	90044		21 50 12.2	+25 35 46	+0.001	0.00	7.3	0.5	3.4	F5 V	−11	60 s		
207552	164686		21 50 12.9	−16 50 41	+0.002	0.00	6.4	1.1	0.2	K0 III		170 s		
207611	145696		21 50 13.8	−0 36 22	−0.008	−0.03	7.8	0.5	4.0	F8 V		57 s		
207583	164689		21 50 23.1	−16 11 48	−0.004	−0.14	7.80	0.8	5.2	G5 V	−24	33 s		
207377	247196		21 50 24.1	−58 18 19	+0.011	−0.10	7.8	0.6	3.2	G5 IV		84 s		
207782	71756		21 50 26.6	+39 26 32	+0.001	−0.01	7.77	0.22	1.7	A3 V		140 s	15375	m
207771	71757		21 50 30.1	+37 35 24	−0.001	−0.03	8.0	1.1	0.2	K0 III		350 s		
207756	71754		21 50 30.7	+32 39 28	−0.004	−0.04	6.8	1.1	0.2	K0 III	−26	210 s		
207855	33800		21 50 32.0	+50 59 56	−0.002	−0.03	7.9	0.4	2.6	F0 V		110 s		
207719	90047		21 50 32.3	+21 56 35	+0.002	−0.02	7.1	1.1	0.2	K0 III		240 s		
207653	127103		21 50 32.6	+0 45 10	+0.003	−0.03	7.8	1.1	−0.1	K2 III		380 s		
207627	145698		21 50 35.2	−8 58 57	+0.004	−0.03	6.7	1.1	0.2	K0 III		200 s		
207740	90050		21 50 36.0	+28 46 03	−0.004	0.00	8.00	0.70	5.2	G5 V		35 s		

HD	SAO	Star Name	α 2000	δ 2000	μ(α)	μ(δ)	V	B-V	M_v	Spec	RV	d(pc)	ADS	Notes
207839	51363		21 50 36.9	+49 26 16	-0.016	-0.01	7.7	1.1	0.2	K0 III		100 mx		
205879	258910		21 50 38.8	-83 57 38	+0.001	0.00	7.7	-0.2	0.4	B9.5 V		280 s		
207428	247201		21 50 39.8	-55 09 55	+0.004	0.00	7.6	1.4	0.2	K0 III		180 s		
208391	10179		21 50 40.8	+77 46 07	-0.004	+0.01	8.0	1.1	0.2	K0 III		320 s		
207603	190677		21 50 46.1	-27 55 57	+0.002	0.00	7.40	0.00	0.4	B9.5 V		250 s	15365	m
207241	255087	o Ind	21 50 47.1	-69 37 47	-0.004	0.00	5.53	1.37	-0.4	M0 III	+20	150 s		
207415	247202		21 50 53.5	-59 38 42	+0.002	-0.04	7.7	0.7	0.2	K0 III		310 s		
206240	258914	λ Oct	21 50 54.2	-82 43 09	+0.035	-0.03	5.29	0.75	4.4	G0 V	-11	11 s		
207642	190680		21 50 56.2	-20 11 19	+0.001	+0.01	7.9	1.6	-0.5	M2 III		490 s		
207990	19677		21 50 57.9	+61 37 02	+0.002	+0.03	7.50	0.1	1.4	A2 V	-29	160 s	15390	m
207757	107436		21 51 01.8	+12 37 32	0.000	0.00	7.6				-17	28 mn		AG Peg, v
207707	145703		21 51 03.2	-7 54 27	+0.005	+0.01	7.9	0.5	4.0	F8 V		60 s		
207857	71767		21 51 04.7	+39 32 12	0.000	+0.01	6.17	-0.08	-1.2	B8 III p	0	290 s		
207687	164693		21 51 05.7	-10 02 16	+0.006	+0.02	7.4	0.6	4.4	G0 V		39 s		
207197	257966		21 51 07.9	-72 08 23	+0.005	0.00	7.8	1.1	-0.1	K2 III		380 s		
207899	51369		21 51 12.2	+44 26 27	+0.004	0.00	7.5	1.1	0.2	K0 III		290 s		
207816	90052		21 51 18.8	+22 51 36	+0.001	+0.01	8.0	1.1	0.2	K0 III		360 s		
207907	51370		21 51 19.6	+44 21 26	+0.006	-0.09	7.7	1.1	-0.1	K2 III		110 mx		
207692	190685		21 51 24.5	-23 16 15	+0.025	-0.09	6.85	0.49	3.4	F5 V	-49	37 mx		
207773	145706		21 51 26.9	-2 14 34	0.000	-0.03	8.0	1.1	0.2	K0 III		360 s		
208105	19683		21 51 28.9	+64 43 18	-0.002	+0.04	7.8	1.1	0.2	K0 III		300 s		
207694	190687		21 51 31.4	-27 23 56	+0.002	0.00	7.9	1.2	0.2	K0 III		260 s		
207796	127114		21 51 33.5	+0 46 14	-0.001	-0.01	6.8	1.1	0.2	K0 III		210 s		
207840	107445		21 51 34.0	+19 49 36	-0.001	+0.02	5.77	-0.10	-0.9	B6 V p	-20	220 s	15383	m
208133	19686		21 51 37.1	+65 45 10	-0.001	-0.01	6.4	0.2		A m	+4		15407	m
207859	107450		21 51 40.0	+19 18 25	-0.002	0.00	6.90	0.4	3.0	F2 V	+16	60 s	15386	m
207952	51374		21 51 41.5	+47 08 15	+0.001	+0.01	7.9	0.1	0.6	A0 V		270 s		
207760	164697		21 51 41.7	-18 37 23	+0.010	-0.08	6.1	1.0	3.0	F2 V	-42	17 s		
207828	107447		21 51 42.4	+12 02 09	+0.002	0.00	7.6	0.1	1.7	A3 V		160 s		
207908	71774		21 51 44.3	+31 54 43	0.000	-0.01	7.5	1.1	-0.1	K2 III	-12	320 s		
207365	257969		21 51 44.8	-70 21 08	-0.017	-0.06	7.7	0.5	3.4	F5 V		71 s		
208106	19685		21 51 47.3	+61 56 34	-0.002	0.00	7.48	0.14	-1.7	B3 V	-24	470 s		
--	71778		21 51 48.2	+40 06 01	+0.002	+0.01	8.0	1.0						
208075	33816		21 51 48.2	+59 55 50	-0.002	+0.01	7.9	0.1	0.6	A0 V		260 s		
208141	19687		21 51 48.4	+64 54 00	+0.006	0.00	7.20	1.1	0.2	K0 III	-20	230 s	15408	m
207954	51377		21 51 51.0	+41 38 56	-0.010	-0.02	7.8	0.8	3.2	G5 IV		83 s		
207861	107452		21 51 51.8	+11 05 29	+0.006	+0.03	6.9	0.2	2.1	A5 V		92 s		
207966	51378		21 51 52.8	+42 20 36	-0.015	-0.32	7.80	0.8	3.2	G5 IV		24 ts	15400	m
207607	247208		21 51 53.6	-54 39 07	+0.004	-0.04	7.3	0.7	0.6	A0 V		79 s		
207991	51382		21 51 55.2	+48 26 14	-0.002	+0.02	6.88	1.60	-4.4	K5 Ib	+36	1300 s		
207862	127121		21 51 56.3	+9 04 46	0.000	-0.02	8.00	0.2	2.1	A5 V	-8	150 s		m
207955	71779		21 51 58.1	+38 25 48	+0.001	+0.01	8.0	1.1	0.2	K0 III		350 s		
207763	213350		21 51 58.4	-30 19 00	-0.001	+0.04	7.9	0.3	2.6	F0 V		79 s		
207618	247209		21 51 58.6	-55 46 44	+0.002	+0.01	7.1	1.3	0.2	K0 III		160 s		
208063	33817		21 52 00.2	+55 47 32	+0.001	+0.02	6.62	-0.03		A1 pe	-6			
208095	33819		21 52 00.8	+55 47 48	+0.001	+0.01	5.34	-0.11	-0.6	B7 V	-7	150 s	15405	m
207967	71782		21 52 02.3	+38 57 16	+0.007	+0.02	7.7	1.1	0.2	K0 III		220 mx		
208022	51387		21 52 04.3	+48 32 56	+0.003	0.00	7.7	0.0	0.4	B9.5 V		270 s		
207726	230873		21 52 06.6	-42 58 47	+0.003	+0.01	7.5	1.8	-0.1	K2 III		140 s		
207575	255093		21 52 09.5	-62 03 11	+0.006	-0.11	7.0	0.5	4.0	F8 V		39 s		
208742	10190		21 52 12.8	+79 33 07	+0.002	+0.02	6.7	1.4	-0.3	K5 III	-16	240 s		
208375	10183		21 52 17.2	+72 29 25	+0.004	+0.02	7.2	0.0	0.4	B9.5 V		200 mx		
207932	90059		21 52 18.0	+21 16 23	+0.001	+0.03	7.0	1.6	-0.5	M4 III		310 s		
207956	90061		21 52 20.5	+24 00 46	0.000	0.00	7.5	0.1	1.4	A2 V	0	170 s		AW Peg, m,v
208185	19694		21 52 20.8	+63 06 04	0.000	0.00	7.37	0.11	-2.5	B2 V	-16	640 s	15417	m
207888	145716		21 52 21.0	-3 10 30	+0.001	-0.02	6.6	0.1	0.6	A0 V		160 s		
207765	230875		21 52 21.4	-41 24 45	+0.001	-0.01	6.8	1.4	0.2	K0 III		120 s		
207977	71786		21 52 21.4	+34 22 50	+0.001	0.00	7.4	1.1	0.2	K0 III		270 s		
208173	19693		21 52 24.6	+60 19 48	-0.004	+0.10	7.5	1.1	0.2	K0 III		100 mx		
207933	107460		21 52 25.1	+12 32 50	0.000	+0.03	7.9	1.1	0.2	K0 III		340 s		
208107	33825		21 52 26.8	+54 40 59	+0.003	+0.02	7.0	1.1	0.2	K0 III		210 s		
207821	190697		21 52 29.0	-27 18 56	+0.001	-0.01	7.9	0.7	0.2	K0 III		340 s		
207736	230874		21 52 29.3	-49 35 11	+0.001	+0.01	7.8	0.5	2.6	F0 V		79 s		
207889	145718		21 52 29.5	-9 25 04	0.000	-0.03	7.3	0.5	3.4	F5 V		62 s		
207978	90065	15 Peg	21 52 29.7	+28 47 36	-0.005	-0.06	5.53	0.42	3.4	F5 V	+19	28 ts		
208134	33828		21 52 30.7	+55 02 25	+0.002	+0.02	7.29	0.12	0.0	B8.5 V		230 s		
208218	19698		21 52 35.1	+62 42 46	0.000	+0.02	6.80	0.24	-4.4	B1 III	-22	920 s		
207891	164705		21 52 41.4	-17 03 59	+0.001	0.00	8.0	1.1	0.2	K0 III		360 s		
207920	145721		21 52 44.7	-3 59 32	+0.001	+0.02	6.7	0.9	3.2	G5 IV	+7	50 s		
208509	10186		21 52 46.8	+73 42 09	+0.008	+0.03	6.6	0.1	0.6	A0 V		130 mx		
208076	71793		21 52 47.9	+38 50 08	-0.002	-0.01	7.3	1.1	0.2	K0 III		250 s		
207715	247212		21 52 48.1	-57 20 03	0.000	+0.03	7.5	1.5	0.2	K0 III		150 s		
207980	107463		21 52 49.9	+13 29 19	+0.001	-0.03	7.6	1.1	0.2	K0 III		300 s		
208007	107465		21 52 58.3	+15 55 55	+0.005	+0.02	7.2	1.1	-0.1	K2 III		250 mx		
207936	164711		21 53 00.8	-10 33 46	+0.002	-0.04	6.8	1.1	0.2	K0 III		210 s		

HD	SAO	Star Name	α 2000	δ 2000	μ(α)	μ(δ)	V	B-V	M_v	Spec	RV	d(pc)	ADS	Notes
208057	90075	16 Peg	21ʰ 53ᵐ 03.6ˢ	+25° 55' 30"	+0ˢ.001	0".00	5.08	-0.17	-1.7	B3 V	-12	230 s		
208121	51401		21 53 04.1	+41 36 13	+0.002	-0.01	7.4	0.1	1.4	A2 V		150 s		
208410	19710		21 53 05.0	+68 29 54	+0.002	0.00	7.4	0.1	0.6	A0 V		220 s		
208219	33832		21 53 07.0	+56 12 45	+0.003	+0.01	6.8	1.1	0.2	K0 III	-15	200 s		
208452	19716		21 53 08.0	+69 41 48	-0.001	+0.01	7.4	0.1	1.4	A2 V		150 s		
208411	19712		21 53 08.2	+68 06 26	+0.001	-0.01	7.5	0.9	-2.1	G8 II	-2	670 s		m
--	--		21 53 08.5	+67 36 11			7.60							m
208510	10188		21 53 12.0	+71 59 23	+0.003	0.00	7.0	0.0	0.4	B9.5 V		190 s		
207895	213364		21 53 14.5	-36 53 49	0.000	-0.06	7.1	1.1	3.2	G5 IV		40 s		
207958	164713	51 μ Cap	21 53 17.6	-13 33 07	+0.021	+0.01	5.08	0.37	2.6	F0 V	-22	30 s		
207852	230887		21 53 18.6	-46 50 01	-0.007	-0.03	7.42	0.57	3.6	G0 IV-V		58 s		m
208253	33835		21 53 19.6	+53 59 50	-0.002	-0.01	6.7	0.1	1.4	A2 V		110 s		
208135	71801		21 53 21.3	+36 07 37	0.000	-0.02	7.5	1.4	-0.3	K5 III		350 s		
207969	164714		21 53 21.6	-14 11 22	+0.001	+0.01	8.00	0.2	2.1	A5 V		150 s		m
207180	257971		21 53 21.7	-78 34 45	+0.010	+0.02	7.50	1.1	-0.1	K2 III		330 s		m
208064	107468		21 53 24.7	+11 38 05	-0.003	0.00	8.0	1.1	0.2	K0 III		360 s		
208008	164717		21 53 35.8	-10 18 42	+0.001	0.00	7.00	0.00	0.4	B9.5 V	-11	210 s		m
208108	107474		21 53 37.2	+19 40 06	+0.001	+0.02	5.68	0.01	0.6	A0 V	+6	100 s		
208109	107473		21 53 37.6	+19 10 31	+0.002	+0.04	7.5	1.6	-0.5	M2 III		400 s		
208122	90079		21 53 38.9	+20 23 55	0.000	-0.01	7.5	1.1	0.2	K0 III		280 s		
207970	190714		21 53 41.7	-28 40 15	+0.013	-0.22	7.6	1.3	3.2	G5 IV		36 s		
208392	19718		21 53 47.9	+62 36 54	-0.001	+0.02	7.02	0.28	-3.9	B1 IV	-26	770 s	15434	EM Cep, m,v
208440	19719		21 53 53.0	+62 36 01	-0.002	0.00	7.95	0.07	0.0	B8.5 V	-14	390 s		
208439	19721		21 53 53.7	+63 44 16	+0.003	+0.03	7.54	0.03	0.6	A0 V		230 s		
208174	90085		21 53 54.5	+28 20 31	-0.012	-0.11	6.7	0.1	1.4	A2 V	-7	63 mx		
207971	213374	γ Gru	21 53 55.6	-37 21 54	+0.009	-0.02	3.01	-0.12	-1.2	B8 III	-2	70 s		
208110	127141		21 53 57.6	+6 51 52	+0.005	0.00	6.15	0.80		G0	-10			
207802	255097		21 53 59.3	-63 06 15	+0.004	-0.01	7.8	0.1	0.6	A0 V		250 mx		
208236	71810		21 54 00.4	+36 34 13	-0.001	-0.03	7.8	1.4	-0.3	K5 III		390 s		
208111	145731		21 54 10.3	-4 16 34	+0.004	-0.09	5.71	1.18	-0.1	gK2	-37	140 s		
208125	145732		21 54 11.7	-5 21 10	+0.005	+0.08	7.4	0.5	3.4	F5 V		63 s		
207926	247220		21 54 12.4	-53 07 31	+0.002	-0.03	7.9	0.8	3.4	F5 V		47 s		
208309	51418		21 54 15.5	+47 00 03	-0.005	-0.05	7.8	1.1	0.2	K0 III		280 mx		
208188	107487		21 54 16.8	+18 01 22	-0.001	-0.03	7.89	1.57	-0.3	K5 III		390 s		
208202	107489		21 54 17.3	+19 43 06	-0.002	0.00	6.39	0.97	0.2	K0 III	+4	170 s	15431	m
208033	213378		21 54 18.9	-36 03 42	+0.002	-0.02	6.7	1.2	0.2	K0 III		160 s		
206053	258916		21 54 21.3	-85 02 01	-0.028	+0.01	6.7	1.3	0.2	K0 III		140 s		
208189	107488		21 54 22.0	+12 45 04	+0.001	0.00	6.6	0.0	0.4	B9.5 V		180 s		
208016	230895		21 54 22.1	-41 15 58	0.000	-0.02	7.6	2.0	-0.5	M2 III		160 s		
209111	3658		21 54 26.3	+80 18 32	+0.001	0.00	6.5	1.6	-0.5	M4 III		250 s		
208296	71820		21 54 35.3	+34 30 50	0.000	+0.01	8.0	1.1	-0.1	K2 III		390 s		
208177	145735		21 54 35.7	-3 18 04	+0.001	-0.02	6.20	0.48		F8	-16		15432	m
208176	145736		21 54 35.8	-2 59 49	+0.002	+0.01	7.5	0.1	0.6	A0 V		240 s		
208362	51425		21 54 36.6	+47 09 20	+0.004	+0.06	7.4	0.4	2.6	F0 V		91 s		
208101	213384		21 54 41.5	-34 08 12	-0.009	-0.01	7.8	1.0	3.4	F5 V		34 s		
208394	51427		21 54 41.8	+48 11 56	-0.005	-0.03	7.4	0.1	1.4	A2 V	-25	150 s		
208629	19735		21 54 43.1	+67 45 13	+0.010	+0.01	7.1	0.4	2.6	F0 V		76 s		
208313	71824		21 54 44.9	+32 19 42	+0.016	-0.24	7.6	1.1	0.2	K0 III	-16	26 ts		
207700	257978		21 54 45.2	-73 26 17	-0.044	-0.42	7.43	0.69	4.4	G0 V		28 mx		
208312	71825		21 54 45.5	+32 40 24	0.000	-0.01	7.7	0.1	0.6	A0 V		250 s		
208441	51432		21 54 52.4	+48 33 12	0.000	-0.06	7.8	0.1	0.6	A0 V		130 mx		
208501	33864	13 Cep	21 54 52.9	+56 36 41	-0.001	0.00	5.80	0.73	-5.6	B8 Ib	-15	710 s		
208277	107499		21 54 53.6	+15 13 31	+0.004	-0.03	8.0	0.1	1.7	A3 V		110 mx		
208345	71831		21 54 54.2	+32 20 20	-0.001	+0.01	7.0	1.4	-0.3	K5 III		290 s		
208412	51431		21 54 54.7	+46 14 23	-0.001	-0.01	8.0	1.1	-0.1	K2 III		390 s		
208344	71834		21 54 58.9	+34 46 23	+0.004	+0.01	7.5	0.1	1.4	A2 V		150 mx		
208502	33865		21 54 59.5	+53 56 08	+0.016	+0.09	6.9	0.5	3.4	F5 V	-3	50 s		
208209	164737		21 55 00.1	-15 15 30	-0.001	-0.03	7.0	1.1	0.2	K0 III		230 s		
208582	33872		21 55 07.8	+59 59 53	-0.002	-0.01	8.04	0.30	0.4	B9.5 V		270 s		
207964	255101		21 55 11.4	-61 53 11	+0.008	-0.08	5.90	0.39	1.7	F0 IV		61 s		m
208471	51438		21 55 12.3	+47 27 03	+0.005	+0.01	7.6	0.1	0.6	A0 V		160 mx		
208657	19739		21 55 14.1	+65 43 29	+0.002	+0.01	7.2	0.1	0.6	A0 V		190 s		
208472	51437		21 55 14.1	+44 25 08	-0.001	0.00	7.4	1.1	0.2	K0 III	+17	260 s		
207576	--		21 55 19.4	-76 57 02			7.8	1.6	-0.5	M4 III		450 s		
208606	19738		21 55 20.6	+61 32 32	0.000	+0.01	6.13	1.60	-4.5	G8 Ib	-32	680 s		
208363	90096		21 55 20.7	+20 50 56	+0.001	-0.02	6.9	0.8	3.2	G5 IV		55 s		
208193	213397		21 55 20.8	-33 00 22	-0.001	+0.02	7.4	1.8	-0.3	K5 III		210 s		
208442	71941		21 55 21.0	+36 08 47	-0.003	-0.01	6.9	1.1	0.2	K0 III		210 s		
208415	71840		21 55 24.7	+30 49 46	+0.002	+0.05	7.90	1.1	0.2	K0 III		320 mx		
207870	257979		21 55 27.8	-70 06 58	+0.012	-0.06	6.8	-0.2	0.6	A0 V		58 mx		
208513	51447		21 55 29.5	+44 56 54	0.000	-0.01	7.65	-0.03	0.4	B9.5 V	-16	280 s		
208682	19742		21 55 30.9	+65 19 16	+0.001	+0.01	5.86	-0.06	-3.0	B2 IV e	-15	500 s	15467	m
208712	19744		21 55 31.3	+66 41 48	+0.002	+0.01	7.9	0.4	2.6	F0 V		110 s		
207896	257980		21 55 34.0	-70 04 06	+0.009	-0.04	6.7	0.4	1.4	A2 V		72 s		
208456	71844		21 55 35.0	+34 58 38	+0.002	+0.01	7.1	0.1	1.4	A2 V		140 s		

557

HD	SAO	Star Name	α 2000	δ 2000	μ(α)	μ(δ)	V	B-V	M_v	Spec	RV	d(pc)	ADS	Notes
			h m s	° ′ ″	s	″								
208563	33875		21 55 35.0	+53 14 36	0.000	+0.01	6.7	1.1	0.2	K0 III		190 s		
208473	71846		21 55 37.4	+35 04 45	+0.001	0.00	7.90	0.1	0.6	A0 V		280 s		m
208271	190741		21 55 38.2	−21 08 24	−0.003	−0.03	7.20	1.6	−0.5	M4 III		350 s		m
208349	127160		21 55 38.5	+2 21 38	+0.002	−0.02	7.2	0.1	1.4	A2 V		140 s		
208284	190742		21 55 42.7	−23 03 12	0.000	−0.03	7.9	1.6	−0.3	K5 III		390 s		
208196	230902		21 55 49.1	−45 14 52	0.000	−0.01	7.9	1.1	0.2	K0 III		320 s		
208418	107514		21 55 49.3	+13 50 20	−0.001	−0.02	7.7	1.4	−0.3	K5 III		390 s		
208491	71850		21 55 50.8	+31 15 52	−0.001	−0.02	7.7	1.4	−0.3	K5 III		390 s		
208365	127163		21 55 51.9	+0 18 02	−0.001	+0.02	8.0	0.1	0.6	A0 V		300 s		
208713	19745		21 55 53.6	+63 15 37	−0.004	−0.05	7.2	0.5	3.4	F5 V		55 s		
208215	230903		21 55 55.2	−46 55 39	−0.012	−0.04	6.52	0.46	3.4	F5 V		41 s		
208285	213406		21 55 55.4	−30 36 23	+0.004	−0.02	6.41	0.93		G5				
208666	33882		21 55 56.4	+59 49 59	+0.002	0.00	7.90	0.1	0.6	A0 V		270 s	15476	m
208443	127166		21 55 58.9	+10 05 51	−0.002	+0.01	7.9	1.1	0.2	K0 III		340 s		
208293	213408		21 55 59.7	−30 55 27	+0.001	0.00	8.0	0.5	3.4	F5 V		83 s		m
208184	247229		21 56 03.1	−52 27 49	+0.001	−0.03	6.6	1.2	0.2	K0 III		140 s		
208258	213405		21 56 03.1	−39 18 33	+0.005	−0.06	7.9	0.5	4.0	F8 V		59 s		
208398	145752		21 56 05.2	−5 49 45	0.000	0.00	7.4	1.1	0.2	K0 III		280 s		
208536	71855		21 56 08.7	+34 33 03	−0.009	−0.02	7.8	1.1	0.2	K0 III		170 mx		
208119	255104		21 56 09.7	−60 13 44	+0.006	+0.02	7.4	1.0	3.0	F2 V		30 s		
208548	71856		21 56 11.9	+33 21 57	−0.003	−0.03	6.9	1.1	−0.1	K2 III		250 s		
208420	145754		21 56 13.7	−6 58 50	+0.001	−0.02	7.50	0.00	0.4	B9.5 V		260 s	15459	m
208149	247230		21 56 13.8	−57 53 58	+0.003	+0.02	6.19	0.21	1.7	A3 V		68 s		
208761	19751		21 56 16.0	+62 53 40	+0.001	−0.01	7.63	0.05	−1.0	B5.5 V		480 s		
208714	33890		21 56 18.8	+58 24 45	−0.001	+0.04	8.0	0.4	2.6	F0 V		110 s		
208459	145757		21 56 20.5	−4 45 23	+0.001	0.00	8.0	1.4	−0.3	K5 III		450 s		
208526	—		21 56 21.9	+22 51 17	−0.002	−0.27	7.70	2.0		N7.7	−27			RX Peg, v
208321	213414		21 56 22.6	−37 15 13	−0.001	0.00	5.46	0.08	1.7	A3 V	+28	57 s		
208527	90112		21 56 23.8	+21 14 24	0.000	+0.02	6.5	1.4	−0.3	K5 III	+2	230 s		
208198	247233		21 56 24.5	−55 41 30	−0.002	−0.04	7.4	2.9	0.2	K0 III		19 s		
208261	247235		21 56 25.9	−51 24 47	+0.005	−0.06	7.4	0.4	0.2	K0 III		180 mx		
208744	33894		21 56 27.2	+59 47 42	+0.003	0.00	6.89	0.09	0.6	A0 V		160 s	15481	m
208251	247237		21 56 28.8	−53 02 56	−0.002	−0.02	8.0	0.1	1.7	A3 V		170 s		
208382	190750		21 56 29.6	−25 29 58	+0.005	+0.04	7.6	0.7	3.2	G5 IV		77 s		
208550	90116		21 56 32.2	+21 47 38	+0.001	+0.01	7.8	0.1	1.4	A2 V		190 s		
208551	90117		21 56 35.8	+21 09 41	+0.002	+0.04	7.2	1.1	0.2	K0 III		250 s		
208384	213420		21 56 38.8	−31 23 26	+0.002	−0.09	7.90	0.1	0.6	A0 V		42 mx		m
208816	19753		21 56 39.0	+63 37 33	0.000	+0.01	4.91	1.77		M1 p	−19	48 mn		VV Cep, m,v
208530	127180		21 56 40.9	+4 00 38	0.000	−0.01	7.9	1.4	−0.3	K5 III		430 s		
208323	230906		21 56 47.6	−46 28 48	+0.008	−0.04	7.44	0.39	2.8	F5 IV-V n		87 s		
208585	107529		21 56 53.0	+17 26 22	−0.001	−0.01	7.9	1.1	0.2	K0 III		340 s		
208904	19754		21 56 53.7	+65 35 37	+0.001	+0.01	7.56	0.00	−1.6	B3.5 V	−17	550 s		
208433	213424		21 56 55.4	−35 21 36	+0.001	−0.01	7.60	0.1	0.6	A0 V		250 s		m
208640	71866		21 56 55.6	+31 51 39	−0.003	−0.02	7.7	1.4	−0.3	K5 III		390 s		
208565	107528	17 Peg	21 56 56.1	+12 04 35	−0.002	−0.01	5.54	0.05	1.4	A2 V	+15	67 s		
208217	255107		21 56 56.5	−61 50 46	+0.003	−0.03	7.6	0.1	0.6	A0 V		250 s		
208618	90121		21 56 58.0	+25 16 04	+0.001	0.00	7.9	1.6	−0.5	M2 III		490 s		
208483	190765		21 57 01.9	−26 28 47	−0.002	−0.07	7.7	0.7	3.4	F5 V		53 s		
208435	213427		21 57 02.0	−37 44 50	+0.002	0.00	6.2	0.3	2.6	F0 V		50 s		
208609	107531		21 57 02.0	+17 40 54	0.000	−0.01	7.12	1.52	−0.3	K4 III	−28	270 s		
208727	51477		21 57 02.1	+48 40 05	+0.001	−0.02	6.42	−0.08	0.6	A0 V	−16	150 s		
208434	213428		21 57 02.4	−37 39 33	+0.005	−0.01	7.2	1.4	−0.1	K2 III		210 s		
—	19759		21 57 02.5	+67 59 51	−0.004	+0.01	7.9	2.1						
208930	19757		21 57 03.6	+66 01 44	+0.019	+0.06	7.5	0.8	3.2	G5 IV		71 s		
207883	257983		21 57 04.4	−76 12 40	−0.030	+0.03	6.6	1.0	0.2	K0 III		180 s		
208507	164759		21 57 04.4	−19 11 23	−0.001	+0.01	7.8	0.7	3.2	G5 IV		81 s		
208728	51476		21 57 04.6	+46 35 34	−0.001	+0.02	6.78	1.19	0.2	K0 III	−15	170 s		
208699	51473		21 57 05.6	+42 36 59	−0.002	+0.02	7.7	0.1	0.6	A0 V		260 s		
208360	230909		21 57 08.2	−49 41 51	+0.005	−0.23	7.63	0.67	4.4	G0 V		39 s		
208947	19760		21 57 10.9	+66 09 22	+0.001	+0.01	6.43	−0.05	−2.5	B2 V	+2	520 s		
208658	90126		21 57 12.2	+28 49 24	−0.002	−0.03	8.00	1.1	0.0	K1 III		400 s		
208905	19758		21 57 17.4	+61 17 43	−0.002	0.00	6.98	0.09	−3.5	B1 V p	−22	930 s	15499	m
208785	—		21 57 17.7	+50 29 22	0.000	+0.01	7.60	1.1	−1.3	K3 II-III	−17	510 s		
208487	213432		21 57 19.8	−37 45 52	+0.010	−0.14	7.8	0.1	4.0	F8 V		57 s		
207670	257981		21 57 23.9	−79 43 17	+0.019	−0.06	7.4	0.7	0.2	K0 III		210 mx		
208971	19762		21 57 24.0	+66 08 22	+0.001	+0.02	7.0	1.4	−0.3	K5 III	+11	260 s		
208667	90127		21 57 26.4	+23 36 03	+0.001	+0.01	7.7	1.1	0.2	K0 III		310 s		
208700	90128		21 57 28.9	+29 18 37	−0.002	−0.04	7.2	1.1	−0.2	K3 III		300 s		
208632	127190		21 57 30.7	+4 09 28	−0.001	−0.02	7.00	0.5	3.4	F5 V		53 s		m
208797	51486		21 57 36.6	+47 25 23	−0.002	+0.01	7.9	0.1	0.6	A0 V		270 s		
208716	90133		21 57 38.3	+28 46 16	−0.002	+0.02	8.0	0.1	0.6	A0 V		300 s		
208787	51484		21 57 38.7	+43 13 04	−0.002	−0.02	7.9	0.7	4.4	G0 V		49 s		
208612	145768		21 57 39.1	−8 33 53	0.000	0.00	6.6	0.0	0.4	B9.5 V		170 s		
208702	90131		21 57 39.2	+20 18 09	0.000	−0.01	7.25	0.03		A0				
208669	107543		21 57 41.6	+10 26 27	0.000	−0.03	7.3	0.6	4.4	G0 V		39 s		

HD	SAO	Star Name	α 2000	δ 2000	μ(α)	μ(δ)	V	B-V	M_v	Spec	RV	d(pc)	ADS	Notes
208621	164768		21 57 47.2	-15 07 24	+0.001	-0.02	7.20	0.1	1.4	A2 V		150 s	15489	m
208717	107544		21 57 47.4	+19 52 40	+0.003	-0.04	7.3	0.1	1.4	A2 V		150 s		
209258	10208		21 57 50.9	+74 59 48	-0.002	0.00	6.6	1.4	-0.3	K5 III	-17	220 s		
208835	51488		21 57 51.0	+46 51 52	-0.003	+0.01	7.50	-0.04	0.6	A0 V	+2	240 s		
208450	247244	δ Ind	21 57 55.0	-54 59 34	+0.006	-0.01	4.40	0.28	1.7	F0 IV	+15	35 s		m
209307	10210		21 57 58.1	+76 05 23	+0.004	+0.04	7.9	1.1	-0.1	K2 III		360 s		
--	--		21 58 01.2	+5 56 25	0.000	-0.01	7.3	0.1	1.4	A2 V		150 s		
208861	51489		21 58 01.8	+43 08 59	+0.003	0.00	7.7	0.1	1.4	A2 V		180 s		
208451	247245		21 58 06.2	-56 42 21	+0.004	-0.01	7.3	2.0	-0.5	M2 III		220 s		
208837	71890		21 58 06.3	+38 55 36	+0.001	0.00	7.6	0.0	0.0	B8.5 V		320 s		
208878	51496		21 58 11.1	+43 14 20	+0.001	0.00	7.43	-0.05	0.4	B9.5 V	-22	260 s		
208648	190779		21 58 11.6	-23 13 47	+0.001	+0.03	7.6	1.2	-0.1	K2 III		340 s		
209942	3673		21 58 12.5	+82 52 11	-0.074	-0.03	6.98	0.52		F5	-22	35 mn	15571	m
208703	145778		21 58 13.1	-5 25 29	+0.002	-0.10	6.33	0.37	3.0	dF2	+1	46 s		
208592	230918		21 58 13.4	-40 09 36	+0.003	-0.01	7.7	1.2	0.2	K0 III		290 s		
209943	3675		21 58 19.1	+82 52 17	-0.077	-0.02	7.49	0.70		F5	-18			
208799	90141		21 58 20.5	+22 14 42	-0.002	-0.01	8.0	1.1	0.2	K0 III		360 s		
208916	51501		21 58 23.2	+44 08 46	+0.001	+0.01	7.6	1.1	-0.1	K2 III		330 s		
208704	164779		21 58 24.0	-12 39 53	+0.001	+0.07	7.0	0.8	3.2	G5 IV	+3	58 s		
208789	107553		21 58 27.1	+14 01 22	+0.002	-0.04	6.6	1.1	-0.1	K2 III		220 s		
208776	127201		21 58 28.2	+3 46 36	-0.017	-0.13	6.94	0.60	4.4	G0 V	+25	31 s		
208625	230920		21 58 29.6	-41 43 16	-0.005	-0.01	6.76	1.58	-0.3	K5 III		230 s		
208722	164780		21 58 29.6	-13 15 08	+0.001	-0.01	8.0	0.9	3.2	G5 IV		93 s		
208496	247247		21 58 29.9	-59 00 45	+0.001	+0.02	6.12	0.46		F5				
208817	107554		21 58 33.3	+15 18 27	0.000	+0.02	7.75	1.00	0.2	K0 III		320 s		
208627	230921		21 58 34.6	-44 03 48	0.000	-0.07	6.55	0.90	3.2	G8 IV		46 s		
208800	127205		21 58 36.5	+6 00 50	+0.001	-0.01	7.90	0.1	1.4	A2 V		200 s		m
208614	230922		21 58 37.9	-45 44 36	+0.005	+0.02	7.8	0.6	1.4	A2 V		82 s		
208674	213445		21 58 38.2	-32 21 21	+0.001	-0.01	7.7	0.5	0.6	A0 V		120 s		
208790	127204		21 58 39.4	+0 32 23	+0.002	+0.02	7.9	1.1	-0.1	K2 III		390 s		
208917	71899		21 58 40.5	+36 41 19	-0.003	-0.03	8.0	1.1	0.2	K0 III		350 s		
208906	90147		21 58 40.6	+29 48 46	-0.029	-0.38	6.94	0.51	3.4	F5 V	+8	37 mx		m
208735	190786		21 58 43.6	-21 10 59	+0.001	0.00	6.12	1.65	-0.5	gM4	+3	210 s		
208864	107556		21 58 48.9	+15 03 14	0.000	-0.04	7.9	1.1	-0.1	K2 III		390 s		
209112	19773		21 58 53.3	+62 41 54	0.000	+0.03	5.93	1.67	-0.5	M4 III	-16	190 s		
208801	145784		21 58 54.8	-4 22 24	0.000	-0.25	6.22	1.00	6.3	dK2	-44	16 mn		
208962	51507		21 58 55.7	+41 41 45	+0.002	-0.02	7.60	-0.06	0.4	B9.5 V		280 s		
209102	33939		21 58 57.8	+60 05 02	-0.018	-0.05	8.0	0.5	4.0	F8 V		61 s		
209123	19775		21 58 59.2	+61 39 45	+0.006	+0.01	8.0	0.1	1.7	A3 V		170 s		
208941	71905		21 58 59.4	+32 28 59	-0.001	-0.01	7.8	0.0	0.4	B9.5 V		300 s		
208897	107561		21 58 59.5	+19 01 13	+0.005	-0.03	6.5	1.1	0.2	K0 III		180 s		
208577	255116		21 59 09.6	-60 36 27	-0.001	-0.01	7.9	0.3	3.0	F2 V		94 s		
208951	71909		21 59 11.4	+30 31 57	-0.001	+0.02	7.90	1.1	-0.1	K2 III		400 s		
209005	51513		21 59 13.9	+41 57 33	+0.002	+0.01	7.55	1.91		B9			15520	m
209369	10216	16 Cep	21 59 14.7	+73 10 48	-0.016	-0.15	5.03	0.44	3.4	F5 V	-21	21 s		m
208808	190794		21 59 17.3	-22 52 19	+0.004	+0.02	7.4	0.5	3.4	F5 V	-10	58 s		
208737	213452		21 59 17.6	-38 23 43	+0.004	0.00	5.50	1.00	0.2	K0 III	-10	120 s		
208793	190793		21 59 17.6	-23 49 59	+0.003	-0.01	7.1	0.2	0.6	A0 V		160 s		
209145	19778		21 59 19.4	+60 17 51	-0.001	-0.01	7.61	0.32	-3.5	B1 V		810 s		
208710	230924		21 59 21.1	-46 20 37	+0.008	-0.01	7.58	1.24	-0.2	K3 III		260 mx		
208932	107564		21 59 21.6	+17 07 35	+0.004	-0.01	7.9	1.1	-0.1	K2 III		350 mx		
209006	71913		21 59 22.3	+34 10 51	+0.005	0.00	7.9	0.1	1.4	A2 V		140 mx		
209124	33943		21 59 22.8	+57 39 31	0.000	-0.01	6.59	0.03	0.6	A0 V	-3	150 s		
208757	230926		21 59 24.1	-42 48 37	+0.002	0.00	7.6	2.0	-0.1	K2 III		190 s		
208867	164795		21 59 27.6	-17 59 44	+0.001	0.00	7.9	1.6	-0.5	M4 III		490 s		
208810	190798		21 59 30.5	-29 03 24	-0.003	-0.02	7.10	1.1	0.2	K0 III		240 s		m
208942	107567		21 59 31.5	+10 34 36	0.000	+0.02	7.2	1.4	-0.3	K5 III		320 s		
209146	33948		21 59 34.3	+56 53 18	+0.007	+0.01	7.3	0.4	2.6	F0 V		85 s		
208851	190800		21 59 34.7	-27 37 49	+0.003	0.00	7.50	0.4	2.6	F0 V		96 s	15509	m
208921	145794		21 59 38.8	-6 16 25	0.000	+0.01	8.0	1.1	0.2	K0 III		360 s		
208922	164799		21 59 45.3	-10 18 48	+0.003	-0.06	7.7	0.5	3.4	F5 V		71 s		
209328	19787		21 59 46.9	+67 58 27	+0.010	+0.03	8.00	0.5	3.4	F5 V		81 s	15542	m
208869	190803		21 59 47.6	-28 37 08	-0.004	-0.01	7.3	1.0	3.0	F2 V		32 s		
209026	90155		21 59 49.5	+23 56 27	-0.001	-0.02	7.00	1.6	-0.5	M4 III		320 s	15525	m
208654	255123		21 59 51.9	-60 45 53	+0.005	0.00	7.6	1.5	-0.1	K2 III		230 s		
208886	213466		21 59 55.4	-31 31 31	0.000	0.00	7.1	0.1	0.0	B8.5 V		200 s		
209203	33955		22 00 02.0	+54 06 28	-0.001	-0.01	7.5	0.5	4.0	F8 V		50 s		
209317	19789		22 00 03.2	+65 25 58	-0.001	0.00	7.2	1.4	-0.3	K5 III	-25	290 s		
208782	247260		22 00 03.7	-52 03 27	+0.005	-0.06	7.6	0.5	2.6	F0 V		75 s		
209147	51531		22 00 04.2	+44 33 10	0.000	+0.03	7.8	0.1	1.4	A2 V	-18	180 s		CM Lac, v
209218	33958		22 00 04.5	+55 01 11	0.000	0.00	8.0	0.1	-2.8	A0 II		1000 s		
209059	90159		22 00 06.4	+26 47 03	0.000	+0.02	7.7	0.0	0.4	B9.5 V		290 s		
209192	33956		22 00 07.1	+51 31 06	0.000	0.00	7.8	0.4	2.6	F0 V		100 s		
208813	230929		22 00 07.3	-47 39 24	0.000	+0.01	7.9	1.8	-0.5	M2 III		330 s		
209008	127219	18 Peg	22 00 07.7	+6 43 02	+0.001	0.00	6.00	-0.12	-2.9	B3 III	-7	530 s		

HD	SAO	Star Name	α 2000	δ 2000	μ(α)	μ(δ)	V	B-V	M_v	Spec	RV	d(pc)	ADS	Notes
			h m s	° ′ ″	s	″								
209125	51532		22 00 10.8	+42 35 00	+0.002	−0.02	7.5	0.1	1.4	A2 V		170 s		
209126	71929		22 00 16.9	+34 37 47	0.000	+0.01	6.9	1.4	−0.3	K5 III		260 s		
208979	164803		22 00 18.9	−17 23 10	+0.001	+0.01	7.8	0.2	2.1	A5 V		140 s		
208796	247262		22 00 24.0	−55 52 58	+0.002	+0.02	6.01	−0.10		B8	+3			
209204	51540		22 00 24.0	+49 03 06	0.000	−0.01	7.9	1.1	0.2	K0 III		320 s		
209149	71932		22 00 26.6	+33 00 22	−0.001	+0.07	6.5	0.5	3.4	F5 V	−2	41 s		
208927	230936		22 00 31.9	−42 27 08	+0.004	−0.02	7.3	0.8	1.7	A3 V		47 s		
209219	51542		22 00 36.5	+44 06 20	0.000	+0.01	7.5	1.1	−0.1	K2 III	−28	310 s		
209028	164809		22 00 38.0	−13 44 45	0.000	+0.01	7.50	0.4	2.6	F0 V		96 s	15534	m
209029	164808		22 00 38.7	−14 09 51	+0.009	0.00	7.6	0.5	3.4	F5 V		71 s		
209339	19792		22 00 39.1	+62 29 17	−0.001	+0.01	6.66	0.06	−4.6	B0 IV	−20	1200 s		m
208982	213469		22 00 42.7	−33 30 30	−0.004	−0.04	7.7	2.3	−0.1	K2 III		76 s		
209193	71937		22 00 45.2	+31 27 00	−0.004	−0.03	7.1	0.4	2.6	F0 V	−8	78 s		
209014	190822	12 η PsA	22 00 50.1	−28 27 13	+0.001	+0.01	5.42	−0.09	−0.2	B8 V	−5	130 s	15536	m
209105	127232		22 00 50.8	+7 51 10	+0.002	+0.01	7.8	0.1	0.6	A0 V		270 s		
209236	71944		22 00 52.5	+39 31 34	−0.002	0.00	7.6	1.1	0.2	K0 III		300 s		
209205	71939		22 00 54.7	+31 32 03	0.000	0.00	7.5	0.1	0.6	A0 V	+6	240 s		
209246	51552		22 00 55.5	+43 38 06	+0.001	−0.01	7.70	0.2	2.1	A5 V		130 s	15547	m
209181	90169		22 00 57.8	+21 51 55	+0.001	−0.05	7.8	1.1	0.2	K0 III		320 s		
208992	213472		22 00 59.0	−36 33 22	+0.002	−0.03	6.7	1.0	0.2	K0 III		200 s		
209015	213474		22 00 59.5	−35 27 28	+0.001	−0.02	7.8	0.4	1.4	A2 V		110 s		
209286	51557		22 01 02.1	+48 22 22	+0.001	0.00	7.9	1.4	−0.3	K5 III		410 s		
209128	127235	28 Aqr	22 01 04.8	+0 36 18	+0.001	−0.01	5.58	1.28	−0.3	gK4	+7	150 s		
209080	164811		22 01 04.9	−13 01 39	0.000	−0.11	7.1	0.5	3.4	F5 V		55 s		
208958	230938		22 01 05.2	−47 27 21	−0.001	−0.04	7.5	1.0	0.2	K0 III		260 s		
209166	107587	20 Peg	22 01 05.2	+13 07 12	+0.004	−0.05	5.60	0.34	3.0	F2 V	+7	32 ts	15543	m
209260	71949		22 01 06.5	+39 14 45	+0.001	0.00	7.12	−0.02	0.6	A0 V	−15	200 s	15549	m
209167	127239	19 Peg	22 01 09.0	+8 15 26	−0.001	−0.01	5.65	1.44	−0.3	gK5	−23	160 s		
209223	90171		22 01 11.2	+26 49 48	0.000	0.00	7.80	0.00	0.4	B9.5 V		300 s	15548	m
209069	190829		22 01 12.5	−25 00 39	+0.001	−0.05	7.9	1.6	0.2	K0 III		160 s		
209272	71952		22 01 19.4	+36 25 59	+0.002	+0.01	7.8	0.0	0.4	B9.5 V		290 s		
209273	71950		22 01 20.0	+33 10 10	0.000	0.00	8.0	0.0	0.4	B9.5 V		330 s		
209453	19800		22 01 23.1	+61 58 26	+0.011	+0.05	7.1	1.1	−0.1	K2 III	−6	200 mx		
209454	19802		22 01 25.6	+61 33 21	+0.001	−0.01	7.78	0.16	−2.5	B2 V	−17	760 s		
209131	164817		22 01 26.9	−14 19 36	+0.001	−0.04	8.0	1.1	0.2	K0 III		360 s		
209082	213478		22 01 27.3	−33 50 10	+0.002	−0.09	7.3	1.1	0.2	K0 III		120 mx		
209329	51564		22 01 28.4	+48 11 54	+0.007	+0.01	8.0	0.4	−4.7	F1 Ib p		97 mx		
209154	164819		22 01 32.8	−15 36 44	+0.003	−0.01	7.10	0.8	3.2	G5 IV		60 s	15546	m
209195	145813		22 01 35.3	−6 12 47	+0.001	+0.02	8.0	1.1	0.2	K0 III		360 s		
208998	247274		22 01 36.4	−53 05 36	+0.006	−0.49	7.13	0.57		F8		25 mn		
209342	51566		22 01 42.5	+41 32 19	−0.004	−0.02	7.2	1.1	−0.1	K2 III		280 s		
209224	145817		22 01 46.1	−1 07 43	+0.011	−0.01	7.8	0.5	4.0	F8 V		57 s		
209419	33985		22 01 50.4	+52 52 56	0.000	+0.01	5.78	−0.11	−2.2	B5 III	−22	360 s		
208500	257990		22 01 52.1	−77 39 45	−0.007	0.00	6.41	0.22		A5				
209262	127243		22 01 53.8	+4 46 14	+0.004	−0.09	7.90	0.9	3.2	G5 IV		87 s	15559	m
209288	107599		22 02 01.3	+10 58 26	+0.001	+0.01	6.40	−0.10	−1.1	B5 V		320 s		
209172	213490		22 02 03.9	−34 22 57	−0.004	−0.01	7.7	1.8	−0.3	K5 III		270 s		
209481	33990	14 Cep	22 02 04.0	+58 00 02	−0.001	0.00	5.56	0.06	−4.8	O9 V	−11	890 s		
209174	213489		22 02 04.7	−35 32 49	+0.003	−0.04	7.3	1.6	0.2	K0 III		120 s		
209480	33991		22 02 05.1	+58 02 59	0.000	+0.01	7.56	−0.03		A				
209393	51578		22 02 05.2	+44 20 35	+0.003	+0.03	7.9	0.9	3.2	G5 IV		87 s		
208638	257991		22 02 06.7	−75 40 44	−0.004	−0.03	6.66	1.43		K5				
208937	255130		22 02 06.8	−65 14 21	+0.009	−0.06	6.8	1.1	0.2	K0 III		170 mx		
209841	10230		22 02 12.4	+75 21 16	+0.022	−0.07	8.0	0.9	3.2	G5 IV		89 s		
209227	190841		22 02 13.1	−23 37 01	+0.004	−0.02	7.5	1.0	0.2	K0 III		270 s		
209394	71965		22 02 15.8	+36 58 58	0.000	−0.01	7.20	1.6	−0.5	M2 III	−60	330 s	15566	m
209557	19807		22 02 18.6	+61 05 40	+0.003	0.00	7.90	0.00	0.6	A0 V		290 s	15574	m
209514	33995		22 02 22.8	+55 56 47	+0.002	+0.01	7.24	0.24	−1.3	B6 IV		320 s		
209278	164829	29 Aqr	22 02 26.1	−16 57 54	+0.001	0.00	7.20	0.1	1.4	A2 V		150 s	15562	m
209405	71967		22 02 26.2	+34 22 44	+0.002	0.00	7.7	1.4	−0.3	K5 III		370 s		
209253	213495		22 02 33.0	−32 07 59	+0.001	+0.05	6.7	0.5	3.4	F5 V		41 s		
209321	145825		22 02 33.0	−0 55 06	+0.002	+0.02	7.6	1.1	−0.1	K2 III		340 s		
209142	247284		22 02 36.3	−51 16 57	+0.003	−0.02	7.6	0.2	1.7	A3 V		130 s		
209439	71970		22 02 37.0	+33 22 46	+0.002	+0.01	6.9	0.1	1.7	A3 V	−7	110 s		
209380	107611		22 02 38.3	+15 59 11	0.000	−0.02	6.69	−0.05		B9				
209280	190846		22 02 41.2	−27 22 02	−0.002	−0.02	7.0	1.6	0.2	K0 III		94 s		
209469	51589		22 02 44.1	+42 48 53	+0.001	+0.01	7.22	−0.05	0.4	B9.5 V	−13	230 s		
209242	213496		22 02 47.3	−39 39 22	0.000	+0.01	7.7	1.3	−0.1	K2 III		360 s		
209886	10236		22 02 48.0	+73 49 29	−0.001	+0.02	7.8	1.1	−0.1	K2 III		330 s		
209691	19814		22 02 49.2	+66 03 51	+0.002	+0.02	6.72	0.14	0.0	B8.5 V	−40	160 s		
209470	71979		22 02 50.6	+39 25 38	+0.001	−0.01	8.0	0.1	0.6	A0 V		290 s		
209483	71981		22 02 54.5	+39 33 46	+0.001	+0.01	7.90	0.1	0.6	A0 V		280 s	15576	m
209421	107620		22 02 55.0	+15 46 56	0.000	−0.01	7.3	0.4	3.0	F2 V		72 s		
209457	90194		22 02 55.0	+29 41 16	+0.001	+0.01	7.80	1.4	−0.3	K5 III		420 s		
209406	107619		22 02 55.8	+14 02 21	0.000	0.00	7.9	1.4	−0.3	K5 III		430 s		

HD	SAO	Star Name	α 2000	δ 2000	μ(α)	μ(δ)	V	B-V	M_v	Spec	RV	d(pc)	ADS	Notes
			$22^h 02^m 56^s.5$	$+44°39'00''$										
209515	51595		22 02 56.5	+44 39 00	$-0^s.002$	$-0''.03$	5.60	-0.03		B9 p	-1		15578	m
209360	145833		22 03 00.8	-8 29 59	+0.001	+0.01	7.3	1.1	0.2	K0 III		260 s		
208741	257993		22 03 03.3	-76 07 06	+0.005	-0.07	5.95	0.39	0.6	F3 III		120 s		m
209407	127252		22 03 03.8	+7 38 13	-0.003	-0.04	8.0	0.5	4.0	F8 V		62 s		
209232	247289		22 03 06.4	-50 18 59	-0.001	0.00	7.6	1.6	0.2	K0 III		140 s		
209408	127253		22 03 07.6	+3 10 57	-0.001	-0.01	7.9	1.1	-0.1	K2 III		390 s		
209325	190848		22 03 08.4	-27 03 05	-0.002	-0.01	7.80	0.1	1.4	A2 V		190 s	15570	m
209458	107623		22 03 10.6	+18 53 03	+0.002	-0.02	7.6	0.5	4.0	F8 V		54 s		
209541	51597		22 03 11.1	+41 24 21	+0.001	-0.02	7.6	1.1	-0.1	K2 III		330 s		
209677	19818		22 03 13.6	+60 50 51	-0.002	-0.01	7.90	1.1	0.2	K0 III		310 s	15590	m
213126	3714		22 03 14.2	+88 04 23	+0.060	+0.02	7.4	0.1	1.4	A2 V	-5	160 s		
209484	90195		22 03 14.2	+30 12 12	0.000	+0.02	7.0	0.0	0.4	B9.5 V	-6	210 s		
209396	145836	30 Aqr	22 03 16.3	-6 31 21	+0.003	+0.01	5.54	0.96		G5	+30			
209336	213500		22 03 16.5	-31 26 43	-0.001	-0.02	7.4	1.1		M6 III				
209335	213502		22 03 17.0	-29 54 15	+0.010	-0.01	7.10	0.32	0.6	F0 III n		99 mx		
209409	145837	31 o Aqr	22 03 18.7	-2 09 19	+0.001	-0.01	4.69	-0.06	-0.2	B8 V	+12	95 s		
209459	107625	21 Peg	22 03 18.9	+11 23 11	+0.001	-0.01	5.80	-0.07	-0.3	B9 IV	0	170 s		
209500	90196		22 03 20.1	+29 44 58	+0.001	0.00	7.70	1.4	-0.3	K5 III		400 s		
209636	34003		22 03 20.7	+54 52 48	+0.001	+0.01	7.01	-0.05	0.4	B9.5 V	-12	210 s		
209100	247287	ε Ind	22 03 21.5	-56 47 10	+0.482	-2.53	4.69	1.06	7.0	K5 V	-40	3.4 t		
209517	90197		22 03 24.1	+30 02 15	-0.001	0.00	7.4	0.0	0.4	B9.5 V	+2	250 s		
209464	127256		22 03 24.6	+5 26 26	+0.001	-0.02	7.4	1.4	-0.3	K5 III		350 s		
209612	51603		22 03 27.5	+49 39 55	0.000	+0.02	7.57	-0.06	0.4	B9.5 V		270 s		
209444	145838		22 03 29.9	-5 41 34	+0.004	+0.02	7.6	1.1	0.2	K0 III		290 mx		
209316	230961		22 03 34.0	-46 07 40	+0.001	-0.04	8.0	1.6	-0.1	K2 III		240 s		
209268	247291		22 03 34.9	-55 58 38	-0.028	-0.09	7.00	0.6	4.4	G0 V		33 s		m
209472	127260		22 03 35.3	+4 23 27	+0.001	-0.05	8.0	0.5	3.4	F5 V		82 s		
209663	34009		22 03 38.0	+52 43 02	-0.002	-0.02	8.0	1.1	-0.1	K2 III		370 s		
209386	190856		22 03 38.7	-29 20 48	0.000	0.00	7.4	-0.6	0.6	A0 V		230 s		
209234	255134		22 03 42.4	-60 26 15	+0.017	-0.01	7.8	0.9	4.4	G0 V		30 s		
209723	34014		22 03 42.6	+59 10 01	+0.006	+0.02	7.9	0.5	4.0	F8 V		58 s		
209637	51606		22 03 42.9	+49 46 17	+0.002	+0.02	7.3	1.1	0.2	K0 III	-45	250 s		
209791	19826	17 ξ Cep	22 03 45.7	+64 37 42	+0.029	+0.09	4.29	0.34		A m	-7	37 t	15600	Kurhah, m
209501	127264		22 03 47.3	+4 47 32	-0.003	-0.06	7.8	0.1	0.6	A0 V		84 mx		
209706	34013		22 03 47.9	+55 50 58	+0.005	+0.08	7.79	0.49	3.7	F6 V		64 s		
209640	51609		22 03 51.6	+44 35 26	-0.001	+0.01	7.8	0.1	0.6	A0 V		260 s		
209772	19828	18 Cep	22 03 52.7	+63 07 12	+0.004	+0.06	5.29	1.58		gM5	-4	41 mn		
209744	34016		22 03 53.7	+59 48 53	0.000	0.00	6.71	0.23	-3.5	B1 V	-17	620 s	15601	m
209598	90201		22 03 59.3	+28 20 54	+0.001	0.00	7.00	1.6		M III	-27			TW Peg, v
209490	164840		22 04 05.7	-13 01 12	-0.001	-0.02	7.5	0.1	0.6	A0 V		240 s		
209679	51614		22 04 06.6	+44 20 43	0.000	+0.01	6.6	0.1	1.4	A2 V	+4	110 s		
209504	164842		22 04 12.0	-16 09 46	+0.001	-0.01	8.0	1.1	0.2	K0 III		360 s		
209809	34019		22 04 12.7	+59 52 02	+0.001	0.00	6.97	0.04		A0			15601	m
209810	34022		22 04 16.9	+59 49 49	+0.001	0.00	7.85	0.08		A0			15601	m
209283	255137		22 04 18.2	-61 52 21	+0.010	-0.02	7.7	1.1	0.2	K0 III		290 mx		
209811	34024		22 04 19.8	+58 46 55	-0.001	0.00	7.6	0.1	0.6	A0 V		240 s		
209449	213514		22 04 22.3	-38 52 35	+0.010	-0.04	7.1	0.8	3.2	G5 IV		61 s		
209601	107633		22 04 22.4	+13 38 53	0.000	-0.04	7.60	0.5	3.4	F5 V		69 s	15596	m
209476	213517	13 PsA	22 04 23.8	-29 55 00	+0.001	+0.01	6.47	1.63		K5				
209532	164844		22 04 27.6	-14 26 50	-0.002	-0.01	7.6	1.1	-0.1	K2 III		350 s		
209506	190862		22 04 29.7	-25 53 00	+0.006	+0.02	7.00	1.1	0.2	K0 III		230 s	15592	m
209615	127267		22 04 30.9	+10 14 25	-0.004	-0.01	6.9	1.1	0.2	K0 III		220 s		
209560	145849		22 04 31.3	-8 43 02	0.000	-0.05	7.3	0.4	2.6	F0 V		85 s		
209665	90206		22 04 32.7	+25 39 31	0.000	-0.01	7.21	0.07	0.6	A0 V		190 s		
209693	71998		22 04 34.2	+32 56 32	-0.001	+0.01	6.38	1.10	3.2	G5 IV	-22	27 s	15602	m
209616	127268		22 04 34.9	+8 54 47	+0.001	+0.04	8.0	0.5	4.0	F8 V		62 s		
209522	190864		22 04 36.6	-26 49 20	+0.001	0.00	5.96	-0.14	-1.1	B5 V n		260 s		
209295	255138		22 04 38.2	-64 43 44	+0.004	-0.07	7.3	0.7	1.4	A2 V		53 mx		
209523	190865		22 04 39.2	-28 26 09	+0.001	0.00	7.0	0.7	0.2	K0 III		230 s		
209507	213522		22 04 42.2	-35 41 47	+0.004	+0.01	7.7	1.7	-0.5	M2 III		370 s		
209898	19841		22 04 44.2	+61 20 17	+0.002	+0.03	7.2	0.1	1.4	A2 V		140 s		
209625	145853	32 Aqr	22 04 47.2	-0 54 23	-0.001	-0.05	5.30	0.23		A m	+20	28 mn		
209477	230968		22 04 48.5	-45 45 23	0.000	0.00	7.8	1.2	-0.1	K2 III		380 s		
209813	51628		22 04 56.5	+47 14 06	+0.006	+0.05	6.5	1.1	0.2	K0 III	-23	170 s		
209708	—		22 04 58.6	+22 38 14	+0.001	0.00	7.4	0.1	1.7	A3 V		140 s		
209960	19847	20 Cep	22 05 00.3	+62 47 09	+0.002	+0.06	5.27	1.41	-0.3	K4 III	-21	130 s		
209645	145854		22 05 03.3	-6 18 10	+0.001	-0.01	7.6	0.1	0.6	A0 V		260 s		
209709	107646		22 05 03.4	+14 48 57	0.000	-0.02	6.6	1.6	-0.5	M2 III	-4	270 s		
209792	72010		22 05 06.6	+39 28 58	0.000	+0.01	8.0	0.4	3.0	F2 V		97 s		
209870	34033		22 05 07.2	+51 42 01	-0.002	-0.03	7.2	0.1	1.4	A2 V		140 s		
209975	19849	19 Cep	22 05 08.7	+62 16 48	0.000	+0.01	5.11	0.08	-5.9	O9.5 Ib	-13	1000 s	15624	m
209537	230971		22 05 11.0	-41 33 47	+0.002	0.00	7.6	2.0	-0.1	K2 III		220 s		
209761	90214		22 05 11.2	+26 40 26	+0.002	+0.04	5.78	1.25	0.2	K0 III	-25	93 s		
209710	107649		22 05 11.8	+12 36 18	+0.001	-0.01	7.6	0.1	0.6	A0 V		250 s		
209856	51631		22 05 12.9	+49 18 46	-0.001	-0.01	7.9	0.1	0.6	A0 V		270 s		

HD	SAO	Star Name	α 2000	δ 2000	μ(α)	μ(δ)	V	B-V	M_v	Spec	RV	d(pc)	ADS	Notes
210011	19851		22 05 13.3	+63 34 42	-0.001	0.00	8.0	0.9	3.2	G5 IV		90 s		
209725	107650		22 05 13.9	+14 29 10	-0.001	-0.02	7.6	1.1	-0.1	K2 III		350 s		
209815	72011		22 05 14.5	+39 40 48	-0.001	0.00	7.7	1.4	-0.3	K5 III		390 s		
209857	51632		22 05 16.2	+46 44 41	-0.004	-0.02	6.14	1.62	-0.5	M4 III	-13	210 s		
209944	34038		22 05 18.1	+58 14 04	0.000	-0.03	7.99	1.15	3.2	K0 IV		70 s		
209990	19854		22 05 28.4	+60 43 47	-0.002	-0.01	7.2	1.1	0.2	K0 III	-3	230 s		
209726	127281		22 05 31.9	+0 37 40	-0.001	0.00	7.1	1.1	0.2	K0 III		240 s		
210040	19858		22 05 32.1	+62 15 42	-0.002	0.00	7.74	0.06	1.7	A3 V		160 s		
209775	107656		22 05 32.3	+17 30 37	-0.004	-0.05	7.8	0.4	2.6	F0 V		110 s		
209833	90217	23 Peg	22 05 34.5	+28 57 50	+0.002	-0.01	5.60	-0.05	0.6	A0 V	-12	100 s		
209843	72016		22 05 34.9	+36 23 06	+0.002	+0.01	7.6	0.0	0.4	B9.5 V		270 s		
209659	213537		22 05 35.3	-30 42 28	+0.004	+0.02	7.9	0.6	4.4	G0 V		48 s		
209991	34046		22 05 39.4	+57 07 35	+0.002	+0.04	7.58	0.45	2.2	F6 IV		120 s		m
209747	127285	22 ν Peg	22 05 40.6	+5 03 31	+0.007	+0.11	4.84	1.44	-0.3	K4 III	-16	99 s		
209776	127287		22 05 41.5	+5 58 00	+0.003	-0.01	7.6	1.1	0.2	K0 III		300 s		
209902	51636		22 05 42.9	+41 21 08	0.000	-0.01	7.9	0.0	0.4	B9.5 V		310 s		
209844	—		22 05 43.3	+29 54 22	-0.004	-0.03	8.00						15620	m
209750	145862	34 α Aqr	22 05 46.8	-0 19 11	+0.001	0.00	2.96	0.97	-4.5	G2 Ib	+8	290 s		Sadalmelik, m
209717	164854		22 05 47.9	-18 40 06	+0.001	+0.01	7.6	0.4	1.7	A3 V		98 s		
209932	51640		22 05 50.4	+45 06 45	+0.002	-0.01	6.44	-0.03	0.6	A0 V	-4	150 s		
209529	247303		22 05 50.8	-59 38 10	+0.007	-0.05	5.62	1.46	-0.3	K5 III		150 s		
209961	51645		22 05 51.0	+48 13 54	-0.001	+0.01	6.27	-0.06	-2.5	B2 V	-18	490 s		
209858	90220		22 05 53.9	+27 58 01	+0.004	-0.20	7.82	0.53	4.0	F8 V		58 s		
209933	51644		22 05 54.1	+44 17 16	-0.003	0.00	7.8	1.4	-0.3	K5 III		390 s		
209946	51646		22 05 57.6	+44 27 20	-0.001	+0.02	7.9	0.1	0.6	A0 V		270 s		
209992	34052		22 05 58.3	+53 37 24	0.000	-0.01	7.0	1.1	0.2	K0 III	-11	220 s		
209945	51650		22 06 01.8	+45 00 52	-0.001	-0.01	5.2	1.4	-0.3	K5 III	-23	120 s		
209662	230976		22 06 03.1	-44 32 14	+0.002	+0.01	7.8	1.0	2.6	F0 V		41 s		
209467	255142		22 06 03.9	-65 40 07	-0.001	-0.03	7.7	0.8	3.2	G5 IV		78 s		
209661	230977		22 06 04.1	-43 57 58	+0.008	-0.02	6.93	1.04	0.2	K0 III		210 s		
209779	145866		22 06 05.2	-5 21 29	+0.011	-0.06	7.5	0.6	4.4	G0 V		43 s		
209688	213543	λ Gru	22 06 06.7	-39 32 36	-0.002	-0.12	4.46	1.37	-0.4	M0 III	+39	94 s		
210014	—		22 06 09.4	+50 17 09			7.5	1.4	-0.3	K5 III		330 s		
209845	127297		22 06 10.5	+10 05 37	+0.001	0.00	7.2	0.1	1.4	A2 V		140 s	15630	m
209993	51652		22 06 12.2	+45 14 55	+0.003	-0.01	6.19	0.09		A2	-2			
209767	164860		22 06 13.0	-14 53 46	+0.016	+0.01	7.1	0.5	3.4	F5 V		56 s		
210071	34055		22 06 13.3	+56 20 35	+0.001	-0.01	6.39	-0.10		A0 p	-20			
209741	190886		22 06 13.4	-29 04 16	+0.002	0.00	8.0	-0.2	0.6	A0 V		300 s		
209740	190888		22 06 14.5	-28 42 18	+0.003	0.00	7.44	1.29	-0.1	K2 III		270 s		
210072	34058		22 06 17.4	+55 14 49	-0.001	0.00	7.65	0.30	-2.5	B2 V		550 s		
209468	255143		22 06 23.0	-68 44 32	+0.010	-0.05	7.5	0.1	0.6	A0 V		66 mx		
209937	90228		22 06 24.5	+25 09 05	+0.001	-0.01	7.9	0.0	0.4	B9.5 V		320 s		
209819	164861	33 ι Aqr	22 06 26.1	-13 52 11	+0.003	-0.05	4.27	-0.07	-0.2	B8 V	-10	78 s		
210167	19862		22 06 32.0	+60 39 50	+0.001	+0.07	7.8	0.5	4.0	F8 V		57 s		
209875	127300		22 06 33.1	+1 51 28	+0.024	+0.22	7.25	0.53	4.0	F8 V	-42	28 mx		
209903	127301		22 06 33.4	+5 11 41	+0.002	+0.01	7.6	0.1	0.6	A0 V		260 s		
209920	107664		22 06 33.8	+12 26 43	0.000	+0.01	7.2	0.1	1.4	A2 V		150 s		
209786	213547		22 06 34.9	-31 34 10	+0.001	+0.02	7.6	2.0	-0.5	M2 III		250 s		
209905	127303		22 06 38.6	+2 26 22	+0.001	-0.01	6.52	-0.06		B9				
210100	34060		22 06 39.7	+51 48 20	0.000	-0.01	7.06	-0.09	0.0	B8.5 V		260 s		
210144	34067		22 06 41.5	+53 07 51	-0.060	-0.33	7.79	0.79	5.6	G8 V	-36	29 ts		
209947	107667		22 06 41.8	+16 44 58	0.000	-0.01	7.75	0.32		F0				
210369	10248		22 06 43.1	+72 13 43	-0.004	+0.05	7.6	1.1	-0.1	K2 III		300 s		
209907	145871		22 06 46.2	-6 43 47	0.000	-0.02	7.6	0.1	1.4	A2 V		180 s		
210026	90239		22 06 58.6	+26 37 22	0.000	0.00	7.90	1.1	0.2	K0 III		350 s		
210027	90238	24 ι Peg	22 07 00.5	+25 20 42	+0.022	+0.03	3.76	0.44	3.4	F5 V	-4	13 ts		m
209802	230985		22 07 00.8	-41 07 59	+0.004	-0.01	7.6	1.0	3.2	G5 IV		38 s		
209864	213556		22 07 01.1	-30 56 02	+0.001	-0.05	7.7	1.9	-0.1	K2 III		130 s		
210087	72040		22 07 02.0	+36 05 37	0.000	-0.01	7.90	0.1	1.7	A3 V		170 s	15645	m
209977	107669		22 07 02.3	+11 46 04	-0.002	-0.02	7.30	1.6	-0.5	M1 III	-66	360 s		
210073	72039		22 07 03.1	+34 31 16	+0.009	+0.06	7.00	0.5	3.4	F5 V		53 s	15646	m
210086	72043		22 07 06.0	+37 54 03	+0.003	0.00	7.9	1.1	0.2	K0 III		330 s		
210060	72041		22 07 07.2	+30 19 03	+0.001	-0.02	7.4	1.1	0.2	K0 III	-9	270 s		
210220	34072		22 07 09.4	+58 50 27	-0.003	-0.02	6.32	0.88	0.3	G6 III	-10	160 s		
210145	51667		22 07 09.8	+46 22 06	-0.001	-0.03	8.0	0.1	1.4	A2 V		200 s		
210431	10253		22 07 10.7	+72 22 22	+0.017	+0.04	7.0	0.1	0.6	A0 V		76 mx		
210341	19872		22 07 13.6	+68 18 08	+0.020	+0.04	7.5	0.4	3.0	F2 V		78 s		
210401	10251		22 07 14.3	+70 41 43	+0.002	+0.05	7.14	0.16	-2.6	B3 III-IV		560 s		
209925	190898		22 07 14.9	-22 14 26	+0.001	+0.02	6.9	1.2	3.2	G5 IV		29 s		
209950	164866		22 07 17.9	-10 26 50	0.000	-0.01	6.9	1.6	-0.5	M4 III		300 s		
210532	10256		22 07 20.9	+74 43 58	-0.002	-0.04	7.7	1.6	-0.5	M2 III		390 s		
209951	164867		22 07 25.1	-19 34 12	-0.002	-0.03	7.4	1.4	0.2	K0 III		150 s		
210221	34076		22 07 25.4	+53 18 26	-0.001	-0.01	6.14	0.43	-4.8	A3 Ib	-26	920 s		
210074	107675		22 07 28.5	+19 28 32	+0.009	+0.04	5.8	0.4	2.6	F0 V	-15	43 s		
210090	107676		22 07 29.8	+18 00 02	+0.002	-0.04	6.35	1.64	-0.5	M2 III	-10	220 s		

HD	SAO	Star Name	α 2000	δ 2000	μ(α)	μ(δ)	V	B-V	M$_v$	Spec	RV	d(pc)	ADS	Notes
209653	255146		22^h 07^m 30^s.4	-68° 01' 23"	-0^s.037	-0".04	7.2	0.6	4.4	G0 V		36 s		
210041	127309		22 07 31.4	+9 40 15	+0.003	0.00	7.0	0.0	0.4	B9.5 V		210 s		
210000	164868		22 07 35.1	-14 29 26	+0.004	-0.01	6.6	0.1	0.6	A0 V		160 mx		
209979	190902		22 07 38.2	-20 45 33	+0.002	-0.02	7.4	1.3	0.2	K0 III		190 s		
209788	247314		22 07 39.2	-59 19 13	-0.024	-0.10	7.0	1.3	0.2	K0 III		70 mx		
210075	107677		22 07 39.4	+10 45 13	+0.005	-0.01	7.4	1.1	0.2	K0 III		270 mx		
210043	145882		22 07 41.4	-5 49 43	+0.001	+0.02	7.9	1.1	0.2	K0 III		340 s		
209970	190904		22 07 44.1	-28 08 51	-0.001	+0.01	7.3	0.2	2.6	F0 V		88 s		
209720	255148		22 07 45.1	-65 06 11	-0.002	-0.04	7.7	1.1	-0.1	K2 III		360 s		
210030	164870		22 07 46.1	-10 04 42	-0.001	-0.01	7.4	1.1	0.2	K0 III		280 s		
210208	51671		22 07 48.8	+42 57 10	+0.001	-0.01	7.5	0.0	0.4	B9.5 V	+9	260 s		
210129	90252	25 Peg	22 07 50.1	+21 42 10	-0.003	-0.06	5.78	-0.10	-0.9	B6 V	-52	210 s		
210222	51674		22 07 52.9	+44 12 22	-0.003	0.00	7.4	1.4	-0.3	K5 III		330 s		
209635	258000		22 07 53.6	-70 17 09	+0.020	+0.01	6.9	0.1	4.0	F8 V		37 s		
210130	107680		22 07 58.0	+13 04 09	+0.003	0.00	7.78	0.29		F0	-10			
210322	34091		22 07 58.5	+56 34 44	+0.004	+0.03	7.96	0.29	2.5	A9 V		120 s		
210477	19888		22 08 07.4	+68 31 08	+0.003	0.00	8.0	0.5	3.4	F5 V		79 s		
210170	107682		22 08 08.9	+17 33 30	+0.002	-0.01	7.10	0.05		A0	-16			
210121	145889		22 08 11.7	-3 31 53	+0.002	0.00	7.5	0.0	0.4	B9.5 V		270 s		
--	51675		22 08 12.4	+43 21 40	-0.002	+0.01	7.8	2.0						
209952	230992	α Gru	22 08 13.8	-46 57 40	+0.013	-0.15	1.74	-0.13	-1.1	B5 V	+12	21 mx		[Al Na'ir], m
210081	190911		22 08 14.6	-23 43 56	-0.001	+0.01	7.84	1.01		G5				
210615	10258		22 08 16.0	+72 46 09	-0.005	+0.02	6.95	1.93	-0.3	K7 III e		260 s		DM Cep, v
209829	255151		22 08 16.4	-63 30 20	+0.002	-0.02	7.9	0.4	3.0	F2 V		96 s		
210289	51678		22 08 16.4	+49 47 47	+0.003	-0.02	6.60	1.4	-0.3	K5 III	+17	220 s	15659	m
210094	--		22 08 16.6	-17 49 47			7.8	0.6	4.4	G0 V		48 s		
210550	19893		22 08 16.9	+69 58 58	-0.001	+0.02	7.85	0.04	0.4	B9.5 V		280 s		
210210	90259		22 08 17.0	+25 32 37	-0.003	-0.03	6.11	0.27	2.6	F0 V	+2	50 s		
209953	247322		22 08 17.2	-50 06 19	-0.002	+0.02	8.0	-0.1	2.6	F0 V		120 s		
210308	--		22 08 18.0	+49 12 18			8.00	0.00	0.6	A0 V		300 s	15661	m
210211	90260		22 08 18.7	+24 09 26	-0.004	-0.07	6.6	0.8	3.2	G5 IV		48 s		
210050	213575		22 08 19.1	-34 33 22	+0.001	-0.06	7.0	1.5	-0.5	M2 III		210 mx		
210049	213576	14 μ PsA	22 08 22.8	-32 59 19	+0.006	-0.03	4.50	0.05	1.4	A2 V	+12	42 s		
210066	213577	υ PsA	22 08 25.8	-34 02 38	+0.001	-0.04	4.99	1.48	-0.3	gK5	+20	110 s		
210123	164880		22 08 26.0	-16 32 35	+0.002	+0.01	7.5	0.9	3.2	G5 IV		71 s		
210443	19889		22 08 27.5	+62 43 50	-0.006	-0.02	7.81	0.16	0.6	A0 V		220 s		
210413	19886		22 08 27.7	+61 14 54	-0.007	-0.01	7.8	1.1	-0.1	K2 III		330 s		
210290	72055		22 08 31.3	+39 22 19	+0.003	+0.01	7.80	0.1	0.6	A0 V		270 s	15662	m
210151	164883		22 08 33.9	-11 36 50	+0.002	-0.03	7.9	1.1	0.2	K0 III		340 s		
210433	34101		22 08 35.7	+59 17 23	0.000	+0.01	7.15	-0.04	0.6	A0 V		200 s	15670	m
210051	230999		22 08 39.3	-43 02 31	+0.002	-0.06	7.01	1.20	1.7	K0 III-IV		85 s		
210185	127318		22 08 40.4	+0 20 16	+0.001	-0.02	8.0	1.1	-0.1	K2 III		410 s		
210334	51684		22 08 40.9	+45 44 31	-0.003	+0.03	6.11	0.74		G5	-35			AR Lac, m,v
210353	51686		22 08 41.6	+47 56 01	+0.003	-0.01	6.80	-0.01	0.6	A0 V	-1	170 s		
210111	213583		22 08 42.5	-33 07 32	+0.002	+0.03	6.5	-0.1	1.4	A2 V		100 s		
209941	247325		22 08 43.8	-59 17 07	-0.001	+0.01	7.8	1.7	-0.3	K5 III		310 s		
--	19897		22 08 44.8	+68 10 50	-0.002	-0.03	8.0	1.8						
210212	127320		22 08 45.3	+7 19 54	-0.001	-0.05	8.0	1.4	-0.3	K5 III		450 s		
210478	19892		22 08 45.4	+61 01 20	0.000	-0.01	7.32	0.08	-3.5	B1 V	-42	990 s		
210139	190913		22 08 45.6	-25 46 13	+0.002	-0.03	7.1	0.4	1.4	A2 V		87 s		
210172	164885		22 08 47.3	-13 18 01	+0.007	-0.01	6.9	0.5	3.4	F5 V		51 s		
210264	90266		22 08 50.2	+22 08 19	0.000	-0.01	7.1	0.9	3.2	G5 IV		60 s		
210640	19900		22 08 55.6	+70 14 24	+0.015	-0.03	7.76	0.46	3.4	F5 V		72 s		
209929	255154		22 08 56.6	-63 53 23	-0.002	-0.01	7.5	1.1	-0.1	K2 III		330 s		
210191	164888	35 Aqr	22 08 58.9	-18 31 11	0.000	-0.01	5.7	0.1	-2.5	B2 V	-5	430 s		
210142	213589		22 08 59.5	-36 03 22	-0.003	-0.02	7.3	1.1	0.2	K0 III		230 s		
210479	34108		22 09 00.0	+58 51 07	+0.003	0.00	8.0	1.1	0.2	K0 III		320 s		
210404	51693		22 09 01.5	+48 31 08	+0.009	+0.05	7.3	0.9	3.2	G5 IV		66 s		
210265	127326		22 09 03.9	+9 27 04	0.000	-0.01	7.9	0.1	0.6	A0 V		280 s		
210387	51691		22 09 05.2	+44 51 17	+0.003	0.00	6.77	-0.04		B9	-9			
210267	127327		22 09 09.0	+2 44 02	+0.001	-0.02	6.6	0.1	0.6	A0 V		160 s		
210354	72064	27 Peg	22 09 13.4	+33 10 21	-0.005	-0.06	5.58	1.00	0.3	gG6	-6	99 s	15672	m
210405	51698		22 09 15.0	+44 50 48	+0.001	+0.01	6.73	-0.03		B9	-5		15679	m
210388	72067		22 09 22.3	+35 07 46	+0.003	-0.03	7.3	0.6	4.4	G0 V		38 s		
209840	258002		22 09 24.8	-70 54 16	+0.007	-0.01	6.9	0.7	0.2	K0 III		220 s		
210269	145905		22 09 26.7	-8 11 09	+0.004	+0.05	7.0	0.9	3.2	G5 IV	-46	57 s		
210457	51705		22 09 28.1	+46 55 15	+0.003	0.00	7.8	0.5	4.0	F8 V		56 s		
210244	190922		22 09 28.9	-23 39 32	+0.003	+0.01	6.5	1.2	3.2	G5 IV		25 s		
210277	145906		22 09 29.7	-7 32 55	+0.006	-0.44	6.6	0.6	4.4	G0 V	-24	18 mx		
210193	231001		22 09 34.7	-41 13 28	+0.017	-0.05	7.85	0.66	5.2	G5 V		34 s		
210373	90278		22 09 39.3	+22 14 33	+0.006	0.00	7.6	1.1	0.2	K0 III		250 mx		
210482	51709		22 09 42.0	+43 38 42	-0.002	-0.02	7.8	1.1	-0.1	K2 III		370 s		
210641	19909		22 09 44.5	+64 07 22	+0.001	+0.01	7.1	0.1	1.4	A2 V		130 s		
210496	51711		22 09 45.5	+43 11 12	0.000	+0.01	7.5	1.4	-0.3	K5 III		360 s		
210807	10265	24 Cep	22 09 48.2	+72 20 29	+0.007	+0.01	4.79	0.92	0.3	G8 III	-15	79 s		

HD	SAO	Star Name	α 2000	δ 2000	μ(α)	μ(δ)	V	B-V	M$_v$	Spec	RV	d(pc)	ADS	Notes
			h m s	° ′ ″	s	″								
210271	213599		22 09 55.5	-34 00 53	-0.001	+0.04	5.37	0.24		A5	+2	17 mn		
210512	51714		22 09 56.4	+44 36 33	+0.007	0.00	8.0	1.4	-0.3	K5 III		270 mx		
210204	231005		22 09 57.8	-48 06 26	+0.008	-0.04	6.43	1.39		K2				
210459	72077	29 π Peg	22 09 59.1	+33 10 42	-0.001	-0.02	4.29	0.46	-0.6	F5 II-III	+2	96 s		
210300	190930		22 10 00.0	-28 17 33	+0.002	+0.03	6.5	0.5	1.7	A3 V		52 s		
210391	127338		22 10 00.7	+4 05 55	+0.007	-0.04	7.7	0.6	4.4	G0 V		45 s		
210444	90281		22 10 02.1	+23 07 34	-0.002	+0.01	7.60	0.5	3.4	F5 V		69 s		m
210390	127339		22 10 02.7	+7 57 14	-0.001	0.00	8.00	0.9	0.3	G8 III		350 s		m
209036	258923		22 10 03.6	-83 21 29	+0.017	+0.01	7.4	1.5	0.2	K0 III		140 s		
210513	72083		22 10 06.4	+35 12 51	-0.001	-0.02	6.9	0.1	1.4	A2 V		120 s		
210499	72081		22 10 07.1	+32 20 33	+0.005	-0.01	7.90	1.1	0.2	K0 III		290 mx	15689	m
210302	213602	15 τ PsA	22 10 08.6	-32 32 54	+0.034	+0.02	4.92	0.48	3.4	dF5	-15	22 ts		
210628	34126		22 10 09.9	+56 05 02	-0.001	0.00	6.94	0.09	-0.9	B6 V	-21	290 s		
210616	34124		22 10 11.2	+54 33 29	+0.001	-0.01	7.8	0.1	0.6	A0 V		250 s		
210565	51719		22 10 11.5	+44 04 34	-0.002	-0.01	8.00	1.1	-0.1	K2 III		400 s	15692	m
210418	127340	26 θ Peg	22 10 11.8	+6 11 52	+0.018	+0.03	3.53	0.08	1.4	A2 V	-6	25 ts		Biham
210873	10267		22 10 15.1	+72 06 40	-0.002	-0.02	6.37	-0.06		A0	-3			
210514	72084		22 10 16.7	+32 17 16	0.000	-0.01	7.30	1.6	-0.5	M4 III	-25	360 s		
210460	107706		22 10 18.8	+19 37 01	+0.006	-0.07	6.18	0.69	4.4	G0 V	+40	19 s		
210759	19917		22 10 20.1	+66 18 16	+0.004	0.00	8.0	0.1	0.6	A0 V		270 s		
210419	145914		22 10 20.9	-3 53 39	+0.001	-0.05	6.27	-0.01		A0				
210236	247337		22 10 21.2	-51 57 37	+0.024	-0.10	7.5	0.7	4.0	F8 V		41 s		
210461	107707		22 10 22.0	+14 37 48	+0.002	-0.02	6.33	1.08	0.2	K0 III	-42	150 s	15690	m
210743	19918		22 10 25.9	+64 20 27	-0.007	-0.01	8.01	0.38	0.6	A0 V		210 s		
210483	107708		22 10 26.1	+18 47 46	+0.025	-0.06	7.9	0.7	4.5	G1 V	-72	49 s		
210501	107709		22 10 27.5	+16 42 28	+0.001	-0.02	7.8	0.1	0.6	A0 V		270 s		
210770	19919		22 10 27.6	+65 31 28	-0.004	+0.02	7.66	-0.01	0.6	A0 V		170 mx		
210516	90287	28 Peg	22 10 29.9	+20 58 40	-0.001	-0.01	6.40	0.12	0.1	A4 III	+8	180 s		
210422	164907		22 10 31.6	-10 49 14	+0.003	+0.03	7.0	1.1	0.2	K0 III	+1	230 s		
210349	213608		22 10 32.0	-33 49 38	-0.001	-0.03	7.80	0.6	4.4	G0 V		48 s		m
---	51727		22 10 32.3	+47 08 27	0.000	+0.01	8.0	1.2						
210434	145916		22 10 33.5	-4 16 02	+0.005	0.00	6.01	0.98	0.2	gK0	-18	150 s		
210629	51728		22 10 34.3	+47 54 49	-0.001	-0.01	7.70	0.5	3.4	F5 V		72 s	15698	m
210339	213609		22 10 37.0	-37 44 59	+0.020	-0.07	7.2	0.1	3.4	F5 V		57 s		
210424	164910	38 Aqr	22 10 37.3	-11 33 54	+0.002	+0.01	5.46	-0.12	-1.9	B6 III	+3	290 s		
210502	107712		22 10 37.4	+11 37 29	-0.002	-0.05	5.78	1.58	-0.3	K5 III	+17	150 s		
210884	19922		22 10 38.7	+70 07 58	-0.012	+0.03	5.44	0.40	3.0	F2 V	+1	29 ts	15719	m
210097	255160		22 10 39.6	-68 19 45	+0.022	-0.05	7.0	1.1	0.2	K0 III		160 mx		
209855	258003		22 10 42.4	-75 52 51	+0.012	0.00	6.55	1.18		K2				
210902	10272		22 10 43.7	+71 07 40	+0.011	-0.09	7.7	0.7	4.4	G0 V		44 s		
210272	247339		22 10 49.7	-55 27 26	+0.024	-0.14	7.22	0.66	4.4	G0 V		32 s		
210195	---		22 10 51.0	-64 01 42			8.0	1.1	0.2	K0 III		360 s		
210745	34137	21 ζ Cep	22 10 51.1	+58 12 05	+0.002	+0.01	3.35	1.57	-4.4	K1 Ib	-18	220 s		
210205	255163		22 10 51.2	-63 05 59	+0.017	-0.01	7.9	0.6	4.4	G0 V		50 s		
210594	72095		22 10 51.4	+30 33 12	0.000	-0.01	6.32	0.18		A5	+4			
210055	258004		22 10 51.8	-71 58 36	0.000	-0.02	7.9	0.1	0.6	A0 V		290 s		
210760	34141		22 10 54.8	+57 56 29	+0.007	+0.04	7.42	0.18	1.9	A4 V		120 s	15706	m
210437	190943		22 10 54.9	-24 42 53	+0.003	-0.06	8.0	0.9	0.2	K0 III		190 mx		
210696	51735		22 10 59.3	+49 02 51	-0.001	0.00	7.5	1.4	-0.3	K5 III		350 s		
210808	19921		22 10 59.4	+63 23 59	0.000	+0.01	7.32	0.19	-1.0	B5.5 V		340 s	15712	m
210395	213615		22 11 01.6	-39 32 51	-0.001	-0.08	8.0	0.4	4.4	G0 V		52 s		
210464	190945		22 11 02.2	-21 13 57	+0.009	-0.03	6.09	0.50	3.7	dF6	-13	29 s		
210396	231018		22 11 02.9	-40 16 52	-0.004	-0.04	8.0	0.5	0.2	K0 III		300 mx		
210666	51734		22 11 03.6	+43 01 51	-0.002	-0.01	6.7	1.1	0.2	K0 III		200 s		
210697	51738		22 11 04.5	+48 40 40	0.000	+0.01	6.57	-0.06	0.4	B9.5 V		170 s		
210952	10274		22 11 06.2	+70 58 35	0.000	+0.05	7.9	0.1	1.4	A2 V		190 s		
210682	51737		22 11 09.3	+43 31 41	-0.001	-0.02	7.4	1.1	-0.1	K2 III		310 s		
210715	34143		22 11 09.7	+50 49 24	+0.014	+0.04	5.40	0.15		A2	-8	31 mn	15708	m
210667	72103		22 11 11.8	+36 15 24	+0.003	-0.23	7.27	0.82	5.9	K0 V		19 s		
210646	90303		22 11 14.6	+27 44 13	0.000	0.00	7.1	0.1	0.6	A0 V	+10	200 s		
210761	34145		22 11 14.9	+52 18 45	0.000	0.00	8.00	0.6	-3.3	G1 Ib-II		1300 s		
210303	255166		22 11 16.4	-60 32 52	+0.005	+0.01	7.9	0.2	3.0	F2 V		94 s		
210731	51739		22 11 20.8	+47 08 54	-0.003	+0.02	7.8	0.5	4.0	F8 V		56 s		
210698	72107		22 11 21.4	+39 42 41	-0.002	-0.04	7.40	1.4	-0.3	K5 III	-15	330 s		
210661	90305		22 11 23.0	+29 15 16	+0.001	-0.01	7.5	0.1	1.4	A2 V	-17	170 s		
210274	255165		22 11 23.7	-64 00 50	+0.001	+0.05	7.2	1.6	-0.5	M2 III		350 s		
210524	190953		22 11 26.9	-27 09 09	0.000	-0.03	7.0	0.7	0.2	K0 III		230 s		
210441	231019		22 11 27.0	-43 50 34	+0.005	+0.02	6.60	1.00	3.2	G8 IV		39 s		
210684	72108		22 11 28.3	+32 05 10	+0.004	+0.01	7.80	0.4	2.6	F0 V		110 s	15709	m
210839	34149	22 λ Cep	22 11 30.5	+59 24 53	0.000	-0.01	5.04	0.25		O6.8	-74	42 mn		
210525	213625		22 11 38.0	-34 27 54	+0.003	-0.01	7.00	0.4	3.0	F2 V		63 s		m
210809	34147		22 11 38.5	+52 25 48	-0.001	+0.01	7.56	0.05	-6.1	O9 Ib	-80			
210471	231020		22 11 40.8	-44 44 56	+0.012	-0.09	7.9	0.6	3.4	F5 V		62 s		
210800	51748		22 11 44.1	+49 56 12	+0.013	-0.03	7.8	0.8	3.2	G5 IV		83 s		
210747	72115		22 11 44.7	+38 24 57	0.000	-0.01	7.3	1.6	-0.5	M2 III		340 s		

HD	SAO	Star Name	α 2000	δ 2000	μ(α)	μ(δ)	V	B-V	M$_V$	Spec	RV	d(pc)	ADS	Notes
			h m s	° ′ ″	s	″								
210885	34156		22 11 48.0	+59 43 15	-0.001	-0.01	7.80	0.9	-2.1	G8 II	-4	750 s	15729	m
210855	34151		22 11 48.5	+56 50 21	+0.028	+0.13	5.24	0.51	4.0	F8 V	-19	21 mn		m
210702	107729		22 11 51.1	+16 02 26	-0.001	-0.01	5.95	0.95	0.0	K1 III	+11	160 s		
210686	127363		22 11 54.6	+6 53 47	-0.001	0.00	7.00	0.4	2.6	F0 V		76 s	15716	m
210056	258006		22 11 55.2	-76 06 59	-0.010	-0.04	6.15	1.00		K0				
210732	90315		22 11 56.4	+23 46 45	+0.001	+0.02	7.9	1.6	-0.5	M4 III		490 s		
210905	34158		22 11 56.7	+59 05 05	+0.016	+0.09	6.30	1.13		K0	-28			
210819	51749		22 11 57.6	+50 11 47	+0.003	+0.01	7.30	0.4	2.6	F0 V		85 s	15727	m
210772	72119		22 11 58.4	+37 39 04	-0.005	-0.01	7.80	0.5	4.0	F8 V		57 s	15723	m
210820	51750		22 12 01.6	+47 05 46	+0.001	+0.03	6.65	-0.03	0.6	A0 V		160 s		
210939	19932		22 12 01.7	+60 45 34	-0.001	+0.02	5.35	1.17	0.0	K1 III	-3	110 s		
210571	213631		22 12 02.0	-38 18 10	-0.002	+0.02	7.60	0.4	3.0	F2 V		83 s		m
210788	72120		22 12 02.5	+37 53 14	+0.001	+0.01	8.0	1.1	0.2	K0 III		350 s		
210940	34159		22 12 02.5	+60 05 26	-0.003	+0.01	7.70	1.4	-0.3	K5 III		360 s	15737	m
210980	19934		22 12 04.5	+63 29 09	+0.001	+0.01	7.7	1.1	-0.1	K2 III		320 s		
210857	51754		22 12 07.1	+49 15 34	+0.003	+0.02	7.6	0.4	2.6	F0 V		100 s		
210762	90317		22 12 07.8	+24 57 00	0.000	-0.02	5.92	1.51	0.2	K0 III	-3	69 s		
210717	127365		22 12 08.8	+10 09 20	+0.002	-0.01	8.0	1.1	-0.1	K2 III		410 s		
211014	19935		22 12 12.9	+63 44 18	+0.002	0.00	8.00	0.1	0.6	A0 V		280 s	15742	m
210687	164922		22 12 16.7	-11 03 56	+0.001	0.00	7.3	1.6	-0.5	M2 III		360 s		
210719	127367		22 12 17.5	+2 44 02	+0.001	-0.02	7.1	0.4	2.6	F0 V		78 s		
210876	51759		22 12 19.6	+47 47 55	0.000	-0.02	8.0	1.1	0.2	K0 III		340 s		
210922	34164		22 12 21.0	+55 05 52	+0.001	0.00	7.16	1.36	0.0	K1 III	-13	180 s		
211029	19937		22 12 22.2	+63 17 29	-0.001	0.00	5.79	1.67	-0.5	M2 III	-14	160 s		
210705	164923	39 Aqr	22 12 25.6	-14 11 38	+0.002	-0.04	6.2	0.4	3.0	dF2	+15	43 s		
210801	90321		22 12 26.4	+23 45 32	-0.002	-0.02	7.5	1.1	-0.1	K2 III		330 s		
211374	10285		22 12 29.2	+76 27 42	+0.007	+0.02	7.1	1.1	-0.1	K2 III		250 s		
210622	231028		22 12 29.9	-40 12 10	+0.002	-0.05	6.6	0.4	0.2	K0 III		190 s		
210733	145938		22 12 35.6	-8 00 51	+0.004	-0.04	7.10	0.4	3.0	F2 V		66 s	15725	m
210678	190961		22 12 36.8	-27 05 06	0.000	-0.01	7.4	0.8	0.2	K0 III		280 s		
210572	247354		22 12 38.1	-54 17 27	+0.015	-0.09	8.0	-0.1	4.0	F8 V		62 s		
210752	145939		22 12 43.3	-6 28 08	+0.015	+0.02	7.40	0.52		G0				
210763	145940		22 12 43.6	-4 43 15	-0.004	-0.03	6.39	0.50	3.3	dF4	+2	37 s		
210737	190963		22 12 46.5	-20 53 23	-0.003	-0.03	7.6	0.5	3.4	F5 V		64 s		
210889	72132		22 12 47.7	+34 36 17	+0.002	-0.04	5.33	1.13	-0.1	K2 III	-7	120 s		
211300	10284		22 12 52.7	+73 18 26	+0.005	+0.03	6.08	1.01	-0.9	K0 II-III	+1	200 mx	15764	m
211193	19944		22 12 53.8	+68 59 00	+0.005	+0.01	7.4	0.1	0.6	A0 V		210 s		
210982	51769		22 12 54.2	+49 22 14	-0.001	0.00	7.90	1.4	-0.3	K5 III		410 s	15747	m
210602	247359		22 12 55.9	-55 56 41	+0.006	0.00	7.0	1.5	-0.5	M2 III		320 s		
210739	190967		22 12 57.3	-26 19 40	-0.001	-0.03	6.17	0.17		A2				
211004	34177		22 12 57.7	+51 11 24	+0.002	0.00	8.0	0.9	3.2	G5 IV		90 s		
210657	231031		22 13 00.7	-49 03 14	+0.004	0.00	7.60	0.8	3.2	G5 IV		76 s		m
210923	72139		22 13 02.9	+33 36 12	0.000	+0.06	7.4	0.6	4.4	G0 V		39 s		
210756	190969		22 13 05.1	-26 29 49	+0.005	-0.06	7.63	1.01		K0				
211093	34190		22 13 09.8	+59 17 58	+0.008	+0.03	7.3	0.2	2.1	A5 V		100 s		
210890	107740		22 13 09.8	+18 16 47	0.000	-0.03	6.6	1.1	-0.1	K2 III		220 s		
212710	3721		22 13 10.4	+86 06 29	+0.050	+0.05	5.27	-0.03		A0	+4	19 mn		
212774	3722		22 13 10.7	+86 13 17	+0.034	+0.01	6.7	1.1	0.2	K0 III	-10	200 s		
210925	90327		22 13 11.3	+25 56 26	-0.003	-0.14	6.57	1.03	0.2	K0 III	-61	190 s		
210944	90329		22 13 13.3	+27 19 08	0.000	-0.04	7.20	0.48		F5	+7			
211057	34189		22 13 14.7	+55 18 50	-0.002	+0.01	7.61	0.08	-0.6	B8 IV		350 s		
211047	51775		22 13 18.3	+49 22 29	+0.001	-0.05	7.5	1.1	-0.1	K2 III		320 s		
210860	145946		22 13 20.1	+0 14 32	+0.003	+0.01	7.9	1.4	-0.3	K5 III		450 s		
211070	34194		22 13 20.8	+55 28 16	+0.003	+0.02	7.57	1.09	-0.1	K2 III		340 s		
210955	72143		22 13 21.9	+31 22 57	+0.001	0.00	7.8	1.1	-0.1	K2 III		390 s		
211071	34192		22 13 23.8	+53 09 14	+0.001	+0.01	7.6	0.2	2.1	A5 V		120 s		
210862	145948		22 13 25.1	-1 39 13	+0.008	-0.08	7.2	1.1	-0.1	K2 III		130 mx		
210845	164935		22 13 26.2	-11 55 34	+0.002	-0.02	7.1	0.8	3.2	G5 IV	-8	59 s		
210562	255174		22 13 36.3	-66 46 59	-0.011	-0.02	7.4	1.1	-0.1	K2 III		220 mx		
211094	51780		22 13 37.9	+49 28 17	+0.011	+0.03	7.3	0.5	3.4	F5 V		60 s		
211006	90337		22 13 38.5	+28 36 30	+0.005	0.00	5.89	1.15	-0.1	gK2	-19	160 s		
211125	34201		22 13 41.4	+52 23 31	-0.001	0.00	8.0	0.0	0.0	B8.5 V		360 s		
210563	255175		22 13 43.0	-68 07 36	+0.006	-0.01	6.7	2.0	-0.1	K2 III		70 s		
210726	247363		22 13 43.4	-52 15 11	+0.005	-0.02	7.9	0.3	3.4	F5 V		78 s		
210848	190976		22 13 44.2	-25 10 51	+0.005	+0.02	5.58	0.49	4.0	F8 V	-28	20 s		
210945	127385		22 13 44.2	+7 42 54	+0.001	0.00	7.90	1.1	-0.1	K2 III		400 s	15750	m
211096	51783		22 13 49.1	+45 26 27	+0.007	+0.01	5.5	0.1	0.6	A0 V	-9	94 s		
211242	19948		22 13 49.3	+63 09 45	-0.003	0.00	6.11	-0.08		B9	+12			
210947	127386		22 13 50.2	+0 52 35	+0.004	-0.04	7.9	0.5	4.0	F8 V		60 s		
210926	145954		22 13 51.3	-4 27 05	0.000	0.00	7.2	1.1	0.2	K0 III		260 s		
210795	231040		22 13 52.4	-45 48 22	+0.002	0.00	7.6	0.9	0.2	K0 III		150 s		
211073	72155		22 13 52.5	+39 42 54	+0.003	+0.01	4.49	1.39	-0.2	K3 III	-11	77 s	15758	m
210851	213651		22 13 52.8	-34 45 34	+0.005	+0.02	7.9	-0.2	2.6	F0 V		110 s		
211149	34205		22 13 53.4	+53 16 06	0.000	0.00	7.92	1.60	-0.3	K5 III		380 s		
211386	19956		22 13 58.1	+69 53 54	-0.001	+0.03	7.84	1.10	0.2	K0 III		290 s		

HD	SAO	Star Name	α 2000	δ 2000	μ(α)	μ(δ)	V	B-V	M_v	Spec	RV	d(pc)	ADS	Notes
			h m s	° ′ ″	s	″								
210984	127391		22 13 58.8	+9 28 49	+0.003	+0.03	8.0	1.1	0.2	K0 III		360 s		
210899	190981		22 14 01.5	−24 37 28	0.000	+0.02	7.7	1.8	0.2	K0 III		110 s		
210741	247365		22 14 02.5	−57 13 06	+0.003	0.00	7.5	2.2	−0.5	M2 III		180 s		
210882	213653		22 14 05.3	−30 28 40	+0.001	−0.07	7.7	1.8	−0.1	K2 III		150 s		
211178	34208		22 14 07.3	+51 26 31	0.000	−0.01	8.0	1.1	−0.1	K2 III		370 s		
210988	145957		22 14 14.0	−5 44 46	+0.005	+0.01	7.8	0.5	3.4	F5 V		74 s		
210960	190986	41 Aqr	22 14 17.8	−21 04 27	+0.001	+0.07	5.32	0.80	0.2	K0 III	−24	110 s	15753	m
211076	107756		22 14 18.2	+17 11 22	−0.006	−0.08	6.42	1.21	−0.3	K4 III	−35	180 mx	15763	m
210934	190985	16 λ PsA	22 14 18.6	−27 46 02	+0.002	0.00	5.43	−0.16	−1.2	B8 III	−6	210 s		
211209	34212		22 14 20.2	+52 56 23	−0.002	0.00	7.90	1.1	−0.1	K2 III		360 s		
210972	164942		22 14 20.3	−19 14 56	+0.002	−0.04	7.2	1.0	3.4	F5 V		27 s		
211138	72160		22 14 21.4	+36 49 54	+0.001	0.00	7.5	0.1	0.6	A0 V		230 s		
211018	145960		22 14 21.9	+0 02 56	0.000	−0.01	8.0	1.4	−0.3	K5 III		450 s		
211227	34214		22 14 24.0	+52 30 49	0.000	0.00	7.90	1.1		K2 IV				
211048	127402		22 14 29.0	+7 58 35	+0.001	0.00	6.80	0.1	0.6	A0 V		170 s	15767	m
210385	258012		22 14 32.2	−76 56 32	+0.018	0.00	7.9	0.5	4.0	F8 V		60 s		
211139	90348		22 14 34.5	+29 34 21	+0.003	−0.03	7.60	0.5	3.4	F5 V	−18	69 s	15769	m
211022	164948		22 14 36.9	−15 06 08	−0.002	−0.10	7.1	0.4	3.0	F2 V		65 s		
211038	164949		22 14 38.0	−15 49 07	0.000	−0.36	6.54	0.90	5.6	G8 V	+12	17 mn		
210918	231045		22 14 38.4	−41 22 54	+0.050	−0.79	6.23	0.65	5.2	G5 V	−18	14 mx		
211062	145963		22 14 41.9	−6 24 02	+0.002	−0.01	7.8	0.1	1.4	A2 V		190 s		
211264	51799		22 14 43.0	+49 21 16	−0.001	+0.02	7.60	0.5	4.0	F8 V		52 s	15778	m
211037	164952		22 14 44.2	−15 42 07	0.000	−0.04	7.9	1.1	−0.1	K2 III		390 s		
211211	51797		22 14 44.2	+42 57 14	+0.005	−0.02	5.7	0.1	0.6	A0 V	−38	100 s		
211115	127404		22 14 46.7	+4 16 57	0.000	+0.02	7.0	1.1	0.2	K0 III		230 s		
211024	190995		22 14 47.1	−24 00 16	+0.009	0.00	6.9	0.7	3.4	F5 V		35 s		
211152	90350		22 14 47.8	+24 18 54	+0.004	+0.03	7.0	0.5	3.4	F5 V		52 s		
211153	90349		22 14 48.4	+22 31 25	0.000	0.00	6.70	1.1	0.2	K0 III		200 s	15771	m
211512	19962		22 14 52.8	+70 00 06	+0.005	+0.03	7.83	0.98	1.8	G8 III-IV		150 s		
211099	145968		22 14 54.0	−6 44 07	+0.002	+0.02	7.7	0.0	0.4	B9.5 V		290 s		
210919	247373		22 14 59.3	−52 11 48	+0.001	+0.01	7.5	1.3	0.2	K0 III		200 s		
211336	34227	23 ε Cep	22 15 01.8	+57 02 37	+0.054	+0.05	4.19	0.28	1.7	F0 IV	−1	30 ts		m
211301	34225		22 15 03.0	+51 29 05	0.000	+0.01	7.80	0.1	0.6	A0 V		260 s	15783	m
211274	51806		22 15 03.2	+43 46 25	−0.004	−0.04	6.9	1.1	0.2	K0 III		210 s		
211402	34231		22 15 04.8	+59 06 52	−0.002	−0.03	7.00	0.04	1.4	A2 V		130 s		
211065	190997		22 15 07.5	−25 38 54	+0.001	−0.02	7.5	1.5	−0.3	K5 III		370 s		
211183	127406		22 15 15.4	+7 01 39	+0.001	−0.02	7.2	1.1	0.2	K0 III		250 s		
210742	258017		22 15 17.1	−70 39 10	+0.011	−0.04	7.2	1.1	0.2	K0 III		230 mx		
210963	247376		22 15 18.1	−52 42 03	+0.003	+0.02	8.0	0.0	4.0	F8 V		62 s		
211322	51811		22 15 28.6	+40 52 04	0.000	−0.02	7.9	1.1	0.2	K0 III		330 s		
211338	51814		22 15 29.1	+43 44 55	+0.003	+0.03	7.6	0.1	0.6	A0 V		240 s		
211430	34238		22 15 29.2	+55 49 07	0.000	+0.01	7.47	−0.05	−0.3	B9 IV		360 s		
211052	231052		22 15 30.3	−43 38 06	+0.011	0.00	7.1	0.9	0.2	K0 III		180 mx		
211053	231053		22 15 34.9	−44 27 07	+0.001	−0.01	6.10	1.02	0.2	gK0		150 s		
211088	231055	μ¹ Gru	22 15 36.7	−41 20 48	+0.005	+0.03	4.79	0.80	0.3	gG4	−7	79 s		
211231	127416		22 15 38.4	+4 16 16	−0.003	−0.02	7.4	0.4	2.6	F0 V		91 s		
211472	34243		22 15 54.0	+54 40 21	+0.025	+0.05	7.52	0.81	5.9	K0 V	−7	21 s	15797	m
211376	72190		22 15 54.0	+38 51 26	0.000	+0.01	7.9	1.1	0.2	K0 III		330 s		
211133	231058		22 15 56.4	−42 41 50	−0.001	−0.03	7.3	2.3	−0.3	K5 III		250 s		
211388	72191	1 Lac	22 15 58.1	+37 44 56	+0.001	+0.01	4.13	1.46	−1.3	K3 II-III	−8	100 s		
211287	127420		22 15 59.5	+8 33 00	0.000	+0.01	6.21	0.02		A0	0			
211489	34244		22 15 59.8	+54 49 34	+0.005	+0.04	7.98	0.25	2.8	F1 V	−12	110 s		
211304	107778		22 16 00.5	+11 45 37	+0.002	+0.01	7.0	0.0	0.4	B9.5 V	+6	210 s		
211234	164962		22 16 00.7	−14 26 09	+0.002	+0.01	7.84	1.24	−0.1	gK2	−20	330 s		
211431	51817		22 16 00.8	+45 05 21	+0.001	+0.01	7.2	0.1	1.7	A3 V		120 s		
211236	191009		22 16 07.1	−23 00 35	+0.001	+0.01	7.3	0.3	1.7	A3 V		94 s		
211055	247382		22 16 07.9	−57 34 06	+0.001	−0.02	7.20	1.6	−0.5	M4 III		350 s		m
211458	51821		22 16 15.8	+43 31 03	+0.002	+0.03	8.0	1.1	−0.1	K2 III		390 s		
211540	34254		22 16 15.8	+57 58 31	+0.001	+0.01	7.95	1.95	−4.4	K5 Ib		1700 s		
211406	90374		22 16 16.0	+30 02 52	0.000	−0.02	8.0	0.5	4.0	F8 V		62 s		
211419	72193		22 16 17.3	+33 43 57	+0.009	+0.01	7.20	0.5	4.0	F8 V		44 s	15798	m
211589	19970		22 16 18.6	+62 18 56	0.000	−0.03	7.0	0.1	1.7	A3 V		110 s		
211554	34256		22 16 26.3	+57 13 13	+0.005	+0.01	5.88	0.93	0.3	G8 III	−8	130 s		
211202	231063	μ² Gru	22 16 26.5	−41 37 39	−0.001	−0.01	5.10	0.92	0.3	gG5	+13	86 s		
211045	255183		22 16 28.5	−63 20 19	+0.008	−0.06	7.3	0.5	3.4	F5 V		59 s		
211432	90379		22 16 29.5	+27 48 16	−0.001	+0.01	6.37	0.99	0.2	K0 III	+16	170 s		
211446	72195		22 16 30.1	+32 37 18	0.000	+0.01	7.5	1.4	−0.3	K5 III		370 s		
211642	19973		22 16 32.0	+63 42 45	+0.001	+0.02	7.4	0.9	3.2	G5 IV		67 s		
212664	3725		22 16 33.3	+84 30 16	−0.006	−0.01	7.5	0.1	0.6	A0 V		240 s		
211356	145989		22 16 33.5	−1 35 47	−0.002	−0.01	6.15	0.19		A2				
211291	191016		22 16 37.3	−25 53 54	+0.003	−0.01	6.15	1.11		K0				
211420	107786		22 16 40.0	+12 55 13	+0.001	+0.01	7.5	1.1	0.2	K0 III		290 s		
211433	90381		22 16 40.0	+22 53 47	−0.003	−0.01	6.8	0.1	0.6	A0 V	+5	180 s		
211460	90382		22 16 40.9	+29 10 24	+0.012	+0.03	6.7	0.9	3.2	G5 IV	−39	51 s		
211360	164973		22 16 45.2	−11 38 57	+0.004	−0.03	7.2	0.9	3.2	G5 IV		64 s		

HD	SAO	Star Name	α 2000	δ 2000	μ(α)	μ(δ)	V	B–V	M_v	Spec	RV	d(pc)	ADS	Notes
			$22^h 16^m 46^s.5$	$-30°29'20''$	$-0^s.001$	$+0''.01$								
211294	213691		22 16 46.5	-30 29 20	-0.001	+0.01	7.8	0.5	3.2	. G5 IV		85 s		
211361	164974	42 Aqr	22 16 47.9	-12 49 53	+0.001	+0.01	5.34	1.14	0.2	K0 III	+13	84 s		
211474	90385		22 16 47.9	+30 05 29	0.000	0.00	7.6	0.0	0.0	B8.5 V	-34	330 s		
211391	145991	43 θ Aqr	22 16 49.9	-7 47 00	+0.008	-0.02	4.16	0.98	1.8	G8 III-IV	-15	26 s		Ancha
211392	145992		22 16 52.4	-9 02 24	-0.003	-0.01	5.79	1.16	-0.2	gK3	+12	160 s		
211497	72200		22 16 54.6	+33 34 20	-0.001	0.00	8.0	1.1	0.2	K0 III		360 s		
211363	191019		22 16 54.8	-20 01 32	+0.007	-0.06	7.4	0.3	1.4	A2 V		51 mx		
211380	164979		22 16 56.4	-14 39 24	+0.001	-0.01	7.14	0.47		F5	+19			
211364	191021		22 16 59.7	-23 08 24	+0.006	-0.05	6.3	0.9	3.2	G5 IV		33 s		
211643	34268		22 17 00.9	+56 10 36	+0.005	-0.01	7.08	0.11		A m				
211349	213694		22 17 04.0	-34 12 53	+0.002	-0.01	8.0	1.1	3.2	G5 IV		59 s		
211434	145993	44 Aqr	22 17 06.3	-5 23 14	0.000	+0.03	5.75	0.88	0.3	gG4	+7	120 s		
211367	213696		22 17 09.9	-34 45 42	+0.002	-0.06	6.8	1.0	0.2	K0 III		170 mx		
211515	90388		22 17 10.2	+22 23 15	-0.002	-0.01	7.5	0.1	1.4	A2 V		170 s		
211746	19982		22 17 11.9	+66 07 44	+0.001	+0.02	7.03	0.12	0.6	A0 V	-14	160 s		
211261	247391		22 17 14.5	-54 19 18	+0.005	-0.03	7.3	0.3	2.6	F0 V		83 s		
211476	107794		22 17 15.0	+12 53 55	+0.058	+0.10	7.04	0.60	4.7	G2 V	-30	31 ts		
211500	107796		22 17 16.3	+15 02 48	-0.001	-0.06	7.1	1.1	0.2	K0 III		240 s		
211395	191022		22 17 17.5	-27 23 23	+0.006	-0.06	7.1	1.2	0.2	K0 III		160 mx		
211555	90395		22 17 24.7	+26 23 13	+0.003	+0.03	7.3	1.1	0.2	K0 III		260 s		
211622	51841		22 17 25.1	+45 16 46	-0.001	-0.02	8.0	1.4	-0.3	K5 III		430 s		
--	51842		22 17 26.8	+49 11 46	-0.012	+0.09	7.9	0.5						
211463	164984		22 17 27.0	-15 58 47	+0.002	+0.01	7.1	1.1	0.2	K0 III		240 s		
211660	51844		22 17 28.4	+49 07 37	+0.004	+0.01	6.60	1.1	0.2	K0 III		190 s	15814	m
211516	127434		22 17 29.6	+5 08 39	0.000	0.00	7.7	1.6	-0.5	M2 III		440 s		
211450	191026		22 17 31.2	-23 06 51	0.000	+0.01	7.5	1.1	0.2	K0 III		260 s		
211867	19986		22 17 31.8	+70 03 30	-0.004	-0.05	7.22	1.02	0.3	G8 III		220 s		
211425	213706		22 17 37.9	-31 45 59	+0.004	+0.01	7.1	0.6	3.2	G5 IV		60 s		
211694	34278		22 17 38.3	+51 19 05	-0.001	+0.01	7.60	-0.08	0.0	B8.5 V		330 s		
211662	51848		22 17 44.7	+42 46 58	+0.017	+0.08	8.0	1.1	0.2	K0 III		86 mx		
211224	255184		22 17 45.3	-65 01 19	+0.007	-0.10	7.9	0.1	1.7	A3 V		35 mx		
211606	90396		22 17 47.0	+26 56 12	-0.003	0.00	6.7	1.4	-2.3	K5 II	-10	640 s		
211645	72214		22 17 49.7	+39 01 33	+0.004	+0.02	7.37	1.22	0.2	G9 III	-25	180 s		
210853	258020	ψ Oct	22 17 49.9	-77 30 42	-0.016	+0.01	5.51	0.31	1.3	A6 IV		60 s		
211702	51852		22 17 50.9	+49 49 48	-0.002	-0.01	8.0	1.1	-0.1	K2 III		390 s		
211483	213709		22 17 55.7	-31 09 05	-0.003	-0.10	7.7	1.8	-0.1	K2 III		120 mx		
211624	107805		22 18 00.1	+19 02 25	+0.004	-0.02	7.9	1.1	0.2	K0 III		300 mx		
211299	255187		22 18 02.1	-62 48 42	+0.003	-0.01	7.60	0.5	3.4	F5 V		69 s		m
211607	107807		22 18 04.0	+13 57 01	+0.006	-0.01	7.0	1.1	0.2	K0 III		230 mx		
211575	146004		22 18 04.1	-0 14 16	-0.003	-0.06	6.39	0.44		F5				
--	72221		22 18 09.2	+39 58 07	+0.004	+0.03	7.9	1.2						
211547	191033		22 18 11.4	-22 17 52	+0.004	+0.03	7.6	0.0	1.7	A3 V		160 s		
211833	19991	25 Cep	22 18 12.6	+62 48 16	+0.006	+0.02	5.75	1.26	0.0	K1 III	-2	100 s		
211415	247400		22 18 15.4	-53 37 40	+0.048	-0.66	5.37	0.60	4.5	G1 V	-14	14 ts		m
211506	213713		22 18 16.9	-36 35 48	+0.004	-0.05	7.2	0.2	2.6	F0 V		82 s		
212150	10312		22 18 20.1	+76 29 18	+0.002	+0.02	6.6	0.1	0.6	A0 V	-18	150 s		
211880	19993		22 18 27.6	+63 13 23	-0.001	0.00	7.34	0.32		B0.5 V				m
211569	191039		22 18 28.8	-23 19 53	+0.004	-0.05	7.3	0.7	3.4	F5 V		45 s		
211628	146010		22 18 29.5	-1 07 21	+0.004	+0.05	7.6	1.1	-0.1	K2 III		230 mx		
211416	255193	α Tuc	22 18 30.1	-60 15 35	-0.008	-0.04	2.86	1.39	-0.2	K3 III	+42	35 s		
211600	191043		22 18 34.0	-20 44 25	-0.001	-0.04	7.7	1.5	-0.3	K5 III		390 s		
211822	34294		22 18 37.5	+52 39 22	-0.001	+0.01	7.4	0.6	0.4	G2 III	-27	230 s		
211571	213721		22 18 38.4	-30 40 13	+0.001	+0.03	8.0	0.5	1.4	A2 V		100 s		
211523	231074		22 18 40.5	-46 08 13	+0.004	0.00	7.5	0.6	0.2	K0 III		280 s		
211684	127442		22 18 41.2	+7 51 56	-0.002	0.00	8.0	1.1	-0.1	K2 III		410 s		
211732	90409		22 18 44.8	+25 46 27	+0.002	0.00	7.71	0.22	1.7	A3 V		140 s		
211783	51865		22 18 45.7	+42 08 33	-0.002	-0.02	7.8	1.1	-0.1	K2 III		360 s		
--	107818		22 18 46.2	+19 17 10	0.000	0.00	7.9							
211718	107819		22 18 46.8	+19 57 44	+0.005	-0.05	7.0	1.1	-0.1	K2 III		180 mx		
211317	255192		22 18 49.4	-68 18 48	+0.006	-0.19	7.10	0.8	3.2	G5 IV		54 mx		m
211704	127445		22 18 51.7	+7 21 04	0.000	+0.01	8.0	1.4	-0.3	K5 III		450 s		
211733	107820		22 18 52.3	+16 15 29	+0.002	+0.01	6.9	0.1	0.6	A0 V	-27	190 s		
211797	72228		22 18 56.0	+37 46 09	+0.005	+0.05	6.08	0.30	2.6	dF0	+7	48 s	15828	m
211676	164996	45 Aqr	22 19 00.5	-13 18 18	+0.005	-0.01	5.95	1.06	0.2	gK0	+30	130 s		
212027	20006		22 19 01.9	+68 25 54	-0.004	-0.06	7.5	1.1	0.2	K0 III		260 s		
211854	51870		22 19 02.0	+46 00 19	-0.001	+0.01	7.9	1.1	0.2	K0 III		340 s		
212237	10315		22 19 05.7	+76 00 49	+0.008	+0.03	7.4	0.1	0.6	A0 V		130 mx		
211799	90415		22 19 06.9	+28 50 51	+0.002	+0.03	7.2	0.5	4.0	F8 V	-23	43 s		
211811	72230		22 19 13.2	+33 30 24	+0.002	+0.01	7.3	1.1	0.2	K0 III		260 s		
211836	72231		22 19 18.4	+34 32 16	-0.004	-0.06	7.0	1.1	-0.1	K2 III		260 s		
211687	191053		22 19 19.9	-28 12 30	+0.002	+0.01	6.8	2.2	-0.3	K5 III		100 s		
211655	231081		22 19 21.7	-40 17 02	0.000	+0.01	8.0	1.3	-0.3	K5 III		450 s		
211785	107827		22 19 23.3	+14 04 12	0.000	0.00	7.9	0.1	1.4	A2 V		200 s		
211800	107829		22 19 23.5	+15 32 45	0.000	-0.01	7.1	1.4	-0.3	K5 III	-1	310 s		
211786	107828		22 19 24.9	+12 27 36	+0.005	-0.09	8.0	0.9	3.2	G5 IV		93 s		

HD	SAO	Star Name	α 2000	δ 2000	μ(α)	μ(δ)	V	B-V	M$_v$	Spec	RV	d(pc)	ADS	Notes
			h m s	° ′ ″	s	″								
211971	34314		22 19 25.6	+60 08 52	0.000	0.00	6.90	1.00	-5.0	A2 Ib	-17	690 s		
211738	164998		22 19 34.3	-16 42 15	+0.003	-0.07	7.4	0.5	4.0	F8 V		49 s		
211823	90417		22 19 34.8	+21 01 26	-0.002	-0.03	7.7	1.1	0.2	K0 III		320 s		
211982	34316		22 19 40.1	+56 12 15	+0.004	+0.04	7.28	1.12	0.0	K1 III	-17	280 s		
211537	255197		22 19 41.3	-63 16 00	-0.001	+0.03	7.8	0.5	3.4	F5 V		75 s		
211972	34315		22 19 44.5	+52 52 53	+0.002	+0.01	7.73	1.57	-0.3	K5 III		370 s		
211753	191060		22 19 47.7	-27 23 19	-0.004	+0.01	7.6	1.3	0.2	K0 III		190 s		
212332	10323		22 19 48.3	+76 07 26	+0.013	0.00	7.8	0.1	1.4	A2 V		180 s		
211802	146021		22 19 49.0	-4 03 59	0.000	-0.02	7.8	0.1	1.7	A3 V		170 s		
211947	51880		22 19 49.6	+46 54 02	-0.002	+0.01	8.0	1.4	-0.3	K5 III		420 s		
211884	90423		22 19 53.8	+25 43 25	-0.002	-0.04	7.60	1.4	-0.3	K5 III		380 s		
211973	34319		22 19 54.9	+52 00 25	+0.010	-0.01	8.0	0.4	3.0	F2 V		95 s		
211974	51884		22 19 58.4	+47 28 21	0.000	0.00	8.0	0.1	1.4	A2 V		180 s		
210967	258928	ε Oct	22 20 01.3	-80 26 23	+0.020	-0.04	5.10	1.47		M6 III	+12	280 mx		
211538	255198		22 20 01.7	-66 53 07	+0.005	-0.03	7.4	0.4	2.6	F0 V		91 s		
212043	34325		22 20 06.0	+56 55 05	0.000	+0.01	6.56	-0.06	-3.7	B6 II		1100 s		
211838	146023	46 ρ Aqr	22 20 11.7	-7 49 16	+0.001	0.00	5.37	-0.06	-0.2	B8 V	-9	130 s		
211905	107842		22 20 21.1	+11 02 21	+0.002	+0.01	7.90	1.1	0.2	K0 III		350 s		m
212044	34327		22 20 22.6	+51 51 39	0.000	-0.01	6.98	0.04	-3.5	B1 V nnpe	-14	940 s		
211768	231086		22 20 22.9	-43 30 10	+0.001	-0.05	7.6	1.3	0.2	K0 III		210 s		
212003	72247		22 20 26.1	+35 07 11	-0.041	+0.10	7.50	0.4	2.6	F0 V		25 mx	15848	m
211924	127453	30 Peg	22 20 27.4	+5 47 22	+0.001	0.00	5.37	-0.02	-2.2	B5 III	-8	270 s	15847	m
212002	72248		22 20 28.1	+37 17 55	0.000	0.00	6.9	1.1	0.2	K0 III		220 s		
211889	146026		22 20 29.1	-0 03 55	+0.006	+0.05	8.0	0.5	3.4	F5 V		82 s		
212265	20019		22 20 31.5	+69 21 45	-0.019	-0.07	7.9	0.5	3.4	F5 V		76 s		
212106	34333		22 20 34.4	+54 50 34	+0.020	-0.05	7.42	0.49	3.4	F5 V		61 s		
211726	247410		22 20 36.1	-57 30 36	+0.003	-0.01	6.3	1.9	-0.3	K5 III		120 s		
212072	51898		22 20 38.3	+49 12 31	0.000	+0.02	7.5	1.4	-0.3	K5 III		350 s		
212136	34338		22 20 38.3	+58 24 37	-0.005	+0.04	6.53	0.95	0.3	G5 III		160 mx		
212310	10325		22 20 39.5	+71 41 31	+0.004	0.00	7.9	0.1	1.4	A2 V		180 s		
212071	34335		22 20 39.5	+50 58 51	+0.001	+0.01	6.5	1.1	-0.1	K2 III	-9	200 s		
211960	107846		22 20 42.2	+14 01 55	+0.001	-0.03	6.6	1.1	0.2	K0 III		190 s		
211984	90434		22 20 42.6	+23 03 40	-0.002	0.00	7.9	1.1	0.2	K0 III		340 s		
211863	213751		22 20 44.3	-34 31 00	-0.006	-0.03	6.9	1.1	3.2	G5 IV		33 s		
212107	51902		22 20 51.2	+46 03 45	+0.001	-0.04	7.5	0.5	3.4	F5 V		64 s		
211935	146032		22 20 51.3	-8 45 55	+0.002	-0.01	7.4	1.1	0.2	K0 III		280 s		
211976	127460		22 20 55.7	+8 11 12	+0.003	+0.03	6.17	0.44	3.3	dF4	+10	37 s		
212266	20021		22 20 55.8	+65 18 19	0.000	+0.01	8.0	0.1	1.4	A2 V		200 s		
212278	20023		22 20 57.5	+66 57 55	+0.008	+0.03	7.40	0.2	2.1	A5 V		110 s	15870	m
212047	90437		22 20 59.9	+26 56 07	+0.001	0.00	6.4	1.6	-0.5	M2 III	-4	240 s		
212120	51904	2 Lac	22 21 01.4	+46 32 12	+0.002	+0.01	4.57	-0.10	-1.3	B6 IV	-10	150 s	15862	m
212046	72253		22 21 01.6	+31 18 35	-0.001	-0.02	7.70	1.1	0.2	K0 III		320 s	15858	m
212183	34346		22 21 03.6	+55 59 10	+0.002	+0.01	8.01	-0.04	-0.6	B9 III-IV		520 s		
211729	255205		22 21 03.7	-62 32 54	+0.023	-0.04	6.7	0.5	4.4	G0 V		29 s		
212267	20024		22 21 07.7	+63 12 52	-0.004	-0.02	8.04	0.30		A0				
212139	51906		22 21 07.9	+48 14 24	0.000	-0.01	7.8	0.1	0.6	A0 V		260 s		
211511	258024		22 21 11.6	-74 58 03	-0.001	0.00	7.8	1.6	-0.5	M2 III		450 s		
212097	90440	32 Peg	22 21 19.2	+28 19 50	+0.001	+0.01	4.81	0.00	-0.2	B8 V	+8	90 s	15863	m
211866	247418		22 21 22.0	-52 14 55	+0.009	-0.03	7.9	0.3	2.6	F0 V		110 s		
211849	247417		22 21 22.7	-55 45 51	+0.008	+0.04	7.6	0.8	4.0	F8 V		37 s		
212022	146041		22 21 23.8	-6 14 35	+0.002	0.00	7.5	0.1	1.4	A2 V	-11	170 s		
211878	247419		22 21 24.2	-51 06 04	-0.002	+0.03	7.9	0.6	3.4	F5 V		62 s		
212098	90441		22 21 26.6	+21 02 54	+0.001	+0.01	7.9	0.1	0.6	A0 V		290 s		
212074	107852		22 21 27.8	+14 53 49	-0.002	+0.02	8.0	1.1	-0.1	K2 III		410 s		
212075	107853		22 21 28.6	+14 22 13	-0.001	0.00	7.0	0.1	1.7	A3 V	-4	110 s		
212076	107854	31 Peg	22 21 30.9	+12 12 19	+0.001	+0.01	5.01	-0.13	-2.5	B2 V	+10	310 s		
212010	191083	47 Aqr	22 21 35.3	-21 35 54	-0.001	-0.08	5.13	1.07	-0.1	K2 III	+49	110 s		
212061	146044	48 γ Aqr	22 21 39.2	-1 23 14	+0.009	+0.01	3.84	-0.05	0.6	A0 V	-15	28 ts	15864	Sadachbia, m
212079	127475		22 21 40.9	+0 38 37	-0.002	-0.01	8.0	0.1	0.6	A0 V		300 s		
212439	10331		22 21 42.4	+70 15 29	-0.003	-0.01	7.9	0.1	0.6	A0 V		260 s		
212391	20035		22 21 45.7	+66 42 22	+0.004	+0.01	6.70				-2		15881	m
212081	146045		22 21 46.2	-6 10 53	0.000	-0.01	8.0	1.6	-0.5	M2 III	+11	510 s		
212222	51918		22 21 50.7	+42 04 42	+0.001	0.00	6.41	-0.08	-1.1	B5 V	-18	300 s		
212351	20031		22 21 50.8	+62 15 40	-0.004	-0.04	8.0	1.4	-0.3	K5 III		390 s		
212365	20032		22 21 51.3	+63 08 24	+0.001	-0.02	6.9	0.1	0.6	A0 V		170 s		
212212	51919		22 21 52.7	+40 39 57	+0.001	+0.02	6.70	1.1	-0.1	K2 III		220 s	15874	m
212034	191087		22 21 55.0	-25 21 53	+0.001	+0.02	7.6	0.7	0.2	K0 III		310 s		
212247	51922		22 21 58.1	+43 44 42	-0.003	-0.02	8.0	1.1	-0.1	K2 III	-23	390 s		
212083	165023		22 21 59.6	-19 26 08	-0.004	-0.06	7.9	0.7	4.4	G0 V		42 s		
211942	247424		22 22 03.9	-53 51 51	+0.002	-0.02	7.3	0.6	0.2	K0 III		260 s		
212035	213767		22 22 04.2	-34 29 21	-0.001	+0.02	7.3	0.5	3.0	F2 V		56 s		
212186	107860		22 22 05.1	+15 39 06	+0.002	0.00	6.69	-0.01		A0	-1			
212123	165025		22 22 11.4	-10 50 54	-0.003	-0.02	7.9	1.1	0.2	K0 III		340 s		
212280	72275		22 22 32.6	+30 21 27	+0.001	-0.01	7.51	0.70	2.8	G0 IV		75 s	15883	m
212224	127486		22 22 33.8	+9 56 25	+0.002	-0.04	8.00	0.4	2.6	F0 V		120 s		m

HD	SAO	Star Name	α 2000	δ 2000	μ(α)	μ(δ)	V	B−V	M$_V$	Spec	RV	d(pc)	ADS	Notes
			h m s	° ′ ″	s	″								
212226	127485		22 22 34.2	+6 28 23	+0.001	−0.02	8.00	0.1	1.7	A3 V		180 s	15880	m
212018	247426		22 22 36.2	−56 09 11	+0.002	−0.03	7.80	0.1	1.7	A3 V		170 s		m
213047	3737		22 22 37.7	+81 56 21	+0.027	+0.07	7.0	0.4	3.0	F2 V		64 s		
212289	72277		22 22 39.7	+30 44 44	0.000	−0.01	7.86	1.23	−2.2	K1 II		1000 s		
212087	231105	π¹ Gru	22 22 43.7	−45 56 52	+0.001	−0.01	6.62	2.01		S5.7				m,v
213593	3742		22 22 48.0	+85 03 43	+0.028	+0.02	7.2	1.1	0.2	K0 III		250 s		
212315	72280		22 22 49.8	+33 35 44	+0.003	−0.03	7.6	0.5	4.0	F8 V		53 s		
212334	72284		22 22 50.1	+36 39 33	0.000	+0.06	6.5	1.1	0.2	K0 III		180 mx		
212687	10342		22 22 50.7	+75 00 16	+0.016	0.00	8.0	1.1	0.2	K0 III		320 s		
212130	231110		22 22 58.0	−44 03 43	+0.013	−0.01	8.0	0.6	2.6	F0 V		31 s		
212454	34383		22 23 00.0	+57 17 04	+0.002	+0.01	6.16	−0.13		B8				
212495	20042		22 23 00.0	+62 25 12	−0.001	+0.05	6.04	0.05	0.6	A0 V	−15	110 s		
212455	34382		22 23 01.2	+55 13 58	−0.002	−0.01	7.87	0.38	−6.3	B5 Iab	−59	3600 s		
212165	231112		22 23 06.4	−43 24 18	+0.006	+0.01	7.9	1.6	3.2	G5 IV		33 s		
212466	34387		22 23 06.7	+55 57 49	−0.001	+0.01	6.53	2.30	−8.0	G8 Ia		1500 s		RW Cep, v
212132	231111	π² Gru	22 23 07.8	−45 55 42	+0.023	−0.05	5.62	0.36	1.7	F0 IV p		57 mx		
212291	127490		22 23 09.0	+9 27 42	+0.021	+0.06	7.92	0.68	5.3	dG6	−8	33 s		
212231	191101		22 23 14.4	−25 50 35	+0.029	−0.11	7.86	0.61		G2		21 mn		
212317	127491		22 23 14.9	+5 38 48	0.000	−0.04	7.50	0.1	1.4	A2 V		170 s	15889	m
212270	165032		22 23 20.3	−14 57 02	−0.004	−0.15	7.1	0.5	3.4	F5 V		55 s		
212180	231116		22 23 24.4	−46 40 08	−0.001	+0.01	6.92	0.06		A2				
212424	51950		22 23 25.8	+42 03 33	0.000	0.00	7.9	0.1	1.4	A2 V		190 s		
212318	127494		22 23 27.5	+0 36 30	−0.001	−0.01	7.02	0.20	1.4	A2 V		110 s		
212378	107877		22 23 27.8	+17 39 21	0.000	−0.02	7.4	1.1	−0.1	K2 III		320 s		
212271	191105	49 Aqr	22 23 30.7	−24 45 45	+0.008	−0.01	5.53	0.99	0.2	K0 III	−11	120 s		
212320	146062		22 23 32.0	−7 11 40	0.000	+0.01	5.93	1.00	0.3	gG6	−14	120 s		
212337	146063		22 23 32.8	−2 18 05	−0.002	−0.03	7.8	0.5	4.0	F8 V		57 s		
212497	34393		22 23 33.1	+51 37 06	+0.002	0.00	7.7	1.4	−0.3	K5 III		350 s		
212496	34395	3 β Lac	22 23 33.4	+52 13 44	−0.002	−0.18	4.43	1.02	0.2	G9 III	−10	66 s		
212468	51957		22 23 33.8	+45 21 00	+0.003	+0.01	7.60	0.8	3.2	G5 IV		75 s	15900	m
212511	34397		22 23 37.1	+51 11 44	0.000	−0.01	7.3	0.1	1.7	A3 V		130 s		
212395	90462	33 Peg	22 23 39.5	+20 50 54	+0.024	−0.01	6.20	0.5	3.4	F5 V	−23	34 ts	15896	m
−−	107880		22 23 40.5	+19 11 17	−0.001	−0.03	8.0							
212937	10351		22 23 41.2	+78 14 37	−0.003	+0.02	6.5	−0.1		B9				
212566	34405		22 23 50.0	+58 32 05	+0.003	−0.01	7.08	0.31	2.6	F0 V		75 s		
212487	72296		22 23 54.0	+38 34 25	+0.021	+0.12	6.22	0.49	4.0	F8 V	+5	43 mn		
212105	255210		22 23 55.6	−65 21 14	−0.002	0.00	7.5	0.4	2.6	F0 V		95 s		
212470	72295		22 23 56.2	+31 15 44	0.000	+0.02	7.4	1.6	−0.5	M4 III	−5	380 s		
212488	72298		22 23 58.0	+35 06 32	+0.002	−0.01	8.0	0.1	0.6	A0 V		300 s		
212591	34407		22 23 59.6	+59 43 50	+0.014	−0.02	7.8	1.1	0.2	K0 III		160 mx		
212442	107884		22 24 00.5	+15 16 54	+0.003	0.00	6.8	0.1	0.6	A0 V	+5	170 s		
212359	191113		22 24 06.4	−26 22 25	+0.005	−0.02	8.0	−0.1	2.6	F0 V		120 s		
212404	146067	51 Aqr	22 24 06.7	−4 50 13	+0.002	0.00	5.78	−0.04		A0	+6		15902	m
212396	165042		22 24 07.4	−10 11 52	−0.003	−0.02	7.2	1.1	0.2	K0 III		250 s		
212417	146068		22 24 10.3	−2 03 43	+0.001	−0.01	7.8	0.1	1.7	A3 V		160 s		
212556	51967		22 24 12.6	+48 49 19	+0.002	0.00	7.9	0.1	1.4	A2 V		190 s		
212775	10349		22 24 14.5	+71 18 05	+0.007	+0.03	7.6	0.1	1.7	A3 V		130 mx		
212399	191115		22 24 15.9	−21 35 50	−0.001	0.00	7.4	1.6	−0.1	K2 III		180 s		
212308	231120		22 24 22.2	−48 51 38	+0.004	−0.03	7.4	0.8	0.2	K0 III		280 s		
212500	107888		22 24 24.9	+16 15 46	+0.001	−0.02	7.1	0.4	3.0	F2 V	−37	66 s		
212430	165044	50 Aqr	22 24 26.9	−13 31 46	+0.003	+0.01	5.76	0.97	3.2	G5 IV	−21	23 s		
212545	72303		22 24 27.2	+35 26 01	−0.001	−0.01	7.7	0.0	−6.3	B5 Iab		6300 s		
212328	231121		22 24 27.6	−48 48 59	+0.003	−0.02	7.7	1.3	−0.1	K2 III		230 s		
212593	51970	4 Lac	22 24 30.8	+49 28 35	−0.001	0.00	4.57	0.09	−6.5	B9 Iab	−26	1500 s		
212825	10352		22 24 32.7	+71 02 49	−0.003	−0.01	7.2	0.5	0.4	B9.5 V		110 s		
212432	191117		22 24 33.7	−23 22 01	0.000	+0.02	7.2	0.5	0.4	B9.5 V		110 s		
212474	146074		22 24 35.6	−1 11 12	+0.007	+0.04	6.7	1.1	0.2	K0 III	−32	140 mx		
211998	258033	ν Ind	22 24 36.8	−72 15 20	+0.286	−0.68	5.29	0.65	4.4	G0 V	+21	14 s		m
212385	213795		22 24 37.6	−39 07 37	+0.011	0.00	6.9	0.1		A m				
212408	213798		22 24 38.6	−33 16 13	−0.001	−0.08	7.8	0.3	4.4	G0 V		48 s		
212386	231123		22 24 39.0	−41 26 25	+0.012	−0.03	6.70	0.5	3.4	F5 V		46 s		m
212210	255214		22 24 39.6	−65 37 21	+0.003	−0.03	7.2	0.1	0.6	A0 V		210 s		
212730	20051		22 24 41.1	+63 34 14	+0.002	+0.02	7.5	1.1	0.2	K0 III		260 s		
212955	10357		22 24 42.2	+74 51 01	−0.006	+0.02	8.0	0.9	3.2	G5 IV	−4	89 s		
212517	127511		22 24 43.3	+9 48 36	+0.001	+0.01	7.7	1.4	−0.3	K5 III		390 s		
212448	191119		22 24 46.3	−26 51 35	−0.002	−0.02	7.1	1.0	−0.1	K2 III		280 s		
212665	34418		22 24 46.8	+54 23 43	0.000	+0.04	8.0	1.1	3.2	K0 IV		89 s		
212526	127513		22 24 51.0	+8 44 12	−0.001	0.00	7.9	1.1	0.2	K0 III		340 s		
212330	247441		22 24 56.2	−57 47 51	+0.018	−0.34	5.32	0.67	4.4	dG0	+8	17 ts		m
212652	51974		22 24 58.8	+45 47 01	0.000	−0.03	7.9	1.1	−0.1	K2 III		380 s		
212611	72313		22 25 02.4	+32 47 28	+0.003	−0.01	6.9	0.1	1.4	A2 V		130 s		
212595	90486		22 25 02.5	+26 36 40	−0.002	−0.02	7.9	0.1	0.6	A0 V		280 s		
212691	34425		22 25 03.2	+51 15 22	+0.005	+0.03	6.8	1.6	−0.5	M2 III		280 s		
212668	51975		22 25 05.4	+44 32 20	−0.005	+0.01	8.0	0.1	0.6	A0 V		120 mx		
212464	213803		22 25 06.2	−37 49 47	+0.002	−0.05	7.3	1.2	0.2	K0 III		190 s		

HD	SAO	Star Name	α 2000	δ 2000	μ(α)	μ(δ)	V	B-V	M_v	Spec	RV	d(pc)	ADS	Notes
			h m s	° ' "	s	"								
213085	10364		22 25 08.7	+77 25 57	-0.001	-0.02	7.3	0.1	1.7	A3 V		130 s		
212211	258034		22 25 10.5	-70 25 54	+0.026	-0.06	5.78	0.38	1.1	F0 III-IV n		84 s		
212669	51976		22 25 12.0	+42 01 54	-0.003	+0.01	7.8	1.4	-0.3	K5 III		390 s		
212571	127520	52 π Aqr	22 25 16.4	+1 22 39	+0.001	+0.01	4.66	-0.03	-4.1	B0 V	+4	450 s		
212423	247443		22 25 16.6	-51 23 56	-0.002	-0.08	7.4	1.3	3.2	G5 IV		33 s		
213000	10360		22 25 20.7	+73 36 10	-0.012	-0.02	8.00	0.4	2.6	F0 V		120 s		m
213021	10363		22 25 23.3	+74 49 42	-0.005	+0.01	8.0	1.1	0.2	K0 III	-23	320 s		
212712	34430		22 25 23.9	+50 24 06	0.000	+0.01	7.2	0.0	0.4	B9.5 V		220 s		
212638	90494		22 25 29.7	+22 34 27	+0.004	0.00	7.60	0.1	1.4	A2 V		150 mx	15924	m
212560	165062		22 25 31.1	-15 35 56	-0.001	+0.02	7.8	0.5	3.4	F5 V		74 s		
212670	107906		22 25 40.5	+18 26 40	+0.002	+0.03	6.27	1.22	0.2	K0 III	+22	180 s		
--	--		22 25 40.9	-37 34 11			7.80B							T Gru, v
212587	165065		22 25 43.9	-15 21 26	-0.003	-0.01	7.9	0.9	3.2	G5 IV		88 s		
212790	34439		22 25 46.2	+53 49 00	0.000	+0.01	7.4	1.1	0.2	K0 III	-34	260 s		
212882	20060		22 25 46.9	+63 19 43	+0.005	+0.01	6.90	1.56	-0.5	M4 III	+8	300 s		
212600	191131		22 25 47.8	-20 14 13	-0.001	0.00	7.10	0.5	4.0	F8 V		42 s	15926	m
212976	20066		22 25 49.6	+69 53 42	-0.002	+0.01	8.0	1.1	0.2	K0 III	+1	320 s		
212168	258036		22 25 51.0	-75 00 55	+0.016	+0.03	6.04	0.64	3.0	G3 IV		41 s		m
212810	34441		22 25 52.5	+53 56 41	-0.007	+0.01	7.4	0.4	2.6	F0 V	-15	89 s		
212538	231134		22 25 56.3	-45 06 49	+0.007	-0.03	7.70	1.1	0.2	K0 III		240 mx		m
213417	3746		22 26 00.2	+80 42 06	+0.010	+0.02	6.7	1.1	0.2	K0 III		190 s		
213022	10366		22 26 00.4	+70 46 17	+0.002	+0.02	5.47	1.20	0.2	K0 III	-17	86 s		
212734	90501		22 26 04.1	+25 55 39	-0.002	0.00	7.05	0.06	0.6	A0 V	-4	170 s		
212750	90502		22 26 06.6	+28 31 05	-0.003	+0.01	7.2	1.1	0.2	K0 III		260 s		
212672	146083		22 26 06.7	-5 10 40	+0.002	0.00	7.26	0.28		F0				
212617	191138		22 26 09.6	-29 04 56	+0.003	-0.02	8.0	0.3	3.0	F2 V		100 s		T PsA, q
212643	191144		22 26 10.5	-23 40 57	+0.001	+0.01	6.2	0.1	0.6	A0 V	-15	120 s		
212633	191140		22 26 10.6	-28 00 46	+0.001	-0.02	7.6	0.9	0.2	K0 III		300 s		
213011	20069		22 26 12.7	+69 02 37	-0.005	-0.02	7.8	0.4	2.6	F0 V		100 s		
212695	146084		22 26 14.3	-2 47 20	-0.003	-0.08	7.1	0.5	3.4	F5 V		54 s		
212898	34452		22 26 26.2	+54 06 13	-0.002	-0.01	7.60	0.1	0.6	A0 V	-4	230 s	15941	m
212871	51998		22 26 26.6	+46 33 42	+0.003	+0.01	7.7	0.1	0.6	A0 V		250 s		
212542	247446		22 26 33.5	-58 52 41	-0.001	-0.03	7.9	1.9	-0.3	K5 III		240 s		
212697	165077	53 Aqr	22 26 34.0	-16 44 29	+0.018	-0.01	6.60	0.6	4.4	G0 V	-3		15934	m
212698	165078	53 Aqr	22 26 34.1	-16 44 33	+0.015	0.00	6.4	0.6	4.4	G0 V	-6	19 ts	15934	m
212717	146087		22 26 34.6	-7 22 39	+0.005	-0.02	6.9	0.4	2.6	F0 V		73 s		
212522	255218		22 26 35.1	-62 32 22	+0.002	-0.02	7.7	1.1	-0.1	K2 III		360 s		
212754	127529	34 Peg	22 26 37.2	+4 23 37	+0.020	+0.05	5.75	0.52	3.4	dF5	-18	28 ts	15935	m
212741	165079		22 26 41.5	-11 13 42	+0.003	-0.01	7.1	0.1	1.7	A3 V		120 s		
213403	10375	28 Cep	22 26 42.4	+78 47 09	-0.006	-0.04	5.8	0.1	1.4	A2 V	-6	73 s		
212858	72343		22 26 43.2	+36 46 12	+0.019	+0.12	8.0	0.5	4.0	F8 V		55 mx		
212832	90511		22 26 44.4	+22 48 43	0.000	-0.02	7.80	0.1	1.4	A2 V		190 s		m
212883	72344		22 26 45.4	+37 26 38	+0.001	0.00	6.46	-0.13	-2.5	B2 V	-7	590 s	15942	m
212833	90512		22 26 46.3	+21 52 11	-0.002	0.00	7.9	1.1	0.2	K0 III		340 s		
212757	146088		22 26 46.9	-8 30 39	0.000	+0.01	8.0	0.1	1.4	A2 V		210 s		
213208	10373		22 26 53.5	+73 34 28	-0.001	-0.02	8.0	0.1	0.6	A0 V		270 s		
212723	213818		22 26 58.1	-35 40 33	-0.001	-0.04	8.0	1.4	0.2	K0 III		220 s		
212986	34460		22 26 59.1	+56 26 01	+0.003	+0.01	6.4	-0.1		B8				
212901	72347		22 27 00.6	+30 57 26	-0.002	-0.01	7.6	1.1	-0.1	K2 III		340 s		
212940	52011		22 27 01.4	+45 47 20	+0.001	-0.04	7.3	0.1	1.4	A2 V		150 s		
212771	165086		22 27 03.0	-17 15 49	-0.005	-0.09	7.6	0.8	3.2	G5 IV		76 s		
213087	20075	26 Cep	22 27 05.2	+65 07 57	0.000	0.00	5.46	0.37		B0.5 Ib	-15			
212837	146092		22 27 06.5	-1 18 36	+0.004	-0.02	6.8	0.6	4.4	G0 V		31 s		
212861	127535		22 27 09.5	+8 55 27	-0.003	-0.02	7.9	0.5	4.0	F8 V		60 s		
213048	20074		22 27 10.5	+61 54 46	-0.006	-0.05	6.6	1.1	0.2	K0 III		180 s		
212581	255222	δ Tuc	22 27 19.9	-64 58 00	+0.012	0.00	4.48	-0.03	-0.2	B8 V	+12	76 s		
213099	20079		22 27 20.3	+63 57 25	+0.003	+0.05	7.9	0.1	0.6	A0 V		260 s		
212942	72355		22 27 20.3	+33 30 53	-0.001	-0.01	7.5	0.4	2.6	F0 V		98 s		
212661	247451		22 27 21.1	-58 00 04	-0.001	0.00	7.10	0.1	1.4	A2 V		140 s		m
212708	231145		22 27 24.5	-49 21 58	+0.035	-0.34	7.48	0.73		G5		19 mn		
212978	72358		22 27 26.3	+39 48 36	0.000	-0.01	6.14	-0.14	-2.5	B2 V	-17	520 s		
212301	258040		22 27 30.0	-77 43 06	+0.019	-0.10	7.9	0.5	4.0	F8 V		60 s		
212977	52020		22 27 30.0	+42 00 50	-0.003	-0.01	7.9	0.5	4.0	F8 V		59 s		
212800	213829		22 27 31.0	-31 22 24	+0.004	-0.06	7.8	1.6	-0.1	K2 III		160 mx		
212801	213830		22 27 31.3	-31 33 18	+0.005	-0.06	7.6	0.6	3.2	G5 IV		77 s		
213036	34469		22 27 32.9	+51 58 36	0.000	+0.01	8.00	0.6	0.6	G0 III p		280 s		
212852	191156		22 27 39.0	-26 25 07	0.000	+0.01	7.2	0.4	2.6	F0 V		69 s		
--	--		22 27 42.9	+4 31 25	+0.002	+0.02	7.9	0.5	4.0	F8 V		61 s		
213050	34471		22 27 43.3	+51 29 36	-0.002	+0.01	7.3	0.1	-2.8	A0 II		940 s		
212988	72366		22 27 46.0	+31 50 25	+0.003	+0.04	5.98	1.45	-0.1	K2 III	+1	110 s		
213012	72370		22 27 47.0	+37 49 20	0.000	0.00	7.9	1.1	0.2	K0 III		330 s		
212890	191158		22 27 50.5	-21 52 11	+0.001	-0.01	8.0	1.3	-0.5	M4 III		510 s		
212943	127540	35 Peg	22 27 51.4	+4 41 44	+0.005	-0.31	4.79	1.05	0.2	K0 III	+54	77 s		m
213290	10378		22 27 53.0	+70 48 04	+0.003	-0.01	7.8	0.1	0.6	A0 V		250 s		
213089	34475		22 27 54.0	+51 34 32	0.000	-0.01	7.57	0.12	0.6	A0 V		220 s		

HD	SAO	Star Name	α 2000	δ 2000	μ(α)	μ(δ)	V	B-V	M$_v$	Spec	RV	d(pc)	ADS	Notes
			^{h} ^{m} ^{s}	$^{\circ}$ ' "	^{s}	"								
213061	52028		22 28 03.2	+44 07 15	-0.001	+0.02	6.80	1.4	-0.3	K5 III		260 s	15968	m
212989	107935		22 28 07.3	+12 14 56	+0.015	+0.01	7.08	0.90		K0		10 mn	15962	m
213025	90527		22 28 10.9	+27 01 08	-0.003	-0.02	6.5	0.9	0.3	G8 III		180 s		
212915	191161		22 28 11.1	-27 12 30	+0.001	-0.05	8.0	0.9	0.2	K0 III		360 s		
213014	107941		22 28 11.3	+17 15 48	+0.001	-0.01	7.5	0.9	0.2	G9 III	-40	290 s	15967	m
213026	90528		22 28 12.5	+24 52 12	0.000	-0.01	7.5	1.1	0.2	K0 III		290 s		
212895	213843		22 28 13.2	-39 05 21	+0.004	-0.01	7.4	1.9	-0.3	K5 III		200 s		
213027	90529		22 28 13.7	+24 47 33	0.000	0.00	6.7	0.0	0.4	B9.5 V		180 s		
213062	72373		22 28 15.6	+37 21 35	+0.001	-0.01	8.0	0.5	4.0	F8 V		62 s		
213091	72374		22 28 18.6	+37 08 22	+0.001	-0.03	7.8	0.0	0.4	B9.5 V		300 s		
213242	20088		22 28 19.5	+64 05 08	+0.003	-0.01	6.29	1.08	0.2	K0 III	-27	150 s		
213159	34484		22 28 25.4	+51 18 22	0.000	0.00	7.73	0.00	0.4	B9.5 V		290 s		
213141	52040		22 28 29.0	+48 32 36	-0.002	+0.02	7.70	1.1	0.2	K0 III		300 s	15973	m
213188	34488		22 28 29.6	+52 51 55	+0.002	+0.02	7.70	1.1	-0.1	K2 III		330 s		
212870	247460		22 28 29.8	-51 47 18	0.000	-0.01	7.00	1.4	-0.3	K5 III		290 s		m
213232	34493		22 28 34.4	+58 32 25	+0.003	+0.02	7.9	0.2		A5 p				
213176	52042		22 28 35.5	+49 27 38	+0.001	+0.01	8.0	1.1	-0.1	K2 III		390 s		
213225	34492		22 28 35.6	+56 55 16	+0.004	+0.01	8.0	1.1	0.2	K0 III		320 s		
212728	255227		22 28 37.6	-67 29 21	+0.028	-0.07	5.55	0.20	1.7	A3 V		36 mx		
212953	213850	ν Gru	22 28 39.0	-39 07 55	+0.003	-0.16	5.47	0.95	-0.4	K9 III	+11	76 mx		m
213269	20091		22 28 41.1	+61 27 21	-0.006	+0.03	7.4	1.1	0.2	K0 III	+6	190 mx		
212963	213851		22 28 42.1	-33 46 15	-0.001	+0.02	7.9	-0.4	2.6	F0 V		110 s		
213556	10392		22 28 45.2	+76 14 08	-0.004	+0.01	7.9	1.4	-0.3	K5 III	-13	380 s		
213005	191168		22 28 49.0	-22 04 13	+0.008	+0.02	7.3	0.2	3.0	F2 V		74 s		
213051	146107	55 ζ¹ Aqr	22 28 49.5	-0 01 13	+0.012	+0.02	4.53	0.40	0.6	F2 III	+29	30 mx	15971	m
213052	146108	55 ζ² Aqr	22 28 49.9	-0 01 12	+0.014	+0.05	4.31			F2 IV	+25		15971	m
212878	255229		22 28 55.5	-60 03 14	-0.002	+0.01	6.8	1.3	0.2	K0 III		150 s		
213191	52048		22 29 01.0	+40 18 58	+0.003	+0.02	7.5	1.6		M5 III	-60			S Lac, v
213177	90542		22 29 03.3	+29 47 43	+0.002	-0.01	7.86	1.09	-2.1	K0 II		980 s		
213243	34500		22 29 03.9	+51 30 30	+0.004	+0.02	7.5	0.1	1.7	A3 V		140 s		
213066	146111		22 29 04.7	-9 44 29	+0.002	-0.09	7.6	0.7	4.4	G0 V		45 s		
213143	90541		22 29 06.2	+21 23 36	-0.001	-0.01	7.7	0.2	2.1	A5 V		130 s		
213019	213857		22 29 07.3	-36 58 13	+0.005	0.00	7.5	1.2	0.2	K0 III		210 s		
213178	90543		22 29 07.7	+29 01 46	+0.002	+0.02	7.16	1.17	0.2	K0 III		200 s		
213119	127544	36 Peg	22 29 07.8	+9 07 44	+0.004	-0.02	5.58	1.55	-0.3	gK5	-30	140 s		
213307	34506	27 δ Cep	22 29 09.1	+58 24 15	+0.001	+0.01	6.31	-0.03	-0.6	B7 V	-21		15987	m
213179	90544		22 29 10.1	+26 45 48	+0.002	0.00	5.79	1.25	-0.1	K2 III	-45	140 s		
213306	34508	27 δ Cep	22 29 10.1	+58 24 55	+0.002	+0.01	3.75	0.60	-4.6	F8 Ib var	-17	410 s	15987	m,v
213042	213859		22 29 15.0	-30 01 06	+0.017	-0.80	7.66	1.10	8.0	K5 V	+5	10 ts		
213009	231154	δ¹ Gru	22 29 15.9	-43 29 45	+0.003	0.00	3.97	1.03	0.3	gG5	+5	43 s		m
213405	20100		22 29 17.2	+65 06 47	-0.003	-0.01	7.95	0.47		B0.5 V				
213210	90545		22 29 19.9	+27 25 42	+0.003	0.00	7.68	0.30		F0				
213480	20104		22 29 21.6	+69 25 22	+0.006	-0.01	8.0	1.4	-0.3	K5 III		400 s		
213322	34511		22 29 22.1	+54 14 46	0.000	+0.01	6.78	-0.02	-1.6	B3.5 V	-10	380 s		
212806	--		22 29 23.5	-69 25 30			7.8	1.2	0.2	K0 III		240 s		
213074	191178		22 29 25.6	-28 39 35	0.000	-0.02	7.90	1.1	0.2	K0 III		350 s	15978	m
213502	10394		22 29 30.6	+70 26 09	+0.012	+0.02	7.3	0.1	1.4	A2 V		110 mx		
213310	52055	5 Lac	22 29 31.7	+47 42 25	0.000	0.00	4.36	1.68	-2.4	M0 II	-4	230 s		
213076	213862		22 29 32.0	-33 51 02	-0.003	+0.03	7.5	1.2	0.2	K0 III		200 s		
213234	90547		22 29 32.3	+29 59 19	-0.002	-0.02	8.00	0.1	1.7	A3 V		180 s		m
213258	72395		22 29 36.8	+35 56 17	+0.001	-0.04	7.8	0.1	1.7	A3 V		170 s		
213245	72393		22 29 37.3	+32 49 44	-0.001	+0.02	7.7	1.4	-0.3	K5 III		390 s		
213045	247465		22 29 43.7	-49 51 57	0.000	0.00	8.0	0.8	2.6	F0 V		57 s		
213272	72399		22 29 43.8	+35 43 32	-0.002	-0.03	6.5	0.1	0.6	A0 V	-2	160 s		
213481	20107		22 29 44.6	+66 28 39	-0.001	-0.01	8.0	0.0	0.0	B8.5 V		360 s		
213080	231161	δ² Gru	22 29 45.3	-43 44 58	-0.001	+0.01	4.11	1.57		M4	+2	27 mn		m
213503	20110		22 29 45.8	+68 13 16	-0.003	-0.01	7.8	0.8	3.2	G5 IV		82 s		
213135	191182		22 29 45.9	-27 06 27	+0.010	-0.01	5.95	0.33	2.6	F0 V		45 s		
--	10404		22 29 50.3	+80 03 12	-0.003	-0.02	8.0	2.4						
212610	258042		22 29 50.8	-78 12 46	+0.002	+0.01	6.7	2.0	-0.1	K2 III		70 s		
213083	231163		22 29 51.6	-46 04 29	0.000	+0.01	7.20	0.9	3.2	G5 IV		48 s		m
213169	191186		22 29 52.3	-23 51 20	-0.004	-0.04	8.0	0.5	3.4	F5 V		75 s		
213798	10402	29 ρ Cep	22 29 52.7	+78 49 28	0.000	-0.02	5.5	0.1	1.4	A2 V	+1	64 s		
213388	34519		22 29 53.4	+52 25 00	+0.002	+0.03	6.60	0.8	0.3	G5 III		170 s	15993	m
213235	127551	37 Peg	22 29 57.8	+4 25 54	-0.002	-0.14	5.48	0.38	0.7	F5 III	+1	85 mx	15988	m
213571	20113		22 30 00.3	+70 10 17	+0.003	0.00	7.16	-0.03	-1.0	B5.5 V	-18	400 s		
213198	165123		22 30 01.4	-12 54 54	+0.011	0.00	6.40	0.32	2.6	F0 V	-11	58 s		
213323	72406	38 Peg	22 30 01.6	+32 34 21	+0.003	-0.01	5.50	-0.10	0.2	B9 V	-16	120 s		
213389	52073		22 30 06.3	+49 21 22	-0.003	-0.04	6.50	1.1	-0.1	K2 III	+4	200 s		m
213340	72407		22 30 08.1	+31 12 57	-0.003	-0.01	6.6	0.1	0.6	A0 V		160 s		
213247	146126		22 30 09.7	+0 08 18	-0.001	0.00	7.70	0.1	0.6	A0 V		260 s	15990	m
213391	52071		22 30 12.5	+40 57 57	-0.004	-0.03	7.50	0.5	3.4	F5 V		66 s	15996	m
213406	52076		22 30 14.3	+46 25 33	-0.001	-0.02	7.9	1.1	-0.1	K2 III		390 s		
213315	90555		22 30 14.4	+22 28 01	+0.001	0.00	8.00	0.35		F0			15992	m
213155	231168		22 30 15.4	-42 17 27	+0.002	+0.01	6.92	-0.04		A0				

HD	SAO	Star Name	α 2000	δ 2000	μ(α)	μ(δ)	V	B-V	M_v	Spec	RV	d(pc)	ADS	Notes
			h m s	° ′ ″	s	″								
213236	165127	56 Aqr	22 30 17.2	−14 35 09	+0.002	−0.04	6.37	−0.04	−	A0				
213035	255235		22 30 17.7	−62 16 56	−0.001	+0.02	7.9	1.1	0.2	K0 III		350 s		
213471	34531		22 30 18.6	+57 13 32	+0.001	0.00	6.7	0.7	4.4	G0 V	−61	29 s	16001	m
213316	107959		22 30 25.8	+10 40 39	+0.003	−0.01	7.9	1.1	0.2	K0 III		360 s		
213251	165129		22 30 26.6	−16 20 51	+0.005	+0.01	7.2	0.8	3.2	G5 IV		63 s		
213278	146128		22 30 27.7	−4 49 00	+0.006	+0.06	7.9	0.9	3.2	G5 IV		88 s		
213420	52079	6 Lac	22 30 29.1	+43 07 25	−0.001	0.00	4.51	−0.09	−3.0	B2 IV	−8	280 s		
213530	20115		22 30 32.6	+61 37 26	+0.002	+0.01	7.60	0.1	0.6	A0 V		230 s	16011	m
213293	146130		22 30 32.7	−8 06 53	+0.004	0.00	7.70	0.4	2.6	F0 V		110 s	15994	m
213320	165134	57 σ Aqr	22 30 38.7	−10 40 41	0.000	−0.03	4.82	−0.06	0.0	A0 IV	+11	92 s		
213342	127563		22 30 40.5	+4 19 54	+0.001	−0.03	8.0	1.1	0.2	K0 III		360 s		
213220	231170		22 30 42.9	−44 05 39	+0.003	+0.01	6.93	0.07		A0				
213483	52091		22 30 46.2	+46 31 44	−0.001	+0.01	8.0	0.0	0.4	B9.5 V		320 s		
213295	191195		22 30 49.1	−24 10 06	0.000	−0.01	7.6	1.5	0.2	K0 III		160 s		
213296	191196	ζ PsA	22 30 53.5	−26 04 25	+0.003	−0.07	6.5	1.1	0.2	K0 III		150 s		
213393	127569		22 30 56.9	+8 25 36	−0.001	−0.02	7.9	1.1	0.2	K0 III		340 s		
213230	231172		22 30 58.5	−48 18 58	+0.002	−0.03	7.6	0.4	1.4	A2 V		100 s		
213240	231175		22 31 00.2	−49 26 01	−0.014	−0.20	6.80	0.61	3.1	G4 IV		55 s		m
213298	213879		22 31 01.1	−31 01 33	+0.005	−0.04	7.3	0.4	1.7	A3 V		72 mx		
213531	—		22 31 05.5	+50 21 53	−0.001	+0.01	8.0	0.0	0.4	B9.5 V		320 s		
213519	52097		22 31 05.5	+45 08 43	−0.018	+0.05	7.7	0.9	3.2	G5 IV		56 mx		
213425	127572		22 31 09.0	+9 19 12	0.000	−0.02	7.8	0.4	2.6	F0 V		110 s		
213462	90564		22 31 12.8	+23 58 34	+0.002	+0.01	7.8	1.1	−0.1	K2 III		380 s		
213379	191201		22 31 13.2	−22 59 48	−0.001	−0.02	7.9	0.4	4.4	G0 V		51 s		
213557	34541		22 31 14.8	+50 52 20	+0.001	0.00	7.80	0.1	1.4	A2 V		190 s	16020	m
213520	72427		22 31 16.5	+39 14 37	−0.002	−0.06	7.71	0.21		A7				
213558	34542	7 α Lac	22 31 17.3	+50 16 57	+0.014	+0.02	3.77	0.01	1.4	A2 V	−4	30 ts	16021	m
213428	146136		22 31 18.1	−2 54 40	−0.001	−0.03	6.16	1.08		K0				
213429	146135		22 31 18.2	−6 33 18	+0.011	−0.10	6.14	0.56		F8		24 t		
213642	20126		22 31 18.7	+63 49 06	0.000	+0.03	7.4	0.9	3.2	G5 IV		68 s		
−−	34545		22 31 25.0	+50 35 43	−0.001	+0.02	7.8	1.8	−0.3	K5 III		270 s		
213303	247470		22 31 28.4	−50 56 14	+0.007	+0.02	7.4	0.8	3.2	G5 IV		67 s		
213398	213883	17 β PsA	22 31 30.1	−32 20 46	+0.005	−0.01	4.29	0.01	0.6	A0 V	+6	53 s		m
−−	213884		22 31 30.4	−32 21 17	+0.004	−0.02	7.8							
213544	52105		22 31 31.2	+42 13 18	−0.004	−0.03	7.7	0.1	1.4	A2 V		180 s		
213534	90568		22 31 34.1	+29 32 34	−0.003	−0.02	6.3	0.2	2.1	A5 V	+2	70 s		
211539	258932	υ Oct	22 31 37.2	−85 58 03	−0.035	+0.06	5.77	1.02	0.2	K0 III		130 s		
213399	231182		22 31 40.5	−39 58 44	0.000	−0.03	7.8	1.4	0.2	K0 III		260 s		
213464	165147	58 Aqr	22 31 41.1	−10 54 21	+0.005	−0.04	6.4	0.4	2.6	F0 V	+4	58 s		
213560	72434		22 31 43.5	+34 13 41	0.000	0.00	7.9	0.5	4.0	F8 V		61 s		
213559	72435		22 31 43.7	+34 39 54	+0.002	−0.01	7.9	0.5	3.4	F5 V		80 s		
213412	213887		22 31 45.2	−35 17 28	+0.006	−0.02	6.6	1.1	0.2	K0 III		150 s		
213363	231184		22 31 47.8	−46 34 49	+0.002	−0.03	7.24	0.46		F8				
213509	127578		22 31 47.9	+7 47 28	+0.002	−0.01	8.0	0.1	0.6	A0 V		300 s		
213522	107978		22 31 49.1	+13 06 16	+0.005	+0.05	7.9	0.4	3.0	F2 V		94 s		
213413	231186		22 31 58.2	−43 32 52	+0.002	0.00	7.7	1.0		Ma				
213596	72442		22 32 06.5	+33 22 38	+0.004	+0.02	8.0	0.1	0.6	A0 V		150 mx		
213222	255241		22 32 08.9	−69 06 28	+0.023	−0.02	7.3	0.5	3.0	F2 V		58 s		
213573	107980		22 32 09.3	+15 49 30	+0.003	0.00	7.9	1.4	−0.3	K5 III		450 s		
213457	231188		22 32 11.4	−43 15 46	0.000	+0.02	6.91	0.98		G5				
214035	10414		22 32 16.0	+76 13 35	−0.006	0.00	5.7	0.1	0.6	A0 V	−22	100 s		
213730	34558		22 32 16.1	+59 29 36	+0.025	−0.01	8.0	0.9	3.2	G5 IV		90 s		
213720	34555		22 32 18.6	+54 02 15	+0.004	+0.03	6.4	0.9	0.3	G8 III	−14	160 s		
213831	20139		22 32 19.0	+67 37 00	+0.001	+0.03	7.8	1.1	−0.1	K2 III		340 s		
213659	52115		22 32 21.2	+40 36 54	−0.002	−0.03	8.01	0.14		A1				
213574	127583		22 32 21.7	+2 35 11	−0.002	−0.02	7.7	1.1	−0.1	K2 III		360 s		
213810	20137		22 32 23.0	+65 17 36	−0.006	−0.03	8.0	1.1	−0.1	K2 III		370 s		
213490	213893		22 32 23.9	−35 02 54	−0.001	−0.03	7.9	1.4	0.2	K0 III		190 s		
213660	72446		22 32 26.2	+39 46 48	0.000	0.00	5.80	0.1	1.7	A3 V	+5	66 s	16031	m
213618	107984		22 32 27.3	+19 37 25	+0.002	+0.04	7.2	1.4	−0.3	K5 III		320 s		
213643	72447		22 32 31.6	+33 48 53	+0.005	+0.04	7.9	0.1	0.6	A0 V		95 mx		
213575	146142		22 32 33.9	−6 28 02	+0.020	+0.03	7.0	0.6	4.4	G0 V		33 s		
213617	107986	39 Peg	22 32 35.3	+20 13 48	+0.011	+0.03	6.3	0.4	2.6	F0 V	−19	55 s		
213661	72448		22 32 35.6	+34 13 44	0.000	0.00	7.90	0.2	2.1	A5 V		150 s	16034	m
213619	107988		22 32 39.2	+13 02 43	+0.008	+0.05	6.6	0.4	3.0	F2 V	+11	53 s		
213620	127589		22 32 46.1	+3 00 17	−0.004	−0.10	8.0	1.1	−0.1	K2 III		190 mx		
213644	107989		22 32 46.7	+15 51 46	+0.001	+0.01	6.3	1.1	0.2	K0 III	−28	170 s		
214071	10417		22 32 51.1	+74 04 27	−0.001	+0.03	7.2	1.1	−0.1	K2 III		260 s		
213776	52127		22 32 53.5	+49 23 16	+0.002	+0.01	7.73	0.41		F5			16046	m
213721	72452		22 32 57.8	+35 26 18	0.000	0.00	7.5	0.1	0.6	A0 V		240 s		
213759	52131		22 32 59.7	+46 04 33	−0.001	+0.01	8.0	1.4	−0.3	K5 III		420 s		
213442	255247	ν Tuc	22 32 59.8	−61 58 56	+0.006	−0.02	4.81	1.61		M5	−3	21 mn		
213612	165156		22 33 01.2	−17 30 05	−0.013	−0.03	7.9	0.9	3.2	G5 IV		88 mx		
213973	20150		22 33 02.6	+69 54 49	+0.022	+0.07	6.0	0.2	2.1	A5 V	−2	59 s	16057	m
213908	20146		22 33 03.9	+63 52 48	+0.001	+0.02	7.8	1.1	0.2	K0 III		300 s		

HD	SAO	Star Name	α 2000	δ 2000	μ(α)	μ(δ)	V	B-V	M_v	Spec	RV	d(pc)	ADS	Notes
			^{h} ^{m} ^{s}	$^\circ$ $'$ $''$	^{s}	$''$								
213989	20151		22 33 07.3	+68 30 17	0.000	-0.01	7.7	0.1	1.7	A3 V		150 s	16058	m
213746	72455		22 33 09.0	+31 24 13	+0.005	-0.10	7.9	0.9	3.2	G5 IV		88 s		
213479	255248		22 33 10.9	-60 25 58	+0.001	+0.05	7.3	0.7	3.2	G5 IV		66 s		
213722	107995		22 33 13.0	+17 47 06	+0.004	+0.01	7.75	0.56		F8				
214019	10418		22 33 16.9	+70 22 26	+0.010	+0.03	6.34	0.1		A0	-19	49 mn	16062	m
213600	213903		22 33 16.9	-35 08 18	0.000	-0.01	7.2	1.4	-0.1	K2 III		220 s		
213812	52136		22 33 19.0	+46 06 56	-0.005	0.00	7.8	0.1	1.7	A3 V		150 mx		
213786	72458		22 33 22.4	+38 07 37	+0.003	-0.02	7.6	1.1	0.2	K0 III		300 s		
213517	255253		22 33 23.7	-59 59 31	0.000	+0.01	7.90	0.93	0.3	G8 III		330 s		
213909	34572		22 33 26.2	+55 55 43	+0.001	+0.01	7.5	1.4	-0.3	K5 III		320 s		
213493	255252		22 33 30.5	-63 28 18	+0.012	+0.01	7.5	1.6	0.2	K0 III		130 s		
213628	213905		22 33 30.6	-35 26 42	+0.028	-0.15	7.80	0.72	5.6	G8 V		29 ts		
213655	191223		22 33 33.4	-29 39 54	+0.002	+0.01	7.5	0.6	2.6	F0 V		61 s		
213802	72463		22 33 34.0	+31 17 02	+0.001	-0.06	7.9	0.7	4.4	G0 V		49 s		
213835	52140		22 33 34.3	+40 49 08	0.000	+0.01	6.80	1.1	0.2	K0 III		200 s		m
213568	247487		22 33 36.6	-58 12 51	-0.001	+0.01	7.9	0.6	2.6	F0 V		72 s		
--	52145		22 33 37.0	+47 57 31	+0.002	0.00	8.0	1.9						
213605	247488		22 33 37.1	-51 23 09	-0.005	-0.02	7.8	2.0	0.2	K0 III		84 s		
213871	52143		22 33 37.4	+46 33 57	+0.001	+0.01	7.3	0.1		A0 p	-10			
213930	34574		22 33 40.3	+56 37 30	+0.009	+0.05	5.71	0.96	0.2	K0 III	-11	130 s		
213854	52146		22 33 43.0	+43 29 08	0.000	+0.02	7.8	1.4	-0.3	K5 III		390 s		
213682	213910		22 33 48.7	-38 08 24	+0.005	-0.01	7.9	2.1	-0.3	K5 III		180 s		
214007	20153		22 33 53.0	+61 46 42	+0.002	+0.02	6.5	0.1	1.4	A2 V		100 s		
213872	72468		22 33 53.7	+36 48 00	0.000	-0.06	7.9	0.1	1.7	A3 V		140 mx		
213764	146159		22 33 54.2	-4 53 07	+0.001	+0.01	8.0	1.1	0.2	K0 III		360 s		
213658	231192		22 33 54.3	-44 41 24	0.000	-0.02	6.79	0.95	1.7	K0 III-IV		100 s		
214051	20154		22 33 54.4	+65 49 53	+0.009	+0.02	7.6	1.1	0.2	K0 III		280 s	16072	
214072	20155		22 33 57.9	+65 13 23	-0.004	0.00	7.8	0.0	0.4	B9.5 V		270 s		
213714	213916		22 34 01.7	-35 36 13	0.000	+0.02	7.7	1.4	-0.1	K2 III		270 s		
213789	146160	60 Aqr	22 34 02.7	-1 34 27	+0.002	-0.03	5.89	0.99	0.3	gG6	-8	120 s		m
213728	213918		22 34 04.9	-32 08 30	0.000	+0.01	6.6	0.5	0.0	B8.5 V		96 s		
213780	146161		22 34 06.5	-9 36 31	-0.001	0.00	6.7	1.1	0.2	K0 III	+5	200 s		
213873	108005		22 34 17.8	+17 09 12	+0.002	0.00	7.8	1.1	0.2	K0 III		330 s		
213669	247490		22 34 18.5	-54 17 53	+0.003	-0.02	7.6	0.0	0.6	A0 V		230 s		
213891	108007		22 34 24.9	+13 24 36	0.000	-0.02	7.7	1.1	0.2	K0 III		320 s		
213770	213924		22 34 26.8	-34 40 31	-0.002	0.00	6.9	1.1	3.2	G5 IV		33 s		
213976	52155		22 34 29.9	+40 46 29	-0.002	0.00	7.02	-0.11	-3.0	B1.5 V	-17	880 s		m
214088	34580		22 34 32.3	+58 09 47	-0.002	-0.01	7.50	1.4	-0.3	K5 III		330 s		
213920	108009		22 34 32.5	+15 41 56	+0.002	0.00	7.9	1.1	0.2	K0 III		340 s		
213977	72481		22 34 33.7	+37 06 12	0.000	+0.03	8.0	1.4	-0.3	K5 III		450 s		
213893	127611		22 34 35.8	+0 35 43	-0.002	-0.07	6.71	1.52	-0.3	K5 III	-88	250 s		
214037	52157		22 34 36.1	+48 45 23	-0.001	+0.01	7.6	0.5	3.4	F5 V		70 s		
213947	90607		22 34 36.4	+26 35 52	0.000	0.00	7.2	1.1	-0.1	K2 III		290 s		
214008	52156		22 34 37.0	+45 00 18	+0.001	0.00	7.3	0.1	0.6	A0 V		220 s		
213874	146166		22 34 38.0	-3 18 24	-0.002	0.00	8.0	0.4	3.0	F2 V		98 s		
213845	191235	59 υ Aqr	22 34 41.5	-20 42 30	+0.016	-0.14	5.20	0.44	3.1	F3 V	-2	25 s		
213992	90613		22 34 47.2	+29 57 28	-0.001	+0.01	7.2	1.1	-0.2	K3 III	+7	310 s		
213785	231200		22 34 47.6	-45 47 06	+0.003	0.00	7.78	0.36		F0				
213994	90615		22 34 57.6	+22 18 05	+0.001	0.00	7.90	1.1	0.2	K0 III		350 s	16081	m
214165	20161		22 35 00.0	+60 49 35	-0.017	-0.01	7.1	0.4	3.0	F2 V	+3	64 s		
214023	72490		22 35 02.3	+30 48 12	+0.003	-0.02	7.35	1.34	-0.2	K3 III		300 s	16085	m
--	52161		22 35 09.5	+48 48 18	+0.003	+0.01	7.9	1.2						
213591	258051		22 35 10.3	-72 33 06	-0.034	+0.10	7.4	0.6	4.4	G0 V		40 s		
214010	108019		22 35 12.1	+14 36 36	+0.001	-0.02	7.70	1.1	-0.1	K2 III		360 s	16087	m
214097	52162		22 35 15.3	+45 09 13	-0.002	0.00	7.0	1.4	-0.3	K5 III		280 s		
214098	52163		22 35 17.8	+43 40 53	-0.001	+0.02	7.9	0.0	0.0	B8.5 V		360 s		
214089	72496		22 35 18.6	+32 49 17	-0.007	-0.06	7.9	0.5	4.0	F8 V		60 s		
213998	146181	62 η Aqr	22 35 21.2	-0 07 03	+0.006	-0.05	4.02	-0.09	-0.2	B8 V	-8	46 mx		
213775	255258		22 35 25.5	-61 38 31	+0.008	-0.04	7.7	0.8	3.2	G5 IV		78 s		
213402	258049		22 35 25.9	-78 46 18	+0.014	0.00	6.15	1.38	0.2	gK0		84 s		
213830	247498		22 35 26.1	-55 11 02	+0.002	+0.01	7.69	1.02		K0				
214112	72500		22 35 27.3	+39 35 10	+0.001	+0.02	7.24	0.28	2.6	F0 V		83 s		
213882	247502		22 35 32.9	-51 36 14	-0.002	0.00	6.7	1.5	0.2	K0 III		110 s		
213940	213933		22 35 34.7	-38 43 27	-0.001	+0.03	7.90	0.4	3.0	F2 V		96 s		m
213986	191246		22 35 36.3	-23 59 28	+0.003	0.00	5.97	0.99	0.2	K0 III	-3	140 s		
214057	108021		22 35 38.7	+10 19 39	0.000	0.00	7.8	1.1	-0.1	K2 III		390 s		
214238	34602		22 35 39.1	+56 51 40	-0.017	-0.04	7.64	0.69		G0			16097	m
214393	20171		22 35 40.2	+70 14 34	+0.002	+0.01	7.2	1.1	0.2	K0 III	-9	230 s		
214099	90625		22 35 40.3	+25 26 43	-0.001	+0.03	7.9	1.1	0.2	K0 III		350 s		
214127	72504		22 35 42.3	+32 47 22	+0.003	+0.01	7.3	0.1	1.4	A2 V		160 s		
214470	10425	31 Cep	22 35 45.8	+73 38 36	+0.040	+0.03	5.08	0.39	-0.6	F4 II-III	0	39 mx		
214113	108026		22 35 47.0	+19 42 59	0.000	0.00	6.7	1.1	0.2	K0 III		200 s		
214003	213940		22 35 47.6	-30 20 08	+0.003	+0.03	7.7	1.7	-0.5	M2 III		360 s		
214028	165178		22 35 48.6	-17 27 37	-0.002	-0.04	6.7	1.4	-0.3	K5 III	-8	260 s		
214128	90628		22 35 49.5	+20 16 27	+0.001	-0.11	6.7	1.1	0.2	K0 III	-33	100 mx		

573

HD	SAO	Star Name	α 2000	δ 2000	μ(α)	μ(δ)	V	B−V	M_v	Spec	RV	d(pc)	ADS	Notes
214279	34605		22^{h}35^{m}51^s.6	+56°04'12"	−0^s.001	−0".01	6.38	0.12	1.4	A2 V	−2	91 s		
214167	72509	8 Lac	22 35 52.2	+39 38 03	0.000	0.00	6.46	−0.14	−3.5	B1 V e	−11	870 s	16095	m
213884	247505		22 35 52.7	−57 53 02	+0.010	−0.01	6.23	0.20	1.7	A3 V n		81 s		
214240	52171		22 35 53.2	+50 04 16	0.000	+0.01	6.29	−0.05	−1.7	B3 V	−15	340 s		
214046	191252		22 35 53.7	−20 56 05	−0.001	−0.01	7.53	1.46	−0.3	K4 III		330 s		m
214080	165181		22 36 06.2	−16 23 16	0.000	−0.13	6.83	−0.13	−5.7	B1 Ib	0	2900 s		
213741	258052		22 36 06.7	−70 57 50	+0.006	+0.01	7.7	1.1	0.2	K0 III		320 s		
213941	247509		22 36 07.5	−54 36 39	+0.042	−0.26	7.59	0.66	5.2	G5 V		30 s		
214199	72518		22 36 07.7	+36 45 46	+0.004	+0.01	7.4	1.4	−0.3	K5 III		350 s		
214200	72517		22 36 07.7	+35 34 38	0.000	−0.06	6.10	1.00	0.2	K0 III	−16	150 s		
213942	247510		22 36 08.0	−55 11 30	0.000	+0.01	7.4	1.3	0.2	K0 III		190 s		
214511	10429		22 36 08.8	+72 52 53	+0.022	+0.06	7.56	0.48	3.7	dF6	−12	35 ts	16111	m
214225	72521		22 36 12.1	+39 20 16	−0.001	+0.01	7.7	1.1	0.2	K0 III		300 s		
214132	146188		22 36 16.8	−7 39 52	+0.004	+0.01	7.2	0.4	2.6	F0 V		82 s		
214226	90634		22 36 21.5	+27 07 20	0.000	+0.01	7.3	0.4	2.6	F0 V		89 s		
213742	258053		22 36 22.1	−72 42 46	+0.028	−0.03	6.6	1.0	0.2	K0 III		170 mx		
214263	72525		22 36 22.2	+37 50 33	0.000	0.00	6.85	−0.13	−2.5	B2 V	−20	710 s		
214369	34614		22 36 27.3	+58 25 35	−0.002	+0.02	7.76	1.89	−8.0	K0 Ia pe		6200 s		W Cep, v
214182	127631		22 36 28.1	+6 30 43	+0.001	−0.01	7.7	1.1	0.2	K0 III		310 s		
214085	231211		22 36 29.1	−40 34 58	+0.004	−0.07	6.28	0.12	1.9	A4 V		51 mx		
214122	213948		22 36 35.2	−31 39 50	−0.003	−0.04	5.82	1.09	0.2	K0 III		110 s		m
214203	108036		22 36 36.2	+11 41 49	−0.001	+0.02	6.40	0.00	1.4	A2 V	−8	100 s		
214227	108037		22 36 37.5	+13 27 52	−0.001	−0.02	7.8	1.1	−0.1	K2 III		380 s		
214265	90640		22 36 38.1	+27 47 03	+0.005	+0.01	7.10	1.1	0.2	K0 III	−19	240 s	16103	m
214066	231213		22 36 38.6	−48 18 23	+0.001	0.00	6.6	1.9	−0.5	M2 III		180 s		
214121	213950		22 36 38.9	−31 38 29	−0.002	+0.01	7.3	1.8	0.2	K0 III		89 s		
214418	34618		22 36 39.4	+59 35 47	−0.009	−0.06	8.0	0.9	3.2	G5 IV		90 s		
214183	127632		22 36 39.9	+0 21 51	0.000	0.00	7.81	0.13		F0				
214245	108038		22 36 41.8	+13 09 21	+0.003	+0.01	7.00	1.57	−0.3	K5 III	−30	260 s		
214471	20176		22 36 41.8	+65 15 28	−0.005	+0.06	7.7	1.1	0.2	K0 III		150 mx		
214094	231214		22 36 42.4	−43 28 16	+0.024	−0.01	6.76	0.53	3.7	F6 V		38 s		
214296	72533		22 36 44.7	+30 47 28	−0.004	−0.02	7.7	0.1	0.6	A0 V		260 s		
214172	191265		22 36 46.7	−21 05 22	0.000	+0.03	8.0	−0.2	0.6	A0 V		300 s		
214410	34619		22 36 47.4	+54 25 24	+0.001	0.00	7.42	−0.05	0.4	B9.5 V		250 s		
213956	—		22 36 48.0	−64 23 03			7.8	1.7	−0.1	K2 III		190 s		
214313	72535		22 36 48.6	+35 39 09	0.000	0.00	6.30	1.38	−0.3	K5 III	+10	210 s		
214267	127634		22 36 53.8	+7 46 57	+0.004	−0.03	7.6	0.4	3.0	F2 V		85 s		
214150	231217		22 36 58.6	−40 35 28	+0.003	−0.07	5.86	0.06		A2	+15			m
214247	146201		22 37 00.7	−7 34 00	+0.012	−0.18	7.8	0.4	2.6	F0 V		44 mx		
214067	247512		22 37 03.6	−59 33 33	−0.033	+0.08	7.5	0.9	4.4	G0 V		26 s		
214298	108043		22 37 04.6	+12 34 38	−0.002	−0.01	6.4	1.4	−0.3	K5 III	−19	220 s		
214248	165198		22 37 06.2	−11 43 52	+0.007	−0.02	7.7	1.1	0.2	K0 III		220 mx		
214333	108045		22 37 10.1	+17 16 58	+0.003	+0.01	7.9	0.5	3.4	F5 V		78 s		
214350	90651		22 37 12.4	+23 28 25	+0.002	+0.01	7.2	0.2	2.1	A5 V		100 s		
214710	10440		22 37 12.8	+75 22 19	+0.011	+0.01	5.79	1.57	−0.3	K5 III	−7	150 s		
214421	52190		22 37 14.8	+42 48 39	+0.002	+0.02	7.0	1.4	−0.3	K5 III		280 s		
214301	146205		22 37 17.8	−6 04 03	−0.002	−0.03	7.9	1.1	−0.1	K2 III		390 s		
214174	231219		22 37 19.0	−44 39 54	0.000	0.00	7.93	1.25		K0				
214606	20179		22 37 19.0	+69 12 30	+0.003	−0.01	7.5	0.6	4.4	G0 V	+12	41 s		
214454	34628	9 Lac	22 37 22.3	+51 32 43	−0.006	−0.10	4.63	0.24	1.5	A7 IV	+12	41 s		
214432	72547		22 37 28.5	+39 26 20	−0.001	0.00	7.59	−0.11	−1.7	B3 V		690 s		
214304	191273		22 37 29.7	−20 22 44	+0.002	−0.06	7.3	0.4	1.7	A3 V		58 mx		
214068	255265		22 37 31.8	−65 33 59	+0.002	−0.01	7.9	1.1	0.2	K0 III		350 s		
214398	90655		22 37 33.2	+24 00 05	−0.002	0.00	7.0	0.1	1.7	A3 V		110 s		
214584	20182		22 37 38.8	+63 46 11	+0.010	+0.02	7.0	1.1	0.2	K0 III	−19	210 s		
213928	—		22 37 42.4	−73 26 03			8.0	0.5	3.4	F5 V		82 s		
214306	191276		22 37 42.5	−29 02 41	+0.004	+0.01	7.9	0.6	3.4	F5 V		68 s		
214376	146210	63 κ Aqr	22 37 45.2	−4 13 41	−0.005	−0.11	5.03	1.14	−0.1	K2 III	+8	110 s		m
214488	52200		22 37 46.5	+45 24 30	−0.002	0.00	7.9	1.1	0.2	K0 III		330 s		
214291	231223		22 37 47.4	−39 51 25	−0.003	+0.01	6.80	0.5	4.0	F8 V		36 s		m
214435	108052		22 37 52.2	+14 59 10	+0.002	−0.16	7.9	0.5	4.0	F8 V		60 s		
214458	90664		22 37 55.6	+29 55 26	+0.005	−0.02	7.40	1.1	−0.1	K2 III		290 mx		
214489	72553		22 37 56.4	+38 56 58	+0.018	+0.10	7.79	0.53	4.0	F8 V		57 s		
214276	247524		22 38 02.0	−52 46 12	−0.001	+0.01	7.8	2.2	−0.3	K5 III		150 s		
214341	213969		22 38 02.1	−33 28 13	+0.005	−0.03	7.60	0.1	1.7	A3 V		89 mx		m
214504	72554		22 38 02.1	+33 13 36	0.000	+0.01	7.8	0.1	1.7	A3 V		160 s		
214308	231225		22 38 04.4	−46 42 34	+0.010	−0.06	7.69	0.43	2.1	F5 IV		93 mx		
214505	72555		22 38 05.0	+32 14 38	+0.011	+0.03	7.9	1.4	−0.3	K5 III		130 mx		
214524	52204		22 38 09.1	+41 07 15	+0.004	0.00	7.49	−0.04	0.0	B8.5 V		180 mx		
214675	20186		22 38 10.6	+66 08 39	+0.001	0.00	7.3	0.1	0.6	A0 V		200 s		
214385	191284		22 38 10.9	−27 26 37	+0.034	+0.02	7.88	0.63	4.7	G2 V		42 s		
214557	52209		22 38 11.3	+45 49 39	−0.012	−0.18	7.1	0.5	4.0	F8 V	−36	42 s		
214405	191286		22 38 12.0	−25 39 24	+0.004	−0.05	7.8	0.6	2.6	F0 V		69 s		
214558	52211		22 38 17.3	+45 10 59	0.000	0.00	6.5	0.5	4.0	F8 V	−4	31 s		
214564	52212		22 38 19.9	+46 43 00	0.000	+0.02	8.0	1.1	−0.1	K2 III		390 s		

HD	SAO	Star Name	α 2000	δ 2000	μ(α)	μ(δ)	V	B-V	M$_v$	Spec	RV	d(pc)	ADS	Notes
214448	146216		22^h 38^m 22.0^s	-7° 53′ 52″	+0.005^s	0.00″	6.23	0.78		G0			16130	m
214438	191288		22 38 24.1	-22 44 20	0.000	-0.02	7.7	1.6	-0.3	K5 III		340 s		
214607	52216		22 38 26.5	+49 32 25	-0.002	-0.03	8.0	1.4	-0.3	K5 III		420 s		
214491	127652		22 38 26.8	+9 15 30	+0.002	0.00	7.7	1.1	-0.1	K2 III		370 s		
214526	90668		22 38 28.6	+21 13 39	-0.001	+0.02	7.8	1.1	-0.1	K2 III		390 s		
214367	231228		22 38 31.2	-48 40 52	0.000	-0.01	7.8	1.6	-0.1	K2 III		240 s		
214588	52217		22 38 33.4	+44 40 08	+0.001	-0.01	7.0	0.1	0.6	A0 V		190 s		
214460	165211		22 38 34.4	-14 04 00	+0.002	+0.03	7.60	0.1	1.7	A3 V		150 s		m
214665	34651		22 38 37.7	+56 47 45	+0.006	-0.02	5.50	1.6	-0.5	M4 III	+8	150 s	16140	m
214576	72560		22 38 37.8	+33 32 43	-0.004	-0.01	7.8	0.5	3.4	F5 V		74 s		
214734	20190	30 Cep	22 38 38.8	+63 35 04	-0.001	-0.02	5.19	0.06		A2	+11	27 mn		
214566	90671		22 38 40.9	+29 42 41	+0.002	-0.01	8.0	0.1	0.6	A0 V		300 s		
214462	191292		22 38 44.5	-28 44 52	-0.003	0.00	6.47	1.04		K0				
214390	247527		22 38 45.7	-53 43 53	+0.003	+0.01	7.90	0.4	2.6	F0 V		120 s		m
214608	52218		22 38 47.2	+44 18 51	+0.022	+0.07	6.83	0.55	4.4	dG0	-19	31 s	16138	m
214546	127658		22 38 50.0	+4 31 49	+0.003	0.00	6.9	0.1	1.7	A3 V		110 s		
214484	213981		22 38 51.4	-33 04 53	+0.001	+0.02	5.66	0.05	1.4	A2 V p	+4	71 s		
214567	108064	40 Peg	22 38 52.5	+19 31 20	-0.003	-0.10	5.82	0.92	3.2	G5 IV	-20	27 s		m
214465	213982		22 38 55.6	-38 28 39	0.000	-0.02	6.6	0.7	-0.1	K2 III		220 s		
214466	213983		22 38 56.5	-39 09 19	+0.001	-0.01	7.0	0.5	0.2	K0 III		230 s		
214485	213985		22 39 02.6	-37 02 36	-0.010	-0.02	7.9	0.6	3.4	F5 V		62 s		
214652	72569		22 39 04.4	+37 22 32	-0.001	0.00	6.84	-0.11	-2.5	B2 V	-13	680 s	16143	m
214764	34659		22 39 05.3	+58 55 39	+0.010	0.00	6.9	0.4	2.6	F0 V		71 s		
214441	247531		22 39 08.3	-52 41 32	+0.001	-0.01	6.65	0.35		F0				m
214680	72575	10 Lac	22 39 15.6	+39 03 01	0.000	0.00	4.88	-0.20	-4.8	O9 V	-10	780 s	16148	m
214572	165217		22 39 15.9	-10 01 40	-0.003	0.00	7.2	0.6	4.4	G0 V	+12	36 s		
214697	52230		22 39 22.9	+40 36 01	+0.004	0.00	8.0	1.1	-0.1	K2 III		390 s		
215319	3769		22 39 24.3	+81 23 31	+0.005	+0.01	6.9	0.2	2.1	A5 V	+11	90 s		
214486	247535		22 39 27.7	-50 40 55	-0.003	-0.02	7.7	2.0	-0.5	M2 III		270 s		
214343	255272		22 39 30.7	-68 03 37	+0.021	-0.01	6.9	1.2	0.2	K0 III		180 mx		
214716	72578		22 39 31.1	+37 21 14	-0.002	0.00	6.7	1.1	0.2	K0 III		200 s		
214713	72579		22 39 31.7	+37 44 37	+0.002	0.00	7.8	0.1	1.4	A2 V		190 s		
211207	—		22 39 32.5	-87 19 52			7.5	1.1	0.2	K0 III		290 s		
214712	72580		22 39 32.7	+38 44 55	-0.003	0.00	6.98	0.29	2.6	F0 V		75 s		
214780	34664		22 39 34.1	+51 09 21	+0.020	+0.06	6.8	0.5	3.4	F5 V		47 s		
214714	72581		22 39 34.1	+37 35 34	+0.001	0.00	6.03	0.86	3.2	G5 IV	-7	33 s	16154	m
214615	165218		22 39 34.4	-12 36 54	+0.016	-0.15	7.76	0.78	5.7	dG9	-11	26 ts	16145	m
214767	52239		22 39 39.4	+42 12 01	+0.001	-0.04	7.5	0.1	1.7	A3 V		140 s		
214740	72586		22 39 42.6	+31 08 44	-0.001	-0.01	8.0	1.1	-0.1	K2 III		420 s		
214599	191308		22 39 43.8	-28 19 31	+0.006	-0.03	6.31	1.01	3.2	K0 IV		46 s	16149	m
214600	191309		22 39 46.1	-28 20 55	+0.006	-0.05	7.11	0.47	3.4	F5 V		46 s	16149	m
214698	108078	41 Peg	22 39 46.8	+19 40 52	0.000	-0.01	6.10	0.00	1.4	A2 V	-11	87 s		
214669	127671		22 39 49.4	+3 14 33	+0.001	+0.01	7.8	1.4	-0.3	K5 III		410 s		
214683	127672		22 39 50.5	+4 06 59	+0.011	+0.13	7.7	1.1	-0.1	K2 III		86 mx		
214654	165221		22 39 55.2	-14 33 08	+0.003	-0.19	7.9	0.5	4.0	F8 V		52 mx		
214757	90686		22 39 55.4	+22 52 37	-0.004	-0.01	7.4	1.1	0.2	K0 III		280 s		
214563	231238		22 39 56.6	-47 44 41	+0.001	-0.01	7.9	1.5	0.2	K0 III		130 s		
214623	213995		22 39 57.9	-35 10 29	+0.002	-0.01	7.2	0.8	0.2	K0 III		250 s		
213615	258937		22 40 00.3	-83 15 09	+0.039	-0.04	6.8	0.1	2.1	A5 V		65 mx		
214657	165223		22 40 02.5	-19 11 55	+0.001	-0.05	7.30	0.5	4.0	F8 V		46 s	16155	m
214821	52246		22 40 04.6	+41 18 51	+0.003	-0.01	7.3	1.1	0.2	K0 III		260 s		
214806	52247		22 40 05.6	+40 40 13	0.000	-0.01	7.0	1.6	-0.5	M2 III		300 s		
214645	213997		22 40 06.6	-29 56 47	0.000	-0.01	7.34	0.96	1.7	K0 III-IV		130 s		
214573	231240		22 40 06.9	-49 35 52	+0.008	-0.05	7.3	1.5	0.2	K0 III		130 s		
214686	146230		22 40 07.8	-9 21 34	+0.011	+0.04	6.8	0.7	4.4	G0 V	-39	30 s		
214722	146232		22 40 12.2	-6 32 05	-0.001	-0.04	7.2	0.6	4.4	G0 V		36 s		
214879	34681		22 40 17.5	+51 32 53	0.000	+0.03	7.9	0.4	2.6	F0 V		110 s		
214878	34682		22 40 18.2	+53 50 46	0.000	-0.01	5.93	0.93	-0.2	B8 V	-6	42 s		
214690	214000		22 40 22.2	-30 39 32	-0.008	-0.20	5.87	1.30	-0.2	K3 III	+79	160 s		
214786	127678		22 40 27.0	+4 34 30	+0.007	-0.12	7.0	1.1	0.2	K0 III		100 mx		
214825	108089		22 40 30.5	+20 02 29	+0.003	-0.01	7.3	0.4	3.0	F2 V		71 s		
214868	52251	11 Lac	22 40 30.7	+44 16 35	+0.009	+0.02	4.46	1.33	-0.2	K3 III	-10	84 s		
—	—		22 40 33.2	-61 33 15			7.70							T Tuc, v
214772	165229		22 40 39.1	-16 46 16	+0.004	-0.01	7.6	1.1	0.2	K0 III		300 s		
214748	191318	18 ε PsA	22 40 39.2	-27 02 37	+0.002	0.00	4.17	-0.11	-0.2	B8 V	+3	75 s		
214749	191319		22 40 43.2	-29 40 30	+0.031	-0.03	7.83	1.13	8.0	K5 V		10 ts		
214906	52253		22 40 43.8	+43 29 37	+0.006	+0.03	8.0	1.4	-0.3	K5 III		240 mx		
214810	146239		22 40 47.9	-3 33 15	0.000	-0.04	6.31	0.52		G0				m
214539	255281		22 40 48.0	-67 41 20	-0.003	-0.03	7.22	0.02	0.2	B9 V		230 s		
214632	247544		22 40 48.9	-57 25 20	+0.008	0.00	5.97	1.46	-0.3	K4 III		170 s		
—	90699		22 40 52.3	+20 49 24	0.000	+0.02	8.0	1.8	-0.1	K2 III		170 s		
214705	231245		22 40 52.5	-45 30 23	+0.009	-0.07	7.6	1.1	3.2	G5 IV		38 s		
214850	108094		22 40 52.5	+14 32 58	+0.018	+0.15	5.71	0.72	4.9	dG3	-10	18 ts	16173	m
215065	20205		22 40 54.5	+66 31 24	+0.035	+0.39	7.5	0.9	3.2	G5 IV	-46	70 s		
214759	214006		22 40 55.2	-31 59 24	+0.029	+0.04	7.39	0.80	5.6	G8 V		21 s		

HD	SAO	Star Name	α 2000	δ 2000	μ(α)	μ(δ)	V	B-V	M$_v$	Spec	RV	d(pc)	ADS	Notes
			h m s	° ' "	s	"								
214882	90700		22 40 59.8	+25 12 37	−0.005	−0.03	7.3	0.4	2.6	F0 V		88 s		
214920	72616		22 41 01.4	+37 04 11	0.000	−0.05	7.8	0.5	3.4	F5 V		74 s		
214976	34694		22 41 02.6	+51 45 31	0.000	+0.01	6.7	1.1	−0.1	K2 III		230 s		
214707	247546		22 41 02.8	−51 27 34	0.000	−0.04	7.7	2.4	−0.5	M4 III		170 s		
214929	72617		22 41 03.7	+36 15 49	−0.001	−0.02	7.0	1.4	−0.3	K5 III		290 s		
214813	191323		22 41 04.2	−23 31 16	+0.004	−0.03	7.1	0.9	4.0	F8 V		26 s		
214883	108097		22 41 06.7	+15 29 07	−0.002	−0.03	7.7	1.1	0.2	K0 III		320 s		
214946	52262		22 41 09.5	+45 00 31	+0.002	+0.02	7.1	0.2	2.1	A5 V	−15	100 s		
214708	247547		22 41 15.2	−57 01 43	−0.002	0.00	7.3	1.5	−0.3	K5 III		340 s		
214852	191328		22 41 22.6	−25 21 02	+0.013	0.00	7.3	0.9	3.4	F5 V		32 s		
215025	52266		22 41 22.7	+50 05 34	+0.001	+0.02	7.8	0.1	0.6	A0 V		270 s		
214873	165233		22 41 25.4	−12 13 49	0.000	−0.02	6.8	1.1	−0.1	K2 III		240 s		
214930	90709		22 41 25.5	+23 50 48	−0.001	−0.01	7.40	−0.13	−1.7	B3 V	−53	650 s		
215250	10464		22 41 26.1	+73 15 24	+0.005	0.00	7.7	0.1	0.6	A0 V		240 s		
214923	108103	42 ζ Peg	22 41 27.6	+10 49 53	+0.005	−0.01	3.40	−0.09	0.0	B8.5 V	+7	48 s	16182	Homam, m
214993	72627	12 Lac	22 41 28.5	+40 13 32	−0.001	0.00	5.25	−0.14	−3.6	B2 III	−15	560 s		DD Lac, m,v
214979	72625		22 41 31.3	+30 57 57	+0.005	+0.01	6.4	1.4	−0.3	K5 III	−35	220 s		
215398	10469		22 41 33.1	+77 36 56	+0.002	+0.02	7.2	0.1	1.4	A2 V		140 s		
215066	34703		22 41 34.1	+54 23 12	+0.002	0.00	7.82	0.03	0.4	B9.5 V		280 s		
215030	52270		22 41 35.9	+41 32 57	+0.013	+0.07	6.0	0.9	0.2	G9 III	−14	92 mx		
214994	90717	43 o Peg	22 41 45.3	+29 18 27	0.000	−0.02	4.79	−0.01	1.2	A1 V	+9	52 s		
215012	90718		22 41 49.4	+22 23 22	+0.003	+0.02	7.3	0.1	0.6	A0 V		220 s		
214902	214019		22 41 54.0	−30 01 42	−0.001	+0.03	7.8	1.7	3.2	G5 IV		24 s		
214995	108109		22 41 57.3	+14 30 59	+0.006	−0.02	5.90	1.11	0.2	K0 III	−26	120 s		
215041	90721		22 42 02.4	+22 06 01	−0.003	−0.03	7.4	1.1	0.2	K0 III		270 s		
214983	146251		22 42 05.8	−5 06 08	+0.005	−0.05	7.10	1.6	−0.5	M2 III		230 mx		m
214982	146252		22 42 06.4	−3 28 26	+0.002	−0.02	7.8	0.1	0.6	A0 V		270 s		
215015	108112		22 42 08.1	+11 05 07	+0.003	−0.02	7.6	0.4	3.0	F2 V		84 s		
215089	52282		22 42 08.5	+42 23 46	+0.006	+0.01	8.0	1.1	0.2	K0 III		280 mx		
215074	72638		22 42 09.7	+33 13 06	−0.001	0.00	7.9	1.4	−0.3	K5 III		430 s		
215031	127698		22 42 09.9	+8 04 29	−0.001	−0.01	7.8	0.8	3.2	G5 IV		84 s		
215251	20210		22 42 16.8	+64 52 20	−0.002	+0.01	7.23	1.20	4.0	F8 V		15 s		
214965	191336		22 42 18.2	−27 22 58	0.000	−0.01	7.7	1.5	−0.3	K5 III		390 s		
215149	52286		22 42 18.2	+49 15 20	+0.003	+0.01	7.9	0.1	1.4	A2 V		190 s		
215159	34713		22 42 20.6	+53 54 32	0.000	0.00	6.12	1.62	−0.1	K2 III	+9	94 s		
215043	127699		22 42 21.0	+5 10 25	+0.002	0.00	7.6	0.1	1.7	A3 V		150 s		
214966	191337	19 PsA	22 42 22.0	−29 21 40	+0.002	−0.01	6.17	1.54	−0.5	M2 III	−9	220 s		
215178	34718		22 42 23.9	+54 14 54	+0.002	−0.01	7.50	0.00	0.4	B9.5 V		240 s	16203	m
215209	34720		22 42 26.4	+58 31 34	+0.002	0.00	7.9	0.1	0.6	A0 V		260 s		
215047	165243		22 42 32.0	−12 33 36	−0.003	−0.01	7.8	0.1	0.6	A0 V		270 s		
215439	10474		22 42 35.9	+74 19 51	+0.019	−0.05	8.0	0.4	3.0	F2 V		95 s		
214953	231257		22 42 36.7	−47 12 38	+0.001	−0.32	5.98	0.58	4.5	G1 V		19 ts		m
214691	—		22 42 36.8	−73 13 47			8.0	0.8	3.2	G5 IV		90 s		
215321	20211		22 42 38.8	+65 40 53	+0.002	+0.01	7.91	0.08	0.6	A0 V		260 s		
214952	231258	β Gru	22 42 39.9	−46 53 05	+0.014	−0.01	2.11	1.62	−2.4	M3 II	+2	53 mx		
214987	231260		22 42 43.0	−44 14 53	+0.004	+0.04	6.07	0.98		K1 IV		410 mx		
215093	146261		22 42 48.9	+0 13 55	0.000	+0.01	6.98	0.32	2.6	F0 V	−16	72 s		
215252	34725		22 42 51.8	+54 45 27	+0.002	+0.02	7.3	1.1	−0.1	K2 III	+7	280 s		
215128	127705		22 42 52.8	+4 58 07	+0.002	+0.01	7.3	1.1	−0.1	K2 III		300 s		
215181	72650		22 42 54.0	+37 26 51	+0.003	0.00	7.3	0.1	1.4	A2 V		150 s		
215866	3782		22 42 55.3	+82 29 37	−0.001	+0.01	7.9	1.1	0.2	K0 III		360 s		
215191	72652		22 42 55.4	+37 48 10	+0.001	0.00	6.43	−0.09	−3.5	B1 V	−18	830 s		
215110	127706		22 42 57.9	+0 24 07	+0.011	−0.15	7.74	0.86	3.2	G5 IV	−9	53 mx		
215163	90731		22 42 57.9	+27 59 19	+0.002	−0.01	7.9	0.0	0.4	B9.5 V		320 s		
215129	127707		22 42 58.0	+1 13 03	−0.001	−0.01	7.00	0.1	1.7	A3 V	−6	120 s	16201	m
215182	90734	44 η Peg	22 43 00.0	+30 13 17	+0.001	−0.02	2.94	0.86	−0.9	G2 II-III	+4	53 s	16211	Matar, m
215097	165248		22 43 01.1	−10 06 10	0.000	+0.01	7.2	1.1	0.2	K0 III	+14	250 s		
215081	191349		22 43 01.6	−21 39 26	+0.002	+0.01	7.2	0.5	0.2	K0 III	+7	260 s		
215833	3781		22 43 02.4	+81 53 46	−0.004	−0.01	7.7	1.1	0.2	K0 III		310 s		
215114	146271		22 43 03.3	−8 18 42	+0.001	0.00	6.45	0.17		A2	+5		16208	m
215371	20214		22 43 04.3	+65 20 10	+0.001	0.00	6.74	−0.01	−2.6	B2.5 IV	−23	580 s		
215242	52296		22 43 04.3	+47 10 07	+0.001	0.00	6.40	0.47		B9 comp	−17		16214	m
215971	3784		22 43 04.8	+83 16 15	+0.006	+0.01	7.5	0.0	0.0	B8.5 V		240 mx		
215183	90735		22 43 05.9	+23 01 57	−0.002	−0.02	7.8	1.1	−0.1	K2 III		380 s		
215143	146273	67 Aqr	22 43 14.0	−6 57 47	+0.002	−0.01	6.41	−0.04		B9				
215322	34734		22 43 15.2	+56 24 31	+0.010	+0.05	6.9	0.4	2.6	F0 V		70 s		
215085	231263		22 43 16.7	−40 00 34	+0.003	−0.04	7.8	1.0	0.2	K0 III		260 s		
215192	108129		22 43 18.2	+17 15 04	+0.003	−0.03	8.0	0.4	3.0	F2 V		98 s		
215383	20215		22 43 25.5	+62 40 18	+0.006	+0.02	6.9	0.9	3.2	G5 IV		54 s		
215104	231265	ρ Gru	22 43 29.9	−41 24 52	+0.002	−0.09	4.85	1.03	0.2	gK0	+29	83 s		m
215120	231266		22 43 29.9	−40 07 44	+0.002	−0.02	7.3	2.3	−0.5	M4 III		290 s		
215324	52303		22 43 31.0	+46 01 34	+0.018	−0.01	7.40	0.5	3.4	F5 V	−33	44 mx	16220	m
215290	72666		22 43 35.0	+32 49 21	−0.001	+0.01	7.30	1.6	−0.4	M0 III	−23	350 s		
215167	165252	66 Aqr	22 43 35.1	−18 49 49	−0.002	−0.02	4.69	1.37	−0.3	K4 III	+22	100 s		
215335	52308		22 43 38.7	+44 31 49	+0.001	0.00	7.3	1.1	0.2	K0 III		250 s		

HD	SAO	Star Name	α 2000	δ 2000	μ(α)	μ(δ)	V	B-V	M$_v$	Spec	RV	d(pc)	ADS	Notes
215274	90742		22^{h}43^{m}40.3	+30°05'32"	+0.018	+0.02	8.03	0.67	5.2	G5 V		37 s		
215344	52310		22 43 41.3	+47 24 43	-0.001	+0.01	6.9	1.1	0.2	K0 III		210 s		
215243	108131		22 43 42.5	+10 56 21	+0.001	-0.16	6.4	0.5	3.4	F5 V	-2	41 s		
215037	255297		22 43 42.7	-60 43 27	-0.004	-0.02	7.3	0.7	2.6	F0 V		49 s		
215730	—		22 43 50.6	+78 31 06	+0.005	+0.01	7.50	0.1	0.6	A0 V		220 s	16243	m
215257	127714		22 43 50.7	+3 53 13	+0.011	+0.34	7.6	0.5	4.0	F8 V		45 mx		
215309	127717		22 44 04.8	+6 44 10	+0.004	-0.08	7.8	0.5	3.4	F5 V		74 s		
215359	72675		22 44 05.1	+39 27 56	0.000	-0.02	5.95	1.48	-0.3	K5 III	-27	180 s	16228	m
215373	52317	13 Lac	22 44 05.3	+41 49 09	-0.001	+0.01	5.08	0.96	0.2	K0 III	+13	95 s	16227	m
215500	20223		22 44 06.0	+64 34 16	+0.012	-0.28	7.80	0.9	5.6	G8 V	-42	27 s		
215867	3785		22 44 06.6	+80 26 14	+0.011	+0.02	7.1	1.1	0.2	K0 III		220 s		
215485	20221		22 44 07.6	+61 30 42	-0.001	0.00	7.7	1.1	-0.1	K2 III		330 s		
215361	90752		22 44 15.9	+20 56 14	+0.003	+0.03	7.7	1.1	0.2	K0 III		320 s		
215121	255299		22 44 16.4	-60 29 58	+0.001	+0.03	6.5	1.2	3.4	F5 V		15 s		
215280	165258		22 44 17.2	-14 36 42	+0.001	0.00	7.6	0.9	3.2	G5 IV		77 s		
215427	52320		22 44 17.2	+49 24 45	-0.001	+0.02	7.2	1.4	-0.3	K5 III	-26	300 s		
215442	34743		22 44 19.4	+51 27 00	+0.014	+0.09	7.4	0.5	4.0	F8 V		47 s		
214917	258063		22 44 20.5	-74 49 14	-0.013	-0.02	7.8	0.5	3.4	F5 V		75 s		
215296	165259		22 44 21.2	-14 40 33	+0.001	+0.01	7.2	1.1	0.2	K0 III		250 s		
215703	10483		22 44 23.6	+74 25 49	-0.001	+0.08	8.0	1.4	-0.3	K5 III		150 mx		
215298	191364		22 44 25.3	-23 45 55	+0.002	-0.01	8.0	0.4	0.6	A0 V		160 s		
215283	214053		22 44 28.0	-31 07 47	+0.001	-0.02	7.7	1.3	4.4	G0 V		16 s		
215385	90754		22 44 28.9	+21 57 07	+0.001	+0.01	7.9	1.4	-0.3	K5 III		430 s		
215362	108140		22 44 29.2	+14 17 25	+0.001	-0.01	7.5	1.1	0.2	K0 III		300 s		
215471	34745		22 44 29.7	+53 45 23	-0.001	-0.02	7.4	1.1	-0.1	K2 III	-16	290 s		
215131	255300		22 44 33.9	-63 57 14	-0.001	0.00	7.10	0.1	0.6	A0 V		200 s		m
215533	20226		22 44 34.7	+62 11 22	+0.006	+0.01	8.0	1.1	-0.1	K2 III		370 s		
215158	255301		22 44 37.2	-62 32 55	+0.006	-0.03	6.7	1.8	0.2	K0 III		63 s		
215206	247562		22 44 39.6	-57 27 28	0.000	+0.02	7.4	1.2	0.2	K0 III		220 s		
215171	255303		22 44 40.3	-60 07 03	+0.008	-0.04	7.80	0.8	3.2	G5 IV		83 s		m
215317	231272		22 44 48.8	-40 46 04	-0.005	-0.03	7.2	0.7	-0.1	K2 III		280 s		
215518	34757		22 44 49.0	+52 31 02	0.000	+0.02	6.6	1.1	-0.1	K2 III	+5	210 s		
215239	247564		22 44 50.0	-56 49 37	+0.004	0.00	7.1	1.2	0.2	K0 III		170 s		
215445	90757		22 44 50.3	+24 22 42	+0.001	-0.01	7.30	1.1	0.2	K0 III		260 s		
215575	20229		22 44 51.2	+60 48 53	+0.004	+0.02	8.02	0.27	0.0	B8.5 V		210 mx		
215473	72682		22 44 52.8	+39 12 04	+0.002	-0.01	6.42	1.50	-0.3	K5 III		220 s		
215402	146292		22 45 00.2	-0 43 56	+0.009	+0.01	7.9	0.5	3.4	F5 V		79 s		
215588	34761		22 45 03.4	+58 08 49	-0.008	-0.13	6.5	0.5	3.4	F5 V		42 s		
215459	108150		22 45 10.3	+11 41 10	0.000	-0.01	8.0	0.4	3.0	F2 V		98 s		
215547	52334		22 45 15.6	+47 12 13	+0.001	0.00	7.6	1.6	-0.5	M4 III		400 s		
215606	34768		22 45 19.3	+57 08 25	+0.002	0.00	7.9	0.2		A m				
215449	146296		22 45 20.5	-9 38 40	+0.001	0.00	7.40	0.2	2.1	A5 V		120 s	16238	m
215475	127735		22 45 21.5	+3 41 22	+0.002	-0.03	8.0	1.1	-0.1	K2 III		410 s		
215476	127737		22 45 22.5	+1 45 09	-0.008	-0.04	7.4	1.1	0.2	K0 III		190 mx		
215566	52336		22 45 24.2	+44 45 08	-0.001	-0.01	7.0	0.0	0.4	B9.5 V	-22	210 s		
206553	258931		22 45 27.8	-88 49 07	+0.058	-0.04	6.57	0.28	1.5	A7 IV		96 s		
215451	165275		22 45 27.9	-15 35 01	+0.002	-0.04	7.1	1.4	-0.3	K5 III		300 s		
215510	108154	45 Peg	22 45 28.0	+19 22 00	-0.002	+0.06	6.25	1.05	0.2	K0 III	-22	130 mx		
215452	191382		22 45 32.2	-25 14 12	+0.006	+0.02	6.9	0.1	0.6	A0 V		91 mx		
215607	34772		22 45 32.2	+53 22 29	+0.002	0.00	7.5	1.1	-0.1	K2 III		300 s		
215452	191382		22 45 32.2	-25 14 12	+0.006	+0.02	6.9	0.1	0.6	A0 V		91 mx		
215607	34772		22 45 32.2	+53 22 29	+0.002	0.00	7.5	1.1	-0.1	K2 III		300 s		
215591	52338		22 45 32.6	+48 09 41	+0.001	-0.01	7.9	0.1	0.6	A0 V		270 s		
215436	214058		22 45 33.5	-32 06 24	+0.003	-0.02	7.8	2.2	-0.1	K2 III		92 s		
215549	72692		22 45 34.2	+30 26 33	-0.020	-0.35	6.38	0.92		K1 III-IV	-1	22 mn	16248	m
215618	34773		22 45 34.6	+51 19 22	0.000	0.00	7.8	1.4	-0.3	K5 III		390 s		
215633	34775		22 45 36.0	+52 53 25	-0.002	-0.01	7.0	1.1	0.2	K0 III	-6	220 s		
215855	10491		22 45 37.1	+74 33 14	-0.006	-0.02	7.6	1.4	-0.3	K5 III		340 s		
215369	247570	η Gru	22 45 37.7	-53 30 00	+0.004	+0.02	4.85	1.18	0.2	gK0	+28	62 s		m
214469	258940		22 45 38.0	-83 44 30	+0.015	+0.01	7.9	0.4	2.6	F0 V		100 s		
215267	255307		22 45 38.0	-66 06 01	+0.004	-0.03	7.9	1.1	0.2	K0 III		340 mx		
215405	231278		22 45 40.5	-46 32 51	-0.003	-0.01	5.51	1.32	-0.2	K3 III	+42	130 s		
215466	191384		22 45 42.0	-26 58 44	0.000	-0.01	7.9	1.7	-0.1	K2 III		200 s		
215524	146300		22 45 49.5	-0 55 59	+0.001	0.00	7.8	1.1	0.2	K0 III		330 s		
215550	108159		22 45 53.9	+11 12 01	+0.002	0.00	7.5	0.8	3.2	G5 IV		72 s		
215619	72701		22 46 01.7	+35 40 20	+0.012	+0.09	7.7	0.5	3.4	F5 V		71 s		
215770	20236		22 46 03.0	+65 17 41	+0.002	0.00	7.60	0.8	3.2	G5 IV		74 s		m
214846	258941	β Oct	22 46 03.1	-81 22 54	-0.028	+0.01	4.15	0.20	2.6	dF0	+24	20 s		
215468	231284		22 46 04.0	-45 50 32	+0.007	+0.01	8.02	0.37		F5				
215424	247572		22 46 04.2	-55 03 49	-0.006	+0.01	6.5	1.3	0.2	K0 III		130 s		
215714	34785		22 46 05.0	+58 04 20	0.000	-0.05	8.00	0.5	4.0	F8 V		62 s	16260	m
215635	72702		22 46 06.0	+38 29 58	-0.001	+0.02	7.9	1.1	-0.1	K2 III		390 s		
215684	34783		22 46 06.3	+52 06 17	0.000	0.00	8.00	0.4	2.6	F0 V		120 s	16259	m
215456	231285		22 46 07.8	-48 58 44	+0.021	-0.04	6.62	0.63	3.6	G0 IV-V		38 s		
215664	52348		22 46 10.0	+44 32 46	+0.013	+0.04	5.8	0.4	2.6	F0 V	-10	44 s		

HD	SAO	Star Name	α 2000	δ 2000	μ(α)	μ(δ)	V	B-V	M$_v$	Spec	RV	d(pc)	ADS	Notes
215675	52349		22^h 46^m 12.7^s	+42° 51' 37"	+0.002	+0.01	7.6	1.1	-0.1	K2 III		330 s		
215553	165285		22 46 14.0	-11 09 59	-0.001	-0.03	6.6	1.6	-0.5	M2 III		260 s		
215704	52353		22 46 20.3	+50 12 38	+0.026	+0.03	7.86	0.80	0.2	K0 III		78 mx		
215705	52352		22 46 20.9	+46 12 55	+0.001	-0.01	6.7	1.4	-0.3	K5 III		250 s		
215208	258066		22 46 22.4	-75 25 20	-0.010	+0.01	7.4	0.0	1.4	A2 V		160 s		
215504	231287		22 46 28.1	-49 41 09	+0.009	-0.27	6.48	1.14	1.7	K0 III-IV		40 mx		
215638	127747		22 46 28.4	+7 34 59	+0.004	+0.01	6.9	0.4	3.0	F2 V		59 s		
215757	34793		22 46 29.4	+54 52 18	+0.001	-0.01	6.85	0.07	0.6	A0 V		160 s		
215665	90775	47 λ Peg	22 46 31.7	+23 33 56	+0.004	-0.01	3.95	1.07	-0.9	G8 II-III	-4	33 ts		
215517	247579		22 46 41.3	-51 03 01	0.000	-0.01	7.6	1.7	-0.5	M4 III		370 s		
215648	108165	46 ξ Peg	22 46 41.4	+12 10 22	+0.016	-0.49	4.19	0.50	3.8	F7 V	-5	14 ts	16261	m
--	34801		22 46 42.3	+53 38 22	+0.001	0.00	7.8	1.7	-0.3	K7 III		310 s		
215545	231290		22 46 43.5	-46 56 23	+0.006	+0.01	6.56	0.30	0.6	F0 III		160 s		m
215676	108166		22 46 44.7	+14 59 42	-0.003	-0.01	7.8	1.1	0.2	K0 III		330 s		
215690	90778		22 46 44.7	+25 48 12	-0.001	0.00	7.9	0.1	0.6	A0 V		290 s		
215601	214072		22 46 47.6	-31 52 21	0.000	-0.17	7.8	0.7	4.4	G0 V		40 s		
215585	231293		22 46 48.4	-42 32 00	+0.002	-0.02	8.00	1.1	0.2	K0 III		210 s		m
215408	255312		22 46 53.4	-69 31 40	+0.005	-0.02	7.8	1.2	-0.1	K2 III		360 s		
215616	214074		22 46 54.1	-33 10 57	+0.003	+0.01	6.8	0.6	3.2	G5 IV		52 s		
215760	72716		22 46 55.7	+39 25 02	-0.003	+0.04	7.5	0.5	4.0	F8 V		50 s		
--	20241		22 46 57.3	+60 45 13	+0.001	0.00	8.0	2.2	0.2	K0 III		66 s		
215733	108170		22 47 02.3	+17 13 59	-0.001	-0.02	7.34	-0.13	-5.1	B1 II	-23	2900 s		
215532	255314		22 47 06.6	-60 04 34	+0.011	-0.05	7.9	0.8	4.4	G0 V		37 s		
215627	231296		22 47 09.0	-41 41 31	+0.003	0.00	6.84	1.28	-0.2	K3 III		260 s		
215641	214078		22 47 09.4	-32 40 30	+0.022	-0.07	7.5	0.6	3.2	G5 IV		67 mx		
215848	34815		22 47 11.3	+54 45 07	+0.002	+0.01	7.78	0.08	0.0	B8.5 V		320 s		
215708	146307		22 47 12.5	-2 42 35	+0.007	-0.03	7.6	0.5	3.4	F5 V		68 s		
215643	214079		22 47 15.8	-39 27 25	+0.004	-0.05	4.20	0.51	3.8	F7 V		12 s		
215869	34818		22 47 18.6	+53 22 53	-0.003	-0.03	6.7	0.1	1.4	A2 V		120 s		
215669	214081		22 47 19.0	-34 09 40	-0.005	+0.05	6.28	1.16		K0				
215793	72720		22 47 20.0	+31 02 16	+0.003	+0.01	8.0	0.1	1.4	A2 V		210 s		
215696	165289		22 47 20.5	-16 08 46	+0.026	-0.05	7.33	0.68		G8	-30	18 mn	16265	m
215762	108176		22 47 21.6	+14 07 23	0.000	-0.03	7.0	1.1	-0.1	K2 III		270 s		
215907	34824		22 47 23.0	+58 28 58	+0.001	+0.01	6.36	-0.04	0.6	A0 V	+4	140 s		
216035	10499		22 47 23.0	+71 32 56	-0.006	+0.01	7.9	0.4	2.6	F0 V		110 s		
215856	52364		22 47 24.5	+46 12 28	+0.005	+0.01	8.0	0.0	0.4	B9.5 V		160 mx		
215709	165290		22 47 25.1	-17 33 00	+0.008	+0.02	6.9	0.5	3.4	F5 V		50 s		
215657	231298		22 47 26.7	-44 57 53	-0.001	+0.03	7.22	0.60	3.9	G3 IV-V		45 s		
215678	214083		22 47 28.2	-34 49 32	0.000	+0.01	7.2	1.0	0.2	K0 III		250 s		
216446	3794		22 47 28.8	+83 09 14	+0.012	+0.05	4.74	1.26	-0.2	K3 III	-31	97 s	16294	m
215777	127763		22 47 28.8	+7 42 32	+0.002	0.00	7.6	0.2	2.1	A5 V		130 s		
215778	127764		22 47 29.6	+5 53 16	+0.005	-0.09	7.7	0.5	4.0	F8 V		54 s		
215749	146310		22 47 29.6	-1 47 20	0.000	-0.01	7.5	1.1	-0.1	K2 III		340 s		
215763	127762		22 47 29.9	+2 54 16	+0.005	-0.01	8.00	0.5	4.2	F9 V	-23	58 s		
216520	3796		22 47 30.9	+83 41 50	-0.098	+0.13	7.5	1.1	-0.1	K2 III		57 mx		
215505	255315		22 47 32.1	-69 28 28	-0.005	+0.03	7.0	0.3	2.6	F0 V		75 s		
215721	165293	68 Aqr	22 47 33.0	-19 36 48	-0.007	-0.20	5.26	0.94	0.3	G7 III	+23	73 mx		
215870	52365		22 47 34.0	+47 38 27	-0.001	+0.02	8.0	1.1	-0.1	K2 III		390 s		
215562	255316		22 47 35.4	-65 33 39	-0.005	-0.02	6.50	1.1	0.2	K0 III		180 s		m
215766	165298	69 Aqr	22 47 42.6	-14 03 23	+0.002	-0.01	5.80	-0.05		B9	+15		16268	m
215724	214087		22 47 46.9	-38 13 20	-0.006	-0.07	6.71	0.48	2.8	G0 IV		61 s		
215896	52370		22 47 47.8	+47 48 45	+0.001	+0.01	7.8	0.1	0.6	A0 V		260 s		
215908	52372		22 47 48.1	+49 10 44	-0.001	+0.03	8.0	1.1	0.2	K0 III		340 s		
215812	146315		22 47 50.1	-4 13 44	-0.014	-0.30	6.66	0.65	5.2	G5 V	-23	20 s	16270	m
216014	20247		22 47 52.9	+65 03 44	0.000	+0.01	6.86	0.30		B0.5 III	-21			AH Cep, m,v
215940	34835		22 47 54.6	+51 58 46	+0.004	+0.01	7.9	0.1	1.7	A3 V		170 s		
215782	191412		22 47 56.1	-25 54 43	+0.009	-0.10	6.30	0.91	0.3	G5 III		74 mx		
215660	247586		22 47 57.4	-58 41 07	+0.004	-0.02	6.9	1.5	-0.1	K2 III		170 s		
215711	247589		22 48 03.9	-51 30 44	+0.005	+0.02	7.5	0.4	0.2	K0 III		280 s		
215768	214093		22 48 07.1	-37 45 24	+0.018	+0.01	7.8	0.5	4.4	G0 V		49 s		
215953	52377		22 48 07.9	+49 34 59	+0.004	+0.01	7.20	1.6	-0.5	M3 III	-55	160 mx		
215860	127771		22 48 09.2	+5 52 43	+0.001	-0.05	7.9	0.5	3.4	F5 V		78 s		
215943	72732		22 48 10.8	+37 25 00	-0.005	-0.05	5.90	1.03	0.2	gK0	-25	130 s		
216720	3799		22 48 12.7	+84 46 33	+0.033	-0.01	7.0	1.1	0.2	K0 III	+2	230 s		
215682	255321		22 48 21.3	-61 41 03	+0.011	+0.01	6.37	1.06		K0				
215955	72737		22 48 27.2	+31 06 01	-0.001	-0.01	7.70	0.00	0.4	B9.5 V	+8	290 s	16278	m
215852	191417		22 48 27.8	-27 33 50	-0.001	-0.09	7.0	1.2	4.4	G0 V		14 s		
215818	231307		22 48 30.1	-43 26 46	-0.001	0.00	7.07	0.99		G5				
215874	165308	70 Aqr	22 48 30.1	-10 33 21	+0.002	+0.01	6.2	0.4	2.6	F0 V	-6	51 s		
215617	258068		22 48 32.0	-70 56 00	+0.004	0.00	7.8	1.4	-0.3	K5 III		420 s		
215789	247593	ε Gru	22 48 33.1	-51 19 01	+0.012	-0.06	3.49	0.08	1.4	A2 V	0	25 s		
215991	72740		22 48 36.1	+33 26 34	0.000	-0.06	7.9	1.1	-0.1	K2 III		290 mx		
216004	52384		22 48 39.2	+40 42 46	+0.005	+0.03	7.9	1.1	-0.1	K2 III		300 mx		
215945	127776		22 48 40.4	+7 50 42	0.000	-0.01	7.8	1.1	0.2	K0 III		330 s		
215863	214099		22 48 40.6	-34 46 24	-0.002	0.00	7.5	0.3	0.6	A0 V		160 s		

HD	SAO	Star Name	α 2000	δ 2000	μ(α)	μ(δ)	V	B-V	M_v	Spec	RV	d(pc)	ADS	Notes
216102	20253		22h 48m 43.9s	+62° 56′ 18″	+0.001	−0.05	6.06	1.20	0.2	K0 III	−27	110 s		
216057	34845		22 48 47.6	+54 24 54	−0.002	+0.01	6.0	−0.1		B8				
216017	72750		22 48 49.9	+38 31 37	+0.001	−0.01	8.0	1.6	−0.5	M2 III		510 s		
215916	165313		22 48 53.6	−16 59 42	0.000	+0.02	7.9	1.4	−0.3	K5 III		430 s		
215769	255324		22 48 54.9	−63 43 06	−0.001	+0.01	6.7	−0.6	0.4	B9.5 V		180 s		
215977	127779		22 48 56.0	+4 17 33	+0.004	−0.08	7.5	0.5	3.4	F5 V		65 s		
215935	191422		22 49 00.8	−24 55 25	+0.001	−0.08	8.0	0.7	2.6	F0 V		66 s		
216172	20260		22 49 01.5	+68 34 16	+0.023	+0.08	7.20	0.5	3.4	F5 V	+3	45 ts	16291	m
216065	52390		22 49 02.9	+49 34 00	+0.002	+0.02	7.7	0.1	1.4	A2 V		180 s		
215891	231313		22 49 05.0	−39 52 24	+0.006	−0.02	7.6	2.0	0.2	K0 III		180 s		
216046	90808		22 49 14.8	+23 53 39	−0.001	−0.01	7.8	1.1	−0.1	K2 III		380 s		
215950	214108		22 49 16.7	−31 12 14	0.000	−0.02	8.0	0.2	0.6	A0 V		210 s		
215981	191428		22 49 17.0	−23 05 32	−0.001	−0.02	6.9	1.1	−0.1	K2 III		260 s		
215729	258070		22 49 17.3	−70 20 52	+0.003	+0.04	6.34	0.07		A2				
216106	34852		22 49 17.6	+50 58 35	0.000	−0.15	6.8	0.7	4.4	G0 V		30 s		
216092	52394		22 49 17.7	+47 55 50	+0.002	+0.02	7.85	−0.07	−3.5	B1 V		1500 s		
216068	72757		22 49 22.3	+32 31 48	−0.001	+0.02	8.0	0.1	0.6	A0 V		300 s		
216018	165316		22 49 26.4	−11 20 56	+0.002	+0.03	7.9	0.2	2.1	A5 V		150 s		
215966	214112		22 49 28.2	−34 59 04	+0.002	0.00	8.0	−0.3	0.4	B9.5 V		330 s		
216048	108204		22 49 32.2	+10 28 44	+0.006	−0.03	6.54	0.29		F0 n	−8			
216032	165321	71 τ Aqr	22 49 35.3	−13 35 33	−0.001	−0.03	4.01	1.57	−0.4	M0 III	+1	74 s		m
216227	20267		22 49 36.1	+66 33 14	+0.001	0.00	7.01	0.06	0.4	B9.5 V		210 s	16298	m
216120	52400		22 49 38.7	+42 37 51	+0.004	0.00	7.2	0.1	1.4	A2 V		140 s		
216508	3798		22 49 39.8	+80 22 04	+0.006	0.00	7.2	1.1	0.2	K0 III		240 s		
216228	20268	32 ι Cep	22 49 40.6	+66 12 02	−0.011	−0.12	3.52	1.05	0.0	K1 III	−12	39 ts		
215631	258069		22 49 40.8	−77 03 02	+0.024	+0.02	6.73	0.11	2.4	A7 V		74 s		
215905	247603		22 49 41.0	−58 55 48	+0.005	−0.02	7.2	1.3	−0.1	K2 III		250 s		
216122	52401		22 49 41.1	+40 30 56	+0.002	−0.02	7.6	0.5	4.0	F8 V		52 s	16293	m
216174	34858		22 49 46.1	+55 54 11	+0.010	+0.05	5.43	1.17	0.0	K1 III	−36	110 s		
216110	90814		22 49 47.2	+28 56 41	+0.003	+0.04	8.0	0.5	3.4	F5 V		82 s		
216189	34859		22 49 47.5	+54 50 48	+0.001	−0.01	7.53	0.02	0.0	B8.5 V		320 s		
216367	10514		22 49 54.7	+75 09 45	+0.004	+0.01	7.80	0.1	0.6	A0 V		250 s		m
216052	191434		22 49 55.9	−20 17 07	+0.001	−0.01	7.4	0.3	1.4	A2 V		110 s		
216175	52407		22 49 56.9	+50 01 02	+0.016	+0.07	8.0	0.9	3.2	G5 IV		90 mx		
216042	214119		22 49 58.9	−32 48 19	−0.004	−0.01	6.33	0.31	1.9	F2 IV		77 s		m
216131	90816	48 μ Peg	22 50 00.0	+24 36 06	+0.011	−0.04	3.48	0.93	0.2	K0 III	+14	45 s		Sadalbari
215985	247605		22 50 00.4	−55 14 06	0.000	−0.03	7.0	1.6		Ma				
215906	255328		22 50 04.7	−66 03 04	−0.014	−0.03	8.0	0.5	3.4	F5 V		82 s		
—	34863		22 50 09.6	+53 55 46	0.000	0.00	8.0	1.6	−0.1	K2 III		220 s		
216206	34862		22 50 10.0	+50 40 37	+0.001	+0.01	6.21	1.14	−4.5	G4 Ib	−9	1200 s		
216054	231323		22 50 10.3	−41 29 25	−0.033	−0.21	7.77	0.74	5.2	G5 V		29 s		
216011	247608		22 50 13.0	−55 54 54	−0.001	0.00	7.6	1.9	−0.1	K2 III		130 s		
216191	72767		22 50 14.1	+36 42 15	+0.006	−0.06	7.9	1.1	−0.1	K2 III		160 mx		
216200	52412	14 Lac	22 50 21.6	+41 57 13	+0.001	0.00	5.92	0.08	−2.3	B3 IV	−14	310 s		
215573	258946	ξ Oct	22 50 22.8	−80 07 27	+0.010	−0.02	5.35	−0.15		B8	+16			
216128	165327		22 50 23.9	−14 03 34	0.000	−0.02	8.0	0.4	2.6	F0 V		120 s		
216024	247610		22 50 28.0	−59 31 33	+0.002	0.00	7.5	0.4	2.6	F0 V		78 s		
216178	127794		22 50 38.2	+4 04 06	+0.001	−0.05	7.2	1.1	0.2	K0 III		250 s		
216201	108211		22 50 38.9	+19 08 27	+0.004	−0.02	6.4	1.1	0.2	K0 III	−39	180 s		
216180	146346		22 50 42.3	−0 34 40	0.000	+0.01	7.4	0.1	1.4	A2 V		160 s		
216275	34870		22 50 46.0	+52 03 42	+0.014	+0.18	7.1	0.6	4.4	G0 V		35 s		
216182	146347		22 50 46.1	−7 18 42	−0.002	−0.01	7.2	1.1	0.2	K0 III		260 s		
216350	20277		22 50 48.1	+66 44 18	+0.009	0.00	7.3	1.1	−0.1	K2 III	−1	270 s		
216257	52417		22 50 48.8	+44 56 18	+0.001	0.00	8.0	1.1	0.2	K0 III		340 s		
216219	108214		22 50 51.9	+18 00 08	+0.003	+0.01	7.4	0.9	3.2	G5 IV		69 s		
216282	34871		22 50 53.8	+51 36 46	−0.001	−0.01	7.5	1.4	−0.3	K5 III		350 s		
216043	255333		22 50 54.5	−62 48 52	−0.001	+0.01	7.8	1.0	4.4	G0 V		28 s		
216149	214134		22 51 02.0	−39 09 25	+0.002	−0.01	5.42	1.43	−0.4	M0 III	+27	150 s		
217157	3808		22 51 02.1	+85 22 25	+0.005	+0.11	5.90	1.32	−0.3	K5 III	−30	100 mx		
216306	34875		22 51 03.7	+52 16 06	−0.003	−0.02	7.3	0.4	2.6	F0 V		86 s		
216276	72779		22 51 05.5	+34 23 03	+0.004	+0.02	7.2	0.4	3.0	F2 V		69 s		
216284	52424		22 51 06.0	+43 14 16	−0.001	−0.03	7.7	1.1	0.2	K0 III		300 s		
216114	247617		22 51 06.4	−56 28 45	0.000	−0.03	7.5	1.0	0.2	K0 III		280 s		
216099	247618		22 51 11.9	−59 05 50	+0.010	+0.01	7.4	0.1	2.6	F0 V		90 s		
216321	52426		22 51 13.4	+48 44 03	+0.003	+0.04	6.9	0.0	0.4	B9.5 V	−16	190 s		
216209	191443		22 51 17.4	−23 45 54	−0.001	+0.02	7.5	0.8	0.2	K0 III		300 s		
216210	191444	21 PsA	22 51 20.8	−29 32 11	0.000	−0.01	5.97	0.91	0.2	K0 III		140 s		
216380	20281		22 51 22.5	+61 41 49	+0.015	+0.05	5.60	0.78	1.8	G8 III-IV	+2	59 s	16317	m
216285	90833		22 51 26.4	+26 23 28	−0.001	+0.01	6.93	0.22		A3	−5		16314	m
216260	127801		22 51 27.4	+5 00 26	+0.005	+0.02	8.0	1.1	0.2	K0 III		240 mx		
216425	20283		22 51 29.2	+62 58 21	+0.001	−0.03	6.8	0.9	3.2	G5 IV		52 s		
216261	146358		22 51 33.1	−1 48 48	−0.002	−0.02	7.90	0.1	0.6	A0 V		290 s		m
216411	34881		22 51 33.5	+59 00 32	−0.003	+0.01	7.20	0.60	−6.6	B1 Ia	−44	2000 s		
217158	3809		22 51 33.9	+85 02 48	+0.022	+0.02	7.10	1.6	−0.5	M4 III		330 s		AR Cep, v
216323	72786		22 51 33.9	+32 48 56	+0.003	−0.01	7.1	0.1	1.4	A2 V		140 s		

HD	SAO	Star Name	α 2000	δ 2000	μ(α)	μ(δ)	V	B-V	M$_V$	Spec	RV	d(pc)	ADS	Notes
216296	127806		22^{h}51^{m}41^s.4	+4°47′09″	+0^s.003	0″.00	7.2	0.4	2.6	F0 V		83 s		
216331	90838		22 51 44.5	+30 02 41	−0.001	0.00	7.90	0.9	−2.1	G5 II		1000 s		
216169	255336		22 51 44.8	−59 52 52	−0.004	+0.01	6.3	1.3	0.2	K0 III		120 s		
216394	34883		22 51 44.8	+53 01 40	−0.001	−0.02	7.2	0.1	0.6	A0 V		200 s		
216237	214144		22 51 45.2	−35 53 18	+0.001	+0.01	7.40	0.1	1.7	A3 V		140 s		m
216308	108216		22 51 45.9	+15 05 09	+0.001	+0.01	6.9	0.1	0.6	A0 V	−10	180 s		
216413	34885		22 51 48.1	+53 11 29	+0.008	+0.03	7.90	0.2	2.1	A5 V		110 mx	16323	m
216369	52433		22 51 49.2	+41 18 46	+0.001	−0.01	7.07	0.01	0.6	A0 V	−18	200 s	16321	m
216396	52435		22 51 50.7	+46 44 08	+0.001	−0.01	7.3	1.1	0.2	K0 III		250 s		
216448	34895		22 52 00.4	+57 43 02	−0.011	−0.19	8.02	1.03		K0			16326	m
216397	52436	15 Lac	22 52 01.9	+43 18 45	+0.010	+0.03	4.94	1.56	−0.4	gM0	−17	120 s	16325	m
216270	214146		22 52 09.0	−38 57 05	+0.002	+0.02	7.3	1.7	0.2	K0 III		110 s		
216187	255339		22 52 09.8	−63 11 19	+0.003	−0.04	6.12	1.03	0.2	K0 III		140 s		m
216290	214147		22 52 10.5	−34 38 03	+0.002	−0.02	7.8	−0.1	1.4	A2 V		190 s		
216281	231332		22 52 18.3	−47 34 31	+0.002	0.00	7.7	−0.3	3.4	F5 V		72 s		
216384	108222		22 52 20.7	+10 24 54	+0.005	0.00	6.8	0.5	3.4	F5 V	+7	47 s		
216383	108224		22 52 21.7	+15 52 56	+0.002	−0.02	7.0	1.1	−0.1	K2 III		260 s		
216385	127810	49 σ Peg	22 52 24.0	+9 50 09	+0.035	+0.05	5.16	0.48	2.3	F7 IV	+12	26 mx		
216301	214149		22 52 25.7	−39 18 01	0.000	0.00	7.9	1.4	0.2	K0 III		210 s		
216291	231334		22 52 26.2	−45 59 29	−0.001	−0.02	7.25	1.46		K0				
216325	214152		22 52 26.9	−31 50 04	+0.001	−0.06	7.7	1.8	−0.5	M2 III		220 mx		
216292	231333		22 52 27.3	−47 59 36	+0.005	−0.03	7.7	0.7	0.2	K0 III		280 mx		
216582	20291		22 52 30.1	+67 59 38	+0.015	+0.08	7.8	0.5	3.4	F5 V		72 s		
216532	20286		22 52 30.2	+62 26 26	−0.003	0.00	8.01	0.55		O8				
216336	214153	22 γ PsA	22 52 31.4	−32 52 32	−0.002	−0.02	4.46	−0.04	0.6	A0 V	+17	59 s		m
216487	52442		22 52 32.3	+47 45 37	0.000	0.00	7.9	0.1	1.4	A2 V		190 s		
216400	127811		22 52 32.6	+3 33 08	0.000	0.00	6.9	0.1	0.6	A0 V		180 s		
216451	72802		22 52 33.7	+35 20 17	+0.001	+0.01	7.1	0.1	1.4	A2 V		140 s		
216419	127813		22 52 34.0	+7 39 25	−0.003	−0.08	8.0	0.4	3.0	F2 V		98 s		
216510	34912		22 52 35.1	+56 31 14	+0.001	+0.01	7.70	1.1	−3.3	K2 Ib-II		1100 s		
216509	34911		22 52 35.3	+56 42 36	+0.012	+0.05	7.3	0.7	4.4	G0 V		38 s		
216357	165351		22 52 35.5	−19 02 16	−0.006	−0.02	7.0	0.9	4.0	F8 V		25 s		
216417	127815		22 52 36.0	+10 13 14	+0.012	−0.08	7.3	0.5	3.4	F5 V		60 s		
216386	146362	73 λ Aqr	22 52 36.6	−7 34 47	0.000	+0.04	3.74	1.64	−0.5	M2 III	−9	71 s		
216500	34909		22 52 37.9	+50 42 18	−0.003	−0.01	7.3	1.6	−0.5	M2 III		340 s		
216401	146363		22 52 39.8	−2 37 34	+0.001	+0.01	7.9	0.9	3.2	G5 IV		88 s		
216533	34915		22 52 41.8	+58 48 24	+0.005	+0.01	7.91	0.08		A2 p				
216361	191460		22 52 41.9	−29 10 53	+0.001	0.00	7.7	0.4	3.2	G5 IV		78 s		
216606	20295		22 52 42.1	+67 59 26	+0.014	+0.07	7.40	0.4	3.0	F2 V		74 s		m
216572	20292		22 52 45.6	+60 54 59	−0.002	−0.01	7.50	0.52		A2 p			16334	m
216402	165353		22 52 46.4	−10 03 31	+0.017	+0.03	6.8	0.5	4.0	F8 V		36 s		
216511	52450		22 52 49.2	+46 32 55	+0.001	0.00	6.7	0.0	0.4	B9.5 V	−15	180 s		
216272	255344		22 52 50.8	−62 48 57	+0.004	+0.02	7.7	0.8	3.2	G5 IV		78 s		
216403	191463		22 52 51.5	−25 12 25	−0.002	0.00	7.9	1.2	0.2	K0 III		270 s		
216523	34917		22 52 52.1	+50 24 43	+0.002	−0.01	6.4	0.0		B9				
216512	72806		22 52 53.8	+38 36 58	−0.001	0.00	7.0	1.6	−0.5	M2 III		320 s		
216489	108231		22 53 02.1	+16 50 28	−0.001	−0.02	5.64	1.12	0.0	K1 III	−12	130 s		
216595	34921		22 53 03.7	+60 06 04	+0.002	+0.01	6.01	1.75	−0.1	K2 III	−7	75 s		
216432	191464		22 53 06.3	−25 08 53	−0.004	−0.01	7.7	0.7	4.0	F8 V		45 s		
216502	90852		22 53 09.1	+26 58 39	0.000	−0.02	7.80	1.1	−0.1	K2 III		380 s		
216538	72812		22 53 11.1	+40 10 02	0.000	−0.01	6.20	−0.08	−1.9	B6 III	+7	380 s		
216560	52454		22 53 11.9	+45 41 36	+0.001	0.00	8.0	1.1	0.2	K0 III		340 s		
216406	231341		22 53 15.2	−45 08 51	+0.005	0.00	6.85	1.12		K0				
216316	255345		22 53 16.0	−63 45 56	+0.010	−0.06	7.5	0.5	4.0	F8 V		50 s		
216349	247630		22 53 18.0	−57 51 15	+0.004	−0.02	7.5	1.3	−0.1	K2 III		290 s		
216471	191469		22 53 26.6	−28 24 58	+0.001	−0.03	7.6	1.1	0.2	K0 III		270 s		
216503	146371		22 53 27.3	−5 59 17	−0.005	−0.04	6.8	0.1	1.4	A2 V		120 s		
216494	165359	74 Aqr	22 53 28.5	−11 37 00	+0.001	0.00	5.80	−0.08		B9				
216443	214162		22 53 30.7	−39 25 11	+0.006	+0.02	7.2	1.2	0.2	K0 III		200 s		
216550	90857		22 53 31.1	+23 24 42	+0.001	+0.01	7.9	1.1	0.2	K0 III		340 s		
216607	34929		22 53 32.9	+51 31 51	+0.001	−0.01	8.0	1.1	−0.1	K2 III		390 s		
216562	72815		22 53 34.9	+30 45 46	+0.005	+0.01	7.50	0.1	1.4	A2 V	−27	170 s	16339	m
216435	231343	τ¹ Gru	22 53 37.7	−48 35 54	+0.022	−0.07	6.04	0.62	3.0	G3 IV		32 ts		
216885	10535		22 53 38.0	+78 54 06	−0.002	0.00	7.7	1.4	−0.3	K5 III		350 s		
216608	52465		22 53 39.9	+44 44 57	−0.002	−0.01	5.80	0.0	0.6	A0 V	+12	110 s	16345	m
216540	146375		22 53 44.1	−4 39 25	0.000	+0.01	7.6	1.1	0.2	K0 III		300 s		
216586	90859		22 53 48.7	+28 37 34	+0.001	−0.04	7.60	1.1	0.0	K1 III		330 s		
216565	127826		22 53 52.0	+1 50 36	0.000	−0.01	7.0	1.4	−0.3	K5 III		290 s		
216991	10543		22 53 55.1	+80 14 25	−0.003	+0.03	7.8	1.1	−0.1	K2 III	−12	340 s		
216712	34933		22 53 58.4	+58 54 26	0.000	+0.01	8.0	1.1	0.2	K0 III		320 s		
216631	72822		22 54 02.0	+34 11 56	−0.005	−0.22	7.5	0.5	4.0	F8 V		49 s		
216553	165365		22 54 05.5	−19 10 33	+0.006	−0.02	6.5	2.1	−0.3	K5 III		110 s		
216848	10536		22 54 06.3	+74 36 44	−0.002	−0.01	7.6	0.5	3.4	F5 V		68 s		
216646	52471		22 54 06.8	+40 22 37	+0.009	+0.04	5.81	1.13	0.2	K0 III	−6	110 s		
216567	165366		22 54 07.0	−12 11 26	+0.003	−0.04	7.1	1.1	0.2	K0 III	+5	240 s		

HD	SAO	Star Name	α 2000	δ 2000	μ(α)	μ(δ)	V	B-V	M_v	Spec	RV	d(pc)	ADS	Notes
			$22^h54^m07^s.1$	$+19°53'32''$	$-0^s.006$	$-0''.12$								
216625	108240		22 54 07.1	+19 53 32	-0.006	-0.12	7.2	0.5	4.0	F8 V	+6	43 s		
216632	90864		22 54 11.2	+28 00 59	-0.005	-0.04	7.68	0.50	4.0	F8 V		55 s	16352	m
216556	214175		22 54 11.5	-30 33 26	-0.003	+0.03	7.9	1.5	0.2	K0 III		180 s		
216886	10541		22 54 14.4	+76 20 18	+0.013	-0.01	7.90	0.1	1.7	A3 V		160 s	16371	m
216730	34937		22 54 16.8	+56 19 59	-0.003	0.00	7.0	0.8	3.2	G5 IV		57 s		
216635	108242		22 54 18.0	+17 47 42	+0.002	-0.03	6.56	1.04		K0				
216580	191481		22 54 19.0	-24 58 48	+0.004	+0.06	7.2	0.9	0.2	K0 III		240 mx		
217159	3812		22 54 19.5	+81 56 07	+0.006	0.00	7.6	1.1	0.2	K0 III		300 s		
216684	52475		22 54 21.1	+43 31 44	+0.001	+0.02	7.76	-0.02	-0.2	B8 V		390 s		
217382	3816		22 54 24.4	+84 20 47	+0.060	+0.03	4.71	1.43	-0.3	K4 III	+3	99 s		
216326	258082		22 54 28.1	-74 51 53	+0.005	+0.01	7.5	0.4	3.0	F2 V		79 s		
216977	10548		22 54 30.7	+77 50 35	-0.008	0.00	8.0	1.1	0.2	K0 III		330 s		
216637	146382	78 Aqr	22 54 34.0	-7 12 17	-0.001	-0.03	6.19	1.28	-0.2	gK3	+9	190 s		
216731	52478		22 54 35.0	+47 16 57	+0.001	+0.01	8.0	1.4	-0.3	K5 III		420 s		
216672	108246		22 54 35.5	+16 56 30	+0.001	0.00	6.4	1.9		S5.1	+11			
216627	165375	76 δ Aqr	22 54 38.8	-15 49 15	-0.003	-0.02	3.27	0.05	-0.2	A2 III	+18	30 ts		Skat
216437	258084	ρ Ind	22 54 39.4	-70 04 26	-0.008	+0.07	6.05	0.66	4.5	G1 V		18 s		
216696	90872		22 54 40.2	+24 23 14	+0.001	+0.01	7.9	1.6	-0.5	M4 III		470 s		
216462	—		22 54 41.3	-67 36 10			7.60	0.4	2.6	F0 V		100 s		m
216673	108247		22 54 42.6	+11 29 17	+0.001	-0.05	8.0	1.1	0.2	K0 III		360 s		
216722	72833		22 54 43.6	+35 59 04	+0.010	+0.04	7.6	1.1	0.2	K0 III		140 mx		
216733	52480		22 54 44.0	+42 30 39	+0.003	+0.01	7.83	0.52	1.4	A2 V		99 s		
216688	127833		22 54 44.1	+7 15 28	-0.001	-0.02	7.80	1.1	-0.1	K2 III		380 s		m
216716	72831		22 54 44.6	+31 28 44	0.000	0.00	7.5	0.1	0.6	A0 V	-8	240 s		
216640	165376	77 Aqr	22 54 45.2	-16 16 19	-0.016	-0.09	5.56	1.14	-0.1	K2 III	-36	36 mx		
216664	191488		22 54 54.6	-22 22 08	+0.004	-0.20	6.7	0.9	3.4	F5 V		25 s		
216887	10546		22 54 54.9	+71 32 34	-0.001	0.00	8.0	1.1	0.2	K0 III		320 s		
216518	255351		22 54 55.1	-65 19 49	+0.001	0.00	8.0	1.4	-0.3	K5 III		460 s		
216723	90875		22 54 55.8	+28 00 34	+0.002	+0.01	7.3	0.9	0.3	G8 III	-15	250 s		
216701	127836	1 Psc	22 54 59.3	+1 03 53	+0.001	+0.01	6.11	0.20		A3	+13			
216724	108255		22 55 00.8	+19 33 35	-0.001	-0.02	7.8	1.6	-0.5	M4 III		470 s		
216756	72838		22 55 02.4	+37 04 36	+0.007	+0.01	5.91	0.42	3.0	F2 V	-28	37 s		
216497	255350		22 55 04.1	-69 06 15	+0.018	-0.02	7.5	1.1	3.0	F2 V		30 s		
216734	127838		22 55 06.9	+9 27 34	0.000	-0.01	8.00	1.1	-0.1	K2 III		420 s	16364	m
216718	146388		22 55 10.8	-4 59 16	+0.002	0.00	5.72	0.88	0.3	gG7	-9	120 s	16365	m
216643	231352		22 55 11.0	-46 40 45	+0.006	-0.07	7.4	1.4	0.2	K0 III		140 mx		
216797	52491		22 55 13.3	+46 22 20	-0.004	-0.01	7.9	0.1	0.6	A0 V		260 mx		
216735	127839	50 ρ Peg	22 55 13.5	+8 48 58	+0.005	+0.02	4.90	0.00	1.2	A1 V	-10	55 s		
216666	214182		22 55 14.7	-36 23 19	+0.004	0.00	6.3	1.4	0.2	K0 III		96 s		
216656	231353		22 55 15.9	-48 29 31	-0.022	+0.06	6.67	0.46		G5	+6	20 mn		m
216655	231354		22 55 15.9	-48 27 57	-0.023	+0.06	7.03	0.57		G0	+3			m
216725	146389		22 55 21.0	-5 41 21	0.000	+0.03	7.7	1.4	-0.3	K5 III		390 s		
216925	10550		22 55 24.5	+70 17 54	+0.006	+0.04	7.7	0.2	2.1	A5 V		110 mx		
216679	231356		22 55 30.1	-46 19 39	+0.001	+0.03	7.52	0.39		F2				
216727	191496		22 55 30.6	-20 08 20	+0.004	0.00	6.5	1.4	-0.1	K2 III		160 s		
216774	108261		22 55 34.1	+18 11 47	+0.002	+0.01	7.8	0.2	2.1	A5 V		140 s		
216786	108262		22 55 40.1	+18 52 31	+0.002	-0.01	8.0	1.6	-0.5	M2 III		510 s		
216898	20323		22 55 42.1	+62 18 22	-0.003	-0.01	8.02	0.53		O8				
216831	72851		22 55 44.4	+36 21 06	+0.002	+0.01	5.60	-0.05	-1.6	B7 III	+1	250 s	16376	m
217049	10555		22 55 46.5	+72 50 37	+0.004	+0.02	8.01	0.08		A0			16384	m
216851	52497		22 55 46.9	+43 33 34	-0.001	+0.01	8.02	0.04	-1.7	B3 V n	-20	780 s		
216850	52498		22 55 46.9	+44 18 30	+0.001	-0.01	6.8	1.1	0.2	K0 III		200 s		
216777	146393		22 55 49.9	-7 49 22	+0.038	-0.06	8.02	0.64	5.3	G6 V	-24	28 ts		m
216761	214187		22 55 51.2	-31 37 59	-0.003	0.00	6.10	1.36	0.2	K0 III		79 s		
216742	231360		22 55 54.0	-41 05 50	+0.005	+0.02	6.8	1.3	0.2	K0 III		150 s		
216888	34960		22 55 54.5	+54 15 14	+0.007	+0.05	8.0	1.1	0.2	K0 III		250 mx		
216992	20329		22 55 55.7	+68 14 57	+0.001	+0.02	8.0	0.0	0.0	B8.5 V		350 s		
216743	231361		22 55 56.0	-42 33 10	+0.003	0.00	7.26	0.16	1.2	A1 V n		160 s		
216854	52502		22 55 56.7	+41 20 07	+0.008	-0.02	7.32	0.49	3.4	F5 V		56 s		
216763	214189	23 δ PsA	22 55 56.8	-32 32 23	+0.001	+0.03	4.21	0.97	0.3	gG4	-12	49 s		m
216519	258085		22 55 57.5	-74 59 43	+0.005	0.00	7.8	1.1	0.2	K0 III		330 s		
216945	20328		22 55 58.1	+62 25 54	+0.001	-0.03	6.54	1.82	-0.1	gK2	-21	74 s		
216878	52506		22 56 02.1	+46 41 53	+0.004	+0.04	8.0	0.9	3.2	G5 IV		91 s		
216912	34967		22 56 03.2	+58 11 46	+0.002	+0.02	7.07	-0.06	0.6	A0 V		200 s		
216993	—		22 56 03.9	+66 43 00			8.00	0.1	0.6	A0 V		270 s	16385	m
216870	72859		22 56 08.1	+35 41 41	-0.004	+0.01	7.8	1.4	-0.3	K5 III		420 s		
216928	34972		22 56 10.5	+56 27 02	+0.001	+0.01	7.24	0.09	0.4	B9.5 V		220 s		
216771	214192		22 56 11.4	-38 04 37	+0.002	0.00	7.6	1.1	-0.1	K2 III		350 s		
217085	10560		22 56 13.1	+72 50 14	+0.001	-0.01	7.38	0.22		A2			16393	m
216916	52512	16 Lac	22 56 23.4	+41 36 14	0.000	0.00	5.59	-0.14	-3.0	B2 IV	-7	500 s	16381	EN Lac, m,v
216879	90881		22 56 23.5	+22 57 22	+0.003	+0.01	7.40	0.4	2.6	F0 V		91 s		m
216803	214197		22 56 23.8	-31 33 57	+0.026	-0.16	6.48	1.10	7.0	K5 V	+6	7.8 t		
216864	108268		22 56 24.7	+13 03 17	+0.007	-0.01	7.8	0.8	3.2	G5 IV		84 s		
216946	52516		22 56 25.8	+49 44 01	0.000	0.00	4.95	1.78	-4.4	K0 Ib	-10	320 s		
216130	—		22 56 30.5	-83 21 14			7.9	0.5	3.4	F5 V		79 s		

HD	SAO	Star Name	α 2000	δ 2000	μ(α)	μ(δ)	V	B-V	M_V	Spec	RV	d(pc)	ADS	Notes
217035	20332		22ʰ56ᵐ30.7ˢ	+62°52′08″	0ˢ.000	+0″.01	7.75	0.46	-4.1	B0 V		920 s	16394	m
216872	127851		22 56 32.0	+6 48 19	+0.010	-0.01	7.7	0.5	4.0	F8 V		54 s		
216960	52519		22 56 36.9	+47 22 18	-0.001	+0.03	7.8	1.1	0.2	K0 III		310 s		
216959	52520		22 56 38.7	+48 06 38	-0.001	+0.03	7.8	1.1	-0.1	K2 III		360 s		
216929	72865		22 56 40.5	+34 20 36	+0.001	+0.01	7.9	0.1	0.6	A0 V		290 s		
217294	10563		22 56 41.5	+78 29 50	-0.043	-0.04	7.98	0.92	3.2	G8 IV	-22	86 s	16407	m
217086	20335		22 56 46.8	+62 43 38	-0.003	0.00	7.65	0.63		O5				m
216823	231364	τ³ Gru	22 56 47.6	-47 58 09	-0.003	0.00	5.70	0.22		A m				
216900	108275		22 56 51.3	+11 50 54	+0.004	0.00	6.60	0.1	1.7	A3 V	+8	96 s	16389	m
217062	34988		22 56 56.0	+59 57 42	+0.002	0.00	7.18	-0.04	0.4	B9.5 V		230 s		
216930	108277		22 57 00.2	+14 25 07	+0.001	-0.01	7.9	1.6	-0.5	M2 III		480 s		
216997	72871		22 57 01.1	+36 30 21	+0.004	0.00	8.0	0.1	1.4	A2 V		170 mx		
217071	34990		22 57 02.6	+56 29 26	-0.001	-0.02	7.4	0.4	2.6	F0 V		88 s		
217050	52526		22 57 04.4	+48 41 03	+0.001	0.00	5.43	-0.09		B2 p	-11			EW Lac, v
216824	247653		22 57 05.8	-57 24 04	+0.009	-0.05	7.3	1.4		M6 III		190 mx		
216907	191516		22 57 06.4	-20 20 38	+0.001	-0.04	7.9	1.6		M5 III	-58			S Aqr, v
216931	146402		22 57 06.7	-3 14 42	+0.002	+0.03	6.60	0.1	0.6	A0 V		160 s	16392	m
216949	108278		22 57 07.9	+14 00 51	+0.001	+0.01	7.8	0.4	2.6	F0 V		110 s		
216998	72872		22 57 08.5	+34 54 44	-0.001	-0.01	7.8	0.5	3.4	F5 V		74 s		
217013	72874		22 57 15.3	+32 11 58	-0.001	+0.02	7.9	1.1	-0.1	K2 III		390 s		
216953	146404		22 57 17.1	-4 48 36	-0.001	+0.01	6.31	0.94	0.3	gG6	-9	150 s		
216847	247655		22 57 21.9	-56 52 29	+0.003	+0.03	6.9	0.2	1.4	A2 V		99 s		
217127	34998		22 57 25.4	+57 00 45	0.000	0.00	7.8	0.1	0.6	A0 V		250 s		
217073	52529		22 57 25.6	+43 00 47	+0.003	+0.02	6.90	1.4	-0.3	K5 III		260 s	16401	m
216983	146405		22 57 25.8	-3 09 27	+0.003	-0.13	7.9	0.5	4.0	F8 V		59 s		
217089	52530		22 57 26.6	+46 49 23	-0.004	+0.02	7.6	0.9	3.2	G5 IV		76 s		
217088	52531		22 57 27.0	+47 01 04	-0.002	+0.02	8.0	0.1	1.4	A2 V		200 s		
217014	90896	51 Peg	22 57 27.8	+20 46 08	+0.015	+0.06	5.49	0.67	5.0	G4 V	-31	13 ts		
217019	127860		22 57 32.6	+3 48 37	+0.004	+0.04	6.28	1.12	0.0	K1 III	+11	170 s		
217037	108286		22 57 35.1	+10 16 22	0.000	-0.01	7.9	0.5	4.0	F8 V		60 s		
217090	72882		22 57 36.8	+39 23 19	+0.004	0.00	7.7	1.1	-0.1	K2 III		360 s		
216968	191526		22 57 37.0	-20 16 28	+0.002	+0.03	7.6	1.2	0.2	K0 III		230 s		
216956	191524	24 α PsA	22 57 38.9	-29 37 20	+0.026	-0.16	1.16	0.09	2.0	A3 V	+7	6.7 t		Fomalhaut
217101	72883		22 57 40.5	+39 18 32	-0.001	+0.01	6.18	-0.15	-2.8	B2 IV-V	-16	590 s		
216867	255357		22 57 40.8	-61 51 40	+0.003	-0.01	8.0	1.4	-0.3	K5 III		460 s		
216985	191527		22 57 42.7	-27 57 58	+0.004	0.00	7.7	0.7	0.2	K0 III		310 s		
217224	20345		22 57 47.1	+68 24 27	-0.001	+0.01	7.8	0.0	0.0	B8.5 V		330 s		
217004	191529		22 57 47.8	-26 06 28	+0.011	-0.26	7.37	0.68		G0		22 mn	16400	m
217295	10566		22 57 48.0	+73 08 00	-0.005	-0.03	6.6	1.1	0.2	K0 III		180 s		
216988	214208		22 57 53.3	-37 56 07	+0.007	-0.01	7.3	0.6	0.2	K0 III		260 s		
217075	108292		22 57 56.2	+13 36 41	+0.006	+0.04	7.90	0.4	3.0	F2 V		96 s	16403	m
216989	231370		22 57 58.3	-45 09 30	+0.009	-0.03	7.72	0.37	2.6	F0 V		98 s		
216910	247657		22 57 59.2	-59 05 22	+0.011	-0.04	6.7	0.7	2.6	F0 V		38 s		
217115	90903		22 58 06.2	+21 30 48	+0.002	-0.03	7.6	1.6	-0.5	M4 III		420 s		
217118	108297		22 58 12.4	+14 43 16	-0.002	-0.09	8.0	1.4	-0.3	K5 III		280 mx		
217107	146412		22 58 15.2	-2 23 42	-0.001	0.00	6.16	0.74		G5				
216838	258092		22 58 22.8	-73 46 35	-0.003	+0.04	7.7	1.1	0.2	K0 III		320 s		
217131	146415		22 58 23.5	-1 24 37	+0.006	0.00	6.37	0.35		F2	-14			
216924	--		22 58 26.2	-65 22 00			8.0	0.5	3.4	F5 V		82 s		
217183	72894		22 58 26.8	+35 16 57	-0.001	-0.08	7.0	1.4	-0.3	K5 III		230 mx		
217165	127869		22 58 29.7	+9 49 33	+0.008	+0.01	7.7	0.6	4.4	G0 V		47 s		
217047	247662		22 58 29.9	-51 07 39	-0.006	+0.02	7.82	0.84	5.3	G6 V		26 s		
217297	20349		22 58 33.3	+63 42 25	+0.002	+0.01	7.41	0.32	-3.5	B1 V	-10	760 s		
217096	214215		22 58 34.9	-35 31 23	0.000	-0.11	6.13	0.58	2.4	F8 IV		51 s		
217166	127870		22 58 35.0	+9 21 26	+0.027	-0.14	6.43	0.64	4.5	dG1	-27	28 ts	16417	m
217226	35010		22 58 35.1	+50 41 53	-0.001	-0.01	7.5	1.4	-0.3	K5 III		350 s		
217084	231376		22 58 36.1	-45 31 09	+0.006	+0.06	7.67	0.64	3.2	G5 V		69 s		
217209	72897		22 58 37.2	+39 47 33	-0.002	-0.03	7.9	0.2	2.1	A5 V		150 s		
217312	20351		22 58 39.3	+63 04 37	-0.006	-0.01	7.40	0.39	-4.6	B0 IV		390 mx		
217187	127873		22 58 42.2	+5 20 32	+0.001	+0.03	7.9	0.9	3.2	G5 IV		88 s		
217186	127874		22 58 42.4	+7 20 23	-0.003	-0.07	6.33	0.06		A0	-1			
217227	52551		22 58 45.7	+43 50 21	+0.001	+0.01	7.16	-0.06	-2.5	B2 V	-14	710 s		
217150	191543		22 58 46.6	-23 26 58	-0.001	-0.02	7.9	0.8	3.2	G5 IV		85 s		
217188	146421		22 58 52.7	-0 18 58	+0.003	+0.02	7.3	1.1	0.2	K0 III		260 s		
217229	90914		22 58 58.3	+27 44 46	0.000	+0.02	8.00	0.5	3.4	F5 V		83 s	16424	m
217241	72900		22 58 58.9	+34 04 04	+0.002	+0.01	7.6	0.1	0.6	A0 V		250 s		
217348	35026		22 59 08.8				6.43	0.03		B9				
217691	3825		22 59 09.6	+80 20 39	+0.014	+0.05	7.40	1.1	-0.1	K2 III		290 s		m
217314	35022		22 59 10.1	+52 39 16	-0.004	+0.03	6.29	1.42	-0.1	K2 III	+28	130 s		
217232	108307	52 Peg	22 59 11.7	+11 43 44	+0.002	-0.04	5.75	0.32	2.6	F0 V	+20	41 s	16428	m
217315	35023		22 59 12.9	+52 18 11	+0.001	-0.01	6.9	0.0	0.4	B9.5 V		190 s		
217172	231380		22 59 14.7	-45 11 21	+0.004	+0.01	7.29	0.14		A0				
217110	255364		22 59 16.9	-60 09 23	0.000	+0.01	7.4	1.0	-0.1	K2 III		310 s		
216976	--		22 59 17.7	-71 49 58			7.4	1.6	-0.5	M2 III		390 s		
217247	127880		22 59 18.8	+4 21 31	0.000	-0.06	7.3	0.5	3.4	F5 V		61 s		

HD	SAO	Star Name	α 2000	δ 2000	μ(α)	μ(δ)	V	B-V	M_v	Spec	RV	d(pc)	ADS	Notes
			h m s	° ′ ″	s	″								
217523	10575		22 59 25.1	+72 42 40	-0.013	-0.13	7.9	0.5	3.4	F5 V		76 s		
217264	127881	2 Psc	22 59 27.3	+0 57 46	+0.006	-0.07	5.43	0.98	0.0	gK1	-13	120 s	16431	m
217335	52567		22 59 30.4	+43 54 50	+0.002	-0.01	7.7	1.1	-0.1	K2 III		340 s		
217251	165425		22 59 35.5	-13 04 15	-0.001	0.00	6.2	1.1	-0.1	K2 III	+13	180 s		
217236	191550		22 59 35.7	-29 27 44	0.000	+0.01	5.51	0.25		A5 n	0			
217155	255370		22 59 44.2	-62 46 19	+0.011	-0.02	7.3	0.5	3.4	F5 V		59 s		
217255	214225		22 59 45.8	-31 07 35	+0.002	+0.02	7.9	2.1	-0.5	M2 III		230 s		
217276	165426		22 59 46.7	-16 23 53	+0.003	-0.17	8.00	0.6	4.4	G0 V	+52	53 s		
217286	146434		22 59 49.3	-2 26 25	+0.004	+0.01	8.0	0.5	3.4	F5 V		82 s		
217339	108316		22 59 52.4	+13 51 44	+0.004	+0.03	7.8	0.4	2.6	F0 V		110 s		
217352	127885		22 59 59.3	+5 09 36	+0.004	+0.01	7.0	1.1	0.2	K0 III		230 s		
217475	35037		23 00 00.6	+57 46 21	-0.002	-0.04	7.3	0.4	2.6	F0 V		86 s		
217259	231382		23 00 02.5	-46 29 32	+0.006	-0.02	8.00	0.4	3.0	F2 V		100 s		m
217476	35039		23 00 04.9	+56 56 44	-0.001	+0.01	5.00	1.42	-8.0	G0 Ia	-58	1700 s		
217303	191554		23 00 05.6	-25 09 51	+0.003	-0.07	5.65	1.25	0.2	K0 III	-35	81 s		
217260	247673		23 00 08.6	-51 42 04	+0.002	+0.01	7.5	0.1	2.1	A5 V		120 s		
217386	108321		23 00 09.4	+13 55 30	+0.001	+0.02	7.7	1.1	-0.1	K2 III		360 s		
217326	191561		23 00 14.7	-27 05 42	+0.001	+0.05	7.6	2.0	-0.1	K2 III		110 s		
217357	191563		23 00 15.9	-22 31 28	-0.066	+0.07	7.89	1.39	8.5	Ma	+17	7.7 t		
217343	191562		23 00 19.3	-26 09 12	+0.010	-0.14	7.4	0.7	4.4	G0 V		32 s		
217376	146438		23 00 19.7	-8 52 51	+0.001	-0.03	6.9	0.4	3.0	F2 V		60 s		
217307	231387		23 00 23.8	-41 09 05	-0.004	-0.04	7.0	0.8	0.2	K0 III		230 s		
217358	191565		23 00 24.4	-25 37 36	+0.006	-0.03	6.4	1.1	0.2	K0 III		94 s		
217427	108327		23 00 32.2	+13 28 07	+0.002	-0.02	7.7	1.1	-0.1	K2 III		360 s		
217491	52587		23 00 34.3	+45 22 30	+0.001	+0.01	6.5	0.1	1.4	A2 V	-4	100 s		
217526	52589		23 00 36.5	+49 23 28	-0.004	-0.01	7.9	0.1	1.7	A3 V		170 s		
217428	146443	3 Psc	23 00 37.7	+0 11 09	+0.003	+0.02	6.21	0.89	0.3	gG4	-16	140 s		
217541	52591		23 00 41.3	+49 57 21	+0.002	-0.01	7.8	0.1	1.7	A3 V		160 s		
217477	72924		23 00 42.4	+31 05 00	+0.002	+0.01	6.60	-0.04		A p	+1		16443	m
217459	127894		23 00 42.7	+3 00 42	+0.001	-0.08	5.83	1.34	-0.3	gK4	+19	170 s		
217363	231388		23 00 47.2	-49 27 08	+0.009	-0.03	7.3	0.0	1.7	A3 V		65 mx		
217478	90939		23 00 49.0	+21 22 54	+0.003	+0.02	7.5	1.1	0.2	K0 III		290 s		
217568	35053		23 00 50.8	+54 19 07	+0.002	-0.01	7.7	1.1	-0.1	K2 III		340 s		
217446	165434		23 00 51.7	-16 52 51	+0.002	0.00	8.0	0.9	3.2	G5 IV		93 s		
217364	247680	ζ Gru	23 00 52.6	-52 45 15	-0.007	-0.01	4.12	0.98	0.3	G5 III	-1	50 s		
217543	72929		23 00 54.6	+38 42 29	0.000	0.00	6.54	-0.11	-1.7	B3 V p	-16	450 s		
217447	165435		23 00 55.2	-19 20 48	0.000	+0.02	7.6	0.8	3.2	G5 IV		78 s		
217380	231390		23 00 55.8	-48 56 37	+0.006	-0.03	7.6	1.6	-0.1	K2 III		200 s		
--	52596		23 00 57.0	+46 58 42	-0.004	-0.04	8.0	1.5						
217417	214247		23 01 00.6	-37 21 32	0.000	+0.02	7.8	1.6	1.4	A2 V		23 s		
217346	247681		23 01 00.8	-58 26 16	+0.009	0.00	7.20	0.1	1.7	A3 V		130 s		m
217810	10589		23 01 00.8	+76 07 22	+0.007	+0.01	7.50	0.9	3.2	G5 IV		71 s	16454	m
217511	90944		23 01 01.9	+22 23 30	+0.011	+0.04	7.6	0.5	3.4	F5 V		68 s		
217832	10590		23 01 07.2	+76 52 11	+0.009	+0.02	7.8	0.1	1.4	A2 V		150 mx		
217403	247683		23 01 07.3	-50 57 01	+0.008	0.00	5.68	1.42	-0.1	gK2	+8	100 s		
217418	231393		23 01 07.8	-45 12 18	+0.005	-0.04	7.7	0.6	-0.1	K2 III		230 mx		
217992	3830		23 01 14.9	+80 46 54	+0.014	+0.03	6.90	0.1	1.7	A3 V		110 s	16469	m
217587	52599		23 01 15.3	+44 11 56	+0.001	+0.03	7.3	0.1	1.7	A3 V	+5	130 s		
217484	191581		23 01 19.3	-28 51 13	+0.005	-0.01	5.55	1.35	0.2	K0 III		63 s		
217559	108334		23 01 21.1	+14 52 14	-0.003	-0.01	7.1	1.1	0.2	K0 III		240 s		
217498	191585		23 01 22.9	-22 47 27	+0.003	-0.02	6.3	0.6	1.4	A2 V		44 s		
217531	146447	81 Aqr	23 01 23.5	-7 03 40	-0.001	-0.01	6.21	1.41		K2	-2			
217533	165440		23 01 24.8	-14 16 18	+0.003	-0.04	7.4	1.1	0.2	K0 III		270 s		
217672	35060		23 01 27.1	+58 15 52	+0.001	+0.01	8.0	0.0	0.0	B8.5 V		360 s		
217634	52604		23 01 28.1	+46 49 49	-0.002	-0.01	7.8	0.4	2.6	F0 V		110 s		
217673	35062		23 01 30.5	+57 06 20	-0.001	0.00	6.20	1.50	-2.2	K2 II	-6	390 s		
217563	146451		23 01 31.6	-4 42 41	+0.001	+0.01	5.94	1.00		K0				
217730	20371		23 01 32.3	+63 17 31	-0.002	-0.01	7.25	1.44	-0.1	K2 III	-21	180 s		
217600	72935		23 01 32.6	+32 36 46	-0.001	0.00	7.7	1.6	-0.5	M4 III		430 s		
217577	108337		23 01 33.0	+19 16 13	+0.012	-0.04	8.00	0.6	4.7	G2 V	-2	46 s		
217648	52606		23 01 33.4	+49 02 34	-0.001	+0.04	7.7	1.4	-0.3	K5 III		370 s		
217635	52605		23 01 33.7	+45 30 37	+0.002	+0.02	7.9	1.1	0.2	K0 III		330 s		
217516	191590		23 01 34.1	-22 25 09	+0.007	-0.03	7.4	0.4	4.0	F8 V		48 s		
217588	90958		23 01 34.8	+23 04 24	-0.001	0.00	7.9	0.5	3.4	F5 V		79 s		
217618	72937		23 01 36.7	+36 59 07	+0.004	-0.04	7.6	1.1	0.2	K0 III		240 mx		
217578	127907		23 01 39.2	+4 27 40	+0.005	-0.10	7.2	0.6	4.4	G0 V		36 s		
217711	35072		23 01 41.4	+60 11 50	+0.003	0.00	7.4	1.1	-0.1	K2 III	-15	290 s		
217781	20376		23 01 42.4	+67 47 42	+0.013	+0.04	8.00	0.5	4.0	F8 V		62 s	16459	m
217590	127908		23 01 43.4	+3 31 51	+0.002	-0.05	6.5	0.9	3.2	G5 IV		46 s		
217522	231398		23 01 46.6	-44 50 26	-0.009	-0.03	7.7	-0.1	2.1	A5 V		110 mx		
217694	35070		23 01 48.6	+50 51 00	-0.004	+0.01	7.40	1.4	-0.3	K4 III	-81	330 s		
217649	72940		23 01 51.2	+34 36 41	+0.001	-0.07	6.7	0.4	2.6	F0 V	0	67 s		
217580	146456		23 01 51.4	-3 50 56	+0.027	-0.22	7.46	0.95	7.4	K4 V	-46	15 ts		
217695	52610		23 01 53.0	+47 57 21	0.000	+0.03	7.5	0.1	0.6	A0 V	-3	240 s		
217636	108344		23 01 54.4	+19 50 15	+0.002	-0.02	7.2	1.1	-0.1	K2 III		280 s		

HD	SAO	Star Name	α 2000	δ 2000	μ(α)	μ(δ)	V	B−V	M_v	Spec	RV	d(pc)	ADS	Notes
217453	255378		23 01 55.0	−63 05 08	+0.005	+0.02	7.0	1.3	−0.1	K2 III		230 s		
217675	52609	1 o And	23 01 55.1	+42 19 34	+0.002	0.00	3.6	0.1	1.4	B6 p	−14	35 mn		m,v
217605	146458		23 01 55.2	−2 41 09	0.000	0.00	7.9	1.4	−0.3	K5 III		430 s		
217796	20381		23 02 05.2	+62 30 42	−0.001	−0.03	8.00	0.2	2.1	A5 V		140 s	16466	m
217627	165450		23 02 09.7	−14 58 33	+0.001	0.00	8.0	1.6	−0.5	M2 III		510 s		
217661	108352		23 02 11.2	+18 32 07	+0.001	+0.01	8.0	1.1	−0.1	K2 III		410 s		
217731	52616		23 02 11.2	+44 34 23	+0.001	−0.01	6.4	1.1	0.2	K0 III	−9	170 s		
217488	255381		23 02 15.9	−64 17 54	+0.005	−0.05	7.60	0.4	2.6	F0 V		100 s		m
217768	35082		23 02 20.2	+53 48 35	+0.003	+0.03	7.7	0.2	2.1	A5 V		130 s		
217715	90964		23 02 20.9	+23 20 25	+0.001	+0.01	6.8	0.1	0.6	A0 V		170 s		
––	35084		23 02 23.9	+58 14 14	+0.001	+0.02	7.8	2.4	−0.3	K5 III		130 s		
217679	127916		23 02 24.2	+0 26 22	−0.004	+0.08	7.6	0.5	4.0	F8 V		52 s		
217817	35086		23 02 25.1	+59 51 09	+0.001	0.00	7.01	−0.01	−1.6	B3.5 V	−33	430 s		
217595	231402		23 02 27.3	−45 18 14	−0.001	0.00	7.20	0.44	3.4	F5 V		58 s		
217597	231403		23 02 31.3	−49 41 48	+0.020	−0.21	7.74	0.86		G5				
217701	146465	82 Aqr	23 02 32.4	−6 34 27	−0.001	−0.03	6.15	1.58		Ma	−8			
217754	72949		23 02 32.9	+31 46 51	0.000	−0.01	6.57	0.35	2.6	F0 V	−17	58 s		
217642	214261		23 02 33.8	−36 25 16	+0.002	−0.03	6.47	0.94	0.2	K0 III		180 s		m
217732	108356		23 02 35.0	+16 13 59	0.000	+0.03	6.8	0.4	0.6	F0 III		170 s		
217782	52623	2 And	23 02 36.1	+42 45 28	+0.005	0.00	5.10	0.09		A2	+2	21 t	16467	m
217860	35090		23 02 37.2	+59 36 19	0.000	0.00	7.2	0.2	2.1	A5 V		100 s		
217684	165456		23 02 37.7	−18 32 30	+0.003	−0.01	6.80	0.1	0.6	A0 V		170 s	16465	m
217872	20384		23 02 39.1	+63 20 28	+0.004	+0.01	6.9	1.1	0.2	K0 III	−10	210 s		
217849	35094		23 02 41.0	+59 33 29	+0.001	+0.02	8.0	1.1	−0.1	K2 III		360 s		
217615	247696		23 02 41.6	−53 47 04	0.000	−0.02	7.5	1.6	3.2	G5 IV		23 s		
217833	35092		23 02 43.6	+55 14 12	+0.001	+0.02	6.60	−0.08		B9			16474	m
217614	247697		23 02 44.0	−52 57 55	+0.006	+0.01	6.9	1.1	3.2	G5 IV		34 s		
217703	191604		23 02 44.1	−20 52 14	−0.004	−0.11	5.97	0.94		G5				
217783	90969		23 02 44.1	+23 07 50	+0.001	−0.01	7.2	0.5	3.4	F5 V		59 s		
217811	52626		23 02 45.0	+44 03 32	0.000	0.00	6.39	0.00	−2.5	B2 V	−8	460 s	16472	m
217835	35093		23 02 46.0	+53 14 47	+0.003	+0.01	8.0	1.1	−0.1	K2 III		390 s		
217010	258955		23 02 48.9	−82 42 28	+0.190	−0.10	7.8	0.8	3.2	G5 IV		45 mx		
217670	231409		23 02 56.4	−47 50 54	+0.002	0.00	6.7	0.3	0.0	B8.5 V		130 s		
217785	108362		23 02 58.2	+16 27 29	+0.001	+0.01	7.9	0.1	1.4	A2 V		200 s		
217813	90973		23 03 04.8	+20 55 06	−0.008	−0.03	6.7	0.6	4.4	G0 V	−3	28 s		
217863	52633		23 03 05.6	+48 49 49	+0.003	−0.01	7.8	1.4	−0.3	K5 III		390 s		
217786	146477		23 03 08.1	−0 25 47	−0.006	−0.17	7.8	0.5	4.2	F9 V	+7	52 s		
217646	255386		23 03 10.3	−62 22 58	+0.005	0.00	8.00	0.1	0.6	A0 V		250 mx		m
217745	214270		23 03 10.5	−31 08 36	−0.001	−0.01	7.6	0.5	2.6	F0 V		75 s		
217726	231412		23 03 11.4	−47 15 17	+0.008	−0.06	7.8	−0.1	3.4	F5 V		77 s		
217686	247699		23 03 15.8	−55 10 07	+0.019	−0.06	7.6	0.4	3.0	F2 V		77 s		
217852	72961		23 03 17.5	+31 41 22	+0.002	+0.01	7.5	0.4	3.0	F2 V		81 s		
217864	72962		23 03 18.7	+38 17 55	−0.002	−0.02	7.1	1.1	0.2	K0 III		240 s		
217944	35103		23 03 21.4	+58 33 53	+0.008	+0.02	6.43	0.90	3.2	G5 IV	+15	37 s		
217943	20393		23 03 23.5	+60 26 43	+0.001	0.00	6.66	−0.01		B5	−17		16481	m
217920	35105		23 03 27.3	+53 18 09	+0.002	+0.01	7.6	1.1	0.2	K0 III		290 s		
217766	231414		23 03 28.6	−43 04 44	+0.002	−0.13	7.77	0.54	4.0	F8 V		55 s		
217792	214275	π PsA	23 03 29.6	−34 44 58	+0.006	+0.09	5.11	0.29	1.7	F0 IV	−14	48 s		
217777	214273		23 03 29.7	−38 26 00	0.000	+0.01	7.1	1.7	−0.3	K5 III		230 s		
217727	247701		23 03 31.5	−56 29 57	−0.001	+0.02	7.4	0.3	3.4	F5 V		64 s		
218029	20398		23 03 32.7	+67 12 34	+0.004	+0.02	5.24	1.26	−0.2	K3 III	−7	120 s		
217874	90978		23 03 35.2	+23 22 43	+0.001	+0.01	7.7	0.1	1.4	A2 V		180 s		
217825	191612		23 03 38.1	−26 48 46	+0.003	+0.03	6.78	1.62		M5 III				
217807	231415		23 03 42.4	−40 48 58	+0.004	+0.03	6.9	1.9	−0.5	M2 III		170 s		
217826	214278		23 03 44.0	−30 26 41	+0.003	+0.03	6.9	0.7	3.2	G5 IV		54 s		
217906	90981	53 β Peg	23 03 46.3	+28 04 58	+0.014	+0.14	2.42	1.67	−1.4	M2 II-III	+9	54 s	16483	Scheat, m,v
217924	90982		23 03 50.4	+21 35 57	−0.001	−0.03	7.3	0.6	4.4	G0 V		39 s		
217839	214279		23 03 51.7	−32 06 38	+0.002	−0.03	8.0	1.1	−0.3	K5 III		460 s		
217891	127934	4 β Psc	23 03 52.5	+3 49 12	+0.001	0.00	4.53	−0.12		B5 pe	0			
217867	191615		23 03 54.4	−25 45 22	0.000	+0.01	8.0		0.2	K0 III				
217877	146482		23 03 57.1	−4 47 43	+0.021	+0.03	6.68	0.58	4.4	G0 V		28 s		
217795	247707		23 03 58.9	−55 41 52	+0.005	−0.03	7.6	0.4	1.4	A2 V		87 mx		
217842	231419		23 03 59.4	−41 28 43	0.000	+0.07	5.79	1.08	0.2	K0 III		120 s		
217926	127937		23 04 00.8	+6 37 00	0.000	+0.02	6.41	0.39		F2	+4			
218168	10607		23 04 02.0	+74 28 41	+0.026	+0.06	8.0	0.9	3.2	G5 IV		90 s		
218066	20401		23 04 02.0	+63 23 48	0.000	−0.01	7.64	0.40	−3.5	B1 V	−10	750 s		CW Cep, m,v
217882	191618		23 04 03.6	−22 42 36	+0.003	+0.01	7.7	1.3	0.2	K0 III		200 s		
217927	127938		23 04 05.1	+3 17 35	+0.007	−0.05	7.8	0.4	3.0	F2 V		89 s		
217894	––		23 04 09.7	−22 13 44			7.5	1.1	0.2	K0 III		240 s		
218031	52649	3 And	23 04 10.8	+50 03 08	+0.017	+0.17	4.65	1.06	0.2	K0 III	−35	71 s		
––	191622		23 04 11.2	−22 14 04	0.000	−0.01	7.1	1.1	0.2	K0 III		240 s		
217982	108375		23 04 19.8	+11 10 17	+0.002	−0.01	7.8	1.1	0.2	K0 III		330 s		
218042	52651		23 04 24.6	+41 16 28	−0.001	0.00	7.7	0.1	0.6	A0 V		260 s		
218091	35121		23 04 25.0	+56 27 25	0.000	+0.01	7.96	1.90	−0.5	M2 III		320 s		
218139	20405		23 04 32.6	+63 14 28	+0.001	+0.05	6.7	1.1	0.2	K0 III		190 s		

HD	SAO	Star Name	α 2000	δ 2000	μ(α)	μ(δ)	V	B-V	M_v	Spec	RV	d(pc)	ADS	Notes
			$23^h04^m36.1^s$	$+31°18'29''$	$+0.005^s$	$-0.01''$								
218043	72981		23 04 36.1	+31 18 29	+0.005	-0.01	6.8	0.4	3.0	F2 V	-8	58 s		
217902	247711	κ Gru	23 04 39.4	-53 57 55	+0.006	-0.11	5.1	2.3	-0.3	K5 III	+18	42 s		
218033	127949		23 04 39.8	+8 51 12	0.000	-0.04	7.35	1.24	-0.3	K5 III		340 s		
218095	52655		23 04 40.5	+49 19 48	+0.005	+0.03	8.0	0.1	1.4	A2 V		140 mx		
218045	108378	54 α Peg	23 04 45.5	+15 12 19	+0.004	-0.04	2.49	-0.04	0.2	B9 V	-4	31 ts		Markab
218003	191631		23 04 50.0	-27 08 10	-0.002	-0.01	6.8	0.0	1.7	A3 V		110 s		
217831	255395		23 04 52.0	-68 49 13	+0.008	+0.07	5.52	0.36	0.6	F2 III p	+4	96 s		
218151	35130		23 04 54.9	+56 41 25	0.000	-0.04	6.7	0.1	0.6	A0 V		160 s		
218058	127953		23 04 57.0	+4 36 00	+0.002	0.00	7.7	1.1	-0.1	K2 III		360 s		
218079	108382		23 05 00.5	+18 37 54	+0.009	+0.03	7.8	0.4	2.6	F0 V		110 s		
218099	90993		23 05 03.1	+24 31 20	-0.004	+0.02	7.2	0.5	3.4	F5 V		56 s		
217988	231425		23 05 03.4	-43 01 36	-0.012	-0.03	7.71	0.98		K3 IV		160 mx		m
218152	52663		23 05 04.6	+49 03 00	+0.004	-0.05	7.8	0.7	4.4	G0 V		49 s		
218209	20408		23 05 06.0	+68 25 02	+0.107	+0.17	7.48	0.65	5.3	G6 V	-18	29 ts		
218187	35133		23 05 06.0	+58 43 58	0.000	+0.02	7.1	1.1	0.2	K0 III	-29	220 s		
218101	108383		23 05 06.1	+16 33 47	-0.013	-0.19	6.44	0.83	3.2	G8 IV	-27	45 s		
218060	146498	83 Aqr	23 05 09.6	-7 41 38	+0.008	+0.02	5.43	0.30	3.0	F2 V	-13	31 s	16497	m
218061	165481		23 05 12.0	-17 04 45	-0.002	-0.03	6.30	1.1	0.2	K0 III		170 s		m
217989	247718		23 05 15.0	-53 43 57	+0.004	-0.03	7.9	1.9	0.2	K0 III		100 s		
218048	191635		23 05 15.2	-24 21 00	-0.001	+0.01	7.0	1.3	0.2	K0 III		150 s		
218103	127960		23 05 17.5	+1 18 25	+0.002	-0.02	6.39	0.94	0.2	G9 III	-12	170 s		
218081	146500		23 05 18.3	-7 45 17	+0.001	0.00	7.5	0.9	0.3	G8 III	-24	280 s		
218071	165482		23 05 21.8	-16 53 53	+0.003	+0.01	7.0	0.8	3.2	G5 IV		58 s		
218074	191638		23 05 25.4	-22 29 12	+0.003	+0.01	7.7	1.7	-0.5	M4 III		380 s		
218142	108388		23 05 25.6	+10 26 53	+0.002	-0.01	7.5	0.4	2.6	F0 V		94 s		
217990	247719		23 05 27.1	-58 53 57	+0.005	+0.02	7.6	0.4	2.1	A5 V		97 s		
218084	214302		23 05 28.1	-30 48 42	+0.001	-0.01	7.8	2.3	-0.3	K5 III		130 s		
218154	90999		23 05 28.8	+24 39 12	-0.002	0.00	7.0	0.1	0.6	A0 V	+9	190 s		
218153	91000		23 05 29.0	+26 00 34	+0.003	0.00	7.90	0.9	-2.1	G8 II	-82	1000 s		
218170	91001		23 05 29.6	+28 59 21	+0.006	+0.03	7.40	1.6	-0.5	M2 III	-57	210 mx		
218155	108392		23 05 32.9	+14 57 33	+0.002	0.00	6.79	0.03		A0	+12			
218172	108395		23 05 35.3	+20 14 29	+0.002	+0.04	7.2	0.5	3.4	F5 V		59 s		
218087	231432		23 05 42.7	-43 04 42	-0.002	0.00	7.2	1.2	0.2	K0 III		190 s		
218220	52672		23 05 47.1	+43 53 05	0.000	-0.02	7.5	1.4	-0.3	K5 III		350 s		
218198	72990		23 05 49.4	+35 16 14	0.000	0.00	8.0	1.1	0.2	K0 III		370 s		
218040	255402		23 05 49.6	-61 11 34	+0.010	-0.03	7.5	1.6	-0.3	K5 III		210 mx		
217987	214301		23 05 52.0	-35 51 12	+0.557	+1.33	7.34	1.50	9.6	M2 V	+10	3.6 t		Lacaille 9352
218173	146505		23 05 52.3	-7 56 12	+0.001	0.00	6.9	0.1	0.6	A0 V		180 s		
218174	165489		23 05 53.5	-10 26 16	0.000	-0.01	7.1	0.1	1.4	A2 V		140 s		
218323	20421		23 05 56.5	+64 17 46	-0.002	+0.05	7.63	0.60		B0.5 II				
218136	214306		23 05 57.2	-39 25 11	+0.008	-0.05	8.0	0.7	0.2	K0 III		160 mx		
218342	20422		23 06 09.0	+63 12 48	+0.002	+0.01	7.40	0.41	-4.6	B0 IV	-13	1000 s		
218257	52677		23 06 09.4	+41 47 24	-0.003	-0.01	7.6	0.9	3.2	G5 IV		76 s	16516	m
218259	72992		23 06 12.2	+36 49 14	0.000	0.00	6.7	1.1	0.2	K0 III		200 s		
218234	108399		23 06 13.1	+18 59 25	0.000	0.00	7.5	0.9	0.3	G8 III	+14	280 s		
218466	10620		23 06 16.6	+75 34 53	+0.004	0.00	6.7	0.0	0.4	B9.5 V		170 s		
218210	146509		23 06 17.1	-9 36 24	+0.006	-0.08	7.3	1.1	0.2	K0 III		120 mx		
218235	108400		23 06 18.0	+18 31 03	+0.016	+0.06	6.13	0.44	3.0	F2 V	-12	37 s		
218299	52678		23 06 21.6	+44 27 08	-0.001	+0.01	7.9	0.1	0.6	A0 V		270 s		
218166	231439		23 06 21.7	-47 47 07	-0.003	+0.01	8.00	1.1	0.2	K0 III		160 s		m
218289	72993		23 06 23.3	+38 34 11	+0.002	+0.02	7.9	1.1	0.2	K0 III		340 s		
218314	52680		23 06 25.6	+42 35 33	0.000	0.00	7.80	1.4	-0.3	K5 III		390 s	16513	m
218375	20424		23 06 28.5	+61 26 24	+0.004	-0.02	6.70						16514	m
218315	52682		23 06 30.6	+41 29 01	+0.001	-0.02	7.9	1.1	0.2	K0 III		340 s		
218261	108402		23 06 31.7	+19 54 39	+0.021	+0.01	6.30	0.49	4.0	F8 V	-5	29 s		
218344	35142		23 06 32.0	+51 04 38	0.000	+0.01	7.42	-0.11	-2.5	B2 V	-13	890 s		
218325	52685		23 06 32.9	+46 55 27	+0.001	+0.02	7.71	0.06		B3				m
218376	35147	1 Cas	23 06 37.6	+59 25 12	+0.001	+0.01	4.85	-0.03	-4.4	B1 III	-9	570 s		
218324	52688		23 06 36.8	+46 56 35	0.000	-0.01	8.0	0.0	0.4	B9.5 V		320 s		
218326	52687		23 06 37.1	+42 39 26	+0.003	-0.01	7.8	0.0	0.4	B9.5 V		280 s		
218205	231443		23 06 38.9	-43 30 13	-0.005	-0.01	4.50	0.6	4.4	G0 V		11 s		m
218292	108403		23 06 38.9	+10 32 34	0.000	-0.03	6.8	1.6		M5 III	+20			R Peg, v
218240	191651	86 Aqr	23 06 40.8	-23 44 36	+0.005	0.00	4.47	0.90	0.2	gG9	+15	71 s	16511	m
218303	127971		23 06 46.4	+5 02 10	0.000	+0.02	7.8	1.4	-0.3	K5 III		410 s		
218227	231444	θ Gru	23 06 52.6	-43 31 14	-0.004	-0.02	4.28	0.42	2.2	F6 IV	+10	26 s		m
218242	214313	υ Gru	23 06 53.5	-38 53 32	+0.003	+0.02	5.61	0.01	0.6	A0 V n	+16	100 s		m
218355	73004		23 06 55.3	+38 01 00	+0.008	+0.04	7.9	0.9	3.2	G5 IV		88 s		
218217	247735		23 06 56.9	-52 46 50	+0.005	+0.02	7.8	1.4	0.2	K0 III		200 s		
218283	214314		23 06 57.6	-30 02 49	+0.002	+0.04	6.6	0.1	3.0	F2 V		53 s		
218329	127976	55 Peg	23 07 00.1	+9 24 34	+0.001	-0.01	4.52	1.57	-0.5	M2 III	-5	100 s		
218366	73006		23 07 02.2	+35 03 10	-0.002	0.00	7.4	0.1	0.6	A0 V		230 s		
218365	73007		23 07 04.8	+35 38 11	+0.004	0.00	6.5	1.1	0.2	K0 III		180 s		
218393	52701		23 07 06.1	+50 11 32	+0.001	0.00	7.02	0.39		B pe	-15			
218356	91019	56 Peg	23 07 06.5	+25 28 06	0.000	-0.03	4.76	1.34	-2.1	K0 II p	-27	240 s		
218255	231445		23 07 09.3	-49 36 24	+0.002	-0.01	6.3	1.5	0.2	K0 III		86 s		

HD	SAO	Star Name	α 2000	δ 2000	μ(α)	μ(δ)	V	B-V	M_V	Spec	RV	d(pc)	ADS	Notes
			$23^h07^m10^s.0$	$+52°48'59''$	$+0^s.001$	$+0''.01$								
218416	35151		23 07 10.0	+52 48 59	+0.001	+0.01	6.11	1.05	0.2	K0 III	+5	140 s		
218440	35152		23 07 10.2	+59 43 39	0.000	+0.01	6.40	−0.01	−2.6	B2.5 IV	−5	490 s		
218331	146515		23 07 11.5	−7 41 41	+0.001	−0.05	7.4	0.1	0.6	A0 V		230 s		
218266	231446		23 07 11.7	−45 50 34	+0.003	−0.03	7.7	1.4	0.2	K0 III		310 s		
218268	247738		23 07 13.7	−50 41 13	−0.003	−0.02	5.83	0.48	3.4	F5 V		29 s		
218269	247739		23 07 14.5	−50 41 11	−0.005	−0.01	5.83	0.48		F5				m
218394	52705		23 07 14.6	+40 33 02	−0.002	+0.01	7.9	0.4	2.6	F0 V		110 s		
218582	10624		23 07 17.1	+74 43 56	+0.003	0.00	7.8	1.1	0.2	K0 III		300 s		
218306	214320		23 07 17.7	−34 51 55	+0.001	+0.04	7.8	2.5	−0.1	K2 III		60 s		
218407	52707		23 07 17.9	+46 04 06	0.000	0.00	6.66	−0.05	−2.5	B2 V	−15	580 s		
218270	247741		23 07 20.4	−52 36 10	+0.001	+0.01	7.4	1.0	3.2	G5 IV		50 s		
218737	3845		23 07 20.8	+80 38 16	+0.013	+0.02	7.7	0.0	0.4	B9.5 V		120 mx		
218535	10623		23 07 22.1	+70 39 40	+0.007	−0.01	7.70	1.1	0.2	K0 III		290 s	16525	m
218358	165504		23 07 24.2	−11 48 23	−0.001	+0.01	8.0	0.1	1.4	A2 V		210 s		
218337	214323		23 07 25.2	−34 12 36	−0.003	0.00	7.6	0.6	2.1	A5 V		70 s		
218381	91021		23 07 25.4	+20 34 54	−0.001	−0.03	6.70	1.1	0.2	K0 III		200 s	16520	m
218417	52709		23 07 25.8	+43 23 26	+0.001	−0.02	7.9	0.1	0.6	A0 V		270 s		
218468	35156		23 07 25.9	+58 59 24	+0.002	+0.01	7.3	1.1	0.2	K0 III	−12	240 s		
218395	73010		23 07 27.5	+32 49 33	−0.002	+0.01	6.00	0.12	1.7	A3 V	−1	69 s	16519	m
218396	91022		23 07 28.6	+21 08 03	+0.008	−0.05	5.99	0.26	2.1	A5 V	−12	30 mx		
218765	3846		23 07 35.1	+80 34 10	−0.002	−0.02	7.2	0.1	1.4	A2 V		140 s		
218452	52711	4 And	23 07 39.1	+46 23 14	−0.001	−0.03	5.33	1.41	−0.3	K5 III	−6	130 s	16526	m
218428	91025		23 07 39.9	+30 03 17	−0.002	0.00	7.2	0.0	0.4	B9.5 V	+3	230 s		
218536	20438		23 07 41.8	+65 37 01	−0.003	−0.01	6.9	0.1	1.4	A2 V		120 s		
218470	52713	5 And	23 07 45.2	+49 17 45	+0.015	+0.14	5.70	0.44	3.4	F5 V	−2	29 s		
218409	146521		23 07 45.4	−0 17 46	−0.001	0.00	7.4	1.1	0.2	K0 III		270 s		
218537	20439		23 07 47.5	+63 38 01	0.000	0.00	6.26	−0.02	0.6	A0 V	−36	140 s	16530	m
218453	73019		23 07 47.7	+39 47 47	+0.004	+0.03	7.70	0.4	3.0	F2 V		87 s	16527	m
218454	73016		23 07 47.7	+30 26 24	−0.001	−0.01	7.50	1.4	−2.3	K4 II	−21	910 s		
218430	108415		23 07 48.0	+12 40 24	−0.001	−0.01	7.90	0.4	2.6	F0 V		120 s	16524	m
218322	247749		23 07 50.3	−56 08 36	+0.002	−0.01	7.7	1.9	−0.5	M2 III		310 s		
218455	91029		23 07 53.0	+29 41 33	+0.001	−0.01	8.0	1.1	−0.1	K2 III		410 s		
218658	10629	33 π Cep	23 07 53.7	+75 23 16	+0.003	−0.02	4.41	0.80	0.4	G2 III	−19	59 s	16538	m
218472	73021		23 07 54.9	+31 27 33	+0.001	−0.01	7.30	0.1	1.4	A2 V	−2	150 s	16528	m
218560	20441		23 07 57.0	+64 13 21	+0.001	+0.01	6.21	1.10	0.2	K0 III	−28	140 s		
218441	127986		23 08 03.1	+4 52 27	0.000	+0.01	7.8	1.1	−0.1	K2 III		380 s		
218525	52717		23 08 12.2	+44 33 42	+0.002	0.00	6.56	0.17		A0	+2			
218444	165508		23 08 18.6	−16 45 01	−0.002	−0.04	7.9	1.1	0.2	K0 III		340 s		
218434	191674		23 08 20.8	−28 49 24	−0.004	−0.03	5.60	0.88	0.2	K0 III		120 s		
218108	258105		23 08 23.4	−79 28 51	+0.031	−0.03	6.12	0.14	2.4	A7 V n		56 s		
218573	35164		23 08 23.4	+57 10 48	−0.001	+0.01	7.8	0.7	4.4	G0 V		47 s		
218488	146526		23 08 25.9	−0 29 59	+0.006	−0.03	7.4	0.5	3.4	F5 V		63 s		
218538	91037		23 08 28.3	+29 09 41	+0.001	+0.02	7.5	0.1	1.4	A2 V	−8	170 s		
218288	258106		23 08 35.6	−73 35 11	+0.007	−0.01	6.15	1.42		K0				
218379	255411		23 08 36.2	−63 52 18	+0.020	−0.19	8.00	0.6	4.4	G0 V		45 mx		m
218392	255413		23 08 37.7	−59 44 11	+0.010	−0.05	7.60	0.4	3.0	F2 V		83 s		m
218527	127993	5 Psc	23 08 40.8	+2 07 40	+0.010	+0.11	5.40	0.91	3.2	G8 IV	−18	26 s		
218586	73035		23 08 45.9	+39 10 04	−0.001	−0.01	7.37	1.58	−0.3	K5 III		300 s		
218550	108426		23 08 47.7	+10 57 30	+0.003	−0.01	7.60	0.45		F5			16539	m
218522	191676		23 08 51.3	−25 49 49	−0.013	−0.03	6.8	0.7	3.4	F5 V		32 s		
218609	73036		23 08 56.8	+38 54 55	0.000	0.00	7.34	0.04	0.6	A0 V		210 s		
218694	20450		23 08 56.8	+66 04 00	+0.002	+0.02	8.0	0.1	1.7	A3 V		170 s		
218672	20447		23 08 58.0	+62 54 33	−0.002	0.00	7.1	1.1	−0.1	K2 III	+1	250 s		
218673	20449		23 08 58.5	+62 48 57	−0.002	0.00	7.6	1.1	−0.1	K2 III		310 s		
218507	231458		23 09 03.7	−45 30 44	+0.004	−0.02	7.6	1.2	0.2	K0 III		270 s		
218696	20452		23 09 08.8	+64 00 00	−0.008	+0.05	7.9	0.1	0.6	A0 V		68 mx		
218610	91043		23 09 09.5	+26 54 48	0.000	+0.02	7.60	1.1	−0.1	K2 III	−13	350 s		
218723	20456		23 09 16.6	+65 12 41	+0.005	0.00	6.68	−0.04	−1.0	B5.5 V	−12	130 mx		
218674	52742		23 09 16.7	+49 39 03	+0.001	0.00	6.74	0.00	−2.3	B3 IV	−4	510 s		
218612	127999		23 09 22.3	+2 08 39	+0.001	+0.00	8.0	1.1	−0.1	K2 III		410 s		
218624	108432		23 09 22.4	+18 44 09	0.000	−0.01	6.60	0.00	0.4	B9.5 V		170 s	16547	m
218594	191683	88 Aqr	23 09 26.6	−21 10 21	+0.004	+0.04	3.66	1.22	0.2	K0 III	+21	33 s		
218779	20464		23 09 28.2	+68 25 42	−0.001	−0.01	7.9	0.0	0.4	B9.5 V		290 s		
218634	128001	57 Peg	23 09 31.3	+8 40 38	0.000	+0.01	5.12	1.47		M4 s	−4	40 mn		m
218660	91049		23 09 32.4	+29 40 11	+0.002	+0.02	6.7	1.1	0.0	K1 III	+9	220 s		
218662	91047		23 09 33.5	+23 35 20	0.000	+0.01	8.0	0.1	1.4	A2 V		210 s		
218753	35186	2 Cas	23 09 43.9	+59 19 59	−0.001	+0.01	5.70	0.33	−2.1	A5 II	−12	310 s	16556	m
218497	255418		23 09 44.5	−67 52 36	0.000	−0.06	6.6	0.0	3.0	F2 V	*	53 s		
218619	191686		23 09 44.5	−28 05 19	+0.001	+0.01	5.87	1.31	0.2	K0 III		78 s		
218714	73048		23 09 48.7	+37 41 15	+0.006	−0.01	7.8	0.4	3.0	F2 V		89 s		
218639	165522		23 09 49.6	−14 30 39	+0.005	−0.02	6.42	0.01	0.6	A0 V		100 mx		
218725	73049		23 09 52.3	+39 28 10	0.000	+0.01	6.5	0.1	0.6	A0 V		150 s		
218640	191687	89 Aqr	23 09 54.6	−22 27 27	+0.002	0.00	4.69	0.65	4.4	G0 V	−5	13 mn		
218829	20466		23 09 55.5	+67 14 33	+0.006	+0.03	6.7	0.1	0.6	A0 V		140 mx		
218687	108437		23 09 57.0	+14 25 33	−0.008	−0.10	7.80	0.6	4.4	G0 V		48 s	16553	m

HD	SAO	Star Name	α 2000	δ 2000	μ(α)	μ(δ)	V	B−V	M$_V$	Spec	RV	d(pc)	ADS	Notes
218738	52753		23^{h}09^{m}57^s.1	+47°57′30″	+0^s.015	0″.00	7.91	0.90		G5				m
218630	231464		23 09 57.1	−42 51 38	−0.030	0.00	5.81	0.48	3.7	F6 V		26 s		
218843	20467		23 09 57.5	+68 37 53	+0.001	−0.01	8.0	0.1	0.6	A0 V		270 s		
218739	52754		23 09 58.3	+47 57 35	+0.011	+0.01	7.17	0.64	4.4	dG0	−8	32 s	16557	m
218700	128007	58 Peg	23 10 01.3	+9 49 19	−0.001	−0.01	5.39	−0.08		B8 n	+9			
218741	73051		23 10 02.4	+36 50 55	+0.002	+0.01	7.70	1.1	−0.1	K2 III		360 s	16558	m
218766	52756		23 10 05.6	+49 59 26	−0.001	+0.02	7.4	1.1	0.2	K0 III	+9	260 s		
218717	128008		23 10 06.4	+7 21 46	−0.001	0.00	7.2	0.0	0.4	B9.5 V		230 s		
218682	191693		23 10 07.2	−28 56 45	+0.001	0.00	8.0	0.8	0.2	K0 III		360 s		
218865	20469		23 10 08.2	+68 49 31	−0.001	−0.01	7.8	1.4	−0.3	K5 III		360 s		
218742	73052		23 10 08.8	+33 46 03	+0.002	−0.04	6.8	1.6	−0.5	M4 III		290 s		
218803	35193		23 10 09.3	+57 26 55	0.000	−0.01	7.30	1.1	−0.1	K2 III		280 s	16560	m
218655	231465		23 10 09.6	−40 35 30	+0.003	−0.04	5.83	1.62	−0.5	M4 III		190 s		
218558	255420		23 10 11.6	−66 51 27	+0.036	+0.04	6.47	0.95		G5				
218606	247764		23 10 15.6	−59 22 10	+0.005	+0.01	7.4	1.1	0.2	K0 III		280 s		
218683	214357		23 10 16.4	−35 23 57	−0.001	−0.01	7.9	1.9	−0.3	K5 III		240 s		
218767	73054		23 10 17.0	+32 29 14	+0.001	+0.01	7.03	−0.04		B9	−2		16561	m
218669	231467		23 10 19.0	−42 45 12	+0.002	+0.02	7.6	0.6	3.2	G5 IV		77 s		
218718	165530		23 10 20.6	−13 38 40	+0.004	+0.03	7.2	1.1	−0.1	K2 III		280 mx		
218790	52759		23 10 21.2	+49 01 06	+0.025	+0.05	7.30	0.6	4.4	G0 V		38 s	16562	m
218670	231468	ι Gru	23 10 21.4	−45 14 48	+0.013	−0.02	3.90	1.02	0.2	K0 III	−4	53 s		
218684	231470		23 10 23.8	−40 02 00	+0.003	−0.01	7.3	1.1	0.2	K0 III		240 s		
218730	146541		23 10 24.6	−7 48 44	+0.010	−0.13	7.7	0.6	4.4	G0 V		47 s		
218866	20470		23 10 24.9	+64 31 49	+0.013	−0.03	7.2	0.5	4.0	F8 V		43 s		
218631	255421		23 10 27.0	−60 33 50	+0.008	0.00	6.7	0.7	1.4	A2 V		46 s		
218804	52761	6 And	23 10 27.0	+43 32 40	−0.018	−0.18	5.94	0.44	2.1	F5 IV	−43	58 s		
218768	108442		23 10 28.7	+10 48 59	+0.002	0.00	7.4	1.1	0.2	K0 III		270 s		
218844	52765		23 10 37.6	+46 22 40	+0.001	0.00	7.9	0.1	0.6	A0 V		270 s		
218769	146543		23 10 39.4	−5 57 38	0.000	0.00	7.0	0.0	0.0	B8.5 V		250 s		
218748	214363		23 10 40.6	−32 10 42	+0.003	+0.01	7.9	0.9	2.6	F0 V		47 s		
218806	91061		23 10 42.2	+26 31 23	+0.002	0.00	7.20	0.1	1.4	A2 V		150 s	16567	m
218792	108443		23 10 42.4	+17 35 39	+0.002	−0.02	5.71	1.34	−0.3	gK5	+2	160 s		
218524	258110		23 10 46.1	−76 27 49	−0.003	0.00	7.9	1.6	−0.5	M4 III		470 s		
218759	191703		23 10 46.4	−29 31 30	−0.003	−0.05	6.51	0.27	2.6	F0 V		61 s		
218868	52768		23 10 49.8	+45 30 46	−0.011	−0.25	7.20	1.1	0.2	K0 III		67 mx		m
218775	214366		23 10 55.2	−33 37 24	+0.001	+0.03	7.8	1.3	0.2	K0 III		230 s		
218852	73062		23 10 58.1	+31 09 03	0.000	−0.01	7.5	0.9	0.3	G8 III	+5	280 s		
218878	52769		23 11 01.2	+43 39 22	−0.001	−0.01	7.3	0.1	0.6	A0 V		210 s		
—	91066		23 11 05.1	+25 22 01	0.000	−0.02	8.0	1.6	−0.1	K2 III		220 s		
218915	35206		23 11 06.8	+53 03 31	0.000	+0.02	7.20	0.02	−6.2	O9 I	−72	3100 s		
218711	255423		23 11 10.6	−63 11 25	+0.013	−0.02	7.6	1.1	−0.1	K2 III		240 mx		
218880	91069		23 11 10.7	+30 02 35	+0.001	+0.02	6.9	1.1	0.2	K0 III		220 s		
218853	128019		23 11 14.0	+5 00 15	+0.001	0.00	6.90	1.6	−0.5	M4 III	−6	300 s		m
218925	35210		23 11 23.4	+50 46 17	+0.006	+0.02	6.8	0.5	4.0	F8 V		37 s		
218949	52780		23 11 38.8	+47 05 55	−0.022	−0.17	7.1	0.9	3.2	G5 IV		61 s		
218918	128022	59 Peg	23 11 44.1	+8 43 12	−0.001	0.00	5.16	0.13	1.4	A2 V	+10	51 s		
218935	91080	60 Peg	23 11 49.0	+26 50 50	−0.014	−0.11	6.17	0.94	3.2	K0 IV	−10	39 s		m
218965	73075		23 11 55.2	+32 39 57	+0.003	−0.02	7.3	1.1	0.2	K0 III		270 s		
218983	52783		23 11 57.7	+44 20 06	+0.003	−0.06	7.9	0.1	0.6	A0 V		110 mx		
218928	165551		23 11 59.5	−11 56 01	+0.002	−0.02	7.80	1.1	0.2	K0 III		330 s	16579	m
218999	52784		23 11 59.5	+49 12 49	+0.001	0.00	7.5	0.1	0.6	A0 V		230 s		
219063	20483		23 12 01.2	+64 43 10	−0.002	0.00	7.32	−0.03	−1.0	B5.5 V	+14	420 s		
218890	231486		23 12 02.8	−43 59 15	0.000	+0.03	7.3	1.0	0.2	K0 III		230 s		
219015	52785		23 12 05.6	+49 14 25	−0.008	0.00	8.0	1.1	0.2	K0 III		300 mx		
219396	3865		23 12 09.4	+82 23 41	+0.034	+0.09	7.6	0.6	4.4	G0 V		43 s		
218559	258961		23 12 11.9	−80 54 46	+0.007	−0.01	6.41	1.50		K2				
218862	255424		23 12 15.7	−61 29 14	−0.008	+0.01	7.6	0.5	3.4	F5 V		68 s		
218861	255425		23 12 17.1	−61 19 59	−0.008	−0.03	7.6	0.9	3.2	G5 IV		75 s		
219049	35227		23 12 20.5	+52 13 48	−0.001	0.00	7.5	0.4	2.6	F0 V		94 s		
219028	73086		23 12 24.9	+36 58 04	+0.004	+0.01	6.8	1.1	0.2	K0 III		210 s		
218957	191719		23 12 30.0	−23 49 10	+0.007	−0.01	7.9	0.4	3.4	F5 V		81 s		
219080	52787	7 And	23 12 32.9	+49 24 23	+0.009	+0.10	4.52	0.29	2.6	F0 V	+13	23 ts		
219126	20488		23 12 33.6	+64 47 58	+0.001	−0.09	7.32	0.01	0.6	A0 V		76 mx		
219029	108456		23 12 36.4	+10 57 09	−0.001	−0.02	7.9	0.8	3.2	G5 IV		89 s		
219018	128034		23 12 38.5	+2 41 10	+0.005	−0.14	7.72	0.63		G5			16591	m
—	52790		23 12 41.8	+45 35 20	+0.012	+0.04	8.0	1.4						
219135	35235		23 12 52.9	+56 32 11	−0.002	−0.01	7.6	0.7	−4.5	G0 Ib		1600 s		
219052	146563		23 12 58.4	−9 34 14	+0.003	−0.01	6.9	1.1	0.2	K0 III		220 s		
219034	214386		23 12 59.2	−32 19 03	+0.002	−0.06	7.60	1.1	0.2	K0 III		180 mx		m
219066	146565		23 13 00.1	+0 01 52	0.000	+0.01	7.70	0.9	0.3	G6 III	0	300 s		
219109	52798		23 13 01.3	+42 03 44	−0.001	−0.01	6.7	0.1	1.4	A2 V		120 s		
219209	20496		23 13 02.0	+67 04 35	+0.004	+0.02	7.71	1.91	−0.5	M2 III		280 s		
219110	91095		23 13 03.8	+29 26 30	−0.002	−0.03	6.35	0.93		K0	+4			
219081	128037		23 13 04.8	+8 57 49	−0.001	+0.01	7.6	1.6	−0.5	M4 III		420 s		
219127	73090		23 13 06.2	+40 00 10	+0.007	0.00	7.60	0.28	2.4	dA8 n	0	110 s	16599	m

HD	SAO	Star Name	α 2000	δ 2000	μ(α)	μ(δ)	V	B−V	M_V	Spec	RV	d(pc)	ADS	Notes
			$23^h 13^m 06{.}^s5$	$-55°06'23''$	$+0{.}^s005$	$-0{.}''03$								
218993	247779		23 13 06.5	−55 06 23	+0.005	−0.03	7.5	0.7	1.4	A2 V		65 s		
219136	52801		23 13 12.4	+41 28 58	+0.002	0.00	7.7	1.4	−0.3	K5 III		410 s		
219023	231493		23 13 14.9	−49 37 08	0.000	−0.01	6.80	0.8	3.2	G5 IV		53 s		m
219085	165562		23 13 16.3	−18 22 09	+0.002	+0.02	7.7	0.3	3.4	F5 V		73 s		
219134	35236		23 13 16.9	+57 10 07	+0.255	+0.30	5.56	1.01	6.4	K3 V	−18	6.8 t		Bradley 3077, m
219069	214389		23 13 18.3	−30 03 16	+0.002	+0.02	7.7	1.5	0.2	K0 III		160 s		
219139	108463		23 13 26.3	+11 03 55	−0.001	+0.01	5.82	1.01	0.2	K0 III	+16	130 s	16603	m
219114	128043		23 13 28.2	+0 55 50	−0.001	+0.01	7.4	0.4	2.6	F0 V		90 s		m
219150	128046		23 13 40.8	+2 12 13	+0.001	+0.04	8.00	0.4	3.0	F2 V		100 s	16607	m
219371	10658		23 13 47.1	+74 33 13	+0.012	+0.06	8.0	1.1	0.2	K0 III		250 mx		
219151	165571		23 13 48.0	−13 23 40	−0.001	−0.01	7.40	1.1	−0.1	K2 III		320 s	16608	m
219172	108468		23 13 48.1	+15 22 03	+0.001	−0.07	7.6	0.6	4.4	G0 V		44 s		
219224	52813		23 13 49.1	+44 35 52	+0.001	−0.01	7.9	0.5	3.4	F5 V		77 s		
219106	—		23 13 51.2	−47 11 24			7.7	1.1	0.2	K0 III		290 s		
219225	52814		23 13 54.9	+40 47 55	+0.002	−0.04	7.8	1.1	−0.1	K2 III		390 s		
219196	108472		23 13 59.1	+19 38 02	+0.002	0.00	6.5	1.1	−0.1	K2 III		210 s		
219212	91107		23 13 59.6	+24 14 00	0.000	+0.01	8.00	0.1	0.6	A0 V		300 s	16612	m
219188	128051		23 14 00.5	+4 59 49	+0.002	0.00	6.9	0.1		B0.5 III	+48			m
219748	3876		23 14 04.5	+84 14 33	+0.007	+0.01	7.8	0.0	0.4	B9.5 V		300 s		
219200	146580		23 14 06.2	−2 38 06	−0.002	−0.02	7.50	0.1	1.4	A2 V		170 s	16613	m
219077	255435		23 14 06.5	−62 42 00	+0.070	−0.42	6.12	0.79	3.2	G5 IV		38 s		
—	35249		23 14 06.5	+57 57 12	0.000	+0.01	8.0	2.2	−0.3	K7 III		180 s		
219175	146577		23 14 07.3	−8 55 28	+0.037	−0.04	7.08	0.61	4.2	F9 V	−32	28 mx	16611	m
219179	165573		23 14 07.5	−19 41 42	+0.003	+0.03	7.5	−0.1	1.4	A2 V		170 s		
219048	258119		23 14 07.8	−70 03 31	+0.033	+0.08	7.0	0.6	4.4	G0 V		33 s		
219290	35251		23 14 14.1	+50 37 04	+0.004	0.00	6.3	0.1	0.6	A0 V	−14	88 mx		
219238	108478		23 14 15.9	+19 45 56	0.000	+0.01	7.8	0.1	0.6	A0 V		280 s		
219202	165574		23 14 17.3	−16 54 33	0.000	−0.01	7.9	1.1	0.2	K0 III		340 s		
219215	146585	90 φ Aqr	23 14 19.2	−6 02 56	+0.003	−0.19	4.22	1.56	−0.5	M2 III	0	75 mx		
219291	91111		23 14 21.5	+29 46 18	−0.001	−0.01	6.42	0.45	3.4	F5 V	+9	40 s		m
219307	52825		23 14 23.0	+43 28 35	+0.003	+0.02	8.0	0.4	3.0	F2 V		96 s		
219306	52829		23 14 26.1	+43 41 52	+0.007	−0.10	7.5	0.5	3.4	F5 V		67 s		
219159	247791		23 14 28.8	−57 14 13	+0.002	−0.02	7.8	1.5	−0.3	K5 III		410 s		
219292	91112		23 14 30.4	+20 26 33	+0.002	+0.01	7.20	1.1	−0.1	K2 III		290 s		m
219308	73109		23 14 32.6	+35 00 58	−0.002	−0.03	7.7	1.1	−0.1	K2 III		360 s		
219310	91113		23 14 36.3	+24 06 11	+0.008	+0.01	6.4	1.1	0.2	K0 III	−27	170 mx		
219485	10664		23 14 37.1	+74 13 53	+0.011	+0.01	5.7	0.1	0.6	A0 V	−3	100 s		
219279	165578		23 14 40.0	−10 41 19	0.000	−0.03	6.40	1.4	−0.3	K5 III		220 s	16618	m
219256	146587		23 14 41.8	−9 41 18	+0.006	−0.02	7.5	0.1	1.7	A3 V		74 mx		
219259	191750		23 14 44.3	−25 33 49	−0.001	−0.04	8.0	1.1	0.2	K0 III		330 s		
219295	146589		23 14 49.6	−7 42 12	0.000	−0.03	8.0	1.6	−0.5	M2 III		510 s		
—	—		23 14 51.6	+43 19 53	−0.002	+0.01	7.8	0.2	2.1	A5 V		140 s		
219314	165581		23 14 52.9	−16 14 22	0.000	−0.03	7.6	1.1	0.2	K0 III		310 s		
219459	20510		23 14 54.5	+65 25 29	−0.007	−0.06	7.1	0.4	3.0	F2 V		66 s		
219361	91118		23 14 57.7	+28 04 15	+0.002	−0.01	7.12	0.04		A0	+2			
219263	231512		23 14 58.5	−41 06 20	+0.010	−0.11	5.77	1.18	−0.1	K2 III	+26	77 mx		
219384	52834		23 14 58.8	+41 08 35	−0.001	+0.02	7.7	1.1	−0.1	K2 III		360 s		
219397	52835		23 15 01.2	+46 31 28	0.000	−0.01	7.0	0.1	0.6	A0 V		190 s		
219249	247795		23 15 09.7	−56 43 46	+0.030	+0.09	7.97	0.70	5.2	G5 V		35 s		
219496	20515		23 15 11.8	+65 00 44	0.000	+0.01	7.7	0.5	3.4	F5 V		71 s		
219386	108493		23 15 18.7	+12 26 16	+0.006	+0.03	7.8	0.4	2.6	F0 V		110 s		
219339	191760		23 15 19.1	−24 51 10	−0.001	−0.03	7.26	0.17	2.1	A5 V		110 s		
219363	165586		23 15 20.7	−11 33 56	−0.002	−0.02	7.5	0.1	1.4	A2 V		170 s		
219364	165585		23 15 20.8	−14 00 55	+0.002	+0.01	7.5	1.1	0.2	K0 III	+9	290 s		
219418	91126		23 15 22.9	+25 40 20	0.000	+0.02	6.7	0.8	0.3	G5 III	+39	190 s		
219301	247798		23 15 23.2	−56 41 30	+0.004	−0.02	6.7	0.1	2.6	F0 V		66 s		
219401	128067		23 15 25.2	+6 10 45	+0.007	+0.04	8.0	1.1	0.2	K0 III		160 mx		
219342	231518		23 15 27.5	−43 36 17	−0.003	−0.09	8.0	0.9	4.0	F8 V		41 s		
219523	20518		23 15 27.9	+64 16 00	−0.004	0.00	7.19	−0.02	−1.0	B5.5 V	−14	330 mx		
219302	247800		23 15 29.9	−59 41 38	−0.004	+0.01	7.70	1.1	−0.1	K2 III		360 s		m
219402	146593		23 15 34.1	−3 29 47	−0.001	0.00	5.55	0.06		A2	+11			
219684	10677		23 15 36.1	+79 13 58	−0.009	0.00	7.8	0.1	0.6	A0 V		260 s		
219404	165592		23 15 36.4	−11 37 55	+0.001	+0.01	8.0	1.4	−0.3	K5 III		450 s		
219586	10671		23 15 37.5	+70 53 18	+0.003	+0.01	5.6	0.1	1.7	A3 V	+12	59 s		
219667	10674		23 15 39.6	+76 47 14	+0.002	0.00	7.3	1.1	0.2	K0 III		240 s		
219420	128069		23 15 40.0	+1 18 30	+0.012	−0.04	6.8	0.5	3.4	F5 V		48 s		
219446	91129		23 15 40.9	+24 25 21	−0.001	−0.02	7.9	1.1	0.2	K0 III		350 s		
219476	73118		23 15 41.5	+31 40 31	+0.012	+0.04	7.6	0.5	4.0	F8 V		52 s		
—	35269		23 15 42.1	+58 02 36	+0.001	+0.01	8.0	2.1	−0.1	K2 III		120 s		
219409	214410		23 15 44.5	−29 50 52	+0.012	−0.02	6.52	1.08	0.0	K1 III		140 mx		
219477	91130	61 Peg	23 15 46.1	+28 14 52	+0.001	0.00	6.5	0.9	3.2	G5 IV	+4	45 s		
219449	146598	91 ψ¹ Aqr	23 15 53.4	−9 05 16	+0.025	−0.01	4.21	1.11	1.8	K0 III	−25	30 ts		m
219537	35271		23 15 51.4	+56 40 34	+0.001	+0.01	7.85	0.02	0.6	A0 V		280 s		m
219452	165595		23 15 52.4	−13 11 02	+0.005	−0.01	7.3	0.4	2.6	F0 V		87 s		
219449	146598		23 15 53.3	−9 05 16	+0.025	−0.01	4.23	1.10	0.2	K0 III	−26	52 mx	16633	m

HD	SAO	Star Name	α 2000	δ 2000	μ(α)	μ(δ)	V	B−V	M_v	Spec	RV	d(pc)	ADS	Notes
219487	91133		23^{h}15^{m}57.7	+24°46'16"	+0.007	+0.01	6.60	0.40	3.0	F2 V	+5	51 s		
219461	146600		23 15 58.2	−1 25 25	+0.002	+0.01	7.1	1.1	0.2	K0 III		240 s		
219512	73122		23 16 00.5	+35 23 59	−0.004	−0.06	6.9	0.4	2.6	F0 V		72 s		
219456	214418		23 16 04.1	−35 40 59	+0.001	+0.02	7.9	1.5	−0.1	K2 III		260 s		
219392	247806		23 16 04.3	−51 52 20	−0.003	−0.01	6.9	2.0	−0.3	K5 III		140 s		
219499	108501		23 16 07.8	+18 15 27	+0.002	−0.03	7.7	1.1	−0.1	K2 III		360 s		
219395	255441		23 16 10.1	−59 46 51	−0.010	−0.01	7.9	0.3	3.4	F5 V		78 s		
219766	10679		23 16 13.6	+78 09 02	−0.007	0.00	7.4	0.2	2.1	A5 V		110 s		
219588	—		23 16 19.8	+50 29 18	+0.005	+0.02	7.6	0.1	0.6	A0 V		160 mx		
219458	247809		23 16 25.0	−55 31 37	+0.005	+0.02	7.0	0.2	3.4	F5 V		53 s		
219599	35280		23 16 25.5	+50 40 17	0.000	0.00	7.7	1.4	−0.3	K5 III		380 s		
219634	20531		23 16 26.6	+61 57 48	−0.004	0.00	6.53	0.23		B8	−8			
219504	191779		23 16 29.2	−23 13 30	−0.004	−0.01	7.1	0.4	2.1	A5 V		76 s		
219505	191778		23 16 30.3	−27 04 53	+0.003	−0.02	7.2	1.9	−0.3	K5 III		190 s		
219612	35283		23 16 31.8	+55 11 27	+0.014	+0.02	8.0	0.4	2.6	F0 V		120 s		
219542	146605		23 16 35.2	−1 35 09	+0.012	+0.03	7.59	0.66	5.0	dG4	−17	32 s	16642	m
219507	231532		23 16 39.5	−44 29 21	0.000	0.00	5.92	1.06	0.2	K0 III		130 s		
219519	191783		23 16 39.5	−28 41 03	+0.003	−0.01	7.2	0.8	−0.1	K2 III		290 s		
219623	35285		23 16 42.2	+53 12 49	+0.012	−0.23	5.54	0.52	4.0	F8 V	−25	26 ts		m
219564	146609		23 16 43.9	−1 49 49	+0.004	−0.04	8.0	1.1	0.2	K0 III		200 mx		
219553	191787		23 16 49.5	−21 12 11	+0.001	−0.04	7.9	0.5	0.2	K0 III		340 s		
219531	231533		23 16 49.6	−41 11 40	−0.002	+0.02	6.47	1.08	0.2	gK0		160 s		
219576	146612	92 χ Aqr	23 16 50.8	−7 43 36	−0.001	−0.01	5.06	1.60		gM5	−15	29 mn		q
219651	35288		23 16 51.3	+57 38 52	−0.001	−0.02	7.4	0.1	0.6	A0 V		210 s		
219626	73138		23 16 56.1	+37 57 44	0.000	0.00	7.8	1.4	−0.3	K5 III		410 s		
219482	255446		23 16 57.6	−62 00 05	+0.025	−0.03	5.66	0.51	4.0	F8 V	−9	22 s		
219627	73137		23 16 57.8	+30 22 26	−0.002	−0.02	7.8	0.1	1.4	A2 V		190 s		
219533	247811		23 16 58.2	−51 19 11	−0.007	−0.12	7.9	0.9	4.4	G0 V		32 s		
219578	165602		23 16 58.3	−18 52 27	−0.003	−0.03	7.0	1.1	0.2	K0 III		200 s		
219592	146614		23 16 58.9	−7 09 38	+0.008	+0.06	6.7	0.4	2.6	F0 V		66 s		
219580	191792		23 17 08.5	−28 26 16	+0.003	−0.08	6.6	1.3	−0.5	M2 III		150 mx		
219604	165605		23 17 09.0	−18 09 56	+0.006	−0.02	7.9	1.1	0.2	K0 III		250 mx		
219615	128085	6 γ Psc	23 17 09.7	+3 16 56	+0.051	+0.02	3.69	0.92	0.3	G8 III	−14	48 s		
219605	—		23 17 11.7	−20 40 17			8.0	0.1	2.6	F0 V		120 s		
219628	146615		23 17 14.9	−4 26 09	−0.002	−0.04	7.9	1.1	−0.1	K2 III		400 s		
219668	52865		23 17 16.5	+45 09 51	+0.009	−0.06	6.43	1.07	0.2	gK0	−38	150 s		
219841	10684		23 17 18.8	+75 17 57	+0.005	+0.01	6.5	0.1	1.4	A2 V	−8	99 s		
219654	91146		23 17 19.0	+29 52 24	+0.001	0.00	7.90	1.6	−0.5	M1 III	+4	480 s		
219536	255449		23 17 20.5	−62 46 53	+0.004	−0.01	7.6	1.1	−0.1	K2 III		350 s		
219332	258125		23 17 23.4	−79 26 02	+0.005	+0.01	7.8	0.4	2.1	A5 V		100 s		
219571	247814	γ Tuc	23 17 25.6	−58 14 08	−0.003	+0.09	3.99	0.40	0.6	F0 III	+18	45 s		
219639	165607		23 17 28.0	−16 10 20	0.000	0.00	6.5	0.0	0.0	B8.5 V		200 s		
219610	231536		23 17 29.2	−45 13 56	+0.003	+0.05	8.00	0.6	4.4	G0 V		53 s		m
219646	165608		23 17 29.3	−13 38 38	+0.009	−0.12	7.8	0.8	3.2	G5 IV		64 mx		
—	20540		23 17 35.5	+60 38 59	+0.001	+0.01	8.0	2.1	−0.3	K5 III		210 s		
219670	108513		23 17 35.7	+16 42 06	+0.005	+0.03	8.0	1.1	0.2	K0 III		220 mx		
219675	108515		23 17 37.1	+18 18 18	−0.004	−0.02	7.00	0.4	2.6	F0 V	+13	76 s	16650	m
219657	146617		23 17 37.5	−1 31 18	+0.017	−0.08	7.87	0.67	3.2	G5 IV	−47	86 s	16649	m
219811	20545		23 17 39.2	+69 39 00	+0.008	+0.04	7.8	1.4	−0.3	K5 III		360 s		
219659	165609		23 17 39.9	−11 42 48	+0.002	0.00	6.34	0.05	0.6	A0 V		130 s		
219712	73149		23 17 41.6	+34 02 49	−0.001	0.00	7.0	0.5	3.4	F5 V	+1	53 s		
219699	73148		23 17 42.0	+31 01 50	+0.003	+0.01	7.2	0.4	2.6	F0 V	−27	83 s		
219734	52871	8 And	23 17 44.5	+49 00 55	+0.004	+0.01	4.85	1.67	−0.5	gM2	−8	100 s	16656	m
219685	108516		23 17 46.0	+16 32 53	+0.002	−0.01	8.0	1.1	−0.1	K2 III		410 s		
219736	73151		23 17 53.5	+30 27 28	−0.004	−0.04	7.2	1.1	0.0	K1 III	−4	270 s		
219688	146620	93 ψ² Aqr	23 17 54.1	−9 10 57	+0.001	−0.01	4.39	−0.15	−1.1	B5 V	−6	130 s		
219749	52876		23 17 55.9	+45 29 20	+0.003	−0.01	6.30	−0.03		B9 p				
219702	165614		23 17 56.3	−13 47 42	−0.001	0.00	6.8	1.1	0.2	K0 III	−9	210 s		
219889	10686		23 17 56.8	+73 41 18	−0.016	−0.02	7.2	1.1	0.2	K0 III		240 s		
219728	128095		23 17 59.6	+9 30 13	0.000	0.00	7.6	1.1	0.2	K0 III		310 s		
219665	231538		23 18 00.5	−48 06 50	0.000	−0.02	7.3	0.4	3.0	F2 V		66 s		
219631	255453		23 18 00.6	−61 00 13	+0.007	−0.08	6.70	0.9	3.2	G5 IV		50 s		m
219977	10689		23 18 00.8	+79 52 26	+0.004	−0.01	7.9	0.1	1.4	A2 V		190 s		
219693	231539	∮ Gru	23 18 09.7	−40 49 29	+0.012	−0.12	5.53	0.44	2.6	dF0	+14	34 s		
219779	91157		23 18 15.2	+27 36 01	+0.003	−0.03	7.4	0.5	3.4	F5 V		64 s		
219769	108529		23 18 18.6	+16 59 07	+0.004	+0.01	7.8	0.2	2.1	A5 V		140 s		
219644	255455		23 18 19.7	−67 26 16	+0.003	+0.02	6.13	1.35		K0				
219815	52881	9 And	23 18 23.2	+41 46 25	−0.001	−0.01	5.9	0.1	1.7	A3 V	−4	69 s		AN And, m,v
219813	52882		23 18 23.4	+47 15 41	+0.001	−0.02	7.80	0.00	0.4	B9.5 V		290 s	16661	m
219780	108533		23 18 24.8	+18 59 47	0.000	0.00	8.0	0.9	3.2	G5 IV		93 s		
219585	258126		23 18 27.5	−74 30 57	+0.001	0.00	8.0	0.3	4.0	F8 V		64 s		
219800	91161		23 18 28.8	+27 36 14	−0.002	−0.05	7.1	1.1	0.2	K0 III		240 s		
219743	191808		23 18 29.4	−28 28 38	+0.004	−0.04	8.0	1.7	0.2	K0 III		140 s		
219758	165617		23 18 29.5	−18 50 29	+0.001	+0.02	6.9	1.1	−0.3	K5 III		280 s		
219697	255456		23 18 35.4	−60 60 00	0.000	0.00	7.1	2.1	0.2	K0 III		52 s		

HD	SAO	Star Name	α 2000	δ 2000	μ(α)	μ(δ)	V	B-V	M_v	Spec	RV	d(pc)	ADS	Notes
			$23^h\,18^m\,36\overset{s}{.}6$	$-58°18'22''$	$+0\overset{s}{.}028$	$-0\overset{''}{.}15$								
219709	247822		23 18 36.6	-58 18 22	+0.028	-0.15	7.34	0.70	4.7	G2 V		31 s		m
219916	20554	34 o Cep	23 18 37.4	+68 06 42	+0.010	+0.02	4.75	0.84	0.2	K0 III	-18	81 s	16666	
219844	73161		23 18 38.9	+33 43 39	+0.002	-0.02	7.2	1.4	-0.3	K5 III		320 s		
219944	10690		23 18 41.6	+71 19 26	+0.002	-0.02	7.8	1.1	-0.1	K2 III		330 s		
219843	73162		23 18 41.9	+36 05 26	+0.003	+0.03	7.3	0.4	2.6	F0 V		87 s		
219682	---		23 18 42.1	-65 18 16			7.9	1.1	0.2	K0 III		350 s		
219761	231542		23 18 45.8	-47 26 07	+0.001	+0.01	6.60	-0.06	0.6	A0 V		160 s		
219828	108536		23 18 46.5	+18 38 45	-0.001	+0.02	8.00	0.6	2.8	G0 IV	-23	110 s		
219784	214444	γ Scl	23 18 49.3	-32 31 55	+0.002	-0.06	4.41	1.13	0.3	G8 III	+16	47 s		
219793	191813		23 18 51.5	-25 56 33	+0.002	+0.02	8.0	1.1	-0.1	K2 III		410 s		
219794	191814		23 18 53.5	-28 44 28	+0.005	-0.01	7.73	0.24		A m				
219829	128108		23 18 54.3	+5 24 20	+0.032	-0.10	8.00	0.83	5.9	K0 V	-14	26 s	16665	m
219901	35321		23 18 54.5	+54 32 20	-0.002	+0.01	7.7	1.1	-0.1	K2 III		340 s		
219804	214445		23 18 54.8	-30 33 12	+0.003	-0.05	6.94	1.09	0.2	K0 III		200 s		
219890	---		23 18 56.6	+48 29 46			7.50	0.2	2.1	A5 V		120 s	16673	m
219832	146635	95 ψ³ Aqr	23 18 57.5	-9 36 39	+0.003	0.00	4.98	-0.02	0.6	A0 V	-10	75 s	16671	m
219833	165622		23 19 02.1	-12 10 13	+0.002	+0.02	7.24	0.00	0.6	A0 V		210 s		
219891	52892		23 19 02.2	+45 08 14	+0.005	+0.01	6.5	0.1	1.4	A2 V	+7	100 s		
219823	191816		23 19 03.1	-28 23 15	+0.001	+0.03	6.7	1.9	-0.3	K5 III		150 s		
219904	52896		23 19 05.8	+43 07 57	-0.003	+0.01	7.7	0.1	0.6	A0 V		250 s		
219834	165625	94 Aqr	23 19 06.5	-13 27 32	+0.020	-0.10	5.08	0.80	3.2	G5 IV	+10	23 s	16672	m
219917	52899		23 19 07.9	+48 55 11	+0.013	+0.02	7.70	0.1	1.4	A2 V		60 mx		m
219572	258127		23 19 08.1	-79 28 22	+0.008	+0.01	6.33	0.91		K0				
219892	128113		23 19 20.0	+7 12 54	+0.003	-0.01	7.7	1.1	0.2	K0 III		310 s		
219825	247829		23 19 23.2	-53 25 11	0.000	0.00	7.5	1.7	-0.3	K5 III		280 s		
219978	20558		23 19 23.2	+62 44 22	-0.004	-0.01	6.72	2.30	-4.4	K5 Ib	-19	330 mx		
219877	146639	96 Aqr	23 19 23.8	-5 07 28	+0.013	-0.02	5.55	0.39	2.6	dF0	-9	36 s	16676	m
219859	214450		23 19 23.8	-36 07 47	-0.002	-0.04	7.9	0.5	0.2	K0 III		340 s		
219879	165628		23 19 23.9	-18 04 31	-0.001	-0.01	5.93	1.52	0.2	gK0	+5	52 s		
219926	73170		23 19 24.2	+35 49 30	+0.001	-0.01	6.5	0.0	0.4	B9.5 V		170 s		
219860	214451		23 19 25.6	-39 09 30	+0.004	-0.02	7.1	-0.1	1.4	A2 V		98 mx		
220140	10697		23 19 26.8	+79 00 14	+0.073	+0.09	7.70	0.9	5.7	G9 V	-17	25 s		m
219927	73171		23 19 27.2	+34 47 36	+0.001	0.00	6.10	-0.08	-1.6	B5 IV	-1	310 s		
219945	52907	11 And	23 19 29.5	+48 37 31	+0.002	+0.06	5.44	1.03	0.2	K0 III	+11	110 s		
220086	10696		23 19 40.3	+73 55 27	-0.004	-0.03	7.80	0.4	2.6	F0 V		110 s	16683	m
219920	165632		23 19 40.5	-15 46 54	+0.002	+0.01	7.5	0.9	3.2	G5 IV		71 s		
219937	91173		23 19 41.1	+23 49 30	+0.001	-0.04	7.9	0.1	0.6	A0 V		280 s		
219962	52912		23 19 41.4	+48 22 52	+0.021	+0.04	6.32	1.12	-0.1	K2 III	+23	82 mx		m
220031	20560		23 19 41.5	+66 24 58	-0.004	+0.01	7.8	0.0	0.4	B9.5 V		280 s		
219885	231550		23 19 42.1	-41 05 01	+0.003	0.00	7.7	0.2	2.1	A5 V		130 s		
219912	214459		23 19 43.0	-33 42 30	+0.004	-0.04	6.37	1.30		K2				m
219952	73174		23 19 44.2	+39 17 19	-0.001	-0.03	7.60	1.05	-0.1	K2 III		350 s		
219949	128119		23 19 48.0	+7 58 55	-0.001	-0.03	7.0	1.1	0.2	K0 III		230 s		
220016	35331		23 19 49.4	+59 38 00	0.000	0.00	7.90	0.04	-1.7	B3 V		640 s		
220007	35330		23 19 49.6	+57 14 37	+0.001	0.00	7.20	1.1	-0.1	K2 III	+2	260 s	16681	m
219981	52914	10 And	23 19 52.2	+42 04 41	+0.003	+0.01	5.79	1.48	-0.1	K2 III	+3	98 s		
219980	52915		23 19 52.5	+42 57 52	+0.006	+0.03	7.5	0.1	0.6	A0 V		110 mx		
219914	231553		23 19 59.4	-47 32 24	+0.001	+0.01	6.7	0.8	0.2	K0 III		200 s		
220057	20562		23 20 00.4	+61 09 01	0.000	+0.02	6.93	0.03	-3.0	B2 IV	-41	700 s		
219989	52917		23 20 01.0	+41 45 17	+0.002	0.00	7.3	0.1	0.6	A0 V		220 s		
220072	20564		23 20 02.9	+64 50 55	+0.005	+0.01	7.2	0.4	2.6	F0 V		81 s		
219964	108548		23 20 03.7	+19 18 42	+0.001	-0.02	7.9	1.1	0.2	K0 III		340 s		
220073	20566		23 20 09.9	+62 37 04	+0.004	+0.02	7.68	0.42		F5				
219992	91179		23 20 11.5	+23 05 30	0.000	+0.01	6.8	1.1	-0.1	K2 III	-2	240 s		
220074	20567		23 20 14.2	+61 58 12	0.000	-0.01	6.5	1.4	-0.3	K5 III	-35	220 s		
219983	146645		23 20 15.7	-3 55 08	+0.019	-0.09	6.6	0.4	3.0	F2 V	-14	42 mx		
220008	128125		23 20 18.1	+6 52 20	-0.006	+0.02	7.80	0.8	5.0	G4 V	-15	36 s		
220009	128126	7 Psc	23 20 20.4	+5 22 52	+0.005	-0.06	5.05	1.20	-0.1	K2 III	+38	100 mx		
220102	20568		23 20 20.7	+60 16 29	0.000	0.00	6.63	0.63	-2.0	F5 II	-24	410 s		
219959	231555		23 20 25.0	-43 08 59	+0.006	-0.02	8.0	0.4	0.2	K0 III		250 mx		
220033	128130		23 20 33.0	+8 55 07	+0.005	-0.03	7.2	1.6		M5 III	+5	250 mx		S Peg, v
220130	20572		23 20 34.4	+62 12 48	0.000	0.00	6.39	1.61	-0.3	K5 III	-23	190 s	16690	m
219853	258130		23 20 37.3	-75 37 57	-0.010	+0.05	7.1	0.4	3.0	F2 V		66 s		
220043	91184		23 20 37.9	+21 11 23	-0.001	-0.03	7.1	1.1	-0.1	K2 III		280 s		
220061	91186	62 τ Peg	23 20 38.1	+23 44 25	+0.002	0.00	4.60	0.17	1.2	A5 IV	+16	47 s		
220044	108554		23 20 40.0	+19 50 27	-0.001	0.00	7.1	0.1	1.7	A3 V		120 s		
220035	146652		23 20 40.7	-5 54 29	-0.007	-0.05	6.17	1.07		G5				
220104	52925		23 20 42.2	+44 34 49	+0.001	-0.02	7.9	1.1	-0.1	K2 III		370 s		
220105	52927		23 20 43.9	+44 06 59	-0.002	-0.02	6.10	0.1	1.7	A3 V	-2	75 s	16685	m
220078	108555		23 20 45.1	+15 02 53	+0.003	-0.03	7.6	0.1	1.4	A2 V	-16	170 s		
220088	73187	63 Peg	23 20 49.4	+30 24 54	+0.006	-0.06	5.59	1.50	-0.3	K5 III	-19	150 s		
220003	247838		23 20 49.8	-50 18 24	+0.003	-0.07	6.05	0.42	0.7	F6 III		98 mx		m
220062	128132		23 20 50.2	+2 27 23	-0.002	0.00	7.90	0.5	3.4	F5 V		79 s	16687	m
220063	165638		23 20 52.7	-10 31 56	+0.004	0.00	7.8	0.8	3.2	G5 IV		84 s		
220117	73190	12 And	23 20 53.1	+38 10 56	+0.011	-0.06	5.77	0.46	3.4	F5 V	-9	29 s		m

HD	SAO	Star Name	α 2000	δ 2000	μ(α)	μ(δ)	V	B-V	M$_v$	Spec	RV	d(pc)	ADS	Notes
220066	165639		23^{h}20^{m}56.5^s	-18°59'18"	+0.007	+0.04	7.9	0.6	4.4	G0 V		50 s		
220091	108560		23 20 57.6	+17 15 09	+0.007	+0.03	6.6	0.4	2.6	F0 V	-19	62 s		
220065	—		23 20 57.8	-18 32 11			7.20	0.9	3.2	G5 IV		63 s	16688	m
220167	20577		23 21 06.9	+60 28 17	-0.009	0.00	7.17	1.14		K1 IV	-65			
220119	128135		23 21 11.1	+2 11 50	-0.002	-0.03	8.0	1.1	0.2	K0 III		360 s		
220208	20580		23 21 11.6	+64 44 28	-0.002	+0.01	7.40	0.02	0.4	B9.5 V		250 s		
220109	146656		23 21 11.8	-8 40 28	+0.003	0.00	7.40	0.1	1.4	A2 V		160 s		m
220123	165643		23 21 14.9	-13 16 10	+0.001	+0.01	7.2	1.1	-0.1	K2 III		300 s		
220096	191840		23 21 15.3	-26 59 13	-0.001	-0.01	5.64	0.82	5.0	G4 V		11 s		
220148	73196		23 21 17.3	+38 34 56	0.000	-0.01	7.4	1.1	0.2	K0 III		280 s		
220134	146659		23 21 21.3	-4 40 22	-0.002	-0.03	6.7	0.1	0.6	A0 V		170 s		
220168	73197		23 21 23.0	+36 30 04	+0.006	+0.01	7.6	1.1	-0.1	K2 III		270 mx		
220112	231562		23 21 28.4	-44 54 37	+0.006	-0.04	7.6	1.3	-0.1	K2 III		200 mx		
220143	165646		23 21 28.6	-16 41 19	+0.002	+0.01	8.0	0.5	3.4	F5 V		84 s		
220182	52935		23 21 36.4	+44 05 53	+0.059	+0.22	7.36	0.80	6.1	K1 V	+2	18 s		
220006	258131		23 21 37.0	-71 34 26	-0.003	+0.01	7.6							
220253	20583		23 21 38.3	+62 49 29	+0.005	+0.02	7.8	1.1	0.2	K0 III		300 s		
220241	35351		23 21 42.0	+51 17 02	+0.001	+0.01	8.0	1.4	-0.3	K5 III		420 s		
220211	91206		23 21 49.1	+20 38 16	0.000	-0.02	7.3	1.6	-0.5	M2 III		370 s		
220172	165651		23 21 50.6	-9 45 42	-0.002	-0.01	7.68	-0.19	-1.7	B3 V n	+13	750 s		
220189	165652		23 21 51.4	-12 26 57	0.000	-0.01	7.9	1.1	0.2	K0 III		340 s		
220222	73205	64 Peg	23 21 54.7	+31 48 45	0.000	0.00	5.32	-0.11	-1.7	B3 V	+2	230 s	16702	m
220242	91208		23 21 58.0	+26 36 32	-0.007	-0.07	6.62	0.37	3.0	F2 V	+10	52 s		
220314	20585		23 22 02.4	+64 31 11	-0.004	0.00	7.95	0.05	0.6	A0 V		300 s		
220274	52942		23 22 02.5	+48 47 45	-0.003	+0.03	7.4	1.1	0.2	K0 III		270 s		
220254	91211		23 22 04.8	+28 41 41	+0.007	-0.07	6.6	0.4	3.0	F2 V	-26	53 s		
220300	35357		23 22 10.3	+56 20 54	0.000	0.00	7.92	0.15	0.0	B8.5 V		320 s		
220204	231569		23 22 13.9	-43 11 13	0.000	+0.02	7.7	1.1	0.2	K0 III		310 s		
220265	91212		23 22 13.9	+23 33 06	+0.002	-0.01	7.3	1.1	0.2	K0 III		260 s		
220158	255469		23 22 25.3	-65 15 44	+0.007	0.00	7.8	0.4	2.6	F0 V		110 s		
220288	91217		23 22 28.5	+25 55 06	+0.001	-0.02	6.34	1.54	-0.2	K3 III	+22	130 s	16707	m
220369	35361		23 22 32.3	+60 08 01	0.000	0.00	5.56	1.68	-2.3	K3 II	-12	260 s		
220278	165658	97 Aqr	23 22 39.1	-15 02 21	+0.007	+0.02	5.20	0.20		A3 n	-12	16 mn	16708	m
220318	91220	65 Peg	23 22 40.3	+20 49 43	+0.001	-0.01	6.20	-0.05	-0.1	B9.5 IV	-14	180 s		
220370	35363		23 22 41.1	+54 13 22	-0.005	+0.01	7.9	1.1	-0.1	K2 III		370 s		
220218	255473		23 22 41.5	-61 25 31	+0.003	+0.03	8.0	0.8	3.2	G5 IV		90 s		
220383	35364		23 22 47.2	+51 02 51	0.000	0.00	7.7	1.1	-0.1	K2 III		340 s		
220334	91222		23 22 48.6	+20 33 32	+0.022	-0.01	6.62	0.60	4.4	dG0	-22	26 s	16713	m
220295	191853		23 22 50.8	-23 27 08	+0.009	-0.06	7.3	0.3	2.6	F0 V		83 s		
220337	128150		23 22 53.0	+2 49 07	+0.001	0.00	6.9	0.1	0.6	A0 V		190 s		
220262	247846		23 22 54.2	-57 37 29	0.000	+0.01	7.6	0.5	3.2	G5 IV		77 s		
220263	255474		23 22 56.8	-60 03 21	+0.013	0.00	6.09	1.60		M1				
220321	191858	98 Aqr	23 22 58.0	-20 06 02	-0.009	-0.09	3.97	1.10	0.2	K0 III	-7	49 s		
220363	108580	66 Peg	23 23 04.4	+12 18 50	+0.002	-0.01	5.08	1.31	-0.2	K3 III	-4	110 s	16715	m
220339	165664		23 23 04.6	-10 45 52	+0.030	+0.26	7.81	0.88	6.3	K2 V	+36	19 ts		
220264	255476		23 23 09.4	-64 53 36	-0.012	-0.07	7.6	1.1	0.2	K0 III		160 mx		
220350	191861		23 23 12.8	-24 36 58	-0.004	-0.05	7.99	0.84		G0				
220376	146678		23 23 17.4	-9 23 06	+0.002	-0.01	7.8	0.5	3.4	F5 V		77 s		
220353	231583		23 23 18.2	-41 36 04	-0.001	+0.02	7.2	0.3	2.6	F0 V		80 s		
220444	35371		23 23 19.3	+52 06 10	-0.002	0.00	8.0	1.1	0.2	K0 III		340 s		
220330	247851		23 23 22.0	-55 33 11	-0.002	0.00	6.4	1.4	0.2	K0 III		110 s		
—	—		23 23 30.5	+0 08 59			6.50							m
220406	128156		23 23 31.8	+0 17 29	+0.004	+0.01	6.31	1.61		K2				m
220485	35378		23 23 38.3	+54 01 52	+0.002	+0.01	6.8	0.1	1.4	A2 V		120 s		
220358	255480		23 23 40.1	-61 01 02	+0.001	0.00	7.5	0.8	0.6	A0 V		74 s		
220400	214513		23 23 44.3	-31 06 39	0.000	+0.01	7.5	1.9	-0.5	M2 III		250 s		
220424	165668		23 23 44.4	-19 06 36	0.000	-0.02	8.0	1.1	0.2	K0 III		320 s		
220436	146683		23 23 45.1	-8 27 37	0.000	-0.02	7.20	1.1	0.2	K0 III	-9	250 s	16725	m
220401	231587		23 23 45.2	-43 07 28	+0.001	+0.02	6.10	1.46		K0				
220460	73223		23 23 47.3	+32 31 53	+0.018	+0.04	6.69	0.45	3.4	F5 V	+10	46 s		
220426	214515		23 23 47.5	-33 22 35	-0.007	+0.06	8.0	1.1	4.0	F8 V		30 s		
220550	20602		23 23 51.5	+65 20 36	+0.008	-0.02	7.4	0.4	3.0	F2 V		74 s		
220453	165670		23 23 52.2	-17 47 23	-0.001	0.00	8.0	0.5	4.0	F8 V		62 s		
220391	247853		23 23 52.3	-53 48 55	+0.005	-0.03	7.12	0.26	2.4	A7 V		77 mx		
220392	247854		23 23 54.2	-53 48 30	+0.007	-0.02	6.15	0.26	0.5	A7 III		130 s		m
220637	10727		23 23 55.4	+77 30 45	-0.013	+0.04	8.0	0.1	1.4	A2 V	-22	190 s		
220428	231590		23 23 58.6	-43 36 00	+0.001	+0.02	7.8	1.0	0.2	K0 III		330 s		
220455	191872		23 24 01.3	-27 16 52	0.000	0.00	8.00	0.00	0.4	B9.5 V		330 s	16728	m
220501	91233		23 24 03.2	+29 40 23	0.000	+0.01	7.1	0.1	0.6	A0 V	+2	200 s		
220466	191873		23 24 03.8	-21 46 29	-0.005	-0.08	6.60	0.5	3.4	F5 V	+25	44 s	16727	m
220476	146686		23 24 06.3	-7 33 03	+0.011	0.00	7.9	0.9	3.2	G5 IV		88 s		
220465	165672		23 24 07.6	-18 41 15	+0.010	+0.07	6.19	1.02	0.3	G5 III		88 mx		
220562	35386		23 24 07.8	+57 32 08	+0.001	0.00	6.80	0.26	-1.0	B5.5 V	-4	230 s	16731	m
220524	52978		23 24 08.8	+41 36 46	-0.001	0.00	6.6	1.6	-0.5	M2 III		260 s		
220440	247858		23 24 12.9	-51 53 29	+0.001	-0.03	5.6	2.0	-0.3	K5 III		73 s		

HD	SAO	Star Name	α 2000	δ 2000	μ(α)	μ(δ)	V	B-V	M_v	Spec	RV	d(pc)	ADS	Notes
			$23^h 24^m 14^s.4$	$+35°19'51''$	$-0^s.001$	$0''.00$								
220537	73229		23 24 14.4	+35 19 51	-0.001	0.00	7.7	1.1	0.2	K0 III		320 s		
220563	52981		23 24 14.6	+50 02 36	0.000	-0.01	8.0	0.0	0.4	B9.5 V		320 s		
220512	128160		23 24 16.5	+3 42 55	+0.001	-0.03	6.80	1.1	-0.1	K2 III	-14	240 s	16730	m
220527	108593		23 24 18.3	+18 58 57	+0.001	0.00	8.0	1.1	-0.1	K2 III		410 s		
220539	91239		23 24 18.8	+23 45 35	-0.007	-0.01	7.2	1.1	0.2	K0 III		250 s		
220367	258135		23 24 19.2	-72 09 54	+0.001	-0.03	6.9	0.5	4.0	F8 V		37 s		
221142	3904		23 24 20.8	+86 25 04	-0.020	+0.02	6.60	0.4	2.6	F0 V	-12	63 s	16759	m
220529	128162		23 24 21.6	+6 11 12	+0.002	-0.01	7.2	0.1	0.6	A0 V		210 s		
220506	214525		23 24 26.4	-37 12 04	-0.001	-0.01	7.35	0.10	1.4	A2 V		140 s		
220564	73233		23 24 30.8	+30 42 44	+0.004	+0.03	7.2	0.4	3.0	F2 V		71 s		
220575	52987		23 24 34.9	+41 06 46	+0.001	0.00	6.5	0.1	0.6	A0 V	-3	150 s		
220638	20611		23 24 35.6	+63 51 12	-0.006	-0.06	6.7	1.1	0.2	K0 III		190 s		
220554	146693		23 24 38.8	-1 44 28	+0.012	-0.02	7.9	0.9	3.2	G5 IV		88 s		
220616	35394		23 24 40.3	+53 24 43	+0.001	0.00	7.9	0.1	0.6	A0 V		270 s		
220507	247863		23 24 41.8	-52 42 08	-0.003	-0.17	7.8	0.0	4.4	G0 V		49 s		
220598	73239		23 24 43.2	+36 21 45	0.000	+0.01	7.02	-0.13	-1.6	B3.5 V	-20	520 s		
220582	91245		23 24 45.1	+25 29 48	0.000	+0.01	7.2	0.0	0.4	B9.5 V	+5	230 s		
220583	91246		23 24 46.4	+23 28 49	+0.002	+0.02	8.0	1.1	-0.1	K2 III		410 s		
220652	20614	4 Cas	23 24 50.1	+62 16 58	+0.001	-0.01	4.97	1.68	-0.3	gK5	-37	86 s		m
220599	73241	67 Peg	23 24 50.7	+32 23 05	+0.001	0.00	5.57	-0.11	-0.8	B9 III	+18	190 s		
220653	35401		23 24 55.5	+56 06 25	-0.001	+0.03	7.6	1.1	0.2	K0 III		270 s		
220639	35400		23 24 55.5	+55 05 09	-0.001	-0.01	7.5	0.9	-0.9	G8 II-III		450 s		
220579	191885		23 24 57.9	-22 30 10	+0.001	-0.03	7.7	1.5	0.2	K0 III		170 s		
220624	52991		23 25 02.5	+42 13 32	+0.003	-0.01	7.9	0.5	3.4	F5 V		78 s		
---	258138		23 25 05.0	-73 16 23	-0.003	+0.04	7.6							
220458	---		23 25 08.0	-73 17 06			8.0	0.8	3.2	G5 IV		90 s		
220606	214534		23 25 14.1	-36 11 08	+0.003	0.00	7.9	0.1	1.4	A2 V		180 s		
220572	247867		23 25 19.2	-56 50 57	+0.010	-0.02	5.59	1.07	0.2	K0 III	-19	110 s		
220520	258139		23 25 20.3	-70 54 37	-0.012	0.00	7.7	1.1	0.2	K0 III		320 s		
220668	73245		23 25 20.3	+36 09 53	-0.001	+0.01	7.7	0.1	0.6	A0 V		260 s		
220657	91253	68 υ Peg	23 25 22.7	+23 24 15	+0.014	+0.04	4.40	0.61	2.4	F8 IV	-11	22 s		
220621	214536		23 25 26.6	-30 00 26	0.000	-0.01	7.9	1.2	-0.1	K2 III		390 s		
220711	35406		23 25 29.4	+55 25 30	+0.001	-0.01	8.0	1.1	-0.1	K2 III		400 s		
220719	35407		23 25 35.8	+52 58 47	-0.002	-0.01	6.8	1.6	-0.5	M2 III	+4	280 s		
220781	20622		23 25 38.7	+68 57 26	-0.004	+0.04	7.0	1.1	0.2	K0 III	-47	210 s		
220694	91256		23 25 39.3	+24 57 14	+0.001	-0.01	6.8	0.1	1.7	A3 V		110 s		
220769	20623		23 25 43.4	+66 55 19	0.000	+0.04	7.6	0.1	0.6	A0 V		240 s		
220747	35408		23 25 44.5	+53 21 44	0.000	-0.01	7.9	1.4	-0.3	K5 III		400 s		
220721	53008		23 25 45.5	+45 20 54	+0.004	-0.01	8.0	0.6	4.4	G0 V		51 s		
220770	20625		23 25 50.1	+61 26 06	-0.004	+0.01	7.82	0.78	-4.8	A5 Ib		170 mx		
220748	53014		23 25 57.6	+45 23 06	+0.004	-0.04	7.8	0.8	3.2	G5 IV		82 s		
220750	73250		23 26 00.9	+39 20 23	+0.002	-0.01	7.01	-0.02	0.6	A0 V	-4	190 s		
220704	191900	99 Aqr	23 26 02.5	-20 38 31	-0.004	-0.05	4.39	1.47	-0.3	K5 III	+16	87 s		
220737	128179		23 26 08.4	+8 55 46	+0.001	0.00	7.6	1.1	0.2	K0 III		310 s		
220841	10732		23 26 11.7	+70 41 09	+0.006	+0.04	6.70	0.1	1.4	A2 V	-15	46 mx	16754	m
220717	191903		23 26 12.1	-27 25 27	+0.007	+0.03	6.6	1.2	0.2	K0 III		150 s		
220739	165692		23 26 19.6	-12 57 03	+0.001	0.00	7.8	0.1	1.4	A2 V		190 s		
220806	35414		23 26 20.4	+53 10 01	-0.001	-0.01	7.8	1.4	-0.3	K5 III		380 s		
220819	20626		23 26 20.8	+61 05 14	0.000	0.00	6.59	0.34	-2.0	F0 II	+1	500 s		
220700	255483		23 26 27.0	-63 14 13	-0.006	+0.02	7.80	0.1	1.7	A3 V		170 s		m
220773	128181		23 26 27.1	+8 38 41	+0.001	-0.19	7.1	0.7	4.4	G0 V	-42	29 mx		
220807	53019		23 26 27.4	+40 23 13	0.000	-0.14	6.7	0.9	3.2	G5 IV		50 s		
220782	108606		23 26 28.8	+17 30 35	-0.001	0.00	7.8	1.1	-0.1	K2 III		380 s		
220793	---		23 26 30.9	+31 44 50	+0.001	+0.01	8.0	0.1	0.6	A0 V		300 s		
220774	146707		23 26 34.2	-6 36 28	+0.004	-0.02	7.2	1.1	-0.1	K2 III		290 s		
220766	191907		23 26 35.3	-21 44 27	+0.004	+0.01	6.60	1.1	0.2	K0 III	+11	190 s	16753	m
220868	20630		23 26 35.3	+64 18 17	-0.003	0.00	8.0	0.1	1.7	A3 V		170 s		
220729	247874	o Gru	23 26 36.4	-52 43 18	+0.004	+0.13	5.52	0.40	1.9	F3 IV	+18	51 s		
220822	53026		23 26 38.0	+42 33 19	-0.002	-0.07	7.9	0.5	4.0	F8 V		60 s		
220834	53028		23 26 43.2	+41 02 44	+0.003	-0.04	7.7	0.2	2.1	A5 V		130 s		
220796	128184		23 26 43.5	+2 28 42	-0.001	+0.02	6.8	1.1	0.2	K0 III		210 s		
220758	247875		23 26 51.1	-59 38 40	+0.011	-0.06	7.8	0.3	3.4	F5 V		77 s		
220679	258143		23 26 54.9	-73 50 13	+0.009	+0.01	7.8	0.9	3.2	G5 IV		82 s		
220846	91271		23 26 55.7	+25 24 10	-0.001	0.00	7.5	0.2	2.1	A5 V		120 s		
220825	128186	8 κ Psc	23 26 55.8	+1 15 20	+0.006	-0.09	4.94	0.03		A2 p	-3	30 t		m
220811	165698		23 26 56.6	-15 14 52	-0.001	-0.01	7.10	0.1	1.7	A3 V		120 s	16758	m
220871	53035		23 26 59.9	+42 12 19	+0.006	+0.03	7.8	0.8	3.2	G5 IV		84 s		
221525	3916		23 27 01.1	+87 18 27	+0.112	+0.02	5.58	0.23	2.6	F0 V	-11	39 s		
220812	191909		23 27 01.1	-22 43 24	+0.003	+0.01	7.3	1.1	0.2	K0 III		210 s		
221344	3912		23 27 04.7	+86 00 29	+0.002	0.00	7.8	0.9	3.2	G5 IV		83 s		
220759	255486		23 27 07.1	-66 34 53	-0.006	-0.02	6.45	1.47		K0				
220885	53039	13 And	23 27 07.2	+42 54 43	+0.008	+0.02	5.6	0.0		B9	-9			
220718	258144		23 27 07.3	-72 52 51	-0.017	+0.03	8.0	0.7	4.4	G0 V		53 s		
220802	247880		23 27 08.9	-50 09 26	+0.003	+0.01	6.20	-0.08		B9	-1			
220780	255488		23 27 09.1	-62 44 15	-0.002	0.00	6.6	2.1	0.2	K0 III		40 s		

HD	SAO	Star Name	α 2000	δ 2000	μ(α)	μ(δ)	V	B-V	M_v	Spec	RV	d(pc)	ADS	Notes
220908	35430		23^{h}27^{m}09.9	+54°51'59"	+0.010	+0.03	8.0	0.5	4.0	F8 V		62 s		
220803	247881		23 27 10.6	-50 16 49	+0.010	-0.04	7.70	1.1	-0.1	K2 III		150 mx		m
220859	146712		23 27 14.4	-2 38 13	+0.010	-0.08	7.6	1.1	0.2	K0 III		94 mx		
220858	128188	9 Psc	23 27 14.6	+1 07 21	+0.003	-0.03	6.25	1.02	0.3	gG7	-7	130 s		
220790	247882		23 27 14.8	-58 28 34	+0.006	+0.08	5.63	0.98	0.2	gK0	-11	120 s		
220974	10737		23 27 16.4	+70 21 36	+0.023	0.00	5.6	0.1	1.4	A2 V	-3	68 s		
220940	20640		23 27 17.0	+65 37 16	-0.001	-0.07	7.00	1.1	0.2	K0 III		200 mx	16764	m
221100	10740		23 27 18.1	+79 47 42	+0.010	+0.01	7.5	0.1	1.4	A2 V		160 mx		
220888	128190		23 27 18.3	+8 04 00	+0.001	+0.01	7.7	1.1	0.2	K0 III		310 s		
220910	53043		23 27 19.4	+43 51 49	0.000	0.00	8.0	1.4	-0.3	K5 III	-14	420 s		
220909	53044		23 27 20.3	+45 22 04	-0.003	-0.03	7.7	0.1	0.6	A0 V		250 s		
220963	20641		23 27 25.1	+63 55 44	0.000	-0.01	7.4	1.1	0.2	K0 III	-6	250 s		
220876	165705		23 27 26.0	-12 55 49	0.000	-0.03	7.4	1.6	-0.5	M2 III	-19	390 s		
220881	191919		23 27 33.3	-27 16 40	+0.009	+0.03	7.44	0.28	0.6	F0 III		140 mx		
220894	214557		23 27 36.8	-33 37 40	-0.024	-0.09	7.4	0.6	4.0	F8 V		45 s		
220933	91278	69 Peg	23 27 40.1	+25 10 02	+0.002	-0.03	5.90	-0.06	-0.6	A0 V	-16	200 s		
220949	53047		23 27 41.4	+44 47 07	-0.001	-0.02	7.5	0.8	3.2	G5 IV		70 s		
221071	10742		23 27 42.8	+74 07 12	+0.013	+0.04	7.50	0.4	2.6	F0 V		93 s	16775	m
220924	128194		23 27 43.3	+1 53 12	+0.005	0.00	7.4	0.5	3.4	F5 V		64 s		
221488	3917		23 27 43.5	+86 33 33	-0.044	+0.05	7.3	0.5	3.4	F5 V		60 s		
220951	73272		23 27 49.5	+32 57 32	-0.003	-0.02	7.4	0.2	2.1	A5 V	-9	120 s		
220896	247888		23 27 50.4	-50 20 38	-0.001	0.00	7.20	0.91		G5				m
220999	35441		23 27 54.0	+59 41 43	-0.007	-0.01	7.56	0.24	0.5	A7 III	-26	260 s		
220935	165707		23 27 56.3	-15 17 59	+0.001	-0.02	6.8	1.4	-0.3	K5 III		260 s		
220954	128196	10 θ Psc	23 27 57.9	+6 22 44	-0.008	-0.04	4.28	1.07	0.0	K1 III	+6	72 s		
221036	20646		23 27 59.8	+63 00 53	-0.009	+0.01	8.0	0.9	3.2	G5 IV		89 s		
220929	214561		23 28 00.5	-35 32 40	0.000	+0.01	6.32	1.20		K2				
219765	258970	τ Oct	23 28 04.0	-87 28 57	+0.032	+0.01	5.49	1.27	-0.1	K2 III		110 s		
220965	128197		23 28 04.2	+8 37 54	-0.002	+0.01	7.9	1.1	-0.1	K2 III		390 s		
220957	165708		23 28 05.0	-11 26 59	+0.007	-0.02	6.5	0.6	4.4	G0 V		26 s		
220938	214562		23 28 06.4	-30 50 14	-0.003	-0.02	7.8	1.7	0.2	K0 III		130 s		
220958	165710		23 28 07.0	-12 12 44	-0.002	0.00	7.6	1.6	-0.5	M2 III		420 s		
221038	20650		23 28 10.2	+61 28 11	-0.002	-0.02	8.05	0.17	-2.0	A7 II		1000 s		
221039	35444		23 28 11.1	+60 03 28	+0.003	+0.02	7.41	1.05	-2.1	K0 II		800 s		
221023	35443		23 28 12.1	+51 21 51	+0.001	+0.02	7.7	1.1	0.2	K0 III		300 s		
220989	108625		23 28 12.8	+18 32 18	0.000	0.00	7.9	0.4	2.6	F0 V		110 s		
220918	247889		23 28 13.2	-56 26 11	+0.010	-0.05	7.80	0.4	3.0	F2 V		91 s		m
221072	20651		23 28 17.8	+63 23 26	-0.003	-0.02	8.0	0.5	3.4	F5 V		79 s		
221024	53057		23 28 20.8	+47 16 28	-0.001	+0.02	8.0	1.4	-0.3	K5 III		420 s		
221011	108627		23 28 23.4	+19 53 06	-0.001	-0.02	6.7	0.4	2.6	F0 V	-14	66 s		
220978	191931		23 28 25.2	-25 25 13	-0.001	0.00	6.9	0.7	1.7	A3 V		51 s		
221057	53059		23 28 27.2	+45 41 34	+0.002	-0.02	7.8	1.1	0.2	K0 III		320 s		
221026	128201		23 28 31.5	+9 25 38	+0.006	+0.01	7.9	0.5	3.4	F5 V		78 s		
221074	53063		23 28 33.2	+48 38 40	0.000	+0.02	7.8	0.4	2.6	F0 V		110 s		
--	73282		23 28 40.1	+39 41 23	+0.001	-0.01	8.0	1.8						
221102	53065		23 28 45.3	+42 24 41	-0.002	0.00	7.7	0.1	0.6	A0 V		260 s		
221043	146725		23 28 46.0	-0 50 09	-0.010	-0.12	7.60	0.5	4.0	F8 V		53 s	16781	m
221044	146724		23 28 46.6	-5 23 21	0.000	-0.01	7.1	1.1	-0.1	K2 III		270 s		
220986	247893		23 28 49.2	-59 26 13	+0.004	-0.01	7.0	1.7	0.2	K0 III		82 s		
221029	214573		23 28 50.4	-31 23 21	-0.005	-0.06	7.9	1.6	0.2	K0 III		160 s		
221124	35454		23 28 51.8	+53 40 02	+0.006	+0.03	7.1	1.1	0.2	K0 III	-28	230 s		
221215	10754		23 28 57.6	+75 13 31	-0.004	0.00	6.6	0.1	0.6	A0 V	-17	150 s		
221081	146729		23 29 00.5	-9 15 58	-0.004	-0.02	6.18	1.44		K0				
221051	231622		23 29 00.6	-44 29 52	+0.003	-0.01	6.43	1.18	0.2	gK0		140 s		
221006	255497		23 29 00.9	-63 06 38	+0.007	-0.01	5.68	-0.17		A pe	+15			
221114	108636		23 29 02.1	+16 00 45	0.000	0.00	7.0	0.1	1.4	A2 V	+3	130 s		
221113	91295		23 29 05.5	+23 02 52	-0.003	-0.09	6.4	0.9	0.2	G9 III	+20	150 mx		
--	--		23 29 05.7	+59 46 15			7.80	0.00	-1.9	B6 III		680 s		
221083	191940		23 29 06.2	-28 15 52	-0.003	-0.05	7.8	0.2	3.4	F5 V		74 s		
221115	108638	70 Peg	23 29 09.1	+12 45 38	+0.004	+0.03	4.55	0.94	0.3	G8 III	-15	70 s		
221168	35461		23 29 12.0	+55 08 12	0.000	+0.01	7.8	1.1	0.2	K0 III		310 s		
221069	247899		23 29 16.6	-54 30 05	+0.001	+0.02	6.5	1.2	0.2	K0 III		130 s		
221160	53072		23 29 16.7	+42 21 36	-0.001	0.00	7.7	0.0	0.4	B9.5 V		280 s		
220522	258973		23 29 18.1	-85 42 33	+0.020	0.00	7.70	1.1	0.2	K0 III		320 s		m
221133	91298		23 29 20.9	+25 48 44	+0.001	+0.01	7.90	1.1	-0.1	K2 III	-26	400 s		
221099	247901		23 29 25.4	-52 40 44	+0.003	0.00	7.0	0.5	3.2	G5 IV		58 s		
221147	146733		23 29 26.8	-1 47 28	-0.001	-0.02	6.5	1.1	0.2	K0 III	+12	180 s		
221169	53074		23 29 26.8	+46 27 53	+0.002	-0.01	6.7	0.1	1.4	A2 V		110 s		
221070	255500		23 29 28.4	-60 44 55	+0.003	+0.01	7.2	1.1	3.4	F5 V		23 s		
221170	73293		23 29 28.7	+30 25 58	0.000	-0.05	7.66	1.12	3.0	G2 IV	-119	44 s		
221146	146735		23 29 30.2	-1 02 09	-0.001	-0.03	7.1	0.6	4.4	G0 V	-14	35 s		
221187	35469		23 29 31.1	+51 41 08	-0.002	+0.01	8.00	1.6	-0.5	M2 III		470 s		m
221148	146736		23 29 32.0	-4 31 58	+0.012	-0.22	6.25	1.09	-0.2	K3 III	-25	36 mx		
221188	53076		23 29 36.9	+48 22 52	+0.003	+0.01	7.8	0.1	0.6	A0 V		260 s		
221203	53077		23 29 39.2	+49 06 27	+0.001	+0.02	8.0	1.1	0.2	K0 III		340 s		

HD	SAO	Star Name	α 2000	δ 2000	μ(α)	μ(δ)	V	B–V	M_V	Spec	RV	d(pc)	ADS	Notes
221190	73298		23ʰ29ᵐ45.8ˢ	+31°42'21"	0.000ˢ	−0.02"	7.70	0.1	1.7	A3 V		160 s		m
221234	53083		23 29 50.7	+43 56 26	−0.001	−0.01	7.9	0.1	0.6	A0 V		270 s		
221237	35476		23 29 52.1	+58 32 54	+0.002	0.00	7.08	0.02	1.2	A1 V	−4	180 s	16795	m
221217	73299		23 29 52.4	+35 00 51	−0.001	0.00	7.9	1.4	−0.3	K5 III		430 s		
221204	91306		23 29 54.2	+24 46 21	−0.001	−0.01	7.2	1.1	0.2	K0 III		260 s		
221218	91307		23 30 01.8	+21 32 58	+0.001	−0.02	7.1	0.0	0.4	B9.5 V		220 s		
221253	35478		23 30 01.8	+58 32 56	+0.002	+0.01	4.91	−0.12	−1.7	B3 V	−16	180 s	16795	AR Cas, m, v
221246	53088		23 30 07.3	+49 07 59	+0.003	0.00	6.3	1.1	−0.1	K2 III	+6	190 s		
221235	128211		23 30 13.3	+5 00 46	0.000	+0.03	7.3	1.1	0.2	K0 III		260 s		
221226	191954		23 30 15.8	−21 52 59	+0.004	+0.03	7.5	0.5	3.0	F2 V		68 s		
221302	20667		23 30 16.3	+68 14 34	0.000	+0.02	7.8	1.1	0.2	K0 III		290 s		
221248	128214		23 30 21.8	+6 25 31	0.000	0.00	7.9	0.1	0.6	A0 V		280 s		
221264	73306		23 30 26.1	+30 49 54	+0.005	−0.03	7.28	0.50	3.4	dF5	−21	72 mn	16800	m
221271	108650		23 30 31.5	+16 33 49	+0.001	0.00	7.3	0.9	3.2	G5 IV		66 s		
221325	20669		23 30 31.9	+64 23 33	+0.005	+0.02	7.3	0.2	2.1	A5 V		110 s		
221241	231637		23 30 33.9	−41 59 21	−0.007	−0.09	6.7	1.2	0.2	K0 III		150 s		
––	231638		23 30 33.9	−41 59 21	−0.005	−0.08	6.8	1.4	0.2	K0 III		120 s		
221405	10764		23 30 34.0	+77 53 40	−0.001	+0.01	7.1	0.1	0.6	A0 V	−1	190 s		
221240	214591		23 30 34.1	−36 57 28	+0.003	0.00	7.7	0.3	2.1	A5 V		110 s		
221293	73308		23 30 39.6	+38 39 44	+0.003	+0.02	6.05	0.99	0.2	K0 III	−9	150 s		
221257	191958		23 30 39.9	−24 11 45	+0.004	+0.04	7.5	0.4	4.4	G0 V	−25	42 s		
221273	128215		23 30 40.3	+0 52 37	−0.002	−0.03	7.5	1.1	−0.1	K2 III		330 s		
221272	128216		23 30 40.6	+5 14 58	−0.001	0.00	7.2	0.1	1.7	A3 V		120 s	16803	m
221279	91311		23 30 43.4	+22 02 50	+0.002	0.00	7.6	0.4	2.6	F0 V		100 s		
221304	73309		23 30 46.8	+33 50 09	0.000	−0.01	7.3	1.1	−0.1	K2 III		300 s		
221334	20673		23 30 49.0	+62 17 19	+0.002	0.00	7.65	0.03		A0				
221282	165730		23 30 54.2	−15 58 58	+0.006	−0.04	7.9	0.5	4.0	F8 V		60 s		
221308	146748		23 31 00.9	−6 17 18	−0.001	−0.03	6.39	1.26		K0				
221231	255503		23 31 02.7	−69 04 35	+0.035	−0.11	7.20	0.7	4.4	G0 V		36 s		m
221327	108654		23 31 04.9	+18 46 44	−0.001	−0.01	7.60	0.1	1.7	A3 V	−3	150 s	16808	m
221318	128223		23 31 06.5	+2 21 48	+0.006	−0.04	7.1	0.5	3.4	F5 V		55 s		
221345	73311	14 And	23 31 17.2	+39 14 11	+0.025	−0.08	5.22	1.02	0.2	K0 III	−59	97 s		
221269	255506		23 31 18.8	−64 22 17	+0.020	+0.01	7.9	0.2	4.0	F8 V		61 s		
221377	35501		23 31 19.6	+52 24 39	+0.012	−0.02	7.40	0.5	3.4	F5 V		63 s	16810	m
221287	247912		23 31 20.3	−58 12 35	+0.025	0.00	7.5	0.7	4.0	F8 V		41 s		
221348	128225		23 31 20.6	+9 45 42	+0.002	+0.01	7.9	1.1	−0.1	K2 III		390 s		
221354	35498		23 31 22.0	+59 09 56	+0.141	+0.11	6.74	0.83	5.9	K0 V	−25	15 s		
221393	35503		23 31 23.2	+59 27 45	−0.001	0.00	7.29	1.71	−0.3	K5 III		250 s		
221355	108659		23 31 24.5	+16 13 05	+0.004	−0.04	7.9	0.5	4.0	F8 V		59 s	16812	m
221378	53114		23 31 25.4	+49 30 31	+0.002	+0.04	7.4	1.1	0.2	K0 III		270 s		
221323	231642		23 31 26.7	−44 50 37	+0.003	−0.01	6.02	1.02	0.2	K0 III		140 s		
221315	247914		23 31 28.4	−56 08 32	−0.005	+0.01	7.2	1.4	0.2	K0 III		150 s		
221364	91318		23 31 29.9	+28 39 58	+0.001	−0.01	6.6	1.1	0.2	K0 III	−5	190 s		
221356	146752		23 31 31.3	−4 05 14	+0.012	−0.18	6.49	0.54	4.4	dG0	−11	26 ts		
221380	73315		23 31 32.0	+33 41 55	−0.005	−0.01	8.0	0.1	1.4	A2 V		190 mx		
221829	3926		23 31 34.1	+86 10 41	0.000	−0.01	7.2	0.2	2.1	A5 V	−23	110 s		
221412	35509		23 31 37.4	+54 08 28	+0.001	−0.01	7.7	0.0	0.0	B8.5 V		330 s		
221537	10772		23 31 39.4	+77 49 11	+0.012	+0.02	7.02	0.05		A0	−4			
221395	91320		23 31 41.4	+24 06 41	+0.002	−0.01	7.9	1.1	0.2	K0 III		340 s		
221357	191970	100 Aqr	23 31 41.8	−21 22 10	0.000	+0.04	6.29	0.33	2.6	F0 V	−8	51 s		
221394	91321		23 31 42.9	+28 24 13	−0.002	0.00	6.41	0.01		A0 pe	−6			
221439	35512		23 31 43.5	+59 05 33	+0.001	0.00	7.36	1.34	−4.4	K0 Ib		2200 s		
221343	247915		23 31 44.0	−53 46 12	+0.011	−0.05	7.8	0.7	3.2	G5 IV		84 s		
221366	165738		23 31 44.9	−14 48 23	−0.001	+0.02	8.0	1.1	−0.1	K2 III		410 s		
221385	191971		23 31 48.1	−21 14 59	−0.005	−0.02	7.3	0.7	2.6	F0 V		52 s		
221422	53118		23 31 50.2	+44 04 23	−0.001	+0.03	7.1	1.1	−0.1	K2 III		260 s		
221398	165740		23 31 51.0	−11 56 54	+0.001	+0.01	7.9	1.1	0.2	K0 III		340 s		
221370	231645		23 31 51.4	−41 45 09	−0.001	+0.01	6.7	1.0	0.2	K0 III		200 s		
221373	231643		23 31 53.2	−46 43 18	+0.010	+0.02	7.9	0.4	3.0	F2 V		30 s		
221387	191972		23 31 55.8	−25 44 38	0.000	+0.04	6.6	0.8	0.2	K0 III		190 s		
221399	165741		23 31 56.1	−14 56 12	+0.003	−0.02	7.9	1.1	−0.1	K2 III		390 s		
221442	73319		23 31 57.3	+38 18 23	−0.001	−0.02	7.60	1.1	−0.1	K2 III		350 s		m
221409	146756	13 Psc	23 31 57.4	−1 05 09	+0.001	+0.02	6.38	1.18	0.0	gK1	−23	170 s		
221440	53122		23 31 57.5	+46 47 05	+0.002	0.00	7.9	0.1	0.6	A0 V		270 s		
221214	258975		23 31 57.8	−80 49 42	0.000	+0.01	7.4	1.9	0.2	K0 III		78 s		
221459	53123		23 32 00.1	+45 53 18	+0.001	−0.04	7.6	1.1	−0.1	K2 III		320 s		
221360	247918		23 32 04.7	−59 00 14	−0.001	−0.02	7.8	1.3	0.2	K0 III		230 s		
221512	20685		23 32 11.3	+65 44 22	+0.009	+0.02	6.6	0.4	2.6	F0 V		63 s		
221427	165746		23 32 12.4	−11 32 42	0.000	−0.03	6.7	1.4	−0.3	K5 III		250 s		
221445	128231		23 32 12.6	+7 05 06	−0.002	−0.05	6.81	0.49	3.7	dF6	−12	41 s	16819	m
221431	214608		23 32 18.2	−33 40 29	0.000	+0.07	7.9	1.4	2.1	A5 V		27 s		
221491	73325		23 32 24.5	+34 57 09	+0.001	0.00	6.65	−0.06		A0	+12			
221479	108669		23 32 24.6	+17 24 15	+0.004	−0.02	7.20	0.4	3.0	F2 V		69 s	16821	m
221513	35523		23 32 24.6	+53 17 51	−0.001	0.00	8.0	0.1	0.6	A0 V		280 s		
221472	165754		23 32 25.6	−10 59 56	−0.001	+0.01	6.7	0.8	3.2	G5 IV		51 s		

HD	SAO	Star Name	α 2000	δ 2000	μ(α)	μ(δ)	V	B−V	M_v	Spec	RV	d(pc)	ADS	Notes
221410	255511		23ʰ32ᵐ27ˢ.9	−63°39′19″	+0ˢ.001	−0″.04	7.3	1.1	0.2	K0 III		260 s		
221493	91333		23 32 28.9	+23 50 38	0.000	−0.01	6.5	1.4	−0.3	K5 III		230 s		
221583	10774		23 32 29.4	+72 00 03	−0.001	−0.01	6.7	0.1	1.4	A2 V		110 s		
221515	53132		23 32 32.4	+48 22 23	+0.002	+0.04	8.0	0.0	0.4	B9.5 V		310 s		
221455	247921		23 32 32.8	−50 06 57	+0.003	0.00	7.6	0.5	1.4	A2 V		90 s		
—	20688		23 32 34.8	+64 08 00	−0.005	+0.01	7.4	1.9						
221538	35526		23 32 34.9	+53 41 12	+0.001	+0.01	7.0	1.1	0.2	K0 III	−19	220 s		
221494	108671		23 32 35.5	+10 58 40	+0.005	0.00	7.8	0.1	1.7	A3 V		120 mx		
221484	—		23 32 37.2	−17 36 55			7.9	1.1	0.2	K0 III		340 s		
221473	231655		23 32 40.5	−45 07 19	+0.003	0.00	7.00	0.1	1.4	A2 V		130 s		m
221568	35532		23 32 47.4	+57 54 20	+0.002	0.00	7.61	0.11		A0 p				
221573	53136		23 32 53.4	+47 00 44	−0.002	−0.03	7.9	0.9	3.2	G5 IV		85 s		
221585	20692		23 32 53.5	+63 09 21	+0.060	+0.05	7.44	0.74	3.2	G8 IV	+9	36 mx		
221530	146765		23 32 55.2	−3 01 00	0.000	−0.02	7.0	0.1	0.6	A0 V		190 s		
221588	53137		23 32 56.4	+46 07 24	+0.006	−0.01	6.8	1.6	−0.5	M2 III		270 s		
221507	214615	β Scl	23 32 58.0	−37 49 07	+0.007	+0.02	4.37	−0.09		B9 p	+2			
221587	—		23 33 01.7	+50 21 18	+0.001	+0.01	8.0	1.1	0.2	K0 III		340 s		
221611	20696		23 33 05.4	+63 07 22	−0.014	0.00	8.0	0.9	3.2	G5 IV		89 s		
221561	146771		23 33 11.9	−6 24 03	+0.008	+0.01	7.8	0.5	4.0	F8 V		57 s		
221625	—		23 33 13.0	+57 24 31	−0.013	−0.03	7.40						16828	m
221564	165761		23 33 13.8	−12 36 33	−0.001	−0.02	7.08	0.15	0.6	A0 V		160 s		
221565	191988	101 Aqr	23 33 16.4	−20 54 52	−0.001	+0.02	4.71	0.02		A0 n	+15	25 mn		m
221420	258154		23 33 19.4	−77 23 07	+0.005	+0.01	5.81	0.68		K0				
221639	20699		23 33 20.4	+60 24 40	+0.012	+0.04	7.18	0.92	6.1	K1 V	0	15 s		m
221599	128246		23 33 22.9	+4 38 23	0.000	+0.01	8.0	1.4	−0.3	K5 III		450 s		
221613	53142		23 33 23.8	+42 50 48	+0.021	+0.18	7.14	0.58	4.4	G0 V		35 s		
221697	10780		23 33 25.0	+74 13 34	−0.004	+0.01	8.0	1.1	0.2	K0 III	−24	330 s		
221578	214622		23 33 25.8	−31 17 22	+0.001	+0.01	7.2	1.2	0.2	K0 III		180 s		
221615	91340	71 Peg	23 33 27.9	+22 29 55	+0.001	−0.01	5.32	1.60	−0.5	M4 III	+3	150 s		
221605	91341		23 33 28.0	+24 20 34	+0.006	+0.03	6.9	0.2	2.1	A5 V		91 s		
221600	146774		23 33 28.3	−4 24 05	+0.001	−0.01	7.1	0.5	3.4	F5 V		55 s		
221642	53144		23 33 29.0	+41 38 16	+0.007	+0.03	7.1	0.9	3.2	G5 IV		61 s		
221552	247924		23 33 29.9	−57 03 12	+0.004	+0.03	7.5	0.5	2.6	F0 V		72 s		
221627	108677		23 33 34.7	+17 49 09	+0.019	+0.04	6.70	0.7	4.4	G0 V	−7	29 s	16830	m
221670	20701		23 33 35.1	+60 28 07	−0.002	0.00	7.34	0.99	5.6	G8 V	+2	16 s		m
221671	35546		23 33 36.1	+60 03 04	+0.001	−0.01	7.61	0.08	−2.8	A0 II	−13	1100 s		
221644	91344		23 33 40.0	+29 52 30	+0.006	−0.02	7.9	0.5	3.4	F5 V		78 s		
221629	146776		23 33 41.2	−2 14 41	−0.001	−0.03	7.8	0.1	1.4	A2 V		190 s		
221661	53147		23 33 42.6	+45 03 30	−0.003	+0.02	6.24	0.98	3.2	G5 IV	+7	30 s		
221609	214624		23 33 45.8	−36 15 46	+0.011	+0.05	7.20	0.4	2.6	F0 V		83 s		m
221672	53149		23 33 49.2	+46 40 52	+0.002	−0.04	6.8	1.1	0.2	K0 III		200 s		
221662	91349		23 33 55.4	+20 50 27	0.000	−0.02	6.06	1.73	−0.5	M2 III	+5	170 s		
221673	73341	72 Peg	23 33 57.0	+31 19 31	+0.004	−0.01	4.98	1.38	−0.3	K4 III	−24	110 s	16836	m
221681	53151		23 33 58.0	+43 54 16	+0.003	+0.02	6.9	0.0	0.4	B9.5 V		190 s		
221647	191996		23 34 00.0	−24 51 35	+0.004	+0.03	7.5	0.8	4.0	F8 V		36 s		
221665	165770		23 34 06.6	−14 13 45	+0.003	0.00	8.0	1.1	0.2	K0 III		370 s		
221774	10785		23 34 07.4	+72 54 52	+0.007	−0.01	7.3	0.1	1.4	A2 V		140 s		
221675	146780	14 Psc	23 34 08.9	−1 14 51	+0.007	−0.01	5.87	0.30		A m	−3			
221711	35554		23 34 09.1	+55 29 20	−0.001	−0.01	7.53	−0.09	−1.6	B3.5 V	−5	600 s		
221701	128260		23 34 25.0	+5 28 13	+0.008	0.00	6.8	0.4	3.0	F2 V		58 s		
221668	255518		23 34 32.8	−60 52 06	+0.002	+0.01	7.2	2.0	0.2	K0 III		63 s		
221638	258158		23 34 34.3	−70 57 46	+0.012	+0.09	7.6	0.5	4.0	F8 V		52 s		
221742	73342		23 34 34.3	+34 21 22	+0.007	+0.01	7.3	1.1	0.2	K0 III		220 mx		
221756	73346	15 And	23 34 37.3	+40 14 11	−0.002	−0.04	5.59	0.10	−0.2	A2 III	+13	140 s		
221758	73345	73 Peg	23 34 38.1	+33 29 51	0.000	+0.03	5.63	1.03	0.2	K0 III	−3	120 s		
221757	73347		23 34 41.0	+37 48 01	+0.009	+0.11	7.3	0.5	4.0	F8 V		46 s		
221766	73349		23 34 44.1	+36 12 17	+0.002	+0.02	7.5	1.1	0.2	K0 III		290 s		
221781	35563		23 34 46.4	+52 36 37	+0.001	0.00	7.9	0.0	0.4	B9.5 V		300 s		
221776	73351		23 34 46.6	+38 01 26	0.000	+0.01	6.18	1.58	−0.3	K5 III	−12	180 s	16843	m
221745	165780		23 34 49.2	−15 14 45	+0.003	−0.08	5.96	1.36	0.2	K0 III		73 s		
221784	53162		23 34 49.9	+43 12 06	+0.001	0.00	7.3	0.1	0.6	A0 V		220 s		
—	53163		23 34 50.1	+46 51 38	−0.003	+0.01	8.0	2.2				130 s		
221736	231672		23 34 50.2	−42 41 05	+0.004	−0.01	6.9	−0.2	1.4	A2 V		240 s		
221750	214634		23 34 54.0	−35 05 06	+0.002	−0.04	7.1	0.7	0.2	K0 III		210 s		
221785	73352		23 34 58.9	+31 50 08	−0.001	+0.01	8.0	0.1	1.4	A2 V				
221861	10790		23 34 58.9	+71 38 32	+0.001	+0.01	5.84	1.80	−4.4	K0 Ib	−3	500 s		
221786	91358		23 34 59.4	+24 25 35	−0.004	0.00	6.6	1.1	0.2	K0 III		190 s		
221738	247933		23 35 01.9	−56 49 31	−0.005	0.00	6.9	0.4	3.0	F2 V		53 s		
221777	146786		23 35 03.9	−7 40 38	0.000	+0.03	7.32	1.30	−0.3	K4 III	+9	330 s		
221760	231675	ι Phe	23 35 04.4	−42 36 54	+0.004	+0.01	4.71	0.08		A p	+19	25 mn		m
221800	108693		23 35 05.4	+16 28 48	+0.002	−0.05	8.0	1.1	−0.1	K2 III		410 s		
221862	20713		23 35 09.3	+67 29 31	+0.006	+0.01	7.4	1.1	0.2	K0 III	−6	250 s		
221813	108694		23 35 10.7	+12 39 44	−0.002	−0.04	7.8	0.1	1.7	A3 V		160 s		
221740	255520		23 35 12.6	−64 41 22	+0.003	+0.01	7.40	0.4	2.6	F0 V		91 s		m
221801	146789		23 35 14.5	−3 51 15	+0.004	+0.03	7.5	0.4	3.0	F2 V		79 s		

HD	SAO	Star Name	α 2000	δ 2000	μ(α)	μ(δ)	V	B-V	M_V	Spec	RV	d(pc)	ADS	Notes
221814	108696		23^{h}35^{m}15.8^s	+10° 57′ 31″	0.000	-0.01	7.4	1.1	0.2	K0 III		270 s		
221848	—		23 35 19.0	+50 18 43	0.000	+0.01	7.1	0.0	0.4	B9.5 V		210 s		
221849	53166		23 35 22.2	+44 07 25	-0.005	+0.01	7.8	0.1	1.7	A3 V		160 s		
221851	73357		23 35 25.4	+31 09 40	-0.016	-0.28	7.9	0.8	3.2	G5 IV		63 mx		
221832	128269		23 35 26.7	+8 31 11	0.000	-0.01	6.8	1.4	-0.3	K5 III		270 s		
221863	35571		23 35 27.0	+52 17 29	+0.009	+0.03	7.0	1.1	0.2	K0 III	+2	200 mx		
221825	165786		23 35 27.6	-10 33 19	+0.002	0.00	7.2	1.4	-0.3	K5 III		320 s		
221815	165785		23 35 27.8	-16 35 01	-0.001	-0.03	7.6	1.1	0.2	K0 III		310 s		
221833	128270		23 35 28.4	+1 18 47	-0.003	-0.02	6.6	1.1	0.2	K0 III	+7	190 s		
221830	73358		23 35 28.8	+31 01 03	+0.043	+0.26	6.85	0.60	4.7	G2 V	-103	27 s		
221850	73359		23 35 31.0	+32 11 50	+0.002	-0.02	6.6	1.1	0.2	K0 III		190 s		
221835	146795		23 35 31.8	-7 27 52	0.000	+0.02	6.39	0.88	0.3	gG5	+5	170 s		
221853	128272		23 35 36.0	+8 22 58	+0.005	-0.03	7.3	0.4	2.6	F0 V		89 s		
221866	108698		23 35 37.3	+14 35 38	+0.005	+0.01	7.6	0.1	1.7	A3 V		150 mx		
221839	192020		23 35 40.1	-27 29 24	+0.009	-0.13	6.64	0.56	2.8	G0 IV		57 mx	16850	m
221856	165787		23 35 41.7	-15 17 58	-0.001	-0.04	7.3	1.1	0.2	K0 III		260 s		
221913	35577		23 35 42.4	+51 15 51	-0.002	+0.02	7.20	1.6	-0.5	M1 III	-19	330 s		
221934	20718		23 35 45.2	+67 01 15	-0.003	+0.03	8.0	0.1	1.4	A2 V		190 s		
221843	214642		23 35 48.2	-38 57 13	+0.005	0.00	7.6	1.1	0.2	K0 III		250 s		
221946	20720		23 35 53.7	+66 51 36	+0.007	-0.04	7.6	0.7	4.4	G0 V		44 s		
221904	91368		23 35 55.2	+27 51 54	-0.001	+0.01	7.3	0.0	0.4	B9.5 V		240 s		
221905	91367		23 35 56.0	+24 33 40	+0.002	+0.02	6.5	1.6	-0.5	M2 III	-12	250 s		
221890	146799		23 36 00.1	-8 45 58	-0.002	-0.03	7.4	0.5	3.4	F5 V		62 s		
221881	214645		23 36 00.8	-30 38 48	+0.004	-0.03	7.5	0.6	3.4	F5 V		52 s		
221914	108707		23 36 05.8	+18 26 34	+0.048	+0.22	7.64	0.70	5.2	G5 V	-25	30 s		
221907	192030		23 36 09.5	-26 52 34	-0.002	+0.01	6.6	0.5	1.4	A2 V		57 s		
221948	53177		23 36 12.1	+42 38 43	-0.001	0.00	7.8	1.1	0.2	K0 III		330 s		
221939	128278		23 36 14.9	+6 51 39	-0.001	-0.01	7.8	1.1	0.2	K0 III		330 s		
221950	128281	16 Psc	23 36 23.1	+2 06 08	-0.007	+0.06	5.68	0.44	2.6	dF0	+39	37 s		
222169	3938		23 36 23.9	+83 12 15	+0.006	+0.01	7.7	1.1	0.2	K0 III		310 s		
221972	91372		23 36 27.9	+20 39 55	+0.003	-0.05	7.8	0.4	3.0	F2 V		89 s		
222016	20725		23 36 28.3	+67 50 56	0.000	-0.02	7.2	0.1	0.6	A0 V		200 s		
221970	73368		23 36 30.4	+32 54 15	-0.001	+0.01	6.35	0.46	3.4	F5 V	-1	38 s		
221910	255522		23 36 31.6	-62 45 18	+0.005	+0.01	7.8	0.1	0.6	A0 V		260 mx		
221992	53179		23 36 31.7	+45 30 55	+0.007	-0.07	8.0	0.9	3.2	G5 IV		91 s		
221943	231692		23 36 33.8	-44 53 34	+0.008	-0.02	6.92	0.19	0.5	A7 III		120 mx		
221961	192033		23 36 41.2	-25 14 16	+0.006	+0.05	7.7	0.5	3.2	G5 IV		81 s		
222028	20726		23 36 46.3	+64 32 17	-0.002	+0.02	8.0	1.1	-0.1	K2 III		360 s		
222018	53187		23 36 51.7	+43 18 31	+0.001	-0.02	7.8	1.1	-0.1	K2 III		380 s		
222008	108715		23 36 58.2	+11 00 54	+0.004	0.00	7.9	0.2	2.1	A5 V		140 s		
221967	255525		23 37 03.2	-60 50 08	+0.010	0.00	7.2	0.5	2.6	F0 V		62 s		
222004	214659		23 37 05.2	-31 52 16	+0.004	+0.01	6.52	1.26	0.2	K0 III		120 s		m
222063	35600		23 37 06.3	+53 27 10	+0.001	+0.01	7.7	0.0	0.4	B9.5 V		280 s		
222033	73377		23 37 06.5	+30 40 42	+0.015	+0.09	7.18	0.63	4.4	G0 V	-13	33 s		
222077	35603		23 37 12.1	+55 52 35	+0.001	-0.02	7.5	1.4	-0.3	K5 III		340 s		
222067	91384		23 37 18.6	+22 14 26	-0.001	-0.02	7.7	1.4	-0.3	K5 III		390 s		
222039	165800		23 37 20.8	-13 20 18	+0.001	0.00	7.9	1.1	0.2	K0 III		340 s		
222078	91385		23 37 20.8	+22 29 23	+0.003	-0.01	7.9	1.1	0.2	K0 III		340 s		
222006	247952		23 37 22.4	-58 10 11	+0.006	-0.01	7.2	1.6	0.2	K0 III		110 s		
222025	247953		23 37 23.6	-51 43 59	-0.003	-0.02	7.8	1.1	0.2	K0 III		300 s		
222104	53197		23 37 27.1	+49 21 53	-0.006	+0.04	8.0	0.5	3.4	F5 V		81 s		
222089	73383		23 37 27.9	+35 48 54	+0.005	-0.02	7.0	1.1	-0.1	K2 III		270 s		
222109	53202		23 37 31.9	+44 25 45	+0.001	-0.01	6.30	-0.06		B9	-11		16877	m
222107	53204	16 λ And	23 37 33.7	+46 27 30	+0.015	-0.42	3.82	1.01	1.8	G8 III-IV	+7	24 mx		m,v
222093	165804		23 37 39.4	-13 03 37	+0.002	+0.03	5.65	1.03	3.2	G5 IV	-13	20 s	16878	m
222098	108729	74 Peg	23 37 39.6	+16 49 32	+0.009	0.00	6.20	0.00	1.4	A2 V	-26	91 s		m
222091	165806		23 37 41.5	-9 58 30	-0.005	-0.02	7.9	1.1	-0.1	K2 III		320 mx		
222096	231706		23 37 49.6	-45 34 07	-0.005	0.00	8.0	1.5	-0.5	M4 III		480 mx		
222095	231707		23 37 50.8	-45 29 33	+0.007	-0.01	4.74	0.08	1.4	A2 V	+10	45 s		
222100	192042		23 37 50.8	-24 52 43	-0.002	-0.06	7.8	0.0	3.0	F2 V		91 s		
222133	108732	75 Peg	23 37 56.6	+18 24 03	+0.003	+0.02	5.40	-0.03	1.2	A1 V	-16	69 s		m
—	192044		23 37 57.6	-27 15 14	+0.003	+0.01	8.0	0.9	3.2	G5 IV		93 s		
222081	247957		23 37 57.9	-58 11 20	+0.015	+0.08	7.5	0.5	3.0	F2 V		68 s		
222143	53210		23 37 58.5	+46 11 59	+0.035	0.00	6.5	0.9	3.2	G5 IV	-1	44 mx		
222155	53211		23 38 00.3	+48 59 48	+0.021	-0.11	7.11	0.65	4.4	G0 V		31 s		
222172	35622		23 38 01.2	+51 14 33	+0.001	+0.01	8.0	0.0	0.4	B9.5 V		320 s		
222125	165808		23 38 02.9	-15 05 42	+0.006	-0.11	6.40	1.08	0.2	K0 III		91 mx		
222174	53213		23 38 03.5	+43 04 41	-0.003	0.00	6.9	1.1	0.2	K0 III		220 s		
222173	53216	17 ι And	23 38 08.0	+43 16 05	+0.003	0.00	4.29	-0.10	-0.2	B8 V	-1	79 s		
222076	258165		23 38 08.1	-70 54 12	-0.014	+0.03	7.47	1.03	0.2	K0 III		270 s		
222177	91392		23 38 10.0	+20 34 09	+0.001	-0.01	7.3	0.4	3.0	F2 V		74 s		
222157	146815		23 38 12.2	-8 37 40	+0.002	-0.02	6.8	0.9	3.2	G5 IV		53 s		
222120	247959		23 38 12.4	-56 27 06	-0.001	-0.01	8.0	0.8	-0.5	M2 III		510 s		
222189	108737		23 38 17.1	+17 03 47	+0.003	-0.02	7.8	1.1	0.2	K0 III		330 s		
222218	35631		23 38 22.1	+58 39 15	-0.002	-0.01	6.94	1.02	0.0	K1 III	0	240 s		

HD	SAO	Star Name	α 2000	δ 2000	μ(α)	μ(δ)	V	B−V	M_v	Spec	RV	d(pc)	ADS	Notes
222060	258166		23^{h}38^{m}24^s.1	−76°52′12″	+0^s.027	−0″.03	6.00	0.90		K0				
—	—		23 38 24.5	+42 31 07			6.80						16890	m
222207	53225		23 38 25.9	+42 30 44	−0.004	+0.03	6.8	0.0	0.4	B9.5 V	−18	86 mx		
222165	—		23 38 29.2	−46 28 48			6.02	1.18	−4.5	G8 Ib		1300 s		
222221	73398		23 38 33.9	+35 02 01	+0.001	−0.02	7.10	0.1	1.4	A2 V		140 s	16894	m
222151	—		23 38 38.4	−64 02 48			7.9	1.1	0.2	K0 III		280 s		
222212	192056		23 38 41.4	−20 52 02	0.000	+0.01	7.9	0.5	4.4	G0 V		51 s		
222167	255531		23 38 43.4	−62 53 07	+0.006	+0.01	6.62	0.93	3.2	G6 IV		42 s		
222275	20745		23 38 48.7	+62 08 06	−0.003	0.00	6.58	0.55	−2.3	A3 II	−34	350 s	16898	m
222249	91399		23 38 49.3	+23 18 31	+0.002	−0.02	7.0	0.5	3.4	F5 V		54 s		
222265	35637		23 38 52.0	+54 26 34	0.000	+0.02	7.5	1.1	0.2	K0 III	−38	270 s		
222303	20748		23 38 53.0	+66 29 33	−0.006	−0.01	7.9	0.1	0.6	A0 V		260 s		
222250	108746		23 38 53.3	+12 11 41	−0.001	0.00	7.6	0.1	1.7	A3 V		150 s		
222226	231715		23 38 55.5	−45 36 42	+0.007	−0.01	7.00	0.30	1.7	F0 IV		120 s		
222268	108747		23 39 00.1	+17 00 51	−0.001	0.00	7.8	1.1	−0.1	K2 III		380 s		
222293	35640		23 39 01.3	+52 15 45	0.000	0.00	6.6	1.6		M6 III				SV Cas, v
222304	35642	18 And	23 39 08.2	+50 28 18	−0.002	0.00	5.30	−0.06		B9	+9			
222278	128304		23 39 10.3	+6 12 05	+0.001	−0.01	7.8	0.2	2.1	A5 V		140 s		
222386	10814		23 39 10.3	+75 17 34	+0.003	+0.02	5.95	0.13		A2	+3			
222325	35644		23 39 14.2	+58 45 03	0.000	0.00	8.0	1.1	0.2	K0 III		330 s		
221763	258978		23 39 15.2	−86 23 57	+0.060	−0.03	7.9	0.0	2.6	F0 V		120 s		
222281	192063		23 39 15.8	−21 55 23	+0.003	+0.06	7.8	0.5	2.6	F0 V		89 s		
222404	10818	35 γ Cep	23 39 20.7	+77 37 57	−0.020	+0.16	3.21	1.03		K1 IV	−42	16 t		Errai
222387	10817		23 39 21.0	+74 00 10	−0.002	+0.01	5.98	0.89	3.2	G5 IV	+9	31 s		
222284	214690		23 39 23.5	−33 29 34	−0.001	+0.02	7.8	1.2	0.2	K0 III		260 s		
222287	231719		23 39 27.8	−46 38 16	+0.003	+0.04	6.09	0.24		A3				
222317	91405		23 39 30.7	+28 14 48	+0.022	+0.24	7.04	0.67	5.2	G5 V	−4	23 s		
222366	35650		23 39 34.9	+58 58 15	+0.004	+0.03	7.45	0.86	5.9	K0 V		20 s		
222237	258167		23 39 37.0	−72 43 20	+0.031	−0.73	7.09	0.99	6.9	K3 V		10 ts		
222355	53246		23 39 37.6	+43 19 58	+0.004	+0.01	7.6	0.5	3.4	F5 V		71 s		
222272	255540		23 39 40.8	−66 15 16	+0.007	−0.06	7.3	1.1	0.2	K0 III		180 mx		
222332	192068		23 39 41.5	−22 32 01	+0.003	−0.02	7.29	0.12	1.2	A1 V		140 s		
222345	165818	102 ω¹ Aqr	23 39 46.9	−14 13 18	+0.004	−0.03	5.00	0.24		A5	−2	18 mn		
222448	10822		23 39 48.4	+75 45 08	−0.011	0.00	7.2	0.1	1.9	F2 IV	+1	110 s		
222367	73418		23 39 50.1	+34 57 21	+0.003	−0.03	7.2	0.1	1.4	A2 V		140 s		
222358	146834		23 39 50.4	−5 32 49	−0.002	−0.02	7.8	1.1	−0.1	K2 III		380 s		
222335	214692		23 39 51.1	−32 44 35	+0.010	−0.28	7.19	0.81	6.1	K1 V		17 s		
222407	20756		23 39 52.9	+63 43 31	+0.007	0.00	6.85	0.10		A2	−17		16911	m
222377	128309		23 39 54.9	+9 40 38	+0.007	−0.01	5.97	0.20		A2	0			
222362	192070		23 39 56.7	−28 56 34	+0.002	−0.02	7.5	1.3	0.2	K0 III		190 s		
222368	128310	17 ι Psc	23 39 56.9	+5 37 35	+0.025	−0.43	4.13	0.51	3.8	F7 V	+5	13 ts		m
222363	231726		23 39 59.6	−47 19 34	+0.002	+0.02	7.00	0.9	3.2	G5 IV		58 s		
222348	247973		23 40 00.0	−51 37 00	−0.002	+0.01	7.5	1.1	0.2	K0 III		260 s		
222390	91410		23 40 00.8	+27 30 55	+0.003	+0.03	6.66	1.06	0.0	K1 III	−12	220 s		
222392	108755		23 40 02.1	+14 48 38	0.000	0.00	8.0	0.1	1.7	A3 V		180 s		
222399	73422		23 40 02.6	+37 39 09	−0.001	−0.08	6.53	0.35	2.6	F0 V	−16	57 s	16913	m
222391	91412		23 40 03.3	+26 48 07	+0.011	+0.04	7.52	0.57	0.6	G0 III	−2	99 mx		
222416	53259		23 40 12.0	+45 04 28	+0.002	+0.02	7.5	0.0	0.4	B9.5 V	−19	250 s		
222396	192073		23 40 13.4	−28 21 35	−0.007	−0.09	7.8	0.7	3.4	F5 V		52 s		
222419	108760		23 40 16.3	+18 12 52	+0.003	0.00	7.2	1.1	0.2	K0 III		250 s		
222402	192076		23 40 19.4	−26 53 18	0.000	+0.01	7.7	1.1	3.2	G5 IV		51 s		
222439	53264	19 κ And	23 40 24.4	+44 20 02	+0.008	−0.01	4.14	−0.08	−0.2	B8 V	−9	60 mx	16916	m
222412	192077		23 40 26.1	−26 12 07	+0.007	−0.15	7.60	0.44	2.2	F6 IV		45 mx		
222440	91415		23 40 30.9	+24 41 42	+0.003	+0.04	7.80	0.5	3.4	F5 V		76 s		m
222433	214701	μ Scl	23 40 38.0	−32 04 24	−0.007	−0.05	5.31	0.97	0.2	K0 III	+14	110 s		
222455	128319		23 40 40.0	+0 24 56	−0.006	−0.04	7.40	1.17	−0.2	K3 III	−2	220 mx	16919	m
222451	73428		23 40 40.4	+36 43 15	+0.019	+0.03	6.23	0.39	3.4	F5 V	0	37 s		
−−	−−		23 40 42.6	+0 12 38			7.70						16919	m
222472	35659		23 40 43.1	+55 08 57	+0.001	0.00	7.9	0.0	0.4	B9.5 V		300 s		
222490	35660		23 40 46.1	+52 33 39	+0.001	+0.02	7.6	0.9	3.2	G5 IV		74 s		
222465	146842		23 40 48.5	−7 54 51	+0.004	−0.04	7.1	0.5	4.0	F8 V		42 s		
222437	247975		23 40 50.5	−56 25 15	−0.006	−0.05	7.2	0.4	0.6	A0 V		110 s		
222491	73433		23 41 00.3	+35 58 19	+0.001	0.00	7.7	0.1	0.6	A0 V		270 s		
222514	35665		23 41 00.6	+57 50 20	−0.003	−0.01	7.24	0.17		A m	+9			
222467	214705		23 41 01.0	−37 52 06	+0.004	−0.05	7.5	1.5	−0.1	K2 III		170 mx		
222499	53267		23 41 04.5	+41 51 03	+0.006	+0.02	6.8	1.1	−0.1	K2 III		240 s		
222516	53270		23 41 05.2	+46 13 07	−0.012	−0.03	7.80	0.5	3.4	F5 V		75 s		m
222485	192083		23 41 06.8	−24 09 37	0.000	−0.01	6.60	1.57	−0.3	gK5		220 s		
222480	214707		23 41 08.2	−32 04 14	+0.009	+0.05	7.13	0.66	3.2	G5 IV		61 s		
222493	165828		23 41 08.7	−11 40 50	+0.004	+0.01	5.89	1.00	0.2	gG9	−11	130 s		
222568	20776		23 41 11.0	+68 21 35	0.000	−0.01	7.69	0.41	−3.9	B1 IV		880 s		
222569	20775		23 41 13.4	+65 07 03	+0.005	+0.01	7.7	1.1	0.2	K0 III		290 s		
222598	10836		23 41 13.5	+75 34 57	−0.004	−0.05	8.0	1.1	−0.1	K2 III	−5	260 mx		
222529	73436		23 41 17.2	+32 33 39	0.000	−0.01	7.40	0.1	0.6	A0 V		230 s	16928	m
222555	53274		23 41 20.0	+46 51 35	0.000	−0.02	7.2	0.0	0.4	B9.5 V	−9	220 s		

HD	SAO	Star Name	α 2000	δ 2000	μ(α)	μ(δ)	V	B-V	M_v	Spec	RV	d(pc)	ADS	Notes
			$23^h41^m26^s.8$	$+49°30'45"$	$0^s.000$	$-0".02$								
222570	53276		23 41 26.8	+49 30 45	0.000	-0.02	6.3	0.1	1.7	A3 V	-6	83 s		
222508	231740		23 41 28.5	-41 34 50	+0.010	0.00	7.81	0.48	3.8	F7 V		63 s		m
222556	53277		23 41 29.4	+40 33 16	-0.002	-0.07	7.1	0.5	3.4	F5 V		54 s		
222572	91420		23 41 33.8	+24 13 26	+0.003	0.00	7.5	0.1	1.4	A2 V		170 s		
222547	165834	103 Aqr	23 41 34.3	-18 01 38	-0.003	-0.07	5.34	1.57	0.2	K0 III	+25	39 s		
222536	214713		23 41 36.6	-37 26 01	+0.001	-0.01	8.0	1.1	3.2	G5 IV		59 s		
222538	247980		23 41 38.2	-50 25 27	-0.001	-0.01	7.8	1.7	0.2	K0 III		130 s		
222562	—		23 41 42.0	-19 45 45			7.8	2.0	-0.5	M2 III		280 s		
222574	165836	104 Aqr	23 41 45.6	-17 48 59	+0.001	+0.01	4.82	0.82	-2.0	G0 II	+3	220 s		m
222582	146849		23 41 51.4	-5 59 09	-0.009	-0.12	7.9	0.9	3.2	G5 IV		88 s		
222629	20784		23 41 53.2	+68 40 19	0.000	+0.01	7.88	0.00	0.6	A0 V		290 s		
222618	35682		23 41 54.3	+57 15 36	0.000	0.00	6.3	0.9	0.3	G8 III	-12	150 s		
222602	128335		23 41 56.5	+7 15 02	-0.002	-0.04	5.89	0.10		A0	+1			
222576	231744		23 41 57.0	-42 16 08	+0.003	-0.04	7.11	1.02	0.0	K1 III		260 s		
222593	192096		23 42 02.0	-29 37 31	+0.002	-0.05	7.7	2.0	-0.1	K2 III		120 s		
222603	128336	18 λ Psc	23 42 02.6	+1 46 48	-0.009	-0.15	4.50	0.20	2.4	A7 V	+12	26 s		
222551	—		23 42 06.5	-69 26 45			8.0	0.9	3.2	G5 IV		80 s		
222630	53290		23 42 08.8	+47 03 48	-0.002	0.00	8.0	1.1	0.2	K0 III		340 s		
222693	10838		23 42 09.1	+74 07 47	+0.004	+0.02	8.0	0.0	0.4	B9.5 V		190 mx		
222595	247988		23 42 12.5	-53 25 23	+0.022	+0.02	7.9	0.9	3.2	G5 IV		74 s		
222642	53291		23 42 14.3	+44 45 20	+0.007	-0.02	6.9	0.4	2.6	F0 V	+5	72 s		
222641	53292		23 42 14.6	+44 59 31	-0.001	-0.01	6.6	1.4	-0.3	K5 III	-10	230 s		
222670	20791		23 42 20.6	+64 30 56	0.000	0.00	6.56	1.88	-0.5	M2 III	-3	170 s	16940	m
222657	53296		23 42 24.8	+49 09 07	-0.001	+0.03	8.0	0.1	0.6	A0 V		280 s		
222617	255548		23 42 25.6	-61 03 43	+0.007	0.00	7.1	1.4	0.2	K0 III		140 s		
222643	165841		23 42 27.7	-15 26 52	+0.001	0.00	5.28	1.37	-0.3	K4 III	+7	130 s		
222659	108780		23 42 31.1	+18 39 59	+0.001	-0.02	7.40	1.1	0.2	K0 III		280 s		m
222682	20793		23 42 31.5	+61 40 47	+0.007	0.00	6.40	1.24	-0.1	K2 III	-16	170 s		m
222638	247991		23 42 33.9	-57 28 41	+0.003	-0.01	7.6	0.8	0.6	A0 V		77 s		
222694	35691		23 42 40.6	+51 25 40	+0.002	+0.02	7.4	0.1	0.6	A0 V		220 s		
222661	165842	105 ω² Aqr	23 42 43.2	-14 32 42	+0.007	-0.06	4.49	-0.04	0.4	B9.5 V	+3	43 mx	16944	m
222683	108782		23 42 43.7	+16 20 09	+0.006	+0.02	6.4	1.1	0.2	K0 III	-2	180 s		
222664	214723		23 42 44.9	-31 14 09	-0.008	-0.19	7.8	1.5	4.4	G0 V		13 s		
222821	3954		23 42 45.4	+81 18 03	+0.001	+0.01	7.9	0.1	0.6	A0 V		270 s		
222676	165843		23 42 47.6	-11 19 44	-0.003	-0.03	7.30	1.1	0.2	K0 III		260 s	16946	m
222684	128341		23 42 49.0	+9 53 07	+0.003	-0.02	8.0	1.1	0.2	K0 III		360 s		
222699	192101		23 43 02.1	-29 02 39	+0.001	-0.02	7.1	1.7	-0.1	K2 III		130 s		
222748	35698		23 43 05.0	+51 56 23	+0.005	+0.02	6.5	1.1	0.2	K0 III		180 s		
222762	35700		23 43 05.4	+53 09 09	-0.001	+0.01	6.5	0.0	0.0	B8.5 V		190 s		
222688	231749		23 43 05.9	-46 18 44	-0.001	+0.02	6.63	0.91	3.2	G8 IV		47 s		m
222701	214729		23 43 07.0	-36 47 04	+0.001	+0.02	7.5	0.5	2.1	A5 V		74 s		
222731	73459		23 43 07.3	+34 44 49	+0.003	-0.07	7.3	1.1	0.2	K0 III		180 mx		
222770	35702		23 43 13.3	+52 14 54	+0.003	+0.02	7.6	0.1	0.6	A0 V		250 s		
222668	258173		23 43 16.6	-70 49 05	+0.011	+0.06	7.50	0.9	3.2	G5 IV		72 s		m
222754	146862		23 43 18.5	-0 42 29	-0.001	0.00	8.0	0.9	3.2	G5 IV		93 s		
222764	108789	77 Peg	23 43 22.2	+10 19 54	0.000	+0.02	5.06	1.68		Ma	-34	17 mn		
222739	214735		23 43 22.3	-33 04 57	0.000	-0.02	7.1	0.9	0.2	K0 III		240 s		
222794	35706		23 43 26.9	+58 04 49	+0.048	+0.49	7.11	0.66	4.7	G2 V	-67	29 s		m
222690	258174		23 43 31.9	-74 12 51	+0.020	+0.05	7.9	0.4	3.0	F2 V		96 s		
222768	192110		23 43 32.3	-22 54 08	-0.010	-0.06	7.7	1.5	0.2	K0 III		150 mx		
—	35709		23 43 34.0	+59 57 08	0.000	+0.01	7.7	2.6	-0.5	M2 III		120 s		
222809	53318		23 43 36.2	+47 20 26	0.000	0.00	7.9	1.4	-0.3	K5 III		400 s		
222743	255552		23 43 39.7	-63 57 40	+0.028	-0.10	7.5	0.6	4.4	G0 V		42 s		
222811	91450		23 43 42.0	+21 56 25	0.000	-0.01	7.9	1.6	-0.5	M2 III		490 s		
222800	165849		23 43 49.3	-15 17 04	+0.002	-0.02	6.70	1.6		M7 pe	-22			R Aqr, v
222780	255554		23 43 49.5	-61 43 47	+0.003	0.00	7.4	1.6	-0.5	M2 III		390 s		
222801	214741		23 43 51.6	-33 25 39	-0.004	-0.01	6.9	0.5	3.4	F5 V		45 s		
222804	231754		23 43 53.7	-45 27 46	+0.004	-0.06	6.81	1.22	-0.2	K3 III		150 mx		
222857	20807		23 43 55.3	+66 33 06	+0.002	-0.02	7.7	0.0	0.4	B9.5 V		260 s		
222669	258983		23 43 57.2	-82 30 29	-0.068	-0.07	7.7	0.8	4.4	G0 V		35 s		
222843	108798		23 43 58.9	+17 30 42	-0.001	-0.01	7.9	0.5	3.4	F5 V		78 s		
222842	91457	78 Peg	23 43 59.2	+29 21 42	+0.005	-0.03	4.93	0.95	0.2	K0 III	-7	88 s	16957	m
222803	231756		23 44 00.9	-45 04 58	+0.029	+0.02	6.09	0.98	3.2	G8 IV		33 s		
222854	53325		23 44 05.1	+42 27 48	-0.005	-0.02	7.7	1.1	0.2	K0 III		310 s		
222837	231758		23 44 11.2	-40 41 10	+0.003	0.00	7.7	1.4	-0.1	K2 III		270 s		
222820	255557		23 44 11.9	-64 24 17	+0.003	+0.03	5.72	1.40	-2.3	K3 II		400 s		
222847	165854	106 Aqr	23 44 11.9	-18 16 37	+0.002	0.00	5.24	-0.08	-0.2	B8 V	+14	120 s		
222838	231759		23 44 15.4	-43 01 28	0.000	+0.05	7.78	0.21		A m				
222858	73470		23 44 16.4	+36 20 21	+0.008	+0.04	7.5	0.5	4.0	F8 V		49 s		
222830	255558		23 44 17.4	-64 20 22	+0.004	+0.01	6.8	1.3	-0.3	K5 III		270 s		
222860	128354		23 44 18.0	+0 42 45	-0.002	-0.03	7.97	0.54		F8	+5			
222849	248001		23 44 19.1	-54 26 08	+0.001	+0.01	7.3	1.7	-0.5	M2 III		320 s		
222887	35721		23 44 22.6	+55 12 23	-0.001	-0.01	7.3	1.1	-0.1	K2 III	-16	280 s		
222886	35723		23 44 24.3	+57 45 33	-0.001	+0.01	8.0	1.1	0.2	K0 III		320 s		
222805	258178		23 44 25.3	-70 29 26	+0.050	+0.06	6.07	0.91	3.2	G8 IV		35 s		

HD	SAO	Star Name	α 2000	δ 2000	μ(α)	μ(δ)	V	B-V	M_v	Spec	RV	d(pc)	ADS	Notes
			h m s	° ′ ″	s	″								
222870	165857		23 44 28.5	-14 25 12	0.000	-0.02	8.00	0.43	3.0	F2 V		92 s		
222872	192116		23 44 28.7	-26 14 47	-0.005	-0.01	6.17	0.50		F5			16963	m
222915	35724		23 44 29.0	+55 29 29	0.000	+0.01	7.7	0.0	0.4	B9.5 V		270 s		
222900	53332		23 44 32.0	+46 22 49	0.000	+0.01	7.60	0.1	0.6	A0 V	-8	240 s	16965	m
222878	146877		23 44 32.4	-3 10 31	0.000	-0.01	7.3	0.5	4.0	F8 V		46 s		
222806	258179		23 44 40.4	-78 47 30	+0.017	0.00	5.75	1.11	0.2	gK0		100 s		
222916	53336		23 44 40.7	+46 15 56	+0.002	+0.01	8.0	1.6	-0.5	M2 III		460 s		
222922	53337		23 44 45.4	+43 44 47	+0.006	+0.02	6.8	0.1	0.6	A0 V	+4	100 mx		
222932	35728		23 44 48.1	+55 47 59	+0.002	0.00	6.6	0.8	3.2	G5 IV	+9	48 s		
222919	128358		23 44 48.7	+7 11 30	+0.001	-0.01	6.9	1.1	0.2	K0 III		220 s		
222923	73474		23 44 48.9	+34 43 32	+0.005	+0.02	8.0	0.5	3.4	F5 V		82 s		
222933	—		23 44 55.0	+50 35 15	-0.001	-0.02	8.0	1.1	0.2	K0 III		340 s		
222958	20818		23 44 56.5	+69 45 17	-0.002	-0.01	7.13	0.06	0.0	B8.5 V		240 s		
222928	146885		23 44 59.8	-0 39 38	0.000	+0.01	7.30	1.4	-0.3	K5 III	+8	330 s		
222940	91464		23 45 03.8	+21 23 23	-0.001	-0.02	6.9	1.1	-0.1	K2 III		260 s		
222946	108810		23 45 12.0	+15 02 36	-0.002	-0.02	7.8	0.1	0.6	A0 V		270 s		
222975	20821		23 45 17.4	+62 10 22	+0.006	-0.05	8.0	0.9	3.2	G5 IV		89 s		
222962	128365		23 45 21.5	+10 10 48	+0.001	+0.01	6.6	0.1	1.7	A3 V	+11	94 s		
222950	165862		23 45 22.0	-13 40 55	+0.003	+0.01	7.9	0.4	2.6	F0 V		120 s		
222978	91467		23 45 29.2	+20 24 55	+0.002	-0.01	6.80	1.1	0.2	K0 III		210 s	16970	m
222957	248008		23 45 33.9	-58 16 44	-0.004	-0.04	7.80	0.1	1.7	A3 V		170 s		m
222995	108813		23 45 37.1	+13 09 10	-0.002	-0.02	7.2	0.4	3.0	F2 V		70 s		
222971	248009		23 45 42.0	-52 14 06	-0.001	-0.02	7.8	1.4	0.2	K0 III		200 s		
223019	91470		23 45 46.0	+26 20 07	-0.003	-0.04	7.80	1.1	-0.2	K3 III	-13	400 s		
223017	73487		23 45 51.6	+33 17 26	+0.002	-0.01	7.3	0.1	0.6	A0 V		220 s		
223006	165864		23 45 52.2	-14 45 17	0.000	+0.01	7.37	1.52	-0.1	K2 III		170 s		
222996	214759		23 45 52.5	-32 55 42	-0.001	-0.05	7.9	0.2	1.4	A2 V		160 s		
223037	53354		23 45 55.0	+50 09 43	-0.001	+0.01	8.0	0.1	1.4	A2 V		200 s		
223022	165865		23 45 58.9	-13 27 16	-0.004	-0.01	7.9	1.1	0.2	K0 III		340 s		
223043	20829		23 46 00.7	+62 40 11	+0.002	-0.02	7.72	0.04		A0				
223024	165867	107 Aqr	23 46 00.8	-18 40 41	+0.009	+0.03	5.29	0.28	2.1	F0 IV-V comp	-2	43 s	16979	m
223011	231769		23 46 01.0	-40 10 57	+0.008	-0.03	6.31	0.20	2.4	A7 V		55 mx		
223057	20830		23 46 01.2	+63 19 03	-0.002	+0.02	7.78	0.10	0.6	A0 V	-2	230 s		
223047	53355	20 ψ And	23 46 01.9	+46 25 13	+0.001	0.00	4.95	1.11	-4.5	G5 Ib	-25	740 s		m
223023	165869		23 46 02.6	-15 08 05	+0.002	-0.05	8.02	1.55	-0.3	K5 III		460 s		m
223046	35742		23 46 04.0	+50 39 47	+0.004	+0.02	7.70	1.1	0.2	K0 III		300 s	16980	m
223032	165870		23 46 04.7	-17 08 11	+0.002	-0.03	7.9	1.4	-0.3	K5 III		430 s		
223070	20832		23 46 07.8	+60 28 24	+0.001	+0.01	7.1	1.1	0.2	K0 III	-35	220 s	16982	m
223061	128372		23 46 18.1	+9 47 05	-0.004	-0.19	7.9	0.7	4.4	G0 V		51 s		
223050	165872		23 46 18.4	-13 21 52	+0.001	0.00	8.00	0.9	3.2	G5 IV		91 s		m
223073	73495		23 46 21.7	+34 11 48	-0.001	-0.03	7.8	1.1	-0.1	K2 III		390 s		
223082	91475		23 46 23.3	+23 27 01	0.000	+0.01	7.9	1.6	-0.5	M2 III		490 s		
223075	128374	19 Psc	23 46 23.3	+3 29 13	-0.002	-0.02	5.04	2.60		C5 II	-11	63 mn		TX Psc, v
223094	91476		23 46 25.3	+28 42 13	-0.002	-0.01	6.92	1.65	-0.3	K5 III	+20	230 s		
223189	3969		23 46 29.1	+81 22 42	-0.004	+0.02	8.0	0.2	2.1	A5 V		140 s		
223066	231774		23 46 32.6	-44 49 44	-0.001	-0.03	7.6	0.9	0.2	K0 III		300 s		
223065	231773		23 46 32.8	-41 34 54	+0.024	-0.85	7.21	0.20	1.4	A2 V	-15	120 s		SX Phe, v
223109	20837		23 46 33.4	+64 18 08	0.000	+0.04	7.4	0.9	3.2	G5 IV		68 s		
223084	146898		23 46 33.9	-8 59 48	+0.012	-0.06	7.3	0.6	4.4	G0 V		38 s		
223096	128375		23 46 34.4	+0 31 48	-0.002	-0.02	7.4	0.9	3.2	G5 IV	+1	68 s		
223055	—		23 46 35.2	-64 34 42			8.0	0.5	3.4	F5 V		82 s		
223128	20838		23 46 36.5	+66 46 57	+0.001	+0.01	5.95	-0.04	-3.0	B2 IV	-14	510 s		
223112	53362		23 46 41.2	+42 01 24	-0.002	-0.01	8.0	1.1	-0.1	K2 III		410 s		
223113	73501		23 46 43.6	+37 30 28	+0.001	-0.03	8.0	1.4	-0.3	K5 III		470 s		
223115	146901		23 46 50.6	-8 53 48	+0.002	-0.01	7.3	0.8	3.2	G5 IV		67 s		
223100	248015		23 46 51.2	-55 48 35	+0.003	+0.02	6.4	2.5	-0.5	M2 III		70 s		
223131	108830		23 46 52.2	+17 31 45	0.000	0.00	8.0	1.4	-0.3	K5 III		450 s		
223138	91482		23 46 52.2	+28 25 12	0.000	-0.01	7.40	1.6	-0.5	M2 III	-4	380 s		m
223151	35755		23 46 52.6	+54 51 40	0.000	-0.01	7.3	0.1	0.6	A0 V		220 s		
223136	53366		23 46 54.4	+48 48 08	+0.003	+0.04	7.4	0.1	0.6	A0 V		220 s		
223152	35757		23 46 56.1	+51 13 44	+0.001	0.00	7.5	0.0	-1.0	B5.5 V	-4	460 s		
223153	53368		23 46 56.4	+46 13 04	+0.002	-0.01	8.0	1.1	0.2	K0 III		340 s		
223125	248016		23 47 00.6	-55 24 31	+0.009	+0.01	6.9	0.8	3.4	F5 V		30 s		
223173	35761		23 47 01.7	+57 27 05	0.000	0.00	5.51	1.65	-2.3	K3 II	-6	260 s		
223165	35763	5 τ Cas	23 47 03.3	+58 39 07	+0.008	+0.06	4.87	1.11	0.0	K1 III	-21	92 s		
223091	258184		23 47 06.4	-73 05 57	+0.025	+0.01	7.7	0.5	3.4	F5 V		71 s		
223170	165880		23 47 15.7	-11 54 39	-0.004	-0.08	5.73	1.08	0.2	K0 III	+11	110 s		
223145	248018	σ Phe	23 47 15.8	-50 13 35	+0.001	-0.01	5.18	-0.19		B5 n	+11			
223171	231781		23 47 20.7	-48 16 33	-0.038	-0.22	6.88	0.66		G5		21 mn		
223148	255567		23 47 23.0	-68 23 39	+0.007	-0.04	6.89	0.46		F2				
223134	258186		23 47 23.4	-71 18 24	+0.016	-0.04	7.90	1.1	0.2	K0 III		170 mx		m
223209	20850		23 47 25.7	+64 08 42	+0.001	+0.01	7.87	0.03		B9				
223211	91490		23 47 29.1	+25 34 47	-0.002	-0.01	7.10	1.1	-0.2	K3 III	-19	290 s	17005	m
223228	53373		23 47 29.7	+49 17 38	0.000	+0.02	7.50	0.00	0.4	B9.5 V		250 s	17004	m
223193	214776		23 47 31.4	-38 09 22	+0.007	+0.01	7.9	0.3	3.2	G5 IV		86 s		

HD	SAO	Star Name	α 2000	δ 2000	μ(α)	μ(δ)	V	B-V	M$_v$	Spec	RV	d(pc)	ADS	Notes
223212	108837		23^{h}47^{m}31^s.9	+16°56'15"	-0^s.002	-0".02	7.8	0.4	2.6	F0 V		110 s		m
223229	53374		23 47 33.0	+46 49 58	+0.001	0.00	6.07	-0.14	-2.3	B3 IV	-24	450 s	17006	m
223230	53375		23 47 35.3	+46 01 04	0.000	-0.01	8.0	1.1	-0.1	K2 III		380 s		
223210	73509		23 47 35.9	+34 31 17	+0.002	-0.01	7.6	0.1	0.6	A0 V		250 s		
223042	258984		23 47 38.1	-83 51 43	+0.017	+0.03	7.8	1.0	0.2	K0 III		340 s		
223216	146908		23 47 38.3	-5 49 33	+0.002	-0.01	8.0	0.5	4.0	F8 V		62 s		
223215	146909		23 47 39.0	-4 27 42	+0.002	+0.01	7.5	0.1	1.4	A2 V		160 s		
223188	255569		23 47 39.8	-65 14 28	+0.001	0.00	7.4	0.2	2.1	A5 V		120 s		
223235	146911		23 47 41.7	-0 45 44	-0.003	-0.03	7.20	1.1	0.2	K0 III		250 s	17007	m
223265	10859		23 47 42.5	+71 29 46	+0.001	+0.01	7.6	1.1	0.2	K0 III		280 s		
223223	248020		23 47 47.5	-52 12 17	-0.002	+0.02	7.4	1.4	3.2	G5 IV		29 s		
223249	108841		23 47 51.7	+17 03 22	+0.002	0.00	7.80	0.1	0.6	A0 V		280 s	17009	m
223238	128385		23 47 52.1	+4 10 31	+0.023	-0.04	7.69	0.64	4.7	G2 V	-16	37 s		
223274	20853		23 47 54.5	+67 48 25	+0.002	+0.01	5.04	-0.01		A0	+10	26 mn		
223226	255571		23 47 55.8	-66 34 04	-0.014	+0.01	7.0	0.3	2.6	F0 V		75 s		
223253	165889		23 47 56.2	-13 55 01	-0.002	-0.09	7.5	1.1	0.2	K0 III		190 mx		
223252	146915	20 Psc	23 47 56.3	-2 45 42	+0.006	+0.01	5.49	0.94	0.3	G8 III	-7	110 s		m
223275	—		23 48 02.9	+50 18 12	0.000	+0.01	8.0	0.4	0.4	B9.5 V		320 s		
223288	20854		23 48 06.2	+63 49 03	+0.005	0.00	7.41	0.08		A0			17010	m
223256	248023		23 48 06.7	-50 53 28	-0.005	-0.03	7.00	1.1	0.2	K0 III		230 s		m
223271	165890		23 48 09.2	-16 41 56	-0.008	-0.04	7.6	0.5	3.4	F5 V		68 s		
223277	128388		23 48 12.0	+8 14 45	+0.002	-0.01	6.6	0.4	2.6	F0 V		64 s		
223292	128389		23 48 14.3	+7 09 47	-0.002	-0.03	7.1	0.9	3.2	G5 IV		60 s		
223291	128390		23 48 18.4	+8 11 20	+0.002	+0.01	8.0	1.1	-0.1	K2 III		410 s		
223298	53387		23 48 18.4	+49 51 54	-0.002	-0.03	7.9	1.1	0.2	K0 III		330 s		
223321	35781		23 48 23.1	+54 41 55	+0.003	0.00	7.8	1.1	0.2	K0 III		310 s		
223323	91500		23 48 31.2	+25 39 14	-0.010	-0.02	7.0	0.4	3.0	F2 V		63 s		
223329	35787		23 48 31.5	+55 38 58	0.000	-0.01	7.6	0.0	-1.0	B5.5 V		480 s		
223311	146919		23 48 32.3	-6 22 50	0.000	-0.02	6.07	1.45	-0.3	gK4	-21	180 s		
223330	53392		23 48 33.4	+42 11 35	-0.002	-0.03	7.7	1.1	-0.1	K2 III		360 s		
223332	91503		23 48 34.9	+28 22 14	0.000	-0.01	7.40	1.4	-2.3	K5 II	+11	870 s		
223331	73523		23 48 35.2	+36 16 27	-0.001	-0.03	7.60	0.6	4.4	G0 V	+7	44 s	17019	m
223358	20866		23 48 38.8	+64 52 35	+0.002	-0.01	6.41	0.06	0.6	A0 V	-3	130 s	17020	m
223361	53396		23 48 47.2	+44 23 50	+0.006	-0.01	7.8	0.2	2.1	A5 V		99 mx		
223328	231788		23 48 47.8	-47 38 59	+0.003	+0.02	6.7	1.7	0.2	K0 III		72 s		
223346	128393		23 48 49.1	+2 12 51	0.000	-0.04	6.46	0.44		F2	-25			
223385	20869	6 Cas	23 48 50.0	+62 12 53	-0.001	0.00	5.43	0.67	-7.6	A3 Ia	-46	2000 s	17022	m
223362	73525		23 48 50.7	+31 43 42	+0.001	+0.02	8.0	0.1	1.4	A2 V		210 s		
223386	35794		23 48 53.8	+59 58 44	+0.006	+0.01	6.34	-0.01		A0	-16			
223352	192167	δ Scl	23 48 55.4	-28 07 49	+0.008	-0.10	4.51	0.00	0.6	A0 V	+14	30 mx	17021	m
223341	231789		23 48 55.8	-46 50 33	+0.003	0.00	7.0	1.5	-0.3	K5 III		290 s		
223421	35798		23 49 11.9	+58 57 47	+0.005	-0.01	6.33	0.40	3.0	dF2	+30	44 s		
223425	91514		23 49 18.1	+22 51 17	-0.003	-0.02	7.1	0.1	0.6	A0 V		200 s		
223408	192174		23 49 19.4	-27 51 14	-0.005	-0.03	7.1	0.1	4.0	F8 V		41 s		
223424	91516		23 49 19.7	+27 02 03	-0.004	-0.01	7.80	1.1	0.2	K0 III	-1	330 s		
223399	255574		23 49 21.7	-60 04 22	+0.003	+0.10	7.6	0.8	4.4	G0 V		32 s		
223429	192176		23 49 25.8	-21 36 51	0.000	+0.01	7.1	0.9	0.2	K0 III		240 s		
223438	128401	21 Psc	23 49 27.3	+1 04 34	0.000	-0.02	5.77	0.16		A3	+5			
223457	35804		23 49 28.8	+55 39 33	+0.002	+0.02	7.9	1.1	0.2	K0 III		330 s		
223428	165905		23 49 31.4	-15 51 40	+0.003	-0.02	6.24	1.22	0.2	K0 III		110 s		
223400	255576		23 49 32.0	-66 15 27	+0.056	-0.07	7.4	0.5	4.0	F8 V		45 mx		
223430	214797		23 49 32.2	-31 24 09	-0.003	+0.02	8.0	1.7	0.2	K0 III		140 s		
223461	91522	79 Peg	23 49 39.2	+28 50 33	+0.005	+0.03	5.9	0.1	1.7	A3 V	-4	69 s		
223460	73535		23 49 40.8	+36 25 31	0.000	-0.05	5.90	0.79	3.2	G5 IV	+1	34 s		
223444	255578		23 49 44.5	-62 50 22	+0.005	-0.03	6.59	1.48		K0				
223466	192180		23 49 49.4	-25 19 53	-0.002	-0.02	6.42	0.12	1.7	A3 V		84 s	17029	m
223486	91524		23 49 50.4	+27 40 52	+0.005	+0.01	7.40	0.4	2.6	F0 V		91 s	17030	m
223501	20881		23 49 53.0	+62 12 51	-0.002	0.00	7.79	0.05		B2.5 e	-38			
223480	231800		23 49 54.3	-42 18 05	-0.002	-0.04	7.9	0.2	2.6	F0 V		110 s		
223452	255579		23 49 54.7	-65 38 30	+0.001	-0.03	7.1	0.4	3.0	F2 V		66 s		
223470	255580		23 49 58.2	-61 08 09	+0.004	0.00	8.0	1.0	-0.5	M4 III		510 s		
223509	165909		23 50 05.6	-11 06 10	-0.009	-0.01	6.7	0.4	3.0	F2 V		56 s		
223521	73541		23 50 05.9	+35 46 17	-0.001	-0.03	7.8	0.5	4.0	F8 V		59 s		
223531	91529		23 50 10.5	+24 51 29	-0.004	+0.01	7.6	0.1	0.6	A0 V		220 mx		
223532	108864		23 50 14.2	+17 38 46	-0.003	0.00	7.8	0.4	3.0	F2 V		89 s		
223524	165911		23 50 14.5	-9 58 27	+0.009	+0.08	6.0	1.1	3.2	K0 IV	-18	36 s		
223515	192188		23 50 14.8	-29 24 07	+0.014	+0.02	7.70	0.9	3.2	G5 IV		79 s		m
223568	20885		23 50 21.0	+66 54 29	+0.003	+0.03	7.2	0.4	3.0	F2 V		69 s		
223552	35823		23 50 22.1	+51 37 18	+0.013	-0.01	6.44	0.37	3.0	F2 V	-21	49 s	17032	m
223578	20887		23 50 24.0	+63 44 37	-0.005	-0.01	7.2	1.1	-0.1	K2 III	-32	260 s		
223569	35824		23 50 24.7	+57 09 41	0.000	+0.02	7.90	0.00	0.4	B9.5 V		290 s		m
223546	192192		23 50 26.0	-22 14 35	+0.003	0.00	8.0	1.4	-0.3	K5 III		450 s		
223542	165912		23 50 26.7	-14 34 46	-0.001	+0.01	6.69	1.37	0.2	K0 III		100 s		
223543	165913		23 50 27.5	-14 59 01	+0.001	0.00	7.49	0.20		A m				
223581	35826		23 50 28.3	+59 36 37	0.000	0.00	7.8	0.1	0.6	A0 V		250 s		

HD	SAO	Star Name	α 2000	δ 2000	μ(α)	μ(δ)	V	B-V	M_V	Spec	RV	d(pc)	ADS	Notes
			$23^h50^m30\overset{s}{.}1$	$+17°36'41''$	$+0\overset{s}{.}001$	$+0\overset{''}{.}01$								
223554	108866		23 50 30.1	+17 36 41	+0.001	+0.01	7.4	0.0	0.4	B9.5 V		250 s		
223541	165914		23 50 30.5	-13 06 38	0.000	0.00	7.1	1.1	0.2	K0 III		240 s		
223570	53411		23 50 30.9	+43 25 09	0.000	0.00	7.80	0.1	0.6	A0 V		280 s	17037	m
223559	165915		23 50 33.1	-14 24 07	+0.002	-0.03	5.72	1.51	0.2	K0 III	-58	51 s		
223529	255582		23 50 33.6	-69 05 34	+0.014	-0.04	7.3	1.5	-0.1	K2 III		180 mx		
223561	192194		23 50 34.0	-20 14 00	+0.002	-0.02	7.3	0.5	1.4	A2 V		82 s		
223551	248037		23 50 34.3	-51 42 19	+0.008	-0.09	7.62	0.76		G5				m
223549	231804		23 50 35.4	-47 22 38	0.000	+0.03	6.5	1.0	0.2	K0 III		180 s		
223582	35828		23 50 37.8	+54 11 54	0.000	-0.03	7.20	0.5	3.4	F5 V		57 s	17038	m
223583	53415		23 50 44.0	+41 49 38	+0.016	+0.05	7.8	0.7	4.4	G0 V		49 s		
223586	—		23 50 45.9	-18 17 39			7.0	1.1	0.2	K0 III		230 s		
223537	258986		23 50 55.5	-79 53 55	-0.029	+0.02	8.0	0.7	4.4	G0 V		43 s		
223616	53418		23 50 57.9	+41 09 50	+0.001	+0.01	7.3	1.1	-0.1	K2 III		300 s		
223615	35835		23 50 59.2	+51 23 53	+0.002	0.00	7.9	1.1	0.2	K0 III		330 s		
223624	20896		23 50 59.3	+63 59 06	0.000	+0.01	6.82	0.04	1.2	A1 V		130 s		
223600	231806		23 51 00.6	-48 39 34	-0.001	-0.01	6.9	1.2	0.2	K0 III		170 s		
223605	231807		23 51 02.6	-41 36 00	-0.002	0.00	7.6	1.6	-0.1	K2 III		200 s		
223617	128417		23 51 06.7	+2 14 16	0.000	0.00	7.2	0.8	3.2	G5 IV		64 s		
223626	35836		23 51 08.6	+51 20 30	+0.001	-0.01	8.0	1.1	-0.1	K2 III		390 s		
223635	53424		23 51 17.4	+43 29 54	-0.002	-0.01	7.8	0.5	4.0	F8 V		57 s		
223636	73556		23 51 18.6	+40 11 56	+0.019	-0.05	6.7	0.5	4.0	F8 V		35 s		
223641	214816		23 51 20.6	-36 02 06	-0.018	-0.15	7.8	0.6	4.4	G0 V		47 s		
223637	128421	80 Peg	23 51 21.0	+9 18 49	-0.001	-0.06	5.79	1.66		Ma	-9			
223640	165918	108 Aqr	23 51 21.1	-18 54 32	+0.001	0.00	5.18	-0.14		A0 p	+13	16 mn		
223633	231811		23 51 21.7	-42 22 22	+0.004	-0.13	7.60	0.46	2.8	F5 IV-V		57 mx		
223642	231813		23 51 26.8	-41 46 22	+0.005	0.00	7.9	1.2	3.2	G5 IV		48 s		
223660	53426		23 51 26.9	+47 45 16	+0.001	+0.01	7.70	0.00	0.4	B9.5 V		270 s	17047	m
223655	214820		23 51 28.0	-32 25 48	+0.003	+0.02	7.80	0.4	2.6	F0 V		110 s		
223672	53427		23 51 33.0	+42 04 58	+0.001	0.00	7.70	0.2	2.1	A5 V		130 s	17050	m
223661	53429		23 51 33.1	+47 29 16	0.000	0.00	7.8	0.4	2.6	F0 V		110 s		
223666	214824		23 51 39.1	-34 41 30	+0.005	-0.02	6.7	0.1	3.2	G5 IV		49 s		
223685	128424		23 51 40.2	+4 41 52	-0.001	+0.03	7.0	1.1	-0.1	K2 III		260 s		
223677	231814		23 51 42.9	-40 49 41	0.000	-0.07	6.7	0.9	0.2	K0 III		180 mx		
223678	231815		23 51 46.4	-42 40 23	+0.002	0.00	7.2	1.0	0.2	K0 III		250 s		
223691	214827		23 51 48.4	-33 27 09	-0.003	+0.07	7.9	0.7	3.2	G5 IV		88 s		
223705	53430		23 51 50.8	+42 54 35	-0.001	-0.01	6.9	1.1	0.2	K0 III		220 s		
223718	73565		23 51 52.3	+37 53 29	-0.006	-0.07	7.80	0.5	3.4	F5 V	-20	76 s	17054	m
223707	128425		23 51 53.4	+4 44 43	+0.001	+0.01	6.7	0.1	1.4	A2 V		110 s		
223700	214829		23 51 54.7	-33 07 18	+0.003	+0.02	6.9	0.7	0.2	K0 III		220 s		
223717	53431		23 51 55.7	+40 44 41	0.000	-0.02	7.5	0.0	0.0	B8.5 V		310 s		
223731	10874		23 51 57.3	+77 35 58	+0.080	-0.09	6.55	0.44	3.4	F5 V	+1	43 s		
223719	128427	22 Psc	23 51 57.6	+2 55 49	+0.001	-0.01	5.55	1.53	-2.3	K4 II	0	340 s		
223716	53433		23 51 58.6	+42 13 43	+0.005	+0.02	7.8	1.1	0.2	K0 III		330 s		
223702	248044		23 52 02.6	-59 12 06	+0.007	-0.05	7.5	1.0	0.2	K0 III		170 mx		
223647	258989	γ¹ Oct	23 52 06.7	-82 01 08	-0.020	-0.01	5.11	0.92	0.3	G7 III	+15	92 s		
223734	53434		23 52 07.5	+50 06 52	+0.007	+0.02	7.6	1.1	0.2	K0 III		280 mx		
223723	214833		23 52 08.2	-31 04 32	+0.018	+0.01	7.8	2.1	4.4	G0 V		48 s		
223712	248045		23 52 09.5	-58 48 22	+0.001	-0.03	7.6	0.9	3.2	G5 IV		75 s		
223724	248046		23 52 12.3	-52 19 58	+0.001	0.00	7.4	1.5	0.2	K0 III		140 s		
223767	20902		23 52 16.5	+61 52 40	-0.004	+0.02	7.23	0.63	-6.9	A4 Iab		150 mx		
223755	91548		23 52 23.2	+21 40 15	-0.003	-0.01	6.11	1.59	-0.5	M2 III	-5	210 s		
223778	10879		23 52 24.9	+75 32 41	+0.082	+0.06	6.39	0.98	6.9	K3 V	+1	11 ts	17062	m
223753	248047		23 52 27.4	-56 38 35	-0.004	0.00	7.6	0.8	0.6	A0 V		77 s		
223754	248048		23 52 28.4	-58 33 00	0.000	+0.01	7.9	-0.1	1.7	A3 V		170 s		
223768	108878	81 ø Peg	23 52 29.2	+19 07 13	0.000	-0.03	5.08	1.60	-0.5	gM2	-8	130 s		
223774	165933		23 52 29.8	-14 15 04	-0.007	0.00	5.87	1.28	0.2	K0 III	+2	82 s		
223781	108879	82 Peg	23 52 37.0	+10 56 51	-0.002	+0.01	5.4	0.1	1.7	A3 V	-3	54 s		
223765	255590		23 52 38.5	-68 01 25	+0.012	+0.02	7.2	1.1	0.2	K0 III		250 s		
223788	35856		23 52 39.6	+54 16 08	0.000	-0.04	7.8	0.5	3.4	F5 V		73 s		
223785	165934		23 52 39.8	-18 33 41	+0.002	+0.01	6.81	0.09	1.4	A2 V n		120 s		
223792	91552		23 52 40.0	+21 44 33	-0.001	-0.02	6.7	1.1	0.2	K0 III	-3	200 s		
223786	192211		23 52 40.8	-24 59 06	+0.001	+0.03	6.8	1.3	-0.1	K2 III		200 s		
223783	165935		23 52 41.3	-16 22 27	-0.001	+0.02	7.7	1.6	-0.5	M2 III		430 s		
223800	165936		23 52 47.4	-12 01 00	-0.001	-0.01	7.5	1.6	-0.5	M2 III		410 s		
223777	258192		23 52 48.5	-72 23 59	+0.008	0.00	7.60	0.4	3.0	F2 V		83 s		m
223819	53441		23 52 49.8	+48 28 56	+0.002	+0.02	7.3	0.1	1.7	A3 V		130 s		
223648	258990		23 52 50.1	-85 53 45	+0.013	+0.02	7.9	0.4	0.0	B8.5 V		190 s		
223807	146953		23 52 50.4	-8 59 48	+0.004	-0.02	5.75	1.17	0.2	gK0	-18	95 s		
223820	53444		23 52 53.2	+46 29 32	+0.002	-0.01	7.7	1.1	0.2	K0 III		290 s		
223802	248052		23 52 53.4	-58 50 19	+0.005	+0.01	7.6	0.2	1.4	A2 V		140 s		
223810	192213		23 52 55.3	-21 28 50	-0.003	-0.02	8.0	1.5	3.4	F5 V		18 s		
223825	146954	24 Psc	23 52 55.4	-3 09 20	+0.005	-0.04	5.93	1.07	0.2	gG9	-6	120 s		m
223848	73582		23 52 56.4	+36 57 17	-0.007	-0.07	6.7	0.6	4.4	G0 V		29 s		
223839	108883		23 52 59.8	+11 55 28	+0.003	-0.02	7.30						17079	m
223835	53445		23 53 00.0	+41 20 45	+0.001	+0.01	7.20	1.6	-0.5	M2 III	-8	350 s		m

601

HD	SAO	Star Name	α 2000	δ 2000	μ(α)	μ(δ)	V	B-V	M_V	Spec	RV	d(pc)	ADS	Notes
223837	108884		23^{h}53^{m}00^s.8	+17°53'58"	+0^s.005	-0".03	6.7	1.1	-0.1	K2 III		230 s		
223847	35860		23 53 02.2	+59 25 19	+0.008	+0.02	7.86	1.05	0.3	G7 III	-15	250 mx		
223855	128436	25 Psc	23 53 04.6	+2 05 27	+0.001	-0.01	6.28	-0.01		A2	+5			
223866	20915		23 53 04.7	+60 42 17	+0.001	+0.01	6.67	1.56		K2			17080	m
223854	128437		23 53 06.7	+2 19 51	+0.003	+0.04	8.02	0.50		F5				
223869	91557		23 53 07.2	+25 59 54	+0.007	-0.05	7.70	1.1	0.0	K1 III	+17	160 mx		
223843	248053		23 53 08.5	-54 21 58	+0.005	+0.02	7.5	1.6	0.2	K0 III		130 s		
223868	73585		23 53 10.8	+31 54 35	+0.002	-0.02	7.4	0.4	3.0	F2 V		76 s		
223879	73587		23 53 11.9	+35 03 08	-0.004	-0.04	7.9	1.4	-0.3	K5 III		350 mx		
223860	165941		23 53 13.5	-11 00 53	+0.009	-0.09	7.9	1.1	0.2	K0 III		87 mx		
223893	73588		23 53 16.4	+35 39 01	0.000	-0.07	7.7	1.1	-0.1	K2 III		210 mx		
223884	192218		23 53 20.6	-24 13 45	+0.003	-0.01	6.2	0.1	1.7	A3 V		80 s		
223899	20917		23 53 21.5	+66 46 15	+0.003	-0.02	8.0	1.4	-0.3	K5 III		390 s		
223892	53451		23 53 21.9	+45 17 18	+0.001	+0.01	8.0	0.1	0.6	A0 V		290 s		
223915	35868		23 53 33.3	+51 56 16	+0.001	0.00	8.0	0.0	0.0	B8.5 V		380 s		
223916	35869		23 53 34.2	+51 31 22	+0.006	+0.01	7.00	1.1	0.2	K0 III		220 s		m
223920	192222		23 53 38.3	-20 31 04	0.000	-0.04	7.5	1.2	0.2	K0 III		200 s		
223913	255594		23 53 39.6	-65 56 51	-0.009	+0.07	6.70	0.5	4.0	F8 V		35 s		m
223934	192224		23 53 44.1	-29 23 49	+0.004	0.00	7.50	1.4	-0.3	K5 III		360 s		m
223932	165947		23 53 45.8	-18 21 54	+0.001	0.00	7.4	1.0	3.2	G5 IV	-20	52 s		
223833	258992		23 53 49.2	-83 00 27	+0.083	+0.02	6.7	2.2	-0.3	K5 III		98 s		
223960	20923		23 53 49.9	+60 51 14	0.000	+0.01	6.90	0.71	-7.1	A0 Ia	-48	2700 s		
223943	128449		23 53 52.1	+0 35 45	0.000	-0.02	7.8	1.6	-0.5	M2 III		460 s		
223952	53459		23 53 52.3	+44 33 21	0.000	0.00	7.4	1.1	0.2	K0 III		270 s		
223954	108894		23 53 54.0	+17 59 29	+0.001	-0.05	7.15	1.17		K0				
223961	108895		23 53 56.2	+12 00 03	+0.004	0.00	7.8	1.4	-0.3	K5 III		360 mx		
223957	214844		23 53 56.9	-30 21 26	-0.004	-0.03	7.8	0.3	3.4	F5 V		75 s		
223969	35875		23 54 02.3	+56 29 21	+0.006	-0.01	7.4	1.1	-0.1	K2 III	-8	310 s		
223963	146962		23 54 02.9	-9 17 24	+0.004	0.00	7.15	1.62	-0.5	M1 III	-34	300 mx		
223971	73597		23 54 04.1	+39 16 57	+0.006	+0.02	6.62	0.67	4.0	F8 V		25 s	17087	m
223974	128451		23 54 05.0	+4 52 16	0.000	-0.04	7.9	1.1	0.2	K0 III		340 s		
223981	53462		23 54 10.9	+44 52 03	-0.004	0.00	7.8	0.0	0.4	B9.5 V		290 s		
223967	248057		23 54 12.7	-59 32 32	+0.004	+0.01	7.0	0.2	0.6	A0 V		150 s		
223987	20927		23 54 13.1	+61 36 21	+0.001	-0.02	7.56	0.50	-5.7	B1 Ib	-45	1800 s		
223989	128454		23 54 19.3	+3 40 40	+0.004	-0.05	7.8	1.1	0.2	K0 III		180 mx		
223991	192231		23 54 21.2	-27 02 32	+0.002	+0.03	6.35	0.20		A m			17090	m
224014	35879	7 ρ Cas	23 54 22.9	+57 29 58	0.000	0.00	4.54	1.22	-8.0	F8 Ia	-43	1500 s		q
223997	53465		23 54 23.2	+48 55 47	+0.002	0.00	7.8	0.1	0.6	A0 V		260 s		
223999	53466		23 54 23.7	+46 57 59	-0.001	0.00	8.0	0.9	3.2	G5 IV		91 s		
223998	53467		23 54 25.7	+48 37 53	+0.001	+0.03	7.7	0.1	0.6	A0 V		250 s		
224022	231842		23 54 38.4	-40 18 01	+0.032	+0.03	6.03	0.57	4.0	F8 V		23 s		
224037	146971		23 54 38.4	-1 56 47	-0.001	-0.02	7.7	1.4	-0.3	K5 III		400 s		
224055	20932		23 54 41.6	+61 50 19	-0.004	-0.02	7.17	0.70	-6.8	B3 Ia	-42	1900 s		
224062	146973		23 54 46.5	+0 06 34	-0.003	-0.01	5.61	1.59		Mb	-2			
224060	108902		23 54 47.4	+18 44 49	+0.002	+0.01	7.60	1.1	-0.2	K3 III	-11	360 s		
224098	10893		23 54 49.1	+74 24 37	-0.001	0.00	6.60	0.00	0.4	B9.5 V	-13	160 s		m
224083	91574		23 54 56.1	+29 28 32	+0.001	0.00	6.60	0.00	0.4	B9.5 V		170 s		m
224011	258199		23 54 56.7	-78 30 18	+0.029	-0.02	7.9	0.8	3.2	G5 IV		86 s		
224084	91577		23 55 01.3	+29 27 03	+0.001	0.00	7.6	1.1	0.2	K0 III		300 s		
224093	146975		23 55 03.7	-9 27 39	+0.005	+0.01	6.8	0.4	2.6	F0 V		70 s		
224085	91578		23 55 03.9	+28 38 01	+0.044	+0.04	7.37	1.01	5.9	K0 V	-20	21 ts		
224099	35887		23 55 04.9	+54 35 27	+0.002	0.00	7.8	0.0	0.0	B8.5 V		340 s		
224100	53477		23 55 07.1	+47 43 44	+0.003	+0.02	7.1	1.1	0.2	K0 III		230 s		
224104	146977		23 55 07.5	-4 40 04	-0.001	0.00	7.9	1.1	0.2	K0 III		350 s		
224103	128466	26 Psc	23 55 07.7	+7 04 16	+0.001	-0.01	6.21	-0.07	0.2	B9 V	+17	160 s		
224107	192235		23 55 10.3	-21 22 47	+0.002	-0.03	7.30	1.1	0.2	K0 III		260 s		m
224116	35891		23 55 14.7	+58 05 21	+0.003	+0.08	7.39	1.09	-0.1	K2 III		120 mx		
224139	10899		23 55 15.3	+74 52 26	-0.003	0.00	7.8	1.1	0.2	K0 III		310 s		
224113	214860		23 55 16.4	-31 55 17	+0.001	+0.01	6.10	-0.08	-1.6	B5 IV	-12	320 s		
224112	214861		23 55 16.4	-31 53 03	0.000	+0.01	6.83	-0.08		A				
224127	73612		23 55 22.5	+32 53 30	-0.001	-0.02	7.6	1.1	0.2	K0 III		300 s		
224128	91582		23 55 22.8	+25 57 18	0.000	0.00	6.6	1.4	-0.3	K5 III	-15	240 s		
224124	248060		23 55 23.3	-50 06 49	-0.003	+0.01	7.9	0.2	0.2	K0 III		340 s		
224135	165951		23 55 25.5	-17 49 30	+0.001	+0.13	7.70	0.5	3.4	F5 V		72 s	17101	m
224140	53485		23 55 32.1	+48 56 07	+0.005	+0.04	8.0	1.1	-0.1	K2 III		330 mx		
224156	128469		23 55 32.3	+3 30 05	-0.014	-0.29	7.71	0.76	3.2	G5 IV		58 mx		
224152	35897		23 55 32.4	+52 44 08	+0.007	+0.03	6.7	1.1	0.2	K0 III	-1	190 s		
224150	35898		23 55 32.4	+59 24 57	+0.002	+0.02	7.3	1.1	0.2	K0 III	-13	240 s		
224165	53486		23 55 33.4	+47 21 21	0.000	-0.01	6.00	1.16	-4.5	G8 Ib	-17	1300 s		
224151	35899		23 55 33.5	+57 24 44	-0.001	0.00	6.00	0.21		B0.5 II	-26	13 mn		V373 Cas, v
224143	192241		23 55 34.8	-20 57 12	0.000	+0.03	7.91	0.64		G0				
224155	128472		23 55 37.5	+8 13 24	0.000	0.00	6.79	-0.02	0.6	A0 V	-2	170 s		
224166	53487		23 55 37.9	+46 21 30	+0.002	-0.01	6.96	-0.07		B9	-18			
224167	53488		23 55 40.5	+43 18 11	-0.001	0.00	7.9	0.1	1.4	A2 V		200 s		
224203	10902		23 55 41.0	+77 55 13	-0.002	+0.02	7.8	1.1	0.2	K0 III		300 s		

HD	SAO	Star Name	α 2000	δ 2000	μ(α)	μ(δ)	V	B-V	M_V	Spec	RV	d(pc)	ADS	Notes
224164	248063		23^{h}55^{m}43.5^s	-53°49'08"	+0.003^s	-0.01"	7.1	0.6	2.1	A5 V		57 s		
224171	146981		23 55 44.3	-9 03 42	+0.004	-0.04	7.3	0.8	3.2	G5 IV		67 s		
224172	165956		23 55 45.4	-13 09 09	-0.003	-0.07	6.8	0.9	3.2	G5 IV		52 s		
224175	53489		23 55 47.5	+48 13 13	0.000	-0.06	6.8	1.1	0.2	K0 III		200 s		
224173	165957		23 55 48.6	-13 57 59	+0.011	+0.04	7.3	0.5	3.4	F5 V		59 s		
224186	108911		23 55 54.6	+15 13 48	+0.007	-0.02	6.5	1.6	-0.5	M4 III	+4	210 mx		
224204	91585		23 55 58.3	+21 09 56	+0.002	0.00	6.6	0.2	2.1	A5 V		80 s		
224208	165960		23 56 01.2	-12 29 15	+0.003	-0.03	8.0	0.4	2.6	F0 V		120 s		
224216	73618		23 56 04.6	+31 04 41	0.000	-0.05	7.6	0.4	2.6	F0 V		98 s		
224225	192250		23 56 06.9	-21 59 31	+0.001	+0.03	7.24	1.64		M3	-5			
224211	248065		23 56 07.1	-52 47 17	+0.003	-0.02	7.2	0.7	3.2	G5 IV		62 s		
224235	73619		23 56 07.4	+33 29 16	+0.001	0.00	6.9	0.0	0.4	B9.5 V	+13	200 s		
224213	255603		23 56 08.5	-62 52 18	0.000	+0.04	7.30	0.1	1.7	A3 V		130 s		m
224228	214868		23 56 10.3	-39 03 08	+0.017	-0.18	7.9	1.1	0.2	K0 III		46 mx		
224233	35900		23 56 11.2	+59 46 04	+0.026	+0.29	7.67	0.63		G0				
224250	128479		23 56 19.4	+10 00 44	0.000	0.00	7.4	0.0	1.4	A2 V		160 s		
224259	108917		23 56 19.7	+20 12 32	+0.001	+0.01	8.0	1.1	-0.1	K2 III		410 s		
224257	35907		23 56 24.8	+55 59 26	0.000	+0.01	7.98	-0.06	-4.6	B0 IV		2600 s		
224273	20946		23 56 25.9	+66 45 01	+0.002	+0.02	7.9	1.1	0.2	K0 III		310 s		
224309	3994		23 56 27.5	+83 11 29	+0.026	+0.02	6.4	0.1	0.6	A0 V	-13	60 mx		V Cep, q
224283	192252		23 56 29.8	-24 44 13	0.000	+0.03	6.3	0.9	3.2	G5 IV		36 s		
224274	35908		23 56 30.9	+53 42 46	+0.001	+0.01	7.8	0.8	3.2	G5 IV		82 s		
224285	192254		23 56 32.0	-29 30 39	+0.005	-0.05	7.7	1.4	-0.3	K5 III		160 mx		
—	35910		23 56 33.5	+53 57 37	+0.001	+0.01	7.7	1.4	-0.1	K2 III		270 s		
224271	248068		23 56 34.4	-55 41 43	+0.001	+0.03	7.4	1.8	0.2	K0 III		92 s		
224303	91595		23 56 41.4	+22 38 53	-0.001	0.00	6.15	1.60	-0.5	M2 III	+1	210 s		
224296	231859		23 56 42.6	-42 11 43	+0.012	-0.05	7.91	0.42	2.8	F5 IV-V		76 mx		
224299	248074		23 56 43.6	-57 08 55	+0.008	-0.01	7.4	0.5	0.2	K0 III		280 s		
224315	128487		23 56 47.1	+4 43 30	+0.003	+0.01	7.40	0.4	2.6	F0 V		91 s	17111	m
224320	35913		23 56 55.5	+55 50 25	0.000	0.00	7.00	1.1	0.2	K0 III	-5	220 s	17114	m
224336	146990		23 57 01.2	-7 21 16	-0.002	-0.03	7.9	0.9	3.2	G5 IV		88 s		
224345	91598		23 57 01.8	+27 34 26	+0.002	+0.01	7.5	0.0	0.4	B9.5 V		260 s		
224363	10908		23 57 02.0	+79 45 25	+0.001	-0.02	7.9	0.4	2.6	F0 V		110 s		
224342	53511		23 57 03.5	+42 39 30	0.000	0.00	5.97	0.69	3.4	F5 V	-7	22 s		
224323	248076		23 57 04.1	-58 07 56	+0.006	+0.01	7.4	2.1	0.2	K0 III		60 s		
224346	128492		23 57 05.6	+3 04 15	-0.002	-0.02	7.6	1.4	-0.3	K5 III		390 s		
224350	192262		23 57 08.1	-26 37 25	+0.005	+0.02	6.3	1.2	0.2	K0 III		130 s		
224355	35917		23 57 08.3	+55 42 21	-0.002	-0.01	5.55	0.49	3.4	F5 V	+13	25 s		
224349	192264		23 57 09.7	-20 50 01	-0.005	0.00	6.6	1.3	0.2	K0 III		120 s		
224371	108925		23 57 14.3	+15 36 35	-0.003	-0.01	7.7	1.1	0.2	K0 III		320 s		
224364	20954		23 57 19.4	+61 01 34	-0.001	0.00	6.59	1.65	-0.4	M0 III	-77	220 s	17119	m
224361	255606		23 57 19.5	-62 57 23	+0.011	+0.03	5.97	0.11		A2 p				
224360	231863		23 57 19.5	-46 06 37	+0.012	+0.03	7.70	0.45	3.4	F5 V		72 s		
224380	53518		23 57 21.8	+48 16 51	-0.001	-0.01	7.5	0.1	0.6	A0 V	-4	230 s		
224381	53519		23 57 22.5	+41 36 29	0.000	+0.01	8.0	0.9	3.2	G5 IV		93 s		
224402	10910		23 57 23.1	+76 18 53	+0.007	+0.01	7.8	0.1	0.6	A0 V	+7	250 s		
224377	255607		23 57 27.5	-61 45 44	+0.004	+0.01	7.8	0.0	0.4	B9.5 V		270 mx		
224386	214876		23 57 28.6	-36 42 22	+0.001	-0.03	7.4	0.9	0.2	K0 III		270 s		
224403	20957		23 57 32.1	+61 59 28	-0.004	-0.01	7.68	0.14	-0.3	B9 IV		300 s		
224362	258996	γ² Oct	23 57 32.7	-82 10 12	-0.015	-0.02	5.73	1.05	0.3	gG8	+27	100 s		
224383	146997		23 57 33.4	-9 38 53	+0.031	-0.16	7.88	0.63	4.7	G2 V	-31	29 mx		
224404	35922		23 57 33.4	+60 01 25	+0.002	0.00	6.47	0.01		B8				
224392	255609	η Tuc	23 57 35.1	-64 17 55	+0.014	-0.06	5.00	0.06		A2 n	+33	13 mn		
224393	255608		23 57 35.4	-65 46 30	+0.044	-0.01	8.0	0.7	4.4	G0 V		53 s		
224406	91608		23 57 36.5	+20 19 50	+0.002	+0.01	8.0	0.9	3.2	G5 IV		93 s		
224410	214878		23 57 38.4	-33 11 19	+0.009	+0.04	6.9	1.0	3.4	F5 V		23 s		
224427	91611	84 ψ Peg	23 57 45.4	+25 08 29	-0.002	-0.03	4.66	1.59	-0.5	M3 III	-4	110 s		
224429	108931		23 57 46.0	+11 28 28	+0.001	0.00	6.62	0.01	0.4	B9.5 V nn		180 s	17125	m
224425	35927		23 57 47.0	+57 08 24	-0.002	0.00	7.30	0.22	1.4	A2 V		120 s		
224431	147000		23 57 48.4	-5 27 22	+0.008	+0.07	7.8	0.5	4.0	F8 V		57 s		
224423	255610		23 57 52.1	-67 00 24	-0.010	-0.03	7.1	0.4	3.0	F2 V		66 s		
224444	214883		23 57 59.0	-38 57 15	+0.002	+0.02	7.13	1.57	-0.3	K4 III		250 s		
—	—		23 58 03.6	+24 20 22	-0.004	-0.22	7.5	0.9	3.2	G5 IV		70 mx	17131	
224465	35934		23 58 06.6	+50 26 51	-0.005	+0.25	6.8	0.9	3.2	G5 IV	+3	34 mx		
224464	248082		23 58 13.0	-57 11 48	+0.018	+0.02	7.4	0.6	4.0	F8 V		45 s		
224468	165971		23 58 15.8	-11 27 30	-0.006	-0.03	7.9	0.9	3.2	G5 IV		88 s		
224474	108939		23 58 18.9	+19 41 37	+0.006	+0.01	7.4	1.1	-0.1	K2 III		230 mx		
224481	165972	1 Cet	23 58 21.0	-15 50 51	+0.005	-0.01	6.26	1.08	0.3	gG8	+4	120 s		
224492	73640		23 58 21.0	+35 00 48	+0.002	+0.02	6.80	0.1	1.7	A3 V		110 s	17136	m
224472	255612		23 58 21.7	-63 00 10	+0.001	+0.06	6.5	1.1	0.2	K0 III		180 s		
—	—		23 58 22.8	+51 23 40	+0.004	-0.03	5.50							R Cas, v
224491	73641		23 58 22.9	+38 57 48	+0.002	-0.02	8.0	1.1	-0.1	K2 III		410 s		
224494	108941		23 58 26.6	+13 49 21	+0.002	+0.01	8.0	0.1	1.4	A2 V		210 s		
224519	20968		23 58 30.4	+60 41 14	+0.001	0.00	7.99	1.12	0.2	K0 III		310 s		
224505	248085		23 58 31.1	-57 16 51	+0.004	0.00	6.7	0.4	4.4	G0 V		29 s		

HD	SAO	Star Name	α 2000	δ 2000	μ(α)	μ(δ)	V	B-V	M$_v$	Spec	RV	d(pc)	ADS	Notes
224514	192278		23ʰ 58ᵐ 31.2ˢ	-24° 10' 08"	+0.003ˢ	+0.04"	7.98	0.27		A2				
224518	20970		23 58 33.3	+69 20 39	-0.002	+0.01	7.9	0.4	2.6	F0 V		110 s		
224512	165975		23 58 34.3	-14 07 33	-0.001	-0.01	8.00	0.4	2.6	F0 V		120 s		m
224533	147008	27 Psc	23 58 40.2	-3 33 22	-0.003	-0.07	4.86	0.93	0.2	G9 III	0	86 s	17137	m
224541	35942		23 58 41.9	+53 23 13	+0.005	-0.01	7.7	1.1	0.2	K0 III		300 s		
224529	214892		23 58 42.3	-38 13 31	+0.007	0.00	7.8	0.1	3.0	F2 V		92 s		
224559	53540		23 58 46.3	+46 24 47	+0.002	0.00	6.54	-0.09	-2.3	B3 IV	-1	520 s		
224544	73650		23 58 49.2	+32 22 55	+0.001	+0.01	6.52	-0.11	-1.3	B6 IV e	-6	370 s		
224543	73649		23 58 50.0	+33 44 33	+0.021	-0.13	7.84	0.66	4.4	G0 V		43 s		
224538	255614		23 58 51.7	-61 35 11	+0.013	-0.01	7.8	0.5	3.4	F5 V		75 s		
224576	73651		23 58 53.0	+39 06 40	+0.002	-0.04	7.25	0.33	3.0	F2 V		71 s		
224554	248087	π Phe	23 58 55.6	-52 44 45	+0.006	+0.07	5.13	1.13	0.0	K1 III	-14	100 s		
224563	128509		23 58 56.1	+4 16 24	0.000	0.00	7.8	1.4	-0.3	K5 III		410 s		
224571	20973		23 59 00.3	+66 48 54	+0.012	+0.02	7.9	0.5	3.4	F5 V		76 s		
224572	35947	8 σ Cas	23 59 00.4	+55 45 18	+0.001	0.00	4.88	-0.07	-3.5	B1 V	-13	400 s	17140	m
224583	248090		23 59 04.2	-56 34 31	-0.001	+0.01	7.3	1.6		M5 III	+10			S Phe, v
224593	147011		23 59 07.7	-8 29 17	+0.002	-0.04	7.9	0.2	2.1	A5 V		150 s		
224602	53549		23 59 08.9	+41 12 06	+0.007	+0.01	7.9	0.6	4.4	G0 V		51 s		
224596	231874		23 59 09.6	-42 14 14	+0.004	+0.02	6.77	0.21	2.4	A7 V n		75 s		
224611	35952		23 59 11.8	+53 33 47	-0.003	-0.02	7.0	0.1	1.4	A2 V		130 s		
224612	35954		23 59 14.5	+50 31 45	0.000	-0.01	7.4	0.9	3.2	G5 IV		68 s		
224617	128513	28 ω Psc	23 59 18.5	+6 51 48	+0.010	-0.11	4.01	0.42		F4	+2	26 mn		
224626	91637		23 59 26.9	+20 24 51	+0.001	0.00	7.8	0.2	2.1	A5 V		140 s		
224624	35959		23 59 27.7	+57 40 16	+0.003	-0.02	7.20	0.14	0.6	A0 V		170 s		m
224630	192294		23 59 27.8	-29 29 06	0.000	0.00	5.62	1.60		K5				
224619	192293		23 59 28.2	-20 02 07	+0.036	-0.30	7.46	0.74	5.5	G7 V	+22	25 s		
224636	73656		23 59 29.0	+33 43 29	-0.005	-0.08	6.60				-5		17149	m
224687	4006		23 59 30.9	+86 42 23	+0.042	+0.01	6.7	0.1	0.6	A0 V	-17	52 mx		
224639	147015		23 59 31.2	-2 50 39	+0.002	-0.01	7.3	0.4	2.6	F0 V		86 s		
224648	35963		23 59 32.6	+50 49 57	0.000	0.00	7.2	0.0	0.4	B9.5 V		220 s		
224642	214899		23 59 34.2	-31 41 18	+0.001	+0.04	7.4	0.6	3.0	F2 V		55 s		
224638	147016		23 59 34.9	-1 51 01	+0.011	+0.02	7.5	0.4	2.6	F0 V		96 mx		
224656	73657		23 59 39.5	+37 48 00	+0.002	0.00	7.80	0.00	0.4	B9.5 V		300 s		m
224661	147017		23 59 40.4	-5 53 35	+0.002	-0.05	6.8	0.9	3.2	G5 IV	+13	52 s		
224657	91639		23 59 43.1	+21 17 21	+0.005	-0.06	7.6	0.4	2.6	F0 V		100 s		
224662	192298		23 59 43.7	-24 38 42	0.000	0.00	7.8	0.9	0.2	K0 III		330 s		
224669	20985		23 59 43.7	+62 59 30	-0.004	0.00	7.89	1.67	-0.2	K3 III		240 s		
224666	248094		23 59 45.6	-50 59 56	-0.001	-0.03	7.6	0.0	0.2	K0 III		300 s		
224677	147018		23 59 46.3	-0 16 48	-0.001	-0.02	6.9	1.6	-0.5	M2 III	-31	310 s		
224674	108948		23 59 47.5	+11 16 27	0.000	+0.03	6.60	0.1	1.4	A2 V		110 s		m
224673	108950		23 59 51.1	+11 40 25	0.000	+0.01	7.2	1.1	-0.1	K2 III		290 s		
224689	128521		23 59 54.2	+5 57 22	+0.003	-0.05	7.6	1.1	0.2	K0 III		210 mx		
224686	255619	ε Tuc	23 59 54.9	-65 34 38	+0.009	-0.02	4.50	-0.08	0.0	B8.5 V	+11	67 mx		
224698	10925		23 59 55.3	+72 36 56	-0.006	-0.04	8.0	1.1	0.2	K0 III		320 s		